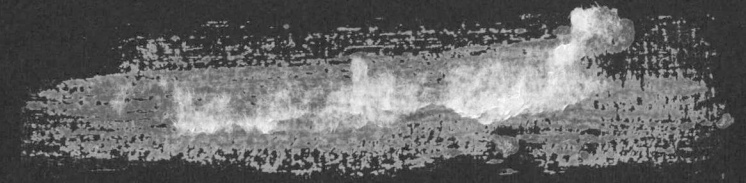

Bretherick's Handbook of **Reactive Chemical Hazards**

Sixth Edition

Volume 1

Bretherick's Handbook of

4566111

Reactive Chemical Hazards

Sixth Edition — Volume 1

Edited by
P G Urben
Akzo-Nobel

Compiler
M J Pitt
Department of Chemical Engineering
University of Leeds

OXFORD AUCKLAND BOSTON JOHANNESBURG MELBOURNE NEW DELHI

Butterworth-Heinemann Ltd
Linacre House, Jordan Hill, Oxford OX2 8DP

A member of the Reed Elsevier plc group

First published 1975
Second edition 1979
Reprinted 1979, 1981, 1984
Third edition 1985
Reprinted 1987, 1989
Fourth edition 1990
Fifth edition 1995
Sixth edition 1999

© Butterworth-Heinemann 1999

British Library Cataloguing in Publication Data
A record for this title is available from the British Library

Library of Congress Cataloguing in Publication Data
A record for this title is available from the Library of Congress

ISBN 0 7506 3605 X

Typeset by Laser Words, Madras, India
Printed and bound in Great Britain by Clays Ltd, St Ives plc

Foreword

It is now commonplace to say that the growth of knowledge of every sort is exponential and that keeping abreast with it becomes more difficult every year. As a result research can often become re-search, discovering again what is already described somewhere in the literature, perhaps in a little-known journal or little-read book.

In many fields the worst result is a waste of resources on repetitive work but when safety is concerned the lack of access to available knowledge can have tragic results. Those of us who have been working in the safety field for many years have seen the same accidents repeat themselves with distressing regularity. We welcome, therefore, every attempt to bring together scattered information on any aspect of safety and make it readily accessible.

In preparing the first four editions of this book Leslie Bretherick, almost unaided, produced a masterly summary of available information on reactive chemicals and their reactions. It was a remarkable achievement for one man, especially when we remember that the earlier editions were prepared in his spare time! Now that he has retired a team of editors, led by Peter Urben, has carried on the good work. They have increased the length of this edition by adding data on about 200 new compounds in Part 1, making the total nearly 5000, and 30 new groups in Part 2, making the total about 650. Leslie's high standard has been maintained.

It is a tribute to Leslie that very little of the new information in the 5th and this edition is old stuff that he missed; almost all comes from new publications. The entries on new groups range from acrylates to yeast passing through drums, fumes, mists and pnictides on the way. Familiar accidents continue to occur and new entries have been added on well-known hazards, such as nitric acid and azides.

In reviewing an earlier edition, I compared Leslie with those immortals, Beilstein and Perry, and said that he would become as well-known. Today I would go further and say that he has done more. He, and his successors, have not just made knowledge available but have saved lives and prevented injuries and damage and will continue to do so for many years.

Knowledge is effective only if it is used. Many coffee table books and cookery books are used to decorate the living room and kitchen rather than add to our knowledge. Some engineers are known to buy books to impress their visitors or perhaps with good intentions that are never fulfilled. As an author I don't complain but I hope your copy of this book does not undergo this fate. Look up the entries on all the compounds and groups of compounds you use or are thinking of using (now, not after your accident). Also, look up the entries on substances you used in the past and think of what might have happened.

The editors have done their bit; the publishers have done their bit by making the data available in book form and on CD-ROM. It is now up to you. As a bonus you will find the data fascinating to browse through, you will come across many facts that you never knew before, or had forgotten, and you will be amused by the deadpan humour of a few entries such as those on air, environmentalism, safety literature, sunspots and superiors.

TREVOR KLETZ

April 1999

Preface to the Sixth Edition

Bretherick's Handbook remains broadly similar to the previous editions but older readers will notice some changes. There are, of course, some hundreds of additional entries and much supplementary information in existing entries. This is the second edition for which I have been responsible and readers will still regret the absence of Leslie Bretherick, who had to withdraw from compilation because of worsening sight but remains a support and stay. The bulk is still his work, which is an indication of his immense labours laying the foundations, when accidents were less often reported and databases harder to compile than they now are. The present editor and his assistants have a far easier task continuing the work.

The change in the cyclic structures, now drawn in the more contemporary notation for which Hampden Data Services are to be thanked continues. The book is also now available as an electronic database, with all the improved ease of searching for cross-references or related materials which that brings. In future, it is proposed to supplement this rather more frequently than the new editions of the book will appear.

The present edition includes the literature until the end of 1998. There are few organisational changes from Leslie Bretherick's model, although no new references to safety data sheets are given. There are ever proliferating series of these, it would be impossible to read all and invidious to distinguish some, while others appear to have been compiled by (mis)information (ne)scientists in the library rather than by chemists skilled and experienced in lab and plant.

Once again, we request all users to inform us of any hazards of which they are aware and of which we are not, as also of any errors they find (regretfully, I must admit that some will certainly have escaped detection). Thanks are given to all those who have contributed to this and previous editions.

I have been valiantly assisted in the compilation and evaluation of data by Dr Martin Pitt of the University of Sheffield. I am also indebted to the staff of Butterworth-Heinemann and Hampden Data Services for their ability to make my files manifest as book and database.

My employers, Courtaulds plc, (recently subsumed into Akzo-Nobel) have generously allowed me time to undertake this work which, while both yet existed, benefited greatly from their library and from the Courtaulds Library at the University of Warwick. But, above all, thanks are due to Leslie Bretherick, not only for assistance and counsel but because, without him, the whole would be inconceivable. We again wish him a long and active (semi)retirement.

P. URBEN

March 1999

Preface to the First Edition

Although I had been aware during most of my career as a preparative chemist of a general lack of information relevant to the reactive hazards associated with the use of chemicals, the realisation that this book needed to be compiled came soon after my reading *Chemistry & Industry* for June 6th, 1964. This issue contained an account of an unexpected laboratory explosion involving chromium trioxide and acetic anhydride, a combination which I knew to be extremely hazardous from close personal experience 16 years previously.

This hazard had received wide publicity in the same journal in 1948, but during the intervening years had apparently lapsed into relative obscurity. It was then clear that currently existing arrangements for communicating 'well-known' reactive chemical hazards to practising chemists and students were largely inadequate. I resolved to try to meet this obvious need for a single source of information with a logically arranged compilation of available material. After a preliminary assessment of the overall problems involved, work began in late 1964.

By late 1971, so much information had been uncovered but remained to be processed that it was apparent that the compilation would never be finished on the spare-time basis then being used. Fortunately I then gained the support of my employers, the British Petroleum Company, Ltd., and have now been able to complete this compilation as a supporting research objective since January 1972.

The detailed form of presentation adopted has evolved steadily since 1964 to meet the dual needs for information on reactive chemical hazards in both specific and general terms, and the conflicting practical requirements of completeness and brevity. A comprehensive explanation of how this has been attempted, with suggestions on using this Handbook to best advantage, is given in the Introduction.

In an attempt to widen the scope of this work, unpublished information has been sought from many sources, both by published appeals and correspondence. In this latter area, the contribution made by a friend, the late Mr A. Kruk-Schuster, of Laboratory Chemicals Disposal Company, Ltd., Billericay, has been outstanding. During 1965–1968 his literature work and global letter campaign to 2000 University chemistry departments and industrial institutions yielded some 300 contributions.

The coverage attempted in this Handbook is wide, but is certainly incomplete because of the difficulties in retrieving relevant information from original literature when it does not appear in the indices of either primary or abstract journals. Details of such new material known to users of this Handbook and within the scope given in the Introduction will be welcomed for inclusion in supplementary or revised editions of this work.

L.B

October 1974

Contents

Volume 1

Introduction

Aims of the Handbook

This compilation has been prepared and revised to give access to a wide and up-to-date selection of documented information to research students, practising chemists, safety officers and others concerned with the safe handling and use of reactive chemicals. This will allow ready assessment of the likely potential for reaction hazards which may be associated with an existing or proposed chemical compound or reaction system.

A secondary, longer-term purpose is to present the information in away which will, as far as possible, bring out the causes of, and interrelationships between, apparently disconnected facts and incidents. This is designed to encourage an increased awareness of potential chemical reactivity hazards in school, college and university teaching laboratories, and to help to dispel the relative ignorance of such matters which is still in evidence in this important area of safety training during the formative years of technical education.

Others involved in a more general way with the storage, handling, packing, transport and distribution of chemicals, or emergencies related thereto, are likely to find information of relevance to their activities.

Scope and source coverage

This Handbook includes all information which had become available to the editor by January 1999 on the reactivity hazards of individual elements or compounds, either alone or in combination. Appropriate source references are included to give access to more expansive information than that compressed into the necessarily abbreviated text entries.

A wide variety of possible sources of published information has been scanned to ensure maximum coverage. Primary sources have largely been restricted to journals known to favour or specialise in publication of safety matters, and the textbook series specialising in synthetic and preparative procedures.

Secondary sources have been a fairly wide variety of both specialised and general textbooks and encyclopaedic collections (notably those of Mellor, Sidgwick, Pascal and Bailar in the inorganic area, Houben-Weyl in the organic and organometallic areas, and Kirk-Othmer in the industrial area). Section 50 of *Chemical Abstracts*, the CAS selection *Chemical Hazards, Health, & Safety*, the *Universities' Safety Association Safety News*, the CIA CISHC *Chemical Safety Summary*, (publication of which ceased in 1986 after 56 years), and the IChE *Loss Prevention Bulletin* have been rich sources, together with the more recent RSC *Laboratory HazardsBulletin* and *Chemical Hazards in Industry*. Additionally, various safety manuals, compilations, summaries, data sheets and case histories have been used, and fuller details of all the sources used are set out in Appendix 1. References in the text to textbooks are characterised by absence of the author's initials after the surname.

More recently, some reports have been picked from the Internet, when two of the three following conditions obtained: the editor finds the report credible; it represents a hazard not already present in the handbook; or the source is authoritative. Information on toxic hazards has been specifically excluded because it is available elsewhere in many well-ordered and readily usable forms.

However, it should be remembered that many of the compounds included in this Handbook show high reactivity of one sort or another toward other materials, so may in general terms be expected to be reactive even in brief contact with animal organisms or tissue (including yours), with possible toxic effects, either acute or chronic. Also, no attempt has been made to include details of all flammable or combustible materials capable of burning explosively when mixed with air and ignited, nor of any incidents related to this most frequent cause of accidents, such information again being available elsewhere.

However, to focus attention on the potential hazards always associated with the use of flammable and especially highly flammable substances, some 560 gases and liquids with flash points below 25°C and/or autoignition temperature below 225°C have been included in the text, their names prefixed with a dagger. The numerical values of the fire hazard-related properties of flashpoint, autoignition temperature and explosive (flammability) limits in air where known are given in the tabular Appendix 2. Those elements or compounds which ignite on exposure to air are included in the text, but not in the Table.

General arrangement

The information presented on reactive hazards is of two main types, specific or general, and these types of information have been arranged differently in their respective separate volumes 1 and 2.

FOR CROSS REFERENCES IN CAPITALS, PAGE NUMBERS REFER TO VOLUME 2.

Specific information on instability of individual chemical compounds, and on hazardous interactions of elements and/or compounds, is contained in the main formula-based Volume 1 of the Handbook. For an example of an unstable compound,
see Ethyl perchlorate
For an example of a hazardous interaction between 2 compounds,
see Nitric acid: Acetone
or 2 separate examples involving the same compound,

see Nitric acid: Acetone, or: Ethanol

and one involving 3 compounds,

see Hydrogen peroxide: Nitric acid, Thiourea

General information relating to classes or groups of elements or compounds possessing similar structural or hazardous characteristics is contained in the smaller alphabetically based Volume 2.

See ACYL NITRATES

PYROPHORIC METALS

References in the text to these general classes or groups of materials is always in small capitals to differentiate them from references to specific chemicals, the names of which are given in normal roman typeface.

Some individual materials of variable composition (substances) and materials which cannot conveniently be formulated and placed in Volume 1 are also included in this general section.

See BLEACHING POWDER, CELLULOSE NITRATE

Both theoretical and practical hazard topics, some indirectly related to the main theme of this book, are also included.

See DISPOSAL, EXPLOSIBILITY

GAS CYLINDERS, OXYGEN ENRICHMENT

Several topics which bring together incidents involving a common physical cause or effect but different types of chemicals are now included in Volume 2.

See CATALYTIC IMPURITY INCIDENTS

GAS EVOLUTION INCIDENTS

Specific chemical entries (Volume 1)

A single unstable compound of known composition is placed in the main first volume and is located on the basis of its empirical molecular formula expressed in the Hill system used by *Chemical Abstracts* (C and H if present, then all other element symbols alphabetically). The use of this indexing basis permits a compound to be located if its structure can be drawn, irrespective of whether a valid name is known for it. A representation of the structure of each compound is given on the third bold title line while the name of the compound appears as the first bold title line. References to the information source are given, followed by a statement of the observed hazard, with any relevant explanation. Cross-reference to similar compounds, often in a group entry, completes the entry. *See* Trifluoroacetyl nitrite p. 244.

Where two or more elements or compounds are involved in a reactive hazard, and an intermediate or product of reaction is identifiable as being responsible for the hazard, both reacting substances are normally cross-referred to the identified product. The well-known reaction of ammonia and iodine to give explosive nitrogentriodide-ammonia is an example of this type. The two entries

Ammonia: Halogens

Iodine: Ammonia

are referred back to the main entry under the identified material

Nitrogen triodide-ammonia

No attempt has been made, however, to list all combinations of reactants which can lead to the formation of a particular main entry compound.

In a multi-reactant system where no identification of an unstable product was possible, one of the reactants had to be selected as primary reactant to prepare and index the main entry, with the other material(s) as secondary reactant(s). No strictly logical basis of choice for this is obvious.

However, it emerged during the compilation phase that most two component reaction hazard systems of this type involve a fairly obvious oxidant material as one of the reactants. Where this situation was recognised, the oxidant has normally been selected as primary (indexing) reactant, with the other as secondary reactant, following the colon.

See Potassium permanganate: Acetic acid, etc.

In the markedly fewer cases where an obvious reducant has been involved as one reactant, that was normally selected as primary reactant.

See Lithium tetrahydroaluminate: 3,5-Dibromocyclopentene

In the relatively few cases where neither (or none) of the reactants can be recognised as an oxidant or reducant, the choice was made which appeared to give the more informative main entry text.

See Chloroform: Acetone, etc.

Where some hazard has been noted during the preparation of a specific compound, but without it being possible to identify a specific cause, an entry for that compound states 'Preparative hazard', and back-refers to the reactants involved in the preparation.

See Sulfur dioxide

Occasionally, departures from these considerations have been made where such action appeared advantageous in bringing out a relationship between formally unrelated compounds or hazards. In all multi-component cases, however, the secondary reactants (except air and water) appear as formula entries back-referred to the main entry text, so that the latter is accessible from either primary or secondary reactants.

See Dimethyl sulfoxide: Acyl halides (main entry)

Acetyl chloride: Dimethyl sulfoxide (back reference)

Grouping of Reactants

There are advantages to be gained in grouping together elements or compounds showing similar structure or reactivity, because this tends to bring out the relationships between structure and activity more clearly than separate treatment. This course has been adopted widely for primary reactants (see next heading), and for secondary reactants where one primary reactant has been involved separately with a large number of secondary materials. Where possible, the latter have been collected together under a suitable general group title indicative of the composition or characteristics of those materials.

See Chlorine: Hydrocarbons

Hydrogen peroxide: Metals, Metal oxides, Metal salts

Hydrogen sulfide: Oxidants

This arrangement means, however, that some practice will be necessary on the user's part in deciding into what group an individual secondary reactant falls before the longer-term advantages of the groupings become apparent. The formal group titles listed in Volume 2, Appendix 3, and classified in Appendix 4, will be of use in this connection. However, it should be noted that sometimes informal group titles are used which do not appear in these Appendices.

General group entries (Volume 2)

In some cases literature references relating to well-defined groups of hazardous compounds or to hazard topics have been found, and these are given, with a condensed version of relevant information at the beginning of the topic or group entry, under a suitable bold title, the latter being arranged in alphabetical order in Volume 2.

Cross references to related group or sub-group entries are also included, with a group list of the names and serial (not page) numbers of the chemicals appearing in Volume 1 which lie within the structural or functional scope of the group entry title. Compounds which are closely similar to, but not in strict conformity with, the group definition are indicated by a prefixed asterisk.

The group entries thus serve as sub-indexes for each structurally based group of hazardous compounds. Conversely, each individual compound entry is back-referred to the group entry, and thence to all its strict structural analogues and related congeners included in Volume 1 of this Handbook. Note that these group lists of chemicals are now in alphabetical (not formula) order, and give the serial-number (not page number) for the chemical.

These features should be useful in attempts to estimate the stability or reactivity of a compound or reaction system which does not appear in this Handbook. The effects on stability or reactivity of changes in the molecular structure to which the destabilising or reactive group(s) is attached are in some cases discussed in the group entry. Otherwise such information may be gained from comparison of the information available from the individual compound entries listed collectively (now in alphabetical order, with serial number) in the group entry.

Care is, however, necessary in extrapolating from the described properties of compounds to others in which the user of this Handbook may be interested. Due allowance must be made for changes in elemental reactivity up or down the columns of the Periodic Table, and for the effects of variation in chain length, branching and point of group-attachment in organic systems. Purity of materials, possible catalytic effects (positive or negative) of impurities, and scale of operations may all have a direct bearing upon a particular reaction rate. These and other related matters are dealt with in more detail in the following Introductory Chapter.

Nomenclature

With the direct encouragement and assistance of the Publishers, an attempt has been made to use chemical names which conform to recent recommendations of IUPAC. While this has not been an essential part of the compilation, because each title name has the corresponding structural and molecular formula adjacent, it seems none the less desirable to minimise possible confusion by adopting the unambiguous system of nomenclature presented in the IUPAC publications.

Where the IUPAC name for a compound is very different from a previously used recent trivial name, the latter is included as a synonym in parentheses (and in single quotes where no longer an acceptable name). Generally, retained trivial names have not been used as main entry titles, but they have been used occasionally in the entry texts. Rarely, on the grounds of brevity, names not conforming strictly to IUPAC principles but recommended for chemicals used in industry in BS 2474: 1983 have been used. The prefix mixo-,to represent the mixtures of isomers sometimes used as industrial materials, is a case in point.

Some of the rigidly systematic names selected by the Association for Science Education for their nomenclature list in 1985 from the IUPAC possibilities, and some of the systematic indexing names used by *Chemical Abstracts* since 1972, are given as synonyms in the Index of Chemical Names (Appendix 4). This should assist those coming into industry and research with a command of those nomenclature systems but who may be unfamiliar with the current variety of names used for chemicals. The inclusion where possible of the CAS Registry Number for each title compound should now simplify the clarification of any chemical name or synonym problems, by reference to the Registry Handbook or other CAS source.

In connection with the group titles adopted for the alphabetically ordered Volume 2, it has been necessary in some cases to devise groupnames (particularly in the inorganic field) to indicate in a very general way the chemical structures involved in various classes, groups or sub-groups of compounds.

For this purpose, all elements have been considered either as METALS or NON-METALS and of the latter, HALOGENS, HYDROGEN, NITROGEN, OXYGEN, and SULFUR were selected as specially important. Group names have then been coined from suitable combinations of these, such as the simple

METAL OXIDES, NON-METAL SULFIDES,

N-HALOGEN COMPOUNDS, NON-METAL HYBRIDES,

METAL NON-METALLIDES, COMPLEX HYBRIDES

or the more complex

METAL OXOHALOGENATES

AMMINECHROMIUM PEROXOCOMPLEXES

OXOSALTS OF NITROGENOUS BASES

METALOXONON-METALLATES

Organic group entries are fairly conventional, such as

HALOALKENES

NITROARL COMPOUNDS

DIAZONIUM SALTS

Where necessary, such group names are explained in the appropriate group entry, of which a full listing is given in Volume 2, Appendix 3, and a classified listing in Appendix 4.

Cross reference system

The cross-reference system adopted in this Handbook plays a large part in providing maximum access to, and use of, the rather heterogeneous collection of information herein. The significance of the five types of cross-reference which have been used is as follows.

See... refers to a directly related item.

See also... refers to an indirectly related item.

See other... refers to listed strict analogues of the compound etc.

See related... refers to listed related compounds(congeners) or groups not strictly analogous structurally.

See entry... points to a, or the relevant, reference in Volume 2.

Information content of individual entries

A conscious effort has been made throughout this compilation to exclude all fringe information not directly relevant to the involvement of chemical reactivity in the various incidents o observations, with just enough detail present to allow the reader to judge the relevance or otherwise of the quoted reference(s) to his or her particular reactivity problems or interests.

It must be stressed that this book can do no more than to serve as a guide to much more detailed information available via the quoted references. It cannot relieve the student, the chemist and their supervisors of their moral and now legal obligation to themselves and to their co-workers, to equip themselves with the fullest possible information from the technical literature resources which are widely available, *before* attempting any experimental work with materials known, or suspected, to be hazardous or potentially so. It could be impossible for you *after* the event.

THE ABSENCE OF A MATERIAL OR A COMBINATION OF MATERIALS FROM THIS HANDBOOK CANNOT BE TAKEN TO IMPLY THAT NO HAZARD EXISTS. LOOK THENFOR ANALOGOUS MATERIALS USING THE GROUP ENTRY SYSTEM AND THE INDEXES THERETO.

One aspect which, although it is excluded from most entry texts, is nevertheless of vital importance, is that of the potential for damage, injury or death associated with the various materials and reaction systems dealt with in this Handbook.

Though some of the incidents have involved little or no damage (*see* CAN OF BEANS), others have involved personal injuries, often of unexpected severity (*see* SODIUM PRESS), and material damage is often immense. For example, the incident given under Perchloric acid: Cellulose derivatives, (reference 1) involved damage to 116 buildings and a loss approaching $ 3M at 1947 values. The death-toll associated with reactive chemical hazards has ranged from 1 or 2 (see Tetrafluoroethylene: Iodine pentafluoride) to some 600 with 2000 injured in the incident at Oppau in 1921 (*see* Ammonium nitrate, reference 4), and now to several thousand with more than 100,000 injured by methyl isocyanate fumes at Bhopal in 1984 (reference 7).

This sometimes vast potential for destruction again emphasises the need to gain the maximum of detailed knowledge *before* starting to use an unfamiliar chemical or reaction system.

Reactive Chemical Hazards

CROSS REFERENCES IN CAPITALS REFER TO PAGE NUMBERS IN VOLUME 2.

This introductory chapter seeks to present an overview of the complex subject of reactive chemical hazards, drawing attention to the underlying principles and to some practical aspects of minimising such hazards. It also serves in some measure to correlate some of the topic entries in the alphabetically arranged Volume 2 of the Handbook.

Basics

All chemical reactions implicitly involve energy changes (energy of activation + energy of reaction), for these are the driving force. The majority of reactions liberate energy as heat (occasionally as light or sound) and are termed exothermic. In a minority of reactions, energy is absorbed into the products, when both the reaction and its products are described as endothermic.

All reactive hazards involve the release of energy in quantities or at rates too high to be absorbed by the immediate environment of the reacting system, and material damage results. The source of the energy may be an exothermic multi-component reaction, or the exothermic decomposition of a single unstable (often endothermic) compound.

All measures to minimise the possibility of occurrence of reactive chemical hazards are therefore directed at controlling the extent and rate of release of energy in a reacting system. In an industrial context, such measures are central to modern chemical engineering practice. Some of the factors which contribute to the possibility of excessive energy release, and appropriate means for their control, are now outlined briefly, with references to examples in the text.

Kinetic Factors

The rate of an exothermic chemical reaction determines the rate of energy release. so factors which affect reaction kinetics are important in relation to possible reaction hazards. The effects of proportions and concentrations of reactants upon reaction rate are governed by the Law of Mass Action, and there are many examples where changes in proportion and/or concentration of reagents have transformed an established uneventful procedure into a violent incident. For examples of the effect of increase in proportion,

see 2-Chloronitrobenzene: Ammonia
   Sodium 4-nitrophenoxide

For the effect of increase in concentration upon reaction velocity,

see Dimethyl sulfate: Ammonia
   Nitrobenzene: Alkali (reference 2)

The effects of catalysts (which effectively reduce the energy of activation), either intentional or unsuspected, is also relevant in this context. Increase in the concentration of a catalyst (normally used at 1–2%) may have a dramatic effect on reaction velocity.

See Trifluoromethanesulfonic acid: Acyl chlorides, etc.
   2-Nitroanisole: Hydrogen
   HYDROGEN CATALYSTS

The presence of an unsuspected contaminant or catalytic impurity may affect the velocity or change the course of reaction. For several examples,

See CATALYTIC IMPURITY INCIDENTS

In the same context, but in opposite sense, the presence of inhibitors (negative catalysts, increasing energy of activation) may seriously interfere with the smooth progress of a reaction. An inhibitor may initiate an induction period which can lead to problems in establishing and controlling a desired reaction. For further details and examples,

See INDUCTION PERIOD INCIDENTS

Undoubtedly the most important factor affecting reaction rates is that of temperature. It follows from the Arrhenius equation that the rate of reaction will increase exponentially with temperature. Practically, it is found that an increase of 10°C in reaction temperature often doubles or trebles the reaction velocity.

Because most reactions are exothermic, they will tend to accelerate as reaction proceeds unless the available cooling capacity is sufficient to prevent rise in temperature. Note that the exponential temperature effect accelerating the reaction will exceed the (usually) linear effect of falling reactant concentration in decelerating the reaction. When the exotherm is large and cooling capacity is inadequate, the resulting accelerating reaction may proceed to the point of loss of control (runaway), and decomposition, fire or explosion may ensue.

The great majority of incidents described in the text may be attributed to this primary cause of thermal runaway reactions. The scale of the damage produced is related directly to the size, and more particularly to the rate, of energy release.

See RUNAWAY REACTIONS

Reactions at high pressure may be exceptionally hazardous owing to the enhanced kinetic energy content of the system.

See HIGH-PRESSURE REACTION TECHNIQUES

Although detailed consideration of explosions is outside the scope of this Handbook, three levels of intensity of explosion (i.e. rates of fast energy release) can be discerned and roughly equated to the material damage potential.

Deflagration involves combustion of a material, usually in presence of air. In a normal liquid pool fire, combustion in an open situation will normally proceed without explosion. Mixtures of gases or vapours with air within the explosive limits which are subsequently ignited will burn at normal flame velocity (a few m/s) to produce a 'soft' explosion, with minor material damage, often limited to scorching by the moving flame front. Injuries to personnel may well be more severe.

If the mixture (or a dust cloud) is confined, even if only by surface irregularities or local partial obstructions, significant pressure effects can occur. Fuel-air mixtures near to stoicheiometric composition and closely confined will develop pressures of several bar within milliseconds, and material damage will be severe. Unconfined vapour explosions of large dimensions may involve higher flame velocities and significant pressure effects, as shown in the Flixborough disaster.

See DUST EXPLOSION INCIDENTS
   PRESSURE INCREASE IN EXOTHERMIC DECOMPOSITION
   VAPOUR CLOUD EXPLOSIONS

Detonation is an extreme form of explosion where the propagation velocity becomes supersonic in gaseous, liquid or solid states. The temperatures and particularly pressures associated with detonation are higher by orders of magnitude than in deflagration. Energy release occurs in a few microseconds and the resulting shattering effects are characteristic of detonation. Deflagration may accelerate to detonation if the burning material and geometry of confinement are appropriate (endothermic compounds, long narrow vessels or pipelines).

See Acetylene (reference 9)
ENDOTHERMIC COMPOUNDS
EXPLOSIONS
UNIT PROCESS INCIDENTS

Factors of importance in preventing such thermal runaway reactions are mainly related to the control of reaction velocity and temperature within suitable limits. These may involve such considerations as adequate heating and particularly cooling capacity in both liquid and vapour phases of a reaction system; proportions of reactants and rates of addition (allowing for an induction period); use of solvents as diluents and to reduce viscosity of the reaction medium; adequate agitation and mixing in the reactor; control of reaction or distillation pressure; use of an inert atmosphere.

See AGITATION INCIDENTS

In some cases it is important not to overcool a reaction system, so that the energy of activation is maintained.

See Acetylene: Halogens (reference 1)

Adiabatic Systems

Because process heating is expensive, lagging is invariably applied to heated process vessels to minimise heat loss, particularly during long-term hot storage. Such adiabatic or near-adiabatic systems are potentially hazardous if materials of limited thermal stability, or which possess self-heating capability, are used in them. Insufficiently stabilised bulk-stored monomers come into the latter category.

See 1,2,4,5-Tetrachlorobenzene: Sodium hydroxide, Solvent
   POLYMERISATION INCIDENTS

Reactivity vs. Composition and Structure

The ability to predict reactivity and stability of chemical compounds from their composition and structure is as yet limited, so the ability accurately to foresee potential hazards during preparation, handling and processing of chemicals and their mixtures is also restricted. Although some considerable progress has been made in the use of computer programs to predict hazards, the best available approach for many practical purposes appears to be an initial appraisal based on analogy with, or extrapolation from, data for existing compounds and processes. This preliminary assessment should be supplemented with calorimetric instrumental examination, then bench-scale testing procedures for thermal stability applied to realistic reaction mixtures and processing conditions. A wide range of equipment and techniques is now available for this purpose.

See ACCELERATING RATE CALORIMETRY
    ASSESSMENT OF REACTIVE CHEMICAL HAZARDS
    COMPUTATION OF REACTIVE CHEMICAL HAZARDS
    DIFFERENTIAL SCANNING CALORIMETRY
    DIFFERENTIAL THERMAL ANALYSIS
    MAXIMUM REACTION HEAT
    REACTION SAFETY CALORIMETRY

It has long been recognised that instability in single compounds, or high reactivity in combinations of different materials, is often associated with particular groupings of atoms or other features of molecular structure, such as high proportions or local concentrations of oxygen or nitrogen. Full details of such features associated with explosive instability are collected under the heading EXPLOSIBILITY.

An approximate indication of likely instability in a compound may be gained from inspection of the empirical molecular formula to establish stoicheiometry.

See HIGH-NITROGEN COMPOUNDS
    OXYGEN BALANCE

Endothermic compounds, formed as the energy-rich products of endothermic reactions, are thermodynamically unstable and may be liable to energetic decomposition with low energy of activation.

See ENDOTHERMIC COMPOUNDS

Reaction Mixtures

So far as reactivity between different compounds is concerned, some subdivision can be made on the basis of the chemical types involved. Oxidants (electron sinks) are undoubtedly the most common chemical type to be involved in hazardous incidents, the other components functioning as fuels or other electron sources. Air (21% oxygen) is the most widely dispersed oxidant, and air-reactivity may lead to either short- or long-term hazards.

Where reactivity of a compound is very high, oxidation may proceed so fast in air that ignition occurs.

See PYROPHORIC MATERIALS

Slow reaction with air may lead to the longer-term hazard of peroxide formation.

See AUTOXIDATION

PEROXIDATION INCIDENTS

PEROXIDATION IN SOLVENTS

PEROXIDISABLE COMPOUNDS

Oxidants more concentrated than air are of greater hazard potxential, and the extent of involvement of the common oxidants

Perchloric acid

Chlorine

Nitric acid

Hydrogen peroxide

Sulfuric acid

METAL CHLORATES

may be judged from the large number of incidents in the text involving each of them, as well as other OXIDENTS, p. 271.

At the practical level, experimental oxidation reactions should be conducted to maintain in the reacting system a minimum oxygen balance consistent with other processing requirements. This may involve adding the oxidant slowly with appropriate mixing and cooling to the other reaction materials to maintain the minimum effective concentration of oxidant for the particular reaction. It will be essential to determine by a suitable diagnostic procedure that the desired reaction has become established, to prevent build-up of unused oxidant and a possible approach to the oxygen balance point.

See OXYGEN BALANCE

Reducants (rich electron sources) in conjunction with reducible materials (electron acceptors) feature rather less frequently than oxidants in hazardous incidents.

See REDUCANTS

Interaction of potent oxidants and reducants is invariably highly energetic and of high hazard potential.

See Dibenzoyl peroxide: Lithium tetrahydroaluminate

Hydrazine: Oxidants

REDOX REACTIONS

ROCKET PROPELLANTS

Similar considerations apply to those compounds which contain both oxidising and reducing functions in the same molecular structure.

See REDOX COMPOUNDS

Water is, after air, one of the most common reagents likely to come into contact with reactive materials, and several classes of compounds will react violently, particularly with restricted amounts of water.

See WATER-REACTIVE COMPOUNDS

Most of the above has been written with deliberate processing conditions in mind, but it must be remembered that the same considerations will apply, and perhaps to a greater degree, under the uncontrolled reaction conditions prevailing when accidental contact of reactive chemicals occurs in storage or transit.

Adequate planning is therefore necessary in storage arrangements to segregate oxidants from fuels and reducants, and fuels and combustible materials from compressed gases and water-reactive compounds. This will minimise the possibility of accidental contact and violent reaction arising from faulty containers or handling operations, and will prevent intractable problems in the event of fire in the storage areas.

See SAFE STORAGE OF CHEMICALS

Unexpected sources of ignition may lead to ignition of flammable materials during chemical processing or handling operations.

See FRICTIONAL IGNITION OF GASES
  IGNITION SOURCES
  SELF-HEATING AND IGNITION INCIDENTS
  STATIC INITIATION INCIDENTS

Protective Measures

The need to provide protective measures will be directly related to the level of potential hazards which may be assessed from the procedures outlined above. Measures concerned with reaction control are frequently mentioned in the following text, but details of techniques and equipment for personal protection, though usually excluded from the scope of this work, are obviously of great importance.

Careful attention to such detail is necessary as a second line of defence against the effects of reactive hazards. The level of protection considered necessary may range from the essential and absolute minimum of effective eye protection, via the safety screen, fume cupboard or enclosed reactor, up to the ultimate of a remotely controlled and blast-resistant isolation cell (usually for high-pressure operations). In the absence of facilities appropriate to the assessed level of hazard, operations must be deferred until such facilities are available.

Volume 1

Specific Chemicals

(Elements and Compounds arranged in formula order)

EXPLANATORY NOTES

This volume gives detailed information on the hazardous properties of individual chemicals, either alone or in combination with other compounds. The items are arranged in order of the empirical formula (at right of second bold title line) which corresponds to the chemical name, or a synonym within parentheses, used as the first line bold title of each main entry; (nomenclature is now rather promiscuous since the systematisers have contrived to give many materials two or even three new names since the first edition was published, while the name used where chemicals are handled, as opposed to in lecture rooms, will be something else again). The 3 part number within square brackets at the left of the second title line is the CAS registry number, now being widely used to provide are liable basis for establishing equivalence between differing chemical names and trade names for the same chemical compound (but note that one compound, within the terms of this work, may have numerous CAS numbers by virtue of isotopic composition, undefined stereo- and regio-chemistry or variant solvation levels). Lack of content within the square brackets indicates that a registry number has not yet been located, and (ion) after the number indicates that the main ion only has been located, rather than the specific title salt. Where possible, a linear or graphical representation of the structure of the title compound is given at the centre of the third title line, otherwise the reader is referred to one of the several lettered pages preceeding on which the corresponding cyclic structural formula is set out.

A † prefixed to the chemical name indicates the existence of tabulated information on fire-related properties in Appendix 2. The † prefix is also appended to the entry (and any synonym) in the index in Appendix 4 of the chemicals appearing as title lines. Immediately under the title lines some references to sources of general safety-related data concerning use and handling precautions for the title chemical are given. The references to the series of *MCA Safety Data Sheets* are given in parentheses because the whole series was withdrawn in 1980, apparently on grounds other than obsolescence of the technical content. Since these data sheets are no longer available, alternative references are given where possible to the Data Sheets available from the National Safety Council *(NSC)*, Chicago; the Fire Protection Association *(FPA)*, London; to the appropriate page of 'Handling Chemicals Safely 1980' *(HCS 1980)*, published in Holland; or to the Laboratory Hazard

Data Sheet series published by the Royal Society of Chemistry *(RSC)*, now in Cambridge.

No new data sheets have been included since 1990, since distinction between the proliferation of sources would be invidious, and even supposedly author-itative bodies are putting out sheets which have evidently been compiled by (mis)information scientists at their computers, unchecked by anyone who has ever seen, smelt, or handled, the material in question. The first reference(s) and data given under the title lines refer to the hazards of the title material alone, or in the presence of air, unless stated otherwise. Where other (secondary) chem-icals are involved with the title compound in a reactive incident, the name(s) follows in roman characters under the bold title entry. As in previous editions of this Handbook, where these secondary chemicals are described in group terms (e.g. Polynitroaryl compounds), reference to the alphabetical group entries now in Volume 2 may suggest other analogous possibilities of hazards. References to original or abstract literature then follow, and sufficient of the relevant information content is given to allow a general picture of the nature and degree of hazard to be seen.

Two features relevant to entries for pairs of reactive chemicals arise from the work of Prof. T. Yoshida in developing a method for the calculation of maximum reaction heats (MRH) possible for binary (or ternary) mixtures of chemicals, and the publication of his tabulated results. Where available for combinations existing in this text, these data are given opposite the name of the secondary chemical in the form MRH 2.9/22. This means that the calculated reaction heat is maximal at 2.9 kJ/g in a mixture containing 22% wt of the secondary reactant with 78% of the main (bold title) compound. The second feature is the inclusion of the secondary entry 'Other reactants' under which the extent of the information available in Yoshida's book for some 240 title compounds is given. More detail onthe origin of these figures is given in Volume 2 under the entry MAXIMUM REACTION HEAT.

All temperatures in the text are expressed in degrees Celsius; pressures in bars, mbars or Pa; volumes in m^3, litres or ml; and energy as joules, kJ or MJ. Where appropriate, attention is drawn to closely similar or related materials or events by *See* or *See also* cross-references. Finally, if a title compound is a member of one of the general classes or groups in Volume 2, it is related to those by a *See other* cross-reference. If the compound is not strictly classifiable, a *See related* cross-reference establishes a less direct link to the group compound index lists in Volume 2, such compounds being prefixed in the lists by an asterisk. In relatively few cases, literature references (or further references) for individual compounds are in the alphabetical entries in Volume 2, and a *See entry* cross-reference leads to that entry with the literature reference. An alphabetical index of the chemical names used as bold titles in Volume 1, together with synonyms, is given in Appendix 4.

Details of corrections of typographical or factual errors, or of further items for inclusion in the text, will be welcomed, and a page which can be photocopied for this purpose will be found at the back of the book.

0001. Silver
[7440-22-4]

Ag

Acetylenic compounds   MRH Acetylene 8.70/99+
See ACETYLENIC COMPOUNDS

Aziridine
See Aziridine: Silver

Bromine azide
See Bromine azide

3-Bromopropyne
See 3-Bromopropyne: Metals

Carboxylic acids
Koffolt, J. H., private comm., 1965
Silver is incompatible with oxalic or tartaric acids, since the silver salts decompose on heating. Silver oxalate explodes at 140°C, and silver tartrate loses carbon dioxide.
See other METAL OXALATES

Chlorine trifluoride   MRH 1.42/36
See Chlorine trifluoride: Metals

Copper, Ethylene glycol
See Ethylene glycol: Silvered copper wire

Electrolytes, Zinc
Britz, W. K. *et al., Chem. Abs.*, 1975, **83**, 150293
Causes of spontaneous combustion and other hazards of silver–zinc batteries were investigated.

Ethanol, Nitric acid
Luchs, J. K., *Photog. Sci. Eng.*, 1966, **10**, 334
Action of silver on nitric acid in presence of ethanol may form the readily detonable silver fulminate.
See Nitric acid: Alcohols
See also SILVER-CONTAINING EXPLOSIVES

Ethylene oxide   MRH 3.72/99+
See Ethylene oxide: Reference 4

Ethyl hydroperoxide
See Ethyl hydroperoxide: Silver

Hydrogen peroxide   MRH 1.59/99+
See Hydrogen peroxide: Metals

Iodoform
  Grignard, 1935, Vol. 3, 320
  In contact with finely divided (reduced) silver, incandescence occurs.

Other reactants
  Yoshida, 1980, 103
  MRH values for 7 combinations, largely with catalytically susceptible materials, are given.

Ozonides
  See OZONIDES

Peroxomonosulfuric acid
  See Peroxomonosulfuric acid: Catalysts

Peroxyformic acid   MRH 5.69/100
  See Peroxyformic acid: Metals
  See other METALS

0002. Silver−aluminium alloy
  [11144-29-9]   Ag−Al

                          Ag−Al

  1. Popov, E. I. *et al., Chem. Abs.,* 1977, **87**, 205143
  2. Popov, E. I. *et al., Chem. Abs.,* 1980, **94**, 35622
  Combustion and explosion hazards of the powdered alloy used in batteries were studied. Increase in silver content leads to higher values of the ignition temperature and COI [1,2].
  See other ALLOYS, SILVER COMPOUNDS

0003. Silvered copper
  [37218-25-0]   Ag−Cu

                          Ag−Cu

Ethylene glycol
  See Ethylene glycol: Silvered copper wire
  See related ALLOYS

0004. Silver−thorium alloy
  [[12785-36-3] (1:2)]   Ag−Th

                          Ag−Th

  See entry PYROPHORIC ALLOYS

0005. Silver tetrafluoroborate
[14104-20-2]   AgBF₄

$$Ag[BF_4]$$

Preparative hazard
1. Meerwein, H. *et al., Arch. Pharm.*, 1958, **291**, 541–544
2. Lemal, D. M. *et al., Tetrahedron Lett.*, 1961, 776–777
3. Olah, G. A. *et al., J. Inorg. Nucl. Chem.*, 1960, **14**, 295–296
Experimental directions must be followed exactly to prevent violent spontaneous explosions during preparation of the salt from silver oxide and boron trifluoride etherate in nitromethane, according to the earlier method [1]. The later method [3] is generally safer than that in [2].
See other SILVER COMPOUNDS

0006. Silver tetrafluorobromate
[35967-89-6]   AgBrF₄

$$AgBrF_4$$

See entry METAL POLYHALOHALOGENATES *See other* SILVER COMPOUNDS

0007. Silver bromate
[7783-89-3]   AgBrO₃

$$AgOBrO_2$$

Other reactants
Yoshida, 1980, 133
MRH values for 16 combinations with oxidisable materials are given.

Sulfur compounds   MRH Sulfur 2.0/12
1. Taradoire, F., *Bull. Soc. Chim. Fr.*, 1945, **12**, 94–95
2. Pascal, 1960, Vol. 13.1, 1004
The bromate is a powerful oxidant, and unstable mixtures with sulfur ignite at 73–75°C, and with disulfur dibromide on contact [1]. Hydrogen sulfide ignites on contact with the bromate [2].
See other METAL OXOHALOGENATES, SILVER COMPOUNDS

0008. Silver chloride
[7783-90-6]   AgCl

$$AgCl$$

Aluminium
See Aluminium: Silver chloride

Ammonia
1. Mellor, 1941, Vol. 3, 382
2. Kauffmann, G. B., *J. Chem. Educ.*, 1977, **54**, 132
3. Ranganathan, S. *et al., J. Chem. Educ.*, 1976, **53**, 347

Exposure of ammoniacal silver chloride solutions to air or heat produces a black crystalline deposit of 'fulminating silver', mainly silver nitride, with silver diimide and silver amide also possibly present [1]. Attention is drawn [2] to the possible explosion hazard in a method of recovering silver from the chloride by passing an ammoniacal solution of the chloride through an ion exchange column to separate the $Ag(NH_3)^+$ ion, prior to elution as the nitrate [3]. It is essential to avoid letting the ammoniacal solution stand for several hours, either alone or on the column [2].
See Silver nitride
See other METAL HALIDES, SILVER COMPOUNDS

0009. Silver azide chloride
[67880-13-1]

N_3AgCl

$AgClN_3$

Frierson, W. J. *et al., J. Amer. Chem. Soc.,* 1943, **65**, 1698
It is shock sensitive when dry.
See other METAL AZIDE HALIDES, SILVER COMPOUNDS

0010. Silver chlorite
[7783-91-7]

$AgOClO$

$AgClO_2$

Alone, or Iodoalkanes
  Levi, G. R., *Gazz. Chim. Ital. [2],* 1923, **53**, 40
  The salt is impact-sensitive, cannot be finely ground, and explodes at 105°C. Attempts to react silver chlorite with iodo-methane or -ethane caused explosions, immediate in the absence of solvents, or delayed in their presence.

Hydrochloric acid, or Sulfur
  Mellor, 1941, Vol. 2, 284
  It explodes in contact with hydrochloric acid or on rubbing with sulfur.

Non-metals
  Pascal, 1960, Vol. 16, 264
  Finely divided carbon, sulfur or red phosphorus are oxidised violently by silver chlorite.
  See other CHLORITE SALTS, SILVER COMPOUNDS

0011. Silver chlorate
[7783-92-8]

$AgOClO_2$

$AgClO_3$

Sorbe, 1968, 126
An explosive compound and powerful oxidant.

Ethylene glycol   MRH 2.68/17
    See Ethylene glycol: Oxidants

Other reactants
    Yoshida, 1980, 69
    MRH values for 17 combinations, largely with oxidisable materials, are given.
    See other METAL CHLORATES, SILVER COMPOUNDS

0012. Silver perchlorate
[7783-93-9]   **AgClO$_4$**
                            AgOClO$_3$

    1. Anon., *Angew. Chem. (Nachr.)*, 1962, **10**, 2
    It melts without decomposition although the enthalpy of conversion to silver chlo-
    ride and oxygen appears to be about −0.5 kJ/g. An explosion while grinding the
    salt (which had not been in contact with organic materials) has been reported [1].
    A powerful oxidant.

Acetic acid   MRH 2.80/22
    Mellor, 1956, Vol. 2, Suppl.1, 616
    The salt solvated with acetic acid is impact sensitive.
    See Aromatic compounds, below

Alkynes, Mercury
    Comyns, A. E. *et al., J. Amer. Chem. Soc.*, 1957, **79**, 4324
    Concentrated solutions of the perchlorate in 2-pentyne or 3-hexyne (complexes are
    formed) explode on contact with mercury.
    See METAL ACETYLIDES

Aromatic compounds   MRH Aniline 3.47/11, toluene 3.51/9
    1. Sidgwick, 1950, 1234
    2. Brinkley, S. R., *J. Amer. Chem. Soc.*, 1940, **62**, 3524
    3. Peone, J. *et al., Inorg. Synth.*, 1974, **15**, 69
    4. Stull, 1977, 22
    Silver perchlorate forms solid complexes with aniline, pyridine, toluene, benzene
    and many other aromatic hydrocarbons [1]. A sample of the benzene complex
    exploded violently on crushing in a mortar. The ethanol complex also exploded
    similarly, and unspecified perchlorates dissolved in organic solvents were observed
    to explode [2]. Solutions of the perchlorate in benzene are said to be dangerously
    explosive [3], but this may be in error for the solid benzene complex. The energy
    released on decomposition of the benzene complex has been calculated as 3.4 kJ/g,
    some 75% of that for TNT [4].

Carbon tetrachloride, Hydrochloric acid
    491M, 1975, 368

Silver perchlorate and carbon tetrachloride in presence of a little hydrochloric acid produce trichloromethyl perchlorate, which explodes at 40°C.
See Trichloromethyl perchlorate

1,2-Diaminoethane
491M, 1975, 368
Dropwise addition of the amine to the salt led to an explosion (possibly initiated by heat liberated by complex formation).

Diethyl ether
1. Heim, F., *Angew. Chem.*, 1957, **69**, 274
After crystallisation from ether, the material exploded violently on crushing in a mortar. It had been considered stable previously, since it melts without decomposition [1].

Dimethyl sulfoxide
Ahrland, S. *et al., Acta. Chem. Scand. A*, 1974, **28**, 825
The crystalline complex solvated with 2DMSO explodes with extreme violence if rubbed or scratched.
See Dimethyl sulfoxide: Metal oxosalts

Ethanol   MRH 3.30/13
See Aromatic compounds, above

Other reactants
Yoshida, 1980, 81
MRH values for 20 combinations with oxidisable materials are given.

1,4-Oxathiane
Barnes, J. C. *et al., J. Chem. Soc. Pak.*, 1982, **4**, 103–113
The perchlorate forms complexes with 2, 3 or 4 mols of oxathiane which explode on heating.

Tetrachlorosilane, or Tetrabromosilane, or Titanium tetrachloride, and Diethyl ether
Schmeisser, M., *Angew. Chem.*, 1955, **67**, 499
Reaction gives explosive volatile organic perchlorates, probably ethyl perchlorate.
See ALKYL PERCHLORATES

Tetrasulfur tetraimide
See Tetrasulfurtetraimide–silver(I) perchlorate
See other METAL PERCHLORATES, OXIDANTS, SOLVATED OXOSALT INCIDENTS

0013. Silver fluoride
[7775-41-9]   **AgF**
$$AgF$$

Calcium hydride
See Calcium hydride: Silver halides

Non-metals
Mellor, 1941, Vol. 3, 389
Boron reacts explosively when ground with silver fluoride; silicon reacts violently.

Titanium
Mellor, 1941, Vol. 7, 20
Interaction at 320°C is incandescent.
See other METAL HALIDES, SILVER COMPOUNDS

0014. Silver difluoride
[7783-95-1] AgF_2

$$AgF_2$$

Boron, Water
Tulis, A. J. *et al., Proc. 7th Symp. Explos. Pyrotechnics*, 1971, **3**(4), 1–12
Mixtures of boron and silver difluoride function as detonators when contacted with water.

Dimethyl sulfoxide
See Iodine pentafluoride: Dimethyl sulfoxide

Hydrocarbons, or Water
Priest, H. F. *Inorg. Synth.*, 1950, **3**, 176
It reacts even more vigorously with most substances than does cobalt fluoride.
See other METAL HALIDES, SILVER COMPOUNDS

0015. Silver amide
[65235-79-2] AgH_2N

$$AgNH_2$$

Brauer, 1965, Vol. 2, 1045
Extraordinarily explosive when dry.
See Nitrogen triiodide–silver amide
See other *N*-METAL DERIVATIVES, SILVER COMPOUNDS

0016. Silver *N*-nitrosulfuric diamidate
[ ] $AgH_2N_3O_4S$

$$AgN(NO_2)SO_2NH_2$$

Sorbe, 1968, 120
The silver salt of the nitroamide is explosive.
See other *N*-METAL DERIVATIVES, *N*-NITRO COMPOUNDS, SILVER COMPOUNDS

0017. Silver phosphinate
[ ] AgH_2O_2P

$$AgP(H)(O)OH$$

Luchs, J. K., *Photog. Sci. Eng.*, 1966, **10**, 335

Explosive, but less sensitive than the azide or fulminate.
See other METAL PHOSPHINATES, SILVER COMPOUNDS

0018. Diamminesilver permanganate
[ ] $AgH_6MnN_2O_4$

$$[(H_3N)_2Ag]MnO_4$$

Pascal, 1960, Vol. 16, 1062
It may explode on impact or shock.
See other AMMINEMETAL OXOSALTS, SILVER COMPOUNDS

0019. Dihydrazinesilver nitrate
[31247-72-0] $AgH_8N_5O_3$

$$(H_4N_2)_2AgNO_3$$

Gall, H. *et al., Z. Anorg. Chem.*, 1932, **206**, 376
The salt explodes at $-1.5°C$.
See other AMMINEMETAL OXOSALTS, SILVER COMPOUNDS

0020. Silver iodate
[7783-97-3] $AgIO_3$

$$AgOIO_2$$

Metals   MRH Potassium 1.50/25, sodium 1.92/35
  See Potassium: Oxidants
     Sodium: Iodates

Other reactants
  Yoshida, 1980, 194
  MRH values for 16 combinations with oxidisable materials are given.

Tellurium
  Pascal, 1960, Vol. 13.2, 1961
  Interaction is violent.
  See other METAL HALOGENATES, SILVER COMPOUNDS

0021. Silver permanganate
[7783-98-4] $AgMnO_4$

$$AgMnO_4$$

Sulfuric acid
  491M, 1975, 369
  The moist salt exploded during drying over the concentrated acid in a vacuum
  desiccator, (presumably owing to formation of traces of manganese heptoxide
  from reaction with sulfuric acid vapour).
  See other OXIDANTS, SILVER COMPOUNDS

0022. Silver nitrate
[7761-88-8]   AgNO₃

AgNO₃

HCS 1980, 822

Acetaldehyde
Luchs, J. K., *Photog. Sci. Eng.*, 1966, **10**, 336
Aqueous silver nitrate reacts with acetaldehyde to give explosive silver fulminate.

Acetylene and derivatives
Mellor, 1946, Vol. 5, 854
Silver nitrate (or other soluble salt) reacts with acetylene in presence of ammonia to
form silver acetylide, a sensitive and powerful detonator when dry. In the absence
of ammonia, or when calcium acetylide is added to silver nitrate solution, explosive
double salts of silver acetylide and silver nitrate are produced. Mercury(I) acetylide
precipitates silver acetylide from the aqueous nitrate.
See 1,3-Butadiyne, and Buten-3-yne, both below
See METAL ACETYLIDES

Acrylonitrile
See Acrylonitrile: Silver nitrate

Aluminium
Laing, M., *J. Chem. Educ.*, 1994, **71**, 270
It is warned that a mixture of aluminium powder and silver nitrate is potentially
as dangerous as that with magnesium, both being capable of producing >8 kJ/g.
See Magnesium, Water; below

Ammonia   MRH 1.46/29
1. *MCA Case History No. 2116*
2. *CISHC Chem. Safety Summ.*, 1976, **47**, 31
3. MacWilliam, E. A. *et al.*, *Photogr. Sci. Eng.*, 1977, **21**, 221–224
A bottle containing Gomari tissue staining solution (ammoniacal silver nitrate)
prepared 2 weeks previously exploded when disturbed. The solution must be
prepared freshly each day, and discarded immediately after use with appropriate
precautions [1]. A large quantity of ammoniacal silver nitrate solution exploded
violently when disturbed by removing a glass rod [2]. However, it has now been
shown that neither the solid precipitated during addition of ammonia to the nitrate,
nor the redissolved complex, is sensitive to initiation by very severe shocks. This
was so for fresh or aged solutions. The solids produced by total evaporation at
95°C or higher would explode only at above 100 kg cm shock force. A pH value
above 12.9 is essential for separation of explosive precipitates, and this cannot be
attained by addition of ammonia alone [3].
See SILVER-CONTAINING EXPLOSIVES (reference 2)
Silver(I) oxide: Ammonia, etc.

9

Ammonia, Ethanol
MCA Case History No. 1733
A silvering solution exploded when disturbed. This is a particularly dangerous mixture, because both silver nitride and silver fulminate could be formed.
See Ethanol, below

Ammonia, Sodium carbonate
Vasbinder, H., *Pharm. Weekblad*, 1952, **87**, 861–865
A mixture of the components in gum arabic solution (marking ink) exploded when warmed.

Ammonia, Sodium hydroxide
1. Milligan, T. W. *et al., J. Org. Chem.*, 1962, **27**, 4663
2. *MCA Case History No. 1554*
3. Morse, J. R., *School Sci. Rev.*, 1955, **37**(131), 147
4. Baldwin, J., *School Sci. Rev.*, 1967, **48**(165), 586
5. MacWilliam, E. A. *et al., Photogr. Sci. Eng.*, 1977, **21**, 221–224
6. Anon., *Univ. Safety Assoc. Safety News*, 1977, (8), 15–16
During preparation of an oxidising agent on a larger scale than described [1], addition of warm sodium hydroxide solution to warm ammoniacal silver nitrate with stirring caused immediate precipitation of black silver nitride which exploded [2]. Similar incidents had been reported previously [3], including one where explosion appeared to be initiated by addition of Devarda's alloy (Al–Cu–Zn) [4]. The explosive species separates at pH values above 12.9, only produced when alkali is added to ammoniacal silver solutions, or when silver oxide is dissolved with ammonia [5]. The Sommer & Market reagent mixture used to identify cellulose derivatives led to a severe explosion [6].
See Silver nitride, *also* Ammonia: Silver compounds
See also SILVERING SOLUTIONS, TOLLENS' REAGENT

Arsenic
Mellor, 1941, Vol. 3, 470
A finely divided mixture with excess nitrate ignited when shaken out onto paper.

1,3-Butadiyne
See 1,3-Butadiyne: Silver nitrate

Buten-3-yne
See Buten-3-yne: Silver nitrate

Chlorine trifluoride
See Chlorine trifluoride: Metals, etc.

Chlorosulfuric acid
Mellor, 1941, Vol. 3, 470
Interaction is violent, nitrosulfuric acid being formed.

Disilver ketenide
See Disilver ketenide–silver nitrate

Ethanol   MRH 2.55/12
1. Tully, J. P., *Ind. Eng. Chem. (News Ed.)*, 1941, **19**, 250
2. Luchs, J. K., *Photog. Sci. Eng.*, 1966, **10**, 334
3. Garin, D. L. *et al., J. Chem. Educ.*, 1970, **47**, 741
4. Perrin, D. D. *et al., Chem. Brit.*, 1986, **22**, 1084; *Chem. Eng. News*, 1987, **65**(2), 2

Reclaimed silver nitrate crystals, damp with the alcohol used for washing, exploded violently when touched with a spatula, generating a strong smell of ethyl nitrate [1]. The explosion was attributed to formation of silver fulminate (which is produced on addition of ethanol to silver nitrate solutions). Ethyl nitrate may also have been involved. Alternatives to avoid ethanol washing of recovered silver nitrate are discussed [2], including use of 2-propanol [3]. Another case of explosion during filtration of silver nitrate purified by progressive dilution with ethanol of its aqueous solution has been reported. Initiation was by agitation of the slurry on a glass frit with a spatula [4].
See Silver fulminate

Magnesium, Water   MRH 6.94/30
1. Marsden, F., private comm., 1973
2. Lyness, D. J. *et al., School Sci. Rev.*, 1953, **35**(125), 139

An intimate mixture of dry powdered magnesium and silver nitrate may ignite explosively on contact with a drop of water [1,2].
See other REDOX REACTIONS

Non-metals   MRH Carbon 2.46/10, phosphorus (y) 3.89/18, sulfur 1.67/16
Mellor, 1941, Vol. 3, 469–473
Under a hammer blow, a mixture with charcoal ignites, while mixtures with phosphorus or sulfur explode, the latter violently.

Other reactants
Yoshida, 1980, 194
MRH values for 19 combinations with oxidisable materials are given.

Phosphine
Mellor, 1941, Vol. 3, 471
Rapid passage of gas into a conc. nitrate solution caused an explosion, or ignition of a slower gas stream. The explosion may have been caused by rapid oxidation of the precipitated silver phosphide derivative by the co-produced nitric acid or dinitrogen tetraoxide.

Phosphonium iodide
See Phosphonium iodide: Oxidants

Polymers
491M, 1975, 366

To assess suitability of plastic storage containers for distribution of silver nitrate, behaviour under fire exposure conditions of various polymers in contact with the salt was examined. All polymers tested burned vigorously.

Silver acetylide
See Silver acetylide–silver nitrate

Thiophene
Southern, T., private communication, 1990
A black solid is produced from these two reagents under influence of ultrasound (but not otherwise) which explodes violently on warming. It is apparently not silver acetylide.

Titanium
Shanley, E. S., *Chem. Eng. News*, 1990, **68**(16), 2
A titanium-containing sludge from a nitric acid bath was separated, before completely dry it exploded, killing a workman. Investigation showed the dry sludge to be a powerful explosive sensitive to heat, friction and impact, composed of about 60:40 silver nitrate:titanium.
See Titanium: Nitric acid
See other METAL NITRATES, SILVER COMPOUNDS

0023. Silver azide
[13863-88-2]

AgN$_3$

$$AgN_3$$

1. Mellor, 1940, Vol. 8, 349; 1967, Vol. 8, Suppl. 2, 47
2. Gray, P. *et al., Chem. & Ind.*, 1955, 1255
3. Kabanov, A. A. *et al., Russ. Chem. Rev.*, 1975, **44**, 538
4. Ryabykh, S. M. *et al., Chem. Abs.*, 1984, **100**, 194549

As a heavy metal azide, it is considerably endothermic ($\Delta H_f^\circ +279.5$ kJ/mol, 1.86 kJ/g). While pure silver azide explodes at 340°C [1], the presence of impurities may cause explosion at 270°C. It is also impact-sensitive and explosions are usually violent [2]. Its use as a detonator has been proposed. Application of an electric field to crystals of the azide will detonate them, at down to −100°C [3], and it may be initiated by irradiation with electron pulses of nanosecond duration [4].
See other CATALYTIC IMPURITY INCIDENTS, IRRADIATION DECOMPOSITION INCIDENTS

Ammonia
Mellor, 1940, Vol. 8, 349
Solutions in aqueous ammonia become explosive around 100°C.

Chlorine azide
See Chlorine azide: Ammonia, etc.

12

Halogens
Mellor, 1940, Vol. 8, 336
Silver azide, itself a sensitive compound, is converted by ethereal iodine into the less stable and explosive compound, iodine azide. Similarly, contact with nitrogen-diluted bromine vapour gives bromine azide, often causing explosions.
See Silver azide chloride

Metal oxides, or Metal sulfides
Kurochin, E. S. *et al., Chem. Abs.*, 1974, **83**, 201390
Pure silver azide explodes at 340°C, but presence of below 10% of copper(I) or (II) oxides or sulfides, copper(I) selenide or bismuth(III) sulfide reduces the detonation temperature to 235°C. Concentrations of 10% of copper(II) oxide, copper(I) selenide or sulfide further reduced it to 200, 190 and 170°C, respectively.

Photosensitising dyes
Aleksandrov, E. *et al., Chem. Abs.*, 1974, **81**, 31755
In a study of dye-sensitised silver azide, it was found that many dyes caused explosions in the initial stages.

Sulfur dioxide
Mellor, 1940, Vol. 8, 349
Mixtures of the slightly soluble azide with liquid sulfur dioxide became explosive at elevated temperatures.
See other ENDOTHERMIC COMPOUNDS, METAL AZIDES, SILVER COMPOUNDS

0024. Silver trisulfurpentanitridate

[ ]   AgN$_5$S$_3$

(Structure unknown)

See 1,3,5-Trichlorotrithiahexahydro-1,3,5-triazine: Ammonia
See other N–S COMPOUNDS, SILVER COMPOUNDS

0025. Silver(II) oxide

[1301-96-8]   AgO

AgO

Hydrogen sulfide
See Hydrogen sulfide: Metal oxides
See other METAL OXIDES, SILVER COMPOUNDS

0026. Silver sulfide

[21548-73-2]   AgS

AgS

Potassium chlorate
See Potassium chlorate: Metal sulfides
See other METAL SULFIDES, SILVER COMPOUNDS

0027. Silver hexahydrohexaborate(2−)

[ ] $Ag_2B_6H_6$

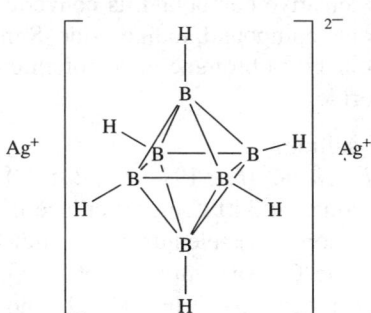

Bailar, 1973, Vol. 1, 808
It is a detonable salt.
See other SILVER COMPOUNDS, COMPLEX HYDRIDES

0028. Silver perchlorylamide
[25870-02-4] Ag_2ClNO_3

$$Ag_2NClO_3$$

See entry PERCHLORYLAMIDE SALTS *See other* SILVER COMPOUNDS

0029. Tetrasulfurtetraimide–silver perchlorate
[64867-41-0] $Ag_2Cl_2H_4N_4O_8S_4$

Nabi, S. N., *J. Chem. Soc., Dalton Trans.*, 1977, 1156
The complex detonates violently at 120°C.
See related METAL PERCHLORATES, N–S COMPOUNDS
See other SILVER COMPOUNDS

0030. Silver imide
[ ] Ag_2HN

$$Ag_2NH$$

Bailar, 1973, Vol. 3, 101
It explodes very violently when dry.
But see Silver(I) oxide: Ammonia (reference 2)
See other N-METAL DERIVATIVES, SILVER COMPOUNDS

0031. Silver hyponitrite
[7784-04-5] $Ag_2N_2O_2$

AgON:NOAg

See Hyponitrous acid
See other METAL OXONON-METALLATES, SILVER COMPOUNDS

0032. Silver(I) oxide
[20667-12-3] Ag_2O

Ag_2O

Aluminium
See Copper(II) oxide: Metals

Ammonia
1. Vasbinder, H., *Pharm. Weekblad*, 1952, **87**, 861–865
2. MacWilliam, E. A., *Photogr. Sci. Eng.*, 1977, **21**, 221–224
The clear solution, obtained by centrifuging a solution of the oxide in aqueous ammonia which had been treated with silver nitrate until precipitation started, exploded on two occasions after 10–14 days' storage in closed bottles in the dark. This was ascribed to slow precipitation of amorphous silver imide, which is very explosive even when wet [1]. When silver oxide is dissolved in ammonia solution, an extremely explosive precipitate (probably Ag_3N_4) will separate. The explosive behaviour is completely inhibited by presence of colloids or ammonium salts (acetate, carbonate, citrate or oxalate). Substitution of methylamine for ammonia does not give explosive materials [2].
See Silver nitrate: Ammonia, etc., *also* Ammonia: Silver compounds

Ammonia or Hydrazine, Ethanol
Silver oxide and ammonia or hydrazine slowly form explosive silver nitride and, in presence of alcohol, silver fulminate may also be produced.
See entries SILVER-CONTAINING EXPLOSIVES, SILVERING SOLUTIONS

Boron trifluoride etherate, Nitromethane
See Silver tetrafluoroborate

Carbon monoxide
Mellor, 1941, Vol. 3, 377
Carbon monoxide is exothermically oxidised over silver oxide, and the temperature may attain 300°C.

Chlorine, Ethylene
See Ethylene: Chlorine

Dichloromethylsilane
See Dichloromethylsilane: Oxidants

Hydrogen sulfide
See Hydrogen sulfide: Metal oxides

15

Magnesium
Mellor, 1941, Vol. 3, 378
Oxidation of magnesium proceeds explosively when warmed with silver oxide in a sealed tube.

Metal sulfides
Mellor, 1941, Vol. 3, 376
Mixtures with gold(III) sulfide, antimony sulfide or mercury(II) sulfide ignite on grinding.

Nitroalkanes
See NITROALKANES: metal oxides

Non-metals
Mellor, 1941, Vol. 3, 376–377
Selenium, sulfur or phosphorus ignite on grinding with the oxide.

Potassium–sodium alloy
See Potassium–sodium alloy: Metal oxides

Seleninyl chloride
See Seleninyl chloride: Metal oxides

Selenium disulfide
Mellor, 1941, Vol. 3, 377
A mixture may ignite under impact.
See other METAL OXIDES, SILVER COMPOUNDS

0033. Silver peroxide
[25455-73-6] Ag_2O_2
$$AgOOAg$$

Poly(isobutene)
Mellinger, T., *Arbeitsschutz*, 1972, 248
Mixtures of silver peroxide with 1% of polyisobutene exploded on 3 separate occasions. Use of a halogenated polymer was safe.
See other METAL PEROXIDES

0034. Silver osmate
[ ] Ag_2O_4Os
$$Ag_2OsO_4$$

Sorbe, 1968, 126
Explodes on impact or heating.
See other HEAVY METAL DERIVATIVES, SILVER COMPOUNDS

16

0035. Silver sulfide
[21548-73-2] Ag_2S

AgSAg

Potassium chlorate
  See Potassium chlorate: Metal sulfides
  See other METAL SULFIDES, SILVER COMPOUNDS

0036. Disilver pentatin undecaoxide
[ ] $Ag_2Sn_5O_{11}$

(Structure unknown)

Mellor, 1941, Vol. 7, 418
The compound 'silver beta-stannate' is formed by long contact between solutions
of silver and stannous nitrates, and loses water on heating, then decomposing
explosively.
  See other METAL OXIDES, SILVER COMPOUNDS

0037. Silver peroxochromate
[ ] Ag_3CrO_8

Sulfuric acid
  Riesenfeld, E. H. *et al., Ber.*, 1914, **47**, 548
  In attempts to prepare 'perchromic acid', a mixture of silver (or barium) peroxo-
  chromate and 50% sulfuric acid prepared at $-80°C$ reacted explosively on warming
  to about $-30°C$.
  See other PEROXOACID SALTS, SILVER COMPOUNDS

0038. Silver nitride
[20737-02-4] Ag_3N

Ag_3N

Hahn, H. *et al., Z. Anorg. Chem.*, 1949, **258**, 77
Very sensitive to contact with hard objects, exploding when moist. An extremely
sensitive explosive when dry, initiable by friction, impact or heating. The impure
product produced by allowing ammoniacal silver oxide solution to stand seems
even more sensitive, often exploding spontaneously in suspension.
  See Silver chloride: Ammonia
  See also SILVER-CONTAINING EXPLOSIVES, SILVERING SOLUTIONS, TOLLENS'
  REAGENT
  See other CATALYTIC IMPURITY INCIDENTS, *N*-METAL DERIVATIVES, SILVER
  COMPOUNDS

0039. Silver 2,4,6-tris(dioxoselena)perhydrotriazine-1,3,5-triide ('Silver triselenimidate')

[ ] $Ag_3N_3O_6Se_3$

See Selenium difluoride dioxide: Ammonia
See other N-METAL DERIVATIVES, SILVER COMPOUNDS

0040. Trisilver tetranitride

[ ] Ag_3N_4

$$Ag_3N_4$$

See Silver(I) oxide: Ammonia (reference 2)
See other N-METAL DERIVATIVES, SILVER COMPOUNDS

0041. Tetrasilver orthodiamidophosphate

[ ] $Ag_4H_3N_2O_3P$

$$(AgO)_3P(NH_2)NHAg$$

Sulfuric acid
  Mellor, 1940, Vol. 8, 705
  Ignites with sulfuric acid.
See other N-METAL DERIVATIVES, SILVER COMPOUNDS

0042. Tetrasilver diimidotriphosphate

[ ] $Ag_4H_3N_2O_8P_3$

$$HOP:O(OAg)NHP:O(OAg)N(Ag)P:O(OAg)OH$$

Alone, or Sulfuric acid
  Mellor, 1940, Vol. 8, 705; 1971, Vol. 8, Suppl. 3, 787
  The dry material explodes on heating, and ignites in contact with sulfuric acid.
  The molecule contains one N−Ag bond.
See other N-METAL DERIVATIVES, SILVER COMPOUNDS

0043. Tetrasilver diimidodioxosulfate

[ ] $Ag_4N_2O_2S$

$$(AgN:)_2S(OAg)_2$$

  Nachbauer, E. *et al., Angew. Chem. (Intern. Ed.)*, 1973, **12**, 339

18

The dry salt explodes on friction or impact.
See other N-METAL DERIVATIVES, SILVER COMPOUNDS

0044. Pentasilver orthodiamidophosphate
[ ] $Ag_5H_2N_2O_3P$
$$(AgO)_3P(NHAg)_2$$

Alone, or Sulfuric acid
   Stokes, H. N., *Amer. Chem. J.*, 1895, **17**, 275
   Explodes on heating, friction or contact with sulfuric acid.
   See other N-METAL DERIVATIVES, SILVER COMPOUNDS

0045. Pentasilver diimidotriphosphate
[ ] $Ag_5H_2N_2O_8P_3$
$$HOP:O(OAg)N(Ag)P:O(OAg)N(Ag)P:O(OAg)OH$$

Alone, or Sulfuric acid
   Mellor, 1940, Vol. 8, 705; 1971, Vol. 8, Suppl. 3, 787
   The salt is explosive and may readily be initiated by friction, heat or contact with
   sulfuric acid. The molecule contains two N−Ag bonds.

0046. Pentasilver diamidophosphate
[ ] $Ag_5N_2O_2P$
$$AgOP:O(NAg_2)_2$$

   1. Bailar, 1973, Vol. 8, 455
   2. Mellor, 1940, Vol. 8, 705
   The salt contains 85% silver and four N−Ag bonds [1], and detonates readily on
   friction, heating or contact with sulfuric acid [2].
   See other N-METAL DERIVATIVES, SILVER COMPOUNDS

0047. Heptasilver nitrate octaoxide
[12258-22-9] Ag_7NO_{11}
$$(Ag_3O_4)_2.AgNO_3$$

Alone, or Sulfides, or Non-metals
   Mellor, 1941, Vol. 3, 483−485
   The crystalline product produced by electrolytic oxidation of silver nitrate (and
   possibly as formulated) detonates feebly at 110°C. Mixtures with phosphorus and
   sulfur explode on impact, hydrogen sulfide ignites on contact and antimony trisulfide ignites when ground with the salt.
   See related METAL NITRATES *See other* SILVER COMPOUNDS

0048. Aluminium
[7429-90-5]   Al
$$Al$$

HCS 1980, 135 (powder)

1. *Haz. Chem. Data*, 1975, 44
2. *Dust Explosion Prevention: Aluminum Powder*, NFPA Standard Codes 65, 651, both 1987
3. Popov, E. I. *et al., Chem. Abs.*, 1975, **82**, 61411
4. Ida, K. *et al., Japan Kokai*, 1975, 75 13 233
5. Steffens, H. D. *et al., Chem. Abs.*, 1979, **90**, 209354
6. Scherbakov, V. K. *et al., Chem. Abs.*, 1979, **91**, 76322
7. Barton, J. A. *et al., Chem. Brit.*, 1986, **22**, 647–650
8. May, D. C. *et al., J. Haz. Mat.*, 1987, **17**(1), 61–88
9. *See entry* DUST EXPLOSION INCIDENTS (reference 22)

Finely divided aluminium powder or dust forms highly explosive dispersions in air [1], and all aspects of prevention of aluminium dust explosions are covered in 2 recent US National Fire Codes [2]. The effects on ignition properties of impurities introduced by recycled metal used to prepare dust were studied [3]. Pyrophoricity is eliminated by surface coating aluminium powder with polystyrene [4]. Explosion hazards involved in arc and flame spraying of the powder are analysed and discussed [5], and the effect of surface oxide layers on flammability was studied [6]. The causes of a severe explosion in 1983 in a plant producing fine aluminium powder are analysed, and improvements in safety practices discussed [7]. A number of fires and explosions involving aluminium dust arising from grinding, polishing and buffing operations are discussed, and precautions detailed [8]. Atomised and flake aluminium powders attain maximum explosion pressures of 5.7 and 8.6 bar, respectively, both with maximum rates of pressure rise exceeding 1.36 kbar/s [9].

Air, Hydrocarbons

Lamnevik, S., *Chem. Abs.*, 1982, **96**, 202034
Presence of dispersed aluminium powder in propane–air or butane–air mixtures enhances the detonative properties of stoichiometric mixtures.

Air, Water

1. Turetzky, M., private comm., 1987
2. *MARS Database*, 1998, short report 028

Fine aluminium dust and chippings from precision surface machining operations were air-transported to a cyclone collection system with outlet bag filters for the dust, and the cyclone contents were discharged into metal scrap bins. Sparks from a portable grinding machine ignited aluminium dust on a maintenance platform, and use of a carbon dioxide extinguisher on the fire transferred it into a metal scrap bin below. Heat from this larger fire activated the automatic water sprinkler system, and contact of the water (89% oxygen) with the burning metal liberated hydrogen, which, after mixing with air, exploded. This primary explosion created an aluminium dust cloud which exploded forcefully, creating larger dust clouds and eventually involving the contents of the other scrap bins and 2 of the dust filter bags, leading to at least 4 separate tertiary explosions in all. Structural damage to adjacent buildings and vehicles was extensive. The dry collection system was replaced by a wet one [1].

A different scenario involving these three occurred when lightning struck an aluminium foundry. It is supposed that this dispersed molten metal droplets in air, which then exploded with the estimated force of 200 kg TNT, causing damage which allowed remaining molten metal to fall into the wet casting pit, producing a second explosion of half the power of the first [2].

Aluminium halides, Carbon oxides
Guntz, A. *et al., Compt. rend.*, 1897, **124**, 187–190
Aluminium powder burns when heated in carbon dioxide, and presence of aluminium chloride or aluminium iodide vapour in carbon monoxide or carbon dioxide accelerates the reaction to incandescence.

Ammonium nitrate   MRH 8.70/75
Mellor, 1946, Vol. 5, 219
Mixtures with the powdered metal are used as an explosive, sometimes with the addition of carbon or hydrocarbons, or other oxidants.
See Ammonium nitrate: Metals

Ammonium nitrate, Calcium nitrate, Formamide
See Calcium nitrate: Aluminium, etc.

Ammonium peroxodisulfate
See Ammonium peroxodisulfate: Aluminium

Antimony or Arsenic
Matignon, C., *Compt. rend.*, 1900, **130**, 1393
Powdered aluminium reacts violently on heating with antimony or arsenic.

Antimony trichloride
Matignon, C., *Compt. rend.*, 1900, **130**, 1393
The metal powder ignites in antimony trichloride vapour.

Arsenic trioxide, Sodium arsenate, Sodium hydroxide
MCA Case History No.1832
An aluminium ladder was used (instead of the usual wooden one) to gain access to a tank containing the alkaline arsenical mixture. Hydrogen produced by alkaline attack on the ladder generated arsine, which poisoned the three workers involved.

Barium peroxide
See Barium peroxide: Metals

Bismuth
Mellor, 1947, Vol. 9, 626
The finely divided mixture of metals produced by hydrogen reduction of co-precipitated bismuth and aluminium hydroxides is pyrophoric.

Butanol

Luberoff, B. J., private comm., 1964

Butanol, used as a solvent in an autoclave preparation at around 100°C, severely attacked the aluminium gasket, liberating hydrogen which caused a sharp rise in pressure. Other alcohols would behave similarly, forming the aluminium alkoxide.
See other CORROSION INCIDENTS

Calcium oxide, Chromium oxide, Sodium chlorate   MRH 10.7/67

Nolan, 1983, Case History 174

In a thermite process to produce chromium metal, the mechanically-mixed ingredients were ignited in a large crucible and the reaction proceeded smoothly. When the mixer broke down, manual mixing was used but gave poorer dispersion of the constituents. An explosion after ignition is attributed to a high local concentration of sodium chlorate and aluminium powder in the mixture.
See other THERMITE REACTIONS

Carbon, Chlorine trifluoride

See Chlorine trifluoride: Metals, etc.

Carbon dioxide

See Carbon dioxide: Metals

Carbon dioxide, Sodium peroxide

See Sodium peroxide: Metals, etc.

Carbon disulfide

Matignon, C., *Compt. rend.*, 1900, **130**, 1391

Aluminium powder ignites in carbon disulfide vapour.

Carbon tetrachloride, Methanol

See Methanol: Carbon tetrachloride, Metals

Catalytic metals

1. *491M*, 1975, 28
2. Kirk-Othmer (3rd ed.), 1982, Vol. 19, 495

At the m.p. of aluminium (600°C) an aluminium-sheathed palladium thermocouple formed an alloy with a flash and an exotherm to 2800°C [1]. The use of thin layers of palladium or platinum on aluminium foil or wire as igniters derives from the intense heat of alloy formation, which is sufficient to melt the intermetallic compounds [2].

Chloroformamidinium nitrate

See Chloroformamidinium nitrate (reference 2)

Copper, Sulfur

Donohue, P. C., US Pat. 3 932 291, 1973

During preparation of aluminium copper(I) disulfide from the elements in an air-free silica tube at 900–1000°C, initial heating must be slow to prevent explosion of the tube by internal pressure of unreacted sulfur vapour.

Copper(I) oxide

Haws, L. D. *et al., Proc. 6th Int. Pyrotech. Semin.*, 1978, 209–222

Technical aspects of the use of aluminium–copper(I) oxide thermite devices are detailed.

See Copper(I) oxide: Aluminium

Copper(II) oxide   MRH 3.29/82

1. Anon., *Chem. Age.*, 1932, **27**, 23
2. Scamaton, W. B. *et al., Chem. Abs.*, 1980, **92**, 8484
3. Peterson, W. S., *Light Met.* (Warrendale, PA), 1987, 699–701

A mixture of aluminium powder and hot copper oxide exploded violently during mixing with a steel shovel on an iron plate. The frictional mixing initiated the thermite-like mixture [1]. Such mixtures are now used in electro-explosive devices [2]. Two cases of violent explosions after adding scrap copper to molten aluminium are discussed. In both cases, when some undissolved copper with adhering aluminium and oxide dross was removed from the furnace, the explosions occurred outside the melting furnace [3].

See Metal oxides, etc., below

Diborane

See Diborane: Metals

Dichloromethane

Piotrowski, A. M. *et al., J. Org. Chem.*, 1988, **53**, 2829–2835

Conditions are described for the safe reaction of aluminium powder with dichloromethane to give bis(dichloroalumino)methane in high yield. Derivatives of this compound are effective methyleneating agents for ketones.

See Halocarbons, below

Diethyl ether

Murdock, T. O., Ph.D. Thesis, Univ. N. Dakota, 1977 (*Diss. Abs. B*, 1978, **39**, 1291)

The aluminium–solvent slurry produced by metal atom/solvent co-condensation at −196°C is so reactive that oxygen is abstracted from the solvent ether as the mixture is allowed to melt. Hydrocarbon solvents are more suitable (but halocarbon solvents would react explosively).

Disulfur dibromide

See Disulfur dibromide: Metals

Drain cleaner, Cola

1. Utterbuck, P., Internet, 1997
2. Editor's comments

A student mixed aluminium foil and drain cleaner in a soft drink bottle, which started emitting gas. Another student carried the bottle outside and was claimed to have been overcome by the toxic fumes [1]. Most drain cleaners are alkalis, so that aluminium will dissolve to produce hydrogen. The bleach that it is is suggested may have been present will produce no toxic fumes in alkali, and one would be surprised to find arsenic or antimony compounds present. If the collapse was not

purely hysterical, the remaining, though remote, possibility would be phosphine. The soft drink the bottle had contained was one of the many perhaps best described as impure dilute phosphoric acid [2].
See Arsenic trioxide, etc., above

Explosives
1. Stettbacher, A., *Chem. Abs.*, 1944, **38**, 4445$_4$
2. Muraour, H., *Chem. Abs.*, 1944, **38**, 4445$_7$
The addition of substantial amounts (up to 32%) of aluminium powder to conventional explosives enhances the energy release by up to 100% [1], involving high temperature reduction of liberated carbon dioxide and water by the metal [2].

Formic acid
Matignon, C., *Compt. rend.*, 1900, **130**, 1392
The metal reduces the acid (itself a reducant) with incandescence
1. Anon., *Angew. Chem.*, 1950, **62**, 584
2. Anon., *Chem. Age*, 1950, **63**, 155
3. Anon., *Chem. Eng. News*, 1954, **32**, 258
4. *Pot. Incid. Rep.*, ASESB, 1968, **39**
5. Anon., *Chem. Eng. News*, 1955, **33**, 942
6. Eiseman, B. J., *J. Amer. Soc. Htg. Refr. Air Condg. Eng.*, 1963, **5**, 63
7. Eiseman, B. J., *Chem. Eng. News*, 1961, **39**(27), 44
8. Laccabue, J. R., *Fluorolube–Aluminium Detonation Point: Report 7E.1500*, San Diego, Gen. Dynamics, 1958
9. Atwell, V. J., *Chem. Eng. News*, 1954, **32**, 1824
10. ICI Mond Div., private comm., 1973
11. Anon., *Ind. Acc. Prev. Bull. RoSPA*, 1953, **21**, 60
12. Wendon, G. W., private comm., 1973
13. Coffee, R. D., *Loss Prev.*, 1971, **5**, 113
14. Schwab, R. F., ibid., 113
15. Corley, R., ibid., 114
16. Heinrich, H. J., *Arbeitsschutz*, 1966, 156–157
17. Lamoroux, A. *et al.*, *Mém. Poudres*, 1957, **39**, 435–445
18. Hamstead, A. C. *et al.*, *Corrosion*, 1958, **14**, 189t–190t
19. Hartmann, I., *Ind. Eng. Chem.*, 1948, **40**, 756
20. *MCA Case History No. 2160*
21. Arias, A., *Ind. Eng. Chem., Prod. Res. Dev.*, 1976, **15**, 150
22. Wendon, G. W., *Ind. Eng. Chem., Prod. Res. Dev.*, 1977, **16**, 112
23. Arias, A., *Ind. Eng. Chem., Prod. Res. Dev.*, 1977, **16**, 112
24. Archer, W. L., Paper 47, 1–11, presented at *Corrosion/78*, Houston, March 6–10, 1978
25. Anon. *Univ. Safety Assoc. Newsletter*, 1982, (16), 5
26. Stull, 1977, 22
27. Cardillo, P., *et al.*, *J. Loss Prevention*, 1992, **5**(2), 81
MRH Chloromethane 2.46/85, bromomethane 1.33/91, chloroform 3.84/82, carbon tetrachloride 4.15/91, carbon tetrafluoride 6.48/71, dichlorodifluoromethane

5.39/77, trichlorofluoromethane 4.76/79, trichloroethylene 4.10/83, trichlorofluoroethane 4.98/77, 1,2-dichloropropene 3.30/86, o-dichlorobenzene 2.67/89

Heating aluminium powder with carbon tetrachloride, chloromethane or carbon tetrachloride–chloroform mixtures in closed systems to 152°C may cause an explosion, particularly if traces of aluminium chloride are present [1]. A mixture of carbon tetrachloride and aluminium powder exploded during ball-milling [2], and it was later shown that heavy impact would detonate the mixture [3]. Mixtures with fluorotrichloroethane and with trichlorotrifluoroethane will flash or spark on heavy impact [4]. A virtually unvented aluminium tank containing a 4:1:2 mixture of o-dichlorobenzene, 1,2-dichloroethane and 1,2-dichloropropane exploded violently 7 days after filling. This was attributed to formation of aluminium chloride which catalysed further accelerating attack on the aluminium tank [5]. An analysis of the likely course of the Friedel-Crafts reaction and calculation of the likely heat release (29.1 kJ/mol) has been published [26].

In a dichlorodifluoromethane system, frictional wear exposed fresh metal surfaces on an aluminium compressor impellor, causing an exothermic reaction which melted much of the impellor. Later tests showed similar results, decreasing in order of intensity, with: tetrafluoromethane; chlorodifluoromethane; bromotrifluoromethane; dichlorodifluoromethane; 1,2-difluorotetrafluoroethane; 1,1,2-trichlorotrifluoroethane [6]. In similar tests, molten aluminium dropped into liquid dichlorodifluoromethane burned incandescently below the liquid [7]. Aluminium bearing surfaces under load react explosively with polytrifluoroethylene greases or oils. The inactive oxide film will be removed from the metal by friction, and hot spots will initiate reaction [8]. An attempt to scale up the methylation of 2-methylpropane with chloromethane in presence of aluminium chloride and aluminium went out of control and detonated, destroying the autoclave. The preparation had been done on a smaller scale on 20 previous occasions without incident [9].

Violent decomposition, with evolution of hydrogen chloride, may occur when 1,1,1-trichloroethane comes into contact with aluminium or its alloys with magnesium [10]. Aluminium-dusty overalls were cleaned by immersion in trichloroethylene. During subsequent drying, violent ignition occurred. This was attributed to presence of free hydrogen chloride in the solvent, which reacted to produce aluminium chloride [10]. This is known to catalyse polymerisation of trichloroethylene, producing more hydrogen chloride and heat. The reaction is self-accelerating and can develop a temperature of 1350°C [11]. Trichloroethylene cleaning baths must be kept neutral with sodium carbonate, and free of aluminium dust. Halocarbon solvents now available with added stabilisers (probably amines) show a reduced tendency to react with aluminium powder [12]

Aluminium powder undergoes an exothermic and uncontrollable reaction with dichloromethane above 95°C under appropriate pressure [13]. Several cases of violent reaction between aluminium and trichloroethylene or tetrachloroethylene in vapour degreasers have been noted [14]. Chloromethane in liquefied storage diffused 70 m along a nitrogen inerting line into the pressure regulator. Interaction with aluminium components of the regulator formed alkylaluminium compounds which ignited when the regulator was dismantled [15]. An explosion in an aluminium degreasing plant using tetrachloroethylene was attributed to overheating of residues on the heating coils. Subsequent tests showed that simultaneous presence of water and aluminium chloride in an aluminium powder–tetrachloroethylene mixture lowered the initiation temperature to below 250°C. Presence of cutting oils reduced it below 150°C, and a temperature of 300°C was reached within 100 min [16]

Reaction of aluminium powder with hexachloroethane in alcohol is not initially violent, but may become so [17]. An aluminium transfer pipe failed after a few hours' service carrying refined 1,2-dichloropropane in warm weather. The corrosive attack was simulated and studied under laboratory conditions [18]. In dichlorodifluoromethane vapour, aluminium dust ignited at 580°C, and suspensions of the dust in the vapour gave strong explosions when sparked [19]. A fire occurred at a liquid outlet from a 40 m³ mild steel tanker of chloromethane. This was traced to the presence of trimethylaluminium produced by interaction of chloromethane and (unsuspected) aluminium baffle plates in the tanker [20]. A proposal to prepare pure aluminium chloride by ball-milling aluminium in carbon tetrachloride [21] was criticised [22] as potentially hazardous. Possible modifications (use of inert solvents, continuously fed mills, etc.) to improve the procedure were suggested [23]. The reaction mechanism of attack and inhibition in aluminium–1,1,1-trichloroethane systems has been investigated [24]. An attempt to clean a motor assembly containing an aluminium alloy gearwheel by soaking overnight in 1,1,1-trichloroethane led to gross degradation of the assembly [25].

A tabulation of theoretically hazardous halocarbon/aluminium combinations, and estimates of safe dilutions, calculated by the CHETAH program is reported [27].
See Bromomethane: Metals
See Dichloromethane, above
See COMPUTATION OF REACTIVE CHEMICAL HAZARDS, *See other* CORROSION INCIDENTS, METAL–HALOCARBON INCIDENTS

Halogens   MRH Bromine 1.96/90, chlorine 5.23/80
1. Mellor, 1946, Vol. 2, 92, 135; Vol. 5, 209
2. Azmathulla, S. *et al., J. Chem. Educ.*, 1955, **32**, 447; *School Sci. Rev.*, 1956, **38**(134), 107
3. Hammerton, C. M., ibid, 1957, **38**(136), 459
Aluminium powder ignites in chlorine without heating, and foil reacts vigorously with liquid bromine at 15°C, and incandesces on warming in the vapour [1]. The metal and iodine react violently in the presence of water, either as liquid, vapour or that present in hydrated salts [2]. Moistening a powdered mixture causes incandescence and will initiate a thermite mixture [3].
See Iodine: Aluminium, Diethyl ether

Hydrochloric acid or Hydrofluoric acid
Kirk-Othmer, 1963, Vol. 1, 952
The metal is attacked violently by the aqueous acids.

Hydrogen chloride   MRH 3.05/80
Batty, G. F., private comm., 1972

26

Erroneous use of aluminium instead of alumina pellets in a hydrogen chloride purification reactor caused a vigorous exothermic reaction which distorted the steel reactor shell.

Interhalogens
See Bromine pentafluoride: Acids, etc.
    Chlorine fluoride: Aluminium
    Iodine chloride: Metals
    Iodine heptafluoride: Metals
    Iodine pentafluoride: Metals

Iron, Water
Chen, W. Y. et al., Ind. Eng. Chem., 1955, 47(7), 32A
A sludge of aluminium dust (containing iron and sand) removed from castings in water was found, during summer weather, to undergo sudden exotherms to 95°C with hydrogen evolution. Similar effects with aluminium-sprayed steel plates exposed to water were attributed to electrolytic action, as addition of iron filings to a slurry of aluminium in water caused self heating and hydrogen evolution to occur.
See Water, below
See other CORROSION INCIDENTS

Mercury(II) salts
Woelfel, W. C., J. Chem. Educ., 1967, 44, 484
Aluminium foil is unsuitable as a packing material in contact with mercury(II) salts in presence of moisture, when vigorous amalgamation ensues.
See Aluminium amalgam (reference 2)
See other CORROSION INCIDENTS

Metal nitrates, Potassium perchlorate, Water
Johansson, S. R. et al., Chem. Abs., 1973, 78, 18435
A pyrotechnic mixture of aluminium powder with potassium perchlorate, barium nitrate, potassium nitrate and water exploded after 24 h storage under water. Tests revealed the exothermic interaction of finely divided aluminium with nitrate and water to produce ammonia and aluminium hydroxide. Under the conditions prevailing in the stored mixture, the reaction would be expected to accelerate, finally involving the perchlorate as oxidant and causing ignition of the mixture.

Metal nitrates, Sulfur, Water
Anon., Chem. Eng. News, 1954, 32, 258
Aluminium powder, barium nitrate, potassium nitrate, sulfur and vegetable adhesives, mixed to a paste with water, exploded on 2 occasions. Laboratory investigation showed initial interaction of water and aluminium to produce hydrogen. It was supposed that nascent hydrogen reduced the nitrates present, increasing the alkalinity and thence the rate of attack on aluminium, the reaction becoming self-accelerating. Cause of ignition was unknown. Other examples of interaction of aluminium with water are known.
See Iron, Water, above; Water, below
See other CORROSION INCIDENTS

Metal oxides or Oxosalts or Sulfides
1. Mellor, 1946, Vol. 5, 217
2. Price, D. J. *et al., Chem. Met. Eng.*, 1923, **29**, 878
MRH Chromium trioxide 6.15/79, iron(II) oxide 3.01/80, iron(II)(III) oxide 3.38/76, calcium sulfate 5.73/65, sodium carbonate 2.92/67, sodium nitrate 8.32/65, sodium sulfate 5.27/66

Many metal oxo-compounds (nitrates, oxides and particularly sulfates) and sulfides are reduced violently or explosively (i.e. undergo 'thermite' reaction) on heating an intimate mixture with aluminium powder to a suitably high temperature to initiate the reaction. Contact of massive aluminium with molten salts may give explosions [1]. Application of sodium carbonate to molten (red hot) aluminium caused an explosion [2].
See Sodium sulfate, below
See Iron(III) oxide: Aluminium, *See other* MOLTEN SALT BATHS, THERMITE REACTIONS

Methanol
Médard, L. *Mém. Poudres*, 1951, **33**, 490–503 (Engl. translation HSE 11270, 1986 available from HSE/LIS)
The explosive nature of mixtures of aluminum or magnesium with methanol or water is detailed.
See also Magnesium: Methanol

Niobium oxide, Sulfur
491M, 1975, 28
A mixture caused a serious fire.
See Metal oxides, etc., above

Nitro compounds, Water
Hajek, V. *et al., Research*, 1951, **4**, 186–191
Dry mixtures of picric acid and aluminium powder are inert, but addition of water causes ignition after a delay dependent upon the quantity added. Other nitro compounds and nitrates are discussed in this context.

Non-metal halides   MRH Phosphorus pentachloride 2.76/83
1. Matignon, C., *Compt. rend.*, 1900, **130**, 1393
2. Lenher, V. *et al., J. Amer. Chem. Soc.*, 1926, **48**, 1553
3. Berger, E., *Compt. rend.*, 1920, **170**, 29
Powdered aluminium ignites in the vapour of arsenic trichloride or sulfur dichloride, and incandesces in phosphorus trichloride vapour [1]. Above 80°C, aluminium reacts incandescently with diselenium dichloride [2]. The powder ignites in contact with phosphorus pentachloride [3].

Non-metals
Matignon, C., *Compt. rend.*, 1900, **130**, 1393–1394
Powdered aluminium reacts violently with phosphorus, sulfur or selenium, and a mixture of powdered metal with red phosphorus exploded when severely shocked.

Oleic acid
 de Ment, J., *J. Chem. Educ.*, 1956, **36**, 308
 Shortly after mixing the two, an explosion occurred, but this could not be repeated.
 The acid may have been peroxidised.

Other reactants
 Yoshida, 1980, 29
 MRH values calculated for 44 combinations, largely with oxidants, are given.

Oxidants   MRH values show % of oxidant
 1. Kirshenbaum, 1956, 4, 13
 2. Mellor, 1947, Vol. 2, 310
 3. Annikov, V. E. *et al., Chem. Abs.*, 1976, **85**, 145389
 4. Nakamura, H. *et al., Chem. Abs.*, 1985, **102**, 97779
 Mixtures of aluminium powder with liquid chlorine, dinitrogen tetraoxide or tetranitromethane are detonable explosives, but not as powerful as aluminium—liquid
 oxygen mixtures, some of which exceed TNT in effect by a factor of 3 to 4 [1].
 Mixtures of the powdered metal and various bromates may explode on impact,
 heating or friction. Iodates and chlorates act similarly [2]. Detonation properties of
 gelled slurries of aluminium powder in aqueous nitrate or perchlorate salt solutions
 have been studied [3]. Reactions of aluminium powder with potassium chlorate or
 potassium perchlorate have been studied by thermal analysis [4].
 For other combinations,
 See Halogens, above; Oxygen, below
      Ammonium peroxodisulfate: Aluminium, etc.
      Nitryl fluoride: Metals
      Potassium chlorate: Metals   MRH 9.20/70
      Potassium perchlorate: Aluminium, or: Metal powders   MRH 9.96/67
      Sodium chlorate: Aluminium, Rubber   MRH (no rubber) 10.71/67
      Sodium nitrate: Aluminium   MRH 8.32/65
      Sodium peroxide: Aluminium, etc., or: Metals   MRH 4.56/81
      Zinc peroxide: Alone, or Metals
 See other REDOX REACTIONS

Oxygen
 Phillips, B. R. *et al., Combust. Flame*, 1979, **35**(3), 249–258
 Metal powders or fibres ignite in an oxygen-fed resonance tube.

Oxygen, Water
 Aleshim, M. A. *et al., Chem. Abs.*, 1979, **91**, 76312
 Injection of oxygen is used to initiate the metal—water reaction in rocket propulsion
 systems.

Paint
 See Zinc: Paint primer base

Palladium
 See Catalytic metals, above

Polytetrafluoroethylene
1. Pittaluga, F. *et al., Termotechnica*, 1981, **25**, 332–334
2. Schwartz, A., Internet, 1996
A mixture of aluminium powder and PTFE has been evaluated as an underwater missile propellant [1]. It is suggested that using PTFE tape to lubricate and seal aluminium screw threads is inadvisable [2].
See Halocarbons, above

2-Propanol
1. Wilds, A. L., *Org. React.*, 1944, **2**, 198
2. Muir, G. D., private comm., 1968
3. Anon., *Safety Digest Univ. Safety Assoc.* 1994, (50), 21
Dissolution of aluminium in 2-propanol to give the isopropoxide is rather exothermic, but often subject to an induction period similar to that in preparation of Grignard reagents [1]. Only small amounts of aluminium should be present until reaction begins [2]. A canister of Video Display Unit Cleaner exploded in a cupboard. This is thought to have been because of faulty interior lacquer which allowed the propanol containing cleaner to attack the aluminium can, generating heat and hydrogen [3].
See Magnesium: Methanol
See Butanol; Methanol; both above

Silicon steel
Partington, 1967, 418
Aluminium is added (0.1%) to molten steel to remove dissolved oxygen and nitrogen to prevent blowholes in castings. It reacts very violently with silicon steels.

Silver chloride   MRH 0.67/94
Anon., *Chem. Eng. News*, 1954, **32**, 258
An intimate mixture of the two powders may lead to reaction of explosive violence, unless excess aluminium is present (but note low MRH value).

Sodium acetylide
See Sodium acetylide: Metals

Sodium dithionite, Water
See Sodium dithionite: Aluminium, Water

Sodium diuranate
Gray, L. W., *Rept. DP-1485*, Richmond (Va), NTIS, 1978
During outgassing of scrap uranium–aluminium cermet reactor cores, powerful exotherms led to melting of 9 cores. It was found that the incident was initiated by reactions at 350°C between aluminium powder and sodium diuranate, which released enough heat to initiate subsequent exothermic reduction of ammonium uranyl hexafluoride, sodium nitrate, uranium oxide and vanadium trioxide by aluminium, leading to core melting.

Sodium hydroxide
1. *MCA Case History No. 1115*
2. *MCA Case History No. 1888*

In an incident involving corrosive attack on aluminium by sodium hydroxide solution, the vigorous evolution of hydrogen was noticed before a tank trailer (supposed to be mild steel) had perforated [1]. Corrosion caused failure of an aluminium coupling between a pressure gauge and a pump, causing personal contamination [2].

Sodium sulfate   MRH 5.27/66
Kohlmeyer, E. J., *Aluminium*, 1942, **24**, 361–362
The violent explosion experienced when an 8:3 molar mixture of metal powder and salt was heated to 800°C was attributed to thermal dissociation (at up to 3000°C) of the metal sulfide(s) formed as primary product(s).
See other THERMITE REACTIONS

Steel
1. *MCA Case Histories Nos. 2161, 2184* (same incident)
2. Bailey, J. C., *Met. Mater.*, 1978, (Oct.), 26–27
A joint between a mild steel valve screwed onto an aluminium pipe was leaking a resin–solvent mixture, and when the joint was tightened with a wrench, a flash fire occurred. This was attributed to generation of sparks by a thermite reaction between the rusted steel valve and the aluminium pipe when the joint was tightened [1]. Hazards involved in the use of aluminium-sprayed steel are reviewed [2].

Sulfur   MRH 4.81/64
Read, C. W. W., *School Sci. Rev.*, 1940, **21**(83), 977
The violent interaction of aluminium powder with sulfur on heating is considered to be too dangerous for a school experiment.

Uranium compounds
See Sodium diuranate, above

Water
1. *MCA Case History No. 462*
2. Shidlovskii, A. A., *Chem. Abs.*, 1947, **41**, 1105d
3. Bamberger, M. *et al., Z. Angew. Chem.*, 1913, **26**, 353–355
4. Gibson, 1969, 2
5. Pittaluga, F. *et al., Termotechnica*, 1977, **31**, 306–312
6. Sukhov, A. V., *Chem. Abs.*, 1978, **88**, 193939
7. Tompa, A. S. *et al., J. Haz. Mat.*, 1978, **2**, 197–215
8. Meguro, T. *et al., Chem. Abs.*, 1980, **90**, 8505
9. Lemmon, A. W., *Light Met.* (New York), 1980, 817–836
10. Boxley, G. *et al., Light Met.* (New York), 859–868
11. Epstein, S. G. *et al., Light Met.* (Warrendale Pa), 1987, 693–698; *ibid.*, 1993, 845
12. Epstein, S. G., *Light Met.* (Warrendale, Pa), 1995, 885
13. Ikata, K., *Chem. Abs.*, 1993, **118**, 127778
14. Breault, R. *et al., Light Met.* (Warrendale, Pa), 1995, 903
15. Hughes, D. T. *et al., IMechE Conf. Trans.*, 1995, (2), 73
16. Taylor, F. R. *et al., Rept. ARAED-TR-87022*, 1–26, 1988;
Cans of aluminium paint contaminated with water contained a considerable pressure of hydrogen from interaction of finely divided metal and moisture [1]. Mixtures of

powdered aluminium and water can be caused to explode powerfully by initiation with a boosted detonator [2]. A propellant explosive composed of aluminium and water has been patented [13]. During granulation of aluminium by pouring the molten metal through a sieve into water, a violent explosion occurred. This was attributed to steam trapped in the cooling metal [3], but see references [9,10]. Moist finely divided aluminium may ignite in air [4]. The aluminium–steam reaction was evaluated for power generation purposes, a peak steam temperature of over 1000°C being attained [5]. The combustion of aluminium in steam has been studied in detail [6]. In a study of reaction of metal powder with water at 100–110°C in presence of various salts, pH values above 9.5 increased the rate of hydrogen evolution [7]. Hazards of storing aluminium powders and pastes in contact with water under alkaline or acid conditions were studied [8]. Investigations of the explosion mechanism when molten aluminium contacts water have been described [9,10,11], and plant incidents since 1940 are presented [11,12]. A study of hazards of transport of molten aluminium in possibly wet 'torpedoes' is available [14]. Investigation of aluminium alloyed with lithium or calcium revealed a more energetic explosion of the melt with water, but did not conclusively demonstrate easier initiation [15]. The use of organic coatings on aluminium or magnesium powder in pyrotechnic compositions prevents reaction with atmospheric moisture and problems resulting from hydrogen evolution [16].
See other CORROSION INCIDENTS, MOLTEN METAL EXPLOSIONS

Zinc
MCA Case History No. 1722
Ball-milling aluminium–zinc (not stated if alloy or mixture) with inadequate inerting arrangements led to fires during operation or discharge of the mill.
See other METALS, REDUCANTS

0049. Aluminium–cobalt alloy (Raney cobalt alloy)
[37271-59-3] 50:50; [12043-56-0] Al$_5$Co; [73730-53-7] Al$_2$Co   Al–Co
                                      Al–Co

The finely powdered Raney cobalt alloy is a significant dust explosion hazard.
See entry DUST EXPLOSION INCIDENTS (reference 22)

0050. Aluminium–copper–zinc alloy (Devarda's alloy)
[8049-11-4]   Al–Cu–Zn
                        Al–Cu–Zn

See DEVARDA'S ALLOY
      Silver nitrate: Ammonia, etc.
See other ALLOYS

0051. Aluminium amalgam (Aluminium–mercury alloy)
[[12003-69-9] (1:1)]   Al–Hg
                     Al–Hg

1. Neely, T. A. *et al., Org. Synth.*, 1965, **45**, 109
2. Calder, A. *et al., Org. Synth.*, 1975, **52**, 78

The amalgamated aluminium wool remaining from preparation of triphenylaluminium will rapidly oxidise and become hot on exposure to air. Careful disposal is necessary [1]. Amalgamated aluminium foil may be pyrophoric and should be kept moist and used immediately [2].
See other ALLOYS

0052. Aluminium–lithium alloy
[103760-93-6] (7:1 atom ratio, 3.7% Li)   Al–Li
Al–Li

Water
Jacoby, J. E. *et al.*, Can. Pat. CA 1 225 8166, 1987 (*Chem. Abs.*, 1988, **108**, 42436)
The molten alloy (2–3% Li) explodes violently if cooled by contact with water, but not if cooled by ethylene glycol containing less than 25% of water.
See other ALLOYS

0053. Aluminium–magnesium alloy
[12042-38-5] (1:3)   Al–Mg
Al–Mg

The finely powdered alloy is a significant dust explosion hazard.
See entry DUST EXPLOSION INCIDENTS (reference 22)

Barium nitrate
See Barium nitrate, Aluminium–magnesium alloy

Iron(III) oxide, Water
Maischak, K. D. *et al., Neue Huette*, 1970, **15**, 662–665
Accidental contact of the molten alloy (26% Al) with a wet rusty iron surface caused violent explosions with brilliant light emission. Initial evolution of steam, causing fine dispersion of the alloy, then interaction of the fine metals with rust in a 'thermite' reaction, were postulated as likely stages. Direct interaction of the magnesium (74%) with steam may also have been involved.
See other THERMITE REACTIONS

Water
Long, G. C., *Spectrum* (Pretoria), 1980, **18**, 30
In a school demonstration, an alloy pencil sharpener body (ignited in a flame) continues to burn vigorously in steam.
See other ALLOYS

0054. Aluminium–magnesium–zinc alloy
[ ]   Al–Mg–Zn
Al–Mg–Zn

Rusted steel
Yoshino, H. *et al., Chem. Abs.*, 1966, **64**, 14017j

Impact of an alloy containing 6% Al and 3% Zn with rusted steel caused incendive sparks which ignited LPG–air mixtures in 11 out of 20 attempts.
See IGNITION SOURCES, THERMITE REACTIONS

0055. Aluminium–nickel alloys (Raney nickel alloys)
[12635-29-9]   Al–Ni

Al–Ni

HCS 1980, 807

Water
Anon., *Angew. Chem. (Nachr.)*, 1968, **16**, 2
Heating moist Raney nickel alloy containing 20% aluminium in an autoclave under hydrogen caused the aluminium and water to interact explosively, generating 1 k bar pressure of hydrogen.
See also HYDROGENATION CATALYSTS *See other* ALLOYS

0056. Aluminium–titanium alloys
[12003-96-2] (1:1)   Al–Ti

Al–Ti

Oxidants
Mellor, 1941, Vol. 7, 20–21
Alloys ranging from Al_3Ti_2 to Al_4Ti have been described, which ignite or incandesce on heating in chlorine; or bromine or iodine vapour; (or hydrogen chloride); or oxygen.
See other ALLOYS

0057. Tetraiodoarsonium tetrachloroaluminate
[124687-11-2]   AlAsCl₄I₄

$AsI_4^+ AlCl_4^-$

Klapötke, T. *et al., Angew. Chem. (Int.)*, 1989, **101**(12), 1742
Decomposes spontaneously at −78°C. Low temperature photolysis led to violent explosions on warming.
See other IODINE COMPOUNDS

0058. Aluminium tetrahydroborate
[16962-07-5]   AlB₃H₁₂

$Al[BH_4]_3$

1. Schlessinger, H. I. *et al., J. Amer. Chem. Soc.*, 1940, **62**, 3421
2. Badin, F. J. *et al., J. Amer. Chem. Soc.*, 1949, **71**, 2950

The vapour is spontaneously flammable in air [1], and explodes in oxygen, but only in presence of traces of moisture [2].

Alkenes, Oxygen
Gaylord, 1956, 26
The tetrahydroborate reacts with alkenes and, in presence of oxygen, combustion is initiated even in absence of moisture. Butene explodes after an induction period, while butadiene explodes immediately.
See other INDUCTION PERIOD INCIDENTS

Dimethylaminoborane
Burg, A. B. *et al., J. Amer. Chem. Soc.*, 1951, **73**, 953
Reaction with dimethylaminoborane or dimethylaminodiborane gives an oily mixture which ignites in air and reacts violently with water.

Water
Semenenko, K. N. *et al., Russ. Chem. Rev.*, 1973, 4
Interaction at ambient temperature is explosive.
See other COMPLEX HYDRIDES

0059. Aluminium tetraazidoborate
[67849-01-8] AlB_3N_{36}
$$Al[B(N_3)_4]_3$$

Mellor, 1967, Vol. 8, Suppl. 2, 2
A very shock sensitive explosive, containing nearly 90 wt% of nitrogen.
See other HIGH-NITROGEN COMPOUNDS, NON-METAL AZIDES

0060. Aluminium bromide
[7727-15-3] $AlBr_3$
$$AlBr_3$$

Dichloromethane
Kramer, G. M. *et al., Acc. Chem. Res.*, 1986, **19**(3), 78–84
Solutions of aluminium bromide in dichloromethane used as a catalyst in hydride-transfer equilibrium experiments should be kept cold, as a potentially dangerous exothermic halide exchange reaction occurs on warming.

Water
Nicholson, D. G. *et al., Inorg. Synth.*, 1950, **3**, 35
The anhydrous bromide should be destroyed by melting and pouring slowly into running water. Hydrolysis is very violent and may destroy the container if water is added to it.
See other METAL HALIDES

0061. Aluminium dichloride hydride diethyl etherate
[13497-97-7] (solvent-free) $AlCl_2H.C_4H_{10}O$

$$Cl_2HAl.OEt_2$$

Dibenzyl ether

Marconi, W. *et al., Ann. Chim.*, 1965, **55**, 897

During attempted reductive cleavage of the ether with aluminium dichloride hydride etherate an explosion occurred. Peroxides may have been present in the susceptible ether.

See other PEROXIDATION INCIDENTS
See related COMPLEX HYDRIDES

0062. Aluminium chloride
[7446-70-0] $AlCl_3$

$$AlCl_3$$

(*MCA SD-62*, 1956); *NSC 435*, 1978; *HCS 1980*, 130; *RSC Lab. Hazards Data Sheet No 72*, 1988

1. Popov, P. V., *Chem. Abs.*, 1947, **41**, 6723d
2. Kitching, A. F., *School Sci. Rev.*, 1930, **12**(45), 79
3. Bailar, 1973, Vol. 1, 1019

Long storage of the anhydrous salt in closed containers caused apparently spontaneous decomposition and occasional explosion on opening [1,2]. This seems likely to have arisen from slow diffusion of moisture (MW 18) in through the closure and pressurisation of the container by the liberated HCl which because of its higher MW (36.5) would diffuse out at a slower rate. The need is emphasised for care in experiments in which the chloride is heated in sealed tubes. High internal pressure may be generated, not only by its vapour pressure and pressure of desorbed hydrogen chloride, but also by the near doubling in volume which occurs when the chloride melts to the monomer [3].

Alkenes

Jenkins, P. A., private. comm., 1975

Mixtures of C_4 alkene isomers (largely isobutene) are polymerised commercially in contact with low levels of aluminium chloride (or other Lewis acid) catalysts. The highly exothermic runaway reactions occasionally experienced in practice are caused by events leading to the production of high local levels of catalyst. Rapid increases in temperature and pressure of 160°C and 18 bar, respectively, have been observed experimentally when alkenes are brought into contact with excess solid aluminium chloride. The runaway reaction appears to be more severe in the vapour phase, and a considerable amount of catalytic degradation contributes to the overall large exotherm.

See other POLYMERISATION INCIDENTS, RUNAWAY REACTIONS

Aluminium oxide, Carbon oxides

See Aluminium: Aluminium halides, Carbon oxides

Aluminium, Sodium peroxide
See Sodium peroxide: Aluminium, etc.

Benzene, Carbon tetrachloride
Nolan, 1983, Case history 105
Triphenylmethyl chloride was manufactured by a Friedel-Crafts reaction of benzene and carbon tetrachloride with excess aluminium chloride. Owing to an operating fault, all the carbon tetrachloride was added to the other reactants without agitation but with water cooling. When the agitator was started, two explosions followed. The first was from the sudden exotherm and gas evolution on mixing the reactants, and this damaged the heat exchangers and allowed ingress of cooling water, leading to a second explosion on contact with the aluminium chloride.
See GAS EVOLUTION INCIDENTS (reference 2); and Water, below
See other AGITATION INCIDENTS, GAS EVOLUTION INCIDENTS

Benzoyl chloride, Naphthalene
Clar, E. *et al., Tetrahedron*, 1974, **30**, 3296
During preparation of 1,5-dibenzoylnaphthalene, addition of aluminium chloride to a mixture of benzoyl chloride and naphthalene must be effected above the m.p. of the mixture to avoid a violent reaction.

Ethylenimine, Substituted anilines
See Ethylenimine, Aluminium chloride, etc.

Ethylene oxide
See Ethylene oxide: Contaminants

Nitrobenzene
1. Riethman, J. *et al., Chem. Ing. Tech.*, 1974, **48**, 729
2. Eigenmann, K. *et al., Proc. Int. Symp. Simulation*, 436, Zurich, Acta Press, 1975
3. Finck, P. *et al., Proc. Symp. Safety Chem. Ind.*, 17–25, Mulhouse, 1978
4. *See entry* SELF-ACCELERATING REACTIONS
5. Nolan, 1983, Case history 104
Mixtures of nitrobenzene and aluminium chloride are thermally unstable and may lead to explosive decomposition. Subsequent to an incident involving rupture of a 4000 l vessel, the decomposition reaction was investigated and a 3-stage mechanism involving formation and subsequent polymerisation of 2- and 4-chloronitrosobenzene was proposed [1]. Further chemical and thermodynamic work [2,3] on the thermal degradation of nitrobenzene–aluminium chloride addition compounds formed in Friedel-Crafts reactions shows that it is characterised by a slow multi-step decomposition reaction above 90°C, which self-accelerates with high exothermicity producing azo- and azoxy-polymers. Simulation of the original incident suggested that the pressure in the reactor probably increased from 3 to 40 bar in 5 s [4]. Nitrobenzene was added to a 2.2 kl reactor in readiness for a Friedel-Crafts reaction, but it leaked away through a faulty bottom valve. Aluminium chloride and a reactant, hexamethyltetralin were added, but when the agitator stopped (solid complex formation), the absence of the nitrobenzene solvent was realised. More nitrobenzene was being sucked in to the reactor when the exotherm from the reaction with aluminium chloride pressurised

the reactor, boiled and ignited the limited amount of nitrobenzene, and shattered a glass fitting [5].

See other AGITATION INCIDENTS, POLYMERISATION INCIDENTS

Nitrobenzene, Phenol
Anon., *Chem. Eng. News*, 1953, **31**, 4915
Addition of aluminium chloride to a large volume of recovered nitrobenzene containing 5% of phenol caused a violent explosion. Experiment showed that mixtures containing all 3 components reacted violently at 120°C.

Nitromethane
See Aluminium chloride–Nitromethane

Other reactants
Yoshida, 1980, 52
MRH values calculated for 12 combinations with oxidants are given.

Oxygen difluoride
See Oxygen difluoride: Halogens, etc.

Phenyl azide
See Phenyl azide: Lewis acids

Perchlorylbenzene
See Perchlorylbenzene: Aluminium chloride

Sodium tetrahydroborate
See Sodium tetrahydroborate: Aluminium chloride, etc.

Water
1. Anon., *Ind. Eng. Chem. (News Ed.)*, 1934, **12**, 194
2. Anon., *Loss Prev. Bull.*, 1980, (035), 9
An unopened bottle of anhydrous aluminium chloride erupted when the rubber bung with which it was sealed was removed. The accumulation of pressure was attributed to absorption of moisture by the anhydrous chloride before packing (but see comment following text of masthead reference 2). The presence of an adsorbed layer of moisture in the bottle used for packing may have contributed. Reaction with water is violently exothermic [1]. The unsuspected presence of 100 kg of the chloride in a vessel led to bursting of a glass vent when the vessel was being flushed with water [2].

See other GLASS INCIDENTS, GAS EVOLUTION INCIDENTS
See other METAL HALIDES

0063. Aluminium chloride–nitromethane
[3495-54-3] $AlCl_3.CH_3NO_2$
$$AlCl_3.CH_3NO_2$$

An alkene
Cowen, F. M. *et al., Chem. Eng. News*, 1948, **26**, 2257
A gaseous alkene was passed into a cooled autoclave containing the complex, initially with agitation, and later without. Later, when the alkene was admitted to a pressure of

5.6 bar at 2°C, a slight exotherm occurred, followed by an explosion. The autoclave contents were completely carbonised. Mixtures of ethylene, aluminium chloride and nitromethane had exploded previously, but at 75°C.
See Ethylene: Aluminium chloride, Catalysts

Carbon monoxide, Phenol
Webb, H. F., *Chem. Eng. News*, 1977, **55**(12), 4
An attempt to formylate phenol by heating a mixture with nitromethane and aluminium chloride in an autoclave under carbon monoxide at 100 bar pressure at 110°C led to a high-energy explosion after 30 min.
See related NITROALKANES

0064. Trihydrazinealuminium perchlorate
[85962-45-4] $AlCl_3H_{12}N_6O_{12}$
$$[(H_4N_2)_3Al] [ClO_4]_3$$

Ind. Inst. Sci., Indian Pat. IN 150422, 1982
Useful as a propellant oxidant.
See other AMMINEMETAL OXOSALTS, REDOX COMPOUNDS

0065. Aluminium chlorate
[15477-33-5] $AlCl_3O_9$
$$Al(ClO_3)_3$$

Sidgwick, 1950, 428
During evaporation, its aqueous solution evolves chlorine dioxide, and eventually explodes.
See other METAL CHLORATES

0066. Aluminium perchlorate
[14452-39-2] $AlCl_3O_{12}$
$$Al(ClO_4)_3$$

Dimethyl sulfoxide
See Dimethyl sulfoxide: Metal oxosalts

Hydrazine
See Trihydrazinealuminium perchlorate
See other METAL PERCHLORATES

0067. Caesium hexahydroaluminate(3−)
[53436-80-9] $AlCs_3H_6$
$$Cs_3[AlH_6]$$

See Potassium hexahydroaluminate(3−)
See other COMPLEX HYDRIDES

0068. Copper(I) tetrahydroaluminate
[62126-20-9]   AlCuH₄

$$Cu[AlH_4]$$

Aubry, J. *et al.*, *Compt. rend.*, 1954, **238**, 2535
The unstable hydride decomposed at −70°C, and ignited on contact with air.
See other COMPLEX HYDRIDES

0069. Lithium tetradeuteroaluminate
[14128-54-2]   AlD₄Li

$$Li[AlD_4]$$

Leleu, Cahiers, 1977, (86), 99
It may ignite in moist air.
See related COMPLEX HYDRIDES

0070. Aluminium hydride
[7784-21-6]   AlH₃

$$AlH_3$$

1. Mirviss, S. B. *et al.*, *Ind. Eng. Chem.*, 1961, **54**(1), 54A
2. Gibson, 1968, 66
It is very unstable and has been known to decompose spontaneously at ambient temperature with explosive violence. Its complexes (particularly the diethyl etherate) are considerably more stable [1]. The hydride ignites in air with or without oxygen enrichment [2].
See Aluminium hydride–trimethylamine
Aluminium hydride–diethyl ether

Carbon dioxide, Methyl ethers
Barbaras, G. *et al.*, *J. Amer. Chem. Soc.*, 1948, **70**, 877
Presence of carbon dioxide in solutions of the hydride in dimethyl or bis(2-methoxyethyl) ether can cause a violent decomposition on warming the residue from evaporation. Presence of aluminium chloride tends to increase the vigour of decomposition to explosion. Lithium tetrahydroaluminate may behave similarly, but is generally more stable.
See Lithium tetrahydroaluminate (reference 8)

Carbon dioxide, or Sodium hydrogen carbonate
Thompson, B. T. *et al.*, *Polyhedron*, 1983, **2**, 619–621
At elevated temperatures, the hydride reduces carbon dioxide or sodium hydrogencarbonate to methane and ethane. The latter are probably the explosive reaction products produced when carbon dioxide extinguishers are used on LAH fires.
See other GAS EVOLUTION INCIDENTS

Tetrazole derivatives
Fetter, N. R. *et al.*, US Pat. 3 396 170, 1968
The 1:1 complexes arising from interaction of the hydride (as a complex with ether or trimethylamine) and various tetrazole derivatives are explosive. Tetrazoles mentioned are 2-methyl-, 2-ethyl-, 5-ethyl-, 2-methyl-5-vinyl-, 5-amino-2-ethyl-, 1-alkyl-5-amino-, and 5-cyano-2-methyl-tetrazole.
See TETRAZOLES
See other METAL HYDRIDES

0071. Alane−N,N-dimethylethylamine ((N,N-Dimethylethanamine)trihydro-aluminium)
[124330-23-0] $AlH_3 \cdot C_4H_{11}N$
$$H_3Al.Me_2NEt$$

Frigo, D. M., *et al., Chem. Mater.*, 1994, **6**(2), 190
Not generally pyrophoric, unless dropped on vermiculite, the complex decomposes slowly in the liquid phase to generate considerable pressures of hydrogen. It appears much more stable in the vapour phase.
See related METAL HYDRIDES

0072. Aluminium hydride−trimethylamine
[17013-07-9] $AlH_3 \cdot C_3H_9N$
$$H_3Al.NMe_3$$

Water
Ruff, J. K., *Inorg. Synth.*, 1967, **9**, 34
It ignites in moist air and is explosively hydrolysed by water.
See related METAL HYDRIDES

0073. Aluminium hydride−diethyl ether
[26351-01-9] $AlH_3 \cdot C_4H_{10}O$
$$H_3Al.OEt_2$$

Water
Schmidt, D. L. *et al., Inorg. Synth.*, 1973, **14**, 51
Interaction of the solid with water or moist air is violent and may be explosive.
See related METAL HYDRIDES

0074. Aluminium hydroxide
[21645-51-2] AlH_3O_3
$$Al(OH)_3$$

Chlorinated rubber
See CHLORINATED RUBBER: Metal oxides or hydroxides

0075. Lithium tetrahydroaluminate (Lithium aluminium hydride)
[16853-85-3]   AlH₄Li

$$Li[AlH_4]$$

HCS 1980, 593; *RSC Lab. Hazard Data Sheet No. 5,* 1982

1. Augustine, 1968, 12
2. Gaylord, 1956, 37
3. Sakaliene, A. *et al., Chem. Abs.*, 1970, **73**, 19115
4. *MCA Case History No. 1832*
5. Walker, E. R. H., *Chem. Soc. Rev.*, 1976, **5**, 36
6. Brendel, G., *Chem. Eng. News*, 1979, **57**(36), 5
7. Anon., *Lab. Haz. Bull.*, 1985, (12), item 852
8. Green, M. L. H., private comm., 1986
9. Schatzschneider, U., Internet, 1996

Care is necessary in handling this powerful reducant, which may ignite if lumps are pulverised with a pestle and mortar, even in a dry box [1]. An actual explosion destroying the mortar, has been claimed as the result of attempting to grind down large pellets, though contributory factors seem probable [9]. A rubber mallet is recommended for breaking up lumps [2]. The explosive thermal decomposition of the aluminate at 150–170°C is due to its interaction with partially hydrolysed decomposition products [3]. A spilled mixture with ether ignited after the ether evaporated [4]. Sodium bis(2-methoxyethoxy)dihydroaluminate, which is of similar reducing capability to lithium tetrahydroaluminate, is safer in that it does not ignite in moist air or oxygen and is stable at 200°C [5]. A detailed comment on the latter states that the commercial (crystalline) lithium compound is not pyrophoric, even in contact with moist air of high humidity [6]. An attempt to decontaminate a polythene bag dusted with residual lithium tetrahydroaluminate by immersion in ethyl acetate caused a fire. Two alternative methods are proposed, one for light coatings of dust, which should be immersed completely in a large volume of water behind screens until reaction ceases. The other method for larger amounts involves suspension of the residue in an inert solvent (light petroleum — *flammable*), and dropwise treatment with ethyl acetate until reaction ceases. After standing, the treatment is repeated with ethanol, and then water [7]. Following the investigation of a laboratory explosion, precautions essential for the safe use of lithium aluminium hydride (LAH) have been defined [8]. The measures given below are to prevent overheating of the hydride and its dissociation to finely divided aluminium, which can then undergo thermite-like reactions with compounds or solvents containing combined oxygen or halogen.

a. All apparatus and reactants should be perfectly dry, and reactions should be run rigorously under nitrogen, with the reaction temperature below 60°C at all times.

b. Order of addition is important. Always first add the hydride to the solvent in the nitrogen-purged apparatus, before adding the other reactant last.

c. The hydride should never be allowed to form a crust above the level of the liquid or to settle to the bottom, so efficient but gentle stirring is absolutely essential.

d. To prevent local overheating of the reaction vessel, heating mantles should never be used: always use an oil bath as heat source.

e. After reduction has been effected, destroy excess LAH by slow and careful addition of dry ethyl acetate (preferably diluted with inert solvent), again under nitrogen and keeping the temperature below 60°C. All LAH reactions should be carried out behind suitable protective screens.

See Ethyl acetate, below

Alkyl benzoates

1. Tados, W. *et al., J. Chem. Soc.,* 1954, 2353
2. Field, B. O. *et al., J. Chem. Soc.,* 1955, 1111

Application of a method for reducing benzaldehydes to the corresponding alcohols with a fourfold excess of the aluminate [1] to alkyl benzoates proved to be difficult to control and frequently dangerous [2].

Bis(2-methoxyethyl) ether

1. Watson, A. R., *Chem. & Ind.,* 1964, 665
2. Adams, R. M., *Chem. Eng. News,* 1953, **31**, 2334
3. Barbaras, G. *et al., J. Amer. Chem. Soc.,* 1948, **70**, 877
4. *MCA Case History No. 1494*
5. Author's comment, 1987

The peroxide-free ether, being dried by distillation at 162°C under inert atmosphere at ambient pressure, exploded violently when the heating bath temperature had been raised to 200°C towards the end of distillation. This was attributed to local overheating of an insulating crust of hydride in contact with oxygen-containing organic material [1]. Two previous explosions were attributed to peroxides [2] and the high solubility of carbon dioxide in such ethers [3]. Stirring during distillation would probably prevent crust formation. Alternatively, drying could be effected with a column of molecular sieve or activated alumina. During distillation of the solvent from the aluminate at 100°C at ambient pressure, the flask broke and the contents ignited explosively. The aluminate decomposes at 125–135°C [4]. It seems probable that at least some of these incidents are better explained as arising from formation of finely divided aluminium from overheating of the hydride, and a thermite-like reaction with the oxygen-containing solvent [5]. See initial reference 8 above for precautions to avoid this.

See Aluminium hydride: Carbon dioxide

Boron trifluoride diethyl etherate

1. Scott, R. B., *Chem. Eng. News,* 1967, **45**(28), 7; **45**(21), 51
2. Shapiro, I. *et al., J. Amer. Chem. Soc.,* 1952, **74**, 90

Use of lumps of the solid aluminate, rather than its ethereal solution, and of peroxide-containing etherate [1], rather than the peroxide-free material specified [2], caused an explosion during the attempted preparation of diborane.

Dibenzoyl peroxide   MRH 4.39/85
    See Dibenzoyl peroxide: Lithium tetrahydroaluminate

3,5-Dibromocyclopentene
    Johnson, C. R. *et al., Tetrahedron Lett.*, 1964, **45**, 3327
    Preparation of the 4-bromo compound by partial debromination of crude 3,5-dibromo-
    cyclopentene by addition of its ethereal solution to the aluminate in ice-cold ether
    is hazardous. Explosions have occurred on 2 occasions about 1 h after addition of
    dibromide.

1,2-Dimethoxyethane
    1. *MCA Case History No. 1182*
    2. Hoffman, K. A. *et al., Org. Synth.*, 1968, **48, 62**
    The finely powdered aluminate was charged through a funnel into a nitrogen-purged
    flask. When the solvent was added through the same funnel, ignition occurred,
    possibly due to local absence of purge gas in the funnel caused by turbulence [1].
    Distillation of the solvent from the solid must not be taken to dryness, to avoid
    explosive decomposition of the residual aluminate [2].

Dioxane
    1. Anon., private comm., 1976
    2. Author's comments, 1987
    Dioxane was purified by distillation from the complex hydride in a glass still, and
    when the residue was cooling down, a severe explosion and fire occurred [1]. This
    may have been caused by ingress of air into the cooling dioxane vapour (flammability
    limits 2–22%), then subsequent oxidative heating of finely divided hydride deposited
    on the upper parts of the still on contact with air to above the rather low autoignition
    temperature of dioxane (180°C). Nitrogen purging will render the operation safe.
    However, the alternative explanation of a thermite-like reaction with dioxane, arising
    from overheating of the hydride (initial reference 8 above) may be equally probable. If
    this were the case, nitrogen purging alone would not necessarily render the operation
    safe [2]. See reference 8 above for preventive measures.

Ethyl acetate
    1. Bessant, K. H. C., *Chem. & Ind.*, 1957, 432
    2. Yardley, J. T., *Chem. & Ind.*, 1957, 433
    Following a reductive dechlorination in ether, a violent explosion occurred when
    ethyl acetate was added to decompose excess aluminate [1]. Ignition was attributed
    to the strongly exothermic reaction occurring when undiluted (and reducible) ethyl
    acetate contacts the solid aluminate. Addition of a solution of ethyl acetate in inert
    solvent or of a moist unreactive solvent to destroy excess reagent is preferable [2].
    See reference 1–7, above

Fluoroamides
    1. Karo. W., *Chem. Eng. News*, 1955, **33**, 1368
    2. Reid, T. S. *et al., Chem. Eng. News*, 1951, **29**, 3042

44

The reduction of amides of fluorocarboxylic acids with the tetrahydroaluminate appears generally hazardous at all stages. During reduction of *N*-ethylheptafluorobutyramide in ether, violent and prolonged gas evolution caused a fire. Towards the end of reduction of trifluoroacetamide in ether, solid separated and stopped the stirrer. Attempts to restart the stirrer by hand caused a violent explosion [1]. During decomposition by water of the reaction complex formed by interaction with tetrafluorsuccinamide in ether, a violent explosion occurred. Reaction complexes similarly obtained from trifluoroacetic acid, heptafluorobutyramide and octafluoroadipamide also showed instability, decomposing when heated. General barricading of all reductions of fluorocompounds with lithium tetrahydroaluminate is recommended [2].
See other AGITATION INCIDENTS, GAS EVOLUTION INCIDENTS

Hydrogen peroxide
See Hydrogen peroxide: Lithium tetrahydroaluminate

Nitrogen
1. Metts, L., *Chem. Eng. News*, 1981, **59**(31), 3
2. Grossman, M. I., *Chem. Eng. News*, 1981, **59**(41), 57
3. Deberitz, J., *Chem. Eng. News*, 1982, **60**(11), 3
The incident which originally described the glowing residue from combustion in air of a pellet of hydride as being able to burn under nitrogen [1] was later interpreted as 'ignition of the hydride in a nitrogen atmosphere' [2]. It is later stressed that the hydride itself does not react directly with nitrogen under any foreseeable conditions of normal use. After combustion, the residue of hot aluminium and lithium metals would be expected to react with nitrogen [3].

Nitromethane
See Nitromethane: Lithium tetrahydroaluminate

Pyridine
Augustine, 1968, 22–23
Addition of the aluminate (0.5 g) to pyridine (50 ml) must be effected very slowly with cooling. Addition of 1 g portions may cause a highly exothermic reaction.

Tetrahydrofuran
Moffett, R. B., *Chem. Eng. News*, 1954, **32**, 4328
The solvent had been dried over the aluminate and then stored over calcium hydride for 2 years 'to prevent peroxide formation'. Subsequent addition of more aluminate caused a strong exotherm and ignition of liberated hydrogen. Calcium hydride does not prevent peroxide formation in solvents.
See other PEROXIDATION INCIDENTS

1,2,3,4-Tetrahydro-1-naphthyl hydroperoxide
See 1,2,3,4-Tetrahydro-1-naphthyl hydroperoxide: Lithium tetrahydroaluminate

Water
1. Leleu, *Cahiers*, 1977, (86), 100
2. Anon., *Fire Prevention*, 1986, (191), 43
Interaction is very vigorous, and with limited water, incandescent [1]. An account of a serious fire caused by inadvertent addition of lithal to water [2]. In the editor's experience ignition of evolved hydrogen will occur with quantities as slight as 200 mg.
See other GAS EVOLUTION INCIDENTS
See other COMPLEX HYDRIDES

0076. Sodium tetrahydroaluminate (Sodium aluminium hydride)
[13770-96-2] AlH_4Na

$$Na[AlH_4]$$

Tetrahydrofuran
Del Giudia, F. P. *et al., Chem. Eng. News*, 1961, **39**(40), 57
During synthesis from its elements in tetrahydrofuran, a violent explosion occurred when absorption of hydrogen had stopped. This was attributed to deposition of solid above the liquid level, overheating and reaction with solvent to give butoxyaluminium hydrides. Vigorous stirring and avoiding overheating are essential.

Water
Gibson, 1969, 85
It may ignite and explode in contact with water.
See other COMPLEX HYDRIDES

0077. Potassium hexahydroaluminate(3−)
[17083-63-5] AlH_6K_3

$$K_3[AlH_6]$$

Ashby, E. C., *Chem. Eng. News*, 1969, **47**(1), 9
A 20 g sample, prepared and stored in a dry box for several months, developed a thin crust of oxidation/hydrolysis products. When the crust was disturbed, a violent explosion occurred, later estimated as equivalent to 230 g TNT. A weaker explosion was observed with potassium tetrahydroaluminate. The effect was attributed to superoxidation of traces of metallic potassium, and subsequent interaction of the hexahydroaluminate and superoxide after frictional initiation. Precautions advised include use of freshly prepared material, minimal storage in a dry diluent under an inert atmosphere and destruction of solid residues. Potassium hydrides and caesium hexahydroaluminate may behave similarly, as caesium also superoxidises in air.
See other COMPLEX HYDRIDES

0078. Aluminium phosphinate
[24704-64-1] $AlH_6O_6P_3$

$$Al[P(H)(O)OH]_3$$

Mellor, 1971, Vol. 8, Suppl. 3, 623

It decomposes at around 220°C liberating spontaneously flammable phosphine.
See other METAL PHOSPHINATES

0079. Aluminium iodide
[7784-23-8] AlI_3

$$AlI_3$$

Alone, or Water
  Bailar, 1973, Vol. 1, 1023
  It reacts violently with water, and on heating produces flammable vapour which may
  explode if mixed with air and ignited.

Aluminium, Carbon oxides
  See Aluminium: Aluminium halides, Carbon oxides
  See other METAL HALIDES

0080. Aluminium–lanthanum–nickel alloy
[66459-02-7] $AlLaNi_4$

$$AlLaNi_4$$

Mendelsohn, M. H. *et al., Inorg. Synth.*, 1083, **22**, 99
The alloy powder (used in hydrogen-storage systems) may ocasionally be pyrophoric
after hydriding–dehydriding operations, igniting when placed on a combustible
surface (e.g. weighing paper).
See other ALLOYS, PYROPHORIC ALLOYS

0081. Lithium tetraazidoaluminate
[67849-02-9] $AlLiN_{12}$

$$Li[Al(N_3)_4]$$

Mellor, 1967, Vol. 8, Suppl. 2, 2
A shock-sensitive explosive.
See related METAL AZIDES *See other* HIGH-NITROGEN COMPOUNDS

0082. Aluminium azide
[39108-14-0] AlN_9

$$Al(N_3)_3$$

Brauer, 1963, Vol. 1, 829
Containing 82.3% of nitrogen, it may be detonated by shock.
See other HIGH-NITROGEN COMPOUNDS, METAL AZIDES

0083. Aluminium phosphide
[20859-73-8]   AlP

AlP

HCS 1980, 133

Mineral acids
1. Wang. C. C. *et al., J. Inorg. Nucl. Chem.*, 1963, **25**, 327
2. Mellor, 1971, Vol. 8, Suppl. 3, 306
Evolution of phosphine is slow in contact with water or alkali, but explosively violent
in contact with dilute mineral acids [1]. However, reports of violent interaction with
concentrated or dilute hydrochloric acid, and of explosive reaction with 1:1 aqua
regia, have been questioned [2].
See other GAS EVOLUTION INCIDENTS
See other METAL NON-METALLIDES

0084. Aluminium copper(I) sulfide
[12003-23-5] $Al_2Cu_2S_4$

$(S:AlSCu-)_2$

Preparative hazard
See Aluminium: Copper, Sulfur
See other METAL SULFIDES

0085. Magnesium tetrahydroaluminate
[17300-62-8] Al_2H_8Mg

$Mg[AlH_4]_2$

Gaylord, 1956, 25
It is similar to the lithium salt.
See other COMPLEX HYDRIDES

0086. Manganese(II) tetrahydroaluminate
[65776-39-8] Al_2H_8Mn

$Mn[AlH_4]_2$

Aubry, J. *et al., Compt. rend.*, 1954, **238**, 2535
The unstable hydride decomposes at $-80°C$ and ignites in contact with air.
See other COMPLEX HYDRIDES

0087. Aluminium oxide ('Alumina')
[1344-28-1] Al_2O_3

Al_2O_3

HCS 1980, 132

Chlorine trifluoride
See Chlorine trifluoride: Metals, etc.

Ethylene oxide
See Ethylene oxide: Contaminants

Halocarbons, Heavy metals
Burbidge, B. W., unpublished information, 1976
It is known that alumina is chlorinated exothermically at above 200°C by contact with halocarbon vapours, and hydrogen chloride, phosgene etc. are produced. It has now been found that a Co/Mo−alumina catalyst will generate a substantial exotherm in contact with vapour of carbon tetrachloride or 1,1,1-trichloroethane at ambient temperature in presence of air. In absence of air, the effect is less intense. Two successive phases appear to be involved: first, adsorption raises the temperature of the alumina; then reaction, presumably metal-catalysed, sets in with a further exotherm.

Oxygen difluoride
See Oxygen difluoride: Adsorbents

Sodium nitrate
See Sodium nitrate: Aluminium, etc.

Vinyl acetate
See Vinyl acetate: Desiccants
See other METAL OXIDES

0088. Dialuminium octavanadium tridecasilicide
[ ] $Al_2Si_{13}V_8$
                              (Complex structure)

Hydrofluoric acid
Sidgwick, 1950, 833
The silicide reacts violently with the aqueous acid.
See other METAL NON-METALLIDES

0089. Cerium(III) tetrahydroaluminate
[65579-06-8] Al_3CeH_{12}
                              $Ce[AlH_4]_3$

Aubry, J. et al., Compt. rend., 1954, **238**, 2535
The unstable hydride decomposes at −80°C, and ignites in contact with air.
See other COMPLEX HYDRIDES

0090. Americium trichloride
[13464-46-5]   AmCl₃

AmCl$_3$

AmCl$_3$

MCA Case History No. 1105
A multi-wall shipping container, holding 400 ml of a solution of americium chloride in a polythene bottle and sealed for over 3 months, exploded. The reason could have been a slow pressure build-up of radiolysis products. Venting and other precautions are recommended.
See Radon: Water
See other GAS EVOLUTION INCIDENTS, METAL HALIDES

0091. Argon
[7440-37-1]   Ar

Ar

Liquid nitrogen
  1. Baker, K., *Chem. Brit.*, 1979, **15**, 65
  2. Janiaut, H., *Actual Chim.*, 1986, (1–2), 59–61
The presence of argon in apparatus cooled by liquid nitrogen can lead to hazardous situations with the possibility of explosion. This is because argon solidifies at −189°C and has so low a vapour pressure at −196°C that the solid may survive exposure to high vacuum for a considerable period, and subsequently evaporate on warming with generation of high pressure. Use of solid CO_2–solvent baths is recommended for argon [1]. A phenylhydrazone was vacuum-degassed and frozen in liquid nitrogen, probably during freeze-drying. Vacuum was broken by slow admission of argon (some of which would condense and solidify), and the flask, still under liquid nitrogen, was sealed. During subsequent warming to ambient temperature, the flask exploded violently from internal gas pressure [2].
See other GAS EVOLUTION INCIDENTS

Magnesium perchlorate
See Magnesium perchlorate: Argon
See other NON-METALS

0092. Arsenic
[7440-38-2]   As

As

NSC 499, 1979 (As and its inorganic compounds); *RSC Lab. Hazard Data Sheet No. 62*, 1987 (As and compounds)

Müller, W. J., *Z. Angew. Chem.*, 1914, **27**, 338
An explosive variety (or compound) of arsenic was produced as a surface layer on the exposed iron surfaces of a corroded lead-lined vessel which contained 35% sulfuric

acid with a high arsenic content. It exploded on friction or ignition, and contained no hydrogen, but variable small amount of iron and lead. It may have been analogous to explosive antimony.

Bromine azide
See Bromine azide

Halogens or Interhalogens
1. Mellor, 1946, Vol. 2, 92
2. Mellor, 1956, Vol. 2, Suppl. 1, 379
The finely powdered element inflames in gaseous chlorine or liquid chlorine at −33°C [1]. The latter is doubtful [2].
See Bromine trifluoride: Halogens, etc.
 Bromine pentafluoride: Acids, etc.
 Chlorine trifluoride: Metals, etc.
 Iodine pentafluoride: Metals, etc.

Metals
Mellor, 1940, Vol. 4, 485–486; 1942, Vol. 15, 629; 1937, Vol. 16, 161
Palladium or zinc and arsenic react on heating with evolution of light and heat, and platinum with vivid incandescence.
See Aluminium: Antimony, etc.

Nitrogen trichloride
See Nitrogen trichloride: Initiators

Oxidants
See Chromium trioxide: Arsenic
 Dichlorine oxide: Oxidisable materials
 Nitrosyl fluoride: Metals, etc.
 Potassium dioxide: Metals
 Potassium permanganate: Antimony, etc.
 Silver nitrate:Arsenic
 Sodium peroxide: Non-metals

Rubidium acetylide
See Rubidium acetylide: Non-metals
See other NON-METALS

0093. Difluoroperchloryl hexafluoroarsenate
 [39003-82-2] $AsClF_8O_2$
$$[F_2ClO_2] [AsF_6]$$

See entry DIFLUOROPERCHLORYL SALTS

0094. Arsenic trichloride
[7784-34-1]   AsCl$_3$

AsCl$_3$

HCS 1980, 165

Aluminium
See Aluminium: Non-metal halides

Hexafluoroisopropylideneaminolithium
See Hexafluoroisopropylideneaminolithium: Non-metal halides
See other NON-METAL HALIDES

0095. Arsenic pentafluoride
[7784-36-3]   AsF$_5$

AsF$_5$

Benzene, Potassium methoxide
Kolditz, L. *et al., Z. Anorg. Chem.*, 1965, **341**, 88–92
Interaction of the pentafluoride and methoxide proceeded smoothly in trichlorotrifluoroethane at 30–40°C, whereas in benzene as solvent repeated explosions occurred.

1,3-Butadiyne
See 1,3-Butadiyne: Arsenic pentafluoride

Krypton difluoride
See Krypton difluoride: Arsenic pentafluoride
See other NON-METAL HALIDES

0096. Fluorokrypton hexafluoroarsenate
[50859-36-2]   AsF$_7$Kr

FKr$^+$F$_6$As$^-$

See Krypton difluoride: Arsenic pentafluoride
See XENON COMPOUNDS

0097. Monofluoroxonium hexafluoroarsenate
[ ]   AsF$_7$H$_2$O

(H$_2$OF)$^+$(AsF$_6$)$^-$

Minkwitz, R. *et al., Angew. Chem. Int.*, 1990, **29**(6), 689
The above salt is a powerful oxidant, reacting explosively with methanol even at −40°C. The antimonate is similar.
See other OXIDANTS

0098. Difluoroammonium hexafluoroarsenate
[56533-30-3] AsF_8H_2N
$$F_2NH_2[AsF_6]$$

Christe, K. O., *Inorg. Chem.*, 1975, **14**, 2821–2824
Solutions of this and the hexafluoroantimonate salt in hydrogen fluoride, kept for
extended periods between −50 and +50°C, burst the Kel-F or Teflon FEP containers.
This was attributed to excess pressure of hydrogen fluoride and nitrogen arising from
decomposition of the salts. The variable rates of decomposition indicated catalysis
by trace impurities. The salts also decompose exothermally after a short period at
ambient temperature.
See other GAS EVOLUTION INCIDENTS, *N*-HALOGEN COMPOUNDS

0099. Trifluoroselenium hexafluoroarsenate
[59544-89-7] AsF_9Se
$$F_3Se[AsF_6]$$

Water
Bartlett, N. *et al., J. Chem. Soc.*, 1956, 3423
Violent interaction.
See related NON-METAL HALIDES

†0100. Arsine
[7784-42-1] AsH_3
$$AsH_3$$

HCS 1980, 167 (cylinders); *RSC Lab. Hazard Data Sheet No. 51*, 1986

1. Rüst, 1948, 301
2. Kayser, J. C., *Proc. 11th Int. Symp. Prev. Occup. Risks Chem. Ind.*, 289–315,
   Heidelberg, ISSA, 1987
Arsine is strongly endothermic, and can be detonated by suitably powerful initiation
[1]. The potential hazards involved in setting up and operating a plant for manufacture
of arsine for semi-conductor use have been studied [2].

Chlorine
See Chlorine: Non-metal hydrides

Nitric acid
See Nitric acid: Non-metal hydrides
See other ENDOTHERMIC COMPOUNDS, NON-METAL HYDRIDES

0101. Arsine–boron tribromide
[65313-32-8] $AsH_3 \cdot BBr_3$

$$H_3As.BBr_3$$

Oxidants
1. Stock, A., *Ber.*, 1901, **34**, 949
2. Mellor, 1939, Vol. 9, 57
Unlike arsine, the complex ignites on exposure to air or oxygen, even at below 0°C
[1]. It is violently oxidised by nitric acid [2].
See related NON-METAL HYDRIDES, NON-METAL HALIDES

0102. Trisilylarsine
[15100-34-6] AsH_9Si_3

$$(H_3Si)_3As$$

Leleu, *Cahiers*, 1977, (88), 363
Ignites in air.
See related SILANES

0103. Triazidoarsine (Arsenous triazide)
[167771-41-7] AsN_9

$$(N_3)_3As$$

See entry Tetraazidoarsonium hexafluoroarsenate
See other NON-METAL AZIDES

0104. Tetraazidoarsonium hexafluoroarsenate
[171565-26-7] $As_2F_6N_{12}$

$$(N_3)_4As^+ \ AsF_6^-$$

Tornieporth-Oetting, I. C. *et al., Angewand. Chem. (Int.)*, 1995, **34**(5), 511
Klapoetke, T. M. *et al., J. Chem. Soc., Dalton Trans.*, 1995, (20), 3365
As might be expected, it is violently explosive, as is the precursor triazidoarsine; any
other salts of this hitherto unknown arsonium ion may also prove so.
See other NON-METAL AZIDES

0105. Arsenic trioxide
[1327-53-3] As_2O_3

$$As_2O_3$$

(*MCA SD-60*, 1956); *HCS 1980*, 166

Various reagents
See Chlorine trifluoride: Metals, etc.
        Hydrogen fluoride: Oxides
        Sodium nitrate: Arsenic trioxide, etc.
        Zinc: Arsenic trioxide
See other NON-METAL OXIDES

0106. Arsenic pentaoxide
    [1303-28-2] As_2O_5

$$As_2O_5$$

Bromine pentafluoride
    See Bromine pentafluoride: Acids, etc.
    See other NON-METAL OXIDES

0107. Platinum diarsenide
    [12044-52-9] As_2Pt

$$PtAs_2$$

Preparative hazard
    See Platinum: Arsenic
    See other METAL NON-METALLIDES

0108. Arsenic trisulfide
    [1303-33-9] As_2S_3

$$As_2S_3$$

Oxidants
    See Chloric acid: Metal sulfides
        Potassium chlorate: Arsenic trisulfide
    See other NON-METAL SULFIDES

0109. Tetraarsenic tetrasulfide
    [12279-90-2] As_4S_4

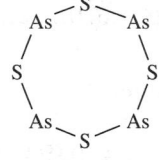

Oxidants
    See Chloric acid: Metal sulfides
        Chlorine: Sulfides
        Potassium nitrate: Metal sulfides
    See other NON-METAL SULFIDES

0110. Gold
    [7440-57-5]   **Au**

$$Au$$

Analytical hazard
    See Gold(III) chloride, next below

55

Hydrogen peroxide
See Hydrogen peroxide: Metals
See other METALS

0111. Gold(III) chloride
[13453-07-1]   AuCl$_3$

$$AuCl_3$$

Ammonia and derivatives
1. Mellor, 1941, Vol. 3, 582–583
2. Sidgwick, 1950, 178
3. *491M*, 1975, 194

Action of ammonia or ammonium salts on gold chloride, oxide or other salts under a wide variety of conditions gives explosive or 'fulminating' gold [1]. Of uncertain composition but containing Au–N bonds, this is a heat-, friction- and impact-sensitive explosive when dry, similar to the related mercury and silver compounds [2]. In an attempt to precipitate finely divided gold from its solution in aqua regia (effectively gold chloride solution), ammonia solution was added instead of ammonium oxalate. The precipitated 'gold' subsequently exploded when heated in a furnace with other metals to prepare an alloy [3].
See FULMINATING GOLD

Potassium cyanate
Selig, W. S., *Microchem. J.*, 1990, **41**(3), 386
A precipitate, explosive when touched after drying, which appeared from a neutralised solution stood overnight with a large excess of potassium cyanate, was described as a fulminate. This is unlikely, though fulminating gold is probable, cyanate hydrolysing to release ammonia (see above). Existence of explosive cyanate complexes is also conceivable.
See other GOLD COMPOUNDS, METAL HALIDES

0112. Gold(III) hydroxide–ammonia
[ ]   2AuH$_3$O$_3$.3H$_3$N

$$2Au(OH)_3.3NH_3$$

1. Sorbe, 1958, 63
2. Ephraim, 1939, 463

Explosive gold, formed from the hydroxide and ammonia, is formulated as above [1]. Dry heating forms the equally explosive Au$_2$O$_3$.3NH$_3$, then 3Au$_2$O.4NH$_3$, while heating with water forms the more explosive Au$_2$O$_3$.2NH$_3$.
See FULMINATING GOLD *See other* GOLD COMPOUNDS

0113. Sodium triazidoaurate(?)
[ ]   AuN$_9$Na

$$Na[Au(N_3)_3]$$

Rodgers, G. T., *J. Inorg. Nucl. Chem.*, 1958, **5**, 339–340

56

The material (of unknown structure, analysing as $Au_{1.5}N_9Na$ and possibly the impure title compound, explodes at 130°C.
See other GOLD COMPOUNDS, METAL AZIDES

0114. Bis(dihydroxygold)imide
[ ] $Au_2H_5NO_4$

$$HN[Au(OH)_2]_2$$

Mellor, 1940, Vol. 8, 259
An explosive compound.
See other GOLD COMPOUNDS, *N*-METAL DERIVATIVES

0115. Gold(III) oxide
[1303-58-8] Au_2O_3

$$Au_2O_3$$

Ammonium salts
  See Gold(III) chloride: Ammonia, etc.
  See other ENDOTHERMIC COMPOUNDS, GOLD COMPOUNDS

0116. Gold(III) sulfide
[1303-61-3] Au_2S_3

$$Au_2S_3$$

Silver oxide
  See Silver(I) oxide: Metal sulfides
  See other GOLD COMPOUNDS, METAL SULFIDES

0117. Gold(I) nitride–ammonia
[ ] $Au_2N_3.H_3N$

$$Au_2N_3.NH_3$$

Raschig, F., *Ann.*, 1886, **235**, 349
An explosive compound, probably present in fulminating gold, produced from action of ammonia on gold(I) oxide.
See FULMINATING GOLD *See other* GOLD COMPOUNDS, *N*-METAL DERIVATIVES

0118. Gold(III) nitride trihydrate
[ ] $Au_3N_2.3H_2O$

$$Au_3N_2.3H_2O \text{ or } (HOAu)_3N.NH_3$$

Mellor, 1940, Vol. 8, 101

Very explosive when dry.
See other GOLD COMPOUNDS, *N*-METAL DERIVATIVES

0119. Boron
[7440-42-8]

B

B

Bailar, 1973, Vol. 1, 692
Many of the previously described violent reactions of boron with a variety of reagents are ascribed to the use of impure or uncharacterised 'boron'. The general impression of the reactivity of pure boron, even when finely divided, is one of extreme inertness, except to highly oxidising agents at high temperatures.

Ammonia
Mellor, 1940, Vol. 8, 109
Boron incandesces when heated in dry ammonia, hydrogen being evolved.

Dichromates, Silicon
Howlett, S. *et al., Thermochim. Acta*, 1974, **9**, 213–216
The mechanism of ignition and combustion of pyrotechnic mixtures of boron with potassium dichromate and/or sodium dichromate in presence or absence of silicon are discussed.

Rubidium acetylide
See Rubidium acetylide: Non-metals

Halogens or Interhalogens
1. Mellor, 1941, Vol. 2, 92
2. Bailar, 1973, Vol. 1, 690–691
Boron ignites in gaseous chlorine or fluorine at ambient temperature, attaining incandescence in fluorine [1]. Powdered boron reacts spontaneously with the halogens from fluorine to iodine at 20, 400, 600 and 700°C respectively [2].
See Bromine trifluoride: Halogens, etc.
    Bromine pentafluoride: Acids, etc.
Chlorine trifluoride: Boron-containing materials
Iodine pentafluoride: Metals, etc.

Metals
Mellor, 1946, Vol. 3, 389; Vol. 5, 15
Explosive interaction when boron and lead fluoride or silver fluoride are ground together at ambient temperature.

Oxidants
See Halogens or Interhalogens, above
    Calcium chromate: Boron
    Dinitrogen oxide: Boron

Lead(II) oxide: Non-metals
Lead(IV) oxide: Non-metals
Nitric acid: Non-metals
Nitrogen oxide: Non-metals
Nitrosyl fluoride: Metals, etc.
Nitryl fluoride: Non-metals
Oxygen difluoride: Non-metals
Potassium nitrite: Boron
Potassium nitrate: Non-metals
Silver difluoride: Boron, etc.
Sodium peroxide: Non-metals

Water
1. Bailar, 1973, Vol. 1, 691
2. Shidlovskii, A. A., *Chem. Abs.*, 1963, **59**, 11178
Interaction of powdered boron and steam may become violent at red heat [1]. The highly exothermic reactions with water might become combustive or explosive processes at sufficiently high temperatures and pressures [2].
See other NON-METALS

0120. Boron bromide diiodide
[14355-21-6] $BBrI_2$

$$BrBI_2$$

Water
Mellor, 1946, Vol. 5, 136
Interaction is violent, as for the tribromide or triiodide.
See other NON-METAL HALIDES

0121. Dibromoborylphosphine
[30641-57-7] (polymeric form) BBr_2H_2P

$$Br_2BPH_2$$

491M, 1975, 146
Ignites in air.
See related NON-METAL HALIDES, NON-METAL HYDRIDES

0122. Boron tribromide
[10294-33-4] BBr_3

$$BBr_3$$

Sodium
See Sodium: Non-metal halides (reference 7)

Tungsten trioxide
Levason, W. *et al., J. Chem. Soc., Dalton Trans.*, 1981, 2501–2507

Tungsten oxide tetrabromide was prepared by condensing a little of the bromide onto the oxide at $-196°C$, then allowing slow warming by immersion of the container in an ice bath. Omission of the ice bath or use of large amounts of bromide may lead to explosions.

Water
1. *BCISC Quart. Safety Summ.*, 1966, **37**, 22
2. Anon., *Lab. Pract.*, 1966, **15**, 797

Boron halides react violently with water, and particularly if there is a deficiency of water, a violent explosion may result. It is therefore highly dangerous to wash glass ampoules of boron tribromide with water under any circumstances. Following a serious accident, experiment showed that an ampoule of boron tribromide, when deliberately broken under water, caused a violent explosion, possibly a detonation. Only dry non-polar solvents should be used for cleaning or cooling purposes [1]. Small quantities of boron tribromide may be destroyed by cautious addition to a large volume of water, or water containing ice [2].

See other GLASS INCIDENTS
See other NON-METAL HALIDES

0123. Difluoroperchloryl tetrafluoroborate
[38682-34-7] $BClF_6O_2$
$$[F_2ClO_2] [BF_4]$$

See entry DIFLUOROPERCHLORYL SALTS

0124. Dichloroborane
[10325-39-0] BCl_2H
$$Cl_2BH$$

Bailar, 1973, Vol. 1, 742
Ignites in air.
See related NON-METAL HALIDES

0125. Dichlorodisilylaminoborane
[25573-61-9] $BCl_2H_6NSi_2$
$$Cl_2BN(SiH_3)_2$$

491M, 1975, 149
Ignites in air.
See related NON-METAL HALIDES, NON-METAL HYDRIDES

0126. Boron azide dichloride
[ ] BCl_2N_3
$$N_3BCl_2$$

1. Anon., *Angew. Chem. (Nachr.)*, 1970, **18**, 27
2. Paetzold, P. I., *Z. Anorg. Chem.*, 1963, **326**, 47

A hard crust of sublimed material exploded when crushed with a spatula [1]. Explosions on sublimation or during solvent removal were known previously [2].
See related NON-METAL AZIDES

0127. Boron trichloride
[10294-34-5] BCl_3

$$BCl_3$$

HCS 1980, 215 (cylinder)

Aniline
Jones, R. G., *J. Amer. Chem. Soc.*, 1939, **61**, 1378
In absence of cooling or diluent, interaction is violent.

Dinitrogen tetraoxide
See Dinitrogen tetraoxide: Boron trichloride

Hexafluoroisopropylideneaminolithium
See Hexafluoroisopropylideneaminolithium: Non-metal halides

Methanol
Anon., *Safety Digest Univ. Safety Assoc.*, 1989 **34**, 16
On addition of boron trichloride, dropwise, to methanol an explosion and fire resulted. The academic reporter appears unaware that reaction of these was reported in 1834, and attributes the explosion to water in his methanol. Since the chloride is a gas, ability to add it dropwise is puzzling. Had a phonetic error meant PCl_3 was employed the fire would be intelligible, since phosphine would be formed as a pyrophoric byproduct.

Phosphine
Mellor, 1946, Vol. 5, 132
Interaction is energetic.

Triethylsilane
Matteson, D. S. *et al.*, *J. Org. Chem.*, 1990, **55**, 2274
A pressure build-up, septum expulsion and combustion was experienced on mixing these reagents at $-78°C$.
See other NON-METAL HALIDES

0128. Dicobalt boride
[12045-01-1] BCo_2

$$Co_2B$$

Heinzmann, S. W. *et al.*, *J. Amer. Chem. Soc.*, 1982, **104**, 6801 (footnote 22)
The boride precipitated from sodium borohydride and cobalt(II) chloride in methanol becomes pyrophoric after vacuum drying. It can safely be stored solvent-moist.
See other METAL NON-METALLIDES, PYROPHORIC MATERIALS

0129. Boron trifluoride
      [7637-07-2]   BF₃
$$BF_3$$

HCS 1980, 216 (cylinder)

Alkali metals, or Alkaline earth metals (not magnesium)
   Merck Index, 1976, 175
   Interaction hot causes incandescence.

Alkyl nitrates
   See ALKYL NITRATES: Lewis acids
   See other NON-METAL HALIDES

0130. Tetrafluoroboric acid
      [16872-11-0]   BF₄H
$$H[BF_4]$$

Acetic anhydride
   See Acetic anhydride: Tetrafluoroboric acid
   See other INORGANIC ACIDS

0131. Nitronium tetrafluoroborate
      [13826-86-3]   BF₄NO₂
$$NO_2^+ \ [BF_4]^-$$

Tetrahydrothiophene-1,1-dioxide
   See entry NITRATING AGENTS

0132. Dioxygenyl tetrafluoroborate
      [12228-13-6]   BF₄O₂
$$O_2^+ \ [BF_4]^-$$

Organic materials
   Goetschel, C. T. *et al., J. Amer. Chem. Soc.*, 1969, **91**, 4706
   It is a very powerful oxidant, addition of a small particle to small samples of benzene
   or 2-propanol at ambient temperature causing ignition. A mixture prepared at −196°C
   with either methane or ethane exploded when the temperature was raised to −78°C.
   See other OXIDANTS

0133. Tetrafluoroammonium tetrafluoroborate
      [15640-93-4]   BF₈N
$$[NF_4]^+ \ [BF_4]^-$$

2-Propanol
   Goetschel, C. T. *et al., Inorg. Chem.*, 1972, **11**, 1700

When the fluorine used for synthesis contained traces of oxygen, the solid behaved as a powerful oxidant (causing 2-propanol to ignite on contact) and it also exploded on impact. Material prepared from oxygen-free fluorine did not show these properties, which were ascribed to the presence of traces of dioxygenyl tetrafluoroborate (above).

See other CATALYTIC IMPURITY INCIDENTS, N-HALOGEN COMPOUNDS

0134. Poly[borane(1)]
[13766-26-2] $(BH)_n$

$$(BH)_n$$

Bailar, 1973, Vol. 1, 740
Ignites in air.
See other BORANES

0135. Borane
[13283-31-3] BH_3

$$BH_3$$

The monomeric borane is extremely endothermic (ΔH_f° +105.5 kJ/mol, 7.62 kj/g) and on formation apparently immediately dimerises to diborane (or higher boranes). It is usually stabilised as the monomer by the formation of various complexes with N, O, P or S donor molecules and many of these are available commercially.
See other BORANES, ENDOTHERMIC COMPOUNDS

0136. Borane–dimethylsulfide
[13292-87-0] $BH_3.C_2H_6S$

$$(CH_3)_2S.BH_3$$

Clinton, F. L., *Chem. Eng. News*, 1992, **70**(35), 5
Bottles of this complex sometimes pressurise during ambient temperature storage. The cause, which it is hoped has now been eliminated, may be traces of thiols reacting to form hydrogen. Cold storage, careful opening and regular venting are desirable.
See other BORANES

0137. Borane–bis(2,2-dinitropropylhydrazine)
[ ] $BH_3.2C_3H_8N_4O_4$

$$BH_3.2H_2NNHCH_2C(NO_2)_2Me$$

Gao, F. *et al.*, *Youji Huaxue*, 1984, (2), 123–124 (Ch.)
It decomposes rapidly at ambient temperature and ignites or explodes within 1–2 h.
See related BORANES, POLYNITROALKYL COMPOUNDS *See other* REDOX COMPOUNDS

0138. Borane–tetrahydrofuran
[14044-65-6] $BH_3.C_4H_8O$

$$C_4H_8O.BH_3$$

1. Bruce, M. I., *Chem. Eng. News*, 1974, **52**(41), 3
2. Hopps, H., *Chem. Eng. News*, 1974, **52**(41), 3
3. Gaines, D. F. *et al., Inorg. Chem.*, 1963, **2**, 526
4. Kollonitsch, J., *Chem. Eng. News*, 1974, **52**(47), 3

A glass bottle containing a 1M solution of the complex in THF exploded after 2 weeks in undisturbed laboratory storage out of direct sunlight at 15°C [1]. The problem of pressure build-up during storage of such commercial solutions (which are stabilised with 5 mol% of sodium tetrahydroborate) at above 0°C had been noted previously, and was attributed to presence of moisture in the original containers [2]. However, by analogy with the known generation of hydrogen in tetrahydoborate–diborane–bis(2-methoxyethyl) ether systems [3], it is postulated that the tetrahydroborate content may in fact destabilise the borane–THF reagent, with generation of hydrogen pressure in the closed bottle [4]. Storage at 0°C and opening bottles behind a screen are recommended [2].

See other GLASS INCIDENTS, GAS EVOLUTION INCIDENTS
See related BORANES

0139. Borane–pyridine
[110-51-0] $BH_3.C_5H_5N$

$$C_5H_5N.BH_3$$

1. Brown, H. C. *et al., J. Amer. Chem. Soc.*, 1956, **78**, 5385
2. Baldwin, R. A. *et al., J. Org. Chem.*, 1961, **26**, 3550

Decomposition was rapid at 120°C/7.5 mbar [1], and sometimes violent on attempted distillation at reduced pressure [2].

See related BORANES

0140. Borane–phosphorus trifluoride
[14391-39-6] $BH_3.F_3P$

$$H_3B.PF_3$$

Mellor, 1971, Vol. 8, Suppl. 3, 442
The unstable gas ignites in air.
See related BORANES, NON-METAL HALIDES

0141. Borane–ammonia
[17596-45-1] $BH_3.H_3N$

$$H_3B.NH_3$$

Sorbe, 1968, 56
It may explode on rapid heating.
See related BORANES

64

0142. Borane–hydrazine
[14391-40-9] $BH_3.H_4N_2$

$$H_3B.N_2H_4$$

Gunderloy, F. C., *Inorg. Synth.*, 1967, **9**, 13
It is shock-sensitive and highly flammable, like the bis(borane) adduct.
See related BORANES

0143. Bis(borane)–hydrazine
[13730-91-1] $(BH_3)_2.H_4N_2$

$$H_3B.NH_2-H_2N.BH_3$$

See Borane–hydrazine *See related* BORANES

0144. Boric acid
[10043-35-3] BH_3O_3

$$B(OH)_3$$

HCS 1980, 214

Acetic anhydride
  See Acetic anhydride: Boric acid

Potassium
  See Potassium: Oxidants
  See other INORGANIC ACIDS

0145. Lithium tetrahydroborate (Lithium borohydride)
[16949-15-8] BH_4Li

$$Li[BH_4]$$

Water
  Gaylord, 1965, 22
  Contact with limited amounts of water, either as liquid or that present as moisture in cellulose fibres, may cause ignition after a delay.
  See other COMPLEX HYDRIDES

0146. Ammonium peroxoborate
[17097-12-0] BH_4NO_3

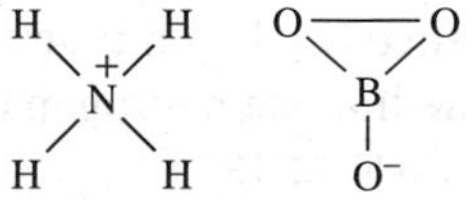

Menzel, H. *et al.*, *Österr. Chem. Z.*, 1925, **28**, 162
Explosive decomposition under vacuum.
See other PEROXOACID SALTS

0147. Sodium tetrahydroborate (Sodium borohydride)
[16940-66-2]   **BH₄Na**

$$Na[BH_4]$$

HCS 1980, 830; *RSC Lab. Hazard Data Sheet No. 39*, 1985

1. le Noble, W. J., *Chem. Eng. News*, 1983, **61**(19), 2
2. le Noble, W. J., private comm., 1983

A several year-old 100 ml glass bottle which had originally contained 25 g of the complex hydride exploded while being opened a few hours after the previous opening to remove a portion [1]. No plausible explanation can be deduced, and a dust explosion seems unlikely in view of the large initiation energy required [2].
See other GLASS INCIDENTS

Acetic acid, Dichloromethane, Methanol
  Ward, D. E. *et al., Tetrahedron Lett.*, 1988, **29**, 517–520
  In a general method for the selective reduction of ketones in presence of conjugated enones, this is effected by the tetrahydroborate in 1:1 methanol–dichloromethane at 75°C. In favourable cases the reaction is carried out at 20°C in dichloromethane containing a little acetic acid. It should be noted that addition of acetic acid to sodium tetrahydroborate in methanol–dichloromethane leads to vigorous evolution of much hydrogen.
  See other GAS EVOLUTION INCIDENTS

Acids
  Bailar, 1973, Vol. 1, 768
  Interaction of sodium and other tetrahydroborates with anhydrous acids (fluorophosphoric, phosphoric or sulfuric) to generate diborane is very exothermic, and may be dangerously violent with rapid mixing. Safer methods of making diborane are detailed.

Alkali
1. Anon., *Angew. Chem. (Nachr.)*, 1960, **8**, 238
2. Volpers, *Proc. 1st Int. Symp. Prev. Occup. Risks Chem. Ind.*, 188–193, Heidelberg, ISSA, 1970
3. Mikheeva, V. I. *et al., Chem. Abs.*, 1969, **71**, 85064

A large volume of alkaline tetrahydroborate solution spontaneously heated and decomposed, liberating large volumes of hydrogen which burst the container. Decomposition is rapid when pH is below 10.5 [1]. A more detailed account of the investigation was published [2]. Dry mixtures with sodium hydroxide containing 15–40% of tetrahydroborate liberate hydrogen explosively at 230–270°C [3].
See other GAS EVOLUTION INCIDENTS

Aluminium chloride, Bis(2-methoxyethyl) ether
1. de Jongh, H. A. P., *Chem. Eng. News*, 1977, **55**(31), 31
2. Brown, H. C., *Chem. Eng. News*, 1977, **55**(35), 5

Addition of a 4% solution of sodium tetrahydroborate in diglyme containing 0.09% of water to a 27% solution of aluminium chloride in the same solvent led to a violent explosion, attributed to formation and ignition of hydrogen. The ignition source arose from contact of the hydroborate solution with the solid chloride, as demonstrated experimentally. Nitrogen purging is essential for all hydride reductions [1], and also for hydroboration, organoborane, Grignard and organometallic reactions generally [2]. Previous work had shown that clear solutions of the sodium tetrahydroborate–aluminium chloride reagent did not ignite in dry air, but the solid-containing reagent could lead to ignition [2].

Charcoal
   Collin, P. A., Sidgwick, C., *Chemistry in Britain*, 1992, **28**(4), 324
   Mixtures of charcoal and borohydride are liable to autoignition in air, the probability being higher if the charcoal has been exposed to a damp atmosphere.
   See other PYROPHORIC MATERIALS

Diborane, Bis(2-methoxyethyl) ether
   See Borane–tetrahydrofuran (reference 3)

Dimethylformamide
   1. Yeowell, A. *et al., Chem. Eng. News*, 1979, **57**(39), 4
   2. Schwartz, J. *et al., J. Org. Chem.*, 1993, **58**(18), 5005
   3. Thiokol-Ventron announcement in *Cambrian News*, 1–2, May 1980, K & K Greeff Chemicals
   Hot solutions of the tetrahydroborate (15.7% wt) in DMF will undergo a violent runaway thermal decomposition, the solid residue attaining a temperature of 310°C. The induction period depends on temperature, and is 45 h at 62, and 45 m at 90°C. In a plant-scale incident, an 83 kg batch led to a violent explosion, ascribed to spontaneous ignition of trimethylamine (AIT 190°C) produced by reduction of the solvent [1]. Further investigation confirmed these results, whether technical, reagent grade or redistilled dry solvent were used, the induction period being independent of added water, amine, metal salts or borate. However, addition of formic (or acetic) acid significantly reduced the induction period, and if the solvent containing 2.6% of formic acid is added to the hydride at ambient temperature, immediate and violent decomposition ensues, involving formation of sodium formyloxytrihydroborate. The latter is responsible for reduction of the amide solvent to trimethylamine. Traces of formic acid and dimethylamine are present in the commercial solvent from hydrolysis, and the latter is catalysed by water, acid or base. Hot solutions of the hydride in DMF above 2M concentration will soon begin to to undergo hydrolysis/reduction reactions, both of which produce formic acid and/or di- or tri-methylamine, so the decomposition is autocatalytic and soon accelerates out of control, causing gross and violent reduction of the solvent. Further investigation is reported and the greater reduction powers of the intermediate exploited [2]. Dimethylacetamide does not react violently with sodium tetrahydroborate, even at 4.7M concentration, and should be considered as a substitute solvent, particularly at higher temperatures [3].
   See other GAS EVOLUTION INCIDENTS, INDUCTION PERIOD INCIDENTS

Glycerol
Epshtein, N. A. *et al., Chem. Abs.*, 1987, **106**, 55145
Contact of a drop of glycerol with a flake of sodium tetrahydroborate leads to ignition, owing to thermal decomposition of the latter at above 200°. Other glycols and methanol also react exothermally, but do not ignite.

Palladium
See Palladium: Sodium tetrahydroborate

Ruthenium salts
Cusumano, J. A., *Nature*, 1974, **247**, 456
Use of borohydride solutions to reduce ruthenium salt solutions to the metal or an alloy gave solid products (possibly hydrides), which when dry, exploded violently in contact with water or when disturbed by a spatula. Hydrazine appears to be a safe reducant for ruthenium salt solutions.

Sulfuric acid
Pascal, 1961, Vol. 6, 337
Ignition may occur if the mixture is not cooled.
See other COMPLEX HYDRIDES

0148. Boron azide diiodide
[68533-38-0] BI_2N_3

$$N_3BI_2$$

Dehnicke, K., *Angew. Chem. (Intern. Ed.)*, 1979, **18**, 510
Decomposes explosively in contact with water.
See related NON-METAL AZIDES, NON-METAL HALIDES

0149. Boron diiodophosphide
[12228-28-3] BI_2P

$$BPI_2$$

Chlorine
See Chlorine: Phosphorus compounds

Metals
Mellor, 1947, Vol. 8, 845
It ignites in contact with mercury vapour or magnesium powder.
See related NON-METAL HALIDES

0150. Boron triiodide
[13517-10-7] BI_3

$$BI_3$$

Though a moderately endothermic compound (ΔH°_f +70.8 kJ/mol), its high MW gives it the rather low specific energy content of 0.18 kJ/g.

Ammonia
Mellor, 1945, Vol. 5, 136
Strong exotherm on contact.

Phosphorus
Mellor, 1946, Vol. 5, 136
Warm red or white phosphorus reacts incandescently.

Water
1. Moissan, H., *Compt. rend.*, 1892, **115**, 204
2. Unpublished information
Violent reaction [1], particularly with limited amounts of water [2].
See other IODINE COMPOUNDS, NON-METAL HALIDES

0151. Lithium tetraazidoborate
[ ]   BLiN$_{12}$

$$Li[B(N_3)_4]$$

Wiberg, E. *et al.*, *Z. Naturforsch.*, 1954, **9B**, 499
Highly explosive, sensitive to heat, impact and friction (e.g. of a spatula when removing solid from a flask).
See related NON-METAL AZIDES *See other* HIGH-NITROGEN COMPOUNDS

0152. Boron nitride
[10043-11-5]   BN

BN

Peroxydisulfuryl difluoride
See Tetra(boron nitride) fluorosulfate

Sodium peroxide
See Sodium peroxide: Boron nitride
See related NON-METALS

0153. Boron triazide (Triazidoborane)
[21844-15-1]   BN$_9$

$$B(N_3)_3$$

1. Anon., *Angew. Chem. (Nachr.)*, 1970, **18**, 27
2. Skillern, K. R. *et al.*, *Inorg. Chem.*, 1977, **16**, 3001
3. Miller, N. E. *et al.*, *Inorg. Chem.*, 1964, **3**, 1064
A sample of the vacuum distilled pyridine complex exploded in a heated capillary sampling tube [1]. Detonation of the trimethylamine complex [2], at or near 200°C [3] is also noted.

Diethyl ether, or Water
Wiberg, E. *et al.*, *Z. Naturforsch.*, 1954, **9B**, 498

The highly explosive material detonated in contact with ether vapour or water at ambient temperature, or with ether at −35°C (probably initiated by the heat of coordination to O).

See other NON-METAL AZIDES

0154. Sodium borate hydrogen peroxidate
[56892-92-5] $BNaO_2.H_2O_2$

$$O:BONa.H_2O_2$$

HCS 1980, 853

See entry CRYSTALLINE HYDROGEN PEROXIDATES
See Sodium peroxoborate (next below)

0155. Sodium peroxoborate
[7632-04-4] $BNaO_3$

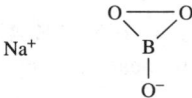

1. Anon., *Angew. Chem.*, 1963, **65**, 41
2. Castrantas, 1965, 5

The true peroxoborate has been reported to detonate on light friction [1]. The common 'tetrahydrate' is not a peroxoborate, but $NaBO_2.H_2O_2.3H_2O$, and while subject to catalytic decomposition by heavy metals and their salts, or easily oxidisable foreign matter, it is relatively stable under mild grinding with other substances [2].

See also CRYSTALLINE HYDROGEN PEROXIDATES
See other CATALYTIC IMPURITY INCIDENTS, PEROXOACID SALTS

0156. Boron phosphide
[20205-91-8]   BP

$$BP$$

Oxidants
  See Nitric acid: Non-metals
      Sodium nitrate: Boron phosphide
  See related NON-METALS

0157. Beryllium tetrahydroborate
[17440-85-6] B_2BeH_8

$$Be[BH_4]_2$$

Air, or Water
1. Mackay, 1966, 169
2. Semenenko, K. N. *et al.*, *Russ. Chem. Rev.*, 1973, 4

It ignites vigorously and often explodes in air [1], or on contact with cold water [2].
See other COMPLEX HYDRIDES

0158. Calcium tetrahydroborate
[17068-96-0] B_2CaH_8

$$Ca[BH_4]_2$$

Köster, R. *et al., Inorg. Synth.*, 1977, **17**, 18
Like other complex hydrides, it reacts vigorously with protic materials and ignition
may occur.

Tetrahydrofuran
Köster, R. *et al., Inorg. Synth.*, 1977, **17**, 18
Dissolution of the 90% pure hydride in dry THF is extremely exothermic.
See other COMPLEX HYDRIDES

0159. Bromodiborane
[23834-96-0] B_2BrH_5

Drake, J. E., *Inorg. Synth.*, 1978, **18**, 146
May ignite violently in air.
See other HALOBORANES

0160. Chlorodiborane
[17927-57-0]   . B_2ClH_5

Bailar, 1973, Vol. 1, 778
A gas above −11°C, it ignites in air.
See other HALOBORANES

0161. Diboron tetrachloride
[13701-67-2] B_2Cl_4

$$Cl_2BBCl_2$$

Wartik, T. *et al., Inorg. Synth.*, 1967, **10**, 125
Sudden exposure to air may cause explosion.

Dimethylmercury
Wartik, T. *et al., Inorg. Chem.*, 1971, **10**, 650

The reaction, starting at −63°C under vacuum, exploded violently on 2 occasions after 23 uneventful runs.

See other HALOBORANES, NON-METAL HALIDES

0162. Diboron tetrafluoride
[13965-73-6] B_2F_4

$$F_2BBF_2$$

Metal oxides
Holliday, A. K. *et al., J. Chem. Soc.*, 1964, 2732
Mixtures with mercury(II) oxide and manganese dioxide prepared at −80°C ignited at 20° and reacted violently at 15°C, respectively. Copper(II) oxide reacted vigorously at 25°C without ignition.

Oxygen
Trefonas, L. *et al., J. Chem. Phys.*, 1958, **28**, 54
The gas is extremely explosive in presence of oxygen.
See other HALOBORANES, NON-METAL HALIDES

0163. Potassium hypoborate
[63706-85-4] $B_2H_2K_2O_4$

$$K^+ \ O^- - B \underset{O}{\overset{O}{<}} B - O^- \ K^+$$

with H attached to top O and bottom O

Mellor, 1946, Vol. 5, 38
As a reducant stronger than the phosphinate, it may be expected to interact more vigorously with oxidants.
See Potassium phosphinate: Air, etc.
See other REDUCANTS

0164. Sodium hypoborate
[16903-32-5] $B_2H_2Na_2O_4$

$$NaOB(OH)B(OH)ONa$$

Mellor, 1946, Vol. 5, 38
As a reducant stronger than sodium phosphinate, it may be expected to interact more vigorously with oxidants.
See Sodium phosphinate: Oxidants
See other REDUCANTS

0165. Iododiborane
[20436-27-5] B_2H_5I

$$H-B(I)(H)-H-B(H)(I)-H$$

Drake, J. E., *Inorg. Synth.*, 1978, **18**, 146
May ignite violently in air.
See other HALOBORANES

†0166. Diborane(6)
[19287-45-7] B_2H_6

$$H_2B(\mu-H)_2BH_2$$

(*MCA SD-84*, 1961); *FPA H81*, 1978; *HCS 1980*, 363 (cylinder)

Preparative hazard
1. Mellor, 1946, Vol. 5, 36
2. Schlessinger, H. I. *et al., Chem. Rev.*, 1942, **31**, 8
3. *Hydrides of Boron and Silicon*, Stock, A. E., Ithaca, Cornell Univ. Press, 1933
4. Follet, M., *Chem. & Ind.*, 1986, 123–128
5. Hariguchi, S. *et al., Chem. Abs.*, 1989, **110**, 62931

The endothermic gas (ΔH°_f (g) +31.3 kJ/mol, 1.12 kJ/g) usually ignites in air unless dry and free of impurities [1], and ignition delays of 3–5 days, followed by violent explosions, have been experienced [2]. Explosion followed spillage of liquid diborane [3]. Problems in industrial-scale use of diborane are discussed, and the advantages of 2 new solid and non-flammable diborane complexes with 1,2-bis(*tert*-butylthio)ethane and 1,4-bis(benzylthio)butane, and the pyrophoric complex with dimethyl sulfide are outlined [4]. Flammability limits and explosion pressures of diborane in mixtures with air, nitrogen or helium, and with hydrogen and air, have been studied. Limits for diborane–air mixtures were 0.84 to 93.3 vol%, and presence of nitrogen or helium was not effective in suppressing explosion, the limiting oxygen value being 1.3 vol%. Explosion pressure in air was 11 bar max. at 10% diborane, with a rate of rise of 1200 bar/s. Presence of hydrogen increases the explosive range. Extraordinary precautions are required for industrial use of diborane [5].
See GAS HANDLING (reference 2)
See Sodium tetrahydroborate: Acids
See Benzene, or Moisture; Oxygen, both below

Ammonia
See Diammineboronium tetrahydroborate

Benzene, or Moisture
Simons, H. P. *et al., Ind. Eng. Chem.*, 1958, **50**, 1665, 1659
Effects of presence of moisture or benzene vapour in air on the spontaneously explosive reaction have been studied.

Chlorine
See Chlorine: Non-metal hydrides

Dimethyl sulfoxide
See Dimethyl sulfoxide: Boron compounds

Halocarbons
Haz. Chem. Data, 1975, 114
Diborane reacts violently with halocarbon liquids used as vaporising fire-extinguishants.

Metals
Haz. Chem. Data, 1975, 114
Interaction with aluminium or lithium gives complex hydrides which may ignite in air.

Nitrogen trifluoride
See Nitrogen trifluoride: Diborane

Octanal oxime, Sodium hydroxide
Augustine, 1968, 78
Addition of sodium hydroxide solution during work-up of a reaction mixture of oxime and diborane in THF is very exothermic, a mild explosion being noted on one occasion.

Oxygen
Whatley, A. I. *et al., J. Amer. Chem. Soc.*, 1954, **76**, 1997–1999
Mixtures at 105–165°C exploded spontaneously after an induction period dependent on temperature and composition.
See title reference 2

Tetravinyllead
Houben-Weyl, 1975, Vol. 13.3, 253
Interaction is explosively violent at ambient temperature.
See other BORANES, ENDOTHERMIC COMPOUNDS

0167. Diammineboronium tetrahydroborate
[23777-63-1] $B_2H_{12}N_2$

$$[BH_2(NH_3)_2]^+[BH_4]^-$$

1. Sidgwick, 1950, 354
2. Bailar, 1973, Vol. 1, 924
The diammine complex of diborane (formulated as above), though less reactive than diborane, ignites on heating in air [1,2].
See related BORANES *See other* COMPLEX HYDRIDES

74

0168. Magnesium boride
[12007-25-9] B_2Mg_3

$$Mg_3B_2$$

Acids
   Mellor, 1946, Vol. 5, 25
   The crude product containing some silicide evolves, in contact with hydrochloric or
   sulfuric acids, a mixture of borane and silane which may ignite.
   See other METAL NON-METALLIDES

0169. Diboron oxide
[12505-77-0] B_2O_2

$$O:BB:O$$

1. Merck, 1976, 1365
2. Halliday, A. K. *et al., Chem. Rev.*, 1962, **62**, 316
Probably polymeric at ambient temperature [1], at 400°C, traces of water react to
cause a violent eruption and incandescence [2].
See other NON-METAL OXIDES

0170. Boron trioxide
[1303-86-2] B_2O_3

$$O:BOB:O$$

Bromine pentafluoride
   See Bromine pentafluoride: Acids, etc.
   See other NON-METAL OXIDES

0171. Thallium(I) peroxodiborate
[ ] $B_2O_7Tl_2.H_2O$

$$TlOB(O_2)OB(O_2)OTl.H_2O$$

Bailar, 1973, Vol. 1, 1154
It liberates oxygen at 18°C and explodes on further warming.
See other PEROXOACID SALTS

0172. Boron trisulfide
[12007-33-9] B_2S_3

$$S:BSB:S$$

Chlorine
   See Chlorine: Sulfides

Water
   Partington, 1967, 415
   Hydrolysis of the sulfide is violent.
   See other NON-METAL SULFIDES

0173. 1,3,5-Trichloro-2,4,6-trifluoroborazine
[56943-26-1] $B_3Cl_3F_3N_3$

Water
Elter, G. *et al., Angew. Chem. (Intern. Ed.)*, 1975, **14**, 709
Hydrolysis of this *N*-chloro-*B*-fluoro compound is explosively violent.
See other N-HALOGEN COMPOUNDS, NON-METAL HALIDES

0174. *B*-1,3,5-Trichloroborazine
[26445-82-9] $B_3Cl_3H_3N_3$

Water
Niedenzu, K. *et al., Inorg. Synth.*, 1967, **10**, 141
Hydrolysis of the *B*-chloro compound is violent.
See related NON-METAL HALIDES

0175. Triboron pentafluoride
[15538-67-7] B_3F_5
$$B_3F_5$$

Air, or Water
Timms, P. L., *J. Amer. Chem. Soc.*, 1967, **89**, 1631
It reacts explosively with air or water.

Tetrafluoroethylene
Timms, P. L., *J. Amer. Chem. Soc.*, 1967, **89**, 1631
The pentafluoride catalyses polymerisation of tetrafluoroethylene smoothly below
−100°C, but explosively above that temperature.
See other NON-METAL HALIDES
See other POLYMERISATION INCIDENTS

76

0176. Borazine
[6569-51-3] $B_3H_6N_3$

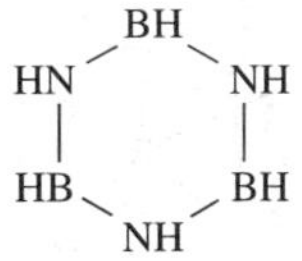

$$
\begin{array}{c}
\text{BH} \\
\text{HN} \quad \text{NH} \\
\text{HB} \quad \text{BH} \\
\text{NH}
\end{array}
$$

Niedenzu, K. *et al., Inorg. Synth.*, 1967, **10**, 144
Samples sealed into ampoules exploded when stored in daylight, but not in the dark.
See other IRRADIATION DECOMPOSITION INCIDENTS
See related NON-METAL HYDRIDES

0177. Sodium octahydrotriborate
[12007-46-4] B_3H_8Na

$$Na[B_3H_8]$$

Solvents
Dewkett, W. J. *et al., Inorg. Synth.*, 1974, **15**, 116
Air should not be drawn through solutions of the compound in ether, or through its
solid complex with dioxane, because such materials have occasionally ignited in air.
See other COMPLEX HYDRIDES

0178. Uranium(III) tetrahydroborate
[ ] $B_3H_{12}U$

$$U(BH_4)_3$$

Semenenko, K. N. *et al., Russ. Chem. Rev.*, 1973, 4
It ignites in air and explodes on heating, unlike the U(VI) analogue.
See other COMPLEX HYDRIDES

0179. Tetraboron tetrachloride
[17156-85-3] B_4Cl_4

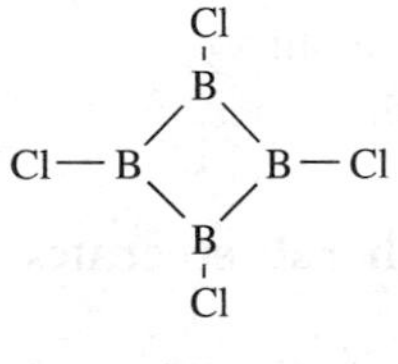

$$
\begin{array}{c}
\text{Cl} \\
\text{B} \\
\text{Cl—B} \quad \text{B—Cl} \\
\text{B} \\
\text{Cl}
\end{array}
$$

Urry, G. *et al., Inorg. Chem.*, 1963, **2**, 398
Ignites in air.
See other HALOBORANES

0180. Tetra(boron nitride) fluorosulfate
[68436-99-7]

$B_4FN_4O_3S$

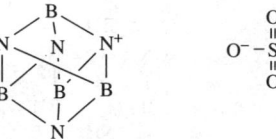

Bartlett, N. *et al., J. Chem. Soc., Chem. Comm.*, 1978, 201

Apparently thermodynamically unstable, because when the peroxodisulfuryl difluoride−boron nitride reaction mixture was heated to 40°C, detonations occurred.

0181. Tetraborane(10)
[18283-93-7]

B_4H_{10}

Oxidants

1. Mellor, 1946, Vol. 5, 36
2. Bailar, 1973, Vol. 1, 790

It ignites in air or oxygen, and explodes with conc. nitric acid [1]. The pure compound is stated not to ignite in air [2].

See Nitric acid: Non-metal hydrides
See other BORANES

0182. Hafnium(IV) tetrahydroborate
[25869-93-6]

$B_4H_{16}Hf$

$Hf(BH_4)_4$

Gaylord, 1956, 58

Violent ignition on exposure to air.
See other COMPLEX HYDRIDES

0183. Uranium(IV) tetrahydroborate etherates
[65579-07-9]

$B_4H_{16}U{\cdot}nR_2O$

$U(BH_4)_4{\cdot}(OR_2)_n$

Water

Rietz, R. R. et al., *Inorg. Chem.*, 1978, **17**, 654, 658

In contact with water, the adduct with dimethyl ether ignites and that of diethyl ether often explodes, as does invariably the bis-THF adduct.
See other COMPLEX HYDRIDES

0184. Zirconium(IV) tetrahydroborate
[23840-95-1] $B_4H_{16}Zr$

$$Zr(BH_4)_4$$

Gaylord, 1956, 58
Violent ignition on exposure to air.
See other COMPLEX HYDRIDES

0185. Sodium tetraborate
[1330-43-4] $B_4Na_2O_7.5H_2O$

$$NaO[B(OH)O]_4Na.3H_2O$$

Zirconium
   See Zirconium: Oxygen-containing compounds
   See related METAL OXONON-METALLATES

0186. 1-Bromopentaborane(9)
[23753-67-5] B_5BrH_8

Air, or Hexamethylenetetramine
   Remmel, R. J. *et al., Inorg. Synth.*, 1979, **19**, 248
   It ignites in air and reacts explosively with hexamine above 90°C.

Dimethyl ether, Potassium hydride
   See Potassium hydride: Bromopentaborane(9), Dimethyl ether
   See other HALOBORANES

0187. 1,2-Dibromopentaborane(9)
[ ] $B_5Br_2H_7$

Remmel, R. J. *et al., Inorg. Synth.*, 1979, **19**, 248

It may detonate above ambient temperature.
See other HALOBORANES

†0188. Pentaborane(9)
   [19624-22-7] B_5H_9

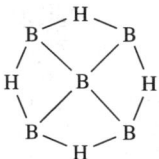

(*MCA SD-84*, 1961); *NSC 508*, 1979 (Boron hydrides)
The endothermic liquid ignites spontaneously in air if impure.

Ammonia
   See Diammineboronium heptahydrotetraborate

Pentacarbonyliron, Pyrex glass
   Shore, S. G. *et al., Inorg. Chem.*, 1979, **18**, 670
   In the preparation of 2-(tricarbonylferra)hexaborane(10) by co-pyrolysis of the reactants in a hot-cold Pyrex tube reactor, the latter was severely etched and weakened,
   sometimes splintering. At 230° a maximum cumulative service life of 4 months was
   observed, and at 260°C the reactor was replaced at the first signs of etching, usually
   after 6 runs.
   See other GLASS INCIDENTS

Oxygen
   Pentaborane, Tech. Bull. LF202, Alton (Il.), Olin Corp. Energy Div., 1960
   Reaction of pentaborane with oxygen is often violently explosive.

Reactive solvents
   1. Cloyd, 1965, 35
   2. Miller, V. R. *et al., Inorg. Synth.*, 1964, **15**, 118–122
   Pentaborane is stable in inert hydrocarbon solvents but forms shock-sensitive solutions in most other solvents containing carbonyl, ether or ester functional groups
   and/or halogen substituents [1]. The later reference gives detailed directions for
   preparation and handling of this exceptionally reactive compound, including a list of
   26 solvents and compounds rated as potentially dangerous in the presence of pentaborane. When large quantities are stored at low temperature in glass, a phase change
   involving expansion of the solid borane may rupture the container [2].
   See other GLASS INCIDENTS
   See Dimethyl sulfoxide: Boron compounds

Tris(difluoroamino)fluoromethane
   See Tris(difluoroamino)fluoromethane: Pentaborane(9)
   See other BORANES, ENDOTHERMIC COMPOUNDS

0189. Pentaborane(11)
[18433-84-6]

$$B_5H_{11}$$

Kit and Evered, 1960, 69
Ignites in air.
See other BORANES

0190. Diammineboronium heptahydrotetraborate (Pentaborane(9) diammoniate)
[28965-70-0]

$$B_5H_{15}N_2$$

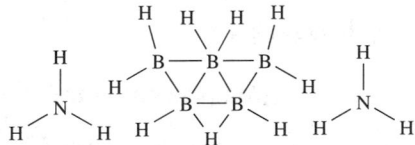

Kodama, G., *J. Amer. Chem. Soc.*, 1970, **92**, 3482
The diammoniate of pentaborane(9) decomposes spectacularly on standing at ambient temperature.
See related BORANES

0191. Hexaborane(10)
[23777-80-2]

$$B_6H_{10}$$

(Complex Structure)

Remmel, R. J. *et al., Inorg. Synth.*, 1979, **19**, 248
It ignites in air.
See other BORANES

0192. Hexaborane(12)
[28375-94-2]

$$B_6H_{12}$$

(Complex Structure)

Mellor, 1946, Vol. 5, 36
It is unstable and ignites in air.
See other BORANES

0193. Lanthanum hexaboride
[12008-21-8]

$$B_6La$$

$$LaB_6$$

See entry REFRACTORY POWDERS

0194. Caesium lithium tridecahydrononaborate
[[12430-27-2] (ion)] $B_9CsH_{13}Li$
(Complex structure)

Siedle, A. R. *et al., Inorg. Chem.*, 1974, **13**, 2737
It ignites in air.
See other COMPLEX HYDRIDES

0195. Disodium tridecahydrononaborate(2−)
[119391-53-6] $B_9H_{13}Na_2$
(Complex structure)

See Sodium tetradecahydrononaborate
See other COMPLEX HYDRIDES

0196. Sodium tetradecahydrononaborate
[70865-40-6] $B_9H_{14}Na$
(Complex structure)

Getman, T. D. *et al, Inorg. Chem.*, 1989, **28**(8), 1507
Explosions occurred during the preparation and handling of this compound and the
disodium salt.
See other COMPLEX HYDRIDES

0197. 1,10-Bis(diazonio)decaboran(8)ate
[ ] $B_{10}H_8N_4$
(Complex structure)

The precursor is explosive.
See Ammonium decahydrodecaborate(2−): Nitrous acid
See related DIAZONIUM SALTS

0198. Decaborane(14)
[17702-41-9] $B_{10}H_{14}$
(Complex Structure)

Ether, or Halocarbons, or Oxygen
    1. *MCA SD-84*, 1961
    2. Hawthorne, M. F., *Inorg. Synth.*, 1967, **10**, 93–94
    3. Shore, S. G., *Chem. Abs.*, 1986, **105**, 202122
It forms impact-sensitive mixtures with ethers (dioxane, etc.) and halocarbons (carbon
tetrachloride) and ignites in oxygen at 100°C [1,2]. An improved and safer synthesis
of decaborane from pentaborane is given [3].
See Dimethyl sulfoxide: Boron compounds
See other BORANES

0199. Ammonium decahydrodecaborate(2—)
  [12008-61-6] $B_{10}H_{18}N_2$
$$(NH_4)_2B_{10}H_{10}$$

Nitrous acid
  Knoth, W. H., *J. Amer. Chem. Soc.*, 1964, **86**, 115
  Interaction of the $B_{10}H_{10}(2-)$ anion with excess nitrous acid gives an inner diazonium salt (of unknown structure, possibly containing a nitronium ion) which is highly explosive in the dry state. It is readily reduced wet to the non-explosive 1,10-bis(diazonio)decaboran(8)ate inner salt.
  See related BORANES

0200. Barium
  [7440-39-3]   Ba
Ba

  HCS 1980, 171

Air
  See Oxidising gases, below

Halocarbons
  1. *Serious Accid. Ser.*, 1952, 23 and Suppl., Washington, USAEC
  2. Anon., *Ind. Res.*, 1968, (9), 15
  3. *Pot. Incid. Rep. 39*, ASESB, 1968
  4. Stull, 1977, 25
  A violent reaction occurred when cleaning lump metal under carbon tetrachloride [1]. Finely divided barium, slurried with trichlorotrifluoroethane, exploded during transfer owing to frictional initiation [2]. Granular barium in contact with fluorotrichloromethane, carbon tetrachloride, 1,1,2-trichlorotrifluoroethane, tetrachloroethylene or trichloroethylene is suceptible to detonation [3]. Thermodynamic calculations indicated a heat of decomposition of 2.60 kJ/g of mixture and a likely adiabatic temperature approaching 3000°C, accompanied by a 30-fold increase in pressure [4].
  See other FRICTIONAL INITIATION INCIDENTS, METAL–HALOCARBON INCIDENTS

Interhalogens
  See Bromine pentafluoride: Acids, etc.
      Iodine heptafluoride: Metals

Oxidising gases
  Kirk-Othmer, 1964, Vol. 3, 78
  The finely divided metal may ignite or explode in air or other oxidising gases.

Water
  Sidgwick, 1950, 844
  Interaction is more violent than with calcium or strontium, but less so than with sodium.
  See other METALS, PYROPHORIC METALS

0201. Barium tetrafluorobromate
[35967-90-9] $BaBr_2F_8$

$$Ba(BrF_4)_2$$

See entry METAL POLYHALOHALOGENATES

0202. Barium bromate
[13967-90-3] $BaBr_2O_6$

$$Ba(BrO_3)_2$$

Hackspill, L. *et al., Compt. rend.*, 1930, **191**, 663
Thermal decomposition with evolution of oxygen is almost explosive at 300°C.

Disulfur dibromide
See Disulfur dibromide: Oxidants

Metals   MRH Aluminium 6.44/22, magnesium 6.40/26
See METAL HALOGENATES: metals

Other reactants
Yoshida, 1980, 186
MRH values calculated for 28 combinations with oxidisable materials are given.

Sulfur   MRH 2.26/12
Taradoire, F., *Bull. Soc. Chim. Fr.*, 1945, **12**, 94, 447
Mixtures are unstable and may ignite 2–11 days after preparation if kept at ambient
temperature, or immediately at 91–93°C. Presence of moisture (as water of crystalli-
sation) accelerates ignition.
See other METAL HALOGENATES

0203. Barium perchlorylamide
[28815-10-3] $BaClNO_3$

$$Ba:NClO_3$$

See entry PERCHLORYLAMIDE SALTS *See other* N-METAL DERIVATIVES

0204. Barium chlorite
[14674-74-9] $BaCl_2O_4$

$$Ba(ClO_2)_2$$

Solymosi, F. *et al., Chem. Abs.*, 1968, **68**, 51465
When heated rapidly, barium chlorite decomposes explosively at 190°C, and the lead
salt at 112°C.

Dimethyl sulfate
Pascal, 1960, Vol. 16.1, 264

84

The sulfate ignites in contact with the unheated chlorite, presumably owing to formation of very unstable methyl chlorite.
See other CHLORITE SALTS, OXIDANTS

0205. Barium chlorate
[13477-00-4]   BaCl$_2$O$_6$
$$Ba(ClO_3)_2$$

HCS 1980, 175

Other reactants
Yoshida, 1980, 72
MRH values calculated for 19 combinations with oxidisable materials are given.
See other METAL CHLORATES

0206. Barium perchlorate
[13465-95-7]   BaCl$_2$O$_8$
$$Ba(ClO_4)_2$$

HCS 1980, 180

Alcohols
Kirk-Othmer, 1964, Vol. 5, 75
Distillation of mixtures with $C_1 - C_3$ alcohols gives the highly explosive alkyl perchlorates. Extreme shock-sensitivity is still shown by n-octyl perchlorate.
See ALKYL PERCHLORATES

Calcium 2,4-pentanedionate
Hamid, I. *et al., Thermochim. Acta*, 1986, **101**, 189
If the ratio of the calcium chelate salt to the trihydrated barium salt in a mixture is above 2:1, the thermal decomposition during DTA/TG analysis may be explosive.
See other METAL PERCHLORATES

0207. Barium hydride
[13477-09-3]   BaH$_2$
$$BaH_2$$

Air, or Oxygen
1. Gibson, 1969, 74
2. Mellor, 1941, Vol. 3, 650
The finely divided hydride ignites in air [1], and coarser material when heated in oxygen [2].

Metal halogenates
See METAL HALOGENATES: Metals, etc. *See other* METAL HYDRIDES

0208. Barium hydroxide
[17194-00-2] BaH_2O_2

$$Ba(OH)_2$$

HCS 1980, 178

Chlorinated rubber
  See CHLORINATED RUBBER: Metal oxides or hydroxides
  See other INORGANIC BASES

0209. Barium amidosulfate
[13770-86-0] $BaH_4N_2O_6S_2$

$$Ba(OSO_2NH_2)_2$$

Metal nitrates or nitrites
  See entry METAL AMIDOSULFATES

0210. Barium phosphinate ('Barium hypophosphite')
[14871-79-5] $BaH_4O_4P_2$

$$Ba[OP(O)H_2]_2$$

Potassium chlorate
  See Potassium chlorate: Reducants
  See other METAL PHOSPHINATES, REDUCANTS

0211. Barium iodate
[10567-69-8] BaI_2O_6

$$Ba(IO_3)_2$$

Other reactants
  Yoshida, 1980, 383
  MRH values calculated for 27 combinations with oxidisable materials are given.
  See other METAL HALOGENATES

0212. Barium nitrate
[10022-31-8] BaN_2O_6

$$Ba(NO_3)_2$$

HCS 1980, 179

Aluminium, Potassium nitrate, Potassium perchlorate, Water
  See Aluminium: Metal nitrates, etc.

Aluminium–magnesium alloy   MRH 4.93/22 or 5.31/27
  Tomlinson, W. R. *et al., J. Chem. Educ.*, 1950, **27**, 606
  An intimate mixture of the finely divided components, once widely used as a photo-
  flash composition, is readily ignitable and extremely sensitive to friction or impact.

Other reactants
   Yoshida, 1980, 202
   MRH values calculated for 15 combinations with oxidisable materials are given.
   See other METAL NITRATES

0213. Barium nitridoosmate
   [25395-83-9] $BaN_2O_6Os_2$

$$Ba(N{:}OsO_3)_2$$

   Mellor, 1942, Vol. 15, 728
   Explodes at 150°C, like the ammonium salt.
   See other N-METAL DERIVATIVES

0214. Barium azide
   [18810-58-7] BaN_6

$$Ba(N_3)_2$$

   HCS 1980, 173

1. Fagan, C. P. *J. and Proc. R. Inst. Chem.*, 1947, 126
2. Ficheroulle, H. *et al., Mém. Poudres*, 1956, **33**, 7
3. Gyunter, P. L. *et al., Chem. Abs.*, 1943, **37**, 1270₉
4. Verneker, V. R-P. *et al., J. Phys. Chem.*, 1968, **72**, 778–783
5. Stull, 1977, 10

The material is impact-sensitive when dry and is supplied and stored damp with ethanol. It is used as a saturated solution and it is important to prevent total evaporation, or the slow growth of large crystals which may become dried and shock-sensitive. Lead drains must not be used, to avoid formation of the detonator, lead azide. Exposure to acid conditions may generate explosive hydrazoic acid [1]. It has been stated that barium azide is relatively insensitive to impact but highly sensitive to friction [2]. Strontium, and particularly calcium azides show much more marked explosive properties than barium azide. The explosive properties appear to be closely associated with the method of formation of the azide [3]. Factors which affect the sensitivity of the azide include surface area, solvent used and ageing. Presence of barium metal, sodium or iron ions as impurities increases the sensitivity [4]. Though not an endothermic compound (ΔH°_f −22.17 kJ/mol, 0.1 kJ/g), it may thermally decompose to barium nitride, rather than to the elements, when a considerable exotherm is produced (98.74 kJ/mol, 0.45 kJ/g of azide) [5].
See other ENDOTHERMIC COMPOUNDS, METAL AZIDES

0215. Barium oxide
   [1304-28-5] BaO

$$BaO$$

Dinitrogen tetraoxide
   See Dinitrogen tetraoxide: Barium oxide

Hydroxylamine
See Hydroxylamine: Oxidants

Sulfur trioxide
See Sulfur trioxide: Metal oxides

Triuranium octaoxide
See Triuranium octaoxide: Barium oxide

Water
Bailar, 1973, Vol. 1, 638
Interaction is so vigorous and exothermic as to be a potential ignition source and fire hazard.
See other METAL OXIDES

0216. Barium peroxide
[1304-29-6] BaO_2

$$BaO_2$$

HSC 1980, 181

Acetic anhydride
Rüst, 1948, 337
The peroxide was substituted for (unavailable) potassium permanganate in a process for purifying the crude anhydride in an open vessel. After several operations, when only minor explosions occurred, a violent explosion and fire occurred. Acetyl peroxide would be produced.

Calcium–silicon alloy
Smolin, A. O. *et al., Chem. Abs.*, 1976, **85**, 194906
Combustion of silico-calcium in a mixture with barium peroxide (title only translated).

Delay compositions
Stupp, J., *Chem. Abs.*, 1975, **82**, 113746
The spontaneous ignition of the peroxide in (unspecified) tracer-ignition delay compositions is described.

Hydrogen sulfide
See Hydrogen sulfide: Metal oxides

Hydroxylamine
See Hydroxylamine: Oxidants

Metals
Pascal, 1958, Vol. 4, 775
Powdered aluminium or magnesium ignite in intimate contact with the peroxide.
See also FLASH POWDER

Non-metal oxides
Pascal, 1958, Vol. 4, 773–774

The heated peroxide attains incandescence in a rapid stream of carbon dioxide or sulfur dioxide.

Organic materials, Water
Koffolt, J. H., private comm., 1966
Contact of barium peroxide and water will readily produce a temperature and a local oxygen concentration high enough to ignite many organic compounds.

Peroxyformic acid
See Peroxyformic acid: Metals, etc.

Propane
Hoffmann, A. B. *et al., J. Chem. Educ.*, 1974, **51**, 419 (footnote 7)
Heating barium peroxide under gaseous propane at ambient pressure caused a violent exothermic reaction which deformed the glass container.
See other GLASS INCIDENTS

Selenium
Johnson, L. B., *Ind. Eng. Chem.*, 1960, **52**, 241–244
Powdered mixtures ignite at 265°C.

Wood
1. Dupré, A., *J. Soc. Chem. Ind.*, 1897, **16**, 492
2. Anon., *Jahresber.*, 1987, 66
Friction of the peroxide between wooden surfaces ignited the latter [1]. A granulated and dried priming composition, largely consisting of barium peroxide, was being sieved by hand to break up agglomerated lumps. Towards the end of the operation, ignition occurred. This was caused by rubbing contact of the priming mixture with the sloping plywood frame of the sieve as the composition was pressed through [2].
See other FRICTIONAL INITIATION INCIDENTS
See other METAL PEROXIDES, OXIDANTS

0217. Barium sulfate
[7727-43-7] BaO_4S

$$BaSO_4$$

Aluminium
See Aluminium: Metal oxides, or Oxosalts

Phosphorus
See Phosphorus: Metal sulfates
See other METAL OXONON-METALLATES

0218. Barium sulfide
[21109-95-5]   **BaS**

BaS

HCS 1980, 182

Dichlorine oxide
  See Dichlorine oxide: Oxidisable materials

Oxidants
  Mellor, 1941, Vol. 3, 745
  Barium sulfide explodes weakly on heating with lead dioxide or potassium chlorate,
  and strongly with potassium nitrate. Calcium and strontium sulfides are similar.

Phosphorus(V) oxide
  Pascal, 1958, Vol. 4, 832
  Interaction is violent, attaining incandescence.
  See other METAL SULFIDES

0219. Barium nitride
[12047-79-9] Ba_3N_2

Ba_3N_2

Air, or Water
  Sorbe, 1968, 34
  It reacts violently with air or water.
  See Barium azide (reference 5)
  See other N-METAL DERIVATIVES

0220. Beryllium
[7440-41-7]   **Be**

Be

HCS 1980, 207; *RSC Lab. Hazards Data Sheet No. 67*, 1988 (Be and compounds)

Carbon dioxide, Nitrogen
  See Carbon dioxide: Metals, Nitrogen

Halocarbons
  Pot. Incid. Rep. 39, ASESB, 1968
  Mixtures of powdered beryllium with carbon tetrachloride or trichloroethylene will
  flash on heavy impact.
  See other METAL–HALOCARBON INCIDENTS

Halogens
  Pascal, 1958, Vol. 4, 22
  Warm beryllium incandesces in fluorine or chlorine.

Phosphorus
  See Phosphorus: Metals
  See other METALS

0221. Beryllium chloride
  [7787-47-5]   BeCl$_2$
                                    BeCl$_2$

Sulfur nitrides
  See Disulfur dinitride: Metal chlorides

Tetrasulfur tetranitride: Metal chlorides
  See other METAL HALIDES

0222. Beryllium perchlorate
  [13597-95-0]   BeCl$_2$O$_8$
                               Be(ClO$_4$)$_2$

Laran, R. J., US Pat. 3 157 464, 1964
A powerful oxidant, insensitive to heat or shock and useful in propellant and igniter systems.
  See other METAL PERCHLORATES

0223. Beryllium fluoride
  [7787-49-7]   BeF$_2$
                                    BeF$_2$

Magnesium
  See Magnesium: Beryllium fluoride
  See other METAL HALIDES

0224. Beryllium hydride
  [7787-52-2]   BeH$_2$
                                    BeH$_2$

Methanol, or Water
  1. Barbaras, G. D., *J. Amer. Chem. Soc.*, 1951, **73**, 48
  2. Brendel, G. J. *et al., Inorg. Chem.*, 1978, **17**, 3589
Reaction of the ether-containing hydride with methanol or water is violent, even at −196°C [1]. A crystalline modification produced by pressure-compaction of the hydride reacts slowly with moist air or water [2].
  See other METAL HYDRIDES

0225. Beryllium oxide
[1304-56-9]   BeO

BeO

HCS 1980, 208

Magnesium
See Magnesium: Metal oxides
See other METAL OXIDES

0226. Bismuth
[7440-69-9]   Bi

Bi

Aluminium
See Aluminium: Bismuth

Oxidants
See Ammonium nitrate: Metals
Bromine pentafluoride: Acids, etc.
Chloric acid: Metals, etc.
Iodine pentafluoride: Metals
Nitric acid: Metals
Nitrosyl fluoride: Metals
Perchloric acid: Bismuth
See other METALS

0227. Bismuth pentafluoride
[7787-62-4]   BiF₅

BiF₅

Water
von Wartenberg, H., *Z. Anorg. Chem.*, 1940, **224**, 344
It reacts vigorously with water, sometimes igniting.
See other METAL HALIDES

0228. Bismuthic acid (Bismuth oxide hydroxide)
[22750-47-6]   BiHO₃

HOBiO₂ (? Bi₂O₅.H₂O)

Hydrofluoric acid
Mellor, 1939, Vol. 9, 657
Interaction of the solid acid with 40% hydrofluoric acid is violent, ozonised oxygen
being evolved.
See related INORGANIC ACIDS *See other* OXIDANTS

0229. Bismuth amide oxide

[ ]   BiH$_2$NO

$$H_2NBi{:}O$$

Watt, G. W. *et al., J. Amer. Chem. Soc.*, 1939, **61**, 1693
The solid, prepared in liquid ammonia, explodes when free of ammonia and exposed to air.
See other N-METAL DERIVATIVES

0230. Bismuth nitride

[12232-97-2]   BiN

$$BiN$$

Alone, or Water
1. Fischer, F. *et al., Ber.*, 1910, **43**, 1471
2. Franklin, E. C., *J. Amer. Chem. Soc.*, 1905, **27**, 847

Very unstable, exploded on shaking [1] or heating, or in contact with water or dilute acids [2].
See other N-METAL DERIVATIVES, METAL NON-METALLIDES

0231. Plutonium bismuthide

[12010-53-6]   BiPu

$$PuBi$$

Williamson, G. K., *Chem. & Ind.*, 1960, 1384
Extremely pyrophoric.
See other ALLOYS, PYROPHORIC MATERIALS

0232. Dibismuth dichromium nonaoxide ('Bismuth chromate')

[37235-82-8]   Bi$_2$Cr$_2$O$_9$

$$Bi_2O_3.2CrO_3, \text{ possibly } (O{:}Bi)_2Cr_2O_7$$

Hydrogen sulfide
Pascal, 1960, Vol. 13.1, 1025
The gas may ignite on contact with the 'chromate'.
See other METAL OXIDES, METAL OXOMETALLATES

0233. Bismuth trioxide

[1304-76-3]   Bi$_2$O$_3$

$$Bi_2O_3$$

Chlorine trifluoride
See Chlorine trifluoride: Metals, etc.

93

Potassium
See Potassium: Metal oxides

Sodium
See Sodium: Metal oxides
See other METAL OXIDES

0234. Bismuth trisulfide
[1345-07-9]

Bi_2S_3

Bi_2S_3

Preparative hazard
Glatz, A. C. *et al., J. Electrochem. Soc.*, 1963, **110**, 1231
Possible causes of explosions in direct synthesis are discussed.
See Sulfur: Metals
See other METAL SULFIDES

0235. Bromine perchlorate
[32707-10-1]

$BrClO_4$

$BrClO_4$

Schack, C. J. *et al., Inorg. Chem.*, 1971, **10**, 1078
It is shock-sensitive.

Perfluorobutadiene
Schack. C. J. *et al., Inorg. Chem.*, 1975, **14**, 151 (footnote 8)
During adduct formation, the perchlorate must be present in excess to prevent formation of a mono-adduct, which may well be explosive.
See Chlorine perchlorate: Chlorotrifluoroethylene
See other GLASS INCIDENTS *See other* HALOGEN OXIDES

0236. Caesium hexafluorobromate
[26222-92-4]

$BrCsF_6$

$Cs[BrF_6]$

See entry METAL POLYHALOHALOGENATES

0237. Caesium bromoxenate
[ ]

$BrCsO_3Xe$

$Cs[BrXeO_3]$

Water
Jaselskis, B. *et al., J. Amer. Chem. Soc.*, 1969, **91**, 1875
Aqueous solutions are extremely unstable and caution is required if isolation of the compound is contemplated.
See other XENON COMPOUNDS

94

0238. Bromine fluoride
[13863-59-7]   BrF

BrF

Sidgwick, 1950, 1149
Chemically it behaves like the other bromine fluorides, but is more reactive.

Hydrogen
Pascal, 1960, Vol. 16.1, 412
Hydrogen ignites in the fluoride at ambient temperature.
See other INTERHALOGENS

0239. Bromyl fluoride
[22585-64-4]   BrFO$_2$

O$_2$BrF

Water
Bailar, 1973, Vol. 2, 1388
Hydrolysis may proceed explosively.
See other HALOGEN OXIDES

0240. Perbromyl fluoride
[25251-03-0]   BrFO$_3$

O$_3$BrF

Fluoropolymers
Johnson, K. G. *et al., Inorg. Chem.*, 1972, **11**, 800
It is considerably more reactive than perchloryl fluoride, and attacks glass and the
usually inert polytetrafluoroethylene and polychlorotrifluoroethylene.
See other CORROSION INCIDENTS, GLASS INCIDENTS, HALOGEN OXIDES

0241. Bromine trifluoride
[7787-71-5]   BrF$_3$

BrF$_3$

1. Davis, R. A. *et al., J. Org. Chem.*, 1962, **32**, 3478
2. Musgrave, W. K. R., *Advan. Fluorine Chem.*, 1964, **1**, 12
The hazards and precautions in use of this very reactive fluorinating agent are outlined
[1]. Contact with rubber, plastics or other organic materials may be explosively
violent and reaction with moisture is very vigorous [2].

Ammonium halides
Sharpe, A. G. *et al., J. Chem. Soc.*, 1948, 2137
Explosive reaction.

Antimony(III) chloride oxide
    Mellor, 1956, Vol. 2, Suppl. 1, 166
    Interaction is violent, even more so than with antimony trioxide.

Carbon monoxide
    Mellor, 1956, Vol. 2, Suppl. 1, 166
    At temperatures rather above 30°C, explosions occurred.

Carbon tetrachloride
    Dixon, K. R. *et al., Inorg. Synth.*, 1970, **12**, 233
    Excess bromine trifluoride may be destroyed conveniently in a hood by slow addition
    to a large volume of the solvent, interaction being vigorous but not dangerous.
    See Solvents, below

Halogens, or Metals, or Non-metals, or Organic materials
    1. Mellor, 1941, Vol. 2, 113; 1956, Vol. 2, Suppl. 1, 164–167
    2. 'Chlorine Trifluoride Tech. Bull.', Morristown, Baker & Adamson, 1970
    Incandescence is caused by contact with bromine, iodine, arsenic, antimony (even
    at −10°C); powdered molybdenum, niobium, tantalum, titanium, vanadium; boron,
    carbon, phosphorus or sulfur [1]. Carbon tetraiodide, chloromethane, benzene or
    ether ignite or explode on contact, as do organic materials generally. Silicon also
    ignites [2].
    See Uranium, below

2-Pentanone
    Stevens, T. E., *J. Org. Chem.*, 1961, **26**, 1629 (footnote 11)
    During evaporation of solvent hydrogen fluoride, an exothermic reaction between
    residual ketone and bromine trifluoride set in and accelerated to explosion.

Potassium hexachloroplatinate
    Dixon, K. R. *et al., Inorg. Synth.*, 1970, **12**, 233–237
    Interaction of the reagents in bromine as diluent to produce the trichlorotrifluoro-
    and then hexafluoro-platinates is so vigorous that increase in scale above 1 g of salt
    is not recommended.

Pyridine
    Kirk-Othmer, 1966, Vol. 9, 592
    The solid produced by action of bromine trifluoride on pyridine in carbon tetrachloride
    ignites when dry. 2-Fluoropyridine reacts similarly.

Silicone grease
    Sharpe, A. G. *et al., J. Chem. Soc.*, 1948, 2136
    As it reacts explosively in bulk, the amount of silicone grease used on joints must be
    minimal.

Solvents
    1. Sharpe, A. G. *et al., J. Chem. Soc.*, 1948, 2135
    2. Simons, J. H., *Inorg. Synth.*, 1950, **3**, 185

96

Bromine trifluoride explodes on contact with acetone or ether [1], and the frozen solid at −80°C reacts violently with toluene at that temperature [2].
See Halogens, etc., above

Tin(II) chloride
Mellor, 1956, Vol. 2, Suppl. 1, 165
Contact causes ignition.

Uranium, Uranium hexafluoride
Johnson, K. *et al., 6th Nucl. Eng. Sci. Conf.*, New York, 1960. Reprint Paper No. 23
Uranium may ignite or explode during dissolution in bromine trifluoride, particularly when high concentrations of the hexafluoride are present. Causative factors are identified.
See Halogens, etc. above

Water
1. Mellor, 1941, Vol. 2, 113
2. *491M*, 1975, 73
Interaction is violent, oxygen being evolved [1], and even at −50°C, reaction with 6Nhydrochloric acid is explosive [2].
See other INTERHALOGENS

0242. Tetrafluoroammonium perbromate
[25483-10-7] BrF_4NO_4

$$F_4N^+ \ BrO_4^-$$

Christe, K. O. *et al., Inorg. Chem.*, 1980, **19**, 1495
Solutions in hydrogen fluoride at −78°C exploded when isolation of the salt was attempted.
See other N-HALOGEN COMPOUNDS, OXOSALTS OF NITROGENOUS BASES

0243. Bromine pentafluoride
[7789-30-2] BrF_5

$$BrF_5$$

Acetonitrile
1. Meinert, H. *et al., Z. Chem.*, 1969, **9**, 190
2. Stein, L., *Chem. Eng. News*, 1984, **62**(28), 4
Although solutions of bromine pentafluoride in acetonitrile were reported stable at ambient temperature [1], it has been found that a 9% solution in the anhydrous solvent prepared at −196°C decomposed violently, bursting the container, about 1 h after attaining ambient temperature [2].
See Hydrogen-containing materials, below

Acids, or Halogens, or Metal halides, or Metals, or Non-metals, or Oxides
1. Mellor, 1956, Vol. 2, Suppl. 1, 172
2. Sidgwick, 1950, 1158

Contact with the following at ambient or slightly elevated temperatures is violent, ignition often occurring: strong nitric acid or sulfuric acids; chlorine (explodes on heating), iodine; ammonium chloride, potassium iodide; antimony, arsenic, boron powder, selenium, tellurium; aluminium powder, barium, bismuth, cobalt powder, chromium, iridium powder, iron powder, lithium powder, manganese, molybdenum, nickel powder, rhodium powder, tungsten, zinc; charcoal, red phosphorus, sulfur, arsenic pentoxide, boron trioxide, calcium oxide, carbon monoxide, chromium trioxide, iodine pentoxide, magnesium oxide, molybdenum trioxide, phosphorus pentoxide, sulfur dioxide or tungsten trioxide [1,2].

Hydrogen-containing materials
1. Mellor, 1956, Vol. 2, Suppl. 1, 172
2. Braker, 1980, 56
Contact with the following materials, containing combined hydrogen, is likely to cause fire or explosion: acetic acid, ammonia, benzene, ethanol, hydrogen, hydrogen sulfide, methane; cork, grease, paper, wax, etc. The carbon content further contributes to the observed reactivity [1]. Chloromethane reacts with explosive violence [2].
See Acetonitrile, above

Perchloryl perchlorate
See Perchloryl perchlorate: Bromine pentafluoride

Water
Sidgwick, 1950, 1158
Contact with water causes a violent reaction or explosion, oxygen being evolved.
See other INTERHALOGENS

0244. Potassium hexafluorobromate
[32312-22-4] BrF_6K

$$K[BrF_6]$$

See entry METAL POLYHALOHALOGENATES

0245. Rubidium hexafluorobromate
[32312-23-5] BrF_6Rb

$$Rb[BrF_6]$$

See entry METAL POLYHALOHALOGENATES

0246. Bromogermane
[13569-43-2] $BrGeH_3$

$$BrGeH_3$$

Preparative hazard
See Bromine: Germane
See related METAL HYDRIDES

0247. Hydrogen bromide
[10035-10-6]   BrH

HBr

HCS 1980, 545 (cylinder gas), 538 (48% solution)

Preparative hazard
See Bromine: Phosphorus

Ammine-1,2-diaminoethanediperoxochromium(IV)
See Ammine-1,2-diaminoethanediperoxochromium(IV): Hydrogen bromide

Fluorine
See Fluorine: Hydrogen halides

Ozone
See Ozone: Hydrogen bromide
See other INORGANIC ACIDS, NON-METAL HALIDES, NON-METAL HYDRIDES

0248. Bromic acid
[7789-31-3]   BrHO$_3$

HOBrO$_2$

In contact with oxidisable materials, reactions are similar to those of the metal bromates.
See entry METAL HALOGENATES
See other INORGANIC ACIDS, OXOHALOGEN ACIDS

0249. Bromamine (Bromamide)
[14519-10-9]   BrH$_2$N

BrNH$_2$

Jander, J. *et al., Z. Anorg. Chem.*, 1958, **296**, 117
The isolated material decomposes violently at $-70°C$, while an ethereal solution is stable for a few hours at that temperature.
See other N-HALOGEN COMPOUNDS

†0250. Bromosilane
[13465-73-1]   BrH$_3$Si

BrSiH$_3$

Ward, L. G. L., *Inorg. Synth.*, 1968, **11**, 161
Ignites in air (gas above 2°C).
See other HALOSILANES

0251. Ammonium bromide
[12124-97-9]   BrH₄N

$$NH_4Br$$

Bromine trifluoride
See Bromine trifluoride: Ammonium halides
See related METAL HALIDES

0252. Ammonium bromate
[13483-59-5]   BrH₄NO₃

$$NH_4OBrO_2$$

1. Sorbe, 1968, 129
2. Shidlovskii, A. A. *et al., Chem. Abs.*, 1968, **69**, 78870
It is a combustible and explosive salt which is very friction-sensitive [1], and may explode spontaneously [2].

Other reactants
Yoshida, 1980, 180
MRH values calculated for 16 combinations with oxidisable materials are given.
See other OXOSALTS OF NITROGENOUS BASES

0253. Poly(dimercuryimmonium bromate)
[ ]   (BrHg₂NO₃)ₙ

$$(Hg{=}N^+{=}HgBrO_3^-)_n$$

Sorbe, 1967, 97
Highly explosive.
See entry POLY(DIMERCURYIMMONIUM) COMPOUNDS

0254. Iodine bromide
[7789-33-5]   BrI

$$IBr$$

Metals
1. Mellor, 1961, Vol. 2, Suppl. 2, 452; 1963, Suppl. 3, 1563
2. Pascal, 1963, Vol. 8.3, 308
A mixture with sodium explodes under a hammer blow, while potassium explodes strongly under the molten bromide [1]. Tin reacts violently with the bromide [2].

Phosphorus
Mellor, 1963, Vol. 2, Suppl. 3, 264
Phosphorus reacts violently with the molten bromide.
See other INTERHALOGENS, IODINE COMPOUNDS

0255. Potassium bromate
[7758-01-2]   BrKO₃
$$KOBrO_2$$

HCS 1980, 760

Anon., Personal communication, 1999
A factory using potassium bromate regularly in small portions found that it caked in storage and had to be broken up. A largely full, but year old, polythene lined fibre-board drum thereof was broken by stabbing the contents with a, possibly rusty, knife and the lumps needed for use removed by gloved hand. The 1.5 kg so removed was observed to be fizzing and was sprayed with water. Minutes later, the 18 kg remaining in the drum deflagrated brightly, described as "sunrise behind the reactor", and ignited a sack of monomer nearby. Water extinguished the fire in a few minutes. Although bromates are thermodynamically unstable with respect to bromides, and there may have been slight surface contamination with organics, propagation of a surface reaction is hard to explain. The packaging would be a substantial fuel, but mixing with that is very poor. The exact cause remains mysterious. The business has decided to buy bromate in smaller lots and to subdivide those into individual 1.5 kg charges when received and before caking can occur. The operators have decided to no longer break up caked chemicals by jumping on the polythene liners containing them.

Aluminium, Dinitrotoluene
Yoshinaga, S. *et al., Chem. Abs.*, 1980, **92**, 44080
The mixture reacts violently at 290°C, with enormous gas evolution and is used to fracture concrete. The mechanism was studied.
See other GAS EVOLUTION INCIDENTS

Azoformamide
Vidal, F. D. *et al., Bakers Dig.*, 1979, **53**(3), 16−18
A mixture of the oxidant and blowing agent (2:1) is used as a dough improver. Potential problems of incompatibility during tableting operations were overcome by incorporating hydrated salts into the tableting formulation.

Ceric ammonium nitrate, Malonic acid, Water
Bartmess, J. *et al., Chem. Eng. News*, 1998, **76**(24), 4
Subsequent to a fire in a teaching laboratory, it was discovered that a mixture of equal weights of the three dry solids, itself stable, reacted violently when wetted with up to two parts of water and was capable of igniting paper. All components (which exhibit an oscillating chemical reaction in solution) were necessary for this effect.

Disulfur dibromide
See Disulfur dibromide: Oxidants

Non-metals   MRH Sulfur 2.55/14
1. Taradoire, F., *Bull. Soc. Chim. Fr.*, 1945, **12**, 94, 466
2. Pascal, 1961, Vol. 6, 440

Mixtures with sulfur are unstable, and may ignite some hours after preparation, depending on the state of subdivision and atmospheric humidity [1]. Selenium reacts violently with aqueous solutions of the oxidant [2].

Other reactants
Yoshida, 1980, 181–182
MRH values calculated for 28 combinations with oxidisable materials are given.
See other METAL OXOHALOGENATES, OXIDANTS

0256. Bromine azide
[13973-87-0] BrN_3

$$BrN_3$$

1. Hargittai, M. et al., Angew. Chem. (Int.), 1993, 32, 759
2. Mellor, 1940, Vol. 8, 336
3. Tornieporth-Oetting, I. C. et al., Angewand. Chem. (Int.), 1995, 34(5), 511
The solid, liquid and vapour are all very shock-sensitive [1]. The liquid explodes on contact with arsenic; sodium, silver foil; or phosphorus. Explosion is likely to be triggered by pressure fluctuations of around 10 Pa [3]. Concentrated solutions in organic solvents may explode on shaking [2].
See other ENDOTHERMIC COMPOUNDS, HALOGEN AZIDES

0257. Sodium bromate
[7789-38-0] $BrNaO_3$

$$NaOBrO_2$$

HCS 1970, 831

Fluorine
See Fluorine: Sodium bromate

Grease
1. MCA Case History No. 874
2. Stull, 1977, 28–29
A bearing assembly from a sodium bromate crusher had been degreased at 120°C, and while still hot the sleeve was hammered to free it. The assembly exploded violently, probably because of the presence of a hot mixture of sodium bromate and a grease component (possibly a sulfurised derivative). It is known that mixtures of bromates and organic or sulfurous matter are heat- and friction-sensitive [1]. The energy of decomposition of the likely components has been calculated as 1.93 kJ/g, with an explosion temperature above 2000°C [2].

Other reactants
Yoshida, 1980, 184–185
MRH values calculated for 29 combinations with oxidisable materials are given.
See other METAL OXOHALOGENATES

0258. Bromine dioxide
[21255-83-4] BrO_2

$$BrO_2$$

Brauer, 1963, Vol. 1, 306
Unstable unless stored at low temperatures, it may explode if heated rapidly.
See other HALOGEN OXIDES

0259. Bromine trioxide
[32062-14-4] BrO_3

$$BrO_3 \text{ or } Br_2O_6$$

1. Lewis, B. *et al., Z. Elektrochem.*, 1929, **35**, 648–652
2. Pflugmacher, A. *et al., Z. Anorg. Chem.*, 1955, **279**, 313
The solid produced at −5°C by interaction of bromine and ozone is only stable
at −80°C or in presence of ozone, and decomposition may be violently explosive
in presence of trace impurities [1]. The structure may be the dimeric bromyl
perbromate, analogous to Cl_2O_6 [2].
See other CATALYTIC IMPURITY INCIDENTS, HALOGEN OXIDES

0260. Thallium bromate
[14550-84-6] BrO_3Tl

$$TlOBrO_2$$

Pascal, 1961, Vol. 6, 950
It decomposes explosively around 140°C.
See other HEAVY METAL DERIVATIVES, METAL OXOHALOGENATES

0261. Bromine
[7726-95-6] Br_2

$$Br_2$$

(*MCA SD-49*, 1968); *NSC 313*, 1979; *FPA H61*, 1977; *HCS 1980*, 218
RSC Lab. Hazard Data Sheet No. 24, 1984

Berthelot, J. *et al., J. Chem. Educ.*, 1986, **63**(11), 1011
The stable complex of bromine with tetrabutylammonium bromide is safer and
more easily handled than bromine itself.

Acetone   MRH 0.46/16
Levene, P. A., *Org. Synth.*, 1943, Coll. Vol. 2, 89
During bromination of acetone to bromoacetone, presence of a large excess of
bromine must be avoided to prevent sudden and violent reaction.
See Carbonyl compounds, below

Acetylene   MRH 8.70/100
  See Acetylene: Halogens

Acrylonitrile   MRH 2.84/100
  See Acrylonitrile: Halogens

Alcohols
  1. Muir, G. D., *Chem. Brit.*, 1972, **8**, 16
  2. Bush, E. L., private comm., 1968
  3. Desty, D. H., private comm., 1986
  Reaction with methanol may be vigorously exothermic. A mixture of bromine
  (9 ml) and methanol (15 ml) boiled in 2 m and in a previous incident such a
  mixture had erupted from a measuring cylinder [1]. The exotherm with industrial
  methylated spirits (ethanol containing 5% methanol) is much greater, and addition
  of 10 ml of bromine to 40 ml of IMS rapidly causes violent boiling [2]. A further
  case of ejection of a methanol solution of bromine from a measuring cylinder was
  described [3].
  See other HALOGENATION INCIDENTS

Aluminium, Dichloromethane
  Nolan, 1983, Case History 26
  Bromochloromethane was being prepared in a 400 l reactor by addition of liquid
  bromine to dichloromethane in presence of aluminium powder (which would form
  some aluminium bromide to catalyse the halogen exchange reaction). The reaction
  was started and run for 1.5 h, stopped for 8 h, then restarted with addition of
  bromine at double the usual rate for 2.5 h, though the reaction did not appear to
  be proceeding. Soon afterwards a thermal runaway occurred, shattering the glass
  components of the reactor.
  See other GLASS INCIDENTS, HALOGENATION INCIDENTS

Ammonia
  Mellor, 1967, Vol. 8, Suppl. 2, 417
  Interaction at normal or elevated temperatures, followed by cooling to −95°C,
  gives an explosive red oil.
  See Nitrogen tribromide hexaammoniate

Boron
  See Boron: Halogens
  See Phosphorus: Halogens

3-Bromopropyne
  See 3-Bromopropyne, (reference 3)
  See Chlorine: 3-Chloropropyne

Carbonyl compounds
  MCA SD-49, 1968

104

Organic compounds containing active hydrogen atoms adjacent to a carbonyl group (aldehydes, ketones, carboxylic acids) may react violently in unmoderated contact with bromine.
See Acetone, above

Chlorotrifluoroethylene, Oxygen
Haszeldine, R. N. *et al., J. Chem. Soc.*, 1959, 1085
Addition of bromine to the gas-phase mixture initiated an explosion, but see Oxygen: Halocarbons (reference 3).

Copper(I) hydride
See Copper(I) hydride: Halogens

Diethyl ether
1. Tucker, H., private comm., 1972
2. Anon., *Safety Digest Univ. Safety Assoc.*, 1989, **34**, 14
Shortly after adding bromine to ether the solution erupted violently (or exploded softly). Photocatalytic bromination may have been involved [1]. Spontaneous ignition occurred on addition of ether to impure bromine [2].
See Tetrahydrofuran, below; or Chlorine: Diethyl ether

Diethylzinc
Houben-Weyl, 1973, Vol. 13.2a, 757
Interaction without diluents may produce dangerous explosions. Even with diluents (ether), interaction of dialkylzincs with halogens is initially violent at 0 to −20°C.

Dimethylformamide
Tayim, H. A. *et al., Chem. & Ind.*, 1973, 347
Interaction is extremely exothermic, and under confinement in an autoclave the internal temperature and pressure exceeded 100°C and 135 bar, causing failure of the bursting disc. The product of interaction is dimethylhydroxymethylenimmonium bromide, and the explosive decomposition may have involved formation of *N*-bromodimethylamine, carbon monoxide and hydrogen bromide.
See N-HALOGEN COMPOUNDS

Ethanol, Phosphorus
Read, C. W. W., *School Sci. Rev.*, 1940, **21**(83), 967
The vigorous interaction of ethanol, phosphorus and bromine to give bromoethane is considered too dangerous for a school experiment.

Fluorine
See Fluorine: Halogens

Germane
Swiniarski, M. F. *et al., Inorg. Synth.*, 1974, **15**, 157–160

During the preparation of mono- or di-bromogermane, either the scale of operation or the rate of addition of bromine must be closely controlled to prevent explosive reaction occurring.
See Non-metal hydrides, below

Hydrogen
Mellor, 1956, Vol. 2, Suppl. 1, 707
Combination is explosive under appropriate temperature and pressure conditions.

Isobutyrophenone
MCA Guide, 1972, 307
Bromine was added dropwise at 20–31°C to a solution of the ketone in carbon tetrachloride. The completed reaction mixture was cooled in ice, but exploded after 15 m.
See other HALOGENATION INCIDENTS

Metal acetylides and carbides
Several of the mono- and di-alkali metal acetylides and copper acetylides ignite at ambient temperature or on slight warming, with either liquid or vapour. The alkaline earth, iron, uranium and zirconium carbides ignite in the vapour on heating.
See Calcium acetylide: Halogens
    Caesium acetylide: Halogens
    Dicopper(I) acetylide: Halogens
    Iron carbide: Halogens
    Lithium acetylide: Halogens
    Rubidium acetylide: Halogens
    Strontium acetylide: Halogens
    Uranium dicarbide: Halogens
    Zirconium dicarbide: Halogens

Metal azides
Mellor, 1940, Vol. 8, 336
Nitrogen-diluted bromine vapour passed over silver azide or sodium azide formed bromine azide, and often caused explosions.

Metals   MRH Aluminium 1.96/10
1. Staudinger, H., *Z. Elektrochem.*, 1925, **31**, 549
2. Mellor, 1941, Vol. 2, 469; 1963, Vol. 2, Suppl. 2.2, 1563, 2174
3. *MCA SD-49*, 1968
4. Mellor, 1941, Vol. 7, 260
5. Mellor, 1939, Vol. 3, 379
6. Hartgen, C. *et al., J. Chem. Soc., Dalton Trans.*, 1980, 70
Lithium is stable in contact with dry bromine, but heavy impact will initiate explosion, while sodium in contact with bromine needs only moderate impact for initiation [1]. Potassium ignites in bromine vapour and explodes violently in contact with liquid bromine, and rubidium ignites in bromine vapour [2]. Aluminium,

mercury or titanium react violently with dry bromine [3]. Warm germanium ignites in bromine vapour [4], and antimony ignites in bromine vapour and reacts explosively with the liquid halogen [5]. During preparation of praseodymium bromide, accidental contact of liquid bromine with small particles of praseodymium led to a violent explosion [6].

See Gallium: Halogens

Nitromethane

Rochat, A. C., private comm., 1990.

A solution of bromine (116 g) in nitromethane (300 ml) was employed in an attempt to brominate a pigment (30 g). On heating the mix in an autoclave, runaway commenced at 70°C, soon shattering the vessel. Nitromethane being an explosive of low oxygen balance, the potential energy certainly, and the sensitivity probably, will be increased by bromine or bromination. Bromonitromethane salts and formaldehyde adducts thereof are intermediates in manufacture of some disinfectants; isolation of the first is considered most unsafe in the industry, the second are thought to need careful handling.

See Nitromethane, Chloronitromethane.

Non-metal hydrides

1. Stock, A. *et al., Ber.*, 1917, **50**, 1739
2. Sujishi, S. *et al., J. Amer. Chem. Soc.*, 1954, **76**, 4631
3. Geisler, T. C. *et al., Inorg. Chem.*, 1972, **11**, 1710
4. Merck, 1976, 955

Interaction of silane and its homologues with bromine at ambient temperature is explosively violent [1] and temperatures of below −30°C are necessary to avoid ignition of the reactants [2]. Ignition of disilane at −95°C and of germane at −112°C emphasises the need for good mixing to dissipate the large exotherm [3]. Phosphine reacts violently with bromine at ambient temperature [4].

See Ethylphosphine: Halogens, or Phosphine: Halogens

Other reactants

Yoshida, 1980, 179

MRH values for 10 combinations with reactive materials are given.

Oxygen, Polymers

Groome, I. J., *Chem. Brit.*, 1983, **19**, 644−665

Bromination of polymers should be effected at temperatures below 120°C to avoid the possibility of explosion likely at higher temperatures if traces of oxygen were present in the highly unsaturated pyrolysis products.

Oxygen difluoride

See Oxygen difluoride: Halogens

Ozone

See Ozone: Bromine

Phosphorus
1. Bandar, L. S. *et al., Zh. Prikl. Khim.*, 1966, **39**, 2304
2. 'Leaflet No. 2', Inst. of Chem., London, 1939
During preparation of hydrogen bromide by addition of bromine to a suspension of red phosphorus in water, the latter must be freshly prepared to avoid the possibility of explosion. This is due to formation of peroxides in the suspension on standing and subsequent thermal decomposition [1]. In the earlier description of such an explosion, action of bromine on boiling tetralin was preferred to generate hydrogen bromide [2], which is now available in cylinders.

Rubber
Pascal, 1960, Vol. 16.1, 371
Bromine reacts violently in contact with natural rubber, but more slowly with some synthetic rubbers.

Sodium hydroxide
MCA Case History No. 1636
A bucket containing 25% sodium hydroxide solution was used to catch and neutralise bromine dripping from a leak. Lack of stirring allowed a layer of unreacted bromine to form below the alkali. Many hours later, a violent eruption occurred when the layers were disturbed during disposal operations. Continuous stirring is essential to prevent stratification of slowly reacting mutually insoluble liquids, especially of such differing densities.

Tetracarbonylnickel
See Tetracarbonylnickel: Bromine

Tetrahydrofuran
Tinley, E. J., private comm., 1983
Rapid addition of bromine to the dried solvent to make a 10% solution caused a vigorous reaction with gas evolution. As this happened in a newly installed brightly illuminated fume cupboard lined with a reflective white finish, photocatalysed bromination of the solvent may have been involved, as has been observed in chlorine–ether systems.
See Diethyl ether, above; Chlorine: Diethyl ether

Tetraselenium tetranitride
See Tetraselenium tetranitride: Alone, or Halogens

Trialkyl boranes
Coates, 1967, Vol. 1, 199
The lower homologues tend to ignite in bromine or chlorine.

Trimethylamine
Bohme, H. *et al., Chem. Ber.*, 1951, **84**, 170–181
The 1:1 adduct (presumably *N*-bromotrimethylammonium bromide) decomposes explosively when heated in a sealed tube.

Trioxygen difluoride
See 'Trioxygen difluoride': Various materials

Tungsten, Tungsten trioxide

Tillack, J., *Inorg. Synth.*, 1973, **14**, 116–120

During preparation of tungsten(IV) dibromide oxide, appropriate proportions of reactants are heated in an evacuated sealed glass ampoule to 400–500°C. Initially only one end should be heated to prevent excessive pressure bursting the ampoule.

See other GLASS INCIDENTS

See other HALOGENS, OXIDANTS

0262. Calcium bromide
[7789-41-5] Br_2Ca

$$CaBr_2$$

Potassium

See Potassium: Metal halides

See other METAL HALIDES

0263. Cobalt(II) bromide
[7789-43-7] Br_2Co

$$CoBr_2$$

Sodium

See Sodium: Metal halides

See other METAL HALIDES

0264. Tetraamminecopper(II) bromate
[ ] $Br_2CuH_{12}N_4O_6$

$$(NH_3)_4Cu^{2+} \ 2BrO_3^-$$

Rammelsberg, C. F., *Pogg. Ann*, 1842, **55**, 63

A bright blue solid detonating at 140°C or if struck.

See other AMMINEMETAL OXOSALTS

0265. Copper(I) bromide
[7787-70-4] Br_2Cu_2

$$BrCuCuBr$$

tert-Butyl peroxybenzoate, Limonene

See tert-Butyl peroxybenzoate: Copper(I) bromide, etc.

See other METAL HALIDES, REDUCANTS

0266. Iron(II) bromide
[7789-46-0] Br_2Fe

$$FeBr_2$$

Potassium

See Potassium: Metal halides

Sodium
See Sodium: Metal halides
See other METAL HALIDES

0267. Dibromogermane
[13769-36-3] Br_2GeH_2

Br_2GeH_2

Preparative hazard
See Bromine: Germane
See related METAL HALIDES, METAL HYDRIDES

0268. *N,N*-Bis(bromomercurio)hydrazine
[ ] $Br_2H_2Hg_2N_2$

$(BrHg)_2NNH_2$

Hofmann, K. A. *et al., Ann.*, 1899, **305**, 217
An explosive compound.
See other MERCURY COMPOUNDS, *N*-METAL DERIVATIVES

0269. Mercury(II) bromide
[7789-47-1] Br_2Hg

$HgBr_2$

HCS 1980, 614

Indium
Clark, R. J. *et al., Inorg. Synth.*, 1963, **7**, 19–20
Interaction at 350°C is so vigorous that it is unsafe to increase the scale of this
preparation of indium bromide.
See other MERCURY COMPOUNDS, METAL HALIDES

0270. Mercury(II) bromate
[[26522-91-8] (dihydrate)] Br_2HgO_6

$Hg(BrO_3)_2$

Janz, 1976, Table 2, 7
It deflagrates around 155°C.
See other MERCURY COMPOUNDS, METAL OXOHALOGENATES, OXIDANTS

0271. Mercury(I) bromate
[13465-33-3] $Br_2Hg_2O_6$

$O_2BrOHgHgOBrO_2$

Hydrogen sulfide
Pascal, 1960, Vol. 13.1, 1004
Contact of the gas with the solid oxidant causes ignition.
See other MERCURY COMPOUNDS, METAL HALOGENATES

0272. N,N'-Dibromosulfurdiimide
[ ]   Br₂N₂S

$$BrN=S=NBr$$

Seppelt, K. *et al.*, *Angew. Chem. (Int.)*, 1969, **8**, 771
This is more shock sensitive than the iodo-analogue, which explodes both on melting and impact. The chloro-compound is not known.
See other N-HALOGEN COMPOUNDS, N–S COMPOUNDS

0273. Titanium diazide dibromide
[32006-07-8]   Br₂N₆Ti

$$Ti(N_3)_2Br_2$$

Dehnicke, K., *Angew. Chem. (Intern. Ed.)*, 1979, **18**, 507: *Chem. Ztg.*, 1982, **106**, 187–188
A highly explosive solid, (possibly polymeric).
See other METAL AZIDE HALIDES

0274. Sulfinyl bromide (Thionyl bromide)
[507-16-4]   Br₂OS

$$O:SBr_2$$

1. Hodgson, P. K. G., private comm., 1981
2. Beattie, T. R., *Chem. Eng. News*, 1982, **60**(28), 5
Opening an unused but old commercial ampoule of the ice-cooled bromide led to eruption of the decomposed contents [1]. Similar occurrences on 2 occasions were reported later [2], though on both occasions hexabromonaphthalene was identified in the residue (suggesting contamination of the bromide by naphthalene as an additional source of internal pressure). Sulfinyl bromide is of limited stability, decomposing to sulfur, sulfur dioxide and bromine, and so should be stored under refrigeration and used as soon as possible.
See other GAS EVOLUTION INCIDENTS, NON-METAL HALIDES

0275. Seleninyl bromide
[7789-51-7]   Br₂OSe

$$O:SeBr_2$$

Metals
Mellor, 1947, Vol. 10, 912
Sodium and potassium react explosively (the latter more violently), and zinc dust ignites, all in contact with the liquid bromide.
See Sodium: Non-metal halides

Phosphorus
Mellor, 1947, Vol. 10, 912

Red phosphorus ignites, and white phosphorus explodes, in contact with the liquid bromide.

See other NON-METAL HALIDES

0276. Bromine bromate (Dibromine trioxide)
[ ] Br_2O_3
$$BrOBrO_2$$

Kuschel, R. *et al., Angew. Chem. (Int.)*, 1993, **32**(11), 1632
Orange needles which decompose above $-40°C$ and detonate if warmed rapidly to $0°C$. The isomeric bromous anhydride would not be expected to be much more stable.

See other HALOGEN OXIDES

0277. Dibromine pentoxide
[ ] Br_2O_5
$$O_2BrOBrO_2$$

Leopold, D. *et al., Angew. Chem. (Int.)*, 1994, **33**(9), 975
The yellow powder decomposes above $-40°C$, sometimes by detonation.
See other HALOGEN OXIDES

0278. Lead bromate
[34018-28-5] Br_2O_6Pb
$$Pb(OBrO_2)_2$$

Sidgwick, 1950, 1227
An explosive salt.
See Lead acetate–lead bromate
See other HEAVY METAL DERIVATIVES, METAL HALOGENATES

0279. Zinc bromate
[14519-07-4] Br_2O_6Zn
$$Zn(OBrO_2)_2$$

See entry METAL HALOGENATES

0280. Sulfur dibromide
[14312-20-0] Br_2S
$$SBr_2$$

Nitric acid
See Nitric acid: Sulfur halides

112

Sodium
See Sodium: Non-metal halides
See other NON-METAL HALIDES

0281. Silicon dibromide sulfide
[13520-74-6] Br_2SSi

$S:SiBr_2$

Water
Bailar, 1973, Vol. 1, 1415
Hydrolysis of the sulfide is explosive.
See related NON-METAL HALIDES, NON-METAL SULFIDES

0282. Disulfur dibromide
[13172-31-1] Br_2S_2

BrSSBr

HCS 1980, 873

Metals
1. Mellor, 1947, Vol. 10, 652
2. Pascal, 1960, Vol. 13.2, 1162
Thin sections of potassium or sodium usually ignite in the liquid bromide. Iron
at about 650°C ignites and incandesces in the vapour [1]. Interaction with finely
divided aluminium or antimony is violent [2].

Oxidants
Taradoire, F., *Bull. Soc. Chim. Fr.*, 1945, **12**, 95
Interaction with moist barium bromate is very violent, and mixtures with potassium
bromate and water (3–4%) ignite at 20°C. In absence of water, ignition occurs at
125°C. Silver bromate also deflagrates.
See Silver bromate: Sulfur compounds
      Nitric acid: Sulfur halides

Phosphorus
See Phosphorus: Non-metal halides
See other NON-METAL HALIDES

0283. Poly(dibromosilylene)
[14877-32-9] $(Br_2Si)_n$

$(SiBr_2)_n$

Oxidants
Brauer, 1963, Vol. 1, 688
Ignites in air at 120°C and reacts explosively with oxidants such as nitric acid.
See related NON-METAL HALIDES

0284. Titanium dibromide
[13783-04-5] Br_2Ti

$$TiBr_2$$

Gibson, 1969, 60–61
It may ignite in moist air.
See other METAL HALIDES, PYROPHORIC MATERIALS

0285. Zirconium dibromide
[24621-17-8] Br_2Zr

$$ZrBr_2$$

Air, or Water
Pascal, 1963, Vol. 9, 558
It ignites in air, reacts violently with water and incandesces in steam.
See other METAL HALIDES, PYROPHORIC MATERIALS

0286. Iron(III) bromide
[10031-26-2] Br_3Fe

$$FeBr_3$$

Potassium
See Potassium: Metal halides

Sodium
See Sodium: Metal halides
See other METAL HALIDES

†0287. Tribromosilane
[7789-57-3] Br_3HSi

$$Br_3SiH$$

Schumb, W. C., *Inorg. Synth.*, 1939, **1**, 42
It usually ignites when poured in air (generating an extended liquid/air interface).
This ignition is remarkable in a compound containing over 89% of bromine.
See other HALOSILANES

0288. Molybdenum azide tribromide
[68825-98-9] Br_3MoN_3

$$Mo(N_3)Br_3$$

Dehnicke, K., *Angew. Chem. (Intern. Ed.)*, 1979, **18**, 510
It is highly explosive.
See other METAL AZIDE HALIDES

0289. Indium bromide
[13469-09-3] Br_3In

$InBr_3$

Preparative hazard
See Mercury(II) bromide: Indium
See other METAL HALIDES

0290. Nitrogen tribromide hexaammoniate (Tribromamine hexaammoniate)
[ ] $Br_3N.6H_3N$

$NBr_3.6NH_3$

Alone, or Non-metals
Mellor, 1967, Vol. 8, Suppl. 2, 417; 1940, Vol. 8, 605
The compound, formed by condensation of its vapour at −95°C explodes suddenly at −67°C. Prepared in another way, it is stable under water but explodes violently in contact with phosphorus or arsenic.
See other N-HALOGEN COMPOUNDS

0291. Tribromamine oxide (Nitrosyl tribromide)
[13444-89-8] Br_3NO

$Br_3N:O$

Sodium−antimony alloy
Mellor, 1940, Vol. 8, 621
The powdered alloy ignites when dropped into the vapour.
See other N-HALOGEN COMPOUNDS, N-OXIDES

0292. Vanadium tribromide oxide
[13520-90-6] Br_3OV

$O:VBr_3$

Water
Bailar, 1973, Vol. 3, 508
The bromide (and analogous chloride) is violently hygroscopic.
See related METAL HALIDES

0293. Phosphorus tribromide
[7789-60-8] Br_3P

PBr_3

HCS 1980, 744

Calcium hydroxide, Sodium carbonate
Seager, J. F., *Chem. Brit.*, 1976, **12**, 105

During disposal of the tribromide by a recommended procedure involving slow addition to a mixture of soda ash and dry slaked lime, a violent reaction, accompanied by flame, occurred a few seconds after the first drop. Cautious addition of the bromide to a large volume of ice water is suggested for disposal.

Oxidants
See Chromyl chloride: Non-metal halides
Oxygen (Gas): Phosphorus tribromide
Ruthenium(VIII) oxide: Phosphorus tribromide

3-Phenylpropanol
Taylor, D. A. H., *Chem. Brit.*, 1974, **10**, 101–102
During dropwise addition of the bromide to the liquid alcohol, the mechanical stirrer stopped, presumably allowing a layer of the dense tribromide to accumulate below the alcohol. Later manual shaking caused an explosion, probably owing to the sudden release of gaseous hydrogen bromide on mixing.
See other AGITATION INCIDENTS, GAS EVOLUTION INCIDENTS

Potassium
See Potassium: Non-metal halides

Sodium, Water
See Sodium: Non-metal halides

Sulfur acids
Dillon, K. B. *et al., J. Chem. Soc., Dalton Trans.*, 1979, 885–887
The tribromide is initially insoluble in 100% sulfuric acid, 25% oleum or fluorosulfuric acid, but violent exotherms occurred after contact for 11, 4 and 5 hours respectively. 65% Oleum reacts violently on contact, and chlorosulfuric acid vigorously after shaking.
See Phosphorus trichloride: Sulfur acids
Phosphorus triiodide: Sulfur acids
See other INDUCTION PERIOD INCIDENTS

1,1,1-Tris(hydroxymethyl)methane
1. Derfer, J. M. *et al., J. Amer. Chem. Soc.*, 1949, **71**, 175
2. Farber, S. *et al., Synth. Comm.*, 1974, **4**, 243
Interaction to form the corresponding tribromomethyl compound is extremely hazardous, even using previously specified precautions [1]. Several fires occurred in the effluent gases, and in reaction residues exposed to air [2], doubtless owing to phosphine or its derivatives.

Water
Mellor, 1940, Vol. 8, 1032
Interaction with warm water is very rapid and may be violent with limited quantities.
See other NON-METAL HALIDES

0294. Tungsten tetrabromide oxide
[13520-77-9] Br_4OW

$$O:WBr_4$$

Preparative hazard
See Boron tribromide: Tungsten trioxide
See related METAL HALIDES

0295. Selenium tetrabromide
[7789-65-3] Br_4Se

$$SeBr_4$$

Trimethylsilyl azide
See Trimethylsilyl azide: Selenium halides
See other NON-METAL HALIDES

0296. Tellurium tetrabromide
[10031-27-3] Br_4Te

$$TeBr_4$$

Ammonia
Sorbe, 1968, 154
Interaction gives a mixture of 'tritellurium' tetranitride and tellurium bromide nitride, which explodes on heating.
See Tetratellurium tetranitride
See other METAL HALIDES

0297. Tungsten azide pentabromide
[ ] Br_5N_3W

$$N_3WBr_5$$

Extremely explosive.
See entry METAL AZIDE HALIDES

0298. Carbon
[7440-44-0] (Amorphous) C_n
[7782-42-5] (Graphite) C_n
[99685-96-8;115383-22-7] (Fullerene) C_{60}, C_{70}
[7782-40-3] (Diamond) C_n
[126487-10-3] (Cyclooctadecanonayne) C_{18}

$$C$$

1. Anon., *Fire Prot. Assoc. J.*, 1964, 337
2. Cameron, A. *et al.*, *J. Appl. Chem.*, 1972, **22**, 1007

3. Zav'yalov, A. N. *et al., Chem. Abs.*, 1976, **84**, 166521
4. Suzuki, E. *et al., Chem. Abs.*, 1980, **93**, 78766
5. Beever, P. F., *Runaway Reactions*, 1981, Paper 4/X, 1–9
6. Van Liempt, J. H. M., *PT-Procestech.* (Neth.), 1988, **43**(9), 45–47
7. Anon., *Loss Prev. Bull.*, 1992, **105**, 15
8. Diederich, F. *et al, Angewand. Chem. (Int.),* 1994, **33**(9), 997
9. Editor's comments
Activated carbon exposed to air is a potential fire hazard because of its very high surface area and adsorptive capacity. Freshly prepared material may heat spontaneously in air, and presence of water accelerates this. Spontaneous heating and ignition may occur if contamination by drying oils or oxidising agents occurs [1]. The spontaneous heating effect has been related to the composition and method of preparation of activated carbon, and the relative hazards may readily be assessed [2]. Free radicals present in charcoal are responsible for auto-ignition effects, and charcoal may be stabilised for storage and transport without moistening by treatment with hot air at 50°C [3]. The causes of accidents involving activated carbon have been surveyed [4]. Fires originating in paper-bagged cargoes of active carbon in holds of ships travelling through tropical waters have been investigated by isothermal methods. Use of oxygen-impermeable plastic bags to limit oxidation and moisture uptake was proposed as a solution to the problem [5]. The potential hazards arising from use of activated carbon for various purposes are reviewed. Hazards usually arise from self-heating from adsorption of vapours or gases (especially oxygen), and may lead to autoignition, fire or explosions, including carbon dust explosions [6]. A fire in an activated charcoal odour abatement system, essentially caused by scale-up without considering the reduced heat loss thus occasioned, is reported [7].

Fullerenes may be expected to be at least as autoxidisable as charcoal (graphite) when suitably finely divided. Heats of formation (solid, w.r.t. graphite): C_{60} 3.2 kJ/g; C_{70} 3.0 kJ/g. They have hitherto proved surprisingly stable kinetically [8].The cyclic C_{18}, when available in substantial quantity, will surely prove capable of spontaneous combustion and will probably be explosive [9].

Generation of pyrophoric carbon
See Barium acetylide
See also HIGH SURFACE-AREA SOLIDS, PETROLEUM COKE

Alkali metals
1. Bailar, 1973, Vol. 1, 443
2. Werner, H. *et al., Fullerene Sci. Technol.*, 1993, **1**(2), 199
Graphite in contact with liquid potassium, rubidium or caesium at 300°C gives intercalation compounds (C_8M) which ignite in air and may react explosively with water [1]. 'Fullerene black' — probably a finely divided and distorted graphite — impregnated with potassium explodes spontaneously in air [2].
See Potassium: Carbon
    Sodium: Non-metals

Chlorinated paraffins, Lead(IV) oxide, Manganese(IV) oxide
See Lead(IV) oxide: Carbon black, etc.

Dibenzoyl peroxide
See Dibenzoyl peroxide: Charcoal

1,4-Diazabicyclo[2.2.2]octane
Hardman, J. S. *et al., Fuel*, 1980, **59**, 213–214
Activated carbon showed an auto-ignition temperature in flowing air of 452–518°C. Presence of 5% of the base ('triethylenediamine') adsorbed on the carbon reduced the AIT to 230–260°C. At high air flow rates an exotherm was seen at 230–260°, but ignition did not then occur until 500°C.

Iron(II) oxide, Oxygen (Liquid)
See Oxygen (Liquid): Carbon, Iron(II) oxide

Lithium, Lithium tetrachloroaluminate, Sulfinyl chloride
See Lithium: Carbon, etc.

Metal salts
MCA Case History No. 1094
Dry metal-impregnated charcoal catalyst was being added from a polythene bag to an aqueous solution under nitrogen. Static so generated ignited the charcoal dust and caused a flash fire. The risk was eliminated by adding a slurry of catalyst in water from a metal container.
See Cobalt(II) nitrate: Carbon
See other STATIC INITIATION INCIDENTS

Molybdenum(IV) oxide
See Molybdenum(IV) oxide: Graphite

2-Nitrobenzaldehyde
See 2-Nitrobenzaldehyde: Active carbon

Nitrogen oxide, Potassium hydrogen tartrate
See Nitrogen oxide: Carbon, Potassium hydrogen tartrate

Other reactants
Yoshida, 1980, 225
MRH values calculated for 25 combinations with oxidants are given.

Oxidants   MRH values show % of carbon
Carbon has frequently been involved in hazardous reactions, particularly finely divided or high-porosity forms exhibiting a high ratio of surface area to mass (up to 2000 m²/g). It then functions as an unusually active fuel which possesses adsorptive and catalytic properties to accelerate the rate of energy release involved in combustion reactions with virtually any oxidant. Less active forms of carbon will ignite or explode on suitably intimate contact with oxygen, oxides, peroxides, oxosalts, halogens, interhalogens and other oxidising species. Individual combinations are found under the entries listed below.

Ammonium perchlorate: Carbon   MRH 6.19/85
Bromine pentafluoride: Acids, etc.
Bromine trifluoride: Halogens, etc.
Chlorine trifluoride: Metals, etc.
Cobalt(II) nitrate: Carbon   MRH 3.68/86
Dichlorine oxide: Carbon, or: Oxidisable materials   MRH 5.14/90
Fluorine: Non-metals
Hydrogen peroxide: Carbon   MRH 6.19/15
Iodine heptafluoride: Carbon
Iodine(V) oxide: Non-metals
Nitrogen oxide: Non-metals   MRH 7.82/16
Nitrogen trifluoride: Charcoal
Oxygen (Liquid): Charcoal   MRH (Gas) 8.95/27
Oxygen difluoride: Non-metals
Ozone: Charcoal, Potassium iodide
Peroxyformic acid: Non-metals   MRH 5.69/tr.
Peroxyfuroic acid: Alone, etc,
Potassium chlorate: Charcoal, etc.   MRH 4.52/87
Potassium dioxide: Carbon
Potassium nitrate: Non-metals
Potassium permanganate: Non-metals   MRH 2.59/7
Silver nitrate: Non-metals   MRH 2.47/10
Sodium nitrate: Non-metals
Sodium peroxide: Non-metals   MRH 2.30/7
'Trioxygen difluoride': Various materials
Zinc nitrate: Carbon   MRH 3.86/87

Potassium hydroxide
Hejduk, J., *Chem. Abs.*, 1989, **110**, 107229
Analytical decomposition of powdered diamond by fusion with potassium
hydroxide may become explosive. This can be avoided by fusion with a potassium
carbonate–sodium carbonate mixture, followed by addition of small portions of
potassium nitrite or nitrate.

Sodium hydrogen carbonate
See Sodium hydrogen carbonate: Carbon, Water

Sodium tetrahydroborate
See Sodium tetrahydroborate

Turpentine
US Environmental Protection Agency, *Alert EPA 550-F-97-002e*, 1997
Fires have been caused when using activated carbon to de-odorise crude sulfate
turpentine. This is another case of adsorption forming hot spots then exposed to air.

Unsaturated oils
1. Bahme, C. W., *NFPA Quart.*, 1952, **45**, 431
2. von Schwartz, 1918, 326

Unsaturated (drying) oils, like linseed oil, etc., will rapidly heat and ignite when distributed on active carbon, owing to the enormous increase in surface area of the oil exposed to air, and in the rate of oxidation, probably catalysed by metallic impurities [1]. A similar, but slower, effect occurs on fibrous materials such as cotton waste [2].

See other CATALYTIC IMPURITY INCIDENTS
See other NON-METALS

0299. Silver cyanide
[506-64-9]   CAgN

$$AgC{\equiv}N$$

Phosphorus tricyanide
See Phosphorus tricyanide (reference 2)

Fluorine
See Fluorine: Metal salts
See other ENDOTHERMIC COMPOUNDS, METAL CYANIDES, SILVER COMPOUNDS

0300. Silver cyanate
[3315-16-0]   CAgNO

$$AgOC{\equiv}N$$

Sorbe, 1968, 125
It explodes on heating.
See related METAL CYANIDES (AND CYANO COMPLEXES) *See other* SILVER COMPOUNDS

0301. Silver fulminate
[5610-59-3]   CAgNO

$$AgC{\equiv}N{\rightarrow}O$$

1. Urbanski, 1967, Vol. 3, 157
2. Collins, P. H. *et al., Propellants, Explos.*, 1978, **3**, 159–162
Silver fulminate is dimeric and rather endothermic (ΔH°_{f} +361.5 kJ/dimol, 1.21 kJ/g). It is readily formed from silver or its salts, nitric acid and ethanol, and is a much more sensitive and powerful detonator than mercuric fulminate [1]. The properties and applications have been reviewed [2].

Hydrogen sulfide
Boettger, A., *J. Pr. Chem.*, 1868, **103**, 309
Contact with hydrogen sulfide at ambient temperature initiates violent explosion of the fulminate.
See other ENDOTHERMIC COMPOUNDS, METAL FULMINATES, SILVER COMPOUNDS

0302. Silver trinitromethanide
[25987-94-4] $CAgN_3O_6$

$$AgC(NO_2)_3$$

Witucki, E. F. *et al., J. Org. Chem.*, 1972, **37**, 152
The explosive silver salt may be replaced with advantage by the potassium salt in
the preparation of 1,1,1-trinitroalkanes.
See other POLYNITROALKYL COMPOUNDS, SILVER COMPOUNDS

0303. Silver azidodithioformate
[74093-43-9] $CAgN_3S_2$

$$AgSC(S)N_3$$

Sorbe, 1968, 126
The tetrahydrated salt explodes on the slightest friction.
See Azidodithioformic acid
See related ACYL AZIDES *See other* SILVER COMPOUNDS

0304. Silver trichloromethanephosphonate
[ ] $CAg_2Cl_3O_3P$

$$(AgO)_2P(O)CCl_3$$

Yakubovich, A. Ya. *et al., Chem. Abs.*, 1953, **47**, 2685i
It explodes on heating.
See other SILVER COMPOUNDS

0305. Disilver cyanamide
[3884-87-0] CAg_2N_2

$$Ag_2NC{\equiv}N$$

1. Chrétien, A. *et al., Compt. rend.*, 1951, **232**, 1114
2. Deb, S. K. *et al., Trans. Faraday Soc.*, 1959, **55**, 106–113
3. Cradock, S., *Inorg. Synth.*, 1974, **15**, 167
During pyrolysis to silver (via silver dicyanamide), initial heating must be slow to
avoid explosion [1]. High intensity illumination will also cause explosive decomposition of a confined sample [2]. Safety precautions for preparation and subsequent
use of the explosive salt are detailed [3].
See other IRRADIATION DECOMPOSITION INCIDENTS
See other N-METAL DERIVATIVES, SILVER COMPOUNDS

0306. Disilver diazomethanide
[54086-40-7] CAg_2N_2

$$Ag_2CN_2$$

Blues, E. T. *et al., J. Chem. Soc., Chem. Comm.*, 1974, 466–467

122

Both the disilver derivative and its precursory dipyridine complex are highly explosive and extremely shock-sensitive when dry.
See other DIAZO COMPOUNDS, SILVER COMPOUNDS

0307. Diazido(trifluoromethyl)arsine
[157951-76-3] $CAsF_3N_6$

$$CF_3As(N_3)_2$$

Ang, H. G. *et al., Inorg. Chem.*, 1994, **33**(20), 4425
A vapour phase study of this azide showed it sometimes to explode when exposed to heat.
See other AZIDES

0308. Gold(I) cyanide
[506-65-0] CAuN

$$AuC \equiv N$$

Magnesium
   See Magnesium: Metal cyanides
   See other GOLD COMPOUNDS, METAL CYANIDES

0309. Carbon tetraboride
[12069-32-8] CB_4

$$CB_4$$

Chlorine trifluoride
   See Chlorine trifluoride: Boron-containing materials
   See related NON-METALS

0310. Bromotrichloromethane
[75-62-7] $CBrCl_3$

$$BrCCl_3$$

Ethylene
   See Ethylene: Bromotrichloromethane
   See other HALOALKANES

0311. Bromotrifluoromethane
[75-63-8] $CBrF_3$

$$BrCF_3$$

Aluminium
   See Aluminium: Halocarbons
   See other HALOALKANES

0312. Bromine(I) trifluoromethanesulfonate
[70142-16-4]   CBrF$_3$O$_3$S

$$Br^+ \ F_3CSO_2O^-$$

Readily oxidisable materials
Katsuhara, Y. *et al., J. Org. Chem.*, 1980, **45**, 2442
Contact of this oxidant with readily oxidisable materials may lead to explosions.
See Chlorine(I) trifluoromethanesulfonate
See related HYPOHALITES

0313. Cyanogen bromide
[506-68-3]   CBrN

$$N{\equiv}CBr$$

HCS 1980, 338

1. Grossman, M. I., *Chem. Eng. News*, 1980, **58**(35), 43
2. Harvey, W. R. *et al., Chem. Eng. News*, 1981, **59**(5), 49
3. Crocker, H. P., *Chem. Eng. News*, 1981, **59**(5), 49

Cyanogen bromide is moderately endothermic (ΔH°_f (g) +50 kJ/mol, 0.47 kJ/g) and shows evidence of instability. The plastic cap of a bottle stored in a laboratory for several years on a high shelf, occasionally at 31°C, shattered and drove fragments into the shelf above [1]. This instability was confirmed, and a procedure outlined to obviate the use of the bromide in autoanalysis by generating cyanogen chloride on demand from Chloramine-T and potassium cyanide [2]. A 50 wt% solution of the bromide in chloroform is a stable and convenient form for use [3].
See Cyanogen chloride
See other CYANO COMPOUNDS, ENDOTHERMIC COMPOUNDS

0314. Tribromonitromethane
[464-10-8]   CBr$_3$NO$_2$

$$Br_3CNO_2$$

Sorbe, 1968, 40
It is used as an explosive.
See related NITROALKANES

0315. Carbon tetrabromide (Tetrabromomethane)
[558-13-4]   CBr$_4$

$$CBr_4$$

HCS 1980, 283

Hexacyclohexyldilead
See Hexacyclohexyldilead: Halocarbons

124

Lithium
  See Lithium: Halocarbons
  See other HALOALKANES

0316. Calcium cyanamide
[156-62-7]   CCaN$_2$

$$Ca:NC\equiv N$$

Water
  Pieri, M., *Chem. Abs.*, 1952, **46**, 8335i
  Absorption of water during handling or storage of technical calcium cyanamide
  may cause explosions, owing to liberation of acetylene from the calcium carbide
  content (up to 2%). Precautions are discussed.
  See other CYANO COMPOUNDS, *N*-METAL DERIVATIVES

0317. Calcium carbonate
[471-34-1]   CCaO$_3$

$$CaCO_3$$

  HCS 1980, 265

Fluorine
  See Fluorine: Metal salts
  See other INORGANIC BASES

0318. (Chlorocarbonyl)imidosulfur difluoride
[53654-97-0]   CClF$_2$NOS

$$ClCO.N{=}SF_2$$

  Mews, R. *et al., Inorg. Synth.*, 1986, **24**, 14–16
  The title compound, particularly if impure, rapidly forms explosive decomposition
  products when stored at ambient temperature.
  See other ACYL HALIDES, N–S COMPOUNDS

0319. Chloroperoxytrifluoromethane
[32755-26-3]   CClF$_3$O$_2$

$$ClOOCF_3$$

Tetrafluoroethylene
  Ratcliffe, C. T. *et al., J. Amer. Chem. Soc.*, 1971, **93**, 3887–3888
  The peroxy compound initiated explosive polymerisation of tetrafluoroethylene
  when a mixture prepared at −196°C warmed to −110°C.
  See other POLYMERISATION INCIDENTS
  See related HYPOHALITES *See other* ORGANIC PEROXIDES

0320. Chlorine(I) trifluoromethanesulfonate
[65597-24-2]   CClF$_3$O$_3$S

$$Cl^+ \ F_3CSO_2O^-$$

Readily oxidisable materials
1. Katsuhara, Y. *et al., J. Amer. Chem. Soc.*, 1979, **101**, 1040
2. Katsuhara, Y. *et al., J. Org. Chem.*, 1980, **45**, 2442
Contact of the oxidant with readily oxidisable materials must be controlled and at low temperatures to prevent explosions.
See Bromine(I) trifluoromethanesulfonate
See related HYPOHALITES

0321. Trifluoromethyl perchlorate
[52003-45-9]   CClF$_3$O$_4$

$$F_3COClO_3$$

Preparative hazard
1. Schack, C. J. *et al., Inorg. Chem.*, 1974, **13**, 2375
2. Schack, C. J. *et al., Inorg. Nucl. Chem. Lett.*, 1974, **10**, 449
Though apparently not explosively unstable [1], its synthesis [2] was occasionally accompanied by deflagrations.
See other ALKYL PERCHLORATES

0322. Trifluoromethanesulfenyl chloride
[421-17-0]   CClF$_3$S

$$F_3CSCl$$

Chlorine fluorides
Sprenger, G. H. *et al., J. Fluorine Chem.*, 1976, **7**, 335
In the preparation of trifluoromethylsulfur trifluoride, the chloride must be dissolved in a fully halogenated solvent to prevent explosion during treatment with chlorine fluoride or chlorine trifluoride.
See other HALOGENATION INCIDENTS

Hexafluoroisopropylideneaminolithium
See Hexafluoroisopropylideneaminolithium: Non-metal halides
See other ACYL HALIDES

0323. Cyanogen chloride
[506-77-4]   CClN

$$N{\equiv}CCl$$

HCS 1980, 339

Preparative hazard
Good, R. J., private comm., 1979

126

Crude cyanogen chloride (endothermic, prepared from hydrogen cyanide and chlorine) may trimerise violently to cyanuric chloride, catalysed by traces of hydrogen chloride or ammonium chloride.

See other CATALYTIC IMPURITY INCIDENTS
See Ammonium chloride
See Cyanogen bromide
See other CYANO COMPOUNDS, ENDOTHERMIC COMPOUNDS

0324. Chlorosulfonyl isocyanate
[1189-71-5]   CClNO$_3$S

$$ClSO_2N:C:O$$

Water
Graf, R., *Org. Synth.*, 1973, Coll. Vol. 5, 228
Interaction is violent.
See related CYANO COMPOUNDS
See other N–S COMPOUNDS, ACYL HALIDES

0325. Triazidocarbenium perchlorate (Triazidomethylium perchlorate)
[ ]   CClN$_9$O$_4$

$$(N_3)_3C^+ \; {}^-OClO_3$$

Petrie, M. A. *et al.*, *J. Amer. Chem. Soc.*, 1997, **119**, 8802
All salts of this cation were friction and shock sensitive, too sensitive for actual use as explosives. The most sensitive though, with estimated energy of 5.4 kJ/g, not the most powerful, was the perchlorate. Even though prepared on low milligram scale this invariably detonated spontaneously during work-up or soon after. The more powerful dinitramide (5.8 kJ/g) decomposed non-explosively on standing.
See other ORGANIC AZIDES

0326. Dichlordifluoromethane (Freon 12)
[75-71-8]   CCl$_2$F$_2$

$$Cl_2CF_2$$

Aluminium
See Aluminium: Halocarbons

Magnesium
See Magnesium: Halocarbons

Water
See LIQUEFIED GASES: WATER *See other* HALOALKANES

0327. Phosphoryl dichloride isocyanate
[870-30-4]   CCl$_2$NO$_2$P

$$Cl_2P(O)N:C:O$$

Preparative hazard
See N-Carbomethoxyiminophosphoryl chloride
See related CYANO COMPOUNDS, NON-METAL HALIDES

0328. Dichlorodinitromethane
[1587-41-3]   CCl$_2$N$_2$O$_4$

$$Cl_2C(NO_2)_2$$

Hocking, M. B. *et al., Chem. & Ind.*, 1976, 952
It exploded during attempted distillation at atmospheric pressure, but was distilled uneventfully at 31°C/13 mbar.
See other POLYNITROALKYL COMPOUNDS

0329. Carbonyl dichloride (Phosgene)
[75-44-5]   CCl$_2$O

$$O:CCl_2$$

(*MCA SD-95*, 1967); *FPA H105*, 1981; *HCS 1980*, 740

Anon., *Sichere Chemiarb.*, 1988, **40**(3), 34
Trichloromethyl chloroformate (diphosgene) is used as a safe substitute for highly toxic phosgene gas. The latter is generated in situ by addition of catalytic amounts of tertiary amines or amides, or active carbon. Diphosgene also disproportionates to 2 equivalents of phosgene on heating above 250°C.

N,N-Dimethylbenzeneamine, Water
Anon., *J. Loss Prev.*, 1994, **7**(3), 257
The aniline was being phosgenated in toluene as solvent, the reaction ran wild and ejected more than 3 tonnes of reactor contents. This is believed to have been due to water contamination, possibly as ice. An initial charge of only part of the phosgene failed to show the exotherm anticipated if water was present, however this may not have been enough, nor was the thermocouple immersed in the solvent.
See Diprotium monoxide

Hexafluoroisopropylideneaminolithium
See Hexafluoroisopropylideneaminolithium: Non-metal halides

Other reactants
Yoshida, 1980, 341
MRH values for 18 combinations with a range of materials are given.

Potassium   MRH 4.76/45
See Potassium: Non-metal halides

Sodium azide
See tert-Butyl azidoformate (reference 4)
See related ACYL HALIDES

0330. Trichlorofluoromethane (Freon 11)
[75-69-4]   CClF$_3$

Cl$_3$CF

HCS 1980, 921

Metals
See Aluminium: Halocarbons
   Barium: Halocarbons
   Lithium: Halocarbons
   METAL–HALOCARBON INCIDENTS
See other HALOALKANES

0331. Trichloronitromethane ('Chloropicrin')
[76-06-2]   CCl$_3$NO$_2$

Cl$_3$CNO$_2$

HCS 1980, 312

Anon., *Chem. Eng. News*, 1972, **50**(38), 13
Tests showed that above a critical volume, bulk containers of trichloronitromethane can be shocked into detonation. Containers below 700 kg content will now be the maximum size as against rail-tanks previously used.

Aniline
Jackson, K. E., *Chem. Rev.*, 1934, **14**, 269
Reaction at 145°C with excess aniline is violent.

3-Bromopropyne
BCISC Quart. Safety Summ., 1968, **39**, 12
An insecticidal mixture in a rail-tank exploded with great violence during pump-transfer operations, possibly owing to the pump running dry and overheating. Both components of the mixture are explosive and the mixture was also found to be shock- and heat-sensitive.

Sodium hydroxide
Scholtz, S., *Explosivstoffe*, 1963, **11**, 159, 181
During destruction of chemical warfare ammunition, pierced shells containing chloropicrin reacted violently with alcoholic sodium hydroxide.

Sodium methoxide
Ramsey, B. G. *et al.*, *J. Amer. Chem. Soc.*, 1966, **88**, 3059

During addition of the nitrocompound in methanol to sodium methoxide solution, the temperature must not be allowed to fall much below 50°C. If this happens, excess nitro compound will accumulate and cause a violent and dangerous exotherm.
See related NITROALKANES

0332. Carbon tetrachloride (Tetrachloromethane)
[56-23-5]   CCl₄

$$CCl_4$$

(*MCA SD-3*, 1963); *HCS 1980*, 284; *RSC Lab. Hazard Data Sheet No. 59*, 1987

Aluminium oxide, Heavy metals
See Aluminium oxide: Halocarbons, etc.

Aluminium chloride, Triethylaluminium
See Triethyldialuminium trichloride: Carbon tetrachloride

Boranes
See BORANES: carbon tetrachloride

Calcium disilicide
See Calcium disilicide: Carbon tetrachloride

Calcium hypochlorite
See Calcium hypochlorite: Carbon tetrachloride

Chlorine trifluoride   MRH 1.21/44
See Chlorine trifluoride: Carbon tetrachloride

Decaborane(14)
See Decaborane(14): Ether, etc.

1,11-Diamino-3,6,9-triazaundecane ('Tetraethylenepentamine')
  1. Hudson, F. L., private comm., 1973
  2. Collins, R. F., *Chem. & Ind.*, 1957, 704
A mixture erupted vigorously one hour after preparation [1]. Interaction (not vigorous) of amines and halocarbons at ambient temperature had been recorded previously [2]. The presence of 5 basic centres in the viscous amine would be expected to enhance exothermic effects.

Dibenzoyl peroxide, Ethylene
See Ethylene: Carbon tetrachloride

Dibenzoyl peroxide: Carbon tetrachloride, Ethylene
See also WAX FIRE

Dimethylformamide   MRH 1.30/28
  1. 'DMF Brochure' Billingham, ICI, 1965
  2. Cardillo, P. *et al., Ann. Chim.* (Rome), 1984, 74, 129–133
There is a potentially dangerous reaction of carbon tetrachloride with
dimethylformamide in presence of iron. The same occurs with 1,2,3,4,5,6-
hexachlorocyclohexane, but not with dichloromethane or 1,2-dichloroethane under
the same conditions [1]. A quantitative study of the reaction by DSC and ARC
techniques shows that in a 1:1 wt. mixture with carbon tetrachloride in absence of
iron, an exothermic reaction sets in below 100°C. Under adiabatic conditions, the
heat release (207.6 J/g) would take a runaway reaction to over 240°C. In presence
of 3% of iron powder, the same mixture shows 2 exotherms, one at 56°C (108 J/g)
and the second at 94°C (275 J/g), a final adiabatic temperature exceeding 285°C
being possible [2]. Dimethylacetamide behaves similarly but more so.
See N,N-Dimethylacetamide: Halogenated compounds
See other CATALYTIC IMPURITY INCIDENTS

Dinitrogen tetraoxide
  See Dinitrogen tetraoxide: Halocarbons

Ethanol (or methanol), Potassium sulfide
  See Potassium dithioformate
  See Sodium: Methanol and Sodium methoxide, both below

Fluorine
  See Fluorine: Halocarbons

Metals   MRH values show % of carbon tetrachloride
  See Aluminium: Halocarbons   MRH 4.14/81
       Barium: Halocarbons   MRH 4.18/36
       Beryllium: Halocarbons
       Lithium: Halocarbons   MRH 8.24/85
       Potassium: Halocarbons   MRH 5.19/50
       Potassium–sodium alloy: Halocarbons
       Sodium: Halocarbons   MRH 6.10/63
       Uranium: Carbon tetrachloride
       Zinc: Halocarbons   MRH 2.43/54
       METAL–HALOCARBON INCIDENTS

Methanol
  See Methanol: Carbon tetrachloride, Metals

Potassium *tert*-butoxide
  See Potassium *tert*-butoxide: Acids, etc.

Other reactants
  Yoshida, 1980, 148–149
  MRH values for 23 combinations with a range of materials are given.

See 2-Propen-1-ol: Carbon tetrachloride
See other HALOALKANES

0333. Tetrachlorotrifluoromethylphosphorane
[1066-48-4] CCl_4F_3P

$$Cl_4PCF_3$$

Tetramethyllead
Yap, N. T. *et al., Inorg. Chem.*, 1979, **18**, 1304
Interaction under vacuum to form methyltrichlorotrifluoromethylphosphorane is hazardous, violent explosions having occurred twice after 30 min at ambient temperature.
See related ALKYLHALOPHOSPHINES

0334. Trichloromethyl perchlorate
[ ] CCl_4O_4

$$Cl_3COClO_3$$

Sidgwick, 1950, 1236
An extremely explosive liquid, only capable of preparation in minute amounts. Oxygen balance is 100%
See Silver perchlorate: Carbon tetrachloride
See other ALKYL PERCHLORATES

0335. Triazidomethylium hexachloroantimonate
[19708-47-5] CCl_6N_9Sb

$$[C(N_3)_3]^+ [SbCl_6]^-$$

Müller, U. *et al., Angew. Chem.*, 1966, **78**, 825
The salt is sensitive to shock or rapid heating.
See related ORGANIC AZIDES

0336. Dideuterodiazomethane
[14621-84-2] CD_2N_2

$$D_2CN_2$$

Gassman, P. G. *et al., Org. Synth.*, 1973, **53**, 38–43
The explosive properties will be similar to those of diazomethane, for which precautions are extensively summarised.
See other DIAZO COMPOUNDS

0337. Poly(carbon monofluoride)
[25136-85-0]

$(CF)_n$

(Complex structure)

Preparative hazard
See Fluorine: Graphite

Hydrogen
Bailar, 1973, Vol. 1, 1269
Above 400°C in hydrogen, deflagration and flaming of the polymer occurs, the vigour depending on the fluorine content. Rapid heating to 500°C in an inert atmosphere causes explosive deflagration.
See related HALOALKANES

0338. Fluorocarbonylperoxonitrate
[ ]

$CFNO_5$

FC(O)OONO₂

Scheffler, D. *et al., Inorg. Chem.*, 1997, **36**(3), 339
Preparation has been achieved; working with more than millimolar quantities is not advised. It is an explosive oxidant in contact with organics and can scarcely be safe in their absence.
See also Trifluoromethylperoxonitrate
See other PEROXYESTERS

0339. Azidocarbonyl fluoride (Fluorocarbonyl azide)
[23143-88-6]

CFN_3O

F.CO.N₃

Mack, H.-G. *et al., J. Mol. Struct.*, *1994*, **291**(2-3), 197
It is advised that the explosive gas be handled only in millimolar quantities and with due precautions.
See other ACYL AZIDES

0340. Fluorotrinitromethane
[1840-42-2]

CFN_3O_6

FC(NO₂)₃

Preparative hazard
See Fluorine: Trinitromethane

Nitrobenzene
Zotov, E. V. *et al., Chem. Abs.*, 1980, **93**, 49732
Detonation characteristics of liquid explosive mixtures with nitrobenzene were studied.
See other POLYNITROALKYL COMPOUNDS

0341. Fluorodinitromethyl azide
[17003-82-6] CFN_5O_4

$$F(NO_2)_2CN_3$$

Unstable at ambient temperature.
See entry FLUORODINITROMETHYL COMPOUNDS
See other ORGANIC AZIDES, POLYNITROALKYL COMPOUNDS

0342. Difluorodiazirine
[693-85-6] CF_2N_2

Craig, N. C. *et al., Spectrochim. Acta*, 1979, **35A**, 895
Explosive decomposition occurred during purification by GLC and when running the gas phase laser Raman spectrum.
See other IRRADIATION DECOMPOSITION INCIDENTS, DIAZIRINES

0343. Carbonyl difluoride
[353-50-4] CF_2O

$$O{:}CF_2$$

Hexafluoroisopropyldeneaminolithium
See Hexafluoroisopropylideneaminolithium: Non-metal halides
See related ACYL HALIDES

0344. Difluoro-*N*-fluoromethanimine
[338-66-9] CF_3N

$$F_2C{:}NF$$

Ginsberg, V. A. *et al., Zh. Obsch. Khim.*, 1967, **37**, 1413
The gas boils at $-60°C$ and explodes in contact with flame.
See other *N*-HALOGEN COMPOUNDS

0345. Nitrosotrifluoromethane
[334-99-6] CF_3NO

$$O{:}NCF_3$$

1. Spaziante, P. M., *Intern. Rev. Sci.: Inorg. Chem. Ser. 1*, 1972, **3**, 141
2. Banks, R. E. *et al., J. Chem. Soc., Perkin Trans. 1*, 1974, 2534–2535

Suggestions of untoward hazard inherent in the preparation of nitrosotrifluoromethane by pyrolysis of trifluoroacetyl nitrite [1] are discounted in the later reference, which gives full details of the equipment and procedure that had been used uneventfully during the previous decade [2].
See other NITROSO COMPOUNDS

0346. Trifluoromethyl peroxonitrate
[50311-48-3] CF_3NO_4

$$F_3COONO_2$$

Hohorst, F. A. *et al., Inorg. Chem.*, 1974, **13**, 715
A small sample exploded under a hammer blow.
See other PEROXYESTERS

0347. Trifluoromethyl azide (Azidotrifluoromethane)
[3802-95-7] CF_3N_3

$$F_3CN_3$$

Schack. C. J., *J. Fluorine Chem.*, 1981, **18**, 584
Potentially hazardous, it must be handled with caution.
See other ORGANIC AZIDES

0348. Trifluoromethylsulfonyl azide
[3855-45-6] $CF_3N_3O_2S$

$$F_3CSO_2N_3$$

Cavender, C. J. *et al., J. Org. Chem.*, 1972, **37**, 3568
An explosion occurred when the azide separated during its preparation in the absence of solvent.
See other ACYL AZIDES

0349. Carbon tetrafluoride (Tetrafluoromethane)
[75-73-0] CF_4

$$CF_4$$

Aluminium
See Aluminium: Halocarbons

Oxygen, Hydrogen
See HALOALKANES
See other HALOALKANES, FLUOROCARBONS

0350. Perfluoroformamidine
[14362-70-0]   CF$_4$N$_2$

$$FC(:NF)NF_2$$

Koshar, R. J. *et al., J. Org. Chem.*, 1967, **32**, 3859, 3862
It explodes on shock or phase change.
See other N, N, N″-TRIFLUOROAMIDINES

0351. Tetrafluorodiaziridine
[17224-09-8]   CF$_4$N$_2$

Firth, W. C., *J. Org. Chem.*, 1968, **33**, 3489, 3491
An explosive oxidant, readily initiated by shock or phase changes during condensation or evaporation, especially around −160°C. Handle only small quantities at
moderately low temperatures.
See other N-HALOGEN COMPOUNDS

0352. Tetrafluorourea
[10256-92-5]   CF$_4$N$_2$O

$$O:C(NF_2)_2$$

Acetonitrile
Fraser, G. W. *et al., Chem. Comm.*, 1966, 532
The solution of the difluoroamide in acetonitrile prepared at −40°C must not be
kept at ambient temperature, since difluorodiazene is formed.
See Acetonitrile: *N*-Fluoro compounds
See other DIFLUOROAMINO COMPOUNDS

0353. Trifluoromethyl hypofluorite
[373-91-1]   CF$_4$O

$$F_3COF$$

Hydrocarbons
Allison, J. A. *et al., J. Amer. Chem. Soc.*, 1959, **81**, 1089–1091
In absence of nitrogen as diluent, interaction with acetylene, cyclopropane or ethyl-
ene is explosive on mixing, and solutions in benzene explode on sparking or UV
irradiation.
See other IRRADIATION DECOMPOSITION INCIDENTS

136

Hydrogen-containing solvents
1. *Catalogue G-7*, 10, Gainesville (Fa.), Peninsular Chem Research, 1973
2. Barton, D. H. R. *et al., Chem. Comm.*, 1968, 804
3. Robins, M. J. *et al., J. Amer. Chem. Soc.*, 1976, **89**, 7389 (footnote 24)
Contact of the extremely reactive compound with hydrogen-containing solvents or conventional plastics tubing, even at −80°C, is undesirable [1]. Fully halogenated solvents are preferred, and some general precautions are described [2]. Violent explosions have occurred on introducing the gas directly into methanol [3].

Lithium
Porter, R. S. *et al., J. Amer. Chem. Soc.*, 1957, **79**, 5625
Interaction set in at about 170°C with a sufficient exotherm to melt the glass container.
See other GLASS INCIDENTS

Polymers
Barton, D. H. R. *et al., J. Org. Chem.*, 1972, **37**, 329
It is a powerful oxidant and only all-glass apparatus, free of polythene, PVC, rubber or similar elastomers should be used. Appreciable concentrations of the gas in oxidisable materials should be avoided.

Pyridine
Barton, D. H. R. *et al., Chem. Comm.*, 1968, 804 (footnote)
Use of pyridine as an acid-acceptor in reactions involving trifluoromethyl hypofluorite is discouraged, as a highly explosive by-product is formed.
See other HYPOHALITES, OXIDANTS

0354. Trifluoromethanesulfinyl fluoride
[812-12-4] CF_4OS

$$F_3CS(O)F$$

Hexafluoroisopropylideneaminolithium
See Hexafluoroisopropylideneaminolithium: Non-metal halides
See other ACYL HALIDES

0355. Difluoromethylene dihypofluorite
[16282-67-0] CF_4O_2

$$F_2C(OF)_2$$

Haloalkenes
Hohorst, F. A. *et al., Inorg. Chem.*, 1968, **7**, 624
Attempts to react the oxidant with *trans*-dichloroethylene or tetrafluoroethylene at ambient temperature in absence of diluent caused violent explosions. The oxidant should not be allowed to contact organic or easily oxidised material without adequate precautions.
See other HYPOHALITES

0356. Xenon(II) fluoride trifluoromethanesulfonate
[39274-39-0]   CF$_4$O$_3$SXe

$$FXeOSO_2CF_3$$

Wechsberg, M. *et al., Inorg. Chem.,* 1972, **11**, 3066
Unless a deficiency of xenon fluoride was used in the preparation at 0°C or below,
the title product exploded violently on warming to ambient temperature.
See other XENON COMPOUNDS

0357. *N,N*-Difluorotrifluoromethylamine (Pentafluoromethanamine)
[335-01-3]   CF$_5$N

$$F_2NCF_3$$

Schack, C. J., *J. Fluorine Chem.,* 1981, **18**, 584
Potentially hazardous, handle with caution.
See other DIFLUOROAMINO COMPOUNDS

0358. 3-Difluoroamino-1,2,3-trifluorodiaziridine
[17224-08-7]   CF$_5$N$_3$

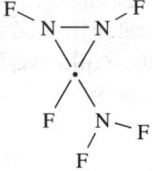

Firth, W. C., *J. Org. Chem.,* 1968, **33**, 3489, 3491
An explosive oxidant, readily initiated by shock or phase change during condensation or evaporation. Only small samples should be used and at moderately low
temperatures.
See other DIFLUOROAMINO COMPOUNDS, *N*-HALOGEN COMPOUNDS

0359. Pentafluoroguanidine
[10051-06-6]   CF$_5$N$_3$

$$FN:C(NF_2)_2$$

Zollinger, J. L. *et al., J. Org. Chem.,* 1973, **38**, 1070–1071
This, and several of its adducts with alcohols, are shatteringly explosive compounds, frequently exploding during phase changes at low temperatures, or on
friction or impact.

Liquid fuels
Scurlock, A. C. *et al.,* US Pat. 3 326 732, 1967

This compound with multiple N–F bonding is useful as an oxidant in propellant technology.

See related N, N, N′-TRIFLUOROAMIDINES
See other DIFLUOROAMINO COMPOUNDS, N-HALOGEN COMPOUNDS

0360. Difluorotrifluoromethylphosphine oxide
[19162-94-8]   CF$_5$OP

$$F_3CP(O)F_2$$

Preparative hazard
See Dinitrogen tetraoxide: Difluorotrifluoromethylphosphine
See related ALKYLHALOPHOSPHINES

0361. Difluorotrifluoromethylphosphine
[1112-04-5]   CF$_5$P

$$F_2PCF_3$$

Dinitrogen tetraoxide
See Dinitrogen tetraoxide: Difluorotrifluoromethylphosphine
See other ALKYLHALOPHOSPHINES

0362. Bis(difluoroamino)difluoromethane
[4394-93-8]   CF$_6$N$_2$

$$(F_2N)_2CF_2$$

Koshar, R. J. *et al., J. Org. Chem.*, 1966, **31**, 4233
Explosions occurred during the handling of this material, especially during phase transitions. Use of protective equipment is recommended for preparation, handling and storage, even on the microscale.
See other DIFLUOROAMINO COMPOUNDS

0363. Trifluoromethylsulfur trifluoride
[374-10-7]   CF$_6$S

$$F_3CSF_3$$

Preparative hazard
See Trifluoromethanesulfenyl chloride: Chlorine fluorides
See other ALKYLNON-METAL HALIDES

0364. Tris(difluoroamino)fluoromethane
[14362-68-6]   CF$_7$N$_3$

$$(F_2N)_3CF$$

Koshar, R. J. *et al., J. Org. Chem.*, 1967, **32**, 3859, 3862
A shock-sensitive explosive, especially in the liquid state (b.p. 5.6°C).

Pentaborane(9)

Marcellis, A. W. *et al.*, US Pat. 4 376 665, 1983

Mixtures prepared at cryogenic temperatures, then allowed to warm, are air-sensitive and powerful explosives.

See other DIFLUOROAMINO COMPOUNDS

0365. Iron carbide
[12011-67-5]   CFe₃

$$Fe_3C$$

Halogens

Mellor, 1946, Vol. 5, 898

Incandesces in chlorine below 100°C, and in bromine at that temperature.

See other METAL NON-METALLIDES

0366. Silver tetrazolide
[13086-63-0]   CHAgN₄

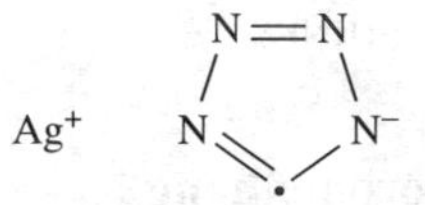

Thiele, J., *Ann.*, 1892, **270**, 59

It explodes on heating.

See other N-METAL DERIVATIVES, SILVER COMPOUNDS, TETRAZOLES

0367. Fluorohydrocyanokrypton(II) hexafluoroarsenate
[ ]   CHAsF7KrN

$$HCN^+\text{-}KrF^-AsF_6$$

Schrobilgen, G, J., *J. Chem. Soc., Chem. Comm.*, 1988, 863

A white solid precipitating from hydrogen cyanide and fluorokrypton in hydrogen fluoride is attributed this structure. It decomposes rapidly above −50°C, usually with violent detonation.

See other XENON COMPOUNDS

0368. Bromoform (Tribromomethane)
[75-25-2]   CHBr₃

$$HCBr_3$$

Acetone, Potassium hydroxide

1. Willgerodt, C., *Ber.*, 1881, **14**, 2451
2. Weizmann, C. *et al.*, *J. Amer. Chem. Soc.*, 1948, **70**, 1189

Interaction in presence of powdered potassium hydroxide (or other bases) is violently exothermic [1], even in presence of diluting solvents [2].
See Chloroform: Acetone, etc.

Cyclic poly(ethylene oxides) ('Crown ethers'), Potassium hydroxide
Le Goaller, R. *et al., Synth. Comm.*, 1982, **12**, 1163–1169
Crown ethers promote dihalocarbene formation from chloroform or bromoform and potassium hydroxide. However, in absence of diluent dichloromethane, dropwise addition of bromoform to the base in cyclohexane led to explosions.

Metals
See Potassium: Halocarbons
Lithium: Halocarbons
METAL–HALOCARBON INCIDENTS
See other HALOALKANES

0369. Chlorodifluoromethane (Freon 22)
[75-45-6] $CHClF_2$
$ClCHF_2$

HCS 1980, 668 (cylinder)

Sand, J. R., *ASHRAE J.*, 1982, **24**(5), 38–40
At elevated pressures, mixtures of the 'refrigerant 22' gas with 50% of air are combustible (though ignition is difficult) and a 6–8-fold pressure increase may occur in closed systems if ignition occurs.

Aluminium
See Aluminium: Halocarbons
See other HALOALKANES

0370. Tetrazole-5-diazonium chloride
[27275-90-7] $CHClN_6$

Shevlin, P. B. *et al., J. Amer. Chem. Soc.*, 1977, **99**, 2628
The crystalline diazonium salt will detonate at the touch of a spatula. An ethereal solution exploded violently after 1 h at −78°C, presumably owing to separation of the solid salt.

See 5-Aminotetrazole: Nitrous acid
See other DIAZONIUM SALTS, TETRAZOLES

0371. 1-Dichloroaminotetrazole
[68594-17-2] $CHCl_2N_5$

Karrer, 1950, 804
1-Dichloroaminotetrazole and its 5-derivatives are extremely explosive, as expected in an *N,N*-dichloro derivative of a high-nitrogen nucleus.
See other N-HALOGEN COMPOUNDS, TETRAZOLES

0372. Chloroform (Trichloromethane)
[67-66-3] $CHCl_3$
$HCCl_3$

(*MCA SD-89*, 1962); *HCS 1980*, 301; *RSC Lab. Hazard Data Sheet No. 44*, 1986

Acetone, Alkali
1. Willgerodt, C., *Ber*, 1881, **14**, 258
2. King, H. K., *Chem. & Ind.*, 1970, 185
3. Hodgson, J. F., *Chem. & Ind.*, 1970, 380
4. Ekely, J. B. *et al., J. Amer. Chem. Soc.*, 1924, **46**, 1253
5. Grew, E. L., *Chem. & Ind.*, 1970, 491
6. Grant, D. H., *Chem. & Ind.*, 1970, 919

Chloroform and acetone interact vigorously and exothermally in presence of solid potassium hydroxide or calcium hydroxide to form 1,1,1-trichloro-2-hydroxy-2-methylpropane [1], and a laboratory incident involving the bursting of a solvent residues bottle was attributed to this reaction. Addition of waste chloroform to a Winchester containing acetone, ether, and petroleum ether chromatography solvent residues led to a vigorous exothermic reaction, the effects of which would be greatly augmented by evolution of vapour of the much lower-boiling solvents in the narrow-necked bottle [2]. No reaction whatsoever occurs in the absence of base, even at 150°C under pressure [1], and the mechanism of the reaction was indicated [3]. Other haloforms and lower ketones react similarly in presence of base [4], but other halocarbons with a less activated hydrogen atom(e.g. dichloromethane) do not undergo the reaction, though there is an exotherm (physical effect) on mixing with acetone [5]. A minor eruption (or sudden boiling) of a chloroform–acetone mixture in new glassware may have been caused by surface alkali [6].

See Bromoform: Acetone, etc.
See other CATALYTIC IMPURITY INCIDENTS, GLASS INCIDENTS

Bis(dimethylamino)dimethylstannane
See Bis(dimethylamino)dimethylstannane: Chloroform

Dinitrogen tetraoxide
See Dinitrogen tetraoxide: Halocarbons

Ethanol (or methanol), Potassium sulfide
See Potassium dithioformate
See Sodium, Methanol; and Sodium methoxide, below

Fluorine
See Fluorine: Halocarbons

Metals
MRH Aluminium 3.85/18, magnesium 5.35/23, potassium 4.94/50, sodium 5.81/37
Davis, T. L. *et al., J. Amer. Chem. Soc.*, 1938, **60**, 720–722
The mechanism of the explosive interaction on impact of chloroform with sodium or potassium has been studied.
See other METAL–HALOCARBON INCIDENTS

Nitromethane
See Nitromethane: Haloforms

Other reactants
Yoshida, 1980, 114
MRH values calculated for 14 combinations with a range of materials are given.

Potassium *tert*-butoxide
See Potassium *tert*-butoxide: Acids, etc.

Sodium, Methanol
Unpublished information, 1948
During attempted preparation of trimethyl orthoformate, addition of sodium to an inadequately cooled chloroform–methanol mixture caused a violent explosion.
See Ethanol (or methanol), Potassium sulfide, above
See Sodium methoxide, below

Sodium hydroxide, Methanol
1. *MCA Case History No. 498*
2. *MCA Case History No. 1913*
A chloroform–methanol mixture was put into a drum contaminated with sodium hydroxide. A vigorous reaction set in and the drum burst. Chloroform normally reacts slowly with sodium hydroxide owing to the insolubility of the latter. The presence of methanol (or other solubiliser) increases the rate of reaction by increasing the degree of contact between chloroform and alkali [1]. Addition of chloroform

143

to a 4:1 mixture of methanol and 50 w/v% sodium hydroxide solution caused the drum to burst [2].

Sodium methoxide
1. *MCA Case History No. 693*
2. Kaufmann, W. E. *et al., Org. Synth.*, 1944, Coll. Vol. 1, 258
For the preparation of methyl orthoformate, solid sodium methoxide, methanol and chloroform were mixed together. The mixture boiled violently and then exploded [1]. The analogous preparation of ethyl orthoformate [2] involves the slow addition of sodium or sodium ethoxide solution to a chloroform–ethanol mixture. The explosion was caused by the addition of the solid sodium methoxide as one portion.
See Sodium, etc., above
See Ethanol (or methanol), Potassium sulfide, above

Triisopropylphosphine
See Triisopropylphosphine: Chloroform
See other HALOALKANES

0373. Fluorodiiodomethane
[1493-01-2]   CHFI$_2$
$$FCHI_2$$

Preparative hazard
See Mercury(I) fluoride: Iodoform
See other HALOALKANES

0374. Fluorodinitromethane
[7182-87-8]   CHFN$_2$O$_4$
$$FCH(NO_2)_2$$

Potentially explosive.
See entry FLUORODINITROMETHYL COMPOUNDS
See other POLYNITROALKYL COMPOUNDS

0375. Trifluoromethanesulfonic acid ('Triflic acid')
[1493-13-6]   CHF$_3$O$_3$S
$$F_3CSO_3H$$

Acyl chlorides, Aromatic hydrocarbons
Effenberger, F. *et al., Angew. Chem. (Intern. Ed.)*, 1972, **11**, 300
Addition of catalytic amounts (1%) of the acid (stronger even than perchloric acid) to mixtures of acyl chlorides and aromatic hydrocarbons causes more or less violent evolution of hydrogen chloride, depending on the reactivity of the Friedel-Crafts components.
See other CATALYTIC IMPURITY INCIDENTS, GAS EVOLUTION INCIDENTS

Perchlorate salts
Dixon, N. E. *et al., Inorg. Synth.*, 1986, **24**, 245
Because trifluoromethanesulfonic acid is a stronger acid than perchloric acid, under no circumstances should perchlorate salts be used with the neat acid, because the hot anhydrous perchloric acid so formed represents an extreme explosion hazard, especially in contact with transition metal complexes (or with organic materials).
See Perchloric acid: Dehydrating agents
See other ORGANIC ACIDS

0376. Iodoform (Triiodomethane)
[75-47-8] CHI_3

$$HCI_3$$

Acetone
See Chloroform: Acetone, etc.

Hexamethylenetetramine
Sorbe, 1968, 137
The 1:1 addition complex exploded at 178°C.

Mercury(I) fluoride
See Mercury(I) fluoride: Iodoform

Silver
See Silver: Iodoform
See other ENDOTHERMIC COMPOUNDS, HALOALKANES

0377. Potassium dinitromethanide
[32617-22-4] $CHKN_2O_4$

$$KCH(NO_2)_2$$

Grakauskas, V. *et al., J. Org. Chem.*, 1978, **43**, 3486
Alkali-metal salts of of dinitromethane are sensitive to impact and should be handled remotely behind shields, in small quantity and with great care.
See other POLYNITROALKYL COMPOUNDS

0378. Potassium dithioformate
[30962-16-4] $CHKS_2$

$$KSCS.H$$

Preparative hazard
1. Martin, K., *Chem. Brit.*, 1988, **24**, 427–428
2. Engler, K., *Z. Allgem. Anorg. Chem.*, 1972, **389**, 145
3. Yoshida, 1980, 387

Potassium dithioformate is prepared by interaction of chloroform with potassium sulfide in ethanol, via a carbene reaction, possibly involving the 3 main stages

$CHCl_3 + K_2S \rightarrow :CCl_2 + KSH + KCl$

$:CCl_2 + KSH \rightarrow KSCHCl_2$

$KSCHCl_2 + K_2S \rightarrow KSC(S)H + 2KCl$

The reaction has been found unpredictable in practice, 2 out of 3 attempts leading to eruption or explosion of the flask contents [1]. However, using a modified version of a published procedure [2], with methanol as solvent, the reaction can be performed without incident, provided that the working scale is restricted to one third of that published (and initially one twelfth until one gains experience of the reaction); that a flanged reaction flask is used; and that the rate of addition of chloroform to the methanolic potassium sulfide is carefully controlled to 4–5 drops/s. This is essential, for if too slow the reaction stops, and if too fast it becomes uncontrollable [1]. Though no figures for the more reactive chloroform are given, the MRH value for a mixture of potassium sulfide with 41 wt% of carbon tetrachloride is 2.05 kJ/g [3].

0379. Lithium diazomethanide
[67880-27-7]   CHLiN$_2$

$$LiCHN_2$$

Müller, E. *et al., Chem. Ber.*, 1954, **87**, 1887

Alkali metal salts of diazomethane are very explosive when exposed to air in the dry state, and should be handled, preferably wet with solvent, under an inert atmosphere.

See other DIAZO COMPOUNDS

†0380. Hydrogen cyanide
[74-90-8]   CHN

$$HC \equiv N$$

(*MCA SD-67*, 1961); *FPA H94*, 1980; *HCS 1980*, 547 (neat), 542 (solutions)

1. *MCA SD-67*, 1961
2. Wöhler, L. *et al., Chem. Ztg.*, 1926, **50**, 761, 781
3. Gause, E. H. *et al., J. Chem. Eng. Data*, 1960, **5**, 351
4. Salomone, G., *Gazz. Chim. Ital. [I]*, 1912, **42**, 617–622
5. Anon., *Jahresber.*, 1979, 70–71
6. Bond, J., *Loss Prev. Bull.*, 1991, **101**, 3

Hydrogen cyanide is highly endothermic and of low MW (ΔH°_f (g) +130.5 kJ/mol, 4.83 kJ/g). A comprehensive guide to all aspects of industrial handling of anhydrous hydrogen cyanide and its aqueous solutions states that the anhydrous liquid is stable at or below room temperature if it is inhibited with acid (e.g. 0.1% sulphuric acid) [1]. Presence of alkali favours explosive polymerisation [2]. In absence of inhibitor, exothermic polymerisation occurs, and if the temperature attains 184°C, explosively rapid polymerisation occurs [3]. A 100 g sample of 95–96% material stored in a glass bottle shielded from sunlight exploded after 8 weeks [4]. The explosive polymerisation of a 33 kg cylinder was attributed to lack of sufficient phosphoric acid

as stabiliser [5]. A tank containing 4 or 5 tonnes of hydrogen cyanide exploded with the force of several kg of TNT, leading to an HCN fire. This was easily extinguished, the clear-up thereafter being impeded by the toxicity of the unburnt cyanide. The explosion was attributed to build up of polymer (the reaction is autocatalytic because of ammonia generation), insufficient stabiliser (oxalic acid), lower than usual purity (<93%) and possibly to mercury contamination. It was recommended that HCN should not be stored if overwet, that tanks should be regularly emptied to inspect for polymer, and that stabiliser should be added before running to storage [6].
See Mercury(II) cyanide: Hydrogen cyanide
See other GLASS INCIDENTS, POLYMERISATION INCIDENTS

Alcohols, Hydrogen chloride
See Hydrogen chloride: Alcohols, etc.

Hypochlorites
See CYANIDES

Other reactants
Yoshida, 1980, 136
MRH values for 15 combinations, mainly with oxidants, are given
See other CYANO COMPOUNDS ENDOTHERMIC COMPOUNDS, ORGANIC ACIDS

0381. Fulminic acid (Hydrogen cyanide *N*-oxide)
[506-85-4]   CHNO

$$HC\equiv N \rightarrow O$$

1. Sorbe, 1968, 72
2. Wentrup, C. *et al., Angew. Chem. (Intern. Ed.)*, 1979, **18**, 467
It is fairly stable as an ethereal solution, but the isolated acid is explosively unstable, and sensitive to heat, shock or friction [1]. In a new method of preparation of the acid or its salts, pyrolysis of 4-oximato-3-substituted-isoxazol-5(4*H*)-ones or their metal salts must be conducted with extreme care under high vacuum to prevent explosive decomposition [2].
See Silver 3-methylisoxazolin-4,5-dione-4-oximate
Sodium 3-methylisoxazolin-4,5-dione-4-oximate
Sodium 3-phenylisoxazolin-4,5-dione-4-oximate
METAL FULMINATES
See other *N*-OXIDES

0382. Monosodium cyanamide
[17292-62-5]   CHN$_2$Na

$$NaNHC\equiv N$$

tert-Butyl hypochlorite
See Cyanonitrene
See other CYANO COMPOUNDS, *N*-METAL DERIVATIVES

0383. Sodium diazomethanide
[67880-28-8] CHN_2Na

$NaCHN_2$

See Lithium diazomethanide
See other DIAZO COMPOUNDS

0384. Sodium dinitromethanide
[25854-41-5] CHN_2NaO_4

$NaCH(NO_2)_2$

Grakauskas, V. *et al., J. Org. Chem.*, 1978, **43**, 3486
Alkali-metal salts of dinitromethane are sensitive to impact and should be handled
remotely behind shields in small quantity, and with great care.
See other POLYNITROALKYL COMPOUNDS

0385. Trinitromethane ('Nitroform')
[517-25-9] CHN_3O_6

$(O_2N)_3CH$

1. Marans, N. S. *et al., J. Amer. Chem. Soc.*, 1950, **72**, 5329
2. Stull, 1977, 20
Explosions occurred during distillation of this polynitro compound [1]. Though not
an endothermic compound, it is of positive oxygen balance and the heat of decomposition (2.80 kJ/g) would give an adiabatic decomposition temperature exceeding
2200°C and a 40-fold increase in pressure [2].
See entry OXYGEN BALANCE

Divinyl ketone
Graff, M. *et al., J. Org. Chem.*, 1968, **33**, 1247
One attempted reaction of trinitromethane with impure ketone caused an explosion
at refrigerator temperature.

2-Propanol
MCA Case History No. 1010
Frozen mixtures of trinitromethane −2-propanol (9:1) exploded during thawing.
The former (of positive oxygen balance) dissolves exothermally in the alcohol, the
heat effect increasing directly with the concentration above 50% w/w. Traces of
nitric acid may also have been present.
See other POLYNITROALKYL COMPOUNDS

0386. Azidodithioformic acid
[4472-06-4] CHN_3S_2

$N_3C(S)SH$

1. Mellor, 1947, Vol. 8, 338
2. Smith, G. B. L., *Inorg. Synth.*, 1939, **1**, 81

The isolated acid or its salts are shock- and heat-sensitive explosives [1]. Safe preparative procedures have been detailed. The heavy metal salts, though powerful detonators, are too sensitive for practical use [2].

See Carbon disulfide: Metal azides
See Bis(azidothiocarbonyl) disulfide
See other ACYL AZIDES, ORGANIC ACIDS

0387. 5-Nitrotetrazole
[55011-46-6] CHN_5O_2

1. Jenkins, J. M., *Chem. Brit.*, 1970, **6**, 401
2. Bates, L. R. *et al.*, US Pat. 4 094 879, 1978
3. Koldobskii, G. I. *et al., Russ. J. Org. Chem.*, 1997, **33**(12), 1771

An acidified solution of the sodium salt was allowed to evaporate during 3 days and spontaneously exploded 2 weeks later. The nature of the explosive species, possibly the *aci*-tetrazolic acid, was being sought [1]. The silver and mercury salts are explosive [2]. The chemical properties of the free nitrotetrazole have been studied. The sodium salt tetrahydrate loses water above 50°C, greatly increasing its friction and shock sensitivity [3].

See other C-NITRO COMPOUNDS, TETRAZOLES

0388. 5-Azidotetrazole
[35038-46-1] CHN_7

Alone, or Acetic acid, or Alkali

1. Thiele, J. *et al., Ann.*, 1895, **287**, 238
2. Lieber, E. *et al., J. Amer. Chem. Soc.*, 1951, **73**, 1313

Though explosive, it (and its ammonium salt) are much less sensitive to impact or friction than its sodium or potassium salts [1]. A small sample of the latter exploded violently during vacuum filtration. The parent compound explodes spontaneously even in acetone (but not in ethanol or aqueous) solution if traces of acetic acid are

149

present [2]. The salts are readily formed from diaminoguanidine salts and alkali nitrites. The ammonium salt explodes on heating, and the silver salt is violently explosive even when wet [1]. The sodium salt is also readily formed from cyanogen azide.

See Sodium 5-azidotetrazolide
See other ORGANIC AZIDES, TETRAZOLES

0389. Triazidomethane
[187585-04-6]   CHN$_9$

$$CH(N_3)_3$$

Hassner, A., private comm., 1986
Hassner, A. *et al., J. Org. Chem.*, 1990, **55**, 2304
The azide form of the quaternary ammonium ion exchange resin IR-400 (which exhibits low friction-sensitivity) reacts with bromoform, analogously to dichloro- and dibromo-methane, to form the highly explosive triazidomethane. Solutions of above 50% concentration explode in contact with a pipette or on injection into a GLC inlet port.

See other ORGANIC AZIDES

0390. Sodium hydrogen carbonate
[144-55-8]   CHNaO$_3$

$$NaOCO.OH$$

HCS 1980, 828

Carbon, Water
CISHC Chem. Safety Summ., 1978, **49**, 33
A mixture was being stirred and steam heated when power failure interrupted stirring, and heating was turned off for a hour before power was restored. When stirring was restarted, the hot contents of the pan erupted immediately. Carbon dioxide is evolved from warm aqueous solutions of the base, and absence of stirring and presence of the carbon adsorbent would lead to non-equilibrium retention of the gas, which would be released instantaneously on stirring.

See other AGITATION INCIDENTS, GAS EVOLUTION INCIDENTS

2-Furaldehyde
See 2-Furaldehyde
See other INORGANIC BASES, METAL OXONON-METALLATES

0391. Silver nitroureide
[[74386-96-2] (ion)]   CH$_2$AgN$_3$O$_3$

$$AgN(NO_2)CO.NH_2$$

See Nitrourea
See other N-METAL DERIVATIVES, N-NITRO COMPOUNDS, SILVER COMPOUNDS

0392. Silver 5-aminotetrazolide
[[50577-64-5] (ion)] CH_2AgN_5

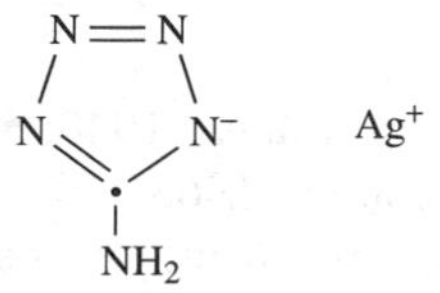

$$\underset{\underset{NH_2}{|}}{\overset{N=N}{\underset{N}{\diagup}}} \quad Ag^+$$

Thiele, J., *Ann.*, 1892, **270**, 59
Similar to silver tetrazolide, it explodes on heating.
See other N-METAL DERIVATIVES, SILVER COMPOUNDS, TETRAZOLES

0393. Cyanoborane oligomer
[60633-76-3] $(CH_2BN)_n$
(-BHCH=N-)$_n$

Gyori, B. *et al., J. Organomet. Chem.*, 1984, **262**, C7
It explodes on mechanical shock, and decomposes violently on heating under nitrogen to 230°C. The scale of preparation may need to be limited.

Sodium chlorate
  See Sodium chlorate: Cyanoborane oligomer
  See related BORANES, CYANO COMPOUNDS

0394. Bis(difluoroboryl)methane
[55124-14-6] $CH_2B_2F_4$
$(F_2B)_2CH_2$

Air, or Water
  Maraschin, N. J. *et al., Inorg. Chem.*, 1975, **14**, 1856
  Highly reactive, it explodes on exposure to air or water.
  See related ALKYLHALOBORANES

0395. Dibromomethane
[74-95-3] CH_2Br_2
Br_2CH_2

Potassium
  See Potassium: Halocarbons
  See other HALOALKANES

0396. Chloronitromethane
[1794-84-9] CH_2ClNO_2
                              $ClCH_2NO_2$

Preparative hazard
   1. Seigle, L. W. *et al., J. Org. Chem.*, 1940, **5**, 100
   2. Libman, D. D., private comm., 1968
   Chlorination of nitromethane following the published general method [1] gave a
   product which decomposed explosively during distillation at 95 mbar [2]. A b.p.
   of 122°C/1 bar is quoted in the literature.
   See Sodium *aci*-nitromethanide: Carbon disulfide, Chlorine
   See related NITROALKANES

†**0397. Dichloromethane**
   [75-09-2] CH_2Cl_2
                              Cl_2CH_2

(*MCA SD-86*, 1962); *NSC 474*, 1979; *HCS 1980*, 648

   1. *RSC Lab. Hazard Data Sheet No. 3*, 1982
   2. Downey, J. R., *Chem. Eng. News*, 1983, **61**(8), 2
   3. Kelling, R. A., *Chem. Eng. News*, 1990, **68**(17), 2
   Previously thought to be non-flammable except at elevated temperature or pressure
   or in oxygen-enriched air [1], it is in fact flammable in the range 12–19% in
   ambient air, given a sufficiently high level of ignition energy. Though it has no
   measurable flash point, it is calculated that flammable regions may exist above
   −9°C [2]. A surprisingly violent burst of a half full separating funnel occurred
   when dichloromethane was shaken with water, air also being present [3].
   See FLASH POINTS (reference 19)

Air, Methanol
   Coffee, R. D. *et al., J. Chem. Eng. Data*, 1972, **17**, 89–93
   Dichloromethane, previously considered to be non-flammable except in oxygen,
   becomes flammable in air at 102°C/1 bar, at 27°C/1.7 bar or at 27°C/1 bar in
   presence of less than 0.5 vol% of methanol (but see reference 2 above). Other
   data are also given.

Aluminium
   See Aluminium: Dichloromethane
   See Metals, below

Aluminium bromide
   See Aluminium bromide: Dichloromethane

Azides
   See Quaternary ammonium azides, below; Sodium azide, below

1,2-Diaminoethane
1. Nolan, 1983, Case history 145
2. Heskey, W. A., *Chem. Eng. News*, 1986, **64**(21), 2
3. Laird, T., *Chem. & Ind.*, 1986, 139

Dichloromethane was being distilled from its mixture with the amine at a bath temperature of 30°C when an exothermic reaction led to deflagration [1]. Heat of reaction, $\Delta H = -343$ kJ/mol (4.04 kJ/g) of dichloromethane [2]. Reaction of amines with dichloromethane at ambient temperature is common, exothermic in conc. solutions, and involves formation of an *N*-chloromethyl quaternary salt. Concentrating dichloromethane solutions of amines to low volumes should therefore be avoided [3].

Dimethyl sulfoxide, Perchloric acid
See Perchloric acid: Dichloromethane, Dimethyl sulfoxide

Dinitrogen pentaoxide
See NITRATING AGENTS

Dinitrogen tetraoxide
See Dinitrogen tetraoxide: Halocarbons

Metals
See Aluminium: Halocarbons
Lithium: Halocarbons
Sodium: Halocarbons
METAL–HALOCARBON INCIDENTS

Nitric acid
See Nitric acid: Dichloromethane

Potassium *tert*-butoxide
See Potassium *tert*-butoxide: Acids, etc.

Quaternary ammonium azides
1. Hassner, A. *et al., Angew. Chem. (Intern. Edn.)*, 1986, **25**, 479–480
2. Demer, F. R., *Lab. Safety Notes* (Univ. of Arizona), 1991(Spring), 2
3. Wood, W., private comm., 1986
4. Bretherick, L., *Chem. Eng. News*, 1986, **64**(51), 2

Quaternary ammonium azides will displace halogens in a synthesis of alkyl azides. Dichloromethane has been used as a solvent, although this can slowly form diazidomethane which may be concentrated by distillation during work-up, thereafter easily exploding [1]. An accident attributed to this cause is described, and acetonitrile recommended as a preferable solvent, supported polymeric azides, excess of which can be removed by filtration are also preferred in place of the tetrabutylammonium salt [2]. A similar explosion was previously recorded when the quaternary azide was generated in situ from sodium azide and a phase transfer catalyst in a part aqueous system [3,4].

Sodium azide
See Quaternary ammonium azides, above
See other HALOALKANES

0398. 1,1-Difluorourea
[1510-31-2] $CH_2F_2N_2O$

$$F_2NCO.NH_2$$

Parker, C. O. *et al., Inorg. Synth.*, 1970, **12**, 309
Concentrated aqueous solutions of difluorourea decompose above −20°C with evolution of tetrafluorohydrazine and difluoramine, both explosive gases.
See other DIFLUOROAMINO COMPOUNDS, GAS EVOLUTION INCIDENTS

0399. Trifluoromethylphosphine
[420-52-0] CH_2F_3P

$$F_3CPH_2$$

491M, 1975, 428

Ignites in air.
See other ALKYLPHOSPHINES

0400. Diiodomethane
[75-11-6] CH_2I_2

$$I_2CH_2$$

Alkenes, Diethylzinc
See Diethylzinc: Alkenes, Diiodomethane

Metals
See Copper−zinc alloys: Diiodomethane, etc.
      Lithium: Halocarbons
      Potassium: Halocarbons
      METAL−HALOCARBON INCIDENTS
See other ENDOTHERMIC COMPOUNDS, HALOALKANES

0401. Methylenedilithium
[21473-42-1] CH_2Li_2

$$H_2CLi_2$$

491M, 1975, 234

Ignites in air.
See other ALKYLMETALS

154

0402. Methylenemagnesium
[25382-52-9]   CH$_2$Mg

$$CH_2Mg$$

Ziegler, K. *et al., Z. Anorg. Chem.*, 1955, **282**, 345
The polymeric form ignites in air.
See other ALKYLMETALS

0403. Sodium *aci*-nitromethanide
[25854-38-0]   CH$_2$NNaO$_2$

$$H_2C=N(O)ONa$$

Mcycr, V., *et al., Ber.*, 1894, **27**, 1601; 3407
Sodium nitromethanoate is relatively stable when solvated. The dry salt is a sensitive and powerful explosive which may be detonated by warming to 100°C, by a strong blow or contact with traces of water. The potassium salt is even more sensitive. Lecture demonstrations of these properties are described.
See Nitromethane: Sodium hydride

Carbon disulfide, Chlorine
  1. Stirling, C. J. M., *Chem. Brit.*, 1986, **22**, 524
  2. Stirling, C. J. M., private comm., 1986
During the preparation of chloronitromethane by adding portions of dry sodium *aci*-nitromethanide to chlorine (40 mol of each) dissolved in carbon disulfide, a violent explosion occurred when the addition was half-complete. Similar reactions using bromine had been executed uneventfully many times previously [1]. No certain explanation has emerged, but the sodium salt is known to be explosively unstable, and mixtures of carbon disulfide vapour and air are of course extremely flammable and explosive. Contact of the dry salt with traces of chlorine above its carbon disulfide solution may have led to an exotherm and ignition of the vapour–air mixture in the flask [2].

Mercury(II) chloride, Acids
Nef, J. U., *Ann.*, 1894, **280**, 263, 305
Interaction gives mercury nitromethanide, which is converted by acids to mercury fulminate.
See Mercury(II) fulminate

Nitric oxide, Base
Traube, W., *Annalen*, 1898, **300**, 107
A slurry of sodium nitromethanoate in excess sodium ethoxide solution forms a (presumably even more) explosive salt on treatment with nitric oxide. It is claimed the parent acid is $O_2NCH_2N_2O_2H$.
See Nitrogen oxide

1,1,3,3-Tetramethyl-2,4-cyclobutanedione
See Sodium 1,3-dihydroxy-1,3-bis(*aci*-nitromethyl)-2,2,4,4-tetramethylcyclobutandiide

Water
1. Nef, J. U., *Ann.*, 1894, **280**, 273
2. Jung. M. E. *et al.*, *J. Chem. Soc., Chem. Comm.*, 1987, 753 (footnore)
The *aci*-sodium salt, normally crystallising with one molecule of ethanol and stable, will explode if moistened with water. This is due to liberation of heat and conversion to sodium fulminate [1]. After several uneventful similar operations, during the destruction of excess sodium salt by pouring water onto it, a violent explosion occurred [2]. A safer procedure would be to add the salt in small portions to a bulk of stirred ice-water.
See also Sodium 1,3-dihydroxy-1,3-bis(*aci*-nitromethyl)-2,2,4,4-tetramethylcyclobutandiide
See other aci-NITRO SALTS

0404. Cyanamide
[420-04-2]   CH$_2$N$_2$
$$N\equiv CNH_2$$

1. Anon., *Fire Prot. Assoc. J.*, 1966, 243
2. Anon., *Sichere Chemiearbeit*, 1976, **28**, 63
Cyanamide is endothermic (ΔH°_f +58.8 kJ/mol, 1.40 kJ/g), thermally unstable and needs storage under controlled conditions. Contact with moisture, acids or alkalies accelerates the rate of decomposition, and at temperatures above 40°C thermal decomposition is rapid and may become violent. A maximum storage temperature of 27°C is recommended [1]. Commercial cyanamide is stabilised with boric acid, phosphoric acid, sodium dihydrogen phosphate etc., but vacuum distillation produces a neutral (unstabilised) distillate, which immediately may decompose spontaneously. Small-scale storage tests showed that unstabilised cyanamide was 47% decomposed after 18 days at 20°C and 75% decomposed after 29 days at 30°C, whereas stabilised material showed only 1% decomposition under each of these conditions. Larger-scale tests with 1–2 kg unstabilised samples led to sudden and violent exothermic polymerisation after storage for 14 days at ambient temperature. If small samples of unstabilised cyanamide are required, they are best prepared by freezing out from aqueous solutions of the stabilised material. Such small samples should be used immediately or stored under refrigeration [2].

1,2-Phenylenediamine salts
Sawatari, K. *et al.*, Japan Kokai, 76 16 669, 1976 (*Chem. Abs.*, 1976, **85**, 63069)
During the preparation of 2-aminobenzimidazoles, reaction conditions are maintained below 90°C to prevent explosive polymerisation of cyanamide.

Water
Pinck, L. A. *et al.*, *Inorg. Synth.*, 1950, **3**, 41
Evaporation of aqueous solutions to dryness is hazardous, owing to the possibility of explosive polymerisation in conc. solution.
See other POLYMERISATION INCIDENTS
See other CYANO COMPOUNDS, ENDOTHERMIC COMPOUNDS

0405. Diazirine
   [157-22-2] CH_2N_2

1. Graham, W. H., *J. Org. Chem.*, 1965, **30**, 2108
2. Schmitz, E. *et al.*, *Chem. Ber.*, 1962, **95**, 800

This cyclic isomer of diazomethane is also a gas (b.p., $-14°C$) which explodes on heating. Several homologues are also thermally unstable [1,2].
See other DIAZIRINES

0406. Diazomethane
   [334-88-3] CH_2N_2

$$H_2C = N^+ = N^-$$

RSC Lab. Hazard Data Sheet No. 4, 1982

1. Eistert, B., in *Newer Methods of Preparative Organic Chemistry*, 517–518, New York, Interscience, 1948
2. de Boer, H. J. *et al.*, *Org. Synth.*, 1963, Coll. Vol. 4, 250
3. Gutsche, C. D., *Org. React.*, 1954, **8**, 392–393
4. Zollinger, H., *Azo and Diazo Chemistry*, 22, London, Interscience, 1961
5. Fieser, L. *et al.*, *Reagents for Organic Synthesis*, Vol. 1, 191, New York, Wiley, 1967
6. Horàk, V. *et al.*, *Chem. & Ind.*, 1961, 472
7. Ruehle, P. H. *et al.*, *Chem. & Ind.*, 1979, 255–256
8. Bowes, C. M., *Univ. Safety Assoc. Safety News*, 1981, **15**, 18–19
9. Hashimoto, N. *et al.*, *Chem. Pharm. Bull.*, 1981, **29**, 1475–1478; 1982, **30**, 119–124
10. Barnes, C. J. *et al.*, *J. Assoc. Off. Anal. Chem.*, 1982, **65**, 273–274
11. Black, H. T., *Aldrichimica Acta*, 1983, **16**(1), 3–10
12. Asyama, T. *et al.*, *Chem. Abs.*, 1984, **101**, 171306
13. Shiga, S., *Chem. Abs.*, 1987, **107**, 117404
14. Carlton, L. *et al.*, *S. African J. Chem.*, 1987, **40**, 203–204
15. Archibald, T. G. *et al.*, US Patent 5,817,778, 1998

Diazomethane is a highly endothermic small molecule ($\Delta H_f°$ (g) +192.5 kJ/mol, 4.58 kJ/g) which boils at $-23°C$ and the undiluted liquid or concentrated solutions may explode if impurities or solids are present [1], including freshly crystallised products [2]. Gaseous diazomethane, even when diluted with nitrogen, may explode at elevated temperatures (100°C or above), or under high-intensity lighting, or if rough surfaces are present [1]. Ground glass apparatus or glass-sleeved stirrers are therefore undesirable when working with diazomethane. Explosive intermediates may also be formed during its use as a reagent, but cold dilute solutions have frequently been used uneventfully [1]. Further safety precautions have been detailed

[2,3,4]. Many precursors for diazomethane generation are available [5], including the stable water soluble intermediate *N*-nitroso-3-methylaminosulfolane [6]. Many of the explosions observed are attributed to uncontrolled or unsuitable conditions of contact between concentrated alkali and undiluted nitroso precursors [1].

A large scale generator and procedure for safe preparation and use of the reagent on 1.1 mol scale are detailed [7]. Insolubility of the substrate in ether led to omission of ether from a microgenerator chamber and use of dioxan to dissolve the substrate in the outer chamber. During addition of alkali to nitrosomethylguanidine (30 mg) in 0.125 ml water, the apparatus exploded. This was attributed to absence of ether vapour to dilute the diazomethane [8]. Trimethylsilyldiazomethane is presented as a safe and effective substitute for the hazardous parent compound [9], and two cheap and safe storage containers for small amounts of ethereal solutions of the latter are described [10]. The available data has recently been summarised with 116 references [11]. The preparation of trimethylsilyldiazomethane and its application as a safe substitute for diazomethane are reviewed [12]. Automated equipment for esterification of small samples of acidic materials for GLC analysis has been developed [13]. A convenient method for preparation of small amounts of pure diazomethane involves entrainment in a stream of helium and condensation as a solid, m.p. $-145°C$, on a liquid nitrogen-cooled cold finger. The solid may be kept almost indefinitely at LN temperature [14]. Procedures for small industrial scale (50 mole) preparation and use in solution have been patented [15].
See also GLASS INCIDENTS

Alkali metals
Eistert, B. in *Newer Methods of Preparative Organic Chemistry*, 518, New York, Interscience, 1948
Contact of diazomethane with alkali metals causes explosions.
See Lithium diazomethanide

Calcium sulfate
Gutsche, C. D., *Org. React.*, 1954, **8**, 392
Calcium sulfate is an unsuitable desiccant for drying tubes in diazomethane systems. Contact of diazomethane vapour and the sulfate causes an exotherm which may lead to detonation. Potassium hydroxide is a suitable desiccant.

Dimethylaminodimethylarsine, Trimethyltin chloride
Krommes, P. *et al., J. Organomet. Chem.*, 1976, **110**, 195–200
Interaction in ether to produce diazomethyldimethylarsine is accompanied by violent foaming, and eye protection is essential.
See other DIAZO COMPOUNDS, ENDOTHERMIC COMPOUNDS

0407. Isocyanoamide ('Isodiazomethane')
[4702-38-9] CH_2N_2
:C=NNH$_2$

1. Müller, E. *et al., Chem. Ber.*, 1954, **87**, 1887
2. Müller, E. *et al., Ann.*, 1968, **713**, 87

This unstable liquid begins to decompose at 15°C and explodes exothermically at 35–40°C [1], but may be handled safely in ether solution [2].

See related CYANO COMPOUNDS

0408. N-Nitromethanimine
[67400-85-2]

$$H_2C{=}NNO_2$$

$CH_2N_2O_2$

Zhao, X. *et al., J. Chem. Phys.*, 1988, **88**, 801

This nitrimine, the nominal monomer of the cyclic nitramine high explosives RDX and HMX, may be involved in their detonation and can be formed from them by pyrolysis.

See 1,3,5-Trinitrohexahydro-1,3,5-triazine

See other N–NITRO COMPOUNDS

0409. Nitrooximinomethane ('Methylnitrolic acid')
[625-49-0]

$$O_2NCH{=}NOH$$

$CH_2N_2O_3$

Sorbe, 1968, 147

An unstable and explosive crystalline solid, formally a nitro-oxime.

See related NITROALKANES, OXIMES

0410. Dinitromethane
[625-76-3]

$$(O_2N)_2CH_2$$

$CH_2N_2O_4$

1. Sorbe, 1968, 148
2. Bedford, C. D. *et al., J. Org. Chem.*, 1979, **44**, 635

It explodes at 100°C [1], and attempted distillation of more than 1 g at 30–35°C/1.5 mbar led to a violent explosion [2].

See other POLYNITROALKYL COMPOUNDS

0411. Tetrazole
[288-94-8]

CH_2N_4

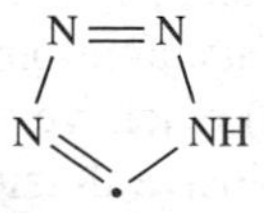

1. Benson, F. R., *Chem. Rev.*, 1947, **41**, 5
2. Stull, 1977, 22

3. *DOC 5*, 1982, 5032
4. Anon., *IST Sci. Tech.*, 1987, (4), 9

It explodes above its m.p., 155°C [1]. It is highly endothermic (ΔH°_f +237.2 kJ/mol, 3.39 kJ/g) with a heat of decomposition (3.27 kJ/g) which would give an adiabatic product temperature of some 1950°C and a 22-fold increase in pressure in a closed container [2]. Its solutions are also explosive when shocked [3]. An explosion during sublimation of tetrazole at ambient pressure was caused by overheating [4].

See other ENDOTHERMIC COMPOUNDS, TETRAZOLES

0412. Lead methylenebis(nitramide) (*N,N'*-Dinitromethanediamine, lead(II) salt)
[86202-43-9] $CH_2N_4O_4Pb$

$$Pb(NNO_2)_2CH_2$$

See Methylenebis(nitramine)
See other HEAVY METAL DERIVATIVES, *N*-NITRO COMPOUNDS

0413. 5-Amino-1,2,3,4-thiatriazole
[6630-99-5] CH_2N_4S

Lieber, E. *et al., Inorg. Synth.*, 1960, **6**, 44
It decomposes with a slight explosion in a capillary tube at 136°C.
See other HIGH-NITROGEN COMPOUNDS, N–S COMPOUNDS

0414. Diazidomethane
[107585-03-5] CH_2N_6

$$CH_2(N_3)_2$$

1. Hassner, A. *et al., Angew. Chem. (Intern. Ed.)*, 1986, **25**, 479–480
2. Hassner, A. *et al., J. Org. Chem.*, 1990, **55**, 2304

The azide form of the quaternary ammonium ion exchange resin IR-400 (which exhibits low friction-sensitivity) reacts very slowly with dichloromethane or dibromomethane to produce the explosive diazidomethane [1]. Solutions of above 70% concentration will explode if a pipette is inserted or on injection into a GLC inlet port [2].

See other ORGANIC AZIDES

0415. 5-*N*-Nitroaminotetrazole
[18558-16-4]   CH₂N₆O₂

$$CH_2N_6O_2$$

1. Lieber, E. *et al., J. Amer. Chem. Soc.*, 1951, **73**, 2328
2. O'Connor, T. E. *et al., J. Soc. Chem. Ind.*, 1949, **68**, 309

It and its monopotassium salt explode at 140°C and the diammonium salt explodes at 220°C after melting [1]. The disodium salt explodes at 207°C [2].

Amminemetals
Complexes with several ammine derivatives of metals are explosive.
See entry NITRAMINE–METAL COMPLEXES
See other *N*-NITRO COMPOUNDS, TETRAZOLES

†0416. Formaldehyde
[50-00-0]   CH₂O

O:CH₂

(*MCA SD-1*, 1960); *NSC 342*, 1982; *FPA H54*, 1977; *HCS 1980*, 506

van den Brink, M. J., *Chem. Mag.* (Rijswijk), 1982, 428
Pure formaldehyde, prepared by vacuum depolymerisation of paraldehyde, was collected as a solid at −189°C. When the flask was transferred to a Cardice–ethanol bath, the contents began to repolymerise exothermally and ignited.
See other POLYMERISATION INCIDENTS
See Glyoxal

Acrylonitrile
See Acrylonitrile, Formaldehyde

Hydrogen peroxide   MRH 6.44/69
See Hydrogen peroxide: Oxygenated compounds

Magnesium carbonate hydroxide
BCISC Quart. Safety Summ., 1965, **36**(143), 44
During neutralisation of the formic acid present in formaldehyde solution by shaking with the basic carbonate in a screw-capped bottle, the latter burst owing to pressure of liberated carbon dioxide. Periodical release of pressure should avoid this.
See other GAS EVOLUTION INCIDENTS, NEUTRALISATION INCIDENTS

Nitromethane

See Nitromethane: Formaldehyde

Other reactants
 Yoshida, 1980, 346
 MRH values for 14 combinations with oxidants are given, and are all high.

Peroxyformic acid   MRH 5.69/100
 See Peroxyformic acid: Organic materials

Phenol
 1. Taylor, H. D. *et al., Major Loss Prevention in Process Industries*, (Symp. Ser.
    No. 34), 46, London, IChE, 1971
 2. Anon., *Chem. Eng. News*, 1992, **70**(10), 7
 3. Anon. *Chem. Eng. News*, 1997, **75**(38), 11
 4. Starkie, A. *et al., Chem. in Brit.*, 1996, **32**(2), 35
 5. Gustin, J.-L. *et al., J. Loss Prev.*, 1993, **6**(2), 103

At least 9 cases of catalysed plant-scale preparations of phenol–formaldehyde resin
which ran away with sudden pressure development and failure of bursting disks or
reactors are briefly mentioned [1]. No details of process conditions are given. A
destructive incident of this type, which must have developed the force of some kg
of TNT, is reported without real detail [2]. Another fatal and destructive explosion
when preparing phenol formaldehyde resins is reported [3]. The editor calculates
from analogues that this reaction could have above 1 kJ/g reagents, depending upon
stoichiometry and concentration, and mostly associated with the second step, dehy-
drative condensation of the initial hydroxymethylphenol to dihydroxybiphenyls.
A photograph of the aftermath of a runaway polymerisation is reproduced [4]. A
very thorough study of hazards of the phenol/formaldehyde polymerisation reaction,
leading to calculations of emergency vent size, is given [5].
See other POLYMERISATION INCIDENTS, RUNAWAY REACTIONS

Potassium permanganate
 See Potassium permanganate: Formaldehyde

Sodium hydroxide
 1. Ashby, E. C. *et al., J. Amer. Chem. Soc.*, 1993, **115**, 1171
 2. Chrisope, D. R. *et. al., Chem. Eng. News.*, 1995, **73**(2), 2
 3. Kapoor, S. *et al., J. Phys. Chem.*, 1995, **99**(18), 6857

In place of the well known Cannizzaro reaction (which is significantly exothermic)
formaldehyde, at lower concentrations, produces hydrogen with alkalis, leading to
possible pressurisation and ignition [1]. It has been demonstrated that this modified Cannizzaro reaction giving hydrogen can operate in real commercial situations
(500 ppm formaldehyde, 1.5% sodium hydroxide). Caution and ventilated or inerted
headspaces when storing alkaline formaldehyde containing products are advisable
[2]. A detailed study of the kinetics and mechanism of hydrogen evolution has been
published [3].

Water, Methanol, Air

Anon., *Jahresbericht*, 1994, 73

An explosion is recorded, consequent upon welding a ladder to the wall of a half-full tank, which had not been inerted with nitrogen and was vented to atmosphere, containing a 37% formaldehyde solution stabilised with 5.6% methanol at 72°C. The flash point of this mixture in air is about 71°C, autoignition temperature 420°C: the head space ignited, killing two workers, injuring nine and blowing off the top of the tank.

See other ALDEHYDES, REDUCANTS

0417. Paraformaldehyde
[9002-81-7] $(CH_2O)_n$

$$HOCH_2(OCH_2)_n OCH_2OH$$

The dry finely powdered linear polymer is a significant dust explosion hazard.
See entry DUST EXPLOSION INCIDENTS (reference 22)
See other ALDEHYDES, REDUCANTS

0418. Formic acid (Methanoic acid)
[64-18-6] CH_2O_2

$$HCO.OH$$

FPA H86, 1979; *HCS 1980*, 508

1. *BCISC Quart. Safety Summ.*, 1973, **44**, 18
2. Falconer, J. *et al., Proc. 2nd Int. Conf. Solid. Surf.*, 1974, 525
3. Anon., *Jahresber.*, 1981, 74
4. Reynolds, R. J., *DNA Repair*, 1981, **1**(A), 11–21, NY, Dekker

The slow decomposition in storage of 98–100% formic acid with liberation of carbon monoxide led to rupture of the sealed glass containers. In absence of gas leakage, a full 2.5 l bottle would develop a pressure of over 7 bar during 1 year at 25°C [1]. Explosive decomposition of formic acid on a clean nickel (1.1.0) surface was studied, using deuteroformic acid [2]. A full 1 l bottle of 96% formic acid burst when the ambient temperature fell to −6°C overnight and the contents froze and expanded. Gas pressure from previous partial decomposition may also have contributed [3]. Hydrolysis of precipitated DNA with 97% acid at 180°C in sealed glass tubes is potentially hazardous because of rapid gas formation [4].

See other GAS EVOLUTION INCIDENTS, GLASS INCIDENTS

Aluminium
See Aluminium: Formic acid

Hydrogen peroxide   MRH 3.30/42
See Peroxyformic acid (references 5,6)

Hydrogen peroxide
See Hydrogen peroxide: Oxygenated compounds

4-Hydroxy-3-methoxybenzaldehyde, Thallium(III) nitrate
See Thallium(III) nitrate: Formic acid, etc.

Nitric acid   MRH 2.64/36
See Nitric acid: Formic acid

Nitric acid, Urea
See Nitric acid: Formic acid, Urea

Nitromethane   MRH 5.69/100
See Nitromethane: Acids

Other reactants
Yoshida, 1980, 100
MRH values calculated for 16 combinations, largely with oxidants, are given.

Palladium–carbon catalyst
Freifelder, 1971, 188
Addition of dry catalyst to 98% formic acid used as a hydrogenation solvent can be extremely hazardous, because hydrogen is released by decomposition of the acid. Addition of acid to the water-wetted catalyst is safer.
See other GAS EVOLUTION INCIDENTS, HYDROGENATION INCIDENTS

Phosphorus pentaoxide
Muir, G. D., private comm., 1968
Attempted dehydration of 95% acid to anhydrous formic acid caused rapid evolution of carbon monoxide.
See other GAS EVOLUTION INCIDENTS

Sodium hypochlorite
See Sodium hypochlorite: Formic acid
See other ORGANIC ACIDS, REDUCANTS

0419. Formaldehyde oxide polymer (Poly(dioxymethylene))
[107981-97-5] $(CH_2O_2)_n$

$$(H_2C{=}O^+{-}O^-)_n$$

Lapalme, R. *et al., Can. J. Chem.,* 1979, **57**, 3272
The dry peroxide polymer (formed in ozonisation of vinyl acetate, and possibly a dioxirane polymer) is a shock-sensitive explosive, which needs care in handling.
See other POLYPEROXIDES

0420. Peroxyformic acid (Methaneperoxoic acid)
[107-32-4] CH_2O_3

HCO.OOH

1. Greenspan, F. P., *J. Amer. Chem. Soc.,* 1946, **68**, 907
2. D'Ans, J. *et al., Ber.,* 1915, **48**, 1136
3. Weingartshofer, A. *et al., Chem. Eng. News,* 1952, **30**, 3041

164

4. Swern, 1970, Vol. 1, 337
5. Isard, A. *et al., Chemical Tech.*, 1974, **4**, 380
6. Anon., *Chem. Eng. News*, 1950, **28**, 418
7. Stull, 1977, 19

Peroxyformic acid solutions are unstable and undergo a self-accelerating exothermic decomposition at ambient temperature [1]. An 80% solution exploded at 80–85°C [2]. A small sample of the pure vacuum-distilled material cooled to below −10°C exploded when the flask was moved [3]. Though the acid has occasionally been distilled, this is an extremely dangerous operation (it is formally the redox compound formyl hydroperoxide) [4]. During preparation of the acid by a patented procedure involving interaction of formic acid with hydrogen peroxide in the presence of meta-boric acid, an explosion occurred which was attributed to spontaneous separation of virtually pure peroxyformic acid [5]. Following an incident in which a 1:1 mixture of formic acid and 90% hydrogen peroxide had exploded violently when handled after use as an oxidant [6], it was calculated that the (redox) decomposition exotherm (1.83 kJ/g of acid) would attain an adiabatic reaction temperature of 1800°C with a 22-fold increase in pressure in a closed container [7].

Chlorine
See Chlorine: Chloromethane

Metals, or Metal oxides
D'Ans, J. *et al., Ber.*, 1915, **48**, 1136
Violence of reaction depends on concentration of acid and scale and proportion of reactants. The following observations were made with additions to 2–3 drops of ca. 90% acid. Nickel powder, becomes violent; mercury, colloidal silver and thallium powder readily cause explosions; zinc powder causes a violent explosion immediately. Iron powder is ineffective alone, but a trace of manganese dioxide promotes deflagration. Barium peroxide, copper(I) oxide, impure chromium trioxide, iridium dioxide, lead dioxide, manganese dioxide and vanadium pentoxide all cause violent decomposition, sometimes accelerating to explosion. Lead(II) oxide, lead(II),(IV) oxide and sodium peroxide all cause an immediate violent explosion.

Non-metals
D'Ans, J. *et al., Ber.*, 1915, **48**, 1136
Impure carbon and red phosphorus are oxidised violently, and silicon, promoted by traces of manganese dioxide, is oxidised with ignition.

Organic materials
1. D'Ans, J. *et al., Ber.*, 1915, **48**, 1136
2. Anon., *Chem. Eng. News*, 1950, **28**, 418
3. Shanley, E. S., *Chem. Eng. News*, 1950, **28**, 3067
4. Öhlenschlager, A. *et al., Chem. Eng. News*, 1987, **65**(15), 2

Formaldehyde, benzaldehyde and aniline react violently with 90% performic acid [1]. An unspecified organic compound was added to the acid (preformed from formic acid and 90% hydrogen peroxide), and soon after the original vigorous reaction had subsided, the mixture exploded violently [2]: (see references 6,7 above). Reaction with alkenes is vigorously exothermic, and adequate cooling is necessary [3]. Reactions with performic acid can be more safely accomplished by slow addition of hydrogen peroxide to a solution of the compound in formic acid, so that the peroxyacid is used as it is formed. Adequate safety screens should be used with all peracid preparations [3]. After the hydroxylation of *trans*-2-pentanoic acid with performic acid according to a published general procedure, while formic acid and water were being distilled off under vacuum at 45°C, a violent explosion occurred. This was attributed to decomposition of unreacted performic acid [4].

Sodium nitrate
D'Ans, J. *et al.*, *Ber.*, 1915, **48**, 1139
The salt may lead to explosive decomposition of the peroxy acid.
See other PEROXYACIDS, REDOX COMPOUNDS

0421. Methylsilver
[75993-65-6]   CH₃Ag

$$H_3CAg$$

Thiele, H., *Z. Elektrochem.*, 1943, **49**, 426
Prepared at −80°C, the addition compound with silver nitrate decomposes explosively on warming to −20°5C.
See other ALKYLMETALS

0422. Silver nitroguanidide
[ ]   CH₃AgN₄O₂

$$AgN(NO_2)C(:NH)NH_2$$

See Nitroguanidine
See other SILVER COMPOUNDS

0423. Methylaluminium diiodide
[2938-46-7]   CH₃AlI₂

$$H_3CAlI_2$$

Nitromethane
See Nitromethane: Alkylmetal halides
See other ALKYLALUMINIUM HALIDES

0424. Dichloromethylarsine
[593-89-5]   CH₃AsCl₂

$$Cl_2AsCH_3$$

Chlorine
See Chlorine: Dichloro(methyl)arsine
See other ALKYLNON-METAL HALIDES

0425. Methylborylene
[62785-41-5] CH_3B

MeB:

Preparative hazard
See Dibromomethylborane: Sodium–potassium alloy (next below)
See related ALKYLBORANES

0426. Dibromomethylborane
[17933-16-2] CH_3BBr_2

MeBBr$_2$

Nöth, H. *et al., Inorg. Synth.*, 1983, **22**, 223
It may ignite in air if warm, or if the heat of oxidation or hydrolysis cannot be dissipated.
See other SELF-HEATING AND IGNITION INCIDENTS

Sodium–potassium alloy
Van der Kerk, S. M. *et al., Polyhedron*, 1983, **2**, 1337
Use of potassium graphite or caesium graphite to generate methylborylene is uneventful, while use of sodium–potassium alloy (1:5 mol) caused an explosion in 2 out of 5 attempts.
See other ALKYLHALOBORANES

0427. Methaneboronic anhydride–pyridine complex
[79723-21-0] $CH_3BO.C_5H_5N$

$H_3CB=O.C_5H_5N$

1. Mattesson, D. S. *et al., J. Org. Chem.*, 1964, **29**, 3399
2. Mattesson, D. S. *et al., Organometallics*, 1982, **1**, 27
The published procedure [1] for preparing the complex failed to note that the organic extracts will ignite spontaneously if exposed to air for a few seconds during workup. Inert blanketing with argon is essential throughout, and the preparation is hazardous on the scale described. Ignition probably arises from by-product trimethylborane [2].
See other AUTOIGNITION INCIDENTS
See related ACID ANHYDRIDES

0428. Methylbismuth oxide (Methyloxobismuthine)
[ ] CH_3BiO

$H_3CBi=O$

Marquardt, A., *Ber.*, 1887, **20**, 1522
It ignites on warming in air.
See related ALKYLMETALS

†0429. Bromomethane (Methyl bromide)
[74-83-9]   CH₃Br

H₃CBr

(*MCA SD-35*, 1968); *HCS 1980*, 641

1. *MCA SD-35*, 1968
2. *MCA Case History No. 746*
Though bromomethane is used as a fire extinguishant, it does in fact form difficultly
flammable mixtures between 10 and 15% (13.5–14.5% but also 8.6–20% have been
noted), the limits in oxygen, or under pressure, being wider [1,2].
See entry FLASH POINTS (reference 19)

Aluminium
Lambert, P. G., *Chem. and Ind.*, 1990, (18), 562
Specially bottled methyl bromide, in an aluminium cylinder under nitrogen pressure,
disgorged a black sludge when liquid was discharged. Inversion and venting, to blow
free the line, gave a burst of flame from the vent. It is presumed that corrosion of
the cylinder produced pyrophoric aluminium alkyls.
See Trimethylaluminium; Aluminium: halocarbons
See other CORROSION INCIDENTS, METAL–HALOCARBON INCIDENTS

Dimethyl sulfoxide
See Trimethylsulfoxonium bromide

Ethylene oxide
See Ethylene oxide: Air, Bromomethane

Metals
MCA Case History No. 746 and addendum
Metallic components of zinc, aluminium and magnesium (or their alloys) are unsuitable for service with bromomethane because of the formation of pyrophoric Grignard-type compounds. The Case History attributes a severe explosion to ignition of
a bromomethane–air mixture by pyrophoric methylaluminium bromides produced
by corrosion of an aluminium component.
See other CORROSION INCIDENTS, METAL–HALOCARBON INCIDENTS
See other HALOALKANES

0430. *N,N*-Dibromomethylamine
[10218-83-4]   CH₃Br₂N

Br₂NCH₃

Cooper, J. C. *et al.*, *Explosivstoffe*, 1969, **17**(6), 129–130
Like the dichloro analogue, it appears to be more sensitive to impact or shock than
N-chloromethylamine.
See other N-HALOGEN COMPOUNDS

0431. Methylcadmium azide
[7568-37-8]   CH$_3$CdN$_3$
                              MeCdN$_3$

Dehnicke, K. *et al., J. Organomet. Chem.*, 1966, **6**, 298
Surprisingly it is thermally stable to 300°C, (cadmium azide is very heat sensitive)
but is hygroscopic and very readily hydrolysed to explosive hydrogen azide.
See related METAL AZIDES

†0432. Chloromethane (Methyl chloride)
[74-87-3]   CH$_3$Cl
                               H$_3$CCl

(*MCA SD-40*, 1970); *FPA H68*, 1979; *HCS 1980*, 644

Aluminium
      See Aluminium: Halocarbons

Aluminium chloride, Ethylene
      See Ethylene: Aluminium chloride

Interhalogens
      See Bromine trifluoride: Halogens, etc.
            Bromine pentafluoride: Hydrogen-containing materials

Metals
      MCA SD-40, 1970
      In presence of catalytic amounts of aluminium chloride, powdered aluminium and
      chloromethane interact to form pyrophoric trimethylaluminium. Chloromethane may
      react explosively with magnesium, or potassium, sodium or their alloys. Zinc prob-
      ably reacts similarly to magnesium.
      See Aluminium: Halocarbons
            Sodium: Halocarbons
            METAL–HALOCARBON INCIDENTS
      See other CATALYTIC IMPURITY INCIDENTS

Other reactants
      Yoshida, 1980, 366
      MRH values calculated for 13 combinations with oxidants are given.
      See other HALOALKANES

0433. Methylmercury perchlorate
[40661-97-0]   CH$_3$ClHgO$_4$
                              MeHgClO$_4$

Anon., *Angew. Chem. (Nachr.)*, 1970, **18**, 214

On rubbing with a glass rod, a sample exploded violently. As the explosion could not be reproduced with a metal rod, initiation by static electricity was suspected.
See other STATIC INITIATION INCIDENTS
See related METAL PERCHLORATES

0434. Methyl hypochlorite
[593-78-2] CH_3ClO

MeOCl

Sandmeyer, T., *Ber.*, 1886, **19**, 859
The liquid could be gently distilled (12°C) but the superheated vapour readily and violently explodes, as does the liquid on ignition.
See other HYPOHALITES

0435. Methanesulfinyl chloride
[676-85-7] CH_3ClOS

MeS(O)Cl

Douglass, I. B. *et al., J. Org. Chem.*, 1964, **29**, 951
A sealed ampoule burst after shelf storage for several months. Store only under refrigeration.
See other ACYL HALIDES

0436. Methyl perchlorate
[17043-56-0] CH_3ClO_4

MeOClO₃

Sidgwick, 1950, 1236
The high explosive instability is due in part to the covalent character of the alkyl perchlorates, and also to the excess of oxygen in the molecule over that required to combust completely the other elements present (i.e. a positive oxygen balance).
See other ALKYL PERCHLORATES

0437. *N,N*-Dichloromethylamine
[7651-91-4] CH_3Cl_2N

Cl₂NMe

Calcium hypochlorite, or Sodium sulfide, or Water
   1. Bamberger, E. *et al., Ber.*, 1895, **28**, 1683
   2. Okon, K. *et al., Chem. Abs.*, 1960, **54**, 17887
A mixture with water exploded violently on warming [1]. Contact with solid sodium sulfide or distillation over calcium hypochlorite also caused explosions [2].
See other N-HALOGEN COMPOUNDS

0438. Dichloromethylphosphine
[676-83-5] CH_3Cl_2P

$$MePCl_2$$

Bis(2-hydroxyethyl)methylphosphine
Piccinni-Leopardi, C. *et al., J. Chem. Soc., Perkin Trans. 2*, 1986, 91
Cyclo-condensation of the 2 components gives 2,6-dimethyl-1,3-dioxa-2,6-diphos-
phacyclooctane, and its subsequent distillation (65–68°C/0.07 mbar) must be effec-
ted with bath temperatures below 120°C to avoid explosion.
See other ALKYLHALOPHOSPHINES

†0439. Methyltrichlorosilane
[75-79-6] CH_3Cl_3Si

$$MeSiCl_3$$

See other ALKYLHALOSILANES

0440. Methylcopper
[1184-53-8] CH_3Cu

$$MeCu$$

1. Coates, 1960, 348
2. Ikariya, T. *et al., J. Organomet. Chem.*, 1974, **72**, 146
The dry solid is very impact-sensitive and may explode spontaneously on being
allowed to dry out at room temperature [1]. Methylcopper decomposes explosively
at ambient temperature, and violently in presence of a little air [2].
See other ALKYLMETALS

†0441. Fluoromethane (Methyl fluoride)
[593-53-3] CH_3F

$$FCH_3$$

See other HALOALKANES

0442. Methyl hypofluorite (Fluoroxymethane)
[36336-08-0] CH_3FO

$$CH_3OF$$

Appelman, E. *et al., J. Amer. Chem. Soc.*, 1991, **113**(7), 2648
This compound, the first known alkyl hypofluorite, was found to be stable but reactive
at −120°C, but samples sometimes exploded on warming to cardice temperatures
(−80°C?).

171

0443. Xenon(II) fluoride methanesulfonate

[ ] CH_3FO_3SXe

$$FXeOSO_2CH_3$$

Wechsberg, M. *et al., Inorg. Chem.*, 1972, **11**, 3066
The solid explodes on warming from 0°C to ambient temperature.
See other XENON COMPOUNDS

0444. Silyl trifluoromethanesulfonate

[2923-28-6] $CH_3F_3O_3SSi$

$$H_3SiOSO_2CF_3$$

Bassindale, A. R. *et al., J. Organomet. Chem.*, 1984, **271**, C2
It cannot be isolated pure by distillation at ambient pressure as it disproportionates to spontaneously flammable silane.
See related SULFUR ESTERS

0445. Iodomethane (Methyl iodide)

[74-88-4] CH_3I

$$H_3CI$$

HCS 1980, 562

Oxygen
    Antonik, S., *Bull. Soc. Chim. Fr.*, 1982, **3–4**, 128–133
    Second stage ignition during oxidation/combustion of iodomethane in oxygen at 300–500°C was particularly violent, occasionally causing fracture of the apparatus, and was attributed to formation and decomposition of a periodic species.

Silver chlorite
    See Silver chlorite: Iodoalkanes

Sodium
    1. Anon., *J. Chem. Educ.*, 1966, **43**, A236
    2. Braidech, M. M., *J. Chem. Educ.*, 1967, **44**, A324
    The first stage of a reaction involved the addition of sodium dispersed in toluene to a solution of adipic ester in toluene. The subsequent addition of iodomethane (b.p. 42°C) was too fast and vigorous boiling ejected some of the flask contents. Exposure of sodium particles to air caused ignition, and a violent toluene–air explosion followed [1]. When a reagent as volatile and reactive as iodomethane is added to a hot reaction mixture, controlled addition, and one or more wide-bore reflux condensers are essential. A similar incident involving benzene was also reported [2].
    See Sodium: Halocarbons

Trialkylphosphines
    Houben-Weyl, 1963, Vol. 12.1, 79

In the absence of a solvent, quaternation of trialkylphosphines with methyl iodide may proceed explosively.
See other HALOALKANES

0446. Methylmagnesium iodide (Iodomethylmagnesium)
[917-64-6]   **CH₃IMg**

$$MeMgI$$

Thiophosphoryl chloride
See Tetramethyldiphosphane disulfide

Vanadium trichloride
See Vanadium trichloride: Methylmagnesium iodide
See other ALKYLMETAL HALIDES, GRIGNARD REAGENTS

0447. Methylzinc iodide (Iodomethylzinc)
[18815-73-1]   **CH₃IZn**

$$MeZnI$$

Nitromethane
See Nitromethane: Alkylmetal halides
See other ALKYLMETAL HALIDES

0448. Methylpotassium
[17814-73-2]   **CH₃K**

$$H_3CK$$

Weiss, E. *et al., Angew. Chem. (Intern. Ed.)*, 1968, **7**, 133
The dry material is highly pyrophoric.
See other ALKYLMETALS

0449. Potassium methanediazoate (Potassium methyldiazeneoxide)
[19416-93-4]   **CH₃KN₂O**

$$H_3CN{=}NOK$$

Water
Hantzsch, A. et al., *Ber.*, 1902, **35**, 901
Interaction with water is explosively violent.
See related ARENEDIAZOATES
See other N–O COMPOUNDS

173

0450. Potassium methoxide
[865-33-8]   CH₃KO

$$KOCH_3$$

It may ignite in moist air.
See entry METAL ALKOXIDES

Arsenic pentafluoride, Benzene
See Arsenic pentafluoride: Benzene, etc.

0451. Potassium methylselenide
[54196-34-8]   CH₃KSe

$$MeSeK$$

2-Nitroacetophenone
See 2-Nitroacetophenone: Potassium methylselenide
See related METAL ALKOXIDES

0452. Methyllithium
[917-54-4]   CH₃Li

$$MeLi$$

Sidgwick, 1950, 71
Ignites and burns brilliantly in air. The commercial solution in diethyl ether is also
pyrophoric.
See ALKALI-METAL DERIVATIVES OF HYDROCARBONS
See other ALKYLMETALS

0453. Formamide
[75-12-7]   CH₃NO

$$HCO.NH_2$$

HCS 1980, 507; *RSC Lab. Hazards Data Sheet No. 58, 1987*

Aluminium, Ammonium nitrate, Calcium nitrate
See Calcium nitrate: Aluminium, etc.

Di-*tert*-butyl hyponitrite
See Di-*tert*-butyl hyponitrite: Formamides

Iodine, Pyridine, Sulfur trioxide
Anon., *J. Chem. Educ.*, 1973, **50**, A293
Bottles containing a modified Karl Fischer reagent with formamide replacing methanol developed gas pressure during several months and burst. No reason was apparent,
but slow formation of sulfuric acid, either by absorption of external water or by

dehydration of some of the formamide to hydrogen cyanide, and liberation of carbon monoxide from the formamide seems a likely sequence.
See other GAS EVOLUTION INCIDENTS

Organozinc compounds
Mistryukov, E. A. *et al., Mendeleev Comm.*, 1993, (6) 242
Adding formamide to a reaction mixture of allylzinc bromide and a ketone as a proton source for a Barbier reaction in ether or tetrahydrofuran gave uncontrollable explosion-like reaction. The reaction proceeded smoothly when starting from zinc dust and allyl bromide, with both formamide and the ketone in situ.

0454. *N*-Hydroxydithiocarbamic acid
[66427-01-8] CH_3NOS_2
                        HONHC(S)SH

Voigt, C. W. *et al., Z. Anorg. Chem.*, 1977, **437**, 233–236
The free acid is unstable and may decompose explosively at sub-zero temperatures.
See other N–O COMPOUNDS, ORGANIC ACIDS

†0455. Methyl nitrite
[624-91-9] CH_3NO_2
                        MeON:O

NSC 693, 1982; *HCS 1980*, 687

Rüst, 1948, 285
Explodes on heating, more powerfully than the ethyl homologue.

Other reactants
Yoshida, 1980, 14
MRH values calculated for 16 combinations, largely with oxidants, are given.
See other ALKYL NITRITES, N–O COMPOUNDS

0456. Nitromethane
[75-52-5] CH_3NO_2
                        MeNO₂

1. *HCS 1980*, 687
2. McKitterick, D. S. *et al., Ind. Eng. Chem. (Anal. Ed.)*, 1938, **10**, 630
3. Makovky, A. *et al., Chem. Rev.*, 1958, **58**, 627
4. Travis, J. R., *Los Alamos Rept. DC 6994*, Washington, USAEC, 1965
5. Sorbe, 1968, 148
6. Lieber, C. O., *Chem. Ing. Tech.*, 1978, **50**, 695–697
7. Perche, A. *et al., J. Chem. Res.*, 1979, (S) 116; (M) 1555–1578
8. Lewis, D. J., *Haz. Cargo Bull.*, 1983, **4**(9), 36–38

9. Stull, 1977, 20
10. Anon., *Jahresber.*, 1984, 66–67
11. Coetzee, J. F. *et al., Pure Appl. Chem.*, 1986, **58**, 1541–1545
12. Brewer, K. R., *J. Org. Chem.*, 1988, **53**, 3776–3779
13. Piermanini, G. J. *et al., J. Phys. Chem.*, 1989, **93**(1), 457–462
14. Dick, J. J., *J. Phys. Chem.*, 1993, **97**(23), 6193
15. Ullmann, 1991, A 17, 405/6

Hazardous properties and handling procedures are summarised [1]. Conditions under which it may explode by detonation, heat or shock were determined. It was concluded that it is potentially very explosive and precautions are necessary to prevent its exposure to severe shock or high temperatures in use [2]. Later work, following two rail tank explosions, showed that shock caused by sudden application of gas pressure, or sudden forced flow through restrictions, could detonate the liquid. The stability and decomposition of nitromethane relevant to use as a rocket fuel are also reviewed [3]. The role of discontinuities in the initiation of shock-compressed nitromethane has been evaluated experimentally [4]. It explodes at about 230°C [5]. The role of pressure waves producing local resonance heating in bubbles and leading to detonation has been studied [6], as have the effects of nitrogen oxide, nitrogen dioxide and formaldehyde upon pyrolytic decomposition of nitromethane [7]. Bulk shipment in rail tanks is now permitted for solutions in appropriate diluents [8].

The heat of decomposition (238.4 kJ/mol, 3.92 kJ/g) has been calculated to give an adiabatic product temperature of 2150°C accompanied by a 24-fold pressure increase in a closed vessel [9]. During research into the Friedel-Crafts acylation reaction of aromatic compounds (components unspecified) in nitrobenzene as solvent, it was decided to use nitromethane in place of nitrobenzene because of the lower toxicity of the former. However, because of the lower boiling point of nitromethane (101°C, against 210°C for nitrobenzene), the reactions were run in an autoclave so that the same maximum reaction temperature of 155°C could be used, but at a maximum pressure of 10 bar. The reaction mixture was heated to 150°C and maintained there for 10 minutes, when a rapidly accelerating increase in temperature was noticed, and at 160°C the lid of the autoclave was blown off as decomposition accelerated to explosion [10]. Impurities present in the commercial solvent are listed, and a recommended purification procedure is described [11]. The thermal decomposition of nitromethane under supercritical conditions has been studied [12]. The effects of very high pressure and of temperature on the physical properties, chemical reactivity and thermal decomposition of nitromethane have been studied, and a mechanism for the bimolecular decomposition (to ammonium formate and water) identified [13]. Solid nitromethane apparently has different susceptibility to detonation according to the orientation of the crystal, a theoretical model is advanced [14]. Nitromethane actually finds employment as an explosive [15].

See Acetone below, or Haloforms below

Acetone
Varob'ev, A. A., *Chem. Abs.*, 1981, **94**, 49735
Mixtures can be detonated (title only translated).

Acids, or Bases
Makovky, A. *et al., Chem. Rev.*, 1958, **58**, 631
Addition of bases or acids to nitromethane renders it susceptible to initiation by a detonator. These include aniline, diaminoethane, iminobispropylamine, morpholine, methylamine, ammonium hydroxide, potassium hydroxide, sodium carbonate, and formic, nitric, sulfuric or phosphoric acids.
See 1,2-Diaminoethane, *N*,2,4,6-Tetranitro-*N*-methylamine, below

Alkalis
Ullmann, 1991, A 17, 405/6
It will form *aci*-nitromethanoate salts, which are sensitive explosives when dry, and may form methazonates (salts of nitroacetaldehyde oxime) which are still more explosive.
See Sodium *aci*-nitromethanide, 2-Nitroacetaldehyde oxime

Alkylmetal halides
Traverse, G., US Pat. 2 775 863, 1957
Contact with R_mMX_n (R is methyl, ethyl; M is aluminium, zinc; X is bromide, iodide) causes ignition. Diethylaluminium bromide, dimethylaluminium bromide, ethylaluminium bromide iodide, methylzinc iodide and methylaluminium diiodide are claimed as specially effective.

Aluminium powder   MRH 9.33/23
Kato, Y. *et al., Chem. Abs.*, 1979, **91**, 159867
Increase in concentration of aluminium powder in a mixture with poly(methyl methacrylate) increases the sensitivity to detonation.

Aluminium chloride
See Aluminium chloride–nitromethane

Aluminium chloride, Ethylene
See Ethylene: Aluminium chloride

Ammonium salts, Organic solvents
Runge, W. F. *et al.*, US Pat. 3 915 768, 1975
Presence of, for example, 5% of methylammonium acetate and 5% of methanol sensitises nitromethane to shock-initiation.

Bis(2-aminoethyl)amine
1. Runge, W. F. *et al.*, US Pat. 3 798 902, 1974
2. Walker, F. E., *Acta Astronaut.*, 1979, **6**, 807–813
Explosive solutions of nitromethane in dichloromethane, sensitised by addition of 10–12% of the amine, retained their sensitivity at −50°C [1]. Presence of 0–5% of the triamine considerably increases detonation sensitivity of nitromethane [2].

Bromine
See Bromine: Nitromethane

Boron trifluoride etherate, Silver oxide
See Silver tetrafluoroborate

Calcium hypochlorite
See Calcium hypochlorite: Nitromethane

Carbon Disulphide
Editor's comments
Salts of the reaction product, nitroethanedithioic acid, are intermediates in manufacture of pharmaceuticals. They are explosive and are to be avoided if not isolated. The free acid will surely be unstable, but not necessarily explosive.

1,2-Diaminoethane, *N*,2,4,6-Tetranitro-*N*-methylaniline
MCA Case History No. 1564
During preparations to initiate the explosion of nitromethane sensitised by addition of 20% of the diamine, accidental contact of the liquid mixture with the solid 'tetryl' detonator caused ignition of the latter.
See Acids, or Bases, above

Formaldehyde
Noland, W. E., *Org. Synth.*, 1961, **41**, 69
Interaction of nitromethane and formaldehyde in presence of alkali gives not only 2-nitroethanol, but also di- and tri-condensation products. After removal of the 2-nitroethanol by vacuum distillation, the residue must be cooled before admitting air into the system to prevent a flash explosion or violent fume-off.

Haloforms
Presles, H. N. *et al., Acta Astronaut.*, 1976, **3**, 531–540
Mixtures with chloroform or bromoform are detonable.

Hydrazine, Methanol
Forshey, D. R. *et al., Explosivstoffe*, 1969, **17**(6), 125–129
Addition of hydrazine strongly sensitises nitromethane and its mixtures with methanol to detonation.

Hydrocarbons
1. Watts, C. E., *Chem. Eng. News*, 1952, **30**, 2344
2. Makovky, A. *et al., Chem. Rev.*, 1958, **58**, 631
Nitromethane may act as a mild oxidant, and should not be heated with hydrocarbons or readily oxidisable materials under confinement [1]. Explosions may occur during cooling of such materials heated to high temperatures and pressures [2]. Mixtures of nitromethane and solvents which are to be heated above the b.p. of nitromethane should first be subjected to small-scale explosive tests [2].

Hydrocarbons, Oxygen
See Oxygen: Hydrocarbons, Promoters

Lithium perchlorate   MRH 5.69/43
1. Titus, J. A., *Chem. Eng. News*, 1971, **49**(23), 6
2. 'Nitroparaffin Data Sheet TDS 1' New York, Comm. Solvents Corp., 1965
3. Egly, R. S., *Chem. Eng. News*, 1973, **51**(6), 30

Explosions which occurred at the auxiliary electrode during electro-oxidation reactions in nitromethane–lithium perchlorate electrolytes, may have been caused by lithium fulminate. This could have been produced by formation of the lithium salt of nitromethane and subsequent dehydration to the fulminate [1], analogous to the known formation of mercury(II) fulminate [2]. This explanation is not considered tenable, however [3].

Lithium tetrahydroaluminate
1. Nystrom, R. F. *et al., J. Amer. Chem. Soc.*, 1948, **70**, 3739
2. Wollweber, H. *et al.*, private comm., 1988

Addition of nitromethane to ethereal lithium tetrahydroaluminate solution at ambient temperature gave an explosively violent rection [1], and this was confirmed when addition of 0.5 ml of dry nitromethane to 10 ml of reducant solution led, after 30 s, to a violent explosion which pulverised the flask [2]. The violence of the explosion suggests that lithium *aci*-nitromethanide (or possibly lithium fulminate) may have been involved.
See Sodium hydride, below
See other aci-NITRO SALTS

Metal oxides
See NITROALKANES: metal oxides

Molecular sieve
1. Wollweber, H., private comm., 1979
2. Bretherick, L., *Chem. & Ind.*, 1979, 532

Nitromethane was dried and stored in a flask over 13X (large-pore) molecular sieve, and when a further portion of freshly activated sieve was added after several weeks, the contents erupted and ignited, breaking the flask [1]. This was attributed to slow formation of sodium *aci*-nitromethanide from the zeolitic sodium ions, which then decomposed in the exotherm arising from adsorption of nitromethane on the freshly activated sieve. Use of a small-pore (3A or 4A) sieve which would exclude nitromethane from the internal channels and greatly reduce the contact with sodium ions should avoid this problem. The possibility of other problems arising from slow release or exchange of ionic species under virtually anhydrous conditions during long term use of molecular sieves was noted [2].
See other MOLECULAR SIEVE INCIDENTS

Nickel
Benziger, J. B., *Appl. Surface Sci.*, 1984, **17**, 309–323
Decomposition of nitromethane on the (111) face of nickel has been studied.

Nitric acid   MRH 6.19/38
See Nitric acid: Nitromethane

Other reactants
Yoshida, 1980, 278
MRH values calculated for 17 combinations, largely with oxidants, are given.

Silver nitrate
Luchs, J. K., *Photog. Sci. Eng.*, 1966, **10**, 336
Aqueous silver nitrate may react to form silver fulminate.

Sodium hydride
1.Pearson, A. J. *et al.*, *J. Organomet. Chem.*, 1980, **202**, 178
2.Johnson, B. F. G. *et al.*, *J. Organomet Chem.*, 1981, **204**, 221–228
Generation of sodium *aci*-nitromethanide by adding de-oiled sodium hydride powder to nitromethane in THF becomes violent at 40°C [1], and if the solvent is omitted,
too fast addition of the hydride leads to a series of small explosions [2].
See Sodium *aci*-nitromethanide

Trimethylsilyl iodide
Voronkov, M. G. *et al.*, *Zh. Obshch. Khim.*, 1989, **59**(5), 1055. (*Chem. Abs.*, **112**,
77304)
Reaction of the silane with nitromethane is explosive, probably by intermediacy of
fulminic acid (a dehydration product of nitromethane).

Uronium perchlorate
See Uronium perchlorate: Organic materials
See other NITROALKANES, *C*-NITRO COMPOUNDS

0457. Methyl nitrate
[598-58-3] CH_3NO_3

MeONO$_2$

1. Kit and Evered, 1960, 268
2. Black. A. P. *et al.*, *Org. Synth.*, 1943, Coll. Vol. 2, 412
3. Goodman, H. *et al.*, *Combust. Flame*, 1972, **19**, 157
4. Griffiths, J. F. *et al.*, *Combust. Flame*, 1982, **45**, 54–66
5. Stull, 1977, 20

It has high shock- and thermal sensitivity, exploding at 65°C, and is too sensitive
for use as a rocket mono-propellant [1]. Conditions during preparation of the ester
from methanol and mixed nitric–sulfuric acids are fairly critical, and explosions
may occur if it is suddenly heated or distilled in presence of acid [2]. Spontaneous
ignition or explosion of the vapour at 250–316°C in presence of gaseous diluents
[3], and the mechanism of exothermic decomposition and ignition [4] have been
studied. The rather high decomposition exotherm (153.8 kJ/mol, 4.42 kJ/g) would
raise the products to an adiabatic temperature approaching 2600°C, with a 31-fold
pressure increase in a closed system [5].

Other reactants
Yoshida, 1980, 207

Of the MRH values calculated for 11 combinations, that with magnesium was most energetic at 8.70/38.

See other ALKYL NITRATES

†0458. Thallium(I) methanediazoate (Thallium *N*-nitrosomethylamide)
[113925-83-0] CH_3N_2OTl

TlON=NMe

1. Keefer, L. K. *et al., J. Amer. Chem. Soc.*, 1988, **110**, 2800–2806, and footmote 20
2. Burns, M. E., private comm., 1985

It appears to be a stable covalent highly crystalline compound (unlike other metal methanediazoates). Alkanediazoates are easily converted to diazoalkanes, so should be regarded as capable of detonation. (Though named by the author as a methanediazoate, it is indexed and registered in *CA* as a nitrosomethylamide salt) [1]. A sample of the freshly synthesised compound was dissolved in dichlorodideuteromethane and sealed into an NMR tube. Four days later, when the tube was being opened for recovery of the sample, the tube exploded. This was attributed to diazomethane formation, possibly from reaction with traces of moisture sealed into the tube [2].

See other HEAVY METAL DERIVATIVES, N–O COMPOUNDS

0459. Methyl azide
[624-90-8] CH_3N_3

MeN$_3$

1. *MCA Case History No. 887*
2. Boyer, J. H. *et al., Chem. Rev.*, 1954, **54**, 323
3. Burns, M. E. *et al., Chem. Eng. News*, 1984, **62**(2), 2

The product, prepared by interaction of sodium azide with dimethyl sulfate and sodium hydroxide, exploded during concurrent vacuum distillation. The explosion was attributed to formation and co-distillation with the product of hydrogen azide, owing to excursion of the pH to below 5 during the preparation. Free hydrogen azide itself is explosive, and it may also have reacted with mercury in a manometer to form the detonator mercuric azide [1]. Methyl azide is stable at ambient temperature, but may detonate on rapid heating [2]. A further explosion during the preparation as above led to incorporation of bromothymol blue in the reaction system to give a visible indication of pH in the generation flask [3].

Mercury

Currie, C. L. *et al., Can. J. Chem.*, 1963, **41**, 1048

Presence of mercury in methyl azide markedly reduces the stability towards shock or electric discharge.

Methanol

Grundman, C. *et al., Angew. Chem.*, 1950, **62**, 410

In spite of extensive cooling and precautions, a mixture of methyl azide, methanol and dimethyl malonate exploded violently while being sealed into a Carius tube. The vapour of the azide is very easily initiated by heat, even at low concentrations.
See other ORGANIC AZIDES

0460. Nitrourea
[556-89-8] $CH_3N_3O_3$

$$O_2NNHCO.NH_2$$

1. Urbanski, 1967, Vol. 3, 34
2. Medard, L., *Mém. Poudres*, 1951, **33**, 113–123; (English transl., HSE 11292, available from HSE/LIS, 1986)

A rather unstable explosive material, insensitive to heating or impact, which gives mercuric and silver salts which are rather sensitive to impact [1]; further detailed data are available [2].
See other N-NITRO COMPOUNDS

0461. 5-Aminotetrazole
[4418-61-5] CH_3N_5

Nitrous acid
1. Elmore, D. T., *Chem. Brit.*, 1966, **2**, 414
2. Thiele, J., *Ann.*, 1892, **270**, 59
3. Gray, E. J. *et al., J. Chem. Soc., Perkin Trans. 1*, 1976, 1503
4. Cawkill, E. *et al., J. Chem. Soc., Perkin Trans. 1*, 1979, 727

Diazotised 5-aminotetrazole is unstable under the conditions recommended for its use as a biochemical reagent. While the pH of the diazotised material (the cation of which contains 87% nitrogen) at 0°C was being reduced to 5 by addition of potassium hydroxide, a violent explosion occurred [1]. This may have been caused by a local excess of alkali causing the formation of the internal salt, 5-diazoniotetrazolide, which will explode in concentrated solution at 0°C [2]. The diazonium chloride is also very unstable in concentrated solution at 0°C. Small-scale diazotisation (2 g of amine) and susequent coupling at pH 3 with ethyl cyanoacetate to prepare ethyl 2-cyano-(1*H*-tetrazol-5-ylhydrazono)acetate proceeded uneventfully, but on double the scale a violent explosion occurred [3]. The importance of adequate dilution of the reaction media to prevent explosions during diazotisation is stressed [4].
See DIAZONIUM SALTS
See other HIGH-NITROGEN COMPOUNDS, TETRAZOLES

182

0462. 5-Hydrazino-1,2,3,4-thiatriazole
[99319-30-9]

CH_3N_5S

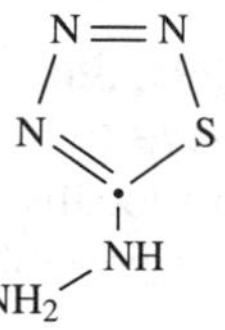

Martin, D., *Z. Chem.*, 1985, **25**(4), 136–137

The explosive title compound was not isolated after preparation from thiocarbazide and nitrous acid, but was condensed with various ketones to give the hydrazones.

See other HIGH-NITROGEN COMPOUNDS, N–S COMPOUNDS
See related TRIAZOLES

0463. Methylsodium
[18356-02-0]

CH_3Na

H_3CNa

Schlenk, W. *et al.*, *Ber.*, 1917, **50**, 262

Ignites immediately in air. The tendency to ignition decreases with ascent of the homologous series of alkylsodiums.

See other ALKYLMETALS

0464. Sodium methoxide
[124-41-4]

CH_3NaO

$NaOCH_3$

1. *HCS 1980*, 850
2. *See entry* THERMOCHEMISTRY AND EXOTHERMIC DECOMPOSITION (reference 2)

It may ignite in moist air. Hazardous properties and handling procedures are summarised [1]. Energy of decomposition (in range 410–460 °C) measured as 0.77 kJ/g [2].

Chloroform
See Chloroform: Sodium methoxide

4-Chloronitrobenzene
See 4-Chloronitrobenzene: Sodium methoxide

Perfluorocyclopropene
See FLUORINATED CYCLOPROPENYL METHYL ETHERS
See other METAL ALKOXIDES

0465. Methanesulfonyl azide
[1516-70-7] $CH_3N_3O_2S$

$$MeSO_2N_3$$

Brown, R. *et al., Chem. Brit.*, 1994, **30**(6), 470
Gram quantities of the azide have twice detonated, once during distillation, the second time also possibly due to localised overheating. It is stated that there was hitherto only a general warning of possible explosivity.
See other ACYL AZIDES

†0466. Methane
[74-82-8] CH_4

$$CH_4$$

FPA H59, 1977; *HCS 1980*, 627 (cylinder)

Halogens, or Interhalogens
See Bromine pentafluoride: Hydrogen-containing materials
   Chlorine: Hydrocarbons
   Chlorine trifluoride: Methane
   Fluorine: Hydrocarbons
   Iodine heptafluoride: Carbon, etc.

Other reactants
   Yoshida, 1980, 361
   MRH values calculated for 12 combinations with oxidants are given.

Oxidants
See Dioxygen difluoride: Various materials
   Dioxygenyl tetrafluoroborate: Organic materials
   Oxygen (Liquid): Hydrocarbons, or: Liquefied gases
   'Trioxygen difluoride': Various materials
See also LIQUEFIED NATURAL GAS

0467. Chloroformamidinium nitrate
[75524-40-2] $CH_4ClN_3O_3$

$$ClC(:NH)N^+H_3\ NO_3^-$$

Alone, or Amines, or Metals
   1. Sauermilch, W., *Explosivstoffe*, 1961, **9**, 71–74
   2. Sorbe, 1968, 144
It is powerfully explosive, and also an oxidant which reacts violently with ammonia or amines, and causes explosive ignition of wet magnesium powder [1], or of powdered aluminium or iron [2].
See Chloroformamidinium chloride: Oxyacids, etc. (next below)
See other OXOSALTS OF NITROGENOUS BASES

0468. Chloroformamidinium chloride
[29671-92-9] $CH_4Cl_2N_2$

$$ClC(:NH)N^+H_3Cl^-$$

Oxyacids and salts

Sauermilch, W., *Explosivstoffe*, 1961, **9**, 71–74, 256

The title compound ('cyanamide dihydrochloride') reacts with perchloric acid to form the perchlorate salt, and with ammonium nitrate, the nitrate salt, both being highly explosive, and the latter also an oxidant.

0469. Chloroformamidinium perchlorate (Methanimidamide monoperchlorate)
[40985-41-9] $CH_4Cl_2N_2O_4$

$$ClC(:NH)N^+H_3ClO_4^-$$

See item above *See other* PERCHLORATE SALTS OF NITROGENOUS BASES

†0470. Dichloromethylsilane
[75-54-7] CH_4Cl_2Si

$$Cl_2SiHMe$$

Oxidants

1. Müller, R. *et al., J. Prakt. Chem.*, 1966, **31**, 1–6
2. Sorbe, 1968, 128

The pure material is not ignited by impact, but it is in presence of potassium permanganate or lead(II),(IV) oxide [1], or by copper oxide or silver oxide, even under an inert gas [2].

See Trichlorosilane

See other ALKYLHALOSILANES

0471. Tetrakis(hydroxymercurio)methane
[ ] $CH_4Hg_4O_4$

$$C(HgOH)_4$$

Breitinger, D. K. *et al., Z. Naturforsch. B*, 1984, **39B**, 123

Intensive and heated high vacuum drying of the title compound (Hofmann's base) over phosphorus(V) oxide may lead to an explosion. Unheated drying appears satisfactory and safer.

See other MERCURY COMPOUNDS

0472. Potassium methylamide
[[54448-39-4] (ion)] CH_4KN

$$KNHCH_3$$

Makhija, R. C. *et al., Can. J. Chem.*, 1971, **49**, 807

Extremely hygroscopic and pyrophoric; may explode on contact with air.

See other N-METAL DERIVATIVES

185

0473. Formylhydrazine
[624-84-0] CH_4N_2O

HCO.NHNH₂

See 4-Amino-4H-1,2,4-triazole
See other HIGH-NITROGEN COMPOUNDS

0474. Methyldiazene
[26981-93-1] CH_4N_2

MeN=NH

Oxygen
Ackermann, M. N. *et al., Inorg. Chem.*, 1972, **11**, 3077
Interaction on warming a mixture from −196°C rapidly to ambient temperature is explosive.
See related AZO COMPOUNDS

0475. Urea (Carbamide)
[57-13-6] CH_4N_2O

O:C(NH₂)₂

NSC 691, 1980; *HCS 1980*, 949

Preparative hazard for ¹⁵N-labelled compound.
See Oxygen (Liquid): Ammonia, etc.

Chromyl chloride
See Chromyl chloride: Urea

Dichloromaleic anhydride, Sodium chloride
See Dichloromaleic anhydride: Sodium chloride, etc.

Metal hypochlorites
491M, 1975, 213
Urea reacts with sodium hypochlorite or calcium hypochlorite to form explosive nitrogen trichloride.
See Chlorine: Nitrogen compounds

Nitrosyl perchlorate
See Nitrosyl perchlorate: Organic materials

Oxalic acid
See Oxalic acid: Urea

Phosphorus pentachloride
See Phosphorus pentachloride: Urea

Sodium nitrite
See Sodium nitrite: Urea

Titanium tetrachloride
See Titanium tetrachloride: Urea
See other ORGANIC BASES

0476. Urea hydrogen peroxidate
[124-43-6] $CH_4N_2O.H_2O_2$

$$O:C(NH_2)_2.H_2O_2$$

MCA Case History No. 719

The contents of a screw-capped brown glass bottle spontaneously erupted after 4 years' storage at ambient temperature. All peroxides should be kept in special storage and checked periodically.
See other CRYSTALLINE HYDROGEN PEROXIDATES, GLASS INCIDENTS

0477. Hydroxyurea
[127-07-1] $CH_4N_2O_2$

$$H_2NC(=O)NHOH$$

Cardillo, P. *et al., Chem. Eng. News*, 1998, **76**(22), 6
An industrial reactor containing a concentrate of this in water was left over the weekend, at an initial temperature possibly around 50°C. On the Tuesday morning it exploded. Investigation of the thermal stability of this drug in pure form showed it to be thermally unstable from 85°C as the solid, 70°C in aqueous solution. The decomposition is probably autocatalytic, gas evolving and with an enthalpy of 2.2 kJ/g. Aged samples were still less stable, decomposing after a variable induction period (from an initial 50°C, 80 hours). Gas evolution combined with temperature elevation to boiling point explains the burst reactor.
See other N–O COMPOUNDS

0478. *N*-Nitromethylamine
[598-57-2] $CH_4N_2O_2$

$$O_2NNHCH_3$$

Diels, O. *et al., Berichte*, 1913, **46**, 2006.
A tiny quantity exploded with the utmost force when strongly heated in a capillary tube.

Amminemetals
Complexes with several ammine derivatives of metals are explosive.
See entry NITRAMINE–METAL COMPLEXES

Sulfuric acid
Urbanski, 1967, Vol. 3, 16
The nitroamine is decomposed explosively by conc. sulfuric acid.
See other N-NITRO COMPOUNDS

0479. Ammonium thiocyanate
[1762-95-4]

$$NH_4SC\equiv N$$

CH$_4$N$_2$S

HCS 1980, 152

Other reactants
Yoshida, 1980, 227
MRH values calculated for 13 combinations, 12 with oxidants, are given.

Potassium chlorate
See Potassium chlorate: Metal thiocyanates
See related CYANO COMPOUNDS

0480. Thiourea
[62-56-6]

$$S:C(NH_2)_2$$

CH$_4$N$_2$S

Acrylaldehyde
See Acrylaldehyde: Acids, etc.

Hydrogen peroxide, Nitric acid
See Hydrogen peroxide: Nitric acid, etc.

Potassium chlorate
See Potassium chlorate: Thiourea
See other ORGANIC BASES

0481. Nitrosoguanidine
[674-81-7]

$$O:NNHC(NH)NH_2$$

CH$_4$N$_4$O

Henry, R. A. *et al., Ind. Eng. Chem.*, 1949, **41**, 846–849
The compound is normally stored and transported as a water-wet paste which slowly decomposes at elevated ambient temperatures, evolving nitrogen. Precautions are described to prevent pressure build-up in sealed containers.
See other NITROSO COMPOUNDS

0482. Nitroguanidine
[556-88-7]

$$O_2NNHC(NH)NH_2$$

CH$_4$N$_4$O$_2$

1. Urbanski, 1967, Vol. 3, 31
2. McKay, A. F., *Chem. Rev.*, 1952, **51**, 301
Nitroguanidine is difficult to detonate, but its mercury and silver complex salts are much more impact-sensitive [1]. Many nitroguanidine derivatives have been considered as explosives [2].
See other N-NITRO COMPOUNDS

188

0483. Methylenebis(nitramine)
[14168-44-6] $CH_4N_4O_4$
$$H_2C(NHNO_2)_2$$

Glowiak, B., *Chem. Abs.*, 1960, **54**, 21761e
It is a powerful and sensitive explosive which explodes at 217°C, and the lead salt at 195°C.
See other N-NITRO COMPOUNDS

†0484. Methanol
[67-56-1] CH_4O
$$H_3COH$$

(*MCA SD-22*, 1970); *NSC 407*, 1979; *FPA H42*, 1975; *HCS 1980*, 628; *RSC Lab. Hazard Data Sheet No. 25*, 1984

Ferris, T. V., *Loss Prev.*, 1974, **8**, 15–19
Explosive behaviour on combustion of methanol–air mixtures at 1.8 bar and 120°C was studied, with or without addition of oxygen and water.

Acetyl bromide
See Acetyl bromide: Hydroxylic compounds

Alkylaluminium solutions
MCA Case History No. 1778
Accidental use of methanol in place of hexane to rinse out a hypodermic syringe used for a dilute alkylaluminium solution caused a violent reaction which blew the plunger out of the barrel.
See ALKYLALUMINIUM DERIVATIVES: alcohols

Beryllium hydride
See Beryllium hydride: Methanol

Boron trichloride
See Boron trichloride

Carbon tetrachloride, Metals
Kuppers, J. R., *J. Electrochem. Soc.*, 1978, **125**, 97–98
The rapid autocatalytic dissolution of aluminium, magnesium or zinc in 9:1 methanol–carbon tetrachloride mixtures is sufficiently vigorous to be rated as potentially hazardous. Dissolution of zinc powder is subject to an induction period of 2 h, which is eliminated by traces of copper(II) chloride, mercury(II) chloride or chromium(III) bromide.
See other INDUCTION PERIOD INCIDENTS

Chloroform, Sodium
See Chloroform: Sodium, Methanol

Chloroform, Sodium hydroxide
See Chloroform: Sodium hydroxide, Methanol

Cyanuric chloride
See 2,4,6-Trichloro-1,3,5-triazine: Methanol

Dichloromethane
See Dichloromethane: Air, etc.

Diethylzinc
See Diethylzinc: Methanol

Hydrogen, Raney nickel catalyst
Klais, O., *Hazards from Pressure, IChE Symp. Ser. No. 102*, 25–36, Oxford, Pergamon, 1987
During hydrogenation of an unspecified substrate in methanol solution under hydrogen at 100 bar with Raney nickel catalyst, a sudden temperature increase led to hydrogenolysis of methanol to methane, and the pressure increase led to an overpressure accident. Such incidents may be avoided by control of agitation, limiting the amount of catalyst, and checking thermal stability of starting materials and end products beforehand.
See other GAS EVOLUTION INCIDENTS

Metals
See Aluminium: Methanol
Magnesium: Methanol
Potassium (Slow oxidation)

Other reactants
Yoshida, 1980, 360
MRH values calculated for 17 combinations, largely with oxidants, are given.

Oxidants   MRH values show % of oxidant
See Barium perchlorate: Alcohols
Bromine: Alcohols
Chlorine: Methanol
Chromium trioxide: Alcohols
Hydrogen peroxide: Oxygenated compounds   MRH 5.98/76
Lead perchlorate: Methanol
Nitric acid: Alcohols (reference 6)   MRH 5.31/70
Sodium hypochlorite: Methanol   MRH 2.47/90

Phosphorus(III) oxide
See Tetraphosphorus hexaoxide: Organic liquids

Potassium *tert*-butoxide
See Potassium *tert*-butoxide: Acids, etc.

Water
MCA Case History Nos. 1822, 2085

190

Static discharge ignited the contents of a polythene bottle being filled with a 40:60 mixture of methanol and water at 30°C, and a later similar incident in a plastics-lined metal tank involved a 30:70 mixture.

0485. Poly[oxy(methyl)silylene]
[9004-73-3]

$(CH_4OSi)_n$

$$\left[\begin{array}{c} Me \\ | \\ ----Si—O---- \\ | \\ H \end{array} \right]_n$$

Preparative hazard

Crivello, J. V. *et al., Chem. Abs.*, 1990, **113**, 41011z; U.S. Pat. 4,895,967.
A procedure for cyclisation of the linear polymer to small oligomers without risk of explosion is described.

1-Allyloxy-2,3-epoxypropane

Anon., *Eur. Chem. News*, 1995, **63**(1656), 27
The two compounds were accidentally mixed in a reactor because of similar packaging. They reacted uncontrollably, rupturing the reactor, and the evolved hydrogen ignited, killing one worker and injuring four. If as is probable a base catalyst was present depolymerisation/disproportionation to oxomethylsilane and methylsilane is also possible, with an increase in both vapour pressure and flammability.
See other SILANES

0486. Methyl hydroperoxide
[3031-73-0]

CH_4O_2

MeOOH

Alone, or Phosphorus(V) oxide, or Platinum

1. Rieche, A. *et al., Ber.*, 1929, **62**, 2458, 2460
2. Hasegawa, K. *et al., Proc. Int. Pyrotech. Semin.*, 1994, (19th), 684

The hydroperoxide is violently explosive and shock-sensitive, especially on warming; great care is necessary in handling. It explodes violently in contact with phosphorus pentaoxide, and a 50% aqueous solution decomposed explosively on warming with spongy platinum. The barium salt is dangerously explosive when dry.
See other ALKYL HYDROPEROXIDES

0487. Hydroxymethyl hydroperoxide
[15932-89-5]

CH_4O_3

HOCH$_2$OOH

Rieche, A., *Ber.*, 1931, 64, 2328; 1935, 68, 1465
It explodes on heating, but is friction-insensitive. Higher homologues are not explosive.
See other 1-OXYPEROXY COMPOUNDS

0488. Methanesulfonic acid
[75-75-2] CH_4O_3S
$$H_3CSO_2OH$$

Ethyl vinyl ether
See Ethyl vinyl ether: Methanesulfonic acid

Hydrogen fluoride
See Hydrogen fluoride: Methanesulfonic acid
See other ORGANIC ACIDS

†0489. Methanethiol
[74-93-1] CH_4S
$$H_3CSH$$

FPA H93, 1980; *HCS 1980*, 655

Mercury(II) oxide
See Mercury(II) oxide: Methanethiol
See other ALKANETHIOLS, ALKYLNON-METAL HYDRIDES

0490. Methanetellurol
[25284-83-7] CH_4Te
$$H_3CTeH$$

Hamada, K. *et al., Synth. React. Inorg. Metal-Org. Chem.*, 1977, **7**, 364 (footnote 1)
It ignites in air and explodes with oxygen at ambient temperature.
See other ALKYLNON-METAL HYDRIDES

0491. Uronium perchlorate (Urea perchlorate)
[18727-07-6] $CH_5ClN_2O_5$
$$H_2NCO.N^+H_3ClO_4^-$$

Aromatic nitro compounds
   1. Shimio, K. *et al., Chem. Abs.*, 1975, **83**, 134504
   2. Fujiwara, S. *et al.*, Japan Kokai, 74 134 812, 1974
Conc. aqueous solutions of the urea salt will dissolve solid or liquid aromatic nitro compounds (e.g. picric acid [1], or nitrobenzene [2]) to give high velocity explosives.
See next item below

Organic materials
Kusakabe, M. *et al., Chem. Abs.*, 1980, **92**, 79028
Liquid mixtures with detonable materials (picric acid, nitromethane) were extremely powerful explosives, and those with nitrobenzene or dimethylformamide less so.
See other PERCHLORATE SALTS OF NITROGENOUS BASES

0492. Fluoromethylsilane

192

[10112-08-0]   CH₅FSi

$$FCH_2SiH_3$$

Chlorine, Dichloromethane
 See Chlorine: Fluoromethylsilane, Dichloromethane
 See other SILANES: halogens

†0493. Methylamine (Methanamine)
   [74-89-5]   CH₅N

$$MeNH_2$$

(*MCA SD-57*, 1965); *HCS 1980*, 637

Nitromethane
 See Nitromethane: Acids, etc.

Other reactants
 Yoshida, 1980, 362
 MRH values calculated for 13 combinations with oxidants are given.
 See other ORGANIC BASES

0494. Uronium nitrate (Urea nitrate)
   [124-47-0]   CH₅N₃O₄

$$H_2NCO.N^+H_3\ NO_3^-$$

1. Kirk-Othmer, 1970, Vol. 21, 38
2. Markalous, F. *et al.*, Czech Pat. 152 080, 1974
3. Lazarov, S. *et al.*, *Chem. Abs.*, 1973, **79**, 94169
4. Medard, L., *Mém. Poudres*, 1951, **33**, 113–123
(English transl. HSE 11292 available from HSE/LIS, 1986)
The nitrate decomposes explosively when heated [1]. Prepared in the presence of phosphates, the salt is much more stable, even when dry [2]. The manufacture and explosive properties of urea nitrate and its mixtures with other explosives are discussed in detail [3]. Explosive properties of the nitrate are detailed [4].

Heavy metals
 Kotzerke, *Proc. 1st Int. Symp. Prev. Occ. Risks Chem. Ind.*, 201–207, Heidelberg, ISSA, 1970
 During evaporation of an alkaline aqueous solution of the nitrate, decomposition led to gas evolution and a pressure explosion occurred. This was attributed to the use of recovered alkali containing high levels of lead and iron, which were found to catalyse the thermal decomposition of the nitrate. Precautions to prevent recurrence are detailed.
 See other GAS EVOLUTION INCIDENTS, OXOSALTS OF NITROGENOUS BASES

0495. 1-Amino-3-nitroguanidine
[18264-75-0] $CH_5N_5O_2$

$$H_2NNHC(NH)NHNO_2$$

Lieber, E. *et al., J. Amer. Chem. Soc.*, 1951, **73**, 2328
It explodes at the m.p., 190°C.
See other HIGH-NITROGEN COMPOUNDS, *N*-NITRO COMPOUNDS

0496. Hydrazinium trinitromethanide (Hydrazinium nitroformate)
[14913-74-7] $CH_5N_5O_6$

$$N_2H_5^+; \ ^-C(NO_2)_3$$

1. Anan, T. *et al., Chem. Abs.*, 1995, **123**, 174351
2. Schoyer, H. F. R. *et al., J. Propul. Power.*, 1995, **11**(4), 856

The compound was examined as a propellant explosive, it proved easily detonable and more sensitive than ammonium perchlorate. A formulation with hydroxyl terminated polybutadiene binder ignited spontaneously at room temperature [1]. Other workers have found it more tractable [2].
See other HYDRAZINIUM SALTS POLYNITROALKYL COMPOUNDS

0497. Methylbis(η^2-peroxo)rhenium oxide hydrate
[ ] CH_5O_6Re

Herrmann, W. A. *et al., Angew. Chem. (Int.)*, 1993, **32**, 1157
This, the active species in methylrheniumtrioxide catalysed peroxidations by hydrogen peroxide, is fairly stable in solution but explosive when isolated.
See other ORGANOMETALLIC PEROXIDES

0498. Methylphosphine
[5931-54-4] CH_5P

$$MePH_2$$

Houben-Weyl, 1963, Vol. 12.1, 69
Primary lower-alkylphosphines readily ignite in air.
See other ALKYLPHOSPHINES

0499. Methylstibine
[23362-09-6] CH_5Sb

$$MeSbH_2$$

Sorbe, 1968, 28
Alkylstibines decompose explosively on heating or shock.
See other ALKYLMETAL HYDRIDES

0500. Methylammonium chlorite
[15875-44-2] CH_6ClNO_2

$$MeN^+H_3ClO_2^-$$

Levi, G. R., *Gazz. Chim. Ital. [1]*, 1922, **52**, 207
A conc. solution caused a slight explosion when poured onto a cold iron plate.
See other CHLORITE SALTS, OXOSALTS OF NITROGENOUS BASES

0501. Methylammonium perchlorate
[16875-44-2] CH_6ClNO_4

$$MeN^+H_3ClO_4^-$$

1. Kasper, F., *Z. Chem.*, 1969, **9**, 34
2. Kempson, R. M., *Chem. Abs.*, 1980, **92**, 61259
The semi-crystalline mass exploded when stirred after standing overnight. The
preparation was based on a published method used uneventfully for preparation
of ammonium, dimethylammonium and piperidinium perchlorates [1]. Its use in
explosives and propellants has been surveyed [2].
See other PERCHLORATE SALTS OF NITROGENOUS BASES

0502. Guanidinium perchlorate
[10308-84-6] $CH_6ClN_3O_4$

$$H_2NC(NH)N^+H_3ClO_4^-$$

1. Davis, 1943, 121
2. Schumacher, 1960, 213
3. Fujiwara, S. *et al., Chem. Abs.*, 1980, **92**, 200443
4. Udupa, M. R., *Propellants, Explos., Pyrotech.*, 1983, **8**, 109–111
Unusually sensitive to initiation and of high explosive power [1], it decomposes
violently at 350°C [2]. The explosive properties have been determined [3], and
thermal decomposition at 275–325° was studied in detail [4].
See DIFFERENTIAL THERMAL ANALYSIS (reference 1)

Iron(III) oxide
Isaev, R. N. *et al., Chem. Abs.*, 1970, **73**, 132626
Addition of 10% of iron oxide reduces the thermal stability of the salt.
See other PERCHLORATE SALTS OF NITROGENOUS BASES

†0503. Methylhydrazine
[60-34-4] CH_6N_2

$$MeNHNH_2$$

491M 1975, 259
When extensively exposed to air (as a thin film, or absorbed on fibrous or porous
solids), it may ignite.

Dicyanofurazan
See Dicyanofurazan: Nitrogenous bases

195

Oxidants

Kirk-Othmer, 1966, Vol. 11, 186

A powerful reducing agent and fuel, hypergolic with many oxidants such as dinitrogen tetraoxide or hydrogen peroxide.

See ROCKET PROPELLANTS *See other* ORGANIC BASES, REDUCANTS

0504. Ammonium *aci*-nitromethanide

[ ] $CH_6N_2O_2$

$$H_4N^+H_2C{=}N(O)O^-$$

Watts, C. E., *Chem. Eng. News*, 1952, **30**, 2344

The isolated salt is a friction-sensitive explosive.

See other aci-NITRO SALTS

0505. Methylammonium nitrite

[68897-47-2] $CH_6N_2O_2$

$$MeN^+H_3 \ NO_2^-$$

Hara, Y., *Chem. Abs.*, 1982, **97**, 55056

It is unstable at ambient temperature, and is formed on contact of the nitrate with sodium or potassium nitrites.

See other NITRITE SALTS OF NITROGENOUS BASES, OXOSALTS OF NITROGENOUS BASES

0506. Methylammonium nitrate (Methanaminium nitrate)

[1941-24-8] $CH_6N_2O_3$

$$MeN^+H_3 \ NO_3^-$$

1. Biasutti, 1981, 145
2. Miron, Y., *J. Haz. Mat.*, 1980, **3**, 301–321
3. Hara, Y., *Chem. Abs.*, 1982, **97**, 55056

Rail tanks of 86% aqueous solutions or slurries of the salt exploded, apparently during pump-transfer operations [1]. The course and mechanism or thermal decomposition has been investigated. Traces of rust or copper powder catalyse and accelerate the decomposition, so corrosion prevention is an important aspect of safety measures [2]. It is of higher thermal stability than the chlorate salt, or the nitrite, which decomposes at ambient temperature [3].

See other CATALYTIC IMPURITY INCIDENTS *See other* OXOSALTS OF NITROGENOUS BASES

0507. Aminoguanidine

[79-17-4] CH_6N_4

$$H_2NNHC(NH)NH_2$$

Kurzer, F. *et al.*, *Chem. & Ind.*, 1962, 1585

All the oxoacid salts are potentially explosive, including the wet nitrate.

See Aminoguanidinium nitrate
See other HIGH-NITROGEN COMPOUNDS, ORGANIC BASES

0508. Carbonic dihydrazide
[497-18-7] CH_6N_4O

$$O:C(NHNH_2)_2$$

Sorbe, 1968, 74
It explodes on heating.

Nitrous acid
    Curtius, J. *et al., Ber.*, 1894, **27**, 55
    Interaction forms the highly explosive carbonic diazide.
    See Carbonic diazide
    See other HIGH-NITROGEN COMPOUNDS

0509. Guanidinium nitrate
[52470-25-4] $CH_6N_4O_3$

$$H_2NC(NH)N^+H_3\ NO_3^-$$

1. Davis, T. L., *Org. Synth.*, 1944, Coll. Vol. 1, 302
2. Smith, G. B. L., and Schmidt, M. T., *Inorg. Synth.*, 1939, **1**, 96, 97
3. Schopf, C. *et al., Angew. Chem.*, 1936, **49**, 23
4. Udupa, M. R., *Thermochim. Acta*, 1982, **53**, 383–385
5. Medard, L. *Mém. Poudres*, 1951, **33**, 113–123 (English transl. HSE 11292 available from HSE/LIS, 1986)
6. Indian Pat. IN 157 293, Projects and Developments (India) Ltd., 1986 (*Chem. Abs.*, 1986, **105**, 210788)

According to an O.S. amendment sheet, the procedure as described [1] is dangerous because the reaction mixture (dicyanodiamide and ammonium nitrate) is similar in composition to commercial blasting explosives. This probably also applies to similar earlier preparations [2]. An earlier procedure which involved heating ammonium thiocyanate, lead nitrate and ammonia demolished a 50 bar autoclave [3]. TGA and DTA studies show that air is not involved in the thermal decomposition [4]. Explosive properties of the nitrate are detailed [5]. An improved process involves catalytic conversion at 90–200°C of a molten mixture of urea and ammonium nitrate to give 92% conversion (on urea) of guanidinium nitrate, recovered by crystallisation. Hazards of alternative processes are listed [6].
See other OXOSALTS OF NITROGENOUS BASES

0510. Methylsilane
[992-94-9] CH_6Si

$$MeSiH_3$$

Mercury, Oxygen
    Stock, A. *et al., Ber.*, 1919, **52**, 706

It does not ignite in air, but explodes if shaken with mercury in oxygen.
See other ALKYLSILANES

0511. Methylhydrazinium nitrate
[29674-96-2] $CH_7N_3O_3$

$$H_2NMeN^+H_2 \ NO_3^-$$

Lawton, E. A. *et al., J. Chem. Eng. Data*, 1984, **29**, 358
Impact sensitivity is comparable to that of ammonium perchlorate.
See other OXOSALTS OF NITROGENOUS BASES

0512. Aminoguanidinium nitrate
[10308-82-4] $CH_7N_5O_3$

$$H_2NNHC(NH)N^+H_3 \ NO_3^-$$

Koopman, H., *Chem. Weekbl.*, 1957, **53**, 97
An aqueous solution exploded violently during evaporation on a steam bath. Nitrate
salts of many organic bases are unstable and should be avoided.
See Aminoguanidine
See other HIGH-NITROGEN COMPOUNDS, OXOSALTS OF NITROGENOUS BASES

0513. Methyldiborane
[23777-55-1] CH_8B_2

$$H_2B:H_2:BHMe$$

Bunting, *Inorg. Synth.*, 1979, 19, 237
It ignites explosively in air.
See other ALKYLBORANES

0514. Diaminoguanidinium nitrate
[10308-83-5] $CH_8N_6O_3$

$$(H_2NNH)_2C{=}N^+H_2NO_3^-$$

Violent decomposition at 260°C.
See entry DIFFERENTIAL THERMAL ANALYSIS (reference 1)
See other OXOSALTS OF NITROGENOUS BASES

0515. Triaminoguanidinium perchlorate
[4104-85-2] $CH_9ClN_6O_4$

$$(H_2NNH)_2C{=}NN^+H_3ClO_4^-$$

Violent decomposition at 317°C.
See entry DIFFERENTIAL THERMAL ANALYSIS (reference 1)
See other PERCHLORATE SALTS OF NITROGENOUS BASES

0516. *O*-Methyl-*N*,*N*-disilylhydroxylamine
[ ] CH_9NOSi_2

$$(H_3Si)_2NOMe$$

Mitzel, N. W. *et al., J. Amer. Chem. Soc.*, 1996, **118**(11), 2664
Reasonably stable at ambient temperature, the compound explodes if heated to 200°C
See other SILANES, N–O COMPOUNDS

0517. Triaminoguanidinium nitrate
[4000-16-2] $CH_9N_7O_3$

$$(H_2NNH)_2C{=}NN^+H_3NO_3^-$$

Violent decomposition at 230°C.
See entry DIFFERENTIAL THERMAL ANALYSIS (reference 1)
See other OXOSALTS OF NITROGENOUS BASES

0518. Caesium cyanotridecahydrodecaborate(2−)
[71250-00-5] $CH_{13}B_{10}Cs_2N$

$$Cs_2[B_{10}H_{13}C{\equiv}N]$$

Hydrochloric acid
    Schultz, R. V. *et al., Inorg. Chem.*, 1979, **18**, 2883
    Addition of the salt to conc. hydrochloric acid is extremely exothermic.
    See other CYANO COMPOUNDS

0519. Pentaamminethiocyanatocobalt(III) perchlorate
[15663-42-0] $CH_{15}Cl_2CoN_6O_8S$

$$[(H_3N)_5CoSC{\equiv}N][ClO_4]_2$$

Explodes at 325°C; medium impact-sensitivity.
See entry AMMINEMETAL OXOSALTS (reference 3)

199

0520. Pentaamminethiocyanatoruthenium(III) perchlorate
[38139-15-0] $CH_{15}Cl_2N_6O_8RuS$

$$[(H_3N)_5RuSC{\equiv}N]\ [ClO_4]_2$$

Armour, J. N., private comm., 1969
After washing with ether, 0.1 g of the complex exploded violently when touched with a spatula.
See other AMMINEMETAL OXOSALTS, REDOX COMPOUNDS

0521. Hafnium carbide
[12069-85-1]   CHf

$$HfC$$

See entry REFRACTORY POWDERS

0522. Poly(diazomethylenemercury) ('Mercury diazocarbide')
[31724-50-2] $(CHgN_2)_n$

$$[HgCN_2]_n$$

Houben-Weyl, 1974, Vol. 13.2b, 19
Interaction of diazomethane with bis(trimethylsilylamino)mercury gives the polymeric explosive solid.
See other DIAZO COMPOUNDS, MERCURY COMPOUNDS

0523. Mercury(I) cyanamide
[72044-13-4] CHg_2N_2

$$Hg_2NC{\equiv}N$$

Deb, S. K. *et al., Trans. Faraday Soc.*, 1959, **55**, 106–113
Relatively large particles explode on heating rapidly to 325°C, or under high-intensity illumination when confined.
See other IRRADIATION DECOMPOSITION INCIDENTS, MERCURY COMPOUNDS, *N*-METAL DERIVATIVES

0524. Iodine isocyanate
[3607-48-5]   CINO

$$IN{=}C{=}O$$

Rosen, S. *et al., Anal. Chem.*, 1966, **38**, 1394
On storage, solutions of iodine isocyanate gradually deposit a touch-sensitive, mildly explosive solid (possibly cyanogen peroxide).
See related *N*-HALOGEN COMPOUNDS *See other* IODINE COMPOUNDS

0525. Carbon tetraiodide (Tetraiodomethane)
[507-25-5] CI_4

$$CI_4$$

Bromine trifluoride
See Bromine trifluoride: Halogens, etc.

Lithium
See Lithium: Halocarbons
See other HALOALKANES

0526. Potassium cyanide
[151-50-8]   **CKN**

$$KC{\equiv}N$$

HCS 1980, 764

Mercury(II) nitrate
See Mercury(II) nitrate: Potassium cyanide

Nitrogen trichloride
See Nitrogen trichloride: Initiators

Perchloryl fluoride
See Perchloryl fluoride: Calcium acetylide, etc.

Sodium nitrite
See Sodium nitrite: Metal cyanides
See entry METAL CYANIDES (AND CYANO COMPLEXES) (reference 1)

0527. Potassium cyanide–potassium nitrite
[ ]   **CKN.KNO$_2$**

$$KC{\equiv}N.KNO_2$$

Sorbe, 1968, 68
The double salt (a redox compound) is explosive.
See other REDOX COMPOUNDS

0528. Potassium cyanate
[590-28-3]   **CKNO**

$$KOC{\equiv}N$$

Water
Prinsky, M. L. *et al.*, *J. Loss Prev. Process Ind.*, 1990, **3**, 345

A drum of 30% solution in water exploded an hour after filling at 50°C, despite having a vent. Calorimetry demonstrated an exothermic, autocatalytic hydrolysis to ammoniacal potassium bicarbonate. In theory, a pressure exceeding 30 bar is obtainable. Aqueous solutions are unstable even at room temperature. Similar hydrolysis may account for an explosive product with Gold(III) chloride.
See Gold(III) chloride
See related METAL CYANATES, METAL CYANIDES

0529. Potassium thiocyanate
[333-20-0]   CKNS

$$KSC{\equiv}N$$

HCS 1980, 778

Calcium chlorite
See Calcium chlorite: Potassium thiocyanate

Other reactants
Yoshida, 1980, 228
MRH values calculated for 13 combinations with oxidants are given.

Perchloryl fluoride
See Perchloryl fluoride: Calcium acetylide, etc.
See related METAL CYANIDES

0530. Potassium trinitromethanide
[14268-23-6] CKN_3O_6

$$KC(NO_2)_3$$

1. Sandler, S. R. *et al., Organic Functional Group Preparations*, 433, New York, Wiley, 1968
2. Shulgin, A. T., private comm., 1968

This intermediate, produced by action of alkali on tetranitromethane, must be kept damp and used as soon as possible with great care, as it may be explosive [1]. Material produced as a by-product in a nitration reaction using tetranitromethane was washed with acetone. It exploded very violently after several months' storage [2].
See other POLYNITROALKYL COMPOUNDS

0531. Potassium carbonate
[584-08-7] CK_2O_3

$$KOCO.OK$$

Carbon
Druce, J. G. F., *School Sci. Rev.*, 1926, **7**(28), 261

Potassium metal prepared by the old process of distilling an intimate mixture of the carbonate and carbon contained some 'carbonylpotassium' (actually potassium benzenehexoxide), and several explosions with old samples of potassium may have involved this compound (or, perhaps more likely, potassium superoxide).
See Potassium (Slow oxidation)

Chlorine trifluoride
See Chlorine trifluoride: Metals, etc.

Magnesium
See Magnesium: Potassium carbonate
See other INORGANIC BASES, METAL OXONON-METALLATES

0532. Lithium thiocyanate
[556-65-0] $\qquad$ CLiNS

$$LiSC\equiv N$$

1,3-Dioxolane
Rao, B. L. M. *et al., J. Electrochem. Soc.*, 1980, **127**, 2333–2335
Thermal decomposition of the thiocyanate in air at 650°C yields 0.01% of volatile cyanide, but presence of moisture or the solvent increases cyanide evolution to 0.5–1.5 wt%.
See related METAL CYANIDES

0533. Lithium carbonate
[554-13-2] $\qquad$ CLi_2O_3

$$LiOCO.OLi$$

HCS 1980, 594

Fluorine
See Fluorine: Metal salts
See other INORGANIC BASES, METAL OXONON-METALLATES

0534. Magnesium carbonate hydroxide
[39409-82-0] $\qquad$ $CMgO_3.\frac{1}{4}H_2MgO_2$

$$4MgCO_3.Mg(OH)_2$$

Formaldehyde
See Formaldehyde: Magnesium carbonate hydroxide
See other INORGANIC BASES, METAL OXONON-METALLATES

0535. Carbonyl(pentasulfur pentanitrido)molybdenum
[ ] $\qquad$ $CMoN_5OS_5$

$$O:CMo(S_5N_5)$$

Wynne, K. J. *et al., J. Inorg. Nucl. Chem.*, 1968, **30**, 2853

A shock- and heat-sensitive amorphous compound of uncertain structure. A sample exploded violently when scraped from a glass sinter, under impact from a hammer, on sudden heating to 260°C or when added to conc. sulfuric acid.

See related CARBONYLMETALS, N–S COMPOUNDS

0536. Sodium cyanide
[143-33-9]   CNNa

$$NaC\equiv N$$

HCS 1980, 836, 837 (30% solution)

Benzyl chloride
See Benzyl chloride: Methanol, Sodium cyanide, etc.

Ethyl chloroacetate
See Ethyl chloroacetate: Sodium cyanide

Oxidants
See entry METAL CYANIDES (AND CYANO COMPLEXES) (reference 1)

0537. Sodium fulminate
[15736-98-8]   CNNaO

$$NaC\equiv N\rightarrow O$$

Smith, 1966, Vol. 2, 99
Even sodium fulminate detonates when touched lightly with a glass rod.
See other METAL FULMINATES

0538. Sodium thiocyanate
[540-72-7]   CNNaS

$$NaSC\equiv N$$

Other reactants
Yoshida, 1980, 229
MRH values calculated for 14 combinations with a variety of reagents are given.

Sodium nitrate
See Sodium nitrate: Sodium thiocyanate

Sodium nitrite   MRH 3.51/72
See Sodium nitrite: Sodium thiocyanate

Sulfuric acid
See Sulfuric acid: Sodium thiocyanate
See related METAL CYANIDES

0539. Thallium fulminate
[20991-79-1]   CNOTl

$$TlC{\equiv}N{\rightarrow}O$$

Boddington, T. *et al., Trans. Faraday Soc.*, 1969, **65**, 509
Explosive, even more shock- and heat-sensitive than mercury(I) fulminate.
See other METAL FULMINATES

0540. Cyanonitrene
[1884-64-6]   CN$_2$

$$N{\equiv}CN{:}$$

Hutchins, M. G. K. *et al., Tetrahedron Lett.*, 1981, **22**, 4599–4602
Interaction of sodium hydrogen cyanamide and *tert*-butyl hypochlorite in methanol
at −50°C to ambient temperature forms cyanonitrene. Isolation gave an orange
dimeric residue which exploded on two occasions.
See related CYANO COMPOUNDS

0541. Nitrosyl cyanide
[4343-68-4]   CN$_2$O

$$O{=}NC{\equiv}N$$

Alone, or Nitrogen oxide
1. Kirby, G. W., *Chem. Soc. Rev.*, 1977, 9
2. Keary, C. M. *et al., J. Chem. Soc., Perkin Trans. 2*, 1978, 243
Direct preparation of the gas is potentially hazardous, and explosive decomposition of the impure gas in the condensed state (below −20°C) has occurred. A safe
procedure involving isolation of the 1:1 adduct with 9,10-dimethylanthracene is
preferred. The impure gas contains nitrogen oxide and it is known that nitrosyl
cyanide will react with the latter to form an explosive compound [1]. The need
to handle this compound of high explosion risk in small quantities, avoiding
condensed states, is stressed [2].
See other CYANO COMPOUNDS, NITROSO COMPOUNDS

0542. Sulfinylcyanamide
[16438-87-3]   CN$_2$OS

$$O{:}S{:}NC{\equiv}N$$

Alone, or Ethanol, or Water
Scherer, O. J. *et al., Angew. Chem. (Intern. Ed.)*, 1967, **6**, 701–702
The crude highly unsaturated material decomposes explosively if stored in a
refrigerator or on attempted distillation under water-pump vacuum (moisture?).
However, it is distillable under oil-pump vacuum and is then stable in refrigerated
storage. It decomposes explosively with traces of water or ethanol, and polymerises
in daylight at ambient temperature.
See other CYANO COMPOUNDS, N–S COMPOUNDS

0543. Thallium(I) azidodithiocarbonate (Thallium(I) dithiocarbonazidate)

[ ] CN_3S_2Tl

$$TlSC(S)N_3$$

Sulfuric acid
    Bailar, 1973, Vol. 1, 1155
    The highly unstable explosive salt is initiated by contact with sulfuric acid.
    See Azidodithioformic acid
    See related ACYL AZIDES

0544. Cyanogen azide
[764-05-6] CN_4

$$N{\equiv}CN_3$$

1. Marsh, F. D., *J. Amer. Chem. Soc.*, 1964, **86**, 4506
2. Marsh, F. D., *J. Org. Chem.*, 1972, **37**, 2966
3. Coppolino, A. P., *Chem. Eng. News*, 1974, **52**(25), 3
4. Mardell, L., *Chem. in Brit.*, 1995, **31**(8), 591

It detonates with great violence when subjected to mild mechanical, thermal or electrical shock [1]. Special precautions are necessary to prevent separation of the azide from solution and to decontaminate equipment and materials [2]. The azide is not safe in storage at −20°C as previously stated. A sample exploded immediately after removal from refrigerated (dark) storage [3]. A student was performing a preparation involving in situ generation of cyanogen azide in acetonitrile. Some unexpected white needles crystallised from the mixture, were filtered, dried on the filter (probably a glass sinter) and scraped off, this caused an explosion removing the student's thumb. It is hypothesised that water contamination, or excess sodium azide, had induced formation of diazidomethylenecyanamide or sodium 5-azidotetrazolide, which had crystallised [4].
See Diazidomethylenecyanamide

Sodium hydroxide
    Marsh, F. D., *J. Org. Chem.*, 1972, **37**, 2966
    Interaction with 10% alkali forms sodium 5-azidotetrazolide, which is violently explosive if isolated.
    See 5-Azidotetrazole
    See related HALOGEN AZIDES

0545. Dinitrodiazomethane
[25240-93-1] CN_4O_4

$$(O_2N)_2CN_2$$

Alone, or Sulfuric acid
    Schallkopf, V. *et al.*, *Angew. Chem. (Intern. Ed.)*, 1969, **8**, 612

It explodes on impact, rapid heating or contact with sulfuric acid.
See other DIAZO COMPOUNDS, POLYNITROALKYL COMPOUNDS

0546. Tetranitromethane
[509-14-8] CN_4O_8

$$C(NO_2)_4$$

1. Liang, P., *Org. Synth.*, 1955, Coll. Vol. 3, 804
2. *MCA Guide*, 1972, 310 (reference 9)

During its preparation from fuming nitric acid and acetic anhydride, strict temperature control and rate of addition of anhydride are essential to prevent a runaway violent reaction [1]. An explosion occurred during preparation in a steel tank [2]. It should not be distilled, as explosive decomposition may occur [1].
See Nitric acid: Acetic anhydride

Aluminium
See Aluminium: Oxidants

Amines
Gol'binder, A. A., *Chem. Abs.*, 1963, **59**, 9730b
Mixtures of amines, for example aniline, with tetranitromethane (19.5%) ignite in 35−55 s, and will proceed to detonation if the depth of liquid is above a critical value.

Aromatic nitro compounds
1. Urbanski, 1964, Vol. 1, 592
2. Zotov, E. V. *et al., Chem. Abs.*, 1982, **98**, 5965

Mixtures with nitrobenzene, 1- or 4-nitrotoluene, 1,3-dinitrobenzene or 1-nitronaphthalene were found to be high explosives of high sensitivity and detonation velocities [1]. Those with nitrobenzene are spark-detonable [2].

Dienes
Eberson, L. *et al., Acta Chem. Scand.*, 1998, **52**(4), 450
Tetranitromethane adds across various non-conjugated dienes to give cyclic products. The products with norbornadiene and cycloocta-1,5-diene are subject to spontaneous violent decomposition.

Ferrocene
Jaworska-Augustyniak, A., *Transition Met. Chem.*, 1979, **4**, 207−208
Contact of tetranitromethane with ferrocene under various conditions leads to violent explosions, though not if the latter is dissolved in methanol or cyclohexane. Other ferrocene derivatives react violently, but not the dimethyl compound.

Hydrocarbons
1. Stettbacher, A., *Z. Ges. Schiess. Sprengstoffw.*, 1930, **25**, 439
2. Hager, K. F., *Ind. Eng. Chem.*, 1949, **41**, 2168

3. Tschinkel, J. G., *Ind. Eng. Chem.*, 1956, **48**, 732
4. Stettbacher, A., *Tech. Ind. Schweiz. Chem. Ztg.*, 1941, **24**, 265–271
5. Anon., *Chem. Ztg.*, 1920, **44**, 497
6. Zotov, E. V. *et al,., Chem. Abs.*, 1980, **93**, 49732, 49734
When mixed with hydrocarbons in approximately stoicheiometric proportions, a sensitive highly explosive mixture is produced which needs careful handling [1,2]. The use of such mixtures as rocket propellants has also been investigated [3]. Explosion of only 10 g of a mixture with toluene caused 10 deaths and 20 severe injuries [4]. The mixture contained excess toluene in error [5]. Detonation characteristics of mixtures with benzene, toluene (and nitrobenzene) were studied [6].

Pyridine
Isaacs, N. S. *et al., Tetrahedron Lett.*, 1982, **23**, 2802
Mixtures occasionally have exploded several hours after preparation.

Sodium ethoxide
Macbeth, A. K., *Ber.*, 1913, **46**, 2537–2538
Addition of the last of several portions of the ethoxide solution caused the violent explosion of 30 g of tetranitromethane (possibly involving formation of sodium trinitromethanide).
See Potassium trinitromethanide

Toluene, Cotton
Winderlich, R., *J. Chem. Educ.*, 1950, **27**, 669
A demonstration mixture, to show combustion of cellulose by combined oxygen, exploded with great violence soon after ignition.
See Hydrocarbons, above
See other OXIDANTS, POLYNITROALKYL COMPOUNDS

0547. Sodium 5-nitrotetrazolide
[67312-43-0] CN_5NaO_2

1. Spear, R. J., *Aust. J. Chem.*, 1982, **35**, 10
2. Redman, L. D. *et al., Chem. Abs.*, 1984, **100**, 194540

Extremely sensitive to mechanical initiation, it should be treated as a primary explosive [1]. A batch of the dihydrate exploded spontaneously [2].
See other C-NITRO COMPOUNDS, TETRAZOLES

0548. 5-Diazoniotetrazolide
[13101-58-1] CN_6

Thiele, J., *Ann.*, 1892, **270**, 60
Conc. aqueous solutions of this internal diazonium salt explode at 0°C.
See other DIAZONIUM SALTS, TETRAZOLES

0549. Disodium tetrazole-5-diazoate
[ ] CN_6Na_2O

Sulfuric acid
Thiele, J., *Ann.*, 1893, **273**, 148–149, 151
It decomposes non-explosively on heating, but ignites in contact with conc. sulfuric acid.
See other TETRAZOLES

0550. Carbonic diazide (Carbonyl azide)
[14435-92-8] CN_6O
$$O{:}C(N_3)_2$$

1. Chapman, L. E. *et al.*, *Chem. & Ind.*, 1966, 1266
2. Kesting, W., *Ber.*, 1924, **57**, 1321
3. Curtius, T. *et al.*, *Ber.*, 1894, **27**, 2684
It is a violently explosive solid, which should be used only in solution, and on a small scale [1]. It exploded violently under ice-water [2], or on exposure to bright light [3].
See other ACYL AZIDES, IRRADIATION DECOMPOSITION INCIDENTS

0551. Sodium 5-azidotetrazolide

[35038-45-0] **CN_7Na**

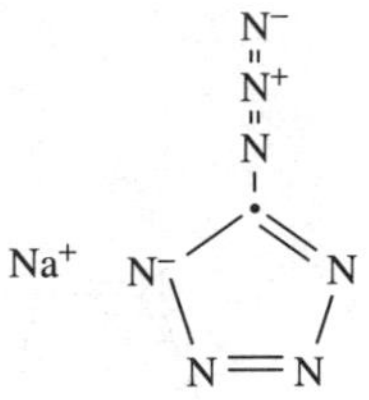

Marsh, F. D., *J. Org. Chem.*, 1972, **37**, 2967, 2969

Readily formed from cyanogen bromide and sodium azide in aqueous solution at 0°C, it is extremely sensitive to friction, heat or pressure. A dry sample under vacuum at 1 mbar will detonate on rapid admission of air.

See 5-Azidotetrazole

See other FRICTIONAL INITIATION INCIDENTS *See other* HIGH-NITROGEN COMPOUNDS, *N*-METAL DERIVATIVES, TETRAZOLES

0552. Sodium carbonate

[497-19-8] **CNa_2O_3**

NaOCO.ONa

HCS 1980, 832

Acidic reaction liquor

Anon., *Loss Prev. Bull.*, 1978, (023), 138

Use of sodium carbonate to neutralise the acid produced during methylation with dimethyl sulfate in a 2-phase system led to severe frothing, and the agitator was turned off to allow it to abate. When agitation was restarted, violent foaming from mixing of the separated acid and alkaline layers ejected 2 t of the reactor contents (remedy worse than ailment!) Use of magnesium oxide to neutralise acid without gas evolution was recommended.

See other AGITATION INCIDENTS, GAS EVOLUTION INCIDENTS, NEUTRALISATION INCIDENTS

Ammonia, Silver nitrate

See Silver nitrate: Ammonia, etc.

An aromatic amine, A chloronitro compound

MCA Case History No. 1964

An unspecified process had been operated for 20 years using synthetic sodium carbonate powder (soda-ash) to neutralise the hydrogen chloride as it was formed by interaction of the amine and chloro compound in a non-aqueous (and probably flammable) solvent in a steel reactor. Substitution of the powdered sodium carbonate by the crystalline sodium carbonate–sodium hydrogencarbonate double

salt ('trona, natural soda') caused a reduction in the rate of neutralisation, the reaction mixture became more acid, and attack on the steel vessel led to contamination by iron. These changed conditions initiated exothermic side reactions, which eventually ran out of control and caused failure of the reactor. Subsequent laboratory work confirmed this sequence and showed that presence of dissolved iron(III) was necessary to catalyse the side reactions.
See Sulfuric acid, below
See other CATALYTIC IMPURITY INCIDENTS, NEUTRALISATION INCIDENTS

2,4-Dinitrotoluene
See 2,4-Dinitrotoluene: Alkali

Fluorine
See Fluorine: Metal salts

Hydrogen peroxide
See CRYSTALLINE HYDROGEN PEROXIDATES

Lithium
See Lithium: Sodium carbonate

Phosphorus pentaoxide
See Tetraphosphorus decaaoxide: Inorganic bases

Sodium sulfide, Water
See SMELT: Water

Sulfuric acid
MCA Case History No. 888
Lack of any mixing arrangements caused stratification of strong sulfuric acid and (probably) sodium carbonate solutions in the same tank. When gas evolution caused intermixing of the layers, a violent eruption of the tank contents occurred.
See An aromatic amine, etc., above
See other AGITATION INCIDENTS, GAS EVOLUTION INCIDENTS

2,4,6-Trinitrotoluene
See 2,4,6-Trinitrotoluene: Added impurities
See other INORGANIC BASES, METAL OXONON-METALLATES

0553. Sodium carbonate hydrogen peroxidate ('Sodium percarbonate')
[15630-89-4] $CNa_2O_3.1.5H_2O_2$
$$NaOCO.ONa.1.5H_2O_2$$

See entry CRYSTALLINE HYDROGEN PEROXIDATES

0554. Sodium monoperoxycarbonate
[4452-58-8]   CNa$_2$O$_4$

NaOCO.OONa

Acetic anhydride
See Sodium carbonate hydrogen peroxidate: Acetic anhydride
See other PEROXOACID SALTS

†0555. Carbon monoxide
[630-08-0]   CO

CO

NSC 415, 1976; FPA H57, 1977 (cylinder); HCS 1980, 282

Aluminium, Aluminium halides
See Aluminium: Aluminium halides, Carbon oxides

Copper(I) perchlorate
See Copper(I) perchlorate, Alkenes, etc.

Dinitrogen oxide
See Dinitrogen oxide: Carbon monoxide

Fluorine, Oxygen
See Bis(fluoroformyl) peroxide

Hydrogen, Oxygen
See Oxygen: Carbon monoxide, Hydrogen

Interhalogens   MRH values show % of interhalogen
  See Bromine pentafluoride: Acids, etc.   MRH 2.47/71
     Bromine trifluoride: Carbon monoxide   MRH 3.05/tr.
     Iodine heptafluoride: Carbon, etc.

Metal oxides
  See Caesium oxide: Halogens, etc.
     Iron(III) oxide: Carbon monoxide
     Silver(I) oxide: Carbon monoxide

Metals
  See Potassium: Non-metal oxides
     Sodium: Non-metal oxides

Other reactants
  Yoshida, 1980, 42
  MRH values calculated for 11 combinations with oxidants are given.

Oxidants
  See Chlorine dioxide: Carbon monoxide
    Oxygen (Liquid): Liquefied gases
    Peroxodisulfuryl difluoride: Carbon monoxide
  See other NON-METAL OXIDES

†0556. Carbonyl sulfide (Carbon oxide sulfide)
[463-58-1]   COS

$$O:C:S$$

See related NON-METAL OXIDES, NON-METAL SULFIDES

0557. Carbon dioxide
[124-38-9] CO_2

$$O:C:O$$

NSC 682, 1980; *HCS 1980*, 279 (cylinder) 280 (solid)

Acrylaldehyde
  See Acrylaldehyde: Acids

Aluminium, Aluminium halides
  See Aluminium: Aluminium halides, Carbon oxides

Aziridine
  See Aziridine: Acids

Barium peroxide
  See Barium peroxide: Non-metal oxides

Caesium oxide
  See Caesium oxide: Halogens, etc.

Flammable materials
  1. *BCISC Quart. Safety Summ.*, 1973, **44**(174), 10
  2. Leonard, J. T. *et al., Inst. Phys. Conf. Ser.*, 1975, **27** (Static Electrif.), 301–310
  3. Anon., *Chem. Eng. News*, 1976, **54**(30), 18
  4. Collocott, S. J. *et al., IEE Proc. part A*, 1980, **127**, 119–120
  5. Moffat, M., *Fire Prot.*, 1980, (7), 9
  6. Kletz, T. A., *Loss Prev. Bull.*, 1984, (056), 9
Dangers attached to the use of carbon dioxide in fire prevention and extinguishing systems in confined volumes of air and flammable vapours are discussed. The main hazard arises from the production of very high electrostatic charges, due to the presence of small frozen particles of carbon dioxide generated during expanding discharge of the compressed gas. Highly incendive sparks (5–15 mJ at 10–20 kV) readily may be produced. Nitrogen is preferred as an inerting gas for enclosed

213

volumes of air and flammable gases or vapours [1]. The electrostatic hazard in discharge of carbon dioxide in tank fire-extinguishing systems is associated with the plastics distribution horns, and their removal eliminated the hazard [2]. Alternatively, a metal sleeve may be fitted inside the plastics horn [3]. The accumulation of charge during discharge of a portable extinguisher has also been studied and found potentially hazardous [4]. The latter hazard is minimised in a statement to clarify the overall situation, which emphasizes the need for avoiding the use of liquid carbon dioxide for inerting purposes [5]. Previous accidents involving carbon dioxide as the cause of explosions are reviewed [6].

Hydrazine, Stainless steel
See Hydrazine: Carbon dioxide, Stainless steel

Metal hydrides
See Aluminium hydride: Carbon dioxide
Lithium tetrahydroaluminate: Bis(2-methoxyethyl) ether

Metal acetylides
See Monolithium acetylide–ammonia: Gases
Monopotassium acetylide: Non-metal oxides
Rubidium acetylide: Non-metal oxides
Sodium acetylide: Non-metal oxides

Metals
1. Hartmann, I., Ind. Eng. Chem., 1948, 40, 756
2. Rhein, R. A., Rept. No. CR-60125, Washington, NASA, 1964
Dusts of magnesium, zirconium, titanium and some magnesium–aluminium alloys [1], and (when heated) of aluminium, chromium and manganese [2], when suspended in carbon dioxide atmospheres are ignitable and explosive, and several bulk metals will burn in the gas.
See Metals, Nitrogen below
Aluminium: Aluminium halides, Carbon oxides
Lithium: Non-metal oxides
Magnesium: Carbon dioxide, etc.
Potassium: Non-metal oxides
Potassium–sodium alloy: Carbon dioxide
Sodium: Non-metal oxides
Titanium: Carbon dioxide
Uranium: Carbon dioxide

Metals, Nitrogen
Rhein, R. A., Rept. No. CR-60125, Washington, NASA, 1964
Powdered beryllium, cerium, cerium alloys, thorium, titanium, uranium and zirconium ignite on heating in mixtures of carbon dioxide and nitrogen; ignition temperatures were determined. Aluminium, chromium, magnesium and manganese ignite only in the absence of nitrogen.

214

Metals, Sodium peroxide
See Sodium peroxide: Metals, Carbon dioxide
See other NON-METAL OXIDES

0558. Lead carbonate
[598-63-0]   CO₃Pb

$$PbCO_3$$

HCS 198, 586

Fluorine
See Fluorine: Metal salts

0559. Carbon sulfide
[2944-05-0]   CS

$$C\cdot S$$

Pearson, T. G., *School Sci. Rev.*, 1938, **20**(78), 189
The gaseous radical polymerises explosively.
See other NON-METAL SULFIDES

†0560. Carbon disulfide
[75-15-0]   CS₂

$$S:C:S$$

(*MCA SD-12*, 1967); *NSC 341*, 1977; *FPA H34*, 1975; *HCS 1980*, 281; *RSC Lab. Hazard Data Sheet No. 7*, 1982

1. Thiem, F., *Chem. Technik*, 1979, **31**, 313–314
2. Briggs, A. G. *et al., Chem. Brit.*, 1982, **18**, 27
3. Bothe, H. *et al., Loss Prevention and Safety Promotion in the Process Industries* Vol. II (Mewis, J. J., Pasman, H. J. & De Rademaker, E. E. Eds.), 327, Amsterdam, Elsevier, 1995

Carbon disulfide vapour, alone or mixed with nitrogen, did not decompose explosively in the range 0.4–2.1 bar/88–142°C when initiated with high energy sparks or a hot wire at 700–900°C. The endothermic sulfide (ΔH°_f (l) +115.3 kJ/mol, 1.52 kJ/g) will, however, decompose explosively to its elements with mercury fulminate initiation [1]. A screening jacket filled with carbon disulfide was used to surround the reaction tube used in flash photolysis experiments. When the quartz lamp was discharged, some vapour of the disulfide which had leaked out ignited in the radiation flash and exploded. Care is also necessary with the vapour and other UV sources [2]. Ignition of vapour air mixes by laser has been studied [3].
See other IRRADIATION DECOMPOSITION INCIDENTS

Air, Rust
1. Mee, A. J., *School Sci. Rev.*, 1940, **22**(86), 95
2. Dickens, G. A., *School Sci. Rev.*, 1950, **31**(114), 264

3. Anon., *J. Roy. Inst. Chem.*, 1956, **80**, 664
4. Thompson, E., private comm., 1981
Disposal of 2 l of the solvent into a rusted iron sewer caused an explosion. Initiation of the solvent–air mixture by rust was suspected [1]. A hot gauze falling from a tripod into a laboratory sink containing some carbon disulfide initiated two explosions [2]. It is a very hazardous solvent because of its extreme volatility and flammability. The vapour or liquid has been known to ignite on contact with steam pipes, particularly if rusted [3]. When a winchester of the solvent fell off a high shelf and broke behind a rusted steel cupboard, ignition occurred [4].

Alkali metals
1. Steimecke, G. *et al., Phos. Sulfur*, 1979, **7**, 49
2. Moradpour, A. *et al., J. Chem. Educ.*, 1986, **62**, 1016
3. Staudinger, H., *Z. Angew. Chem.*, 1922, **35**, 659
Attempts to follow a published procedure for the preparation of 1,3-dithiole-2-thione-4,5-dithiolate salts [1], involving reductive coupling of carbon disulfide with alkali metals, have led to violent explosions with potassium metal, but not with sodium [2]. However, mixtures of carbon disulfide with potassium–sodium alloy, potassium, sodium, or lithium are capable of detonation by shock, though not by heating. The explosive power decreases in the order given above, and the first mixture is more shock-sensitive than mercury fulminate [3].
See Metals, below

Amines
1. Anon., Personal communication, 1991
2. Editor's comments.
Carbon disulphide was used as an extraction solvent when analysing epoxy resins. On one occasion, adding to a hardener produced a vigorous fume-off leaving a residue looking like sulfur [1]. Amines and complexes thereof are used as hardeners, and the reaction with, especially, polyamines to give dithiocarbamates is surprisingly exothermic [2].

Chlorine, Sodium *aci*-nitromethanide
See Sodium *aci*-nitromethanide: Carbon disulfide, etc.

Halogens
See Chlorine: Carbon disulfide
    Fluorine: Sulfides

Metal azides
Mellor, 1947, Vol. 8, 338
Carbon disulfide and aqueous solutions of metal azides interact to produce metal azidodithioformates, most of which are explosive with varying degrees of power and sensitivity to shock or heat.
See Azidodithioformic acid, *also* Sodium azide: Carbon disulfide

See Alkali metals, above
      Aluminium: Carbon disulfide
      Sodium: Sulfides, etc.
      Zinc: Carbon disulfide

Other reactants
  Yoshida, 1980, 281
  MRH values calculated for 16 combinations, largely with oxidants, are given.

Oxidants
  See Halogens, above
      Dinitrogen tetraoxide: Carbon disulfide
      Nitrogen oxide: Carbon disulfide
      Permanganic acid: Organic materials

Phenylcopper–triphenylphosphine complexes
  Camus, A. *et al., J. Organomet. Chem.*, 1980, **188**, 390
  The bis- or tris-complexes of phenylcopper with triphenylphosphine react violently
and exothermically with carbon disulfide, even at 0°C. Suitable control procedures
are described.
  See other ENDOTHERMIC COMPOUNDS, NON-METAL SULFIDES

0561. Titanium carbide
[12070-08-5]   CTi

TiC

MCA Case History No. 618

A violent, and apparently spontaneous, dust explosion occurred while the finely
ground carbide was being removed from a ball-mill. Static initiation seems a likely
possibility.
See entries DUST EXPLOSION INCIDENTS, REFRACTORY POWDERS
See other METAL NON-METALLIDES

0562. Uranium carbide
[12070-09-6]   CU

CU

Schmitt, C. R., *J. Fire Flamm.*, 1971, **2**, 163
The finely divided carbide is pyrophoric.
See other METAL NON-METALLIDES, PYROPHORIC MATERIALS

0563. Tungsten carbide
[12070-12-1]   CW

WC

Fluorine
See Fluorine: Metal acetylides and carbides

Nitrogen oxides
Mellor, 1946, Vol. 5, 890
At about 600°C, the carbide ignites and incandesces in dinitrogen mono- or tetra-oxides.
See other METAL NON-METALLIDES

0564. Ditungsten carbide
[12070-13-2]   CW$_2$

W$_2$C

Oxidants
Mellor, 1946, Vol. 5, 890
The carbide burns incandescently at red heat in contact with dinitrogen mono- or tetra-oxides.
See Fluorine: Metal acetylides and carbides
See other METAL NON-METALLIDES

0565. Zirconium carbide
[12070-14-3]   CZr

ZrC

See entry REFRACTORY POWDERS

0566. Silver chloroacetylide
[ ]   C$_2$AgCl

AgC≡CCl

Kirk-Othmer, 1964, Vol. 5, 204
Explosive, even when wet or under water.
See related HALOACETYLENE DERIVATIVES, METAL ACETYLIDES
See other SILVER COMPOUNDS

0567. Silver cyanodinitromethanide
[12281-65-1]   C$_2$AgN$_3$O$_4$

AgC(CN)(NO$_2$)$_2$

Parker, C. O. *et al.*, *Tetrahedron*, 1962, **17**, 86

218

A sample in a m.p. capillary exploded at 196°C, shattering the apparatus.
See other POLYNITROALKYL COMPOUNDS, SILVER COMPOUNDS

0568. Silver acetylide
[7659-31-6] C_2Ag_2

$$AgC{\equiv}CAg$$

1. Miller, 1965, Vol. 1, 486
2. Bailar, 1973, Vol. 3, 102
3. Reynolds, R. J. *et al., Analyst,* 1971, **96**, 319
4. Koehn, J., *Chem Abs.,* 1979, **91**, 7164

Silver acetylide is a more powerful detonator than the copper derivative, but both will initiate explosive acetylene-containing gas mixtures [1]. It decomposes violently when heated to 120–140°C [2]. Formation of a deposit of this explosive material was observed when silver-containing solutions were aspirated into an acetylene-fuelled atomic absorption spectrometer. Precautions to prevent formation are discussed [3]. The effect of ageing for 16 months on the explosive properties of silver and copper acetylides has been studied. Both retain their hazardous properties for many months, and the former is the more effective in initiating acetylene explosions [4].
See other METAL ACETYLIDES, SILVER COMPOUNDS

0569. Silver acetylide–silver nitrate
[15336-58-0] $C_2Ag_2.AgNO_3$

$$AgC{\equiv}CAg.AgNO_3$$

1. Baker, W. E. *et al., Chem. Eng. News,* 1965, **43**(49), 46
2. Hogan, V. D. *et al., OTS Rept. AD 419 625,* Washington, US Dept. Comm., 1960
3. Benham, R. A., *Chem. Abs.,* 1981, **94**, 142109

The dry complex is exploded by high-intensity light pulses, or by heat or sparks. As a slurry in acetone, it is stable for a week if kept dark [1]. The complex explodes violently at 245°C/1 bar and at 195°C/1.3 mbar [2], or on exposure to light [3].
See related METAL ACETYLIDES
See other IRRADIATION DECOMPOSITION INCIDENTS, SILVER COMPOUNDS

0570. Disilver ketenide
[27278-01-4] C_2Ag_2O

$$Ag_2C{=}C{=}O$$

1. Blues, E. T. *et al., Chem. Comm.,* 1970, 699
2. Bryce-Smith, D. *et al., J. Chem. Soc., Perk. Trans. 2,* 1993, (9), 1631.

This and its pyridine complex explode violently if heated or struck [1] (but see below). More data on the stability of disilver ketenide and its complexes is contained in [2].

See other SILVER COMPOUNDS

0571. Disilver ketenide–silver nitrate
[ ] $C_2Ag_2O.AgNO_3$

$$Ag_2C=C=O.AgNO_3$$

1. Bryce-Smith, D., *Chem. & Ind.*, 1975, 154
2. Blues, E. T. *et al., Chem. Comm.*, 1970, 701
The red complex is more dangerously explosive than the ketenide itself, now described as mildly explosive in the dry state when heated strongly or struck [1]. It was previously formulated as a 2:1 complex [2].

See other SILVER COMPOUNDS

0572. Silver oxalate
[533-51-7] $C_2Ag_2O_4$

$$AgOCO.CO.OAg$$

1. Sidgwick, 1950, 126
2. Anon., *BCISC Quart. Safety Summ.*, 1973, **44**, 19
3. Kabanov, A. A. *et al., Russ. Chem. Rev.*, 1975, **44**, 538
4. MacDonald, J. Y. *et al., J. Chem. Soc.*, 1925, **127**, 2625
5. Waltenberger-Razniewska, M. *et al., Chem. Abs.*, 1982, **96**, 203117
6. Ullmann, 1993, Vol. A24, 161
Above 140°C its exothermic decomposition to metal and carbon dioxide readily becomes explosive [1]. A 1 kg batch which had been thoroughly dried at 50°C exploded violently when mechanical grinding in an end-runner mill was attempted [2]. Explosions have been experienced when drying the oxalate as low as 80°C [6]. It is a compound of zero oxygen balance. The explosion temperature of the pure oxalate is lowered appreciably (from 143 to 122°C) by application of an electric field [3]. The salt prepared from silver nitrate with excess of sodium oxalate is much less stable than that from excess nitrate [4]. Decomposition at 125°C in glycerol prevents explosion in the preparation of silver powder [5].

See ELECTRIC FIELDS
See other METAL OXALATES, SILVER COMPOUNDS

0573. Gold(I) acetylide
[70950-00-4] C_2Au_2

$$AuC\equiv CAu$$

1. Mellor, 1946, Vol. 5, 855
2. Matthews, J. A. *et al., J. Amer. Chem. Soc.*, 1900, **22**, 110

Produced by action of acetylene on sodium gold thiosulfate (or other gold salts), the acetylide is explosive and readily initiated by light impact, friction or rapid heating to 83°C [1]. This unstable detonator is noted for high brisance [2].
See other GOLD COMPOUNDS, METAL ACETYLIDES

0574. Digold(I) ketenide
[54086-41-8] C_2Au_2O

$$Au_2C=C=O$$

Blues, E. T. *et al., J. Chem. Soc., Chem. Comm.*, 1974, 513–514
The ketenide is shock-sensitive when dry, and it or its complexes with tertiary heterocyclic bases (pyridine, methylpyridines, 2,6-dimethylpyridine or quinoline) explode when heated above 100°C.
See other GOLD COMPOUNDS

0575. Barium acetylide
[12070-27-8] $(C_2Ba)_n$

$$(-BaC{\equiv}C-)_n$$

Air, or Hydrogen, or Hydroxylic compounds
Masdupay, E. *et al., Compt. rend.*, 1951, **232**, 1837–1839
It is much more reactive than the diacetylide, and ignites in contact with water or ethanol in air. It may incandesce on heating to 150°C under vacuum or hydrogen, the product from the latter treatment being very pyrophoric owing to the presence of pyrophoric carbon.

Halogens
Mellor, 1946, Vol. 5, 862
Barium acetylide incandesces with chlorine, bromine or iodine at 140, 130 or 122°C respectively.

Selenium
Mellor, 1946, Vol. 5, 862
A mixture incandesces when heated to 150°C.
See other METAL ACETYLIDES

0576. Barium thiocyanate
[2092-17-3] $C_2BaN_2S_2$

$$Ba(SC{\equiv}N)_2$$

Potassium chlorate
See Potassium chlorate: Metal thiocyanates

Sodium nitrite
See Sodium nitrite: Potassium thiocyanate
See related METAL CYANIDES

0577. Barium 5,5′-azotetrazolide
[87478-73-7] C_2BaN_{10}

See entry METAL AZOTETRAZOLIDES *See other* TETRAZOLES

0578. Bromochloroacetylene
[25604-70-0] C_2BrCl

$$BrC{\equiv}CCl$$

Houben-Weyl, 1977, Vol. 5.2a, 604 (footnote 6)
Highly explosive and unstable.
See other HALOACETYLENE DERIVATIVES

†0579. Bromotrifluoroethylene
[598-73-2] C_2BrF_3

$$BrFC{=}CF_2$$

Oxygen
    See Oxygen (Gas): Halocarbons
    See other HALOALKENES

0580. Lithium bromoacetylide
[ ] C_2BrLi

$$LiC{\equiv}CBr$$

See Sodium bromoacetylide (next below); Lithium chloroacetylide; Lithium trifluoropropynide; Sodium chloroacetylide
See other HALOACETYLENE DERIVATIVES, METAL ACETYLIDES

0581. Sodium bromoacetylide
[ ] C_2BrNa

$$NaC{\equiv}CBr$$

Cradock, S. *et al., J. Chem. Soc., Dalton Trans.*, 1978, 760
The dry material (perhaps the ultimate 'carbene precursor') is an extremely shock-sensitive explosive.
See Lithium bromoacetylide; Lithium chloroacetylide; Lithium trifluoropropynide; Sodium chloroacetylide
See other HALOACETYLENE DERIVATIVES, METAL ACETYLIDES

222

0582. Dibromoacetylene
[624-61-3] C_2Br_2

$$BrC{\equiv}CBr$$

Rodd, 1951, Vol. 1A, 284
It ignites in air and explodes on heating.
See other HALOACETYLENE DERIVATIVES

0583. Oxalyl dibromide
[15219-34-8] $C_2Br_2O_2$

$$BrCO.CO.Br$$

Potassium
See Potassium: Oxalyl dihalides
See other ACYL HALIDES

0584. Hexabromoethane
[594-73-0] C_2Br_6

$$Br_3CCBr_3$$

Hexacyclohexyldilead
See Hexacyclohexyldilead: Halocarbons
See other HALOALKANES

0585. Calcium acetylide (Calcium carbide)
[75-20-7] $(C_2Ca)_n$

$$(-CaC{\equiv}C-)_n$$

HCS 1980, 264

1. Anon., *Ind. Safety Bull.*, 1940, **8**, 41
2. Déribère, M., *Chem. Abs.*, 1936, **30**, 8169.8
3. Orlov, V. N., *Chem. Abs.*, 1975, **83**, 168138

Use of a steel chisel to open a drum of carbide caused an incendive spark which ignited traces of acetylene in the drum. The non-ferrous tools normally used for this purpose should be kept free from embedded ferrous particles [1]. If calcium carbide is warm when filled into drums, absorption of the nitrogen from the trapped air may enrich the oxygen content up to 28%. In this case, less than 3% of acetylene (liberated by moisture) is enough to form an explosive mixture, which may be initiated on opening the sealed drum. Other precautions are detailed [2]. Use of carbon dioxide to purge carbide drums, and of brass or bronze non-sparking tools to open them are advocated [3].

Calcium hypochlorite
See Calcium hypochlorite

Halogens
Mellor, 1946, Vol. 5, 862
The acetylide incandesces with chlorine, bromine or iodine at 245, 350 or 305°C, respectively. Strontium and barium acetylides are more reactive.

Hydrogen chloride   MRH 3.97/54
Mellor, 1946, Vol. 5, 862
Incandescence on warming; strontium and barium acetylides are similar.

Iron(III) chloride, Iron(III) oxide
Partington, 1967, 372
The carbide is an energetic reducant. A powdered mixture with iron oxide and chloride burns violently when ignited, producing molten iron.
See other THERMITE REACTIONS

Lead difluoride
Mellor, 1946, Vol. 5, 864
Incandescence on contact at ambient temperature.

Magnesium
Mellor, 1940, Vol. 4, 271
A mixture incandesces when heated in air.

Methanol
Unpublished observations, 1951
Interaction of calcium carbide with methanol to give calcium methoxide is very vigorous, but subject to an induction period of variable length. Once reaction starts, evolution of acetylene gas is very fast, and a wide-bore condenser and adequate ventilation are necessary.
See other GAS EVOLUTION INCIDENTS, INDUCTION PERIOD INCIDENTS

Other reactants
Yoshida, 1980, 223
MRH values calculated for 8 combinations with various reagents are given.

Perchloryl fluoride
See Perchloryl fluoride: Calcium acetylide

Selenium
See Selenium: Metal acetylides

Silver nitrate
Luchs, J. K., *Photog. Sci. Eng.*, 1966, **10**, 334
Addition of calcium acetylide to silver nitrate solution precipitates silver acetylide, a highly sensitive explosive. Copper salt solutions would behave similarly.
See METAL ACETYLIDES

Sodium peroxide   MRH 3.01/86
    See Sodium peroxide: Calcium acetylide

Sulfur   MRH 4.35/33
    See Sulfur: Metal acetylides

Tin(II) chloride
    Mellor, 1941, Vol. 7, 430
    A mixture can be ignited with a match, and reduction to metallic tin proceeds with
    incandescence.

Water
    1. Jones, G. W. *et al., Explosions in Med. Press. Generators, Invest. Rept. 3775*,
       Washington, US Bur. Mines, 1944
    2. Moll, H., *Chem. Weekbl.*, 1933, **30**, 108
    At the end of a generation run, maximum temperature and high moisture content of
    acetylene may cause the finely-divided acetylide to overheat and initiate explosion
    of pressurised gas [1]. During analysis of technical carbide by addition of water,
    the explosive mixture formed in the unpurged reaction vessel exploded, ignited
    either by excessive local temperature or possibly by formation of crude phosphine
    from the phosphide present as impurity in the carbide [2].
    See other METAL ACETYLIDES, REDUCANTS

0586. Bis(trifluoromethyl)cadmium
[3327-66-1] C_2CdF_6

$$F_3CCdCF_3$$

Eujen, R. *et al., J. Organomet. Chem.*, 1995, **503**(2), C51
The uncomplexed compound decomposes to difluorocarbene at $-5°$ and explodes
violently on warming to room temperature.
See METAL DERIVATIVES OF ORGANOFLUORINE COMPOUNDS

0587. Bis(trifluoromethyl)cadmium-1,2-dimethoxyethane adduct
[33327-66-1] $C_2CdF_6 \cdot C_4H_{10}O_2$

$$(F_3C)_2Cd.glyme$$

Ontiveros, C. D. *et al., Inorg. Synth.*, 1986, **24**, 57
It may ignite in air if not thoroughly dry and pure.
See related ALKYLMETALS

0588. Cadmium cyanide
[542-83-6] C_2CdN_2

$$Cd(C{\equiv}N)_2$$

It is moderately endothermic ($\Delta H_f^°$ (s) $+163.2$ kJ/mol, 0.99 kJ/g)

Magnesium
See Magnesium: Metal cyanides
See other ENDOTHERMIC COMPOUNDS, METAL CYANIDES

0589. Cadmium fulminate
[42294-95-6] $C_2CdN_2O_2$

$$Cd(C\equiv N\rightarrow O)_2$$

It is a moderately endothermic compound ($\Delta H°_f$ (s) +163.2 kJ/mol, 0.80 kJ/g).
See entries ENDOTHERMIC COMPOUNDS, METAL FULMINATES

0590. Cerium carbide
[12012-32-7] C_2Ce

$$CeC_2$$

See Lanthanum carbide

†0591. Chlorotrifluoroethylene
[79-38-9] C_2ClF_3

$$ClFC=CF_2$$

Bromine, Oxygen
See Bromine: Chlorotrifluoroethylene, Oxygen

Chlorine perchlorate
See Chlorine perchlorate: Chlorotrifluoroethylene

1,1-Dichloroethylene
See 1,1-Dichloroethylene: Chlorotrifluoroethylene

Ethylene
See Ethylene: Chlorotrifluoroethylene

Oxygen
See Oxygen (Gas): Halocarbons (references 3,4)
See other HALOALKENES

0592. Poly(chlorotrifluoroethylene)
[9002-83-9] $(C_2ClF_3)_n$

$$(\text{-}CF_2\text{-}CFCl\text{-})_n$$

Aluminium
See Aluminium: Halocarbons
See related HALOALKANES

0593. 3-Chloro-3-trifluoromethyldiazirine
[58911-30-1] $C_2ClF_3N_2$

$$N=N$$

Cl   CF$_3$

Moss. R. A. *et al., J. Amer. Chem. Soc.*, 1981, **103**, 6168–6169
The potentially explosive compound was used safely as a solution in pentane.
See other DIAZIRINES

0594. Trifluoroacetyl hypochlorite
[65597-25-3] $C_2ClF_3O_2$

$F_3CCO.OCl$

1. Tari, I. *et al., Inorg. Chem.*, 1979, **18**, 3205–3208
2. Schack, C. J. *et al.*, US Pat. Appl. 47,588, 1979
3. Gard, G. L. *et al., Inorg. Chem.*, 1965, **4**, 594
4. Rozen, S. *et al., J. Org. Chem.*, 1980, **45**, 677

Thermally unstable at 22°C and explosive in the gas phase at pressures above 27–62 mbar [1], there is also an explosion hazard during distillation and trapping at −78°C [2]. The crude (80%) material prepared by action of 3% fluorine in nitrogen on sodium trifluoroacetate [3] has, however, been used synthetically for 3 years without mishap [4].
See other ACYL HYPOHALITES

0595. 2-Chloro-1,1-bis(fluorooxy)trifluoroethane
[72985-56-9] $C_2ClF_5O_2$

$ClCF_2CF(OF)_2$

Sekiya, A. *et al., Inorg. Chem.*, 1980, **19**, 1330
Extremely explosive, detonation of a 1 mmol sample in a vacuum line led to 'impressive damage'.
See other BIS(FLUOROOXY)PERHALOALKANES

0596. *N*-Chlorobis(trifluoromethanesulfonyl)imide
[91742-17-5] $C_2ClF_6NO_4S_2$

$ClN(O_2SCF_3)_2$

Foropoulos, J. *et al., Inorg. Chem.*, 1984, **23**, 3721
During preparation of the title compound from the imide and chlorine fluoride, initially at −111°C with slow warming during 12 h to 0°C, the scale should be limited to 40 mmol or below. Forceful explosions occurred on several occasions when this scale was exceeded, but care is necessary on any scale of working.
See other *N*-HALOGEN COMPOUNDS

0597. Bis(trifluoromethyl)chlorophosphine
[650-52-2] C_2ClF_6P

$$(F_3C)_2PCl$$

491M, 1975, 60
It ignites in air.
See other ALKYLHALOPHOSPHINES

0598. Chloroiodoacetylene (Chloroiodoethyne)
[25604-71-1] C_2ClI

$$ClC{\equiv}CI$$

Houben-Weyl, 1977, Vol. 5.2a, 604 (footnote 6)
Highly unstable and explosive.
See other HALOACETYLENE DERIVATIVES

0599. Lithium chloroacetylide (Lithium chloroethynide)
[6180-21-8] C_2ClLi

$$LiC{\equiv}CCl$$

1. Cadiot, P., *BCISC Quart. Safety Summ.*, 1965, **36**(142), 27
2. Sinden, B. B. *et al., J. Org. Chem.*, 1980, **45**, 2773
During its synthetic use in liquid ammonia, some of the salt separated as a crust
owing to evaporation of solvent, and exploded violently. Such salts are stable
in solution, but dangerous in the solid state. Evaporation of ammonia must be
prevented or made good until unreacted salt has been decomposed by addition
of ammonium chloride [1]. An improved preparation of the lithium derivative is
available [2].
See Lithium bromoacetylide; Lithium trifluoropropynide; Sodium bromoacetylide;
Sodium chloroacetylide

Methyl chloroformate
Acheson, R. M. *et al., J. Chem. Res.; Synop.*, 1986, (10), 378–379
An attempt to prepare methyl chloropropiolate from chloroethynyllithium and
methyl chloroformate led to an explosion.
See other HALOACETYLENE DERIVATIVES, METAL ACETYLIDES

0600. Cyanoformyl chloride
[4474-17-3] C_2ClNO

$$N{\equiv}CCO.Cl$$

Water
Appel, R. *et al., Angew. Chem. (Intern. Ed.)*, 1983, **22**, 785
Reaction with water is violent, forming triazinetricarboxylic acid.
See other ACYL HALIDES, CYANO COMPOUNDS

0601. Sodium chloroacetylide (Sodium chloroethynide)

[ ] C_2ClNa

$$NaC{\equiv}CCl$$

Viehe, H. G., *Chem. Ber.*, 1959, **92**, 1271

Though stable and usable in solution, the sodium salt, like the lithium and calcium salts, is dangerously explosive in the solid state.

See Lithium bromoacetylide; Lithium chloroacetylide; Sodium bromoacetylide; Lithium trifluoropropynide

See other HALOACETYLENE DERIVATIVES, METAL ACETYLIDES

0602. Dichloroacetylene (Dichloroethyne)

[7572-29-4] C_2Cl_2

$$ClC{\equiv}CCl$$

1. Wotiz, J. H. *et al., J. Org. Chem.*, 1961, **26**, 1626
2. Ott, E., *Ber.*, 1942, **75**, 1517
3. Kirk-Othmer, 1964, Vol. 5, 203–205
4. Siegel, J. *et al., J. Org. Chem.*, 1970, **35**, 3199
5. Riemschneider, R. *et al., Ann.*, 1961, **640**, 14
6. *MCA Case History No. 1989*
7. Kende, A. S. *et al., Synthesis*, 1982, 455
8. Kende. A. S. *et al., Tetrahedron Lett.*, 1982, **23**, 2369, 2373
9. Denis, J. N. *et al., J. Org. Chem.*, 1987, **52**, 3461–3462

Dichloroacetylene is rather endothermic (ΔH°_f (g) +149.4 kJ/mol, 1.57 kJ/g) and a heat-sensitive explosive gas which ignites in contact with air. However, its azeotrope with diethyl ether (55.4% dichloroacetylene) is not explosive and is stable to air [1,2]. It is formed on catalysed contact between acetylene and chlorine, or sodium hypochlorite at low temperature; or by the action of alkali upon polychloro-ethane and -ethylene derivatives, notably trichloroethylene [3]. A safe synthesis has been described [4]. Ignition of a 58 mol% solution in ether on exposure to air of high humidity, and violent explosion of a conc. solution in carbon tetrachloride shortly after exposure to air have been reported. Stirring the ethereal solution with tap water usually caused ignition and explosion [5]. Dichloroacetylene had been collected without incident on 6 previous occasions as a dilute solution. When the cooling system was modified by lowering the temperature in the water separator, liquid chloroacetylene separated there and exploded when a stopcock was turned [6]. A safer low temperature preparation of material free of trichloroethylene has been described [7,8]. A practical high-yielding synthesis of dichloroacetylene involves addition of a catalytic amount of methanol to a mixture of trichloroethylene and potassium hydride in THF [9].

See Tetrachloroethylene: Sodium hydroxide

   Trichloroethylene: Alkali, or: Epoxides

See Chlorocyanoacetylene

See other HALOACETYLENE DERIVATIVES

0603. Chlorodifluoroacetyl hypochlorite
[68674-44-2] $C_2Cl_2F_2O_2$

ClCF$_2$CO.OCl

Unstable above 22°C and explosive in gas phase at pressures above 27–63 mbar.
See entry ACYL HYPOHALITES (reference 1)

0604. 1,2-Dichlorotetrafluoroethane
[76-14-2] $C_2Cl_2F_4$

ClCF$_2$CF$_2$Cl

Aluminium
  See Aluminium: Halocarbons
  See other HALOALKANES

0605. Oxalyl dichloride (Oxalyl chloride)
[79-37-8] $C_2Cl_2O_2$

ClCO.CO.Cl

Alumina
  Kropp. P. J. *et al., J. Amer. Chem. Soc.*, 1993, **115**(8), 3071
  Vigorous gas evolution (probably CO — Ed.) was produced when oxalyl chloride
  was adsorbed on alumina.

Dimethyl sulfoxide
  See Dimethyl sulfoxide: Acyl halides

Potassium
  See Potassium: Oxalyl dihalides
  See other ACYL HALIDES

0606. 1,1,2-Trichlorotrifluoroethane
[76-13-1] $C_2Cl_3F_3$

Cl$_2$CFCF$_2$Cl

Metals
  See Aluminium: Halocarbons
       Barium: Halocarbons
       Lithium: Halocabons
       Samarium: 1,1,2-Trichlorotrifluoroethane
       Titanium: Halocarbons
  See other HALOALKANES

230

0607. Trichloroacetonitrile
[545-06-2] C_2Cl_3N

$$Cl_3CC \equiv N$$

Ammonium chloride, Sodium azide
  See 5-Trichloromethyltetrazole
  See other CYANO COMPOUNDS

0608. Sodium trichloroacetate
[650-51-1] $C_2Cl_3NaO_2$

$$Cl_3CCO.ONa$$

HCS 1980, 864

1. Doyle, W. H., *Loss Prev.*, 1969, **3**, 15
Energy of decomposition (in range 150–250 °C) was measured as 0.23 kJ/g by DSC, and T_{ait24} was determined as 119°C by adiabatic Dewar tests, with an apparent energy of activation of 140 kJ/mol.
  Bags of the salt ignited in storage (cause unknown) [1].
See entry THERMOCHEMISTRY AND EXOTHERMIC DECOMPOSITION (reference 2)

0609. Tetrachloroethylene
[127-18-4] C_2Cl_4

$$Cl_2C = CCl_2$$

(*MCA SD-24*, 1971); *HCS 1980*, 887; *RSC Lab. Hazard Data Sheet No. 6*, 1982

Aluminium
  See Aluminium: Halocarbons

Aluminium, Zinc oxide
  Katz, S. *et al., Chem. Abs.*, 1981, **94**, 33106
  The mixture is burned militarily to produce dense smoke.

Dinitrogen tetraoxide
  See Dinitrogen tetraoxide: Halocarbons

Metals
  See Barium: Halocarbons
       Lithium: Halocarbons

Sodium hydroxide
  Mitchell, P. R., private comm., 1973
  The presence of 0.5% of trichloroethylene as impurity in tetrachloroethylene during unheated drying over solid sodium hydroxide caused generation of dichloroacetylene. After subsequent fractional distillation, the volatile fore-run exploded.

See Dichloroacetylene; *also* Trichloroethylene: Alkali
See other HALOALKENES

Zinc
J. A. Young, private communication, 1997
Distillation of tetrachloroethylene (formerly a dry-cleaning solvent) in new galvanised steel equipment produces traces of dichloroacetylene. This is toxic and may cause ill-health in those exposed. Should the acetylene chance to be concentrated (as in the sub-entry above), it is also very explosive. The galvanised metal becomes passivated in a few days and the effect was not found with other steels.
See Dichloroacetylene

0610. 3-Chloro-3-trichloromethyldiazirine
   [35295-56-5] $C_2Cl_4N_2$

$$
\begin{array}{c}
\text{N}=\!=\!=\text{N} \\
\diagdown\diagup \\
\diagup\diagdown \\
\text{Cl}\quad\text{CCl}_3
\end{array}
$$

Liu, M. T. H., *Chem. Eng. News*, 1974, **52**(36), 3
An ampoule of the compound exploded when scored with a file, indicating high shock-sensitivity.
See other DIAZIRINES

0611. Hexachloroethane
   [67-72-1] C_2Cl_6

$$Cl_3CCCl_3$$

HCS 1980, 530

Metals
   See Aluminium: Halocarbons
         Zinc: Halocarbons
   See other HALOALKANES

0612. Chromyl isothiocyanate
   [ ] $C_2CrN_2O_2S_2$

$$O_2Cr(N=C=S)_2$$

Forbes, G. S. *et al.*, *J. Amer. Chem. Soc.*, 1943, **65**, 2273
The salt, prepared in carbon tetrachloride solution, exploded feebly several times as the reaction proceeded. Oxidation of thiocyanate by chromium(VI) is postulated.
See related METAL CYANIDES

0613. Chromyl isocyanate
[ ] $C_2CrN_2O_4$

$$O_2Cr(N{=}C{=}O)_2$$

Forbes, G. S. *et al., J. Amer. Chem. Soc.*, 1943, **65**, 2273
The product is stable during unheated vacuum evaporation of its solutions in carbon
tetrachloride. Evaporation with heat at ambient pressure led to a weak explosion
of the salt.
See related METAL CYANIDES

0614. Caesium acetylide
[22750-56-7] C_2Cs_2

$$CsC{\equiv}CCs$$

Mineral acids
  Mellor, 1946, Vol. 5, 848
  It ignites in contact with gaseous hydrogen chloride or its conc. aqueous solution,
  and explodes with nitric acid.

Iron(III) oxide
  Mellor, 1946, Vol. 5, 849
  Incandescence on warming.

Halogens
  Mellor, 1946, Vol. 5, 848
  It burns in all four halogens at or near ambient temperature.

Lead dioxide
  See Lead(IV) oxide: Metal acetylides

Non-metals
  Mellor, 1946, Vol. 5, 848
  Mixtures with boron or silicon react vigorously on heating.
  See other METAL ACETYLIDES

0615. Copper(II) acetylide
[12540-13-5] $(C_2Cu)_n$

$$(\text{-}CuC{\equiv}C\text{-})_n$$

1. Urbanski, 1967, Vol. 3, 228
2. Bond, J., *Hazards from Pressure, IChE Symp. Ser. No. 102*, 37–44, Oxford,
   Pergamon, 1987
3. Anon., *Jahresber.*, 1987, 64; 1988, 140 (same incident)
Copper(II) acetylide (black or brown) is much more sensitive to impact, friction
or heat than copper(I) acetylide (red-brown) which is used in electric fuses or

detonators [1]. An explosion during maintenance work on a copper condenser in a methanol plant was attributed to copper acetylide deposits in the heat exchanger [2]. During internal inspection of an air-purged 15 year old 330 m^3 liquefied gas cylinder, it was noticed that there was a 2–3 cm layer of rust-like deposit at the bottom. As the inspection proceded and the layer was disturbed, small flames were seen, and then a detonation occurred. The deposit was found to contain 20% of copper but less than 1% of sulfur, so was thought to be a mixture of rust and copper acetylide, friction-sensitive when dry [3].

See Silver acetylide (reference 4)

See other METAL ACETYLIDES

0616. Copper(II) cyanide
[544-92-3] C_2CuN_2

$$Cu(C≡N)_2$$

Magnesium
    See Magnesium: Metal cyanides
    See other METAL CYANIDES (AND CYANO COMPLEXES)

0617. Copper(II) fulminate
[22620-90-2] $C_2CuN_2O_2$

$$Cu(CN→O)_2$$

See entry METAL FULMINATES

0618. Copper(II) thiocyanate
[15192-76-4] $C_2CuN_2S_2$

$$Cu(SCN)_2$$

Tudela, D., *J. Chem. Educ.*, 1993, **70**(2), 174

On addition to boiling water, black cupric thiocyanate converts to the white cuprous salt with violent gas evolution. It is suggested that the gas is hydrogen cyanide, though carbon dioxide seems probable.

See related METAL CYANATES

0619. Dicopper(I) acetylide
[1117-94-8] $(C_2Cu_2)_n$

$$(-CuC≡CCu-)_n$$

1. Mellor, 1946, Vol. 5, 851, 852
2. Rutledge, 1968, 84–85
3. Morita, S., *J. Soc. High Press. Gas Ind.*, 1955, **19**, 167–176
4. Bond, J., *Loss Prev. Bull.*, 1995, **121**, 17

Readily formed from copper or its compounds and acetylene, it detonates on impact or heating above 100°C. If warmed in air or oxygen, it explodes on subsequent contact with acetylene [1]. Explosivity of the precipitate increases with acidity of the salt solutions, while the stability increases in the presence of reducing agents (formaldehyde, hydrazine, or hydroxylamine). The form with a metallic lustre was the most explosive acetylide made. Catalysts with the acetylide supported on a porous solid are fairly stable [2]. The ignition temperature of the pure red acetylide is 260–270°C. On exposure to air or oxygen, it is converted to black copper(II) acetylide, which ignites and explodes at 100°C [3]. It may be formed inadvertently in copper chemical plant; an account of such an incident and a procedure for making such equipment safe (treatment with warm 5% hydrogen peroxide, followed by sodium diethyldithiocarbamate of the same concentration) is reported [4].

Halogens
Mellor, 1946, Vol. 5, 852
Ignition occurs on contact with chlorine, bromine vapour or finely divided iodine.

Silver nitrate
Mellor, 1946, Vol. 5, 853
Contact with silver nitrate solution transforms copper(I) acetylide into a sensitive and explosive mixture of silver acetylide and silver.

Talc
Chambionnat, A., *Chem. Abs.*, 1951, **45**, 7791e
The effect of adding talc as an inert desensitiser upon the sensitivity of the acetylide has been studied.
See other METAL ACETYLIDES

0620. Copper(I) cyanide
[544-92-3] $C_2Cu_2N_2$
$$N\equiv CCuCuC\equiv N$$

3-Methoxy-2-nitrobenzoyl chloride
See 3-Methoxy-2-nitrobenzoyl chloride
See other METAL CYANIDES

0621. Dicopper(I) ketenide
[[41084-90-6] (ion)] C_2Cu_2O
$$Cu_2C{=}C{=}O$$

Blues, E. T. et al., J. Chem. Soc., Chem. Comm., 1973, 921
The dry compound is mildly explosive.
See other HEAVY METAL DERIVATIVES

0622. Copper(I) oxalate
[53421-36-6] $C_2Cu_2O_4$

CuOCO.CO.OCu

Sidgwick, 1950, 126
It explodes feebly on heating.
See other METAL OXALATES

0623. Difluoroethyne (Difluoroacetylene)
[689-99-6] C_2F_2

FC≡CF

1. Feast, W. J., Personal communication, 1993
2. Bürger, H. *et al., J. Chem. Soc., Chem. Comm.*, 1991, 456
Like other substituted acetylenes, this is violently unstable. Attempts to prepare
poly(difluoroacetylene) thence usually end in explosion [1]. Difluoroacetylene
decomposes slowly even at liquid nitrogen temperatures [2].
See other HALOACETYLENE DERIVATIVES

0624. 2-(Difluoroamino)-2,2-dinitroacetonitrile
[180570-35-8] $C_2F_2N_4O_4$

NCC(NO₂)₂NF₂

See 1,1-DINITRODIFLUORAMINES, *See other* N-HALOGEN COMPOUNDS, POLYNITRO-
ALKYL COMPOUNDS

0625. Bis(fluoroformyl) peroxide
[692-74-0] $C_2F_2O_4$

FCO.OOCO.F

1. Talbot, R. L., *J. Org. Chem.*, 1968, **33**, 2095
2. Czerepinski, R. *et al., Inorg. Chem.*, 1968, **7**, 109
Several explosions occurred during the preparation, which involves charging carbon
monoxide into a mixture of fluorine and oxygen [1]. It has been known to decompose or explode at elevated temperatures, and all samples should be maintained
below 30°C and well shielded [2].
See other DIACYL PEROXIDES

0626. Trifluoroacetyl nitrite
[667-29-8] $C_2F_3NO_3$

F₃CCO.ON:O

1. Taylor, C. W. *et al., J. Org. Chem.*, 1962, **27**, 1064
2. Gibbs, R. *et al., J. Chem. Soc., Perkin Trans. 2*, 1972, 1340
3. Banks, R. E. *et al., J. Chem. Soc. (C)*, 1956, 1350–1353
4. Banks, R. E. *et al., J. Chem. Soc., Perkin Trans. 1*, 1974, 2535
Though much more stable than acetyl nitrite even at 100°C, the vapour of trifluoroacetyl nitrite will explode at 160–200°C unless diluted with inert gas to below

about 50 vol% concentration. Higher perfluorohomologues are more stable [1]. A detailed examination of the explosion parameters has been made [2]. This and higher polyfluoroacyl nitrites tend to explode above 140°C at ambient pressure, and handling of large quantities should be avoided [3], especially during pyrolysis [4].
See other ACYL NITRITES

0627. Trifluoroacetyl azide
[23292-52-6] $C_2F_3N_3O$

$F_3CCO.N_3$

Sprenger, G. H. *et al., Inorg. Chem.*, 1973, **12**, 2891
It explodes on exposure to mechanical or thermal shock. Care is necessary during preparation to eliminate hydrogen chloride from the precursory acid chloride, to prevent formation of hydrogen azide.
See other ACYL AZIDES

†0628. Tetrafluoroethylene
[116-14-3] C_2F_4

$F_2C=CF_2$

1. Graham, D. P., *J. Org. Chem.*, 1966, **31**, 956
2. Muller, R. *et al., Paste u. Kaut.*, 1967, **14**, 903
3. Kirk Othmer, 1994, Vol. 11, 624; Ullmann, 1988, **A11**, 396

A terpene inhibitor is usually added to the monomer to prevent spontaneous polymerisation, and in its absence, the monomer will spontaneously explode at pressures above 2.7 bar. The inhibited monomer will explode if ignited [1]. Explosion under thermal initiation is now held to be a disproportionation, that to tetrafluoromethane and carbon gives 3.2 kJ/g, the same energy as black powder [3]. Liquid tetrafluoroethylene, being collected in a liquid nitrogen-cooled trap open to air, formed a peroxidic polymer which exploded [2].
See other POLYMERISATION INCIDENTS

Air, Hexafluoropropene
Dixon, G. D., private comm., 1968
A mixture of the two monomers and air sealed in an ampoule formed a gummy peroxide during several weeks. The residue left after opening the ampoule exploded violently on warming.
See other POLYPEROXIDES

Chloroperoxytrifluoromethane
See Chloroperoxytrifluoromethane: Tetrafluoroethylene

Difluoromethylene dihypofluorite
See Difluoromethylene dihypofluorite: Haloalkenes

Dioxygen difluoride
See Dioxygen difluoride: Various materials

Iodine pentafluoride, Limonene
MCA Case History No. 1520
Accidental contamination of a tetrafluoroethylene gas supply system with iodine pentafluoride caused a violent explosion in the cylinders. Exothermic reaction of the limonene inhibitor with the contaminant present in the gas cylinders may have depleted the inhibitor and initiated explosive polymerisation.
See other POLYMERISATION INCIDENTS

Metal alkoxides
See Methyl trifluorovinyl ether

Oxygen
See Oxygen (Gas): Tetrafluoroethylene

Sulfur trioxide
See Sulfur trioxide: Tetrafluoroethylene

Triboron pentafluoride
See Triboron pentafluoride: Tetrafluoroethylene
See other HALOALKENES

0629. Poly(tetrafluoroethylene)
[9002-84-0] $(C_2F_4)_n$
$(-CF_2F_2C-)_n$

Boron, or Magnesium, or Titanium
Shidlovskii, A. A. *et al., Chem. Abs.*, 1978, **88**, 193944
The combustion of compressed mixtures of the polymer with the finely powdered elements under nitrogen or argon has been studied, reactivity decreasing in the order B, Mg, Ti.
Aluminium: Polytetrafluoroethylene

Fluorine
See Fluorine: Polymeric materials

Metal hydrides
Andreev, B. M. *et al., Chem. Abs.*, 1987, **106**, 162978
In a study of thermal stability and hydrogen sorption characteristics of a series of sorbent tablets composed of hydride-forming metals dispersed in polymers under a 50% hydrogen in argon atmosphere, it was found that tablets of 80% palladium in PTFE, and 80% of 1:5 atom lanthanum−nickel alloy in PTFE could not be used above 247°C because of explosive decomposition of the PTFE.

Oxygen
See FLUOROCARBONS
See related HALOALKANES

0630. (Difluoroamino)difluoroacetonitrile
[5131-88-4]   C₂F₄N₂

F₂NCF₂CN

Hydrazine
  John, E. O. *et al., Inorg. Chem.*, 1988, **27**, 3100–3104
  The major product of reaction of the nitrile with anhydrous hydrazine in THF
  at −196°, followed by slow warming to 25°C for 2 h, is 2-difluoroamino-2,2-
  difluoro-1-iminoethylhydrazine, F₂NCF₂C(:NH)NHNH₂, which is stable only in
  solution. Evaporation of solvent led on each occasion to mild explosions, caused
  by decomposition of the product with considerable gas evolution, for which a
  mechanism is suggested.
  See other GAS EVOLUTION INCIDENTS
  See other CYANO COMPOUNDS, DIFLUOROAMINO COMPOUNDS

0631. 3-Fluoro-3-(trifluoromethyl)-3*H*-diazirine
[11713-32-0]   C₂F₄N₂

Dailey, W. P., *Tetrahedron Lett.*, 1987, **28**, 5801–5804
As it is potentially explosive in neat liquid or solid forms, it is best handled in
solution.
See other DIAZIRINES, *N*-HALOGEN COMPOUNDS

0632. Tetrafluorooxirane (Tetrafluoroethylene oxide)
[694-17-7]   C₂F₄O

F₂C — CF₂
    \ /
     O

Preparative hazard
  Eleuterio, H. S., *J. Macromol. Sci., Chem.*, 1972, **6**, 1027
  Formation from tetrafluoroethylene and oxygen may be explosive. It rearranges to
  trifluoroacetyl fluoride above its b.p., −63.5°C.
  See Oxygen: Tetrafluoroethylene
  See other 1,2-EPOXIDES

0633. Trifluoroacetyl hypofluorite
[359-46-6]   C₂F₄O₂

F₃CCO.OF

Alone, or Potassium iodide, Water
  Cady, G. H. *et al., J. Amer. Chem. Soc.*, 1953, **75**, 2501–2502

The gas explodes on sparking and often during preparation or distillation. Unless much diluted with nitrogen, it explodes on contact with aqueous potassium iodide.

See other ACYL HYPOHALITES

0634. Xenon(II) fluoride trifluoroacetate
[25710-89-8] $C_2F_4O_2Xe$

$$F_3CCO.OXeF$$

Jha. N. K., *RIC Rev.*, 1971, **4**, 157
It explodes on thermal or mechanical shock.
See other XENON COMPOUNDS

0635. Tetrafluorooxathietane-2,2-dioxide (Tetrafluoroethane sultone)
[697-18-7] $C_2F_4O_3S$

$$F_2C\underset{CF_2}{\overset{O}{\diagup\diagdown}}SO_2$$

Preparative hazard
See Sulfur trioxide: Tetrafluoroethylene

0636. Pentafluoroethyllithium (Perfluoroethyllithium)
[91935-83-0] C_2F_5Li

$$CF_3CF_2Li$$

Roddick, D. M., *Chem. Eng. News*, 1997, **75**(40), 6
This reagent, prepared in situ from butyllithium and chloropentafluoroethane as an ether solution, has several times decomposed violently on addition of a few drops of a chlorophosphine. This is attributed to running the preparation too cool ($< -90°C$), so that it had not proceeded. However:
See other METAL DERIVATIVES OF ORGANOFLUORINE COMPOUNDS

0637. Perfluoro-*N*-cyanodiaminomethane
[16408-94-9] $C_2F_5N_3$

$$F_2NCF_2N(F)C\equiv N$$

See Fluorine: Sodium dicyanamide
See other CYANO COMPOUNDS, DIFLUOROAMINO COMPOUNDS

0638. Bis(trifluoromethyl) nitroxide
[2154-71-4] C_2F_6NO

$$(F_3C)_2NO^•$$

Benzoyl azide
Banks, R. E. *et al.*, *J. Chem. Soc., Perkin Trans. 1*, 1981, 455–456
Interaction on warming from $-196°C$ in absence of solvent led to a violent explosion.

240

Platinum hexafluoride
See Platinum hexafluoride: Bis(trifluoromethyl) nitroxide
See other N–O COMPOUNDS

0639. Bis(trifluoromethyl)phosphorus(III) azide
[1479-48-7]   C$_2$F$_6$N$_3$P

$$(F_3C)_2PN_3$$

1. Allcock, H. R., *Chem. Eng. News*, 1968, **46**(18), 70
2. Tesi, G. *et al., Proc. Chem. Soc.*, 1960, 219
Explosive, but stable if stored cold [1]. Previously it was found to be unpredictably
unstable, violent explosions having occurred even at −196°C [2].
See other NON-METAL AZIDES

†**0640. *N*-(Trifluoromethylsulfinyl)trifluoromethylimidosulfinyl azide**
[81341-46-0]   C$_2$F$_6$N$_4$O$_2$S$_2$

$$F_3CS(O)N=S(O)(N_3)CF_3$$

Bechtold, T. *et al., J. Fluorine Chem.*, 1981–2, **19**, 385
An explosive liquid.
See other ACYL AZIDES, N–S COMPOUNDS

0641. 1,1-Bis(fluorooxy)tetrafluoroethane
[16329-92-3]   C$_2$F$_6$O$_2$

$$F_3CCF(OF)_2$$

1. Thompson, P. G. *et al., J. Amer. Chem. Soc.*, 1967, **89**, 2203
2. Sekiya, A. *et al., Inorg. Chem.*, 1980, **19**, 1329–1330
Though reported to be stable up to 200°C in metal [1], samples in glass at 22°C
exploded during vaporisation at ambient pressure, or when a sealed tube was
broken. Higher homologues are less stable [2].
See other BIS(FLUOROOXY)PERHALOALKANES, GLASS INCIDENTS

0642. Bis(trifluoromethyl)trioxide
[1718-18-9]   C$_2$F$_6$O$_3$

$$CF_3OOOCF_3$$

Gobbato, K. I. *et al., Angew. Chem. (Int.).* 1995, **34** 20, 2244
This unstable gas (b.p. −16°C) has never been reported as exploding, possibly
because experimenters have always exercised the greatest caution. It is recommended that this continue. As well as being thermodynamically unstable, the
immediate decomposition products being trifluoromethyl peroxide and oxygen,
it should also be a strong oxidant and therefore a greater danger in presence of
combustible organics.
See ORGANIC PEROXIDES

0643. Bis(trifluoromethyl) sulfide
[371-78-8] C_2F_6S

$$(F_3C)_2S$$

Chlorine fluorides
Sprenger, G. H. *et al., J. Fluorine Chem.*, 1976, **7**, 335
In the preparation of bis(trifluoromethyl)sulfur difluoride, explosions have occurred
when chlorine mono- or tri-fluorides were treated with 2–3 mmol of the sulfide in
absence of solvent. Because chlorine trifluoride is known to deprotonate solvents
with formation of explosive carbene species, only fully halogenated solvents are
suitable as diluents.
See other ALKYLNON-METALS

0644. Bis(trifluoromethyl) disulfide
[372-64-5] $C_2F_6S_2$

$$F_3CSSCF_3$$

Chlorine fluorides
Sprenger, G. H. *et al., J. Fluorine Chem.*, 1976, **57**, 335
In the preparation of bis(trifluoromethyl)sulfur difluoride, the presence of fully
halogenated solvents is essential to prevent explosion during treatment of the
disulfide with chlorine mono- or tri-fluoride.
See related ALKYLNON-METALS

0645. Fluorobis(trifluoromethyl)phosphine oxide
[34005-83-9] C_2F_7OP

$$(F_3C)_2P(O)F$$

Preparative hazard
See Fluorobis(trifluoromethyl)phosphine, next below
See related ALKYLHALOPHOSPHINES

0646. Fluorobis(trifluoromethyl)phosphine
[1426-40-0] C_2F_7P

$$(F_3C)_2PF$$

Mahler, W., *Inorg. Chem.*, 1979, **18**, 352
Admission of air to effect oxidation to the phosphine oxide must be slow to prevent
ignition.
See other ALKYLHALOPHOSPHINES

0647. Perfluoro-1-aminomethylguanidine
[16408-93-8] $C_2F_8N_4$

$$F_2NCF_2NFC(N:F)NF_2$$

See Fluorine: Cyanoguanidine
See other N,N,N'-TRIFLUOROAMIDINES

0648. Bis(trifluoromethyl)sulfur difluoride
[30341-38-9] C_2F_8S

$$(F_3C)_2SF_2$$

Preparative hazard
See Bis(trifluoromethyl) sulfide: Chlorine fluorides
    Bis(trifluoromethyl) disulfide: Chlorine fluorides
See related ALKYLNON-METAL HALIDES

0649. C,N-Perfluoro-N-aminomethyltriaminomethane
[16408-92-7] $C_2F_{10}N_4$

$$F_2NCF_2NFCF(NF_2)_2$$

See Fluorine: Cyanoguanidine
See other DIFLUOROAMINO COMPOUNDS

0650. Monosilver acetylide
[13092-75-6] C_2HAg

$$AgC{\equiv}CH$$

Anon., Chem. Trade J., 1966, **158**, 153
A poorly stoppered dropping bottle of silver nitrate solution absorbed sufficient acetylene from the atmosphere in an acetylene plant laboratory to block the dropping tube. A violent detonation occurred on moving the dropping tube.
See Silver nitrate: Acetylene
See other METAL ACETYLIDES

0651. Bromoacetylene
[593-61-3] C_2HBr

$$BrC{\equiv}CH$$

1. Tanaka, R. et al., J. Org. Chem., 1971, **36**, 3856
2. Hucknall, D. J. et al., Chem. & Ind., 1972, 116
It is dangerous and may burn or explode in contact with air, even when solid at −196°C [1]. Procedures for safe generation, transfer and storage, all under nitrogen, are described [2].
See other HALOACETYLENE DERIVATIVES

0652. Chloroacetylene
[593-63-5] C_2HCl

$$ClC{\equiv}CH$$

Tanaka, R. et al., J. Org. Chem., 1971, **36**, 3856

It is notably endothermic (ΔH°_f (g) +213 kJ/mol, 3.52 kJ/g). Chloroacetylene will burn and may explode in contact with air, and its greater volatility makes it more dangerous than bromoacetylene. Procedures for safe generation, handling and storage, all under nitrogen, are described.

See Trichloroethylene: Alkali
See other ENDOTHERMIC COMPOUNDS, HALOACETYLENE DERIVATIVES

0653. Difluoroacetyl hypochlorite
[71359-63-2] $C_2HClF_2O_2$

$$F_2CHCO.OCl$$

See entry ACYL HYPOHALITES (reference 1)

0654. Chloratomercurio(formyl)methylenemercury(II)
[ ] $C_2HClHg_2O_4$

$$O_2ClOHgC(CO.H){:}Hg$$

Whitmore, 1921, 154
An extremely sensitive crystalline solid, which explodes on shaking with its crystallisation liquor, or upon gentle mixing when dry with copper oxide (for analysis).
See other MERCURY COMPOUNDS

0655. 5-Chloro-1,2,3-thiadiazole
[4113-57-9] C_2HClN_2S

1. Yoshida, T., private comm., 1982
2. Ouchi, H. *et al., Kogyo Kayaku*, 1984, **45**, 73 (*Chem. Abs.*, 1985, **102**, 187475)
Two recent industrial-scale accidents which involved explosive decomposition of this intermediate caused great damage. The compound has considerable explosive potential, for though it lacks oxygen in its structure, it is of rather low oxygen balance [1]. A comprehensive examination using several hazard evaluation techniques revealed it as an extremely dangerous compound. DSC and the drop-hammer test on the solid do not indicate hazard potential, but ARC at 126°C under adiabatic conditions give a corrected time-to-explosion of 40 min. The impact-sensitivity was determined as 2 kg cm [2].
See entry OXYGEN BALANCE *See other* N–S COMPOUNDS

†0656. **Trichloroethylene**
[79-01-6] C_2HCl_3

$$Cl_2C=CHCl$$

(*MCA SD-14*, 1956); *NSC 389*, 1979; *HCS 1980*, 919; *RSC Lab. Hazard Data Sheet No. 13*, 1983

1. Author's comments, 1988
2. Santon, R. C. *et al., Proc. 6th Intl. Sympos. Loss Prev. Safety Prom. Proc. Ind.*, 1989, Oslo, Norweg. Soc C. Eng.

Though usually considered to be non-flammable in use, it has unusually wide flammability limits, and with an AIT of 410°C it can be ignited by a flame or other intense source. The most recent Data Sheet is devoted to all aspects of laboratory use [1]. The limits of flammability have been determined more accurately as 8.5 vol%/25°C to 14.5%/35°, widening to 6.0–47% at 150°C. Intermediate values are also given [2].
See FLASH POINTS (reference 19)

Alkali
1. Fabian, F., private comm., 1960
2. *ABCM Quart. Safety Summ.*, 1956, **27**, 17
3. Anon., *Jahresber.*, 1978, 70

An emulsion, formed during extraction of a strongly alkaline liquor with trichloroethylene, decomposed with evolution of the spontaneously flammable gas, dichloroacetylene [1]. This reaction could also occur if alkaline metal-stripping preparations were used in conjunction with trichloroethylene degreasing preparations, some of which also contain amines as inhibitors, which could also cause the same reaction [2]. Apparently accidental contact of the solvent with potassium hydroxide solution led to generation of flames in the charging port of a stirred reactor [3].
See Tetrachloroethylene: Sodium hydroxide

Aluminium
See Aluminium: Halocarbons

Epoxides
Dobinson, B. *et al., Chem. & Ind.*, 1972, 214

1-Chloro-2,3-epoxypropane, the mono- and di-2,3-epoxypropyl ethers of 1,4-butanediol, and 2,2-bis[4(2′,3′-epoxypropoxy)phenyl]propane can, in presence of catalytic quantities of halide ions, cause dehydrochlorination of trichloroethylene to dichloroacetylene, which causes minor explosions when the mixture is boiled under reflux. A mechanism is discussed.
See Alkali, above

Metals
See Aluminium: Halocarbons
    Barium: Halocarbons
    Beryllium: Halocarbons

245

Lithium: Halocarbons
Magnesium: Halocarbons
Titanium: Halocarbons

Oxidants
See Dinitrogen tetraoxide: Halocarbons
Oxygen (Gas): Halocarbons
Oxygen (Liquid): Halocarbons
Perchloric acid: Trichloroethylene

Water
Brade, C., *Chem. Tech.* (Berlin), 1952, **4**, 506–507
A violent explosion occurred during routine recovery of wet trichloroethylene by distillation at ambient pressure. This was attributed to hydrolytic formation of hydrochloric acid and concentration of the latter, leading to corrosion and blocking of the vent line which caused pressure development and a consequential increase in distillation temperature of the vessel contents. The corrosion products may have catalysed the exothermic decomposition reaction which led to explosion. The vessel contained much carbonised residue.
See other CORROSION INCIDENTS *See other* HALOALKENES

0657. 5-Trichloromethyltetrazole
[2925-21-5] $C_2HCl_3N_4$

1. Beck, W. *et al., Chem. Eng. News*, 1984, **62**(10), 39
2. Finnegan, W. G. *et al., J. Amer. Chem. Soc.*, 1958, **80**, 3908
3. Howe, R. K. *et al., Chem. Eng. News*, 1983, **61**(3), 4
4. Burke, L. A., *Chem. Eng. News*, 1983, **61**(17), 2

The later publication [1] reveals that the title compound is in fact a relatively stable compound. The previously attempted preparation of the then unknown compound from trichloroacetonitrile, sodium azide and ammonium chloride (0.14:0.42:0.2 mol) by an analogous established method [2], but at lower initial temperature because of the exothermic reaction, gave, after vacuum evaporation of solvent, an oily product. When sampled with a pipette, this evolved gas and then exploded violently. It was thought that an azidomethyltetrazole may have been formed by displacement of chloro-substituent(s) by the excess azide employed [3]. An alternative hypothesis which involved isomerisation of the title compound to the open chain azidoazomethine [4] was discounted, because no trace of this could be detected [1].
See other TETRAZOLES

0658. Trichloroacetic acid
    [76-03-9] $C_2HCl_3O_2$

$$Cl_3CCO.OH$$

Copper, Dimethyl sulfoxide
See Dimethyl sulfoxide: Copper, etc.

Methanol, Water
1. Mabbott, D. J., Personal communication, 1997
2. Ullmann, 1986, Vol. A6, 545.
3. Merck Index, 1989, 1515
A bottle labelled as 12% acid in aqueous methanol, and probably two years forgotten, exploded in storage, breaking adjacent bottles[1]. Trichloracetic acid is known to be more unstable, with respect to carbon dioxide and chloroform, in aqueous solution than pure. The reaction usually requires either heat or base catalysis [2]. Storage of trichloroacetic acid at less than 30% concentration is not advised [3]. Hydrolysis of the trichlorogroup is also conceivable, which would yield intermediate oxalyl monochloride, which habitually breaks down to give carbon monoxide, dioxide and hydrogen chloride.
See other ORGANIC ACIDS

0659. Trichloroperoxyacetic acid
    [7796-16-9] $C_2HCl_3O_3$

$$Cl_3CCO.OOH$$

Swern, D., *Chem. Rev.*, 1945, **45**, 10
Very unstable, forming mainly the toxic gaseous products phosgene, chlorine, carbon monoxide and hydrogen chloride.
See other GAS EVOLUTION INCIDENTS, PEROXYACIDS

0660. Pentachloroethane
    [76-01-7] C_2HCl_5

$$Cl_3CCHCl_2$$

Potassium
See Potassium: Halocarbons
See other HALOALKANES

0661. Monocaesium acetylide
    [30180-52-0] C_2HCs

$$CsC \equiv CH$$

Mellor, 1946, Vol. 5, 849–850
It reacts similarly to the dicaesium compound.
See Caesium acetylide
See other METAL ACETYLIDES

0662. Fluoroacetylene
[2713-09-9]   C₂HF

$$FC \equiv CH$$

Alone, or Bromine
Middleton, W. J. *et al., J. Amer. Chem. Soc.*, 1959, **81**, 803–804
It is notably endothermic ($\Delta H°_f$ (g) +125 kJ/mol, 2.84 kJ/g), and liquid fluoroacetylene is treacherously explosive close to its b.p., −80°C. The gas does not
ignite in air and is not explosive. Ignition occurred in contact with a solution of
bromine in carbon tetrachloride. The mercury and silver salts were stable to impact,
but the latter exploded on heating, whereas the former decomposed violently.
See other ENDOTHERMIC COMPOUNDS, HALOACETYLENE DERIVATIVES

†0663. Trifluoroethylene
[359-11-5]   C₂HF₃

$$F_2C{=}CHF$$

Hulbert, J. D. *et al., Chem. Eng. News* 1997, **75**(51), 6
Trifluoroethene at pressures above 14 bar, and possibly below, deflagrates with
tenfold pressure rises when initiated by an exploding wire. The tests were on
air-free and inhibited commercial material.
Ethylene, Tetrafluoroethylene
See other HALOALKENES

0664. 2,2,2-Trifluorodiazoethane
[371-67-5]   C₂HF₃N₂

$$F_3CCHN_2$$

Fields, R. *et al., J. Fluorine Chem.*, 1979, **13**, 154
The stabilising effect of the trifluoromethyl group is insufficient to prevent explosions, either of the compound itself, or of unstable reaction intermediates formed at
low temperatures, especially in absence of solvent. Energy released on explosion
is estimated at 50–100% TNT eqivalent.

Ozone
See Ozone: Trifluoroethylene
See other DIAZO COMPOUNDS

0665. Trifluoroacetic acid
[76-05-1]   C₂HF₃O₂

$$F_3CCO.OH$$

HCS 1980, 935

Aromatic hydrocarbons, Hydrogen peroxide
See Hydrogen peroxide: Aromatic hydrocarbons, Trifluoroacetic acid

Lithium tetrahydroaluminate
See Lithium tetrahydroaluminate: Fluoroamides
See other ORGANIC ACIDS

0666. Peroxytrifluoroacetic acid
[359-48-8] $C_2HF_3O_3$

$$F_3CCO.OOH$$

Sundberg, R. J. *et al., J. Org. Chem.*, 1968, **33**, 4098
This extremely powerful oxidant must be handled and used with great care.

4-Iodo-3,5-dimethylisoxazole
Plepys, R. A. *et al.*, US Pat. 3 896 140, 1975
Interaction to produce 3,5-dimethyl-4-bis(trifluoroacetoxy)iodoisoxazole yields a
detonable by-product, believed to be iodine pentaoxide contaminated with organic
material.
See Iodine(V) oxide: Non-metals
See other OXIDANTS, PEROXYACIDS

0667. 3,3,5-Trifluoro-1,2,4-trioxolane (Trifluoroethylene ozonide)
[86013-87-8] $C_2HF_3O_3$

Explosive.
See Ozone: Trifluoroethylene
See other OZONIDES

0668. 1,1,2,2-Tetrafluoro-1-(fluoroxy)ethane (1,1,2,2-Tetrafluoroethyl hypofluorite)
[ ] C_2HF_5O

$$HCF_2CF_2OF$$

Randolph, B. B. *et al., J. Fluorine Chem.*, 1993, **64**(1-2), 129
The hypofluorite exploded on attempted purification. A variety of related com-
pounds were also unstable, if not explosive.
See other HYPOHALITES

0669. Poly(dimercuryimmonium acetylide)
[ ] $(C_2HHg_2N)_n$

$$(Hg=N^+=HgC_2H^-)_n$$

Sorbe, 1968, 97
Highly explosive.
See other POLY(DIMERCURYIMMONIUM) COMPOUNDS

0670. Iodoacetylene
[14545-08-5]   C₂HI

$$IC{\equiv}CH$$

Sorbe, 1968, 65
It explodes above 85°C.
See other HALOACETYLENE DERIVATIVES

0671. Monopotassium acetylide
[1111-63-3]   C₂HK

$$KC{\equiv}CH$$

Chlorine
  Mellor, 1946, Vol. 5, 849
  It ignites in chlorine.

Non-metal oxides
  Mellor, 1946, Vol. 5, 849
  Interaction with sulfur dioxide at ambient temperature, or with carbon dioxide on
  warming, causes incandescence.
  See other METAL ACETYLIDES

0672. Dipotassium *aci*-nitroacetate
[19419-98-8]   C₂HK₂NO₄

$$KOCO.CH{=}N(O)OK$$

Water
  1. Lyttle, D. A., *Chem. Eng. News*, 1949, **27**, 1473
  2. Whitmore, F. C., *Org. Synth.*, 1941, Coll. Vol. 1, 401
  An inhomogeneous mixture of the dry salt with a little water exploded violently
  after 30 min [1]. This was probably owing to exothermic decarboxylation generating the *aci*-salt of nitromethane, which is explosively unstable. The decomposition of sodium nitroacetate proceeds exothermically above 80°C [2].
  See other aci-NITRO SALTS

0673. Monolithium acetylide
[1111-64-4]   C₂HLi

$$LiC{\equiv}CH$$

Mellor, 1946, Vol. 5, 849
It reacts similarly to the dilithium salt.
See Lithium acetylide
See other METAL ACETYLIDES

0674. Monolithium acetylide–ammonia
[ ] $C_2HLi.H_3N$

$$LiC{\equiv}CH.NH_3$$

Gases, or Water
Mellor, 1946, Vol. 5, 849
The complex ignites in contact with carbon dioxide, sulfur dioxide, chlorine or water.
See related METAL ACETYLIDES

0675. Diazoacetonitrile
[13138-21-1] C_2HN_3

$$N_2CHC{\equiv}N$$

1. Phillips, D. D. *et al., J. Amer. Chem. Soc.*, 1956, **78**, 5452
2. *MCA Case History No. 2169*
3. Roelants, F. *et al., Tetrahedron*, 1978, **34**, 2231
4. Author's comments

Removal of a rubber stopper from a flask of a concentrated solution in methylene chloride initiated an explosion, probably through friction on solvent-free material between the flask neck and bung. Handling only in dilute solution is recommended [1]. An explosion also occurred during operation of the glass stopcock of a dropping funnel while 20 ml of the apparently undiluted nitrile was being added to a reaction mixture [2]. All attempts to isolate the nitrile from its solutions in chlorinated solvents caused explosions [3]. By inference from the thermodynamic properties of diazomethane and acetonitrile, diazoacetonitrile should be strongly endothermic, with heat of formation exceeding 200 kJ/mol, 3 kJ/g [4].
See other DIAZO COMPOUNDS, FRICTIONAL INITIATION INCIDENTS, GLASS INCIDENTS

0676. Dinitroacetonitrile
[921-22-2] $C_2HN_3O_4$

$$(O_2N)_2CHC{\equiv}N$$

See entry POLYNITROALKYL COMPOUNDS

0677. 3-Diazo-3*H*-1,2,4-triazole
[64781-78-8] C_2HN_5

The precursory 1,2,4-triazole-3-diazonium nitrate is extremely unstable, even in ice-cold aqueous solution, and no attempt was made to isolate the title diazo compound from its solution, prepared by cautious treatment of the cold diazonium nitrate solution with dilute aqueous potassium hydroxide with cooling.

See entry DIAZOAZOLES

See other DIAZO COMPOUNDS, HIGH-NITROGEN COMPOUNDS, TRIAZOLES

0678. 4-Diazo-1,2,3-triazole
[85807-68-7] C_2HN_5

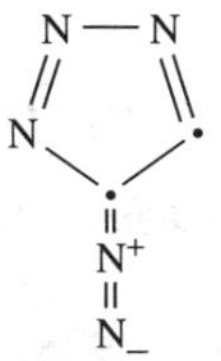

See entry DIAZOAZOLES

See other DIAZO COMPOUNDS, HIGH-NITROGEN COMPOUNDS, TRIAZOLES

0679. Diazoacetyl azide
[19932-64-0] C_2HN_5O

$$N_2CHCO.N_3$$

Neunhöffer, G. *et al., Ann.*, 1968, **713**, 97–98

The material may be distilled cautiously at 20–21°/0.27 mbar but is preferably used in solution as prepared. Either the solid (below 7°C) or the liquid explodes with great violence on impact or friction.

See other ACYL AZIDES, DIAZO COMPOUNDS

0680. 1,2,4-Triazole-3-diazonium nitrate
[59104-93-7] $C_2HN_6O_3$

$$N{=}CHN{=}NCN_2^+NO_3^-$$

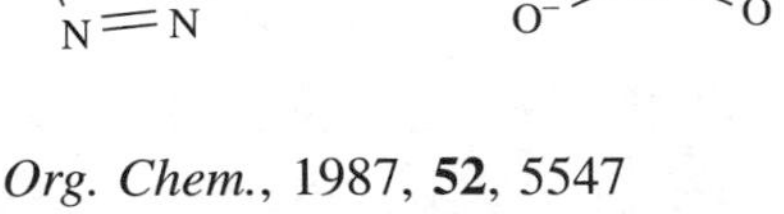

Magee, H. L. *et al., J. Org. Chem.*, 1987, **52**, 5547

The diazonium nitrate is extremely unstable and may only be prepared as an aqueous solution of limited life at below 0°C.

See other DIAZONIUM SALTS, HIGH-NITROGEN COMPOUNDS, TRIAZOLES

0681. Sodium 5-(dinitromethyl)tetrazolide
[2783-96-2]

C$_2$HN$_6$NaO$_4$

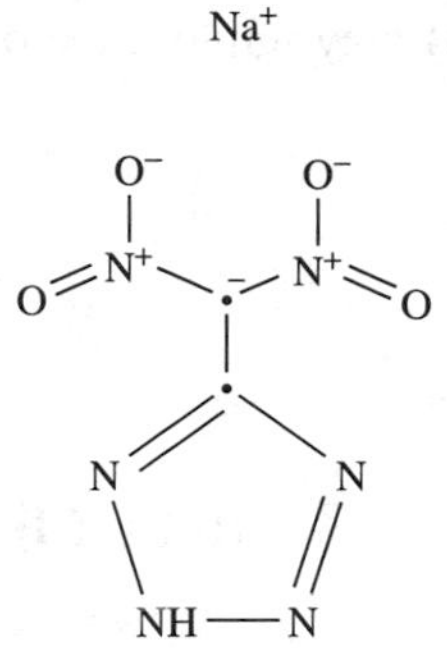

Na$^+$

Einberg, F., *J. Org. Chem.*, 1964, **29**, 2021
It explodes violently at the m.p., 160°C, and may be an *aci*-nitro salt.
See other POLYNITROALKYL COMPOUNDS, TETRAZOLES

0682. Sodium 5(5′-hydroxytetrazol-3′-ylazo)tetrazolide
[ ]

C$_2$HN$_{10}$NaO

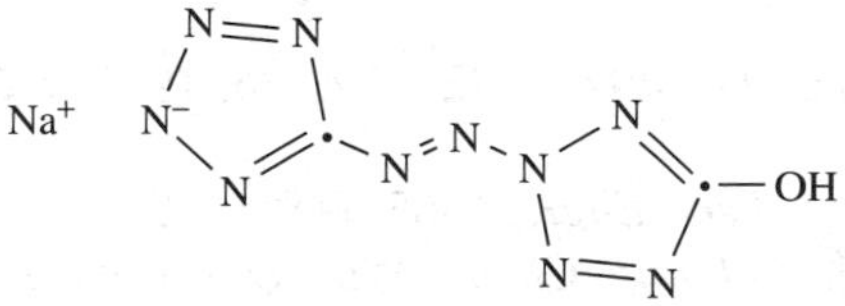

Thiele, J. *et al., Ann.*, 1893, **273**, 150
The salt explodes very violently on heating.
See other AZO COMPOUNDS, TETRAZOLES

0683. 1,1,1-Triazidodinitroethane
[56522-42-0]

C$_2$HN$_{11}$O$_4$

$$(N_3)_3CCH(NO_2)_2$$

Frankel, M. B. et al., UK Pat. Appl. 2 123 829, 1984
An energetic compound (with 63% N and of low oxygen balance).
See other POLYNITROALKYL COMPOUNDS, ORGANIC AZIDES

0684. Monosodium acetylide
[1066-26-8]

C$_2$HNa

$$NaC{\equiv}CH$$

1. Mellor, 1946, Vol. 5, 849
2. Greenlee, K. W., *Inorg. Synth.*, 1947, **2**, 81
3. Houben-Weyl, Vol. 13.1, 280

It reacts similarly to the disodium salt [1]. If heated to 150°C, it decomposes extensively, evolving gas which ignites in air owing to presence of pyrophoric carbon. The residual carbon is also highly reactive [2]. The dry powder (from solution in liquid ammonia) may ignite if exposed to air as an extended layer, e.g. on filter paper [3].

See Sodium acetylide
See other METAL ACETYLIDES

0685. Monorubidium acetylide
[20720-71-2] C_2HRb

$$RbC{\equiv}CH$$

Mellor, 1946, Vol. 5, 849
It reacts similarly to the dirubidium salt.
See Rubidium acetylide
See other METAL ACETYLIDES

†0686. Acetylene (Ethyne)
[74-86-2] C_2H_2

$$HC{\equiv}CH$$

(*MCA SD-7*, 1957); *NSC 494*, 1982; *FPA H6*, 1972; *HCS 1980*, 110, 111

1. Mayes, H. A., *Chem. Engr.*, 1965, (185), 25
2. Nedwick, J. J., *Ind. Eng. Chem., Proc. Des. Dev.*, 1962, **1**, 137; Tedeschi, R. J. *et al.*, ibid., 1968, **7**, 303
3. Miller, 1965, Vol. 1, 506
4. Kirk-Othmer, 1963, Vol. 1, 195–202
5. Rimarski, W., *Angew. Chem.*, 1929, **42**, 933
6. Foote, C. S., private comm., 1965
7. Sutherland, M. E. *et al.*, *Chem. Eng. Progr.*, 1973, **69**(4), 48–51
8. Williams, A. *et al.*, *Chem. Rev.*, 1970, **70**, 270–271
9. Menge, R. *et al.*, *Brandschutz, Explosionschutz*, 1987, **16**, 109
10. Yantovskii, S. A., *Chem. Abs.*, 1964, **61**, 13117f
11. Landesman, Ya. M. *et al.*, *Chem. Abs.*, 1974, **81**, 151352
12. Glikin, M. A. *et al.*, *Chem. Abs.*, 1975, **83**, 100289
13. Savitskaya, L. M. *et al.*, *Chem. Abs.*, 1978, **88**, 157743
14. Viallon, J., *Loss Prev. Safety Prom. Proc. Ind.*, 121–122, Frankfurt, Dechema, 1978
15. Lietze, D. *et al.*, *Chem. Ing. Tech.*, 1978, **50**, 808–809 (summary of 21 page paper)
16. Code of Practice for Acetylene Pipelines, Paris, Euro. Comm. Indust. Gases, 1978
17. Grosse-Wortmann, H. *et al.*, *Chem. Ing. Tech.*, 1981, **53**, 461–463
18. Lietze, D., *Chem. Abs.*, 1986, **105**, 48048

19. Ashmore, F. S. *et al., Fire Prev.*, 1987, (204), 37–39; *Loss Prev. Bull.*, 1988, (080), 15–18
20. Marcks, G., *Proc. 11th Int. Symp. Prev. Occup. Risks Chem. Ind.*, 623–649, Heidelberg, ISSA, 1987
21. Dorfinger, S., *ibid.*, 653–663
22. Schildberg, H. P. *et al., Chem. Ing. Tech.*, 1994, **66**(10), 1389

The extremely endothermic gas (ΔH_f° (g) +226.7 kJ/mol, 8.72 kJ/g) may decompose explosively in absence of air. Addition of up to 30% of miscible diluent did not sufficiently desensitise explosive liquid acetylene to permit of its transportation. The solid has similar properties [1]. However, safe techniques for the use of liquid acetylene in reaction systems at pressures up to 270 bar have been described [2]. Solid acetylene in admixture with liquid nitrogen at −181°C is sensitised by presence of grit (carborundum) and may readily explode on impact [3]. The hazards of preparing and using acetylene have been adequately described, including many reports of explosions in acetylene generators and distribution systems (formerly including domestic lighting installations) where faulty pressure control has allowed pressures to approach 1.4 bar, when explosion occurs in presence of moisture [3,4,5]. A glass flask which had contained acetylene was left open to atmosphere for a week, and then briefly purged with air. It still contained enough acetylene to form an explosive mixture [6]. The explosive decomposition of acetylene in absence of air readily escalates to detonation in tubular vessels. This type of explosive decomposition has been experienced in a 7 mile acetylene pipeline system [7].

In a review on combustion and oxidation of acetylene, the factors affecting spontaneous ignition at low temperatures are discussed [8]. The autoignition temperature may be as low as 85°C when mixed with 1–2% of nitric oxide (NO) [9]. The explosive decomposition of acetylene and conditions necessary for spontaneous ignition to occur have been reviewed [10,11,12]. Use of C_1–C_4 alcohols, C_3–C_4 ketones, C_1 and C_4 acids and C_3–C_4 diols or hydrocarbons to retard explosive decomposition was studied comparatively. The retardation mechanism involves thermal and chemical factors, related to heat content of diluent and to structure, respectively. Practical systems are detailed [13]. Acetylene at 170°C is somewhat stabilised by presence of acetic acid [14]. It is destabilised by potassium hydroxide [22]. The course of explosive decomposition of the gas in pipelines has been studied experimentally [15]. The EU Code of Practice divides pipeline systems into 3 'working ranges', depending on internal pressure and diameter of the pipe, and specifies suitable design and construction criteria for each range [16]. The relationship between initiation energy and decomposition has been studied [17]. The explosibility of acetylene, alone or admixed with air, hydrogen or ethylene has been reviewed, including boundary limits for deflagration/detonation transformation [18]. Accidental local heating to 185°C or above of part of the wall (as little as 6 sq. cm may be enough) of a cylinder containing acetylene may lead to the development of an extremely dangerous situation. At this temperature, exothermic and self-sustaining decomposition, or polymerisation, of (endothermic) acetylene may set in, and if not stopped by rapid and effective cooling (large volume water spray), the cylinder may explode without warning. Flame flash-back into a cylinder from

a wrongly adjusted and/or damaged welding or cutting torch can cause the same effect. Methods of identifying such unstable cylinders and of emergency control are detailed [19]. The hazards associated with installations of banks of acetylene cylinders connected to a high-pressure manifold are discussed [20], and a detailed account of an accident involving explosion of two such overheated cylinders is given [21].
See Magnesium: Barium carbonate, Water

Bleaching powder
See Oxidants, below

Cobalt   MRH 8.70/tr.
Mellor, 1942, Vol. 14, 513
Finely divided (pyrophoric) cobalt decomposes acetylene on contact, becoming incandescent.

Copper   MRH 8.70/tr.
Anon., *ABCM Quart. Safety Summ.*, 1946, **17**, 24
Rubber-covered electric cable, used as a makeshift handle in the effluent pit of an acetylene plant, formed copper acetylide with residual acetylene and the former detonated when disturbed and initiated explosion of the latter. All heavy metals must be rigorously excluded from locations where acetylene may be present.
See METAL ACETYLIDES

Copper(I) acetylide   MRH 8.70/tr.
See Dicopper(I) acetylide

Dinitrogen oxide
See ATOMIC ABSORPTION SPECTROSCOPY

Ethylene, Hydrogen
See Hydrogen (Gas): Acetylene, Ethylene

Gaseous additives
1. Konovalov, E. N. *et al., Chem. Abs.*, 1975, **82**, 111548
2. Detz, C. M., *Combust. Flame*, 1979, **34**, 187–193; US Pat. 4 161 495, 1979
Addition of 0.3–3.5 wt% of propane or butane reduces the explosion hazards of acetylene–air mixtures [1]. Out of 10 gaseous additives tested for effect on the threshold temperature for initiation of explosive decomposition of acetylene by a heated wire at 2–22 bar, nitrogen oxide, hydrogen bromide, hydrogen chloride, hydrogen iodide and vinyl bromide showed stabilising effects, and sulfur dioxide a mild destabilising effect. The patent specifies use of one or more of these radical scavenging agents at levels of 0.5–10% [2].

Halogens   MRH Chlorine or Iodine 8.70/tr.
1. Muir, G. D., private comm., 1968
2. von Schwartz, 1918, 142, 321

3. *MCA SD-7*, 1957
4. Humphreys, V., *Chem. & Ind.*, 1971, 681–682
5. Sokolova, E. I. *et al., Chem. Abs.*, 1970, **72**, 71120
6. Davenport, A. P., *School Sci. Rev.*, 1973, **55**(191), 332
7. Long, G. C., *Spectrum* (Pretoria), 1980, **18**, 30
8. Nolan, 1983, Case History 22
9. Nolan, 1983, Case History 23

Tetrabromoethane is made by passing acetylene into bromine in carbon tetrachloride at reflux. The rate of reaction falls off rapidly below reflux temperature, and if the rate of addition of acetylene is insufficient to maintain the temperature, high concentrations of unreacted acetylene build up, with the possibility of a violent delayed reaction [1]. In absence of a diluent, reaction may be explosive [2]. Mixtures of acetylene and chlorine may explode upon initiation by sunlight or other UV source, or at high temperature, sometimes very violently [2]. Interaction with fluorine is very violent [3] and with iodine possibly explosive [2]. Dilution of equimolar mixtures of chlorine and acetylene with 55 mol% of nitrogen or 70% air prevented spontaneous explosion. At higher dilutions, sparking did not initiate explosion [4]. Explosive interaction of chlorine and acetylene in the dark is initiated by presence of oxygen at between 0.1 and about 40 vol%. The reaction is inhibited by inert gases or oxygen at higher concentrations [5]. Safe techniques for demonstrating explosive combination of acetylene and chlorine have been described [6,7], the latter involving dropping a small lump of calcium carbide into acidified sodium hypochlorite solution. Tetrachloroethane is manufactured by reacting excess chlorine with acetylene at 100°C in presence of ferric chloride. On one occasion the temperature fell to 60°C (low chlorine flow?)and there was an explosion. It was suggested that monochloroacetylene had formed and decomposed, initiating an acetylene/chlorine or gas/air explosion. A number of such explosions have occurred, particularly during start-up [8]. Another explosion was attributed to damp ferric chloride [9].

Heavy metals and salts
See METAL ACETYLIDES

Liquid nitrogen
Houseman, T. H. *et al., J. Labelled Compd. Radiopharm.*, 1978, **14**, 164
Liquid nitrogen should not be used as a trap coolant with acetylene, owing to the explosive nature of liquid or solid acetylene (title reference 5 above).

Mercury(II) salts, Nitric acid, (Sulfuric acid)
See Nitric acid: Acetylene, Mercury(II) salts

Nitric acid, Perchloric acid
See Perchloric acid: Acetylene, Nitric acid

Other reactants
Yoshida, 1980, 15

MRH values calculated for 10 combinations with materials catalysing the decomposition of acetylene are given.

Oxidants
Since acetylene is endothermic (values above) and effectively a reducing agent, its reactions with oxidants are usually violent or explosive if uncontrolled.
See Dinitrogen oxide or Halogens, both above; Oxygen, below
    Calcium hypochlorite: Acetylene
    Nitric acid: Acetylene, Mercury(II) salts
    Nitrogen oxide: Dienes, Oxygen
    Ozone: Acetylene
    Trifluoromethyl hypofluorite: Hydrocarbons

Oxides of nitrogen
See Nitrogen oxide: Dienes, Oxygen

Oxygen
    1. Fowles, G. *et al., School Sci. Rev.*, 1940, **22**(85), 6; 1962, **44**(152), 161; 1963, **44**(154), 706; 1964, **45**(156), 459
    2. Kiyama, R. *et al., Rev. Phys. Chem. Japan*, 1953, **23**, 43–48
    3. Moye, A., *Chem. Ztg.*, 1922, **46**, 69
    4. Ivanov, U., *Chem. Abs.*, 1976, **84**, 57968
The explosion of acetylene–oxygen mixtures in open vessels is a very dangerous experiment (stoicheiometric mixtures detonate with great violence, completely shattering the container). Full precautions are essential for safety [1]. When a mixture of acetylene and oxygen (54:46) at 270°C/10.9 bar was compressed in 0.7 s to 56.1 bar, the resulting explosion attained a pressure of several kbar. In other tests rapid compression of acetylene or its mixtures with air caused no explosions [2]. Previously, passage of acetylene into liquid air to deliberately generate a paste of solid acetylene and liquid oxygen, 'by far the most powerful of explosives', had been proposed [3]. Acetylene had been collected for teaching purposes over water in a pneumatic trough. Later, oxygen was collected in the same way without changing the water, and the sample exploded violently when exposed to a glowing splint. Acetylene remaining dissolved in the water had apparently been displaced by the oxygen stream, the lower explosive limit for acetylene being only 2.5% in air, and less in oxygen [4].

Potassium   MRH 8.70/tr.
    Berthelot, M., *Bull. Soc. Chim. Fr. [2]*, 1866, **5**, 188
    Molten potassium ignites in acetylene, then explodes.

Potassium hydroxide
    Schildberg, H. P. *et al., Loss Prevention and Safety Promotion in the Process Industries*, (Mewis, Pasman and De Rademaker, eds.), Vol. 1, 401, Amsterdam, Elsevier, 1995

At temperatures above 100°C and pressures above 2 bar, potassium hydroxide powder can initiate self-propagating decomposition of acetylene. More detail on limits is given.
See other ALKYNES

0687. Poly(acetylene)
[25067-58-7] $(C_2H_2)_n$

$$(-CH=CH-)_n$$

Feast, W. J. *et al., J. Chem. Soc., Chem. Comm.*, 1985, 202–203
The second stage of an improved synthesis of poly(acetylene), which involves disproportionation of a soluble polymer by heating a thin film at 75°C to give 1,2-bis(trifluoromethyl)benzene and poly(acetylene), must not be done in bulk because the reaction then becomes explosive. The earlier synthesis by direct polymerisation of acetylene was considerably more dangerous
See Poly([7,8-bis(trifluoromethyl)tetracyclo[4.2.0.0^{2,8}.0^{5,7}]octane-3,4-diyl]-1,2-ethenediyl)

Iodine
See Iodine: Poly(acetylene)
See related DIENES

Perchlorate-doping
See PERCHLORATE-DOPED CONDUCTING POLYMERS
See related DIENES

0688. Silver 1,2,3-triazolide
[ ] $C_2H_2AgN_3$

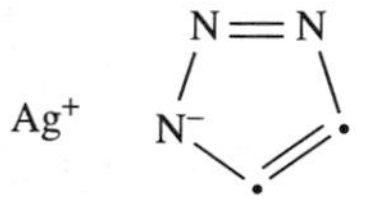

See 1,2,3-Triazole
See other N-METAL DERIVATIVES, SILVER COMPOUNDS, TRIAZOLES

0689. Silver dinitroacetamide
[[26163-27-9] (ion)] $C_2H_2AgN_3O_5$

$$AgC(NO_2)_2CO.NH_2$$

Parker, C. O. *et al., Tetrahedron*, 1962, **17**, 108
The sodium salt explodes at 160°C and the potassium and silver salts at 159 and 130°C, respectively.
See other POLYNITROALKYL COMPOUNDS, SILVER COMPOUNDS

0690. Silver 1,3-di(5-tetrazolyl)triazenide
[ ] $C_2H_2AgN_{11}$

Ag^+

See 1,3-Di(5-tetrazolyl)triazene
See other SILVER COMPOUNDS, TETRAZOLES, TRIAZENES

0691. 2-Iodosovinyl chloride
[ ] C_2H_2ClIO

$$ClCH{=}CHI{:}O$$

Alone, or Water
Thiele, J. *et al., Ann.,* 1909, **369**, 131
It explodes at 63°C and contact with water disproportionates it to the more explosive iodyl compound (next below).
See other IODINE COMPOUNDS

0692. 2-Iodylvinyl chloride
[ ] $C_2H_2ClIO_2$

$$ClCH{=}CHI({:}O)_2$$

Thiele, J. *et al., Ann.,* 1909, **369**, 131
Explodes violently on impact, friction or heating to 135°C.
See other IODINE COMPOUNDS

0693. Azidoacetyl chloride
[30426-58-5] $C_2H_2ClN_3O$

$$N_3CH_2CO.Cl$$

1. Wieland, T. *et al., Chem. Ber.,* 1960, **93**, 1236
2. Prodan, K. A. *et al., Chem. Eng. News,* 1981, **59**(14), 59
3. Hoppe, D. *et al., Ann.,* 1980, 1519

Using a published procedure [1] for preparing azidoacetyl chloride, but at 1.7-fold higher concentration, azidoacetic acid and sulfinyl chloride were slowly heated up to reflux. After 40 min. decomposition started and accelerated to a violent explosion [2]. The cause seems likely to have been the higher initial concentration of azido acid, which would have produced a higher concentration (87%) of acid chloride in excess sulfinyl chloride, and at a significantly higher reflux temperature, than the original method (about 50%). A relatively safe way of preparing the explosively unstable material involves a low temperature partial distillation procedure, but full precautions are recommended [3].

260

See Azidoacetic acid
See other ORGANIC AZIDES

0694. Sodium chloroacetate
[3296-62-3]   C₂H₂ClNaO₂

ClCH₂CO.ONa

Energy of decomposition (in range 130–220°C) was measured as 0.29 kJ/g by DSC, and T_{ait24} was determined as 96°C by adiabatic Dewar tests, with an apparent energy of activation of 112 kJ/mol.
See entry THERMOCHEMISTRY AND EXOTHERMIC DECOMPOSITION (reference 2)

†0695. 1,1-Dichloroethylene (Vinylidene chloride)
[75-35-4]   C₂H₂Cl₂

Cl₂C=CH₂

HCS 1980, 378

1. Reinhardt, R. C., *Chem. Eng. News*, 1947, **25**, 2136
2. *MCA Case History No. 1172*
3. *MCA Case History No. 1693*
4. Harmon, 1974, iv, 2.21
5. Warren, H. S. *et al.*, *Chem. Abs.*, 1978, **89**, 219959

When stored at between −40° and +25°C in the absence of inhibitor and in presence of air, vinylidene chloride rapidly absorbs oxygen with formation of a violently explosive peroxide. The latter initiates polymerisation, producing an insoluble polymer which adsorbs the peroxide. Separation of this polymer in the dry state must be avoided, since if more than 15% of peroxide is present, the polymer may be detonable by slight shock or heat. Hindered phenols are suitable inhibitors to prevent peroxidation [1]. The earlier Case History describes an explosion during handling of a pipe used to transfer the chloride [2]. Two further cases of formation of the polymeric peroxide in bottles of vinylidene chloride in refrigerated storage, and subsequent explosions during handling or disposal, are recorded [3]. The monomer is normally handled at −10°C in absence of light or water, which tend to promote self-polymerisation [4]. Hazards from explosions when the chloride is stored or handled in presence of air are discussed [5].
See entries POLYMERISATION INCIDENTS, POLYPEROXIDES, VIOLENT POLYMERISATION

Chlorotrifluoroethylene
Raasch, M. S. *et al.*, *Org. Synth.*, 1962, **42**, 46
Condensation of the reactants at 180°C under pressure to give 1,1,2-trichloro-2,3,3-trifluorocyclobutane was effected smoothly several times in a 1 l autoclave. Scaling up to a 3 l preparation led to uncontrollable polymerisation which distorted the larger autoclave.

261

Ozone
'Vinylidene Chloride Monomer', 9, Midland (Mich.), Dow Chemical Co., 1968
The reaction products formed with ozone are particularly dangerous.
See other OZONIDES

Perchloryl fluoride
See Perchloryl fluoride: Hydrocarbons, etc.
See other HALOALKENES, PEROXIDISABLE COMPOUNDS

†0696. *cis*-1,2-Dichloroethylene
[156-60-5] $C_2H_2Cl_2$
                    HCCl=CHCl

See other HALOALKENES

†0697. *trans*-1,2-Dichloroethylene
[156-59-2] $C_2H_2Cl_2$
                    ClCH=CHCl

1. Anon., *BCISC Quart. Safety Summ.*, 1964, **35**, 37
2. Anon., *Sichere Chemiearb.*, 1964, **16**, 35
Under appropriate conditions, dichloroethylene, previously thought to be non-flammable, can cause a fire hazard [1]. Addition of a hot liquid to the cold solvent caused sudden emission of sufficient vapour to cause a flame to flash back 12 m from a fire. Although the bulk of the solvent did not ignite, various items of paper and wood in the room were ignited by the transient flame [2].

Alkalies
1. Anon., *Fire Accid. Prev.*, 1956, **42**, 28
2. Thron, H., *Chem. Ztg.*, 1924, **48**, 142
1,2-Dichloroethylene in contact with solid caustic alkalies or their conc. solutions will form chloroacetylene which ignites in air [1]. Distillation of ethanol containing 0.25% of the halocarbon with aqueous sodium hydroxide gave a product which ignited in air [2].
See Trichloroethylene: Alkali
    1,1,2,2-Tetrachloroethane: Alkalies

Difluoromethylene dihypofluorite
See Difluoromethylene dihypofluorite: Haloalkenes

Dinitrogen tetraoxide
See Dinitrogen tetraoxide: Halocarbons
See other HALOALKENES

262

0698. Trichloroacetaldehyde oxime
[1117-99-3] $C_2H_2Cl_3NO$

$$Cl_3CCH{=}NOH$$

Alkali
Rodd, 1965, Vol. 1C, 46
The oxime decomposes explosively with alkali, forming hydrogen cyanide, hydrogen chloride and carbon dioxide.
See other GAS EVOLUTION INCIDENTS, OXIMES

0699. 1,1,2,2-Tetrachloroethane
[79-34-5] $C_2H_2Cl_4$

$$Cl_2CHCHCl_2$$

(*MCA SD-34*, 1949); *HCS 1980*, 886

Alkalis
MCA SD-34, 1949
It is not an inert solvent, and on heating with solid potassium hydroxide or other base, hydrogen chloride is eliminated and chloro- or dichloro-acetylene, which ignite in air, are formed.

Metals
See Potassium: Halocarbons
     Sodium: Halocarbons
See METAL–HALOCARBON INCIDENTS
See other HALOALKANES

†0700. 1,1-Difluoroethylene (Vinylidene fluoride)
[75-38-7] $C_2H_2F_2$

$$F_2C{=}CH_2$$

HCS 1980, 409

Hydrogen chloride
Jensen, J. H., *Chem. Eng. News*, 1981, **59**(14), 3, 59
As well as the normal addition reaction, an extremely exothermic decomposition reaction may occur, particularly at high vessel loadings. At loadings of 0.8 ml of 1:1 mixture per ml, the violent reaction, catalysed by iron(III) chloride, initiates at −40°C and will attain pressures above 0.7 kbar at the rate of 14 kbar/s. At 0.5 ml loading density, a maximum pressure of 68 bar, attained at 114 bar/s, was observed.
See entries CATALYTIC IMPURITY INCIDENTS, PRESSURE INCREASE IN EXOTHERMIC DECOMPOSITION

Ozone
  See Ozone: 1,1-Difluoroethylene
  See other HALOALKENES

0701. 2-(Difluoroamino)-2,2-dinitroacetamide
[180570-38-1] $C_2H_2F_2N_4O_5$

$$F_2NC(NO_2)_2CONH_2$$

See 1,1-DINITRODIFLUORAMINES, *See other* N-HALOGEN COMPOUNDS, POLYNITRO-
ALKYL COMPOUNDS

0702. Difluoroacetic acid
[381-73-7] $C_2H_2F_2O_2$

$$F_2CHCO.OH$$

Caesium fluoride, Fluorine
  See Fluorine: Caesium fluoride, etc.
  See other ORGANIC ACIDS

0703. 1,1-Difluoroethylene ozonide (3,3-Difluoro-1,2,4-trioxolane)
[69932-17-8] $C_2H_2F_2O_3$

  See Ozone: 1,1-Difluoroethylene
  See other OZONIDES

0704. Trifluoroacetamide
[354-38-1] $C_2H_2F_3NO$

$$F_3CCO.NH_2$$

Lithium tetrahydroaluminate
  See Lithium tetrahydroaluminate: Fluoroamides

0705. Sodium 2,2,2-trifluoroethoxide
[420-87-1] $C_2H_2F_3NaO$

$$NaOCH_2CF_3$$

Schmutz, J. L. *et al., Inorg. Chem.*, 1975, **14**, 2437
During preparation from sodium and excess alcohol in ether, attempts to remove
last traces of the alcohol by warming under vacuum led to explosions.
  See related METAL ALKOXIDES

0706. 1,1,2,2-Tetrafluoroethane
[811-97-2] $C_2H_2F_4$

$$F_2CHCHF_2$$

Schröder, V. *et al. Loss Prevention and Safety Promotion in the Process Industries* Vol II (Mewis, J. J., Pasman, H. J. & De Rademaker, E. E. Eds.), 339, Amsterdam, Elsevier, 1995
Although not flammable under normal conditions in air, it is at elevated pressure, or with oxygen enrichment. Explosive limits are reported and reviewed.
See also FLUOROCARBONS

0707. Mercury(II) *aci*-dinitromethanide
[ ] $C_2H_2HgN_4O_8$

$$[O_2NCH{=}N(O)O]_2Hg$$

Urbanski, 1967, Vol. 3, 158
This salt of *aci*-dinitromethane shows detonator properties.
See other MERCURY COMPOUNDS, *aci*-NITRO SALTS

0708. 1,2-Bis(hydroxomercurio)-1,1,2,2-bis(oxydimercurio)ethane ('Ethane hexamercarbide')
[67536-44-1] $C_2H_2Hg_6O_4$

Hofmann, K. A., *Ber.*, 1898, **31**, 1904
The compound, formulated as above, explodes very violently at 230°C.
See other MERCURY COMPOUNDS

0709. Potassium dinitroacetamide
[50650-92-5] $C_2H_2KN_3O_5$

$$KC(NO_2)_2CO.NH_2$$

See Silver dinitroacetamide
See other POLYNITROALKYL COMPOUNDS

0710. Diazoacetaldehyde
[6832-13-9] $C_2H_2N_2O$

$$N_2CHCO.H$$

Arnold, Z., *Chem. Comm.*, 1967, 299

It may be distilled out continuously as formed at 40°C/13 mbar, but readily detonates with great violence if overheated.
See other ALDEHYDES, DIAZO COMPOUNDS

0711. 2-Nitroethanonitrile (Nitroacetonitrile)
[ ] $CH_2N_2O_2$
$$N\equiv CCH_2NO_2$$

Koenig, L. *et al.*, Personal communication, 1997
Made by dehydration of nitroacetaldehyde oxime, this is an explosion hazard. It is improbable that the nitrile is not, itself, detonable.
See 2-Nitroacetaldehyde oxime
See other *C*-NITRO COMPOUNDS

0712. Sodium dinitroacetamide
[[26163-27-9] (ion)] $C_2H_2N_3NaO_5$
$$NaC(NO_2)_2CO.NH_2$$

See Silver dinitroacetamide
See other POLYNITROALKYL COMPOUNDS

0713. Sodium 2,2,2-trinitroethanide
[ ] $C_2H_2N_3O_6Na$
$$(O_2N)_3CCH_2Na$$

See *N,N'*-Bis(trinitroethyl)urea
See other POLYNITROALKYL COMPOUNDS

0714. Azidoacetonitrile
[57707-64-9] $C_2H_2N_4$
$$N_3CH_2C\equiv N$$

Freudenberg, K. *et al., Ber.*, 1932, **65**, 1188
With over 51% nitrogen content, it is, as expected, explosive, sensitive to impact or heating to 250°C.
See other CYANO COMPOUNDS, HIGH-NITROGEN COMPOUNDS, ORGANIC AZIDES

0715. 1,2,4,5-Tetrazine
[290-96-0] $C_2H_2N_4$

Sulfuric acid
  Sorbe, 1968, 139

266

The solid base decomposes violently in contact with conc. acid.
See other HIGH-NITROGEN COMPOUNDS, ORGANIC BASES

0716. 3-Nitro-1,2,4-triazolone (Oxynitrotriazole)
[932-64-9] $C_2H_2N_4O_3$

Ullmann, 1987, A10, 162
Finds employment as an high explosive (oxygen balance −40%).
See EXPLOSIVES, OXYGEN BALANCE
See other HIGH-NITROGEN COMPOUNDS, NITRO COMPOUNDS, TRIAZOLES

0717. 3-Azido-1,2,4-triazole
[21041-85-0] $C_2H_2N_6$

Denault, G. C. *et al., J. Chem. Eng. Data*, 1968, **13**, 514
Samples exploded during analytical combustion.
See other ORGANIC AZIDES, TRIAZOLES

0718. 3-Amino-6-nitro-1,2,4,5-tetrazine-2,4-dioxide
[153757-95-0] $C_2H_2N_6O_4$

See 3,6-Diamino-1,2,4,5-tetrazine-1,4-dioxide
See other HIGH-NITROGEN COMPOUNDS, *See also* AMINE OXIDES

267

0719. 5-(Diazomethylazo)tetrazole
[ ] $C_2H_2N_8$

Mayants, A. G. *et al., Zh. Org. Khim.*, 1991, **27**(11), 2450; *Chem. Abs.*, 1992, **116**, 234937

The compound, produced as an hydrate from dilute acid hydrolyses of disodium azotetrazolide, exploded on heating for analytical purposes.

See other AZO COMPOUNDS, DIAZO COMPOUNDS, TETRAZOLES

0720. 1,2-Diazidocarbonylhydrazine
[67880-17-5] $C_2H_2N_8O_2$

N₃CO.NHNHCO.N₃

Kesting, W., *Ber.*, 1924, **57**, 1321

Similar in properties to lead or silver azide, it explodes on heating or impact.

See other ACYL AZIDES, HIGH-NITROGEN COMPOUNDS

0721. 5,5'-Tetrazolyl disulfide (5,5'-Dithiobis[1H]tetrazole)
[72848-18-1] $C_2H_2N_8S_2$

Christensen, T. B. *et al., Sulfur Lett.*, 1993, **16**(5-6), 205

In a paper devoted to tetrazole chemistry, the explosive nature of this compound excited comment.

See other TETRAZOLES; *See also* HIGH NITROGEN COMPOUNDS

0722. Ketene (Ethenone)
[463-51-4]
$$H_2C=C=O$$

C_2H_2O

Hydrogen peroxide
See Hydrogen peroxide: Ketene

†0723. Ethanedial (Glyoxal)
[107-22-2]
$$HCO.CO.H$$

$C_2H_2O_2$

HCS 1980, 522

Preparative hazard
See Nitric acid: 2,4,6-Trimethyltrioxane

Alone, or Water
1. Merck, 1983, 647
2. van den Brink, M. J., *Chemisch Magazine* (Rijswijk), 1982, (7–8), 428
A powerfully reducing and reactive solid (m.p., 15°C) of high vapour pressure (b.p., 50°C). Mixtures with air may explode, and contact with water causes violent polymerisation [1]. More usually encountered polymeric or as a hydrate. Like formaldehyde, pure glyoxal may polymerise exothermally and ignite in storage [2].

Nitric acid, Water
See Nitric acid: Glyoxal
See other ALDEHYDES, REDUCANTS

0724. Glyoxylic acid (Oxoacetic acid)
[298-12-4]
$$O=CHCOOH$$

$C_2H_2O_3$

Preparative hazard
See Nitric acid: Glyoxal
See other ORGANIC ACIDS *See related* ALDEHYDES

0725. Oxalic acid (Ethanedioic acid)
[144-62-7]
$$HOCO.CO.OH$$

$C_2H_2O_4$

NSC 406, 1978; *HCS 1980*, 706

CHETAH, 1990, 184
Surprisingly, even the dihydrate is apparently detonated by a 50 g tetryl booster, the anhydrous acid is thermally less stable and thus probably more sensitive. There is, however, no history of explosion.

269

Other reactants
Yoshida, 1980, 178
MRH values calculated for 5 combinations are given.

Silver
See Silver: Carboxylic acids

Sodium chlorite
See Sodium chlorite: Oxalic acid

Urea
von Bentzinger, R. *et al., Praxis Naturwiss. Chem.*, 1987, **36**(8), 41–42
During the qualitative analysis of an 'unknown' mixture of oxalic acid, urea and
a little charcoal, a sample was put in an ignition tube under a layer of granular
copper oxide, topped by glass wool and a little anhydrous copper sulfate (to detect
evolved water). The top end of the tube was drawn down obliquely into a capillary
so that the pyrolysis gases could be passed into barium hydroxide solution (to
detect carbon dioxide). It was not noticed that the end of the capillary was sealed,
and when the sample was heated in a flame, an unexpectedly violent explosion
followed. Later experiments showed that there was no explosion if oxalic acid
alone, or urea alone, was heated with copper oxide in a sealed tube, but that a
mixture of the two always led to a violent explosion. The same also happened in
absence of copper oxide. Further work (DSC, and measurement of volumes of gas
liberated) led to the conclusions that the reactions involved were:-
$$2CO.(NH_2)_2 + 3(CO.OH)_2 \rightarrow 5CO_2 + 3CO + 4NH_3 + H_2O$$
and that a gas pressure approaching 100 bar would be produced in the accidentally
sealed ignition tube at 570°C.
See other GAS EVOLUTION INCIDENTS, GLASS INCIDENTS, REDOX REACTIONS
See other ORGANIC ACIDS

0726. Lead carbonate–lead hydroxide (Basic lead carbonate)
[598-63-0] $C_2H_2O_8Pb_3$
$$2PbCO_3.Pb(OH)_2$$

Fluorine
See Fluorine: Metal salts
See other INORGANIC BASES

†0727. Bromoethylene (Vinyl bromide)
[593-60-2] C_2H_3Br
$$BrCH=CH_2$$

1. *HCS 1980*, 956
2. *DOC 5*, 1982, 799
The highly flammable gas reacts violently with oxidants [1] and may polymerise
in sunlight [2].
See other HALOALKENES

0728. Acetyl bromide
[506-96-7] C_2H_3BrO

$$H_3CCO.Br$$

HCS 1980, 108

Hydroxylic compounds
Merck, 1976, 11
Interaction with water, methanol or ethanol is violent, hydrogen bromide being evolved.
See other ACYL HALIDES

0729. Acetyl hypobromite
[4254-22-2] $C_2H_3BrO_2$

$$H_3CCO.OBr$$

1. Reilly, J. J. *et al., J. Org. Chem.*, 1974, **39**, 3992
2. Skell, P. S. *et al., J. Amer. Chem. Soc.*, 1983, **105**, 4000, 4007
The instability of the hypobromite is noted, and protective shielding is recommended for work with more than 5 g quantities [1]. Though stable in fluorotrichloromethane solution at −23°C under nitrogen in the dark, the isolated solid exploded at −23°C at below 1.2 mbar [2].
See other ACYL HYPOHALITES

†0730. Chloroethylene (Vinyl chloride)
[75-01-4] C_2H_3Cl

$$ClCH=CH_2$$

(*MCA SD-56*, 1972); *HCS 1980*, 957; *RSC Lab. Hazards Data Sheet No. 56*, 1987
1. *MCA Case History No. 625*
2. Harmon, 1974, 2.20, 4.74−4.77
3. *MCA Case History No. 1551*
4. Scali, C. *et al., Proc. 3rd Int. Symp. Loss Prev. Safety Prom. Proc. Ind.*, 1350−1358, Basle, SSCI, 1980
5. Anon., *Eur. Chem. News*, 1993, **60**(1589), 26
Discharge of a spray of vapour and liquid under pressure from a cylinder into a fume hood generated static and ignited the vapour. Discharge of the vapour phase only did not lead to ignition [1]. Vinyl chloride tends to self-polymerise explosively if peroxidation occurs, and several industrial explosions have been recorded [2]. Accidental exposure of the recovered monomer to atmospheric oxygen for a long period caused formation of an unstable polyperoxide which initiated an explosion. Suitable precautions are discussed, including the use of 20−30% aqueous sodium hydroxide solution to destroy the peroxide [3]. Problems of stability and control of batch reactors for the suspension polymerisation of vinyl chloride have been studied in relation to process kinetics and plant heat transfer coefficients [4]. A polymerisation reaction ran beyond control, generating back pressure which caused the monomer to leak through a valve, forming a vapour cloud which ignited and exploded [5].

See 1,1-Dichloroethylene
See VIOLENT POLYMERISATION
See other POLYMERISATION INCIDENTS, POLYPEROXIDES, STATIC INITIATION INCIDENTS, VAPOUR CLOUD EXPLOSIONS

Oxides of nitrogen
Povey, R., private comm., 1979
An explosion in a valve in a liquid monomer line was ascribed to traces of oxides of nitrogen remaining after the valve had been passivated by treatment with nitric acid.

Oxygen
Tartari, V., *Proc. 34th Ital. Chem. Congr. Safety Chem. Proc.*, 51–60, Milan, 1983
The stability of the monomer to heat in the presence (and absence) of oxygen has been studied by ARC. The intensity of the exotherm at 90–100°C depends on the partial pressure of oxygen.
See other HALOALKENES, PEROXIDISABLE COMPOUNDS

†0731. 1-Chloro-1,1-difluoroethane
[75-68-3] $C_2H_3ClF_2$

$$ClCF_2CH_3$$

Difficult to ignite vapour
See other HALOALKANES
See FLASH POINTS (reference 19)

0732. Ethenylmagnesium chloride (Vinylmagnesium chloride, Chloroethenylmagnesium)
[3536-96-7] C_2H_3ClMg

$$CH_2{=}CHMgCl$$

Rittmeyer, P. *et al., Z. Naturforsch., B: Chem. Sci.*, 1993, **48**(9), 1223
An incident during concentration of a tetrahydrofuran solution on pilot plant led to a study of the concentration dependent exothermic decomposition of the reagent and generation of safety rules.
See other GRIGNARD REAGENTS

0733. 3-Chloro-3-methyldiazirine
[4222-21-3] $C_2H_3ClN_2$

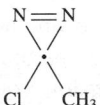

1. Liu, M. T. H., *Chem. Eng. News*, 1974, **53**(36), 10
2. Jones, W. E. *et al., J. Photochem.*, 1976, **5**, 233–239
3. Frey, H. M. *et al., J. Chem. Soc., Faraday Trans. 1*, 1977, **73**, 2011

Extremely shock-sensitive and violently explosive; initiation has been caused by prolonged freezing at $-196°C$, or by sawing a stopcock off a metal trap containing trace amounts [1]. It may be stored safely at $-80°C$ [2]. Detonative explosion during trap-to-trap distillation of purified material is noted [3].
See other DIAZIRINES

0734. 3-Chloro-3-methoxydiazirine
[4222-27-9] $C_2H_3ClN_2O$

Smith, N. P. *et al., J. Chem. Soc., Perkin Trans.* 2, 1979, 213
Unpredictably explosive as the neat liquid, it may be handled safely in solution. All distillation traps must contain solvent.
See other DIAZIRINES

†0735. Acetyl chloride (Ethanoyl chloride)
[75-36-5] C_2H_3ClO

$H_3CCO.Cl$

FPA H65, 1978; *HCS 1980*, 109

Preparative hazard
See Phosphorus trichloride: Carboxylic acids

Dimethyl sulfoxide
See Dimethyl sulfoxide: Acyl halides

Other reactants
Yoshida, 1980, 51
MRH values calculated for 13 combinations with oxidants are given.

Water
1. *Haz. Chem. Data*, 1975, 30
2. Anon., *CISHC Chem. Safety Summ.*, 1986, **57**(225), 3
3. Anon., personal communication 1997
Interaction is violent [1]. A polythene-lined drum which had previously contained acetyl chloride was inspected and appeared empty, but when it was washed out with water a violent eruption occurred. This was attributed to trapping of residual acetyl chloride between the split liner and the drum wall [2]. A very similar but more violent event, rupturing the drum and sending part of it 30 m. with sufficient force to puncture another drum which it hit, occurred in Scotland in 1997 [3].
See other ACYL HALIDES

273

†0736. **Methyl chloroformate (Methyl carbonochloridate)**
[79-22-1] $C_2H_3ClO_2$
$$H_3COCO.Cl$$

HCS 1980, 645

See other ACYL HALIDES

0737. **Peroxyacetyl perchlorate**
[66955-43-9] $C_2H_3ClO_6$
$$MeCO.OOClO_3$$

Yakovleva, A. A. *et al., Chem. Abs.*, 1979, **91**, 210847
The peroxyester of +9.1% oxygen balance exploded on detonation or friction.
See Peroxypropionyl perchlorate
See related ALKYL PERCHLORATES, PEROXYESTERS

0738. **1,1-Dichloro-1-fluoroethane**
[1717-00-6] $C_2H_3Cl_2F$
$$MeCCl_2F$$

1. Anon., *Sichere Chemiearbeit*, 1993, **45**(8), 105
2. Bretherick, L., *Chem. Eng. News*, 1988, **66**(50), 2
An 'empty' drum which had contained this exploded when cut into by a grinder. It is presumably more flammable than might be anticipated [1]. Like trichloroethane the vapour is difficult to ignite and may not show a flash point, but can form explosive mixtures with air [2].
See other HALOALKANES

0739. *N,N*-**Dichloroglycine**
[58941-14-3] $C_2H_3Cl_2NO_2$
$$Cl_2NCH_2CO.OH$$

Vit, J. *et al., Synth. Comm.*, 1976, **6**(1), 1−4
It explodes at 65°C, so solutions must be evaporated at below 40°C.
See other N-HALOGEN COMPOUNDS, ORGANIC ACIDS

†0740. **1,1,1-Trichloroethane**
[71-55-6] $C_2H_3Cl_3$
$$Cl_3CCH_3$$

(*MCA SD-90*, 1965); *HCS 1980*, 918; *RSC Lab. Hazard Data Sheet No. 1*, 1982

1. Wrightson, I. *et al., Loss Prev. Bull.*, 1988, (083), 21−23
2. Santon, R. C. *et al., Proc. 6th Int. Sympos. Loss Prev. Safety Prom. Process Ind.*, 1989, Oslo, Norweg. Soc. Chem. Eng.

Although no flash point is measurable by standard tests, the vapour can be ignited by a high-energy source, the limits being 8.0–10.5 vol% in air at 25°C, with an autoignition temperature of 500°C. A 1050 m^3 storage tank containing 70 t of the inhibited solvent, normally blanketted with nitrogen to exclude moisture, had developed gas leaks in the roof, and the purge system had been switched to manual control. Previous withdrawal of 1150 t had drawn in air, which had not been effectively displaced by manual purging. Arc-welding to repair the roof leaks led to an explosion which separated the roof from the tank for 65% of the 36 m circumference. Only one other explosion, when the end of an 'empty' 200 l drum was cut with an oxy-acetylene torch, has been reported [1]. New work has extended downwards the temperature limits for ignition to 10°C for lel of 8.0 vol%, and 22° for uel of 14.2 vol%. These values diverge with increasing temperature to 6% and 15.5%, respectively, at 100°C [2].
See FLASH POINTS (reference 19)

Aluminium oxide, Heavy metals
See Aluminium oxide: Halocarbons, etc.

Dinitrogen tetraoxide
See Dinitrogen tetraoxide: Halocarbons

Inhibitors
See FLASH POINTS (reference 19)

Metals
See Aluminium: Halocarbons
   Magnesium: Halocarbons
   Potassium: Halocarbons
   Potassium–sodium alloy: Halocarbons
   METAL–HALOCARBON INCIDENTS

Molecular sieve
1. Salmon, J. *et al., Chem. & Ind.*, 1995, 814
2. Wilson, W., *Chem. & Ind.*, 1995, 854
3. Bretherick, L., personal communication
4. Winterton, N. *et al., Chem. & Ind.*, 1995, 902
5. Whitmore, M. W. *et al., Chem. & Ind.*, 1995, 942

Some unstabilised trichloroethane had stood over molecular sieve (uncertain which grade) for a period of some years. The sieve was filtered out and the bottle, still containing some sieve, purged with nitrogen, after which it exploded when moved. The remaining sieve deflagrated on addition to a bucket of water [1]. It is suggested that the sieve catalysed dehydrochlorination to chloroacetylene, which is explosive [2]. This seems improbable since the acetylene is also volatile and not very polar, thus should not adhere to the dried sieve. An alternative explanation is that dehydrochlorination occured, giving vinylidene chloride (1,1-dichloroethene) which polymerised with atmospheric oxygen as a comonomer, leading to a polyperoxide which would adhere to the sieve [3,4]. Another sample of ancient trichloroethane, stored over sieves, showed no instability or peroxidation [5].

Oxygen
See Oxygen (Gas): Halocarbons
        Oxygen (Liquid): Halocarbons
See other HALOALKANES

0741. 1,1,2-Trichloroethane
[79-00-5] $C_2H_3Cl_3$

$$Cl_2CHCH_2Cl$$

HCS 1980, 917

Potassium
See Potassium: Halocarbons
See other HALOALKANES

0742. Methyltrifluoromethyltrichlorophosphorane
[69517-30-2] $C_2H_3Cl_3F_3P$

$$F_3CP(Me)Cl_3$$

Preparative hazard
See Tetrachlorotrifluoromethylphosphorane: Tetramethyllead

0743. N-Carbomethoxyiminophosphoryl chloride
[25147-05-1] $C_2H_3Cl_3NO_2P$

$$MeOCO.N=PCl_3$$

Mellor, 1971, Vol. 8, Suppl. 3, 589
This intermediate (or its ethyl homologue), produced during the preparation of phosphoryl dichloride isocyanate from interaction of phosphorus pentachloride and methyl (or ethyl) carbamate, is unstable. Its decomposition to the required product may be violent or explosive unless moderated by presence of a halogenated solvent.
See related NON-METAL HALIDES

0744. 2,2,2-Trichloroethanol
[115-20-8] $C_2H_3Cl_3O$

$$Cl_3CCH_2OH$$

Sodium hydroxide
MCA Case History No. 1574
Accidental contact of 50% sodium hydroxide solution with residual trichloroethanol in a pump caused an explosion. This was confirmed in laboratory experiments.

Chlorohydroxyacetylene, or the isomeric chloroketene, or chlorooxirene may have been formed by elimination of hydrogen chloride.

See related HALOALKANES

0745. 2,2,2-Trichloro-1,1-ethanediol (Chloral hydrate)
[302-17-0] $C_2H_3Cl_3O_2$

$$Cl_3CCH(OH)_2$$

1. Marvel, C. S. *et al., Org. Synth.*, 1941, Coll. Vol. 1, 377
2. Goodwin, B., *Chem. Brit.*, 1988, **24**, 336
3. Gandy, R. *et al., Chem. Brit.*, 1988, **24**, 336

Following a published procedure for converting substituted anilines to isatins by reaction with chloral hydrate and hydroxylamine [1], it was noticed that at the end of the first stage (formation of an isonitrosoacetanilide), the odour of hydrogen cyanide was present, and this was confirmed by a Prussian blue test [2]. In related work, concentrations of 100–200 ppm of hydrogren cyanide were found [3]. A mechanism for its formation from chloral hydrate and hydroxylamine was proposed [2], and the need for appropriate precautions was stressed [2,3].

See other GAS EVOLUTION INCIDENTS

†0746. Trichlorovinylsilane
[74-94-5] $C_2H_3Cl_3Si$

$$Cl_3SiCH=CH_2$$

It may ignite in air.

Water
491M, 1975, 426
It reacts violently with water or moist air.
See related ALKYLHALOSILANES

†0747. Fluoroethylene (Vinyl fluoride)
[75-02-5] C_2H_3F

$$FCH=CH_2$$

HCS 1980, 958

Ozone
See Fluoroethylene ozonide
See other HALOALKENES

0748. 1-Fluoro-1,1-dinitroethane
[13214-58-9] $C_2H_3FN_2O_4$

$$FC(NO_2)_2CH_3$$

See entry FLUORODINITROMETHYL COMPOUNDS

0749. 2(?)-Fluoro-1,1-dinitroethane
[68795-10-8] $C_2H_3FN_2O_4$

$$FCH_2CH(NO_2)_2$$

MCA Case History No. 784
During a prolonged fractional distillation of the material at 75°C/50 mbar, an exothermic decomposition began. As a remedial measure (!), air was admitted to the hot decomposing residue, causing a violent explosion. Admission of nitrogen and/or cooling the residue before admitting air might have avoided the incident.
See related FLUORODINITROMETHYL COMPOUNDS

0750. 2-Fluoro-2,2-dinitroethanol
[17003-75-7] $C_2H_3FN_2O_5$

$$FC(NO_2)_2CH_2OH$$

1. Cochoy, R. E. *et al., J. Org. Chem.*, 1972, **37**, 3041
2. Cochoy, R. E. *et al., J. Org. Chem.*, 1990, **55**(4) 1401
It is a potentially explosive vesicant, from which a series of esters, expected to be explosive, was prepared [1]. During concentration of a dichloromethane solution, there is a sensitivity maximum between 30% solution and the pure compound [2].
See other FLUORODINITROMETHYL COMPOUNDS, POLYNITROALKYL COMPOUNDS

0751. Acetyl hypofluorite
[78948-09-1] $C_2H_3FO_2$

$$H_3CCO.OF$$

1. Adam, M. J., *Chem. & Eng. News*, 1985, **63**(7), 2
2. Appelman, E. *et al., J. Amer. Chem. Soc.*, 1985, **107**, 6516–6518
A minor explosion in the use of gaseous acetyl hypofluorite was caused by inadvertent contact of the gas inlet tube with a cooling bath, which condensed liquid hypofluorite which subsequently exploded on contact with organic material or air [1]. More detail on the isolation of this compound of limited stability, particularly in the liquid phase, is given [2].
See other ACYL HYPOHALITES, OXIDANTS

0752. Fluoroethylene ozonide (3-Fluoro-1,2,4-trioxolane)
[60553-18-6] $C_2H_3FO_3$

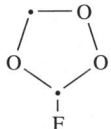

1. Mazur, W. *et al., J. Org. Chem.*, 1979, **44**, 3183
2. Hillig, K. W. *et al., J. Amer. Chem. Soc.*, 1982, **104**, 991

During ozonolysis of vinyl fluoride an explosive solid residue is produced, and the volatile ozonide, trapped at $-63°C$, may explode spontaneously or during removal by syringe [1]. During the cryogenic distillation of the ozonide (formulated as a trioxolane), several explosions occurred, and the explosive reaction residue was destroyed by digestion with 5% potassium iodide solution for 24 h [2].

See other OZONIDES

0753. 2-(Difluoroamino)-2,2-dinitroethane
[121832-04-0] $C_2H_3F_2N_3O_4$

$$F_2NC(NO_2)_2CH_3$$

See 1,1-DINITRODIFLUORAMINES, *See other* N-HALOGEN COMPOUNDS, POLYNITRO-ALKYL COMPOUNDS

†0754. 1,1,1-Trifluoroethane
[420-46-2] $C_2H_3F_3$

$$F_3CCH_3$$

See other HALOALKANES

0755. Pentafluorosulfur peroxyacetate
[60672-60-8] $C_2H_3F_5O_3S$

$$MeCO.OOSF_5$$

Hopkinson, M. J. *et al., J. Fluorine Chem.*, 1976, **7**, 505
The peroxyester and its homologues, like other fluoroperoxy compounds, are potentially explosive and may detonate on thermal or mechanical shock.

See related FLUORINATED PEROXIDES AND SALTS, PEROXYESTERS

0756. Potassium 1-nitroethane-1-oximate
[3454-11-3] $C_2H_3KN_2O_3$

$$O_2NC(Me)=NOK$$

Rodd, 1965, Vol. 1B, 98
Isolated salts of 'nitrolic acids', produced by action of nitrous acid on 1-nitroalkanes, are explosive.

See related NITROALKANES, OXIMES

0757. Vinyllithium
[917-57-7] C_2H_3Li

$$H_2C=CHLi$$

Juenge, E. C. *et al., J. Org. Chem.*, 1961, **26**, 564

When freshly prepared, it is violently pyrophoric but on storage it becomes less reactive and slow to ignite in air, possibly owing to polymerisation.
See related ALKYLMETALS

†**0758. Acetonitrile (Ethanenitrile)**
[75-05-8] C_2H_3N

$$H_3CC\equiv N$$

NSC 683, 1981; *FPA H108*, 1981; *HCS 1980*, 104
RSC Lab. Hazard Data Sheet No. 31, 1984

A moderately endothermic compound (ΔH_f° (l) +53.1 kJ/mol, 1.29 kJ/g) which, though reactive, does not exhibit inherent instability under normal conditions.

Chlorine fluoride, Fluorine
    See Fluorine: Acetonitrile, etc.

2-Cyano-2-propyl nitrate
    See NITRATING AGENTS

Dinitrogen tetraoxide, Indium
    See Dinitrogen tetraoxide: Acetonitrile, Indium

Diphenyl sulfoxide, Trichlorosilane
    See Trichlorosilane: Acetonitrile, etc.

N-Fluoro compounds
    Fraser, G. W. *et al., Chem. Comm.*, 1966, 532
    Nitrogen–fluorine compounds are potentially explosive in contact with acetonitrile.
    See Tetrafluorourea: Acetonitrile

Iron(III) perchlorate
    See Iron(III) perchlorate: Acetonitrile

Lanthanide perchlorate
    Forsberg, J. H., *Chem. Eng. News*, 1984, **62**(6), 33
    A solution of an unspecified lanthanide perchlorate in acetonitrile detonated while being heated under reflux.
    See other METAL PERCHLORATES, SOLVATED OXOSALT INCIDENTS

Nitric acid   MRH 6.15/77
    See Nitric acid: Acetonitrile

Other reactants
    Yoshida, 1980, 19
    MRH values calculated for 14 combinations with oxidants are given.

Perchloric acid
See Perchloric acid: Acetonitrile

Sulfuric acid, Sulfur trioxide   MRH Sulfuric acid 1.76/54
    Lee, S. A., private comm., 1972
    A mixture of acetonitrile and sulfuric acid on heating (or self-heating) to 53°C
    underwent an uncontrollable exotherm to 160°C in a few seconds. The presence
    of 28 mol% of sulfur trioxide reduces the initiation temperature to about 15°C.
    Polymerisation of the nitrile is suspected.
    See other CYANO COMPOUNDS, ENDOTHERMIC COMPOUNDS

0759. Methyl isocyanide
   [593-75-9] C_2H_3N

                              $H_3CN:C$

1. Lemoult, M. D., *Compt. rend.*, 1906, **143**, 902
2. Stein, A. R., *Chem. Eng. News*, 1968, **46**(45), 7–8
3. Collister, J. L., *Diss. Abs. Intl. B*, 1977, **37**, 4479
4. Pritchard, H. O., *Can. J. Chem.*, 1979, **57**, 2677–2679
It is very endothermic (ΔH°_f (g) +150.2 kJ/mol, 3.66 kJ/g). A sample exploded
when heated in a sealed ampoule [1], and during redistillation at 59°C/1 bar, a
drop of liquid fell back into the dry boiler flask and exploded violently [2]. The
explosive decomposition has been studied in detail [3], and existing data on thermal
explosion parameters have been re-examined and discrepancies eliminated [4].
See Ethyl isocyanide
See related CYANO COMPOUNDS *See other* ENDOTHERMIC COMPOUNDS

0760. Glycolonitrile (Hydroxyacetonitrile)
   [107-16-4] C_2H_3NO

                          $HOCH_2C{\equiv}N$

1. Anon., *BCISC Quart. Safety Summ.*, 1964, **35**, 2
2. Gaudry, R., *Org. Synth.*, 1955, Coll. Vol. 3, 436
3. Anon., *Chem. Eng. News*, 1966, **44**(49), 50
4. Parker, A. J. *et al.*, *Aust. J. Chem.*, 1978, **31**, 1187
A year-old bottled sample, containing syrupy phosphoric acid as stabiliser and
which already showed signs of tar formation, exploded in storage. The pressure
explosion appeared to be due to polymerisation, after occlusion of inhibitor in the
tar, in a container in which the stopper had become cemented by polymer [1].
A similar pressure explosion occurred when dry redistilled nitrile, stabilised with
ethanol [2], polymerised after 13 days [3]. The spontaneous and violent decomposition of the nitrile on standing for more than a week is usually preceded by
formation of a red polymer [4].
See other POLYMERISATION INCIDENTS

Alkali

Sorbe, 1968, 146

Traces of alkali promote violent polymerisation.

See other CYANO COMPOUNDS

†0761. Methyl isocyanate
[624-83-9]

$$H_3CN{=}C{=}O$$

C₂H₃NO

HCS 1980, 570

1. Anon., *CISHC Chem. Safety Summ.*, 1979, **50**, 93
2. Author's comments
3. Anon., *Chem. & Ind.*, 1985, 202
4. Anon., *Loss Prev. Bull.*, 1985, (063), 1–8
5. Worthy, W., *Chem. Eng. News*, 1985, **63**(6), 27–33
6. D'Silva, T. D. J., *J. Org. Chem.*, 1986, **51**, 3781–3788
7. Bowander, B. *et al.*, *J. Haz. Mat.*, 1988, **19**, 237–269

Bottled samples had been supplied 3 years previously packed in vermiculite in sealed cans. The caps had disintegrated in storage allowing the volatile contents to escape into the external can. When this was pierced, a jet of isocyanate sprayed out uncontrollably. Improved caps are now used for this material [1].

There was a disastrous release of methyl isocyanate at Bhopal in December 1984 which caused over 2000 fatalities. Although making it in a continuous process, the plant stored several weeks supply of the isocyanate. An exothermic reaction in one of the buried 13 m³ methyl isocyanate storage vessels raised the internal temperature to above the boiling point (39°C). A cooling system existed but appears to have been turned off at the time. The continuing exotherm caused a further substantial increase in pressure, which was eventually relieved through the emergency alkali scrubber and flare stack system. Unfortunately the scrubbing system proved inadequate to deal with the rapid vapour release, and the flare was not alight, so that of the 35 t of isocyanate in the tank, some 23 t were released during the 2 h the safety valve was open.

The origin of the exothermic reaction was not clear at the time, but it was known that alkyl isocyanates would trimerise to trialkyl isocyanurates in the presence of various catalysts, would react with amines to form alkylureas, and were hydrolysed by water in presence of catalytic amounts of acids or bases, with evolution of carbon dioxide [2]. The key role of water has now been confirmed [3]. The Union Carbide report on the probable causes and course of the Bhopal disaster has been summarised [4,5]. Analysis of the residue in the distorted tank led to proposal of a mechanism, involving both water and chloroform as contaminants, to account for the detailed composition of the residue [6]. A detailed analysis of the many factors and failings which led to the ingress of water into the storage tank and to the subsequent disastrous large-scale vapour release has been presented [7]. Shortly after the event fatalities were estimated as 2,500, but such tolls do not shrink in retelling and figures up to 8,000 have since been given.

Bhopal exhibits several features of maladministration of chemical industry familiar to many who work in it on other continents. Construction of a fine chemical plant on a green field site soon causes local planners to regard the surrounding area as suitable for cheap high-density housing. As Trevor Kletz has reiterated in numerous books, add-on safety features, of which there were many, are a menace. If multiple, any one will be turned off as soon as it causes trouble or expense, and reliance be placed on the others which may be ill-designed and will, in any case, be later turned off in full confidence that the first is still working.

See related ORGANIC ISOCYANATES

0762. Acetyl nitrite
[5813-49-0] $C_2H_3NO_3$

$$H_3CCO.ON:O$$

Francesconi, L. *et al., Gazz. Chim. Ital. [I],* 1895, **34**, 439
An unstable liquid, decomposed by light, of which the vapour is violently explosive on mild heating.

See other ACYL NITRITES

0763. Nitroacetaldehyde (Nitroethanal)
[44397-85-5] $C_2H_3NO_3$

$$O_2NCH_2CHO$$

Beilstein, Vol. 1, **III**, 2678
Decomposes on standing, explodes on contact with flame.

See other C-NITRO COMPOUNDS

0764. Poly(vinyl nitrate) (Poly(ethenyl nitrate))
[ ] $(C_2H_3NO_3)_n$

$$(-CH_2CH(ONO_2)-)_n$$

Durgapal, U. C. *et al., Propellants, Explosives, Pyrotechnics,* 1995, **20**, 64
This compound, made by mild nitration of polyvinyl alcohol, is an explosive as shock sensitive as tetryl.

See other ALKYL NITRATES

0765. Acetyl nitrate
[591-09-3] $C_2H_3NO_4$

$$H_3CCO.ONO_2$$

1. Pictet, A. *et al., Ber.,* 1907, **40**, 1164
2. Bordwell, F. G. *et al., J. Amer. Chem. Soc.,* 1960, **82**, 3588
3. Konig, W., *Angew. Chem.,* 1955, **667**, 517
4. Wibaut, J. P., *Chem. Weekbl.,* 1942, **39**, 534

Acetyl nitrate, readily formed above 0°C from acetic anhydride and conc. nitric acid, is thermally unstable and its solutions may decompose violently above 60°C (forming tetranitromethane, a powerful oxidant). The pure nitrate explodes violently on rapid heating to above 100°C [1]. Isolation before use as a nitrating agent at −10°C is not necessary, but the mixture must be preformed at 20–25° before cooling to −10°C to avoid violent reactions [2]. Spontaneous explosions of pure isolated material a few days old had been reported previously [3]. Spontaneous detonation of freshly distilled nitrate has also been recorded, as well as explosion on touching with a glass rod [4].

Ethyl 3,4-dihydroxybenzenesulfonate, Oleum
Dohmen, E. A. M. F. *et al., Chem. Weekbl.*, 1942, **39**, 447–448
Attempted nitration of the sulfonate in 20% oleum led to a violent explosion, probably from decomposition of the nitrate.

Mercury(II) oxide
Chrétien, A. *et al., Compt. rend.*, 1945, **220**, 823
Acetyl nitrate explodes when mixed with red mercuric oxide, or other 'active' (catalytically) oxides.
See other ACYL NITRATES

0766. Peroxyacetyl nitrate (Acetyl nitro peroxide)
[2278-22-0] $C_2H_3NO_5$
$$H_3CCO.OONO_2$$

1. Stephens, E. R. *et al., J. Air Poll. Control Ass.*, 1969, **19**, 261–264
2. Stephens, E. R., *Anal. Chem.*, 1964, **36**, 928
3. Louw, R. *et al., J. Amer. Chem. Soc.*, 1975, **97**, 4396
Accidental production of the liquid nitrate during overcooled storage at 0°C of a rich mixture of the vapour with helium is thought to have caused a violent explosion [1]. Mixtures of up to 0.1 vol% at 7 bar stored at 10°C or above are quite safe. During dilution with nitrogen of vapour samples in evacuated bulbs, slow pressurisation is necessary to avoid explosion [2]. It is extremely explosive and may only be handled in high dilution with nitrogen. The propionyl analogue is similarly explosive, but higher homologues less so [3].
See related ACYL NITRATES, PEROXYESTERS

0767. Mercaptoacetonitrile (Acetonitrilethiol)
[54524-31-1] C_2H_3NS
$$HSCH_2C\equiv N$$

Mathias, E. *et al., J. Chem. Soc., Chem. Comm.*, 1981, 569–570
Great caution is necessary when handling conc. solutions or the solvent-free nitrile. Phosphoric acid stabilises it to some extent, but sudden foaming polymerisation

may occur. Gaseous decomposition products may burst the container, even when kept in refrigerated storage.
See other CYANO COMPOUNDS, GAS EVOLUTION INCIDENTS

0768. 1,2,3-Triazole
[27070-49-1] $C_2H_3N_3$

Baltzer, O. *et al., Ann.*, 1891, **262**, 320, 322
The vapour readily explodes if superheated (above 200°C), and the silver derivative explodes on heating.
See other TRIAZOLES

0769. 1,2,4-Triazole
[288-88-0] $C_2H_3N_3$

Energy of decomposition (in range 280–420°C) measured as 1.1 kJ/g.
See entry THERMOCHEMISTRY AND EXOTHERMIC DECOMPOSITION (reference 2)
See other TRIAZOLES

0770. Vinyl azide
[7570-25-4] $C_2H_3N_3$

$$H_2C=CHN_3$$

Wiley, R. H. *et al., J. Org. Chem.*, 1957, **22**, 995
A sample contained in a flask detonated when the ground joint was rotated. Literature statements that it is surprisingly stable are erroneous.
See other ORGANIC AZIDES

0771. Acetyl azide
[24156-53-4] $C_2H_3N_3O$

$$H_3CCO.N_3$$

Smith, 1966, Vol. 2, 214
It is treacherously explosive.
See other ACYL AZIDES

0772. Azidoacetaldehyde
[67880-11-9] $C_2H_3N_3O$

$$N_3CH_2CO.H$$

Forster, M. O. *et al., J. Chem. Soc.*, 1908, **93**, 1870
It decomposed with vigorous gas evolution below 80°C at 5 mbar. A preparative
reaction mixture of chloroacetaldehyde hydrate and sodium azide had previously
exploded mildly on heating in absence of added water.
See other 2-AZIDOCARBONYL COMPOUNDS

0773. 5-Methoxy-1,2,3,4-thiatriazole
[19155-52-3] $C_2H_3N_3OS$

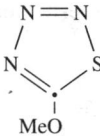

Jensen, K. A. *et al., Acta. Chem. Scand.*, 1964, **18**, 825
It explodes at ambient temperature, and its higher homologues are unstable.
See related TRIAZOLES

0774. Azidoacetic acid
[18523-48-3] $C_2H_3N_3O_2$

$$N_3CH_2CO.OH$$

Iron, or Iron salts
Borowski, S. J., *Chem. Eng. News*, 1976, **54**(44), 5
The pure acid is insensitive to shock at up to 175°C, and not explosively unstable at
up to 250°C. In contrast, the presence of iron or its salts leads to rapid exothermic
decomposition of the acid at 25°C, and explosion at 90°C or lower under strong
illumination with visible light.
See other CATALYTIC IMPURITY INCIDENTS, IRRADIATION DECOMPOSITION INCI-
DENTS
Azidoacetyl chloride *See other* 2-AZIDOCARBONYL COMPOUNDS, ORGANIC ACIDS

0775. 1,1,1-Trinitroethane
[595-86-8] $C_2H_3N_3O_6$

$$(O_2N)_3CCH_3$$

Mukhametshin, F. M., Russ. Pat. 515 378, 1976
Safety aspects of the preparation from halotrinitromethanes or tetranitromethane
by treatment with methyl iodide are improved by use of an aprotic solvent (DMF,
DMSO or HMPA) in diethyl ether or carbon tetrachloride at 30–60°C.
See other POLYNITROALKYL COMPOUNDS

0776. 2,2,2-Trinitroethanol
[918-54-7] $C_2H_3N_3O_7$

$$(O_2N)_3CCH_2OH$$

1. Marans, N. S. *et al., J. Amer. Chem. Soc.*, 1950, **72**, 5329
2. Cochoy, R. E. *et al., J. Org. Chem.*, 1972, **37**, 3041
3. Shiino, K., *Chem. Abs.*, 1978, **88**, 104580

It is a moderately shock-sensitive explosive which has exploded during distillation [1]. A series of its esters, expected to be explosive, was prepared [2], and explosive ether and ester derivatives are reviewed [3].

See other POLYNITROALKYL COMPOUNDS

0777. 4-Nitroamino-1,2,4-triazole
[52096-16-9] $C_2H_3N_5O_2$

It explodes at the m.p., 72°C.

See entry N-AZOLIUM NITROIMIDATES *See other* TRIAZOLES

0778. 1,3-Di(5-tetrazolyl)triazene
[56929-36-3] $C_2H_3N_{11}$

Hofman, K. A. *et al., Ber.*, 1910, **43**, 1869–1870

The barium salt explodes weakly on heating, and the copper salt and silver salt strongly on heating or friction.

See other TETRAZOLES, TRIAZENES

0779. Sodium acetate
[127-09-3] $C_2H_3NaO_2$

$$H_3CCO.ONa$$

HCS 1980, 824

Diketene
See Diketene: Acids, etc.

Potassium nitrate
See Potassium nitrate: Sodium acetate
See other METAL OXONON-METALLATES

0780. Sodium peroxyacetate (Sodium ethaneperoxoate)
[64057-57-4]   C₂H₃NaO₃

H₃CCO.OONa

Humber, L. G., *J. Org. Chem.*, 1959, **24**, 1789
A sample of the dry salt exploded at room temperature.
See other PEROXOACID SALTS, METAL OXONON-METALLATES

†**0781. Ethylene (Ethene)**
[74-85-1]   C₂H₄

H₂C=CH₂

FPA H17, 1973; *HCS 1980*, 458 (cryogenic liquid), 477 (cylinder gas)

1. Waterman, H. I. *et al., J. Inst. Petr. Tech.*, 1931, **17**, 505–510
2. Stull, 1977, 15–16
3. Conrad, D. *et al., Chem. Ing. Tech.*, 1975, **47**, 265 (summary of full paper)
4. Luft, G. *et al., Chem. Ing. Tech.*, 1978, **50**, 620–622
5. Lawrence, W. W. *et al., Loss Prev.*, 1967, **1**, 10–12
6. McKay, F. F. *et al., Hydrocarbon Proc.*, 1977, **56**(11), 487–494
7. Tanimoto, S., *Safety in Polyethylene Plants*, 1974, **2**, 14–21
8. Anon., *Oil Gas J.*, 1982, (1), 54
9. Watanabe, H. *et al., Chem. Abs.*, 1983, **98**, 218185
10. Brown, D. G., *Proc. 4th Int. Symp. Loss Prev. Safety Prom. Proc. Ind.*, Vol. 1, M1–M10, Rugby, IChE, 1983
11. Britton, L. G., *Process Safety Progr.*, 1994, **13**(3), 128
12. Britton, L. G. *et al., Plant/Oper. Progr.*, 1986, **5**, 238–251
13. Gebauer, M. *et al.*, E. Ger. Pats DD 251 260–2, *Chem. Abs.*, 1988, **109**, 55464–6
14. Britton, L. G., *Process Safety Progr.*, 1996, **15**(3), 128
15. Black, D. E., *Int. Conf. Workshop Risk Anal. Process Saf.*, 1997, 461

Explosive decomposition in absence of air occurred at 350°C under a pressure of 170 bar [1]. An experimental investigation of the decomposition of ethylene (which is moderately endothermic, H°_f (g) +52.3 kJ/mol, 2.18 kJ/g) at 68 bar with high energy initiation (hot wire plus guncotton) gave results which were in close agreement with those calculated from thermodynamic data (exotherm of 4.33 kJ/g, flame temperature 1360°C, 6-fold pressure increase) [2]. The limiting pressures and temperatures for explosive decomposition of ethylene with electric initiation were determined in the ranges 100–250 bar and 120–250°C. Limiting conditions are much lower for high energy (exploding wire) initiation [3]. The study of spontaneous and initiated decomposition has been extended up to 5000 bar and 450°C, in the context of high pressure polymerisation of ethylene [4]. Previously, reaction parameters including effect of pipe size and presence of nitrogen as inert diluent upon the propagation of explosive decomposition had been studied using thermite initiation: 10 vol% of nitrogen markedly interferes with propagation [5].

Sudden pressurising with ethylene of part of an air-containing pipeline system from 1 to 88.5 bar led to adiabatic compressive heating and autoignition of the ethylene. The decomposition flame slowly propagated upstream into the main 30 cm pipeline and subsequent rupture of the latter occurred at an estimated wall temperature of 700–800°C. The AIT for ethylene in air at 1 bar is 492°C, but this falls with increase in pressure and could lie between 204 and 371°C at 68–102 bar, and the ethylene decomposition reaction may start at 315–371° within these pressure limits. Detonation was not, however, observed. Precautions against such incidents in ethylene installations are discussed [6]. In a published symposium mainly devoted to engineering aspects of large-scale polyethylene manufacture, this paper describes an extensive large-scale field experiment to determine the blast and destructive effects when various ethylene–air mixtures were detonated to simulate a compressor house explosion [7]. A fire and explosion in an ethylene separator was attributed to ingress of air which formed a catalytically active species which initiated exothermic polymerisation eventually leading to pipe rupture [8]. The decomposition of ethylene at 100–400 bar/20–149°C has been studied in detail [9]. An account of a decomposition incident at a high density polythene plant which was considered safe because it operated with ethylene diluted in a solvent is available [15].

In a review of low level ethylene decomposition since its discovery in 1842, the main factors are recognised as pressure–temperature–ignition relationships, heat balance in reaction zone and presence of impurities or contaminants. Case histories are discussed and safety practices and general recommendations are presented [10]. In a survey of incidents on high pressure ethylene plant, most appeared to involve overheating on molecular sieves in driers from polymerisation under conditions of reduced flow. Large pored sieves are not advisable [11]. In a review of the stability of ethylene at elevated pressures, the factors of thermal runaway, catalysis, influence of contaminants, and adsorptive heating, compressive ignition, and direct ignition are discussed, the effects of decomposition being described in terms of pressure and propagation [12]. A series of 3 patents describe start-up procedures to avoid explosive decomposition of ethylene during high-pressure polymerisation at 1600 bar using peroxide initiation. The procedures involve use of ethylene–hydrogen mixtures (6–30:1) with compression in several stages, interspersed with circulation through the reactor [13]. Another study of the slow deflagrative decomposition of ethene at around 300 bar to mostly methane, hydrogen and carbon is reported. Addition of hydrogen facilitates ignition, but not propagation, and increases the proportion of methane in the product. Hot spots, thence initiation, may develop over finely divided iron, starting from hydrogen reduction of rust from as low as 70°C [14].

Acetylene, Hydrogen
See Hydrogen (Gas): Acetylene, etc.

Air, Chlorine
Fujiwara, T. *et al.*, Jap. Pat., 74 32 841, 1974

During oxychlorination of ethylene to 1,2-dichloroethane, excess hydrogen chloride is used to maintain the reaction mixture outside the explosive limits.

Air, Polyethylene
Fonin, M. F. *et al., Chem. Abs.*, 1984, **101**, 111718
The solubility of ethylene in freshly prepared polyethylene, and its diffusion out of the latter were studied in relation to the formation of explosive ethylene–air mixtures in storage. Explosive mixtures may be formed, because the solubility of ethylene in its polymer (e.g. 1130 ppm w/w at 30°C) considerably exceeds the concentration (30 ppm at 30°C) necessary to exceed the lower explosive limit above the gas-containing polymer in closed storage, and the diffusion coefficient is also 30% higher than for aged polymer samples.

Aluminium chloride, Catalysts
Waterman, H. I. *et al., J. Inst. Pet.*, 1947, **33**, 254
Mixtures of ethylene and aluminium chloride, initially at 30–60 bar, rapidly heat and explode in presence of supported nickel catalysts, methyl chloride or nitromethane.
See Aluminium chloride: Alkenes

Aluminium chloride, Nitromethane
See Aluminium chloride–nitromethane: An alkene

Boron trioxide, Oxygen
Ts'olkovskii, T. I. *et al., Chem. Abs.*, 1979, **90**, 8511
AITs for ethylene–oxygen mixtures at 1 bar in stainless steel were reduced by 30–40°C by coating the vessel walls with boron trioxide.

Bromotrichloromethane   MRH 1.21/90
Elsner, H. *et al., Angew. Chem.*, 1962, **74**, 253
Following a literature method for preparation of 1-bromo-3,3,3-trichloropropane, the reagents were being heated at 120°C/51 bar. During the fourth preparation, a violent explosion occurred.
See Halocarbons, below

Carbon tetrachloride   MRH 1.55/85
Zakaznov, V. F. *et al., Khim. Prom.*, 1968, **8**, 584
Mixtures of ethylene and carbon tetrachloride can be initiated to explode at temperatures betwen 25 and 105°C and pressures of 30–80 bar, causing a six-fold pressure increase. At 100°C and 61 bar, explosion initiated in the gas phase propagated into the liquid phase. Increase of halocarbon conc. in the gas phase decreased the limiting decomposition pressure.
See Halocarbons, below, *also* Dibenzoyl peroxide: Carbon tetrachloride, etc.

Chlorine
491M, 1975, 103

Interaction is explosive when catalysed by sunlight or UV irradiation, or in presence of mercury(I) oxide, mercury(II) oxide or silver oxide.
See Chlorine: Hydrocarbons

Chlorotrifluoroethylene
Colombo, P. *et al., J. Polymer Sci.*, 1963, **B1**(8), 435–436
Mixtures containing ratios of about 20:1 and 12:1 of ethylene:haloalkene undergoing polymerisation under gamma irradiation at 308 krad/h exploded violently after a total dose of 50 krad. Dose rate and haloalkene conc. were both involved in the initiation process.
See Halocarbons, below

Copper
Dunstan, A. E. *et al., J. Soc. Chem. Ind.*, 1932, **51**, 1321
Polymerisation of ethylene in presence of metallic copper becomes violent above a pressure of 54 bar at about 400°C, much carbon being deposited.

Halocarbons
See Bromotrichloromethane, and Carbon tetrachloride, and Chlorotrifluoroethylene, all above; Tetrafluoroethylene, below

Lithium
See Lithium: Ethylene

Molecular sieve
See MOLECULAR SIEVE INCIDENTS: Ethylene

Other reactants
Yoshida, 1980, 46
The MRH values calculated and given for 5 combinations are noted above and below.

Oxides of nitrogen
See Nitrogen oxide: Dienes, Oxygen

Ozone
See Ozone: Combustible gases, or: Ethylene

Steel-braced tyres
Anon., *CISHC Chem. Safety Summ.*, 1974–5, **45–46**, 2–3
Two hours after a road tanker had crashed, causing the load of liquid ethylene to leak, one of the tyres of the tanker burst and ignited the spill, eventually causing the whole tanker to explode. The tyre failed because it froze and became embrittled, and it is known that such failure of steel-braced tyres gives off showers of sparks. This could therefore be a common ignition source in cryogenic transportation spillage incidents.
See other IGNITION SOURCES

Tetrafluoroethylene

Coffman, D. D. *et al., J. Amer. Chem. Soc.*, 1949, **71**, 492

A violent explosion occurred when a mixture of tetrafluoroethylene and excess ethylene was heated at 160°C and 480 bar. Traces of oxygen must be rigorously excluded. Other alkenes reacted smoothly.

See Oxygen (Gas): Tetrafluoroethylene, *also* Halocarbons, above

Trifluoromethyl hypofluorite

See Trifluoromethyl hypofluorite: Hydrocarbons

Vinyl acetate

Albert J. *et al., Chem. Eng. Process*, 1998, **37**(1), 55

At pressures above 25 Mpa and temperatures above 250°C, vinyl acetate admixture destabilises ethylene and increases the maximum explosion pressure from its decomposition.

See other ALKENES

0782. Poly(ethylene)

[9002-88-4] $(C_2H_4)_n$

Complex structure

Experimental investigation of poly(ethylene) powder explosions.

See entry DUST EXPLOSION INCIDENTS (reference 18)

0783. 1-Bromoaziridine

[19816-89-8] C_2H_4BrN

Graefe, A. F., *J. Amer. Chem. Soc.*, 1958, **80**, 3940

The compound is very unstable and always decomposes, sometimes explosively, during or shortly after distillation.

See other AZIRIDINES, *N*-HALOGEN COMPOUNDS

0784. *N*-Bromoacetamide

[79-15-2] C_2H_4BrNO

$H_3CCO.NHBr$

'Organic Positive Bromine Compounds', Brochure, Boulder (Co.) Arapahoe Chemicals Inc., 1962

It tends to decompose rapidly at elevated temperatures in presence of moisture and light.

See other *N*-HALOGEN COMPOUNDS

0785. 1,2-Dibromoethane (Ethylene dibromide)
[106-93-4]   C₂H₄Br₂

$$BrCH_2CH_2Br$$

HCS 1980, 365

Magnesium
   See Magnesium: Halocarbons
   See other HALOALKANES

0786. 1-Chloroaziridine
[25167-31-1]   C₂H₄ClN

1. Davies, C. S., *Chem. Eng. News*, 1964, **42**(8), 41
2. Graefe, A. F., *Chem. Eng. News*, 1958, **36**(43), 52
A sample of redistilled material exploded after keeping in an amber bottle at ambient temperature for 3 months [1]. A similar sample had exploded very violently when dropped [2].
See other AZIRIDINES, *N*-HALOGEN COMPOUNDS

0787. Chloroacetaldehyde oxime
[51451-05-9]   C₂H₄ClNO

$$ClCH_2CH=NOH$$

1. Brintzinger, H. *et al.*, *Chem. Ber.*, 1952, **85**, 345
2. Anon., *Chem. Eng. News*, 1993, **71**(34), 6
Vacuum distillation of the product at 61°C/27 mbar must be interrupted when a solid separates from the residue, to avoid an explosion [1]. A serious explosion occurred in a cooling loop containing the oxime and hydrochloric acid, during shutdown to clear a blockage elsewhere in a pesticide plant [2].
See Bromoacetone oxime
See other OXIMES
See related ALDEHYDES

0788. Chloroacetamide
[79-07-2]   C₂H₄ClNO

$$ClCH_2CO.NH_2$$

Energy of decomposition (in range 170–300°C) measured as 0.67 kJ/g.
See entry THERMOCHEMISTRY AND EXOTHERMIC DECOMPOSITION (reference 2)

293

0789. *N*-Chloroacetamide
[598-49-2]   C$_2$H$_4$ClNO

MeCO.NHCl

Muir, G. D., private comm., 1968
It has exploded during desiccation of the solid or during concentration of its
chloroform solution. It may be purifed by pouring a solution in acetone into water
and air-drying the product.
See other N-HALOGEN COMPOUNDS

†**0790. 1,1-Dichloroethane**
[75-34-3]   C$_2$H$_4$Cl$_2$

Cl$_2$CHCH$_3$

HCS 1980, 494

See other HALOALKANES

†**0791. 1,2-Dichloroethane (Ethylene dichloride)**
[107-06-2]   C$_2$H$_4$Cl$_2$

ClCH$_2$CH$_2$Cl

(*MCA SD-18*, 1971); *NSC 350*, 1977; *FPA H77*, 1979; *HCS 1980*, 482

Chlorine
See Chlorine: 1,2-Dichloroethane

3-Dimethylaminopropylamine
491M, 1975, 171
After the amine was charged into a tank containing some residual wet halocarbon,
a violent explosion occurred later. Subsequent investigation showed it was an
extremely hazardous combination (as dehydrochlorination of the solvent would
give acetylene).

Dinitrogen tetraoxide
See Dinitrogen tetraoxide: Halocarbons

Metals
See Aluminium: Halocarbons
     Potassium: Halocarbons

Nitric acid
See Nitric acid: 1,2-Dichloroethane
See other HALOALKANES

0792. Azo-*N*-chloroformamidine
[502-98-7] $C_2H_4Cl_2N_6$

$$H_2NC(:NCl)N=NC(:NCl)NH_2$$

Braz, G. I. *et al., Chem. Abs.*, 1946, **40**, 2267.9
Decomposition is explosive at 155°C, and accelerated by contact with metals.
See other AZO COMPOUNDS, *N*-HALOGEN COMPOUNDS

†0793. Bis(chloromethyl) ether
[542-88-1] $C_2H_4Cl_2O$

$$ClCH_2OCH_2Cl$$

HCS 1980, 377; *RSC Lab. Hazards Data Sheet No. 46*, 1986

0794. 1,1-Dichloroethyl hydroperoxide
[90584-32-0] $C_2H_4Cl_2O_2$

$$MeCCl_2OOH$$

Gäb, S. *et al., J. Org. Chem.*, 1984, **49**, 2714
Samples of the oil, either neat or as conc. solutions, rapidly decomposed exother-
mally on warming to ambient temperature. Although no explosions were experi-
enced, caution is advised.
See Ozone: trans-2,3-Dichloro-2-butene
See other ALKYL HYDROPEROXIDES

0795. Ethylene diperchlorate
[52936-25-1] $C_2H_4Cl_2O_8$

$$(-CH_2OClO_3)_2$$

Schumacher, 1960, 214
A highly sensitive, violently explosive material, capable of detonation by addition
of a few drops of water.
See other ALKYL PERCHLORATES

0796. Bis(chloromethyl)thallium chloride
[ ] $C_2H_4Cl_3Tl$

$$(ClCH_2)_2TlCl$$

Yakubovich, A. Ya. *et al., Dokl. Akad. Nauk SSSR*, 1950, **3**, 957
An explosive solid of low stability.
See related ALKYLMETAL HALIDES

0797. 1,2-Bis(dichlorophosphino)ethane
[28240-69-9] $C_2H_4Cl_4P_2$

$$Cl_2PC_2H_4PCl_2$$

Chatt, J. *et al., J. Chem. Soc., Dalton Trans.*, 1985, 1131–1136
A modified method of preparation from ethylene, phosphorus trichloride and phosphorus in a stainless autoclave gives a highly pyrophoric reaction residue.
See related ALKYLHALOPHOSPHINES, PYROPHORIC MATERIALS

0798. Ethylenedicaesium
[65313-36-2] $C_2H_4Cs_2$

$$CsCH_2CH_2Cs$$

Leleu, *Cahiers*, 1977, (88), 366
Reacts violently with water.
See related ALKYLMETALS

0799. Hydroxycopper(II) glyoximate
[63643-78-7] $C_2H_4CuN_2O_3$

$$HOCuON{=}CHCH{=}NOH$$

Morpurgo, G. O. *et al., J. Chem. Soc., Dalton Trans.*, 1977, 746
The complex loses weight up to 140°C, then explodes.
See related OXIMES *See other* HEAVY METAL DERIVATIVES

0800. 2-Fluoro-2,2-dinitroethylamine
[18139-02-1] $C_2H_4FN_3O_4$

$$FC(NO_2)_2CH_2NH_2$$

Adolph, H. G. *et al., J. Org. Chem.*, 1969, **34**, 47
Outstandingly explosive among fluorodinitromethyl compounds, samples stored neat at ambient temperatures regularly exploded within a few hours. Occasionally, concentrated solutions in dichloromethane have decomposed violently after long storage.
See other FLUORODINITROMETHYL COMPOUNDS

†0801. 1,1-Difluoroethane
[75-37-6] $C_2H_4F_2$

$$F_2CHCH_3$$

HCS 1980, 408 (cylinder)

See other HALOALKANES

0802. 1,2-Bis(difluoroamino)ethanol
[13084-47-4] $C_2H_4F_4N_2O$

$$F_2NCH_2CH(NF_2)OH$$

Reed. S. F., *J. Org. Chem.*, 1967, **32**, 2894
It is slightly more impact-sensitive than glyceryl trinitrate.
See other DIFLUOROAMINO COMPOUNDS

0803. 1,2-Bis(difluoroamino)-*N*-nitroethylamine
[18273-30-8] $C_2H_4F_4N_4O_2$

$$F_2NCH_2CH(NF_2)NHNO_2$$

Tyler, W. E., US Pat. 3 344 167, 1967
The crude product tends to explode spontaneously on storage, though the triple-distilled material appears stable on prolonged storage. Generally, such nitroamines are unstable and explode at 75°C or above.
See other CATALYTIC IMPURITY INCIDENTS, DIFLUOROAMINO COMPOUNDS, *N*-NITRO COMPOUNDS

0804. Mercury(II) formohydroxamate
[ ] $C_2H_4HgN_2O_4$

$$Hg(ONHCO.H)_2 \text{ or } Hg(ON{=}CHOH)_2$$

Urbanski, 1967, Vol. 3, 158
It possesses detonator properties.
See other MERCURY COMPOUNDS, N–O COMPOUNDS

0805. Bis(iodomethyl)zinc
[14399-53-2] $C_2H_4I_2Zn$

$$ICH_2ZnCH_2I$$

Dichloromethane
Charette, A.*Chem. Eng. News*, 1995, **73**(6), 2
An explosion was twice experienced when working with a suspension of this compound in dichloromethane at near ambient temperature.
See Copper-zinc alloys: Diiodomethane, Ether
See also MAGNETIC STIRRERS

0806. Potassium 1-nitroethoxide
[ ] $C_2H_4KNO_3$

$$KOCH(NO_2)CH_3$$

Meyer, V. *et al., Ber.*, 1872, **5**, 1032

The solid exploded on heating, and an aqueous solution sealed into a Carius tube exploded violently when the tube was cracked open with a red-hot glass rod. (The *aci*-salt may have been formed).
See other C-NITRO COMPOUNDS

0807. 2-Nitroethanol, sodium salt
[2406-51-1] $C_2H_4NNaO_3$

$$HOCH_2CH{=}NO_2^-Na^+$$

Beilstein, **H1**, 339
Deflagrates at 120°C.
See other C-NITRO COMPOUNDS

0808. 3-Methyldiazirine
[765-31-1] $C_2H_4N_2$

Schmitz, E. *et al., Chem. Ber.*, 1962, **95**, 795
The gas explodes on heating.
See other DIAZIRINES

0809. 2-Nitroacetaldehyde oxime ('Methazonic acid')
[5653-21-4] $C_2H_4N_2O_3$

$$O_2NCH_2CH{=}NOH$$

1. Sorbe, 1968, 149
2. Cooke, F., *Chem. Eng. News*, 1981, **59**(34), 3
3. Hoglen, D. K. *et al.*, Personal communication, 1997
4. Editor's comments

The oxime (an isomer of 'ethylnitrolic acid') decomposes gradually at ambient temperature, but explosively above 110°C [1]. The product from treatment of nitromethane with strong sodium hydroxide solution at 50°C, followed by acidification and ether extraction, gave, after vacuum evaporation, a residue which exploded when air was admitted. On this occasion, a weekend elapsed between alkali and acid treatments, and it seems possible that formation of fulminic acid derivatives may have occurred [2]. A 75 g sample of methazonic acid, from a preparation which had transiently overheated to 70°C, during the addition of nitromethane to sodium hydroxide solution, was stored for a week at −15°C, then allowed to warm to room temperature. Twelve hours later it spontaneously detonated, destroying a fume hood. There is no evident cause for this explosion [3]. It seems sensible to apply Occam's Razor, assume that methazonic acid is itself explosive, and cease to invoke impurities [4]. The salts of methazonic acid are also explosive.

See Nitromethane: Acids, or Bases; Alkalis
See other OXIMES
See related ALDEHYDES

0810. 1-Nitro-1-oximinoethane ('Ethylnitrolic acid')
[600-26-0] $C_2H_4N_2O_3$

$$H_3CC(NO_2)=NOH$$

Sorbe, 1968, 147
An explosive solid, formally a nitro-oxime.
See other OXIMES

0811. Ethylidene dinitrate
[55044-04-7] $C_2H_4N_2O_6$

$$H_3CCH(ONO_2)_2$$

Kacmarek, A. J. *et al., J. Org. Chem.*, 1975, **40**, 1853
An analytical sample exploded in the hot zone of the combustion apparatus.
See Dinitrogen pentaoxide: Acetaldehyde
See other ALKYL NITRATES

0812. 4-Amino-4*H*-1,2,4-triazole
[584-13-4] $C_2H_4N_4$

Preparative hazard
Tomann, J. *et al., Chem. Prum.*, 1987, **37**, 489–492
The title compound (with 66.5% nitrogen content) is prepared by condensing formylhydrazine (2 mols, with elimination of $2H_2O$) by heating to 170°C. During a pilot production run in a 500 l reactor, an explosion destroyed the vessel. The heat of decomposition of the compound was determined by thermal analysis as 1.5 kJ/g, with an energy of activation of 91 kJ/mol.
See other HIGH-NITROGEN COMPOUNDS, TRIAZOLES

0813. Cyanoguanidine ('Dicyanodiamide')
[461-58-5] $C_2H_4N_4$

$$N\equiv CNHC(:NH)NH_2$$

HCS 1980, 386

Oxidants
Baumann, J., *Chem. Ztg.*, 1920, **44**, 474
Mixtures of cyanoguanidine with ammonium nitrate, potassium chlorate, etc., were
formerly proposed for use as powerful explosives.
See other CYANO COMPOUNDS, HIGH-NITROGEN COMPOUNDS

0814. 2-Methyltetrazole
[16681-78-0]   C$_2$H$_4$N$_4$

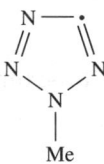

Aluminium hydride
See Aluminium hydride: Tetrazole derivatives
See other TETRAZOLES

0815. Azoformaldoxime
[74936-21-3]   C$_2$H174N$_4$O$_2$
$$HON=CHN=NCH=NOH$$

Armand, J. *et al., J. Chem. Res.; Synop.*, 1980, 304; *Microf.* 1980, 3853–3869
The dioxime exploded on melting, as did the cyclised isomeric triazole derivative,
1-hydroxy-2-hydroxyamino-1,3,4-triazole.
See other AZO COMPOUNDS, N–O COMPOUNDS, OXIMES

0816. Azoformamide (Azodicarboxamide)
[123-77-3]   C$_2$H$_4$N$_4$O$_2$
$$(H_2NCO.N:)_2$$

Whitmore, M. W. *et al., J. Loss Prev.*, 1993, **6**(3), 169
An 850 kg batch of a slightly doped form of azodicarbonamide exploded violently,
with a TNT equivalence of 3.3 kg, 5 minutes after sampling at the end of drying.
The probable initial temperature was 65°C, the lowest self accelerating decomposition temperature 90°C, and such decomposition is not explosive. Full explosibility
tests, including detonability, had shown no hazard. Further study demonstrated that
slightly contained azodicarboxamide, thermally initiated at the bottom of a column
or conical vessel could explode even at the 5 kg scale. The above TNT equivalence corresponds to decomposition of 4% of the available charge. The cause of
the presumptive hot spot is unknown.
    Energy of decomposition (in range 160–230°C) measured as 1.36 kJ/g by DSC,
and T$_{ait24}$ was determined as 120°C by adiabatic Dewar tests, with an apparent
energy of activation of 131 kJ/mol.

See entries BLOWING AGENTS, PRESSURE INCREASE IN EXOTHERMIC DECOMPOSI-
TION
See entry THERMOCHEMISTRY AND EXOTHERMIC DECOMPOSITION (reference 2)

Potassium bromate
See Potassium bromate: Azoformamide
See other AZO COMPOUNDS

0817. 1-Hydroxy-2-hydroxylamino-1,3,4-triazole
[76002-04-5] $C_2H_4N_4O_2$

Armand, J. *et al., J. Chem Res.; Synop.*, 1980, 304; *Microf.*, 1980, 3853–3869
The triazole exploded on melting, as did the isomeric azoformaldoxime, from
which it is prepared by cyclisation.
See other HIGH-NITROGEN COMPOUNDS, N–O COMPOUNDS, TRIAZOLES

0818. 1,1-Diazidoethane
[67880-20-0] $C_2H_4N_6$

$$(N_3)_2CHCH_3$$

Forster, M. O. *et al., J. Chem. Soc.*, 1908, **93**, 1070
The extreme instability and explosive behaviour of this diazide caused work on
other *gem*-diazides to be abandoned.
See other ORGANIC AZIDES

0819. 1,2-Diazidoethane
[629-13-0] $C_2H_4N_6$

$$N_3C_2H_4N_3$$

Alone, or Sulfuric acid
Forster, M. O. *et al., J. Chem. Soc.*, 1908, **93**, 1070
Though less unstable than the 1,1-isomer (next above), it explodes on heating, and
irreproducibly in contact with sulfuric acid.
See other ORGANIC AZIDES

0820. Azidocarbonylguanidine
[54567-24-7] $C_2H_4N_6O$

$$N_3CO.NHC(:NH)NH_2$$

Thiele, J. *et al., Ann.*, 1898, **303**, 93
It explodes violently on rapid heating.
See related ACYL AZIDES

0821. 3,6-Diamino-1,2,4,5-tetrazine-1-oxide
[153757-93-8]

$C_2H_4N_6O$

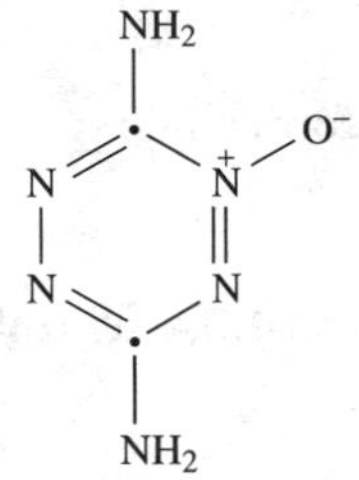

See 3,6-Diamino-1,2,4,5-tetrazine-1,4-dioxide
See other HIGH-NITROGEN COMPOUNDS, See also AMINE OXIDES

0822. 3,6-Diamino-1,2,4,5-tetrazine-1,4-dioxide
[153757-93-8]

$C_2H_4N_6O_2$

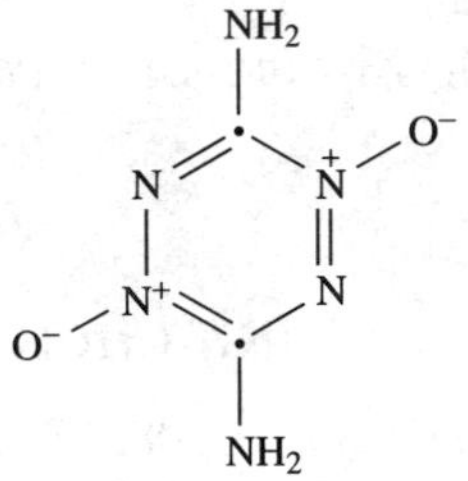

Coburn, M. D. et al., J. Heterocycl. Chem., 1993, **30**(6), 1593
This, the mono-oxide and the 3-amino-6-nitro-2,4-dioxide are all explosives to be handled with care. All are thermally stable well above 100°C.
See other HIGH-NITROGEN COMPOUNDS, See also AMINE OXIDES

0823. Ammonium 3,5-dinitro-1,2,4-triazolide
[76556-13-3]

$C_2H_4N_6O_4$

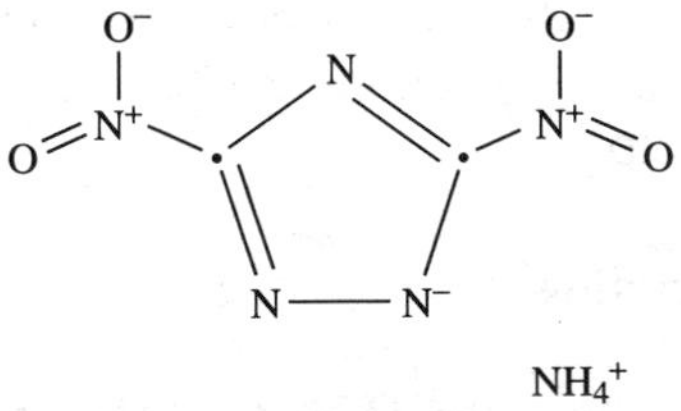

Selig, W., Propellants Explos., 1981, **6**, 96–98
It is an explosive.
See other TRIAZOLES

0824. 1,3-Diazido-2-nitroazapropane (N,N-Bis(azidomethyl)nitric amide)
[67362-62-3] $C_2H_4N_8O_2$

$$(N_3CH_2)_2NNO_2$$

Flanagan, J. E. *et al.*, US Pat. 4 085 123, 1978
It is useful as a high energy plasticiser for propellants.
See other N-NITRO COMPOUNDS, ORGANIC AZIDES

0825. Azo-N-nitroformamidine
[53144-64-2] $C_2H_4N_8O_4$

$$(O_2NNHC(:NH)N:)_2$$

Wright, G. F., *Can. J. Chem.*, 1952, **30**, 64
Explosive decomposition at 165°C.
See other AZO COMPOUNDS, N-NITRO COMPOUNDS

0826. 5,5′-Hydrazotetrazole
[74999-19-2] $C_2H_4N_{10}$

1. Thiele, J. *et al.*, *Ann.*, 1898, **303**, 66
2. Reddy, G. O., *J. Haz. Mat.*, 1984, **9** 291–303
It explodes without melting when heated [1]. The title compound and its barium, lead(II), and mercury(II) salts were studied by DTA and DSC techniques. The lead salt is the least stable and most powerful primary explosive of the 4 compounds [2].
See other HIGH-NITROGEN COMPOUNDS, TETRAZOLES

0827. 1,6-Bis(5-tetrazolyl)hexaaza-1,5-diene
[68594-19-4] $C_2H_4N_{14}$

Hofman, K. A. *et al.*, *Ber.*, 1911, **44**, 2953
This very high-nitrogen compound (87.5%) explodes violently on pressing with a glass rod, or on heating to 90°C.
See other HIGH-NITROGEN COMPOUNDS, TETRAZOLES

†0828. **Acetaldehyde (Ethanal)**
[75-07-0]   C₂H₄O

H₃CCO.H

(*MCA SD-43*, 1952); *FPA H29*, 1974; *HCS 1980*, 97; *RSC Lab. Hazards Data Sheet No. 38*, 1985

1. White, A. G. *et al., J. Soc. Chem. Ind.*, 1950, **69**, 206
2. Vervalin, 1973, 90–91
3. Garbuzynk, T. A., *Chem. Abs.*, 1978, **89**, 131835; 1989, **111**, 117216b
4. Nalbandyan, A. B. *et al., Prog. Astronaut. Aeronaut.*, 1988, **113**, 58; *Chem. Abs.*, 1989, **111**, 22912; *Khim. Fiz.*, 1988, **7**(12), 1709; *Chem. Abs.*, 1989, **111**, 56791

(The MCA Data Sheet describes acetaldehyde as extremely or violently reactive with: acid anhydrides, alcohols, halogens, ketones, phenols, amines, ammonia, hydrogen cyanide or hydrogen sulfide). Mixtures of 30–60% of acetaldehyde vapour with air (or 60–80% with oxygen) may ignite on surfaces at 176 and 105°C, respectively, owing to formation and subsequent violent decomposition of peroxyacetic acid [1]. Acetaldehyde vapour leaking into a building equipped only with flameproof electrical equipment nevertheless ignited, possibly on contact with rusted steel, corroded aluminium or hot steam lines [2]. The minimum AIT (130°C) for acetaldehyde in air mixtures is at 55–57 vol% of aldehyde. The effect of presence of acetone or methane upon AIT was studied, they inhibit ignition during the oxidation to peroxyacetic acid; the AIT is surface sensitive and may be as low as 80°C in quartz [3]. The mechanism of this ignition has been studied [4], it is catalysed by peroxy-acids on the surface of the reactor.
See other CATALYTIC IMPURITY INCIDENTS

Acetic acid   MRH 1.13/tr.
MCA Case History No. 1764
A drum contaminated with acetic acid was filled with acetaldehyde. The ensuing exothermic polymerisation reaction caused a mild eruption lasting several hours.

Cobalt acetate, Oxygen
1. Phillips, B. *et al. J. Amer. Chem. Soc.*, 1957, **79**, 5982
2. Bloomfield, G. F. *et al., J. Soc. Chem. Ind.*, 1935, **54**, 129T
Oxygenation of acetaldehyde in presence of cobalt acetate at −20°C caused precipitation of 1-hydroxyethyl peroxyacetate (acetaldehyde hemi-peracetate), which exploded violently on stirring [1]. Ozone or UV light also catalyses the autoxidation [2].

Desiccants, Hydrogen peroxide
See Hydrogen peroxide: Acetaldehyde, Desiccants

Dinitrogen pentaoxide
See Dinitrogen pentaoxide: Acetaldehyde

Halocarbons
Jones, J. C. *et al., Combust. Flame*, 1983, **52**, 211–213

Cool flame behaviour of acetaldehyde is apparently eliminated by *tert*-butyl bromide, and reduced by methyl iodide.

Hydrogen peroxide   MRH 6.27/79
See Hydrogen peroxide: Oxygenated compounds

Mercury(II) oxosalts
Sorbe, 1968, 97
Some of the products of interaction of acetaldehyde and mercury(II) chlorate or mercury(II) perchlorate are highly explosive and extremely shock-sensitive.
See Chloratomercurio(formyl)methylenemercury(II)

Metals
Sorbe, 1968, 103
Impure material will polymerise readily in presence of trace metals (iron) or acids.
See Sulfuric acid, below
See other CATALYTIC IMPURITY INCIDENTS

Other reactants
Yoshida, 1980, 17
MRH values calculated for 18 combinations, largely with oxidants, are given.

Oxygen   MRH 8.66/64
1. *MCA Case History No. 117*
2. Armstamyan, A. M. *et al., Chem. Abs.,* 1984, **101**, 194603
Oxygen leaked into the free space in an acetaldehyde storage tank normally purged with nitrogen. Accelerating exothermic oxidation led to detonation [1]. The self-ignition temperature of acetaldehyde–oxygen mixtures depends on dimensions of the reactor and the partial pressure of peracetic acid accumulated on the walls. Spontaneous ignition temperatures of 71–73°C were observed [2].

Silver nitrate
See Silver nitrate: Acetaldehyde

Sulfuric acid   MRH 1.34/43
Sorbe, 1968, 103
Acetaldehyde is polymerised violently by the conc. acid.
See Metals, above
See other POLYMERISATION INCIDENTS *See other* ALDEHYDES, PEROXIDISABLE COMPOUNDS

†0829. Ethylene oxide (Oxirane)
[75-21-8]   C₂H₄O

(*MCA SD-38*, 1971); *FPA H32*, 1974; *HCS 1980*, 486; *RSC Lab. Hazard Data Sheet No. 2*, 1982

1. *Guidelines for Bulk Handling of Ethylene Oxide, London, CIA, 1983*
2. Hess, L. G. *et al., Ind. Eng. Chem.*, 1950, **42**, 1251
3. Burgoyne, J. H. *et al., Proc. 1st Symp. Chem. Proc. Hazards*, Symp. Ser. No. 7, 30–36, London, IChE, 1960; *Proc. 3rd Symp. Chem. Proc. Hazards*, Symp. Ser. No. 25, 1–7, London, IChE, 1967
4. *MCA SD-38*, 1971
5. *MCA Case History No. 1666*
6. Harmon, 1974, 2.9
6a Anon., *Loss Prev. Bull.*, 1991, **100**, 1
6b Mellin, B. E., *ibid.*, 13
6c Viera, G. A. *et al., Chem. Eng. Progress*, 1993, **89**(8), 66
7. Heuser, S. G. B., *Chem. Brit.*, 1977, **13**, 317
8. Pogany, G. A., *Chem. & Ind.*, 1979, 16–21
9. Pesetsky, B. *et al., Loss Prev.*, 1980, **13**, 123–131
10. Pesetsky, B. *et al., Loss Prev.*, 1980, **13**, 132–141
11. Chen, L. D. *et al., Combust. Flame*, 1981, **40**, 13–28
12. *See entry* THERMOCHEMISTRY AND EXOTHERMIC DECOMPOSITION (reference 2)
13. Loeffler U., *Prax. Sicherheitstech.*, 1997, **4**, 31
14. Britton, L. G., *Plant/Oper. Progr.*, 1990, **9**(2), 75

The detailed code of practice [1] serves as a comprehensive guide to the industrial use of ethylene oxide to replace the US Safety Data Sheet withdrawn in 1980. Ethylene oxide vapour may readily be initiated into explosive decomposition in absence of air, and storage and handling requirements were detailed [2]. Explosive decomposition may be suppressed by many diluents, not necessarily inert [3]. Metal fittings containing copper, silver, mercury or magnesium should not be used in ethylene oxide service, since traces of acetylene in the oxide could produce metal acetylides capable of detonating the vapour [4]. Ethylene oxide exposed to heating and subsequently cooled (e.g. by exposure to fire conditions) may continue to polymerise exothermically, leading to container pressurisation and explosion. A mechanism consistent with several observed incidents is proposed [5]. Presence of hot spots in processing plants is identified as a particular hazard [6]. These can be generated by the polymerisation/autoxidation of leaks absorbed onto lagging material, with devastating results [6a, 6b, 13], or by catalysis of decomposition by γ-ferric oxide and γ-FeO(OH) [6c]. Appropriate column packing can inhibit propagation of ethylene oxide decomposition from hot spots in ethylene oxide stills [13]. Combustion properties, including ignition of leaks onto packing are reviewed thoroughly in [14]. The autoignition temperature may be as low as 140°C in presence of rust. Aerosols, detonated rather than ignited, are used in fuel/air weapons, giving a TNT equivalence of 2.4 kg/kg. Under conditions of ignition and deflagration TNT equivalence is more like zero.

Hazards attendant on use of ethylene oxide in steriliser chambers arise from difficulties in its subsequent removal by evacuation procedures, owing to its ready absorption or adsorption by the treated material. Even after 2 evacuation cycles the oxide may still be present. Safety is ensured by using the oxide diluted with up to 90% of Freon or carbon dioxide. If high concentrations of oxide are used, an inert gas purge between cycles is essential [7]. The main factors in safe handling and use on laboratory or small pilot plant scales have been identified [8]. Liquid phase decomposition, previously thought not possible, was observed under conditions of temperature and pressure outside the normal working range with different submerged igniter wires, and a mechanism is discussed [9]. The use of methane as a diluent for ethylene oxide has been studied in detail [10]. The decomposition threshold temperature for ethylene oxide has been related to pressure, ignition source and geometry of the containment vessel, pressure having the greatest effect [11]. The energy of decomposition (in range 320–490°C) has been revised to 1.516 kJ/g [12].

See entry THERMOCHEMISTRY AND EXOTHERMIC DECOMPOSITION (reference 2)
See other POLYMERISATION INCIDENTS
See also VIOLENT POLYMERISATION; INSULATION

Air, Bromomethane
Baratov, A. N. *et al.*, p.24 of BLL translation (628.74, issued 1966) of Russian book on 'Fire Prevention and Firefighting Symposium'. The addition of bromomethane to ethylene oxide (used for germicidal sterilising) to reduce the risk of explosion is relatively ineffective, the inhibiting concentration being 31.2, as against 5.8 for hexane and 13.5 vol.% for hydrogen.

Air, Ethylene
See Oxygen, below

Alkanethiols, or An alcohol
1. Meigs, D. P., *Chem. Eng. News*, 1942, **20**, 1318
2. Personal experience (PGU)
Autoclave reactions involving ethylene oxide with alkanethiols or an (unspecified) alcohol went out of control and exploded violently. Similar previous reactions had been uneventful [1]. An arenethiol was being reacted with ethylene oxide under catalysis by a fraction of a percent of sodium hydroxide (solid) dissolved in the thiol to which the oxirane was slowly charged. After an initial exotherm a white solid precipitated, the exotherm died away and later resumed, with dissolution of the solid, the reaction then running out of control from the backlog of charged oxirane [2].

Ammonia
1. *MCA Case History No. 792*
2. Troyan, J. E. *et al., Loss Prev.*, 1968, **2**, 125–130
3. Stull, 1977, 26–27
Accidental contamination by aqueous ammonia of an ethylene oxide feed tank containing 22 t caused violent polymerisation which ruptured the tank and led to a devastating vapour cloud explosion [1,2]. The close similarity to other base-catalysed incidents was stressed [3].
See Contaminants, and Trimethylamine, both below
See other POLYMERISATION INCIDENTS, RUNAWAY REACTIONS, VIOLENT POLYMERISATION

Contaminants
1. Gupta, A. K., *J. Soc. Chem. Ind.*, 1949, **68**, 179
2. Stull, 1977, 26–27
3. Freeder, B. G. *et al.*, *J. Loss Prev. Process Ind.*, 1988, **1**, 164–168

Accidental contamination of a 90 kg cylinder of ethylene oxide with a little sodium hydroxide solution led to explosive failure of the cylinder over 8 hours later [1]. Based on later studies of the kinetics and heat release of the polycondensation reaction, it was estimated that after 8 hours and 1 min, some 12.7% of the oxide had condensed with an increase in temperature from 20 to 100°C. At this point the heat release rate was calculated to be 2.1 MJ/min, and 100 s later the temperature and heat release rate would be 160° and 1.67 MJ/s respectively, with 28% condensation. Complete reaction would have been attained some 16 s later at a temperature of 700°C [2]. Precautions designed to prevent explosive polymerisation of ethylene oxide are discussed, including rigid exclusion of acids; covalent halides, such as aluminium chloride, iron(III) chloride, tin(IV) chloride; basic materials like alkali hydroxides, ammonia, amines, metallic potassium; and catalytically active solids such as aluminium oxide, iron oxide, or rust [1]. A comparative study of the runaway exothermic polymerisation of ethylene oxide and of propylene oxide by 10 wt% of solutions of sodium hydroxide of various concentrations has been done using ARC. Results below show onset temperatures/corrected adiabatic exotherm/maximum pressure attained; and heat of polymerisation for the least (0.125 M) and most (1 M) concentrated alkali solutions used as catalysts.

EO 55.0, 22,5°C/439, 415°/40.5, 44.6 bar; 1.06, 1.0 kJ/g.

PO 91, 53°/451, 452°/26.6, 20.9 bar; 1.09, 1.1 kJ/g

The more reactive ethylene oxide clearly shows much higher self-heating rates and hazard potential than its higher homologue, though more detailed work is needed to quantify the differences [3].

See Ammonia, above; Trimethylamine, below

See other CATALYTIC IMPURITY INCIDENTS, POLYCONDENSATION REACTION INCIDENTS, POLYMERISATION INCIDENTS

Dinitrogen pentaoxide
See Dinitrogen pentaoxide: Strained ring heterocycles

Glycerol
Anon., *Loss Prev. Bull.*, 1979, (028), 92–93

Glycerol was to be ethoxylated at 115–125°C in a circulating reaction system with separate reactor, heat exchanger and catalyst units. The valve at the base of the reactor was still closed, but an inoperative flow indicator failed to indicate absence of circulation and a total of 3 tonnes of the oxide, plus glycerol, was charged to the reactor. Upon subsequent opening of the valve, the reaction mixture passed through the heater, now at 200°C, and a runaway reaction developed, the reactor burst and an explosion followed.

See other RUNAWAY REACTIONS

Iron(III) hexacyanoferrate(4−) ('Iron blue pigment')
See Iron(III) hexacyanoferrate(4−): Ethylene oxide

Lagging materials

Hilado, C. J., *J. Fire Flamm.*, 1974, **5**, 321–326

The self-heating temperature (effectively the AIT) of a 50:50 mixture of ethylene oxide and air is reduced from 456°C on passage through various thermal insulation (lagging) materials to 251–416°C, depending on the particular material (of which 13 were tested).

Magnesium perchlorate

See Magnesium perchlorate: Ethylene oxide

4-Methoxybenzenelithium

Taylor, D. A. H., personal comunication, 1993

Oxirane was added to the organolithium (probably in ether solution) at a temperature perhaps −40°C. The mixture was then allowed to warm up, went out of control and distributed itself widely. Presumably too much ethylene oxide had accumulated because of sluggish reaction at the low temperature.

See Alkanethiols, above

3-Nitroaniline

Anon., *Angew. Chem. (Nachr.)*, 1958, **70**, 150

Interaction of the two compounds in an autoclave at 150–160°C is described as safe in Swiss Pat. 171 721. During careful repetition of the reaction with stepwise heating, an autoclave exploded at 130°C.

Nitrogen

Grosse-Wortmann, H., *Chem. Ing. Tech.*, 1970, **42**, 85–86

Concentration boundaries for explosive decomposition of the oxide and its mixtures with nitrogen at elevated temperatures and pressures are presented graphically.

Other reactants

Yoshida, 1980, 47

MRH values calculated for 13 combinations with oxidants are given.

Polyhydric alcohol, Propylene oxide

Vervalin, 1973, 82

A polyether-alcohol, prepared by co-condensation of ethylene oxide and propylene oxide with a polyhydric alcohol, was stored at above 100°C and exposed to air via a vent line. After 10–15 h, violent decomposition occurred, rupturing the vessel. It was subsequently found that exothermic oxidation of the product occurred above 100°C, and that at 300°C a rapid exothermic reaction set in, accompanied by vigorous gas evolution.

See other GAS EVOLUTION INCIDENTS

Oxygen

Griffiths, R. F. *et al.*, *Proc. 18th Int. Combust. Symp.*, 1981, 893–901

If rapid compression of dilute ethylene oxide vapours occurs, ignition and/or decomposition may occur. Thus 5% of oxide in argon will ignite if compressed more than 11-fold, and presence of traces of oxygen enhances the effect. Even 1% or less of ethylene in the oxide in air may be hazardous if modest but rapid compression occurs.

Pyridine, Sulfur dioxide
Nolan, 1983, Case History 151
Ethylene sulfite is prepared from ethylene oxide, pyridine, and sulfur dioxide in excess to prevent polymerisation of ethylene oxide. Use of a deficiency of sulfur dioxide led to rupture of a reactor from that cause.

Sodium hydroxide
See Contaminants, above

Sucroglyceride
1. Slade, R. C., *Chem. Brit.*, 1977, **13**, 317
2. Parker, K. J., private comm., 1977
During ethoxylation of the glyceride by stirring and heating to 120°C with excess ethylene oxide in an autoclave fitted with a 300 bar bursting disc [1], exothermic reaction occurred, usually causing an increase in temperature to 190°C (even with ice cooling applied) when large excesses were used. In one such attempt, an operating error led to the initial ethylene oxide charge being distilled out of the autoclave via an open vent valve, and the dry residue being heated to a temperature considerably above the 100°C indicated by the (unwetted) pocketed thermocouple. After cooling, more oxide was added and the preparation was continued. At an indicated temperature of 126°C and a pressure reading above 10 bar, it was decided to cool the autoclave to moderate the accelerating reaction. At this point, the rapidly increasing pressure and temperature caused the autoclave lid bolts to stretch and relieve the explosive reaction [2]. It is surmised that during the 'false start' of the reaction, a layer of the sucroglyceride became distributed as a paste round the heated autoclave wall, and was then overheated, forming an insulating layer on the autoclave wall. After more ethylene oxide was added and heating restarted, the indicated liquid temperature (126°C) would be far below that of the heated autoclave wall behind the insulating layer of thermally degraded solid. Dissolution of the inner layers of solid in the hot ethylene oxide may have exposed a hot spot which initiated thermal decomposition. Additionally or alternatively, local concentrations of alkali arising from thermal decomposition of potassium soaps present as impurity in the sucroglyceride may have initiated exothermic polymerisation of the oxide.
See entry PRESSURE INCREASE IN EXOTHERMIC DECOMPOSITION

Trimethylamine
1. Anon., *BCISC Quart. Safety Summ.*, 1966, **37**, 44
2. Anon., *Chem. Trade J.*, 1956, **138**, 1376
Accidental contamination of a large ethylene oxide feed-cylinder by reaction liquor containing trimethylamine caused the cylinder to explode 18 h later. Contamination was possible because of a faulty pressure gauge and suck-back of froth from above the liquid level [1]. A similar incident had occurred previously [2].
See Ammonia, and Contaminants, both above

Water
Vanderwater, R. G., *Chem. Eng. Progress*, 1989, **85**(12), 16
A supposedly empty tankwagon, which may have still held 12 tonnes of oxirane, was filled with water as the first stage of a cleaning process and the valves then closed. Several hours later, an explosion shattered the tank, sending parts up to 600 metres distance. It appears that filling will not have given complete mixing but left an oxirane layer on top of the water, which would be heated, but not mixed, by reaction with water at the interface, possibly aided by catalysis at sites of corrosion. There should have been enough water to act as a heat sink had it mixed. Explosion appears to have taken place within the supernatant ethylene oxide layer.

See other CORROSION INCIDENTS
See other 1,2-EPOXIDES

0830. Cyclic poly(ethylene oxides) ('Crown ethers')
[[294-93-9] ($n = 4$) [33100-27-5] ($n = 5$) [17455-13-9] ($n = 6$)] $(C_2H_4O)_n$

Bromoform, Potassium hydroxide
See Bromoform: Cyclic poly(ethylene oxides), etc.

0831. Poly(vinyl alcohol)
[9002-89-5] $(C_2H_4O)_n$

$$[-CH_2CH(OH)-]_n$$

Energy of decomposition (in range 125–430°C) measured as 0.59 kJ/g.
See entry THERMOCHEMISTRY AND EXOTHERMIC DECOMPOSITION (reference 2)

†0832. Thiolacetic acid
   [507-09-5]   C₂H₄OS

$$H_3CCO.SH$$

See other ORGANIC ACIDS

0833. Acetic acid (Ethanoic acid)
   [64-19-7]   C₂H₄O₂

$$H_3CCO.OH$$

(*MCA SD-41*, 1951); *NSC 410*, 1979; *FPA H120*, 1983; *HCS 1980*, 99

Acetaldehyde
   See Acetaldehyde: Acetic acid

Acetic anhydride, Water
   See Acetic anhydride: Acetic acid, Water

5-Azidotetrazole
   See 5-Azidotetrazole: Alone, or Acetic acid

Other reactants
   Yoshida, 1980, 121
   MRH values calculated for 15 combinations are given.

Oxidants   MRH values show % of oxidant
   See Bromine pentafluoride: Hydrogen-containing materials
         Chromium trioxide: Acetic acid   MRH 2.17/12
         Hydrogen peroxide: Acetic acid   MRH 5.10/69
         Potassium permanganate: Acetic acid   MRH 2.34/84
         Sodium peroxide: Acetic acid   MRH 2.80/84

Phosphorus trichloride
   See Phosphorus trichloride: Carboxylic acids (references 1,2)

Potassium *tert*-butoxide
   See Potassium *tert*-butoxide: Acids
   See other ORGANIC ACIDS

†0834. Methyl formate (Methyl methanoate)
   [107-31-3]   C₂H₄O₂

$$MeOCO.H$$

HCS 1980, 653

Methanol, Sodium methoxide
   Pond, D. M., *Chem. Eng. News*, 1982, **60**(37), 43

312

The crude product of reaction of methanol and carbon monoxide at 100°C/70 bar in presence of 0.5% of sodium methoxide was discharged after cooling into a storage bottle, which burst 4 h later. This was attributed to extreme instability of the ester in presence of the base, leading to the reverse reaction with vigorous evolution of carbon monoxide. Immediate neutralisation of the reaction mixture would prevent the decomposition, which also occurs with ethyl formate and base.
See other GAS EVOLUTION INCIDENTS

0835. Poly(ethylidene peroxide)
[ ] $(C_2H_4O_2)_n$

(-CHMeOO-)$_n$

Rieche, A. *et al., Angew. Chem.*, 1936, **49**, 101
The highly explosive polyperoxide is present in peroxidised diethyl ether and has been responsible for many accidents during distillation of the solvent.
See other POLYPEROXIDES

0836. Ethylene ozonide (1,2,4-Trioxolane)
[289-14-5] $C_2H_4O_3$

1. Harries, C. *et al., Ber.*, 1909, **42**, 3305
2. Briner, E. *et al., Helv. Chim. Acta*, 1921, **12**, 154
3. Criegee, R., *Angew. Chem.*, 1958, **65**, 398–399
It explodes very violently on heating, friction or shock [1]. Stable at 0°C, but often decomposes explosively at ambient temperature [2]. Pouring the liquid from one vessel to another initiated a violent explosion [3].
See other OZONIDES

0837. Peroxyacetic acid (Ethaneperoxoic acid)
[79-21-0] $C_2H_4O_3$

MeCO.OOH

FPA H89, 1980; *HCS 1980*, 720 (40% solution in acetic acid)

1. Swern, D., *Chem. Rev.*, 1949, **45**, 7
2. Davies, 1961, 56
3. Phillips, B. *et al., J. Org. Chem.*, 1959, **23**, 1823
4. Anon., *Angew. Chem. (Nachr.)*, 1957, **5**, 178
5. Smith, I. C. P., private comm., 1973
6. *MCA Case History No. 1804*
7. Slattery, G. H., US Pat. 4 137 256, 1979

It is insensitive to impact but explodes violently at 110°C [1]. The solid acid has exploded at −20°C [2]. Safe procedures (on basis of detonability experiments) for preparation of anhydrous peracetic acid solutions in chloroform or ester solvents have been detailed [3]. However, a case of explosion on impact has been recorded [4]. During vacuum distillation, turning a ground glass stopcock in contact with the liquid initiated a violent explosion, but the grease may have been involved as well as friction [5]. During a pilot-scale preparation of a solution of the acid by adding acetic anhydride slowly to 90% hydrogen peroxide in dichloromethane, an explosion of great violence occurred. Subsequent investigation revealed that during the early stages of the addition a 2-phase system existed, which was difficult to detect because of closely similar densities and refractive indices. The peroxide-rich phase became extremely shock-sensitive when between 10 and 30% of the anhydride had been added [6]. A safer process for the vapour phase oxidation of acetaldehyde to peroxyacetic acid has been claimed [7].

See Hydrogen peroxide: Acetic acid, or: Acetic anhydride, or: Vinyl acetate
See PEROXYACIDS (references 5,6) See other GLASS INCIDENTS

Acetic anhydride
1. MCA Case History No. 1795
2. Augustine 1969, Vol. 2, 217
During an attempt to prepare an anhydrous 25% solution of peroxyacetic acid in acetic acid by dehydrating a water-containing solution with acetic anhydride, a violent explosion occurred. Mistakes in the operational procedure allowed heated evaporation to begin before the anhydride had been hydrolysed. Acetyl peroxide could have been formed from the anhydride and peroxyacid, and the latter may have detonated and/or catalysed violent hydrolysis of the anhydride [1]. A technique for preparing the anhydrous acid in dichloromethane without acetyl peroxide formation has been described [2].

See Acetic anhydride: Water

5-p-Chlorophenyl-2,2-dimethyl-3-hexanone
Gillespie, J. S. et al., Tetrahedron, 1975, 31, 5
During the attempted preparation of 3-p-chlorophenylbutanoic acid by addition of the ketone (2 g mol) to peracetic acid (50%) in acetic acid at 65–70°C, a serious explosion occurred.

Ether solvents
Augustine, 1971, Vol. 2, 159
Ether solvents, such as THF, diethyl ether etc. are unsuitable solvents for peracetic acid oxidations, as interaction of the acid with the peroxidisable solvent is violent.
See other PEROXIDATION INCIDENTS

Metal chlorides
Wienhofer, E., Chem. Ztg., 1980, 104, 146
Addition of chloride ions (as solid calcium chloride, potassium chloride or sodium chloride) to aqueous solutions containing 40% of peroxyacetic acid and 1% of

acetic acid leads to a violently exothermic decomposition reaction. Chlorine is evolved, most of the liquid evaporates and the residue (often red coloured) deflagrates.

See other GAS EVOLUTION INCIDENTS

3-Methyl-3-buten-1-yl tetrahydropyranyl ether
See TETRAHYDROPYRANYL ETHER DERIVATIVES

1-Octene
1. Swern, D. et al., *J. Amer. Chem. Soc.*, 1946, **68**, 1506
2. van den Brink, M. J., private comm., 1983

Following a published procedure [1], octene was treated with a solution of peroxyacetic acid in acetic acid for 8 h to form the epoxide, but the reaction mixture was then allowed to stand uncooled overnight. Next morning, when a 3μl sample was injected into a heated GLC injection port, the syringe shattered. This was attributed to formation of diacetyl peroxide during the overnight standing, and its subsequent explosion in the heated port [2].

Paper
See other PACKAGING INCIDENTS
See other PEROXYACIDS

0838. Hydroperoxymethyl formate (Formyloxymethyl hydroperoxide)
[ ] $C_2H_4O_4$

$$HOOCH_2OCH=O$$

Thamm, J. *et al., Chem. Phys. Lett.*, 1996, **258**(1,2), 155

This compound was shown to be an intermediate in ozonolysis of some ethenes. In concentrated form it is very explosive; minimal handling and working with strong gloves behind a blast screen are recommended.

See other ALKYL HYDROPEROXIDES

0839. 2,4-Dithia-1,3-dioxane-2,2,4,4-tetraoxide ('Carbyl sulfate')
[503-41-3] $C_2H_4O_6S_2$

N-Methyl-4-nitroaniline
1. Deucker, W. *et al., Chem. Ing. Tech.*, 1973, **45**, 1040–1041
2. Wooton, D. L. *et al., J. Org. Chem.*, 1974, **39**, 2112

The title cyclic ester (1 mol) and substituted aniline (>2 mol) condense when heated in nitrobenzene solution, initially at 45°C and eventually at 75°C under vacuum in a sealed vessel, to give the ester-salt, *N*-methyl-4-nitroanilinium 2-(*N*-methyl-*N*-4-nitrophenylaminosulfonyl)ethyl sulfate. The reaction technique had been used uneventfully during 8 years, but the 3 m³ reactor was violently ruptured when the reaction ran away and developed an internal pressure exceeding 70 bar.

Subsequent DTA investigation showed that an exothermic reaction set in above 75°C after an induction period depending on the initial temperature and concentration of reactants, which attained nearly 300°C, well above the decomposition point of the cyclic ester component (170°C). The reaction conditions used could have permitted local over-concentration and overheating effects to occur, owing to slow dissolution of the clumped solid ester and aniline in the nitrobenzene solvent [1]. Crude 'carbyl sulfate' contains excess sulfur trioxide [2].
See DIFFERENTIAL THERMAL ANALYSIS (reference 4)
See other INDUCTION PERIOD INCIDENTS, RUNAWAY REACTIONS

†0840. Thiirane (Ethylene sulfide)
[420-12-2] C_2H_4S

Acids
Sorbe 1968, 122
The sulfide may polymerise violently in presence of acids, especially when warm.
See related 1,2-EPOXIDES

0841. Ethylaluminium bromide iodide
[32673-60-2] C_2H_5AlBrI
$$EtAl(Br)I$$

Nitromethane
See Nitromethane: Alkylmetal halides
See other ALKYLALUMINIUM HALIDES

0842. Ethylaluminium dibromide
[2386-62-1] $C_2H_5AlBr_2$
$$EtAlBr_2$$

See entry ALKYLALUMINIUM HALIDES

0843. Ethylaluminium dichloride
[563-43-9] $C_2H_5AlCl_2$
$$EtAlCl_2$$

HCS 1980, 465

See entry ALKYLALUMINIUM HALIDES

316

0844. Ethylaluminium diiodide
[2938-73-0] $C_2H_5AlI_2$

$EtAlI_2$

491M, 1975, 153

It ignites in air.
See other ALKYLALUMINIUM HALIDES

0845. Dichloroethylborane
[1739-53-3] $C_2H_5BCl_2$

$EtBCl_2$

491M, 1975, 149

It ignites in air.
See other ALKYLHALOBORANES

†0846. Bromoethane (Ethyl bromide)
[74-96-4] C_2H_5Br

$EtBr$

FPA H98, 1981; *HCS 1980*, 472

Preparative hazard
See Bromine: Ethanol, etc.
See other HALOALKANES

0847. Ethylmagnesium bromide
[925-90-6] C_2H_5BrMg

$EtMgBr$

Water
    Nolan, 1983, Case History 108
    Ethylmagnesium bromide was prepared as usual on a large scale in one reactor,
    and then transferred by nitrogen pressurisation to another. The glass transfer line
    had not been completely dried; the ethane evolved on contact of the Grignard
    reagent with moisture overpressurised the line and it burst.
    See other GAS EVOLUTION INCIDENTS, GLASS INCIDENTS
    See other GRIGNARD REAGENTS

†0848. Chloroethane (Ethyl chloride)
[75-00-3] C_2H_5Cl

$EtCl$

(*MCA SD-50*, 1953); *FPA H114*, 1982; *HCS 1980*, 474

Potassium
See Potassium: Halocarbons
See other HALOALKANES

0849. 2-Chloro-*N*-hydroxyacetamidine (Chloroacetamide oxime)
[3272-96-6] $C_2H_5ClN_2O$
$$ClCH_2C(NH_2):NOH$$

Maytum, D. M., *Chem. Brit.*, 1978, **14**, 382
During vacuum evaporation at 60°C of ethyl acetate from a solution of the oxime, a violent explosion occurred. The oxime was found to be thermally unstable above 60°C, and at the m.p., 90°C, an exotherm of over 100°C occurred accompanied by rapid gas evolution.
See related OXIMES *See other* GAS EVOLUTION INCIDENTS

†0850. Chloromethyl methyl ether
[107-30-2] C_2H_5ClO
$$ClCH_2OCH_3$$

HCS 1980, 302

See related HALOALKANES

0851. Ethyl hypochlorite
[624-85-1] C_2H_5ClO
$$EtOCl$$

Alone, or Copper
Sandmeyer, T., *Ber.*, 1885, **18**, 1768
Though distillable slowly (at 36°C), ignition or rapid heating of the vapour causes explosion, as does contact of copper powder with the cold liquid.
See other HYPOHALITES

0852. Ethyl perchlorate
[22750-93-2] $C_2H_5ClO_4$
$$EtOClO_3$$

Sidgwick, 1950, 1235
Reputedly the most explosive substance known, it is very sensitive to impact, friction and heat. It is readily formed from ethanol and perchloric acid.
See other ALKYL PERCHLORATES

318

0853. 2,2-Dichloroethylamine
[5960-88-3] $C_2H_5Cl_2N$

$$Cl_2CHCH_2NH_2$$

Roedig, A. *et al., Chem. Ber.*, 1966, **99**, 121
A violent explosion ocurred during evaporation of an ethereal solution at 260 mbar from a bath at 80–90°C. No explosion occurred when the bath temperature was limited to 40–45°C, or during subsequent distillation at 60–64°C/76 mbar. (Aziridine derivatives may have been formed by dehydrohalogenation.)

†0854. Trichloroethylsilane
[115-21-9] $C_2H_5Cl_3Si$

$$Cl_3SiEt$$

See other ALKYLHALOSILANES

†0855. Fluoroethane (Ethyl fluoride)
[353-36-6] C_2H_5F

$$EtF$$

See other HALOALKANES

0856. Ethyl fluorosulfate
[371-69-7] $C_2H_5FO_3S$

$$EtOSO_2F$$

Kumar, R. C. *et al., Inorg. Chem.*, 1984, **23**, 3113–3114
The compound, prepared by by co-condensation of diethyl sulfite (3.5 mmol) and chlorine fluoride (10 mmol) at −196°C, followed by slow warming to −78, then −20°, is unstable. Trap-to-trap distillation must be effected with great care, as violent explosions occurred (even on this small scale) when cryogenic cooling was removed from the traps. Scaling up is not recommended.
See other SULFUR ESTERS

0857. 2-Hydroxyethylmercury(II) nitrate
[51821-32-0] $C_2H_5HgNO_4$

$$HOC_2H_4HgNO_3$$

Whitmore, 1921, 110
It decomposes with a slight explosion on heating.
See other MERCURY COMPOUNDS *See other* ORGANOMETALLIC NITRATES

0858. Iodoethane (Ethyl iodide)
[75-03-6] C_2H_5I

EtI

HCS 1980, 561

Preparative hazard
See Iodine: Ethanol, etc.

Silver chlorite
See Silver chlorite: Iodoalkanes
See other HALOALKANES

0859. Ethylmagnesium iodide
[10467-10-4] C_2H_5IMg

EtMgI

Ethoxyacetylene
Jordan, C. F., *Chem. Eng. News*, 1966, **44**(8), 40
A stirred mixture of ethoxyacetylene with ethylmagnesium iodide in ether exploded when the agitator was turned off. The corresponding bromide had been used without incident in earlier smaller-scale attempts.
See other AGITATION INCIDENTS, ALKYLMETAL HALIDES, GRIGNARD REAGENTS

0860. 2-Iodoethanol
[624-76-0] C_2H_5IO

IC$_2$H$_4$OH

Trimethylstibine
See 2-Hydroxyethyltrimethylstibonium iodide
See related HALOALKANES

0861. Potassium ethoxide
[917-58-8] C_2H_5KO

KOEt

1. *See entry* METAL ALKOXIDES
2. Anon. private comm., 1984
It may ignite in moist air [1]. The dry powder is very dusty, and suspensions in dry air have a low minimum ignition energy, 17.1 mJ. After an induction period depending on the temperature and humidity of the atmosphere, a sample confined in a Dewar flask decomposed rapidly, with an exotherm of 200°C. Heats of solution in ethanol and in water are 0.63 and 0.65 kJ/g, respectively. [2].
See other INDUCTION PERIOD INCIDENTS

0862. Ethyllithium
[811-49-4] C_2H_5Li

EtLi

Sorbe, 1968, 82
It ignites in air.
See other ALKYLMETALS

†0863. Aziridine (Ethylenimine)
[151-56-4] C_2H_5N

NH

HCS 1980, 169

Preparative hazard
Energy of decomposition (in range 130–390°C) measured as 2.02 kJ/g.
See 2-Chloroethylammonium chloride: Alkali
See entry THERMOCHEMISTRY AND EXOTHERMIC DECOMPOSITION (reference 2)

Acids
'Ethylenimine', Brochure 125-521-65, Midland (Mich.), Dow Chemical Co., 1965
It is very reactive chemically and subject to aqueous auto-catalysed exothermic polymerisation, which may be violent if uncontrolled by dilution, slow addition or cooling. It is normally stored over solid caustic alkali, to minimise polymerisation catalysed by presence of carbon dioxide.
See other CATALYTIC IMPURITY INCIDENTS, POLYMERISATION INCIDENTS

Aluminium chloride, Substituted anilines
Lehman, D. *et al., Chem. Abs.*, 1983, **98**, 125559
In the preparation of a series of substituted phenylethylenediamines, it is essential to add the reagents to an aromatic solvent at 30–80°C in the order aniline, then aluminium chloride, then ethylenimine to prevent uncontrollable exothermic reaction.

Chlorinating agents
Graefe, A. F. *et al., J. Amer. Chem. Soc.*, 1958, **80**, 3939
It gives the explosive 1-chloroaziridine on treatment with, e.g. sodium hypochlorite solution.
See 1-Chloroaziridine

Dinitrogen pentaoxide
See Dinitrogen pentaoxide: Strained ring heterocycles

Silver
'Ethylenimine' Brochure 125-521-65, Midland (Mich.), Dow Chemical Co., 1965
Explosive silver derivatives may be formed in contact with silver or its alloys,
including silver solder, which is therefore unsuitable in handling equipment.
See other AZIRIDINES

0864. Vinylamine (Etheneamine)
[593-67-9] C_2H_5N

$$H_2C=CHNH_2$$

Isoprene
Seher, A., *Ann.*, 1952, **575**, 153–161
Attempts at interaction led to explosions.

†0865. Acetaldehyde oxime (Hydroximinoethane)
[107-29-9] C_2H_5NO

$$MeCH=NOH$$

HCS 1980, 98

See other OXIMES

0866. *N*-Methylformamide
[123-39-7] C_2H_5NO

$$MeNHCO.H$$

HCS 1980, 670

See other APROTIC SOLVENTS

†0867. Ethyl nitrite
[109-95-5] $C_2H_5NO_2$

$$EtON:O$$

Haz. Chem. Data, 1975, 161

It may decompose explosively around 90°C.
See other ALKYL NITRITES

0868. Methyl carbamate
[598-55-0] $C_2H_5NO_2$

$$MeOCO.NH_2$$

Phosphorus pentachloride
See N-Carbomethoxyiminophosphoryl chloride

0869. Nitroethane
[79-24-3] $C_2H_5NO_2$

EtNO$_2$

HCS 1980, 682

1. Kirk-Othmer, 1996, Vol. 17, 216
2. *Beilsteins Handbuch der organischen Chemie*, **H**(1), 101, Springer, Berlin, 1918
Although it is certainly less explosive, being detonable only if both hot and heavily
confined [1] , many reference works recommend precautions similar to those for
nitromethane. The sodium salt explodes on heating [2]
See Nitromethane

Metal oxides
See NITROALKANES: Metal oxides

Other reactants
Yoshida, 1980, 271
MRH values calculated for 13 combinations with oxidants are given.
See other NITROALKANES

†0870. Ethyl nitrate
[625-58-1] $C_2H_5NO_3$

EtONO$_2$

Hydrocarbons, Oxygen
See Oxygen: Hydrocarbons, Promoters

Lewis acids
See ALKYL NITRATES (references 2,3)

Other reactants
Yoshida, 1980, 191
MRH values calculated for 7 combinations are given.
See other ALKYL NITRATES

0871. 2-Nitroethanol
[625-48-9] $C_2H_5NO_3$

O$_2$NC$_2$H$_4$OH

Anon., *ABCM Quart. Safety Summ.*, 1956, **27**, 24
An explosion occurred towards the end of vacuum distillation of a relatively small
quantity of 2-nitroethanol. This was attributed to the presence of peroxides, but
the presence of traces of alkali seems a possible alternative cause.
See other C-NITRO COMPOUNDS

0872. Ethyl azide
[871-31-8] $C_2H_5N_3$

EtN₃

1. Boyer, J. H. *et al.*, *Chem. Rev.*, 1954, **54**, 32
2. Koch, E., *Angew. Chem. (Nachr.)*, 1970, **18**, 26
3. Burns, M. E. *et al.*, *Chem. Eng. News*, 1985, **63**(50), 2
4. Nielsen, C. J. *et al.*, *Spectrochim. Acta A*, 1988, **44A**, 409–422

Though stable at room temperature, it may detonate on rapid heating [1]. A sample stored at −55°C exploded after a few minutes' exposure at ambient temperature, possibly owing to development of internal pressure in the stoppered vessel. It will also explode if dropped from 1 m in a small flask on to a stone floor [2]. An explosion occurred during laboratory preparation of ethyl azide when bromomethyl blue indicator was added to the reaction mixture to show the pH. This was attributed either to the acidic nature of the indicator, or to exposure of the azide vapour to the ground glass joint. Precautions recommended are similar to those when using diazomethane, and include small scale working, no ground joints, protection from intense light, and dilution of the product with solvent [3]. A safe synthesis is described, and IR, NMR and Raman spectral data were recorded in vapour, liquid and solid states [4].

See other GLASS INCIDENTS, ORGANIC AZIDES

0873. 2-Azidoethanol
[1517-05-1] $C_2H_5N_3O$

HOC₂H₄N₃

1. Appleby, I. C., *Chem. & Ind.*, 1986, 337
2. Appleby, I. C., private comm., 1987

During the preparation of 2-azidoethanol from a stirred mixture of 2-bromoethanol (14.6 mol) and sodium azide (15.4 mol) heated on a steam bath, a violent explosion occurred after 100 min. The preparation had been carried out previously without mishap. The need for care in handling azides of low MW is stressed [1]. Later detailed studies showed that the most probable cause of the explosion was the extraordinarily high mechanical and thermal sensitivity of the compound, with initiation by vibration from the agitator [2].

See other ORGANIC AZIDES

0874. Imidodicarbonic diamide (Biuret)
[108-19-0] $C_2H_5N_3O_2$

H₂NCO.NHCO.NH₂

Chlorine
See Chlorine: Nitrogen compounds

0875. *N*-Methyl-*N*-nitrosourea
[684-93-5] $C_2H_5N_3O_2$

$$H_2NCO.N(N:O)Me$$

1. Arndt, F., *Org. Synth.*, 1943, Coll. Vol. 2, 462
2. Anon., *Angew. Chem. (Nachr.)*, 1957, **5**, 198
3. Anon., *Jahresber.*, 1978, 70–71
4. *See entry* SELF-ACCELERATING REACTIONS

This must be stored under refrigeration to avoid sudden decomposition after storage at ambient temperature or slightly above (30°C) [1]. More stable precursors for generation of diazomethane are now available. Material stored at 20°C exploded after 6 months [2]. Explosive decomposition and ignition of solvent ether occured when 50 g of the compound in a nickel spoon was being added to 40% potassium hydroxide solution to generate diazomethane [3]. DSC examination of several *N*-nitroso compounds showed that most underwent self accelerating decomposition, some with extraordinary velocity. The time to maximum rate for the urea falls from 1200 min at 40° to near zero at 80°C, when the maximum heat flow is 8.7 J/s/g [4].
See Diazomethane
See other NITROSO COMPOUNDS

0876. 1-Methyl-3-nitro-1-nitrosoguanidine
[70-25-7] $C_2H_5N_5O_3$

$$MeN(N:O)C(:NH)NHNO_2$$

1. Eisendrath, J. N., *Chem. Eng. News*, 1953, **31**, 3016
2. Aldrich advertisement, *J. Org. Chem.*, 1974, **39**(7), cover iv

Formerly used as a diazomethane precursor, this material will detonate under high impact, and a sample exploded when melted in a sealed capillary tube [1]. Although the crude product from the aqueous nitrosation is pyrophoric, recrystallised material is stable (though powerfully mutagenic) [2].
See other *N*-NITRO COMPOUNDS, NITROSO COMPOUNDS

0877. Ethylsodium
[676-54-0] C_2H_5Na

$$EtNa$$

Houben-Weyl, 1970, Vol. 13.1, 384
The powder ignites in air.
See other ALKYLMETALS

0878. Sodium ethoxide
[141-52-6] C_2H_5NaO

$$NaOEt$$

May ignite in moist air.
See entry METAL ALKOXIDES

0879. Sodium dimethylsulfinate
[58430-57-2] C_2H_5NaOS

$$Na^+ Me_2SO^-$$

4-Chlorotrifluoromethylbenzene
See 4-Chlorotrifluoromethylbenzene: Sodium dimethylsulfinate
See Dimethyl sulfoxide: Sodium hydride

0880. Sodium 2-hydroxyethoxide
[7388-28-5] $C_2H_5NaO_2$

$$NaOC_2H_4OH$$

Polychlorobenzenes
Milnes, M. H., *Nature*, 1971, **232**, 395–396
It was found that the monosodium salt of ethylene glycol decomposed at about 230°C, during an investigation of the 1968 Coalite explosion during the preparation of 2,3,5-trichlorophenol by hydrolysis of 1,2,4,5-tetrachlorobenzene with sodium hydroxide in ethylene glycol at 180°C. Decomposition in the presence of polychlorobenzenes was exothermic and proceeded rapidly and uncontrollably, raising the temperature to 410°C, and liberating large volumes of white vapours, one of which was identified as ethylene oxide. The decomposition proceeded identically in presence of the tetrachlorobenzene or the 3 isomeric trichlorobenzenes (and seems also likely to occur in the absence of these or the chlorophenols produced by hydrolysis). The latter view has now been confirmed.
See Sodium hydroxide: Glycols

†0881. Ethane
[74-84-0] C_2H_6

$$H_3CCH_3$$

FPA H37, 1975; *HCS 1980*, 456 (cylinder)

Chlorine
See Chlorine: Hydrocarbons

Dioxygenyl tetrafluoroborate
See Dioxygenyl tetrafluoroborate: Organic materials

0882. Dimethylaluminium bromide
[3017-85-4] C_2H_6AlBr

$$Me_2AlBr$$

Nitromethane
See Nitromethane: Alkylmetal halides
See other ALKYLALUMINIUM HALIDES

326

0883. Dimethylaluminium chloride
[1184-58-3] C_2H_6AlCl

$$Me_2AlCl$$

Air, or Water
  Gaines, D. F. *et al., Inorg. Chem.*, 1974, **15**, 203–204
  It ignites in air and reacts violently with water.
  See other ALKYLALUMINIUM HALIDES

0884. Chlorodimethylarsine
[557-89-1] C_2H_6AsCl

$$ClAsMe_2$$

Leleu, *Cahiers*, 1977, (88), 362
  It ignites in air, even at 0°C.
  See other ALKYLNON-METAL HALIDES

0885. Dimethylfluoroarsine
[420-23-5] C_2H_6AsF

$$Me_2AsF$$

Leleu, *Cahiers*, 1977, (88), 363
  It ignites in air.
  See other ALKYLNON-METAL HALIDES

0886. Dimethyliodoarsine
[676-75-5] C_2H_6AsI

$$Me_2AsI$$

Millar, I. T. *et al., Inorg. Synth.*, 1960, **6**, 117
  It ignites when heated in air.
  See other ALKYLNON-METAL HALIDES

0887. Bromodimethylborane
[5158-50-9] C_2H_6BBr

$$Me_2BBr$$

Katz, H. E., *J. Org. Chem.*, 1985, 50, **50** 28
  It is pyrophoric in air.
  See other ALKYLHALOBORANES, PYROPHORIC MATERIALS

0888. Azidodimethylborane
[2306-67-0] $C_2H_6BN_3$

$$N_3BMe_2$$

Anon., *Angew. Chem. (Nachr.)*, 1970, **18**, 27

A sample exploded on contact with a warm sampling capillary.
See other NON-METAL AZIDES

0889. Barium methyl hydroperoxide

[ ] $C_2H_6BaO_4$

$$Ba(OOMe)_2$$

See Methyl hydroperoxide
See related ALKYL HYDROPEROXIDES

0890. Dimethylberyllium

[506-63-8] C_2H_6Be

$$Me_2Be$$

Air, or Carbon dioxide, or Water
Coates, 1967, Vol. 1, 106
It ignites in moist air or in carbon dioxide, and reacts explosively with water.
See other ALKYLMETALS

0891. Dimethylberyllium-1,2-dimethoxyethane

[ ] $C_2H_6Be.C_4H_{10}O_2$

$$Me_2Be.MeOC_2H_4OMe$$

Houben-Weyl, 1973, Vol. 13.2a, 38
It ignites in air.
See related ALKYLMETALS

0892. Chloroethylbismuthine

[65313-33-9] C_2H_6BiCl

$$ClBi(Et)H$$

Leleu, *Cahiers*, 1977, (88), 364
It ignites in air.
See related ALKYLMETAL HALIDES, ALKYLMETAL HYDRIDES

0893. Dimethylbismuth chloride

[ ] C_2H_6BiCl

$$Me_2BiCl$$

Sidgwick, 1950, 781

328

It ignites when warm in air.
See other ALKYLMETAL HALIDES

0894. Dimethylcadmium
[506-82-1] C_2H_6Cd

Me$_2$Cd

1. Egerton, A. *et al., Proc. R. Soc.*, 1954, **A225**, 429
2. Davies, A. G., *Chem. & Ind.*, 1958, 1177
3. Sidgwick, 1950, 268

On exposure to air, dimethylcadmium peroxide is formed as a crust [1] which explodes on friction [2]. Ignition of dimethylcadmium may occur if a large area:volume ratio is involved, as when it is dripped onto filter paper [3]. It is a mildly endothermic compound ($\Delta H°_f$ (l) +67.8 kJ/mol, 0.47 kJ/g).
See other ALKYLMETALS, ENDOTHERMIC COMPOUNDS

0895. *N*-Chlorodimethylamine
[1585-74-6] C_2H_6ClN

ClNMe$_2$

Antimony chlorides
Weiss, W. *et al., Z. Anorg. Chem.*, 1977, **433**, 207–210
The crystalline addition compounds formed with antimony trichloride and antimony pentachloride in dichloromethane at −78°C tend to explode at ambient temperature, or when shocked or heated.
See other N-HALOGEN COMPOUNDS

0896. 2-Chloroethylamine
[689-98-5] C_2H_6ClN

ClC$_2$H$_4$NH$_2$

It may polymerise explosively.
See 2-Chloroethylammonium chloride: Alkali *also* 2-Chloro-1,4-phenylenediamine
See other POLYMERISATION INCIDENTS

0897. Methyl iminioformate chloride
[15755-09-6] C_2H_6ClNO

MeOCH(:N$^+$H$_2$) Cl$^-$

Preparative hazard
See Hydrogen chloride: Alcohols, Hydrogen cyanide

0898. 2-Aza-1,3-dioxolanium perchlorate (Ethylenedioxyammonium perchlorate)
[ ] $C_2H_6ClNO_6$

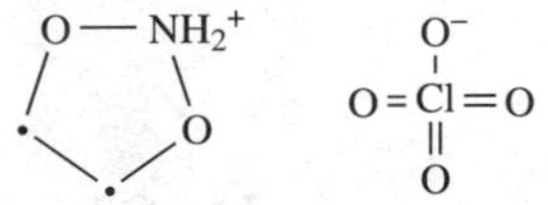

MCA Case History No. 1622
A batch of this sensitive compound (low oxygen balance) exploded violently, probably during recrystallisation.
See other PERCHLORATE SALTS OF NITROGENOUS BASES, N–O COMPOUNDS

0899. Chlorodimethylphosphine
[811-62-1] C_2H_6ClP

$$ClPMe_2$$

1. Staendeke, H. *et al.*, *Angew. Chem. (Intern. Ed.)*, 1973, **12**, 877
2. Parshall, G. W., *Inorg. Synth.*, 1974, **15**, 192–193
It ignites in contact with air [1], and handling procedures are detailed [2].
See other ALKYLHALOPHOSPHINES

0900. Dimethylantimony chloride
[18380-68-2] C_2H_6ClSb

$$Me_2SbCl$$

Sidgwick, 1950, 777
It ignites at 40°C in air.
See other ALKYLMETAL HALIDES

0901. Dimethyl *N,N*-dichlorophosphoramidate
[29727-86-4] $C_2H_6Cl_2NO_3P$

$$(MeO)_2P(O)NCl_2$$

Preparative hazard
See Chlorine: Dimethyl phosphoramidate
See other N-HALOGEN COMPOUNDS, PHOSPHORUS ESTERS

†0902. Dichlorodimethylsilane
[75-78-5] $C_2H_6Cl_2Si$

$$Cl_2SiMe_2$$

HCS 1980, 428

Water
Leleu, *Cahiers*, 1977, (88), 370
Violent reaction with water.
See other ALKYLHALOSILANES

330

†0903. Dichloroethylsilane
[1789-58-8] $C_2H_6Cl_2Si$

Cl_2SiHEt

See other ALKYLHALOSILANES

0904. (Dimethyl ether)oxodiperoxochromium(VI)
[ ] $C_2H_6CrO_6$

Schwarz, R. *et al., Ber.*, 1936, **69**, 575
The blue solid explodes powerfully at −30°C.
See related AMMINECHROMIUM PEROXOCOMPLEXES

0905. Copper(II) glycinate nitrate (Aquaglycinatonitratocopper)
[94791-14-7] $C_2H_6CuN_2O_6$

Davies, H. O. *et al., J. Chem. Soc. Chem. Comm.*, 1992, (3), 226
The mixed salt explodes on heating to 167°C, first losing water.
See related METAL NITRATES

†0906. 2,2,2,4,4,4-Hexafluoro-1,3-dimethyl-1,3,2,4-diazadiphosphetidine
[3880-04-4] $C_2H_6F_6N_2P_2$

tert-Butyllithium
See tert-Butyllithium: 2,2,2,4,4,4-Hexafluoro-1,3,2,4-diazadiphosphetidine

0907. Dimethylmercury
[593-74-8]   C₂H₆Hg

$$Me_2Hg$$

It is mildly endothermic (ΔH°_f (l) +75.3 kJ/mol, 0.33 kJ/g).

Diboron tetrachloride
    See Diboron tetrachloride: Dimethylmercury
    See other ALKYLMETALS, ENDOTHERMIC COMPOUNDS, MERCURY COMPOUNDS

0908. Dimethylmagnesium
[2999-74-8]   C₂H₆Mg

$$Me_2Mg$$

Air, or Water
    Gilman, H. *et al., J. Amer. Chem. Soc.*, 1930, **52**, 5049
    Contact with moist air usually caused ignition of the dry powder, and water always ignited the solid or its ethereal solution.
    See other ALKYLMETALS, DIALKYLMAGNESIUMS

0909. Dimethylmanganese
[33212-68-9]   C₂H₆Mn

$$Me_2Mn$$

1. Bailar, 1973, Vol. 3, 851; Vol. 4, 792
2. *491M 1975*, 159
It is a readily explosive powder, (probably polymeric) [1], which also ignites in air [2].
See other ALKYLMETALS

0910. Azomethane (Dimethyldiazene)
[503-28-6]   C₂H₆N₂

$$H_3CN{=}NCH_3$$

Allen, A. O. *et al., J. Amer. Chem. Soc.*, 1935, **57**, 310–317
The slow thermal decomposition of gaseous azomethane becomes explosive above 341°C/250 mbar and 386°C/24 mbar.
See other AZO COMPOUNDS

0911. Disodium *N,N'*-dimethoxysulfonyldiamide
[ ]   C₂H₆N₂Na₂O₄S

$$O_2S[N(Na)OMe]_2$$

Goehring, 1957, 87

The dry solid readily explodes, as do many *N*-metal hydroxylamides.
See other N-METAL DERIVATIVES, N–O COMPOUNDS

0912. Acetohydrazide
[1068-57-1] $C_2H_6N_2O$

$$MeCO.NHNH_2$$

Energy of decomposition (in range 460–510°C) measured as 1.02 kJ/g.
See entry THERMOCHEMISTRY AND EXOTHERMIC DECOMPOSITION (reference 2)

0913. Dimethyl hyponitrite (Dimethoxydiazene)
[29128-41-4] $C_2H_6N_2O_2$

$$MeON{=}NOMe$$

Mendenhall, G. D. *et al., J. Org. Chem.*, 1975, **40**, 1646
Samples of the ester exploded violently during low-temperature distillation, and
on freezing by liquid nitrogen. The undiluted liquid is exceptionally unpredictable,
but addition of mineral oil before vaporisation improved safety aspects.
See other DIALKYL HYPONITRITES

0914. Dimethyltin dinitrate
[40237-34-1] $C_2H_6N_2O_6Sn$

$$Me_2Sn(NO_3)_2$$

Sorbe, 1968, 159
It decomposes explosively on heating.
See other ORGANOMETALLIC NITRATES

0915. *N*-Azidodimethylamine
[2156-66-3] $C_2H_6N_4$

$$N_3NMe_2$$

Bock, H. *et al., Angew. Chem.*, 1962, **74**, 327
It is rather explosive.
See related ORGANIC AZIDES

0916. *N,N'*-Dinitro-1,2-diaminoethane
[505-71-5] $C_2H_6N_4O_4$

$$(O_2NNHCH_2{-})_2$$

1. Urbanski, 1967, Vol. 3, 20
2. Stull, 1977, 20

This powerful but relatively insensitive explosive decomposes violently at 202°C, and gives lead and silver salts which are highly impact sensitive [1]. Though not endothermic (ΔH°_f −103.3 kJ/mol), as a bis-nitramine it has a rather high heat of decomposition (3.91 kJ/g) which it is calculated would attain an adiabatic decomposition temperature over 2250°C, with a 60-fold pressure increase in a closed system [2].

See other N-NITRO COMPOUNDS

Amminemetals

Complexes with several ammine derivatives of metals are explosive.

See entry NITRAMINE–METAL COMPLEXES

0917. *S*, *S*-Dimethylpentasulfur hexanitride
[71901-54-7] $C_2H_6N_6S_5$

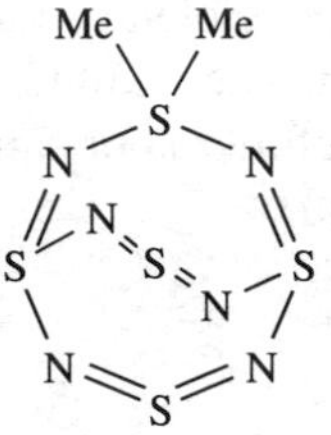

Sheldrick, W. S. *et al., Inorg., Chem.*, 1980, **19**, 339

Powerfully explosive like the non-methylated compound, breaking the quartz combustion tubes during microanalysis.

See other N–S COMPOUNDS

0918. Diazidodimethylsilane
[4774-73-6] $C_2H_6N_6Si$

$$(N_3)_2SiMe_2$$

Anon., *Angew. Chem. (Nachr.)*, 1970, **18**, 26–27

A 3 year old sample exploded violently on removing the ground stopper.

See other FRICTIONAL INITIATION INCIDENTS, NON-METAL AZIDES

†0919. Dimethyl ether (Oxybismethane)
[115-10-6] C_2H_6O

MeOMe

FPA H113, 1983, *HCS 1980*, 429

†0920. Ethanol (Ethyl alcohol)
[64-17-5] C_2H_6O
$$H_3CCH_2OH$$

NSC 391, 1968; *FPA H2*, 1972; *HCS 1980*, 639; *RSC Lab. Hazard Data Sheet No. 50*, 1986

Acetic anhydride, Sodium hydrogensulfate
   See Acetic anhydride: Ethanol, etc.

Acetyl bromide
   See Acetyl bromide: Hydroxylic compounds

Ammonia, Silver nitrate
   See Silver nitrate: Ammonia, Ethanol

Disulfuric acid, Nitric acid
   See Nitric acid: Alcohols, Disulfuric acid

Dichloromethane, Sulfuric acid, (A nitrate or nitrite?)
   See Sulfuric acid, Dichloromethane, etc.

Disulfuryl difluoride
   See Disulfuryl difluoride: Ethanol

Magnesium perchlorate
   See Magnesium perchlorate: Ethanol

Nitric acid, Silver
   See Silver: Ethanol, Nitric acid

Other reactants
   Yoshida, 1980, 43
   MRH values calculated for 18 combinations with oxidants are given.

Oxidants   MRH values show % of oxidant
   See Barium perchlorate: Alcohols
      Bromine pentafluoride: Hydrogen-containing materials
      Calcium hypochlorite: Hydroxy compounds   MRH 2.30/93
      Chloryl perchlorate: Organic matter
      Chromium trioxide: Alcohols   MRH 2.55/90
      Chromyl chloride: Organic solvents
      Dioxygen difluoride: Various materials
      'Fluorine nitrate': Organic materials
      Hydrogen peroxide: Alcohols   MRH 6.19/21
      Iodine heptafluoride: Organic solvents
      Nitric acid: Alcohols   MRH 5.56/76

335

Nitrosyl perchlorate: Organic materials
Perchloric acid: Alcohols   MRH 6.28/79
Permanganic acid: Organic materials
Peroxodisulfuric acid: Organic liquids
Potassium dioxide: Ethanol
Potassium perchlorate: Ethanol
Potassium permanganate: Ethanol, etc.   MRH 2.68/91
Ruthenium(VIII) oxide: Organic materials
Silver perchlorate: Aromatic compounds   MRH 3.30/87
Sodium peroxide: Hydroxy compounds   MRH 2.43/91
Uranium hexafluoride: Aromatic hydrocarbons, etc.
Uranyl perchlorate: Ethanol
    See N-HALOMIDES: Alcohols

Phosphorus(III) oxide
    See Tetraphosphorus hexaoxide: Organic liquids

Platinum
    See Platinum: Ethanol

Potassium
    See Potassium: Slow oxidation

Potassium *tert*-butoxide
    See Potassium *tert*-butoxide: Acids, etc.

Silver nitrate
    See Silver nitrate: Ethanol

Silver oxide
    See Silver(I) oxide: Ammonia, etc.

Sodium
    See Sodium: Ethanol

Tetrachlorosilane
    See Tetrachlorosilane: Ethanol

†0921. Dimethyl sulfoxide (Sulfinylbismethane)
    [69-68-5] C_2H_6OS
                        Me$_2$S(:O)

HCS 1980, 435; *RSC Lab. Hazard Data Sheet No. 11*, 1983

1. Brogli, F. *et al., Proc. 3rd Int. Symp. Loss Prev. Safety Prom. Proc. Ind.*, 681–682, Basle, SSCI, 1980
2. Cardillo, P. *et al., Chim. e Ind.* (Milan), 1982, **44**, 231–234

3. Santosusso, T. M. *et al., Tetrahedron Lett.*, 1974, 4255–4258
4. *See entry* THERMOCHEMISTRY AND EXOTHERMIC DECOMPOSITION (reference 2)
Two instances of used DMSO decomposing exothermally while being kept at
150°C prior to recovery by vacuum distillation were investigated. Traces of alkyl
bromides lead to a delayed, vigorous and strongly exothermic reaction ($Q =$
0.85 kJ/g) at 180°C. Addition of zinc oxide as a stabiliser extends the induction period and markedly reduces the exothermicity [1]. ARC examination shows
that exothermic decomposition sets in by a radical mechanism at 190°C, just above
the b.p., 189°C. The proposed retardants, sodium carbonate and zinc oxide, do not
affect the decomposition temperature, and a maximum decomposition pressure of
60 bar was attained (at up to 4 bar/min) at the low sample loading of 18 w/v% in
the bomb [2]. The thermolytic degradation of the sulfoxide to give acidic products
which catalyse further decomposition had been discussed previously [3]. T_{ait24} was
determined as 213°C by adiabatic Dewar tests, with an apparent energy of activation of 243 kJ/mol. At elevated temperatures (200°C) DSC shows decomposition
to be both faster and more energetic when chloroform or sodium hydroxide is
present.
See other INDUCTION PERIOD INCIDENTS, SELF-ACCELERATING REACTIONS

Acid anhydrides
See Trifluoroacetic anhydride, below

Acids, or Bases
Hall, J., *Loss Prev. Bull.*, 1993, (114), 2
These may catalyse exothermic decomposition. A runaway of a mixture with 4-
nitrobenzenesulfonic acid, possibly from as low as 60°C (though a steam valve
may have been leaking) is reported by Hall.
See reference 2, next below

Acyl halides, or Non-metal halides   MRH Thionyl chloride 1.63/60
1. Buckley, A., *J. Chem. Educ.*, 1965, **42**, 674
2. Allan, G. G. *et al., Chem. & Ind.*, 1967, 1706
3. Mancuso, A. J. *et al., J. Org. Chem.*, 1979, **44**, 4148–4150
In absence of diluent or other effective control of reaction rate, the sulfoxide
reacts violently or explosively with the following: acetyl chloride, benzenesulfonyl chloride, cyanuric chloride, phosphorus trichloride, phosphoryl chloride,
tetrachlorosilane, sulfur dichloride, disulfur dichloride, sulfuryl chloride or thionyl
chloride [1]. These violent reactions are explained in terms of exothermic polymerisation of formaldehyde produced under a variety of conditions by interaction
of the sulfoxide with reactive halides, acidic or basic reagents [2]. Oxalyl chloride reacts explosively with DMSO at ambient temperature, but controllably in
dichloromethane at −60°C [3].
See Carbonyl diisothiocyanate, and Dinitrogen tetraoxide, and Hexachlorocyclotriphosphazine, and Sodium hydride, all below
See Perchloric acid: Sulfoxides

Allyl trifluoromethanesufonates
See ALLYL TRIFLUOROMETHANESULFONATES: Alone, or Aprotic solvents

Boron compounds
Shriver, 1969, 209
DMSO forms an explosive mixture with nonahydrononaborate(2−) ion and with diborane. It is probable that other boron hydrides and hydroborates behave similarly.

Bromides
1. Hall, J., *Loss Prev. Bull.*, 1993, (114), 2
2. Aida, T., *et al., Bull. Chem. Soc. Jap.*, 1976, **49**, 117
A review of runaway decomposition reactions of DMSO [1], several involving bromides, some alkaline and under vacuum, from temperatures possibly as low as 130°C. A bromide/bromine catalysed decomposition reaction is known [2].

4(4′-Bromobenzoyl)acetanilide
491M, 1975, 74
An explosion occurred after a solution of the anilide in DMSO had been held at 100°C for 30 min.

Bromomethane
See Trimethylsulfoxonium bromide

Carbonyl diisothiocyanate
Bunnenberg, R. *et al., Chem. Ber.*, 1981, **114**, 2075–2086
Interaction of the solvent with the strong electrophile is explosive.
See Acyl halides, above

Copper, Trichloroacetic acid
1. Giessemann, B. W., *Chem. Eng. News*, 1981, **59**(28), 4
2. Dunne, T. G. *et al., Chem. Eng. News*, 1981, **59**(50), 4
The acid was added to copper wool and rinsed down with DMSO. Within 20 s, the contents of the flask were ejected and the neck was distorted by intense heat [1]. Adding the copper wool to a solution of the acid in the solvent gave an exothermic but controlled reaction [2]. (Formation of a reactive carbene species by dehydrohalogenation seems a remote possibility.)
See Acyl halides, above

Dichloromethane, Perchloric acid
See Perchloric acid: Dichloromethane, Dimethyl sulfoxide

Dinitrogen tetraoxide
Buckley, A., *J. Chem. Educ.*, 1965, **42**, 674
Interaction may be violent or explosive.
See Acyl halides
See Nitric acid: Dimethyl sulfoxide

Hexachlorocyclotriphosphazine
Pierce, T. *et al., Lab. Haz. Bull.*, 1984, (4), item 215

Addition of about 100 mg of hexachlorocyclotriphosphazine (also containing some phenoxylated products) to 0.5 ml of DMSO in an NMR tube led to an exothermic reaction which ejected the tube contents. General precautions are suggested.
See Acyl halides, above

Iodine pentafluoride
Lawless, E. M., *Chem. Eng. News*, 1969, **47**(13), 8
Interaction is explosive, after a delay, in either tetrahydrothiophene-1,1-dioxide (sulfolane) or trichlorofluoromethane as solvent, on 0.15 g mol scale, though not on one tenth this scale.

Magnesium perchlorate   MRH 5.52/78
See Magnesium perchlorate: Dimethyl sulfoxide

Metal alkoxides
MCA Case History No. 1718
Addition of potassium *tert*-butoxide or of sodium isopropoxide to the solvent led to ignition of the latter. This was attributed to presence of free metal in the alkoxides, but a more likely explanation seems to be that of direct interaction between the powerful bases and the sulfoxide.
See Acids, or Bases, above (reference 2)
   Potassium *tert*-butoxide: Acids, etc.

Metal oxosalts
   1. Martin, 1971, 435
   2. Dehn, H., Brit. Pat. 1 129 777, 1968
   3. Sandstrom, M. *et al., Acta Chem. Scand.*, 1978, **A32**, 610
Mixtures of the sulfoxide with metal salts of oxoacids are powerful explosives. Examples are aluminium perchlorate, sodium perchlorate and iron(III) nitrate [1]. The water in hydrated oxosalts (aluminium perchlorate, iron(III) perchlorate, iron(III) nitrate) may be partially or totally replaced by dimethyl (or other) sulfoxide to give solvated salts useful as explosives [2]. Metal nitrates and perchlorates solvated with DMSO are generally powerfully explosive, and under certain conditions a violent reaction is easily triggered [3]. Several other explosions involving perchlorates and the sulfoxide have been reported.
See Chromium(III) perchlorate . 6 dimethyl sulfoxide
   Magnesium perchlorate: Dimethyl sulfoxide
   Mercury(II) perchlorate . 6(or 4)dimethyl sulfoxide
   Silver perchlorate: Dimethyl sulfoxide
See Perchloric acid: Sulfoxides

Nitric acid
See Nitric acid: Dimethyl sulfoxide

Non-metal halides
See Acyl halides, etc., above

Other reactants
Yoshida, 1980, 171
MRH values calculated for 6 combinations are given.

Perchloric acid   MRH 6.19/77
See Perchloric acid: Sulfoxides

Periodic acid
See Periodic acid: Dimethyl sulfoxide

Phosphorus(III) oxide
See Tetraphosphorus hexaoxide: Organic liquids

Potassium
Houben-Weyl 1970, Vol. 13.1, 295
Interaction of potassium 'sand' and dimethyl sulfoxide is violent in the absence of
a diluent, and leads to partial decomposition of the potassium dimethylsulfinate.
THF is a suitable diluent.
See Metal alkoxides, above

Potassium permanganate   MRH 3.01/14
See Potassium permanganate: Dimethyl sulfoxide

Silver difluoride
Lawless, E. M., *Chem. Eng. News*, 1969, **47**(13), 8
Interaction is violent.

Sodium
Hall, J., *Loss Prev. Bull.*, 1993, (114), 2
A preparation of dimsyl sodium from sodium and DMSO, on 14 litre scale, overheated, then exploded causing two fatalities. It has not proved possible to duplicate
the runaway.

Sodium hydride
1. French, F. A., *Chem. Eng. News*, 1966, **44**(15), 48
2. Olson, G. A., *Chem. Eng. News*, 1966, **44**(24), 7
3. Russell, G. A. *et al., J. Org. Chem.*, 1966, **31**, 248
4. Batchelor, J. F., private comm., 1976
5. Anon., *Loss Prev. Bull.*, 1980, (030), 161
6. Allan, G. G. *et al., Chem. & Ind.*, 1967, 1706
7. Itoh, M. *et al., Lab. Haz. Bull.* 1985(5), item 304
Two violent pressure-explosions occurred during preparations of dimethylsulfinyl
anion on 3–4 g mol scale by reaction of sodium hydride with excess solvent. In
each case, the explosion occurred soon after separation of a solid. The first reaction
involved addition of 4.5 g mol of hydride to 18.4 g mol of sulfoxide, heated to
70°C [1], and the second 3.27 and 19.5 g mol respectively, heated to 50°C [2].
A smaller scale reaction at the original lower hydride concentration [3], did not

explode, but methylation was incomplete. Explosions and fire occurred when the reaction mixture was overheated (above 70°C) [4]. Reaction of 1 g mol of hydride with 0.5 l of sulfoxide at 80°C led to an exotherm to 90°C with explosive decomposition [5]. These and similar incidents are explicable in terms of exothermic polymerisation of formaldehyde produced from sulfoxide by reaction with the hydride base [6]. The heat of reaction was calculated and determined experimentally. Thermal decomposition of the solution of hydride is not very violent, but begins at low temperatures, with gas evolution [7].
See other GAS EVOLUTION INCIDENTS
See Sodium dimethylsulfinate

Sulfur trioxide   MRH 2.51/59
See Sulfur trioxide: Dimethyl sulfoxide

Trifluoroacetic anhydride
Sharma, A. K. *et al., Tetrahedron Lett.*, 1974, 1503–1506
Interaction of the sulfoxide with some acid anhydrides or halides may be explosive. The highly exothermic reaction with trifluoroacetic anhydride was adequately controlled in dichloromethane solution at below −40°C.
See Acyl halides, above
See other APROTIC SOLVENTS

0922. Poly(dimethylsiloxane)
[9016-00-6] $(C_2H_6OSi)_n$

$$(-SiMe_2O-)_n$$

See SILICONE OIL

0923. Dimethyl peroxide
[690-02-8] $C_2H_6O_2$

$$MeOOMe$$

1. Rieche, A. *et al., Ber.*, 1928, **61**, 951
2. Baker, G. *et al., Chem. & Ind.*, 1964, 1988
Extremely explosive, heat- and shock-sensitive as liquid or vapour [1]. During determination of the impact-sensitivity of the confined material, rough handling of the container caused ignition. The material should only be handled in small quantity and with great care [2].
See other DIALKYL PEROXIDES

0924. Ethylene glycol (1,2-Ethanediol)
[107-21-1] $C_2H_6O_2$

$$HOC_2H_4OH$$

FPA H47, 1976; HCS 1980, 483; RSC Lab. Hazards Data Sheet No. 69, 1988

Dimethyl terephthalate, Titanium butoxide
See Dimethyl terephthalate: Ethylene glycol, Titanium butoxide

Other reactants
Yoshida, 1980, 48
MRH values calculated for 15 combinations with oxidants are given.

Oxidants   MRH values show % of oxidant
1. Oouchi, H. *et al., Chem. Abs.*, 1982, **97**, 8605
2. Inoue, Y. *et al., Chem. Abs.*, 1987, **107**, 120353
In a study of hypergolic ignition of ethylene glycol by oxidants, chromium trioxide, potassium permanganate and sodium peroxide caused ignition on contact at ambient temperature, and ammonium dichromate, silver chlorate, sodium chlorite and uranyl nitrate at 100°C [1]. Results from an exothermicity test in which 1 g portions each of a glycol and an oxidising agent were mixed at 100°C in a Dewar flask were compared with those from DTA experiments [2].
See Chromium trioxide: Ethylene glycol   MRH 2.47/84
    Perchloric acid: Glycols and their ethers   MRH 5.69/70
    Potassium dichromate: Ethylene glycol   MRH 1.46/89
See Sodium hypochlorite: Ethanediol

Phosphorus pentasulfide
See Tetraphosphorus decasulfide: Alcohols

Silvered copper wire
1. Downs, W. R., *TN D-4327*, Washington, NASA Tech. Note, 1968
2. Stevens, H. D., *Chem. Abs.*, 1975, **83**, 134485
Contact of aqueous ethylene glycol solutions with d.c.-energised silvered copper wires causes ignition of the latter. Bare copper or nickel- or tin-plated wires were inert and silver-plated wire can be made so by adding benzotriazole as a metal deactivator to the coolant solution [1]. This problem of electrical connector fires in aircraft has been studied in detail to identify the significant factors [2].

Sodium hydroxide
See Sodium hydroxide: Glycols, and Sodium 2-hydroxyethoxide

Sodium hypochlorite
See Sodium hypochlorite: Ethanediol

0925. Ethyl hydroperoxide
[3031-74-1] $C_2H_6O_2$
                                  EtOOH

Baeyer, A. *et al., Ber.*, 1901, **34**, 738
It explodes violently on superheating; the barium salt is heat- and impact-sensitive.

Hydriodic acid
Sidgwick, 1950, 873
The concentrated acid (a reducer) is oxidised explosively.
See other REDOX REACTIONS

Silver
Sidgwick, 1950, 873
Finely divided ('molecular') silver decomposes the hydroperoxide, sometimes explosively.
See other ALKYL HYDROPEROXIDES

0926. Hydroxymethyl methyl peroxide
[ ] $C_2H_6O_3$

HOCH₂OOMe

Rieche, A. *et al., Ber.*, 1929, **62**, 2458
Violently explosive, impact-sensitive when heated.
See other 1-OXYPEROXY COMPOUNDS

0927. Dimethyl sulfite
[616-42-2] $C_2H_6O_3S$

(MeO)₂S:O

Phosphorus (III) oxide
See Tetraphosphorus hexaoxide: Organic liquids
See other SULFUR ESTERS

0928. Bis(hydroxymethyl) peroxide
[17088-73-2] $C_2H_6O_4$

HOCH₂OOCH₂OH

Wieland, H., *et al., Ber.*, 1930, **63**, 66
Highly explosive and very friction-sensitive. Higher homologues are more stable.
See other 1-OXYPEROXY COMPOUNDS

†0929. Dimethyl sulfate
[77-78-1] $C_2H_6O_4S$

(MeO)₂SO₂

(*MCA SD-19*, 1966); *HCS 1980*, 434

Ammonia
1. Lindlar, H., *Angew. Chem.*, 1963, **75**, 297
2. Claesson, P. *et al., Ber.*, 1880, **13**, 1700

A violent reaction occurred which shattered the flask when litre quantities of dimethyl sulfate and conc. aqueous ammonia were accidentally mixed. Use dilute ammonia in small quantities to destroy dimethyl sulfate [1]. Similar incidents were noted previously with ammonia and other bases [2].
See Tertiary bases, below
See other AMINATION INCIDENTS

Barium chlorite
See Barium chlorite: Dimethyl sulfate

Other reactants
Yoshida, 1980, 395
MRH values calculated for 13 combinations with oxidants are given.

Tertiary bases
Sorbe, 1968, 123
In absence of diluent, quaternation of some tertiary organic bases may proceed explosively.
See Ammonia, above

Unnamed material
MCA Case History No. 1786
The product of methylating an unnamed material at 110°C was allowed to remain in the reactor of a pilot plant. and after 80 min the reactor exploded. This was ascribed to exothermic decomposition of the mixture above 100°C, and subsequent acceleration and boiling decomposition at 150°C.
See other SULFUR ESTERS

0930. Dimethyl selenate
[6918-51-0] $C_2H_6O_4Se$

$$(MeO)_2SeO_2$$

Sidgwick, 1950, 977
Dimethyl selenate, like its ethyl and propyl homologues, can be distilled under reduced pressure, but explodes at around 150°C under ambient pressure, though less violently than the lower alkyl nitrates.
See related SULFUR ESTERS

0931. Dimethanesulfonyl peroxide
[1001-62-3] $C_2H_6O_6S_2$

$$MeSO_2OOSO_2Me$$

Haszeldine, R. N. *et al., J. Chem. Soc.*, 1964, 4903
A sample heated in a sealed tube exploded violently at 70°C. Unconfined, the peroxide decomposed (sometimes explosively) immediately after melting at 79°C.
See other DIACYL PEROXIDES

†0932. **Dimethyl sulfide (Thiobismethane)**
[75-18-3]   C₂H₆S

$$Me_2S$$

Dibenzoyl peroxide
  See Dibenzoyl peroxide: Dimethyl sulfide

1,4-Dioxane, Nitric acid
  See Nitric acid: Dimethyl sulfide, 1,4-Dioxane

Other reactants
  Yoshida, 1980, 389
  MRH values calculated for 14 combinations with oxidants are given.

Oxygen
  See Oxygen (Gas): Dimethyl sulfide

Xenon difluoride
  See Xenon difluoride: Dimethyl sulfide

†0933. **Ethanethiol**
[75-08-1]   C₂H₆S

$$EtSH$$

HCS 1980, 497

Other reactants
  Yoshida, 1980, 44
  MRH values calculated for 14 combinations with oxidants are given.
  See other ALKANETHIOLS

†0934. **Dimethyl disulfide**
[624-92-0]   C₂H₆S₂

$$MeSSMe$$

0935. **Dimethylzinc**
[544-97-8]   C₂H₆Zn

$$Me_2Zn$$

Air, or Oxygen
  1. Egerton, A. *et al., Proc. R. Soc.*, 1954, **A225**, 429
  2. Frankland, E., *Phil. Trans. R. Soc.*, 1852, 417
  It ignites in air (owing to peroxide formation) [1], and explodes in oxygen [2]. It
  is slightly endothermic (ΔH°_{f} (l) +25.1 kJ/mol, 0.26 kJ/g).

2,2-Dichloropropane
Houben-Weyl, 1973, Vol. 13.2a, 767
Uncontrolled reaction is explosive.
See DIALKYLZINCS: Alkyl chlorides

Ozone
See Ozone: Alkylmetals

Water
Leleu, *Cahiers*, 1977, (88), 371
Reaction is explosively violent.
See other ALKYLMETALS, DIALKYLZINCS, ENDOTHERMIC COMPOUNDS

0936. Dimethylaluminium hydride
[865-37-2] C_2H_7Al

$$Me_2AlH$$

Houben-Weyl, 1970, Vol. 13.4, 58
Slight contact with air or moisture causes ignition.
See other ALKYLALUMINIUM ALKOXIDES AND HYDRIDES

0937. Dimethylarsine
[593-57-7] C_2H_7As

$$Me_2AsH$$

1. von Schwartz, 1918, 322
2. Sidgwick, 1950, 762
It inflames in air [1], even at 0°C [2].
See other ALKYLNON-METAL HYDRIDES

0938. 1,1-Dimethyldiazenium perchlorate
[[53534-20-6] (ion)] $C_2H_7ClN_2O_4$

$$Me_2N^+{=}NHClO_4^-$$

McBride, W. E. *et al.*, *J. Amer. Chem. Soc.*, 1957, **79**, 576 footnote 16
The salt is impact sensitive.
See other PERCHLORATE SALTS OF NITROGENOUS BASES

0939. 1-Methyl-3-nitroguanidinium perchlorate
[ ] $C_2H_7ClN_4O_6$

$$MeH_2N^+C(:NH)NHNO_2ClO_4^-$$

McKay, A. F. *et al.*, *J. Amer. Chem. Soc.*, 1947, **69**, 3029
The salt is sensitive to impact, exploding violently.
See other N-NITRO COMPOUNDS, PERCHLORATE SALTS OF NITROGENOUS BASES

0940. 2-Chloroethylammonium chloride (2-Chloroethylamine hydrochloride)
[870-24-6] $C_2H_7Cl_2N$

$$ClC_2H_4N^+H_3Cl^-$$

Alkali
  1. Wystrach, V. P. *et al., J. Amer. Chem. Soc.*, 1955, **77**, 5915
  2. Wystrach, V. P. *et al., Chem. Eng. News*, 1956, **34**, 1274
The procedure described previously [1] for the preparation of ethylenimine is erroneous. 2-Chloroethylammonium chloride must be added with stirring as a 33% solution in water to strong sodium hydroxide solution. Addition of the solid salt to the alkali caused separation in bulk of 2-chloroethylamine, which polymerised explosively. Adequate dilution and stirring, and a temperature below 50°C are all essential [2].
See other POLYMERISATION INCIDENTS

†0941. Dimethylamine (*N*-Methylmethanamine)
[124-40-3] C_2H_7N

$$Me_2NH$$

(*MCA SD-57*, 1955); *HCS 1980*, 421 (40% solution)

Acrylaldehyde
  See Acrylaldehyde: Acids, etc.

4-Chloroacetophenone
  See 4-Chloroacetophenone: Dimethylamine

Fluorine   MRH 10.6/77
  See Fluorine: Nitrogenous bases

Maleic anhydride
  See Maleic anhydride: Bases, etc.

Other reactants
  Yoshida, 1980, 169
  MRH values calculated for 14 combinations with oxidants are given.
  See other ORGANIC BASES

†0942. Ethylamine (Ethanamine)
[75-04-7] C_2H_7N

$$EtNH_2$$

FPA H67, 1978; *HCS 1980*, 467 (cylinder)

Cellulose nitrate
  See CELLULOSE NITRATE: Amines
  See other ORGANIC BASES

347

0943. 2-Hydroxyethylamine (Ethanolamine)
[141-43-5] C_2H_7NO

$$HOC_2H_4NH_2$$

HCS 1980, 457

Carbon dioxide, Pyrophoric deposits
1. Bond, J., *Hazards from Pressure*, IChE Symp. Ser. No. 102, 37–44, Oxford, Pergamon, 1987
2. Bond, J., *Loss Prev. Bull.*, 1995, **121**, 18

Ethanolamine has a large affinity for carbon dioxide, and is used to remove it from process gas streams. A storage tank used to hold the amine had a steel steam coil to warm the contents, and the coil developed a pyrophoric deposit, probably from contact with sulfur-containing impurities. When part of the coil was not covered by the amine, the surface deposit caused a fire which ignited some of the ethanolamine in the closed tank until the air was exhausted and the fire went out. The combined effect of cooling of the hot gases in the tank and of the rapid absorption of the carbon dioxide produced in the fire caused a significant pressure reduction and the tank collapsed [1]. It was also suggested that the pyrophoric material might have been tris(hydroxyethylamine)iron(0), a rather improbable species, or tris(aminoethoxy)iron(III) trihydrate which is perhaps more probable, though its postulated decomposition to pyrophoric iron is less so [2].
See PYROPHORIC IRON–SULFUR COMPOUNDS

Cellulose nitrate
See CELLULOSE NITRATE: Amines

N,N-Dimethyl-N,N-dinitrosoterephthalamide
See N,N'-Dimethyl-N,N'-dinitrosoterephthalamide: 2-Hydroxyethylamine

Other reactants
Yoshida, 1980, 370
MRH values calculated for 15 combinations are given.
See other ORGANIC BASES

0944. *O*-(2-Hydroxyethyl)hydroxylamine
[3279-95-6] $C_2H_7NO_2$

$$HOC_2H_4ONH_2$$

Sulfuric acid
Campbell, H. F., *Chem. Eng. News*, 1975, **53**(49), 5
After 30 min at 120°C under vacuum, an equimolar mixture with the conc. acid exploded violently. Salts of unsubstituted hydroxylamine are thermally unstable.
See HYDROXYLAMINIUM SALTS *See other* N–O COMPOUNDS

0945. 1,3-Dimethyltriazene ('Diazoaminomethane')
[3585-32-8] $C_2H_7N_3$

$$MeN=NNHMe$$

Dimroth, O., *Ber.*, 1906, **39**, 3910
A drop explodes sharply in contact with flame.
See other TRIAZENES

0946. 1,2-Dimethylnitrosohydrazine
[101672-10-0] $C_2H_7N_3O$

$$MeN(N:O)NHMe$$

Smith, 1966, Vol. 2, 459
The liquid deflagrates on heating.
See other NITROSO COMPOUNDS

0947. 1-Methyl-3-nitroguanidinium nitrate
[ ] $C_2H_7N_5O_5$

$$MeN^+H_2C(:NH)NHNO_2NO_3^-$$

McKay, A. F. *et al., J. Amer. Chem. Soc.*, 1947, **69**, 3029
The salt could be exploded by impact between steel surfaces.
See other N-NITRO COMPOUNDS, OXOSALTS OF NITROGENOUS BASES

0948. Dimethylphosphine
[676-59-5] C_2H_7P

$$Me_2PH$$

1. Houben-Weyl, 1963, Vol. 12.1, 69
2. Parshall, G. W., *Inorg. Synth.*, 1968, **11**, 158
3. Trenkle, A. *et al., Inorg. Synth.*, 1982, **21**, 180–181
Secondary lower-alkylphosphines readily ignite in air [1]; preparative precautions are detailed [2,3].
See other ALKYLPHOSPHINES

0949. Ethylphosphine
[593-68-0] C_2H_7P

$$EtPH_2$$

Houben-Weyl, 1963, Vol. 12.1, 69
Primary lower-alkylphosphines readily ignite in air.

Halogens, or Nitric acid
  von Schwartz, 1918, 324–325

It explodes on contact with chlorine, bromine or fuming nitric acid, and inflames with conc. acid.

See other ALKYLPHOSPHINES

0950. Dimethylammonium perchlorate
[14488-49-4] $C_2H_8ClNO_4$

$$Me_2N^+H_2ClO_4^-$$

1. Gore, P. H., *Chem. Brit.*, 1976, **12**, 205
2. Menzer, M. *et al., Z. Chem.*, 1977, **17**, 344
3. Gore, P. H. *et al., Synth. Commun.*, 1980, **10**, 320

A violent explosion occurred during vacuum evaporation of an aqueous mixture of excess dimethylamine and perchloric acid [1], and a similar incident occurred when the moist solid from a like preparation was moved [2]. Use of this explosive salt as an intermediate for the preparation of 3-dimethylamino-2-propenylidenedimethylimmonium perchlorate should be avoided, the methylsulfate or triflate salts being preferred [3].

See other PERCHLORATE SALTS OF NITROGENOUS BASES

0951. 2-Hydroxyethylaminium perchlorate (Ethanolamine perchlorate)
[38092-76-1] $C_2H_8ClNO_5$

$$HOC_2H_4N^+H_3ClO_4^-$$

Annikov, V. E. *et al., Chem. Abs.*, 1983, **99**, 160860

Ethanolamine perchlorate was outstanding amongst other nitrogenous perchlorate salts for its capacity for detonation as an aqueous solution.

See other PERCHLORATE SALTS OF NITROGENOUS BASES

0952. Dimethyl phosphoramidate
[2697-42-9] $C_2H_8NO_3P$

$$(MeO)_2P(:O)NH_2$$

Chlorine
  See Chlorine: Dimethyl phosphoramidate
  See other PHOSPHORUS ESTERS

0953. 1,2-Diaminoethane (Ethylenediamine)
[107-15-3] $C_2H_8N_2$

$$H_2NC_2H_4NH_2$$

Cellulose nitrate
  See CELLULOSE NITRATE: Amines

Diisopropyl peroxydicarbonate
  See Diisopropyl peroxydicarbonate: Amines, etc.

350

Nitromethane
See Nitromethane: Acids, etc., and: 1,2-Diaminoethane, etc.

Other reactants
Yoshida, 1980, 50
MRH values calculated for 13 combinations with oxidants are given.

Silver perchlorate
See Silver perchlorate: 1,2-Diaminoethane
See other ORGANIC BASES

†0954. 1,1-Dimethylhydrazine
[57-14-7] $C_2H_8N_2$

Me_2NNH_2

HCS 1980, 431

Energy of decomposition (in range 280–380°C) measured as 1.15 kJ/g.
See entry THERMOCHEMISTRY AND EXOTHERMIC DECOMPOSITION (reference 2)

Dichloromethane
See Dichloromethane: 1,2-Diaminoethane

Dicyanofurazan
See Dicyanofurazan: Nitrogenous bases

Oxidants
1. Kirk-Othmer, 1966, Vol. 11, 186
2. Wannagat, U. *et al., Monatsh.*, 1969, **97**, 1157–1162
3. *491M*, 1975, 158
A powerful reducing agent and fuel, hypergolic with many oxidants, such as dinitrogen tetraoxide, hydrogen peroxide and nitric acid [1]. The ignition delay with fuming nitric acid was determined as 8 ms, explosion also occurring [2]. When spread (as a thin film, or absorbed on porous or fibrous material) to expose a large surface to air, ignition may occur [3].
See Hexanitroethane: 1,1-Dimethylhydrazine
See also ROCKET PROPELLANTS
See other ORGANIC BASES, REDUCANTS

†0955. 1,2-Dimethylhydrazine
[540-73-8] $C_2H_8N_2$

MeNHNHMe

See other ORGANIC BASES, REDUCANTS

0956. 1-Hydroxyethylidene-1,1-diphosphonic acid
[2809-21-4] $C_2H_8O_7P_2$

$$MeCOH[P(O)(OH)_2]_2$$

'Data Sheet ADPA 60A', Oldbury, Albright & Wilson, 1977
The anhydrous acid decomposes, often violently, at above 200°C, to give phosphine, phosphoric acid and other products.
See other ORGANIC ACIDS

0957. 1,2-Diphosphinoethane
[5518-62-7] $C_2H_8P_2$

$$H_2PC_2H_4PH_2$$

Taylor, R. C. *et al., Inorg. Chem.*, 1973, **14**, 10–11
It ignites in air.
See other ALKYLPHOSPHINES

0958. 2-Aminoethylammonium perchlorate
[25682-07-9] $C_2H_9ClN_2O_4$

$$H_2NC_2H_4N^+H_3ClO_4^-$$

Hay, R. W. *et al., J. Chem. Soc., Dalton Trans.*, 1975, 1467
Solutions of the perchlorate salts of 1,2-diaminoethane and other amines must be evaporated without heating to avoid the risk of violent explosions.
See other PERCHLORATE SALTS OF NITROGENOUS BASES

0959. Dimethyl hydrazidophosphate (Dimethylphosphorohydrazine)
[58816-51-8] $C_2H_9N_2O_3P$

$$(MeO)_2P(O)NHNH_2$$

Anon., *Sichere Chemiearbeit*, 1990, **42**, 56
The compound decomposed explosively upon vacuum distillation at 150°C. Caution is advised with similar hydrazides.
See related PHOSPHORUS ESTERS

†0960. 1,1-Dimethyldiborane
[16924-32-6] $C_2H_{10}B_2$

$$Me_2B:H_2:BH_2$$

See other ALKYLBORANES

†0961. 1,2-Dimethylborane
[17156-88-6] $C_2H_{10}B_2$

$$MeHB:H_2:BHMe$$

See other ALKYLBORANES

0962. *B*-Chlorodimethylaminodiborane
[ ] $C_2H_{10}B_2ClN$

Burg, A. B. *et al., J. Amer. Chem. Soc.*, 1949, **71**, 3454
It ignites in air.
See related BORANES, HALOBORANES

0963. 2,2-Dimethyltriazanium perchlorate
[20446-69-9] $C_2H_{10}ClN_3O_4$

$$Me_2N^+(NH_2)_2ClO_4^-$$

Rubtsov, Yu. I. *et al., Chem. Abs.*, 1981, **95**, 117860
It decomposes rapidly above 122°C.
See other PERCHLORATE SALTS OF NITROGENOUS BASES

0964. 1,2-Ethylenebis(ammonium) perchlorate
[15718-71-5] $C_2H_{10}Cl_2N_2O_8$

$$(-CH_2N^+H_3)_2 \ 2ClO_4^-$$

Lothrop, W. C. *et al., Chem. Rev.*, 1949, **44**, 432
An explosive which appreciably exceeds the power and brisance of TNT.
See other PERCHLORATE SALTS OF NITROGENOUS BASES

0965. Aqua-1,2-diaminoethanediperoxochromium(IV)
[17168-82-0] $C_2H_{10}CrN_2O_5$

1. Childers, R. F. *et al., Inorg. Chem.*, 1968, **7**, 749
2. House, D. A. *et al., Inorg. Chem.*, 1966, **5**, 840
The monohydrate is light-sensitive and explodes at 96–97°C if heated at 2°/min
[1]. It effervesces vigorously on dissolution in perchloric acid [2].
See other AMMINECHROMIUM PEROXOCOMPLEXES

0966. 1,2-Diammonioethane nitrate (Ethylenediamine dinitrate)
[20829-66-7] $C_2H_{10}N_4O_6$
$$[H_3N^+C_2H_4N^+H_3]\ [NO_3^-]_2$$

Dobratz, B. M., *Chem. Abs.*, 1984, **100**, 88163
Ethylenediamine dinitrate was formerly used as a military explosive.
See other OXOSALTS OF NITROGENOUS BASES

0967. Dimethylaminodiborane
[23273-02-1] $C_2H_{11}B_2N$

Keller, P. G., *Inorg. Synth.*, 1977, **17**, 34
It ignites in air, like the parent diborane.
See related BORANES

0968. Ammine-1,2-diaminoethanediperoxochromium(IV)
[17168-82-0] $C_2H_{11}CrN_3O_4$
$$[H_3NCr(C_2H_8N_2)(O_2)_2]$$

House, D. A. *et al.*, *Inorg. Chem.*, 1967, **6**, 1077
The monohydrate is potentially explosive at 25°C and decomposes or explodes at
115° during slow or moderate heating.

Hydrogen bromide
Hughes, R. G. *et al.*, *Inorg. Chem.*, 1968, **7**, 74
Interaction must be slow with cooling to prevent explosion.
See other AMMINECHROMIUM PEROXOCOMPLEXES

0969. Tetraamminedithiocyanatocobalt(III) perchlorate
[36294-69-6] (ion) $C_2H_{12}ClCoN_6O_4S_2$
$$[(H_3N)_4Co(SCN)_2]\ ClO_4$$

Tomlinson, W. R. *et al.*, *J. Amer. Chem. Soc.*, 1949, **71**, 375
Explodes at 335°C, medium impact-sensitivity.
See other AMMINEMETAL OXOSALTS

354

0970. Tetraamminebis(5-nitro-2*H*-tetrazolato)cobalt(1$^+$) perchlorate

[117412-28-9] $C_2H_{12}ClCoN_{14}O_8$

Fronabarger, J. W., *et al., Proc. Int. Pyrotech. Semin.*, 1996, 22nd, 645
It is being developed as an initiating explosive.
See other AMMINEMETAL OXOSALTS, TETRAZOLES

0971. Guanidinium dichromate

[27698-99-3] $C_2H_{12}Cr_2N_6O_7$

$$[H_2NC(:N^+H_2)NH_2]_2Cr_2O_7^{2-}$$

Ma, C., *J. Amer. Chem. Soc.*, 19512, **73**, 1333–1335
Heating causes orderly decomposition, but under confinement, a violent explosion.
See other DICHROMATE SALTS OF NITROGENOUS BASES

0972. Ammonium 1,2-ethylenebis(nitramide)

[3315-89-7] $C_2H_{12}N_6O_4$

$$(N^+H_4)_2 (CH_2N^-NO_2)_2$$

Violent decomposition occurred at 191°C.
See entry DIFFERENTIAL THERMAL ANALYSIS (reference 1)
See other N-NITRO COMPOUNDS

0973. Ethylpentaborane(9)

[28853-06-7] $C_2H_{13}B_5$

Leleu, *Cahiers*, 1977, (88), 365
It ignites in air.
See other ALKYLBORANES

0974. 2(5-Cyanotetrazole)pentaamminecobalt(III) perchlorate

[70247-32-4] $C_2H_{16}Cl_3CoN_{10}O_{12}$

$[C_2N_5HCo(NH_3)_5]\ 3ClO_4$

Loyola, V. M. *et al., Chem. Abs.*, 1981, **95**, 222342

It is a practical explosive of zero oxygen balance, but much less sensitive to accidental explosion than conventional detonators.

See other AMMINEMETAL OXOSALTS, TETRAZOLES

0975. Mercury(II) acetylide

[37297-87-3] $(C_2Hg)_n$

$(-HgC\equiv C-)_n$

Bailar, 1973, Vol. 3, 314–315

Possibly polymeric, it explodes when heated or shocked.

See other MERCURY COMPOUNDS, METAL ACETYLIDES

0976. Mercury(II) cyanide

[592-04-1] C_2HgN_2

$Hg(C\equiv N)_2$

Ullmann, A8, 176

It is a moderately endothermic compound but of high MW (ΔH_f° (s) +261.5 kJ/mol, 1.035 kJ/g). It may explode if detonated, although it is used as an inerting diluent for the more sensitive and more powerful oxycyanide, $Hg(CN)_2 \cdot HgO$.

Fluorine

See Fluorine: Metal salts

Hydrogen cyanide

Wöhler, L. *et al., Chem. Ztg.*, 1926, **50**, 761

The cyanide is a friction- and impact-sensitive explosive, and may initiate detonation of liquid hydrogen cyanide. Other heavy metal cyanides are similar.

Magnesium

See Magnesium: Metal cyanides

Sodium nitrite

See Sodium nitrite: Metal cyanides

See other ENDOTHERMIC COMPOUNDS, MERCURY COMPOUNDS, METAL CYANIDES

0977. Mercury(II) cyanate

[3021-39-4] $C_2HgN_2O_2$

$Hg(OC\equiv N)_2$

Biasutti, 1981, 53

Two explosions during the crushing of the cyanate are recorded.

See related METAL CYANIDES (AND CYANO COMPLEXES) *See other* MERCURY COM-POUNDS

0978. Mercury(II) fulminate
[628-86-4] $C_2HgN_2O_2$

$$Hg(C\equiv N \rightarrow O)_2$$

NSC 309, 1978

Alone, or Sulfuric acid
1. Urbanski, 1967, Vol. 3, 135, 140
2. Carl, L. R., *J. Franklin Inst.*, 1945, **240**, 149

Mercury fulminate, readily formed by interaction of mercury(II) nitrate, nitric acid and ethanol, is endothermic (ΔH°_f (s) +267.7 kJ/mol, 0.94 kJ/g) and was a very widely used detonator. It may be initiated when dry by flame, heat, impact, friction or intense radiation. Contact with sulfuric acid causes explosion [1]. The effects of impurities on the preparation and decomposition of the salt have been described [2].

See other ENDOTHERMIC COMPOUNDS, MERCURY COMPOUNDS, METAL FULMI-NATES

0979. Mercury(II) thiocyanate
[592-85-8] $C_2HgN_2S_2$

$$Hg(SCN)_2$$

HCS 1980, 619

Unpublished information, 1950

It is a moderately endothermic and thermally unstable compound of high MW (ΔH°_f (s) +200.8 kJ/mol, 0.63 kJ/g). A large batch of the damp salt became over-heated in a faulty steam drying oven, and decomposed vigorously, producing an enormous 'Pharaoh's serpent'.

See related METAL CYANIDES (AND CYANO COMPLEXES) *See other* ENDOTHERMIC COMPOUNDS, MERCURY COMPOUNDS

0980. Mercury 5,5′-azotetrazolide
[87552-19-0] C_2HgN_{10}

See entry METAL AZOTETRAZOLIDES *See other* MERCURY COMPOUNDS, TETRA-ZOLES

0981. Mercury(II) 5-nitrotetrazolide
[60345-95-1] $C_2HgN_{10}O_4$

1. Gilligan, W. H. *et al., Rept.AD-A036086*, Richmond (Va.), USNTIS, 1976
2. Farncomb, R. E. *et al., Chem. Abs.*, 1977, **87**, 203672
3. Redman, L. D. *et al. Chem. Abs.*, 1984, **100**, 194540

It is a promising replacement for lead azide in detonators [1], and a remotely controlled procedure for the preparation of this oxygen-balanced compound is described [2]. Preparative methods have been assessed for safety features [3].
See other MERCURY COMPOUNDS, *C*-NITRO COMPOUNDS, TETRAZOLES

0982. Mercury(II) oxalate
[3444-13-1] C_2HgO_4

(-OCO.CO.OHg-)$_n$

1. Anon., *ABCM Quart. Safety Summ.*, 1953, **24**, 30, 45
2. Muir, G. D., private comm., 1968

When dry, the oxygen-balanced salt explodes readily on percussion, grinding or heating to 105°C. This instability is attributed to presence of impurities (nitrate, oxide, or basic oxalate) in the product [1]. It is so thermally unstable that storage is inadvisable [2].
See other CATALYTIC IMPURITY INCIDENTS, MERCURY COMPOUNDS, METAL OXALATES
See entry OXYGEN BALANCE

0983. Dimercury dicyanide oxide (Mercury(II) oxycyanide)
[1335-31-5] C_2Hg_2O

NCHgOHgCN or Hg(CN)$_2$.HgO

1. Merck Index, 1983, 840; *ibid.* 1989, 5781
2. Kast, H. *et al., Chem. Abs.*, 1922, **16**, 4065
3. *MCA Guide*, 1972, 309

The compound is explosive when pure, and sensitive to impact or heat. It is stabilised for commerce by the presence of a major excess of mercury(II) cyanide [1]. Several explosive incidents have been described [2], most involving friction [3].
See related METAL CYANIDES (AND CYANO COMPLEXES) *See other* FRICTIONAL INITIATION INCIDENTS, MERCURY COMPOUNDS

0984. Thallium(I) iodacetylide

[ ]   C$_2$ITl

$$TlC\equiv CI$$

Fawcett, H. H., private comm., 1984
The acetylide, produced as an unexpected major product, is a shock- and friction-sensitive explosive. A few mg exploded when touched with a spatula, and the remaining 2 g sample detonated violently when destroyed by a controlled explosion.
See other HALOACETYLENE DERIVATIVES, METAL ACETYLIDES

0985. Diiodoacetylene (Diiodoethyne)

[624-74-8]   C$_2$I$_2$

$$IC\equiv CI$$

1. Anon., *Sichere Chemiearb.*, 1955, **7**, 55
2. Vaughn, T. H. *et al.*, *J. Amer. Chem. Soc.*, 1932, **54**, 789
3. Taylor, G. N., *Chem. Brit.*, 1981, **17**, 107
Pure recrystallised material exploded while being crushed manually in a mortar. The decomposition temperature is 125°C, and this may have been reached locally during crushing [1]. Explosion on impact, on heating to 84°C, and during attempted distillation at 98°C/5 mbar had been recorded previously [2]. During an attempt to repeat the published preparation [2], a relatively large amount of the very sensitive explosive, nitrogen triiodide (ibid.), was isolated [3].
See other HALOACETYLENE DERIVATIVES

0986. Tetraiodoethylene

[513-92-8]   C$_2$I$_4$

$$I_2C=CI_2$$

Iodine pentafluoride
See Iodine pentafluoride: Tetraiodoethylene
See other HALOALKENES, IODINE COMPOUNDS

0987. Potassium acetylide

[22754-96-7]   C$_2$K$_2$

$$KC\equiv CK$$

Water
Bahme, 1972, 80
Contact with limited amounts of water may cause ignition and explosion of evolved acetylene.
See other METAL ACETYLIDES

0988. Potassium dinitrooxalatoplatinate(2−)
  [15213-49-7] $C_2K_2N_2O_8Pt$

$$K^+ \qquad\qquad O_2N \diagdown \begin{array}{c} O \\ \diagup \end{array} \overset{2-}{Pt} \begin{array}{c} O \\ \diagdown O \end{array} \overset{\displaystyle O}{\underset{\displaystyle O}{}}$$

$$K^+ \qquad\qquad O_2N$$

Vèzes, M. *Compt. rend.*, 1897, **125**, 525
The salt decomposes violently at 240°C.
See related METAL OXALATES *See other* PLATINUM COMPOUNDS

0989. Potassium 1,1,2,2-tetranitroethanediide
  [32607-31-1] $C_2K_2N_4O_8$
                           $K_2[(O_2N)_2CC(NO_2)_2]$

1. Borgardt, F. G. *et al., J. Org. Chem.*, 1966, **31**, 2806, 2810
2. Griffin, T. S. *et al., J. Org. Chem.*, 1980, **45**, 2882
This anhydrous salt, and the mono- and di-hydrates of the analogous lithium and sodium salts, are all very impact-sensitive. The potassium salt, an intermediate in the preparation of hexanitroethane [1], was not allowed to become dry during isolation, but after precipitation was washed with methanol and dichloromethane and used wet with the latter for the succeeding nitration stage [2].
See other POLYNITROALKYL COMPOUNDS

0990. Potassium acetylene-1,2-dioxide (Potassium ethynediolate)
  [2851-55-0] $C_2K_2O_2$
                           $KOC\equiv COK$

Air, or Halocarbons, or Halogens, or Protic compounds
  Taylor, C. K., *Chem. Abs.*, 1983, **99**, 32105
Produced by action of carbon monoxide on potassium in liquid ammonia at −50°C, the yellow powder burns explosively in contact with air, halocarbons, halogens, alcohols, water and any material with acidic hydrogen. Analogous metal derivatives are reviewed.
See Potassium benzenehexoxide
See other ACETYLENIC COMPOUNDS

0991. Lanthanum carbide
  [12071-15-7] C_2La
                           LaC_2

Borlas, R. A. *et al., Chem. Abs.*, 1976, **84**, 182199

'On the flammability and explosiveness of lanthanum carbide and cerium carbide powders' (title only translated).
See other METAL NON-METALLIDES

0992. Lithium acetylide
[1070-75-3] C_2Li_2

$$LiC{\equiv}CLi$$

Halogens
    Mellor, 1946, Vol. 5, 848
    It burns brilliantly when cold in fluorine or chlorine, but must be warm before ignition occurs in bromine or iodine vapours.

Lead(II) oxide
    See Lead(II) oxide: Metal acetylides

Non-metals
    Mellor, 1946, Vol. 5, 848
    It burns vigorously in phosphorus, selenium or sulfur vapours.
    See other METAL ACETYLIDES

0993. Lithium ethynediolate
[88906-07-4] $C_2Li_2O_2$

$$LiOC{\equiv}COLi$$

Weiss, E., *Angew. Chem. (Int.)*, 1993, **32**(11), 1518
Too shock sensitive to obtain crystallographic data.
See Lithium benzenehexoxide
See other ACETYLENIC COMPOUNDS

0994. Lithium 1,1,2,2-tetranitroethanediide
[ ] $C_2Li_2N_4O_8$

$$Li_2[(O_2N)_2CC(NO_2)_2]$$

See Potassium 1,1,2,2-tetranitroethanediide
See other POLYNITROALKYL COMPOUNDS

0995. Dicarbonylmolybdenum diazide
[68348-85-6] $C_2MoN_6O_2$

$$(CO)_2Mo(N_3)_2$$

Alone, or Water
    Dehnicke, K., *Angew. Chem. (Intern. Ed.)*, 1979, **18**, 513

This and the tungsten analogue (both homopolymeric) are extremely sensitive, exploding on the slightest mechanical stress, or violently in contact with traces of water.
See related CARBONYLMETALS, METAL AZIDES

†0996. Dicyanogen (Ethanedinitrile)
[460-19-5] C_2N_2

$$N\equiv CC\equiv N$$

Oxidants
Dicyanogen is extremely endothermic (ΔH°_f (g) +307.9 kJ/mol, 5.91 kJ/g) and the potential energy of mixtures with powerful oxidants may be released explosively under appropriate circumstances.
See Dichlorine oxide: Dicyanogen
     Fluorine: Halogens
     Oxygen (Liquid): Liquefied gases
     Ozone: Dicyanogen
     ROCKET PROPELLANTS
See other CYANO COMPOUNDS, ENDOTHERMIC COMPOUNDS

0997. Nickel(II) cyanide
[557-19-7] C_2N_2Ni

$$Ni(CN)_2$$

It is a somewhat endothermic compound (ΔH°_f (s) +113.4 kJ/mol, 1.02 kJ/g), though not notably thermally unstable.

Magnesium
See Magnesium: Metal cyanides
See other ENDOTHERMIC COMPOUNDS, METAL CYANIDES

0998. Dicyanogen N,N'-dioxide
[4331-98-0] $C_2N_2O_2$

$$(-C\equiv N \rightarrow O)_2$$

Grundmann, C. J., *Angew. Chem.*, 1963, **75**, 450; *Ann.*, 1965, **687**, 194
The solid decomposes at −45°C under vacuum, emitting a brilliant light before exploding.
See related CYANO COMPOUNDS, HALOGEN OXIDES

0999. Lead(II) cyanide
[592-09-2] C_2N_2Pb

$$Pb(CN)_2$$

Magnesium
See Magnesium: Metal cyanides
See other METAL CYANIDES

362

1000. Lead(II) thiocyanate
[592-87-0] $C_2N_2PbS_2$

$$Pb(SCN)_2$$

Urbanski, 1967, Vol. 3, 230
The explosive properties of lead thiocyanate have found limited use.
See related METAL CYANIDES (AND CYANO COMPLEXES) *See other* HEAVY METAL
DERIVATIVES

1001. Thiocyanogen
[505-14-6] $C_2N_2S_2$

$$N{\equiv}CSSC{\equiv}N$$

Söderbäck, E., *Ann.*, 1919, **419**, 217
Low-temperature storage is necessary, as it polymerises explosively above its m.p.,
$-2°C$ ($-7°C$ is also recorded).
See other POLYMERISATION INCIDENTS
See related CYANO COMPOUNDS

1002. Sulfur thiocyanate
[57670-85-6] $C_2N_2S_3$

$$S(SC{\equiv}N)_2$$

Lecher, H. *et al., Ber.*, 1922, **55**, 1481
It decomposes explosively (but harmlessly) on storage at ambient temperature.
See related CYANO COMPOUNDS *See other* N–S COMPOUNDS

1003. Disulfur thiocyanate
[ ] $C_2N_2S_4$

$$(-SSC{\equiv}N)_2$$

Lecher, H. *et al., Ber.*, 1922, **55**, 1485
It melts at $-3.5°C$ and decomposes explosively (but harmlessly) after turning yellow.
See related CYANO COMPOUNDS *See other* N–S COMPOUNDS

1004. Zinc cyanide
[557-21-1] C_2N_2Zn

$$Zn(CN)_2$$

It is modestly endothermic ($\Delta H^°_f$ (s) $+77$ kJ/mol, 0.65 kJ/g) but not notably thermally unstable.

Magnesium
   See Magnesium: Metal cyanides
   See other ENDOTHERMIC COMPOUNDS, METAL CYANIDES

1005. Dicyanodiazene (Azocarbonitrile)
[1557-57-9] C_2N_4

$$N\equiv CN=NC\equiv N$$

1. Marsh, F. D. *et al., J. Amer. Chem. Soc.*, 1965, **87**, 1819
2. Ittel, S. D. *et al., Inorg. Chem.*, 1975, **14**, 1183

The solid explodes when mechanically shocked or heated in a closed vessel [1]. Preparative methods are hazardous because of the need to heat the explosive precursor, cyanogen azide [1,2].

See Cyanonitrene *See other* AZO COMPOUNDS, CYANO COMPOUNDS

1006. Disodium dicyanodiazenide
[ ] $C_2N_4Na_2$

$$Na_2[NC{\cdot}N=NC{\cdot}N]$$

Marsh, F. D. *et al., J. Amer. Chem. Soc.*, 1965, **87**, 1820

The radical anion salt is an explosive powder.

See related AZO COMPOUNDS, CYANO COMPOUNDS

1007. Sodium 1,1,2,2-tetranitroethanediide
[4415-23-0] $C_2N_4Na_2O_8$

$$Na_2[(O_2N)_2CC(NO_2)_2]$$

See Potassium 1,1,2,2-tetranitroethanediide
See other POLYNITROALKYL COMPOUNDS

1008. 3,4-Dinitrofurazan-2-oxide
[153498-61-4] $C_2N_4O_6$

Godovikova, T. L., *Mendeleev Comm.*, 1993, (5) 209

This oxygen rich explosive is claimed to be the first pernitro heterocycle known.

See other FURAZAN *N*-OXIDES, *C*-NITRO COMPOUNDS

1009. Trinitroacetonitrile
[630-72-8] $C_2N_4O_6$

$$(O_2N)_3CCN$$

1. Schischkow, A., *Ann. Chim. [3]*, 1857, **49**, 310
2. Parker, C. O. *et al., Tetrahedron*, 1962, **17**, 79, 84

It explodes if heated quickly to 220°C [1]. It is also a friction- and impact-sensitive explosive, which may be used conveniently in carbon tetrachloride solution to minimise handling problems [2].
See other CYANO COMPOUNDS, POLYNITROALKYL COMPOUNDS

1010. Tetranitroethylene
[13223-78-4]

$C_2N_4O_8$

$$(O_2N)_2C{=}C(NO_2)_2$$

Baum, K, *et al., J. Org. Chem.*, 1985, **50**, 2736–2739
This exceptionally reactive dienophile (at least one order of magnitude greater than tetracyanoethylene) reacted normally with dilute solutions of unsaturated compounds in dichloromethane, but sometimes explosively in absence of solvent.
See other NITROALKENES

1011. Thiocarbonyl azide thiocyanate
[ ]

$C_2N_4S_2$

$$N_3C({:}S)SC{\equiv}N$$

CHETAH, 1990, 182
Shock sensitive

Ammonia, or Hydrazine
Audrieth, L. F. *et al., J. Amer. Chem. Soc.*, 1930, **52**, 2799–2805
The unstable (undoubtedly endothermic) compound reacts explosively with ammonia gas, and violently with conc. hydrazine solutions.
See related CYANO COMPOUNDS *See other* ORGANIC AZIDES

1012. Disodium 5-tetrazolazocarboxylate
[68594-24-1]

$C_2N_6Na_2O_2$

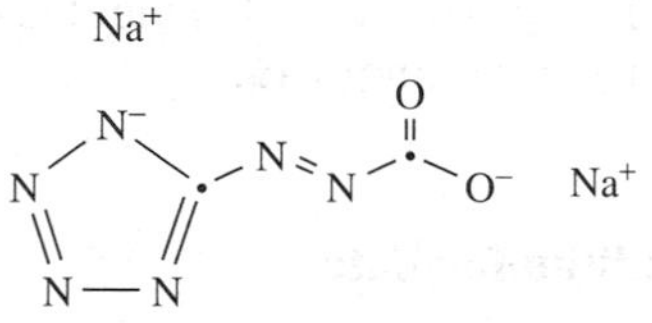

Thiele, J. *et al., Ann.*, 1895, **287**, 238
It is explosive.
See other AZO COMPOUNDS, TETRAZOLES

1013. Dicarbonyltungsten diazide
[68379-32-8]

$C_2N_6O_2W$

$$(CO)_2W(N_3)_2$$

Dehnicke, K., *Angew. Chem. (Intern. Ed.)*, 1979, **18**, 513

This and the molybdenum analogue (both homopolymeric) are extremely sensitive, exploding on the slightest mechanical stress, or violently in contact with traces of water.

See related CARBONYLMETALS, METAL AZIDES

1014. Hexanitroethane
[918-37-6] $C_2N_6O_{12}$

$$(O_2N)_3CC(NO_2)_3$$

1. Loewenschuss, A. *et al., Spectrochim. Acta*, 1974, **30A**, 371–378
2. Sorbe, 1968, 149

Grinding the solid to record its IR spectrum was precluded on safety grounds [1]. It decomposes explosively above 140°C [2].

Boron
Finnerty, A. E. *et al., Chem. Abs.*, 1979, **91**, 177486
There is an autocatalytic reaction at ambient temperature in this new explosive combination.

1,1-Dimethylhydrazine
Noble, P. *et al., Am. Inst. Aeron. Astronaut. J.*, 1963, **1**, 395–397
Hexanitroethane is a powerful oxidant and hypergolic with dimethylhydrazine or other strong organic bases.

Organic compounds
Will, M., *Ber.*, 1914, **47**, 961–965
Though relatively insensitive to friction, impact or shock, it can be detonated. With hydrogen containing organic compounds, this oxygen-rich compound (+200% oxygen balanced) forms powerfully explosive mixtures. The addition compound with 2-nitroaniline ($C_8H_6N_8O_{14}$, −27% balance) is extremely explosive.

See other POLYNITROALKYL COMPOUNDS

1015. Palladium(II) azidodithioformate
[29149-89-1] $C_2N_6PdS_4$

$$Pd(SCS.N_3)_2$$

Fehlhammer, W. P. *et al., Z. Naturforsch.*, 1983, **38B**, 547
In the first stage of the preparation of bis(triphenylphosphine)palladium(II) isothiocyanate, a large deficiency of aqueous palladium nitrate must be added with rapid stirring to an excess of sodium azidodithioformate solution to avoid the precipitation of explosive palladium(II) azidodithioformate.

See Azidodithioformic acid
See other ACYL AZIDES

1016. Bis(azidothiocarbonyl) disulfide

[ ] $C_2N_6S_4$

$$N_3C(:S)SSC(:S)N_3$$

Smith, G. B. L., *Inorg Synth.*, 1939, **1**, 81

This compound, readily formed by iodine oxidation of azidodithioformic acid or its salts, is a powerful explosive. It is sensitive to mechanical impact or heating to 40°C, and slow decomposition during storage increases the sensitivity. Preparative precautions are detailed.

See Azidodithioformic acid
See other ACYL AZIDES

1017. Diazidomethylenecyanamide

[67880-22-2] C_2N_8

$$(N_3)_2C{=}NC{\equiv}N$$

Marsh, F. D., *J. Org. Chem.*, 1972, **37**, 2967

This explosive solid may be produced during preparation of cyanogen azide.
See other CYANO COMPOUNDS, HIGH-NITROGEN COMPOUNDS, ORGANIC AZIDES

1018. Disodium 5,5′-azotetrazolide

[41463-64-3] $C_2N_{10}Na_2$

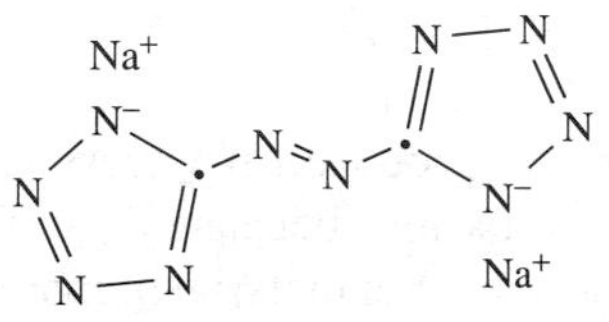

See entry METAL AZOTETRAZOLIDES *See other* TETRAZOLES

1019. Disodium 5,5′-azoxytetrazolide

[ ] $C_2N_{10}Na_2O$

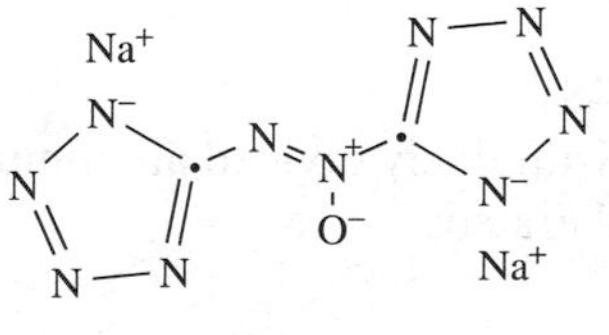

Thiele, J. *et al.*, *Ann.*, 1893, **273**, 151

Insensitive to friction or impact, but explodes violently on heating in a melting point tube or on analytical combustion, destroying the apparatus in both cases.
See related METAL AZOTETRAZOLIDES *See other* N-OXIDES, TETRAZOLES

1020. Lead 5,5'-azotetrazolide
[87489-52-9] $C_2N_{10}Pb$

See entry METAL AZOTETRAZOLIDES *See other* TETRAZOLES

1021. Diazidomethyleneazine
[ ] C_2N_{14}

$$(N_3)_2C=NN=C(N_3)_2$$

Houben-Weyl, 1965, Vol. 10.3, 793
This very explosive bis-*gem*-diazide contains over 89% of nitrogen.
See other HIGH-NITROGEN COMPOUNDS, ORGANIC AZIDES

1022. Sodium acetylide (Sodium ethynide)
[2881-62-1] C_2Na_2

$$NaC{\equiv}CNa$$

1. Opolsky, S., *Bull. Acad. Cracow*, 1905, 548
2. Houben-Weyl, 1977, Vol. 5.2a, 361
A brown explosive form is produced if excess sodium is used in preparation of thiophene homologues — possibly because of sulfur compounds [1]. As normally produced, it is a dry stable solid, but material prepared from acetylene and sodium–oil dispersions ignites in air [2].

Halogens
Mellor, 1946, Vol. 5, 848
Sodium acetylide burns in chlorine and (though not stated) probably also in fluorine, and in contact with bromine and iodine on warming.

Metals
Mellor, 1946, Vol. 5, 848
Trituration in a mortar with finely divided aluminium, iron, lead or mercury may be violent, carbon being liberated.

Metal salts
Mellor, 1946, Vol. 5, 848
Rubbing in a mortar with some metal chlorides or iodides may cause incandescence or explosion. Sulfates are reduced, and nitrates would be expected to behave similarly.

Non-metal oxides
  von Schwartz, 1918, 328
  Sodium acetylide incandesces in carbon dioxide or sulfur dioxide.

Oxidants
  Mellor, 1946, 5, 848
  Ignites on warming in oxygen, and incandesces at 150°C in dinitrogen pentaoxide.
  See Halogens, above

Phosphorus
  See Phosphorus: Metal acetylides

Water
  1. Mellor, 1946, Vol. 5, 848
  2. Davidsohn, W. E., *Chem. Rev.*, 1967, **67**, 74
  Excess water is necessary to prevent explosion [1]; the need for care in handling is
  stressed [2].
  See other METAL ACETYLIDES

1023. Sodium ethynediolate
[2611-42-9] $C_2Na_2O_2$

$$NaOC\equiv CONa$$

  See Sodium: Non-metal oxides
  See other ACETYLENIC COMPOUNDS

1024. Sodium peroxydicarbonate
[3312-92-6] $C_2Na_2O_6$

$$NaOCO.OOCO.ONa$$

Acetic anhydride
  See Sodium carbonate hydrogen peroxide: Acetic anhydride
  See other PEROXOACID SALTS

1025. Rubidium acetylide
[22754-97-8] C_2Rb_2

$$RbC\equiv CRb$$

Acids
  Mellor, 1946, Vol. 5, 848
  With conc. hydrochloric acid ignition occurs, and contact with nitric acid causes
  explosion.

Halogens
  Mellor, 1946, Vol. 5, 848
  It burns in all four halogens.

Metal oxides
Mellor, 1946, Vol. 5, 848–850
Iron(III) oxide and chromium(III) oxide react exothermally, and lead oxide explosively. Copper oxide and manganese dioxide react at 350°C incandescently.

Non-metal oxides
Mellor, 1946, Vol. 5, 848
Warming in carbon dioxide, nitrogen oxide or sulfur dioxide causes ignition.

Non-metals
Mellor, 1946, Vol. 5, 848
It reacts vigorously with boron or silicon on warming, ignites with arsenic, and burns in sulfur or selenium vapours.
See other METAL ACETYLIDES

1026. Strontium acetylide
[12071-29-3] $(C_2Sr)_n$
$$(-SrC{\equiv}C-)_n$$

Halogens
Mellor, 1946, Vol. 5, 862
Strontium acetylide incandesces with chlorine, bromine or iodine at 197, 174 and 182°C, respectively.
See other METAL ACETYLIDES

1027. Thorium dicarbide
[12674-40-7] C_2Th
$$ThC_2$$

Non-metals, or Oxidants
Mellor, 1946, Vol. 5, 862
Contact with selenium or sulfur vapour causes the heated carbide to incandesce. Contact of the carbide with molten potassium chlorate, potassium nitrate or even potassium hydroxide causes incandescence.
See other METAL NON-METALLIDES

1028. Uranium dicarbide
[12071-33-9] C_2U
$$UC_2$$

1. Mellor, 1946, Vol. 5, 890
2. Sidgwick, 1950, 1071
Uranium carbide emits brilliant sparks on impact, ignites on grinding in a mortar [1], or on heating in air to 400°C [2].

Halogens
Mellor, 1946, Vol. 5, 891
Incandesces in warm fluorine, in chlorine at 300°C and weakly in bromine at 390°C.

Hydrogen chloride
See Hydrogen chloride: Metal acetylides or carbides

Nitrogen oxide
See Nitrogen oxide: Metal acetylides or carbides

Water
1. Sidgwick, 1950, 1071
2. Mellor, 1946, Vol. 5, 890–891
Interaction with hot water is violent [1], and the carbide ignites in steam at dull red heat [2].
See other METAL NON-METALLIDES

1029. Zirconium dicarbide
[12070-14-3] C_2Zr

$$ZrC_2$$

Halogens
Mellor, 1946, Vol. 5, 885
It ignites in cold fluorine, and in chlorine, bromine and iodine at 250, 300 and 400°C, respectively.
See other METAL NON-METALLIDES

1030. Silver trifluoropropynide
[ ] C_3AgF_3

$$AgC{\equiv}CCF_3$$

Henne, A. L. *et al., J. Amer. Chem. Soc.*, 1951, **73**, 1042
Explosive decomposition on heating.
See other HALOACETYLENE DERIVATIVES, SILVER COMPOUNDS

1031. Aluminium carbide
[1299-86-1] C_3Al_4

$$Al_4C_3$$

Oxidants
Mellor, 1946, Vol. 5, 872
Incandescence on warming with lead dioxide or potassium permanganate.
See other METAL NON-METALLIDES

1032. Potassium 1,3-dibromo-1,3,5-triazine-2,4-dione-6-oxide
[15114-46-2] $C_3Br_2KN_3O_3$

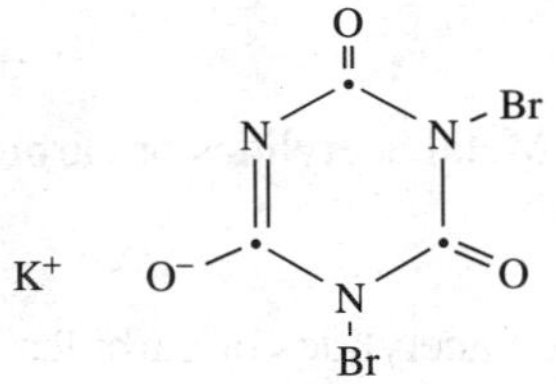

It may be expected to show similar properties to the chloro-analogue.
See Sodium 1,3-dichloro-1,3,5-triazine-2,4-dione-6-oxide
See other N-HALOIMIDES

1033. 1-Chlorodifluoromethyl-1-trifluoromethyldioxirane
[35356-48-3] $C_3ClF_5O_2$

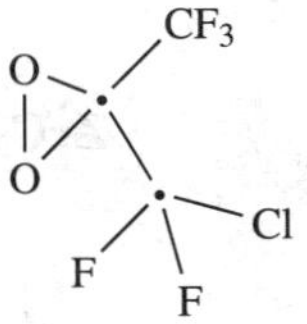

See Bis(trifluoromethyl)dioxirane
See other CYCLIC PEROXIDES

1034. Pentafluoropropionyl hypochlorite
[71359-61-0] $C_3ClF_5O_2$

$$F_5C_2CO.OCl$$

Tari, I. *et al., Inorg. Chem.*, 1979, **18**, 3205–3208
Thermally unstable at 22°C, and explosive in the gas phase at pressures above
27–62 mbar.
See other ACYL HYPOHALITES

1035. Heptafluoroisopropyl hypochlorite
[22675-68-9] C_3ClF_7O

$$(F_3C)_2CFOCl$$

Schack, C. J. *et al., J. Amer. Chem. Soc.*, 1969, **91**, 2904

Material condensed at −95°C may suddenly decompose completely and vaporise.
See other HYPOHALITES

1036. Chlorocyanoacetylene (Chloropropynenitrile)
[2003-31-8]   C₃ClN

$$ClC\equiv CC\equiv N$$

Hashimoto, N. *et al., J. Org. Chem.*, 1970, **35**, 675
Avoid contact with air at elevated temperatures because of its low (unstated) ignition
temperature. Burns moderately in the open, but may explode in a nearly closed vessel.
Presence of mono- and di-chloroacetylenes as impurities increases the flammability
hazard, which may be reduced by addition of 1% of ethyl ether.
See other HALOACETYLENE DERIVATIVES

1037. Sodium 1,3-dichloro-1,3,5-triazine-2,4-dione-6-oxide
[2893-78-9]   C₃Cl₂N₃NaO₃

'Fi-Clor 60S', Brochure NH/FS/67.4, Loughborough, Fisons, 1967
This compound (sodium dichloroisocyanurate), used in chlorination of swimming
pools, is a powerful oxidant and indiscriminate contact with combustible materials
must be avoided. Ammonium salts and other nitrogenous materials are incompatible
in formulated products. The dibromo analogue, used for the same purpose, will
behave similarly.

Preparative hazard
See 1,3,5-Trichloro-1,3,5-triazinetrione: Cyanuric acid, etc.

Calcium hypochlorite
1. Jones, C., INTERNET: CatJones@FC. KERN.ORG
2. Kirk Othmer, 1998, Vol. 25, 592
Although calcium hypochlorite is used in conjunction with sodium dichloroisocyanurate (usually as the dihydrate) for swimming pool maintenance, they must not
be mixed undiluted. Noxious fumes and violent explosion or fire can result [1].
Although the organic compound is, at first sight, at its highest oxidation level, it
contains disguised ammonia, and nitrogen trichloride may appear on reaction with
hypochlorites [2].
See other N-HALOIMIDES

1038. 2,4,6-Trichloro-1,3,5-triazine (Cyanuric chloride)
[108-77-0] $C_3Cl_3N_3$

HCS 1980, 925

1. Anon., *Loss Prev. Bull.*, 1979, (025), 21–22
2. *See entry* SELF-ACCELERATING REACTIONS

Of the factors associated with the high reactivity of cyanuric chloride (high exother-micity, rapid hydrolysis in presence of water-containing solvents, acid catalysed reactions, liberation of up to 3 mol hydrogen chloride/mol of chloride, formation of methyl chloride gas with methanol, formation of carbon dioxide from bicarbonates), several were involved in many of the incidents recorded [1] (and given below). The acid catalysed self acceleration and high exothermicity are rated highest [2]. It is also a mildly endothermic compound (ΔH°_f (s) +91.6 kJ/mol, 0.49 kJ/g).

Acetone, Water
Anon., *Loss Prev. Bull.*, 1979, (025), 20
The chloride was to be purified by dissolution in dry acetone, but in error, acetone containing 40% of water was used. The acid-catalysed exothermic hydrolysis reac-tion of the chloride accelerated to runaway, and gas and vapour evolution ruptured the vessel, leading to fire and explosion.
See Methanol, *also* Water, both below
See other GAS EVOLUTION INCIDENTS, SELF-ACCELERATING REACTIONS

Allyl alcohol, Sodium hydroxide, Water
Anon., *Loss Prev. Bull.*, 1974, (001), 11; 1979, (029), 21
When aqueous sodium hydroxide was added to a mixture of the chloride and alcohol at 28°C instead of the normal 5°C, a rapidly accelerating reaction led to rupture of the bursting disc and a gasket, and subsequently to a flash-fire and explosion.
See Methanol, *also* Water, both below

2-Butanone, Sodium hydroxide, Water
Anon., *Loss Prev. Bull.*, 1979, (025), 21
The total product of a batch containing a chlorotriazine, water, sodium hydroxide and 2-butanone had not been discharged from the 11,000 l vessel when a further 1.5 t of cyanuric chloride was charged ready for the next batch. The ensuing rapid

exothermic hydrolysis led to eruption of the reactor contents, ignition and explosion of butanone vapour, and damage of 3 M dollars.

Dimethylformamide

Anon., *BCISC Quart. Safety Summ.*, 1960, **35**, 24

Cyanuric chloride reacts vigorously and exothermically with DMF after a deceptively long induction period. The 1:1 adduct initially formed decomposes above 60°C with evolution of carbon dioxide and formation of a dimeric unsaturated quaternary ammonium salt. Dimethylformamide is appreciably basic and is not a suitable solvent for acyl halides.

See other INDUCTION PERIOD INCIDENTS

Dimethyl sulfoxide

See Dimethyl sulfoxide: Acyl halides

2-Ethoxyethanol

Anon., *Loss Prev. Bull.*, 1979, (025), 20

Alcoholysis of the chloride on the plant scale was effected at 40°C (with brine cooling) by adding portions to the alcohol alternately with finely crystalline disodium phosphate to neutralise the hydrogen chloride produced. On one occasion, use of coarsely crystalline sodium phosphate (of low surface area) reduced the rate of neutralisation, the mixture became acid, and a runaway exotherm to 170°C developed leading to eruption of vessel contents. On another occasion, accidental addition of sodium sulfate instead of phosphate led to a similar situation beginning to develop, but an automatic pH alarm allowed remedial measures to be instituted successfully.

See other NEUTRALISATION INCIDENTS

Methanol

Anon., *ABCM Quart. Safety Summ.*, 1960, **31**, 40

Cyanuric chloride dissolved in methanol reacted violently and uncontrollably with the solvent. This was attributed to the absence of an acid acceptor to prevent the initially acid catalysed (and later auto-catalysed) exothermic reaction of all 3 chlorine atoms simultaneously.

Methanol, Sodium hydrogen carbonate

Anon., *Loss Prev. Bull.*, 1979, (025), 19

A crust of residual cyanuric chloride left in a reactor from a previous batch reacted with the methanol (usually charged first) to form hydrogen chloride. When the base was added (usually before the chloride), vigorous evolution of carbon dioxide expelled some of the solvent. In a second incident, accidentally doubling the charge of cyanuric chloride but not the base, led to the development of free acid (which auto-catalyses the reaction with methanol), and a runaway reaction developed causing violent boiling of the solvent, methyl chloride evolution and damage to the plant.

See Methanol, above *See other* RUNAWAY REACTIONS

Sodium azide
See 2,4,6-Triazido-1,3,5-triazine (reference 2)

Water
MCA Case History No. 1869
A reaction mixture containing the chloride and water, held in abeyance before processing, developed a high internal pressure in the containing vessel. Hydrolysis (or alcoholysis) of the chloride becomes rapidly exothermic above 30°C.
See Allyl alcohol, *also* Methanol, both above
See related ACYL HALIDES *See other* ENDOTHERMIC COMPOUNDS

1039. 1,3,5-Trichloro-1,3,5-triazinetrione ('Trichloroisocyanuric acid') [87-90-1]

$C_3Cl_3N_3O_3$

HCS 1980, 920

Combustible materials
'Fi-Clor 91' Brochure, Loughborough, Fisons, 1967
This compound, used in chlorination of swimming pools, is a powerful oxidant, and indiscriminate contact with combustible materials must be avoided. The tribromo analogue would be similar.

Cyanuric acid, Sodium hydroxide
Brennan, J. P., US Pat. 4 118 570, 1978
Formation of sodium 1,3-dichloro-1,3,5-triazine-2,4-dione-6-oxide ('sodium dichloroisocyanurate') from an aqueous slurry of the trione, cyanuric acid and sodium hydroxide is accompanied by evolution of nitrogen trichloride. Nitrogen purging will keep the concentration of the explosive gas below the lower explosive limit of 5–6%.

Water
Young, J. A., private comm., 1986
If mixed with a small amount of water, the conc. solution (with pH around 2) may explode, owing to evolution of nitrogen trichloride. It is believed that hydrolysis leads to formation of hypochlorous acid and dichloro-*s*-triazinetrione, and the protonated acid then attacks the C=N bonds in the triazine ring leading to formation of

chloramines and nitrogen trichloride. The dichloro compound is stable to acid in absence of hypochlorous acid.

See other N-HALOIMIDES

†1040. Trichloromethyl carbonate (Triphosgene)
[32315-10-9] $C_3Cl_6O_3$

$$(CCl_3O)_2CO$$

Hollingsworth, M. D., *Chem. Eng. News*, 1993, **70**(28), 2

A package of this, consisting of a wax-coated glass jar in vermiculite within a sealed can, was found to be pressurised when opened. The cap of the jar had cracked, and it was assumed that residual water on the vermiculite had reacted to form hydrogen chloride, phosgene and carbon dioxide. Similar mishaps had been experienced elsewhere. Less packing would have given rise to less danger; the supplier has, however, supplemented the four layers initially present with a fifth and a recommendation not to store long and then under refrigeration. (The editor would predict that there are possible contaminants which could rearrange this compound slowly to phosgene in absence of water).

See other PACKAGING INCIDENTS

1041. Tetrachloropropadiene
[18608-30-5] C_3Cl_4

$$Cl_2C{=}C{=}CCl_2$$

Tetracarbonylnickel
  See Dicarbonyl-η-trichloropropenyldinickel chloride dimer
  See other HALOALKENES

1042. Tetrachloroethylene carbonate
[22432-68-4] $C_3Cl_4O_3$

Tributylamine
  1. Fagenburg, D. R., *Chem. Eng. News*, 1982, **60**(33), 45
  2. Melby, L. R., *Chem. Eng. News*, 1982, **60**(40), 2

Interaction of the two compounds led to the evolution of a toxic gas thought to be chlorine [1]. It is the far more poisonous phosgene, arising from the known base-catalysed disproportionation of the carbonate to oxalyl chloride and phosgene, which occurs even at ambient temperature [2]. (The editor knows that amides, too, catalyse this rearrangement and suspects that Lewis acids will also)

See other GAS EVOLUTION INCIDENTS

1043. 2,4,6-Tris(dichloroamino)-1,3,5-triazine (Hexachloromelamine)
[2428-04-8] $C_3Cl_6N_6$

As a trifunctional dichloroamino compound, it is probably more reactive and less stable than the monochloroamino analogue.

Acetone, or Bases
491M, 1975, 197
Addition of acetone, ammonia, aniline or diphenylamine to the oxidant causes rapid exothermic reactions, with or without flame, and large amounts under confinement would explode. The trichloro analogue is similar, but less vigorous.
See 2,4,6-Tris(chloroamino)-1,3,5-triazine (next above)
See other N-HALOGEN COMPOUNDS

1044. Potassium tricyanodiperoxochromate(3−)
[65521-60-0] $C_3CrK_3N_3O_4$

$$K_3[(CN)_3Cr(O_2)_2]$$

Bailar, 1973, Vol. 4, 167
A highly explosive material, with internal redox features.
See other PEROXOACID SALTS, REDOX COMPOUNDS

1045. Trifluoromethyliodine(III) isocyanate
[127510-65-0] $C_3F_3IN_2O_2$

$$CF_3I(NCO)_2$$

Naumann, D. *et al., J. Fluor. Chem.*, 1989, **45**(3), 401
This compound explosively decomposes when warmed up to room temperature.
See other IODINE COMPOUNDS, ORGANIC ISOCYANATES

1046. Lithium trifluoropropynide
[14856-86-1] C_3F_3Li

$$LiC{\equiv}CCF_3$$

See Lithium chloroacetylide
See other HALOACETYLENE DERIVATIVES, METAL ACETYLIDES

1047. 2-Trifluoroacetyl-1,3,4-dioxazalone
[87050-94-0] $C_3F_3NO_3$

Middleton, W. J., *J. Org. Chem.*, 1983, **48**, 3845
A sample contained in a dropping funnel at 25°C exploded forcefully. Though not sensitive to mechanical shock, it could be detonated by a hot wire when unconfined, or if rapidly heated to 220°C when confined in a stainless steel capillary.
See 1,3,4-DIOXAZOLONES

1048. Dicyanoiodonium triflate
[ ] $C_3F_3IN_2O_3S$

$$(NC)_2I^+CF_3SO_3^-$$

Steng, P. J. *et al., Tetrahedron Lett.*, 1992, **33**(11), 1419
This iodinating reagent is storable under nitrogen at -20°C; it may explode after contact with air. It decomposes in minutes at ambient temperatures.
See other CYANO COMPOUNDS, IODINE COMPOUNDS

1049. Trifluoroacryloyl fluoride
[667-49-2] C_3F_4O

$$F_2C=CFCO.F$$

Sodium azide
Middleton, W. J., *J. Org. Chem.*, 1973, **38**, 3294
The product of interaction was an unidentified highly explosive solid.
See other ACYL HALIDES

1050. *O*-Trifluoroacetyl-*S*-fluoroformyl thioperoxide
[27961-70-2] $C_3F_4O_3S$

$$F_3CCO.OSCO.F$$

Anon., *Angew. Chem. (Nachr.)*, 1970, **18**, 378
It exploded spontaneously in a glass bomb closed with a PTFE-lined valve. No previous indications of instability had been noted during distillation, pyrolysis or irradiation.
See related DIACYL PEROXIDES

†1051. **Hexafluoropropene**
[116-15-4] C_3F_6

$$F_2C=CFCF_3$$

Air, Tetrafluoroethylene
See Tetrafluoroethylene: Air, Hexafluoropropene

Grignard reagents
Dinowski, W., *J. Fluorine Chem.*, 1981, **18**, 25–30
Reaction of various substituted phenylmagnesium bromides with hexafluoropropene under pressure at ambient temperature had been effected on the 0.2–0.6 g mol scale without incident. An attempt to scale up to 0.8 g mol with phenylmagnesium bromide led to an explosion.

Oxygen, Oxygen difluoride
See Oxygen difluoride: Hexafluoropropene, etc.
See other HALOALKENES

1052. **Hexafluoroisopropylideneaminolithium**
[31340-36-0] C_3F_6LiN

$$(F_3C)_2C=NLi$$

Non-metal halides
Swindell, R. F., *Inorg. Chem.*, 1972, **11**, 242
Interaction of the lithium derivative with a range of chloro- and fluoro-derivatives of arsenic, boron, phosphorus, silicon and sulfur during warming to 25°C tended to be violently exothermic in absence of solvent. Thionyl chloride reacted with explosion.
See Hexafluoroisopropylideneamine: Butyllithium See other N-METAL DERIVA-
TIVES

1053. **Bis(trifluoromethyl)cyanophosphine**
[431-97-0] C_3F_6NP

$$(F_3C)_2PC\equiv N$$

491M, 1975, 60

It ignites in air.
See related ALKYLHALOPHOSPHINES, CYANO COMPOUNDS

1054. **Pentafluoropropionyl fluoride**
[422-61-7] C_3F_6O

$$F_5C_2CO.F$$

Fluorinated catalysts
Sorbe, 1968, 2

The acid fluoride may decompose explosively in contact with fluorinated catalysts.
See other ACYL HALIDES

1055. Bis(trifluoromethyl)dioxirane
[35357-46-1] $C_3F_6O_2$

Ando, W. (Ed.), *Organic Peroxides*, 195, Wiley, New York, 1992.
Formed by fluorine oxidation of the dilithium salt of hexafluoroacetone hydrate, it is unstable and explosive. The chloropentafluoro homologue is similar. Proponents of their use as reagents claim that the dimethyl and methyltrifluoromethyl analogues are not explosive; this seems improbable, especially since the less stressed lower dioxetanes (homodioxiranes) are all dangerous.
See other CYCLIC PEROXIDES

1056. Pentafluoropropionyl hypofluorite
[ ] $C_3F_6O_2$
$$F_5C_2CO.OF$$

Menefee, A. *et al., J. Amer. Chem. Soc.*, 1954, **76**, 2020
Less stable than its lower homologue, the hypofluorite explodes on sparking, on distillation at ambient pressure (b.p., 2°C), though not at below 0.13 bar.
See Fluorine: Caesium heptafluoropropoxide
See other ACYL HYPOHALITES

1057. Trifluoroacetyl trifluoromethanesulfonate
[68602-57-3] $C_3F_6O_4S$
$$F_3CCO.OSO_2CF_3$$

Forbus, T. R. *et al., J. Org. Chem.*, 1979, **44**, 313
The mixed anhydride reacts almost explosively with water.
See other ACID ANHYDRIDES, SULFUR ESTERS

1058. 1-Fluoroiminohexafluoropropane
[78343-32-5] C_3F_7N
$$FN=CFC_2F_5$$

Fluorine
See Fluorine: 1- or 2-Fluoroiminohexafluoropropane
See other N-HALOGEN COMPOUNDS

1059. 2-Fluoroiminohexafluoropropane
[2802-70-2] C_3F_7N

$$FN{=}C(CF_3)_2$$

Fluorine
  See Fluorine: 1- or 2-Fluoroiminohexafluoropropane
  See other N-HALOGEN COMPOUNDS

1060. Heptafluoropropyl hypofluorite
[2203-57-8] C_3F_8O

$$F_7C_3OF$$

1. Sorbe, 1968, 62
2. *MCA Case History No. 1045*
An explosive compound [1], and the preparation is hazardous, involving the fluorination of caesium perfluoropropoxide [2].
See Fluorine: Caesium heptafluoropropoxide
See other HYPOHALITES

1061. 1,1-Bis(fluorooxy)hexafluoropropane
[72985-54-7] $C_3F_8O_2$

$$(FO)_2CFC_2F_5$$

Explosive, less stable than its lower homologue.
See entry BIS(FLUOROOXY)PERHALOALKANES

1062. 2,2-Bis(fluorooxy)hexafluoropropane
[16329-93-4] $C_3F_8O_2$

$$(FO)_2C(CF_3)_2$$

It is especially prone to explosive decomposition.
See entry BIS(FLUOROOXY)PERHALOALKANES

1063. Tris(trifluoromethyl)phosphine
[432-04-2] C_3F_9P

$$(F_3C)_3P$$

Oxygen
  Mahler, W., *Inorg. Chem.*, 1979, **18**, 352

Contact with oxygen at 25°C/0.5 bar causes ignition of the phosphine.
See related ALKYLPHOSPHINES

1064. Sodium tricarbonylnitrosylferrate
[25875-18-7] $C_3FeNNaO_4$

$$[(CO)_3FeNO]^-Na^+$$

Davies, S. G. *et al., J. Organomet. Chem.*, 1990, **386**(2), 195
This compound is air sensitive and decomposes explosively. Use of the more stable
tetrabutylammonium salt as a reagent for carbonylation of alkyl halides is recommended.

1065. Propioloyl chloride (Propynoyl chloride)
[50277-65-1] C_3HClO

$$HC{\equiv}CCO.Cl$$

Balfour, W. J. *et al., J. Org. Chem.*, 1974, **39**, 726
This chloride, purified by distillation at 58–60°C/1 bar usually ignites spontaneously
in air owing to presence of chloroacetylene (formed by decarbonylation), but vacuum
distillation at cryogenic temperatures prevents formation of the impurity.
See other ACYL HALIDES, ACETYLENIC COMPOUNDS

1066. 3,3,3-Trifluoropropyne
[661-54-1] C_3HF_3

$$HC{\equiv}CCF_3$$

Haszeldine, R. N., *J. Chem. Soc.*, 1951, 590
It tends to explode during analytical combustion, and the copper(I) and silver derivatives decomposed violently (with occasional explosion) on rapid heating.
See other HALOACETYLENE DERIVATIVES

†1067. Hexafluoroisopropylideneamine (2-Iminohexafluoropropane)
[1645-75-6] C_3HF_6N

$$(F_3C)_2C{=}NH$$

Butyllithium
Swindell, R. F. *et al., Inorg. Chem.*, 1972, **11**, 242
The exothermic reaction which set in on warming the reagents in hexane to 0°C
sometimes exploded if concentrated solutions of butyllithium (above 2.5 M) were
used, but not if diluted (to about 1.2 M) with pentane.

1068. Potassium hydrogen diazirine-3,3-dicarboxylate

[ ] $C_3HKN_2O_4$

See Diazirine-3,3-dicarboxylic acid
See other DIAZIRINES

1069. 4-Azidocarbonyl-1,2,3-thiadiazole

[ ] C_3HN_5OS

Pain, D. L. *et al. J. Chem. Soc.*, 1965, 5167
The azide is extremely explosive in the dry state, though the oxygen balance is −54%.
See OXYGEN BALANCE *See other* ACYL AZIDES

1070. Diazopropyne
[17761-23-8] $C_3H_2N_2$
Wierlacher, S. *et al., Chem. Phys. Lett.*, 1994, **222**(4), 319
This compound is highly explosive and has detonated violently when in use on laboratory scale.
See other ACETYLENIC COMPOUNDS, DIAZO COMPOUNDS

1071. Silver malonate
[57421-56-4] $C_3H_2Ag_2O_4$

$$CH_2(CO.OAg)_2$$

Sorbe, 1968, 126
It explodes on heating.
See other SILVER COMPOUNDS

1072. 2,2-Dinitro-2-fluoroethoxycarbonyl chloride
[31841-79-9] $C_3H_2ClFN_2O_6$

$$(O_2N)_2CFCH_2OCO.Cl$$

Potassium picrate
Becuwe, A. G. *et al.*, US Pat. 4 145 361, 1979
Interaction to give the (explosive) ester, 2,2-dinitro-2-fluoroethyl 2,4,6-trinitrophenyl carbonate must be effected at below 50°C to prevent explosion.
See other ACYL HALIDES, FLUORODINITROMETHYL COMPOUNDS

1073. 3-Chloro-1-iodopropyne
[109-71-7] C_3H_2ClI

$$IC{\equiv}CCH_2Cl$$

Houben-Weyl, 1977, Vol. 5.2a, 610
Distillation at 47°C/5 mbar should only be taken to a 33% residue; less may lead to violent decomposition, as may admission of air to the hot residue.
See other HALOACETYLENE DERIVATIVES

†1074. 2-Chloroacrylonitrile
[920-37-6] C_3H_2ClN

$$H_2C{=}CClCN$$

See other CYANO COMPOUNDS *See related* HALOALKENES

1075. Cyanoacetyl chloride
[16130-58-8] C_3H_2ClNO

$$N{\equiv}CCH_2CO.Cl$$

491M, 1975, 140
After treating cyanoacetic acid with phosphorus pentachloride, volatile materials were distilled off and the crude chloride was left in a stoppered flask. After 24 h the flask exploded.
See other ACYL HALIDES, CYANO COMPOUNDS

1076. Sodium diformylnitromethanide hydrate
[34461-00-2] $C_3H_2NNaO_4.H_2O$

$$Na^+O_2NC(CO.H)_2.H_2O \text{ or } NaON(O){=}C(CO.H)CH(OH)_2$$

Fanta, P. E., *Org. Synth.*, 1962, Coll. Vol. 4, 844
The monohydrate, possibly the *gem*-dihydroxy compound and an *aci*-nitro salt, is an impact-sensitive solid and must be carefully handled with precautions.
See related C-NITRO COMPOUNDS, *aci*-NITRO SALTS

1077. Diisocyanomethane
[ ] $C_3H_2N_2$

$$C\equiv NCH_2N\equiv C$$

Buschmann, J. *et al., Angew. Chem. (Int.)*, 1997, **36**(21), 2372
A clear crystalline solid at $-55°C$, this compound melts, with browning at $-15°C$, and decomposes explosively at $-10°C$, leaving a polymer.
See other DIISOCYANIDE LIGANDS

†1078. Malononitrile (Propanedinitrile)
[109-77-3] $C_3H_2N_2$

$$CH_2(C\equiv N)_2$$

Energy of decomposition (in range $180–270°C$) measured as 1.65 kJ/g.
See entry THERMOCHEMISTRY AND EXOTHERMIC DECOMPOSITION (reference 2)

Alone, or Bases
1. Personal experience
2. 'Malononitrile' Brochure, p. 11, Basle, Lonza Ltd., 1974
3. Anon., *CISHC Chem. Safety Summ.*, 1978, **49**, 29
It may polymerise violently on heating at $130°C$, or in contact with strong bases at lower temperatures [1]. The stability of the molten nitrile decreases with increasing temperature and decreasing purity, but no violent decomposition below $100°C$ has been recorded [2]. However, a partially filled drum of malononitrile stored in an oven at $70–80°C$ for 2 months exploded violently [3].
See other CYANO COMPOUNDS

†1079. Diisocyanatomethane (Methylene diisocyanate)
[4747-90-4] $C_3H_2N_2O_2$

$$CH_2(C=N=O)_2$$

Dimethylformamide
'DMF Chemical Uses', Brochure, Wilmington, Du Pont, 1967
Contact with the basic solvent causes violent polymerisation of the isocyanate.
See other POLYMERISATION INCIDENTS *See related* CYANO COMPOUNDS

1080. Diazirine-3,3-dicarboxylic acid
[76429-98-6] $C_3H_2N_2O_4$

Schustov, G. V. *et al., Angew. Chem. (Intern. Ed.)*, 1981, **20**, 200

While the dipotassium salt is stable for over 2 months at 20°C, the monopotassium salt exploded within a few hours at 0°C. Both salts exploded when triturated.
See other DIAZIRINES, ORGANIC ACIDS

1081. Diazomalonic acid
[59348-62-8]   C₃H₂N₂O₄

$$N_2C(CO.OH)_2$$

1. *491M*, 1975, 144
2. *DOC 5*, 1982, 1588

During attempted distillation of the impure acid at 4 mbar, the (involatile) sample exploded [1]. The free acid is unknown, but the diethyl ester has exploded during low pressure distillation [2].
See other DIAZO COMPOUNDS, ORGANIC ACIDS

1082. 2-Diazo-2*H*-imidazole
[50846-98-5]   C₃H₂N₄

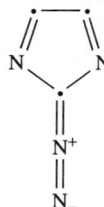

As an isolated solid, the compound is shock-sensitive and should be freshly prepared only as a solution for immediate use.
See entry DIAZOAZOLES
See other DIAZO COMPOUNDS, HIGH-NITROGEN COMPOUNDS

1083. Cyanohydrazonoacetyl azide
[115057-44-8]   C₃H₂N₆O

$$N_3CO.C(:NNH_2)CN$$

Kobolov, M. Yu. *et al., Chem. Abs.*, 1988, **109**, 20908
It explodes on melting.
See other ACYL AZIDES, CYANO COMPOUNDS, HIGH-NITROGEN COMPOUNDS

1084. Bis(1,2,3,4-thiatriazol-5-ylthio)methane
[ ]   C₃H₂N₆S₄

Pilgram, K. *et al., Angew. Chem.*, 1965, **77**, 348

This compound (which is 'sulfur balanced'), and its three longer chain homologues, explodes loudly with a flash on impact, or on heating to the m.p.

See OXYGEN BALANCE *See other* N–S COMPOUNDS

†1085. Propiolaldehyde (Propynal)
[624-67-9] C_3H_2O

$$HC \equiv CCO.H$$

Bases
1. Sauer, J. C., *Org. Synth.*, 1963, Coll. Vol. 4, 814
2. Makula, D. *et al.*, *L'Actual. Chim.*, 1983, (6), 31–34

The acetylenic aldehyde undergoes vigorous polymerisation in presence of alkalies and, with pyridine, the reaction is almost explosive [1]. The ground glass stopper of a 1 l brown bottle containing 750 ml of the aldehyde in toluene (2:1) and stored for 8 years was found to be seized. During attempts to open it an explosion occurred, leaving 250 ml of a polymeric residue in the bottle. DSC experiments and modeling calculations led to the conclusion that polymerisation had been initiated at ambient temperature by flakes of (alkaline) soda glass and peroxidic material (which had seized the stopper) falling into the solution of aldehyde. It is recommended that propynal should not be stored in glass, and only as a dilute (10%) solution [2] and for strictly limited periods.

See Acrylaldehyde
See other ACETYLENIC COMPOUNDS, ALDEHYDES, GLASS INCIDENTS, POLYMERISATION INCIDENTS

1086. Propiolic acid (Propynoic acid)
[471-25-0] $C_3H_2O_2$

$$HC \equiv CCO.OH$$

Ammonia, Heavy metal salts
Baudrowski, E., *Ber.*, 1882, **15**, 2701

Interaction of the acid with ammoniacal solutions of copper(I) or silver salts gives precipitates which explode on warming or impact. The structures are not given, but may be amminemetal acetylide salts.

See METAL ACETYLIDES *See other* ACETYLENIC COMPOUNDS, ORGANIC ACIDS

†1087. 1,3-Dioxol-4-en-2-one (Vinylene carbonate)
[872-36-6] $C_3H_2O_3$

Archer, M., private comm., 1979

388

A 7 year old screw capped sample burst in storage. Peroxide formation seems a less likely cause than slow hydrolysis and carbon dioxide evolution, though both are possibilities.

See related GAS EVOLUTION INCIDENTS, PEROXIDISABLE COMPOUNDS

1088. Silver 3-hydroxypropynide

[ ] C_3H_3AgO

$$AgC≡CCH_2OH$$

Karrer, P., *Organic Chemistry*, 111, London, Elsevier, 4th Engl. Edn., 1950
The silver salt is explosive.

See other METAL ACETYLIDES, SILVER COMPOUNDS

1089. Aluminium formate

[7360-53-4] $C_3H_3AlO_6$

$$Al(OCO.H)_3$$

Anon., *ABCM Quart. Safety Summ.*, 1939, **10**, 1
An aqueous solution of aluminium formate was being evaporated over a low flame. When the surface crust was disturbed, an explosion occurred. This seems likely to have been due to thermal decomposition of the solid, liberation of carbon monoxide and ignition of the latter admixed with air.

†1090. 3-Bromopropyne (Propargyl bromide)

[106-96-7] C_3H_3Br

$$HC≡CCH_2Br$$

HCS 1980, 784, (785 in toluene)

1. Coffee, R. D. *et al., Loss Prev.*, 1967, **1**, 6–9
2. Driedger, P. E. *et al., Chem. Eng. News*, 1972, **50**(12), 51
3. Brown, J. P,. 1991, personal communication
4. Forshey, *Fire Technol.*, 1969, **5**, 100–111

This liquid acetylenic endothermic compound (ΔH°_f estimated as 230–270 kJ/mol, ~2 kJ/g) may be decomposed by mild shock, and when heated under confinement, it decomposes with explosive violence and may detonate. Addition of 20–30 wt% of toluene makes the bromide insensitive in laboratory impact and confinement tests [1]. More recently, it was classed as extremely shock-sensitive [2]. It can be ignited by impact derived from the 'liquid-hammer' effect of accidental pressurisation of the aerated liquid, and will then undergo sustained (monopropellant) burning decomposition. Propargyl bromide, added dropwise to bromine, exploded as it neared the halogen [3]. The chloro analogue is similar, but less readily ignited [4].

See other ENDOTHERMIC COMPOUNDS

Metals
Dangerous Substances, 1972, Sect. 1, 27

389

There is a danger of explosion in contact with copper, high-copper alloys, mercury or silver (arising from metal acetylide formation).
See METAL ACETYLIDES

Trichloronitromethane
See Trichloronitromethane: 3-Bromopropyne
See other HALOACETYLENE DERIVATIVES

1091. 2,4,6-Tris(bromoamino)-1,3,5-triazine (Tribromomelamine)
[22755-34-6] $C_3H_3Br_3N_6$

Vona, J. A. *et al., Chem. Eng. News*, 1952, **30**, 1916
Bromination with this and similar *N*-halogen compounds may become violent or explosive after an induction period as long as 15 min. Small scale preliminary experiments, designed to avoid the initial presence of excess brominating agent, are recommended.
See other INDUCTION PERIOD INCIDENTS

Allyl alcohol
Vona, J. A. *et al., Chem. Eng. News*, 1952, **30**, 1916
The components reacted violently 15 min after mixing at ambient temperature. This seems likely to have been a radical-initiated polymerisation of the alcohol (possibly peroxidised) in absence of diluent.
See other *N*-HALOGEN COMPOUNDS

†**1092. 3-Chloropropyne (Propargyl chloride)**
[624-65-7] C_3H_3Cl

$$HC{\equiv}CCH_2Cl$$

HCS 1980, 786 (787 in toluene)

Doyle, W. H., *Loss Prev.*, 1969, **3**, 15
Pumping the liquid against a closed valve caused the pump to explode, which detonated the contents of the reservoir tank-car.

Ammonia
Anon., *Sichere Chemiearb.*, 1956, **8**(6), 45
Interaction of 3-chloropropyne and liquid ammonia under pressure in a static steel bomb had been used several times to prepare the amine. On one occasion the usual

slow exothermic reaction did not occur, and the bomb was shaken mechanically. The increased reaction rate led to a rapid exothermic reaction, followed by an explosion. Other cases of instability in propyne derivatives are known.

See 3-Propynol, and 3-Propynethiol
See other AMINATION INCIDENTS

Chlorine
See Chlorine: 3-Chloropropyne
See other HALOACETYLENE DERIVATIVES

†1093. Acryloyl chloride (2-Propenoyl chloride)
[814-68-6] $\qquad$ C_3H_3ClO

$$H_2C=CHCO.Cl$$

1. Pyriadi, T. M., *Chem. Eng. News*, 1985, **63**(44), 4
2. Griffith, T. E., *Chem. Eng. News*, 1985, **63**(50), 2
3. Author's comment, 1986

A 500 ml bottle of the acid chloride stabilised with 0.05% of phenothiazine was shipped to a hot climate without refrigeration, and was stored on arrival in a fume cupboard for 2 days at temperatures approaching 50°C. The material polymerised, bursting the bottle and forming a solid foam [1]. The label recommended storage at 4°C. Another manufacturer proposed mislabeling or contamination by moisture to explain the incident [2], but the presence of polymeric foam appears to confirm that polymerisation at the extremely high ambient storage temperature was the primary cause, perhaps with some thermal decomposition and evolution of hydrogen chloride assisting in foam formation [3].

See other ACYL HALIDES, POLYMERISATION INCIDENTS

1094. 1,3-Dithiolium perchlorate
[3706-77-2] $\qquad$ $C_3H_3ClO_4S_2$

1. Ferraris, J. P. *et al., Chem. Eng. News*, 1974, **52**(37), 3
2. Klingsberg, E., *J. Amer. Chem. Soc.*, 1964, **86**, 5292
3. Leaver, D. *et al., J. Chem. Soc.*, 1962, 5109
4. Wudl, F. *et al., J. Org. Chem.*, 1974, **39**, 3608–3609; *Inorg. Chem.*, 1979, **19**, 28–29
5. Melby, L. R. *et al., J. Org. Chem.*, 1974, **39**, 2456

The salt (an intermediate in the preparation of 1,4,5,8-tetrahydro-1,4,5,8-tetrathiafulvalene) exploded violently during removal from a glass frit with a Teflon-clad spatula [1]. Previous references to the salt exploding at 250°C [2] or melting at

264°C [3] had been made. Use of a salt alternative to the perchlorate is urged [1]. Safer methods suitable for small-scale [4] and large-scale [5] preparations have been described.

See other NON-METAL PERCHLORATES

†1095. 2,4,6-Tris(chloroamino)-1,3,5-triazine (Trichloromelamine)
[7673-09-8] $C_3H_3Cl_3N_6$

Acetone, or Bases
491M, 1975, 197
Addition of acetone, ammonia, aniline or diphenylamine to the oxidant causes, after a few seconds delay, a rapid reaction accompanied by smoke and flame.
See 2,4,6-Tris(bromoamino)-1,3,5-triazine
See other N-HALOGEN COMPOUNDS

1096. Methyl trichloroacetate
[598-99-2] $C_3H_3Cl_3O_2$

$$MeOCO.CCl_3$$

Trimethylamine
Anon., *Angew. Chem. (Nachr.)*, 1962, **10**, 197
A stirred uncooled mixture in an autoclave reacted violently, the pressure developed exceeding 400 bar. Polymerisation of a reactive species produced by dehydrochlorination of the ester or a decomposition product seems a possibility.
See Sodium hydride: Ethyl 2,2,3-trifluoropropionate
See other POLYMERISATION INCIDENTS

1097. Trichloromethyl peroxyacetate
[94089-34-6] $C_3H_3Cl_3O_3$

$$CCl_3OOCO.CH_3$$

Turner, W. V. *et al.*, *J. Org. Chem.*, 1992, **57**(5), 1610
This compound has caused several explosions in the author's laboratory.
See other PEROXYESTERS

1098. 1-Propynylcopper(I)
[30645-13-7] C_3H_3Cu

$$MeC{\equiv}CCu$$

Houben-Weyl, 1977, Vol. 5.2a, 570

Preparation and use of this very explosive compound are detailed.
See other METAL ACETYLIDES

†1099. 3,3,3-Trifluoropropene
[677-21-4] $C_3H_3F_3$
$$H_2C=CHCF_3$$

See other HALOALKENES

†1100. 1,1,1-Trifluoroacetone
[421-50-1] $C_3H_3F_3O$
$$F_3CCO.CH_3$$

1101. Methyl trifluorovinyl ether
[3823-94-7] $C_3H_3F_3O$
$$MeOCF=CF_2$$

1. Dixon, S., US Pat. 2 917 548, 1959
2. Anderson, A. W., *Chem. Eng. News*, 1976, **54**(16), 5
3. O'Neill, G. J., US Pat. 4 127 613, 1978

Methyl trifluorovinyl ether, b.p. 10.5–12.5°C, prepared from tetrafluoroethylene and sodium methoxide [1], has considerable explosive potential. On ignition, it decomposes more violently than acetylene and should be treated with extreme caution [2]. Other trifluorovinyl ethers are similarly available from higher alkoxides [1], and although not tested for instability, should be handled carefully. Presence of fluorohaloalkanes boiling lower than the ether stabilises the latter against spark-initiated decomposition in both fluid phases [3].
See related HALOALKENES

1102. Methyltrifluoromethyldioxirane
[115464-59-0] $C_3H_3F_3O_2$

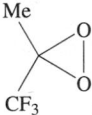

See Dimethyldioxirane, Bis(trifluoromethyl)dioxirane
See other CYCLIC PEROXIDES

1103. Trifluoromethyl peroxyacetate
[33017-08-2] $C_3H_3F_3O_3$
$$F_3COOCO.CH_3$$

Bernstein, P. A. *et al.*, *J. Amer. Chem. Soc.*, 1971, **93**, 3885

A 1 g sample cooled to $-196°C$ exploded violently when warmed in a bath at $22°C$.
See other PEROXYESTERS

1104. 1,1-Bis(difluoroamino)-2,2-difluoro-2-nitroethyl methyl ether
[30957-47-2] $C_3H_3F_6N_3O_3$

$$O_2NCF_2C(NF_2)_2OMe$$

Ross, D. L. *et al., J. Org. Chem.*, 1970, **35**, 3096–3098
A shock-sensitive explosive.
See other DIFLUOROAMINO COMPOUNDS

1105. 3-Iodopropyne
[659-86-9] C_3H_3I

$$HC{\equiv}CCH_2I$$

Whiting, M. C., *Chem. Eng. News*, 1972, **50**(23), 86–87
It explodes during distillation at $180°C$.
See other HALOACETYLENE DERIVATIVES

1106. Potassium 1-tetrazolacetate
[51286-83-0] $C_3H_3KN_4O_2$

Eizember, R. F. *et al., J. Org. Chem.*, 1974, **39**, 1792–1793
During oven-drying, kg quantities of the salt exploded violently. Investigation showed that self-propagating and extremely rapid decomposition of a cold sample can be initiated by local heating to over $200°C$ by a flint spark, prolonged static spark or flame. The sodium salt could only be initiated by flame, and the free acid is much less sensitive.
See other TETRAZOLES

†1107. Acrylonitrile (Propenenitrile)
[107-13-1] C_3H_3N

$$H_2C{=}CC{\equiv}N$$

(*MCA SD-31*, 1964); *FPA H51*, 1976; *HCS 1980*, 116; *RSC Lab. Hazards Data Sheet No. 48*, 1986
1. *MCA SD-31*, 1964
2. Harmon, 1974, 2.3
3. Taylor, B. J., *Univ. Safety Assoc. Safety News*, 1972, (1), 5–6
4. Bond, J., *Loss Prev. Bull.*, 1985, (065), 26

The monomer is sensitive to light, and even when inhibited (with aqueous ammonia) it will polymerise exothermally at above 200°C [1]. It must never be stored uninhibited, or adjacent to acids or bases [2]. Polymerisation of the monomer in a sealed tube in an oil bath at 110°C led to a violent explosion. It was calculated that the critical condition for runaway thermal explosion was exceeded by a factor of 15 [3]. Runaway polymerisation in a distillation column led to an explosion and fire [4].
See other POLYMERISATION INCIDENTS

Acids   MRH Nitric acid 6.32/78, sulfuric acid 2.26/48
1. 'Acrylonitrile', London, British Hydrocarbon Chemicals Ltd., 1965
2. Kaszuba, F. J., *J. Amer. Chem. Soc.*, 1945, **67**, 1227
3. Shirley, D. A., *Preparation of Organic Intermediates*, 3, New York, Wiley, 1951
4. Kaszuba, F. J., *Chem. Eng. News*, 1952, **30**, 824
5. Anon., *Loss Prev. Bull.*, 1978, (024), 167
6. Bond, J., *Loss Prev. Bull.*, 1985, (065), 26
Contact of strong acids (nitric or sulfuric) with acrylonitrile may lead to vigorous reactions. Even small amounts of acid are potentially dangerous, as these may neutralise the aqueous ammonia present as polymerisation inhibitor and leave the nitrile unstabilised [1]. Precautions necessary in the hydrolysis of acrylonitrile [2] are omitted in the later version of the procedure [3]. It is essential to use well-chilled materials (acrylonitrile, diluted sulfuric acid, hydroquinone, copper powder) to avoid eruption and carbonisation. A really wide-bore condenser is necessary to contain the vigorous boiling of unhydrolysed nitrile [4]. Leaking valves or operator error caused contamination of a 1000 l dosing vessel of the nitrile with conc. sulfuric acid, and an explosion ensued [5]. Contamination of a drum of acrylonitrile by nitric acid residues in the filling hose led to a slowly accelerating polymerisation reaction which burst the drum after 10 days [6].
See Bromine, etc., below

Bases   MRH Potassium or sodium hydroxides 2.84/tr.
1. *MCA SD-31*, 1964
2. Castaneda Hernandez, H. R. *et al.*, *Chem. Abs.*, 1987, **107**, 78274
Acrylonitrile polymerises violently in contact with strong bases, whether stabilised or unstabilised [1]. Alkaline hydrolysis of acrylonitrile is exothermic and violent, especially when the temperature is above 60°C, the pressure is above atmospheric, and when heating at 60°C is prolonged above 10 mins. Polymerisation does not induce a violent reaction at 40–50°C at a concentration of 3% of sodium hydroxide in water [2].

Benzyltrimethylammonium hydroxide, Pyrrole
MCA Guide, 1972, 299
To catalyse the cyanoethylation of pyrrole, 3 drops of the basic catalyst solution were added to the reaction mixture of pyrrole (30%) in the nitrile. An exotherm developed and base-catalysed polymerisation of the nitrile accelerated to explosion.
See Bases, above, *also* Tetrahydrocarbazole, etc., below
See other POLYMERISATION INCIDENTS

Formaldehyde
Personal experience
The reaction between acrylonitrile and formaldehyde (as paraformaldehyde or tri-oxane), under strong acid catalysis (usually sulphuric) and most often in presence of catalytic quantities of acetic anyhydride, to produce triacrylohexahydrotriazine, is inclined to violent exotherm after an induction period. The runaway can be uncontrollable on sub-molar scale. It may be due to acrylate polymerisation or to increasing reactivity of the formaldehyde equivalent due to progressive de-oligomerisation. Procedures claimed to prevent the risk have been described in the literature but do not seem reliable.

Halogens   MRH 2.84/tr. Bromine
1. *MCA Case History No. 1214*
2. Towell, G., 1989, Personal communication
Bromine was being added in portions to acrylonitrile with ice cooling, with intermediate warming to 20°C between portions. After half the bromine was added, the temperature increased to 70°C; then the flask exploded. This was attributed either to an accumulation of unreacted bromine (which would be obvious) or to violent polymerisation [1]. The latter seems more likely, catalysed by hydrogen bromide formed by substitutive bromination. Chlorine produces similar phenomena, even if the flask stays intact. The runaway is preceded by loss of yellow colouration and accompanied by formation of 3-chloroacrylonitrile and derivatives. It can be suppressed by presence of bases [2].
See Acids, above

Initiators
1. Zhulin, V. M. *et al., Dokl. Akad. Nauk SSSR*, 1966, **170**, 1360–1363
2. Biesenberger, J. A. *et al., Polymer Eng. Sci.*, 1976, **16**, 101–116
At pressures above 6000 bar, free radical polymerisation sometimes proceeded explosively [1]. The parameters were determined in a batch reactor for thermal runaway polymerisation of acrylonitrile initiated by azoisobutyronitrile, dibenzoyl peroxide or di-*tert*-butyl peroxide [2].
See other POLYMERISATION INCIDENTS
See VIOLENTPOLYMERISATION

Other reactants
Yoshida, 1980, 11
MRH values calculated for 17 combinations, mainly with oxidants, are given

Silver nitrate
Anon., *ABCM Quart. Safety Summ.*, 1962, **33**, 24
Acrylonitrile containing undissolved solid silver nitrate is liable, on long standing, to polymerise explosively and ignite. This was attributed to the slow deposition of a thermally insulating layer of polymer on the solid nitrate, which gradually gets hotter and catalyses rapid polymerisation. Photocatalysed decomposition of the salt with formation of traces of nitric acid may have been involved in the initial stages.
See other POLYMERISATION INCIDENTS

Tetrahydrocarbazole, Benzyltrimethylammonium hydroxide

Anon., *BCISC Quart. Safety Summ.*, 1968, **39**, 36

Cyanoethylation of 1,2,3,4-tetrahydrocarbazole initiated by the quaternary base had been effected smoothly on twice a published scale of working. During a further fourfold increase in scale, the initiator was added at 0°C, and shortly after cooling had been stopped and heating begun, the mixture exploded. A smaller proportion of initiator and very slow warming to effect reaction are recommended (to avoid rapid polymerisation of the nitrile by the base).

See Benzyltrimethylammonium hydroxide, etc., and Bases, both above

See other CYANO COMPOUNDS

See related HALOALKENES

1108. Poly(acrylonitrile)
[25014-41-9] $(C_3H_3N)_n$

$$(-CH_2CH(CN)-)_n$$

Methacrylate polymer, Nitric acid

See Nitric acid: Acrylonitrile–methacrylate copolymer

1109. Vinyl isocyanide (Isocyanoethene)
[14668-82-7] C_3H_3N

$$H_2C=CHN=C:$$

1. Matteson, D. S. *et al., J. Amer. Chem. Soc.*, 1968, **90**, 3765
2. Bailey, R., *Chem. Abs.*, 1968, **69**, 113029

The molar heat of formation of this endothermic compound ($+230-250$ kJ, 4.5 kJ/g) is comparable with that of buten-3-yne (vinylacetylene). While no explosive decomposition of the isocyanide has been reported, the possibility should be borne in mind [1]. It is stable at $-15°C$, but isomerises to acrylonitrile and polymerises at ambient temperature [2].

See related CYANO COMPOUNDS

See other ENDOTHERMIC COMPOUNDS

†1110. Isoxazole
[288-14-2] C_3H_3NO

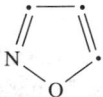

Sealed samples decompose exothermally above 136°C.

See entry ISOXAZOLES

See other N–O COMPOUNDS

†1111. Oxazole
[288-42-6]

C_3H_3NO

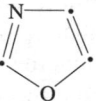

See other N–O COMPOUNDS

1112. 2-Thioxo-4-thiazolidinone ('Rhodanine')
[141-84-4]

$C_3H_3NOS_2$

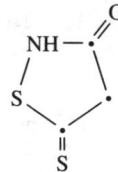

Merck, 1983, 1180
It may explode on rapid heating.
See other N–S COMPOUNDS

1113. Cyanoacetic acid
[372-09-8]

$C_3H_3NO_2$

$$N{\equiv}ICCH_2CO.OH$$

Furfuryl alcohol
See Furfuryl alcohol: Acids
See other CYANO COMPOUNDS, ORGANIC ACIDS

1114. 3-Azidopropyne (Propargyl azide)
[14989-89-0]

$C_3H_3N_3$

$$HC{\equiv}CCH_2N_3$$

Almlof, J. *et al., J. Mol. Struct.*, 1987, **160**, 1
It is extremely explosive and must be handled with utmost care. Explosions during synthesis (100 mmol, 8.1 g) wrecked a ventilated hood. (The energy of decomposition of this endothermic acetylenic azide probably exceeds 250 kJ/mol, 3 kJ/g.)
See other ACETYLENIC COMPOUNDS, ENDOTHERMIC COMPOUNDS, ORGANIC AZIDES

1115. 1,3,5-Triazine
[290-87-9]

$C_3H_3N_3$

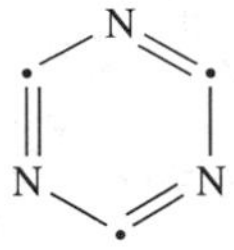

Nitric acid, Trifluoroacetic anhydride
See Nitric acid: Triazine, etc.
See other ORGANIC BASES

†1116. 2-Carbamoyl-2-nitroacetonitrile ('Fulminuric acid')
[475-08-1]

$C_3H_3N_3O_3$

$$H_2NCO.CH(NO_2)C\equiv N$$

Sorbe, 1968, 74
It explodes on heating.
See other CYANO COMPOUNDS, *C*-NITRO COMPOUNDS

1117. 2-Amino-5-nitrothiazole
[121-66-4]

$C_3H_3N_3O_2S$

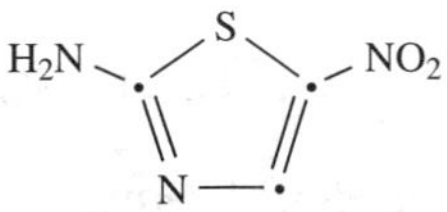

CHETAH, 1990, 183
Can be exploded with a detonator, not by mechanical shock.

Preparative hazard
See Nitric acid: 2-Aminothiazole, Sulfuric acid

†1118. 2,4,6-Trihydroxy-1,3,5-triazine (Cyanuric acid)
[108-80-5]

$C_3H_3N_3O_3$

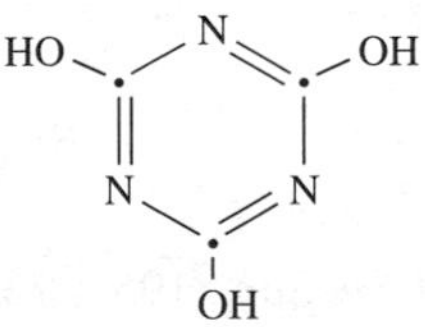

Chlorine
See Chlorine: Nitrogen compounds

1119. Sodium 1-tetrazolacetate
[51286-84-1] $C_3H_3N_4NaO_2$

See Potassium 1-tetrazolacetate
See other TETRAZOLES

1120. 5-Cyano-2-methyltetrazole
[91511-39-6] $C_3H_3N_5$

Aluminium hydride
See Aluminium hydride: Tetrazole derivatives
See other TETRAZOLES

1121. 2,3-Diazidopropiononitrile
[101944-90-5] $C_3H_3N_7$

$$N_3CH_2CH(N_3)CN$$

Owston, P. G. *et al., J. Chem. Res.; Synop.*, 1985, (11), 352–353
It is unstable and potentially explosive: routine use of a fume cupboard and safety screens is recommended.
See other CYANO COMPOUNDS, HIGH-NITROGEN COMPOUNDS, ORGANIC AZIDES

1122. 2-Hydroxy-4,6-bis(nitroamino)-1,3,5-triazine
[19899-80-0] $C_3H_3N_7O_5$

Atkinson, J. R., *J. Amer. Chem. Soc.*, 1951, **73**, 4443–4444
The explosive nitration product of 'melamine' was identified as the title compound of low oxygen balance, which is easily detonated on impact.
See other N-NITRO COMPOUNDS

1123. Sodium methoxyacetylide
[ ]   C$_3$H$_3$NaO

$$NaC \equiv COMe$$

Houben-Weyl, 1970, Vol. 13.1, 649
It may ignite in air.

Brine
    Jones, E. R. H. *et al., Org. Synth.*, 1963, Coll. Vol. 4, 406
    During addition of saturated brine at −20°C to the sodium derivative at −70°C, minor explosions occur. These may have been caused by particles of sodium igniting the liberated methoxyacetylene.
See other METAL ACETYLIDES

†1124. Propadiene (Allene)
[463-49-0]   C$_3$H$_4$

$$H_2C = C = CH_2$$

HCS 1980, 779 (cylinder)

1. Bondor, A. M. *et al., Khim. Prom.*, 1965, **41**, 923
2. Forshey, D. R., *Fire Technol.*, 1969, **5**, 100–111
The diene is endothermic (ΔH_f° (g) +192.1 kJ/mol, 4.80 kJ/g) and the pure gas can decompose explosively under a pressure of 2 bar [1], but this is also quoted as the upper limiting pressure for flame propagation during sustained (monopropellant) burning at 25°C [2].
See Propyne (below)

Oxides of nitrogen
    See Nitrogen oxide: Dienes, Oxygen
    See other DIENES, ENDOTHERMIC COMPOUNDS

†1125. Propyne
[74-99-7]   C$_3$H$_4$

$$HC \equiv CCH_3$$

HCS 1980, 633 (cylinder)

1. *MCA Case History No. 632*
2. Fitzgerald, F., *Nature*, 1960, **186**, 386–387
3. Hurden, D., *J. Inst. Fuel.*, 1963, **36**, 50–54
4. Stull, 1977, 16
The liquid material in cylinders (which contains around 30% of propadiene) is not shock-sensitive, but a wall temperature of 95°C (even very localised) accompanied by pressures of about 3.5 bar, will cause a detonation to propagate from the hot spot [1]. Induced decomposition of the endothermic hydrocarbon leads to flame

propagation in absence of air above minimum pressures of 3.4 and 2.1 bar at 20 and 120°C, respectively [2]. Application as a monopropellant and possible hazards therefrom (including formation of explosive copper propynide) have been discussed [3]. Although the pure material is highly endothermic (ΔH_f° (g) +185.4 kJ/mol, 4.64 kJ/g), the commercial mixture with propadiene and propane (MAPP gas) is comparable with ethylene for handling requirements and potential hazard [4].

Silver nitrate
Grignard, 1935, Vol. 3, 175
Interaction in ammoniacal solution gives the silver derivative which ignites around 150°C.
See other ALKYNES, ENDOTHERMIC COMPOUNDS

1126. 3-Bromo-1,1,1-trichloropropane
[13749-37-6] $C_3H_4BrCl_3$

$$BrC_2H_4CCl_3$$

Preparative hazard
See Ethylene: Bromotrichloromethane
See other HALOALKANES

†1127. 1-Chloro-3,3,3-trifluoropropane
[460-35-5] $C_3H_4ClF_3$

$$ClC_2H_4CF_3$$

See other HALOALKANES

1128. 2-Chloro-*N*,*N*,*N*'-trifluoropropionamidine
[25238-02-2] $C_3H_4ClF_3N_2$

$$MeCHClC(:NF)NF_2$$

Ross, D. L. *et al.*, *J. Org. Chem.*, 1970, **35**, 3096–3097
A shock-sensitive explosive.
See other *N*,*N*,*N''*-TRIFLUOROAMIDINES

1129. *N*-Chloro-3-aminopropyne (*N*-Chloropropargylamine)
[103698-31-3] C_3H_4ClN

$$CH{\equiv}CCH_2NHCl$$

It explodes at ambient temperature.
See entry N-HALOGEN COMPOUNDS (reference 4)
See other ACETYLENIC COMPOUNDS, N-HALOGEN COMPOUNDS

1130. 2-Chloro-1-cyanoethanol (3-Chlorolactonitrile)
[33965-80-9] C_3H_4ClNO

$$ClCH_2CH(CN)OH$$

Scotti, F. *et al.*, *J. Org. Chem.*, 1964, **29**, 1800

Distillation (at 110°C/4 mbar) is hazardous, since slight overheating may cause explosive decomposition to 2-chloroacetaldehyde and hydrogen cyanide.
See other CYANO COMPOUNDS

1131. *N*-(2-Chloroethyl)-*N*-nitrosocarbamoyl azide
[60784-40-9] $C_3H_4ClN_5O_2$

$$ClC_2H_4N(N:O)CO.N_3$$

Eisenbrand, G., Ger. Offen. 2 659 862, 1978
It should not be isoated from solution, as it is potentially explosive.
See other ACYL AZIDES, NITROSO COMPOUNDS

†1132. 2,3-Dichloropropene
[78-88-6] $C_3H_4Cl_2$

$$H_2C=CClCH_2Cl$$

See other HALOALKENES

1133. 2,2,3,3-Tetrafluoropropanol
[76-37-9] $C_3H_4F_4O$

$$F_2CHCF_2CH_2OH$$

Potassium hydroxide, or Sodium
Bagnall, R. D., private comm., 1972
Attempted formation of sodium tetrafluoropropoxide by adding the alcohol to sodium (40 g) caused ignition and a fierce fire which melted the flask. This was attributed to alkoxide-induced elimination of hydrogen fluoride, and subsequent exothermic polymerisation. In an alternative preparation of the potassium alkoxide by adding the alcohol to solid potassium hydroxide, a vigorous exotherm occurred. This was not seen when the base was added slowly to the alcohol.
See other POLYMERISATION INCIDENTS

1134. 1-Hydroxyimidazole *N*-oxide
[35321-46-1] $C_3H_4N_2O_2$

Flynn, A. P., *Chem. Brit.*, 1984, **20**, 30

On DSC examination it exhibits a large decomposition exotherm at 200–250°C, and is a sensitive detonating and deflagrating explosive.
See 1-Methoxyimidazole *N*-oxide
See other N–O COMPOUNDS, *N*-OXIDES

†1135. 3-Diazopropene (Vinyldiazomethane)
[2032-04-4] $C_3H_4N_2$

$$H_2C=CHCHN_2$$

Salomon, R. G. *et al., J. Org. Chem.*, 1975, **40**, 758
Potentially explosive, it should be stored in solution at 0°C and shielded from light.
See other DIAZO COMPOUNDS

1136. 3-Aminoisoxazole
[1750-42-1] $C_3H_4N_2O$

Sealed samples decompose exothermally above 115°C.
See entry ISOXAZOLES *See other* N–O COMPOUNDS

1137. 4-Amino-3-isoxazolidinone
[68-39-3] $C_3H_4N_2O_2$

Sealed samples decompose exothermally above 78°C.
See entry ISOXAZOLES *See other* N–O COMPOUNDS

1138. Methyl diazoacetate
[6832-16-2] $C_3H_4N_2O_2$

$$MeOCO.CHN_2$$

Searle, N. E., *Org. Synth.*, 1963, Coll. Vol. 4, 426
This ester must be handled with particular caution as it explodes with extreme violence on heating.
See other DIAZO COMPOUNDS

1139. 3-Nitro-2-isoxazoline
[1121-14-8] $C_3H_4N_2O_3$

CHETAH, 1990, 188
Shock sensitive
See other C-NITRO COMPOUNDS, N–O COMPOUNDS

1140. 2-Aminothiazole
[96-50-4] $C_3H_4N_2S$

MCA Case History No. 1587
Drying 2-aminothiazole in an oven without forced air circulation caused development of hot spots and eventual ignition. It has a low auto-ignition temperature and
will ignite after 3.5 h at 100°C.

Nitric acid
See Nitric acid: 2-Aminothiazole

Nitric acid, Sulfuric acid
See Nitric acid: 2-Aminothiazole, Sulfuric acid

†1141. Imidazoline-2,4-dithione ('Dithiohydantoin')
[5789-17-3] $C_3H_4N_2S_2$

Pouwels, H., *Chem. Eng. News*, 1975, **53**(49), 5
A 70 g sample, sealed into a brown glass ampoule, exploded after storage at
ambient temperature for 17 years. This was attributed to slow decomposition
and gas generation (perhaps initiated by traces of alkali in the ampoule
glass).

See Pyrimidine-2,4,5,6-(1*H*,3*H*)-tetraone
See other GLASS INCIDENTS

1142. 1,3-Dinitro-2-imidazolidinone
[2536-18-7]   C₃H₄N₄O₅

Violent decomposition occurred at 238°C.
See entry DIFFERENTIAL THERMAL ANALYSIS(DTA) (reference 1) *See other* N-NITRO
COMPOUNDS

1143. 1,3-Diazidopropene
[22750-69-2]   C₃H₄N₆

$$N_3CH=CHCH_2N_3$$

Forster, M. O. *et al., J. Chem. Soc.*, 1912, **101**, 489
A sample exploded while being weighed.
See other ORGANIC AZIDES

1144. Ammonium 2,4,5-trinitroimidazolide
[63839-60-1]   C₃H₄N₆O₆

Coburn, M. D. US Pat. 4 028 154, 1977
The ammonium salt is an explosive comparable to RDX but of higher thermal
stability.
See related POLYNITROARYL COMPOUNDS

†1145. Acrylaldehyde (Propenal)
[107-02-8]   C₃H₄O

$$H_2C=CHCO.H$$

(*MCA SD-85*, 1961); *NSC 436*, 1978; *NFPA H49*, 1976; *HCS 1980*, 113; *RSC Lab.
Hazards Data Sheet No. 70*, 1988

Energy of decomposition (in range 70–380°C) measured as 0.864 kJ/g by DSC, and T_{ait24} was determined as 117°C by adiabatic Dewar tests, with an apparent energy of activation of 214 kJ/mol.
See entry THERMOCHEMISTRY AND EXOTHERMIC DECOMPOSITION (reference 2)

Acids, or Bases
1. *MCA SD-85*, 1961
2. Hearsey, C. J., private comm., 1973
3. Personal experience (PGU)
4. Catalogue note, Hopkin & Williams, 1973
5. Bond, J., *Loss Prev. Bull.*, 1985, (065), 26

Acrylaldehyde (acrolein) is very reactive and will polymerise rapidly, accelerating to violence, in contact with strong acid or basic catalysts. Normally an induction period, shortened by increase in contamination, water content or initial temperature, precedes the onset of polymerisation. Uncatalysed polymerisation sets in at 200°C in the pure material [1]. Exposure to weakly acidic conditions (nitrous fumes, sulfur dioxide, carbon dioxide), some hydrolysable salts, or thiourea will also cause exothermic and violent polymerisation. A 2 year old sample stored in a refrigerator close to a bottle of dimethylamine exploded violently, presumably after absorbing enough volatile amine (which penetrates plastics closures) to initiate polymerisation [2]. Measurement of acrolein in a plastic measuring cylinder previously used for triethylamine (and allowed to evaporate dry) gave rise to violent polymerisation and a lacquered fume cupboard within less than a minute [3]. The stabilising effect of the added hydroquinone may cease after a comparatively short storage time. Such unstabilised material could polymerise explosively [4]. Violent polymerisation in a 250 kl storage tank led to an explosion and widespread damage [5].
See other ALDEHYDES, CATALYTIC IMPURITY INCIDENTS, INDUCTION PERIOD INCIDENTS, PEROXIDISABLE COMPOUNDS, POLYMERISATION INCIDENTS

Trienes
Anon., 1989, Personal communication
Calorimetric investigation of a Diels Alder reaction between propenal and a triene, which had caused problems on scale-up, showed, after the exotherm due to the Diels Alder reaction, and from a temperature a litle above 200°C, a second, more exothermic, reaction with a very fast pressure rise which burst the ARC can employed. This is presumably aromatisation of the alkenylcyclohexenealdehyde first formed, with probable liberation of hydrogen and carbon monoxide.
See ACCELERATING RATE CALORIMETRY

†1146. Methoxyacetylene (Ethynyl methyl ether)
[6443-91-0] C_3H_4O

$$MeOC{\equiv}CH$$

See Sodium methoxyacetylide: Brine
See other ACETYLENIC COMPOUNDS

†1147. 3-Propynol (Propargyl alcohol)
[107-19-7]   C₃H₄O

$$HC{\equiv}CCH_2OH$$

Alkalies
Anon., *Angew. Chem. (Nachr.)*, 1954, **2**, 209
If propynol and similar acetylenic compounds are dried with alkali before distil-
lation, the residue may explode (probably owing to acetylenic salt formation).
Sodium sulfate is recommended as a suitable desiccant.

Mercury(II) sulfate, Sulfuric acid, Water
1. Reppe, W. *et al., Ann.*, 1955, **596**, 38
2. Nettleton, J., private comm., 1972
Following the published procedure [1], hydroxyacetone was being prepared on half
the scale by treating propargyl alcohol as a 30% aqueous solution with mercury
sulfate and sulfuric acid (6 g and 0.6 g/mol of alcohol, respectively). On stirring and warming the mixture to 70°C a violent exothermic eruption occurred.
Quartering the scale of operations to 1 g mol and reducing the amount of acid to
0.37 g/mol gave a controllable reaction at 70°C. Adding the alcohol to the other
reactants at 70°C is an alternative possibility to avoid the suspected protonation
and polymerisation of the propargyl alcohol.
See other POLYMERISATION INCIDENTS

Phosphorus pentaoxide
491M, 1975, 321
Addition of the (acidic) oxide to the alcohol caused ignition.

Sulfuric acid
1. Sauer, J. C., *Org. Synth.*, 1963, Coll. Vol. 4, 813
2. Crouse, D. M., *Chem. Eng. News*, 1979, **57**(41), 4
3. Hart, E. V., *Chem. Eng. News*, 1980, **58**(2), 4
During addition of 56% aqueous acid to the aqueous 33% alcohol, using a published
procedure for conversion to propiolaldehyde [1], a violent explosion was attributed
to lack of adequate cooling [2]. This attribution is disputed, and contamination,
possibly with a heavy metal ion, is postulated [3]. The published procedure is rated
not hazardous.
See Mercury(II) sulfate, etc., above
See other ACETYLENIC COMPOUNDS

†1148. Acrylic acid (Propenoic acid)
[79-10-7]   C₃H₄O₂

$$H_2C{=}CHCO.OH$$

FPA H95, 1980; *HCS 1980*, 115; *RSC Lab. Hazard Data Sheet No. 68*, 1988

1. *Haz Chem. Data*, 1975, 34
2. Anon., *RoSPA OS & H Bull.*, 1976, (10), 3

3. Anaorian, M. *et al.*, *Eur. Polym. J.*, 1981, **17**, 823–838
4. Bond, J., *Loss Prev. Bull.*, 1985, (065), 26
5. Kurland, J. J. *et al.*, *Plant/Oper. Progr.*, 1987, **6**, 203–207
6. Yassin, A. A. *et al.*, *Chem. & Ind.*, 1985, 275
7. Levy, L. B., *Plant/Oper. Progr.*, 1987, **6**, 188–189; *ibid.*, 1989, **8**(2), 105
8. Wamper, F. M., *Plant/Oper. Progr.*, 1988, **7**, 183–189

Acrylic acid is normally supplied as the inhibited monomer, but because of its relatively high freezing point (14°C) it often partly solidifies, and the solid phase (and the vapour) will then be free of inhibitor which remains in the liquid phase. Even the uninhibited acid may be stored safely below the m.p., but such material will polymerise exothermically at ambient temperature, and may accelerate to a violent or explosive state if confined. Narrow vents may become blocked by polymerisation of uninhibited vapour [1]. A 17 m^3 tank trailer of glacial acrylic acid was being warmed by internal coils containing water from an unmonitored steam and water mixer to prevent the acid freezing in sub-zero temperatures. The extremely violent explosion which occurred later probably involved explosive polymerisation accelerated by both the latent heat liberated when the acid froze, and the uncontrolled deliberate heating [2]. The detailed mechanism of auto-accelerating polymerisation of the acid in hexane–methanol solution, which becomes explosive, has been studied [3]. Contamination of a 220 kl cargo of acrylic acid from an adjacent storage tank in a marine tanker led to an escalating polymerisation which could not be controlled, and the cargo could not be off-loaded. Eventually the tank had to be cut out of the ship [4]. It appears that the contaminant leaking into the acrylic acid was ethylidenenorbornene, which reacted with the traces of oxygen necessary to activate the inhibitor in the acrylic acid, leading to onset of polymerisation [5]. Hexamethylenetetramine is an effective stabiliser for acrylic acid in both liquid and vapour phases, preventing polymerisation at concentrations of 50–100 ppm [6]. Dimerisation of acrylic acid is temperature and solvent dependent and cannot be controlled by inhibitors, but vinyl polymerisation can be inhibited by conventional radical-trapping compounds, such as hydroquinone or phenothiazine, both of which need access of some oxygen for activation [7]. The dimerisation process of moist acrylic acid has been further investigated, and an empirical equation permits prediction of the extent of dimerisation in storage at different temperatures [8].

See other POLYMERISATION INCIDENTS

Initiator, Water
Anon., *Sichere Chemiearb.*, 1978, **30**, 3
During the experimental large scale continuous polymerisation of acrylic acid in aqueous solution in presence of an initiator and a moderator, failure of one of the feed pumps led to an unusually high concentration of monomer in solution. This led to runaway polymerisation which burst a glass vent line and the escaping contents ignited and led to an explosion and fire.

See other GLASS INCIDENTS, POLYMERISATION INCIDENTS

Other reactants
Yoshida, 1980, 8

MRH values calculated for 13 combinations with oxidants are given.
See VIOLENT POLYMERISATION *See other* ORGANIC ACIDS

†1149. Vinyl formate (Ethenyl methanoate)
[692-45-5]   C₃H₄O₂

$$H_2C=CHOCO.H$$

1150. Pyruvic acid (2-Oxopropanoic acid)
[127-17-3]   C₃H₄O₃

$$MeCO.CO.OH$$

1. Anon., *Sichere Chemiearb.*, 1977, **29**, 87
2. Wolff, L., *Ann.*, 1901, **317**, 2

A bottle of analytical grade material exploded in laboratory storage at 25°C (undoubtedly from internal pressure of carbon dioxide). Pure material, protected from light and air, is only stable on a long term basis if kept refrigerated. Otherwise slow decomposition and decarboxylation occurs [1], possibly accelerated by enzymic catalysis from ingress of airborne yeasts. At ambient temperature, the acid dimerises and dehydrates to 2-oxo-4-carboxyvalerolactone [2].
See other GAS EVOLUTION INCIDENTS, ORGANIC ACIDS

1151. Malonic acid (Propanedioic acid)
[141-82-2]   C₃H₄O₄

$$CH_2(COOH)_2$$

Although undoubtedly capable of generating carbon dioxide by thermal decomposition, the editor has heard of no pressurisation problems arising from this.

Potassium bromate, Ceric ammonium nitrate, Water
See Potassium bromate: Ceric ammonium nitrate, Malonic acid, Water

1152. 3-Propynethiol
[27846-30-6]   C₃H₄S

$$HC{\equiv}CCH_2SH$$

1. Sato, K. *et al., Chem. Abs.*, 1956, **53**, 5112b
2. Brandsma, 1971, 179

When distilled at ambient pressure, it polymerised explosively. It distils smoothly under reduced pressure at 33–35°C/127 mbar [1]. The polymer produced on exposure to air may explode on heating. Presence of a stabiliser is essential during handling or storage under nitrogen at −20°C [2].
See other ACETYLENIC COMPOUNDS, POLYMERISATION INCIDENTS

410

†1153. 3-Bromo-1-propene (Allyl bromide)
[106-95-6] C_3H_5Br

$$H_2C=CHCH_2Br$$

HCS 1980, 124

See other ALLYL COMPOUNDS, HALOALKENES

1154. Propionyl hypobromite
[82198-80-9] $C_3H_5BrO_2$

$$EtCO.OBr$$

Skell, P. S. *et al., J. Amer. Chem. Soc.*, 1983, **105**, 4000
As an isolated solid it is unpredictably explosive.
See other ACYL HYPOHALITES

1155. Allylzinc bromide (Bromo-2-propenylzinc)
[18925-10-5] C_3H_5BrZn

$$CH_2=CHCH_2ZnBr$$

Formamide, Ketones
  See Formamide: Organozinc compounds
  See also GRIGNARD REAGENTS

†1156. 1-Chloro-1-propene
[590-21-6] C_3H_5Cl

$$ClCH=CHMe$$

See other HALOALKENES

†1157. 2-Chloropropene
[557-98-2] C_3H_5Cl

$$H_2C=CHClMe$$

See other HALOALKENES

†1158. 3-Chloropropene (Allyl chloride)
[107-05-1] C_3H_5Cl

$$H_2C=CHCH_2Cl$$

(*MCA SD-99*, 1973); *HCS 1980*, 125; *RSC Lab. Hazards Data Sheet No. 28*, 1984
Piccinini, N. *et al., Plant/Operations Progr.*, 1982, **1**, 69–74
An operability analysis to identify potential risks in an allyl chloride manufacturing
plant has been published.

Aromatic hydrocarbons, Ethylaluminium chlorides
491M, 1975, 22
Friedel-Crafts alkylation of benzene or toluene by allyl chloride in presence of ethylaluminium chlorides is vigorous even at $-70°C$, and explosions have occurred.
See Lewis acids, etc., next below

Lewis acids, Metals
MCA SD-99, 1973
Contact with aluminium chloride, boron trifluoride, sulfuric acid etc., may cause violently exothermic polymerisation. Organometallic products of contact of the chloride with aluminium, magnesium, zinc (or galvanised metal) may produce similar results.
See other ALLYL COMPOUNDS, HALOALKENES

1159. Ethyl *N*-chloro-*N*-sodiocarbamate
[17510-52-0] $C_3H_5ClNNaO_2$
ClN(Na)CO.OEt

1. Herranz, E. *et al.*, *J. Amer. Chem. Soc.*, 1978, **100**, 3598, footnote 8
2. Scully, F. E., *Tetrahedron Lett.*, 1990, **31**(37), 5261
It may decompose (non-explosively) exothermally with gas evolution, and the scale of preparation should be limited to 0.2 g mol [1]. Samples drying in desiccators several times shattered them. Use in solution is recommended [2].
See other *N*-HALOGEN COMPOUNDS

1160. 1-Chloro-2,3-propylene dinitrate
[2612-33-1] $C_3H_5ClN_2O_6$
ClCH$_2$CH(ONO$_2$)CH$_2$ONO$_2$

Sorbe, 1968, 54
An explosive syrup.
See related ALKYL NITRATES

†1161. Chloroacetone (Chloro-2-propanone)
[78-95-5] C_3H_5ClO
ClCH$_2$CO.Me

HCS 1980, 291

1. Allen, C. F. H. *et al.*, *Ind. Eng. Chem. (News Ed.)*, 1931, **9**, 184
2. Ewe, G. E., *Ind. Eng. Chem. (News Ed.)*, 1931, **9**, 229
Two separate incidents involved explosive polymerisation of chloroacetone stored in glass bottles under ambient conditions for extended periods were reported [1,2].
See Bromoacetone oxime

†1162. 1-Chloro-2,3-epoxypropane (Chloromethyloxirane, Epichlorohydrin)
[106-89-8]   **C₃H₅ClO**

HCS 1980, 454

Energy of decomposition (in range 375–500°C) measured as 0.5 kJ/g.
See entry THERMOCHEMISTRY AND EXOTHERMIC DECOMPOSITION (reference 2)

Aniline
Hearfield, F., *Chem. Abs.*, 1980, **92**, 115649
In a review of thermal instability in chemical reactors, the explosive interaction of
'epichlorohydrin' with aniline is detailed.

Catalyst, Heterocyclic nitrogen compounds
Anon., *Sichere Chemiearb.*, 1985, **37**, 81
The epoxide and the nitrogen compound were mixed and stirred in a 250 l reactor
fitted for reflux and protected by a rupture disk. After addition of the catalyst,
subsequent heating to the required temperature seems to have been effected by
direct application of steam to the jacket, rather than the usual hot water. The too-
rapid heating led to a violently exothermic reaction which ruptured the disk and
stretched the lid clamping bolts.

Contaminants
1. TV News item, ITN, 5th January, 1984
2. Sittig, 1981, 293
A road tanker was loaded with 1-chloro-2,3-epoxypropane and then driven
250 miles overnight to the delivery point. On arrival, the contents were found
to have self heated (undoubtedly from polymerisation initiated by some unknown
contaminant) to the boiling point (115°C at ambient pressure) and soon afterwards
the relief valve lifted and discharged large volumes of vapour. Cooling with
water sprays eventually restored thermal control over the remaining tanker
contents [1]. The material is incompatible with strong acids, caustic alkalies,
zinc, aluminium, aluminium chloride or iron(III) chloride, all of which catalyse
exothermic polymerisation [2].
See other CATALYTIC IMPURITY INCIDENTS, POLYMERISATION INCIDENTS

Isopropylamine
Barton, N. *et al., Chem. & Ind.*, 1971, 994
With slow mixing and adequate cooling, smooth condensation to 1-chloro-3-
isopropylamino-2-propanol occurs. With rapid mixing and poor cooling, a variable

413

induction period with slow warming precedes a rapid, violent exotherm (to 350°C in 6 s). Other primary and secondary amines behave similarly. Moderating effects of water and the nature of the products are discussed.
See N-Substituted aniline, below
See other AMINATION INCIDENTS, INDUCTION PERIOD INCIDENTS

Potassium *tert*-butoxide
See Potassium *tert*-butoxide: Acids, etc.

Sodium chlorite
Anon. *Chemical Engineer*, 1996, (620), 4; *ibid*, 1996, (621), 7
A tanker of aqueous 'sodium chlorite' was part charged to a tank of epichlorohydrin before it was reported it was mis-labelled. There was a very violent subsequent explosion, which blew down the wall of an adjacent factory. (The editor suspects it was, in fact, sodium hypochlorite, commonly known simply as hypochlorite, which sounds like epichlorohydrin. Sodium hypochlorite would be expected to be the more reactive of the two, generating an alkyl hypochlorite.)
See HYPOHALITES

N-Substituted aniline (unspecified)
Schierwater, F.-W., *Major Loss Prevention*, 1971, 47
Interaction is exothermic and the mixture was normally maintained at 60°C by stirring and cooling. Malfunction caused a temperature increase to 70°C and cooling capacity was insuffecnt to regain control. The temperature steadily increased to 120°C, when explosive decomposition occurred. This was attributed to thermal instability of the reaction system and inadequate pressure relief arrangements.
See Isopropylamine, above

Sulfuric acid
Leleu, *Cahiers*, 1974, (75), 276
Interaction is violent.

Trichloroethylene
See Trichloroethylene: Epoxides
See other 1,2-EPOXIDES

†1163. Propionyl chloride (Propanoyl chloride)
[79-03-8] C_3H_5ClO
EtCO.Cl

HCS 1980, 794

Preparative hazard
See Phosphorus trichloride: Carboxylic acids

Diisopropyl ether
Koenst, W. M. B., *J. Haz. Mat.*, 1981, **4**, 291–298

A facile exothermic reaction, catalysed by traces of zinc chloride and iron(III) chloride (4 and 40 ppm, respectively) to produce isopropyl propionate and 2-chloropropane (b.p., 35°C) led to pressure build up and bursting of a closed galvanised drum after 24 h. Similar reactions are thermodynamically possible with other acid chlorides and ethers (particularly if secondary or tertiary alkyl ethers), so such mixtures should only be prepared immediately prior to use.

See Sulfinyl chloride: Esters, Metals
See other CATALYTIC IMPURITY INCIDENTS, GAS EVOLUTION INCIDENTS
See other ACYL HALIDES

†1164. Ethyl chloroformate (Ethyl carbonochloridate)
[541-41-3] $C_3H_5ClO_2$

$$EtOCO.Cl$$

HCS 1980, 476

See other ACYL HALIDES

†1165. Methoxyacetyl chloride (2-Methoxyethanoyl chloride)
[38870-89-2] $C_3H_5ClO_2$

$$MeOCH_2CO.Cl$$

Anon., *Jahresber.*, 1979, 75
A sample of redistilled material in a screw capped bottle exploded 15 weeks after capping. This was attributed to development of internal pressure (probably of hydrogen chloride arising from hydrolysis, and perhaps also of chloromethane from scission). (The editor has also known methoxyacetic mixed anhydrides to decarbonylate, releasing carbon monoxide)
See other ACYL HALIDES, GAS EVOLUTION INCIDENTS

1166. Methoxycarbonylmethyl perchlorate
[95407-67-3] $C_3H_5ClO_6$

$$MeOCO.CH_2OClO_3$$

See entry ALKYL PERCHLORATES (reference 6)

1167. Peroxypropionyl perchlorate
[66955-44-0] $C_3H_5ClO_6$

$$EtCO.OOClO_3$$

Yakarka, A. A. *et al., Chem. Abs.*, 1979, **91**, 210847
The oxidant exploded on detonation or friction.
See related ACID ANHYDRIDES
See other NON-METAL PERCHLORATES, ORGANIC PEROXIDES

†**1168.** *N,N*-**Dichloro-β-alanine (2-Dichloroaminopropanoic acid)**
[58941-15-4] $C_3H_5Cl_2NO_2$

$$Cl_2NCH(Me)CO.OH$$

Vit, J. *et al., Synth. Commun.*, 1976, **6**(1), 1–4
It is thermally unstable above 95°C.
See other N-HALOGEN COMPOUNDS

1169. Allyl phosphorodichloridite
[1498-47-1] $C_3H_5Cl_2OP$

$$H_2C=CHCH_2OPCl_2$$

Kamler, M. *et al., Can. J. Chem.*, 1985, **63**, 825–826
The title compound is formed by reaction of allyl alcohol with phosphorus trichloride and is reported to polymerise explosively if the material, when being purified by distillation, is taken down to less than half its bulk.
See other ALLYL COMPOUNDS, PHOSPHORUS ESTERS

1170. *N*-**Carboethoxyiminophosphoryl chloride**
[3356-63-6] $C_3H_5Cl_3NO_2P$

$$EtOCO.N=PCl_3$$

See N-Carbomethoxyiminophosphoryl chloride
See related NON-METAL HALIDES

†**1171. 1-Fluoro-2,3-epoxypropane (Fluoromethyloxirane)**
[503-09-3] C_3H_5FO

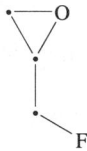

See other 1,2-EPOXIDES

1172. *N,N,N′*-**Trifluoropropionamidine**
[21372-60-1] $C_3H_5F_3N_2$

$$EtC(:NF)NF_2$$

Ross, D. L. *et al., J. Org. Chem.*, 1970, **35**, 3096–3097
A shock-sensitive explosive.
See other N, N, N″-TRIFLUOROAMIDINES

1173. Allylmercury(II) iodide
[2845-00-3] C_3H_5HgI

$$H_2C=CHCH_2HgI$$

Potassium cyanide
Whitmore, 1921, 122
Among the products of interaction, a minor one is an explosive liquid.
See other ALLYL COMPOUNDS, MERCURY COMPOUNDS

†1174. 3-Iodopropene (Allyl iodide)
[556-56-9] C_3H_5I

$$H_2C=CHCH_2I$$

See other ALLYL COMPOUNDS, HALOALKENES, IODINE COMPOUNDS

1175. Potassium 1,1-dinitropropanide
[30533-63-2] $C_3H_5KN_2O_4$

$$KC(NO_2)_2Et$$

Bisgrove, D. E. *et al., Org. Synth.*, 1963, Coll. Vol. 4, 373
The potassium salt of 1,1-dinitropropane, isolated as a by-product during preparation of 3,4-dinitro-3-hexene, is a hazardous explosive.
See NITROALKANES: alkali metals *See other* POLYNITROALKYL COMPOUNDS

†1176. Potassium *O*-ethyl dithiocarbonate ('Potassium ethyl xanthate')
[140-89-6] $C_3H_5KOS_2$

$$KSC(:S)OEt$$

Diazonium salts
See DIAZONIUM SULFIDES AND DERIVATIVES

1177. Allyllithium
[3052-45-7] C_3H_5Li

$$H_2C=CHCH_2Li$$

Leleu, *Cahiers*, 1977, (88), 367
As usually prepared, it is a pyrophoric solid.
See other ALKYLMETALS, ALLYL COMPOUNDS

†1178. Ethyl isocyanide (Isocyanoethane)
[624-79-3] C_3H_5N

$$EtN=C:$$

Lemoult, M. P., *Compt. rend.*, 1906, **143**, 903

This showed a strong tendency to explode while being sealed into glass ampoules.
See Methyl isocyanide
See related CYANO COMPOUNDS

†1179. Propiononitrile (Propanonitrile)
[107-12-0]   C$_3$H$_5$N

EtCN

Preparative hazard
See Nitric acid: Aliphatic amines

N-Bromosuccinimide
See *N*-Bromosuccinimide: Propiononitrile
See other CYANO COMPOUNDS

†1180. Acrylamide (Propenamide)
[79-06-1]   C$_3$H$_5$NO

H$_2$C=CHCO.NH$_2$

HCS 1980, 114

Bretherick, 1981, 165
It may polymerise with violence on melting at 86°C.
See other POLYMERISATION INCIDENTS

†1181. 2-Cyanoethanol (3-Hydroxypropanenitrile)
[109-78-4]   C$_3$H$_5$NO

N≡CC$_2$H$_4$OH

HCS 1980, 479

Acids, or Bases
HCS 1980, 479
Reacts violently with mineral acids, amines or inorganic bases, probably because of
dehydration to acrylonitrile and subsequent catalysed polymerisation of the latter.
See other POLYMERISATION INCIDENTS *See other* CYANO COMPOUNDS

1182. 2,3-Epoxypropionaldehyde oxime (Oxiranecarboxaldehyde oxime)
[67722-96-7]   C$_3$H$_5$NO$_2$

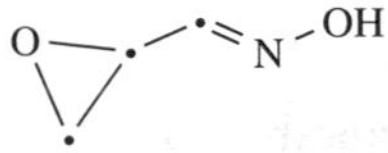

Payne, G. B., *J. Amer. Chem. Soc.*, 1959, **81**, 4903
The residue from distillation at 48–49°C/1.3 mbar polymerised violently, and the
distilled material polymerised explosively after 1–2 h at ambient temperature.

See 2,3-Epoxypropionaldehyde 2,4-dinitrophenylhydrazone
See other 1,2-EPOXIDES, OXIMES, POLYMERISATION INCIDENTS
See related ALDEHYDES

1183. 2-Nitropropene
[4749-28-4] $C_3H_5NO_2$

$$H_2C{=}C(NO_2)CH_3$$

Miyashita, M. *et al., Org. Synth.*, 1981, **60**, 101–103
Vacuum distillation must be effected at below 80°C to avoid fume-offs, particularly if air be admitted to the warm residue. The lachrymatory material also polymerises in contact with alkalies.
See other NITROALKENES, POLYMERISATION INCIDENTS

†1184. Nitroacetone (1-Nitro-2-propanone)
[10230-68-9] $C_3H_5NO_3$

$$O_2NCH_2COMe$$

1. Tegeler, J. J., *Chem. Eng. News*, 1987, **65**(1), 4
2. Author's comment, 1987

Nitroacetone (237 g) was generated in dichloromethane–ethyl acetate solution by treating its dicyclohexylamine *aci*-salt with sulfuric acid. Towards the end of concentration of the solution by rotary vacuum evaporation at 40–50°C, a small explosion occurred. Extreme caution and small scale handling of α-nitroketones are urged [1]. If exact neutralisation of the dicyclohexylamine salt had not been achieved, presence of excess sulfuric acid (or of some undecomposed *aci*-salt) may have reduced the stability of the nitroacetone [2].
See Nitromethane: Acids, or Bases;
See also aci-NITRO SALTS
See related NITROALKANES

1185. Propionyl nitrite
[28128-14-5] $C_3H_5NO_3$

$$EtCO.ON{:}O$$

See entry ACYL NITRITES

†1186. 2,3-Epoxypropyl nitrate (Oxiranemethanol nitrate)
[6659-62-7] $C_3H_5NO_4$

Urbanski, 1965, Vol. 2, 129
Shock-sensitive and explodes at 200°C.
See related ALKYL NITRATES, 1,2-EPOXIDES

1187. Peroxypropionyl nitrate
[5796-89-4] $C_3H_5NO_5$

$$EtCO.OONO_2$$

1. Stephens, E. R., *Anal. Chem.*, 1964, **36**, 928–929
2. Louw, R. *et al., J. Amer. Chem. Soc.*, 1975, **97**, 4396
Like the lower homologues, it is extremely explosive [1], and may only be handled in high dilution with air or nitrogen. Higher homologues are less explosive [2].
See related ACID ANHYDRIDES, ACYL NITRATES, PEROXYESTERS

†1188. 3-Azidopropene (Allyl azide)
[821-13-6] $C_3H_5N_3$

$$H_2C=CHCH_2N_3$$

1. Anon., *Saf. Digest*, (Univ. Saf. Assoc.), 1985, (13), 2–3
2. Forster, M. O., *J. Chem. Soc.*, 1908, **93**, 1174
An explosion occurred in the final stages of evaporation of ether from allyl azide [1]. Peroxides in the ether may possibly have been involved, but the azide itself is of limited thermal stability, though apparently capable of distillation at 76°C/1 bar. The vapour is weakly explosive [2].
See other ALLYL COMPOUNDS, ORGANIC AZIDES

1189. 1-Methyl-1,2,3-triazole
[16681-65-5] $C_3H_5N_3$

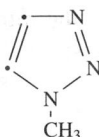

Anon., private comm., 1980
An explosion during the preparation of the triazole was attributed to local over-heating or presence of impurities.
See other HIGH-NITROGEN COMPOUNDS, TRIAZOLES

†1190. Azidoacetone (Azido-2-propanone)
[4504-27-2] $C_3H_5N_3O$

$$N_3CH_2CO.Me$$

1. Spauschus, H. O. *et al., J. Amer. Chem. Soc.*, 1951, **73**, 209
2. Forster, M. O. *et al., J. Chem. Soc.*, 1908, **93**, 72
A small sample exploded after storage in the dark for 6 months [1]. The freshly prepared material explodes when dropped on to a hotplate and burns brilliantly [2].
See other 2-AZIDOCARBONYL COMPOUNDS

1191. Glycidyl azide (Azidomethyloxirane)
[80044-09-3]

$C_3H_5N_3O$

[80044-10-6] polymer$(-CH_2CHOHCH_2(N_3)-)_n$

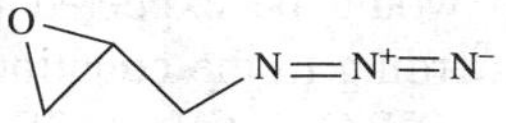

Nazare, A. N. *et al., J. Energ. Mater.*, 1992, **10**(1), 43

A review on the derived polymer as a propellant component, which use depends upon the energy of the azide. The monomer would be expected to have several kJ/g, including oxirane ring strain.

See 1,2-EPOXIDES, ORGANIC AZIDES

1192. 5-Amino-3-methylthio-1,2,4-oxadiazole
[55864-39-6]

$C_3H_5N_3OS$

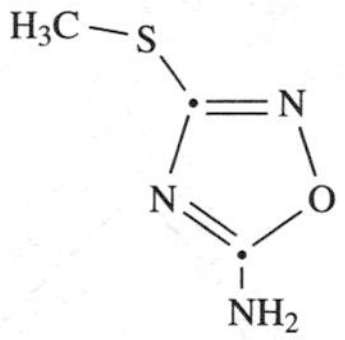

Wittenbrook, L. S., *J. Heterocycl. Chem.*, 1975, **12**, 38

It undergoes rapid and moderately violent decomposition at the m.p., 97–99°C.

See other N–O COMPOUNDS

†1193. Ethyl azidoformate (Ethyl carbonazidate)
[817-87-8]

$C_3H_5N_3O_2$

EtOCO.N$_3$

Forster, M. O. *et al., J. Chem. Soc.*, 1908, **93**, 91

It is liable to explode if boiled at ambient pressure (at 114°C). The methyl ester (b.p. 103°C) can behave similarly.

See other ACYL AZIDES

1194. 2-Ammoniothiazole nitrate
[57530-25-3]

$C_3H_5N_3O_3S$

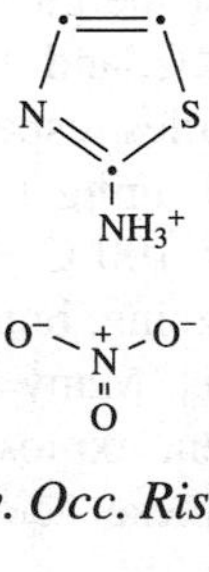

1. Rebsch, M., *Proc. Int. Symp. Prev. Occ. Risks Chem. Ind.*, 485–486, Heidelberg, ISSA., 1976
2. Soldarna Marlor, L., *Chem. Abs.*, 1977, **87**, 203834

During air-pressurised discharge of a hot 53% aqueous solution of the nitrate salt from a reaction vessel via a filter press, a violent explosion occurred. The nitrate salt begins to decompose below 100°C, and at the likely internal temperature of 142°C, decomposition would be expected to be very rapid, involving much gas/vapour generation according to the equation below.

$$4C_3H_5N_3O_3S \rightarrow 12C + 10H_2O + 6N_2 + 3S + SO_2$$

Air-pressurisation would have prevented the water from boiling and absorbing the exotherm to moderate the thermal explosion. Decomposition of the solution may have been accelerated by the presence of impurities, or by the solid salt splashed onto the heated vessel wall [1]. There is an independent account of the incident [2].

See other GAS EVOLUTION INCIDENTS

1195. 3,3-Dinitroazetidine
[129660-17-9] $C_3H_5N_3O_4$

Hiskey, M. A. *et al., J. Energ. Mater.,* 1993, **11**(3), 157.
Dinitroazetidine salts were studied as explosives: nitrate, dinitroimidate, 2,4-dinitroimidazolate, 4,4',5,5'-tetranitro-2,2'-biimidazolate, 5-nitro-1,2,4-triazolonate and 3,5-dinitrotriazolate.

See other C-NITRO COMPOUNDS, STRAINED-RING COMPOUNDS

†1196. Glyceryl trinitrate (1,2,3-Propanetriyl nitrate)
[55-63-0] $C_3H_5N_3O_9$

$$O_2NOCH_2CH(ONO_2)CH_2ONO_2$$

1. Urbanski, 1965, Vol. 2, 47, 53–54
2. Bond, J., *Loss Prev. Bull.,* 1983, (050), 18

The relatively high sensitivity of the pure liquid material to initiation by shock or heating (it explodes at 215–218°C, and decomposes energetically above 145°C) is enhanced by presence of impurities, especially nitric acid, arising from imperfect washing, hydrolysis or thermal aging (e.g. heating to 50°C). Intense UV radiation will explode a sample at 100°C. The sensitivity of the viscous liquid to shock is slightly reduced by freezing, but is greatly diminished by absorption onto porous solids, as in dynamite [1]. Many incidents, some dating back to 1870, are summarised, all involving violent explosions when cans or buckets of nitroglycerine were dropped. Some of these detonations initiated the detonation of bulk material nearby [2].

See other ALKYL NITRATES, IRRADIATION DECOMPOSITION INCIDENTS

†1197. Cyclopropane
[75-19-4] C_3H_6

HCS 1980, 351 (cylinder)

Stull, 1977, 16
It is fairly endothermic and of low MW ($\Delta H^°_f$ (g) +82.3 kJ/mol, 1.96 kJ/g), and
a minor constituent of MAPP gas.
See Propyne
See other ENDOTHERMIC COMPOUNDS, STRAINED-RING COMPOUNDS

†1198. Propene (Propylene)
[115-07-1] C_3H_6

$$H_2C=CHCH_3$$

(*MCA SD-59*, 1956); *FPA H75*, 1978; *HCS 1980*, 788 (cylinder)

1. Russell, F. R. *et al., Chem. Eng. News*, 1952, **30**, 1239
2. Stull, 1977, 16
Several sets of values for flammability limits in air are quoted; 1.4–7.1, 2.4–11.1,
2.7–36.0 vol%.
   Propene at 955 bar and 327°C was being subjected to further rapid compression.
At 4.86 kbar explosive decomposition occurred, causing a pressure surge to
10 kbar or above. Decomposition to carbon, hydrogen and methane must have
occurred to account for this pressure. Ethylene behaves similarly at much lower
pressure, and cyclopentadiene, cyclohexadiene, acetylene and a few aromatic
hydrocarbons have been decomposed explosively [1]. It is mildly endothermic
($\Delta H^°_f$ (g) +20.4 kJ/mol, 0.49 kJ/g) and a minor constituent of MAPP gas [2].

Lithium nitrate, Sulfur dioxide
   Pitkethly, R. C., private comm., 1973
   A mixture under confinement in a glass pressure bottle at 20°C polymerised explosively, the polymerisation probably being initiated by access of light through the
clear glass container. Such alkene–sulfur dioxide co-polymerisations will not occur
above a ceiling temperature, different for each alkene.
   See other GLASS INCIDENTS, POLYMERISATION INCIDENTS

Other reactants
   Yoshida, 1980, 321
   MRH values calculated for 13 combinations with oxidants are given.

Oxides of nitrogen
   See Nitrogen oxide: Dienes, Oxygen

423

Trifluoromethyl hypofluorite
See Trifluoromethyl hypofluorite: Hydrocarbons

Water
See LIQUEFIED GASES: water
See other ALKENES

1199. Cyanodimethylarsine
[683-45-4]

C_3H_6AsN

$$N\equiv CAsMe_2$$

491M, 1975, 140
It ignites in air.
See related ALKYLNON-METAL HALIDES

1200. Dimethylgold selenocyanate
[42494-76-1]

C_3H_6AuNSe

$$Me_2AuSeC\equiv N$$

Stocco, F. *et al., Inorg. Chem.*, 1971, **10**, 2640
Very shock-sensitive and explodes readily when precipitated from aqueous solution. Crystals obtained by slow evaporation of a carbon tetrachloride extract were less sensitive.
See other GOLD COMPOUNDS *See related* METAL CYANATES

1201. Bromoacetone oxime (1-Bromo-2-oximinopropane)
[62116-25-0]

C_3H_6BrNO

$$BrCH_2C(Me)=NOH$$

Forster, M. O *et al., J. Chem. Soc.*, 1908, **93**, 84
It decomposes explosively during distillation.
See Chloroacetaldehyde oxime, *and* Chloroacetone
See other OXIMES

†1202. *N*-Chloroallylamine (*N*-Chloro-3-aminopropene)
[82865-33-6]

C_3H_6ClN

$$H_2C=CHCH_2NHCl$$

It explodes at ambient temperature.
See entry N-HALOGEN COMPOUNDS (reference 4)
See other ALLYL COMPOUNDS

†1203. 1,1-Dichloropropane
[78-99-9] $C_3H_6Cl_2$

$$Cl_2CHEt$$

See other HALOALKANES

†1204. 1,2-Dichloropropane
[78-87-5] $C_3H_6Cl_2$

$$ClCH_2CHClCH_3$$

HCS 1980, 383

Aluminium
See Aluminium: Halocarbons (reference 18)

Aluminium, 1,2-Dichlorobenzene, 1,2-Dichloroethane
See Aluminium: Halocarbons (reference 5)
See other HALOALKANES

1205. 2,2-Dichloropropane
[594-20-7] $C_3H_6Cl_2$

$$H_3CCCl_2CH_3$$

Dimethylzinc
See Dimethylzinc: 2,2-Dichloropropane
See other HALOALKANES

1206. 1-Chloro-2-propyl perchlorate
[58426-27-0] $C_3H_6Cl_2O_4$

$$ClCH_2CH(Me)OClO_3$$

See entry ALKYL PERCHLORATES (reference 6)

1207. 3-Chloro-2-hydroxypropyl perchlorate
[101672-07-5] $C_3H_6Cl_2O_5$

$$ClCH_2CH(OH)CH_2OClO_3$$

Hofmann, K. A. *et al.*, *Ber.*, 1909, **42**, 4390
It explodes violently on shaking in a capillary.
See other ALKYL PERCHLORATES

†1208. **Dichloromethylvinylsilane**
[124-70-9] $C_3H_6Cl_2Si$

$$H_2C=CHSi(Me)Cl_2$$

See related ALKYLHALOSILANES

1209. **Dichloromethylenedimethylammonium chloride**
[33843-02-3] $C_3H_6Cl_3N$

$$Cl_2C=N^+Me_2Cl^-$$

1. Bader, A. R., private comm., 1971
2. Senning, A., *Chem. Rev.*, 1965, **65**, 388
The compound does not explode on heating [1]; the earlier reference [2] is in error.

1210. **Potassium *O*-propionohydroxamate**
[71939-10-1] $C_3H_6KNO_2$

$$EtCO.NHOK$$

Anon., *Actual. Chim. (Fr.)*, 1983, **8**, 49–50
The solid explodes on attempted drying. DTA indicates that the rapid decomposition at 70–74°C to potassium hydroxide and ethyl isocyanate, accompanied by an exotherm of 1.37 kJ/g, is involved in the explosion.
See other N–O COMPOUNDS

1211. **Dimethylthallium fulminate**
[ ] C_3H_6NOTl

$$Me_2TlC\equiv N \rightarrow O$$

Beck, W. *et al., J. Organomet. Chem.*, 1965, **3**, 55
Highly explosive, unlike the diphenyl analogue.
See related METAL FULMINATES

1212. **3-Aminopropiononitrile**
[151-18-8] $C_3H_6N_2$

$$H_2NC_2H_4C\equiv N$$

1. Shner, V. F., *Zhur. Prikl. Khim.*, 1966, **39**, 2386 (Engl. transl.)
2. *MCA Guide*, 1972, 309
3. Merck, 1976, 65
Vacuum distilled material kept dark for 6 months was found to have largely polymerised to a yellow solid, which exploded 15 days later [1]. A bottle of uninhibited distilled material exploded after shelf storage for several months [2]. It may be stored for several months out of contact with air under refrigeration. In contact with ambient air, it polymerises slowly, but rapidly in contact with acidic materials [3].
See other CYANO COMPOUNDS, POLYMERISATION INCIDENTS

1213. 2-Ethyltetrazole
[82944-28-3]   C₃H₆N₄

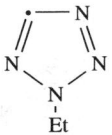

Aluminium hydride
See Aluminium hydride: Tetrazole derivatives
See other TETRAZOLES

1214. 5-Ethyltetrazole
[50764-78-8]   C₃H₆N₄

Aluminium hydride
See Aluminium hydride: Tetrazole derivatives
See other TETRAZOLES

1215. Azidoacetone oxime
[101672-04-2]   C₃H₆N₄O

$$N_3CH_2C(Me){=}NOH$$

Forster, M. O. *et al., J. Chem. Soc.,* 1908, **93**, 83
During distillation at 84°C/2.6 mbar, the large residue darkened and finally exploded violently.
See Bromoacetone oxime
See other 2-AZIDOCARBONYL COMPOUNDS

1216. 2,2-Diazidopropane
[85620-95-7]   C₃H₆N₆

$$Me_2C(N_3)_2$$

Al-Khalil, S. I. *et al., J. Chem. Soc., Perkin Trans. 1,* 1986, 555–565
The liquid azide, b.p. 28–30°C/0.6 mbar, exploded during attempted analytical combustion.
See other ORGANIC AZIDES

1217. 1,3,5-Trinitrosohexahydro-1,3,5-triazine
[13980-04-6] $C_3H_6N_6O_3$

Sulfuric acid
Urbanski, 1967, Vol. 3, 122
It decomposes explosively in contact with conc. sulfuric acid.
See other NITROSO COMPOUNDS

1218. 5-Nitro-2-(nitroimino)hexahydro-1,3,5-triazine (1,4,5,6-Tetrahydro-
N,5-dinitrotriazin-2-amine)
[130400-13-4] $C_3H_6N_6O_4$

Dagley, I. J., *et al., J. Energ. Mater.*, 1995, **13**(1 & 2), 35
This nitroguanidine analogue is more sensitive than RDX by the drop-weight
impact test.
See other N-NITRO COMPOUNDS

1219. 1,3,5-Trinitrohexahydro-1,3,5-triazine (Cyclotrimethylenetrinitramine)
[121-82-4] $C_3H_6N_6O_6$

Botcher, T. R. *et al., J. Amer. Chem. Soc.*, 1992, **114**(2), 8303
Some studies relating to the explosion mechanism of this military explosive (RDX
or Cyclonite) and its higher homologue HMX.

Calcium hydroxide, Water
Iida, M. *et al., Chem. Abs.*, 1988, **109**, 112968
The self-heating and decomposition of the explosive and aqueous alkali was studied by DSC in a sealed capsule and in a larger scale furnace test. A rapid exothermic decomposition reaction can be initiated at 100°C or below, and may lead to spontaneous ignition and then deflagration or detonation.
See other N-NITRO COMPOUNDS
See also NITROAROMATIC–ALKALI HAZARDS

†1220. Acetone (2-Propanone)
[67-64-1] C_3H_6O

$$H_3CCO.CH_3$$

(*MCA SD-87*, 1962); *NSC 398*, 1982; *FPA H1*, 1972; *HCS 1980*, 102; *RSC Lab. Hazard Data Sheet No. 21*, 1984

Bromoform
See Bromoform: Acetone, etc.

Carbon, Air
Boiston, D. A., *Brit. Chem. Eng.*, 1968, **13**, 85
(In the absence of other ignition source), fires in plant to recover acetone from air with active carbon are due to the bulk surface effect of oxidative heating when air flow is too low to cool effectively.

Chloroform
See Chloroform: Acetone, etc.

2-Methyl-1,3-butadiene
See 2-Methyl-1,3-butadiene: Acetone

Nitric acid, Sulfuric acid   MRH Nitric 5.86/77, Sulfuric 1.05/53
Fawcett, H. H., *Ind. Eng. Chem.*, 1959, **51**(4), 89A
Acetone will be oxidised with explosive violence if brought into contact with the mixed (nitrating) acids, particularly under confinement.
See Oxidants, below

Other reactants
Yoshida, 1980, 20
MRH values calculated for 17 combinations with oxidants are given.

Oxidants   MRH values show % of oxidant
See Nitric acid, Sulfuric acid, above
Bromine: Acetone

Bromine trifluoride: Solvents
Chromium trioxide: Acetone   MRH 2.59/90
Chromyl chloride: Organic solvents
Dioxygen difluoride: Various materials
Hydrogen peroxide: Acetone   MRH 6.36/81
Hydrogen peroxide: Ketones, or: Oxygenated compounds
Nitric acid: Acetone   MRH 5.86/77
Nitrosyl chloride: Acetone, etc.
Nitrosyl perchlorate: Organic materials
Oxygen (Gas): Acetylene, Acetone
Peroxomonosulfuric acid: Acetone

Peroxyacetic acid
  See 5-Bromo-4-pyrimidinone

Potassium *tert*-butoxide
  See Potassium *tert*-butoxide: Acids, etc.

Sulfur dichloride
  See Sulfur dichloride: Acetone

Thiotrithiazyl perchlorate
  See Thiotrithiazyl perchlorate: Organic solvents

†1221. Methyl vinyl ether (Methoxyethene)
[107-25-5] C_3H_6O
MeOCH=CH$_2$

HCS 1980, 664 (cylinder)

Acids
  1. Braker, 1980, 487
  2. 'MVE' Brochure, Billingham, ICI, 1962
  Methyl vinyl ether is rapidly hydrolysed by contact with dilute acids to form
  acetaldehyde, which is more reactive and has wider flammability limits than the
  ether [1]. Presence of base is essential during storage or distillation of the ether
  to prevent rapid acid-catalysed homopolymerisation, which is not prevented by
  antioxidants. Even mildly acidic solids (calcium chloride or some ceramics) will
  initiate exothermic polymerisation [2].
  See other POLYMERISATION INCIDENTS

Halogens, or Hydrogen halides
  Braker, 1980, 487
  Addition reactions with bromine, chlorine, hydrogen bromide or hydrogen chloride
  are very vigorous and may be explosive if uncontrolled.
  See other PEROXIDISABLE COMPOUNDS

†1222. Oxetane
[503-30-0]

C_3H_6O

Dinitrogen pentaoxide
See Dinitrogen pentaoxide: Strained ring heterocycles
See other STRAINED-RING COMPOUNDS

†1223. 2-Propen-1-ol (Allyl alcohol)
[107-18-6]

C_3H_6O

$$H_2C{=}CHCH_2OH$$

HCS 1980, 122

Energy of decomposition (in range 360–500°C) measured as 0.69 kJ/g.
See entry THERMOCHEMISTRY AND EXOTHERMIC DECOMPOSITION (reference 2)

Alkali, 2,4,6-Trichloro-1,3,5-triazine   MRH Sodium hydroxide 1.21/tr.
See 2,4,6-Trichloro-1,3,5-triazine: Allyl alcohol, etc.

Carbon tetrachloride   MRH 1.42/73
491M, 1975, 21
Interaction gives an unstable mixture of halogenated C_4 epoxides which exploded
during distillation.

Other reactants
Yoshida, 1980, 24
MRH values calculated for 15 combinations, mainly with oxidants, are given.

Sulfuric acid   MRH 1.42/53
Senderens, J. B., *Compt. rend.*, 1925, **181**, 698–700
During preparation of diallyl ether by dehydration of the alcohol with sulfuric
acid, a violent explosion may occur (possibly involving peroxidation and certainly
polymerisation).
See other PEROXIDATION INCIDENTS, POLYMERISATION INCIDENTS

2,4,6-Tris(bromoamino)-1,3,5-triazine
See 2,4,6-Tris(bromoamino)-1,3,5-triazine: Allyl alcohol
See other ALLYL COMPOUNDS

†1224. Propionaldehyde (Propanal)
[123-38-6]

C_3H_6O

$$EtCO.H$$

HCS 1980, 791

Methyl methacrylate
See Methyl methacrylate: Propionaldehyde
See other ALDEHYDES, PEROXIDISABLE COMPOUNDS

†1225. Propylene oxide (Methyloxirane)
[75-56-9]   C₃H₆O

FPA H88, 1980; *HCS 1980*, 800

1. *CHETAH*, 1990, 183
2. Kulik, M. K., *Science*, 1967, **155**, 400
3. Smith, R. S., *Science*, 1967, **156**, 12
4. Pogany, G. A., *Chem. & Ind.*, 1979, 16–21
5. *See entry* THERMOCHEMISTRY AND EXOTHERMIC DECOMPOSITION (reference 2)
The liquid can be exploded by a detonator, though not by mechanical shock
[1]. Use of propylene oxide as a biological sterilant is hazardous because of
ready formation of explosive mixtures with air (2.8–37%). Commercially available mixtures with carbon dioxide, though non-explosive, may be asphyxiant and
vesicant [1]. Such mixtures may be ineffective, but neat propylene oxide vapour
may be used safely, provided that it is removed by evacuation using a water-jet
pump [2]. The main factors involved in the use and safe handling on a laboratory
scale have been discussed [3]. The energy of decomposition (in range 340–500°C)
has been measured as 1.114 kJ/g [4].

Epoxy resin
Sheaffer, J., *CHAS Notes*, 1981, **1**(4), 5
Mixing of propylene oxide and epoxy resin in a waste bottle led to an explo-
sion, probably owing to the polymerisation of the oxide catalysed by the amine
accelerator in the resin.
See Sodium hydroxide, below

Ethylene oxide, Polyhydric alcohol
See Ethylene oxide: Polyhydric alcohol, etc.

Other reactants
Yoshida, 1980, 322
MRH values calculated for 13 combinations with oxidants are given.

Oxygen
1. Hassan, A. *et al.*, *Combust. Sci. Technol.*, 1983, **35**, 215–230
2. Peng, J., *Proc. Int. Pyrotech, Semin.*, 1991, (17th), Vol. 1, 353

The transition of deflagration to detonation in mixtures was studied with respect to mixing ratio, pressure and spark energy [1]. A study of TNT equivalences in propylene oxide fuel/air explosives is made [2].

Sodium hydroxide
1. *MCA Case History No. 31*
2. Freeder, B. G. *et al., J. Loss Prev. Process Ind.*, 1988, **1**, 164–168
A drum of crude product containing unreacted propylene oxide and sodium hydroxide catalyst exploded and ignited, probably owing to base-catalysed exothermic polymerisation of the oxide [1]. A comparative ARC study of the runaway exothermic polymerisation of ethylene oxide and the less reactive propylene oxide in presence of sodium hydroxide solutions, as typical catalytically active impurities, has been done. The results suggest that the hazard potential for propylene oxide is rather less than that for the lower homologue, though more detailed work is needed to quantify the difference [2].
See Ethylene oxide: Contaminants
See other CATALYTIC IMPURITY INCIDENTS, POLYMERISATION INCIDENTS
See other 1,2-EPOXIDES

1226. Allyl hydroperoxide (2-Propenyl hydroperoxide)
[6069-42-7] $C_3H_6O_2$

$$H_2C=CHCH_2OOH$$

1. Dykstra, S. *et al., J. Amer. Chem. Soc.*, 1957, **79**, 3474
2. Seyfarth, H. E. *et al., Angew. Chem.*, 1965, **77**, 1078
When impure, the material is unstable towards heat or light and decomposes to give an explosive residue. The pure material is more stable to light, but detonates on heating or in contact with solid alkalies [1]. Preparation by action of oxygen on diallylzinc gives improved yields, but there is a risk of explosion. The peroxide is also impact-sensitive if sand is admixed [2].
See other ALKYL HYDROPEROXIDES, ALLYL COMPOUNDS

1227. Dimethyldioxirane
[74087-85-7] $C_3H_6O_2$

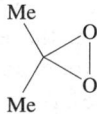

1. Murray, R. W., private communication, 1992
This material, and homologues, are being recommended as oxidants, they are prepared in situ and very dilute. Thus far, the material has shown itself surprisingly stable during isolation and pyrolysis (decomposes slowly at ambient temperatures and quietly on heating) and even explosivity testing [1]. However several

433

known oligomers, presumably of lower (strain) energy, are dangerously sensitive explosives. Considerable caution during work-up of reactions is recommended.

See Bis(trifluoromethyl)dioxirane
See Hydrogen Peroxide: Acetone
See other CYCLIC PEROXIDES

†1228. 1,3-Dioxolane
[646-06-0]   C₃H₆O₂

$$C_3H_6O_2$$

Lithium perchlorate
See Lithium perchlorate: 1.3-Dioxolane

1229. 2,3-Epoxypropanol (Oxiranemethanol)
[556-52-5]   C₃H₆O₂

$$C_3H_6O_2$$

HCS 1980, 518

1. Anon., *Jahresber.*, 1982, 67–68; *Sichere Chemiearb.*, 1984, **36**, 80
2. Anon, *Loss Prevention Bull.*, 1989, **88**, 17
3. Cardillo, P. *et al., J. Loss Prev. Process Ind.*, 1991, **4**(4), 242

Off-specification 'glycidol' from a production unit was collected in a 40 t storage tank and recovered occasionally by redistillation. As no stabiliser is known, the contents were water-cooled to prevent self-heating and polymerisation, and pump-circulated to ensure homogeneity. A 9 t quantity of the stored material was found to be quite warm, and soon after, the tank exploded, causing 3 fatalities. The contents of the tank had been subjected to storage tests at 60°C on 200 l scale, when decomposition but no runaway was seen, presumably the drums were not adiabatic enough. It is known that the epoxy compound will undergo explosive decomposition in presence of strong acids or bases, salts (aluminium chloride, iron(III) chloride, tin(IV) chloride), or metals (copper or zinc), but these were not likely to have been present. It was concluded that the water cooling had been turned off, and that the energy dissipated by the circulation pump would have raised the temperature of the tank contents to 50°C under the prevailing weather conditions, the latter conclusion being confirmed experimentally [1]. The energy of decomposition (in the range 130–450°C) has been measured by DSC as 1.365 kJ/g, and T_{ait24} was determined as the low figure of 65°C by adiabatic Dewar tests, with an apparent energy of activation of 71 kJ/mol. A detailed study of the thermochemistry of glycidol, including CHETAH calculations and calorimetry (ARC and DSC), showed a two stage reaction; polymerisation from 60°C, followed by a decomposition with pressure generation from 280°C. Polymerisation was catalysed by NaOH and triethanolamine but not, apparently, by HCl or triethylamine [3].

See entry THERMOCHEMISTRY AND EXOTHERMIC DECOMPOSITION
See other 1,2-EPOXIDES
See also CALORIMETRY; COMPUTATION OF REACTIVE CHEMICAL HAZARDS

†1230. Ethyl formate (Ethyl methanoate)
[109-94-4] $C_3H_6O_2$

EtOCO.H

HCS 1980, 488

Base
See Methyl formate: Methanol, Sodium methoxide

1231. Hydroxyacetone (Propan-1-ol-2-one)
[116-09-6] $C_3H_6O_2$

HOCH₂CO.CH₃

Preparative hazard
See 3-Propynol: Mercury(II) sulfate, etc.

†1232. Methyl acetate (Methyl ethanoate)
[79-20-9] $C_3H_6O_2$

MeOCO.CH₃

FPA H124, 1983; *HCS 1980*, 426

Other reactants
Yoshida, 1980, 123
MRH values calculated for 13 combinations with oxidants are given.

†1233. Dimethyl carbonate
[616-38-6] $C_3H_6O_3$

MeOCO.OMe

HCS 1980, 426

Other reactants
Yoshida, 1980, 224
MRH values calculated for 13 combinations with oxidants are given.

435

Potassium *tert*-butoxide
See Potassium *tert*-butoxide: Acids, etc.

1234. Lactic acid (2-Hydroxypropanoic acid)
[598-82-3] $C_3H_6O_3$

MeCHOHCO.OH

Hydrofluoric acid, Nitric acid
See Nitric acid: Hydrofluoric acid, Lactic acid
See other ORGANIC ACIDS

1235. 3-Methyl-1,2,4-trioxolane (Propene ozonide)
[38787-96-1] $C_3H_6O_3$

Briner, E. *et al., Helv. Chim. Acta*, 1929, **12**, 181
Stable at 0°C but often explosively decomposes at ambient temperature.
See other CYCLIC PEROXIDES, OZONIDES

1236. Peroxypropionic acid (Propaneperoxoic acid)
[4212-43-5] $C_3H_6O_3$

EtCO.OOH

1. Swern, D., *Chem. Rev.*, 1949, **45**, 9
2. Phillips, B. *et al., J. Org. Chem.*, 1959, **23**, 1823
More stable than its lower homologues, it merely deflagrates on heating. Higher
homologues appear to be still more stable [1]. Safe procedures (on the basis of
detonability experiments) for preparation of anhydrous solutions of peroxypropionic acid in chloroform or ethyl propionate have been described [2].

Organic solvents
Manly, T. D., *Chem. Brit.*, 1982, **18**, 341
A violent explosion occurred during the distillation of solvents that had contained
the peroxyacid. Although the solvents had been treated by addition of sodium
sulfite to reduce the acid, dipropionyl peroxide (which is not reduced by sulfite,
and which may have been produced by the sulfite treatment) remained in solution
and subsequently decomposed on heating. If the solvent is added to the sulfite
solution, the diacyl peroxide is not formed.
See DIACYL PEROXIDES See other PEROXYACIDS

436

1237. 1,3,5-Trioxane
[110-88-3]   C₃H₆O₃

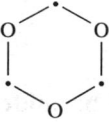

HCS 1980, 942
MCA Case History No. 1129

Use of an axe to break up a large lump of trioxane caused ignition and a vigorous
fire. Peroxides may have been involved.

Acrylonitrile
   See Acrylonitrile: Formaldehyde

Hydrogen peroxide, Lead
   See Hydrogen peroxide: Lead, Trioxane

Oxygen (Liquid)
   See Oxygen (Liquid): 1,3,5-Trioxane
   See PEROXIDISABLE COMPOUNDS

1238. 3-Thietanol-1,1-dioxide
[22524-35-2]   C₃H₆O₃S

Preparative hazard
   See Hydrogen peroxide: Acetic acid, 3-Thietanol

†1239. 2-Propene-1-thiol (Allyl mercaptan)
[870-23-5]   C₃H₆S
                                        H₂C=CHCH₂SH

   See other ALLYL COMPOUNDS

1240. Diazomethyldimethylarsine
[59871-26-0]   C₃H₇AsN₂
                                        N₂CHAsMe₂

Preparative hazard
   See Diazomethane: Dimethylaminodimethylarsine
   See other DIAZO COMPOUNDS

†1241. 1-Bromopropane (Propyl bromide)
[106-94-5] C_3H_7Br

$$BrCH_2Et$$

See other HALOALKANES

†1242. 2-Bromopropane (Isopropyl bromide)
[75-26-3] C_3H_7Br

$$MeCHBrMe$$

See other HALOALKANES

†1243. 1-Chloropropane (Propyl chloride)
[540-54-5] C_3H_7Cl

$$ClCH_2Et$$

See other HALOALKANES

†1244. 2-Chloropropane (Isopropyl chloride)
[75-29-6] C_3H_7Cl

$$MeCHClMe$$

HCS 1980, 578

See other HALOALKANES

1245. *S*-Carboxymethylisothiouronium chloride (Carboxymethyl
carbamimoniothioate chloride)
[5425-78-5] $C_3H_7ClN_2O_2S$

$$HOCO.CH_2SC(:N^+H_2)NH_2Cl^-$$

Chlorine
See Chlorine: Nitrogen compounds (reference 4)
See related ORGANIC ACIDS

†1246. Chloromethyl ethyl ether (Chloromethoxyethane)
[3188-13-4] C_3H_7ClO

$$ClCH_2OEt$$

See related HALOALKANES

1247. Isopropyl hypochlorite
[53578-07-7] C_3H_7ClO

$$Me_2CHOCl$$

1. Chattaway, F. D. *et al.*, *J. Chem. Soc.*, 1923, **123**, 3001
2. Mellor, 1956, Vol. 2, Suppl. 1, 550

Of extremely low stability; explosions occurred during its preparation if cooling was inadequate [1], or on exposure to light [2].

See other HYPOHALITES, IRRADIATION DECOMPOSITION INCIDENTS

1248. 3-Chloro-1,2-propanediol
[96-24-2] $C_3H_7ClO_2$

$$ClCH_2CHOHCH_2OH$$

Perchloric acid
   See Perchloric acid: Glycols, etc. (reference 1)

1249. Propyl perchlorate
[22755-14-2] $C_3H_7ClO_4$

$$PrOClO_3$$

See entry ALKYL PERCHLORATES

1250. 2-Propyl perchlorate
[52936-53-1] $C_3H_7ClO_4$

$$Me_2CHOClO_3$$

See entry ALKYL PERCHLORATES (reference 6)

1251. Propylcopper(I)
[18365-12-3] C_3H_7Cu

$$PrCu$$

Houben-Weyl 1970, Vol. 13.1, 737
Small quantities only should be handled because of the danger of explosion.
See other ALKYLMETALS

†1252. 2-Iodopropane (Isopropyl iodide)
[75-30-9] C_3H_7I

$$MeCHIMe$$

See other HALOALKANES, IODINE COMPOUNDS

1253. Propyllithium
[2417-93-8] C_3H_7Li

$$PrLi$$

Leleu, *Cahiers*, 1977, (88), 368
It ignites in air.
See other ALKYLMETALS

†1254. 3-Aminopropene (Allylamine)
[107-11-9]   C₃H₇N

$$H_2C{=}CHCH_2NH_2$$

HCS 1980, 123

See other ALLYL COMPOUNDS

†1255. Azetidine
[503-29-7]   C₃H₇N

Dinitrogen pentaoxide
   See Dinitrogen pentaoxide: Strained ring heterocycles
   See other STRAINED-RING COMPOUNDS

†1256. Cyclopropylamine
[765-30-0]   C₃H₇N

†1257. 2-Methylaziridine (Propyleneimine)
[75-55-8]   C₃H₇N

Acids
   Inlow, R. O. *et al., J. Inorg. Nucl. Chem.*, 1975, **37**, 2353
   Like the lower homologue ethyleneimine, it may polymerise explosively if exposed
to acids or acidic fumes, so it must always be stored over solid alkali.
   See other POLYMERISATION INCIDENTS

1258. Acetone oxime (2-Hydroxyiminopropane)
[127-06-0]   C₃H₇NO

$$Me_2C{=}NOH$$

Energy of decomposition (in range 200–420°C) measured as 1.5 kJ/g.

440

See entry THERMOCHEMISTRY AND EXOTHERMIC DECOMPOSITION (reference 2)
See other OXIMES

1259. Dimethylformamide
[68-12-2]   C₃H₇NO

$$Me_2NCO.H$$

HCS 1980, 430; *RSC Lab. Hazard Data Sheet No. 17*, 1983

This powerful aprotic solvent is not inert and reacts vigorously or violently with
a range of materials.
See endo-2,5-Dichloro-7-thiabicyclo[2.2.1]heptane: Dimethylformamide
     Diisocyanatomethane: Dimethylformamide
     Lithium azide: Alkyl nitrates, etc.
     Tetraphosphorus decaoxide: Organic liquids
     Sodium hydride: Dimethylformamide
     Sodium tetrahydroborate: Dimethylformamide
     Sodium: Dimethylformamide
     Sulfinyl chloride: Dimethylformamide
     Triethylaluminium: Dimethylformamide
     2,4,6-Trichloro-1,3,5-triazine: Dimethylformamide

Allyl trifluoromethanesulfonates
     See ALLYL TRIFLUOROMETHANESULFONATES: alone, or aprotic solvents

Halocarbons
     See Carbon tetrachloride: Dimethylformamide
          1,2,3,4,5,6-Hexachlorocyclohexane: Dimethylformamide
     See N,N-Dimethylacetamide: Halogenated compounds, etc.
     See other CATALYTIC IMPURITY INCIDENTS

Other reactants
     Yoshida, 1980, 172
     MRH values calculated for 13 combinations with oxidants are given.

Oxidants
     Pal, B. C. *et al., Chem. Eng. News*, 1981, **59**(17), 47
     Explosions involving chromium trioxide or potassium permanganate and dimethyl-
     formamide are explained in terms of DMF being, like formic acid, a reducing agent.
     Dimethylacetamide is recommended as a safer, less reducing substitute solvent for
     DMF. For other combinations listed
     See Bromine: Dimethylformamide
          Chlorine: Dimethylformamide
          Chromium trioxide: Dimethylformamide
          Magnesium nitrate: Dimethylformamide
          Potassium permanganate: Dimethylformamide
          Uronium perchlorate: Organic materials
     See other APROTIC SOLVENTS

1260. S -Methylthioacetohydroximate
[13749-94-5]   C₃H₇NOS

$$Me(MeS)C=NOH$$

Preparative hazard
   Oren, Z. *et al., Israeli Patent IL 89641,* 1995; *Chem. Abs.,* 1996, **124**, 260401g
   It is claimed that a prior method of manufacture from acetonitrile, methanethiol
   and hydroxylamine hydrochloride gave uncontrolled reaction and explosion. A
   safer preparation starting from acetaldoxime is claimed.

1261. Ethyl carbamate
[51-79-6]   C₃H₇NO₂

$$EtOCO.NH_2$$

Phosphorus pentachloride
   See N-Carbomethoxyiminophosphoryl chloride

†1262. Isopropyl nitrite
[541-42-4]   C₃H₇NO₂

$$Me_2CHON:O$$

See other ALKYL NITRITES

1263. 1-Nitropropane
[108-03-2]   C₃H₇NO₂

$$O_2NCH_2Et$$

Metal oxides
   See NITROALKANES: metal oxides

Other reactants
   Yoshida, 1980, 276
   MRH values calculated for 14 combinations, mainly with oxidants, are given.
   See other NITROALKANES

1264. 2-Nitropropane
[79-46-9]   C₃H₇NO₂

$$O_2NCHMe_2$$

HCS 1980, 689; *RSC Lab. Hazard Data Sheet No. 10,* 1983

Carbon, Hopcalite
   'Tech. Data Sheet No. 7', New York, Comml. Solvents Corp., 1968

The heat of adsorption of 2-nitropropane is very high, so carbon-containing respirators should not be used in high vapour concentrations. Also, if Hopcalite catalyst (co-precipitated copper(II) oxide and manganese (IV) oxide) is present in the respirator cartridge, ignition may occur.

Amines, Heavy metal oxides
'Tech Data Sheet No. 7', New York, Comml. Solvents Corp., 1968
Contact with amines and mercury oxide or silver oxide may lead to formation of unstable salts analogous to those of nitromethane.
See other NITROALKANES

†1265. Propyl nitrite
[543-67-9] $C_3H_7NO_2$

$$PrON:O$$

Anon., private comm., 1985
Bottles of isopropyl nitrite decomposed during refrigerated storage, developing enough internal pressure to burst the containers and blow open the refrigerator door. Other alkyl nitrites lower than amyl nitrite may be expected to behave similarly.
See other ALKYL NITRITES

†1266. Isopropyl nitrate
[1712-64-7] $C_3H_7NO_3$

$$Me_2CHONO_2$$

1. 'Tech. Bull. No. 2', London, ICI Ltd., 1964
2. Brochet, C., *Astronaut. Acta*, 1970, **15**, 419–425
3. Beeley, P. *et al.*, *Chem. Abs.*, 1977, **87**, 120000
4. Beeley, P. *et al.*, *Comb. Flame*, 1980, **39**, 255–268, 269–281

The self-sustaining exothermic decomposition/combustion characteristics of this fuel of negative oxygen balance (−68%) renders it suitable as a rocket monopropellant. Storage and handling precautions are detailed [1]. Conditions for detonation of the nitrate, an explosive of low sensitivity, have been investigated [2]. The pure vapour ignites spontaneously at very low temperatures and pressures, and other aspects were studied in detail [3], including spontaneous ignition on compression [4].

Lewis acids
See ALKYL NITRATES: lewis acids *See other* ALKYL NITRATES

†1267. Propyl nitrate
[627-13-4] $C_3H_7NO_3$

$$PrONO_2$$

Other reactants
Yoshida, 1980, 204
MRH Values calculated for 11 combinations are given.

Solvents

Breslow, B. A., US Pat. 4 042 032, 1977

The shock-sensitive nitrate is desensitised by 1–2% of propane, butane, chloroform, dimethyl ether or diethyl ether.

See THERMAL EXPLOSIONS *See other* ALKYL NITRATES

1268. 5-Amino-2-ethyl-2*H*-tetrazole
[95112-14-4] $C_3H_7N_5$

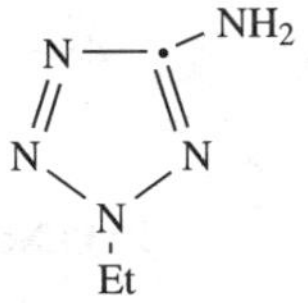

Aluminium hydride

See Aluminium hydride: Tetrazole derivatives

See other TETRAZOLES

1269. Propylsodium
[15790-54-2] C_3H_7Na

PrNa

Houben-Weyl, 1970, Vol. 13.1, 384

The powder ignites in air.

See other ALKYLMETALS

1270. Sodium isopropoxide
[683-60-3] C_3H_7NaO

$NaOCHMe_2$

Dimethyl sulfoxide

See Dimethyl sulfoxide: Metal alkoxides

See other METAL ALKOXIDES

†1271. Propane
[74-98-6] C_3H_8

$H_3CCH_2CH_3$

FPA H10, 1973 (cylinder); *HCS 1980*, 780 (cylinder)

Barium peroxide

See Barium peroxide: Propane

See also CRYOGENIC LIQUIDS (reference 10), LIQUEFIED PETROLEUM GASES

444

1272. Ethyliodomethylarsine
[65313-31-7] C_3H_8AsI

$$EtAs(I)Me$$

Leleu, *Cahiers*, 1977, (88), 363
It ignites in air.
See other ALKYLNON-METAL HALIDES

1273. Ethyl iminioformate chloride
[16694-46-5] C_3H_8ClNO

$$EtOC(:N^+H_2)H\ Cl^-$$

Preparative hazard
See Hydrogen chloride: Alcohols, Hydrogen cyanide

1274. Isopropyldiazene
[26981-95-3] $C_3H_8N_2$

$$Me_2CHN=NH$$

Chlorine
Abendroth, H. J., *Angew. Chem.*, 1959, **71**, 340
The crude chlorination product of 'isoacetone hydrazone' (presumably an *N*-chloro
compound) exploded violently during drying at 0°C.
See N-HALOGEN COMPOUNDS *See related* AZO COMPOUNDS

1275. 3-Hydrazinopropanenitrile (3-Cyanoethylhydrazine)
[353-07-1] $C_3H_7N_3$

$$N\equiv CCH_2CH_2NHNH_2$$

Jasiewicz, M. L., *Chem. & Ind.*, 1996, (3), 70
In the late stage of work up of a sample prepared from acrylonitrile and hydrazine
hydrate and stripped of water by a procedure involving dichloromethane, the
distillation flask pressurised and burst, shattering the front of the fume cupboard.
Previously, slight pressurisation had once been observed. Involving both a nitrile
and a hydrazino moiety, this molecule cannot be thermodynamically stable but it
has not given previous problems.
See Dichloromethane: 1,2-Diaminoethane, 3-Aminopropionitrile
See CYANO COMPOUNDS

1276. *N,N'*-Dinitro-*N*-methyl-1,2-diaminoethane
[10308-90-4] $C_3H_8N_4O_4$

$$MeN(NO_2)C_2H_4NHNO_2$$

Violent decomposition occurred at 210°C.

See entry DIFFERENTIAL THERMAL ANALYSIS (reference 1) See other N-NITRO COMPOUNDS

1277. 2,2-Dinitropropylhydrazine

[ ] $C_3H_8N_4O_4$

$$MeC(NO_2)_2CH_2NHNH_2$$

Borane
See Borane bis(2,2-dinitropropylhydrazine)
See other POLYNITROALKYL COMPOUNDS

†1278. Ethyl methyl ether (Methoxyethane)

[540-67-0] C_3H_8O

$$EtOMe$$

HCS 1980, 498

†1279. Propanol (1-Hydroxypropane)

[71-23-8] C_3H_8O

$$PrOH$$

HCS 1980, 797

Other reactants
Yoshida, 1980, 315
MRH values calculated for 13 combinations with oxidants are given.

Potassium tert-butoxide
See Potassium tert-butoxide: Acids, etc.

†1280. 2-Propanol (Isopropanol)

[67-63-0] C_3H_8O

$$Me_2CHOH$$

(MCA SD-98, 1972); FPA H45, 1976; HCS 1980, 783
1. Renfrew, M. M., J. Chem. Educ., 1983, **60**(9), A229
2. Bonafede, J. D., J. Chem. Educ., 1984, **61**, 632
3. Redemann, C. E., J. Amer. Chem. Soc., 1942, **64**, 3049
4. Mirafzal, G. A. et al., J. Chem. Educ., 1988, **65**(9), A226–229
5. Bohanon, J. T., Chem. Eng. News, 1989, **67**(1), 4
Several explosions have occurred during laboratory distillation of isopropanol [1,2,4,5], some with a sample stored for 5 years in a part empty can [5]. No cause was apparent, but presence of traces of ketone(s) promoting peroxidation is a probability. Previously, the presence of 0.36 M peroxide had been reported

in a 99.5% pure sample of isopropanol stored for several months in a partially full clear glass bottle in strong daylight [3]. The reformation of peroxides in deperoxided isopropanol 'within a few days' had been noted [2]. It appears that the tertiary H on the 2-position is susceptible to autoxidation, and that 2-propanol must be classed as peroxidisable. 2-Hydroperoxy-2-hydroxypropane has, in fact, been isolated from photocatalysed oxidation of isopropanol.

See PEROXIDISABLE COMPOUNDS (reference 7)

See 2-Butanone, below; also Oxygen (Gas): Alcohols

Aluminium
See Aluminium: 2-Propanol

Aluminium isopropoxide, Crotonaldehyde
Wagner-Jauregg, T., Angew. Chem., 1939, 53, 710
During distillation of 2-propanol recovered from the reduction of crotonaldehyde with isopropanol/aluminium isopropoxide, a violent explosion occurred. This was attributed to peroxidised diisopropyl ether (a possible by-product) or to peroxidised crotonaldehyde. An alternative or additional possibility is that the isopropanol may have contained traces of a higher secondary alcohol (e.g. 2-butanol) which would be oxidised during the Meerwein-Ponndorf reduction procedure to 2-butanone. The latter would then effectively sensitise the isopropanol or other peroxidisable species to peroxidation.

See 2-Butanone, next below; also Diisopropyl ether

2-Butanone
1. Bathie, H. M., Chem. Brit., 1974, 10, 143
2. Unpublished observations, 1974
3. Pitt, M. J., Chem. Brit., 1974, 10, 312
4. Schenk, G. O. et al., Angew. Chem., 1958, 70, 504

Distillation to small volume of a small sample of a 4-year-old mixture of the alcohol with 0.5% of the ketone led to a violent explosion, and the presence of peroxides was subsequently confirmed [1]. Pure alcohols which can form stable radicals (secondary branched structures) may slowly peroxidise to a limited extent under normal storage conditions (isopropanol to 0.0015 M in brown bottle, subdued light during 6 months; to 0.0009 M in dark during 5 years) [2]. The presence of ketones markedly increases the possibility of peroxidation by sensitising photochemical oxidation of the alcohol. Acetone (produced during autoxidation of isopropanol) is not a good sensitiser, but the presence of even traces of 2-butanone in isopropanol would be expected to accelerate markedly peroxidation of the latter. Treatment of any mixture or old sample of a secondary alcohol with tin(II) chloride and then lime before distillation is recommended [3]. The product of photosensitised oxidation is 2-hydroperoxy-2-propanol [4].

See 2-Butanol

Hydrogen peroxide
Baratov, A. N. et al., Chem. Abs., 1978, 88, 39537

Addition of a small amount of hydrogen peroxide may reduce sharply the AIT of the alcohol (455°C), probably to that of the hydroperoxide.

Other reactants
Yoshida, 198, 41
MRH values calculated for 13 combinations with oxidants are given.

Oxidants   MRH values show % of oxidant
See Barium perchlorate: Alcohols
        Chromium trioxide: Alcohols   MRH 2.47/91
        Dioxygenyl tetrafluoroborate: Organic materials
        Hydrogen peroxide: Oxygenated compounds   MRH 6.32/84
        Oxygen (Gas): Alcohols
        Sodium dichromate: 2-Propanol, etc.
        Trinitromethane: 2-Propanol

Potassium *tert*-butoxide
See Potassium *tert*-butoxide: Acids, etc.

†1281. Dimethoxymethane
[109-87-5] $C_3H_8O_2$

$$CH_2(OMe)_2$$

HCS 1980, 419

Oxygen
Molem, M. J. *et al., Chem. Abs.*, 1975, **83**, 45495
The nature of the reaction and the products in methylal–oxygen mixtures during cool flame or explosive oxidation were studied.
See other PEROXIDISABLE COMPOUNDS

1282. Ethyl methyl peroxide
[70299-48-8] $C_3H_8O_2$

$$EtOOMe$$

Rieche, A., *Ber.*, 1929, **62**, 218
Shock-sensitive as a liquid or vapour, it explodes violently on superheating.
See other DIALKYL PEROXIDES

1283. Isopropyl hydroperoxide (2-Hydroperoxypropane)
[3031-75-2] $C_3H_8O_2$

$$Me_2CHOOH$$

Medvedev, S. *et al., Ber.*, 1932, **65**, 133
It explodes just above the b.p., 107–109°C.
See other ALKYL HYDROPEROXIDES

1284. 2-Methoxyethanol (Ethylene glycol monomethyl ether)
[109-86-4] $C_3H_8O_2$

$$MeOC_2H_4OH$$

HCS 1980, 799

491M, 1975, 256
It forms explosive peroxides.
See other PEROXIDISABLE COMPOUNDS

1285. Propylene glycol (1,2-Propanediol)
[57-55-6] $C_3H_8O_2$

$$MeCHOHCH_2OH$$

HCS 1980, 782

Hydrofluoric acid, Nitric acid, Silver nitrate
See Nitric acid: Hydrofluoric acid, Propylene glycol, etc.

1286. Glycerol (1,2,3-Propanetriol)
[56-81-5] $C_3H_8O_3$

$$HOCH_2CHOHCH_2OH$$

HCS 1980, 516

Ethylene oxide
See Ethylene oxide: Glycerol

Other reactants
Yoshida 1980, 448
MRH values calculated for 16 combinations, mainly with oxidants, are given.

Oxidants   MRH values show % of oxidant
Unpublished comments, 1978
The violent or explosive reactions exhibited by glycerol in contact with many solid
oxidants are due to its unique properties of having three centres of reactivity, of
being a liquid which ensures good contact, and of high boiling point and viscosity
which prevents dissipation of oxidative heat. The difunctional, less viscous liquid
glycols show similar but less extreme behaviour.
See Calcium hypochlorite: Hydroxy compounds   MRH 2.13/88
     Chlorine: Glycerol
     Hydrogen peroxide: Organic compounds (reference 2)   MRH 5.56/71
     Nitric acid: Glycerol
     Nitric acid: Glycerol, Hydrochloric acid
     Nitric acid: Glycerol, Hydrofluoric acid

Nitric acid: Glycerol, Sulfuric acid   MRH Nitric 4.89/65, Sulfuric 0.9/26,
Perchloric acid, Glycerol, etc.
Potassium permanganate   MRH 2.76/85
Sodium peroxide: Hydroxy compounds   MRH 2.72/85

Phosphorus triiodide
See Phosphorus triiodide: Hydroxylic compounds

Sodium hydride
See Sodium hydride: Glycerol

Sodium tetrahydroborate
See Sodium tetrahydroborate: Glycerol

1287. 2,2-Bis(hydroperoxy)propane
[2614-76-8] $C_3H_8O_4$

$$Me_2C(OOH)_2$$

1. Hutton, W. *et al.*, *Chem. Eng. News*, 1984, **62**(38), 4
2. Sauer, M. C. V., *J. Phys. Chem.*, 1972, **76**, 1283–1288
3. Hutton, W., private comm., 1984
4. Schwoegler, E. J., Chem. Eng. News, 1985, 63(1), 6
5. Bodner, G. M., J. Chem. Educ., 1985, 62, 1105

The thermal instability of the bis-hydroperoxide ('peroxyacetone') has long been recognised and used in demonstrations of the violent combustion of peroxides. In thin layers a fireball is produced, while ignition of lumps or samples confined in a deflagrating spoon was likely to produce detonation. The compound had been used several times for such demonstrations without indication of undue sensitivity to shock or normal careful handling procedures. When one sample, stored in an open jar for 48 hours, was being sprinkled from it as usual for disposal by burning, it detonated violently [1]. Though exposure to acid fumes might have led to formation of the extremely sensitive trimeric acetone peroxide [2], no such exposure was likely. It was concluded that the peroxide may have become unusually dry and sensitive during its exposure to atmosphere in a hood, and that it should no longer be used for such demonstrations. [3]. Hutton's interpretation is incorrect, because 2,2-dihydroperoxypropane is not a solid, but a liquid. Most probably the material was the solid trimeric acetone peroxide [4]. Several other incidents, usually involving injury, are described [5].
See 3,3,6,6,9,9-Hexamethyl-1,2,4,5,7,8-hexoxonane

†1288. Ethyl methyl sulfide
[624-89-5] C_3H_8S

$$EtSMe$$

†1289. Propanethiol
[107-03-9]

C_3H_8S

PrSH

Calcium hypochlorite
See Calcium hypochlorite: Organic sulfur compounds
See other ALKANETHIOLS

†1290. 2-Propanethiol
[75-33-2]

C_3H_8S

Me$_2$CHSH

See other ALKANETHIOLS

1291. Trimethylaluminium
[75-24-1]

C_3H_9Al

Me$_3$Al

HCS 1980, 938

Kirk-Othmer 1963, Vol. 2, 40
Extremely pyrophoric, ignition delay 13 ms in air at 235°C/0.16 bar.

Dichlorodi-μ-chlorobis(pentamethylcyclopentadienyl)dirhodium
(Structure III, p. S-2)
See Dichlorodi-μ-chlorobis(pentamethylcyclopentadienyl)dirhodium: Air, Alkyl-metals

Other reactants
Yoshida, 1980, 257
MRH values calculated for 15 combinations, largely with oxidants, are given.

Tetrafluorobenzene-1,4-diol
Spence, R., *Chem. Eng. News*, 1996, **74**(21), 4
The dried product of reaction of approximately equimolar quantities of the diol and methyl aluminium exploded when crushed with a spatula.
See METAL DERIVATIVES OF ORGANOFLUORINE COMPOUNDS
See entry TRIALKYLALUMINIUMS
See other ALKYLMETALS

1292. Trimethyldialuminium trichloride (Aluminium chloride–trimethyl-aluminium complex)
[12542-85-7]

$C_3H_9Al_2Cl_3$

Me$_3$Al.AlCl$_3$

HCS 1980, 636

See entry ALKYLALUMINIUM HALIDES

1293. Ethylmethylarsine
[689-93-0] C_3H_9As

EtAsHMe

Leleu, *Cahiers*, 1977, (88), 362
It ignites in air.
See other ALKYLNON-METAL HYDRIDES

1294. Trimethylarsine
[593-88-4] C_3H_9As

Me₃As

Air, or Halogens
  Sidgwick, 1950, 762, 769
  It inflames in air, and interaction with halogens is violent.
See other ALKYLNON-METALS

1295. Trimethylborane
[593-90-8] C_3H_9B

Me₃B

Stock, A. *et al., Ber.*, 1921, **54**, 535
The gas ignites in air.
See Chlorine: Trialkylboranes
See other ALKYLBORANES

†1296. Trimethyl borate
[121-43-7] $C_3H_9BO_3$

(MeO)₃B

1297. 2-(Tricarbonylferra)hexaborane(10)
[75952-58-8] $C_3H_9B_5FeO_3$

(Complex Structure)

Preparative hazard
  See Pentaborane(9): Pentacarbonyliron, Pyrex glass

1298. Trimethylbismuthine
[593-91-9] C_3H_9Bi

Me₃Bi

1. Coates, 1967, Vol. 1, 536
2. Sorbe, 1968, 157

452

It ignites in air, but is unreactive with water [1], and explodes at 110°C [2].
See TRIALKYLBISMUTHS *See other* ALKYLMETALS

1299. Trimethylsulfoxonium bromide
[25596-24-1]   C₃H₉BrOS

$$Me_3S^+O\ Br^-$$

1. Scaros, M. G. *et al., Chem. Brit.*, 1973, **9**, 523; private comm., 1974
2. Scaros, M. G. *et al., Proc. 7th Conf. Catal. Org. Synth. 1978*, 1980, 301–309
3. Dryden, H. L. *et al.*, US Pat. 4 141 920, 1979

Preparation of the title compound by interaction of dimethyl sulfoxide and bromo-methane, sealed into a resin-coated glass bottle and heated at 65°C, led to an explo-sion after 120 h. The isolated salt thermally decomposes above 180°C to produce formaldehyde and a residue of methanesulfonic acid. In solution in dimethyl sulfoxide, the salt begins to decompose after several hours at 74–80°C (after exposure to light), the exothermic reaction accelerating with vigorous evolution of vapour (including formaldehyde and dimethyl sulfide), the residue of methanesul-fonic acid finally attaining 132°C. Some white solid, probably poly-formaldehyde, is also produced. The explosion seems likely to have been a pressure-burst of the container under excessive internal pressure from the decomposition products. A safe procedure has been developed in which maximum reaction temperatures and times of 65°C and 55 h are used. Once prepared, the salt should not be redis-solved in dimethyl sulfoxide. In contrast, trimethylsulfoxonium iodide appears simply to dissolve on heating rather than decomposing like the bromide [1]. The decomposition mechanism involves isomerisation of some of the product to dimethylmethoxysulfonium bromide, which then degrades to formaldehyde, dimethyl sulfide and hydrogen bromide, and the latter catalyses further decom-position reactions [2]. Use of orthoformate or orthocarbonate esters as bromine and hydrogen bromide scavengers, and a cooling/pressure venting system avoids the possibility of exothermic decomposition of the reaction mixture [3].
See other GAS EVOLUTION INCIDENTS, GLASS INCIDENTS

1300. *N*-Bromotrimethylammonium bromide(?)
[ ]   C₃H₉Br₂N

$$BrN^+Me_3\ Br^-$$

See Bromine: Trimethylamine
See other *N*-HALOGEN COMPOUNDS

1301. Trimethylsilyl chlorochromate
[102488-24-4]   C₃H₉ClCrO₃Si

$$Me_3SiOCrO_2Cl$$

Aizpurna, J. M. *et al., Tetrahedron Lett.*, 1983, **24**, 4367

The oxidant appears safe when prepared and used in dichloromethane solution, but attempted isolation by distillation led to a violent explosion.
See other OXIDANTS

1302. Trimethylsilyl perchlorate
[18204-79-0]   C₃H₉ClO₄Si

$$Me_3SiOClO_3$$

See entry ORGANOSILYL PERCHLORATES *See other* NON-METAL PERCHLORATES

1303. Trimethylsulfonium chloride
[3086-29-1]   C₃H₉ClS

$$Me_3S^+Cl^-$$

Preparative hazard
Byrne, B. *et al., Tetrahedron Lett.*, 1986, **27**, 1233–1236
During its preparation by heating dimethyl sulfide and methyl chloroformate in a glass pressure bottle at 80°C, it is essential to interrupt the reaction after 4 h, cool and release the internal pressure of carbon dioxide, before recapping the bottle and heating for a further 22 h. This avoids the possibility of excessive pressure build-up and failure of the cap seal.
See other GAS EVOLUTION INCIDENTS, GLASS INCIDENTS

†1304. Chlorotrimethylsilane
[75-77-4]   C₃H₉ClSi

$$ClSiMe_3$$

HCS 1980, 941

Preparative hazard
See Chlorine: Antimony trichloride, Tetramethylsilane

Hexafluoroisopropylideneaminolithium
See Hexafluoroisopropylideneaminolithium: Non-metal halides

Water
Leleu, *Cahiers*, 1977, (88), 370
It reacts violently with water.
See other ALKYLHALOSILANES

1305. Trimethylgallium
[1445-79-0]   C₃H₉Ga

$$Me_3Ga$$

Air, or Water
Sidgwick, 1950, 461

454

It ignites in air, even at $-76°C$, and reacts violently with water.
See other ALKYLMETALS

†1306. Iodotrimethylsilane
[16029-98-4]   C₃H₉ISi

$$Me_3SiI$$

See other ALKYLHALOSILANES, IODINE COMPOUNDS

1307. Trimethylindium
[3385-78-2]   C₃H₉In

$$Me_3In$$

1. Bailar, 1973, Vol. 1, 1116
2. Houben-Weyl, 1970, Vol. 13.4, 351

Trimethylindium and other lower-alkyl derivatives ignite in air [1], including the 2:1 complex with diethyl ether [2].
See other ALKYLMETALS

†1308. Isopropylamine (2-Propanamine)
[75-31-0]   C₃H₉N

$$Me_2CHNH_2$$

FPA H72, 1978; *HCS 1980*, 577

1-Chloro-2,3-epoxypropane
See 1-Chloro-2,3-epoxypropane: Isopropylamine

Perchloryl fluoride
See Perchloryl fluoride: Nitrogenous bases
See other ORGANIC BASES

†1309. Propylamine (Propanamine)
[107-10-8]   C₃H₉N

$$PrNH_2$$

HCS 1980, 796

Triethynylaluminium
See Triethynylaluminium: Diethyl ether, etc.
See other ORGANIC BASES

†1310. Trimethylamine (*N,N*-Dimethylmethanamine)
[75-50-3]   C₃H₉N

$$Me_3N$$

(*MCA SD-57*, 1955); *HCS 1980*, 939 (anhydrous), 940 (40% solution)

Bromine
See Bromine: Trimethylamine

Ethylene oxide   MRH 3.72/100
See Ethylene oxide: Trimethylamine

Other reactants
Yoshida, 1980, 256
MRH values calculated for 14 combinations, largely with oxidants, are given.
See other ORGANIC BASES

1311. 1-Amino-2-propanol
[78-96-6] C_3H_9NO

$$H_2NCH_2CHOHCH_3$$

Cellulose nitrate
See CELLULOSE NITRATE: amines

2,4-Hexadienal
See 2,4-Hexadienal: 1-Amino-2-propanol
See other ORGANIC BASES

†1312. 2-Methoxyethylamine
[109-85-3] C_3H_9NO

$$MeOC_2H_4NH_2$$

Energy of decomposition (in range 140–200°C) measured as 0.23 kJ/g.
See entry THERMOCHEMISTRY AND EXOTHERMIC DECOMPOSITION (reference 2)
See other AMINOMETHOXY COMPOUNDS

1313. Trimethylamine oxide
[1184-78-7] C_3H_9NO

$$Me_3N \rightarrow O$$

1. *Inorg. Synth.*, 1989, **25**, 107
2. Ringel, C., *Z. Chem.*, 1969, **9**, 188
3. Editor's comments
Energy of decomposition (in range 160–280°C) measured as 1.455 kJ/g.
   The assessors experienced an explosion while drying the oxide in ethyl ether. Rather drastic precautions are recommended in handling it [1]. A preparation, allowed to stand for a week rather than the day specified, exploded during concentration [2]. Amine oxides from the standard preparation are inclined to retain hydrogen peroxide of 'hydration' unless it is destroyed during work-up. The peroxidate (or diperoxidate) of dimethylamine oxide would be expected to be far more dangerous than the oxide itself [3].
See entry THERMOCHEMISTRY AND EXOTHERMIC DECOMPOSITION (reference 2)
See other N-OXIDES

†1314. Trimethylsilyl azide (Azidotrimethylsilane)
[4648-54-8]   C₃H₉N₃Si

$$Me_3SiN_3$$

1. Sundermeyer, W., *Chem. Ber.*, 1963, **96**, 1293
2. Thayer, J. S. *et al., Inorg. Chem.*, 1964, **3**, 889
3. West, R. *et al., Chem. Eng. News,* 1984, **62**(24), 4

A standard literature method [1,2] had been used frequently and uneventfully to prepare the azide from trimethylsilyl chloride and sodium azide in presence of aluminium chloride as catalyst. A recent duplication led to a violent detonation during distillation of the product, and this was attributed to carry-over of traces of aluminium azides into the distillation flask. Precautions are detailed [3].

Rhenium hexafluoride
1. Fawcett, J. *et al., J. Chem. Soc., Chem. Comm.*, 1982, 958
2. Fawcett, J. *et al., J. Chem. Soc., Dalton Trans.*, 1987, 567–571

Interaction at −60°C in 1,1,2-trichloroethane to give rhenium nitride tetrafluoride is violent and an explosion is possible even with repeated cooling with liquid nitrogen. The final warming to 0°C is also hazardous [1]. If the reactants are mixed at −196°C, at any stage of warming, the reaction can become extremely vigorous and detonate, even with maximum cooling control [2].

Selenium halides
Kennett, F. A. *et al., J. Chem. Soc., Dalton Trans.*, 1982, 851–857

In attempted preparation of poly(selenium nitride), the black solid formed by interaction with selenium tetrabromide in acetonitrile exploded violently within 1 min at 0°C. The solid produced from diselenium tetrachloride in acetonitrile exploded at around 100°C, and in dichloromethane the product exploded in contact with a nickel spatula.

Tungsten hexafluoride
See Azidopentafluorotungsten
See related NON-METAL AZIDES

1315. Trimethyl phosphite
[121-45-9]   C₃H₉O₃P

$$(MeO)_3P$$

Magnesium perchlorate
See Magnesium perchlorate: Trimethyl phosphite

Trimethylplatinum(IV) azide tetramer
See Dodecamethyltetraplatinum(IV)azide: Trimethyl phosphite
See other PHOSPHORUS ESTERS

1316. Trimethyl thiophosphate (Trimethyl phosphorothioate)
[152-18-1] $C_3H_9O_3PS$

$$(MeO)_3P(:S)$$

Chlorine
See Chlorine: Trimethyl thionophosphate
See other PHOSPHORUS ESTERS

1317. Titanium(III) methoxide
[7245-18-3] $C_3H_9O_3Ti$

$$Ti(OMe)_3$$

Bailar, 1973, Vol. 3, 384
A pyrophoric solid.
See other METAL ALKOXIDES

1318. Trimethyl phosphate
[512-56-1] $C_3H_9O_4P$

$$(MeO)_3P(:O)$$

Anon., *ABCM Quart. Safety Summ.*, 1953, **25**, 3
The residue from a large scale atmospheric pressure distillation of trimethyl phosphate exploded violently. This was attributed to rapid decomposition of the ester, catalysed by the acidic degradation products, with evolution of gaseous hydrocarbons. It is recommended that only small batches of alkyl phosphates should be vacuum distilled and in presence of magnesium oxide to neutralise any acid by-products, and to suppress the acid catalysed reaction.
See other CATALYTIC IMPURITY INCIDENTS, PHOSPHORUS ESTERS

1319. Trimethylphosphine
[594-09-2] C_3H_9P

$$Me_3P$$

1. Personal experience
2. Labinger, J. A. *et al., Inorg. Synth.*, 1978, **18**, 63
It may ignite in air [1]. It is readily regenerated by heating its air-stable complex with silver iodide [2], which became commercially available in 1983.
See other ALKYLPHOSPHINES

1320. Trimethylantimony (Trimethylstibine)
[594-10-5] C_3H_9Sb

$$Me_3Sb$$

Von Schwartz, 1918, 322
It ignites in air.

Halogens
Sidgwick, 1950, 777
Interaction is violent.

2-Iodoethanol
See 2-Hydroxyethyltrimethylstibonium iodide
See other ALKYLMETALS

1321. Trimethylthallium
[3003-15-4]   C$_3$H$_9$Tl

$$Me_3Tl$$

Sidgwick, 1950, 463
It ignites in air, and is liable to explode violently above 90°C.

Diethyl ether
Leleu, *Cahiers*, 1977, (88), 371
The complex with ether explodes at 0°C.
See other ALKYLMETALS

1322. Trimethylammonium perchlorate
[15576-35-9]   C$_3$H$_{10}$ClNO$_4$

$$Me_3N^+H \ ClO_4^-$$

Jain, S. R. *et al., Chem. Abs.*, 1982, **96**, 220110
The salt burns more readily as a propellant than does ammonium perchlorate.
See other PERCHLORATE SALTS OF NITROGENOUS BASES

†1323. Trimethylhydroxylammonium perchlorate (Trimethylamine oxide perchlorate)
[22755-36-8]   C$_3$H$_{10}$ClNO$_5$

$$Me_3N^+OH \ ClO_4^-$$

Hofmann, K. A. *et al., Ber.*, 1910, **43**, 2624
The salt, formed from trimethylamine oxide and perchloric acid, explodes on heating, or under a hammer blow.
See other PERCHLORATE SALTS OF NITROGENOUS BASES, REDOX COMPOUNDS

1324. Dihydrazinecadmium(II) malonate (Bis(hydrazine)[propanedioato(2−)] cadmium)
[159793-64-3]   C$_3$H$_{10}$CdN$_4$O$_4$

See HYDRAZINE METAL MALONATES AND SUCCINATES

†1325. 1,2-Diaminopropane (1,2-Propanediamine)
[78-90-0] $C_3H_{10}N_2$
$$H_2NCH_2CHNH_2Me$$

HCS 1980, 798

See other ORGANIC BASES

†1326. 1,3-Diaminopropane (1,3-Propanediamine)
[109-76-2] $C_3H_{10}N_2$
$$H_2NC_3H_6NH_2$$

HCS 1980, 361

See other ORGANIC BASES

1327. S-Ethylisothiouronium hydrogen sulfate (Ethyl carbamimoniothioate hydrogen sulfate)
[22722-03-8] $C_3H_{10}N_2O_4S_2$
$$EtSC(:N^+H_2)NH_2\ HSO_4^-$$

Chlorine
See Chlorine: Nitrogen compounds (reference 4)

1328. Dihydrazinenickel(II) malonate (Bis(hydrazine)[propanedioato(2−)]nickel)
[159793-62-1] $C_3H_{10}N_4NiO_4$

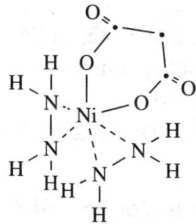

See HYDRAZINE METAL MALONATES AND SUCCINATES

1329. Trimethylplatinum hydroxide
[14477-33-9] $C_3H_{10}OPt$
$$Me_3PtOH$$

Höchstetter, M. N. *et al., Inorg. Chem.*, 1969, **8**, 400
The compound (originally reported as tetramethylplatinum) and usually obtained in admixture with an alkoxy derivative, detonates on heating.
See other PLATINUM COMPOUNDS

1330. Trimethylsilyl hydroperoxide
[18230-75-6] $C_3H_{10}O_2Si$

$$Me_3SiOOH$$

Buncel, E. *et al., J. Chem. Soc.*, 1958, 1551
It decomposes rapidly above 35°C and may have been involved in an explosion
during distillation of the corresponding peroxide, which is stable to 135°C.
See related ORGANOMETALLIC PEROXIDES

1331. Propylsilane
[13154-66-0] $C_3H_{10}Si$

$$PrSiH_3$$

Leleu, *Cahiers*, 1977, (88), 369
It ignites in air.
See other ALKYLSILANES

1332. Trimethylgermylphosphine
[20519-92-0] $C_3H_{11}GeP$

$$Me_3GePH_2$$

Oxygen
Dahl, A. R. *et al., Inorg. Chem.*, 1975, **14**, 1093
A solution in chloroform at −45°C reacts rapidly with oxygen, and at oxygen pres-
sures above 25 mbar combustion occurs. Dimethyldiphosphinogermane behaves
similarly.
See related ALKYLMETALS, PHOSPHINES

1333. Di(hydroperoxy)trimethylantimony(V)
[ ] $C_3H_{11}O_4Sb$

$$Me_3Sb(OOH)_2$$

Dodd, M. *et al., Appl. Organomet. Chem.*, 1992, **6**(2), 207
This compound, prepared from trimethylstibine and hydrogen peroxide, exploded
violently when heated under vacuum.
See other ORGANOMETALLIC PEROXIDES

1334. Trimethyldiborane
[21107-27-7] $C_3H_{12}B_2$

$$Me_2B{:}H_2{:}BHMe$$

Leleu, *Cahiers*, 1977, (88), 366
It ignites in air.
See other ALKYLBORANES

1335. 2,4,6-Trimethylborazine (*B*-Trimethylborazine)
[5314-85-2] $C_3H_{12}B_3N_3$

Nitryl chloride
See Nitryl chloride: *B*-Trimethylborazine
See related BORANES

1336. Aqua-1,2-diaminopropanediperoxochromium(IV) dihydrate
[17185-68-1] $C_3H_{12}CrN_2O_5.2H_2O$
$$[H_2O(C_3H_{10}N_2)Cr(O_2)_2].2H_2O$$

House, D. A. *et al., Inorg. Chem.*, 1966, **5**, 40
Several preparations of the dihydrate exploded spontaneously at 20–25°C, and the
aqua complex explodes at 88–90°C during slow heating.
See other AMMINECHROMIUM PEROXOCOMPLEXES

1337. Beryllium tetrahydroborate–trimethylamine
[ ] $C_3H_{17}B_2BeN$
$$Be(BH_4)_2.NMe_3$$

Air, or Water
Burg, A. B. *et al., J. Amer. Chem. Soc.*, 1940, **62**, 3427
The complex ignites in contact with air or water.
See other COMPLEX HYDRIDES

1338. Cadmium hydrazinium tris(hydrazinecarboxylate) trihydrate
[ ] $C_3H_{20}CdN_8O_9$
Dhas, N. A. *et al., Chem. Abs.*, 1994, **121**, 314276d
The hydrazinecarboxylate decomposed explosively on heating to 120°C.

1339. Triiodocyclopropenium iodide
[99796-78-8] C_3I_4

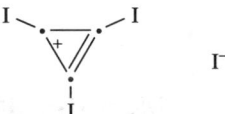

Weiss, R. *et al., Angew. Chem. (Intern. Ed.)*, 1986, **25**, 103
This first example of an ionic halocarbon is extremely explosive, not only when
dry, but also upon dissolution in polar solvents (ethanol, acetonitrile). Only 1 mmol

of the salt should be prepared in a single experiment, and only samples moist with dichloromethane should be used or handled.

See HALOCARBONS
See other IODINE COMPOUNDS

1340. Dipotassium diazirine-3,3-dicarboxylate
[76429-97-5]   C$_3$K$_2$N$_2$O$_4$

See Diazirine-3,3-dicarboxylic acid
See other DIAZIRINES

1341. Oxopropanedinitrile (Carbonyl dicyanide)
[1115-12-4]   C$_3$N$_2$O

$$O:C(CN)_2$$

Water
Martin, E. L., *Org. Synth.*, 1971, **51**, 70
The nitrile reacts explosively with water.
See related ACYL HALIDES *See other* CYANO COMPOUNDS

1342. Carbonyl diisothiocyanate
[6470-09-3]   C$_3$N$_2$OS$_2$

$$O:C(N=C=S)_2$$

Dimethyl sulfoxide
See Dimethyl sulfoxide: Carbonyl diisothiocyanate
See related CYANO COMPOUNDS

1343. Phosphorus tricyanide
[1116-01-4]   C$_3$N$_3$P

$$P(C\equiv N)_3$$

1. Absolom, R. *et al., Chem. & Ind.*, 1967, 1593
2. Smith, T. D. *et al., Chem. & Ind.*, 1967, 1969
3. Mellor, 1971, Vol. 8, Suppl. 3, 583

During vacuum sublimation at around 100°C, explosions occurred on 3 occasions. These were attributed to slight leakage of air into the sublimer. The use of joints remote from the heating bath, a high-melting grease, and a cooling liquid other than water (to avoid rapid evolution of hydrogen cyanide in the event of breakage) are recommended, as well as working in a fume cupboard [1]. However, the later publication suggests that the silver cyanide used in the preparation could be the cause of the explosions. On long keeping, it discolours and may then contain the nitride (fulminating silver). Silver cyanide precipitated from slightly acidic solution and stored in a dark bottle is suitable [2]. Phosphorus tricyanide ignites in air if touched with a warm glass rod, and reacts violently with water.

See related NON-METAL HALIDES AND THEIR OXIDES
See other CYANO COMPOUNDS

1344. Diazomalononitrile (Diazodicyanomethane)
[1618-08-2] C_3N_4

$$N_2C(C\equiv N)_2$$

Ciganek, E., *J. Org. Chem.*, 1965, **30**, 4200
Small (mg) quantities melt at 75°C, but larger amounts explode. It also has borderline sensitivity towards static electricity and must be handled with full precautions.
See other CYANO COMPOUNDS, DIAZO COMPOUNDS

1345. 5-Cyano-4-diazo-4*H*-1,2,3-triazole
[16968-06-2] C_3N_6

The crystalline solid explodes at the m.p., 125–6°C.
See entry DIAZOAZOLES
See other DIAZO COMPOUNDS, HIGH-NITROGEN COMPOUNDS, TRIAZOLES

1346. Cyanodiazoacetyl azide
[115057-40-4] C_3N_6O

$$NCC(N_2)CO.N_3$$

Kobolov, M. Yu. *et al.*, *Chem. Abs.*, 1988, **109**, 22908
It explodes on melting.
See other ACYL AZIDES, CYANO COMPOUNDS, DIAZO COMPOUNDS, HIGH-NITROGEN COMPOUNDS

1347. Diazidomalononitrile (Diazidodicyanomethane)
[67880-21-1]

C_3N_8

$$(N_3)_2C(C{\equiv}N)_2$$

MCA Case History No. 820

An ethereal solution of some 100 g of the crude nitrile was allowed to spontaneously evaporate and crystallise. The crystalline slurry so produced exploded violently without warning. Previously such material had been found not to be shock-sensitive to hammer blows, but dry recrystallised material was very shock-sensitive. Traces of free hydrogen azide could have been present, and a metal spatula had been used to stir the slurry, so metal azides could have been formed.
See other CYANO COMPOUNDS, ORGANIC AZIDES

1348. 2,4,6-Triazido-1,3,5-triazine
[5637-83-2]

C_3N_{12}

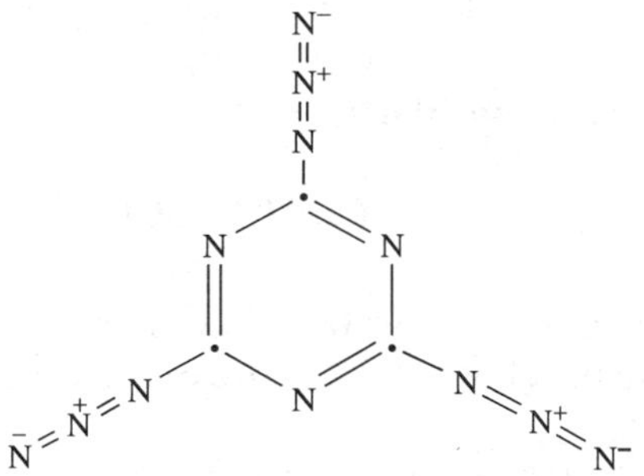

1. Ott, *Ber.*, 1921, **54**, 183
2. Simmonds, R. J. *et al.*, *J. Chem. Soc., Perkin Trans. 1*, 1982, 1824
This polyazide (82.3% N) explodes on impact, shock or rapid heating to 170–180°C [1]. When preparing 2,4-diazido-6-dimethylamino-1,3,5-triazine from the 2,4-dichloro compound and sodium azide, there is the possibility of forming the triazido derivative, which detonates violently when touched. Reactions involving organic chlorides and excess sodium azide are extremely dangerous [2].
See other HIGH-NITROGEN COMPOUNDS, ORGANIC AZIDES

†1349. Propadiene-1,3-dione ('Carbon suboxide')
[504-64-3]

C_3O_2

$$O{=}C{=}C{=}C{=}O$$

See related ACID ANHYDRIDES

1350. Propadienedithione (Carbon subsulfide)
[627-34-9] C_3S_2

$$S=C=C=C=S$$

Mellor, 1940, Vol. 6, 88
A red oil decomposing to a black polymer, explosively if heated rapidly to 100–120°C. Probably highly endothermic.
See other ENDOTHERMIC COMPOUNDS, NON-METAL SULFIDES

1351. 1,4-Dibromo-1,3-butadiyne
[36333-41-2] C_4Br_2

$$BrC\equiv CC\equiv CBr$$

Petterson, R. C. *et al., Chem. Eng. News*, 1977, **55**(48), 36
An explosion temperature has been published, but a solvent-free sample standing unheated under nitrogen exploded after 1 h.
See other HALOACETYLENE DERIVATIVES

1352. Heptafluorobutyryl hypochlorite
[71359-62-1] $C_4ClF_7O_2$

$$C_3F_7CO.OCl$$

Tari, I. *et al., Inorg. Chem.*, 1979, **18**, 3205–3208
It is thermally unstable above 22°C and explosive in the gas phase at pressures above 27–62 mbar.
See entry ACYL HYPOHALITES (reference 1)

1353. 1,4-Dichloro-1,3-butadiyne
[51104-87-1] C_4Cl_2

$$ClC\equiv CC\equiv CCl$$

Sorbe, 1968, 50
It explodes above 70°C.
See other HALOACETYLENE DERIVATIVES

1354. Copper(I) chloroacetylide
[ ] $C_4Cl_2Cu_2$

$$ClC\equiv CCuCuC\equiv CCl$$

Kirk-Othmer, 1964, Vol. 5, 204
Explosive.
See other HALOACETYLENE DERIVATIVES, HEAVY METAL DERIVATIVES

1355. 4,5-Dichloro-3,3,4,5,6,6-hexafluoro-1,2-dioxane
[ ] $C_4Cl_2F_6O_2$

Haszeldine, R. N. *et al., J. Chem. Soc.*, 1959, 1089
Formed in the controlled oxidation of chlorotrifluoroethylene, it explodes violently on heating.
See other CYCLIC PEROXIDES

1356. Mercury bis(chloroacetylide)
[64771-59-1] C_4Cl_2Hg

$$(ClC{\equiv}C)_2Hg$$

Whitmore, 1921, 119
It explodes fairly violently above the m.p., 185°C.
See other HALOACETYLENE DERIVATIVES, MERCURY COMPOUNDS, METAL ACETYLIDES

1357. 3,4-Dichloro-2,5-dilithiothiophene
[29202-04-8] $C_4Cl_2Li_2S$

Gilman, H., private comm., 1971
The dry solid is explosive, though relatively insensitive to shock.
See other ORGANOLITHIUM REAGENTS

1358. Tetracarbonylmolybdenum dichloride
[15712-13-7] $C_4Cl_2MoO_4$

$$(OC)_4MoCl_2$$

Newton, W. E. *et al., Inorg. Synth.*, 1978, **18**, 54
If stored at ambient temperature, the complex decomposes to a dark pyrophoric powder.
See related CARBONYLMETALS

1359. Dichloromaleic anhydride
[1122-17-4]

$C_4Cl_2O_3$

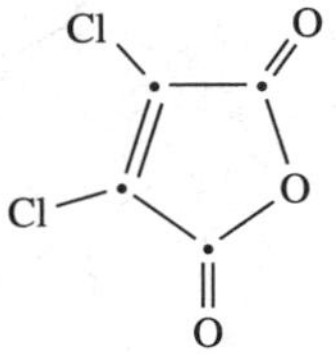

Sodium chloride, Urea
  Bott, D. C., private comm., 1980
  To prepare dichloromaleimide, the 3 components are melted together with stirring. At 118°C the vigorous exothermic reaction sets in, and rapid ice cooling must then be applied to prevent explosion.
  See other ACID ANHYDRIDES

1360. 1,1,4,4-Tetrachlorobutatriene
[19792-18-8]

C_4Cl_4

$$Cl_2C{=}C{=}C{=}CCl_2$$

Heinrich, B. *et al., Angew. Chem. (Intern. Ed.)*, 1968, **7**, 375
  One of the by-products formed by heating this compound is an explosive polymer.
  See other HALOALKENES *See related* DIENES

1361. Bis(trichloroacetyl) peroxide
[2629-78-9]

$C_4Cl_6O_4$

$$Cl_3CCO.OOCO.CCl_3$$

1. Swern, 1971, Vol. 2, 815
2. Miller, W. T., *Chem. Abs.*, 1952, **46**, 7889e
  Pure material explodes on standing at ambient temperature [1], but the very shock-sensitive solid may be stored safely in trichlorofluoromethane solution at −20°C [2].
  See other DIACYL PEROXIDES

1362. Tetracarbon monofluoride
[12774-81-1]

C_4F

$$C_4F$$

Bailar, 1973, Vol. 1, 1271
  Though generally inert, rapid heating causes deflagration.
  See related HALOALKANES, NON-METAL HALIDES

1363. 1,1,4,4-Tetrafluorobutatriene
[2252-95-1] C_4F_4

$$F_2C{=}C{=}C{=}CF_2$$

Martin, E. L. *et al., J. Amer. Chem. Soc.*, 1959, **81**, 5256
In the liquid state, it explodes above $-5°C$.
See other HALOALKENES *See related* DIENES

1364. Perfluorobutadiene
[87-68-3] C_4F_6

$$F_2C{=}CFCF{=}CF_2$$

Feast, W. J., personal communication, 1993.
Attempts to use this material as a monomer are not recommended as reaction is uncontrollable and it is liable to explode.

Bromine perchlorate
  See Bromine perchlorate: Perfluorobutadiene
  See other DIENES, HALOALKENES

1365. Bis(trifluoroacetoxy)xenon
[21961-22-8] $C_4F_6O_4Xe$

$$Xe(OOCCF_3)_2$$

Kirk Othmer, 1980, Vol. 12, 293
Detonates easily and should be handled with extreme care.
See other XENON COMPOUNDS

1366. Trifluoroacetic anhydride
[407-25-0] $C_4F_6O_3$

$$F_3CCO.OCO.CF_3$$

Dimethyl sulfoxide
  See Dimethyl sulfoxide: Trifluoroacetic anhydride

Nitric acid, 1,3,5-Triacetylhexahydro-1,3,5-triazine
  See Nitric acid: 1,3,5-Triacetylhexahydrotriazine

Nitric acid, 1,3,5-Triazine
  See Nitric acid: 1,3,5-Triazine, etc.
  See other ACID ANHYDRIDES

1367. Bis(trifluoroacetyl) peroxide
[383-73-3] $C_4F_6O_4$

$$F_3CCO.OOCO.CF_3$$

Swern, 1971, Vol. 2, 815

Pure material explodes on standing at ambient temperature.
See other DIACYL PEROXIDES

1368. Heptafluorobutyryl nitrite
[663-25-2] $C_4F_7NO_3$

$$C_3F_7CO.ON:O$$

Banks, R. E. *et al., J. Chem. Soc. (C)*, 1966, 1351
This may be explosive like its lower homologues, and suitable precautions are
neccessary during heating or distillation.
See other ACYL NITRITES

1369. Heptafluorobutyryl hypofluorite
[ ] $C_4F_8O_2$

$$C_3F_7CO.OF$$

Cady, G. H. *et al., J. Amer. Chem. Soc.*, 1954, **76**, 2020
The vapour slowly decomposes at ambient temperature, but will decompose explo-
sively on spark initiation.
See other ACYL HYPOHALITES

1370. Perfluoro-*tert*-nitrosobutane (Tris(trifluoromethyl)nitrosomethane)
[354-93-8] C_4F_9NO

$$(F_3C)_3CN:O$$

Nitrogen oxides
  Sterlin, S. R. *et al.*, Russ. Pat. 482 432, 1975
The danger of explosions during oxidation of the nitroso compound to perfluoro-
tert-butanol with nitrogen oxides and subsequent hydrolysis, was reduced by working in a flow system at 160–210°C with 8–10% of nitrogen oxides in air, and
using conc. sulfuric acid for hydrolysis.
See other NITROSO COMPOUNDS

1371. Decafluorobutyramidine
[41409-50-1] $C_4F_{10}N_2$

$$C_3F_7C(:NF)NF_2$$

Ross, D. L. *et al., J. Org. Chem.*, 1970, **35**, 3096–3097
A shock-sensitive explosive
See other N,N,N''-TRIFLUOROAMIDINES

1372. Decafluoro-2,5-diazahexane 2,5-dioxyl
[36525-64-1] $C_4F_{10}N_2O_2$

$$F_3CN(O\cdot)C_2F_4N(O\cdot)CF_3$$

Preparative hazard

Banks, R. E. *et al., J. Chem. Soc., Perkin Trans. 1*, 1974, 2535−2536
During an increased-scale preparation of the dioxyl by permanganate oxidation of the hydrolysate of a nitrosotrifluoromethane−tetrafluoroethylene−phosphorus trichloride adduct, an impurity in the dioxyl, trapped out at −96°C (? −196°C) and <2.5 mbar, caused a violent explosion to occur when the trap content was allowed to warm up. A procedure to eliminate the hazard is detailed.
See related N-OXIDES

1373. Perfluoro-*tert*-butyl peroxyhypofluorite
[66793-67-7] $C_4F_{10}O_2$

$$(F_3C)_3COOF$$

Yu, S.-L. *et al., Inorg. Chem.*, 1978, **17**, 2485
It explodes when warmed to 22°C in a sealed tube with a liquid phase present, but is apparently stable in the gas phase alone.
See related HYPOHALITES *See other* ORGANIC PEROXIDES

1374. 1,1,4,4-Tetrakis(fluoroxy)hexafluorobutane
[22410-18-0] $C_4F_{10}O_4$

$$(FO)_2CFC_2F_4CF(OF)_2$$

Sekiya, A. *et al., Inorg. Chem.*, 1980, **19**, 1329
It explodes energetically at −20°C.
See other BIS(FLUOROOXY)PERHALOALKANES

1375. Di[bis(trifluoromethyl)phosphido]mercury
[ ] $C_4F_{12}HgP_2$

$$[(F_3C)_2P]_2Hg$$

Grobe, J. *et al., Angew. Chem. (Intern. Ed.)*, 1972, **11**, 1098
It ignites in air.
See related ALKYLPHOSPHINES *See other* MERCURY COMPOUNDS

1376. Sodium tetracarbonylferrate(2−)
[14878-31-0] $C_4FeNa_2O_4$

$$Na_2[Fe(CO)_4]$$

1. Collman, J. P., *Accts. Chem. Res.*, 1975, **8**, 343
2. Finke, R. G. *et al., Org. Synth.*, 1979, **59**, 102−112

It is extremely oxygen-sensitive and ignites in air [1]. The salt solvated with 1.5 mols of dioxane is similarly reactive [2].

See related CARBONYLMETALS

1377. Germanium isocyanate (Tetraisocyanatogermane)
[4756-66-5] $C_4GeN_4O_4$

$$Ge(N=C=O)_4$$

Water
Johnson, D. H., *Chem. Rev.*, 1951, **48**, 287
The rate of exothermic hydrolysis becomes dangerously fast above 80°C.
See related CYANO COMPOUNDS, METAL CYANATES

1378. Dichloromaleimide (3,4-Dichloro-2,5-pyrrolidinedione)
[1193-54-0] $C_4HCl_2NO_2$

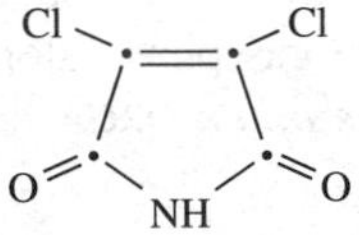

Preparative hazard
See Dichloromaleic anhydride: Sodium chloride, Urea

1379. Poly(1-pentafluorothio-1,2-butadiyne)
[84864-30-2] $(C_4HF_5S)_n$

See Structure

Kovacina, T. A. *et al., Ind. Eng. Chem., Prod. Res. Dev.*, 1983, **22**, 170–172
The insoluble polymer formed in the liquid phase was pressure- and shock-sensitive and should not be heated above 25°C. Cutting with a razor or breaking the polymer causes detonation and ignition.
See related HALOALKENES, DIENES

1380. Perfluoro-*tert*-butanol
[2378-02-1] C_4HF_9O

$$(F_3C)_3COH$$

Preparative hazard
See Perfluoro-*tert*-nitrosobutane: Nitrogen oxides

1381. 1-Iodo-1,3-butadiyne
[6088-91-1] C_4HI

$$IC{\equiv}CC{\equiv}CH$$

1. Schluhbach, H. H. *et al.*, *Ann.*, 1951, **573**, 118, 120
2. Kloster-Jensen, E., *Tetrahedron*, 1966, **22**, 969

Crude material exploded violently at 35°C during attempted vacuum distillation, and temperatures below 30°C are essential for safe handling [1]. A sample of pure material exploded on scratching under illumination [2].

See other HALOACETYLENE DERIVATIVES, IRRADIATION DECOMPOSITION INCIDENTS

1382. Potassium hydrogen acetylenedicarboxylate
[928-84-1] C_4HKO_4

$$KOCO.C{\equiv}CCO.OH$$

Energy of decomposition (in range 170–360°C) measured as 0.775 kJ/g by DSC, and T_{ait24} was determined as 202°C by adiabatic Dewar tests, with an apparent energy of activation of 155 kJ/mol.

See entry THERMOCHEMISTRY AND EXOTHERMIC DECOMPOSITION (reference 2)
See other ACETYLENIC COMPOUNDS *See related* ORGANIC ACIDS

1383. Cyanoform (Tricyanomethane)
[454-50-2] C_4HN_3

$$HC(C{\equiv}N)_3$$

1. Stull, 1977, 19
2. Rodd, 1976, **IE**, 333

As expected for a tricyano compound, it is highly endothermic (ΔH_f° (s) +348.1 kJ/mol, 3.78 kJ/g), so its stability seems dubious [1]. However, because of isomerisation to the acidic dicyano-enimine $(NC)_2C{=}C{=}NH$, it is considerably stabilised and the silver and sodium salts surprisingly are described as stable [2].

See other CYANO COMPOUNDS, ENDOTHERMIC COMPOUNDS

1384. 2-Azido-3,5-dinitrofuran
[70664-49-2] $C_4HN_5O_5$

Sitzmann, M. E., *J. Heterocycl. Chem.*, 1979, **16**, 477–480

Very sensitive to heat and impact.
See other ORGANIC AZIDES, POLYNITROARYL COMPOUNDS

†1385. 1,3-Butadiyne
[460-12-8] C_4H_2

$$HC{\equiv}CC{\equiv}CH$$

1. Armitage, J. B. *et al., J. Chem. Soc.*, 1951, 44
2. Mushii, R. Y. *et al., Khim. Prom.*, 1963, **39**, 109
3. Moshkovich, F. B. *et al., Khim. Prom.*, 1965, **41**, 137–139
4. Anon., *Chem. Age*, 1951, **64**(1667), 955–958
5. Schilling, H. *et al.*, Ger. Pat. 860 212, 1956

Potentially very explosive, it may be handled and transferred by low temperature distillation. It should be stored at −25°C to prevent decomposition and formation of explosive polymers [1]. The critical pressure for explosion is 0.04 bar, but presence of 15–40% of diluents (acetylene, ammonia, carbon dioxide or nitrogen) will raise the critical pressure to 0.92 bar [2]. Further data on attenuation by inert diluents of the explosive decomposition of the diyne are available [3]. During investigation of the cause of a violent explosion in a plant for separation of higher acetylenes, the most important finding was to keep the concentration of 1,3-butadiyne below 12% in its mixtures. Methanol is a practical diluent [4]. The use of butane (at 70 mol%) or other diluents to prevent explosion of 1,3-butadiyne when heated under pressure has been claimed [5]. It polymerises rapidly above 0°C.

Arsenic pentafluoride
Russo, P. J. *et al., J. Chem. Soc., Chem. Comm.*, 1982, 53–54
Polymerisation of the diyne by arsenic pentafluoride at low temperature may be explosive if too rapid mixing in the gas phase, or too rapid warming of the solids from −196°C, is permitted for 200 mg quantities.
See other POLYMERISATION INCIDENTS

Silver nitrate
Grignard, 1935, Vol. 3, 181
Reaction with ammoniacal silver nitrate gives a very explosive friction-sensitive silver salt.
See METAL ACETYLIDES
See other ALKYNES, PEROXIDISABLE COMPOUNDS

1386. Poly(butadiyne)
[61565-16-0] $(C_4H_2)_n$

$$(\text{-CH}{=}\text{CHC}{\equiv}\text{C-})_n$$

Preparative hazard
See 1,3-Butadiyne: Arsenic pentafluoride
See related ALKYNES

1387. 1,2-Dibromo-1,2-diisocyanatoethane polymers
[51877-12-4] $(C_4H_2Br_2N_2O_2)_{2 \text{ or } 3}$

$$(O{=}N{=}CCHBrCHBrC{=}N{=}O)_{2 \text{ or } 3}$$

2-Phenyl-2-propyl hydroperoxide
See 2-Phenyl-2-propyl hydroperoxide: 1,2-Dibromo-1,2-diisocyanatoethane polymers
See related CYANO COMPOUNDS, ORGANIC ISOCYANATES

1388. 4-Nitrothiophene-2-sulfonyl chloride
[40358-04-1] $C_4H_2ClNO_4S_2$

Libman, D. D., private comm., 1968
After distilling the chloride up to 147°C/6 mbar, the residue decomposed vigorously.
See other NITROACYL HALIDES

1389. 1,1,2,3-Tetrachloro-1,3-butadiene
[1637-31-6] $C_4H_2Cl_4$

$$Cl_2C{=}CClCCl{=}CH_2$$

It autoxidises to an unstable peroxide.
See 3,3,4,5-Tetrachloro-3,6-dihydro-1,2-dioxin
See other DIENES, HALOALKENES, PEROXIDISABLE COMPOUNDS

1390. 3,3,4,5-Tetrachloro-3,6-dihydro-1,2-dioxin
[ ] $C_4H_2Cl_4O_2$

Akopyan, A. N. *et al., Chem. Abs.*, 1975, **82**, 111537
This autoxidation product of 1,1,2,3-tetrachloro-1,3-butadiene exploded during attempted vacuum distillation.
See other CYCLIC PEROXIDES

1391. Perfluorosuccinic acid (Tetrafluorobutanedioic acid)
[377-38-8] $C_4H_2F_4O_2$

$$(HOCO.CF_2-)_2$$

Caesium fluoride, Fluorine
See Fluorine: Caesium fluoride, etc.
See other ORGANIC ACIDS

1392. Heptafluorobutyramide
[662-50-0] $C_4H_2F_7NO$

$$F_7C_3CO.NH_2$$

Lithium tetrahydroaluminate
See Lithium tetrahydroaluminate: Fluoroamides

1393. Iron(II) maleate
[7705-12-6] $C_4H_2FeO_4$

$$Fe(-OCO.CH:)_2$$

Schwab, R. F. *et al., Chem. Eng. Prog.*, 1970, **66**(9), 53
The finely divided maleate, a by-product of phthalic anhydride manufacture, is
subject to rapid aerial oxidation above 150°C, and has been involved in plant fires.
See other HEAVY METAL DERIVATIVES, PYROPHORIC MATERIALS

1394. Potassium diethynylpalladate(2−)
[ ] $C_4H_2K_2Pd$

$$K_2[Pd(C\equiv CH)_2]$$

Air, or Water
Immediately pyrophoric in air, explosive decomposition with aqueous reagents;
the sodium salt is similar.
See entry COMPLEX ACETYLIDES

1395. Potassium diethynylplatinate(2−)
[ ] $C_4H_2K_2Pt$

(Complex Structure)

Air, or Water
Pyrophoric in air, explosive decomposition with water; the sodium salt is similar.
See entry COMPLEX ACETYLIDES

1396. Manganese(II) bis(acetylide)
[ ] C_4H_2Mn

$$Mn(C\equiv CH)_2$$

Bailar, 1973, Vol. 3, 855

A highly explosive compound.
See other METAL ACETYLIDES

1397. Fumarodinitrile
[766-42-1]   C₄H₂N₂

$$N{\equiv}CCH{=}CHC{\equiv}N \text{ (trans)}$$

Energy of decomposition (in range 340–380°C) measured as 0.454 kJ/g.
See entry THERMOCHEMISTRY AND EXOTHERMIC DECOMPOSITION (reference 2)
See other CYANO COMPOUNDS

†1398. Pyrimidine-2,4,5,6-(1*H*,3*H*)-tetrone ('Alloxan')
[50-71-5]   C₄H₂N₂O₄

1. Wheeler, A. S., and Bogert, M. T., *J. Amer. Chem,. Soc.*, 1910, **32**, 809
2. Gartner, R. A., *J. Amer. Chem. Soc.*, 1911, **33**, 85
3. Franklin, E. C., *J. Amer. Chem. Soc.*, 1901, **23**, 1362
4. Sorbe, 1968, 140

The tetrone slowly decomposes during storage at ambient temperature with generation of carbon dioxide. Two separate pairs of incidents involving bursting of bottles [1], and of pressure generation [2,3], were reported. It also explodes at temperatures above 170°C [4].
See Imidazoline-2,4-dithione
See other GAS EVOLUTION INCIDENTS

1399. 3-Diazoniopyrazolide-4-carboxamide
[102613-59-2]   C₄H₂N₄O

Cheng, C. C. *et al.*, *J. Med. Chem.*, 1986, **29**, 1546
Though the preparative method is suitable for large scale synthesis, the product must be handled with care. On rapid heating, the internal diazonium salt decomposes violently at 155–160°C accompanied by a sharp sound. The dried powder should not be scratched with a metal spatula or ground finely.
See other DIAZONIUM SALTS, HIGH-NITROGEN COMPOUNDS

1400. 2,6-Diazidopyrazine
[74273-75-9]

$C_4H_2N_8$

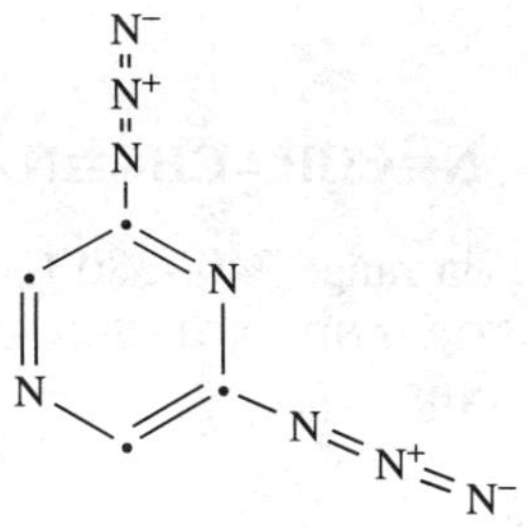

Shaw, J. T. *et al., J. Heterocycl. Chem.*, 1980, **17**, 14
It is heat- and impact-sensitive, exploding at 200°C, or under a hammer-blow.
See other HIGH-NITROGEN COMPOUNDS, ORGANIC AZIDES

1401. 3,3′-Azo-(1-nitro-1,2,4-triazole)
[104364-15-0]

$C_4H_2N_{10}O_4$

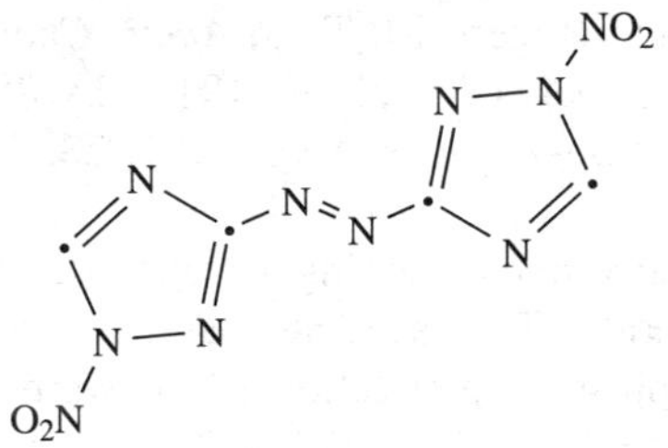

Lee, K. Y., *Energy Res. Abstr.*, 1985, **10**(17), 34672
A candidate for high-energy propellant applications.
See other AZO COMPOUNDS, HIGH-NITROGEN COMPOUNDS, *N*-NITRO COMPOUNDS,
TRIAZOLES

1402. Poly(furan-2,5-diyl)
[51325-04-3]

$(C_4H_2O)n$

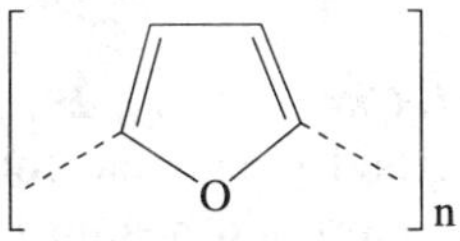

Perchlorate ions
See entry PERCHLORATE-DOPED CONDUCTING POLYMERS

1403. Acetylenedicarboxaldehyde (1,4-Dioxobut-2-yne)
[21251-20-7] $C_4H_2O_2$

$$OHCC \equiv CCHO$$

Gorgues, A. *et al., Tetrahedron*, 1990, **46**(8), 2817
The solid explodes, flamelessly, even under nitrogen, at its melting point, $-11°C$.
The vapour was handled at reduced pressure up to $60°C$.
See other ACETYLENIC COMPOUNDS, ALDEHYDES

1404. Maleic anhydride (2,5-Dihydrofuran-2,5-dione)
[108-31-6] $C_4H_2O_3$

(*MCA SD-88*, 1962); *HCS 1980*, 605; *RSC Lab. Hazards Data Sheet No. 53*, 1983

Vaughn, C. B. *et al., J. Loss Prevention*, 1993, **6**(1), 61
Energy of decomposition (in range $260-370°C$) measured as 0.92 kJ/g.
A fire within a maleic acid plant left pipework glowing red hot. Investigation of possible sources of ignition showed that the anhydride, spread on insulation materials, undergoes runaway autoxidation/decomposition from $190°C$.
See also INSULATION, THERMOCHEMISTRY AND EXOTHERMIC DECOMPOSITION (reference 2)

Bases, or Cations
1. Vogler, C. E. *et al., J. Chem. Eng. Data*, 1963, **8**, 620
2. Davie, W. R., *Chem. Eng. News.*, 1964, **42**(8), 41
3. *MCA Case History No. 622*
4. *MCA Case History No. 2032*
Maleic anhydride decomposes exothermically, evolving carbon dioxide, in the presence of alkali- or alkaline earth-metal or ammonium ions, dimethylamine, triethylamine, pyridine or quinoline, at temperatures above $150°C$ [1]. Sodium ions and pyridine are particularly effective, even at concentrations below 0.1%, and decomposition is rapid [2]. An industrial incident involved gas-rupture of a large insulated tank of semi-solid maleic anhydride which had been contaminated with sodium hydroxide. Use of additives to reduce the sensitivity of the anhydride has been described [3]. Accidental transfer of an aqueous solution of sodium 2-benzothiazolethiolate into a bulk storage tank of the anhydride led to eventual explosive destruction of the tank [4].
See other CATALYTIC IMPURITY INCIDENTS, GAS EVOLUTION INCIDENTS

1-Methylsilacyclopenta-2,4-diene
See 1-Methylsilacyclopenta-2,4-diene: Dienophiles
See other ACID ANHYDRIDES

1405. Acetylenedicarboxylic acid
[142-45-0] $C_4H_2O_4$

$$HOCO.C{\equiv}CCO.OH$$

Energy of decomposition (in range 140–270°C) measured as 1.355 kJ/g.
See entry THERMOCHEMISTRY AND EXOTHERMIC DECOMPOSITION (reference 2)

Copper salts
Zamudio, W. *et al., Bol. Soc. Chil. Quim.*, 1994, **39**(4), 339
Aqueous solutions of copper acetylenedicarboxylates decompose thermally to precipitate explosive copper acetylides. Presumably other elements readily forming explosive acetylides might do likewise.
See Copper(II) acetylide
See also METAL ACETYLIDES
See other ACETYLENIC COMPOUNDS, ORGANIC ACIDS

1406. Maleic anhydride ozonide
[101672-17-7] $C_4H_2O_6$

Briner, E. *et al., Helv. Chim. Acta*, 1937, **20**, 1211
It explodes on warming to −40°C.
See other OZONIDES

1407. Poly(thiophene)
[25233-34-5] $(C_4H_2S)_n$

Perchlorate ions
See entry PERCHLORATE-DOPED CONDUCTING POLYMERS

480

1408. Silver buten-3-ynide
[15383-68-3] C_4H_3Ag

$$AgC{\equiv}CCH{=}CH_2$$

Alone, or Ammonia, or Nitric acid
 1. Willstätter, R. *et al., Ber.*, 1913, **46**, 535–538
 2. Grignard, 1935, Vol. 3, 187
The silver salt, which deflagrates on heating, explodes if moistened with fuming nitric acid [1], or with ammonia solution [2].
See other METAL ACETYLIDES, SILVER COMPOUNDS

1409. Silver 3-methylisoxazolin-4,5-dione-4-oximate
[70247-51-7] $C_4H_3AgN_2O_3$

Wentrup, C. *et al., Angew. Chem. (Intern. Ed.)*, 1979, **18**, 467
It explodes on rapid heating. On slow heating under vacuum, it gives dangerously explosive silver fulminate.
See Fulminic acid (reference 2), *also* Sodium 3-methylisoxazolin-4,5-dione-4-oximate
See related OXIMES *See other* N–O COMPOUNDS, SILVER COMPOUNDS

1410. 5-Bromo-4-pyrimidinone
[19808-30-1] $C_4H_3BrN_2O$

Preparative hazard
 Dietsche, T. J. *et al., Chem. Eng. News*, 1995, **78**(28), 4
 A literature procedure whereby bromopyrimidine is oxidised by excess peroxyacetic acid in acetone, with sulfuric acid catalysis, was being scaled up. The crude product from the fourth batch at two molar scale was filtered out and allowed to dry to dry in the sintered glass funnel over the weekend. An explosion occurred when it was scraped out to complete purification on the Monday. This was considered due to acetone peroxides, which had probably concentrated locally by wicking or sublimation.
 See CYCLIC PEROXIDES, SINTERED GLASS

1411. 1-Chloro-3,3-difluoro-2-methoxycyclopropene
[59034-34-3] $C_4H_3ClF_2O$

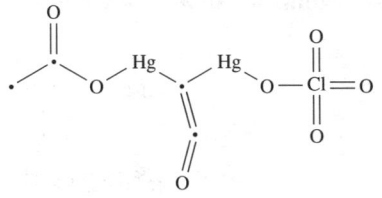

See entry FLUORINATED CYCLOPROPENYL METHYL ETHERS

1412. Acetoxymercurio(perchloratomercurio)ethenone
[73399-68-5] $C_4H_3ClHg_2O_7$

Blues, E. T. *et al., J. Chem. Soc., Chem. Comm.*, 1979, 1043
It is dangerously explosive.
See related METAL PERCHLORATES *See other* MERCURY COMPOUNDS

1413. Acetoxydimercurio(perchloratodimercurio)ethenone
[73399-71-0] $C_4H_3ClHg_4O_7$
$$AcOHgHgC(:C:O)HgHgOClO_3$$
Blues, E. T. *et al., J. Chem. Soc., Chem. Comm.*, 1979, 1043
It is dangerously explosive.
See related METAL PERCHLORATES *See other* MERCURY COMPOUNDS

1414. 1,3,3-Trifluoro-2-methoxycyclopropene
[59034-32-1] $C_4H_3F_3O$

See entry FLUORINATED CYCLOPROPENYL METHYL ETHERS

1415. Potassium 3-methylfurazan-4-carboxylate 2-oxide
[37895-51-5] $C_4H_3KN_2O_4$

Gasco, A. *et al., Tetrahedron Lett.,* 1974, 630 (footnote 2)

The dried salt explodes violently on heating, impact or friction.

See other FURAZAN *N*-OXIDES

1416. 3-Cyanopropyne (1-Butyne-4-nitrile)
[2235-08-7] C_4H_3N

$$HC\equiv CCH_2C\equiv N$$

Reddy, G. S. *et al., J. Amer. Chem. Soc.,* 1961, **83**, 4729

After preparation from interaction of 3-bromopropyne with copper(I) cyanide and filtration from copper salts, an explosion occurred during distillation of the evaporated filtrate at 45–60°C/66 mbar. This was attributed to explosion of some dissolved copper acetylide(s). After refiltration the product was again distilled at 45–48°C/53 mbar without incident, and it appeared to be stable, unlike true haloalkynes. However it is undoubtedly an endothemic compound with its two triple bonds.

See related HALOACETYLENE DERIVATIVES
See other ACETYLENIC COMPOUNDS, CYANO COMPOUNDS

1417. Poly(pyrrole)
[30604-81-0] $(C_4H_3N)_n$

Perchlorate ions
See entry PERCHLORATE-DOPED CONDUCTING POLYMERS

1418. Maleimide
[541-59-3] C_4H_3NO

Energy of decomposition (in range 140–300°C) measured as 0.49 kJ/g.
See entry THERMOCHEMISTRY AND EXOTHERMIC DECOMPOSITION (reference 2)

1419. Sodium 3-methylisoxazolin-4,5-dione-4-oximate
[70247-50-6] $C_4H_3N_2NaO_3$

Wentrup, C. *et al., Angew. Chem. (Intern. Ed.)*, 1979, **18**, 467
It explodes on rapid heating, like the analogous silver salt and 3-phenyl derivative.
See related OXIMES *See other* N–O COMPOUNDS

1420. 5-(Prop-2-ynyloxy)-1,2,3,4-thiatriazole
[ ] $C_4H_3N_3OS$

Banert, K. *et al., Angew. Chem. (Int.)*, 1992, **31**(7), 866
Decomposes in solution at 0°C but can be handled, with caution even sublimed,
as an explosive crystalline solid, m.p. 66°C.
See other ACETYLENIC COMPOUNDS, N–S COMPOUNDS

1421. 2-Amino-3,5-dinitrothiophene
[2045-70-7] $C_4H_3N_3O_4S$

Fast flame propagation occurs on heating the powder moderately.
See entry HIGH RATE DECOMPOSITION

1422. 3-Diazoniopyrazolide-4-carboxamide
[102613-59-2] $C_4H_3N_5O$

Cheng, C. C. *et al., J. Med. Chem.*, 1986, **29**, 1546
Though the preparative method is suitable for large scale synthesis, the product
must be handled with care. On rapid heating, the internal diazonium salt decomposes violently at 155–160°C accompanied by a sharp sound. The dried powder
should not be scratched with a metal spatula or ground finely.
See other DIAZONIUM SALTS, HIGH-NITROGEN COMPOUNDS

†1423. Buten-3-yne (Vinylacetylene)
[689-97-4]   C$_4$H$_4$

$$HC \equiv CCH = CH_2$$

1. Hennion, G. F. *et al., Org. Synth.*, 1963, Coll. Vol. 4, 684
2. Rutledge, 1968, 28
3. Strizhevskii, I. I. *et al., Chem. Abs.*, 1973, **78**, 98670
4. Vervalin, 1973, 38–43

It must be stored out of contact with air to avoid the formation of explosive (peroxidic) compounds [1]. The sodium salt is a safe source of butenyne for synthetic use [2]. Explosive properties of the liquid and gaseous hydrocarbon alone, or diluted with decahydronaphthalene, were determined [3]. Mechanical failure of a compressor circulating the gaseous hydrocarbon through a reaction system led to local overheating and explosive decomposition, which propagated throughout the large plant [4]. As anticipated for a multiply unsaturated hydrocarbon, vinylacetylene is strongly endothermic (ΔH_f° (g) +143.5 kJ/mol, 2.76 kJ/g).
See other ENDOTHERMIC COMPOUNDS

1,3-Butadiene
1. Jarvis, H. C., *Chem. Eng. Progr.*, 1971, **67**(6), 41–44
2. Freeman, R. H. *et al., Chem. Eng. Progr.*, 1971, **67**(6), 45–51
3. Keister, R. G. *et al., Loss Prev.*, 1971, **5**, 67–75
4. Carver, F. W. S. *et al., Proc. 5th Symp. Chem. Proc. Haz.*, No. 39a, 99–116, London, IChE, 1975

Unusual conditions in the main fractionation column separating product butadiene from by-product butenyne (vinylacetylene, thought to be safe at below 50 mol% concentration) caused the concentration of the latter to approach 60% in part of the column as fractionation proceeded. Explosive decomposition, possibly initiated by an overheated unstable organic material derived from sodium nitrite, destroyed the column and adjacent plant [1,2]. Subsequent investigation showed that all mixtures of butenyne and butadiene can reproducibly be caused to react exothermically and then decompose explosively at appropriately high heating rates under pressure. Butadiene alone will behave similarly at higher heating rates and pressures [3]. Another study of the detailed mechanism and type of decomposition of C$_4$ hydrocarbons containing butenyne showed that both were dependent on energy of initiation and time of application, as well as on the composition and phases present [4].
See Sodium nitrite: 1,3-Butadiene

Oxygen
1. Dolgopolskii, I. M. *et al., Chem., Abs.*, 1958, **52**, 19904g
2. Gosa, V. *et al., Rev. Chim.* (Bucharest), 1981, **32**, 1103–1105
3. Ionescu, N. I. *et al., Rev. Chim.* (Bucharest), 1982, **33**, 1021–1026

The rate of absorption of oxygen by liquid butenyne increased with time, and eventually a yellow liquid phase separated. After evaporation of excess hydrocarbon, the yellow peroxidic liquid was explosive. Presence of 5% of chloroprene

485

increased the rate of absorption 5–6-fold, and of 2% of water decreased the rate by 50%, but residues were explosive in each case [1]. Explosive combustion in admixture with oxygen has been studied [2], and the effects of presence of nitrogen upon explosion parameters were determined [3].

Silver nitrate
Grignard, 1935, Vol. 3, 187
An explosive silver salt is formed.
See Silver buten-3-ynide
See other ALKYNES, PEROXIDISABLE COMPOUNDS

1424. Barium 1,3-di(5-tetrazolyl)triazenide
[ ] $C_4H_4BaN_{22}$

See 1,3-Di(5-tetrazolyl)triazene
See other HIGH-NITROGEN COMPOUNDS, *N*-METAL DERIVATIVES, TETRAZOLES, TRIAZENES

1425. *N*-Bromosuccinimide (1-Bromo-2,5-pyrrolidinedione)
[128-08-5] $C_4H_4BrNO_2$

Aniline, or Diallyl sulfide, or Hydrazine hydrate
See N-HALOIMIDES: alcohols, etc.

Dibenzoyl peroxide, 4-Toluic acid
1. Tcheou, F. K. *et al., J. Chinese Chem. Soc.*, 1950, **17**, 150
2. Mould, R. W., private comm., 1975
Thirtyfold increase in scale of a published method [1] for radical-initiated side-chain bromination of the acid in carbon tetrachloride led to violent reflux and eruption of the flask contents through the condenser [2].
See other HALOGENATION INCIDENTS

Propiononitrile
MCA Guide, 1972, 309

After refluxing for 24 hours at 105°C, a mixture exploded, possibly due to dehydrohalogenation of the bromonitrile to acrylonitrile, and polymerisation of the latter.

See other N-HALOGEN COMPOUNDS, N-HALOIMIDES

1426. 2-Chloro-2-propenyl trifluoromethanesulfonate
[62861-56-7] $C_4H_4ClF_3O_3S$

$$H_2C=CClCH_2OSO_2CF_3$$

See entry ALLYL TRIFLUOROMETHANESULFONATES *See other* SULFUR ESTERS

1427. N-Chlorosuccinimide (1-Chloro-2,5-pyrrolidinedione)
[128-09-6] $C_4H_4ClNO_2$

Energy of decomposition (in range 160–380°C) measured as 1.61 kJ/g.
See entry THERMOCHEMISTRY AND EXOTHERMIC DECOMPOSITION (reference 2)

Alcohols, or Benzylamine
See N-HALOIMIDES: alcohols, etc.

Dust
Boscott, R. J., private comm., 1968
Smouldering in a stored drum of the chloroimide was attributed to dust contamination.
See other N-HALOGEN COMPOUNDS, N-HALOIMIDES

1428. 1,4-Dichloro-2-butyne
[821-10-3] $C_4H_4Cl_2$

$$ClCH_2C\equiv CCH_2Cl$$

1. Ford, M. *et al., Chem. Eng. News*, 1972, **50**(3), 67
2. Herberz, T., *Chem. Ber.*, 1952, **85**, 475–482
3. *CHETAH*, 1990, 182 and 188

Preparation of the acetylenic dichloride by conversion of the diol in pyridine with neat thionyl chloride is difficult to control, and hazardous on a large scale. Use of dichloromethane as diluent and operation at −30°C renders the preparation reproducible and safer [1]. During distillation at up to 110°C/7–8 mbar, slight overheating of the residue to 120°C caused explosive decomposition [2]. Sensitive to detonator, sometimes to mechanical shock [3].
See other HALOACETYLENE DERIVATIVES

1429. Copper(II) 1,3-di(5-tetrazolyl)triazenide
[32061-49-7]

C$_4$H$_4$CuN$_{22}$

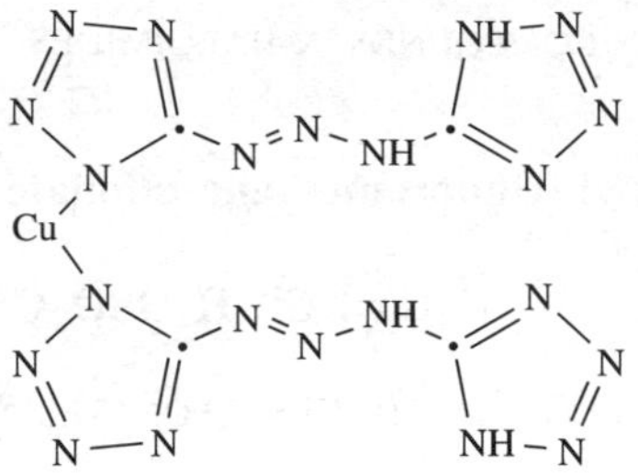

See 1,3-Di(5-tetrazolyl)triazene
See other HIGH-NITROGEN COMPOUNDS, *N*-METAL DERIVATIVES, TETRAZOLES, TRIAZENES

1430. Tetrafluorosuccinamide
[377-37-7]

C$_4$H$_4$F$_4$N$_2$O$_2$

$$(-CF_2CO.NH_2)_2$$

Lithium tetrahydroaluminate
See Lithium terahydroaluminate: Fluoroamides

1431. Poly[bis(2,2,2-trifluoroethoxy)phosphazene]
[28212-50-2]

(C$_4$H$_4$F$_6$NO$_2$P)$_n$

$$[(F_3CCH_2O)_2P=N]_n$$

Allcock, H. R. *et al., Macromolecules*, 1974, **7**, 284–290
Sealed tubes containing the linear P=N polymer exploded after heating at 300°C for 24–30 h (or after shorter times at higher temperatures), owing to pressure build-up from formation of the more volatile cyclic tri- and tetra-mers.

1432. 2-Cyano-1,2,3-tris(difluoroamino)propane
[16176-02-6]

C$_4$H$_4$F$_6$N$_4$

$$F_2NC(CN)(CH_2NF_2)_2$$

Reed, S. F., *J. Org. Chem.*, 1968, **33**, 1864
The 95% pure material is shock-sensitive.
See other CYANO COMPOUNDS, DIFLUOROAMINO COMPOUNDS

1433. Succinodinitrile (1,4-Butanedinitrile)
[110-61-2]

C$_4$H$_4$N$_2$

$$(-CH_2C\equiv N)_2$$

1. Kniess, H., private comm., 1983
2. Berthold, W. *et al., Proc. 4th Int. Symp. Loss Prev. Safety Prom. Proc. Ind.,* Vol. 3, C1–C4, Rugby, IChE, 1983

After vacuum distillation of several t of the nitrile at 150°C, lack of a receiver necessitated hot storage of the product at 80°C for 46 h, and the drainage line blocked with solid. Though the bulk of the material was still liquid, heating to 195°C did not clear the line, and shortly afterwards decomposition occurred. It was subsequently found that the distillation residue had also decomposed at 175°C, 2 h after distillation ended [1]. DTA showed that the nitrile is unusual in that no heat is evolved during the induction period of around 33 h/200°C, or 1.5 h/280°C, and the rapid onset and narrow peak width suggested a self-accelerating decomposition process, with $Q = 700$ J/g. No stabilising additives were found, but addition of 1% of hydrogen cyanide reduced the induction period for the nitrile to 20% of its value, while 1% of potassium cyanide reduced it to 2%. This supports the proposed mechanism of cyanide ion-catalysed decomposition to hydrogen cyanide and acrylonitrile, followed by their simultaneous polymerisation, again by self-accelerating processes. Adiabatic storage tests on 150 g samples gave the same induction period and a decomposition peak of 400°C/40 bar [2].

See entry SELF-ACCELERATING REACTIONS
See other CYANO COMPOUNDS, INDUCTION PERIOD INCIDENTS

1434. α-Diazo-γ-thiobutyrolactone
[79472-87-0] $C_4H_4N_2OS$

See α-Bromo-γ-thiobutyrolactone
See other DIAZO COMPOUNDS

1435. Maleic hydrazide (1,2-Dihydropyridazine-3,6-dione)
[123-33-1] $C_4H_4N_2O_2$

Energy of decomposition (in range 280–300°C) measured as 0.59 kJ/g by DSC, and T_{ait24} was determined as 261°C by adiabatic Dewar tests, with an apparent energy of activation of 257 kJ/mol.

489

1436. 2-Amino-4,6-dihydroxy-5-nitropyrimidine
[80466-56-4]

$C_4H_4N_4O_4$

Madeley, J. P., personal communication 1989

A 200 g sample, prepared by addition of substrate to mixed acid, followed by quenching into water, filtration and washing, decomposed vigorously at a late stage of drying in a vacuum oven, which was pressurised and the seal forced open. This behaviour might have been due to inadequate washing and residual sulfuric acid.

See other C-NITRO COMPOUNDS

1437. 2,3-Diazido-1,3-butadiene
[91686-86-1]

$C_4H_4N_6$

$$H_2C=CN_3CN_3=CH_2$$

Mielsen, C. J. *et al., J. Mol. Struct.*, 1986, **147**, 217–229

Although it is extremely explosive, like other polyunsaturated azides, it was possible to run gaseous electron diffraction and IR spectra at cryogenic temperatures. Sample decomposition occurred during laser Raman spectral determination.

See other DIENES, ENDOTHERMIC COMPOUNDS, HIGH-NITROGEN COMPOUNDS, ORGANIC AZIDES

1438. Succinoyl diazide
[40428-75-9]

$C_4H_4N_6O_2$

$$(\text{-CH}_2\text{CO.N}_3)_2$$

1. Maclaren, J. A., *Chem. & Ind.*, 1971, 3952
2. France, A. D. G. *et al., Chem. & Ind.*, 1962, 2065

This intermediate for the preparation of ethylene diisocyanate exploded violently during isolation [1]. Previously the almost dry solid had exploded violently on stirring with a spatula [2].

See other ACYL AZIDES, FRICTIONAL INITIATION INCIDENTS

†1439. Furan
[110-00-9]

C_4H_4O

See other APROTIC SOLVENTS, PEROXIDISABLE COMPOUNDS

490

1440. 1,2-Cyclobutanedione
[33689-28-0] $C_4H_4O_2$

Dennis, J. M. *et al., Org. Synth.*, 1981, **60**, 18–19
The dione must be stored cold and handled out of direct light to prevent polymerisation.

†1441. Diketene (4-Methylene-2-oxetanone)
[674-82-8] $C_4H_4O_2$

HCS 1980, 416

1. Vervalin, 1973, 86
2. Zdenek, F. *et al.*, Czech Pat. 156 584, 1975
3. *See entry* PRESSURE INCREASE IN EXOTHERMIC DECOMPOSITION

Diketene residues in a tank trailer awaiting incineration decomposed violently on standing, blew off the dome cover and ignited [1]. The risk of autoignition during the exothermic dimerisation of diketene to dehydroacetic acid is eliminated by operating in a non-flammable solvent [2]. It produces a moderate pressure rise on exothermic decomposition at 125°C [3].

Acids, or Bases, or Sodium acetate
1. *Haz. Chem. Data*, 1975, 136
2. Laboratory Chemical Disposal Co. Ltd., confid. information, 1968

Presence of mineral or Lewis acids, or bases including amines, will catalyse violent polymerisation of this very reactive dimer, accompanied by gas evolution [1]. Sodium acetate is sufficiently basic to cause violent polymerisation at 0.1% concentration when added to diketene at 60°C [2].

See other CATALYTIC IMPURITY INCIDENTS, GAS EVOLUTION INCIDENTS, POLYMERISATION INCIDENTS

†1442. Methyl propiolate (Methyl propynoate)
[922-67-8] $C_4H_4O_2$

$$MeOClO.C{\equiv}CH$$

Octakis(trifluorophosphine)dirhodium
See Octakis(trifluorophosphine)dirhodium: Acetylenic esters
See other ACETYLENIC COMPOUNDS

1443. Succinic anhydride
[108-30-5] $C_4H_4O_3$

$$CH_2CH_2CO.OCO$$

Sodium hydroxide
1. Hoekstra, M. C., *Synth. Comm.*, 1984, **14**, 1379–1380
2. Kirk Othmer, 1997, (4th edn.), Vol. 22, 1076
Succinic anhydride is dimerised to 1,6-dioxaspiro [4.4] nonane-2,7-dione by heating with sodium hydroxide. Modification of an existing procedure by adding further sodium hydroxide after the initial reaction led to a severe exothermic reaction after heating for some 30 h which fused the glass flask to the heating mantle, probably at a temperature approaching 550°C. The reason for this was not known [1]. At elevated temperatures and under influence of alkali, succinic acid condenses decarboxylatively beyond the dimeric spiroacetal, sometimes explosively. Contamination of the anhydride with base is to be avoided [2].
See other GLASS INCIDENTS
See other ACID ANHYDRIDES

1444. 2,3,7-Trioxabicyclo[2,2,1]hept-5-ene
[6824-18-6] $C_4H_4O_3$

Lorencak, P. *et al., J. Phys. Chem.*, 1989, **93**(6), 2276
This, the ozonide of cyclobutadiene (and also a likely peroxide of furan), may be stored in liquid nitrogen. It can explode with considerable force on warming.
See other CYCLIC PEROXIDES, OZONIDES

1445. 3,6-Dioxo-1,2-dioxane (Succinyl peroxide)
[ ] $C_4H_4O_4$

CHETAH, 1990, 183
This cyclic peroxide, readily formed from succinate derivatives and peroxides, is claimed to be more stable than its molecular formula might indicate. It is, however, shock sensitive.
See other DIACYL PEROXIDES, CYCLIC PEROXIDES

1446. Fumaric acid (*trans*-2-Butene-1,4-dioic acid)
[110-17-8] $C_4H_4O_4$

$$HOCO.CH=CHCO.OH$$

Energy of decomposition (in range 290–340°C) measured as 0.925 kJ/g.
See entry THERMOCHEMISTRY AND EXOTHERMIC DECOMPOSITION (reference 2)
See other ORGANIC ACIDS

1447. Dihydroxymaleic acid (Dihydroxybutenedioic acid)
[526-84-1] $C_4H_4O_6$

$$(:C(OH)CO.OH)_2$$

Axelrod, B., *Chem. Eng. News*, 1955, **33**, 3024
A glass ampoule of the acid exploded during storage at ambient temperature. It is
unstable and slowly loses carbon dioxide, even at 4°C.
See other GAS EVOLUTION INCIDENTS, GLASS INCIDENTS, ORGANIC ACIDS

†1448. Thiophene
[110-02-1] C_4H_4S

HCS 1980, 899

Nitric acid
  See Nitric acid: Thiophene

N-Nitrosoacetanilide
  See *N*-Nitrosoacetanilide: Thiophene

1449. Ethynyl vinyl selenide
[101672-11-1] C_4H_4Se

$$HC\equiv CSeCH=CH_2$$

Brandsma, L. *et al., Rec. Trav. Chim.*, 1962, **81**, 539
Distillable at 30–40°C/80 mbar, it decomposes explosively on heating at ambient
pressure.
See other ACETYLENIC COMPOUNDS

1450. α-Bromo-γ-thiobutyrolactone (3-Bromo-2-oxotetrahydrothiophene)
[20972-64-9]   C₄H₅BrOS

Miller, G. A. *et al., J. Org. Chem.*, 1981, **46**, 4751–4753
In the preparation *via* diazotisation and bromination of homocysteine thiolactone, presence of unreacted diazo intermediate led to an explosion during distillation of the product.
See DIAZO COMPOUNDS

†1451. 2-Chloro-1,3-butadiene (Chloroprene)
[126-99-8]   C₄H₅Cl

$$H_2C=CClCH=CH_2$$

1. Bailey, H. C., in *Oxidation of Organic Compounds-I*, ACS No. 75 (Gould, R. F., Ed.), 138–149, Washington, ACS, 1968
2. Bailey, H. C., private comm., 1974

Chloroprene monomer will autoxidise very rapidly with air, and even at 0°C it produces an unstable peroxide (a mixed 1,2- and 1,4-addition copolymer with oxygen), which effectively will catalyse exothermic polymerisation of the monomer. The kinetics of autoxidation have been studied [1]. It forms 'popcorn polymer' at a greater rate than does butadiene [2].
See other DIENES, HALOALKENES, PEROXIDISABLE COMPOUNDS, POLYMERISATION INCIDENTS

Preparative hazard
See 2-Chloro-1-nitro-4-nitroso-2-butene

1452. 2-Chloro-1-nitro-4-nitroso-2-butene
[58675-02-8]   C₄H₅ClN₂O₃

$$O_2NCH_2C(Cl)=CHCH_2NO$$

Simmons, H. E. *et al., Chem. Eng. News*, 1995, **73**(32), 4
A worker was injured by an explosion when cutting into a "decontaminated" steel pipe removed from a storage tank in a chloroprene plant. The pipe was found to be coated by a deposit of this compound, existing as a dimer. Investigation showed it to be unstable from 80°C, deflagrating with an energy of 2.5 kJ/g (about that of black powder). It is thought it may be capable of true detonation. It was formed by reaction of chloroprene with dinitrogen trioxide, generated during distillation of the chloroprene by decomposition of an ill-defined nitrosated additive used for vapour phase stabilisation against polymerisation reactions. Although originally present

494

at negligible concentration the nitroso-dimer, being very insoluble in chloroprene, accumulated in the pipe. Warning is given that reaction of alkenes with nitrogen oxides, to form similar products, is a general reaction.
See Nitrogen dioxide: Alkenes

1453. 4-Chloro-1-methylimidazolium nitrate
[[4897-21-6] (base)] $C_4H_5ClN_3O_3$

Personal experience
Towards the end of evaporation of a dilute aqueous solution, the residue decomposed violently. Nitrate salts of many organic bases are thermally unstable and should be avoided.
See other OXOSALTS OF NITROGENOUS BASES

1454. 1-Chloro-1-buten-3-one
[7119-27-9] C_4H_5ClO

$$ClCH{=}CHCO.CH_3$$

Pohland, A. E. *et al., Chem. Rev.*, 1966, **66**, 164
If prepared from vinyl chloride, this unstable ketone will decompose almost explosively within one day, apparently owing to the presence of some *cis*-isomer.
See related HALOALKENES

1455. 4-Chloro-2-butynol
[13280-07-4] C_4H_5ClO

$$ClCH_2C{\equiv}CCH_2OH$$

1. Taylor, S. K. *et al., J. Org. Chem.*, 1983, **48**, 595
2. Harmon, D. P. G. *Chem. Brit.* 1998, (7), 19
It exploded during distillation [1]. Another explosion during distillation of a crude sample is reported. It is claimed that there were no previous warnings of the hazard [2].
See other HALOACETYLENE DERIVATIVES

1456. Ethyl oxalyl chloride (Ethyl chloroglyoxylate)
[4755-77-5] $C_4H_5ClO_3$

$$EtO.OCCO.Cl$$

Urben, P. G., Personal experience

A drum of clear and bright material which had been vacuum distilled pressurised on standing overnight. Investigation revealed a crack in the heat exchanger which had been leaking water into the oxalyl chloride, forming partial anhydrides. These are unstable and decompose to give carbon monoxide and dioxide, other oxalyl halides will behave similarly. On another occasion, a vacuum distillation went suddenly to positive pressure on heating to about 70°C. This was catalysis by traces of iron or aluminium chlorides, which induce break up to, probably, carbon monoxide, carbon dioxide and ethyl chloride.

See Oxalyl dichloride

See other ACYL HALIDES, CATALYTIC IMPURITY INCIDENTS, GAS EVOLUTION INCIDENTS

1457. 2-Methyl-2-propenoyl chloride (Methacryloyl chloride)
[920-46-7] C_4H_5ClO

$$CH_2{=}C(CH_3)CO.Cl$$

Frazier, J., *Chem. Eng. News*, 1996, **74**(23), 4

A five year old bottle, originally stabilised with phenothiazine, was found to have polymerised explosively in storage. The resultant black goo was not acidic, suggesting something more complicated than an acrylate polymerisation. It is recommended that this monomer be stored under refrigeration.

See VIOLENT POLYMERISATION

1458. 4-Fluoro-4,4-dinitrobutene
[19273-49-5] $C_4H_5FN_2O_4$

$$FC(NO_2)_2CH_2CH{=}CH_2$$

See entries FLUORODINITROMETHYL COMPOUNDS, NITROALKENES

1459. Bis(2-fluoro-2,2-dinitroethyl)amine
[18139-03-2] $C_4H_5F_2N_5O_8$

$$(FC(NO_2)_2CH_2)_2NH$$

1. Gilligan, W. H., *J. Org. Chem.*, 1972, **37**, 3947
2. Gilligan, W. H., *J. Chem. Eng. Data*, 1982, **27**, 97–99

The amine and several derived amides are explosives, and need appropriate care in handling [1]. Several other derivatives, most of them explosive, are prepared and described [2].

See other FLUORODINITROMETHYL COMPOUNDS, ORGANIC BASES

†1460. Ethyl trifluoroacetate
[383-63-1] $C_4H_5F_3O_2$

$$EtOCO.CF_3$$

Diethyl succinate, Sodium hydride

See Sodium hydride: Diethyl succinate, etc.

496

1461. Prop-2-enyl trifluoromethanesulfonate
[41029-45-2] $C_4H_5F_3O_3S$

$$H_2C=CHCH_2OSO_2CF_3$$

It explodes at 20°C.
See entry ALLYL TRIFLUOROMETHANESULFONATES *See other* SULFUR ESTERS

1462. Potassium hydrogen tartrate
[868-14-4] $C_4H_5KO_6$

KOCO.CHOHCHOHCO.OH

Carbon, Nitrogen oxide
See Nitrogen oxide: Carbon, etc.
See related ORGANIC ACIDS

1463. Cyanocyclopropane (Cyclopropanecarbonitrile)
[5500-21-0] C_4H_5N

Preparative hazard
See 4-Chlorobutyronitrile: Sodium hydroxide
See other CYANO COMPOUNDS, STRAINED-RING COMPOUNDS

†1464. 1-Cyanopropene (2-Butenonitrile)
[627-26-9] C_4H_5N

$$N\equiv CCH=CHCH_3$$

See other CYANO COMPOUNDS

†1465. 3-Cyanopropene (3-Butenonitrile)
[109-75-1] C_4H_5N

$$H_2C=CHCH_2C\equiv N$$

See other ALLYL COMPOUNDS, CYANO COMPOUNDS

1466. Pyrrole
[109-97-7] C_4H_5N

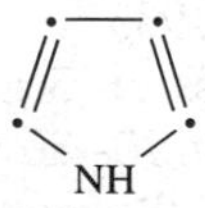

2-Nitrobenzaldehyde
See 2-Nitrobenzaldehyde: Pyrrole

1467. 1-Cyano-2-propen-1-ol (2-Hydroxy-3-butenonitrile)
[5809-59-6] C_4H_5NO

$$N{\equiv}CCHOHCH{=}CH_2$$

Unpublished information
It shows a strong tendency to exothermic polymerisation of explosive violence
in presence of light and air above 25–30°C. Presence of an inhibitor and inert
atmosphere are essential for stable storage.
See other CYANO COMPOUNDS

1468. 5-Methylisoxazole
[576-44-6] C_4H_5NO

Sealed samples decompose exothermally above 171°C.
See entry ISOXAZOLES *See other* N–O COMPOUNDS

1469. *N*-Hydroxysuccinimide
[6066-82-6] $C_4H_5NO_3$

Energy of decomposition (in range 180–370°C) measured as 1.51 kJ/g.
See entry THERMOCHEMISTRY AND EXOTHERMIC DECOMPOSITION (reference 2)
See other N–O COMPOUNDS

1470. Methyl isocyanoacetate
[39687-95-1] $C_4H_5NO_2$

$$:C{:}NCH_2CO.OMe$$

Heavy metals
Naef, H. *et al.*, private comm., 1983
Crude material prepared in glass on 3 g mol scale was distilled uneventfully at
40°C/0.067 mbar from a bath at 70–80°C. A 30 mol batch prepared in a glass-
lined vessel with a stainless steel thermo-probe (and later found to contain 15 ppm
of iron) decomposed very violently during distillation at 75°C/13 mbar from a bath
at 130°C. Thermal analysis showed that the stability of the methyl (and ethyl) ester

was very sensitive to traces of heavy metals (iron, copper, chromium, etc.) and was greatly reduced. Addition of traces of hydrated iron(II) sulfate led to explosive decomposition at 25°C.

See related CYANO COMPOUNDS *See other* CATALYTIC IMPURITY INCIDENTS

1471. Allyl isothiocyanate (3-Isothiocyanatopropene)
[57-06-7] $\qquad$ C₄H₅NS

$$H_2C=CHCH_2N{:}C{:}S$$

Anon., *Ind. Eng. Chem. (News Ed.)*, 1941, **19**, 1408

A routine preparation by interaction of allyl chloride and sodium thiocyanate in an autoclave at 5.5 bar exploded violently at the end of the reaction. Peroxides were not present or involved and no other cause could be found, but extensive decomposition occurred when allyl isothiocyanate was heated to 250°C in glass ampoules. Exothermic polymerisation seems a likely possibility.

See related CYANO COMPOUNDS, ORGANIC ISOCYANATES
See other ALLYL COMPOUNDS

1472. 2-Azido-1,3-butadiene
[01686-88-3] $\qquad$ C₄H₅N₃

$$H_2C=CN_3CH=CH_2$$

Schel, S. A. *et al.*, *J. Mol. Struct.*, 1986, **147**(3–4), 203–215

Although it is highly explosive, like other polyunsaturated azides, it was possible to record spectral data under the following conditions: gaseous electron diffraction; IR spectra of matrix-isolated species in argon at 15°K; of amorphous and crystalline solids at 90°K and Raman spectra of the liquid at 240°K.

See other DIENES, ENDOTHERMIC COMPOUNDS, HIGH-NITROGEN COMPOUNDS, ORGANIC AZIDES

1473. Azido-2-butyne
[105643-77-4] $\qquad$ C₄H₅N₃

$$N_3CH_2C{\equiv}CCH_3$$

Nielsen, C. J. *et al.*, *J. Mol. Struct.*, 1987, **16**(1–2), 41–56

Precautions appropriate to a potentially explosive compound were observed during synthesis and handling. Electron diffraction studies on the vapour, and IR, NMR and Raman spectral studies on the vapour, liquid and solid (−183°C) phases were effected without incident.

See other ACETYLENIC COMPOUNDS, ENDOTHERMIC COMPOUNDS, HIGH-NITROGEN COMPOUNDS, ORGANIC AZIDES

1474. Iminobisacetonitrile
[628-87-5] $\qquad$ C₄H₅N₃

$$NCCH_2NHCH_2CN$$

Anon., *Loss. Prev. Bull.*, 1993, (109), 7

Iminobisacetonitrile crystallised in a holding tank of a production plant. Cleaning was by filling with water and heating to dissolve the nitrile. It was heated to a supposed 75°C. An hour later the tank exploded causing severe structural damage. The pyrometer used to measure the temperature was found inaccurate. The nitrile decomposes exothermically from perhaps as low as 70°C in alkaline water (135°C dry) to a variety of products including HCN.
See other CYANO COMPOUNDS

1475. 2-Methyl-4-nitroimidazole
[696-23-1] $C_4H_5N_3O_2$

Zmojdzin, A. *et al.*, Brit. Pat. 1 418 538, 1974
2-Methylimidazole is difficult to nitrate, and use of conventional reagents under forcing conditions (excess sulfuric/nitric acids, high temperature) involves a high risk of cleavage and violent runaway oxidation. A new and safe process involves the use of nitration liquor from a previous batch (and optional use of excess nitric acid), both of which moderate the nitration reaction, in conjunction with balancing amounts of sulfuric acid, which tends to accelerate the nitration of 2-methylimidazole. Accurate control of the highly exothermic reaction is readily effected.
See other NITRATION INCIDENTS

Nitric acid, Sulfuric acid
Zmojdzin, A. *et al.*, Fr. Demande 2 220 523, 1974
The danger of explosion during further nitration with nitrating acid is eliminated by addition of excess nitric acid as the reaction proceeds.
See NITRATION INCIDENTS *See related* NITROARYL COMPOUNDS

1476. 5-Aminoisoxazole-3-carbonamide
[3445-52-1] $C_4H_5N_3O_2$

Sealed samples decompose exothermally above 155°C.
See entry ISOXAZOLES *See other* N–O COMPOUNDS

1477. 1-Hydroxyimidazole-2-carboxaldoxime 3-oxide
[35967-34-3] $C_4H_5N_3O_3$

Hayes, K. J., *J. Heterocycl. Chem.*, 1974, **11**, 615
It is an explosive solid, containing three types of N–O bonds.
See other N–O COMPOUNDS, OXIMES

1478. Sodium ethoxyacetylide
[73506-39-5] C_4H_5NaO

$$NaC\equiv COEt$$

1. Jones, E. R. H. *et al., Org. Synth.*, 1963, Coll. Vol. 4, 405
2. Brandsma, 1971, 120
It is extremely pyrophoric, apparently even at $-70°C$ [1], and may explode after prolonged contact with air [2].
See other METAL ACETYLIDES, PEROXIDISABLE COMPOUNDS

†1479. 1,2-Butadiene
[590-19-2] C_4H_6

$$H_2C{=}C{=}CHCH_3$$

Highly endothermic, $\Delta H°_f$ (g) +165.5 kJ/mol, 3.06 kJ/g.

Cyclopentadiene (or its dimer)
Donike, W., Ger. Offen. DE 3 617 461, 1987 (*Chem. Abs.*, 1988, **109**, 210597)
In the Diels-Alder condensation of the 2 neat endothermic dienes to give 5-ethylidene- and 5-methyl-6-methylene-bicyclo[2.2.1]hept-2-ene, there is a serious risk of explosive decomposition arising from local overheating of the reactor walls. This hazard is eliminated by the presence of various hydrocarbons and their mixtures as diluents.
See other DIENES, ENDOTHERMIC COMPOUNDS

†1480. 1,3-Butadiene
[106-99-0] C_4H_6

$$H_2C{=}CHCH{=}CH_2$$

(*MCA SD-55*, 1954); *FPA H63*, 1977; *HCS 1980*, 223 (cylinders); *RSC Lab. Hazards Data Sheet No. 45*, 1986

1. Scott, D. A., *Chem. Eng. News*, 1940, **18**, 404
2. Hendry, D. G. *et al.*, *Ind. Eng. Chem., Prod. Res. Dev.*, 1968, **7**, 136, 1145
3. Mayo, F. R. *et al.*, *Prog. Rept. No. 40*, Sept. 1971, Stamford Res. Inst. Project PRC-6217
4. Keister, R. G. *et al.*, *Loss Prev.*, 1971, **5**, 69
5. Penkina, O. M. *et al.*, *Chem. Abs.*, 1975, **83**, 29457
6. Bailey, H. C., private comm., 1974
7. Miller, G. H. *et al.*, *J. Polymer Sci. C*, 1964, 1109–1115
8. Vervalin, 1964, 335–338, 358–359
9. Harmon, 1974, 2.5–2.6
10. Alexander, D. S., *Ind. Eng. Chem.*, 1959, **51**, 733–738

Solid butadiene at below −113°C will absorb enough oxygen at reduced pressure to make it explode violently when allowed to melt. The peroxides formed on long contact with air are explosives sensitive to heat or shock, but may also initiate polymerisation [1]. The hazards associated with peroxidation of butadiene are closely related to the fact that the polyperoxide is insoluble in butadiene and progressively separates. If local concentrations build up, self-heating from the (initially slow) spontaneous decomposition begins, and when a large enough mass of peroxide accumulates, explosion occurs. Critical mass at 27°C is a 9 cm sphere; the size decreases rapidly with increasing temperature [2]. Although isoprene and styrene also readily peroxidise, the peroxides are soluble in the monomers, and the degree of hazard is correspondingly less [3].

Butadiene is rather endothermic (ΔH_f° (g) +111.9 kJ/mol, 2.07 kJ/g) and will decompose explosively if heated under pressure at 30–40°C/min to exceed critical temperatures of 200–324°C and critical pressures of 1.0–1.2 kbar simultaneously [4]. A more recent study of the explosion properties of butadiene at elevated temperatures and pressures has been published (but title only translated) [5]. Phenolic antioxidants (e.g. *tert*-butylcatechol at 0.02 wt.%) are effective in stabilising butadiene against autoxidation in clean storage at moderate ambient temperatures. The presence of rust and/or water in steel storage cylinders rapidly consumes the antioxidant [6]. In prolonged storage, butadiene will (even when very pure and in sealed glass containers) produce 'popcorn' polymer [7]. Growth of this involves continued diffusion of monomer into an existing polymeric matrix which continues to increase in bulk. The expansion eventually may rupture the container [6]. Case histories of two industrial explosions involving peroxide formation with air have been detailed [8], and the literature relating to violent polymerisation of butadiene has been reviewed [9]. A detailed account of the experimental investigation following a large-scale explosion in 1951 of a partially full 159 m³ butadiene storage vessel is given. The involvement of peroxide formation in the explosion was established, as was the effectiveness of the presence of aqueous sodium hydroxide in destroying peroxide and preventing explosion [10].

See Poly(1,3-butadiene peroxide): Butadiene
See other SELF-HEATING AND IGNITION INCIDENTS, PEROXIDATION INCIDENTS
See entries POLYMERISATION INCIDENTS, VIOLENT POLYMERISATION
See other GLASS INCIDENTS

Aluminium tetrahydroborate
See Aluminium tetrahydroborate: Alkenes, etc.

Boron trifluoride etherate, Phenol
MCA Case History No. 790
The hydrocarbon–phenol reaction, catalysed by the etherate, was being run in petroleum ether solution in a sealed pressure bottle. The bottle burst, possibly owing to exothermic polymerisation of the diene.

Buten-3-yne
See Buten-3-yne: 1,3-Butadiene

Cobalt
See reference 7 above

Crotonaldehyde
See Crotonaldehyde: Butadiene

Ethanol, Iodine, Mercury oxide
See 2-Ethoxy-1-iodo-3-butene

Other reactants
Yoshida, 1980, 296
MRH values calculated for 13 combinations with oxidants are given.

Oxides of nitrogen, Oxygen   MRH Nitrogen oxide 8.74/85
Vervalin, 1973, 63–65
An explosion and fire occurred in the pipework of a vessel in which dilute butadiene was stored under an 'inert' atmosphere, generated by the combustion of fuel gas in a limited air supply. The 'inert' gas, which contained up to 1.8% of oxygen and traces of oxides of nitrogen, reacted in the vapour phase over an extended period to produce concentrations of gummy material containing up to 64% of butadiene peroxide and 4.2% of a butadiene–nitrogen oxide complex. The deposits eventually decomposed explosively.
See Nitrogen oxide: Dienes, Oxygen

Oxygen
MCA Case History No. 303
A leaking valve allowed butadiene to accumulate in a pipeline exposed to an inerting gas containing up to 2% of oxygen. Peroxide formed and initiated 'popcorn' polymerisation which burst the pipeline.
See reference 6 above

Sodium nitrite
See Sodium nitrite: 1,3-Butadiene
See other DIENES, ENDOTHERMIC COMPOUNDS, PEROXIDISABLE COMPOUNDS

†1481. 1-Butyne
[107-00-6] C_4H_6

$$HC{\equiv}CEt$$

Highly endothermic (ΔH°_f (g) +166.1 kJ/mol, 3.07 kJ/g).
See other ALKYNES, ENDOTHERMIC COMPOUNDS

†1482. 2-Butyne
[503-17-3] C_4H_6

$$MeC{\equiv}CMe$$

HCS 1980, 253

Highly endothermic (ΔH°_f (g) +148.0 kJ/mol, 2.74 kJ/g).
See other ALKYNES, ENDOTHERMIC COMPOUNDS

†1483. Cyclobutene
[822-35-3] C_4H_6

See other ALKENES

1484. *cis*-Poly(butadiene)
[9003-17-2] $(C_4H_6)_n$

$$(CH_2CH{=}CHCH_2)_n$$

Sedov, V. V. *et al., Chem. Abs.*, 1976, **85**, 109747
Stereoregular *cis*-poly(butadiene) compositions may explode when heated at 337–427°C/1–0.01 µbar, presumably owing to exothermic cyclisation.
See related DIENES

1485. Dimethylaminobis(trifluoromethyl)borane (*N,N*-Dimethyl-1,1-bis(trifluoromethyl)boranamine)
[105224-90-6] $C_4H_6BF_6N$

$$(CF_3)_2BNMe_2$$

Primary alkynes
Bürger, H. *et al., Main Group Met. Chem.*, 1995, **18**(5), 235

The adducts of primary alkynes with the borane may explode when heated to 90°C.
See other ACETYLENIC COMPOUNDS

1486. Barium acetate
[543-80-6] $C_4H_6BaO_4$

$$Ba(OCO.Me)_2$$

Copper(II) oxide, Yttrium oxide
1. Fleischer, N., *Chem. Eng. News*, 1988, **66**(15), 2
2. Author's comments
3. Castrillon, J., *Chem. Eng. News*, 1988, **66**(25), 2
A pelleted mixture containing barium acetate, copper(II) oxide and yttrium oxide, 3 g in all in a quartz tube, was heated in a furnace, and a small explosion occurred during the early stages, 'from formation of pyrolysis products'. It was suggested that the tube should in future be purged, preferably with an oxygen-containing gas [1]. This suggestion is hazardous, however, as the pyrolysis product is acetone. Only inert gas should be used for purging [2]. Pyrolysis of barium acetate gives high yields of acetone, and is catalysed by the metal oxides present in the mixture [3].
See Cadmium propionate
See other HEAVY METAL DERIVATIVES

1487. 4-Chlorobutyronitrile (4-Chlorobutanonitrile)
[628-20-6] C_4H_6ClN

$$Cl[CH_2]_3C\equiv N$$

Sodium hydroxide
Brogli, F. *et al., Runaway Reactions*, 1981, Paper 3/M, 5–6, 10
Unstable plant-scale operation in the catalysed cyclisation by sodium hydroxide to cyclopropanecarbonitrile was investigated using a bench scale calorimeter. Crust formation on the reactor wall, which caused the erratic operation, was eliminated by using liquid alkali instead of solid.
See other CYANO COMPOUNDS

1488. *N*-Chloro-5-methyl-2-oxazolidinone
[25480-76-6] $C_4H_6ClNO_2$

Walles, W. E., US Pat. 3 850 920, 1974
It exploded at 160°C.
See other N-HALOGEN COMPOUNDS

1489. N-Chloro-3-morpholinone
[33744-03-5] $C_4H_6ClNO_2$

Walles, W. E., US Pat 3 850 920, 1974
It exploded at 115°C.
See other N-HALOGEN COMPOUNDS

1490. N-Chloro-4,5-dimethyltriazole
[72040-09-6] $C_4H_6ClN_3$

Gallagher, T. C. *et al., J. Chem. Soc., Chem. Comm.*, 1979, 420
The solid (or a concentrated solution) decomposes vigorously on standing at
ambient temperature and needs careful handling.
See other N-HALOGEN COMPOUNDS, TRIAZOLES

1491. 1,1-Dichloroethyl peroxyacetate
[59183-18-5] $C_4H_6Cl_2O_3$

$$MeCO.OOCCl_2Me$$

Griesbaum, K. *et al., J. Amer. Chem. Soc.*, 1976, **98**, 2880
A friction- and heat-sensitive viscous liquid, it exploded violently during attempted
injection into a gas chromatograph.
See other PEROXYESTERS

1492. 3,6-Dichloro-3,6-dimethyltetraoxane
[59183-17-4] $C_4H_6Cl_2O_4$

Griesbaum, K. *et al., J. Amer. Chem. Soc.*, 1976, **98**, 2880

Extremely sensitive to shock, heat or minor friction, it has exploded very violently when touched with a glass rod or spatula.
See other CYCLIC PEROXIDES

1493. Chromium(II) acetate
[628-52-4]   C₄H₆CrO₄

$$Cr(OAc)_2$$

1. Ocone, L. R. *et al., Inorg. Synth.*, 1966, **8**, 129–131
2. Young, G. C., *J. Chem. Educ.*, 1988, **65**, 918–919
The anhydrous salt is pyrophoric in air, or chars slowly in lower oxygen concentrations [1]. A safe and convenient synthesis of the dimeric monohydrate is described [2].
See Bis(2,4-pentanedionato)chromium
See other PYROPHORIC MATERIALS

1494. Chromyl acetate
[4112-22-5]   C₄H₆CrO₆

$$O_2Cr(OAc)_2$$

Preparative hazard
See Chromium trioxide: Acetic anhydride
See other OXIDANTS

1495. 3,3-Dinitroazetidinium trifluoromethanesulfonate
[ ]   C₄H₆F₃N₃O₇S

Hiskey, M. A. *et al., J. Heterocycl. Chem.*, 1992, **29**(7), 1855
The *gem*-dinitro compound is shock sensitive, deflagrating under a heavy hammer blow.
See related POLYNITROALKYL COMPOUNDS

1496. 1,2-Bis(difluoroamino)ethyl vinyl ether
[13084-45-2]   C₄H₆F₄N₂O

$$F_2NCH_2CH(NF_2)OCH=CH_2$$

Reed, S. F., *J. Org. Chem.*, 1967, **32**, 2894
It is slightly less impact-sensitive than glyceryl nitrate.
See other DIFLUOROAMINO COMPOUNDS

1497. Di-1,2-bis(difluoroaminoethyl) ether
[13084-46-3] $C_4H_6F_8N_4O$

$$[F_2NCH_2CH(NF_2)]_2O$$

Reed, S. F., *J. Org. Chem.*, 1967, **32**, 2894
It is slightly more impact-sensitive than glyceryl nitrate.
See other DIFLUOROAMINO COMPOUNDS

1498. Divinylmagnesium
[6928-74-1] C_4H_6Mg

$$(H_2C{=}CH)_2Mg$$

Houben-Weyl, 1973, Vol. 13.2a, 203
The solid may ignite in air.
See related ALKYLMETALS

1499. 1-Methylimidazole
[616-47-7] $C_4H_6N_2$

Osmium(VIII) oxide
See Osmium(VIII) oxide: 1-Methylimidazole

1500. 2-Amino-4-methyloxazole
[35629-70-0] $C_4H_6N_2O$

Hydrogen peroxide
See Hydrogen peroxide: 2-Amino-4-methyloxazole
See related ISOXAZOLES

1501. 3-Amino-5-methylisoxazole
[1072-67-9] $C_4H_6N_2O$

Cardillo, P., *J. Loss Prev. Proc. Ind.*, 1988, **1**, 46–48; *Chem. Eng. News*, 1988, **66**(1), 4

508

The molten compound (100 kg) at 80°C was put into a 200 l drum for storage, but the closed drum exploded after 2 h. Analysis showed the presence of 5-amino-3-methylisoxazole as the main impurity. Thermal stability of the title compound, of the isomeric impurity, and of a mixture of the former with 5% of the latter was investigated by DSC and ARC. Very high self-heating rates were seen, and while the pure 3-amino compound decomposed from 140–160°C, the impure mixture and the prepared mixture of isomers started to decompose at 62°C, the decomposition exotherms being above 1.9 kJ/g in all cases, accompanied by gas evolution. The rapid rate of pressure rise and the final pressure (above 170 bar) which were observed explain the violent bursting of the drum of exothermically decomposing material. The thermal stability of the isomeric 5-amino-3-methylisoxazole was lower than that of the title compound (though the decomposition exotherm was lower at 1.42 kJ/g), and the presence of a few % of the 5-amino compound led to violent exothermic decomposition of the molten compound at 80°C.

See entry ISOXAZOLES *See other* N–O COMPOUNDS

1502. 5-Amino-3-methylisoxazole
[14678-02-5] $C_4H_6N_2O$

See 3-Amino-5-methylisoxazole (next above)
See entry ISOXAZOLES *See other* N–O COMPOUNDS

1503. Ethyl diazoacetate
[623-73-4] $C_4H_6N_2O_2$

$$N_2CHCO.OEt$$

1. Searle, N. E., *Org. Synth.*, 1962, Coll. Vol. 4, 426
2. Soos, R. *et al., Chem. Abs.*, 1984, **101**, 230032

It is explosive, and distillation, even under reduced pressure as described, may be dangerous [1]. A Hungarian patent describes a safe procedure for *in-situ* generation of the ester, azeotropic dehydration and subsequent metal-catalysed reaction with 1,3-dienes to give alkyl cyclopropanecarboxylates [2].

Tris(dimethylamino)antimony
Krommes, P. *et al., J. Organomet. Chem.*, 1975, **97**, 63, 65
Interaction at ambient temperature, either initially or on warming from −20°C or below, becomes explosively violent.
See other DIAZO COMPOUNDS

509

1504. 3-Ethyl-4-hydroxy-1,2,5-oxadiazole (4-Ethyl-3-furazanone)
[34529-29-8] $C_4H_6N_2O_2$

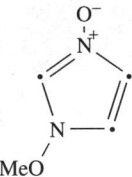

Sodium hydroxide
   Barker, M. D., *Chem. & Ind.*, 1971, 1234
   The sodium salt obtained by vacuum evaporation at 50°C of an aqueous alcoholic
   mixture of the above ingredients exploded violently when disturbed.
   See related FURAZAN *N*-OXIDES *See other* N–O COMPOUNDS

1505. 1-Methoxyimidazole *N*-oxide
   [90052-22-5] $C_4H_6N_2O_2$

   Flynn, A. P., *Chem. Brit.*, 1984, **20**, 30
   Attempted removal of water from a 3 g sample of the hydrate by
   vacuum distillation at 150–180°C/0.027 mbar caused a violent explosion as
   distillation started. DSC examination showed a large decomposition exotherm at
   140–180°C.
   See 1-Methoxy-3,4,5-trimethylpyrazole *N*-oxide, *also* 1-Hydroxyimidazole *N*-
   oxide
   See other N-OXIDES

1506. 2-Cyano-2-propyl nitrate
   [40561-27-1] $C_4H_6N_2O_3$
                                    $Me_2C(CN)ONO_2$

   Freeman, J. P., *Org. Synth.*, 1963, **43**, 85
   It is a moderately explosive material of moderate impact-sensitivity.
   See NITRATING AGENTS
   See other ALKYL NITRATES, CYANO COMPOUNDS, N–O COMPOUNDS

510

†1507. 1,2:3,4-Diepoxybutane
[298-18-0] $C_4H_6O_2$

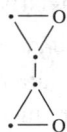

See other 1,2-EPOXIDES

1508. 1,1-Dinitro-3-butene
[10229-09-1] $C_4H_6N_2O_4$

$(O_2N)_2CHCH_2CH{=}CH_2$

Witucki, E. F. *et al., J. Org. Chem.*, 1972, **37**, 152
A fume-off during distillation of this compound (b.p., 63–65°C/2 mbar) illustrates
the inherent instabilty of this type of compound.
See other NITROALKENES, POLYNITROALKYL COMPOUNDS

1509. 2,3-Dinitro-2-butene
[28103-68-6] $C_4H_6N_2O_4$

$MeC(NO_2){=}C(NO_2)Me$

Bisgrove, D. E. *et al., Org. Synth.*, 1962, Coll. Vol. 4, 374
Only one explosion has been recorded during vacuum distillation at
135°C/14 mbar.
See other NITROALKENES, POLYNITROALKYL COMPOUNDS

1510. Diacetatoplatinum(II) nitrate
[ ] $C_4H_6N_2O_{10}Pt$

$[MeC{:}O_2{:}Pt{:}O_2{:}CMe]\ [NO_3]_2$

Preparative hazard
See Nitric acid: Acetic acid, etc.
See other PLATINUM COMPOUNDS

1511. 2-Methyl-5-vinyltetrazole
[15284-39-6] $C_4H_6N_4$

Aluminium hydride
See Aluminium hydride: Tetrazole derivatives
See other TETRAZOLES

1512. N,N'-Dimethyl-N,N'-dinitrosooxamide
[7601-87-8] $C_4H_6N_4O_4$

$$(MeN(N:O)CO.-)_2$$

Preussmann, R., *Angew. Chem.*, 1963, **75**, 642

Removal of the solvent carbon tetrachloride (in which nitrosation had been effected) at ambient, rather than reduced, pressure caused a violent explosion at the end of distillation. Lowest possible temperatures should be maintained in the preparation. Other precursors seem more suitable as sources of diazomethane.

See Diazomethane
See other NITROSO COMPOUNDS

1513. N,N'-Dimethyl-N,N'-dinitrooxamide
[14760-99-7] $C_4H_6N_4O_6$

$$(MeN(NO_2)CO.-)_2$$

Allenby, O. C. W. *et al.*, *Can. J. Res.*, 1947, **25B**, 295–300

It is an explosive.

See other N-NITRO COMPOUNDS

1514. Octahydro-2,5-bis(nitroimino)imidazo[4,5-d]imidazole (1.3a,4,6a-Tetrahydro-N,N'-dinitroimidazo[4,5-d]imidazole-2,5-diamine)
[139394-51-7] $C_4H_6N_8O_4$

Dagley, I. J. *et al.*, *J. Energ. Mater.*, 1995, **13**(1 & 2), 35

This nitroguanidine analogue is more sensitive than RDX by the drop-weight impact test.

See other N-NITRO COMPOUNDS

†1515. 1-Buten-3-one (Methyl vinyl ketone)
[78-94-4] C_4H_6O

$$H_2C=CHCO.Me$$

HCS 1980, 665

Haz. Chem. Data, 1975, 207

The uninhibited monomer polymerises on exposure to heat or sunlight. The inhibited monomer may also polymerise if heated sufficiently (by exposure to fire) and lead to rupture of the containing vessel.

See other POLYMERISATION INCIDENTS

Other reactants

Yoshida, 1980, 368

MRH values calculated for 13 combinations with oxidants are given.

†1516. Crotonaldehyde (2-Butenal)
 [4170-30-3]   C₄H₆O

MeCH=CHCO.H

HCS 1980, 330

Butadiene
  1. Greenlee, K. W., *Chem. Eng. News*, 1948, **26**, 1985
  2. Hanson, E. S., *Chem. Eng. News*, 1948, **26**, 2551
  3. Watson, K. M., *Ind. Eng. Chem.*, 1943, **35**, 398
  An autoclave without a bursting disk and containing the two poorly mixed reactants
  was wrecked by a violent explosion which occurred on heating the autoclave to
  180°C [1]. This was attributed to not allowing sufficient free space for liquid
  expansion to occur [2]. The need to calculate separate reactant volumes under
  reaction conditions for all autoclave preparations is stressed [3].
  See 1,3-Butadiene

Ethyl acetoacetate
  Soriano, D. S. *et al., J. Chem. Educ.*, 1988, **65**, 637
  A procedure using a phase-transfer catalyst is employed to prevent the rapid poly-
  merisation of crotonaldehyde during the Robinson annulation reaction.
  See related POLYMERISATION INCIDENTS

Nitric acid
  See Nitric acid: Crotonaldehyde
  See other ALDEHYDES, PEROXIDISABLE COMPOUNDS

†1517. 2,3-Dihydrofuran
 [1191-91-7]   C₄H₆O

See other PEROXIDISABLE COMPOUNDS

1518. Dimethylketene (2-Methyl-1-propene-1-one)
 [598-26-5]   C₄H₆O

Me₂C=C=O

  1. Smith, G. W. *et al., Org. Synth.*, 1962, Coll. Vol. 4, 348
  2. Staudinger, H., *Ber.*, 1925, **58**, 1079–1080
  Dimethylketene rapidly forms an extremely explosive peroxide when exposed to
  air at ambient temperatures. Drops of solution allowed to evaporate may explode.
  Inert atmosphere should be maintained above the monomer [1]. The peroxide is

513

polymeric and very sensitive, exploding on friction at −80°C. Higher homologues are very unstable and unisolable [2].

See other FRICTIONAL INITIATION INCIDENTS

Oxygen
See Poly(peroxyisobutyrolactone)
See other POLYPEROXIDES

†1519. Divinyl ether (1,1′-Oxybisethene)
[109-93-3]   C₄H₆O

$$(H_2C=CH)_2O$$

Anon., *Chemist & Druggist*, 1947, **157**, 258
The presence of *N*-phenyl-1-naphthylamine as inhibitor considerably reduces the development of peroxide in the ether.

Nitric acid
See Nitric acid: Divinyl ether
See other PEROXIDISABLE COMPOUNDS

†1520. 3,4-Epoxybutene (Vinyloxirane)
[930-22-3]   C₄H₆O

See other 1,2-EPOXIDES

†1521. Ethoxyacetylene (Ethoxyethyne)
[927-80-0]   C₄H₆O

$$EtOC≡CH$$

Jacobs, T. L. *et al., J. Amer. Chem. Soc.*, 1942, **64**, 223
Small samples rapidly heated in sealed tubes to around 100°C exploded.

Ethylmagnesium iodide
See Ethylmagnesium iodide: Ethoxyacetylene
See other ACETYLENIC COMPOUNDS

†1522. Methacrylaldehyde (2-Methylpropenal)
[78-85-3]   C₄H₆O

$$H_2C=CMeCO.H$$

See other ALDEHYDES, PEROXIDISABLE COMPOUNDS

514

1523. 3-Methoxypropyne (Methyl propargyl ether)
[627-41-8]

$$HC{\equiv}CCH_2OMe$$

C$_4$H$_6$O

Brandsma, 1971, 13
It explodes on distillation at its atmospheric b.p., 61°C.
See other ACETYLENIC COMPOUNDS

†1524. Allyl formate (3-Propenyl methanoate)
[1838-59-1]

$$H_2C{=}CHCH_2OCO.H$$

C$_4$H$_6$O$_2$

See other ALLYL COMPOUNDS

†1525. Butane-2,3-dione
[431-03-8]

$$MeCO.CO.Me$$

C$_4$H$_6$O$_2$

HCS 1980, 228

1526. 2-Butyne-1,4-diol
[110-65-6]

$$HOCH_2C{\equiv}CCH_2OH$$

C$_4$H$_6$O$_2$

Alkalies, or Halide salts, or Mercury salts and acids
Kirk-Othmer, 1960, Suppl. Vol. 2, 45
The pure diol may be distilled unchanged, but traces of alkali or alkaline earth
hydroxides or halides may cause explosive decomposition during distillation. In
presence of strong acids, mercury salts may cause violent decomposition of the diol.
See other ACETYLENIC COMPOUNDS, CATALYTIC IMPURITY INCIDENTS

1527. Butyrolactone (2(3*H*)-Dihydrofuranone)
[96-48-0]

C$_4$H$_6$O$_2$

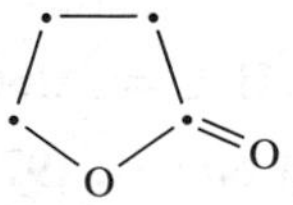

Butanol, 2,4-Dichlorophenol, Sodium hydroxide
Anon., *CISHC Chem. Safety Summ.*, 1977, **48**, 3
In an altered process to prepare 2,4-dichlorophenoxybutyric acid, the lactone was
added to the other components, and soon after, the reaction temperature reached
165°C, higher than the usual 160°C. Application of cooling failed to check the

515

thermal runaway, and soon after reaching 180°C the vessel began to fail and an explosion and fire occurred.

See other RUNAWAY REACTIONS

1528. Dimethyl azoformate (Dimethyl diazenedicarboxylate)
[2446-84-6] $C_4H_6N_2O_4$

MeOCO.N=NCO.OMe

1. US Pat. 3 347 845, 1967
2. Kauer, J. C., *Org. Synth.*, 1962, Coll. Vol. 4, 412
3. Young, J. A., private communication, 1993
4. Diels, O. *et al., Berichte*, 1913, **46**, 2006

It is shock-sensitive [1]. An undergraduate student was preparing a 200 g batch following the Organic Syntheses procedure [2]. The first stage quality is believed to have been poor. The final product exploded during vacuum distillation, costing her the sight of an eye [3]. Since the first man to describe it compared it with guncotton, when heated under slight confinement [4], it might be considered unsuitable for undergraduates. The editor suspects that hydrazoic acid reported as an hydrolysis product may indicate that procedure [2] can produce methyl azidoformate as a byproduct. Diels [4] also demonstrated formation of another very unstable compound from its further reaction with nitric acid, which is the oxidant used in the second stage of preparation [2].

See Ethyl azidoformate
See Diethyl azoformate
See other AZO COMPOUNDS

†1529. 1,2:3,4-Diepoxybutane
[298-18-0] $C_4H_6O_2$

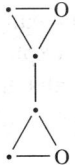

See other 1,2-EPOXIDES

1530. Methacrylic acid (2-Methylpropenoic acid)
[79-41-4] $C_4H_6O_2$

$H_2C=CMeCO.OH$

1. Anon., *CISHC Chem. Safety Summ.*, 1979, **50**, 34
2. Bond, J., *Loss Prev. Bull.*, 1991, **101**, 1
3. Anderson, S. E. *et al., Plant/Oper. Progress*, 1992, **11**(3), 151
4. Nicolson, A., *Plant/Oper. Progress.*, 1991, **10**(3), 171

A drum of the uninhibited acid (m.p., 16°C) which had been stored outside under winter conditions was transferred into a warm room to liquefy the acid. Later, exothermic polymerisation led to bulging of the drum and leakage of the acid vapour [1]. A more serious accident involved a railtank of acid-washed crude technical methacrylic acid (liable to throw down a dilute H_2SO_4 layer) which exploded, throwing debris 300 metres, some 20 hrs after a pressure relief valve was seen to lift, despite being sprayed with water meanwhile [2,3]. It was shipped in unlined steel without corrosion tests because that had previously been used for purified inhibited product and it contained negligible stabiliser. A study of the complex interactions of oxygen (inhibitory at low but not at high concentrations) and stabilisers is reported [3]. Phenothiazine seems a better inhibitor than the usual phenols.

See Acrylic acid
See other CORROSION INCIDENTS
See other ORGANIC ACIDS, CORROSION INCIDENTS, POLYMERISATION INCIDENTS

†1531. Methyl acrylate (Methyl propenoate)
[96-33-3] $C_4H_6O_2$

$$MeOCO.CH=CH_2$$

(*MCA SD-79*, 1960); *HCS 1980*, 634

1. *MCA Case History No. 2033*
2. Harmon, 1974, 2.11
3. *See entry* SELF-ACCELERATING REACTIONS

A 4 l glass bottle of the ester (sealed and partly polymerised old stock) exploded several hours after being brought from storage into a laboratory. An inhibitor had originally been present, but could have been consumed during prolonged storage, and peroxides may have initiated exothermic polymerisation of the remaining monomer [1]. The monomer is normally stored and handled inhibited and at below 10°C, but not under inert atmosphere, because traces of oxygen are essential to the inhibition process [2]. DSC investigation showed that the temperature-dependent induction period (425 min/125°C, 25/165, 6/185, in which the stabiliser becomes consumed) is followed by autocatalytic and rapidly accelerating polymerisation, with $Q = 0.89$ kJ/g [3].

See VIOLENT POLYMERISATION
See other GLASS INCIDENTS, INDUCTION PERIOD INCIDENTS, POLYMERISATION INCIDENTS

Other reactants
Yoshida, 1980, 10
MRH values calculated for 13 combinations with oxidants are given.

†1532. Vinyl acetate (Ethenyl ethanoate)
[108-05-4] $C_4H_6O_2$

$$H_2C=CHOCO.Me$$

(*MCA SD-75*, 1970); *FPA H36*, 1975; *HCS 1980*, 955

1. Harmon, 1974, 2.19
2. Levy, L. B., *Process Safety Progress, 1993,* **12** (1), 47
3. Gustin, J. L. *et al., Chem. Abs.*, 1998, **129**, 264630g

The monomer is volatile and tends to self-polymerise, and is therefore stored and handled cool and inhibited, with storage limited to below 6 months. Several industrial explosions have been recorded [1]. Unlike acrylic monomers, oxygen is not involved in stabilisation and is detrimental at higher temperatures [2]. The polymerisation has been modelled and causes of accidents proposed [3].

Air, Water
MCA SD-75, 1970
Vinyl acetate is normally inhibited with hydroquinone to prevent polymerisation. A combination of too low a level of inhibitor and warm, moist storage conditions may lead to spontaneous polymerisation. This process involves autoxidation of acetaldehyde (a normal impurity produced by hydrolysis of the monomer) to a peroxide which initiates exothermic polymerisation as it decomposes. In bulk, this may accelerate to a dangerous extent. Other peroxides or radical sources will initiate the exothermic polymerisation.

Desiccants
MCA SD-75, 1970
Vinyl acetate vapour may react vigorously in contact with silica gel or alumina.

Dibenzoyl peroxide, Ethyl acetate
Vervalin, 1973, 81
Polymerisation of the ester with dibenzoyl peroxide in ethyl acetate accelerated out of control and led to discharge of a large volume of vapour which ignited and exploded.
See other POLYMERISATION INCIDENTS

Ethylene
See Ethylene: Vinyl acetate

Hydrogen peroxide   MRH 6.19/78
See Hydrogen peroxide: Vinyl acetate

Other reactants
Yoshida, 1980, 122
MRH values calculated for 15 combinations with oxidants are given.

Oxygen
Barnes, C. E. *et al., J. Amer. Chem. Soc.*, 1950, **72**, 210
The unstabilised polymer exposed to oxygen at 50°C generated an ester–oxygen interpolymeric peroxide which, when isolated, exploded vigorously on gentle heating.
See other POLYPEROXIDES

Ozone
See Vinyl acetate ozonide

Toluene

MCA Case History No. 2087
The initial exotherm of a solution polymerisation of the ester in boiling toluene in a $10 m^3$ reactor was too great for the cooling and vent systems, and the reaction began to accelerate out of control. Failure of a gasket released a quantity of the flammable reaction mixture which became ignited and destroyed the containing building.
See other POLYMERISATION INCIDENTS, PEROXIDISABLE COMPOUNDS

1533. Poly(1,3-butadiene peroxide)
[28655-95-0] $(C_4H_6O_2)_n$

Complex structure

Butadiene
1. Alexander, D. S., *Ind. Eng. Chem.*, 1959, **51**, 733
2. Hendry, D. G. *et al., Ind. Eng. Chem., Prod. Res. Dev.*, 1968, **7**, 136, 145
3. *ASTM D1022-76*
4. Bailey, H. C., private comm., 1974
5. *CISHC Chem. Safety Summ.*, 1980, **51**, 39
6. *Sichere Chemiearb.*, 1980, **32**, 6–8
7. *MCA Case History No. 2270*
A violent explosion in a partially full butadiene storage sphere was traced to butadiene peroxide. The latter had been formed by contact with air over a long period, and had eventually initiated the exothermic polymerisation of the sphere contents. A tank monitoring and purging system was introduced to prevent a recurrence [1]. The polyperoxide (formed even at 0°C) is a mixed 1,2- and 1,4-addition copolymer of butadiene with oxygen [2], effectively a dialkyl peroxide. Its concentration in butadiene is very seriously underestimated by a standard method applicable to hydroperoxides [3], which indicates only 5% of the true value. Alternative methods of greater reliability are likely to form the basis of a revised standard method currently under consideration [4]. A bolt falling into an 'empty' rail tanker during maintenance work led to a serious explosion. Other rail tankers used to convey unstabilised butadiene were later found to contain several kg of the gelatinous polyperoxide [5]. Peroxide formation in uninhibited monomer is very rapid, air-saturated monomer giving in 24 h 124 ppm of peroxide/18°C and 460 ppm/50°C. The peroxide ($n = 7-9$) has very low solubility in the monomer, and separates as a lower layer which is very shock-sensitive and powerfully explosive [6]. Solidified peroxide plugging a tube condenser decomposed explosively during drilling operations to clear the tubes [7].
See 1,3-Butadiene
See other POLYPEROXIDES

1534. Acetic anhydride
[108-24-7] $C_4H_6O_3$

MeCO.OCO.Me

(*MCA SD-15*, 1962); *HCS 1980*, 101

The principal reaction hazard attached to use of acetic anhydride is the possibility of rapid and exothermic acid-catalysed hydrolysis unless the conditions prevailing (temperature, agitation, order of mixing, proportion of water) are such as to promote smooth and progressive hydrolysis with adequate heat removal. The examples below illustrate these factors.

Acetic acid, Water
MCA Case History No. 1865
Erroneous addition of aqueous acetic acid into a tank of the anhydride caused violent exothermic hydrolysis of the latter.
See Water, below

Ammonium nitrate, Hexamethylenetetraminium acetate, Nitric acid
See Nitric acid: Acetic anhydride, etc.

Barium peroxide
See Barium peroxide: Acetic anhydride

Boric acid
1. Lerner, L. M., *Chem. Eng. News*, 1973, **51**(34), 42
2. *Experiments in Organic Chemistry*, 281, Fieser, L. F., Boston, Heath, 1955
Attempted preparation of acetyl borate by slowly heating a stirred mixture of the anhydride and solid acid led to an eruptive explosion at 60°C [1]. The republished procedure being used [2] omitted the reference to a violent reaction mentioned in the German original. Modifying the procedure by adding portions of boric acid to the hot stirred anhydride should give a smoother reaction.

Bromohydroxybiphenyl, Pyridine, Water
Nolan, 1983, Case History 36
Bromohydroxybiphenyl was *O*-acetylated with excess acetic anhydride in pyridine. The excess anhydride was to be hydrolysed by addition of water, but this was done without proper control using a hosepipe. The hydrolysis reaction ran away, causing boiling and evaporation of the reactor contents.
See Acetic acid, above; Water, below

N-tert-Butylphthalamic acid, Tetrafluoroboric acid
Boyd, G. V. *et al., J. Chem. Soc., Perkin Trans. 1*, 1978, 1346
Interaction to give *N-tert*-butylphthalisoimidium tetrafluoroborate was very violent, possibly because of exothermic hydrolysis of the anhydride by the 40% aqueous tetrafluoroboric acid.
See Tetrafluoroboric acid, below

Chromic acid
1. Dawkins, A. E., *Chem. & Ind.*, 1956, 196
2. Baker, W., *Chem. & Ind.*, 1956, 280

520

Addition of acetic anhydride to a solution of chromium trioxide in water caused violent boiling [1], due to the acid-catalysed exothermic hydrolysis of the anhydride [2].

Chromium trioxide   MRH 2.38/84
See Chromium trioxide: Acetic anhydride

1,3-Diphenyltriazene
See 1,3-Diphenyltriazene: Acetic anhydride

Ethanol, Sodium hydrogen sulfate
Staudinger, H., *Angew. Chem.*, 1922, **35**, 657
Accidental presence of the acid salt vigorously catalysed a large scale preparation of ethyl acetate, causing violent boiling and emission of vapour which became ignited and exploded.
See other CATALYTIC IMPURITY INCIDENTS

Glycerol, Phosphoryl chloride
Bellis, M. P., *Hexagon Alpha Chi Sigma* (Indianapolis), 1949, **40**(10), 40
Violent acylation occurs in catalytic presence of phosphoryl chloride, because the high viscosity of the mixture in absence of solvent prevents mixing and dissipation of the high heat of reaction.

Hydrochloric acid, Water
1. Vogel, 1957, 572–573
2. Batchelor, J. F., private comm., 1976
Crude dimethylaniline was being freed of impurities by treatment with acetic anhydride according to a published procedure [1]. However, three times the recommended proportion of anhydride was used, and the reaction mixture was ice cooled before addition of diluted hydrochloric acid to hydrolyse the excess anhydride. Hydrolysis then proceeded with explosive violence.

Hydrogen peroxide   MRH 5.52/71
See Hydrogen peroxide: Acetic anhydride

Hypochlorous acid
See Hypochlorous acid: Acetic anhydride

Metal nitrates
1. Davey, W. *et al., Chem. & Ind.*, 1948, 814
2. Collman, J. P. *et al., Inorg. Synth.*, 1963, **7**, 205–207
3. James, B. D., *J. Chem. Educ.*, 1974, **51**(9), 568
Use of mixtures of metal nitrates with acetic anhydride as a nitrating agent may be hazardous, depending on the proportions of reactants and on the cation; copper nitrate or sodium nitrate usually cause violent reactions [1]. An improved procedure for the use of the anhydride–copper(II) nitration mixture [2] has been further modified [3] to improve safety aspects.

Nitric acid   MRH 4.93/66
See Nitric acid: Acetic anhydride

Other reactants
Yoshida, 1980, 356
MRH values calculated for 16 combinations with oxidants are given.

Perchloric acid, Water
1. Anon., *ABCM Quart. Safety Summ.*, 1954, **25**, 3
2. Turner, H. S. *et al., Chem. & Ind.*, 1965, 1933
3. McG. Tegart, W. J., *The Electrolytic and Chemical Polishing of Metals in Research and Industry*, London, Pergamon, 2nd Ed., 1959

Anhydrous solutions of perchloric acid in acetic acid are prepared by using acetic anhydride to remove the diluting water from 72% perchloric acid. It is essential that the anhydride is added slowly to the aqueous perchloric–acetic acid mixture under conditions where it will react readily with the water, i.e. at about 10°C. Use of anhydride cooled in a freezing mixture caused delayed and violent boiling to occur. Full directions for the preparation are given [1]. A violent explosion occurred during the preparation of an electropolishing solution by addition of perchloric acid solution to a mixture of water and acetic anhydride. The cause of the explosion, the vigorously exothermic acid-catalysed hydrolysis of acetic anhydride, is avoided if water is added last to the mixture produced by adding perchloric acid solution to acetic anhydride [2]. The published directions [3] are erroneous and insufficiently detailed for safe working.
See Perchloric acid: Acetic anhydride, or: Dehydrating agents

Peroxyacetic acid
See Peroxyacetic acid: Acetic anhydride

Polyphosphoric acid, Water
Montfort, B. *et al., Bull. Soc. Chim. Fr.*, 1987, 848–854
2-Aryloxy-1,2-diarylethanones can be cyclodehydrated to diarylbenzofurans by heating with sodium acetate and acetic anhydride in polyphosphoric acid. Quenching the hot reaction mixture with water leads to initially violent acid-catalysed hydrolysis of the excess anhydride.

Potassium permanganate   MRH 2.55/86
See Potassium permanganate: Acetic acid, etc.

Sodium percarbonate
See Sodium carbonate hydrogen peroxidate: Acetic anhydride

Tetrafluoroboric acid
1. Lichtenberg, D. W. *et al., J. Organomet. Chem.*, 1975, **94**, 319
2. Wudl, F. *et al., Inorg. Synth.*, 1979, **19**, 29
Dehydration of the aqueous 48% acid by addition to the anhydride is rather exothermic, and caution is advised [1]. Operation at 0°C led to an explosion [2].
See Chromic acid, *also* Hydrochloric acid, *also* Perchloric acid, all above

522

4-Toluenesulfonic acid, Water
1. Jones,B. J. *et al., Clin. Chem.*, 1955, **1**, 305
2. Pearson, S. *et al., Anal. Chem.*, 1953, **25**, 813–814

Caution is advised [1] to prevent explosions when using an analytical method involving sequential addition of acetic acid, aqueous 4-toluenesulfonic acid and acetic anhydride to serum [2]. It is difficult to see why this should happen, unless the anhydride were all added before the sulfonic acid solution.
See Water, below

Water
1. Leigh, W. R. D. *et al., Chem. & Ind.*, 1962, 778
2. Benson, G., *Chem. Eng. News*, 1947, **25**, 3458

Accidental slow addition of water to a mixture of the anhydride and acetic acid (85:15) led to a violent, large scale explosion. This was simulated closely in the laboratory, again in the absence of mineral-acid catalyst [1]. If unmoderated, the rate of acid-catalysed hydrolysis of (water insoluble) acetic anhydride can accelerate to explosive boiling [2]. Essentially the same accident, fortunately with no injuries or fatalities this time, was repeated in 1990.
See other ACID ANHYDRIDES

1535. Peroxycrotonic acid (Peroxy-2-butenoic acid)
[5813-77-4] $C_4H_6O_3$

$$MeCH=CHCO.OOH$$

Vasilina, T. U. *et al., Chem. Abs.*, 1974, **81**, 151446
The decomposition of the acid has been mentioned in a safety context, but the details were not translated.
See other PEROXYACIDS

1536. Poly(peroxyisobutyrolactone) (Poly(dimethylketene peroxide))
[67772-28-5] $(C_4H_6O_3)_n$

$$(-OCMe_2CO.O-)_n$$

Turro, N. J. *et al., J. Amer. Chem. Soc.*, 1978, **100**, 5580
Autoxidation of dimethylketene with oxygen in ether at −20°C gives the poly(peroxylactone) which as a dry solid is liable to undergo unpredictable and violent detonation.
See Dimethylketene
See related PEROXYESTERS
See other POLYPEROXIDES

1537. Diacetyl peroxide
[110-22-2] $C_4H_6O_4$

$$MeCO.OOCO.Me$$

1. Shanley, E. S., *Chem. Eng. News*, 1949, **27**, 175
2. Kuhn, L. P., *Chem. Eng. News*, 1948, **26**, 3197

3. Kharasch, M. S. *et al., J. Amer. Chem. Soc.*, 1948, **70**, 1269
4. Moore, C. G., *J. Chem. Soc.*, 1951, 236
Acetyl peroxide may readily be prepared and used in ethereal solution. It is essential to prevent separation of the crystalline peroxide even in traces, since, when dry, it is shock-sensitive and a high explosion risk [1]. Crystalline material, separated and dried deliberately, detonated violently [2]. The commercial material, supplied as a 30% solution in dimethyl phthalate, is free of the tendency to crystallise and is relatively safe. It is, however, a powerful oxidant [1]. Precautions necessary for the preparation and thermolysis of the peroxide have been detailed [3,4].
See Fluorine: Sodium acetate

Other reactants
Yoshida, 1980, 134
MRH values calculated for 15 combinations with oxidisable elements and compounds are given.
See other DIACYL PEROXIDES

1538. *cis*-1,4-Dioxenedioxetane (2,5,7,8-Tetraoxa[4.2.0]bicyclooctane)
[59261-17-5] $C_4H_6O_4$

Wilson, T. *et al., J. Amer. Chem. Soc.*, 1976, **98**, 1087
A small sample of the solid peroxide exploded on warming to ambient temperature. Storage at $-20°C$ in solution appears safe.
See other CYCLIC PEROXIDES

1539. Poly(vinyl acetate peroxide)
[ ] $(C_4H_6O_4)_n$

(-CH(OAc)CH$_2$OO-)$_n$

See Vinyl acetate: Oxygen
See other POLYPEROXIDES

1540. Lead acetate–lead bromate
[ ] $C_4H_6O_4Pb.Br_2O_6Pb$

Pb(OAc)$_2$.Pb(BrO$_3$)$_2$

1. Berger, A., *Arbeits-Schutz.*, 1934, **2**, 20
2. Leymann,–, *Chem. Fabrik*, 1929, 360–361
The compound (formulated as the double salt, (fuel + oxidant), rather than the mixed salt) may be formed during the preparation of lead bromate from lead acetate and

potassium bromate in acetic acid, and is explosive and very sensitive to friction. Although lead bromate is stable up to 180°C, it is an explosive salt [1]. Further details of the incident are available [2].
See related METAL OXOHALOGENATES

1541. Palladium(II) acetate
[375-31-3] $C_4H_6O_4Pd$

$$Pd(OCO.Me)_2$$

Phenylacetylene
See Phenylacetylene: Palladium(II) acetate
See related PLATINUM COMPOUNDS

1542. Monoperoxysuccinic acid
[3504-13-0] $C_4H_6O_5$

$$HOCO.C_2H_4CO.OOH$$

1. Lombard, R. *et al., Bull. Soc. Chim. Fr.*, 1963, **12**, 2800
2. Castrantas, 1965, 16
It explodes in contact with flame [1] and is weakly shock-sensitive [2].
See other PEROXYACIDS

1543. Vinyl acetate ozonide (3-Acetoxy-1,2,4-trioxolane)
[101672-23-5] $C_4H_6O_5$

Kirk-Othmer, 1970, Vol. 21, 320
The ozonide formed by vinyl acetate is explosive when dry.
See other OZONIDES

1544. Dimethyl peroxydicarbonate
[15411-45-7] $C_4H_6O_6$

$$MeOCO.OOCO.OMe$$

Explodes on heating to 55–60°C, or readily under a hammer blow, and more powerfully than dibenzoyl peroxide.
See entry PEROXYCARBONATE ESTERS

1545. Tartaric acid (2,3-Dihydroxybutanedioic acid)
[87-69-4] $C_4H_6O_6$

$$(-CHOHCO.OH)_2$$

HCS 1980, 883

Silver
  See Silver: Carboxylic acids
  See other ORGANIC ACIDS

1546. 2-Butyne-1-thiol
[101672-05-3] C_4H_6S

$$EtC≡CSH$$

Brandsma, 1971, 180
Presence of a stabiliser is essential during handling or storage at −20°C under nitrogen. Exposure to air leads to formation of a polymer which may explode on heating.
See other ACETYLENIC COMPOUNDS

1547. Divinylzinc
[1119-22-8] C_4H_6Zn

$$(H_2C=CH)_2Zn$$

491M, 1975, 164

It ignites in air.
See related ALKYLMETALS

†1548. 1-Bromo-2-butene
[4784-77-4] C_4H_7Br

$$BrCH_2CH=CHMe$$

See other ALLYL COMPOUNDS, HALOALKENES

†1549. 4-Bromo-1-butene
[5162-44-7] C_4H_7Br

$$H_2C=CHCH_2CH_2Br$$

Chloromethylphenylsilane, Chloroplatinic acid
  See Chloromethylphenylsilane: 4-Bromobutene, etc.
  See other HALOALKENES

1550. 1,1,1-Tris(bromomethyl)methane (3-(Bromomethyl)-1,3-dibromopropane)
[62127-48-4] $C_4H_7Br_3$

$$(BrCH_2)_3CH$$

Preparative hazard
See Phosphorus tribromide: 1,1,1-Tris(hydroxymethyl)methane

†**1551. 2-Chloro-2-butene**
[4461-41-0] C_4H_7Cl

$$MeCH{=}CClMe$$

See other HALOALKENES

†**1552. 3-Chloro-1-butene**
[563-52-0] C_4H_7Cl

$$H_2C{=}CHCHClMe$$

See other HALOALKENES

†**1553. 3-Chloro-2-methyl-1-propene**
[563-47-3] C_4H_7Cl

$$H_2C{=}CMeCH_2Cl$$

See other HALOALKENES

1554. *N*-Chloro-4-methyl-2-imidazolinone
[55341-15-6] $C_4H_7ClN_2O$

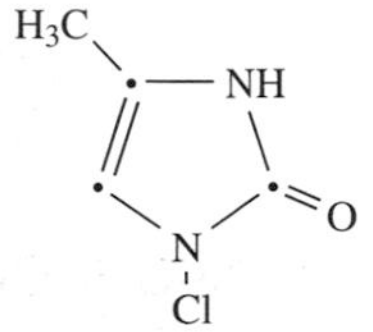

Walles, W. E., US Pat. 3 850 920, 1974
It exploded after several hours at ambient temperature.
See other *N*-HALOGEN COMPOUNDS

†**1555. Butyryl chloride (Butanoyl chloride)**
[141-75-3] C_4H_7ClO

$$PrCO.Cl$$

See other ACYL HALIDES

1556. 1-Chloro-2-butanone
[616-27-3] C_4H_7ClO

$$ClCH_2CO.Et$$

Tilford., C. H. private comm., 1965
An amber bottle of stabilised material spontaneously exploded.
See Chloroacetone

†1557. 2-Chloroethyl vinyl ether (2-Chloroethoxyethene)
[110-75-8] C_4H_7ClO

$$ClC_2H_4OCH{=}CH_2$$

See other PEROXIDISABLE COMPOUNDS

†1558. Isobutyryl chloride (2-Methylpropanoyl chloride)
[79-30-1] C_4H_7ClO

$$Me_2CHCO.Cl$$

See other ACYL HALIDES

1559. Ethyl chloroacetate
[105-39-5] $C_4H_7ClO_2$

$$EtOCO.CH_2Cl$$

HCS 1980, 475

Sodium cyanide
Brown, E. A. B., private comm., 1982
A well-established procedure (1 g mol, 20 runs) for preparation of ethyl cyanoacetate by heating the reactants together suddenly erupted out of control.

†1560. Isopropyl chloroformate
[108-23-6] $C_4H_7ClO_2$

$$Me_2CHOCO.Cl$$

1. *491M*, 1975, 223
2. Young, J. A., Chem. Health Saf., 1998, **5** (5), 4
A sample exploded in refrigerated storage. It is known that if iron salts are present during the preparation from 2-propanol and phosgene, catalysed thermal decompositon may occur [1]. A 2 litre glass bottle, presumably less than a year old and stored unrefrigerated at eye-level, exploded during stocktaking, splashing the eyes and face of a researcher [2]. Iron is usually present in glass, especially brown glass, while most eye protection is designed primarily to stop fast particles and shields poorly against slow liquids.
See other ACYL HALIDES, CATALYTIC IMPURITY INCIDENTS

1561. Isopropylisocyanide dichloride
[29119-58-2] $C_4H_7Cl_2N$

$$Me_2CHN{=}CCl_2$$

Iron(III) chloride, Metal oxides, or Water
Fuks, R. *et al., Tetrahedron*, 1973, **29**, 297–298
Interaction of the 3:2 complex with iron(III) chloride and calcium oxide, mercury oxide or silver oxide was usually too violent for preparative purposes, but zinc oxide was satisfactory. Reaction with water was violent.
See related CYANO COMPOUNDS

1562. 1-Fluoro-1,1-dinitrobutane
[19273-47-3] $C_4H_7FN_2O_4$

$$FC(NO_2)_2Pr$$

See entry FLUORODINITROMETHYL COMPOUNDS

†1563. Butyronitrile (Butanonitrile)
[109-74-0] C_4H_7N

$$PrCN$$

See other CYANO COMPOUNDS

1564. (Dimethylamino)acetylene (*N,N*-Dimethylethynamine)
[24869-88-3] C_4H_7N

$$Me_2NC{\equiv}CH$$

Water
Brandsma, 1971, 139
The amine reacts extremely vigorously with neutral water.
See other ACETYLENIC COMPOUNDS, ORGANIC BASES

†1565. Isobutyronitrile (2-Methylpropanonitrile)
[78-82-0] C_4H_7N

$$Me_2CHCN$$

See other CYANO COMPOUNDS

1566. 2-Cyano-2-propanol
[75-86-5] C_4H_7NO

$$Me_2C(CN)OH$$

HCS 1980, 103

Sulfuric acid
See Sulfuric acid: 2-Cyano-2-propanol
See other CYANO COMPOUNDS

1567. 3,6-Dihydro-1,2,2H-oxazine
[3686-43-9]

C_4H_7NO

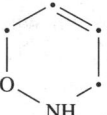

Nitric acid
See Nitric Acid: 3,6-Dihydro-1,2,2*H*-oxazine
See other N–O COMPOUNDS

1568. 2-Methylacryaldehyde oxime
[28051-68-5]

C_4H_7NO

$$H_2C=CMeCH=NOH$$

Mowry, D. T. *et al., J. Amer. Chem. Soc.*, 1947, **69**, 1831
The viscous distillation residue may decompose explosively if overheated. A polymeric peroxide may have been involved.
See other OXIMES, POLYPEROXIDES

†1569. 2-Methyl-2-oxazoline
[1120-64-5]

C_4H_7NO

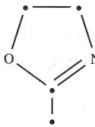

See other N–O COMPOUNDS

1570. 2,3-Butanedione monoxime
[57-71-6]

$C_4H_7NO_2$

$$MeCO.C(:NOH)Me$$

Muir, G. D., private comm., 1968
Distillation from a mantle heated flask at 6.5 mbar caused explosions on several occasions. Distillation from a steam heated flask at below 1.3 mbar appeared safe.
See other OXIMES

530

1571. N-Hydroxymethylacrylamide (N-Hydroxymethyl-2-propenamide)
[924-42-5] $C_4H_7NO_2$

$$HOCH_2NHCO.CH{=}CH_2$$

491M, 1975, 213

Excessive heat, smoke, flames and crackling noises were emitted from stored fibre drums of the monomer, some unopened. Polymerisation may have been initiated by minor contaminants (perhaps as vapours), and/or by excessively warm storage conditions.
See other POLYMERISATION INCIDENTS

1572. 4-Nitro-1-butene
[32349-29-4] $C_4H_7NO_2$

$$O_2NC_2H_4CH{=}CH_2$$

Witucki, E. F. *et al., J. Org. Chem.*, 1972, **37**, 152
A fume-off during distillation of one sample of this compound (b.p., 55°C/25 mbar) illustrates the inherent instability of this type of compound.
See other NITROALKENES

1573. Butyryl nitrate
[101672-06-4] $C_4H_7NO_4$

$$PrCO.ONO_2$$

Francis, F. E., *Ber.*, 1906, **39**, 3798
It detonates on heating.
See other ACYL NITRATES

1574. Butyryl peroxonitrate
[5796-89-4] $C_4H_7NO_5$

$$PrCO.OONO_2$$

Stephens, E. R., *Anal. Chem.*, 1964, **36**, 928
It is highly explosive, like its lower homologues.
See related PEROXYESTERS

1575. 4-Hydroxy-3,5-dimethyl-1,2,4-triazole
[35869-74-0] $C_4H_7N_3O$

Armand, J. *et al., J. Chem. Res. (M)*, 1980, 3868

It explodes on melting at 122°C.
See other N–O COMPOUNDS, TRIAZOLES

1576. 1,3-Bis(5-amino-1,3,4-triazol-2-yl)triazene
[3751-44-8] $C_4H_7N_{11}$

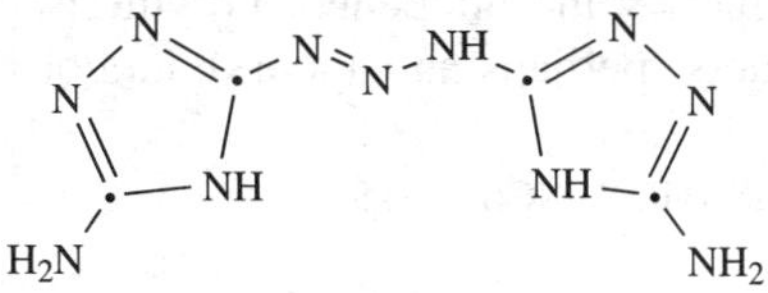

Hauser, M., *J. Org. Chem.*, 1964, **29**, 3449
It explodes at the m.p., 187°C.
See other HIGH-NITROGEN COMPOUNDS, TRIAZOLES

†1577. 1-Butene
[106-98-9] C_4H_8

$$H_2C{=}CHEt$$

HCS 1980, 229 (cylinder)

Aluminium tetrahydroborate
See Aluminium tetrahydroborate: Alkenes, etc.
See other ALKENES

†1578. *cis*-2-Butene
[624-64-6] C_4H_8

$$MeCH{=}CHMe$$

See other ALKENES

†1579. *trans*-2-Butene
[590-18-1] C_4H_8

$$MeCH{=}CHMe$$

See other ALKENES

†1580. Cyclobutane
[287-23-0] C_4H_8

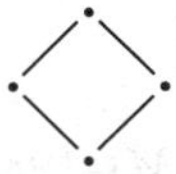

†1581. Methylcyclopropane
[594-11-6]   C₄H₈

See other STRAINED-RING COMPOUNDS

†1582. 2-Methylpropene (Isobutene)
[115-11-7]   C₄H₈

$$H_2C{=}CMe_2$$

FPA H108, 1981; *HCS 1980*, 566

See other ALKENES

1583. Poly(isobutene)
[9003-27-4]   (C₄H₈)ₙ

$$(\text{-}CMe_2CH_2\text{-})_n$$

Silver peroxide
See Silver peroxide: Poly(isobutene)

1584. *N*-Chloropyrrolidine
[19733-68-7]   C₄H₈ClN

It explodes at ambient temperature.
See entry N-HALOGEN COMPOUNDS (reference 4)

1585. *N*,*N*-Dimethyl-2-chloroacetamide
[2675-89-0]   C₄H₈ClNO

$$ClCH_2CO.N(Me)_2$$

Energy of decomposition (in range 180–440°C) measured as 0.59 kJ/g.
See entry THERMOCHEMISTRY AND EXOTHERMIC DECOMPOSITION (reference 2)

1586. 4-Morpholinesulfenyl chloride
[2958-89-6]

C_4H_8ClNOS

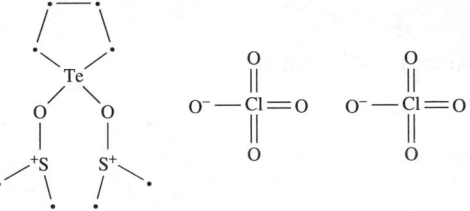

MCA Case History No. 1806

The product from chlorination of 4,4'-dithiodimorpholine exploded violently after vacuum stripping of the solvent carbon tetrachloride. As it was a published process that previously had been operated uneventfully, no cause was apparent. It seems remotely possible that an unstable *N*-chloro derivative could have been produced by N−S bond cleavage if chlorination conditions had differed from those previously employed.
See other N–S COMPOUNDS

†1587. *mixo*-Dichlorobutane
[26761-81-9]

$C_4H_8Cl_2$

$$ClCH_2CHClEt + ClCH_2CH_2CHClMe$$

See other HALOALKANES

1588. 1,1-Bis(dimethyl sulfoxide)telluracyclopentane diperchlorate
[100242-83-9]

$C_4H_8Cl_2O_8.2C_2H_6OS$

Srivastava, T. *et al., Polyhedron*, 1985, **4**, 1223–1229
The complex explodes at its m.p., 116°C.
See other ORGANOMETALLIC PERCHLORATES

1589. Bis(2-chloroethyl)sulfide (1,1'-Thiobis(2-chloroethane))
[505-60-2]

$C_4H_8Cl_2S$

$$(ClC_2H_4)_2S$$

Bleaching powder
See BLEACHING POWDER

1590. *N*-Chloro-bis(2-chloroethyl)amine
[63915-60-6]

$C_4H_8Cl_3N$

$$ClN(C_2H_4Cl)_2$$

Butters, M. *et al.*, UK Pat. Appl. GB 2 165 244, 1986

The *N*-chloro derivative is potentially explosive when neat, often exploding during distillation at 60°C/2.6 mbar.

See other N-HALOGEN COMPOUNDS

1591. Bis(1-chloroethylthallium chloride) oxide

[ ]   C₄H₈Cl₄OTl₂

$$(MeCHClTlCl)_2O$$

Yakubovich, A. Ya. *et al.*, *Doklad. Akad. Nauk SSSR*, 1950, **73**, 957–958
An explosive solid of low stability.
See related ALKYLMETALS, METAL HALIDES, METAL OXIDES

1592. 2-Methyl-1-nitratodimercurio-2-nitratomercuriopropane

[ ]   C₄H₈Hg₃N₂O₆

$$O_3NHgHgCH_2CMe_2HgNO_3$$

Whitmore, 1921, 116
The compound, prepared from isobutene and mercury(II) nitrate, explodes on impact at 80°C.
See other ORGANOMETALLIC NITRATES

1593. 1,1'-Biaziridinyl

[4388-03-8]   C₄H₈N₂

Oxygen
Graefe, A. F., *J. Amer. Chem. Soc.*, 1958, **80**, 3941
It exploded violently during analytical combustion in oxygen.
See other AZIRIDINES

1594. 3-Propyldiazirine

[70348-66-2]   C₄H₈N₂

Schmitz, E. *et al.*, *Ber.*, 1962, **95**, 800
It exploded on attempted distillation from calcium chloride at about 75°C.
See other DIAZIRINES

1595. Butane-2,3-dione dioxime
[95-45-4]

$C_4H_8N_2O_2$

MeC(:NOH)C(:NOH)Me

Energy of decomposition (in range 220–360°C) measured as 1.98 kJ/g.
See entry THERMOCHEMISTRY AND EXOTHERMIC DECOMPOSITION (reference 2)
See other OXIMES

1596. 1-Methylamino-1-methylthio-2-nitroethene (*N*-Methyl-1-(methylthio)-2-nitroethenamine)
[61832-41-5]

$C_4H_8N_2O_2S$

MeNHC(SMe)=CHNO$_2$

Preparative hazard
1. Koshy, A. *et al., J. Hazard. Mater.*, 1995, 42(3), 215
Goulding, J. P. in *Safety and Runaway Reactions*, Smeder, B & Mitchison, N. (Eds.),
EUR 17723 EN, European Commission, 1997, p31
This pharmaceutical intermediate, itself unstable, is prepared via intermediates which
are actually explosive. A calorimetric study of risks is reported [1]. The byproducts
are also hazardous, a study into an explosion and fire consequent upon isopropanol
(solvent) recovery from a manufacturing process is reported. This was done by distillation, leaving a concentrated residue, more than one distillation being performed
before the residue was discharged for incineration. A fortnight's accumulated residue,
as about a 25% solution or suspension in isopropanol was left in the reactor for a
week. This was probably below minimum stir volume and below the level of the
temperature probe. During this week, there was a power cut which appears to have
stopped the flow of coolant to the reactor; the coolant coils may in any event not have
dipped far into the residue. Two days later, shortly preceded by some indications
of modest warming, the explosion occurred, rupturing storage tanks, igniting their
contents and thus causing considerable secondary damage. Calorimetry suggested
that the deposited solids became unsafe at 40°C, the stirred bulk from about 65°C. It
was concluded that both the design and operation of the recovery process had been
inadequate. (Elsewhere in this book, an account of another explosion and destructive
sequence of secondary fires, when treating wastes from a slightly different manufacturing process, is to be found. In this other case neither product nor byproducts
were directly responsible — Ed.)
See Nitromethane; Carbon disulfide
See other C-NITRO COMPOUNDS

1597. Ethyl *N*-methyl-*N*-nitrosocarbamate
[615-53-2]

$C_4H_8N_2O_3$

EtOCO.N(NO)Me

1. Hartman, W. W., *Org. Synth.*, 1943, Coll. Vol. 2, 465
2. Druckrey, H. *et al., Nature*, 1962, **195**, 1111

The material is unstable and explodes if distilled at ambient pressure [1], or may become explosive if stored above 15°C [2]. Its use for preparation of diazomethane has been superseded by more stable intermediates.

See Diazomethane
See other NITROSO COMPOUNDS

1598. *trans*-4-Hydroperoxy-5-hydroxy-4-methylimidazolin-2-one
[85576-52-9] $C_4H_8N_2O_4$

Preparative hazard
See Hydrogen peroxide: 2-Amino-4-methyloxazole
See related ALKYL HYDROPEROXIDES

1599. 2,2′-Oxybis(ethyl nitrate)
[693-21-0] $C_4H_8N_2O_7$

$$O(C_2H_4ONO_2)_2$$

1. Kit and Evered, 1960, 268
2. Toso, R. *et al., Chem. Abs.*, 1988, **108**, 24140

A powerful explosive, sensitive to vibration and mechanical shock, too heat-sensitive for a rocket propellant [1]. In the industrial preparation of this explosive by the continuous nitration of diethylene glycol by nitric–sulfuric acid mixtures, the main factor limiting the overall safety of the process is the instability of the spent acid mixture, and handling of the latter is the most important specific problem. Close control of the acid composition and of the glycol:acid ratio reduces the nitric acid and water contents of the organic phase in the reaction mixture, and so increases the stability of the organic phase. Continuous addition of oleum to the spent acid coming from the separator increases its stability and permits safe handling and storage [2].
See other NITRATION INCIDENTS, ALKYL NITRATES

1600. *N*-Allylthiourea
[101-57-9] $C_4H_8N_2S$

$$H_2C{=}CHCH_2NHCS.NH_2$$

Energy of decomposition (in range 170–270°C) measured as 0.549 kJ/g.
See entry THERMOCHEMISTRY AND EXOTHERMIC DECOMPOSITION (reference 2)
See other ALLYL COMPOUNDS

1601. Azo-N-methylformamide
[18880-14-3] $\qquad$ $C_4H_8N_4O_2$

MeNHCO.N=NCO.NHMe

Smith, R. F. *et al., J. Org. Chem.*, 1975, **40**, 1855
Thermolysis is vigorously exothermic, the temperature increasing from 176 to 259°C during decomposition.
See other AZO COMPOUNDS, BLOWING AGENTS

1602. Bis(2-nitratoethyl)nitric amide
[4185-47-1] $\qquad$ $C_4H_8N_4O_8$

$O_2NN(C_2H_4ONO_2)_2$

Chromium compounds
$\quad$ Aleksandrov, V. V. *et al., Combust. Flame*, 1979, **35**, 1–15
$\quad$ The burning rate of the solid explosive is increased by various chromium compounds.
$\quad$ *See related* ALKYL NITRATES, N-NITRO COMPOUNDS

1603. 2,2-Diazidobutane
[90329-44-5] $\qquad$ $C_4H_8N_6$

$MeC(N_3)_2Et$

Kirchmeyer, S. *et al., Synthesis*, 1983, 501
Potentially explosive (a *gem*-diazide).
See other ORGANIC AZIDES

1604. 1,1′-Oxybis-2-azidoethane
[24345-74-2] $\qquad$ $C_4H_8N_6O$

$O(CH_2CH_2N_3)_2$

1. Rankine A., *Safety Digest Univ. Safety Assn.*, 1992, **43**, 22
2. Peacock, R. D., *Chemistry in Britain*, 1992, **28** (4), 327

An explosion was experienced during isolation of supposed diamine produced by reduction of the diazide with hydrogen sulfide. This was attributed to incomplete reduction [1]. An explosion, apparently during reduction, of a closely related compound by this method has earlier been reported [2].
See 1,2-Bis(2-azidoethoxy)ethane
See other ORGANIC AZIDES

1605. 1,3,5,7-Tetranitroperhydro-1,3,5,7-tetrazocine
[2691-41-0] $\qquad$ $C_4H_8N_8O_8$

The military explosive HMX. Violent decomposition occurred at 279°C.

See entry DIFFERENTIAL THERMAL ANALYSIS (reference 1)
See 1,3,5-Trinitrohexahydro-1,3,5-triazine
See other N-NITRO COMPOUNDS

†1606. 2-Butanone (Ethyl methyl ketone)
[78-93-3]   C$_4$H$_8$O
$$MeCO.Et$$

(*MCA SD-83*, 1961); *FPA H16*, 1973; *HCS 1980*, 650; *RSC Lab. Hazard Data Sheet
No. 18*, 1983

Chloroform, Alkali
  See Chloroform: Acetone, etc.

Hydrogen peroxide   MRH ,
  MRH 6.36/83

Nitric acid   MRH 5.77/79
  See Hydrogen peroxide: Ketones, etc.

Other reactants
  Yoshida, 1980, 365
  MRH values calculated for 13 combinations with oxidants are given.

Potassium *tert*-butoxide
  See Potassium *tert*-butoxide: Acids, etc.

2-Propanol
  See 2-Propanol: 2-Butanone

†1607. Butyraldehyde (Butanal)
[123-72-8]   C$_4$H$_8$O
$$PrCO.H$$

(*MCA SD-78*, 1960); *FPA H70*, 1978; *HCS 1980*, 254

See other ALDEHYDES, PEROXIDISABLE COMPOUNDS

†1608. Cyclopropyl methyl ether (Methoxycyclopropane)
[540-47-6]   C$_4$H$_8$O

MeO

See other PEROXIDISABLE COMPOUNDS, STRAINED-RING COMPOUNDS

†1609. 1,2-Epoxybutane (Ethyloxirane)
[106-88-7]   C₄H₈O

HCS 1980, 455

Dinitrogen pentaoxide
   See Dinitrogen pentaoxide: Oxygen heterocycles
   See other 1,2-EPOXIDES, STRAINED-RING COMPOUNDS

†1610. Ethyl vinyl ether (Ethoxyethene)
[109-92-2]   C₄H₈O

                              EtOCH=CH₂

Methanesulfonic acid
   Eaton, P. E. *et al. J. Org. Chem.*, 1972, **37**, 1947
   Methanesulfonic acid is too powerful a catalyst for *O*-alkylation with the vinyl ether,
   causing explosive polymerisation of the latter on multimolar scale. Dichloroacetic
   acid is a satisfactory catalyst on the 3 g mol scale.

Other reactants
   Yoshida, 1980, 288
   MRH values calculated for 13 combinations with oxidants are given.
   See other PEROXIDISABLE COMPOUNDS

†1611. Isobutyraldehyde (2-Methylpropanal)
[78-84-2]   C₄H₈O

                              Me₂CHCO.H

(*MCA SD-78*, 1960)

See other PEROXIDISABLE COMPOUNDS

†1612. Tetrahydrofuran
[109-99-9]   C₄H₈O

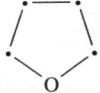

FPA H60, 1977; *HCS 1980*, 891; *RSC Lab. Hazard Data Sheet No. 12*, 1983

1. *Recommended Handling Procedures*, THF Brochure FC3-664, Wilmington, Du Pont, 1964
2. Anon., *Org. Synth.*, 1966, **46**, 105
3. Schurz, J. *et al., Angew. Chem.*, 1956, **68**, 182
4. Davies, A. G., *J. R. Inst. Chem.*, 1956, **80**,386
5. Halonbrenner, R. *et al., Chem. Eng. News*, 1978, **56**(6), 3
6. Coates, J. S., *Chem. Eng. News*, 1978, **56**(6), 3
7. Wissink, H. G., *Chem. Eng. News*, 1997, **75**(9), 6
8. Schwartz, A. M., *Chem. Eng. News*, 1978, **56**(24), 88
9. Hanson, E. S., *Chem. Eng. News*, 1978, **56**(24), 88
10. Nagano, H., *Chem. Abs.*, 1980, **92**, 163833

Like many other ethers and cyclic ethers, in absence of inhibitors tetrahydrofuran is subject to autoxidation on exposure to air, when initially the 2-hydroperoxide forms. This tends to decompose smoothly when heated, but if allowed to accumulate for a considerable period, it becomes transformed into other peroxidic species which will decompose violently [1]. Commercial material is supplied stabilised with a phenolic antioxidant which is effective under normal closed storage conditions in preventing the formation and accumulation of peroxide. Procedures for testing for the presence of peroxides and also for their removal are detailed [1,2,4]. The last of these references recommends the use of copper(I) chloride for removal of trace amounts of peroxide. If more than trace amounts are present, the peroxidised solvent (small quantities only) should be discarded by dilution and flushing away with water [2]. An attempt to remove peroxides by shaking with solid ferrous sulfate before distillation did not prevent explosion of the distillation residue [3]. Alkali treatment to destroy peroxides [1] appears not to be safe [2]. (See Caustic alkalies, below). Distillation or alkali treatment of stabilised THF removes the involatile antioxidant and the solvent must be restabilised or stored under nitrogen to prevent peroxide formation during storage, which should not exceed a few days' duration in the absence of stabiliser [2]. The use of lithium tetrahydroaluminate is only recommended for drying THF which is peroxide-free and is not grossly wet [2]. A violent explosion during reflux of the solvent with calcium hydride [5] was attributed to cleavage of the cyclic ether by overheated excess hydride [6]. Other metal hydrides (and some other strong bases) can also cause decomposition to ethene and enolate salts of acetaldehyde [7], pressurising containers by gas emission. The use of sodium benzophenone ketyl [8], or of an activated alumina column [9] to remove moisture and peroxides are detailed. Peroxides in THF may be destroyed by passage through activated carbon at 20–66°C with contact time >2 min [10].

See 2-Tetrahydrofuryl hydroperoxide
See also ETHERS (references 9,10)

2-Aminophenol, Potassium dioxide
   See Potassium dioxide: 2-Aminophenol, Tetrahydrofuran

Borane
   See Borane–tetrahydrofuran

Bromine
  See Bromine: Tetrahydrofuran

Calcium hydride
  See (references [5,6])

Caustic alkalies
  1. *NSC Newsletter, Chem. Sect.*, 1964(10); 1967(3)
  2. Anon., *Org. Synth.*, 1966, **46**, 105
  It is not safe to store quantities of THF which have been freed of the phenolic inhibitor
  (e.g. by alkali treatment) since dangerous quantities of peroxides may build up in
  prolonged storage. Peroxidised materials should not be dried with sodium hydroxide
  or potassium hydroxide, as explosions may occur [1,2].

Diisobutylaluminium hydride
  See Diisobutylaluminium hydride: Tetrahydrofuran

Lithium tetrahydroaluminate
  See Lithium tetrahydroaluminate: Tetrahydrofuran

Metal halides
  See Hafnium tetrachloride: Tetrahydrofuran
      Titanium tetrachloride: Tetrahydrofuran
      Zirconium tetrachloride: Tetrahydrofuran

Other reactants
  Yoshida, 1980, 238
  MRH values calculated for 12 combinations with oxidants are given.

Oxygen, strong base
  Butora, G. *et al., J. Amer. Chem. Soc.*, 1997, **119**(33), 7694
  As part of a synthesis, a rather dilute THF solution of an heterocycle, converted into
  a dianion with sec-butyllithium, was oxidised by bubbling gaseous oxygen through
  at −78 C. In the light of above entries, formation of peroxides is not surprising;
  however, the reaction tended to inflame or explode while still warming to ambient
  temperature for treatment with deperoxidising reagents. The technique is not recommended and an electrochemical alternative, far superior both in yield and safety, was
  devised.

Sodium tetrahydroaluminate
  See Sodium tetrahydroaluminate: Tetrahydrofuran

Sulfinyl chloride
  See Sulfinyl chloride: Tetrahydrofuran
  See other APROTIC SOLVENTS, PEROXIDISABLE COMPOUNDS

1613. 1,4-Oxathiane
[15980-15-1]

C₄H₈OS

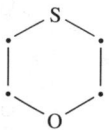

Metal perchlorates
See Silver perchlorate: 1,4-Oxathiane
    Copper(I) perchlorate: 1,4-Oxathiane

1614. Butyric acid (Butanoic acid)
[107-92-6]

C₄H₈O₂

PrCO.OH

HCS 1980, 255

Chromium trioxide
See Chromium trioxide: Butyric acid
See other ORGANIC ACIDS

1615. 3,3-Dimethyl-1,2-dioxetane
[32315-88-1]

C₄H₈O₂

Adam, W. *et al.*, *J. Amer. Chem. Soc.*, 1994, **116**(17), 7581
This molecule sometimes detonated spontaneously even below 0°C. The trimethyl and tetramethyl homologues should not be handled above 0°C and the precursor bromohydroperoxides are also hazardous.
See other DIOXETANES, STRAINED-RING COMPOUNDS; CYCLIC PEROXIDES

†1616. 1,3-Dioxane
[505-22-6]

C₄H₈O₂

See other PEROXIDISABLE COMPOUNDS

†**1617. 1,4-Dioxane**
  [123-91-1]   **C₄H₈O₂**

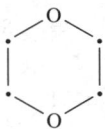

HCS 1980, 442; *RSC Lab. Hazard Data Sheet No. 29*, 1984

1. Dasler, W. *et al., Ind. Eng. Chem. (Anal. Ed.)*, 1946, **18**, 52

Like other monofunctional ethers but more so because of the four susceptible hydrogen atoms, dioxane exposed to air is susceptible to autoxidation with formation of peroxides which may be hazardous if distillation (causing concentration) is attempted. Because it is water-miscible, treatment by shaking with aqueous reducants (iron(II) sulfate, sodium sulfide, etc.) is impracticable. Peroxides may be removed, however, under anhydrous conditions by passing dioxane (or any other ether) down a column of activated alumina. The peroxides (and any water) are removed by adsorption onto the alumina, which must then be washed with methanol or water to remove them before the column material is discarded [1]. The heat of decomposition of dioxane has been determined (130–200°C) as 0.165 kJ/g.

See entry THERMOCHEMISTRY AND EXOTHERMIC DECOMPOSITION (reference 2)

Decaborane(14)
  See Decaborane(14): Ether, etc.

Nickel
  Mozingo, R., *Org. Synth.*, 1955, Coll. Vol. 3, 182
  Dioxane reacts with Raney nickel catalyst almost explosively above 210°C.

Nitric acid, Perchloric acid   MRH Nitric acid 5.81/74
  See Perchloric acid: Dioxane, Nitric acid

Other reactants
  Yoshida, 1980, 150
  MRH values calculated for 14 combinations with oxidants are given.

Sulfur trioxide   MRH 2.26/38
  See Sulfur trioxide: Diethyl ether, etc.

Triethynylaluminium
  See Triethynylaluminium: Dioxane
  See other APROTIC SOLVENTS, PEROXIDISABLE COMPOUNDS

†**1618. Ethyl acetate (Ethyl ethanoate)**
  [141-78-6]   **C₄H₈O₂**
                                   EtOCO.CH₃

  (*MCA SD-51*, 1953); *FPA H8*, 1973; *HCS 1980*, 462; *RSC Lab. Hazards Data Sheet No. 41*, 1985

1. Anon., *Loss. Prev. Bull.*, 1977, (013), 16

2. Bond, J., *Loss Prev. Bull.*, 1994, (119), 9

During rotary evaporation of solvent (from an ethyl acetate extract of a fermentation culture) at 55°C/50 mbar, the flask exploded. (It seems likely that some unsuspected peroxidised material may have been present in the extract, or perhaps the flask was scratched or cracked, and imploded [1].) Several fires and explosions suggest that ethyl acetate may be a worse generator of static electricity than its measured conductivity would indicate [2].

See Hydrogen peroxide: Ethyl acetate

See other STATIC INITIATION INCIDENTS

Lithium tetrahydroaluminate
See Lithium tetrahydroaluminate: Ethyl acetate

Potassium *tert*-butoxide
See Potassium *tert*-butoxide: Acids, etc.

†1619. Isopropyl formate
[625-55-8] $C_4H_8O_2$

$$Me_2CHOCO.H$$

†1620. Methyl propionate
[554-12-1] $C_4H_8O_2$

$$MeOCO.Et$$

HCS 1980, 659

†1621. Propyl formate
[110-74-7] $C_4H_8O_2$

$$PrOCO.H$$

Potassium *tert*-butoxide
See Potassium *tert*-butoxide: Acids, etc.

1622. Tetrahydrothiophene-1,1-dioxide (Sulfolane)
[126-33-0] $C_4H_8O_2S$

Nitronium tetrafluoroborate
See NITRATING AGENTS *See other* APROTIC SOLVENTS

1623. *trans*-2-Butene ozonide (3,5-Dimethyl-1,2,4-trioxolane)
[16187-15-8] $C_4H_8O_3$

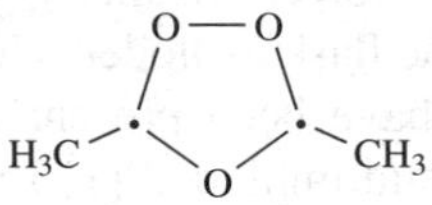

See trans-2-Hexene ozonide
See other OZONIDES

1624. 2-Tetrahydrofuryl hydroperoxide
[4676-82-8] $C_4H_8O_3$

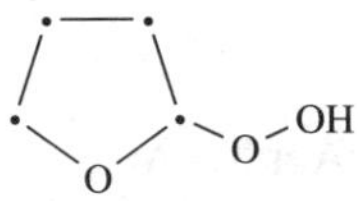

1. Rein, H., *Angew. Chem.*, 1950, **62**, 120
2. Criegee, R., *Angew. Chem.*, 1950, **62**, 120
This occurs as the first product of the ready autoxidation of tetrahydrofuran, and is relatively stable [1]. It readily changes, however, to a highly explosive polyalkylidene peroxide, which is responsible for the numerous explosions observed on distillation of peroxidised tetrahydrofuran [2].
See Tetrahydrofuran
See other ALKYL HYDROPEROXIDES

1625. 3,6-Dimethyl-1,2,4,5-tetraoxane
[cis [102502-32-9]; trans [118171-56-5]] $C_4H_8O_4$

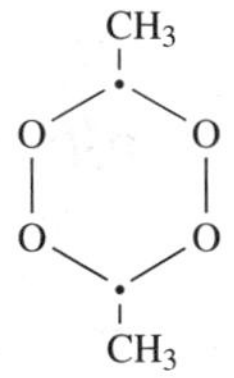

Rieche, A. *et al., Ber.*, 1939, **72**, 1933
This dimeric 'ethylidene peroxide' is an extremely shock-sensitive solid which explodes violently at the slightest touch. Extreme caution in handling is required.
See other CYCLIC PEROXIDES

1626. 1-Hydroxyethyl peroxyacetate
[7416-48-0] $C_4H_8O_4$

MeCH(OH)OOCO.Me

An explosive low-melting solid, readily formed during autoxidation of acetaldehyde.

546

See Acetaldehyde: Cobalt acetate, etc.
See other PEROXYESTERS

†1627. Tetrahydrothiophene
[110-01-0]

C_4H_8S

Hydrogen peroxide
See Hydrogen peroxide: Tetrahydrothiophene

1628. Acetyldimethylarsine
[21380-82-5]

C_4H_9AsO

AcAsMe$_2$

DOC 5, 1982, 38

It ignites in air.
See related ALKYLNON-METALS

1629. Butyldichloroborane
[14090-22-3]

$C_4H_9BCl_2$

BuBCl$_2$

Air, or Water
Niedenzu, K. et al., Inorg. Synth., 1967, 10, 126
It ignites on prolonged exposure to air; hydrolysis may be explosive.
See other ALKYLHALOBORANES

1630. But-2-en-2-ylboronic acid
[[125261-72-5] E -]
[[125261-73-6] Z -]

$C_4H_9BO_2$

CH$_3$CH=C(CH$_3$).B(OH)$_2$

Trombini, C. et al. J. Chem. Soc. Perk. 1, 1989, 1025
The compound is pyrophoric when dry.
See other PYROPHORIC COMPOUNDS

†1631. 1-Bromobutane (n -Butyl bromide)
[109-65-9]

C_4H_9Br

BrCH$_2$Pr

Bromobenzene, Sodium
See Sodium: Halocarbons (reference 7)
See other HALOALKANES

†1632. 2-Bromobutane (*sec*-Butyl bromide)
[78-76-2] C_4H_9Br
                          MeCHBrEt

See other HALOALKANES

†1633. 1-Bromo-2-methylpropane (Isobutyl bromide)
[78-77-3] C_4H_9Br
                        $BrCH_2CHMe_2$

See other HALOALKANES

†1634. 2-Bromo-2-methylpropane (*tert*-Butyl bromide)
[507-19-7] C_4H_9Br
                           $BrCMe_3$

See other HALOALKANES

†1635. 2-Bromoethyl ethyl ether
[592-55-2] C_4H_9BrO
                        BrC_2H_4OEt

1636. 1,1-Dimethylethyldibromamine (*tert*-Butyldibromamine)
[51655-36-8] $C_4H_9Br_2N$
                         Me_3CNBr_2

Kille, G. *et al.*, *Ind. Chem. Libr.*, 1991, **3**(1), 75; *Chem. Abs.*, 1992, **116620**(11), 105572
Differential scanning calorimetry (DSC) showed this to be prone to highly exothermic decomposition (100 J/g) at ambient temperatures. Solutions are a little more stable.
See other N-HALOGEN COMPOUNDS
See also DIFFERENTIAL SCANNING CALORIMETRY

†1637. 1-Chlorobutane (*n*-Butyl chloride)
[109-69-3] C_4H_9Cl
                          $ClCH_2Pr$

HCS 1980, 297

See other HALOALKANES

†1638. 2-Chlorobutane (*sec*-Butyl chloride)
[78-86-4] C_4H_9Cl
                          MeCHClEt

See other HALOALKANES

†1639. 1-Chloro-2-methylpropane (Isobutyl chloride)
[513-36-0]   C₄H₉Cl

$$ClCH_2CHMe_2$$

See other HALOALKANES

†1640. 2-Chloro-2-methylpropane (*tert*-Butyl chloride)
[507-20-0]   C₄H₉Cl

$$ClCMe_3$$

HCS 1980, 304

See other HALOALKANES

1641. Butylmagnesium chloride
[693-04-9]   C₄H₉ClMg

$$BuMgCl$$

Water
Nolan, 1983, Case history 109
Butylmagnesium chloride was prepared by adding a mixture of ether, butyl chloride and butyl bromide to magnesium, then cyclohexane was added prior to adding stannic chloride to form tetrabutyltin. However, the cyclohexane contained water and this reacted with the Grignard reagent to liberate butane which ruptured the bursting disk and ignited.
See other GAS EVOLUTION INCIDENTS, GRIGNARD REAGENTS

1642. *tert*-Butyl hypochlorite
[507-40-4]   C₄H₉ClO

$$Me_3COCl$$

1. Lewis, J. C., *Chem. Eng. News*, 1962, **40**(43), 62
2. Teeter, H. M. *et al., Org. Synth.*, 1962, Coll. Vol. 4, 125
3. Mintz, M. J. *et al., Org. Synth.*, 1973, Coll. Vol. 5, 185

This material, normally supplied or stored in sealed ampoules and used as a paper chromatography spray reagent, is photo-sensitive. Exposure to UV light causes exothermic decomposition to acetone and chloromethane. Ampoules have burst because of pressure build-up after exposure to fluorescent or direct day light. Store cool and dark, and open ampoules with personal protection. The material also reacts violently with rubber [1,2]. It should not be heated to above its boiling point [3]. There is also a preparative hazard.
See Chlorine: *tert*-Butanol

Sodium hydrogen cyanamide
See Cyanonitrene
See other HYPOHALITES, IRRADIATION DECOMPOSITION INCIDENTS

549

1643. 2(2-Hydroxyethoxy)ethyl perchlorate
[ ] $C_4H_9ClO_6$

$$HOC_2H_4OC_2H_4OClO_3$$

Hofmann, K. A. *et al., Ber.*, 1909, **42**, 4390
It explodes violently on heating in a capillary.
See other ALKYL PERCHLORATES

1644. *tert*-Butyl peroxophosphoryl dichloride
[ ] $C_4H_9Cl_2O_3P$

$$Me_3COOP(O)Cl_2$$

The product decomposed violently after isolation.
See entry tert-BUTYL PEROXOPHOSPHATE DERIVATIVES

1645. *N*-Fluoro-*N*-nitrobutylamine
[14233-86-4] $C_4H_9FN_2O_2$

$$BuN(F)NO_2$$

Grakauskas, V. *et al., J. Org. Chem.*, 1972, **37**, 334
A sample exploded on vacuum distillation at 60°C, though not at 40°C/33 mbar.
See other N-HALOGEN COMPOUNDS, *N*-NITRO COMPOUNDS

1646. *tert*-Butyldifluorophosphine
[29149-32-4] $C_4H_9F_2P$

$$Me_3CPF_2$$

Stelzer, O. *et al., Inorg. Synth.*, 1978, **18**, 174
It ignites in air.
See other ALKYLHALOPHOSPHINES

†1647. 2-Iodobutane (*sec*-Butyl iodide)
[513-48-4] C_4H_9I

$$MeCHIEt$$

See other HALOALKANES

†1648. 1-Iodo-2-methylpropane (Isobutyl iodide)
[513-38-2] C_4H_9I

$$ICH_2CHMe_2$$

See other HALOALKANES

†1649. 2-Iodo-2-methylpropane (*tert*-Butyl iodide)
[558-17-8]   C₄H₉I

Me₃CI

See other HALOALKANES

1650. Potassium *tert*-butoxide
[865-47-4]   C₄H₉KO

KOCMe₃

Fieser, 1967, Vol. 1, 911
This extremely powerful base may ignite if exposed to air (or oxygen) at elevated temperatures.
See other PYROPHORIC MATERIALS

Acids, or Reactive solvents
Manwaring, R. *et al., Chem. & Ind.*, 1973, 172
Contact of 1.5 g portions of the solid butoxide with drops of the liquid (l) or with the vapours (v) of the reagents below caused ignition after the indicated period (min).
Acetic acid, v, 3; sulfuric acid, l, 0.5
Methanol, l, 2; ethanol, v, 7; propanol, l, 1; isopropanol, l, 1
Ethyl acetate, v, 2; butyl acetate, v, 2; propyl formate, v, 4; dimethyl carbonate, l, 1; diethyl sulfate, l, 1
Acetone, v, 4, l, 2; 2-butanone, v, 1, l, 0.5; 4-methyl-2-butanone, v, 3
Dichloromethane, l, 2; chloroform, v, 2, l, 0; carbon tetrachloride, l, 1; 1-chloro-2,3-epoxypropane, l, 1
The potentially dangerous reactivity with water, acids or halocarbons was already known, but that arising from contact with alcohols, esters or ketones was unexpected. Under normal reaction conditions, little significant danger should exist where excess of solvent will dissipate the heat, but accidental spillage of the solid butoxide could be hazardous.

Dimethyl sulfoxide
See Dimethyl sulfoxide: Metal alkoxides
See other METAL ALKOXIDES

1651. Butyllithium
[109-72-8]   C₄H₉Li

BuLi

Air, or Carbon dioxide, or Water
1. (*MCA SD-91*, 1966)
2. Dezmelyk, E. W. *et al., Ind. Eng. Chem.*, 1961, **53**(6), 56A
This reactive liquid is now normally supplied commercially as a solution (up to 25 wt.%) in pentane, hexane or heptane, because it reacts with ether and ethereal solutions must be stored under refrigeration. Reaction with atmospheric oxygen or

water vapour is highly exothermic and solutions of 20% concentration will ignite on exposure to air at ambient temperature and about 70% relative humidity. Solutions of above 25% concentration will ignite at any humidity. Contact with liquid water will cause immediate ignition of solutions in these highly flammable solvents [1]. Though contact with carbon dioxide causes an exothermic reaction, it is less than with atmospheric oxygen, so carbon dioxide extinguishers are effective on butyllithium fires. Full handling, operating and disposal procedures are detailed [2].

Styrene
See Styrene: Butyllithium

Water
See Air, etc., above
See other ALKYLMETALS

1652. *tert*-Butyllithium
[594-19-4]   C₄H₉Li

$$Me_3CLi$$

Turbitt, T. D. *et al., J. Chem. Soc., Perkin Trans.* 2, 1974, 183
The 2 M solution in heptane ignites in air.

Copper(I) iodide, Tetrahydrofuran
Kurzen, H. *et al., J. Org. Chem.*, 1985, **50**, 3222
t-Butyllithium in presence of copper(I) iodide in THF is used for alkylation. If an excess of the lithium derivative is present, introduction of oxygen or air can cause a violent explosion, even at −78°C.

2,2,2,4,4,4-Hexafluoro-1,3-dimethyl-1,3,2,4-diazadiphosphetidine
Harris, R. *et al., J. Chem. Soc., Dalton Trans.*, 1976, 23
Interaction in hexane to produce the 2,4-di-*tert*-butyl derivative often starts only after an induction period and may then proceed very violently. Careful temperature control is imperative.
See other INDUCTION PERIOD INCIDENTS *See other* ALKYLMETALS

†1653. Pyrrolidine
[123-75-1]   C₄H₉N

Benzaldehyde, Propionic acid

See Benzaldehyde: Propionic acid, etc.

1654. 2-Butanone oxime (2-Oximinobutane)
[96-29-7] C_4H_9NO

MeC(:NOH)Et

HCS 1980, 652

Tyler, L. J., *Chem. Eng. News*, 1974, **52**(35), 3
Investigation following two violent explosions involving the oxime or its derivatives
showed that it may be distilled at 152°C at ambient pressure only if highly purified.
Presence of impurities, especially acidic impurities (e.g. the oxime hydrochloride)
drastically lowers the temperature at which degradation occurs.
See 2-Butanone oxime hydrochloride
See other CATALYTIC IMPURITY INCIDENTS

Sulfuric acid
MCA Guide, 1972, 300
The oxime was hydrolysed with aqueous sulfuric acid and the 2-butanone liberated was distilled out at 15 mbar from a bath at 110–115°C. Soon after release of
vacuum and bath removal, the residue (crude hydroxylaminium sulfate) decomposed
violently.
See HYDROXYLAMINIUM SALTS *See other* OXIMES

1655. Butyraldehyde oxime (1-Hydroxyiminobutane)
[151-00-8] C_4H_9NO

PrCH=NOH

1. Anon., *Sichere Chemiearb.*, 1966, **18**(3), 20
2. Kröger, *Proc. 1st Int. Symp. Prev. Occ. Risks Chem. Ind.*, 180–187, Heidelberg,
   ISSA, 1970
A large batch exploded violently (without flame) during vacuum distillation at
90–100°C/20–25 mbar. Since the distilled product contained up to 12% butyronitrile, it was assumed that the the oxime had undergone the Beckman rearrangement
to butyramide and then dehydrated to the nitrile. The release of water into a system
at 120°C would generate excessive steam pressure which the process vessel could
not withstand. The rearrangement may have been catalysed by metallic impurities
[1]. This hypothesis was confirmed in a detailed study, which identified lead oxide
and rust as active catalysts for the rearrangement and dehydration reactions [2].
See Ethyl 2-formylpropionate oxime: Hydrogen chloride
      Sulfuric acid: Cyclopentanone oxime
See other CATALYTIC IMPURITY INCIDENTS, OXIMES

1656. *N,N* -Dimethylacetamide
[127-19-5] C_4H_9NO

Me$_2$NAc

HCS 1980, 420

Halogenated compounds

1. 'DMAC Brochure A-79931', Wilmington, Du Pont, 1969
2. Cardillo, P. *et al., 3rd Euro. Symp. Therm. Anal. Calorim.*, Interlaken, 1984 (private comm.,)

The tertiary amide acts as a dehydrohalogenating agent, and reaction with some highly halogenated compounds (carbon tetrachloride, hexachlorocyclohexane) is very exothermic and may become violent, particularly if iron is present [1]. A detailed investigation by ARC shows that 1:1 w/w mixtures with carbon tetrachloride are most reactive and that 2 exotherms are evident. One starting at 91° and ending at 97°C is relatively weak (40 J/g) while the second, starting at 97° and ending at 172° has its maximum rate at 147°C, with an exotherm of 508 J/g, which under adiabatic runaway conditions would give a final temperature exceeding 450°C. A considerable release of gas generates a maximum pressure of 12.8 bar at 172 °C. Presence of 1% of iron powder initiates the first exotherm at 71°C and increases it to 60 J/g, the second exotherm being marginally reduced by presence of iron. The exothermic effect is much greater than in the similar reactions of DMF [2].

See Carbon tetrachloride: Dimethylformamide
See 3,4-Dichloronitrobenzene
See other CATALYTIC IMPURITY INCIDENTS, GAS EVOLUTION INCIDENTS

Other reactants
Yoshida, 1980, 165
MRH values calculated for 13 combinations with oxidants are given.
See other APROTIC SOLVENTS

1657. Morpholine (Tetrahydro-1,4-oxazine)
[110-91-8]   C$_4$H$_9$NO

HCS 1980, 671; RSC Lqab. Hazards Data Sheet No. 73, 1988

Cellulose nitrate
See CELLULOSE NITRATE: amines

Nitromethane
See Nitromethane: Acids, etc.
See other ORGANIC BASES

†1658. Butyl nitrite
[544-16-1]   C$_4$H$_9$NO$_2$

BuON:O

See other ALKYL NITRITES

†1659. *tert*-Butyl nitrite
[540-80-7] $C_4H_9NO_2$

$$Me_3CON{:}O$$

Lopez, F., *Chem. Eng. News*, 1992, **70**(51), 2
DSC showed that exothermic decomposition (−1200 J/g) begins at about 110°C. A falling hammer test indicated shock induced decomposition. Considerable caution is advised when employing this useful reagent.
See DIFFERENTIAL SCANNING CALORIMETRY
See other ALKYL NITRITES

1660. *tert*-Nitrobutane (2-Methyl-2-nitropropane)
[594-70-7] $C_4H_9NO_2$

$$O_2NCMe_3$$

Preparative hazard
Mee, A. J., *School Sci. Rev.*, 1940, **22**(85), 96
A sample exploded during distillation.
See Potassium permanganate: Acetone, *tert*-Butylamine
See other NITROALKANES

1661. Butyl nitrate
[928-45-0] $C_4H_9NO_3$

$$BuONO_2$$

Lewis acids
See ALKYL NITRATES: lewis acids

Other reactants
Yoshida, 1980, 203
MRH values calculated for 10 combinations with various reagents are given.
See other ALKYL NITRATES

1662. Ethyl 2-nitroethyl ether (2-Nitroethoxyethane)
[31890-52-5] $C_4H_9NO_3$

$$EtOC_2H_4NO_2$$

Cann, P. F. *et al.*, *J. Chem. Soc., Perkin Trans. 2*, 1974, 817–819
Addition of diphenyl ether as an inert diluent to the crude ether before distillation at 0.13 mbar is essential to prevent violent explosion of the residue after it has cooled.
See related NITROALKANES

1663. *tert*-Butyl peroxynitrate (1,1-Dimethylethyl peroxynitrate)
[ ] $C_4H_9N_4$

$$Me_3COONO_2$$

Like other low molecular weight peroxyesters, or alkylnitrates, this is a material to avoid isolating even when you can make it.

Preparative hazard
See tert-Butylhydroperoxide: Toluene, Dinitrogen pentaoxide
See other PEROXYESTERS

1664. Tris(hydroxymethyl)nitromethane
(2-Hydroxymethyl-2-nitropropane-1,3-diol)
[126-11-4] $C_4H_9NO_5$

$$(HOCH_2)_3CNO_2$$

Hydrogen, Nickel
Ostis, K. *et al., Chem. Abs.*, 1975, **82**, 36578
Catalytic hydrogenation of the title compound to the amine with hydrogen at
5–10 bar is described as hazardous, and an electrochemical reduction process is
recommended.
See other CATALYTIC NITRO REDUCTION PROCESSES, HYDROGENATION INCIDENTS
See related NITROALKANES

†1665. Cyanotrimethylsilane
[7677-24-9] C_4H_9NSi

$$Me_3SiCN$$

See related ALKYLHALOSILANES *See other* CYANO COMPOUNDS

1666. 1-Azidobutane (Butyl azide)
[7332-00-5] $C_4H_9N_3$

$$Me(CH_2)_3N_3$$

See Butyl toluenesulfonate
See AZIDES

1667. Butylsodium
[3525-44-8] C_4H_9Na

$$BuNa$$

Preparative hazard
See Sodium: 1-Chlorobutane *See other* ALKYLMETALS

†1668. Butane
[106-97-8] C_4H_{10}

$$MeC_2H_4Me$$

FPA H15, 1973 (cylinder); *HCS 1980*, 224 (cylinder)

See LIQUEFIED PETROLEUM GASES

†1669. Isobutane (2-Methylpropane)
[75-28-5] C_4H_{10}

$$Me_2CHCH_3$$

FPA H119, 1983 (cylinder); *HCS 1980*, 565

1670. Diethylaluminium bromide
[760-19-0] $C_4H_{10}AlBr$

$$Et_2AlBr$$

Nitromethane
See Nitromethane: Alkylmetal halides
See entry ALKYLALUMINIUM HALIDES

1671. Diethylaluminium chloride
[96-10-6] $C_4H_{10}AlCl$

$$Et_2AlCl$$

HCS 1980, 390

Chlorine azide
Houben-Weyl 1970, Vol. 13.4, 76
Interaction produces distillable ethylaluminium azide chloride, but the residue is explosive.
See other ALKYLALUMINIUM HALIDES

1672. Diethylgold bromide
[26645-10-3] $C_4H_{10}AuBr$

$$Et_2AuBr$$

Sorbe, 1968, 63
It explodes at 70°C.
See other ALKYLMETAL HALIDES, GOLD COMPOUNDS

1673. Chlorodiethylborane
[5314-83-0] $C_4H_{10}BCl$

$$ClBEt_2$$

Koster, R. *et al., Inorg. Synth.*, 1974, **15**, 149–152
It ignites in air; preparative hazard also described.
See other ALKYLHALOBORANES

1674. Boron trifluoride diethyl etherate
[109-63-7] $C_4H_{10}BF_3O$

$$Et_2O \rightarrow BF_3$$

HCS 1980, 217

Lithium tetrahydroaluminate
  See Lithium tetrahydroaluminate: Boron trifluoride diethyl etherate
  See related NON-METAL HALIDES

1675. Diethylberyllium
[542-63-2] $C_4H_{10}Be$

$$Et_2Be$$

Air, or Water
  Coates, 1967, Vol. 1, 106
  Ignites in air, even when containing ether, and reacts explosively with water.
  See other ALKYLMETALS

1676. Diethylbismuth chloride
[65313-34-0] $C_4H_{10}BiCl$

$$Et_2BiCl$$

Gilman, H. *et al., J. Org. Chem.*, 1939, **4**, 167
It ignites in air.
See other ALKYLMETAL HALIDES

1677. Diethylcadmium
[592-02-9] $C_4H_{10}Cd$

$$Et_2Cd$$

1. Krause, E., *Ber.*, 1918, **50**, 1813
2. Houben-Weyl, 1973, Vol. 13.2a, 903 (footnote 3), 908
The vapour decomposes explosively at 180°C. Exposure to ambient air produces
white fumes which turn brown and then explode violently [1]. Moderate quantities
decompose explosively when heated rapidly to 130, not 180°C as stated previously.
This, apart from allylcadmium derivatives, is the only pyrophoric dialkylcadmium
[2] (but *see* Dimethylcadmium). It is mildly endothermic (ΔH°_f (l) +61.5 kJ/mol,
0.36 kJ/g).
See other ALKYLMETALS, ENDOTHERMIC COMPOUNDS

1678. 2-Butanone oxime hydrochloride (2-Hydroxyliminiobutane chloride)
[4154-69-2] $C_4H_{10}ClNO$

$$MeC(:N^+HOH)Et\ Cl^-$$

Tyler, L. J., *Chem. Eng. News*, 1974, **52**(35), 3

The salt undergoes violent degradation at 50–70°C, and its presence in trace quantities may promote degradation of the oxime.
See related OXIMES

1679. Diethylaminosulfinyl chloride
[34876-30-7] $C_4H_{10}ClNOS$

$$Et_2NS(:O)Cl$$

Preparative hazard
See Sulfinyl chloride: Bis(dimethylamino) sulfoxide
See other N–S COMPOUNDS

1680. Morpholinium perchlorate
[35175-75-8] $C_4H_{10}ClNO_5$

Udupa, M. R., *Thermochim. Acta*, 1980, **38**, 241–143
It decomposes exothermally at 230°C.
See other PERCHLORATE SALTS OF NITROGENOUS BASES

1681. Diethyl phosphorochloridate (Diethyl chlorophosphate)
[814-49-3] $C_4H_{10}ClO_3P$

$$(EtO)_2P(O)Cl$$

Silbert, L. A., *J. Org. Chem.*, 1971, **36**, 2162
Presence of hydrogen chloride as impurity causes an uncontrollable exothermic reaction during preparation of diethyl phosphate from the title compound.
See related PHOSPHORUS ESTERS *See other* CATALYTIC IMPURITY INCIDENTS

1682. Diethylthallium perchlorate
[[22392-07-0] (ion)] $C_4H_{10}ClO_4Tl$

$$Et_2TlClO_4$$

Cook, J. R. *et al.*, *J. Inorg. Nucl. Chem.*, 1964, **26**, 1249
It explodes at the m.p., 250°C.
See related METAL PERCHLORATES

†1683. Dichlorodiethylsilane
[1719-53-5] $C_4H_{10}Cl_2Si$

$$Et_2SiCl_2$$

See other ALKYLHALOSILANES

1684. Diethylaminosulfur trifluoride
[38078-09-0] $C_4H_{10}F_3NS$

Et_2NSF_3

Alone, or Water
1. Cochran, J., *Chem. Eng. News*, 1979, **57**(12), 4, 74
2. Middleton, W. J., *Chem. Eng. News*, 1979, **57**(21), 43
3. Middleton, W. J. *et al.*, *J. Org. Chem.*, 1980, **45**, 2884 (footnote 3)
4. Middleton, W. J. *et al.*, *Org. Synth.*, 1977, **57**, 50
5. Breckenridge, R. J. *et al.*, *Chem. Brit.*, 1982, **18**, 619
6. Suckling, C. J., private comm., 1982
7. Pivawer, P. M., private comm., 1983

Explosive decomposition occurred when stripping solvent or attempting vacuum distillation. Exothermic decomposition begins at about 50°C and is violent above 90°C. Contact with water causes explosive decomposition [1]. Later comments suggest that the compound may be distilled safely at pressures below 13 mbar and handled safely in solution, or in bulk below 50°C [2,3]. While repeating an earlier procedure [4], two explosions occurred. One was when admitting air to the residue from vacuum distillation, and the second immediately prior to redistillation of the cold product from vacuum stripping of solvent [5]. This was attributed to formation of some impurities related to a dialkylaminofluorosulfur imine [6]. During preliminary distillation to remove sulfur tetrafluoride, etc., the line from the cold trap to the alkaline scrubber must end in an inverted funnel to avoid blockage with solid [7].

See S-Diethylamino(methylimino)sulfur(IV) fluoride
See other N–S COMPOUNDS *See related* NON-METAL HALIDES

1685. Potassium diethylamide
[2245-68-3] $C_4H_{10}KN$

$KNEt_2$

Pez, G. P. *et al.*, *Pure Appl. Chem.*, 1986, **57**, 1925
The solid is highly pyrophoric and will react violently even with limited exposure to air
See other N-METAL DERIVATIVES, PYROPHORIC MATERIALS

1686. Lithium diethylamide
[816-43-3] $C_4H_{10}LiN$

$LiNEt_2$

1. Fischer, E. O. *et al.*, *Inorg. Synth.*, 1979, **19**, 165
2. Pez, G. P. *et al.*, *Pure Appl. Chem.*, 1986, **57**, 1924
Extremely pyrophoric in air [1], and will react violently upon even brief contact [2].
See other N-METAL DERIVATIVES, PYROPHORIC MATERIALS

1687. Diethylmagnesium
[557-18-6]

Et_2Mg

$C_4H_{10}Mg$

Air, or Water

Gilman, H. *et al., J. Amer. Chem. Soc.*, 1930, **52**, 5048

Contact with moist air usually caused ignition of the dry powder, and water always ignited the solid or its ethereal solution.

Carbon dioxide

Merck, 1983, 453

It will glow and ignite in carbon dioxide.

See other ALKYLMETALS, DIALKYLMAGNESIUMS

1688. Sodium diethylamide
[34992-80-8]

$NaNEt_2$

$C_4H_{10}NNa$

Pez, G. P. *et al., Pure Appl. Chem.*, 1986, **57**, 1925

The solid is highly pyrophoric and will react violently even on limited exposure to air.

See other N-METAL DERIVATIVES, PYROPHORIC MATERIALS

1689. Piperazine (Hexahydropyrazine)
[110-85-0]

$C_4H_{10}N_2$

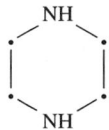

HCS 1980, 756

Dicyanofurazan

See Dicyanofurazan: Nitrogenous bases

See other ORGANIC BASES

1690. Diethyl hyponitrite (Diethoxydiazene)
[4549-46-6]

$EtON=NOEt$

$C_4H_{10}N_2O_2$

See entry DIALKYL HYPONITRITES

1691. N-Nitrosoethyl-2-hydroxyethylamine
[13147-25-6] $C_4H_{10}N_2O_2$

$$EtN(N:O)C_2H_4OH$$

Preussman, R., *Chem. Ber.*, 1962, **95**, 1572
It decomposed explosively during attempted distillation at 18 mbar from a bath at 170°C, but was distilled with slight decomposition at 103–105°/0.4 mbar.
See other NITROSO COMPOUNDS

1692. Diethyllead dinitrate
[17498-10-1] $C_4H_{10}N_2O_6Pb$

$$Et_2Pb(NO_3)_2$$

Potts, D. *et al., Can. J. Chem.*, 1969, **47**, 1621
The salt is unstable above 0°C and explodes on heating.
See related ALKYLMETALS, HEAVY METAL DERIVATIVES

1693. N-Butylamidosulfuryl azide
[13449-22-4] $C_4H_{10}N_4O_2S$

$$BuNHSO_2N_3$$

Shozda, R. J. *et al., J. Org. Chem.*, 1967, **32**, 2876
It exploded during analytical combustion.
See other ACYL AZIDES

1694. 1-Butanol
[78-83-1] $C_4H_{10}O$

$$HOCH_2Pr$$

FPA H64, 1977; *HCS 1980*, 236

Aluminium   MRH 1.17/11
  See Aluminium: Butanol

Chromium trioxide   MRH 2.55/92
  See Chromium trioxide: Alcohols

Other reactants
  Yoshida, 1980, 297
  MRH values calculated for 15 combinations, largely with oxidants, are given.

†1695. 2-Butanol
[78-92-2] $C_4H_{10}O$

$$MeCH(OH)Et$$

HCS 1980, 237

1. Posdev, V. S. *et al., Chem. Abs.*, 1977, **86**, 194347
2. Peterson, D., *Chem. Eng. News*, 1981, **59**(19), 3
3. Sharpless, T. W., *J. Chem. Educ.*, 1984, **61**, 476
4. Doyle, R. R., *Chem. Eng. News*, 1986, **64**(7), 4: *J. Chem. Educ.*, 1986, **63**(20), 186

A powerful explosion which occurred during distillation of a 10-year-old sample of the alcohol was attributed to presence of peroxy compounds formed by autoxidation, possibly involving 2-butanone as an effective photochemical sensitiser [1]. After a later explosion, it was found that the sample being distilled contained 12% of peroxide [2]. A further incident involved a 12-year old sample which exploded at the end of distillation, and which also contained a high level of peroxide. Several other stock alcohols were found to contain much lower levels of peroxide than the 2-butanol, and recommendations on clean-up or disposal, depending on the level of peroxide, are made [3]. A further report of an explosion at the end of laboratory distillation confirms the potential for peroxide formation on prolonged storage of 2-butanol [4].
See 2-Propanol: 2-Butanone

Chromium trioxide
See Chromium trioxide: Alcohols
See other PEROXIDISABLE COMPOUNDS

†1696. *tert*-Butanol
[75-65-0]   C$_4$H$_{10}$O

Me$_3$COH

HCS 1980, 238

Potassium−sodium alloy
See Potassium−sodium alloy: *tert*-Butanol

†1697. Diethyl ether (Ethoxyethane)
[60-29-7]   C$_4$H$_{10}$O

EtOEt

(*MCA SD-29*, 1965); *NSC 396*, 1968; *FPA H40*, 1975; *HCS 1980*, 487; *RSC Lab. Hazards Data Sheet No. 48*, 1965

1. Criegee, R. *et al., Angew. Chem.*, 1936, **49**, 101
2. Criegee, R. *et al., Angew. Chem.*, 1958, **70**, 261
3. Mallinckrodt, E. *et al., Chem. Eng. News*, 1955, **33**, 3194
4. Ray, A., *J. Appl. Chem.*, 1955, 188
5. Anon., *Chemist & Druggist*, 1947, **157**, 258
6. Feinstein, R. N., *J. Org. Chem.*, 1959, **24**, 1172
7. Davies, A. G., *J. R. Inst. Chem.*, 1956, **80**, 386
8. Anon., *Lab. Accid. in Higher Educ.*, Item 27, HSE, Barking, 1987

The hydroperoxide initially formed by autoxidation of ether is not particularly explosive, but on standing and evaporation, polymeric 1-oxyperoxides are formed which are dangerously explosive, even below 100°C [1]. Numerous laboratory explosions have been caused by evaporation of peroxidised ether [2]. Formation of peroxide in stored ether may be prevented by presence of sodium diethyldithiocarbamate (0.05 ppm), [4], which probably deactivates traces of metals which catalyse peroxidation [2]; of pyrogallol (1 ppm) [4]; or by larger proportions (5 to 20 ppm) of other inhibitors [5]. Once present in ether, peroxides may be detected by the iodine–starch test, and removed by percolation through anion exchanger resin [6] or activated alumina [7], which leaves the ether dry, or by shaking with aqueous ferrous sulfate or sodium sulfite solutions. Many other methods have been described [7]. A small quantity of ether stored in an unmodified domestic refrigerator in a biochemistry laboratory led to an explosion, with ignition by the thermostat contacts [8].

See 1-OXYPEROXY COMPOUNDS
See other PEROXIDISABLE COMPOUNDS, POLYPEROXIDES

Boron triazide
    See Boron triazide: Diethyl ether, etc.

Halogens, or Interhalogens
    See Bromine trifluoride: Halogens, etc., or: Solvents
        Bromine pentafluoride: Hydrogen-containing materials
        Bromine: Diethyl ether
        Chlorine: Diethyl ether
        Iodine heptafluoride: Organic solvents

Other reactants
    Yoshida, 1980, 140
    MRH values calculated for 13 combinations with oxidants are given.

Oxidants	MRH values show % of oxidant
See Halogens, or Interhalogens, above	
Chromyl chloride: Organic solvents	
'Fluorine nitrate': Organic materials	
Hydrogen peroxide: Diethyl ether	MRH 6.53/84
Iodine(VII) oxide: Diethyl ether	
Lithium perchlorate: Diethyl ether	
Nitric acid: Diethyl ether	MRH 5.94/79
Nitrosyl perchlorate: Organic materials	
Nitryl perchlorate: Organic solvents	
Ozone: Diethyl ether	
Perchloric acid: Diethyl ether	
Permanganic acid: Organic materials	
Peroxodisulfuric acid: Organic liquids	
Silver perchlorate: Diethyl ether	
Sodium peroxide: Organic liquids	MRH 2.55/92
Liquid Air: Diethyl ether	

Peat soils

Walkley, A., *Austral. Chem. Inst. J.*, 1939, **6**, 310

Explosions occurred during the extraction of fats and waxes from the soils with ether, as well as when heating the extract at 100°C. Although the latter is scarcely surprising (the ether contained 230 ppm of peroxides), the former observation is unusual.

Sulfur, or Sulfur compounds

See Sulfonyl chloride: Diethyl ether

Sulfur: Diethyl ether

Thiotrithiazyl perchlorate: Organic solvents

Uranyl nitrate

See Uranyl nitrate: Diethyl ether

Wood pulp extracts

Durso, D. F., *Chem. Eng. News*, 1957, **35**(34), 91, 115

Ethereal extracts of pulp exploded during or after concentration by evaporation. Although the ether used for the extraction previously had been freed from peroxides by treatment with cerium(III) hydroxide, the ethereal extracts had been stored for 3 weeks before concentration was effected. (During this time the ether and/or extracted terpenes would be expected to again form peroxides, but no attempt seems to have been made to test for, or to remove them before distillation was begun).

See other PEROXIDISABLE COMPOUNDS

1698. *tert*-Butyl hydroperoxide (2-Hydroperoxy-2-methylpropane) [75-91-2]

$C_4H_{10}O_2$

Me₃COOH

1. Milas, N. A. *et al., J. Amer. Chem. Soc.*, 1946, **68**, 205
2. Castrantas, 1965, 15
3. Sharpless, K. B. *et al., Aldrichimica Acta*, 1979, **12**, 63–73
4. Verhoeff, J., *Runaway Reactions*, 1981, Paper 3/T, 1–12
5. Hill, J. H. *et al., J. Org. Chem.*, 1983, **48**, 3607–3608

Though relatively stable, explosions have been caused by distillation to dryness [1], or attempted distillation at ambient pressure [2]. In a comprehensive review of the use of the hydroperoxide as a selective metal-catalysed oxygenator for alkenes and alkynes, attention is drawn to several potential hazards in this application. One specific hazard to be avoided stems from the fact that Lucidol TBHP-70 contains 19% of di-*tert*-butyl peroxide which will survive the catalysed reaction and may lead to problems in the work-up and distillation [3]. A thorough investigation of the stability and explosive properties of the 70% solution in water has been carried out [4]. The anhydrous peroxide as a solution in toluene may now readily be prepared azeotropically, and the solutions are stable in storage at ambient temperature. This solution is now a preferred method for using the anhydrous hydroperoxide [5].

See 1,2-Dichloroethane, below,

See also THERMAL EXPLOSIONS

Preparative hazard
See Hydrogen peroxide: *tert*-Butanol, Sulfuric acid

Acids
1. Sharpless, K. B. *et al., Aldrichimica Acta*, 1979, **12**, 71
2. Andreozzi, R. *et al., J. Haz. Mat.*, 1988, **17**, 305–313
Not even traces of strong acids should be added to high strength solutions of the hydroperoxide [1]. The thermal stability and mechanism of cleavage of mixtures with 4-toluenesulfonic acid have been studied under adiabatic conditions, and there is potential for development of a thermal runaway [2].

1,2-Dichloroethane
Hill, J. H. *et al., J. Org. Chem.*, 1983, **48**, 3607
Solutions of the hydroperoxide in halogenated solvents, and especially dichloro-ethane are much less stable than in toluene (reference 5 above). On a large scale, the azeotropic drying of solutions of the hydroperoxide in dichloroethane may present a thermal hazard.

Molecular sieves
Sharpless, K. B. *et al., Aldrichimica Acta*, 1979, **12**, 71
Use of the 4A sieve to dry the hydroperoxide (6% water) had been used routinely without incident, but inadvertent use of the larger-pored 13X sieve gave a high exotherm in the packed bed which soon led to autoignition of the hydroperoxide. This was attributed to the heat of adsorption occurring within the larger sieve pores (10 Angstroms in 13X), but caution is advised when using sieve to dry any high strength (90%) solutions of the hydroperoxide, and an azeotropic method is preferred (see reference 5 above).
See Nitromethane: Molecular sieve
See other MOLECULAR SIEVE INCIDENTS

Toluene, Dinitrogen pentaoxide
1. Owens, K. A. *et al., J. Loss Prevention*, 1995, **8**(5), 296
2. Editor's comments
Toluene was used, in place of tetrachloromethane, as solvent when preparing t-Butyl peroxynitrate. An explosion ensued, attributed to trinitrotoluene formation and detonation [1]. Since TNT is not easy to form and difficult to explode, this seems improbable. The intended product will certainly be a more dangerous explosive than TNT, as can be the expected byproduct, an unreacted solution of anhydrous nitric acid in toluene. Most explosions in nitration systems are due to free radical oxidation of an organic substrate or solvent by the nitrating agent; this is quite probable here. Though nitrogen pentoxide is itself more an ionic nitration reagent than a radical oxidant, all peroxy species are radical generators par excellence, and commonly used to initiate free radical reaction sequences [2].

Transition metal salts
Sharpless, K. B. *et al., Aldrichimica Acta*, 1979, **12**, 71

Transition metal salts known to be good autoxidation catalysts (of cobalt, iron or manganese) should never be added to high-strength solutions of the hydroperoxide to avoid the near-certainty of vigorous evolution of a large volume of oxygen.
See other ALKYL HYDROPEROXIDES

1699. Diethyl peroxide
[628-37-5] $C_4H_{10}O_2$

$$EtOOEt$$

1. Baeyer, A. *et al., Ber.*, 1900, **33**, 3387
2. Castrantas, 1965, 15
3. Baker, G. *et al., Chem. & Ind.*, 1964, 1988
4. Gray, P. *et al., Proc. R. Soc.*, 1971, **A325**, 175

While it is acknowledged as rather explosive, the stated lack of shock-sensitivity at ambient temperature [1] is countered by its alternative description as shock-sensitive [2] and detonable [3]. The vapour explodes above a certain critical pressure, at temperatures above 190°C [4].
See other DIALKYL PEROXIDES

†1700. 1,1-Dimethoxyethane
[534-15-6] $C_4H_{10}O_2$

$$(MeO)_2CHCH_3$$

HCS 1980, 417

See other PEROXIDISABLE COMPOUNDS

†1701. 1,2-Dimethoxyethane
[110-71-4] $C_4H_{10}O_2$

$$MeOC_2H_4OMe$$

HCS 1980, 418

Phifer, L. H., *CHAS Notes*, 1982, **1**(3), 2
A seven-year-old peroxidised sample exploded during distillation.

Lithium tetrahydroaluminate
  See Lithium tetrahydroaluminate: 1,2-Dimethoxyethane
  See other PEROXIDISABLE COMPOUNDS

1702. 2-Ethoxyethanol
[110-80-5] $C_4H_{10}O_2$

$$EtOC_2H_4OH$$

HCS 1980, 459

Hydrogen peroxide, Poly(acrylamide) gel, etc.
See Hydrogen peroxide: 2-Ethoxyethanol, etc.
See other PEROXIDISABLE COMPOUNDS

†1703. Methyl propyl ether (1-Methoxypropane)
[557-17-5] $C_4H_{10}O_2$

MeOPr

HCS 1980, 660

1704. Zinc ethoxide
[3851-22-7] $C_4H_{10}O_2Zn$

$Zn(OEt)_2$

Nitric acid
Solomartin, V. S. et al., Chem. Abs., 1977, **87**, 210678
In the analysis of diethylzinc, a 1 ml sample is cooled to −196°C and treated with 2 ml of ethanol to give the ethoxide. During subsequent conversion to zinc nitrate (prior to pyrolysis to the oxide) by treatment with 3 ml of 30% nitric acid, cooling must be continued to avoid an explosion hazard.
See other METAL ALKOXIDES

1705. Bis(2-hydroxyethyl) ether (Diethylene glycol)
[111-46-6] $C_4H_{10}O_3$

$(HOC_2H_4)_2O$

HCS 1980, 397

Other reactants
Yoshida, 1980, 144
MRH values calculated for 14 combinations, largely with oxidants, are given.

Sodium hydroxide
See Sodium hydroxide: Glycols

1706. 1-Hydroxy-3-butyl hydroperoxide (3-Hydroperoxy-1-butanol)
[ ] $C_4H_{10}O_3$

$HOC_2H_4CH(Me)OOH$

Rieche, A., Ber., 1930, **63**, 2462
It explodes on heating.

1707. 2-Methoxyprop-2-yl hydroperoxide (2-Methoxy-1-methylethyl hydroperoxide)
[10027-74-4] $C_4H_{10}O_3$

$$Me_2(MeO)COOH$$

Dussault, P., *Chem. Eng. News*, 1993, **71**(32), 2
This reagent, made by methanolic ozonolysis of tetramethylethylene, may deflagrate if isolated.
See Hydrogen peroxide: Acetone
See related 1-OXYPEROXY COMPOUNDS

†1708. Trimethyl orthoformate (Trimethoxymethane)
[149-73-5] $C_4H_{10}O_3$

$$(MeO)_3CH$$

Preparative hazard
See Chloroform: Sodium, Methanol, or: Sodium methoxide

1709. Diethyl sulfite
[623-81-4] $C_4H_{10}O_3S$

$$(EtO)_2S{:}O$$

Chlorine fluoride
See Ethyl fluorosulfate
See other SULFUR ESTERS

1710. Diethyl sulfate
[64-67-5] $C_4H_{10}O_4S$

$$(EtO)_2SO_2$$

HCS 1980, 406

3,8-Dinitro-6-phenylphenanthridine, Water
Hodgson, J. F., *Chem. & Ind.*, 1968, 1399
Accidental ingress of water to the heated mixture liberated sulfuric acid or its half ester which caused a violent reaction with generation of a large volume of solid black foam.
See 4-Nitroaniline-2-sulfonic acid, *also* Sulfuric acid: Nitroaryl bases, etc.

Iron, Water
Siebeneicher, K., *Angew. Chem.*, 1934, **47**, 105
Moisture in a sealed iron drum of the ester caused hydrolysis to sulfuric acid, leading to corrosion of the metal and development of a high internal pressure of hydrogen, which ruptured the drum.
See other CORROSION INCIDENTS, GAS EVOLUTION INCIDENTS

569

Potassium *tert*-butoxide
  See Potassium *tert*-butoxide: Acids, etc.
  See other SULFUR ESTERS

1711. Zinc ethylsulfinate
  [24308-92-7] $C_4H_{10}O_4S_2Zn$

$$Zn[OS(:O)Et]_2$$

Preparative hazard
  See Diethylzinc: Sulfur dioxide
  See related SULFUR ESTERS

†1712. Butanethiol
  [109-79-5] $C_4H_{10}S$

$$BuSH$$

Nitric acid
  See Nitric acid: Alkanethiols
  See other ALKANETHIOLS

†1713. 2-Butanethiol
  [513-53-1] $C_4H_{10}S$

$$MeEtCHSH$$

See other ALKANETHIOLS

†1714. Diethyl sulfide (3-Thiapentane)
  [352-93-2] $C_4H_{10}S$

$$EtSEt$$

HCS 1980, 407

†1715. 2-Methylpropanethiol
  [513-44-0] $C_4H_{10}S$

$$Me_2CHCH_2SH$$

Calcium hypochlorite
  See Calcium hypochlorite: Organic sulfur compounds
  See other ALKANETHIOLS

†1716. 2-Methyl-2-propanethiol
  [75-66-1] $C_4H_{10}S$

$$Me_3CSH$$

See other ALKANETHIOLS

1717. Diethyl telluride
[627-54-3]

Et$_2$Te

C$_4$H$_{10}$Te

Ellern, 1968, 24–25
It ignites in air.
See other ALKYLMETALS

1718. Diethylzinc
[557-20-0]

Et$_2$Zn

C$_4$H$_{10}$Zn

Aluminium Alkyls and other Organometallics, 5, New York, Ethyl Corp., 1967

It is immediately pyrophoric in air.

Alkenes, Diiodomethane
Houben-Weyl, 1973, Vol. 13.2a, 845
During preparation of cyclopropane derivatives, it is important to add the diiodomethane slowly to a solution of diethylzinc in the alkene. Addition of diethylzinc to an alkene–diiodomethane mixture may be explosively violent.

Halogens
See Bromine: Diethylzinc
Chlorine: Diethylzinc

Methanol
Houben-Weyl, 1973, Vol. 13.2a, 855
Interaction is explosively violent and ignition ensues.

Nitro compounds
Leleu, *Cahiers*, 1977, (88), 371
Interaction is usually violent.

Non-metal halides
Leleu, *Cahiers*, 1977, (88), 372
Interaction with arsenic trichloride or phosphorus trichloride is violent, forming pyrophoric triethylarsine or triethylphosphine.

Ozone
See Ozone: Alkylmetals

Sulfur dioxide
Houben-Weyl, 1973, 13.2a, 709
During preparation of zinc ethylsulfinate, addition of diethylzinc to liquid sulfur dioxide at −15°C leads to an explosively violent reaction. Condensation of sulfur dioxide into cold diethylzinc leads to a controllable reaction on warming.

Water
Ito, Y. *et al.*, *Org. Synth.*, 1979, **59**, 113, 116
It reacts violently with water, but a procedure for destroying excess reagent with ice-water is detailed.
See other ALKYLMETALS, DIALKYLZINCS

1719. Diethylaluminium hydride
[871-27-2]   C$_4$H$_{11}$Al

$$Et_2AlH$$

HCS 1980, 391

See entry ALKYLALUMINIUM ALKOXIDES AND HYDRIDES

1720. Diethylarsine
[692-42-2]   C$_4$H$_{11}$As

$$Et_2AsH$$

Sidgwick, 1950, 762
It ignites in air, even at 0°C.
See other ALKYLNON-METAL HYDRIDES

1721. Diethylgallium hydride (Diethylhydrogallium)
[93481-56-2]   C$_4$H$_{11}$Ga

$$Et_2GaH$$

Air, or Water
Eisch, J. J., *J. Amer. Chem. Soc.*, 1962, **84**, 3835
It ignites in air, and reacts violently with water.
See other ALKYLMETAL HYDRIDES

1722. Trimethylsilylmethyllithium
[1872-00-0]   C$_4$H$_{11}$LiSi

$$Me_3SiCH_2Li$$

Tessier-Youngs, C. *et al.*, *Inorg. Synth.*, 1986, **24**, 92
It ignites spontaneously in air and reacts violently with water, like the aluminium analogue.
See related ALKYLMETALS, ALKYLSILANES

†1723. Butylamine (1-Aminobutane)
[109-73-9]   C$_4$H$_{11}$N

$$BuNH_2$$

FPA H100, 1982; *HCS 1980*, 239

Other reactants
Yoshida, 1980, 305
MRH values calculated for 13 combinations with oxidants are given.

Perchloryl fluoride
See Perchloryl fluoride: Nitrogenous bases
See other ORGANIC BASES

†1724. 2-Butylamine (2-Aminobutane)
[13952-84-6]

$C_4H_{11}N$

MeEtCHNH$_2$

See other ORGANIC BASES

†1725. *tert*-Butylamine (2-Amino-2-methylpropane)
[75-64-9]

$C_4H_{11}N$

Me$_3$CNH$_2$

HCS 1980, 240

2,2-Dibromo-1,3-dimethylcyclopropanoic acid
See 2,2-Dibromo-1,3-dimethylcyclopropanoic acid: *tert*-Butylamine
See other ORGANIC BASES

†1726. Diethylamine
[109-89-7]

$C_4H_{11}N$

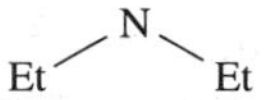

(*MCA SD-97*, 1971); *FPA H85*, 1979; *HCS 1980*, 392

Cellulose nitrate
See CELLULOSE NITRATE: amines

Dicyanofurazan
See Dicyanofurazan: Nitrogenous bases

Other reactants
Yoshida, 1980, 139
MRH values calculated for 13 combinations with oxidants are given.

Sulfuric acid
See Sulfuric acid: Diethylamine
See other ORGANIC BASES

†1727. Ethyldimethylamine
[598-56-1]

$C_4H_{11}N$

EtNMe$_2$

See other ORGANIC BASES

†1728. Isobutylamine (2-Methypropylamine)
[78-81-9]
$C_4H_{11}N$

$$Me_2CHCH_2NH_2$$

See other ORGANIC BASES

1729. N-2-Hydroxyethyldimethylamine (2-Dimethylaminoethanol)
[108-01-0]
$C_4H_{11}NO$

$$HOC_2H_4NMe_2$$

HCS 1980, 422

Cellulose nitrate
See CELLULOSE NITRATE: amines

Other reactants
Yoshida, 1980, 167
MRH values calculated for 14 combinations with oxidants are given.
See other ORGANIC BASES

1730. 3-Methoxypropylamine
[5332-73-0]
$C_4H_{11}NO$

$$MeOC_3H_6NH_2$$

Energy of decomposition (in range 150–550°C) measured as 0.57 kJ/g.
See entry THERMOCHEMISTRY AND EXOTHERMIC DECOMPOSITION (reference 2)
See other AMINOMETHOXY COMPOUNDS

1731. Tris(hydroxymethyl)methylamine
[77-86-1]
$C_4H_{11}NO_3$

$$(HOCH_2)_3CNH_2$$

Preparative hazard
See Tris(hydroxymethyl)nitromethane: Hydrogen, etc.
See other ORGANIC BASES

1732. Dimethyl ethanephosphonite
[15715-42-1]
$C_4H_{11}O_2P$

$$(MeO)_2PEt$$

Arbuzov, B. A. *et al., Chem. Abs.*, 1953, **47**, 3226a
It ignites in air when absorbed onto filter paper.
See other PHOSPHORUS ESTERS, PYROPHORIC MATERIALS

1733. Diethyl phosphite
[762-04-9] $C_4H_{11}O_3P$

$$(EtO)_2P(:O)H$$

4-Nitrophenol
    Phifer, L. H., private comm., 1982
    Interaction in absence of solvent in a stirred flask heated by a regulated mantle led
    to a runaway reaction and explosion.
    See other PHOSPHORUS ESTERS

1734. Diethylphosphine
[627-49-6] $C_4H_{11}P$

$$Et_2PH$$

    Houben-Weyl 1963, Vol. 12.1, 69
    Secondary lower-alkylphosphines readily ignite in air.
    See other ALKYLPHOSPHINES

1735. Ethyldimethylphosphine
[1605-51-2] $C_4H_{11}P$

$$EtPMe_2$$

    Smith, J. F., private comm., 1970
    May ignite in air.
    See other ALKYLPHOSPHINES

1736. Tetramethyldiarsane (Tetramethyldiarsine)
[471-35-2] $C_4H_{12}As_2$

$$Me_2AsAsMe_2$$

    Sidgwick, 1950, 770
    Inflames in air.
    See other ALKYLNON-METALS

1737. Bis(dimethylarsinyl) oxide
[503-80-0] $C_4H_{12}As_2O$

$$(Me_2As)_2O$$

    von Schwartz, 1918, 322
    Ignites in air.
    See related ALKYLNON-METALS

1738. Bis(dimethylarsinyl) sulfide
CAS591-10-6C$_4$H$_{12}$As$_2$S

$$(Me_2As)_2S$$

von Schwartz, 1918, 322
Ignites in air.
See related ALKYLNON-METALS

1739. Tetramethyldigold diazide
[22653-19-6]   C$_4$H$_{12}$Au$_2$N$_6$

$$Me_2Au:(N_3)_2:AuMe_2$$

Beck, W. *et al., Inorg. Chim. Acta*, 1968, **2**, 468
The dimeric azide is extremely sensitive and may explode under water if touched.
See other GOLD COMPOUNDS, METAL AZIDES

1740. Lithium tetramethylborate
[2169-38-2]   C$_4$H$_{12}$BLi

$$Li[Me_4B]$$

491M, 1975, 238
It may ignite in moist air.
See related ALKYLNON-METALS

1741. Dihydrazinecadmium(II) succinate (Bis(hydrazine)[butanedioato(2−)] cadmium)
[159793-67-6]   C$_4$H$_{12}$CdN$_4$O$_4$

See HYDRAZINE METAL MALONATES AND SUCCINATES

1742. Tetramethylammonium chlorite
[67922-18-3]   C$_4$H$_{12}$ClNO$_2$

$$Me_4N^+ClO_2^-$$

Levi, *Gazz. Chim. Ital. [1]*, 1922, **52**, 207
The dry solid explodes on impact.
See other CHLORITE SALTS, OXOSALTS OF NITROGENOUS BASES

1743. 1,2-Dichlorotetramethyldisilane
[39437-99-5]   C$_4$H$_{12}$Cl$_2$Si$_2$

$$Me_2ClSiSiClMe_2$$

Rich, J. D., *Chem. Eng. News*, 1990, **68**(13), 2

Explosions were experienced in the still-head when distilling this compound. The cause is not very clear and may have to do with impurities.
See Chloropentamethyldisilane,
See other ALKYLHALOSILANES

1744. Lithium tetramethylchromate(II)
[ ] $C_4H_{12}CrLi$

$$Li_2[Me_4Cr]$$

Kurras, E. *et al., J. Organomet. Chem.*, 1965, **4**, 114–118
Isolated as a dioxane complex, it ignites in air.
See other ALKYLMETALS

1745. Tetramethylammonium monoperchromate
[ ] $C_4H_{12}CrNO_5$

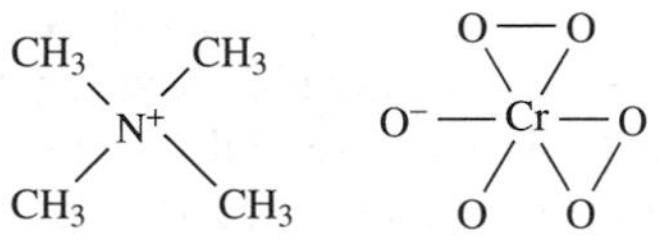

Mellor, 1943, Vol. 11, 358
It explodes on moderate heating or in contact with sulfuric acid. Now thought probably to have the structure shown (and corresponding empirical formula) it has in the intervening period been considered to be a dimer, a pentaperoxydichromate.
See Tetramethylammonium pentaperoxodichromate
See other OXOSALTS OF NITROGENOUS BASES, PEROXOACID SALTS, PEROXOCHROMIUM COMPOUNDS, QUATERNARY OXIDANTS

1746. 1,2-Diamino-2-methylpropaneoxodiperoxochromium(VI)
[ ] $C_4H_{12}CrN_2O_5$

$$[(C_4H_{12}N_2)Cr(O)(O_2)_2]$$

House, D. A. *et al., Inorg. Chem.*, 1967, **6**, 1078 (footnote 6)
This blue precipitate is formed intermediately during preparation of 1,2-diamino-2-methylpropaneaquadiperoxochromium(IV) monohydrate, and its analogues are dangerously explosive and should not be isolated without precautions.
See other AMMINECHROMIUM PEROXOCOMPLEXES

1747. Tetramethylammonium pentafluoroxenoxide
[161535-67-6] $C_4H_{12}F_5NOXe$

$$Me_4N^+ \ ^-OXeF_5$$

Christe, K. O. *et al., Inorg. Chem.*, 1995, **34**(7), 1868
Not surprisingly, this salt is highly explosive.
See other XENON COMPOUNDS, *See also* QUATERNARY OXIDANTS

1748. Tetramethyldigallane
[65313-37-3] $C_4H_{12}Ga_2$

$$Me_2GaGaMe_2$$

Leleu, *Cahiers*, 1977, (88), 366
It ignites in air.
See other ALKYLMETALS

1749. Tetramethylammonium periodate (N,N,N-Trimethylmethaminium periodate)
[55999-69-4] $CH_{12}IO_4$

$$Me_4N^+ INO_{4-}$$

1. Wagner, R. I. *et al., Inorg. Chem.*, 1997, **36**(12), 2564
2. Malpass, J. R. *et al., Chem. Brit.*, 1998, **34**(12), 18
3. Cullis, P. M., Personal communication, 1998

Explodes violently on heating above 240°C [1]. A I4g sample exploded while being transferred to a cold, empty, unagitated flask via a polythene powder funnel, injuring the worker. The isolated salt was essentially neutral, so excess periodic acid cannot be involved. Subsequent attempts to explode the material deliberately have failed, but it is felt that static electricity may be the cause [2], [3].
See other QUATERNARY OXIDANTS

1750. Tetramethylammonium diazidoiodate(I)
[68574-13-0] $C_4H_{12}IN_7$

$$Me_4N^+[I(N_3)_2]^-$$

Dehnicke, K., *Angew. Chem. (Intern. Ed.)*, 1979, **18**, 512
Highly explosive crystals.
See related NON-METAL AZIDES

1751. Tetramethylammonium superoxide
[3946-86-9] $C_4H_{12}NO_2$

$$Me_4N^+O_2^-$$

1. McElroy, A. O. *et al., Inorg. Chem.*, 1964, **3**, 1798
2. Foote, C. S., *Photochem. Photobiol.*, 1978, **28**, 718–719

During the preparation by solid phase interaction of tetramethylammonium hydroxide and potassium superoxide [1] by tumbling for several days in a rotary evaporator flask, a violent explosion occurred [2]. This may have been caused by ingress of grease or other organic material leading to contact with potassium superoxide, a powerful oxidant.
See Potassium dioxide: Organic materials
See other QUATERNARY OXIDANTS

578

1752. N -(Diethylphosphinoyl)hydroxylamine
[ ] $C_4H_{12}NO_2P$

Et₂P(O)NHOH

Harger, M. J. P. *et al., Tetrahedron Lett.*, 1991, **32**(36), 4769
Described as a hazardously unstable oil.
See other N–O COMPOUNDS

1753. Tetramethylammonium ozonate (*N,N,N* -Trimethylmethanaminium ozonide)
[78657-29-1] $C_4H_{12}NO_3$

$Me_4N^+O_3^-$

Preparative hazard
See Ozone: Tetramethylammonium hydroxide
See other QUATERNARY OXIDANTS

1754. Tetrakis(hydroxymethyl)phosphonium nitrate
[24748-25-2] $C_4H_{12}NO_7P$

$(HOCH_2)_4P^+NO_3^-$

Calamari, T. A. *et al., Chem. Eng. News*, 1979, **57**(12), 74
A violent explosion ocurred during prolonged azeotropic drying at 105–110°C of a 75 wt% benzene solution of the salt. Traces of a nitrate ester may have been formed from a slight excess of nitrate ion.
See related OXOSALTS OF NITROGENOUS BASES, REDOX COMPOUNDS

†1755. 2-Dimethylaminoethylamine
[108-00-9] $C_4H_{12}N_2$

$Me_2NC_2H_4NH_2$

See other ORGANIC BASES

1756. N -2-Hydroxyethyl-1,2-diaminoethane (2-(2-Aminoethyl)aminoethanol)
[111-41-1] $C_4H_{12}N_2O$

$HOC_2H_4NHC_2H_4NH_2$

HCS 1980, 555

Cellulose nitrate
See CELLULOSE NITRATE: amines *See other* ORGANIC BASES

1757. Bis(dimethylamino) sulfoxide
[3768-60-3] $C_4H_{12}N_2OS$

$Me_2NS(:O)NMe_2$

Sulfinyl chloride
See Sulfinyl chloride: Bis(dimethylamino) sulfoxide
See other N–S COMPOUNDS

1758. Seleninyl bis(dimethylamide)
[2424-09-1] $C_4H_{12}N_2OSe$

$$SeO(NMe_2)_2$$

Bailar, 1973, Vol. 2, 975
Vacuum distillation at temperatures around 50–60°C leads to explosive decomposition.
See related N-METAL DERIVATIVES, N–S COMPOUNDS

1759. Tetramethyl-2-tetrazene
[6130-87-6] $C_4H_{12}N_4$

$$Me_2NN{=}NNMe_2$$

1. Houben-Weyl, 1967, Vol. 10.2, 830
2. Stull, 1977, 22
It explodes above its b.p., 130°C/1 bar [1], and is strongly endothermic ($\Delta H^°_f$ (l) +226.4 kJ/mol, 1.98 kJ/g) [2].
See other ENDOTHERMIC COMPOUNDS, TETRAZENES

1760. Dihydrazinenickel(II) succinate (Bis(hydrazine)[butanedioato(2−)]nickel)
[45101-14-2] $C_4H_{12}N_4NiO_4$

See HYDRAZINE METAL MALONATES AND SUCCINATES

1761. Bis(dimethylstibinyl) oxide
[ ] $C_4H_{12}OSb_2$

$$(Me_2Sb)_2O$$

Sidgwick, 1950, 777
It ignites in air.
See related ALKYLMETALS

1762. 1,1-Diethoxy-1,3-disiladioxetane
[ ] $C_4H_{12}O_4Si_2$

491M, 1975, 150

It ignites in air.
See related NON-METAL HYDRIDES

1763. Diethylhydroxytin hydroperoxide
[ ] $C_4H_{12}O_3Sn$

$$Et_2Sn(OH)OOH$$

Bailar, 1973, Vol. 2, 68
It is explosive.
See other ORGANOMETALLIC PEROXIDES

†1764. Tetramethoxysilane (Tetramethyl silicate)
[681-84-5] $C_4H_{12}O_4Si$

$$(MeO)_4Si$$

HCS 1980, 662

Metal hexafluorides
Jacob, E., *Angew. Chem. (Intern. Ed.)*, 1982, **21**, 143, (footnote 6)
During the preparation of the hexamethoxides of rhenium, molybdenum and tungsten by co-condensation with excess tetramethoxysilane on a cold surface, simultaneous co-condensation is necessary to avoid the danger of explosion present when sequential condensation of the reactants is employed. In the latter case, the high concentrations of hexafluoride at the interface leads to violent reaction with the silane.
See related SILANES

1765. Tetramethyldiphosphane (Tetramethyldiphosphine)
[3676-91-3] $C_4H_{12}P_2$

$$Me_2PPMe_2$$

1. Kardosky, G. *et al., Inorg. Synth.*, 1973, **14**, 15
2. Goldsberry, R. *et al., Inorg. Synth.*, 1972, **13**, 28
It ignites in air [1], and unreacted tetramethyldiphosphane residues may also ignite [2].
See Bis(tetramethyldiphosphane disulfide)cadmium perchlorate
See other ALKYLPHOSPHINES

1766. Tetramethyldiphosphane disulfide
[3676-97-9] $C_4H_{12}P_2S_2$

$$Me_2P(S)P(S)Me_2$$

Preparative hazard
1. Parshall, G. W., *Org. Synth.*, 1975, Coll. Vol. 5, 1016
2. Butter, A. S. *et al., Inorg. Synth.*, 1974, **15**, 186
3. Davies, S. G., *Chem. Brit.*, 1984, **20**, 403
4. Bercaw, J. E., *Chem. Eng. News*, 1984, **62**(18), 4
5. Burt, R. J. *et al., J. Organomet. Chem.*, 1979, **182**, 203

6. Bercaw, J. E. *et al., Inorg. Synth.*, 1985, **25**, 199

7. Parshall, G. W., *Org. Synth.*,1985, **63**, 226

The published preparation of the disulfide from thiophosphoryl chloride (4.28 mole) and methylmagnesium iodide (4.0 mole) [1,2] is hazardous. On the stated scale and at the recommended rate of addition of the 47% solution of thiophosphoryl chloride in ether, it is difficult to maintain the temperature at 0–5°C. Even at below one tenth of the recommended addition rate, after half of the solution had been added during 9 hours, the temperature rose uncontrollably from −5 to +60 °C in 10 s and a violent explosion followed, pulverising the apparatus [3]. A closely similar accident was simultaneously reported [4]. An alternative method is recommended [5], the product of which, though air-sensitive, does not ignite in air as previously stated. In the light of recent experience, the procedure in [1] is preferred to that in [2] because more dilute solutions are used in the former method [6]. To avoid explosions in the preparation of the title compound from methylmagnesium bromide and thiophosphoryl chloride in ether, the chloride should be redistilled and reaction vessel cooling should be with ice–salt rather than acetone–carbon dioxide, to assist in maintaining the reaction temperature in the range 0–5°C. If the temperature drops below −5°C, stop the addition and cautiously warm to 0–5°C [7].

See related ALKYLPHOSPHINES, NON-METAL SULFIDES

1767. Tetramethyllead (Tetramethylplumbane)
[75-74-1] $C_4H_{12}Pb$

Me$_4$Pb

HCS 1980, 895

Sidgwick, 1950, 463
It is liable to explode violently above 90°C.

Tetrachlorotrifluoromethylphosphorane
See Tetrachlorotrifluoromethylphosphorane: Tetramethyllead
See other ALKYLMETALS

1768. Tetramethylplatinum
[22295-11-0] $C_4H_{12}Pt$

Me$_4$Pt

Gilman, H. *et al., J. Amer. Chem. Soc.*, 1953, **75**, 2065
It explodes weakly on heating.
See Trimethylplatinum hydroxide
See other ALKYLMETALS, PLATINUM COMPOUNDS

1769. Tetramethyldistibane (Tetramethyldistibine)
[41422-43-9] $C_4H_{12}Sb_2$

Me$_2$SbSbMe$_2$

Sidgwick, 1950, 779

It ignites in air.
See other ALKYLMETALS

†1770. Tetramethylsilane
[75-76-3]

$$Me_4Si$$

$C_4H_{12}Si$

Chlorine
See Chlorine: Antimony trichloride, etc.
See other ALKYLSILANES

†1771. Tetramethyltin (Tetramethylstannane)
[594-27-4]

$$Me_4Sn$$

$C_4H_{12}Sn$

Dinitrogen tetraoxide
See Dinitrogen tetraoxide: Tetramethyltin
See other ALKYLMETALS

1772. Tetramethyltellurium(IV)
[123311-08-0]

$$Me_4Te$$

$C_4H_{12}Te$

Gedridge, R. W. *et al., Organometallics*, 1989, **8**(12), 2817
Extremely pyrophoric, sometimes explosively so.
See other ALKYLMETALS

1773. Bis(2-aminoethyl)aminesilver nitrate
[ ]

$$[HN(C_2H_4NH_2)_2Ag]\ NO_3$$

$C_4H_{13}AgN_4O_3$

Tennhouse, G. J. *et al., J. Inorg. Nucl. Chem.*, 1966, **28**, 682–684
The complex explodes at 200°C.
See other AMMINEMETAL NITRATES, SILVER COMPOUNDS

1774. 1,1,3,3-Tetramethylsiloxalane
[ ]

$$(Me)_2SiHOAl(Me)_2$$

$C_4H_{13}AlOSi$

This compound, which exists as the dimer, explodes on contact with water.
See ALKYLALUMINIUM DERIVATIVES, *See related* ALKYLSILANES

1775. Bis(2-aminoethyl)aminecobalt(III) azide
[26493-63-0]

$$[HN(C_2H_4NH_2)_2Co]\ [N_3]_3$$

$C_4H_{13}CoN_{12}$

Druding, L. F. *et al., Acta Cryst.*, 1974, **B30**, 2386

The shock-sensitivity and dangerously explosive nature is stressed.
See AMMINECOBALT(III) AZIDES *See related* METAL AZIDES

1776. Bis(2-aminoethyl)aminediperoxochromium(IV)
[59419-71-5] $C_4H_{13}CrN_3O_4$

1. House, D. A. *et al., Inorg. Chem.*, 1966, **5**, 840
2. Ghosh, S. K. *et al., J. Organomet. Chem.*, 1988, **340**(1), 59

The monohydrate explodes at 109–110°C during slow heating [1]. The solid complex exploded when manipulated on a glass sinter with a metal spatula [1]. Preparation on small scale and use of plastic equipment is recommended [2].

See other AMMINECHROMIUM PEROXOCOMPLEXES

1777. Bis(2-aminoethyl)amine (Diethylenetriamine)
[111-40-0] $C_4H_{13}N_3$

$$HN(C_2H_4NH_2)_2$$

HCS 1980, 399

Cellulose nitrate
  See CELLULOSE NITRATE: amines

Nitromethane
  See Nitromethane: Bis(2-aminoethyl)amine

Other reactants
  Yoshida, 1980, 147
  MRH values calculated for 13 combinations with oxidants are given.
  See other ORGANIC BASES

1778. Tetramethyldialuminium dihydride
[33196-65-5] $C_4H_{14}Al_2$

Wiberg, W. E. *et al., Angew. Chem.*, 1939, **52**, 372
It ignites and burns explosively in air.
See other ALKYLALUMINIUM ALKOXIDES AND HYDRIDES

1779. Tetramethyldiborane
[21482-59-7] $C_4H_{14}B_2$

$$H_3C\diagdown_{B}\diagup^{H}\diagdown_{B}\diagup^{CH_3}$$
$$H_3C\diagup\quad\diagdown_{H}\quad\diagdown^{CH_3}$$

Leleu, *Cahiers*, 1977, (88), 365
It ignites in air.
See other ALKYLBORANES

1780. Diacetatotetraaquocobalt (Cobalt(II) acetate tetrahydrate)
[6147-53-1] $C_4H_{14}CoO_8$

$$Co(OAc)_2.4H_2O$$

Widmann, G. *et al., Thermochim. Acta*, 1988, **134**, 451–455
Energy of decomposition (270–340°C) was determined by DSC as 2.42 kJ/g, peaking at 328°C. The exotherm is largely from combustion of the evolved carbon
monoxide and acetone in air.
See other HEAVY METAL DERIVATIVES, REDUCANTS

1781. 1,2-Diamino-2-methylpropaneaquadiperoxochromium(IV)
[17168-83-1] $C_4H_{14}CrN_2O_5$

House, D. A. *et al., Inorg. Chem.*, 1967, **6**, 1078
The monohydrate exploded at 83–84°C during slow heating, and is potentially explo-
sive at 20–25°C. There is also a preparative hazard.
See 1,2-Diamino-2-methylpropaneoxodiperoxochromium(VI)
See other AMMINECHROMIUM PEROXOCOMPLEXES

1782. Tetramethylammonium amide
[13422-81-6] $C_4H_{14}N_2$

$$Me_4N^+NH_2^-$$

Ammonia
Musker, W. K., *J. Org. Chem.*, 1967, **32**, 3189

During the preparation, the liquid ammonia used as solvent must be removed completely at $-45°C$. The compound decomposes explosively at ambient temperature in presence of ammonia.

See Sodium: Ammonia, Aromatic hydrocarbons
See related N-METAL DERIVATIVES

1783. Tetramethyldisiloxane
[3277-26-7] $C_4H_{14}OSi_2$

$$HMe_2SiOSiMe_2H$$

Allylphenols, Platinum
See SILANES: olefins, platinum

Oxygen, Sodium hydroxide, Water
See Oxygen (Gas): Tetramethyldisiloxane
See related ALKYLSILANES

1784. Bis(1,2-diaminoethane)dinitrocobalt(III) perchlorate
[[14781-32-9] (c-) [14781-33-0] (t-)] $C_4H_{16}ClCoN_6O_8$

$$[(C_2H_8N_2)_2Co(NO_2)_2] ClO_4$$

Seel, F. *et al., Z. Anorg. Chem.*, 1974, **408**, 281
The *cis-* and *trans*-isomers are both dangerously explosive compounds.
See other AMMINEMETAL OXOSALTS

1785. Tetrakis(thiourea)manganese(II) perchlorate
[50831-29-3] $C_4H_{16}Cl_2MnN_8O_8S_4$

$$[(CH_4N_2S)_4Mn] [ClO_4]_2$$

Karnaukhov, A. S. *et al., Chem. Abs.*, 1973, **79**, 140195
The complex decomposes explosively at 257°C.
See related AMMINEMETAL OXOSALTS

1786. Bis(1,2-diaminoethane)dichlorocobalt(III) chlorate
[26388-78-3] $C_4H_{16}Cl_3CoN_4O_3$

$$[(C_2H_8N_2)_2CoCl_2] ClO_3$$

It explodes at 320°C.
See entry AMMINEMETAL OXOSALTS (reference 2)

1787. Bis(1,2-diaminoethane)dichlorocobalt(III) perchlorate
[14932-06-0] $C_4H_{16}Cl_3CoN_4O_4$

$$[(C_2H_8N_2)_2CoCl_2] ClO_4$$

It explodes at 300°C; low impact-sensitivity.
See entry AMMINEMETAL OXOSALTS (reference 2)

1788. *cis*-Bis(1,2-diaminoethane)dinitrocobalt(III) iodate
[ ] $C_4H_{16}CoIN_6O_7$

$$[(C_2H_8N_2)_2Co(NO_2)_2] IO_3$$

Lobanov, N. I., *Zh. Neorg. Khim.*, 1959, **4**, 151
It dissociates explosively on heating.
See other AMMINEMETAL OXOSALTS

1789. Diazido-bis(1,2-diaminoethane)ruthenium(II) hexafluorophosphate
[30459-28-0] $C_4H_{16}F_6N_{10}PRu$

$$[(C_2H_8N_2)_2Ru(N_3)_2] PF_6$$

Kane-Maguire, L. A. P. *et al., Inorg. Synth.*, 1970, **12**, 24
Small explosions were observed on scratching the crystals with a metal spatula.
See related METAL AZIDES

1790. 2,4,6,8-Tetramethylcyclotetrasiloxane
[2370-88-9] $C_4H_{16}O_4Si_4$

See 2,4,6,8-Tetraethenyl-2,4,6,8-tetramethylcyclotetrasiloxane
See other SILANES

1791. Bis(1,2-diaminoethane)hydroxooxorhenium(V) perchlorate
[19267-68-6] $C_4H_{17}Cl_2N_4O_{10}Re$

$$[(C_2H_8N_2)_2Re(OH)O] [ClO_4]_2$$

Murmann, R. K. *et al., Inorg. Synth.*, 1966, **8**, 174–175
It explodes violently when dried at above ambient temperature and is also shock-sensitive. Several explosions occurred during analytical combustion.
See other AMMINEMETAL OXOSALTS

1792. Pentaamminepyrazineruthenium(II) perchlorate
[41481-90-7] $C_4H_{19}Cl_2N_7O_8Ru$

$$[(H_3N)_5RuC_4H_4N_2] [ClO_4]_2$$

Creutz, C. A., private comm., 1969

587

After ether washing, 30 mg of the salt exploded violently when disturbed.
See other AMMINEMETAL OXOSALTS

1793. Bis(1,2-diaminoethane)diaquacobalt(III) perchlorate
[55870-36-5] $C_4H_{20}Cl_3CoN_4O_{14}$

$$[(C_2H_8N_2)_2Co(H_2O)_2] [ClO_4]_3$$

Seel, F. *et al., Z. Anorg. Chem.*, 1974, **408**, 281
During evaporation of a solution of the complex salt, a very violent explosion occurred.
See other AMMINEMETAL OXOSALTS

1794. Tetraammine-2,3-butanediimineruthenium(III) perchlorate
[56370-81-1] $C_4H_{20}Cl_3N_6O_{12}Ru$

$$[(H_3N)_4RuC_4H_8N_2] [ClO_4]_3$$

Evans, I. P. *et al., J. Amer. Chem. Soc.*, 1977, **98**, 8042
It is explosive.
See other AMMINEMETAL OXOSALTS

1795. Bis(dimethylaminoborane)aluminium tetrahydroborate
[39047-21-7] $C_4H_{22}AlB_3N_2$

$$(Me_2NBH_3)_2Al\ BH_4$$

Air, or Water
Burg, A. B. *et al., J. Amer. Chem. Soc.*, 1951, **73**, 957
The impure oily product ignites in air and reacts violently with water.
See other COMPLEX HYDRIDES *See related* BORANES

1796. 1,1′-Azo-1,2-dicarbadecaborane(14)
[ ] $C_4H_{22}B_{20}N_2$

Complex structure

See entry AZOCARBABORANES

1797. Potassium tetracyanomercurate(II)
[591-89-9] $C_4HgK_2N_4$

$$K_2[Hg(CN)_4]$$

Ammonia
Pieters, 1957, 30
Contact may be explosive.
See Potassium hexacyanoferrate(III): Ammonia
See other MERCURY COMPOUNDS *See related* METAL CYANIDES

1798. μ-1,2-Bis(cyanomercurio)ethanediylidenedimercury(II)

[ ] $C_4Hg_4N_2$

$$NCHgC(:Hg)C(:Hg)HgCN$$

Whitmore, 1921, 128
It explodes slightly on heating.
See related METAL CYANIDES *See other* MERCURY COMPOUNDS

1799. 1,4-Diiodo-1,3-butadiyne

[53214-97-4] C_4I_2

$$IC{\equiv}CC{\equiv}CI$$

Sorbe, 1968, 66
It explodes at about 100°C.
See other HALOACETYLENE DERIVATIVES

1800. Potassium tetracyanotitanate(IV)

[75038-71-0] $C_4K_4N_4Ti$

$$K_4[Ti(CN)_4]$$

Water
Nicholls, D. *et al., J. Chem. Soc., Chem Comm.*, 1974, 635–636
Interaction is violent.
See other METAL CYANIDES

†1801. Dicyanoacetylene (2-Butynedinitrile)

[1071-98-3] C_4N_2

$$N{\equiv}CC{\equiv}CC{\equiv}N$$

1. Kirshenbaum, A. D. *et al., J. Amer. Chem. Soc.*, 1956, **78**, 2020
2. Robinson, Rodd, 1965, Vol. ID, 338, 366
3. Ciganek, E. *et al., J. Org. Chem.*, 1968, **33**, 542

As anticipated for a conjugated multiple triply bound structure, it is extremely endothermic (ΔH_f° (g) +626.4 kJ/mol, 8.22 kJ/g), and may decompose explosively to carbon powder and nitrogen. When burnt in oxygen a flame temperature exceeding 4700°C is attained [1], and it ignites in air at 130°C [2]. It is potentially explosive in the pure state or in conc. solutions, but is fairly stable in dilute solution [3]. It rapidly polymerises with water, weak bases, alcohols, etc. [2].
See related HALOACETYLENE DERIVATIVES
See other CYANO COMPOUNDS, ENDOTHERMIC COMPOUNDS

1802. Sodium tetracyanatopalladate(II)

[ ] $C_4N_4Na_2O_4Pd$

$$Na_2[Pd(OCN)_4]$$

Bailar, 1973, Vol. 3, 1288

It explodes on heating or impact.

See related METAL CYANATES

1803. Dicyanofurazan
[55644-07-0] C_4N_4O

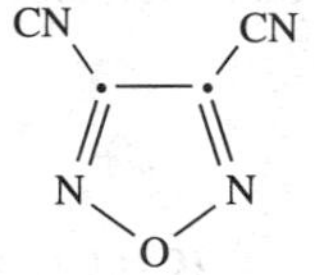

Homewood, R. H. *et al.*, US Pat. 3 832 249, 1974
Although highly endothermic (ΔH°_f (s) +456.3 kJ/mol, 3.80 kJ/g), it is a stable and relatively insensitive but powerful explosive.

Nitrogenous bases
Denson, D. B. *et al.*, US Pat. 3 740 947, 1973
Contact of dicyanofurazan, or its *N*-oxide (dicyanofuroxan), with hydrazine, mono- or di-methylhydrazine, piperidine, piperazine, diethylamine or their mixtures is instantaneously explosive.
See other CYANO COMPOUNDS, ENDOTHERMIC COMPOUNDS, N–O COMPOUNDS

1804. Dicyanofurazan *N*-oxide (Dicyanofuroxan)
[55644-07-0] $C_4N_4O_2$

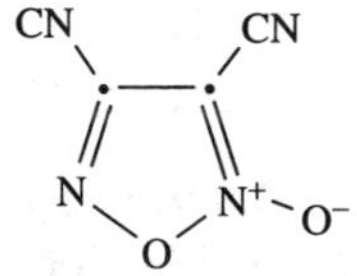

It is highly endothermic (ΔH°_f (s) +565.5 kJ/mol, 3.43 kJ/g).

Nitrogenous bases
Denson, D. B. *et al.*, US Pat. 3 740 947, 1973
Contact of dicyanofurazan *N*-oxide or the parent furazan with hydrazine, mono- or di-methylhydrazine, piperidine, piperazine, diethylamine or their mixtures is instantaneously explosive.
See other CYANO COMPOUNDS, ENDOTHERMIC COMPOUNDS, FURAZAN *N*-OXIDES

†1805. Tetracarbonylnickel (Nickel tetracarbonyl)
[13463-39-3] C_4NiO_4
$$(CO)_4Ni$$

HCS 1980, 885

Bromine
  Blanchard, A. A. *et al., J. Amer. Chem. Soc.*, 1926, **48**, 872
  The two compounds interact explosively in the liquid state, but smoothly as vapour.

Dinitrogen tetraoxide
  See Dinitrogen tetraoxide: Tetracarbonylnickel

Mercury, Oxygen
  Mellor, 1946, Vol. 5, 955
  A mixture of the dry carbonyl and oxygen will explode on vigorous shaking with mercury (presumably catalysed by mercury or its oxide).

Oxygen, Butane
  1. Egerton, A. *et al., Proc. R. Soc.*, 1954, **A225**, 427
  2. Badin, E. J. *et al., J. Amer. Chem. Soc.*, 1948, **70**, 2055
  The carbonyl on exposure to atmospheric oxygen produces a deposit which becomes peroxidised and may ignite. Mixtures with air or oxygen at low partial and total pressures explode after a variable induction period [1]. Addition of the carbonyl to a butane–oxygen mixture at 20–40°C caused explosive reaction in some cases [2].
  See other INDUCTION PERIOD INCIDENTS

Tetrachloropropadiene
  See Dicarbonyl-η-trichloropropenyldinickel chloride dimer
  See other CARBONYLMETALS

1806. Hexafluoroglutaryl dihypochlorite
[71359-64-3] $C_5Cl_2F_6O_4$

$$F_2C(CF_2CO.OCl)_2$$

  Tari, I. et al., *Inorg. Chem.*, 1979, **18**, 3205–3208
  Explosive above −10°C.
  See other ACYL HYPOHALITES

1807. Tetrachlorodiazocyclopentadiene
[21572-61-2] $C_5Cl_4N_2$

  Abel, E. W. *et al., Inorg. Chim. Acta*, 1980, **44**(3), L161–163

Conversion to octachloronaphthalene by heating at 150°C in absence of solvent was explosive.

See other DIAZO COMPOUNDS

1808. Hexachlorocyclopentadiene
[77-47-4] C_5Cl_6

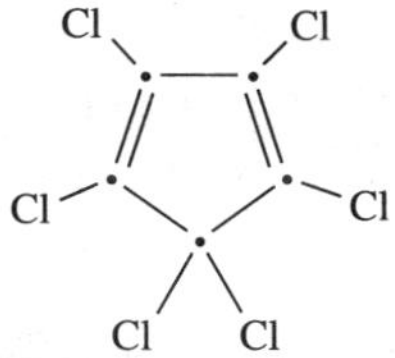

Sodium
   See Sodium: Halocarbons
   See other HALOALKENES

1809. Potassium azidopentacyanocobaltate(3−)
[14705-99-8] $C_5CoK_3N_8$

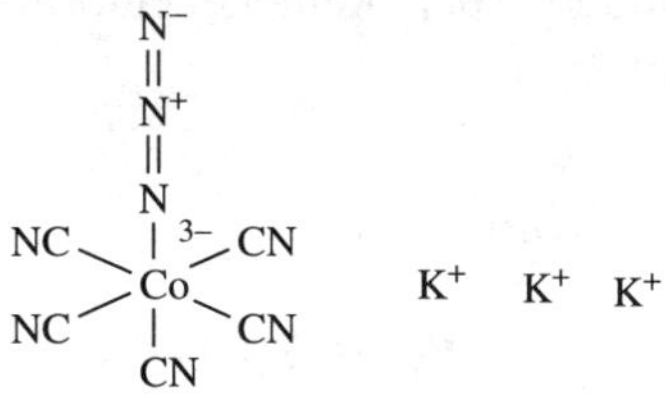

Huan, J. *et al., Thermochim. Acta*, 1988, **130**, 77−85
Thermal decomposition under hydrogen of a series of pentacyanocobaltate complexes (CN-, NO₂-, NO- or N₃-ligands) revealed that the latter complex is the most exothermic by far. Presence of iron powder suppresses hydrogen cyanide formation.

See related METAL AZIDES, METAL CYANIDES

1810. Potassium pentacyanodiperoxochromate(5−)
[ ] $C_5CrK_5N_5O_4$

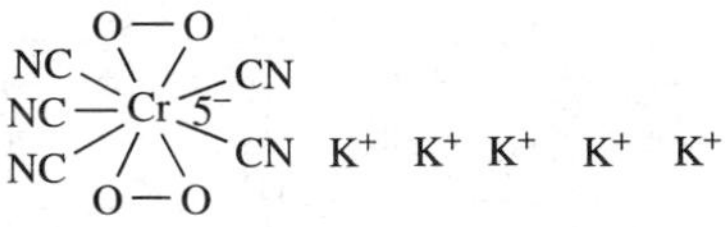

Bailar, 1973, Vol. 4, 167
A highly explosive material, with internal redox features.

See related AMMINECHROMIUM PEROXOCOMPLEXES *See other* REDOX COMPOUNDS

1811. Caesium pentacarbonylvanadate(3−)
[78937-12-9] $C_5Cs_3O_5V$

Air, or Hydroxy compounds
Ellis, J. F. *et al., J. Amer. Chem. Soc.*, 1981, **103**, 6101
Pyrophoric in air, deflagrates under nitrogen on scratching with a metal spatula, and reacts explosively with alcohols or water.
See related CARBONYLMETALS

1812. 2-Heptafluoropropyl-1,3,4-dioxazolone
[87050-95-1] $C_5F_7NO_3$

Middleton, W. J., *J. Org. Chem.*, 1983, **48**, 3845
It could not be detonated by hot wire initiation, but a sample exploded during distillation at ambient pressure at 102°C.
See 1,3,4-DIOXAZOLONES *See other* N–O COMPOUNDS

1813. Sodium pentacyanonitrosylferrate(2−)
[14402-89-2] $C_5FeN_6Na_2O$

Sodium nitrite
See Sodium nitrite: Metal cyanides
See related METAL CYANIDES

†1814. **Pentacarbonyliron (Iron pentacarbonyl)**
[13463-40-6]   C₅FeO₅

$(OC)_5Fe$

Merck, 1983, 736
Pyrophoric in air.

Acetic acid, Water
Braye, E. H. *et al., Inorg. Synth.*, 1966, **8**, 179
A brown pyrophoric powder is produced if the carbonyl is dissolved in acetic acid
containing above 5% of water.

Nitrogen oxide
See Nitrogen oxide: Pentacarbonyliron

Pentaborane(9), Pyrex glass
See Pentaborane(9): Pentacarbonyliron, etc.

Transition metal halides, Zinc
Lawrenson, M. J., private comm., 1970
The preparation of carbonylmetals by treating a transition metal halide either with
carbon monoxide and zinc, or with iron pentacarbonyl is well-known and smooth.
However, a violent eruptive reaction occurs if a methanolic solution of a cobalt
halide, a rhodium halide or a ruthenium halide is treated with both zinc and iron
pentacarbonyl.
See other CARBONYLMETALS

1815. **1,3-Pentadiyn-1-ylsilver**
[ ]   C₅H₃Ag

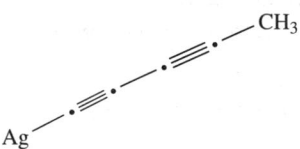

Alone, or Sulfuric acid
Schluhbach, H. H. *et al., Ann.*, 1950, **568**, 155
Very sensitive to impact or friction, and explodes when moistened with sulfuric acid.
See other METAL ACETYLIDES, SILVER COMPOUNDS

1816. 2-Chloro-3-pyridinediazonium tetrafluoroborate
[70682-07-4] $C_5H_3BClF_4N_3$

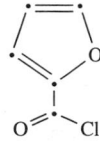

Verde, C., Span. Pat. ES 548 695, 1986
The diazonium salt decomposes violently above 25°C or when dry.
See other DIAZONIUM TETRAHALOBORATES

1817. 2-Furylchlorodiazirine
[ ] $C_5H_3ClN_2$

Khasanova, T. *et al., J. Amer. Chem. Soc.*, 1998, **120**(1), 233
This previously unknown compound is, not surprisingly, detonable.
See other DIAZIRINES

1818. Furoyl chloride
[1300-32-9] $C_5H_3ClO_2$

Preparative hazard
    Wrigley, T. C., private comm., 1979
    A sample of freshly distilled material exploded violently during overnight storage.
    Self-acylation or polymerisation catalysed by hydrogen chloride may have been
    involved.
    See Phosphorus trichloride: Carboxylic acids
    See also 2-HALOMETHYL-FURANS OR -THIOPHENES
    See other ACYL HALIDES, POLYCONDENSATION REACTION INCIDENTS

1819. 1,3-Pentadiyn-1-ylcopper
   [115609-87-5] C_5H_3Cu

Schluhbach, H. H. *et al., Ann.*, 1950, **568**, 155
Explodes on impact or friction.
See other METAL ACETYLIDES

1820. Lithium 3-(1,1,2,2-tetrafluoroethoxy)propynide
   [82906-06-7] $C_5H_3F_4LiO$

Von Werner, K., *J. Fluorine Chem.*, 1982, **20**, 215–216
The explosion during work-up of a substantial amount of the lithiated fluorocompound emphasises the need for great care and full precautions in handling such materials.
See related HALOACETYLENE DERIVATIVES *See other* METAL ACETYLIDES

1821. 2-Furoyl azide
   [20762-98-5] $C_5H_3N_3O_2$

Dunlop, 1953, 544
The azide explodes violently on heating in absence of a solvent or diluent.
See other ACYL AZIDES

1822. 2-Hydroxy-3,5-dinitropyridine
[2980-33-8] $C_5H_3N_3O_5$

Glowiak, B., *Chem. Abs.*, 1963, **58**, 498h
Various heavy metal salts show explosive properties, and the lead salt might be useful as an initiating explosive.
See other POLYNITROARYL COMPOUNDS

1823. 2-Formylamino-3,5-dinitrothiophene
[ ] $C_5H_3N_3O_5S$

Fast flame propagation on heating the powder moderately.
See entry HIGH RATE DECOMPOSITION *See other* POLYNITROARYL COMPOUNDS

1824. Methyl 3,3-diazido-2-cyanoacrylate
[82140-87-2] $C_5H_3N_7O_2$

Saalfrank, R. W. *et al., Angew. Chem. (Intern. Ed.)*, 1987, **26**, 1161 (footnote 14)
The crystalline solid decomposes explosively around 70°C, so working scale was limited to 0.5 g.
See other CYANO COMPOUNDS, HIGH-NITROGEN COMPOUNDS, ORGANIC AZIDES

1825. 1,3-Pentadiyne
   [4911-55-1]   C₅H₄

Brandsma, 1971, 7, 36
It explodes on distillation at ambient pressure.
See other ACETYLENIC COMPOUNDS

1826. 3-Pyridinediazonium tetrafluoroborate
   [586-92-5]   C₅H₄BF₄N₃

1. Johnson, E. P. *et al., Chem. Eng. News*, 1967, **45**(44), 44
2. Roe, A. *et al., J. Amer. Chem. Soc.*, 1947, **69**, 2443

A sample, air-dried on aluminium foil, exploded spontaneously and another sample exploded on heating to 47°C [1]. The earlier reference describes the instability of the salt above 15°C if freed from solvent, and both the 2- and 4-isomeric salts were found to be very unstable and incapable of isolation [2].
See other DIAZONIUM TETRAHALOBORATES

1827. 3-Bromopyridine
   [626-55-1]   C₅H₄BrN

Acetic acid, Hydrogen peroxide
   See Hydrogen peroxide: Acetic acid, *N*-Heterocycles

1828. 3-Bromopyridine N-oxide
  [2402-97-3] C_5H_4BrNO

Br
(structure of 3-bromopyridine N-oxide)
N^+—O^-

Preparative hazard
  See Hydrogen peroxide: Acetic acid, N-Heterocycles

1829. 2-Chloropyridine N-oxide
  [2402-95-1] C_5H_4ClNO

Cl
(structure of 2-chloropyridine N-oxide)
N^+—O^-

1. Kotoyori, T., private comm., 1983
2. *MARS Database*, 1998, short report 025
A 500 kg batch had been prepared in glacial acetic acid using 35% aqueous hydrogen peroxide with 2% sulfuric acid as catalyst. When the reaction was complete, solvents were removed by vacuum distillation, steam was turned off and the reaction vessel jacket filled with water. After standing overnight, violent decomposition of the contents occurred, causing the safety valve to lift, even although the vessel was vented via its condenser/receiver system. An external fire ensued, and carbonised residue was found inside the reactor. It was surmised that as the product began to solidify inwards from the outer cooled reactor wall, the inner liquid material effectively went into adiabatic storage. The compound is of limited thermal stability and begins to decompose exothermally at 90–100°C, and salt formation from liberated hydrogen chloride will tend to accelerate the process. The hot weather (and possibly a leak of steam into the jacket) may have exaggerated the adiabatic effect. The necessity to assess thermal stabilities of reactants and products, to minimise heating times and to maximise cooling rates, and to monitor and supervise processes is stressed [1]. Another report of a process explosion during manufacture of this, using maleic anhydride as catalyst and without a solvent, was attributed to adding a shot of 70% hydrogen peroxide slightly faster than normal. Since the reaction was almost complete and no product remained in the reactor afterwards, it is possible that the amine oxide was also implicated [2]
See other N-OXIDES

1830. 2-Chloro-5-chloromethylthiophene
[23784-96-5]   C₅H₄Cl₂S

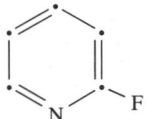

Rosenthal, N. A., *J. Amer. Chem. Soc.*, 1951, **73**, 590
Storage at ambient temperature may lead to explosively violent decomposition.
See other 2-HALOMETHYL-FURANS OR -THIOPHENES

1831. 2-Fluoropyridine
[372-48-5]   C₅H₄FN

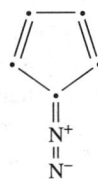

Bromine trifluoride
See Bromine trifluoride: Pyridine

1832. Diazocyclopentadiene
[1192-27-4]   C₅H₄N₂

$$C_5H_4Cl_2S$$

1. Ramirez, F., *J. Org. Chem.*, 1958, **23**, 2036–2037
2. Wedd, A. G., *Chem. & Ind.*, 1970, 109
3. Aarons, L. J. *et al.*, *J. Chem. Soc., Faraday Trans. 2*, 1974, **70**, 1108
4. DeMore, W. B. *et al.*, *J. Amer. Chem. Soc.*, 1959, **81**, 5875
5. Weil, T., *J. Org. Chem.*, 1963, **28**, 2472
6. Regitz, M. *et al.*, *Tetrahedron*, 1967, **23**, 2706
7. Gillespie, R. J. *et al.*, *J. Chem. Soc., Perkin Trans. 1*, 1979, 2626
8. Reimer, K. J. *et al.*, *Inorg. Synth.*, 1980, **20**, 189

A violent explosion occurred after distillation at 47–49°C/60–65 mbar [1],and also
during distillation at 48–53°C/66 mbar; use of solutions of undistilled material was
recommended [2]. The compound exploded on one occasion in the solid state after
condensation at −196°C, probably owing to fortuitous tribomechanical shock [3]. A
similar explosion on chilling a sample in liquid nitrogen had been noted previously
[4]. Of two simplified methods of isolating the material without distillation [4,5],

the former appears to give higher yields. A further explosion during distillation has been reported [7], but solutions in pentane are considered convenient and safe [8].

See other DIAZO COMPOUNDS

1833. 4-Nitropyridine *N*-oxide
[1124-33-0] $C_5H_4N_2O_3$

$$O_2NC_5H_4N{:}O$$

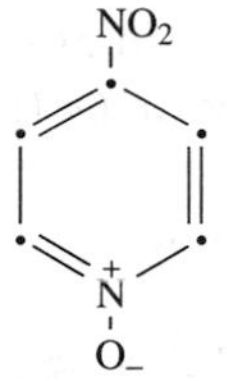

Diethyl 1,4-dihydro-2,6-dimethylpyridine-3,5-dicarboxylate
Kurbatova, A. S. *et al., Chem. Abs.*, 1978, **88**, 120946
Dehydrogenation of the 'Hantzsch' ester in a melt with the oxide at 130–140°C proceeded explosively.

See other N-OXIDES

1834. 1,2-Dihydropyrido[2,1-e]tetrazole
[ ] $C_5H_4N_4$

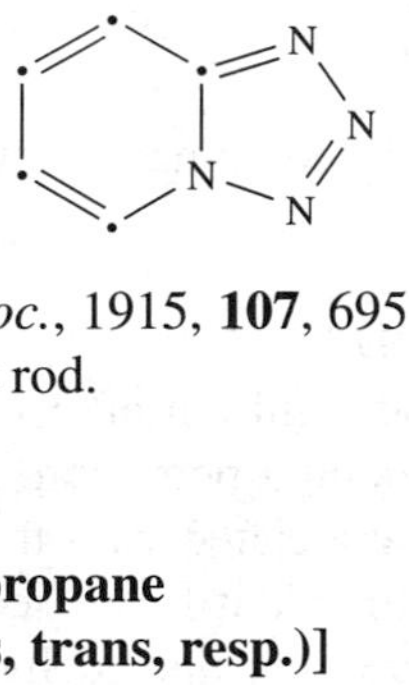

Fargher, R. G. *et al., J. Chem. Soc.*, 1915, **107**, 695
Explodes on touching with a hot rod.

See other TETRAZOLES

1835. 1,2-Bis(azidocarbonyl)cyclopropane
[[68979-48-6; 114752-52-2] (cis, trans, resp.)] $C_5H_4N_6O_2$

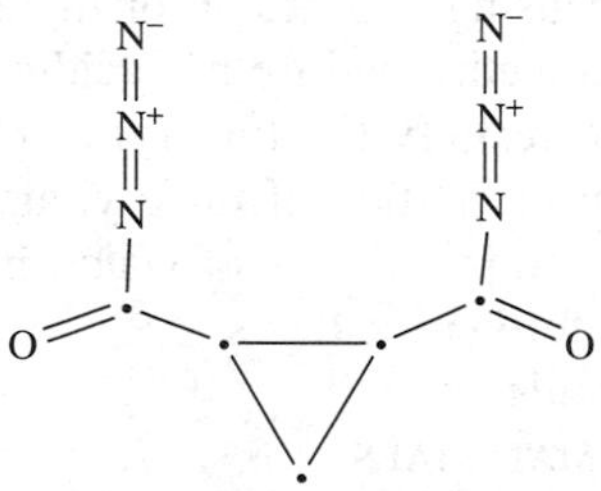

1. Landgrebe, J. A., *Chem. Eng. News*, 1981, **59**(17), 47
2. Majchrzak, M. W. *et al., Synth. Comm.*, 1981, **11**, 493–495
3. Johnson, R. E., *Chem. Eng. News*, 1988, **66**(13), 2

Trituration of the crude *cis*-azide with hexane at ambient temperature caused detonation to occur. Isolation of low molecular weight carbonyl azides should be avoided, or extreme precautions taken. A similar incident with the analogous cyclobutane diazide was reported [1]. During use as an intermediate in preparation of the 1,2-diamine, the explosive diazide was never isolated or handled free of solvent [2]. The *trans* isomer (0.25 mole) was prepared by the latter technique, but as a 45 w/v% solution in toluene, rather than as the recommended 10% solution. When a magnetic spin-bar was introduced, the solution detonated. Initiation may have involved friction from adventitious presence of iron filings adhering to the magnetic bar [3].

See other ACYL AZIDES, FRICTIONAL INITIATION INCIDENTS

1836. 2-Furaldehyde (Furfural)
[98-01-1] $C_5H_4O_2$

HCS 1980, 511

Other reactants
Yoshida, 1980, 313
MRH values calculated for 13 combinations with oxidants are given.

Sodium hydrogen carbonate
Unpublished information, 1977
Several cases of spontaneous ignition after exposure to air of fine coke particles removed from filter strainers on a petroleum refinery furfural extraction unit have been noted. This has been associated with the use of sodium hydrogen carbonate (bicarbonate) injected into the plant for pH control, which produced a pH of 10.5 locally. This would tend to resinify the aldehyde, but there is also the possibility of a Cannizzaro reaction causing conversion of the aldehyde to furfuryl alcohol and furoic acid. The latter, together with other acidic products of autoxidation of the aldehyde, would tend to resinify the furfuryl alcohol. Pyrolysis GLC showed the presence of a significant proportion of furfuryl alcohol-derived resins in the coke. The latter is now discarded into drums of water, immediately after discharge from the strainers, to prevent further incidents.
See Furfuryl alcohol: Acids
See other PYROPHORIC MATERIALS

Sodium hypochlorite
See Sodium hypochlorite: Furfuraldehyde
See other ALDEHYDES, PEROXIDISABLE COMPOUNDS

1837. Peroxyfuroic acid
[5797-06-8]

C₅H₄O₄

$$C_5H_4O_4$$

Alone, or Metal salts, or Organic materials
Milas, N. A. *et al., J. Amer. Chem. Soc.*, 1934, **56**, 1221
Thermal decomposition of the acid becomes violently explosive at 40°C. Intimate contact with finely divided metal salts (mainly halides), ergosterol, pyrogallol or animal charcoal led to explosive decomposition (often violent) in 18 out of 28 experiments.
See other PEROXYACIDS

1838. Silver cyclopropylacetylide
[ ]

$$C_5H_5Ag$$

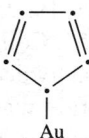

Slobodin, Ya. M. *et al., Chem. Abs.*, 1952, **46**, 10112e
It explodes on heating.
See other METAL ACETYLIDES, SILVER COMPOUNDS

1839. Cyclopentadienylgold(I)
[21254-73-9]

$$C_5H_5Au$$

Huttel, R. *et al., Angew. Chem. (Intern. Ed.)*, 1967, **6**, 862
It is sensitive to friction and heat, often deflagrating on gentle warming.
See other ALKYLMETALS, GOLD COMPOUNDS

1840. 2-Bromomethylfuran
[4437-18-7]

$$C_5H_5BrO$$

Dunlop, 1953, 231

It is very unstable, and the liberated hydrogen bromide accelerates further decomposition to explosive violence.

See other 2-HALOMETHYL-FURANS OR -THIOPHENES

1841. 2-Chloromethylfuran
[615-88-9] C_5H_5ClO

Anon., *ABCM Quart. Safety Summ.*, 1962, **33**, 2

A small sample of freshly prepared and distilled material, when stored over a weekend, exploded violently owing to polymerisation or decomposition (probably arising from the facile hydrolysis with traces of moisture and liberation of hydrogen chloride). Material should be prepared and used immediately, but if brief storage is inevitable, refrigeration is essential.

See other 2-HALOMETHYL-FURANS OR -THIOPHENES

1842. 2-Chloromethylthiophene
[617-88-9] C_5H_5ClS

1. Bergeim, F. H., *Chem. Eng. News*, 1952, **30**, 2546
2. Meyer, F. C., *Chem. Eng. News*, 1952, **30**, 3352
3. Wiberg, K. B., *Org. Synth.*, 1955, Coll. Vol. 3, 197

The material is unstable and gradually decomposes, even when kept cold and dark, with liberation of hydrogen chloride which accelerates the decomposition [1]. If kept in closed containers, the pressure increase may cause an explosion [2]. Amines stabilise the material, which may then be kept cold in a vented container for several months [3].

See other GAS EVOLUTION INCIDENTS, 2-HALOMETHYL-FURANS OR -THIOPHENES

1843. Poly(cyclopentadienyltitanium dichloride)
[35398-20-0] $(C_5H_5Cl_2Ti)_n$

(Complex structure)

Lucas, C. R. *et al., Inorg, Synth.*, 1976, **16**, 238

It is pyrophoric in air.

See other ORGANOMETALLICS, PYROPHORIC MATERIALS

604

1844. Oxodiperoxopyridinechromium *N*-oxide
[38293-27-5]   C$_5$H$_5$CrNO$_6$

1. Westland, A. D. *et al., Inorg. Chem.*, 1980, **19**, 2257
2. Daire, E. *et al., Nouv. J. Chim.*, 1984, **8**, 271–274
This chromium complex, and the molybdenum and tungsten analogues with two *N*-oxide ligands, were all explosive [1]. The chromium complex has exploded during desiccation [2].
See Triphenylphosphine oxide-oxodiperoxochromium(VI)
See other AMMINECHROMIUM PEROXOCOMPLEXES

1845. 1-Iodo-3-penten-1-yne
[ ]   C$_5$H$_5$I

Vaughn, J. A. *et al., J. Amer. Chem. Soc.*, 1934, **56**, 1208
After distillation at 72°C/62 mbar, the residue always exploded if heating were continued.
See other HALOACETYLENE DERIVATIVES

1846. Potassium cyclopentadienide
[30994-24-2]   C$_5$H$_5$K

Angus, P. L. *et al., Inorg. Synth.*, 1977, **17**, 177
The dry powder is very air-sensitive and pyrophoric.
See other ALKYLMETALS, PYROPHORIC MATERIALS

605

1847. Potassium 2,5-dinitrocyclopentanonide
[26717-79-3] $C_5H_5KN_2O_5$

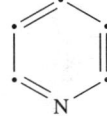

Wieland, H. *et al., Ann.*, 1928, **461**, 304
The salt (possibly an *aci*-nitro salt) explodes at 154–158°C and is less stable than
the di-potassium salt (241–245°C).
See other C-NITRO COMPOUNDS

†1848. Pyridine
[110-86-1] C_5H_5N

NSC 310, 1978; *FPA H58*, 1977; *HCS 1980*, 802; *RSC Lab. Hazard Data Sheet
No. 8*, 1983

Formamide, Iodine, Sulfur trioxide
  See Formamide: Iodine, etc.

Maleic anhydride
  See Maleic anhydride: Bases, etc.

Other reactants
  Yoshida, 1980, 292
  MRH values calculated for 15 combinations with oxidants are given.

Oxidants	MRH values show % of oxidant
See Bromine trifluoride: Pyridine	MRH 3.05/75
Chromium trioxide: Pyridine	MRH 3.05/75
Dinitrogen tetraoxide: Heterocyclic bases	MRH 7.07/78
Fluorine: Nitrogenous bases	
Trifluoromethyl hypofluorite: Pyridine	

See other ORGANIC BASES

1849. Pyridine *N*-oxide
[694-59-7]   C₅H₅NO

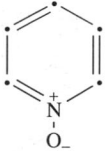

Energy of decomposition (in range 250–380°C) measured as 1.076 kJ/g by DSC, and T_{ait24} was determined as 205°C by adiabatic Dewar tests, with an apparent energy of activation of 124 kJ/mol.
See entry THERMOCHEMISTRY AND EXOTHERMIC DECOMPOSITION (reference 2)

Hexamethyldisilane, Tetrabutylammonium fluoride
   Vorbrueggen, H. *et al., Tetrahedron Lett.*, 1983, **24**, 5337
   Slow addition of hexamethyldisilane in THF to pyridine *N*-oxide and tetrabutylammonium fluoride in THF effected smooth reduction to pyridine, while addition of undiluted fluoride led to an explosion on two occasions.
   See other N-OXIDES

1850. Methylnitrothiophene
   [ ]   C₅H₅NO₂S
                              (Unstated isomer)

Preparative hazard
   See Nitric acid: Methylthiophene
   See other C-NITRO COMPOUNDS

1851. 2-(*N*-Nitroamino)pyridine *N*-oxide
   [85060-25-9]   C₅H₅N₃O₃

Talik, T. *et al., Chem. Abs.*, 1983, **98**, 160554
Powerfully explosive.
See other N-NITRO COMPOUNDS, N-OXIDES

607

1852. 4-(N-Nitroamino)pyridine N-oxide (N-Nitro-4-pyridinamine 1-oxide)
[85060-27-1] $C_5H_5N_3O_3$

Talik, T. *et al., Chem. Abs.*, 1983, **98**, 160554
Powerfully explosive.
See other N-NITRO COMPOUNDS, N-OXIDES

1853. Ethyl 4-diazo-1,2,3-triazole-5-carboxylate
[85807-69-8] $C_5H_5N_5O_2$

See entry DIAZOAZOLES
See other DIAZO COMPOUNDS, HIGH-NITROGEN COMPOUNDS, TRIAZOLES

1854. 1,1,1,3,5,5,5-Heptanitropentane
[20919-99-7] $C_5H_5N_7O_{14}$

Klager, K. *et al., Propellants, Explos., Pyrotech.*, 1983, **8**, 25–28
This explosive compound has a +12% oxygen balance so can function as an oxidant.
See other OXIDANTS, POLYNITROALKYL COMPOUNDS

1855. Cyclopentadienylsodium
[4984-82-1] C_5H_5Na

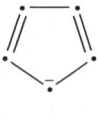

Na$^+$

King, R. B. *et al., Inorg. Synth.*, 1963, **7**, 101
The solid obtained by evaporation of the air-sensitive solution is pyrophoric in air.
See ALKALI-METAL DERIVATIVES OF HYDROCARBONS

Ammonium hexanitrocerate
See Sodium nitrate: Tetracyclopentadienylcerium

Lead(II) nitrate
See Lead(II) nitrate: Cyclopentadienylsodium
See other ALKYLMETALS, ORGANOMETALLICS

1856. Bicyclo[2.1.0]pent-2-ene
[5164-35-2] C_5H_6

Andrist, A. H. *et al., Org. Synth.*, 1976, **55**, 18
A purified undiluted sample was reported to explode.
See other STRAINED-RING COMPOUNDS

†1857. Cyclopentadiene
[542-92-7] C_5H_6

HCS 1980, 348

1. Kirk-Othmer, 1965, Vol. 6, 690
2. Raistrick, B. *et al., J. Chem. Soc.*, 1939, 1761–1773
3. Starkie, A. *et al., Chem. in Brit.*, 1996, **32**(2), 35
Dimerisation is highly exothermic, the rate increasing rapidly with temperature, and
may cause rupture of a closed uncooled container. The monomer may largely be
prevented from dimerising by storage at −80°C or below [1]. The polymerisation of
the undiluted diene may become explosive within the range 0−40°C and at pressures
above 2300 bar. The effect of diluents was also studied [2]. A polymerisation was

conducted with insufficient solvent, leading to a runaway and an explosion which killed three workers and destroyed plant up to 300m away [3].

See Propene
See other POLYMERISATION INCIDENTS

Nitric acid
See Nitric acid: Hydrocarbons (reference 9)

Oxides of nitrogen
See Nitrogen oxide: Dienes, Oxygen
Dinitrogen tetraoxide: Hydrocarbons (reference 10)

Oxygen
Hock, H. *et al., Chem. Ber.*, 1951, **84**, 349
Exposure of the diene to oxygen gives peroxidic products containing some monomeric, but largely polymeric, peroxides, which explode strongly on contact with a flame.
See other POLYPEROXIDES

Oygen, Ozone
See Ozone: Dienes, Oxygen

Potassium hydroxide
Wilson, P. J. *et al., Chem. Rev.*, 1944, **34**, 8
Contact with the ethanolic base causes vigorously exothermic resin formation.
See other POLYMERISATION INCIDENTS

Sulfuric acid
See Sulfuric acid: Cyclopentadiene
See other DIENES

†1858. 2-Methyl-1-buten-3-yne
[78-80-8] C_5H_6
$$H_2C{=}CMeC{\equiv}CH$$

See other ALKYNES

1859. Cyclopentadiene–silver perchlorate
[ ] $C_5H_6AgClO_4$
$$C_5H_6 \cdot AgClO_4$$

Ulbricht, T. L. V., *Chem. & Ind.*, 1961, 1570

The complex explodes on heating.
See related METAL PERCHLORATES *See other* SILVER COMPOUNDS

1860. 3,5-Dibromocyclopentene
[1890-04-6] $C_5H_6Br_2$

Lithium tetrahydroaluminate
See Lithium tetrahydroaluminate: 3,5-Dibromocyclopentene
See other HALOALKENES

1861. 4,4-Dibromo-3,5-dimethylpyrazole
[84691-20-3] $C_5H_6Br_2N_2$

Adam, W. A. *et al., J. Org. Chem.*, 1994, **59**(23), 7067
This compound decomposed in a few days at −18°C. On warming to ambient it
decomposed rapidly, exploding on one occasion.

1862. Pyridinium chlorochromate
[26299-14-9] $C_5H_6ClCrNO_3$

1. Glaros, G., *J. Chem. Educ.*, 1978, **55**, 410
2. Singh, J. H. *et al., Chem. & Ind.*, 1986, 751

611

It is a stable and safe oxidant which can replace oxodiperoxodipyridine-chromium(VI) in oxidation of alcohols to aldehydes [1]. The quinolinium analogue has also found similar application [2].

See Oxodiperoxodipyridinechromium(VI)

See other OXIDANTS, QUATERNARY OXIDANTS

1863. Pyridinium perchlorate
[15598-34-2] $C_5H_6ClNO_4$

1. Kuhn, R. *et al., Chem. Ztg.*, 1950, **74**, 139
2. Anon., *Sichere Chemiearb.*, 1963, **15**(3), 19
3. Arndt, F. *et al., Chem. Ztg.*, 1950, **74**, 140
4. Schumacher, 1960, 213
5. Zacherl, M. K., *Mikrochemie*, 1948, **33**, 387–388
6. Anon., Private communication, 1991

It can be detonated on impact, but is normally considered a stable intermediate (m.p. 288°C), suitable for purification of pyridine [1]. Occasionally explosions have occurred when the salt was disturbed [2], which have been variously attributed to presence of ethyl perchlorate, ammonium perchlorate, or chlorates. A sample was being dried in a glass vacuum desiccator. Over-enthusiastic release of vacuum distributed some of the salt onto the lid. The grinding action as the lid was slid off initiated explosion of this, shattering the lid [6]. A safer preparative modification is described [3]. It explodes on heating to above 335°C, or at a lower temperature if ammonium perchlorate is present [4]. A violent explosion which occurred during the final distillation according to the preferred method [3] was recorded [5].

See other PERCHLORATE SALTS OF NITROGENOUS BASES

1864. 5-Chloro-1,3-dimethyl-4-nitro-1*H*-pyrazole
[13551-73-0] $C_5H_6ClN_3O_2$

See NITRATION INCIDENTS (reference 8) *See other* C-NITRO COMPOUNDS

1865. 1,3-Dichloro-5,5-dimethyl-2,4-imidazolidindione
[118-52-5] $C_5H_6Cl_2N_2O_2$

Xylene
1. Anon., *Chem. Trade J.*, 1951, **129**, 136
2. *See entry* SELF-ACCELERATING REACTIONS
An attempt to chlorinate xylene with the 'dichlorohydantoin' caused a violent explosion [1]. The haloimide undergoes immediate self accelerating decomposition in presence of solvents. Safe conditions (including lower temperatures and progressive addition of reagent to match its consumption) can be developed for its use [2].
See other N-HALOIMIDES

1866. 3,3-Difluoro-1,2-dimethoxycyclopropene
[59034-33-2] $C_5H_6F_2O_2$

See entry FLUORINATED CYCLOPROPENYL METHYL ETHERS

1867. 4,4-Bis(difluoroamino)-3-fluoroimino-1-pentene
[33364-51-1] $C_5H_6F_5N_3$

Parker, C. O. *et al.*, *J. Org. Chem.*, 1972, **37**, 922
Samples of this and related poly-difluoroamino compounds exploded during analytical combustion.
See other DIFLUOROAMINO COMPOUNDS, N-HALOGEN COMPOUNDS

1868. 4-Iodo-3,5-dimethylisoxazole
[10557-85-4] C_5H_6INO

Peroxytrifluoroacetic acid
See Peroxytrifluoroacetic acid: 4-Iodo-3,5-dimethylisoxazole
See other N–O COMPOUNDS, IODINE COMPOUNDS

1869. 1,3-Diisocyanopropane
[72399-85-0] $C_5H_6N_2$

See entry DIISOCYANIDE LIGANDS

1870. Glutarodinitrile
[544-13-8] $C_5H_6N_2$

Energy of decomposition (in range 200–340°C) measured as 0.098 kJ/g.
See entry THERMOCHEMISTRY AND EXOTHERMIC DECOMPOSITION (reference 2)
See other CYANO COMPOUNDS

1871. 4,5-Cyclopentanofurazan-N-oxide (Trimethylenefuroxan)
[54573-23-8] $C_5H_6N_2O_2$

Barnes, J. F. et al., J. Chem. Res. (S), 1979, 314–315
It decomposes explosively on attempted distillation at 150°C at ambient pressure.
See other FURAZAN N-OXIDES

614

1872. Furan-2-amidoxime
[50892-99-4] $C_5H_6N_2O_2$

Pearse, G. A., *Chem. Brit.*, 1984, **20**, 30
Attempted vacuum distillation of the oxime at a pot temperature of 100°C led to an explosion as distillation began. When heated at ambient pressure to above 65°C, a rapid exotherm occurs with gas evolution.
See other GAS EVOLUTION INCIDENTS, OXIMES

1873. Pyridinium nitrate
[543-53-3] $C_5H_6N_2O_3$

Goldstein, N. *et al., J. Chem. Phys.*, 1979, **70**, 5073
Like the perchlorate salt, pyridinium nitrate explodes on heating, but not with shock or friction.
See Pyridinium perchlorate
See other OXOSALTS OF NITROGENOUS BASES

1874. Glutaryl diazide
[64624-44-8] $C_5H_6N_6O_2$

Curtiss, T. *et al., J. Prakt. Chem.*, 1900, **62**, 196
A very small sample exploded sharply on heating.
See other ACYL AZIDES

615

1875. N,N'-Bis(2,2,2-trinitroethyl)urea
[918-99-0] $C_5H_6N_8O_{13}$

$$O_2N-\overset{\overset{\displaystyle NO_2}{|}}{\underset{\underset{\displaystyle NO_2}{|}}{C}}-CH_2-NH-\overset{\overset{\displaystyle }{}}{\underset{\underset{\displaystyle O}{\|}}{C}}-NH-CH_2-\overset{\overset{\displaystyle NO_2}{|}}{\underset{\underset{\displaystyle NO_2}{|}}{C}}-NO_2$$

Sodium hydroxide
   Biasutti, G. S., private comm., 1981
   A pilot plant for manufacture of the title compound (an explosive of zero oxygen
   balance) was decommissioned and treated with concentrated alkali, but not com-
   pletely drained. After 4 years, an explosion occurred, attributed to the hydrolytic
   formation of trinitroethanol or its sodium salt(s).
   See 2,2,2-Trinitroethanol
   See other POLYNITROALKYL COMPOUNDS

1876. 2,4-Diazido-6-dimethylamino-1,3,5-triazine
[83297-43-2] $C_5H_6N_{10}$

Preparative hazard
   See 2,4,6-Triazido-1,3,5-triazine
   See other ORGANIC AZIDES

1877. Divinyl ketone (1,4-Pentadien-3-one)
[890-28-4] C_5H_6O

Trinitromethane
   See Trinitromethane: Divinyl ketone

616

†1878. 2-Methylfuran
[534-22-5] C_5H_6O

1879. 2-Penten-4-yn-3-ol
[101672-20-2] C_5H_6O

Brandsma, 1971, 73
The residue from distillation at 20 mbar exploded vigorously at above 90°C.
See other ACETYLENIC COMPOUNDS

1880. Furfuryl alcohol (2-Furanemethanol)
[98-00-0] $C_5H_6O_2$

Acids
1. Tobie, W. C., *Chem. Eng. News*, 1940, **18**, 72
2. *MCA Case History No. 858*
3. Dunlop, 1953, 214, 221, 783
4. Golosova, L. V., *Chem. Abs.*, 1975, **83**, 194098
A mixture of the alcohol with formic acid rapidly self-heated, then reacted violently
[1]. A stirred mixture with cyanoacetic acid exploded violently after application of
heat [2]. Contact with acids causes self-condensation of the alcohol, which may be
explosively violent under unsuitable physical conditions. The general mechanism
has been discussed [3]. The explosion hazards associated with the use of acidic
catalysts to polymerise furfuryl alcohol may be avoided by using as catalyst the
condensation product of 1,3-phenylenediamine and 1-chloro-2,3-epoxypropane [4].
See Nitric acid: Alcohols (reference 6)
See other POLYMERISATION INCIDENTS

Hydrogen peroxide
See Hydrogen peroxide: Alcohols

Sulfur tetrafluoride, Triethylamine
See Sulfur tetrafluoride: 2-(Hydroxymethyl)furan, Triethylamine

617

1881. Methyl 2-butynoate
[23326-27-4]

$C_5H_6O_2$

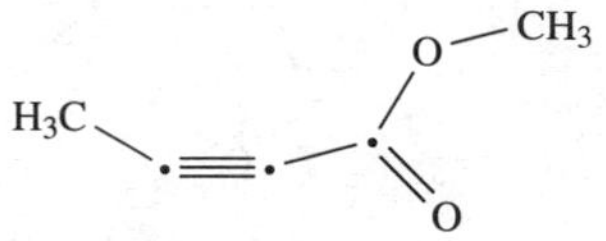

Cochran, J., *Chem. Eng. News* 1994, **72**(46), 4
Attempts to manufacture above the gram scale by several different routes have often resulted in violent decomposition during vacuum distillation. Distillation of more than a few grams demands caution.
See other ACETYLENIC COMPOUNDS

†1882. 2-Methylthiophene
[554-14-3]

C_5H_6S

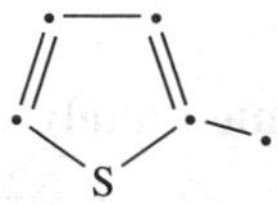

Nitric acid
See Nitric acid: Methylthiophene

1883. 2-Propynyl vinyl sulfide
[21916-66-5]

C_5H_6S

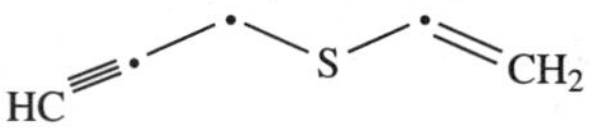

Brandsma, 1971, 7, 182
It decomposes explosively above 85°C.
See other ACETYLENIC COMPOUNDS

1884. 4-Bromocyclopentene
[1781-66-4]

C_5H_7Br

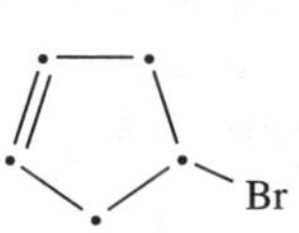

Preparative hazard
See Lithium tetrahydroaluminate: 3,5-Dibromocyclopentene
See other HALOALKENES

1885. 3-Chlorocyclopentene
[96-40-2]

C_5H_7Cl

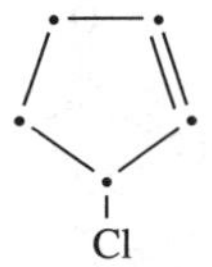

MCA Guide, 1972, 305
A 35 g portion decomposed (or polymerised) explosively after storage for 1 day.
See other HALOALKENES

1886. Ethyl 2,2,3-trifluoropropionate
[28781-86-4]

$C_5H_7F_3O_2$

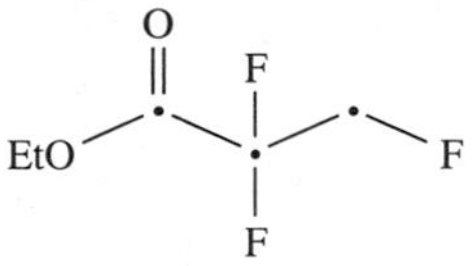

Sodium hydride
    See Sodium hydride: Ethyl 2,2,3-trifluoropropionate

†1887. 1-Methylpyrrole
[96-54-8]

C_5H_7N

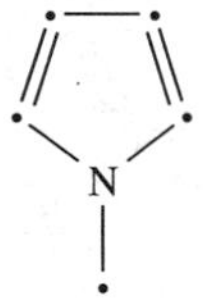

1888. 3,5-Dimethylisoxazole
[300-87-8]

C_5H_7NO

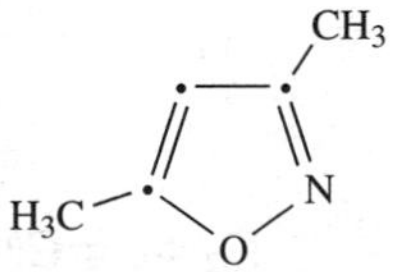

Sealed samples decompose exothermally above 210°C.
See entry ISOXAZOLES *See other* N–O COMPOUNDS

619

1889. Ethyl cyanoacetate
[105-56-6]

$C_5H_7NO_2$

Preparative hazard
    See Ethyl chloroacetate: Sodium cyanide
    See other CYANO COMPOUNDS

1890. Ethyl 2-azido-2-propenoate
[81852-50-8]

$C_5H_7N_3O_2$

Kakimoto, M. *et al., Chem. Lett.*, 1982, 525–526
Attempted distillation caused a severe explosion; such materials should be purified without heating.
See other 2-AZIDOCARBONYL COMPOUNDS

†1891. Cyclopentene
[142-29-0]

C_5H_8

HCS 1980, 350

See other ALKENES

†1892. 3-Methyl-1,2-butadiene
[598-25-4]

C_5H_8

$$H_2{=}C{=}CMe_2$$

It is rather endothermic ($\Delta H°_f$ (g) +129.7 kJ/mol, 1.90 kJ/g).
See other DIENES, ENDOTHERMIC COMPOUNDS

†1893. 2-Methyl-1,3-butadiene (Isoprene)
   [78-79-5] C_5H_8
                        $H_2C{=}CMeCH{=}CH_2$

FPA H112, 1982; *HCS 1980*, 573

Kirk-Othmer, 1967, Vol. 12, 79
It is somewhat endothermic, ΔH°_f (g) +75.7 kJ/mol, 1.11 kJ/g. In absence of
inhibitors, isoprene absorbs atmospheric oxygen to form peroxides which do not
separate from solution. Although the solution is not detonable, the gummy peroxidic
polymer obtained by evaporation can be detonated by impact under standard
conditions.
See 1,3-Butadiene (reference 2)

Acetone
   Lokhmacheva, I. K. *et al., Chem. Abs.*, 1975, **82**, 63856
   Prevention of peroxidation of isoprene–acetone mixtures, and other hazards
   involved in the industrial preparation of synthetic citral, are discussed.

Other reactants
   Yoshida, 1980, 40
   MRH values calculated for 14 combinations with oxidants are given.

Oxygen, Ozone
   See Ozone: Dienes, Oxygen

Ozone
   Loveland, J. W., *Chem. Eng. News*, 1956, **34**(3), 292
   Isoprene (1 g) dissolved in heptane was ozonised at −78°C. Soon after cooling
   was stopped, a violent explosion, followed by a lighter one, occurred. This was
   attributed to high concentrations of peroxides and ozonides building up at the rather
   low temperature employed. Operation at a higher temperature would permit the
   ozonides and peroxides to decompose, so avoiding high concentrations in the reaction
   mixture.
   See other OZONIDES

Vinylamine
   See Vinylamine: Isoprene
   See other DIENES, ENDOTHERMIC COMPOUNDS

†1894. 3-Methyl-1-butyne
   [598-23-2] C_5H_8
                        $HC{\equiv}CCHMe_2$

It is considerably endothermic, ΔH°_f (g) +136.4 kJ/mol, 2.0 kJ/g.
See other ALKYNES, ENDOTHERMIC COMPOUNDS

†1895. 1,2-Pentadiene
[591-95-7] C_5H_8

$$H_2C=C=CHEt$$

It is considerably endothermic, ΔH_f° (g) +145.6 kJ/mol, 2.14 kJ/g.
See other DIENES, ENDOTHERMIC COMPOUNDS

†1896. 1,3-Pentadiene
[504-60-9] C_5H_8

$$H_2C=CHCH=CHMe$$

It is somewhat endothermic, ΔH_f° (g) +77.8 kJ/mol, 1.14 kJ/g.
See other DIENES, ENDOTHERMIC COMPOUNDS

†1897. 1,4-Pentadiene
[591-93-5] C_5H_8

$$H_2C=CHCH_2CH=CH_2$$

It is somewhat endothermic, ΔH_f° (g) +105.4 kJ/mol, 1.55 kJ/g.
See other DIENES, ENDOTHERMIC COMPOUNDS

†1898. 1-Pentyne
[627-19-0] C_5H_8

$$HC\equiv CPr$$

It is somewhat endothermic, ΔH_f° (g) +144.3 kJ/mol, 2.12 kJ/g.
See other ALKYNES, ENDOTHERMIC COMPOUNDS

†1899. 2-Pentyne
[627-21-4] C_5H_8

$$MeC\equiv CEt$$

It is somewhat endothermic, ΔH_f° (g) +128.9 kJ/mol, 1.89 kJ/g.

Mercury, Silver perchlorate
  See Silver perchlorate: Alkynes, etc.
  See other ALKYNES, ENDOTHERMIC COMPOUNDS

1900. Tetrakis(N,N-dichloroaminomethyl)methane

[ ] $C_5H_8Cl_8N_4$

Lévy, R. S. *et al., Mém. Poudres*, 1958, **40**, 109
Shock- and heat-sensitive, it is a brisant more powerful than mercury fulminate.
See other N-HALOGEN COMPOUNDS

1901. 2,2-Bis[(nitrooxy)methyl]propane-1,3-diol dinitrate

[78-11-5] $C_5H_8N_4O_{12}$

$$C(CH_2ONO_2)_4$$

This heart drug (-14.3% oxygen balance) enjoys larger scale use as the explosive PETN (PentaErythritolTetraNitrate).
See other ALKYL NITRATES

1902. 3,3-Bis(azidomethyl)oxetane

[17607-20-4] $C_5H_8N_6O$

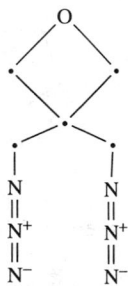

Frankel, M. B. *et al., J. Chem. Eng. Data*, 1981, **26**, 219
Explosive, of considerable sensitivity.
See other ORGANIC AZIDES

1903. Ethyl 2,3-diazidopropionate
  [85590-62-1] $C_5H_8N_6O_2$

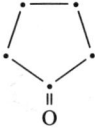

Owston, P. G. *et al., J. Chem. Res., Synop.*, 1985, 352–353
It is unstable and potentially explosive: routine use of a fume cupboard and safety
screens is recommended.
See other 2-AZIDOCARBONYL COMPOUNDS, ORGANIC AZIDES

†**1904. Allyl vinyl ether (3-Ethenoxypropene)**
  [3917-15-5] C_5H_8O

$$H_2C=CHCH_2OCH=CH_2$$

See other ALLYL COMPOUNDS

1905. Cyclopentanone
  [120-92-3] C_5H_8O

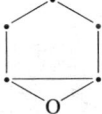

Hydrogen peroxide, Nitric acid
  See Hydrogen peroxide: Ketones, etc.

†**1906. Cyclopentene oxide (6-Oxabicyclo[3.1.0]hexane)**
  [285-67-6] C_5H_8O

See other 1,2-EPOXIDES

624

†1907. Cyclopropyl methyl ketone
[765-43-5]   C₅H₈O

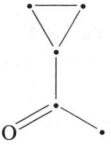

†1908. 2,3-Dihydropyran
[110-87-2]   C₅H₈O

HCS 1980, 410

1909. 1-Ethoxy-2-propyne
[628-33-1]   C₅H₈O

Brandsma, 1971, 172
Distillation of a 1 kg quantity at 80°C/1 bar led to a violent explosion. As the
compound had not been stored under nitrogen during the 3 weeks since preparation,
peroxide formation was suspected.
See other ACETYLENIC COMPOUNDS, PEROXIDISABLE COMPOUNDS

†1910. 2-Methyl-3-butyn-2-ol
[115-19-5]   C₅H₈O

$$Me_2C(OH)C{\equiv}CH$$

Sulfur tetrafluoride
See Sulfur tetrafluoride: 2-Methyl-3-butyn-2-ol
See other ACETYLENIC COMPOUNDS

†1911. Methyl isopropenyl ketone (2-Methylbuten-3-one)
[814-78-8]   C₅H₈O

$$MeCO.CMe{=}CH_2$$

†**1912. Allyl acetate**
  [591-87-7] $C_5H_8O_2$

$$H_2C{=}CHCH_2OCO.Me$$

Energy of decomposition (in range 170–470°C) measured as 0.45 kJ/g.
See entry THERMOCHEMISTRY AND EXOTHERMIC DECOMPOSITION (reference 2)
See other ALLYL COMPOUNDS

†**1913. Ethyl acrylate (Ethyl propenoate)**
  [140-88-5] $C_5H_8O_2$

$$EtOCO.CH{=}CH_2$$

FPA H73, 1978; *HCS 1980*, 464

MCA Case History No. 1759
Inhibited monomer was transferred from a steel drum into a 4 l clear glass bottle
exposed to sunlight in a laboratory in which the ambient temperature was temporarily
higher than usual. Exothermic polymerisation set in and caused the bottle to burst.
Precautions recommended included increase in inhibitor concentration tenfold (to
200 ppm) for laboratory-stored samples, and use of metal or brown glass containers.
See other POLYMERISATION INCIDENTS

†**1914. Methyl crotonate (Methyl 2-butenoate)**
  [623-43-8] $C_5H_8O_2$

$$MeOCO.CH{=}CHMe$$

†**1915. Methyl methacrylate (Methyl 2-methylpropenoate)**
  [80-62-6] $C_5H_8O_2$

$$MeOCO.CMe{=}CH_2$$

FPA H62, 1977; *HCS 1980*, 656

1. Harmon, 1974, 2.13
2. Barnes, C. E. *et al., J. Amer. Chem. Soc.*, 1950, **72**, 210
3. Bond, J., *Loss Prev. Bull.*, 1985, (065), 26
4. Huschke, D., *Chem. Processing* (Chicago), 1988, **51**(1), 136–137

The monomer tends to self polymerise and this may become explosive. It must be
stored inhibited [1]. Exposure of the pure (unstabilised) monomer to air at ambient
temperature for 2 months generated an ester–oxygen interpolymer, which exploded
on evaporation of the surplus monomer at 60°C (but not at 40°C) [2]. Polymerisation,
probably initiated by rust, in a drum of the monomer led to development of overpressure which sheared off the base and propelled the drum into the roof of the building
[3]. Some oxygen must be present in the nitrogen used to inert storage tanks, to activate the stabiliser and prevent gelling of the monomer. Use of an oxygen-selective
permeable membrane to produce enriched nitrogen from compressed air for this
purpose is described [4].
See Initiators, below;

See also VIOLENT POLYMERISATION
See other POLYPEROXIDES

Dibenzoyl peroxide
See Dibenzoyl peroxide: Methyl methacrylate

Initiators
1. Schulz, G. V. et al., Z. Elektrochem., 1941, **47**, 749–761
2. Biesenberg, J. S. et al., J. Polym. Eng. Sci., 1976, **16**, 101–116
Polymerisation of methyl methacrylate initiated by oxygen or peroxides proceeds with a steady increase in velocity during a variable induction period, at the end of which a violent 90°C exotherm occurs. This was attributed to an increase in chain branching, and not to a decrease in heat transfer arising from the increasing viscosity [1]. The parameters were determined in a batch reactor for thermal runaway polymerisation of methyl methacrylate, initiated by azoisobutyronitrile, dibenzoyl peroxide or di-*tert*-butyl peroxide [2].

Other reactants
Yoshida, 1980, 359
MRH values calculated for 13 combinations with oxidants are given.

Propionaldehyde
CISHC Chem. Safety Summ., 1983, **54**(216), 498
Following an incident in which a drum containing bulked drainings (from other drums awaiting reconditioning) fumed and later exploded after sealing, it was found that methyl methacrylate and propionaldehyde can, under certain conditions of mixing, lead to a rapid exothermic reaction. Precautions are discussed.
See other POLYMERISATION INCIDENTS

†1916. Isopropenyl acetate
[108-72-5] $C_5H_8O_2$

$$H_2C{=}C(Me)OCO.Me$$

†1917. Methyl cyclopropanecarboxylate
[2868-37-3] $C_5H_8O_2$

MeO   O

See other STRAINED-RING COMPOUNDS

†1918. Vinyl propionate
[105-38-4] $C_5H_8O_2$

$$H_2C{=}CHOCO.Et$$

1919. Poly(methyl methacrylate peroxide)
[ ] $(C_5H_8O_4)_n$

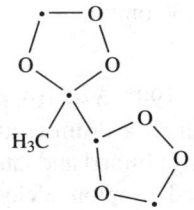

See Methyl methacrylate (reference 2) *also*
See POLYPEROXIDES

1920. Isoprene diozonide (3-(3-Methyl-1,2,4-trioxolan-3-yl)-1,2,4-trioxolane)
[[118112-41-7] [118112-40-6] (diastereoisomers)] $C_5H_8O_6$

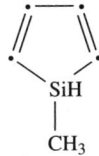

See 2-Methyl-1,3-butadiene: Ozone
See other OZONIDES

1921. 1-Methylsilacyclopenta-2,4-diene (1-Methylsilole)
[73132-51-5] C_5H_8Si

Dienophiles
   Barton, T. J., *J. Organomet. Chem.*, 1979, **179**, C19
   Thermal cracking at 150°C of the dimer (mixed isomers) of the silole in presence of reactive dienophiles (maleic anhydride, tetracyanoethylene or dimethyl
   acetylenedicarboxylate) inevitably produced violent explosions arising from exothermic Diels-Alder reactions.

628

1922. 2,2,2-Tris(bromomethyl)ethanol
[1522-92-5]

$C_5H_9Br_3O$

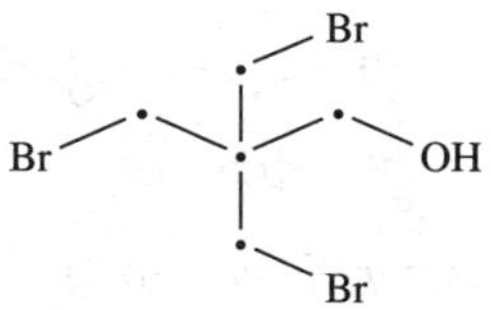

Ethyl acetoacetate, Zinc
  See Ethyl acetoacetate: 2,2,2-Tris(bromomethyl)ethanol, Zinc

†1923. Chlorocyclopentane
[930-28-9]

C_5H_9Cl

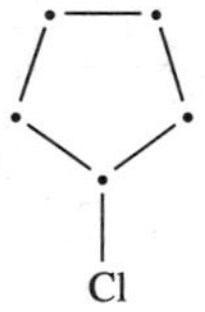

See other HALOALKANES

1924. *N*-(Chlorocarbonyloxy)trimethylurea
[52716-12-8]

$C_5H_9ClN_2O_3$

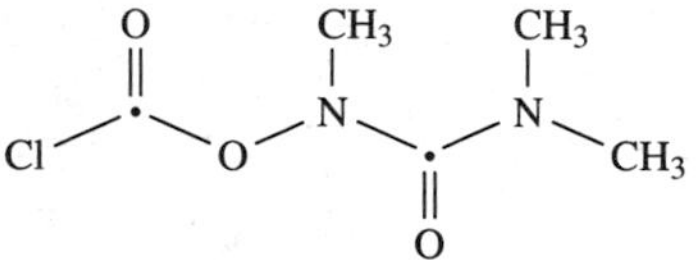

Groebner, P. *et al.*, *Euro. J. Med. Chem., Chim. Ther.*, 1974, **9**, 32–34
A small sample decomposed explosively during vacuum distillation.
See other ACYL HALIDES, N–O COMPOUNDS

†1925. Pivaloyl chloride (Trimethylacetyl chloride)
[3282-30-2]

C_5H_9ClO

$$Me_3CCO.Cl$$

See other ACYL HALIDES

1926. *tert*-Butyl chloroperoxyformate
[56139-33-4] $C_5H_9ClO_3$

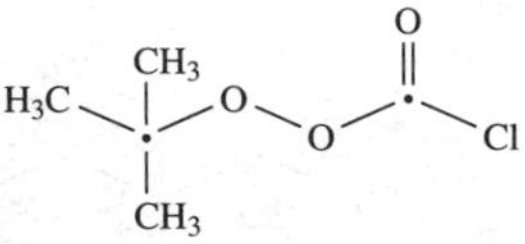

1. Bartlett, P. D. *et al., J. Amer. Chem. Soc.*, 1963, **85**, 1858
2. Hanson, P., *Chem. Brit.*, 1975, **11**, 418

The peroxyester is stable in storage at −25°C, and samples did not explode on friction, impact or heating. However, a 10 g sample stored at ambient temperature heated spontaneously, exploded and ignited [1]. A sample stored at −30°C exploded during manipulation at 0–5°C [2].

See other ACYL HALIDES, PEROXYESTERS

1927. 5-Trichloromethyl-1-trimethylsilyltetrazole
[72385-44-5] $C_5H_9Cl_3N_4Si$

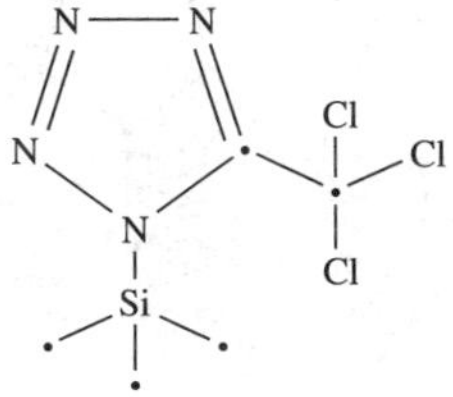

Lazukina, L. A. *et al., J. Org. Chem. (USSR)*, 1980, **15**, 2009
It explodes at 80–90°C.
See other TETRAZOLES

1928. 2(2-Iodoethyl)-1,3-dioxolane
[83665-55-8] $C_5H_9IO_2$

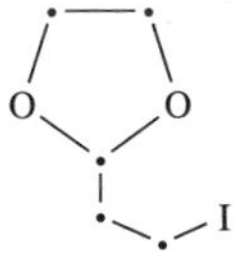

Stowell, J. C. *et al., J. Org. Chem.*, 1983, **48**, 5382
Chromatographic purification on alumina is recommended because the compound decomposes with some violence on distillation at 55°C/1.6 mbar.
See other IODINE COMPOUNDS

†1929. Pivalonitrile (Trimethylacetonitrile)
[630-18-2] C_5H_9N

$$Me_3CCN$$

See other CYANO COMPOUNDS

630

† 1930. 1,2,3,6-Tetrahydropyridine
[694-05-3]   C$_5$H$_9$N

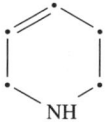

See other ORGANIC BASES

† 1931. Butyl isocyanate
[111-36-4]   C$_5$H$_9$NO

$$BuN=C=O$$

Energy of decomposition (in range 160–450°C) measured as 0.55 kJ/g.
See entry THERMOCHEMISTRY AND EXOTHERMIC DECOMPOSITION (reference 2)
See other ORGANIC ISOCYANATES

1932. Cyclopentanone oxime
[1192-28-5]   C$_5$H$_9$NO

Sulfuric acid
   1. Brookes, F. R., private comm., 1968
   2. Peacock, C. J., private comm., 1978
Heating the oxime with 85% sulfuric acid to prepare 2-piperidone caused eruption
of the stirred flask contents. Benzenesulfonyl chloride in alkali is a less vigorous
reagent [1]. A similar reaction using 70% acid and methanol solvent proceeded
uneventfully until vacuum distillation to remove volatiles had been completed at
90°C (bath)/27 mbar when the dark residue exploded [2].
See other OXIMES

1933. 2-Ethylacryladehyde oxime (2-Methylenebutanal oxime)
[99705-27-8]   C$_5$H$_9$NO

Marvel, C. S. *et al., J. Amer. Chem. Soc.*, 1950, **72**, 5408

An explosion during distillation was recorded, possibly attributable to peroxide formation (or to Beckmann rearrangement).
See other OXIMES, PEROXIDISABLE COMPOUNDS

1934. 2-Piperidone
[675-20-7] C_5H_9NO

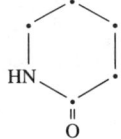

Preparative hazard
See Cyclopentanone oxime: Sulfuric acid

1935. Pivaloyl azide (Trimethylacetyl azide)
[4981-48-0] $C_5H_9N_3O$

Bühler, A. *et al., Helv. Chim. Acta*, 1943, **26**, 2123
The azide exploded very violently on warming, and on one occasion, on standing.
See other ACYL AZIDES

1936. *tert*-Butyl azidoformate
[1070-19-5] $C_5H_9N_3O_2$

1. Carpino, L. A. *et al., Org. Synth.*, 1964, **44**, 15
2. Sakai, K. *et al., J. Org. Chem.*, 1971, **36**, 2387
3. Insalaco, M. A. *et al., Org. Synth.*, 1970, **50**, 12
4. Yajima, H. *et al., Chem. Pharm. Bull.*, 1970, **18**, 850–851

5. Fenlon, W. J., *Chem. Eng. News*, 1976, **54**(22), 3
6. Koppel, H. C., *Chem. Eng. News*, 1976, **54**(39), 5
7. Aldrich advertisement, *J. Chem. Soc., Chem. Comm.*, 1976, (23), i
8. Anon., *Org. Synth.*, 1977, **57**, 122
9. Aldrich advertisement, *J. Org. Chem.*, 1979, **44**(3), cover iv
10. Feyen, P., *Angew. Chem. (Intern. Ed.)*, 1977, **16**, 115
Explosion during distillation at 74°C/92 mbar had been recorded on only one out of several hundred occasions [1], but an alternative procedure avoiding distillation has been described [2]. Although distillation at 56–71°C/53 mbar has been accomplished on many occasions, the structure of the compound suggests that it should be considered as potentially explosive [3]. It is important in the preparation to eliminate all traces of phosgene before adding sodium azide to avoid formation of the dangerously explosive carbonyl azide [4]. The undiluted compound is a shock-sensitive explosive, which is thermally unstable between 100 and 135°C and autoignites at 143°C [5]. A safer reagent for *tert*-butoxycarbonylation of amino groups was then made available [6,7], and other safer alternatives are also available [8], including di-*tert*-butyl dicarbonate [9]. An explosion during evaporation of an ethereal solution of the azide at 50°C has also been reported [10].
See other ACYL AZIDES

1937. 1,1,1-Tris(azidomethyl)ethane
[31044-86-7]

$C_5H_9N_9$

Hydrogen, Palladium catalyst.
Zompa, L. J. *et al.*, *Org. Prep. Proced. Int.*, 1974, **6**, 103–106
There is an explosion hazard during the palladium-catalysed hydrogenation of the tris-azide in ethanol at 2 bar to the tris-amine.
See other CATALYTIC NITRO REDUCTION PROCESSES, HYDROGENATION INCIDENTS, ORGANIC AZIDES

1938. Dimethyl-1-propynylthallium
[ ] C_5H_9Tl

Nast, R. *et al., J. Organomet. Chem.*, 1966, **6**, 461
It explodes on heating and is sensitive to stirring or impact.
See other ALKYLMETALS, METAL ACETYLIDES

†1939. Cyclopentane
[287-92-3] C_5H_{10}

HCS 1980, 349

Other reactants
Yoshida, 1980, 154
MRH values calculated for 13 combinations with oxidants are given.

†1940. 1,1-Dimethylcyclopropane
[1630-94-0] C_5H_{10}

†1941. Ethylcyclopropane
[1191-96-4] C_5H_{10}

†1942. 2-Methyl-1-butene
[563-46-2] C_5H_{10}

$$H_2C{=}C(Me)Et$$

†1943. 2-Methyl-2-butene
[513-35-9] C_5H_{10}

$$Me_2C{=}CHMe$$

†1944. 3-Methyl-1-butene
[563-45-1]

C_5H_{10}

$$H_2C=CHCHMe_2$$

†1945. Methylcyclobutane
[598-61-8]

C_5H_{10}

†1946. 1-Pentene
[109-76-1]

C_5H_{10}

$$H_2C=CHPr$$

HCS 1980, 718

†1947. 2-Pentene
[646-04-8]

C_5H_{10}

$$MeCH=CHEt$$

1948. *N*-Chloropiperidine
[2156-71-0]

$C_5H_{10}ClN$

Claxton, G. P. *et al., Org. Synth.*, 1977, **56**, 120
To avoid rapid spontaneous decomposition of the compound, solvent ether must only
be partially distilled out of the extracted product, and from a water bath maintained
at below 60°C.
See other N-HALOGEN COMPOUNDS

1949. *N*-Perchlorylpiperidine
[768-34-3]

$C_5H_{10}ClNO_3$

Alone, or Piperidine

Gardner, D. L. *et al., J. Org. Chem.*, 1964, **29**, 3738–3739

It is a dangerously sensitive oil which has exploded on storage, heating or contact with piperidine. Absorption in alumina was necessary to desensitise it to allow non-explosive analytical combustion.

See other PERCHLORYL COMPOUNDS

1950. Dichloromethylenediethylammonium chloride
[33842-02-3] $C_5H_{10}Cl_3N$

1. Bader, A. R., private comm., 1971
2. Viehe, H. G. *et al., Angew. Chem. (Intern. Ed.)*, 1971, **10**, 573

The compound does not explode on heating [1], the published statement [2] is in error.

1951. *N*-(2-Cyanoethyl)ethylhydroxylamine
[ ] $C_5H_{10}N_2O$

$$EtN(OH)CH_2CH_2CN$$

Krueger H., *Chem. Abs.*, 1998, **128**, 92406z

Although this compound would undoubtedly be unstable, it here appears to be a misprint for the isomeric ethanolamine entered directly below.

See 3-(2-Hydroxyethylamino)propionitrile

1952. 3-(2-Hydroxyethylamino)propionitrile (*N*-(2-Cyanoethyl)ethanolamine)
[33759-44-3] $C_5H_{10}N_2O$

$$HOCH_2CH_2NHCH_2CH_2CN$$

Krueger H., *Prax. Sicherheitstech.*, 1997, **4**, 83

A storage tank containing this compound self-heated, with evolution of ammonia, then burst with generation of polymer. The reaction appears to be autocatalytic generation of 2-vinyloxazoline, which then polymerises. Thermal analysis showed it unsafe for storage at 50°C, although storable at room temperature. Other homologues became increasingly stable as the distance between hydroxyl and amino groups increases.

See other CYANO COMPOUNDS

1953. 2,4-Dinitropentane isomers
[109745-04-2] [109745-05-3] (R*,S* and R*,R* (±), resp.) $C_5H_{10}N_2O_4$

Wade, P. A. *et al., J. Amer. Chem. Soc.*, 1987, **109**, 5052–5055
A grossly impure sample of the (±) mixed isomers decomposed vigorously when distilled at 130°C. During steam distillation of the R*,S* isomer, overheating of the still pot caused vigorous decomposition: destruction of the pot residues with nitric acid was also vigorous for both.
See other POLYNITROALKYL COMPOUNDS

1954. 3,7-Dinitroso-1,3,5,7-tetraazabicyclo[3.3.1]nonane
[101-25-7] $C_5H_{10}N_6O_2$

1. *MCA Case History No. 841*
2. *See entry* HIGH RATE DECOMPOSITION
3. Hancyk, B. *et al., Chem. Abs.*, 1988, **109**, 212215
A cardboard drum of the blowing agent 'dinitrosopentamethylenetetramine' (as a 40% dispersion in fine silica) ignited when roughly handled in storage [1]. It exhibits fast flame propagation on moderate heating, either neat or mixed with 20% chalk [2]. Mixtures with mineral dust (1:4) eliminated neither fire nor explosion hazards, and mixtures with liquid wax (1:4) eliminated explosion hazard but increased the flammability hazard. The neat blowing agent is highly sensitive to explosion if impacted or dropped. Thermal analysis showed that presence of desensitising agents had no effect on thermal stability and course of decomposition of the title compound, which at above 180°C undergoes self-accelerating decomposition [3]. The recently calculated value of 78°C for the critical ignition temperature is in close agreement with the previous value of 75°C.
See entry CRITICAL IGNITION TEMPERATURE

Other reactants
Yoshida, 1980, 158
MRH values calculated for 11 mixtures with other materials (usually present in catalytic proportions) are given.
See other BLOWING AGENTS, NITROSO COMPOUNDS

637

†1955. Allyl ethyl ether (3-Ethoxypropene)
[557-31-3] $C_5H_{10}O$

$$H_2C{=}CHCH_2OEt$$

HCS 1980, 127

Anon., *ABCM Quart. Safety Summ.*, 1963, **34**, 7
A commercial sample was distilled without being tested for peroxides and exploded
towards the end of distillation. It was later shown to contain peroxides.
See other PEROXIDISABLE COMPOUNDS

†1956. Ethyl propenyl ether (1-Ethoxypropene)
[928-55-2] $C_5H_{10}O$

$$EtOCH{=}CHMe$$

See other PEROXIDISABLE COMPOUNDS

†1957. Isopropyl vinyl ether
[926-65-8] $C_5H_{10}O$

$$Me_2CHOCH{=}CH_2$$

See other PEROXIDISABLE COMPOUNDS

†1958. Isovaleraldehyde
[590-86-3] $C_5H_{10}O$

$$Me_2CHCH_2CO.H$$

See other ALDEHYDES, PEROXIDISABLE COMPOUNDS

†1959. 3-Methyl-2-butanone
[563-80-4] $C_5H_{10}O$

$$MeCO.CHMe_2$$

†1960. 2-Methyl-3-buten-2-ol
[115-18-4] $C_5H_{10}O$

$$H_2C{=}CHCMe_2OH$$

†1961. 2-Methyltetrahydrofuran
[96-47-9] $C_5H_{10}O$

See other PEROXIDISABLE COMPOUNDS

†1962. 2-Pentanone
[107-87-9]
$C_5H_{10}O$

MeCO.Pr

FPA H110, 1982; *HCS 1980*, 661

Bromine trifluoride
See Bromine trifluoride: 2-Pentanone

Other reactants
Yoshida 1980, 369
MRH values calculated for 13 combinations with oxidants are given.

†1963. 3-Pentanone
[96-22-0]
$C_5H_{10}O$

$Et_2C{:}O$

HCS 1980, 401

Hydrogen peroxide   MRH ,
MRH 6.44/84

Nitric acid   MRH 5.82/79
See Hydrogen peroxide: Ketones, etc.

Other reactants
Yoshida, 1980, 338
MRH values calculated for 13 combinations with oxidants are given.

†1964. 4-Penten-1-ol
[821-09-0]
$C_5H_{10}O$

$H_2C{=}CH[CH_2]_3OH$

†1965. Tetrahydropyran
[142-68-7]
$C_5H_{10}O$

See other PEROXIDISABLE COMPOUNDS

†1966. Valeraldehyde (Pentanal)
  [110-62-3]

$C_5H_{10}O$

$$BuCO.H$$

See ALDEHYDES

†1967. Butyl formate
  [592-84-7]

$C_5H_{10}O_2$

$$BuOCO.H$$

HCS 1980, 247

†1968. 3,3-Dimethoxypropene
  [6044-68-4]

$C_5H_{10}O_2$

$$H_2C=CHCH(OMe)_2$$

See other ALLYL COMPOUNDS, PEROXIDISABLE COMPOUNDS

†1969. 2,2-Dimethyl-1,3-dioxolane
  [2916-31-6]

$C_5H_{10}O_2$

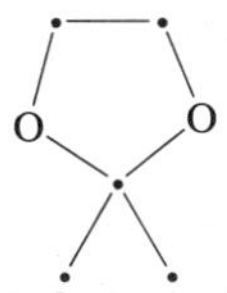

†1970. Ethyl propionate (Ethyl propanoate)
  [105-37-3]

$C_5H_{10}O_2$

$$EtOCO.Et$$

HCS 1980, 499

†1971. Isobutyl formate
  [542-55-2]

$C_5H_{10}O_2$

$$Me_2CHCH_2OCO.H$$

†1972. Isopropyl acetate
  [108-21-4]

$C_5H_{10}O_2$

$$Me_2CHOCO.Me$$

640

†1973. 2-Methoxyethyl vinyl ether
[1663-35-0] $C_5H_{10}O_2$

$$MeOC_2H_4OCH=CH_2$$

See other PEROXIDISABLE COMPOUNDS

†1974. Methyl butyrate
[623-42-7] $C_5H_{10}O_2$

$$MeOCO.Pr$$

†1975. 4-Methyl-1,3-dioxane
[1120-97-4] $C_5H_{10}O_2$

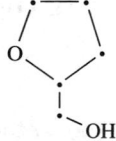

†1976. Methyl isobutyrate
[547-63-7] $C_5H_{10}O_2$

$$MeOCO.CHMe_2$$

†1977. Propyl acetate
[109-60-4] $C_5H_{10}O_2$

$$PrOCO.Me$$

HCS 1980, 795

1978. Tetrahydrofurfuryl alcohol
[97-99-4] $C_5H_{10}O_2$

HCS 1980, 892

3-Nitro-*N*-bromophthalimide
See N-HALOIMIDES: alcohols

Other reactants
    Yoshida, 1980, 239
    MRH values calculated for 13 combinations with oxidants are given.

1979. Trimethyl-1,2-dioxetane
[22688-10-6]   C₅H₁₀O₂

$C_5H_{10}O_2$

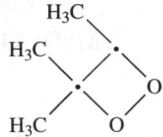

See 3,3-Dimethyl-1,2-dioxetane
See other DIOXETANES, STRAINED-RING COMPOUNDS; CYCLIC PEROXIDES

1980. Tellurane-1,1-dioxide
[ ] $C_5H_{10}O_2Te$

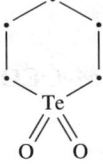

Alone, or Acids
    Morgan, G. T. *et al., J. Chem. Soc.*, 1928, 321
    It exploded on rapid heating, and decomposed violently in contact with nitric or
    sulfuric acids.
    See related ALKYLMETALS

†1981. Diethyl carbonate
[105-58-8] $C_5H_{10}O_3$

$$(EtO)_2C{:}O$$

HCS 1980, 396

Norton, C., *Chemical Engineer*, 1995, (595), 6
A drum containing distillation wastes of ethyl carbonate, ethyl bromoacetate and
another bromoester (not intelligibly named) burst spontaneously, releasing a cloud
which caused considerable lachrymation and respiratory irritation in nearby housing.
Carbon dioxide was doubtless generated, but whether by water contamination or a
previously unknown catalytic reaction of the known contents is unclear to the editor.

1982. *trans*-2-Pentene ozonide (3-Ethyl-5-methyl-1,2,4-trioxolane)
[16187-03-4] $C_5H_{10}O_3$

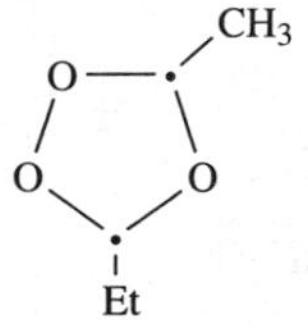

See *trans*-2-Hexene ozonide
See other OZONIDES

1983. Allyldimethylarsine
[691-35-0] $C_5H_{11}As$

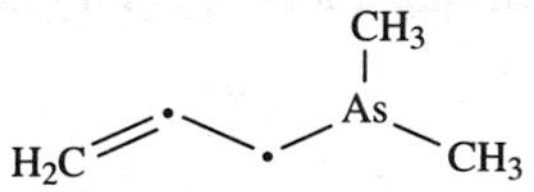

Ellern, 1968, 24–25
It ignites in air if exposed on filter paper.
See related ALKYLNON-METALS *See other* ALLYL COMPOUNDS

†1984. 1-Bromo-3-methylbutane
[107-82-4] $C_5H_{11}Br$

$$BrCH_2CH_2CHMe_2$$

See other HALOALKANES

†1985. 2-Bromopentane
[107-81-3] $C_5H_{11}Br$

$$MeCHBrPr$$

See other HALOALKANES

†1986. 1-Chloro-3-methylbutane
[107-84-6] $C_5H_{11}Cl$

$$ClCH_2CH_2CHMe_2$$

See other HALOALKANES

†1987. 2-Chloro-2-methylbutane
[594-36-5] $C_5H_{11}Cl$

$$Me_2CClEt$$

See other HALOALKANES

†**1988. 1-Chloropentane**
[543-59-9]

$C_5H_{11}Cl$

ClCH₂Bu

See other HALOALKANES

1989. 2-Chlorovinyltrimethyllead
[ ]

$C_5H_{11}ClPb$

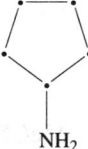

Steingross, W. *et al., J. Organomet. Chem.*, 1966, **6**, 109
It may explode violently during vacuum distillation at 102°C/62.5 mbar.
See related ALKYLMETALS

†**1990. 2-Iodopentane**
[637-97-8]

$C_5H_{11}I$

MeCHIPr

See other HALOALKANES

†**1991. Cyclopentylamine**
[1003-03-8]

$C_5H_{11}N$

See other ORGANIC BASES

†**1992. *N*-Methylpyrrolidine**
[120-94-5]

$C_5H_{11}N$

See other ORGANIC BASES

644

†1993. Piperidine
  [110-89-4]   C₅H₁₁N

HCS 1980, 757

Dicyanofurazan
  See Dicyanofurazan: Nitrogenous bases

N-Nitrosoacetanilide
  See *N*-Nitrosoacetanilide: Piperidine

Other reactants
  Yoshida, 1980, 291
  MRH values calculated for 13 combinations with oxidants are given.

N-Perchlorylpiperidine
  See *N*-Perchlorylpiperidine: Alone, etc.
  See other ORGANIC BASES

1994. *N*-*tert*-Butylformamide
  [2425-74-3]   C₅H₁₁NO

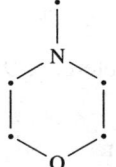

Di-*tert*-butyl hyponitrite
  See *trans*-Di-*tert*-butyl hyponitrite: Formamides

†1995. 4-Methylmorpholine
  [109-02-4]   C₅H₁₁NO

HCS 1980, 657

See other ORGANIC BASES

†1996. Isopentyl nitrite
[110-46-3] $C_5H_{11}NO_2$

$$Me_2CHC_2H_4ON{:}O$$

See other ALKYL NITRITES

1997. *N*-Methylmorpholine oxide (4-Methylmorpholine-4-oxide)
[7529-22-8] $C_5H_{11}NO_2$
[70187-32-5], [85489-61-8], [80913-66-2], [80913-65-1] (hydrates);
[95650-61-6] (cellulose complex)

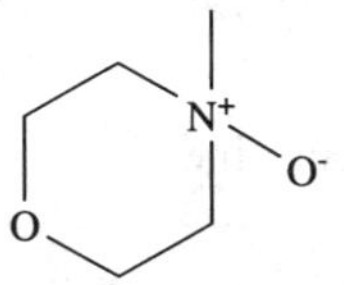

Cellulose
Peguy, A. *et al.*, *J. Appl. Poly, Science*, 1990, **40**(3/4), 429
Solutions of cellulose in wet methylmorpholine oxide can undergo exothermic reaction to the point of explosion if confined at elevated temperatures from about 120°C or if otherwise heated to 180°C. The reaction is catalysed by some metals, notably copper.
See other CATALYTIC IMPURITY INCIDENTS, *N*-OXIDES

†1998. Pentyl nitrite
[463-04-7] $C_5H_{11}NO_2$

$$BuCH_2ON{:}O$$

Sorbe, 1968, 146
It explodes on heating to above 250°C.
See other ALKYL NITRITES

†1999. 2,2-Dimethylpropane (Neopentane)
[463-82-1] C_5H_{12}

$$Me_4C$$

Preparative hazard
See Aluminium: Halocarbons (reference 9)

†2000. 2-Methylbutane (Isopentane)
[78-78-4] C_5H_{12}

$$Me_2CHEt$$

HCS 1980, 572

†2001. Pentane
[109-66-0] C_5H_{12}

BuMe

FPA H55, 1977; *HCS 1980*, 716

2002. *N*-Bromotetramethylguanidine
[6926-40-5] $C_5H_{12}BrN_3$

Papa, A. J., *J. Org. Chem.*, 1966, **31**, 1426
The material is unstable even at 0°C, and explodes if heated above 50°C at ambient pressure.
See other N-HALOGEN COMPOUNDS

2003. 1-Chloro-3-dimethylaminopropane
[109-54-6] $C_5H_{12}ClN$

Lithium, Sodium
See Lithium: 1-Chloro-3-dimethylaminopropane, etc.

2004. *N*-Chlorotetramethylguanidine
[6926-39-2] $C_5H_{12}ClN_3$

Papa, A. J., *J. Org. Chem.*, 1961, **31**, 1426
The material is unstable even at 0°C, and explodes if heated above 50°C at ambient pressure.
See other N-HALOGEN COMPOUNDS

2005. Tetramethylammonium azidocyanoiodate(I)
[68574-17-4]

C$_5$H$_{12}$IN$_5$

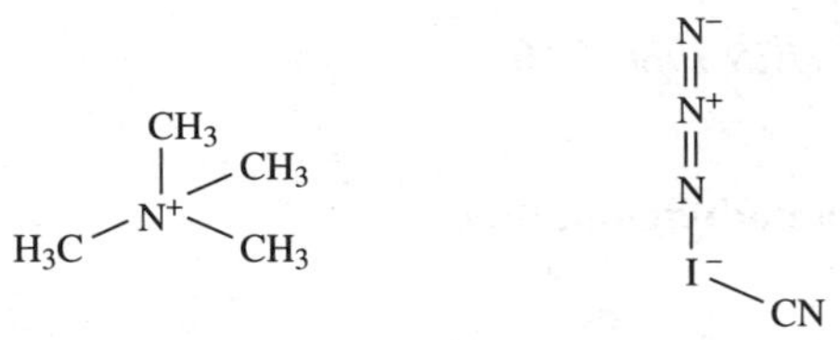

Dehnicke, K., *Angew. Chem. (Intern. Ed.)*, 1979, **18**, 512
This, and the cyanato and selenocyanato (pseudohalogen) analogues are explosive in the solid state, but may be handled with comparative safety. They are also sensitive to laser light.
See related METAL AZIDES

2006. Tetramethylammonium azidocyanatoiodate(I)
[68574-15-2]

C$_5$H$_{12}$IN$_5$O

See Tetramethylammonium azidocyanoiodate(I) (next above)
See related METAL AZIDES, METAL CYANATES

2007. Tetramethylammonium azidoselenocyanatoiodate(I)
[ ]

C$_5$H$_{12}$IN$_5$Se

See Tetramethylammonium azidocyanoiodate(I)
See related METAL AZIDES, METAL CYANATES

2008. Dimethylthallium *N*-methylacetohydroxamate

[ ] $C_5H_{12}NO_2Tl$

Schwering, H. U. *et al., J. Organomet. Chem.*, 1975, **99**, 22

It exploded below 160°C, unlike the other dialkylmetal derivatives, which showed high thermal stability.

See other N–O COMPOUNDS

2009. *O*, *O*-Dimethyl *S*-methylcarbamoylmethyl phosphorodithioate

[60-51-5] $C_5H_{12}NO_3PS_2$

1. Matos, E. *et al., Chem. Abs.*, 1983, **98**, 94955
2. Ludovisi, G. *et al.* in *Safety and Runaway Reactions*, Smeder, B. & Mitchison, N. (Eds.), EUR 17723 EN, European Commission, 1997, p11

Accidental exothermic decomposition of the insecticide Dimethoate in a 6000 l vessel led to gross contamination; (no further details of circumstances available from author) [1]. A second incident of bulk decomposition in cyclohexanone solution (concentration not given) is slightly better reported. This was a formulation solution; the details of the processing are not clear from the report but appear to have involved azeotroping the dimethoate dry, then discharging the warm solution (no temperatures given) to the insulated, uncooled, storage vessel which burst. It is claimed that out of specification material had been processed at above previous temperatures. In any event, there was a runaway exotherm, loss of containment, fire, the attentions of regulatory authorities and closure of the factory. The manufacturer's safety study went beyond the then legal requirements, but, as so often, concentrated on toxicity while neglecting the reactivity which could cast that toxicity abroad. A laboratory sample of the liquid residue of the accident itself later burst its container [2].

See other PHOSPHORUS ESTERS

†2010. Butyl methyl ether

[628-28-4] $C_5H_{12}O$

BuOMe

649

†2011. *tert*-Butyl methyl ether
[1634-04-4]

$C_5H_{12}O$

Me₃COMe

1. Hage, T., *Chem. Abs.*, 1987, **107**, 21833
2. Pearson, H., *Chem. Abs.*, 1988, **108**, 115254

tert-Butyl methyl ether, now manufactured in bulk as a gasoline component, is a less hazardous extracting solvent than diethyl ether, as it scarcely forms peroxides, is less volatile (b.p. 55°C) and has a narrower flammable range (2.5–15%) in air [1]. However, fires involving it are difficult to extinguish with foam, as no film-forming effect is shown [2].

Sulfuric acid
Anon., *Eur. Chem. News*, 1995, **64**(1678), 5

An explosion and fire, costing the life of one worker, occurred when sulphuric acid entered a distillation during scale-up (in glass) of a carotenoid synthesis. This caused a sudden rise in temperature and pressure, bursting the distillation column and an adjacent vessel containing methyl t-butyl ether. Reading between the lines: this was the solvent being distilled off; acid catalysis will cause it to revert to methanol and butene. Large scale glass apparatus may burst with overpressures as low as 0.2 bar.

†2012. Ethyl isopropyl ether
[625-54-7]

$C_5H_{12}O$

EtOCHMe₂

See other PEROXIDISABLE COMPOUNDS

†2013. Ethyl propyl ether
[628-32-0]

$C_5H_{12}O$

EtOPr

2014. Isopentanol (3-Methylbutanol)
[123-51-3]

$C_5H_{12}O$

FPA H123, 1983; *HCS 1980*, 642

Hydrogen trisulfide
See Hydrogen trisulfide: Pentanol

†2015. *tert*-Pentanol (1,1-Dimethylpropanol)
[75-85-4] $C_5H_{12}O$

$$Me_2C(OH)Et$$

HCS 1980, 395

†2016. Diethoxymethane
[462-95-3] $C_5H_{12}O_2$

$$(EtO)_2CH_2$$

†2017. 1,1-Dimethoxypropane
[4744-10-9] $C_5H_{12}O_2$

$$(MeO)_2CHEt$$

†2018. 2,2-Dimethoxypropane
[77-76-9] $C_5H_{12}O_2$

$$(MeO)_2CMe_2$$

Becerra, R. *et al., An. Quim., Ser. A*, 1983, **79**, 163–167
The relationship between the cool flames observed at 210°C and subsequent explosions were studied.
See entry COOL FLAMES

Metal perchlorates
1. Dickinson, R. C. *et al., Chem. Eng. News*, 1970, **48**(28), 6
2. Cramer, R., *Chem. Eng. News*, 1970, **48**(45), 7
3. Mikulski, S. M., *Chem. Eng. News*, 1971, **49**(4), 8
During dehydration of manganese(II) perchlorate [1] or nickel(II) perchlorate [2] with dimethoxypropane, heating above 65°C caused violent explosions, probably involving oxidation by the anion [1] (possibly of the methanol liberated by hydrolysis). Triethyl orthoformate is recommended as a safer dehydrating agent [2] (but methanol would still be liberated).
See Dimethyl sulfoxide: Metal oxosalts
See also METAL PERCHLORATES (reference 2)

2019. 2-(2-Methoxyethoxy)ethanol
[111-77-3] $C_5H_{12}O_3$

Calcium hypochlorite
See Calcium hypochlorite: Hydroxy compounds

2020. Pentanesulfonic acid
[35452-30-3]

$C_5H_{12}O_3S$

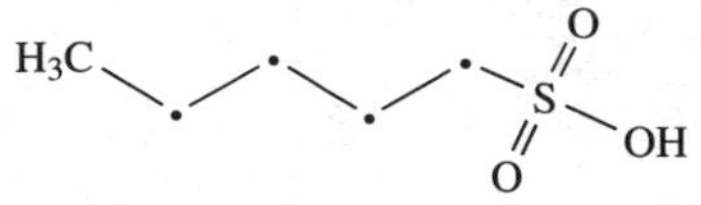

Preparative hazard
See Nitric acid: Alkanethiols (reference 3)
See other ORGANIC ACIDS

2021. Pentaerythritol
[115-77-5]

$C_5H_{12}O_4$

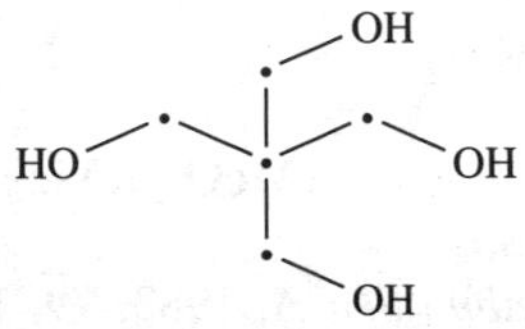

Thiophosphoryl chloride
See Thiophosphoryl chloride: Pentaerythritol

†2022. Tetramethyl orthocarbonate (Tetramethoxymethane)
[1850-14-2]

$C_5H_{12}O_4$

$$(MeO)_4C$$

†2023. 2-Methylbutane-2-thiol
[1679-09-0]

$C_5H_{12}S$

$$Me_2C(SH)Et$$

See other ALKANETHIOLS

†2024. 3-Methylbutanethiol
[541-31-1]

$C_5H_{12}S$

$$Me_2CHC_2H_4SH$$

See other ALKANETHIOLS

†2025. Pentanethiol
  [110-66-7] $C_5H_{12}S$

                              BuCH$_2$SH

Nitric acid
    See Nitric acid: Alkanethiols (reference 3)
    See other ALKANETHIOLS

2026. *N,N,N′,N′*-Tetramethylformamidinium perchlorate
  [2506-80-1] $C_5H_{13}ClN_2O_4$

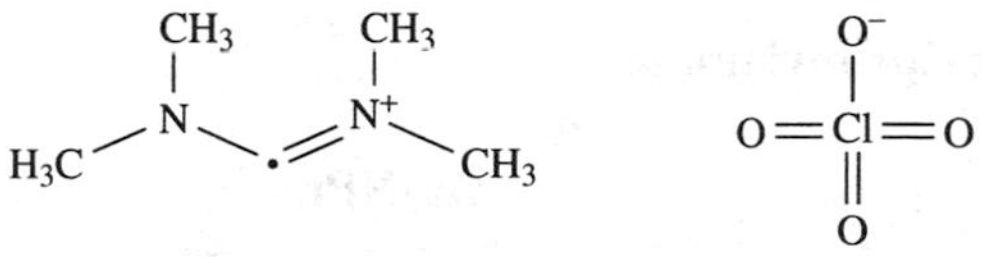

Menzer, M. *et al., Z. Chem.*, 1977, **17**, 344
The undeuterated salt was refluxed with deuterium oxide to effect deuterium
exchange, then the excess was evaporated off under vacuum. Towards the end
of the evaporation the moist residue exploded violently. The presence of D is not
likely to affect instability of the structure.
See other PERCHLORATE SALTS OF NITROGENOUS BASES

2027. *S*-Diethylamino(methylimino)sulfur(IV) fluoride
  [ ] $C_5H_{13}FNS$

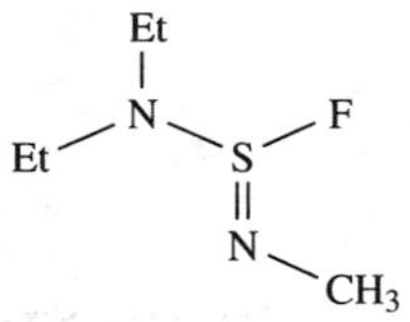

von Halasz, A. *et al., Chem. Ber.*, 1971, **104**, 1250, 1253
Care is necessary in the preparation, because if coloured impurities are present, the
compound may polymerise violently and very exothermically at ambient temperature.
See other N–S COMPOUNDS

†2028. 1,1-Dimethylpropylamine
  [594-39-8] $C_5H_{13}N$

                              EtCMe$_2$NH$_2$

See other ORGANIC BASES

†2029. 1,2-Dimethylpropylamine
[598-74-3] $C_5H_{13}N$

$$Me_2CHCHMeNH_2$$

See other ORGANIC BASES

†2030. 2,2-Dimethylpropylamine (Neopentylamine)
[5813-64-9] $C_5H_{13}N$

$$Me_3CCH_2NH_2$$

See other ORGANIC BASES

†2031. *N,N*-Dimethylpropylamine
[926-63-6] $C_5H_{13}N$

$$Me_2NPr$$

See other ORGANIC BASES

†2032. Isopentylamine
[107-85-7] $C_5H_{13}N$

$$Me_2CHC_2H_4NH_2$$

See other ORGANIC BASES

†2033. *N*-Methylbutylamine
[110-68-9] $C_5H_{13}N$

$$BuNHMe$$

See other ORGANIC BASES

†2034. Pentylamine
[110-58-7] $C_5H_{13}N$

$$BuCH_2NH_2$$

See other ORGANIC BASES

2035. Bis(2-hydroxyethyl)methylphosphine
[53490-67-8] $C_5H_{13}O_2P$

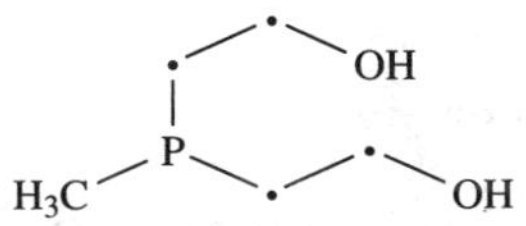

Dichloromethylphosphine
   See Dichloromethylphosphine: Bis(2-hydroxyethyl)methylphosphine
   See related ALKYLPHOSPHINES

2036. Diethylmethylphosphine
[1605-58-9]

Et₂PMe

C₅H₁₃P

Et
|
Et — P
      \
       CH₃

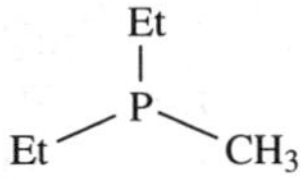

Personal experience
May ignite in air with long exposure.
See other ALKYLPHOSPHINES

2037. 2-Hydroxyethyltrimethylstibonium iodide
[81924-93-8]

C₅H₁₄IOSb

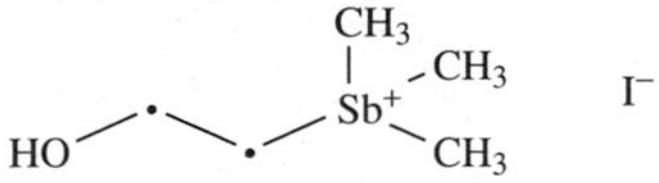

$$CH_3$$

HO—•—•—Sb⁺(CH₃)(CH₃)   I⁻

Meyer, E. M. *et al., Biochem. Pharm.*, 1981, **30**, 3003–3005
An attempt to prepare this antimony analogue of choline by heating trimethylstibine with iodoethanol in bis(2-methoxyethyl) ether in a sealed tube at 150°C led to an explosion.

†2038. 2-Dimethylamino-*N*-methylethylamine
[142-25-6]

C₅H₁₄N₂

Me₂NC₂H₄NHMe

2039. 3-Dimethylaminopropylamine
[109-55-7]

C₅H₁₄N₂

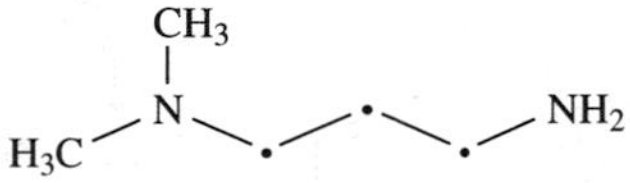

CH₃
|
H₃C—N—•—•—NH₂

HCS 1980, 423

Cellulose nitrate
   See CELLULOSE NITRATE: amines

1,2-Dichloroethane
   See 1,2-Dichloroethane: 3-Dimethylaminopropylamine

Other reactants
   Yoshida, 1980, 168
   MRH values calculated for 14 combinations with oxidants are given.

2040. Pentamethylbismuth
[148739-67-7] $C_5H_{15}Bi$

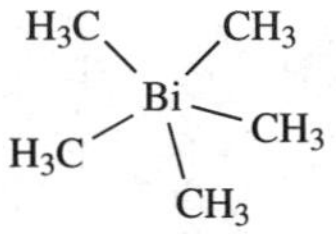

$$H_3C \quad CH_3$$
$$H_3C-Bi-CH_3$$
$$CH_3$$

Seppelt, K. *et al., Angew. Chem. (Int.)*, 1994, **33**(9), 976
This compound, formed at −90°C is unstable in solution and the solid explodes on rapid warming to room temperature.
See other ALKYL METALS

2041. Lithium 2,2-dimethyltrimethylsilylhydrazide
[13529-75-4] $C_5H_{15}LiN_2Si$

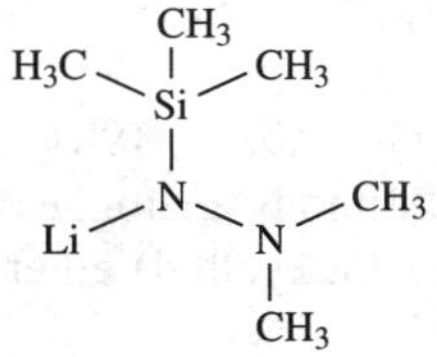

$$CH_3$$
$$H_3C-Si-CH_3$$
$$N$$
$$Li-N \quad N-CH_3$$
$$CH_3$$

Oxidants
See entry SILYLHYDRAZINES

2042. Lithium pentamethyltitanate–bis(2,2′-bipyridine)
[50662-24-3] $C_5H_{15}LiTi.2C_{10}H_8N_2$

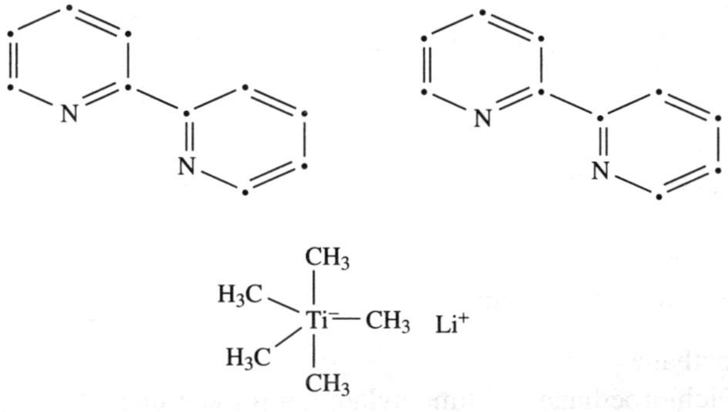

$$H_3C \quad CH_3$$
$$H_3C-Ti\!=\!CH_3 \quad Li^+$$
$$CH_3$$

Houben-Weyl, 1975, Vol. 13.3, 288
Friction-sensitive, it decomposes explosively.
See related ALKYLMETALS

†2043. Dimethylaminotrimethylsilane
[2083-91-2] $C_5H_{15}NSi$

$$Me_2NSiMe_3$$

Xenon difluoride
   See Xenon difluoride: Silicon–nitrogen compounds
   See related ALKYLSILANES

2044. 1,1,1-Tris(aminomethyl)ethane
[15995-42-3] $C_5H_{15}N_3$

Preparative hazard
   See 1,1,1-Tris(azidomethyl)ethane: Hydrogen, etc.

2045. Dimethyltrimethylsilylphosphine
[26464-99-3] $C_5H_{15}PSi$

Air, or Water
   Goldsberry, R. *et al., Inorg. Chem.*, 1972, **13**, 29–32
   It ignites in air and is hydrolysed to dimethylphosphine, also spontaneously
   flammable.
   See related ALKYLPHOSPHINES, ALKYLSILANES

2046. Pentamethyltantalum
[53378-72-6] $C_5H_{15}Ta$

Mertis, K. *et al., J. Organomet. Chem.*, 1975, **97**, C65
It should be handled with extreme caution, even in absence of air, because dangerous
explosions have occurred on warming frozen samples or during transfer operations.
See other ALKYLMETALS

2047. 1,2-Dimethyl-2-trimethylsilylhydrazine
[13271-93-7]

$C_5H_{16}N_2Si$

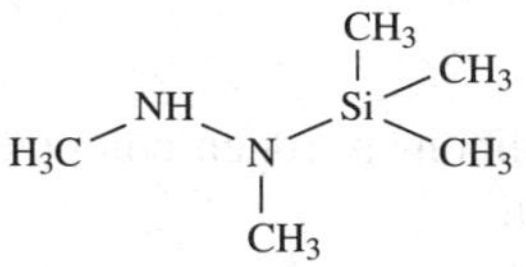

Oxidants
See entry SILYLHYDRAZINES

2048. Pentaamminepyridineruthenium(II) perchlorate
[19482-31-6]

$C_5H_{20}Cl_2N_6O_8Ru$

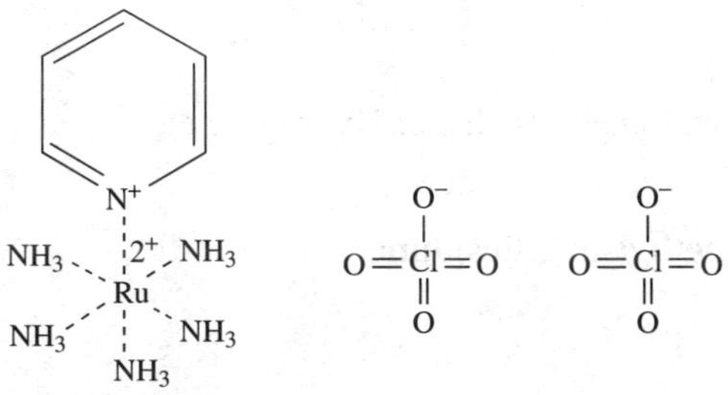

Creutz, C. A., private comm., 1969
The dry salt exploded on touching.
See other AMMINEMETAL OXOSALTS

2049. Potassium pentacarbonylvanadate(3−)
[78937-14-1]

$C_5K_3O_5V$

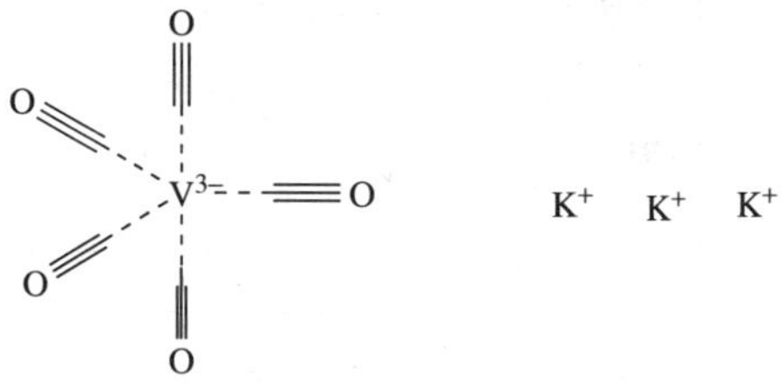

Alone, or Poly(chlorotrifluoroethylene)
Ellis, J. E. *et al., J. Amer. Chem. Soc.*, 1981, **103**, 6101
It is a pyrophoric and treacherously shock-sensitive explosive, especially as the dry
black form. It will deflagrate under nitrogen or argon when scratched with a spatula,
and ignites in contact with poly(chlorotrifluoroethylene), (Fluorolube). The caesium
and rubidium salts are not shock-sensitive but will deflagrate on friction and are

pyrophoric. An insoluble by-product from the preparation will explode if treated with limited amounts of water or alcohol.

See related CARBONYLMETALS

2050. 2-Diazonio-4,5-dicyanoimidazolide (Diazodicyanoimidazole)
[40953-35-3] C_5N_6

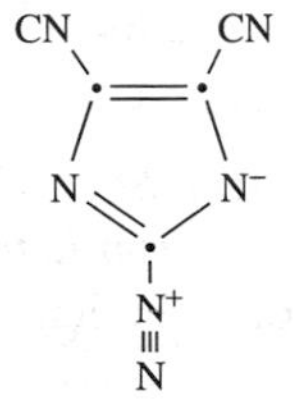

1. Sheppard, W. A. *et al.*, *J. Amer. Chem. Soc.*, 1973, **95**, 2696
2. Sheppard, W. A. *et al.*, *J. Org. Chem.*, 1979, **44**, 1718

It is highly shock-sensitive when dry and explodes above 150°C [1]. It must not be scraped or rubbed, or used in above 2 g portions. The crystalline 1:1 complex with 18-crown-6 ether is significantly less hazardous [2].

See other CYANO COMPOUNDS, DIAZONIUM SALTS, HIGH-NITROGEN COMPOUNDS

2051. Sodium pentacarbonylrhenate
[33634-75-2] C_5NaO_5Re

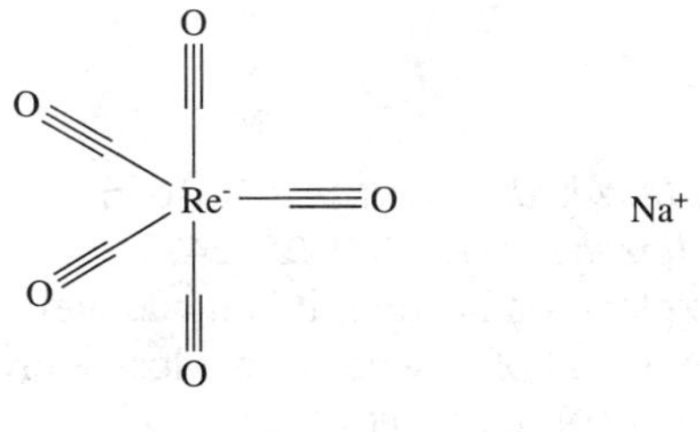

Bailar, 1973, Vol. 3, 955

This and other alkali-metal salts are pyrophoric.

See related CARBONYLMETALS

2052. Silver 1,3,5-hexatriynide
[ ] C_6Ag_2

Hunsman, W., *Chem. Ber.*, 1950, **83**, 216

The silver salt may be handled moist, but when dry it explodes violently on touching with a glass rod.

See other METAL ACETYLIDES, SILVER COMPOUNDS

2053. Pentafluorophenylaluminium dibromide
[4457-90-3] $C_6AlBr_2F_5$

Air, or Water
1. Chambers, R. D. *et al., J. Chem. Soc. (C)*, 1967, 2185
2. Chambers, R. D. *et al., Tetrahedron Lett.*, 1965, 2389
It ignites in air [1], and explodes violently on rapid heating to 195°C or during uncontrolled hydrolysis [2].
See other HALO-ARYLMETALS

2054. 2,5-Diazido-3,6-dichlorobenzoquinone
[26157-96-0] $C_6Cl_2N_6O_2$

1. Gilligan, W. H., *Tetrahedron Lett.*, 1978, 1676
2. Winkelman, E., *Tetrahedron*, 1969, **25**, 2427
While it does not explode on heating, it is moderately impact-sensitive [1], contradicting an earlier report [2] of absence of explosive properties.
See other 2-AZIDOCARBONYL COMPOUNDS

2055. Hexachlorobenzene (Perchlorobenzene)
[118-74-1] C_6Cl_6

Calcium hydride
See Calcium hydride: Hexachlorobenzene

2056. Hexacarbonylchromium
[13007-92-6]   C₆CrO₆

$$C_6CrO_6$$

Weast, 1972, B-83
It explodes at 210°C.
See other CARBONYLMETALS

2057. 1,3,5-Trifluorotrinitrobenzene
[1423-11-6] $C_6F_3N_3O_6$

Nucleophiles
Koppes, W. M. *et al., J. Chem. Soc., Perkin Trans. 1*, 1981, 1817
The title compound readily undergoes substitution reactions with a variety of N,
O, C and X nucleophiles. Many of the products of such reactions, especially with
hydrazine and its derivatives, are explosive.
See other POLYNITROARYL COMPOUNDS

2058. Pentafluoroiodosylbenzene
[14353-90-3] C_6F_5IO

1. Collmann, J. P., *Chem. Eng. News*, 1985, **63**(11), 2
2. Groves, J. T. *et al., Chirality*, 1998, **10**, 106

A 12 g sample decomposed explosively on heating under vacuum at 60°C, though it has been described as stable up to 210°C when it decomposes violently. Avoid heating it to remove volatile impurities [1]. In later safety warnings it has been described as able to detonate spontaneously [2].

See other IODINE COMPOUNDS

2059. Pentafluorophenyllithium
[1076-44-4] C_6F_5Li

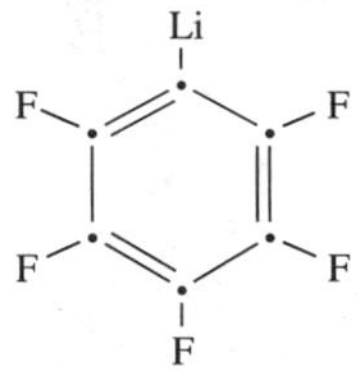

Krafft, T. E. USA Pat. 5,679,289, 1996

A procedure is described for safe preparation and use of this compound in bulk. It depends upon keeping inventories low.

Deuterium oxide

Kinsella, E. *et al., Chem. & Ind.*, 1971, 1017

During addition of an ethereal solution of deuterium oxide (containing some peroxide) to a suspension of the organolithium reagent in pentane, a violent explosion occurred. This may have been initiated by the peroxide present, but probably involved elimination of lithium fluoride.

See other HALO-ARYLMETALS
See other FLUORINATED ORGANOLITHIUM COMPOUND

†2060. Hexafluorobenzene
[392-56-3] C_6F_6

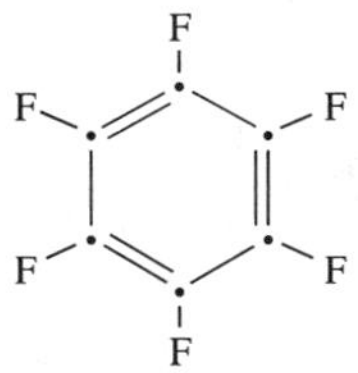

Metals

See HALOARENEMETAL π-COMPLEXES

Silane

Koga, Y. *et al., J. Phys. Chem.*, 1987, **91**(2), 298–305

Reaction of hexafluorobenzene and silane under single pulse irradiation of a MW CO_2 laser was explosive.

See other IRRADIATION DECOMPOSITION INCIDENTS

2061. Trifluoromethyl 3-fluorocarbonylhexafluoroperoxybutyrate
[32750-98-4] $C_6F_{10}O_4$

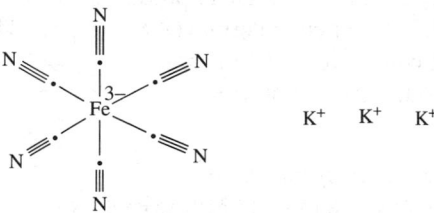

Bernstein, P. A. *et al., J. Amer. Chem. Soc.*, 1971, **93**, 3885
A small sample exploded at 70°C.
See other PEROXYESTERS

2062. Perfluorohexyl iodide
[355-43-1] $C_6F_{13}I$

Sodium
See Sodium: Halocarbons (reference 6)

2063. Potassium hexacyanoferrate(III) ('Potassium ferricyanide')
[13746-66-2] $C_6FeK_3N_6$

HCS 1980, 765

Ammonia
1. Pieters, 1957, 30
2. Sidgwick, 1950, 1359
Contact may be explosive [1], possibly owing to rapid oxidation of ammonia by alkaline 'ferricyanide' [2].
See Potassium tetracyanomercurate(II)
See other AMINATION INCIDENTS

Chromium trioxide
See Chromium trioxide: Potassium hexacyanoferrate(III)

Hydrochloric acid
Ephraim, 1939, 303
Treatment of the complex salt with acid liberates the corresponding complex ferri-cyanic acid, an oxidant which is rather endothermic (ΔH°_{f} (aq) +640.5 kJ/mol, 2.96 kJ/g), and which forms solid complexes with ether, etc.

Sodium nitrite
See Sodium nitrite: Metal cyanides
See other METAL CYANIDES (AND CYANO COMPLEXES)

2064. Potassium hexacyanoferrate(II) ('Potassium ferrocyanide') [13943-58-3] $C_6FeK_4N_6$

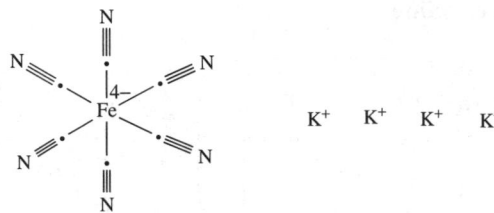

HCS 1980, 766

Copper(II) nitrate
See Copper(II) nitrate: Potassium hexacyanoferrate(II)

Hydrochloric acid
Ephraim, 1939, 303
Treatment of the complex salt with acid produces the corresponding complex ferro-cyanic acid, which is rather endothermic (ΔH°_{f} (aq) +534.7 kJ/mol, 2.48 kJ/g), and which forms solid complexes with ether, etc.
See other ENDOTHERMIC COMPOUNDS

Sodium nitrite
See Sodium nitrite: Metal cyanides
See other METAL CYANIDES (AND CYANO COMPLEXES)

2065. Iron(III) oxalate [2944-66-3] $C_6Fe_2O_{12}$

Fe^{3+} Fe^{3+}

Weinland, R. *et al., Z. Anorg. Chem.*, 1929, **178**, 219

The salt, probably complex, decomposes at 100°C.
See other METAL OXALATES

2066. 4-Chloro-2,5-dinitrobenzenediazonium 6-oxide
[ ] $C_6HClN_4O_5$

See Nitric acid: 4-Chloro-2-nitroaniline
See other ARENEDIAZONIUM OXIDES

2067. 3,4-Difluoro-2-nitrobenzenediazonium 6-oxide
[ ] $C_6HF_2N_3O_3$

Finger, G. C. *et al., J. Amer. Chem. Soc.*, 1951, **73**, 148
Diazotisation of 3,4,6-trifluoro-2-nitroaniline under differing conditions gave either this isomer or the 3,6-difluoro 4-oxide (next below), both of which exploded on heating, ignition or impact.
See other ARENEDIAZONIUM OXIDES

2068. 3,6-Difluoro-2-nitrobenzenediazonium 4-oxide
[ ] $C_6HF_2N_3O_3$

Finger, G. C. *et al., J. Amer. Chem. Soc.*, 1951, **73**, 148
Diazotisation of 3,4,6-trifluoro-2-nitroaniline under differing conditions gave either this isomer or the 3,4-difluoro 6-oxide (next above), both of which exploded on heating, ignition or impact.
See other ARENEDIAZONIUM OXIDES

2069. 1,2,3,-4,5-Pentafluorobicyclo[2.2.0]hexa-2,5-diene
[21892-31-9] C_6HF_5

1. Rotajezak, E., *Roczn. Chem.*, 1970, **44**, 447–449
2. Widmann, G. *et al., Thermochim. Acta*, 1988, **134**, 451–455

Both isomers were stored as frozen solids, for as liquids they are more explosive than the hexafluoro 'Dewar benzene' analogue [1]. Energy of decomposition (320–540°C) was determined by DSC as 4.60 kJ/g.

See other HALOALKENES, STRAINED-RING COMPOUNDS

2070. 1,2,3,-5,6-Pentafluorobicyclo[2.2.0]hexa-2,5-diene
[28663-71-0] C_6HF_5

1. Rotajezak, E., *Roczn. Chem.*, 1970, **44**, 447–449
2. Widmann, G. *et al., Thermochim. Acta*, 1988, **134**, 451–455

Both isomers were stored as frozen solids, for as liquids they are more explosive than the hexafluoro 'Dewar benzene' analogue [1]. Energy of decomposition (320–540°C) was determined by DSC as 4.60 kJ/g.

See other HALOALKENES, STRAINED-RING COMPOUNDS

2071. Lead 2,4,6-trinitroresorcinoxide ('Lead styphnate')
[15245-44-0] $C_6HN_3O_8Pb$

1. *MCA Case History No. 957*
2. Okazaki, K. *et al., Chem. Abs.*, 1977, **86**, 57638
3. Biasutti, 1981, 63, 73, 75, 86, 108

Three beakers containing lead styphnate were being heated in a laboratory oven to dry the explosive salt. When one of the beakers was moved, all 3 detonated. Other heavy metal salts of polynitrophenols are dangerously explosive when dry [1]. The desensitising effect of presence of water upon the friction sensitivity of this priming explosive was studied. There is no effect up to 2% content and little at 5%. Even at 20% water content it is still as sensitive as dry pentaerythritol tetranitrate [2]. In one incident in 1927, 20 kg of the dry salt exploded while being poured into paper bags. Initiation by static sparks in a very dry atmosphere was supposed. In another incident, 1 kg exploded during drying in thin layers in an oven held below 65°C. In a third incident, sifting of the dry material also led to the explosion of adjacent wet material, though in a fourth incident, the latter did not also explode. The other incidents involved explosions during drying or handling operations [3].

See LEAD SALTS OF NITRO COMPOUNDS

See other HEAVY METAL DERIVATIVES, METAL NITROPHENOXIDES, POLYNITROARYL COMPOUNDS

2072. 2,3,5-Trinitrobenzenediazonium-4-oxide

[ ] $C_6HN_5O_7$

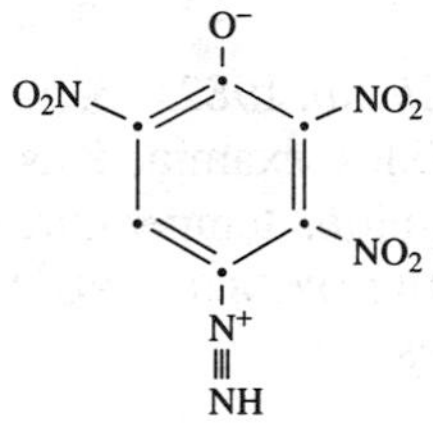

Meldola, R. *et al., J. Chem. Soc.*, 1909, **95**, 1383

It is extremely explosive.

See other ARENEDIAZONIUM OXIDES

2073. 3,4-Bis(1,2,3,4-thiatriazol-5-ylthio)maleimide

[1656-16-2] $C_6HN_7O_2S_4$

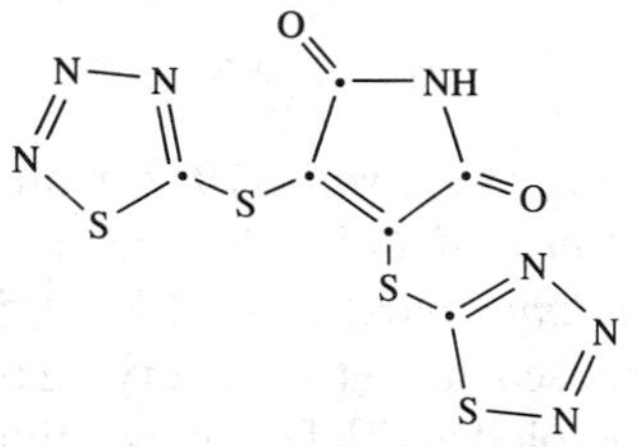

Pilgram, K. *et al., Angew. Chem.*, 1965, **77**, 348

This compound of low oxygen balance explodes loudly on impact or at the m.p.

See other HIGH-NITROGEN COMPOUNDS, N–S COMPOUNDS

2074. 1,3,5-Hexatriyne

[3161-99-7] C_6H_2

1. Hunsman, W., *Chem. Ber.*, 1950, **83**, 213–217
2. Armitage, J. B. *et al., J. Chem. Soc.*, 1952, 2013
3. Bjornov, E. *et al., Spectrochim. Acta*, 1974, **30A**, 1255–1256

It polymerises slowly at −20°C, but rapidly at ambient temperature in air to give a friction-sensitive explosive solid, probably a peroxide [1]. Although stable under vacuum at −5°C for a limited period, exposure at 0°C to air led to violent explosions soon afterwards [2]. Explosive hazards have again been stressed [3].
See other ALKYNES

2075. Silver 2-azido-4,6-dinitrophenoxide
[82177-80-8] $C_6H_2AgN_5O_5$

Spear, R. J. *et al., Aust. J. Chem.*, 1982, **35**, 5
Of a series of 17 compounds examined as stab-initiation sensitisers, the title compound had the lowest ignition temperature of 122°C.
See other METAL NITROPHENOXIDES, ORGANIC AZIDES, POLYNITROARYL COMPOUNDS, SILVER COMPOUNDS

2076. 2,6-Dibromobenzoquinone-4-chloroimide
[537-45-1] $C_6H_2Br_2ClNO$

1. Hartmann, W. W. *et al., Org. Synth.*, 1943, Coll. Vol. 2, 177
2. Horber, D. F. *et al., Chem. & Ind.*, 1967, 1551
3. Taranto, B. J., *Chem. Eng. News*, 1967, **45**(24), 54
The chloroimide decomposed violently on a drying tray at 60°C [1], and a bottle accidentally heated to 50°C exploded [2]. Decomposition of the heated solid occurs after a temperature-dependent induction period. Unheated material eventually exploded after storage at ambient temperature [3].
See 2,6-Dichlorobenzoquinone-4-chloroimide
See other N-HALOGEN COMPOUNDS, INDUCTION PERIOD INCIDENTS

2077. 1,5-Dichloro-2,4-dinitrobenzene
[3698-83-7] $C_6H_2Cl_2N_2O_4$

Dimethyl sulfoxide, Potassium fluoride
Koch-Light Laboratories Ltd., private comm., 1976
A mixture of the (highly activated) dichloro compound with potassium fluoride in the solvent was heated to reflux to effect replacement of chlorine by fluorine. The reaction accelerated out of control and exploded, leaving much carbonised residue. Analogous reactions had been effected uneventfully on many previous occasions.
See Dimethyl sulfoxide: Acyl halides, etc.
See other HALOARYL COMPOUNDS, POLYNITROARYL COMPOUNDS

2078. 2,6-Dichlorobenzoquinone-4-chloroimide
[101-38-2] $C_6H_2Cl_3NO$

Taranto, B. J., *Chem. Eng. News*, 1967, **45**(52), 54
The heated solid material decomposed violently after a temperature-dependent induction period. Unheated material eventually exploded after storage at ambient temperature.
See 2,6-Dibromobenzoquinone-4-chloroimide
See other N-HALOGEN COMPOUNDS, INDUCTION PERIOD INCIDENTS

2079. 1,2,4,5-Tetrachlorobenzene
[95-94-3] $C_6H_2Cl_4$

Sodium hydroxide, Solvent
1. Anon., *Chem. Eng. News*, 1983, **61**(23), 44
2. *MCA Case History No. 620*
3. Hofmann, H. T., *Arch. exptl. Pathol. Pharmakol.*, 1957, **232**, 228–230
4. Cardillo, P. *et al., J. Haz. Mat.*, 1984, **9**, 221–234
5. Sambeth, J. A., *Chimia*, 1982, **36**, 128–132
6. Anon., *ABCM Quart. Safety Summ.*, 1952, **23**, 5
7. May, G., *Brit. J. Ind. Med.*, 1973, **30**, 276–277
8. Milnes, M. H., *Nature*, 1971, **232**, 396
9. Calnan, A. B., private comm., 1972
10. Theofanous, T. G., Statement at special meeting of Society of Chemical Industry's Industrial Health & Safety Group, London, 23rd March, 1981; *Chem. & Ind.*, 1981, 197; *Nature*, 1981, **291**, 640

11. Theofanous, T. G., *Chem. Engrg. Sci.*, 1983, **38**, 1615
12. Salomon, C. M., *Chimia*, 1982, **36**, 133–139
13. Künzi, H., *Chimia*, 1982, **36**, 162–168
14. Cardillo, P. *et al., Chim. e Ind.* (Milan), 1983, **65**, 611–615
15. Theofanous, T. G., *Chem. Engrg. Sci.*, 1983, **38**, 1631
16. Cardillo, P. *et al., Chim. e Ind. (Milan)*, 1982, **64**, 781–784
17. Marshall, V. C., *Loss Prevention Bull.*, 1992, **104**, 15

Eight serious accidents occurred during the commercial preparation of 2,4,5-trichlorophenol by alkaline partial hydrolysis of 1,2,4,5-tetrachlorobenzene during the period 1949 to 1976 [1], involving two distinct processes, each with their particular potential hazards. The earlier process used methanolic alkali under autogenous pressure to effect the hydrolysis, and on 2 occasions around 1949 the reaction at 125°C went out of control, one attaining 400°C [2]. One of the principal hazards of this pressurised process is associated with the fact that no reflux is possible, so the reaction exotherm must be removed by applied cooling in a batch reactor, (or by controlled flow rate in a continuous reactor) to prevent undue temperature rise. A second potential hazard resides in the fact that if the sodium hydroxide normally used as base (and in excess) is not completely dissolved in the methanol before the temperature of onset of hydrolysis (165–180°C) is reached, the hydrolysis exotherm (108.4 kJ/mol of tetrachlorobenzene) is increased by the solution exotherm (perhaps 30–90 kJ/mol) of undissolved alkali, and a runaway reaction, accompanied by a rapid and uncontrollable pressure increase may develop. In a further incident in 1953, the explosion was associated with the post-reaction stage during distillation of methanol from the reaction mixture [3]. Under these conditions a virtually solid crust of the sodium phenoxide is formed and may become overheated [4]. Four explosions occurred during the commercial use of the methanol process [5].

The later process used ethylene glycol as solvent to effect hydrolysis with sodium hydroxide and operated essentially at or near atmospheric pressure. Hydrolysis was effected at between 140 and 180°C during several hours, and the excess glycol (that portion which did not condense to relatively involatile diethylene glycol) might or might not be recovered by vacuum distillation, depending on the process design. When another solvent was present in the reaction system (dichlorobenzene to dissolve sublimed tetrachlorobenzene from the condenser in the Coalite system, or xylene to assist in water and glycol removal in the Seveso system), some inherent temperature control by reflux during the hydrolysis reaction was possible. Violent decomposition could, however, occur during the subsequent solvent distillation phase in the absence of effective temperature control. A laboratory residue from vacuum stripping using electric heating (without knowledge of the liquid temperature) exploded when the vapour temperature had reached 160°C [6]. In the Coalite plant incident in 1968, where the hydrolysis was run at about 180°C in a reaction vessel heated by circulating oil at 300°C, failure of the manually regulated oil heating system led to an uncontrollable temperature increase during 50 m to above 250°C, when a violent explosion occurred [7]. Subsequent work showed that an accelerating exothermic decomposition sets in at 230°C, capable of attaining 410°C. This involves sodium hydroxide and ethylene glycol (or diethylene glycol) and produces hydrogen, thus pressurisation, and sodium carbonate [8,16].

The Seveso accident in 1976 also involved the glycol-based process, but differed fundamentally from the 1968 incident. While the latter apparently involved a thermal runaway initiated during the hydrolysis reaction by application of excessive heat by the faulty hot oil system [7], the process design adopted by Icmesa at Seveso featured heating the reaction vessel by steam at 12 bar (192°C if saturated) to ensure a minimum 40°C safety margin below the known decomposition temperature of 230°C [5]. At Seveso the exothermic hydrolysis reaction had been completed, but not the ensuing xylene–glycol solvent distillation, before the weekend shutdown, when the liquid temperature was 158°C. The bursting disk set at 3.8 bar failed some 7.5 h after processing operations, agitation and heating had been terminated [5]. The accident was followed by an unprecedented amount of investigational work; giving a highly probable explanation if several factors and new information are taken into account.

It was found that the steam supply to the reactor was often superheated (just prior to shutdown to 330°C) [10]. Although this degree of superheat would not grossly increase the temperature of the inner reactor wall in contact with the liquid (or the bulk liquid temperature) [11], it seems probable that any reaction material splashed onto and dried out at the top of the coil-heated wall would have become heated to a much higher temperature. Further detailed work on the thermal stability of the mixture showed that a previously unsuspected very slow exothermic decomposition existed, beginning at 180°C and proceeding at an appreciable rate only above 200°C, so that the exotherm was insufficient to heat the contents of the reactor from the last recorded temperature of 158°C to the decomposition temperature of 230°C in 7.5 h [12,13,14]. It was concluded that an alternative (effectively an external) source of heat was necessary to account for the observed effect, and the residual superheat from the steam at 330°C seems to have been that source.

Any superheating effect would be localised near the steam inlet at the top of the vessel, and there would not have been enough excess heat to raise the temperature of the whole reactor contents by more than a few °C. A new physicochemical process involving conduction, convection and radiation was proposed, and verified experimentally, which concentrated the available excess heat into the top surface layer of a few cm (a few % of the total depth) of the reactor contents, so that the critical temperature of 230°C would be attained in that top layer [11]. The exothermic decomposition reaction, (perhaps catalysed by any thermal degradation products falling into the top layer from the overheated top of the reactor) would then have propagated rapidly downwards into the reactor contents (some 6 t), leading to runaway reaction, pressure build-up and bursting disk failure with release of the reactor contents to atmosphere [11,15].

A publication summarises all the then available technical evidence related to the Seveso accident, and recommends operational criteria to ensure safety in commercial processes to produce trichlorophenol [4]. All the plant scale incidents were characterised [1] by the subsequent occurrence of chloracne arising from the extremely toxic and dermatitic compound 2,3,7,8-tetrachlorodibenzodioxin (structure IX, p. S-3), formed during the thermal runaway reaction and dispersed in the ensuing explosion. It is also extremely resistant to normal chemical decontamination procedures, and after the 1968 explosion, further cases occurred after transient contact with plant which had been decontaminated and allowed to weather for 3 years, and which appeared free of 'dioxin' [7,9]. The consequences at Seveso: 447 cases of chemical burns (NaOH); and 179 of chloracne, only 34 with both [17].

See Sodium hydroxide: Glycols
See other GAS EVOLUTION INCIDENTS, RUNAWAY REACTIONS
See other HALOARYL COMPOUNDS

2080. 1,5-Difluoro-2,4-dinitrobenzene
[327-92-4] $C_6H_2F_2N_2O_4$

Preparative hazard
See 1,5-Dichloro-2,4-dinitrobenzene: Dimethyl sulfoxide, etc.
See other HALOARYL COMPOUNDS, POLYNITROARYL COMPOUNDS

†2081. 1,2,4,5-Tetrafluorobenzene
[327-54-8] $C_6H_2F_4$

See other HALOARYL COMPOUNDS

2082. Tetrafluorobenzene-1,4-diol
[771-63-1] $C_6H_2F_4O_2$

See Trimethylaluminium: Tetrafluorobenzene-1,4-diol

2083. Sodium 3-hydroxymercurio-2,6-dinitro-4-*aci*-nitro-2,5-cyclohexadienonide
[ ]   C₆H₂HgN₃NaO₅

Hantzsch, A. *et al., Ber.*, 1906, **39**, 1111
This salt of the mono-*aci p*-quinonoid form of 3-hydroxymercurio-2,4,6-trinitro-phenol explodes on rapid heating.
See entry aci-NITROQUINONOID COMPOUNDS

2084. 3,4,5-Triiodobenzenediazonium nitrate
[68596-99-6]   C₆H₂I₃N₃O₃

Kalb, L. *et al., Ber.*, 1926, **59**, 1867
Unstable on warming, explodes on heating in a flame.
See other DIAZONIUM SALTS

2085. Potassium picrate (Potassium 2,4,6-trinitrophenoxide)
[573-83-1]   C₆H₂KNO₇

2,2-Dinitro-2-fluoroethoxycarbonyl chloride
See 2,2-Dinitro-2-fluoroethoxycarbonyl chloride: Potassium picrate
See other PICRATES

2086. Sodium picrate (Sodium 2,4,6-trinitrophenoxide)
[73771-13-8] $C_6H_2N_3NaO_7$

1. Coningham, H., *H. M. Insp. Expl. Rept. 236*, London, 1919
2. Nakagawa, K. *et al., Bull. Chem. Soc. Japan*, 1987, **60**, 2038, footnote

A large scale explosion involving initiation of wet sodium picrate by impact was investigated [1]. A small sample heated to above 250°C exploded with sufficient violence to destroy the DTA thermocouple [2].

See other METAL NITROPHENOXIDES, PICRATES

2087. 3,5-Dinitrobenzenediazonium 2-oxide (5,7-Dinitrobenzoxa-1,2,3-diazole)
[4682-03-5] $C_6H_2N_4O_5$
[87-31-0] (oxadiazole form)

It appears uncertain which structure this explosive has, quite possibly both, since it can be made either by diazotisation or nucleophilic cyclisation.

See N,2,3,5-Tetranitroaniline
See other ARENEDIAZONIUM OXIDES

2088. 4,6-Dinitrobenzenediazonium 2-oxide
[ ] $C_6H_2N_4O_5$

Urbanski, 1967, Vol. 3, 204

This priming explosive, as sensitive as mercury fulminate, is much more powerful than metal-containing initiators.

See other ARENEDIAZONIUM OXIDES

2089. 4,6-Dinitrobenzofurazan *N*-oxide
[5128-28-9] $C_6H_2N_4O_6$

Norris, W. P. *et al., Aust. J. Chem.*, 1983, **36**, 306
A powerful high explosive, of comparable sensitivity to dry picric acid.

Cysteine
  See 4-(2′-Ammonio-2′-carboxyethylthio)-5,7-dinitro-4,5-dihydro-benzofurazanide *N*-oxide

Furan, or *N*-Methylindole, or *N*-Methylpyrrole
  Halle, J. C. *et al., Tetrahedron Lett.*, 1983, **24**, 2258
  The mono- or bis-furoxan adducts are explosive, and work on them was discontinued.

Ketones
  Terrier, F. *et al., J. Org. Chem.*, 1981, **46**, 3538
  The adducts formed with acetone, cyclopentanone, cyclopentanedione, 2,4-pentane-dione or 3-methyl-2,4-pentanedione (See Structure str01, p. S-1, R=CH₃CO.CH₂, etc.) form insoluble potassium salts (M=K), but these are highly explosive and shock-sensitive and should not be isolated, but handled only in solution.

Nucleophilic reagents
  Norris, W. P. *et al., Aust. J. Chem.*, 1983, **36**, 297–309
  Reaction of the title compound with a series of nucleophilic oxygen reagents (potassium hydrogencarbonate in water or methanol, potassium hydroxide in methanol), or nitrogen reagents (ammonia, hydroxylamine or hydrazine hydrate) gives a series of mono- or bi-nuclear 7-substituted Meisenheimer complexes (structure II, p. S-4, R=OH, OMe, NH₂, NHOH or NHNH₂). All of these are explosives sensitive to initiation by impact, friction or electrostatic discharge, which need due care in handling.
  See other FURAZAN *N*-OXIDES, POLYNITROARYL COMPOUNDS

2090. 3,5-Dinitro-4-hydroxybenzenediazonium 2-oxide
[29906-35-2] $C_6H_2N_4O_6$

Dynamit Nobel, Belg. Pat. 874 187, 1979
The barium, lead and potassium salts (on the 4-position) are sensitive detonators.
See other ARENEDIAZONIUM OXIDES, POLYNITROARYL COMPOUNDS

2091. 2,3,4,6-Tetranitrophenol
[641-16-7] $C_6H_4N_2O_9$

Sorbe, 1968, 152
Exceptionally explosive.
See other POLYNITROARYL COMPOUNDS

2092. Picryl azide (2,4,6-Trinitrophenyl azide)
[1600-31-3] $C_6H_2N_6O_6$

Schrader, E., *Ber.*, 1917, **50**, 778
Explodes weakly under impact, but not on heating.
See other ORGANIC AZIDES, POLYNITROARYL COMPOUNDS

676

2093. Pentanitroaniline
[21958-87-5] $C_6H_2N_6O_{10}$

Koppes, W. M. *et al., J. Chem. Soc., Perkin Trans. 1*, 1981, 1816
It is a very sensitive explosive.
See other POLYNITROARYL COMPOUNDS

2094. 2,4-Hexadiyne-1,6-dioic acid
[ ] $C_6H_2O_4$

Sorbe, 1968, 109
Explodes on heating.
See other ACETYLENIC COMPOUNDS, ORGANIC ACIDS

2095. Triethynylaluminium
[61204-16-8] C_6H_3Al

Diethyl ether, or Dioxane, or Trimethylamine
1. Chini, P. *et al., Chim. e Ind.* (Milan), 1962, **44**, 1220
2. Houben-Weyl, 1970, Vol. 13.4, 159
The residue from sublimation of the complex with dioxane is explosive, and the
complex should not be dried by heating. The trimethylamine complex may also

explode on sublimation [1]. Triethynylaluminium or its complex with diethyl ether may decompose explosively on heating. Sublimation is not therefore advised as a purification method [2].

See other METAL ACETYLIDES

2096. Triethynylarsine
[687-78-5] C_6H_3As

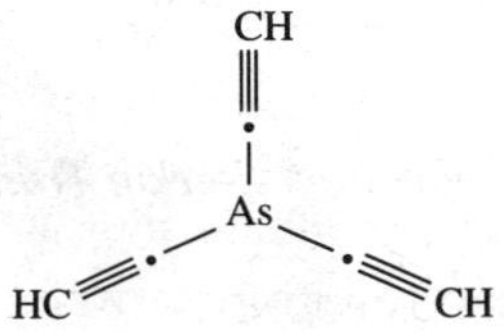

Voskuil, W. *et al., Rec. Trav. Chim.*, 1964, **83**, 1301
It explodes on strong friction.
See related METAL ACETYLIDES

2097. 6-Bromo-2,4-dinitrobenzenediazonium hydrogen sulfate
[65036-47-7] $C_6H_3BrN_4O_8S$

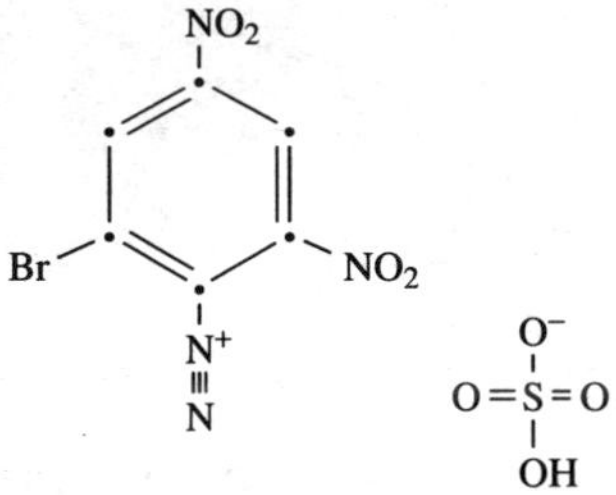

See entry DIAZONIUM SULFATES
See other POLYNITROARYL COMPOUNDS

2098. 1-Chloro-2,4-dinitrobenzene
[97-00-7] $C_6H_3ClN_2O_4$

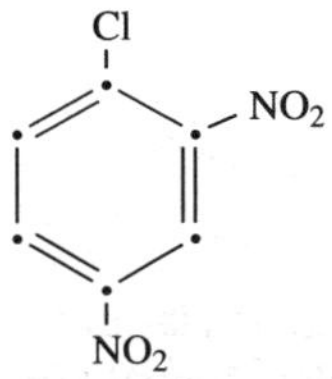

HCS 1980, 299

Sorbe, 1968, 56

T_{ait24} was determined as 250°C by adiabatic Dewar tests, with an apparent energy of activation of 185 kJ/mol.

It has been used as an explosive.

See 1-Fluoro-2,4-dinitrobenzene

See entry THERMOCHEMISTRY AND EXOTHERMIC DECOMPOSITION (reference 2)

Ammonia
1. Uhlmann, P. W., *Chem. Ztg.*, 1914, **38**, 389–390
2. Wells, F. B. *et al., Org. Synth.*, 1943, Coll. Vol. 2, 221–222

During the preparation of 2,4-dinitroaniline by a well-established procedure involving heating the reactants in a direct-fired autoclave (170°C and 40 bar were typical conditions), a sudden increase in temperature and pressure preceded a violent explosion [1]. An alternative process avoiding the use of a sealed vessel is now used [2].

See 2-Chloronitrobenzene: Ammonia
See POLYNITROARYL COMPOUNDS: bases, or salts
See other AMINATION INCIDENTS

Hydrazine hydrate
491M, 1975, 201

In absence of diluent, the reaction is sufficiently exothermic and violent to shatter the flask.

Other reactants
Yoshida, 1980, 157

MRH values calculated for 7 combinations with materials catalysing its decomposition are given.

See other HALOARYL COMPOUNDS, POLYNITROARYL COMPOUNDS

2099. 2,4-Dinitrobenzenesulfenyl chloride
[528-76-7] $C_6H_3ClN_2O_4S$

Kharasch, N. *et al., Org. Synth.*, 1964, **44**, 48

During removal of solvent under vacuum, the residual chloride must not be overheated, as it may explode.

See related NITROACYL HALIDES *See other* POLYNITROARYL COMPOUNDS

2100. 2,4-Dinitrobenzenesulfonyl chloride
[1656-44-6] $C_6H_3ClN_2O_6S$

Preparative hazard
   See Bis(2,4-dinitrophenyl) disulfide: Chlorine
   See other NITROACYL HALIDES, POLYNITROARYL COMPOUNDS

2101. 2,6-Dinitro-4-perchlorylphenol
[ ] $C_6H_3ClN_2O_8$

Gardner, D. M. *et al., J. Org. Chem.*, 1963, **28**, 2652
This analogue of picric acid is dangerously explosive and very shock-sensitive.
   See other PERCHLORYL COMPOUNDS, POLYNITROARYL COMPOUNDS

2102. 6-Chloro-2,4-dinitrobenzenediazonium hydrogen sulfate
[68597-05-7] $C_6H_3ClN_4O_8S$

   See entry DIAZONIUM SULFATES
   See other POLYNITROARYL COMPOUNDS

680

2103. Dichloronitrobenzene

[?] $C_6H_3Cl_2NO_2$

$$Cl_2C_6H_3NO_2$$

Aniline

Anon., *Chem. Hazards Ind.*, 1992, (Feb.), 171

A reaction with an unspecified isomer of dichloronitrobenzene ran wild on overheating.
See other RUNAWAY REACTIONS *See other* NITROARYL COMPOUNDS

2104. 2,4,-Dichloronitrobenzene

[611-06-3] $C_6H_3Cl_2NO_2$

Acetic acid, Potassium fluoride, Dimethylacetamide

Mooney, D. G., *Hazards XI*, 38132, Symp. Ser. 124, Rugby (UK), IChE, 1991
van Reijendam, J. W. *et al., J. Loss Prev. Proc. Ind.*, 1992, **5**(4), 211

A Halex process, exchanging the chlorine for fluorine in solvent dimethylacetamide, on 15,000l scale with above a tonne of substrate, showed a runaway exotherm from 160°C, followed by rupture of the vessel and formation of a fireball by the contents. Fragments and damage extended up to 500m distance. Examination suggested that a complex chain of events, starting with large scale contamination of a previous batch by water during work up, had produced recycled dimethylacetamide contaminated with acetic acid. Acetate then acted as nucleophile, subsequently eliminating ketene to produce a phenol, itself another nucleophile producing diphenyl ethers. The limited exotherm produced by these reactions was enough to take the mix to 250°C, where a more exothermic reaction, probably involving the oxidant nitro group, sets in. Presumably similar reactions could take place with other halonitrobenzenes.
See 5-Fluoro-2-nitrophenol
See other RUNAWAY REACTIONS *See other* HALOARYL COMPOUNDS, NITROARYL COMPOUNDS

2105. 3,4-Dichloronitrobenzene

[99-54-7] $C_6H_3Cl_2NO_2$

Catalyst, Hydrogen

See 3,4-Dichlorophenylhydroxylamine
See other HALOARYL COMPOUNDS, NITROARYL COMPOUNDS

2106. 2,6-Dichloro-4-nitrobenzenediazonium hydrogen sulfate
   [68597-06-8] $C_6H_3Cl_2N_3O_6S$

See entry DIAZONIUM SULFATES

2107. 2,4,5-Trichlorophenol
   [95-95-4] $C_6H_3Cl_3O$

Preparative hazard
   See 1,2,4,5-Tetrachlorobenzene: Sodium hydroxide, Solvent
   See other HALOARYL COMPOUNDS

2108. 1-Fluoro-2,4-dinitrobenzene
   [70-34-8] $C_6H_3FN_2O_4$

HCS 1980, 505

1. Halpern, B. D., *Chem. Eng. News*, 1951, **29**, 2666
2. Muir, G. D., private comm., 1969
3. Reinheimer, J. D. *et al., J. Amer. Chem. Soc.*, 1961, **83**, 837

The residue left from conversion of the chloro to the fluoro compound exploded during distillation at 1.3 mbar [1]. Reheating the residue from distillation caused violent decomposition [2]. Traces of an *aci*-nitroquinonoid species (structure str02, p. S1-S7) may have been formed in the chlorodinitrobenzene–potassium fluoride reaction system [3].

See related aci-NITROQUINONOID COMPOUNDS

Ether peroxides
Shafer, P. R., private comm., 1967
When air was admitted after vacuum evaporation of a (peroxidic) ether solution of the dinitro compound (5 g), a violent explosion occurred. Exploding ether peroxide may have initiated the dinitro compound.

See other HALOARYL COMPOUNDS, POLYNITROARYL COMPOUNDS

2109. 2,4-Difluoronitrobenzene
[446-35-5] $C_6H_3F_2NO_2$

Sodium hydroxide, Dioxane, Water
See 5-Fluoro-2-nitrophenol
See other NITROARYL COMPOUNDS

†2110. 1,2,4-Trifluorobenzene
[367-23-7] $C_6H_3F_3$

See other HALOARYL COMPOUNDS

2111. Sodium 2-hydroxymercurio-6-nitro-4-*aci*-nitro-2,5-cyclohexadienonide
[ ] $C_6H_3HgN_2NaO_6$

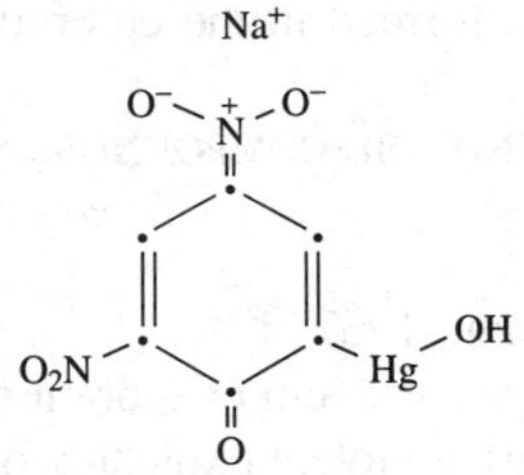

Na$^+$

O$^-$\N$^+$/O$^-$

O$_2$N
Hg—OH
O

Hantzsch, A. *et al., Ber.*, 1906, **39**, 1113
This salt of the mono-*aci- p*-quinonoid form of 2-hydroxymercurio-4,6-dinitrophenol
explodes violently on strong heating.
See entry *aci*-NITROQUINONOID COMPOUNDS
See other MERCURY COMPOUNDS, POLYNITROARYL COMPOUNDS

2112. Potassium 1,3,5-trinitrobenzene
[ ] $C_6H_3KN_3O_6$

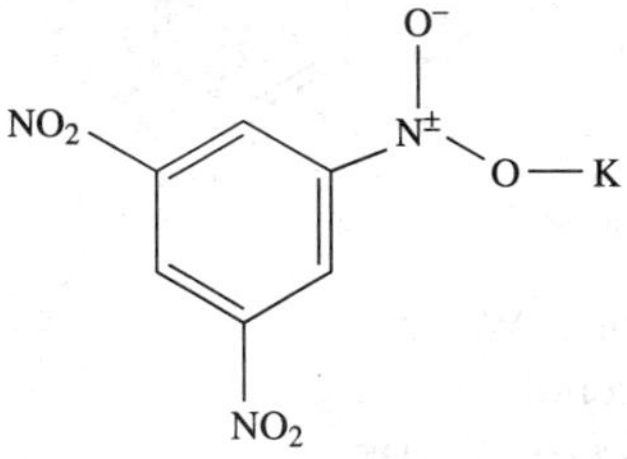

NO$_2$

O$^-$
N$^\pm$
O—K

NO$_2$

See entry NITROAROMATIC–ALKALI HAZARDS; potassium radical salts

2113. Potassium 6-*aci*-nitro-2,4-dinitro-2,4-cyclohexadieniminide
[[12244-59-6] (ion)] $C_6H_3KN_4O_6$

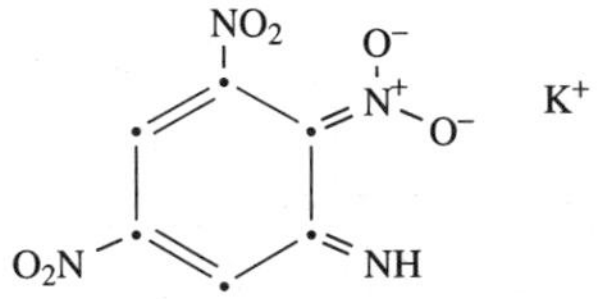

NO$_2$
O$^-$
N$^+$
O$^-$
K$^+$

O$_2$N
NH

Green, A. G. *et al., J. Chem. Soc.*, 1913, **103**, 508–513
This salt of the *aci-o*-iminoquinonoid form of 2,4,6-trinitroaniline explodes violently
at 110°C.
See entry *aci*-NITROQUINONOID COMPOUNDS
See other POLYNITROARYL COMPOUNDS

2114. Potassium 4-hydroxy-5,7-dinitro-4,5-dihydrobenzofurazanide 3-oxide
[[57891-85-7] (ion)] $C_6H_3KN_4O_7$

1. Boulton, A. J. *et al., J. Chem. Soc.*, 1965, 5414
2. Spear, R. J., *Aust. J. Chem.*, 1982, **35**, 4
3. Spear, R. J. *et al., Prop. Explos., Pyrotech.*, 1983, **8**, 95–98

The complex formed between 4,6-dinitrobenzofurazan 3-oxide and potassium hydroxide has been evaluated as an explosive [1]. The structure is now believed to be that of a Meisenheimer complex [2]. The analogous sodium, silver and barium salts are also explosive, the latter being less sensitive to initiation [3].

See other FURAZAN *N*-OXIDES, POLYNITROARYL COMPOUNDS

2115. Sodium 2,4-dinitrophenoxide
[38892-09-0] $C_6H_3N_2NaO_5$

1. Hickson, B., *Chem. Age*, 1926, **14**, 522
2. Uhlmann, P. W., *Chem. Ztg.*, 1914, **38**, 388–390

A violent explosion occurred during centrifugal dehydration of an aqueous paste of the sodium salt [1], which was known to be explosive [2]. The instability may be associated with the presence of some of the *aci-o-* or *p*-quinonoid salt.

See other METAL NITROPHENOXIDES, NITROAROMATIC–ALKALI HAZARDS, POLYNI-TROARYL COMPOUNDS

2116. 1,3,5-Trinitrobenzene
[99-35-4] $C_6H_3N_3O_6$

Alkylmetallates, or Arylmetallates
Artamkina, G. M. *et al., J. Organomet. Chem.*, 1979, **182**, 185

The THF-containing complexes formed between trinitrobenzene and the lithium or potassium salts of trimethyl-, triethyl- or triphenyl-germanate, -silanate or -stannate decompose explosively on heating, though trinitrobenzene–potassium trimethyl-stannate decomposes explosively at ambient temperature.

Methanol, Potassium hydroxide
See Potassium 4-methoxy-1-*aci*-nitro-3,5-dinitro-2,5-cyclohexadienide
See other POLYNITROARYL COMPOUNDS

2117. Trinitrosophloroglucinol (2,4,6-Trinitrosobenzene-1,3,5-triol)
[ ] $C_6H_3N_3O_6$

The isomeric cyclohexanehexone 1,3,5-trioxime is identified as [14378-99-5].

Heavy metal compounds
Heavy metal salts (possibly of the oxime form) may be hazardous.
See Lead(II) trinitrosobenzene-1,3,5-trioxide
See other NITROSO COMPOUNDS See related OXIMES

2118. Picric acid (2,4,6-Trinitrophenol)
[88-89-1] $C_6H_3N_3O_7$

NSC 351, 1979; FPA H128, 1983; HCS 1980, 755; RSC Lab. Hazards Data Sheet No. 34, 1985

Alone, or Heavy metals, or Bases
1. Cooper-Key, A., *Home Office Rept. 211*, 9, HMSO, 1914
2. Kirk-Othmer, 1965, Vol. 8, 617
3. Urbanski, 1964, Vol. 1, 518
4. Garey, H. E., *Chem. Eng. News*, 1979, **57**(41), 51
5. Ventrone, T. A., *CHAS Notes*, 1982, **1**(3), 1–2; (4), 4

686

Picric acid, in common with several other polynitrophenols, is an explosive material in its own right and is usually stored as a water-wet paste. Several dust explosions of dry material have been reported [1]. It forms salts with many metals, some of which (lead, mercury, copper or zinc) are rather sensitive to heat, friction or impact. The salts with ammonia and amines, and the molecular complexes with aromatic hydrocarbons, etc. are, in general, not so sensitive [2]. Contact of picric acid with concrete floors may form the friction-sensitive calcium salt [3]. Contact of molten picric acid with metallic zinc or lead forms the metal picrates which can detonate the acid. Picrates of lead, iron, zinc, nickel, copper, etc. should be considered dangerously sensitive. Dry picric acid has little effect on these metals at ambient temperature. Picric acid of sufficient purity is of the same order of stability as TNT, and is not considered unduly hazardous in regard to sensitivity [4]. Details of handling and disposal procedures have been collected and summarised [5].
See Sodium picrate, *also* Ammonium picrate

Aluminium, Water
See Aluminium: Nitro compounds, etc.

Uronium perchlorate
See Uronium perchlorate: Organic materials
See other ORGANIC ACIDS, POLYNITROARYL COMPOUNDS

2119. Trinitroresorcinol (2,4,6-Trinitrobenzene-1,3-diol)
[82-71-3] $C_6H_3N_3O_8$

Sorbe, 1968, 152
The compound ('styphnic acid', +6.7% oxygen balance) and its salts are very explosive.
See other ORGANIC ACIDS, POLYNITROARYL COMPOUNDS

2120. Trinitrophloroglucinol (2,4,6-Trinitrobenzene-1,3,5-triol)
[4328-17-0] $C_6H_3N_3O_9$

1. Sorbe, 1968, 153
2. DeFusco, A. A., *Org. Prep. Proced. Int.*, 1982, **14**, 393–395

It explodes on heating [1], and an improved preparation is described [2]. The compound has +20% oxygen balance, so can function as an overall oxidant. It would also be expected to form explosive metal salts.
See other POLYNITROARYL COMPOUNDS

2121. N,2,3,5-Tetranitroaniline
[57284-58-7] $C_6H_3N_5O_8$

1. Mudge, P. R. et al., J. Chem. Soc., Chem. Comm., 1975, 569
2. Meldola, R. et al., J. Chem. Soc., 1905, **87**, 1204, 1205
Thermal rearrangement to the internal salt, 3,5-dinitrobenzenediazonium 2-oxide, proceeds explosively in absence of solvent [1]. The oxide itself explodes violently at 190°C [2].
See other N-NITRO COMPOUNDS, POLYNITROARYL COMPOUNDS

2122. 2,3,4,6-Tetranitroaniline
[3698-54-2] $C_6H_3N_5O_8$

$$(O_2N)_4C_6HNH_2$$

Preparative hazard
See 3-Nitroaniline: Nitric acid, Sulfuric acid
See other POLYNITROARYL COMPOUNDS

2123. N,2,4,6-Tetranitroaniline
[4591-46-2] $C_6H_3N_5O_8$

Olsen, R. E. et al., ACS 54, 1966, 50
Impure tetryl must be recrystallised with care as it may deflagrate at only 50°C.
See other N-NITRO COMPOUNDS, POLYNITROARYL COMPOUNDS

2124. Triethynylphosphine
[687-80-9] C_6H_3P

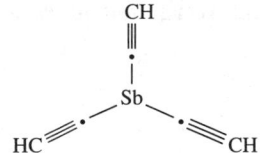

Voskuil, W. *et al., Rec. Trav. Chim.*, 1964, **83**, 1301
It explodes on strong friction, and on standing it decomposes and may then explode spontaneously.
See related ALKYLPHOSPHINES *See other* ACETYLENIC COMPOUNDS

2125. Triethynylantimony
[687-81-0] C_6H_3Sb

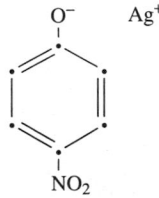

Voskuil, W. *et al., Rec. Trav. Chim.*, 1964, **83**, 1301
It explodes on strong friction.
See other METAL ACETYLIDES

2126. Silver 4-nitrophenoxide
[86255-25-6] $C_6H_4AgNO_3$

Spiegel, L. *et al., Ber.*, 1906, **39**, 2639
A sample decomposed explosively after intensive drying at 110°C. The *aci*-nitro-*p*-quinonoid salt may have been formed.
See related aci-NITROQUINONOID COMPOUNDS
See other METAL NITROPHENOXIDES, SILVER COMPOUNDS

2127. Silver benzo-1,2,3-triazole-1-oxide
[93693-72-2]

$C_6H_4AgN_3O$

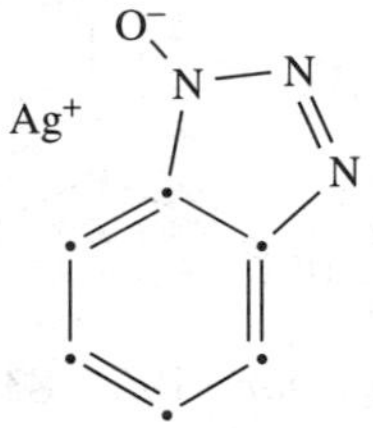

Deorha, D. S. *et al.*, *J. Indian Chem. Soc.*, 1964, **41**, 793
It is unsuitable as a gravimetric precipitate, as it explodes on heating rather than smoothly giving weighable silver.
See other N–O COMPOUNDS, SILVER COMPOUNDS, TRIAZOLES

2128. 2-Nitrobenzenediazonium tetrachloroborate
[ ]

$C_6H_4BCl_4N_3O_2$

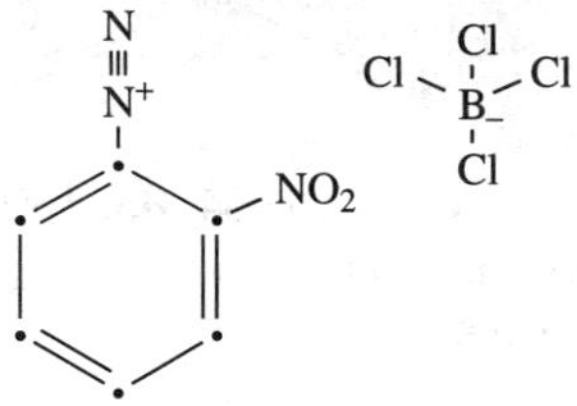

See entry DIAZONIUM TETRAHALOBORATES

2129. 3-Bromophenyllithium
[6592-86-1]

C_6H_4BrLi

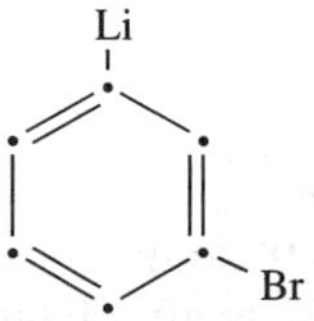

See entry ORGANOLITHIUM REAGENTS *See other* HALO-ARYLMETALS

2130. 4-Bromophenyllithium
  [22480-64-4] C_6H_4BrLi

Anon., *Angew. Chem. (Nachr.)*, 1962, **10**, 65
It explodes if traces of oxygen are present in the inert atmosphere needed for the preparation.
See entry ORGANOLITHIUM REAGENTS *See other* HALO-ARYLMETALS

2131. 4-Bromobenzenediazonium salts
  [17333-82-2] (ion) $C_6H_4BrN_2Z$

Hydrogen sulfide
  See DIAZONIUM SULFIDES AND DERIVATIVES (reference 2) *See other* DIAZONIUM
  SALTS

2132. 2-Bromo-4,6-dinitroaniline
  [1817-73-8] $C_6H_4BrN_3O_4$

This and the 4-bromo-2,6-dinitro analogue show local decomposition on moderate heating.
See entry HIGH RATE DECOMPOSITION

2133. 4-Bromo-2,6-dinitroaniline
[62554-90-9] $C_6H_4BrN_3O_4$

This and the 2-bromo-4,6-dinitro analogue show local decomposition on moderate heating.
See entry HIGH RATE DECOMPOSITION

†**2134. 2-Chlorofluorobenzene**
[348-51-6] C_6H_4ClF

See other HALOARYL COMPOUNDS

†**2135. 3-Chlorofluorobenzene**
[625-98-9] C_6H_4ClF

See other HALOARYL COMPOUNDS

†**2136. 4-Chlorofluorobenzene**
[352-33-0] C_6H_4ClF

See other HALOARYL COMPOUNDS

692

2137. 4-Chlorobenzenediazonium triiodide
[68596-93-0]

$C_6H_4ClI_3N_2$

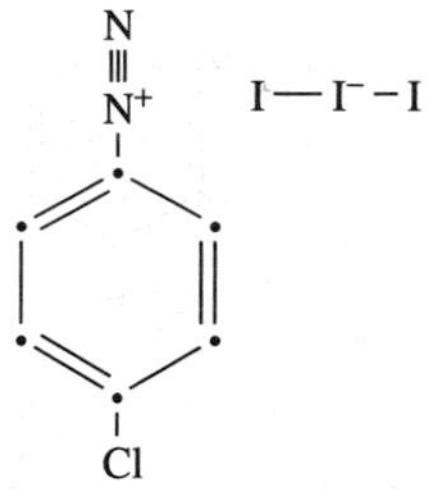

See entry DIAZONIUM TRIIODIDES

2138. 3-Chlorophenyllithium
[25077-87-6]

C_6H_4ClLi

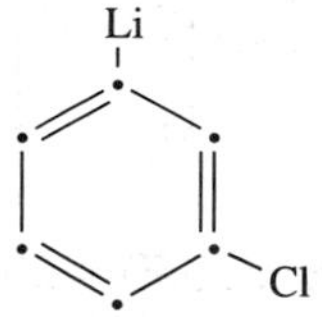

See entry ORGANOLITHIUM REAGENTS *See other* HALO-ARYLMETALS

2139. 4-Chlorophenyllithium
[14774-78-8]

C_6H_4ClLi

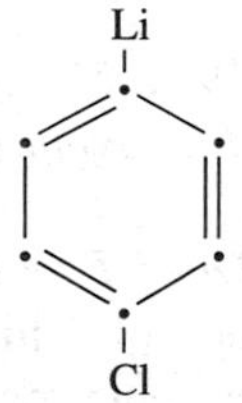

Anon., *Angew. Chem. (Nachr.)*, 1962, **10**, 65
It explodes if traces of oxygen are present in the inert atmosphere needed for the preparation.
See entry ORGANOLITHIUM REAGENTS *See other* HALO-ARYLMETALS

2140. 1,4-Benzoquinone-4-chloroimine
[637-61-6] C_6H_4ClNO

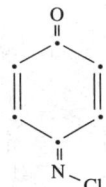

Sorbe, 1968, 112
It explodes on heating, like the bis-chloroimine.
See other N-HALOGEN COMPOUNDS

2141. 2-Chloronitrobenzene
[88-73-3] $C_6H_4ClNO_2$

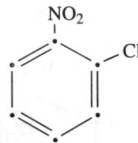

Energy of decomposition (in range 350–450°C) measured as 1.83 kJ/g by DSC, and T_{ait24} was determined as 264°C by adiabatic Dewar tests, with an apparent energy of activation of 173 kJ/mol.
See entry THERMOCHEMISTRY AND EXOTHERMIC DECOMPOSITION (reference 2)

Alkalis, Alcohols
See 4-Nitrochlorobenzene, next below

Ammonia
1. *MCA Case History No. 1624*
2. Vincent, G. C., *Loss. Prev.*, 1971, **5**, 46–52
3. Jacobson, O. F., *Proc. 3rd Int. Sympos. Loss Prev. Safety Prom. Proc. Ind.*, 417–421, SSCI, Basle, 1982
4. Gordon, M. D. *et al., Plant/Oper. Progr.*, 1982, **1**, 27–33
During the large-scale preparation of 2-nitroaniline at 160–180°C/30–40 bar in a jacketed autoclave, several concurrent processing abnormalities (excess chloro compound, too little ammonia solution, failure to apply cooling or to vent the autoclave and non-failure of a rupture-disk) led to a runaway reaction and pressure-explosion of the vessel [1]. A full analysis of the incident is published [2]. A retrospective hazard and operability study revealed that 6 simultaneous fault conditions had been involved [3], and the thermal stability of the amination process for manufacturing 2-nitroaniline was studied using mathematical modelling techniques [4]. The results of modelling the process abnormalities closely followed the actual events [1,2].
See other AMINATION INCIDENTS

694

Methanol, Sodium hydroxide

Klais, O. *et al., Prax. Sicherheitstech.*, 1997, **4**, 41

Klais, O. *et al.* in *Safety and Runaway Reactions*, Smeder, B & Mitchison, N. (Eds.), EUR 17723 EN, European Commission, 1997

A very large scale reaction to produce 2-nitroanisole from chloronitrobenzene and methanolic sodium hydroxide ran out of control and painted the town orange. This was attributed to reduction of the nitro group at temperatures above 100°C, a far more exothermic reaction than intended. This temperature was reached because the methanolic alkali was charged without agitation, and, the reaction not starting, the batch was heated to 90°C. The agitation was only then switched on.

See also 4-Nitrochlorobenzene, next below

See other AGITATION INCIDENTS

See other HALOARYL COMPOUNDS, NITROARYL COMPOUNDS

2142. 4-Chloronitrobenzene
[100-00-5] $C_6H_4ClNO_2$

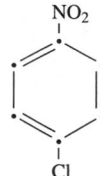

HCS 1980, 308

1. Doyle, W. H., *Loss. Prev.*, 1969, **3**, 16

A steam-heated still used to top the crude material exploded

Energy of decomposition (in range 300–450°C) measured as 2.05 kJ/g by DSC, and T_{ait24} was determined as 275°C by adiabatic Dewar tests, with an apparent energy of activation of 172 kJ/mol.

See entry THERMOCHEMISTRY AND EXOTHERMIC DECOMPOSITION (reference 2)

Alkalis, Alcohols

Anderson, H. L. *et al., Thermochim. Acta*, 1996, 271, 195

Anderson, H. L. *et al., Loss Prevention and Safety Promotion in the Process Industries*, Vol II, (Mewis, J. J., Pasman, H. J. & De Rademaker, E. E. Eds.), 387, Amsterdam, Elsevier, 1995

A calorimetric study of reaction with sodium or potassium hydroxides in ethanol or 2-propanol is given. At starting temperatures below 70°C the product is the appropriate nitrophenyl ether; above that temperature, reduction of the nitro groups may come into play, to give much more energy and a variety of other products. This reaction is inhibited by oxygen. There is potential for runaway if such reactions are operated industrially with poor temperature control. The editor suspects that the stimulus for this study was an accident which sprayed the German environment with 2-nitroanisole.

See Sodium methoxide; below

695

Potassium hydroxide

Mixtures with potassium hydroxide (1:1.5 mol) deflagrate readily at a rate of 1.3 cm/min.

See entries PRESSURE INCREASE IN EXOTHERMIC DECOMPOSITION (reference 3), NITROAROMATIC–ALKALI HAZARDS

See other DEFLAGRATION INCIDENTS

Sodium methoxide

Anon., *ABCM Quart. Safety Summ.*, 1944, **15**, 15

Addition of the chloro compound to a solution of sodium methoxide in methanol caused an unusually exothermic reaction to occur. The lid of the 450 l vessel was blown off, and a fire and explosion followed. No cause for the unusual vigour of the reaction was found.

See Alkalis, Alcohols; above

See other HALOARYL COMPOUNDS, NITROARYL COMPOUNDS

2143. 3-Nitroperchlorylbenzene
[20731-44-6] $C_6H_4ClNO_5$

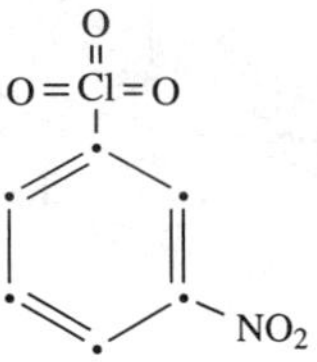

McCoy, G., *Chem. Eng. News*, 1960, **38**(4), 62

The nitration product of perchloryl benzene is explosive, comparable in shock-sensitivity with lead azide, with a very high propagation rate.

See other NITROARYL COMPOUNDS, PERCHLORYL COMPOUNDS

2144. 2-Chloro-5-nitrobenzenesulfonic acid
[96-73-1] $C_6H_4ClNO_5S$

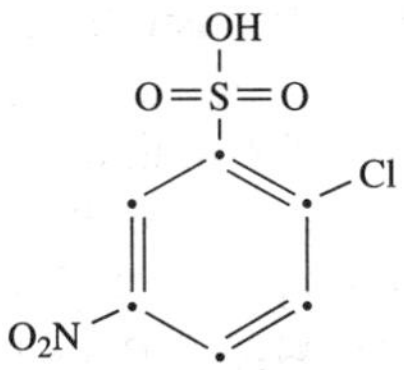

Preparative hazard

Exothermic decomposition of the acid at 150°C ($Q = 0.708$ kJ/g) is time-delayed, may be faster in a closed than an open system, and is dangerous.

See Sulfuric acid: 4-Chloronitrobenzene

696

See entry SELF-ACCELERATING REACTIONS
See other HALOARYL COMPOUNDS, NITROARYL COMPOUNDS, ORGANIC ACIDS

2145. 4-Chloro-3-nitrobenzenesulfonic acid
[71799-32-1] $C_6H_4ClNO_5S$

Négyesi, G., *Process Safety Progress*, 1996, **15**(1), 42
This compound shows signs of impact and friction sensitivity.
See other NITROARYL COMPOUNDS

2146. 4-Hydroxy-3-nitrobenzenesulfonyl chloride
[ ] $C_6H_4ClNO_5S$

Vervalin, C. H., *Hydrocarbon Proc.*, 1976, **55**(9), 321–322
The chloride, produced by interaction of 2-nitrophenol and chlorosulfuric acid at 4°C, decomposed violently during discharge operations from the 2000 l vessel, leaving a glowing residue. It was subsequently found that accelerating exothermic decomposition sets in at 24–27°C.
See related SULFONATION INCIDENTS *See other* NITROACYL HALIDES

2147. Dicarbonylpyrazinerhodium(I) perchlorate
[ ] $C_6H_4ClN_2O_6Rh$

Uson, R. *et al.*, *J. Organomet. Chem.*, 1981, **220**, 109

This (notably) and other polynuclear complexes exploded violently on heating during microanalysis.

See related AMMINEMETAL OXOSALTS

2148. 2-Chlorobenzenediazonium salts
[17333-83-4] (ion) $C_6H_4ClN_2Z$

Potassium 2-chlorothiophenoxide
See DIAZONIUM SULFIDES AND DERIVATIVES (reference 9) *See other* DIAZONIUM SALTS

2149. 3-Chlorobenzenediazonium salts
[17333-84-5] (ion) $C_6H_4ClN_2Z$

Potassium thiophenoxide, or Sodium polysulfide
See DIAZONIUM SULFIDES AND DERIVATIVES (reference 9) *See other* DIAZONIUM SALTS

2150. 1-Chlorobenzotriazole
[21050-95-3] $C_6H_4ClN_3$

Hopps, H. B., *Chem. Eng. News*, 1971, **49**(30), 3
Spontaneous ignition occurred during packing operations.
See other N-HALOGEN COMPOUNDS

2151. 3-Nitrobenzenediazonium chloride
[2028-76-4] $C_6H_4ClN_3O_2$

Potassium *O*-ethyldithiocarbamate
See DIAZONIUM SULFIDES AND DERIVATIVES *See other* DIAZONIUM SALTS

2152. 4-Chlorobenzenesulfonyl azide
[4547-68-6] $C_6H_4ClN_3O_2S$

N-Methyltetrahydro-β-carboline
Bailey, A. S. *et al., J. Chem. Soc., Perkin Trans. 1*, 1980, 1513
In absence of solvent chloroform, interaction at ambient temperature is explosive.
See other ACYL AZIDES

2153. 2-Chloro-4,6-dinitroaniline
[3531-19-9] $C_6H_4ClN_3O_4$

The solid amine can be caused to detonate by heating, or by a powerful initiating
charge, and fast local decomposition occurs on heating moderately.
See entry HIGH RATE DECOMPOSITION

699

Nitrosylsulfuric acid

Bersier, P. *et al., Chem. Ing. Tech.*, 1971, **43**, 1311–1315

Diazotisation of the potentially explosive 2-chloro-4,6-dinitroaniline in the third batch of a new process (at a higher than usual concentration in 40% nitrosylsulfuric acid) led to a violent explosion soon after the temperature had been increased to 50°C. Subsequent DSC work showed that the temperature at which thermal decomposition of the diazonium sulfate solution or suspension sets in is inversely proportional to the concentration of amine, falling from 160°C at 0.3 mmol/g to 80°C at 2 mmol/g. Thermal stability of 17 other diazonium derivatives was similarly investigated.

See other DIAZONIUM SULFATES *See other* POLYNITROARYL COMPOUNDS

2154. 4-Chloro-2,6-dinitroaniline
[5388-62-5] $C_6H_4ClN_3O_4$

CHETAH , 1990, 183

Can be caused to detonate by 50 g tetryl booster.

Preparative hazard

See Nitric acid: 4-Chloro-2-nitroaniline

Nitrosylsulfuric acid

Anon., *Angew. Chem. (Nachr.)*, 1970, **18**, 62

During large-scale diazotisation of the amine, severe local overheating is thought to have caused the explosion observed. The effect could not be reproduced in the laboratory.

See DIAZOTISATION *See other* POLYNITROARYL COMPOUNDS

2155. 3-Nitrobenzenediazonium perchlorate
22751-24-2 $C_6H_4ClN_3O_6$

Schumacher, 1960, 205

Explosive, very sensitive to heat or shock.
See other DIAZONIUM PERCHLORATES

2156. 1,2-Dichlorobenzene
[95-50-1] $C_6H_4Cl_2$

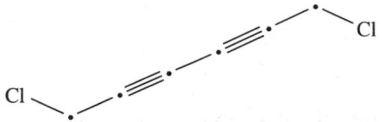

Aluminium, Halocarbons
See Aluminium: Halocarbons (reference 5)

Other reactants
Yoshida, 1980, 156
MRH values calculated for 13 combinations with oxidants are given.
See other HALOARYL COMPOUNDS

2157. 1,6-Dichloro-2,4-hexadiyne
[16260-59-6] $C_6H_4Cl_2$

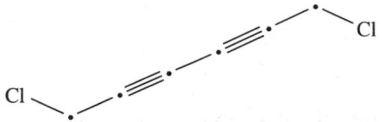

1. Driedger, P. E. *et al., Chem. Eng. News*, 1972, **50**(12), 51
2. Whiting, M. C., *Chem. Eng. News*, 1972, **50**(23), 86
3. Ford, M *et al., Chem. Eng. News*, 1972, **50**(30), 67
4. Walton, D. R. M., private comm., 1987
5. Anon., *Inst. Univ. Saf. Officers Bull.*, 1987, (8), 9

It is extremely shock-sensitive, a 4.0 kg cm shock causing detonation in 50% of
test runs (cf. 3.5 kg cm for propargyl bromide; 2.0 kg cm for glyceryl nitrate). The
intermediate bis-chlorosulfite involved in the preparation needs low temperatures to
avoid vigorous decomposition. The corresponding diiodo derivative was expected
to be similarly hazardous [1], and this has been confirmed [2]. Improvements in
preparative techniques (use of dichloromethane solvent at −30°C) to avoid violent
reaction have also been described [3]. An attempt to distill the compound (b.p.
55–58°C/0.6 mbar, equivalent to about 230°C/1 bar) at atmospheric pressure from
a heating mantle led to a violent explosion [4]. The compound involved was erroneously given as 1,6-dichloro-2,4-hexadiene [5].
See other HALOACETYLENE DERIVATIVES

2158. 4-Nitrophenylphosphorodichloridate
[777-52-6] $C_6H_4Cl_2O_4P$

Frommer, M. A. *et al.*, 1990, US Pat. 4910190; *Chem. Abs.*, 1991, **114**, 122705b
This compound may decompose violently during distillation.
See other NITROARYL COMPOUNDS
See related NON-METAL HALIDES (AND THEIR OXIDES)

2159. Benzoquinone 1,4-bis(chloroimine) (1,4-Bis(chlorimido)-2,5-cyclohexadiene)
[637-70-7] $C_6H_4Cl_2N_2$

Sorbe, 1968, 112
It explodes on heating, like the mono-chloroimine.
See other N-HALOGEN COMPOUNDS

2160. Benzene-1,4-bis(diazonium perchlorate)
[43008-25-9] $C_6H_4Cl_2N_4O_8$

Hofman, K. A. *et al., Ber.*, 1910 **43**, 2624
Described by Hofman as the most powerfully explosive compound of which he
knew. Since it is of better oxygen balance than the aromatic high explosives, and
the diazonium unit is also high energy, this is quite probable.
See DIAZONIUM PERCHLORATES

2161. Dichlorophenol mixed isomers
[25167-81-1] $C_6H_4Cl_2O$

HCS 1980, 381 (2,4-Dichlorophenol)

Laboratory Chemical Disposal Co. Ltd., confidential information, 1968
During vacuum fractionation of the mixed dichlorophenols produced by partial hydrolysis of trichlorobenzene, rapid admission of air to the receiver caused the column contents to be forced down into the boiler at 210°C, and a violent explosion ensued.
See other HALOARYL COMPOUNDS

2162. 2,4-Hexadiynylene chlorosulfite
[ ] $C_6H_4Cl_2O_4S_2$

Driedger, P. E. *et al., Chem. Eng. News*, 1972, **50**(12), 51
This is probably an intermediate in the preparation of the 1,6-dichloro compound from 2,4-hexadiyne-1,6-diol and thionyl chloride in DMF. The reaction mixture must be kept at a low temperature to avoid vigorous decomposition and charring. This may be due to interaction with the solvent.
See 2,4-Hexadiynylene chloroformate *also* Dimethylformamide
See other ACETYLENIC COMPOUNDS, ACYL HALIDES

2163. 2,3,4-Trichloroaniline
[634-67-3] $C_6H_4Cl_3N$

Energy of decomposition (in range 220–400°C) measured as 0.21 kJ/g.

See entry THERMOCHEMISTRY AND EXOTHERMIC DECOMPOSITION (reference 2)
See other HALOANILINES

2164. Cobalt(II) chelate of 1,3-bis(*N*-nitrosohydroxylamino)benzene
[27662-11-9] $C_6H_4CoN_4O_4$

Battei, R. S. *et al., J. Inorg. Nucl. Chem.*, 1970, **32**, 1525
The cobalt(II) chelate of 'dicupferron' explodes on heating at 205°C.
See Cobalt(II) chelate of bi(1-hydroxy-3,5-diphenylpyrazol-4-yl *N*-oxide)
See other N–O COMPOUNDS

2165. Dicopper(I) 1,5-hexadiynide
[86425-12-9] $C_6H_4Cu_2$

Chow, Y. L. *et al., Can. J. Chem.*, 1983, **61**, 795–800
It exploded when dried in a desiccator.
See other METAL ACETYLIDES

2166. 4-Fluorophenyllithium
[1493-23-8] C_6H_4FLi

See entry ORGANOLITHIUM REAGENTS *See other* HALO-ARYLMETALS

2167. 5-Fluoro-2-nitrophenol
[446-36-6] $C_6H_4FNO_3$

Preparative hazard
Robinson, N., *Chem. Brit.*, 1987, **23**, 837

Following a patented procedure for the conversion of 2,4-dinitrochlorobenzene to 5-chloro-2-nitrophenol, 2,4-difluoronitrobenzene was treated with sodium hydroxide in hot aqueous dioxane containing a phase transfer catalyst. On the small scale, the reaction and isolation of 5-fluoro-2-nitrophenol, including vacuum distillation, were uneventful. On the 20 l scale, vacuum distillation of combined batches of the crude product led to onset of decomposition at 150°C, which could not be controlled, and the residue erupted with explosive violence and a small fire ensued. Thermal examination of fresh small-scale crude material has shown that it is capable of highly exothermic decomposition, with onset of the exotherm at 150°C (ARC). It was then realised that difficulty in controlling the reaction temperature had been experienced on the 20 l scale. It is recommended that this procedure and purification should not be attempted on so large a scale.

See other NITROARYL COMPOUNDS

†2168. 1,3-Difluorobenzene
[372-18-9] $C_6H_4F_2$

See other HALOARYL COMPOUNDS

†2169. 1,4-Difluorobenzene
[540-36-3] $C_6H_4F_2$

See other HALOARYL COMPOUNDS

2170. Octafluoroadipamide
[355-66-8] $C_6H_4F_8N_2O_2$

Lithium tetrahydroaluminate
See Lithium tetrahydroaluminate: Fluoroamides

2171. Sodium 2-hydroxymercurio-4-*aci*-nitro-2,5-cyclohexadienonide
[ ] $C_6H_4HgNNaO_4$

Hantzsch, A. *et al., Ber.*, 1906, **39**, 1115
This salt of the *aci-p*-quinonoid form of 2-hydroxymercurio-4-nitrophenol explodes
on heating.
See other MERCURY COMPOUNDS, *aci*-NITROQUINONOID COMPOUNDS

2172. 1,2-Diiodobenzene
[615-42-9] $C_6H_4I_2$

Rüst, 1948, 302
A small sample in a sealed glass capillary exploded violently at 181°C, breaking the
surrounding oil bath.
See other HALOARYL COMPOUNDS, IODINE COMPOUNDS

706

2173. 1,6-Diiodo-2,4-hexadiyne
[44750-17-6] $C_6H_4I_2$

See 1,6-Dichloro-2,4-hexadiyne
See other HALOACETYLENE DERIVATIVES

2174. Potassium 4-nitrophenoxide
[1124-31-8] $C_6H_4KNO_3$

1. L. Bretherick, Personal experience
2. Thorpe's Dictionary of Applied Chemistry; 4th Edn., London, 1940, **IV**, 475
A sample of this potassium salt exploded after long storage in a non-evacuated
desiccator [1]. This may have been caused by formation of some tautomeric *aci*-
quinonoid salt by prolonged desiccation. This compound had earlier been described
as exploding on percussion [2].
See 4-Nitrophenol: Potassium hydroxide
See other METAL NITROPHENOXIDES, NITROARYL COMPOUNDS, *aci*-NITROQUINONOID
COMPOUNDS

2175. Potassium 4-nitrobenzeneazosulfonate
[ ] $C_6H_4KN_3O_5S$

491M, 1975, 345
During examination of a 10 g batch of the crystalline material on a filter paper
using a hand lens, it exploded. This was attributed to sudden rearrangement of the

metastable syn- to the anti-form (perhaps triggered by light energy concentrated by the lens).

See other AZO COMPOUNDS, N–S COMPOUNDS

2176. Potassium 4-hydroxyamino-5,7-dinitro-4,5-dihydrobenzofurazanide 3-oxide
[86341-95-9] $C_6H_4KN_5O_7$

Norris, W. P. *et al., Chem. Abs.*, 1983, **99**, 40662

Sensitivity to mechanical, electrostatic and thermal shock is typical of a primary explosive.

See other FURAZAN *N*-OXIDES

2177. Lithium 4-nitrothiophenoxide
[78350-94-4] $C_6H_4LiNO_2S$

1. Julia, M., *Chem. & Ind.*, 1981, 130; *Chem. Eng. News*, 1981, **59**(14), 3
2. Bretherick, L., *Chem. & Ind.*, 1981, 202; *Chem. Eng. News*, 1981, **59**(31), 3

A vacuum dried sample of the salt exploded violently when exposed to air [1]. It was postulated that the *aci*-quinonoid species may have been the unstable species [2].

See other NITROAROMATIC–ALKALI HAZARDS, *aci*-NITROQUINONOID COMPOUNDS
See related METAL NITROPHENOXIDES

2178. 1,3-Dilithiobenzene
[2592-85-0] $C_6H_4Li_2$

See entry ORGANOLITHIUM REAGENTS *See other* ARYLMETALS

708

2179. 1,4-Dilithiobenzene
[7279-42-0] $C_6H_4Li_2$

See entry ORGANOLITHIUM REAGENTS *See other* ARYLMETALS

2180. N,N,4-Trilithioaniline
[ ] $C_6H_4Li_3N$

Gilman, H. *et al., J. Amer. Chem. Soc.*, 1949, **71**, 2934
Highly explosive in contact with air.
See other N-METAL DERIVATIVES, ORGANOMETALLICS

2181. Sodium 4-nitrosophenoxide
[823-87-0] $C_6H_4NNaO_2$

Sorbe, 1968, 86
The dry solid tends to ignite.
See other NITROSO COMPOUNDS

2182. Sodium 2-nitrothiophenoxide
[22755-25-5]

$C_6H_4NNaO_2S$

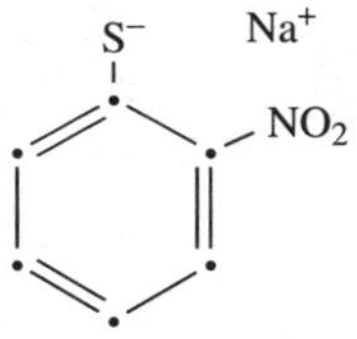

Davies, H. J., *Chem. & Ind.*, 1966, 257

The material exploded when the temperature of evaporation of the slurry was increased by adding a higher-boiling solvent. This may have been caused by presence of some of the tautomeric *aci*-nitro-*o*-thioquinone. Such salts are thermally unstable.

See other NITROAROMATIC–ALKALI HAZARDS, *aci*-NITROQUINONOID COMPOUNDS
See related METAL NITROPHENOXIDES,

2183. Sodium 4-nitrophenoxide
[824-78-2]]

$C_6H_4NNaO_3$

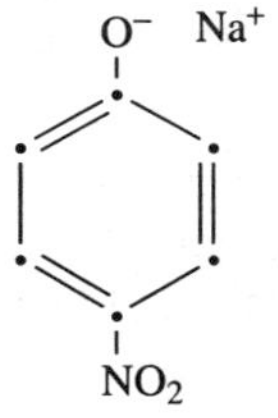

1. Semiganowski, N., *Z. Anal. Chem.*, 1927, **72**, 29
2. Baltzer, private comm., 1976

An ammoniacal solution of 4-nitroaniline containing ammonium chloride was being treated with 50% sodium hydroxide solution to displace ammonia. In error, double the amount required to give the usual 10% of free alkali was added, and during the subsequent degassing operation (heating to 130°C under pressure, followed by depressuring to vent ammonia), complete conversion to sodium 4-nitrophenoxide occurred unwittingly [1]. This solid product, separated by centrifuging, was then heated to dry it, when it decomposed violently and was ejected through the vessel opening like a rocket [2]. Some of the tautomeric *aci*-nitroquinonoid salt may have been produced during the drying operation.

See other METAL NITROPHENOXIDES, NITROAROMATIC–ALKALI HAZARDS, *aci*-NITRO-QUINONOID COMPOUNDS

2184. Sodium 3-nitrobenzenesulfonate
[127-68-4]

$C_6H_4NNaO_5S$

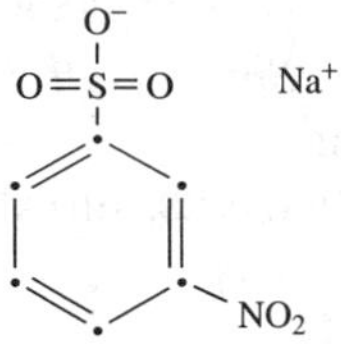

1. Grewer, T. *et al., Hazards from Pressure*, IChE Symp. Ser. No. 102, 1–9, Oxford, Pergamon, 1987
2. Grewer, T. *et al., Exothermic Decomposition*, Technical Report 01VD 159/0329 for Federal German Ministry for Res. Technol., Bonn. 1986

It ignites at 335°C and the deflagration rate of the solid is 4.5 cm/min [1]. T_{ait24} was determined as 305°C by adiabatic Dewar tests, with an apparent energy of activation of 441 kJ/mol [2].

See 3-Nitrobenzenesulfonic acid
See other DEFLAGRATION INCIDENTS, NITROARYL COMPOUNDS

2185. Disodium 4-nitrophenylphosphate
[4264-83-9]

$C_6H_4NNa_2O_6P$

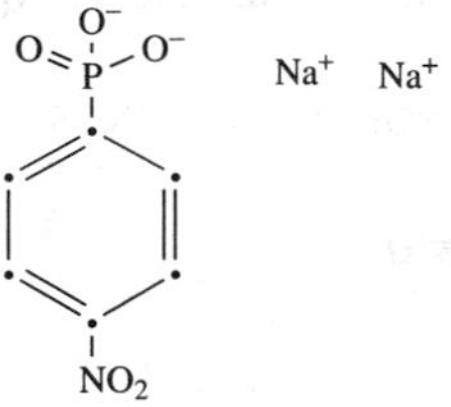

Preparative hazard
See Nitric acid: Disodium phenyl orthophosphate
See other NITROARYL COMPOUNDS

2186. Thallium(I) 2-nitrophenoxide
[32707-02-1]

$C_6H_4NO_3Tl$

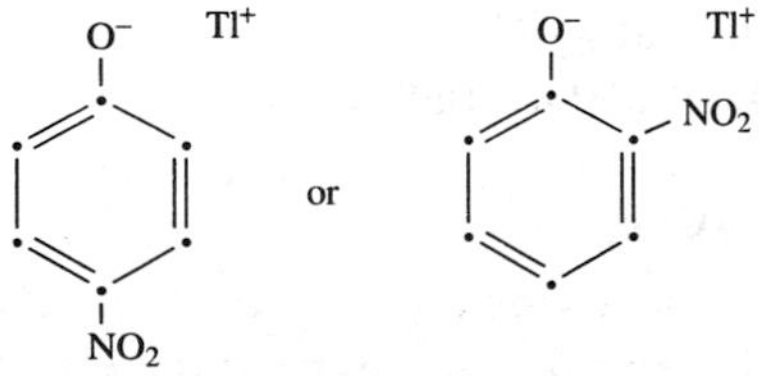

Langhans, A., *Z. Ges. Schiess.-Sprengstoff W.*, 1936, **31**, 359, 361
Both solids explode on heating.
See other HEAVY METAL DERIVATIVES, METAL NITROPHENOXIDES, NITROARYL COMPOUNDS

2187. Thallium(I) 4-nitrophenoxide
[32707-03-2]   C₆H₄NO₃Tl

$$TlOC_6H_4NO_2$$

Langhans, A., *Z. Ges. Schiess.-Sprengstoff W.*, 1936, **31**, 359, 361
Both solids explode on heating.
See other HEAVY METAL DERIVATIVES, METAL NITROPHENOXIDES, NITROARYL COMPOUNDS

2188. Sodium 1,4-bis(*aci*-nitro)-2,5-cyclohexadienide
[ ]   C₆H₄N₂Na₂O₄

Meisenheimer, J. *et al., Ber.*, 1906, **39**, 2529
The sodium and potassium salts of the bis-*aci*-quinonoid form of 1,4-dinitrobenzene deflagrate on heating.
See other *aci*-NITROQUINONOID COMPOUNDS

2189. Benzenediazonium-4-oxide
[6925-01-5]   C₆H₄N₂O

Hantzsch, A. *et al., Ber.*, 1896, **29**, 1530
It decomposes violently at 75°C.
See other ARENEDIAZONIUM OXIDES

2190. Benzofurazan-*N*-oxide
[480-96-6]   C₆H₄N₂O₂

CHETAH , 1990, 188
Shock sensitive (probably not very).
See other FURAZAN *N*-OXIDES

2191. Benzo-1,2,3-thiadiazole 1,1-dioxide
[37150-27-9] $C_6H_4N_2O_2S$

1. Wittig, G. *et al., Chem. Ber.*, 1962, **95**, 2718
2. Wittig, G. *et al., Org. Synth.*, 1967, **47**, 6
The solid explodes at 60°C, on impact or friction, or sometimes spontaneously. Solutions decompose below 10°C to give benzyne [1]. Full preparative and handling details are given [2].
See other N–S COMPOUNDS

2192. Benzenediazonium-2-sulfonate
[612-31-7] $C_6H_4N_2O_3S$

Franklin, E. C., *Amer. Chem. J.*, 1899, **20**, 459
The internal diazonium salt explodes on contact with flame or on percussion.
See related DIAZONIUM CARBOXYLATES *See other* DIAZONIUM SALTS

2193. Benzenediazonium-4-sulfonate
[305-80-6] $C_6H_4N_2O_3S$

1. Wichelhaus, H., *Ber.*, 1901, **34**, 11
2. Anon., *Sichere Chemiearb.*, 1959, **11**, 59
3. Anon., *Sichere Chemiearb.*, 1978, **30**, 77
This internal salt of diazotised sulfanilic acid (Pauly's reagent) exploded violently on touching when thoroughly dry [1]. Use of a metal spatula to remove a portion of a refrigerated sample of the solid salt caused a violent explosion [2], and a similar incident was reported 19 years later [3]. All solid diazo compounds must be stored only in small quantity under refrigeration in loosely plugged containers. Handle gently with non-metallic spatulae using personal protection [2].

See related DIAZONIUM CARBOXYLATES *See other* DIAZONIUM SALTS

2194. 1,2-Dinitrobenzene
[528-29-0] $C_6H_4N_2O_4$

Nitric acid
See Nitric acid: Nitroaromatics
See other POLYNITROARYL COMPOUNDS

2195. 1,3-Dinitrobenzene
[99-65-0] $C_6H_4N_2O_4$

Anon., *Sichere Chemiearbeit*, 1993, **45**(7), 82

Heat of decomposition was determined as 4.6 kJ/g (c.f. MRH calculated value of 5.02). T_{ait24} was determined as 284°C by adiabatic Dewar tests, with an apparent energy of activation of 284 kJ/mol.

A vessel containing dinitrobenzene for drying, after washing out styphnic and picric acid byproducts in alkali, retained a heel of some hundreds of litres, with a residue of rust, after emptying. This was cleaned out with low pressure steam at 130°C every two years. During cleaning, the tank burst and burnt out. It was supposed that residual nitrophenol salts had accumulated on the rust, which was shown to thermally sensitise them, and deflagrated, initiating the dinitrobenzene.

See entry THERMOCHEMISTRY AND EXOTHERMIC DECOMPOSITION (references 1,2)

Nitric acid
   See Nitric acid: Nitroaromatics

Other reactants
   Yoshida, 1980, 160
   MRH values calculated for 7 combinations with materials catalysing its decomposition are given.

Tetranitromethane
   See Tetranitromethane: Aromatic nitro compounds
   See other POLYNITROARYL COMPOUNDS

2196. 1,4-Dinitrobenzene
   [100-25-4] $C_6H_4N_2O_4$

Nitric acid
   See Nitric acid: Nitroaromatics
   See other POLYNITROARYL COMPOUNDS

2197. 2,4-Dinitrophenol
   [51-28-5] $C_6H_4N_2O_5$

NSC 690, 1981; *HCS 1980*, 439

T_{ait24} was determined as 189°C by adiabatic Dewar tests, with an apparent energy of activation of 144 kJ/mol. Dinitrophenol is now classified as an explosive in the UK, and is normally available from laboratory suppliers wetted with 15% water, as is picric acid.
See entry THERMOCHEMISTRY AND EXOTHERMIC DECOMPOSITION (reference 2)

715

Bases

1. Uhlmann, P. W., *Chem. Ztg.*, 1914, **38**, 389–390
2. *Thorpe's Dictionary of Applied Chemistry* ; London, 1940, **IV**, 477

Dinitrophenol forms explosive (*aci*-quinonoid ?) salts with alkalies or ammonia, and should not be heated with them in closed vessels [1]. Dinitrophenol forms explosive salts with most metals [2].

See Sodium 2,4-dinitrophenoxide
See other NITROAROMATIC–ALKALI HAZARDS, *aci*-NITROQUINONOID COMPOUNDS
See other POLYNITROARYL COMPOUNDS

2198. 4,6-Dinitro-1,3-benzenediol (4,6-Dinitroresorcinol)
[616-74-0] $C_6H_4N_2O_6$

Preparative hazard

Schmitt., R. J. *et al., J. Org. Chem.*, 1988, **53**, 5568–5569

The reaction conditions necessary to obtain a good yield of the title compound (a difficult isomer), and to avoid hazards during the nitration of resorcinol , are critical and strict adherence to those specified is essential. The necessary 80% white fuming nitric acid must be completely free from oxides of nitrogen and nitrous acid, and procedures for this are detailed. Then the temperature during addition of the diacetate must be kept between −10 and 0°C by regulating the rate of addition. The alternative use of 80% sulfuric acid as solvent for the 80% nitric acid (5 equiv.) is preferred as more reliable, but both methods have led to violent exothermic decomposition, accompanied by fume-off, after an induction period. In any event, the explosive 2,4,6-trinitroresorcinol ('styphnic acid') is produced as a by-product.

See other INDUCTION PERIOD INCIDENTS, NITRATION INCIDENTS, POLYNITROARYL COMPOUNDS

2199. Benzo-1,2,3-thiadiazole
[273-77-8] $C_6H_4N_2S$

Jordis, U. *et al., Phos. Sulfur*, 1984, **19**, 279

716

Purification by extraction and then distillation led to violent decomposition. Steam distillation avoided this problem.

See other N–S COMPOUNDS

2200. Sodium 6-*aci*-nitro-4-nitro-2,4-cyclohexadieniminide

[ ] $C_6H_4N_3NaO_4$

Green, A. G. *et al.*, *J. Chem. Soc.*, 1913, **103**, 508–513

The sodium and potassium salts of the mono-*aci*-*o*-quinonoid form of 2,4-dinitro-aniline readily deflagrate on heating.

See other aci-NITROQUINONOID COMPOUNDS, POLYNITROARYL COMPOUNDS

2201. 2-Nitrobenzenediazonium salts

[25910-37-6] (ion) $C_6H_4N_3O_2Z$

Sodium disulfide, or Sodium polysulfide

See DIAZONIUM SULFIDES AND DERIVATIVES

2202. 4-Nitrobenzenediazonium salts

[14368-49-1] (ion) $C_6H_4N_3O_2Z$

Hydrogen sulfide, or Sodium sulfide, or Potassium *O*, *O*-diphenyl phosphorodithioate

See DIAZONIUM SULFIDES AND DERIVATIVES

2203. Sodium 3,5-bis(*aci*-nitro)cyclohexene-4,6-diiminide

[ ] $C_6H_4N_4Na_2O_4$

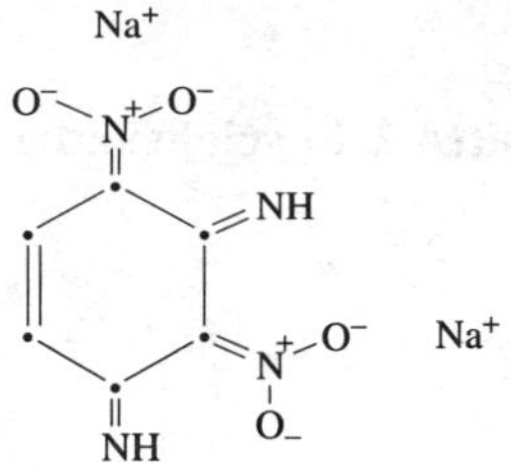

Alone, or Water

Meisenheimer, J. *et al., Ber.*, 1906, **39**, 2538

The disodium salt of the bis-*aci-o*-quinonoid form of 2,4-dinitro-1,3-phenylene-diamine deflagrates violently in contact with heat or moisture.

See other aci-NITROQUINONOID COMPOUNDS

2204. 4-Nitrobenzenesulfinyl azide

[53852-44-1] $C_6H_4N_4O_3S$

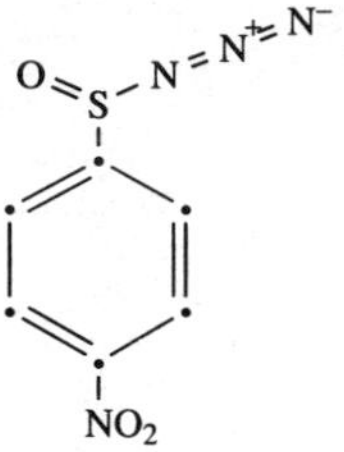

See entry SULFINYL AZIDES *See other* ACYL AZIDES

2205. 4-Nitrobenzenediazonium nitrate

[42238-29-9] $C_6H_4N_4O_5$

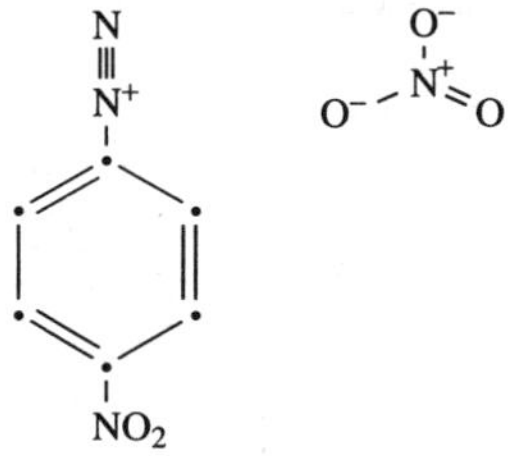

Bamberger, E., *Ber.*, 1895, **28**, 239

The isolated dry salt explodes on heating, but not on friction.

See other DIAZONIUM SALTS

2206. 2,4-Dinitrobenzenediazonium hydrogen sulfate
[64445-49-4] $C_6H_4N_4O_8S$

See entry DIAZONIUM SULFATES *See other* POLYNITROARYL COMPOUNDS

2207. 1,3-Diazidobenzene
[13556-50-8] $C_6H_4N_6$

Acids
Forster, M. O. *et al., J. Chem. Soc.*, 1907, **91**, 1953
The diazide ignites and explodes mildly with concentrated acid.
See other ORGANIC AZIDES

2208. 1,4-Diazidobenzene
[2294-47-5] $C_6H_4N_6$

Griess, P., *Ber.*, 1888, **21**, 1561
The diazide explodes very violently on heating.
See other ORGANIC AZIDES

2209. 4-Nitrobenzenediazonium azide
[68560-64-3] $C_6H_4N_6O_2$

Hantzsch, A., *Ber.*, 1903, **36**, 2058
The dry azide salt exploded violently with a brilliant flash.
See other DIAZONIUM SALTS, NON-METAL AZIDES

2210. Benzene-1,3-bis(sulfonyl azide)
[4547-69-7] $C_6H_4N_6O_4S_2$

Balabanov, G. P. *et al., Chem. Abs.*, 1969, **70**, 59427
The title compound was the most explosive of a series of 7 substituted benzenesul-
fonyl azides.
See other ACYL AZIDES

2211. 1,4-Di(1,2,3,4-thiatriazol-5-yloxy)but-2-yne
[ ] $C_6H_4N_6O_2S_2$

Banert, K. *et al., Angew. Chem. (Int.)*, 1992, **31**(7), 866

720

An explosive crystalline solid.
See other ACETYLENIC COMPOUNDS, N–S COMPOUNDS

2212. 2-Ethynylfuran
[18649-64-4] C_6H_4O

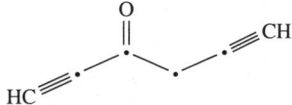

Alone, or Nitric acid
Dunlop, 1953, 77
It explodes on heating or in contact with conc. nitric acid.
See other ACETYLENIC COMPOUNDS

2213. 1,5-Hexadiyn-3-one
[66737-76-7] C_6H_4O

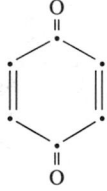

Dowd, P. *et al., Synth. Commun.*, 1978, **8**, 20
The reactive material should be vacuum distilled at ambient temperature to prevent
the near-explosive decomposition which occurs if it is heated.
See other ACETYLENIC COMPOUNDS

2214. 1,4-Benzoquinone (Cyclohexadiene-1,4-dione)
[106-51-4] $C_6H_4O_2$

Anon., *Loss Prev. Bull.*, 1977, (013), 4
Drums containing moist quinone self-heated, generating smoke and internal pressure.
Investigation showed that the moist material showed an exotherm at 40–50°C and
decomposed at 60–70°C, possibly owing to presence of impurities. Drums exposed
to direct sunlight soon attain a surface temperature of 50–60°C.
See other SELF-HEATING AND IGNITION INCIDENTS

Other reactants
Yoshida, 1980, 102
MRH values calculated for 13 combinations with oxidants are given.

2215. 2,2′-Bi-1,3-dithiole (1,4,5,8-Tetrahydro-1,4,5,8-tetrathiafulvalene)
[31366-25-3] $C_6H_4S_2$

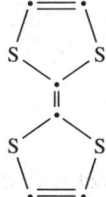

Preparative hazard
See 1,3-Dithiolium perchlorate

2216. 6-Fulvenoselone
[72443-10-8] C_6H_4Se

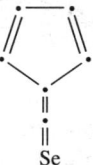

Schulz, R. *et al., Angew. Chem. (Intern. Ed.)*, 1980, **19**, 69
It polymerises explosively at −196°C.
See other POLYMERISATION INCIDENTS

2217. Phenylsilver
[5274-48-6] C_6H_5Ag

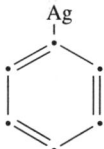

1. Houben-Weyl, 1970, Vol. 13.1, 774
2. Krause, E. *et al., Ber.*, 1923, **56**, 2064
The dry solid explodes on warming to room temperature, or on light friction or
stirring [1]. It is more stable wet with ether [2].
See other ARYLMETALS, SILVER COMPOUNDS

2218. Silver phenoxide
[61514-68-9] C_6H_5AgO

Macomber, R. S. *et al., Synth. React. Inorg. Met.-Org. Chem.*, 1977, **7**, 111–122
The dried solid product from interaction of lithium phenoxide and silver perchlorate
in benzene (probably largely silver phenoxide) exploded on gentle heating. Other
silver alkoxide derivatives were unstable.
See Silver perchlorate
See other SILVER COMPOUNDS

2219. Silver benzeneselenonate
[39254-49-4] $C_6H_5AgO_3Se$

See Benzeneselenonic acid
See other SILVER COMPOUNDS

2220. 4-Hydroxy-3,5-dinitrobenzenearsonic acid
[6269-50-7] $C_6H_5AsN_2O_8$

Phillips, M. A., *Chem. & Ind.*, 1947, 61

A heated, unstirred water-wet sludge of this analogue of picric acid exploded violently, apparently owing to local overheating in the containing flask.
See other ORGANIC ACIDS, POLYNITROARYL COMPOUNDS

2221. Phenylgold
[ ]   C₆H₅Au

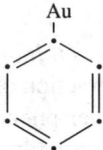

De Graaf, P. W. J. *et al., J. Organomet. Chem.*, 1977, **141**, 353
The impure solid at −70°C exploded when touched with a spatula.
See other ARYLMETALS, GOLD COMPOUNDS

2222. Dichlorophenylborane
[873-51-8]   C₆H₅BCl₂

Gerwantch, U. W. *et al., Inorg. Synth.*, 1983, **22**, 207
It is pyrophoric when hot, and extremely sensitive to hydrolysis.
See related ALKYLHALOBORANES

2223. Benzenediazonium tetrafluoroborate
[369-57-3]   C₆H₅BF₄N₂

Caesium fluoride, Difluoroamine
See Difluoramine: Benzenediazonium tetrafluoroborate, etc.
See other DIAZONIUM TETRAHALOBORATES

2224. Bromobenzene
[106-86-1]

C_6H_5Br

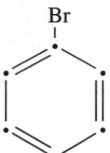

HCS 1980, 667

Bromobutane, Sodium
   See Sodium: Halocarbons (reference 7)

Other reactants
   Yoshida, 1980, 323
   MRH values calculated for 13 combinations with oxidants are given.
   See other HALOARYL COMPOUNDS

2225. Phenylmagnesium bromide
[100-58-3]

C_6H_5BrMg

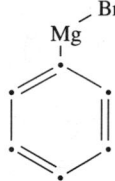

Chlorine
   See Chlorine: Phenylmagnesium bromide

Hexafluoropropene
   See Hexafluoropropene: Grignard reagents

Water
   Wosnick, J., INTERNET, 1996
   An overconcentrated ether solution of the reagent was poured down a sink. Reaction
   was violent.
   See other GRIGNARD REAGENTS

2226. 2,4-Dibromoaniline
[615-57-6]

$C_6H_5Br_2N$

Energy of decomposition (in range 150–310°C) measured as 0.24 kJ/g.
See entry THERMOCHEMISTRY AND EXOTHERMIC DECOMPOSITION (reference 2)
See other HALOANILINES

2227. Benzenediazonium tribromide
[19521-84-7]

$C_6H_5Br_3N_2$

Sorbe, 1968, 40
An explosive compound.
See other DIAZONIUM SALTS

2228. Chlorobenzene
[108-90-7]

C_6H_5Cl

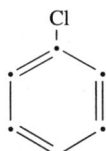

Other reactants
Yoshida, 1980, 113
MRH values calculated for 13 combinations, largely with oxidants, are given.

Phosphorus trichloride, Sodium
See Sodium: Chlorobenzene, etc.

Sodium
See Sodium: Halocarbons (reference 8)
See other HALOARYL COMPOUNDS

2229. Sodium *N*-chlorobenzenesulfonamide ('Chloramine B')
[127-52-6] $C_6H_5ClNNaO_2S$

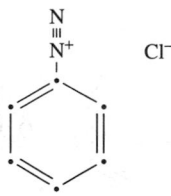

Piotrowski, T., *Chem. Abs.*, 1984, **101**, 60040
In a study of explosive potential of 20 pharmaceutical products, 'Chloramine B' was
found to present the greatest fire hazard, decomposing explosively at 185°C.
See other N-HALOGEN COMPOUNDS, N-METAL DERIVATIVES

2230. Benzenediazonium chloride
[100-34-5] $C_6H_5ClN_2$

1. Hantzsch, A. *et al., Ber.*, 1901, **34**, 3338
2. Grewer, T., *Chem. Ing. Tech.*, 1975, **47**, 233–234
3. See entry THERMOCHEMISTRY AND EXOTHERMIC DECOMPOSITION
The dry salt was more or less explosive, depending on the method of preparation
[1]. DTA examination of a 15% solution of the diazonium salt in hydrochloric acid
showed an exothermic decomposition reaction with the exotherm peak at 65°C.
Adiabatic decompositon of the solution at 20°C took 80 min to attain the maximum

temperature of 80°C [2]. The heat of decomposition Q was determined as 1.5 kJ/g [3].
See Benzenediazonium tetrachlorozincate

Potassium *O*-methyldithiocarbonate
See Thiophenol
See also DIAZONIUM SULFIDES AND DERIVATIVES
See other DIAZONIUM SALTS

2231. *N*-Chloro-4-nitroaniline
[59483-61-3] $C_6H_5ClN_2O_2$

1. Goldschmidt, S. *et al., Ber.*, 1922, **55**, 2450
2. Pyetlewski, L. L., *Rept. AD A028841*, 24, Richmond (Va.), NTIS, 1976
Though stable at −70°C, it soon decomposes explosively at ambient temperature
[1], and violently on heating slightly [2].
See other *N*-HALOGEN COMPOUNDS, NITROARYL COMPOUNDS

2232. Benzenediazonium perchlorate
[114328-95-9] $C_6H_5ClN_2O_4$

Alone, or Sulfuric acid
1. Vorlander, D., *Ber.*, 1906, **39**, 2713−2715
2. Hofmann, K. A. *et al., Ber.*, 1906, **39**, 3146−3148
The salt explodes on friction, impact or contact with sulfuric acid when still wet with
liquor [1], and is frightfully explosive when dry [2].
See other DIAZONIUM PERCHLORATES

2233. Phenylphosphonic azide chloride
[ ] $C_6H_5ClN_3OP$

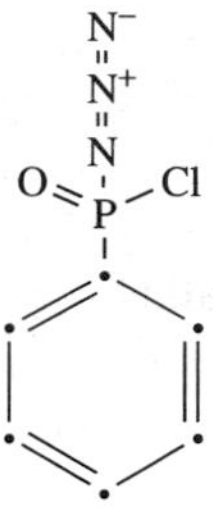

$$N^-$$
$$\|$$
$$N^+$$
$$\|$$
$$N$$
$$O = \overset{|}{P} - Cl$$

Utvary, K., *Inorg. Nucl. Chem. Lett.*, 1965, **1**, 77–78
A violent explosion occurred during distillation at 80–88°C/0.46 mbar of mixtures containing the azide chloride and the diazide
See Phenylphosphonic diazide
See other ACYL AZIDES

2234. Benzenesulfinyl chloride
[4972-29-6] C_6H_5ClOS

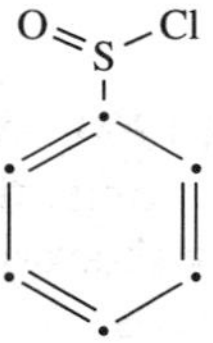

$$O = S - Cl$$

Anon., *Chem. Eng. News*, 1957, **35**, 57
A glass bottle, undisturbed for several months, exploded, probably from internal gas pressure arising from photolytic or hydrolytic decomposition.
See other ACYL HALIDES, GAS EVOLUTION INCIDENTS

2235. Benzenesulfonyl chloride
[98-09-9] $C_6H_5ClO_2S$

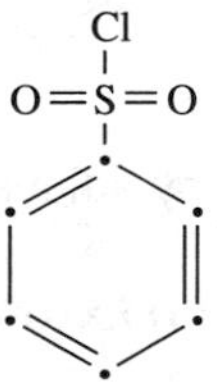

$$Cl$$
$$|$$
$$O = S = O$$

HCS 1980, 188

Anon., *Chem. Eng. News*, 1957, **35**(47), 57
A glass container exploded after shelf storage for several months, (probably from internal gas pressure arising from slow hydrolysis by atmospheric or adsorbed moisture).
See other GAS EVOLUTION INCIDENTS

Dimethyl sulfoxide
    See Dimethyl sulfoxide: Acyl halides
    See other ACYL HALIDES

2236. Perchlorylbenzene
[5390-07-8] $C_6H_5ClO_3$

Aluminium chloride
    Bruce, W. F., *Chem. Eng. News*, 1960, **38**(4), 63
    A mixture is quiescent for some time, then suddenly explodes. Many perchloryl aromatics appear to be inherently shock-sensitive.
    See other PERCHLORYL COMPOUNDS

2237. 2,3-Dichloroaniline
[608-27-5] $C_6H_5Cl_2N$

T_{ait24} was determined as 212°C by adiabatic Dewar tests, with an apparent energy of activation of 119 kJ/mol.
See entry THERMOCHEMISTRY AND EXOTHERMIC DECOMPOSITION (reference 2)
See other HALOANILINES

2238. 2,4-Dichloroaniline
[554-00-7] $C_6H_5Cl_2N$

T_{ait24} was determined as 211°C by adiabatic Dewar tests, with an apparent energy
of activation of 129 kJ/mol.
See entry THERMOCHEMISTRY AND EXOTHERMIC DECOMPOSITION (reference 2)
See other HALOANILINES

2239. 2,5-Dichloroaniline
[98-82-9] $C_6H_5Cl_2N$

T_{ait24} was determined as 216°C by adiabatic Dewar tests, with an apparent energy
of activation of 126 kJ/mol.
See entry THERMOCHEMISTRY AND EXOTHERMIC DECOMPOSITION (reference 2)
See other HALOANILINES

2240. 2,6-Dichloroaniline
[608-31-1] $C_6H_5Cl_2N$

T_{ait24} was determined as 217°C by adiabatic Dewar tests, with an apparent energy
of activation of 227 kJ/mol.

See entry THERMOCHEMISTRY AND EXOTHERMIC DECOMPOSITION (reference 2)
See other HALOANILINES

2241. 3,4-Dichloroaniline
[95-76-1] $C_6H_5Cl_2N$

HCS 1980, 374

T_{ait24} was determined as 187°C by adiabatic Dewar tests, with an apparent energy of activation of 146 kJ/mol.
See entry THERMOCHEMISTRY AND EXOTHERMIC DECOMPOSITION (reference 2)

Preparative hazard
Kotoyori, T., private comm., 1983
During pilot-scale vacuum distillation from a mild steel still of the crude hydrogenation product of 3,4-dichloronitrobenzene, normal conditions had been established as 160–170°C/43–63 mbar with steam heating at 194°C. While the seventh 8 t batch was being distilled, distillation conditions were 187°C/135 mbar, and later 194°C/165 mbar, at which point heating was stopped but the temperature and pressure kept rising. The vessel (without a pressure relief) was isolated for 2.5 h while the vacuum line was examined for blockage. Before vacuum could be reapplied, the vessel ruptured at the lid flange with eruption of white smoke and black tar. Several contributory factors for the decomposition were established later. 3,4-Dichloroaniline undergoes dehydrochlorination-polymerisation reactions, forming hydrochloride salts, at temperatures above 180°C in presence of ferric chloride, and the salts dissociate above 260°C with violent liberation of hydrogen chloride gas. Leaks of water into the still from a condenser, and of air from a hardened and cracked gasket, tended to increase the rate of formation of ferric chloride by hydrolysis of the aniline and oxidation of iron dissolved from the vessel, as well as causing increases in distillation temperature and pressure. The level of dissolved iron in the residue after the accident was around 1% (80 kg Fe), owing to the fact that distillation residues were not removed after each batch was completed. Recommended precautionary measures included fitting a pressure relief, distilling under alkaline conditions, better gasket maintenance to prevent air leaks, and removal of the residue after each batch.
See 3,4-Dichlorophenylhydroxylamine
See other HALOANILINES

2242. N,N-Dichloroaniline
[70278-00-1] $C_6H_5Cl_2N$

Goldschmidt, S., *Ber.*, 1913, **46**, 2732
The oil prepared at −20°C explodes while warming to ambient temperature.
See other N-HALOGEN COMPOUNDS

2243. 3,4-Dichlorophenylhydroxylamine
[33175-34-7] $C_6H_5Cl_2NO$

1. Tong, W. R. *et al.*, *Loss Prev.*, 1977, **11**, 71–75
2. MacNab, J. I., *Runaway Reactions*, 1981, 3/S, 1–12
A violent explosion during the plant-scale catalytic hydrogenation of 3,4-dichloronitrobenzene to the aniline (heat of hydrogenation is 523 kJ/mol, 2.72 kJ/g) was traced to thermal decomposition of one of the reaction intermediates, 3,4-dichlorophenylhydroxylamine. The latter is normally present to the extent of about 5%, but contamination of the nitro compound by unseparated wash-water containing 2% of nitrite ion had slowed the reaction and increased the proportion of hydroxylamine to 35%. Investigation of the thermal stability of the reaction mixture as hydrogenation proceeded showed that the minimum autodecomposition temperature of 260°C coincided with the maximum concentration of the hydroxylamine at 60% of the theoretical hydrogen uptake. The pure hydroxylamine begins to decompose at 80°C. Shortly before the explosion occurred, the heat input had been increased to offset the slower reaction rate and had triggered the decomposition. The sequence was confirmed experimentally, and maximum rates of increase of 6.7°C/s and 36 bar/s were recorded for the decomposition [1]. The intial temperature for exothermic decomposition of this reduction intermediate, alone or diluted with the nitro or amino compounds, was 80°C or below, with exotherms up to 60°C [2].
See 2-Chloro-5-methylphenylhydroxylamine, *also* 3,4-Dichloroaniline
See other N–O COMPOUNDS, CATALYTIC NITRO REDUCTION PROCESSES

2244. *N,N* -Dichlorobenzenesulfonamide
[473-34-7] $C_6H_5Cl_2NO_2S$

Energy of decomposition (in range 160–260°C) measured as 0.92 kJ/g.
See entry THERMOCHEMISTRY AND EXOTHERMIC DECOMPOSITION (reference 2)
See other N-HALOGEN COMPOUNDS

2245. Phenylphosphonyl dichloride
[824-72-6] $C_6H_5Cl_2OP$

Koch-Light Ltd., private comm., 1976
A bottle of the dichloride exploded violently while on a bench. No cause was established, but diffusive ingress of moisture, or an intramolecular Friedel-Crafts reaction,
both of which would form free hydrogen chloride, seem possible contributory factors.
See related ACYL HALIDES, NON-METAL HALIDES

2246. Phenylvanadium(V) dichloride oxide
[28597-01-5] $C_6H_5Cl_2OV$

Houben-Weyl, 1975, Vol. 13.3, 365
Thermal decomposition may be explosive.
See related ARYLMETALS, METAL HALIDES

2247. Phenyliodine(III) chromate ('[(Chromyldioxy)iodo]benzene')
[ ]   C₆H₅CrIO₄

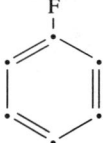

Sidgwick, 1950, 1250
It explodes at 66°C.
See other IODINE COMPOUNDS

†2248. Fluorobenzene
[462-06-6]   C₆H₅F

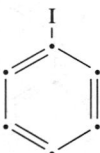

See other HALOARYL COMPOUNDS

2249. Iodobenzene
[591-50-4]   C₆H₅I

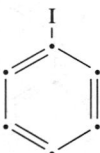

Sorbe, 1968, 66
It explodes on heating above 200°C.
See other HALOARYL COMPOUNDS, IODINE COMPOUNDS

2250. Benzenediazonium iodide
[57965-32-9] $C_6H_5IN_2$

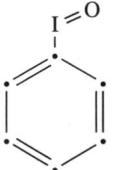

See entries DIAZONIUM SALTS, DIAZONIUM TRIIODIDES

2251. Phenyliodine(III) nitrate
[58776-08-2] $C_6H_5IN_2O_6$

Sorbe, 1968, 66
It explodes above 100°C.
See other IODINE COMPOUNDS

2252. Iodosylbenzene
[536-80-1] C_6H_5IO

1. Saltzman, A. *et al., Org. Synth.*, 1963, **43**, 60
2. Banks, D. F., *Chem. Rev.*, 1966, **66**, 255
It explodes at 210°C [1], and homologues generally explode on melting [2].
See other IODINE COMPOUNDS

2253. Iodylbenzene
[696-33-3] $C_6H_5IO_2$

1. Sharefkin, J. G. *et al., Org. Synth.*, 1963, **43**, 65
2. Banks, D. F., *Chem. Rev.*, 1966, **66**, 259

It explodes at 230°C [1]. Extreme care should be used in heating, compressing or grinding iodyl compounds, as heat or impact may cause detonation of homologues [2].
See other IODINE COMPOUNDS

2254. Benzenediazonium triiodide
[68596-95-2] $C_6H_5I_3N_2$

See entry DIAZONIUM TRIIODIDES

2255. Potassium nitrobenzene
[ ] $C_6H_5KNO_2$

See entry NITROAROMATIC–ALKALI HAZARDS; potassium radical salts

2256. Potassium 3,5-dinitro-2(1-tetrazenyl)phenoxide
[70324-35-5] $C_6H_5KN_6O_5$

Kenney, J. F., US Pat. 2 728 760, 1952

It is explosive.

See other HIGH-NITROGEN COMPOUNDS, POLYNITROARYL COMPOUNDS

2257. Potassium benzenesulfonylperoxosulfate

[ ] $C_6H_5KO_7S_2$

Davies, 1961, 65
It explodes on friction or warming.
See other DIACYL PEROXIDES

2258. Potassium citrate tri(hydrogen peroxidate)

[31266-90-1] $C_6H_5K_3O_7.3H_2O_2$

Anon., *Jahresber.*, 1978, 75
Treatment of the citrate monohydrate with a 20% excess of 70% hydrogen peroxide gave a crystalline product which, after vacuum drying at 50°C corresponded approximately to the tri(hydrogen peroxidate). It decomposed violently when disturbed with a porcelain spatula.
See other CRYSTALLINE HYDROGEN PEROXIDATES

2259. Phenyllithium

[591-51-5] C_6H_5Li

Preparative hazard
See Lithium: Bromobenzene

738

Titanium tetraethoxide

Gilman, H. *et al., J. Org. Chem.*, 1945, **10**, 507

The product of interaction at 0°C (of unknown composition) ignited in air and reacted violently with water.

See other ARYLMETALS

2260. 1,4-Benzoquinone monoimine (2,5-Cyclohexadienone-4-imine)
[3009-34-5]   C₆H₅NO

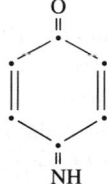

Willstätter, R. *et al., Ber.*, 1906, **37**, 4607

The solid decomposes with near-explosive violence.

See related DIENES

2261. Nitrosobenzene
[586-86-9]   C₆H₅NO

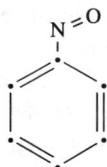

Energy of decomposition (in range 190–380°C) measured as 0.823 kJ/g.

See entry THERMOCHEMISTRY AND EXOTHERMIC DECOMPOSITION (reference 2)

See other NITROSO COMPOUNDS

2262. Nitrobenzene
[98-95-3]   C₆H₅NO₂

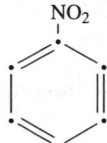

(*MCA SD-21*, 1967); *HCS 1980*, 681; 32RSC Lab Hazards Data Sheet No. 77, 1988

739

CHETAH, 1990, 182

It is marginally detonable; requiring a 50 g tetryl booster.

The heat of decomposition of nitrobenzene was determined (360–490°C) as 1.76 kJ/g by DSC, and T_{ait24} was as 280°C by adiabatic Dewar tests, with an apparent energy of activation of 216 kJ/mol.

See entry THERMOCHEMISTRY AND EXOTHERMIC DECOMPOSITION (reference 2)

Alkali
1. Hatton, J. P., private comm., 1976
2. Anon., *Sichere Chemiearb.*, 1975, **27**, 89; Bilow, W., private comm., 1976
3. Wohl, A., *Ber.*, 1899, **32**, 3846; 1901, **34**, 2444
4. Anon., *ABCM Quart. Safety Summ.*, 1953, **24**, 42
5. Bretherick, L., *Chem. & Ind.*, 1976, 576
6. *CHSO Rept. Y4648-181*, Washington (DC), OSHA, 1981

Heating a mixture of nitrobenzene, flake sodium hydroxide and a little water in an autoclave led to an explosion [1]. During the technical-scale preparation of a warm solution of nitrobenzene in methanolic potassium hydroxide (flake 90% material), accidental omission of most of the methanol led to an accelerating exothermic reaction which eventually ruptured the 6 m³ vessel. Laboratory investigation showed that no exothermic reaction occurred between potassium hydroxide and nitrobenzene, either alone, or with the full amount (3.4 vol.) of methanol, but that it did if only a little methanol were present. The residue was largely a mixture of azo- and azoxy-benzene [2]. The violent conversion of nitrobenzene to (mainly) 2-nitrophenol by heating with finely powdered and anhydrous potassium hydroxide had been described earlier [3]. Accidental substitution of nitrobenzene for aniline as diluent during large-scale fusion of benzanthrone with potassium hydroxide caused a violent explosion [4]. Related incidents involving aromatic nitro compounds and alkali are also discussed [5]. Drying wet nitrobenzene in presence of flake sodium hydroxide (probably by distilling out the nitrobenzene–water azeotrope) led to separation of finely divided solid base which remained in contact with the hot dry nitrobenzene under essentially adiabatic conditions. A violent explosion subsequently blew out a valve and part of the 10 cm steel pipeline. These conditions, when simulated in an ARC test run, also led to thermal runaway and explosion [6].

See other NITROAROMATIC–ALKALI HAZARDS

Aluminium chloride
See Aluminium chloride: Nitrobenzene, or :Nitrobenzene, Phenol

Aniline, Glycerol, Sulfuric acid
See Quinoline

Other reactants
Yoshida, 1980, 277
MRH values calculated for 2 combinations with oxidants are given.

Oxidants   MRH values show % of oxidant
See Dinitrogen tetraoxide: Nitrobenzene
     Fluorodinitromethane: Nitrobenzene

Nitric acid: Nitroaromatics, or: Nitrobenzene
Peroxodisulfuric acid: Organic liquids
Sodium chlorate: Nitrobenzene   MRH 4.94/12
Tetranitromethane: Aromatic nitro compounds   MRH 7.36/76
Uronium perchlorate: Organic materials

Phosphorus pentachloride
See Phosphorus pentachloride: Nitrobenzene

Potassium
See Potassium: Nitrogen-containing explosives

Sulfuric acid
1. *MCA Case History No. 678*
2. Silverstein, J. L. *et al., Loss Prev.*, 1981, **14**, 78
Nitrobenzene was washed with dilute (5%) sulfuric acid to remove amines, and
became contaminated with some tarry emulsion that had formed. After distillation,
the hot tarry acidic residue attacked the iron vessel with hydrogen evolution, and an
explosion eventually occurred. It was later found that addition of the nitrobenzene
to the diluted acid did not give emulsions, while the reversed addition did. A final
wash with sodium carbonate solution was added to the process [1]. During hazard
evaluation of a continuous adiabatic process for manufacture of nitrobenzene, it was
found that the latter with 85% sulfuric acid gave a violent exotherm above 200°C,
and with 69% acid a mild exotherm at 150–170°C [2].

Sulfuric acid, Sulfur trioxide
See 3-Nitrobenzenesulfonic acid

Tin(IV) chloride
Unpublished observations, 1948
Mixtures of nitrobenzene and tin(IV) chloride undergo exothermic decomposition
with gas evolution above 160°C.
See other GAS EVOLUTION INCIDENTS
See other NITROARYL COMPOUNDS

2263. 2-Nitrosophenol (1,2-Benzoquinone monoxime)
[13168-78-0] ([637-62-7]) $C_6H_5NO_2$

Energy of decomposition (in range 80–320°C) measured as 1.214 kJ/g by DSC, and
T_{ait24} was determined as the very low figure of 53°C by adiabatic Dewar tests, with
an apparent energy of activation of 462 kJ/mol.
See entry THERMOCHEMISTRY AND EXOTHERMIC DECOMPOSITION (reference 2)
See other DEFLAGRATION INCIDENTS

Alone, or Acids

Baeyer, A. *et al., Ber.*, 1902, **35**, 3037

It explodes on heating or in contact with concentrated acids (the latter possibly involving Beckmann rearrangement and/or polymerisation of the oxime form, which is effectively a 1,3-diene).

See other NITROSO COMPOUNDS, OXIMES
See related DIENES

2264. 4-Nitrosophenol (1,4-Benzoquinone monoxime)
[637-62-7] ([104-91-6])   C₆H₅NO₂

Alone, or Acids, or Alkalies

1. Milne, W. D., *Ind. Eng. Chem.*, 1919, **11**, 489
2. Kuznetsov, V., *Chem. Abs.*, 1940, **34**, 3498i
3. Anon., *Sichere Chemiearb.*, 1980, 3532, 6
4. Klais, O. *et al., Proc. 4th Int. Symp. Loss Prev. Safety Prom. Proc. Ind.*, Vol. 3, C24–C34, Rugby, IChE, 1983
5. Grewer, T. *et al., Hazards from Pressure*, IChE Symp. Ser. No. 102, 1–9, Oxford, Pergamon, 1987

Stored barrels heated spontaneously and caused a fire. Contamination of the bulk material by acid or alkali may cause ignition [1]. The tendency to spontaneous ignition was found to be associated with the presence of nitrates in the sodium nitrite used in preparation. A diagnostic test for this tendency is to add sulfuric acid, when impure material will effervesce or ignite. The sodium salt is more suitable for storage purposes [2] (*but see* Sodium 4-nitrosophenoxide). Compressed material has a greater tendency to self-ignition than the loose solid, and a 20–30 kg quantity of centrifuged material ignited in a drum located near a steam heater [3]. Material for laboratory use is now supplied wet with 30–40% water. The heat of decomposition was determined as 1.2 kJ/g, and the pressure rise occurring during exothermic decomposition is low [4]. However, the deflagration rate for the solid of 8 cm/min is relatively high [5].

See entry THERMOCHEMISTRY AND EXOTHERMIC DECOMPOSITION
See other NITROSO COMPOUNDS
See related OXIMES

2265. 2-Nitrophenol
[88-75-5] $C_6H_5NO_3$

Energy of decomposition (in range 260–365°C) measured as 2.15 kJ/g by DSC, and T_{ait24} was determined as 182°C by adiabatic Dewar tests, with an apparent energy of activation of 135 kJ/mol.
See entry THERMOCHEMISTRY AND EXOTHERMIC DECOMPOSITION (reference 2)

Chlorosulfuric acid
  See 4-Hydroxy-3-nitrobenzenesulfonyl chloride

Potassium hydroxide
  491M, 1975, 342
  The molten phenol reacts violently with commercial 85% potassium hydroxide pellets (possibly involving formation of the *aci-o*-nitroquinonoid salt).
  See NITROAROMATIC–ALKALI HAZARDS *See other* NITROARYL COMPOUNDS

2266. 3-Nitrophenol
[554-81-7] $C_6H_5NO_3$

T_{ait24} was determined as 213°C by adiabatic Dewar tests, with an apparent energy of activation of 148 kJ/mol.
See entry THERMOCHEMISTRY AND EXOTHERMIC DECOMPOSITION (reference 2)
See other NITROARYL COMPOUNDS

2267. 4-Nitrophenol
[100-02-7] $C_6H_5NO_3$

HCS 1980, 688

1. Grewer, T. *et al.*, *Proc. 3rd Int. Symp. Loss Prev. Safety Prom. Proc. Ind.*, 657–664, Basle, SSCI, 1980
2. Grewer, T. *et al.*, *Exothermic Decomposition*, Technical Report 01VD 159/0329 for Federal German Ministry for Res. Technol., Bonn. 1986

There is a high rate of pressure increase during exothermic decomposition [1]. Energy of decomposition (in range 240–440°C) measured as 1.58 kJ/g by DSC, and T_{ait24} was determined as 185°C by adiabatic Dewar tests, with an apparent energy of activation of 117 kJ/mol [2].

Diethyl phosphite
See Diethyl phosphite: 4-Nitrophenol

Potassium hydroxide
Solid mixtures of 4-nitrophenol with potassium hydroxide (1:1.5 mol) readily deflagrate, at the rapid rate of 30 cm/min.
See entries METAL NITROPHENOXIDES, PRESSURE INCREASE IN EXOTHERMIC DECOMPOSITION (reference 3)
See Potassium 4-nitrophenoxide
See other DEFLAGRATION INCIDENTS
See other NITROARYL COMPOUNDS

2268. 3-Nitrobenzenesulfonic acid
[984-47-5] $C_6H_5NO_5S$

Sulfuric acid, Sulfur trioxide
1. *MCA Case History No. 1482*
2. *MCA Case History No. 944*
3. *See entry* SELF-ACCELERATING REACTIONS
4. Nolan, 1983, Case history 102
5. Bergroth, K., *Loss Prev. Bull.*, 1993, (109), 1

The acid is prepared by sulfonation of nitrobenzene with oleum, and the reaction product consists essentially of a hot solution of the acid in sulfuric acid. A completed 270 l batch exploded violently after hot storage at ~150°C for several hours. An exotherm develops at 145°C, and the acid is known to decompose at ~200°C [1]. A similar incident arose from water leaking from a cooling coil into the fuming sulfuric acid reaction medium, which caused an exotherm to over 150°C and subsequent violent decomposition [2]. Detailed examination of the thermal decomposition of the acid shows that it is much slower for the isolated acid than for the reaction mass, and that the concentration of sulfur trioxide in the oleum used for sulfonation bears directly upon the decomposition rate. Care must be exercised when interpreting short-term DSC test results, because the self-acceleration features are not immediately evident. A Q value of 1 kJ/g of reaction mass after sulfonation with 25% oleum was found experimentally [3]. A 2000 l batch of the acid prepared with 20% excess of 65% oleum had been cooled to 70°C prior to being blown over into a tank for treatment with sodium carbonate. Application of air pressure at 2.75 bar to the reactor failed to transfer the product, so the transfer lines were isolated and steamed at 2 bar to clear them. On reconnecting the lines, the condensed water reacted with the excess oleum, leading to exothermic decomposition of the acid, rupture of the vessel and a fire [4]. An explosion in a plant resulted because of crystallisation from the oleum sulfonation mixture in a control valve, which failed. Manual operation led to steam flowing back from the quench to a nitration-product holding tank, which overheated and exploded [5].

See other NITROARYL COMPOUNDS, ORGANIC ACIDS, SULFONATION INCIDENTS

2269. Benzotriazole
[95-14-7] $C_6H_5N_3$

1. Anon., *Chem. Eng. News*, 1956, **34**, 2450
2. *CHETAH*, 1990, 182
3. Stull, 1977, 22–23

A 1 t batch decomposed exothermally, then detonated at 220°C, during distillation at 160°C/2.5 mbar. No cause was found, and similar batches had previously distilled satisfactorily. The multiple N–N bonding would tend to cause instability in the molecule, particularly in presence of heavy metals, but these were absent in this case [1]. It is shock sensitive (probably not very) [2]. Benzotriazole is an endothermic compound (ΔH°_f (s) +249.8 kJ/mol, 2.1 kJ/g) and this energy on release would attain an adiabatic decomposition temperature approaching 1100°C, with an 18 bar pressure increase in the closed system [3].

See other ENDOTHERMIC COMPOUNDS, HIGH-NITROGEN COMPOUNDS, TRIAZOLES

2270. 1,2,4-Triazolo[4,3-a]pyridine silver nitrate
[ ] $C_6H_5N_3 \cdot AgNO_3$

Fargher, R. G. *et al.*, *J. Chem. Soc.*, 1915, **107**, 696

The base [274-80-6] forms an addition compound with silver nitrate which decomposes at 228°C in a m.p. capillary, but explodes when heated in larger amount.
See other AMMINEMETAL NITRATES, SILVER COMPOUNDS

2271. Phenyl azide (Azidobenzene)
[622-37-7] $C_6H_5N_3$

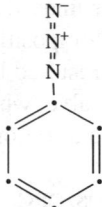

1. Lindsay, R. O. *et al., Org. Synth.*, 1955, Coll. Vol. 3, 710
2. Grewer, T. *et al., Exothermic Decomposition*, Technical Report 01VD 159/0329 for Federal German Ministry for Res. Technol., Bonn. 1986
3. Lee, S. W. *et al., Inorg. Chem.*, 1988, **27**, 1213–1219

Though distillable at considerably reduced pressure, phenyl azide explodes when heated at ambient pressure, and occasionally at lower pressures [1]. The energy of decomposition has been determined (130–260°C) as 1.36 kJ/g by DSC, and T_{ait24} was determined as 87°C by adiabatic Dewar tests, with an apparent energy of activation of 119 kJ/mol [2]. It is best stored as a 3M solution in toluene. Exposure to heavy metals should be avoided [3].

Lewis acids
Boyer, J. H. *et al., Chem. Rev.*, 1954, **54**, 29
The ready decomposition of most aryl azides with sulfuric acid and Lewis acids may be vigorous or violent, depending on structure and conditions. In absence of a diluent (carbon disulfide), phenyl azide and aluminium choride exploded violently.
See other ORGANIC AZIDES

2272. 1-Hydroxybenzotriazole
[2592-95-2] $C_6H_5N_3O$

Anon., *CISHC Chem. Safety Summ.*, 1978, **49**, 52–53
A sample of the compound (which can isomerise to two *N*-oxide forms) was inadvertently heated, probably to well over 160°C, during vacuum drying and exploded violently. Several salts have been reported as powerfully explosive.

Copper(II) nitrate
See Copper(II) nitrate: Hydroxybenzotriazole, Ethanol
See other HIGH-NITROGEN COMPOUNDS, N–O COMPOUNDS, TRIAZOLES

2273. Benzenesulfinyl azide
[21230-20-6]

C$_6$H$_5$N$_3$OS

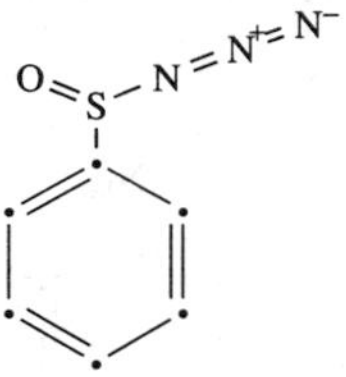

See entry SULFINYL AZIDES *See other* ACYL AZIDES

2274. Benzenesulfonyl azide
[938-10-3]

C$_6$H$_5$N$_3$O$_2$S

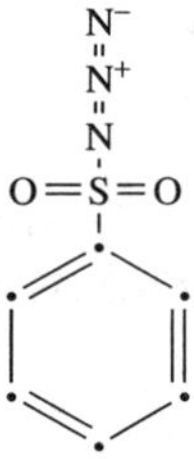

Dermer, O. C. *et al., J. Amer. Chem. Soc.*, 1955, **77**, 71
Whereas the pure azide decomposes rapidly but smoothly at 105°C, the crude material explodes violently on heating.
See other ACYL AZIDES

2275. Benzenediazonium nitrate
[619-96-7]

C$_6$H$_5$N$_3$O$_3$

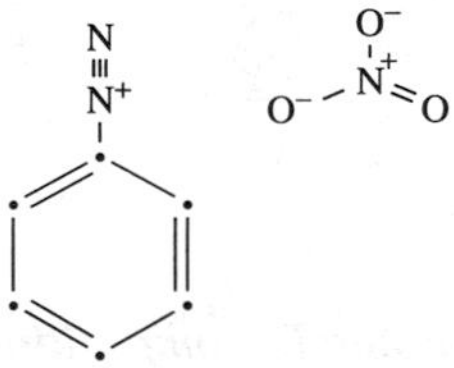

1. Urbanski, 1967, Vol. 3, 201
2. Stull, 1977, 19

Although not a practical explosive, the isolated salt is highly sensitive to friction and impact, and explodes at 90°C [1]. It is, like most diazonium salts, endothermic, with ΔH°_f (s) +200.8 kJ/mol, 1.2 kJ/g [2].

See other DIAZONIUM SALTS, ENDOTHERMIC COMPOUNDS

2276. Dinitroanilines (: all isomers except 3,5-)

2,3- [602-03-9]; 2,4- [97-02-9]; 2,5- [619-18-1]; 2,6- [606-22-4]; $C_6H_5N_3O_4$
3,4- [610-41-3]

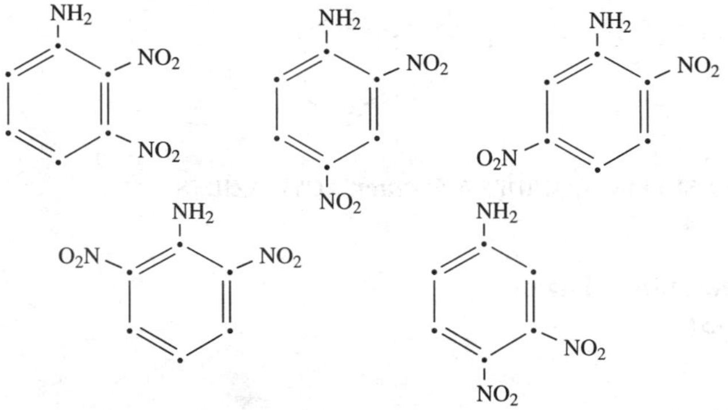

Chlorine, Hydrochloric acid
See Hydrochloric acid: Chlorine, Dinitroanilines
See other POLYNITROARYL COMPOUNDS

2277. 2,4-Dinitroaniline

[97-02-9] $C_6H_5N_3O_4$

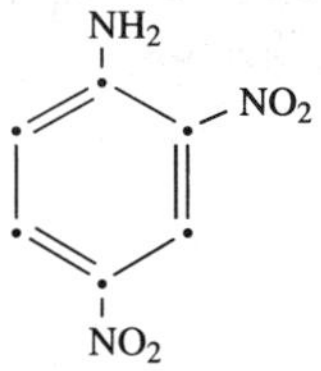

HCS 1980, 437

Preparative hazard
1. Sorbe, 1968, 151
2. *CHETAH*, 1990, 183
3. Grewer, T. *et al., Exothermic Decomposition*, Technical Report 01VD 159/0329 for Federal German Ministry for Res. Technol., Bonn. 1986

It is a compound with high fire and explosion hazards [1], showing fast flame propagation on heating the powder moderately. Not sensitive to mechanical shock,

it can be exploded by a detonator [2]. T_{ait24} was determined as 217°C by adiabatic Dewar tests, with an apparent energy of activation of 129 kJ/mol [3].
See 1-Chloro-2,4-dinitrobenzene: Ammonia
See entry HIGH RATE DECOMPOSITION

Charcoal
Leleu, *Cahiers*, 1980, (99), 278
A mixture with powdered charcoal smoked above 280°C and ignited at 350°C.
See other POLYNITROARYL COMPOUNDS

2278. 2-Amino-4,6-dinitrophenol
[96-91-3]

$C_6H_5N_3O_5$

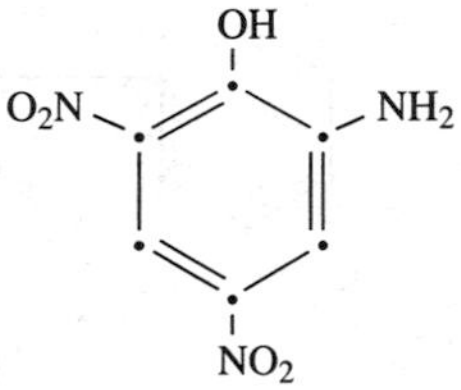

Sorbe, 1968, 152
This partially reduced derivative of picric acid, 'picramic acid', explodes very powerfully when dry.
See other POLYNITROARYL COMPOUNDS

2279. *O*-(2,4-Dinitrophenyl)hydroxylamine
[17508-17-7]

$C_6H_5N_3O_5$

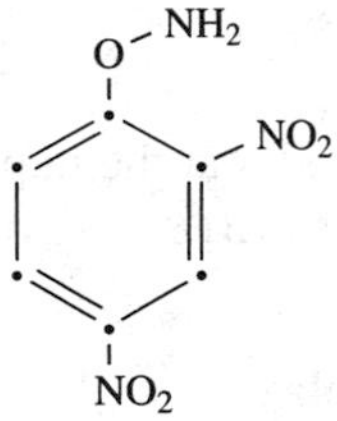

Potassium hydride
Radhakrishna, A. S. *et al., J. Org. Chem.*, 1979, **44**, 4838
Interaction in THF gives 2,4-dinitrophenol and ammonia, and on one occasion a violent detonation occurred; potassium dinitrophenoxide may have been involved.
See Sodium 2,4-dinitrophenoxide
See other N–O COMPOUNDS, POLYNITROARYL COMPOUNDS

2280. 2-Acetylamino-3,5-dinitrothiophene
[51249-08-2] $C_6H_5N_3O_5S$

Fast flame propagation on heating the powder moderately.
See entry HIGH RATE DECOMPOSITION *See other* POLYNITROARYL COMPOUNDS

2281. Benzotriazolium 1-nitroimidate
[ ] $C_6H_5N_5O_2$

A dangerous impact-sensitive solid which explodes at the m.p., 74°C.
See entry N-AZOLIUM NITROIMIDATES

2282. Benzotriazolium 2-nitroimidate
[52096-20-5] $C_6H_5N_5O_2$

A dangerous solid which explodes at the m.p., 84°C.
See entry N-AZOLIUM NITROIMIDATES

2283. 1-(Nitroamino)benzotriazole
[52096-18-1] $C_6H_5N_5O_2$

This is the structure assigned by *Chem. Abs.* to the claimed benzotriazolium nitroimidate below. The editor suspects they are right and that the nitroamine structure should also be considered for the 2-isomer. The *aci*-nitro structures are also possible.

See Benzotriazolium 1-nitroimidate
See also N-NITRO COMPOUNDS

2284. Phenylphosphonic diazide

[ ]   C$_6$H$_5$N$_6$OP

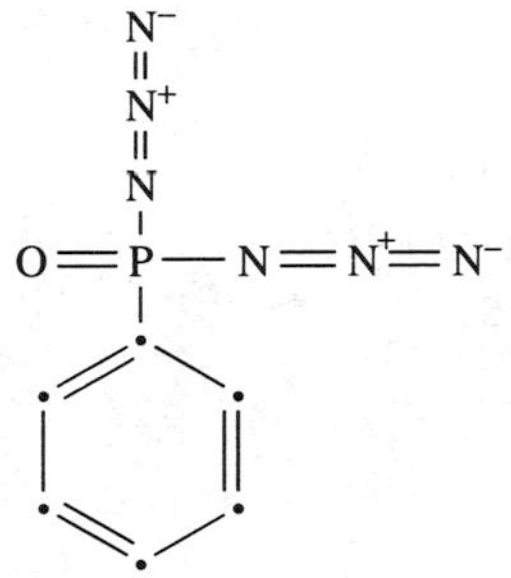

$$
\begin{array}{c}
N^- \\
\| \\
N^+ \\
\| \\
N \\
| \\
O{=}P{-}N{=}N^+{=}N^-
\end{array}
$$

Baldwin, R. A., *J. Org. Chem.*, 1965, **30**, 3866

Although a small sample had been distilled at 72–74°C/0.13 mbar without incident, it decomposed vigorously on exposure to flame, and the liquid was impact-sensitive, exploding vigorously. The material can be prepared and handled safely in pyridine solution.

See other ACYL AZIDES

2285. Phenylthiophosphonic diazide

[ ]   C$_6$H$_5$N$_6$PS

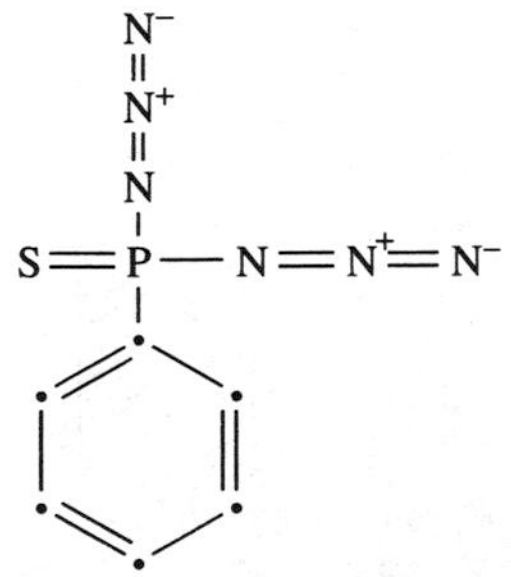

$$
\begin{array}{c}
N^- \\
\| \\
N^+ \\
\| \\
N \\
| \\
S{=}P{-}N{=}N^+{=}N^-
\end{array}
$$

Baldwin, R. A., *J. Org. Chem.*, 1965, **30**, 3866

During an attempt to distil crude material at 80°C/0.13 mbar, a violent explosion occurred. The material can be prepared and handled safely in pyridine solution.

See other ACYL AZIDES

2286. Phenylthallium diazide
[ ] $C_6H_5N_6Tl$

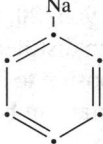

Pascal, 1961, Vol. 6, 1019
It decomposes violently at 200°C.
See related METAL AZIDES

2287. Phenylsodium
[1623-99-0] C_6H_5Na

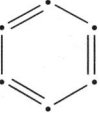

Preparative hazard
Sorbe, 1968, 86
Suspensions containing phenylsodium may ignite and burn violently in moist air.
See Sodium: Halocarbons (reference 8)
See other ARYLMETALS

†2288. Benzene
[71-43-2] C_6H_6

(*MCA SD-2*, 1960); *NSC 308*, 1979; *FPA H18*, 1973; *HCS 1980*, 187; *RSC Lab. Hazard Data Sheet No. 20*, 1984

Arsenic pentafluoride, Potassium methoxide
See Arsenic pentafluoride: Benzene, etc.

Diborane
See Diborane: Benzene, etc.

Hydrogen, Raney nickel
Anon., *Loss Prev. Bull.*, 1979, (028), 91–92
Hydrogenation of benzene to cyclohexane was effected in a fixed bed reactor at 210–230°C, but a fall in conversion was apparent. Increasing the bed temperature by 10°C and the hydrogen flow led to a large increase in reaction rate which the interbed cooling coils could not handle, and an exotherm to 280°C developed, with a hot spot of around 600°C which bulged the reactor wall.
See other HYDROGENATION INCIDENTS

Interhalogens   MRH value shows % of interhalogen
See Bromine trifluoride: Halogens, etc.
Bromine pentafluoride: Hydrogen-containing materials   MRH 3.85/73
Iodine pentafluoride: Benzene
Iodine heptafluoride: Organic solvents

Other reactants
Yoshida, 1980, 334
MRH values calculated for 17 combinations with oxidants are given.

Oxidants   MRH values show % of oxidant
See Dioxygen difluoride: Various materials
Dioxygenyl tetrafluoroborate: Organic materials
Nitric acid: Hydrocarbons   MRH 8.15/82
Nitryl perchlorate: Organic solvents
Oxygen (Liquid): Hydrocarbons   MRH 9.83/75
Ozone: Aromatic compounds, or: Benzene, etc.   MRH Benzene 12.05/75
Permanganic acid: Organic materials
Peroxodisulfuric acid: Organic liquids
Peroxomonosulfuric acid: Aromatic compounds
Silver perchlorate   MRH 3.51/91
Sodium peroxide: Organic liquids, etc.   MRH 2.76/93
Ozone: Aromatic compounds, or: Benzene, etc.   MRH Benzene 12.05/75

Uranium hexafluoride
See Uranium hexafluoride: Aromatic hydrocarbons

2289. Benzvalene
[659-85-8] C_6H_6

Katz, T. J. *et al., J. Amer. Chem. Soc.*, 1971, **93**, 3782

When isolated, this highly strained valence tautomer of benzene exploded violently when scratched. It may be handled safely in solution in ether.

See other STRAINED-RING COMPOUNDS

2290. 1,3-Hexadien-5-yne
[10420-90-3] C_6H_6

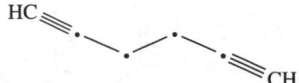

Brandsma, 1971, 132

Distillation of this rather unstable material at normal pressure involves risk of an explosion.

See other ALKYNES, DIENES

†2291. 1,5-Hexadien-3-yne (Divinylacetylene)
[821-08-9] C_6H_6

$$H_2C{=}CHC{\equiv}CCH{=}CH_2$$

1. Nieuwland, J. A. *et al., J. Amer. Chem. Soc.*, 1931, **53**, 4202
2. Cupery, M. E. *et al., J. Amer. Chem. Soc.*, 1934, **56**, 1167–1169
3. Handy, C. T. *et al., J. Org. Chem.*, 1962, **27**, 41

The extreme hazards involved in handling this highly reactive material are stressed. Freshly distilled material rapidly polymerises at ambient temperature to produce a gel and then a hard resin. These products can neither be distilled nor manipulated without explosions ranging from rapid decomposition to violent detonation. The hydrocarbon should be stored in the mixture with catalyst used to prepare it, and distilled out as required [1]. The dangerously explosive gel is a peroxidic species not formed in absence of air, when some 1,2-di(3-buten-1-ynyl)cyclobutane is produced by polymerisation [2]. The dienyne reacts readily with atmospheric oxygen, forming an explosively unstable polymeric peroxide. Equipment used with it should be rinsed with a dilute solution of a polymerisation inhibitor to prevent formation of unstable residual films. Adequate shielding of operations is essential [3].

See other ALKYNES, DIENES, POLYPEROXIDES

2292. 1,5-Hexadiyne
[628-16-0] C_6H_6

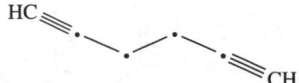

Grignard, 1935, Vol. 3, 181

At the end of distillation at 86°C/1 bar, the last drop ignited. It explodes if heated to 100–120°C, and the copper(I) salt also ignites.
See other ALKYNES

2293. 2,4-Hexadiyne
[2809-69-0] C_6H_6

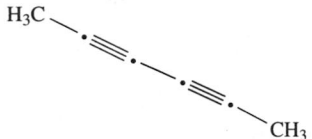

1. Bohlman, F., *Chem. Ber.*, 1951, **84**, 785–794
2. Armitage, J. B., *J. Chem. Soc.*, 1951, 45
Filtration through alumina, which prevents explosion on subsequent distillation at 55°C/13 μbar [1], or at ambient pressure [2], presumably removes peroxidic impurities.
See other ALKYNES

2294. Prismane
[650-42-0] C_6H_6

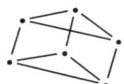

Katz, T. J. *et al., J. Amer. Chem. Soc.*, 1973, **95**, 2739
This tetracyclic isomer of benzene is an explosive liquid.
See other STRAINED-RING COMPOUNDS

2295. Iodosylbenzene tetrafluoroborate
[120312-56-3] C_6H_6BF 4IO

$$PhIOH^+BF_4^-$$

van Look G., *Chem. Eng. News*, 1989 **67**(30), 2
A compound with tetrafluoroboric acid, previously claimed to be a useful safe oxidant, proved not to be on preparation at 0.5 molar scale. It exploded spontaneously while drying in vacuo at room temperature. *Chem. Abs.* (**112** 98095p) identifies it with this title compound, which was claimed to be too unstable to explode by the original users. It seems more likely that the explosive was the "anhydride", oxybisphenyliodonium tetrafluoroborate. Several related compounds have also proved to be explosive.
See Oxybisphenyliodonium bistetrafluoroborate, also Iodosylbenzene
See other IODINE COMPOUNDS

2296. 4-Bromoaniline
[106-40-1]

C_6H_6BrN

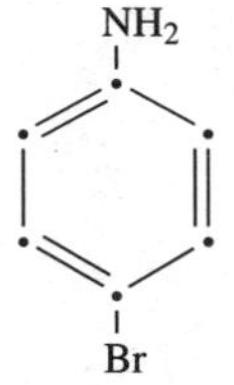

HCS 1980, 220

Energy of decomposition (in range 220–270°C) measured as 0.428 kJ/g.
See entry THERMOCHEMISTRY AND EXOTHERMIC DECOMPOSITION (reference 2)

Nitrous acid
See 4-Bromobenzenediazonium salts
See other HALOANILINES

2297. 2-Bromo-4-methylpyridine *N*-oxide
[17117-12-3]

C_6H_6BrNO

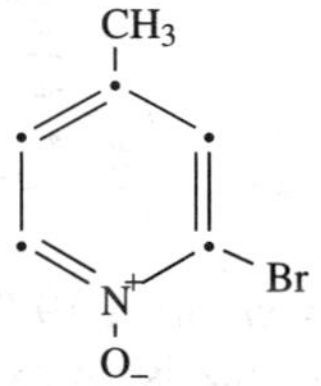

Preparative hazard
See 3-Chloroperoxybenzoic acid: 2-Bromo-4-methylpyridine
See other N-OXIDES

2298. 2,3-Dibromo-5,6-epoxy-7,8-dioxabicyclo[2.2.2]octane
[56411-66-6]

$C_6H_6Br_2O_3$

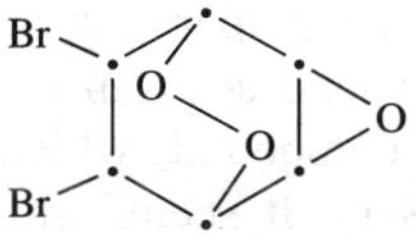

Sasaoka, M. *et al., J. Org. Chem.*, 1979, **44**, 373 (footnote 10)
It explodes on heating.
See other CYCLIC PEROXIDES, 1,2-EPOXIDES

2299. (Hydroxy)(oxo)(phenyl)-λ^3-iodanium perchlorate (Iodylbenzene perchlorate)
[101672-14-4]

$C_6H_6ClIO_6$

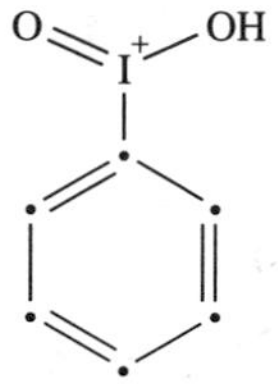

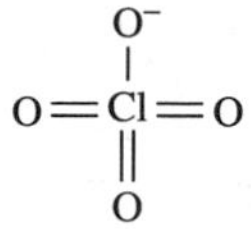

Masson, I., *J. Chem. Soc.*, 1935, 1674
A small sample exploded very violently while still damp.
See other IODINE COMPOUNDS, NON-METAL PERCHLORATES

2300. 2-Chloroaniline
[95-51-2]

C_6H_6ClN

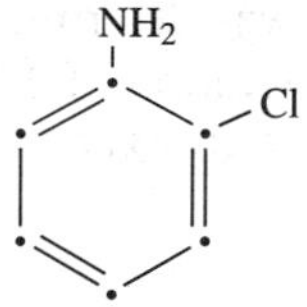

HCS 1980, 295

Heat of decomposition was determined as 0.41 kJ/g.
See entry THERMOCHEMISTRY AND EXOTHERMIC DECOMPOSITION

Nitrous acid
  See 2-Chlorobenzenediazonium salts
  See other HALOANILINES

2301. 3-Chloroaniline
[108-42-9]

C_6H_6ClN

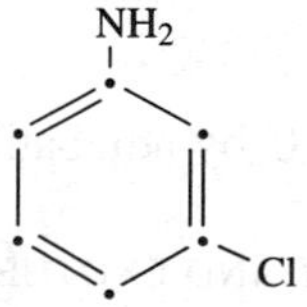

Heat of decomposition was determined as zero.
See entry THERMOCHEMISTRY AND EXOTHERMIC DECOMPOSITION

Nitrous acid
  See 3-Chlorobenzenediazonium salts
  See other HALOANILINES

2302. 4-Chloroaniline
[106-47-8]

C_6H_6ClN

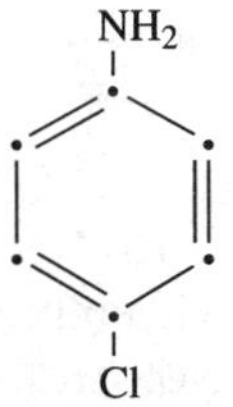

HCS 1980, 294

Energy of exothermic decomposition in range 210–400°C was measured as 0.63 kJ/g by DSC, and T_{ait24} was determined as 200°C by adiabatic Dewar tests, with an apparent energy of activation of 124 kJ/mol.
See entry THERMOCHEMISTRY AND EXOTHERMIC DECOMPOSITION (reference 2)

Nitrous acid
  See DIAZONIUM SULFIDES AND DERIVATIVES (references 8,9) *See other* HALOANILINES

2303. 4-Chloro-2-aminophenol
[95-85-2]

C_6H_6ClNO

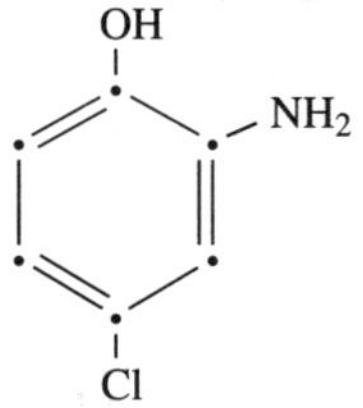

T_{ait24} was determined as 110°C by adiabatic Dewar tests, with an apparent energy of activation of 140 kJ/mol.
See entry THERMOCHEMISTRY AND EXOTHERMIC DECOMPOSITION (reference 2)
See other HALOANILINES

2304. 4-Aminobenzenediazonium perchlorate
[ ] $C_6H_6ClN_3O_4$

Hofmann, K. A. *et al., Ber.*, 1910, **43**, 2624
Extremely explosive.
See other DIAZONIUM PERCHLORATES

2305. 2,5-Bis(chloromethyl)thiophene
[28569-48-4] $C_6H_6Cl_2S$

Griffing, J. M. *et al., J. Amer. Chem. Soc.*, 1948, **70**, 3417
It polymerises at ambient temperature and must therefore be stored cold.
See other 2-HALOMETHYL-FURANS OR -THIOPHENES

2306. 1,2,3,4,5,6-Hexachlorocyclohexane
[608-73-1] $C_6H_6Cl_6$

Dimethylformamide
'DMF Brochure', Billingham, ICI, 1965
There is a potentially dangerous reaction of hexachlorocyclohexane with DMF
in presence of iron. The same occurs with carbon tetrachloride, but not with
dichloromethane or 1,2-dichloroethane under the same conditions.
See N,N-Dimethylacetamide: Halogenated compounds
See related HALOALKANES
See other CATALYTIC IMPURITY INCIDENTS

759

2307. *N*-Ethylheptafluorobutyramide
[70473-76-6] **$C_6H_6F_7NO$**

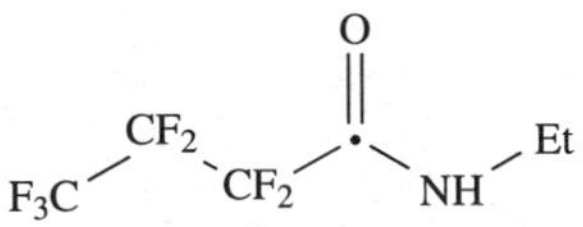

Lithium tetrahydroaluminate
See Lithium tetrahydroaluminate: Fluoroamides

2308. Potassium bis(propynyl)palladate
[ ] **$C_6H_6K_2Pd$**

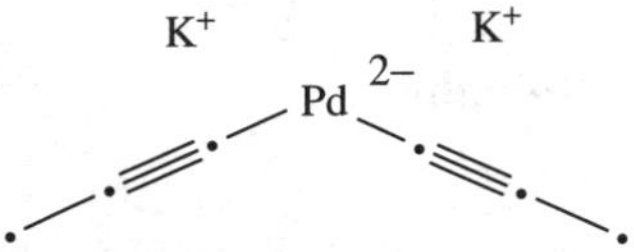

Air, or Water
Immediately pyrophoric in air, and explosive decomposition with aqueous reagents. The sodium salt is similar.
See entry COMPLEX ACETYLIDES

2309. Potassium bis(propynyl)platinate
[ ] **$C_6H_6K_2Pt$**

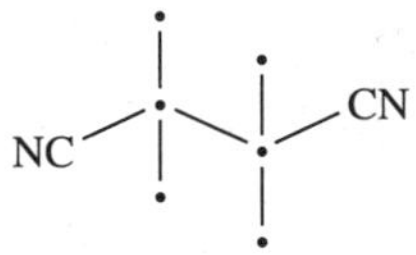

Air, or Water
Pyrophoric in air, explosive decomposition with water, and the sodium salt is similar.
See entry COMPLEX ACETYLIDES

2310. 1,4-Benzoquinone diimine (1,4-Diimido-2,5-cyclohexadiene)
[4377-73-5] $C_6H_6N_2$

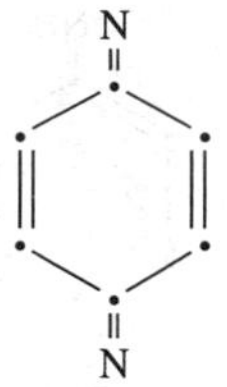

Acids

Willstätter, R. *et al., Ber.*, 1906, **37**, 4607

The unstable solid decomposes explosively in contact with conc. hydrochloric acid or sulfuric acid.

See related DIENES

2311. 1,4-Dicyano-2-butene
[1119-69-3] $C_6H_6N_2$

MCA Case History No. 1747

Overheating in a vacuum evaporator initiated accelerating polymerisation-decomposition of (endothermic) dicyanobutene, and the rapid gas evolution eventually caused pressure-failure of the process equipment.

See other CYANO COMPOUNDS, ENDOTHERMIC COMPOUNDS, GAS EVOLUTION INCIDENTS, POLYMERISATION INCIDENTS

2312. Bis(acrylonitrile)nickel(0)
[12266-58-9] $C_6H_6N_2Ni$

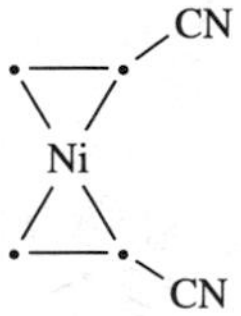

Schrauzer, G. N., *J. Amer. Chem. Soc.*, 1959, **81**, 5310

Pyrophoric in air.

See other CYANO COMPOUNDS, PYROPHORIC MATERIALS

2313. 2-Nitroaniline
[88-74-4] $C_6H_6N_2O_2$

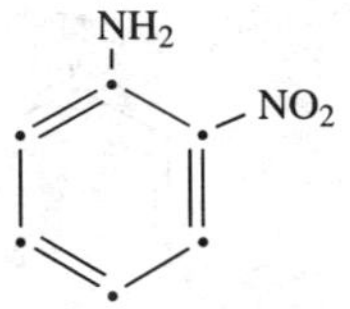

NH$_2$
NO$_2$

Energy of decomposition (in range 280–380°C) measured as 1.81 kJ/g by DSC, and T_{ait24} was determined as 223°C by adiabatic Dewar tests, with an apparent energy of activation of 171 kJ/mol.
See entry THERMOCHEMISTRY AND EXOTHERMIC DECOMPOSITION (reference 2)

Preparative hazard
Duck, M. W. *et al., Plant/Oper. Progr.*, 1982, **1**, 19–26
Exothermic decomposition of 2-nitroaniline in chemical processes was studied by DSC and ARC techniques. The stability in reaction mixtures was markedly less than for the pure, isolated compound.
See 2-Chloronitrobenzene: Ammonia

Hexanitroethane
See Hexanitroethane: Organic compounds

Magnesium, Nitric acid
See Nitric acid: Magnesium, 2-Nitroaniline

Nitrous acid
See 2-Nitrobenzenediazonium salts

Sulfuric acid
See Sulfuric acid: Nitroaryl bases
See other NITROARYL COMPOUNDS

2314. 3-Nitroaniline
[99-09-2] $C_6H_6N_2O_2$

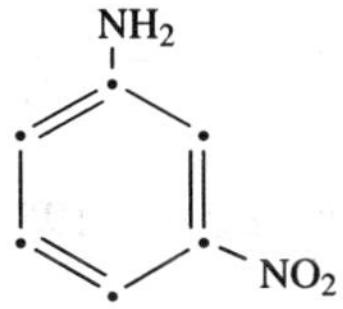

NH$_2$
NO$_2$

Hakl, J. *et al., Runaway Reactions*, 1981, Paper 3/L, 4, 10

762

Energy of decomposition (in range 280–380°C) measured as 1.882 kJ/g by DSC, and T_{ait24} was determined as 213°C by adiabatic Dewar tests, with an apparent energy of activation of 149 kJ/mol.

Initial exothermic decomposition occurs at 247°C.

See entry THERMOCHEMISTRY AND EXOTHERMIC DECOMPOSITION (reference 2)

Ethylene oxide
See Ethylene oxide: 3-Nitroaniline

Nitric acid, Sulfuric acid
Vanderah, D. J., *J. Energ. Mater.*, 1990, **58**(5), 378
Nitration to 2,3,4,6-tetranitroaniline proved unsafc in mixed acid at 30–50°C. Efficient and safe nitration at 20–30°C was obtained by dissolving the substrate in sulfuric acid, then adding mixed acid (made with oleum) thereto.
See other NITRATION INCIDENTS
See other NITROARYL COMPOUNDS

2315. 4-Nitroaniline
[100-01-6] $C_6H_6N_2O_2$

NH_2

NO_2

HCS 1980, 680

Energy of decomposition (in range 280–380°C) measured as 1.882 kJ/g by DSC, and T_{ait24} was determined as 217°C by adiabatic Dewar tests, with an apparent energy of activation of 160 kJ/mol.
See entry THERMOCHEMISTRY AND EXOTHERMIC DECOMPOSITION (reference 2)

Nitrous acid
See 4-Nitrobenzenediazonium salts

Sodium hydroxide
See Sodium 4-nitrophenoxide

Sulfuric acid
See Sulfuric acid: Nitroaryl bases
See other NITROARYL COMPOUNDS

2316. 2-Amino-5-nitrophenol
[121-88-0] $C_6H_6N_2O_3$

Nitrous acid

Anon., *CISHC Chem. Safety Summ.*, 1978, **49**, 53

An explosion in a poorly maintained and infrequently cleaned rubber lined vessel used for diazotisation of 2-amino-5-nitrophenol was attributed to decomposition of traces of a diazonium derivative (possibly 4-nitrobenzenediazonium-2-oxide) which had accumulated and dried out on the underside of the lid.

See ARENEDIAZONIUM OXIDES *See other* NITROARYL COMPOUNDS

2317. 3-Methyl-4-nitropyridine *N*-oxide
[1074-98-2] $C_6H_6N_2O_3$

Preparative hazard

1. Ross, W. C. J., *J. Chem. Soc. C*, 1966, 1816
2. Kuilen, A. v. d. *et al.*, *Chem. Eng. News*, 1989, **67**(5), 2
3. Baumgarten, H. E., *J. Amer. Chem. Soc.*, 1957, **79**, 3145

An explosion occurred during attempted preparation of the title compound following a published procedure. This involved conversion of 3-methylpyridine to the *N*-oxide by treatment with hydrogen peroxide in glacial acetic acid, vacuum evaporation of the volatiles, and then addition of sulfuric and nitric acids, followed by slight warming to effect nitration [1]. Soon after warming the final mixture, a vigorous reaction set in, followed by an explosion. This was attributed to the presence of residual conc. peracetic acid in the *N*-oxide, and its subsequent detonation by the nitration exotherm [2]. An alternative procedure for preparing *N*-oxides precludes formation of peracetic acid [3].

See other N-OXIDES, NITRATION INCIDENTS

2318. Benzenediazonium hydrogen sulfate
[36211-73-1] $C_6H_6N_2O_4S$

Sorbe, 1968, 122
The salt explodes at about 100°C, but not on impact.
See other DIAZONIUM SULFATES

2319. 4-Nitroaniline-2-sulfonic acid
[96-75-3] $C_6H_6N_2O_5S$

See Sulfuric acid: Nitroaryl bases
See other NITROARYL COMPOUNDS, ORGANIC ACIDS

2320. η^6-Benzeneruthenium(II) dinitrate
[ ] $C_6H_6N_2O_6Ru$

Bennett, M. A. *et al., Organometallics*, 1992, **11**(9), 3069
This compound is a shock sensitive explosive.
See other ORGANOMETALLIC NITRATES

2321. 2,4-Dinitrophenylhydrazine
[119-26-6] $C_6H_6N_4O_4$

This spot-test reagent is now supplied water-wet since it has been found to be detonable and legally classifiable as an explosive. The editor has not heard of any problems in use.

See other POLYNITROARYL COMPOUNDS

2322. Ammonium picrate
[131-74-8] $C_6H_6N_4O_7$

Unpublished data.

A small sample, isolated incidentally during decomposition of a picrate with ammonia, exploded during analytical combustion. The presence of traces of metallic picrates (arising from metal contact) increases heat sensitivity. The salt may also explode on impact.

See other PICRATES, POLYNITROARYL COMPOUNDS

766

2323. *cis*-1,2-Bis(azidocarbonyl)cyclobutane
[ ] $C_6H_6N_6O_2$

Landgrebe, J. A., *Chem. Eng. News*, 1981, **59**(17), 47
Explosion during preparation is reported for this and its cyclopropane analogue.
See other ACYL AZIDES

2324. 1,3,5-Triaminotrinitrobenzene
[3058-38-0] $C_6H_6N_6O_6$

Hydroxylaminium perchlorate
See Hydroxylaminium perchlorate: 1,3,5-Triaminotrinitrobenzene
See other POLYNITROARYL COMPOUNDS

2325. Hexanitrohexaazaisowurtzitane (Octahydro-1,3,4,7,8,10-hexanitro-5,2,6-(iminomethenimino-1*H* -imidazo[4,5-*b*]pyrazine))
[135285-90-4] $C_6H_6N_{12}O_{12}$

Borman, S., *Chem. Eng. News*, 1994, **72**(3), 18
This nitramine is claimed to be the most energetic molecular explosive known to man (which the editor doubts).
See other N-NITRO COMPOUNDS

2326. Di(1-propynyl) ether

[ ] C_6H_6O

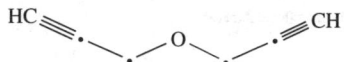

Brandsma, L. *et al., Rec. Trav. Chim.*, 1962, **81**, 510
It is unstable, exploding at ambient temperature.
See other ACETYLENIC COMPOUNDS

2327. Di(2-propynyl) ether (Dipropargyl ether)

[6921-27-3] C_6H_6O

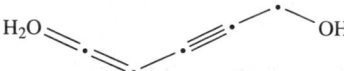

Guide for Safety, 1972, 302
Distillation of the ether in a 230 l still led to an explosion, attributable to peroxidation.
See other ACETYLENIC COMPOUNDS, PEROXIDISABLE COMPOUNDS

2328. 4,5-Hexadien-2-yn-1-ol

[2749-79-3] C_6H_6O

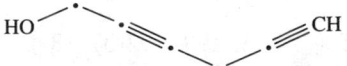

Brandsma, 1971, 8, 54
Dilution with white mineral oil before distillation is recommended to prevent explosion of the concentrated distillation residue.
See other ACETYLENIC COMPOUNDS, DIENES, PEROXIDISABLE COMPOUNDS

2329. 2,5-Hexadiyn-1-ol

[28255-99-4] C_6H_6O

Houben-Weyl, 1977, Vol. 5.2a, 480

The residue after vacuum distillation (at 59–62°C/0.7 mbar) may explode on strong heating or on admission of air, unless mineral oil diluent is present.
See other ACETYLENIC COMPOUNDS, PEROXIDISABLE COMPOUNDS

2330. Phenol (Hydroxybenzene)
[108-95-2]   C$_6$H$_6$O

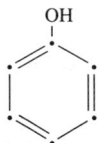

(*MCA SD-4*, 1964); *NSC 405*, 1978; *FPA H82*, 1979; *HCS 1980*, 725; *RSC Lab. Hazards Data Sheet No. 42*, 1985

Aluminium chloride, Nitrobenzene
  See Aluminium chloride: Nitrobenzene, etc.

Aluminium chloride–nitromethane complex, Carbon monoxide
  See Aluminium chloride–nitromethane: Carbon monoxide, etc.

Formaldehyde
  See Formaldehyde: Phenol

Other reactants
  Yoshida, 1980, 295
  MRH values calculated for 15 combinations, largely with oxidants, are given.

Peroxodisulfuric acid
  See Peroxodisulfuric acid: Organic liquids

Peroxomonosulfuric acid
  See Peroxomonosulfuric acid: Aromatic compounds

Sodium nitrate, Trifluoroacetic acid
  See Sodium nitrate: Phenol, etc.

Sodium nitrite
  See Sodium nitrite: Phenol

2331. 1,2-Benzenediol (Pyrocatechol)
[120-80-9]   C$_6$H$_6$O$_2$

HCS 1980, 285

Nitric acid   MRH 5.73/74
  See Nitric acid: Pyrocatechol

Other reactants
  Yoshida, 1980, 89
  MRH values calculated for 13 combinations with oxidants are given.

2332. 1,3-Benzenediol (Resorcinol)
   [108-46-3] $C_6H_6O_2$

Nitric acid
  See Nitric acid: Resorcinol

2333. 1,4-Benzenediol (Hydroquinone)
   [123-31-9] $C_6H_6O_2$

HCS 1980, 554

Other reactants
  Yoshida, 1980, 287
  MRH values calculated for 13 combinations, largely with oxidants, are given.

Oxygen
  See Oxygen (Gas): 1,4-Benzenediol

Sodium hydroxide   MRH 1.09/tr.
  491M, 1975, 385
  Accidental mixing of the hot crude hydroquinone with conc. sodium hydroxide
  solution led to extensive exothermic decomposition.
  See other REDUCANTS

2334. 1,4-Benzenediol–oxygen complex (Hydroquinone–oxygen clathrate)
[20471-57-2] $C_6H_6O_2.O_2$

Preparative hazard
See Oxygen: 1,4-Benzenediol

2335. Benzeneseleninic acid
[6996-92-5]   · $C_6H_6O_2Se$

Preparative hazard
See Hydrogen peroxide: Diphenyl diselenide

Hydrazine derivatives
Back, T. G. *et al., J. Org,. Chem.*, 1981, **46**, 1568
The acid or its anhydride react very vigorously with hydrazine derivatives in absence of solvent. A solid mixture of the acid with benzohydrazide reacted violently after an induction period of several minutes.
See other INDUCTION PERIOD INCIDENTS *See other* ORGANIC ACIDS

2336. endo-2,3-Epoxy-7,8-dioxabicyclo[2.2.2]oct-5-ene (Oxepin-3,6-endoperoxide)
[39597-90-5] $C_6H_6O_3$

Foster, C. H. *et al., J. Org. Chem.*, 1975, **40**, 3744

Evaporation of (peroxide-contacted) solvent ether from 9.8 g of crude peroxide led to an explosion. Subsequent preparations were effected on a smaller scale.
See other CYCLIC PEROXIDES, 1,2-EPOXIDES

2337. Benzeneperoxyseleninic acid
[62865-97-6] $C_6H_6O_3Se$

Syper, L. *et al., Tetrahedron*, 1987, **43**, 207–213
It explodes at the m.p., 52°C.
See other PEROXYACIDS

2338. Benzeneselenonic acid
[39254-48-3] $C_6H_6O_3Se$

Stöcker, M. *et al., Ber.*, 1906, **39**, 2917
It explodes feebly at 180°C but the silver salt is more explosive.
See other ORGANIC ACIDS

2339. 2,3:5,6-Diepoxy-7,8-dioxabicyclo[2.2.2]octane
[56411-67-7] $C_6H_6O_4$

Sasaoka, M. *et al., J. Org. Chem.*, 1979, **44**, 373 (footnote 10)
It explodes when heated.
See other CYCLIC PEROXIDES, 1,2-EPOXIDES

2340. Dimethyl acetylenedicarboxylate (Dimethyl 2-butynedioate)
[762-42-5] $C_6H_6O_4$

Energy of decomposition (in range 150–430°C) measured as 1.224 kJ/g by DSC, and T_{ait24} was determined as 86°C by adiabatic Dewar tests, with an apparent energy of activation of 91 kJ/mol.
See entry THERMOCHEMISTRY AND EXOTHERMIC DECOMPOSITION (reference 2)

1-Methylsilacyclopenta-2,4-diene
  See 1-Methylsilacyclopenta-2,4-diene: Dienophiles

Octakis(trifluorophosphine)dirhodium
  See Octakis(trifluorophosphine)dirhodium: Acetylenic esters
  See other ACETYLENIC COMPOUNDS

2341. Benzeneperoxysulfonic acid
[ ] $C_6H_6O_4S$

Preparative hazard
  See Hydrogen peroxide: Benzenesulfonic anhydride
  See other PEROXYACIDS

2342. *E*-Propene-1,2,3-tricarboxylic acid (Aconitic acid)
[4023-65-8] $C_6H_6O_6$

Hydrogen peroxide
  See Hydrogen peroxide: Aconitic acid
  See other ORGANIC ACIDS

773

2343. Benzene triozonide (Ozobenzene)
[71369-69-2] $C_6H_6O_9$

(Uncertain structure)

1. Harries, C. *et al., Ber.*, 1904, **37**, 3431
2. *DOC 5*, 1982, Vol. 4, 4471

The product of ozonising benzene exhaustively is extremely explosive at the slightest touch [1]. The structure is not well defined as previously thought, but is probably polymeric [2].

See Ozone: Aromatic compounds
See other OZONIDES

2344. Benzenethiol (Thiophenol)
[108-98-5] C_6H_6S

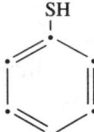

HCS 1980, 189

Preparative hazard

1. Leuckart, R., *J. Prakt. Chem.*, 1890, **41**, 179
2. Graesser, R., private comm., 1968

During preparation of thiophenol by addition of a cold solution of potassium *O*-methyldithiocarbonate to a cold solution of benzenediazonium chloride, a violent explosion accompanied by an orange flash occurred [1]. This was attributed to the formation and decomposition of bis(benzenediazo) disulfide. A preparation in which the diazonium solution was added to the 'xanthate' solution proceeded smoothly [2].

See also DIAZONIUM SULFIDES AND DERIVATIVES

2345. 4-Bromo-1,2-diaminobenzene
[1575-37-7] $C_6H_7BrN_2$

T_{ait24} was determined as the rather low value of 70°C by adiabatic Dewar tests, with an apparent energy of activation of 133 kJ/mol.

See related HALOANILINES

774

2346. 2-Bromomethyl-5-methylfuran
[57846-03-4]   C₆H₇BrO

C_6H_7BrO

Dunlop, 1953, 261
It distils with violent decomposition at 70°C/33 mbar.
See other 2-HALOMETHYL-FURANS OR -THIOPHENES

2347. 4-Chloro-1,2-benzenediamine
[95-83-0] $C_6H_7ClN_2$

Energy of decomposition (in range 160–260°C) measured as 0.78 kJ/g by DSC, and T_{ait24} was determined as 100°C by adiabatic Dewar tests, with an apparent energy of activation of 140 kJ/mol.
See entry THERMOCHEMISTRY AND EXOTHERMIC DECOMPOSITION (reference 2)
See related HALOANILINES

2348. 4-Chloro-1,3-benzenediamine
[5131-60-2] $C_6H_7ClN_2$

Energy of decomposition (in range 200–250°C) measured as 1.13 kJ/g by DSC, and T_{ait24} was determined as 1006°C by adiabatic Dewar tests, with an apparent energy of activation of 121 kJ/mol.

See entry THERMOCHEMISTRY AND EXOTHERMIC DECOMPOSITION (reference 2)
See related HALOANILINES

2349. 2-Chloro-1,4-benzenediamine
[615-66-7] $C_6H_7ClN_2$

Frankenberg, P. E., *Chem. Eng. News*, 1982, **60**(3), 97

Violently explosive decomposition at the start of distillation at 165°C/33 mbar of a 2 kg sample was attributed to autocatalytic decomposition which can occur at 125°C in small samples. Increased lability of the 2-chloro substituent when hydrochloride formation has occurred is postulated as a possible cause. Caution on heating any halogenated amine is advised.

See 2-Chloroethylamine
See other HALOANILINES

2350. 4-Nitroanilinium perchlorate
[15873-50-4] $C_6H_7ClN_2O_6$

Jain, S. R. *et al., Combust. Flame*, 1979, **35**, 289–295

Of a series of substituted anilinium perchlorate salts, that of 4-nitroaniline showed the lowest thermal stability (DTA peak temperature 201°C) and highest shock-sensitivity.

See other NITROARYL COMPOUNDS, PERCHLORATE SALTS OF NITROGENOUS BASES

2351. 2,4-Dinitrophenylhydrazinium perchlorate
[104503-71-1] $C_6H_7ClN_4O_8$

Anon., *Angew. Chem. (Nachr.)*, 1967, **15**, 78
Although solutions of the perchlorate are stable as prepared, explosive decomposition may occur during concentration by heated evaporation.
See also HYDRAZINIUM SALTS
See other POLYNITROARYL COMPOUNDS, PERCHLORATE SALTS OF NITROGENOUS BASES

2352. 2-Chloromethyl-5-methylfuran
[52157-57-0] C_6H_7ClO

Dunlop, 1953, 261
It is even more unstable than its lower homologue, 2-chloromethylfuran.
See other 2-HALOMETHYL-FURANS OR -THIOPHENES

2353. 3-Methoxycarbonylpropen-2-yl trifluoromethanesulfonate
[62861-57-8] $C_6H_7F_3O_5S$

See entry ALLYL TRIFLUOROMETHANESULFONATES

2354. Aniline (Benzeneamine)
[62-53-3] C_6H_7N

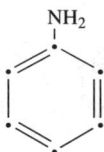

(*MCA SD-17*, 1963); *NSC 409*, 1979; *FPA H80*, 1979; *HCS 1980*, 156; *RSC Lab. Hazard Data Sheet No. 23*, 1984

Anilinium chloride
   Anon., *Chem. Met. Eng.*, 1922, **27**, 1044
   Heating a mixture of the components in an autoclave at 240–260°C/7.6 bar to produce diphenylamine led to a violent explosion, following a sudden increase in pressure to 17 bar. The equipment and process had been used previously and uneventfully.

Benzenediazonium-2-carboxylate
   See Benzenediazonium-2-carboxylate
       Benzenediazonium-2-carboxylate: Aniline, etc.

Boron trichloride
   See Boron trichloride: Aniline

1-Chloro-2,3-epoxypropane
   See 1-Chloro-2,3-epoxypropane: Aniline

Dibenzoyl peroxide
   See Dibenzoyl peroxide: Aniline

Nitromethane   MRH 5.69/99+
   See Nitromethane: Acids, etc.

Nitrous acid
   See DIAZONIUM SALTS

Other reactants
   Yoshida, 1980, 22
   MRH values calculated for 20 combinations with oxidants are given.

Oxidants   MRH values show % of oxidant
   See Diisopropyl peroxydicarbonate: Amines, etc.
       'Fluorine nitrate': Organic materials
       Fluorine: Nitrogenous bases

778

Hydrogen peroxide: Organic compounds (reference 2)   MRH 6.44/83
Nitric acid: Aromatic amines   MRH 5.98/81
Nitrosyl perchlorate: Organic materials
Ozone: Aromatic compounds   MRH 11.5/72
Perchloric acid: Aniline, etc.   MRH 6.65/82
Perchloryl fluoride: Nitrogenous bases
Peroxodisulfuric acid: Organic liquids
Peroxomonosulfuric acid: Aromatics
Peroxyformic acid: Organic materials
Sodium peroxide: Organic liquids, etc.   MRH 2.76/93
N-HALOIMIDES: Alcohols, Amines

Tetranitromethane
   See Tetranitromethane: Amines

Trichloronitromethane   MRH 2.38/64
   See Trichloronitromethane: Aniline

2355. 2-Methylpyridine (2-Picoline)
[109-06-8] C_6H_7N

HCS 1980, 754

Hydrogen peroxide, Iron(II) sulfate, Sulfuric acid
   See Hydrogen peroxide: Iron(II) sulfate, etc.
   See other ORGANIC BASES

2356. N-Phenylhydroxylamine (N-Hydroxybenzeneamine)
[100-65-2] C_6H_7NO

Preparative hazard
   See Zinc: Nitrobenzene
   See Phenylhydroxylaminium chloride *See other* N–O COMPOUNDS, ORGANIC BASES

2357. 4-Oximino-4,5,6,7-tetrahydrobenzofurazan *N*-oxide
[ ] $C_6H_7N_3O_3$

Preparative hazard
See Sulfinyl chloride: 1,2,3-Cyclohexanetrione trioxime, etc.
See other FURAZAN *N*-OXIDES, OXIMES

2358. 5,6-Dinitro-2-dimethylaminopyrimidinone
[ ] $C_6H_7N_5O_5$

Tennant, G. *et al., J. Chem. Soc., Chem. Comm.*, 1982, 61
It decomposed violently on melting at 190°C.
See related POLYNITROARYL COMPOUNDS

2359. Ethyl 2-cyano-2-(1-*H*-tetrazol-5-ylhydrazono)acetate
[61033-15-6] $C_6H_7N_7O_2$

Preparative hazard
See 5-Aminotetrazole (reference 3)
See other CYANO COMPOUNDS, TETRAZOLES

2360. Phenylphosphine
[638-21-1]   **C$_6$H$_7$P**

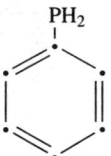

Haszeldine, R. N. *et al., J. Fluorine Chem.*, 1977, **10**, 38
It is pyrophoric in air.
See related ALKYLPHOSPHINES

†2361. 1,3-Cyclohexadiene
[592-57-4]   **C$_6$H$_8$**

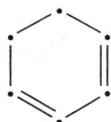

1. Bodendorf, K., *Arch. Pharm.*, 1933, **271**, 11
2. Hock, H. *et al., Chem. Ber.*, 1951, **84**, 349
Cyclohexadiene autoxidises slowly in air, but the residual (largely polymeric)
peroxide explodes very violently on ignition [1]. The monomeric peroxide has
also been isolated [2].
See Propene
See other DIENES, POLYPEROXIDES

†2362. 1,4-Cyclohexadiene
[628-41-1]   **C$_6$H$_8$**

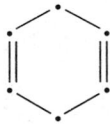

See Propene
See other DIENES

781

2363. Methylcyclopentadiene
[26519-91-5]   C₆H₈

C_6H_8

CH₃

Bis(pentafluorophenyl)ytterbium
  See Bis(pentafluorophenyl)ytterbium: Methylcyclopentadiene
  See other DIENES

2364. 2,2-Dibromo-1,3-dimethylcyclopropanoic acid
[72957-64-3] $C_6H_8Br_2O_2$

H₃C
              Br

       Br

H₃C
         OH
    O

tert-Butylamine
  Baird, M. S. *et al., J. Chem. Soc., Perkin Trans. 1*, 1979, 2324
  Interaction is highly exothermic.
  See other ORGANIC ACIDS

2365. Anilinium chloride
[142-04-1] C_6H_8ClN

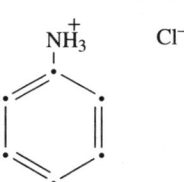

NH₃⁺   Cl⁻

HCS 1980, 157

Aniline
  See Aniline: Anilinium chloride

2366. Phenylhydroxylaminium chloride
[22755-09-5] C_6H_8ClNO

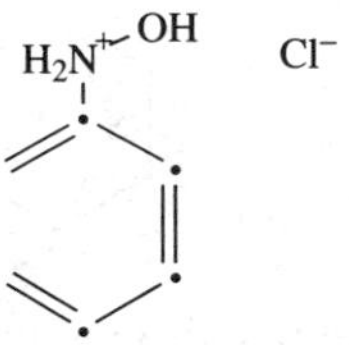

MCA Guide, 1972, 299

A brown glass bottle containing 700 g of the salt exploded forcefully after shelf storage for 2 weeks.
See HYDROXYLAMINIUM SALTS *See other* N–O COMPOUNDS

2367. Anilinium perchlorate
[14796-11-3] $C_6H_8ClNO_4$

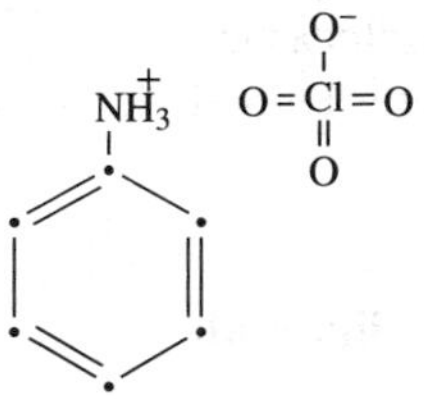

Metal oxides
    Rao, M. V. *et al., Combust. Flame*, 1980, **39**, 63–68
    The effect of metal oxides in sensitising the thermal decomposition and explosion of the salt is in the order: manganese dioxide > copper oxide > nickel oxide.
    See other AMINIUM PERCHLORATES, PERCHLORATE SALTS OF NITROGENOUS BASES

2368. 2-Azatricyclo[2.2.1.0^{2,6}]hept-7-yl perchlorate
[98566-27-9] $C_6H_8ClNO_4$

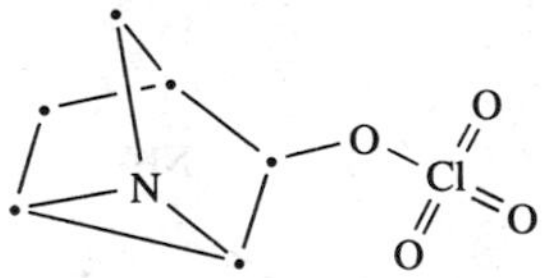

Durrant, M. L. *et al., J. Chem. Soc., Chem. Comm.*, 1985, 1879
A crystalline sample exploded without warning in the freezer compartment of a refrigerator.
See related ALKYL PERCHLORATES, PERCHLORATE SALTS OF NITROGENOUS BASES, STRAINED-RING COMPOUNDS

783

2369. endo-2,5-Dichloro-7-thiabicylo[2.2.1]heptane
  [6522-40-3] $C_6H_8Cl_2S$

Dimethylformamide, Sodium tetrahydroborate
1. Corey, E. J. *et al., J. Org. Chem.*, 1969, **34**, 1234 (footnote 5)
2. Paquette, L. A., private comm., 1975
3. Brown, H. C. *et al., J. Org. Chem.*, 1962, **27**, 1928–1929
To effect reduction to the parent heterocycle, a solution of the dichloro compound in DMF was being added to a hot solution of sodium tetrahydroborate in the same solvent, when a violent explosion occurred [1,2]. This may have arisen either from interaction of the dichloro compound with the solvent, or from the known instability of hot solutions of the tetrahydroborate in DMF. Use of aqueous diglyme as an alternative solvent [3] would only be applicable to this and other hydrolytically stable halides [2].
See Dimethylformamide: Halocarbons
        Sodium tetrahydroborate: Dimethylformamide
See HALOCARBONS

2370. *N,N,N′,N′*-Tetrachloroadipamide
  [ ] $C_6H_8Cl_4N_2O_2$

Water
  See Dichloramine *See other* N-HALOGEN COMPOUNDS

2371. 1,4-Diaminobenzene (*p*-Phenylenediamine)
  [106-50-3] $C_6H_8N_2$

The finely powdered base is a significant dust explosion hazard.

See entry DUST EXPLOSION INCIDENTS   (reference 22)
See other ORGANIC BASES

2372. 1,4-Diisocyanobutane
[929-25-9] $C_6H_8N_2$

See entry DIISOCYANIDE LIGANDS *See related* CYANO COMPOUNDS

2373. Phenylhydrazine
[100-63-0] $C_6H_8N_2$

HCS 1980, 735

et al., Exothermic Decomposition, Technical Report 01VD 159/0329 for Federal German Ministry for Res. Technol., Bonn. 1986
Energy of decomposition (in range 280–375°C) measured as 0.662 kJ/g by DSC, and T_{ait24} was determined as 227°C by adiabatic Dewar tests, with an apparent energy of activation of 165 kJ/mol.
   Heat of decomposition was determined as 0.65 kJ/g.
See entry THERMOCHEMISTRY AND EXOTHERMIC DECOMPOSITION

Lead(IV) oxide   MRH 0.46/95
   See Lead(IV) oxide: Nitrogen compounds

Other reactants
   Yoshida, 1980, 294
   MRH values calculated for 13 combinations with oxidants are given.

Perchloryl fluoride
   See Perchloryl fluoride: Nitrogenous bases

α-Phenylazo hydroperoxide
   See α-Phenylazo hydroperoxide: Phenylhydrazine
   See other ORGANIC BASES, REDUCANTS

2374. 2,3-Diazabicyclo[2.2.2]octa-2,5-diene N-oxide
[37436-17-2] $C_6H_8N_2O$

Preparative hazard
   See Hydrogen peroxide: 4-Methyl-2,4,6-triazatricyclo[5.2.2.0^{2,6}]undeca-8-ene-3,5-dione, etc.
   See other N-OXIDES

2375. 2-Diazocyclohexanone
[3242-56-6] $C_6H_8N_2O$

Regitz, M. *et al.*, *Org. Synth.*, 1971, **51**, 86
It may explode on being heated, and a relatively large residue should be left during vacuum distillation to prevent overheating.
See other DIAZO COMPOUNDS

2376. Butane-1,4-diisocyanate
[4538-37-8] $C_6H_8N_2O_2$

Zlobin, V. A. *et al.*, *Chem. Abs.*, 1987, **107**, 39144
Conditions for the safe preparation of the diisocyanate from adipoyl chloride and sodium azide in acetonitrile–toluene mixtures were established.
See other ORGANIC ISOCYANATES

2377. 2-Buten-1-yl diazoacetate
[14746-03-3] $C_6H_8N_2O_2$

Blankley, C. J. *et al.*, *Org. Synth.*, 1969, **49**, 25
Precautions are necessary during vacuum distillation of this potentially explosive material.
See other DIAZO COMPOUNDS

2378. 2,5-Dimethylpyrazine 1,4-dioxide
[6890-38-6] $C_6H_8N_2O_2$

Preparative hazard
    See Hydrogen peroxide: Acetic acid, *N*-Heterocycles
    See other N-OXIDES

2379. Anilinium nitrate
[542-15-4] $C_6H_8N_2O_3$

Nitric acid
    See Nitric acid: Anilinium nitrate
    See other OXOSALTS OF NITROGENOUS BASES

2380. Isosorbide dinitrate (1,4:3,6-Dianhydro-D-glucitol dinitrate)
[87-33-2] $C_6H_8N_2O_8$

Reddy, G. O. *et al., J. Hazard. Mater.*, 1992, **32**(1), 87
This heart drug is detonable when dry but non-explosive with 30% of water.
See other ALKYL NITRATES

2381.
Dodecahydro-4,8-dinitro-2,6-bis(nitroimino)diimidazo[4,5-*b*:4′,5′-*e*]pyrazine
[134927-04-1] [160693-88-2] (stereoisomer) $C_6H_8N_{12}O_8$

Dagley, I. J. *et al., J. Energ. Mater.*, 1995, **13**(1 & 2), 35
This nitroguanidine analogue is more sensitive than RDX to friction, but equivalent
by the drop-weight impact test.
See other *N*-NITRO COMPOUNDS

†2382. 2,5-Dimethylfuran
[625-86-5] C_6H_8O

2383. 2,4-Hexadienal
[142-83-6] C_6H_8O

1-Amino-2-propanol
Steele, A. B. *et al., Chem. Engrg.*, 1959, **66**(8), 166

788

Following a plant incident in which the crude aldehyde had polymerised during distillation, it was found that the presence of 2% of isopropanolamine (a probable impurity) led to an explosive polymerisation reaction. It is conceivable that aldehydic peroxides may also have been involved.

See other PEROXIDISABLE COMPOUNDS, POLYMERISATION INCIDENTS

2384. 3-Methyl-2-penten-4-yn-1-ol
[105-29-3] C_6H_8O

1. Lorentz, F., *Loss Prev.*, 1967, **1**, 1–5
2. Fesenko, G. V. *et al., Chem. Abs.*, 1975, **83**, 47605

Temperature control during pressure hydrogenation of *cis*- or *trans*-isomers is essential, since at 155°C violent decomposition to carbon, hydrogen and carbon monoxide with development of over 1 kbar pressure will occur. The material should not be heated above 100°C, particularly if acid or base is present, to avoid exothermic polymerisation [1]. The *cis*-isomer is readily cyclised to 2,3-dimethylfuran, which promotes fire and explosion hazards. These were measured for the *cis*- and *trans*-isomers, and for *trans*-3-methyl-1-penten-4-yn-3-ol [2].

See other GAS EVOLUTION INCIDENTS, HYDROGENATION INCIDENTS, POLYMERISATION INCIDENTS

Sodium hydroxide
1. *MCA Case History No. 363*
2. Silver, L., *Chem. Eng. Progr.*, 1967, **63**(8), 43

Presence of traces of sodium hydroxide probably caused formation of the acetylenic sodium salt, which exploded during high-vacuum distillation in a metal still [1]. A laboratory investigation which duplicated the explosion, without revealing the precise cause, was also reported [2].

See other ACETYLENIC COMPOUNDS

2385a. 1,2-Cyclohexanedione
[765-87-7] $C_6H_8O_2$

Preparative hazard
See Nitric acid: 4-Methylcyclohexanone

2385b. 2,4-Hexadienoic acid (Sorbic acid)
[110-44-1] $C_6H_8O_2$

$$Me(CH=CH)_2CO.OH$$

The dry finely powdered acid is a significant dust explosion hazard.
See entry DUST EXPLOSION INCIDENTS (reference 22)
See other ORGANIC ACIDS

2386. Poly(1,3-cyclohexadiene peroxide)
[ ] $(C_6H_8O_2)_n$

(Complex structure)

See 1,3-Cyclohexadiene
See other POLYPEROXIDES

2387. 2,5-Dimethyl-2,5-dihydrothiophene-2,5-endoperoxide
[67711-62-0] $C_6H_8O_2S$

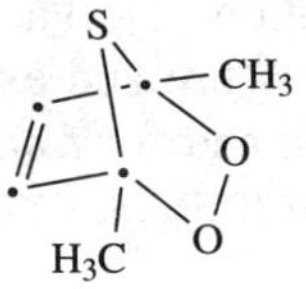

Golnick, K. *et al., Tetrahedron Lett.*, 1984, **25**, 4291–4294
The endoperoxide of 2,5-dimethylthiophene (effectively a thiaozonide) when neat decomposes violently at ambient temperature.
See other CYCLIC PEROXIDES
See related OZONIDES

2388. 1,4-Dimethyl-2,3,7-trioxabicyclo[2.2.1]hept-5-ene (2,5-Dimethyl-2,5-dihydrofuran-2,5-endoperoxide)
[45722-89-2] $C_6H_8O_3$

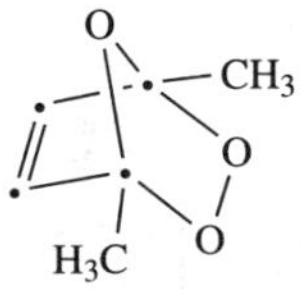

Gollnick, K. *et al., Angew. Chem. (Intern. Ed.)*, 1983, **22**, 726
The solvent-free peroxide always explodes if amounts above 100 mg are handled.
See other CYCLIC PEROXIDES

790

2389. Citric acid (2-Hydroxypropane-1,2,3-tricarboxylic acid)
[77-92-9]

$C_6H_8O_7$

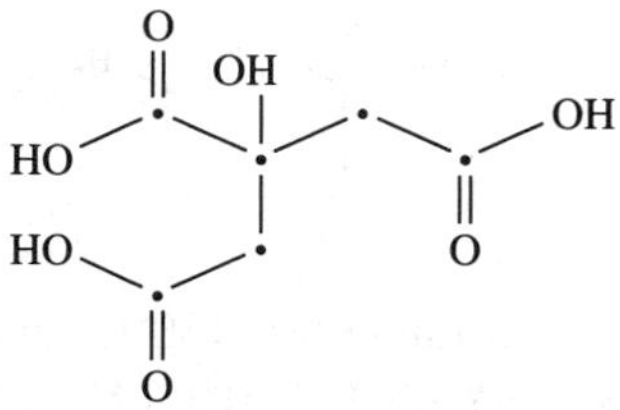

Metal nitrates
  See METAL NITRATES: citric acid *See other* ORGANIC ACIDS

2390. Triacetyl borate
[4887-24-5]

$C_6H_9BO_6$

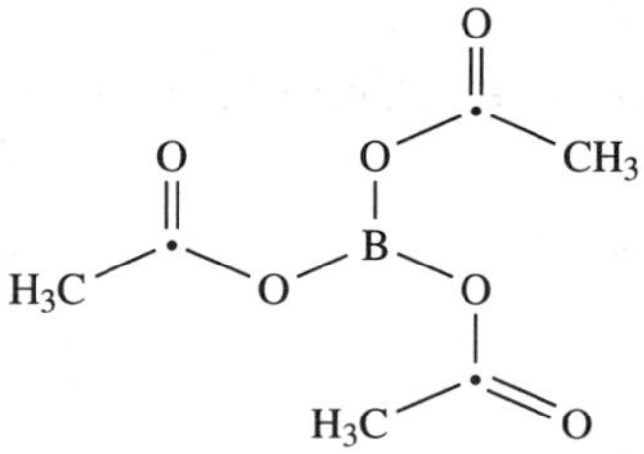

Preparative hazard
  See Acetic anhydride: Boric acid

2391. Trivinylbismuth
[65313-35-1]

C_6H_9Bi

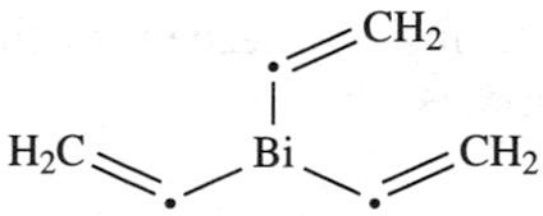

Coates, 1967, Vol. 1, 538
It ignites in air.
See related ALKYLMETALS, TRIALKYLBISMUTHS

2392. 2-Bromocyclohexanone
[822-85-5] C_6H_9BrO

Energy of decomposition (in range 80–130°C) measured as 0.29 kJ/g.
See entry THERMOCHEMISTRY AND EXOTHERMIC DECOMPOSITION (reference 2)

2393. 1-Iodohexa-2,4-diene
[ ] C_6H_9I

Hayashi, T., *et al., J. Chem. Soc., Chem. Comm.*, 1990, (19), 1362
This compound undergoes explosive decomposition when freed of solvent. Homo-
logues are also unstable.
See other DIENES, HALOALKENES, IODINE COMPOUNDS

2394. Iodine triacetate
[6540-76-7] $C_6H_9IO_6$

1. Schutzenberger, P., *Compt. rend.*, 1861, **52**, 135
2. Oldham, F. W. H. *et al., J. Chem. Soc.*, 1941, 368
The acetate explodes at 140°C [1], and higher homologues decompose at about
120°C, the violence decreasing with ascent of the homologous series [2].
See other IODINE COMPOUNDS

2395. 3,3-Dimethyl-1-nitro-1-butyne (*tert*-Butylnitroacetylene)
[22691-91-4] $C_6H_9NO_2$

Amines
Jäger, V. *et al., Angew. Chem. (Intern. Ed.)*, 1969, **8**, 273

Reaction with primary, secondary or tertiary amines proceeds explosively with ignition in absence of a solvent.

See other ACETYLENIC COMPOUNDS *See related* NITROALKENES

2396. 1,4,3,6-Dianhydroglucitol 2-nitrate (Isosorbide mononitrate)
[16016-20-0] $C_6H_9NO_6$

Négyesi, G., *Process Safety Progress*, 1996, **15**(1), 42
Lead block tests show this to be a feeble explosive, it is not impact sensitive but shows some decomposition on friction.
See Isosorbide dinitrate
See other ALKYL NITRATES

2397. Bis(2-cyanoethyl)amine
[111-94-4] $C_6H_9N_3$

Anon., *BCISC Quart. Safety Summ.*, 1967, **38**, 42
Some 18-month old bottles of 'iminodipropionitrile' exploded, probably owing to slow hydrolysis and release of ammonia.
See other CYANO COMPOUNDS, GAS EVOLUTION INCIDENTS

2398. 3,6,9-Triazatetracyclo[6.1.0.0^{2,4}.0^{5,7}]nonane (*cis*-Benzene triimine)
[52851-26-0] $C_6H_9N_3$

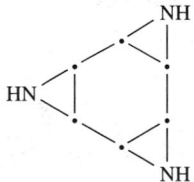

Schwesinger, R. *et al.*, *Angew. Chem. (Intern. Ed.)*, 1973, **12**, 989
It explodes if rapidly heated to 200°C.
See other STRAINED-RING COMPOUNDS

2399. Ammonium _N_ -nitrosophenylaminooxide ('Cupferron')
    [135-20-6] $C_6H_9N_3O_2$

Thorium salts
   Pittwell, L. R., _J. R. Inst. Chem._, 1956, **80**, 173
   Solutions of this reagent are destabilised by the presence of thorium ions. If a
   working temperature of $10-15°C$ is much exceeded, the risk of decomposition, not
   slowed by cooling and accelerating to explosion, exists. Titanium and zirconium
   salts also cause slight destabilisation, but decomposition temperatures are then 35
   and 40°C, respectively.
   See other NITROSO COMPOUNDS, N–O COMPOUNDS

2400. 1,2,3-Cyclohexanetrione trioxime
    [3570-93-2] $C_6H_9N_3O_3$

Sulfinyl chloride
   See Sulfinyl chloride: 1,2,3-Cyclohexanetrione trioxime, etc.
   See other OXIMES

2401. 1,3,5-Cyclohexanetrione trioxime
    [622-22-7] $C_6H_9N_3O_3$

   Baeyer, A., _Ber._, 1886, **19**, 160

The oxime of phloroglucinol explodes rather violently at 155°C.
See other OXIMES

2402. Lead(IV) acetate azide
[ ] $C_6H_9N_3O_6Pb$

See Lead(IV) azide
See related METAL AZIDES

2403. 3(2,3-Epoxypropyloxy)2,2-dinitropropyl azide
[76828-34-7] $C_6H_9N_5O_6$

Frankel, M. B. *et al., J. Chem. Eng. Data*, 1981, **26**, 219
The monomeric azide and several precursors are explosives of moderate to considerable sensitivity to initiation by impact, shock, friction or heat.
See other 1,2-EPOXIDES, ORGANIC AZIDES, POLYNITROALKYL COMPOUNDS

2404. Ethyl sodioacetoacetate
[19232-39-4] $C_6H_9NaO_3$

2-Iodo-3,5-dinitrobiphenyl
See 2-Iodo-3,5-dinitrobiphenyl: Ethyl sodioacetoacetate

2405. Trivinylantimony
[5613-68-3] C_6H_9Sb

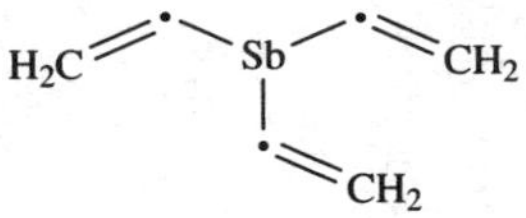

Leleu, *Cahiers*, 1977, (88), 362
It ignites in air.
See related ALKYLMETALS

†2406. Cyclohexene
[110-83-8] C_6H_{10}

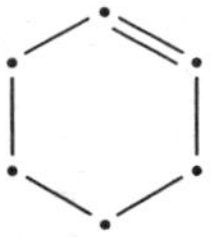

HCS 1980, 346

See other ALKENES

2407. 2,3-Dimethyl-1,3-butadiene
[513-81-5] C_6H_{10}

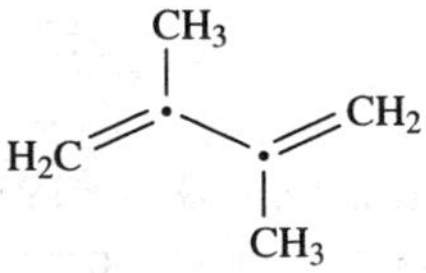

Bodendorf, K., *Arch. Pharm.*, 1933, **271**, 33
A few mg of the autoxidation residue, a largely polymeric peroxide, exploded
violently on ignition.
See other POLYPEROXIDES

Oxygen, Ozone
See Ozone: Dienes, Oxygen

Thiazyl fluoride
See Thiazyl fluoride: Alkylbutadienes
See other DIENES

796

2408. 3,3-Dimethyl-1-butyne
[917-92-0] C_6H_{10}

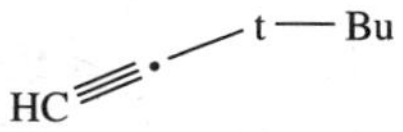

$$HC\equiv\!\!\cdot\!-t-Bu$$

Preparative hazard

See 2,2-Dichloro-3,3-dimethylbutane: Sodium hydroxide
See other ALKYNES

†2409. 1,3-Hexadiene
[592-48-3] C_6H_{10}

$$H_2C{=}CHCH{=}CHEt$$

See other DIENES

†2410. 1,4-Hexadiene
[592-45-0] C_6H_{10}

$$H_2C{=}CHCH_2CH{=}CHMe$$

See other DIENES

†2411. 1,5-Hexadiene
[592-42-7] C_6H_{10}

$$H_2C{=}CHC_2H_4CH{=}CH_2$$

See other DIENES

†2412. *cis*-2-*trans*-4-Hexadiene
[5194-50-3] C_6H_{10}

$$MeCH{=}CHCH{=}CHMe$$

See other DIENES

†2413. *trans*-2-*trans*-4-Hexadiene
[5194-51-4] C_6H_{10}

$$MeCH{=}CHCH{=}CHMe$$

See other DIENES

†2414. 1-Hexyne
[693-02-7] C_6H_{10}

$$HC{\equiv}CBu$$

See other ALKYNES

†2415. 3-Hexyne
[928-49-4]   C$_6$H$_{10}$

Mercury, Silver perchlorate
   See Silver perchlorate: Alkynes, etc.
   See other ALKYNES

†2416. 2-Methyl-1,3-pentadiene
[1118-58-7]   C$_6$H$_{10}$

$$H_2C=CMeCH=CHMe$$

See other DIENES

†2417. 4-Methyl-1,3-pentadiene
[926-56-7]   C$_6$H$_{10}$

$$H_2C=CHCH=CMe_2$$

See other DIENES

2418. **Cadmium propionate**
[16986-83-7]   C$_6$H$_{10}$CdO$_4$

Anon., *Chem. Age*, 1957, **77**, 794
The salt exploded during drying at 60–100°C in an electric oven, presumably because of overheating which caused pyrolysis to 3-pentanone and ignition of its vapour.
See other HEAVY METAL DERIVATIVES

2419. **1-Chloro-1-nitrosocyclohexane**
[695-64-7]   C$_6$H$_{10}$ClNO

Labaziewicz, H. *et al.*, *J. Chem. Soc., Perkin Trans. 1*, 1979, 2928

There is a risk of explosion if the pressure rises during its vacuum distillation.
See other NITROSO COMPOUNDS

2420. *trans*-2-Chlorocyclohexyl perchlorate
[81971-84-8] $C_6H_{10}Cl_2O_4$

See entry ALKYL PERCHLORATES (reference 6)

2421. Bis(2-fluoro-2,2-dinitroethoxy)dimethylsilane
[73526-98-4] $C_6H_{10}F_2N_4O_{10}Si$

Neale, R. S., *Ind. Eng,. Chem., Prod. Res. Dev.*, 1980, **19**, 634
This substituted silane (intended as an explosive plasticiser), though relatively
stable to heat and impact of a hammer, is exceedingly sensitive to compressive
shock. Great care must therefore be exercised in syringing samples of the viscous
liquid.
See other ALKYLSILANES, FLUORODINITROMETHYL COMPOUNDS

2422. 3,3-Pentamethylenediazirine
[930-82-5] $C_6H_{10}N_2$

1. Schmitz, E. *et al., Org. Synth.*, 1965, **45**, 85
2. Schmitz, E. *et al., Chem. Ber.*, 1961, **94**, 2168

Small-scale preparation is recommended [1], in view of a previous explosion through superheating of the liquid during distillation at 109°C [2].
See other DIAZIRINES

2423. *tert*-Butyl diazoacetate
[35059-50-8]

$C_6H_{10}N_2O_2$

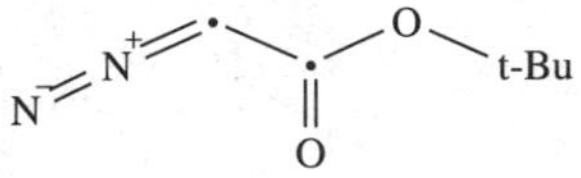

Regitz, M. et al., Org. Synth., 1968, **48**, 36
Vacuum distillation of the compound is potentially hazardous.
See other DIAZO COMPOUNDS

2424. *N*-Nitroso-6-hexanelactam
[35784-01-1]

$C_6H_{10}N_2O_2$

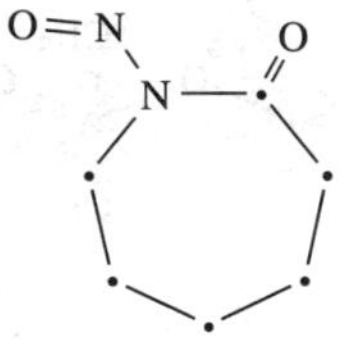

Preparative hazard
See 6-Hexanelactam: Acetic acid, etc.
See other NITROSO COMPOUNDS

2425. Diethyl azoformate (1,2-Bisethoxycarbonyldiazene)
[1972-28-7]

$C_6H_{10}N_2O_4$

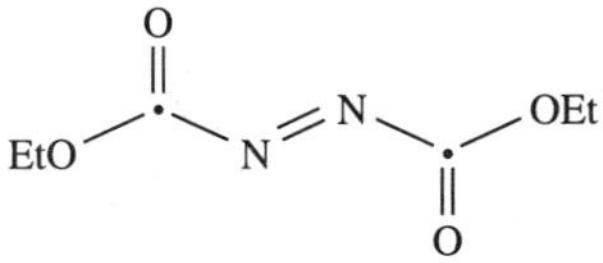

1. US Pat. 3 347 845, 1967
2. Kauer, J. C., *Org. Synth.*, 1963, Coll. Vol. 4, 412
3. Houben Weyl, 1967, 10/2, 809
Shock-sensitive [1], it explodes on heating in a (sealed?) capillary, like its lower homologue. The crude product prepared by chlorine oxidation of the corresponding

hydrazide may detonate violently during distillation. This is thought due to either excess chlorine or inadequate washing leaving chloramine byproducts [3]. However, the editor believes the alternative preparation of azodicarboxylates by fuming nitric acid oxidation of the hydrazide [2] can give azide and nitramine impurities; crude methyl ester from this route has also exploded during distillation.
See Dimethyl azoformate
See other AZO COMPOUNDS

2426. *N,N'*-Diacetyl-*N,N'*-dinitro-1,2-diaminoethane
[922-89-4] $C_6H_{10}N_4O_6$

Violent decomposition occurred at 142°C.
See entry DIFFERENTIAL THERMAL ANALYSIS (reference 1) *See other* N-NITRO COMPOUNDS

2427. *trans*-1,4,5,8-Tetranitro-1,4,5,8-tetraazadecahydronaphthalene
[83673-81-8] $C_6H_{10}N_8O_8$

Willer, R. L., *Propellants, Explos. Pyrotech.*, 1983, **8**, 65
A new explosive compound.
See other N-NITRO COMPOUNDS

2428. Butoxyacetylene
[3329-56-4] $C_6H_{10}O$

$$BuOC{\equiv}CH$$

Jacobs, T. L. *et al.*, *J. Amer. Chem. Soc.*, 1942, **64**, 223
Small samples rapidly heated in sealed tubes to around 100°C exploded.
See other ACETYLENIC COMPOUNDS

2429. Cyclohexanone
[108-94-1] $C_6H_{10}O$

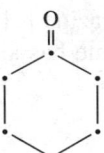

HCS 1980, 344; *RSC Lab. Hazards Data Sheet No. 80*, 1989

Hydrogen peroxide, Nitric acid
  See Hydrogen peroxide: Ketones, etc.

Nitric acid
  See Nitric acid: 4-Methylcyclohexanone

Other reactants
  Yoshida, 1980, 152
  MRH values calculated for 14 combinations with oxidants are given.

2430. Diethylketene (2-Ethyl-1-butene-1-one)
[24264-08-2] $C_6H_{10}O$

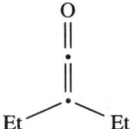

Sorbe, 1968, 118
Like the lower homologue, it readily forms explosive peroxides with air at ambient
temperature.
See other PEROXIDISABLE COMPOUNDS

†2431. Diallyl ether (Di-2-propenyl ether)
[557-40-4] $C_6H_{10}O$
$$(H_2C{=}CHCH_2)_2O$$

HCS 1980, 126

Preparative hazard
  MCA Case History No. 412
  The ether was left exposed to air and sunlight for 2 weeks before distillation, and
  became peroxidised. During distillation to small bulk, a violent explosion occurred.

See 2-Propen-1-ol: Sulfuric acid
See *other* ALLYL COMPOUNDS, PEROXIDATION INCIDENTS, PEROXIDISABLE
COMPOUNDS

2432. 2-Hexenal
   [6728-26-3] $C_6H_{10}O$

$$PrCH{=}CHCO.H$$

Nitric acid
   See Nitric acid: 2-Hexenal
   See *other* PEROXIDISABLE COMPOUNDS

†2433. 2-3-Methyl-2-methylenebutanal (Isopropylacryladehyde)
   [4417-80-5] $C_6H_{10}O$

$$Me_2CHC({=}CH_2)CO.H$$

See *other* PEROXIDISABLE COMPOUNDS

2434. 1-Allyloxy-2,3-epoxypropane (Allyl glycidyl ether, 2-Propenyloxy-
   methyloxirane)
   [106-92-3] $C_6H_{10}O_2$

Peroxide test strips
   Anon., *Jahresber.*, 1987, 65
   Peroxide test strips may not be sufficiently reliable as the sole means of testing
   for presence of peroxides in a wide range of susceptible liquids.
      A sample of the readily peroxidisable allyl ether, tested with a peroxide test
   strip gave a negative test result, so was charged into a distillation apparatus, but
   there was a detonation during distillation. It was recommended that an additional
   test (potassium iodide–acetic acid) should be used unless it is certain that the test
   strips will give a positive result with a particular peroxide-containing liquid.
   See PEROXIDES IN SOLVENTS, PEROXIDISABLE COMPOUNDS

Poly[oxy(methyl)silylene]
   See Poly[oxy(methyl)silylene]: 1-Allyloxy-2,3-epoxypropane
   See *other* ALLYL COMPOUNDS, 1,2-EPOXIDES

2435. 2-Cyclohexenyl hydroperoxide
[4845-05-0] $C_6H_{10}O_2$

Leleu, *Cahiers*, 1973, (71), 226
Accidental contact of a mixture containing the peroxide with the hot top of the freshly sealed glass container led to an explosion.
See other ALKYL HYDROPEROXIDES

†2436. Ethyl crotonate
[623-70-1] $C_6H_{10}O_2$

$$MeCH=CHCO.OEt$$

†2437. Ethyl cyclopropanecarboxylate
[4606-07-9] $C_6H_{10}O_2$

See other STRAINED-RING COMPOUNDS

†2438. Ethyl methacrylate
[97-63-2] $C_6H_{10}O_2$

$$H_2C=CMeCO.OEt$$

†2439. Vinyl butyrate
[123-20-6] $C_6H_{10}O_2$

$$PrCO.OCH=CH_2$$

2440. Ethyl acetoacetate (Ethyl 3-oxobutanoate)
[141-97-9] $C_6H_{10}O_3$

HCS 1980, 463

2,2,2-Tris(bromomethyl)ethanol, Zinc

1. Reeder, J. A., U S Pat. 3 578 619, 1971
2. *491M*, 1975, 167

The patent describes the formation of complex metal chelates by treatment of the ketoester simultaneously with an alcohol and a metal to effect trans-esterification and chelate formation by distilling out the by-product ethanol [1]. This process was being applied to produce the zinc chelate of 2-tris(bromomethyl)ethyl acetoacetate, and when 80% of the ethanol had been distilled out (and the internal temperature had increased considerably), a violent decomposition occurred [2]. This presumably involved interaction of a bromine substituent with excess zinc to form a Grignard-type reagent, and subsequent exothermic reaction of this with one or more of the bromo or ester functions present.

See related GRIGNARD REAGENTS

2441. Adipic acid (Hexane-1,6-dioic acid)
[124-04-9] $C_6H_{10}O_4$

$$HO.OC(CH_2)_4CO.OH$$

Kaiser, M. A., *Plant/Oper. Progr.*, 1991, **10**(2), 100

After two minor dust explosions in an industrial adipic acid dryer, evidence was obtained that adipic acid forms an iron complex capable of both decarboxylation/dehydration of adipic acid to cyclopentanone and of catalysing air oxidation, giving exotherms from as low as 135°C.

See other CATALYTIC IMPURITY INCIDENTS, SELF-HEATING AND IGNITION INCIDENTS, ORGANIC ACIDS

2442. Dipropionyl peroxide
[3248-28-0] $C_6H_{10}O_4$

Swern, 1971, Vol. 2, 815

Pure material explodes on standing at ambient temperature.

Other reactants

Yoshida, 1980, 317

MRH values calculated for 13 combinations, largely with materials catalysing its decomposition, are given.

See other DIACYL PEROXIDES

805

2443. Diallyl sulfate
[27063-40-7] $C_6H_{10}O_4S$

$$H_2C{=}\!\!\cdot\!\!\diagdown\!\!\diagup^{O}{\diagdown}SO_2{\diagup}^{O}{\diagdown}\!\!\cdot\!\!\diagup{=}CH_2$$

von Braun, J. *et al., Ber.*, 1917, **50**, 293
The explosive decomposition during distillation may well have been caused by
polymerisation initiated by acidic decomposition products.
See other ALLYL COMPOUNDS, POLYMERISATION INCIDENTS, SULFUR ESTERS

2444. Diethyl dicarbonate (Ethyl pyrocarbonate)
[1609-47-8] $C_6H_{10}O_5$

EtOCO.OCO.OEt

Anon., *Univ. Safety Assoc. Safety Digest*, 1991, **42**, 7
Unopened bottles of this, all originating from the same batch and well within the
manufacturer's expiry date, exploded while warming to room temperature. This
was attributed to water contamination leading to hydrolysis and carbon dioxide
formation, without supporting evidence — a number of other contaminants, both
nucleophiles and acids, might catalyse carbon dioxide formation (Editor).
See other GAS EVOLUTION INCIDENTS

2445. 2-Hydroxy-2-methylglutaric acid
[503-49-1] $C_6H_{10}O_5$

$$HO{\diagup}^{O}{\diagdown}\!\!\cdot\!\!\diagdown\!\!\diagup^{CH_3}_{OH}{\diagdown}^{O}{\diagup}OH$$

Preparative hazard
See 4-Hydroxy-4-methyl-1,6-heptadiene: Ozone
See other ORGANIC ACIDS

2446. Diethyl peroxydicarbonate
[14666-78-5] $C_6H_{10}O_6$

$$EtO{\diagdown}\!\!\underset{O}{\overset{\displaystyle}{\diagup}}\!\!^{O}{\diagdown}^{O}{\diagup}\!\!\overset{\displaystyle O}{\diagdown}OEt$$

An oil of extreme instability when impure, and sensitive to heat or impact,
exploding more powerfully than dibenzoyl peroxide.
See entry PEROXYCARBONATE ESTERS

806

2447. Diallyl sulfide
[592-88-1]   C₆H₁₀S

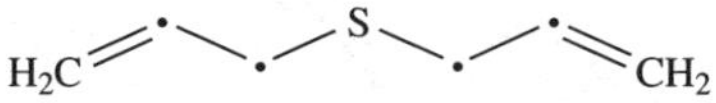

$$H_2C = \cdot \diagdown \cdot \diagup S \diagdown \cdot \diagup \cdot = CH_2$$

N-Bromosuccinimide

See N-HALOIMIDES: alcohols, etc. See other ALLYL COMPOUNDS

2448. N,N,N'-Trifluorohexanamidine
[31330-22-0]   C₆H₁₁F₃N₂

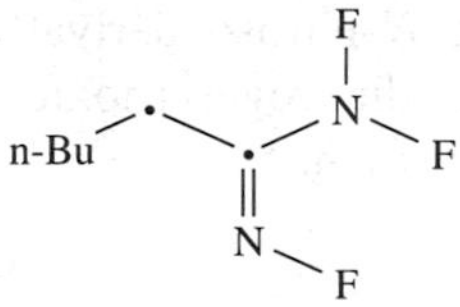

Ross, D. L. *et al., J. Org. Chem.*, 1970, **35**, 3096–3097
A shock-sensitive explosive.
See other N,N,N'-TRIFLUOROAMIDINES

2449. 2-Ethoxy-1-iodo-3-butene
[13957-21-6]   C₆H₁₁IO

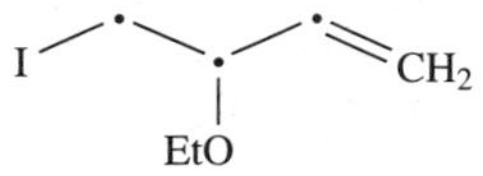

Trent, J. *et al., Chem. Eng. News*, 1966, **44**(33), 7
During a 1 g mol-scale preparation by a published method from butadiene, ethanol, iodine and mercury oxide, a violent explosion occurred while ethanol was being distilled off at 35°C under slight vacuum. The cause of the explosion could not be established, and several smaller-scale preparations had been uneventful.
See related HALOALKENES *See other* IODINE COMPOUNDS

†2450. Diallylamine
[124-02-7]   C₆H₁₁N

$$(H_2C{=}CHCH_2)_2NH$$

HCS 1980, 356

See other ALLYL COMPOUNDS, ORGANIC BASES

2451. Caprolactam (6-Hexanelactam)
[105-60-2] $C_6H_{11}NO$

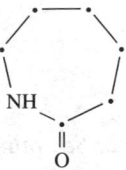

HCS 1980, 258

Acetic acid, Dinitrogen trioxide
Huisgen, R. *et al., Ann.*, 1952, **575**, 174–197
During preparation of the *N*-nitroso derivative from the lactam in acetic acid
solution, the treatment with dinitrogen trioxide must be very effectively cooled to
prevent explosive decomposition.

2452. Cyclohexanone oxime
[100-64-1] $C_6H_{11}NO$

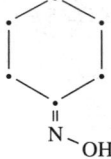

HCS 1980, 345

Ullmann, 1986, **A5** 34
It can decompose violently, for unknown reasons, during vacuum distillation (probably exothermic Beckmann rearrangement and subsequent polymerisation — Editor)

Oleum (fuming sulfuric acid)
Ale, B. J. M., *Proc. 3rd Intl. Symp. Loss Prev. Safety Prom. Proc. Ind.*, Vol. 4,
1541–1550, Basle, SSCI, 1980
During startup of the Beckmann rearrangement stage in the caprolactam process, a
mixture of caprolactam and oleum is preheated to operating temperature, then oxime
feed and cooling are begun. On one occasion, the exothermic rearrangement ($Q =$
230.8 kJ/mol, 2.04 kJ/g of oxime) failed to start, and the temperature decreased. On
application of steam to the cooler to start the reaction, the reactor vessel burst. At
temperatures above 150–160°C, oleum oxidises the oxime with evolution of carbon
dioxide and sulfur trioxide. An investigation of the operating stability of the reactor
system and transient behaviour during startup allowed safe starting procedures to be
clearly defined.
See other GAS EVOLUTION INCIDENTS, OXIMES

2453. 2-Isopropylacrylaldehyde oxime

[ ] $C_6H_{11}NO$

Marvel, C. S. *et al., J. Amer. Chem. Soc.*, 1950, **72**, 5408–5409
Hydroquinone must be added to the oxime before distillation, to prevent formation and subsequent violent decomposition of a peroxide (there being 2 susceptible hydrogen atoms in the molecule).
See other OXIMES, PEROXIDISABLE COMPOUNDS

2454. Ethyl 2-formylpropionate oxime

[ ] $C_6H_{11}NO_3$

Hydrogen chloride (?)
  Loftus, F., private comm., 1972
  The formyl ester and hydroxylamine were allowed to react in ethanol, which was then removed by vacuum evaporation at 40°C. The dry residue decomposed violently 5 min later. Analysis of the residue suggested that the oxime had undergone an exothermic Beckmann rearrangement, possibly catalysed by traces of hydrogen chloride in the reaction residue.
  See Butyraldehyde oxime
  See other OXIMES
  See other CATALYTIC IMPURITY INCIDENTS

2455. 5-Methyl-1(1-methylethyl)-1,2,3-azadiphosphole (3(1-Methylethyl)-4-methyldiphosphazole)

126330-30-1 $C_6H_{11}NP_2$

Gueth, W., *New J. Chem.* 1989, **13**(4–5), 309

Extremely pyrophoric
See other PYROPHORIC MATERIALS, *See related* ALKYLPHOSPHINES

2456. Diallyl phosphite (Di-2-propenyl phosphonite)
[23679-20-1]

$C_6H_{11}O_3P$

$$H_2C=\!\!\!=\!\!\!\bullet\!\!\!-\!\!\!\bullet\!\!\!-\!\!O\!\!-\!\!P\!\!-\!\!O\!\!-\!\!\bullet\!\!\!-\!\!\!\bullet\!\!\!=\!\!\!=\!\!CH_2$$
$$\overset{\parallel}{O}$$

Houben-Weyl, 1964, Vol. 12.2, 22
The ester is liable to explode during distillation unless thoroughly dry allyl alcohol is used for the preparation, and if more than 60% of the product is distilled over. Acid-catalysed polymerisation is probably involved.
See other ALLYL COMPOUNDS, PHOSPHORUS ESTERS, POLYMERISATION INCIDENTS

†2457. Cyclohexane
[110-82-7]

C_6H_{12}

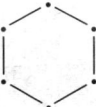

(*MCA SD-68*, 1957); *FPA H25*, 1974; *HCS 1980*, 342; *RSC Lab. Hazards Data Sheet No. 54*, 1986

Dinitrogen tetraoxide
See Dinitrogen tetraoxide: Hydrocarbons

Other reactants
Yoshida, 1980, 153
MRH values calculated for 14 combinations with oxidants are given.

†2458. Ethylcyclobutane
[4806-61-5]

C_6H_{12}

ET

†2459. 1-Hexene
[592-41-6]

C_6H_{12}

$$H_2C=CHBu$$

See other ALKENES

810

†2460. 2-Hexene
[592-43-8]   C$_6$H$_{12}$

$$\text{MeCH=CHPr}$$

See other ALKENES

†2461. Methylcyclopentane
[96-37-7]   C$_6$H$_{12}$

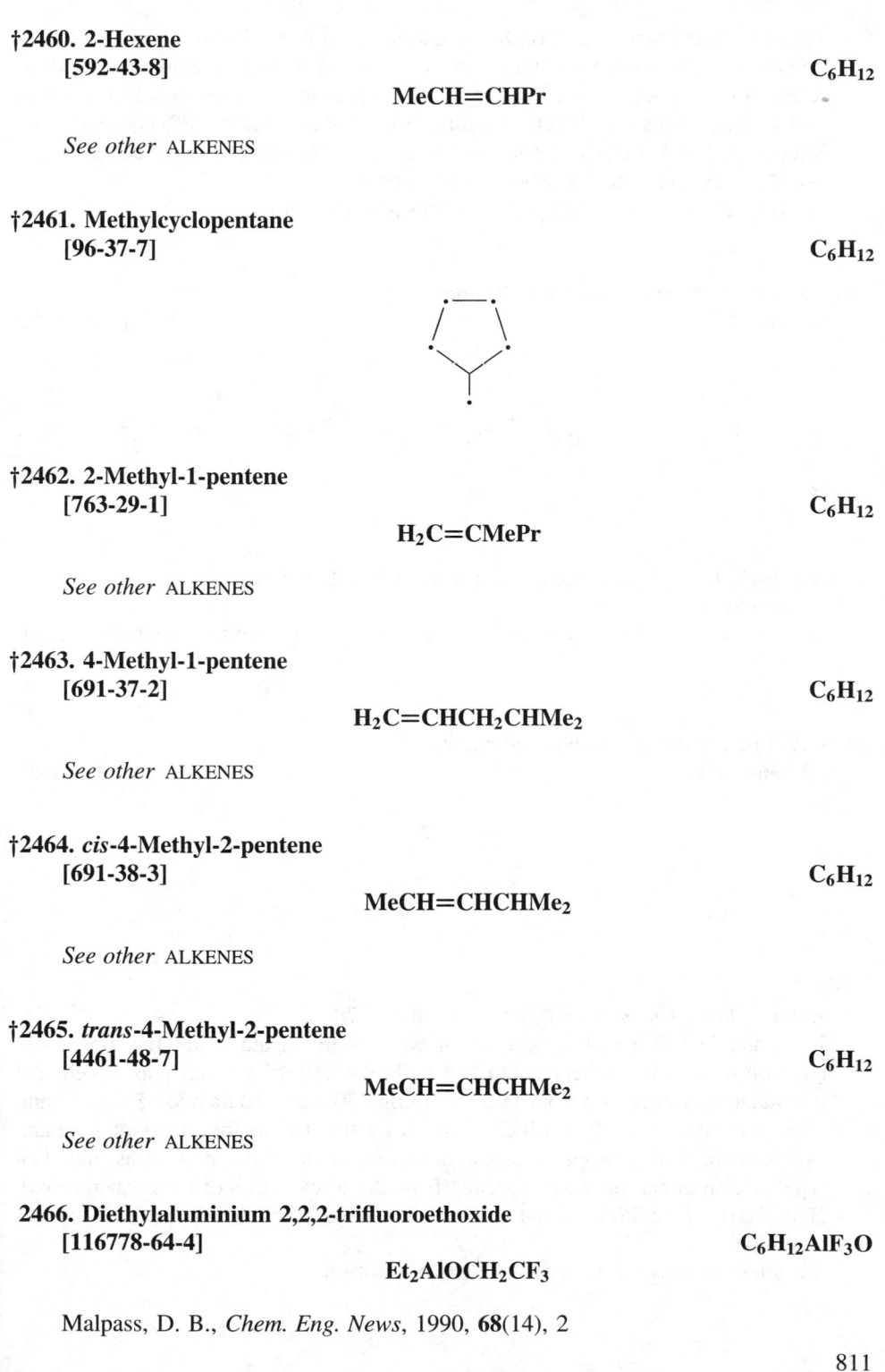

†2462. 2-Methyl-1-pentene
[763-29-1]   C$_6$H$_{12}$

$$\text{H}_2\text{C=CMePr}$$

See other ALKENES

†2463. 4-Methyl-1-pentene
[691-37-2]   C$_6$H$_{12}$

$$\text{H}_2\text{C=CHCH}_2\text{CHMe}_2$$

See other ALKENES

†2464. *cis*-4-Methyl-2-pentene
[691-38-3]   C$_6$H$_{12}$

$$\text{MeCH=CHCHMe}_2$$

See other ALKENES

†2465. *trans*-4-Methyl-2-pentene
[4461-48-7]   C$_6$H$_{12}$

$$\text{MeCH=CHCHMe}_2$$

See other ALKENES

2466. Diethylaluminium 2,2,2-trifluoroethoxide
[116778-64-4]   C$_6$H$_{12}$AlF$_3$O

$$\text{Et}_2\text{AlOCH}_2\text{CF}_3$$

Malpass, D. B., *Chem. Eng. News*, 1990, **68**(14), 2

This compound was found, on investigation of an explosion in a syringe during transfer, to have a decomposition energy equivalent to that of commercial explosives. Slow decomposition, even at room temperature, becomes explosive above 100°C. Shock sensitive. Stable as hydrocarbon solution below 25% concentration. Recommended that this and related compounds be handled only in such solution.
See FLUORINATED ORGANOLITHIUM COMPOUND
See related ALKYLALUMINIUM ALKOXIDES AND HYDRIDES

2467. Bis(dimethylarsinyldiazomethyl)mercury
[63382-64-9] $C_6H_{12}As_2HgN_4$

Glozbach, E. *et al., J. Organomet. Chem.*, 1977, **132**, 359
It is explosive.
See related ALKYLNON-METALS *See other* DIAZO COMPOUNDS, MERCURY COMPOUNDS

2468. *N*-Ethyl-*N*-propylcarbamoyl chloride
[98456-61-2] $C_6H_{12}ClNO$

Water
Anon., *CISHC Chem. Safety Summ.*, 1978, **49**, 78
The chloride (60 l) and 2 volumes of water were mixed ready for subsequent addition of alkali to effect controlled hydrolysis. Before alkali was added, the internal temperature rose from 15 to 25° during 30 min, and then to 35°C in 5 min, when gas was suddenly evolved. This was attributed to the effect of liberated hydrochloric acid causing autocatalytic acceleration of the hydrolysis and then rapid release of carbon dioxide arising from decarboxylation of the carbamic acid. Hydrolysis by addition of the chloride to excess alkali would prevent the gas evolution.
See other ACYL HALIDES, GAS EVOLUTION INCIDENTS

2469. 2,2-Dichloro-3,3-dimethylbutane
[594-84-3] $C_6H_{12}Cl_2$

$$\underset{H_3C}{\overset{Cl}{\underset{}{}}}\overset{Cl}{\underset{t\text{-Bu}}{C}}$$

Sodium hydroxide
1. Bartlett, P. D. *et al., J. Amer. Chem. Soc.*, 1942, **64**, 543
2. Kocienski, P. J., *J. Org. Chem.*, 1974, **39**, 3285–3286
A previous method [1] of preparing 3,3-dimethylbutyne by dehydrochlorination of the title compound in a sodium hydroxide melt is difficult to control and hazardous on the large scale. Use of potassium *tert*-butoxide as base in DMSO is a high-yielding, safe and convenient alternative method of preparation of the alkyne [2].
See Dimethyl sulfoxide: Metal alkoxides
See other HALOALKANES

2470. 1,6-Hexanediyl perchlorate
[95407-64-0] $C_6H_{12}Cl_2O_8$

See entry ALKYL PERCHLORATES (reference 6)

2471. Hexamethylenetetramine tetraiodide
[12001-65-9] $C_6H_{12}I_4N_4$

Merck, 1976, 779
The hexamine–iodine complex deflagrates at 138°C.
See other IODINE COMPOUNDS

2472. (Oxodiperoxy(1,3-dimethyl-2,4,5,6-tetrahydro-2-1H)-pyrimidinone) molybdenum
[128568-96-7] $C_6H_{12}MoN_2O_6$

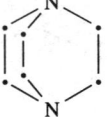

Paquette, L. A. *et al., Chem. Eng. News*, 1992, **70**(37), 2
A well dried sample of this complex exploded violently when pushed through a funnel with a steel spatula. The pyridine adduct, which was being prepared, is claimed to be a useful, safe, oxidant, replacing the hexamethylphosphoramide complex of MoO_5 (ibid.).
See other METAL PEROXIDES

2473. 1,4-Diazabicyclo[2.2.2]octane
[280-57-9] $C_6H_{12}N_2$

An exceptionally strong base.

Carbon
  See Carbon: 1,4-Diazabicyclo[2.2.2]octane

Cellulose nitrate
  See CELLULOSE NITRATE: amines

Hydrogen peroxide
  See next below *See other* ORGANIC BASES

2474. 1,4-Diazabicyclo[2.2.2]octane hydrogen peroxidate
[38910-25-8] $C_6H_{12}N_2.H_2O_2$

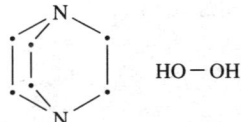

Davies, A. G. *et al., J. Chem. Soc., Perkin Trans. 2*, 1981, 1512

A sample of the 1:1 complex exploded while being dried overnight in a desiccator.
See other CRYSTALLINE HYDROGEN PEROXIDATES

2475. 1,3-Dimethylhexahydropyrimidone (Dimethylpropyleneurea) [7226-23-5] $C_6H_{12}N_2O$

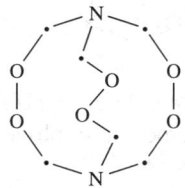

Chromium trioxide
See Chromium trioxide: 1,3-Dimethylhexahydropyrimidone

2476. 1,6-Diaza-3,4,8,9,12,13-hexaoxabicyclo[4.4.4]tetradecane ('Hexamethylenetriperoxydiamine') [283-66-9] $C_6H_{12}N_2O_6$

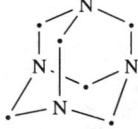

Alone, or Bromine, or Sulfuric acid
Leulier, *J. Pharm. Chim.*, 1917, **15**, 222–229
The compound, precipitated by interaction of hexamethylene tetramine and acidic 30% hydrogen peroxide, is a heat- and shock-sensitive powerful explosive when dry, much more shock sensitive than mercury fulminate. It explodes in contact with bromine or sulfuric acid.
See other CYCLIC PEROXIDES

2477. Hexamethylenetetramine (1,3,5,7-Tetraazatricyclo[3.3.1.1³,⁷]decane) [100-97-0] $C_6H_{12}N_4$

Finely powdered dry hexamine is a significant dust explosion hazard.
See entry DUST EXPLOSION INCIDENTS (reference 22)

Acetic acid, Acetic anhydride, Ammonium nitrate, Nitric acid
See Nitric acid: Acetic anhydride, Hexamethylenetetramine acetate

1-Bromopentaborane(9)
   See 1-Bromopentaborane(9): Air, or Hexamethylenetetramine

Hydrogen peroxide
   See 1,6-Diaza-3,4,8,9,12,13-hexaoxabicyclo[4.4.4]tetradecane

Iodine
   See Hexamethylenetetramine tetraiodide

Iodoform
   See Iodoform: Hexamethylenetetramine
   See other ORGANIC BASES

2478. 4a,8a,9a,10a-Tetraaza-2,3,6,7-tetraoxaperhydroanthracene
   [262-38-4] $C_6H_{12}N_4O_4$

Katritzky, A. R. *et al., J. Chem. Soc., Perkin Trans. 2*, 1979, 1136
The bis-peroxy tricyclic heterocycle explodes at 120°C.
See other CYCLIC PEROXIDES

2479. 1,2-Bis(2-azidoethoxy)ethane
   [59559-04-7] $C_6H_{12}N_6O_2$

Ethanol, Hydrogen sulfide
   Anon., *Lab. Accid. Higher Educ.* Item 32, HSE, Barking, 1987
   During the reduction of the azide in ethanol with hydrogen sulfide (no regulation
   other than main cylinder valve) under reflux (with insufficient flow of condenser
   water), the mixture apparently overheated, then exploded violently.
   See other ORGANIC AZIDES

816

2480. 1,4-Diazabicyclo[2.2.2]octane 1,4-bis(nitroimidate)
[51470-74-7] $C_6H_{12}N_6O_4$

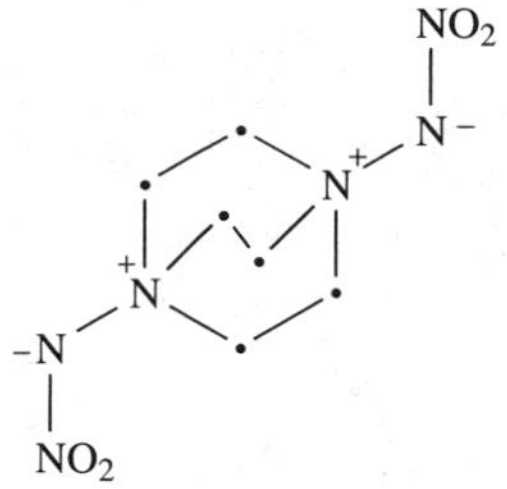

See N-AZOLIUM NITROIMIDATES

2481. Bis(2-azidoethoxymethyl)nitramine
[88487-87-6] $C_6H_{12}N_8O_4$

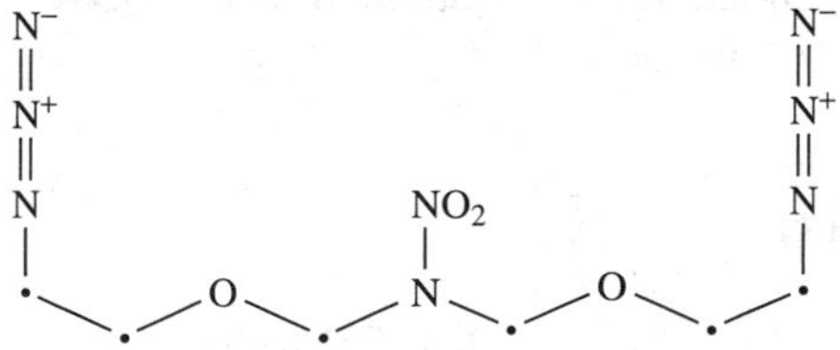

See entry ENERGETIC COMPOUNDS See other N-NITRO COMPOUNDS, ORGANIC AZIDES

2482. Tris(2-azidoethyl)amine
[84928-99-4] $C_6H_{12}N_{10}$

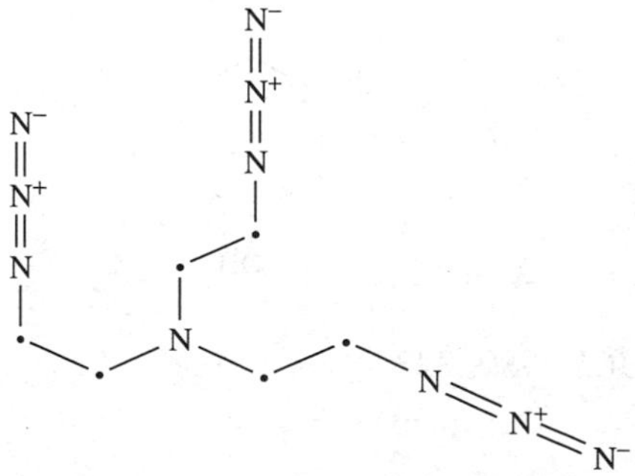

See entry ENERGETIC COMPOUNDS See other ORGANIC AZIDES

817

2483. 1,1,3,3,5,5-Tris-spiro(N,N'-dinitroethylenediamino)cyclotriphosphazene (1,4,8,11,14,17-Hexanitro-1,4,6,8,11,12,14,17,18-nonaaza-5λ^5,7λ^5,13λ^5-triphosphatatrispiro[4.1.4.1.4.1]octadeca-5,7(12),13(18)-triene)
[155793-91-2] $C_6H_{12}N_{15}O_{12}P_3$

Paritosh, D. R. *et al., Chem. Abs.*, 1994, **121**, 35869j; U.S. Pat. Appl. 42,229
Paritosh, D. R. *et al., Phosphorus, Sulfur, Silicon Relat. Elem.*, 1994, **90**(1–4), 175
This and several other compounds with fewer nitro groups and partial replacement of the spiro rings by halogen are patented as explosives.
See other N-NITRO COMPOUNDS

†2484. Butyl vinyl ether
[111-34-2] $C_6H_{12}O$

$$BuOCH=CH_2$$

Glikin, M. A. *et al., Chem. Abs.*, 1978, **89**, 30115
'Explosion hazard of industrial synthesis of butyl vinyl ether' (title only translated).
See other PEROXIDISABLE COMPOUNDS

2485. Cyclohexanol
[108-93-0] $C_6H_{12}O$

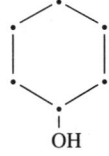

FPA H91, 1980; *HCS 1980*, 343

Other reactants
  Yoshida, 1980, 151
  MRH values calculated for 12 combinations with oxidants are given.

Oxidants MRH values show % of oxidant

See Chromium trioxide: Alcohols MRH 2.55/92

Nitric acid: Alcohols MRH 5.86/81

†2486. 3,3-Dimethyl-2-butanone

[75-97-8] **$C_6H_{12}O$**

MeCO.CMe_3

†2487. 2-Ethylbutanal

[97-96-1] **$C_6H_{12}O$**

$Et_2CHCO.H$

See other PEROXIDISABLE COMPOUNDS

†2488. 2-Hexanone

[591-78-6] **$C_6H_{12}O$**

MeCO.Bu

HCS 1980, 643

†2489. 3-Hexanone

[589-38-8] **$C_6H_{12}O$**

EtCO.Pr

†2490. Isobutyl vinyl ether

[109-53-5] **$C_6H_{12}O$**

$Me_2CHCH_2OCH{=}CH_2$

†2491. 2-Methylpentanal

[123-15-9] **$C_6H_{12}O$**

$Me_2CHC_2H_4CO.H$

†2492. 3-Methylpentanal

[15877-57-3] **$C_6H_{12}O$**

$EtCHMeCH_2CO.H$

†2493. 2-Methyl-3-pentanone

[565-69-5] **$C_6H_{12}O$**

$Me_2CHCO.Et$

†2494. 3-Methyl-2-pentanone

[565-61-7] **$C_6H_{12}O$**

MeCOCHMeEt

†2495. 4-Methyl-2-pentanone
[108-10-1] $C_6H_{12}O$

MeCO.CH$_2$CHMe$_2$

FPA H97, 1981; *HCS 1980*, 576

1. Smith, G. H., *Chem. & Ind.*, 1972, 291
2. Bretherick, L., *Chem. & Ind.*, 1972, 363
4-Methyl-2-pentanone had not been considered prone to autoxidation, but an explosion during prolonged and repeated aerobic hot evaporation of the solvent [1] was attributed to formation and explosion of a peroxide [2].

Other reactants
Yoshida, 1980, 363
MRH values calculated for 13 combinations with oxidants are given.

Potassium *tert*-butoxide
  See Potassium *tert*-butoxide: Acids, etc.
  See other PEROXIDISABLE COMPOUNDS

†2496. Butyl acetate
[123-86-4] $C_6H_{12}O_2$

BuOCO.Me

FPA H116, 1982; *HCS 1980*, 233

Potasium *tert*-butoxide
  See Potassium *tert*-butoxide: Acids, etc.

†2497. 2-Butyl acetate
[105-46-4] $C_6H_{12}O_2$

MeCO.OCHMeEt

HCS 1980, 234

†2498. 2,6-Dimethyl-1,4-dioxane
[10138-17-7] $C_6H_{12}O_2$

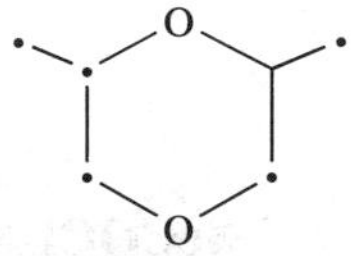

†2499. Ethyl isobutyrate
[97-62-1] $C_6H_{12}O_2$

EtOCO.CHMe$_2$

†2500. 2-Ethyl-2-methyl-1,3-dioxolane
[126-39-6] $C_6H_{12}O_2$

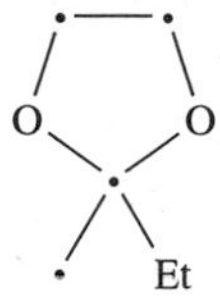

†2501. 4-Hydroxy-4-methyl-2-pentanone
[123-42-2] $C_6H_{12}O_2$

MeCO.CH$_2$CH(OH)Me$_2$

HCS 1980, 355

†2502. Isobutyl acetate
[110-19-0] $C_6H_{12}O_2$

Me$_2$CHCH$_2$OCO.Me

HCS 1980, 567

†2503. Isopentyl formate
[110-45-2] $C_6H_{12}O_2$

Me$_2$CHC$_2$H$_4$OCO.H

HCS 1980, 564

†2504. Isopropyl propionate
[637-78-5] $C_6H_{12}O_2$

Me$_2$CHOCO.Et

†2505. Methyl isovalerate
[556-24-1] $C_6H_{12}O_2$

MeOCO.CH$_2$CHMe$_2$

†2506. Methyl pivalate
[598-98-1] $C_6H_{12}O_2$

MeOCO.CMe$_3$

†2507. Methyl valerate
[624-24-8]

$C_6H_{12}O_2$

MeOCO.Bu

2508. Tetramethyl-1,2-dioxetane
[35856-82-7]

$C_6H_{12}O_2$

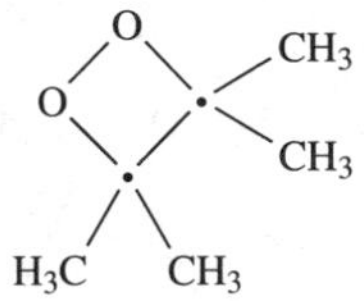

Kopecky, K. R. *et al., Can. J. Chem.*, 1975, **53**, 1107
Several explosions occurred in vacuum-sealed samples kept at ambient tempera-
ture. Storage under air at $-20°C$, or as solutions up to 2 Mol appear safe.
See 3,3-Dimethyl-1,2-dioxetane
See DIOXETANES *See other* CYCLIC PEROXIDES

2509. *tert*-Butyl peroxyacetate
[107-71-1]

$C_6H_{12}O_3$

Me₃COOCO.Me

1. Martin, J. H., *Ind. Eng. Chem.*, 1960, **52**(4), 49A
2. Castrantas, 1965, 16
3. *CHETAH*, 1990, 188

The peroxyester explodes with great violence when rapidly heated to a critical
temperature. Previous standard explosivity tests had not shown this behaviour.
The presence of benzene (or preferably a less toxic solvent) as diluent prevents
the explosive decomposition, but if the solvent evaporates, the residue is dangerous
[1]. The pure ester is also shock-sensitive and detonable, but the commercial 75%
solutions are not [2]. However, a 75% benzene solution has been exploded with a
detonator, though not by mechanical shock [3].
See other PEROXYESTERS

2510. Cyclobutylmethyl methanesulfonate
[22524-45-4]

$C_6H_{12}O_3S$

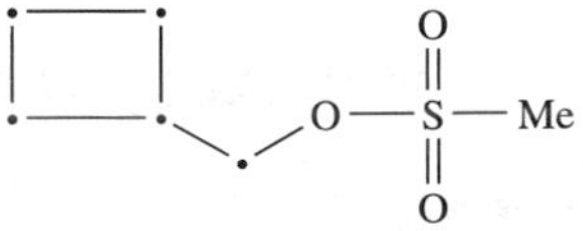

Explodes during distillation at $81-89°C/0.8$ mbar.
See entry SULFONIC ACID ESTERS *See other* SULFUR ESTERS

822

2511. 2-Ethoxyethyl acetate
[111-15-9] $C_6H_{12}O_3$

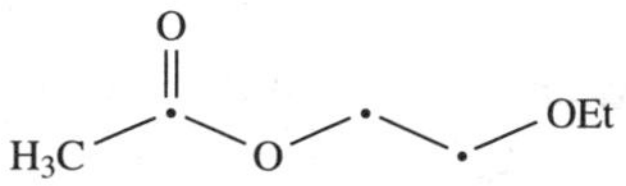

HCS 1980, 461

1. Anon., private comm., 1975
2. Author's comments

Mild explosions have occurred at the end of technical-scale distillations of 'ethyl glycol acetate' in a copper batch still about 20 min after the kettle heater was shut off. It is possible that air was drawn into the distillation column as it cooled, creating a flammable mixture, but no mechanism for ignition could be established. Oxidation studies did not indicate the likely formation of high peroxide concentrations in the liquid phase of this ether–ester [1]. It seems remotely possible that small amounts of dioxane (formed by dehydration of traces of ethylene glycol) or of acetaldehyde (arising from traces of ethanol or from cracking of the ester, catalysed perhaps by copper), and both having very low autoignition temperatures, may have been involved in the incidents [2].

†2512. Isobutyl peroxyacetate
[55153-36-1] $C_6H_{12}O_3$

$$Me_2CHCH_2OOCO.Me$$

See other PEROXYESTERS

2513. *trans*-2-Hexene ozonide (3-Methyl-5-propyl-1,2-4-trioxolane)
[ ] $C_6H_{12}O_3$

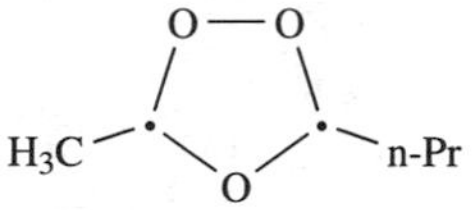

Loan, L. D. *et al., J. Amer. Chem. Soc.*, 1965, **87**, 41

Attempts to get C and H analyses by combustion of the ozonides of 2-butene, 2-pentene and 2-hexene caused violent explosions, though oxygen analyses were uneventful.

See other OZONIDES

2514. Peroxyhexanoic acid
[5106-46-7] $C_6H_{12}O_3$

$$BuCH_2CO.OOH$$

Swern, D., *Chem. Rev.*, 1949, **45**, 10

Fairly stable at ambient temperature, it explodes and ignites on rapid heating.
See other PEROXYACIDS

†2515. 2,4,6-Trimethyltrioxane
[123-63-7]

$C_6H_{12}O_3$

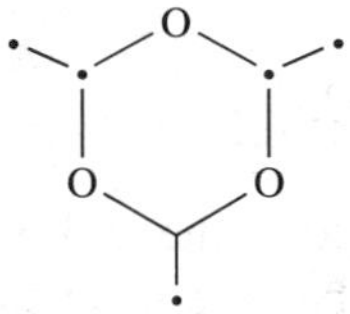

Nitric acid
    See Nitric acid: 2,4,6-Trimethyltrioxane

2516. Tetramethoxyethylene
[1069-12-1]

$C_6H_{12}O_4$

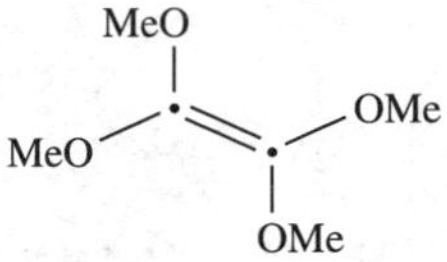

Preparative hazard
    See 1,2,3,4-Tetrachloro-7,7-dimethoxy-5-phenylbicyclo[2.2.1]hepta-2,5-diene

2517. 3,3,6,6-Tetramethyl-1,2,4,5-tetraoxane
[1073-91-2]

$C_6H_{12}O_4$

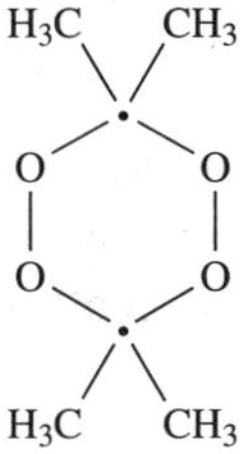

Baeyer, A. *et al., Ber.*, 1900, **33**, 858
Action of hydrogen peroxide or permonosulfuric acid on acetone produces this dimeric acetone peroxide, which explodes violently on impact, friction or rapid heating.
See other CYCLIC PEROXIDES

824

2518. Glucose
[50-99-7] $C_6H_{12}O_6$

Energy of decomposition (in range 224–330°C) measured as 0.406 kJ/g.
See entry THERMOCHEMISTRY AND EXOTHERMIC DECOMPOSITION (reference 2)

Alkali
  See SUGARS

Potassium nitrate, Sodium peroxide
  See Sodium peroxide: Glucose, Potassium nitrate
  See other REDUCANTS

2519. Bis(dimethylthallium)acetylide
[ ] $C_6H_{12}Tl_2$

Nast, R. *et al., J. Organomet. Chem.*, 1966, **6**, 461
Extremely explosive, heat- and friction-sensitive.
See related ALKYLMETALS, METAL ACETYLIDES

2520. Hexyl perchlorate
[52936-24-0] $C_6H_{13}ClO_4$

See entry ALKYL PERCHLORATES (reference 6)

†2521. Cyclohexylamine
[108-91-8]

$C_6H_{13}N$

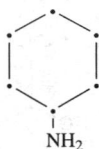

HCS 1980, 347

Nitric acid
    See Nitric acid: Cyclohexylamine *See other* ORGANIC BASES

†2522. 1-Methylpiperidine
[626-67-5]

$C_6H_{13}N$

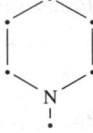

See other ORGANIC BASES

†2523. 2-Methylpiperidine
[109-05-7]

$C_6H_{13}N$

See other ORGANIC BASES

†2524. 3-Methylpiperidine
[626-56-2]
See other ORGANIC BASES

$C_6H_{13}N$

†2525. 4-Methylpiperidine
[626-58-4]

$C_6H_{13}N$

See other ORGANIC BASES

†2526. Perhydroazepine (Hexamethyleneimine)
[111-49-9] $C_6H_{13}N$

See other ORGANIC BASES

2527. *N*-Methylpiperidine *N*-oxide
[17206-00-7] $C_6H_{13}NO$

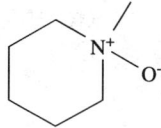

Cope, A. C. *et al., J. Amer. Chem. Soc.*, 1960, **82**, 4656
This amine oxide is thermally surprisingly stable, decomposing explosively at 215°C.
See other N-OXIDES

2528. *N*-Butyl-*N*-2-azidoethylnitramine
[84928-98-3] $C_6H_{13}N_5O_2$

$$BuN(NO_2)C_2H_4N_3$$

See entry ENERGETIC COMPOUNDS
See other N-NITRO COMPOUNDS, ORGANIC AZIDES

†2529. 2,2-Dimethylbutane
[75-83-2] C_6H_{14}

$$Me_3CEt$$

†2530. 2.3-Dimethylbutane
[79-29-8] C_6H_{14}

$$CH_3CHMeCHMe_2$$

†2531. Hexane
[110-54-3] C_6H_{14}

$$EtC_2H_4Et$$

FPA H83, 1979; *HCS 1980*, 533 (covers isomers); *RSC Lab. Hazards Data Sheet No. 63*, 1987

Dinitrogen tetraoxide
See Dinitrogen tetraoxide: Hydrocarbons

Other reactants
Yoshida, 1980, 326
MRH values calculated for 13 combinations with oxidants are given.

†2532. Isohexane (2-Methylpentane)
[107-83-5] C_6H_{14}

$$Me_2CHPr$$

†2533. 3-Methylpentane
[96-14-0] C_6H_{14}

$$EtCHMeEt$$

2534. Chlorodipropylborane
[22086-53-9] $C_6H_{14}BCl$

Leleu, *Cahiers*, 1977, (88), 365
It ignites in air
See other ALKYLHALOBORANES

2535. Diisopropylberyllium
[15721-33-2] $C_6H_{14}Be$

$$(Me_2CH)_2Be$$

Water
Coates, G. E. *et al., J. Chem. Soc.*, 1954, 22
Uncontrolled reaction with water is explosive.
See other ALKYLMETALS

2536. 1,4-Bis(isothiouronio)-2-butene dichloride
[67405-47-4] $C_6H_{14}Cl_2N_4S_2$

Harrp, D. N. *et al., Org. Prep. Proceed. Int.*, 1978, **10**, 13
Crystallisation from 95% ethanol is strongly exothermic.

2537. Dipropylmercury
[628-85-3] $C_6H_{14}Hg$

$$Pr_2Hg$$

Iodine
Whitmore, 1921, 100
Interaction is violent.
See other ALKYLMETALS, MERCURY COMPOUNDS

2538. Diisopropyl hyponitrite (Bis(2-propyloxy)diazene)
[82522-47-2]; *E* **-[86886-16-0]** $C_6H_{14}N_2O_2$

$$Me_2CHON{=}NOCHMe_2$$

Ogle, C. A. *et al., J. Org. Chem.*, 1983, **48**, 3729
It detonated violently when deliberately struck.
See other DIALKYL HYPONITRITES

2539. Dipropyl hyponitrite (Bis(propyloxy)diazene)
[ ] $C_6H_{14}N_2O_2$

$$PrON{=}NOPr$$

See entry DIALKYL HYPONITRITES

†2540. Butyl ethyl ether
[628-81-9] $C_6H_{14}O$

$$BuOEt$$

HCS 1980, 473

See other PEROXIDISABLE COMPOUNDS

†2541. *tert***-Butyl ethyl ether**
[637-92-3] $C_6H_{14}O$

$$Me_3COEt$$

†2542. Diisopropyl ether
[108-20-3] $C_6H_{14}O$

$$Me_2CHOCHMe_2$$

FPA H101, 1981; *HCS 1980*, 579

1. Anon., *Chem. Eng. News*, 1942, **20**, 1458
2. *MCA Case Histories Nos. 603, 1607*

3. Douglass, I. B., *J. Chem. Educ.*, 1963, **40**, 469
4. Rieche, A. *et al., Ber.*, 1942, **75**, 1016
5. Hamstead, A. C. *et al., J. Chem. Eng. Data.*, 1960, **5**, 583; *Ind. Eng. Chem.*, 1961, **53**(2), 63A
6. Walton, G. C., *CHAS Notes*, 1988, **6**, 3

There is a long history of violent explosions involving peroxidised diisopropyl ether, with initiation by disturbing a drum [1], unscrewing a bottle cap [2], or accidental impact [3]. This ether, with two susceptible hydrogen atoms adjacent to the oxygen link, is extremely readily peroxidised after only a few hours' exposure to air, initially to a dihydroperoxide which disproportionates to dimeric and trimeric acetone peroxides. These separate from solution and are highly explosive [4]. Presence of a crystalline solid in a sample of the ether should be a cause for great concern, no attempt being made to open the container. Professional assistance should be sought for the safe disposal of substantially peroxidised samples. It has been reported that the ether may be inhibited completely against peroxide formation for a considerable time by addition of *N*-benzyl-4-aminophenol at 16 ppm, or by addition of diethylenetriamine, triethylenetetramine or tetraethylenepentamine at 50 ppm [5]. Further information on the extreme hazards of peroxidised diisopropyl ether is presented, with directions for safe disposal in a remote location by controlled explosion. The dry crystalline peroxide is photo-sensitive and will detonate on exposure to sunlight. Four additional literature references are included [6].
See 3,3,6,6-Tetramethyl-1,2,4,5-tetraoxane, *and* 3,3,6,6,9,9-Hexamethyl-1,2,4,5,7,8-hexoxonane
See also ETHERS

Propionyl chloride
See Propionyl chloride: Diisopropyl ether
See other PEROXIDATION INCIDENTS, PEROXIDISABLE COMPOUNDS

†2543. Dipropyl ether
[111-43-3] $C_6H_{14}O$

PrOPr

See other PEROXIDISABLE COMPOUNDS

†2544. 2-Methyl-2-pentanol
[590-36-3] $C_6H_{14}O$

$Me_2C(OH)Pr$

†2545. 1,1-Diethoxyethane (Diethyl acetal)
[105-57-7] $C_6H_{14}O_2$

$(EtO)_2CHMe$

Mee, A. J., *School Sci. Rev.*, 1940, **22**(86), 95

A peroxidised sample exploded violently during distillation.
See other PEROXIDISABLE COMPOUNDS

†2546. 1,2-Diethoxyethane
[629-14-1] $C_6H_{14}O_2$

EtOC$_2$H$_4$OEt

HCS 1980, 400

See other PEROXIDISABLE COMPOUNDS

2547. Dipropyl peroxide
[29914-92-9] $C_6H_{14}O_2$

PrOOPr

Swern, 1972, Vol. 3, 21
Unexpected explosions have occurred with dipropyl peroxides.
See other DIALKYL PEROXIDES

2548. 2,6-Dimethyl-1,3-dioxa-2,6-diphosphacyclooctane
[79251-56-2; 79251-57-3] *cis-*, *trans-*, resp. $C_6H_{14}O_2P_2$

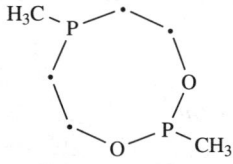

Piccini-Leopardi, C. *et al., J. Chem. Soc., Perkin Trans 2*, 1986, 85–92
The isomers could not be separated by distillation at 65–68°C/0.07 mbar, and bath
temperatures above 120°C caused explosions.
See related ALKYLPHOSPHINES, PHOSPHORUS ESTERS

2549. Bis-(2-methoxyethyl) ether
[111-96-6] $C_6H_{14}O_3$

MeO ⁀ O ⁀ OMe

'Diglyme Properties and Uses', Brochure, London, ICI, 1967
Details of properties, applications and safe handling are given. Light, heat and
oxygen (air) promote formation of potentially explosive peroxides. These may be
removed by stirring with a suspension of iron oxide in aqueous alkali.

Metal hydrides
   See Aluminium hydride: Carbon dioxide, etc.
      Lithium tetrahydroaluminate: Bis(2-methoxyethyl) ether
   See other PEROXIDISABLE COMPOUNDS

2550. 1-Pentyl methanesulfonate
[7958-20-3]   C₆H₁₄O₃S

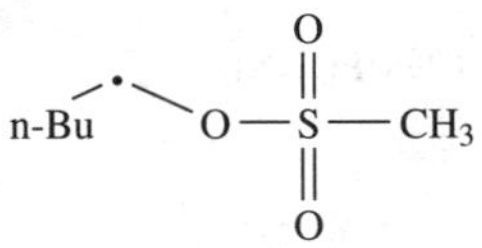

It decomposes vigorously at 185°C.
See entry SULFONIC ACID ESTERS *See other* SULFUR ESTERS

2551. Dipropylzinc
[628-91-1]   C₆H₁₄Zn
                          PrZnPr

Ellern, 1968, 24
It ignites in air if exposed on an extended surface.
See other DIALKYLZINCS

2552. Dipropylaluminium hydride
[2036-15-9]   C₆H₁₅Al
                          Pr₂AlH

HCS 1980, 444

See entry ALKYLALUMINIUM ALKOXIDES AND HYDRIDES

†2553. Triethylaluminium
[97-93-8]   C₆H₁₅Al
                          Et₃Al

HCS 1980, 929

It ignites in air.
See TRIALKYLALUMINIUMS

Alcohols   MRH Methanol, 1.45/46
   Mirviss, S. B. *et al., Ind. Eng. Chem.*, 1961, **53**(1), 54A

Interaction of methanol, ethanol and 2-propanol with the undiluted trialkylaluminium is explosive, while *tert*-butanol reacts vigorously.
See ALKYLALUMINIUM DERIVATIVES

Carbon tetrachloride   MRH 1.51/85
Reineckel, H., *Angew. Chem. (Intern. Ed.)*, 1964, **3**, 65
A mixture prepared at 0°C with a 3:1 molar excess of halocarbon exploded violently soon after removal of the ice bath. Formation of a 1:1 chlorine-bridged adduct was assumed.
See ALKYLALUMINIUM DERIVATIVES: halocarbons

Dimethylformamide   MRH 1.38/66
'DMF Brochure', Billingham, ICI, 1965
A mixture of the amide solvent and triethylaluminium explodes when heated.

Other reactants
Yoshida, 1980, 248
MRH values calculated for 14 combinations, largely with oxidants, are given.

Tris(pentafluorophenyl)boron
Pohlmann, J. L. W. *et al., Z. Naturforsch.*, 1965, **20b** (1), 5
A mixture detonated on warming to 70°C.
See METAL DERIVATIVES OF ORGANOFLUORINE COMPOUNDS
See other ALKYLMETALS, TRIALKYLALUMINIUMS

2554. Diethylethoxyaluminium
[1586-92-1] $C_6H_{15}AlO$

Dangerous Loads, 1972
The pure material, or solutions of above 20% concentration, ignite in air.
See other ALKYLALUMINIUM ALKOXIDES AND HYDRIDES

2555. Triethoxydialuminium tribromide
[65232-69-1] $C_6H_{15}Al_2Br_3O_3$

Air, or Ethanol, or Water
Anon., *Chem. Eng. Progr.*, 1966, **62**(9), 128

833

It ignites in air and explodes with ethanol and water.
See related ALKYLALUMINIUM ALKOXIDES AND HYDRIDES, METAL HALIDES

2556. Triethyldialuminium trichloride
[12075-68-2]

$C_6H_{15}Al_2Cl_3$

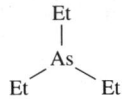

HCS 1980, 466

Carbon tetrachloride
Reineckel, H., *Angew. Chem. (Intern. Ed.)*, 1964, **3**, 65
A mixture exploded when warmed to room temperature.
See other ALKYLALUMINIUM HALIDES *See related* METAL HALIDES

2557. Triethylarsine
[617-75-4]

$C_6H_{15}As$

Sidgwick, 1950, 762
It inflames in air.
See other ALKYLNON-METALS

2558. Triethylphosphinegold nitrate
[14243-51-7]

$C_6H_{15}AuNO_3P$

Coates, G. E. *et al., Aust. J. Chem.*, 1966, **19**, 541
After thorough desiccation, the crystalline solid exploded spontaneously.
See other GOLD COMPOUNDS *See related* ORGANOMETALLIC NITRATES

2559. Triethylborane
[97-94-9] $C_6H_{15}B$

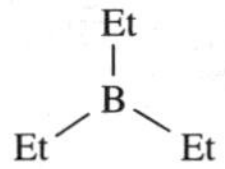

$$Et_2B-Et \text{ (structure: B with three Et groups)}$$

HCS 1980, 931

Meerwein, H., *J. Prakt. Chem.*, 1937, **147**, 240
It ignites on exposure to air.

Halogens, or Oxygen
Coates, 1960, 84
It ignites in contact with chlorine or bromine, and explodes in oxygen.

Triethylaluminium
Hurd, D. T., *J. Amer. Chem. Soc.*, 1948, **70**, 2053
Mixtures with triethylaluminium have been used as hypergolic igniters in rocket propulsion systems.
See other ALKYLBORANES

†2560. Triethyl borate
[150-46-9] $C_6H_{15}BO_3$

2561. Triethylbismuth
[617-77-6] $C_6H_{15}Bi$

Et₃Bi

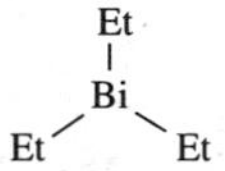

Coates, 1967, Vol. 1, 537
It ignites in air, and explodes at about 150°C, before distillation begins.
See other TRIALKYLBISMUTHS

2562. Triethylsilyl perchlorate
[18244-91-2] $C_6H_{15}ClO_4Si$

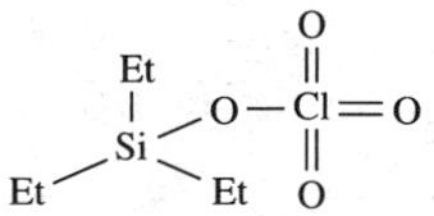

1. Wannagat, U. *et al.*, *Z. Anorg. Chem.*, 1950, **302**, 185–198
2. Barton, T. J. *et al.*, *J. Org. Chem.*, 1978, **43**, 3651

Several trialkyl- or triaryl-silyl perchlorates explode on heating [1]. A syringe used to inject a sample of the title perchlorate ester into a reaction flask exploded soon after addition was complete [2].

See other ORGANOSILYL PERCHLORATES

2563. Triethylgallium
[1115-99-7] $C_6H_{15}Ga$

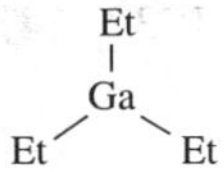

Air, or Water

Dennis, L. M. *et al., J. Amer. Chem. Soc.*, 1932, **54**, 182

The compound ignites in air. Breaking a bulb containing 0.2 g under cold water shattered the container. The monoetherate behaved similarly under 6 M nitric acid

See other ALKYLMETALS

2564. Triethylindium
[923-34-2] $C_6H_{15}In$

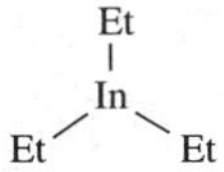

Leleu, *Cahiers*, 1977, (88), 367

It ignites in air.

See other ALKYLMETALS

†2565. Butylethylamine
[13360-63-9] $C_6H_{15}N$

BuNHEt

See other ORGANIC BASES

†2566. Diisopropylamine
[108-18-9] $C_6H_{15}N$

$(Me_2CH)_2NH$

See other ORGANIC BASES

†2567. 1,3-Dimethylbutylamine
[108-09-8] $C_6H_{15}N$

$Me_2CHCH_2CHMeNH_2$

See other ORGANIC BASES

†2568. Dipropylamine
[142-84-7] $C_6H_{15}N$

$$Pr_2NH$$

See other ORGANIC BASES

†2569. Triethylamine
[121-44-8] $C_6H_{15}N$

$$Et_3N$$

FPA H92, 1980; *HCS 1980*, 930; *RSC Lab. Hazard Data Sheet No. 27*, 1984

2-Bromo-2,5,5-trimethylcyclopentanone, Potassium hydroxide
See 2-Bromo-2,5,5-trimethylcyclopentanone: Potassium hydroxide, etc.

Dinitrogen tetraoxide
See Dinitrogen tetraoxide: Triethylamine

Maleic anhydride
See Maleic anhydride: Bases, etc.

Other reactants
Yoshida, 1980, 247
MRH Values calculated for 13 combinations with oxidants are given.
See other ORGANIC BASES

2570. Triethylamine hydrogen peroxidate
[ ] $C_6H_{15}N.4H_2O_2$

See entry CRYSTALLINE HYDROGEN PEROXIDATES

2571. 2-Diethylammonioethyl nitrate nitrate
[ ] $C_6H_{15}N_3O_6$

1. Barbière, J., *Bull. Soc. Chim. Fr.*, 1944, **11**, 470–480
2. Fakstorp, J. *et al.*, *Acta Chem. Scand.*, 1951, **5**, 968–969

Repetition of the original preparation [1] involving interaction of diethyl-laminoethanol with fuming nitric acid, followed by vacuum distillation of excess acid, invariably caused explosions during this operation. A modified procedure without evaporation is described [2].

See related ALKYL NITRATES, OXOSALTS OF NITROGENOUS BASES

2572. Diethyl ethanephosphonite
[2651-85-6]   C₆H₁₅O₂P

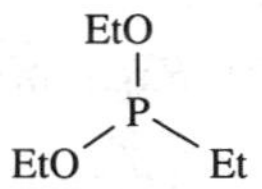

Arbuzov, B. A. *et al., Chem. Abs.*, 1953, **47**, 3226a
Absorbed on to filter paper it ignites in air.
See other PHOSPHORUS ESTERS *See related* ALKYLPHOSPHINES

2573. Triethylphosphine
[554-70-1]   C₆H₁₅P

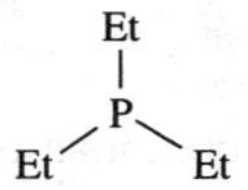

Oxygen
  Engler, C. *et al., Ber.*, 1901, **34**, 365
  Action of oxygen at low temperature on the phosphine produces an explosive product.

Palladium(II) perchlorate
  See Perchloratotris(triethylphosphine)palladium(II) perchlorate
  See other ALKYLPHOSPHINES

2574. Triethylantimony
[617-85-6]   C₆H₁₅Sb

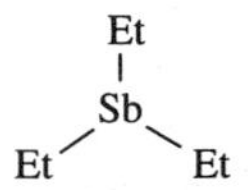

von Schwartz, 1918, 322

838

It ignites in air.

See other ALKYLMETALS

2575. Sodium dihydrobis(2-methoxyethoxy)aluminate
[22722-98-1] $C_6H_{16}AlNaO_4$

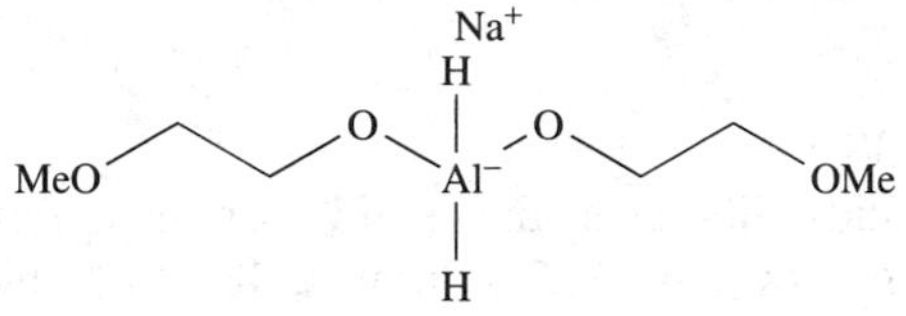

Preparative hazard
1. Casensky, B. *et al., Inorg. Synth.*, 1978, **18**, 150
2. Strauf, O. *et al., J. Organomet. Chem. Library 15*, 1985

The preparation is effected by heating sodium, aluminium powder and 2-methoxyethanol in toluene at 150°C under 70bar pressure of hydrogen in an autoclave. The reaction is exothermic, and hydrogen is first evolved by the dissolving metals, initially increasing the pressure, before absorption of hydrogen into the complex hydride occurs to reduce the pressure. Care in controlling temperature and the volume of reagents charged to the autoclave is necessary to avoid hazards. The solid complex hydride is not pyrophoric with air or water, but concentrated solutions (above 70%) may ignite when exposed to air on dry textiles [1]. It is, however a novel and versatile reducing agent which can with advantage (including that of increased safety in use) replace other complex hydrides [2].

See other COMPLEX HYDRIDES, GAS EVOLUTION INCIDENTS

2576. Bis(methyl 1-methylhydrazinocarbodithioate N^2,S')(perchlorato-O,O') copper(1+) perchlorate
[67870-97-7] $C_6H_{16}Cl_2CuN_4O_8S_4$

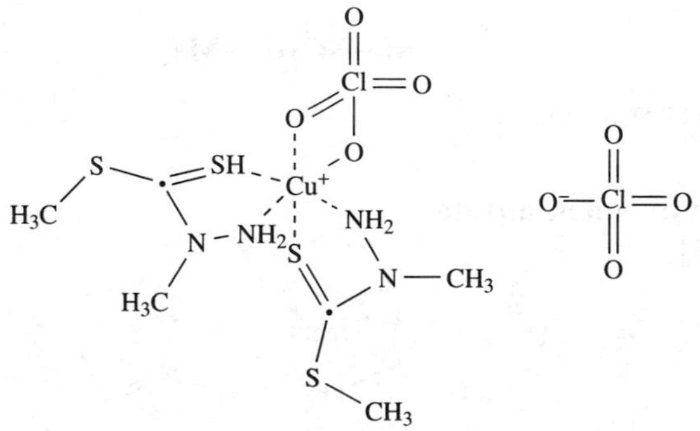

See Bis-O, N[(N'-pent-2-en-2-oxy-4-ylidene)-N,S-dimethyldithiocarbazate] copper(II) perchlorate

839

2577. Ammonium hexacyanoferrate(II)
[14481-29-9] $C_6H_{16}FeN_{10}$

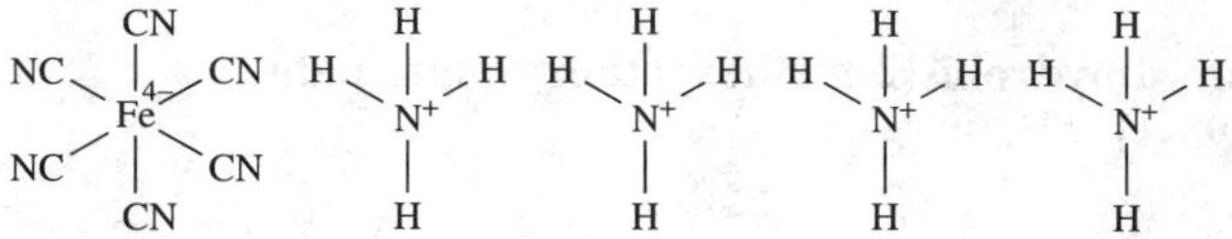

Hydrochloric acid
Ephraim, 1939, 303
Contact with acids liberates the solid complex ferrocyanic acid which is endothermic (ΔH°_{f} (aq) +534.7 kJ/mol, 2.48 kJ/g), and forms complexes with diethyl ether etc.

Metal nitrates
See Cobalt(II) nitrate: Ammonium hexacyanoferrate
Copper(II) nitrate: Ammonium hexacyanoferrate
See related ENDOTHERMIC COMPOUNDS, METAL CYANIDES (AND CYANO COMPLEXES)

2578. Lithium triethylsilylamide
[13270-95-6] $C_6H_{16}LiNSi$

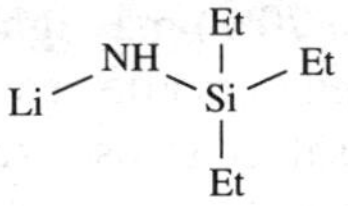

Oxidants
It is hypergolic with fluorine or fuming nitric acid, and explodes with ozone.
See SILYLHYDRAZINES *See other* N-METAL DERIVATIVES

†2579. 1,2-Bis(dimethylamino)ethane
[110-18-9] $C_6H_{16}N_2$

$$Me_2NC_2H_4NMe_2$$

See other ORGANIC BASES

2580. Triethylammonium nitrate
[27096-31-7] $C_6H_{16}N_2O_3$

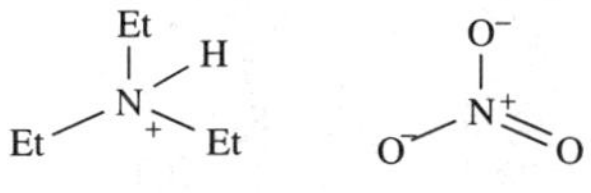

Dinitrogen tetraoxide
See Dinitrogen tetraoxide: Triethylammonium nitrate
See other OXOSALTS OF NITROGENOUS BASES

840

2581. Diazido(*N*,*N*,*N*,*N* -tetramethylethanediamine)palladium
[158668-93-0] $C_6H_{16}N_8Pd$

Kim, Y. *et al., Bull. Korean Chem. Soc.*, 1994, **15**(8), 690
Exploded during melting point determination.
See AZIDES

†2582. Diethoxydimethylsilane
[78-62-6] $C_6H_{16}O_2Si$

$$(EtO)_2SiMe_2$$

See related ALKYLSILANES

2583. Triethyltin hydroperoxide
[ ] $C_6H_{16}O_2Sn$

Hydrogen peroxide
Aleksandrov, Y. A. *et al., Zh. Obsch. Khim.*, 1965, **35**, 115
While the hydroperoxide is stable at ambient temperature, the addition compound
which it readily forms with excess hydrogen peroxide is not, decomposing violently.
See other ORGANOMETALLIC PEROXIDES

2584. Triethoxysilane
[998-30-1] $C_6H_{16}O_3Si$

$$(EtO)_3SiH$$

1. Xin, S. *et al., Can. J. Chem.*, 1990, **68**, 471
2. Buchwald, S. L., *Chem. Eng. News*, 1993, **71**(13), 2
This compound can disproportionate, generating silane (SiH_4) which is liable to be
pyrophoric. Presumably other alkoxy silanes can do likewise [1]. Its use as a reducing
agent for the preparation of alcohols from esters is considered safer in air than under

an inert atmosphere, which latter permits accumulation of silane to hazardous levels and later fire or explosion. [2]
See related SILANES

2585a. Ethylenebis(dimethylphosphine)
   [23936-60-9] $C_6H_{16}P_2$

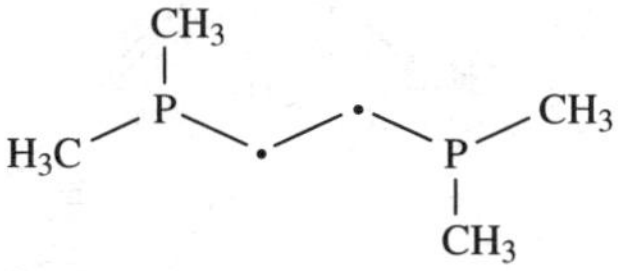

1. Burt, R. J., *J. Organomet. Chem.*, 1979, **182**, 205
2. Butter, S. A. *et al.*, *Inorg. Synth.*, 1974, **15**, 190
Though air sensitive [1], it is not pyrophoric as reported previously [2].
See other ALKYLPHOSPHINES

2585b. Triethylsilane
   [617-86-7] $C_6H_{16}Si$

Et₃SiH

Boron trichloride
   See Boron trichloride: Triethylsilane
   See other ALKYLSILANES

2586. 1-Ethyl-1,1,3,3-tetramethyltetrazenium tetrafluoroborate
   [65651-91-4] $C_6H_{17}BF_4N_4$

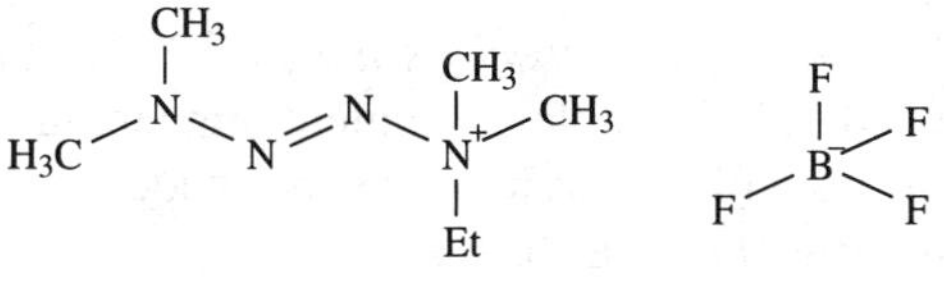

Stöldt, E. *et al., Angew. Chem. (Intern,. Ed.)*, 1978, **17**, 203
An explosive salt.
See other PERCHLORATE SALTS OF NITROGENOUS BASES, TETRAZENES

2587. 3,3′-Iminobispropylamine (Norspermidine)
   [56-18-8] $C_6H_{17}N_3$
                                    HN(CH₂CH₂CH₂NH₂)₂

Editor's comments, 1993

Various secondary sources of safety data are now listing this as an explosive. I can find no primary source for this classification, which seems very improbable. Simple minded use of many computational hazard prediction procedures would show thermodynamically that this compound, like most lower amines, could hypothetically convert to alkane, ammonia and nitrogen with sufficient energy (about 3 kJ/g) to count as an explosion hazard. This reaction is not known to happen. (Simple minded thermodynamicists would rate this book, or computer, and its reader as a severe hazard in an air environment.) Like other bases, iminobispropylamine certainly sensitises many nitro-explosives to detonation. It is used experimentally to study the effect, which may have found technical exploitation and, garbled, could have led to description of the amine as itself an explosive.

See C-NITRO COMPOUNDS, COMPUTATION OF REACTIVE CHEMICAL HAZARDS

2588. Triethyldiborane
[62133-36-2] $C_6H_{18}B_2$

Leleu, *Cahiers*, 1977, (88), 365
It ignites in air.
See other ALKYLBORANES

2589. 1,2-Bis(ethylammonio)ethane perchlorate
[53213-78-8] $C_6H_{18}Cl_2N_2O_8$

Sunderlin, K. G. R., *Chem. Eng. News*, 1974, **52**(31), 3
It exploded mildly under an impact of 11.5 kgm.
See other PERCHLORATE SALTS OF NITROGENOUS BASES

2590. Lithium hexamethylchromate(3−)
[14931-97-6] (tris-dioxane complex) $C_6H_{18}CrLi_3$

Kurras, E. *et al.*, *J. Organomet. Chem.*, 1965, **4**, 114–118

Isolated as a dioxane complex, it ignites in air.
See related ALKYLMETALS

2591. Bis(trimethylsilyl) chromate
[1746-09-4] $C_6H_{18}CrO_4Si$

1. Anon., *ABCM Quart. Safety Summ.*, 1960, **31**, 16
2. Schmidt, M. *et al., Angew. Chem.*, 1958, **70**, 704

Small quantities can be distilled at about 75°C/1.3 mbar, but larger amounts are liable to explode violently owing to local overheating [1]. An attempt to prepare an analogous poly(dimethylsilyl) chromate by heating a polydimethylsiloxane with chromium trioxide at 140°C exploded violently after 20min at this temperature [2].
See related ALKYLSILANES, METAL OXIDES

2592. Hexamethylerbium–hexamethylethylenediaminelithium complex
[66862-11-1] $C_6H_{18}Er.3C_2H_{16}N_2Li$

Schumann, H. *et al., Angew. Chem. (Intern. Ed.)*, 1981, **20**, 120–121
It ignites in air, as do the analogous derivatives of lutetium and other lanthanoids
See related ALKYLMETALS

2593. Bis(trimethylsilyl)mercury
[4656-04-6] $C_6H_{18}HgSi_2$

Eisch, J. J. *et al., J. Organomet. Chem.*, 1979, **171**, 149

844

Momentary exposure of traces of the product on a stopper to air usually causes ignition.

See related ALKYLSILANES *See other* MERCURY COMPOUNDS

2594. Lithium bis(trimethylsilyl)amide
[4039-32-1] $C_6H_{18}LiNSi_2$

Amonoo-Neizer, E. H. *et al., Inorg. Synth.*, 1966, **8**, 21
It is unstable in air and ignites when compressed.
See other N-METAL DERIVATIVES

2595. Dilithium 1,1-bis(trimethylsilyl)hydrazide
[15114-92-8] $C_6H_{18}Li_2N_2Si_2$

It ignites in air.
See entry SILYLHYDRAZINES *See other* N-METAL DERIVATIVES, METAL HYDRAZIDES

2596. Molybdenum hexamethoxide
[135840-44-7] $C_6H_{18}MoO_6$

Preparative hazard
See Tetramethoxysilane: Metal hexafluorides
See other METAL ALKOXIDES

2597. Bis(dimethylamino)dimethylstannane
[993-74-8] $C_6H_{18}N_2Sn$

$$\underset{\substack{| \\ CH_3}}{H_3C-\underset{|}{N}-Sn}\overset{\substack{CH_3 \quad CH_3 \\ | \qquad |}}{\underset{|}{}}\overset{}{N}-CH_3$$

Chloroform
Randall, E. W. *et al.*, *Inorg. Nucl. Chem. Lett.*, 1965, **1**, 105
Gentle heating of a 1:1 chloroform solution of the stannane led to a mild explosion.
Similar explosions had been experienced at the conclusion of fractional distillation of
other bis(dialkylamino)stannanes where the still pot temperature had risen to about
200°C.
See other N-METAL DERIVATIVES *See related* ALKYLMETALS

2598. Tris(dimethylamino)antimony
[7289-92-1] $C_6H_{18}N_3Sb$

$$H_3C-\overset{\substack{CH_3 \quad CH_3 \\ | \qquad |}}{N}-Sb-\overset{}{N}-CH_3$$
$$H_3C-\overset{|}{N}-CH_3$$

Ethyl diazoacetate
See Ethyl diazoacetate: Tris(dimethylamino)antimony
See other N-METAL DERIVATIVES

2599. N,N'-Bis(2-aminoethyl)1,2-diaminoethane (Triethylenetetramine)
[112-24-3] $C_6H_{18}N_4$

$$H_2N\diagdown\diagup\diagdown_{NH}\diagdown\diagup\diagdown_{NH}\diagdown\diagup\diagdown_{NH_2}$$

Cellulose nitrate
See CELLULOSE NITRATE: amines

Other reactants
Yoshida, 1980, 251
MRH values calculated for 13 combinations with oxidants are given.
See other ORGANIC BASES

2600. 2,4,6,8,9,10-Hexamethylhexaaza-1,3,5,7-tetraphosphaadamantane
[10369-17-2] $C_6H_{18}N_6P_4$

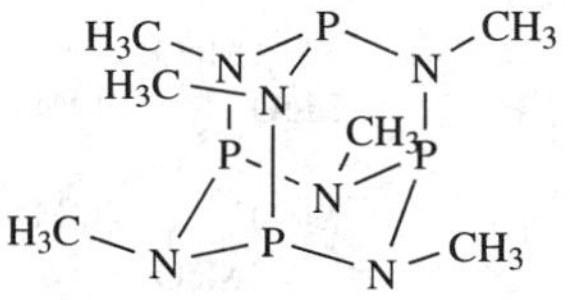

Oxidants

Verdier, F., *Rech. Aerosp.*, 1970, (137), 181–189

Among other solid P–N compounds examined, the title compound ignited immediately on contact with nitric acid, hydrogen peroxide or dinitrogen trioxide.

See ROCKET PROPELLANTS

†2601. Bis(trimethylsilyl) oxide
[107-46-0] $C_6H_{18}OSi_2$

$$(Me_3Si)_2O$$

See related ALKYLSILANES

2602. Bis(trimethylsilyl) peroxomonosulfate
[23115-33-5] $C_6H_{18}O_5SSi_2$

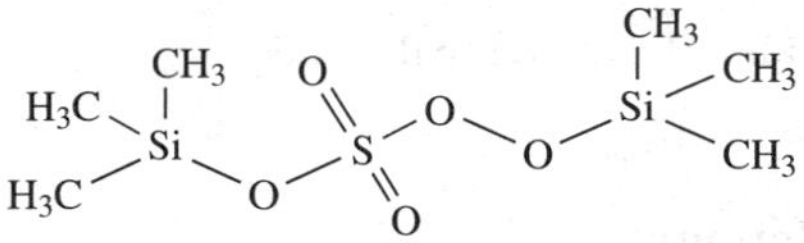

Blaschette, A. *et al., Angew. Chem. (Intern. Ed.)*, 1969, **8**, 450

It is stable at $-30°C$, but on warming to ambient temperature it decomposes violently evolving sulfur trioxide.

See other GAS EVOLUTION INCIDENTS, PEROXYESTERS, SULFUR ESTERS

2603. Rhenium hexamethoxide
[ ] $C_6H_{18}O_6Re$

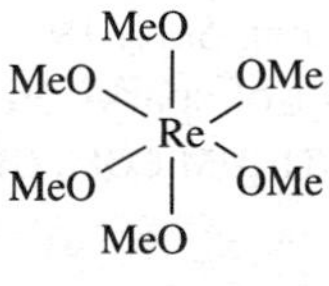

Preparative hazard

See Tetramethoxysilane: Metal hexafluorides

See other METAL ALKOXIDES

2604. Tungsten hexamethoxide
[35869-33-1]

$C_6H_{18}O_6W$

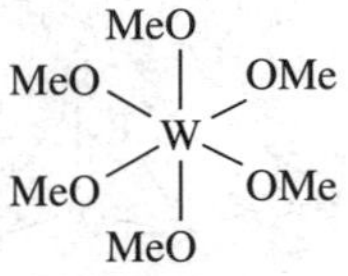

$$
\begin{array}{c}
\text{MeO} \\
\text{MeO} \diagdown \mid \diagup \text{OMe} \\
\text{W} \\
\text{MeO} \diagup \mid \diagdown \text{OMe} \\
\text{MeO}
\end{array}
$$

Preparative hazard
See Tetramethoxysilane: Metal hexafluorides
See other METAL ALKOXIDES

2605. (Dimethylsilylmethyl)trimethyllead
[ ]

$C_6H_{18}PbSi$

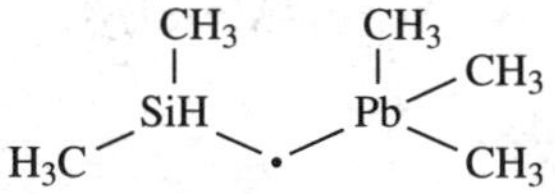

$$
\begin{array}{c}
\quad \text{CH}_3 \qquad\quad \text{CH}_3 \\
\quad\ \mid \qquad\qquad\ \mid \quad \diagup \text{CH}_3 \\
\quad \text{SiH} \qquad\quad \text{Pb} \\
\text{H}_3\text{C} \diagup \quad \bullet \diagdown \qquad\ \diagdown \text{CH}_3
\end{array}
$$

Schmidbaur, H., *Chem. Ber.*, 1964, **97**, 270
It decomposes slowly above 100°C, and with explosive violence in presence of oxygen.
See related ALKYLMETALS, ALKYLSILANES

2606. Hexamethyldiplatinum
[4711-74-4]

$C_6H_{18}Pt_2$

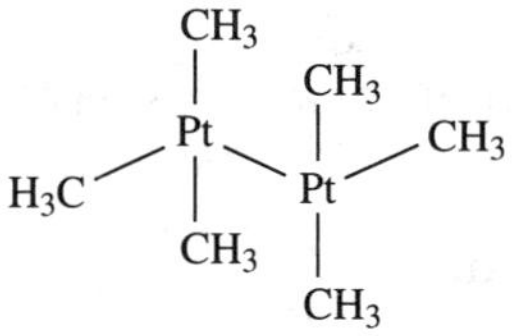

$$
\begin{array}{c}
\qquad \text{CH}_3 \\
\qquad\ \mid \qquad \text{CH}_3 \\
\qquad \text{Pt} \diagdown \quad\ \mid \quad \diagup \text{CH}_3 \\
\text{H}_3\text{C} \diagup \ \mid \ \ \text{Pt} \\
\qquad \text{CH}_3 \ \ \mid \\
\qquad\qquad\ \text{CH}_3
\end{array}
$$

Gilman, H. *et al.*, *J. Amer. Chem. Soc.*, 1953, **75**, 2065
It explodes sharply in a shower of sparks on heating.
See other ALKYLMETALS, PLATINUM COMPOUNDS

2607. Hexamethylrhenium
[56090-02-9]

$C_6H_{18}Re$

$$\begin{array}{c} CH_3 \\ H_3C \diagdown \;\; | \;\; \diagup CH_3 \\ Re \\ H_3C \diagup \;\; | \;\; \diagdown CH_3 \\ CH_3 \end{array}$$

1. Mertis, K. *et al., J. Organomet. Chem.*, 1975, **97**, C65
2. Mertis, K. *et al., J. Chem. Soc., Dalton Trans.*, 1976, 1489
3. Edwards, P. G. *et al., J. Chem. Soc., Dalton Trans.*, 1980, 2474

It should be handled with extreme caution, even in absence of air, because dangerous explosions have occurred on warming frozen samples, or during transfer operations [1]. It is unstable above −20°C, and on one occasion admission of nitrogen to the solid under vacuum caused a violent explosion [2]. It should be prepared only as required and in small amounts, and in use, oxygen and moisture must be rigorously excluded [3].

See other ALKYLMETALS

2608. Hexamethyldisilane
[1450-14-2]

$C_6H_{18}Si_2$

$$\begin{array}{c} CH_3 \\ H_3C \diagdown \;\; | \\ Si \diagdown \quad CH_3 \\ H_3C \diagup \quad Si \diagup \\ \quad | \diagdown CH_3 \\ CH_3 \end{array}$$

Pyridine *N*-oxide, Tetrabutylammonium fluoride
See Pyridine *N*-oxide: Hexamethyldisilane, Tetrabutylammonium fluoride
See other ALKYLSILANES

2609. Hexamethyltungsten
[36133-73-0]

$C_6H_{18}W$

$$\begin{array}{c} CH_3 \\ H_3C \diagdown \;\; | \;\; \diagup CH_3 \\ W \\ H_3C \diagup \;\; | \;\; \diagdown CH_3 \\ CH_3 \end{array}$$

Galyer, L. *et al., J. Organomet. Chem.*, 1975, **85**, C37

Several unexplained and violent explosions during preparation and handling indicate that the compound must be handled with great care and as potentially explosive, particularly during vacuum sublimation.

See other ALKYLMETALS

†**2610. Hexamethyldisilazane**
    [999-97-3] $C_6H_{19}NSi_2$

Me₃SiNHSiMe₃

See related ALKYLSILANES

2611. Bis(trimethylsilyl) phosphonite
    [30148-50-6] $C_6H_{19}O_2PSi_2$

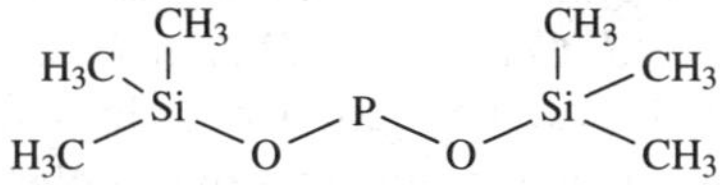

Hata, T. *et al., Chem. Lett.*, 1977, 1432
It ignites in air.
See other PHOSPHORUS ESTERS, PYROPHORIC MATERIALS

2612. *N,N*-Bis(trimethylsilyl)aminoborane
    [73452-31-0] $C_6H_{20}BNSi_2$

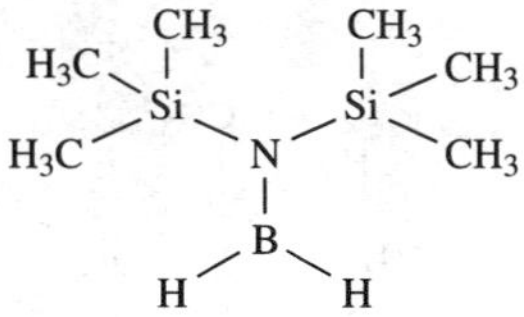

Wisian-Nielsen, P., *J. Inorg. Nucl. Chem.*, 1979, **41**, 1546
It ignites in direct contact with air.
See related ALKYLSILANES, BORANES

2613. Bis(1,2-diaminopropane)-*cis*-dichlorochromium(III) perchlorate
    [59598-02-6] $C_6H_{20}Cl_3CrN_4O_4$

[(C₃H₁₀N₂)₂CrCl₂] ClO₄

Perchloric acid
    Anon., *Chem. Eng. News*, 1963, **41**(27), 47
    A mixture of the complex and 20 volumes of 70% acid was being stirred at 22°C when
    it exploded violently. Traces of ether or ethanol may have been present. Extreme
    care must be exercised when concentrated perchloric acid is contacted with organic
    materials with agitation or without cooling.
    See other AMMINEMETAL OXOSALTS

2614. (Benzenesulfinato-*S*)pentaamminecobalt(III) trichloro(perchlorato)-stannate(II)

[64825-48-5] $C_6H_{20}Cl_4CoN_5O_6SSn$

$$[(NH_3)_5CoS(O_2)Ph] [Cl_3SnClO_4]$$

Elder, R. C. *et al., Inorg. Chem.*, 1978, **17**, 427
It explodes at temperatures around 150°C. Thermal decomposition appears to involve an internal redox reaction in the anion leading to formation of trichloro(chlorato)oxostannate(IV).
See other AMMINEMETAL OXOSALTS, REDOX COMPOUNDS
See related METAL PERCHLORATES, REDOX REACTIONS

2615. *cis*-Bis(trimethylsilylamino)tellurium tetrafluoride
[86045-52-5] $C_6H_{20}F_4N_2Si_2Te$

Hartl, H. *et al., Inorg. Chem.*, 1983, **22**, 2183–2184
The colourless solid explodes mildly at 100°C, but a 50 g sample on keeping for 6 months in a stoppered cylinder was converted to a yellow extremely explosive solid, probably a hydrolysis product.
See other AMMINEMETAL HALIDES
See related ALKYLSILANES, METAL HALIDES

2616. 1,2-Bis(trimethylsilyl)hydrazine
[692-56-8] $C_6H_{20}N_2Si_2$

Oxidants
It is hypergolic with fluorine or fuming sulfuric acid.
See entry SILYLHYDRAZINES

2617. Hexamethyltrisiloxane
[1189-93-1] $C_6H_{20}O_2SI_3$

$$H-\underset{\underset{H_3C}{|}}{\overset{\overset{H_3C}{|}}{Si}}-O-\underset{\underset{CH_3}{|}}{\overset{\overset{CH_3}{|}}{Si}}-O-\underset{\underset{CH_3}{|}}{\overset{\overset{CH_3}{|}}{Si}}-H$$

See entry SILANES; olefins, platinum
See related ALKYLSILANES

2618. Isonicotinamidepentaammineruthenium(II) perchlorate
[31279-70-6] $C_6H_{21}Cl_2N_7O_9Ru$

$$[C_6H_6N_2ORu(NH_3)_5]\,[ClO_4]_2$$

Stanbury, D. M. *et al., Inorg Chem.*, 1980, **19**, 519
It is explosive.
See other AMMINEMETAL OXOSALTS

2619. Tris(1,2-diaminoethane)chromium(III) perchlorate
[15246-55-6] $C_6H_{24}Cl_3CrN_6O_{12}$

$$[(C_2H_8N_2)_3Cr]\,[ClO_4]_3$$

Olander, D. E., *Chem. Eng. News*, 1984, **62**(10), 2, 39
It is a treacherously sensitive, very brisant primary explosive (while its cobalt
analogue is a well behaved secondary explosive).
See other AMMINEMETAL OXOSALTS

2620. Hexaureagallium(III) perchlorate
[31332-72-6] $C_6H_{24}Cl_3GaN_{12}O_{18}$

$$[(H_2NCO.NH_2)_6Ga]\,[ClO_4]_3$$

Lloyd, D. J. *et al., J. Chem. Soc.*, 1943, 76
It decomposes violently when heated above its m.p., 179°C.
See related AMMINEMETAL OXOSALTS

2621. Tris(1,2-diaminoethane)ruthenium(III) perchlorate
[67187-50-2] $C_6H_{24}Cl_3N_6O_{12}Ru$

$$[(C_2H_8N_2)_3Ru]\,[ClO_4]_3$$

Lavallee, U. *et al., Inorg. Chem.*, 1978, **17**, 2218
An attempt to prepare the perchlorate salt led to an explosion.
See other AMMINEMETAL OXOSALTS

2622. Tris(1,2-diaminoethane)cobalt(III) nitrate
[6865-68-5] $C_6H_{24}CoN_9O_9$

$$[(C_2H_8N_2)_3Co] [NO_3]_3$$

Tomlinson, W. R. *et al., J. Amer. Chem. Soc.*, 1949, **71**, 375
No explosion on heating, but medium impact-sensitivity.
See other AMMINEMETAL NITRATES

2623. Hexaureachromium(III) nitrate
[22471-42-7] $C_6H_{24}CrN_{15}O_{15}$

$$[(H_2NCO.NH_2)_6Cr] [NO_3]_3$$

Tomlinson, W. R. *et al., J. Amer. Chem. Soc.*, 1949, **71**, 375
Explodes at 265°C, medium impact-sensitivity.
See related AMMINEMETAL NITRATES

2624. 1,1′-Azo-[2-methyl-1,2-dicarbadecaborane(14)]
[73469-45-1] $C_6H_{26}B_{20}N_2$
Complex Structure

Preparative hazard
Totari, T. *et al., J. Chem. Soc., Chem. Commun.*, 1979, 1051
The reaction mixture from permanganate oxidation of the aminocarbaborane in liquid
ammonia must be diluted with toluene at −60°C to avoid detonation when the
ammonia is subsequently evaporated.
See AZOCARBABORANES *See related* BORANES

2625. Potassium cyclohexanehexone 1,3,5-trioximate
[ ] $C_6K_3N_3O_6$

Alone, or Acids
Benedikt, R., *Ber.*, 1878, **11**, 1375
The salt explodes on heating above 130°C or if moistened with sulfuric or nitric
acids. The lead salt also explodes violently on heating.
See related OXIMES
See other N–O COMPOUNDS

853

2626. Potassium benzenehexoxide
[3264-86-6]

$C_6K_6O_6$

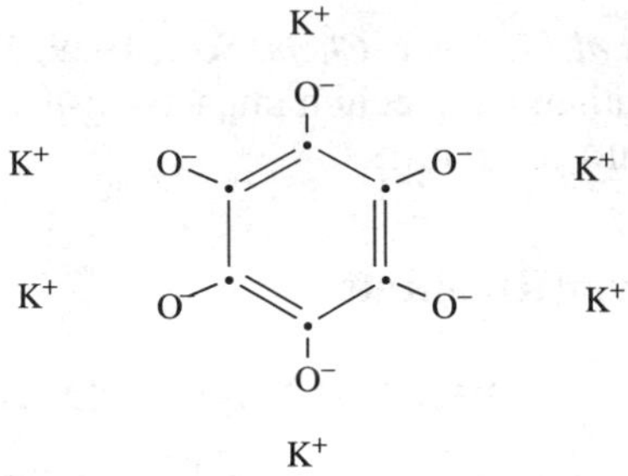

Air, or Oxygen, or Water
1. Sorbe, 1968, 68, 69
2. Mellor, 1963, Vol. 2, Suppl. 2.2, 1567–1568; 1946, Vol. 5, 951
3. Weiss, E., *Angew. Chem. (Int.)*, 1993, **32**(11), 1518

The compound 'potassium carbonyl' is now thought to be the hexameric potassium salt of hexahydroxybenzene, and reacts with moist air to give a very explosive product [1]. It reacts violently with oxygen, and explodes on heating in air or in contact with water [2]. There are probably two compounds described, the original preparation from carbon monoxide and liquid potassium may well give the aromatic structure. X-ray crystallography shows the product from potassium and carbon monoxide in liquid ammonia to be dipotassium ethynediolate [3].
See Potassium: Non-metal oxides (reference 4)

2627. Lithium benzenehexoxide
[101672-16-6]

$C_6Li_6O_6$

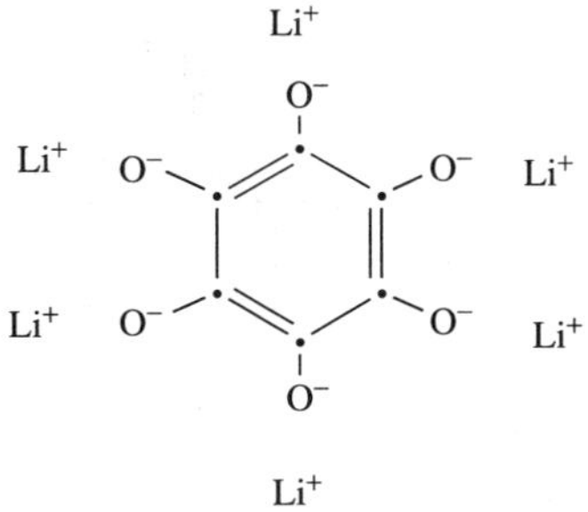

Water
1. Mellor, 1961, Vol. 2, Suppl. 2.1, 88
2. Weiss, E., *Angew. Chem. (Int.)*, 1993, **32**(11), 1518

Like the potassium compound, it explodes with water [1]. It is probable that this compound is actually dilithium ethynediolate [2].

2628. Hexacarbonylmolybdenum
[13939-06-5]

$$(CO)_6Mo$$

C_6MoO_6

Diethyl ether
  Owen, B. B. *et al., Inorg. Synth.*, 1950, **3**, 158
  Solutions of hexacarbonylmolybdenum in ether have exploded after extended storage.
See other CARBONYLMETALS

2629. Tetracyanoethylene
[670-54-2]

C_6N_4

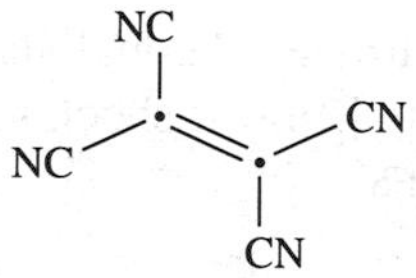

1-Methylsilacyclopenta-2,4-diene
  See 1-Methylsilacyclopenta-2,4-diene: Dienophiles
  See other CYANO COMPOUNDS, ENDOTHERMIC COMPOUNDS

2630. 3(3-Cyano-1,2,4-oxadiazol-5-yl)-4-cyanofurazan 2-(or 5-) oxide
[56092-91-2]

$C_6N_6O_3$

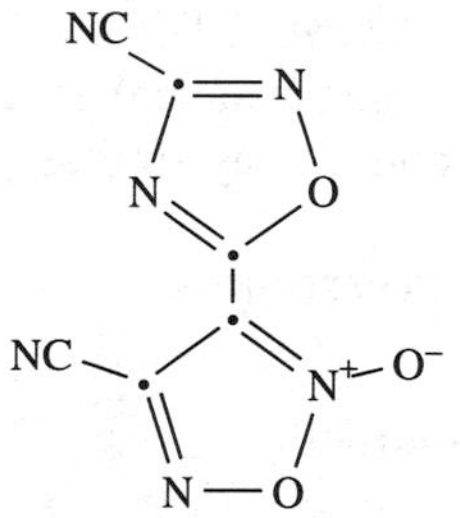

Hydrazine derivatives
  Lupton, E. C. *et al., J. Chem. Eng. Data*, 1975, **20**, 136
  The compound is highly endothermic ($\Delta H°_f$ (s) +565.5 kJ/mol, 2.77 kJ/g), and interaction with hydrazines is explosive.
See Dicyanofurazan: Nitrogenous bases
See other CYANO COMPOUNDS, ENDOTHERMIC COMPOUNDS, FURAZAN *N*-OXIDES

2631. Benzotri(furazan N-oxide) (Hexanitrosobenzene)
[3470-17-5]

$C_6N_6O_6$

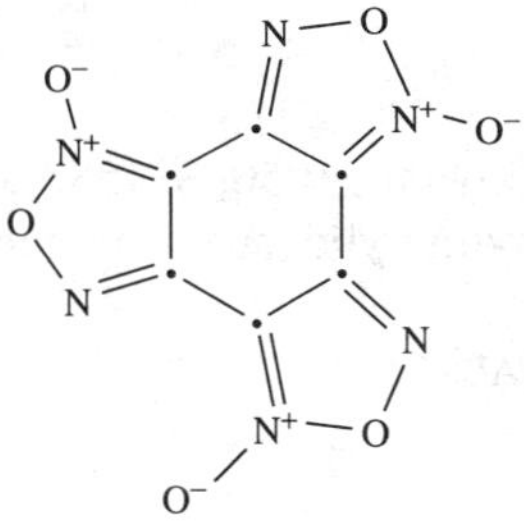

1. Selig, W., *Z. Anal. Chem.*, 1978, **289**, 44–46
2. McGuire, R. R., *Chem. Abs.*, 1978, **89**, 131857
3. *CHETAH*, 1990, 182

It is explosive [1], and the limited thermal stability is reduced by traces of triazidotrinitrobenzene as impurity [2]. It is shock sensitive [3].

See other FURAZAN N-OXIDES

2632. Hexanitrobenzene
[13232-74-1]

$C_6N_6O_{12}$

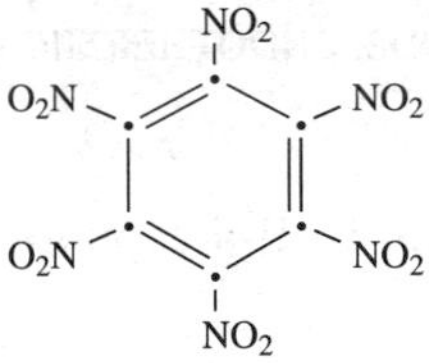

Nielsen, A. T. *et al., J. Org. Chem.*, 1979, **44**, 1181

The compound (of zero oxygen balance) is a powerful explosive, as is pentanitrobenzene. As a nitrocarbon, it has similar properties to tetranitromethane and hexanitroethane.

See other POLYNITROARYL COMPOUNDS

2633. Tetraazido-1,4-benzoquinone
[22826-61-5]

$C_6N_{12}O_2$

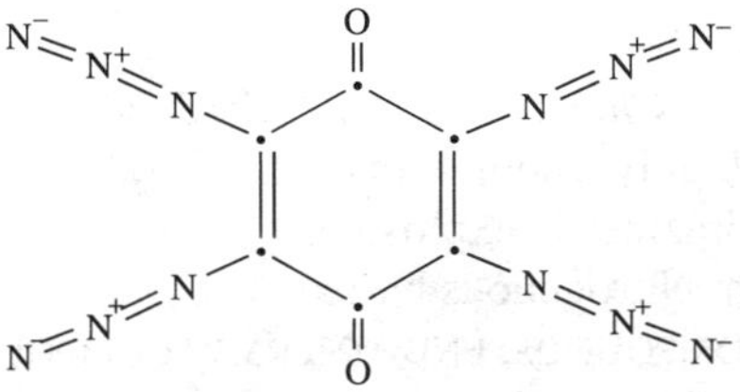

1. Friess, K. *et al., Ber.*, 1923, **56**, 1304

2. Gilligan, W. H. *et al., Tetrahedron Lett.*, 1978, 1675–1676
3. Winkelmann, E., *Tetrahedron*, 1969, **25**, 2427

Extremely explosive, sensitive to heat, friction and impact. A single crystal heated on a spatula exploded and bent it badly [1]. It explodes at 120°C, and the impact-sensitivity was too high to be measured, extreme friction sensitivity causing explosions when samples were being arranged on sandpaper. Quantities above 100–200 mg are too dangerous to prepare and handle under ordinary laboratory conditions [2], the 24 g scale previously described [3] being dangerously excessive.

See other 2-AZIDOCARBONYL COMPOUNDS, HIGH-NITROGEN COMPOUNDS

2634. Sodium benzenehexoxide
[101672-22-4] $C_6Na_6O_6$

See Sodium: Non-metal oxides (reference 4)

2635. Hexacarbonylvanadium
[14024-00-1] C_6O_6V

$$(CO)_6V$$

1. Pruett, R. L. *et al., Chem. & Ind.*, 1960, 119
2. Bailar, 1973, Vol. 3, 529

The pyrophoric compound previously regarded as 'dodecacarbonyl divanadium' [1] is now known to be the monomeric hexacarbonylvanadium [2].

See other CARBONYLMETALS

2636. Hexacarbonyltungsten
[14040-11-0] C_6O_6W

$$(CO)_6W$$

Preparative hazard

Hurd, D. T. *et al., Inorg. Chem.*, 1957, **5**, 136

It is dangerous to attempt the preparation from tungsten hexachloride, aluminium powder and carbon monoxide in an autoclave of greater than 0.3 l capacity.

See other CARBONYLMETALS

857

2637. 2-Chloro-1,3-dinitro-5-trifluoromethylbenzene

[393-75-9] $C_7H_2ClF_3N_2O_4$

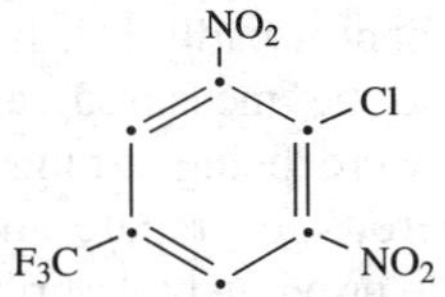

Nazarov, V. P. *et al., Chem. Abs.*, 1980, **93**, 2387

Mathematical modelling was used to develop the recommendation to use a cascade of 3 reactors for the highly exothermic dinitration of 4-chloro-trifluoromethylbenzene.

See other NITRATION INCIDENTS, POLYNITROARYL COMPOUNDS

2638. 5-Nitro-2-picryltetrazole

[82177-75-1] $C_7H_2N_8O_8$

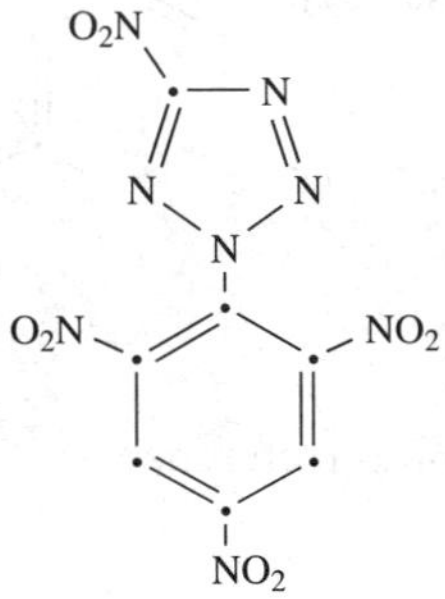

Spear, R. J. *et al., Chem. Abs.*, 1983, **98**, 91969

Powerfully explosive but of low impact-sensitivity.

See other POLYNITROARYL COMPOUNDS, TETRAZOLES

2639. 4-Iodobenzenediazonium-2-carboxylate

[ ] $C_7H_3IN_2O_2$

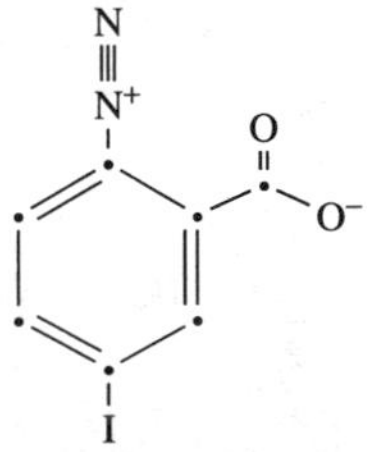

Gommper, R. *et al., Chem. Ber.*, 1968, **101**, 2348

It is a highly explosive solid.

See other DIAZONIUM CARBOXYLATES

858

2640. 2,4,6-Trinitrobenzoic acid
 [129-66-8] $C_7H_3N_3O_8$

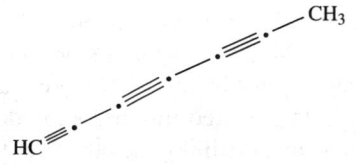

Preparative hazard
 See Sodium dichromate: Sulfuric acid, Trinitrotoluene

Heavy metals
 Kranz, A. *et al., Chem. Age* (London), 1925, **13**, 392
 All heavy metal salts prepared were explosive on heating or impact.
 See other POLYNITROARYL COMPOUNDS

2641. Hepta-1,3,5-triyne
 [66486-68-8] C_7H_4

Cook, C. L. *et al., J. Chem. Soc.*, 1952, 2890
 It explodes very readily (in absence of air) above 0°C, and the distillation residues
 also exploded on admission of air.
 See other ALKYNES

2642. Silver 3,5-dinitroanthranilate
 [58302-42-4] $C_7H_4AgN_3O_6$

Gupta, S. P. *et al., Def. Sci. J.*, 1975, **25**, 101–106

It explodes at 394°, and the copper salt at 371°C, each after a 10 s delay. They are not impact-sensitive.

See other POLYNITROARYL COMPOUNDS, SILVER COMPOUNDS

2643. 2-, 3- or 4-Trifluoromethylphenylmagnesium bromide
[395-47-1, 402-26-6, 402-51-7, resp.] $C_7H_4BrF_3Mg$

1. Appleby, J. C., *Chem. & Ind.*, 1971, 120
2. 'Benzotrifluorides Catalog 6/15' West Chester (Pa)., Marshallton Res. Labs., 1971
3. Taylor, R. *et al.*, private comm., 1973
4. Moore, *et al., Chem. Eng. News*, 1997, **75** (11), 6

During a 2 kg preparation under controlled conditions, the 3-isomer exploded violently [1]. The same had occurred previously in a 1.25 kg preparation 30 min after all magnesium had dissolved. A small preparation of the 2-isomer had also exploded. It was found that 3-preparations in ether decomposed violently at 75° and in benzene–THF at 90°C. Molar preparations or less have been done several times uneventfully when temperatures below 40°C were maintained, but caution is urged [2]. The 3- and 4-reagents prepared in ether exploded when the temperature was raised by adding benzene and distilling the ether out [3]. Twentyfive years after the establishment of a hazard, a sample of the 4-isomer, which had been all but isolated by evaporation of ether, exploded; the experimenters sought to blame hypothetical residues of highly active magnesium [4].

See other GRIGNARD REAGENTS, HALO-ARYLMETALS, FLUORINATED ORGANO-LITHIUM COMPOUNDS

2644. 4-Bromobenzoyl azide
[14917-59-0] $C_7H_4BrN_3O$

Curtiss, T. *et al., J. Prakt. Chem.*, 1898, **58**, 201

860

It explodes violently above its m.p., 46°C.

See other ACYL AZIDES

2645. 3-Bromo-3(4-nitrophenyl)-3*H*-diazirine
[115127-49-6]   C₇H₄BrN₃O₂

Tetrabutylammonium fluoride
  Terpinski, J. *et al., Magn. Reson. Chem.*, 1987, **25**, 924
  Direct introduction of the bromodiazirine into molten tetrabutylammonium fluoride
  (at 63°C) to effect halogen exchange leads to an explosion. Use of acetonitrile at
  0°C as solvent gives a low yield of the required fluorodiazirine.
  See other DIAZIRINES, *N*-HALOGEN COMPOUNDS, NITROARYL COMPOUNDS

2646. 4-Chlorotrifluoromethylbenzene
[98-56-6]   C₇H₄ClF₃

Sodium dimethylsulfinate
  Meschino, J. A. *et al., J. Org. Chem.*, 1971, **36**, 3637, footnote 9
  Interaction with dimethylsulfinate anion (from the sulfoxide and sodium hydride)
  at −5°C is very exothermic, and addition of the chloro compound must be slow to
  avoid violent eruption.
  See other HALOARYL COMPOUNDS

2647. 2-Chloro-5-trifluoromethylbenzenediazonium hydrogen sulfate
[29362-18-3] (ion) $C_7H_4ClF_3N_2O_4S$

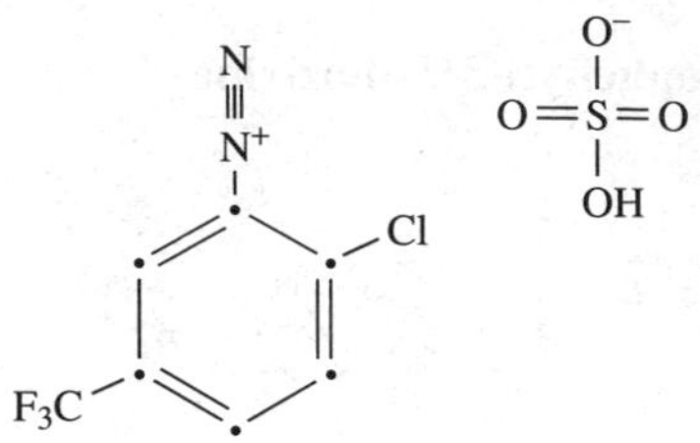

See entry DIAZONIUM SULFATES

2648. 4-Chlorophenyl isocyanate
[104-12-1] C_7H_4ClNO

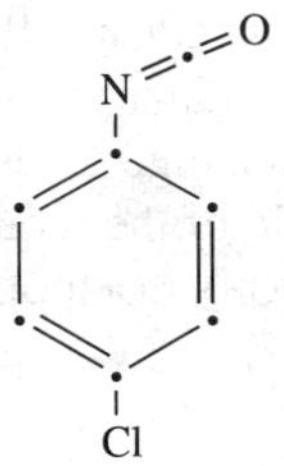

Preparative hazard
See 4-Chlorobenzoyl azide
See other ORGANIC ISOCYANATES

2649. 2-Chloro-5-nitrobenzaldehyde
[6361-21-3] $C_7H_4ClNO_3$

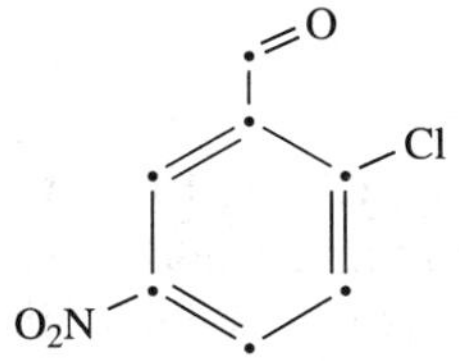

Initial decomposition temperature by ARC was 156°C.
See entry NITROBENZALDEHYDES See other NITROARYL COMPOUNDS

2650. 2-Chloro-6-nitrobenzaldehyde
[6361-22-4]

$C_7H_4ClNO_3$

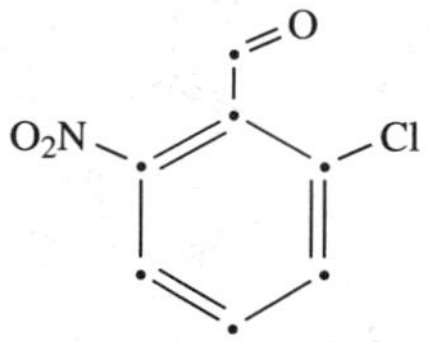

Initial decomposition temperature by ARC was 146°C.
See entry NITROBENZALDEHYDES
See other ALDEHYDES, NITROARYL COMPOUNDS

2651. 4-Chloro-3-nitrobenzaldehyde
[16588-34-4]

$C_7H_4ClNO_3$

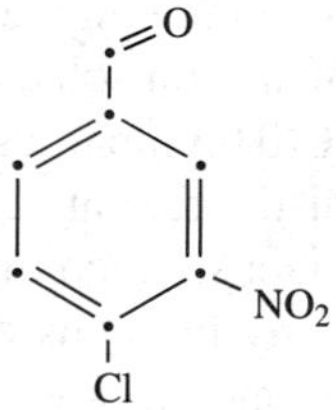

Initial decomposition temperature by ARC was 116°C.
See entry NITROBENZALDEHYDES
See other ALDEHYDES, NITROARYL COMPOUNDS

2652. 5-Chloro-2-nitrobenzaldehyde
[6628-86-0]

$C_7H_4ClNO_3$

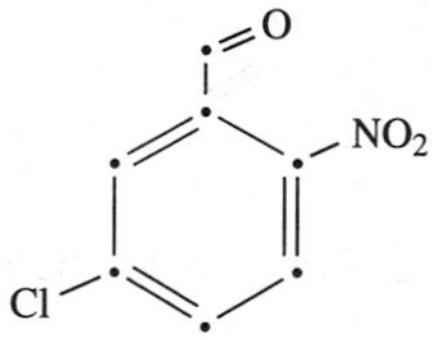

Initial decomposition temperature by DSC was 240°C.
See entry NITROBENZALDEHYDES
See other ALDEHYDES, NITROARYL COMPOUNDS

2653. 2-Nitrobenzoyl chloride
[610-14-0]

$C_7H_4ClNO_3$

1. Cook, N. C. *et al., Chem. Eng. News*, 1945, **23**, 2394
2. Bonner, W. D. *et al., J. Amer. Chem. Soc., 1946,* **68**, 344; *Chem. & Ind.*, 1948, 89
3. *MCA Case History No. 1915*
4. Lockemann, G. *et al., Ber.*, 1947, **80**, 488
5. *CHETAH*, 1990, 189

The hot material remaining after vacuum stripping of solvent up to 130°C decomposed with evolution of gas and then exploded violently 50 min after heating had ceased. Further attempts to distil the acid chloride even in small amounts at below 1.3 mbar caused exothermic decomposition at 110°C. It was, however, possible to flash-distil the chloride in special equipment [1]. Two later similar publications recommend use in solution of the unisolated material [2]. Smaller scale distillation of the chloride at 94–95°C/0.03 mbar had been uneventful, but a 1.2 mol scale preparation exploded during distillation at 128°C/1.2 mbar [3], even in presence of phosphorus pentachloride, previously recommended to reduce the danger of explosion during distillation [4]. Many previous explosions had been reported, and the need for adequate purity of intermediates was stressed. It is shock sensitive [5].
See other GAS EVOLUTION INCIDENTS, NITROACYL HALIDES

2654. 4-Chlorobenzoyl azide
[14848-01-2]

$C_7H_4ClN_3O$

1. Cobern, D. *et al., Chem. & Ind.*, 1965, 1625; 1966, 375
2. Tyabji, M. M., *Chem. & Ind.*, 1965, 2070

A violent explosion occurred during vacuum distillation of 4-chlorophenyl isocyanate, prepared by Curtius reaction from the azide. It was found by IR spectroscopy that this isocyanate (as well as others prepared analogously) contained some unchanged azide, to which the explosion was attributed. The use of IR spectroscopy to check for absence of azides in isocyanates is recommended before distillation [1]. Subsequently, the explosion was attributed to free hydrogen azide, produced by hydrolysis of the unchanged acyl azide [2].
See other ACYL AZIDES

2655. 2-Trifluoromethylphenyllithium
[49571-31-5] $C_7H_4F_3Li$

$$F_3CC_6H_4Li$$

(Preparative hazard for 2-)
See entry ORGANOLITHIUM REAGENTS *See other* HALO-ARYLMETALS

2656. 3-Trifluoromethylphenyllithium
[368-49-0] $C_7H_4F_3Li$

See entry ORGANOLITHIUM REAGENTS *See other* HALO-ARYLMETALS

2657. 4-Trifluoromethylphenyllithium
[2786-01-8] $C_7H_4F_3Li$

See entry ORGANOLITHIUM REAGENTS *See other* HALO-ARYLMETALS

2658. Sodium 2-benzothiazolylthiolate
[26249-01-4] $C_7H_4NNaS_2$

Maleic anhydride
See Maleic anhydride: Bases, etc.

2659. Benzenediazonium-2-carboxylate
[17333-86-7] $C_7H_4N_2O_2$

1. Yaroslavsky, S., *Chem. & Ind.*, 1965, 765
2. Sullivan, J. M., *Chem. Eng. News*, 1971, **49**(16), 5
3. Embree, H. D., *Chem. Eng. News*, 1971, **49**(30), 3
4. Mich, T. K. *et al., J. Chem. Educ.*, 1968, **45**, 272
5. Matuszak, C. A., *Chem. Eng. News*, 1971, **49**(24), 39
6. Logullo, F. M. *et al., Org. Synth.*, 1973, Coll. Vol. 5, 54–59
7. Stiles, R. M. *et al.*, 1963, **85**, 1795, footnote 30a
8. Rabideau, P. W., *Org. Prep. Proced. Int.*, 1986, **18**, 113–116
The isolated internal salt is explosive, and should only be handled in small amounts [1]. This has been amply confirmed [2,3]. The salt may be precipitated during the generation of benzyne by diazotisation of anthranilic acid at ambient temperature; some heating is essential for complete decomposition [4]. During decomposition of the isolated salt, too-rapid addition to hot solvent caused a violent explosion [5]. The published procedure [6] must be closely followed for safe working. It was claimed that, contrary to earlier beliefs, diazonium carboxylate hydrohalide salts are also shock-sensitive explosives [7], but this is now firmly refuted. The hydrochloride salt cannot be detonated by impact, but if deliberately ignited by flame contact, it will flash. Thus, when a 30 g quantity was ignited, it burned away very rapidly without explosion. The hydrochloride appears to have other practical advantages, in that the yield of dibenzobarrelene, formed from benzyne and naphthalene is much higher when the hydrochloride is used as benzyne precursor, rather than the internal salt [8].

Aniline, or Isocyanides
Huisgen, R. *et al., Chem. Ber.*, 1965, **98**, 4104
It reacts explosively with aniline, and violently with aryl isocyanides.
See other AMINATION INCIDENTS

1-Pyrrolidinylcyclohexene
Kuehne, M. E., *J. Amer. Chem. Soc.*, 1962, **84**, 841
Decomposition of benzenediazonium-2-carboxylate in the enamine at 40°C caused a violent explosion.
See other DIAZONIUM CARBOXYLATES

2660. Benzenediazonium-3-carboxylate
[17333-87-8]

$C_7H_4N_2O_2$

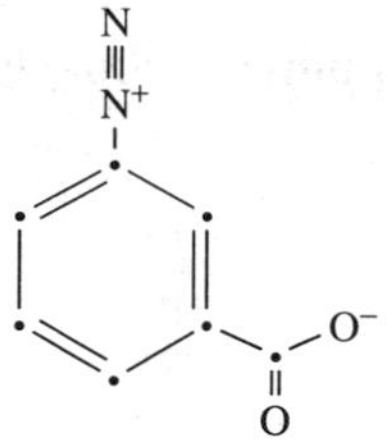

Potassium O, O-diphenyl dithiophosphate
See DIAZONIUM SULFIDES AND DERIVATIVES (reference 10)
See other DIAZONIUM CARBOXYLATES

2661. Benzenediazonium-4-carboxylate
[17333-88-9]

$C_7H_4N_2O_2$

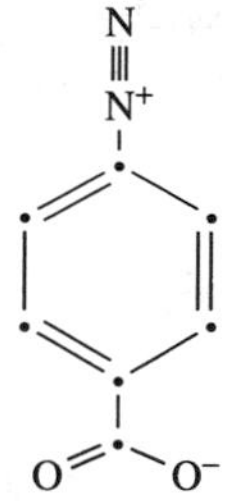

Potassium O, O-diphenyl dithiophosphate
See DIAZONIUM SULFIDES AND DERIVATIVES (reference 10)
See other DIAZONIUM CARBOXYLATES

2662. 2-Nitrobenzonitrile
[612-24-8]

$C_7H_4N_2O_2$

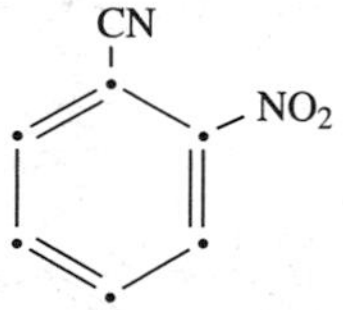

1. Miller, C. S., *Org. Synth.*, 1955, Coll. Vol. 3, 646
2. Partridge, M. W., private comm., 1968
When the published method [1] for preparing the 4-isomer is used to prepare 2-nitrobenzonitrile, a moderate explosion often occurs towards the end of the reaction period. (This may be owing to formation of nitrogen trichloride as a by-product.) Using an alternative procedure, involving heating 2-chloronitrobenzene with copper(I) cyanide in pyridine for 7 h at 160°C, explosions occurred towards the end of the heating period in about 20% of the preparations [2].

See 4-Chloro-3-nitrotoluene: Copper(I) cyanide, Pyridine
See other CYANO COMPOUNDS, NITROARYL COMPOUNDS

2663. 4-Hydroxybenzenediazonium-3-carboxylate
[68596-89-4] $C_7H_4N_2O_3$

1. Auden, W., *Chem. News*, 1899, **80**, 302
2. Puxeddu, E., *Gazz. Chim. Ital.*, 1929, **59**, 14

It explodes at either 155° [1] or 162°C [2], probably depending on the heating rate.
See other DIAZONIUM CARBOXYLATES

2664. 3-Nitrobenzoyl nitrate
[101672-19-9] $C_7H_4N_2O_6$

Francis, F. E., *J. Chem. Soc.*, 1906, **89**, 1

The nitrate explodes if heated rapidly, like benzoyl nitrate.
See other ACYL NITRATES, NITROARYL COMPOUNDS

2665. 3,5-Dinitrosalicylic acid (2-Hydroxy-3,5-dinitrobenzoic acid)
[609-99-4] $C_7H_4N_2O_7$

The editor would be surprised if it proved impossible to detonate this analytical reagent, the carboxylation product of 2,4-dinitrophenol, especially in contact with bases. There is no record of problems in use.

See other POLYNITROARYL COMPOUNDS
See 2,4-Dinitrophenol; 2,4-Dinitrophenylhydrazine

2666. 3,5-Dinitro-2-methylbenzenediazonium-4-oxide
[70343-16-7]

$C_7H_4N_4O_5$

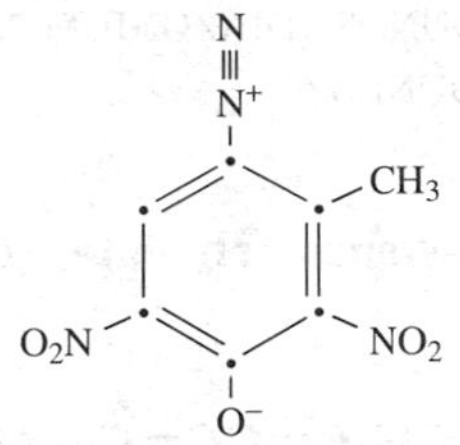

Nielsen, A. T. *et al., J. Org. Chem.*, 1979, **44**, 2503–2504
A very shock-sensitive explosive.
See other ARENEDIAZONIUM OXIDES, POLYNITROARYL COMPOUNDS

2667. 3,5-Dinitro-6-methylbenzenediazonium-2-oxide
[70343-16-7]

$C_7H_4N_4O_5$

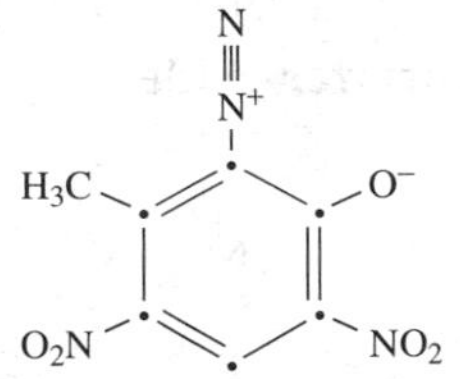

Nielsen, A. T. *et al., J. Org. Chem.*, 1979, **44**, 2503–2504
A very shock-sensitive explosive which melts at 152°C, then decomposes explosively.
See other ARENEDIAZONIUM OXIDES, POLYNITROARYL COMPOUNDS

2668. 2-Cyano-4-nitrobenzenediazonium hydrogen sulfate
[68597-10-4]

$C_7H_4N_4O_6S$

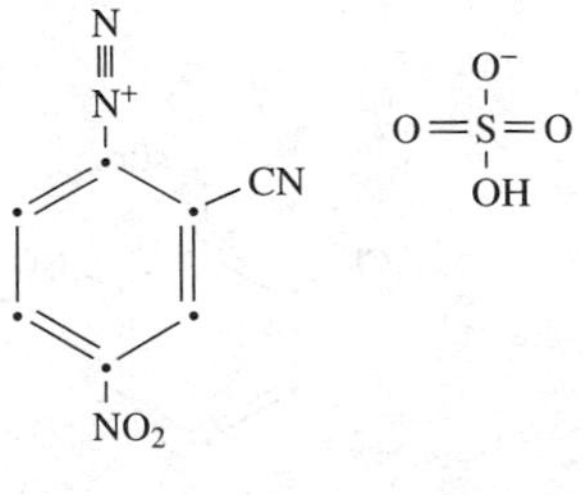

Sulfuric acid
Grewer, T., *Chem. Ing. Tech.*, 1975, **47**, 233–234

DTA examination of a 35% solution of the diazonium salt in sulfuric acid showed 3 exotherms, corresponding to hydrolysis of the nitrile group (peak at 95°), decomposition of the diazonium salt (peak at 160°) and loss of the nitro group (large peak at 240°C). Adiabatic decomposition of the solution from 50°C also showed 3 steps, with induction periods of around 30, 340 and 380 min, respectively.

See THERMAL STABILITY OF REACTION MIXTURES AND SYSTEMS
See other CYANO COMPOUNDS, DIAZONIUM SULFATES, INDUCTION PERIOD INCIDENTS, NITROARYL COMPOUNDS

2669. Poly(2,4-hexadiyne-1,6-ylene carbonate) (Poly(oxycarbonyl-2,4-hexadiyne-1,6-diyl))
[63354-77-8] $(C_7H_4O_3)_n$

$$(-CH_2C{\equiv}CC{\equiv}CH_2OC(O)O-)_n$$

Kuehling, S. *et al., Macromolecules*, 1993, **23**(19), 4192
This polymer, prepared by copper catalysed air oxidation of dipropynyl carbonate, exploded readily on warming or manipulation. The corresponding adipate and sebacate esters are non-explosive. The authors wonder if peroxide formation is involved; such explanation seems otiose.
See other ACETYLENIC COMPOUNDS

2670. Silver 3-cyano-1-phenyltriazen-3-ide
[70324-20-8] $C_7H_5AgN_4$

Bretschneider, H., *Monats.*, 1950, **81**, 981
It is explosive.
See other CYANO COMPOUNDS, SILVER COMPOUNDS, TRIAZENES

2671. 6-Chloro-2-nitrobenzyl bromide
[56433-01-3] $C_7H_5BrClNO_2$

See entry NITROBENZYL COMPOUNDS (reference 2)
See other BENZYL COMPOUNDS, NITROARYL COMPOUNDS

2672. 2,6-Dinitrobenzyl bromide
[3013-38-5] $C_7H_5BrN_2O_4$

Potassium phthalimide
Reich, S. *et al., Bull. Soc. Chim. Fr. [4]*, 1917, **21**, 119
Reaction temperature in absence of solvent must be below 130–135°C to avoid
explosive decomposition.
See entry NITROBENZYL COMPOUNDS *See other* POLYNITROARYL COMPOUNDS

2673. Phenylchlorodiazirine
[4460-46-2] $C_7H_5ClN_2$

1. Wheeler, J. J., *Chem. Eng. News*, 1970, **48**(30), 10
2. Manzara, A. P., *Chem. Eng. News*, 1974, **52**(42), 5
3. Padwa, A. *et al., Org. Synth.*, 1981, **60**, 53–57
The neat material is about three times as shock-sensitive as glyceryl nitrate, and
should not be handled undiluted [1]. It exploded during vacuum distillation at
3.3 mbar from a bath at 140°C. Impact- and spark-sensitivities were determined,
and autoignition occurred after 30 s at 107°C [2]. An explosion occurred during
distillation at a pot temperature of 80°C. Full handling precautions are detailed.
Dilution of neat material with solvent eliminates shock-sensitivity [3].
See other AUTOIGNITION INCIDENTS

Phenylacetylene
See 3-Chloro-1,3-diphenylcyclopropene
See other DIAZIRINES

2674. N-Chloro-5-phenyltetrazole
[65037-43-6]   C$_7$H$_5$ClN$_4$

Gallacher, T. C. *et al., J. Chem. Soc., Chem. Comm.*, 1979, 420
Thermal decomposition is vigorous, and was explosive on one occasion.
See other N-HALOGEN COMPOUNDS, TETRAZOLES

2675. Benzoyl chloride
[98-88-4]   C$_7$H$_5$ClO

Aluminium chloride, Naphthalene
See Aluminium chloride: Benzoyl chloride, etc.

Dimethyl sulfoxide   MRH 1.30/tr.
See Dimethyl sulfoxide: Acyl halides

Other reactants
Yoshida, 1980, 58
MRH values calculated for 13 combinations with oxidants are given.
See other ACYL HALIDES

2676. 3-Chloroperoxybenzoic acid
[937-14-4]   C$_7$H$_5$ClO$_3$

2-Bromo-4-methylpyridine
1. Brand, W. W., *Chem. Eng. News*, 1978, **56**(24), 88
2. Shanley, E. S., *Chem. Eng. News*, 1978, **56**(29), 43

Vacuum evaporation of the product of unheated conversion of the pyridine to its N-oxide with 5% excess of the peroxy acid in chloroform gave a residue which decomposed violently [1]. This was attributed to the relative stability of the peroxy acid in the cold pure state, which when concentrated and finally heated with other materials underwent accelerating decomposition [2].

See other PEROXYACIDS

2677. 1,3-Benzodithiolium perchlorate
[32283-21-9] $C_7H_5ClO_4S_2$

1. Pelter, A., *Chem. Abs.*, 1981, **95**, 97643 (ref. there to *Tetrahedron Lett.* original is incorrect)
2. Degari, I. *et al., J. Chem. Soc., Perkin Trans. 1*, 1976, 1886
3. Anon., *Univ. Safety Assoc. Safety News*, 1978, (10), 11–12

Highly friction sensitive, with considerable explosion hazard [1]. Repeating a published method ([2], where explosion at the m.p., 180°C is noted), the product was dried in a desiccator. When the lid (ground glass flange) was slid off, the friction initiated explosion of a little of the compound deposited on the flange, and a severe explosion resulted [3].

See other GLASS INCIDENTS, NON-METAL PERCHLORATES

2678. 2-Chloro-4-nitrobenzyl chloride
[50274-95-8] $C_7H_5Cl_2NO_2$

See entry NITROBENZYL COMPOUNDS (reference 2)
See other BENZYL COMPOUNDS, NITROARYL COMPOUNDS

2679. 4-Chloro-2-nitrobenzyl chloride
[938-71-6] $C_7H_5Cl_2NO_2$

See entry NITROBENZYL COMPOUNDS (reference 2)
See other BENZYL COMPOUNDS, NITROARYL COMPOUNDS

2680. α-(Difluoroamino)-α,α-dinitrotoluene ((Difluoroaminodinitromethyl)
benzene, N,N -Difluoro-α,α-dinitrobenzenemethanamine)
[157066-32-5] $C_7H_5F_2N_3O_4$

See 1,1-DINITRODIFLUORAMINES, See other N-HALOGEN COMPOUNDS, POLYNITRO-
ALKYL COMPOUNDS

2681. 2-Iodylbenzoic acid (1-Hydroxy-1,2-benziodoxol-3-one-1-oxide)
[64297-64-9] (Acid) $C_7H_5IO_4$
[131-62-4] (Iodoxolone)

Plumb, J. B., Chem. Eng. News, 1990, **68**(29), 3
The pure dry acid explodes at 233°C, and violently if confined, possibly <200°C.
It is also impact-sensitive, exploding under a hammer blow, or under impact of
a 534 g steel ball falling from a height of 1 m. Several of its salts (ammonium,
potassium, sodium, silver, barium, calcium and magnesium) are also explosive. The
acid is precursor to the 'Dess-Martin periodinane' mild oxidant, and is produced
when the latter is treated with water.

874

Calcium 2-iodylbenzoate
See 1,1,1-Triacetoxy-1,2-benziodoxol-3-one
See other IODINE COMPOUNDS, ORGANIC ACIDS

2682. Potassium phenyldinitromethanide
[2918-51-8] $C_7H_5KN_2O_4$

Rohman, A. *et al., J. Org. Chem.*, 1976, **41**, 124
It is unwise to use more than 0.02 mol of potassium salts of dinitroalkanes as they are explosive.
See related aci-NITRO SALTS, POLYNITROALKYL COMPOUNDS

2683. Potassium 2,4,6-trinitrotoluene
[ ] $C_7H_5KN_3O_6$

See entry NITROAROMATIC–ALKALI HAZARDS; potassium radical salts

2684. Potassium *O–O*-benzoylmonoperoxosulfate
[ ] $C_7H_5KO_5S$

Alone, or Sulfuric acid
    Willstätter, R. *et al., Ber.*, 1909, **42**, 1839

The anhydrous salt explodes on grinding, and the monohydrate on heating to 70–80°C, or in contact with sulfuric acid.

See other DIACYL PEROXIDES, PEROXOACID SALTS

2685. Phenyl isocyanate
[103-71-9] C_7H_5NO

Energy of decomposition (in range 240–460°C) measured as 0.41 kJ/g.

See entry THERMOCHEMISTRY AND EXOTHERMIC DECOMPOSITION (reference 2)

See other ORGANIC ISOCYANATES

2686. 2-Nitrobenzaldehyde
[552-89-6] $C_7H_5NO_3$

1. Grewer, T. *et al., Exothermic Decomposition*, Technical Report 01VD 159/0329 for Federal German Ministry for Res. Technol., Bonn. 1986
2. Grewer, T. *et al., Proc. 3rd Int. Symp. Loss Prev. Safety Prom. Proc. Ind.,* 657–664, Basle, SSCI, 1980
3. Grewer, T. *et al., Hazards from Pressure*, IChE Symp. Ser. No. 102, 1–9, Oxford, Pergamon, 1987
4. Anon., *Loss Prev. Bull.,* 1989, (087), 27–29

Energy of decomposition (in range 220–370°C) measured as 2.124 kJ/g by DSC, and T_{ait24} was determined as 123°C by adiabatic Dewar tests, with an apparent energy of activation of 116 kJ/mol [1]. Initial decomposition temperature by ARC was 176°C. A very high rate of pressure increase was observed in exothermic decomposition, and further work on homogeneous decomposition under confinement has been reported [2,3]. Exothermic decomposition during vacuum distillation of an atypical 600 kg batch of 2-nitrobenzaldehyde containing other isomers, and which accelerated to explosion, has been reported and discussed [4].

See entry NITROBENZALDEHYDES

Active carbon

Mixtures with active carbon (1.5:1 mol) deflagrate at 1.25 cm/min.

See entry PRESSURE INCREASE IN EXOTHERMIC DECOMPOSITION (reference 3)

Pyrrole

Sorrell, T. N., *Inorg. Synth.*, 1980, **20**, 162

In the preparation of 5,10,15,20-tetrakis(2-nitrophenyl)porphyrin by dropwise addition of the aldehyde in acetic acid, the exothermic reaction may become so violent that solvent sprays out of the condenser. However, a low yield may ensue if the reaction is insufficiently vigorous, so a careful compromise is required.

See other ALDEHYDES, NITROARYL COMPOUNDS

2687. 3-Nitrobenzaldehyde
[99-61-6] $C_7H_5NO_3$

1. Lange, J. *et al., Chem. & Ind.*, 1967, 1424
2. Klais, O. *et al., Proc. 4th Int. Symp. Loss Prev. Safety Prom. Proc. Ind.*, Vol. 3, C24–C34, Rugby, IChE, 1983
3. Grewer, T. *et al., Exothermic Decomposition*, Technical Report 01VD 159/0329 for Federal German Ministry for Res. Technol., Bonn. 1986

Delays in working up the crude product caused violent explosions during attempted vacuum distillation. An alternative method of crystallisation is described [1]. There is a very high rate of pressure increase in exothermic decomposition [2]. Energy of decomposition (in range 180–420°C) measured as 2.19 kJ/g by DSC, and T_{ait24} was determined as 147°C by adiabatic Dewar tests, with an apparent energy of activation of 168 kJ/mol [3]. The initial decomposition temperature by ARC was 166°C.

See entry NITROBENZALDEHYDES
See other ALDEHYDES, NITROARYL COMPOUNDS

2688. 4-Nitrobenzaldehyde
[555-16-5] $C_7H_5NO_3$

1. Grewer, T. *et al., Exothermic Decomposition*, Technical Report 01VD 159/0329 for Federal German Ministry for Res. Technol., Bonn. 1986

Energy of decomposition (in range 220–450°C) measured as 1.602 kJ/g by DSC, and T_{ait24} was determined as 171°C by adiabatic Dewar tests, with an apparent energy of activation of 202 kJ/mol [1],The initial decomposition temperature by ARC was 226°C. A very high rate of pressure increase was observed in exothermic decomposition.

See entry PRESSURE INCREASE IN EXOTHERMIC DECOMPOSITION
See entry NITROBENZALDEHYDES
See other ALDEHYDES, NITROARYL COMPOUNDS

2689. Benzoyl nitrate
[6786-32-9] $C_7H_5NO_4$

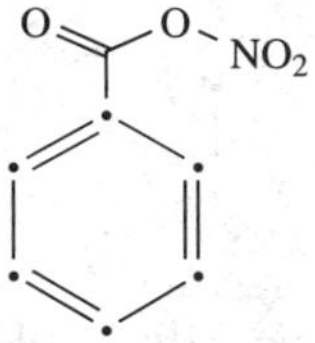

1. Francis, F. E., *J. Chem. Soc.*, 1906, **89**, 1
2. Ferrario, E., *Gazz. Chim. Ital. [1]*, 1901, **40**, 99

An unstable liquid which is capable of distillation under reduced pressure, but which explodes violently on rapid heating at ambient pressure [1]. It may also explode on exposure to light [2].

See other IRRADIATION DECOMPOSITION INCIDENTS

Water
Francis, F. E., *J. Chem. Soc.*, 1906, **89**, 1

Reactivity of benzoyl nitrate towards moisture is so great that attempted filtration through an undried filter paper causes explosive decomposition (possibly involving cellulose nitrate?).

See other ACYL NITRATES

2690. 3-Hydroxy-4-nitrobenzaldehyde
[704-13-2] $C_7H_5NO_4$

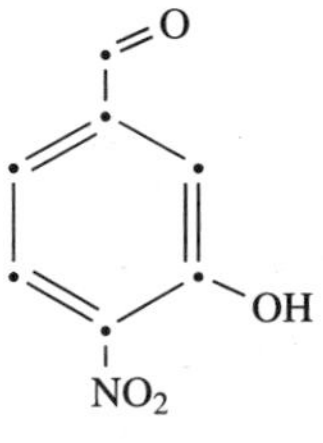

Initial decomposition temperature by DSC was 200°C.

See entry NITROBENZALDEHYDES *See other* ALDEHYDES, NITROARYL COMPOUNDS

878

2691. 4-Hydroxy-3-nitrobenzaldehyde
[3011-34-5]

$C_7H_5NO_4$

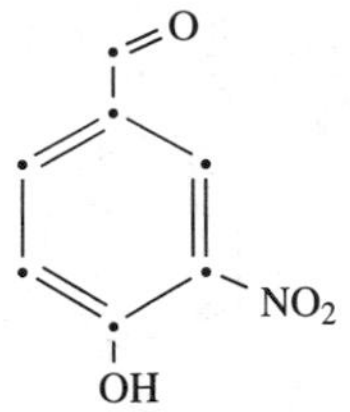

Initial decomposition temperature by DSC was 200°C.

See entry NITROBENZALDEHYDES *See other* ALDEHYDES, NITROARYL COMPOUNDS

2692. 5-Hydroxy-2-nitrobenzaldehyde
[42454-06-8]

$C_7H_5NO_4$

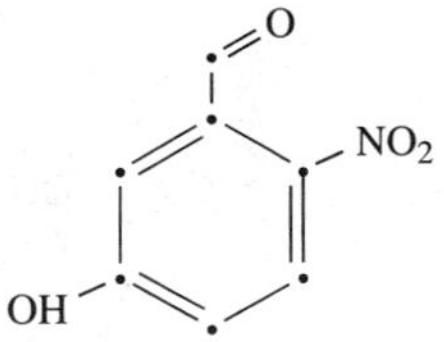

Initial decomposition temperature by DSC was 175°C.

See entry NITROBENZALDEHYDES *See other* NITROARYL COMPOUNDS

2693. 2-Nitrobenzoic acid
[552-16-9]

$C_7H_5NO_4$

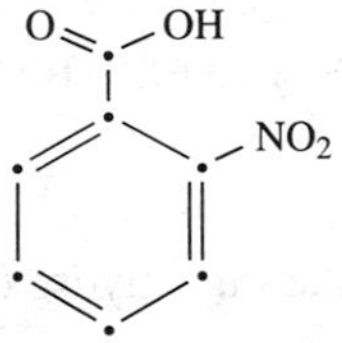

Energy of decomposition (in range 230–450°C) measured as 1.72 kJ/g by DSC, and T_{ait24} was determined as 189°C by adiabatic Dewar tests, with an apparent energy of activation of 156 kJ/mol.

See entry THERMOCHEMISTRY AND EXOTHERMIC DECOMPOSITION (reference 2)

See other NITROARYL COMPOUNDS, ORGANIC ACIDS

2694. 3-Nitrobenzoic acid
  [121-92-6] $C_7H_5NO_4$

Energy of decomposition (in range 320–410°C) measured as 1.734 kJ/g by DSC, and T_{ait24} was determined as 245°C by adiabatic Dewar tests, with an apparent energy of activation of 151 kJ/mol.

See entry THERMOCHEMISTRY AND EXOTHERMIC DECOMPOSITION (reference 2)
See other NITROARYL COMPOUNDS, ORGANIC ACIDS

2695. 4-Nitrobenzoic acid
  [62-23-7] $C_7H_5NO_4$

Energy of decomposition (in range 320–360°C) measured as 2.234 kJ/g by DSC, and T_{ait24} was determined as 247°C by adiabatic Dewar tests, with an apparent energy of activation of 131 kJ/mol.

See entry THERMOCHEMISTRY AND EXOTHERMIC DECOMPOSITION (reference 2)

Potassium hydroxide
  Mixtures of the acid with potassium hydroxide (1:2 mol) readily deflagrated, at a rate of 5.1 cm/min.

See entry PRESSURE INCREASE IN EXOTHERMIC DECOMPOSITION (reference 3)
See other DEFLAGRATION INCIDENTS, NITROARYL COMPOUNDS, ORGANIC ACIDS

2696. 2,5-Pyridinedicarboxylic acid
[100-26-5]

$C_7H_5NO_4$

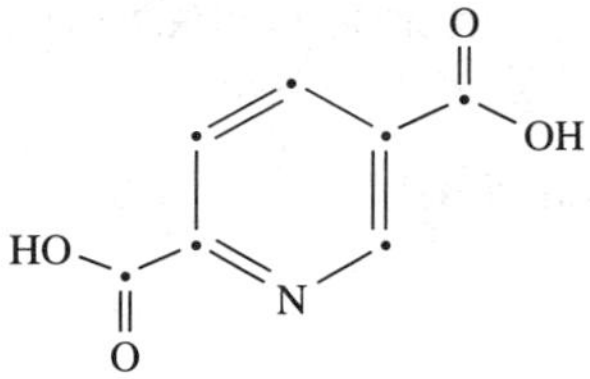

Preparative hazard
See Nitric acid: 5-Ethyl-2-methylpyridine
See other ORGANIC ACIDS

2697. 4-Azidobenzaldehyde
[24173-36-2]

$C_7H_5N_3O$

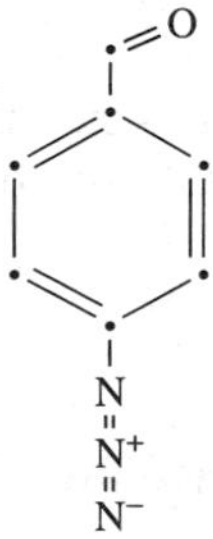

Lahti, P. M., *Chem. Eng. News*, 1998, **76** (28), 8
Very shock sensitive; explodes in drop-weight test at 11 kg-cm.
Energy of decomposition (in range 110–220°C) measured as 1.67 kJ/g.
See entry THERMOCHEMISTRY AND EXOTHERMIC DECOMPOSITION (REFERENCE 2)
See other ALDEHYDES, ORGANIC AZIDES

2698. Benzoyl azide
[582-61-6]

$C_7H_5N_3O$

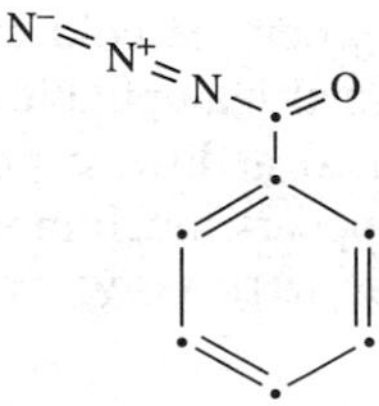

Bergel, F., *Angew. Chem.*, 1927, **40**, 974

The sensitivity towards heat of this explosive compound is increased by previous compression, confinement and presence of impurities. Crude material exploded violently between 120 and 165°C.

See other ACYL AZIDES, CATALYTIC IMPURITY INCIDENTS

2699. Carboxybenzenesulfonyl azide
[56743-33-0] $C_7H_5N_3O_4S$

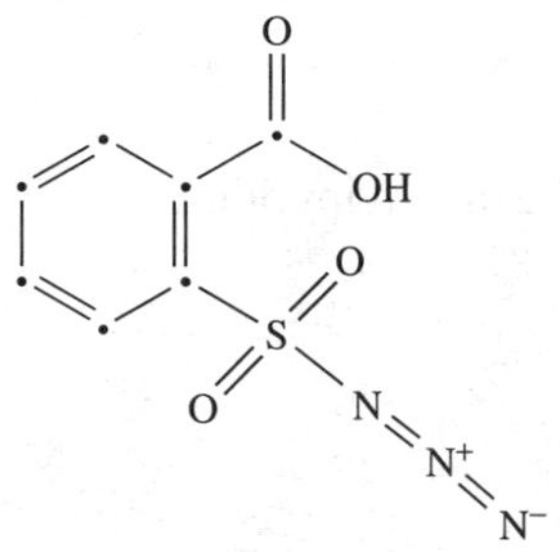

Anon., *Res. Discl.*, 1975, **134**, 44

The azide (structure not stated) is not impact-sensitive, but decomposes explosively at 120°C. Blending with 75% of polymer as diluent eliminated the explosive decomposition.

See other ACYL AZIDES

2700. 2-Nitrophenylsulfonyldiazomethane
[49558-46-5] $C_7H_5N_3O_4S$

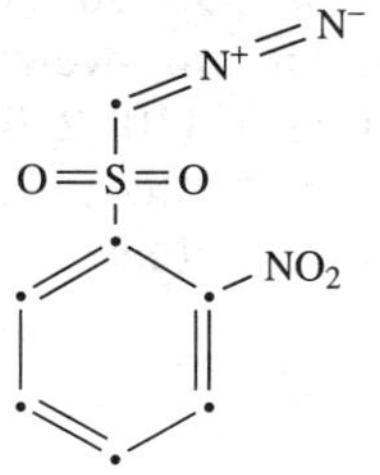

Wagenaar, A. *et al., J. Org. Chem.*, 1974, **39**, 411−413

Severe explosions occurred when the cold crystalline solid was allowed to warm to ambient temperature. Normal manipulation of dilute solutions appears feasible, but the solid must be handled at lowest possible temperatures with full safety precautions. The 4-isomer appears much more stable.

See other DIAZO COMPOUNDS, NITROARYL COMPOUNDS

882

2701. 2,4,6-Trinitrotoluene
[118-96-7] $C_7H_5N_3O_6$

NSC 314, 1981

Heat of decomposition was determined as 5.1 kJ/g.
See entry THERMOCHEMISTRY AND EXOTHERMIC DECOMPOSITION (REFERENCE 1)

Added impurities
Haüptli, H., private comm., 1972
During investigation of effect of 1% of added impurities on the thermal explosion temperature of TNT (297°C), it was found that fresh red lead, sodium carbonate and potassium hydroxide reduced the explosion temperatures to 192, 218, and 192°C, respectively.
See Potassium: Nitrogen-containing explosives
See other CATALYTIC IMPURITY INCIDENTS

Bases   MRH Sodium hydroxide 5.40/tr.
Capellos, C. *et al., Chem. Abs.*, 1982, **96**, 88003
The formation of *aci*-quinonoid transient intermediates by electronic excitation has been identified spectroscopically. Proton abstractors (bases) such as sodium hydroxide, potassium iodide or tetramethylammonium octahydrotriborate induce deflagration in molten TNT.
See NITROAROMATIC–ALKALI HAZARDS

Metals, Nitric acid
Kovache, A. *et al., Mé* m. Poudres, 1952, **34**, 369–378
Trinitrotoluene in contact with nitric acid and lead or iron produces explosive substances which may readily be ignited by shock, friction or contact with nitric or sulfuric acids. Such materials have been involved in industrial explosions.
See HEAVY METAL DERIVATIVES

Other reactants
Yoshida, 1980, 254
MRH values calculated for 3 combinations with catalytically active materials are given.

Potassium hydroxide
Copisarow, M., *Chem. News*, 1915, **112**, 283–284

TNT and potassium hydroxide in methanol will interact even at −65°C to give explosive *aci*-nitro salts (presumably *o*-quinonoid, or possibly Meisenheimer complexes). The explosion temperature is lowered to 160°C by the presence of a little potassium hydroxide.
See entry aci-NITROQUINONOID COMPOUNDS

Sodium carbonate, Water
Iida, M. *et al., Chem. Abs.*, 1997, **127**, 222627q
TNT reacts vigorously with sodium carbonate, in presence of water at 100°C and above. The solid residue resulting ignites on contact with strong acid.
See 2,4,5-Trinitrotoluene; next below
See NITROAROMATIC–ALKALI HAZARDS

Sodium dichromate, Sulfuric acid
See Sodium dichromate: Sulfuric acid, Trinitrotoluene
See other POLYNITROARYL COMPOUNDS

2702. 2,4,5-Trinitrotoluene
[610-25-3] $C_7H_5N_3O_6$

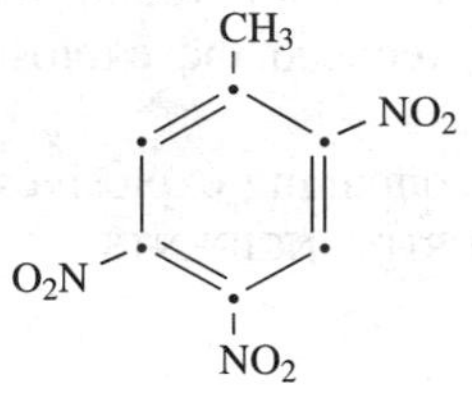

Sodium carbonate
Biasutti, 1981, 78
Sodium carbonate was added to a heated tank of washed TNT to neutralise excess acidity. It is known that the assym. isomer of TNT reacts with sodium carbonate to form unstable compounds, and decomposition led to a fire and detonation of 5 tonnes of explosive.
See other NITROAROMATIC–ALKALI HAZARDS, POLYNITROARYL COMPOUNDS

2703. 3-Methyl-2,4,6-trinitrophenol
[602-99-3] $C_7H_5N_3O_7$

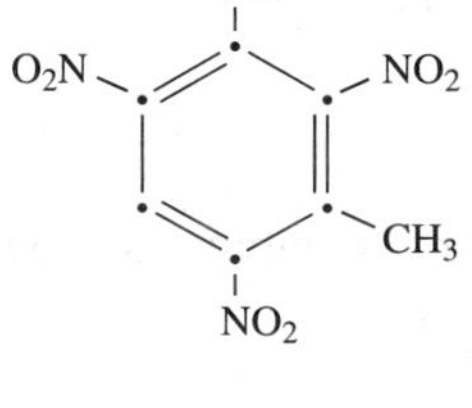

Sorbe, 1968, 153

884

It explodes above 150°C.
See other POLYNITROARYL COMPOUNDS

2704. Benzimidazolium 1-nitroimidate
[52096-22-7] $C_7H_5N_4O_2$

It explodes at the m.p., 169°C.
See entry N-AZOLIUM NITROIMIDATES

2705. 1-Nitro-3-(2,4-dinitrophenyl)urea
[22751-18-4] $C_7H_5N_5O_7$

McVeigh, J. L *et al., J. Chem. Soc.,* 1945, 621
It is an explosive of lower power, but of greater impact- or friction-sensitivity than picric acid, so should not be stored in bottles with ground-in stoppers.
See other N-NITRO COMPOUNDS, POLYNITROARYL COMPOUNDS

2706. N,2,4,6-Tetranitro-N -methylaniline (Tetryl)
[479-45-8] $C_7H_5N_5O_8$

Hydrazine
See Hydrazine: Oxidants

885

Trioxygen difluoride
See 'Trioxygen difluoride': Various materials
See other N-NITRO COMPOUNDS, POLYNITROARYL COMPOUNDS

2707. 1-Heptene-4,6-diyne
[ ] C_7H_6

1. Armitage, J. B. et al., J. Chem. Soc., 1952, 1994
2. CHETAH, 1990, 182
It decomposes explosively with great ease [1] but was shown to be exploded by a detonator, not by mechanical shock [2].
See other ALKYNES, ALLYL COMPOUNDS

2708. 2-Azidomethylbenzenediazonium tetrafluoroborate
[59327-98-9] $C_7H_6BF_4N_5$

Trichloroacetonitrile
Kreher, R. et al., Tetrahedron Lett., 1976, 4260
Thermal decomposition of the salt in nitriles produces 2-substituted quinazolines.
In trichloroacetonitrile it proceeded explosively.
See other DIAZONIUM TETRAHALOBORATES, ORGANIC AZIDES

2709. 5-Fluoro-2-methylphenylmagnesium bromide (Bromo(5-fluoro-2-methylphenyl)magnesium)

[ ] C_7H_6BrFMg

Preparative Hazard
Jasiewicz, M. L. *Chem. & Ind.*, 1995, 246

A considerable explosion resulted from preparation of this compound (which presumably exists with several tightly complexed solvent molecules) in limited tetrahydrofuran under reflux. Investigation showed that there is an exotherm at 130°C, giving 440 J/g, followed by a more violent one at 300°C giving 690 J/g. Using a large excess of solvent no problems are observed.

See other GRIGNARD REAGENTS

2710. 2-Nitrobenzyl bromide
[3958-60-9] $C_7H_6BrNO_2$

1. Biasutti, 1981, 114
2. Cardillo, P. *et al., Rivista Combust.*, 1982, **36**, 304
3. Cardillo, P. *et al., Proced. 34th Ital. Chem. Congr., 'Safety of Chemical Processes'*, 71–82, Milan, 1983

During the first large scale preparation of the bromide from 2-nitrotoluene, the reactor exploded violently, though hundreds of runs had been done uneventfully in a pilot scale plant [1]. Following a thermal explosion during drying of the compound at a slightly elevated temperature, its thermal stability has been determined by TGA, DSC and ARC techniques. The latter shows that exothermic decomposition begins at 100°C and proceeds in two stages with abundant gas evolution. The former techniques indicate onset of decomposition at 130° and 112°C, respectively [2]. Further work on the 3- and 4-isomers showed them to be marginally more stable than the 2-nitro compound.

See 2-Nitrobenzyl chloride
See other GAS EVOLUTION INCIDENTS, NITROBENZYL COMPOUNDS, NITROARYL COMPOUNDS

2711. 2-Chloro-4-nitrotoluene
[121-86-8]

$C_7H_6ClNO_2$

Other reactants
Yoshida, 1980, 112
MRH values calculated for 13 combinations with oxidants are given.

Sodium hydroxide   MRH 3.39/tr.
1. Anon., *ABCM Quart. Safety Summ.*, 1962, **33**, 20
2. *MCA Case History No. 907*
The residue from vacuum distillation of crude material (contaminated with sodium hydroxide) exploded after showing signs of decomposition. Experiment showed that 2-chloro-4-nitrotoluene decomposes violently when heated at 170–200°C in presence of alkali. Thorough water washing is therefore essential before distillation is attempted [1]. A similar incident was reported with mixed chloronitrotoluene isomers [2].
See other NITROAROMATIC–ALKALI HAZARDS
See other NITROARYL COMPOUNDS

2712. 4-Chloro-3-nitrotoluene
[890-60-1]

$C_7H_6ClNO_2$

Copper(I) cyanide, Pyridine
1. Hub, L., *Proc. Chem. Process. Haz. Symp. VI* (Manchester, 1977), No. 49, 39–46, Rugby, IChE, 1977
2. Hub, L., private comm., 1977
Industrial preparation of 4-cyano-3-nitrotoluene by heating the reaction components at around 170°C for 6 h led to an explosion in 1976. Subsequent investigation by DSC showed that the cyano compound in presence of the starting materials exhibited an exotherm at 180°C. After 6 h reaction, this threshold temperature fell to 170°C. Isothermal use of a safety calorimeter showed that a large exotherm occurred during the first hour of reaction and that, in absence of strong cooling,

the reaction accelerated and the vessel contents were ejected by the vigour of the decomposition [2].

See -nitrobenzonitrile, REACTION SAFETY CALORIMETRY
See other NITROARYL COMPOUNDS

2713. 2-Nitrobenzyl chloride
[612-23-7]
$C_7H_6ClNO_2$

Cardillo, P. *et al.*, *Proc. 34th Ital. Chem. Congr.*, *'Safety of Chem. Processes'*, 71–82, Milan, 1983
Following an explosion during drying of 2-nitrobenzyl bromide, the thermal stability of all six isomeric nitrobenzyl halides was investigated by TGA, DSC and ARC techniques. All the halides decomposed exothermally with abundant gas evolution, and the 2-nitro chloride is less thermally stable than the 3- or 4-isomers.
See 2-Nitrobenzyl bromide
See other BENZYL COMPOUNDS, NITROBENZYL COMPOUNDS, NITROARYL COMPOUNDS

2714. 4-Nitrobenzyl chloride
[100-14-1]
$C_7H_6ClNO_2$

$$O_2NC_6H_4CH_2Cl$$

T_{ait24} was determined as 189°C by adiabatic Dewar tests, with an apparent energy of activation of 124 kJ/mol.
See entry THERMOCHEMISTRY AND EXOTHERMIC DECOMPOSITION (reference 2)
See other NITROARYL COMPOUNDS, NITROBENZYL COMPOUNDS

2715. 2-Chloro-5-nitrobenzyl alcohol
[80866-80-4]
$C_7H_6ClNO_3$

See entry NITROBENZYL COMPOUNDS (reference 3)
See other BENZYL COMPOUNDS, NITROARYL COMPOUNDS

2716. 4-Chloro-2-nitrobenzyl alcohol
[22996-18-5]

$C_7H_6ClNO_3$

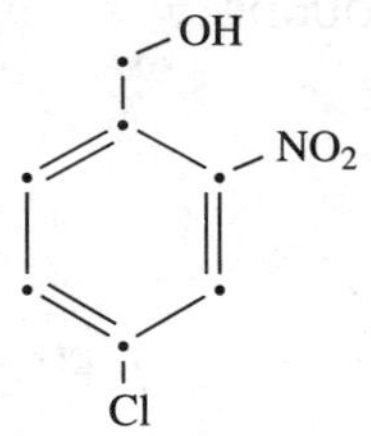

See entry NITROBENZYL COMPOUNDS (reference 3)
See other BENZYL COMPOUNDS, NITROARYL COMPOUNDS

2717. 4-Chloro-3-nitrobenzyl alcohol
[55912-20-4]

$C_7H_6ClNO_3$

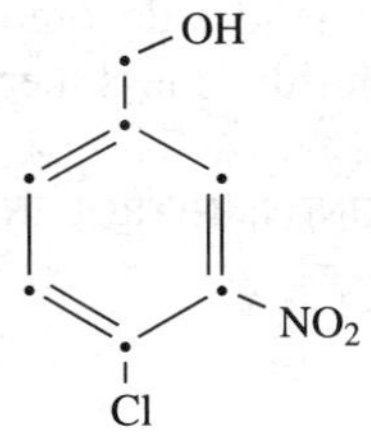

See entry NITROBENZYL COMPOUNDS (reference 3)
See other BENZYL COMPOUNDS, NITROARYL COMPOUNDS

2718. 5-Chloro-2-nitrobenzyl alcohol
[73033-58-6]

$C_7H_6ClNO_3$

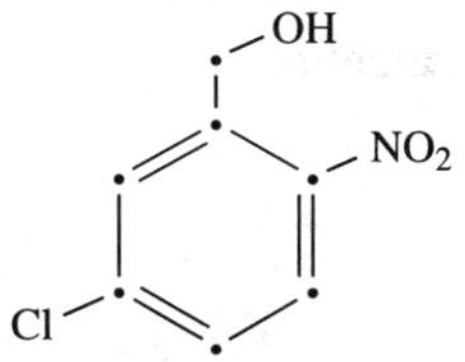

See entry NITROBENZYL COMPOUNDS (reference 3)
See other BENZYL COMPOUNDS, NITROARYL COMPOUNDS

890

2719. 4-Chloro-2-methylbenzenediazonium salts
  [27165-08-8] (ion) $\qquad$ $C_7H_6ClN_2^+ Z^-$

Sodium hydrogen sulfide, or Sodium disulfide, or Sodium polysulfide
  See entry DIAZONIUM SULFIDES AND DERIVATIVES (references 1–4)
  See other DIAZONIUM SALTS

2720. Sodium 4-chloro-2-methylphenoxide
  [52106-86-2] $\qquad$ C_7H_6ClNaO

Anon., private comm., 1984
A dry solution of the sodium salt in n-butanol was usually prepared by azeotropic drying. Use of excessively wet recovered butanol led to complete removal of the butanol with the water and heating of the dry salt at 200°C, when rapid decomposition occurred, leaving a glowing carbonised residue.
See 4-Chloro-2-methylphenol: Sodium hydroxide

2721. Potassium 2,6-dinitrotoluene
  [ ] $\qquad$ $C_7H_6KN_4O_4$

The o- and m-isomers are similar.
See entry NITROAROMATIC–ALKALI HAZARDS; potassium radical salts

891

2722. Potassium 4-methoxy-1-*aci*-nitro-3,5-dinitro-2,5-cyclohexadienonide
[1270-21-9] $C_7H_6KN_3O_7$

1. Meisenheimer, J., *Ann.*, 1902, **323**, 221
2. Millor, H. D. *et al.*, *Sidgwick's Organic Chemistry of Nitrogen*, 396, Oxford, Clarendon Press, 3rd edn, 1966

The product of interaction of trinitrobenzene and conc. aqueous potassium hydroxide in methanol is explosive, and analyses as the hemihydrate of a hemiacetal of the *aci-p*-quinonoid form of picric acid [1], and/or the mesomeric *o*-forms [2].

See other *aci*-NITROQUINONOID COMPOUNDS, POLYNITROARYL COMPOUNDS

2723. Thallium *aci*-phenylnitromethanide
[53847-48-6] $C_7H_6NO_2Tl$

McKillop, A. *et al., Tetrahedron*, 1974, **30**, 1369

No attempt should be made to isolate or dry this compound, as it is treacherously explosive.

See other HEAVY METAL DERIVATIVES, *aci*-NITRO SALTS

2724. Benzimidazole (Benzodiazole)
[51-17-2] $C_7H_6N_2$

CHETAH, 1990, 188
Shock sensitive (which seems unlikely but not impossible).
See Benzotriazole

2725. Phenyldiazomethane
[766-91-6] $C_7H_6N_2$

Schneider, M. *et al.*, *Tetrahedron*, 1976, **36**, 621
The need for care in distillation of phenyldiazomethane is stressed. Previously it has been used as prepared in solution.
See other DIAZO COMPOUNDS

2726. 2,4-Dinitrotoluene
[121-14-2] $C_7H_6N_2O_4$

(*MCA SD-93*, 1966); *NSC 658*, 1976

1. Grewer, T. *et al.*, *Exothermic Decomposition*, Technical Report 01VD 159/0329 for Federal German Ministry for Res. Technol., Bonn. 1986
2. *MCA SD-93*, 1966
3. Biasutti, 1981, 143
4. Bond, J., *Loss Prev. Bull.*, 1983, 050, 18
5. Anon., *Loss Prevention Bull.*, 1989, **88**, 13

T_{ait24} was determined as 181°C by adiabatic Dewar tests, with an apparent energy of activation of 153 kJ/mol [1]. The commercial material, containing some 20% of the 2,6-isomer, decomposes at 250°, but at 280°C decomposition becomes self-sustaining. Prolonged heating below these temperatures may also cause some decomposition, and the presence of impurities may decrease the decomposition temperatures [2]. Although not considered to be explosive, several cases of detonation in manufacture or storage have been reported [3], including an explosion when a jug of dinitrotoluene was dropped into a machine [4]. An isolated pipe filled with crude dinitrotoluene at a temperature between 125 and 135°C ran away to

893

detonation over 3.5 hours. Having been separated from spent acid, but not washed, it contained traces of both sulfuric and nitric acids. Investigation showed that this combination destabilised the material more than either separately. A sample (2 phase) of approximate composition dinitrotoluene 45%, nitric acid 5%, sulfuric acid 40%, water 10% showed an exothermic reaction on ARC as low as 66°C [5].

Alkali
Bateman, T. L. *et al., Loss. Prev.*, 1974, **8**, 117–122
Dinitrotoluene held at 210°C (rather than the 125° intended) for 10 days in a 50 mm steam-heated transfer pipeline exploded. Subsequent tests showed decomposition at 210°C (producing a significant pressure rise) in 1 day, and presence of sodium carbonate (but not rust) reduced the induction period. A maximum handling temperature of 150°C was recommended, (when the induction period was 32 days, or 14 days for alkali-contaminated material).
See other CATALYTIC IMPURITY INCIDENTS, INDUCTION PERIOD INCIDENTS, NITRO-AROMATIC–ALKALI HAZARDS

Nitric acid   MRH 4.56/tr.
See Nitric acid: Nitroaromatics

Other reactants
Yoshida, 1980, 159
MRH values calculated for 10 combinations, largely with catalytically active materials, are given.

Sodium oxide
Anon., *Rept. of Explos. Div.*, Ottawa, Dept. of Mines, 1929
Admixture of the two solids caused a rapid reaction and fire.
See other NITROAROMATIC–ALKALI HAZARDS
See other POLYNITROARYL COMPOUNDS

2727. 5-Phenyltetrazole
[3999-10-8] $C_7H_6N_4$

Wedekind, E., *Ber.*, 1898, **31**, 948
It explodes on attempted distillation.
See other TETRAZOLES

894

2728. *N*-Azidocarbonylazepine

[ ] $C_7H_6N_4O$

Chapman, L. E. *et al., Chem. & Ind.,* 1966, 1266
It explodes on attempted distillation.
See other ACYL AZIDES

2729. *N*-Phenyl-1,2,3,4-thiatriazolamine (Phenylaminothia-2,3,4-triazole)

[13078-30-3] $C_7H_6N_4S$

CHETAH, 1990, 182
Shock sensitive.
See other HIGH NITROGEN COMPOUNDS, N–S COMPOUNDS

2730. Phenyldiazidomethane

[17122-97-3] $C_7H_6N_6$

Barash, L. *et al., J. Amer. Chem. Soc.,* 1967, **89**, 3391
A sample of the *gem*-diazide detonated on contact with a hot glass surface during
ampoule sealing operations.
See other ORGANIC AZIDES

†2731. **Benzaldehyde**
[100-52-7]   C$_7$H$_6$O

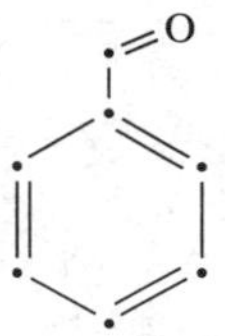

HCS 1980, 191

Other reactants
   Yoshida, 1980, 333
   MRH values calculated for 14 combinations with oxidants are given.

Peroxyformic acid   MRH 5.69/99+
   See Peroxyformic acid: Organic materials

Propionic acid, Pyrrolidine
   Kaufman, J. A., private comm., 1987
   An explosion occurred when benzaldehyde, pyrrolidine (pyrrole ?) and propionic
   acid were heated to form porphins.
   See 2-Nitrobenzaldehyde: Pyrrole
   See other ALDEHYDES, REDUCANTS

2732. **Benzoic acid**
[65-85-0]   C$_7$H$_6$O$_2$

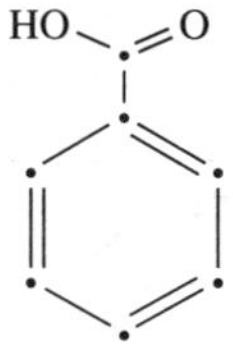

HCS 1980, 191

The finely powdered dry acid is a significant dust explosion hazard.
See entry DUST EXPLOSION INCIDENTS (reference 22)

Other reactants
   Yoshida, 1980, 32
   MRH values calculated for 12 combinations with oxidants are given.

Oxygen
   Oxygen (Gas): Benzoic acid *See other* ORGANIC ACIDS

2733. Peroxybenzoic acid
[93-59-4] $C_7H_6O_3$

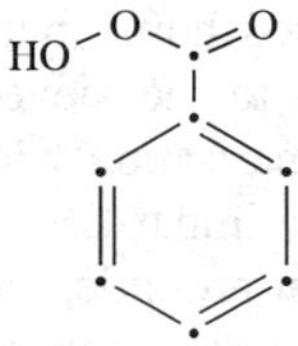

1. Baeyer, A. *et al., Ber.*, 1900, **33**, 1577
2. Silbert, L. S. *et al., Org. Synth.*, 1963, **43**, 95

It explodes weakly on heating [1]. A chloroform solution of peroxybenzoic acid exploded during evaporation of solvent. Evaporation should be avoided or conducted with full precautions [2].

See other PEROXYACIDS

2734. Silver *N*-perchlorylbenzylamide
[89521-43-7] $C_7H_7AgClNO_3$

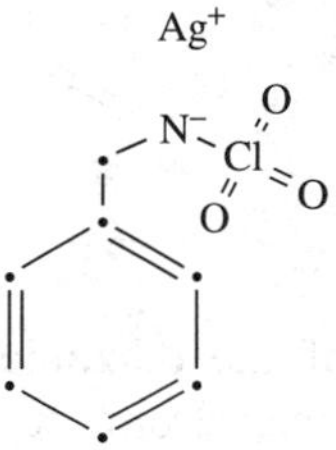

Hennrichs, W. *et al., Z. Anorg. Chem.*, 1983, **50** 6, 203–207
It explodes on impact or on heating above 105°C.
See entry PERCHLORYLAMIDE SALTS (reference 2)
See other *N*-METAL DERIVATIVES, SILVER COMPOUNDS

2735. Benzyl bromide
[100-39-0] C_7H_7Br

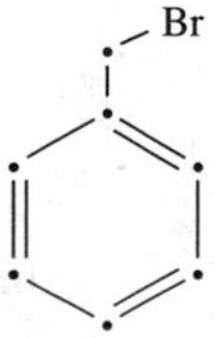

HCS 1980, 200

Molecular sieve

Harris, T. D., *Chem. Eng. News*, 1979, **57**(12), 74

Redistilled bromide was stored over 4A molecular sieve previously activated at 300°C. After 8 days the closed bottle burst from internal pressure of hydrogen bromide. This was attributed to the sieve catalysing a Friedel-Craft type intermolecular condensation-polymerisation with liberation of hydrogen bromide.

See Benzyl chloride: Catalytic impurities

Benzyl alcohol: Hydrogen bromide, Iron

See other GAS EVOLUTION INCIDENTS, MOLECULAR SIEVE INCIDENTS, POLYCONDENSATION REACTION INCIDENTS

See other BENZYL COMPOUNDS

2736. 2-Toluenediazonium bromide
[54514-12-4]   C₇H₇BrN₂

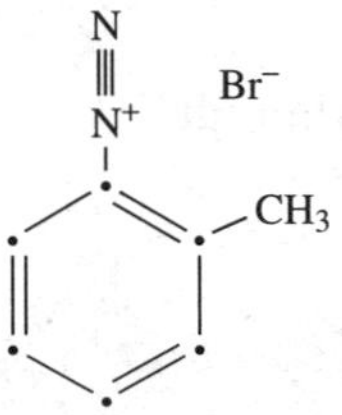

Copper powder

Vogel, 1956, 606–607

During the decomposition of the diazonium solution by warming with copper bronze to give 2-bromotoluene, strong cooling must be applied immediately to prevent the nitrogen evolution becoming violent and ejecting the flask contents.

See other DIAZONIUM SALTS, GAS EVOLUTION INCIDENTS

2737. Methyl 4-bromobenzenediazoate
[67880-26-6]   C₇H₇BrN₂O

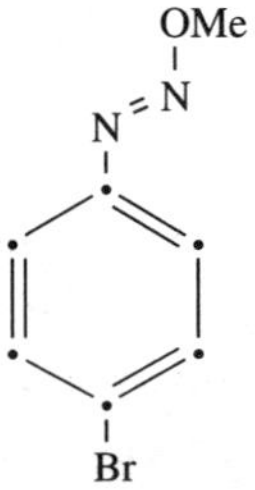

Bamberger, E., *Ber.*, 1895, **28**, 233

It explodes on heating.

See other ARENEDIAZOATES, N–O COMPOUNDS

898

2738. Benzyl chloride
[100-44-7] C_7H_7Cl

HCS 1980, 202

Oehme, F., *Chem. Tech.* (Berlin), 1952, **4**, 404
During distillation of the technical chloride, hydrogen chloride was evolved, air was bubbled through to remove it, and there was an explosion. This was attributed to a polycondensation (intermolecular Friedel-Craft) reaction, possibly catalysed by oxidation products.
Energy of decomposition (in range 290–370°C) measured as 0.14 kJ/g.
See entry THERMOCHEMISTRY AND EXOTHERMIC DECOMPOSITION (reference 2)

Catalytic impurities
Naef, H., private comm., 1980
The chloride is usually (but not always) stabilised in storage by addition of aqueous alkali or anhydrous amines as acid acceptors. A 270 kg batch which was not stabilised polymerised violently when charged into a reactor. Contact of the chloride (slightly hydrolysed and acidic) with rust led to formation of ferric chloride which catalysed an intermolecular Friedel-Craft reaction to form polybenzyls with evolution of further hydrogen chloride. Contact of unstabilised benzyl chloride with aluminium, iron or rust should be avoided to obviate the risk of polycondensation.
See Benzyl bromide: Molecular sieve
Benzyl alcohol: Hydrogen bromide, Iron
1,2-Bis(chloromethyl)benzene: Catalytic impurities
See other GAS EVOLUTION INCIDENTS, POLYCONDENSATION REACTION INCIDENTS

Other reactants
Yoshida, 1980, 331
MRH values for 13 combinations with oxidants are given, many values being high.

Sodium acetate, Pyridine
Bast,–, *Proc. 1st Int. Symp. Prev. Occ. Risks Chem. Ind.*, 194–199, Heidelberg, ISSA, 1970
Benzyl acetate was prepared by addition of benzyl chloride (containing 0.6% pyridine as stabiliser) to preformed sodium acetate at 70°, followed by heating at 115°, then finally up to 135°C to complete the reaction. On one occasion, gas began to be evolved at the end of the dehydration phase, and the reaction accelerated to a violent explosion, rupturing the 25 mm thick cast iron vessel. This was attributed to presence of insufficient pyridine to maintain basicity, dissolution of iron by the

acidic mixture, and catalysis by ferric chloride of a Freidel-Craft type polycondensation reaction to polybenzyls, with evolution of hydrogen chloride, which at 130°C
would produce an overpressure approaching 100 bar. Previously the chloride had
been supplied in steel drums containing 10% sodium carbonate or 3% sodium
hydroxide solutions as stabilisers, but the packing was changed to polythene-lined
kegs and the stabiliser omitted. However, higher pyridine concentrations effectively
stabilise the chloride in steel drums.
See Catalytic impurities, above

Sodium cyanide
Nolan, 1983, Case history 192
Benzyl cyanide was prepared from the chloride and sodium cyanide in aqueous
methanol. Inadequate cooling of the reactor led to a fire.
See other BENZYL COMPOUNDS, GAS EVOLUTION INCIDENTS, POLYCONDENSATION
REACTION INCIDENTS

2739. Sodium *N*-chloro-4-toluenesulfonamide
[127-65-1] $C_7H_7ClNNaO_2S$

1. Klundt, I. L., *Chem. Eng. News*, 1977, **55** (49), 56
2. Kirk Othmer, (4th Edn), 1993, Vol IV, 921
The anhydrous salt ('Chloramine-T') explodes at 175°C, but a small quantity
exploded after storage in a bottle at ambient temperature [1]. The trihydrate may
explode as low as 130°C, the anhydrous compound is less stable still [2].

Calcium carbonate, Isonitriles
Sorbe, 1968, 137
Mixtures explode when warmed on the steam-bath in presence of calcium
carbonate.

Isonitriles, Anthranilates, Tetraethylammonium chloride
Bossio, R. *et al.*, *J. Heterocycl. Chem.*, 1995, **32** (4), 1115
A one pot synthesis of heterocycles involving the above reagents and catalyst
sometimes showed an induction period followed by violent reaction.
See other N-HALOGEN COMPOUNDS

900

2740. 2-Toluenediazonium perchlorate
[69597-04-6] $C_7H_7ClN_2O_4$

Hofmann, K. A. *et al.*, *Ber.*, 1906, **39**, 3146
The solid salt is explosive even when wet.
See other DIAZONIUM PERCHLORATES

2741. 4-Chloro-2-methylphenol
[1570-64-5] C_7H_7ClO

Sodium hydroxide
Anon., *ABCM Quart. Safety Summ.*, 1957, **28**, 39
A large quantity (700 kg) of the chlorophenol, left in contact with conc. sodium
hydroxide solution for 3 days, decomposed, attaining red heat and evolving fumes
which ignited explosively. Although this could not be reproduced under laboratory
conditions, it is believed that exothermic hydrolysis to the hydroquinone (possibly
with subsequent aerobic oxidation to the quinone) occurred, the high viscosity of
the liquid preventing dissipation of heat.
See Sodium 4-chloro-2-methylphenoxide
See other HALOARYL COMPOUNDS

2742. Tropylium perchlorate
[25230-72-2] $C_7H_7ClO_4$

1. Ferrini, P. G. *et al.*, *Angew. Chem.*, 1962, **74**, 488
2. Barltrop, J., *Chem. Brit.*, 1980, **16**, 452
Pressing the salt through a funnel with a glass rod caused a violent explosion [1],
and grinding 19 g of the dry salt in a mortar led to the same result.
See other NON-METAL PERCHLORATES

2743. 2-Tolylcopper
[20854-03-9]

C_7H_7Cu

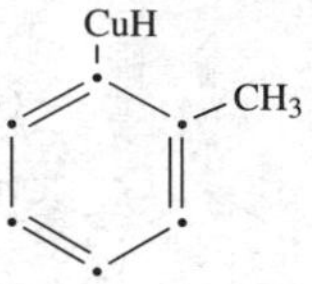

Camus, A. *et al., J. Organomet. Chem.*, 1968, **14**, 442–443
The solid 2-isomer usually exploded strongly on exposure to oxygen at 0°C, or weakly on heating above 100°C under vacuum.
See other ARYLMETALS

2744. 3-Tolylcopper
[20854-05-1]

C_7H_7Cu

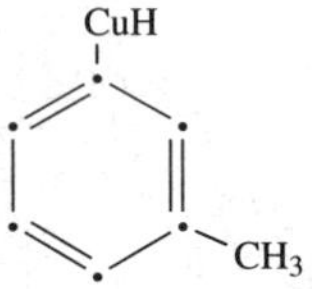

Camus, A. *et al., J. Organomet. Chem.*, 1968, **14**, 442–443
The solid 3-isomers usually exploded strongly on exposure to oxygen at 0°C, or weakly on heating above 100°C under vacuum.
See other ARYLMETALS

2745. 4-Tolylcopper
[5588-74-9]

C_7H_7Cu

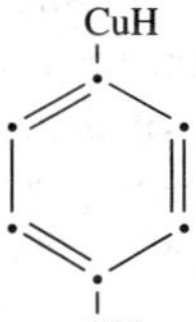

Camus, A. *et al., J. Organomet. Chem.*, 1968, **14**, 442–443
The solid 4-isomers usually exploded strongly on exposure to oxygen at 0°C, or weakly on heating above 100°C under vacuum.
See other ARYLMETALS

2746. Benzyl fluoride (α-Fluorotoluene)
[25496-08-6]

C_7H_7F

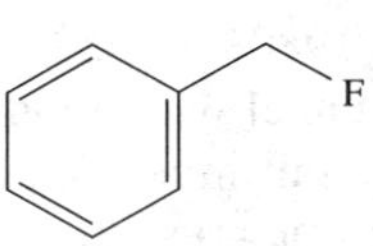

Szucs, S. S., *Chem. Eng. News.*, 1990 **68**(34), 2

A bottle exploded on storage, coating the surroundings with polymer. It is considered that glass has catalytic properties for the Friedel Crafts polymerisation reaction.

See other BENZYL COMPOUNDS, POLYCONDENSATION REACTION INCIDENTS

†2747. 2-Fluorotoluene
[952-52-3] C_7H_7F

See other HALOARYL COMPOUNDS

†2748. 3-Fluorotoluene
[352-70-5] C_7H_7F

See other HALOARYL COMPOUNDS

†2749. 4-Fluorotoluene
[352-32-9] C_7H_7F

See other HALOARYL COMPOUNDS

2750. 4-Iodotoluene
[624-31-7] C_7H_7I

Sorbe, 1968, 66

It explodes above 200°C.
See other HALOARYL COMPOUNDS, IODINE COMPOUNDS

2751. 2-Toluenediazonium iodide
[54514-15-7]

$C_7H_7IN_2$

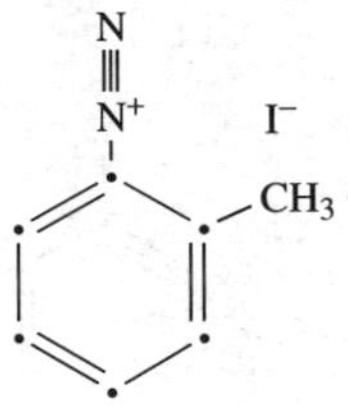

See 3-Toluenediazonium salts
See also DIAZONIUM TRIIODIDES
See other DIAZONIUM SALTS

2752. 3-Toluenediazonium iodide
[54514-15-7]

$C_7H_7IN_2$

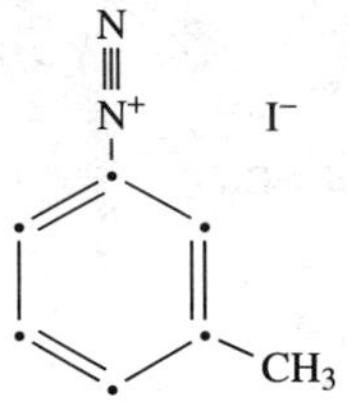

See 3-Toluenediazonium salts
See also DIAZONIUM TRIIODIDES
See other DIAZONIUM SALTS

2753. 4-Iodosyltoluene
[69180-59-2]

C_7H_7IO

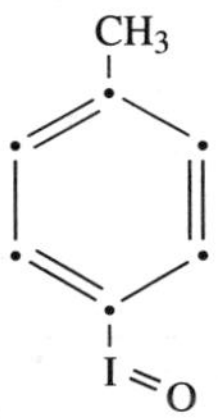

Sorbe, 1968, 66
It explodes at above 175°C
See other IODINE COMPOUNDS

904

2754. 4-Iodyltoluene
[16825-72-2] $C_7H_7IO_2$

Sorbe, 1968, 66
It explodes at above 200°C.
See other IODINE COMPOUNDS

2755. 4-Iodylanisole
[16825-74-4] $C_7H_7IO_3$

Leymann,–, *Chem. Fabrik.*, 1928, 705
A 1.5 kg quantity of pure material, prepared as a disinfectant, exploded with great
violence when a door was opened, possibly during drying operations. It explodes
at 225°C.
See other IODINE COMPOUNDS

2756. 4-Toluenediazonium triiodide
[68596-94-1] $C_7H_7I_3N_2$

See entry DIAZONIUM TRIIODIDES

905

2757. 2-Methoxybenzenediazonium triiodide
[68596-92-9] $C_7H_7I_3NO$

See entry DIAZONIUM TRIIODIDES

2758. 4-Methoxybenzenediazonium triiodide
[68596-91-8] $C_7H_7I_3NO$

See entry DIAZONIUM TRIIODIDES

2759. Vinylpyridine
[1337-81-1] C_7H_7N

$$H_2C=CHC_5H_4N$$

Aminium nitrites
    Karakuleva, G. I. *et al.*, Russ. Pat. 396 349, 1973
    Nitrite salts inhibit the sometimes explosive spontaneous polymerisation of 'N-vinylpyridine derivatives'.
    See other ORGANIC BASES, POLYMERISATION INCIDENTS

2760. Benzaldehyde oxime
[932-90-1]   C$_7$H$_7$NO

Energy of decomposition (in range 160–350°C) measured as 1.37 kJ/g by DSC, and T$_{ait24}$ was determined as 108°C by adiabatic Dewar tests, with an apparent energy of activation of 96 kJ/mol.
See entry THERMOCHEMISTRY AND EXOTHERMIC DECOMPOSITION (reference 2)
See other OXIMES

2761. 1-Nitrocyclohepta-1,3,5-triene
[156545-33-4]   C$_7$H$_7$NO$_2$

Burnett, I. J. *et al., J. Chem. Soc., Chem Comm.*, 1994, (10) 1187.
This compound can decompose explosively if stored at room temperature.
See other C-NITRO COMPOUNDS

2762. *mixo*-Nitrotoluene
[1321-12-6]   C$_7$H$_7$NO$_2$

1. Staudinger, H., *Z. Angew. Chem.*, 1922, **35**, 657–659
2. Harris, G. F. P. *et al., Runaway Reactions*, 1981, Paper 4/W, 1–11
3. Anon., *Euro. Chem. News*, 1992, 3558(1548), 27; *ibid.*, 1993, **59**(1562), 22; *Chem. Brit.*, 1993, **29**(1), 15

907

4. Anon., *Chem. & Ind.*, 1994, 485
5. The Fire at Hickson and Welch Ltd., HSE Books, Sudbury UK, 1994
The combined residues (300–400 kg) from several vacuum distillations at up to 180°C/40 mbar of a mixture of the 2- and 4-isomers exploded several hours after distillation had been completed. It appeared probable that admission of air to cool the hot residue initiated autoxidative heating, possibly involving nitrogen oxide and catalysis by iron, which eventually led to explosion of the residue, which probably contained polynitro compounds [1]. Explosions have occurred during fractional distillation to separate mixed nitrotoluene isomers when excessive heating was employed, or when the materials were held at more moderate temperatures but for long periods. The factors involved were identified by a series of systematic thermal stability tests on various mixtures, some including dinitrotoluenes. It was established that decomposition would not occur on the large scale if still temperatures were limited to 190°C [2]. An explosion occurred, killing five workers, when raking out (with metal rakes) the 40 cm accumulated residues from thirty years nitrotoluene distillations. It appears to have been initiated by runaway decomposition from hot-spots developed when a steam coil, possibly at 165°C, was used to soften the still tars, which were not in contact with the thermocouple, by heating to a specified 80°C. There is a suggestion that nitrocresol byproducts were involved [3,4]. A full report on this event has ben published [5].
See other NITROARYL COMPOUNDS

2763. 2-Nitrotoluene
[88-72-2]

$C_7H_7NO_2$

HCS 1980, 690

T_{ait24} was determined as 232°C by adiabatic Dewar tests, with an apparent energy of activation of 162 kJ/mol.
See entry THERMOCHEMISTRY AND EXOTHERMIC DECOMPOSITION (REFERENCE 2)

Alkali
1. Anon., *Loss Prev. Bull.*, 1977, (013), 2
2. Hauptli, H., private comm., 1976
1,2-Bis(2-nitrophenyl)ethane (2,2'-dinitrobibenzyl) is prepared industrially by action of alkali on 2-nitrotoluene, and the highly exothermic reaction is controlled by the rate of addition of the nitro compound to the other cooled reaction components, to keep the temperature at 5–10°C. During one batch, failure of the agitator caused increase in temperature and at 15° the temperature monitor stopped addition of nitrotoluene and increased cooling, the temperature rose to 20° and the

building was evacuated. After the exotherm had subsided, the agitator was used very intermittently to assist cooling and mixing. At 12°C full agitation was restored and processing continued to completion [1]. Similar caution is necessary during the preparation of other analogous nitrobibenzyls [2].

See other AGITATION INCIDENTS

Sodium hydroxide

Anon., *ABCM Quart. Safety Summ.*, 1945, **16**, 2

Crude 2-nitrotoluene, containing some hydrochloric and acetic acids, was charged into a vacuum still with flake sodium hydroxide to effect neutralisation prior to distillation. An explosion occurred later. Similar treatments had been used uneventfully previously, using the weaker bases lime or sodium carbonate for neutralisation. It seems likely that the explosion involved formation and violent decomposition of the sodium salt of an aci-nitro species, possibly of quinonoid type.

See ACI-NITROQUINONOID COMPOUNDS, NITROAROMATIC–ALKALI HAZARDS
See other NITROARYL COMPOUNDS

2764. 4-Nitrotoluene
[99-99-0]   C₇H₇NO₂

HCS 1980, 691

1. Grewer, T. *et al., Exothermic Decomposition*, Technical Report 01VD 159/0329 for Federal German Ministry for Res. Technol., Bonn. 1986
2. Widmann, G. *et al., Thermochim. Acta*, 1988, **134**, 451–455
3. Anon., *ABCM Quart. Safety Summ.*, 1938, **9**, 65; 1939, **10**, 2

Energy of decomposition (in range 300–420°C) was measured as 1.74 kJ/g by DSC, and T_{ait24} was determined as 232°C by adiabatic Dewar tests, with an apparent energy of activation of 162 kJ/mol [1]. Energy of decomposition of pure 4-nitrotoluene in the range 240–460°C was determined by micro DSC as 3.52 kJ/g, peaking at 409°C [2]. The residue from large-scale vacuum distillation of 4-nitrotoluene was left to cool after turning off heat and vacuum, and after 8 h a violent explosion occurred. This was variously attributed to: presence of an excessive proportion of dinitrotoluenes in the residue, which would decompose on prolonged heating; presence of traces of alkali in the crude material, which would form unstable *aci*-nitro salts during distillation; ingress of air to the hot residue on shutting off the vacuum supply, and subsequent accelerating oxidation of the

909

residue. All these factors may have contributed, and several such incidents have occurred [3].

Sodium

Schmidt, J., *Ber.*, 1899, **32**, 2970

One of the products as treating 4-nitrotoluene in ether with sodium is a dark brown sodium derivative which ignites in air.

See other NITROAROMATIC–ALKALI HAZARDS

Sulfuric acid

1. Hunt, J. K., *Chem. Eng. News*, 1949, **27**, 2504
2. Chem. Abs., 1949, **43**, 8681a
3. McKeand, G., private comm., 1974
4. McKeand, G., *Chem. & Ind.*, 1974, 425

Solutions of 4-nitrotoluene in 93% sulfuric acid decompose very violently if heated to 160°C. This happened on plant-scale when automatic temperature control failed [1], but the temperature was erroneously abstracted as 135°C [2]. The explosion temperature of 160°C for the mixture (presumably containing a high proportion of 4-nitrotoluene-2-sulfonic acid) is 22°C lower than that observed for onset of decomposition when 4-nitrotoluene and 93% sulfuric acid are heated at a rate of 100°C/h [3]. Mixtures of 4-nitrotoluene with 98% acid or 20% oleum begin to decompose at 180 and 190°C, respectively [3,4]. Thereafter, decomposition accelerates (190–224° in 14 min, 224–270 in 1.5 min) until eruption occurs with evolution of much gas [4].

See also 3-NITROBENZENESULFONIC ACID: SULFURIC ACID

Sulfuric acid, Sulfur trioxide

Vervalin, H. C., *Hydrocarbon Process.*, 1976, **55**(9), 323

During sulfonation of 4-nitrotoluene at 32°C with 24% oleum in a 2000 l vessel, a runaway decomposition reaction set in and ejected the contents as a carbonaceous mass. The thermal decomposition temperature was subsequently estimated as 52°C (but see above).

See other SULFONATION INCIDENTS

Tetranitromethane

See Tetranitromethane: Aromatic nitro compounds

See other NITROARYL COMPOUNDS

2765. Benzyl nitrate

[15285-42-4] $C_7N_7NO_3$

1. Nef, J. U., *Ann.*, 1899, **309**, 172
2. Verbal comment on presentation by Verhoeff, J., *Proc. 4th Int. Symp. Loss Prev. Safety Prom. Proc. Ind.*, Vol. 3, C5–13, Rugby, IChE, 1983

It decomposed explosively at 180–200°C [1], but there is a pressure band between ambient and 64 bar in which thermal explosion does not occur [2].

Lewis acids
See entry ALKYL NITRATES *See other* BENZYL COMPOUNDS

2766. 3-Methyl-4-nitrophenol
[2581-34-2] $C_7H_7NO_3$

Dartnell, R. C. *et al., Loss Prev.*, 1971, **5**, 53–56; *MCA Case History No. 1649*
A batch of 8 t of material accumulated in storage at 154°C during 72 h decomposed explosively. Stability tests showed that thermal instability developed when 3-methyl-4-nitrophenol is stored molten at temperatures above 140°C. Decomposition set in after 14 h at 185° or 45 h at 165°, with peak temperatures of 593 and 521°C, respectively. In a closed vessel, a peak pressure of 750 bar was attained, with a maximum rate of increase of 40 kbar/s. Thermal degradation involves an initially slow exothermic free radical polymerisation process, followed by a rapid and violently exothermic decomposition at take-off.
See THERMAL STABILITY OF REACTION MIXTURES AND SYSTEMS
See entry PRESSURE INCREASE IN EXOTHERMIC DECOMPOSITION
See other NITROARYL COMPOUNDS

2767. 4-Methyl-2-nitrophenol
[119-33-5] $C_7H_7NO_3$

Sodium hydroxide, Sodium carbonate, Methanol
MCA Case History No. 701
Failure to agitate a large-scale mixture of the reagents caused an eruption owing to action of the exotherm when mixing did occur.
See entry NITROAROMATIC–ALKALI HAZARDS *See other* AGITATION INCIDENTS

2768. 2-Nitroanisole
[91-23-6] $C_7H_7NO_3$

Hydrogen
1. Carswell, T. S., *J. Amer. Chem. Soc.*, 1931, **53**, 2417
2. Adkins, H., *J. Amer. Chem. Soc.*, 1931, **53**, 2808
Catalytic hydrogenation on 400 g scale at 34 bar under excessively vigorous conditions (250°C, 12% catalyst, no solvent) caused the thin autoclave without bursting disk to rupture [1]. Under more appropriate conditions the hydrogenation is safe [2].
See other CATALYTIC NITRO REDUCTION PROCESSES

Sodium hydroxide, Zinc
Anon., *Angew. Chem. (Nachr.)*, 1955, **3**, 186
In the preparation of 2,2′-dimethoxyazoxybenzene, solvent ethanol was distilled out of the mixture of 2-nitroanisole, zinc and sodium hydroxide before reaction was complete. The exothermic reaction continued unmoderated, and finally exploded.
See other NITROAROMATIC–ALKALI HAZARDS *See other* NITROARYL COMPOUNDS

2769. 2-Nitrobenzyl alcohol
[612-25-9] $C_7N_7NO_3$

See entry NITROBENZYL COMPOUNDS (reference 3)
See other BENZYL COMPOUNDS, NITROARYL COMPOUNDS

2770. 3-Nitrobenzyl alcohol
[619-25-0] $C_7H_7NO_3$

See entry NITROBENZYL COMPOUNDS (reference 3)
See other BENZYL COMPOUNDS, NITROARYL COMPOUNDS

2771. 4-Nitrobenzyl alcohol
   [619-37-8] $C_7H_7NO_3$

See entry NITROBENZYL COMPOUNDS (reference 3)
See other BENZYL COMPOUNDS, NITROARYL COMPOUNDS

2772. 2-Methyl-5-nitrobenzenesulfonic acid
   [121-03-9] $C_7H_7NO_5S$

Preparative hazard
   See 4-Nitrotoluene: Sulfuric acid, or: Sulfuric acid, Sulfur trioxide
   See other NITROARYL COMPOUNDS, ORGANIC ACIDS

2773. 4-Methyl-3-nitrobenzenesulfonic acid
   [97-06-3] $C_7H_7NO_5S$

Hakl, J., *Runaway Reactions*, 1981, paper 3/L, 5
Initial exothermic decomposition sets in at 170°C.
See other NITROARYL COMPOUNDS, ORGANIC ACIDS

2774. 3-Toluenediazonium salts
[14604-29-6] (ion)

$$C_7H_7N_2^+Z^-$$

Ammonium sulfide, or Hydrogen sulfide
See entry DIAZONIUM SULFIDES AND DERIVATIVES

Potassium *O*-ethyl dithiocarbonate
1. Tarbell, D. S. *et al., Org. Synth.*, 1955, Coll. Vol. 3, 810
2. Parham, W. E. *et al., Org. Synth.*, 1967, **47**, 107
During interaction of the diazonium sulfide and the *O*-ethyl dithiocarbonate ('xanthate') solutions, care must be taken to ensure that the intermediate diazonium dithiocarbonate decomposes to 2-thiocresol as fast as it is formed [1]. This can be assured by presence of a trace of nickel in the solution to effect immediate catalytic decomposition [2]. When the 2 solutions were mixed cold and then heated to effect decomposition, a violent explosion occurred.
See other DIAZONIUM SULFIDES AND DERIVATIVES

Potassium iodide
Trumbull, E. R., *J. Chem. Educ.*, 1971, **48**, 640
Addition of potassium iodide solution to diazotised 3-toluidine was accommpanied on 3 occasions by an eruption of the beaker contents. This was not observed with the 2- and 4-isomers or other aniline derivatives.
See DIAZONIUM TRIIODIDES
See other DIAZONIUM SALTS

2775. 2-Toluenediazonium salts
[17333-74-3] (ion)

$$C_7H_7N_2^+Z^-$$

Ammonium sulfide, or Hydrogen sulfide
See entry DIAZONIUM SULFIDES AND DERIVATIVES
See other DIAZONIUM SALTS

2776. 4-Toluenediazonium salts
[14597-45-6] (ion) $C_7H_7N_2^+ Z^-$

Ammonium sulfide, or Hydrogen sulfide
See entry DIAZONIUM SULFIDES AND DERIVATIVES *See other* DIAZONIUM SALTS

2777. 7-Azidobicyclo[2.2.1]hepta-2,5-diene
[104172-76-1] $C_7H_7N_3$

Hoffmann, R. W. *et al., Chem. Ber.*, 1986, **119**, 3309
It is explosive, and when distilled at 41°C/1.3 mbar, the bath temperature must not exceed 100°C.
See other DIENES, ORGANIC AZIDES

2778. 5-Nitrosalicylhydrazide
[946-32-7] $C_7H_7N_3O_4$

Khadikar, P. V. *et al., Thermochim Acta*, 1986, (99), 109–118
Several heavy metal complexes of the hydrazide ($ML_2.2H_2O$) exploded around 250°C during thermal analysis.
See related HEAVY METAL DERIVATIVES *See other* NITROARYL COMPOUNDS

2779. Benzyl azide
[622-79-7]
$C_7N_7N_3$

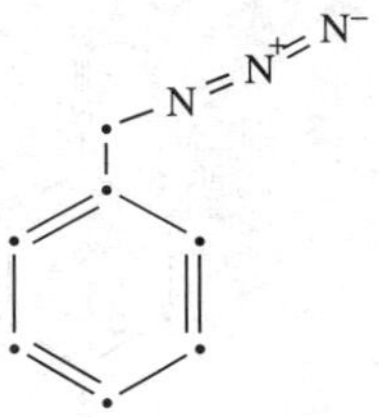

Sorbe, 1968, 138
A particularly heat-sensitive explosive oil.

Bis(trifluoromethyl) nitroxide
  See Bis(trifluoromethyl) nitroxide: Benzoyl azide
  See other BENZYL COMPOUNDS, ORGANIC AZIDES

2780. 4-Toluenesulfinyl azide
[40560-76-7]
$C_7H_7N_3OS$

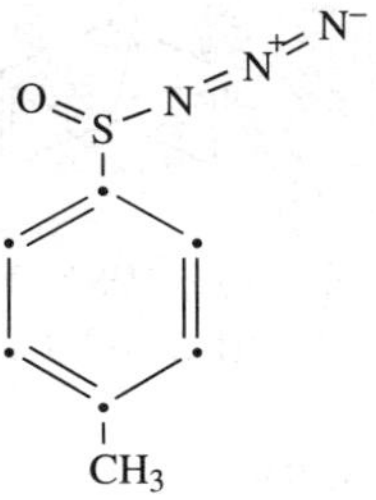

See entry SULFINYL AZIDES *See other* ACYL AZIDES

2781. 4-Toluenesulfonyl azide
[941-55-9]
$C_7H_7N_3O_2S$

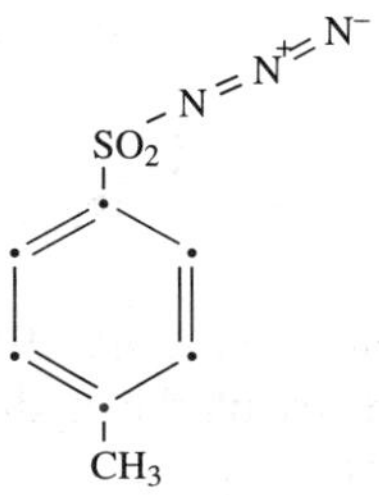

1. Rewicki, D. *et al., Angew. Chem. (Intern. Ed.)*, 1972, **11**, 44

916

2. Roush, W. R. *et al., Tetrahedron. Lett.*, 1974, 1391–1392
3. Spencer, H., *Chem. Brit.*, 1981, **17**, 106

The impure compound, present as a majority in the distillation residue from preparation of 1-diazoindene, will explode if the bath temperature exceeds 120°C [1]. A polymer-bound sulfonyl azide reagent has been described as safer in use than the title compound [2]. Distillation of the azide at 130–132°C/~0.5 mbar from an oil bath at 145°C led to darkening, then violent explosion of the 30–40% residue [3].

See other ACYL AZIDES

2782. Methyl 2-nitrobenzenediazoate
[62375-91-1] $C_7H_7N_3O_3$

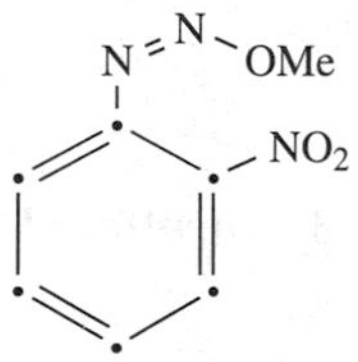

Bamberger, E., *Ber.*, 1895, **28**, 237

It explodes violently on heating, or on disturbing after 24 h confinement in a sealed tube at ambient temperature. The 4-nitro analogue is more stable.

See other ARENEDIAZOATES

2783. 3-Amino-2,5-dinitrotoluene
[65321-68-8] $C_7H_7N_3O_4$

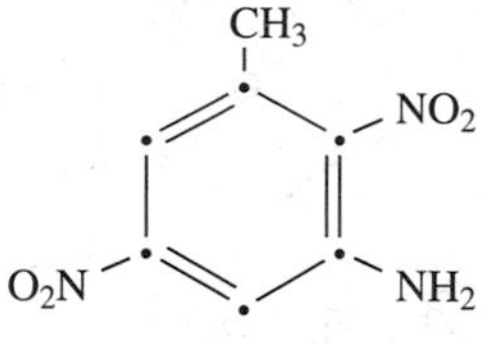

Preparative hazard

See 2,5-Dinitro-3-methylbenzoic acid: Oleum, Sodium azide
See other POLYNITROARYL COMPOUNDS

2784. 2-Methanesulfonyl-4-nitrobenzenediazonium hydrogen sulfate
[68597-08-0] $C_7H_7N_3O_8S_2$

See entry DIAZONIUM SULFATES

2785. 4-Methylaminobenzene-1,3-bis(sulfonyl azide)
[87425-02-3] $C_7H_7N_7O_4S_2$

Lenox, R. S., US Pat. 4 393 006, 1983
It detonates on melting.
See other ACYL AZIDES

918

2786. Tris(2,2,2-trinitroethyl) orthoformate
  [14548-59-5] $C_7H_7N_9O_{21}$

See entry TRINITROETHYL ORTHOESTERS
See other POLYNITROALKYL COMPOUNDS

2787. Benzylsodium
  [1121-53-5] C_7H_7Na

Leleu, *Cahiers*, 1977, (88), 370
It ignites in air.
See related ALKYLMETALS *See other* BENZYL COMPOUNDS, ORGANOMETALLICS

2788. Sodium 4-methylphenoxide (Sodium 4-cresolate)
  [22113-51-5] (ion) C_7H_7NaO

May & Baker Ltd, private comm., 1968

919

Sodium 4-methylphenoxide solution was dehydrated azeotropically with chlorobenzene, and the filtered solid was dried in an oven, where it soon ignited and glowed locally. This continued for 30 min after it was removed from the oven. A substituted potassium phenoxide, prepared differently, also ignited on heating. Finely divided and moist alkali phenoxides may be prone to vigorous oxidation (or perhaps reaction with carbon dioxide) when heated in air.

†2789. Bicyclo[2.2.1]hepta-2,5-diene
[121-46-0]

C_7H_8

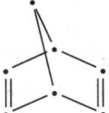

HCS 1980, 209

See related DIENES

†2790. 1,3,5-Cycloheptatriene
[544-25-2]

C_7H_8

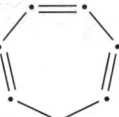

HCS 1980, 341

See related DIENES

†2791. Toluene (Methylbenzene)
[108-88-3]

C_7H_8

(*MCA SD-63*, 1956); *NSC 204*, 1968; *FPA H11*, 1973; *HCS 1980*, 906; *RSC Lab. Hazard Data Sheet No. 15*, 1983

Other reactants
  Yoshida, 1980, 261
  MRH Values calculated for 15 combinations with oxidants are given.

Oxidants   MRH values show % of oxidant
  See Bromine trifluoride: Solvents   MRH 2.97/80
    1,3-Dichloro-5,5-dimethyl-2,4-imidazolidindione: Xylene
    Dinitrogen tetraoxide: Hydrocarbons   MRH 7.15/81
    Nitric acid: Hydrocarbons   MRH 6.07/82
    Tetranitromethane: Hydrocarbons
    Uranium hexafluoride: Aromatic hydrocarbons

Sulfur dichloride
  See Sulfur dichloride: Toluene

2792. Ethyl 4-bromo-1,1,3-trioxoisothiazoleacetate
[126623-64-1]   C₇H₈BrNO₅S

Ethyl diazoacetate
  Burri, K. F., *Helv. Chim. Acta*, 1989, **72**(6), 1416
  It is suggested that both adducts of these compounds are explosive.
  See Ethyl 6-ethoxycarbonyl-3,4-dihydro-1,1,3-trioxo-2-pyrazolo [3,4-d]isothiazole-
  2-acetate
  See Ethyl diazoacetate
  See other N–S COMPOUNDS

2793. 2-Chloro-5-methylaniline
[95-81-8]   C₇H₈ClN

Preparative hazard
  See 2-Chloro-5-methylphenylhydroxylamine
  See other HALOANILINES

2794. 4-Chloro-2-methylaniline
[95-69-2]

C_7H_8ClN

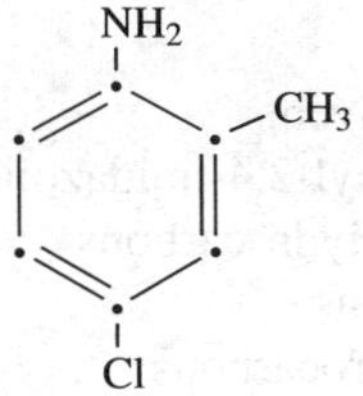

See 4-Chloro-2-methylbenzenediazonium salts

Copper(II) chloride
Kotoyori, T. *et al., J. Haz. Mat.*, 1976, 252–262
At the end of vacuum distillation of 4-chloro-2-methylaniline, the 1300 l still burst owing to internal gas pressure from an unexpected decompositon reaction. Subsequent investigation showed that copper(I) oxide present in the still charge would react with hydrogen chloride and air leaking in to produce copper(II) chloride. The latter is an effective catalyst for decomposition of the haloaniline at 239°C to produce hydrogen chloride. The physical conditions of the installation were compatible with this mechanism, and enough chloroaniline was present to exceed the bursting pressure of the vessel by a factor of 3. It is concluded that halogenated aromatic amines should be distilled under alkaline conditions in absence of air.
See CATALYTIC IMPURITY INCIDENTS See other HALOANILINES

2795. 3-Chloro-4-methoxyaniline
[5345-54-0]

C_7H_8ClNO

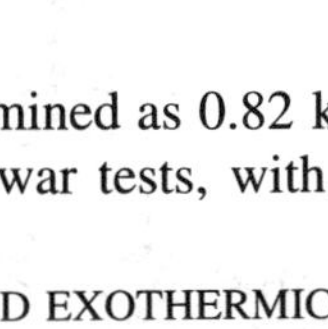

Heat of decomposition was determined as 0.82 kJ/g by DSC, and T_{ait24} was determined as 158°C by adiabatic Dewar tests, with an apparent energy of activation of 135 kJ/mol.
See entry THERMOCHEMISTRY AND EXOTHERMIC DECOMPOSITION (reference 2)
See other AMINOMETHOXY COMPOUNDS, HALOANILINES

2796. 2-Chloro-5-methylphenylhydroxylamine
[65039-20-5] C_7H_8ClNO

Rondevstedt, C. S., *Chem. Eng. News*, 1977, **55**(27), 38; *Synthesis*, 1977, 851–852
In an attempt to reduce 2-chloro-5-methylnitrobenzene to the aniline by treatment
with excess hydrazine and Pd/C catalyst, the hydroxylamine was unexpectedly
produced as major product. During isolation of the product, after removal of solvent
it was heated to 120°C under vacuum and exploded fairly violently. Many aryl
hydroxylamines decompose violently when heated above 90–100°C, especially in
presence of acids. GLC is not suitable as an analytical diagnostic for arylhydroxyl-
amines because of this thermal decomposition.
See 3,4-Dichlorophenylhydroxylamine
See other CATALYTIC NITRO REDUCTION PROCESSES, N–O COMPOUNDS

2797. *N*-Chlorotoluene-4-sulfonamide
[144-86-5] $C_7H_8ClNO_2S$

Energy of exothermic decomposition in range 110–330°C was measured as
1.36 kJ/g.
See entry THERMOCHEMISTRY AND EXOTHERMIC DECOMPOSITION (reference 2)

2798. Benzohydrazide
[613-94-5] $C_7H_8N_2O$

Benzeneseleninic acid
See Benzeneseleninic acid: Hydrazine derivatives

923

2799. Methyl benzenediazoate (1-Phenyl-2-methoxydiazene)
[66217-76-3]
C₇H₈N₂O

$$C_7H_8N_2O$$

Bamberger, E., *Ber.*, 1895, **28**, 228
It explodes on heating or after storage for 1 h in a sealed tube at ambient temperature.
See other ARENEDIAZOATES

2800. *N*-Methyl-4-nitroaniline
[100-15-2]
$$C_7H_8N_2O_2$$

2,4-Dithia-1,3-dioxane-2,2,4,4-tetraoxide ('Carbyl sulfate')
See 2,4-Dithia-1,3-dioxane-2,2,4,4-tetraoxide: *N*-Methyl-4-nitroaniline
See other NITROARYL COMPOUNDS

2801. 2-Methyl-5-nitroaniline
[99-55-8]
$$C_7H_8N_2O_2$$

Hakl, J., *Runaway Reactions*, 1981, Paper 3/L, 5
Initial exothermic decomposition sets in at 150°C.
See other NITROARYL COMPOUNDS

2802. 2-Methoxy-5-nitroaniline
[99-59-2] $C_7H_8N_2O_3$

Preparative hazard
See 2-Methoxyanilinium nitrate: Sulfuric acid
See other AMINOMETHOXY COMPOUNDS, NITROARYL COMPOUNDS

2803. Ammonium 3-methyl-2,4,6-trinitrophenoxide
[58696-86-9] $C_7H_8N_4O_7$

Biasutti, 1981, 63
An old store of the explosive material exploded violently with no apparent cause. A
similar incident had been recorded previously. Some isomeric *aci*-nitroquinonoid
salt may have been formed.
See aci-NITROQUINONOID COMPOUNDS *See other* POLYNITROARYL COMPOUNDS

2804. 3-Phenyl-1-(5-tetrazolyl)-1-tetrazene
[ ] $C_7H_8N_8$

Thiele, J., *Ann.*, 1892, **270**, 60

It explodes on warming, as do *C*-substituted homologues.
See other HIGH-NITROGEN COMPOUNDS, TETRAZENES, TETRAZOLES

2805. Methylenebis(3-nitramino-4-methylfurazan)
[ ] $C_7H_8N_8O_6$

Willer, R. L. *et al., J. Heterocycl. Chem.*, 1992, **29**(7), 1835
An explosive of moderate sensitivity, thermally decomposing (non-explosively)
from 120°C. The 4-nitro analogue was also explosive and decomposed from 90°C.
See other N-NITRO COMPOUNDS, HIGH NITROGEN COMPOUNDS

2806. Benzyl alcohol
[100-51-6] C_7H_8O

HCS 1980, 198

Other reactants
Yoshida, 1980, 329
MRH values calculated for 13 combinations with oxidants are given.

Sulfuric acid   MRH 1.13/48
See Sulfuric acid: Benzyl alcohol

Hydrogen bromide, Iron
Iselin, E. *et al., Chem. Ing. Tech.*, 1972, **44**, 462
Benzyl alcohol contaminated with 1.4% of hydrogen bromide and 1.1% of dis-
solved iron(II) polymerises exothermally above 100°C. Bases inhibit the polymeri-
sation reaction. In a laboratory test, alcohol containing 1% of HBr and 0.04% of
Fe polymerised at about 150° with an exotherm to 240°C. Formation and iron-
catalysed poly-condensation of benzyl bromide seems to have been implicated.
See Benzyl bromide: Molecular sieve, or: Catalytic impurities
See other BENZYL COMPOUNDS, POLYCONDENSATION REACTION INCIDENTS

2807. 2-Acetyl-3-methyl-4,5-dihydrothiophen-4-one
[ ] $C_7H_8O_2S$

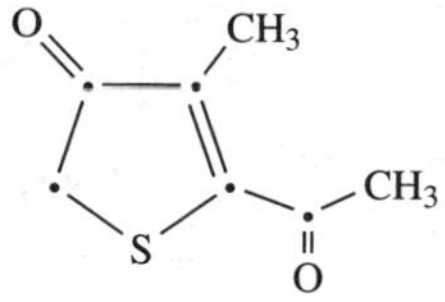

MCA Guide, 1972, 307

An explosion occurred during vacuum distillation. The highly conjugated compound could have been susceptible to peroxidation as well as capable of condensative polymerisation.
See other PEROXIDATION INCIDENTS, POLYCONDENSATION REACTION INCIDENTS

2808. 4-Toluenesulfonic acid
[104-15-4] $C_7H_8O_3S$

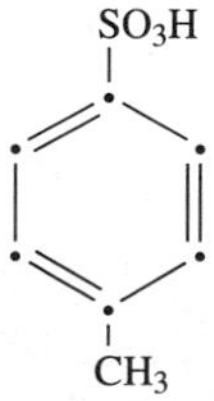

HCS 1980, 908

Acetic anhydride, Water
See Acetic anhydride: 4-Toluenesulfonic acid, etc.
See other ORGANIC ACIDS

2809. 3-Thiocresol
[108-40-7] C_7H_8S

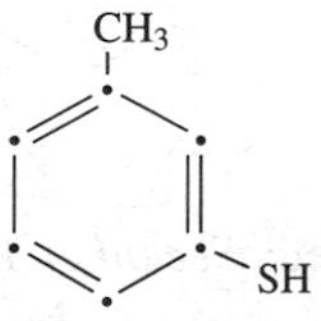

Preparative hazard
See 3-Toluenediazonium salts: Potassium *O*-ethyl dithiocarbonate

2810. Chloromethylphenylsilane
[1631-82-9]

C₇H₉ClSi

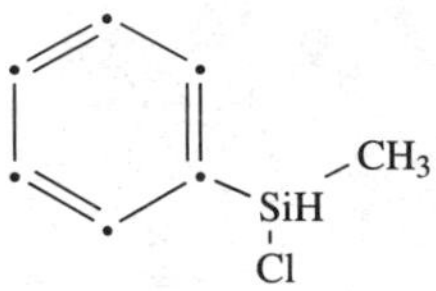

C_7H_9ClSi

4-Bromobutene, Chloroplatinic acid
Brook, A. G. *et al., J. Organomet. Chem.*, 1975, **87**, 265
Interaction of chloromethylphenylsilane and 4-bromobutene, catalysed by chloroplatinic acid at 100°C, is extremely exothermic (to 165°C) and is accompanied by vigorous frothing.
See other ALKYLHALOSILANES

2811. 2-Deuterobicyclo[2.2.1]hept-2-ene
[694-94-0]

C_7H_9D

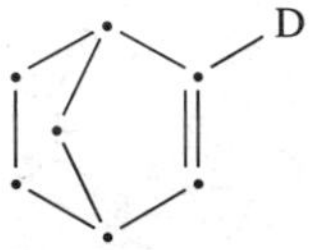

Preparative hazard
Shackelford, S. A., *Tetrahedron Lett.*, 1977, 4268, note 12
The metallation reaction involved in the preparation proved hazardous and unpredictable.
See related ALKENES

2812. Benzylamine
[100-46-9]

C_7H_9N

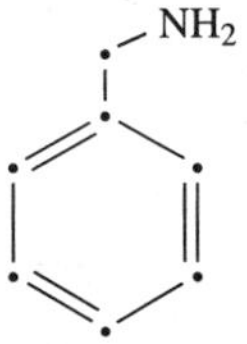

N-Chlorosuccinimide
See N-HALOIMIDES: alcohols, etc. *See other* BENZYL COMPOUNDS, ORGANIC BASES

928

2813. 2-Toluidine
   [95-53-4] C_7H_9N

$$CH_3$$

HCS 1980, 910

See 2- or 4-Toluenediazonium salts

Nitric acid   MRH 6.11/81
   See Nitric acid: Aromatic amines

Other reactants
   Yoshida, 1980, 260
   MRH values calculated for 13 combinations with oxidants are given.

2814. 3-Toluidine
   [108-44-1] C_7H_9N

$$CH_3$$

See 3-Toluenediazonium salts

2815. 4-Toluidine
   [106-49-0] C_7H_9N

$$CH_3$$

See 2- or 4-Toluenediazonium salts

2816. 2-Methoxyaniline (*o*-Anisidine)
  [90-04-0]   C₇H₉NO

Energy of decomposition (in range 190–480°C) measured as 0.387 kJ/g.
See entry THERMOCHEMISTRY AND EXOTHERMIC DECOMPOSITION (reference 2)
See other AMINOMETHOXY COMPOUNDS

2817. 3-Methoxyaniline (*m*-Anisidine)
  [536-90-3]   C₇H₉NO

Energy of decomposition (in range 410–500°C) measured as 0.518 kJ/g.
See entry THERMOCHEMISTRY AND EXOTHERMIC DECOMPOSITION (reference 2)
See other AMINOMETHOXY COMPOUNDS

2818. 4-Methoxyaniline (*p*-Anisidine)
  [100-94-9]   C₇H₉NO

Energy of decomposition (in range 340–490°C) measured as 0.429 kJ/g.
See entry THERMOCHEMISTRY AND EXOTHERMIC DECOMPOSITION (reference 2)
See other AMINOMETHOXY COMPOUNDS

930

2819. 2,6-Dimethylpyridine *N*-oxide
   [1073-23-0]   C₇H₉NO

Phosphoryl chloride
   See Phosphoryl chloride: 2,6-Dimethylpyridine *N*-oxide
   See other N-OXIDES

2820. 4(1-Hydroxyethyl)pyridine *N*-oxide
   [34277-48-0]   C₇H₉NO₂

Taylor, R., *J. Chem. Soc., Perkin Trans. 2*, 1975, 279
The compound exploded during vacuum distillation, possibly owing to the presence
of traces of peracetic acid.
See other N-OXIDES

2821. *O*-4-Toluenesulfonylhydroxylamine
   [52913-14-1]   C₇H₉NO₃S

Carpino, L. A., *J. Amer. Chem. Soc.*, 1960, **82**, 3134
It deflagrated spontaneously on attempted isolation and drying at ambient temper-
ature.
See *O*-Mesitylenesulfonylhydroxylamine
See other N–O COMPOUNDS

2822. 4-Chloro-2,6-diamino-N-methylaniline
[48114-68-7] $C_7H_{10}ClN_3$

T_{ait24} determined as 148°C in adiabatic Dewar tests with apparent energy of activation of 123 kJ/mol.

See entry THERMOCHEMISTRY AND EXOTHERMIC DECOMPOSITION (reference 2)
See other HALOANILINES

2823. N-(Methylphenylphosphinoyl)hydroxylamine
[94193-96-3] $C_7H_{10}NO_2P$

Harger, M. J. P., *J. Chem. Soc., Perkin Trans. 1*, 1987, 685
A sample decomposed vigorously after several hours at warm ambient temperature. The ethyl and isopropyl homologues appeared stable, but all (including the derived *O*-sulfonic acids) were treated as potentially unstable and were only heated briefly during crystallisation, and were stored at −20°C.
See related HYDROXYLAMINIUM SALTS *See other* N–O COMPOUNDS

2824. Dicyclopropyldiazomethane
[16102-24-2] $C_7H_{10}N_2$

Ensslin, H. M *et al., Angew. Chem. (Intern. Ed.)*, 1967, **6**, 702–703

Storable in ether at $-50°C$, but violent decomposition of undiluted material sets in at $-15°C$.
See other DIAZO COMPOUNDS, STRAINED-RING COMPOUNDS

2825. 1,5-Diisocyanopentane
[97850-58-3]   C$_7$H$_{10}$N$_2$

See entry DIISOCYANIDE LIGANDS *See related* CYANO COMPOUNDS

2826. 4-Methoxy-1,3-phenylenediamine
[615-05-4]   C$_7$H$_{10}$N$_2$O

Energy of decomposition (in range $260-420°C$) measured as 0.42 kJ/g by DSC, and T_{ait24} was determined as $207°C$ by adiabatic Dewar tests, with an apparent energy of activation of 114 kJ/mol.
See entry THERMOCHEMISTRY AND EXOTHERMIC DECOMPOSITION (reference 2)
See other AMINOMETHOXY COMPOUNDS

2827. 4-Toluenesulfonylhydrazide
[877-66-7]   C$_7$H$_{10}$N$_2$O$_2$S

1. Hansell, D. P. *et al.*, *Chem. & Ind.*, 1975, 464

While drying in a tray dryer at 60°C, a batch decomposed fairly exothermically but without fire. As a blowing agent, it is designed to have limited thermal stability [1]. One of the recently calculated values of 87° and 78°C for induction periods of 7 and 60 days, respectively, for critical ignition temperature coincides with the previous value of 87°/7 days. Autocatalytic combustion is exhibited.
See other BLOWING AGENTS

2828. Diethyl diazomalonate
[5256-74-6] $C_7H_{10}N_2O_4$

DOC 5, 1982, Vol. 2, 1588

It may explode during distillation at low pressure.
See other DIAZO COMPOUNDS

2829. 2-Methoxyanilinium nitrate
[ ] $C_7H_{10}N_2O_4$

1. Vervalin, C. H., *Hydrocarb. Process.*, 1976, **55**(9), 321
2. *See entry* HIGH RATE DECOMPOSITION
Use of a screw feeder to charge a reactor with the 2-anisidine salt led to ignition of the latter. It was known that the salt would decompose exothermically above 140°C, but later investigation showed that lower-quality material could develop an exotherm above 46°C under certain conditions [1]. Fast flame propagation occurs on moderately heating anisidine nitrate powder [2].

Sulfuric acid
Vervalin, C. H., *Hydrocarb. Process.*, 1976, **55**(9), 321
In a process for preparation of 2-methoxy-5-nitroaniline the anisidine salt was added to stirred sulfuric acid. An accidental deficiency of the latter prevented

934

proper mixing and dissipation of the heat of solution, and local decomposition spread through the entire content of the 2000 l vessel, attaining red heat.

See other OXOSALTS OF NITROGENOUS BASES, SULFONATION INCIDENTS
See related AMINOMETHOXY COMPOUNDS

2830. Methyl 3-methoxycarbonylazocrotonate
[63160-33-8] $C_7H_{10}N_2O_4$

Sommer, S., *Tetrahedron Lett.*, 1977, 117
Attempted high-vacuum fractional distillation of an isomeric mixture (10% *cis-*) led to explosive decomposition. The 'diene' polymerises if stored at 20°C.
See other AZO COMPOUNDS

2831. 5-*tert*-Butyl-3-diazo-3*H*-pyrazole
[62072-11-1] $C_7H_{10}N_4$

The light-sensitive and piezosonic crystalline solid explodes at 97°C.
See entry DIAZOAZOLES
See other DIAZO COMPOUNDS, HIGH-NITROGEN COMPOUNDS

2832. Benzylsilane
[766-06-3] $C_7H_{10}Si$

491M, 1975, 58

It ignites in air.
See related ALKYLSILANES *See other* BENZYL COMPOUNDS

2833. Cyclopentadienyldimethylborane

[ ]

C$_7$H$_{11}$B

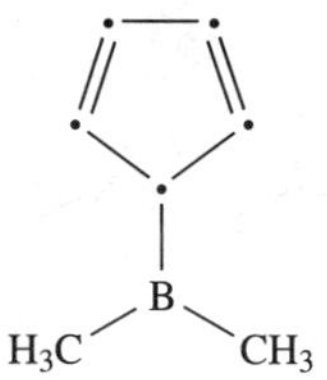

Herberich, G. E. *et al.*, *Organometallics*, 1996, **15**(1), 58
Spontaneously inflames in air.
See other ALKYLBORANES

2834. Lithium 1-heptynide

[42017-07-2]

C$_7$H$_{11}$Li

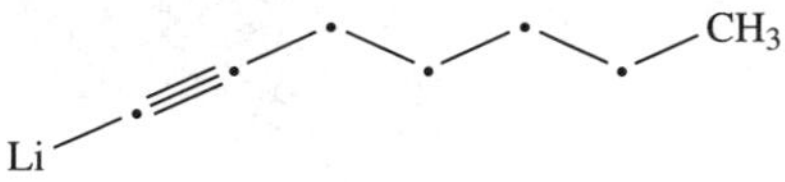

Ammonia, Iodine
See Iodine: Ammonia, Lithium 1-heptynide
See other METAL ACETYLIDES

2835. 3-Methyl-4-nitro-1-buten-3-yl acetate

[61447-07-2]

C$_7$H$_{11}$NO$_4$

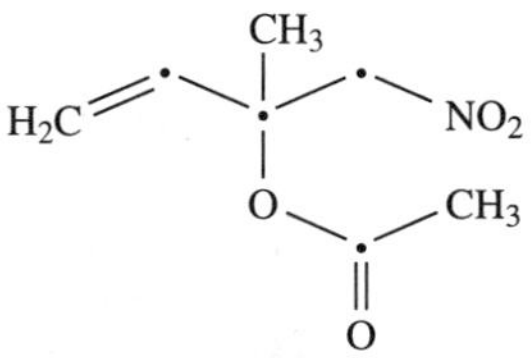

Wehrli, P. A. *et al.*, *J. Org. Chem.*, 1977, **42**, 2940
This and the isomeric nitroester (below), both produced in the nitroacetoxylation
of isoprene, are of limited thermal stability and it is recommended that neither be

936

heated above 100°C, either neat or in solution. Vacuum distillation of the nitro esters should be limited to 1 g portions, as decomposition 'fume-offs' have been observed.

See related NITROALKENES

2836. 3-Methyl-4-nitro-2-buten-1-yl acetate
[1447-06-1] $C_7H_{11}NO_4$

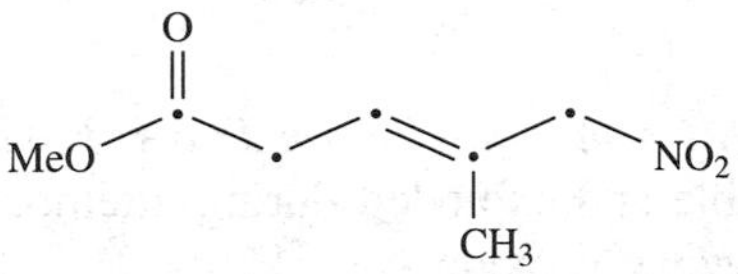

See 3-Methyl-4-nitro-1-buten-3-yl acetate
See related NITROALKENES

†2837. Cycloheptene
[628-92-2] C_7H_{12}

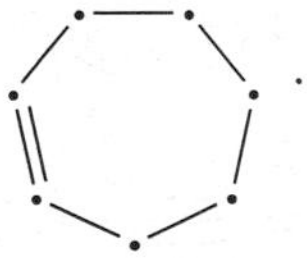

See related ALKENES

†2838. 1-Heptyne
[628-71-7] C_7H_{12}

$$HC\equiv CC_5H_{11}$$

See other ALKYNES

†2839. 4-Methylcyclohexene
[591-47-9] C_7H_{12}

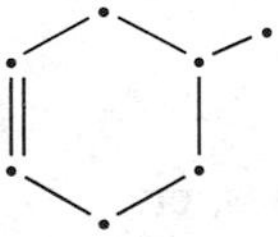

See related ALKENES

2840. 1-Methoxy-3,4,5-trimethylpyrazole *N*-oxide
     [39753-42-9] $C_7H_{12}N_2O_2$

Boyle, F. T. *et al., J. Chem. Soc., Perkin Trans. 1*, 1973, 167
It is thermally unstable and exploded during attempted distillation.
See 1-Methoxyimidazole *N*-oxide
See other N–O COMPOUNDS, *N*-OXIDES

2841. 2-Heptyn-1-ol
     [1002-36-4] $C_7H_{12}O$

Brandsma, 1971, 69
Dilution of the alcohol with white mineral oil before vacuum distillation is recom-
mended to avoid the possibility of explosion of the undiluted distillation residue.
See other ACETYLENIC COMPOUNDS

2842. 3-Methylcyclohexanone
     [591-24-2] $C_7H_{12}O$

Hydrogen peroxide, Nitric acid
  See Hydrogen peroxide: Ketones, etc.

2843. 4-Methylcyclohexanone
[589-92-4] $C_7H_{12}O$

Nitric acid
See Nitric acid: 4-Methylcyclohexanone

2844. 4-Ethoxy-2-methyl-3-butyn-2-ol
[2041-76-1] $C_7H_{12}O_2$

Brandsma, 1971, 78
Traces of acid adhering to glassware are sufficient to induce explosive decomposition of the alcohol during distillation, and must be neutralised by pre-treatment with ammonia gas. Low pressures and temperatures are essential during distillation. Explosions during distillation using water pump vacuum and bath temperatures above 115°C were frequent.
See ETHOXYETHYNYL ALCOHOLS
See other ACETYLENIC COMPOUNDS, CATALYTIC IMPURITY INCIDENTS

2845. 1,2-Dimethylcyclopentene ozonide (1,5-Dimethyl-6,7,8-trioxabicyclo[3.2.1] octane)
[19987-14-5] $C_7H_{12}O_3$

Criegee, R. *et al., Ber.*, 1953, **86**, 3
Though the ozonide appeared stable to small-scale high-vacuum distillation, the residue exploded violently at 130°C.
See other OZONIDES

2846. *N*-Cyano-2-bromoethylbutylamine
[ ] $C_7H_{13}BrN_2$

$$BuN(CN)C_2H_4Br$$

Anon., *BCISC Quart. Safety Summ.*, 1964, **35**, 23
A small sample heated to 160°C decomposed exothermically, reaching 250°C. It
had previously been distilled at 95°/0.65 mbar without decomposition.
See *N*-Cyano-2-bromoethylcyclohexylamine
See other CYANO COMPOUNDS

2847. 3-Trimethylsilylprop-2-enyl trifluoromethanesulfonate
[62905-84-4] $C_7H_{13}F_3O_3SSi$

See entry ALLYL TRIFLUOROMETHANESULFONATES *See other* ALLYL COMPOUNDS,
SULFUR ESTERS

2848. *N*-Dimethylethyl-3,3-dinitroazetidine (*N*-*tert*-Butyl-3,3-dinitroazetidine)
[125735-38-8] $C_7H_{13}N_3O_4$

Archibald, T. G. *et al., J. Org. Chem.*, 1990, **55**, 2920
The above compound, prepared as an intermediate to 1,3,3-trinitroazetidine for
study as an explosive, decomposed explosively when heated above 120°C for
distillation. Precautions when handling nitroazetidines are counselled.
See related POLYNITROALKYL COMPOUNDS *See other* STRAINED RING COM-
POUNDS

†2849. Cycloheptane
[291-64-5] C_7H_{14}

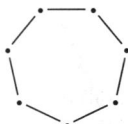

†2850. Ethylcyclopentane
[1640-89-7] C_7H_{14}

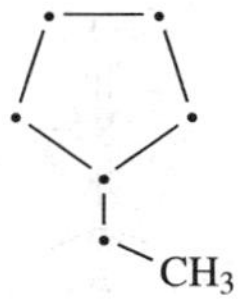

†2851. 1-Heptene
[592-76-7] C_7H_{14}

$$H_2C=CHC_5H_{11}$$

See other ALKENES

†2852. 2-Heptene
[592-77-8] C_7H_{14}

$$MeCH=CHBu$$

See other ALKENES

†2853. 3-Heptene
[592-78-9] C_7H_{14}

$$EtCH=CHPr$$

See other ALKENES

†2854. Methylcyclohexane
[108-87-2] C_7H_{14}

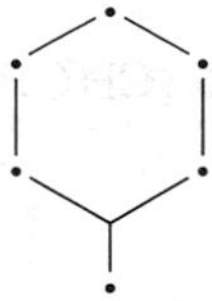

HCS 1980, 646

†2855. 2,3,3-Trimethylbutene
[594-56-9] C_7H_{14}

$$H_2C=CMeCMe_3$$

See other ALKENES

2856. Dimethylmethyleneoxosulfanenickel(0) diethylene complex
[ ] $C_7H_{14}NiOS$

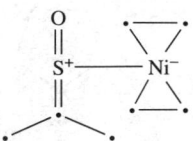

Pörschke, K. R., *Chem. Ber.*, 1987, **120**, 425–427
The complex decomposes violently at 0°C, evolving ethane, cyclopropane and methane.
See other GAS EVOLUTION INCIDENTS, HEAVY METAL DERIVATIVES

†2857. **2,4-Dimethyl-3-pentanone**
[565-80-0] $C_7H_{14}O$

$$Me_2CHCO.CHMe_2$$

†2858. **3,3-Diethoxypropene**
[3054-95-3] $C_7H_{14}O_2$

$$H_2C=CHCH(OEt)_2$$

See other PEROXIDISABLE COMPOUNDS

†2859. **Ethyl isovalerate**
[108-64-5] $C_7H_{14}O_2$

$$Me_2CHCH_2CO.OEt$$

†2860. **Isobutyl propionate**
[540-42-1] $C_7H_{14}O_2$

$$Me_2CHCH_2OCO.Et$$

†2861. **Isopentyl acetate**
[123-92-2] $C_7H_{14}O_2$

$$MeCO.OC_2H_4CHMe_2$$

NSC 208, 1977; *FPA H117*, 1982; *HSC 1980*, 563

†2862. **Isopropyl butyrate**
[638-11-9] $C_7H_{14}O_2$

$$Me_2CHOCO.Pr$$

†2863. Isopropyl isobutyrate
[617-50-5] $C_7H_{14}O_2$

$$Me_2CHOCO.CHMe_2$$

†2864. Pentyl acetate
[628-63-7] $C_7H_{14}O_2$

$$MeCO.OC_5H_{11}$$

HCS 1980, 153

†2865. 2-Pentyl acetate
[53496-15-4] $C_7H_{14}O_2$

$$MeCO.OCH(Me)Pr$$

HCS 1980, 154

2866. 4(2,3-Epoxypropoxy)butanol
[4711-95-9] $C_7H_{14}O_3$

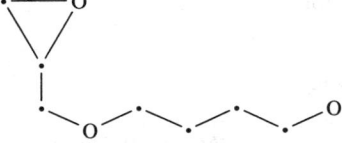

Trichloroethylene
See Trichloroethylene: Epoxides
See other 1,2-EPOXIDES

2867. Cyclopentylmethyl methanesulfonate
[73017-74-2] $C_7H_{14}O_3S$

$$C_5H_9CH_2OSO_2.Me$$

It decomposes vigorously at 128°C.
See entry SULFONIC ACID ESTERS *See other* SULFUR ESTERS

†2868. 2,6-Dimethylpiperidine
[504-03-0] $C_7H_{15}N$

See other ORGANIC BASES

943

†2869. 1-Ethylpiperidine
[766-09-6] $C_7H_{15}N$

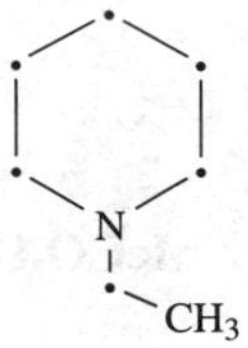

See other ORGANIC BASES

†2870. 2-Ethylpiperidine
[1484-80-6] $C_7H_{15}N$

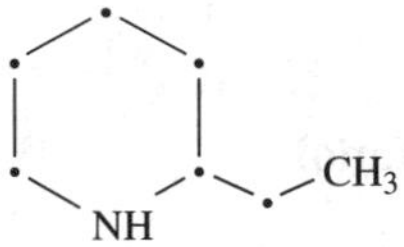

See other ORGANIC BASES

†2871. 2,3-Dimethylpentane
[565-59-3] C_7H_{16}

$$Me_2CHCHMeEt$$

†2872. 2,4-Dimethylpentane
[108-08-7] C_7H_{16}

$$Me_2CHCH_2CHMe_2$$

†2873. Heptane
[142-82-5] C_7H_{16}

$$Me[CH_2]_5Me$$

HCS 1980, 524 (includes isomers)

†2874. 2-Methylhexane
[591-76-4] C_7H_{16}

$$Me_2CHBu$$

†2875. 3-Methylhexane
[589-34-4] C_7H_{16}

$$EtCHMePr$$

944

†2876. 2,2,3-Trimethylbutane
  [464-06-2]   C₇H₁₆

Me₃CCHMe₂

2877. 3-Diethylaminopropylamine
  [14642-66-1]   C₇H₁₈N₂

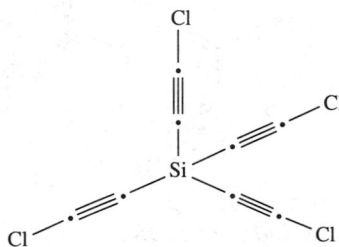

Cellulose nitrate
  See CELLULOSE NITRATE: amines *See other* ORGANIC BASES

2878. *N*-*tert*-Butyl-*N*-trimethylsilylaminoborane
  [73452-32-1]   C₇H₂₀BNSi

t-BuN(Me₃Si)BH₂

Wisian-Nielsen, P. *et al., J. Inorg. Nucl. Chem.*, 1979, **41**, 1546
It ignites in direct contact with air.
See related BORANES, ALKYLSILANES

2879. Tetrakis(chloroethynyl)silane
  [ ]   C₈Cl₄Si

Viehe, H. G., *Chem. Ber.*, 1959, **92**, 3075
Though thermally stable, it exploded violently on grinding (with potassium
bromide) or on impact.
See other FRICTIONAL INITIATION INCIDENTS, HALOACETYLENE DERIVATIVES

2880. Potassium octacyanodicobaltate(8−)
[23705-25-1] $C_8Co_2K_8N_8$

$$K_8[Co_2(CN)_8]$$

Bailar, 1973, Vol. 4, 175
A very unstable, pyrophoric material.
See other METAL CYANIDES (AND CYANO COMPLEXES), PYROPHORIC MATERIALS

2881. Octacarbonyldicobalt
[10210-68-1] $C_8Co_2O_8$

$$(OC)_3Co:(CO)_2:Co(CO)_3$$

Wender, I. *et al., Inorg. Synth.*, 1957, **5**, 191
When the carbonyl is prepared (by rapid cooling) as a fine powder, it decomposes
in air to give pyrophoric dodecacarbonyltetracobalt.
See other CARBONYLMETALS

2882. Caesium graphite
[12079-66-2] C_8Cs

$$CsC_8$$

See Carbon: Alkali metals
See other METAL NON-METALLIDES

2883. 3-Bromo-2,7-dinitrobenzo[b]thiophene-5-diazonium-4-oxide
[ ] $C_8HBrN_4O_5S$

Brown, I. *et al., Chem. & Ind.*, 1962, 982
It is an explosive compound.
See Nitric acid: Benzo[*b*]thiophene derivatives
See other ARENEDIAZONIUM OXIDES

2884. *N*-Bromo-3-nitrophthalimide
   [ ]   C$_8$H$_3$BrN$_2$O$_4$

Tetrahydrofurfuryl alcohol
   See N-HALOIMIDES: alcohols *See other* N-HALOGEN COMPOUNDS, NITROARYL
   COMPOUNDS

2885. 1,3-Bis(trifluoromethyl)-5-nitrobenzene
   [328-75-6]   C$_8$H$_3$F$_6$NO$_2$

Preparative hazard
   See Nitric acid: 1,3-Bis(trifluoromethyl)benzene
   See other NITROARYL COMPOUNDS

2886. 4-Nitro-1-picryl-1,2,3-triazole
   [31123-28-1]   C$_8$H$_3$N$_7$O$_8$

Storm, C. B. *et al., J. Phys. Chem.*, 1989, **93**, 1000–1007
Though the 1- and 2-picryl isomers are similarly explosive, the former has a
much higher impact-sensitivity than the latter. This is attributed to the former

947

having a shorter N2 to N3 interatomic distance, longer adjacent bonds, and a lower activation energy for elimination of molecular nitrogen from the 2,3 positions.

See 5-nitro-2-picryltetrazole, 1-Picryl-1,2,3-triazole

See other POLYNITROARYL COMPOUNDS, TRIAZOLES

2887. Silver isophthalate
[57664-97-8] $C_8H_4Ag_2O_4$

Fields, E. K., *J. Org. Chem.*, 1976, **41**, 918–919

Decomposition of the salt at 375°C under nitrogen, with subsequent cooling under hydrogen, gives a black carbon-like polymer containing metallic silver which ignites at 25°C on exposure to air.

See related PYROPHORIC MATERIALS *See other* SILVER COMPOUNDS

2888. Isophthaloyl chloride
[99-63-8] $C_8H_4Cl_2O_2$

Preparative hazard
    See 1,3-Bis(trichloromethyl)benzene: Oxidants

Methanol
    Morrell, S. H., private comm., 1968
    Violent reaction occurred between isophthaloyl chloride and methanol when they were accidentally added in succession to the same waste solvent bottle.
    See other ACYL HALIDES

2889. Terephthaloyl chloride (1,4-Benzenedicarbonyl chloride)
[100-20-9] $C_8H_4Cl_2O_2$

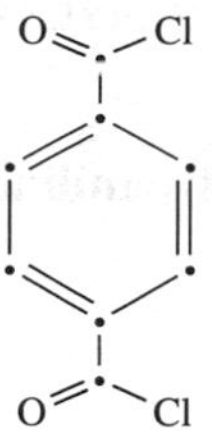

Water

Anon., *Loss Prev. Bull.*, 1986, (071), 9

Hydrolysis of 7 t of the acid chloride by water in a herbicide processing unit ran
out of control, gas evolution eventually rupturing the reactor violently.

See other ACYL HALIDES, GAS EVOLUTION INCIDENTS

2890. 2,4-Hexadiynylene chloroformate
[ ] $C_8H_4Cl_2O_4$

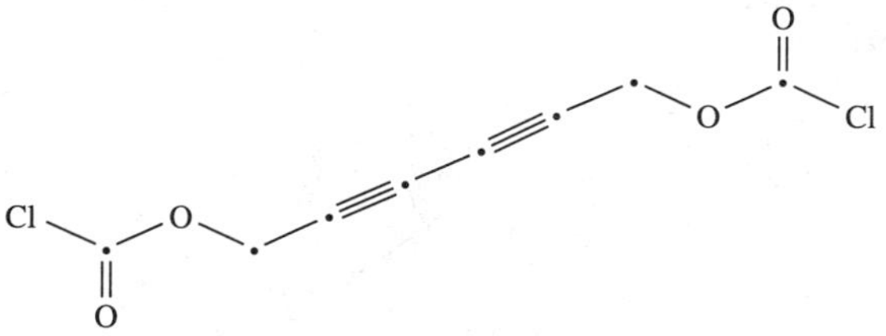

Driedger, P. E. *et al., Chem. Eng. News*, 1972, **50**(12), 51

One sample exploded with extreme violence at 15°C/0.2 mbar, while another
smaller sample had been distilled uneventfully at 114–115°C/0.2 mbar.

See other ACYL HALIDES, ACETYLENIC COMPOUNDS

2891. 1,3-Bis(trichloromethyl)benzene
[881-99-2] $C_8H_4Cl_6$

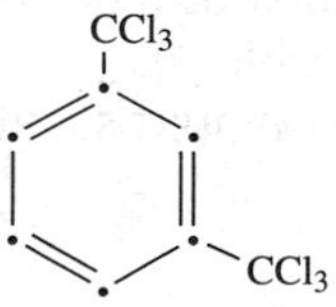

Oxidants

Rondevstedt, C. S., *J. Org. Chem.*, 1976, **41**, 3574–3577

Heating the bis(trichloromethyl)benzene with potassium nitrate, selenium dioxide or sodium chlorate to effect conversion to the bis(acyl chloride) led to eruptions at higher temperatures, and was too dangerous to pursue.
See related HALOALKANES, HALOARYL COMPOUNDS

2892. *N*-Trifluoroacetoxy-2,4-dinitroaniline
[127526-90-3] $C_8H_4F_3N_3O_6$

Bryant, I. R. *et al., Austr. J. Chem.*, 1989, **42**(12), 2275
The crude compound was isolated, began to evolve trifluoroacetic acid and, after 15 minutes, exploded.
See other N–O COMPOUNDS

2893. 1-Picryl-1,2,3-triazole
[18922-71-9] $C_8H_4N_6O_6$

Storm, C. B. *et al., J. Phys. Chem.*, 1989, **93**, 1000–1007
Though the 1- and 2-picryl isomers are similarly explosive, the former has a much higher impact-sensitivity than the latter. This is attributed to the former having a shorter N2 to N3 interatomic distance, longer adjacent bonds, and a lower activation energy for elimination of molecular nitrogen from the 2,3 positions.
See 4-Nitro-1-picryl-1,2,3-triazole
See other POLYNITROARYL COMPOUNDS, TRIAZOLES

2894. 1,3-Bis(trifluoromethyl)benzene
[402-31-3]   C₈H₄F₆

$C_8H_4F_6$

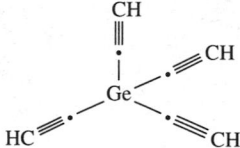

Nitric acid, Sulfuric acid
See Nitric acid: 1,3-Bis(trifluoromethyl)benzene, etc.
See related HALOALKANES, HALOARYL COMPOUNDS

2895. Tetraethynylgermanium
[4531-35-5] C_8H_4Ge

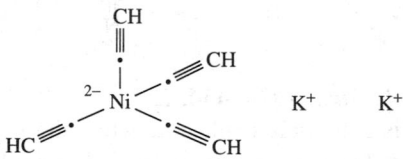

1. Chokiewicz, W. *et al., Compt. rend.*, 1960, **250**, 866
2. Shikkiev, I. A. *et al., Doklad. Akad. Nauk SSSR*, 1961, **139**, 1138 (Engl. transl., 830)

It explodes on rapid heating [1], or friction [2].
See other METAL ACETYLIDES

2896. Potassium tetraethynylnickelate(2—)
[65664-23-5] $C_8H_4K_2Ni$

Air, or Water
Vigorously pyrophoric in air, it may explode on contact with drops of water.
See entry COMPLEX ACETYLIDES

2897. Potassium tetraethynylnickelate(4−)
[ ] $C_8H_4K_4Ni$

The diammoniate (and that of the sodium salt) explodes on exposure to impact, friction or flame.

See entry COMPLEX ACETYLIDES

2898. Phthaloyl diazide
[50906-29-1] $C_8H_4N_6O_2$

Lindemann, H. *et al., Ann.*, 1928, **464**, 237

The *symm*-diazide is extremely explosive, while the *asymm*-isomer (*gem*-diazido-isobenzofuranone) is less so, melting at 56°C, but exploding on rapid heating.

See other ACYL AZIDES

2899. Phthalic anhydride
[85-44-9]   C$_8$H$_4$O$_3$

HCS 1980, 752

Preparative hazard
1. Kratochvil, V., *Chem. Abs.*, 1974, **80**, 3215
2. English transl. by D. S. Rosenberg, 10.2.69 of German transl. by S. Vedrilla (Chemiebau) of 'Low Temperature Pyrophoric Compounds in Process for Preparing Phthalic Anhydride', a Japanese paper in *Anzen Kogyo Kogaku*, 1968
3. Schwab, R. F. *et al., Chem. Eng. Progr.*, 1970, **66**(9), 49–53

Investigation of an explosion in a phthalic anhydride plant showed that naphthoquinones (by-products from naphthalene oxidation) reacted with iron(III) salts of phthalic, maleic or other acids (corrosion products) to form labile materials. The latter were found to undergo exothermic oxidation and self-ignition at around 200°C. Process conditions to minimise hazards are discussed, with 27 references [1]. Fires and explosions in the condenser sections of phthalic anhydride plants were traced to the presence of deposits of thermally unstable basic iron salts of maleic or phthalic acids and iron sulfides derived from sulfur in the naphthalene feedstock. A small-scale laboratory test procedure was developed to measure the exotherm point and maximum temperature rise in various mixtures of materials found to be present in the condensing section. A basic iron(III) maleate hydrate showed an exotherm point of 165°C and temperature rise of 369°C, changed by pretreatment with hydrogen sulfide to 164 and 392°C, respectively. The corresponding figures for basic iron(III) phthalate were 236 and 477°C, both unchanged by presulfiding treatment. The part played by acid-corrosion of the mild steel condensers in formation of the unstable pyrophoric compounds was also investigated. It was concluded that the origins of the many mishaps which have occurred soon after plant start-up were explicable in terms of these experimental findings [2]. Other hazards in phthalic anhydride production units have also been discussed [3].
See PYROPHORIC IRON–SULFUR COMPOUNDS *See other* CORROSION INCIDENTS

Other reactants
Yoshida, 1980, 357
MRH Values calculated for 11 combinations with oxidants are given.

Oxidants	MRH values show % of oxidant
See Copper(II) oxide: Phthalic acid, etc.	MRH 2.09/tr.
Nitric acid: Phthalic anhydride	MRH 5.52/71
Sodium nitrite: Phthalic acid, etc.	MRH 2.09/tr.

See other ACID ANHYDRIDES

2900. Phthaloyl peroxide (2,3-Benzodioxin-1,4-dione)
[4733-52-2] $C_8H_4O_4$

Jones, M. *et al., J. Org. Chem.*, 1971, **36**, 1536
Detonable by impact or by melting at 123°C. It may be a linear polymer, rather
than the cyclic structure implied by the alternative name.
See other DIACYL PEROXIDES

2901. Tetraethynyltin
[16413-88-0] C_8H_4Sn

Jenkner, H., Ger. Pat. 115 736, 1961
It explodes on rapid heating.
See other METAL ACETYLIDES

2902. 2,4-Dinitrophenylacetyl chloride
[109799-62-4] $C_8H_5ClN_2O_5$

Bonner, T. G., *J. R. Inst. Chem.*, 1957, **81**, 596

954

Explosive decomposition occurred during attempted vacuum distillation, attributed either to the presence of some trinitro compound in the unpurified dinitrophenylacetic acid used, or to the known instability of *o*-nitro acid chlorides. A previous publication (ibid., 407) erroneously described the decomposition of 2,4-dinitrobenzoyl chloride.
See other NITROACYL HALIDES

2903. Trifluoroacetyliminoiodobenzene
[94286-09-6] $C_8H_5F_3INO$

$$F_3CCO.N{=}IPh$$

Mansuy, D. *et al., J. Chem. Soc., Chem. Comm.*, 1984, 1163
The solid undergoes explosive decomposition at 100°C.
See other N-HALOGEN COMPOUNDS, IODINE COMPOUNDS

2904. 3-Benzocyclobutenylpotassium
[ ] C_8H_5K

Perchloryl fluoride
   See Perchloryl fluoride: Benzocyclobutene, etc.
   See related ARYLMETALS

2905. 3-Nitrophenylacetylene
[3034-94-4] $C_8H_5NO_6$

Sorbe, 1968, 152
The compound (isomer not stated) and its metal derivatives are explosive.
See other ALKYNES, NITROARYL COMPOUNDS

2906. 3-Nitrophthalic acid
[603-11-2] $C_8H_5NO_6$

Preparative hazard
    See Nitric acid: Phthalic anhydride, etc.
    See other NITROARYL COMPOUNDS, ORGANIC ACIDS

2907. Nitroterephthalic acid
[603-11-2] $C_8H_5NO_6$

Preparative hazard
    See Nitric acid: Sulfuric acid, Terephthalic acid
    See other NITROARYL COMPOUNDS, ORGANIC ACIDS

2908. 3-Sodio-5-(5′-nitro-2′-furfurylideneamino)imidazolidin-2,4-dione
[54-87-5] $C_8H_5N_4NaO_5$

Boros, L. et al., Ger. Offen., 2 328 927, 1974
The crystalline *N*-sodium salt is explosive.
See other N-METAL DERIVATIVES

956

2909. 3-Diazo-5-phenyl-3*H*-1,2,4-triazole
[80670-36-6]

$C_8H_5N_5$

It is a light-sensitive and highly shock-sensitive solid, which can be weighed or stored in the dark at ambient temperature as the dry solid, but it decomposes in solution in a few hours.

See entry DIAZOAZOLES

See other DIAZO COMPOUNDS, HIGH-NITROGEN COMPOUNDS, TRIAZOLES

2910. 4-Diazo-5-phenyl-1,2,3-triazole
[64781-77-7]

$C_8H_5N_5$

The crystalline solid explodes at the m.p., 133 − 5°C

See entry DIAZOAZOLES

See other DIAZO COMPOUNDS, HIGH-NITROGEN COMPOUNDS, TRIAZOLES

2911. Sodium phenylacetylide
[1004-22-4]

C_8H_5Na

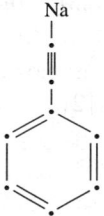

Nef, J. U., *Ann.*, 1899, **308**, 264

The ether-moist powder ignites in air.

See ALKALI-METAL DERIVATIVES OF HYDROCARBONS *See other* METAL ACETYLIDES

2912. Phenylacetylene
[536-74-3] C_8H_6

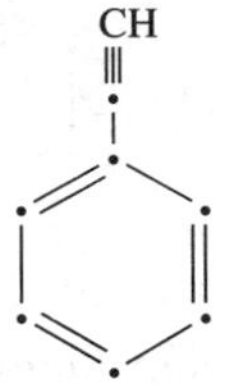

Palladium(II) acetate

Trumbo, D. L. *et al., J. Polym. Sci., A, Polymer Chem.*, 1987, **25**, 1027–1034
Among palladium(II) salts used to polymerise phenylacetylene, the acetate led to rapid and very exothermic polymerisation, sometimes leading to explosion.
See other POLYMERISATION INCIDENTS

Perchloric acid

See Perchloric acid: Phenylacetylene
See other ALKYNES

2913. 3-Methyl-2-nitrobenzoyl chloride
[50424-93-6] $C_8H_6ClNO_3$

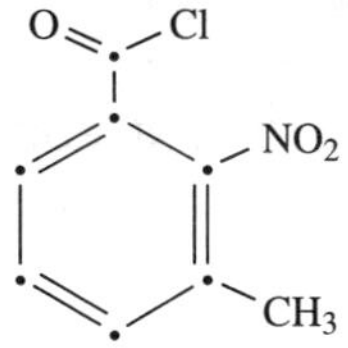

1. Dahlbom, R., *Acta Chem. Scand.*, 1960, **14**, 2049
2. Anon., *Loss Prev. Bull.*, 1977, (013), 16

Attempted distillation caused an explosion [1], and a 200 g sample deflagrated at 120–130°C under high vacuum [2].
See other NITROACYL HALIDES

2914. 2-Nitrophenylacetyl chloride
[22751-23-1] $C_8H_6ClNO_3$

Hayao, S., *Chem. Eng. News*, 1964, **42**(13), 39
Distillation of solvent chloroform from the preparation caused the residue to explode violently on 2 occasions. Previous publications had recommended use of solutions of the chloride, in preference to isolated material.
See other NITROACYL HALIDES

2915. 3-Methoxy-2-nitrobenzoyl chloride
[15865-57-3] $C_8H_6ClNO_4$

1. Goodwin, B., *Chem. Brit.*, 1987, **23**, 1180–1181
2. Bretherick, L., *Chem. Brit.*, 1988, **24**, 125
During an attempt to prepare 3-methoxy-2-nitrobenzoyl cyanide by heating the title chloride with copper(I) cyanide to over 200°C, the reaction mixture had decomposed vigorously at around 60°C. This was attributed to thermal instability of the intended product [1], but available evidence suggests that instability of the title nitroacyl halide is a much more likely cause. However, there is also the possibility of a reaction involving demethylation of the methoxynitrobenzoyl chloride by traces of hydrogen chloride, and of a subsequent polycondensation reaction of the hydroxynitroacyl chloride so formed and of its subsequent exothermic decomposition, analogous to that of the 4-methoxy-3-nitro isomer. The presence of the copper salt would be expected to catalyse either of the possibilities of exothermic reaction [2].
See 4-Methoxy-3-nitrobenzoyl chloride
See entry NITROACYL HALIDES

2916. 4-Methoxy-3-nitrobenzoyl chloride
[10397-28-1] $C_8H_6ClNO_4$

Preparative hazard

Grewer, T. *et al., Chem. Ing. Tech.*, 1977, **49**, 562–563

Preparation of the acid chloride by an established procedure involved heating the acid with sulfinyl chloride, finally at 100°C for 4 h, when evolution of hydrogen chloride and sulfur dioxide was complete (and the vessel appears then to have been isolated from the gas absorption system). An apparently normally completed batch led to a double explosion with rupture of the process vessel. Subsequent thermoanalytical investigation revealed that the nitroacyl chloride is demethylated by hydrogen chloride produced in the primary reaction, methyl chloride being formed as well as the hydroxyacyl chloride. The latter condenses to form a poly(nitrophenylcarboxylic ester) and hydrogen chloride, which can lead to further demethylation. The sequence of exothermic reactions leads to accelerating decomposition, which becomes violent at about 350–380°C, with generation of pressure up to 40 bar by the volatiles produced. The double explosion was attributed to ignition of chloromethane, followed by that of the decomposing poly(nitro ester). Preventive measures are discussed.

See other GAS EVOLUTION INCIDENTS, NITROACYL HALIDES, POLYCONDENSATION REACTION INCIDENTS

2917. 4-Cyano-3-nitrotoluene
[26830-95-5] $C_8H_6N_2O_2$

Preparative hazard

See 4-Chloro-3-nitrotoluene: Copper(I) cyanide, Pyridine

See other CYANO COMPOUNDS, NITROARYL COMPOUNDS

960

2918. 1,4-Dinitropentacyclo[4.2.0.0^{2,5}.0^{3,8}.0^{4,7}]octane
[87830-30-6]

$C_8H_6N_2O_4$

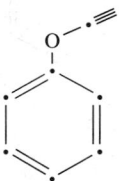

See entry CUBANES

2919. 2,5-Dinitro-3-methylbenzoic acid
[70343-15-6]

$C_8H_6N_2O_6$

Oleum, Sodium azide
Nielsen, A. T. *et al., J. Org. Chem.*, 1979, **44**, 2504
Addition of sodium azide to a solution of the acid in 20% oleum at 5–10°C to produce 3-amino-2,5-dinitrotoluene must be slow (0.1 g portions during 1 h) to avoid explosion.
See other ORGANIC ACIDS, POLYNITROARYL COMPOUNDS

2920. Phenoxyacetylene
[4279-76-9]

C_8H_6O

Jacobs, T. L. *et al., J. Amer. Chem. Soc.*, 1942, **64**, 223
Small samples heated rapidly in sealed tubes to around 100°C exploded.
See other ACETYLENIC COMPOUNDS

2921. *cis*-3,4-Diethynylcyclobut-3-ene-1,2-diol
  [125358-28-3] $C_8H_6O_2$

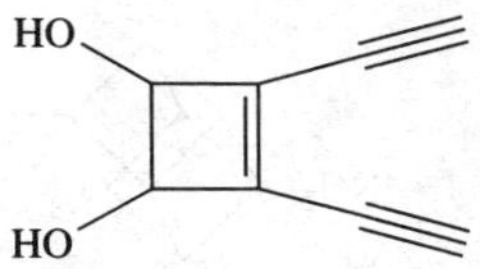

Diederich, F., *et al., J. Amer. Chem. Soc.*, 1990, **112**, 1616
The title compound explodes on heating to 55°C. The ethylene ketal of the corresponding diketone explodes at 118°C.
See 11,12-Diethynyl-1,4,7,10-tetraoxadispiro[4.0.4.2]dodec-11-ene
See other ACETYLENIC COMPOUNDS

†2922. Phenylglyoxal
  [1074-12-0] $C_8H_6O_2$

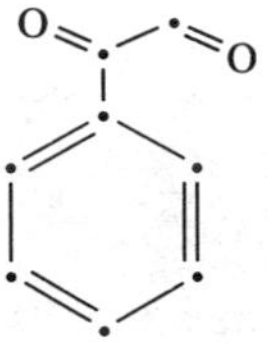

2923. Phthalic acid
  [88-99-3] $C_8H_6O_4$

$$1,2\text{-}C_6H_4(CO.OH)_2$$

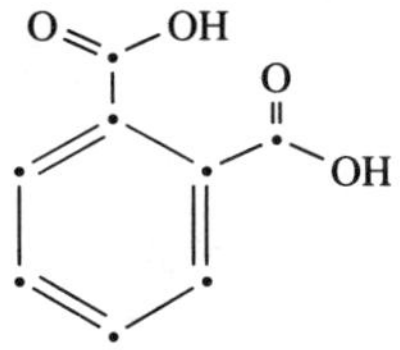

HCS 1980, 753

Other reactants
  Yoshida, 1980, 298
  MRH values calculated for 12 combinations with oxidants are given.

Sodium nitrite   MRH 3.64/79
  See Sodium nitrite: Phthalic acid
  See other ORGANIC ACIDS

962

2924. Terephthalic acid
[100-21-0]

$C_8H_6O_4$

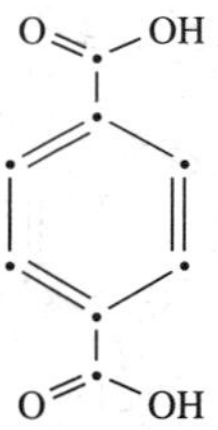

Preparative hazard
See Nitric acid: Hydrocarbons (reference 6)
See other ORGANIC ACIDS

2925. Diperoxyterephthalic acid
[1711-42-8]

$C_8H_6O_6$

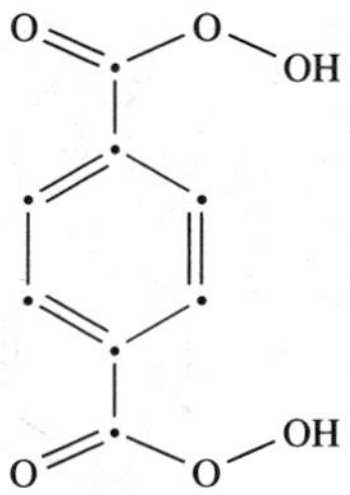

Baeyer, A. *et al., Ber.*, 1901, **34**, 762
It explodes when heated or struck.
See other PEROXYACIDS

2926. 4-Bromomethylbenzoic acid
[6232-88-8]

$C_8H_7BrO_2$

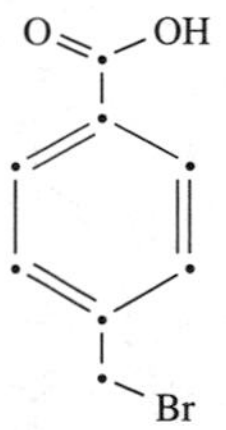

Preparative hazard
See N-Bromosuccinimide: Dibenzoyl peroxide, etc.
See other BENZYL COMPOUNDS, ORGANIC ACIDS

2927. Sodium 4-chloroacetophenone oximate
[1956-39-4] (oxime) $C_8H_7ClNNaO$

van Dijk, J. *et al., J. Med. Chem.*, 1977, **20**, 1201, 1205
Admission of air to the evacuated container of the dry oximate led to explosion on 2 occasions: nitrogen purging is advised.
See related OXIMES *See other* N–O COMPOUNDS

2928. Chloro-(4-methoxyphenyl)diazirine
[4222-26-8] $C_8H_7ClN_2O$

Griller, D. *et al., J. Amer. Chem. Soc.*, 1982, **104**, 5549
A cold sample decomposed violently on rapid warming to ambient temperature.
See other DIAZIRINES

2929. 4-Chloroacetophenone
[99-91-2] C_8H_7ClO

Dimethylamine
Lundstedt, T. *et al., Acta Chem. Scand., Ser. B*, 1984, **B38**, 717–719

964

During optimisation of a preparative procedure, heating mixtures of 4-chloroaceto-phenone and dimethylamine (4.22:1) at 234°C led to two explosions. Use of a safety screen is advised.

See other AMINATION INCIDENTS

2930. 4-Methoxybenzoyl chloride (Anisoyl chloride)
[100-07-2] $C_8H_7ClO_2$

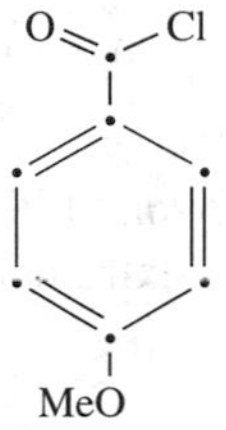

1. Carroll, D. W., *Chem. Eng. News*, 1960, **38**(34), 40
2. Ager, J. H. *et al.*, *Chem. Eng. News*, 1960, **38**(43), 5

Two incidents involving explosion of bottles of chloride stored at ambient temper-ature are described [1,2]. Safe preparation with storage at 5°C is detailed [2].

See other ACYL HALIDES

2931. Benzyl chloroformate
[501-53-1] $C_8H_7ClO_2$

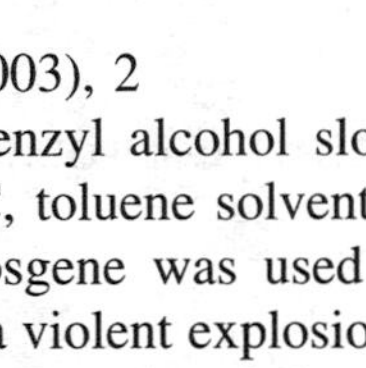

Iron salts

Anon., *Loss Prev. Bull.*, 1975, (003), 2

The ester is made by adding benzyl alcohol slowly to a preformed solution of phosgene in toluene at 12–16°C, toluene solvent finally being distilled off under vacuum. When discoloured phosgene was used (probably containing iron salts from corrosion of the cylinder), a violent explosion occurred during the distillation phase, presumably involving iron-catalysed decomposition of the chloroformate ester.

See other CATALYTIC IMPURITY INCIDENTS, CORROSION INCIDENTS

Water

See ARYL CHLOROFORMATES: water *See related* ACYL HALIDES

2932. Tricarbonyl(1-methylpyrrole)chromium(0) (Tricarbonyl[(1,2,3,4,5-η)-1-methyl-1H-pyrrole]chromium)
[33506-43-3] $C_8H_7CrNO_3$

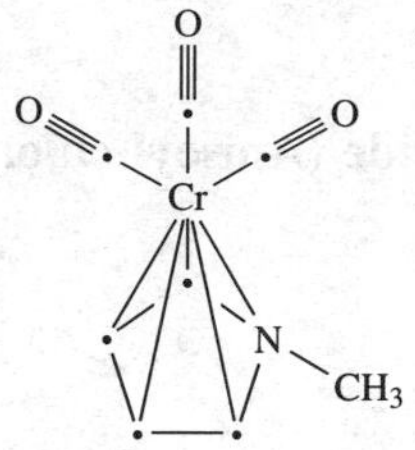

Goti, A. *et al., J. Organomet. Chem.*, 1994, **470**(1–2), C4
This half-sandwich compound is extremely pyrophoric.
See other ORGANOMETALLICS

2933. Copper 1,3,5-octatrien-7-ynide
[   ] C_8H_7Cu

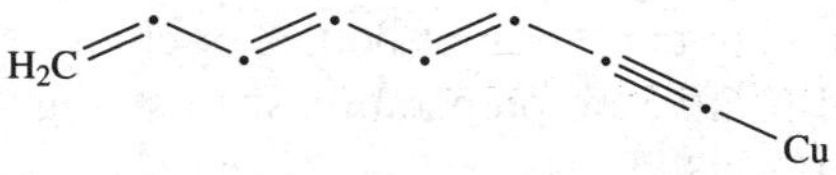

Georgieff, K. K. *et al., J. Amer. Chem. Soc.*, 1954, **76**, 5495
It deflagrates on heating in air.
See other METAL ACETYLIDES

2934. 1-Fluoro-1,1-dinitro-2-phenylethane
[22692-30-4] $C_8H_7FN_2O_4$

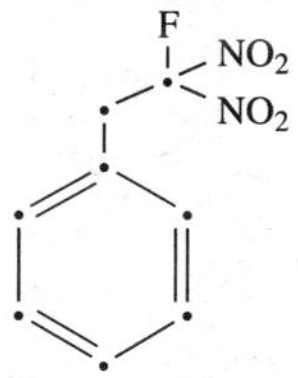

See entry FLUORODINITROMETHYL COMPOUNDS (reference 4)

2935. Phenylacetonitrile
[140-29-4] C_8H_7N

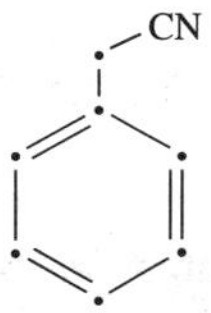

HCS 1980, 203

966

Sodium hypochlorite
See Sodium hypochlorite: Phenylacetonitrile
See other BENZYL COMPOUNDS, CYANO COMPOUNDS

2936. 2-Nitroacetophenone
[614-21-1] $C_8H_7NO_3$

Potassium methylselenide
Leitem, L. *et al., Compt. rend. C*, 1974, **278**, 276
Interaction in dimethylformamide was explosive.
See other NITROARYL COMPOUNDS

2937. 3-Methoxy-2-nitrobenzaldehyde
[53055-05-3] $C_8H_7NO_4$

$$MeOC_6H_3(NO_2)CO.H$$

Initial decomposition temperature by DSC was 245°C.
See entry NITROBENZALDEHYDES
See other ALDEHYDES, NITROARYL COMPOUNDS

2938. 4-Nitrophenylacetic acid
[104-03-0] $C_8H_7NO_4$

$$O_2NC_6H_4CH_2CO.OH$$

Acetic anhydride, Pyridine
1. Batchelor, J. F., 1992, Personal communication
2. Smith, J., *J. Amer. Chem. Soc.*, 1953, **75**, 1134
The distillation residue from 0.8 molar preparation of 2-acetoxy-1(4-nitrophenyl)
prop-1-ene from reaction of the above followed by distillation, according to [2],

erupted on being left to cool, spraying a fume cupboard with tar. The preparation had previously been uneventful at smaller scale [1]. The reaction is not high yielding, indicating much residue and a pot temperature perhaps in excess of 200°C to distil the product.

See other NITROARYL COMPOUNDS

2939. Methyl 2-azidobenzoate
[16714-23-1] $C_8H_7N_3O_2$

491M, 1975, 257

It exploded during distillation.
See other ORGANIC AZIDES

2940. 2-Methyl-5-nitrobenzimidazole
[1792-40-1] $C_8H_7N_3O_2$

Safe preparation
See Nitric acid: 2-Methylbenzimidazole, etc.
See other NITROARYL COMPOUNDS

2941. 3,5-Dinitro-2-toluamide
[148-01-6] $C_8H_7N_3O_5$

1. *The Explosion at the Dow Chemical Factory, King's Lynn 27th June 1976*, London, Health & Safety Executive, HMSO, 1977
2. Grossel, S. S., *J. Loss Prev. Proc. Ind.*, 1988, **1**, 62–74

A 1.3 t batch of off-spec. material had been redried in a rotating double cone vacuum drier heated by steam to around 120–130°C for 23 h. It was left in the drier after heating had been discontinued (though cooling was not applied) and air had been allowed in, to await discharge the next working day. After 2 h, a violent explosion occurred causing one fatality and much destruction. Later investigation of the stability of the compound, notably by ARC techniques, showed that the excessively long drying time had probably caused the onset of thermal decomposition while heat was still being applied, and that further storage under virtually adiabatic conditions would have led to the steady acceleration of the exothermic decomposition to the point of explosion. At this point, it was estimated that a temperature in excess of 800°C had been attained in the drier at detonation, when the shock wave would generate a pressure approaching 800 bar. DSC techniques had shown exotherms at 273° for unheated material, and of 248–259 for material which had been preheated using conditions thought to be present in the drier [1]. The finely powdered amide is also a significant dust explosion hazard, maximum explosion pressures of 10.4 bar and a maximum rate of rise above 680 bar/s having been determined [2].

See other POLYNITROARYL COMPOUNDS
See DRYING, THERMOCHEMISTRY AND EXOTHERMIC DECOMPOSITION

2942. Pentacyclo[4.2.0.0^{2,5}.0^{3,8}.0^{4,7}]octane
[277-10-1]

C$_8$H$_8$

See entry CUBANES

†2943. 1,3,5,7-Cyclooctatetraene
[629-20-9]

C$_8$H$_8$

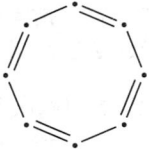

The tetraene is endothermic by some 2.5 kJ/g; which is about the energy of explosion of old-fashioned black powder.

Dimethyl acetylenedicarboxylate
1. Bianchi, G., *Chem. in Brit.*, 1996, **32**(8), 22
In an attempt to obtain a 1:1 Diels Alder adduct, small quantities were sealed into a steel bomb and incubated in an oven at 140°C. After a time there was a violent explosion, converting the oven to an irregular perforated sphere [1]. The editor

suspects that the bursting bomb released flammable vapours that ignited in the cubic oven, distorting the shapnel punctured body.

See Lithium perchlorate: Diethyl ether

Oxygen
    See Oxygen: Cyclooctatetraene
    See related DIENES

2944. 1,3,7-Octatrien-5-yne
[16607-77-5]
C_8H_8

$$H_2C \diagup\!\!\!=\!\!\diagdown\diagup\!\!\!=\!\!\diagdown\!\!\!\equiv\!\!\!\diagdown\diagup\!\!=CH_2$$

1. Nieuwland, J. A. *et al., J. Amer. Chem. Soc.*, 1931, **53**, 4201
2. Dolgopolski, I. M. *et al., Chem. Abs.*, 1948, **42**, 4517e
3. Georgieff, K. K. *et al., J. Amer. Chem. Soc.*, 1954, **76**, 5495

This tetramer of acetylene decomposes violently on distillation at 156°C, but not at reduced pressure [1]. It polymerises on standing to a solid detonable by shock [2]. Explosions occurred during attempted analytical combustion [3].

See related ALKYNES, DIENES

2945. Styrene (Ethenylbenzene)
[100-42-5]
C_8H_8

(*MCA SD-37*, 1971); *NSC 627*, 1982; *FPA H44*, 1976; *HCS 1980*, 866
RSC Lab. Hazard Data Sheet No. 22, 1984

1. Bond, J., *Loss Prev. Bull.*, 1985, (065), 25
2. *MCA SD-37*, 1971
3. Harmon, 1974, 2.17
4. Grewer, T. *et al., Exothermic Decomposition*, Technical Report 01VD 159/0329 for Federal German Ministry for Res. Technol., Bonn. 1986

The autocatalytic exothermic polymerisation reaction exhibited by styrene was involved in a plant-scale incident where accidental heating caused violent ejection of liquid and vapour from a storage tank [1]. Polymerisation becomes self-sustaining above 95°C [2]. The monomer has been involved in several plant-scale explosions, and must be stored at below 32°C, and for less than 3 months [3]. Styrene is a moderately endothermic compound (ΔH°_f (l) +147.7 kJ/mol,

1.42 kJ/g), and the energy of decomposition (110–300°C) has been measured as 0.67 kJ/g [4].
See other ENDOTHERMIC COMPOUNDS

Air, Polymerising styrene
Cullis, C. F. *et al., Combust. Flame*, 1978, **31**, 1–5
Introduction of polymerising liquid styrene into a heated styrene vapour–air mixture can, despite the temporary cooling effect, cause ignition to occur under conditions where it would not normally do so. This effect had been noted previously in plant incidents.
See other POLYMERISATION INCIDENTS

Alkali-metal–graphite compounds
Williams, N. E., private comm., 1968
Interlaminar compounds of sodium or potassium in graphite will ionically polymerise styrene (and other monomers) smoothly. The occasional explosions experienced were probably due to rapid collapse of the layer structure and release of very finely divided metal.

Butyllithium
Roper, A. N. *et al., Br. Polym. J.*, 1975, **7**, 195–203
Thermal explosion which occurred during fast anionic polymerisation of styrene, catalysed by butyllithium, was prevented by addition of low MW polystyrene before the catalyst.
See other POLYMERISATION INCIDENTS

Chlorine, Iron(III) chloride
See Chlorine: Iron(III) chloride, Monomers

Dibenzoyl peroxide
Sebastian, D. H. *et al., Polym. Eng. Sci.*, 1976, **16**, 117–123
The conditions were determined for runaway/non-runaway polymerisation of styrene in an oil-heated batch reactor at 3 bar, using dibenzoyl peroxide as initiator at 3 concentrations. Results are presented diagrammatically.
See other POLYMERISATION INCIDENTS

Initiators
Biesenberger, J. A. *et al., Polym. Eng. Sci.*, 1976, **16**, 101–116
The parameters were determined in a batch reactor for thermal runaway polymerisation of styrene, initiated by azoisibutyronitrile, dibenzoyl peroxide or di-*tert*-butyl peroxide.
See other POLYMERISATION INCIDENTS

Other reactants
Yoshida, 1980, 218
MRH values calculated for 14 combinations with oxidants are given.

Oxygen   MRH 9.75/73
Barnes, C. E. *et al., J. Amer. Chem. Soc.*, 1950, **72**, 210

Exposure of unstabilised styrene to oxygen at 40–60°C generated a styrene–oxygen interpolymeric peroxide which, when isolated, exploded violently on gentle heating.
See other POLYPEROXIDES

Trichlorosilane
See Trichlorosilane: Styrene
See related ALKENES

2946. 1,2-Bis(chloromethyl)benzene
[612-12-4]   C₈H₈Cl₂

$C_8H_8Cl_2$

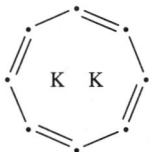

Catalytic impurities
Naef, H., private comm., 1980
The molten material, after holding for 4 h at 78°C in a stainless steel vessel, underwent a thermal runaway reaction and 500 kg erupted through the vent line. It was later found that addition of 0.1% of rust to the hot material led to an accelerating self-condensation Friedel-Craft reaction, catalysed by iron(III) chloride, which led to formation of poly-benzyls accompanied by evolution of hydrogen chloride.
See Benzyl bromide: Molecular sieve
    Benzyl chloride: Catalytic impurities
See other BENZYL COMPOUNDS, GAS EVOLUTION INCIDENTS, POLYCONDENSATION REACTION INCIDENTS

2947. Dipotassium μ-cyclooctatetraene
[59391-85-4]   C₈H₈K₂

$C_8H_8K_2$

Air, or Oxygen
1. Starks, D. F. *et al., Inorg. Chem.*, 1974, **13**, 1307
2. Evans, W. J. *et al., J. Org. Chem.*, 1981, **46**, 3927
The dry solid is violently pyrophoric, exploding on contact with oxygen [1], or air [2]. The solid produced by passing oxygen into a THF solution of the potassium complex must be kept wet with solvent, as it is shock-sensitive when dry [2].
See other ORGANOMETALLICS

2948. Dilithium μ-cyclooctatetraene
   [37609-69-1]   C₈H₈Li₂

Spencer, J. L., *Inorg. Synth.*, 1979, **19**, 214
The solid (but not an ether solution) is pyrophoric in air.
See other ORGANOMETALLICS

2949. *N*-Nitrosoacetanilide
   [938-81-8]   C₈H₈N₂O₂

Piperidine
   Huisgen, R., *Ann.*, 1951, **573**, 163–181
   The dry anilide explodes on contact with a drop of piperidine.

Thiophene
   Bamberger, E., *Ber.*, 1887, **30**, 367
   A mixture prepared at 0°C exploded when removed from the cooling bath.
   See other NITROSO COMPOUNDS

2950. 4-Nitroacetanilide
   [104-04-1]   C₈H₈N₂O₃

Sulfuric acid
   See Sulfuric acid: Nitroaryl bases, etc.
   See other NITROARYL COMPOUNDS

2951. Sodium 4,4-dimethoxy-1-*aci*-nitro-3,5-dinitro-2,5-cyclohexadienide
[12275-58-0] $C_8H_8N_3NaO_8$

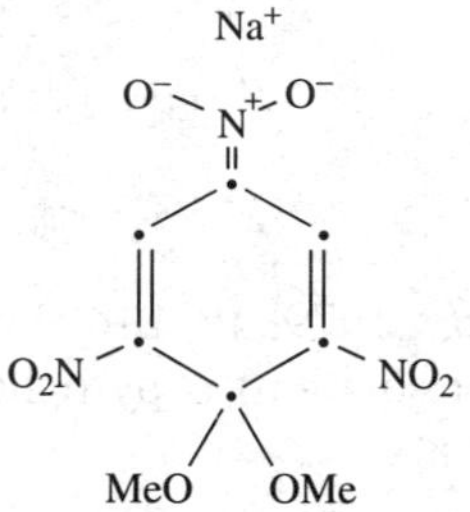

Jackson, C. L. *et al., Amer. Chem. J.*, 1898, **20**, 449

The compound is produced from interaction of trinitroanisole and sodium methoxide and explodes with great violence on heating in a free flame. It should probably be formulated as shown, a dimethyl acetal of the *aci-p*-quinonoid form of picric acid.

See other aci-NITROQUINONOID COMPOUNDS, POLYNITROARYL COMPOUNDS

2952. 5-Amino-3-phenyl-1,2,4-triazole
[4922-98-9] $C_8H_8N_4$

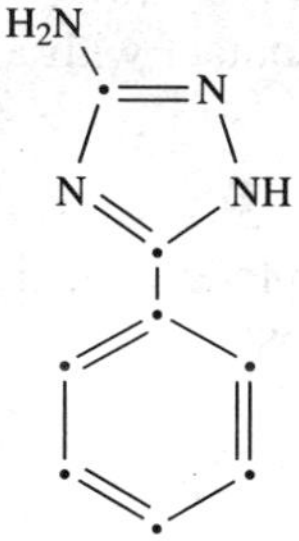

Nitrous acid

Polya, J. B., *Chem. & Ind.*, 1965, 812

The solution, obtained by conventional diazotisation of the amine, contained some solid which was removed by filtration on a glass frit. The solid, thought to be precipitated diazonium salt (possibly an internal salt), exploded violently when disturbed with a metal spatula.

See 5-Aminotetrazole

See other DIAZONIUM SALTS, DIAZOAZOLES, TRIAZOLES

974

2953. 1,1-Diazidophenylethane
[87272-16-0] $C_8H_8N_6$

The structure shows a central carbon bonded to H3C, a phenyl ring, and two azide groups (one drawn vertically with N⁻=N⁺=N⁻, and one horizontal N=N⁺=N⁻).

Kinlager, S. *et al., Synthesis*, 1983, 501
The *gem*-diazide is potentially explosive.
See other ORGANIC AZIDES

2954. 1,2-Diazido-1-phenylethane
[22710-73-2] $C_8H_8N_6$

Preparative hazard
See Manganese(III) azide: Styrene
See other ORGANIC AZIDES

2955. 7,8-Dioxabicyclo[4.2.2]-2,4,7-decatriene
[83931-97-8] $C_8H_8O_2$

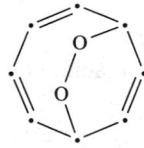

Preparative hazard
See Oxygen (Gas): Cyclooctatetraene
See other CYCLIC PEROXIDES

2956. 4-Methoxybenzaldehyde (Anisaldehyde)
    [123-11-5]   **$C_8H_8O_2$**

Preparative hazard
    See Oxygen (Gas): 4-Methoxytoluene
    See ALDEHYDES

2957. Poly(styrene peroxide)
    [33873-53-9]   **$(C_8H_8O_2)_n$**

$$(-OOCH(Ph)CH_2-)_n$$

    See Styrene: Oxygen
    See other POLYPEROXIDES

2958. 4-Hydroxy-3-methoxybenzaldehyde (Vanillin)
    [121-33-5]   **$C_8H_8O_3$**

Formic acid, Thallium(III) nitrate
    See Thallium(III) nitrate: Formic acid, etc.
    See other ALDEHYDES

2959. 4-Methylbenzyl chloride
[104-82-5] C_8H_9Cl

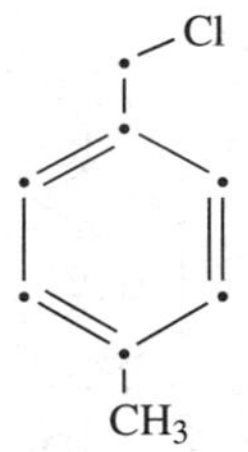

Cl
|
(benzene ring structure)
|
CH_3

Iron

Hakl, J., *Runaway Reactions*, 1981, Paper 3/L, 4
Initial exothermic decomposition sets in at 55°C in presence of 0.02% of iron.
See other BENZYL COMPOUNDS, CATALYTIC IMPURITY INCIDENTS

2960. Dimethyl 2-chloro-4-nitrophenyl thionophosphate
[2643-84-5] $C_8H_9ClNO_5PS$

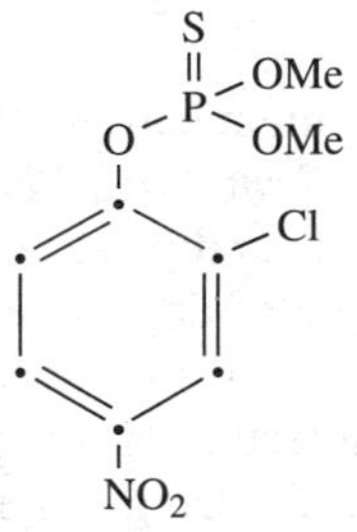

(structure: S double-bonded to P, with OMe, OMe, O linked to benzene ring bearing Cl and NO₂)

NO_2

McPherson, J. B. *et al.*, *J. Agric. Food Chem.*, 1956, **4**, 42
Thermal decomposition of a 1.5 g sample at about 270°C leads to effervescence
after 20 s, with later deflagration and combustion, but a 5 g sample ignites immediately during decomposition. The induction period increases with decreasing
temperature, to 384 s at 174°C. The 3-chloro analogue behaves similarly.
See other INDUCTION PERIOD INCIDENTS, PHOSPHORUS ESTERS

2961. Dimethyl 3-chloro-4-nitrophenyl thionophosphate
[500-28-7] $C_8H_9ClNO_5PS$

See Dimethyl 2-chloro-4-nitrophenyl thionophosphate (next above)
See other INDUCTION PERIOD INCIDENTS, PHOSPHORUS ESTERS

2962. 4-Methoxybenzyl chloride (Anisyl chloride)
[824-94-2] C_8H_9ClO

1. Bryant, R. J., personal communication 1990
2. Harger, M. J. P. *et al., J. Chem. Soc., Perk. I*, 1997, (4), 527

This compound is usually made and used in situ. A 5 g sample was allowed to sit in a screw top vial for a couple of days, one morning it was found to have blown a hole in the lid and extruded a 20 ml plume of polymeric foam [1]. An attempt to dry anisyl chloride over molecular sieves pressurised the container, presumably with hydrogen chloride evolved during polymerisation [2].
See other BENZYL COMPOUNDS, POLYCONDENSATION REACTION INCIDENTS

2963. 2,4-Dimethylbenzenediazonium triiodide
[ ] $C_8H_9I_3N_2$

See entry DIAZONIUM TRIIODIDES

2964. Potassium 1-phenylethanediazoate
[ ] $C_8H_9KN_2O$

$$PhCHMeN{=}NOK$$

White, E. H. *et al., J. Amer. Chem. Soc.*, 1992, **114**(221), 8023
The *syn*-isomer decomposed exothermically to styrene when handled in a nitrogen atmosphere.
See related DIAZONIUM SALTS *See other* N–O COMPOUNDS

2965. 5(1,1-Dinitroethyl)-2-methylpyridine
[ ] $C_8H_9N_3O_4$

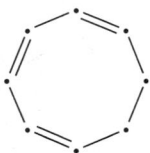

See Nitric acid: 5-Ethyl-2-methylpyridine
See other POLYNITROALKYL COMPOUNDS

2966. 1,3,5-Cyclooctatriene
[1871-52-9] C_8H_{10}

See Diethyl acetylenedicarboxylate: 1,3,5-Cyclooctatriene
See related DIENES

2967. 6,6-Dimethylfulvene
[2175-91-9] C_8H_{10}

Air, or Ether
Engler, C. *et al., Ber.*, 1901, **34**, 2935

Dimethylfulvene readily peroxidised to give an insoluble (polymeric?) peroxide which exploded violently on heating to 130°C, and also caused ether used to triturate it to ignite.

See 1,4-Epidioxy-1,4-dihydro-6,6-dimethylfulvene

See other DIENES, POLYPEROXIDES

†2968. Ethylbenzene
[100-41-4] C_8H_{10}

FPA H99, 1981; *HCS 1980*, 470; *RSC Lab. Hazards Safety Data Sheet No. 74*, 1988

†2969. *mixo*-Xylene
[1330-20-7] C_8H_{10}

NSC 204, 1968; *FPA H76*, 1978; *RSC Lab. Hazard Data Sheet No. 16*, 1983

†2970. *o*-Xylene
[95-47-6] C_8H_{10}

Gorecki, J. *et al.*, *Arch. Combust.*, 1988, **8**, 123–127

The aerobic oxidation of *o*-xylene to phthalic acid or anhydride at elevated temperatures is industrially important, and the flammability limits and explosion parameters at 350°C under a range of pressures have been redetermined.
See also OXIDATION PROCESSES

Other reactants
Yoshida, 1980, 101
MRH values calculated for 13 combinations with oxidants are given.

†2971. *m*-Xylene
[108-38-3]   C$_8$H$_{10}$

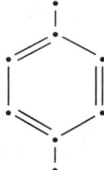

HCS 1980, 962

Sodium
See Sodium: Xylene

†2972. *p*-Xylene
[106-42-3]   C$_8$H$_{10}$

HCS 1980, 963

Acetic acid, Air
Shraer, B. I., *Khim. Prom.*, 1970, **46**(10), 747–750
In liquid phase aerobic oxidation of *p*-xylene in acetic acid to terephthalic acid, it is important to eliminate the inherent hazards of this fuel–air mixture. Effects of temperature, pressure and presence of steam on the explosive limits of the mixture have been studied.
See OXIDATION PROCESSES

1,3-Dichloro-5,5-dimethyl-2,4-imidazolidindione
See 1,3-Dichloro-5,5-dimethyl-2,4-imidazolidindione: Xylene

Nitric acid
See Nitric acid: Hydrocarbons

2973. 4-Bromodimethylaniline
[586-77-6] $C_8H_{10}BrN$

Anon., *Chem. Eng. News*, 1961, **39**(13), 37
During vacuum distillation, self-heating could not be controlled, and proceeded to explosion. Internal dehydrohalogenation to a benzyne seems a possibility.
See other HALOANILINES

2974. 2-Bromo-3,5-dimethoxyaniline
[70277-99-5] $C_8H_{10}BrNO_2$

491M, 1975, 74

During laboratory-scale fractional distillation of a mixture of the aniline and its mono- and di-brominated derivatives, an explosion occurred, attributed to the lability of the bromine substituent.
See other AMINOMETHOXY COMPOUNDS, HALOANILINES

2975. 2-Chloro-*N*-(2-hydroxyethyl)aniline
[94-87-1] $C_8H_{10}ClNO$

CISHC Chem. Safety Summ., 1977, **48**, 25; Shaw, A. W., private comm., 1978

During the vacuum fractional distillation of bulked residues (7.2 t containing 30–40% of the bis(hydroxyethyl) derivative, and up to 900 ppm of iron) at 210–225°C/445–55 mbar in a mild steel still, a runaway decomposition set in and accelerated to explosion. Laboratory work on the material charged showed that exothermic decomposition on the large scale would be expected to set in around 210–230°C, and that the induction time at 215°C of 12–19 h fell to 6–9 h in presence of mild steel. Quantitative work in sealed tubes showed a maximum rate of pressure rise of 45 bar/s, to a maximum developed pressure of 200 bar. The thermally induced decomposition produced primary amine, hydrogen chloride, ethylene, methane, carbon monoxide and carbon dioxide.

See related HALOANILINES

See other GAS EVOLUTION INCIDENTS, INDUCTION PERIOD INCIDENTS

2976. 2(3-Chlorophenoxy)ethylamine
[6488-00-2] $C_8H_{10}ClNO$

Energy of exothermic decomposition in range 270–360°C was measured as 0.27 kJ/g

See entry THERMOCHEMISTRY AND EXOTHERMIC DECOMPOSITION (reference 2)

2977. *N*-(4-Chloro-2-nitrophenyl)-1,2-diaminoethane
[59320-16-0] $C_8H_{10}ClN_3O_2$

See N-(2-Nitrophenyl)-1,2-diaminoethane

See other HALOANILINES, NITROARYL COMPOUNDS

2978. Bis(ethoxycarbonyldiazomethyl)mercury
[20539-85-9] $C_8H_{10}HgN_4O_4$

Buchner, E., *Ber.*, 1895, **28**, 217
It decomposes with foaming on melting at 104°C, and will explode under a
hammer-blow.
See other DIAZO COMPOUNDS, MERCURY COMPOUNDS

2979. Dimethyl 4-nitrophenyl thionophosphate
[298-00-0] $C_8H_{10}NO_5PS$

McPherson, J. B. *et al., J. Agric. Food Chem.*, 1956, **4**, 42–49
Overheating during removal of solvent by distillation from a pilot batch of 'methyl
parathion' led to explosive decomposition, and the course of the 2-stage decomposition was studied. A 1.5 g sample immersed at 270°C decomposes after an
induction period of 54 s and the residue later deflagrates, but a 5 g sample deflagrates during the initial decomposition.
See other INDUCTION PERIOD INCIDENTS, PHOSPHORUS ESTERS

2980. *N,N*-Dimethyl-4-nitrosoaniline
[138-89-6] $C_8H_{10}N_2O$

Energy of exothermic decomposition (in range 120–310°C) measured as 1.92 kJ/g by DSC, and T_{ait24} was determined as 87°C by adiabatic Dewar tests, with an apparent energy of activation of 123 kJ/mol.

See entry THERMOCHEMISTRY AND EXOTHERMIC DECOMPOSITION (reference 2)

Acetic anhydride, Acetic acid
Bain, P. J. S. *et al., Chem. Brit.*, 1971, **7**, 81
An exothermic reaction, sufficiently violent to expel the flask contents, occurs after an induction period of 15–30 s when acetic anhydride is added to a solution of the nitroso compound in acetic acid. 4,4′-Azobis(*N,N*-dimethylaniline), isolated from the reaction tar, may have been formed in a redox reaction, possibly involving an oxime derived from the nitroso compound.

See other INDUCTION PERIOD INCIDENTS, REDOX REACTIONS
See other NITROSO COMPOUNDS

2981. 2-Ethylpyridine-4-carbothioamide
[536-33-4] $C_8H_{10}N_2S$

See FRICTIONAL INITIATION INCIDENTS (reference 3)

2982. 1-Acetyl-4-(4′-sulfophenyl)-3-tetrazene (4-(4-Acetyl-1-tetrazenyl)benzene-sulfonic acid)
[70324-27-5] $C_8H_{10}N_4O_4S$

Houben-Weyl, 1965, Vol. 10.3, 740
It explodes on heating.
See other HIGH-NITROGEN COMPOUNDS, ORGANIC ACIDS, TETRAZENES

2983. Benzeneethanol (Phenethyl alcohol)
[60-12-8] $C_8H_{10}O$

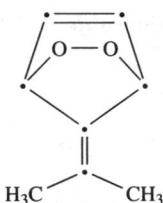

Nitric acid
See Nitric acid: Benzeneethanol, Sulfuric acid

2984. 1,4-Epidioxy-1,4-dihydro-6,6-dimethylfulvene (7-(1-Methylethylidene)-2,3-dioxabicyclo[2.2.1]hept-5-ene)
[51027-90-8] $C_8H_{10}O_2$

Harada, N. *et al., Chem. Lett. (Japan)*, 1973, 1173–1176
When free of solvent it decomposes explosively above −10°C.
See 6,6-Dimethylfulvene: Air, or Ether
See other CYCLIC PEROXIDES

2985. 1-Hydroperoxphenylethane
[3071-32-7] $C_8H_{10}O_2$

Druliner, J. D., *Chem. Eng. News*, 1981, **59**(14), 3
A 9 g sample of the freshly prepared hydroperoxide decomposed after 20 min at ambient temperature, bursting the 20 ml glass container. A 30% solution of the hydroperoxide in ethylbenzene is stable.
See other ALKYL HYDROPEROXIDES

2986. Dicrotonoyl peroxide
[93506-63-9]

$C_8H_{10}O_4$

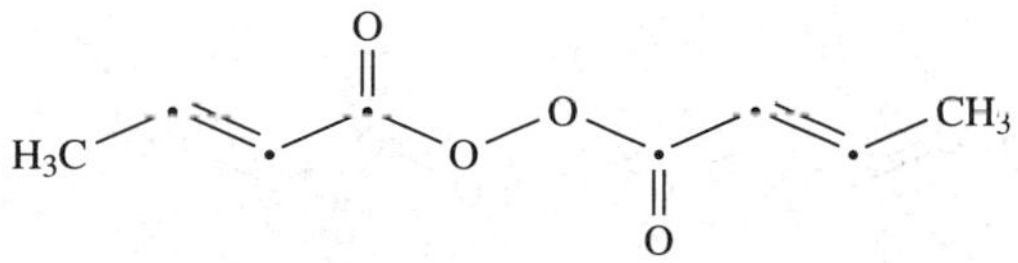

Guillet, J. E. *et al.*, Ger. Pat. 1 131 407, 1962
The diacyl peroxide and its lower homologues, though of relatively high thermal stability, are very shock-sensitive. Replacement of the hydrogen atoms adjacent to the carbonyl groups with alkyl groups renders the materials non-shock-sensitive.
See other DIACYL PEROXIDES

2987. Diethyl acetylenedicarboxylate
[762-21-0]

$C_8H_{10}O_4$

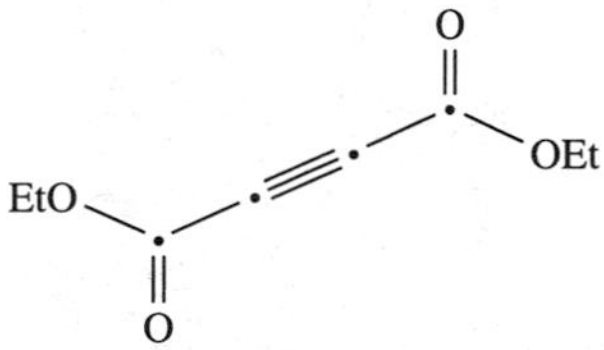

1,3,5-Cyclooctatriene
Foote, C. S., private comm., 1965
A mixture of the reactants being heated at 60°C to effect a Diels-Alder addition exploded. Onset of this vigorously exothermic reaction was probably delayed by an induction period, and presence of a solvent and/or cooling would have moderated it.
See other INDUCTION PERIOD INCIDENTS, ACETYLENIC COMPOUNDS

2988. Ethyl 3,4-dihydroxybenzenesulfonate
[ ]

$C_8H_{10}O_5S$

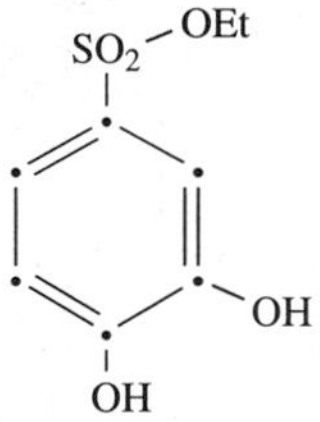

Acetyl nitrate
See Acetyl nitrate: Ethyl 3,4-dihydroxybenzenesulfonate
See other SULFUR ESTERS

987

2989. Diallyl peroxydicarbonate
[34037-79-1]

$C_8H_{10}O_6$

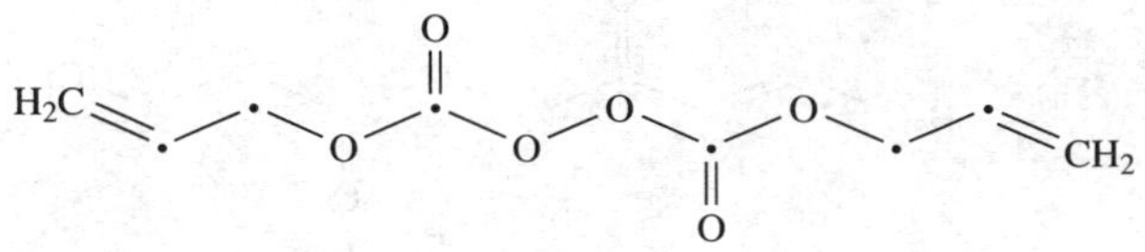

The distilled oil exploded on storage at ambient temperature.
See entry PEROXYCARBONATE ESTERS *See other* ALLYL COMPOUNDS

2990. Bis(3-carboxypropionyl) peroxide
[123-23-9]

$C_8H_{10}O_8$

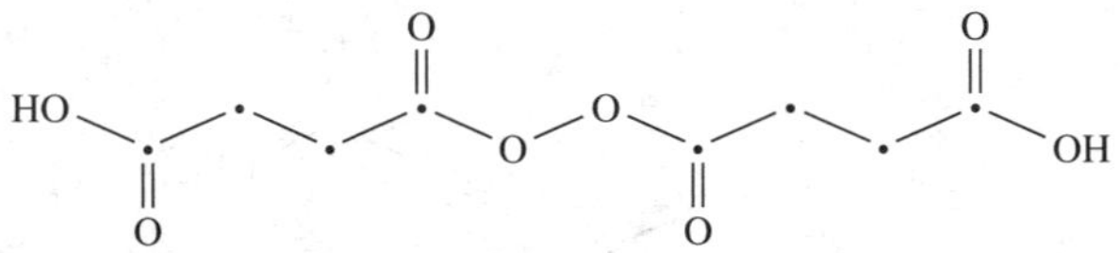

Lombard, R. *et al., Bull. Soc. Chim. Fr.*, 1963, **12**, 2800
It explodes on contact with flame. The commercial dry 95% material ('succinic acid peroxide') is highly hazard-rated.
See other COMMERCIAL ORGANIC PEROXIDES, DIACYL PEROXIDES, ORGANIC ACIDS

2991. Dimethylphenylarsine
[696-26-4]

$C_8H_{11}As$

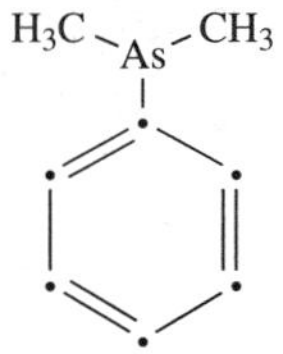

Hydrogen peroxide
 See Hydrogen peroxide: Dimethylphenylarsine
 See related ALKYLNON-METALS

2992. 2,4,6-Trimethylpyrilium perchlorate
[940-93-2] $C_8H_{11}ClO_5$

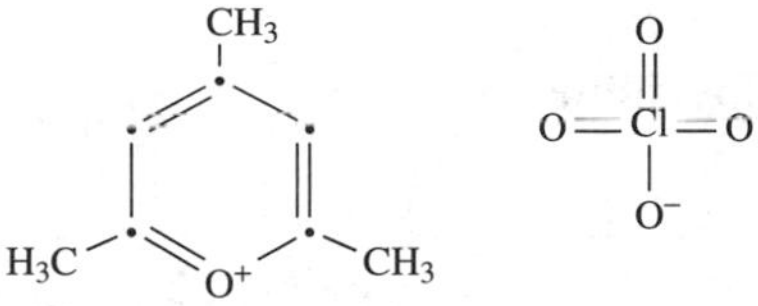

Hafner, K. *et al., Org. Synth.*, 1964, **44**, 102
The crystalline solid is an impact- and friction-sensitive explosive and must be
handled with precautions. These include use of solvent-moist material and storage
in a corked rather than glass-stoppered vessel.
See other GLASS INCIDENTS, NON-METAL PERCHLORATES

2993. *N,N*-Dimethylaniline
[121-69-7] $C_8H_{11}N$

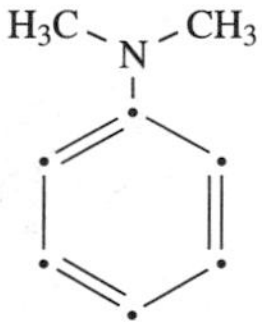

HCS 1980, 425

Dibenzoyl peroxide   MRH 2.26/99+
   See Dibenzoyl peroxide: *N,N*-Dimethylaniline

Diisopropyl peroxydicarbonate
   See Diisopropyl peroxydicarbonate: Amines, etc.

Other reactants
   Yoshida, 1980, 166
   MRH values calculated for 14 combinations with oxidants are given.
   See other ORGANIC BASES

2994. *N*-Ethylaniline
[103-69-5] $C_8H_{11}N$

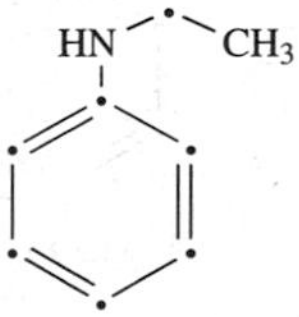

HCS 1980, 469

Nitric acid
   See Nitric acid: Aromatic amines

2995. 5-Ethyl-2-methylpyridine
   [104-90-5] $C_8H_{11}N$

Nitric acid
   See Nitric acid: 5-Ethyl-2-methylpyridine
   See other ORGANIC BASES

2996. 3,3-Dimethyl-1-phenyltriazene
   [7227-91-0] $C_8H_{11}N_3$

1. Baeyer, O. *et al.*, *Ber.*, 1875, **8**, 149
2. Heusler, F., *Ann.*, 1890, **260**, 249
It decomposes explosively on attempted distillation at ambient pressure [1], but may be distilled uneventfully at reduced pressure [2].
See other TRIAZENES

2997. 3-Methoxybenzylamine
   [5071-96-5] $C_8H_{11}NO$

Energy of decomposition (in range 280–430°C) measured as 0.22 kJ/g.

See entry THERMOCHEMISTRY AND EXOTHERMIC DECOMPOSITION (reference 2)
See other AMINOMETHOXY COMPOUNDS, BENZYL COMPOUNDS

2998. 2,4-Dimethoxyaniline
[2735-04-8] $C_8H_{11}NO_2$

Energy of decomposition (in range 160–440°C) measured as 0.55 kJ/g.
See entry THERMOCHEMISTRY AND EXOTHERMIC DECOMPOSITION (reference 2)
See other AMINOMETHOXY COMPOUNDS

2999. 3,5-Dimethoxyaniline
[10272-07-8] $C_8H_{11}NO_2$

Energy of decomposition (in range 320–450°C) measured as 0.83 kJ/g.
See entry THERMOCHEMISTRY AND EXOTHERMIC DECOMPOSITION (reference 2)
See other AMINOMETHOXY COMPOUNDS

3000. 1-(4-Methoxyphenyl)-3-methyltriazene
[53477-43-3] $C_8H_{11}N_3O$

Rondevstedt, C. S. *et al., J. Org. Chem.*, 1957, **22**, 203
The liquid product exploded on attempted vacuum distillation at below 1 mbar.
See other TRIAZENES   *See related* AMINOMETHOXY COMPOUNDS

3001. N-(2-Nitrophenyl)-1,2-diaminoethane
[51138-16-0] $C_8H_{11}N_3O_2$

Doleschall, G. *et al., Tetrahedron*, 1976, **32**, 59

This, and the corresponding 5-chloro and 5-methoxy derivatives, tend to explode during vacuum distillation, and minimal distillation pressures and temperatures are recommended. This instability, (which was not noted for the homologues in which the hydrogen atom of the secondary amino group was replaced by alkyl) may be connected with the possibilty of isomerisation to the *aci*-nitro iminoquinone internal salt species.

See NITROAROMATIC–ALKALI HAZARDS, *aci*-NITROQUINONOID COMPOUNDS
See other NITROARYL COMPOUNDS

3002. Dimethylphenylphosphine oxide
[10311-08-7] $C_8H_{11}OP$

Preparative hazard
See Hydrogen peroxide: Dimethylphenylphosphine
See related ALKYLPHOSPHINES

3003. Dimethylphenylphosphine
[672-66-2] $C_8H_{11}P$

Hydrogen peroxide
See Hydrogen peroxide: Dimethylphenylphosphine
See other ALKYLPHOSPHINES

†3004. 4-Vinylcyclohexene
   [100-40-3] C_8H_{12}

See other PEROXIDISABLE COMPOUNDS

3005. 1-(4-Chlorophenyl)biguanidinium hydrogen dichromate
   [15842-89-4] $C_8H_{12}ClCr_2N_5O_7$

See entry DICHROMATE SALTS OF NITROGENOUS BASES *See other* OXOSALTS OF
NITROGENOUS BASES

3006. 1,6-Diisocyanohexane
   [929-57-7] $C_8H_{12}N_2$

See entry DIISOCYANIDE LIGANDS

3007. Tetramethylsuccinodinitrile
   [3333-52-6] $C_8H_{12}N_2$

Preparative hazard
    Azoisobutyronitrile: Heptane *See other* CYANO COMPOUNDS

3008. 1,6-Diisocyanatohexane
[822-06-0]
$C_8H_{12}N_2O_2$

Alcohols
491M, 1975, 19
Base-catalysed reactions of isocyanates, such as the title compound, with alcohols may be explosively violent in absence of diluting solvents.
See other ORGANIC ISOCYANATES

3009. *tert*-Butyl 2-diazoacetoacetate
[13298-76-5]
$C_8H_{12}N_2O_3$

Regitz, M. *et al., Org. Synth.*, 1968, **48**, 38
During low-temperature crystallisation, scratching to induce seeding of the solution must be discontinued as soon as the sensitive solid separates.
See other DIAZO COMPOUNDS

3010. 6-Aminopenicillanic acid *S*-oxide
[4888-97-5]
$C_8H_{12}N_2O_4S$

Preparative hazard
1. Micetich, R. G., *Synthesis*, 1976, **4**, 264
2. Noponen, A., *Chem. Eng. News*, 1977, **55**(8), 5
3. Micetich, R. G., *Chem. Brit.*, 1977, **13**, 163

According to a published procedure, the sulfur-containing acid is oxidised to the sulfoxide with hydrogen peroxide and precipitiated as the 4-toluenesulfonate salt in the presence of acetone [1]. During subsequent purification, trimeric acetone peroxide was precipitated and exploded violently after filtration [2]. The use of acetone in this and similar preparations involving hydrogen peroxide is not now recommended, and is highly dangerous [3].

See Hydrogen peroxide: Acetone, etc., *also* 3,3,6,6,9,9-Hexamethyl-1,2,4,5,7,8-hexoxonane

3011. Azoisobutyronitrile (AZDN)
[78-67-1] $C_8H_{12}N_4$

1. Lamy, P., *J. Calorim. Anal. Therm.*, 1980, **11**, 2-10-1 – 2-10-11
2. Grewer, T. *et al., Exothermic Decomposition*, Technical Report 01VD 159/0329 for Federal German Ministry for Res. Technol., Bonn. 1986
3. Grewer, T. *et al., Hazards from Pressure*, IChE Symp. Ser. No. 102, 1–9, Oxford, Pergamon, 1987
4. Kilattab, M. A. *et al., Bull. Fac. Sci. Alexandria Univ.*, 1995, **35**(2), 229; *Chem. Abs.*, 1996, **124**, 184211c
5. *CHETAH*, 1990, 187
6. Gladwell, P., *Loss Prev. Bull.*, 1998, 139, 3

AZDN is a much used polymerisation initiator. A mathematical model of thermal dissociation leading to explosion was verified experimentally with solutions of the nitrile in dibutyl phthalate and silicone oil [1]. It shows a very high rate of pressure increase during exothermic decomposition. Energy of decomposition (78–130°C) measured by DSC as 1.3 kJ/g [2], and further work on thermal decomposition under confinement has been done [3]. The recently calculated value of 56°C for the critical ignition temperature is in agreement with the previous value of 54°C and a more recent calorimetric study of runaway hazards [4]. It is shock sensitive [5]. A review of numerous fires and explosions caused by, or involving, AZDN has been published. Several appear to have started below the 40°C temperature now required by transport regulations. Refrigeration is desirable and confinement must be avoided. Solutions are even less stable than the solid and are not normally transported or stored [6].

See PRESSURE INCREASE IN EXOTHERMIC DECOMPOSITION
See entry CRITICAL IGNITION TEMPERATURE
See BLOWING AGENTS

Acetone
Carlisle, P. J., *Chem. Eng. News*, 1949, **27**, 150

During recrystallisation of technical material from acetone, explosive decomposition occurred. Non-explosive decomposition occurred when the nitrile was heated alone, or in presence of methanol.

Ammonium peroxodisulfate
See Ammonium peroxodisulfate: Azoisobutyronitrile
See other AZO COMPOUNDS, BLOWING AGENTS, CYANO COMPOUNDS

Heptane
Scheffold, R. *et al., Helv. Chim. Acta*, 1975, **58**, 60
During preparation of tetramethylsuccinodinitrile by thermal decomposition of 100 g of the azonitrile by slow warming in unstirred heptane, an explosion occurred. Successive addition of small portions of the nitrile to heptane at 90–92°C is a safer, preferred method.

3012. Diazido[(1,2,5,6-η)-1,5-Cyclooctadiene]platinum
[144673-73-4]

$C_8H_{12}N_6Pt$

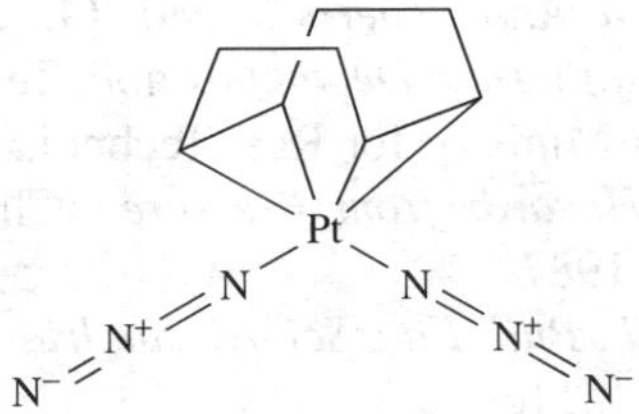

Kim, Y. *et al., Bull. Korean Chem. Soc.*, 1994, **15**(8), 690
Exploded during melting point determination.
See other AZIDES

3013. 1,3,4,6-Tetrakis(2-methyltetrazol-5-yl)-hexaaza-1,5-diene
[83195-98-6]

$C_8H_{12}N_{22}$

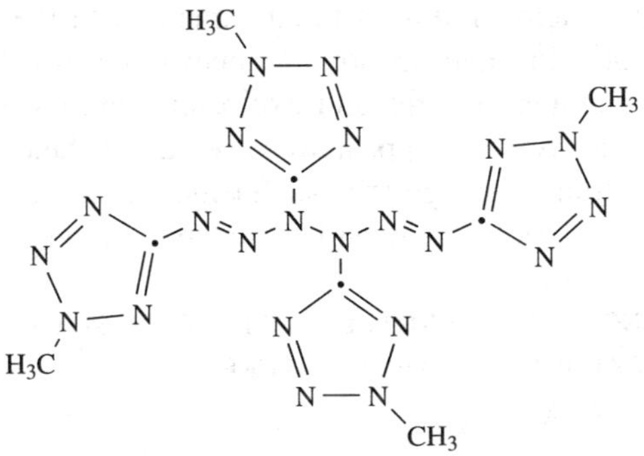

Butler, R. N. *et al., J. Chem. Res., Synop.*, 1982, 183

With 74% N content, it is explosive, sensitive to shock, friction or rapid heating to 121–125°C.

See other HIGH-NITROGEN COMPOUNDS, TETRAZOLES

3014. 3-Ethoxymethylene-2,4-pentanedione
[33884-41-2] $C_8H_{12}O_3$

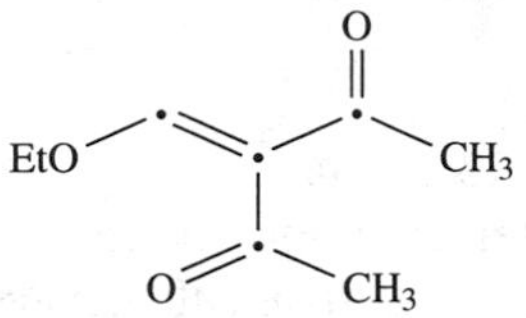

Riley, D. P. *et al., Inorg. Synth.*, 1978, **18**, 37

The residue from distillation of the title compound (prepared from ethyl orthoformate, pentanedione and acetic anhydride) at 165–170°C/13 mbar, is pyrophoric when hot.

3015. *O*–*O*-*tert*-Butyl hydrogen monoperoxymaleate
[1931-62-0] $C_8H_{12}O_5$

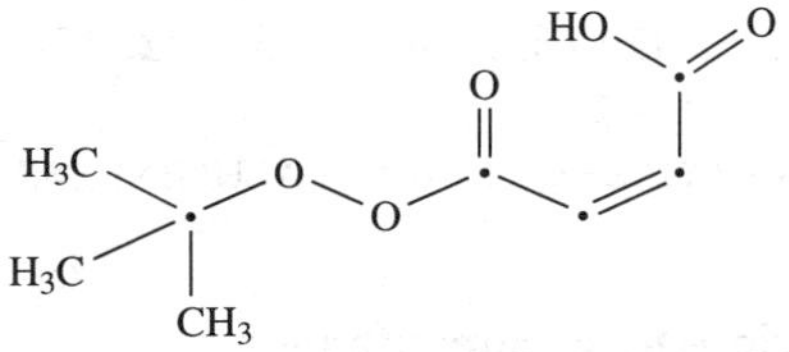

Castrantas, 1965, 17

Slightly shock-sensitive, the commercial dry 95% material is highly hazard-rated.

See other COMMERCIAL ORGANIC PEROXIDES, PEROXYESTERS

3016. Tetravinyllead
[866-87-5] $C_8H_{12}Pb$

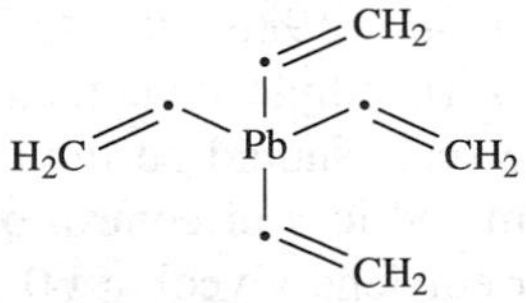

Holliday, A. K. *et al., Chem. & Ind.*, 1968, 1699

997

In the preparation from lead(II) chloride and vinylmagnesium bromide in THF–hexane, violent explosions occurred during isolation of the product by distillation of solvent. This could be avoided by a procedure involving steam distillation of the tetravinyllead, no significant loss of yield by hydrolysis being noted. It is likely to be considerably more endothermic than tetraethyllead.

Diborane
See Diborane: Tetravinyllead

Phosphorus trichloride
Houben-Weyl 1975, Vol. 13.3, 244
In absence of solvent, interaction may be explosive
See related ALKYLMETALS, ENDOTHERMIC COMPOUNDS

3017. 3-Buten-1-ynyldiethylaluminium
[ ] $C_8H_{13}Al$

Petrov, A. A. *et al., Zh. Obsch. Khim.*, 1962, **32**, 1349
It ignites in air.
See other METAL ACETYLIDES, TRIALKYLALUMINIUMS

3018. 2-Bromo-2,5,5-trimethylcyclopentanone
[69167-99-3] $C_8H_{13}BrO$

Potassium hydroxide, Triethylene glycol
Adam, W., *J. Amer. Chem. Soc.*, 1986, **108**, 4559
In the preparation of 2,5,5-trimethylcyclopenten-3-one by dehydrobromination of the title bromoketone, the scale should be limited to below 10 g of the latter, to avoid a vigorous exotherm and loss of control on addition of the bromoketone to potassium hydroxide in triethylene glycol at 60–70°C.
See related HALOALKANES

3019. 1-Phenylbiguanidinium hydrogen dichromate
[15760-45-9] $C_8H_{13}Cr_2N_5O_7$

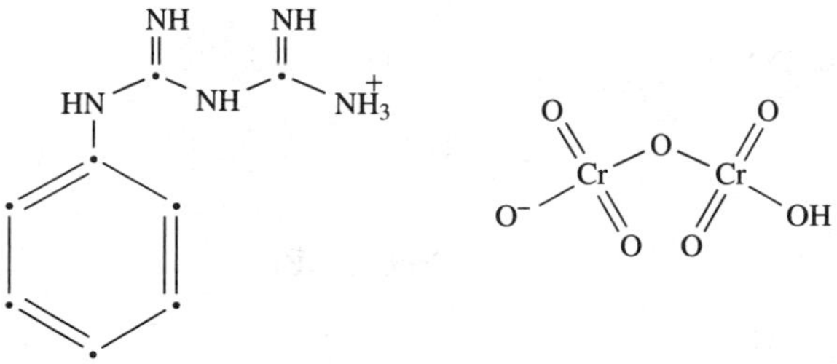

See entry DICHROMATE SALTS OF NITROGENOUS BASES
See other OXOSALTS OF NITROGENOUS BASES

3020. 1-Diethylamino-1-buten-3-yne
[1809-53-6] $C_8H_{13}N$

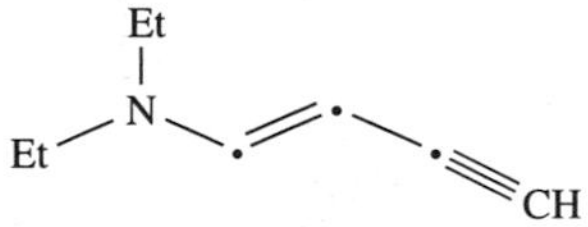

Brandsma, 1971, 163
Dilution of the amine with white mineral oil is recommended before vacuum distillation, to avoid explosive decomposition of the residue.
See other ACETYLENIC COMPOUNDS

†3021. 1,7-Octadiene
[3710-30-3] C_8H_{14}

$$H_2C{=}CH[CH_2]_4CH{=}CH_2$$

See other DIENES

†3022. 1-Octyne
[629-05-0] C_8H_{14}

$$HC{\equiv}C[CH_2]_5Me$$

See other ALKYNES

†3023. 2-Octyne
[2809-67-8] C_8H_{14}

$$MeC{\equiv}C[CH_2]_4Me$$

See other ALKYNES

†3024. 3-Octyne
[15232-76-5] C_8H_{14}

$$EtC{\equiv}CBu$$

See other ALKYNES

†3025. 4-Octyne
[1942-45-6] C_8H_{14}

$$PrC{\equiv}CPr$$

See other ALKYNES

†3026. Vinylcyclohexane
[695-12-5] C_8H_{14}

See related ALKENES

3027. Di-2-butenylcadmium
[7544-40-3] $C_8H_{14}Cd$

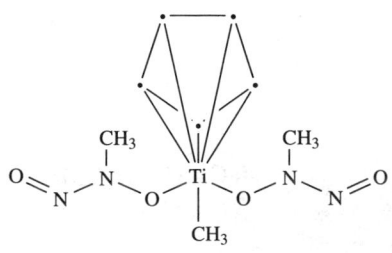

Houben-Weyl, 1973, Vol. 13.2a, 881
When rapidly warmed from −5°C to ambient temperature, it decomposes explosively.
See related ALKYLMETALS, ALLYL COMPOUNDS

3028. μ-Cyclopentadienyl(methyl)-bis-(*N*-methyl-*N*-nitrosohydroxylamino)titanium
[58659-15-7] $C_8H_{14}N_4O_4Ti$

Clark, R. J. H. *et al., J. Chem. Soc., Dalton Trans.*, 1974, 122

The complex was stable under nitrogen or vacuum, but exploded quite readily and decomposed in air.

See other NITROSO COMPOUNDS, N–O COMPOUNDS, ORGANOMETALLICS

3029. 4-Hydroxy-4-methyl-1,6-heptadiene
[25021-40-5] $C_8H_{14}O$

Ozone
1. Rabinowitz, J. L., *Biochem. Prep.*, 1958, **6**, 25
2. Miller, F. W., *Chem. Eng. News*, 1973, **51**(6), 29
Following a published procedure [1], but using 75 g of diene instead of the 12.6 g specified, the ozonisation product exploded under desiccation. Eight previous preparations on the specified scale had been uneventful.
See OZONIDES

3030. Poly(vinyl butyral)
[63148-65-2] $(C_8H_{14}O_2)_n$

Korshova, I. T. *et al.*, *Chem. Abs.*, 1983, **99**, 195868
Ignition and autoignition temperatures, lower explosion limit and oxygen index were determined for the powdered butyral in relation to dispersibilty and viscosity.

3031. Diethyl succinate
[123-25-1] $C_8H_{14}O_4$

HCS 1980, 405

Ethyl trifluoroacetate, Sodium hydride
See Sodium hydride: Diethyl succinate, etc.

3032. Diisobutyryl peroxide
[3437-84-1] $C_8H_{14}O_4$

1. Kharasch, S. *et al., J. Amer. Chem. Soc.*, 1941, **63**, 526
2. *MCA Case History No. 579*

A sample of the peroxide in ether, prepared according to a published procedure [1], was being evaporated to dryness with a stream of air when it exploded violently. Handling the peroxide as a dilute solution at low temperature is recommended [2].
See other DIACYL PEROXIDES

3033. Acetyl cyclohexanesulfonyl peroxide
[3179-56-4] $C_8H_{14}O_5S$

1. 'Code of Practice for Storage of Organic Peroxides', GC14, Widnes, Interox
   Chemicals Ltd, 1970
2. *CHETAH*, 1990, 182

While the commercial material damped with 30% water is not shock- or friction-sensitive, if it dries out it may become a high-hazard material. As a typical low-melting diacyl peroxide, it may be expected to decompose vigorously or explosively on slight heating or on mechanical initiation [1]. It is shock sensitive (presumably when dry) [2].
See other COMMERCIAL ORGANIC PEROXIDES, DIACYL PEROXIDES

3034. Diisopropyl peroxydicarbonate
[105-64-6] $C_8H_{14}O_6$
$$(Me_2CHOCO.O-)_2$$

1. Strong, W. A., *Ind. Eng. Chem.*, 1964, **56**(12), 33
2. 'Bulletin, T. S. 350', Pittsburgh, Pittsburgh Plate Glass Co., 1963

When warmed slightly above its m.p., 10°C, the ester undergoes slow but self-accelerating decomposition, which may become dangerously violent under confinement. Bulk solutions of the peroxyester (45%) in benzene–cyclohexane stored at 5°C developed sufficient heat to decompose explosively after 1 day, and 50–90% solutions were found to be impact-sensitive [1]. The solid is normally stored and transported at below −18°C in loose-topped trays [2].

Amines, or Potassium iodide

Strain, F., *J. Amer. Chem. Soc.*, 1950, **72**, 1254

At 20–30°C, decomposition occurred in 10–30 min. Addition of 1% of aniline, 1,2-diaminoethane or potassium iodide caused instant decomposition, and of dimethylaniline, instant explosion.

See other COMMERCIAL ORGANIC PEROXIDES, DIACYL PEROXIDES, PEROXYCARBONATE ESTERS

3035. Dipropyl peroxydicarbonate
[16066-38-9] $C_8H_{14}O_6$

$$(PrOCO.O-)_2$$

Barter, J. A., US Pat. 3 775 341, 1973

Methylcyclohexane is a suitable solvent to reduce the hazardous properties of the peroxyester.

See other DIACYL PEROXIDES, PEROXYCARBONATE ESTERS

3036. 4-Hydroxy-4-methyl-1,6-heptadiene diozonide
[ ] $C_8H_{14}O_7$

See 4-Hydroxy-4-methyl-1,6-heptadiene: Ozone
See other OZONIDES

3037. Bis(2-methoxyethyl) peroxydicarbonate
[22575-95-7] $C_8H_{14}O_8$

It exploded at 34°C.
See entry PEROXYCARBONATE ESTERS

3038. μ-Cyclopentadienyltrimethyltitanium
[38386-55-9] **$C_8H_{14}Ti$**

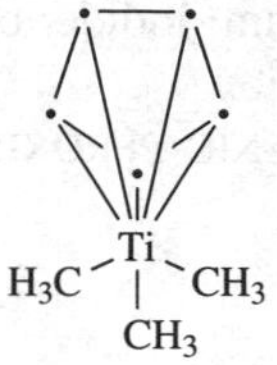

Bailar, 1973, Vol. 3, 395
Pyrophoric at ambient temperature.
See other ALKYLMETALS, ORGANOMETALLICS

†3039. Dimethylcyclohexanes (5 isomers)
[27195-67-1] **C_8H_{16}**

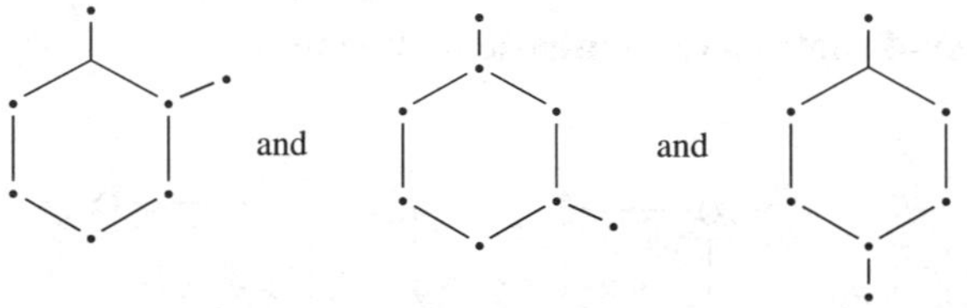

†3040. 1-Octene
[111-66-0] **C_8H_{16}**

$H_2C{=}CHC_6H_{13}$

See other ALKENES

†3041. 2-Octene
[111-67-1] **C_8H_{16}**

$MeCH{=}CH[CH_2]_4Me$

See other ALKENES

†3042. 2,3,4-Trimethyl-1-pentene
[565-76-4] **C_8H_{16}**

$H_2C{=}CMeCHMeCHMe_2$

See other ALKENES

†3043. 2,4,4-Trimethyl-1-pentene
[107-39-1] C_8H_{16}

$$H_2C{=}CMeCH_2CMe_3$$

HCS 1980, 414

See other ALKENES

†3044. 2,3,4-Trimethyl-2-pentene
[565-77-5] C_8H_{16}

$$Me_2C{=}CMeCHMe_2$$

See other ALKENES

†3045. 2,4,4-Trimethyl-2-pentene
[107-40-4] C_8H_{16}

$$Me_2C{=}CHCMe_3$$

See other ALKENES

†3046. 3,4,4-Trimethyl-2-pentene
[598-96-9] C_8H_{16}

$$MeCH{=}CMeCMe_3$$

See other ALKENES

3047. 4,4′-Dithiodimorpholine
[103-34-4] $C_8H_{16}N_2O_2S_2$

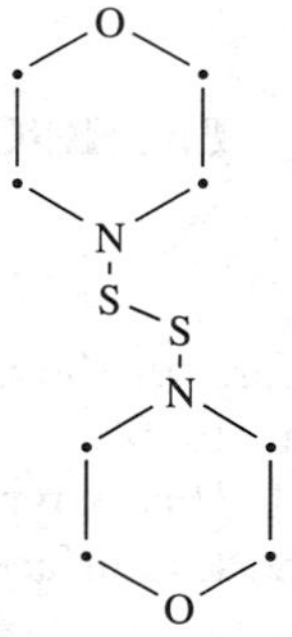

Chlorine
See 4-Morpholinesulfenyl chloride

3048. 3-Azoniabicyclo[3.2.2]nonane nitrate
[10308-93-7] $C_8H_{16}N_2O_3$

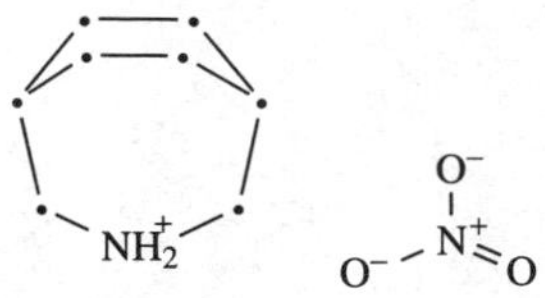

Violent decomposition at 258°C.
See entry DIFFERENTIAL THERMAL ANALYSIS
See other OXOSALTS OF NITROGENOUS BASES

3049. 1,12-Diazido-3,10-dioxa-5,8-dinitrazadodecane
[88487-88-7] $C_8H_{16}N_{10}O_6$

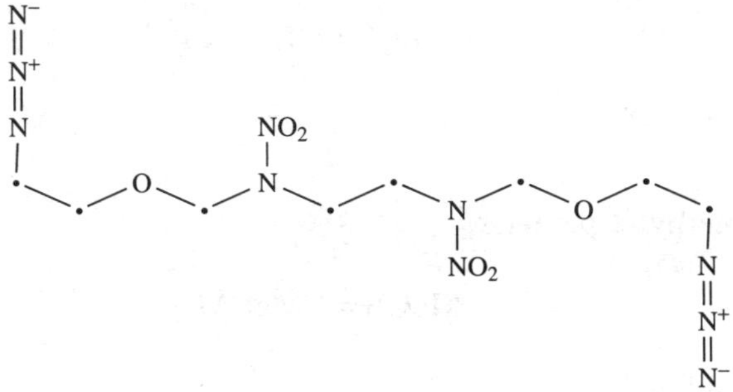

See entry ENERGETIC COMPOUNDS
See other N-NITRO COMPOUNDS, ORGANIC AZIDES

†3050. 2-Ethylhexanal
[123-05-7] $C_8H_{16}O$

BuCHEtCO.H

HCS 1980, 489

1. Steele, A. B. *et al., Chem. Engrg.*, 1969, **66**, 160
2. Urben, P. G., private comm., 1989
The aldehyde ignited in air [1]. This is typical of mid-range aldehydes if sorbed on paper or cloth which increases surface exposure, ignition occurring within 2 hours [2].
See α-Pentylcinnamaldehyde
See other ALDEHYDES, PEROXIDISABLE COMPOUNDS

3051. *O–O-tert*-**Butyl isopropyl monoperoxycarbonate**
[2372-21-6] $C_8H_{16}O_4$

$$Me_3COOCO.OCHMe_2$$

See entry THERMAL EXPLOSIONS *See other* PEROXYCARBONATE ESTERS

3052. 3,6-Diethyl-3,6-dimethyl-1,2,4,5-tetraoxane
[33817-92-4] $C_8H_{16}O_4$

1. Castrantas, 1965, 18
2. Swern, 1970, Vol. 1, 37

By analogy, this dimeric 2-butanone peroxide and the corresponding trimer probably contribute largely to the high shock-sensitivity of the commercial 'MEK peroxide' mixture, which contains these and other peroxides [1,2].
See other CYCLIC PEROXIDES, KETONE PEROXIDES

3053. Octylsodium
[2875-36-7] $C_8H_{17}Na$

Houben-Weyl, 1970, Vol. 13.1, 384
The powder ignites in air.
See other ALKYLMETALS

†3054. 2,3-Dimethylhexane
[584-94-1] C_8H_{18}

$$Me_2CHCHMePr$$

†3055. 2,4-Dimethylhexane
[589-43-5] C_8H_{18}

$$Me_2CHCH_2CHMeEt$$

†3056. 3-Ethyl-2-methylpentane
[609-26-7] C_8H_{18}

$$Me_2CHCHEt_2$$

†3057. 2-Methylheptane
[529-27-8] C_8H_{18}

$$Me_2CHCH_2Bu$$

†3058. 3-Methylheptane
[589-81-1] C_8H_{18}

$$EtCHMeBu$$

†3059. Octane
[111-65-9] C_8H_{18}

$$Me[CH_2]_6Me$$

HCS 1980, 696

†3060. 2,2,3-Trimethylpentane
[564-02-3] C_8H_{18}

$$Me_3CCHMeEt$$

†3061. 2,2,4-Trimethylpentane
[540-84-1] C_8H_{18}

$$Me_3CCH_2CHMe_2$$

HCS 1980, 696

†3062. 2,3,3-Trimethylpentane
[560-21-4] C_8H_{18}

$$Me_2CHCMe_2Et$$

†3063. 2,3,4-Trimethylpentane
[565-75-3] C_8H_{18}

$$Me_2CHCHMeCHMe_2$$

3064. Diisobutylaluminium chloride
[1779-25-5] $C_8H_{18}AlCl$

$$(Me_2CHCH_2)_2AlCl$$

HCS 1980, 411

See entry ALKYLALUMINIUM HALIDES

3065. Chlorodibutylborane
[1730-69-4]

$$ClB(Bu)_2$$

$C_8H_{18}BCl$

491M, 1975, 146

It ignites in air.
See other ALKYLHALOBORANES

3066. Di-*tert*-butyl chromate
[1189-85-1]

$$(Me_3CO)_2CrO_2$$

$C_8H_{18}CrO_4$

1. Anon., *ABCM Quart. Safety Summ.*, 1953, **24**, 2
2. Koenst, W. M. B. *et al., Synth. Comm.*, 1980, **10**, 905–909
Large-scale preparation by addition of *tert*-butanol to chromium trioxide in a full
unstirred flask with poor cooling detonated owing to local overheating. Effective
cooling and stirring are essential [1]. It may safely be prepared by addition of a
40% aqueous solution of chromium trioxide to the alcohol [2].

Valencene
1. Hunter, G. L. K. *et al., J. Food Sci.*, 1965, **30**, 876
2. Wilson, C. W. *et al., J. Agric. Food Chem.*, 1978, **26**, 1430–1432
Oxidation of the sesquiterpene valencene to the 4-en-3-one proceeds explosively
when at sub-ambient temperatures [1]. A safer alternative oxidation procedure is
detailed [2].
See other OXIDANTS

3067. Di-*tert*-butylfluorophosphine
[29146-24-5]

$$(Me_3C)_2PF$$

$C_8H_{18}FP$

Stelzer, O. *et al., Inorg. Synth.*, 1978, **18**, 176
It may ignite in air.
See other ALKYLHALOPHOSPHINES

3068. Dibutylmagnesium
[1191-47-5]

$$Bu_2Mg$$

$C_8H_{18}Mg$

491M, 1975, 147

It ignites in air.
See other ALKYLMETALS

3069. Dibutyl hyponitrite
[86886-17-1] $C_8H_{18}N_2O_2$

$$BuON=NOBu$$

See entry DIALKYL HYPONITRITES

3070. *trans*-Di-*tert*-butyl hyponitrite
[14976-54-6] $C_8H_{18}N_2O_2$

$$Me_3CON=NOCMe_3$$

Mendenhall, G. D. *et al., J. Amer. Chem. Soc.*, 1982, **104**, 5113
It is unstable and shock-sensitive, and needs careful handling.

Formamides
Protasiewicz, J. *et al., J. Org. Chem.*, 1985, **50**, 3222
An equimolar mixture with *N-tert*-butylformamide underwent a sudden exotherm after placing in a bath at 60°C and the tube contents were ejected vigorously as a small cloud. A 1:2 mixture of the ester with formamide emitted a loud pop on heating in a bath at 70°C.
See other DIALKYL HYPONITRITES

†3071. Dibutyl ether
[142-96-1] $C_8H_{18}O$

$$BuOBu$$

HCS 1980, 369

Dasler, W. *et al., Ind. Eng. Chem. (Anal. Ed.)*, 1946, **18**, 52
Peroxides formed in storage are removed effectively by percolation through alumina.
See ETHERS, PEROXIDES IN SOLVENTS *See other* PEROXIDISABLE COMPOUNDS

3072. 1-(1,1-Dimethylethoxy)-2-methylpropane (*tert*-Butyl isobutyl ether)
[33021-02-2] $C_8H_{18}O$

$$Me_3COCH_2CHMe_2$$

Bowen, R. D. *et al., J. Chem. Soc. Perk. 2*, 1990, 147
Despite testing negative for peroxides, a sample exploded during distillation. Precautions are recommended.
See other ETHERS, PEROXIDISABLE COMPOUNDS

3073. 3,5-Dimethyl-3-hexanol
[4209-91-0] $C_8H_{18}O$

Hydrogen peroxide, Sulfuric acid
See Hydrogen peroxide: 3,5-Dimethyl-3-hexanol, etc.

†3074. Di-*tert*-butyl peroxide
[110-05-4] $C_8H_{18}O_2$

Me₃COOCMe₃

$Me_3COOCMe_3$

HCS 1980, 370

Preparative hazard
1. Griffiths, J. F. *et al., Combust. Flame*, 1984, **56**, 135–148
2. Stull, 1977, 21
3. Sime, R. J., *Chem. Eng. News*, 1988, **66**(24), 4
Thermal decomposition is exothermic and self-ignition may result, especially if oxygen is present [1]. Though the heat of exothermic decomposition (1.32 kJ/g) is not exceptionally high, the weakness of the peroxide link tends to ready decomposition, and the energy release, coupled with the high volume of gaseous products (7.6 mol) at the adiabatic maximum of 550°C would give a 21-fold pressure increase in a closed vessel [2]. A routine student experiment on the gas-phase decomposition of the peroxide (0.2 ml) by injection into a 250 ml flask immersed in an oil bath at 155°C. The procedure was identical to a published method except that a mercury manometer was used to measure the pressure in place of a Bourdon gauge, and the procedure had been used on 30 occasions without incident. During a run at 165°C a violent explosion occurred [3].
See Hydrogen peroxide: *tert*-Butanol, etc.
See entry THERMAL EXPLOSIONS
See other DIALKYL PEROXIDES, SELF-HEATING AND IGNITION INCIDENTS

3075. 3,5-Dimethyl-3-hexyl hydroperoxide
[ ] $C_8H_{18}O_2$

Preparative hazard
See Hydrogen peroxide: 3,5-Dimethyl-3-hexanol, etc.
See other ALKYL HYDROPEROXIDES

†3076. Bis(2-ethoxyethyl) ether
[112-36-7]
$C_8H_{18}O_3$

$$(EtOC_2H_4)_2O$$

HCS 1980, 398

See other PEROXIDISABLE COMPOUNDS

†3077. 2-(2-Butoxyethoxy)ethanol (Diethyleneglycol monobutyl ether)
[112-34-5]
$C_8H_{18}O_3$

$$Bu(OC_2H_4)_2OH$$

HCS 1980, 231

3078. Bis(2-hydroperoxy-2-butyl) peroxide (Dioxybis(1-methylpropyl hydroperoxide))
[126-76-1]
$C_8H_{18}O_6$

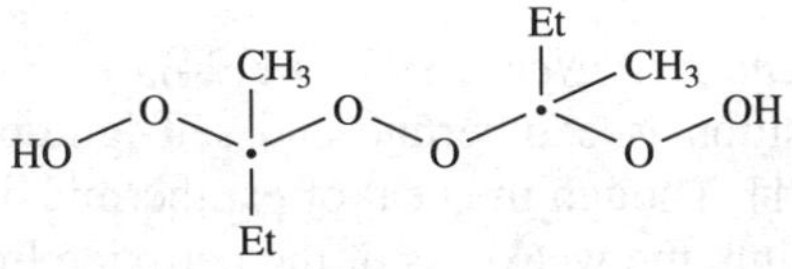

Leleu, *Cahiers*, 1973, (71), 238
The triperoxide, the main constituent of 'MEK peroxide', is explosive in the pure state, but insensitive to shock as the commercial 50% solution in dimethyl phthalate. The solution will explode at about 85°C, and slowly liberates oxygen at ambient temperature.
See other COMMERCIAL ORGANIC PEROXIDES, KETONE PEROXIDES
See related ALKYL HYDROPEROXIDES

3079. Diammonium Aquabis(peroxotartratovanadate)(2-)
[   ]
$C_8H_{18}N_2O_{19}V_2$

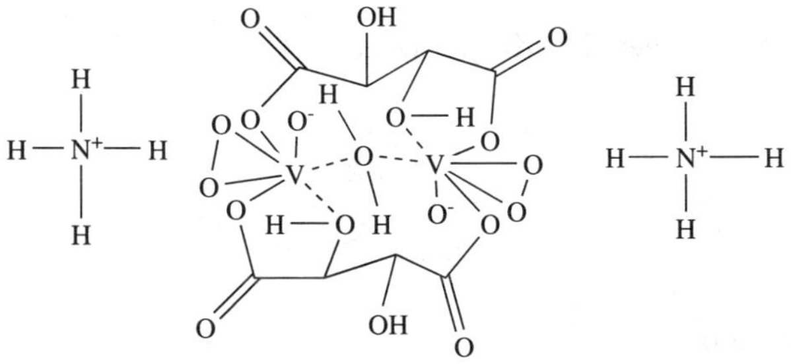

Schwendt, P, *et al.*, *Polyhedron*, 1998, **17**(13-14), 2161

This compound, which exists as a pentahydrate and for which a considerably weirder structure than shown has been inferred from X-ray crystallography on a homologue, may decompose explosively at room temperature. Handling more than 20 mg is discountenanced.

See related PEROXOMOLYBDATES AND TUNGSTATES, PEROXOCHROMIUM COMPOUNDS

3080. Dibutylzinc
[1119-90-0] $C_8H_{18}Zn$

$$Bu_2Zn$$

Leleu, *Cahiers*, 1977, (88), 371
It fumes in air, and may ignite under warm conditions.
See other DIALKYLZINCS, ALKYLMETALS

3081. Diisobutylzinc
[1854-19-9] $C_8H_{18}Zn$

$$(Me_2CHCH_2)_2Zn$$

Gibson, 1969, 180
It ignites in air.
See other DIALKYLZINCS, ALKYLMETALS

3082. Diisobutylaluminium hydride
[1191-15-7] $C_8H_{19}Al$

$$(Me_2CHCH_2)_2AlH$$

1. Mirviss, S. B. *et al., Ind. Eng. Chem.*, 1961, **53**(1), 54A
2. 'Specialty Reducing Agents', Brochure TA-2002/1, New York, Texas Alkyls, 1971

The higher thermal stability of dialkylaluminium hydrides over the corresponding trialkylaluminiums is particularly marked in this case with 2 branched alkyl groups [1]. Used industrially as a powerful reducant, it is supplied as a solution in hydrocarbon solvents. The undiluted material ignites in air unless diluted to below 25% concentration [2].

Tetrahydrofuran
Wissink, H. G., *Chem. Eng. News*, 1997, **75**(9), 6
The reagent can react with this solvent to give gaseous byproducts, capable of bursting glass containers. Initial complex formation is exothermic, so good cooling is needed to prevent decomposition when solutions are prepared, they should not be stored.
See other ALKYLALUMINIUM ALKOXIDES AND HYDRIDES, REDUCANTS

3083. Dibutylamine
[111-92-2]   C$_8$H$_{19}$N

$$Bu_2NH$$

HCS 1980, 367

Cellulose nitrate   MRH 4.81/99+
    See CELLULOSE NITRATE: amines

Other reactants
    Yoshida, 1980, 163
    MRH values calculated for 14 combinations with oxidants are given.
    See other ORGANIC BASES

†3084. Di-2-butylamine
[626-23-3]   C$_8$H$_{19}$N

$$(EtMeCH)_2NH$$

See other ORGANIC BASES

†3085. Diisobutylamine
[110-96-3]   C$_8$H$_{19}$N

$$(Me_2CHCH_2)_2NH$$

See other ORGANIC BASES

3086. Dibutyl hydrogen phosphite
[1809-19-4]   C$_8$H$_{19}$O$_3$P

$$(BuO)_2P(O)H$$

Morrell, S. H., private comm., 1968
The phosphite was being distilled under reduced pressure. At the end of distillation
the air-bleed was opened more fully, when spontaneous combustion occurred inside
the flask, probably of phosphine formed by thermal decomposition.
See other PHOSPHORUS ESTERS

3087. Tetraethyldiarsane
[612-08-8]   C$_8$H$_{20}$As$_2$

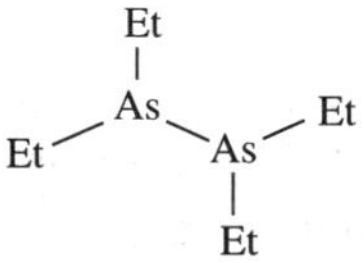

Sidgwick, 1950, 770

It ignites in air.
See other ALKYLNON-METALS

3088. Tetraethylammonium perchlorate
[2567-83-1] $\qquad$ $C_8H_{20}ClNO_4$

Udupa, M. R., *Propellants, Explos. Pyrotech.*, 1982, **7**, 155–157
It explodes at 298°C after an energetic phase transformation (10.4 kJ/mol)at 98°C.
See other QUATERNARY OXIDANTS, PERCHLORATE SALTS OF NITROGENOUS BASES

3089. 2,2′-Azobis(2-amidiniopropane) chloride
[2997-92-4] $\qquad$ $C_8H_{20}Cl_2N_6$

Sodium peroxodisulfate
See 2,2′-Azobis(2-amidiniopropane) peroxodisulfate
See other AZO COMPOUNDS, POLYMERISATION INCIDENTS

3090. Tetraethylammonium periodate
[5492-69-3] $\qquad$ $C_8H_{20}INO_4$

Preparative hazard
See Periodic acid: Tetraethylammonium hydroxide
See also TETRAMETHYLAMMONIUM PERIODATE
See other QUATERNARY OXIDANTS, OXOSALTS OF NITROGENOUS BASES

3091. 2,2′-Azobis(2-amidiniopropane) peroxodisulfate

[ ] $C_8H_{20}N_6O_8S_2$

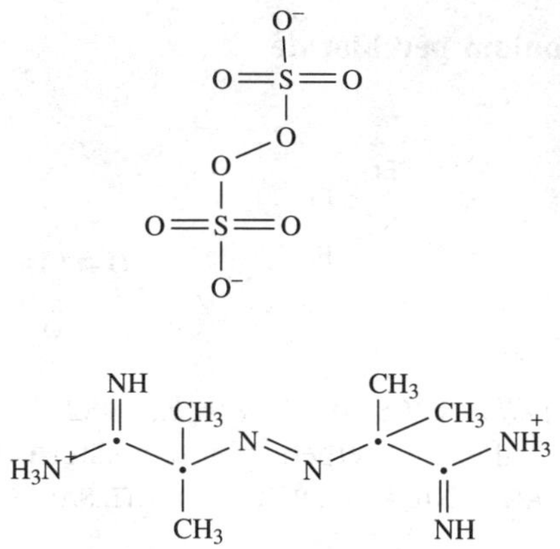

Rausch, D. A., *Chem. Eng. News*, 1988, **66**(5), 2

When aqueous solutions of the polymerisation initiators 2,2′-azobis(2-amidinio-propane) chloride and sodium peroxodisulfate are mixed, the title compound separates as a water insoluble shock-sensitive salt. The shock-sensitivity increases as the moisture level decreases, and is comparable with that of lead azide. Stringent measures should be used to prevent contact of the solutions outside the polymerisation environment. (The instability derives from the high nitrogen (21.4%) and oxygen (31.6%) contents, and substantial oxygen balance, as well as the structural factors present in the salt.)

See other AZO COMPOUNDS, HIGH-NITROGEN COMPOUNDS, OXOSALTS OF NITROGE-NOUS BASES

3092. 1,3,6,8-Tetraazatricyclo[6.2.1.1^{3,6}]dodecane tetranitrate

[10308-95-9] $C_8H_{20}N_8O_{12}$

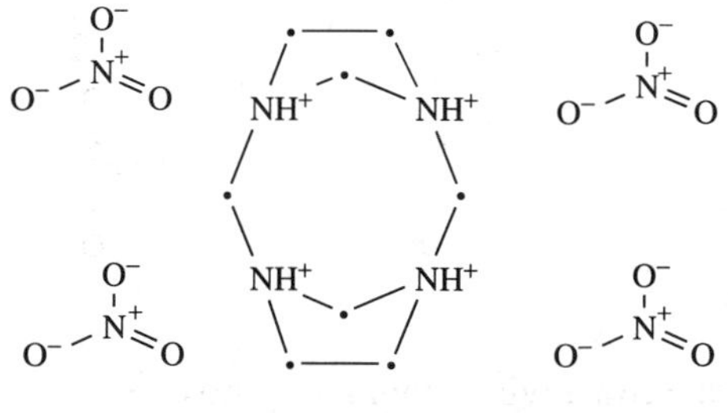

Violent decomposition occurred at 260°C.

See entry DIFFERENTIAL THERMAL ANALYSIS (DTA)

See other OXOSALTS OF NITROGENOUS BASES

1016

3093. Bis(trimethylphosphine)nickel(0)–acetylene complex
[98316-02-0] $C_8H_{20}NiP_2$

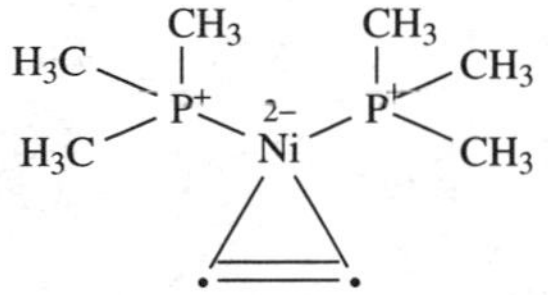

$$H_3C \diagdown \underset{H_3C \diagup}{\overset{CH_3}{\underset{|}{P^+}}} \overset{2-}{\underset{Ni}{}} \underset{}{\overset{CH_3}{\underset{|}{P^+}}} \diagdown \overset{CH_3}{\underset{CH_3}{}}$$

Pörschke, K. R. *et al., Z. Naturforsch. B*, 1985, **40B**, 199–209
The complex with acetylene (1 mol) decomposed explosively at 0°C.
See related COMPLEX ACETYLIDES, ALKYLPHOSPHINES

3094. Ethoxytriethyldiphosphinyl oxide
[ ] $C_8H_{20}O_2P_2$

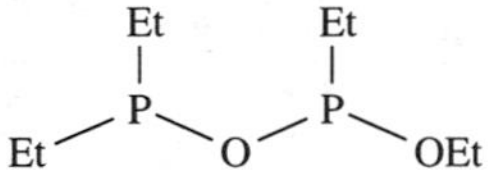

$$\underset{Et}{\overset{Et}{\underset{|}{P}}} \diagdown \underset{O}{} \diagup \underset{}{\overset{Et}{\underset{|}{P}}} \diagdown OEt$$

491M, 1975, 167

It ignites in air.
See related ALKYLPHOSPHINES

3095. Tetraethyllead
[78-00-2] $C_8H_{20}Pb$

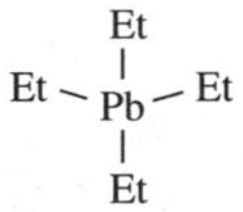

$$Et \diagdown \underset{Et}{\overset{Et}{\underset{|}{Pb}}} \diagup Et$$

HCS 1980, 890

Doyle, F. P., *Loss Prev.*, 1969, **3**, 16
It is mildly endothermic (ΔH_f° (l) +217.5 kJ/mol, 0.56 kJ/g).
    Failure to cover the residue with water after emptying a tank of the compound caused explosive decomposition after several days.

Other reactants
    Yoshida, 1980, 234
    MRH values calculated for 12 combinations with oxidants are given.
See other ALKYLMETALS, ENDOTHERMIC COMPOUNDS

3096. Tetrakis(ethylthio)uranium
[ ] $C_8H_{20}S_4U$

Bailar, 1973, Vol. 5, 416
It ignites in air.
See related METAL ALKOXIDES

3097. Tetraethyltin
[597-64-8] $C_8H_{20}Sn$

Sorbe, 1968, 160
It tends to ignite in air.
See other ALKYLMETALS

3098. 'Tetraethyldiborane'
[12081-54-8] $C_8H_{22}B_2$

Köster, R. *et al., Inorg. Synth.*, 1974, **15**, 141–146
The material (actually an equilibrium mixture of diborane and highly ethylated homologues) ignites in air. The propyl analogue behaves similarly.
See other ALKYLBORANES

3099. Bis(trimethylsilylmethyl)magnesium
[51329-17-0] $C_8H_{22}MgSi_2$

$$H_3C-\underset{H_3C}{\overset{CH_3}{\underset{|}{Si}}}-Mg-\underset{CH_3}{\overset{CH_3}{\underset{|}{Si}}}-CH_3$$

Oxygen, or Water
Andersen, R. A. *et al.*, *Inorg. Synth.*, 1979, **19**, 264
It is sensitive to oxygen and reacts violently with water.
See related ALKYLMETALS, ALKYLSILANES

3100. 3,6,9-Triaza-11-aminoundecanol ('Hydroxyethyltriethylenetetramine')
[38361-85-2] $C_8H_{22}N_4O$

$$HO\diagdown\diagup NH\diagdown\diagup NH\diagdown\diagup NH\diagdown\diagup NH_2$$

HCS 1980, 889

Cellulose nitrate
See CELLULOSE NITRATE: amines *See other* ORGANIC BASES

3101. 1,11-Diamino-3,6,9-triazaundecane ('Tetraethylenepentamine')
[112-57-2] $C_8H_{23}N_5$

$$H_2N\diagdown\diagup NH\diagdown\diagup NH\diagdown\diagup NH\diagdown\diagup NH_2$$

Carbon tetrachloride
See Carbon tetrachloride: 1,11-Diamino-3,6,9-triazaundecane

Cellulose nitrate
See CELLULOSE NITRATE: amines
See other ORGANIC BASES

3102. Bis(tetramethyldiphosphane disulfide)cadmium perchlorate
[71231-59-9] $C_8H_{24}CdCl_2O_8P_4S_4$
$$[(Me_2P(S)P(S)Me_2)_2Cd]\,[ClO_4]_2$$

McQuillan, G. P. *et al.*, *J. Chem. Soc., Dalton Trans.*, 1979, 899

This and the nickel analogue are shock-sensitive and violently explosive. Other metal complexes (except cobalt) decompose slowly liberating the dimethyldiphosphane which tends to ignite when old sample bottles are opened.
See related AMMINEMETAL OXOSALTS

3103. Tetramethylammonium pentaperoxodichromate

[ ] $\qquad$ $C_8H_{24}Cr_2N_2O_{12}$

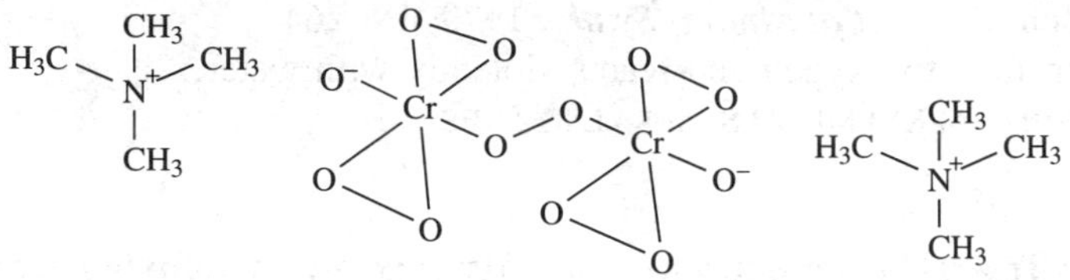

Mellor, 1943, Vol. 11, 358
The structure given is a 1960's vintage re-interpretation of the salt of CrO_5^- given by Mellor.
See Tetramethylammonium monoperchromate

3104. Tetrakis(dimethylamino)titanium

[ ] $\qquad$ $C_8H_{24}N_4Ti$

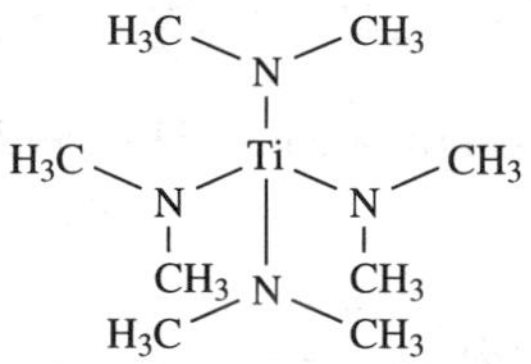

Hydrazine
See Hydrazine: Titanium compounds
See other N-METAL DERIVATIVES

3105. 1,2-Diaminoethanebis(trimethylgold)

[ ] $\qquad$ $C_8H_{26}Au_2N_2$

$$(Me_3Au:NH_2CH_2-)_2$$

Alone, or Nitric acid
Gilman, H. *et al., J. Amer. Chem. Soc.*, 1948, **70**, 550
The solid complex is very sensitive to light and explodes violently on heating in an open crucible. A drop of conc. nitric acid added to the dry compound causes explosion.
See related ALKYLMETALS *See other* GOLD COMPOUNDS

1020

3106. Bis(diethylenetriamine)cobalt(III) perchlorate
[28872-17-5] $C_8H_{26}Cl_3CoN_6O_{12}$

$$[(C_4H_{13}N_3)_2Co]\ [ClO_4]_3$$

It explodes at 325°C; high impact-sensitivity.
See entry AMMINEMETAL OXOSALTS

3107. Potassium graphite
[12081-88-8] C_8K
(Complex Structure)

Water
    Bailar, 1973, Vol. 1, 1276
    It may explode with water.
See Carbon: Alkali metals
See other METAL NON-METALLIDES

3108. Oligo(octacarbondioxide) (Cyclo-*oligo*-butadiynyl-2,3-dioxocyclo-butenylidene)
[123002-86-8] $n = 3$; [123002-87-9] $n = 4$; [ ] $n = 5$ $(C_8O_2)_n$

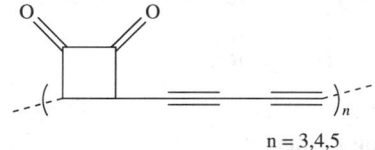

n = 3,4,5

Diederich, F. *et al., J. Amer. Chem. Soc.*, 1991, **113**(2), 495
These compounds exploded violently on heating above 85°C in an attempt to obtain melting points.
See 11,12-Diethynyl-1,4,7,10-tetraoxadispiro[4.0.4.2]dodec-11-ene
See other ACETYLENIC COMPOUNDS

3109. Lithium octacarbonyltrinickelate
[81181-41-5] (ion) $C_8Li_2Ni_3O_8$

$$Li_2[(OC)_8Ni_3]$$

Bailar, 1973, Vol. 3, 1117
A pyrophoric salt.
See related CARBONYLMETALS

3110. Rubidium graphite
[12193-29-2] C_8Rb
(Complex Structure)

See Carbon: Alkali metals
See other METAL NON-METALLIDES

3111. Nonacarbonyldiiron
[20982-74-5]   C₉Fe₂O₉

$$(OC)_3Fe\equiv(CO)_3\equiv Fe(CO)_3$$

Anon., *ABCM Quart. Safety Summ.*, 1943, **14**, 18
Commercial iron carbonyl (fl.p.35°C) has an autoignition temperature in contact
with brass of 93°C, lower than that of carbon disulfide.
See other CARBONYLMETALS

3112. 2-Nonen-4,6,8-triyn-1-al
[ ]   C₉H₄O

Bohlman, F. *et al., Chem. Ber.*, 1963, **96**, 2586
Extremely unstable, explodes after a few minutes at ambient temperature and
ignites at 110°C.
See other ACETYLENIC COMPOUNDS

3113. 2-Nitrophenylpropiolic acid
[530-85-8]   C₉H₅NO₄

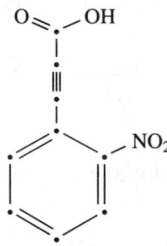

Sorbe, 1968, 152
It explodes above 150°C.
See other ACETYLENIC COMPOUNDS, NITROARYL COMPOUNDS, ORGANIC ACIDS

3114. Sodium 3-phenylisoxazolin-4,5-dione-4-oximate
[70247-49-3] $C_9H_5N_3NNaO_3$

Wentrup, C. *et al., Angew. Chem. (Intern. Ed.)*, 1979, **18**, 467
It explodes on rapid heating, like the 5-methyl analogue and the silver salt.
See Fulminic acid(reference 2)
See other N–O COMPOUNDS
See related OXIMES

3115. 3,5-Dimethyl-4-[*I*,*I*-bis(trifluoroacetoxy)iodo]isoxazole
[57508-69-7] $C_9H_6F_6INO_5$

Preparative hazard
See Peroxytrifluoroacetic acid: 4-Iodo-3,5-dimethylisoxazole
See [*I*,*I*-Bis(trifluoroacetoxy)iodo]benzene
See other IODINE COMPOUNDS

3116. 1-Diazoindene
[35847-40-6] $C_9H_6N_2$

Quinn, S. *et al., Inorg. Chim. Acta*, 1981, **50**, 141–146

The preparation of this is hazardous, and it is recommended that diazoindene always be kept in solution.
See 4-Toluenesulfonyl azide
See other DIAZO COMPOUNDS

3117. 2,4-Diisocyanatotoluene
[584-84-9] $C_9H_6N_2O_2$

(*MCA SD-73*, 1971); *NSC 489*, 1977; *FPA H5*, 1972; *HCS 1980*, 907

Acyl chlorides, or Bases
MCA SD-73, 1971
The diisocyanate may undergo exothermic polymerisation in contact with bases or more than traces of acyl chlorides, sometimes used as stabilisers.

Other reactants
Yoshida, 1980, 259
MRH values calculated for 13 combinations with oxidants are given.

Water
Kirk-Othmer, 1981, Vol. 13, 812
Polythene containers holding the diisocyanate may harden and burst in prolonged storage, because of slow absorption of water vapour through the wall leading to urea deposition from hydrolysis and generation of pressure of liberated carbon dioxide.
See other ORGANIC ISOCYANATES

3118. 3-Diazo-5-phenyl-3H-pyrazole
[62072-18-8] $C_9H_6N_4$

The light-sensitive crystalline solid is shock-sensitive and explodes at 115°C.
See entry DIAZOAZOLES
See other DIAZO COMPOUNDS, HIGH-NITROGEN COMPOUNDS

3119. 5-(4-Diazoniobenzenesulfonamido)thiazole tetrafluoroborate
[41334-40-1] (ion) $C_9H_7BF_4N_4O_2S$

Zuber, F. *et al., Helv. Chim. Acta*, 1950, **33**, 1269–1271
It explodes at 135–137°C.
See entry DIAZONIUM TETRAHALOBORATES

3120. 3-Iodo-1-phenylpropyne
[73513-15-2] C_9H_7I

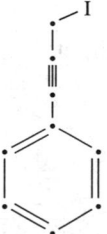

Whiting, M. C., *Chem. Eng. News*, 1972, **50**(23), 86

It detonated on distillation at around 180°C.

See other HALOACETYLENE DERIVATIVES

3121. Quinoline
[91-22-5]

C$_9$H$_7$N

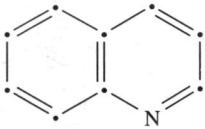

HCS 1980, 805

1. Blumann, A., *Proc. R. Aust. Chem. Inst.*, 1964, **31**, 286
2. *MCA Case History No. 1008*

The traditional unpredictably violent nature of the Skraup reaction (preparation of quinoline and derivatives by treating anilines with glycerol, sulfuric acid and an oxidant, usually nitrobenzene) is attributed to lack of stirring and adequate temperature control in many published descriptions [1]. A reaction on 450 l scale, in which sulfuric acid was added to a stirred mixture of aniline, glycerol, nitrobenzene, ferrous sulfate and water, went out of control soon after the addition. A 150 mm rupture disk blew out first, followed by the manhole cover of the vessel. The violent reaction was attributed to doubling the scale of the reaction, an unusually high ambient temperature (reaction contents at 32°C) and the accidental addition of excess acid. Experiment showed that a critical temperature of 120°C was attained immediately on addition of excess acid under these conditions [2].

Dinitrogen tetraoxide
    See Dinitrogen tetraoxide: Heterocyclic bases

Hydrogen peroxide
    See Hydrogen peroxide: Organic compounds (reference 2)

Linseed oil, Thionyl chloride
    See Sulfinyl chloride: Linseed oil, etc.

Maleic anhydride
    See Maleic anhydride: Bases, etc.
    See other ORGANIC BASES

3122. 3-Phenyl-5-isoxazolone
[7713-79-3]

$C_9H_7NO_2$

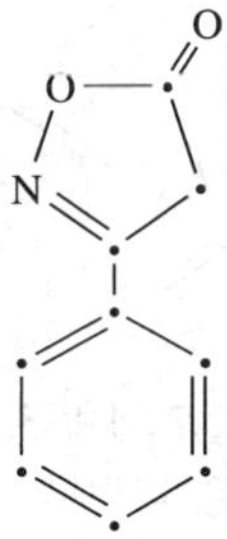

Sealed samples decompose exothermally above 105°C.
See entry ISOXAZOLES
See other N–O COMPOUNDS

3123. Dicarbonyl-π-cycloheptatrienyltungsten azide
[59589-13-8]

$C_9H_7N_3O_2W$

$$(OC)_2W(C_7H_7)N_3$$

Hoch, G. *et al., Z. Naturforsch. B*, 1976, **31B**, 295
It was insensitive to shock, but decomposed explosively at 130°C.
See related CARBONYLMETALS, METAL AZIDES, ORGANOMETALLICS

3124. 3-Methoxy-2-nitrobenzoyldiazomethane
[24115-83-1]

$C_9H_7N_3O_4$

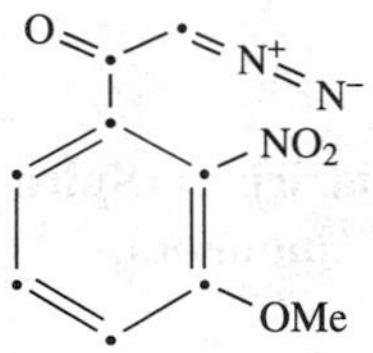

Alone, or Sulfuric acid
Musajo, L. *et al., Gazz. Chim. Ital.*, 1950, **80**, 171–176
The diazoketone explodes at 138–140°C, or on treatment with conc. sulfuric acid.
See other DIAZO COMPOUNDS, NITROARYL COMPOUNDS

3125. Tetracyclo[4.3.0³,⁵.0⁴,⁶]nona-1,7-diene (2,4-Dihydro-1H-1,2,6a-methano-pentalene)
[85861-34-3] C_9H_8

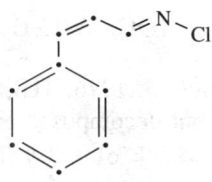

Burger, U. *et al., Helv. Chim. Acta*, 1983, **66**, 60–67
The highly strained hydrocarbon, produced from dilithiopentalene and chlorocar-bene, explodes violently at temperatures as low as −40°C when concentrated. Other isomers are probably similar.
See other DIENES, STRAINED-RING COMPOUNDS

3126. N-Chlorocinnamaldimin (3-Chlorimino-1-phenylpropene)
[ ] C_9H_8ClN

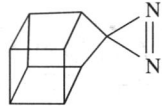

Hauser, C. R. *et al., J. Amer. Chem. Soc.*, 1935, **57**, 570
It decomposes vigorously after storage at ambient temperature for 30 m.
See other N-HALOGEN COMPOUNDS
See related DIENES

3127. Spiro(homocubane-9,9′-diazirine) (Spiro[3H-diazirine-3,9′-pentacyclo[4,3,0,0,²,⁵0,³,⁸0,⁴,⁷]nonane])
[129215-84-5] $C_9H_8N_2$

Platz, M. S. *et al., J. Amer. Chem. Soc.*, 1991, **113**, 4991
A white, dangerously explosive solid, apparently storable at room temperature but decomposing before melting.
See other DIAZIRINES, STRAINED-RING COMPOUNDS

3128. 3,5-Dimethylbenzenediazonium-2-carboxylate
[68596-88-3] $C_9H_8N_2O_2$

Wilcox, C. F. *et al., J. Amer. Chem. Soc.*, 1984, **106**, 7196

It is more sensitive to thermal decomposition than other diazonium carboxylates, and will decompose readily when inadequate stirring, large scale or excessively rapid diazotisation produces too large a temperature rise. There is significant potential for violent eruption of suspensions, or detonation of the filtered solid, and all recommended precautions should be used.

See 2-Carboxy-3,6-dimethylbenzenediazonium chloride
See other DIAZONIUM CARBOXYLATES

3129. 4,6-Dimethylbenzenediazonium-2-carboxylate
[ ] $C_9H_8N_2O_2$

Gommper, R. *et al., Chem. Ber.*, 1968, **101**, 2348
It is a highly explosive solid.
See other DIAZONIUM CARBOXYLATES

3130. 2,3-Epoxypropionaldehyde 2,4-dinitrophenylhydrazone
[ ] $C_9H_8N_4O_5$

Payne, G. B., *J. Amer. Chem. Soc.*, 1959, **81**, 4903

Although it melts at 96–98°C in a capillary, 0.5 g in a test tube decomposed explosively at 100°C.

See 2,3-Epoxypropionaldehyde oxime

See other 1,2-EPOXIDES, POLYNITROARYL COMPOUNDS *See related* HYDRAZONES

3131. 3-(2-Nitrophenyl)-2-hydroxyiminopropanoic acid (*o*-Nitrophenylpyruvic acid oxime)

[27878-36-0] $C_9H_8N_2O_5$

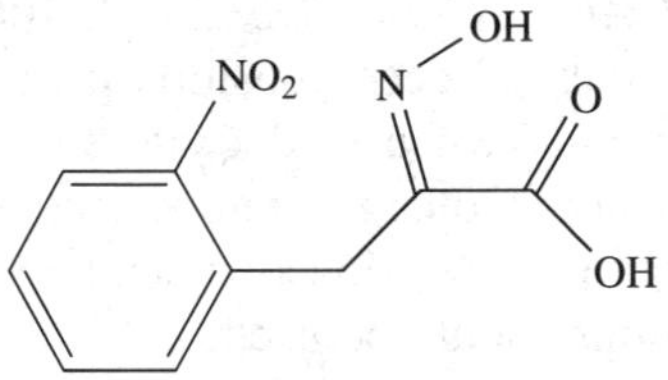

CHETAH, 1990, 187

Shock sensitive

See other NITROARYL COMPOUNDS, ORGANIC ACIDS, OXIMES

3132. Tetrakis(2,2,2-trinitroethyl)orthocarbonate

[14548-58-4] $C_9H_8N_{12}O_{28}$

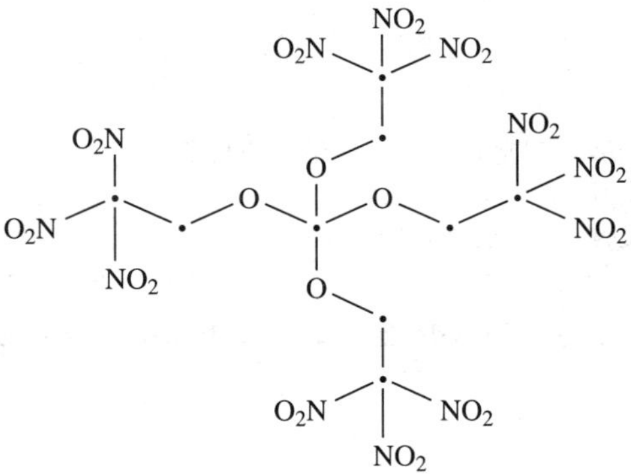

See entry TRINITROETHYL ORTHOESTERS

3133. Benzyloxyacetylene
[40089-12-1] C_9H_8O

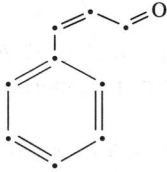

Olsman, H. *et al., Rec. Trav. Chim.*, 1964, **83**, 305
If heated above 60°C during vacuum distillation, explosive rearrangement occurs.
See other ACETYLENIC COMPOUNDS, BENZYL COMPOUNDS

3134. Cinnamaldehyde (3-Phenylpropenal)
[104-55-2] C_9H_8O

HCS 1980, 316

Sodium hydroxide
Morrell, S. H., private comm., 1968
Rags soaked in sodium hydroxide and in the aldehyde overheated and ignited owing to aerobic oxidation when they came into contact in a waste bin.
See other ALDEHYDES, PEROXIDISABLE COMPOUNDS, SELF-HEATING AND IGNITION INCIDENTS

3135. 2-Methyl-3,5,7-octatriyn-2-ol
[ ] C_9H_8O

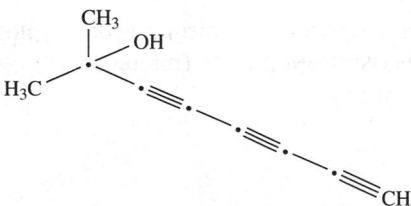

Cook, C. J. *et al., J. Chem. Soc.*, 1952, 2885, 2890

The crude material invariably deflagrated at ambient temperature, and once during drying at 0°C/0.013 mbar.

See other ACETYLENIC COMPOUNDS

3136. 4-Hydroxy-*trans*-cinnamic acid
[7400-08-0] $C_9H_8O_3$

Tanaka, Y. *et al., Proc. 4th Int. Conf. High Press. (1974)*, 704, 1975
During spontaneous solid-state polymerisation at 325°C/8–10 kbar, the acid exploded violently.

See HIGH PRESSURE REACTION TECHNIQUES
See other ORGANIC ACIDS, POLYMERISATION INCIDENTS

3137. *O*-Acetylsalicylic acid
[50-78-2] $C_9H_8O_4$

For experimental investigation of aspirin powder explosions,
See entry DUST EXPLOSION INCIDENTS (reference 18)
See other ORGANIC ACIDS

3138. 2-Carboxy-3,6-dimethylbenzenediazonium chloride
[36794-93-1]   C₈H₉ClN₂O₂

Hart, H. *et al., J. Org. Chem.*, 1972, **37**, 4272

The compound (the hydrochloride of the internal carboxylate salt) appears to be stable for considerable periods at ambient temperature, but explodes on melting at 88°C.

See related DIAZONIUM CARBOXYLATES
See other ORGANIC ACIDS

3139. 2-Phenylethyl isocyanate
[14649-03-7]   C₉H₉NO

Energy of decomposition (in range 220–350°C) measured as 0.50 kJ/g.

See entry THERMOCHEMISTRY AND EXOTHERMIC DECOMPOSITION (reference 2)
See other ORGANIC ISOCYANATES

3140. *mixo*-Nitroindane
[34701-14-9, 7436-07-9] (4- and 5-isomers, resp.)   C₉H₉NO₂

1. Lindner, J. *et al., Ber.*, 1927, **60**, 435
2. Gribble, G. W., *Chem. Eng. News*, 1973, **51**(6), 30, 39

The crude mixture of 4- and 5-nitroindanes produced by mixed acid nitration of indane following a literature method [1] is hazardous to purify by distillation. The warm residue from distillation of 15 mmol at 80°C/1.3 mbar exploded on admission of air, and a 1.3 mol batch exploded as distillation began under the same conditions. Removal of higher-boiling poly-nitrated material before distillation is recommended.
See other NITROARYL COMPOUNDS

3141. 3-Nitropropiophenone
[17408-16-1]   C₉H₉NO₃

Preparative hazard
See Nitric acid: Propiophenone, etc.
See other NITROARYL COMPOUNDS

3142. 4-Nitrophenylpropan-2-one
[5332-96-7]   C₉H₉NO₃

Preparative Hazard
4-Nitrophenylacetic acid. *See other* NITROARYL COMPOUNDS

3143. 2-Isocyanoethyl benzenesulfonate
[57678-14-5]   C₉H₉NO₃S

Matteson, D. S. *et al., J. Amer. Chem. Soc.*, 1968, **90**, 3761

Heating the product under vacuum to remove pyridine solvent caused a moderately forceful explosion.

See other ORGANIC ISOCYANATES, SULFUR ESTERS

3144. 3-Phenylpropionyl azide

[ ] $C_9H_9N_3O$

Curtiss, T. *et al., J. Prakt. Chem.*, 1901, **64**, 297

A sample exploded on a hot water bath.

See other ACYL AZIDES

3145. 1,3,5-Tris(nitromethyl)benzene

[ ] $C_9H_9N_3O_6$

See Nitric acid: Hydrocarbons (reference 8)

See related POLYNITROALKYL COMPOUNDS, POLYNITROARYL COMPOUNDS

3146. 4-(2-Ammonio-2-carboxyethylthio)-5,7-dinitro-4,5-dihydrobenzofurazanide
N-oxide (4,6-Dinitrobenzofurazan *N*-oxide.cysteine complex)

[79263-47-1] $C_9H_9N_5O_8S$

Strauss, M. J. *et al., Tetrahedron Lett.*, 1981, **22**, 1945–1948

When thoroughly dry, the cysteine complex of 5,7-dinitrobenzofurazan *N*-oxide exploded on heating.

See other FURAZAN N-OXIDES

3147. 2-Chloro-1-nitroso-2-phenylpropane
[6866-10-0]   C₉H₁₀ClNO

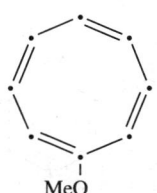

MCA Case History No. 747

A sample of the air-dried material (which is probably the dimer) decomposed vigorously on keeping in a closed bottle at ambient temperature overnight.

See other NITROSO COMPOUNDS

3148. 1,3-Dimethyl-2,1-benzisoxazolium perchlorate
[63609-41-6]   C₉H₁₀ClNO₅

Haley, N. F., *J. Org. Chem.*, 1978, **43**, 1236
The salt, m.p. 152°, explodes at 154°C.

See other PERCHLORATE SALTS OF NITROGENOUS BASES

3149. Methoxy-1,3,5,7-cyclooctatetraene
[7176-89-8]   C₉H₁₀O

Oxygen
Adam, W. *et al., Tetrahedron Lett.*, 1982, **23**, 2837–2840

The evaporated residue from sensitised photochemical oxidation of the polyene ignited spontaneously, on several occasions explosions occurring.

See related DIENES

See other PEROXIDISABLE COMPOUNDS, SELF-HEATING AND IGNITION INCIDENTS

3150. 3-Phenoxy-1,2-epoxypropane (Glycidyl phenyl ether)
[122-60-1] $C_9H_{10}O$

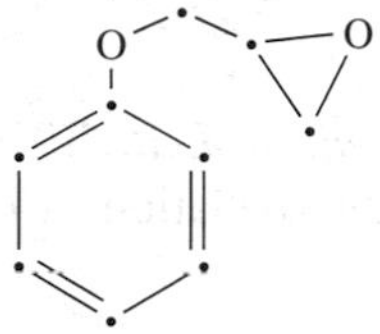

Energy of decomposition (in range 360–450°C) measured as 0.626 kJ/g.

See entry THERMOCHEMISTRY AND EXOTHERMIC DECOMPOSITION (reference 2)

3151. Propiophenone
[93-55-0] $C_9H_{10}O$

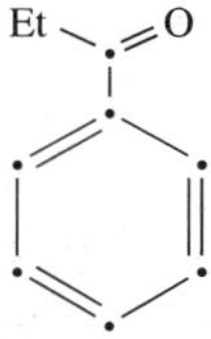

Nitric acid, Sulfuric acid

See Nitric acid: Propiophenone, etc.

3152. 3,5-Dimethylbenzoic acid
[499-06-9] $C_9H_{10}O_2$

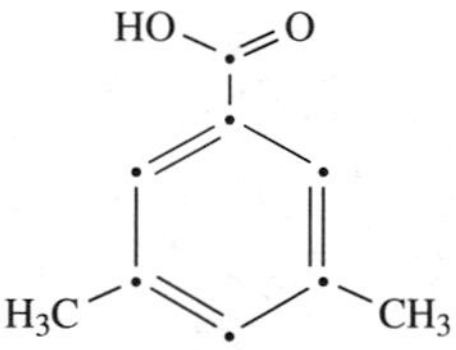

Preparative hazard

See Nitric acid: Hydrocarbons (reference 8)

See other ORGANIC ACIDS

3153. 1-(*cis*-Methoxyvinyl)-1,4-endoperoxy-2,5-cyclohexadiene (Z -1-(2-Methoxyethenyl)-2,3-dioxabicyclo[2.2.2]octa-5,7-diene)

[ ] $C_9H_{10}O_3$

Matsumoto, M. *et al., Tetrahedron. Lett.*, 1979, 1607–1610
The product of sensitised photooxidation of *cis*-2-methoxystyrene is explosive.
See other CYCLIC PEROXIDES

3154. 3- or 4-Methoxy-5,6-benzo-6*H* -1,2-dioxin
[69681-97-6, 69681-98-7, resp.] $C_9H_{10}O_3$

Lerdal, D. *et al., Tetrahedron. Lett.*, 1978, 3227–3228
Both isomers decomposed violently on warming in absence of solvent.
See other CYCLIC PEROXIDES

3155. Allyl benzenesulfonate
[7575-57-7] $C_9H_{10}O_3S$

$$H_2C=CHCH_2OSO_2Ph$$

Dye, W. T. *et al., Chem. Eng. News*, 1950, **28**, 3452
The residue from vacuum distillation at 92–135°C/2.6 mbar darkened, thickened, then exploded after removal of the heat source.
    For precautions, *see* Triallyl phosphate
See other ALLYL COMPOUNDS, SULFUR ESTERS

3156. 2-(4-Bromophenylazo)-2-propyl hydroperoxide
[72447-41-7] $C_9H_{11}BrN_2O_2$

See entry α-PHENYLAZO HYDROPEROXIDES (reference 4)
See related ALKYL HYDROPEROXIDES, AZO COMPOUNDS

3157. 4-Nitroisopropylbenzene
[1817-47-6] $C_9H_{11}NO_2$

High rate of pressure increase during exothermal decomposition.
See entry PRESSURE INCREASE IN EXOTHERMIC DECOMPOSITION
See other NITROARYL COMPOUNDS

3158. Nitromesitylene
[603-71-4] $C_9H_{11}NO_2$

Preparative hazard
See Nitric acid: Hydrocarbons (references 7,8)
See other NITROARYL COMPOUNDS

3159. 2-Azido-2-phenylpropane
     [32366-26-0] $C_9H_{11}N_3$

Energy of exothermic decomposition in range 180–270°C was measured as 1.2 kJ/g
See entry THERMOCHEMISTRY AND EXOTHERMIC DECOMPOSITION (reference 2)
See other ORGANIC AZIDES

3160. 4-Methyl-2,4,6-triazatricyclo[5.2.2.0^{2,6}]undeca-8-ene-3,5-dione (5,8-
     Dihydro-2-methyl-5,8-ethano-1H-[1,2,4]triazolo[1,2a] pyridazine-
     1,3(2H)dione)
     [78790-57-5] $C_9H_{11}N_3O_2$

Hydrogen peroxide, Potassium hydroxide
   See Hydrogen peroxide: 4-Methyl-2,4,6-triazatricyclo[5.2.2.0^{2,6}]undeca-8-ene-3,5-
   dione

3161. 5-(4-Dimethylaminobenzeneazo)tetrazole
     [53004-03-8] $C_9H_{11}N_7$

Thiele, J., *Ann.*, 1892, **270**, 61

It explodes at 155°C.
See other AZO COMPOUNDS, TETRAZOLES

3162. Mesitylene (1,3,5-Trimethylbenzene)
[108-67-3] C_9H_{12}

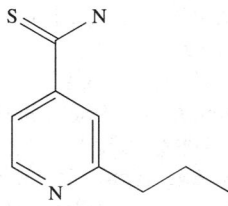

HCS 1980, 621

Nitric acid
See Nitric acid: Hydrocarbons (reference 8)

3163. 2-Propylpyridine-4-carbothioamide (2-Propyl(thioisonicotinamide))
[14222-60-7] $C_9H_{12}N_2S$

See FRICTIONAL INITIATION INCIDENTS (REFERENCE 3)

3164. 2-Methoxyethylbenzene
[3558-60-9] $C_9H_{12}O$

$$C_6H_5CH_2CH_2OMe$$

Nitric acid
See Nitric acid: 2-Methoxyethylbenzene, Sulfuric acid

3165. 3-Phenylpropanol
[122-97-4] $C_9H_{12}O$

Phosphorus tribromide
See Phosphorus tribromide: 3-Phenylpropanol

3166. 2-Phenyl-2-propyl hydroperoxide (Cumyl hydroperoxide)
[80-15-9] $C_9H_{12}O_2$

$$H_3C \underset{|}{\overset{O-OH}{\underset{|}{C}}} CH_3$$

1. Leroux, A., *Mém. Poudres*, 1955, **37**, 49
2. Simon, A. H. *et al., Chem. Ber.*, 1957, **90**, 1024
3. Hulanicki, A., *Chem. Anal.* (Warsaw), 1973, **18**, 723–726
4. *MCA Case History No. 906*
5. Leleu, *Cahiers*, 1973, (71), 226
6. Fleming, J. B. *et al., Hydrocarbon Process.*, 1976, **55**(1), 185–196
7. Redoshkin, B. A. *et al., Chem. Abs.*, 1964, **60**, 14359h
8. Antonovskii, V. L. *et al., Chem. Abs.*, 1964, **60**, 14360c
9. Kletz, T. A., *Loss Prev.*, 1979, **12**, 98
10. Anon., *Loss Prev. Bull.*, 1986, (071), 9
11. Schwab, R. F., *Loss. Prev. Bull.*, 1988, (083), 20–32
12. Anon., *Loss Prev. Bull.*, 1992, **103**, 31
13. Francisco, M. A., *Chem. Eng. News*, 1993, **71**(22), 4
14. Shanley, E. S., *Chem. Eng. News*, 1993, **71**(35), 61

The explosibility of this unusually stable hydroperoxide has been investigated [1]. It is difficult, but not impossible, to induce explosive decomposition [2]. A colorimetric method of analysing mixtures of cumene and its hydroperoxide which involves alkaline decomposition of the latter is safer than the use of acidic conditions, or of reducants, which may cause explosive reactions [3]. During vacuum concentration of the hydroperoxide by evaporation of cumene, a $6.5m^3$ quantity of the 36% material decomposed explosively after storage at 109°C for 5 h. Catalytic decomposition under near-adiabatic conditions may have been involved [4]. At the end of concentration from hexane, after purification, 100 ml of peroxide exploded violently for no apparent reason; isolation of no more than 5 g is recommended [13]. Contact with copper, copper or lead alloys, mineral acids or reducants may lead to violent decomposition [5]. Cobalt also catalyses the decomposition reaction which is autocatalytic. The acid cleavage reaction used industrially is highly exothermic, and safety aspects of commercial phenol/acetone processes are discussed fully [6]. Much detailed kinetic work has also been published [7,8]. Five explosions due to overheating and subsequent decomposition of the concentrated solutions in cumene have been noted. Another explosion arose from reduced flow of acid gas used in the cleavage reaction because the cylinder was outdoors on a cold night. The reduced flow led to spurious control signals which eventually caused the explosion [9]. During a shut-down, a 100 t quantity of hydroperoxide solution in intermediate storage was being kept hot with steam but became overheated, causing cumene to vent. The vapour ignited explosively and ruptured the storage tank [10]. A detailed analysis of the incident is available [11]. A drum of the hydroperoxide exploded, this was attributed to possible contamination with sodium hydroxymethylsulfonate. This might form catalytic acids if oxidised, but the drum had been sampled with a, possibly corroded, stainless steel dip pipe [12].

CHETAH calculations for energy/hazard of an hypothetical decomposition reaction to carbon, water and methane are reported [14].

See THERMAL EXPLOSIONS, COMPUTATION OF REACTIVE CHEMICAL HAZARDS
See other CATALYTIC IMPURITY INCIDENTS, CORROSION INCIDENTS

Charcoal
Leleu, *Cahiers*, 1980, (99), 278
Contact with charcoal powder at ambient temperature gives a strong exotherm (probably owing to traces of heavy metals present).

1,2-Dibromo-1,2-diisocyanatoethane polymers
Lapshin, N. M. *et al., Chem. Abs.*, 1974, **81**, 64013
A mixture of the dimeric and trimeric dibromo compounds with the hydroperoxide in benzene may react vigorously or explode if the solution is heated to concentrate it.

Iodides
Duswalt, A. A., *Chem. Eng. News*, 1990, **68**(6), 2
Following a laboratory incident, cumene hydroperoxide was shown to be explosively decomposed by catalytic sodium iodide.

Other reactants
Yoshida, 1980, 105
MRH values calculated for 15 combinations with oxidisable materials are given.
See other ALKYL HYDROPEROXIDES, COMMERCIAL ORGANIC PEROXIDES

3167. 2,3,4,6-Tetramethylpyrilium perchlorate
[35941-38-9] $C_9H_{13}ClO_5$

Balaban, A. T. *et al., Org. Prep. Proced. Int.*, 1982, **14**, 31–38
It must be handled cautiously.
See other NON-METAL PERCHLORATES

3168. Benzyldimethylamine
[28262-13-7] $C_9H_{13}N$

$$\text{(structure: benzene ring with N(CH}_3\text{)}_2\text{ substituent)}$$

Cellulose nitrate
    See CELLULOSE NITRATE: amines *See other* BENZYL COMPOUNDS, ORGANIC BASES

3169. 2,4,6-Trimethylaniline
[88-05-1] $C_9H_{13}N$

$$\text{(structure: benzene ring with NH}_2\text{ and three CH}_3\text{ groups)}$$

Nitrosyl perchlorate
    See Nitrosyl perchlorate: Organic materials

3170. *O*-Mesitylenesulfonylhydroxylamine
[36016-40-7] $C_9H_{13}NO_3S$

$$\text{(structure: benzene ring with SO}_2\text{-O-NH}_2\text{ and three CH}_3\text{ groups)}$$

1. Tamura, Y. *et al., Tetrahedron Lett.*, 1972, **40**, 4133
2. Ning, R. Y., *Chem. Eng. News*, 1973, **51**(51), 36–37
3. Carpino, L. A., *J. Amer. Chem. Soc.*, 1960, **82**, 3134
4. Ning, R. Y., private comm., 1974

5. Johnson, C. R. *et al., J. Org. Chem.*, **39**, 2459, footnote 15
6. Scopes, D. I. C. *et al., J. Org. Chem.*, 1977, **42**, 376
7. Dillard, R. D. *et al., J. Med. Chem.*, 1980, **23**, 718
8. Fluka advertisement, *J. Org. Chem.*, 1980, **45**(9), cover ii

The title compound was prepared following a published procedure [1] and a dried sample decomposed soon after putting it into an amber bottle for storage, the screw cap being shattered [2]. Although the instability of the compound had been mentioned [3], no suggestion of violent decomposition had previously been made. It seems likely that traces of surface alkali in the soda-glass bottle had catalysed both the formation of the highly active imidogen radical (HN:) from the base-labile compound, and its subsequent exothermic decomposition. Storage in dichloromethane solution appears safe [4]. The small crystals produced by a modified method appear to be safe in storage at 0°C, or in use [5]. Attempted vacuum drying of the compound at ambient temperature led to a mild explosion. Subsequently, a solution of the wet solid in dimethoxyethane was dried over molecular sieve and used in further work [6]. The solid may explode violently when heated to 60°C. Large amounts should be used either in solution or as the crystalline solid. The neat molten liquid form should be avoided, as it evolves heat on crystallisation. It is best to avoid storage and generate the reagent as needed [7]. A stable precursor, *tert*-butyl *N*-mesitylenesulfonyloxycarbamate has become available [8].

See O-4-Toluenesulfonylhydroxylamine

See other CATALYTIC IMPURITY INCIDENTS, GLASS INCIDENTS, N–O COMPOUNDS

3171. 3-Ethyl-1(4-methylphenyl)triazene
[50707-40-9]

$C_9H_{13}N_3$

Anon., *Loss Prev. Bull.*, 1980, (030), 160

A 5 g sample was distilled under vacuum uneventfully from a bulb tube with air-bath heating, but a 50 g portion exploded violently when distilled at 0.13 mbar from an oil-bath at 120°C.

See other TRIAZENES

3172. 2-(2-Aminoethylamino)-5-methoxynitrobenzene
[13556-31-5] $C_9H_{13}N_3O_3$

See N-(2-Nitrophenyl)-1,2-diaminoethane
See other NITROARYL COMPOUNDS

3173. 3,3,6,6-Tetrakis(bromomethyl)-9,9-dimethyl-1,2,4,5,7,8-hexoxonane
[16007-16-2] $C_9H_{14}Br_4O_6$

Schulz, M. *et al., Chem. Ber.,* 1967, **100**, 2245
It explodes on impact or friction, as do the tetrachloro- and 9-ethyl-9-methyl analogues.
See other CYCLIC PEROXIDES

3174. 1-(4-Methyl-1,3-diselenonylidene)piperidinium perchlorate
[53808-72-3] $C_9H_{14}ClNO_4Se_2$

See entry 1-(1,3-DISELENONYLIDENE)PIPERIDINIUM PERCHLORATES
See other PERCHLORATE SALTS OF NITROGENOUS BASES

3175. 6,6-Dimethylbicyclo[3.1.1]heptan-2-one (Nopinone)
[24903-95-5]; S-[7782-63-9]
$C_9H_{14}O$

Preparative hazard
1. Gordon, P. M., *Chem. Eng. News*, 1990, **68**(30), 2
2. Ferreira, J. T. B., *ibid.*, (50), 2
3. Weissman, S., *ibid.*, 1991, **69**(4), 2
Several reports have been received of explosion during work-up of the products obtained from ozonolysis of pinene, despite prior treatment with reducants. A safe alternative procedure is given in [3].
See OZONIDES

3176. 2,6-Dimethyl-2,5-heptadien-4-one diozonide
[ ]
$C_9H_{14}O_7$

Harries, G. *et al.*, *Ann.*, 1910, **374**, 338
'Phorone' diozonide ignites on warming to ambient temperature.
See other OZONIDES

3177. Tri-2-propenylborane (Triallylborane)
[688-61-9]
$C_9H_{15}B$

Borane
Alder, R. W. *et al.*, *J. Chem. Soc., Perk. I*, 1996, (7), 657
The 1:1 adduct is highly pyrophoric, so too is its pyrolysis product. Both are polymeric, diborabicycloundecane is not.
See other ALKYLBORANES

3178. N -Cyano-2-bromoethylcyclohexylamine
[53182-16-4]

$C_9H_{15}BrN_2$

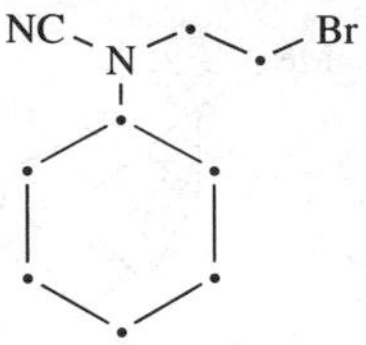

Anon., *BCISC Quart. Safety Summ.*, 1964, **35**, 23

The reaction product from N-cyclohexylaziridine and cyanogen bromide (believed to be the title compound) exploded violently on attempted distillation at 160°C/0.5 mbar.

See N-Cyano-2-bromoethylbutylamine

See other CYANO COMPOUNDS

3179. Tris(ethylthio)cyclopropenium perchlorate
[50744-08-6]

$C_9H_{15}ClO_4S_3$

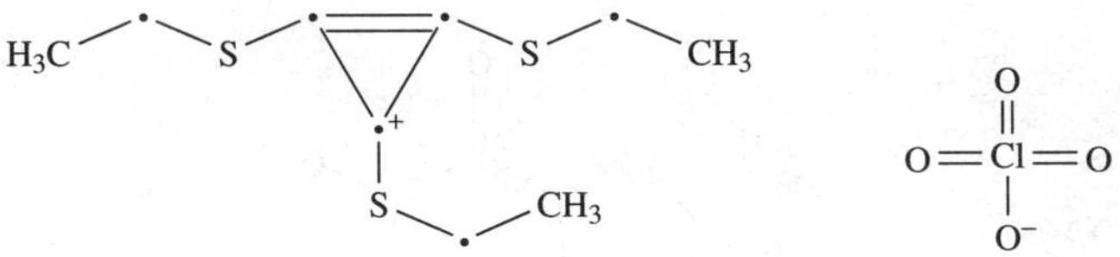

Sunderlin, K. G. R., *Chem. Eng. News*, 1974, **52**(31), 3

Some 50 g of the compound had been prepared by a method used for analogous compounds, and the solvent-free oil had been left to crystallise. Some hours later it exploded with considerable violence. Preparation of this and related compounds by other workers had been uneventful.

See other NON-METAL PERCHLORATES

3180. Triallylchromium
[12082-46-1]

$C_9H_{15}Cr$

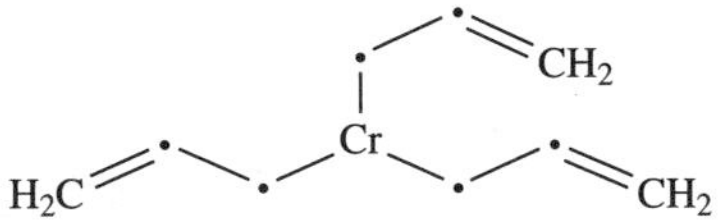

O'Brien, S. *et al., Inorg. Chem.*, 1972, **13**, 79

Triallylchromium and its thermal decomposition products are pyrophoric in air.

See related ALKYLMETALS

See other ALLYL COMPOUNDS

3181. Nitrilotris(oxiranemethane) (Triglycidylamine)
[61014-23-1] $C_9H_{15}NO_3$

McKelvey, J. *et al., J. Org. Chem.*, 1960, **25**, 1424
The crude product polymerises violently during vacuum distillation. The pure is storable cool
See other 1,2-EPOXIDES

3182. 4-Tolylbiguanidium hydrogen dichromate
[15760-46-0] $C_9H_{15}Cr_2N_5O_7$

See entry DICHROMATE SALTS OF NITROGENOUS BASES

3183. 1,3,5-Triacetylhexahydro-1,3,5-triazine
[26028-46-6] $C_9H_{15}N_3O_3$

Nitric acid, Trifluoroacetic anhydride
See Nitric acid: 1,3,5-Triacetylhexahydro-1,3,5-triazine, etc.

3184. Triallyl phosphate
[1624-19-4] $C_9H_{15}O_4P$

1. Dye, W. T. *et al., Chem. Eng. News*, 1950, **28**, 3452
2. Steinberg, G. M., *Chem. Eng. News*, 1950, **28**, 3755
Alkali-washed material, stabilised with 0.25% of pyrogallol, was distilled at 103°C/4 mbar until slight decomposition began. The heating mantle was then removed and the still-pot temperature had fallen below its maximum value of 135°C when the residue exploded violently [1]. The presence of solid alkali [2] or 5% of phenolic inhibitor is recommended, together with low-temperature high-vacuum distillation, to avoid formation of acidic decomposition products, which catalyse rapid exothermic polymerisation.
See other ALLYL COMPOUNDS, CATALYTIC IMPURITY INCIDENTS, PHOSPHORUS ESTERS

3185. 1,4,7-Triazacyclononanetricarbonylmolybdenum hydride perchlorate
[ ] $C_9H_{16}ClMoN_3O_7$
[C_6H_{15}N_3MoH(CO)_3] ClO_4

See 1,4,7-Triazacyclononanetricarbonyltungsten hydride perchlorate (next below)
See related AMMINEMETAL OXOSALTS, CARBONYLMETALS

3186. 1,4,7-Triazacyclononanetricarbonyltungsten hydride perchlorate
[88253-43-4] $C_9H_{16}ClN_3O_7W$
[C_6H_{15}N_3WH(CO)_3] ClO_4

Chaudhuri, P. *et al., Inorg. Chem.*, 1984, **23**, 429
The dry tungsten and molybdenum complexes are extremely explosive when dry; handle no more than 10 mg portions. The corresponding Mo complexes with Br or I replacing H are also explosive.
See related AMMINEMETAL OXOSALTS, CARBONYLMETALS

3187. Azelaic acid
[123-99-9] $C_9H_{16}O_4$

Hydrogen peroxide, Sulfuric acid
 See Hydrogen peroxide: Azelaic acid, Sulfuric acid
 See other ORGANIC ACIDS

3188. Diisopropyl malonate
[13196-64-7] $C_9H_{16}O_4$

$$CH_2(CO.OCHMe_2)_2$$

Sulfinyl chloride
 See Sulfinyl chloride: Diisopropyl malonate

3189. Diperoxyazelaic acid
[1941-79-3] $C_9H_{16}O_6$

Preparative hazard
 See Hydrogen peroxide: Azelaic acid, Sulfuric acid
 See other PEROXYACIDS

†3190. 2,6-Dimethyl-3-heptene
[2738-18-3] C_9H_{18}

$$Me_2CHCH=CHCH_2CHMe_2$$

See other ALKENES

†3191. 1,3,5-Trimethylcyclohexane
[1795-27-3] C_9H_{18}

3192. 1,4-Bis(2-chloroethyl)-1,4-bis(azonia)bicyclo[2.2.1]heptane periodate
[77628-04-7] $C_9H_{18}Cl_2I_2N_2O_8$

Pettit, G. R. *et al., Can. J. Chem.*, 1981, **59**, 218, 220
It exploded violently at about 150°C, or when ignited openly or on gentle impact.
See related PERCHLORATE SALTS OF NITROGENOUS BASES

3193. Dibutylthallium isocyanate
[74637-34-6] $C_9H_{18}NOTl$

$$Bu_2TlN:C:O$$

Srinastava, T. N. *et al., Indian J. Chem. Sect. A*, 1980, **19A**, 480
It explodes at 35°C.
See related ALKYLMETALS, METAL CYANATES *See other* N-METAL DERIVATIVES

3194. Di-*tert*-butyl diperoxycarbonate
[3236-56-4] $C_9H_{18}O_5$

$$(Me_3COO)_2CO$$

A partially decomposed sample exploded violently at 135°C.
See entry PEROXYCARBONATE ESTERS *See other* PEROXYESTERS

3195. 3,3,6,6,9,9-Hexamethyl-1,2,4,5,7,8-hexoxonane
[17088-37-8] $C_9H_{18}O_6$

1. Dilthey, W. *et al., J. Prakt. Chem.*, 1940, **154**, 219
2. Rohrlich, M. *et al., Z. Ges. Schiess u. Sprengstoffw.*, 1943, **38**, 97–99

1052

This trimeric acetone peroxide is powerfully explosive [1], and will perforate a steel plate when heated on it [2].

See 6-Aminopenicillanic acid *S*-oxide
      Hydrogen peroxide: Ketones, Nitric acid
      Ozone: Citronellic acid
See other CYCLIC PEROXIDES

3196. 3,6,9-Triethyl-1,2,4,5,7,8-hexoxonane
[ ] $C_9H_{18}O_6$

Rieche, A. *et al., Ber.*, 1939, **72**, 1938

This trimeric 'propylidene peroxide', formed from propanal and hydrogen peroxide, is an extremely explosive and friction-sensitive oil.
See other CYCLIC PEROXIDES

3197. 2,2,4,4,6,6-Hexamethyltrithiane
[828-26-2] $C_9H_{18}S_3$

Nitric acid
See Nitric acid: 2,2,4,4,6,6-Hexamethyltrithiane

3198. Pivaloyloxydiethylborane
[32970-52-8] $C_9H_{19}O_2$

Air, or Water
    Köster, R. *et al., Inorg. Synth.*, 1983, **22**, 196, 189

It reacts violently with air or water.
See related ACID ANHYDRIDES, ALKYLBORANES

†3199. 3,3-Diethylpentane
[1067-20-5]
C_9H_{20}

$$Et_4C$$

†3200. 2,5-Dimethylheptane
[2216-30-0]
C_9H_{20}

$$Me_2CHC_2H_4CHMeEt$$

†3201. 3,5-Dimethylheptane
[926-82-9]
C_9H_{20}

$$EtMeCHCH_2CHMeEt$$

†3202. 4,4-Dimethylheptane
[1069-19-5]
C_9H_{20}

$$Me_2CPr_2$$

†3203. 3-Ethyl-2,3-dimethylpentane
[16747-33-4]
C_9H_{20}

$$Me_2CHC(Me)Et_2$$

†3204. 3-Ethyl-4-methylhexane
[3074-77-9]
C_9H_{20}

$$Et_2CHCHMeEt$$

†3205. 4-Ethyl-2-methylhexane
[3074-75-7]
C_9H_{20}

$$Me_2CHCH_2CHEt_2$$

†3206. 2-Methyloctane
[3221-61-2]
C_9H_{20}

$$Me_2CH[CH_2]_5Me$$

†3207. 3-Methyloctane
[2216-33-3]
C_9H_{20}

$$Me_2CH[CH_2]_5Me$$

†3208. 4-Methyloctane
   [2216-34-4 resp.] C_9H_{20}

$$Me_2CH[CH_2]_5Me$$

†3209. Nonane
   [111-84-2] C_9H_{20}

$$Me[CH_2]_7Me$$

†3210. 2,2,5-Trimethylhexane
   [3522-94-9] C_9H_{20}

$$Me_3CC_2H_4CHMe_2$$

†3211. 2,2,3,3-Tetramethylpentane
   [7154-79-2] C_9H_{20}

$$Me_3CCMe_2Et$$

†3212. 2,2,3,4-Tetramethylpentane
   [1186-53-4] C_9H_{20}

$$Me_3CCHMeCHMe_2$$

3213. 1-Methoxy-2-methyl-1-(trimethylsilyloxy)-1-butene ([(1-Methoxy-2-methyl-
   1butenyl)oxy]trimethylsilane)
   [84393-12-4] $C_9H_{20}O_2Si$

Bromoanisole, Sodamide
   See Sodamide: Aryl halide, 1-Alkoxy-1-(trimethylsilyloxy)alkenes

3214. Triisopropylaluminium
   [2397-67-3] $C_9H_{21}Al$

$$(Me_2CH)_3Al$$

See entry TRIALKYLALUMINIUMS

3215. Tripropylaluminium
[102-67-0]

Pr_3Al

$C_9H_{21}Al$

See entry TRIALKYLALUMINIUMS

3216. Aluminium isopropoxide
[555-31-2]

$Al(OCHMe_2)_3$

$C_9H_{21}AlO_3$

Crotonaldehyde, 2-Propanol
See 2-Propanol: Aluminium isopropoxide, etc.

Hydrogen peroxide
See Hydrogen peroxide: Aluminium isopropoxide
See other METAL ALKOXIDES

3217. Tripropylborane
[1116-61-6]

Pr_3B

$C_9H_{21}B$

Gibson, 1969, 147
It ignites in air.
See other ALKYLBORANES

3218. Tripropylsilyl perchlorate
[ ]

$Pr_3SiOClO_3$

$C_9H_{21}ClO_4Si$

See entry ORGANOSILYL PERCHLORATES

3219. Tripropylindium
[3015-98-3]

$C_9H_{21}In$

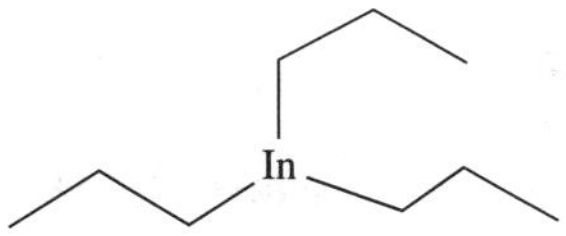

Leleu, *Cahiers*, 1977, (88), 367
It ignites in air.
See other ALKYLMETALS

3220. Triisopropylphosphine
[6476-36-4]

$C_9H_{21}P$

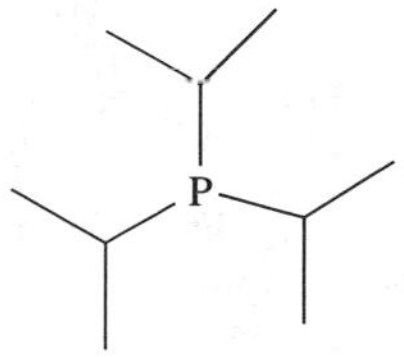

Chloroform, or Oxidants
Catalogue entry, Deutsche Advance Produktion, 1968
Particularly this phosphine reacts, when undiluted, rather vigorously with most peroxides, ozonides, *N*-oxides, and also chloroform. It may be safely destroyed by pouring into a solution of bromine in carbon tetrachloride.
See other ALKYLPHOSPHINES

3221. Tripropylantimony
[5613-69-4]

$C_9H_{21}Sb$

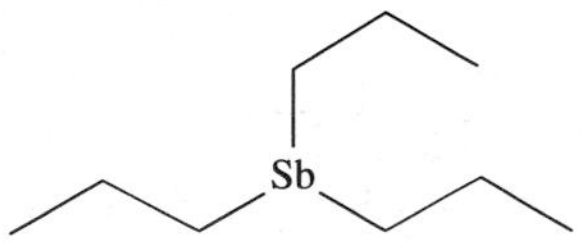

Gibson, 1969, 171
The solid ignites or chars when exposed to air on filter paper.
See other ALKYLMETALS

3222. Tris(2-propylthio)silane
[17891-55-3]

$C_9H_{22}S_3Si$

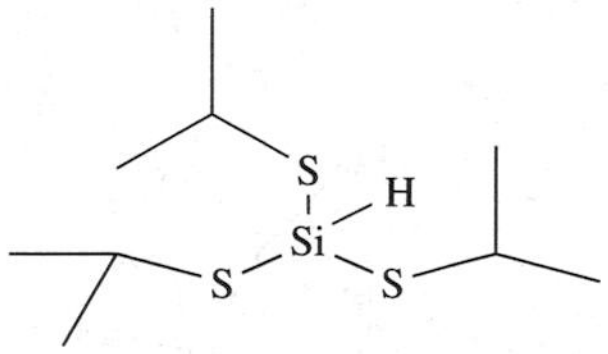

Lambert, J. A., *J. Amer. Chem. Soc.*, 1988, **110**, 2210
The silane, prepared by reduction of tris(2-propylthio)silylenium perchlorate with diisobutylaluminium hydride at −78°C, is subsequently isolated by high-vacuum distillation from the mixture. Heating must be very mild to prevent explosion.
See related ALKYLSILANES

3223. N,N-Bis(diethylboryl)methylamine
[19162-21-1] $C_9H_{23}B_2N$

491M, 1975, 258

It ignites in air.
See related ALKYLBORANES

3224. Tris(dimethylfluorosilylmethyl)borane
[62497-91-0] $C_9H_{24}BF_3Si_3$

Hopper, S. P. *et al., Synth. React. Inorg. Met.-Org. Chem.*, 1976, **6**, 378
The crude or pure material will ignite in air, especially if exposed on filter paper
or a similar extended surface.
See related ALKYLBORANES, ALKYLHALOSILANES

3225. N,N,N′,N′-Tetramethylethane-1,2-diamine, trimethylpalladium(IV)
bromide (Bromotrimethyl-N,N,N′,N′-tetramethylethanediamine-N,N′-
palladium)
[123147-95-5] $C_9H_{25}BrN_2Pd$

De Graaf, W. *et al., Organometallics*, 1990, **9**(5), 1479

Decomposes explosively when heated to 50°C.
See other AMMINEMETAL HALIDES, *See related* PLATINUM COMPOUNDS

3226. *N,N,N,N*-Tetramethylethane-1,2-diamine, trimethylpalladium(IV) iodide
[123147-84-2] $C_9H_{25}IN_2Pd$

De Graaf, W. *et al., Organometallics*, 1989, **8**(12), 2907
Decomposes explosively when brought to room temperature.
See other AMMINEMETAL HALIDES
See related PLATINUM COMPOUNDS

3227. Tris(trimethylsilyl)aluminium
[65343-66-0] $C_9H_{27}AlSi_3$

Rösch, L., *Angew. Chem. (Intern. Ed.)*, 1977, **16**, 480
The crystalline solid ignites in air.
See related ALKYLSILANES

3228. Potassium dinitrogentris(trimethylphosphine)cobaltate(1−)
[63181-09-9] $C_9H_{27}CoKN_2P_3$

$$K[N_2Co(PMe_3)_3]$$

Hammer, R. *et al., Angew. Chem. (Intern. Ed.)*, 1977, **16**, 485
The crystalline solid ignites immediately on exposure to air.
See related ALKYLPHOSPHINES, *N*-METAL DERIVATIVES

3229. Tris(trimethylsilyl)phosphine
[5573-38-3] $C_9H_{27}PSi_3$

Dahl, O., *Acta Chem. Scand. Ser. B*, 1976, **30**, 799
It is pyrophoric.
See related ALKYLSILANES

3230. Tris(trimethylsilyl)hydrazine
[13272-02-1] $C_9H_{28}N_2Si_3$

Oxidants
It is hypergolic with fluorine or fuming nitric acid, and explodes with ozone.
See entry SILYLHYDRAZINES

3231. Tris(trimethylsilyl)silane (1,1,1,3,3,3-Hexamethyl-2-(trimethylsilyl)trisilane)
[1873-77-4] $C_9H_{28}Si_4$

$$(Me_3Si)_3SiH$$

Preparative hazard
Dickhaut, J. *et al., Org. Synth.*, 1992, **70**, 164
The solid residue of the preparation from lithium powder, trimethylsilyl chloride, tetrachlorosilane and methyllithium may be highly pyrophoric. Caution in quenching into dilute acid is advised.
See other ALKYLSILANES

3232. *N,N,N'-***Tris(trimethylsilyl)diaminophosphine**
[63104-54-1] $C_9H_{29}N_2PSi_3$

$$(Me_3Si)_2NPHNHSiMe_3$$

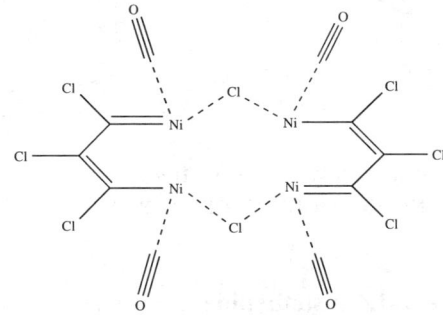

Niecke, E. *et al., Angew. Chem. (Intern. Ed.),* 1977, **16**, 487
The finely divided solid ignites in air.
See related ALKYLSILANES, PHOSPHINES

3233. Dicarbonyl-η-trichloropropenyldinickel chloride dimer
[64297-36-5] $C_{10}Cl_8Ni_4O_4$

Posey, R. G. *et al. J. Amer. Chem. Soc.*, 1977, **99**, 4865 (footnote 13)
Nickel carbonyl and tetrachloropropadiene react in a wide variety of solvents
to give the crystalline title compound which is extremely shock-sensitive, and
explodes if touched when dry.
See related CARBONYLMETALS

3234. 2-Azidoperfluoronaphthalene
[74415-60-4] $C_{10}F_7N_3$

Banks, R. E. *et al., J. Chem. Soc., Chem. Comm.*, 1980, 151

Thermolyses at 300°C/7 mbar to give 2-cyanoperfluoroindene were terminated by explosions.
See other ORGANIC AZIDES

3235. Octatetrayne-1,8-dicarboxylic acid
[40575-26-6]

$C_{10}H_2O_4$

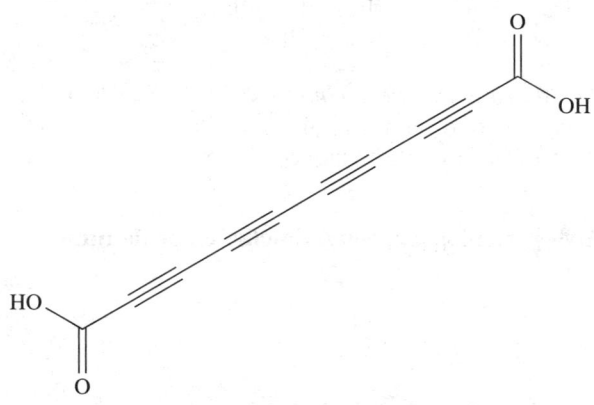

Sorbe, 1968, 151
An explosive solid, extremely heat-sensitive.
See other ACETYLENIC COMPOUNDS, ORGANIC ACIDS

3236. Copper(I) benzene-1,4-bis(ethynide)
[40575-26-6]

$(C_{10}H_4Cu_2)_n$

Royer, E. C. *et al.*, *J. Inorg. Nucl. Chem.*, 1981, **43**, 708
Explosive, and pyrophoric at 200°C.
See other METAL ACETYLIDES

3237. 1,3,6,8-Tetranitronaphthalene
[28995-89-3] $C_{10}H_4N_4O_8$

Sorbe, 1968, 151
An explosive solid, most sensitive to heating.
See other POLYNITROARYL COMPOUNDS

3238. 2,2′-Azo-3,5-dinitropyridine
[55106-91-7] $C_{10}H_4N_8O_8$

Explosive.
See entry POLYNITROAZOPYRIDINES
See other AZO COMPOUNDS, POLYNITROARYL COMPOUNDS

3239. [*I*,*I*-Bis(trifluoroacetoxy)iodo]benzene
[2712-78-9] $C_{10}H_5F_6IO_4$

Increase in pH converts the title compound into an explosive 'dimer'.
See μ-Oxo*I*,*I*-bis(trifluoroacetato-*O*)-*I*,*I*-diphenyldiiodine(III)
See 3,5-Dimethyl-4[*I*,*I*-bis(trifluoroacetoxy)iodo]isoxazole
See other *I*,*I*-(DIBENZOYLPEROXY)ARYLIODINESIODINE COMPOUNDS

3240. 2-Hydroxy-6-nitro-1-naphthalenediazonium-4-sulfonate
    [5366-84-7] $C_{10}H_5N_3O_6S$

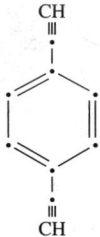

HCS 1980, 362

Fast smouldering propagation occurs on heating the powder moderately.
See entry HIGH RATE DECOMPOSITION
See related DIAZONIUM CARBOXYLATES

3241. 1,4-Diethynylbenzene
    [935-14-8] $C_{10}H_6$

CH
|||

CH

Preparative hazard
    See 1,4-Bis(1,2-dibromoethyl)benzene: Potassium hydroxide
    See other ALKYNES

3242. 1,5-Dinitronaphthalene
    [605-71-0] $C_{10}H_6N_2O_4$

NO$_2$

NO$_2$

Sulfur, Sulfuric acid
    Hub, L., *Proc. Chem. Process Haz. Symp. VI*, (Manchester, 1977), No. 49, 43,
    Rugby, IChE, 1977

For industrial conversion to 5-aminonaphthoquinone derivatives, dinitronaphthalene had been mixed cold with sulfuric acid and sulfur (to form sulfur dioxide), then heated to 120°C on over 100 occasions without incident. When dinitronaphthalene from a different supplier was used, the mixture exploded violently. Investigation in the safety calorimeter showed that an exothermic reaction begins at only 30°C, and that the onset and intensity of the exotherm (up to 400°C) markedly depends on quality of the dinitronaphthalene.

See REACTION SAFETY CALORIMETRY *See other* POLYNITROARYL COMPOUNDS

3243. 6-Quinolinecarbonyl azide
[78141-00-1] $C_{10}H_6N_4O$

Houben-Weyl, 1952, Vol. 8, 682
A sample heated above its m.p. (88°C) exploded violently.
See other ACYL AZIDES

3244. 2,4-Diethynylphenol
[30427-53-3] $C_{10}H_6O$

Kotlyarevskii, I. L. *et al., Chem. Abs.,* 1971, **74**, 53198
Oxidation of the methyl or propyl ethers gave insoluble polymeric solids which exploded on heating.
See other ACETYLENIC COMPOUNDS, POLYPEROXIDES

3245. Di-2-furoyl peroxide
[25639-45-6] $C_{10}H_6O_6$

Castrantas, 1965, 17
It explodes violently on friction and heating.
See other DIACYL PEROXIDES

3246. 1-Naphthalenediazonium perchlorate
[68597-02-4] $C_{10}H_7ClN_2O_4$

Hofmann, K. A. *et al., Ber.*, 1906, **39**, 3146
The salt explodes under light friction or pressure when dry.
See other DIAZONIUM PERCHLORATES

3247. 2-Naphthalenediazonium perchlorate
[69587-03-5] $C_{10}H_7ClN_2O_4$

Hofmann, K. A. *et al., Ber.*, 1906, **39**, 3146
The salt explodes under light friction or pressure when dry.
See other DIAZONIUM PERCHLORATES

3248. 2-Naphthalenediazonium trichloromercurate
[68448-47-5] $C_{10}H_7Cl_3HgN_2$

Nesmeyanov, A. M., *Org. Synth.*, 1943, Coll. Vol. 2, 433
The isolated double salt precipitated by mercury(II) chloride explodes violently if heated during drying.
See other DIAZONIUM SALTS, MERCURY COMPOUNDS

3249. 1-Nitronaphthalene
[86-57-7] $C_{10}H_7NO_2$

Nitric acid, Sulfuric acid
See Nitric acid: 1-Nitronaphthalene, Sulfuric acid

Tetranitromethane
See Tetranitromethane: Aromatic nitro compounds
See other NITROARYL COMPOUNDS

3250. 1-Nitroso-2-naphthol
[131-91-9] $C_{10}H_7NO_2$

1. Starobinskii, V. A. *et al., Chem. Abs.*, 1980, **92**, 7828
2. Grewer, T. *et al., Exothermic Decomposition*, Technical Report 01VD 159/0329 for Federal German Ministry for Res. Technol., Bonn. 1986

3. Grewer, T. *et al., Hazards from Pressure*, IChE Symp. Ser. No. 102, 1–9, Oxford, Pergamon, 1987

4. *CHETAH*, 1990, 187

1-Nitroso-2-naphthol, an intermediate in the preparation of 'Pigment green', undergoes an exothermic rearrangement at 124°C to an unstable material which appears to ignite spontaneously. Pigment green (an iron complex of a bisulfite derivative of the title compound precipitated with sodium carbonate in presence of additives) decomposes at 230°C with a large exotherm, and the instances of ignition observed during drying of the pigment may be connected with the presence of some free nitrosonaphthol in it [1]. Energy of decomposition (in range 110–180°C) measured by DSC as 1.409 kJ/g by DSC, and T_{ait24} was determined as the low figure of 78°C by adiabatic Dewar tests, with an apparent energy of activation of 209 kJ/mol [2]. The solid deflagrates at 0.74 cm/min, and will attain a maximum of 68 bar pressure in a closed system [3]. It is shock sensitive [4].

See other NITROSO COMPOUNDS, SELF-HEATING AND IGNITION INCIDENTS

3251. 1-Naphthalenediazonium salts
[15511-25-8] (ions) $C_{10}H_7N_2Z$

Ammonium sulfide, or Hydrogen sulfide, or Sodium sulfide
See entry DIAZONIUM SULFIDES AND DERIVATIVES

3252. 2-Naphthalenediazonium salts
[36097-38-8] (ions) $C_{10}H_7N_2Z$

Ammonium sulfide, or Hydrogen sulfide, or Sodium sulfide
See entry DIAZONIUM SULFIDES AND DERIVATIVES

3253. Naphthylsodium
[25398-08-7]

$C_{10}H_7Na$

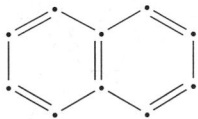

Chlorinated biphenyl
MCA Case History No. 565
A chlorinated biphenyl heat-transfer liquid was added to a burning batch of naphthylsodium to help to extinguish it. An exothermic reaction, followed by an explosion occurred. Sodium is known to react violently with many halogenated materials.
See Sodium: Halocarbons
See other ARYLMETALS

3254. Naphthalene
[91-20-3]

$C_{10}H_8$

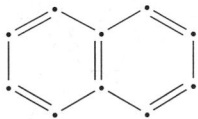

(*MCA SD-58*, 1956); *NSC 370*, 1978; *FPA H69*, 1978; *HCS 1980*, 672
Zavetskii, A. D., *Chem. Abs.*, 1983, **98**, 128923
Fires inside wood-packed benzene scrubbers in coke oven gas plants were attributed to saturation of the wood with naphthalene, and vapour-phase oxidation of the latter to phthalic anhydride, which participates in exothermic free radical chain reactions.

Aluminium chloride, Benzoyl chloride
See Aluminium chloride: Benzoyl chloride, etc.

Dinitrogen pentaoxide
See Dinitrogen pentaoxide: Naphthalene

Other reactants
Yoshida, 1980, 266
MRH values calculated for 13 combinations with oxidants are given.

3255. 2,2′-Bipyridyldichloropalladium(IV) perchlorate
[14871-92-2] (ion) $C_{10}H_8Cl_4N_2O_8Pd$

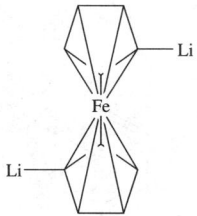

Gray, L. R. *et al., J. Chem. Soc., Dalton Trans.*, 1983, 133–141
Though no explosions occurred during preparation (by addition of 72% perchloric acid to solutions of the complex in conc. nitric/hydrochloric mixtures), this and related amino and phosphino analogues are potentially unstable.
See other AMMINEMETAL OXOSALTS

3256. μ,1,1′-Ferrocenediyldilithium (Dilithioferrocene)
[33272-09-2] $C_{10}H_8FeLi_2$
[174872-99-2] tetramethylethanediamine complex

Anhydrides
Rulkens, R. *et al., J. Amer. Chem. Soc.*, 1997, **119**(45), 10976
A synthetic procedure was devised which involved mixing this organometallic, as its TMEDA complex, with either solid benzenesulphonothioic acid anhydrosulphide $(PhSO_2)_2S$, or the equivalent selenide, at liquid nitrogen temperatures, and adding tetrahydrofuran. The first few drops of THF, which presumably allowed the dog to see the rabbit, produced vigorous interaction, with possible pressure generation. Reacting anhydrides with alkyl metals, in absence of diluents, may surely be expected to be violent.
See other ALKYLMETALS

3257. 1,3-Bis(isocyanomethyl)benzene
[ ] $C_{10}H_8N_2$

See entry DIISOCYANIDE LIGANDS *See related* CYANO COMPOUNDS

3258. 2,2′-Bipyridyl 1-oxide
  [33421-43-1] $C_{10}H_8N_2O$

Preparative hazard
  See Hydrogen peroxide: Acetic acid, *N*-Heterocycles
  See other N-OXIDES

3259. 2-Furaldehyde azine
  [5428-37-5] $C_{10}H_8N_2O_2$

Nitric acid
  See Nitric acid: Aromatic amines (reference 5), and : Hydrazines
  See related HYDRAZONES

3260. 2,2′-Oxybis(iminomethylfuran) mono-*N*-oxide (Dehydrofurfural oxime)
  [ ] $C_{10}H_8N_2O_4$

Ponzio, G. *et al.*, *Gazz. Chim. Ital.*, 1906, **36**, 338–344
It explodes at 130°C.
See other N-OXIDES, N–O COMPOUNDS

3261. Pentacyclo[4.2.0.0^{2,5}.0^{3,8}.0^{4,7}]octane-1,2-dicarboxylic acid (Cubane-1,2-dicarboxylic acid)
[129103-51-1] $C_{10}H_8O_4$

Eaton, P. E. *et al.*, *J. Org. Chem.*, 1990, **55**(22), 5746
Attempts to dehydrate this to the anhydride were unsuccessful; attempting to distil off monomeric products under vacuum at 80°C after a dehydration attempt with ethoxyacetylene led to a violent explosion. This is said to be atypical of cubanes, but:
See other STRAINED-RING COMPOUNDS
See entry CUBANES

3262. Poly(ethylene terephthalate)
[25038-59-9] $(C_{10}H_8O_4)_n$

Preparative hazard
See Dimethyl terephthalate: Ethylene glycol, Titanium butoxide

3263. 2,2′-Bi-5,6-dihydro-1,3-dithiolo[4,5-b][1,4]dithiinylidene
[66946-48-3] $C_{10}H_8S_8$

Preparative hazard
Larsen, J. *et al.*, *Org. Synth.*, 1995, **72**, 265

A procedure avoiding the sometimes explosive reaction of carbon disulfide with alkali metals is detailed.
See Carbon disulfide: Alkali metals

3264. 1-Naphthylamine
[134-32-7] $C_{10}H_9N$

Nitrous acid
See 1-Naphthalenediazonium salts and 2-Naphthalenediazonium salts

3265. 2-Naphthylamine
[91-59-8] $C_{10}H_9N$

Nitrous acid
See 1-Naphthalenediazonium salts and 2-Naphthalenediazonium salts

3266. 4-Acetoxy-3-methoxy-2-nitrobenzaldehyde
[2698-69-3] $C_{10}H_9NO_6$

Preparative hazard
See Nitric acid: 4-Acetoxy-3-methoxybenzaldehyde
See other ALDEHYDES, NITROARYL COMPOUNDS

1073

3267. Dipyridinesilver(I) perchlorate
[2152-13-7]

$[(C_5H_5N)_2Ag]\ ClO_4$

$C_{10}H_{10}AgClN_2O_4$

Acids
Kauffman, G. B. *et al., Inorg. Synth.*, 1960, **6**, 7, 8
Contact with acids, especially hot, must be avoided to prevent the possibility of violent explosion.
See other AMMINEMETAL OXOSALTS, SILVER COMPOUNDS

3268. 1,4-Bis(1,2-dibromoethyl)benzene
[25393-98-0]

$C_{10}H_{10}Br_4$

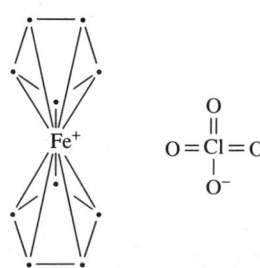

Potassium hydroxide
Kotlyarevskii, I. L. *et al., Chem. Abs.*, 1978, **89**, 23858
Tetra-dehydrobromination of the compound in benzene with ethanolic potassium hydroxide for the commercial production of 1,4-diethynylbenzene is highly exothermic and needs careful control to avoid hazard.
See related HALOALKANES

3269. Ferrocenium perchlorate
[12312-33-3]

$C_{10}H_{10}ClFeO_4$

See 1,3-Di[bis(cyclopentadienyl)iron]-2-propen-1-one: Perchloric acid, etc.
See other ORGANOMETALLIC PERCHLORATES

3270. 1-Chloro-4-(2-nitrophenyl)-2-butene
[7318-34-5] $C_{10}H_{10}ClNO_2$

Acheson, R. M. *et al., J. Chem. Soc., Perkin Trans. 1*, 1978, 1119
It may explode during distillation at 112–114°C/60 mbar.
See other ALLYL COMPOUNDS, NITROARYL COMPOUNDS

3271. Dicyclopentadienylperoxyniobium chloride
[75109-11-4] $C_{10}H_{10}ClNbO_2$

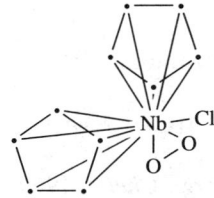

Sala-pala, J. *et al., J. Mol. Catal.*, 1980, **7**, 142–143
The compound and its methyl homologue exploded occasionally when scraped
from the walls of a tube, and invariably when exposed to a Raman laser beam,
even at −80°C.
See other IRRADIATION DECOMPOSITION INCIDENTS, ORGANOMETALLIC PEROX-
IDES

3272. 4-(2,4-Dichlorophenoxy)butyric acid
[94-82-6] $C_{10}H_{10}Cl_2O_3$

Preparative hazard
See γ-Butyrolactone: 2,4-Dichlorophenol, etc.

3273. Dicyclopentadienylchromium (Chromocene)
[1271-24-5] $C_{10}H_{10}Cr$

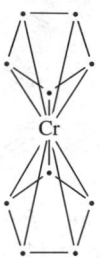

Alcohols
Votinsky, J. *et al., Coll. Czech. Chem. Comm.*, 1979, **44**, 82
Reaction with C_1-C_4 alcohols at 20°C produces pyrophoric dialkoxychromiums and cyclopentadiene.
See METAL ALKOXIDES *See other* ORGANOMETALLICS

3274. Oxodiperoxodipyridinechromium(VI)
[ ] $C_{10}H_{10}CrN_2O_5$
$$[(C_5H_5N)_2CrO(O_2)_2]$$

1. Caldwell, S. H. *et al., Inorg,. Chem.*, 1969, **8**, 151
2. Adams, D. M. *et al., J. Chem. Educ.*, 1966, **43**, 94
3. Collins, J. C. *et al., Org. Synth.*, 1972, **52**, 5–8
4. Wiede, O. F., *Ber.*, 1897, **30**, 2186
5. Glaros, G., *J. Chem. Educ.*, 1978, **55**, 410

This complex, formerly called 'pyridine perchromate' and now finding application as a powerful and selective oxidant, is violently explosive when dry [1]. Use while moist on the day of preparation and destroy any surplus with dilute alkali [2]. Preparation and use of the reagent have been detailed further [3]. The analogous complexes with aniline, piperidine and quinoline may be similarly hazardous [4]. The damage caused by a 1 g sample of the pyridine complex exploding during desiccation on a warm day was extensive. Desiccation of the aniline complex had to be at ice temperature to avoid violent explosion [4]. Pyridinium chlorochromate is commercially available as a safer alternative oxidant of alcohols to aldehydes [5].
See Chromium trioxide: Pyridine
        Dipyridinium dichromate
See other AMMINECHROMIUM PEROXOCOMPLEXES

3275. Bis(cyclopentadienyldinitrosylchromium)
[66649-16-9] $C_{10}H_{10}Cr_2N_4O_4$

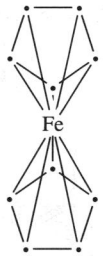

Flitcroft, N. *et al., Chem. & Ind.*, 1969, 201
A small sample exploded violently upon laser irradiation for Raman spectroscopy.
See other IRRADIATION DECOMPOSITION INCIDENTS, ORGANOMETALLICS

3276. Ferrocene
[102-54-5] $C_{10}H_{10}Fe$

Mercury(II) nitrate
  See Mercury(II) nitrate: Ferrocene

Tetranitromethane
  See Tetranitromethane: Ferrocene
  See other ORGANOMETALLICS

3277. Bis(cyclopentadienyl)magnesium
[1284-72-6] $C_{10}H_{10}Mg$

Barber, W. A., *Inorg. Synth.*, 1960, **6**, 15

It may ignite on exposure to air.
See other ORGANOMETALLICS

3278. Bis(cyclopentadienyl)manganese
[1271-27-8] $C_{10}H_{10}Mn$

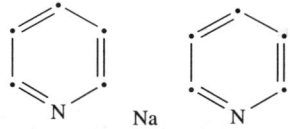

491M, 1975, 248

It ignites in air.
See other ORGANOMETALLICS

3279. Oxodiperoxodi(pyridine *N*-oxide)molybdenum
[73680-23-6] $C_{10}H_{10}MoN_2O_7$

Westland, A. D. *et al., Inorg. Chem.*, 1980, **19**, 2257
The analogous complex of tungsten, and that of chromium with a single *N*-oxide
ligand, were also explosive.
See related AMMINECHROMIUM PEROXOCOMPLEXES

3280. Dipyridinesodium
[101697-88-5] $C_{10}H_{10}N_2Na$

Sidgwick, 1950, 89
The addition product of sodium and pyridine (or its 2-methyl derivative) ignites
in air.
See related *N*-METAL DERIVATIVES

3281. 4a,5,7a,8-Tetrahydro-4,8-methano-4H-indeno[5,6-c]-1,2,5-oxadiazole 1-oxide (Dicyclopentadienefurazan N-oxide)

[ ] $C_{10}H_{10}N_2O_2$

Crosby, J. *et al.*, Br. Pat. 1 474 693, 1977
The solid isomeric mixture of the 1-oxides explodes at 80–85°C, but may be handled safely in solution.
See other FURAZAN N-OXIDES

3282. 4a,5,7a,8-Tetrahydro-4,8-methano-4H-indeno[5,6-c]-1,2,5-oxadiazole 3-oxide (Dicyclopentadienefurazan N-oxide)

[ ] $C_{10}H_{10}N_2O_2$

Crosby, J. *et al.*, Br. Pat. 1 474 693, 1977
The solid isomeric mixture of the 3-oxide explodes at 80–85°C, but may be handled safely in solution.
See other FURAZAN N-OXIDES

3283. Oxodiperoxodi(pyridine N-oxide)tungsten

[ ] $C_{10}H_{10}N_2O_7W$

Westland, A. D. *et al., Inorg. Chem.*, 1980, **19**, 2257

The analogous complex of molybdenum, and that of chromium with a single N-oxide ligand, were also explosive.

See related AMMINECHROMIUM PEROXOCOMPLEXES

3284. N,N'-Dimethyl-N,N'-dinitrosoterephthalamide
[133-55-1] $C_{10}H_{10}N_4O_4$

2-Hydroxyethylamine
MCA Case History No. 2308
Addition of the dry nitrosamide to a flask apparently containing residual ethanolamine led to evolution and ignition of diazomethane and the dry nitrosamide powder.
See Diazomethane
See other NITROSO COMPOUNDS

3285. Bis(cyclopentadienyl)tungsten diazide oxide
[53504-80-6] $C_{10}H_{10}N_6OW$

Anand, S. P., *J. Inorg. Nucl. Chem.*, 1974, **36**, 926
It is highly explosive.
See related METAL AZIDES

3286. Bis(cyclopentadienyl)vanadium diazide
[85836-55-1] $C_{10}H_{10}N_6V$

$$N^- = N^+ = N - V - N = N^+ = N^-$$

Blandy, C. *et al., Tetrahedron. Lett.*, 1983, **24**, 4189–4192
It is explosive in the solid state.
See related METAL AZIDES

3287. Nickelocene
[1271-28-9] $C_{10}H_{10}Ni$

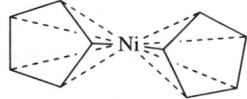

See 1-(1-Methyl-1-phenylethyl)-4-(2-propynyloxy)benzene
See other ORGANOMETALLICS

3288. 2-Indanecarboxaldehyde
[37414-44-1] $C_{10}H_{10}O$

1. Black, T. H., *Chem. Eng. News*, 1988, **66**(32), 2
2. Black, T. H., private comm., 1988
A violent explosion occurred at the start of bulb-tube distillation of a 2 g sample
of the crude aldehyde (b.p. 97°/4 mbar) at 75°C/0.13 mbar [1]. It seems likely that
the crude product could have formed peroxide during the 2 day's storage (in a
closed but not hermetically sealed vessel) between preparation and distillation. The
structure is such that the proton on C-2, which effectively has 2 benzyl substituents
and an adjacent carbonyl function, seems rather susceptible to autoxidation [2].
See other ALDEHYDES, PEROXIDISABLE COMPOUNDS

3289. 1-Oxo-1,2,3,4-tetrahydronaphthalene
  [529-34-0] $C_{10}H_{10}O$

Preparative hazards
  See Chromium trioxide: Acetic anhydride
    Hydrogen peroxide: Acetone, etc.

3290. 4-Acetoxy-3-methoxybenzaldehyde
  [881-68-5] $C_{10}H_{10}O_4$

Nitric acid
  See Nitric acid: 4-Acetoxy-3-methoxybenzaldehyde
  See other ALDEHYDES

3291. 1,3-Diacetoxybenzene (Resorcinol diacetate)
  [108-58-7] $C_{10}H_{10}O_4$

Nitric acid
  See 4,6-Dinitro-1,3-benzenediol

1082

3292. Dimethyl terephthalate
[120-61-6] $C_{10}H_{10}O_4$

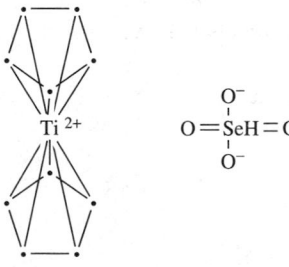

The finely powdered ester is a significant dust explosion hazard.
See entry DUST EXPLOSION INCIDENTS (reference 22)

Ethylene glycol, Titanium butoxide
1. Rix, G. C., private comm., 1977
2. Goodman, I. in *Encyclopaedia of Polymer Science and Technology*, Mark, H. F.
   (Ed.), Vol. 11, 112–113, New York, Interscience, 1969

The polymer had been produced for 12 years by transesterification under nitrogen of the dimethyl ester with ethylene glycol at 250°C in presence of titanium butoxide catalyst. After increasing the heating capacity of the vessel from a half coil to a full coil, 3 incidents of ignition of vapour after opening the vessel were noted [1]. This is attributed to formation and ignition of mixtures of acetaldehyde or dioxane with ingressing air on the hot vessel surfaces. Acetaldehyde is produced by thermal degradation of the polymer [2], and has rather wide flammability limits (4–57%) and is readily ignited on hot or corroded surfaces. Dioxane, readily formed by heating ethylene glycol with acids, metal oxides or other catalysts, also has wide limits (2–22%) and a low autoignition temperature (180°C). It seems likely that if either of these were present, perhaps with residual traces of methanol vapour, migration towards an open vent would produce turbulence and admixture with air.
See related SELF-HEATING AND IGNITION INCIDENTS

3293. Bis(cyclopentadienyl)titanium selenate
[87612-48-4] $C_{10}H_{10}O_4SeTi$

Malek, J. *et al., Z. Chem.*, 1983, **23**, 189

It is explosive, sensitive to heat or friction.
See other ORGANOMETALLICS   *See related* ORGANOMETALLIC NITRATES

3294. Bis(cyclopentadienyl)lead
   [1294-74-2] $C_{10}H_{10}Pb$

$$(C_5H_5)_2Pb$$

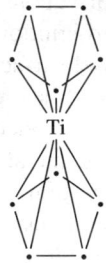

Preparative hazard
   See Lead(II) nitrate: Cyclopentadienylsodium
   See other ORGANOMETALLICS

3295. Bis(cyclopentadienyl)titanium
   [1271-29-9] $C_{10}H_{10}Ti$

Bailar, 1973, Vol. 3, 389
Pyrophoric crystals (but see Polar solvents, below).

Polar solvents
   Watt, G. W. *et al., J. Amer. Chem. Soc.*, 1966, **88**, 1139–1140
   Though not pyrophoric, it reacts very vigorously with oxygen-free water and other
   polar solvents.
   See other ORGANOMETALLICS

3296. Bis(cyclopentadienyl)zirconium
  [12116-83-5] $C_{10}H_{10}Zr$

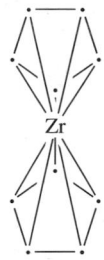

Watt, G. W. *et al., J. Amer. Chem. Soc.*, 1966, **88**, 5926
A pyrophoric solid.
See other ORGANOMETALLICS

3297. 3-(4-Chlorophenyl)butanoic acid
  [1012-17-5] $C_{10}H_{11}ClO_2$

Preparative hazard
  See Peroxyacetic acid: 5-(4'-Chlorophenyl)-2,2-dimethyl-3-hexanone
  See other ORGANIC ACIDS

3298. Dimethyl(phenylethynyl)thallium
  [10158-43-7] $C_{10}H_{11}Tl$

Nast, R. *et al., J. Organomet. Chem.*, 1966, **6**, 461
May explode on stirring, heating or impact.
See other ALKYLMETALS, METAL ACETYLIDES

†3299. Dicyclopentadiene
[77-73-6] $C_{10}H_{12}$

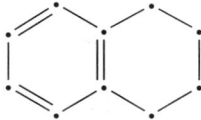

HCS 1980, 387

Ahmed, M., *et al., Plant/Oper. Progr.*, 1991, **10**(3), 143
The compound is capable of exothermic pressure generating decompositions, from
about 170°C.
See other DIENES, GAS EVOLUTION INCIDENTS

3300. Tetrahydronaphthalene
[119-64-2] $C_{10}H_{12}$

HCS 1980, 893; *RSC Lab. Hazards Safety Data Sheet No. 78*, 1989

1. Grignard, 1949, Vol. 17.1, 673
2. Duckworth, D. F., private comm., 1977
3. McDonald, J. S., *Chem. Abs.*, 1960, **54**, 15933
Readily peroxidises in air, and explosions have occurred during large-scale distillations [1], particularly towards the end when sufficiently concentrated [2]. Use
of tetralin as secondary coolant in a molten sodium heat transfer system led to an
explosion due to peroxide formation [3].
See 1,2,3,4-Tetrahydro-1-naphthyl hydroperoxide
See other PEROXIDISABLE COMPOUNDS

3301. 4-Chloro-2,5-diethoxynitrobenzene
[91-43-0] $C_{10}H_{12}ClNO_4$

CHETAH, 1990, 183

Although not sensitive to mechanical shock, this compound can be exploded by a detonator.

See other HALOARYL COMPOUNDS, NITROARYL COMPOUNDS

3302. Ethylphenylthallium(III) acetate perchlorate
[34521-47-6] $C_{10}H_{12}ClO_6Tl$

Preparative hazard
  See Perchloric acid: Ethylbenzene, Thallium triacetate
  See related METAL PERCHLORATES

3303. 6,6′-Dihydrazino-2,2′-bipyridylnickel(II) perchlorate
[53346-21-7] $C_{10}H_{12}Cl_2N_6NiO_8$

Lewis, J. *et al., J. Chem. Soc., Dalton Trans.*, 1977, 738

It detonates violently on heating, but is stable to shocks.

See other AMMINEMETAL OXOSALTS

3304. Dipyridinium dichromate
[20039-37-6] $C_{10}H_{12}Cr_2N_2O_7$

Preparative hazard
  1. Coates, W. M. *et al., Chem. & Ind.,* 1969, 1594
  2. Fieser, M. *et al., Reagents for Organic Synth.,* 1981, Vol. 9, 399
  3. Corey, E. J. *et al., Tetrahedron Lett.,* 1979, 399
  4. Salmon, J., *Chem. & Ind.,* 1982, 616; *Chem. Brit.,* 1982, **18**, 703
  5. Gill, P. *et al., Thermochim. Acta,* 1984, **80**, 193–196

It is known that preparation of the oxidant salt under anhydrous conditions is explosion-prone [1], but during repetition of a supposedly safe preparative method [2], recommended for large-scale use [3], ignition and a violent explosion occurred. Use of more water to dissolve completely the chromium trioxide, and a reaction temperature below 25°C, are recommended [4]. During a study by TGA of the thermal degradation of the salt, too-rapid heating of the samples led to explosions with fire [5].

See other DICHROMATE SALTS OF NITROGENOUS BASES, OXIDANTS

3305. Sodium ethylenediaminetetraacetate
[67401-50-7] $C_{10}H_{12}N_2Na_4O_8$

Sodium hydroxide, Sodium hypochlorite
  See Sodium hypochlorite: Sodium ethylenediaminetetraacetate, etc.

1088

3306. 4,6-Dinitro-2-*sec*-butylphenol
[88-85-7] $C_{10}H_{12}N_2O_5$

Nolan, 1983, Case history 99
A batch became overheated during drying (leaking steam valve) and decomposed violently.
See other POLYNITROARYL COMPOUNDS

3307. *O*, *O*-Dimethyl-*S*-[(4-oxo-1,2,3-benxotriazin-3[4*H*]-yl)methyl] phosphorodithioate (Azinphos-methyl)
[86-50-0] $C_{10}H_{12}N_3O_3PS_2$

Kirschner, E., *Chem. Eng. News*, 1997, **75**(20), 12
A smouldering bag in a pesticide warehouse, believed to be of this, led to an explosion, killing three firemen, and fire which took six days to extinguish (possibly because of caution concerning anti-cholinesterase toxicity). This is a moderately high energy compound by virtue of the triazene function. Other pesticides present were of less energetic structure.
See other TRIAZENES

3308. 2-Dimethylamino-3-methylbenzoyl azide
[161616-67-7] $C_{10}H_{12}N_4O$

Waldron, N. M. *et al.*, *J. Chem. Soc., Chem. Comm.*, 1995, 81

With only 20% azide nitrogen by weight, this azide might be thought safe for cautious isolation. A 1 g sample isolated by evaporation of solvent at −10°C, exploded violently and spontaneously on warming to 0°C.
See other ACYL AZIDES

3309. 3,4-Dimethyl-4-(3,4-dimethyl-5-isoxazolylazo)isoxazolin-5-one
[4100-38-3] $C_{10}H_{12}N_4O_3$

Boulton, A. J. *et al.*, *J. Chem. Soc.*, 1965, 5415
It invariably decomposed explosively if heated rapidly to 100°C, but was stable to impact or friction.
See other AZO COMPOUNDS, ISOXAZOLES, N–O COMPOUNDS

3310. 1,3,5,7-Tetranitroadamantane
[75476-36-7] $C_{10}H_{12}N_4O_8$

Gilbert, E. E. *et al.*, US Pat. 4 329 522, 1982
It shows explosive properties
See related POLYNITROALKYL COMPOUNDS

3311. Isobutyrophenone
[611-70-1] $C_{10}H_{12}O$

Bromine
See Bromine: Isobutyrophenone

3312. 1,2,3,4-Tetrahydro-1-naphthyl hydroperoxide
[771-29-9] $C_{10}H_{12}O_2$

Hock, H. *et al., Ber.*, 1933, **66**, 61
It explodes on superheating the liquid.

Lithium tetrahydroaluminate
Sutton, D. A., *Chem. & Ind.*, 1951, 272
Interaction in ether is vigorously exothermic.
See other REDOX REACTIONS *See related* ALKYL HYDROPEROXIDES

3313. 3-Methoxy-3-methyl-5,6-benzo-6*H*-1,2-dioxin
[69681-99-8, *cis*-; 69682-00-4, *trans*-] $C_{10}H_{12}O_3$

Lerdal, D. *et al., Tetrahedron Lett.*, 1978, 3228
Both isomers decomposed violently on heating in absence of solvent.
See other CYCLIC PEROXIDES

3314. 3-Methoxy-4-methyl-5,6-benzo-6*H*-1,2-dioxin
[69682-01-5] $C_{10}H_{12}O_3$

Lerdal, D. *et al., Tetrahedron Lett.*, 1978, 3228
It decomposed violently on heating in absence of solvent.
See other CYCLIC PEROXIDES

3315. Allyl 4-toluenesulfonate
[4873-09-0]

$C_{10}H_{12}O_3S$

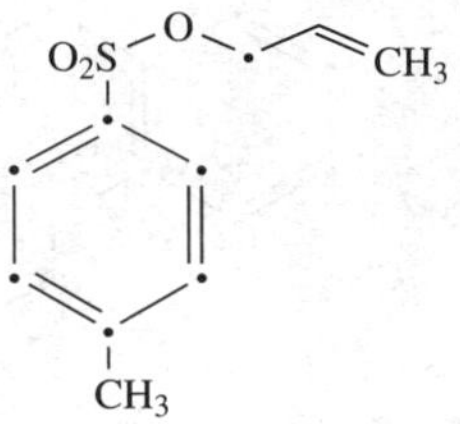

Rüst, 1948, 302
Explosion and charring of the ester during high-vacuum distillation from an oil-bath at 110°C was ascribed to exothermic polymerisation, (probably catalysed by acidic decomposition products).
See other ALLYL COMPOUNDS, POLYMERISATION INCIDENTS, SULFUR ESTERS

3316. 2-Buten-1-yl benzenesulfonate
[20443-67-8]

$C_{10}H_{12}O_3S$

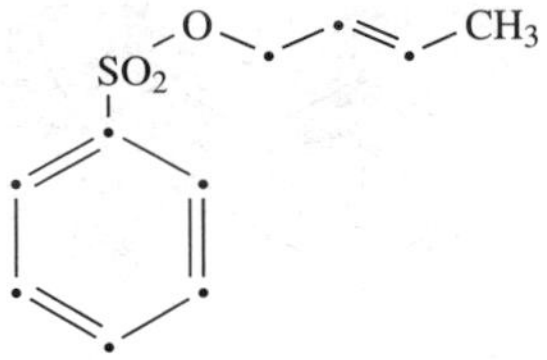

Sorbe, 1968, 122
After evaporation of solvent, the ester may explode in contact with air (possibly owing to exothermic polymerisation catalysed by peroxide formation).
See other POLYMERISATION INCIDENTS, SULFUR ESTERS

3317. 1-Phenylboralane
[1075-17-8]

$C_{10}H_{13}B$

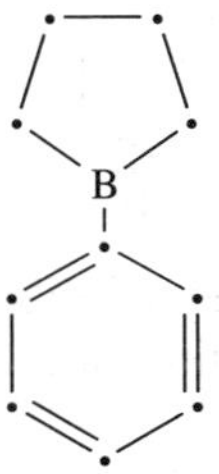

Leleu, *Cahiers*, 1977, (88), 365

It ignites in air.

See related ALKYLBORANES

3018. 2-(4-Chlorophenyl)-1,1-dimethylethyl hydroperoxide
[71356-36-0] $C_{10}H_{13}ClO_2$

Preparative hazard
1. Kochi, J. B. *et al., J. Amer. Chem. Soc.*, 1964, **86**, 5268
2. Parris, M. *et al., Can. J. Chem.*, 1979, **57**, 1233–1234
Preparation by the method used for the 2-phenyl analogue [1] is hazardous [2].
(No further details)
See Hydrogen peroxide: 2-Phenyl-1,1-dimethylethanol, etc.
See other ALKYL HYDROPEROXIDES

3319. 4-*tert*-Butyliodylbenzene
[64297-75-2] $C_{10}H_{13}IO_2$

Ranganathan, S. *et al., Tetrahedron Lett.*, 1985, **26**, 4955–4956
Although a stable crystalline compound (m.p. 217–221°C), its oxidation potential
approximates to that of ozone, and it should only be used in a suitably inert solvent
(chlorobenzene or nitrobenzene).
See other IODINE COMPOUNDS, OXIDANTS

3320. Diethyl trifluoroacetosuccinate
[94633-25-7] $C_{10}H_{13}F_3O_5$

Preparative hazard
See Sodium hydride: Diethyl succinate, etc.

3321. 4-Dimethylaminoacetophenone
[18925-69-4] $C_{10}H_{13}NO$

Preparative hazard
See 4-Chloroacetophenone: Dimethylamine

3322. 2-Formylamino-1-phenyl-1,3-propanediol
[51317-78-3] $C_{10}H_{13}NO_3$

Nitric acid
See Nitric acid: 2-Formylamino-1-phenyl-1,3-propanediol

1094

3323. Butylbenzene
[104-51-8]

C₁₀H₁₄

n-Bu

Preparative hazard
See Sodium: Halocarbons (reference 7)

3324. Bis(cyclopentadienyl)niobium tetrahydroborate
[37298-41-2]

$C_{10}H_{14}BNb$

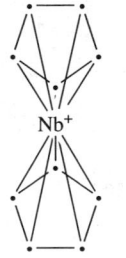

Lucas, C. R., *Inorg. Synth.*, 1976, **16**, 109–110
It is pyrophoric in air, particularly after sublimation.
See other COMPLEX HYDRIDES, ORGANOMETALLICS

3325. Calcium 2,4-pentanedionate
[19372-47-2]

$C_{10}H_{14}CaO_4$

Ca(OCMe=CHCO.Me)₂ bicyclic chelate

Barium perchlorate
See Barium perchlorate: Calcium 2,4-pentanedionate

3326. Bis(2,4-pentanedionato)chromium
[14024-50-1]

$C_{10}H_{14}CrO_4$

Ocone, L. R. *et al., Inorg. Synth.*, 1966, **8**, 131

It ignites in air.

See Chromium(II) acetate
See related METAL ALKOXIDES, ORGANOMETALLICS

3327. Diaquabis(cyclopentadienyl)titanium dichromate
[87612-49-5] $\qquad$ $C_{10}H_{14}Cr_2O_9Ti$

Malek, J. *et al, Z. Chem.*, 1983, **23**, 189
Explosive, sensitive to heat or friction.
See other ORGANOMETALLICS

3328. Diethyl 4-nitrophenyl thionophosphate
[25737-28-4] $\qquad$ $C_{10}H_{14}NO_5PS$

1. Bright, N. H. F. *et al., J. Agr. Food Chem.*, 1950, **1**, 344
2. Gweek, B. F. *et al., Ind. Eng. Chem.*, 1959, **51**, 104–112

Vacuum distillation of 'parathion' at above 100°C is hazardous, frequently leading to violent decomposition [1]. Following a plant explosion, the process design was modified and featured a high degree of temperature sensing and control to avoid a recurrence [2].

Endrin

During the blending of the two insecticides (endrin is a halogenated polycyclic epoxide) into a petroleum solvent, an unexpected exothermic reaction occurred which vaporised some solvent and led to a vapour–air explosion. Faulty agitation was suspected.
See other AGITATION INCIDENTS, NITROARYL COMPOUNDS, PHOSPHORUS ESTERS

3329. Diethyl 4-nitrophenyl phosphate
[311-45-5] $C_{10}H_{14}NO_6P$

Preparative hazard
See Diethyl phosphite: 4-Nitrophenol
See other NITROARYL COMPOUNDS, PHOSPHORUS ESTERS

3330. 1,4-Bis(isocyanomethyl)cyclohexane
[ ] $C_{10}H_{14}N_2$

See entry DIISOCYANIDE LIGANDS

3331. 4-Azido-*N*,*N*-dimethylaniline
[24573-95-3] $C_{10}H_{14}N_4$

491M, 1975, 51

It decomposed violently on attempted distillation.
See other ORGANIC AZIDES

3332. 2-Phenyl-1,1-dimethylethyl hydroperoxide
[1944-83-8]   C$_{10}$H$_{14}$O$_2$

Preparative hazard
See Hydrogen peroxide: 2-Phenyl-1,1-dimethylethanol, etc.
See other ALKYL HYDROPEROXIDES

3333. 2-Chlorobenzylidenemalononitrile ('CS gas')
[2698-41-1]   C$_{10}$H$_{15}$ClN$_2$

The finely powdered nitrile is a significant dust explosion hazard.
See entry DUST EXPLOSION INCIDENTS (reference 22)
See other CYANO COMPOUNDS

3334. *N*-Methoxy-*N*-methylbenzylidenimmonium methylsulfate
[ ]   C$_{10}$H$_{15}$NO$_5$S

Volpiana,–, *Proc. 1st Int. Symp. Prev. Risks Chem. Ind.*, 176–179, Heidelberg, ISSA, 1970

Quaternation of *O*-methylbenzaldoxime was effected by addition to dimethyl sulfate at 110°C with cooling to maintain that temperature for 2 hours more. Cessation of temperature control allowed a slow exotherm to proceed unnoticed, and at 140–150°C a violent decomposition set in.

See related OXIMES *See other* N–O COMPOUNDS

3335. 1-Phenyl-3-*tert*-butyltriazene
[63246-75-3] $C_{10}H_{15}N_3$

Rondevstedt, C. S. *et al., J. Org. Chem.*, 1957, **22**, 203
It exploded on attempted distillation at below 1 mbar.
See other TRIAZENES

3336. Δ3-Carene (3,7,7-Trimethylbicyclo[4.1.0]hept-3-ene)
[13466-78-9] $C_{10}H_{16}$

Nitric acid
　See Nitric acid: Hydrocarbons (reference 15)
　See related ALKENES

3337. Limonene (4-Isopropenyl-1-methylcyclohexene)
[5989-27-5] $C_{10}H_{16}$

HCS 1980, 591

Iodine pentafluoride, Tetrafluoroethylene
　See Tetrafluoroethylene: Iodine pentafluoride, etc.

Sulfur
See Sulfur: Limonene
See other DIENES, PEROXIDISABLE COMPOUNDS

3338. 1,5-*p*-Menthadiene (5-Isopropyl-2-methyl-1,3-cyclohexadiene) [99-83-2]

$C_{10}H_{16}$

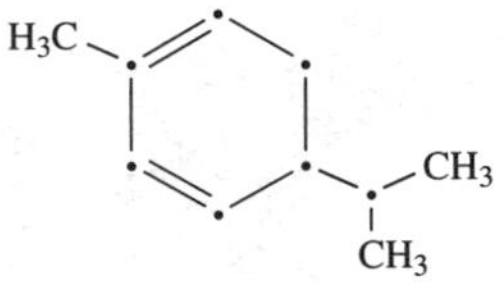

Bodendorf, K., *Arch. Pharm.*, 1933, **271**, 28
The terpene readily peroxidises with air, and the (polymeric) peroxidic residue exploded violently on attempted distillation at 100°C/0.4 mbar.
See other DIENES, PEROXIDISABLE COMPOUNDS, POLYPEROXIDES

3339. 2-Pinene (2,6,6-Trimethylbicyclo[3.1.1]hept-2-ene) [80-56-8]

$C_{10}H_{16}$

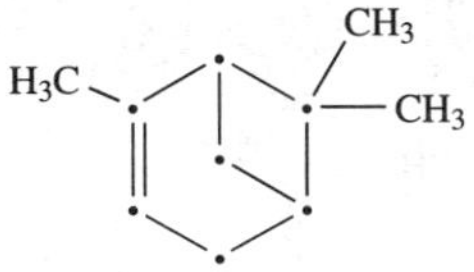

Nitrosyl perchlorate
See Nitrosyl perchlorate: Organic materials
See related ALKENES

3340. 1-(4,5-Dimethyl-1,3-diselenonylidene)piperidinium perchlorate [53808-74-5]

$C_{10}H_{16}ClNO_4Se_2$

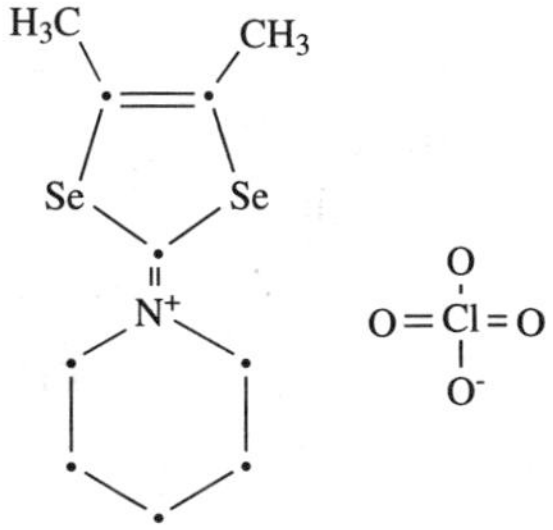

See entry 1-(1,3-DISELENONYLIDENE)PIPERIDINIUM PERCHLORATES
See other PERCHLORATE SALTS OF NITROGENOUS BASES

3341. Sebacoyl chloride
[111-19-3] $C_{10}H_{16}Cl_2O_2$

Hüning, S. *et al., Org. Synth.*, 1963, **43**, 37
During vacuum distillation of the chloride at 173°C/20 mbar, the residue frequently decomposes spontaneously, producing a voluminous black foam.
See other ACYL HALIDES

3342. Benzyltrimethylammonium permanganate
[76710-75-3] $C_{10}H_{16}MnNO_4$

It explodes at 80–90°C.
See entry QUATERNARY OXIDANTS

3343. 1,8-Diisocyanooctane
[ ] $C_{10}H_{16}N_2$

See entry DIISOCYANIDE LIGANDS

3344. Sodium 1,3-dihydroxy-1,3-bis(*aci*-nitromethyl)-2,2,4,4-tetramethylcyclo-butandiide

[ ] $C_{10}H_{16}N_2Na_2O_6$

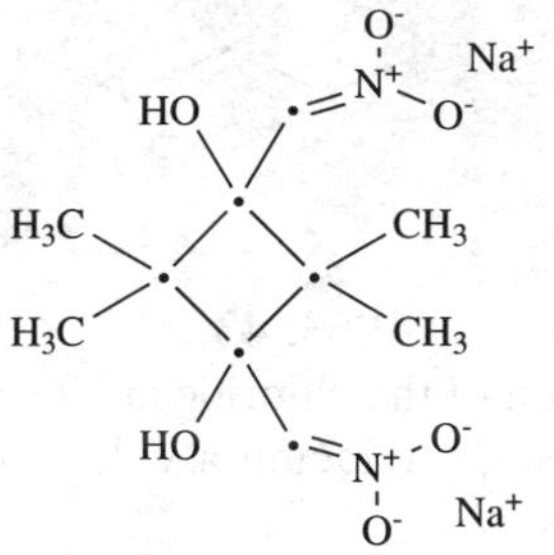

Water

Dauben, H. J., *Org. Synth.*, 1963, Coll. Vol. 4, 223

The dry powdered condensation product of sodium *aci*-nitromethanide (2 mol) with dimethylketene dimer exploded violently when added to crushed ice.

See Sodium *aci*-nitromethanide: Water

See other aci-NITRO SALTS

3345. 2,2′-Azoisovaleronitrile

[70843-45-7] $C_{10}H_{16}N_4$

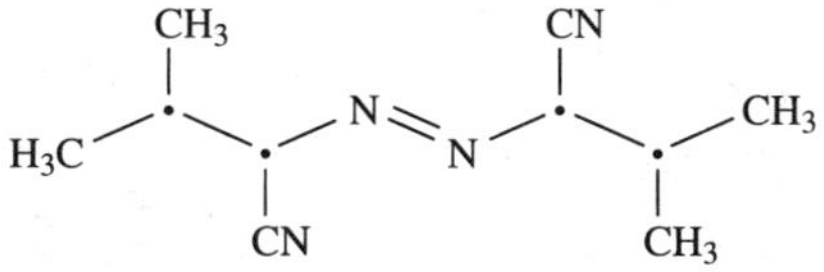

See entry BLOWING AGENTS *See other* AZO COMPOUNDS, CYANO COMPOUNDS

3346. 3,7-Dimethyl-2,6-octadienal (Citral)

[5392-40-5] $C_{10}H_{16}O$

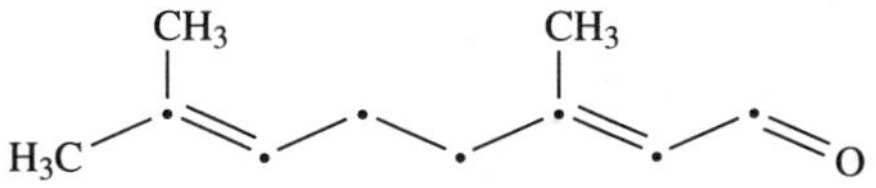

Preparative hazard

See 2-Methyl-1,3-butadiene: Acetone

See related DIENES

1102

3347. 1,4-Epidioxy-2-*p*-menthene (Ascaridole)
[512-85-6] $C_{10}H_{16}O_2$

1. Castrantas, 1965, 15
2. Henry, T. A. *et al., J. Chem. Soc.*, 1921, **119**, 1722
3. Walton, F. P., private comm., 1980
Explosive decomposition occurs on heating from 130 to 150°C [1]. When heated above 160°C, explosive rearrangement occurs with an exotherm to 300°C. An explosion occurred during fractional distillation at 108–110°C/20 mbar. It may be distilled without explosion by a flash evaporation technique [2,3].
See other CYCLIC PEROXIDES

3348. 3-Peroxycamphoric acid (1,2,2-Trimethyl-1-carboxycyclopentane-3-peroxycarboxylic acid)
[31921-03-6] $C_{10}H_{16}O_5$

1. Milas, N. A. *et al., J. Amer. Chem. Soc.*, 1933, **55**, 350
2. Pirkle, W. H. *et al., J. Org. Chem.*, 1977, **42**, 2080–2082
Explosive decomposition occurs at 80–100°C [1], but the isomeric 1-peroxy acid may also have been present in the sample [2].
See other ORGANIC ACIDS, PEROXYACIDS

3349. 7,7,10,10-Tetramethyl-1,2,5,6-tetroxecane-3,4-dione (2,5-Dimethyl-2,5-hexylenebis diperoxyoxalate)
[35551-00-7] $C_{10}H_{16}O_6$

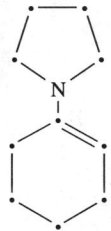

Preparative hazard

Adam, W. *et al., J. Amer. Chem. Soc.*, 1980, **102**, 4803

The preparation is hazardous because a highly explosive peroxidic polymer is formed as the major product. After extraction of the title product with pentane, the residue must be disposed of forthwith to avoid the possibility of detonation of the polymer as it dries, as happened on one occasion.

See other CYCLIC PEROXIDES, PEROXYESTERS

3350. 1-Pyrrolidinylcyclohexene
[1125-99-1] $C_{10}H_{17}N$

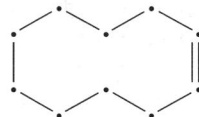

Benzenediazonium-2-carboxylate

See Benzenediazonium-2-carboxylate: 1-Pyrrolidinylcyclohexene

See related ALKENES

3351. *cis*-Cyclododecene
[2198-20-1] $C_{10}H_{18}$

Preparative hazard

See Hydrazine: Metal salts (reference 3)

See other ALKENES

1104

3352. Citronellic acid (3,7-Dimethyl-6-hexenoic acid)
[502-47-6]

$C_{10}H_{18}O_2$

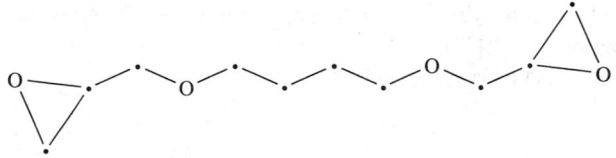

Ozone
See Ozone: Citronellic acid
See other ORGANIC ACIDS

3353. 1,4-Bis(2,3-epoxypropoxy)butane
[2425-79-8]

$C_{10}H_{18}O_4$

Trichloroethylene
See Trichloroethylene: Epoxides
See other 1,2-EPOXIDES

3354. Di-2-methylbutyryl peroxide
[1607-30-3]

$C_{10}H_{18}O_4$

(EtCHMeCO.O-)$_2$

Swern, 1971, Vol. 2, 815
Pure material explodes on standing at ambient temperature.
See other DIACYL PEROXIDES

3355. 1,2-Bis(acetoxyethoxy)ethane (Triethylene glycol diacetate)
[111-21-7]

$C_{10}H_{18}O_6$

Acetic acid
Kolodner, H. J. *et al., Proc. 3rd Int. Symp. Loss Prev. Safety Prom. Proc. Ind.,* 345–355, Basle, SSCI, 1980

When fractional distillation at 150°C/27 mbar of a mixture of the ester with acetic acid (36% mol) was interrupted by admission of air, smoking and gas evolution occurred. This was attributed to cool flame auto-ignition and detailed investigation confirmed this. Nitrogen should always be used to break vacuum, and further precautions are recommended.

See 2-Ethoxyethyl acetate
See COOL FLAMES

3356. *tert*-Butyl diperoxyoxalate
[14666-77-4] $C_{10}H_{18}O_6$

$$(Me_3COOCO.-)_2$$

1. Castrantas, 1965, 17
2. Nutter, D. E. *et al., Chem. Abs.*, 1997, **126**, 263799v

It exploded on removal from a freezing mixture [1]. Now attracts warnings as 'highly explosive' [2].

See other PEROXYESTERS

3357. 3-Buten-1-ynyltriethyllead
[30478-96-7] $C_{10}H_{18}Pb$

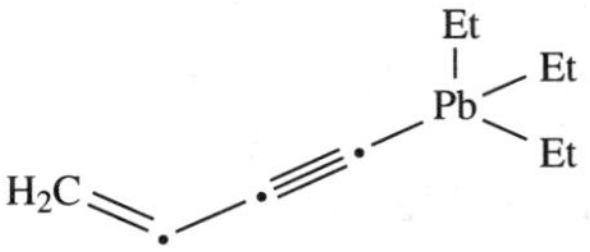

Zavagorodnii, S. V. *et al., Dokl. Akad. Nauk SSSR*, 1962, **143**, 855 (Engl. transl. 268)

It explodes on rapid heating.

See other ALKYLMETALS, METAL ACETYLIDES

3358. Manganese(II) *N,N*-diethyldithiocarbamate
[15685-17-3] $C_{10}H_{20}MnN_2S_4$

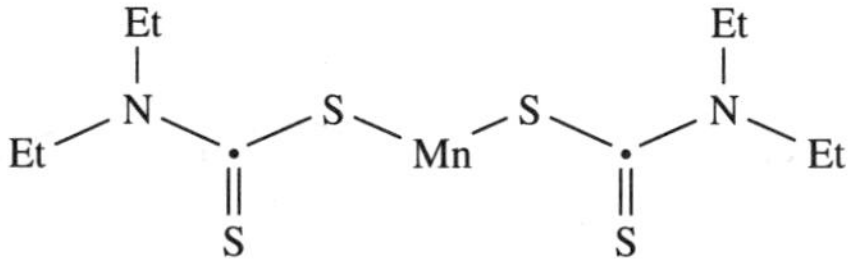

Bailar, 1973, Vol. 3, 872

A pyrophoric solid.

See other PYROPHORIC MATERIALS

3359. 2,2'-Di-*tert*-butyl-3,3'-bioxaziridine
[54222-33-2] $C_{10}H_{20}N_2O_2$

Putnam, S. J. *et al., Chem. Eng. News*, 1958, **36**(23), 46
A sample prepared by an established method and stored overnight at 2°C exploded
when disturbed with a metal spatula.
See 2,2''-Ethylenebis(3-phenyloxaziridine)
See other N–O COMPOUNDS, STRAINED-RING COMPOUNDS

†3360. 2-Ethylhexyl vinyl ether
[103-44-6] $C_{10}H_{20}O$

$$BuCHEtCH_2OCH{=}CH_2$$

†3361. Isopentyl isovalerate
[659-70-1] $C_{10}H_{20}O_2$

$$Me_2CHC_2H_4OCO.CH_2CHMe_2$$

3362. '10-Carbon oxime'
[ ] $C_{10}H_{21}NO$
(Unknown structure)

Knies, H., private comm., 1983
A 10-carbon oxime at 60°C was stored in an insulated container, and within 24 h
the temperature had risen to 125°C, possibly owing to an exothermic Beckmann
rearrangement to an amide. Application of cooling prevented a thermal runaway.
See other OXIMES

3363. Tripropyllead fulminate
[43135-83-7] $C_{10}H_{21}NOPb$

$$Pr_3PbC{\equiv}N \rightarrow O$$

Houben-Weyl, 1975, Vol. 13.3, 101
An extremely explosive solid.
See related ALKYLMETALS, METAL FULMINATES

3364. Decyl nitrite
[1653-57-2]

$C_{10}H_{21}NO_2$

Hakl, J., *Runaway Reactions*, 1981, Paper 3/L, 4
Initial exothermic decomposition occurs at 115°C.
See other ALKYL NITRITES

†3365. Decane
[124-18-5]

$C_{10}H_{22}$

$$Me[CH_2]_8Me$$

†3366. 2-Methylnonane
[34464-38-5]

$C_{10}H_{22}$

$$Me_2CH[CH_2]_6Me$$

3367. Oxodiperoxodipiperidinechromium(VI)
[ ]

$C_{10}H_{22}CrN_2O_5$

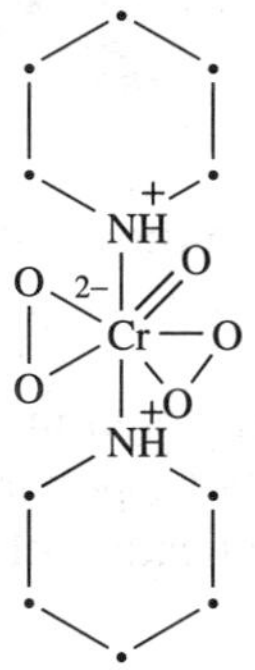

See Oxodiperoxodipyridinechromium(VI)
See other AMMINECHROMIUM PEROXOCOMPLEXES

3368. Diisopentylmercury
[24423-68-5]

$C_{10}H_{22}Hg$

$$(Me_2CHC_2H_4)_2Hg$$

Iodine
  Whitmore, 1921, 103
  Interaction is violent, accompanied by hissing.
  See other ALKYLMETALS, MERCURY COMPOUNDS

†3369. Dipentyl ether
[693-65-2] **$C_{10}H_{22}O$**

$BuCH_2OCH_2Bu$

HCS 1980, 719

3370. Formyl(triisopropyl)silane
[ ] **$C_{10}H_{22}OSi$**

$(Me_2CH)_3SiCO.H$

Soderquist, J. A. *et al., J. Amer. Chem. Soc.*, 1992, **114**(25), 10078
Pyrophoric.
See other ALKYLSILANES

3371. Diisopentylzinc
[21261-07-4] **$C_{10}H_{22}Zn$**

$(Me_2CHC_2H_4)_2Zn$

Gibson, 1969, 181
It ignites in air.
See other ALKYLMETALS, DIALKYLZINCS

3372. 4-Ethoxybutyldiethylaluminium
[65235-78-1] **$C_{10}H_{23}Al$**

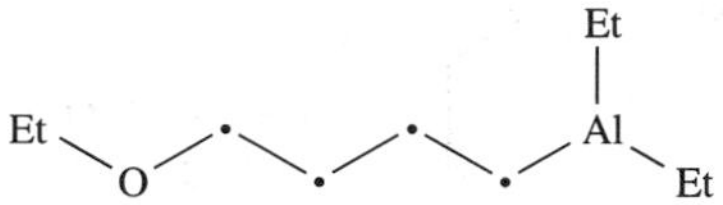

Bahr, G. *et al., Chem. Ber.*, 1955, **88**, 256
It ignites in air.
See related ALKYLALUMINIUM ALKOXIDES AND HYDRIDES

3373. Ethoxydiisobutylaluminium
[15769-72-9] **$C_{10}H_{23}AlO$**

$EtOAl(CH_2CHMe_2)_2$

May ignite in air.
See entry ALKYLALUMINIUM ALKOXIDES AND HYDRIDES

3374. Di(*O–O-tert*-butyl) ethyl diperoxophosphate
[ ] **$C_{10}H_{23}O_6P$**

$(Me_3COO)_2P(O)OEt$

Rieche, A. *et al., Chem. Ber.*, 1962, **95**, 385

The liquid deflagrated soon after isolation.
See other *tert*-BUTYL PEROXOPHOSPHATE DERIVATIVES, PHOSPHORUS ESTERS

3375. *N*-Trimethylsilyl-*N*-trimethylsilyloxyacetoacetamide
[87318-59-0] $C_{10}H_{23}NO_3Si_2$

Oster, T. A. *et al., J. Org. Chem.*, 1983, **48**, 4309
It may decompose spontaneously if too high a distillation temperature is attained.
See other N–O COMPOUNDS
See related ALKYLSILANES

3376. 1,4,8,11-Tetraazacyclotetradecanenickel(II) perchlorate
[15220-72-1] $C_{10}H_{24}Cl_2N_4Ni_2O_8$

Barefield, E. K., *Inorg. Chem.*, 1972, **11**, 2274
It exploded violently during analytical combustion.
See TETRAAZAMACROCYCLANEMETAL PERCHLORATES
See other AMMINEMETAL OXOSALTS

3377. 2,4-Di-*tert*-butyl-2,2,4,4-tetrafluoro-1,3-dimethyl-1,3,2,4-diazadiphosphetidine
[58656-11-4] $C_{10}H_{24}F_4N_2P_2$

Preparative hazard
See Butyllithium: 2,2,2,4,4,4-Hexafluoro-1,3-dimethyl-1,3,2,4-diazadiphospheti-
dine

3378. 1,2-Bis(*tert*-butylphosphino)ethane
[25196-24-9] $C_{10}H_{24}P_2$

$$Me_3CPHC_2H_4PHCMe_3$$

Benn, R. *et al., Z. Naturforsch. B*, 1987, **41B**, 680–691
The title compound and the bis(diisopropylphosphino) homologue were pyrophoric
in air.
See related ALKYLPHOSPHINES

3379. 1,2-Bis(diethylphosphino)ethane
[6411-21-8] $C_{10}H_{24}P_2$

$$Et_2PC_2H_4PEt_2$$

Mays, M. J. *et al., Inorg. Synth.*, 1974, **15**, 2
It ignites in air.
See other ALKYLPHOSPHINES

3380. Potassium [8-(4-nitrophenylthio)undecahydrodicarbaundecaborato]
undecahydrodicarbaundecaboratocobaltate(1−) (Potassium [(7,8,9,10,11-η)-
1,2,3,4,5,6,7,8,9,11-decahydro-10-(4-nitrobenzenethiolato-*S*)-7,8-
dicarbaundecaborato(2−)][(7,8,9,10,11-η)undecahydro-7,8-
dicarbadecaborato(2−)cobaltate(1−)])
[155147-43-6] $C_{10}H_{25}B_{18}CoKNO_2S$

$$[O_2NC_6H_4SC_2B_9H_{10}CoC_2B_9H_{11}]^-K^+$$

See Potassium bis(8-(4-nitrophenylthio)undecahydrodicarbaundecaborato)
cobaltate(1−)
See other BORANES

3381. Tris(trimethylsilyl)methylcaesium
[165610-13-9] $C_{10}H_{27}CsSi_3$

$$(Me_3Si)_3CCs$$

Eaborn, C. *et al., Angewand. Chem. (Int.)*, 1995, **34**(6), 687
This compound was isolated with 3.5 benzenes of crystallisation as moisture sensi-
tive, pyrophoric, yellow plates.
See other ORGANOMETALLICS

3382. Tetramethylbis(trimethysilanoxy)digold
[15529-45-0] $C_{10}H_{30}Au_2O_2Si_2$

Schmidbaur, H. *et al., Inorg. Chem.*, 1966, **5**, 2069
Sublimed crystals decomposed explosively at 120°C.
See related ALKYLMETALS *See other* GOLD COMPOUNDS

3383. *trans*-Diammine(1,4,8,11-tetraazacyclotetradecane)chromium(III) perchlorate
[94484-09-0] $C_{10}H_{30}Cl_3CrN_6O_{12}$

Kane-Maguire, N. A. P. *et al., Inorg. Chem.*, 1985, **24**, 597–605
It exploded on heating. Hexafluorophosphate is suggested as a more stable anion
for this type of salt.
See other AMMINEMETAL OXOSALTS

3384. Pentakis(dimethylamino)tantalum
[ ] $C_{10}H_{30}N_5Ta$

$$Ta(NMe_2)_5$$

Preparative hazard
Chestnut, R. W. *et al., Chem. Eng. News*, 1990, **68**(31), 2
Preparation by reaction of tantalum pentachloride with pure lithium dimethylamide
in pentane is unsafe. Initial non-reaction is followed by explosion during manip-
ulation of the slurry. Presence of ether or dimethylamine gives smooth reaction.
Stepwise displacement of chlorine, starting with the free amine, is also recom-
mended.
See other N-METAL DERIVATIVES

3385. 4,4'-Bipyridyl-bis(pentaammineruthenium(III) perchlorate)
[69042-75-7] $C_{10}H_{38}Cl_6N_{12}O_{24}Ru_2$

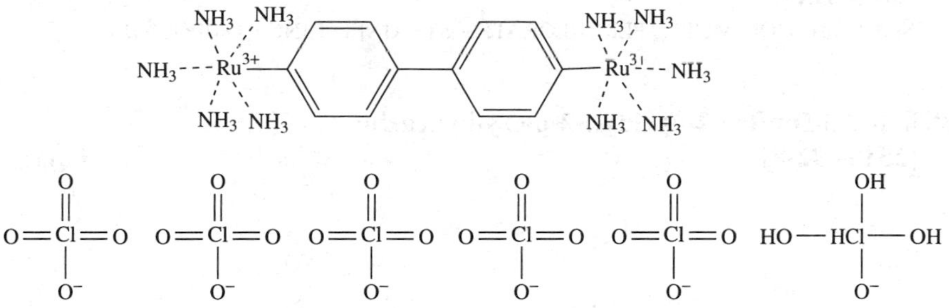

Sutton, J. E. *et al.*, *Inorg. Chem.*, 1979, **18**, 1018

Use of the perchlorate salt gives a cleaner product than the tosylate, but is more hazardous.

See other AMMINEMETAL OXOSALTS

3386. Decacarbonyldirhenium
[14285-68-8] $C_{10}O_{10}Re_2$

$$(OC)_5ReRe(CO)_5$$

Bailar, 1973, Vol. 3, 953

It tends to ignite in air above 140°C.

See other CARBONYLMETALS

3387. 3,5-Dinitro-2-(picrylazo)pyridine
[55106-93-9] $C_{11}H_4N_8O_{10}$

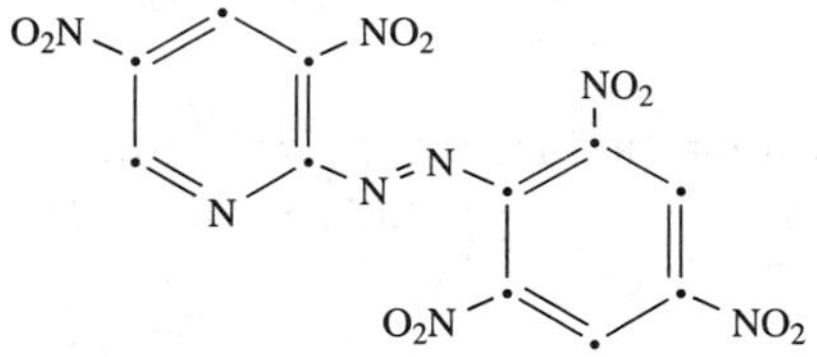

Explosive.

See entry POLYNITROAZOPYRIDINES

3388. 1-Chloro-2,3-di(2-thienyl)cyclopropenium perchlorate
[ ] $C_{11}H_6Cl_2O_4S_2$

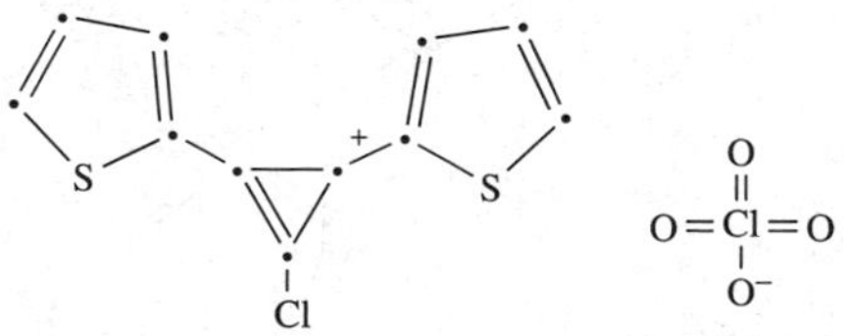

Komitsu, K. *et al.*, *Tetrahedron Lett.*, 1978, 806

When isolated and dried, a violent explosion occurred. The tetrachloroaluminate salt is safer.

See other NON-METAL PERCHLORATES, STRAINED-RING COMPOUNDS

3389. 1-(3,5-Dinitro-2-pyridyl)-2-picrylhydrazine
[55106-92-8] $C_{11}H_6N_8O_{10}$

Explosive.
See entry POLYNITROAZOPYRIDINES

3390. 1-Naphthyl isocyanate
[1984-04-9] $C_{11}H_7NO$

Energy of decomposition (in range 250–380°C) measured as 0.31 kJ/g.
See entry THERMOCHEMISTRY AND EXOTHERMIC DECOMPOSITION (reference 2)
See other ORGANIC ISOCYANATES

3391. (1-Napththyl)-1,2,3,4-thiatriazole
[95322-62-6] $C_{11}H_7N_3S$

CHETAH, 1990, 182
Shock sensitive.
See other N–S COMPOUNDS

3392. 1-(4-Diazoniophenyl)-1,2-dihydropyridine-2-iminosulfinate
[61224-59-7] $C_{11}H_8N_4O_2S$

Hoffmann, H. *et al.*, US Pat. 3 985 724, 1977
Diazotisation of 2-*p*-aminophenylsulfonamidopyridine with subsequent adjustment of pH to 3–6 gives the internal diazonium salt, which decomposes violently at 230°C.
See other DIAZONIUM SALTS, N–S COMPOUNDS
See related DIAZONIUM CARBOXYLATES

3393. 2,4-Diethynyl-5-methylphenol
[32229-04-2] $C_{11}H_8O$

Myasnikova, R. N. *et al.*, *Izv. Akad. Nauk SSSR*, 1970, **11**, 2637
Oxidation of the phenol or its methyl or propyl ethers in pyridine gave insoluble polymers. Those of the ethers exploded on heating.
See other ACETYLENIC COMPOUNDS, POLYPEROXIDES

3394. 1-(2-Naphthyl)-3-(5-tetrazolyl)triazene
[68499-65-0] $C_{11}H_9N_7$

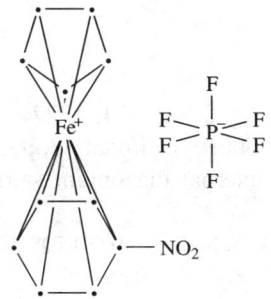

Thiele, J., *Ann.*, 1892, **270**, 54; 1893, **273**, 144
It explodes at 184°C.
See other TETRAZOLES, TRIAZENES

3395. (Nitrobenzene)(cyclopentadienyl)iron(II) hexafluorophosphate
[77486-94-3] $C_{11}H_{10}F_6FeNO_2P$

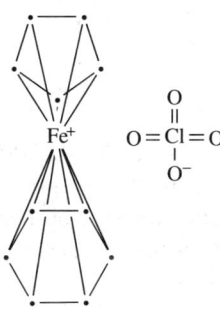

Gassman, P. G. *et al., Organometallics*, 1994, **13**(7), 2890
Heating this complex above 200°C may lead to explosion.
See other ORGANOMETALLICS

3396. η-Benzenecyclopentadienyliron(II) perchlorate
[ ] $C_{11}H_{11}ClFeO_4$

Denning, R. G. *et al., J. Organomet. Chem.*, 1966, **5**, 292

The dry material is shock-sensitive and detonated on touching with a spatula.
See other ORGANOMETALLIC PERCHLORATES

3397. Ethyl α-azido-*N*-cyanophenylacetimidate
[51688-25-6] $C_{11}H_{11}N_5O$

Petersen, H. J., *J. Med. Chem.*, 1974, **17**, 104
The crude imidoester should be used as a concentrated solution in ether. A small solvent-free sample exploded violently.
See related 2-AZIDOCARBONYL COMPOUNDS

3398. 3,3-Dimethyl-1-(3-quinolyl)triazene
[70324-23-1] $C_{11}H_{12}N_4$

Rondevstedt, C. S. *et al., J. Org. Chem.*, 1957, **22**, 201
The crude material decomposes violently if allowed to dry, and purified material explodes at 131°C or during analytical combustion.
See other TRIAZENES

3399. 3-Butyn-1-yl 4-toluenesulfonate
[23418-85-1] $C_{11}H_{12}O_3S$

Eglinton, G. *et al., J. Chem. Soc.*, 1950, 3653

The material could be distilled in small amounts at below 0.01 mbar, but exploded on attempted distillation at 0.65 mbar.

See other ACETYLENIC COMPOUNDS, SULFUR ESTERS

3400. *tert*-Butyl 4-nitroperoxybenzoate
[16166-61-3] $C_{11}H_{13}NO_5$

Criegee, R. *et al., Ann.*, 1948, **560**, 135

This and the 4-nitrobenzoates of homologous *tert*-alkyl hydroperoxides explode in contact with flame.

See other NITROARYL COMPOUNDS, PEROXYESTERS

†3401. *tert*-Butyl peroxybenzoate
[614-45-9] $C_{11}H_{14}O_3$

1. Criegee, R., *Angew. Chem.*, 1953, **65**, 398–399
2. Schnur, R. C., *Chem. Eng. News*, 1981, **59**(19), 3
3. Verhoeff, J., *Experimental Study of the Thermal Explosion of Liquids*, Doctoral thesis, University of Delft, 1983
4. *CHETAH*, 1990, 182

Shortly after interruption of vacuum distillation from an oil-bath at 115°C to change a thermometer, the ester exploded violently and this was attributed to overheating [1]. A commercial sample of undetermined age exploded violently just after vacuum distillation had begun [2]. In an examination of thermal explosion behaviour, the title compound was used as a model compound in autoclave

1118

experiments at low, high or constant pressure. Three stages in the overall process were identified: thermal runaway, initiation, and then explosion, and these are studied and discussed in detail. In the high pressure experiments, maximum rates of pressure rise approaching 1000 kbar/s were observed [3]. It is shock sensitive [4].
See entry PRESSURE INCREASE IN EXOTHERMIC DECOMPOSITION

Copper(I) bromide, Limonene
Wilson, C. W. *et al., J. Agric. Food Chem.*, 1975, **23**, 636
Addition of all the perester in one portion to limonene and catalytic amounts of copper(I) bromide before oxidation had begun, as shown by development of a blue-green colour, led to an explosion.
See other PEROXYESTERS

3402. 1-Chloro-3.phenylpent-1-en-4-yn-3-ol
[33543-69-0] $C_{11}H_9ClO$

Acheson, R. M. *et al., J. Chem. Soc., Perkin Trans. 1*, 1987, 2322. 2326
In spite of very careful temperature control, explosive decomposition always occurred towards the end of vacuum distillation at 133–136°C/1 mbar.
See other ACETYLENIC COMPOUNDS, HALOACETYLENE DERIVATIVES

3403. Ethyl 6-ethoxycarbonyl-3,4-dihydro-1,1,3-trioxo-2-pyrazolo[3,4-d] isothiazole-2-acetate
[126623-77-6] $C_{11}H_{13}N_3O_7S$

Burri, K. F., *Helv. Chim. Acta*, 1989, **72**(6), 1416
It is suggested that this compound, and two precursors, can explode.
See Ethyl 4-bromo-1,1,3-trioxoisothiazoleacetate
See other N–S COMPOUNDS

3404. Cyclobutyl 4-methylbenzenesulfonate
[10437-85-1] $C_{11}H_{14}O_3S$

It decomposes vigorously at 118°C.
See entry SULFONIC ACID ESTERS *See other* SULFUR ESTERS

3405. Methyl spiro[5-bromobicyclo[2.2.1]heptane 2,2′ 1,3-dioxolane]-7-carboxylate
[71155-12-9] $C_{11}H_{15}BrO_4$

Perchloryl fluoride
See Perchloryl fluoride: Methyl 2-bromo-5,5-(ethylenedioxy)[2.2.1]bicycloheptane-7-carboxylate

3406. 2-*tert*-Butyl-3-phenyloxaziridine
 [7731-34-2]

$C_{11}H_{15}NO$

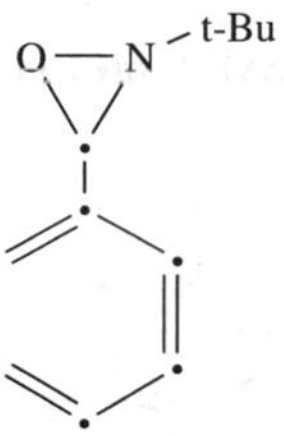

Emmons, W. D. *et al., Org. Synth.*, 1969, **49**, 13
Vacuum distillation of this active oxygen compound is potentially hazardous and precautions are necessary.
See other N–O COMPOUNDS, STRAINED-RING COMPOUNDS

3407. *N*-Phenylazopiperidine
 [16978-76-0]

$C_{11}H_{15}N_3$

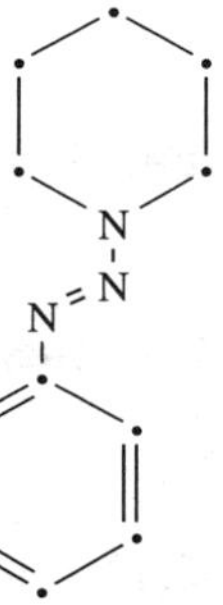

Hydrofluoric acid
 Wallach, O. *et al., Ann.*, 1886, **235**, 258; 1888, **243**, 219
 Interaction to give fluorobenzene is violent and is not suitable for above 10 g quantities.
 See other TRIAZENES

3408. 2-(1,3-Diselena-4,5,6,7-tetrahydroindanylidene)piperidinium perchlorate
 [76371-73-8]

$C_{11}H_{16}ClNO_4Se_2$

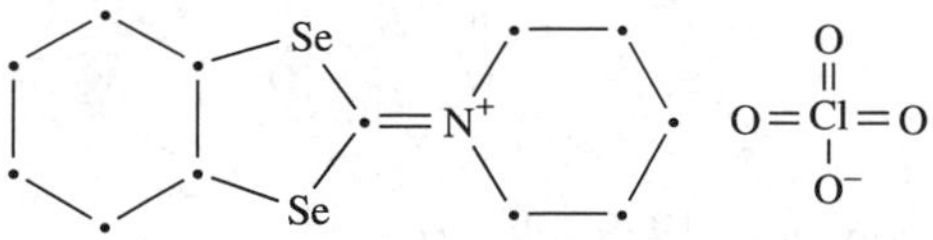

See entry 1-(1,3-DISELENONYLIDENE)PIPERIDINIUM PERCHLORATES
See other PERCHLORATE SALTS OF NITROGENOUS BASES

3409. Butyl toluenesulfonate (Butyl 4-methylbenzenesulfonate)
[778-28-9]

$C_{11}H_{16}O_3S$

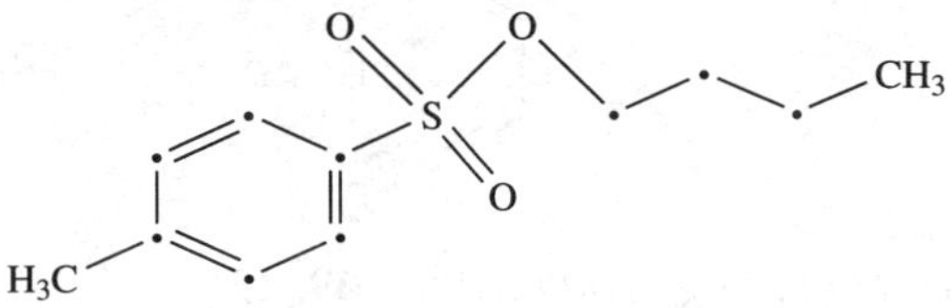

Boyer, J. H. *et al., J. Org. Chem.* 1958, **23**, 1051

Eight grams of the crude ester, which had been prepared from tosyl azide and sodium butoxide, then stored for four weeks at 0°C, was to be distilled at 0.2 mm Hg. On the bath reaching 70°C, there was a violent explosion. This was attributed to a toluenesulphonic acid autocatalysed elimination. However, presence of butyl azide as an impurity seems probable.

See other SULFONIC ACID ESTERS, *See also* AZIDES

3410. Oxodiperoxy(pyridine)(1,3-dimethyl-2,4,5,6-tetrahydro-2-1*H*)-(pyrimidinone)molybdenum
[128575-71-3]

$C_{11}H_{17}MoN_3O_6$

Complex structure

Preparative hazard
See Oxodiperoxy(1,3-dimethyl-2,4,5,6-tetrahydro-2-1*H*)-(pyrimidinone) molybdenum
See related AMMINECHROMIUMPEROXO COMPLEXES

3411. Di-*tert*-butyl diazomalonate
[35207-75-1]

$C_{11}H_{18}N_2O_4$

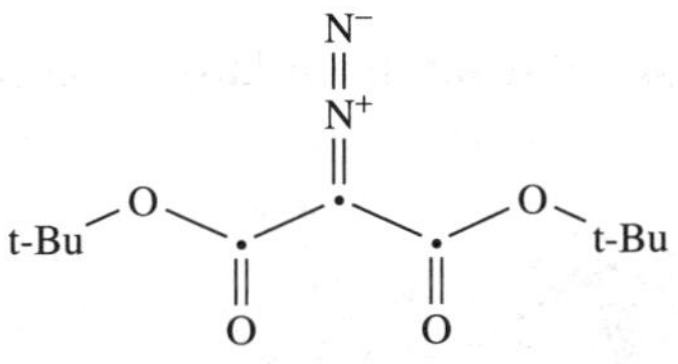

Ledon, H. J., *Org. Synth.*, 1979, **59**, 70

Although no explosions have been encountered during the distillation of 10 g portions under argon at 54–58°C/0.003 mbar, it should be treated as potentially explosive. Contact with rough or metallic surfaces should be avoided.
See other DIAZO COMPOUNDS

3412. 3-Iodo-4-methoxy-4,7,7-trimethylbicyclo[4.1.0]heptane
[65082-64-6] $C_{11}H_{19}IO$

See vic-IODO-ALKOXY COMPOUNDS
See other IODINE COMPOUNDS

3413. 3-Methyl-3-buten-1-ynyltriethyllead
[ ] $C_{11}H_{20}Pb$

Zavgorodnii, S. V. *et al., Doklad. Akad. Nauk SSSR*, 1962, **143**, 855 (Engl. transl. 268)
It explodes on rapid heating.
See other ALKYLMETALS, METAL ACETYLIDES

3414. Diethyl-3-diethylaminopropylaluminium
[73566-71-5] $C_{11}H_{26}AlN$

491M, 19758, 151

It ignites in air.
See related TRIALKYLALUMINIUMS

3415. N,N-Bis(3-aminopropyl)-1,4-diazacycloheptanenickel(II) perchlorate
[70212-12-3] $C_{11}H_{26}Cl_2N_4NiO_8$

Billo, E. J., *Inorg. Nucl. Chem. Lett.*, 1978, **14**, 502
An explosive solid.
See other AMMINEMETAL OXOSALTS

3416. 3-Dibutylaminopropylamine
[30734-83-9] $C_{11}H_{26}N_2$

$$Bu_2N[CH_2]_3NH_2$$

Cellulose nitrate
See CELLULOSE NITRATE: amines *See other* ORGANIC BASES

3417. 1-Trimethylsilyloxy-1-trimethylsilylphosphylidine-2,2-dimethylpropane
(4-(Dimethylethyl)-2,2,6,6-tetramethyl-3-oxa-5-phospha-2,6-disilahept-4-ene)
[78114-26-8] $C_{11}H_{27}OPSi_2$

$$Me_3SiP=C(CMe_3)OSiMe_3$$

Becker, G. *et al., Inorg. Synth.*, 1990, **27**, 249
Extremely oxygen sensitive and pyrophoric.
See other ALKYLPHOSPHINES

3418. Bis(pentafluorophenyl)aluminium bromide
[4457-91-4] $C_{12}AlBrF_{10}$

$$(F_5C_6)_2AlBr$$

Air, or Ether, or Water
1. Chambers, R. D. *et al., J. Chem. Soc. (C)*, 1967, 2185
2. Chambers, R. D. *et al., Tetrahedron Lett.*, 1965, 2389
3. Cohen, C. S. *et al., Advan. Fluorine Chem.*, 1970, **6**, 156
It ignites in air, explodes during uncontrolled hydrolysis [1], and chars during controlled hydrolysis [2]. When isolated as the etherate, attempts to remove solvent ether caused violent decomposition [3].
See other HALO-ARYLMETALS

3419. Bis(hexafluorobenzene)cobalt(0)

[ ] $C_{12}CoF_{12}$

It decomposes explosively at 10°C
See entry HALOARENEMETAL π-COMPLEXES

3420. Dodecacarbonyltetracobalt

[19212-11-4] $C_{12}Co_4O_{12}$

$$(OC)_{12}Co_4$$

Blake, E. J., private comm., 1974
Cobalt catalysts discharged from 'oxo'-process reactors are frequently pyrophoric,
owing to the presence of the carbonylcobalt.
See other CARBONYLMETALS, PYROPHORIC CATALYSTS

3421. Bis(hexafluorobenzene)chromium(0)

[59420-01-8] $C_{12}CrF_{12}$

It decomposes explosively at 40°C.
See entry HALOARENEMETAL π-COMPLEXES

3422. Graphite hexafluorogermanate
[76321-00-0]

$C_{12}GeF_6$

$C_{12}F_6Ge$

McCarron, E. M. *et al.*, *J. Chem. Soc., Chem. Comm.*, 1980, 890–891
The solid product of intercalating graphite with germanium tetrafluoride–fluorine mixtures at 20°C is thought to have an oxidising and fluorinating capability close to that of elemental fluorine itself.
See Fluorine
See other OXIDANTS

3423. Bis(pentafluorophenyl)ytterbium
[66080-22-6]

$C_{12}F_{10}Yb$

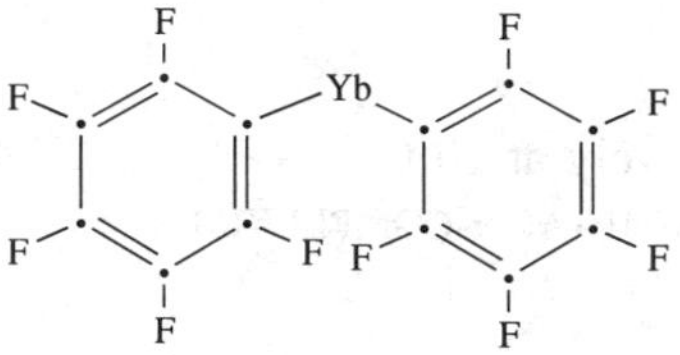

Methylcyclopentadiene
Deacon, G. B. *et al.*, *Aust. J. Chem.*, 1985, **38**, 1763
The THF-solvated product of the ligand exchange reaction between bis(pentafluorophenyl)ytterbium and methylcyclopentadiene was explosive. The expected product, bis(methylcyclopentadiene)ytterbium, obtained by another exchange route, is not explosive.
See other ORGANOMETALLICS

3424. Bis(hexafluorobenzene)iron(0)
[37279-26-8]

$C_{12}F_{12}Fe$

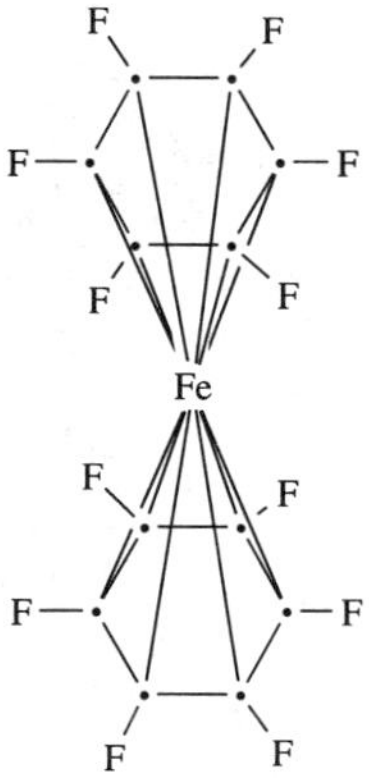

It decomposes explosively at −40°C.
See entry HALOARENEMETAL π-COMPLEXES

3425. Bis(hexafluorobenzene)nickel(0)
  [74167-05-8] $C_{12}F_{12}Ni$

Alone, or Air, or Hydrogen, or Carbon monoxide
or Allyl bromide, or Trifluoromethyl iodide
  Klabunde, K. J. *et al., J. Fluorine Chem.*, 1974, **4**, 114–115
  The complex decomposes explosively on slight provocation (flakes falling into the
  reactor, static charge or uneven warming) and at 70°C with careful heating. At
  ambient temperature, interaction with air or the reagents above is explosive, but
  can be controlled by preliminary cooling to −196°C, followed by controlled slow
  warming.
  See entry HALOARENEMETAL π-COMPLEXES

3426. Bis(hexafluorobenzene)titanium(0)
  [ ] $C_{12}F_{12}Ti$

It decomposes explosively at −50°C.
  See entry HALOARENEMETAL π-COMPLEXES

3427. Bis(hexafluorobenzene)vanadium(0)

[ ] $C_{12}F_{12}V$

It decomposes explosively at 100°C.
See entry HALOARENEMETAL π-COMPLEXES

3428. Dodecacarbonyltriiron

[17658-52-8] $C_{12}Fe_3O_{12}$

$$(OC)_3Fe\equiv(CO)_3\equiv Fe\equiv(CO)_3\equiv Fe(CO)_3$$

King, R. B. *et al., Inorg. Synth.*, 1963, **7**, 195
On prolonged storage, pyrophoric decomposition products are formed.
See other CARBONYLMETALS

3429. Silver hexanitrodiphenylamide

[32817-38-2] (ion) $C_{12}H_4AgN_7O_{12}$

Taylor, C. A. *et al., Army Ordnance*, 1926, **7**, 68–69
It had been evaluated as a detonator.
See other N-METAL DERIVATIVES, POLYNITROARYL COMPOUNDS, SILVER COMPOUNDS

3430. Calcium picrate
[16824-78-5] $C_{12}H_4CaN_6O_{14}$

See Picric acid (reference 3)
See other PICRATES, POLYNITROARYL COMPOUNDS

3431. Bis(2,4,5-trichlorobenzenediazo) oxide
[ ] $C_{12}H_4Cl_6N_4O$

Alone, or Benzene
Kaufmann, T. *et al., Ann.*, 1960, **634**, 77
The dry solid explodes under a hammer-blow, or on moistening with benzene.
See other BIS(ARENEDIAZO) OXIDES, HALOARYL COMPOUNDS

3432. Copper(II) picrate
[33529-09-8] $C_{12}H_4CuN_6O_{14}$

See Picric acid (reference 2)
See other PICRATES, POLYNITROARYL COMPOUNDS

3433. Mercury(II) picrate
[19528-48-4] $C_{12}H_4HgN_6O_{14}$

See Picric acid (reference 2)
See other MERCURY COMPOUNDS, PICRATES, POLYNITROARYL COMPOUNDS

3434. 1,3-Dicyano-2-diazo-2,6-azulene quinone
(2-Diazo-2,6-dihydro-6-oxo-1.3-azulenedicarbonitrile)
[53535-47-0] $C_{12}H_4N_4O$

Nosoe, T. *et al., J. Org. Chem,.* 1995, **60**(18), 5919
The compound sometimes decomposes explosively at 155°C.
See other ARENEDIAZOMIUM OXIDES

3435. Nickel picrate
[63085-06-3] $C_{12}H_4N_6NiO_{14}$

Bernardi, A., *Gazz. Chim. Ital.*, 1930, **60**, 169–171
It deflagrates or explodes very violently on heating.
See other HEAVY METAL DERIVATIVES, PICRATES, POLYNITROARYL COMPOUNDS

3436. Lead(II) picrate
[6477-64-1]

$C_{12}H_4N_6O_{14}Pb$

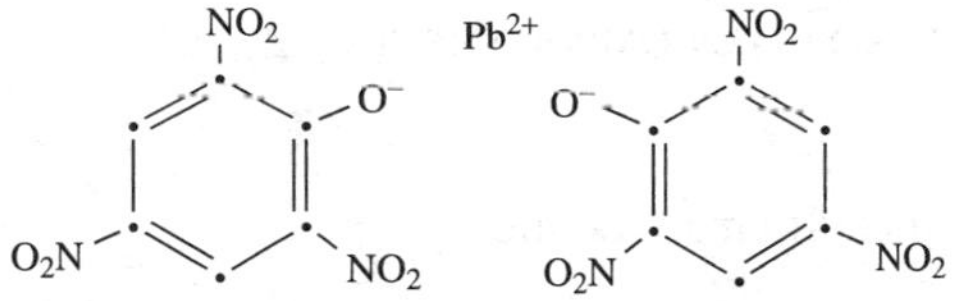

Belcher, R., *J. R. Inst. Chem.*, 1960, **84**, 377

During the traditional qualitative inorganic analytical procedure, samples containing the lead and salicylate radicals can lead to the formation and possible detonation of lead picrate. This arises during evaporation of the filtrate with nitric acid, after precipitation of the copper–tin group metals with hydrogen sulfide. Salicylic acid is converted under these conditions to picric acid, which in presence of lead gives explosive lead picrate. An alternative (MAQA) scheme is described which avoids this possibility.

See Picric acid (reference 2)
See other HEAVY METAL DERIVATIVES, PICRATES, POLYNITROARYL COMPOUNDS

3437. Zinc picrate
[16824-81-0]

$C_{12}H_4N_6O_{14}Zn$

$$Zn(OC_6H_2(NO_2)_3]_2$$

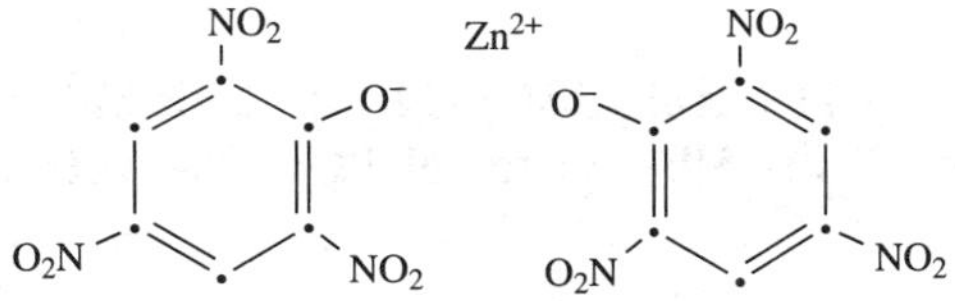

See Picric acid (reference 2)
See other PICRATES, POLYNITROARYL COMPOUNDS

3438. 4,6-Dinitro-1-picrylbenzotriazole
[50892-86-9]

$C_{12}H_4N_8O_{10}$

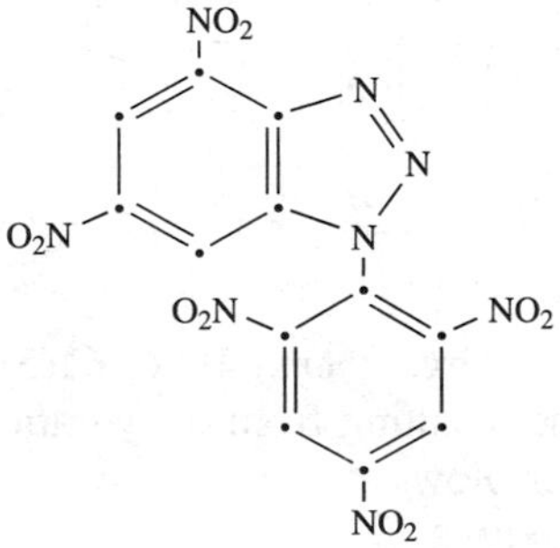

Coburn, M. D., *J. Heterocycl. Chem.*, 1973, **10**, 743–746

The isomeric dinitro derivatives, and some of the mononitro derivatives, are impact-sensitive explosives.

See other POLYNITROARYL COMPOUNDS, TRIAZOLES

3439. 5,7-Dinitro-1-picrylbenzotriazole
[50892-90-5] $C_{12}H_4N_8O_{10}$

Coburn, M. D., *J. Heterocycl. Chem.*, 1973, **10**, 743–746

The isomeric dinitro derivatives, and some of the mononitro derivatives, are impact-sensitive explosives.

See other POLYNITROARYL COMPOUNDS, TRIAZOLES

3440. 5,12-Dioxo-4,4,11,11-tetrahydroxy-14*H*-[1,2,5]oxadiazolo[3,4-*e*] [1,2,5]oxadiazolo[3′,4′:4,5]benzotriazolo[2,1-*a*]benzotriazolo-*b*-ium 1,8-dioxide
[ ] $C_{12}H_4N_8O_{10}$

Subramanian, G. *et al.*, *J. Org. Chem.*, 1996, **61**(5), 1898

This, and the tetraketone resulting from dehydration of the gem-diol groups, both explode under a hammer blow.

See other FURAZAN *N*-OXIDES

3441. 4,6-Dinitro-2-(2,4,6-trinitrophenyl)benzotriazole *N*-oxide
[87604-86-2] $C_{12}H_4N_8O_{11}$

Renfrow, R. A. *et al., Aust. J. Chem.*, 1983, **36**, 1847
It is explosive and a super-electrophile, like 4,6-dinitrobenofurazan *N*-oxide.
See other N-OXIDES, POLYNITROARYL COMPOUNDS, TRIAZOLES

3442. 1,3,6,8-Tetranitrocarbazole
[4543-33-3] $C_{12}H_5N_5O_8$

Sorbe, 1968, 153
It may readily explode when dry, so it should be stored wet with water (10% is normally added to commercial material).
See other POLYNITROARYL COMPOUNDS

3443. 1,3,5-Triethynylbenzene
[17814-74-3] $C_{12}H_6$

Shvartsberg, M. S. *et al., Izv. Akad. Nauk SSSR, Ser. Khim.*, 1963, **110**, 1836
It polymerised explosively on rapid heating and compression.
See other ACETYLENIC COMPOUNDS, POLYMERISATION INCIDENTS

3444. 2,6-Diperchloryl-4,4-diphenoquinone

[ ] $C_{12}H_6Cl_2O_8$

Gardner, D. M. *et al., J. Org. Chem.*, 1963, **28**, 2650
A shock-sensitive explosive.
See other PERCHLORYL COMPOUNDS

3445. Bis(3,4-dichlorobenzenesulfonyl) peroxide

[ ] $C_{12}H_6Cl_4O_6S_2$

Dannley, R. *et al., J. Org. Chem.*, 1966, **31**, 154
The peroxide was too unstable to dry thoroughly; such samples often exploded spontaneously.
See other DIACYL PEROXIDES

3446. Potassium hexaethynylcobaltate(4−)

[ ] $C_{12}H_6CoK_4$

Alone, or Ammonia
> Nast, R. *et al., Z. Anorg. Chem.*, 1955, **282**, 210
> It is moderately stable at below −30°C, is very shock- and friction-sensitive, and explodes violently on contact with water. At ambient temperature, it rapidly forms explosive decomposition products. Its addition compound with ammonia behaves similarly, exploding on contact with air.
> *See other* COMPLEX ACETYLIDES

3447. Iron(III) maleate

[14451-00-4] $C_{12}H_6Fe_2O_{12}$

Iron(III) hydroxide
> Thermally unstable basic iron(III) maleates produced from mixtures of the above may be pyrophoric.
> *See* Phthalic anhydride (reference 2)

Sulfur compounds
> *See* Phthalic anhydride (reference 2)
> *See other* HEAVY METAL DERIVATIVES

3448. Potassium hexaethynylmanganate(3−)

[ ] $C_{12}H_6K_3Mn$

Bailar, 1973, Vol. 3, 855
A highly explosive solid.
See other COMPLEX ACETYLIDES

3449. Bis(2,4-dinitrophenyl) disulfide

[2217-55-2] $C_{12}H_6N_4O_8S_2$

Chlorine
MCA Guide, 1972, 317
The disulfide was chlorinated in wet tetrachloroethylene to give dinitrobenzenesul-
fonyl chloride. When the solvent was being removed by evaporation, the residue
exploded. Some of the less stable sulfenyl chloride may have been present.
See 2,4-Dinitrobenzenesulfenyl chloride
See other NON-METAL SULFIDES, POLYNITROARYL COMPOUNDS

3450. 2-Iodo-3,5-dinitrobiphenyl
[ ] $C_{12}H_7IN_2O_4$

Ethyl sodioacetoacetate
Zaheer, S. H. *et al., J. Indian Chem. Soc.*, 1955, **32**, 491
Interaction of 2-halo-3,5-dinitrobiphenyls with the sodio ester should be limited to
5–6 g of the title compound to avoid explosions experienced with larger amounts.
See other HALOARYL COMPOUNDS, POLYNITROARYL COMPOUNDS

3451. Potassium 6-*aci*-nitro-2,4-dinitro-1-phenylimino-2,4-cyclohexadienide
[[12245-12-4] (ion)] $C_{12}H_7KN_4O_6$

Sudborough, J. J. *et al., J. Chem. Soc.*, 1906, **89T**, 586
The salt of the mono-*aci*-quinonoid form of 2,4,6-trinitrodiphenylamine is explo-
sive, like the *N*-1- and -2-naphthyl analogues.
See other aci-NITROQUINONOID COMPOUNDS

3452. 1,4-Dihydrodicyclopropa[*b*, *g*]naphthalene
[52781-75-6] $C_{12}H_8$

Ippen, J. *et al., Angew. Chem. (Intern. Ed.)*, 1974, **13**, 736

It undergoes explosive decomposition at 132–133°C, and is a shock-sensitive solid which requires handling with caution.

See other STRAINED-RING COMPOUNDS

3453. Bis(bromobenzenesulfonyl) peroxide
[29342-58-3] $C_{12}H_8Br_2O_6S_2$

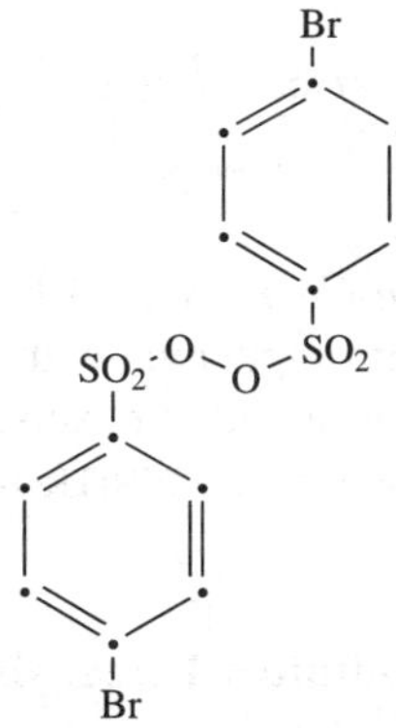

1. Bolte, J. *et al., Tetrahedron Lett.*, 1965, 1529
2. Dannley, R. L. *et al., J. Org. Chem.*, 1966, **31**, 154

The peroxide was too unstable to dry thoroughly; such samples often exploded spontaneously.

See other DIACYL PEROXIDES

3454. 2,2′-Bipyridine *N,N′*-dioxide-dicarbonylrhodium(I) perchlorate
[84578-88-1] $C_{12}H_8ClN_2O_8Rh$

$$[(C_{10}H_8N_2O_2)Rh(CO)_2]\ ClO_4$$

Uson, R. *et al., J. Organomet. Chem.*, 1982, **240**, 433 footnote f

It exploded violently during analytical combustion.

See related AMMINEMETAL OXOSALTS

3455. Thianthrenium perchlorate
[21299-20-7] $C_{12}H_8ClO_4S_2$

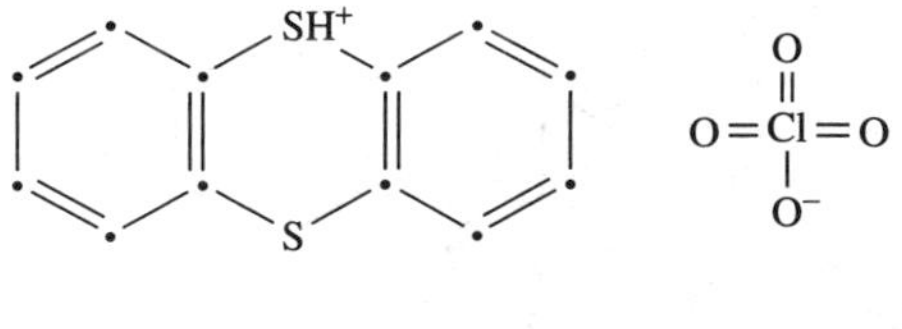

1138

Shine, H. J. *et al., Chem. & Ind.*, 1969, 782; *J. Org. Chem.*, 1971, **36**, 2925
A 1–2 g portion of the freshly prepared suction-dried material exploded violently during transfer from a sintered filter. Initiation may have been caused by friction of transfer or rubbing with a glass rod. Preparation of 50–100 mg quantities only is recommended.

See other FRICTIONAL INITIATION INCIDENTS, NON-METAL PERCHLORATES

3456. Bis(4-chlorobenzenediazo) oxide
[ ] $C_{12}H_8Cl_2N_4O$

Alone, or Benzene
Bamberger, E., *Ber.*, 1896, **29**, 464
More stable than unsubstituted analogues, it may be desiccated at 0°C, but is then extremely sensitive and violently explosive. Contact with benzene (even at 0°C) is violent and the reaction may become explosive.

See other BIS(ARENEDIAZO) OXIDES, HALOARYL COMPOUNDS

3457. 4,4′-Biphenylenebis(diazonium) perchlorate
(1,1′-Biphenyl-4,4′-bisdiazonium perchlorate)
[ ] $C_{12}H_8Cl_2N_4O_8$

Vorlander, D., *Ber.*, 1906, **39**, 2713–2715
The perchlorate derived from tetrazotised benzidine is unstable and explosive.
See other DIAZONIUM PERCHLORATES

3458. Bis(4-chlorobenzenesulfonyl) peroxide
[1886-69-7] $C_{12}H_8Cl_2O_6S_2$

Dannley, R. L. *et al., J. Org. Chem.*, 1966, **31**, 154
The peroxide was too unstable to dry thoroughly; such samples often exploded
spontaneously.
See other DIACYL PEROXIDES

3459. Endrin (1,2,3,4,10,10-Hexachloro-6,7-epoxy-1,4,4a,5,6,7,8,8a-octahydro-endo,endo-1,4:5,8-dimethanonaphthalene))
[72-20-8] $C_{12}H_8Cl_6O$

Diethyl 4-nitrophenyl thionophosphate
See Diethyl 4-nitrophenyl thionophosphate: Endrin
See other 1,2-EPOXIDES

3460. Cobalt(II) picramate
[33393-70-3] $C_{12}H_8CoN_6O_{10}$

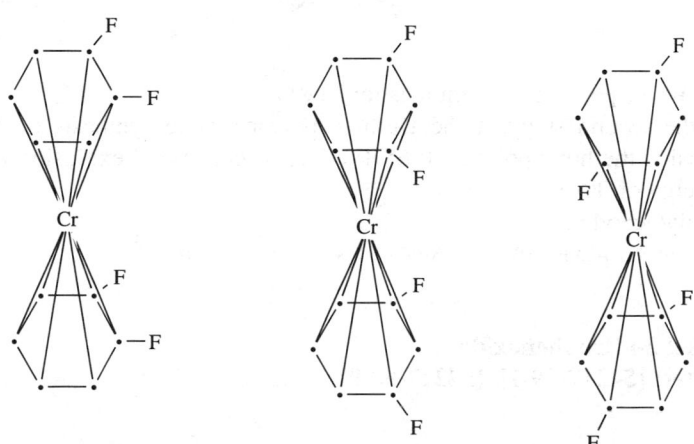

Agrawal, J. P. *et al., Propellants, Explos.*, 1981, **6**, 112–116
The picramate is less stable than the picrate.
See related METAL NITROPHENOXIDES, PICRATES
See other HEAVY METAL DERIVATIVES, POLYNITROARYL COMPOUNDS

3461. Bis(1,*n*-difluorobenzene)chromium(0) isomers
[53504-66-8] [53504-62-4] [39422-94-1] (1,2-, 1,3-, 1,4- resp.) $C_{12}H_8CrF_4$

All isomers unstable and explosive.
See entry HALOARENEMETAL π-COMPLEXES

3462. 3,6-Bis(trifluoromethyl)pentacyclo [6.2.0.0^{2,4}.0^{3,6}.0^{5,7}]dec-9-ene
[24516-71-0], [96570-67-1] homopolymer $C_{12}H_8F_6$

Editor's comments

The monomer is the highest energy stage of the Durham synthesis of polyacetylene; since the next stage, metathesized polymer, is known to be potentially explosive, the monomer is not likely to be absolutely safe.

See Poly(acetylene)

3463. Poly([7,8-bis(trifluoromethyl)tetracyclo [4.2.0.0^{2,8}.0^{5,7}]octane-3,4-diyl]-1,2-ethenediyl)
[102868-70-2]   ($C_{12}H_8F_6)_n$

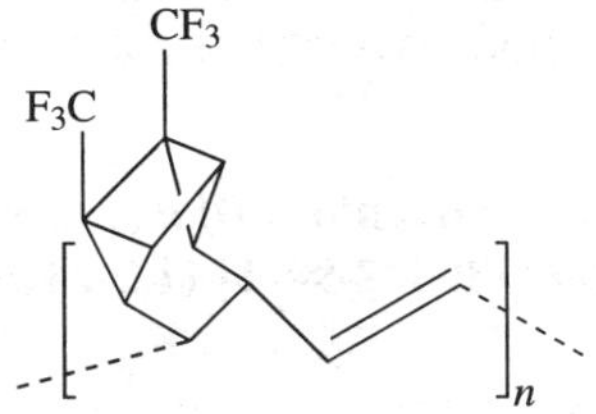

Feast, W. J., personal communication, 1993

This, the second stage of the Durham polyacetylene synthesis (and presumably identical to the homopolymer [96570-67-1]) is capable of explosion with one sixth the energy of TNT.

See Poly(acetylene)
See other STRAINED-RING COMPOUNDS

3464. Nickel 2-nitrophenoxide
hydrates: [54217-99-1], [54217-98-0] $C_{12}H_8N_2NiO_6$

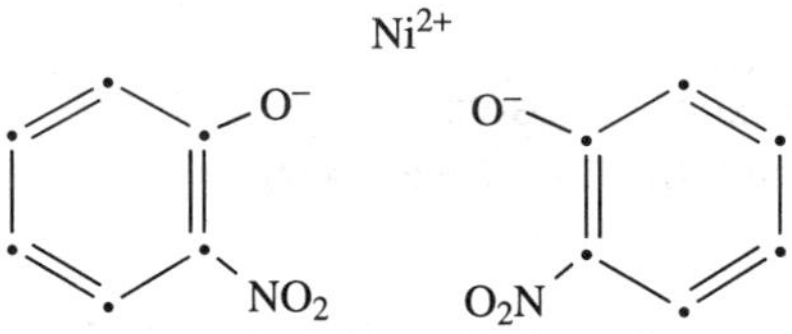

Bernardi, A., *Gazz. Chim. Ital.*, 1930, **60**, 166

It deflagrates violently or explodes on heating.

See other HEAVY METAL DERIVATIVES, METAL NITROPHENOXIDES, NITROARYL COMPOUNDS

1142

3465. Bis(2-nitrophenyl) disulfide
[115-00-6] $C_{12}H_8N_2O_4S_2$

Anon., *Jahresber.*, 1977, 72–74

This is prepared from 2-nitrochlorobenzene and sodium disulfide in ethanol, and the reaction residue after recovery of the ethanol by steam distillation is known to be of limited thermal stability. On one occasion, a container (atypically with closed vents) of bulked residues totalling over 2 t exploded violently while awaiting collection for disposal. It seems remotely possible that an *aci*-nitro thioquinonimine species derived from 2-nitrobenzenethiol may have been involved.

See Sodium 2-nitrothiophenoxide

See other NON-METAL SULFIDES, POLYNITROARYL COMPOUNDS

3466. 2-*trans*-1-Azido-1,2-dihydroacenaphthyl nitrate
[68206-59-5] $C_{12}H_8N_4O_3$

Trahanovsky, W. S. *et al., J. Amer. Chem. Soc.*, 1971, **93**, 5257

Although several other 1-azido-2-nitrato-alkanes appeared thermally stable, the acenaphthane derivative exploded violently on heating (probably during analytical combustion).

See other ALKYL NITRATES, ORGANIC AZIDES

1143

3467. Bis(4-nitrobenzenediazo) sulfide

[ ] $C_{12}H_8N_6O_4S$

Tomlinson, W. R., *Chem. Eng. News*, 1951, **29**, 5473
The dry material is extremely sensitive and can be exploded by very light friction.
It is too sensitive to handle other than as a solution, or as a dilute slurry in excess
solvent, and then only on 1 g scale.
See other BIS(ARENEDIAZO) SULFIDES

3468. 4,4′-Biphenylene-bis-sulfonylazide
[4712-19-0] $C_{12}H_8N_6O_4S_2$

$$N_3SO_2(C_6H_4)_2SO_2N_3$$

CHETAH, 1990, 182
Shock sensitive.
See other ACYL AZIDES

3469. Lead 2-amino-4,6-dinitrophenoxide
[11802-21-2] $C_{12}H_8N_6O_{10}Pb$

Sobela, B. *et al.*, Czech Pat. CS 245 822, 1987
The lead salt is used as a minor component in compositions for low-power electric
detonators.
See other HEAVY METAL DERIVATIVES, POLYNITROARYL COMPOUNDS
See related METAL NITROPHENOXIDES

3470. (2,2-Dichloro-1-fluorovinyl)ferrocene
[58281-24-6] $C_{12}H_9Cl_2FFe$

Okohura, K., *J. Org. Chem.*, 1976, **41**, 1493, footnote 14

Two attempts to purify the crude material by vacuum distillation led to sudden exothermic decomposition of the still contents. Distillation in steam was satisfactory.

See related HALOALKENES *See other* ORGANOMETALLICS

3471. Potassium tricarbonyltris(propynyl)molybdate(3—)
[ ] $C_{12}H_9K_3MoO_3$

$$K_3[(OC)_3Mo(C{\equiv}CCH_3)_3]$$

Houben-Weyl, 1975, Vol. 13.3, 470

It is pyrophoric.

See related CARBONYLMETALS, COMPLEX ACETYLIDES

3472. 4-Nitrodiphenyl ether
[2216-12-8] $C_{12}H_9NO_3$

T_{ait24} was determined as 310°C by adiabatic Dewar tests, with an apparent energy of activation of 314 kJ/mol.

See entry THERMOCHEMISTRY AND EXOTHERMIC DECOMPOSITION (reference 2)

See other NITROARYL COMPOUNDS

3473. Chloromercuriodiphenylgold
[65381-02-4]

$C_{12}H_{10}AuClHg$

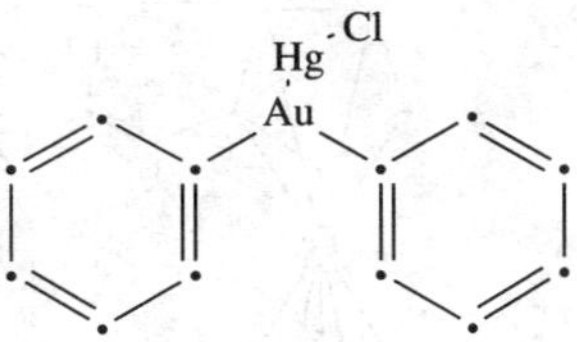

De Graaf, P. W. J. *et al., J. Organomet. Chem.*, 1977, **141**, 346
When dry it explodes violently, even at $-20°C$.
See other GOLD COMPOUNDS, MERCURY COMPOUNDS

3474. 3,5-Dibromo-7-bromomethylene-7,7a-dihydro-1,1-dimethyl-1*H*-azirino[1,2-a]indole
[34489-60-6]

$C_{12}H_{10}Br_3N$

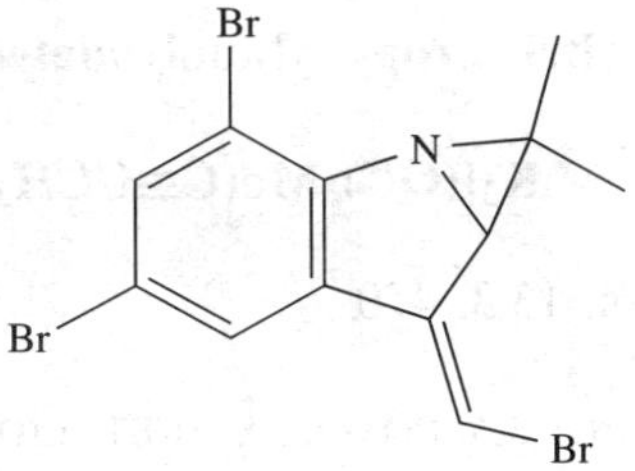

Brown, J. P., *et al., J. Chem. Soc. (C)*, 1971, 3631
A melting point sample exploded at $150°C$. Larger quantities showed signs of autocatalytic decomposition from $-30°C$. This presumably involves release of strain in the aziridine ring.
See other STRAINED-RING COMPOUNDS

3475. Oxybisphenyliodonium bistetrafluoroborate
[119701-48-3]

$C_{12}H_{10}B_2F_8I_2O$

$$(PhI^+)_2O \ 2BF_4^-$$

1. van Look G., *Chem. Eng. News*, 1989 **67**(30), 2
2. Zhdankin, V. V. *et al., J. Org. Chem.*, 1989, **54**, 2607; 2609

An explosion [1] of "iodosylbenzene tetrafluoroborate" attributed by *Chem. Abs.* to another structure is more likely to be of this, in view of the original synthetic papers [2].
See Iodosylbenzene tetrafluoroborate
See other IODINE COMPOUNDS

3476. Benzenediazonium tetrachlorozincate
[15727-43-2]

$C_{12}H_{10}Cl_4N_4Zn$

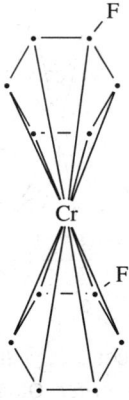

Muir, G. D., *Chem. & Ind.*, 1956, 58

A batch of the 'double salt' exploded, either spontaneously or from slight vibration, after thorough drying under vacuum at ambient temperature overnight. Although dry diazonium salts are known to be light-, heat- and shock-sensitive when dry, the double salts with zinc chloride were thought to be more stable. Presence of traces of solvent reduces the risk of frictional heating and deterioration.

See other DIAZONIUM SALTS

3477. Bis(fluorobenzene)chromium(0)
[42087-90-1]

$C_{12}H_{10}CrF_2$

Unstable, potentially explosive

See entry HALOARENEMETAL π-COMPLEXES

3478. Bis(fluorobenzene)vanadium(0)
[53966-09-9]

$C_{12}H_{10}F_2V$

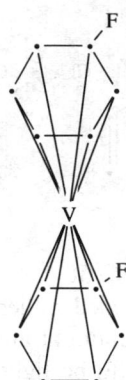

Unstable, potentially explosive.
See entry HALOARENEMETAL π-COMPLEXES

3479. Ferrocene-1,1′-dicarboxylic acid
[1293-87-4]

$C_{12}H_{10}FeO_4$

Phosphoryl chloride
   See Phosphoryl chloride: Ferrocene-1,1′-dicarboxylic acid
   See related ORGANIC ACIDS *See other* ORGANOMETALLICS

3480. Diphenylmercury
[587-85-9]

$C_{12}H_{10}Hg$

Chlorine monoxide, or Sulfur trioxide
1. Dreher, E. *et al.*, *Ann.*, 1870, **154**, 127
2. Otto, R., *J. Prakt. Chem. [2]*, 1870, **1**, 183
Chlorine monoxide reacts violently [1], and sulfur trioxide very violently [2], with diphenylmercury.
See other ARYLMETALS, MERCURY COMPOUNDS

3481. Potassium *O,O*-diphenyl dithiophosphate
[3514-82-7] $C_{12}H_{10}KO_2PS_2$

$$PhO-\underset{\underset{PhO}{|}}{\overset{\overset{S}{||}}{P}}-S^-\quad K^+$$

Arenediazonium salts
See DIAZONIUM SULFIDES AND DERIVATIVES (reference 10)
See related PHOSPHORUS ESTERS

3482. Diphenylmagnesium
[555-55-4] $C_{12}H_{10}Mg$

$$Ph\overset{Mg}{\diagup\diagdown}Ph$$

Air, or Carbon dioxide, or Water
1. Sidgwick, 1950, 234
2. Leleu, *Cahiers*, 1977, (88), 368
It ignites in moist (but not dry) air, reacts violently with water, reaching incandescence [1], and ignites in carbon dioxide [2].
See other ARYLMETALS, DIALKYLMAGNESIUMS

3483. Azobenzene (Diphenyldiazene)
[103-33-3] $C_{12}H_{10}N_2$

$$Ph\diagup\underset{N}{\overset{N}{\diagdown}}\diagup Ph$$

HCS 1980, 170

Heat of decomposition is determined as 0.80 kJ/g.
See entry THERMOCHEMISTRY AND EXOTHERMIC DECOMPOSITION
See other AZO COMPOUNDS

3484. cis-Azobenzene
[17082-12-1]

$C_{12}H_{10}N_2$

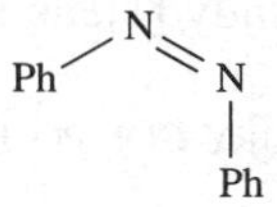

Ph—N=N—Ph

Stull, 1977, 19

It is highly endothermic (ΔH_f° (s) +362.3 kJ/mol, 1.99 kJ/g). The *trans*-isomer [1080-16-6] will be less so.

See other ENDOTHERMIC COMPOUNDS

3485. Azoxybenzene
[495-48-7]

$C_{12}H_{10}N_2O$

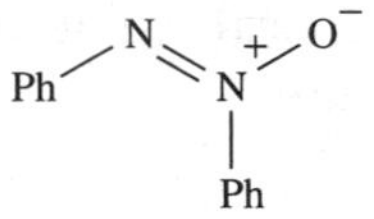

Ph—N=N⁺—O⁻ / Ph

1. Stull, 1977, 18
2. Grewer, T. *et al.*, *Exothermic Decomposition*, Technical Report 01VD 159/0329 for Federal German Ministry for Res. Technol., Bonn. 1986

Azoxybenzene is moderately endothermic (ΔH_f° (s) +152.3 kJ/mol, 0.77 kJ/g). (The isolated stereoisomeric components [21650-65-7] '*cis*-', [20972-43-4] '*trans*-' may differ significantly from these values) [1]. Energy of decomposition in range 245–420°C measured by DSC as 1.303 kJ/g by DSC, and T_{ait24} was determined as 191°C by adiabatic Dewar tests, with an apparent energy of activation of 194 kJ/mol [2].

See other ENDOTHERMIC COMPOUNDS

3486. N-Nitrosodiphenylamine
[86-30-6]

$C_{12}H_{10}N_2O$

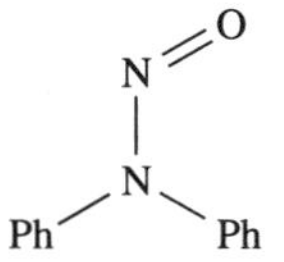

Ph—N—Ph, N=O

Energy of decomposition (in range 300–500°C) measured as 0.65 kJ/g.

See entry THERMOCHEMISTRY AND EXOTHERMIC DECOMPOSITION (reference 2)
See other NITROSO COMPOUNDS

3487. 4-Aminophenylazobenzene
[60-09-3] $C_{12}H_{11}N_3$

Energy of decomposition (in range 145–365°C) measured as 0.522 kJ/g.
See entry THERMOCHEMISTRY AND EXOTHERMIC DECOMPOSITION (reference 2)
See other AZO COMPOUNDS

3488. Diphenylphosphorus(III) azide
[4230-63-1] $C_{12}H_{10}N_3P$

Allcock, H. R. *et al., Macromolecules*, 1975, **8**, 380
The azide monomer, prepared in solution from the chlorophosphine and sodium azide, should not be isolated because it is potentially explosive.
See other NON-METAL AZIDES

3489. Diphenyl azidophosphate
[26386-83-9] $C_{12}H_{10}N_3O_3P$

Nebuloni, M. *et al., J. Calorim., Anal. Therm. Thermodyn.*, 1986, **17**, 398–
Various methods were used to assess the thermal stability and potential hazards in handling the title compound.
See related NON-METAL AZIDES, PHOSPHORUS ESTERS

3490. Bis(benzeneazo) oxide
[ ] $C_{12}H_{10}N_4O$

$$Ph-N=N-O-N=N-Ph$$

Bamberger, E., *Ber.*, 1896, **29**, 460
Extremely unstable, it explodes on attempted isolation from the liquor, or on allowing the latter to warm to −18°C.
See other BIS(ARENEDIAZO) OXIDES

3491. Bis(benzenediazo) sulfide
[22755-07-3] $C_{12}H_{10}N_4S$

$$Ph-N=N-S-N=N-Ph$$

Tomlinson, W. R., *Chem. Eng. News*, 1951, **29**, 5473
The wet solid can be exploded by impact or heating, and explodes while handling in air at ambient temperature. The material is too sensitive to handle other than as a solution or dilute slurry in excess solvent, and then only on 1 g scale.
See other BIS(ARENEDIAZO) SULFIDES

3492. Diphenyl ether
[101-84-8] $C_{12}H_{10}O$

$$Ph-O-Ph$$

HCS 1980, 731

Chlorosulfuric acid
    See Chlorosulfuric acid: Diphenyl ether

3493. Diphenyl sulfoxide
[945-51-7] $C_{12}H_{10}OS$

$$\underset{Ph}{\overset{\displaystyle \overset{O}{\|}}{\underset{}{S}}}\,Ph$$

Acetonitrile, Trichlorosilane
See Trichlorosilane: Acetonitrile, etc.

3494. Diphenylselenone
[10504-99-1] $C_{12}H_{10}O_2Se$

$$\underset{Ph}{\overset{OO}{Se}}\,Ph$$

Krafft, F. *et al., Ber.*, 1896, **29**, 424
It explodes feebly on heating in a test tube.
See related NON-METAL OXIDES

3495. Benzeneseleninic anhydride
[17697-12-0] $C_{12}H_{10}O_3Se_2$

$$\underset{Ph}{Se}\;O\;\underset{Ph}{Se}$$

Preparative hazard
See Hydrogen peroxide: Diphenyl diselenide

Benzhydrazide
See Benzeneseleninic acid: Hydrazine derivatives
See other ACID ANHYDRIDES

3496. 11,12-Diethynyl-1,4,7,10-tetraoxadispiro[4.0.4.2]dodec-11-ene
   [123002-93-7] $C_{12}H_{10}O_4$

See cis-3,4-Diethynylcyclobut-3-ene-1,2-diol
See Oligo(octacarbondioxide)
See other ACETYLENIC COMPOUNDS

3497. Bis(4-hydroxyphenyl) sulfone
   [80-09-1] $C_{12}H_{10}O_4S$

When prepared by thermal dehydration of 4-hydroxybenzenesulfonic acid, the reaction mixture begins to decompose exothermally around 240°C. Decomposition is delayed but still occurs at lower temperatures (160°C), and the presence of iron reduces the time to maximum rate of decomposition. Above 800 ppm of iron, the time to maximum rate is less than the dehydration reaction time, leading to severe control problems. Improved processing conditions were developed.
See entry SELF-ACCELERATING REACTIONS *See other* INDUCTION PERIOD INCIDENTS

3498. Benzenesulfonic anhydride
   [512-35-6] $C_{12}H_{10}O_5S_2$

Hydrogen peroxide
   See Hydrogen peroxide: Benzenesulfonic anhydride
   See other ACID ANHYDRIDES

1154

3499. Dibenzenesufonyl peroxide
[29342-61-8]

$C_{12}H_{10}O_6S_2$

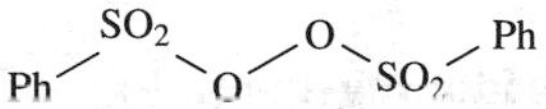

$$Ph-SO_2-O-O-SO_2-Ph$$

Alone, or Nitric acid, or Water
1. Davies, 1961, 65
2. Crovatt, L. W. *et al., J. Org. Chem.*, 1959, **24**, 2032
3. Weinland, R. F. *et al., Ber.*, 1903, **36**, 2702

It explodes at 53–54°C [1], and decomposes somewhat violently after storage overnight at ambient temperature, but may be stored unchanged for several weeks at −20°C [2]. It is also shock-sensitive, and explodes with fuming nitric acid, or addition to boiling water [3].
See other DIACYL PEROXIDES

3500. Diphenyldistibene (Stibobenzene)
[5702-61-4]

$C_{12}H_{10}Sb_2$

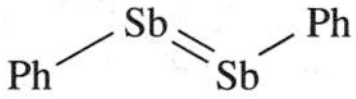

$$Ph-Sb=Sb-Ph$$

Air, or Nitric acid
Schmidt, H., *Ann.*, 1920, **421**, 235
This antimony analogue of azobenzene ignites in air, and is oxidised explosively by nitric acid.
See related ARYLMETALS, AZO COMPOUNDS

3501. Diphenyl diselenide
[1666-13-3]

$C_{12}H_{10}Se_2$

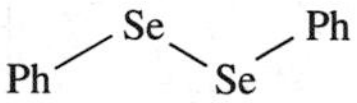

$$Ph-Se-Se-Ph$$

Hydrogen peroxide
See Hydrogen peroxide: Diphenyl diselenide

3502. Diphenyltin
[6381-06-2]

$C_{12}H_{10}Sn$

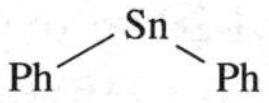

$$Ph-Sn-Ph$$

1155

Nitric acid
Nitric acid: Diphenyltin *See other* ARYLMETALS

3503. Lithium diphenylhydridotungstate(2−)
[ ] $C_{12}H_{11}Li_2W$

Sanny, B., *Z. Anorg. Chem.*, 1964, **329**, 221
It is pyrophoric in air.
See related ARYLMETALS, METAL HYDRIDES

3504. Diphenylamine
[122-39-4] $C_{12}H_{11}N$

Preparative hazard
See Aniline: Anilinium chloride

3505. Sodium 2-allyloxy-6-nitrophenylpyruvate oxime
[80841-08-3] $C_{12}H_{11}N_2NaO_6$

Hydrochloric acid
Lading, P. *et al.*, *Dan. Kemi*, 1981, **62**, 242−244
Conversion of the sodium salt to the free acid by hydrochloric acid treatment led
to an explosion. This was attributed to the thermal instability of the acid, and its
Claisen rearrangement to 3-allyl-2-hydroxynitrotoluene, identified in the residue.
See other ALLYL COMPOUNDS, NITROARYL COMPOUNDS, OXIMES

3506. 1,3-Diphenyltriazene
[136-35-6] $C_{12}H_{11}N_3$

$$Ph-N=N-NH-Ph$$

1. Müller, E. *et al., Chem. Ber.*, 1962, **95**, 1257
2. Grewer, T. *et al., Exothermic Decomposition*, Technical Report 01VD 159/0329 for Federal German Ministry for Res. Technol., Bonn. 1986
3. Grewer, T. *et al., Proc. 4th Symp. Loss Prev. Safety Prom. Proc. Ind.*, Vol. 3, A1–A11, Rugby, IChE, 1983
4. Grossel, S. S., *J. Loss Prev. Proc. Ind.*, 1988, **1**, 62–74
5. Grewer, T. *et al., Hazards from Pressure*, IChE Symp. Ser. No. 102, 1–9, Oxford, Pergamon, 1987

It decomposes explosively at the m.p., 98°C [1], and shows a high rate of pressure increase on exothermic decomposition [2]. The heat of decomposition was determined as 1.34 kJ/g by DSC, and T_{ait24} was determined as 88°C by adiabatic Dewar tests, with an apparent energy of activation of 223 kJ/mol [3]. When finely divided, it also shows significant dust explosion hazards, with a maximum explosion pressure of 7.75 bar, and a maximum rate of pressure rise above 680 bar/s [4]. Further work on homogeneous decomposition under confinement has been reported [5].

Acetic anhydride
Heusler, F., *Ber.*, 1891, **24**, 4160
A mixture exploded with extraordinary violence on warming.
See other TRIAZENES

3507. 1,5-Diphenylpentaazadiene
[ ] $C_{12}H_{11}N_5$

$$Ph-N=N-NH-N=N-Ph$$

Griess, P., *Ann.*, 1866, **137**, 81
The dry solid explodes violently on warming, impact or friction, and *C*-homologues behave similarly.
See other HIGH-NITROGEN COMPOUNDS *See related* TRIAZENES

3508. Diphenylphosphine
[829-85-6] $C_{12}H_{11}P$

Ireland, R. F. *et al., Org. Synth.*, 1977, **56**, 47
If the phosphine is spilled onto a paper towel (or other extended surface), it may ignite in air.
See related ALKYLPHOSPHINES

3509. 1,2-Di(3-buten-1-ynyl)cyclobutane
cis [80605-27-2], trans [80605-36-3] $C_{12}H_{12}$

Cupery, M. E. *et al., J. Amer. Chem. Soc.*, 1934, **56**, 1167
During isolation of the title product from polymerised 1,5-hexadiene-3-yne, high vacuum distillation must be carried only to a limited extent to prevent sudden explosive decomposition of the more highly polymerised residue.
See other ACETYLENIC COMPOUNDS, POLYMERISATION INCIDENTS

3510. Tetraacrylonitrilecopper(I) perchlorate
[ ] $C_{12}H_{12}ClCuN_4O_4$
$$(H_2C{=}CHC{\equiv}N)_4CuClO_4$$

Ondrejovic, G., *Chem. Zvesti*, 1964, **18**, 281
Like the copper(II) analogue, it decomposes explosively on heating
See related AMMINEMETAL OXOSALTS *See other* CYANO COMPOUNDS

3511. Bis(η-benzene)chromium(0)
[1271-54-1]

$C_{12}H_{12}Cr$

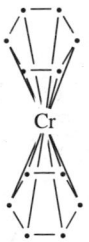

Oxygen
Anon., *Chem. Eng. News*, 1964, **42**(38), 55
The orange-red complex formed with oxygen in benzene decomposes vigorously on friction or heating in air.
See related ARYLMETALS *See other* ORGANOMETALLICS

3512. Bis(η-benzene)iron(0)
[68868-87-1]

$C_{12}H_{12}Fe$

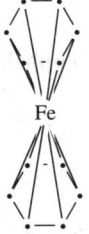

Timms, P. L., *Chem. Eng. News*, 1969, **47**(18), 43
Prepared in the vapour phase at low temperature, the solid explodes at −40°C.
See related ARYLMETALS *See other* ORGANOMETALLICS

3513. Potassium tetrakis(propynyl)nickelate(4−)
[ ]

$C_{12}H_{12}K_4Ni$

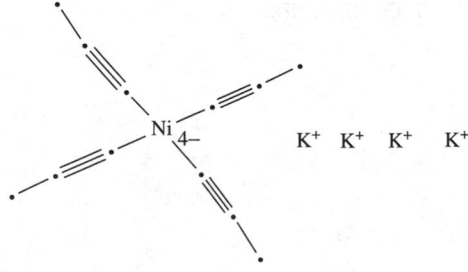

Ammonia
The diammoniate ignites on friction, impact or flame contact.
See entry COMPLEX ACETYLIDES

3514. Bis(η-benzene)molybdenum(0)
[12129-68-9]

$C_{12}H_{12}Mo$

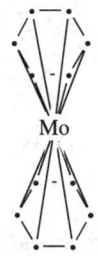

Silverthorn, W. E., *Inorg. Synth.*, 1977, **17**, 56
The dry crude material is highly pyrophoric in air.
See related ARYLMETALS *See other* ORGANOMETALLICS

3515. Dimethyl 4-acetamino-5-nitrophthalate
[52412-56-3]

$C_{12}H_{12}N_2O_7$

Preparative Hazard
See Nitric acid: Dimethyl 4-acetamidophthalate
See other NITROARYL COMPOUNDS

3516. Benzidine (4,4'-Diaminobiphenyl) (4,4'-Diaminobiphenyl)
[92-87-5] $C_{12}H_{12}N_2$

Nitric acid
See Nitric acid: Benzidine

3517. 1,2-Diphenylhydrazine
[530-50-7] $C_{12}H_{12}N_2$

Energy of decomposition (in range 135–405°C) measured as 0.442 kJ/g by DSC, and T_{ait24} was determined as 131°C by adiabatic Dewar tests, with an apparent energy of activation of 139 kJ/mol.
See entry THERMOCHEMISTRY AND EXOTHERMIC DECOMPOSITION (reference 2)

Perchloryl fluoride
See Perchloryl fluoride: Nitrogenous bases
See other REDUCANTS

3518. Amminebarium bis(nitrophenylide)
[91410-67-2] $C_{12}H_{13}BaN_3O_4$

Stevenson, G. R. *et al., J. Org. Chem.*, 1984, **49**, 3444
A 5 day old sample of the crystalline radical complex exploded violently on agitation.
See related N-METAL DERIVATIVES, NITROARYL COMPOUNDS

3519. 1,1-Diphenylhydrazinium chloride
[530-47-2]

$C_{12}H_{13}ClN_2$

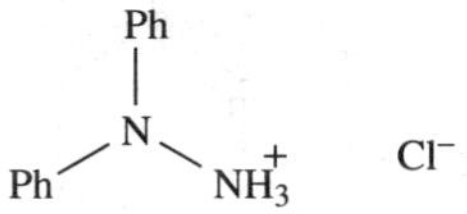

Energy of decomposition (in range 140–200°C) measured as 0.59 kJ/g.
See entry THERMOCHEMISTRY AND EXOTHERMIC DECOMPOSITION (reference 2)
See other HYDRAZINIUM SALTS

3520. Dimethyl 4-acetamidophthalate
[51832-56-5]

$C_{12}H_{13}NO_5$

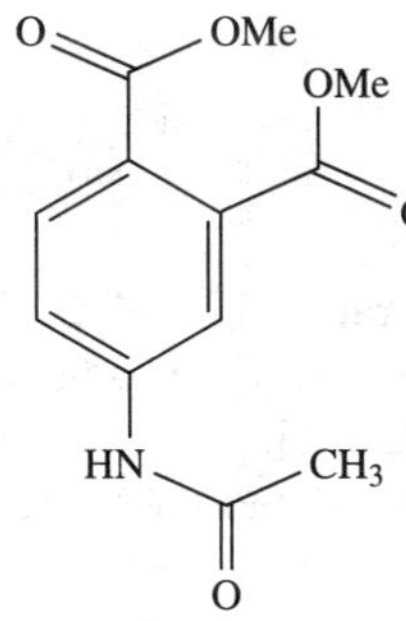

Nitric acid
Anon., *Jahresber.*, 1987, 64
Nitration of the ester with 90% nitric acid was effected according to a literature method. Examination of the reaction mixture by TLC showed that both expected mononitro derivatives had been formed, and it was allowed to stand at ambient temperature overnight. Ice and dichloromethane were then added, and the separated solvent layer was washed with aqueous sodium hydrogen carbonate, dried, and then freed of solvent by evaporation. The viscous oily residue decomposed and ignited in the flask. The preparation was repeated, but worked up immediately and under nitrogen while still cold, and furnished samples of both of the pure isomeric nitro derivatives. Thermal analysis of these showed both to be thermally stable. It was concluded that the decomposition was due to an impurity formed during the 16 h interval between nitration and work-up.
See other NITRATION INCIDENTS

3521. *N*-*tert*-Butylphthalisoimidium tetrafluoroborate
(3-*N*-*tert*-Butyliminioisobenzofuranone tetrafluoroborate)
[69338-55-2]

$C_{12}H_{14}BF_4NO_2$

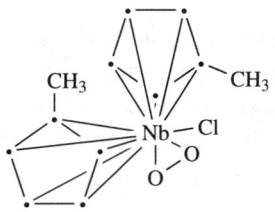

Preparative hazard
See Acetic anhydride: *N*-*tert*-Butylphthalamic acid, etc.

3522. Bis(methylcyclopentadienyl)peroxoniobium chloride
[77625-37-7]

$C_{12}H_{14}ClNbO_2$

Sala-Pala, J. *et al., J. Mol. Catal.*, 1980, **7**, 142
Occasionally the dry product exploded violently when scraped off the glass container. It invariably exploded when irradiated with a Raman laser beam, even at −80°C.
See related ORGANOMETALLIC PEROXIDES *See other* IRRADIATION DECOMPOSITION INCIDENTS

3523. Dianilineoxodiperoxochromium(VI)
[ ]

$C_{12}H_{14}CrN_2O_5$

[(PhNH$_2$)$_2$CrO(O$_2$)$_2$]

See Oxodiperoxodipyridinechromium(VI)
See other AMMINECHROMIUM PEROXOCOMPLEXES

3524. Bis(2-methylpyridine)sodium

[ ] $C_{12}H_{14}N_2Na$

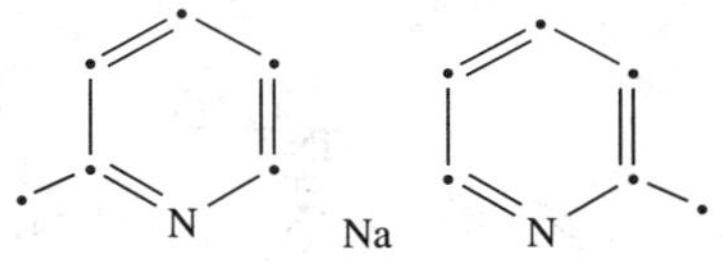

Sidgwick, 1950, 89

The addition product of sodium and 2-methylpyridine (or pyridine) ignites in air.

3525. 4,4′-Oxybis(benzenesulfonylhydrazide)

[80-51-3] $C_{12}H_{14}N_4O_5S_2$

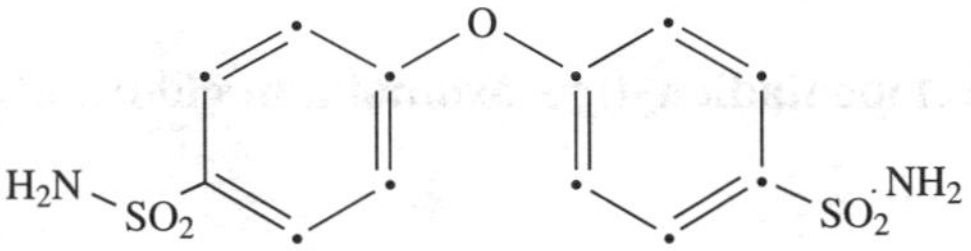

The recently calculated value of 78°C for critical ignition temperature is appreciably lower than the previous value of 90°C.

See entry CRITICAL IGNITION TEMPERATURE (reference 2)

See entry BLOWING AGENTS *See other* N–S COMPOUNDS

3526. Bis(2-hydroxyethyl) terephthalate

[959-26-2] $C_{12}H_{14}O_6$

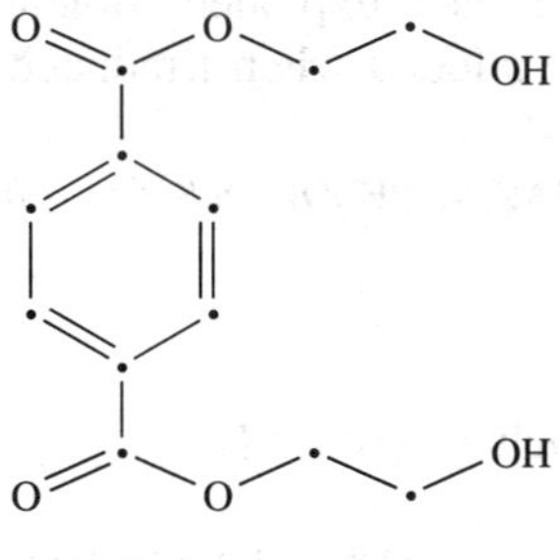

Ethanol, Perchloric acid

See Perchloric acid: Bis(2-hydroxyethyl) terephthalate, etc.

3527. 6-Isopropylidenethiacyclodeca-3,8-diyne

[ ] $C_{12}H_{14}S$

Gleiter, R. *et al.*, *Synthesis*, 1993, (6), 558
The compound exploded during Kugelrohr distillation at about 1 millibar, temperature not stated but probably about 100°C.
See other ACETYLENIC COMPOUNDS

3528. 1,3,4,7-Tetramethylisoindole

[20944-65-4] $C_{12}H_{15}N$

Cobalt
Dolphin, D. *et al.*, *Inorg. Synth.*, 1980, **20**, 157
The scale of the procedure described for preparing [octamethyltetrabenzoporphinato (2−)]cobalt(II) by heating the reagents in an evacuated Carius tube at 390°C must not be increased or the tube will explode. Personal protection is also necessary when opening the sealed tube.
See other ORGANIC BASES

3529. Acetoacet-4-phenetidide

[122-82-7] $C_{12}H_{15}NO_3$

The fine powder is a significant dust explosion hazard.
See entry DUST EXPLOSION INCIDENTS (reference 22)

3530. 2,4,6-Triallyloxy-1,3,5-triazine ('Triallylcyanurate')
[1025-15-6] $C_{12}H_{15}N_3O_3$

Preparative hazard
See 2,4,6-Trichloro-1,3,5-triazine: Allyl alcohol
See other ALLYL COMPOUNDS

3531. 1,3,5-Tris(1-oxoprop-2-enyl)-hexahydro-1,3,5-triazine
(1,3,5-Triacrylylhexahydro-*sym*-triazine)
[959-52-4] $C_{12}H_{15}N_3O_3$

Preparative Hazard
See Acrylonitrile: Formaldehyde

3532. 5-*tert*-Butyl-2,4,6-trinitro-1,3-xylene
[81-15-2] $C_{12}H_{15}N_3O_6$

1166

See Nitric acid: *tert*-Butyl-*m*-xylene, Sulfuric acid
See other POLYNITROARYL COMPOUNDS

3533. 1,3,5-Tris(2,3-epoxypropyl)triazine-2,4,6-trione ('Triglycidylisocyanurate')

[2589-01-7] $C_{12}H_{15}N_3O_6$

Anon., *Loss Prev. Bull.*, 1977, (013), 3

The epoxy resin component is made by a 2-stage process involving reaction of 1-chloro-2,3-epoxypropane (epichlorhydrin) with isocyanuric acid to give the 1,3,5-tris(2-hydroxy-3-chloropropyl) derivative, which is then treated with sodium hydroxide to eliminate hydrogen chloride to form the title compound. One batch contained more than the normal amount of hydrolysable chlorine, and when excess epichlorhydrin was distilled off, the residual material decomposed with explosive violence. It was later established that the abnormal chlorine content was associated with reduced thermal stability, and criteria for hydrolysable chlorine, epoxy content and pH have been set to prevent distillation of off-spec. material.

See other 1,2-EPOXIDES

3534. Pentamethylcyclopentadienyl-bis(thiolatothiatriazyl)rhodium

[71178-86-4] $C_{12}H_{15}N_6RhS_4$

Rigby, W. *et al.*, *J. Chem. Soc., Dalton Trans.*, 1979, 378

The material, which is probably polymeric, may detonate.

See other N–S COMPOUNDS, ORGANOMETALLICS

3535. 2,4,6-Tri(2-acetylhydrazino)-1,3,5-trinitrobenzene
[78925-61-8]

$C_{12}H_{15}N_9O_9$

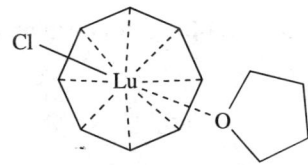

Hydrochloric acid

Lawrence, G. W. *et al., Tetrahedron Lett.*, 1980, **21**, 1618

Hydrolysis of the acetylhydrazide by 12 M hydrochloric acid at ambient temperature gave a solid product which was surprisingly sensitive to a mild hammer-blow. Formation of a diazo structure was suspected.

See other DIAZONIUM SALTS *See other* POLYNITROARYL COMPOUNDS

3536. Chloro($\eta^8$1,3,5,7-cyclooctatetraene)(tetrahydrofuran)lutetium
[96504-50-6]

$C_{12}H_{16}ClLuO$

Wayda, A. L., *Inorg. Synth.*, 1990, **27**, 150

The dry powder may ignite spontaneously in air, partially oxidised samples can explode.

See other ORGANOMETALLICS

3537. Dianilinium dichromate
[101672-09-7] $C_{12}H_{16}Cr_2N_2O_7$

Gibson, G. M., *Chem. & Ind.*, 1966, 553
It is unstable on storage.
See other DICHROMATE SALTS OF NITROGENOUS BASES

3538. Tetrakis(pyrazole)manganese(II) sulfate
[ ] $C_{12}H_{16}MnN_8O_4S$

$$[(C_3H_4N_2)_4Mn]\ SO_4$$

See entry AMMINEMETAL OXOSALTS (reference 9)

3539. 1-Phenylazocyclohexyl hydroperoxide
[950-32-3] $C_{12}H_{16}N_2O_2$

See entry α-PHENYLAZO HYDROPEROXIDES (reference 4)

3540. Tris(3-methylpyrazole)zinc sulfate
[55060-80-5] $C_{12}H_{18}N_6O_4SZn$

$$[(MeC_3H_3N_2)_3Zn]\ SO_4$$

See entry AMMINEMETAL OXOSALTS (reference 9)

3541. 1,2-Bis(2-hydroperoxy-2-propyl)benzene
  [29014-32-2] $C_{12}H_{18}O_4$

Velenskii, M. S. *et al., Chem. Abs.*, 1974, **81**, 108066
This difunctional analogue of cumyl hydroperoxide appears to be no more
hazardous that the latter. Though impact-sensitive, the decomposition was mild
and incomplete.
See other ALKYL HYDROPEROXIDES

3542. 1,4-Bis(2-hydroperoxy-2-propyl)benzene
  [3159-98-6] $C_{12}H_{18}O_4$

Velenskii, M. S. *et al., Chem. Abs.*, 1974, **81**, 108066
This difunctional analogue of cumyl hydroperoxide appears to be no more
hazardous that the latter. Though impact-sensitive, the decomposition was mild
and incomplete.
See other ALKYL HYDROPEROXIDES

3543. 1,10-Di(methanesulfonyloxy)deca-4,6-diyne
  [42404-59-1] $C_{12}H_{18}O_6S_2$

CHETAH, 1990, 189
Shock sensitive. The parent diol is not but shows thermal risks on calorimetry.
See other ACETYLENIC COMPOUNDS, SULFUR ESTERS

3544. 3-Acetoxy-4-iodo-3,7,7-trimethylbicyclo[4.1.0]heptane
[ ] $C_{12}H_{19}IO_2$

See vic-IODO-ALKOXY COMPOUNDS
See other IODINE COMPOUNDS

3545. Tetraallyl-2-tetrazene (Tetra-2-propenyl-2-tetrazene)
[52999-19-6] $C_{12}H_{20}N_4$

Houben-Weyl, 1967, Vol. 10.2, 831
It explodes with great violence when heated above its b.p., 113°C/1 bar.
See other ALLYL COMPOUNDS, TETRAZENES

3546. 3,6-Di(spirocyclohexane)-1,2,4,5-tetraoxane
(7,8,15,16-Tetraoxadispiro[5.2.5.2]hexadecane)
[183-84-6] $C_{12}H_{20}O_4$

Dilthey, W. *et al., J. Prakt. Chem.*, 1940, **154**, 219
This dimeric cyclohexanone peroxide explodes on impact.
See other CYCLIC PEROXIDES

3547. 1-Acetoxy-6-oxo-cyclodecyl hydroperoxide
[ ] $C_{12}H_{20}O_5$

Criegee, R. *et al., Ann.,* 1949, **564**, 15
The crystalline compound prepared at $-70°C$ explodes mildly after a few minutes at ambient temperature.
See related ALKYL HYDROPEROXIDES, 1-OXYPEROXY COMPOUNDS

3548. Tetraallyluranium
[28711-64-0] $C_{12}H_{20}U$

Bailar, 1973, Vol. 5, 405
Only stable below $-20°C$, it ignites in air.
See related ALKYLMETALS *See other* ALLYL COMPOUNDS

3549. 3-Buten-1-ynyldiisobutylaluminium
[18864-05-6] $C_{12}H_{21}Al$

Petrov, A. A. *et al., Zh. Obsch. Khim.,* 1962, **32**, 1349

1172

It ignites in air.
See other METAL ACETYLIDES, TRIALKYLALUMINIUMS

3550. Perhydro-9b-boraphenalene
[16664-33-8] $C_{12}H_{21}B$

Negishi, E.-I. *et al., Org. Synth.*, 1983, **61**, 103–111
The title product and several precursors are highly pyrophoric, and all leaks must be eliminated from vacuum handling equipment.
See related ALKYLBORANES

3551. 6-Azidohexyl 6-azidohexanoate
[84487-84-3] $C_{12}H_{22}N_6O_2$

See entry ENERGETIC COMPOUNDS
See other ORGANIC AZIDES

3552. 1-Butoxyethyl 3-trimethylplumbylpropiolate
[21981-95-3] $C_{12}H_{22}O_3Pb$

$$BuOCHMeOCO.C{\equiv}CPbMe_3$$

Houben-Weyl, 1975, Vol. 13.3, 80
It is explosive.
See related ALKYLMETALS, METAL ACETYLIDES

3553. Bis(1-hydroxycyclohexyl) peroxide (1,1′-Dioxybiscyclohexanol)
[2407-94-5] $C_{12}H_{22}O_4$

1. Stoll, M. *et al., Helv. Chim. Acta*, 1930, **13**, 142
2. *CHETAH*, 1990, 183
Normally stable, it explodes on attempted vacuum distillation [1] and is shock sensitive [2].
See other 1-OXYPEROXY COMPOUNDS

3554. Dihexanoyl peroxide
[2400-59-1] $C_{12}H_{22}O_4$

$$BuCH_2CO.OOCO.CH_2Bu$$

Castrantas, 1965, 17
It explodes at 85°C.
See other DIACYL PEROXIDES

3555. 1-Hydroperoxy-1′-hydroxydicyclohexyl peroxide
(1-[(1-Hydroperoxycyclohexyl)dioxy]cyclohexanol)
[78-18-2] $C_{12}H_{22}O_5$

1. Davies, 1961, 74
2. Cardinale, G. *et al., Tetrahedron*, 1985, **45**, 2901

This appears to be a main constituent of commercial 'cyclohexanone peroxide' together with the bis(hydroperoxy) peroxide (below), and is also known to be hazardous [1]. Prepared from cyclohexanone and 30% hydrogen peroxide in presence of hydrochloric acid catalyst, it must be kept moist after isolation as it may explode on drying out [2].

See entry COMMERCIAL ORGANIC PEROXIDES *See other* 1-OXYPEROXY COMPOUNDS

3556. Bis(1-hydroperoxycyclohexyl) peroxide (Dioxydicyclohexylidenebishydroperoxide) [2699-12-9]

$C_{12}H_{22}O_6$

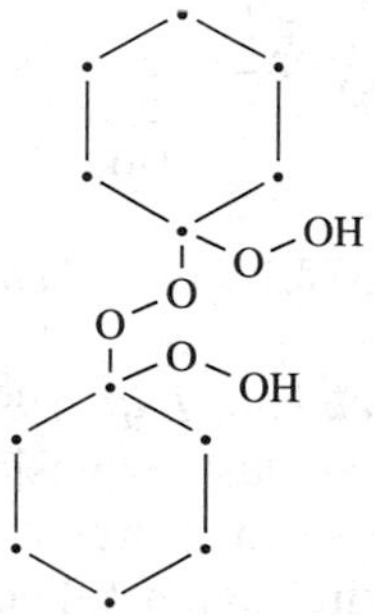

Criegee, R. *et al., Ann.*, 1949, **565**, 17–18

One of the components in 'cyclohexanone peroxide', it explodes violently at elevated temperatures (during vacuum distillation or on exposure to flame).

See entry COMMERCIAL ORGANIC PEROXIDES *See other* 1-OXYPEROXY COMPOUNDS

3557. Lactose [63-42-3]

$C_{12}H_{22}O_{11}$

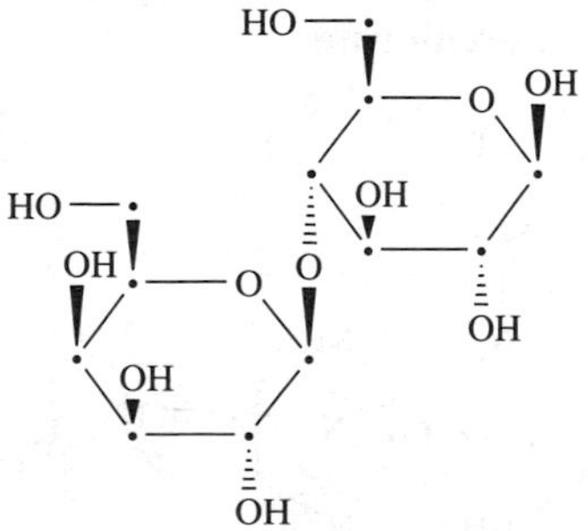

Alkalis
See SUGARS

Oxidants
Treumann, H., *Chem. Abs.*, 1981, **94**, 177551
Explosion hazards of mixtures of lactose monohydrate with potassium chlorate, potassium nitrate or potassium perchlorate are assessed.
See Potassium chlorate: Sugars

3558. Sucrose
[57-50-1] $C_{12}H_{22}O_{11}$

1. Geysen, W. J. *et al.*, *ASTM Spec. Tech. Publ. 958*, 234–242, Philadelphia, ASTM, 1987
2. Beck, H. *Gefahrstoffe-Reinhalt. Luft*, 1996, **56**(10), 369
The investigation of a sugar dust explosion in a Belgian factory is discussed and preventive and protective measures are described in detail [1]. Sugar dust explosions and safety measures against them are reviewed [2].
   Energy of decomposition (in range 280–350°C) measured as 0.48 kJ/g.
See entry THERMOCHEMISTRY AND EYOTHERMIC DECOMPOSITION (reference 2)
See other DUST EXPLOSION INCIDENTS

Nitric Acid
See Nitric Acid: Sucrose

Sulfuric Acid
See Sulfuric Acid: Sucrose

3559. *trans*-Bromoazido(meso-5,12-dimethyl-1,4,8,11-tetraazacyclotetradeca-4,11-diene)cobalt(III) perchlorate
[74190-87-7] $C_{12}H_{24}BrClN_7O_4$

1176

Lawrence, G. A. *et al., Aust. J. Chem.*, 1980, **33**, 274

Perchlorate salts of the azide complexes, and also of the dichloro analogue, may present an explosion hazard.

See related [14] DIENE-N₄COMPLEXES

3560. 2,4,6,8-Tetraethenyl-2,4,6,8-tetramethylcyclotetrasiloxane.
[2554-06-5]   C₁₂H₂₄O₄Si₄

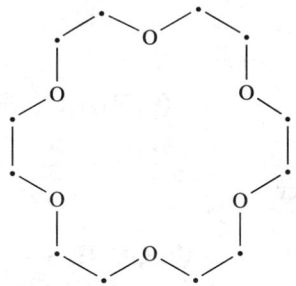

Rubinsztajn, S. *et al., Chem. Eng. News.*, 1994, **72**(17), 4

Hydrosilylation with 2,4,6,8-tetramethylcyclotetrasiloxane has sometimes given uncontrollable exotherms or explosions in the laboratory. Use of safety screens and the lowest possible (<20 ppm) level of platinum catalyst are recommended.

See VINYLSILOXANES

3561. 1,4,7,10,13,16-Hexaoxacyclooctadecane ('18-Crown-6')
[17455-13-9]   C₁₂H₂₄O₆

Preparative hazard

1. Cram, D. J. *et al., J. Org. Chem.*, 1974, **39**, 2445
2. Stott, P. E., *Chem. Eng. News*, 1976, **54**(37), 5
3. Gouw, T. H., *Chem. Eng. News*, 1976, **54**(44), 5
4. Stott, P. E., *Chem. Eng. News*, 1976, **54**(51), 5
5. de Jong, F. *et al., J. R. Neth. Chem. Soc.*, 1981, **100**, 449–452

There is a potential explosion hazard during larger scale operation of the published (small-batch) procedure [1]. This arose during thermal decomposition of the crown ether–potassium chloride complex under reduced pressure, when the crude ether distilled out. The larger batch size involved more extensive heating to complete the decomposition of the complex, and a considerable amount of 1,4-dioxane was produced by cracking and blocked the vacuum pump trap. Admission of air to the overheated residue led to a violent explosion, attributed to autoignition of the dioxane–air mixture. Dioxane has a relatively low AIT (180°C) and rather wide explosive limits. Practical precautions are detailed [2]. Subsequently the observed hazards were attributed to poor distillation procedures [3], but further more detailed information on the experimental procedure was published to refute this view [4]. A safe method of purifying the product without distillation involves formation of complexes with alkaline earth metal alkanedisulfonates [5].

See other APROTIC SOLVENTS

3562. 3,6,9-Triethyl-3,6,9-trimethyl-1,2,4,5,7,8-hexoxonane
[24748-23-0] $C_{12}H_{24}O_6$

See 3,6-Diethyl-3,6-dimethyl-1,2,4,5-tetraoxane
See other CYCLIC PEROXIDES

†3563. Dihexyl ether
[112-58-3] $C_{12}H_{26}O$

$$(C_6H_{13})_2O$$

3564. Di-2-butoxyethyl ether
[112-73-2] $C_{12}H_{26}O_3$

$$(BuOC_2H_4)_2O$$

Nitric acid
See Nitric acid: Di-2-butoxyethyl ether

3565. 2,2-Di(*tert*-butylperoxy)butane
[2167-23-9] $C_{12}H_{26}O_4$

$$(Me_3COO)_2CMeEt$$

Dickey, F. H. *et al., Ind. Eng. Chem.*, 1949, **41**, 1673

The pure material explodes on heating to about 130°C, on sparking or on impact.
See related DIALKYL PEROXIDES

3566. Bis(2-hydroperoxy-4-methyl-2-pentyl) peroxide
[53151-88-5] $C_{12}H_{26}O_6$

$$(Me_2CHCH_2CMe(OOH)O-)_2$$

Leleu, *Cahiers*, 1973, (71), 238
The triperoxide, main component in 'MIBK peroxide', is explosive in the pure
state, but insensitive to shock as the commercial 60% solution in dimethyl phthalate. The solution will explode at about 75°C, and slowly liberates oxygen at
ambient temperatures.
See KETONE PEROXIDES
See related ALKYL HYDROPEROXIDES, DIALKYL PEROXIDES

3567. Dodecanethiol
[1322-36-7] $C_{12}H_{26}S$

HCS 1980, 449

Nitric acid
   See Nitric acid: Alkanethiols
   See other ALKANETHIOLS

†3568. Triisobutylaluminium
[100-99-2] $C_{12}H_{27}Al$

$$(Me_2CHCH_2)_3Al$$

HCS 1980, 937

1. 'Specialty Reducing Agents', Brochure TA-2002/1, New York, Texas Alkyls,
   1971
2. Fischer, T. S. *et al.*, *Int. Annu. Conf. ICT*, 1996, **27th** (Energetic Materials),
   107

Used industrially as a powerful reducant, it is supplied as a solution in hydrocarbon
solvents. The undiluted material is of relatively low thermal stability (decomposing above 50°C) and ignites in air unless diluted to below 25% concentration
[1]. A thermal explosion of the compound has been studied from a theoretical
standpoint [2].
See other TRIALKYLALUMINIUMS

3569. *mixo*-**Tributylborane**
[122-56-5, 1116-39-8] (n-, iso- resp.) $C_{12}H_{27}B$
Bu_3B, $(Me_2CHCH_2)_3B$

1. Hurd, D. T., *J. Amer. Chem. Soc.*, 1948, **70**, 2053
2. Brown, C. A., *Inorg. Synth.*, 1977, **17**, 27
A mixture of the n- and iso-compounds ignited on exposure to air [1]. In bulk, the liquid does not ignite in air, but may do so when absorbed onto a porous surface like paper, or if spread as a thin film [2]. *See other* ALKYLBORANES

3570. Tri-2-butylborane
[1113-78-6] $C_{12}H_{27}B$
$(MeEtCH)_3B$

Brown, C. A., *Inorg. Synth.*, 1977, **17**, 27
It ignites in air.
See other ALKYLBORANES

3571. Tributylbismuth
[3692-81-7] $C_{12}H_{27}Bi$
Bu_3Bi

Air, or Oxygen
Gilman, H. *et al., J. Amer. Chem. Soc.*, 1939, **61**, 1170
It ignites in air and explodes violently in oxygen.
See other TRIALKYLBISMUTHS

3572. Tributylgallium
[15677-44-8] $C_{12}H_{27}Ga$
Bu_3Ga

Kovar, R. A. *et al., Inorg. Chem.*, 1980, **19**, 3265
Tributylgallium and dibutylgallium chloride are probably pyrophoric in air.
See other ALKYLMETALS

3573. Tributylindium
[15676-66-1] $C_{12}H_{27}In$
Bu_3In

Houben-Weyl, 1970, Vol. 13.4, 347
It ignites in air.
See other ALKYLMETALS

3574. Tributyl phosphate
[126-73-8] $C_{12}H_{27}O_4P$

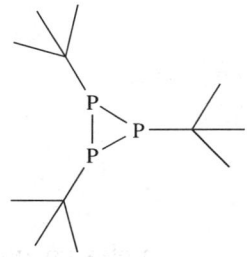

Nitric acid, Nitrates
See NUCLEAR WASTES

†3575. Tributylphosphine
[998-40-3] $C_{12}H_{27}P$

$$Bu_3P$$

Steele, A. B. *et al., Chem. Engrg.*, 1969, **66**(4), 160
It ignites in air.
See other ALKYLPHOSPHINES

3576. Tris(2,2-dimethylethyl)triphosphirane
[61695-12-3] $C_{12}H_{27}P_3$

Baudler, M. *et al., Inorg. Synth.*, 1989, **25**, 1
The compound is reported as pyrophoric when finely divided.
See related ALKYLPHOSPHINES

3577. Tetraisopropylchromium
[38711-69-2] $C_{12}H_{28}Cr$

$$(Me_2CH)_4Cr$$

Müller, J. *et al., Angew. Chem. (Intern. Ed.)*, 1975, **14**, 761
The finely crystalline form is pyrophoric.
See other ALKYLMETALS

3578. 1,10-Phenanthroline-5,6-dione
[27318-90-7]

$C_{12}H_8N_2O_2$

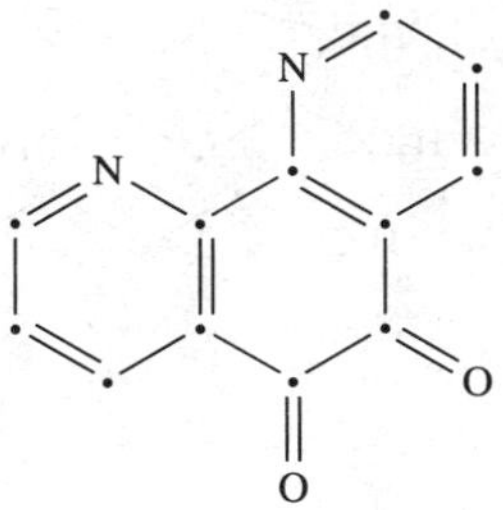

Preparative hazard
  Black, K. J. *et al., Inorg. Chem.*, 1993, **32**(24), 5591–6
  Preparation by oxidising tris(phenanthroline)cobalt(III) tetrafluoroborate with nitric
  acid in sulfuric acid with potassium bromide catalysis is potentially explosive.
  See Nitric acid

3579. Titanium tetraisopropoxide
[546-68-9]

$C_{12}H_{28}O_4Ti$

Ti(OCHMe_2)_4

  HCS 1980, 894

  Hydrazine
    See Hydrazine: Titanium compounds
    See other METAL ALKOXIDES

3580. 'Tetrapropyldiborane'
[22784-01-6]

$C_{12}H_{30}B_2$

(Complex mixture)

  'Tetraethyldiborane' *See other* ALKYLBORANES

3581. Bis(triethyltin) peroxide
[4403-63-8]

$C_{12}H_{30}O_2Sn_2$

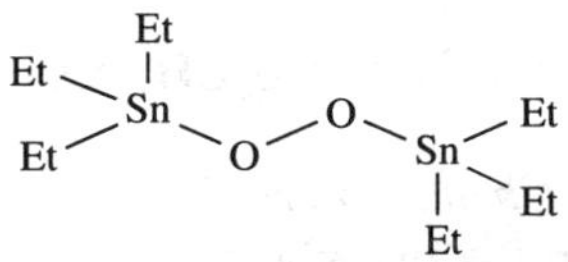

  Sorbe, 1968, 160

1182

Readily decomposed, it explodes at 50°C.
See other ORGANOMETALLIC PEROXIDES

3582. Tetra(3-aminopropanethiolato)trimercury perchlorate
[95247-08-1] $C_{12}H_{32}Cl_2Hg_3O_8S_4$

$$[(H_2NC_3H_6S)_4Hg_3] [ClO_4]_2$$

Barrera, H. *et al.*, *Chem. Abs.*, 1985, **102**, 159407
It decomposed violently but non-explosively.
See other MERCURY COMPOUNDS
See related AMMINEMETAL OXOSALTS

3583. 1,2-Bis(triethylsilyl)hydrazine
[13271-97-1] $C_{12}H_{32}N_2Si_2$

See entry SILYLHYDRAZINES *See related* ALKYLSILANES

3584. Tris(trimethylsilylmethyl)aluminium
[41924-27-0] $C_{12}H_{33}AlSi_3$

Tessier-Youngs, C. *et al.*, *Inorg. Synth.*, 1986, **24**, 92, 94
It burns spontaneously in air and reacts violently with water, as does the congener, bis(trimethylsilylmethyl)aluminium bromide.
See related TRIALKYLALUMINIUMS, ALKYLMETALS, ALKYLSILANES

3585. Tris(trimethylsilylmethyl)indium
[69833-15-4] $C_{12}H_{33}InSi_3$

Kopasz, J. P et al., Inorg. Synth., 1986, **24**, 89
It burns spontaneously in air.
See related TRIALKYLALUMINIUMS, ALKYLMETALS, ALKYLSILANES

3586. Triethylsilyl-1,2-bis(trimethylsilyl)hydrazine
[13272-03-2] $C_{12}H_{34}N_2Si_3$

See entry SILYLHYDRAZINES *See related* ALKYLSILANES

3587. Dodecamethyltetraplatinum(IV) perchlorate (Trimethylplatinum(IV) perchlorate tetramer)
[23411-89-4] (monomer) $C_{12}H_{36}Cl_4O_{16}Pt_4$

Preparative hazard
Neruda, B. et al., J. Organomet. Chem., 1977, **131**, 318
The salt is a heat and shock-sensitive explosive. Attempts to prepare it from
trimethylplatinum hydroxide and perchloric acid led to violent explosions at higher
acid concentrations.
See related ALKYLMETALS, METAL PERCHLORATES *See other* PLATINUM COM-
POUNDS

1184

3588. Tetrakis(trimethylsilyl)diaminodiphosphene
[84521-55-1] $C_{12}H_{36}N_2P_2$

Niecke, E. *et al., Angew. Chem. (Intern. Ed.)*, 1983, **22**, 155
It is pyrophoric in air.
See related ALKYLSILANES, PHOSPHINES

3589. Tris(2,3-diaminobutane)nickel(II) nitrate
[66599-67-5] $C_{12}H_{36}N_8NiO_6$

$$[(MeCH(NH_2)CH(NH_2)Me)_3Ni]\ [NO_3]_2$$

Billo, E. J. *et al., Inorg. Chim. Acta*, 1978, **26**, L10
The complex is potentially explosive.
See other AMMINEMETAL OXOSALTS

3590. Dodecamethyltetraplatinum(IV) azide (Trimethylplatinum(IV) azide tetramer)
[52732-14-6] $C_{12}H_{36}N_{12}Pt_4$

von Dahlen, K. H. *et al., J. Organomet. Chem.*, 1974, **65**, 267
The azide is not shock-sensitive, but detonates violently on rapid heating or exposure to flame.

Trimethyl phosphite
Neruda, B. *et al., J. Organomet. Chem.*, 1976, **111**, 241–248
In the exothermic reaction with trimethyl phosphite to give *cis*-dimethyl-bis(trimethyl phosphito)platinum, the azide must be added to the phosphite in small portions with stirring. Addition of the phosphite to the solid azide led to a violent explosion, probably involving the transitory by-product methyl azide.
See related METAL AZIDES *See other* PLATINUM COMPOUNDS

3591. μ-Peroxobis[ammine(2,2′,2″-triaminotriethylamine)cobalt(III)](4+) perchlorate

[37480-75-4] $C_{12}H_{42}Cl_4Co_2N_{10}O_{18}$

$[H_3N (H_2NC_2H_4)_3NCoOOCoN(C_2H_4NH_2)_3 NH_3] [ClO_4]_4$

1. Yang, C.-H. *et al., Inorg. Chem.*, 1973, **12**, 666
2. Mori, M. *et al., J. Amer. Chem. Soc.*, 1968, **90**, 619

The salt exploded at 220°C [1], and other oxosalts of this type (permanganates, possibly nitrates) are also explosive [2].

See related AMMINECHROMIUM PEROXOCOMPLEXES, AMMINEMETAL OXOSALTS

3592. Di[tris-1,2-diaminoethanecobalt(III)] triperoxodisulfate

[ ] $C_{12}H_{48}Co_2N_{12}O_{24}S_6$

$[(C_2H_8N_2)_3Co]_2 [OSO_2OOSO_2O]_3$

Beacom, S. E., *Nature*, 1959, **183**, 38

It explodes upon ignition, or after UV irradiation and then heating to 120°C.

See other AMMINEMETAL OXOSALTS

3593. Di[tris(1,2-diaminoethanechromium(III)] triperoxodisulfate

[ ] $C_{12}H_{48}Cr_2N_{12}O_{24}S_6$

$[(C_2H_8N_2)_3Cr]_2 [OSO_2OOSO_2O]_3$

Beacom, S. E., *Nature*, 1959, **183**, 38

It explodes upon ignition, or after UV irradiation and then heating to 115°C.

See other AMMINEMETAL OXOSALTS

3594. Lead(II) trinitrosobenzene-1,3,5-trioxide

[ ] $C_{12}N_6O_{12}Pb_3$

Freund, H. E., *Angew. Chem.*, 1961, **73**, 433

An air dried sample of the lead salt of trinitrosophoroglucinol exploded when disturbed, possibly owing to aerobic oxidation to the trinitro compound.

See other HEAVY METAL DERIVATIVES, NITROSO COMPOUNDS

3595. 4,5,11,12-Tetraoxo-14H -[1,2,5]oxadiazolo[3,4-e][1,2,5]oxadiazolo[3′,4′:4,5]
benzotriazolo[2,1-a]benzotriazolo-b -ium 1,8-dioxide
[ ] $C_{12}N_8O_8$

See 5,12-Dioxo-4,4,11,11-tetrahydroxy-14H -[1,2,5]oxadiazolo[3,4-e][1,2,5]
oxadiazolo[3′,4′:4,5]benzotriazolo[2,1-a]benzotriazolo-b-ium 1,8-dioxide
See other FURAZAN N-OXIDES

3596. $N,N′$-dichlorobis(2,4,6-trichlorophenyl)urea
[2899-02-7] $C_{13}H_4Cl_8N_2O$

Ammonia
  Pytlewski, L. L., *Rep. AD-A028841*, 13, Richmond (Va.), USNTIS, 1976
  Contact of gaseous ammonia with the N-chlorourea, either alone, or mixed with
  zinc oxide, leads to ignition. The same could happen in contact with conc. aqueous
  ammonia, solid ammonium carbonate or organic amines.

Dimethyl sulfoxide
  1. Pytlewski, L. L., *Rep. AD-A028841*, 13, Richmond (Va.), USNTIS, 1976
  2. Owens, C. *et al., Phosphorus, Sulfur*, 1976, **2**, 177–180

Violent ignition occurs on mixing [1]. Interaction is explosive, and the products have been identified and a homolytic mechanism proposed for the reaction [2].

1-(4-Nitrophenylazo)-2-naphthol, Zinc oxide
Pytlewski, L. L., *Rep. AD-A028841*, 13, Richmond (Va.), USNTIS, 1976
Spontaneous combustion in storage (occasionally at high ambient temperatures) of clothing impregnation kits containing the three title compounds was investigated. The *N*-chlorourea when heated evolves chlorine to give the isocyanate and a nitrene. Chlorine and the azo-dye react violently and serve as an initiation source of heat. Zinc oxide is converted to the chloride, which catalyses violently exothermic polymerisation of the isocyanate, the main contribution to the total very high exotherm (some 2 MJ/mol), which leads to vigorous smouldering decomposition of the whole mass, temperatures in excess of 315°C being attained. In absence of the dye, heating a mixture of the urea and zinc oxide at 5°/min leads to ignition at 130°C.
See other N-HALOGEN COMPOUNDS

3597. 5-Benzoylbenzenediazonium-2-oxide
[ ] $C_{13}H_8N_2O_2$

Berry, W. L., *J. Org. Chem.*, 1961, **26**, 27
Red crystals, exploding on melting (115°C). It was the highest molecular weight of several homologues to do this. Smaller compounds were often too unstable to isolate, larger are not reported as exploding. Also exploding were: 5-aceto, red, 96°; 5-acetamino, orange, 120°C; 5-phenylmethyl, yellow, temperature unstated.
See other ARENEDIAZONIUM OXIDES

3598. Bis(1-benzo[d]triazolyl) carbonate
[88544-01-8] $C_{13}H_8N_6O_3$

Keller, O. *et al.*, *Chimia*, 1985, **39**(2–3), 63
It is explosive in the dry state, but if prepared at below 60°C, it may be handled wet with 30% of 1,1,2-trichloroethane.
See other HIGH-NITROGEN COMPOUNDS, N–O COMPOUNDS, TRIAZOLES

3599. 1-Bromo-1,2-cyclotridecadien-4,8,10-triyne
[17530-57-3]   C₁₃H₉Br

$C_{13}H_9Br$

Leznoff, C. C. *et al.*, *J. Amer. Chem. Soc.*, 1968, **90**, 731

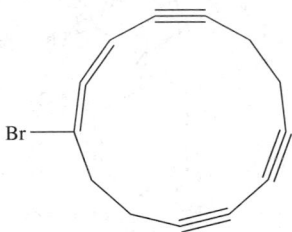

It explodes at 85°C and slowly decomposes in the dark at 0°C.
See other HALOACETYLENE DERIVATIVES

3600. 2-Benzylideneamino-4,6-dinitrophenol
[53088-06-5] $C_{13}H_9N_3O_5$

Metal salts
See entry 2-ARYLIDENEAMINO-4,6-DINITROPHENOL SALTS
See other POLYNITROARYL COMPOUNDS

3601. 2-(4-Nitrophenoxyazo)benzoic acid
[ ] $C_{13}H_9N_3O_5$

Griess, P., *Ber.*, 1884, **17**, 338
It explodes on heating.
See other ARENEDIAZOATES, N–O COMPOUNDS, ORGANIC ACIDS

3602. 1-(2-Nitrophenyl)-5-phenyltetrazole
[55761-75-8] $C_{13}H_9N_5O_2$

Houghton, P. G. *et al., J. Chem. Soc., Chem. Comm.*, 1979, 771
Thermolysis of the undiluted tetrazole is violent, but controlled if dispersed in sand or in solution.
See other NITROARYL COMPOUNDS, TETRAZOLES

3603. Diazidodiphenylmethane
[17421-82-8] $C_{13}H_{10}N_6$

Kirchmeyer, S. et al., Synthesis, 1983, 301
Potentially explosive (a *gem*-diazide).
See other ORGANIC AZIDES

3604. Sodium diphenylketyl
[3463-17-0] $C_{13}H_{10}NaO$

Nitrogen oxide
See Nitrogen oxide: Sodium diphenylketyl

1190

3605. Silver 1-benzeneazothiocarbonyl-2-phenylhydrazide
[12154-56-2]

$C_{13}H_{11}AgN_4S$

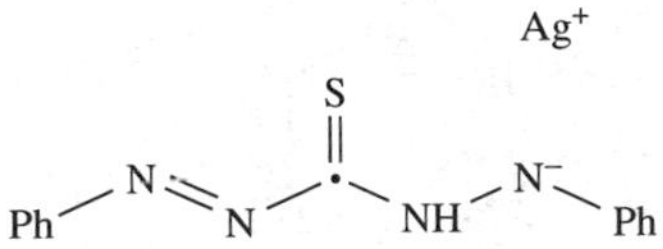

Sorbe, 1968, 126

The silver derivative of dithizone decomposes explosively at higher temperatures.

See other AZO COMPOUNDS, SILVER COMPOUNDS

3606. α-(4-Bromophenylazo)benzyl hydroperoxide
[72437-42-4]

$C_{13}H_{11}BrN_2O_2$

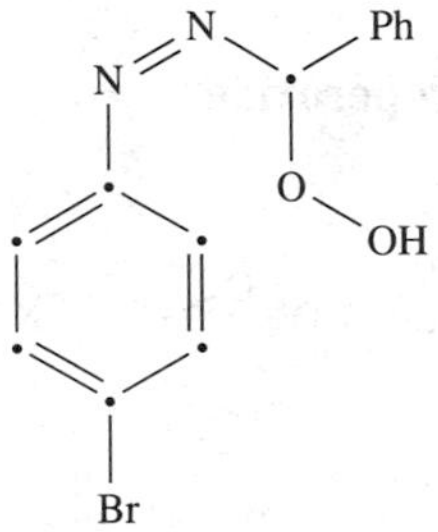

See entry α-PHENYLAZO HYDROPEROXIDES (reference 4)

See related ALKYL HYDROPEROXIDES, AZO COMPOUNDS

3607. α-Phenylazo-4-bromobenzyl hydroperoxide
[83844-92-2]

$C_{13}H_{11}BrN_2O_2$

BrC$_6$H$_4$CH(OOH)N=NPh

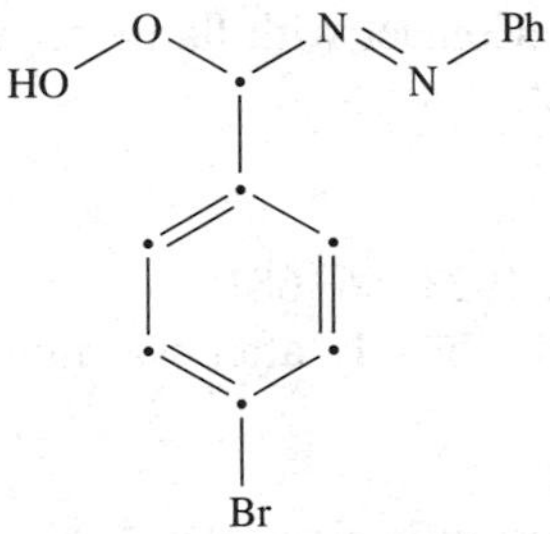

See entry α-PHENYLAZO HYDROPEROXIDES (reference 3)

See related ALKYL HYDROPEROXIDES, AZO COMPOUNDS

3608. α-Phenylazo-4-fluorobenzyl hydroperoxide
  [83844-91-1] $C_{13}H_{11}FN_2O_2$

See entry α-PHENYLAZO HYDROPEROXIDES (reference 3)
See related ALKYL HYDROPEROXIDES, AZO COMPOUNDS

3609. α-Phenylazobenzyl hydroperoxide
  [2829-31-4] $C_{13}H_{12}N_2O_2$

Alone, or Acids
  1. Swern, 1971, Vol. 2, 19
  2. Busch, M. *et al., Ber.*, 1914, **47**, 3277
The phenylhydrazones of benzaldehyde and its homologues, (or of acetone) are
readily autoxidised in solution and rearrange to give the azo-hydroperoxides,
isolable as solids which may explode after a short time on standing, though not
on friction or impact [1]. Contact with flame or with conc. sulfuric or nitric acids
also initiates explosion [2].

Phenylhydrazine
  Bergman, M. *et al., Ber.*, 1923, **56**, 681
The hydroperoxide reacts violently after warming for a few min. with phenylhy-
drazine.
See other REDOX REACTIONS
See entry α PHENYLAZO HYDROPEROXIDES
See related ALKYL HYDROPEROXIDES, AZO COMPOUNDS

3610. 1,1,1-Triacetoxy-1,2-benziodoxol-3-one
[87413-09-0] $C_{13}H_{13}IO_8$

Plumb, J. B., *Chem. Eng. News*, 1990, **68**(29), 3

The title compound is produced by treatment of 2-iodylbenzoic acid with acetic anhydride in acetic acid, and has found wide application as a mild oxidant ('Dess-Martin periodinane') for 1y and 2y alcohols. Although it appears not to be sensitive to impact, unlike the precursor acid, both explode violently when heated under confinement. The oxidant, on treatment with water is hydrolysed back to the explosive 2-iodylbenzoic acid. Forethought and caution are advised before using these explosive materials on any scale of working.

See 2-Iodylbenzoic acid
See other IODINE COMPOUNDS, OXIDANTS

3611. 3-Ethoxycarbonyl-4,4,5,5-tetracyano-3-trimethylplumbyl-4,5-dihydro-3*H* -pyrazole
[52762-90-0] $C_{13}H_{14}N_6O_2Pb$

Houben-Weyl, 1975, Vl. 13.3, 221

It tends to explode on heating.
See related ALKYLMETALS, CYANO COMPOUNDS

1193

3612. 2-(Dimethylaminomethyl)fluoroferrocene
[54747-08-9]   C₁₃H₁₆FFeN

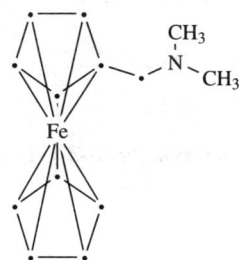

Preparative hazard
See Perchloryl fluoride: 2-Lithio(dimethylaminomethyl)ferrocene
See other ORGANOMETALLICS

3613. Dimethylaminomethylferrocene
[1271-86-9]   C₁₃H₁₇FeN

Nitric acid, Water
See Nitric acid: Dimethylaminomethylferrocene, etc.
See other ORGANOMETALLICS

3614. 2-Methyl-2-[4-(2-methylpropyl)phenyl]oximinoethane
(*p*-Isobutylphenyl-α-methylacetaldehyde oxime)
[58609-72-6]   C₁₃H₁₉NO

P. Cardillo, P., 1992, Personal Communication

On the first full-scale run of a modified process for the nickel catalysed isomerisation of the oxime to the corresponding amide, on 1200 kg scale in toluene solution under reflux in place of the previous water, the reaction overheated (to >180°C), pressurised, and then escaped confinement. Investigation showed the explosion to be purely a runaway reaction caused by a slower start than previously. It was recommended that the oxime be charged in portions, and the solvent changed to the higher-boiling xylene.

See other OXIMES
See related ALDEHYDES

3615. Diethyl 1,4-dihydro-2,6-dimethylpyridine-3,5-dicarboxylate
[1149-23-1] $C_{13}H_{19}NO_4$

4-Nitropyridine *N*-oxide
See 4-Nitropyridine *N*-oxide: Diethyl 1,4-dihydro-2,6-dimethylpyridine-3,5-dicarboxylate

3616. Iron(II) chelate of bis-*N*,*N* -(2-pentanon-4-ylidene)-1,3-diamino-2-hydroxypropane
[22281-49-8] $C_{13}H_{20}FeN_2O_3$

Berenbaum, M. B., US Pat. 3 388 141, 1968
The solid chelate is pyrophoric in air, burning to iron(III) oxide.
See other PYROPHORIC MATERIALS

3617. Benzyltriethylammonium permanganate
[68844-25-7] $C_{13}H_{22}MnNO_4$

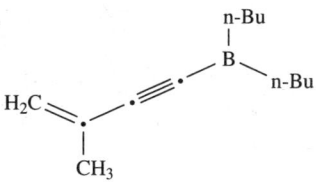

1. Schmidt, H.-J. *et al. Angew. Chem. (Intern. Ed.)*, 1979, **18**, 69
2. Jäger, H. *et al., Angew. Chem. (Intern. Ed.)*, 1979, **18**, 786–787
3. Schmidt, H.-J. *et al., Angew. Chem. (Intern. Ed.)*, 1979, **18**, 787
4. Leddy, B. P. *et al., Tetrahedron Lett.*, 1980, **21**, 2261–2262
5. *See entry* SELF-ACCELERATING REACTIONS
6. Graefe, J. *et al., Angew. Chem. (Intern. Ed.)*, 1983, **23**, 625

The quaternary oxosalt, previously reported as insensitive to hammer-blows or heating to 100°C for 5 min. [1], exploded violently during vacuum drying at 80°C/1.3 mbar. Detailed work has shown that it explodes at 120°C when heated at 4°/min, or at above 50° after various induction periods, e.g. 14 h/60°, 90 min/80°, 25, 28 min/90°, 7 min/100°C [2]. It is also sensitive to heavy blows [2,3]. It explodes in a m.p. capillary at 80–90°C, but may be stabilised by absorption on alumina [4]. An induction period of 80 h/60°C has also been reported [5]. In a further laboratory accident, material dried for 36 h at only 20°C under high vacuum ignited during transfer from a flask to another container, generating flames 1 m long. Since it undergoes violent decomposition in the absence of heating, extreme care and precautions are obviously necessary. It appears that drying conditions may be critical [5].
See other INDUCTION PERIOD INCIDENTS, QUATERNARY OXIDANTS

3618. Dibutyl-3-methyl-3-buten-1-ynlborane
[ ] $C_{13}H_{23}B$

Davidsohn, W. E., *Chem. Rev.*, 1967, **67**, 75
It ignites in air.
See other ACETYLENIC COMPOUNDS, ALKYLBORANES

3619. Tridecanal

[ ] $C_{13}H_{26}O$

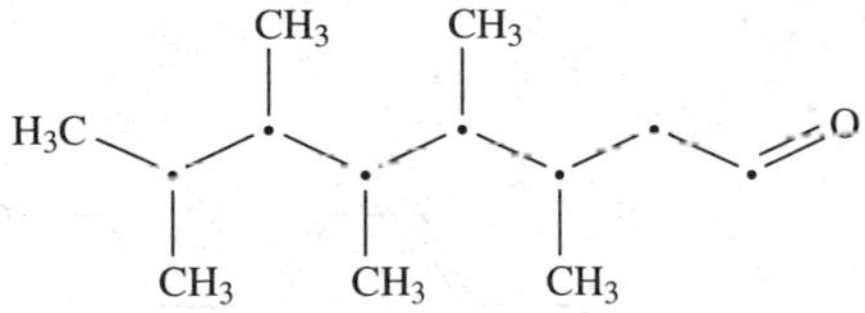

1. Steele, A. B. *et al., Chem. Engrg.*, 1959, **66**(4), 160
2. Urben, P. G., private comm., 1989

Reported to ignite in air [1]. This is typical of medium-range aldehydes, particularly if exposure is increased by sorption on paper or cloth, ignition often occurring within 2 hours [2].

See other ALDEHYDES, PEROXIDISABLE COMPOUNDS

3620. Tris(trimethylsilyl)aluminium etherate ([1,1′-Oxybis[ethane]]tris(trimethyl silyl)aluminium)

[75441-10-0] $C_{13}H_{37}AlOSi_3$

$$(Me_3Si)_3Al.OEt_2$$

Avery, M. A. *et al., J. Amer. Chem. Soc.*, 1992, **114**(3), 974

This compound was pyrophoric even as a relatively dilute pentane solution (Editor's Note: This was a crude preparation probably containing many other organometallics).

See Tris(trimethylsilyl)aluminium

See other ALKYLMETALS, ALKYLSILANES

3621. 1,8-Dihydroxy-2,4,5,7-tetranitroanthraquinone

[ ] $C_{14}H_4N_4O_{12}$

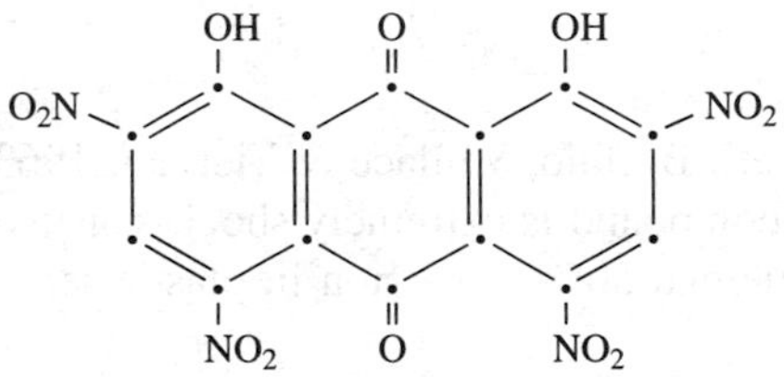

Sorbe, 1968, 153

It is explosive.

See other POLYNITROARYL COMPOUNDS

3622. 2,2′-[1,4-Phenylenebis(azidomethylidyne)]bis(propanedinitrile)

[ ] $C_{14}H_4N_{10}$

Moore, J. A., *et al., Macromolecules*, 1993, **26**(5), 916
It exploded when filtered on a sinter, destroying the apparatus.
See related BENZYL COMPOUNDS, CYANO COMPOUNDS, ORGANIC AZIDES

3623. Bis(2,4-dichlorobenzoyl) peroxide

[133-14-2] $C_{14}H_6Cl_4O_4$

'Lucidol Data Sheet', Buffalo, Wallace & Tiernan, 1963
Whereas the pure compound is extremely shock-sensitive and decomposes rapidly
at 80°C, the commercial 50% dispersion in plasticiser is not shock-sensitive.

Charcoal
   Leleu, *Cahiers*, 1980, (99), 279
   A mixture with a solution of the peroxide in a dialkyl phthalate plasticiser decomposes exothermically but moderately, (probably catalysed by trace heavy metals in the carbon).
   See other COMMERCIAL ORGANIC PEROXIDES, DIACYL PEROXIDES

3624. 1-Nitroanthraquinone
[82-34-8] $C_{14}H_7NO_4$

MARS Database, 1998, short report 007

A runaway decomposition in a melting vessel associated with distillation of the crude product burst the vessel and led to a major fire. The short report does not include detailed circumstances. The decomposition was attributed to the catalytic effects of impurities. However, it seems that nitric acid was also present.

See Nitric acid: most sub-entries

3625. Calcium 2-iodylbenzoate
[59643-77-5] $C_{14}H_8CaI_2O_8$

1. Unpublished information, 1948
2. Merck, 1976, 55

Formulated granules accidentally dried to below normal moisture content exploded violently [1]. The ammonium salt is similarly unstable to heating [2].

See 2-Iodylbenzoic acid
See other IODINE COMPOUNDS

3626. Dicarbonyl(phenanthroline *N*-oxide)rhodium(I) perchlorate
[84578-90-5] $C_{14}H_8ClN_2O_7Rh$

1199

Uson, R. *et al., J. Organomet. Chem.*, 1982, **240**, 433, footnote f
It exploded violently during microanalytical combustion.
See related AMMINEMETAL OXOSALTS

3627. Copper(II) 3,5-dinitroanthranilate
[58302-41-3] $C_{14}H_8CuN_6O_{12}$

See Silver 3,5-dinitroanthranilate
See other POLYNITROARYL COMPOUNDS

3628. Bis(2-azidobenzoyl) peroxide
[20442-99-3] $C_{14}H_8N_6O_4$

1. Leffler, J. E., *Chem. Eng. News*, 1963, **41**(48), 45
2. Hoffman, J., *Chem. Eng. News*, 1963, **41**(52), 5
A small sample of crystalline material on a sintered glass funnel detonated with extreme violence when touched with a metal spatula [1]. Static electrical initiation may have been involved [2].
See related DIACYL PEROXIDES, ORGANIC AZIDES, STATIC INITIATION INCIDENTS

3629. Bis(1-benzo[d]triazolyl) oxalate
[89028-37-5] $C_{14}H_8N_6O_4$

ReftitleKeller, O. *et al., Chimia*, 1985, **39**(2–3), 63
It is explosive in the dry state, but if prepared at below 50°C, it may be handled
wet with 40% of 1,1,2-trichloroethane.
See other HIGH-NITROGEN COMPOUNDS, N–O COMPOUNDS, TRIAZOLES

3630. 9,10-Epidioxyanthracene
[220-42-8] $C_{14}H_8O_2$

Dufraisse, C. *et al., Compt. rend.*, 1935, **201**, 428
It decomposes explosively at 120°C.
See other CYCLIC PEROXIDES

3631. 2,2-Biphenyldicarbonyl peroxide (Dibenzo-d,f-1,2-dioxocin-3,8-dione)
[6109-04-2] $C_{14}H_8O_4$

Ramirez, F., *J. Amer. Chem. Soc.*, 1964, **86**, 4394
It explodes violently on heating to 70°C or on impact, but can be preserved at low
temperature.
See other CYCLIC PEROXIDES, DIACYL PEROXIDES

1201

3632. Anthracene
[120-12-7]   **C₁₄H₁₀**

$$C_6H_4:(CH)_2:C_6H_4$$

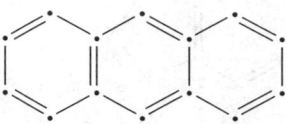

HCS 1980, 159

Fluorine
See Fluorine: Hydrocarbons

Other reactants
Yoshida, 1980, 35
MRH values calculated for 13 combinations with oxidants are given.

3633. 2,3:5,6-Dibenzobicyclo[3.3.0]hexane (9,10 Dewar anthracene)
[79403-75-1]   **C₁₄H₁₀**

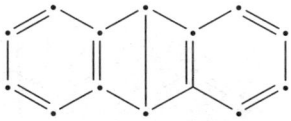

Pritschins, W. *et al., Tetrahedron Lett.*, 1982, **23**, 1153
It reverts violently to anthracene at 73°C in a tube sealed under argon.
See other STRAINED-RING COMPOUNDS

3634. Diphenylethyne (Diphenylacetylene)
[501-65-5]   **C₁₄H₁₀**
                                    PhCCPh

Aluminium chloride, Nitrobenzene
1. Eisch, J. J. *et al., Chem. Eng. News*, 1998, **76**(6), 2
2. Eisch, J. J., Personal communication, 1998.
On adding one drop of nitrobenzene to an equimolar, ten millimolar, mixture of the
other two solids a violent reaction produced gas and carbonaceous material. This
was initially attributed to the oxidative powers of the nitrobenzene [1]. However,
diphenylacetylene is a high energy molecule, ΔH_f° +315 kJ/mole. At least 98%
of the potential chemical energy present will have been the diphenylacetylene.
It is probable that the nitrobenzene merely provided a liquid phase in which the
aluminium chloride could interact with the acetylene, catalysing reaction beyond
the intended azulene dimerisation product[2].
See other ACETYLENIC COMPOUNDS

1202

3635. Phenyl, phenylethynyliodonium perchlorate
[126208-50-2] $C_{14}H_{10}ClIO_4$

$$PhC \equiv CIPh^+ ClO_4^-$$

Stang, P. J., *Chem. Eng. News*, **68**(34), 2
This salt is explosive, which was attributed to the iodonium group.
See other ACETYLENIC COMPOUNDS, IODINE COMPOUNDS, NON-METAL PERCHLO-
RATES

3636. Bis(cyclopentadienyl)hexafluoro-2-butynechromium
[63618-84-8] $C_{14}H_{10}CrF_6$

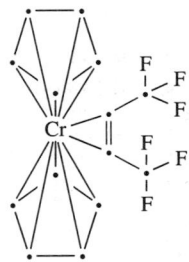

Chisholm, M. H. *et al., Synth. React. Inorg. Metal-Org. Chem.*, 1977, **7**, 283
This adduct of chromocene with hexafluoro-2-butyne decomposes after a few mins.
at ambient temperature, either under nitrogen or vacuum. The process is autocatalytic and violently exothermic, several explosions having occurred.
See other ACETYLENIC COMPOUNDS, ORGANOMETALLICS

3637. Mercury(II) peroxybenzoate
[18918-17-7] $C_{14}H_{10}HgO_6$

Castrantas, 1965, 19
It explodes if heated above its normal decomposition temperature of 100–110°C.
See other MERCURY COMPOUNDS, PEROXOACID SALTS

3638. 1-Benzoyl-1-phenyldiazomethane
[3469-17-8] $C_{14}H_{10}N_2O$

Nenitzescu, C. D. *et al., Org. Synth.*, 1943, Coll. Vol. 2, 497
The material may explode if heated to above 40°C.
See other DIAZO COMPOUNDS

3639. Dibenzoyl peroxide
[94-36-0] $C_{14}H_{10}O_4$

(*MCA SD-81*, 1960); *FPA H56*, 1977; *HCS 1980*, 196 (both latter relate to commercial material dispersed in water or plasticiser)
1. *MCA SD-81*, 1960
2. Lappin, G. R., *Chem. Eng. News*, 1948, **26**, 3518
3. Taub, D., *Chem. Eng. News*, 1949, **27**, 46
4. Nametkin, S. S., *Chem. Abs.*, 1931, **25**, 4127$_8$
5. Nozaki, K. *et al., J. Amer. Chem. Soc.*, 1946, **68**, 1692
6. Fine, D. J. *et al., Combust. Flame*, 1967, **11**, 71–78
7. McCloskey, C. M. *et al., Chem. Abs.*, 1967, **66**, 12613c
8. Uetake, K. *et al., Chem. Abs.*, 1974, **81**, 5175
9. Sekida, O., Jap. Pat. 40 220, 1974

10. Anon., *Sichere Chemiearb.*, 1976, **28**, 49
11. Anon., *Jahresber.*, 1980, 66

The dry material is readily ignited, burns very rapidly and is moderately sensitive to heat, shock, friction or contact with combustible materials. When heated above its m.p. (103–105°C), instantaneous and explosive decomposition occurs without flame, but the decomposition products are flammable. If under confinement (or in large bulk), decomposition may be violently explosive [1]. An explosion which occurred when a screw-capped bottle of the peroxide was opened was attributed to friction initiating a mixture of peroxide and organic dust in the cap threads [2]. Waxed paper tubs are recommended to store this and other sensitive solids [3]. Crystallisation of the peroxide from hot chloroform solution involves a high risk of explosion. Precipitation from cold chloroform solution by methanol is safer [4]. Water- or plasticiser-containing pastes of dibenzoyl peroxide are much safer for industrial use.

The explosive decomposition of the solid has been studied in detail [6]. The effect of moisture upon ignitibility and explosive behaviour under confinement was studied. A moisture content of 3% allowed slow burning only, and at 5% ignition did not occur [7]. Thermal instability was studied using a pressure vessel test, ignition delay time, TGA and DSC, and decomposition products were identified [8]. The presence of acyl chlorides renders dibenzoyl peroxide impact-sensitive [9]. There is a further report of a violent explosion during purification of the peroxide by Soxhlet extraction with hot chloroform [10]. Residual traces of the peroxide in a polythene feed pipe exploded when it was cut with a handsaw [11]. The heat of decomposition has been determined as 1.39 kJ/g. The recently calculated value of 69°C for critical ignition temperature coincides with that previously recorded.

See entry THERMOCHEMISTRY AND EXOTHERMIC DECOMPOSITION

Aniline
Bailey, P. S. *et al., J. Chem. Educ.*, 1975, **52**, 525
Addition of a drop of aniline to 1 g of the peroxide leads to mildly explosive decomposition after a short delay.

N-Bromosuccinimide, 4-Toluic acid
See N-Bromosuccinimide: Dibenzoyl peroxide, etc.

Carbon tetrachloride, Ethylene
Bolt, R. O. *et al., Chem. Eng. News*, 1947, **25**, 1866
Interaction of ethylene and carbon tetrachloride at elevated temperatures and pressures, initiated with benzoyl peroxide as radical source, caused violent explosions on several occasions. Recommended precautions include use of minimum pressure and quantity of initiator, maximum agitation, and presence of water as an inert moderator of high specific heat.
See Ethylene: Carbon tetrachloride

Charcoal
Leleu, *Cahiers*, 1980, (99), 279

At 50°C, a mixture reacts violently, evolving white fumes. (Catalytic decomposition by traces of heavy metals in the charcoal seems likely to be involved.)

N,N-Dimethylaniline
Horner, L. *et al., Chem. Ber.*, 1953, **86**, 1071–1072
The solid peroxide exploded on contact with a drop of dimethylaniline.

Dimethyl sulfide
Pryor, W. A. *et al., J. Org. Chem.*, 1972, **37**, 2885
The rapid decomposition of benzoyl peroxide by dimethyl sulfide is explosive in absence of solvent.

Lithium tetrahydroaluminate   MRH 4.39/15
Sutton, D. A., *Chem. & Ind.*, 1951, 272
One or two attempts to reduce the diacyl peroxide in ether led to a moderately violent explosion.
See other REDOX REACTIONS

Methyl methacrylate
MCA Case History No. 996
Local overheating and ignition occurred when solid benzoyl peroxide was put into a beaker which had been rinsed out with methyl methacrylate. Contact between the peroxide, a powerful oxidant and radical source, and oxidisable or polymerisable materials should only be under controlled conditions.

Vinyl acetate
See Vinyl acetate: Dibenzoyl peroxide, etc.
See other DIACYL PEROXIDES

3640. Bis-3-(2-furyl)acryloyl peroxide
[22978-89-8] $C_{14}H_{10}O_6$

Milas, N. A. *et al., J. Amer. Chem. Soc.*, 1934, **56**, 1219
It explodes violently on heating.
See other DIACYL PEROXIDES

3641. N-(3-Methylphenyl)-2-nitrobenzimidyl chloride
[ ] $C_{14}H_{11}ClN_2O_2$

Preparative hazard
See Phosphorus pentachloride: 3'-Methyl-2-nitrobenzanilide

3642. 1,1-Diphenylethylene
[530-48-3] $C_{14}H_{12}$

Oxygen
Staudinger, H., *Ber.*, 1925, **58**, 1075
Exposure of the alkene to oxygen at ambient temperature and pressure produces
an alkene–oxygen interpolymeric peroxide which explodes lightly on heating. An
attempt to react the alkene with oxygen at 100 bar and 40–50°C caused a violent
explosion in the autoclave.
See other POLYPEROXIDES See related ALKENES

3643. *trans*-1,2-Diphenylethylene (Stilbene)
[103-30-0] $C_{14}H_{12}$

It is mildly endothermic (ΔH°_f (s) +135.4 kJ/mol, 0.75 kJ/g), and the *cis*-isomer
[645-49-8] will be rather more so.
See related ALKENES See other ENDOTHERMIC COMPOUNDS

3644. 5-Chlorotoluene-2-diazonium tetrachlorozincate
[89452-69-0] $C_{14}H_{12}Cl_6N_4Zn$

Anon., *ABCM Quart Safety Summ.*, 1953, **24**, 42
A batch containing only half the normal water content (60%) exploded violently
during ball-milling. Tests later showed the dry material to be shock-sensitive.
See Benzenediazonium tetrachlorozincate
See other DIAZONIUM SALTS

3645. 3′-Methyl-2-nitrobenzanilide
[50623-64-8] $C_{14}H_{12}N_2O_3$

Phosphorus pentachloride
  See Phosphorus pentachloride: 3′-Methyl-2-nitrobenzanilide
  See other NITROARYL COMPOUNDS

3646. 1,2-Bis(2-nitrophenyl)ethane (2,2′-Dinitrobibenzyl)
[16968-19-7] $C_{14}H_{12}N_2O_4$

1208

Preparative hazard
See 2-Nitrotoluene: Alkali
See other POLYNITROARYL COMPOUNDS

3647. 2,2-Diphenyl-1,3,4-thiadiazoline
[79999-60-3] $C_{14}H_{12}N_2S$

Kolwinsch, I. *et al., J. Amer. Chem. Soc.*, 1981, **103**, 7032
The crystalline solid isolated at $-78°C$ suddenly decomposed with nitrogen evolution at $-20°C$.
See related AZO COMPOUNDS, N–S COMPOUNDS

3648. α-(4-Bromophenylazo)phenylethyl α-hydroperoxide
[91364-94-2] $C_{14}H_{13}BrN_2O_2$

See entry α-PHENYLAZO HYDROPEROXIDES (reference 4)
See related AZO COMPOUNDS, ALKYL HYDROPEROXIDES

3649. Barium *N*-perchlorylbenzylamide
[89521-42-6] $C_{14}H_{14}BaCl_2N_2O_6$

It explodes on impact.

See entry PERCHLORYLAMIDE SALTS (reference 2)

See other BENZYL COMPOUNDS, *N*-METAL DERIVATIVES

3650. Dibenzyl phosphorochloridate ('Dibenzyl chlorophosphate')
[538-37-4] $C_{14}H_{14}ClO_3P$

Atherton, F. R. *et al.*, *J. Chem. Soc.*, 1945, 382

It is too unstable to be distilled, and the precursory phosphite also tends to decompose on distillation.

Dibenzyl phosphite *See other* BENZYL COMPOUNDS, PHOSPHORUS ESTERS

3651. Mercury(II) *N*-perchlorylbenzylamide
[89521-44-8] $C_{14}H_{14}Cl_2HgN_2O_6$

It explodes on impact, or on heating above 120°C.

See entry PERCHLORYLAMIDE SALTS (reference 2)

See other BENZYL COMPOUNDS, MERCURY COMPOUNDS, *N*-METAL DERIVATIVES

3652. Bis(1-methylbenzotriazole)cobalt(II) nitrate
[123668-66-6] $C_{14}H_{14}CoN_8O_6$

Zafiropoulos, T. F. *et al.*, *Monatsh.*, 1989, **120**(4), 357
Explodes at 260 C.
See other AMMINEMETAL NITRATES, TRIAZOLES

3653. 2-Azoxyanisole
[13620-57-0] $C_{14}H_{14}N_2O_3$

Preparative hazard
See 2-Nitroanisole: Sodium hydroxide, etc.
See related AZO COMPOUNDS

3654. Bis(toluenediazo) oxide (1,1-Oxybis(4-methylphenyldiazene))
[90238-04-3] $C_{14}H_{14}N_4O$

Alone, or Toluene
Bamberger, E., *Ber.*, 1896, **39**, 452, 458
Extremely unstable, it explodes under its reaction liquor at above −4°C. Very shock- and friction-sensitive, a small sample exploded when dried on a porous tile and set off the moist material some distance away. Contact with toluene, even at −5°C, causes an explosive reaction with flame.
See other BIS(ARENEDIAZO) OXIDES

3655. Dibenzyl ether
[103-50-4]

$C_{14}H_{14}O$

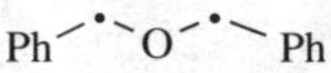

$$Ph\diagup O \diagdown Ph$$

HCS 1980, 204

Aluminium dichloride hydride
See Aluminium dichloride hydride diethyl etherate

Other reactants
Yoshida, 1980, 330
MRH values calculated for 13 combinations with oxidants are given.
See other BENZYL COMPOUNDS, PEROXIDISABLE COMPOUNDS

3656. Di-4-toluenesulfonyl peroxide
[1886-68-6]

$C_{14}H_{14}O_6S_2$

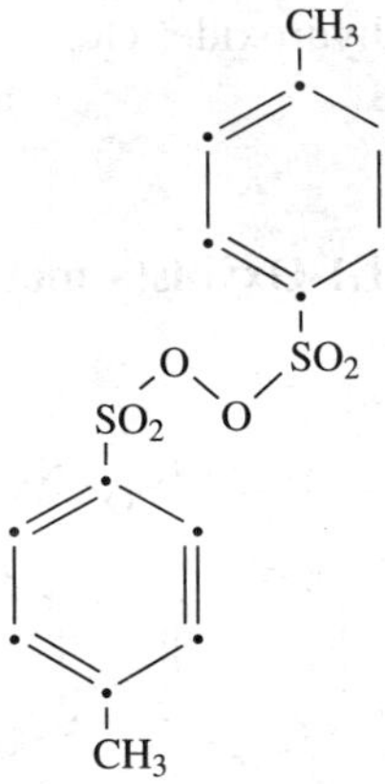

1. Bolte, J. *et al., Tetrahedron Lett.*, 1965, **6**, 1929
2. Dannley, R. L. *et al., J. Org. Chem.*, 1966, **31**, 154
The peroxide was too unstable to dry thoroughly [1]; such samples often exploded spontaneously [2].
See other DIACYL PEROXIDES

3657. 1-Benzyl-3-(4-tolyl)triazene
   [17683-09-9] $C_{14}H_{15}N_3$

White, E. H. *et al., Org. Synth.*, 1973, Coll. Vol. 5, 799
Vacuum sublimation at 90–100°C led to a violent explosion.
See other BENZYL COMPOUNDS, TRIAZENES

3658. Dibenzyl phosphite
   [17176-77-1] $C_{14}H_{15}O_3P$

Atherton, F. R. *et al., J. Chem. Soc.*, 1945, 382; 1948, 1106
It decomposes at 160°C, but prolonged heating at 120°C may have the same effect. Not more than 50 g should be distilled at one time, using high-vacuum conditions (b.p 100–120°C/0.001 mbar) unless a preliminary treatment to remove acidic impurities has been used.
See other BENZYL COMPOUNDS, PHOSPHORUS ESTERS

3659. 1-(4-Phenyl-1,3-diselenolylidene)piperidinium perchlorate
   [53808-76-7] $C_{14}H_{16}ClNO_4Se_2$

See entry 1-(1,3-DISELENONYLIDENE)PIPERIDINIUM PERCHLORATES

3660. 3,3′-Dimethoxy-4,4′-diaminobiphenyl (*o*-Dianisidine)
[119-90-4]

$C_{14}H_{16}N_2O_2$

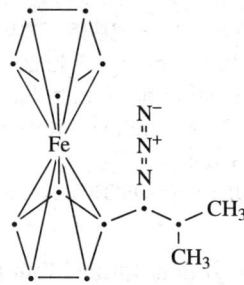

The finely powdered carcinogen is a significant dust explosion hazard.
See entry DUST EXPLOSION INCIDENTS (reference 22)
See other ORGANIC BASES

3661. 1-Ferrocenyl-2-methylpropyl azide
[105017-95-6]

$C_{14}H_{17}FeN_3$

Cushman, M. *et al., J. Org. Chem.*, 1987, **52**, 1517–1521
Heating the title azide at 100°C caused explosive decomposition.
See related ORGANIC AZIDES, ORGANOMETALLICS

3662. 2,10-Dimethyl-1,2,3,4,5,10-hexahydroazepino[3-4,b]indole
    [75142-81-3] $C_{14}H_{18}N_2$

4-Chlorobenzenesulfonyl azide
See 4-Chlorobenzenesulfonyl azide: 2,10-Dimethyl-1,2,3,4.

3663. α-Pentylcinnamaldehyde (3-Phenyl-2-pentylpropenal)
    [122-40-7] $C_{14}H_{18}O$

Anon., *Chem. Trade. J.*, 1937, **100**, 362
This is very prone to spontaneous oxidative heating. A mixture with absorbent cotton attained a temperature of 230°C 4 min after exposure to air.
See other ALDEHYDES, PEROXIDISABLE COMPOUNDS

3664. 5-*p*-Chlorophenyl-2,2-dimethyl-3-hexanone
    [55058-76-9] $C_{14}H_{19}ClO$

Peroxyacetic acid
See Peroxyacetic acid: 5-*p*-Chlorophenyl-2,2-dimethyl-3-hexanone

3665. Bis(2-dimethylaminopyridine *N*-oxide)copper(II) perchlorate
[67517-99-7]   C₁₄H₂₀Cl₂CuN₄O₁₀

West, D. X., *Inorg. Nucl. Chem. Lett.*, 1978, **14**, 155–159
It decomposes explosively at 270°C without prior weight loss.
See related AMMINEMETAL OXOSALTS

3666. 2,6-Di-*tert*-butyl-4-nitrophenol
[728-40-5]   C₁₄H₂₁NO₃

1. *ASESB Expl. Incid. Rept.*, 1961, 24
2. Barnes, T. J. *et al., J. Chem. Soc.*, 1961, 953
A sample of the compound exploded violently after short heating to 100°C.
Although this was attributed to presence of polynitro derivatives [1], the thermal
decomposition of this type of nitro compound is known [2].
See other NITROARYL COMPOUNDS

3667. Dicyclohexylcarbonyl peroxide
[4904-55-6]   C₁₄H₂₂O₄

Castrantas, 1965, 17
Large quantities may explode without apparent reason.
See other DIACYL PEROXIDES

3668. 2,2′-Azobis(2,4-dimethylvaleronitrile)
[4419-11-8] $C_{14}H_{24}N_4$

$$(Me_2CHCH_2MeC(CN)N:)_2$$

Yoshida, T., private comm., 1984
The low-melting solid of limited thermal stability (max. safe storage temperature
20°C, self-heating decomposition detected above 30°C) will explode on heavy
impact (150 kg cm), but is of low sensitivity to ignition. However, 3 cases of
spontaneous decomposition in storage or transportation have been noted.
See entry BLOWING AGENTS *See other* AZO COMPOUNDS, CYANO COMPOUNDS

3669. 1-Acetoxy-1-hydroperoxy-6-cyclodecanone
[ ] $C_{14}H_{24}O_5$

Criegee, R. *et al., Ann.*, 1949, **564**, 9
It explodes on removal from a freezing mixture.
See other 1-OXYPEROXY COMPOUNDS

3670. Bis(dipropylborino)acetylene
[ ] $C_{14}H_{28}B_2$

$$Pr_2BC{\equiv}CBPr_2$$

Hartmann, H. *et al., Z. Anorg. Allgem. Chem.*, 1959, **299**, 179
Both n- and iso-propyl derivatives ignite in air.
See other ACETYLENIC COMPOUNDS, ALKYLBORANES

†3671. Tetradecane
[629-59-4] $C_{14}H_{30}$

$$Me[CH_2]_{12}Me$$

3672. Acetylenebis(triethyllead)
[5120-07-0]

$C_{14}H_{30}Pb_2$

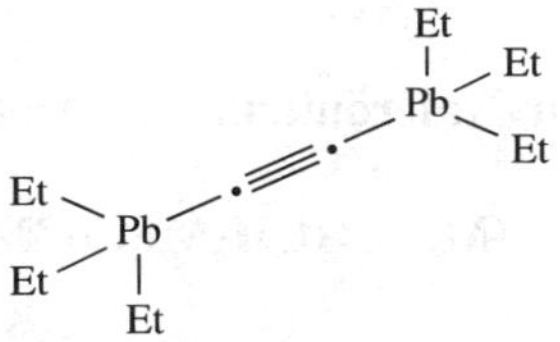

Beerman, C. *et al., Z. Anorg. Chem.*, 1954, **276**, 20
It is very sensitive to heat, oxygen or light, and should not be dried.
See related ALKYLMETALS, METAL ACETYLIDES

3673. 1,2-Bis-(di-2-propylphosphino)ethane
[87532-69-2]

$C_{14}H_{32}P_2$

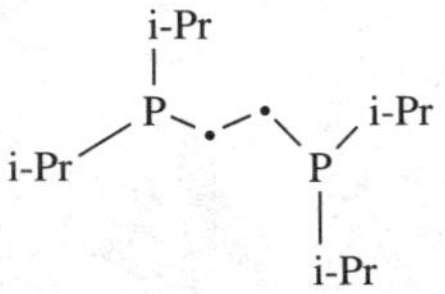

Benn, R. *et al., Z. Naturforsch. B*, 1986, **41** B, 680–691
The title ligand, and the bis(*tert*-butylphosphino) lower homologue are pyrophoric in air.
See related ALKYLPHOSPHINES

3674. Acetylenebis(triethyltin)
[994-99-0]

$C_{14}H_{30}Sn_2$

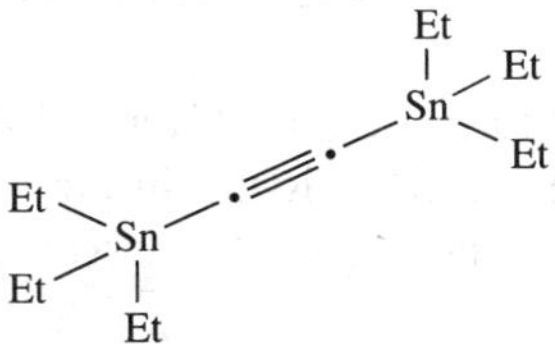

Stannic chloride
  Beerman, C. *et al., Z. Anorg. Chem.*, 1954, **276**, 20
  The product of interaction is highly explosive.
  See related ALKYLMETALS, METAL ACETYLIDES

1218

3675. N,N'-Di-tert-butyl-N,N'-bis(trimethylsilyl)diaminophosphene
[85923-32-6] $C_{14}H_{36}N_2P_2Si_2$

Niecke, E. *et al., Angew. Chem. (Intern. Ed.)*, 1983, **22**, 486
Highly pyrophoric, like the tetrakis(trimethylsilyl) analogue.
See related ALKYLSILANES, PHOSPHINES

3676. Heptakis(dimethylamino)trialuminium triboron pentahydride
[28016-59-3] $C_{14}H_{47}Al_3B_3N_7$
Complex structure

Hall, R. E., *et al., Inorg. Chem.*, 1969, **8**, 270
The crystalline solid ignites in air.
See other COMPLEX HYDRIDES
See related BORANES

3677. 5,6-Diphenyl-1,2,4-triazine-3-diazonium tetrafluoroborate 2-oxide
[98571-91-6] $C_{15}H_{10}BF_4N_5O$

Jovanovic, V., *Heterocycles*, 1985, **23**, 1969–1981
The isolated diazonium salt exploded when exposed to air, and also on impact.
See other DIAZONIUM TETRAHALOBORATES, HIGH-NITROGEN COMPOUNDS, *N*-OXIDES

3678. Pyrocatecholato(2−)(quinolin-8-olato-N,O)-trioxygenido(2−)phosphorus
[82434-14-8] $C_{15}H_{10}NO_6P$

Koenig, M. *et al., Tetrahedron Lett.*, 1982, **23**, 422
The hexavalent phosphorus ozonide, stable at −20°C, exploded in contact with air
or on warming to ambient temperature.
See other OZONIDES
See related PHOSPHORUS ESTERS

3679. 3-Chloro-1,3-diphenylcyclopropene
[20421-00-5] $C_{15}H_{11}Cl$

Preparative hazard
Padwa, A. *et al., J. Org. Chem.*, 1969, **34**, 2732
If the crude title product (from interaction of 3-chloro-3-phenyldiaziridine and
diphenylacetylene) is not purified immediately after preparation, violent decompo-
sition occurs.
See other STRAINED-RING COMPOUNDS

3680. Diphenylcyclopropenylium perchlorate
[37647-36-2] $C_{15}H_{11}ClO_4$

Hughes, R. P. *et al., Organometallics*, 1985, **4**, 1761–1766
It should not be air dried, and as an explosive salt it must be handled with care.
See other NON-METAL PERCHLORATES

3681. 4(4′-Bromobenzoyl)acetanilide
[ ] $C_{15}H_{12}BrNO_2$

Dimethyl sulfoxide
See Dimethyl sulfoxide: 4(4′-Bromobenzoyl)acetanilide

3682. 1,2,3,4-Tetrachloro-7,7-dimethoxy-5-phenylbicyclo[2.2.1]hepta-2,5-diene
[15584-65-3] $C_{15}H_{12}Cl_4O_2$

Hoffmann, R. W. *et al., Tetrahedron*, 1965, **21**, 900
Pyrolysis of the material at 130°C under nitrogen at low pressure to give tetra-methoxyethylene may be explosive if more than 25 g is used.

3683. Tris(cyclopentadienyl)cerium
[1298-53-9]

$C_{15}H_{15}Ce$

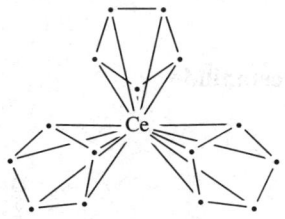

Preparative hazard
See Sodium nitrate: Tris(cyclopentadienyll)cerium
See other ORGANOMETALLICS

3684. Tris(cyclopentadienyl)plutonium
[12216-68-9]

$C_{15}H_{15}Pu$

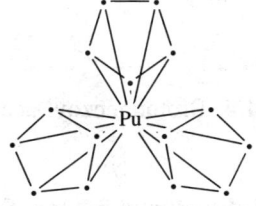

Bailar, 1973, Vol. 5, 407
It ignites in air.
See other ORGANOMETALLICS

3685. Tris(cyclopentadienyl)uranium
[54007-00-0]

$C_{15}H_{15}U$

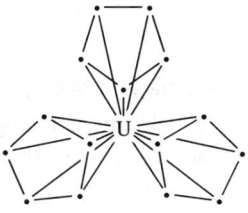

Bailar, 1973, Vol. 5, 407
It ignites in air.
See other ORGANOMETALLICS

3686. 2,2-Bis(4-hydroxyphenyl)propane
[80-05-7]

$C_{15}H_{16}O_2$

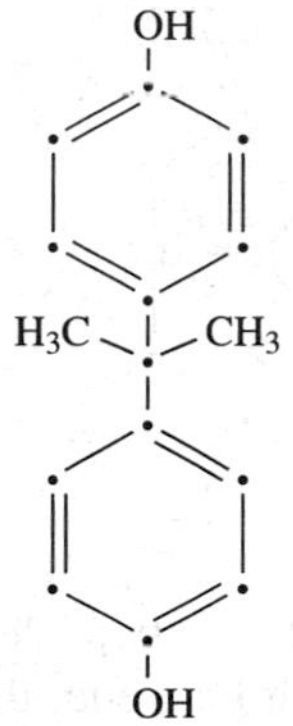

The finely powdered resin component (Bisphenol-A) is a significant dust explosion hazard.

See entry DUST EXPLOSION INCIDENTS (reference 22)

3687. *trans*-Aquadioxo(terpyridine)ruthenium(2+) diperchlorate
[ ]

$C_{15}H_{17}Cl_2N_5O_{10}Ru$

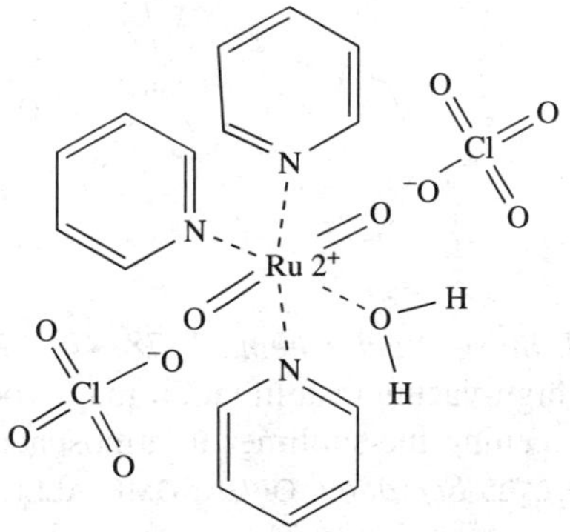

Meyer, T. J. *et al., Inorg. Chem.*, 1992, **31**(8), 1375
The complex sometimes explodes when scraped against a glass frit with a metal spatula.
See other AMMINEMETAL OXOSALTS

3688. Tricyclopentadienyluranium tetrahydroaluminate
[107633-87-4] $C_{15}H_{19}AlU$

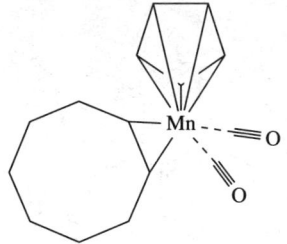

Ossola, S. *et al., J. Organomet. Chem.*, 1986, **310**(1), C1
It ignites in air, and reacts with benzene, dimethoxyethane, THF and toluene.
See related COMPLEX HYDRIDES, HEAVY METAL DERIVATIVES, ORGANOMETALLICS,
PYROPHORIC MATERIALS

3689. *cis*-Dicarbonyl(cyclopentadienyl)cyclooctenemanganese
[49716-47-4] $C_{15}H_{19}MnO_2$

Butler, I. S. *et al., J. Inorg. Nucl. Chem.*, 1978, **40**, 1937
Overheating during high-vacuum sublimation may produce pyrophoric manganese
which explodes on opening the sublimer to atmosphere.
See PYROPHORIC METALS *See other* ORGANOMETALLICS

3690. Tris(2,4-pentanedionato)molybdenum(III)
[14284-90-3] $C_{15}H_{21}MoO_6$
Larson, M. L. *et al., Inorg. Synth.*, 1966, **8**, 153
It rapidly oxidises in air, sometimes igniting.
See related ORGANOMETALLICS

3691. 2,6-Di-*tert*-butyl-4-cresol
[120-37-0] $C_{15}H_{24}O$

The finely powdered anti-oxidant is a significant dust explosion hazard.
See entry DUST EXPLOSION INCIDENTS (reference 22)

3692. *tert*-Butyl 1-adamantaneperoxycarboxylate
[21245-43-2] $C_{15}H_{24}O_3$

Razuvajev, G. A. *et al., Tetrahedron*, 1969, **25**, 4925
It explodes on heating to 90–100°C.
See other PEROXYESTERS

3693. Tri(spirocyclopentane)1,1,4,4,7,7-hexoxonane
[4884-18-8] $C_{15}H_{24}O_6$

Bjorklund, G. H. *et al.*, *Trans. R. Soc. Can. (Sect. III)*, 1950, **44**, 25
A violent explosive, very sensitive to shock, friction and rapid heating.
See Hydrogen peroxide: Ketones, etc.
See other CYCLIC PEROXIDES

3694. 2,4,6-Tris(dimethylaminomethyl)phenol
[90-72-2]

$C_{15}H_{27}N_3O$

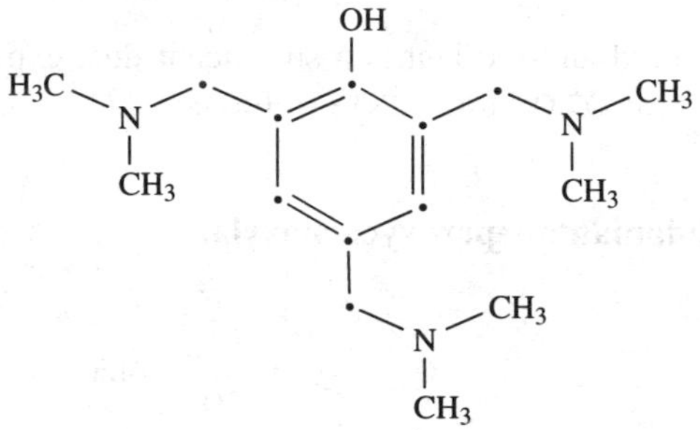

Cellulose nitrate
See CELLULOSE NITRATE: amines *See other* ORGANIC BASES

3695. Hexaethyltrialuminium trithiocyanate
[17548-36-6]

$C_{15}H_{30}Al_3N_3S_3$

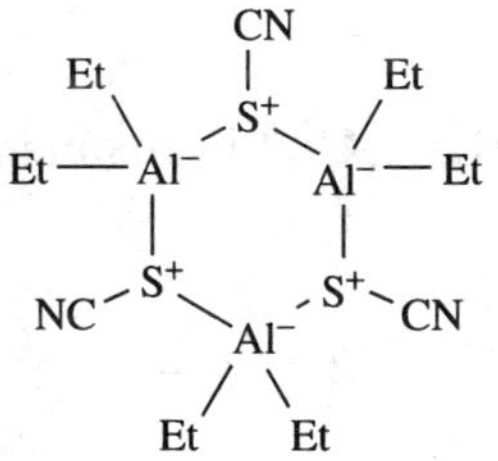

Dehnicke, K., *Angew. Chem. (Intern. Ed.)*, 1967, **6**, 947
On heating at 210° under vacuum it disproportionates explosively, but smoothly at 180°C.
See related ALKYLALUMINIUM HALIDES, METAL CYANATES

3696. 1,2-Bis(triethylsilyl)trimethylsilylhydrazine
[13272-04-3] $C_{15}H_{40}N_2Si_3$

See entry SILYLHYDRAZINES *See related* ALKYLSILANES

3697. 1-(4-Chloro-2-nitrobenzeneazo)-2-hydroxynaphthalene
[6410-13-5] $C_{16}H_{10}ClN_3O_3$

Lead chromate
See Lead chromate: Azo-dyestuffs
See other AZO COMPOUNDS

3698. μ-Oxo-*I,I*-bis(trifluoroacetato-*O*)-*I,I*-diphenyldiodine(III)
[91879-79-7] $C_{16}H_{10}F_6I_2O_5$

Boutin, R. H. *et al., J. Org. Chem.*, 1984, **49**, 4278, 4284
It detonated at 220°C during determination of the m.p.
See [*I,I*-Bis(trifluoroacetoxy)iodo]benzene
See other IODINE COMPOUNDS

3699. Potassium bis(phenylethynyl)palladate(2−)
[66986-75-2] $C_{16}H_{10}K_2Pd$

$$K^+ \quad K^+$$
$$Pd^{2-}$$
$$Ph \overset{\cdot}{\equiv} \cdot \quad \cdot \equiv \cdot Ph$$

Air, or Water

Immediately pyrophoric in air, and explosive decomposition occurs with aqueous reagents; the sodium salt is similar.

See entry COMPLEX ACETYLIDES (reference 2)

3700. Potassium bis(phenylethynyl)platinate(2−)
[ ] $C_{16}H_{10}K_2Pt$

$$K^+ \quad K^+$$
$$Pt^{2-}$$
$$Ph \overset{\cdot}{\equiv} \cdot \quad \cdot \equiv \cdot Ph$$

Air, or Water

Immediately pyrophoric in air, and explosive decomposition occurs with water; the sodium salt is similar.

See entry COMPLEX ACETYLIDES (reference 2)

3701. *mixo*-Dimethoxydinitroanthraquinone
[6407-56-3] (1,5,4,8-) $C_{16}H_{10}N_2O_8$
[53939-55-2] (1,8:4,5-)

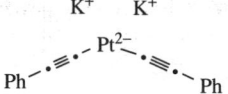

and

1228

Sulfuric acid

Hildreth, J. D., *Chem. & Ind.*, 1970, 1592

During hydrolysis of crude dimethoxy compound by heating in sulfuric acid, a runaway exothermic decomposition occurred causing vessel failure. Experiment showed a threshold decomposition temperature of 150–155°C, and an oxidising effect of the nitro groups, yielding CO and CO_2 above 162°C.

Oleum

White, D. L. *et al.*, Ger. Offen. 2 451 569, 1975

In an alternative method of preparing *mixo*-dihydroxyanthraquinonedisulfonic acid by heating the title compound with oleum to effect simultaneous hydrolysis, denitration and sulfonation of the nucleus, there is the possibility of formation of methyl nitrate from the scission fragments.

See other POLYNITROARYL COMPOUNDS

3702. 1-(2,4-Dinitrobenzeneazo)-2-hydroxynaphthalene (Pigment orange 5)
[3468-63-1] $C_{16}H_{10}N_4O_5$

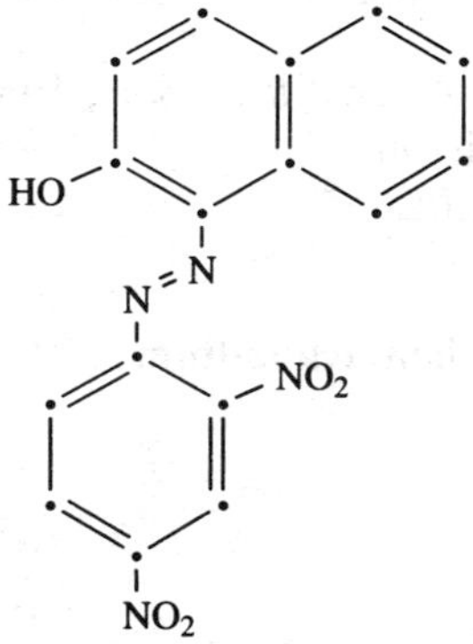

The solid deflagrates at 1.8 cm/min.

See entry PRESSURE INCREASE IN EXOTHERMIC DECOMPOSITION (reference 3)

See other DEFLAGRATION INCIDENTS

Lead chromate

See Lead chromate: Azo-dyestuffs

See other AZO COMPOUNDS, POLYNITROARYL COMPOUNDS

3703. Bis(cyclopentadienyl)pentafluorophenylzirconium hydroxide
[ ] $C_{16}H_{11}F_5OZr$

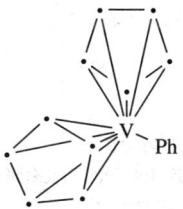

Chaudhari, M. A. *et al., J. Chem. Soc. A*, 1966, 840
Decomposition on heating in air above 260°C is sometimes explosive.
See related HALO-ARYLMETALS *See other* ORGANOMETALLICS

3704. Copper 1,3,5-octatrien-7-ynide
[ ] $C_{16}H_{14}Cu_2$

$$(H[CH=CH]_3C\equiv C)_2Cu_2$$

Georgieff, K. K. *et al., J. Amer. Chem. Soc.*, 1954, **76**, 5495
It deflagrates on heating in air.
See other METAL ACETYLIDES

3705. Bis(cyclopentadienyl)phenylvanadium
[12212-56-5] $C_{16}H_{15}V$

Gibson, 1969, 178
It ignites in air.
See related ARYLMETALS *See other* ORGANOMETALLICS

3706. Bis(cyclooctatetraenyl)cerium (Bis(η^8-1,3,5,7-cyclooctatetraene)cerium)
[37205-27-9] $C_{16}H_{16}Ce$

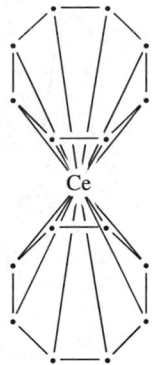

Edelmann, F. T. *et al., Angew. Chem. (Int.)*, 1994, **33**(15/16), 1618
This sandwich compound, nominally the Ce(IV) derivative of cyclooctate-traene(2−) is highly pyrophoric.
See other ORGANOMETALLICS

3707. 2,2′-(1,2-Ethylenebis)3-phenyloxaziridine
[54222-34-3] $C_{16}H_{16}N_2O_2$

Ph

O—N $\cdots$ N—O

Ph

MCA Case History No. 2175

The washed crude product from oxidation of 1,2-dibenzylideneaminoethane in methylene chloride solution exploded violently during vacuum evaporation at a relatively low temperature.
See 2,2′-Di-*tert*-butyl-3,3′-bioxaziridine
See other N–O COMPOUNDS, STRAINED-RING COMPOUNDS

3708. Bis(cyclooctatetraene)uranium(0)
   [11079-26-8] $C_{16}H_{16}U$

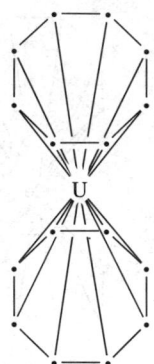

Streitweiser, A. *et al., J. Amer. Chem. Soc.*, 1968, **90**, 7364
It inflames in air.
See other ORGANOMETALLICS

3709. 2-(4-Bromophenyl)-2-propyl 1-(1,1-dimethyl-2-pentyn-4-enyl) peroxide
   [88519-63-5] $C_{16}H_{19}BrO_2$

See 2-(4-Chlorophenyl)-2-propyl 1-(1,1-dimethyl-2-pentyn-4-enyl) peroxide
(below)
See other ACETYLENIC PEROXIDES, DIALKYL PEROXIDES

3710. 1,4,7-Trithia[7]ferrocenophene–acetonitrilecopper(I) perchlorate
[110725-79-6]   C₁₆H₁₉ClCuFe₂NO₄S₃

$C_{16}H_{19}ClCuFe_2NO_4S_3$

Sato, M. *et al., Bull. Chem. Soc. Japan*, 1986, **59**, 3611–3615
It exploded on heating.
See other AMMINEMETAL OXOSALTS, NON-METAL PERCHLORATES

3711. 2-(4-Chlorophenyl)-2-propyl 1-(1,1-dimethyl-2-pentyn-4-enyl) peroxide
[ ] $C_{16}H_{19}ClO_2$

Vilenskaya, M. R. *et al., Zh. Org. Khim., 1981,* **17**, 959–961
The acetylenic peroxide and its bromo analogue explode at 120–130°C.
See other ACETYLENIC PEROXIDES, DIALKYL PEROXIDES

3712. *O–O-tert*-Butyl diphenyl monoperoxophosphate
[20914-03-0] $C_{16}H_{19}O_5P$

Rieche, A. *et al.*, *Chem. Ber.*, 1962, **95**, 385
The material deflagrated as a solid, or decomposed exothermically in its reaction mixture soon after preparation.
*See other tert-*BUTYL PEROXOPHOSPHATE DERIVATIVES, PHOSPHORUS ESTERS

3713. *N*-Methyl-*p*-nitroanilinium 2(*N*-methyl-*N*-*p*-nitrophenylaminosulfonyl) ethylsulfate
[ ] $C_{16}H_{20}N_4O_{10}S_2$

There is a preparative hazard for this reaction product of *N*-methyl-*p*-nitroaniline and 'carbyl sulfate'.
See 2,4-Dithia-1,3-dioxane-2,2,4,4-tetraoxide: *N*-Methyl-4-nitroaniline, etc.
See related NITROARYL COMPOUNDS, SULFUR ESTERS

3714. Dibutyl phthalate
[84-74-2] $C_{16}H_{22}O_4$

Chlorine
See Chlorine: Dibutyl phthalate

3715. Di-*tert*-butyl diperoxyphthalate
[2155-71-7] $C_{16}H_{22}O_6$

Castrantas, 1965, 17
It is shock-sensitive.
See other PEROXYESTERS

3716. Tetrakis(3-methylpyrazole)cadmium sulfate
[55060-82-7] $C_{16}H_{24}CdN_8O_4S$

$$[(C_4H_6N_2)_4Cd]SO_4$$

See entry AMMINEMETAL OXOSALTS (reference 9)

3717. Tetrakis(3-methylpyrazole)manganese(II) sulfate
[55060-79-2] $C_{16}H_{24}MnN_8O_4S$

$$[(C_4H_6N_2)_4Mn]SO_4$$

See entry AMMINEMETAL OXOSALTS (reference 9)

3718. Bis-*O*, *N* [(*N'*-pent-2-en-2-oxy-4-ylidene)-*N*,*S*-dimethyldithiocarbazate] copper(II) perchlorate
[ ] $C_{16}H_{28}Cl_2CuN_4O_{10}S_4$

$$[MeSC(S)N(Me)N=C(Me)CH=C(Me)O]_2Cu\,(ClO_4)_2$$

Preparative hazard
Akbar Ali, M. *et al., J. Inorg. Nucl. Chem.*, 1978, **40**, 452
Attempts to prepare this and the corresponding *S*-benzyl complex gave highly explosive products. It appears that the ligand hydrolysed and that the bis(*N*,*S*-dialkyldithiocarbazate)copper perchlorates were obtained and it was these which exploded.
See related AMMINEMETAL OXOSALTS

3719. Potassium bis(8-(4-nitrophenylthio)undecahydrodicarbaundecaborato) cobaltate(1−)
[155147-44-7] $C_{16}H_{28}B_{18}CoKN_2O_4S_2$

$$(O_2NC_6H_4SC_2B_9H_{10})_2Co^-K^+$$

Knyazev, S. P. *et al., Chem. Abs.*, 1994, **120** 298911d
Nitro compounds of this series are reported as exploding on grinding or on rapid heating to >200°C. The equivalent ferrate and a mono(2-nitrophenyl) substituted bisboratocobaltate were also prepared.
See other BORANES

3720. Potassium bis(8-(4-nitrophenylthio)undecahydrodicarbaundecaborato) ferrate(1−)
[155147-41-4] $C_{16}H_{28}B_{18}FeKN_2O_4S_2$

$$(O_2NC_6H_4SC_2B_9H_{10})_2Fe^-K^+$$

See Potassium bis(8-(4-nitrophenylthio)undecahydrodicarbaundecaborato) cobaltate(1−)
See other BORANES

3721. Bis (2-phenyl-1,2-dicarbadodecaborane(12)-1-yl)diazene
[73399-75-4] $C_{16}H_{30}B_{20}N_2$
Complex structure

See entry AZOCARBABORANES

3722. Bromo-5,7,7,12,14,14-hexamethyl-1,4,8,11-tetraaza-4,11-cyclo-tetra decadieneiron(II) perchlorate
[31122-43-7] $C_{16}H_{32}BrClFeN_4O_4$

See entry [14] DIENE-N₄ COMPLEXES

3723. Iodo-5,7,7,12,14,14-hexamethyl-1,4,8,11-tetraaza-4,11-cyclotetra-decadieneiron(II) perchlorate
[31122-35-7] $C_{16}H_{32}ClFeIN_4O_4$

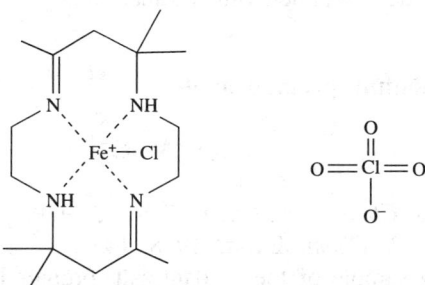

See entry [14] DIENE-N_4 COMPLEXES

3724. Chloro-5,7,7,12,14,14-hexamethyl-1,4,8,11-tetraaza-4,11-cyclotetra-decadieneiron(II) perchlorate
[31122-42-6] $C_{16}H_{32}Cl_2FeN_4O_4$

See entry [14] DIENE-N_4 COMPLEXES

3725. Dichloro-5,7,7,12,14,14-hexamethyl-1,4,8,11-tetraaza-4,11-cyclotetra-decadieneiron(III) perchlorate
[36691-97-1] $C_{16}H_{32}Cl_3FeN_4O_4$

See entry [14] DIENE-N_4 COMPLEXES

1237

†3726. Hexadecane
[544-76-3] $C_{16}H_{34}$

$$Me[CH_2]_{14}Me$$

3727. Hexadecanethiol
[2917-26-2] $C_{16}H_{34}S$

Nitric acid
See Nitric acid: Alkanethiols

3728. Tetrabutylammonium fluoride
[429-41-4] $C_{16}H_{36}FN$

$$Bu_4NF$$

Hexamethyldisilazane, Pyridine *N*-oxide
See Pyridine *N*-oxide: Hexamethyldisilazane, etc.

3729. Tetrabutylammonium permanganate
[35638-41-6] $C_{16}H_{36}MnNO_4$

$$Bu_4N^+MnO_4^-$$

1. Sala, T. *et al., J. Chem. Soc., Chem. Comm.*, 1978, 254
2. Morris, J. A. *et al., Chem. & Ind.*, 1978, 446
Portions of a 20 g sample of the oxidant salt, prepared exactly as the published
description [1], had been used uneventfully during a few days. Subsequently, as
a 2 g portion was tipped out onto glazed paper, it ignited and burned violently,
leading also to ignition of the bottled material [2].
See other QUATERNARY OXIDANTS

3730. Titanium butoxide
[5593-70-4] $C_{16}H_{36}O_6Ti$

$$Ti(OBu)_4$$

HCS 1980, 884

Ethyl terephthalate, Ethylene glycol
See Dimethyl terephthalate: Ethylene glycol, Titanium butoxide
See other METAL ALKOXIDES

3731. Tetrakis(butylthio)uranium

[ ] $C_{16}H_{36}S_4U$

$$(BuS)_4U$$

Bailar, 1973, Vol. 5, 416
It ignites in air.
See related METAL ALKOXIDES, METAL SULFIDES

3732. Tetrabutylammonium hydrogen monoperoxysulfate

[104548-30-3] $C_{16}H_{37}NO_5S$

$$Bu_4N^+O^-SO_2OOH$$

Trost, B. M. *et al., J. Org. Chem.*, 1987, **53**, 532–537
Although no instability was noted, and a small sample was insensitive to a hammer
blow, it should be treated as potentially unstable as an organic oxidant.
See other QUATERNARY OXIDANTS

3733. Tetrakis(diethylphosphino)silane

[ ] $C_{16}H_{40}P_4Si$

Air, or Water
Fritz, G. *et al., Angew. Chem. (Intern. Ed.)*, 1963, **2**, 262
It ignites in air and is extremely sensitive to water.
See related ALKYLPHOSPHINES, ALKYLSILANES

3734. 2,6-Bis(picrylazo)-3,5-dinitropyridine

[55106-96-2] $C_{17}H_5N_{13}O_{16}$

Explosive.
See entry POLYNITROAZOPYRIDINES

3735. 2,6-Bis(2-picrylhydrazino)-3,5-dinitropyridine
[55106-95-1]

$C_{17}H_9N_{13}O_{16}$

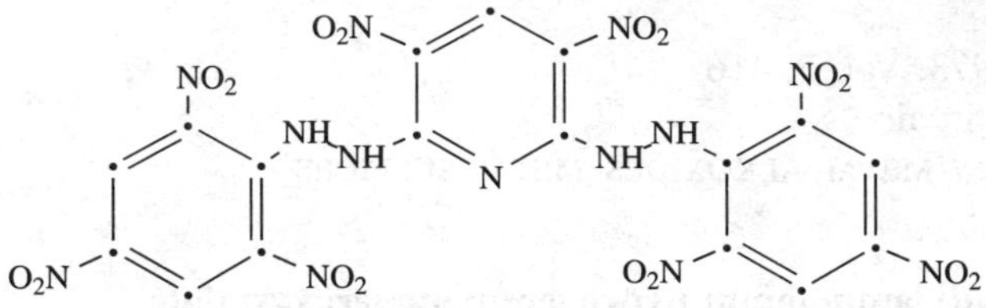

Explosive.
See entry POLYNITROAZOPYRIDINES

3736. Benzanthrone
[82-05-3]

$C_{17}H_{10}O$

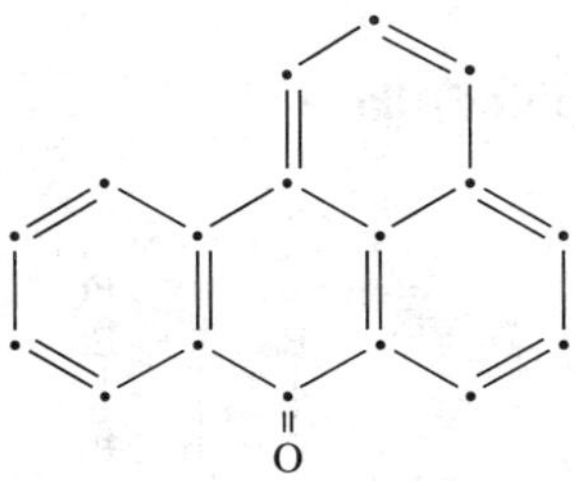

Nitrobenzene, Potassium hydroxide
See Nitrobenzene: Alkali (reference 4)

3737. *S*-7-Methylnonylthiouronium picrate
[ ]

$C_{17}H_{27}N_5O_7S$

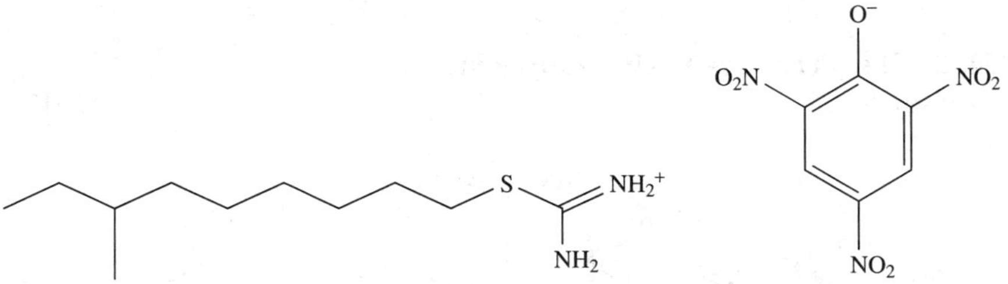

Leese, C. L. *et al. J. Chem. Soc.*, 1950, 2739
The salt explodes if heated rapidly.
See other PICRATES

3738. Tris(pentafluorophenyl)boron
[1109-15-5]

$C_{18}BF_{15}$

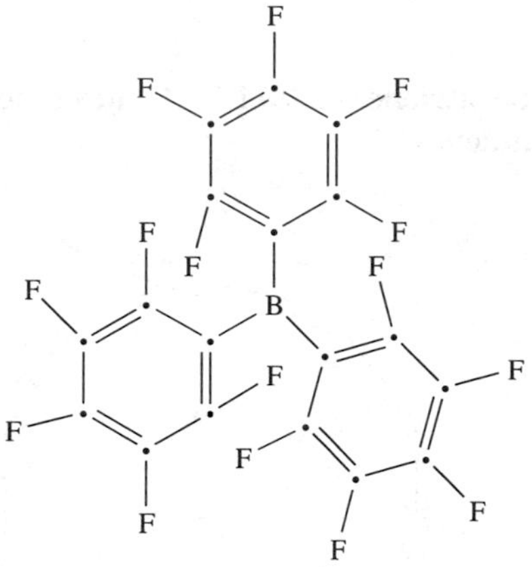

See Triethylaluminium: Tris(pentafluorophenyl)boron
See also METAL DERIVATIVES OF ORGANOFLUORINE COMPOUNDS

3739. Iron(III) hexacyanoferrate(4−)
[14038-43-8]

$C_{18}Fe_7N_{18}$

$$Fe_4[Fe(CN)_6]_3$$

Widmann, G. *et al., Thermochim. Acta*, 1988, **134**, 451–455
The energy of decomposition of Prussian blue in air in the range 210–360°C was determined by DSC as 2.42 kJ/g, peaking at 271°C. When nitrogen was used as inerting gas in the sealed micro-crucible, decomposition was delayed to 250–350°C with a broad peak.

Blown castor oil, Turkey red oil (sulfonated castor oil)
Gärtner, K., *Farben-Ztg.*, 1938, **43**, 1118
A colour mixture containing the 3 components ignited spontaneously. Oxidation products in the air-blown oil may have reacted exothermally with the complex cyanide, a reducant.

Ethylene oxide
1. Wagle, N. G., *Chem. Brit.*, 1977, **13**, 317
2. Heuser, S. G. B., *Chem. Brit.*, 1977, **13**, 436
Interaction of 'iron blue' pigment with ethylene oxide vapour at ambient temperature is highly exothermic and gives a product which ignites in contact with air [1]. The hazards of using undiluted ethylene oxide in sterilisers, and the need to use inert purging between vacuum degassing stages is stressed [2].

Lead chromate
See Lead chromate: Iron(III) hexacyanoferrate(4−)
See other METAL CYANIDES (AND CYANO COMPLEXES)

3740. 1,7,13-Cyclooctadecatriene-3,5,9,11,15,17-hexayne (Hexahydro[18]annulene) [16668-69-2]

$C_{18}H_6$

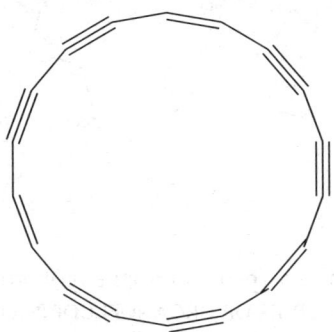

Diederich, F. *et al., Angew. Chem. (Int.),* 1992, **31**(9), 1101
Explodes on heating to 85°C.
See other ALKYNES

3741. Lanthanum picrate

La^{3+}

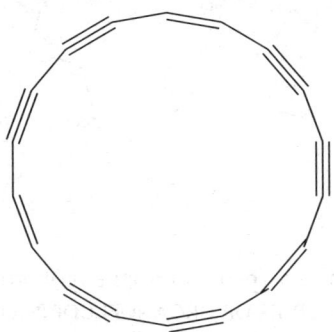

Tucholskii, T., *Rept. AD 633414*, Springfield (Va.), USNTIS, 1966
In a study of the thermal stability of the picrates of group III metals, that of lanthanum occasionally exploded prematurely on heating.
See other PICRATES

3742. Manganese picrate hydroxide
[41570-79-0] $C_{18}H_7Mn_2N_9O_{22}$

Bernardi, A., *Gazz. Chim. Ital.*, 1930, **60**, 169–171

The basic manganese picrate double salt,, and its 2,4-dinitro- and 2-nitro-analogues, all deflagrated violently on heating.

See other HEAVY METAL DERIVATIVES, METAL NITROPHENOXIDES, PICRATES, POLYNI-TROARYL COMPOUNDS

3743. 1,6-Bis(4-chlorophenyl)-2,5-bis(diazo)-1,3,4,6-tetraoxohexane
[76695-68-6] $C_{18}H_8Cl_2N_4O_4$

See 1,6-Diphenyl-2,5-bis(diazo)-1,3,4,6-tetraoxohexane (next below)
See other DIAZO COMPOUNDS

1243

3744. 1,6-Diphenyl-2,5-bis(diazo)-1,3,4,6-tetraoxohexane
[76695-64-2] $C_{18}H_{10}N_4O_4$

Rubin, M. B. *et al.*, *J. Chem. Soc., Perkin Trans. I*, 1980, 2671
The title compound and its 4,4′-chloro-, methyl- and methoxy-derivatives all
decomposed violently at 155–165°C.
See other DIAZO COMPOUNDS

3745. Nickel 2,4-dinitrophenoxide hydroxide
[ ] $C_{18}H_{10}N_6Ni_2O_{16}$

Bernardi, A., *Gazz. Chim. Ital.*, 1930, **60**, 166
It deflagrates very violently on heating.
See other HEAVY METAL DERIVATIVES, METAL NITROPHENOXIDES, POLYNITROARYL
COMPOUNDS

3746. 4,4'-Diphenyl-2,2'-bi(1,3-dithiol)-2'-yl-2-ylium perchlorate
[53213-77-7] $C_{18}H_{12}ClO_4S_4$

Sunderlin, K. G. R., *Chem. Eng. News*, 1974, **52**(31), 3
A few mg exploded violently during determination of the m.p., shattering the apparatus.
See other NON-METAL PERCHLORATES

3747. 9-Phenyl-9-iodafluorene
[32714-73-5] $C_{18}H_{13}I$

Banks, D. F., *Chem. Rev.*, 1966, **66**, 248
It explodes at 105°C.
See other IODINE COMPOUNDS

3748. Oxodiperoxodiquinolinechromium(VI)
[ ] $C_{18}H_{14}CrN_2O_5$

See Oxodiperoxodipyridinechromium(VI)
See other AMMINECHROMIUM PEROXOCOMPLEXES

1245

3749. Triphenylaluminium
[841-76-9] $C_{18}H_{15}Al$

$$\underset{Ph}{\overset{Ph}{\underset{}{\Big|}}} Al \diagdown Ph$$

Water
Neely, T. A. *et al., Org. Synth.*, 1965, **45**, 107
Triphenyl aluminium and its etherate evolved heat and sparks on contact with water.
See other ARYLMETALS

3750. Triphenylsilyl perchlorate
[101652-99-7] $C_{18}H_{15}ClO_4Si$

$$\underset{Ph}{\overset{Ph}{\underset{Ph}{\Big|}}} Si - O - Cl \overset{O}{\underset{O}{\Big|\Big|}} = O$$

See entry ORGANOSILYL PERCHLORATES

3751. Triphenylchromium tetrahydrofuranate
[16462-53-6] $C_{18}H_{15}Cr.3C_4H_8O$

$$Ph_3Cr.3C_4H_8O$$

Diethyl ether
Bailar, 1973, Vol. 4, 974
On warming, or on treatment with ether, the solvated complex give a black pyrophoric material of unknown constitution.
See other ARYLMETALS

3752. Triphenylphosphine oxide-oxodiperoxochromium(VI)
[93228-65-0] $C_{18}H_{15}CrO_6P$

$$[Ph_3P(O)Cr(O_2)_2O]$$

Daire, E. *et al., Nouv. J. Chim.*, 1984, **8**, 271–274
It effectively hydroxylates hydrocarbons but is free of the explosion risk of the analogous pyridine *N*-oxide complex.
See related AMMINECHROMIUM PEROXOCOMPLEXES

1246

3753. Triphenyllead nitrate
[21483-09-8] (ion)

$C_{18}H_{15}NO_3Pb$

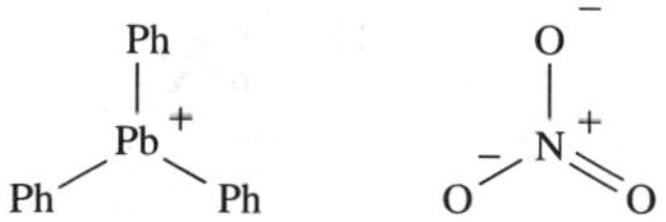

Sulfuric acid

Gilman, H. *et al., J. Org. Chem.*, 1951, **16**, 466
It ignites in contact with conc. acid
See related ARYLMETALS

3754. 1,3,5-Triphenyl-1,4-pentaazadiene
[30616-12-7]

$C_{18}H_{15}N_5$

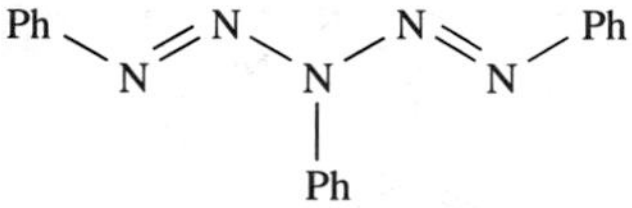

Sorbe, 1968, 141
It explodes at 80°C and is shock-sensitive.
See other HIGH-NITROGEN COMPOUNDS

3755. Triphenylphosphine oxide hydrogen peroxidate
[3319-45-7]

$C_{18}H_{15}OP.0.5H_2O_2$

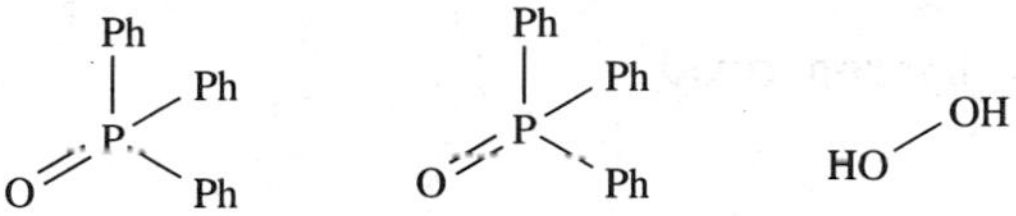

Bradley, D. C. *et al., J. Chem. Soc., Chem. Comm.*, 1974, 4
The 2:1 complex may explode.
See CRYSTALLINE HYDROGEN PEROXIDATES
See related REDOX COMPOUNDS

3756. Triphenylphosphine
[603-35-0]

$C_{18}H_{15}P$

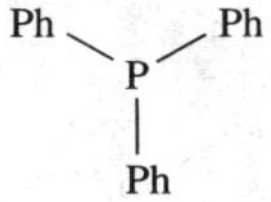

$$Ph \diagdown \underset{|}{\overset{Ph}{P}} \diagup Ph$$

$$Ph$$

Preparative hazard

See Sodium: Chlorobenzene, Phosphorus trichloride

See related ALKYLPHOSPHINES

3757. 1,3-Bis(phenyltriazeno)benzene
[70324-26-4]

$C_{18}H_{16}N_6$

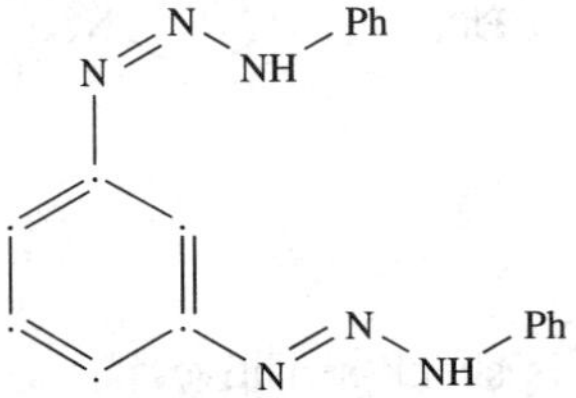

Kleinfeller, H., *J. Prakt. Chem. [2]*, 1928, **119**, 61

It explodes if heated rapidly.

See other TRIAZENES

3758. Triphenyltin hydroperoxide
[4150-34-9]

$C_{18}H_{16}O_2Sn$

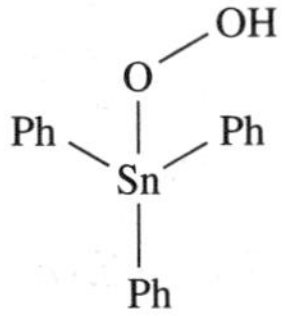

Dannley, R. L. *et al., J. Org. Chem.*, 1965, **30**, 3845

It explodes reproducibly at 75°C.

See other ORGANOMETALLIC PEROXIDES

3759. Sodium hexakis(propynyl)ferrate(4−)

[ ] $C_{18}H_{18}FeNa_4$

Bailar, 1973, Vol. 3, 1025
An explosive complex salt.
See other COMPLEX ACETYLIDES

3760. 1-(1-Methyl-1-phenylethyl)-4-(2-propynyloxy)benzene

[ ] $C_{18}H_{18}O$

Douglas, W. E. *et al., J. Organomet. Chem.*, **444**(1–2), C62
Explosively polymerises when heated >110°C in presence of nickelocene catalyst.
See VIOLENT POLYMERISATION
See other ACETYLENIC COMPOUNDS, POLYMERISATION INCIDENTS

3761. 1,5-Cyclooctadiene-bis(4-chloropyridine *N*-oxide)rhodium(I) perchlorate

[80005-73-8] $C_{18}H_{20}Cl_3N_2O_6Rh$

Uson, R. *et al., J. Organomet. Chem.*, 1981, **217**, 252
The complex, one of a series of substituted pyridine derivatives, exploded violently during microanalytical combustion.
See related AMMINEMETAL OXOSALTS

3762. 1,2-Dihydroperoxy-1,2-bis(benzeneazo)cyclohexane
[71404-13-2]   C₁₈H₂₀N₄O₄

See Oxygen: Cyclohexane-1,2-dione bis(phenylhydrazone)
See other α-PHENYLAZO HYDROPEROXIDES

3763. *O*–*O*-*tert*-Butyl di(4-tolyl) monoperoxophosphate
[ ]   C₁₈H₂₃O₅P

Rieche, A. *et al., Chem. Ber.*, 1962, **95**, 385
The material deflagrated as a solid, or decomposed exothermally in its reaction mixture soon after preparation.
See other *tert*-BUTYL PEROXOPHOSPHATE DERIVATIVES, PHOSPHORUS ESTERS

3764. 2-(1-Methylheptyl)-4,6-dinitrophenyl crotonate
[27661-44-5]

$C_{18}H_{24}N_2O_6$

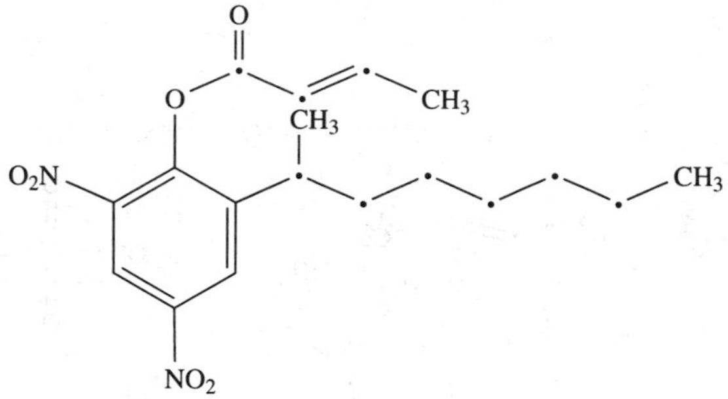

Anon., *CISHC Chem. Safety Summ.*, 1978, **49**, 2
Decomposition and a pressure explosion occurred while a 10 kl tanker with steam-heating coils was being unloaded.
See related PEROXIDISABLE COMPOUNDS
See other POLYNITROARYL COMPOUNDS

3765. 2,2,4-Trimethyldecahydroquinolinium picrate
[ ]

$C_{18}H_{23}N_4O_7$

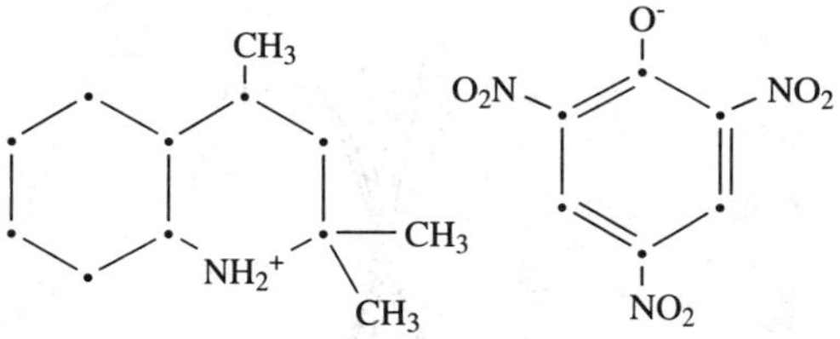

2-(2-Butoxyethoxy)ethanol
Franklin, N. C., private comm., 1967
Evaporation of a solution of the picrate in the diether (b.p. 230°C) caused a violent explosion. The solvent had probably peroxidised during open storage and the residual mixture of peroxide and picrate had exploded as evaporation proceeded. Use of peroxide-free solvent and lower (vacuum) evaporation temperature appeared to be safe.
See other PICRATES

3766. Bis(benzyl 1-methylhydrazinocarbodithioate
N^2,S')(perchlorato-O,O')copper(1+) perchlorate
[67870-99-9] $C_{18}H_{24}Cl_2CuN_4O_8S_4$

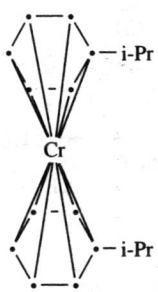

See Bis-O,N[(N'-pent-2-en-2-oxy-4-ylidene)-N,S-dimethyldithiocarbazate]
copper(II) perchlorate

3767. Dicumenechromium(0)
[12001-89-7] $C_{18}H_{24}Cr$

$$(Me_2CHC_6H_5)_2Cr$$

Aldrich catalogue, 1986
It is pyrophoric.
See other ARYLMETALS, ORGANOMETALLICS

1252

3768. **5,6,14,15,20,21-Hexamethyl-1,3,4,7,8,10,12,13,16,17,19,22-dodecaazatetra-cyclo[8.8.4.1^{3,17}.1^{8,12}]tetracosa-4,6,13,15,19,21-hexaene-N^4,N^7,N^{13},N^{16},N^{19},N^{22}cobalt(II) perchlorate** [72644-04-3]

$C_{18}H_{30}Cl_2CoN_{12}O_8$

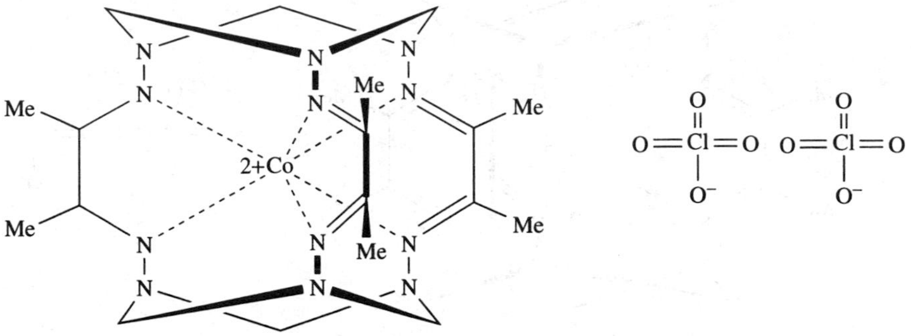

See entry CLATHROCHELATED METAL PERCHLORATES

3769. **5,6,14,15,20,21-Hexamethyl-1,3,4,7,8,10,12,13,16,17,19,22-dodecaazatetra-cyclo[8.8.4.1^{3,17}.1^{8,12}]tetracosa-4,6,13,15,19,21-hexaene-N^4,N^7,N^{13},N^{16},N^{19},N^{22}iron(II) perchlorate** [75516-37-9]

$C_{18}H_{30}Cl_2FeN_{12}O_8$

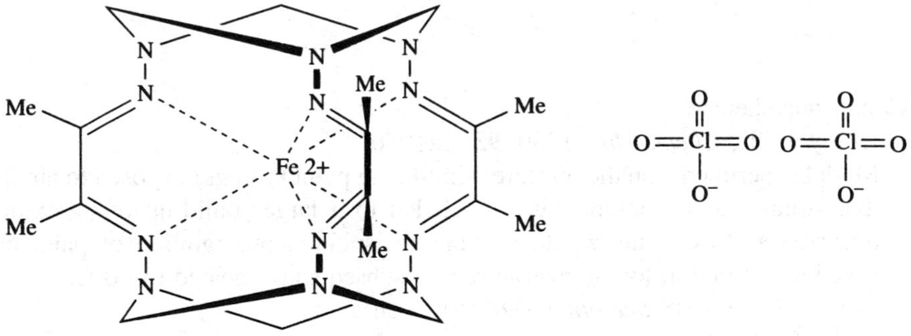

See entry CLATHROCHELATED METAL PERCHLORATES

3770. 5,6,14,15,20,21-Hexamethyl-1,3,4,7,8,10,12,13,16,17,19,22-dodecaazatetra-cyclo[8.8.4.1^{3,17}.1^{8,12}]tetracosa-4,6,13,15,19,21-hexaene-N^4,N^7,N^{13},N^{16},N^{19},N^{22}nickel(II) perchlorate [63128-08-5]

$C_{18}H_{30}Cl_2N_{12}NiO_8$

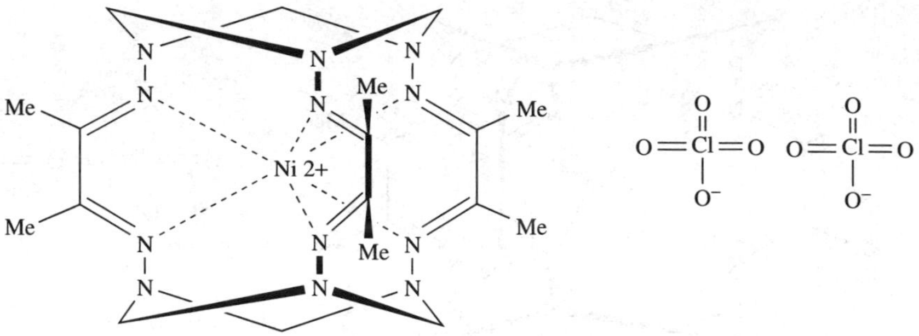

See entry CLATHROCHELATED METAL PERCHLORATES

3771. 9,12,15-Octadecatrienoic acid (Linolenic acid) [463-40-1]

$C_{18}H_{30}O_2$

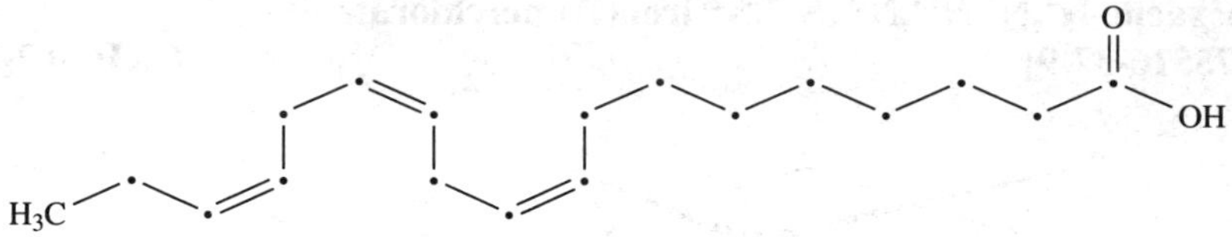

Cobalt naphthenate

Kotoyori, T., *Chem. Abs.*, 1980, **93**, 221950

Model experiments on the mixture (simulating paint tailings) exposed to air showed that storage at low temperature (7°C) led to a larger build-up of peroxide than occurred at 35°C. This would account for spontaneous ignition of paint tailings (residues) stored at low temperature and subsequently heated to 60°C.

See AUTOXIDATION *See other* ORGANIC ACIDS

3772. Oleoyl chloride (*cis*-9-Octadecenoyl chloride) [112-77-6]

$C_{18}H_{33}ClO$

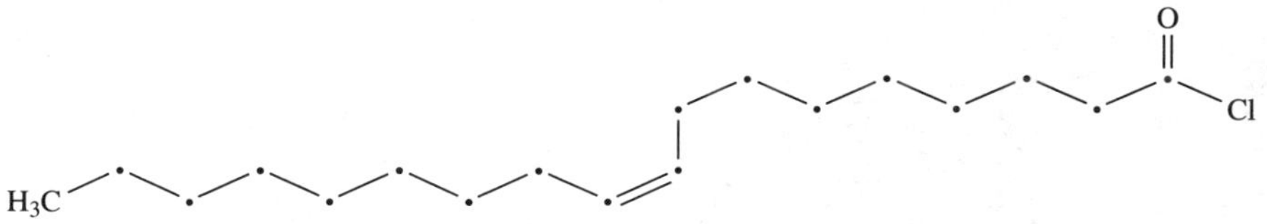

'Acid tars'
Anon., *Jahresbericht*, 1994, 74
The residue of a sample of 'olein chloride' (crude oleoyl chloride) was tipped into a waste container containing 'acid tars' (70–80% waste sulfuric acid). There was a vigorous reaction, with gas evolution, which sprayed the analytical technician involved with the liquid contents of the container, causing burns.

Nitric acid
See Nitric acid: Oleoyl chloride
See other ACYL HALIDES

3773. 1,4-Octadecanolactone
[502-26-1] $C_{18}H_{34}O_2$

Preparative hazard
See Perchloric acid: Oleic acid

3774. *cis*-9-Octadecenoic acid (Oleic acid)
[112-80-1] $C_{18}H_{34}O_2$

HCS 1980, 701

Aluminium
See Aluminium: Oleic acid

Perchloric acid
See Perchloric acid: Oleic acid
See other ORGANIC ACIDS

3775. Bis(dibutylborino)acetylene
[ ] $C_{18}H_{36}B_2$

$$Bu_2BC{\equiv}CBBu_2$$

Hartmann, H. *et al., Z. Anorg. Chem.*, 1959, **299**, 174
Both the n- and iso-derivatives ignite in air.
See other ACETYLENIC COMPOUNDS, ALKYLBORANES

3776. 1,2-Bis(di-*tert*-butylphosphino)ethane
[107783-62-0] $C_{18}H_{40}P_2$

$$(Me_3C)_2PC_2H_4P(CMe_3)_2$$

Benn, R. *et al., Z. Naturforsch. B*, 1986, **41B**, 680–691
The ligand, like its lower homologues, is pyrophoric in air.
See related ALKYLPHOSPHINES

3777. 1,2-Bis(tripropylsilyl)hydrazine
[13271-98-2] $C_{18}H_{44}N_2Si_2$

$$(Pr_3SiNH-)_2$$

See entry SILYLHYDRAZINES *See related* ALKYLSILANES

3778. Perchloratotris(triethylphosphine)palladium(II) perchlorate
[94288-47-8] $C_{18}H_{45}Cl_2O_8P_3Pd$

$$[(Et_2P)_3PdClO_4]\ ClO_4$$

Bruce, D. W. *et al., J. Chem. Soc., Dalton Trans.*, 1984, 2252
It explodes on heating.
See related AMMINEMETAL OXOSALTS, METAL PERCHLORATES, PLATINUM COMPOUNDS

3779. Hexamethylenetetrammonium tetraperoxochromate(V)?
[ ] $C_{18}H_{48}Cr_4N_{12}O_{32}$

House, D. A. *et al., Inorg. Chem.*, 1966, **6**, 1078, footnote 8
Material (believed to be the title compound) recrystallised from water, and washed with methanol to dry by suction, ignited and then exploded on the filter funnel.
See related AMMINECHROMIUM PEROXOCOMPLEXES, AMMINEMETAL OXOSALTS

3780. 3,8-Dinitro-6-phenylphenanthridine
[82921-86-6]

$C_{19}H_{11}N_3O_4$

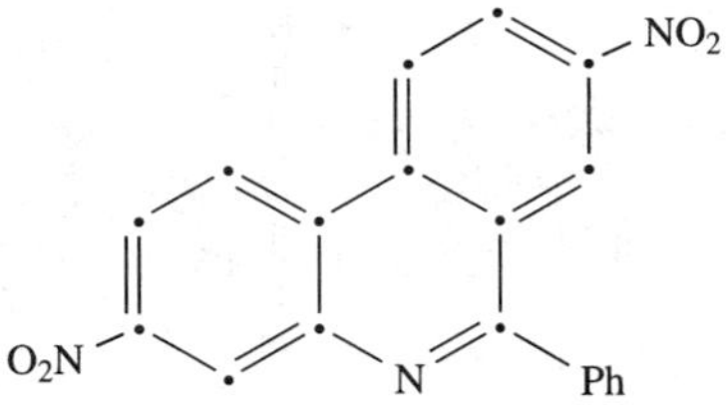

Diethyl sulfate
    See Diethyl sulfate: 3,8-Dinitro-6-phenylphenanthridine
    See other POLYNITROARYL COMPOUNDS

3781. Triphenylmethylpotassium
[1528-27-4]

$C_{19}H_{15}K$

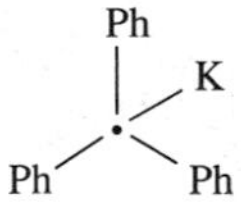

Houben-Weyl, 1970, Vol. 13.1, 269
The dry powder ignites in air.
See other ORGANOMETALLICS

3782. Triphenylmethyl nitrate
[111422-41-4]

$C_{19}H_{15}NO_3$

Lewis acids
    See entry ALKYL NITRATES

3783. Triphenylmethyl azide
[14309-25-2]

$C_{19}H_{15}N_3$

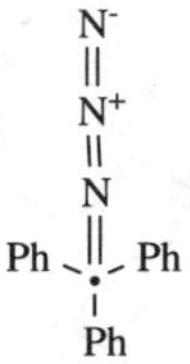

$$N^-$$
$$\parallel$$
$$N^+$$
$$\parallel$$
$$N$$

Ph$\diagdown$ $\overset{\parallel}{\underset{\mid}{C}}$ $\diagup$Ph

Ph

Robillard, J. J., Ger. Offen. 2 345 787, 1974
The unstable explosive azide may be stabilised by adsorption for reprographic purposes.
See other ORGANIC AZIDES

3784. Methyltriphenylphosphonium permanganate
[73335-41-8]

$C_{19}H_{18}MnO_4P$

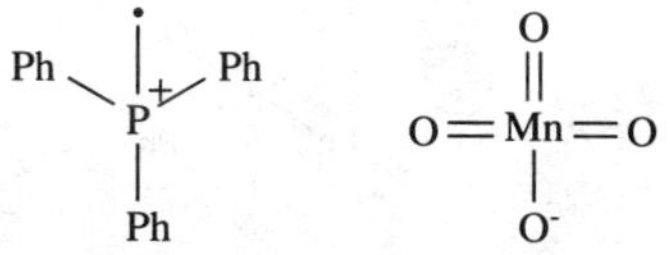

Ph$\diagdown$ $\overset{\bullet}{\underset{\mid}{P^+}}$ $\diagup$Ph

Ph

$$O$$
$$\parallel$$
$$O = Mn = O$$
$$\mid$$
$$O^-$$

1. Reischl, W. *et al., Tetahedron*, 1979, **35**, 1109–1110
2. Leddy, B. P. *et al., Tetrahedron Lett.*, 1980, **21**, 2261–2262
It decomposes explosively above 70° [1], or at 80–90°C [2].
See other QUATERNARY OXIDANTS

3785. Hexadecyltrimethylammonium permanganate (Cetyltrimethylammonium permanganate)
[73257-07-5]

$C_{19}H_{42}MnNO_4$

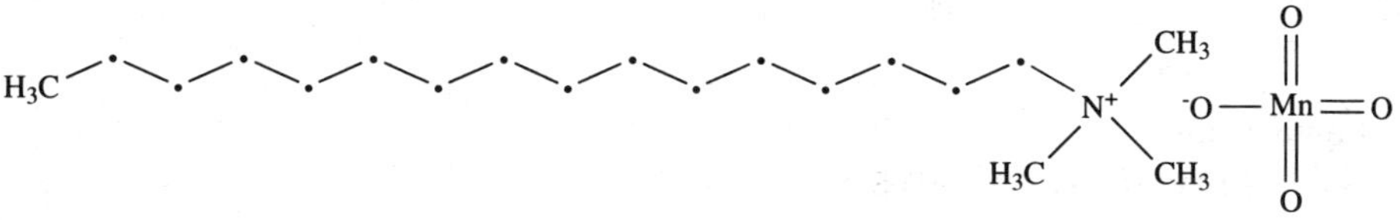

H$_3$C$\diagup\diagdown\diagup\diagdown\diagup\diagdown\diagup\diagdown\diagup\diagdown\diagup\diagdown\diagup\diagdown\diagup$

CH$_3$

N$^+$

H$_3$C   CH$_3$

$^-$O — Mn = O

$$O$$
$$\parallel$$
$$\parallel$$
$$O$$

Dash, S. *et al., Chem. Abs., 1995*, **123**, 227457m
Stable when kept in the dark at room temperature, it decomposes violently at 115°C.
See other QUATERNARY OXIDANTS

1258

3786. Dibenzocyclododeca-1,7-dienetetrayne
   (5,6,7,8,13,14,15,16-Octadehydrodibenzo[a,g]cyclododecene)
   [7203-21-6] $C_{20}H_8$

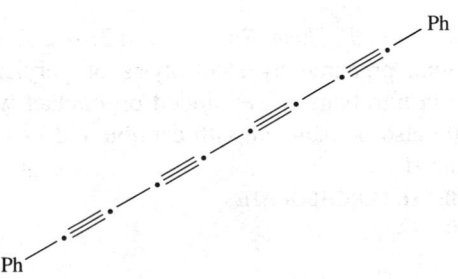

Guo, L. *et al., J. Chem. Soc., Chem. Comm.*, 1994, (3), 243
An unstable synthetic byproduct, attributed the above structure, explodes when
scratched in air, leaving carbon as the residue. It blackens rapidly even in vacuum.
See other ACETYLENIC COMPOUNDS

3787. 2′,4′,5′,7′-Tetrabromo-3′,6′-dihydroxyspiro[isobenzofuran-
   1(3H),9′-[9H]xanthen-3-one] (Eosin)
   [15876-39-8], [122530-01-2] $C_{20}H_8Br_4O_5$

Lead salts, Red lead
   See Dilead(II)lead(IV) oxide: Eosin

3788. 1,8-Diphenyloctatetrayne
   [4572-12-7] $C_{20}H_{10}$

1259

Nakagawa, M., *Chem. Abs.*, 1951, **45**, 7082a

Though stable in the dark at ambient temperature for a year, when exposed to light on a metal plate it decomposed explosively.

See other ALKYNES, IRRADIATION DECOMPOSITION INCIDENTS

3789. *I,I*-Bis(4-nitrobenzoylperoxy)-4-chlorophenyliodine
[56391-49-2]

$C_{20}H_{12}ClIN_2O_{10}$

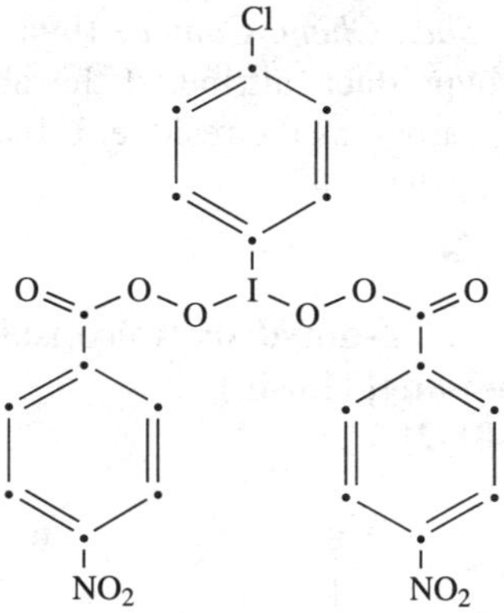

See entry I,I-DI(BENZOYLPEROXY)ARYLIODINES

3790. Perylenium perchlorate
[12576-63-5]

$C_{20}H_{12}Cl_2O_8$

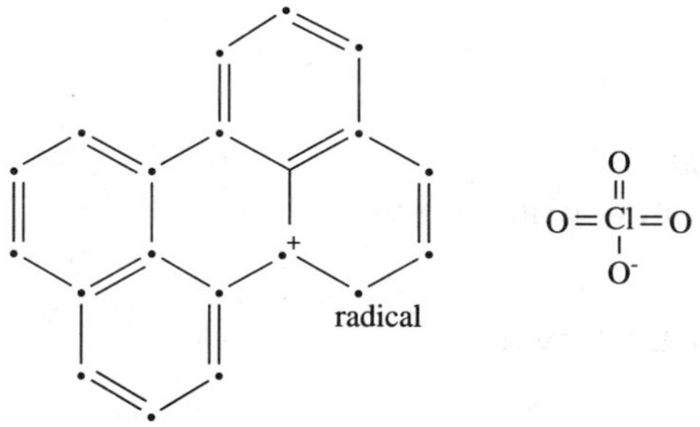

Rosseinsky, D. R. *et al.*, *J. Chem. Soc., Perkin Trans. 2*, 1985, 135–138

The title compound, prepared by electrolysis of perylene and tetrabutylammonium perchlorate in nitrobenzene, exploded on contact with nickel. Co-produced compounds should also be handled with caution and in small amounts, especially in contact with metals.

See other NON-METAL PERCHLORATES

3791. Bis(1,8-naphthyridine)tetracarbonyldirhodium(I) perchlorate
[90635-49-7] $C_{20}H_{12}Cl_2N_4O_{12}Rh_2$

$$[(C_8H_6N_2)_2Rh_2(CO)_4] [ClO_4]_2$$

Camellini, A. T. *et al., J. Chem. Soc., Dalton Trans.*, 1984, 130
Attempted microanalytical combustion of old samples of the salt or its 2-methyl
homologue led to violent explosions.
See related CARBONYLMETALS, AMMINEMETAL OXOSALTS

3792. 11,12-Diethynyl-9,10-dihydro-9,10-ethenoanthracene
[126487-13-6] $C_{20}H_{12}$

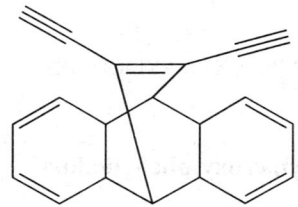

Diederich, F. *et al., Angew. Chem. (Int.)*, 1992, **31**(9), 1101
The crystalline solid has been known to explode spontaneously.
See other ALKYNES

3793. *I,I* -Bis(3-chlorobenzoylperoxy)-4-chlorophenyliodine
[ ] $C_{20}H_{12}Cl_3IO_6$

See entry *I,I*-DI(BENZOYLPEROXY)ARYLIODINES

3794. *I,I*-Bis(3-chlorobenzoylperoxy)phenyliodine

[30242-75-2]

$C_{20}H_{13}Cl_2IO_6$

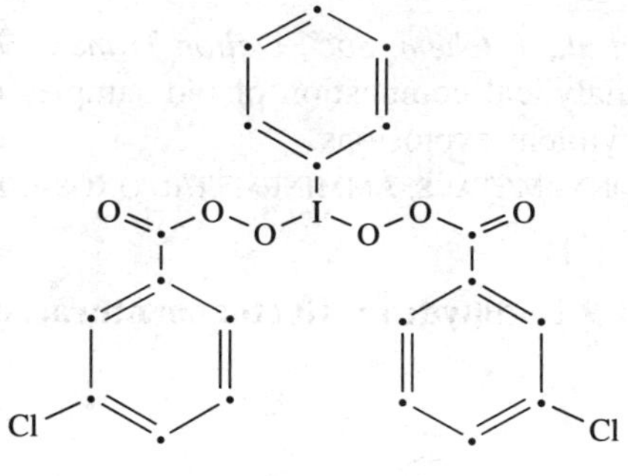

See entry I,I-DI(BENZOYLPEROXY)ARYLIODINES

3795. *I,I*-Bis(4-nitrobenzoylperoxy)phenyliodine

[30030-30-9]

$C_{20}H_{13}IN_2O_{10}$

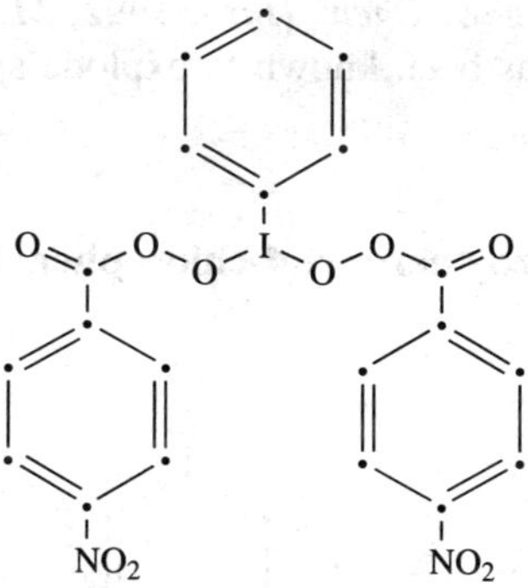

See entry I,I-DI(BENZOYLPEROXY)ARYLIODINES

3796. 1,6-Di(4′-tolyl)-2,5-bis(diazo)-1,3,4,6-tetraoxohexane
[76695-67-5] $C_{20}H_{14}N_4O_4$

See 1,6-Diphenyl-2,5-bis(diazo)-1,3,4,6-tetraoxohexane

3797. 1,6-Di(4′-methoxyphenyl)-2,5-bis(diazo)-1,3,4,6-tetraoxohexane
[77695-66-4] $C_{20}H_{14}N_4O_6$

See 1,6-Diphenyl-2,5-bis(diazo)-1,3,4,6-tetraoxohexane

1263

3798. 2,5-Diphenyl-3,4-benzofuran-2,5-endoperoxide
   (1,4-Epoxy-1,4-dihydro-1,4-diphenyl-2,3-benzodioxin)
   [41337-62-6] $C_{20}H_{14}O_3$

Alone, or Carbon disulfide
   1. Dufraisse, C. *et al., Compt. rend.*, 1946, **223**, 735
   2. Rio, G. *et al., J. Chem. Soc., Chem. Comm.*, 1982, 72
Formally an ozonide, this photo-peroxide explodes at 18°C [1], and its dimerisation, catalysed by carbon disulfide, was exothermic and sometimes explosive at
ambient temperature [2].
See other CYCLIC PEROXIDES, OZONIDES

3799. *cis*-Dichlorobis(2,2′-bipyridyl)cobalt(III) chloride
   [23380-38-3 or 14522-39-5] $C_{20}H_{16}Cl_3CoN_4$

Preparative hazard
   See Chlorine: Cobalt(II) chloride, Methanol

1264

3800. 1,1-Bis(4-nitrobenzoylperoxy)cyclohexane

[ ] $C_{20}H_{18}N_2O_{10}$

Criegee, R. *et al., Ann.*, 1948, **560**, 135
It explodes at 120°C.
See other PEROXYESTERS

3801. Tetrapyridinecobalt(II) chloride

[13985-87-0] $C_{20}H_{20}Cl_2CoN_4$

$$[(C_5H_5N)_4Co]Cl_2$$

Chlorine, Methanol
See Chlorine: Methanol, Tetrapyridinecobalt(II) chloride

3802. *trans*-Dichlorotetrapyridinecobalt(III) chloride

[144077-02-2 or 27883-34-7] $C_{20}H_{20}Cl_3CoN_4$

Preparative hazard
See Chlorine: Methanol, Tetrapyridinecobalt(II) chloride

3803. Oxybis[bis(cyclopentadienyl)titanium]
[51269-39-7]

$C_{20}H_{20}OTi_2$

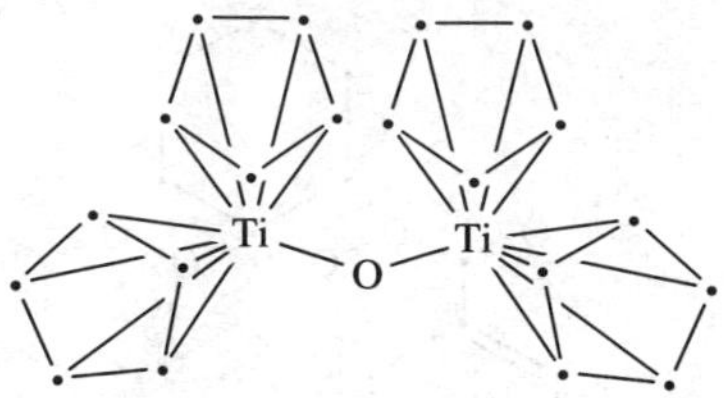

Bottomley, F. *et al.*, *J. Amer. Chem. Soc.*, 1980, **102**, 5241
The dry solid is explosively oxidised in contact with air.
See other ORGANOMETALLICS

3804. 1,1-Bis(benzoylperoxy)cyclohexane
[13213-29-1]

$C_{20}H_{20}O_6$

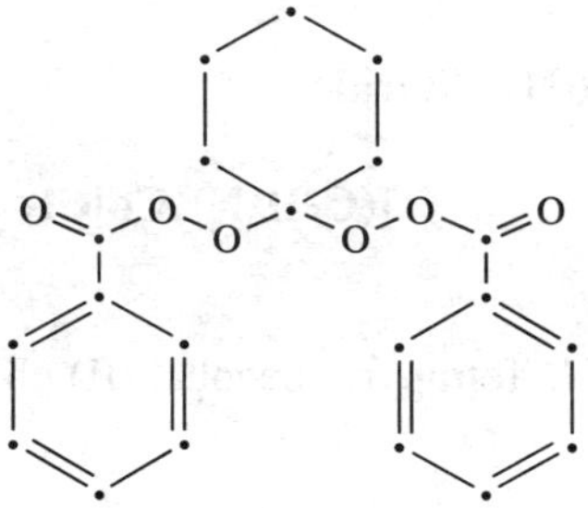

Criegee, R. *et al.*, *Ann.*, 1948, **560**, 135
It explodes sharply in a flame.
See other PEROXYESTERS

3805. Sodium abietate
[14351-66-7]

$C_{20}H_{29}NaO_2$

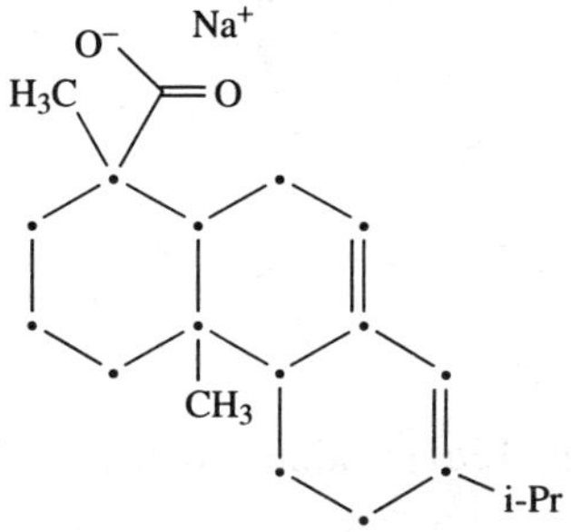

See entry METAL ABIETATES

1266

3806. Dichlorodi-μ-chlorobis(pentamethylcyclopentadienyl)dirhodium
[12354-85-7] $C_{20}H_{30}Cl_4Rh_2$

Air, Alkylmetals
Isobe, K. *et al., J. Chem. Soc., Dalton Trans.*, 1983, 1445
In preparation of di-μ-methylenebis(methyl-pentamethylcyclopentadienyl)dirhodium complexes by aerobic oxidation of a solution of the halocomplex and methyllithium or trimethylaluminium in ether–benzene, the reaction mixture occasionally ignited and burned violently. Full precautions and a working scale below 1 mmol are recommended.
See related ALKYLMETAL HALIDES

3807. Di-3-camphoroyl peroxide
[ ] $C_{20}H_{30}O_8$

Milas, N. A. *et al., J. Amer. Chem. Soc.*, 1933, **55**, 350–351
Explosive decomposition occurred at the m.p., 142°C, or on exposure to flame.
See other DIACYL PEROXIDES

3808. *trans*-Dichlorobis[1,2-phenylenebis(dimethylarsine)]palladium(IV) diperchlorate
[60489-76-1]

$C_{20}H_{32}As_4Cl_4O_8Pd$

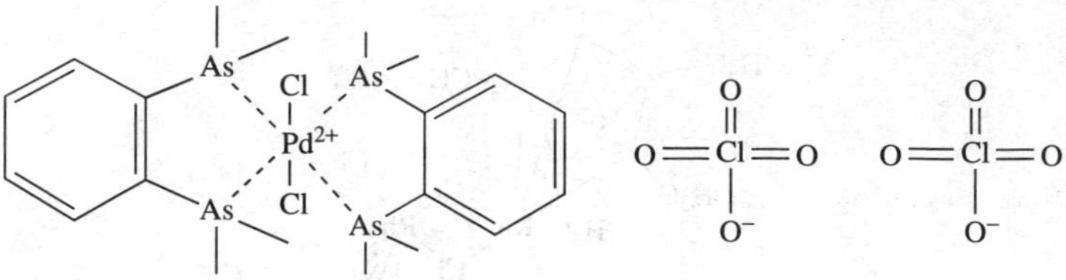

Gray, L. R. *et al., J. Chem. Soc., Dalton Trans.*, 1983, 33–41
Though no explosions occurred during the preparation by addition of 75% perchloric acid to solutions of the complex in conc. nitric/hydrochloric acid mixtures, this and the amino- and phosphino-analogues are potentially unstable.
See related AMMINEMETAL OXOSALTS

3809. Sodium 5,8,11,14-eicosatetraenoate (Sodium arachidonate)
[6610-25-9]

$C_{20}H_{32}NaO_2$

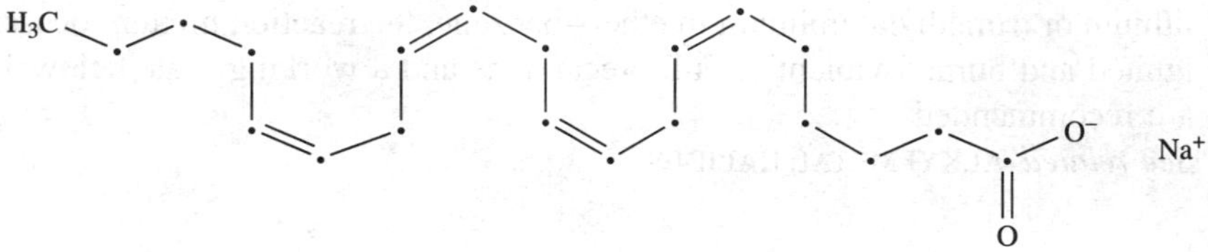

Henderson, J. D. *et al., Chem. Eng. News*, 1983, **61**(33), 2
An opened 100 mg ampoule of the salt was stored dark in a desiccator at −20°C between occasional samplings during a few months. A few minutes after return to storage, the ampoule contents decomposed violently. It was supposed that peroxide formation (and perhaps polymerisation of the polyene) was involved in the incident.
See other PEROXIDISABLE COMPOUNDS

3810. 1,1,6,6-Tetrakis(acetylperoxy)cyclododecane
[ ]

$C_{20}H_{32}O_{12}$

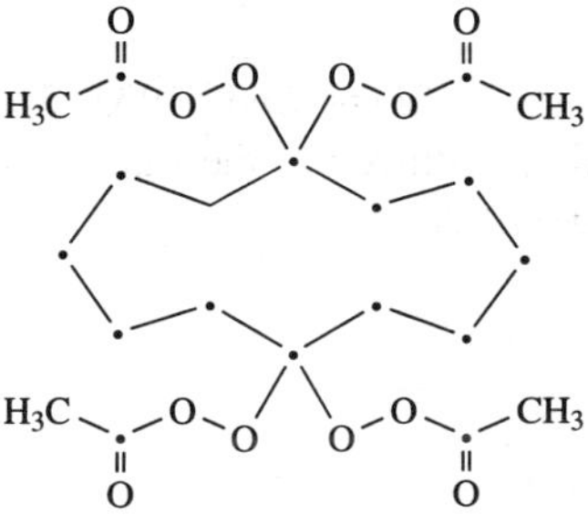

Criegee, R. *et al., Ann.*, 1948, **560**, 135

Weak friction causes strong explosion.
See other PEROXYESTERS

3811. Tributyl(phenylethynyl)lead
[21249-40-0] $C_{20}H_{32}Pb$

$$Bu_3PbC\equiv CPh$$

Houben-Weyl., 1975, Vol. 13.3, 80
Violent decomposition occurred during attempted distillation.
See related ALKYLMETALS, METAL ACETYLIDES

3812. *B*-Iododiisopinocampheylborane (Iodobis(2,6,6-trimethylbicyclo [3.1.1]hept-3-yl)borane)
[116005-09-5] $C_{20}H_{34}BI$

Brown, H. C. *et al., Heteroat. Chem.*, 1995, **6**(2), 117
Pyrophoric, very hygroscopic, fumes in air.
See other ALKYLHALOBORANES

3813. Diacetonitrile-5,7,7,12,14,14-hexamethyl-1,4,8,11-tetraaza-4,11-cyclotetradecadieneiron(II) perchlorate
[50327-56-5] $C_{20}H_{38}Cl_2FeN_6O_8$

See entry [14] DIENE-N$_4$ COMPLEXES

3814. Tetracyanooctaethyltetragold
[ ] $C_{20}H_{40}Au_4N_4$

Burawoy, A. *et al.*, *J. Chem. Soc.*, 1935, 1026
This tetramer of cyanodiethylgold is friction-sensitive and also decomposes explosively above 80°C. The propyl homologue is not friction-sensitive, but also decomposes explosively on heating in bulk.
See other CYANO COMPOUNDS, GOLD COMPOUNDS

3815. Lanthanum 2-nitrobenzoate
[114515-72-9] $C_{21}H_{12}LaN_3O_{12}$

Ferenc, W. *et al.*, *Monatsh. Chem.*, 1987, **118**, 1087–1100
Preparation of the 2-nitrobenzoate salts of yttrium and the lanthanide metals (except praseodymium) as mono- or di-hydrates was studied. All melted and decomposed explosively above 250°C.
See other HEAVY METAL DERIVATIVES, NITROARYL COMPOUNDS

3816. Scandium 3-nitrobenzoate
[111016-97-8] $C_{21}H_{12}N_3O_{12}Sc$

Brzyska, W. *et al., J. Therm. Anal.*, 1987, **32**, 671–678
After loss of the hemi-hydrate water, the anhydrous complex decomposes explosively on further heating.
See other HEAVY METAL DERIVATIVES, NITROARYL COMPOUNDS

3817. Yttrium 4-nitrobenzoate trihydrate
[115627-20-8] $C_{21}H_{12}N_3O_{12}Y.3H_2O$

Ferenc, W. *et al., Monatsh. Chem.*, 1988, **119**, 407–424
The yttrium, lanthanum and other lanthanide salts exploded after dehydration during heating to above 300°C.
See other HEAVY METAL DERIVATIVES, NITROARYL COMPOUNDS

3818. 1,3,5-Tris(4-azidosulfonylphenyl)-1,3,5-triazinetrione
[31328-33-3] $C_{21}H_{12}N_{12}O_9S_3$

Ulrich, H. *et al., J. Org. Chem.*, 1975, **40**, 804

This trimer of 4-azidosulfonylphenylisocyanate melts at 200°C with violent decomposition.

See other ACYL AZIDES

3819. *I,I*-Bis(4-nitrobenzoylperoxy)-2-methoxyphenyliodine
[ ] $C_{21}H_{15}IN_2O_{11}$

See entry *I,I*-DI(BENZOYLPEROXY)ARYLIODINES

3820. *N,N*-Diphenyl-3-phenylpropenylidenimmonium perchlorate
[77664-36-9] $C_{21}H_{18}ClNO_4$

Qureshi, M. *et al., J. Indian Chem. Soc.*, 1980, **57**, 915–917

The immonium salt explodes on heating.

See other PERCHLORATE SALTS OF NITROGENOUS BASES

3821. *I*-(2-Azido-2-phenylvinyl)phenyliodonium 4-toluenesulfonate
[118235-47-5] $C_{21}H_{18}IN_3O_3S$

Kitamura, T. *et al., Tetrahedron Lett.*, 1988, **29**, 1887–1890
It decomposes explosively on melting.
See other IODINE COMPOUNDS, ORGANIC AZIDES

3822. Tribenzylarsine
[5888-61-9]

$C_{21}H_{21}As$

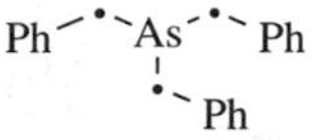

Dondorov, J. *et al., Ber.*, 1935, **68**, 1255
The pure material oxidises slowly at first in air at ambient temperature, but reaction becomes violent through autocatalysis.
See other ALKYLNON-METALS, BENZYL COMPOUNDS

3823. Tri-4-methylphenylsilyl perchlorate
[ ]

$C_{21}H_{21}ClO_4Si$

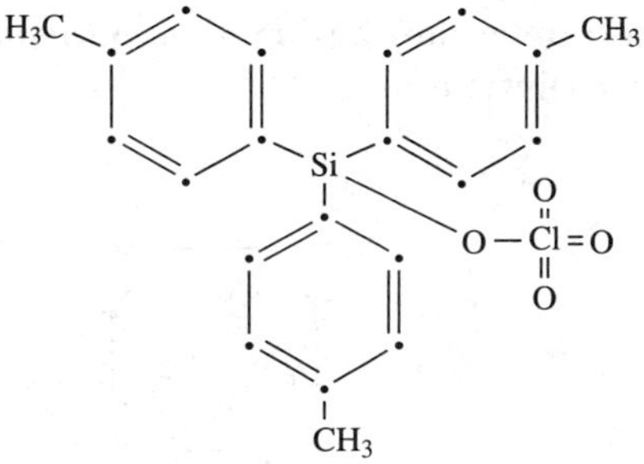

See entry ORGANOSILYL PERCHLORATES

3824. Tri(4-tolyl)ammonium perchlorate
[34729-49-2]

$C_{21}H_{22}ClNO_4$

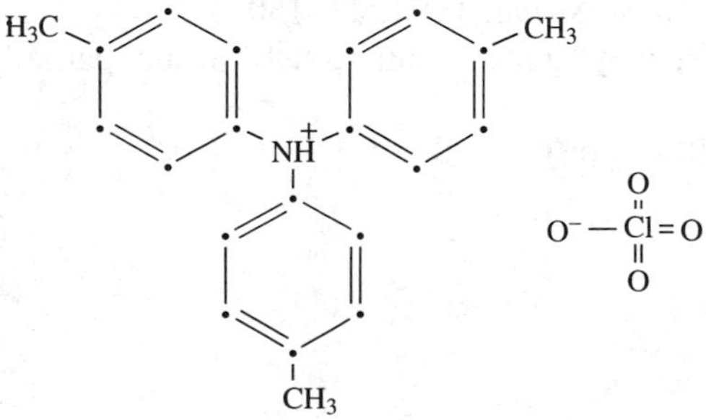

1273

Weitz, E. *et al., Ber.*, 1926, **59**, 2307
It explodes violently when heated above the m.p., 123°C.
See other PERCHLORATE SALTS OF NITROGENOUS BASES

3825. 2,2-Bis[4(2',3'-epoxypropoxy)phenyl]propane
[1675-54-3]

$C_{21}H_{24}O_4$

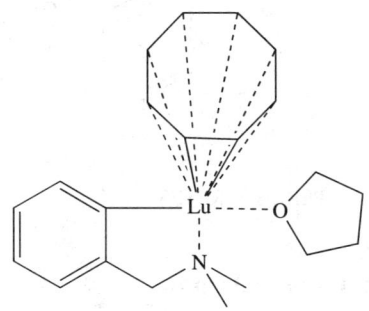

Trichloroethylene
 See Trichloroethylene: Epoxides
 See other 1,2-EPOXIDES

3826. (η^8 1,3,5,7-Cyclooctatetraene)[2-(dimethylaminomethyl)phenyl-
C,N](tetrahydrofuran)lutetium
[84582-81-0]

$C_{21}H_{28}LuNO$

Wayda, A. L., *Inorg. Synth.*, 1990, **27**, 150
The dry powder may ignite spontaneously in air, partially oxidised samples can explode.
See related ORGANOMETALLICS

3827. *N*-Hexadecylpyridinium permanganate
[76710-77-5] $C_{21}H_{38}MnNO_4$

It explodes at 80–90°C.
See entry QUATERNARY OXIDANTS

3828. Acetonitrileimidazole-5,7,7,12,14,14-hexamethyl-1,4,8,11-tetraaza-4,11-cyclotetradecadieneiron(II) perchlorate
[33691-99-3] $C_{21}H_{39}Cl_2FeN_7O_8$

See entry [14] DIENE-N$_4$ COMPLEXES

3829. [1,1′-Oxybis[ethane]]tris(pentafluorophenyl)aluminium (Tris(pentafluorophenyl)aluminium etherate)
[14524-44-8] $C_{22}H_{10}AlF_{15}O$

Pohlmann, J. L. W. *et al., Z. Naturforsch.*, 1965, **20b**(1), 5

This compound sometimes explodes on heating.

See other METAL DERIVATIVES OF ORGANOFLUORINE COMPOUNDS

3830. Bis(cyclopentadienyl)bis(pentafluorophenyl)zirconium
[12097-97-1]

$C_{22}H_{10}F_{10}Zr$

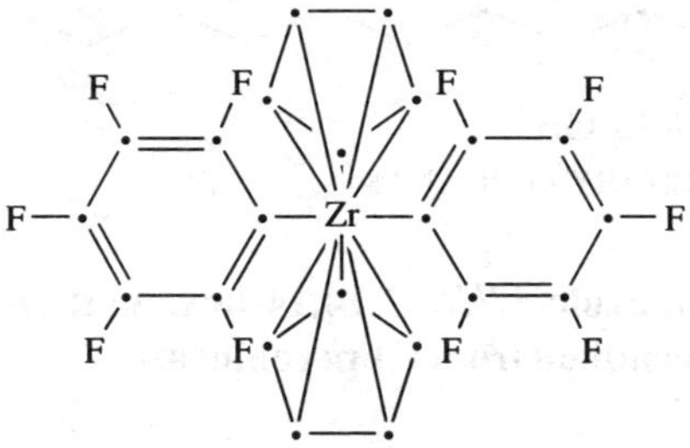

Chaudhari, M. A. *et al., J. Chem. Soc. (A)*, 1966, 838

It explodes in air (but not nitrogen) above the m.p., 219°C.

See other HALO-ARYLMETALS, ORGANOMETALLICS

3831. Di-1-naphthoyl peroxide
[29903-04-6]

$C_{22}H_{14}O_4$

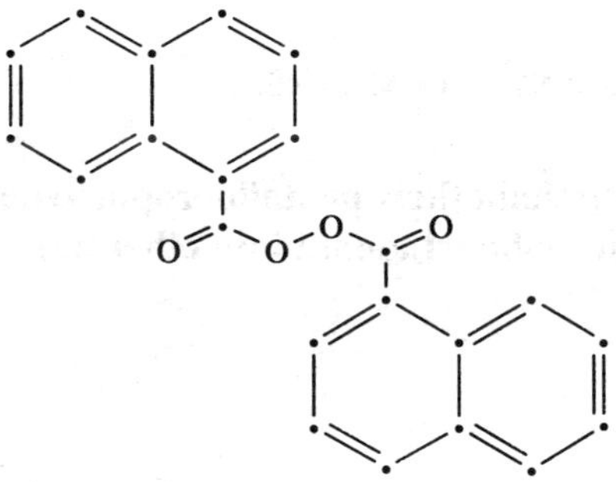

Castrantas, 1965, 17

It explodes on friction.

See other DIACYL PEROXIDES

3832. Tetraspiro[2.2.2.2.2.2.4]docosa-4,9,14,19,21pentyne
[159148-53-5] $C_{22}H_{16}$

Scott, L. T. *et al., J. Amer. Chem. Soc.*, 1994, **116**(22), 10275
This compound melted at 155°C with decomposition, one sample exploded in an oil bath at 192°C.
See also ALKYNES, *See also* ROTANES

3833. Bis(2-methyl-1,8-naphthyridine)tetracarbonyldirhodium(I) perchlorate
[90635-53-3] $C_{22}H_{16}Cl_2N_4O_{12}Rh_2$

See Bis(1,8-naphthyridine)tetracarbonyldirhodium(I) perchlorate
See related CARBONYLMETALS, AMMINEMETAL OXOSALTS

3834. Bis(benzeneiron)–fulvalenediyl complex
[78398-48-8] $C_{22}H_{20}Fe_2$

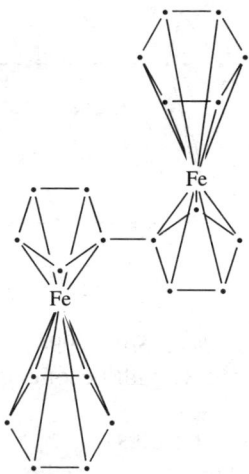

Desbois, M. H. *et al.*, *J. Amer. Chem. Soc.*, 1985, **107**, 5280–5282
Unstable above −10°C, it burns explosively on contact with air.
See other ORGANOMETALLICS

3835. Bis(dicarbonylcyclopentadienyliron)–bis(tetrahydrofuran)magnesium
[55800-08-3] $C_{22}H_{26}Fe_2MgO_6$

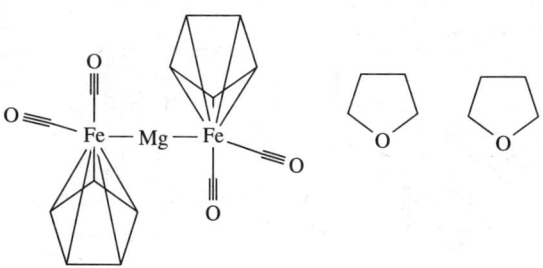

McVicker, G. B., *Inorg. Synth.*, 1976, **16**, 56–58
This, like other transition metal carbonyl derivatives of magnesium, is pyrophoric.
See other CARBONYLMETALS, ORGANOMETALLICS

**3836. 4-[2-(4-Hydrazino-1-phthalazinyl)hydrazino]-4-methyl-2-pentanone
(4-hydrazino-1-phthalazinyl)hydrazonedinickel(II) tetraperchlorate
[9018-45-5]** $\qquad$ $C_{22}H_{28}N_{12}Ni_2Cl_4O_{16}$

Rosen, W., *Inorg. Chem.*, 1971, **10**, 1833

The green polymeric complex isolated directly from the reaction mixture explodes fairly violently at elevated temperatures. The tetrahydrate produced by recrystallisation from aqueous methanol is also moderately explosive when heated rapidly.

See other AMMINEMETAL OXOSALTS

See related TETRAAZAMACROCYCLANEMETAL PERCHLORATES

**3837. Bis(trimethylphosphine)di(3,5-dibromo-2,6-dimethoxyphenyl)nickel
[83459-96-5]** $\qquad$ $C_{22}H_{32}Br_4NiO_4P_2$

Wada, S. *et al.*, *J. Chem. Soc., Dalton Trans.*, 1982, 1446

The complex decomposed explosively at 194°C.

See related ALKYLPHOSPHINES, HALO-ARYLMETALS

3838. 4,4'-(Butadiyne-1,4-diyl)bis(2,2,6,6-tetramethyl-4-hydroxypiperidine-1-oxyl)
[14306-88-8] $C_{22}H_{34}N_2O_4$

Miller, J. S. *et al., Chem. Mater.*, 1990, **2**(1), 60
Both crystalline forms decompose explosively to lower molecular weight products
at 140°C under nitrogen.
See related ALKYNES

3839. 2,4,6-Triphenylpyrilium perchlorate
[1484-88-4] $C_{23}H_{17}ClO_5$

Pelter, A., *Chem. Abs.*, 1981, **95**, 97643 (reference to *Tetrahedron Lett.* original is
incorrect)
Though not as sensitive as 1,3-benzodithiolium perchlorate, this salt is also explosive.
See other NON-METAL PERCHLORATES

3840. 1,3-Di[bis(cyclopentadienyl)iron]-2-propen-1-one
[12171-97-0] $C_{23}H_{20}Fe_2O$

Perchloric acid, Acetic anhydride, Cyclopentanone, Ether, Methanol
Anon., *Chem. Eng. News*, 1966, **44**(49), 50
Condensation of the iron complex with cyclopentanone in perchloric acid–acetic anhydride–ether medium had been attempted. The non-crystalline residue, after methanol washing and drying in air for several weeks, exploded on being disturbed. This was attributed to possible presence of a derivative of ferrocenium perchlorate, a powerful explosive and detonator. However, methyl or ethyl perchlorates alternatively may have been involved.
See Perchloric acid: Acetic anhydride, etc., or: Diethyl ether
See other ORGANOMETALLICS

3841. 3,3,5-Triphenyl-4,4-dimethyl-5-hydroperoxy-4,5-dihydro(3H)pyrazole
(4,5-Dihydro-4,4-dimethyl-3,5,5-triphenyl-3H-pyrazol-3-yl hydroperoxide)
[133610-01-2] $C_{23}H_{22}N_2O_2$

Baumstark, A. L, *et al., J. Heterocyclic Chem.*, 1991, **28**(1), 113
This, the most stable of 5 homologues, the others having one or two aryl substituents replaced by methyl, exploded at 90°C in a melting point determination. It appears that azohydroperoxy compounds are even more explosive than the sum of their parts might indicate.
See related α-PHENYLAZO HYDROPEROXIDES, ORGANIC PEROXIDES

3842. Tetrakis(pentafluorophenyl)titanium
[37759-21-0] $C_{24}F_{20}Ti$

Houben-Weyl, 1975, Vol. 13.3, 300
It explodes at 120–130°C.
See other HALO-ARYLMETALS

3843. 1,7,13,19-Cyclotetracosatetraene-3,5,9,11,15,17,21,23-octayne
(Octahydro[24]annulene)
[30047-26-8]

$C_{24}H_8$

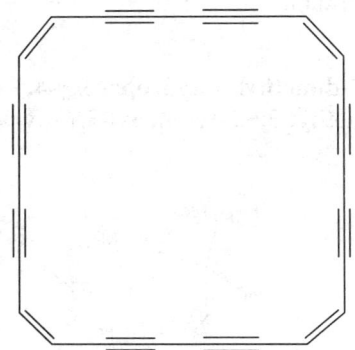

Diederich, F. *et al., Angew. Chem. (Int.)*, 1992, **31**(9), 1101
Explodes on heating.
See other ALKYNES

3844. Iron(III) phthalate
[52118-12-4]

$C_{24}H_{12}Fe_2O_{12}$

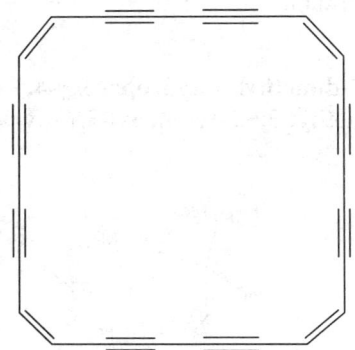

Sulfur compounds
See Phthalic anhydride (reference 2)

3845. 1,5-Dibenzoylnaphthalene
[83-80-7] $C_{24}H_{16}O_2$

Preparative hazard
See Aluminium chloride: Benzoyl chloride, etc.

3846. Aquafluorobis(1,10-phenanthroline)chromium(III) perchlorate
[77812-44-3] $C_{24}H_{18}Cl_2CrFN_4O_{10}$

Delover, M. *et al., J. Chem. Soc., Dalton Trans.*, 1981, 985
Its explosive properties prevented analysis.
See other AMMINEMETAL OXOSALTS

3847. 1(2′-Diazoniophenyl)2-methyl-4,6-diphenylpyridinium diperchlorate
[55358-25-3] $C_{24}H_{19}Cl_2N_3O_8$

Dorofenko, G. N. *et al.*, *Chem. Abs.*, 1975, **82**, 139908

The title compound, its 3'- and 4'-isomers, and their 2,4,6-triphenyl analogues, exploded on heating.

See other DIAZONIUM PERCHLORATES

3848. Tetraphenylarsonium permanganate
[4312-28-1]   C₂₄H₂₀AsMnO₄

It explodes at 120–130°C.

See entry QUATERNARY OXIDANTS

3849. 1,3,6,8-Tetraphenyloctaazatriene
[ ]   C₂₄H₂₀N₈

Wohl, A. *et al.*, *Ber.*, 1900, **33**, 2741

This compound (and several *C*-homologues) is unstable and explodes sharply on heating, impact or friction. In a sealed tube, the explosion is violent.

See other HIGH-NITROGEN COMPOUNDS

3850. Tetraphenyllead
[595-89-1]   C₂₄H₂₀Pb

Potassium amide
See Potassium amide: Tetraphenyllead

Sulfur
Houben-Weyl, 1975, Vol. 13.3, 236
Interaction may be explosive.
See other ARYLMETALS

3851. Bis(benzene)chromium dichromate
[ ] $C_{24}H_{24}Cr_4O_7$

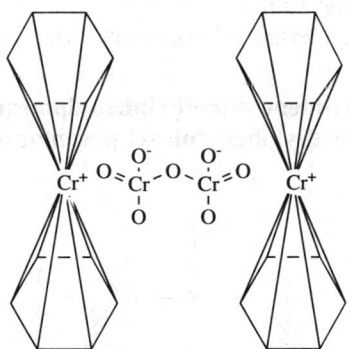

Anon., *Chem. Eng. News*, 1964, **42**(38), 55
This catalyst exists as explosive orange-red crystals.
See other ORGANOMETALLICS *See related* METAL OXOMETALLATES

3852. 3,3,6,6-Tetraphenylhexahydro-3,6-disilatetrazine
[13272-05-4] $C_{24}H_{24}N_4Si_2$

See entry SILYLHYDRAZINES

3853. Bis(*O*-salicylidenaminopropylaziridine)iron(III) perchlorate
[79151-63-6] $C_{24}H_{30}ClFeN_4O_6$

Federer, W. D. *et al., Inorg. Chem.*, 1984, **23**, 3863

Heat- and mildly shock-sensitive, a sample of the complex exploded after hand-grinding in a mortar for 30 min.

See other AZIRIDINES, AMMINEMETAL OXOSALTS

3854. 2-Tetrahydrofuranylidene(dimethylphenylphosphine-trimethyl-phosphine)-2,4,6-trimethylphenylnickel perchlorate

[ ] $C_{24}H_{37}ClNiO_5P_2$

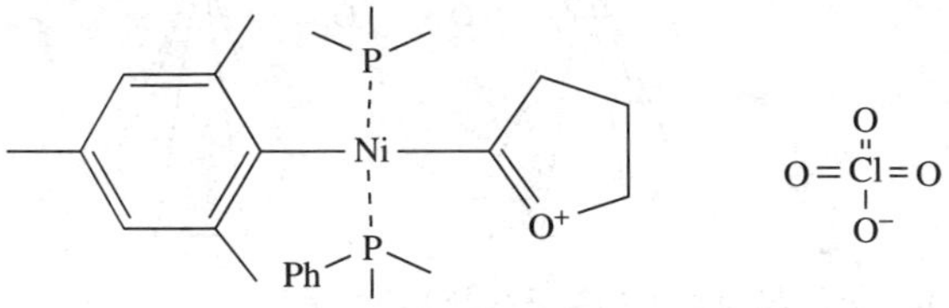

Wada, M. *et al., J. Chem. Soc., Dalton Trans.*, 1982, 794

It explodes on heating, like a tetrahydropyranylidene analogue.

See other ORGANOMETALLIC PERCHLORATES
See related ALKYLPHOSPHINES

3855. Di-μ-methylenebis(methylpentamethylcyclopentadienyl)dirhodium

[80410-45-3] $C_{24}H_{40}Rh_2$

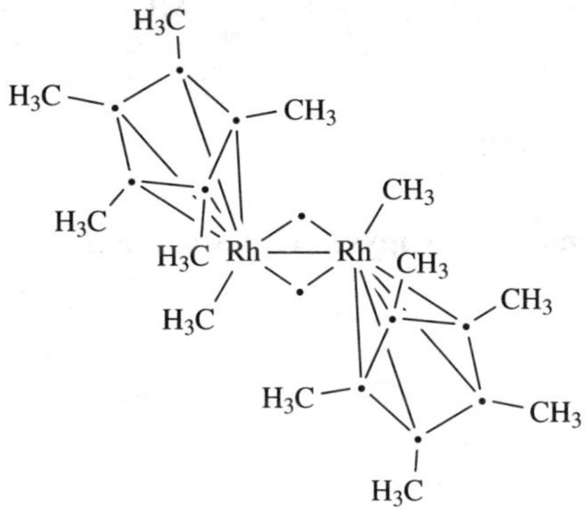

Preparative hazard

See Dichlorodi-μ-chlorobis(pentamethylcyclopentadienyl)dirhodium: Air, Alkyl-metals

See related ALKYLMETALS

3856. Tris(bis-2-methoxyethyl ether)potassium hexacarbonylniobate(1−)
[57304-94-6] $C_{24}H_{42}KNbO_{15}$

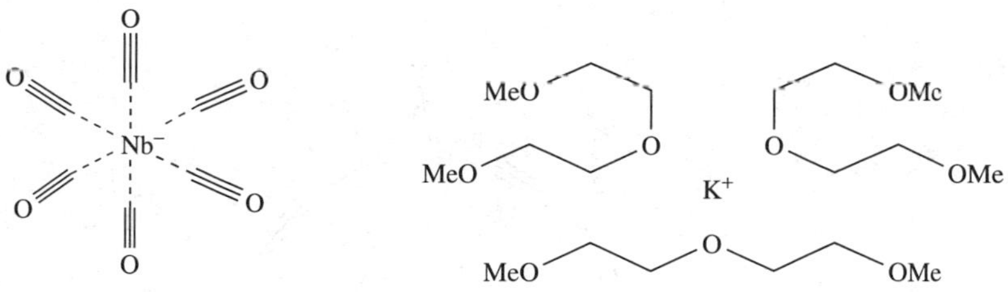

Solvents

Attempts to recover the complex from solutions in solvents other than diglyme led to the formation of pyrophoric solids.

See related CARBONYLMETALS

3857. Didodecanoyl peroxide (Dilauroyl peroxide)
[105-74-8] $C_{24}H_{46}O_4$

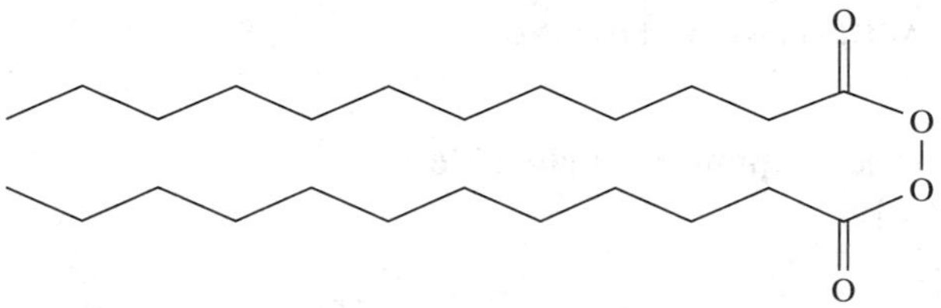

1. *Haz. Chem. Data*, 1975, 137
2. *CHETAH*, 1990, 183

Though regarded as one of the more stable peroxides, it becomes shock-sensitive on heating, and self-accelerating decomposition sets in at 49°C [1]. One of the recently calculated values of 46 and 42°C for induction periods of 7 and 60 days, respectively, for critical ignition temperatures is closely similar to that (45°/7 days) previously recorded. Autocatalytic combustion of the polymerisation initiator is exhibited. Although not ordinarily shock sensitive, it responds to a detonator [2].

Charcoal

Leleu, *Cahiers*, 1980, (99), 279

Mixtures react exothermally at 90°C, sometimes igniting. (Catalysis by trace metals in the carbon may well be involved).

See other DIACYL PEROXIDES

3858. Tetracyclo[20.2.0.0^{6,9}.0^{14,17}]tetracosa-1(22),6(9),14(17)-triene-2,4,10,12,18,20-hexayne-7,8,15,16,23,24-hexone
[123002-86-8]   C$_{24}$O$_6$

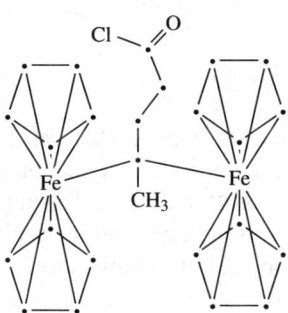

See Oligo(octacarbondioxide)
See other ACETYLENIC COMPOUNDS

3859. 4,4-Diferrocenylpentanoyl chloride
[56386-22-2]   C$_{25}$H$_7$ClFe$_2$O

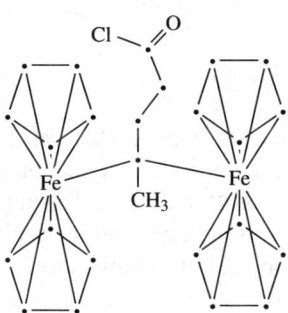

Nielsen, A. T. *et al., J. Org. Chem.*, 1976, **41**, 657
The acid chloride is thermally unstable much above 25°C. At 80–100°C it rapidly evolves hydrogen chloride and forms a black solid.
See other ACYL HALIDES, ORGANOMETALLICS

3860. 1,5-(or 1,8-)Bis(dinitrophenoxy)-4,8-(or 4,5-)dinitroanthraquinone
[ ] $C_{26}H_{10}N_6O_{16}$

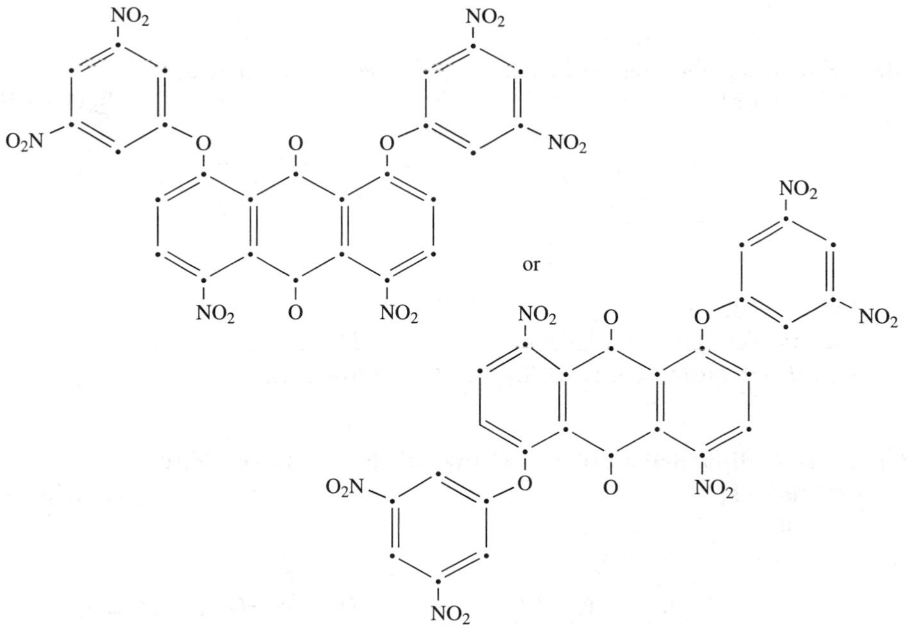

or

Fast flame propagation occurs on heating the powders moderately.
See entry HIGH RATE DECOMPOSITION

3861. Copper bis(1-benzeneazothiocarbonyl-2-phenyl-2-hydrazide)
[12213-13-7] $C_{26}H_{22}CuN_8S_2$

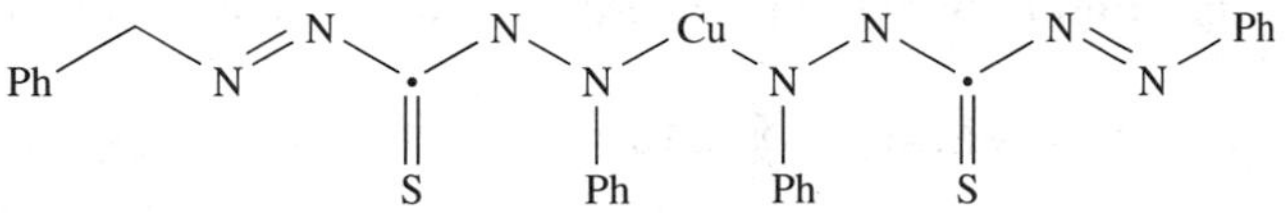

Sorbe, 1968, 80
The copper derivative of dithizone explodes at 150°C.
See other N-METAL DERIVATIVES, METAL HYDRAZIDES

3862. Lead bis(1-benzeneazothiocarbonyl-2-phenyl-2-hydrazide)
[21519-20-0] $C_{26}H_{22}N_8PbS_2$

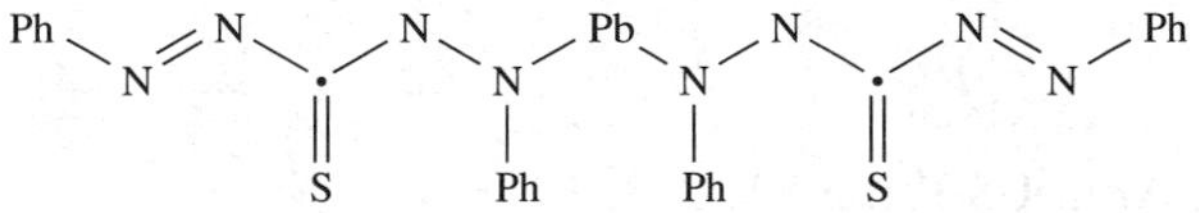

1289

Sorbe, 1968, 36
The lead derivative of dithizone explodes at 215°C.
See other N-METAL DERIVATIVES, METAL HYDRAZIDES

3863. Zinc bis(1-benzeneazothiocarbonyl-2-phenyl-2-hydrazide)
[21790-54-5] $C_{26}H_{22}N_8S_2Zn$

Sorbe, 1968, 159
The zinc derivative of dithizone explodes at 215°C.
See other N-METAL DERIVATIVES, METAL HYDRAZIDES

3864. 1,2-Bis(diphenylphosphino)ethanepalladium(II) perchlorate
[69720-86-1] $C_{26}H_{24}Cl_2O_8P_2Pd$

Davies, J. A. *et al., J. Chem. Soc., Dalton Trans.*, 1980, 2246–2249
Extremely sensitive to thermal and mechanical shock, occasionally detonating spontaneously without external stimulus, it should only be prepared and used in solution. The solid is an extreme hazard, and the acetone-solvated complex is also explosive.
See related AMMINEMETAL OXOSALTS

3865. Didodecyl peroxydicarbonate
[24356-04-5] $C_{26}H_{50}O_6$

D'Angelo, A. J., US Pat. 3 821 273, 1974

It decomposes violently after 90 min at 40°C, while the hexadecyl homologue is stable for a week at 40°C.

See other PEROXYCARBONATE ESTERS

3866. Methylaluminiumbis(pentamethylcyclopentadienyltrimethyliridium)
[ ] $C_{27}H_{51}AlIr_2$

$$MeAl(IrMe_3C_5Me_5)_2$$

Isobe, K. *et al., J. Chem. Soc., Dalton Trans.*, 1984, 932

When the complex in pentane at ambient temperature was deliberately exposed to air to produce pentamethylcyclopentadienyltetramethyliridium, it occasionally ignited. Using proper precautions it is not dangerous, but increase in scale above 0.4 mmol is not advised.

See related ALKYLMETALS *See other* ORGANOMETALLICS

3867. Tetrakis(pyridine)bis(tetracarbonylcobalt)magnesium
[51006-26-9] $C_{28}H_{20}Co_2MgN_4O_8$

$$[(C_5H_5N)_2Co(CO)_4]_2Mg$$

McVicker, G. B., *Inorg. Synth.*, 1976, **16**, 58–59

This, like other transition metal carbonyl derivatives of magnesium, is pyrophoric.

See related CARBONYLMETALS

3868. Bis[1,5-bis(4-methylphenyl)-1,3-pentaazadienato-N3,N5]-(T-4) cobalt
[113634-30-3] $C_{28}H_{28}CoN_{10}$

Schmid, R. *et al.*, *Z. Naturforsch. B*, 1987, **42B**, 911–916
The red complex is explosive.
See other HEAVY METAL DERIVATIVES, HIGH-NITROGEN COMPOUNDS, N-METAL
DERIVATIVES, PENTAAZADIENES

3869. 5,7,7,12,14,14-Hexamethyl-1,4,8,11-tetraaza-4,11,-cyclotetradecadiene-
1,10-phenanthrolineiron(II) perchlorate
[36732-77-1] $C_{28}H_{40}Cl_2FeN_6O_8$

See entry [14] DIENE-N_4 COMPLEXES

3870. 1(2′-, 3′-, or 4′-Diazoniophenyl)-2,4,6-triphenylpyridinium diperchlorate
[55358-23-1, 55358-29-7, 55358-33-3, resp.] $C_{29}H_{21}Cl_2N_3O_8$

Dorofenko, G. N. *et al.*, *Chem. Abs.*, 1975, **82**, 139908
The three diazonium perchlorate isomers, and their 2-methyl-4,6-diphenyl
analogues, all exploded on heating.
See other DIAZONIUM PERCHLORATES

3871. Sodium tris(O,O'-1-oximatonaphthalene-1,2-dione)ferrate (Pigment green B)
[16143-80-9] $C_{30}H_{18}FeN_3NaO_6$

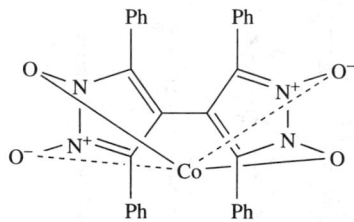

Piotrowski, T., *Chem. Abs.*, 1983, **99**, 159938
The pigment (containing below 1% of inorganic salts) may ignite or explode during milling or drying. The causes of ignition and impact sensitivity were determined and safe handling procedures are proposed.
See related OXIMES *See other* N–O COMPOUNDS

3872. Cobalt(II) chelate of bi(1-hydroxy-3,5-diphenylpyrazol-4-yl N-oxide)
[75027-71-3] $C_{30}H_{20}CoN_4O_4$

Battei, R. S. *et al., J. Inorg. Nucl. Chem.*, 1980, **42**, 494
The title cobalt chelate exploded at 175°C during DTA studies, 30° lower than the Co chelate of 'dicupferron'.
See Cobalt(II) chelate of 1,3-bis(N-nitrosohydroxylamino)benzene
See other N–O COMPOUNDS, N-OXIDES

3873. Tris(2,2′-bipyridine)silver(II) perchlorate

[ ] $C_{30}H_{24}AgCl_2N_6O_8$

Morgan, G. T. *et al., J. Chem. Soc.*, 1930, 2594
It explodes on heating.
See other AMMINEMETAL OXOSALTS

3874. Tris(2,2′-bipyridine)chromium(II) perchlorate
[15388-46-2] $C_{30}H_{24}Cl_2CrN_6O_8$

Holah, D. G. *et al., Inorg. Synth.*, 1967, **10**, 34
It explodes violently on slow heating to 250°C and can be initiated by static sparks,
but not apparently by impact.
See other AMMINEMETAL OXOSALTS, STATIC INITIATION INCIDENTS

3875. Tris(2,2′-bipyridine)chromium(0)
[14751-89-4] $C_{30}H_{24}CrN_6$

Herzog, S. *et al., Z. Naturforsch. B*, 1957, **12B**, 809
It ignites in air.
See other PYROPHORIC MATERIALS

3876. Benzenediazonium tetraphenylborate
[2200-13-7] $C_{30}H_{25}BN_2$

Sakakura, T. *et al., J. Chem. Soc., Perk. Trans. 1*, 1994, (3), 283
On two occasions this salt exploded during drying. The dry material is dangerous
to handle. In view of the high molecular weight, the diazonium function, alone,
represents a low energy per unit mass for explosivity
See other DIAZONIUM SALTS

3877. 2-Tetrahydropyranylidene-bis(dimethylphenylphosphine)-
3,4,6-trimethylphenyl-nickel perchlorate
[82647-51-6] $C_{30}H_{41}ClNiO_5P_2$

Wada, M. *et al., J. Chem. Soc., Dalton Trans.*, 1982, 794
It explodes on heating, like a tetrahydrofuranylidene analogue.
See other ORGANOMETALLIC PERCHLORATES
See related ALKYLPHOSPHINES

3878. 2,12,18,28-Tetramethyl-3,7,11,19,23,27,33,34-octaazatricyclo [27.3.1.1(13,17)]tetratriaconta-1(33),2,11,13,15,17(34)18,27,29,31-decaenetetraaquadiiron(II) tetraperchlorate
[94643-57-9] $C_{30}H_{52}Cl_4Fe_2N_8O_{20}$

Cabral, M. F. *et al., Inorg. Chim. Acta*, 1984, **90**, 170
The dinuclear complex exploded readily on slight friction.
See other POLYAZACAGED METAL PERCHLORATES

3879. Bis(μ_3-methylidyne) triangulo tris(pentamethylcyclopentadienylrhodium)
[83350-11-2] $C_{32}H_{47}Rh_3$

Vasquez de Miguel, A. *et al., Organometallics*, 1982, **1**, 1606

Preparation of hexamethyldialuminium in benzene, necessary for reaction with $(C_5H_5Rh)_2Cl_4$ to give the title compound, is hazardous and should only be attempted by experienced workers.

See other ORGANOMETALLICS

3880. 2,7,12,18,23,28-Hexamethyl-3,7,11,19,23,27,33,34-octaazatricyclo [27.3.1.1(13,17)]tetratriaconta-1(33),2,11,13,15,17(34),18,27,29,31-decaenediaquadichlorodiiron(II) diperchlorate
[94643-57-9] $C_{32}H_{52}Cl_4Fe_2N_8O_{10}$

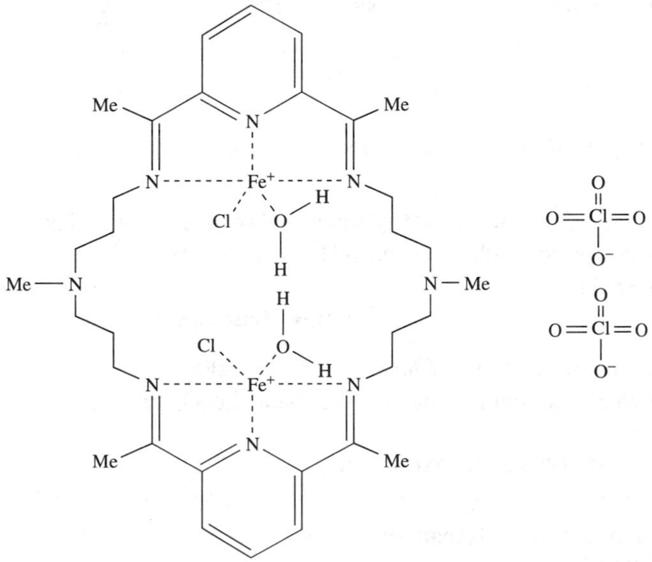

Cabral, M. F. *et al.*, *Inorg. Chim. Acta*, 1984, **90**, 170
The dinuclear complex exploded readily on slight friction.
See other POLYAZACAGED METAL PERCHLORATES

3881. Hydroxobis[5,7,7,12,14,14-hexamethyl-1,4,8,11-tetraaza-4,11-cyclotetradecadieneiron(II)] triperchlorate
[36913-76-5] $C_{32}H_{65}Cl_3Fe_2N_8O_{13}$

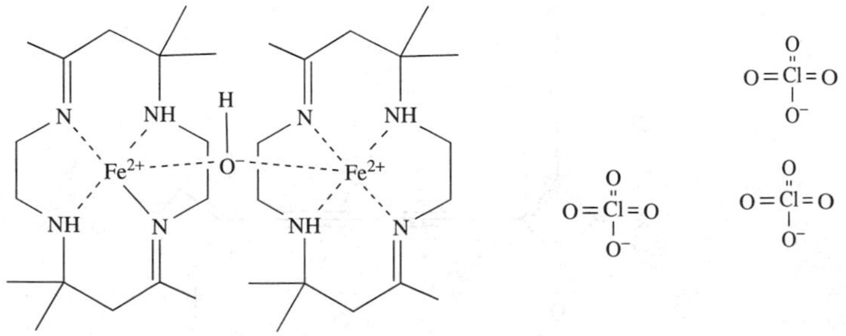

See other [14] DIENE-N₄ COMPLEXES

1297

3882. Oxybis[aqua-5,7,7,12,14,14-hexamethyl-1,4,8,11-tetraaza-4,11-cyclotetradecadieneiron(II)] tetraperchlorate
[36841-23-3]
$C_{32}H_{68}Cl_4Fe_2N_8O_{19}$

See other [14] DIENE-N$_4$ COMPLEXES

3883. Tetrakis(μ_3-2-amino-2-methylpropanolato)tetrakis(μ_2-2.amino-2-methylpropanolato)hexacopper(II) perchlorate
[100113-85-7]
$C_{32}H_{80}Cl_4Cu_6O_{24}$

Complex structure

Muhonen, H. *et al., Inorg. Chem.*, 1986, **25**, 800
The complex, containing a bicapped cubane Cu_6O_8 cluster, may explode if heated when dry.
See other AMMINEMETAL OXOSALTS

3884. Octacarbondioxide tetramer
[123002-87-9]
$C_{32}O_8$

Oligo(octacarbondioxide) *See other* ACETYLENIC COMPOUNDS

3885. Triferrocenylcyclopropenium perchlorate
    [56581-58-9]
                                                          $C_{33}H_{27}ClFe_3O_4$

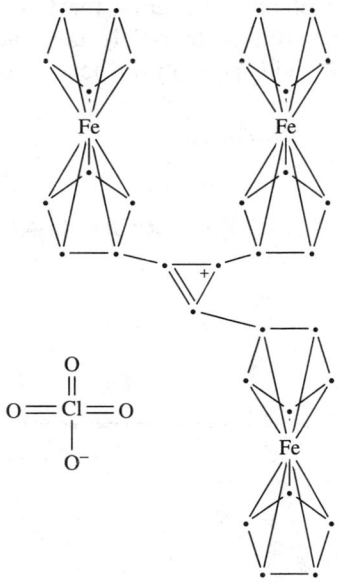

Agranat, I. *et al., Tetrahedron*, 1979, **3** 5, 738
The salt is thermally stable below 171°C, but may explode on friction at ambient
temperature.
See other FRICTIONAL INITIATION INCIDENTS, NON-METAL PERCHLORATES

3886. Tris(4-methoxy-2,2″bipyridine)ruthenium(II) perchlorate
    [ ]
                                                          $C_{33}H_{30}Cl_2N_6O_{11}Ru$

Thomas, R. N., *Univ. Safety Assoc. Safety News*, 1981, **15**, 16–17

Failure of the product (0.5 g) to crystallise out from the aqueous DMF reaction liquor led to vacuum evaporation of the solution at 60–70°C. During evaporation the mixture exploded violently, shattering the fume cupboard sash of toughened glass. The product may well be thermally unstable, but reaction of DMF with excess warm perchloric acid, possibly in near-absence of water, may also have been involved.

See other AMMINEMETAL OXOSALTS

3887. Pentaspiro[2.4.2.4.2.4.2.4.2.4]pentatriaconta-4,6,11,13,18,20,25,27,32,34-decayne
[164077-35-4] $C_{35}H_{20}$

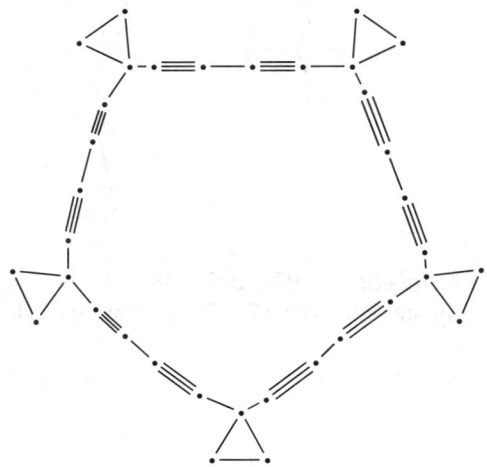

de Meijere, A. *et al., Chem. Eur. J.* 1995, **1**(2), 124

Like others of the series, it is explosive. A sample destroyed the apparatus during melting point determination.

See other ROTANES

1300

3888. 1,2:5,6:11,12:15,16-Tetrabenzocycloconta-1,5,11,15-tetraene-3,7,9,13,17,19-hexayne (1,2:5,6:11,12:15,16-Tetrabenzo-3,7,9,13,17,19-hexahydro[20]annulene)

[ ] $C_{36}H_{16}$

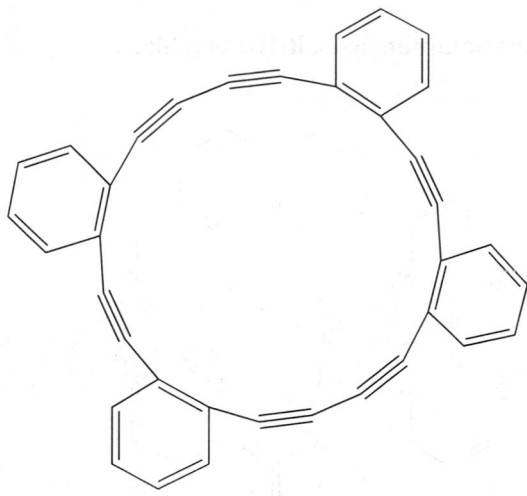

Matzger, A. J. *et al.*, *J. Amer. Chem. Soc.*, 1997, **119**, 2052
Explodes at 245°C to give various forms of carbon, hydrogen and methane and 1 kJ/g of heat.
See other ACETYLENIC COMPOUNDS

3889. Tris(1,10-phenanthroline)ruthenium(II) perchlorate
[27778-27-4] $C_{36}H_{24}Cl_2N_6O_8Ru$

Sulfuric acid

Gillard, R. D. *et al., J. Chem. Soc., Dalton Trans.*, 1974, 1235

Dissolution of the salt in the conc. acid must be very slow with ice-cooling to prevent an explosive reaction.

See other AMMINEMETAL OXOSALTS

3890. Tris(1,10-phenanthroline)cobalt(III) perchlorate
[14516-66-6]

$C_{36}H_{24}Cl_3CoN_6O_{12}$

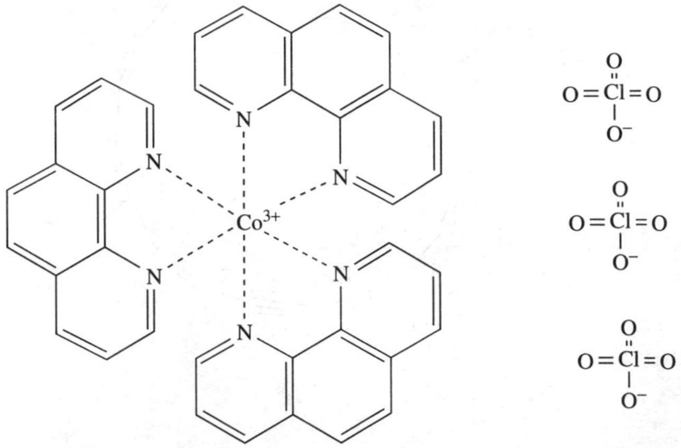

1. Hunt, H. R., *J. Chem. Educ.*, 1977, **54**, 710
2. Kildall, L. K., *J. Chem. Educ.*, 1978, **55**, 476
3. Hunt, H. R., *J. Chem. Educ.*, 1978, **55**, 476

The use of the perchlorate anion to precipitate the cobalt complex to determine the yield [1] is deprecated on the grounds of potential hazard [2], though it was not found possible to cause the salt to detonate by pounding, but it will burn if ignited. A spectroscopic assay method is suggested as an alternative to precipitation [3].

See other AMMINEMETAL OXOSALTS

3891. Hexaphenylhexaarsane
[20738-31-2]

$C_{36}H_{30}As_4$

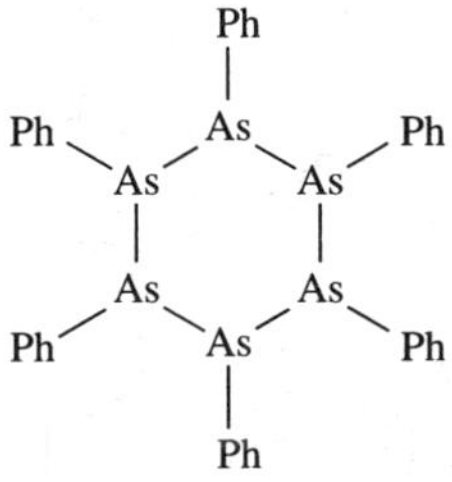

Oxygen

1. Maschmann, E., *Ber.*, 1926, **59**, 1143
2. Blicke, F. F. *et al.*, *J. Amer. Chem. Soc.*, 1930, **52**, 2946–2950

The explosively violent reaction of trimeric 'arsenobenzene' with oxygen [1] was later ascribed to the presence of various catalytic impurities, because the pure compound is stable towards oxygen [2].

See related ALKYLNON-METALS

3892. Lithium hexaphenyltungstate(2−)

[ ] $C_{36}H_{30}Li_2W$

$$Li_2[Ph_6W].3Et_2O$$

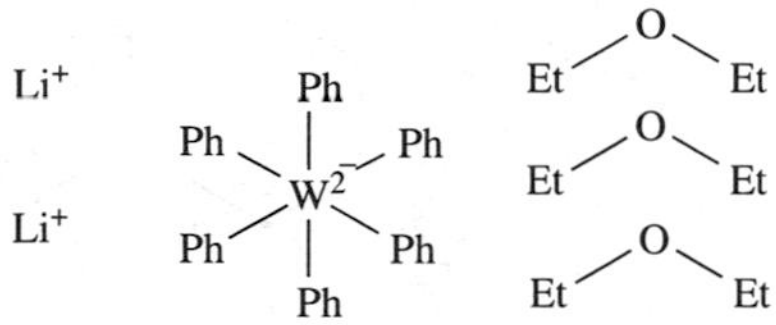

Sarry, B. *et al.*, *Z. Anorg. Chem.*, 1964, **329**, 218

The trietherate is pyrophoric in air.

See related ARYLMETALS

3893. Di[N,N'-Ethylenebis(2-oxidoacetophenoneiminato)copper(II)] oxovanadium(IV) diperchlorate

[93895-68-2] $C_{36}H_{36}Cl_2Cu_2N_4O_{13}V.3H_2O.\frac{1}{2}EtNO_2$

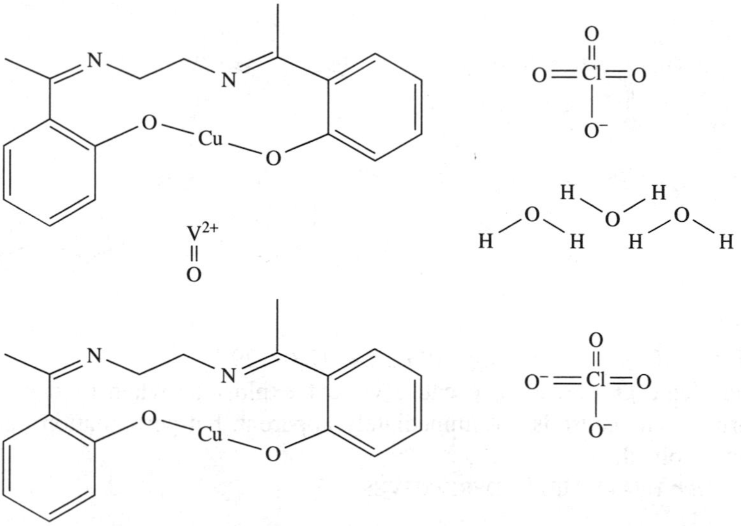

Bencini, A. *et al.*, *Inorg. Chem.*, 1985, **24**, 696

It is thermally unstable and may easily explode.

See related AMMINEMETAL OXOSALTS

3894. Hexacyclohexyldilead
[6713-82-2] $C_{36}H_{66}Pb_2$

Halocarbons
Houben-Weyl, 1975, Vol. 13.3, 168
In absence of other solvent and in presence of air, interaction with carbon tetrabromide is explosive, and with hexabromoethane, more so.
See related ALKYLMETALS

3895. Lead oleate
[1120-46-3] $C_{36}H_{66}O_4Pb$

Mineral oil
Williams, C. G., *Mech. Eng.*, 1932, **54**, 128–129
Lead oleate greases tend to cause violent explosions when used on hot-running bearings. The cause is not immediately apparent, but peroxidation may well have been involved.
See other HEAVY METAL DERIVATIVES

1304

3896. Calcium stearate
[1592-23-0] $C_{36}H_{70}CaO_4$

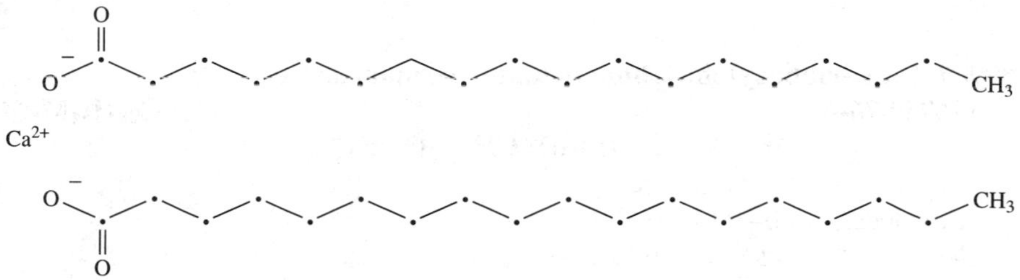

Ca^{2+}

1. Schmutzler, G. *et al.*, *Plaste Kaut.*, 1967, **14**, 827–829
2. Grossel, S. S., *J. Loss Prev. Proc. Ind.*, 1988, **1**, 62–74

Dust explosions and spontaneous ignition hazards for calcium stearate and related plastics additives are detailed and discussed [1]. A maximum pressure increase of 6.6 bar and a maximum rate of rise above 680 bar/s have been recorded [2].

See other DUST EXPLOSION INCIDENTS, AUTOIGNITION INCIDENTS

3897. Zinc stearate
[557-05-1] $C_{36}H_{70}O_4Zn$

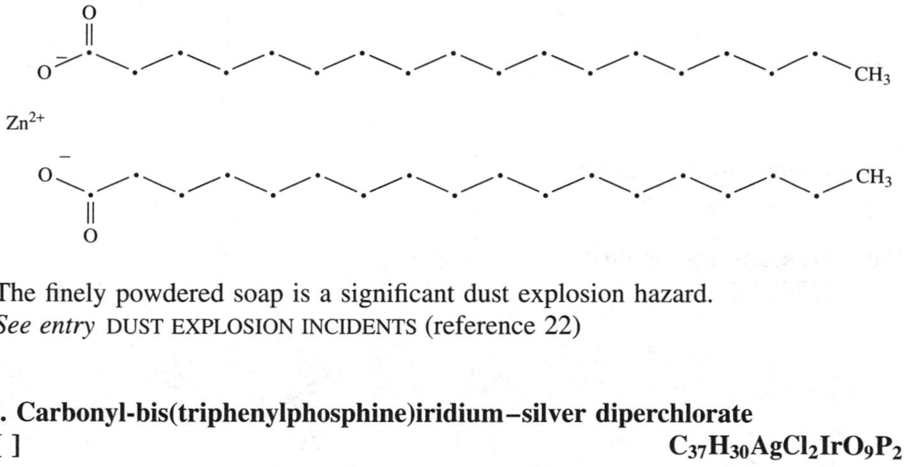

Zn^{2+}

The finely powdered soap is a significant dust explosion hazard.
See entry DUST EXPLOSION INCIDENTS (reference 22)

3898. Carbonyl-bis(triphenylphosphine)iridium–silver diperchlorate
[ ] $C_{37}H_{30}AgCl_2IrO_9P_2$

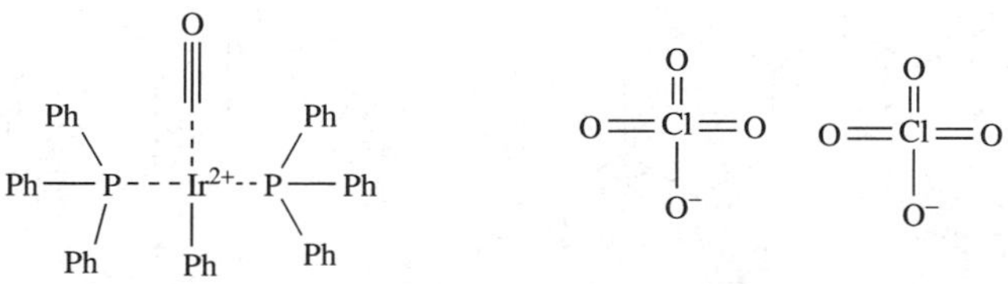

1305

Kuyper, J. *et al.*, *J. Organomet. Chem.*, 1976, **107**, 130
Highly explosive, detonated by heat or shock.
See related AMMINEMETAL OXOSALTS

3899. 1,2-Bis(triphenylphosphonio)ethane permanganate
[76710-76-4] $C_{38}H_{34}Mn_2O_8P_2$

$$(Ph_3P^+CH_2-)_2(MnO_4^-)_2$$

It explodes at 80–90°C.
See entry QUATERNARY OXIDANTS

3900. Calcium abietate
  [13463-98-4] $C_{40}H_{58}CaO_4$

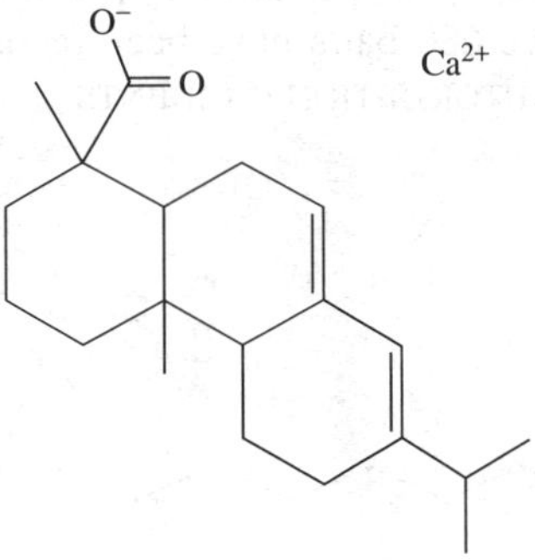

See entry METAL ABIETATES

3901. Manganese abietate
  [54675-76-2] $C_{40}H_{58}MnO_4$

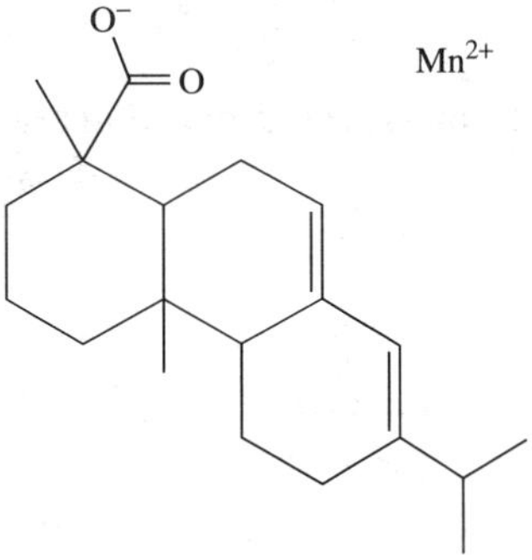

See entry METAL ABIETATES

3902. Lead abietate
[5434-72-0]

$C_{40}H_{58}O_4Pb$

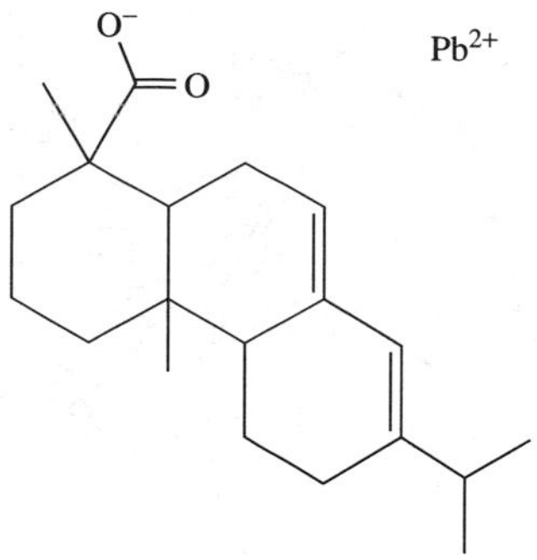

See entry METAL ABIETATES

3903. Zinc abietate
[6798-76-1]

$C_{40}H_{58}O_4Zn$

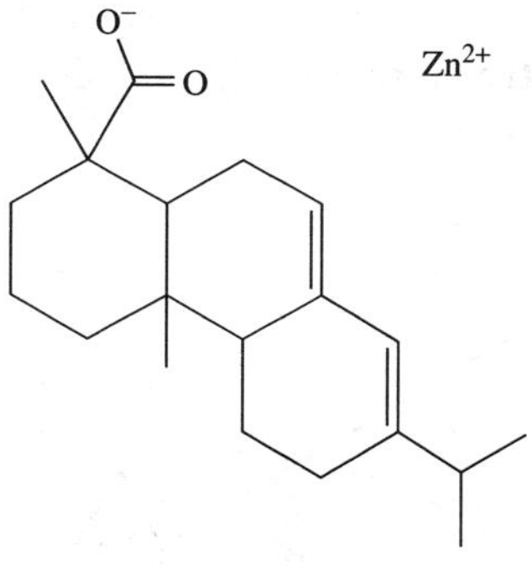

See entry METAL ABIETATES

3904. (Octacarbondioxide) pentamer

[ ] $C_{40}O_{10}$

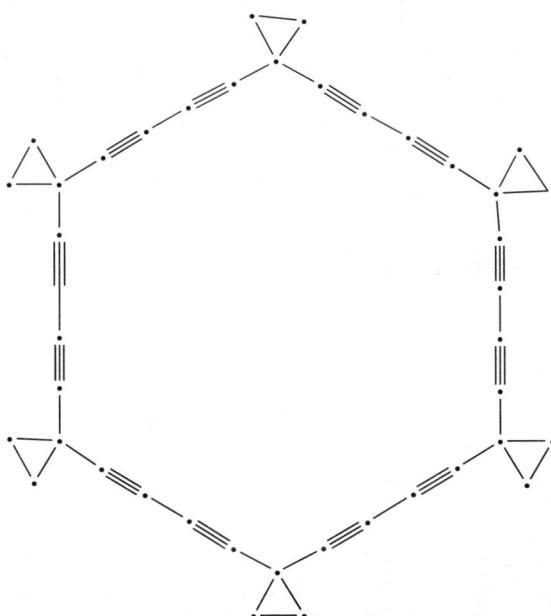

See Oligo(octacarbondioxide)
See other ACETYLENIC COMPOUNDS

3905. Hexaspiro[2.4.2.4.2.4.2.4.2.4.2.4]dotetraconta-
4,6,11,13,18,20,25,27,32,34,39,41-dodecayne
[155797-52-7] $C_{42}H_{24}$

1. de Meijere, A. *et al., Angew. Chem. (Int.)*, 1994, **33**(8), 869
2. de Meijere, A. *et al., Chem. Eur. J.* 1995, **1**(2), 124

The white solid explodes on friction or impact (photo on the cover of this *Angewandte*) and decomposes at 130°C. The instability is associated with the cyclopropane rings as it is claimed to be absent from the dodecamethyl homologue [1]. Calculation indicates 6 kJ/g available energy [2].

See other STRAINED-RING COMPOUNDS, ACETYLENIC COMPOUNDS, ROTANES

3906. Hexakis(pyridine)iron(II) tridecacarbonyltetraferrate(2−)
[23129-50-2] $C_{43}H_{30}Fe_5N_6O_{13}$

$$[(C_5H_5N)_6Fe]^{2+} [(OC)_{13}Fe_4]^{2-}$$

Brauer, 1965, Vol. 2, 1758
Extremely pyrophoric in air.
See related AMMINEMETAL OXOSALTS, CARBONYLMETALS

3907. 5,10,15,20-Tetrakis(2-nitrophenyl)porphine
[22834-73-8] $C_{44}H_{26}N_8O_8$

Preparative hazard
See 2-Nitrobenzaldehyde: Pyrrole
See other NITROARYL COMPOUNDS

3908. Tetrakis(4-N-methylpyridinio)porphinecobalt(III)(5+) perchlorate
    [53149-77-2] $C_{44}H_{36}Cl_5CoN_8O_{20}$

Pasternack, R. F. *et al., J. Inorg. Nucl. Chem.*, 1974, **36**, 600
A sample obtained by precipitation exploded.
See other AMMINEMETAL OXOSALTS

3909. Tetrakis(4-N-methylpyridinio)porphineiron(III)(5+) perchlorate
    [64365-00-0] $C_{44}H_{36}Cl_5FeN_8O_{20}$

Harris, F. L. *et al., Inorg. Chem.*, 1978, **17**, 71
A small sample exploded when scraped with a metal spatula on a glass frit.
See other AMMINEMETAL OXOSALTS

3910. 1,4,8,11,15,18,22,25-Octamethyl-29H,31H -
tetrabenzo[b.g.l.q]porphinato(2—)cobalt(II)
[27662-34-6] $C_{44}H_{36}N_4Co$

Preparative hazard
See 1,3,4,7-Tetramethylisoindole: Cobalt

3911. Oxybis(N,N -dimethylacetamidetriphenylstibonium) perchlorate
[37035-60-2] $C_{44}H_{50}Cl_2N_2O_{11}Sb_2$

Goel, R. G. *et al., Inorg. Chem.*, 1972, **11**, 2143
It exploded on several occasions during handling and attempted analysis.
See related AMMINEMETAL OXOSALTS

3912. Heptaspiro[2.4.2.4.2.4.2.4.2.4.2.4.2.4]nonatetraconta-4,6,11,13,18,20,25,27,32,34,39,41,46,48-tetradecayne
[168848-73-5]

$C_{49}H_{28}$

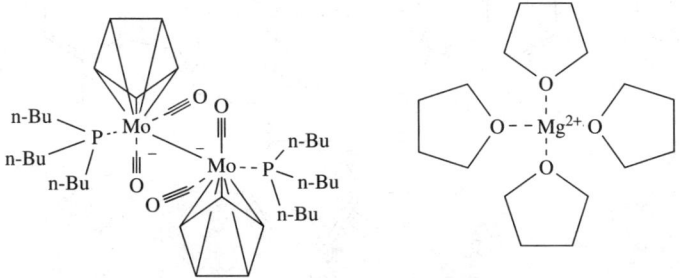

Cyclic

See other ROTANES

3913. Bis[dicarbonyl(cyclopentadienyl)tributylphosphinemolybdenum]–tetrakis(tetrahydrofuran)magnesium
[82148-84-3]

$C_{54}H_{96}MgMo_2O_8P_2$

McVicker, G. B., *Inorg. Synth.*, 1976, **16**, 59–61

This and other transition metal carbonyl derivatives of magnesium is pyrophoric.
See related CARBONYLMETALS *See other* ORGANOMETALLICS

3914. Aluminium stearate
[637-12-7]

$C_{54}H_{105}AlO_6$

The finely powdered soap is a significant dust explosion hazard.
See entry DUST EXPLOSION INCIDENTS (reference 22)

1312

3915. Octaspiro[2.4.2.4.2.4.2.4.2.4.2.4.2.4.2.4]hexapentaconta-4,6,11,13,18,20,25,27,32,34,39,41,46,48,53,55-hexadecayne
[164077-36-5] $C_{56}H_{32}$

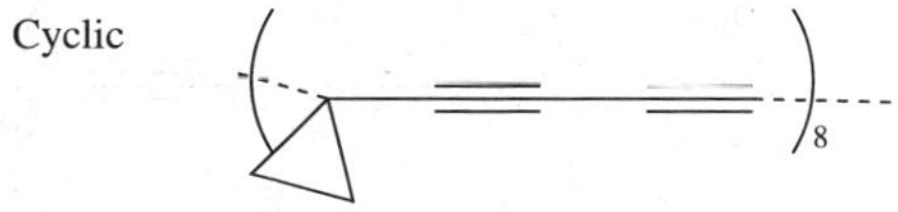

Cyclic

See other ROTANES

3916. Buckminsterfullerene
[99685-96-8] C_{60}
See Carbon

3917. Aluminium abietate
[32454-63-0] $C_{60}H_{87}AlO_6$

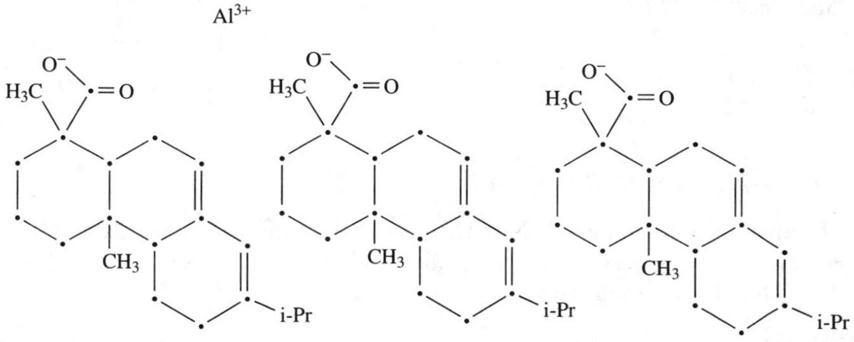

See entry METAL ABIETATES

3918. Nonaspiro[2.4.2.4.2.4.2.4.2.4.2.4.2.4.2.4.2.4]trihexaconta-4,6,11,13,18,20,25,27,32,39,41,46,48,53,55,60,62-octadecayne
[155797-58-3] $C_{63}H_{36}$

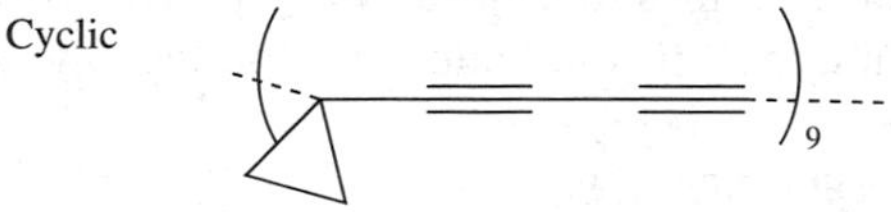

Cyclic

See other ROTANES

3919. Fullerene
[ ] C_{70}
See Carbon

3920. Decaspiro[2.4.2.4.2.4.2.4.2.4.2.4.2.4.2.4.2.4]heptaconta-4,6,11,13,18,20,25,27,32,39,41,46,48,53,55,60,62,67,69-eicosayne [168848-75-7]

$C_{70}H_{40}$

Cyclic

See other ROTANES

3921. Dodecaspiro[2.4.2.4.2.4.2.4.2.4.2.4.2.4.2.4.2.4.2.4]tetraoctaconta-4,6,11,13,18,20,25,27,32,39,41,46,48,53,55,60,62,67,69,74,76,81,83-tetracosayne [168848-76-8]

$C_{84}H_{48}$

Cyclic

See other ROTANES

3922. Calcium
[7440-70-2]

Ca

Ca

FPA H121, 1983; *HCS 1980*, 262

'Product Information Sheet No. 212', Sandwich, Pfizer Chemicals, 1969
Calcium is pyrophoric when finely divided.
See other PYROPHORIC METALS

Alkalies
Merck, 1983, 227
Interaction with alkali-metal hydroxides or carbonates may be explosive.

Ammonia
1. Partington, 1967, 369
2. Gibson, 1969, 369
At ambient temperature, ammonia gas reacts exothermally with calcium, but if warmed the latter becomes incandescent [1]. The metal dissolves unchanged in liquid ammonia, but if the latter evaporates, the finely divided metal is pyrophoric [2].
See other PYROPHORIC METALS

Asbestos cement
Scott, P. J., *School Sci. Rev.*, 1967, **49**(167), 252
Drops of molten calcium falling on to hard asbestos cement sheeting caused a violent explosion which perforated the sheet. Interaction with sorbed water seems likely to have occurred.
See Water, below

1314

Carbon dioxide, Nitrogen
See Carbon dioxide: Metals, Nitrogen

Chlorine fluorides
See Chlorine trifluoride: Metals
    Chlorine pentafluoride: Metals

1,2-Diaminoethane
Benkeser, R. A. *et al., Tetrahedron Lett.*, 1984, **25**, 2089–2092
The use of calcium in 1,2-diaminoethane as a safer substitute for sodium or lithium in liquid ammonia for the improved Birch reduction of aromatic hydrocarbons is described in detail.

Dinitrogen tetraoxide
See Dinitrogen tetraoxide: Metals

Halogens   MRH Chlorine 7.15/64
Mellor, 1941, Vol. 3, 638
Massive calcium ignites in fluorine at ambient temperature, and finely divided (but not massive) calcium ignites in chlorine.

Lead chloride
Mellor, 1941, Vol. 3, 369
Interaction is explosive on warming.

Mercury
Pascal, 1958, Vol. 4, 290
Amalgam formation at 390°C is violent.

Other reactants
Yoshida, 1980, 98
MRH values calculated for 16 combinations, largely with oxidants, are given.

Phosphorus(V) oxide
See Tetraphosphorus decaoxide: Metals

Silicon
Mellor, 1940, Vol. 6, 176–177
Interaction is violently incandescent above 1050°C after a short delay.

Sodium, Mixed oxides
Anon., *BCISC Quart. Safety Summ.*, 1966, **37**(145), 6
An operator working above the charging hole of a sludge reactor was severely burned when a quantity of burning sludge containing calcium and sodium metals and their oxides was ejected. This very reactive mixture is believed to have been ignited by drops of perspiration falling from the operator.
See Water, below

Sulfur   MRH 6.69/44
Mellor, 1941, Vol. 3, 639
A mixture reacts explosively when ignited.

Sulfur, Vanadium(V) oxide
See Vanadium(V) oxide: Calcium, etc.

Water
'Product Information Sheet No. 212', Sandwich, Pfizer Chemicals, 1969
Calcium or its alloys react violently with water (or dilute acids) and the heat of
reaction may ignite evolved hydrogen under appropriate contact conditions.
See ASBESTOS CEMENT, ABOVE *See other* METALS

3923. Calcium chloride
[10043-52-4] $CaCl_2$

$$CaCl_2$$

HCS 1980, 247

Methyl vinyl ether
See Methyl vinyl ether: Acids

Water
1. *MCA Case History No. 69*
2. *MCA Case History No. 2300*
The exotherm produced by adding solid calcium chloride to hot water caused
violent boiling [1], and the same happened 25 years later [2].

Zinc
Anon., *ABCM Quart. Safety Summ.*, 1932, **3**, 35
Prolonged action of calcium chloride solution upon the zinc coating of a galvanised
iron vessel caused slow evolution of hydrogen, which became ignited and exploded.
See other CORROSION INCIDENTS, GAS EVOLUTION INCIDENTS
See other METAL HALIDES

3924. Calcium hypochlorite
[7778-54-3] $CaCl_2O_2$

$$Ca(OCl)_2$$

FPA H46, 1976; *HCS 1980*, 269

1. Sidgwick, 1950, 1217
2. Clancey, V. J., *J. Haz. Mat.*, 1975, **1**, 83–94
3. Cave, R. F., *Chem. in Austr.*, 1978, **45**, 313
4. Uehara, Y. *et al.*, *Chem. Abs.*, 1979, **91**, 159834
5. Uehara, Y. *et al.*, *Chem. Abs.*, 1979, **91**, 180693
6. Uehara, Y. *et al.*, *Chem. Abs.*, 1980, **92**, 220207
7. Clancey, V. J., *Hazards of Pressure*, IChE Symp. Ser. No. 102, 11–23, Oxford, Pergamon, 1987
8. Wejtowicz, J. A., *ibid.*, 323–351
This powerful oxidant is technically of great importance for bleaching and steril-
isation applications, and contact with reducants or combustible materials must be

under controlled conditions. It is present in diluted state in bleaching powder, which is a less powerful oxidant with lower available chlorine content [1]. There have been several instances of mild explosions and/or intense fires on ships carrying cargoes of commercial hypochlorite, usually packed in lacquered steel drums and of Japanese origin. Investigational work was still in progress, but preliminary indications suggested that presence of magnesium oxide in the lime used to prepare the hypochlorite may have led to the presence of magnesium hypochlorite which is known to be of very limited stability. Unlike many other oxidants, the hypochlorite constitutes a hazard in the absence of other combustible materials, because after initiation, local rapid thermal decomposition will spread through the contained mass of hypochlorite as a vigorous fire which *evolves*, rather than consumes, oxygen [2]. Thermal decomposition with evolution of oxygen and chlorine sets in around 175°C and will propagate rapidly through the whole bulk.

Several instances of containers or drums igniting or erupting on reopening after previous use have been reported, and may well have involved formation by slow hydrolysis of a relatively high concentration of dichlorine monoxide in the containers. This may have been caused to decompose from sudden exposure to light, by friction on opening the drum, or by a static spark. In one incident, a dull (gas) explosion was heard immediately before the solid erupted. Dry storage is essential [3]. Extrapolation from small-scale determinations of critical thermal ignition temperatures to commercial dimensions (35.4 or 38.2 cm diameters) gave values of 76.3 and 75°, as compared with experimental values of 77 and 75°C, respectively [4]. Fire and explosion hazards of high strength (70%) calcium hypochlorite were studied, with measurements of thermal properties, ignition temperatures, shock-sensitivity, flame propagation rates and explosive power for mixtures with organic materials [5]. A review of the results, and their comparison with predictions from computer simulation has been published [6]. Further possible causes of the instability of commercial 70% calcium hypochlorite are discussed [7], but it is held that the material, if stored and used correctly, is a relatively stable product in the absence of impurities [8]. The recently calculated value of 75°C for criticial ignition temperature agrees with the previously recorded value of 76°C.

See BLEACHING POWDER

See related FRICTIONAL INITIATION INCIDENTS, IRRADIATION DECOMPOSITION INCIDENTS, STATIC INITIATION INCIDENTS

Acetylene
Rüst, 1948, 338
Contact of acetylene with bleaching powder, etc., may lead (as with chlorine itself) to formation of explosive chloroacetylenes. Application of hot water to free a partially frozen acetylene purification system which contained bleaching powder caused a violent explosion to occur.
See related HALOACETYLENE DERIVATIVES

Acetic acid, Potassium cyanide
Pitt, M. J., private comm., 1984

A jug containing calcium hypochlorite (probably as moist solid) was used as a disposal receptacle for cyanide wastes from student preparations of benzoin. When a little acetic acid residue was inadvertently added, an explosion occurred, attributed to a cyanide–chlorine redox reaction.
See other REDOX REACTIONS

Algaecide
Cave, R. F., *Chem. in Austr.*, 1978, **45**, 314
When algaecide (possibly a conc. solution of a quaternary ammonium salt) was spilt on a drum of swimming pool sterilant, it ignited and flared up immediately.

Ammonium chloride   MRH 1.05/27
1. Morris, D. L., *Science Teach.*, 1968, **35**(6), 4
2. Anderson, M. B., *Science Teach.*, 1968, **35**(9), 4
3. Rai, H., *Chem. News*, 1918, **117**, 253
When calcium hypochlorite was inadvertently used in place of calcium hydroxide to prepare ammonia gas, an explosion occurred [1], attributed to formation of nitrogen trichloride [2]; this was early claimed to be one of the most efficent methods of preparing the trichloride [3].
See Nitrogenous bases, below

Calcium carbide
Saville, B. W., 1988, private communication
A mixture explodes violently on addition of a drop of water.

Carbon
Mellor, 1941, Vol. 2, 262
A confined intimate mixture of hypochlorite and finely divided charcoal exploded on heating.

Carbon tetrachloride
491M, 1975, 84
Use of a carbon tetrachloride extinguisher on a fire in an open container of hypochlorite caused a severe explosion.
See Oxygen (Gas): Halocarbons

Contaminants
1. *MCA Case History No. 666*
2. Cave, R. F., *Chem. in Austr.*, 1971, **45**, 314
The contents of a drum erupted and ignited during intermittent use. This was attributed to contamination of the soldered metal scoop (normally kept in the drum) by oil, grease or water, or all three, and subsequent exothermic reaction with the oxidant [1]. Use of a bucket, possibly contaminated with traces of engine oil, to transport the oxidant led to eruption and ignition of the contents [2].
See Iron oxide, below; *and* Organic matter, below

N,N-Dichloromethylamine
See N,N-Dichloromethylamine: Calcium hypochlorite, etc.

Hydrocarbons
  Young, J. A., *CHAS Notes*, 1992, **X**(5), 3
  'Gasoline' inflames some minutes after being poured onto the hypochlorite. A fire
  consequent upon transporting the two in adjacent containers in a car is reported.
  The editor imagines that non-hydrocarbon fuel additives may also be involved.
  See Organic matter, etc., below

Hydroxy compounds   MRH values below reference
  1. Fawcett, H. H., *Ind. Eng. Chem.*, 1959, **51**(4), 90A
  2. Kirkbride, K. P. *et al.*, *J. Forensic Sci.*, 1991, **36**(3), 902
  MRH values: Glycerol 13/12, phenol 2.26/7, ethanol 2.30/7, methanol 2.34/10
  Contact of the solid hypochlorite with glycerol, diethylene glycol monomethyl
  ether or phenol causes ignition within a few min, accompanied by irritant smoke,
  especially with phenol (formation of chlorophenols). Ethanol may cause an explosion, as may methanol, undoubtedly owing to formation of the alkyl hypochlorites
  [1]. Reaction with polyethylene glycol hydraulic fluid may produce ignition, with
  a fireball. A mechanism for creation of combustible gases is proposed [2].

Iron oxide (rust)
  Gill, A. H., *Ind. Eng. Chem.*, 1924, **16**, 577
  Presence of rust in metal containers has caused many accidental explosions. Metal
  oxides catalyse the oxygen-evolving decomposition of the oxidant.

Nitrogenous bases   MRH Ethylamine 2.92/15
  Kirk-Othmer, 1963, Vol. 2, 105
  Primary aliphatic or aromatic amines react with calcium (or sodium) hypochlorite
  to form *N*-mono- or di-chloroamines which are explosively unstable, but less so
  than nitrogen trichloride.

Nitromethane   MRH 5.69/99+
  Fawcett, H. H., *Trans. Nat. Safety Congr. Chem. Fertilizer Ind.*, Vol. 5, 32,
  Chicago, NSC, 1963
  They interact, after a delay, with extreme vigour.

Oil, Polyethylene film, Water, Wood
  1. Armstrong, B. S., *Loss Prev. Bull.*, 1988, (084), 25–31
  2. Sampson, R. G., private comm., 1989
  Impact damage occurred to one of 48 10 kg polypropylene tubs of calcium
  hypochlorite, each closed with a polythene lid, all supported on a wooden pallet
  and stretch-wrapped round the base with polyethylene film. The spilled material
  was swept up and discarded, and the pallet was moved to another part of the
  warehouse, and some 30 mins. later flames were seen at the base of the pallet
  [1]. Ignition was attributed to contact between residual solid hypochlorite on the
  pallet and lubricant drips on the concrete floor from fork lift traffic. Application
  of a dry powder extinguisher had little effect on the fire, but use of a 40 l water
  extinguisher led to rapid escalation of the fire as contact between hypochlorite

and wood was increased. Flash-over occurred and the fire eventually involved a considerable fraction of the 1400 t of assorted warehouse contents [2].
See Organic matter, etc., below
See other IGNITION SOURCES

Other reactants
Yoshida, 1980, 133
MRH values calculated for 11 combinations with oxidisable materials are given.

Organic matter
1. 'Halane' Information Sheet, Wyandotte Chem. Co., Michigan, 1958
2. Weicherz, J., *Chem. Ztg.*, 1928, **52**, 729
3. Cave, R. F., *Chem. in Austr.*, 1978, **45**, 313
Mixtures of the solid hypochlorite with 1% of admixed organic contaminants are sensitive to heat in varying degree. Wood caused ignition at 176°, while oil caused violent explosion at 135°C [1]. When solid hypochlorite was transferred to a bucket which had been cleaned out with an alcoholic cleaning composition, it ignited violently. This could not be repeated, but it was found that many organic liquids and fibrous materials caused ignition if warmed to 100°C with the oxidant [2]. Heating a little solid oxidant in a lightly greased crucible led to violent ignition around 180°C [3].
See reference 5 above; Contaminants, above; *and* Turpentine, below

Organic matter, Water
1. Tatara, S. *et al.*, Ger. Offen., 2 450 816, 1975
2. Lawrence, S., Internet, 1996
   Calcium hypochlorite containing over 60% active chlorine normally ignites in contact with lubricating oils, but addition of 16–22% of water will prevent this. Examples show the effect of 18% (but not 15%) of added water in preventing ignition when glycerol was dripped onto a hypochlorite (79% active chlorine) containing 2% of oil [1]. Ignition with brake fluid is also reported, in this case reaction is accelerated by cola soft drinks — which are dilute acids rather than just water [2].

Organic sulfur compounds
1. Stephenson, F. G., *Chem. Eng. News*, 1973, **51**(26), 14
2. Laboratory Waste Disposal Manual, 142, Washington, MCA, 1969–1972 edns (withdrawn in 1980)
3. *MCA Guide for Safety in the Chemical Laboratory*, 464, New York, Van Nostrand-Reinhold, 1972
4. *491M*, 1975, 85
5. Wacker, J. V. *et al., Rec. Trav. Chim.*, 1986, **105**(3), 99–107
Contact of the solid oxidant with organic thiols or sulfides may cause a violent reaction and flash fire [1]. This procedure was recommended formerly for treating spills of sulfur compounds [2,3], but is now withdrawn as potentially hazardous. Use of an aqueous solution of up to 15% concentration, or of 5% sodium hypochlorite solution is recommended [1]. Addition of 10 g of oxidant to 5 ml portions of

1-propanethiol or isobutanethiol led to explosions [4]. Application of factorial design techniques to experimental planning gave specific conditions for the safe oxidation of organic sulfides to sulfoxides using calcium hypochlorite or sodium chlorate [5].
See Sulfur, below

Sodium 1,3-dichloro-1,3,5–triazine-2,4-dione-6-oxide
See Sodium 1,3-dichloro-1,3,5–triazine-2,4-dione-6-oxide: Calcium hypochlorite

Sodium hydrogen sulfate, Starch, Sodium carbonate
Anon., *Ind. Eng. Chem. (News Ed.)*, 1937, **15**, 282
Shortly after a mixture of the four ingredients had been compressed into tablets, incandescence and an explosion occurred. This may have been caused by inter-action of the oxidant and starch, accelerated by the acid salt, but may also have involved dichlorine monoxide liberated by the same salt.

Sulfur
Katz, S. A. *et al., Chem. Eng. News*, 1965, **46**(29), 6
Admixture of damp sulfur and 'swimming pool chlorine' caused a violently exo-thermic reaction, and ejection of molten sulfur.

Turpentine
491M, 1975, 85
An explosion occurred soon after the oxidant was put into a can which had contained turpentine but was thought to be empty.
See other METAL HYPOCHLORITES, OXIDANTS

3925. Calcium chlorite
[14674-72-7] $CaCl_2O_4$
$$Ca(ClO_2)_2$$

Potassium thiocyanate
Pascal, 1960, Vol. 16, 264
Mixtures may ignite spontaneously.
See other CHLORITE SALTS, METAL OXOHALOGENATES

3926. Calcium chromate
[13816-48-3] $CaCrO_4$
$$CaCrO_4$$

Boron
1. Rogers, J. W. *et al., Energy Res. Abs.*, 1981, **6**, 20813
2. Rogers, J. W., *Proc. 8th Int. Pyrotech. Seminar*, 1982, 556–573
Pyrotechnic properties of blends with 20% boron were studied [1], and the perfor-mance of several other blends has been assessed [2].
See other METAL OXOMETALLATES, OXIDANTS

3927. Calcium hydride
[7789-78-8]

CaH$_2$

$$CaH_2$$

HCS 1980, 268

Halogens
Mellor, 1941, Vol. 16, 651
Heating the hydride strongly with chlorine, bromine or iodine leads to incandescence.

Hexachlorobenzene
Loiselle, S. *et al., Environ. Sci. Technol.* 1997, **31**(1), 261; *J. Solid State Chem.,* 1997, **129**(2), 263
It was discovered that milling hexachlorobenzene with calcium hydride, to effect dechlorination, could give rise to explosive reaction, the conditions under which this happens were studied.

Manganese dioxide
See Manganese(IV) oxide: Calcium hydride

Metal oxohalogenates
MRH values: Ammonium bromate 4.52/82, ammonium chlorate 6.36/76, ammonium perchlorate 6.53/70, barium bromate 3.97/76, barium chlorate 4.85/72, calcium bromate 5.02/70, calcium chlorate 6.53/62, potassium bromate 4.56/73, potassium chlorate 5.81/56, potassium perchlorate 6.07/63, silver chlorate 4.60/73, sodium bromate 4.85/70, sodium chlorate 6.90/63, sodium perchlorate 6.69/60
1. Mellor, 1946, Vol. 3, 651
2. Yoshida, 1980, 214
Mixtures of the hydride with various bromates, chlorates and perchlorates explode on grinding [1]. MRH values calculated for 14 such combinations are given [2].

Other reactants
See next above

Silver halides
Mellor, 1941, Vol. 3, 389, 651
A mixture of silver fluoride and the hydride incandesces on grinding, and the iodide reacts vigorously on heating.

Tetrahydrofuran
See Tetrahydrofuran: Lithium tetrahydroaluminate
See other METAL HYDRIDES

3928. Calcium hydroxide
[1305-62-0] CaH_2O_2

$$Ca(OH)_2$$

Polychlorinated phenols, Potassium nitrate
   Kozloski, R. P., *Chem. Eng. News*, 1982, **60**(11), 3
   When chlorinated phenols are heated for analytical purposes with calcium hydroxide–potassium nitrate mixtures, chlorinated benzodioxins analogous to the extremely toxic tetrachlorodibenzodioxin may be formed.

Sugars
   See SUGARS

1,3,5-Trinitrohexahydro-1,3,5-triazine
   See 1,3,5-Trinitrohexahydro-1,3,5-triazine: Calcium hydroxide, Water
   See other INORGANIC BASES

3929. Calcium hydroxide *O*-hydroxylamide
[ ] CaH_3NO_2

$$Ca(OH)ONH_2$$

Sorbe, 1968, 43
It explodes on heating (and presumably would be formed if distillation of hydroxylamine from calcium oxide were attempted).
See Hydroxylamine: Metals
See other N–O COMPOUNDS

3930. Calcium bis(*O*-hydroxylamide)
[ ] $CaH_4N_2O_2$

$$Ca(ONH_2)_2$$

See Hydroxylamine: Metals
See other N–O COMPOUNDS

3931. Calcium phosphinate (Calcium hypophosphite)
[7789-79-9] $CaH_4O_4P_2$

$$Ca(OP(O)H_2)_2$$

Other reactants
   Yoshida, 1980, 344
   MRH values calculated for 8 combinations with various reagents are given.

Potassium chlorate   MRH 2.88/49
   See Potassium chlorate: Reducants
   See other METAL PHOSPHINATES, REDUCANTS

3932. Calcium hydrazide
[ ] CaH_6N_4

$$Ca^{2+}(HN^- - NH_2)_2$$

Hydrazine and its Derivatives, Schmidt, F. W., New York, Wiley, 1984, 75
Liable to explode on preparation from the metal, or amide, and hydrazine. Other
authors, however, report a 1:1 hydrazide [28330-64-5] as being fairly stable.
See other METAL HYDRAZIDES

3933. Hexaamminecalcium
[12133-31-2] $CaH_{18}N_6$

$$(NH_3)_6Ca$$

491M, 1975, 83

It ignites in air.
See related N-METAL DERIVATIVES

3934. Calcium permanganate
[10118-76-0] $CaMn_2O_8$

$$Ca(MnO_4)_2$$

Acetic acid   MRH ,
   MRH 2.93/18

Acetic anhydride   MRH 3.14/15
   See Potassium permanganate: Acetic acid, etc.

Cellulose
   Kirk-Othmer, (3rd ed) 1982, Vol. 19, 494
   Calcium permanganate is a more active oxidant than the potassium salt, and ignites
   paper or cotton on contact.

Hydrogen peroxide   MRH 1.59/99+
   See Hydrogen peroxide: Metals, etc.

Other reactants
   Yoshida, 1980, 92
   MRH values calculated for 18 combinations, largely with oxidisable materials, are
   given.
   See other OXIDANTS

3935. Calcium nitrate
[10124-37-5]

$$Ca(NO_3)_2$$

CaN$_2$O$_6$

HCS 1980, 270

Aluminium, Ammonium nitrate, Formamide, Water   MRH Aluminium 9.33/36
1. Wilson, J. F. *et al.*, S. Afr. Pat. 74 03 305, 1974
2. Slykehouse, T. E., Can. Pat. 1 011 561, 1977
A mixture containing 51% of calcium nitrate and 12% ammonium nitrate with 27% formamide and 10% water is detonable at −20°C [1]. Addition of aluminium powder improves performance as a blasting explosive [2].

Ammonium nitrate, Hydrocarbon oil
Haid, A. *et al.*, *Jahresber. Chem. Tech. Reichsanst.*, 1930, **8**, 108–115
The explosibility of the double salt, calcium ammonium nitrate, is enhanced by presence of oil.
See Ammonium nitrate: Organic fuels

Ammonium nitrate, Water-soluble fuels
Clark, W. F. *et al.*, US Pat. 3 839 107, 1970
Up to, or over 40% of the ammonium nitrate content of explosive mixtures with water-soluble organic fuels may be replaced with advantage by calcium nitrate.

Organic materials
Sorbe, 1968, 423
Mixtures of the nitrate with organic materials may be explosive.

Other reactants
Yoshida, 1980, 193
MRH values calculated for 16 combinations with oxidisable materials are given.
See other METAL NITRATES, OXIDANTS

3936. Calcium azide
[19465-88-8]

$$Ca(N_3)_2$$

CaN$_6$

1. Mellor, 1940, Vol. 8, 349
2. Stull, 1977, 10
Calcium, strontium and barium azides are not shock-sensitive, but explode on heating at about 150, 170 and 225 (or 152)°C, respectively. In sealed tubes, the explosion temperatures are higher [1]. Although calcium azide is rather mildly endothermic (ΔH_f° (s) +46 kJ/mol, 0.37 kJ/g), it can decompose much more exothermally to the nitride (189.9 kJ/mol, 1.53 kJ/g) than to the elements [2].
See other ENDOTHERMIC COMPOUNDS, METAL AZIDES

3937. Calcium oxide
    [1305-78-8]   **CaO**

CaO

NSC 241, 1982; *HCS 1980*, 272

Ethanol
    Keusler, V., *Apparatebau*, 1928, **40**, 88–89
    The lime–alcohol residue from preparation of anhydrous alcohol ignited on dis-
    charge from the still and caused a vapour explosion. The finely divided and reactive
    lime may have heated on exposure to atmospheric moisture and caused ignition.
    See Water, below

Hydrogen fluoride
    See Hydrogen fluoride: Oxides

Interhalogens
    See Bromine pentafluoride: Acids, etc.   MRH 3.23/56
        Chlorine trifluoride   MRH 4.06/52

Other reactants
    Yoshida, 1980, 126
    MRH values calculated for 6 combinations with various reagents are given.

Phosphorus(V) oxide
    See Tetraphosphorus decaoxide: Inorganic bases

Water
    1. Mellor, 1941, Vol. 3, 673
    2. Anon., *Fire*, 1935, **28**, 30
    3. Amos, T., *Zentr. Zuckerind.*, 1923, **32**, 103
    4. Anon., *BCISC Quart. Safety Summ.*, 1967, **38**, 15
    5. Anon., *BCISC Quart. Safety Summ.*, 1971, **42**(168), 4
    6. Anon., *Engrg. Digest*, 1908, **4**, 661
    Crystalline calcium oxide reacts imperceptibly slowly with water, but the powdered
    material reacts with explosive violence after a few minutes delay [1]. Quicklime,
    when mixed with $\frac{1}{3}$ its weight of water, will reach 150–300°C (depending on
    quantity) and may ignite combustible material. Occasionally 800–900°C has been
    attained [2]. Moisture present in wooden storage bins caused ignition of the latter
    [3]. A water jet was used unsuccessfully to clear a pump hose blocked with quick-
    lime. On standing, the exothermic reaction generated enough steam to clear the
    blocked hose explosively [4]. Two glass bottles of calcium oxide powder burst
    while on a laboratory shelf, owing to the considerable increase in bulk which
    occurs on hydration. Storage in plastic bottles is recommended, and granular oxide
    should cause fewer problems in this respect than the powder [5]. The powerful
    expansion effect was formerly used in coal-mining operations [6].

See other GLASS INCIDENTS
See other INORGANIC BASES, METAL OXIDES

3938. Calcium peroxide
[1305-79-9] CaO_2

$$CaO_2$$

Chlorinated paraffin
Anon., *Chem. Brit.*, 1997, **33**(6), 7
A fire is reported subsequent to industrial mixing of calcium peroxide and a chlo-
rinated paraffin. Even completely chlorinated hydrocarbons can be oxidised, to
phosgene.
See Oxidisable materials; below

Other reactants
Yoshida, 1980, 84
MRH values calculated for 14 combinations with oxidisable materials are given.

Oxidisable materials   MRH Magnesium 6.02/25
1. Castrantas, 1965, 4
2. Sorbe, 1968, 43
Grinding the peroxide with oxidisable materials may cause fire [1], and the octahy-
drated solid explodes at high temperature [2].

Polysulfide polymers
491M, 1975, 87
The dry peroxide, added to cause cross-linking in liquid polysulfide polymers with
pendant thiol groups, caused sparking or ignition, depending on the scale of the
process.
See other METAL PEROXIDES, OXIDANTS

3939. Calcium sulfate
[7778-18-9] CaO_4S

$$CaSO_4$$

Aluminium
See Aluminium: Metal oxides, or Oxosalts

Diazomethane
See Diazomethane: Calcium sulfate

Phosphorus
See Phosphorus: Metal sulfates
See other METAL OXONON-METALLATES

3940. Calcium peroxodisulfate
[13235-16-0] CaO_8S_2

$$Ca(OSO_2O-)_2$$

Castrantas, 1965, 6
It is shock-sensitive and explodes violently.
See other OXIDANTS, PEROXOACID SALTS

3941. Calcium sulfide
[20548-54-3] CaS

$$CaS$$

HCS 1980, 274

Oxidants
Mellor, 1941, Vol. 3, 745
Alkaline-earth sulfides react vigorously with chromyl chloride, lead dioxide, potassium chlorate (explodes lightly) and potassium nitrate (explodes violently).
See other METAL SULFIDES

3942. Calcium polysulfide
[1332-67-8] (tetrasulfide) CaS_x
[1332-69-9] (pentasulfide)

$$CaS_x$$

Edwards, P. W., *Chem. Met. Eng.*, 1922, **27**, 986
The powdered sulfide, admixed with small amounts of calcium thiosulfate and sulfur, has been involved in several fires and explosions, some involving initiation by static discharges.
See other METAL SULFIDES, STATIC INITIATION INCIDENTS

3943. Calcium silicide
[12013-55-7] $CaSi$

$$CaSi$$

1. Grossel, S. S., *J. Loss Prev. Proc. Ind.*, 1988, **1**, 62–74
2. Yan, J. *et al., Chem. Abs.*, 1987, **107**, 182628
3. Deng, X. *et al., Arch. Combust.*, 1987, **7**(1–2), 19–31

The finely powdered silicide is a significant dust explosion hazard [1]. The lower explosion limit for a calcium–silicon dust cloud of mean particle size 9.7 μm was measured as 79 g/m3, in good agreement with a calculated value [2]. Other dust cloud parameters are presented and related to predictions [3].

Acids
Mellor, 1940, Vol. 6, 177
Interaction is vigorous, and the evolved silanes ignite in air.

1328

Boron
See Boron: Calcium–silicon alloy
See other METAL NON-METALLIDES, PYROPHORIC MATERIALS

3944. Calcium disilicide
[12013-56-8]   CaSi₂

$$CaSi_2$$

Carbon tetrachloride
Zirconium Fire and Explosion Incidents, TID-5365, Washington, USAEC, 1956
Calcium disilicide exploded when milled in the solvent.

Iron(III) oxide
See Iron(III) oxide: Calcium disilicide

Metal fluorides
Berger, E., *Compt. rend.*, 1920, **170**, 29
Calcium disilicide ignites in close contact with alkali metal fluorides, forming
silicon tetrafluoride.

Other reactants
Yoshida, 1980, 99
MRH values calculated for 14 combinations with oxidants are given.

Potassium nitrate
See Potassium nitrate: Calcium silicide
See other METAL NON-METALLIDES

3945. Calcium triperoxochromate
[ ] $Ca_3Cr_2O_{12}$

$$Ca_3[OO)_3Cr]_2$$

Raynolds, J. A. *et al.*, *J. Amer. Chem. Soc.*, 1930, **52**, 1851
A buff solid, prepared by means appropriate to a tetraperoxochromate(3−), it
analysed as the dodecahydrate of this structure. It explodes at 100°C.
See other OXIDANTS, PEROXOACID SALTS

3946. Calcium nitride
[12013-82-0] Ca_3N_2

$$Ca_3N_2$$

von Schwartz, 1918, 322
Spontaneously flammable in air (probably when finely divided and in moist air).

Halogens
Mellor, 1940, Vol. 8, 103

Reaction with incandescence in chlorine gas or bromine vapour.
See other N-METAL DERIVATIVES, NITRIDES

3947. Calcium phosphide
[1305-99-3]   Ca₃P₂

$$Ca_3P_2$$

Dichlorine oxide
See Dichlorine oxide: Oxidisable materials

Oxygen
Van Wazer, 1958, Vol. 1, 145
Calcium and other alkaline-earth phosphides incandesce in oxygen at about 300°C.

Water
Mellor, 1940, Vol. 8, 841
Calcium and other phosphides on contact with water liberate phosphine, which usualy ignites in air, owing to the diphosphane content.
See other METAL NON-METALLIDES

3948. Columbium
[7440-03-1]   Cb

Cb

See Niobium

3949. Cadmium
[7440-43-9]   Cd

Cd

NSC 312, 1978; *HCS 1980*, 257; *RSC Lab. Hazards Safety Data Sheet No. 55*, 1987 (element and compounds)

The finely divided metal is pyrophoric.
See CADMIUM HYDRIDE *See other* PYROPHORIC METALS

Hydrazoic acid
See Cadmium azide (reference 2)

Oxidants
See Nitryl fluoride: Metals
     Ammonium nitrate: Metals

Selenium, or Tellurium
Mellor, 1940, Vol. 4, 480

Reaction on warming powdered cadmium with selenium or tellurium is exothermic, but less vigorous than that of zinc.

See other METALS

3950. Trihydrazinecadmium chlorate
[ ] $CdCl_2H_{12}N_6O_6$

$$[(H_4N_2)_3Cd] [ClO_3]_2$$

Mellor, 1956, Vol. 2, Suppl. 2.1, 592
It explodes on impact.
See other AMMINEMETAL OXOSALTS

3951. Basic trihydrazinecadmium perchlorate
[ ] $Cd_2Cl_2H_{14}N_6O_8$

$$[(H_4N_2)_3Cd] [ClO_4]_2 \cdot Cd(OH)_2 \cdot H_2O$$

Mellor, 1967, Vol. 8, Suppl. 2, 88
The monohydrated basic salt is extremely explosive.
See related AMMINEMETAL OXOSALTS

3952. Cadmium chlorate
[10137-74-3] $CdCl_2O_6$

$$Cd(ClO_3)_2$$

Sulfides
Mellor, 1956, Vol. 2, Suppl. 1, 584
Interaction with copper(II) sulfide is explosive, and with antimony(II) sulfide, arsenic(III) sulfide, tin(II) sulfide and tin(IV) sulfide, incandescent.
See METAL CHLORATES: phosphorus, etc.
also Chloric acid: Metal sulfides
See other METAL CHLORATES, METAL OXOHALOGENATES

3953. Cadmium hydride
[72172-64-6] CdH_2

$$CdH_2$$

Barbaras, G. D. *et al., J. Amer. Chem. Soc.*, 1951, **73**, 4585
The hydride, prepared at $-78°C$, suddenly decomposes during slow warming at $2°C$, leaving a residue of pyrophoric cadmium.
See other METAL HYDRIDES

3954. Cadmium amide
[22750-53-4]   CdH$_4$N$_2$

$$Cd(NH_2)_2$$

Alone, or Water
Mellor, 1940, Vol. 8, 261
When heated rapidly, the amide may explode. Reaction with water is violent.
See other N-METAL DERIVATIVES

3955. Trihydrazinecadmium(II) nitrate
[32434-37-5]   CdH$_{12}$N$_8$O$_6$
See HYDRAZINE METAL NITRATES

3956. Tetraamminecadmium permanganate
[34410-88-3]   CdH$_{12}$Mn$_2$N$_4$O$_8$

$$[H_3N)_4Cd] [MnO_4]_2$$

Mellor, 1942, Vol. 12, 335
It explodes on impact.
See other AMMINEMETAL OXOSALTS

3957. Cadmium azide
[14215-29-3]   CdN$_6$

$$Cd(N_3)_2$$

1. Turney, T. A., *Chem. & Ind.*, 1965, 1295
2. Mellor, 1967, Vol. 8, Suppl. 2.2, 25, 50
3. Schechkov, G. T., *Chem. Abs.*, 1983, **98**, 146010

A solution, prepared by mixing saturated solutions of cadmium sulfate and sodium azide in a 10 ml glass tube, exploded violently several hours after preparation [1]. The dry solid is extremely hazardous, exploding on heating or light friction. A violent explosion occurred with cadmium rods in contact with aqueous hydrogen azide [2]. A DTA study showed a lesser thermal stability than lead azide [3]. It is strongly endothermic (ΔH_f° (s) +451 kJ/mol, 2.32 kJ/g).
See other ENDOTHERMIC COMPOUNDS, METAL AZIDES

3958. Cadmium oxide
[1306-19-0]   CdO

$$CdO$$

HCS 1980, 260

Magnesium
See Magnesium: Metal oxides *See other* METAL OXIDES

3959. Cadmium selenide
[1306-24-7]   CdSe

CdSe

Preparative hazard
See Selenium: Metals
See related METAL SULFIDES

3960. Cadmium nitride
[12380-95-9] Cd_3N_2

Cd_3N_2

Fischer, F. *et al., Ber.*, 1910, **43**, 1469
The shock of the violent explosion caused by heating a sample of the nitride caused an unheated adjacent sample to explode. It is much less endothermic (ΔH_f° (s) +161.5 kJ/mol, 0.44 kJ/g) than the azide.

Acids, or Bases
Mellor, 1964, Vol. 8, Suppl. 2.1, 161
It reacts explosively with dilute acids or bases.

Water
Mellor, 1940, Vol. 8, 261
It explodes on contact.
See other ENDOTHERMIC COMPOUNDS, *N*-METAL DERIVATIVES, NITRIDES

3961. Cerium
[7440-45-1]   Ce

Ce

Alone, or Metals
Mellor, 1945, Vol. 5, 602–603
Cerium or its alloys readily give incendive sparks (pyrophoric particles) on frictional contact, this effect of iron alloys being widely used in various forms of 'flint' lighters. The massive metal ignites and burns brightly at 160°C, and cerium wire burns in a Bunsen flame more brightly than magnesium. Of its alloys with aluminium, antimony, arsenic, bismuth, cadmium, calcium, copper, magnesium, mercury, sodium and zinc, those containing major proportions of cerium are often extremely pyrophoric. The mercury amalgams ignite spontaneously in air without the necessity for frictional generation of small particles. The interaction of cerium with zinc is explosively violent, and with antimony or bismuth, very exothermic.
See other PYROPHORIC ALLOYS

Carbon dioxide, Nitrogen
See Carbon dioxide: Metals, Nitrogen

Halogens
Mellor, 1946, Vol. 5, 603
Cerium filings ignite in chlorine or in bromine vapour at about 215°C.

Phosphorus
See Phosphorus: Metals

Silicon
Mellor, 1946, Vol. 5, 605
Interaction at 1400°C to form cerium silicide is violently exothermic, often destroying the containing vessel.

Water
Uno, T. *et al., Anzen Kagaku*, 1988, **27**(3), 157–161 (*Chem. Abs.*, 1989, **110**, 12857)
The mixed alloy of cerium with lanthanum and other rare-earth metals ('Mischmetall') may ignite spontaneously in contact with aqueous solutions, owing to its oxidation by water and ignition of evolved hydrogen.
See other GAS EVOLUTION INCIDENTS, REDOX REACTIONS
See other METALS, PYROPHORIC METALS

3962. Cerium dihydride
[13569-50-1] CeH_2

$$CeH_2$$

Libowitz, G. G. *et al., Inorg. Synth.*, 1973, **14**, 189–192
The polycrystalline hydride is frequently pyrophoric, particularly at higher hydrogen contents (up to $H_{2.85}$), while monocrystalline material appears to be less reactive to air.
See other METAL HYDRIDES

3963. Cerium trihydride
[13864-02-3] CeH_3

$$CeH_3$$

Muthmann, W. *et al., Ann.*, 1902, **325**, 261
The hydride is stable in dry air, but may ignite in moist air.
See other METAL HYDRIDES

3964. Ammonium hexanitrocerate (Ceric ammonium nitrate, Ammonium cerium(IV) nitrate)
[16774-21-3] $CeH_8N_8O_{12}$

$$(NH_4)_2Ce(NO_3)_6$$

Cyclopentadienylsodium
See Sodium nitrate: Tris(cyclopentadienyl)sodium

Potassium bromate, Malonic acid, Water
　　See Potassium bromate: Ceric ammonium nitrate, Malonic acid, Water.
　　See related METAL NITRATES *See other* OXIDANTS

3965. Cerium nitride
[25764-08-3]　　　　　　　　　　　　　　　　　　　　　　　　CeN

$$CeN$$

Water
　　Mellor, 1940, Vol. 8, 121
　　Contact with water vapour slowly causes incandescence, while a limited amount of water or dilute acid causes rapid incandescence with ignition of evolved ammonia and hydrogen.
　　See other N-METAL DERIVATIVES, NITRIDES

3966. Cerium azide
[ ]　　　　　　　　　　　　　　　　　　　　　　　　　　　CeN_9

$$Ce(N_3)_3$$

　　Mellor, 1940, Vol. 8, 354
　　The precipitate from cerium nitrate and sodium azide is explosive.
　　See other METAL AZIDES

3967. Cerium trisulfide
[12014-93-6]　　　　　　　　　　　　　　　　　　　　　　　Ce_2S_3

$$Ce_2S_3$$

　　Mellor, 1946, Vol. 5, 649
　　It is pyrophoric in ambient air when finely divided.
　　See other METAL SULFIDES, PYROPHORIC MATERIALS

3968. Chromyl azide chloride
[14259-67-7]　　　　　　　　　　　　　　　　　　　　　　$ClCrN_3O_2$

$$CrO_2(N_3)Cl$$

　　An explosive solid.
　　See entry METAL AZIDE HALIDES

3969. Caesium tetrafluorochlorate(1−)
[15321-67-7]　　　　　　　　　　　　　　　　　　　　　　$ClCsF_4$

$$Cs[ClF_4]$$

　　Highly reactive oxidant.
　　See entry METAL POLYHALOHALOGENATES

3970. Caesium chloroxenate
[26283-13-6]   ClCsO₃Xe

$$Cs[ClXeO_3]$$

Jaselskis, B. *et al., J. Amer. Chem. Soc.*, 1967, **89**, 2770
It explodes at 205°C under vacuum.
See other XENON COMPOUNDS

3971. Chlorine fluoride
[7790-89-8]   ClF

$$ClF$$

1. Sidgwick, 1950, 1149
2. 'Product Information Sheet ClF', Tulsa, Ozark-Mahoning Co., 1970
This powerful oxidant reacts with other materials similarly to chlorine trifluoride
or fluorine [1], but more readily than the latter [2].
See Chlorine trifluoride, or Fluorine

Acetonitrile, Fluorine
See Fluorine: Acetonitrile, etc.

Aluminium
Mellor, 1956, Vol. 2, Suppl. 1, 63
Aluminium burns more readily in chlorine fluoride than in fluorine.

Bis(trifluoromethyl) disulfide
See Bis(trifluoromethyl) disulfide: Chlorine fluorides

Bis (trifluoromethyl) sulfide
See Bis(trifluoromethyl) sulfide: Chlorine fluorides

tert-Butanol
Young, D. F. *et al., J. Amer. Chem. Soc.*, 1970, **92**, 2314
A mixture prepared at −196°C exploded at −100°C during slow warming.

N-Chlorosulfinylamine
Kuta, G. S. *et al., Intern. J. Sulfur Chem.*, 1973, **8**, 335–340
Interaction of the reagents in equimolar proportions produced a highly explosive
and strongly oxidising material (possibly an N–F compound).

Diethyl sulfite
See Ethyl fluorosulfate

Fluorocarbon polymers
'Product Information Sheet ClF', Tulsa, Ozark-Mahoning Co., 1970
Chlorine fluoride can probably ignite Teflon and Kel-F at high temperatures, or
under friction or flow conditions.

Phosphorus trifluoride

1. Fox, W. B. *et al., Inorg. Nucl. Chem. Lett.*, 1971, **7**, 861
2. Neilson, R. H. *et al., Inorg. Chem.*, 1975, **14**, 2019

Interaction of the fluorides to produce chlorotetrafluorophosphorane [1] is uncontrollably violent even at −196°C [2]. An improved method of making the phosphorane from phosphorus pentafluoride and boron trichloride is detailed [2].

Tellurium

Mellor, 1943, Vol. 11, 26
Interaction is incandescent.

Trifluoromethanssulfenyl chloride

See Trifluoromethanesulfenyl chloride: Chlorine fluorides

Trifluorosulfur nitride

Waterfeld, A. *et al., Angew. Chem. (Intern. Ed.)*, 1981, **20**, 1017
In the preparation of chloriminosulfur tetrafluoride or its dimer by slowly warming a mixture at −196° to −78°C and then ambient temperature, too rapid warming leads to violent explosions.

Trifluorosulfur nitride

See N,N-Dichloropentafluorosulfanylamine

Water

Pascal, 1960, Vol. 16.1, 189
Interaction with liquid water or that bound in crystalline hydrates is violent.
See other INTERHALOGENS

3972. Fluoronium perchlorate

[ ] $ClFH_2O_4$

$$H_2F^+ClO_4^-$$

Water

Hantzsch, A., *Ber.*, 1930, **63**, 97
This hydrogen fluoride−perchloric acid complex reacts explosively with water.
See other NON-METAL PERCHLORATES

3973. Chloryl hypofluorite

[101672-08-6] $ClFO_3$

$$O_2ClOF$$

1. Hoffman, C. J., *Chem. Rev.*, 1964, **64**, 97
2. Christe, K. O. *et al., Inorg. Chem.*, 1987, **56**, 920

Not then completely purified or characterised, its explosive nature was in contrast to the stability of the isomeric perchloryl fluoride [1]. It is also shock-sensitive [2].
See other HALOGEN OXIDES, HYPOHALITES

3974. Perchloryl fluoride
[7616-94-6]   **ClFO₃**

O₃ClF

1. 'Booklet DC-1819', Philadelphia, Pennsalt Chem. Corp., 1957
2. Anon., *Chem. Eng. News*, 1960, **38**, 62
3. Barton, D. H. R., *Pure Appl. Chem.*, 1970, **21**, 185
4. Sharts, C. M. *et al., Org. React.*, 1974, **21**, 232–234

Procedures relevant to safe handling and use are discussed. Perchloryl fluoride is stable to heat, shock and moisture, but is a powerful oxidiser comparable with liquid oxygen. It forms flammable and/or explosive mixtures with combustible gases and vapours [1,2]. It only reacts with strongly nucleophilic centres, and the by-product, chloric acid is dangerously explosive in admixture with organic compounds [3]. Safety aspects of practical use of perchloryl fluoride have been reviewed [4].

Benzocyclobutene, Butyllithium, Potassium *tert*-butoxide
Adcock, W. *et al., J. Organomet. Chem.*, 1975, **91**, C20
An attempt to convert benzocyclobutenylpotassium (prepared from the reagents above) to the fluoro derivative at −70°C with excess perchloryl fluoride led to a violent explosion when the reaction mixture was removed from the cooling bath after stirring for an hour.

Calcium acetylide, or Potassium cyanide, or Potassium thiocyanate, or Sodium iodide
Kirk-Othmer, 1966, Vol. 9, 602
Unreactive at 25°, these solids react explosively in the gas at 100–300°C.

Charcoal
Inman, C. E. *et al., Friedel-Craft and Related Reactions*, (Olah, G. A., Ed.), Vol. 3, 1508, New York, Interscience, 1964
Adsorption of perchloryl fluoride on charcoal can, like liquid oxygen, produce a powerful 'Sprengel' explosive.

Ethyl 4-fluorobenzoylacetate
Fuqua, S. A. *et al., J. Org. Chem.*, 1964, **29**, 395, footnote 3
Fluorination of the ester with perchloryl fluoride by an established method led to a violent explosion.

Finely divided solids
McCoy, G., *Chem. Eng. News*, 1960, **38**(4), 62
Oxidisable organic materials of high surface to volume ratio (carbon powder, foamed elastomers, lampblack, sawdust) react very violently, even at −78°C, with perchloryl fluoride, which should be handled with the same precautions as liquid oxygen.

Hydrocarbons, or Hydrogen sulfide, or Nitrogen oxide, or Sulfur dichloride, or Vinylidene chloride
  Braker, 1980, 578
  At ambient temperature, perchloryl fluoride is unreactive with the compounds above, but reaction is explosive at 100–300°C, or if the mixtures are ignited.

3α-Hydroxy-5β-androstane-11,17-dione 17-hydrazone
  Nomine, G. *et al., J. Chem. Educ.*, 1969, **46**, 329
  A reaction mixture in aqueous methanol exploded violently at −65°C. (Hydrazine–perchloryl fluoride redox reaction?) A previous reaction at 20°C had been uneventful, and the low-temperature explosion could not be reproduced.

Laboratory materials
  Schlosser, M. *et al., Chem. Ber.*, 1969, **102**, 1944
  Contact of perchloryl fluoride with laboratory greases or rubber tubing, etc., has led to several explosions.

Lithiated compounds
  Houben-Weyl, 1970, Vol. 13.1, 223
  There is a danger of explosion during replacement of a lithium substituent by fluorine using perchloryl fluoride.
  See Benzocyclobutene, etc., above

2-Lithio(dimethylaminomethyl)ferrocene
  Peet, J. H. J. *et al., J. Organomet. Chem.*, 1974, **82**, C57
  Preparation of dimethylaminomethyl-2-fluoroferrocene, by interaction of the lithio compound in THF at −70°C with perchloryl fluoride diluted with helium, unexpectedly exploded violently.

Methyl 2-bromo-5,5-(ethylenedioxy)[2.2.1]bicycloheptane-7-carboxylate
  Wang, C.-L. J. *et al., J. Chem. Soc., Chem. Comm.*, 1976, 468
  In an attempt to improve isomer distribution from the fluorination reaction, one run was cooled to below −40°C, but on quenching with water, a violent explosion occurred.

Nitrogenous bases
  1. Scott, F. L. *et al., Chem. & Ind.*, 1960, 258
  2. Gardner, D. M. *et al., J. Org. Chem.*, 1967, **32**, 1115
  3. Parish, E. J. *et al., J. Org. Chem.*, 1980, **45**, 4036–4037
  Interaction, in presence of diluent below 0°C, with isopropylamine or isobutylamine caused separation of explosive liquids, and with aniline, phenylhydrazine and 1,2-diphenylhydrazine, explosive solids [1]. In absence of diluents, contact with most aliphatic or non-aromatic heterocyclic amines often leads to uncontrolled oxidation and/or explosions [2]. During oxidation of two steroidal dienes in dry pyridine at −35 to −40°C, on one occasion each of the reactions was accompanied by violent explosions [3].
  See 3α-Hydroxy-5β-androstane-11,17-dione 17-hydrazone, above

Podophyllotoxin enolate tetrahydropyranyl ether

Glinski, M. B. *et al., J. Org. Chem.*, 1987, **53**, 2749–2753

Attempted preparation of 2-fluoropodophyllotoxin by treatment of the tetrahydropyranyl ether with perchloryl fluoride led to a violent explosion.

Sodium methoxide, Methanol

Papesch, V., *Chem. Eng. News*, 1959, **36**, 60

Addition of solid methoxide to a reaction vessel containing methanol vapour and
gaseous perchloryl fluoride caused ignition and explosion. This could be avoided by
adding all the methoxide first, or by nitrogen purging before addition of methoxide.

See other HALOGEN OXIDES, PERCHLORYL COMPOUNDS

3975. Chlorine fluorosulfate
[13997-90-5] $ClFO_3S$

$$ClOSO_2F$$

Bailar, 1973, Vol. 2, 1469

Hydrolysis is violent, producing oxygen.

See related ACYL HALIDES, HYPOHALITES

3976. Fluorine perchlorate (Perchloryl hypofluorite)
[10049-03-3] $ClFO_4$

$$O_3ClOF$$

Alone, or Laboratory materials, or Potassium iodide

1. Rohrback, G. H. *et al., J. Amer. Chem. Soc.*, 1949, **69**, 677
2. Christe, K. O. *et al., Inorg. Chem.*, 1982, **21**, 2938

The pure liquid explodes on freezing at −167°C, and the gas is readily initiated
by sparks, flame or contact with grease, dust or rubber tubing. Contact of the gas
with aqueous potassium iodide also caused an explosion [1]. The pure material is
now readily accessible via the relatively stable intermediate, tetrafluoroammonium
perchlorate [2].

Difluorimide–potassium fluoride adduct

Christe, K. O. *et al., Inorg. Chem.*, 1987, **56**, 925

A mixture prepared at −196°C exploded when warmed towards −78°C.

Bis(trifluoromethanesulfonyl)imide

See N-Chlorobis(trifluoromethanesulfonyl)imide

Hydrogen

Hoffman, C. J., *Chem. Rev.*, 1964, **64**, 97

Ignition of the perchlorate occurs in excess hydrogen gas.

See other HALOGEN OXIDES, HYPOHALITES

3977. Xenon(II) fluoride perchlorate
[25582-86-9]   ClFO$_4$Xe

$$XeFClO_4$$

Bartlett, N. *et al., Chem. Comm.*, 1969, 703
Thermodynamically unstable, it explodes readily and sometimes with violence.
See other NON-METAL PERCHLORATES, XENON COMPOUNDS

3978. Nitrogen chloride difluoride
[13637-87-1]   ClF$_2$N

$$ClNF_2$$

Petry, R. C., *J. Amer. Chem. Soc.*, 1960, **82**, 2401
Caution in handling is recommended for this *N*-halogen compound.
See other N-HALOGEN COMPOUNDS

3979. Phosphorus chloride difluoride
[14335-40-1]   ClF$_2$P

$$ClPF_2$$

Hexafluoroisopropylideneaminolithium
See Hexafluoroisopropylideneaminolithium: Non-metal halides
See other NON-METAL HALIDES

3980. Thiophosphoryl chloride difluoride
[2524-02-9]   ClF$_2$PS

$$ClP(S)F_2$$

Mellor, 1971, Vol. 8, Suppl. 3, 536
Mixtures with air explode spontaneously at certain concentrations.
See other AUTOIGNITION INCIDENTS *See related* NON-METAL HALIDES (AND THEIR OXIDES)

3981. Chlorine trifluoride
[7790-91-2]   ClF$_3$

$$ClF_3$$

1. Anon., *J. Chem. Educ.*, 1967, **44**, A1057–1062
2. O'Connor, D. J. *et al., Chem. & Ind.*, 1957, 1155
3. Farrar, R. L. *et al., Some Considerations in Handling Fluorine and Chlorine Fluorides*, Oak Ridge Rept. K/ET-252, 1979, 41pp

Handling procedures for this highly reactive oxidant gas have been detailed [1]. Surplus gas is best burned with town- or natural gas, followed by absorption in alkali [2]. More recent and comprehensive information is available [3].

Mellor, 1956, Vol. 2, Suppl. 1, 157
Strong nitric and sulfuric acids reacted violently, (possibly owing to water content).

Ammonium fluoride, or Ammonium hydrogen fluoride
Gardner, D. M. *et al., Inorg. Chem.*, 1963, **2**, 413
The reaction gases (containing chlorodifluoramine) must be handled at below $-5°C$ to avoid explosion. Ammonium hydrogen fluoride behaves similarly.

Bis(trifluoromethyl) sulfide, or disulfide
See Bis(trifluoromethyl) sulfide: Chlorine fluorides
Bis(trifluoromethyl) disulfide: Chlorine fluorides

Boron-containing materials
Bryant, J. T. *et al., J. Spacecr. Rockets*, 1971, **8**, 192–193
Finely divided boron, tetraboron carbide, and boron–aluminium mixtures will ignite on exposure to the gas.
See Metals, etc., below

Carbon tetrachloride   MRH 1.21/56
Mellor, Vol. 2, Suppl. 1, 156
Chlorine trifluoride will dissolve in carbon tetrachloride at low temperatures without reaction. Such solutions are dangerous, being capable of detonation. If it is used as a solvent for fluorination with the trifluoride, it is therefore important to prevent build-up of high concentrations of the latter.

Chromium trioxide
Mellor, 1943, Vol. 11, 181
Interaction of the two oxidants is incandescent.

Deuterium, or Hydrogen   MRH Hydrogen 6.82/3.0
Haberland, H. *et al., Chem. Phys.*, 1975, **10**, 36
Studies of the interaction of chlorine trifluoride with deuterium or hydrogen atoms in a scattering chamber were accompanied by frequent flashes or explosions, within 3 h if the reactor had been vented, or after 8 h if it had not.

Fluorinated solvents
'Chlorine Trifluoride', Tech. Bull. TA 8522-3, Morristown, N. J., Baker & Adamson Div. of Allied Chemicals Corp., 1968
Bulk surfaces of polytetrafluoroethylene or polychlorotrifluoroethylene are resistant to the liquid or vapour under static conditions, but breakdown and ignition may occur under flow conditions.
See Polychlorotrifluoroethylene, below

Fuels
See ROCKET PROPELLANTS

Halocarbons
   Brower, K. R., *J. Fluorine Chem.*, 1986, **31**, 333–349
   Combinations of liquid chlorine trifluoride with several halocarbons except perfluo-
   rohexane exploded immediately when suddenly mixed at all temperatures between
   25° and −70°C. Poly(chlorotrifluoroethylene), below
   See Carbon tetrachloride, Fluorinated solvents, both above

Hydrocarbons
   Brower, K. R., *J. Fluorine Chem.*, 1986, **31**, 333–349
   Combinations of liquid chlorine trifluoride and several hydrocarbons exploded
   immediately when suddenly mixed at all temperatures between 25° and −70°C.
   See Organic materials, below

Hydrogen-containing materials   MRH Ammonia 5.52/16
Hydrogen sulfide   MRH 4.68/13
   Mellor, 1956, Vol. 2, Suppl. 1, 157
   Explosive reactions occur with ammonia, coal-gas, hydrogen or hydrogen sulfide.

Ice
   See Water, below

Iodine
   Mellor, 1956, Vol. 2, Suppl. 1, 157
   Ignition on contact.

Metals, or Metal oxides, or Metal salts, or Non-metals, or Non-metal oxides

   MRH Iron 5.10/68, potassium 2.90/63, Calcium oxide 4.05/48, magnesium oxide
   3.89/39, manganese dioxide 1.80/48, Calcium oxide 4.05/48, magnesium oxide
   3.89/39, manganese dioxide 1.80/48, Sulfur 4.10/15, silver nitrate 0.5/73
   1. Mellor, 1956, Vol. 2, Suppl. 1, 155–157
   2. Sidgwick, 1950, 1156
   3. Rhein, R. A., *J. Spacecr. Rockets*, 1969, **6**, 1328–1329
   4. Chlorine Trifluoride' Tech. Bull. TA 8532-3, Morristown, N. J., Baker &
      Adamson Div. Allied Chem. Corp., 1968
   Chlorine trifluoride is a hypergolic oxidiser with recognised fuels, and contact with
   the materials following at ambient or slightly elevated temperatures is violent,
   ignition often occurring. The state of subdivision may affect the results.
      Antimony, arsenic, selenium, tellurium, iridium, iron, molybdenum, osmium,
   potassium, rhodium, tungsten; (and when primed with charcoal,) aluminium,
   copper, lead, magnesium, silver, tin, zinc. Interaction of lithium or calcium with
   chlorine tri- or penta-fluorides is hypergolic and particularly energetic.
      Aluminium oxide, arsenic trioxide, bismuth trioxide, calcium oxide, chromic
   oxide, lanthanum oxide, lead dioxide, magnesium oxide, manganese dioxide,
   molybdenum trioxide, phosphorus pentoxide, stannic oxide, sulfur dioxide
   (explodes), tantalum pentoxide, tungsten trioxide, vanadium pentoxide.

Red phosphorus, sulfur; but with carbon, the observed ignition has been attributed to presence of impurities; mercury iodide, potassium iodide, silver nitrate, potassium carbonate.

Methane
Baddiel, C. B. *et al., Proc. 8th Combust. Symp.*, 1960, 1089–1095
The explosive interaction of chlorine trifluoride with methane and its homologues has been studied in detail.

Nitroaryl compounds
Mellor, 1956, Vol. 2, Suppl. 1, 156
Several nitro compounds are soluble in chlorine trifluoride, but the solutions are extremely shock-sensitive. These include trinitrotoluene, hexanitrobiphenyl, hexanitrodiphenyl-amine, -sulfide or -ether. Highly chlorinated compounds behave similarly.

Organic materials   MRH Acetic acid 3.05/19, benzene 5.35/36, ether 5.56/28
Mellor, 1956, Vol. 2, Suppl. 1, 155
Violence of the reaction, sometimes explosive, with e.g., acetic acid, benzene, ether, is associated with both their carbon and hydrogen contents. If nitrogen is also present, explosive fluoroamino compounds may be involved. Fibrous materials–cotton, paper, wood–invariably ignite.

Other reactants
Yoshida, 1980, 129
MRH values calculated for 18 combinations with oxidisable materials are given.

Polychlorotrifluoroethylene
Anon., *Chem. Eng. News*, 1965, **43**(20), 41
An explosion occurred when chlorine trifluoride was being bubbled through the fluorocarbon oil at $-4°C$. Moisture (snow) may have fallen into the mixture, reacted exothermally with the trifluoride and initiated the mixture.
See Water, below

Refractory materials
1. Cloyd, 1965, 58
2. Mellor, 1956, Vol. 2, Suppl. 1, 157
Fibrous or finely divided refractory materials, asbestos, glass wool, sand, or tungsten carbide, may ignite with the liquid and continue to burn in the gas [1]. The presence of adsorbed or lattice water seems necessary for attack on the siliceous materials to occur [2].
See other GLASS INCIDENTS

Ruthenium
Burns, R. C. *et al., J. Inorg. Nucl. Chem.*, 1980, **42**, 1614
Formation of the adduct chlorine trifluoride–ruthenium pentafluoride by reaction with ruthenium metal at ambient temperature is extremely violent.

Selenium tetrafluoride
See Selenium tetrafluoride: Chlorine trifluoride

Trifluoromethanesulfenyl chloride
See Trifluoromethanesulfenyl chloride: Chlorine fluorides

Water
1. Sidgwick, 1950, 1156
2. Mellor, 1956, Vol. 2, Suppl. 1, 156, 158
3. Ruff, O. *et al., Z. Anorg. Chem.*, 1930, **190**, 270
4. Farrar, R. L. *et al.*, USDOE Rept. K/ET-252, 1979, 41

Interaction is violent and may be explosive, even with ice, oxygen being evolved [1]. Part of the water dropped into a flask of the gas was expelled by the violent reaction ensuing [2]. An analytical procedure, involving absorption of chlorine trifluoride into 10% sodium hydroxide solution from the open capillary neck of a quartz ampoule to avoid explosion, was described [3]. Inadvertent collection of chlorine trifluoride and ice in a cryogenic trap led to a small but violent explosion when the trap began to warm up overnight [4].
See other INTERHALOGENS

3982. Chlorine trifluoride oxide
[30708-80-6] ClF_3O

$$ClF_3O$$

Bougon, R. *et al.*, Fr. Pat. 2 110 555, 1972
It is a powerful oxidant, potentially useful in rocketry.
See other HALOGEN OXIDES

3983. Chlorine dioxygen trifluoride
[12133-60-7] ClF_3O_2

$$Cl(O_2)F_3$$

Organic materials
1. Streng, A. G., *Chem. Rev.*, 1963, **63**, 607
2. Christe, K. O. *et al., Inorg. Chem.*, 1973, **12**, 1357

It is a very powerful oxidant, but of low stability, which reacts explosively with organic materials [1]; such combinations should be avoided [2].

3984. Potassium tetrafluorochlorate(1−)
[19195-69-8] ClF_4K

$$K[ClF_4]$$

See entry METAL POLYHALOHALOGENATES

3985. Nitrosyl tetrafluorochlorate
[13815-21-9] ClF_4NO

$$N{:}O[ClF_4]$$

Organic compounds
Sorbe, 1968, 133
The powerful oxidant reacts explosively with many organic compounds.
See other NITROSO COMPOUNDS, OXIDANTS

3986. Tetrafluoroammonium perchlorate
[13706-14-4] ClF_4NO_4

$$N^+F_4ClO_4^-$$

Christe, K. O. *et al., Inorg. Chem.*, 1980, **19**, 1495
It decomposes rapidly at 25°C, giving the very shock-sensitive chloryl hypofluorite.
See other N-HALOGEN COMPOUNDS, PERCHLORATE SALTS OF NITROGENOUS BASES

3987. Chlorotetrafluorophosphorane
[13637-88-2] ClF_4P

$$ClPF_4$$

Preparative hazard
See Chlorine fluoride: Phosphorus trifluoride
See other NON-METAL HALIDES

3988. Rubidium tetrafluorochlorate(1−)
[15321-10-5] ClF_4Rb

$$Rb[ClF_4]$$

See entry METAL POLYHALOHALOGENATES

3989. Chlorine pentafluoride
[13637-63-3] ClF_5

$$ClF_5$$

Metals
See Chlorine trifluoride: Metals

Nitric acid
Christe, K. O., *Inorg. Chem.*, 1972, **11**, 1220
Interaction of anhydrous nitric acid with chlorine pentafluoride vapour at −40°C,
or with the liquid at above −100°C, is very vigorous.

Water
1. Pilipovich, D. *et al.*, *Inorg. Chem.*, 1967, **6**, 1918
2. Christe, K. O., *Inorg Chem.*, 1972, **11**, 1220
Interaction of liquid chlorine pentafluoride with ice at $-100°C$ [1], or of the vapour with water vapour above $0°C$ [2], is extremely vigorous.
See other INTERHALOGENS

3990. Pentafluorosulfur peroxyhypochlorite
[58249-49-3] ClF_5O_2S

$$F_5SOOCl$$

Haloalkenes
Hopkinson, M. J. *et al.*, *J. Org. Chem.*, 1976, **41**, 1408
The peroxyhypochlorite is especially reactive, and the fluoroperoxy compounds produced by its interaction with haloalkenes can detonate when subjected to thermal or mechanical shock. However, no explosions were experienced during this work.
See related HYPOHALITES, FLUORINATED PEROXIDES AND SALTS

3991. Difluoroperchloryl hexafluoroplatinate
[36609-92-4] ClF_8O_2Pt

$$F_2ClO_2 \ [PtF_6]$$

See entry DIFLUOROPERCHLORYL SALTS

3992. Chlorogermane
[13637-65-5] $ClGeH_3$

$$ClGeH_3$$

Ammonia
Johnson, O. H., *Chem. Rev.*, 1951, **48**, 274
Both mono- and di-chlorogermanes react with ammonia to give involatile products (presumably with N—Ge bonds) which explode on heating.
See related METAL HALIDES, METAL HYDRIDES
See other AMINATION INCIDENTS

3993. Hydrogen chloride (Hydrochloric acid)
[7647-01-0] ClH

$$HCl$$

FPA H41, 1975 (gas); *HCS 1980*, 546 (cylinder); *HCS 1980*, 541 (solution)

Preparative hazard
See Sulfuric acid, below

Alcohols, Hydrogen cyanide

1. Pinner, A., *Die Imidoäther*, Berlin, Openheim, 1892
2. Erickson, J. G., *J. Org. Chem.*, 1955, **20**, 1573

Preparation of alkyliminioformate chlorides (imidoester hydrochlorides) by passing hydrogen chloride rapidly into alcoholic hydrogen cyanide proceeds explosively (probably owing to a rapid exotherm), even with strong cooling [1]. Alternative procedures involving very slow addition of hydrogen chloride into a well-stirred mixture kept cooled to ambient temperature, or rapid addition of cold alcoholic hydrogen cyanide to cold alcoholic hydrogen chloride, are free of this hazard [2].

Aluminium   MRH 3.05/20

See Aluminium: Hydrogen chloride
Aluminium–titanium alloys: Oxidants

Caesium telluroacylates

Kato, S. *et al.*, *J. Amer. Chem. Soc.*, 1995, **116**(12), 1262

While generating tellurobenzoic acids by acidifying the corresponding cesium salts with hydrogen chloride in tubes immersed in liquid nitrogen, it was found that explosions sometimes occurred while later allowing the reaction tube to warm. This is due to overcooling; HCl freezes, and therefore becomes unreactive, well above liquid nitrogen temperature, permitting sudden and exothermic reaction during the warming period.

Chlorine, Dinitroanilines

1. Sihlbohm, L., *Acta Chem. Scand.*, 1953, **7**, 1197–1206
2. Harris, G. P. F. *et al.*, *Chem. & Ind.*, 1983, 183

The previously reported cleavage of dinitroanilines to chloronitrodiazonium salts by hydrochloric acid [1] is apparently catalysed by chlorine or other oxidants. The reaction takes place vigorously with copious gas evolution, but at low temperatures (40–25°C or lower) there can be a very long induction period before onset of the vigorous reaction. This could be hazardous on attempting to scale-up laboratory processes. The reactivity of the isomeric dinitroanilines varies, 2,3- being most reactive and 3,5- being unreactive. The cleavage reaction is specific for conc. hydrochloric acid, and has not been observed with hydrobromic acid or 30% sulfuric acid [2].

See other GAS EVOLUTION INCIDENTS, INDUCTION PERIOD INCIDENTS

1,1-Difluoroethylene

See 1,1-Difluoroethylene: Hydrogen chloride

Fluorine   MRH 3.18/35

See Fluorine: Hydrogen halides

Hexalithium disilicide

See Hexalithium disilicide: Acids

Metal acetylides or carbides
    Calcium carbide 3.97/46
    See Caesium acetylide: Mineral acids
        Rubidium acetylide: Acids
        Uranium dicarbide: Hydrogen chloride

Metals
    Anon., *Loss Prev. Bull.*, 1994, (116), 22
    Hydrochloric acid is capable of evolving hydrogen on reaction with many common metals. A magnetic flow meter is reported as exploding after the acid reached the electric amplifier compartment.

Other reactants
    Yoshida, 1980, 55
    MRH values calculated for 11 combinations, largely with oxidants, are given.

Potassium permanganate   MRH 0.75/37
    See Potassium permanganate: Hydrochloric acid

Silicon dioxide
    See Silicon dioxide: Hydrochloric acid

Sodium
    See Sodium: Acids

Sulfuric acid.
    1. *MCA Case History No. 1785*
    2. Libman, D. D., *Chem. & Ind.*, 1948, 728
    3. Smith, G. B. L. *et al., Inorg. Synth.*, 1950, **3**, 132
    Accidental addition of 6500 l of conc. hydrochloric acid to a bulk sulfuric acid storage tank released sufficient hydrogen chloride gas by dehydration to cause the tank to burst violently [1]. Complete dehydration of hydrochloric acid solution releases some 250 volumes of gas. A laboratory apparatus for effecting this safely has been described [2], which avoids the possibility of layer formation in unstirred flask generators [3].
    See other GAS EVOLUTION INCIDENTS

Tetraselenium tetranitride
    See Tetraselenium tetranitride: Alone, or Halogens and derivatives
    See other INORGANIC ACIDS, NON-METAL HALIDES, NON-METAL HYDRIDES

3994. Monopotassium perchlorylamide
   [16971-92-9]   ClHKNO$_3$
                              KNHClO$_3$

Sorbe, 1968, 67

Like the dipotassium salt, it will explode on impact or exposure to flame.
See entry PERCHLORYLAMIDE SALTS *See other* N-METAL DERIVATIVES

3995. Hypochlorous acid
[7790-92-3]   ClHO

HOCl

Acetic anhydride
   Rüst, 1948, 341
   A mixture of the anhydride and hypochlorous acid exploded violently while being
   poured. Some acetyl hypochlorite and/or dichlorine monoxide may have been
   formed.

Alcohols
   Mellor, 1956, Vol. 2, Suppl. 1, 560
   Contact of these, or of chlorine and alcohols, readily forms unstable alkyl hypochlo-
   rites.
   See HYPOHALITES

Ammonia
   Mellor, 1940, Vol. 8, 217
   The violent explosion occurring on contact with ammonia gas is due to formation of
   nitrogen trichloride and its probable initiation by the heat of solution of ammonia.
   See Chloramine (reference 3)

Arsenic
   Mellor, 1941, Vol. 2, 254
   Ignition occurs on contact.
   See other HYPOHALITES, OXOHALOGEN ACIDS

3996. Chloric acid
[7790-93-4]   ClHO$_3$

HOClO$_2$

   Muir, G. D., private comm., 1968
   Aqueous chloric acid solutions decompose explosively if evaporative concentration
   is carried too far.

Cellulose
   Mellor, 1946, Vol. 2, 310
   Filter paper ignites after soaking in chloric acid.

Metal sulfides
   Mellor, 1956, Vol. 2, Suppl. 1, 584
   Copper sulfide explodes with concentrated chloric acid solution (and also with
   cadmium, magnesium or zinc chlorates). Antimony trisulfide or arsenic trisulfide,
   or tin(II) and (IV) sulfides react incandescently with concentrated solutions.

Metals, or Organic materials
Majer, V., *Chemie* (Prague), 1948, **3**, 90–91
The recorded explosions of chloric acid have been attributed to the formation of explosive compounds with antimony, bismuth and iron (including hydrogen also in the latter case). Organic materials (and ammonia) are violently oxidised.

Oxidisable materials
In contact with oxidisable materials, reactions are similar to those of the metal chlorates.
See entry METAL CHLORATES *See other* OXOHALOGEN ACIDS

3997. Chlorosulfuric acid
[7790-94-5] $ClHO_3S$

$$ClSO_2OH$$

FPA H122, 1983; *HCS 1980*, 314

Diphenyl ether
Ehama, T. *et al.*, Jap. Pats. 74 45 034, 74 45 035, 1974
Presence of various nitrogen-containing compounds, or of fatty acids or their derivatives, controls the vigorous interaction of the ether and chlorosulfuric acid at above 40°C, producing higher yields of 4,4′-oxybisbenzenesulfonyl chloride.

Hydrocarbons   MRH Hexane, 0.96/24; Toluene, 1.29/37
1. Anon., *CIHSC Chem. Safety Summ.*, 1986, **57**(226), 65
2. Baker, J. D., *Chem. Eng. News*, 1989, **67**(50), 2
Use of heptane to wash sludge from a chlorosulfuric acid weigh tank led to violent rupture of the 2.5 kl vessel. Laboratory work showed that the two compounds react vigorously with gas evolution (HCl) sufficient to overpressure the tank which was fitted with a 40 mm relief line [1]. Another(?) tank containing chlorosulfonic acid ruptured after the lines to it were washed out with heptane, subsequent investigation showed that chlorosulfonic acid gives reaction, with large gas evolution, when stirred with commercial heptane fractions (and with hexane, but not cyclohexane, in the editor's experience.) Subsequent correspondence sought to attribute the observation to water contamination, which seems unlikely [2].
See other GAS EVOLUTION INCIDENTS, SULFONATION INCIDENTS

Other reactants
Yoshida, 1980, 115–116
MRH values calculated for 21 combinations, mainly with oxidisable materials, are given.

Phosphorus
Heumann, K. *et al., Ber.*, 1882, **15**, 417
The acid is a strong oxidant, and above 25–30°C, interaction is vigorous and accelerates to explosion.

Silver nitrate
See Silver nitrate: Chlorosulfuric acid

Sulfuric acid
Anon., *Loss Prev. Bull.*, 1977, (013), 2–3
Failure to start an agitator before charging 500 kg of 98% sulfuric acid ($d = 1.84$) into 560 kg of chlorosulfuric acid ($d = 1.44$) led to formation of a lower layer of sulfuric acid. Subsequent starting of the agitator led to eruption of part of the contents owing to rapid formation of hydrogen chloride gas from interaction of the chlorosulfuric acid with the water content (2%) in the sulfuric acid.
See other GAS EVOLUTION INCIDENTS
See other ACYL HALIDES, INORGANIC ACIDS, OXIDANTS

3998. Perchloric acid
[7601-90-3] $ClHO_4$
$HOClO_3$

(*MCA SD-11*, 1965); *NSC 311*, 1982; *FPA H53*, 1977; *HCS 1980*, 721

1. *MCA SD-11*, 1965
2. Lazerte, D., *Chem. Eng. News*, 1971, **49**(3), 33
3. Muse, L. A., *J. Chem. Educ.*, 1972, **49**, A463
4. Graf, F. A., *Chem. Eng. Progr.*, 1966, **62**(10), 109
5. Friedmann, W., *Chem. Eng. News*, 1947, **25**, 3458
6. Anon., *ABCM Quart. Safety Summ.*, 1936, **7**, 51
7. Annikov, V. E. *et al.*, *Chem. Abs.*, 1983, **99**, 160860
8. Anon., *Indian Chem. Manuf.*, 1976, **14**(5), 9–13
9. Schilt, A. A., *Perchloric Acid and Perchlorates*, Columbus (Oh.), G. F. Smith Chemical Co., 1979
10. Furr, A. K. (Ed.), *Handbook of Laboratory Safety*. 3rd Edn., 207, CRC Press, Boca Raton (Florida), 1990.

Most of the numerous and frequent hazards experienced with perchloric acid have been associated with either its exceptional oxidising power or the inherent instability of its covalent compounds, some of which form readily. Although the 70–72% acid of commerce behaves when cold as a very strong, but non-oxidising acid, it becomes an extreme oxidant and powerful dehydrator at elevated temperatures (160°C) or when anhydrous [1].
Where an equally strong but non-oxidising acid can be used, trifluoromethanesulfonic acid is recommended [2]. Safe handling procedures have been detailed [3] and an account of safe handling of perchloric acid in large-scale preparation of hydrazinium diperchlorate, with recommendations for materials of construction, has been published [4]. The lack of oxidising power of the unheated 72% aqueous acid was confirmed during analytical treatment of oil sludges on over 100 occasions [5]. A report of two explosions during evaporation of aqueous perchloric acid in a method for estimating potassium infers that the aqueous acid may be unstable. It is more likely that oxidisable material (probably ethanol) was present during the

evaporation [6]. The aqueous acid can be caused to detonate with a booster charge [7]. Safety procedures for using perchloric acid for various purposes have been reviewed [8], and the section on safety aspects in a text [9] is rated by a reviewer as particularly pertinent. Precautions for using perchloric acid and perchlorates are given, and many incidents described in [10].

Acetic anhydride
See Dehydrating agents, below (references 2–4)

Acetic anhydride, Acetic acid,   MRH Acetic acid 4.98/35
Organic materials
  1. Burton, H. *et al., Analyst*, 1955, **80**, 4
  2. Schumacher, 1960, 187, 193
  3. Kuney, J. H., *Chem. Eng. News*, 1947, **25**, 1659
  4. *Tech. Survey No. 2, Fire Hazards and Safeguards for Metalworking Industries*, US Board of Fire Underwriters, 1954

Mixtures of 70% (dihydrated) perchloric acid with enough acetic anhydride will produce a solution of anhydrous perchloric acid in acetic acid/anhydride, which is of high catastrophic potential [1]. Sensitivity to heat and shock depend on the composition of the mixture, and the vapour evolved on heating is flammable [2]. Such solutions have been used for electropolishing operations, and during modifications to an electroplishing process by the unqualified supervisor, a cellulose acetate rack was introduced into a large volume of an uncooled mixture of perchloric acid and acetic anhydride. Dissolution of the rack introduced organic materials (largely cellulose acetate, capable of transesterification to cellulose perchlorate) into the virtually anhydrous acid and caused the whole to explode disastrously, causing enormous damage. This cause was confirmed experimentally [3,4].
See Alcohols; Cellulose and derivatives; Vegetable matter, all below
See Acetic anhydride: Perchloric acid

Acetic anhydride, Carbon tetrachloride, 2-Methylcyclohexanone
  Gall, M. *et al., Org. Synth.*, 1972, **52**, 40
  During acetylation of the enolised ketone, the 70% perchloric acid must be added last to the reaction mixture to provide maximum dilution and cooling effect.

Acetic anhydride, Organic materials, Transition metals
  Hikita, T. *et al., J. Chem. Soc. Japan, Ind. Chem. Sect.*, 1951, **54**, 253–255
  The stability ranges of mixtures of the acid, anhydride and organic materials (ethanol, gelatine) used in electropolishing were studied. Presence of transition metals (chromium, iron, nickel) increases the possibility of explosion. (This is why such mixtures must not be stored after use for etching metals.)

Acetonitrile
  Andrussow, L., *Chim. Ind.* (Paris), 1961, **86**, 542–545
  The latent hazards in storing and handling the explosive mixtures of perchloric acid with acetonitrile or dimethyl ether are discussed.

Acetylene, Nitrous oxide
1. Hobson, J. D., *Chem. Brit.*, 1985, **21**, 632
2. Anon., *Lab. Haz. Bull.*, 1987, (9), item 735
3. Andrew, B. E., *J. Anal. At. Spectrom.*, 1988, **3**, 401–405
A number of explosions have been experienced when using sample solutions containing perchloric acid in atomic absorption spectrometers using acetylene–nitrous oxide flames. Further information is being sought from other AAS users [1]. A warning against using perchloric acid digests in AA instruments using such flames has also been given by an American source [2]. The role of perchlorate accumulation and the effects of instrumental and experimental variables upon such explosions has been studied and preventive measures proposed [3].
See ATOMIC ABSORPTION SPECTROSCOPY

Alcohols   MRH Ethanol 6.27/21
1. Sweasey, D., *Lab. Practice*, 1968, **17**, 915
2. Moureu, H. *et al., Arch. Mal. Prof. Med.*, 1951, **12**, 157–159
3. Michael, A. T. *et al., Amer. Chem. J.*, 1900, **23**, 444
4. Bailar, 1973, Vol. 2, 1451
During digestion of a lipid extract (1 mg) with 72% acid (1.5 ml), a violent explosion occurred. This was attributed to residual traces of the extraction solvent (methanol–chloroform) having formed methyl perchlorate [1]. In the analytical determination of potassium as perchlorate, heating the solid containing traces of ethanol and perchloric acid caused a violent explosion [2]. Contact of drops of anhydrous perchloric acid and ethanol caused immediate violent explosion [3]. Partial esterification of polyfunctional alcohols (ethylene glycol, glycerol, pentaery-thritol) with the anhydrous acid gives extremely explosive liquids which may explode on pouring from one vessel to another [4].
See Glycols and their ethers; *and* Methanol, Triglycerides; both below
See ALKYL PERCHLORATES

Aniline, Formaldehyde   MRH Aniline 6.65/18
Aniline, 86, New York, Allied Chem. Corp., 1964
Aniline reacts with perchloric acid and then formaldehyde to give an explosively combustible condensed resin.

Antimony(III) compounds
Burton, H. *et al., Analyst*, 1955, **80**, 4
Treatment of tervalent compounds of antimony with perchloric acid can be very hazardous.
See Bismuth, below

Azo-pigment, Orthoperiodic acid
1. Anon., *ABCM Quart. Safety Summ.*, 1961, **32**, 125
2. Smith, G. F. *et al., Talanta*, 1960, **4**, 185
During the later stages of the wet oxidation of an azo-pigment with mixed perchloric and orthoperiodic acids, a violent reaction, accompanied by flashes of

light, set in and terminated in an explosion [1]. The general method upon which this oxidation was based is described as hazard-free [2].

Bis(1,2-diaminopropane)-*cis*-dichlorochromium(III) perchlorate
See Bis(1,2-diaminopropane)-*cis*-dichlorochromium(III) perchlorate: Perchloric acid

1,3-Di[bis(cyclopentadienyl)iron]-2-propene-1-one
See 1,3-Di[bis(cyclopentadienyl)iron]-2-propene-1-one: Perchloric acid

Bis(2-hydroxyethyl) terephthalate, Ethanol, Ethylene glycol
491M, 1975, 303
When a solution of the ester in ethanol containing 5% of perchloric acid was evaporated to dryness, there was a violent explosion (probably owing to formation of ethyl perchlorate, perhaps with some ester perchlorate or ethylene glycol perchlorate also being formed). A similar mixture containing ethylene glycol flashed brightly after refluxing for 18 h.
See Alcohols, above; Glycols and their ethers, below

Bismuth
Nicholson, D. G. *et al., J. Amer. Chem. Soc.*, 1935, **57**, 817
Attempts to dissolve bismuth and its alloys in hot perchloric acid carry a very high risk of explosion. At 110°C a dark brown coating is formed, and if left in contact with the acid (hot or cold), explosion occurs sooner or later. The same is true of antimony and its tervalent compounds.

Calcium compounds
Akerman, L. A. et al., *Combustion, Explos., & Shockwaves*, 1987, **23**, 178
Calcium chloride and oxide catalyse the second, violent, stage of ammonium perchlorate decomposition and increase the shock sensitivity of mixtures with sugar.

Carbon   MRH 6.19/17
1. Mellor, 1941, Vol. 2, 380
2. Pascal, 1960, Vol. 16, 300–301
Contact of a drop of the anhydrous acid with wood charcoal causes a very violent explosion [1], and carbon black also reacts violently [2].
See Graphitic carbon, Nitric acid, below

Cellulose and derivatives   MRH Cellulose 5.81/33
1. Schumacher, 1960, 187, 195
2. Harris, E. M., *Chem. Eng.*, 1949, **56**, 116
3. Sutcliffe, *J. Textile Ind.*, 1950, **41**, 196T
Contact of the hot concentrated acid or the cold anhydrous acid with cellulose (as paper, wood fibre or sawdust, etc.) is very dangerous and may cause a violent explosion. Many fires have been caused by long-term contact of diluted acid with wood (especially in fume cupboards) with subsequent evaporation and ignition

[1,2]. Contact of cellulose acetate with 1200 1 of uncooled anhydrous acid in acetic anhydride led to a particularly disastrous explosion [1], and interaction of benzyl cellulose with boiling 72% acid was also explosive [3]. Perchlorate esters of cellulose may have been involved in all these incidents.

See Hydrofluoric acid, Structural materials, below

Charcoal, Chromium trioxide
Randall, W. R., private comm., 1977
A wet-ashing technique used for dissolution of graphite in perchloric acid involved boiling a mixture of 70% perchloric acid and 1% of chromium trioxide as an aqueous solution. This method was later applied to 6–14 mesh charcoal, and after boiling for 30 min the reaction rate increased (foaming) and accelerated to explosion. The charcoal contained traces of extractable tar.

Clay
Atkinson, G. F., *Chem. Eng. News*, 1987, **65**(39), 2
Digestion of clay samples with perchloric acid led to 3 explosions, each of which destroyed a ceramic hotplate used as heat source. There seems a good case for using metal hotplates for reactions involving risk of explosion.

See Hydrogen peroxide: Acetic acid, Jute

Combustible materials
Elliot, M. A. *et al., Rept. Invest. No. 4169*, Washington, US Bur. Mines, 1948
Tests of sensitivity to initiation by heat, impact, shock or ignition sources were made on mixtures of a variety of absorbent materials containing a stoicheiometric amount of 40–70% perchloric acid. Wood meal with 70% acid ignited at 155°C and a mixture of coal and 60% acid which did not ignite below 200°C ignited at 90°C when metallic iron was added. Many of the mixtures were more sensitive and dangerous than common explosives.

Copper dichromium tetraoxide
Solymosi, F. *et al., Proc. 14th Combust. Symp.*, 1309–1316, 1973
The mixed oxide (copper chromite) was the most effective of several catalysts for the vapour-phase decomposition of perchloric acid, decomposition occurring above 120°C.

Dehydrating agents
1. Mellor, 1941, Vol. 2, 373, 380; 1956, Vol. 2, Suppl. 2.1, 598, 603
2. Kuney, J. H., *Chem. Eng. News*, 1947, **25**, 1659
3. Burton, H. *et al., Analyst*, 1955, **80**, 4
4. Schumacher, 1960, 71, 187, 193
5. Wirth, C. M. P., *Lab. Practice*, 1966, **15**, 675
6. Plesch, P. H. *et al., Chem. & Ind.*, 1971, 1043
7. Musso, H. *et al., Angew. Chem.*, 1970, **82**, 46
Although commercial 70–72% perchloric acid (approximating to the dihydrate) itself is stable, incapable of detonation (except by a booster charge) and readily

stored, it may be fairly readily dehydrated to the anhydrous acid by contact with drying agents. This is not safe when stored at ambient temperature, since it slowly decomposes, even in the dark, with accumulation of chlorine dioxide in the solution, which darkens and finally explodes after some 30 days [1,2,5]. The 72% acid (or perchlorate salts) may be converted to the anhydrous acid by heating with sulfuric acid, phosphorus pentoxide or phosphoric acid, or by distillation under reduced pressure [1,4,5]. In contact with cold acetic anhydride, mixtures of the anhydrous acid with excess anhydride and acetic acid are formed, which are particularly dangerous, being sensitive to mechanical shock, heating, or the introduction of organic contaminants [2,3,4]. A solution of the monohydrated acid in chloroform exploded in contact with phosphorus pentoxide [1]. A safer method of preparing anhydrous solutions of perchloric acid in dichloromethane, which largely avoids the risk of explosion, has been described [6]. Further precautions are detailed in an account of an explosion during a similar preparation [7]. Solutions of anhydrous acid of less than 5% concentration in acetic acid or anhydride are relatively stable.

Deoxyribonucleic acid
Cochrane, A. R. G. *et al., School Sci. Rev.*, 1977, **58**, 706–708
Hazards of using perchloric acid to hydrolyse DNA are stressed. Perchloric acid can cause ignition of any organic material, even a considerable time after contact. Other acids to effect hydrolysis are suggested.

Dichloromethane, Dimethyl sulfoxide   MRH Dimethyl sulfoxide 6.19/23
Anon., *Univ. Safety Assoc. Safety News*, 1978, (9), 10
When a syringe used for DMSO and rinsed with dichloromethane was being filled with perchloric acid, a violent explosion resulted.
See Sulfoxides, below

Diethyl ether   MRH 6.61/18
Michael, A. T. *et al., Amer. Chem. J.*, 1900, **23**, 444
The explosions sometimes observed on contact of the anhydrous acid with ether are probably owing to formation of ethyl perchlorate by scission of the ether, (or possibly to formation of diethyloxonium perchlorate).

Dimethyl ether
See Acetonitrile, above

Dioxane, Nitric acid
MCA Guide, 1972, 312–313
Boron trifluoride in aqueous dioxane was being evaporated with nitric acid treatment (3 portions) for analysis. Final addition of perchloric acid with continued heating led to an explosion while unattended. Ring scission of dioxane to diethylene glycol and formation of diethylene glycol nitrate and/or perchlorate may have been involved.
See Nitric acid, Organic matter, below

Ethylbenzene, Thallium triacetate
Uemura, S. *et al., Bull. Chem. Soc., Japan.*, 1971, **44**, 2571
Application of a published method of thallation to ethylbenzene caused a violent explosion. A reaction mixture of thallium triacetate, acetic acid, perchloric acid and ethylbenzene was stirred at 65°C for 5 h, then filtered from thallous salts. Vacuum evaporation of the filtrate at 60°C gave a pasty residue which exploded. This preparation of ethylphenylthallic acetate perchlorate monohydrate had been done twice previously and uneventfully, as had been analogous preparations involving thallation of benzene, toluene, three isomeric xylenes and anisole in a total of 150 runs, where excessive evaporation had been avoided.

Faecal material, Nitric acid
Anon., *Univ. Safety Assoc. Safety News*, 1979, (12), 24
In the determination of metal ions in animal faeces, digestion with 12% perchloric and 56% nitric acids was in progress when an explosion occurred. This was attributed to one of the samples going to dryness on a sand tray heater.
See Nitric acid, Organic matter, below

Fluorine
Rohrbock, G. H. *et al., J. Amer. Chem. Soc.*, 1947, **69**, 677–678
Contact of fluorine and 72% acid at ambient temperature produces a high yield of the explosive gas, fluorine perchlorate.

Fume cupboards
Furr, A. K. (Ed.), *Handbook of Laboratory Safety.* 3rd Edn., 207, CRC Press, Boca Raton (Florida), 1990.
Fume cupboards or their vent lines which have frequently been used with perchloric acid have often spontaneously deflagrated or exploded because of spillage or absorbed vapour.
See Hydrofluoric acid, Structural materials, below; Cellulose and derivatives, above

Glycerol, Lead oxide
MCA Case History No. 799
During maintenance work on casings of fans used to extract perchloric acid fumes, seven violent explosions occurred when flanges sealed with lead oxide–glycerol cement were disturbed. The explosions, attributed to formation of explosive compounds by interaction of the cement with perchloric acid, may have involved perchlorate esters and/or lead salts. Use of an alternative inorganic silicate–hexafluorosilicate cement is recommended.

Glycols and their ethers   MRH Ethylene glycol 5.69/30
1. Schumacher, 1960, 195, 214
2. Comas, S. M. *et al., Metallography*, 1974, **7**, 45–47
Glycols and their ethers undergo violent decomposition in contact with ~70% perchloric acid. This seems likely to involve formation of the glycol perchlorate esters (after scission of ethers) which are explosive, those of ethylene glycol

and 3-chloro-1,2-propanediol being more powerful than glyceryl nitrate, and the former so sensitive that it explodes on addition of water [1]. Investigation of the hazards associated with use of 2-butoxyethanol for alloy electropolishing showed that mixtures with 50–95% of acid at 20°C, or 40–90% at 75°C, were explosive and initiable by sparks. Sparking caused mixtures with 40–50% of acid to become explosive, but 30% solutions appeared safe under static conditions of temperature and concentration.

See Alcohols, above

Graphitic carbon, Nitric acid
Buzzelli, G. *et al., Talanta*, 1977, **24**, 383–385
Dissolution procedures are described for gram samples of graphite or pyrolytic carbon, milligram samples of irradiated fuel particles, and for more readily oxidised forms of carbon, such as charcoal. The first two methods involve heating the samples with mixtures of 70% perchloric and 90% nitric acids (10:1), and must only be used for graphite or pyrolytic carbon. Other forms of carbon must not be oxidised in this way (to avoid explosions), but by a preliminary treatment with nitric acid alone and in portions.

See Carbon, above

Hydrofluoric acid, Structural materials
1. Anon., *Sichere Chemiearb.*, 1983, **35**, 85–86
2. Anon., *Sichere Chemiearb.*, 1984, **36**, 59
A violent explosion in a fume hood, in which inorganic siliceous materials had been digested with perchloric and hydrofluoric acids on a hotplate-heated sandbath during several years, originated from the tiled working surface. This consisted of ceramic tiles set in acid-proof organic resin-bond cement with mineral filler and supported on a 4 cm chipboard base. It appeared that prolonged use of the acids had led to development of hair cracks in the tiled surface, slow penetration of the acids into the underlying chipboard base and progressive formation of cellulose perchlorate and other perchlorate esters from the resin binder beneath the tiled surface. The explosion was probably initiated by impact or friction, perhaps from moving the hotplate on the tiled surface. The significance of such factors in preventing such accidents is discussed [1]. A new piece of equipment should prevent the occurrence of such accidents caused by long-term use of perchloric acid. This is designed to accomodate 8 separate digestion vessels (PTFE for up to 250°C, other material in development for higher temperatures) heated in an aluminium block. Each vessel is connected by a PTFE line into a gas manifold which leads to an alkali-containing absorption vessel to completely neutralise the gases or vapours evolved during acid digestion [2].

See Cellulose and derivatives, above

Hydrogen
1. Dietz, W., *Angew. Chem.*, 1939, **52**, 616
2. Schumacher, 1960, 189
Occasional explosions experienced during use of hot perchloric acid to dissolve steel samples for analysis [1] were attributed to formation of hydrogen–perchloric

acid vapour mixtures and their ignition by steel particles at temperatures as low as 215°C [2].

Hydrogen halides
See Fluoronium perchlorate, *also* Chloronium perchlorate

Iodides
Michael, A. T. *et al., Amer. Chem. J.*, 1900, **23**, 444
The anhydrous acid ignites in contact with sodium iodide or hydriodic acid.

Iron(II) sulfate
Tod, H., private comm., 1968
During preparation of iron(II) perchlorate, a mixture of iron sulfate and perchloric acid was being strongly heated when a most violent explosion occurred. Heating should be gentle to avoid initiating this redox system.
See other REDOX REACTIONS

Ketones   MRH Acetone 6.48/21
Schumacher, 1960, 195
Ketones may undergo violent decomposition in contact with 70% acid.

Methanol, Triglycerides
  1. Mavrikos, P. J. *et al., J. Amer. Oil Chem. Soc.*, 1973, **50**(5), 174
  2. Wharton, H. W., *J. Amer. Oil Chem. Soc.*, 1974, **51**(2), 35–36
Attention is drawn to the hazards involved in the use of perchloric acid in a published method [1] for transesterification of triglycerides with methanol. Alternative acid catalysts and safety precautions are suggested [2].

2-Methylpropene, Metal oxides
Lesnikovich, L. I. *et al., Chem. Abs.*, 1975, **83**, 149760; 1976, 85, 7929
Mixtures of the alkene and perchloric acid vapour (5:1 molar) in nitrogen ignite spontaneously at 250°C. Some metal oxides of low specific surface reduced the ignition temperature below 178°C.

Nitric acid, Organic matter
  1. Anon., *Ind. Eng. Chem. (News Ed.)*, 1937, **15**, 214
  2. Lambie, D. A., *Chem. & Ind.*, 1962, 1421
  3. Mercer, E. R., private comm., 1967
  4. Muse, L. A., *Chem. Eng. News*, 1973, **51**(6), 29–30
  5. Cooke, G. W., private comm., 1967
  6. Kuney, J. H., *Chem. Eng. News*, 1947, **25**, 1659
  7. Balks, R. *et al., Chem. Abs.*, 1939, **33**,2438.9
  8. Rooney, R. C., *Analyst*, 1975, **100**, 471–475
  9. Martinie, G. D. *et al., Anal. Chem.*, 1976, **48**, 70–74
  10. *MCA Case History No. 2145*
  11. Anon., *Univ. Safety Assoc. Safety News*, 1978, (9), 7
  12. Satzger, R. D. *et al., J. Assoc. Off. Anal. Chem.*, 1983, **66**, 985–988

13. Frank, A., *Chem. Abs.*, 1983, **99**, 197442
14. Anon., *Lab. Accid Higher Educ.*, Item 15, HSE, Barking, 1987

The mixed acids have been used to digest organic materials prior to analysis, but several explosions have been reported, including those with vegetable oil [1], milk [2], and calcium oxalate precipitates from plants [3]. To avoid trace metal contamination by homogenising rat carcases in a blender, the whole carcases were dissolved in nitric acid. After separation of fat and addition of perchloric acid (125 ml), evaporation to near-dryness caused a violent explosion [4]. Finely ground plant material in contact with perchloric/nitric acid mixture on a heated sand bath became hot before all the plant material was saturated with acid and it exploded. Subsequent digests left overnight in contact with cold acids proceeded smoothly [5]. Cellulose nitrate and/or perchlorate may have been involved. Treatment with nitric acid before adding perchloric acid was a previously used and well-tried safe procedure [6]. Application to animal tissues of a method previously found satisfactory for plant material caused a violent explosion [7].

A method which involves gradually increasing the liquid temperature by controlled distillation during the digestion of organic matter with the hot mixed acids had been described as safe provided that strict control of the distillation process were observed [4]. During oxidation of large, high-fat samples of animal matter by the the usual technique, the existence of a layer of separated fat on the perchloric acid mixture created a highly hazardous situation. A modified procedure is free of explosion risk [8]. The effectiveness of mixed perchloric and nitric acids in wet oxidation of a wide range of organic materials has been studied. Violent oxidation occasionally occurred, but addition of sulfuric acid prevented any explosions or ignition during digestions [9]. The case history describes an explosion during wet-ashing of a phosphorus-containing polymer. One of the flasks may not have been topped up (i.e. diluted) with nitric acid during the digestion phase [10]. Digestion of vegetable matter with a 4:1 mixture of nitric/perchloric acids led to an explosion [11]. A safer alternative to the use of mixed acids for digestion of bone-meal for lead determination is pressurised dissolution in hydrochloric acid, then stripping voltammetry [12]. Equipment and procedures to minimise the risk of explosion during wet digestion of fat-containing materials are detailed [13]. An untried procedure to digest fresh cows milk with the mixed acids at 180°C led to a violent explosion [14].

Nitric acid, Pyridine, Sulfuric acid
Randall, W. L., *Safety in Handling Hazardous Chemicals*, 8–10, UCID-16610, Lawrence Livermore Lab., Univ. Calif., 1974
Some rare-earth fluoride samples had been wet-ashed incompletely with the three mixed acids and some gave low results. These samples, now containing pyridine, were reprocessed by addition of more acids and slow evaporation on a hot-plate. One of the samples frothed up and then exploded violently. Pyridinium perchlorate seems likely to have been involved.

Nitrogenous epoxides
Harrison, G. E., private comm., 1966

Traces of perchloric acid used as hydration catalyst for ring opening of nitrogenous epoxides caused precipitation of an organic perchlorate salt which was highly explosive. The concentration of acid was less than 1% by vol.
See other PERCHLORATE SALTS OF NITROGENOUS BASES

Oleic acid
1. Swern, D. *et al.*, US Pat. 3 054 804; *Chem. Eng. News*, 1963, **41**(12), 39
2. Anon., *Chem. Eng. News*, 1963, **41**(27), 47
The improved preparation of 1,4-octadecanolactone [1] involves heating oleic acid (or other C_{18} acids) with 70% perchloric acid to 115°C. This is considered to be a potentially dangerous method [2].

Organic materials, Sodium hydrogen carbonate
1. Zderic, J. A. *et al.*, *J. Amer. Chem. Soc.*, 1960, **82**, 446
2. Anon., *Univ. Safety Assoc. Safety News*, 1978, (9), 8–9
Following a published procedure [1], perchloric acid was used as catalyst in preparing a diol ketal, and was neutralised with sodium hydrogencarbonate before workup. Some sodium perchlorate remained dissolved after filtering the reaction mixture, and during subsequent vacuum distillation (with the bath temperature increasing to 200°C) an explosion occurred.

Organophosphorus compounds
Anderson, R. L. *et al.*, *Clin. Chim. Acta*, 1982, **121**, 111–116
The standard method for assaying organophosphorus compounds can be modified to use sulfuric acid to digest the samples and hydrogen peroxide as oxidant in place of perchloric acid.

Other reactants
Yoshida, 1980, 78
MRH values calculated for 19 combinations with oxidisable materials are given.

Phenylacetylene
1. Martheard, J.-P. *et al.*, *Tetrahedron Lett.*, 1982, **23**, 3484
2. Martheard, J.-P. *et al.*, *J. Chem. Res., Synop.*, 1983, 224–225
A mixture of phenylacetylene and the pure (?anhydrous) acid prepared at −180°C exploded when allowed to warm to −78°C, possibly owing to formation of unstable 1-phenylethenyl perchlorate [1]. Reaction of various phenylacetylenes in acetic acid at 40°C is controllable [2].

Phosphine
See Phosphonium perchlorate

Pyridine
See Pyridinium perchlorate

Sodium phosphinate   MRH 3.68/63
Smith, G. F., *Analyst*, 1955, **80**, 16

Although no interaction occurs in the cold, these powerful oxidising and reducing agents violently explode on heating.
See other REDOX REACTIONS

Sulfinyl chloride
Bailar, 1973, Vol. 2, 1442
The anhydrous acid ignites the chloride.

Sulfoxides   MRH Dimethyl sulfoxide 6.19/23
1. Therésa, J. de B., *Anales Soc. Espan. Fis. y Quim.*, 1949, **45B**, 235
2. Graf, F. A., *Chem. Eng. Progr.*, 1966, **62**(10), 109
3. Uemura, S. *et al., Bull. Chem. Soc. Japan*, 1971, **44**, 2571
4. *471M*, 1975, 302
5. Anon., *Jahresber.*, 1990, 64
6. Eigenmann, K. *et al., Angew. Chem. (Intern. Ed.)*, 1975, **14**, 647–648

Lower members of the series of salts formed between organic sulfoxides and perchloric acid are unstable and explosive when dry. That from dibenzyl sulfoxide explodes at 125°C [1]. Dimethyl sulfoxide explodes on contact with 70% perchloric acid solution [2]; one drop of acid added to 10 ml of sulfoxide at 20°C caused a violent explosion [3], and dibutyl sulfoxide behaves similarly [4]. A fatal explosion resulted from mistakenly connecting a DMSO reservoir to an autopipette previously used with perchloric acid [5]. (The editor has met a procedure for methylthiolation of aromatics where DMSO was added to excess 70% perchloric acid; he did not feel justified in trying to scale it up.) Explosions reported seem usually to result from addition to excess sulfoxide. Aryl sulfoxides condense uneventfully with phenols in 70% perchloric acid, but application of these conditions to the alkyl sulfoxide (without addition of the essential phosphoryl chloride) led to a violent explosion [4]. Subsequent investigation showed that mixtures of phenol and perchloric acid are thermally unstable (ester formation?) and may decompose violently, the temperature range depending on composition. DSC measurements showed that sulfoxides alone may decompose violently at elevated temperatures; e.g. dimethyl sulfoxide, 270–355, cyclohexyl methyl sulfoxide, 181–255, or methyl phenyl sulfoxide, 233–286°C, respectively [6].
See Dimethyl sulfoxide: Metal oxosalts

Sulfuric acid, Organic materials
Young, E. G. *et al., Science* (New York), 1946, **104**, 353
Precautions are necessary to prevent explosions when using the mixed acids to oxidise organic materials for subsequent analysis. The sulfuric acid probably tends to dehydrate the 70% perchloric acid to produce the hazardous anhydrous acid.
See Nitric acid, etc., above

Sulfur trioxide
Pascal, 1960, Vol. 16, 301
Interaction of the anhydrous acid and sulfur trioxide is violent and highly exothermic, even in presence of chloroform as diluent, and explosions are frequent.
See Dehydrating agents, above

Trichloroethylene
Prieto, M. A. *et al., Research on the Stabilisation and Characterisation of Highly Concentrated Perchloric Acid*, 36, Whittier, Calif., American Potash and Chem. Corp., 1962
The solvent reacts violently with the anhydrous acid.

Trimethylplatinum hydroxide
See Dodecamethyltetraplatinum(IV) perchlorate

Vegetable matter
Tabatabai, M. A. *et al., Agron. J.*, 1982, **74**, 404–406
As a safer alternative to digestion of vegetable matter with perchloric acid, alkaline oxidation of sulfur compounds to sulfate by sodium hypobromite, and reduction of sulfate to hydrogen sulfide by hydriodic acid/formaldehyde/phosphinic acid is recommended.
See Sulfuric acid, Organic materials, above

Zinc phosphide
Muir, G. D., private comm., 1968
Use of perchloric acid to assist solution of a sample for analysis caused a violent reaction.
See other REDOX REACTIONS
See other INORGANIC ACIDS, OXIDANTS, OXOHALOGEN ACIDS

3999. Mercury(II) amide chloride
[10124-48-8]
ClH_2HgN
$$H_2NHgCl$$

Halogens
Schwarzenbach, V., *Ber.*, 1875, **8**, 1231–1234
Several min after addition of ethanol to a mixture of the amide chloride ('fusible white precipitate') and iodine, an explosion occurs. Addition of the compound to chlorine gas or bromine vapour leads to a delayed violent or explosive reaction. Amminemetal salts behave similarly, and formation of N-halogen compounds is involved in all cases.
See other N-METAL DERIVATIVES, MERCURY COMPOUNDS

4000. Chloramine (Chloramide)
[10599-90-3]
ClH_2N
$$ClNH_2$$

1. Marckwald, W. *et al., Ber.*, 1923, **56**, 1323
2. Coleman, G. H. *et al., Inorg. Synth.*, 1939, **1**, 59
3. Walek, W. *et al., Tetrahedron*, 1976, **32**, 627

The solvent-free material, isolated at −70°C, disproportionates violently (sometimes explosively) at −50°C to ammonium chloride and nitrogen trichloride [1]. Ethereal solutions of chloramine are readily handled [2]. In the preparation of chloramine by reaction of sodium hypochlorite with ammonia, care is necessary to avoid excess chlorine in the preparation of the hypochlorite from sodium hydroxide, because nitrogen trichloride may be formed in the subsequent reaction with ammonia [3].

See other N-HALOGEN COMPOUNDS

4001. Ammonium chloride
[12125-02-9] ClH_4N

$$NH_4Cl$$

HCS 1980, 142

1. Alfenaar, M. *et al.*, Neth. Appl. 73 07 035, 1974
2. Knothe, M., *Schriftenr. GDMB*, 1998, (82), 101

Ammonium chloride is generally inoffensive. However, subjected to oxidation, especially electrochemically, it has the potential to form nitrogen trichloride. Cyanogen halides may be prepared by electrolysis of hydrogen cyanide or its salts mixed with halide salts. If ammonium chloride is used as the halide salt, precautions to prevent formation of explosive nitrogen trichloride are necessary [1]. Nitrogen trichloride is an under-recognised hazard of electrolytic operations in hydrometallurgy, where it may appear at pH <9 from electrolysis of ammonium or amine salts in chloride electrolytes [2].

See ELECTROLYSIS

Interhalogens
See Bromine trifluoride: Ammonium halides
    Bromine pentafluoride: Acids, etc.

Other reactants
Yoshida, 1980, 53
MRH values calculated for 10 combinations with various reagents are given.

Potassium chlorate   MRH 1.46/60
See Potassium chlorate: Ammonium chloride
See related METAL HALIDES

4002. Hydroxylaminium chloride
[5470-11-1] ClH_4NO

$$HON^+H_3 \ Cl^-$$

FPA H66, 1978

Manganese dioxide
Chaterjee, B. P. *et al.*, *Talanta*, 1977, **24**, 180–181

In a method for titrimetric determination of manganese in pyrolusite ore, addition of the powdered ore to a cold 20% solution of the salt causes vigorous decomposition to occur.

See entry HYDROXYLAMINIUM SALTS

4003. Ammonium chlorate
[10192-29-7] ClH_4NO_3

$$N^+H_4 \; ClO_3^-$$

1. Brauer, 1963, Vol. 1, 314
2. Urbanski, 1965, Vol. 2, 476

It occasionally explodes spontaneously, and invariably above 100°C [1]. It will explode after 11 h at 40°C, and after 45 min at 70°C. Ammonium and chlorate salts should not be mixed together [2].

Water

Mellor, 1941, Vol. 2, 339; 1956, Vol. 2, Suppl. 1, 591

A cold saturated solution may decompose explosively after a few days if much excess solid is present. Hot aqueous solutions have exploded during evaporation in steam heated vessels.

See other OXOSALTS OF NITROGENOUS BASES

4004. Ammonium perchlorate
[7790-98-9] ClH_4NO_4

$$N^+H_4 \; ClO_4^-$$

1. *MCA Case History No. 1002*
2. Barret, P., *Cah. Therm.*, 1974, **4**, 13–22
3. Seltzer, R. J., *Chem. Eng. News*, 1988, **66**(32), 7

Materials for a batch of ammonium perchlorate castable propellant were charged into a mechanical mixer. A metal spatula was left in accidentally, and the contents ignited when the mixer was started, owing to local friction caused by the spatula. A tool-listing safety procedure was instituted [1]. The literature on the kinetics of thermal decomposition has been reviewed critically [2].

For an account of the destruction of a plant producing ammonium perchlorate, which was not generally considered to be explosive, see [3]. Below 400°C it behaves as an oxidising agent but decomposition is not violent. The event suggests that tonnage quantities can detonate in some circumstances. The fire was started in spilt waste by sparks from welding and spread to storage areas on asphalt (fuel!); two aluminium bins, each containing five tonnes of perchlorate then detonated.

Aluminium   MRH 9.99/27

Loftus, H. J. *et al., Rept. AD-769283/3GA*, 1–98, Springfield (Va.), USNTIS, 1973

The powdered solid materials have been evaluated as a practical propellant pair.

Galwey, A. K. *et al., Trans. Faraday Soc.*, 1960, **56**, 581

Below 240°C intimate mixtures with sugar charcoal undergo exothermic decomposition, while mild explosions occur above 240°C.

Catalysts
1. Solymosi, F., *Acta Phys. Chem.*, 1974, **20**, 83–103
2. Shadman-Yazdi, F., *Proc. 1973 Iran. Congr. Chem. Eng.*, 1974, **1**, 353–356
3. Glazkova, A. P., *Chem. Abs.*, 1976, **85**, 162846
4. Rastogi, R. P. *et al., J. Appl. Chem. Biotechnol.*, 1978, **28**, 889–894

In a review of the course and mechanism of the catalytic decomposition of ammonium perchlorate, the considerable effects of metal oxides in reducing the explosion temperature of the salt are described [1]. Solymosi's previous work had shown reductions from 440° to about 270° by dichromium trioxide, to 260° by 10 mol% of cadmium oxide and to 200°C by 0.2% of zinc oxide. The effect of various concentrations of 'copper chromite', copper oxide, iron oxide and potassium permanganate on the catalysed combustion of the propellant salt was studied [2]. Similar studies on the effects of compounds of 11 metals and potassium dichromate in particular, have been reported [3]. Presence of calcium carbonate or calcium oxide has a stabilising effect on the salt, either alone or in admixture with polystyrene [4].

Copper   MRH 2.17/57
Anon., *Chem. Engrg.*, 1955, **62**(12), 335

Crystalline ammonium perchlorate ignited in contact with hot copper pipes.

Ethylene dinitrate
MCA Case History No. 1768

Samples of mixtures of ammonium perchlorate and the highly explosive liquid nitrate kept at 60°C ignited after 7 days. Many adverse criticisms of the general planning and execution of the experiments were made.

Impurities
1. Jacobs, P. W. M. *et al., Chem. Rev.*, 1969, **69**, 590
2. Jain, S. R. *et al., Combust. Flame*, 1981, **40**, 113–120

The medium impact-sensitivity of this solid propellant component is greatly increased by co-crystallisation of certain impurities, notably nitryl perchlorate, potassium periodate and potassium permanganate [1]. The presence of certain minimum amounts of mono-, di, tri- or tetra-methylammonium perchlorates in the salt leads to a single step decomposition, at around 290°C for the mono-derivative [2].

See other CATALYTIC IMPURITY INCIDENTS

Metal perchlorates
1. Solymosi, F., *Combust. Flame*, 1966, **10**, 398–399
2. Solymosi, F., *Z. Phys. Chem.*, 1969, **67**, 76–85

3. Patel, K. C. *et al.*, *Combust. Flame*, 1976, **27**, 295–298
4. Al Fakir, M. S., *Progr. Astronaut. Aeronaut.*, 1981, **76**, 5512–564
Admixture of lithium perchlorate [1] or zinc perchlorate [2] leads to decomposition with explosion at 290° or ignition at 240°C, respectively. The role of ammine derivatives of lithium and magnesium perchlorates in catalysing the thermal decomposition of ammonium perchlorate has been studied [3], and lithium perchlorate has a strong catalytic effect on the burning rate [4].

Metals, or Organic materials,   MRH Magnesium 12.97/49
or Sulfur   MRH Sulfur 3.59/27
1. *Haz. Chem. Data*, 1975, 51
2. Klager, K. *et al.*, *Chem. Abs.*, 1982, **97**, 94903
The powdered oxidant functions as an explosive when mixed with finely divided metals, organic materials or sulfur, which increase the shock-sensitivity up to that of picric acid [1]. The hazardous properties of such mixtures increase as the particle size of the oxidant salt decreases [2].

Nitrophenol–formaldehyde polymer
Girdhar, H. L. *et al.*, *Indian J. Technol.*, 1981, **19**, 531–532
The combination (or one with potassium perchlorate) is an explosive propellant.

Other reactants
Yoshida, 1980, 79
MRH values calculated for 19 combinations with oxidisable materials are given.
See other PERCHLORATE SALTS OF NITROGENOUS BASES

4005. Hydroxylaminium perchlorate
[15588-62-2] ClH_4NO_5

$$HON^+H_3\ ClO_4^-$$

1. Rafeev, V. A. *et al.*, *Chem. Abs.*, 1981, **95**, 83166
2. Bokii, V. A. *et al.*, *Chem. Abs.*, 1981, **94**, 49734
The decomposition and combustion of the redox salt [1], and its mechanism [2] were studied.

1,3,5-Triaminotrinitrobenzene
Quong, R., *Chem. Abs.*, 1975, **82**, 113753
A saturated aqueous solution of the perchlorate, and the solid fuel, are individually non-explosive, but form a viable explosive composition on admixture.
See other PERCHLORATE SALTS OF NITROGENOUS BASES, REDOX COMPOUNDS

4006. Phosphonium perchlorate
[101672-21-3] ClH_4O_4P

$$P^+H_4\ ClO_4^-$$

1. Fichter, F. *et al.*, *Helv. Chim. Acta*, 1934, **17**, 222
2. Mellor, 1971, Vol. 8, Suppl. 3, 274

The crystalline salt obtained by action of phosphine on 68% perchloric acid at −20°C is dangerously explosive, sensitive to moist air, increase in temperature, or friction [1] and cannot be dried [2].

See related PERCHLORATE SALTS OF NITROGENOUS BASES
See other REDOX COMPOUNDS

4007. Hydrazinium chloride
[2644-70-4]   ClH₅N₂

$$H_2NN^+H_3Cl^-$$

See entry THERMAL EXPLOSIONS *See other* HIGH-NITROGEN COMPOUNDS

4008. Hydrazinium chlorite
[66326-45-2]   ClH₅N₂O₄

$$H_2NN^+H_3ClO_2^-$$

Levi, G. R., *Gazz. Chim. Ital. [2]*, 1923, **53**, 105–108
It is spontaneously flammable when dry.
See other CHLORITE SALTS, REDOX COMPOUNDS

4009. Ammonium perchlorylamide
[20394-65-4]   ClH₅N₂O₃

$$NH_4NHClO_3$$

See entry PERCHLORYLAMIDE SALTS

4010. Hydrazinium chlorate
[66326-46-3]   ClH₅N₂O₃

$$H_2NN^+H_3ClO_3^-$$

Salvadori, R., *Gazz. Chim. Ital. [2]*, 1907, **37**, 32–40
It explodes violently at its m.p., 80°C.
See other OXOSALTS OF NITROGENOUS BASES, REDOX COMPOUNDS

4011. Hydrazinium perchlorate
[13762-80-6]   ClH₅N₂O₄

$$H_2NN^+H_3ClO_4^-$$

1. Levy, J. B. *et al., ACS 54*, 1966, 55
2. Grelecki, C. J. *et al., ACS 54*, 1966, 73
3. Salvadori, R., *Gazz. Chim. Ital. [2]*, 1907, **37**, 32–40

The deflagration [1] and thermal decomposition [2] of the salt, a component of solid rocket propellants, have been studied. It also explodes on impact [3].

Metal salts, or Metal oxides
1. Shidlovskii, A. A. *et al., Zh. Priklad. Chim.*, 1962, **35**, 756
2. Probhakaran, C. *et al., Def. Sci. J.*, 1982, **31**, 285–291
Presence of 5% of copper(II) chloride caused explosion to occur at 170°C [1]. Of the series of additives copper chromite, copper chloride, nickel oxide, iron oxide, magnesium oxide, the earlier members have the greatest effect in increasing the sensitivity of the perchlorate to heat, impact and friction.
See other PERCHLORATE SALTS OF NITROGENOUS BASES, REDOX COMPOUNDS

4012. Poly(dimercuryimmonium perchlorate)

[ ] $(ClHg_2NO_4)_n$

$$(Hg{=}N^+{=}Hg\ ClO_4^-)_n$$

Sorbe, 1968, 97
Highly explosive.
See other POLY(DIMERCURYIMMONIUM) COMPOUNDS, PERCHLORATE SALTS OF NITROGENOUS BASES

4013. Iodine chloride
[7790-99-0] CII

$$ICl$$

Metals
Mellor, 1940, Vol. 2, 119; 1956, Vol. 2, Suppl. 1, 452; 1963, Vol. 2, Suppl. 2.2, 1563
Mixtures containing sodium explode only on impact, while potassium explodes on contact with the chloride. Aluminium foil ignites after prolonged contact (probably after the surface layer of oxide has been dissolved).

Phosphorus trichloride
Mellor, 1956, Vol. 2, Suppl. 1, 502
Reaction is intensely exothermic.
See other INTERHALOGENS, IODINE COMPOUNDS

4014. Indium(I) perchlorate
[62763-56-8] $ClInO_4$

$$InClO_4$$

Ashraf, M. *et al., J. Chem. Soc., Dalton Trans.*, 1977, 1723
The solvent-free material detonated when crushed with a glass rod. The nitrate (also a redox compound) probably would behave similarly.
See other METAL PERCHLORATES, REDOX COMPOUNDS

4015. Potassium chloride
[7447-40-7]   ClK

$$KCl$$

Potassium permanganate, Sulfuric acid
See Potassium permanganate: Potassium chloride, etc.
See other METAL HALIDES

4016. Potassium chlorite
[14314-27-3] $ClKO_2$

$$KClO_2$$

Other reactants
Yoshida, 1980, 3
MRH values calculated for 18 combinations with oxidisable materials are given.

Sulfur   MRH 3.22/13
Leleu, *Cahiers*, 1974, (74), 137
Interaction is violent.
See other CHLORITE SALTS

4017. Potassium chlorate
[3811-04-9] $ClKO_3$

$$KClO_3$$

HCS 1980, 762

1. Rüst, 1948, 294
2. Biasutti, 1981, 151
3. Anon., *Fire Prevention*, 1986, (186), 48
Although most explosive incidents have involved mixtures of the chlorate with combustible materials, the exothermic decomposition of the chlorate to chloride and oxygen can accelerate to explosion if a sufficient quantity and powerful enough heating are involved. A case history of a fire-heated explosion of a store of 80 t of chlorate is given. The more stable sodium chlorate will also explode under similar conditions [1]. The enthalpy of conversion of potassium chlorate to either chloride or perchlorate is slight, about $\frac{1}{4}$kJ/g, less than is the case for the sodium salt. During sieving of the dry chlorate in a store attached to a fuse igniter plant, a violent explosion of 75 kg of the salt occurred. It seems likely that some oxidisable material was involved in the incident, though the possibility of dropping a drum onto the concrete floor is noted as a potential cause of ignition [2]. An account of an explosion of 16 tonnes during a fire in a warehouse, demolishing 100 metres of the wall is given. The chlorate does not seem to have been isolated from other goods [3].

Agricultural materials
See Sodium chlorate: Agricultural materials

Aluminium,   MRH 9.20/30
Antimony trisulfide   MRH 2.84/39
   Crozier, T. H. *et al., 52nd Ann. Rept. HM Insp. Explos.*, London, Home Office,
   1928
   A pyrotechnic mixture containing the powdered ingredients was found dangerously
   sensitive to frictional initiation and highly explosive.
   See Metals; *and* Metal sulfides; both below

Ammonia, or Ammonium sulfate   MRH Ammonia 4.23/21
   Mellor, 1941, Vol. 2, 702; 1940, Vol. 8, 217
   High concentrations of ammonia in air react so vigorously with potassium chlorate
   as to be dangerous. Mixtures with ammonium sulfate when heated decompose with
   incandescence.

Ammonium chloride
   1. Potjewijd, T., *Pharm. Weekbl.*, 1935, **72**, 68–69
   2. Ellern, 1968, 155
   Addition of ammonium chloride to a drum of weed-killer was suspected as the
   cause of a violent explosion (involving formation of ammonium chlorate) [1].
   Mixtures, used for smoke compositions, are hazardous [2].

Aqua regia, Ruthenium
   Sidgwick, 1950, 1459
   Ruthenium is insoluble in aqua regia, but addition of potassium chlorate causes
   explosive oxidation.

Arsenic trisulfide
   Ganguly, A., *J. Indian Acad. Forensic Sci.*, 1973, **12**, 29–30
   Dry powdered mixtures containing over 30% oxidant exploded under a hammer-
   blow.
   See Metal sulfides, below

Carbon   MRH 4.52/13
   Read, C. W. W., *School Sci. Rev.*, 1941, **22**(87), 341
   Accidental substitution of powdered carbon for manganese dioxide in 'oxygen
   mixture' caused a violent explosion when the mixture was heated.

Cellulose
   1. Watanabe, M. *et al., Chem. Abs.*, 1988, **108**, 207227
   2. Ishida, H. *et al., Chem. Abs.*, 1988, **109**, 56797
   Mixtures of stoichiometric proportions (zero oxygen balance) are a high deflagration hazard and show remarkable pressure increase effects on ignition [1], as well
   as lowest ignition temperatures by ARC [2].
   See PRESSURE INCREASE IN EXOTHERMIC DECOMPOSITION (reference 4)

Charcoal, Potassium nitrate, Sulfur
   Biasutti, 1981, 4

A batch of black powder, modified by addition of potassium chlorate, was being mixed mechanically with added water. A friction-sensitive crust appears to have formed, leading to initiation of the explosion which followed.
See Non-metals, below

Charcoal, Sulfur
Biasutti, 1981, 15, 42, 46
A demonstration by Berthollet in 1788 of replacement of potassium nitrate in gunpowder by the chlorate led to a violent explosion during the crushing operation which caused two fatalities. Later incidents involving factories for chlorate-containing explosives led to widespread destruction.

Cyanides
See METAL CYANIDES (AND CYANO COMPLEXES) (reference 1)

Cyanoguanidine
See Cyanoguanidine: Oxidants

Dinickel trioxide
Mellor, 1942, Vol. 15, 395
Interaction at 300°C is violently exothermic, red heat being attained.

Fabric
Anon., Accidents, 1968, 24
Fabric gloves (wrongly used in place of impervious plastics gloves) became impregnated during handling operations and were subsequently ignited from cigarette ash.
See Cellulose, above

Fluorine
Pascal, 1960, Vol. 16, 316
Interaction at low pressure leads to formation of the explosive gas, fluorine perchlorate.
See Fluorine: Sodium bromate

Gallic acid, Gum tragacanth
Anon., Lab. Accid. Higher Educ., item 7, Barking, HSE, 1987
A 'pyrotechnic device' was being prepared by loading 16 g of a mixture of 73% potassium chlorate, 24% gallic acid and 3% gum tragacanth into a 9 mm copper tube 280 mm long, when it exploded violently and caused shrapnel injuries. Such mixtures are very friction-sensitive.

Hydrocarbons   MRH Hexane 4.77/10, Toluene 4.81/12
1. Bjorkman, P. O., Chem. Abs., 1934, 28, 6311a
2. von Feilitzen, G., Chem. Abs., 1934, 28, 6311b
Mixtures of powdered chlorate and hydrocarbons explode as violently as nitro compound explosives [1]. Porous masses of chlorate impregnated with hydrocarbons are friction-sensitive explosives [2].

Hydrogen iodide
  Mellor, 1941, Vol. 2, 310
  Molten potassium chlorate ignites hydrogen iodide gas.

Manganese dioxide
  1. Mellor MIC, 1961, 333
  2. Partington, 1967, 651
  3. Renfrew, M. M., *J. Chem. Educ.*, 1983, **60**, A229
  When oxygen is generated in the laboratory by heating potassium chlorate with manganese dioxide as decomposition catalyst, the latter must be free of organic matter or an explosion will occur [1]. The mixture (17% wt of manganese dioxide, 'oxygen mixture') must be cautiously heated at 250°C with a small flame to prevent the exothermic decomposition reaction becoming violent [2]. Mixing the components dry with a partially plastic-coated micro-spatula caused a fire [3].
  See other CATALYTIC IMPURITY INCIDENTS

Manganese dioxide, Potassium hydroxide
  Molinari, E. *et al., Inorg. Chem.*, 1964, **3**, 898
  The oxidation of manganese dioxide to manganate by solid alkali–chlorate mixtures becomes explosive above 80–90°C at pressures above 19 kbar.

Metal, Wood
  Davis, G. C., *J. Ind. Eng. Chem.*, 1909, **1**, 118
  Explosions were caused by transportation of metal castings in wooden kegs previously used to store potassium chlorate, impact or friction of the metal causing initiation of the chlorate-impregnated wood.

Metal phosphides
  Mellor, 1940, Vol. 8, 839, 844
  Tricopper diphosphide and trimercury tetraphosphide form impact-sensitive mixtures with potassium chlorate. By analogy, the phosphides of aluminium, magnesium, silver and zinc, etc., would be expected to form similar mixtures with metal halogenates.

Metal phosphinates ('hypophosphites')
  Mellor, 1940, Vol. 8, 881, 883
  Dry mixtures of barium phosphinate and potassium chlorate burn rapidly with a feeble report if unconfined, but even under the slight confinement of enclosing in paper, a sharp explosion occurs. The mixture is readily initiated by sparks, impact or friction. A mixture of calcium phosphinate, potassium chlorate and quartz exploded during mixing. Mixtures of various phosphinates and chlorates have been proposed as explosives, but they are very sensitive to initiation by sparks, friction or shock. Admixture of powdered magnesium causes a brilliant flash on initiating the mixture.
  See Sodium phosphinate: Oxidants
  See other FRICTIONAL INITIATION INCIDENTS, STATIC INITIATION INCIDENTS

Metals   MRH Aluminium 9.20/30, magnesium 9.50/37, iron 3.64/47

1. Mellor, 1941, Vol. 2, 310
2. Anon., *Chem. Eng. News*, 1936, **14**, 451
3. Mellor, 1940, Vol. 4, 480
4. Mellor, 1943, Vol. 11, 163
5. Mellor, 1941, Vol. 7, 20, 116, 260
6. Jackson, H., *Spectrum*, 1969, **7**, 82
7. Ganguly, A., *J. Indian Acad. Forensic Sci.*, 1973, **12**, 29–30

Mixtures of finely divided aluminium, copper, magnesium [1,2] and zinc [3] with potassium chlorate (or other metal halogenates) are explosives and may be initiated by heat, impact or light friction. Chromium incandesces in the molten salt [4] and germanium explodes on heating with potassium chlorate. Titanium explodes on heating, while zirconium gives mild explosions on heating, and ignites when the mixture is impacted [5]. Contact of molten chlorate with steel wool causes violent combustion of the latter [6]. Qualitative experiments showed that hammer impact would explode mixtures with aluminium powder containing over 10% of chlorate [7].

See also FLASH POWDER
See other METAL OXOHALOGENATES, METAL CHLORATES

Metal sulfides   MRH Antimony trisulfide 2.34/39, silver sulfide 1.38/62

1. Mellor, 1941, Vol. 2, 310
2. Mellor, 1939, Vol. 9, 523
3. Mellor, 1941, Vol. 3, 447
4. Rüst, 1948, 335

Many metal sulfides when mixed intimately with metal halogenates form heat-, impact- or friction-sensitive explosive mixtures [1]. That with antimony trisulfide can be initiated by a spark [2] and with silver sulfide a violent reaction occurs on heating [3]. For the preparation of 'oxygen mixture', antimony trisulfide was used in error instead of manganese dioxide, and during grinding, the mixture of sulfide and chlorate exploded very violently [4].

See also METAL HALOGENATES

Metal thiocyanates

1. von Schwartz, 1918, 299–300
2. Anon., *Chem. Age.*, 1936, **35**, 42

Mixtures of thiocyanates with chlorates (or nitrates) are friction- and heat-sensitive, and explode on rubbing, heating to 400°C, or initiation by spark or flame [1]. A violent explosion occurred when a little chlorate was ground in a mortar contaminated with ammonium thiocyanate. A similar larger-scale explosion involving traces of barium thiocyanate is also described [2].

Nitric acid, Organic materials

Asthana, S. S. *et al., Chem. & Ind.*, 1976, 953–954

In a new method of degrading organic matter prior to analysis, small portions (0.3 g only) of chlorate are added to a hot suspension of the organic matter in conc. nitric acid. Use of larger portions of chlorate may lead to explosive oxidation of the organic matter.

Non-metals   MRH Carbon 4.52/13, phosphorus 5.98/23, sulfur 3.18/18
1. Mellor, 1941, Vol. 2, 310
2. Mellor, 1940, Vol. 8, 785–786
3. Mellor, 1946, Vol. 5, 15
4. Ganguly, A., *J. Indian Acad. Forensic Sci.*, 1973, **12**, 29–30
Potassium chlorate (or other metal halogenate) intimately mixed with arsenic, carbon, phosphorus, sulfur or other readily oxidised materials gives friction-, impact- and heat-sensitive mixtures which may explode violently [1]. When potassium chlorate is moistened with a solution of phosphorus in carbon disulfide, it eventually explodes as the solvent evaporates and oxidation proceeds [2]. Boron burns in molten chlorate with dazzling brilliance [3]. Mixtures of the chlorate and finely powdered sulfur containing over 20% of the latter will explode under a hammer-blow [4].
See Sulfur, below
See also METAL CHLORATES, METAL HALOGENATES

Other reactants
Yoshida, 1980, 65–66
MRH values calculated for 24 combinations with oxidisable materials are given.

Phosphorus (red)
1. Haarmann, D. J., *American Fireworks News*, 1985, #51; *ibid.*, 1986, #54,
2. *Explosives and Their Power*, Berthelot, M., (translated and condensed from the French), London, 1892
This combination has, in the past, been the base of various impact sensitive pyrotechnics, described by Haarmann [1] and more recently on the Internet, including reference to a report [2] of a mere 60–70 kg, dispersed in children's toy caps, demolishing a building in an accidental explosion. Red phosphorus is a material of variable composition and reactivity, so unreliability is to be anticipated. There have been passivating components in most of these mixtures, it is suggested that the pure dry powders will often react on contact. For the more reactive white allotrope:
See Non-metals, above

Reducants
Mellor, 1941, Vol. 3, 651; 1940, Vol. 8, 881, 883
Mixtures with calcium hydride or strontium hydride may explode readily, and interaction of the molten chlorate is, of course, violent. A mixture of syrupy sodium phosphinate ('hypophosphite') and the powdered chlorate on heating eventually explodes as powerfully as glyceryl nitrate. Calcium phosphinate mixed with the chlorate and quartz detonates (the latter producing friction to initiate the mixture). Dried mixtures of barium phosphinate and the chlorate are very sensitive and highly explosive under the lightest confinement (screwed up in paper).

Sugars

1. Zaehringer, A. J., *Rocketscience*, 1949, **3**, 64–68
2. Ganguly, A., *J. Indian Acad. Forensic Sci.*, 1973, **12**, 29–30
3. Scanes, F. S., *Combust. Flame*, 1974, **23**, 363–371

A stoicheiometric mixture with sucrose ignites at 159°C, and has been evaluated as a rocket propellant [1]. Dry powdered mixtures with glucose containing above 50% of chlorate explode under a hammer-blow [2]. Pyrotechnic mixtures with lactose begin to react exothermally at about 200°C, when the lactose melts, and carbon is formed. This is then oxidised by the chlorate at about 340°C. The mechanism was studied by DTA [3].

Sugar, Sulfuric acid

Hanson, R. M., *J. Chem. Educ.*, 1976, **53**, 578

Addition of a drop of sulfuric acid to a mound of the chlorate–sugar mixture leads to ignition.

Sulfur   MRH 3.18/18

1. Tanner, H. G., *J. Chem. Educ.*, 1959, **36**, 59
2. Storey, P. D., *Proc. 13th Int. Pyrotech. Semin.*, 1988, 765–784
3. Chapman, D. *et al.*, *J. Pyrotech.*, 1998 (7), 51

The chemistry involved in this explosively unstable system is reviewed [1]. The mechanism of the trigger reactions that initiate the exothermic decomposition of chlorate–sulfur mixtures has been studied. Mixtures containing 1–30% of sulfur can decompose well below the m.p. of sulfur, and addition of sulfur dioxide, the suspected chemical trigger, causes immediate onset of the reaction [2]. Autoignition of stoichiometric mixtures can be as low as 115°C, with frictional sensitivity at 5N, the lowest load the test apparatus permitted. Both were dependent upon the history of the sulphur used [3].
See Non-metals, above

Sulfur, Metal derivatives

McKown, G. L. *et al.*, *Chem. Abs.*, 1977, **87**, 103955

The sensitivity of chlorate–sulfur mixtures to initiation is increased by cobalt and its oxide, greatly so by copper nitride, copper sulfate, and extremely so by copper chlorate. Implications for manufacturing operations are discussed.

Sulfur dioxide   MRH 1.80/56

Mellor, 1947, Vol. 10, 217; 1941, Vol. 2, 311

Contact at temperatures above 60°C causes flashing of the evolved chlorine dioxide. Solutions of sulfur dioxide in ethanol or ether cause an explosion on contact at ambient temperature.

Sulfuric acid

1. Mellor, 1947, Vol. 10, 435
2. Bailar, 1973, Vol. 2, 1362

Addition of potassium chlorate in portions to sulfuric acid maintained at below 60 or above 200°C causes brisk effervescence. At intermediate temperatures,

explosions are caused by the chlorine dioxide produced, and these reach maximum intensity at 120–130°C [1]. Uncontrolled contact of any chlorate with sulfuric acid may be explosive. The need for great caution with this system was stressed by Davy in 1815 [2].
See METAL CHLORATES: acids

Sodium amide   MRH 2.63/24
Mellor, 1940, Vol. 8, 258
A mixture explodes.

Tannic acid
Rüst, 1948, 336
A mixture exploded during grinding in a mortar.
See METAL CHLORATES

Thiourea
1. Soothill, D., *Safety Management*, 1992, **8**(6), 11.
2. Wharton, R. K. *et al., Propellants, Explos., Pyrotech.*, 1993, **18**(2), 77
An explosion is mentioned consequent upon grinding potasssium chlorate in equipment previously used for thiourea. It is claimed the mixture explodes spontaneously [1]. This is an exaggeration, it is very sensitive to friction, shock and heat (ignition <155°C) [2].

Thorium dicarbide
See Thorium dicarbide: Non-metals, etc.
See other METAL CHLORATES, OXIDANTS

4018. Potassium perchlorate
[7778-74-7] $ClKO_4$
$KClO_4$

Aluminium, Aluminium fluoride
1. Hahma, A., *Propellant, Explosives, Pyrotechnics*, 1996, 21, 100
2. Freeman, E. S. *et al., Combust. Flame*, 1966, **80**, 16
Whether mixtures of the perchlorate with aluminium truly detonate is under debate. However, appropriate compositions certainly explode violently even when uncontained, so the deflagration/detonation dispute is essentially academic [1]. Presence of aluminium fluoride increases the ease of ignition of aluminium–perchlorate mixtures, owing to complex fluoride formation [2].

Aluminium, Barium nitrate, Potassium nitrate, Water
See Aluminium: Metal nitrates, etc.

Aluminium powder, Titanium dioxide
Anon., *Fire Prot. Assoc. J.*, 1957, (36), 9

A mixture of the 3 compounds exploded violently during mixing. Previously the mixture had been accidentally ignited by a spark. Aluminium powder is incompatible with oxidants, and its mixture with titanium dioxide is a thermite-like combination.

See Aluminium: Metal oxides
See also THERMITE REACTIONS

Barium chromate, Tungsten and/or Titanium
1. Carrazza, J. A. *et al., Chem. Abs.*, 1975, **82**, 113751
2. Nagaishi, T., *Chem. Abs.*, 1978, **88**, 173088
Mixtures containing 65% of tungsten with organic binders were developed as priming charges for surface flares [1]. Ignition of the dual metal–oxidant mixtures by IR irradiative heating was studied [2].
See other IRRADIATION DECOMPOSITION INCIDENTS

Boron, Magnesium, Silicone rubber
Paulin, C. J. *et al.*, US Pat, 4 101 352, 1978
Ignition of a compressed mixture, applied as a 6 mm layer on electronic components generates enough heat effectively to destroy them.

Combustible materials   MRH Ethylene glycol 4.48/27
Grodzinski, J., *J. Appl. Chem.*, 1958, **8**, 523–528
Explosion temperatures were determined for a wide range of combustible liquid and solid organic materials sealed into glass tubes with potassium perchlorate. The lowest temperatures were shown by mixtures with ethylene glycol (240°), cotton linters (245°) and furfural (270°C).

Ethanol   MRH 4.85/18
Burton, H. *et al., Analyst*, 1955, **80**, 16
Many explosions have been experienced during the gravimetric determination of either perchlorates or potassium as potassium perchlorate by a standard method involving an ethanol extraction. During subsequent heating, formation and explosion of ethyl perchlorate is very probable.

Ferrocenium diamminetetrakis(thiocyanato-*N*)chromate(1−)
Guslev, V. G. *et al., Chem. Abs.*, 1974, **86**, 57615
Presence of 25% of the organometallic salt (ferrocenium reineckate) considerably increases the rate of thermal decomposition of the perchlorate, involving hydrogen cyanide arising from the thiocyanato groups.

Lactose
See Lactose: Oxidants

Metal powders   MRH Aluminium 9.96/33, iron 3.76/51, magnesium 10.25/41,
                                                                  sodium 4.27/45
1. Schumacher, 1960, 210
2. Riffault, M. L., *Chem. Abs.*, 1975, **82**, 127130

3. Ywenkeng, H., *Chem. Abs.*, 1981, **94**, 194474
4. Rittenhouse, C. *et al., Proc. 6th Int. Pyrotech. Semin.*, 1978, 519–536
5. Wild, R. *J. Haz. Mat.*, 1982, **7**, 75–79
6. Munger, A. C. *et al., Chem. Abs.*, 1973, **98**, 91971
7. Parks, R. L., *Chem. Abs.*, 1981, **95**, 100052
8. Sarawadekar, R. G. *et al., Proc. 8th Int. Protech. Semin.*, 1982, 574–587

The mixture of aluminium and/or magnesium powders with potassium perchlorate (a photoflash composition) is very readily ignited and three industrial explosions have occurred. Mixtures of nickel and titanium powders with the oxidant and infusorial earth are very friction-sensitive, causing severe explosions, and easily ignited by very small (static) sparks [1]. A mixture containing 70% of molybdenum powder ignites at 330°C [2], and the rate of combustion of a mixture containing 48% is accelerated by 1% of various additives [3]. The pyrotechnic mixture with aluminium is insensitive to spark discharges but ignited consistently with a hot wire source [4], and blast waves produced on ignition of mixtures with 70% oxidant have been investigated [5]. The mixture with iron powder is sufficiently exothermic after ignition for use in metal welding, and an iron oxide–aluminium–nickel thermite mixture was also developed for the same purpose [6]. Data on detonation and deflagration of the pyrotechnic mixture with titanium were determined in open and closed vessels, and subsequently used to design a containment room for blending the powdered components [7]. The pyrotechnic/explosive perchlorate–tantalum system has been studied [8].

See Aluminium, etc., Boron, etc. and Barium chromate, etc., all above
See other FRICTIONAL INITIATION INCIDENTS, STATIC INITIATION INCIDENTS
See also THERMITE REACTIONS

Other reactants
Yoshida, 1980, 80
MRH values calculated for 18 combinations with oxidisable materials are given.

Potassium hexacyanocobaltate(3−)
Massis, T. M. *et al., Chem. Abs.*, 1977, **87**, 8142
Mixtures serve as gasless pyrotechnic compositions.

Reducants
See Metal powders, above; Titanium hydride, below

Sulfur   MRH 3.39/20
Schumacher, 1960, 211–212
Mixtures of sulfur and potassium perchlorate, used in pyrotechnic devices, can be exploded by moderate impact. All other inorganic perchlorates form such impact-sensitive mixtures.

Titanium hydride
1. Massis, T. M. *et al., Rept. SAND-75-5889*, Richmond (Va.), USNTIS, 1976
2. Collins, L. W., *J. Haz. Mat.*, 1982, **5**, 325–333
3. Parks, R. L., *J. Haz. Mat.*, 1982, **5**, 359–371

The stability of the pyrotechnic mixture has been studied, including the effect of hydrogen content of the hydride [1]. Presence of moisture and impurities adversely affects stability [2], and remote handling facilities for the mixture have been developed [3].

See other METAL PERCHLORATES, OXIDANTS

4019. Potassium perchlorylamide
[15320-25-9] ClK_2NO_3

$$K_2NClO_3$$

Sorbe, 1968, 67
Like the monopotassium salt, it will explode on impact or exposure to flame.
See entry PERCHLORYLAMIDE SALTS *See other* N-METAL DERIVATIVES

4020. Lithium chlorite
[27505-49-3] $ClLiO_2$

$$LiClO_2$$

Janz, 1976, Table 2, p. 10
It disproportionates violently on heating.

Chlorine, 2-Chloroalkyl aryl sulfides
See Chlorine: 2-Chloroalkyl aryl sulfides, etc.
See other CHLORITE SALTS, OXIDANTS

4021. Lithium perchlorate
[7791-03-9] $ClLiO_4$

$$LiClO_4$$

Diethyl ether
1. Urben, P. G., *Chemtech*, 1991, (May), 259
2. Silva, R. A., *Chem. Eng. News*, 1992 **70**(51), 2
3. Handy, S. T. *et al., Synlett.*, 1995, 565
A warning was given that the 5 molar solution in ether used as a solvent for Diels-Alder reactions would lead to explosions [1]. Such a reaction of dimethyl acetylenedicarboxylate and cyclooctatetraene in this solvent exploded very violently on heating. The cyclooctatetraene was blamed, with no supporting evidence [2]. It would appear desirable to find the detonability limits of the mixture with ether before any attempt is made to scale up. A safe alternative to lithium perchlorate/ether as a solvent for Diels-Alder reactions is proposed [3].
See Dimethyl acetylenedicarboxylate, 1,3,3,7-Cyclooctatetraene
See Magnesium perchlorate

1,3-Dioxolane
Newman, G. H. *et al., J. Electrochem. Soc.*, 1980, **127**, 2025–2027

Lithium perchlorate–dioxolane electrolyte systems are unsafe for secondary battery applications, as an explosion occurred during overnight cyclic testing of a Li/TiS$_2$ system. The effect was duplicated under all over-discharge or cell-reversal conditions.

Hydrazine
Rosolovskii, V. Ya., *Chem,. Abs.*, 1969, **70**, 53524
Interaction gives only the dihydrazine complex, which explodes on grinding, but loses hydrazine uneventfully on heating at atmospheric or reduced pressure. The sodium salt is similar.

Nitromethane
See Nitromethane: Lithium perchlorate

Olefins, Electrophiles.
Zefirov, N. S., XIII Int. Symp. Org. Chem. Sulfur, 1988 (lecture; this information not in the later published proceedings).
Electrophilic additions to olefins, especially of XCl species, can give substantial quantities of perchlorate esters as by-products if performed in the presence of lithium perchlorate. These esters are extremely explosive, and may concentrate during distillation of crude products.
See other METAL PERCHLORATES, OXIDANTS

4022. Manganese chloride trioxide
[15605-27-3]   ClMnO$_3$
ClMnO$_3$

Briggs, S. T., *J. Inorg. Nucl. Chem.*, 1968, **30**, 2867–2868
Explosively unstable if isolated as a liquid at ambient temperature, it may be handled safely in carbon tetrachloride solution.
See Manganese dichloride dioxide, Manganese trichloride oxide
See related METAL OXIDES

4023. Nitrosyl chloride
[2696-92-6]   ClNO
O:NCl

It is a moderately endothermic compound (ΔH°_f (g) +52.6 kJ/mol, 0.80 kJ/g).

Acetone, Platinum
Kaufmann, G. B., *Chem. Eng. News*, 1957, **35**(43), 602
A cold sealed tube containing nitrosyl chloride, platinum wire and traces of acetone exploded violently on being allowed to warm up.

Hydrogen, Oxygen
See Nitrogen oxide: Hydrogen, etc.

See other ENDOTHERMIC COMPOUNDS, *N*-HALOGEN COMPOUNDS, NITROSO COM-
POUNDS, OXIDANTS

4024. *N*-Chlorosulfinylimide
[25081-01-0]   CINOS

$$O:S:NCl$$

Anon., *Angew. Chem. (Nachr.)*, 1970, **18**, 318
The ampouled solid exploded violently on melting. Distillation at ambient pressure
and impact tests had not previously indicated instability.

Chlorine fluoride
    See Chlorine fluoride: *N*-Chlorosulfinylamine
    See other *N*-HALOGEN COMPOUNDS, N–S COMPOUNDS

4025. Nitryl chloride
[13444-90-1] $CINO_2$

$$O_2NCl$$

It is feebly endothermic (ΔH°_f (g) +12.1 kJ/mol, 0.15 kJ/g) but a powerful oxidant
gas.

Inorganic materials
    Batey, H. H. *et al., J. Amer. Chem. Soc.*, 1952, **74**, 3408
    Interaction of the chloride with ammonia or sulfur dioxide is very violent, even at
    −75°C, and is vigorous with tin(II) bromide or tin(II) iodide.

Organic matter
    Kaplan, R. *et al., Inorg. Synth.*, 1954, **4**, 54
    It attacks organic matter rapidly, sometimes explosively.

B-Trimethylborazine
    Hirata, T., *Rept, AD-7293939*, Richmond (Va.), USNTIS, 1971
    Interaction is violent in absence of a diluent.
    See other ENDOTHERMIC COMPOUNDS, *N*-HALOGEN COMPOUNDS, OXIDANTS

4026. Nitryl hypochlorite ('Chlorine nitrate')
[14545-72-3] $CINO_3$

$$O_2NOCl$$

Metals, or Metal chlorides
    Bailar, 1973, Vol. 2, 379
    Interaction with most metals and metal chlorides is explosive at ambient temper-
    ature, but controllable at between −40 and −70°C.

Organic materials
Schmeisser, M., *Inorg. Synth.*, 1967, **9**, 129
Not inherently explosive, but it reacts explosively with alcohols, ethers and most organic materials.
See Nitryl hypofluorite
See other HYPOHALITES, OXIDANTS

4027. Nitrosyl perchlorate
[15605-28-4]   ClNO$_5$

$$O:NOClO_3$$

Gerding, H. *et al.*, *Chem. Weekbl.*, 1956, **52**, 282–283
Although stable at ambient temperature, it begins to decompose below 100°C, and at 115–120°C the decomposition becomes a low-order explosion.

Pentaammineazidocobalt(III) perchlorate, Phenyl isocyanate
Burmeister, J. L. *et al.*, *Chem. Eng. News*, 1968, **46**(8), 39
During an attempt to introduce phenyl isocyanate into the Co coordination sphere, a mixture of the 3 components exploded violently when stirring was stopped.

Organic materials
Hofmann, K. A. *et al.*, *Ber.*, 1909, **42**, 2031
As the anhydride of nitrous and perchloric acids, it is a very powerful oxidant. Pinene explodes sharply; acetone and ethanol ignite, then explode; ether evolves gas, then explodes after a few s delay. Small amounts of primary aromatic amines–aniline, toluidines, xylidines, mesidine–ignite on contact, while larger amounts exploded dangerously, probably owing to rapid formation of diazonium perchlorates. Urea ignites on stirring with the perchlorate, (probably for a similar reason).
See other NON-METAL PERCHLORATES, OXIDANTS

4028. Nitronium perchlorate
[12051-08-0]   ClNO$_6$

$$O_2N^+ \; ClO_4^-$$

Albright, Hanson, 1976, 2
The explosively unstable behaviour of stored nitronium perchlorate is attributed to the formation of small equilibrium concentrations of the isomeric covalent nitryl perchlorate ester (below).

1,2-Epoxides
Golding, P. *et al.*, *Tetrahedron Lett.*, 1988, **29**, 2733
Reaction with epoxides gives the dangerously unstable and explosive mixed nitrate–perchlorate diesters, such as 1,2-ethanediyl nitrate perchlorate from ethylene oxide.

See related NITRATION INCIDENTS
See other NON-METAL PERCHLORATES

4029. Nitryl perchlorate
[17495-81-7]   ClNO₆

$$O_2NOClO_3$$

Ammonium perchlorate
See Ammonium perchlorate: Impurities

Organic solvents
1. Spinks, J. W. T., *Chem. Eng. News*, 1960, **38**(15), 5
2. Gordon, W. E. *et al., Can. J. Res.*, 1940, **18B**, 358
Interaction with benzene gave a slight explosion and flash [1], while sharp explosions with ignition were observed with acetone and ether[2].
See other NON-METAL PERCHLORATES, OXIDANTS

4030. Chlorine azide
[13973-88-1]   ClN₃

$$ClN_3$$

Alone, or Ammonia, or Phosphorus, or Silver azide, or Sodium
1. Frierson, W. J. *et al., J. Amer. Chem. Soc.*, 1943, **65**, 1696, 1698
2. Rice, W. J. *et al., J. Chem. Educ.*, 1971, **48**, 659
3. Combourieu, J. *et al., Chem. Abs.*, 1977, **87**, 123313
4. Tornieporth-Oetting, I. C. *et al., Angewand. Chem. (Int.)*, 1995, **34**(5), 511
The undiluted material is extremely unstable, usually exploding violently without cause at any temperature, even as solid at $-100°C$ [1]. Explosion is likely to be triggered by pressure fluctuations of around 10 Pa [4]. It gives an explosive yellow liquid with liquid ammonia; when condensed on to yellow phophorus at $-78°C$ an extremely violent explosion soon occurs. Addition of phosphorus to a solution of the azide in carbon tetrachloride at $0°C$ causes a series of mild explosions if the mixture is stirred, or a violent explosion without stirring. Contact of the liquid or gaseous azide with silver azide at $-78°C$ gave a blue colour, soon followed by explosion, and sodium reacted similarly under the same conditions [1]. When chlorine azide (25 mol %) is used as a thermally activated explosive initiator in a chemical gas laser tube, the partial pressure of azide should never exceed 16 mbar [2]. The explosive decomposition has been studied in detail [3].
See entry HALOGEN AZIDES

4031. Sulfuryl azide chloride
[13449-15-5]   ClN₃O₂S

$$N_3SO_2Cl$$

Preparative hazard
Shozda, R. J. *et al., J. Org. Chem.*, 1967, **32**, 2876

1385

During the preparation of this explosive liquid by interaction of sulfuryl chloride fluoride and sodium azide, traces of chlorine must be eliminated from the former to avoid detonation. The product is nearly as shock-sensitive as glyceryl nitrate and may explode on rapid heating. Solutions (25 wt%) in solvents may be handled safely. The corresponding fluoride is believed to behave similarly.
See other ACYL AZIDES, ACYL HALIDES, NON-METAL AZIDES

4032. Thiotrithiazyl perchlorate

[ ] $ClN_3O_4S_4$

Organic solvents
Goehring, 1957, 74
The precipitated perchlorate salt exploded on washing with acetone or ether.
See other NON-METAL PERCHLORATES, N–S COMPOUNDS

4033. Thiotrithiazyl chloride

[12015-30-4] ClN_3S_4

Alone, or Ammonia
1. Mellor, 1940, Vol. 8, 631–632
2. Bailar, 1973, Vol. 2, 903
The dry chloride, which explodes on heating in air, will rapidly absorb ammonia gas and then explode [1]. The structure of the cation is now known to be a 7-membered ring with only 2 adjacent sulfur atoms. Thiotrithiazyl salts other than the chloride are also explosive [2].
See Thiotrithiazyl nitrate, Thiotrithiazyl perchlorate
See other N–S COMPOUNDS, NON-METAL HALIDES

4034. Cyclopentaazathienium chloride

[88433-73-2] ClN_5S_5

$$N_5S_5^+ \ Cl^-$$

Banister, A. J. *et al., J. Chem. Soc., Chem. Comm.,* 1983, 1017
It detonates on percussion and decomposes rapidly above 90°C to give a residue, largely thiotrithiazyl chloride, which explodes above 170°C.
See other N–S COMPOUNDS

4035. Triazidochlorosilane
[67880-25-5]   ClN₉Si

$$(N_3)_3SiCl$$

It detonates on percussion and decomposes rapidly above 90°C to give a residue, largely thiotrithiazyl chloride, which explodes above 170°C.
See Silicon tetraazide
See other NON-METAL AZIDES
See other N–S COMPOUNDS

4036. Sodium chloride
[7647-14-5]   ClNa

$$NaCl$$

HCS 1980, 834

Dichloromaleic anhydride, Urea
See Dichloromaleic anhydride: Sodium chloride, etc.

Lithium
See Lithium: Sodium carbonate, etc.

Nitrogen compounds
Rosinski, M. *et al., Chem. Abs.*, 1980, **93**, 15512
Electrolysis of sodium chloride in presence of nitrogenous compounds to produce chlorine may lead to formation of explosive nitrogen trichloride. Precautions are detailed.

Water
Aleshin, G. Ya., *Chem. Abs.*, 1982, **96**, 220112
The mechanism of explosion of molten salt at 1100°C in accidental contact with water was studied.
See MOLTEN SALTS *See other* METAL HALIDES

4037. Sodium hypochlorite
[7681-52-9]   ClNaO

$$NaOCl$$

FPA H52, 1976; *HCS 1980*, 213, 847, 848 (5, 10, 15% solns, resp.)

1. Brauer, 1963, Vol. 1, 311
2. Sorbe, 1968, 85
The anhydrous solid obtained by desiccation of the pentahydrate will decompose violently on heating or friction [1,2].

Amines   MRH Aniline 3.10/26
Kirk-Othmer, 1963, Vol. 2, 105

Primary aliphatic or aromatic amines react with sodium hypochlorite (or calcium hypochlorite) to form N-mono- or di-chloroamines which are explosively unstable, but less so than nitrogen trichloride.

Ammonium salts
1. Anon., *Loss Prev. Bull.*, 1978, (022), 116
2. Anon., *Occ. Safety Health*, 1976, **6**(4), 2
3. Anon., *Sichere Chemiearbeit*, 1996, **48**(10), 114
Contact in the drains of an effluent containing the hypochlorite with one containing ammonium salts and acid led to formation of nitrogen trichloride which decomposed explosively. The effect was reproduced under laboratory conditions [1]. Cleaning a brewery tank with an acidified ammonium sulfate cleaning preparation, then sodium hypochlorite solution without intermediate rinsing, led to nitrogen trichloride formation and a violent explosion [2]. The reservoir of a scrubber system whch had previously been employed to scrub ammonia containing off-gas with dilute acid was charged with sodium hypochlorite solution for another process. There was a violent explosion, destroying the reservoir and scrubber assembly. Investigation showed that the reservoir had not been fully cleaned from the previous operation [3].
See Amines, above; Phenylacetonitrile, below

Aziridine
Graefe, A. F. *et al., J. Amer. Chem. Soc.*, 1958, **80**, 3839
Interaction of ethyleneimine with sodium (or other) hypochlorite gives the explosive N-chloro compound.
See 1-Chloroaziridine

Carbonised residue
Dokter, T., *J. Haz. Mat.*, 1985, **10**, 85
A saucepan of vegetable stew had been heated too long and had formed a thick carbonised adherent cake. In an attempt to clean the pan, 1 l of domestic bleach was added and the pan was left to heat on an electric hotplate. Again it was left too long and after all the water had evaporated, the residue exploded violently. This was attributed to formation of sodium chlorate during evaporation, and ignition of the overheated chlorate-impregnated carbonised mass.
See Sodium chlorate: Organic matter

Ethanediol
Bickerton, J., *Chem. Brit.*, 1991, **27**(6), 504
Mixtures of aqueous sodium hypochlorite (presumably the 15% available chlorine commercial product) and ethylene glycol were observed to erupt violently after an induction period of 4 to 8 minutes. Caution is advised in view of the use of glycol as a cooling fluid in industrial reactors.

Formic acid
Khristoskova, S., *Chem. Abs.*, 1983, **98**, 95125

Removal of formic acid from industrial waste streams with sodium hypochlorite solution becomes explosive at 55°C.
See other REDOX REACTIONS

Furfuraldehyde
Birchwood, P. J., *Chem. Brit.*, 1985, **21**, 29
In a variation of the usual reaction conditions for oxidising furfuraldehyde to 2-furoic acid with hypochlorite, the aldehyde was added dropwise to a 10% excess of commercial sodium hypochlorite solution at 20–25°C, but without the inclusion of additional sodium hydroxide. When aldehyde addition was almost complete, a violent explosion occurred. Subsequent investigation showed that the pH of the reaction mixture fell progressively with addition of aldehyde, and at pH 8.5 the reaction mixture erupted violently, the temperature increased by 70°C and the pH fell to 2. Similar results were seen with benzaldehyde, but not with thiophene-2-aldehyde.

'Imidate'
Rayner, P., *Safety Digest Univ. Safety Assoc.*, 1991, **40**, 14
A methyl but-3-enylimidate ester hydrochloride, was charged to a tenfold excess of stirred 14% hypochlorite solution cooled in ice. After 50 min the flask was removed to replenish the ice. Shortly after returning the flask to the icebath a violent explosion shattered both flask and icebath. This was attributed to thermal runaway (although available energy is scarcely sufficient to boil the water in the flask). It seems more likely that trichloroamine was generated by the excess hypochlorite, settled when removed from the magnetic stirrer, and detonated from friction when this restarted. The reaction is said to have been performed many times previously without incident.
See Nitrogen trichloride
See other AGITATION INCIDENTS

Methanol   MRH 2.47/10
ICI Mond Div., private comm., 1968
Several explosions involving methanol and sodium hypochlorite were attributed to formation of methyl hypochlorite, especially in presence of acids or other esterification catalyst.

Other reactants
Yoshida, 1980, 132
MRH values calculated for 17 combinations with oxidisable materials are given.

Phenylacetonitrile
Libman, D. D., private comm., 1968
Use of sodium hypochlorite solution to destroy acidifed benzyl cyanide residues caused a violent explosion, thought to have been due to formation of nitrogen trichloride.

Photographic developer

Anon., *Environment, Safety & Health Bull.*, 1993, **93**(2), 1

As part of an analytical procedure prior to silver recovery, 10 ml commercial hypochlorite solution (15% available chlorine?) was added to highly alkaline waste developer solution. The redox reaction was violent enough to spray the worker, causing alkali burns.

Sodium ethylenediaminetetracetate, Sodium hydroxide

Schierwater, F.-W., *Sichere Chemiearb.*, 1987, **39**, 35

Operating instructions specified that 50% sodium hydroxide solution, sodium hypochlorite solution and sodium EDTA solution were to be added separately by pumping into an off-gas scrubber unit, when the exotherms would be dissipated slowly in the solution tank. An attempt to simplify the operation by premixing the 3 solutions in a drum before pumping the mixture, led to vigorous foaming decomposition. Mixing the conc. alkali with the bleach caused the oxidant to become heated by the heat of dilution of the former, and oxidation of the EDTA component then proceeded exothermically with decomposition (and evolution of carbon dioxide).

See other GAS EVOLUTION INCIDENTS

Water

Hazard Note HN(76)189, Dept. of Health and Social Security, London, 1976

Two 2.5 l bottles of strong sodium hypochlorite solution (10−14% available chlorine) burst in storage, owing to failure of the cap designed to vent oxygen slowly evolved during storage. This normal tendency may have been accelerated by the unusually hot summer. Vent caps should be checked with full personal protection, and material should be stored at 15−18°C and out of direct sunlight, which accelerates decomposition.

See other METAL HYPOCHLORITES

4038. Sodium chlorite
[7758-19-2] $ClNaO_2$

$$NaClO_2$$

FPA H109, 1972; *HCS 1980*, 835

Alone

1. Brauer, 1963, Vol. 1, 312
2. Leleu, *Cahiers*, 1974, (74), 137
3. Mellor, 1956, Vol. 2 (Suppl. II, Pt. 1), 573
4. Ullmann, 1986, A6, 500
5. Simoyi, R. H. *et al.*, *Chem. Eng. News*, 1993, **71**(12), 4

The anhydrous salt explodes on impact [1], and decomposes violently at 200°C [2]. The trihydrate is also percussion sensitive [3] though other sources suggest that in clean, grease- and oil-free equipment the anhydrous salt is shockproof [4]. A bottle of the purified anhydrous salt exploded then 'burnt' on opening, (transition to sodium chloride and oxygen has an enthalpy of 1.2 kJ/g), presumably initiated by friction in the screw cap (most screw caps are combustible). Various contaminants, including iron and ammonium salts, may sensitise sodium chlorite [5].

Acids   MRH Several inorganic acids, all 1.17/tr.
Bailar, 1973, Vol. 2, 1413
Under normal conditions, solutions of sodium (and other) chlorites when acidified do not evolve chlorine dioxide in dangerous amounts. However, explosive concentrations may result if acid is dropped onto solid chlorites.
See Oxalic acid, below

Carbon   MRH 3.60/9
Leleu, *Cahiers*, 1980, (99), 278
A mixture with powdered charcoal ignited above 60°C.

Ethylene glycol   MRH 3.51/17
See Ethylene glycol: Oxidants

Oils   MRH Octane 4.60/22, toluene 4.35/33
Karzhenyak, I. G. *et al., Chem. Abs.*, 1981, **95**, 67122
In presence of oils, the chlorite is friction- and shock-sensitive.

Organic matter   MRH Cellulose 3.76/24
The Diox Process, Newark, N. J., Wallace and Tiernan, 1949
Intimate mixtures of the solid chlorite with finely divided or fibrous organic matter may be explosive and very sensitive to heat, impact or friction.

Other reactants
Yoshida, 1980, 4–7
MRH Values calculated for no fewer than 79 combinations, largely with oxidisable materials, are given.

Oxalic acid
1. *MCA Case History No. 839*
2. Stull, 1977, 20
A bleach solution was being prepared by mixing solid sodium chlorite, oxalic acid, and water, in that order. As soon as water was added, chlorine dioxide was evolved and later exploded. The lower explosive limit of the latter is 10%, and the mixture is photo- and heat-sensitive [1]. It was calculated that the heat of reaction (1.88 kJ/g of dry mixture) would heat the expected products to an adiabatic temperature approaching 1500°C with an 18-fold increase in pressure in a closed vessel [2].

Phosphorus   MRH (Yellow) 6.23/21
Mellor, 1971, Vol. 8, Suppl. 2.2, 645
Red phosphorus and the chlorite react very exothermally in aqueous suspension above 50°C, and there may be a sudden and near-explosive stage in the redox reaction after an induction period.
See other INDUCTION PERIOD INCIDENTS, REDOX REACTIONS

Sodium dithionite
Anon., *Chem. Trade J.*, 1953, **132**, 564
Use of a scoop contaminated with sodium dithionite for sodium chlorite caused ignition of the latter. Materials containing sulfur (dithionite, natural rubber gloves) cause decomposition of sodium chlorite and contact should be avoided.

Sulfur-containing materials   MRH Sulfur 3.64/15, diethyl sulfide 4.27/28
1. Leleu, *Cahiers*, 1974, (74), 138
2. Weber, J. U. *et al., Rec. Trav. Chim.*, 1986, **105**(3), 99–102
Sodium chlorite reacts very violently with organic compounds of divalent sulfur, or with free sulfur (which may ignite), even in presence of water. Contact of the chlorite with rubber vulcanised with sulfur or a divalent sulfur compound should therefore be avoided [1]. Application of factorial design techniques to experimental planning gave specific conditions for the safe oxidation of organic sulfides to sulfoxides using sodium chlorite or calcium hypochlorite [2].
See Sodium dithionite, above

Zinc   MRH 3.60/58
Karzhenyak, I. G. *et al., Chem. Abs.*, 1981, **95**, 64606
Mixtures containing less than 0.1% of zinc dust or other assorted technical materials are not explosive or combustible on impact or friction.
See other CHLORITE SALTS, METAL OXOHALOGENATES, OXIDANTS

4039. Sodium chlorate
[775-09-9] $ClNaO_3$
$$NaClO_3$$

(*MCA SD-52*, 1952); *FPA H7*, 1972; *HCS 1980*, 833

1. *The Fire and Explosion at Braehead Container Depot, Renfrew, 1977*, London, HMSO, 1979
2. *The Fire and Explosion at River Road, Barking, Essex, 1980*, London, HSE, 1980
3. *Anon., New Scientist*, 1982, 595 (1325), 892
Following a violent explosion during a fire involving 67 t of drummed sodium chlorate, experiments confirmed that pure sodium chlorate would decompose explosively under intense fire conditions. An appended report by HM Inspector of Explosives detailed six incidents since 1899 involving explosions during fires in stores holding potassium or sodium chlorate [1]. A later incident involving explosive thermal decomposition of 2.5 t of the salt led to a drastic downward revision of the minimum quantity thought to pose this hazard [2]. A third incident in 1982 again involved explosion of drums of sodium chlorate in a warehouse fire [3]. It is difficult to understand this apparent reluctance to gain benefit from the recorded experience of over 80 years.
See Potassium chlorate

Preparative hazard
Kirk-Othmer, 1994, 4th Edn., **9**, 153
There is a risk of generating hydrogen/chlorine/oxygen mixtures during electrolytic preparation from brine. An explosive limits diagram for this ternary system is given.

Agricultural materials
Reimer, B. *et al., Chem. Technik*, 1974, **26**, 447–447 (condensed paper)
The potential for explosive combustion of mixture of sodium chlorate-based herbicides with other combustible agricultural materials was determined. Initiation temperatures and maximum combustion temperatures were measured for mixtures of sodium (or potassium) chlorate with peat, powdered sulfur, sawdust, urotropine (hexamethylenetetramine), thiuram and other formulated materials. With many combinations, maximum temperature increases of 500–1000°C at rates of 400–1200°/s were recorded for 2 g samples.

Alkenes, Potassium osmate
Lloyd, W. D. *et al., Synthesis*, 1972, 610
The hydroxylation of alkenes to diols with potassium osmate–oxidant mixtures has been described, with either hydrogen peroxide or sodium chlorate as the oxidant. The sodium chlorate method is not applicable where the diol is to be distilled from the mixture, because of the danger of explosive oxidation of the product diol by the chlorate.

Aluminium, Rubber
Olson, C. M., *J. Electrochem. Soc.*, 1949, **116**, 34C
The rubber belt of a bucket elevator, fitted with aluminium buckets and used for transporting solid chlorate, jammed during use. Friction from the rotating drive pulley heated and powdered the jammed belt. A violent explosion consumed all the rubber belt and most of the 90 aluminium buckets. Bronze and steel equipment is now installed.

Ammonium salts, or Metals,   MRH values below references
or Non-metals, or Sulfides
1. *MCA SD-42, 1952*
2. *MCA Case History No. 2019*
MRH Aluminium 10.71/33, iron 4.35/50, magnesium 10.88/40, manganese 5.06/50, sodium 5.56/55, phosphorus 7.32/25, sulfur 4.27/20

Mixtures of the chlorate with ammonium salts, powdered metals, phosphorus, silicon, sulfur or sulfides are readily ignited and potentially explosive [1]. Residues of ammonium thiosulfate in a bulk road tanker contaminated the consignment of dry sodium chlorate subsequently loaded, and exothermic reaction occurred with gas evolution during several hours. Laboratory tests showed that such a mixture could be made to decompose explosively. A reaction mechanism is suggested.
See other GAS EVOLUTION INCIDENTS

Arsenic trioxide
Ellern, 1968, 51
Ignition may occur on contact.

1,3-Bis(trichloromethyl)benzene
See 1,3-Bis(trichloromethyl)benzene: Oxidants

Cyanides
See METAL CYANIDES (reference 1)

Cyanoborane oligomer
Gyori, B. *et al., J. Organomet. Chem.*, 1984, **262**(1), C7
A mixture exploded violently on mild mechanical agitation.

Diols
See Alkenes, etc., above

Grease
Olson, C. M., *J. Electrochem. Soc.*, 1969, **116**, 34C
The greased bearing of a small grinder exposed to chlorate dust exploded violently during cleaning. Fluorocarbon-based greases and armoured bearings are recommended for chlorate service, with full operator protection during cleaning operations.
See Organic matter, below; *also* Sodium bromate: Grease

Leather
MCA Case History No. 1979
Shoes became contaminated with a chlorate weed-killer solution which dried out in wear. A welding spark later fell into the shoe and the front was blown off.

Nitrobenzene   MRH 4.94/88
Hodgson, J. F., private comm., 1973
The combination is powerfully explosive and has been widely used in recent terrorist activities.

Organic matter   MRH Carbon 4.73/12
1. *MCA SD-42*, 1952
2. Cook, W. H., *Can. J. Res.*, 1933, **8**, 509
3. Biasutti, 1981, 30
4. Anon., *Chem. Haz. in Ind.*, 1987, (7), item 1688
5. Anon., *Chem. Haz. in Ind.*, 1987, (10), item 2666
Mixtures of sodium (or other) chlorate with fibrous or absorbent organic materials (charcoal, flour, shellac, sawdust, sugar) are hazardous. If the chlorate concentration is high, the mixtures may be ignited or caused to explode by static sparks, friction or shock [1]. Even at 10–15% concentration, low relative humidity may allow easy ignition and rapid combustion to occur [2]. A fire originating thus led to destruction of part of an explosives factory [3]. Sparks from a grinder ignited newly-cleaned overalls, which had previously been contaminated with sodium chlorate [4]. It was later confirmed that the overalls had been dry-cleaned in perchloroethylene with added water, rather than being laundered in hot water.

The latter cleaning method would have removed completely the sodium chlorate, which is soluble in water but not in halogenated solvents [5].

See Paper, etc; Wood, both below *See also* OXIDANTS AS HERBICIDES
See other FRICTIONAL INITIATION INCIDENTS, STATIC INITIATION INCIDENTS

Osmium
Rogers, D. B. *et al., Inorg. Synth.*, 1972, **13**, 141
During the preparation of osmium dioxide, the ampoule containing the reaction mixture must be cooled during sealing operations to prevent violent reaction occurring. Subsequent heating, first to 300 and then 600°C must also be effected slowly.

Other reactants
Yoshida, 1980, 70–71
MRH values calculated for 33 combinations with oxidisable materials are given.

Paint, Polythene
Lamb, J. A., *Chem. Brit.*, 1979, **15**, 125–126
A violent explosion occurred when the lid was levered off an old paint tin used to contain a polythene bag of sodium chlorate. Subsequent tests showed that low energy input (0.2 J) would cause explosion of unconfined chlorate on a rusty steel surface. The presence of contaminants (sawdust, copper acetate, paint flakes or shredded polythene) increases the sensitivity, and a mixture with paint flakes and polythene caused 100 mg portions to explode under 0.2 J impact in 24 out of 100 attempts.

Paper, Static electricity
1. Ewing, O. R., *Chem. Eng. News*, 1952, **30**, 3210
2. Grimmett, R. E. R., *New Zealand J. Agric.*, 1938, **57**, 224–225
3. Biasutti, 1981, 92
Paper impregnated with sodium chlorate and dried can be ignited by static sparks, but not by friction or impact. Paper bags or card cartons are unsuitable packing materials [1]. A previous incident involving paper sacks which had formerly contained mixed sodium and calcium chlorates had been noted [2]. The tendency of sodium chlorate to deliquesce or effloresce, depending on atmospheric temperature and humidity may lead, if the solid is packed in pervious containers, to the appearance of very fine crystals on the outside of a paper bag or cardboard box, etc. These fine crystals of high specific surface area are very sensitive to friction and initiation of ignition [3].
See other FRICTIONAL INITIATION INCIDENTS, STATIC INITIATION INCIDENTS

Phosphorus   MRH (Yellow) 7.32/25
Anon., *Angew. Chem. (Nachr.)*, 1957, **5**, 78
A mixture of red phosphorus and sodium chlorate exploded violently.

Sodium phosphinate   MRH 2.89/55
See Sodium phosphinate: Oxidants

Sulfuric acid   MRH 1.30/tr.
    1. Anon., *ABCM Quart. Safety Summ.*, 1944, **15**, 3
    2. *MCA Case History No. 282*
Erroneous addition of conc. sulfuric acid to sodium chlorate instead of sodium chloride caused an explosion owing to formation of chlorine dioxide [1]. Accidental contact of 93% acid on clothing previously splashed with sodium chlorate caused immediate ignition [2].

Titanium
    See Titanium: Oxidants
         Titanium: Sodium chlorate, etc.

Triethylene glycol, Wood
    Anon., *CISHC Chem. Safety Summ.*, 1980, **51**, 112
    A wooden pallet which ignited and burned fiercely was found to be contaminated with both the glycol and sodium chlorate to a level of 27%.

Wood
    1. Anon., *ABCM Quart. Safety Summ.*, 1947, **18**, 25
    2. Anon., *BCISC Quart. Safety Summ.*, 1967, **38**, 42
    3. Kletz, T. A., *J. Loss Prev.Proc.Ind.*, 1989, **2**, 117
Various fires and explosions caused by use of wooden containers with chlorates, and precautions necessary during handling and storage, are discussed [1,2]. A wooden pallet burst into flames as it was dragged across ground contaminated with sodium chlorate [3].
    See other FRICTIONAL INITIATION INCIDENTS
    See Organic matter, above
    See other METAL CHLORATES, OXIDANTS

4040. Sodium perchlorate
    [7601-89-0] $ClNaO_4$
                              $NaClO_4$

Hydrazine
    Rosolovskii, V. Ya., *Chem. Abs.*, 1969, **70**, 53524
    Interaction gives only a 1:1 complex which explodes on grinding, but dissociates uneventfully on heating at atmospheric or reduced pressure. The lithium salt is similar.

Other reactants
    Yoshida, 1980, 82
    MRH values calculated for 18 combinations with oxidisable materials are given.

Water
    Moureu, H. *et al.*, *Chem. Abs.*, 1951, **45**, 5929h
    During concentration of an aqueous solution, an explosion occurred. Involvement of an unsuspected impurity seems probable.

1396

Water-soluble fuels   MRH values below reference
  Annikov, V. E. *et al., Chem. Abs.*, 1983, **99**, 73190
  MRH Acetone 4.89/18, ethanol 4.73/18, ethylene glycol 4.35/26
    The detonation and combustion limits of mixtures of sodium perchlorate, water
  and ethylene glycol, glycerol, 1,3-butylene glycol, 2,3-butylene glycol, formamide,
  dimethylformamide, ethanolamine, diaminoethane, acetone, urea and galactose
  have been studied.
  See other METAL PERCHLORATES, OXIDANTS

4041. Antimony(III) chloride oxide
  [7791-08-4]   ClOSb
                                        O:SbCl

Bromine trifluoride
  See Bromine trifluoride: Antimony(III) chloride oxide
  See related METAL HALIDES, METAL OXIDES

4042. Chlorine dioxide
  [10049-04-4] ClO_2
                                        ClO_2

HCS 1980, 289

  1. Sidgwick, 1950, 1203
  2. Mellor, 1941, Vol. 2, 288
  3. Stedman, R. F., *Chem. Eng. News*, 1951, **29**, 5030
  4. Cameron, A. E., *Chem. Eng. News*, 1951, **29**, 3196
  5. Fawcett, H. H., *Chem. Eng. News*, 1951, **29**, 4459
  6. McHale, E. T. *et al., J. Phys. Chem.*, 1968, **72**, 1849
  7. Jansen, M., *et al., Angew. Chemie (Int.)*, 1991, **30**(11), 1510
  8. *Hazards of Chlorine Dioxide*, New York, Natl. Board of Fire Underwriters,
     1950; *Chem. Eng. News*, 1950, **28**, 611
  9. Derby, R. I. *et al., Inorg. Synth.*, 1954, **4**, 152
  10. Bielz, S., Ger. Offen. 2 917 132, 1980
  11. Cowley, G., *Loss Prev. Bull.*, 1993, (113), 1
  12. Simpson, G. D. *et al., Ind. Water Treat.*, 1995, **27**(5), 48

Chlorine dioxide is considerably endothermic (ΔH°_{f} (g) +103.3 kJ/mol, 1.53 kJ/g)
and of limited stability. It is a powerful oxidant and explodes violently on the
slightest provocation as gas or liquid [1]. It is initiated by contact with several
materials (below), on heating rapidly to 100°C or on sparking [2], or by impact as
solid at −100°C [3]. A small sample exploded during vacuum distillation at below
−50°C [4], and it was stated that decomposition by sparking begins to become
hazardous at concentrations of 7–8% in air [3], and that at 10% concentration in
air (0.1 bar partial pressure) explosion may occur from any source of initiation
energy, such as sunlight, heat or electrostatic discharge [5]. A kinetic study of
the decomposition shows that it is explosive above 45°C even in absence of light,
and subject to long induction periods due to formation of intermediate dichlorine

trioxide. UV irradiation greatly sensitises the dioxide to explosion [6]. The solid (A dimer) can be relatively safely handled below −40°C and the gas at pressures below 50 mbar [7]. A guide on fire and explosion hazards in industrial use of chlorine dioxide is available [8], and preparative precautions have been detailed [9]. An improved and safer method for continuous production of chlorine dioxide is claimed [10]. A thorough review has been written, detailing numerous incidents, of hazards attending industrial preparation and use of chlorine dioxide (now much used as a low chlorine bleach). Liquid ClO_2 can separate from aqueous solutions >60 g/l, it is exceedingly shock sensitive. A partial pressure of 130 mbar in air is thought entirely safe, as are aqueous solutions (but not necessarily the head space above them) [11]. The safe use of chlorine dioxide and sodium chlorite has been reviewed [12].

See other ENDOTHERMIC COMPOUNDS, IRRADIATION DECOMPOSITION INCIDENTS, STATIC INITIATION INCIDENTS

Carbon monoxide   MRH 5.35/45
    Mellor, 1941, Vol. 2, 288
    Explosion on mixing.

Fluoramines
    Lawless, 1968, 171
    Interaction with difluoramine or trifluoramine in the gas phase is explosive.

Hydrocarbons   MRH Hexane, 6.61/12, Toluene 6.65/13
    Leleu, *Cahiers*, 1979, (95), 297
    Mixtures with butadiene, ethane, ethylene, methane or propane always explode spontaneously.

Hydrogen
    Mellor, 1941, Vol. 2, 288
    Near-stoicheiometric mixtures detonate on sparking, or on contact with platinum sponge.

Mercury
    Mellor, 1941, Vol. 2, 288
    The gas explodes on shaking with mercury.

Non-metals
    Mellor 1941, Vol. 2, 289; 1956, Vol. 2, Suppl. 1, 532
    Phosphorus, sulfur, sugar or combustible materials ignite on contact and may cause explosion.

Other reactants
    Yoshida, 1980, 268
    MRH values calculated for 19 combinations, largely with oxidisable materials, are given.

Phosphorus pentachloride   MRH 1.55/71
    See Phosphorus pentachloride: Chlorine dioxide

Potassium hydroxide   MRH 1.92/37
    Mellor, 1941, Vol. 2, 289
    The liquid or gaseous oxide will explode in contact with solid potassium hydroxide
    or its conc. solution.

Sodium nitrite, Acid
    Briggs, T. S., *J. Chem. Educ.*, 1991, **68**(11), 938
    An experiment is described which purports to show the detonation of a mixture
    of chlorine dioxide (described as a stable gas!) and nitric oxide formed from the
    above reagents. It seems likely it demonstrates detonation of chlorine dioxide,
    itself, initiated by either the nitrite or nitric oxide.
    See other HALOGEN OXIDES, OXIDANTS

4043. Thallium(I) chlorite
    [40898-91-7] ClO_2Tl

                        TlClO₂

    Bailar, 1973, Vol. 1, 1167
    It is detonable by shock.
    See other CHLORITE SALTS, HEAVY METAL DERIVATIVES

4044. Chlorine trioxide
    [13932-10-0] ClO_3

                        ClO₃

    Schmeisser, M., *Angew. Chem.*, 1955, **67**, 498
    The dimeric form is formulated as chloryl perchlorate.
    See Chloryl perchlorate
    See other HALOGEN OXIDES

4045. Rhenium chloride trioxide (Perrhenyl chloride)
    [42246-25-3] ClO_3Re

                        ClReO₃

Preparative hazard
    See Oxygen: Trirhenium nonachloride
    See related METAL HALIDES, METAL OXIDES

4046. Antimony(III) oxide perchlorate

[ ]   ClO$_5$Sb

O:SbClO$_4$

Mellor, 1956, Vol. 2, Suppl. 1, 613
Decomposes with decrepitation above 60°C.
See related METAL PERCHLORATES

4047. Chlorine

[7782-50-3]   Cl$_2$

Cl$_2$

(*MCA SD-80*, 1970); *NSC 207*, 1982; *FPA H39*, 1975; *HCS 1980*, 288 (cyl);
RSC Lab. Hazard Data Sheet No. 30, 1984

1. Dokter, T., *J. Haz. Mat.*, 1985, **10**, 73–87
2. Anon., *Sichere Chemiearbeit*, 1993, **45**(1), 9

The fire and explosion hazards of chlorine-containing systems have been reviewed
[1]. A rail tanker of chlorine was being emptied to a chlorine gas distribution
network at 7 bar, when air supply to a pneumatic valve failed, closing the system
down. After repair, valve operation caused ignition of the pipework in a chlorine–iron fire, resulting in a leak of chlorine and almost complete destruction of a
valve. Ignition is believed to have occurred because the electrolytic chlorine in the
system contained 0.7% of hydrogen. Condensation of gaseous chlorine trapped
in the pipework (at 0°C) during the shutdown produced a gas phase enriched
with hydrogen to within explosive limits and easily ignited, in turn igniting the
steelwork [2].

See Hydrogen, below; Steel, below
See CHLORINE-CONTAINING SYSTEMS
See other IGNITION SOURCES
See also REPAIR AND MAINTENANCE

Preparative hazard (electrolysis)
1. Anon., *Fire Prevention*, 1986, (189), 43.
2. Kucinski, R. E. *et al.*, *Mod. Chlor-Alkali Technol.*, 1995, 6, 89

A fire and explosions caused by hydrogen from a chloralkali plant seeping through
a defective non-return valve, then spontaneously igniting is reported [1]. A consideration of the risks of producing hydrogen chlorine mixtures, related to the design
of plant in question, claims the ion-exchange membrane plant to be safest [2].

See Sodium chloride: Nitrogen compounds
See Ammonia, below

Acetylene   MRH 8.70/99+
See Acetylene: Halogens

Air, Ethylene
See Ethylene: Air, Chlorine

Alcohols
See tert-Butanol, below
See also HYPOHALITES

Aluminium   MRH 5.23/20
Ann. Rep. Chief Insp. Factories, 1953 (Cmd. 9330), 171, HMSO
Corrosive failure of a vaporiser used in manufacture of aluminium chloride caused
liquid chlorine to contact molten aluminium. A series of explosions occurred.
See Aluminium: Halogens, or: Oxidants
See Metals, below
See other CORROSION INCIDENTS

Amidosulfuric acid
MCA Guide, 1972, 306
Chlorination of aqueous sulfamic acid led to an explosion from formation of
nitrogen trichloride.
See Nitrogen compounds, below (reference 6)

Ammonia
Khattak, M. A., *Engineering Horizons* (Pakistan), 1991, (Aug.), 33
Chlorine reacts with ammonia and compounds to form the treacherously explosive
nitrogen trichloride. A change was made to ammonia as refrigerant in the production of liquid chlorine; some months later minor explosions when transferring the
chlorine culminated in several fatalities when the delivery pipe, in which the less
volatile trichloride had presumably concentrated by partial evaporation of chlorine,
shattered. Further explosions were experienced while decontaminating the plant.
See Preparative hazard, above
See Nitrogen trichloride

Antimony trichloride, Tetramethylsilane
Bush, R. P., Br. Pat. 1 388 991, 1975
Antimony trichloride-catalysed chlorination of the silane to chlorotrimethylsilane
in absence of diluent was explosive at 100°C, but controllable at below 30°C.
See other HALOGENATION INCIDENTS

N-Arylsulfinamides
Johnson, C. R. *et al., J. Org. Chem.*, 1983, **48**, 1
Oxidation by chlorine to the arylimidosulfonyl chlorides is sometimes violent, and
tert-butyl hypochlorite is a milder reagent.

Aziridine   MRH 36.98/99+ (?3.6)
See Aziridine: Chlorinating agents

Benzene   MRH 2.09/27
1. *491M*, 1975, 102
2. Yakushina, E. P. *et al., Chem. Abs.*, 1981, **94**, 3389

A mixture in the vapour phase exploded when exposed to light [1]. Benzene–chlorine mixtures, though combustible, represent a low explosion hazard [2].
See Hydrocarbons, below

Bis(2,4-dinitrophenyl) disulfide
See Bis(2,4-dinitrophenyl) disulfide: Chlorine

Bromine pentafluoride
See Bromine pentafluoride: Acids, etc.

tert-Butanol
1. Bradshaw, C. P. C. *et al., Proc. Chem. Soc.*, 1963, 213
2. Mintz, M. J. *et al., Org. Synth.*, 1969, **49**, 9
Rate of admission of chlorine into the alcohol during preparation of *tert*-butyl hypochlorite must be regulated to keep the temperature below 20°C to prevent explosion [1]. A safer and simpler preparation uses hypochlorite solution in place of chlorine [2].
See tert-Butyl hypochlorite

Butyl rubber, Naphtha
Komarovskii, N. A. *et al., Chem. Abs.*, 1979, **91**, 158791
Chlorination of butyl rubber in naphtha with chlorine–nitrogen mixtures may lead to explosion if N contents below 77% or Cl contents above 16% are used.

Caesium oxide
See Caesium oxide: Halogens

Carbon disulfide   MRH 1.17/99+
MCA Case History No. 971
When liquid chlorine was added to carbon disulfide in an iron cylinder, an explosion occurred, owing to the iron-catalysed chlorination of carbon disulfide to carbon tetrachloride. The operation had been done previously in glassware without incident.
See other CATALYTIC IMPURITY INCIDENTS

Carbon disulfide, Sodium *aci*-nitromethanide
See Sodium *aci*-nitromethanide: Carbon disulfide, etc.

Chlorinated pyridine, Iron powder
1. Anon., *SciQuest*, 1979, **52**(7), 26–27
2. De Haven, E. S., *Plant/Oper. Progr.*, 1990, **9**(2), 131
An explosion occurred during the preparation of iron(III) chloride from iron powder and chlorine gas in a chlorinated pyridine solvent. This was attributed to formation of iron(II) chloride, its interaction with the solvent to give iron(III) chloride, then reduction of the latter by iron to iron(II) chloride. The exotherm and increasing evolution of hydrogen chloride caused the reactor to fail [1].

Another account of what seems to be the same incident attributes the runaway to Friedel Crafts type polymerisation of trichloromethylpyridines largely present in the reactor [2].
See other GAS EVOLUTION INCIDENTS

2-Chloroalkyl aryl sulfides, Lithium perchlorate
Zefirov, N. S. *et al., Chem. Abs.*, 1986, **104**, 168050
Chlorine scission of 2-chloroheptyl 4-nitrophenyl sulfide or of the 2,4-dinitro analogue in presence of lithium perchlorate formed the explosive 2-chloroheptyl perchlorate in 5–7% yield.
See other ALKYL PERCHLORATES, HALOGENATION INCIDENTS

Chloromethane
Dokter, T., *J. Haz. Mat.*, 1985, **10**, 73–87
The autoignition temperature of chloromethane in presence of chlorine is 215°C, some 400° lower than in presence of air.

3-Chloropropyne
Anon., *Loss Prev. Bull.*, 1974, (001), 10
A vigorous explosion during chlorination of 3-chloropropyne in benzene at 0°C over 4 h was attributed to presence of excess chlorine arising from the slow rate of reaction at low temperature.
See Acetylene: Halogens, *also* Hydrocarbons, below (reference 10)
See other HALOGENATION INCIDENTS

Cobalt(II) chloride, Methanol
1. Vlcek, A. A., *Inorg. Chem.*, 1967, **6**, 1425–1427
2. Gillard, R. D. *et al., Chem. & Ind.*, 1973, 77
3. Gillard, R. D., private comm., 1973
During the preparation of *cis*-dichlorobis(2,2′-bipyridyl)cobalt(III) chloride by repeating a published procedure [1], passage of chlorine into an ice-cold solution of cobalt chloride, bipyridyl and lithium chloride in methanol soon caused an explosion followed by ignition of the methanol inside the reaction vessel. The fire could not be extinguished with carbon dioxide, but went out when the flow of chlorine stopped [2]. This is consistent with the formation and self ignition of methyl hypochlorite in the system at a slightly elevated temperature. An alternative route to the cobalt complex not involving chlorine is available[3].
See Methanol, *also* Methanol, Tetrapyridinecobalt(II)chloride, both below
See other HALOGENATION INCIDENTS

Dibutyl phthalate
Statesir, W. A., *Chem. Eng. Progr.*, 1973, **69**(4), 54
A mixture of the ester and liquid chlorine confined in a stainless steel bomb reacted explosively at 118°C.

1,2-Dichloroethane
Steblen, A. V., *Chem. Abs.*, 1981, **95**, 67123

Although some mixtures of the two components will burn, even that with 34% of haloalkane leads only to a 2-fold pressure increase. Such mixtures are comparatively safe.

Dichloro(methyl)arsine
Dillon, K. B. *et al., J. Chem. Soc., Dalton Trans.*, 1976, 1479
In the attempted preparation of tetrachloromethylarsinane ($MeAsCl_4$) by interaction of chlorine with the arsine while slowly warming from $-196°C$, several sealed ampoules exploded at well below $0°C$, probably owing to liberation of chloromethane.
See other GAS EVOLUTION INCIDENTS

Diethyl ether   MRH 12.97/21
1. Harrison, G. E., private comm., 1966
2. Unpublished observation, 1949
Chlorine caused ignition of ether on contact [1]. Exposure of an ethereal solution of chlorine to daylight on removal from a fume cupboard caused a mild photocatalysed explosion [2].
See Bromine: Diethyl ether, or: Tetrahydrofuran
See other IRRADIATION DECOMPOSITION INCIDENTS

Diethylzinc   MRH 2.22/64
Weast, 1979, C-721
Ignition occurs on contact.

Dimethylformamide
Woltornist, A., *Chem. Eng. News*, 1983, **61**(6), 4
After a thermal runaway reaction during chlorination in DMF solution, investigation revealed that saturated solutions of chlorine in DMF are hazardous, and will self-heat and erupt under either adiabatic or non-adiabatic conditions. Principal products are tetramethylformamidinium chloride and carbon dioxide, with dimethylammonium chloride and carbon monoxide in small amounts. A detailed account of the mechanism is to be published.

Dimethyl phosphoramidate
1. Block, H. D. *et al., Angew. Chem. (Intern. Ed.)*, 1971, **10**, 491
2. Zwierak, A. *et al., Tetrahedron*, 1970, **26**, 3521
3. Walker, B. J. *et al., Chem. Brit.*, 1979, **15**, 66
4. Walker, B. J. *et al., J. Chem. Soc., Chem. Comm.*, 1980, 462
In a 1.5 g mol preparation of dimethyl *N,N*-dichlorophosphoramidate by chlorination of the ester, a violent explosion occurred during the period of stirring after the reaction. This did not occur on 0.5 g mol scale, and the longer reaction time may have led to liberation of explosive nitrogen trichloride. Suggested precautions include a working scale below 0.2 g mol and short reaction time. The diethyl homologue appears more stable and may be prepared on a kg scale[1]. Another incident involving explosion while following the same preparative procedure [2] was reported some years later [3,4].

Dioxan
1. Leclerc-Battin, F. *et al., J. Loss Prevention*, 1991, **4**(3), 170
2. Elaissi, A. *et al., Loss Prevention and Safety Promotion in the Process Industries* Vol II (Mewis, J. J., Pasman, H. J. & De Rademaker, E. E. Eds.), 349, Amsterdam, Elsevier, 1995

A study of autoignition of dioxan vapour and chlorine, in the gas phase at rather below 1 atm. Ignition was once observed at 25°C in daylight. In the dark it occurred from 100°C. The maximum pressure rise observed was 26 MPa per sec [1]. Detonation properties of gaseous mixtures of chlorine and dioxane, sometimes diluted with argon, are studied in detail [2]. The combination is more sensitive than most fuel/air mixtures.

Dioxygen difluoride
See Dioxygen difluoride: Various materials

Disilyl oxide
Bailar, 1973, Vol. 1, 1377
Interaction is explosive.

4,4'-Dithiodimorpholine
See 4-Morpholinesulfenyl chloride

Fluorine
See Fluorine: Halogens

Fluoromethylsilane, Dichloromethane
Bürger, H. *et al., Organometallics*, 1993, **12**(12), 4930
Chlorination of the silane in dichloromethane proved uncontrollable even at −100°C. Explosions with decomposition and generation of carbon occurred.

Glycerol
Statesir, W. A., *Chem. Eng. Progr.*, 1973, **69**(4), 54
A mixture of liquid chlorine and glycerol confined in a stainless steel bomb exploded at 70–80°C.

Hexachlorodisilane
Martin, G., *J. Chem. Soc.*, 1914, **105**, 2859
Hexachlorodisilane vapour ignited in chlorine above 300°C; violent explosions sometimes occurred.

Halocarbons
1. Dokter, T., *J. Haz. Mat.*, 1985, **10**, 75–87
2. *See entry* CHLORINE-CONTAINING SYSTEMS
The auto-ignition temperatures of various halogenated hydrocarbons in presence of chlorine are considerably below the corresponding values in air. Examples are (in °C) chloromethane 215 (618 in air); dichloromethane 262 (556); 1,2-dichloropropane 180 (555°C). Flammability limits are usually wider in chlorine

than in air. Gaseous mixtures of chloromethane and chlorine are detonable in the range 12.5–55% [1]. Liquid mixtures of dichloromethane and chlorine may well also be detonable [2].

Hydrocarbons,   MRH Benzene 2.09/27, acetylene 8.70/99+
(Inorganics)

1. Mellor, 1956, Vol. 2, Suppl. 1, 380
2. Eisenlohr, D. H., US Pat. 2 989 571, 1961
3. von Schwartz, 1918, 142, 321
4. von Schwartz, 1918, 324
5. Mamadaliev, Y. G. *et al., Chem. Abs.*, 1937, **31**, 8502^5
6. Brooks, B. T., *Ind. Eng. Chem.*, 1924, **17**, 752
7. Johnson, J. H., *Hydrocarbon Proc. Petr. Ref.*, 1963, **42**(2), 174
8. Statesir, W. A., *Chem. Eng. Progr.*, 1973, **69**(4), 52–54
9. de Oliveria, D. B., *Hydrocarbon Proc.*, 1973, **53**(3), 112–126
10. Kilby, J. L., *Chem. Eng. Progr.*, 1968, **64**(6), 52
11. Lawrence, W. W. *et al., Ind. Eng. Chem., Proc. Des. Develop.*, 1970, **9**, 47–49
12. Kokochashvili, V. I. *et al., Chem. Abs.*, 1976, **84**, 20030
13. Rozlovskii, A. I. *et al., Chem. Abs.*, 1976, **84**, 49418
14. Jenkins, P. A., unpublished information, 1974

Interaction of chlorine with methane is explosive at ambient temperature over yellow mercury oxide [1], and mixtures containing above 20 vol% of chlorine are explosive [2]. Mixtures of acetylene and chlorine may explode on initiation by sunlight, other UV source, or high temperatures, sometimes very violently [3]. Mixtures with ethylene explode on initiation by sunlight, etc., or over mercury, mercury oxide or silver oxide at ambient temperature, or over lead oxide at 100°C [1,4]. Interaction with ethane over activated carbon at 350°C has caused explosions, but added carbon dioxide reduces the risk [5]. Accidental introduction of gasoline into a cylinder of liquid chlorine caused a slow exothermic reaction which accelerated to detonation. This effect was verified [6]. Injection of liquid chlorine into a naphtha–sodium hydroxide mixture (to generate hypochlorite in situ) caused a violent explosion. Several other incidents involving violent reactions of saturated hydrocarbons with chlorine were noted [7].

In a review of incidents involving explosive reactivity of liquid chlorine with various organic auxiliary materials, two involved hydrocarbons. A polypropylene filter element fabricated with zinc oxide filler reacted explosively, rupturing the steel case previously tested to over 300 bar. Zinc chloride derived from the oxide may have initiated the runaway reaction. Hydrocarbon-based diaphragm pump oils or metal-drawing waxes were violently or explosively reactive [8]. A violent explosion in a wax chlorination plant may have involved unplanned contact of liquid chlorine with wax or chlorinated wax residues in a steel trap. Corrosion products in the trap may have catalysed the runaway reaction, but hydrogen (also liberated by corrosion in the trap) may also have been involved [9].

During maintenance work, simultaneous release of chlorine and acetylene from two plants into a common vent line leading to a flare caused an explosion in the line [10]. The violent interaction of liquid chlorine injected into ethane at 80°C/10 bar

becomes very violent if ethylene is also present [11]. The relationship between critical pressure and composition for self-ignition of chlorine–propane mixtures at 300°C was studied, and the tendency is minimal for 60:40 mixtures. Combustion is explosive under some conditions [12]. Precautions to prevent explosions during chlorination of solid paraffin hydrocarbons are detailed [13]. In the continuous chlorination of polyisobutene at below 100°C in absence of air, changes in conditions (increase in chlorine flow, decrease in polymer feed) leading to over-chlorination caused an exotherm to 130°C and ignition [14].
See Benzene, above; Methane, Oxygen, etc., or Synthetic rubber, both below
See also TURPENTINE: halogens

Hydrocarbons, Lewis acids
Howard, W. B., *Loss Prev.*, 1973, **7**, 78
During chlorination of hydrocarbons with Lewis acid catalysis, the catalyst must be premixed with the hydrocarbon before admission of chlorine. Addition of catalyst to the chlorine–hydrocarbon mixture is very hazardous, causing instantaneous release of large volumes of hydrogen chloride.
See Iron(III) chloride, Monomers, below
See other GAS EVOLUTION INCIDENTS, HALOGENATION INCIDENTS

Hydrochloric acid, Dinitroanilines
See Hydrogen chloride: Chlorine, Dinitroanilines

Hydrogen   MRH 2.55/3
  1. Mellor, 1956, Vol. 2, Suppl. 1, 373–375
  2. Weissweiler, A., *Z. Elektrochem.*, 1936, **42**, 499
  3. Eichelberger, W. C. *et al., Chem. Eng. Progr.*, 1961, **57**(8), 94
  4. Wood, J. L., *Loss Prev.*, 1969, **3**, 45–47
  5. Stephens, T. J. R. *et al., Paper 62B*, 65th AIChE Meeting, New York, 1972
  6. de Oliveria, D. B., *Hydrocarbon Proc.*, 1973, **52**(3), 112–126
  7. Antonov, V. N. *et al., Chem. Abs.*, 1974, **81**, 172184
  8. Frolov, U. E., *Chem. Abs.*, 1977, **87**, 119999
  9. Anon., *Fire Prev.*, 1986, (189), 43
Combination of the elements may be explosive over a wide range of physical conditions, with initiation by sparks, radiant energy or catalysis, e.g. by yellow mercuric oxide at ambient temperature [1]. There is a narrow range of concentrations in which the mixture is supersensitive to initiation [2]. Explosion–detonation phenomena in chlorine production cells have been investigated [3,4]. The explosive limits of the mixture vary with the container shape and method of initiation, but are usually within the range 5–89% of hydrogen by volume [1]. After an explosion in a chlorine distillate receiver where hydrogen had been produced by corrosion, no initiation source could be identified or reasonably postulated following a thorough investigation [5]. Several other hydrogen–chlorine explosions without identifiable ignition sources are also mentioned. Hydrogen may have been involved in a severe wax chlorinator explosion [6]. Pressure increase during explosive combustion of mixtures within the critical concentration region of 5–15 vol% of hydrogen was studied, and found less than anticipated [7]. Critical self-ignition temperatures are

397–284°C for mixtures containing 8–18% of hydrogen [8]. A faulty non-return valve allowed hydrogen to leak into a fibreglass-reinforced pipeline containing chlorine, and spontaneous ignition and explosion occurred [9].

Hydrogen, Nitrogen trichloride
See Nitrogen trichloride: Chlorine, Hydrogen

Hydrogen, Other gases
Munke, K., *Chem. Technik*, 1974, **26**, 292–295
Available data on explosibility of chlorine and hydrogen in admixture with air, hydrogen chloride, oxygen or inert gases is discussed and presented as triangular or rectangular diagrams.
See Sodium chlorate: Preparative hazard

Hydrogen(?), Sulfuric acid
Tabata, Y. *et al., Chem. Abs.*, 1984, **100**, 90601
Chlorine gas, produced by electrolysis of brine, was dried by passage through a mist of sulfuric acid in a drying tower, where static charge was generated. An explosion in the tower, (presumably involving hydrogen as fuel) was attributed to initiation by static spark.
See Hydrogen: Oxygen, Sulphuric acid
See other STATIC INITIATION INCIDENTS

Iron(III) chloride, Monomers
1. *MCA Case History No. 2115*
2. *MCA Case History No. 2147*
During chlorination of styrene in carbon tetrachloride at 50°C, a violent reaction occurred when some 10% of the chlorine gas had been fed in. Laboratory examination showed that the eruption was caused by a rapid decomposition reaction catalysed by ferric chloride [1]. Various aromatic monomers decomposed in this way when treated with gaseous chlorine or hydrogen chloride (either neat, or in a solvent) in the presence of steel or iron(III) chloride. Exotherms of 90°C (in 50% solvent) to 200°C (no solvent) were observed, and much gas and polymeric residue was forcibly ejected.
See Hydrocarbons, Lewis acids, above

Mercury(II) oxide
See Mercury(II) oxide: Chlorine

Metal acetylides and carbides
The mono- and di-alkali metal acetylides, copper acetylides, iron, uranium and zirconium carbides all ignite in chlorine, the former often at ambient temperature.
See Caesium acetylide: Halogens
    Dicopper(I) acetylide: Halogens
    Iron carbide: Halogens

Lithium acetylide: Halogens
Monocaesium acetylide
Monorubidium acetylide
Strontium acetylide: Halogens
Uranium dicarbide: Halogens
Zirconium dicarbide: Halogens
See Monolithium acetylide–ammonia: Gases

Metal hydrides   MRH Calcium hydride 5.35/37, Potassium hydride 4.98/52
Mellor, 1941, Vol. 2, 483; Vol. 3, 73
Potassium, sodium and copper hydrides all ignite in chlorine at ambient temperatures.

Metal phosphides   MRH Magnesium phosphide 4.04/39
See Copper(II) phosphide: Oxidants

Metals   MRH values below references
1. Mellor, 1941, Vol. 2, 92, 95; 1956, Vol. 2, Suppl. 1, 380, 469; 1941, Vol. 3,
   638; 1940, Vol. 4, 267, 480; 1941, Vol. 7, 208, 260, 436; 1941, Vol. 9, 379,
   626, 849; 1942, Vol. 12, 312; Vol. 15, 146
2. Hanson, B. H., *Process Eng.*, 1975, (2), 77
3. Davies, D. J., *Process Eng.*, 1975, (2), 77
4. Anon., *Loss Prev. Bull.*, 1975, (006), 1
5. Anon., *Loss Prev. Bull.*, 1991, (097), 9
MRH Aluminium 5.23/20, bismuth 1.21/66, calcium 7.15/36, copper 1.55/47,
germanium 2.51/34, magnesium 6.78/25, manganese 3.85/44, nickel 2.34/45, potassium 5.81/53, sodium 6.98/40, tin 2.47/45, vanadium 3.72/42, zinc 3.05/48

Tin ignites in liquid chlorine at $-34°C$, aluminium powder in the gas at $-20°C$,
while vanadium powder explodes on contact at $0°C$ with the pressurised liquid.
A solution in heptane ignites with powdered copper well below $0°C$. Aluminium,
brass foil, calcium powder, copper foil, iron wire, manganese powder, and potassium all ignite in the dry gas at ambient temperature, as do powdered antimony,
bismuth and germanium sprinkled into the gas, while magnesium, sodium and
zinc ignite in the moist gas. Thorium, tin and uranium ignite and incandesce on
warming (uranium to 150°C), powdered nickel burns at 600°C, while mercury at
$200-300°C$ ignites in a stream of chlorine. Aluminium — titanium alloys also
ignite on heating in chlorine, and niobium ignites on gentle warming [1]. Unlike
most other metals, titanium is not suitable for components in contact with *dry*
gaseous or liquid chlorine, as ignition may occur [2]. Minimum water content to
prevent attack may be from 0.015 to 1.5%, depending on conditions [3]. A statement on ignition of titanium in wet chlorine is incorrect [4]. Titanium washers
were fitted to a liquid chlorine line due to mis-labelling. Within an hour of exposure to chlorine they burnt through in a 'puff of smoke' followed by a chlorine
leak.
See Aluminium, above; Steel, below; *also* Beryllium: Halogens

Methane, Oxygen
Mal'tseva, A. S. *et al., Chem. Abs.*, 1982, **96**, 90922
Ignition in an oxychlorination reactor was attributed to the presence of free space above the packed reaction zone. Other hazards are discussed.

Methanol
Trans. US Natl. Safety Congr., 1937, 273
Passage of chlorine through cold recovered methanol (but not fresh methanol) led to a mild explosion and ignition, formation of methyl hypochlorite apparently being catalysed by an impurity.
See Cobalt(II) chloride, Methanol, above; Methanol, Tetrapyridinecobalt(II) chloride, below
See Bromine: Alcohols
See other CATALYTIC IMPURITY INCIDENTS

Methanol, Tetrapyridinecobalt(II) chloride
Glerup, J. *et al., Acta Chem. Scand., Ser. A*, 1978, **A32**, 673
In preparation of dichlorotetrapyridinecobalt(III) chloride by oxidation of the Co(II) salt in methanol solution with gaseous chlorine, severe explosions (formation of methyl hypochlorite) have occurred. Use of liquid chlorine at $-40°$ *See* Cobalt(II) chloride, Methanol, above

Nitrogen compounds   MRH values below references
1. Mellor, 1941, Vol. 2, 95; 1940, Vol. 8, 99, 288, 313, 607
2. Bowman, W. R. *et al., Chem. & Ind.*, 1963, 979
3. Bainbridge, E. G., *Chem. & Ind.*, 1963, 1350
4. Folkers, K. H. *et al., J. Amer. Chem. Soc.*, 1941, **63**, 3530
5. King, J. F. *et al., Synthesis*, 1980, 285
6. Schierwater, F.-W., *Major Loss Prevention*, 1971, 49
7. Sorbe, 1968, 120
8. Kirk-Othmer, 1964, Vol. 5, 2
9. Dokter, T., *J. Haz. Mat.*, 1985, **10**, 207–224
10. Nothe, M. *et al., Inorg. Chem.*, 1996, **35**(15), 4529
MRH Ammonia 1.88/14, hydrazine 2.43/19, hydroxylamine 3.22/49, calcium nitride 5.90/41

Ammonia—chlorine mixtures are explosive if warmed or if chlorine is in excess, owing to formation of nitrogen trichloride. Hydrazine, hydroxylamine and calcium nitride ignite in chlorine, and nitrogen triiodide may explode on contact with chlorine [1]. During chlorination of impure biuret in water at 20°C, a violent explosion occurred [2]. This was attributed to conversion of the cyanuric acid impurity (3%) to nitrogen trichloride and spontaneous explosion of the latter [3]. During interaction of chlorine and alkylthiouronium salts to give alkanesulfonyl chlorides, the dangerously explosive nitrogen trichloride may be produced if excess chlorine or slow chlorination is used. General precautions are discussed [4]. More recently it has been recommended that prompt working up of the reaction mixture will avoid formation of nitrogen trichloride [5]. Aziridine readily gives the explosive *N*-chloro compound. During chlorination of 2,4,6-triketo-hexahydro-1,3,5-triazine (cyanuric acid), presence of the diaminoketo and aminodiketo analogues as impurities, or of an unusually low pH value, may lead to formation of nitrogen trichloride. UV irradiation may be used to destroy this in a continuous circulation reactor and prevent build-up of dangerous concentrations [6]. Chlorination of amidosulfuric acid [7] or acidic ammonium chloride solutions [8] gives the powerfully explosive oil, nitrogen trichloride. The formation and separation of nitrogen trichloride have been studied in detail [9], [10].

See N-HALOGEN COMPOUNDS

Non-metal hydrides   MRH Diborane 3.59/12, phosphine 2.13/11
    Mellor, 1939, Vol. 9, 55, 396; 1939, Vol. 8, 65; 1940, Vol. 6, 219; 1941, Vol. 5, 37
    Arsine, phosphine and silane all ignite in contact with chlorine at ambient temperature, while diborane and stibine react explosively, the latter also with chlorine water.
    See Ethylphosphine: Halogens

Non-metals
    Mellor, 1941, Vol. 2, 292; 1956, Vol. 2, Suppl. 1, 380; 1943, Vol. 11, 26
    Liquid chlorine at −34°C explodes with white phosphorus, and a solution in heptane at 0°C ignites red phosphorus. Boron, active carbon, silicon and phosphorus all ignite in contact with gaseous chlorine at ambient temperature. Arsenic incandesces on contact with liquid chlorine at −34°C, and the powder ignites when sprinkled into the gas at ambient temperature. Tellurium must be warmed slightly before incandescence occurs.

Other reactants
    1. Yoshida, 1980, 61–62
    2. Gustin, J. J. *et al., Loss Prevention and Safety Promotion in the Process Industries*, Vol I, (Mewis, J. J., Pasman, H. J. & De Rademaker, E. E. Eds.), 157, Amsterdam, Elsevier, 1995
    MRH values calculated for 31 combinations with oxidisable materials are given [1]. Flammability limits and auto-ignition temperatures for a number of mixtures with organic fuels are measured and reviewed. All experimental work reported is in the vapour phase, detonation is common. Precautions for chlorination reactions are considered [2].

Oxygen difluoride
    See Oxygen difluoride: Halogens

Phenylmagnesium bromide
    1. Datta, R. L. *et al., J. Amer. Chem. Soc.*, 1919, **41**, 287
    2. Zakharkin, L. I. *et al., J. Organomet. Chem.*, 1970, **21**, 271

After treatment of the Grignard reagent with chlorine, the solid which separated exploded when shaken [1]. This solid was not seen later, using either diethyl ether or THF as solvent [2]. (The explosive solid could have been magnesium hypochlorite if moisture were present in the chlorine.)
See Magnesium hypochlorite

Phosphorus compounds   MRH values below references
1. Mellor, 1940, Vol. 8, 812, 842, 844–845, 897
2. von Schwartz, 1918, 324
MRH Phosphine 2.13/11, phosphorus trioxide 0.46/43, magnesium phosphide 4.02/39

Borondiiodophosphide, phosphine, phosphorus trioxide and trimercury tetraphosphide all ignite in contact with chlorine at ambient temperature. Trimagnesium diphosphide and trimanganese diphosphide ignite in warm chlorine [1], while ethylphosphine explodes with chlorine [2]. Unheated boron phosphide incandesces in chlorine.
See Trimethyl thionophosphate, below

Polychlorobiphenyl
Statesir, W. A., *Chem. Eng. Progr.*, 1973, **69**(4), 53–54
A mixture of a polychlorobiphenyl process oil and liquid chlorine confined in a stainless steel bomb reacted exothermally between 25 and 81°C.

Silicones
Statesir, W. A., *Chem. Eng. Progr.*, 1973, **69**(4), 53–54
Silicone process oils mixed with liquid chlorine confined in a stainless steel bomb reacted explosively on heating; polydimethysiloxane at 88–118°C, and poly-methyltrifluoropropylsiloxane at 68–114°C. Previously, leakage of a silicone pump oil into a liquid chlorine feed system had caused rupture of a stainless steel ball valve under a pressure surge of about 2 kbar.

Sodium hydroxide
MCA Case History No. 1880
Attempted disposal of a small amount of liquid chlorine by pouring it into 20% sodium hydroxide solution caused a violent reaction leading to personal contamination.

Steel
1. *MCA Case History No. 608*
2. Stephens, T. J. R. *et al., Paper 62B*, 65th AIChE Meeting, New York, 1972
3. Vervalin, 1973, 85
4. Schwarz, E., *Proc. 3rd Int. Symp. Prev. Occ. Risks Chem. Ind.*, 429–434, Heidelberg, ISSA, 1976
5. Fishwick, T., *Loss Prev. Bull.*, 1998, 139, 12
6. Anon., *Loss. Prev. Bull.*, 1983, (052), 7

Chlorine leaking into a steam-heated mild steel pipe caused ignition of the latter at about 250°C [1]. Sheet steel in contact with chlorine usually ignites at 200–250°C, but the presence of soot, rust, carbon or other catalysts may reduce the ignition temperature to 100°C. Dry steel wool ignites in chlorine at only 50°C [2]. Use of a carbon steel inlet pipe in a paraffin chlorination system led to rupture of the pipe and a fire [3]. Presence of an organic contaminant in a steel liquid chlorine vaporiser operating at 50°C led to local exothermic heating to 140–150°C and an iron–chlorine fire ensued. In another incident, admission of chlorine gas into a recently-welded (warm) steel pipe again led to a fire [4]. Another account of the same, or a very similar, vaporiser fire is found in [5]. When steel tanks containing chlorine are exposed to fire conditions, the tank wall above the liquid level may soon attain a temperature at which combustion of steel in chlorine will occur, perforating the tank. Two incidents involving derailed chlorine tank cars are described [6].
See other CORROSION INCIDENTS, IGNITION SOURCES

Sulfides
   Mellor, 1940, Vol. 4, 952; 1946, Vol. 6, 144; 1939, Vol. 9, 270
   Arsenic disulfide, boron trisulfide and mercuric sulfide all ignite in chlorine at ambient temperature, the first only in a rapid stream.

Synthetic rubber
   Murray, R. L., *Chem. Eng. News*, 1948, **26**, 3369
   During interaction of synthetic rubber and liquid chlorine, a violent explosion occurred. It is known that natural and synthetic rubbers will burn in liquid chlorine.
   See Hydrocarbons, above

Tetraselenium tetranitride
   See Tetraselenium tetranitride

Trialkyl boranes
   Coates, 1967, Vol. 1, 199
   The lower homologues tend to ignite in chlorine or bromine.

Trimethyl thionophosphate
   MCA Case History No. 371
   At an early stage in the preparation of methyl parathion, it is supposed that the phosphorus ester was being chlorinated to give dimethyl thionophosphorochloridate. Thermocouple failure indicated a low reaction temperature and the process controller boosted the chlorine feed rate, but when this fault situation was realised, the chlorine flow and agitator were stopped. However, an exothermic runaway reaction developed, eventually leading to a violent explosion.
   See other RUNAWAY REACTIONS

Tungsten dioxide
Mellor, 1943, Vol. 11, 851
Incandescence on warming.

Water
1. Kosharov, P., *Chem. Abs.*, 1940, **34**, 3917q
2. Bray, A. W., private comm., 1982
The statement that 'Mixtures of chlorine and water at certain concentrations are capable of explosion by spark ignition' [1] should read 'Mixtures of chlorine and hydrogen' [2].
See other HALOGENS, OXIDANTS

4048. Cobalt(II) chloride
[7646-79-9] Cl_2Co

$$CoCl_2$$

HCS 1980, 320

Metals
See Potassium: Metal halides
        Sodium: Metal halides
See other METAL HALIDES

4049. Dihydrazinecobalt(II) chlorate
[ ] $Cl_2CoH_8N_4O_6$

$$[(H_4N_2)_2Co]\ [ClO_3]_2$$

Salvadori, R., *Gazz. Chim. Ital. [2]*, 1910, **40**, 9
It explodes powerfully on the slightest impact or friction, or on heating to 90°C.
See other AMMINEMETAL OXOSALTS

4050. Pentaamminephosphinatocobalt(III) perchlorate
[ ] $Cl_2CoH_{17}N_5O_{10}P$

$$[(H_3N)_5CoH_2PO_2]\ [ClO_4]_2$$

Anon., *BCISC Quart. Safety Summ.*, 1965, **36**, 58
When a platinum wire (which may have been hot) was dipped for a flame test into a sintered funnel containing the air-dried complex, detonation occurred. This may have been due to heat and/or friction on a compound containing both strongly oxidising and reducing radicals. Avoid dipping (catalytically active) platinum wire into bulk samples of materials of unknown potential.
See other AMMINEMETAL OXOSALTS, REDOX COMPOUNDS

1414

4051. Cobalt(II) perchlorate hydrates
[13478-33-6, 33827-55-3, 73954-59-3] (6, 4, $2H_2O$, resp.) $Cl_2CoO_8.2H_2O$

Co(ClO$_4$)$_2$.2H$_2$O

1. Salvadori, R., *Gazz. Chim. Ital. [2]*, 1910, **40**, 15–16
2. Robinson, P. J., private comm., 1982
3. Robinson, P. J., *Nature*, 1981, **293**, 696
4. Cook, R. E. *et al., J. Chem. Res. (S)*, 1982, 267 (*M*, 2772–2783)
5. Cook, R. E. *et al., Chem. Brit.*, 1982, **18**, 859

In view of the ready commercial availability and apparent stability of the hexahydrate, it is probable that the earlier report of explosion on impact, and deflagration on rapid heating [1] referred to the material produced by partial dehydration at 100°C, rather than the hexahydrate [2]. The caked crystalline hydrated salt, prepared from aqueous perchloric acid and excess cobalt carbonate with subsequent heated evaporation, exploded violently when placed in a mortar and tapped gently to break up the crystalline mass, when a nearby dish of the salt also exploded [3]. Subsequent investigation revealed the probable cause as heating the solid stable hexahydrate to a temperature (~150°C) at which partial loss of water produced a lower and endothermic hydrate (possibly a trihydrate) capable of explosive decomposition. This hazard may also exist for other hydrated metal perchlorates, and general caution is urged [4,5].

See other METAL PERCHLORATES, SOLVATED OXOSALT INCIDENTS

4052. Chromium(II) chloride
[10049-05-5] Cl_2Cr

CrCl$_2$

Water
1. *MCA Case History No. 1660*
2. Bretherick, L., *Chem. Brit.*, 1976, **12**, 204

An unopened bottle of chromous chloride solution exploded after prolonged storage [1]. This was most likely caused by internal pressure of hydrogen developed by slow reduction of the solvent water by the powerfully reducing Cr(II) ion [2].

See Chromium(II) sulfate

See other GAS EVOLUTION INCIDENTS, METAL HALIDES, REDUCANTS

4053. Pentaamminephosphinatochromium(III) perchlorate
[65901-39-5] $Cl_2CrH_{17}N_5O_{10}P$

[(H$_3$N)$_5$CrH$_2$PO$_2$] [ClO$_4$]$_2$

Coronas, J. M. *et al., Inorg. Chim. Acta*, 1977, **25**, 110

A sample detonated violently on heating.

See other AMMINEMETAL OXOSALTS, REDOX COMPOUNDS

4054. Chromyl chloride
[14977-61-8]   Cl$_2$CrO$_2$

$$O_2CrCl_2$$

1. Sidgwick, 1950, 1004
2. Pitwell, L. R., private comm., 1964

Though a powerful and often violent oxidant of inorganic and organic materials in
absence of a diluent, it has found use as solutions in preparative organic chemistry
for the controlled oxidation of alkyl aromatic derivatives [1]. In such a reaction,
failure of a stirrer during addition of the chloride caused a build-up of unreacted
material, followed by a violent explosion [2].
See other AGITATION INCIDENTS

Ammonia
   Mellor, 1943, Vol. 11, 394
   Contact with ammonia causes incandescence.

Isopropylcyclopropane
   Wang, K. *et al., J. Org. Chem.*, 1997, **62**(13), 4248
   A violent reaction was experienced on adding this hydrocarbon to the chloride at
   room temperature. They can be safely mixed at −78°C.

Non-metal halides
   1. Mellor, 1943, Vol. 11, 394–395
   2. Pascal, 1959, Vol. 14, 153
   A jet of chromyl chloride vapour ignites in the vapour of disulfur dichloride,
   and addition of drops of chromyl chloride to cooled phosphorus trichloride causes
   incandescence and sometimes explosion [1]. Phosphorus tribromide may also ignite
   with the chloride [2].

Non-metal hydrides
   Pascal, 1959, Vol. 14, 153
   Hydrogen sulfide or phosphine may ignite in contact with the chloride.

Non-metals
   Mellor, 1943, Vol. 11, 394–395
   Moist phosphorus explodes in contact with chromyl chloride, while flowers of
   sulfur ignites.

Organic solvents
   Mellor, 1943, Vol. 11, 396
   Acetone, ethanol or ether ignite on contact with the chloride, and turpentine
   behaves similarly.

Sodium azide
   Mellor, 1967, Vol. 8, Suppl. 2.2, 36

Interaction of chromyl chloride and sodium azide to form chromyl azide is explosive in absence of a diluent.

Urea
Pascal, 1959, Vol. 14, 153
Urea ignites in contact with chromyl chloride.

Water
Partington, 1966, 751
Hydrolysis of chromyl chloride is violent.
See related METAL HALIDES
See other OXIDANTS

4055. Chromyl perchlorate
[62597-99-3] Cl_2CrO_{10}

$$CrO_2(ClO_4)_2$$

Alone, or Organic solvents
Schmeisser, M., *Angew. Chem.*, 1955, **67**, 499
A powerful oxidant which may explode violently above 80°C, and which causes organic solvents to ignite on contact. It decomposes on exposure to light, usually slowly, sometimes explosively.
See related METAL PERCHLORATES
See other OXIDANTS

4056. Copper(I) chloride
[7758-89-6] Cl_2Cu_2

$$ClCuCuCl$$

HCS 1980, 337

Preparative hazard
See Copper: Complexing agents, Water

Lithium nitride
See Lithium nitride: Copper(I) chloride
See other METAL HALIDES

4057. Copper(II) perchlorate
[10294-46-9] Cl_2CuO_8

$$Cu(ClO_4)_2$$

Polyfunctional amines
Fabbrizzi, L. *et al.*, *J. Chem. Soc., Dalton Trans.*, 1984, 1496

Complexes formed with the extended chain polyfunctional amines $H_2N[CH_2]_n NH$ $[CH_2]_m NH[CH_2]_n NH_2$ (n or $m = 2$ or 3) may be explosive and must be handled with care.

See related AMMINEMETAL OXOSALTS, [14] DIENE-N_4 COMPLEXES

N-(2-Pyridyl)acylacetamides

Illiopoulos, P. *et al., J. Chem. Soc., Dalton Trans.*, 1986, 437–443

The dihydrated binuclear complexes between copper(II) perchlorate and *N*-(2-pyridyl)-acetoacetamide and -benzoylacetamide, $[Cu_2L_2(ClO_4)_2]$.$2H_2O$, exploded on heating to 260°C.

See related AMMINEMETAL OXOSALTS, [14] DIENE-N_4 COMPLEXES

See other METAL PERCHLORATES

4058. Copper(I) perchlorate

[15061-57-1] $Cl_2Cu_2O_8$

$O_4ClCuCuClO_4$

Alkenes, or Carbon monoxide

1. Ogura, T., *Inorg. Chem.*, 1976, **15**, 2301–2303
2. Thompson, J. S. *et al., Inorg. Chem.*, 1984, **23**, 2814

If preparation of the perchlorate (by stirring copper(II) perchlorate in water with copper powder) is effected under an atmosphere of the ligands ethylene, allene, 1,3-butadiene or carbon monoxide, two types of explosive complexes, L_2.$Cu_2(ClO_4)_2$.nH_2O, (L represents one double bond), are formed. There is a crystalline series with $n = 4$, e.g. $2CO$.$Cu_2(ClO_4)_2$.$4H_2O$, and an amorphous series with $n = 3$, e.g. C_3H_4.$Cu_2(ClO_4)_2$.$3H_2O$. All the complexes are explosive, but unpredictably so. Sometimes a complex would not explode under a hammer-blow, but sometimes it would explode on stirring with a Teflon-coated stirrer bar. On one occasion a 10 mg sample stored in glass exploded, propelling glass fragments 10 m [1]. When the ethylene complex is prepared in methanol solution, it must not be allowed to dry out or it will explode [2].

1,4-Oxathiane

Barnes, J. C. *et al., J. Chem. Soc. Pak.*, 1982, **4**, 103–113

Of the two complexes which the perchlorate forms with oxathiane (1:3, 1:4), the latter explodes on heating.

See other METAL PERCHLORATES, REDOX COMPOUNDS

4059. Dichlorofluoramine

[17417-38-8] Cl_2FN

Cl_2NF

1. Sukornick, B. *et al., Inorg. Chem.*, 1963, **2**, 875
2. Batty, W. E., *Chem. & Ind.*, 1969, 1232

In the liquid phase, it is an extremely friction- and shock-sensitive explosive [1], like the trichloro analogue [2].

See other N-HALOGEN COMPOUNDS

4060. *N,N*-**Dichloropentafluorosulfanylamine**
[22650-46-0] Cl_2F_5NS
$$Cl_2NSF_5$$

Kirk Othmer, 4th Edn., 1993, Vol IV, 917
Shock sensitive and unstable from 80°C.

Preparative hazard
Waterfeld, A. *et al., J. Chem. Soc., Chem. Comm.*, 1982, 839
Reaction of chlorine fluoride with trifluorosulfur nitride is rather hazardous and
may lead to violent explosions. A safer alternative preparation is to use chlorine
and mercuric fluoride in place of chlorine fluoride.
See other N-HALOGEN COMPOUNDS, N–S COMPOUNDS

4061. Iron(II) chloride
[7758-94-3] Cl_2Fe
$$FeCl_2$$

Ozonides
See OZONIDES: metals, etc.
See other METAL HALIDES, REDUCANTS

4062. Iron(II) perchlorate
[13933-23-8] Cl_2FeO_8
$$Fe(ClO_4)_2$$

Preparative hazard
See Perchloric acid: Iron(II) sulfate
See other METAL PERCHLORATES, REDOX COMPOUNDS

4063. Dichloramine (Chlorimide)
[3400-09-7] Cl_2HN
$$Cl_2NH$$

Eckert, P. *et al., Chem. Abs.*, 1951, **45**, 7527g
Dissolving *N,N,N′,N′*-tetrachloroadipamide in boiling water gives a highly explo-
sive oil, probably dichloramine.
See other N-HALOGEN COMPOUNDS

4064. *N,N′*-**Bis(chloromercurio)hydrazine**
[ ] $Cl_2H_2Hg_2N_2$
$$ClHgNHNHHgCl$$

Mellor, 1940, Vol. 4, 874, 881; Vol. 8, 318

It explodes when heated or struck, and the bromo and other analogues are similar.

See Mercury(II) chloride: Hydrazine salts, Base

See other MERCURY COMPOUNDS *See related* METAL HYDRAZIDES

4065. Chloronium perchlorate

[ ] $Cl_2H_2O_4$

$$Cl^+H_2\ ClO_4^-$$

Hantzsch, A., *Ber.*, 1930, **63**, 1789

This hydrogen chloride–perchloric acid complex spontaneously dissociates with explosive violence.

See other NON-METAL PERCHLORATES

†4066. Dichlorosilane

[4109-96-0] Cl_2H_2Si

$$Cl_2SiH_2$$

1. Sharp, K. G. *et al., J. Electrochem. Soc.*, 1982, **129**
2. Britton, L. G., *Plant/Oper. Progr.*, 1990, **9**(1), 16

It is unusual in being flammable with so high a chlorine content (70%), and mixtures with air may detonate if confined when ignited. One case of spontaneous ignition under ambient conditions was observed [1]. It has an unusually low autoignition temperature (43±3°C). A survey of hazards and combustion is found in [2]. There is a risk of combustion with fluorochloromethanes. Autoignition was observed on contact with traces of a mixture of alkali nitrates/nitrite.

See other HALOSILANES

4067. Bis(hydroxylamine)zinc chloride

[33774-96-8] $Cl_2H_6N_2O_2Zn$

$$[(HONH_2)_2Zn]Cl_2$$

Walker, J. E. *et al., Inorg. Synth.*, 1967, **9**, 2

It explodes at 170°C.

See related AMMINEMETAL HALIDES

4068. Hydrazinium diperchlorate

[13812-39-0] $Cl_2H_6N_2O_8$

$$H_3N^+N^+H_32ClO_4^-$$

Grelecki, C. J. *et al., ACS 54*, 1966, 73

The thermal decomposition of this solid rocket propellant component has been studied.

Metal compounds

Probhakaran, C. *et al., Def. Sci. J.*, 1982, **31**, 285–291

Of the series of additives copper chromate, copper chloride, nickel oxide, iron(III) oxide or magnesium oxide, the earlier members have the greatest effect in increasing the sensitivity of the explosive salt towards heat, impact or friction.

See other PERCHLORATE SALTS OF NITROGENOUS BASES

4069. Bis(hydrazine)nickel perchlorate

[ ] $Cl_2H_8N_4NiO_8$

$$[(H_4N_2)_2Ni]\ [ClO_4]_2$$

Maissen, B. *et al., Helv. Chim. Acta*, 1951, **34**, 2084–2085

The salt, known to be explosive when heated dry (but not under a hammer-blow), exploded violently when stirred as a dilute aqueous suspension.

See other AMMINEMETAL OXOSALTS

4070. Bis(hydrazine)tin(II) chloride

[55374-98-6] $Cl_2H_8N_4Sn$

$$[(H_4N_2)_2Sn]Cl_2$$

Mellor, 1941, Vol. 7, 430

It explodes on heating.

See related AMMINEMETAL HALIDES, METAL HYDRAZIDES

4071. Tetraamminebis(dinitrogen)osmium(II) perchlorate

[ ] $Cl_2H_{12}N_8O_8Os$

$$[(H_3N)_4Os(N_2)_2]\ [ClO_4]_2$$

Creutz, C. A., private comm., 1969

The dry salt may explode on touching.

See next entry below *See other* AMMINEMETAL OXOSALTS

4072. Pentaamminedinitrogenosmium(II) perchlorate

[20611-53-4] $Cl_2H_{15}N_7O_8Os$

Bohr, J. D. *et al., Inorg. Chem.*, 1980, **19**, 2417

It can be detonated.

See entry next above *See other* AMMINEMETAL OXOSALTS

4073. Pentaammineazidoruthenium(III) chloride
[28223-30-5] (ion) $Cl_2H_{15}N_8Ru$

Hazardous intermediate
See Pentaamminechlororuthenium chloride: Sodium azide
See related METAL AZIDES

4074. Hexaamminenickel chlorate
[ ] $Cl_2H_{18}N_6NiO_6$

$$[(H_3N)_6Ni] [ClO_3]_2$$

Mellor, 1956, Vol. 2, Suppl. 1, 592
It explodes on impact.
See other AMMINEMETAL OXOSALTS

4075. Hexaamminenickel perchlorate
[14322-50-0] $Cl_2H_{18}N_6NiO_8$

$$[(H_3N)_6Ni] [ClO_4]_2$$

Mellor, 1956, Vol. 2, Suppl. 1, 592
It explodes on impact, but is less sensitive than the chlorate.
See other AMMINEMETAL OXOSALTS

4076. Mercury(II) chloride
[7847-94-7] Cl_2Hg

$$HgCl_2$$

HCS 1980, 615

Hydrazine salts, Base
Hofmann, K. A. *et al., Annalen*, 1899, **305**, 191
A yellowish precipitate was obtained from the chloride and hydrazinium chlorides when basified with sodium acetate. It had N—Hg bonds and was perhaps ClHgNHNHHgCl. It exploded on heating or shock. A similar compound resulted from the bromides.

Sodium *aci*-nitromethanide
See Sodium *aci*-nitromethanide: Mercury(II) chloride

Sodium azide
See Sodium azide: Heavy metals
See other MERCURY COMPOUNDS, METAL HALIDES

4077. Mercury(II) chlorite
[73513-17-4] Cl_2HgO_4

$$Hg(ClO_2)_2$$

Levi, G. R., *Gazz. Chim. Ital. [2]*, 1915, **45**, 161
It is extremely unstable when dry, exploding spontaneously.
See other CHLORITE SALTS, MERCURY COMPOUNDS

4078. Mercury(II) perchlorate
[7616-83-3] Cl_2HgO_8

$$Hg(ClO_4)_2$$

Samuels, G. J. *et al., Inorg. Chem.*, 1980, **19**, 233
A filtered solution, prepared by heating mercuric oxide with a slight excess of perchloric acid, after standing for several months precipitated a little unidentified white solid. This (but not the supernatant liquid) was very shock sensitive, and detonated as it was being rinsed out with water. (Traces of a volatile amine may have been absorbed into the acid liquor to give an amminemercury perchlorate, expected to be explosive).
See other AMMINEMETAL OXOSALTS, MERCURY COMPOUNDS, METAL PER-CHLORATES

4079. Mercury(II) perchlorate . 6 (or 4)dimethyl sulfoxide
[62909-81-3] (.6DMSO) Cl_2HgO_8.6 (or 4)C_2H_6OS

$$[(Me_2S{:}O)_6Hg]\ [ClO_4]_2$$

1. Rayner, P. N. G., *Chem. Brit.*, 1977, **13**, 396; private comm., 1977
2. Cotton, F. A. *et al., J. Amer. Chem. Soc.*, 1960, **82**, 2886
3. Brown, A. J. *et al., J. Chem. Soc., Dalton Trans.*, 1978, 1778
4. Sandstrom, M. *et al., Acta Chem. Scand.*, 1978, **A32**, 610

The hexasolvated compound was being prepared from a mixture of the hydrated salt, methanol and dimethyl sulfoxide but inexplicably mercury(I) oxide came out of solution. This was redissolved by addition of perchloric acid and the solution was refrigerated for 60 h. After addition of molecular sieve to remove the water present, the solution was filtered and the sieve treatment repeated four more times. During the final filtration, a violent explosion occurred in the funnel. The most likely cause was formation of methyl perchlorate promoted by the dehydrating action of the sieve, but dimethyl sulfoxide also forms an unstable salt with perchloric acid, so this also may have been involved [1]. During preparation of the hexasolvate by a published method [2], a violent explosion occurred [3]. The solid tetrasolvated product from treatment of a solution of the perchlorate in methanol–perchloric acid with DMSO at ambient temperature exploded with extreme violence after prolonged desiccation (possibly owing to formation of methyl perchlorate). Materials prepared at $-78°C$ appeared to be free of this hazard, though they will explode under mechanical impact, etc. [4].

See Dimethyl sulfoxide: Metal oxosalts
See other METAL PERCHLORATES
See other MOLECULAR SIEVE INCIDENTS, SOLVATED OXOSALT INCIDENTS

4080. Mercury(I) chlorite
[101672-18-8] $Cl_2Hg_2O_4$

$$Hg_2(ClO_2)_2$$

Levi, G. R., *Gazz. Chim. Ital. [2]*, 1915, **45**, 161
It is extremely unstable when dry, exploding spontaneously.
See other CHLORITE SALTS, MERCURY COMPOUNDS, REDOX COMPOUNDS

4081. Magnesium chloride
[7786-30-3] Cl_2Mg

$$MgCl_2$$

Air, Mild steel, Water
1. Anon., *CISHC Chem. Safety Summ.*, 1984, **55**, 66
2. Haudt, P. R., *Mater. Perform.*, 1986, **25**(9), 49–55
3. Melcher, R. G. *et al.*, *Am. Ind. Hyg. Assoc. J.*, 1987, **48**, 608–612
A large steel evaporator used for magnesium chloride solution was shut down for maintenance. During maintenance operations a fatality occurred from atmospheric oxygen deficiency inside the evaporator. It was found later that the oxygen content in the evaporator fell from the normal 21% to about 1% in under 24 h, and this was confirmed in laboratory tests. This was attributed to very rapid rusting of the steel under warm humid conditions in the presence of traces of magnesium chloride [1]. Further work shows that other salts (calcium bromide, calcium chloride, magnesium sulfate, potassium chloride) behave similarly, and that presence of scale is a contributory factor [2]. Magnetite scale (Fe_3O_4) on mild steel increases the depletion rate by a factor of 10, while the rust formed during the corrosion has little effect [3].
See other CORROSION INCIDENTS

Jute, Sodium nitrate
See Sodium nitrate: Jute, Magnesium chloride
See other METAL HALIDES

4082. Magnesium hypochlorite
[10233-03-1] Cl_2MgO_2

$$Mg(OCl)_2$$

Clancey, V. J., *J. Haz. Mat.*, 1975, **1**, 83–94
The compound is very unstable, and its presence may be one of the causes of the observed explosive and apparently spontaneous decomposition of calcium hypochlorite, if produced from magnesia-containing lime (derived from dolomite).

See Chlorine: Phenylmagnesium bromide
See other METAL HYPOCHLORITES

4083. Magnesium chlorate
[10326-21-3] Cl_2MgO_6

$$Mg(ClO_3)_2$$

Sulfides
   Mellor, 1956, Vol. 2, Suppl. 1, 584
   Interaction with copper(I) sulfide is explosive, and with antimony(III) sulfide,
   arsenic(III) sulfide, tin(II) sulfide or tin(IV) sulfide, incandescent.
   See other METAL CHLORATES

4084. Magnesium perchlorate
[10034-81-8] Cl_2MgO_8

$$Mg(ClO_4)_2$$

Marusch, H., *Chem. Tech.* (Berlin), 1956, **8**, 482–485
The drying agent may contain traces of perchloric acid remaining from manufacturing operations, and owing to the great desiccating power of the salt, the acid
will be anhydrous. Many of the explosions experienced with magnesium perchlorate have their origins in contact of the anhydrous acid with oxidisable materials,
or materials able to form unstable perchlorate esters or salts.

Alkenes
   Heertjes, P. M. *et al., Chem. Weekbl.*, 1941, **38**, 85
   The anhydrous salt which had been used for drying unsaturated hydrocarbons
   exploded on heating to 220°C for reactivation. The need to avoid contact with
   acidic materials is stressed. (Traces of an alkyl perchlorate may conceivably have
   been formed from free perchloric acid).
   See Organic materials, below

Ammonia
   491M, 1975, 244
   Intensive drying of ammonia gas by passing it over the desiccant in a steel drying
   tube led to an exotherm, followed by a violent explosion. (An ammine derivative
   may have been formed).
   See related AMMINEMETAL OXOSALTS

Argon
   Dam, J. W., *Chem. Weekbl.*, 1958, **54**, 277
   An explosion which occurred when argon was being dried obviously involved
   some (unstated) impurity in the system.

Arylhydrazine, Ether
Belcher, R., private comm., 1968
Anhydrous magnesium perchlorate was used to dry thoroughly an ethereal solution of an arylhydrazine. During evaporation of the filtered solution it exploded completely and violently. Magnesium perchlorate is rather soluble in ether and may contain traces of free perchloric acid (probably in the anhydrous form, as the magnesium salt is a powerful dehydrator). It is entirely unsuitable for drying organic solvents. (The hydrazine may have complexed with the magnesium to form a redox salt).
See related AMMINEMETAL OXOSALTS

Cellulose, Dinitrogen tetraoxide, Oxygen
Anon., ABCM Quart. Safety Summ., 1961, 32, 6
Magnesium perchlorate contained in a glass tube between wads of cotton wool was used to dry a mixture of oxygen and dinitrogen tetraoxide. After several days the drying tube exploded violently. It seems probable that the acidic fumes and cotton produced cellulose nitrate, aided by the dehydrating action of the perchlorate.

Dimethyl sulfoxide
1. Tobe, M. L. et al., J. Chem. Soc., 1964, 2991
2. Anon., Chem. Eng. News, 1965, 43(47), 62
3. Dessy, R. E. et al., J. Amer. Chem. Soc., 1964, 86, 28
In the preparation of anhydrous DMSO by a literature method [1], an explosion occurred during distillation from anhydrous magnesium perchlorate [2]. This may have been due to the presence of some free methanesulfonic acid as an impurity in the solvent, which could liberate traces of perchloric acid. It is known that sulfoxides react explosively with 70% perchloric acid, (but several metal perchlorates also form unstable solvates with DMSO) The alternative procedure for drying DMSO with calcium hydride [3] seems preferable, as this would also remove any acidic impurities.
See Dimethyl sulfoxide: Metal oxosalts

Ethanol
Raymond, K. R., private comm., 1983
The desiccant in a drying tube, accidentally exposed to ethanol vapour, was left for several months. The explosion which occurred when the desiccant was scraped out was certainly due to formation of ethyl perchlorate.
See ALKYL PERCHLORATES

Ethylene oxide
491M, 1975, 314
Desiccation of gaseous ethylene oxide led to an explosion, (possibly involving formation of 2-hydroxyethyl perchlorate).

Fluorobutane, Water
Anon., Ind. Eng. Chem. (News Ed.), 1939, 17, 70

A violent explosion followed the use of magnesium perchlorate to dry wet fluorobutane. The latter was presumed to have hydrolysed to give hydrogen fluoride which had liberated perchloric acid, explosively unstable when anhydrous. (This explanation seems unlikely in view of the large disparity between dissociation constants of the two acids). Magnesium perchlorate is unsuitable for drying acidic or flammable materials; calcium sulfate would be suitable.

Organic materials
1. Hodson, R. J., *Chem. & Ind.*, 1965, 1873
2. *MCA Case History No. 243*
The use of the perchlorate as desiccant in a drybag where contamination with organic compounds is possible is considered dangerous [1]. Magnesium perchlorate ('Anhydrone') was inadvertently used instead of calcium sulfate (anhydrite) to dry an unstated reaction product before vacuum distillation. The error was realised and all solid was filtered off. Towards the end of the distillation, decomposition and an explosion occurred, possibly owing to the presence of dissolved magnesium perchlorate, or more probably to perchloric acid present as impurity in the salt [2].

Phosphorus
See Phosphorus: Magnesium perchlorate

Trimethyl phosphite
Jercinovic, L. M. *et al.*, *J. Chem. Educ.*, 1968, **45**, 751
Contact between the salt and ester caused violent explosions on several occasions. (Methyl perchlorate may have been formed).
See related ALKYL PERCHLORATES
See other METAL PERCHLORATES

4085. Manganese(II) chloride
[7773-01-5] Cl_2Mn
$$MnCl_2$$

Zinc
See Zinc: Manganese dichloride
See other METAL HALIDES

4086. Manganese dichloride dioxide
[51819-69-3] Cl_2MnO_2
$$MnCl_2O_2$$

Briggs, T. S., *J. Inorg. Nucl. Chem.*, 1968, **30**, 2867–2868
Explosively unstable if isolated as liquid at ambient temperature, it may be handled safely in carbon tetrachloride solution.
See Manganese chloride trioxide, Manganese trichloride oxide
See related METAL HALIDES, METAL OXIDES

4087. Manganese(II) chlorate
[104813-96-9] Cl_2MnO_6

$$Mn(ClO_3)_2$$

Bailar, 1973, Vol. 3, 835
The hexahydrated salt decomposes explosively above 6°C, producing chlorine dioxide.
See other METAL CHLORATES, REDOX COMPOUNDS

4088. Manganese(II) perchlorate
[13770-16-6] Cl_2MnO_8

$$Mn(ClO_4)_2$$

Sidgwick, 1950, 1285
It explodes at 195°C.

2,2-Dimethoxypropane
See 2,2-Dimethoxypropane: Metal perchlorates
See other METAL PERCHLORATES, REDOX COMPOUNDS

4089. Chloro-1,2,4-triselenadiazolium chloride (3-Chloro-1,3,4-triselenadiazolium chloride)
[ ] $Cl_2N_2Se_3$

1. Chivers, T. *et al., Angew. Chem. (Int.)*, 1992, **31**(11), 1518
2. Chivers, T. *et al., Inorg. Chem.*, 1993, **32**(20), 4391
The black explosive powder previously described as Se_4N_2 is now assigned this structure. It is very sensitive when dry and handling it damp with an hydrocarbon is recommended. There is brown dimer, also explosive [2].
See related NITRIDES, NON-METAL HALIDES, N–S COMPOUNDS

4090. Vanadyl azide dichloride
[ ] Cl_2N_3OV

$$O{:}V(N_3)Cl_2$$

An explosive solid.
See entry METAL AZIDE HALIDES

4091. Di(triselenadiazolium) dichloride

[ ] $Cl_2N_4Se_6$

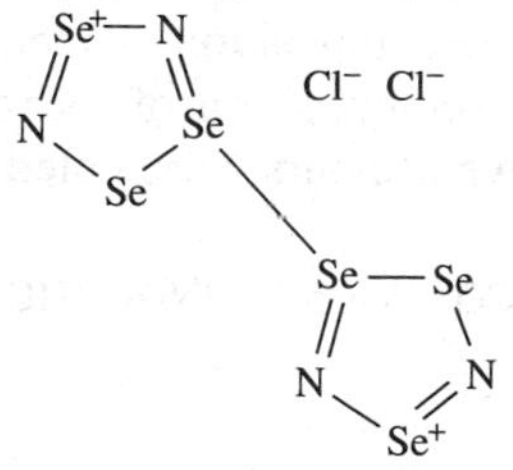

Se⁺—N
N Se   Cl⁻ Cl⁻
Se

Se—Se
N N
Se⁺

Chivers, T. *et al., Inorg. Chem.*, 1993, **32**(20), 4391
A brown explosive powder
See Chloro-1,2,4-triselenadiazolium chloride

4092. Diazidodichlorosilane

[67880-19-7] Cl_2N_6Si

$(N_3)_2SiCl_2$

See Silicon tetraazide *See other* NON-METAL AZIDES

4093. Nickel chlorite

[72248-72-7] Cl_2NiO_4

$Ni(ClO_2)_2$

1. Mellor, 1956, Vol. 2, Suppl. 1, 574
2. Levi, G. R., *Gazz. Chim. Ital. [2]*, 1923, **53**, 245
The dihydrate explodes at 100°C [1], or on percussion, even when wet [2].
See other CHLORITE SALTS

4094. Nickel perchlorate

[13637-71-3] Cl_2NiO_8

$Ni(ClO_4)_2$

2,2-Dimethoxypropane
See 2,2-Dimethoxypropane: Metal perchlorates
See other METAL PERCHLORATES

4095. Dichlorine oxide

[7791-21-1] Cl_2O

Cl_2O

1. Sidgwick, 1950, 1202
2. Jacobs, P. W. M. *et al., Chem. Rev.*, 1969, **69**, 559
3. Pilipovich, D. *et al., Inorg. Chem.*, 1972, **11**, 2190
4. Gray, P. *et al., Combust. Flame*, 1972, **18**, 361–371
5. Cady, G. H., *Inorg. Synth.*, 1957, **5**, 156

It is somewhat endothermic (ΔH°_f (g) +87.5 kJ/mol, 1.0 kJ/g), the liquid may explode on pouring or sparking at 2°C, and the gas readily explodes on rapid heating or sparking [1,2], on adiabatic compression in a U-tube, or often towards the end of slow thermal decomposition. Kinetic data are summarised [3]. The spontaneously explosive decomposition of the gas was studied at 42–86°C, and induction periods up to several hours were noted [4]. Preparative precautions have been detailed [5].
See other ENDOTHERMIC COMPOUNDS, INDUCTION PERIOD INCIDENTS

Alcohols
Gallais, 1957, 677
Alcohols are oxidised explosively.

Carbon
Mellor, 1946, Vol. 5, 824
Addition of charcoal to the gas causes an immediate explosion, probably initiated by the heat of adsorption of the gas on the solid.

Carbon disulfide
Mellor, 1940, Vol. 6, 110
The vapours explode on contact.

Dicyanogen
Brotherton, T. K. et al., Chem. Rev., 1959, 59, 843
Contact causes ignition or explosion.

Diphenylmercury
See Diphenylmercury: Chlorine monoxide, etc.

Ethers
Gallais, 1957, 677
Ethers are oxidised explosively.

Hydrocarbons
Ip, J. K. K. et al., Combust. Flame, 1972, 19, 117–129
The spontaneously explosive interaction of dichlorine oxide with methane, ethane, propane, ethylene or butadiene was investigated at 50–150°C. Self-heating occurs with ethylene, ethane and propane mixtures.
See other SELF-HEATING AND IGNITION INCIDENTS

Nitrogen oxide
Mellor, 1940, Vol. 8, 433
Interaction is explosive.

Oxidisable materials
1. Mellor, 1941, Vol. 2, 241–242; 1946, Vol. 5, 824
2. Cady, G. H., Inorg. Synth., 1957, 5, 156
The heat sensitivity (above) may explain the explosions which occur on contact of many readily oxidisable materials with this powerful oxidant. Such materials include ammonia, potassium; arsenic, antimony; sulfur, charcoal (adsorptive

heating may also contribute); calcium phosphide, phosphine, phosphorus; hydrogen sulfide, antimony sulfide, barium sulfide, mercury sulfide and tin sulfide [1]. Various organic materials (paper, cork, rubber, turpentine, etc.) behave similarly [2]. Mixtures with hydrogen detonate on ignition [1].

Potassium
See Potassium: Non-metal oxides
See other HALOGEN OXIDES, OXIDANTS

4096. Sulfinyl chloride (Thionyl chloride)
[7719-09-7]   Cl₂OS

$$O:SCl_2$$

HCS 1980, 898; *RSC Lab. Hazard Data Sheet No. 26*, 1984

Ammonia   MRH 0.84/16
Foote, C. S., private comm., 1965
Addition of a solution of 4-nitrobenzoyl chloride (1 g) in a large excess (10 ml) of sulfinyl chloride to ice-cold conc. ammonia solution caused a violent explosion. This may certainly be attributed to the instantaneous hydrolysis of the excess sulfinyl chloride by the aqueous ammonia with production of several l of un-neutralised acid gases in a test tube.
See Water, below

Bis(dimethylamino) sulfoxide
Armitage, D. A. *et al., J. Inorg. Nucl. Chem.*, 1974, **36**, 993
Interaction of the chloride with the sulfoxide or its higher homologues to form dialkylaminosulfinyl chlorides causes extensive decomposition, possibly explosive above 80°C.

Chloryl perchlorate
See Chloryl perchlorate: Thionyl chloride

1,2,3-Cyclohexanetrione trioxime, Sulfur dioxide
1. Tokura, N. *et al., Bull. Chem. Soc. Japan*, 1962, **35**, 723
2. Lewis, J. J., *J. Heterocycl. Chem.*, 1975, 12, 601
A previous method of making 4-oximino-4,5,6,7-tetrahydrobenzofurazan by cyclising the oxime with sulfinyl chloride in liquid sulfur dioxide sometimes led to explosive reactions [1]. A new procedure involving aqueous calcium carbonate is quite safe [2].

Diisopropyl malonate
McKeown, R. H., private comm., 1985
During reflux of a mixture to produce malonoyl chloride, vivid sparks were seen in the flask, and the reaction was closed down without mishap. No explanation is apparent, but the diisopropyl ester structure appears likely to be susceptible to autoxidation on storage, and peroxides may possibly have been involved in the phenomenon.
See related PEROXIDATION INCIDENTS

Dimethylformamide

1. Spitulnik, M. J., *Chem. Eng. News*, 1977, **55**(31), 31
2. Joshi, M. S., *Chem. Eng. News*, 1986, **64**(14), 2
3. Zemlicka, J., *Chem. Eng. News*, 1986, **64**(25), 4

Some 200 kg of a mixture of sulfinyl chloride and DMF decomposed vigorously after storage for several h at ambient temperature. This was attributed to the presence of 90 ppm each of iron and zinc in the chloride used for the preparation. Mixtures of the pure components remained unchanged for 48 h, but addition of 200 ppm of iron powder led to exothermic decomposition after stirring for 22 h [1]. During vacuum removal of sulfur dioxide from the reaction mixture in preparation of *N*,*N*-dimethylchloromethyliminium chloride, a sudden exotherm and pressure increase occurred in a 400 l reactor. Study of the reaction in an adiabatic calorimeter showed that a runaway reaction developed within 8 h at 60°C, or within 20 min at 85°C at ambient pressure. Reaction slurry from which part of the sulfur dioxide had been removed decomposed even more rapidly [2]. The use of phosgene in the reaction in place of sulfinyl chloride gives a more controlled reaction which does not need heating [3].

See other GAS EVOLUTION INCIDENTS
See Dimethylformamide: Halocarbons, etc.
See other CATALYTIC IMPURITY INCIDENTS

Dimethyl sulfoxide
See Dimethyl sulfoxide: Acyl halides

Esters, Metals

1. Spagnolo, C. J. *et al.*, *Chem. Eng. News*, 1992, **70**(22), 2
2. Wang, S. S. Y. *et al.*, *Process Safety Progr.*, 1994, **13**(3), 153

It was found that a solution of thionyl chloride in ethyl acetate burst galvanised drums on storage. Further investigation demonstrated an exothermic gas-evolving reaction between the ester and the chloride, initiated by iron or zinc, to give chloroethane, acetyl chloride and sulfur dioxide. A similar reaction is sometimes used synthetically on lactones with Lewis acid catalysis, especially zinc chloride. This is therefore probably a general ester reaction[1]. A more comprehensive, but scarcely more comprehending, study of the ethyl acetate/thionyl chloride reaction demonstrates involvement of zinc. Elemental sulphur is apparently among the products [2].

See Diisopropyl malonate (above)
See Propionoyl chloride: Diisopropyl ether
See Sulfonyl chloride: Organic materials, metals
See other CATALYTIC IMPURITY INCIDENTS, GAS EVOLUTION INCIDENTS

Heterocyclic amines
See HETEROCYCLIC *N*-SULFINYLAMINES

Hexafluoroisopropylideneaminolithium
See Hexafluoroisopropylideneaminolithium: Non-metal halides

Linseed oil, Quinoline   MRH Quinoline 1.63/99+
  1. Fieser, L. F. *et al., Org. Synth.*, 1943, Coll. Vol. 2, 570
  2. Laws, G. F., private comm., 1966
  It is important to add the quinoline and linseed oil, used in purifying the chloride
  [1], to the chloride. If reversed addition is used, a vigorous decomposition may
  occur [2].

Lithium
  See Lithium: Sulfinyl chloride

Other reactants
  Yoshida, 1980, 57
  MRH values calculated for 14 combinations with various reagents are given.

Sodium   MRH 5.27/33
  See Sodium: Non-metal halides

Tetrahydrofuran
  Jackson, R. W. *et al., Chem. Eng. News*, 1988, **66**(27), 2
  Unstirred mixtures of sulfinyl chloride and tetrahydrofuran may react extremely
  exothermally and with gas evolution when heated to 60°C, the products of ring-
  scission being bis(4-chlorobutyl) ether, 1,4-dichlorobutane and sulfur dioxide.
  Some ratios of the two reactants enhance the tendency, and relatively confined
  mixtures may explode. Sufficient heat to initiate the reaction may arise from self-
  mixing of the layered liquids, the chloride being twice as dense as THF.
  See other AGITATION INCIDENTS, GAS EVOLUTION INCIDENTS

Toluene, Ethanol, Water
  Anon., *Univ. Safety Assoc., Safety Newsletter*, 1982–1984
  A solution of the chloride (120 ml) in toluene (750 ml) was treated (apparently
  without effective stirring) with excess sodium bicarbonate solution to destroy it.
  When reaction had ceased, the organic layer was poured into a waste solvent drum.
  Vigorous evolution of sulfur dioxide and hydrogen chloride then ensued from reaction with ethanol (toluene-soluble) in the waste drum. For destruction of solutions
  of sulfinyl chloride in water-insoluble solvents, extremely good agitation is necessary to ensure proper contact with a basic reagent. Ammonia is more soluble
  in toluene than is water, so ammonia solution should be used after bicarbonate
  treatment to ensure complete destruction.

Water
  MCA Case History No. 1808
  Passage of thionyl chloride through a flexible metal transfer hose which was
  contaminated with water or sodium hydroxide solution caused the hose to burst.
  Interaction with water violently decomposes the chloride to hydrogen chloride
  (2 mol) and sulfur dioxide (1 mol), the total expansion ratio from liquid to gas
  being 993:1 at 20°C, so very high pressures may be generated.
  See other ACYL HALIDES, NON-METAL HALIDES

4097. Seleninyl chloride
[7791-23-3]   Cl₂OSe

$$O:SeCl_2$$

Antimony
  Mellor, 1947, Vol. 10, 906
  Powdered antimony ignites on contact with the chloride.

Metal oxides
  Mellor, 1947, Vol. 10, 909
  In contact with silver oxide, light is evolved and sufficient heat to decompose
  some of it. Similar effects were observed with lead(II) oxide, lead(IV) oxide and
  lead (II)(IV) oxide.

Potassium
  Mellor, 1947, Vol. 10, 908
  Potassium explodes violently in contact with the liquid.

Phosphorus
  Mellor, 1947, Vol. 10, 906
  Red phosphorus evolves light and heat in contact with the chloride, while white
  phosphorus explodes.
  See other NON-METAL HALIDES

4098. Lead(II) hypochlorite
[ ]   Cl₂O₂Pb

$$Pb(OCl)_2$$

Hydrogen sulfide
  Pascal, 1960, Vol. 13.1, 1004
  Ignition occurs on contact.
  See other METAL HYPOCHLORITES

4099. Sulfonyl chloride (Sulfuryl chloride)
[7791-25-5]   Cl₂O₂S

$$O_2SCl_2$$

Alkalies
  Brauer, 1963, Vol. 1, 385
  Reaction with alkalies may be explosively violent.

Diethyl ether
  Dunstan, I. *et al.*, *Chem. & Ind.*, 1966, 73
  A solution of sulfuryl chloride in ether vigorously decomposed, evolving hydrogen
  chloride. This was shown to be accelerated by the presence of peroxides. Peroxide-
  free ether should be used, and with care.

Dimethyl sulfoxide
  See Dimethyl sulfoxide: Acyl halides

Dinitrogen pentaoxide
See Dinitrogen pentaoxide: Sulfur dichloride, etc.

Lead dioxide
See Lead(IV) oxide: Non-metal halides

Organic materials, Metals
Anon., personal communication, 1996
Waste ethyl acetate containing sulfuryl chloride and thionyl chloride was stored in a galvanised steel drum. This later burst, although sulfuryl chloride is supplied in such drums they are unsafe with contaminants. The material can chlorinate many active organics, including aromatics, ketones and probably esters. This will release sulphur dioxide and hydrogen chloride, a fairly soluble gas, which will react with zinc and iron to produce hydrogen, an insoluble gas, and zinc or iron chlorides, both catalysts for many reactions including ester cleavage.
See Sulfinyl chloride: Esters, Metals

Phosphorus
See Phosphorus: Non-metal halides
See other ACYL HALIDES, NON-METAL HALIDES

4100. Dichlorine trioxide
[17496-59-2] Cl_2O_3

$$Cl_2O_3$$

1. McHale, E. T. *et al., J. Amer. Chem. Soc.,* 1967, **89**, 2796
2. McHale, E. T. *et al., J. Phys. Chem.,* 1968, **72**, 1849–1856
It is endothermic (ΔH_f° (g) +76.2 kJ/mol, 0.64 kJ/g), very unstable, and the vapour explodes at about 2 mbar and well below 0°C [1]. It has been identified as the intermediate involved in the delayed explosion of chlorine dioxide [2].
See other ENDOTHERMIC COMPOUNDS, HALOGEN OXIDES

4101. Chlorine perchlorate
[27218-16-2] Cl_2O_4

$$ClOClO_3$$

1. Schack, C. J. *et al., Inorg. Chem.,* 1970, **9**, 1387
2. Schack, C. J. *et al., Inorg. Chem.,* 1971, **10**, 1078
A shock-sensitive explosive compound [1]. Bromine perchlorate is also unstable [2].

Chlorotrifluoroethylene
Schack, C. J. *et al., Inorg. Chem.,* 1973, **12**, 897
Interaction of the reactants (pre-mixed at −196°C) during warming to −78°, then ambient temperature, exploded. Progressive addition of the alkene to the perchlorate at −78° was uneventful.

Perfluoroalkyl iodides
Schack, C. J. *et al., Inorg. Chem.,* 1975, **14**, 145–151

Reaction mixtures of chlorine perchlorate with perfluoromethyl iodide, 1,2-diiodoperfluoroethane or 1,3-diiodoperfluoropropane occasionally deflagrated at or below ambient temperature.
See other HALOGEN OXIDES

4102. Lead(II) chlorite
[13453-57-1] Cl_2O_4Pb
$$Pb(ClO_2)_2$$

Alone, or Antimony sulfide, or Sulfur
  Mellor, 1941, Vol. 2, 283
  It explodes on heating above 100°C or on rubbing with antimony sulfide or fine sulfur.

Non-metals
  Pascal, 1960, Vol. 16, 264
  Carbon, red phosphorus, or sulfur are oxidised violently by the chlorite.
  See other CHLORITE SALTS, METAL OXOHALOGENATES

4103. Disulfuryl dichloride
[7791-27-7] $Cl_2O_5S_2$
$$(ClSO_2)_2O$$

Phosphorus
  See Phosphorus: Non-metal halides

Water
  Sveda, M., *Inorg. Synth.*, 1950, **3**, 127
  The pure acid chloride reacts slowly with water, but if more than a few % of chlorosulfuric acid (a usual impurity) is present, the reaction is rapid and could become violent with large quantities.
  See other ACYL HALIDES, NON-METAL HALIDES (AND THEIR OXIDES) *See related* ACID ANHYDRIDES

4104. Chloryl perchlorate
[12442-63-6] Cl_2O_6
$$ClO_2^+ \ ClO_4^-$$

The value of the heat of formation quoted for the monomeric form, chlorine trioxide, indicates considerable endothermicity; ($\Delta H°_f$ (g) +154.8 kJ/mol, 1.85 kJ/g).

Organic matter, or Water
  1. Mellor, 1956, Vol. 2, Suppl. 1, 539
  2. Wechsberg, M. *et al., Inorg. Chem.*, 1972, **11**, 3066

3. Sorbe, 1968, 45
4. Pascal, J. L. *et al., J. Chem. Soc., Dalton Trans.*, 1985, 297
Though the least explosive of the chlorine oxides, being insensitive to heat or
shock, it is a very powerful oxidant and needs careful handling. It violently or
explosively oxidises ethanol, stopcock grease, wood and organic matter generally,
and explodes on contact with water [1]. It explodes on heating [2], and explosions
in contact with organic matter below $-70°C$ have been reported [3]. A later reference indicates that it is shock-sensitive, however [4], and stresses the need for care
in its use as a most effective perchlorylating agent.

Tetrachlorosilane, or Tetrabromosilane
Schmeisser, M., *Angew. Chem.*, 1955, **67**, 499
Reaction gives explosive solids, apparently perchloratosiloxanes.
See Tetraperchloratosilicon

Thionyl chloride
Schmeisser, M. *Angew. Chem.*, 1955, **67**, 499
It is liable to explode on contact with thionyl chloride in absence of inert solvent.
See other ENDOTHERMIC COMPOUNDS, HALOGEN OXIDES, OXIDANTS

4105. Lead(II) chlorate
[10294-47-0] Cl_2O_6Pb

$$Pb(ClO_3)_2$$

Bailar, 1973, Vol. 2, 130
Thermal decomposition may be explosive.
See other METAL CHLORATES

4106. Zinc chlorate
[10361-95-2] Cl_2O_6Zn

$$Zn(ClO_3)_2$$

Sorbe, 1968, 158
The tetrahydrated salt decomposes explosively at 60°C.

Other reactants
Yoshida, 1980, 63–64
MRH values calculated for 30 combinations with oxidisable materials are given.

Sulfides   MRH values below reference
Mellor, 1956, Vol. 2, Suppl. 1, 584
MRH Copper(II) sulfide 2.84/38, antimony(III) sulfide 2.89/41, tin(II) sulfide
3.05/43, tin(IV) sulfide 3.1/37
    Interaction with copper(II) sulfide is explosive, and incandescent with anti-
mony(III) sulfide, arsenic(III) sulfide, tin(II) sulfide and tin(IV) sulfide.
See other METAL CHLORATES

4107. Perchloryl perchlorate (Dichlorine heptaoxide)
[10294-48-1] Cl_2O_7

$$ClO_{3+}ClO_{4-}$$

1. Mellor, 1956, Vol. 2, Suppl. 1, 542
2. Schmeisser, M., *Angew. Chem.*, 1955, **67**, 498
It is rather endothermic (ΔH_f° (g) +265.3 kJ/mol, 1.45 kJ/g) and explodes violently
on impact or rapid heating, but is a less powerful oxidant than other chlorine oxides
[1]. It is now formulated as perchloryl perchlorate [2].

Bromine pentafluoride
Lawless, 1968, 173
Mixtures are shock-sensitive explosives.

Copper phthalocyanine
Osadchaya, L. I. *et al., Chem. Abs.*, 1997, **126**, 309255v
A shock-sensitive explosive complex is formed, which could not be characterised.

Iodine
Kirk-Othmer, 1964, Vol. 5, 5
Explosion on contact.
See other ENDOTHERMIC COMPOUNDS, HALOGEN OXIDES

4108. Lead perchlorate
[13637-76-8] Cl_2O_8Pb

$$Pb(ClO_4)_2$$

Methanol
Willard, H. H. *et al., J. Amer. Chem. Soc.*, 1930, **52**, 2396
A saturated solution of anhydrous lead perchlorate in dry methanol exploded
violently when disturbed. Methyl perchlorate may have been involved.
See ALKYL PERCHLORATES *See other* METAL PERCHLORATES

4109. Tin(II) perchlorate
[25253-54-7] Cl_2O_8Sn

$$Sn(ClO_4)_2$$

Bailar, 1973, Vol. 2, 76
The trihydrated salt decomposes explosively at 250°C.

Aromatic Schiff bases
Dwivedi, B. K. *et al., Indian J. Chem.*, 1987, **26A**, 618–620
Twelve tris-complexes [Sn(ClO$_4$)$_2$.3L] between tin(II) perchlorate and Schiff bases
derived from salicylaldehyde, anisaldehyde, or 2-hydroxy-1-naphthaldehyde and
aromatic amines were investigated for thermal instability.

See related AMMINEMETAL OXOSALTS
See other METAL PERCHLORATES, REDOX COMPOUNDS

4110. Xenon(II) perchlorate (Bis(perchloryloxy)xenon)
[25523-79-9]   Cl$_2$O$_8$Xe

$$Xe(ClO_4)_2$$

1. Wechsberg, M. *et al., Inorg. Chem.*, 1972, **11**, 3066
2. Kirk Othmer, 1980, **Vol. 12**, 293
During preparation from perchloric acid and xenon difluoride at $-50°C$, violent explosions occurred if the reaction mixture was allowed to warm up rapidly [1]. Detonates easily and should be handled with extreme care [2].
See other XENON COMPOUNDS, NON-METAL PERCHLORATES

4111. Uranyl perchlorate
[13093-00-0]   Cl$_2$O$_{10}$U

$$O_2U(ClO_4)_2$$

Ethanol
Erametsa, O., *Suomen Kemist*, 1942, **15B**, 1
Attempted recrystallisation of the salt from ethanol caused an explosion (probably involving ethyl perchlorate).
See ALKYL PERCHLORATES *See other* METAL PERCHLORATES

4112. Lead chloride
[7758-95-4]   Cl$_2$Pb

$$PbCl_2$$

Calcium
See Calcium: Lead chloride *See other* METAL HALIDES

4113. Sulfur dichloride
[10545-99-0]   Cl$_2$S

$$ClSCl$$

(*MCA SD-77*, 1960); *HCS 1980*, 875

Acetone
Fawcett, F. S. *et al., Inorg. Synth.*, 1963, **7**, 121
Acetone is an effective solvent for cleaning traces of sulfur chlorides from reaction vessels, but care is necessary, as the reaction is vigorous if more than a trace is present.

Dimethyl sulfoxide
See Dimethyl sulfoxide: Acyl halides, etc.

Hexafluoroisopropylideneaminolithium
See Hexafluoroisopropylideneaminolithium: Non-metal halides

Metals
See Potassium: Non-metal halides, *and* Sodium: Non-metal halides

Oxidants
See Dinitrogen pentaoxide: Sulfur dichloride
Nitric acid: Sulfur halides
Perchloryl fluoride: Hydrocarbons, etc.

Toluene
1. Caporossi, J. C., *Chem. Eng. News*, 1988, **66**(32), 2
2. Author's comments, 1988
3. Urben, P. G., private comm., 1989
Toluene was added to a closed mild steel storage tank containing sulfur dichloride, and the vessel was subsequently ruptured by overpressurisation. It was later found that the exothermic reaction between toluene and sulfur dichloride was catalysed by iron or iron(III) chloride [1]. Several reactions may have been involved in this chlorination system, many capable of iron catalysis arising from the inevitable formation of traces of iron(III) chloride from contact of sulfur dichloride with mild steel. Initially at low temperatures, exothermic formation of 2- and 4-chlorotoluene with evolution of hydrogen chloride and formation of more iron(III) chloride would occur. As the temperature and concentration of catalyst rose, side-chain chlorination would set in, leading to formation of chloromethyl-, dichloromethyl- and trichloromethyl-benzenes and much more heat and hydrogen chloride. The direct catalysed reaction of toluene with sulfur dichloride to form toluenesulfenyl chloride could also occur [2], with the subsequent possibility of oxidation to the disulfide [3]. It is also known that chloromethylbenzene (benzyl chloride) will undergo a Friedel-Crafts self-condensation reaction, catalysed by iron(III) chloride, to form polybenzyls. Both of these reactions would again liberate hydrogen chloride. With so many possibilities for evolution of hydrogen chloride gas, it is perhaps not surprising that the mild steel vessel ruptured [2].
See Benzyl chloride: Catalytic impurities

Water
MCA SD-77, 1960
Exothermic reaction with water or steam.
See other NON-METAL HALIDES

4114. Disulfur dichloride
[10025-67-9] Cl_2S_2

ClSSCl

(*MCA SD-77*, 1960); *HCS 1980*, 874

Aluminium
See Aluminium: Non-metal halides

Antimony, or Antimony sulfide, or Arsenic sulfide
Mellor, 1947, Vol. 10, 641
Interaction at ambient temperature is surprisingly energetic.

Chromyl chloride
See Chromyl chloride: Non-metal halides

Dimethyl sulfoxide
See Dimethyl sulfoxide: Acyl halides, etc.

Mercury oxide
Mellor, 1947, Vol. 10, 643
Interaction is rapid and very exothermic.

Phosphorus(III) oxide
See Tetraphosphorus decaoxide: Disulfur dichloride

Potassium
Mellor, 1947, Vol. 10, 642
A mixture of potassium and the liquid chloride is shock-sensitive and explodes violently on heating.

Sodium peroxide
See Sodium peroxide: Non-metal halides

Tin
Pascal, 1963, Vol. 8.3, 309
Interaction is violent.

Unsaturated materials
1. Mellor, 1947, Vol. 10, 64–642
2. Huestis, B. L., *Safety Eng.*, 1927, **54**, 95
Alkenes, terpenes and unsaturated glycerides react exothermically, some vigorously [1]. Ignition may occur with some organic materials [2].

Water
MCA SD-77, 1960
As with sulfinyl chloride, the exothermic reaction with limited amounts of water may be dangerously violent under confinement because of rapid gas evolution.
See other GAS EVOLUTION INCIDENTS, NON-METAL HALIDES

4115. Diselenium dichloride
[10025-68-0] Cl_2Se_2

                              ClSeSeCl

Alkali metals or oxides
  See Potassium: Non-metal halides
      Potassium dioxide: Diselenium dichloride
      Sodium: Non-metal halides
      Sodium peroxide: Non-metal halides

Aluminium
  See Aluminium: Non-metal halides

Trimethylsilyl azide
  See Trimethylsilyl azide: Selenium halides
  See other NON-METAL HALIDES

4116. Tin(II) chloride
[7772-99-8] Cl_2Sn

                               $SnCl_2$

  HCS 1980, 900

Bromine trifluoride
  See Bromine trifluoride: Tin(II) chloride

Calcium acetylide   MRH 1.51/25
  See Calcium acetylide: Tin(II) chloride

Hydrazine
  Mellor, 1941, Vol. 7, 430
  Stannous chloride reacts with hydrazine in ethanol to give a precipitate, believed
  the dihydrazino adduct, which explodes on heating.

Hydrogen peroxide   MRH 1.21/15
  See Hydrogen peroxide: Tin(II) chloride

Metal nitrates   MRH Ammonium nitrate 1.21/30
  See METAL NITRATES: esters, etc.

Other reactants
  Yoshida, 1980, 56
  MRH values calculated for 16 combinations, largely with oxidants, are given.
  See other METAL HALIDES, REDUCANTS

4117. Titanium(II) chloride
[10049-06-6]   Cl₂Ti

$$Cl_2Ti$$

TiCl₂

1. Mellor, 1941. Vol. 7, 75
2. Brauer, 1965, Vol. 2, 1187
3. *NSC Data Sheet 485*, 1966

It readily ignites in air, particularly if moist [1,2]. The dichloride on heating under inert atmospheres disproportionates into the tetrachloride and pyrophoric titanium [3].

Water
Gallais, 1957, 431
Interaction at ambient temperature is violent, and water is reduced with evolution of hydrogen.
See other METAL HALIDES, REDUCANTS

4118. Vanadium dichloride
[10580-52-6]   Cl₂V

VCl₂

Platinum, Water
Ephraim, 1954, 261
Vanadium(II) chloride dissolved in water is slowly oxidised by the solvent with evolution of hydrogen. Contact with platinum foil accelerates the reaction to violence.
See other METAL HALIDES, REDUCANTS

4119. Tungsten dichloride
[13470-12-7]   Cl₂W

ClWCl

Preparative hazard
Mussell, R. D. *et al., Inorg. Chem.*, 1990, **29**(19), 3711
Explosions sometimes resulted on opening sealed tubes in which complex mixed chloride salts of this compound had been prepared. There is no obvious source of pressure in the reaction mixture of tungsten hexachloride, sodium and aluminium chlorides and aluminium metal.
See other METAL HALIDES, REDUCANTS

4120. Zinc chloride
[7646-85-7]   Cl₂Zn

ZnCl₂

HCS 1980, 965

Zinc
See Zinc: Zinc chloride
See other METAL HALIDES

4121. Zirconium(II) chloride
[13762-26-0] Cl_2Zr

$$ZrCl_2$$

Sidgwick, 1950, 652
When warm it ignites in air.
See other METAL HALIDES, REDUCANTS

4122. Cobalt(III) chloride
[10241-04-4] Cl_3Co

$$CoCl_3$$

Pentacarbonyliron, Zinc
See Pentacarbonyliron: Transition metal halides, etc.
See other METAL HALIDES

4123. Pentaamminechlorocobalt(III) perchlorate
[15156-18-0] $Cl_3CoH_{15}N_5O_8$

$$[(H_3N)_5CoCl] [ClO_4]_2$$

Explodes at 320°C; high impact sensitivity.
See entry AMMINEMETAL OXOSALTS

4124. Pentaammineaquacobalt(III) chlorate
[13820-81-0] $Cl_3CoH_{17}N_5O_{10}$

$$[(H_3N)_5CoH_2O] [ClO_3]_3$$

Salvadori, R., *Gazz. Chim. Ital. [2]*, 1910, **40**, 9
It is of zero oxygen balance and explodes on impact, or heating to 130°C.
See entry OXYGEN BALANCE See other AMMINEMETAL OXOSALTS

4125. Hexaamminecobalt(III) chlorate
[26156-56-9] $Cl_3CoH_{18}N_6O_9$

$$[(H_3N)_6Co] [ClO_3]_3$$

Friederich, W. *et al., Z. ges. Schiess-Sprengstoffw.*, 1926, **21**, 49
The complex salt is explosive.
See other AMMINEMETAL OXOSALTS

4126. Hexaamminecobalt(III) perchlorate
[13820-83-2] $Cl_3CoH_{18}N_6O_{12}$

$$[(H_3N)_6Co] [ClO_4]_3$$

It explodes at 360°C, is highly impact-sensitive and has +14.2% oxygen balance.
See entries AMMINEMETAL OXOSALTS, OXYGEN BALANCE

4127. Chromium(III) chloride
[10025-73-7] Cl_3Co

$$CoCl_3$$

Lithium, Nitrogen
See Lithium: Metal chlorides, etc.
See other METAL HALIDES

4128. Bis(hydrazine)diperchloratochromium(III) perchlorate
[74311-45-7] $Cl_3CrH_8N_4O_{12}$

$$[(H_4N_2)_2Cr(ClO_4)_2] ClO_4$$

Patil, K. C. *et al., Chem. & Ind.*, 1979, 902
It ignites at 240°C in air and decomposes explosively, but surprisingly is not
sensitive to friction or impact, unlike its nickel analogue. It shows +118% oxygen
balance.
See entry OXYGEN BALANCE *See other* AMMINEMETAL OXOSALTS, REDOX
COMPOUNDS

4129. Hexaamminechromium(III) perchlorate
[15203-80-2] $Cl_3CrH_{18}N_6O_{12}$

$$[(H_3N)_6Cr] [ClO_4]_3$$

Ethanol, Ether
1. Pennington, D. E., *Chem. Eng. News*, 1982, **60**(33), 55
2. Cartwright, R. V., *Chem. Eng. News*, 1983, **61**(6), 4
A second crop of material (+14.2% oxygen balance), precipitated by addition of
excess perchloric acid to the liquor, was collected on a sinter, washed and dried
with ethanol, then ether. Stirring the filter cake led to a violent explosion [1]:
formation of ethyl perchlorate seems to have been the most likely cause. The
advantages of using plastic sintered funnels in reducing friction and the extent of
fragmentation if an explosion occurs were stressed later [2].
See ALKYL PERCHLORATES *See entry* OXYGEN BALANCE
See other AMMINEMETAL OXOSALTS, FRICTIONAL INITIATION INCIDENTS

4130. Chromium(III) perchlorate . 6dimethyl sulfoxide
[59675-70-6] $Cl_3Cr.6C_2H_6OS$

$$Cr(ClO_4)_3.6DMSO$$

Langford, C. H. *et al., J. Chem. Soc., Chem. Comm.*, 1977, 139

The hexasolvated complex may be hazardous with respect to detonation. Other metal perchlorates and nitrates solvated with DMSO are explosive.
See Dimethyl sulfoxide: Metal oxosalts
See other METAL PERCHLORATES, SOLVATED OXOSALT INCIDENTS

4131. Dysprosium perchlorate
[14017-53-9] Cl_3DyO_{12}

$$Dy(ClO_4)_3$$

Birnbaum, E. R. *et al., Inorg. Chem.*, 1973, **12**, 379
The anhydrous salt was stable to 300°C or to impact of a steel hammer, but a mild explosion occurred when a grease-containing sample was disturbed with a metal spatula.
See other FRICTIONAL INITIATION INCIDENTS, METAL PERCHLORATES

4132. Erbium perchlorate
[14017-55-1] Cl_3ErO_{12}

$$Er(ClO_4)_3$$

Acetonitrile
Wolsey, W. C., *J. Chem. Educ.*, 1973, **50**(6), A336–337
The shock-sensitive glassy residue (containing traces of acetonitrile), left after heating the tetrasolvated salt to above 150°C on a vacuum line, exploded violently when scraped with a spatula. (Oxygen balance of salt +700%).
See other FRICTIONAL INITIATION INCIDENTS, METAL PERCHLORATES, SOLVATED OXOSALT INCIDENTS

4133. Iron(III) chloride
[7705-08-0] Cl_3Fe

$$FeCl_3$$

HCS 1980, 501

Chlorine, Monomers
See Chlorine: Iron(III) chloride, Monomers

Ethylene oxide
See Ethylene oxide: Contaminants

Metals
See Potassium: Metal halides
    Sodium: Metal halides
See other METAL HALIDES

4134. Iron(III) perchlorate
[13537-24-1]

Cl_3FeO_{12}

$$Fe(ClO_4)_3$$

Acetonitrile
  Bancroft, G. M. *et al., Can. J. Chem.*, 1974, **52**, 783
  The violent reaction which occurred on dissolution of the anhydrous salt in acetonitrile did not occur with the hydrated salt.

Dimethyl sulfoxide
  See Dimethyl sulfoxide: Metal oxosalts
  See other METAL PERCHLORATES, SOLVATED OXOSALT INCIDENTS

4135. Gallium perchlorate
[17835-81-3]

Cl_3GaO_{12}

$$Ga(ClO_4)_3$$

1. Foster, L. S., *J. Amer. Chem. Soc.*, 1933, **61**, 3123
2. Foster, L. S., *Inorg. Synth.*, 1946, **2**, 28
During preparation by dissolving the metal in 72% perchloric acid, the hexahydrate separates as a crystalline solid. After filtration, the damp crystals must not contact any organic material (filter paper, horn spatula, or recrystallisation solvent) as the adherent perchloric acid is above 72% concentration owing to the hexahydrate formation [1,2].
  See Perchloric acid: Dehydrating agents
  See other METAL PERCHLORATES
  See related SOLVATED OXOSALT INCIDENTS

†4136. Trichlorosilane
[10025-78-2]

Cl_3HSi

$$Cl_3SiH$$

HCS 1980, 924

1. Müller, R. *et al., J. Prakt. Chem.*, 1966, **31**, 1–6
2. Britton, L. G., *Plant/Oper. Progr.*, **9**(1), 16
The pure material is not impact-ignitable in absence of electrostatic charges, but technical material (possibly containing dichlorosilane) is [1]. Hazards are reviewed and an Ait of 182°C established [2].
  See other STATIC INITIATION INCIDENTS

Acetonitrile, Diphenyl sulfoxide
  Benkeser, R. A., *Chem. Eng. News*, 1978, **56**(32), 107
  Use of a single portion of trichlorosilane added to acetonitrile at 10°C to reduce diphenyl sulfoxide led to a violent explosion. The reaction previously had been effected uneventfully in a wide range of other solvents. The explosion was

attributed to use of acetonitrile as solvent and/or the addition of the trichlorosilane as a single portion.

Other reactants
Yoshida, 1980, 252
MRH values calculated for 13 combinations with oxidants are given.

Styrene
Anon., *Chem. Eng. News*, 1994, **72**(26), 6
A drum of these two, apparently freshly mixed as feed for a reaction, pressurised and exploded. The subsequent fire destroyed the plant and injured 31 people. Styrene is polymerisable and the two probably react by Friedel-Crafts type chemistry, if suitably catalysed. It is not reported whether the drum was lined, plain steel, or galvanised.
See other HALOSILANES

4137. Pentaamminechlororuthenium chloride
[18532-87-1] $Cl_3H_{15}N_5Ru$

$$[(H_3N)_5RuCl]Cl_2$$

Sodium azide
Allen, A. D. *et al., Inorg. Synth.*, 1970, **12**, 5–6
During treatment with sodium azide of an intermediate (believed to be pentaammineaquaruthenium(III) derived from the title compound) to produce pentaamminedinitrogenruthenium(II) solutions, a dangerously explosive red solid may be produced. The solid, pentaammineazidoruthenium(III) will, however, decompose on standing to give the desired dinitrogen species.
See Hydrazine: Ruthenium(III) chloride
See other AMMINEMETAL HALIDES *See related* METAL HALIDES

4138. Octaammine-μ-hydroxy[μ-(peroxy-O : O')]dirhodium(3+) perchlorate
[113728-41-9] $Cl_3H_{25}N_8O_{15}Rh_2$

Springborg, J. *et al.*, *Acta Chem. Scand. Sec. A*, 1987, **A41**, 485
The complex may detonate when subjected to mechanical shock.
See other AMMINEMETAL OXOSALTS

4139. Iodine trichloride
[865-44-1]   Cl₃I

ICl₃

Phosphorus
Pascal, 1960, Vol. 16.1, 578
Phosphorus ignites on contact with iodine trichloride.
See other INTERHALOGENS

4140. Iodine(III) perchlorate
[38005-31-1]   Cl₃IO₁₂

I(ClO₄)₃

Christe, K. O. *et al.*, *Inorg. Chem.*, 1972, **11**, 1683
The solid was stable at −45°C but exploded on laser irradiation at low temperature.
See other IODINE COMPOUNDS, IRRADIATION DECOMPOSITION INCIDENTS, NON-
METAL PERCHLORATES

4141. Manganese trichloride oxide
[23097-77-0]   Cl₃MnO

O:MnCl₃

Briggs, T. S., *J. Inorg. Nucl. Chem.*, 1968, **30**, 2867–2868
Explosively unstable if isolated as a liquid at ambient temperature, it may be
handled safely in carbon tetrachloride solution.
See Manganese chloride trioxide, Manganese dichloride dioxide
See related METAL HALIDES, METAL OXIDES

4142. 1,3,5-Trichlorotrithiahexahydro-1,3,5-triazinemolybdenum
[ ]   Cl₃MoN₃S₃

Water
Wynne, K. J. *et al.*, *J. Inorg. Nucl. Chem.*, 1968, **30**, 2851

A 2 g quantity of the product of interaction of hexacarbonylmolybdenum and thiazyl chloride, possibly constituted as shown, reacted explosively with water.
See other N-HALOGEN COMPOUNDS, N–S COMPOUNDS

4143. Nitrogen trichloride (Trichloramine)
[10025-85-1] Cl_3N

$$NCl_3$$

HCS 1980, 686

1. Mellor, 1940, Vol. 8, 598–604; 1967, Vol. 8, Suppl. 2.2, 411
2. Sidgwick, 1950, 705
3. Anon., *ABCM Quart. Safety Summ.*, 1946, **17**, 17
4. Brauer, 1963, Vol. 1, 479
5. Schlessinger, G. G., *Chem. Eng. News*, 1966, **44**(33), 46
6. Kovacic, P. *et al., Org. Synth.*, 1968, **48**, 46
7. Dokter, T., private comm., 1985
8. Dokter, T., *J. Haz. Mat.*, 1985, **10**, 205–222

Contact above 0°C of excess chlorine or a chlorinating agent with aqueous ammonia, ammonium salts or a compound containing a hydrolysable amino-derivative, or electrolysis of ammonium chloride solution produces the highly endothermic ($\Delta H°_f$ (g) +230.1 kJ/mol, 1.91 kJ/g) and explosive nitrogen trichloride as a water-insoluble yellow oil [1,2,3]. It is usually prepared [4] in solution in a solvent, and such solutions in chloroform are reported as stable up to 18% concentration. However, an 18% solution in dibutyl ether exploded on cooling in a refrigerator, and a 12% solution prepared without cooling had decomposed vigorously [5]. In absence of other materials, explosion of the trichloride may be initiated in a wide variety of ways. The solid frozen under vacuum in liquid nitrogen explodes on thawing, and the liquid explodes on heating to 60 or 95°C. Exposure to impact, light or ultrasonic irradiation will cause (or sensitise) detonation [1]. The preparation and synthetic use of solutions of nitrogen trichloride in dichloromethane is described in detail [6]. Trichloramine is surprisingly stable, can be formed at 80°C and above, and has been made in 200 kg lots for possible use as a chlorinating agent [7]. The formation and separation of the trichloride have been described in detail [8].
See Chlorine: Nitrogen compounds, or: Dimethyl phosphoramidate
   Phosphorus pentachloride: Urea
   Sodium chloride: Nitrogen compounds
See other IRRADIATION DECOMPOSITION INCIDENTS

Chlorine, Hydrogen
   Ashmore, P. G., *Nature*, 1953, **172**, 449–450
   Equimolar mixtures of chlorine and hydrogen containing 0.1–0.2% of the trichlo-ride will explode in absence of light if the pressure is below a limiting value dependent on temperature (30 mbar at 20°, 132 mbar at 57°C).

Initiators

1. Mellor, 1940, Vol. 8, 601–604; 1967, Vol. 8, Suppl. 2.2, 412
2. Pascal, 1956, Vol. 10, 264

A wide variety of solids, liquids and gases will initiate the violent and often explosive decomposition of nitrogen trichloride. These include conc. ammonia, arsenic, dinitrogen tetraoxide (above −40°C, with more than 25% solutions of trichloride in chloroform), hydrogen sulfide, hydrogen trisulfide, nitrogen oxide, organic matter (including grease from the fingers), ozone, phosphine, phosphorus (solid, or in carbon disulfide solution), potassium cyanide (solid, or aqueous solution), potassium hydroxide solution or selenium [1]. All four hydrohalide acids will also initiate explosion of the trichloride [2].

See other ENDOTHERMIC COMPOUNDS, *N*-HALOGEN COMPOUNDS

4144. Nitrosylruthenium trichloride
[18902-42-6] Cl_3NORu

$$O:NRuCl_3$$

Sidgwick, 1950, 1486
It decomposes violently at 440°C.
See other NITROSO COMPOUNDS, *N*-METAL DERIVATIVES

4145. 1,3,5-Trichlorotrithiahexahydro-1,3,5-triazine (Thiazyl chloride)
[18428-81-4] $Cl_3N_3S_3$

Bailar, 1973, Vol. 2, 908
It explodes violently on sudden heating.

Ammonia, Silver nitrate
Goehring, 1957, 67
Thiazyl chloride, treated with aqueous ammonia and then silver nitrate, gives a compound AgN_5S_3 (of unknown structure) which is shock-sensitive and explodes violently.

Hexacarbonylmolybdenum
See Carbonyl(pentasulfur pentanitrido)molybdenum
    1,3,5-Trichlorotrithiahexahydro-1,3,5-triazinemolybdenum
See other *N*-HALOGEN COMPOUNDS, N–S COMPOUNDS

4146. Tin azide trichloride

[ ]   Cl₃N₃Sn

$$N_3SnCl_3$$

An explosive solid.
See entry METAL AZIDE HALIDES

4147. Titanium azide trichloride

[ ]   Cl₃N₃Ti

$$N_3TiCl_3$$

An explosive solid.
See entry METAL AZIDE HALIDES

4148. Neodymium perchlorate . 2acetonitrile

[13498-06-1]   Cl₃NdO₁₂.2C₂H₃N

$$Nd(ClO_4)_3.2MeCN$$

1. Forsberg, J. H. *et al., Inorg. Chem.*, 1971, **10**, 2656
2. Eigenbrot, C. W. *et al., Inorg. Chem.*, 1982, **21**, 2867
3. Raymond, K. N., *Chem. Eng. News*, 1983, **61**(49), 4
4. Bretherick, L., *Chem. Eng. News*, 1983, **61**(50), 2

The perchlorate, prepared from neodymium oxide and perchloric acid [1] was purified and isolated as the salt tetrasolvated with acetonitrile [2], which has not been
found to be shock-sensitive. On vacuum drying at 80°C/24h, it is converted to the
disolvated salt which is very shock-sensitive and a 1 g sample exploded violently
in a hand-held flask [3]. Other examples of partially desolvated perchlorate salts
becoming friction- and shock-sensitive are known [4]. (The disolvate is very close
to zero oxygen balance).
See other METAL PERCHLORATES, SOLVATED OXOSALT INCIDENTS

4149. Phosphoryl chloride

[10025-87-3]   Cl₃OP

$$O:PCl_3$$

(*MCA SD-26*, 1968); *HCS 1980*, 751

Carbon disulfide   MRH Carbon disulfide 1.17/99+
Disposal of a benzene solution of phosphoryl chloride into a waste drum containing
carbon disulfide (and other solvents) caused an instantaneous reaction, with evolution of (probably) hydrogen chloride. Presence of a hydroxylic compound seems
likely.

Dimethylaniline
Anon., private comm., 1984

In a chlorination reaction using phosphoryl chloride with dimethylaniline as acid acceptor, the reagents were added all at once. After an induction period, most of the mixture was ejected from the flask. On a smaller scale, the reaction had been uneventful.

Dimethylformamide, 2,5-Dimethylpyrrole
MCA Case History No. 1460
Poor stirring during formylation of 2,5-dimethylpyrrole with the preformed complex of dimethylformamide with phosphoryl chloride caused eruption of the flask contents. Reaction of the complex with a local excess of the pyrrole may have been involved.

2,6-Dimethylpyridine *N*-oxide
1. Kato, T. *et al., J. Pharm. Soc. Japan.*, 1951, **71**, 217
2. Anon., *Chem. Eng. News*, 1965, **43**(47), 40
3. Evans, R. F. *et al., J. Org. Chem.*, 1962, **27**, 1333
Interaction of the reagents in absence of diluent, according to a published procedure [1], caused an explosion [2]. The use of a chlorinated solvent as diluent prevented explosion, confirming an earlier report [3].
See Pyridine *N*-oxide, below

Dimethyl sulfoxide   MRH 1.80/99+
See Dimethyl sulfoxide: Acyl halides

Ferrocene-1,1'-dicarboxylic acid
Marvel, C. S., *Chem. Eng. News*, 1962, **40**(3), 55
An explosion occurred immediately after pouring and capping of the chloride recovered from preparation of ferrocene-1,1'-dicarbonyl chloride. The storage bottle contained phosphoryl chloride recovered from similar preparations and which had been stored for some 3 months. No explanation was apparent.

Other reactants
Yoshida, 1980, 76
MRH values calculated for 10 combinations with various reagents are given.

Pyridine *N*-oxide
MCA Guide 1972, 321
Dropwise addition of the chloride caused a steady exotherm to 60–65°C when a runaway reaction accelerated to explosion.
See 2,6-Dimethylpyridine *N*-oxide, above

Sodium   MRH 4.27/31
See Sodium: Non-metal halides (reference 7)

Water
1. *MCA SD-26*, 1968
2. Unpublished observations

3. *MCA Case History No. 1274*
4. *MCA Case History No. 2286*
5. Anon., *Loss Prev. Bull.*, 1972, (013), 22–23
6. Anon., *Loss Prev. Bull.*, 1980, (038), 27
7. Anon., *Sichere Chemiearb.*, 1985, **37**(11), 130
8. Davies, R., *Loss Prev Bull.*, 1995, 121, 13

The hazards arising from interaction of phosphoryl chloride and water derive from there being often a considerable delay in onset of the exothermic hydrolysis reaction, which may proceed with enough vigour to generate steam and liberate hydrogen chloride gas. Conditions tending to favour delayed or violent reaction include limited quantities of water and/or ice for hydrolysis, lack of stirring, cold or frozen phosphoryl chloride, and reaction in closed or virtually closed containers [1]. A layer of the dense and cold liquid may survive for several minutes under water before violent, almost instantaneous hydrolysis occurs, particularly when disturbed [2]. The Case History describes a violent explosion which occurred when water was added to a drum containing some phosphoryl chloride which had been stored below its freezing point, 2°C [3]. Bursting of a 300 kg drum during filling was attributed to delayed reaction of the chloride with water contamination in the drum [4]. Methods for the safe decontamination of non-returnable drums which have contained the chloride are detailed, and involve thorough draining of the drum and internal inspection before careful application of large amounts of water [5]. When 10 l of the chloride was sucked into a measuring vessel, a violent explosion occurred which ruptured it. It was assumed that water previously used to clean the vessel had not fully drained out [6]. In outdoor storage, rainwater found its way into the space between the outer container and the polythene liner containing solid methylchlorouracil. When this was tipped into phosphoryl chloride in a reactor, some 4–5 l of water also ran in and caused an explosive evolution of hydrogen chloride gas which led to a fatality [7]. An account of an explosion when a transfer hose largely full of phosphoryl chloride was flushed with water is given [8]

See other GAS EVOLUTION INCIDENTS, PLANT CLEANING INCIDENTS

Zinc   MRH 1.34/39
Mellor, 1940, Vol. 8, 1025
Zinc dust ignites in contact with a little phosphoryl chloride, and subsequent addition of water liberates phosphine which ignites.
See other NON-METAL HALIDES (AND THEIR OXIDES)

4150. Antimony trichloride oxide (Antimonyl chloride)
[14459-54-2]   Cl₃OSb
O:SbCl₃

Bromine trifluoride
See Bromine trifluoride: Antimony trichloride oxide
See related METAL HALIDES, METAL OXIDES

1454

4151. Vanadium trichloride oxide (Vanadyl chloride)
[7727-18-6] Cl_3OV

$$O:VCl_3$$

HCS 1980, 953

Bailar, 1973, Vol. 3, 508
The chloride (and analogous bromide) is violently hygroscopic.

Rubidium
See Rubidium: Vanadium trichloride oxide

Sodium
Quam, G. N. *et al.*, in *Safety in the Chemical Laboratory*, (Ed. Steere, N. V.),
Vol. 1, 1967, 82 (expanded articles reprinted from *J. Chem. Educ.*, 1964–67)
In a procedure to purify the chloride oxide with sodium, too-rapid addition of the
metal led to a violent explosion.
See Sodium: Metal halides (reference 2)
See related METAL HALIDES, METAL OXIDES

4152. Vanadyl perchlorate
[63672-69-3] $Cl_3O_{13}V$

$$O:V(ClO_4)_3$$

Alone, or Organic solvents
Schmeisser, M., *Angew. Chem.*, 1955, **67**, 499
A powerful oxidant which may explode above 80°C, and which ignites organic
solvents on contact. It decomposes, sometimes explosively, on exposure to light.
See other METAL PERCHLORATES, OXIDANTS

4153. Phosphorus trichloride
[7719-12-2] Cl_3P

$$PCl_3$$

(*MCA SD-27*, 1972); *HCS 1980*, 749

Carboxylic acids
1. Coghill, R. D., *J. Amer. Chem. Soc.*, 1938, **60**, 488
2. Peacocke, T. A., *School Sci. Rev.*, 1962, **44**(152), 217
3. Scrimgeour, C. M., *Chem. Brit.*, 1975, **11**, 267
4. Taylor, D. A. H. *et al.*, *Chem. Brit.*, 1976, **12**, 105
5. Bretherick, L., *Chem. Brit.*, 1976, **12**, 26
Use of a free flame instead of a heating bath to distil acetyl chloride produced
from acetic acid and phosphorus trichloride caused the residual phosphonic ('phos-
phorous') acid to decompose violently to give spontaneously flammable phosphine

[1]. Two later explosions in the same preparative system after reflux but before distillation from a water-bath may have been due to ingress of air into the cooling flask and ignition of traces of phosphine [2]. During preparation of furoyl chloride, the excess trichloride was distilled off at atmospheric pressure, then vacuum was applied prior to intended distillation at 100°C/100 mbar, and an explosion occurred shortly after [3]. A similar incident occurred during conversion of propionic acid to the chloride [4]. These incidents were attributed to thermal decomposition of the by-product phosphonic acid to give a phosphine–diphosphane mixture and leakage of air into the evacuated system to produce a spontaneously explosive mixture. Sulfinyl chloride is a safer chlorinating agent for carboxylic acids [5].

Chlorobenzene, Sodium
    See Sodium: Chlorobenzene, etc.

Dimethyl sulfoxide
    See Dimethyl sulfoxide: Acyl halides, etc.

Hexafluoroisopropylideneaminolithium
    See Hexafluoroisopropylideneaminolithium: Non-metal halides

Hydroxylamine
    See Hydroxylamine: Phosphorus chlorides

Metals   MRH Potassium 2.80/46, sodium 3.10/33
    Mellor, 1940, Vol. 8, 1006; 1941, Vol. 2, 470
    Potassium ignites in phosphorus trichloride, while molten sodium explodes on contact.
    See Aluminium: Non-metal halides

Nitroheterocycle
    Hallam, S., 1993, personal communication
    A multi-kg mixture of the trichloride and a nitroheterocycle obtained after de-oxygenation of the corresponding N-oxide was being worked up by distillation under vacuum. It burst the glass vessel. Investigation showed that this was due to a higher temperature (85-90°C) than previously employed, permitting the phosphorus trichloride to react vigorously with the product.

Other reactants
    Yoshida, 1980, 59
    MRH values calculated for 8 combinations with various reagents are given.

Oxidants   MRH values show % of oxidant
    See Chromium pentafluoride: Phosphorus trichloride
        Chromyl chloride: Non-metal halides
        Fluorine: Phosphorus halides
        Iodine chloride: Phosphorus trichloride
        Lead(IV) oxide: Non-metal halides   MRH 0.67/69

Nitric acid: Phosphorus halides   MRH 0.63/31
Selenium dioxide: Phosphorus trichloride
Sodium peroxide: Non-metal halides   MRH 2.55/46

Oxygen
Vorob'ev, N. I. *et al., Izv. Akad. Nauk SSSR, Neorg. Mater.*, 1974, **10**, 2039–2042
Interaction to form phosphoryl chloride at 100–700°C was investigated, and the
limiting concentrations to prevent explosions were determined.

Sodium Carbonate
Holmes, W. S., 1989, personal communication
This is often recommended for clearing up spillage. A dangerously vigorous reac-
tion can result (probably depending upon the hydration level of the carbonate?).
Large quantities of dry sand followed by addition to water are preferred.
See Water, below

Sulfur acids
Dillon, K. B. *et al., J. Chem. Soc., Dalton Trans.*, 1979, 883–885
The trichloride is initially insoluble in 100% sulfuric acid, fluorosulfuric acid or
25% oleum, but dissolves very exothermally after delays of 8, 1 and 0.5 h, respec-
tively. Reaction with 65% oleum is immediately violent.
See Phosphorus tribromide: Sulfur acids *and* Phosphorus triiodide: Sulfur acids
See other INDUCTION PERIOD INCIDENTS

Tetravinyllead
See Tetravinyllead: Phosphorus trichloride

Water
1. *MCA SD-27*, 1972
2. Albright and Wilson, (Manufacturer's safety sheet)
3. Coghill, R. D., *J. Amer. Chem. Soc.*, 1938, **60**, 488
4. *MCA Case History No. 520*
5. Anon., *Loss. Prev. Bull.*, 1977, (013), 21–23
Reaction with water is exothermic and immediately violent (unlike phosphoryl
chloride), and is accompanied by evolution of some diphosphane which ignites
[1]. There can sometimes be an induction period, especially if cold. Yellow solids
produced by hydrolysis may ignite spontaneously if dried [2]. Evaporation of the
trichloride from an open beaker on a steam-bath led to ignition, which did not occur
if a hotplate were used as a dry heat source [3]. Interaction of the trichloride and
water in a virtually sealed container caused the latter to burst under the pressure
of hydrogen chloride generated [4]: (1 vol. trichloride gives 830 vol. of hydrogen
chloride at 20°C). Methods for the safe decontamination of non-returnable drums
which have contained the chloride are detailed, and involve careful application of
large amounts of water to the well-drained drum [5].
See other GAS EVOLUTION INCIDENTS *See other* NON-METAL HALIDES (AND THEIR
OXIDES)

4154. Thiophosphoryl chloride
[3982-91-0]   Cl₃PS

$$S:PCl_3$$

Möller, T. *et al., Inorg. Synth.*, 1954, **4**, 72
During the preparation from phosphorus trichloride and sulfur, the quantity and
quality of the aluminium chloride catalyst is critical to prevent the exothermic
reaction going out of control.

Methylmagnesium iodide
See Tetramethyldiphosphane disulfide

Pentaerythritol
MCA Case History No. 1315
On two occasions violent explosions occurred after heating of equimolar propor-
tions of the reagents (for 4 h at 160°C according to a literature method) had been
discontinued. (This suggests spontaneous ignition of traces of phosphine deriva-
tives as air was drawn into the cooling reaction vessel).
See Phosphorus trichloride: Carboxylic acids (reference 5)
See related NON-METAL HALIDES

4155. Rhodium(III) chloride
[10049-07-7]   Cl₃Rh

$$RhCl_3$$

Pentacarbonyliron, Zinc
See Pentacarbonyliron: Transition metal halides
See other METAL HALIDES

4156. Ruthenium(III) chloride
[10049-08-8]   Cl₃Ru

$$RuCl_3$$

Hydrazine
See Hydrazine: Ruthenium(III) chloride

Pentacarbonyliron, Zinc
See Pentacarbonyliron: Transition metal halides
See other METAL HALIDES

4157. Antimony trichloride
[10025-91-9]   Cl₃Sb

$$SbCl_3$$

(*MCA SD-66*, 1957); *HCS 1980*, 164

1458

Aluminium
See Aluminium: Antimony trichloride

N-Chlorodimethylamine
See N-chlorodimethylamine: Antimony chlorides
See other METAL HALIDES

4158. Titanium trichloride
[7705-07-9] Cl_3Ti

$$TiCl_3$$

HCS 1980, 905

Air, or Water
1. Lerner, R. W., *Ind. Eng. Chem.*, 1961, **53**(12). 56A
2. Ingraham, T. K. *et al., Inorg. Synth.*, 1960, **5**, 56
It reacts vigorously with air and/or water (vapour or liquid) and adequate handling precautions are necessary [1]. The finely divided powder is pyrophoric in air [2].
See other METAL HALIDES, PYROPHORIC MATERIALS, REDUCANTS

4159. Vanadium trichloride
[7718-98-1] Cl_3V

$$VCl_3$$

Methylmagnesium iodide
Cotton, F. A., *Chem. Rev.*, 1955, **55**, 560
Reaction of vanadium trichloride and similar halides with Grignard reagents is almost explosively violent under a variety of conditions.
See other METAL HALIDES, REDUCANTS

4160. Zirconium trichloride
[10241-03-9] Cl_3Zr

$$ZrCl_3$$

Water
Pascal, 1963, Vol. 9, 540
Interaction is very violent, hydrogen being evolved.
See other METAL HALIDES, REDUCANTS

4161. Caesium tetraperchloratoiodate
[53078-10-7] Cl_4CsIO_{16}

$$Cs[I(ClO_4)_4]$$

Christe, K. O. *et al., Inorg. Chem.*, 1972, **11**, 683

Though stable at ambient temperature, samples exploded under laser irradiation at low temperatures.
See other IODINE COMPOUNDS, IRRADIATION DECOMPOSITION INCIDENTS, NON-METAL PERCHLORATES

4162. Germanium tetrachloride
[10038-98-9] Cl_4Ge

$$GeCl_4$$

Water
Mellor, 1941, Vol. 7, 270
Interaction is very exothermic, accompanied by crackling if the chloride is dropped into water.
See other METAL HALIDES

4163. Hafnium tetrachloride
[13499-05-3] Cl_4Hf

$$HfCl_4$$

Tetrahydrofuran
Manzer, L. E., *Inorg. Synth.*, 1982, **21**, 137
Addition of the anhydrous chloride directly to THF causes a violent exothermic reaction. Add THF dropwise to a suspension of the chloride in dichloromethane.
See other METAL HALIDES

4164. Molybdenum diazide tetrachloride
[14259-66-6] Cl_4MoN_6

$$(N_3)_2MoCl_4$$

Highly explosive.
See entry METAL AZIDE HALIDES

4165. Chloriminovanadium trichloride
[14989-38-9] Cl_4NV

$$ClN:VCl_3$$

1. Strähle, J. *et al., Angew. Chem.*, 1966, **78**, 450
2. Strähle, J. *et al., Z. Anorg. Chem.*, 1965, **338**, 287
A new method of preparation from vanadium nitride and chlorine [1] is free of the explosion hazards of chlorine azide and vanadium azide tetrachloride present in an earlier method [2].
See other AMMINEMETAL HALIDES, *N*-HALOGEN COMPOUNDS
See related METAL HALIDES

4166. Vanadium azide tetrachloride
[ ] Cl_4N_3V

$$N_3VCl_4$$

Explosive solid.
See entry METAL AZIDE HALIDES

4167. Rhenium tetrachloride oxide
[13814-76-1] Cl_4ORe

$$O:ReCl_4$$

Ammonia
Sidgwick, 1950, 1302
Interaction with gaseous or liquid ammonia is violent.
See related METAL HALIDES, METAL OXIDES
See other AMINATION INCIDENTS

4168. Diphosphoryl chloride
[13498-14-1] $Cl_4O_3P_2$

$$Cl_2P(O)OP(O)Cl_2$$

Water
Mellor, 1971, Vol. 8, Suppl. 3, 505
Hydrolysis is as vigorous as that of phosphorus pentoxide.
See other NON-METAL HALIDES

4169. Tetraperchloratosilicon
[ ] $Cl_4O_{16}Si$

$$Si(OClO_3)_4$$

1. Sorbe, 1968, 127
2. Schmeisser, M., *Angew. Chem.*, 1955, **67**, 499
A highly explosive liquid [1]. Early attempts failed to isolate it but prepared numerous other explosive compounds. Reaction of dichlorine hexoxide with silicon tetrachloride or tetrabromide gave an explosive solid, apparently a perchlorato oligosiloxane. Silver perchlorate and silicon tetrahalides in ether gave explosive volatile organics, perhaps ethyl perchlorate. Replacing ether by acetonitrile as solvent, a solid (di)acetonitrile adduct of the tetraperchlorate precipitated, described as exceptionally explosive even in the smallest quantities [2].
See other NON-METAL PERCHLORATES

4170. Titanium tetraperchlorate
[60580-20-3] $Cl_4O_{16}Ti$

$$Ti(OClO_3)_4$$

Kirk Othmer, 4th. Edn., 1996, Vol. 18, 161
The compound sublimes from 70°C and explodes on heating to 130°C at atmospheric pressure.

Diethyl ether, or Formamide, or Dimethylformamide
Laran, R. J., US Pat. 3 157 464, 1964
Described as insensitive to heat or shock, this powerful oxidant explodes on contact with diethyl ether, and ignites with formamide or DMF.
See other METAL PERCHLORATES, OXIDANTS

4171. Tetrachlorodiphosphane
[13497-91-1] Cl_4P_2

$$Cl_2PPCl_2$$

Besson, A. *et al., Compt. rend.*, 1910, **150**, 102
It oxidises rapidly in air, sometimes igniting.
See other HALOPHOSPHINES, NON-METAL HALIDES

4172. Lead tetrachloride
[13463-30-4] Cl_4Pb

$$PbCl_4$$

1. Bailar, 1973, Vol. 2, 136, 1066
2. Kaufmann, G. B. *et al., Inorg. Synth.*, 1983, **22**, 150
It may disproportionate explosively to lead(II) chloride and chlorine above 100°C [1]. Preparative precautions are detailed [2].

Potassium
See Potassium: Metal halides

Sulfuric acid
Friedrich, H., *Ber.*, 1893, **26**, 1434
It explodes on warming with diluted sulfuric acid or on attempted distillation from the concentrated acid in a stream of chlorine.
See other METAL HALIDES

4173. Tetrachlorosilane (Silicon tetrachloride)
[10026-04-7] Cl_4Si

$$SiCl_4$$

HCS 1980, 821

Dimethyl sulfoxide
See Dimethyl sulfoxide: Acyl halides, etc.

Ethanol, Water
Skinner, S. J., *Loss. Prev.*, 1981, **14**, 178–180
In the preparation of ethyl polysilicate by mixing tetrachlorosilane and industrial methylated spirit containing some water, failure of the agitator is thought to have led to layering of the alcohol over the dense chloride. Evolution of hydrogen chloride led to mixing of the layers, and a greatly increased rate of reaction and self-accelerating gas evolution which burst the reactor.
See other AGITATION INCIDENTS, GAS EVOLUTION INCIDENTS

Sodium
See Sodium: Non-metal halides (reference 8)
See other HALOSILANES, NON-METAL HALIDES

4174. Tin(IV) chloride
[7646-78-8] Cl_4Sn

$SnCl_4$

HCS 1980, 902

Alkyl nitrates
See ALKYL NITRATES: lewis acids

Ethylene oxide
See Ethylene oxide: Contaminants

Nitrobenzene
See Nitrobenzene: Tin(IV) chloride

Turpentine
Mellor, 1941, Vol. 7, 446
Interaction is strongly exothermic and may lead to ignition.
See other METAL HALIDES

4175. Tellurium tetrachloride
[10026-07-0] Cl_4Te

$TeCl_4$

Ammonia
Mellor, 1943, Vol. 11, 58
Interaction with liquid ammonia at $-15°C$ forms tellurium nitride (?), which explodes at 200°C.
See other NON-METAL HALIDES

4176. Titanium tetrachloride
[7550-45-0]   Cl₄Ti

$$TiCl_4$$

HCS 1980, 904

Lithium nitride
See Lithium nitride: Transition metal halides

Sulfur nitrides
See Disulfur dinitride: Metal chlorides
    Tetrasulfur tetranitride: Metal chlorides

Tetrahydrofuran
    Manzer, L. E., *Inorg. Synth.*, 1982, **21**, 135
    Addition of the anhydrous chloride directly to THF causes a violent exothermic
    reaction. Add THF dropwise to a suspension of the chloride in dichloromethane.

Urea
    Ionova, E. A. *et al., Chem. Abs.*, 1966, **64**, 9219b
    The liquid hexaurea complex formed during 6 weeks at 80°C decomposed violently
    at above 90°C. *N*-Halogen compounds may have been formed.
    See other METAL HALIDES

4177. Vanadium tetrachloride
[7632-51-1]   Cl₄V

Lithium nitride
See Lithium nitride: Transition metal halides

4178. Zirconium tetrachloride
[10026-11-6]   Cl₄Zr

$$ZrCl_4$$

Ethanol, or Water
    Rosenheim, A. *et al., Ber.*, 1907, **40**, 811
    Interaction with either is violent.

Lithium, Nitrogen
    See Lithium: Metal chlorides, etc.

Tetrahydrofuran
    Manzer, L. E., *Inorg. Synth.*, 1982, **21**, 136
    Addition of the anhydrous chloride directly to THF causes a violent exothermic
    reaction. Add THF dropwise to a suspension of the chloride in dichloromethane.
    See other METAL HALIDES

4179. Diamminedichloroaminotrichloroplatinum(IV)
[18815-16-2] $Cl_5H_6N_3Pt$

$$[(H_3N)_2Pt(NCl_2)Cl_3]$$

Kukushkin, Yu. N., *Chem. Abs.*, 1958, **52**, 13509i
It tended to decompose violently.
See other N-HALOGEN COMPOUNDS, PLATINUM COMPOUNDS

4180. Molybdenum pentachloride
[10241-05-1] Cl_5Mo

$$MoCl_5$$

Sodium, or Sodium sulfide
  1. Berry, D. H., *Chem. Eng. News*, 1989, **67**(47), 2
  2. Kaner, R. B., *Nature*, 1991, **349**, 510
Placing the chloride in contact with finely divided sodium caused an explosion
after a short induction period [1]. Reaction with sodium sulfide gives molybdenum
disulfide in a vigorous deflagration; autoignition will occur on mixing if the sodium
sulfide is finely divided. Using sintered sulfide, which is recommended, initiation
may take place on local heating to less than 60°C [2].
See Sodium: Metal halides
See other METAL HALIDES

4181. Uranium azide pentachloride
[55042-15-4] Cl_5N_3U

$$N_3UCl_5$$

Explosive solid.
See entry METAL AZIDE HALIDES

4182. Tungsten azide pentachloride
[88495-99-2] Cl_5N_3W

$$N_3WCl_5$$

Extremely explosive.
See entry METAL AZIDE HALIDES

4183. Phosphorus pentachloride
[10026-13-8] Cl_5P

$$PCl_5$$

HCS 1980, 746

Aluminium   MRH 2.76/17
  See Aluminium: Non-metal halides

1465

Carbamates
   See N-Carbomethoxyiminophosphoryl chloride

Chlorine dioxide, Chlorine   MRH Chlorine dioxide 1.55/29
   Mellor, 1941, Vol. 2, 281; 1940, Vol. 8, 1013
   Contact between phosphorus pentachloride and a mixture of chlorine and chlorine
   dioxide (previously considered to be dichlorine trioxide) usually causes explosion,
   possibly owing to formation of the more sensitive chlorine monoxide.

Fluorine
   See Fluorine: Phosphorus halides

Hydroxylamine
   See Hydroxylamine: Phosphorus chlorides

Magnesium oxide
   Mellor, 1940, Vol. 8, 1016
   A heated mixture incandesces brilliantly.

3'-Methyl-2-nitrobenzanilide
   Partridge, M. W., private comm., 1968
   The residue from interaction of the chloride and anilide in benzene after removal
   of solvent and phosphoryl chloride under vacuum exploded violently on admission
   of air.

Nitrobenzene   MRH 4.52/99+
   Unpublished observations
   A solution of phosphorus pentachloride in nitrobenzene is stable at 110° but begins
   to decompose with accelerating violence above 120°C, with evolution of nitrous
   fumes.

Other reactants
   Yoshida, 1980, 60
   MRH values calculated for 8 combinations with varioue reagents are given.

Phosphorus(III) oxide
   Mellor, 1940, Vol. 8, 898
   Interaction is rather violent at ambient temperature.

Sodium   MRH 4.89/35
   See Sodium: Non-metal halides

Urea
   Anon., Angew. Chem. (Nachr.), 1960, 8, 33
   A dry mixture exploded after heating, probably owing to formation of nitrogen
   trichloride. Other chlorinating agents will react similarly with nitrogenous mate-
   rials.
   See Chlorine: Nitrogen compounds

1466

Water
   Mellor, 1940 Vol. 8, 1012
   Interaction with water in limited quantities is violent, and the hydrolysis products
   may themselves react violently with more water.
   See Phosphoryl chloride: Water
   See other NON-METAL HALIDES

4184. Antimony pentachloride
   [7647-18-9] Cl_5Sb

$$SbCl_5$$

   HCS 1980, 160

N-Chlorodimethylamine
   See N-Chlorodimethylamine: Antimony chlorides

Oxygen difluoride
   See Oxygen difluoride: Halogens, etc.

Phosphonium iodide
   See Phosphonium iodide: Oxidants
   See other METAL HALIDES, OXIDANTS

4185. Tantalum pentachloride
   [7721-01-9] Cl_5Ta

$$TaCl_5$$

Lithium dimethylamide
   See Pentakis(dimethylamino)tantalum(V)
   See other METAL HALIDES

4186. Ammonium hexachloroplatinate
   [1332-76-9] $Cl_6H_8N_2Pt$

$$[NH_4]_2 \ [PtCl_6]$$

Potassium hydroxide, Combustible materials
   Mellor, 1942, Vol. 16, 336
   Boiling ammonium hexachloroplatinate with alkali gives a product (possibly potas-
   sium hexahydroxyplatinate) which after drying will explode violently on heating
   alone or with combustible materials.
   See other PLATINUM COMPOUNDS

4187. Potassium hexachloroplatinate
[1307-80-8] Cl_6K_2Pt

$$K_2[PtCl_6]$$

Bromine trifluoride
  See Bromine trifluoride: Potassium hexachloroplatinate
  See other PLATINUM COMPOUNDS

4188. Bis(trichlorophosphoranylidene)sulfamide
[14259-65-5] $Cl_6N_2O_2P_2S$

$$(Cl_3P:N)_2SO_2$$

Cellulose, or Ethanol, or Water
  Vandi, A. et al., Inorg. Synth., 1966, **8**, 119
  It is extremely reactive with water or alcohol, and causes filter paper to ignite.
  See related NON-METAL HALIDES (AND THEIR OXIDES) See other N–S COMPOUNDS

4189. Hexachlorocyclotriphosphazine
[940-71-6] $Cl_6N_3P_3$

Dimethyl sulfoxide
  See Dimethyl sulfoxide: Hexachlorocyclotriphosphazine
  See related NON-METAL HALIDES

4190. Bis(triperchloratosilicon) oxide
[ ] $Cl_6O_{25}Si_2$

$$[ClO_4)_3Si]_2O$$

Schmeisser, M., Angew. Chem., 1955, **67**, 499
This solid decomposition product of tetraperchloratosilicon was so explosive, even
in small amounts, that work was discontinued.
  See related NON-METAL PERCHLORATES

4191. Hexachlorodisilane
[13465-77-5] Cl_6Si_3

$$Cl_3SiSiCl_3$$

  See entry HALOSILANES

Chlorine
  See Chlorine: Hexachlorodisilane

4192. Uranium hexachloride
[13763-23-0] Cl_6U

$$UCl_6$$

Water
  Bailar, 1973, Vol. 5, 189
  Interaction is violent.
  See other METAL HALIDES

4193. Tungsten hexachloride
[13283-01-7] Cl_6W

$$WCl6$$

Sodium sulfide
  Kaner, R. B., *Nature*, 1991, **349**, 510
  Reacts with sodium sulfide even more exothermically than molybdenum pentachloride.
  See Molybdenum pentachloride.
  See other METAL HALIDES

4194. Octachlorotrisilane
[13596-23-1] Cl_8Si_3

$$Cl_3SiSiCl_2SiCl_3$$

See entry HALOSILANES

4195. Tetrazirconium tetraoxide hydrogen nonaperchlorate
[ ] $Cl_9HO_{40}Zr_4$

$$4ZrO(ClO_4)_2.HClO_4$$

Mellor, 1946, Vol. 2, 403
The salt, 'zirconyl perchlorate', explodes if heated rapidly. Later work suggests alternative formulations for the salt.
See other METAL PERCHLORATES

4196. Trirhenium nonachloride
[14973-59-2] Cl_9Re_3

$$Re_3Cl_9$$

Oxygen
  See Oxygen: Trirhenium nonachloride
  See other METAL HALIDES

4197. Decachlorotetrasilane
[13763-19-4]

$$Si_4Cl_{10}$$

$Cl_{10}Si_4$

See entry HALOSILANES

4198. Dodecachloropentasilane
[13596-24-2]

$$Si_5Cl_{12}$$

$Cl_{12}Si_5$

See entry HALOSILANES

4199. Cobalt
[7440-48-4]

Co

$$Co$$

1. Bailar, 1973, Vol. 3, 1056
2. Chadwell, A. J. *et al., J. Phys. Chem.*, 1956, **60**, 1340
Finely divided cobalt is pyrophoric in air [1]. Raney cobalt catalyst appears to be less hazardous than Raney nickel [2].

Acetylene MRH 8.70/99+
See Acetylene: Cobalt

Hydrazinium nitrate
See Hydrazinium nitrate: Alome, or Metals, etc.

Other reactants
Yoshida, 1980, 120
MRH values calculated for 11 combinations, largely with oxidants, are given.

Oxidants MRH value shows % of oxidant
See Ammonium nitrate: Metals MRH 2.55/58
Bromine pentafluoride: Acids, etc.
Nitryl fluoride: Metals

1,3,4,7-Tetramethylisoindole
See 1,3,4,7-Tetramethylisoindole: Cobalt
See other METALS, PYROPHORIC METALS

4200. Cobalt trifluoride
[10026-18-3]

$$CoF_3$$

CoF_3

Hydrocarbons, or Water
Priest, H. F., *Inorg. Synth.*, 1950, **3**, 176

It reacts violently with hydrocarbons or water, and finds use as a powerful fluorinating agent.

Silicon
Mellor, 1956, Vol. 2, Suppl. 1, 64
A gently warmed mixture reacts exothermally, attaining red-heat.
See other METAL HALIDES

4201. Cobalt(III) amide
[ ] CoH_6N_3

$$Co(NH_2)_3$$

Schmitz-Dumont, O. *et al., Z. Anorg. Chem.*, 1941, **284**, 175
Heating converts it to pyrophoric cobalt(III) nitride.
See other N-METAL DERIVATIVES

4202. Diamminenitratocobalt(II) nitrate
[33362-98-0] $CoH_6N_4O_6$

$$[(H_3N)_2CoNO_3]NO_3$$

McPherson, G. L. *et al., Inorg,. Chem.*, 1971, **10**, 1574
A sample of the molten salt exploded at 200°C.
See other AMMINEMETAL OXOSALTS, REDOX COMPOUNDS

4203. Cobalt tris(dihydrogenphosphide)
[ ] CoH_6P_3

$$Co(PH_2)_3$$

Zehr, J., *Staub*, 1962, **22**, 494–508
It ignites in air, especially when finely divided.
See related NON-METAL HYDRIDES, PHOSPHINES

4204. Triamminetrinitrocobalt(III)
[13600-88-9] $CoH_9N_6O_6$

$$[(H_3N)_3Co(NO_2)_3]$$

Explodes at 305°C, medium impact-sensitivity.
See entry AMMINEMETAL OXOSALTS (reference 2)

4205. Trihydrazinecobalt(II) nitrate
[82434-33-1] $CoH_{12}N_8O_6$

$$[(H_4N_2)_2Co] [NO_3]_2$$

Franzen, H. *et al., Z. Anorg. Chem.*, 1908, **60**, 247, 274

It is explexplosive.
See other AMMINEMETAL NITRATES

4206. Ammonium hexanitrocobaltate(3−)
[14652-46-1] $CoH_{12}N_9O_{12}$

$$[NH_4]_3[Co(NO_2)_6]$$

Explodes at 230°C, medium impact-sensitivity.
See entry AMMINEMETAL OXOSALTS (reference 2)

4207. Pentaamminenitratocobalt(III) nitrate
[14404-36-5] $CoH_{15}N_8O_9$

$$[(H_3N)_5CoNO_3] [NO_3]_2$$

Explodes at 310°C, medium impact-sensitivity.
See entry AMMINEMETAL OXOSALTS (reference 2)

4208. Hexaamminecobalt(III) iodate
[14589-65-2] $CoH_{18}I_3N_6O_9$

$$[(H_3N)_6Co] [IO_3]_3$$

Explodes at 355°C, low impact-sensitivity.
See entry AMMINEMETAL OXOSALTS (reference 2)

4209. Hexaamminecobalt(III) permanganate
[22388-72-3] $CoH_{18}Mn_3N_6O_{12}$

$$[(H_3N)_6Co] [MnO_4]_3$$

1. Mellor, 1942, Vol. 12, 336
2. Joyner, T. B., *Can. J. Chem.*, 1969, **47**, 2730
It explodes on heating [1], and is of treacherously high impact-sensitivity [2].
See other AMMINEMETAL OXOSALTS

4210. Hexaamminecobalt(III) nitrate
[10534-86-8] $CoH_{18}N_9O_9$

$$[(H_3N)_6Co] [NO_3]_3$$

Explodes at 295°C, medium impact-sensitivity.
See entry AMMINEMETAL OXOSALTS (reference 2)

4211. Hexahydroxylaminecobalt(III) nitrate
[18501-44-5] (ion) $CoH_{18}N_9O_{15}$

$$[(HONH_2)_6Co] [NO_3]_3$$

Werner, A. *et al., Ber.*, 1905, **38**, 897
It usually explodes during preparation or handling.
See other AMMINEMETAL OXOSALTS

4212. Potassium triazidocobaltate(1−)
[52324-65-9] $CoKN_9$

$$K[(N_3)_3Co]$$

1. Fritzer, H. P. *et al., Angew. Chem. (Intern Ed.)*, 1971, **10**, 829
2. Fritzer, H. P. *et al., Inorg. Nucl. Chem. Lett.*, 1974, **10**, 247–252
The complex azide is highly explosive and must be handled with extreme care.
The analogous potassium and caesium derivatives of zinc azide and nickel azide
deflagrate strongly in a flame and some are shock-sensitive [1]. The potassium
salt alone out of 8 azido-complexes exploded during X-irradiation in an ESCA
study [2].
See other IRRADIATION DECOMPOSITION INCIDENTS
See AZIDE COMPLEXES OF COBALT(III) *See related* METAL AZIDES

4213. Potassium hexanitrocobaltate(3−)
[13782-01-9] $CoK_3N_6O_{12}$

1. Broughton, D. B. *et al., Anal. Chem.*, 1947, **19**, 72
2. Horowitz, O., *Anal. Chem.*, 1948, **20**, 89
3. Tomlinson, W. R. *et al., J. Amer. Chem. Soc.*, 1949, **71**, 375
Evaporation by heating a filtrate from precipitation of 'potassium cobaltinitrite'
caused it to turn purple and explode violently [1]. This was attributed to interaction of nitrite, nitrate, acetic acid and residual cobalt with formation of fulminic or
methylnitrolic acids or their cobalt salts, all of which are explosive [2]. Mixtures
containing nitrates, nitrites and organic materials are potentially dangerous, especially in presence of acidic materials and heavy metals. A later publication confirms
the suggestion of formation of nitro- or nitrito-cobaltate(III) [3].
See related METAL NITRITES

4214. Cobalt(III) nitride
[12139-70-7]

CoN

CoN

Schmitz-Dumont, O., *Angew. Chem.*, 1955, **67**, 231
A pyrophoric powder.
See Cobalt(III) amide
See other NITRIDES, PYROPHORIC MATERIALS

4215. Cobalt(II) nitrate
[10026-22-9]

CoN_2O_6

$Co(NO_3)_2$

HCS 1980, 321

Ammonium hexacyanoferrate(II)
Wolski, W. *et al., Acta Chim.* (Budapest), 1972, **72**, 25–32
Interaction is explosive at 220°C.
See Copper(II) nitrate: Ammonium hexacyanoferrate(II), or: Potassium hexacyano-ferrate(II)

Carbon   MRH 3.68/14
Crowther, J. R., private comm., 1970
Charcoal impregnated with the nitrate exploded lightly during sieving. Possibly a dust or 'black powder' explosion.

Other reactants
Yoshida, 1980, 195
MRH values calculated for 16 combinations with oxidisable materials are given.
See other METAL NITRATES, OXIDANTS

4216. Cobalt(II) azide
[14215-31-7]

CoN_6

$Co(N_3)_2$

Mellor, 1940, Vol. 8, 355
It explodes at 200°C.
See also AMMINEMETAL AZIDES
See other METAL AZIDES

4217. Cobalt(II) oxide
[1307-96-6]

CoO

CoO

Hydrogen peroxide
See Hydrogen peroxide: Metals, etc.
See other METAL OXIDES

4218. Cobalt(II) sulfide
   [1317-42-6]   CoS

$$CoS$$

Brauer, 1965, Vol. 2, 1523
Material dried at 300°C is pyrophoric.
See other METAL SULFIDES, PYROPHORIC MATERIALS

4219. Hexaamminecobalt(III) hexanitrocobaltate(3−)
   [15742-33-3] $Co_2H_{18}N_{12}O_{12}$

$$[(H_3N)_6Co]\ [Co(NO_2)_6]$$

1. Tomlinson, W. R. *et al., J. Amer. Chem. Soc.*, 1949, **71**, 376
2. Shidlovskii, A. A. *et al., Chem. Abs.*, 1977, **87**, 70416
An unstable compound of low impact-sensitivity [1]. In a comparative study of a
series of cobalt complexes ranging from triamminecobalt(III) nitrite to ammonium
hexanitrocobaltate(3−), the title compound burned the fastest [2].
See other AMMINEMETAL OXOSALTS

4220. *trans* Tetraamminediazidocobalt(III)
   ***trans*-diamminetetraazidocobaltate(1−)**
   [54689-17-7] $Co_2H_{18}N_{24}$

$$[(H_3N)_4Co(N_3)_2][(H_3N)_2Co(N_3)_4]$$

Druding, L. F. *et al., Inorg. Chem.*, 1975, **14**, 1365
During its preparation, solutions of the salt, which is a dangerous detonator, must
not be evaporated to dryness. Surprisingly, mixtures with potassium bromide could
be compressed (for IR examination) to 815 bar without decomposition. It contains
71% of nitrogen.
See AMMINEMETAL AZIDES, AZIDE COMPLEXES OF COBALT(III)
See related METAL AZIDES *See other* HIGH-NITROGEN COMPOUNDS

4221. Cobalt(III) oxide
   [1308-04-9] Co_2O_3

$$Co_2O_3$$

(Commercial samples contain some cobalt(II) oxide)

Hydrogen peroxide
   See Hydrogen peroxide: Metals, etc.

Nitroalkanes
   See NITROALKANES: metal oxides *See other* METAL OXIDES

4222. Chromium
[7440-47-3]

Cr

Cr

RSC Lab. Hazards Safety Data Sheet No. 65, 1987 (metal and compounds)

1. Sidgwick, 1950, 1013
2. Uel'skii, A. A. *et al., Chem. Abs.*, 1978, **89**, 116214

Evaporation of mercury from mercury amalgam leaves pyrophoric chromium [1]. Increasing the temperature at which hexacarbonylchromium is thermally decomposed increases the surface area and pyrophoricity of the chromium powder produced [2].

Carbon dioxide
See Carbon dioxide: Metals

Oxidants
See Ammonium nitrate: Metals
Bromine pentafluoride: Acids, etc.
Nitrogen oxide: Metals
Sulfur dioxide: Metals
See other METALS, PYROPHORIC METALS

4223. Copper chromate oxide
[1308-09-4]

$CrCuO_4.2CuO$

$CuCrO_4.2CuO$

Hydrogen sulfide
Pascal, 1960, Vol. 13.1, 1025
The gas may ignite on contact with the basic chromate.
See other METAL OXOMETALLATES

4224. Ammonium fluorochromate
[58501-09-0]

$CrFH_4NO_3$

NH_4FCrO_3

Ribas Bernat, J. G., *Ion* (Madrid), 1975, **35**, 573–576
It decomposes violently at 220°C.
See related OXOSALTS OF NITROGENOUS BASES

4225. Chromyl fluorosulfate
[33497-88-0]

$CrF_2O_8S_2$

$O_2Cr(OSO_2F)_2$

Water
Rochat, W. V. *et al., Inorg. Chem.*, 1969, **8**, 158

1476

Interaction is violent.
See related ACYL HALIDES, OXIDANTS

4226. Chromium pentafluoride
[14884-42-5]   CrF$_5$

$$CrF_5$$

Bailar, 1973, Vol. 2, 1086
It reacts violently in halogen-exchange and oxidation-reduction reactions.

Phosphorus trichloride
O'Donnell, T. A. *et al., Inorg. Chem.*, 1966, **5**, 1435, 1437
Interaction is violent after slight warming.
See other METAL HALIDES, OXIDANTS

4227. Potassium hydroxyoxodiperoxochromate(1−)
[40330-52-7] (ion)   CrHKO$_6$

$$K[HO(O)Cr(O_2)_2]$$

See entry PEROXOCHROMIUM COMPOUNDS

4228. Thallium hydroxyoxodiperoxochromate(1−)
[40330-52-7] (ion)   CrHO$_6$Tl

$$Tl[HO(O)Cr(O_2)_2]$$

See entry PEROXOCHROMIUM COMPOUNDS

4229. Chromic acid
[7738-94-5]   CrH$_2$O$_4$

$$H_2CrO_4$$

HCS 1980, 385 (dichromate/sulfuric acid ± added water)

1. Bryson, W. R., *Chem. Brit.*, 1975, **11**, 377
2. Pitt, M. J., *Chem. Brit.*, 1975, **11**, 456
3. Downing, S., *Chem. Brit.*, 1975, **11**, 456
4. Baker, P. B., *Chem. Brit.*, 1975, **11**, 456
5. Kipling, B., *Chem. Brit.*, 1976, **12**, 169
6. Kaufman, J. A., private comm., 1987

A closed bottle of unused potassium dichromate−sulfuric acid mixture exploded after several months in storage [1]. Previous similar incidents were summarised, and the possibility of the bottle having burst from internal pressure of carbon dioxide arising from trace contamination by carbon compounds was advanced [1,2]. Two further reports of incidents within 1 or 2 days of preparation of the mixture were reported [3,4], the latter involving exothermic precipitation of chromium trioxide. Presence of traces of chloride in the dichromate leads to formation of chromyl chloride, which may be unstable [2,5]. The use of other glass-cleaning agents, and non-storage of chromic acid mixtures is again recommended [2]. Used chromic acid cleaning solution was returned to the bottle and capped; it exploded 2–3 mins. later [6].

See other GAS EVOLUTION INCIDENTS

Acetone
MCA Case History No. 1583
During glass cleaning operations, acetone splashed into a beaker used previously to contain potassium dichromate–sulfuric acid mixture and the solvent ignited. Alcohols behave similarly.

Alcohols, Silica gel
Lou, J.-D. *et al., Chem. & Ind.*, 1987, 531–532
Chromic acid adsorbed on silica gel ($CrO_3 : SiO_2 = 3 : 10$) is a mild and safe reagent to oxidise saturated or unsaturated alcohols in carbon tetrachloride to the corresponding aldehydes.

Oxidisable material
MCA Case History No. 1919
A waste plating solution (containing 22% sulfuric acid and 40% w/v of chromium) was being sucked into an acid-disposal tanker. When 500 l had been transferred, a mild explosion in the tanker blew back through the transfer pump and hose. The oxidisable component in the tanker was not identified.

See other INORGANIC ACIDS, OXIDANTS

4230. Ammonium hydroxyoxodiperoxochromate(1−)
[40330-52-7] (ion) CrH_5NO_6
$$NH_4[HO(O)Cr(O_2)_2]$$

See entry PEROXOCHROMIUM COMPOUNDS

4231. Triamminediperoxochromium(IV)
[17168-85-3] $CrH_9N_3O_4$
$$[(H_3N)_3Cr(O_2)_2]$$

1. Hughes, R. G. *et al., Inorg. Chem.*, 1968, **7, 882**
2. Kauffman, G. B., *Inorg. Synth.*, 1966, **8**, 133
It must be handled with care because it may explode or become incandescent on sudden heating or shock. Heating at 20°/min caused a violent explosion at around 120°C [1]. Preparative and handling precautions have been detailed [2].

See other AMMINECHROMIUM PEROXOCOMPLEXES

4232. Ammonium tetraperoxochromate(3−)
[67165-30-4] $CrH_{12}N_3O_8$

$$[NH_4]_3[Cr(O_2)_4]$$

Alone, or Sulfuric acid
  Mellor, 1943, Vol. 11, 356
  Explodes at 50°C, on impact, or in contact with sulfuric acid.
  See other PEROXOACID SALTS

4233. Pentaamminenitrochromium(III) nitrate
[31255-94-4] $CrH_{15}N_8O_8$

$$[(H_3N)_5CrNO_2][NO_3]_2$$

Mellor, 1943, Vol. 11, 477
  It explodes on heating, and may be impact-sensitive.
  See other AMMINEMETAL OXOSALTS

4234. Hexaamminechromium(III) nitrate
[15263-28-7] $CrH_{18}N_9O_9$

$$[(H_3N)_6Cr][NO_3]_3$$

Moderately impact-sensitive, explodes at 263°C.
  See entry AMMINEMETAL OXOSALTS (reference 2)

4235. Potassium tetraperoxochromate(3−)
[12331-76-9] CrK_3O_8

$$K_3[Cr(O_2)_4]$$

Alone, or Sulfuric acid
  Mellor, 1943, Vol. 11, 356
  Not sensitive to impact, but explodes at 178°C or in contact with sulfuric acid.
  The impure salt is less stable, and explosive.
  See other PEROXOACID SALTS

4236. Lithium chromate
[14307-35-8] $CrLi_2O_4$

$$Li_2CrO_4$$

Zirconium
  de Boer, H. J. *et al.*, *Z. Anorg. Chem.*, 1930, **191**, 113
  During reduction of the chromate to lithium at 450–600°C, a considerable excess
  of zirconium must be used to avoid explosions.
  See other METAL OXOMETALLATES

4237. Chromium nitride
[24094-93-7]   CrN

CrN

Potassium nitrate
  See Potassium nitrate: Chromium nitride
  See other NITRIDES

4238. Chromyl nitrate
[16017-38-2] CrN_2O_8

$$O_2Cr(NO_3)_2$$

Organic materials
  1. Schmeisser, M., *Angew. Chem.*, 1955, **67**, 495
  2. Harris, A. D. *et al., Inorg. Synth.*, 1967, **9**, 87
  Many hydrocarbons and organic solvents ignite on contact with this powerful
  oxidant and nitrating agent [1], which reacts like fuming nitric acid in contact
  with paper, rubber or wood [2].
  See related METAL NITRATES

4239. Chromyl azide
[ ] CrN_6O_2

$$O_2Cr(N_3)_2$$

Preparative hazard
  See Chromyl chloride: Sodium azide
  See related METAL AZIDES

4240. Sodium tetraperoxochromate(3−)
[12206-14-3] $CrNa_3O_8$

$$Na_3[Cr(O_2)_4]$$

Mellor, 1943, Vol. 11, 356
It explodes at 115°C.
See other PEROXOACID SALTS

4241. Chromium(II) oxide
[12018-00-7]   CrO

CrO

1. Férée, J., *Bull. Soc. Chim, Fr.*, 1901, **25**, 620
2. Ellern, 1961, 33
The black powder ignites if ground or heated in air [1]. The oxide obtained by
oxidation of chromium amalgam is pyrophoric [2].
See other METAL OXIDES, PYROPHORIC MATERIALS, REDUCANTS

1480

4242. Chromium trioxide
[1333-82-0]

CrO_3

(*MCA SD-52*, 1944); *FPA H48*, 1976; *HCS 1980*, 315

1. Baker, W., *Chem. & Ind.*, 1956, 280
2. Goertz, A., *Arbeitschutz*, 1935, 323

Presence of nitric acid or nitrates in chromium trioxide may cause oxidation reactions to accelerate out of control, possibly owing to formation of chromyl nitrate. Samples of the oxides should be tested by melting before use, and those evolving oxides of nitrogen should be discarded [1]. A closed container of the pure oxide exploded violently when laid down on its side. This was attributed to unsuspected contamination of the container [2].

See Chromyl nitrate

Acetic acid   MRH 2.22/18

1. Anon., *BCISC Quart. Safety Summ.*, 1966, **37**, 30
2. Anon., *Sichere Chemiearb.*, 1987, **39**, 70

An explosion occurred during initial heating of a large volume of glacial acetic acid being treated with chromium trioxide. This was attributed to violent interaction of solid chromium trioxide and liquid acetic acid on a hot, exposed steam coil, and subsequent initiation of an explosive mixture of acetic acid vapour and air. The risk has been obviated by using a solution of dichromate in sulfuric acid as oxidant, in place of the trioxide. The sulfuric acid is essential, as the solid dichromate, moist with acetic acid obtained by evaporating an acetic acid solution to near-dryness, will explode [1]. Use of a mixture of chromium trioxide and acetic acid as an oxidant shattered a glass reaction vessel [2].

See other GLASS INCIDENTS
See Butyric acid, below

Acetic anhydride   MRH 2.38/16

1. Dawber, J. G., *Chem. & Ind.*, 1964, 973
2. Bretherick, L., *Chem. & Ind.*, 1964, 1196
3. Baker, W., *Chem. & Ind.*, 1956, 280
4. Eck, C. R. *et al.*, *J. Chem. Soc., Chem. Comm.*, 1974, 865
5. *MCA Guide*, 1972, 297

A literature method for preparation of chromyl acetate by interaction of chromium trioxide and acetic anhydride was modified by omission of cooling and agitation. The warm mixture exploded violently when moved [1]. A later publication emphasised the need for cooling, and summarised several such previous occurrences [2]. An earlier reference attributes the cause of chromium trioxide–acetic anhydride oxidation mixtures going out of control to presence of nitric acid or nitrates in the chromium trioxide, and a simple test to check this point is given [3]. Mixtures used as a reagent for the remote oxidation of carboxylic esters are potentially explosive, and must be made up and used at below 25°C under controlled conditions [4]. An attempt to purify the anhydride by warming with 2% w/v of trioxide led to an explosion at 30°C [5].

Acetic anhydride, 3-Methylphenol
Thorne, J. G., private comm., 1978
Addition of the oxide in portions to the reaction mixture (for preparation of 3-hydroxybenzaldehyde or its acetate) at 75–80°C proceeded smoothly, but the final portion caused a large exotherm leading to eruption of the flask contents.

Acetic anhydride, Tetrahydronaphthalene
Peak, D. A., *Chem. & Ind.*, 1949, 14
Use of an anhydride solution of the trioxide to prepare tetralone caused a vigorous fire. This was attributed to use of the more hygroscopic granular trioxide, which is less preferable than the flake type.

Acetone   MRH 2.59/10
Delhez, R., *Chem. & Ind.*, 1956, 931
The use of chromium trioxide to purify acetone is hazardous, ignition on contact occurring at ambient temperature. Methanol behaves similarly when used to reduce the trioxide in preparing hexaaquachromium(III) sulfate.

Acetylene
Grignard, 1935, Vol. 3, 167
Acetylene is oxidised violently.

Alcohols   MRH Ethanol 2.55/10
1. Newth, F. H. *et al.*, *Chem. & Ind.*, 1964, 1482
2. Neumann, H., *Chem. Eng. News*, 1970, **48**(24), 4
3. *491M*, 1975, 126
4. Kishore, K., *Def. Sci. J.*, 1978, **28**, 13–14
When methanol was used to rinse a pestle and mortar which had been used to grind coarse chromium trioxide, immediate ignition occurred due to vigorous oxidation of the solvent. The same occurred with ethanol, 2-propanol, butanol and cyclohexanol. Water is a suitable cleaning agent for the trioxide [1]. For oxidation of *sec*-alcohols in DMF, the oxide must be finely divided, as lumps cause violent local reaction on addition to the solution [2]. Use of methanol to reduce the Cr(VI) oxide to a Cr(III) derivative led to an explosion and fire [3]. The ignitability of the butanols decreases from *n*-through *sec*- to *tert*-butanol [4].
See Dimethylformamide, below

Alkali metals   MRH Potassium 1.46/23, sodium 3.39/21
Mellor, 1943, Vol. 11, 237
Sodium or potassium reacts with incandescence.

Ammonia   MRH Gas 2.51/15, solution 1.63/26
1. Mellor, 1943, Vol. 11, 233
2. Pascal, 1959, Vol. 14, 215
Gaseous ammonia leads to incandescence [1], and the aqueous solution is oxidised very exothermally [2].

Arsenic
  Mellor, 1943, Vol. 11, 234
  Interaction is incandescent.

Bromine pentafluoride
  See Bromine pentafluoride: Acids, etc.

Butyric acid
  Wilson, R. D., *Chem. & Ind.*, 1957, 758
  A mixture of chromium trioxide and butyric acid became incandescent on heating to 100°C.
  See Acetic acid, above

Chlorine trifluoride
  See Chlorine trifluoride: Chromium trioxide

Chromium(II) sulfide
  Mellor, 1943, Vol. 11, 430
  Interaction causes ignition.

Dimethylformamide   MRH 2.64/11
  1. Neumann, H. *Chem. Eng. News*, 1970, **48**(28), 4
  2. Heathcock, C. H., *Chem. Eng. News*, 1981, **59**(8), 9
  During oxidation of a *sec*-alcohol to ketone in cold DMF solution, addition of solid trioxide caused ignition. Addition of lumps of trioxide was later found to cause local ignition on addition to ice-cooled DMF under nitrogen [1]. Addition of 2 g of chromium trioxide to 18 ml of solvent to form a 10 wt% solution caused immediate ignition and ejection of the flask contents [2].
  See Dimethylformamide: Oxidants

1,3-Dimethylhexahydropyrimidone
  Mukhopadhyay, T. *et al., Helv. Chim. Acta*, 1982, **65**, 386 (footnote 9)
  Contact of the trioxide with the aprotic amide solvent ('dimethylpropyleneurea') is always explosive, with fire, whereas this only occurs with hexamethylphosphoramide (below) if the oxide is previously crushed. However the former is much less toxic than the latter solvent.

Ethylene glycol
  See Ethylene glycol: Oxidants

Glycerol   MRH 2.38/17
  Pieters, 1957, 30
  Interaction is violent; the mixture may ignite owing to oxidation of the trihydric alcohol, which is viscous and unable to dissipate the exotherm.
  See Glycerol: Oxidants

Hexamethylphosphoramide
  Cardillo, G. *et al., Synthesis*, 1976, **6**, 394–396

Stirring chromium trioxide (added in small portions) with the unheated solvent leads to the formation of a complex useful for oxidising alcohols to carbonyl derivatives. The trioxide must not be crushed before being added to the solvent, because violent decomposition may then occur.
See 1,3-Dimethylhexahydropyrimidone, above

Hydrazine   MRH 3.01/19
    See Hydrazine: Oxidants (reference 2)

Hydrogen sulfide   MRH 2.05/12
    Mellor, 1943, Vol. 11, 232
    Contact with the heated oxide causes incandescence.

Organic materials,   MRH values below references show % of organic
or Solvents
    1. Leleu, *Cahiers*, 1976, (83), 286–287
    2. Fawcett, H. H., *Ind. Eng. Chem.*, 1959, **51**(4), 90A
    3. Mikhailov, V., *Chem. Abs.*, 1960, **54**, 23331f
    4. Mellor, 1943, Vol. 11, 235
    MRH Aniline 2.59/8, diethyl ether 2.59/8, hexane 2.46/6, phenol 2.51/9, toluene 2.55/7
    Combustible materials may ignite or explode on contact with the oxide. A few drops of oil which fell into a container of the oxide led to an explosion which produced fatal burns [1]. If a few drops of an organic solvent (acetone, 2-butanone, ethanol) contact solid chromium trioxide, a few seconds' delay ensues while some of the oxidant attains the critical temperature of 330°C. Then combustion occurs with enough vigour to raise a fire-ball several feet, and spattering also occurs [2]. Possible ignition hazards were studied for a range of 60 organic liquids and solids in contact with the solid oxide. Hot liquids were added to the oxide; solids were covered with a layer of the oxide. The most dangerous materials were methanol, ethanol, butanol, isobutanol, acetaldehyde, propionaldehyde, butyraldehyde, benzaldehyde, acetic acid, pelargonic acid, ethyl acetate, isopropyl acetate, pentyl acetate, diethyl ether, methyldioxane, dimethyldioxane, acetone and benzylethylaniline. Other materials evolved heat, especially in presence of water. Segregation in storage or transport is essential [3]. Benzene ignites in contact with the powdered oxide [4].
    See 1,3-Dimethylhexahydropyrimidone, or Dimethylformamide, or Glycerol, or Hexamethylphosphoramide, all above

Other reactants
    Yoshida, 1980, 354–355
    MRH values calculated for 28 combinations, largely with oxidisable materials, are given.

Peroxyformic acid
    See Peroxyformic acid: Metals, etc.

Phosphorus   MRH (Yellow) 3.68/16
  Moissan, H., *Ann. Chim. Phys. [6]*, 1885, **5**, 435
  Phosphorus and the molten trioxide react explosively.

Potassium hexacyanoferrate(III)   MRH 2.59/25
  Anon., *BCISC Quart. Safety Summ.*, 1965, **36**(144), 55
  Mixtures of the ferrate ('ferricyanide') and chromium trioxide explode and inflame
  when heated above 196°C. Friction alone is sufficient to ignite violently the mixture
  when ground with silver sand.
  See other FRICTIONAL INITIATION INCIDENTS

Pyridine   MRH 2.63/92
  1. Dauben, W. G. *et al., J. Org. Chem.*, 1969, **34**, 3587
  2. Poos, G. I. *et al., J. Amer. Chem. Soc.*, 1953, **75**, 427
  3. Collins, J. C. *et al., Org. Synth.*, 1972, **52**, 7
  4. Ratcliffe, R. *et al., J. Org. Chem.*, 1970, **35**, 4001
  5. Stensio, K. E., *Acta Chem. Scand.*, 1971, **25**, 1125
  6. *MCA Case History No. 1284*
  During preparation of the trioxide–pyridine complex (a powerful oxidant), lack
  of really efficient stirring led to violent flash fires as the oxide was added to the
  pyridine at −15 to −18°C. Reversed addition of pyridine to the oxide is extremely
  dangerous [1], ignition usually occurring [2]. A later preparation specifies temperature limits of 10–20°C to avoid an excess of unreacted trioxide [3]. A safe method
  of preparing the complex in solution has been described [4], and preparation and
  use of solutions of the isolated complex in dichloromethane [3] or acetic acid [5]
  have been detailed. The Case History gives further information on preparation of
  the complex. Solution of the oxide is not smooth; it first swells, then suddenly
  dissolves in pyridine with an exotherm. This hazard may be eliminated without
  loss of yield by dissolving the trioxide in an equal volume of water before adding
  it to 10 volumes of pyridine. Pulverising chromium trioxide before use is not
  recommended, as this increases its rate of reaction with organic compounds to a
  hazardous level [6].
  See Oxodiperoxopyridinechromium *N*-oxide

Selenium
  Mellor, 1943, Vol. 11, 233
  Interaction is violent.

Sodium   MRH 3.39/21
  See Sodium: Metal oxides
  See other REDOX REACTIONS

Sodium amide
  Mellor, 1943, Vol. 11, 234
  Grinding a mixture leads to violent reaction.

Sulfur   MRH 1.76/14
  Mellor, 1943, Vol. 11, 232

A mixture ignites on warming.
See other METAL OXIDES, OXIDANTS
See related ACID ANHYDRIDES

4243. Lead chromate
[7758-97-6]   CrO₄Pb
$$PbCrO_4$$

HCS 1980, 587

Aluminium, Dinitronaphthalene
Nagaishi, T. *et al.*, *Chem. Abs.*, 1977, **86**, 59602
The considerable energy released by the mixture derives from chromate-catalysed exothermic decomposition of the nitro compound, coupled with a thermite-type reaction of the aluminium and chromate. It is useful for cracking concrete.

Azo-dyestuffs
Anon., *Loss Prev. Bull.*, 1978, (022), 117
Under certain conditions, dry mixtures of lead chromate pigments with the azo-dyes 1-(2′,4′-dinitrobenzeneazo)-2-hydroxynaphthalene (dinitroaniline orange) or 1-(4′-chloro-2′-nitrobenzeneazo)-2-hydroxynaphthalene (chlorinated para-red) may lead to violent explosions during mixing/blending operations.
See N,N″-Dichlorobis(2,4,6-trichlorophenyl)urea: 1-*p*-Nitrobenzeneazo-2-naphthol

Iron(III) hexacyanoferrate(4−)
1. Anon., *Chem. Processing*, 1967, **30**, 118
2. Twitchett, H. J., *Chem. Brit.*, 1977, **13**, 437
During grinding operations, the intimate mixture of pigments was ignited by a spark and burned fiercely [1]. Spontaneous ignition of Brunswick Green pigment (which also contains lead sulfate) soon after grinding was not uncommon, and similar incidents had led to the loss of ships with cargoes of Prussian Blue or Brunswick Green in wooden casks [2].
See Azo-dyestuffs, above
See other SELF-HEATING AND IGNITION INCIDENTS

Sulfur
Jackson, H., *Spectrum*, 1969, **7**(2), 82
The mixture is pyrophoric.

Tantalum
Sarawadekar, R. G. *et al.*, *Thermochim. Acta*, 1983, **70**, 133
The mixture is a pyrotechnic composition.
See other METAL OXOMETALLATES, OXIDANTS

1486

4244. Chromium(II) sulfate

[13825-86-0] CrO_4S

$$CrSO_4$$

Water

1. van Bemmelen, J. M., *Rec. Trav. Chim.*, 1887, **6**, 202–204
2. Bailar, 1973, Vol. 3, 657–658

Crystals of the heptahydrate, damp with surplus water, were sealed into a glass tube and stored in darkness, and after a year the tube exploded. This was attributed to the pressure of hydrogen liberated by reduction of water by the chromium(II) salt [1]. More recent information [2] confirms this hypothesis.

See Chromium(II) chloride *See other* METAL OXONON-METALLATES, REDUCANTS

4245. Chromium(II) sulfide

[12018-06-3]   CrS

$$CrS$$

Chromium trioxide
   See Chromium trioxide: Chromium(II) sulfide

Fluorine
   See Fluorine: Sulfides
   See other METAL SULFIDES

4246. Ammonium dichromate

[7789-09-5] $Cr_2H_8N_2O_7$

$$NH_4[OCrO_2OCrO_2O]NH_4$$

(*MCA SD-45*, 1952); *HCS 1980*, 143

1. Mellor, 1943, Vol. 11, 324
2. Grewer, T. *et al., Exothermic Decomposition*, Technical Report 01VD 159/0329 for Federal German Ministry for Res. Technol., Bonn. 1986
3. Anon., *Chem. Eng. News*, 1986, **64**(4), 6
4. Bretherick, L., *Chem. Eng. News*, 1986, **64**(22), 4

Thermal decomposition of the salt (which is of zero oxygen balance) is initiated by locally heating to 190°C, and flame and sparks spread rapidly through the mass, and if confined, it may become explosive [1]. Under close confinement, the deflagrating salt shows the extraordinarily high rate of pressure increase of 68 kbar/s, attaining a final pressure of 510 bar in about 10 ms, and further study of its deflagration and homogeneous decomposition under confinement have been reported [2]. Energy of decomposition (in range 230–260°C)was measured as 0.76 kJ/g by DSC, and T_{ait24} was determined as 96°C by adiabatic Dewar tests, with an apparent energy of activation of 112 kJ/mol. Following a report of an explosion during heated vacuum drying of a 1 t quantity of the salt [3], attention was drawn [4] to the extreme potential hazard involved in heating the salt under confinement.

See entry THERMOCHEMISTRY AND EXOTHERMIC DECOMPOSITION
See other DEFLAGRATION INCIDENTS

Ethylene glycol
  See Ethylene glycol: Oxidants
  See other DICHROMATE SALTS OF NITROGENOUS BASES

4247. Ammonium pentaperoxodichromate(2−)

[ ] $Cr_2H_8N_2O_{12}$

$$NH_4[O(O_2)_2CrOOCr(O_2)_2O]NH_4$$

Brauer, 1965, Vol. 2, 1392
It explodes at 50°C.
See other PEROXOACID SALTS

4248. Potassium dichromate

[7778-50-9] $Cr_2K_2O_7$

$$K_2[OCrO_2OCrO_2O]$$

FPA H43, 1976

Boron, Silicon
  See Boron: Dichromates, etc.

Ethylene glycol   MRH 1.46/11
  Yoshida, T., private comm., 1982
  Mixing of equal weights at ambient temperature is uneventful, but at 100° an
  exotherm of 170°C occurs.

Hydrazine   MRH 1.59/99+
  Mellor, 1943, Vol. 11, 234
  Explosive interaction.
  See Hydrazine: Oxidants (reference 2)

Hydroxylamine   MRH 3.85/99+
  See Hydroxylamine: Oxidants

Iron   MRH 0.42/34
  Laye, P. G. *et al., Thermochim. Acta*, 1980, **39**, 3567–359
  Pyrotechnic mixtures (1:1 wt) attained a maximum temperature of about 1090°C
  on ignition.
  See Tungsten, below

Sulfuric acid
  See Chromic acid

Tungsten
  Boddington, T. *et al., Combust. Flame*, 1975, **24**, 137–138
  Combustion of a pyrotechnic mixture of the two materials (studied by DTA and
  temperature profile analysis) attains a temperature of about 1700°C in 0.1–0.2 s.
  See other METAL OXOMETALLATES, OXIDANTS

4249. Potassium pentaperoxodichromate

[ ] $Cr_2K_2O_{12}$

$$K_2[(O_2)_2OCrOOCr(O_2)_2O]$$

Mellor, 1943, Vol. 11, 357; Sidgwick, 1950, 1007
The powdered salt explodes above 0°C.
See other PEROXOACID SALTS

4250. Sodium dichromate

[10588-01-9] $Cr_2Na_2O_7$

$$Na_2[OCrO_2OCrO_2O]$$

HCS 1980, 838

Acetic acid, 2-Methyl-2-pentenal
Nolan, 1983, Case history 73
2-Methyl-2-pentenal was oxidised to the acid in a process involving addition over
a period into a mixture of acetic acid and sodium dichromate at 50°C. Addition
of the aldehyde, effected by sucking it into the reactor by application of vacuum,
was entrusted to an inexperienced operator, but was apparently too fast and the
exotherm led to a runaway and eruption of the reactor contents.

Acetic anhydride   MRH 1.84/13
Marszalek, G., private comm., 1973
Addition of the dehydrated salt to acetic anhydride caused an exothermic reaction
which accelerated to explosion. Presence of acetic acid (including that produced
by hydrolysis of the anhydride by the hydrate water) has a delaying effect on the
onset of violent reaction, which occurs where the proportion of anhydride to acid
(after hydrolysis) exceeds 0.37:1, with an initial temperature above 35°C. Mixtures
of dichromate (30 g) with anhydride–acid mixtures (70 g, to give ratios of 2:1,
1:1, 0.37:1) originally at 40°C accelerated out of control after 18, 43 and 120 min,
to 160, 155 and 115°C, respectively.
See other INDUCTION PERIOD INCIDENTS

Boron, Silicon
See Boron: Dichromates, etc.

Ethanol, Sulfuric acid
1. Annable, E. H., *School Sci. Rev.*, 1951, **32**(117), 249
2. Anon., *Education in Science*, 1990, (140), 30
During preparation of acetic acid by acid dichromate oxidation of ethanol according
to a published procedure, minor explosions occurred on two occasions after
refluxing had been discontinued. This possibly may have involved formation of
acetaldehyde (which has an AIT of 140°C) and ingress of air into the reaction
vessel as it cooled [1]. Runaway reactions during small scale oxidation of ethanol
have apparently been experienced by many teachers, poor initial mixing or starting
too cool may be the cause [2].
See 2-Propanol, Sulfuric acid, below

Hydrazine
  Mellor, 1943, Vol. 11, 234
  Interaction is explosive.
  See Hydrazine: Oxidants

Hydroxylamine
  See Hydroxylamine: Oxidants

Organic residues, Sulfuric acid   MRH Acetone 1.97/8
  1. *HCS 1980*, 385
  2. Bradshaw, J. R., *Process Biochem.*, 1970, **5**(11), 19
  3. Anon., *CISHC Chem. Safety Summ.*, 1979, **50**, 95
  The well-known 'chromic acid mixture' of dichromate and sulfuric acid [1] for
  cleaning glassware is by design a powerful oxidant, and contact with large amounts
  of tarry or other organic residues in process vessels should be avoided as it may lead
  to a violent reaction. Further, if solvents are first used to clean glassware roughly
  before acid treatment, traces of readily oxidisable solvents must be removed before
  adding the oxidant mixture. In many cases treatment with a properly formulated
  detergent will ensure adequate cleanliness and avoid possible hazard [2]. Addition
  of 1 l of oxidant mixture to a vessel containing residues of acetone or similar
  solvent led to a violent exothermic reaction [3].
  See Chromic acid: Acetone, *also* Chromium trioxide: Acetic acid

Other reactants
  Yoshida, 1980, 177
  MRH values calculated for 16 combinations with oxidisable materials are given.

2-Propanol, Sulfuric acid
  Cochrane, A., private comms., 1982, 1983
  An established school preparation of 2-propanone (acetone) involves the small-
  scale (and rather exothermic) oxidation of the alcohol with dichromate(VI). It was
  observed in several laboratories that when the acidified dichromate solution was
  added to the alcohol in small portions (1–2 cc) rather than dropwise as specified,
  small sparks or incandescent particles were produced which sometimes survived
  long enough to escape from the neck of the flask. This also happened if the alcohol
  and/or the oxidant solution were diluted with extra water, with old or new samples
  of alcohol, and if air were displaced from the flask by carbon dioxide. It is therefore
  important not to exceed the specified dropwise rate of addition of oxidant solution.
  It is very unusual for glowing particles to be produced from a homogeneous liquid
  reaction system.
  See Ethanol, Sulfuric acid, above

Sulfuric acid, Trinitrotoluene
  Clarke, H. T. *et al.*, *Org. Synth.*, 1941, Coll. Vol. 1, 543–544
  During oxidation of TNT in sulfuric acid to trinitrobenzoic acid, stirring of the
  viscous reaction mixture must be very effective to prevent added portions of solid
  dichromate causing local ignition.

1490

See other AGITATION INCIDENTS, SELF-HEATING AND IGNITION INCIDENTS
See other METAL OXOMETALLATES, OXIDANTS

4251. Chromium(III) oxide
[1308-38-9] Cr_2O_3

$$Cr_2O_3$$

Chlorine trifluoride
See Chlorine trifluoride: Metals, etc.

Copper oxide
See COPPER CHROMITE CATALYST

Lithium
See Lithium: Metal oxides

Nitroalkanes
See NITROALKANES: metal oxides

Rubidium acetylide
See Rubidium acetylide: Metal oxides
See other METAL OXIDES

4252. Ammonium trichromate(2−)
[32390-97-1] $Cr_3H_8N_2O_{10}$

$$[NH_4]_2 \ [O(CrO_2O)_2CrO_2O]$$

Mellor, 1943, Vol. 11, 349
It explodes at 190°C.
See other OXOSALTS OF NITROGENOUS BASES

4253. Ammonium tetrachromate(2−)
[54153-83-2] $Cr_4H_8N_2O_{13}$

$$[NH_4]_2 \ [O(CrO_2O)_3CrO_2O]$$

Mellor, 1943, Vol. 11, 352
It decomposes suddenly at 175°C.
See other OXOSALTS OF NITROGENOUS BASES

4254. Caesium
[7440-46-2]   Cs

$$Cs$$

1. Mellor, 1941, Vol. 2, 468; 1963, Vol. 2, Suppl. 2.2, 2291
2. Pepmiller, P. L. *et al., Rev. Sci. Instr.*, 1985, **56**, 1832–1833

It ignites immediately in air or oxygen [1]. A simple device to open glass vials of caesium and dispense the contents safely gives a good margin of safety to the user [2].

Acids
Pascal, 1957, Vol. 3, 94
Most acids react violently, even when anhydrous.

Halogens
Gibson, 1969, 8
Interaction at ambient temperature is violent with all halogens.

Non-metals
Pascal, 1957, Vol. 3, 94
Interaction with sulfur or phosphorus attains incandescence.

Water
1. Mellor, 1941, Vol. 2, 468; 1963, Vol. 2, Suppl. 2.2, 2328
2. Markowitz, M. M., *J. Chem. Educ.*, 1963, **40**, 633–636
Caesium reacts very violently with cold water and the evolved hydrogen ignites [1]. The reactivity of caesium and other alkali metals with water has been discussed in detail [2].
See other METALS, PYROPHORIC METALS

4255. Caesium fluoride
[13400-13-0]   CsF

CsF

Benzenediazonium tetrafluoroborate, Difluoramine
See Difluoramine: Benzenediazonium tetrafluoroborate, etc.
See other METAL HALIDES

4256. Caesium fluoroxysulfate
[70806-67-6]   CsFO$_4$S

CsOSO$_2$OF

Appelman, E. H., *Inorg. Synth.*, 1986, **24**, 22–23
Caesium fluorooxysulfate is thermodynamically unstable. Mild detonations have occurred occasionally during handling of 100 mg portions of salt, usually when deliberate or violent crushing or scraping occurred. It is likely that detonations were brought about by surface impurities on containers or spatulae. Amounts of 10–20 g have been used routinely over several years without detonations, though great care, shields and gloves were always used. It is also a powerful oxidant and reacts violently with organic solvents, including dimethylformamide, dimethyl sulfoxide, ethylene dichloride and pyridine.
See Rubidium fluoroxysulfate
See other ACYL HYPOHALITES, OXIDANTS

4257. Caesium pentafluorotelluramide
[42081-47-0] $\qquad$ CsF$_5$HNTe

CsNHTeF$_5$

Seppelt, K., *Inorg. Chem.*, 1973, **12**, 2838
Heating must be avoided during the preparation or subsequent drying, as occasional explosions occurred. It exploded immediately upon laser irradiation for Raman spectroscopy.
See other IRRADIATION DECOMPOSITION INCIDENTS, *N*-METAL DERIVATIVES

4258. Caesium hydride
[13772-47-9] $\qquad$ CsH

CsH

Oxygen
    Gibson, 1969, 76
    The unheated hydride ignites in oxygen.
    See other METAL HYDRIDES

4259. Caesium hydrogen xenate
[[73378-56-0], Sesquihydrate] $\qquad$ CsHO$_4$Xe

CsOXeO$_2$OH

Alone, or Alcohols
    Jaselskis, B. *et al., J. Amer. Chem. Soc.*, 1966, **88**, 2150; 1969, **91**, 1874
    It is unstable to friction, thermal or mechanical shock, and may explode on contact with alcohols.
    See other XENON COMPOUNDS

4260. Caesium amide
[22205-57-8] $\qquad$ CsH$_2$N

CsNH$_2$

Pez, G. P. *et al., Pure Appl. Chem.*, 1985, **57**, 1917–1926
Caesium amide is extremely pyrophoric and must be handled in an atmosphere virtually free from oxygen and moisture (each below 5 ppm).

Water
    Mellor, 1940, Vol. 8, 256
    Interaction is incandescent in presence of air.
    See other INORGANIC BASES, *N*-METAL DERIVATIVES

4261. Caesium nitrate
[7789-18-6] $CsNO_3$

$$CsNO_3$$

Xenon hexafluoride
See Xenon hexafluoride: Caesium nitrate

Xenon tetrafluoride
See Xenon tetrafluoride: Caesium nitrate
See other METAL NITRATES

4262. Caesium azide
[22750-57-8] CsN_3

$$CsN_3$$

Sulfur dioxide
Kennet, F. A. et al., J. Chem. Soc., Dalton Trans., 1982, 853
The azide ignites in contact with sulfur dioxide at ambient temperature.
See other METAL AZIDES

4263. Caesium trioxide (Caesium ozonide)
[12053-67-7] CsO_3

$$CsO_3$$

Water
Whaley, T. P. et al., J. Amer. Chem. Soc., 1951, **73**, 79
Reaction of caesium or potassium ozonides with water or aqueous acids is violent, producing oxygen and flashes of light.
See other METAL OXIDES

4264. Caesium oxide
[20281-00-9] Cs_2O

$$Cs_2O$$

Ethanol
Pascal, 1957, Vol. 3, 104
Contact of a little alcohol with the oxide may ignite the solvent.

Halogens, or Non-metal oxides
Mellor, 1941, Vol. 2, 487
Above 150–200°C, incandescence occurs with fluorine, chlorine or iodine. In presence of moisture, contact at ambient temperature with carbon monoxide or carbon dioxide causes ignition, while dry sulfur dioxide causes incandescence on heating.

Water
Pascal, 1957, Vol. 3, 104

Incandescence on contact.
See other METAL OXIDES

4265. Caesium selenide
[31052-46-7] Cs_2Se

CsSeCs

Sidgwick, 1950, 92
When warm it ignites in air.
See related METAL SULFIDES

4266. Caesium nitride
[12134-29-1] Cs_3N

Cs_3N

Alone, or Chlorine, or Non-metals
Mellor, 1940, Vol. 8, 99
It burns in air, and is readily attacked by chlorine, phosphorus or sulfur.
See other NITRIDES

4267. Copper
[7440-50-8]   Cu

Cu

Acetylenic compounds
See ACETYLENIC COMPOUNDS: metals

Aluminium, Sulfur
See Aluminium: Copper, Sulfur

3-Bromopropyne
See 3-Bromopropyne: Metals

Complexing agents, Water
Shah, I. U., *Chem. Eng. News*, 1991, **69**(30), 2
A bottle of 'cuprous chloride' solution prepared by standing cupric chloride in strong hydrochloric acid over excess copper burst on standing. In the presence of some complexing agents, copper can react with aqueous media to form hydrogen. Slow pressurisation by this means explains the above explosion (Editor's comments). The metal is also known to dissolve in cyanides and some amine solutions.

Dimethyl sulfoxide, Trichloroacetic acid
See Dimethyl sulfoxide: Copper, Trichloroacetic acid

Ethylene oxide
See Ethylene oxide (reference 3)

Lead azide
See Lead(II) azide: Copper

Other reactants
Yoshida, 1980, 243
MRH values calculated for 11 combinations with oxidants are given.

Oxidants   MRH Sodium chlorate 2.0/36
Mellor, 1941, Vol. 2, 310
Mixtures of finely divided copper with chlorates or iodates explode on friction,
shock or heating.
See Ammonium nitrate: Metals
  Chlorine: Metals
  Chlorine trifluoride: Metals, etc.
  Fluorine: Metals
  Hydrazinium nitrate: Alone, or Metals
  Hydrogen sulfide: Metals
  Potassium dioxide: Metals
  Sulfuric acid: Copper

Water
Zyszkowski, W., *Int. J. Heat Mass Transf.*, 1976, **19**, 849–868
The vapour explosion which occurs when liquid copper is dumped into water has
been studied.
See MOLTEN METAL EXPLOSIONS *See other* METALS

4268. Copper–zinc alloys
  [12019-27-1]   Cu−Zn
                          Cu−Zn

Alkyl halides
See DIALKYLZINCS

Diiodomethane, Ether
  1. Foote, C. S., private comm., 1965
  2. Repic, O. *et al., Tetrahedron Lett.*, 1982, **23**, 2729–2732
Lack of cooling during preparation of the Simmons-Smith organozinc reagent
caused the reaction to erupt. The possibly pyrophoric nature of organozinc
compounds and the presence of ether presents a severe fire hazard [1]. An
alternative, safer method of activating the zinc for the reaction involves use of
ultrasonic irradiation rather than the copper–zinc couple [2].
See other ALLOYS

4269. Copper iron(II) sulfide
[12015-76-8]

$CuFeS_2$

$$CuFeS_2$$

Ammonium nitrate
See Ammonium nitrate: Copper iron(II) sulfide

Water
Gribin, A. A., *Chem. Abs.*, 1943, **37**, 1272_8
A large dump of copper pyrites ore ignited after heavy rain. The thick layer (6–7 m) and absence of ventilation were contributory factors to the accelerating aerobic oxidation which finally led to ignition.
See other METAL SULFIDES, SELF-HEATING AND IGNITION INCIDENTS

4270. Copper(II) azide hydroxide
[22887-15-6]

$CuHN_3O$

$$N_3CuOH$$

Cirulis, A., *Z. Anorg. Chem.*, 1943, **251**, 332–334
It exploded at 203°C or on impact. The dimeric double salt is similarly sensitive.
See related METAL AZIDES

4271. Lithium dihydrocuprate
[53201-99-3]

CuH_2Li

$$Li[CuH_2]$$

Ashby, E. C. *et al., J. Chem. Soc., Chem. Comm.*, 1974, 157–158
The solid is highly pyrophoric, but stable as a slurry in ether at ambient temperature.
See other COMPLEX HYDRIDES

4272. Copper(II) phosphinate
[34461-68-2]

$CuH_4O_4P_2$

$$Cu(OPHOH)_2$$

Mellor, 1940, Vol. 8, 883: 1971, Vol. 8, Suppl. 3, 623
The solid suddenly explodes at about 90°C, and forms impact-sensitive priming mixtures.
See other METAL PHOSPHINATES, REDOX COMPOUNDS

4273. Lithium pentahydrocuprate(4−)
[64010-65-7]

CuH_5Li_4

$$Li_4[CuH_5]$$

Ashby, E. C. *et al., J. Org. Chem.*, 1978, **43**, 187

Of a series of lithium hydrocuprates Li_nCuH_{n+1} ($n = 1-5$), only the title compound (when solvated with THF) appeared a more powerful reducant than lithium tetrahydroaluminate. Safety precautions similar to those adopted for the complex aluminium hydride seem appropriate.
See other COMPLEX HYDRIDES

4274. Tetraamminecopper(II) sulfate
[14283-05-7] $CuH_{12}N_4O_4S$
$$[(H_3N)_4Cu]SO_4$$

Iodine
See Iodine: Tetraamminecopper(II) sulfate
See other AMMINEMETAL OXOSALTS

4275. Tetraamminecopper(II) nitrite
[39729-81-2] $CuH_{12}N_6O_4$
$$[(H_3N)_4Cu][NO_2]_2$$

Mellor, 1940, Vol. 8, 480
The salt is nearly as shock-sensitive as picric acid. When pure it does not explode on heating, but traces of nitrate cause explosive decomposition.
See other AMMINEMETAL OXOSALTS

4276. Tetraamminecopper(II) nitrate
[31058-64-7] $CuH_{12}N_6O_6$
$$[(H_3N)_4Cu][NO_3]_2$$

Explodes at 330°C, high impact-sensitivity.
See entry AMMINEMETAL OXOSALTS (reference 2)

4277. Tetraamminecopper(II) azide
[70992-03-9] $CuH_{12}N_{10}$
$$[(H_3N)_4Cu][N_3]_2$$

Mellor, 1940, Vol. 8, 348; 1967, Vol. 8, Suppl.2, 26
Explosive on heating or impact.
See related METAL AZIDES

4278. Lithium hexaazidocuprate(4−)
[ ] $CuLi_4N_{18}$
$$Li_4[(N_3)_6Cu]$$

Urbanski, 1967, Vol. 3, 185

It is, like copper(II) azide, an exceptionally powerful initiating detonator, containing 61.5% of nitrogen.

See related METAL AZIDES *See other* HIGH-NITROGEN COMPOUNDS

4279. Copper(II) nitrate

[3251-23-8] $\qquad$ CuN$_2$O$_6$

[10031-43-3]; [13478-38-1] hydrates

Cu(NO$_3$)$_2$

HCS 1980, 336

1. Priestley J., *Experiments and Observations on Different Kinds of Air*, 1790, 3, 203, Birmingham

Although the anhydrous nitrate is relatively stable and can be distilled under partial pressure, as is remarked in most recent text books, these do not tell that the hydrate actually obtainable by purchase or by normal preparations is among the least stable and most oxidising of main valence nitrates. It was early observed that paper contaminated with it ignited easily [1].

Acetic anhydride $\qquad$ MRH 3.56/26

See Acetic anhydride: Metal nitrates

Ammonia, Potassium amide

Sorbe, 1968, 80

Interaction gives an explosive precipitate containing the Cu(I) derivatives Cu$_3$N, Cu$_3$N.nNH$_3$, Cu$_2$NH, CuNK$_2$.NH$_3$.

Ammonium hexacyanoferrate(II)

1. Wolski, J. *et al.*, *Z. Chem.*, 1973, **13**, 95–97
2. Wolski, J. *et al.*, *J. Therm. Anal.*, 1973, **5**, 67

At 220°C interaction is explosive in wet mixtures [1], or in dry mixtures if the nitrate is in excess [2].

See other REDOX REACTIONS

Ammonium nitrate

Mellor, 1923, Vol. III, 284

Evaporation of mixed solutions gives violent decomposition during concentration.

Hydrazine, Ethanol

Hofmann, K. A. *et al., Annalen*, 1899, **305**, 222

A blue crystalline complex, assigned a monohydrazino structure, was obtained from the nitrate and hydrazine hydrate in ethanol. It exploded on heating or, sometimes, after thorough drying in a desiccator, on shaking.

See related AMMINEMETAL NITRATES

1-Hydroxybenzotriazole, Ethanol

Anon., *Safety Digest Univ. Safety Assoc.*, 1992, **44**, 24

A mixture of the above was evaporated to dryness, on scraping the residue with a spatula a 'dust explosion', presumably of only a small part of the charge, resulted, covering the experimenter with powder.

See 1-Hydroxybenzotriazole

See other REDOX REACTIONS

See also AMMINEMETAL NITRATES

Organometallic materials

Buffinger, D. R., *et al., J. Coord. Chem.*, 1988, **19**(1–3), 197

A double salt of the nitrate and the copper(II) salt of an organometallic ligand was found to explode at around 100°C. In view of the oxidising powers of the nitrate, complexes of this nature are not likely to be stable.

See ORGANOMETALLIC NITRATES

Other reactants

Yoshida, 1980, 198

MRH values calculated for 19 combinations with oxidisable materials are given.

Potassium hexacyanoferrate(II)

Wolski, J. *et al., Explosivstoffe*, 1969, **17**(5), 103–110

Interaction is explosive at 220°C.

Tin   MRH 2.68/62

Ellern, 1968, 46

Tin foil in contact with a solution of the nitrate may ignite or give sparks.

See other METAL NITRATES

4280. Copper(II) azide
[14215-30-6]   CuN$_6$

$$Cu(N_3)_2$$

1. Mellor, 1940, Vol. 8, 348; 1967, Vol. 8, Suppl. 2.2, 42–50
2. Urbanski, 1967, Vol. 3, 185
3. Jenkins, C. L. *et al., J, Org. Chem.*, 1971, **36**, 3095–3102
4. Sood, R. K. *et al., Chem. Abs*, 1997, **126**, 48910

The azide is very explosive, even when moist. Loosening the solid from filter paper caused frictional initiation. Explosion initiated by impact is very violent, and spontaneous explosion has also been recorded [1]. It is also an exceptionally powerful initiator [2]. Detonation of the azide when dry has been confirmed [3]. Good crystals are considerably more sensitive to shock, friction and electric discharge than is powder [4].

See Sodium azide: Heavy metals

See other METAL AZIDES

4281. Copper(II) oxide
[1317-38-0]   CuO

CuO

Aluminium   MRH 3.89/18
1. Krah, W., *Chem Abs.*, 1978, **89**, 89652
2. Haws, L. D. *et al., Chem. Abs.*, 1979, **90**, 189294
In demonstrating the use of powdered aluminium to reduce the oxide in a thermite-
type reaction, the mixture must be heated behind a safety screen because of the
small explosion produced [1]. Consolidation of the thermite mixture into a high-
density composite gives chemical heat sources which are safe to handle [2].
See Metals, below
See other THERMITE REACTIONS

Anilinium perchlorate
See Anilinium perchlorate: Metal oxides

Barium acetate, Yttrium oxide
See Barium acetate: Copper(II) oxide, etc.

Boron
Mellor, 1946, Vol. 5, 17
The exothermic reaction on heating a mixture melted the glass container.
See other GLASS INCIDENTS

Dichloromethylsilane
See Dichloromethylsilane: Oxidants

Rubidium acetylide
See Rubidium acetylide: Metal oxides

Hydrogen
Read, C. W. W., *School Sci. Rev.*, 1941, **22**(87), 340
Reduction of the heated oxide in a combustion tube by passage of hydrogen caused
a violent explosion. (The hydrogen may have been contaminated with air.)

Hydrogen sulfide   MRH 0.59/30
See Hydrogen sulfide: Metal oxides
Hydrogen trisulfide: Metal oxides

Metals   MRH values below show % of metal
1. Browne, T. E., *School Sci. Rev.*, 1967, **48**(166), 921
2. Vella, A. J., *J. Chem. Educ.* 1994, **71**(4), 328
An attempted thermite reaction with aluminium powder and copper(II) oxide in
place of iron(III) oxide caused a violent explosion. An anonymous comment
suggests that a greater reaction rate and exothermic effect were involved, and adds
that attempted use of silver oxide would be even more violent [1]. An explosion

1501

resulted from heating a mixture of copper(II) oxide and magnesium in a school laboratory. Avoidance of magnesium thermites in teaching is counselled [2].
See Aluminium, above

Aluminium: Metal oxides   MRH 3.89/18

Magnesium: Metal oxides   MRH 4.30/23

Potassium: Metal oxides

Sodium: Metal oxides   MRH 2.05/37

Other reactants
Yoshida, 1980, 127
MRH values calculated for 7 combinations, largely with oxidisable materials, are given.

Phospham
See Phospham: Oxidants

Phthalic anhydride
'Leaflet No. 5', Inst. of Chem., London, 1940
A mixture of the anhydride and anhydrous oxide exploded violently on heating.

Reducants   MRH Hydrazine 1.17/17
Mellor, 1941, Vol. 3, 137
Interaction with hydroxylamine or hydrazine is vigorous.
See other METAL OXIDES

4282. Copper(II) sulfate
[18939-61-2] CuO_4S

$$CuSO_4$$

HCS 1980, 327

Hydroxylamine
See Hydroxylamine: Copper(II) sulfate
See other METAL OXONON-METALLATES

4283. Copper monophosphide
[12517-41-8] CuP

$$CuP$$

See Copper(II) phosphide *See other* METAL NON-METALLIDES

4284. Copper diphosphide
[12019-11-3] CuP_2

$$CuP_2$$

See Copper(II) phosphide *See other* METAL NON-METALLIDES

4285. Copper(II) sulfide
    [1317-40-4]   **CuS**

CuS

Chlorates
    Mellor, 1956, Vol. 2, Suppl. 1, 584
    Copper(II) sulfide explodes in contact with magnesium chlorate, zinc chlorate or
    cadmium chlorate, or with a conc. solution of chloric acid.
    See METAL HALOGENATES

Other reactants
    Yoshida, 1980, 391
    MRH values calculated for 12 combinations with various reagents are given.
    See other METAL SULFIDES

4286. Copper(I) hydride
    [13517-00-5]   **Cu$_2$H$_2$**

HCuCuH

Gibson, 1969, 77
    The dry hydride ignites in air.

Halogens
    Mellor, 1941, Vol. 3, 73
    Ignition occurs on contact with fluorine, bromine or iodine.
    See other METAL HYDRIDES

4287. Copper(I) azide
    [14336-80-2]   **Cu$_2$N$_6$**

N$_3$CuCuN$_3$

1. Mellor, 1940, Vol. 8, 348; 1967, Vol. 8, Suppl. 2.2, 42–50
    2. Urbanski, 1967, Vol. 3, 185
    It is highly endothermic (ΔH_f° (s) 253.1 kJ/mol, 2.40 kJ/g). One of the more
    explosive metal azides, it decomposes at 205°C [1], and is very highly impact-
    sensitive [2].
    See other ENDOTHERMIC COMPOUNDS, METAL AZIDES

4288. Copper(I) oxide
    [1317-39-1]   **Cu$_2$O**

Cu$_2$O

Aluminium
    Moddeman, W. E. *et al., Chem. Abs.*, 1981, **84**, 159186

The role of surface chemistry in reactivity of this thermite combination has been studied.
See other THERMITE REACTIONS

Lithium nitride
See Lithium nitride: Copper(I) chloride, etc.

Peroxyformic acid
See Peroxyformic acid: Metals, etc.
See other METAL OXIDES

4289. Copper(I) nitride
[1308-80-1]
Cu_3N

$$Cu_3N$$

It is mildly endothermic ($\Delta H^°_f$ (s) +74.5 kJ/mol, 0.36 kJ/g).
Mellor, 1940, Vol. 8, 100
It may explode on heating in air.

Nitric acid
See Nitric acid: Copper(I) nitride
See other N-METAL DERIVATIVES

4290. Copper(II) phosphide
[12134-35-9]
Cu_3P_2

$$Cu_3P_2$$

Oxidants
Mellor, 1940, Vol. 8, 839
The powdered phosphide burns vigorously in chlorine. Mixtures with potassium chlorate explode on impact, and with potassium nitrate, on heating. The monophosphide and diphosphide behave similarly.
See Potassium chlorate: Metal phosphides
See other METAL NON-METALLIDES

†4291. Deuterium oxide
[7789-20-0]
D_2O

$$D_2O$$

1. Anderson, C., *Nature*, 1992, **355**(6356), 102
2. Baum, R., *Chem. Eng. News*, 1992, **70**(26), 8
3. Zhang, X. *et al.*, *Front. Sci. Ser.*, 1993, (4[Frontiers of Cold Fusion]), 381
Electrolysis of heavy water over palladium electrodes is claimed to give inexplicable energy release — 'Cold Fusion'. It can certainly kill [1], even though the mechanism be ignition of an explosive mixture of deuterium and oxygen resulting

from past electrolysis after a platinum recombination catalyst for the off-gases failed and a safety valve blocked. The cell then exploded when the experimenter picked it up; ignition presumably at a catalytic surface exposed by moving liquid [2]. Others hold out for less probable causes [3].

Pentafluorophenyllithium
  See Pentafluorophenyllithium: Deuterium oxide
  See Diprotium monoxide
  See other NON-METAL OXIDES

4292. Europium
[7740-53-1]   Eu

Eu

Bailar, 1973, Vol. 4, 69
Europium is the most reactive lanthanide metal, and may ignite on exposure to air if finely divided.
See LANTHANIDE METALS: oxidants
See other METALS, PYROPHORIC METALS

4293. Europium(II) sulfide
[12020-65-4]   EuS

EuS

Sidgwick, 1950, 454
Pyrophoric in air.
See other METAL SULFIDES, PYROPHORIC MATERIALS

4294. Hydrogen fluoride
[7664-39-3]   FH

HF

(*MCA SD-25*, 1970); Soln. *NSC 459*, 1978; *HCS 1980*, 543; Gas *HCS 1980*, 548

1. Keen, M. J. *et al., Chem. & Ind.*, 1957, 805
2. Braker, 1980, 391
3. Anon., *Lab. Hazards Bull.*, 1991, **11**(09), 556
4. Watson, E. F., *Chem. Eng. News*, 1997, **75**(17), 28, 6

Handling precautions for the gas or anhydrous liquid are detailed [1], and a polythene condenser for disposal of hydrogen fluoride is described [2]. A steel cylinder of the nominally anhydrous material exploded after 42 years. This is attributed to ppm levels of water, which catalyses reaction with the steel to produce hydrogen, slowly increasing the pressure. It is recommended that cylinders should be vented at yearly intervals, and those more than two years old returned to the manufacturer [3]. More warnings against prolonged storage in steel are given [4].

Bismuthic acid
  See Bismuthic acid: Hydrofluoric acid

Cyanogen fluoride
  See Cyanogen fluoride: Hydrogen fluoride

Glycerol, Nitric acid
  See Nitric acid: Glycerol, etc.

Mercury(II) oxide, Organic materials
  Ormston, J., *School Sci. Rev.*, 1944, **26**(98), 32
  During the fluorination of organic materials by passing hydrogen fluoride into a vigorously stirred suspension of the oxide (to form transiently mercury difluoride, a powerful fluorinator), it is essential to use adequate and effective cooling below 0°C to prevent loss of control of the reaction system.

Metal alloys
  Young, J. A., *Chem. Health and Safety*, 1995, **2**(2), 6
  It is reported that an aluminium cleaner containing low concentrations of hydrofluoric acid can generate stibine from antimony containing bearing-metal alloys, to the permanent detriment of the health of nearby workers. Presumably arsine could appear from arsenic containing alloys; both are gases and extremely toxic.

Methanesulfonic acid
  Gramstad, T. *et al., J. Chem., Soc.*, 1956, 173–180
  Electrolysis of a mixture produced oxygen difluoride which exploded.
  See Oxygen difluoride

Nitric acid, Lactic acid
  See Nitric acid: Hydrofluoric acid, Lactic acid

Nitric acid, Propylene glycol
  See Nitric acid: Hydrofluoric acid, Propylene glycol

Other reactants
  Yoshida, 1980, 311
  MRH values calculated for 7 combinations with various reagents are given.

Oxides   MRH Calcium oxide 2.84/59
  Mellor, 1956, Vol. 2, Suppl. 1, 122; 1939, Vol. 9. 101
  Arsenic trioxide and calcium oxide incandesce in contact with liquid hydrogen fluoride.

N-Phenylazopiperidine
  See N-Phenylazopiperidine: Hydrofluoric acid

Phosphorus(V) oxide
  See Tetraphosphorus decaoxide: Hydrogen fluoride

Potassium permanganate
See Potassium permanganate: Hydrofluoric acid

Potassium tetrafluorosilicate(2−)
Mellor, 1956, Vol. 2, Suppl. 1, 121
Contact with liquid hydrogen fluoride causes violent evolution of silicon tetrafluoride. (The same is probably true of metal silicides and other silicon compounds generally.)
See Dialuminium octavanadium tridecasilicide

Sodium   MRH 6.94/53
See Sodium: Acids

Sulfuric acid
See Sulfuric acid: Hydrofluoric acid
See other INORGANIC ACIDS, NON-METAL HALIDES, NON-METAL HYDRIDES

4295. Fluoroselenic acid
[14986-53-9] FHO_3Se
$$FSeO_2OH$$

Cellulose
Bartels, H. *et al.*, *Helv. Chim. Acta*, 1962, **45**, 179
A strong oxidant which reacts violently with filter paper or similar organic matter, igniting it if dry.
See other INORGANIC ACIDS, OXIDANTS

4296. Fluoramine (Fluoramide)
[15861-05-9] FH_2N
$$FNH_2$$

1. Hoffmann, C. J., *et al.*, *Chem. Rev.*, 1962, **62**, 7
2. Minkwitz, R. *et al.*, *Z. Naturforsch. B, Chem. Sci.*, 1988, **43B**, 1478−1480
The impure material is very explosive [1]. The pure material, prepared by a thermolytic method, is stable as a solid at −103°C, but on melting it decomposes to ammonium hydrogen difluoride and nitrogen [2].
See other N-HALOGEN COMPOUNDS

4297. Fluorophosphoric acid
[13537-32-1] FH_2O_3P
$$FP(O)(OH)_2$$

Sodium tetrahydroborate
See Sodium tetrahydroborate: Acids
See other INORGANIC ACIDS

4298. Fluorosilane
[13537-33-2]   FH$_3$Si

FSiH$_3$

Azidogermane
See Azidogermane: Fluorosilane
See other HALOSILANES

4299. Ammonium fluoride
[16099-75-5]   FH$_4$N

NH$_4$F

Chlorine trifluoride
See Chlorine trifluoride: Ammonium fluoride
See related METAL HALIDES

4300. Potassium fluoride hydrogen peroxidate
[32175-44-3]   FK.H$_2$O$_2$

KF.H$_2$O$_2$

See entry CRYSTALLINE HYDROGEN PEROXIDATES

4301. Manganese fluoride trioxide
[22143-20-0]   FMnO$_3$

FMnO$_3$

Alone, or Organic compounds, or Water
Engelbrecht, A. *et al., J. Amer. Chem. Soc.*, 1954, **76**, 2042–2044
It decomposes, usually explosively, above 0°C or in contact with moisture. It is a
powerful oxidant and reacts violently with organic compounds.
See related METAL HALIDES, METAL OXIDES *See other* OXIDANTS

4302. Nitrosyl fluoride
[7789-25-5]   FNO

O:NF

Haloalkene (unspecified)
MCA Case History No. 928
Interaction of a mixture in a pressure vessel at −78°C caused it to rupture when
moved from the cooling bath.

Metals, or Non-metals
1. Schmutzler, R., *Angew. Chem. (Intern. Ed.)*, 1968, **7**, 442
2. Pascal, 1956, Vol. 10, 346

Antimony, bismuth, arsenic, boron, red phosphorus, silicon [1] and tin [2] all react with incandescence.

Oxygen difluoride
Ruff, O. *et al.*, *Z. Anorg. Chem.*, 1932, **208**, 293
Explosion occurs on mixing, even at low temperatures.

Sodium
Mellor, 1940, Vol. 8, 612
Reaction is incandescent.
See other N-HALOGEN COMPOUNDS, NON-METAL HALIDES, OXIDANTS

4303. Nitryl fluoride
[10022-50-1]   FNO₂

$$O_2NF$$

Metals
Aynsley, E. E. *et al.*, *J. Chem. Soc.*, 1954, 1122
When nitryl fluoride is passed at ambient temperature over molybdenum, potassium, sodium, thorium, uranium or zirconium, glowing or white incandescence occurs. Mild warming is needed to initiate similar reactions of aluminium, cadmium, cobalt, iron, nickel, titanium, tungsten, vanadium or zinc, and 200–300°C for lithium or manganese.

Non-metals
Aynsley, E. E. *et al.*, *J. Chem. Soc.*, 1954, 1122
Boron and red phosphorus glow in the fluoride at ambient temperature, while hydrogen explodes at 200–300°C. Carbon and sulfur are also attacked.

Sodium azide
See Sodium azode: Nitryl fluoride
See other N-HALOGEN COMPOUNDS, NON-METAL HALIDES, OXIDANTS

4304. Nitryl hypofluorite ('Fluorine nitrate')
[7789-26-6]   FNO₃

$$O_2NOF$$

1. Schmutzler, R., *Angew. Chem. (Intern. Ed.)*, 1968, **7**, 454
2. Engelbrecht, A., *Monatsh.*, 1964, **95**, 633
3. Lustig, M. *et al.*, *Adv. Fluorine Chem.*, 1973, 7, 189
4. Christe, K. O. *et al.*, *Inorg. Chem.*, 1987, **56**, 920

It is mildly endothermic (ΔH°_f (g) +10.4 kJ/mol, 0.12 kJ/g) but a powerful oxidant.

It is a toxic colourless gas which is dangerously explosive in the gaseous, liquid and solid states [1]. It is produced during electrolysis of nitrogenous compounds in hydrogen fluoride [2]. Later work (perhaps with purer material?) did not show the explosive instability [3]. The shock-sensitivity is confirmed [4].

Gases
Hoffmann, C. J., *Chem. Rev.*, 1964, **64**, 94
Immediate ignition in the gas phase occurs with ammonia, dinitrogen oxide or hydrogen sulfide.

Organic materials
Brauer, 1963, 189
The very powerful liquid oxidant explodes when vigorously shaken, or immediately on contact with alcohol, ether, aniline or grease. It is also sensitive in the vapour or solid state (but see reference 3 above).
See Nitryl hypochlorite
See other ENDOTHERMIC COMPOUNDS, OXIDANTS

4305. Sulfur oxide-(*N*-fluorosulfonyl)imide (Sulfinylsulfamoyl fluoride)
[16829-30-4] FNO_3S_2

$$O:S:NSO_2F$$

Water
Roesky, H. W., *Angew. Chem. (Intern. Ed.)*, 1967, **6**, 711
It reacts explosively with water at ambient temperature, but smoothly at $-20°C$.
See other ACYL HALIDES, N–S COMPOUNDS

4306. Thiazyl fluoride
[18820-63-8]   FNS

$$N\equiv SF$$

Alkylbutadienes
Bludssus, W. *et al., J. Chem. Soc., Chem. Comm.*, 1979, 35–36
Interaction with, for example, 2,3-dimethylbutadiene is explosive.
See other NON-METAL HALIDES, N–S COMPOUNDS

4307. Fluorine azide
[14986-60-8] FN_3

$$FN=N^+=N^-$$

1. Bauer, S. H., *J. Amer. Chem. Soc.*, 1947, **69**, 3104
2. Gholivand, K. *et al., Inorg. Chem.*, 1987, **26**, 2137–2140
This unstable material usually explodes on vaporisation (at $-82°C$) [1]. It is extremely explosive in the liquid and solid states. A safe method has been developed for preparing the pure gas on 20 mg scale. It may be stored safely at $10-20$ mbar/$-80°C$ for several months. Cooling the gas to $-196°C$, or evaporation of the liquid at a fast rate may lead to very violent explosions. Fluorine azide is now described as triazadienyl fluoride [2], as shown above.
See other HALOGEN AZIDES

1510

4308. Fluorothiophosphoryl diazide
[38005-27-5] FN_6PS

$$FP(:S)(N_3)_2$$

O'Neill, S. R. *et al., Inorg. Chem.*, 1972, **11**, 1630
The explosion of the glassy material at $-183°C$ was attributed to crystallisation
of the glass.
See other ACYL AZIDES

4309. Rubidium fluoroxysulfate
[[73347-64-5] (ion)] FO_4RbS

$$RbOSO_2OF$$

Appelman, E. H. *et al., J. Amer. Chem. Soc.*, 1979, **101**, 388
This and the caesium analogue, both powerful oxidants, detonate mildly at 100°C,
evolving oxygen.
See other ACYL HYPOHALITES, OXIDANTS

4310. Fluorine
[7782-41-4] F_2

$$F_2$$

FPA H79, 1979 (cylinder); *HCS 1980*, 503 (cylinder)

1. Kirk-Othmer, 1966, Vol. 9, 506
2. Gall, J. F. *et al., Ind. Eng. Chem.*, 1947, **39**, 262
3. Landau, R. *et al., Ind. Eng. Chem.*, 1947, **39**, 262
4. Turnbull, S. G. *et al., Ind. Eng. Chem.*, 1947, **39**, 286
5. Long, G., *Apparatus for Disposal of Fluorine on a Laboratory Scale*, Harwell,
    AERE, 1956
6. Gordon, J. *et al., Ind. Eng. Chem.*, 1960, **52**(5), 63A
7. Farrar, R. L. *et al., Some Considerations in the Handling of Fluorine and the
    Chlorine Fluorides*, Union Carbide Oak Ridge Rept. K/ET-252 for US DoE,
    1979
8. Ring, R. J. *et al., Review of Fluorine Cells and Production*, Lucas Heights,
    Aus. At. En. Comm., 1973
9. Bailar, 1973, Vol. 2, 1015–1019
10. DesMarteau, D. D. *et al., J. Amer. Chem. Soc.*, 1987, **109**, 7194–7196

Fluorine is the most electronegative and reactive element known, reacting, often
violently, with most of the other elements and their compounds (note the large
MRH values quoted below). Handling hazards and disposal of fluorine on a laboratory scale are adequately described [1,2,3,4,5][6], and a more general review
is also available [7]. Safety practices associated with the use of laboratory- and
industrial-scale fluorine cells and facilities have been reviewed [8]. Equipment
and procedures for the laboratory use of fluorine and volatile fluorides have been

detailed [9]. A new series of N-fluorosulfonimides shows promise as milder and safer reagents than elemental fluorine for aromatic fluorination [10].

Acetonitrile, Chlorine fluoride
Sekiya, A. *et al., Inorg. Chem.*, 1981, **20**, 1–2
When fluorine was condensed onto acetonitrile and chlorine fluoride frozen at −196°C, a small explosion occurred in the reactor.

Acetylene   MRH 11.92/41
See Acetylene: Halogens

Alkanes, Oxygen
Von Elbe, G., US Pat. 3 957 883, 1976
Interaction of propane, butane or 2-methylpropane with fluorine and oxygen produces peroxides. Appropriate reaction conditions are necessary to prevent explosions.

Ammonia   MRH 10.25/23
1. Mellor, 1940, Vol. 8, 216
2. Müller, W., *Umsch. Wiss. Tech.*, 1974, **74**(15), 485–486
Ammonia ignites in contact with fluorine [1], and the anhydrous liquids have been used as a propellant pair [2]. Aqueous ammonia ignites or explodes on contact [1].
See Water, below

Boron nitride
Moissan, 1900, 232
Unheated interaction leads to incandescence.

Caesium fluoride, Fluorocarboxylic acids
Sekiya, A. *et al., Inorg. Chem.*, 1980, **19**, 1329
Low temperature fluorination of fluorocarboxylic acids to give explosive 1,1-bis(fluoroxy)perfluoroalkanes occasionally led to explosive reactions. Thus difluoroacetic acid led to explosion at −195°, and perfluorosuccinic acid at −111° or −20°C.
See BIS(FLUOROOXY)PERHALOALKANES

Caesium heptafluoropropoxide
1. Gumprecht, W. H., *Chem. Eng. News*, 1965, **43**(9), 36
2. *MCA Case History No. 1045*
Fluorination of caesium heptafluoropropoxide at −40°C with nitrogen-diluted fluorine exploded violently after 10 h. This may have been caused by ingress of moisture, formation of some pentafluoropropionyl derivative and conversion of this to pentafluoropropionyl hypofluorite, known to be explosive if suitably initiated. Other possible explosive intermediates are peroxides or peresters.
See Pentafluoropropionyl hypofluorite

Ceramic materials
Farrar, R. L. *et al., US DoE Rept. K/ET-252*, 1979, 9
Even finely divided ceramic materials may be ignited in fluorine.

Covalent halides
1. Mellor, 1940, Vol. 2, 12; 1956, Vol. 2, Suppl. 1, 64; 1940, Vol. 8, 995, 1003, 1013
2. Leleu, *Cahiers*, 1973, (73), 509; 1974, (74), 427
Chromyl chloride at high concentration ignites in fluorine, while phosphorus penta-chloride, phosphorus trichloride and phosphorus trifluoride ignite on contact [1]. Boron trichloride ignites in cold fluorine, and silicon tetrachloride on warming [2].

Cyanogen (Dicyanogen)
Klapötke, T. M. *et al., Angew. Chemie (Int.)*, 1991, **30**(11), 1485
An explosion was experienced on allowing a mix of these reagents, with arsenic pentafluoride, which had been uv irradiated at −196°C, to warm to −85°C.
See Halogens, below

Cyanoguanidine
Rausch, D. A. *et al., J. Org. Chem.*, 1968, **33**, 2522
The products, perfluoro-1-aminomethylguanidine and perfluoro-*N*-aminomethyltri-aminomethane, and by-products of the reaction of fluorine with cyanoguanidine are extremely explosive in gas, liquid and solid states.
See Sodium dicyanamide, below

1- or 2-Fluoriminoperfluoropropane
Sekiya, A. *et al., J. Fluorine Chem.*, 1981, **17**, 463–468
Interaction is explosive.

Graphite
1. Lagow, R. J. *et al., J. Chem. Soc., Dalton Trans.*, 1974, 1270
2. Ruff, O. *et al., Z. Anorg. Chem.*, 1934, **217**, 1
3. Simons, J. H. *et al., J. Amer. Chem. Soc.*, 1939, **61**, 2962–2966
4. Farrar, R. L. *et al., US DoE Rpt. K/ET-252*, 1979, 33–34
5. Kita, Y. *et al., Chem. Abs.*, 1983, **99**, 107531
During interaction at ambient temperature in a bomb to produce poly(carbon monofluoride), admission of fluorine beyond a pressure of 13.6 bar must be extremely slow and carefully controlled to avoid a violently exothermic explosion [1]. Previously it had been shown that explosive interaction of carbon and fluorine was due to the formation and decomposition of the graphite intercalation compound, poly(carbon monofluoride) [2]. Presence of mercury compounds prevents explosion during interaction of charcoal and fluorine [3]. Reaction of surplus fluorine with graphite or carbon pellets was formerly used as a disposal method, but is no longer recommended. Violent reactions observed when an exhausted trap was opened usually involved external impact on the metal trap, prodding the trap contents to empty the trap, or possibly ingress of moist air

[4]. Removal of higher fluorocarbons (above C_4) from the circulating gas stream prevents explosive decomposition of the graphite fluoride [5].

Halocarbons   MRH Carbon tetrachloride 2.34/67, chloroform 3.77/56
1. Mellor, 1956, Vol. 2, Suppl. 1, 198
2. Schmidt, 1967, 82
3. Fletcher, E. A. *et al., Combust. Inst. Paper WSS/CI-67-23*, 1967
4. Moissan, 1900, 241
5. Farrar, R. L. *et al., US DoE Rept. K/ET-252*, 1979, 36–40
The violent or explosive reactions which carbon tetrachloride, chloroform, etc., exhibit on direct local contact with gaseous fluorine [1], can be moderated by suitable dilution, catalysis and diffused contact [2]. Combustion of perfluorocyclobutane–fluorine mixtures was detonative between 9.04 and 57.9 vol% of the halocarbon [3]. Iodoform reacts very violently with fluorine owing to its high iodine content [4]. Explosive properties of mixtures with 1,2-dichlorotetrafluoroethane have been studied [5].
See Poly(tetrafluoroethylene), etc., below

Halogens, or Dicyanogen   MRH values below references
1. Mellor, 1940, Vol. 2, 12
2. Sidgwick, 1950, 1148
3. Tari, I. *et al., Inorg. Chem.*, 1979, **18**, 3205
MRH Bromine 1.38/89, chlorine 1.67/38, iodine 3.89/43
    While bromine, iodine and dicyanogen all ignite in fluorine at ambient temperature [1], a mixture of chlorine and fluorine (containing essential moisture) needs sparking before ignition occurs, though an explosion immediately follows [2]. Heating fluorine and chlorine in a Monel pressure vessel gives contained explosions at around 100°C [3].
See Cyanogen, above

Hexalithium disilicide
Mellor, 1940, Vol. 6, 169
It incandesces on warming in fluorine.

Hydrocarbons   MRH Anthracene 7.57/49, acetylene 11.92/41
1. Mellor, 1940, Vol. 2, 11; 1956, Vol. 2, Suppl.1, 198
2. Sidgwick, 1950, 1117
3. Schmidt, 1967, 82
4. Moissan, 1900, 240–241
Violent explosions occur when fluorine directly contacts liquid hydrocarbons, even at −210 with anthracene or turpentine, or solid methane at −190°C with liquid fluorine. Many lubricants ignite in fluorine [1,2]. Contact and reaction under carefully controlled conditions with catalysis can now be effected smoothly [3]. Gaseous hydrocarbons (town gas, methane) ignite in contact with fluorine, and mixtures with unsaturated hydrocarbons (ethylene, acetylene) may explode on exposure to sunlight. Each bubble of fluorine passed through benzene causes ignition, but a rapid stream may lead to explosion [4].

Hydrofluoric acid
See Hydrogen halides, below

Hydrogen   MRH 13.39/5
1. Mellor, 1940, Vol. 2, 11; 1956, Vol. 2, Suppl.1, 55
2. Sidgwick, 1950, 1102
3. Kirshenbaum, 1956, 46
4. Gmelin, 1980, Fluorine, Suppl. Vol. 2
5. Truby, F. K., *J. Appl. Phys.*, 1978, **49**, 3481–3484
The violently explosive reactions which sometimes occur when the 2 elements
come into contact under conditions ranging from solid fluorine and liquid hydrogen
at −252°C to the mixed gases at ambient temperature [1] are caused by the catalytic
effects of impurities or the physical nature of the walls of the containing vessel
[2]. Even in absence of such impurities, spontaneous explosions still occur in the
range of 75–90 mol% fluorine in the gas phase [3]. Recent data, including the
inhibiting effect of small amounts of oxygen, has been reviewed [4], but presence
of relatively large amounts does not necessarily inhibit explosions [5].

Hydrogen, Oxygen
Getzinger, R. W., *Rept. LA-5659*, Los Alamos Sci. Lab., 1974
The conditions under which mixtures of the gases at 1 bar and ambient temperature
will react non-explosively have been studied.

Hydrogen fluoride, Seleninyl fluoride
Seppelt, K., *Angew. Chem. (Intern. Ed.)*, 1972, **11**, 630
Preparation of pentafluoroorthoselenic acid from the above reagents in an autoclave above ambient temperature caused occasional explosions. A safer alternative
preparation is described.

Hydrogen halides   MRH HBr 2.30/8, HCl 3.18/65, HI 1.97/87
Mellor, 1940 Vol. 2, 12
Hydrogen bromide, hydrogen chloride and hydrogen iodide ignite in contact with
fluorine, and the conc. aqueous solutions, including that of hydrogen fluoride, also
produce flame.

Hydrogen sulfide
See Sulfides, below

Ice
1. English, W. D. *et al.*, *Cryog. Technol.*, 1965, **1**, 260
2. *UK Scientific Mission Report 68/79*, Washington, UKSM, 1968
Mixtures of liquid fluorine and ice are highly impact-sensitive, with a power
comparable to that of TNT. Contact of moist air or water with liquid fluorine
can thus be very hazardous [1,2].
See Water, below

Metal acetylides and carbides
Mellor, 1946, Vol. 5, 849, 885, 890–891

Monocaesium acetylide and caesium acetylide, lithium acetylide and rubidium acetylide, tungsten carbide and ditungsten carbide, and zirconium dicarbide all ignite in cold fluorine, while uranium dicarbide ignites in warm fluorine.

Metal borides
Bailar, 1973, Vol. 1, 729
Interaction frequently attains incandescence.

Metal cyanocomplexes
Moissan, 1900, 228
Potassium hexacyanoferrate(II), lead hexacyanoferrate(III) and potassium hexacyanoferrate(III) become incandescent in fluorine, and the liberated dicyanogen also ignites.

Metal hydrides
Mellor, 1940, Vol. 2, 12, 483; 1956, Vol. 2, Suppl. 1, 56; 1941, Vol. 3, 73
Copper hydride, potassium hydride and sodium hydride all ignite on contact with fluorine at ambient temperature.

Metal iodides
Moissan, 1900, 227
Fluorine decomposes calcium iodide, lead iodide, mercury iodide and potassium iodide at ambient temperature, and the liberated iodine ignites, evolving much heat.

Metal oxides   MRH Calcium oxide 6.15/60
Mellor, 1940, Vol. 2, 13, 469; 1941, Vol. 3, 663; 1942, Vol. 13, 715; 1936, Vol. 15, 380, 399
Oxides of the alkali and alkaline earth metals and nickel(II) oxide incandesce in cold fluorine, and iron(II) oxide when warmed. Nickel(IV) oxide also burns in fluorine.

Metals   MRH Caesium 3.43/28, calcium 15.56/51, copper 3.01/76, lead 3.39/76, lithium 23.51/27, magnesium 17.78/39, manganese 8.91/49, molybdenum 7.57/46, potassium 9.58/68, rubidium 5.19/82, sodium 13.43/55, uranium 6.02/72, zinc 2.45/63
1. Mellor, 1940, Vol. 2, 13, 469; 1941, Vol. 3, 638; 1940, Vol. 4, 267, 476; 1946, Vol. 5, 421; 1939, Vol. 9, 379, 891; 1943, Vol. 11, 513, 730; 1942, Vol. 12, 344; 1942, Vol. 15, 675
2. Schmidt, 1967, 78–80
3. Kirk-Othmer, 1966, Vol. 9, 507–508
The vigour of reaction is greatly influenced by the state of sub-division of the metals involved. Massive calcium, moist magnesium, manganese powder, molybdenum powder, potassium, sodium, rubidium and antimony all ignite in cold fluorine gas. Warm tantalum powder or cold thallium ignites on contact with fluorine. Fine copper wire (as wool) ignites at 121°C, and osmium and tin begin to burn at 100°C, while iron powder (100-mesh, but not 20-mesh) ignites in liquid

fluorine. Titanium will ignite if impacted under the liquid at −188°C and has ignited in presence of catalysts in the gas at −80°C, but in all cases the fluoride film prevents further propagation. Tungsten and uranium powders ignite in the gas without heating, while zinc ignites at about 100°C. Molybdenum, tungsten and Monel wires ignited in atmospheric fluorine at 205, 283 and 396°C (averaged values), respectively. Generally, strongly electropositive metals, or those forming volatile fluorides are attacked the most vigorously.

Metal salts   MRH Potassium dichromate 2.97/66, potassium permanganate 3.18/67, sodium chlorate 2.34/85, sodium nitrate 1.34/56
1. Mellor, 1940, Vol. 2, 13; 1956, Vol. 2, Suppl. 1, 63
2. Schmidt, 1967, 83–84
3. Moissan, 1900, 228–239
4. Pascal, 1960, Vol. 16, 57–60
5. Fichter, F. *et al., Helv. Chim. Acta*, 1930, **13**, 99–102
Calcium carbonate, lead carbonate, basic lead carbonate and sodium carbonate all ignite and burn fiercely in contact with fluorine. Chlorides and cyanides are vigorously attacked by cold fluorine, including lead fluoride and thallium(I) chloride, both of which become molten. Mercury(II) cyanide ignites in fluorine when warmed gently, and silver cyanide reacts explosively when cold [1]. Sodium metasilicate ignites in fluorine [2]. Unheated calcium phosphate [3], sodium thiosulfate and sodium diphosphate [4] all incandesce in contact with fluorine, and barium thiocyanate or mercury thiocyanate ignites [3]. On warming, chromium(III) chloride [4], calcium arsenate or copper borate incandesce, and sodium arsenate ignites [3]. Introduction of fluorine into solutions of silver fluoride, silver nitrate, silver perchlorate or silver sulfate causes violent exothermic reactions to occur, with liberation of ozone-rich oxygen [5].

Metal silicides
Mellor, 1940, Vol. 6, 169, 178
Calcium disilicide readily ignites, and lithium hexasilicide becomes incandescent, when warmed in fluorine.

Miscellaneous materials
1. Mellor, 1940, Vol. 2, 13
2. Schmidt, 1967, 84, 110
3. Lafferty, R. H. *et al., Chem. Eng. News*, 1948, **26**, 3336
Town gas ignites in contact with gaseous fluorine, as does a mixture of lead oxide and glycerol (formerly used as a jointing compound) [1]. Spillage tests involving action of liquid fluorine alone or as a 30% solution in liquid oxygen caused asphalt and crushed limestone to ignite, and coke and charcoal to burn, the latter brilliantly, while JP4 liquid hydrocarbon fuel produced violent explosions and a large fireball. Humus-rich soil also burned with a bright flame [2]. Immersion of various glove materials in liquid fluorine was examined. Cotton exploded violently and Neoprene slightly with ignition, while leather charred but did not ignite [3].

Nitric acid   MRH 0.50/53
  1. Cady, G. H., *J. Amer. Chem. Soc.*, 1934, **56**, 2635
  2. Schmutzler, R., *Angew. Chem. (Intern. Ed.)*, 1968, **8**, 453
Interaction of fluorine with either the concentrated or very dilute acid caused
explosions, while use of 4 N acid did not [1]. Later and safer methods of preparing
nitryl hypofluorite are summarised [2].

Nitrogenous bases   MRH Dimethylamine 10.63/23, pyridine 8.28/45
  Hoffman, C. J., *Chem. Rev.*, 1962, **62**, 12
Aniline, dimethylamine and pyridine incandesce on contact with fluorine.
See Ammonia, and Cyanoguanidine, both above

Non-metal oxides   MRH Carbon monoxide 6.44/42, dinitrogen tetraoxide 1.00/45,
                          nitrogen oxide 2.93/99+, sulfur dioxide 5.10/36
  1. Mellor, 1940, Vol. 2, 11; 1939, Vol. 9, 101
  2. Hoffman, C. J. *et al., Chem. Rev.*, 1962, **62**, 10
  3. Pascal, 1960, Vol. 16, 53, 62
  4. Moissan, 1900, 138
Arsenic trioxide reacts violently and nitrogen oxide ignites in excess fluorine.
Bubbles of sulfur dioxide explode separately on contacting fluorine, while addition of the latter to sulfur dioxide causes an explosion at a certain concentration [1].
Reaction of fluorine with dinitrogen tetraoxide usually causes ignition [2]. Interaction with carbon monoxide may be explosive. Anhydrous silica incandesces in
the gas, and interaction with liquid fluorine at −80°C is explosive [3,4]. Boron
trioxide also incandesces in the gas [3].
See Bis(fluoroformyl) peroxide

Non-metals   MRH Arsenic 7.19/57, silicon 14.73/27, sulfur 8.28/22
  1. Mellor, 1940, Vol. 2, 11, 12: 1956, Vol. 2, Suppl. 1, 60; 1946, Vol. 5, 785, 822;
     1940, Vol. 6, 161; 1939, Vol. 9, 34; 1943, Vol. 11, 26
  2. Schmidt, 1967, 52, 107
  3. Pascal, 1960, Vol. 16, 58
  4. Moissan, 1900, 125–128
Boron, phosphorus (yellow or red), selenium, tellurium and sulfur all ignite in
contact with fluorine at ambient temperature, silicon attaining a temperature above
1400°C [1]. The reactivity shown by various forms of carbon (charcoal, lampblack,
soot) all of which ignite and burn vigorously in fluorine [1] has been reported to be
due to presence of various impurities, moisture and hydrocarbons [1,2]. Carefully
purified carbon (massive graphite) is inert to fluorine at ambient or slightly elevated
temperatures for a short period but may then react explosively [2]. Phosphorus [3]
and sulfur incandesce in liquid fluorine, and sulfur ignites even at −188°C [4].
See Graphite, above

Other reactants
  Yoshida, 1980, 307–310
MRH values calculated for 58 combinations with a very wide variety of other
reagents are given. Many of the values are extremely high.

1518

Oxygenated organic compounds

MRH Acetaldehyde 8.33/19, butanol 9.12/33, dimethylformamide 8.28/43, methanol 8.37/22

1. Pascal, 1960, Vol. 16, 65
2. Moissan, 1900, 242–245
3. Mellor, 1940, Vol. 2, 13

Methanol, ethanol and 3-methylbutanol [1], acetaldehyde, trichloroacetaldehyde [2] and acetone [3] all ignite in contact with gaseous fluorine. Lactic acid, benzoic acid and salicylic acid ignite, while gallic acid becomes incandescent. Ethyl acetate and methyl borate ignite in fluorine [2].

Perchloric acid
See Perchloric acid: Fluorine

Phosphorus halides
See Covalent halides, above

Polymeric materials, Oxygen
Schmidt, 1967, 87

Various polymeric materials were tested statically with both gaseous and liquefied mixtures of fluorine and oxygen containing from 50 to 100% of the former. The materials which burned or reacted violently were: phenol–formaldehyde resins (Bakelite); polyacrylonitrile–butadiene (Buna N); polyamides (Nylon); polychloroprene (Neoprene); polyethylene; polytrifluoropropylmethylsiloxane (LS63); polyvinyl chloride–vinyl acetate (Tygan); polyvinylidene fluoride–hexafluoropropylene (Viton); polyurethane foam. Under dynamic conditions of flow and pressure, the more resistant materials which burned were: chlorinated polyethylenes, polymethyl methacrylate (Perspex); polytetrafluoroethylene (Teflon).

Polytetrafluoroethylene
Appelman, E. H., *Inorg. Chem.*, 1969, **8**, 223

Teflon tubing, when used to conduct fluorine into a reaction mixture, sometimes ignites. Combustion stops when the flow of fluorine is shut off.

Polytetrafluoroethylene, Trichloroethylene
Lawler, A. E., *Advan. Cryog. Eng.*, 1967, **12**, 780–783

Tests showed that Teflon gaskets containing more than 0.35 wt% of sorbed trichloroethylene were potentially hazardous in contact with liquid fluorine.
See Halocarbons, above

Potassium chlorate
See Potassium chlorate: Fluorine

Potassium hydroxide
1. Fichter, F. *et al., Helv. Chim. Acta*, 1927, **10**, 551
2. Kacmarek, A. J. *et al., Inorg. Chem.*, 1962, **1**, 659–661
3. Bailar, 1973, Vol. 2, 789–791

Interaction at $-20°C$ produces potassium trioxide, a spontaneously explosive solid [1]. Later references suggest that these compounds are not spontaneously explosive [2,3].

Purge gases
Farrar, R. L. et al., US DoE Rept. K/ET-252, 1979, 36
Purge gases used with liquid fluorine ($-188°C$ or below) must be scrupulously dry and of low hydrocarbon content (<5 ppm), to prevent formation of ice crystals or solid hydrocarbons.
See Hydrocarbons, and Ice, both above

Sodium acetate   MRH 6.44/35
Mellor, 1956, Vol. 2, Suppl. 1, 562
Application of fluorine to aqueous sodium acetate solution causes an explosion, involving formation of diacetyl peroxide.
See Diacetyl peroxide

Sodium bromate   MRH 1.30/89
Appelman, E. H., Inorg. Chem., 1969, 8, 223
The oxidation of alkaline bromate by fluorine to perbromate is not smooth and small explosions may occur in the vapour above the solution. The reaction should not be run unattended.

Sodium dicyanamide
Rausch, D. A. et al., J. Org. Chem., 1968, 33, 2522
The product, perfluoro-N-cyanodiaminomethane, and many of the by-products from interaction of fluorine and sodium dicyanamide, are extremely explosive in gaseous, liquid and solid states.
See Cyanoguanidine, above

Stainless steel
Stewart, J. W., Proc. 7th Int. Conf. Low Temp. Phys. (Toronto), 1960, 671
During the study of phase transitions of solidified gases at high pressures, solid fluorine reacted explosively with apparatus made from stainless steel.

Sulfides   MRH Antimony(III) sulfide 6.40/44, carbon disulfide 8.24/25, hydrogen
                                                                                                        sulfide 9.29/18
1. Mellor, 1940, Vol. 2, 11, 13; 1940, Vol. 6, 110; 1938, Vol. 9, 522; 1947, Vol. 10, 133; 1943, Vol. 11, 430
2. Moissan, 1900, 231–232
Antimony trisulfide, carbon disulfide vapour, chromium(II) sulfide and hydrogen sulfide all ignite in contact with fluorine at ambient temperature, the solids becoming incandescent [1]. Iron(II) sulfide reacts violently on mild warming, and barium sulfide, potassium sulfide or zinc sulfide all incandesce in the gas, as does molybdenum(III) sulfide at $200°C$ [2].

Trinitromethane
Smith, W. L. et al., Chem. Abs., 1976, 84, 46818

1520

During preparation of fluorotrinitromethane, an instrumental method can be used to avoid occurrence of dangerous over-fluorination of nitroform.

Water
1. Mellor, 1940, Vol. 2, 11
2. Schmidt, 1967, 53, 119

Treatment of liquid air (containing condensed atmospheric moisture) with fluorine give a potentially explosive precipitate, thought to be fluorine hydrate [1]. Contact of liquid fluorine with a bulk of water causes violent explosions. Ice tends to react explosively with fluorine gas after an indeterminate induction period [2].
See other INDUCTION PERIOD INCIDENTS

Xenon, Catalysts
Levec, J. *et al., J. Inorg. Nucl. Chem.*, 1974, **36**, 997–1001

Interaction may be explosive in the presence of finely divided nickel fluoride or silver difluoride, or nickel(III) oxide or silver(I) oxide, or if initiated by local heating. The mechanism is discussed.
See other HALOGENS, OXIDANTS

4311. Difluoramine (Fluorimide)
[10405-27-3]   F₂HN

$$F_2HN$$

$$F_2NH$$

1. Stevens, T. E., *J. Org. Chem.*, 1968, **33**, 2664, 2671
2. Petry, R. C. *et al., J. Org. Chem.*, 1967, **32**, 4034
3. Rosenfeld, D. D. *et al., J. Org. Chem.*, 1968, **33**, 2521
4. Martin, K. J., *J. Amer. Chem. Soc.*, 1965, **87**, 395
5. *MCA Case History No. 768*
6. Parker, C. O. *et al., Inorg. Synth.*, 1970, **12**, 310–311
7. Christe, K. O., *Inorg. Chem.*, 1975, **14**, 2821
8. Christe, K. O. *et al., Inorg. Chem.*, 1987, **26**, 920, 924
9. Chapman, R. D. *et al., J. Org. Chem.*, 1998, **63**(5), 1566

It is a dangerous explosive and must be handled with skill and care and appropriate precautions [1,2]. Explosions have occurred when it was condensed at −196°C [3] or allowed to melt [4], and a glass bulb containing the gas exploded violently when accidentally dropped [5]. Although difluoramine may be condensed safely at −78 or −130° [6], or at −142° [7], liquid nitrogen should not be used to give a trapping temperature of −196°C, as explosions are very likely to occur [6]. The solid adduct with caesium fluoride prepared at −142°C always explodes if allowed to warm towards 0°, and the adduct with rubidium fluoride sometimes exploded [8]. A safer procedure for handling difluoroamine is reported [9].

Benzenediazonium tetrafluoroborate, Caesium fluoride
Baum, K., *J. Org. Chem.*, 1968, **33**, 4333

Use of caesium fluoride as base to effect condensation caused an explosion in absence of solvent. Pyridine or potassium fluoride, and use of dichloromethane gave satisfactory results.

See other N-HALOGEN COMPOUNDS

4312. Mercury(I) fluoride
[13967-25-4] F_2Hg_2

FHgHgF

Iodoform
Andrews, L. *et al., J. Fluorine Chem.*, 1979, **13**, 273–278
Attempted fluorination of iodoform with a 10 year-old sample of mercury(I) fluoride led to an explosion immediately on heating.
See other MERCURY COMPOUNDS, METAL HALIDES

4313. Krypton difluoride
[13773-81-4] F_2Kr

KrF$_2$

$\Delta H°_f$ (g) 60 kJ/mol.

Ullmann, 1991, **A17**, 497
Krypton difluoride is a fluorinating agent some 50 kJ/mol more powerful than fluorine itself. It forms adducts, salts of FKr^+, with high valency fluorides such as AsF_5 and SbF_5, these react explosively with organic compounds.

Arsenic pentafluoride
Christe, K. O. *et al., Inorg. Chem.*, 1983, **22**, 3056
Interaction can lead to spontaneous exothermic decomposition of the fluoride accompanied by a bright flash and gas evolution. Safety precautions are required for this reaction system.
See other NON-METAL HALIDES (AND THEIR OXIDES)
See related XENON COMPOUNDS

4314. Difluorodiazene
[10578-16-2] F_2N_2

FN=NF

Bensoam, J. *et al., Tetrahedron Lett.*, 1977, 2797–2800
An explosion may occur when the diazene vapour is condensed, either to liquid or solid phases.
Both geometrical isomers are mildly endothermic ($\Delta H°_f$ (g) *cis*- +68.35 kJ/mol, 1.04 kJ/g; *trans*- +80.9 kJ/mol, 1.22 kJ/g).

Hydrogen
Kuhn, L. P. *et al., Inorg. Chem.*, 1970, **9**, 60

Explosive interaction occurs above 90°C.
See other ENDOTHERMIC COMPOUNDS, *N*-HALOGEN COMPOUNDS

4315. Phosphorus azide difluoride
[37388-50-4] F_2N_3P

$$N_3PF_2$$

1. O'Neill, S. R. *et al., Inorg. Chem.*, 1972, **11**, 1630
2. Lines, E. L. *et al., Inorg. Chem.*, 1972, **11**, 2270
It is photolytically and thermally unstable and has exploded at 25°C. It is also explosively sensitive to sudden changes in pressure, as occur on expansion into a vacuum or in surging during boiling [1]. It also ignites in air [2].
See related HALOPHOSPHINES, NON-METAL AZIDES, NON-METAL HALIDES

4316. Phosphorus azide difluoride–borane
[38115-19-4] $F_2N_3P.BH_3$

$$N_3PF_2.BH_3$$

Lines, E. L. *et al., Inorg. Chem.*, 1972, **11**, 2270
The liquid complex exploded violently during transfer operations.
See related BORANES, NON-METAL AZIDES, NON-METAL HALIDES

4317. Oxygen difluoride
[7783-41-7] F_2O

$$OF_2$$

It is mildly endothermic (ΔH°_f (g) +23.0 kJ/mol, 0.43 kJ/g).

Adsorbents
1. Metz, F. I., *Chem. Eng. News*, 1965, **43**(7), 41
2. Streng, A. G., *Chem. Eng. News*, 1965, **43**(12), 5
3. Streng, A. G., *Chem. Rev.*, 1963, **63**, 611
Mixtures of silica gel and the liquid difluoride sealed in tubes at 334 mbar exploded above −196°C, presence of moisture rendering the mixture shock-sensitive at this temperature [1]. Reaction of oxygen difluoride with silica, alumina, molecular sieve or similar surface-active solids is exothermic, and under appropriate conditions may be explosive [2]. A quartz fibre can be ignited in the difluoride [3].
See other MOLECULAR SIEVE INCIDENTS

Combustible gases
Streng, A. G., *Chem. Rev.*, 1963, **63**, 612
Mixtures with carbon monoxide, hydrogen and methane are stable at ambient temperatures, but explode violently on sparking. Hydrogen sulfide explodes with oxygen difluoride at ambient temperature and, though interaction is smooth at

−78°C under reduced pressure, the white solid produced exploded violently when cooling was stopped.

Diborane
Rhein, R. A., *Combust. Inst. Paper WSC1-67-10*, 73–80, 1967
Although it reacts slowly at ambient temperature, a mixture of the components which is stable at −195°C could explode during warming to ambient conditions.

Diboron tetrafluoride
Holliday, A. K. *et al., J. Chem. Soc.*, 1964, 2732
Ignition occurred on mixing at, or on warming mixtures to, −80°C.

Halogens, or Metal halides
Streng, A. G., *Chem. Rev.*, 1963, **63**, 611
Mixtures with chlorine, bromine or iodine explode on warming. A mixture with chlorine passed through a copper tube at 300°C exploded with variable intensity. Aluminium chloride explodes in the difluoride, and antimony pentachloride lightly at 150°C.

Hexafluoropropene, Oxygen
Afonso, M. Dos Santos *et al., Chem. Abs.*, 1987, **107**, 175302
The thermal homogeneous chain reaction to give mainly octafluoropropane and hexafluoropropylene oxide becomes explosive above a minimum oxygen pressure of 26 mbar.
See other CATALYTIC IMPURITY INCIDENTS

Metals
Streng, A. G., *Chem. Rev.*, 1963, **63**, 611
Finely divided platinum group metals react on gentle warming, and coarser materials at higher temperatures; aluminium, barium, cadmium, magnesium, strontium, zinc and zirconium evolving light. Lithium, potassium and sodium incandesce brilliantly at 400°C, while tungsten explodes.

Nitrogen oxide
Streng, A. G., *Chem. Rev.*, 1963, **63**, 612
Gaseous mixtures may explode on sparking. The mixed gases slowly react to give a mixture (NO, NOF) which, if liquefied by cooling, will explode on warming.

Nitrosyl fluoride
Hoffman, C. J. *et al., Chem. Rev.*, 1962, **62**, 9
Solid mixture explodes on melting, and the gaseous components ignite on mixing.

Non-metals
1. Sidgwick, 1950, 1136
2. Streng, A. G., *Chem. Rev.*, 1963, **63**, 611
Pressure of the gas must be limited during concentration by contact with cooled charcoal to avoid violent explosions [1]. Red phosphorus ignites when gently

warmed, and powdered boron and silicon generate sparks on heating in the difluoride [2].

Phosphorus(V) oxide
Streng, A. G., *Chem. Rev.*, 1963, **63**, 611
Ignition occurs spontaneously on contact.

Sulfur tetrafluoride
Oberhammer, H. *et al., Inorg. Chem.*, 1978, **17**, 1435
A mixture of the 2 fluorides, used at low temperature to prepare bis(pentafluorosulfur) oxide, is described as possibly explosive.

Water
Streng, A. G., *Chem. Rev.*, 1963, **63**, 610
Presence of water or water vapour in oxygen difluoride is dangerous, the mixture (even when diluted with oxygen) exploding violently on spark ignition, especially at 100°C (i.e. with steam).
See Adsorbents, above
See other ENDOTHERMIC COMPOUNDS, HALOGEN OXIDES, OXIDANTS, OXYGEN FLUORIDES

4318. Sulfinyl fluoride
[7783-42-8]   F$_2$OS

O:SF$_2$

Sodium
See Sodium: Non-metal halides (reference 8)
See other NON-METAL HALIDES

4319. Xenon difluoride oxide
[13780-64-8]   F$_2$OXe

F$_2$Xe:O

Alone, or Mercury, or Fluorides
1. Jacob, E. *et al., Angew. Chem. (Intern. Ed.)*, 1976, **15**, 158–159
2. Gillespie, R. J. *et al., J. Chem. Soc., Chem. Comm.*, 1977, 595–597
Although stable at below −40°C in absence of moisture, it will explode if warmed rapidly (>20°C/h). Explosive decomposition of the solid difluoride oxide at −196°C occurs on contact with mercury, or antimony pentafluoride or arsenic pentafluoride [1]. The fluoride explodes at about 0°C, and also in contact with arsenic pentafluoride in absence of hydrogen fluoride at −78°C [2].
See other NON-METAL HALIDES, XENON COMPOUNDS

4320. Dioxygen difluoride
[7783-44-0]

O_2F_2

F_2O_2

It is mildly endothermic (ΔH°_f (g) +19.8 kJ/mol, 0.28 kJ/g) but a powerful oxidant.

Sulfur trioxide

Solomon, I. J. *et al., J. Chem. Eng. Data*, 1968, **13**, 530
Interaction of the endothermic fluoride with the trioxide is very vigorous, and explosive in absence of solvent.

Various materials

Streng, A. G., *Chem. Rev.*, 1963, **63**, 615
Though not shock-sensitive, it is of limited thermal stability, decomposing below its b.p., −57°C, and explosively in contact with fluorided platinum at −113°C. It is a very powerful oxidant and reacts vigorously or violently with many materials at cryogenic temperatures. It explodes with methane at −194°, with ice at −140°, with solid ethanol at below −130° and with acetone–solid carbon dioxide at −78°C. Ignition or explosion may occur with chlorine, phosphorus trifluoride, sulfur tetrafluoride or tetrafluoroethylene, in the range −130 to −190°C. Even a 2% solution in hydrogen fluoride ignites solid benzene at −78°C.
See other ENDOTHERMIC COMPOUNDS, HALOGEN OXIDES, OXIDANTS, OXYGEN FLUORIDES

4321. Selenium difluoride dioxide
[14984-81-7]

F_2SeO_2

F_2O_2Se

Ammonia

1. Bailar, 1973, Vol. 2, 966
2. Engelbrecht, A. *et al., Monatsh.*, 1962, **92**, 555, 581

Interaction is violent [1], and many of the products and derivatives are both shock- and heat-sensitive explosives [2]. These include the ammonium, potassium, silver and thallium salts of the 'triselenimidate' ion, systematically 2,4,6-tris(dioxoselena)perhydrotriazine-1,3-5-triide.
See other NON-METAL HALIDES

4322. Xenon difluoride dioxide
[13875-06-4]

F_2XeO_2

F_2O_2Xe

Kirk Othmer, 1980, Vol. 12, 292
It is explosively unstable.

Preparative hazard
See Xenon hexafluoride: Water (reference 2)
Xenon tetrafluoride oxide: Caesium nitrate
See other XENON COMPOUNDS

4323. 'Trioxygen difluoride'
[16829-28-0] F_2O_3

$$F_2O_3$$

Bailar, 1973, Vol. 2, 761–763
This material is now considered to be an equimolar mixture of dioxygen difluoride and tetraoxygen difluoride, rather than the title species.

Various materials
Streng, A. G., *Chem. Rev.*, 1963, **63**, 619
Though thermally rather unstable, decomposing above its m.p., −190°C, it appears not to be inherently explosive. It is, however, an extremely potent oxidiser and contact with oxidisable materials causes ignition or explosions, even at −183°C. At this temperature, single drops added to solid hydrazine or liquid methane cause violent explosions, while solid ammonia, bromine, charcoal, iodine, red phosphorus and sulfur react with ignition and/or mild explosion. It is also extremely effective at initiating ignition of combustible materials in liquid oxygen, even at 0.1% concentration, whereas mixtures of ozone and fluorine in liquid oxygen are ineffective. This effect has been examined for use in hypergolic rocket propellant systems. Tetryl detonates spontaneously on contact with the difluoride.
See other HALOGEN OXIDES, OXIDANTS, OXYGEN FLUORIDES

4324. Fluorine fluorosulfate (Fluorosulfuryl hypofluorite)
[13536-85-1] F_2O_3S

$$FSO_2OF$$

1. Cady, G. H., *Chem. Eng. News*, 1966, **44**(8), 40
2. Cady, G. H. *et al.*, *Inorg. Synth.*, 1968, **11**, 155
3. Cady, G. H., *Intra-Sci. Chem. Rep.*, 1971, **5**, 1
4. Lustig, M. *et al.*, *Adv. Fluorine Chem.*, 1973, **7**, 175

The crude fluorosulfate, produced as by-product in preparation of peroxodisulfuryl difluoride, was distilled into a cooled steel cylinder and, on warming to ambient temperature, the cylinder exploded. It decomposes at 200°C, but not explosively [1]. Preparative and handling procedures are detailed [2]. Further warnings on the need to handle with care have been given [3,4].
See other ACYL HYPOHALITES, OXIDANTS *See related* ACYL HALIDES

4325. Difluorotrioxoxenon
[15192-14-0] F_2O_3Xe

$$FXe(O)_3F$$

Kirk Othmer, 1980, **Vol. 12**, 292
Explosively unstable.
See other XENON COMPOUNDS

4326. Disulfuryl difluoride
[13036-75-4] $F_2O_5S_2$

$$FSO_2OSO_2F$$

Ethanol
  Hayek, E., *Monatsh.*, 1951, **82**, 942
  Violent reaction on mixing at ambient temperature.
See other ACYL HALIDES *See related* ACID ANHYDRIDES

4327. Hexaoxygen difluoride
[12191-80-9] F_2O_6

$$O_6F_2$$

Bailar, 1973, Vol. 2, 764
Flashlight illumination or rapid warming of the solid at $-213°$ to $-183°C$ may
lead to explosion.
See other HALOGEN OXIDES, IRRADIATION DECOMPOSITION INCIDENTS, OXYGEN
FLUORIDES

4328. Peroxodisulfuryl difluoride
[13709-32-5] $F_2O_6S_2$

$$FSO_2OOSO_2F$$

Shreeve, J. M. *et al., Inorg. Synth.*, 1963, **7**, 124
A powerful oxidant which ignites organic materials on contact. Preparative and
handling procedures are detailed.

Boron nitride
  See Tetra(boron nitride) fluorosulfate

Carbon monoxide
  Gatti, R. *et al., Z. Phys. Chem.*, 1965, **47**, 323–336
  Interaction proceeds explosively above 20°C.

Dichloromethane
  Kirchmeier, R. L. *et al., Inorg. Chem.*, 1973, **12**, 2889

1528

Equimolar amounts explode while warming together to ambient temperature after initial contact at −183°C. Dilution of the dichloromethane with trichlorofluoromethane prevented explosion at −20°C.
See Fluorine fluorosulfate

N-Fluoroiminosulfur tetrafluoride
See N-Fluoroiminosulfur tetrafluoride: Alone, etc.
See other ACYL HALIDES, DIACYL PEROXIDES, OXIDANTS

4329. Lead(II) fluoride
[7783-46-2] F_2Pb

$$PbF_2$$

Fluorine
See Fluorine: Metal salts *See other* METAL HALIDES

4330. Poly(difluorosilylene)
[30582-57-1] $(F_2Si)_n$

$$(-SiF_2-)_n$$

1. Bailar, 1973, Vol. 1, 1352
2. Perry, D. L. *et al., J. Chem. Educ.*, 1976, **53**, 696–699
Produced by condensation at low temperature, the rubbery polymer ignites in air [1], and preparation, handling and reactions have been detailed [2].
See related HALOSILANES, NON-METAL HALIDES (AND THEIR OXIDES) *SEE OTHER* PYROPHORIC MATERIALS

4331. Tin(II) fluoride
[7783-47-3] F_2Sn

$$SnF_2$$

Magnesium nitrate
See Magnesium nitrate: Tin(II) fluoride
See other METAL HALIDES, REDUCANTS

4332. Xenon difluoride
[13709-36-9] F_2Xe

$$XeF_2$$

Shackelford, S. A., *J. Org. Chem.*, 1979, **44**, 3490–3492
Though a powerful oxidant, the difluoride is not explosively unstable. Safe procedures for the use of xenon difluoride in fluorination reactions are detailed. Residual traces of the fluoride are rapidly destroyed by dichloromethane at ambient temperatures.

Alkylaluminiums
   Bulgakov, R. G., *et al., Chem. Abs.*, 1990, **112**, 7548
   Reaction with diethoxyethylaluminium is explosive in the absence of solvent.
   See ALKYLALUMINIUM ALKOXIDES AND HYDROXIDES, ALKYLMETALS

Combustible materials
   Klimov, B. D. *et al., Chem. Abs.*, 1970, **72**, 85784
   Xenon difluoride (or the tetrafluoride, or their mixtures) could not be caused
   to detonate by impact. Xenon difluoride and xenon tetrafluoride both may
   cause explosion in contact with acetone, aluminium, pentacarbonyliron, styrene,
   polyethylene, lubricants, paper, sawdust, wool or other combustible materials.
   Their vigorous reactions with ethanol, potassium iodate or potassium permanganate
   are not explosive, however.

Dimethyl sulfide
   Forster, A. M., *J. Chem. Soc., Dalton Trans.*, 1984, 2827
   Interaction in absence of a solvent is explosive at ambient temperature.

Silicon–nitrogen compounds
   Gibson, J. A. *et al., Can. J. Chem.*, 1975, **53**, 3050
   Interaction of xenon difluoride and dimethylaminotrimethylsilane in presence or
   absence of solvent became explosive at sub-zero temperatures.
   See Xenon tetrafluoride
   See other NON-METAL HALIDES, XENON COMPOUNDS

†4333. Trifluorosilane
      [13465-71-9] F_3HSi

F_3SiH

   See other HALOSILANES

4334. Iodine dioxide trifluoride
      [25402-50-0] F_3IO_2

O_2IF_3

Organic materials
   Engelbrecht, A. *et al., Angew. Chem. (Intern. Ed.)*, 1969, **8**, 769
   It ignites in contact with flammable organic materials.
   See other HALOGEN OXIDES, IODINE COMPOUNDS, OXIDANTS

4335. Manganese trifluoride
      [7783-53-1] F_3Mn

MnF_3

Glass
   Mellor, 1942, Vol. 12, 344

When heated in contact it attacks glass violently, silicon tetrafluoride being evolved. It is a powerful fluorinating agent.
See other GLASS INCIDENTS, METAL HALIDES

4336. Nitrogen trifluoride
[7783-54-2]   F₃N

$$NF_3$$

Bromotrifluoromethane, Ethylene
  Wyatt, J. R., *Chem. Abs.*, 1982, **97**, 75008
  The presence of the halocarbon extinguishant significantly reduces the severity of the explosive oxidation of ethylene.

Charcoal
  Massonne, J. *et al., Angew. Chem. (Intern. Ed.)*, 1966, **5**, 317
  Adsorption of nitrogen trifluoride on to activated granular charcoal at $-100°C$ caused an explosion, attributed to the heat of adsorption not being dissipated on the porous solid and causing decomposition to nitrogen and carbon tetrafluoride. No reaction occurs at $+100°C$ in a flow system, but incandescence occurs at 150°C.

Chlorine dioxide
  Lawless, 1968, 171
  Interaction in the gas phase is explosive.

Diborane
  Lawless, 1968, 34–35
  No interaction occurred at ambient temperature and at pressures up to 8 bar, but violent explosions occurred at low temperatures in the liquid phase, even in absence of the impurity oxygen difluoride.

Hydrogen-containing materials.
  1. Ruff., O., *Z. Angew. Chem.*, 1929, **42**, 807
  2. Hoffman, C. J. *et al., Chem. Rev.*, 1962, **62**, 4
  Sparking of mixtures with ammonia or hydrogen causes violent explosions, and with steam, feeble ones [1]. Mixtures with ethylene, methane and hydrogen sulfide (also carbon monoxide) explode on sparking [2].

Metals
  Richter, R. F. *et al., Chem. Health & Safety*, 1995, **2**(2), 18
  It is claimed that in storage nitrogen fluoride can slowly react with metals, including stainless steel, to generate tetrafluorohydrazine.
  See Tetrafluoohydrazine, and next below

Tetrafluorohydrazine
  MCA Case History No. 683

A crude mixture of the 2 compounds, kept for 3 days in a stainless steel cylinder, exploded violently during valve manipulation.
See other N-HALOGEN COMPOUNDS

4337. Trifluoroamine oxide
[13847-65-9]   F₃NO

$$F_3N \rightarrow O$$

Gupta, O. D. *et al., Inorg. Chem.*, 1990, **29**(3), 573
A strong oxidant, mixtures with both organic and inorganic compounds are potentially explosive. It is recommended that the synthesis be not scaled up and be performed with due safety precautions.
See AMINE OXIDES
See other N-HALOGEN COMPOUNDS, N–O COMPOUNDS, OXIDANTS

4338. Trifluorosulfur nitride
[15930-75-3]   F₃NS

$$F_3S \equiv N$$

Chlorine fluoride
    See Chlorine fluoride: Trifluorosulfur nitride
    See other NON-METAL HALIDES, NITRIDES, N–S COMPOUNDS

4339. Phosphorus trifluoride
[7783-55-3]   F₃P

$$PF_3$$

Borane
    See Borane–phosphorus trifluoride

Dioxygen difluoride
    See Dioxygen difluoride: Various materials

Fluorine
    See Fluorine: Covalent halides

Hexafluoroisopropylideneaminolithium
    See Hexafluoroisopropylideneaminolithium: Non-metal halides
    See other HALOPHOSPHINES, NON-METAL HALIDES

4340. Thiophosphoryl fluoride
[2404-52-6]   F₃PS

$$S:PF_3$$

Air, or Sodium
    Mellor, 1940, Vol. 8, 1072–1073

In contact with air, the fluoride ignites or explodes, depending on contact conditions. Heated sodium ignites in the gas.
See related NON-METAL HALIDES

4341. Palladium trifluoride
[13842-82-5]   F$_3$Pd

$$PdF_3$$

Hydrogen
  Sidgwick, 1950, 1574
  Contact with hydrogen causes the unheated fluoride to be reduced incandescently.
See other METAL HALIDES

4342. Bis-*N* (imidosulfurdifluoridato)mercury
[23303-78-8]   F$_4$HgN$_2$S$_2$

$$Hg(N{=}SF_2)_2$$

Preparative hazard
  Mews, R. *et al., Inorg. Synth.*, 1986, 24, 14–16
  By-products formed during the preparation of the title compound from (fluorocarbonyl)iminosulfur difluoride, mercury(II) fluoride and bromine often react violently with water: cleaning of equipment should initially be effected with carbon tetrachloride.
See other MERCURY COMPOUNDS, N–S COMPOUNDS

4343. Manganese tetrafluoride
[15195-58-1]   F$_4$Mn

$$MnF_4$$

Petroleum oil
  Sorbe, 1968, 84
  Interaction leads to fire.
See other METAL HALIDES, OXIDANTS

4344. Rhenium nitride tetrafluoride
[84159-29-5]   F$_4$NRe

$$NReF_4$$

Preparative hazard
  See Trimethylsilyl azide: Rhenium hexafluoride
  See other NITRIDES *See related* METAL HALIDES

†4345. Tetrafluorohydrazine
[10036-47-2]

F₂NNF₂ centered...

†4345. Tetrafluorohydrazine
[10036-47-2] F_4N_2
F_2NNF_2

1. Logothetis, A. L., *J. Org. Chem.*, 1966, **31**, 3686, 3689
2. Modica, A. P. *et al., Rept. No. 357-275*, Princetown Univ., 1963
3. Martin, K. J., *J. Amer. Chem. Soc.*, 1965, **87**, 394
4. Richter, M. F. *et al., Chem. Health & Safety*, 1995, **2**(2), 18

General precautions for use of the explosive gas tetrafluorohydrazine and derived reaction products include: reactions on as small a scale as possible and behind a barricade; adequate shielding during work-up of products because explosions may occur; storage of tetrafluorohydrazine at −80°C under 1−2 bar pressure in previously fluorinated Monel or stainless steel cylinders with Monel valves; distillation of volatile difluoroamino products in presence of an inert halocarbon oil to prevent explosions in dry distilling vessels [1]. Light-initiated explosion of the gas has been reported [2], and it explodes on contact with air and combustible vapours, so careful inerting is essential [3]. Disposal of some aged tetrafluorohydrazine cylinders is described, by puncturing the cylinder with explosives in a pit full of lime[4].

Alkenyl nitrates
Reed, S. F. *et al., J. Org. Chem.*, 1972, **37**, 3329

The products of interaction of tetrafluorohydrazine and alkenyl nitrates, bis(difluoroamino)alkyl nitrates, are heat- and impact-sensitive explosives.
See DIFLUOROAMINO COMPOUNDS

Hydrocarbons
Petry, R. C. *et al., J. Org. Chem.*, 1967, **32**, 4034

Mixtures are potentially highly explosive, approaching the energy of hydrogen−oxygen systems.

Hydrocarbons, Oxygen
See Oxygen: Hydrocarbons, Promoters

Hydrogen
Kuhn, L. P. *et al., Inorg. Chem.*, 1970, **9**, 602

Explosive interaction is rather unpredictable, the initiation temperature required (20−80°C) depending on the condition of the vessel wall.

Nitrogen trifluoride
See Nitrogen trifluoride: Tetrafluorohydrazine

Organic materials
Reed, S. F., *J. Org. Chem.*, 1968, **33**, 2634

Mixtures with organic materials in presence of air constitute explosion hazards. Appropriate precautions are essential.

Ozone
Sessa, P. A. *et al., Inorg. Chem.*, 1971, **10**, 2067
When tetrafluorohydrazine was pyrolysed at 310°C to generate NF_2 radicals and the mixture contacted liquid ozone at -196°C, a violent explosion occurred.
See other DIFLUOROAMINO COMPOUNDS, N-HALOGEN COMPOUNDS

4346. Xenon tetrafluoride oxide
[13774-85-1] F_4OXe

$$OXeF_4$$

Preparative hazard
See Xenon hexafluoride: Silicon dioxide
Xenon hexafluoride: Caesium nitrate

Caesium nitrate
Christe, K. O., *Inorg. Chem.*, 1988, **27**, 3763
In the preparation of xenon difluoride dioxide from caesium nitrate and xenon tetrafluoride oxide, the latter must always be used in excess to prevent formation of explosive xenon trioxide.
See Xenon trioxide, below

Graphite, Potassium iodide, Water
Selig, H. *et al., Inorg. Nucl. Chem. Lett.*, 1975, **11**, 75–77
The graphite–xenon tetrafluoride oxide intercalation compound exploded in contact with potassium iodide solution.

Polyacetylene
Selig, H. *et al., J. Chem. Soc., Chem. Comm.*, 1981, 1288
During doping of polyacetylene films, contact with the liquid tetrafluoride led to ignition of the film.

Xenon trioxide
Christe, K. O., *Inorg. Chem.*, 1988, **27**, 3763
Liquid mixtures of the 2 xenon compounds, cooled in liquid nitrogen, tend to flash, then explode after a few seconds.
See other OXIDANTS, XENON COMPOUNDS

4347. Palladium tetrafluoride
[13709-55-2] F_4Pd

$$PdF_4$$

Water
Bailar, 1973, Vol. 3, 1278
Interaction is violent.
See other METAL HALIDES

4348. Platinum tetrafluoride
[13455-15-7]   **F₄Pt**

$$PtF_4$$

Water
    Sidgwick, 1950, 1614
    Interaction is violent.
    See other METAL HALIDES

4349. Rhodium tetrafluoride
[60617-65-4]   **F₄Rh**

$$RhF_4$$

Water
    Bailar, 1973, Vol. 3, 1235
    Interaction is violent.
    See other METAL HALIDES

4350. Sulfur tetrafluoride
[7783-60-0]   **F₄S**

$$SF_4$$

2-(Hydroxymethyl)furan, Triethylamine
    Jenzen, A. F. *et al., J. Fluorine Chem.*, 1988, **38**, 205–208
    Reaction of sulfur tetrafluoride with 2-hydroxymethylfuran in presence of triethy-
    lamine at −50°C is explosive in absence of a solvent.
    See other HALOGENATION INCIDENTS

2-Methyl-3-butyn-2-ol
    Boswell, G. A. *et al., Org. React.*, 1974, **21**, 8
    Interaction at −78°C is explosively vigorous.

Dioxygen difluoride
    See Dioxygen difluoride: Various materials
    See other NON-METAL HALIDES

4351. Selenium tetrafluoride
[13465-66-2]   **F₄Se**

$$SeF_4$$

Chlorine trifluoride
    Olah, G. A. *et al., J. Amer. Chem. Soc.*, 1974, **96**, 927
    The tetrafluoride is prepared by interaction of chlorine trifluoride and selenium in
    selenium tetrafluoride as solvent. The crude tetrafluoride must be substantially free
    from excess chlorine trifluoride to avoid danger during subsequent distillation at
    106°C/1 bar.

Water
  Aynsley, E. E. *et al., J. Chem. Soc.*, 1952, 1231
  Interaction is violent.
  See other NON-METAL HALIDES

4352. Silicon tetrafluoride
[7783-61-1]   F$_4$Si

SiF$_4$

Lithium nitride
  See Lithium nitride: Silicon tetrafluoride

Sodium
  See Sodium: Non-metal halides (reference 8)
  See other NON-METAL HALIDES

4353. Xenon tetrafluoride
[13709-61-0]   F$_4$Xe

XeF$_4$

  1. Shieh, J. C. *et al., J. Org. Chem.*, 1970, **35**, 4022
  2. Malm, J. G. *et al., Inorg. Synth.*, 1966, **8**, 254
  3. Chernick, C. L., *J. Chem. Educ.*, 1966, **43**, 619
  4. Holloway, J. H., *Talanta*, 1967, **14**, 871
  5. Falconer, E. E. *et al., J. Inorg. Nucl. Chem.*, 1967, **29**, 1380
  Moisture converts it to highly shock-sensitive xenon oxides [1]. Precautions necessary for various aspects of its use and application are detailed [2,1,1,5].
  See Xenon trioxide

Flammable materials
  Klimov, B. D. *et al., Chem. Abs.*, 1970, **72**, 85784
  Xenon tetrafluoride (or the difluoride, or their mixtures) could not be caused to detonate by impact. Xenon difluoride and xenon tetrafluoride both may cause explosion in contact with acetone, aluminium, pentacarbonyliron, styrene, polyethylene, lubricants, paper, sawdust, wool or other combustible materials. Their vigorous reactions with ethanol, potassium iodate or potassium permanganate are not explosive, however.
  See other NON-METAL HALIDES, OXIDANTS, XENON COMPOUNDS

4354. Pentafluoroorthoselenic acid
[38989-47-8]   F$_5$HOSe

F$_5$SeOH

Preparative hazard
  See Fluorine: Hydrogen fluoride, Seleninyl fluoride
  See other INORGANIC ACIDS *See related* NON-METAL HALIDES

4355. Iodine pentafluoride
   [7783-66-6]   **F₅I**

IF₅

Benzene
   Ruff, O. *et al., Z. Anorg. Chem.*, 1931, **201**, 245
   Interaction becomes violent above 50°C.

Calcium carbide, or Potassium hydride
   Booth, H. S. *et al., Chem. Rev.*, 1947, **41**, 425
   Both incandesce on contact, the carbide when warmed.

Diethylaminotrimethylsilane
   Oates, G. *et al., J. Chem. Soc., Dalton Trans.*, 1974, 1383
   A mixture exploded at around −80°C. Reactions with other silanes were very
   exothermic.

Dimethyl sulfoxide
   Lawless, E. M., *Chem. Eng. News*, 1969, **47**(13), 8, 109
   Unmoderated reaction with the sulfoxide is violent, and in the presence of diluents
   the reaction may be delayed and become explosively violent. Although small-scale
   reactions were uneventful, reactions involving about 0.15 g mol of the pentafluoride and sulfoxide in presence of trichlorotrifluoromethane or tetrahydrothiophene-
   1,1-dioxide as diluents caused delayed and violent explosions. Silver difluoride and
   other fluorinating agents also react violently with the sulfoxide.
   See Dimethyl sulfoxide: Acyl halides, etc.
   See other INDUCTION PERIOD INCIDENTS

Limonene, Tetrafluoroethylene
   See Tetrafluoroethylene: Iodine pentafluoride, etc.

Metals, or Non-metals
   1. Sidgwick, 1950, 1159
   2. Mellor, 1940, Vol. 2, 114
   3. Booth, H. S. *et al., Chem. Rev.*, 1947, **41**, 424–425
   Contact with boron, silicon, red phosphorus, sulfur, or arsenic, antimony or bismuth
   usually causes incandescence [1]. Solid potassium or molten sodium explode with
   the pentafluoride, and aluminium foil ignites on prolonged contact [2]. Molybdenum and tungsten incandesce when warmed [3].

Organic materials, or Potassium hydroxide
   Pascal, 1960, Vol. 16.1, 582
   The pentafluoride chars and usually ignites organic materials, and interaction with
   potassium hydroxide is violently exothermic.

Tetraiodoethylene
   Mellor, 1956, Vol. 2, Suppl. 1, 176
   Rapid mixing leads to explosion.

1538

Water
Ruff, O. *et al., Z. Anorg. Chem.*, 1931, **201**, 245
Reaction with water or water-containing materials is violent.
See other INTERHALOGENS, IODINE COMPOUNDS

4356. Tetrafluoroiodosyl hypofluorite
[72151-31-6] F_5IO_2

$$F_4I(:O)OF$$

Christe, K. O. *et al., Inorg. Chem.*, 1981, **20**, 2105
Two explosions were encountered in reactions involving the hypofluorite.
See other HALOGEN OXIDES, HYPOHALITES, IODINE COMPOUNDS

4357. *N*-Fluoroiminosulfur tetrafluoride
[74542-20-4] F_5NS

$$FN:SF_4$$

Alone, or Peroxydisulfuryl difluoride
O'Brien, B. A. *et al., Inorg. Chem.*, 1984, **23**, 2189
Samples of the *N*-fluoro compound, either alone or in admixture with the difluoride, exploded forcefully while warming from $-196°C$ to $+22°C$. Sample size should not exceed 3 mmol.
See other *N*-HALOGEN COMPOUNDS, N–S COMPOUNDS

4358. Azidopentafluorotungsten
[75900-58-2] F_5N_3W

$$N_3WF_5$$

Fawcett, J. *et al., J. Chem. Soc., Dalton Trans.*, 1980, 2294–2296
Trimethylsilyl azide reacts with excess tungsten hexafluoride in solvent to give the title compound, which explodes at $63°C$ or if dried in dynamic vacuum. A deficiency of the hexafluoride gave a mixture (presumably containing poly-azides) which exploded at ambient temperature.
See other METAL AZIDE HALIDES

4359. Tripotassium hexafluoroferrate(3−)
[13815-30-0] F_6FeK_3

$$3K^+F_6Fe^{3-}$$

Wright, S. W., *J. Chem. Educ.*, 1994, **71**(3), 251
The anhydrous salt will undergo a 'thermite' reaction with aluminium to produce molten iron.
See other THERMITE REACTIONS

4360. Hydrogen hexafluorophophosphate (Hexafluorophosphoric acid)
[16940-81-1] (anhydrous) F_6HP
$$HPF_6$$

Borosilicate glass
1. Pearse, G. A., *Chem. Brit.*, 1989, **25**, 30
2. Woolf, A. A., *Chem. Brit.*, 1989, **25**, 361
Copper(II) hexafluorophosphate was being prepared by adding 60–65% aqueous
acid to aqueous copper carbonate, followed by evaporation of the reaction mixture
at 80–90°C by heating on a ceramic hotplate. During evaporation, the Pyrex
vessel was dissolved at the liquid level, and the leaking solution also dissolved the
ceramic hotplate. This was attributed to presence of hydrofluoric acid, arising from
hydrolysis of the fluorophosphoric acid. Teflon-lined vessels are recommended for
such operations [1]. Aqueous solutions contain but little of the acid, in hydrolytic
equilibrium with mono- and di-fluorophophoric acids, orthophosphoric acid and
hydrofluoric acid, and such a mixture is clearly incompatible with glass containers,
and warnings have been published. Anhydrous hexafluorophosphate salts may be
prepared in glass by metathesis in anhydrous solvents [2].
See other GLASS INCIDENTS *See other* INORGANIC ACIDS

4361. Azidoiodoiodonium hexafluoroantimonate
[ ] $F_6I_2N_3Sb$
$$N_3I^+I \ SbF_6^-$$

Schleyer, P. von R. *et al., Angew. Chem. (Int.)*, 1992, **31**(10), 1338
This compound, which is more sensitive than iodonium azide, can explode spon-
taneously below −20°C. It is the first known I–N containing cationic species.
See other N-HALOGEN COMPOUNDS, IODINE COMPOUNDS, NON-METAL AZIDES

4362. Iridium hexafluoride
[7783-75-7] F_6Ir
$$IrF_6$$

Silicon
See Silicon: Metal hexafluorides
See other METAL HALIDES, OXIDANTS

4363. Potassium hexafluoromanganate(IV)
[16962-31-5] F_6K_2Mn
$$K_2[MnF_6]$$

Preparative hazard
See Potassium permanganate: Hydrofluoric acid
See related METAL HALIDES

4364. Potassium hexafluorosilicate(2—)
[16871-90-2]   F₆K₂Si

$$K_2[SiF_6]$$

Hydrogen fluoride
   See Hydrogen fluoride: Potassium tetrafluorosilicate(2—)
   See related NON-METAL HALIDES

4365. Molybdenum hexafluoride
[7783-77-9]   F₆Mo

$$MoF_6$$

Tetramethoxysilane
   See Tetramethoxysilane: Metal hexafluorides
   See other METAL HALIDES

4366. Neptunium hexafluoride
[14521-05-2]   F₆Np

$$NpF_6$$

Water
   Bailar, 1973, Vol. 5, 168
   Interaction with water at ambient temperature is violent.
   See other METAL HALIDES, OXIDANTS

4367. Pentafluorosulfur hypofluorite
[15179-32-5]   F₆OS

$$F_5SOF$$

Ruff, J. K., *Inorg. Synth.*, 1968, **11**, 137
It is considered to be potentially explosive.
See other HYPOHALITES

4368. Pentafluoroselenium hypofluorite
[27218-12-8]   F₆OSe

$$F_5SeOF$$

1. Mitra, G. *et al.*, *J. Amer. Chem. Soc.*, 1959, **81**, 2646
2. Smith, J. E. *et al.*, *Inorg. Chem.*, 1970, **9**, 1442
A cooled sample exploded when allowed to dry rapidly [1], but this may have
been owing to impurities, as it did not happen in the later work [2].
See other HYPOHALITES

4369. Pentafluorotellurium hypofluorite
[83314-21-0]   F$_6$OTe

$$F_5TeOF$$

Christe, K. O. *et al., Inorg. Chem.*, 1987, **26**, 920
A shock-sensitive explosive, which needs full handling precautions.
See other HYPOHALITES

4370. Osmium hexafluoride
[13768-38-2]   F$_6$Os

$$OsF_6$$

Organic materials
  Sorbe, 1968, 91
  It causes ignition of paraffin oil and other organic materials.

Silicon
  See Silicon: Metal hexafluorides
  See other METAL HALIDES, OXIDANTS

4371. Platinum hexafluoride
[13693-05-5]   F$_6$Pt

$$PtF_6$$

Bis(trifluoromethyl) nitroxide
  Christe, K. O. *et al., J. Fluorine Chem.*, 1974, **4**, 425
  Interaction of the nitroxide radical and this powerful oxidant was very violent
  during warming from $-196°C$.
  See other METAL HALIDES, OXIDANTS

4372. Plutonium hexafluoride
[13693-06-6]   F$_6$Pu

$$PuF_6$$

Water
  Bailar, 1973, Vol. 5, 168
  Interaction with water at ambient temperature is violent.
  See other METAL HALIDES, OXIDANTS

4373. Rhenium hexafluoride
[10049-17-9]   F$_6$Re

$$ReF_6$$

Silicon
  See Silicon: Metal hexafluorides

Tetramethoxysilane
See Tetramethoxysilane: Metal hexafluorides

Trimethylsilyl azide
See Trimethylsilyl azide: Rhenium hexafluoride
See other METAL HALIDES, OXIDANTS

4374. Sulfur hexafluoride
[2551-62-4] F_6S

$$SF_6$$

Disilane
See Disilane: Non-metal halides, etc.
See other NON-METAL HALIDES

4375. Uranium hexafluoride
[7783-81-5] F_6U

$$UF_6$$

Aromatic hydrocarbons, or Hydroxy compounds
Sidgwick, 1950, 1072
Interaction with benzene, toluene, or xylene is very vigorous, with separation of
carbon, and violent with ethanol or water.
See other METAL HALIDES, OXIDANTS

4376. Tungsten hexafluoride
[7783-82-6] F_6W

$$WF_6$$

Aitchison, K. A. *et al., Chem. Abs.*, 1987, **106**, 147494
Safety procedures for use of tungsten hexafluoride in CVD processing operations
are emphasized.

Tetramethoxysilane
See Tetramethoxysilane: Metal hexafluorides

Trimethylsilyl azide
See Azidopentafluorotungsten
See other METAL HALIDES, OXIDANTS

4377. Xenon hexafluoride
[13693-09-9]

F_6Xe

$$XeF_6$$

Caesium nitrate
  Christe, K. O., *Inorg. Chem.*, 1988, **27**, 3763
  In the preparation of xenon tetrafluoride oxide from the hexafluoride and caesium nitrate, the former must always be used in excess to prevent formation of explosive xenon trioxide.
  See Xenon tetrafluoride oxide: Xenon trioxide

Fluoride donors, Water
  Bailar, 1973, Vol. 1, 317
  Adducts of the hexafluoride with sodium fluoride, potassium fluoride, rubidium fluoride, caesium fluoride or nitrosyl fluoride react violently with water.

Hydrogen
  Malm, J. G. *et al.*, *J. Amer. Chem. Soc.*, 1963, **85**, 110
  Interaction is violent.

Silicon dioxide
  1. McKee, D. E. *et al.*, *Inorg. Chem.*, 1973, **12**, 1722
  2. Aleinikov, N. N. *et al.*, *J. Chromatog.*, 1974, **89**, 367
  Interaction of the yellow hexafluoride with silica to give xenon tetrafluoride oxide must be interrupted before completion (disappearance of colour) to avoid the possibility of formation and detonation of xenon trioxide [1]. An attempt to collect the hexafluoride in fused silica traps at $-20°C$ after separation by preparative gas chromatography failed because of reaction with the silica and subsequent explosion of the oxygen compounds of xenon so produced [2].
  See Xenon trioxide

Water
  1. Gillespie, R. J. *et al.*, *Inorg. Chem.*, 1974, **13**, 2370–2374
  2. Schumacher, G. A. *et al.*, *Inorg. Chem.*, 1984, **23**, 2923
  Although uncontrolled reaction of xenon hexafluoride and moisture produces explosive xenon trioxide, controlled action by progressive addition of limited amounts of water vapour with agitation to a frozen solution of the hexafluoride in anhydrous hydrogen fluoride at $-196°C$ to give xenon oxide tetrafluoride or xenon dioxide difluoride is safe [1]. Controlled hydrolysis in solution in hydrogen fluoride is, however, described as hazardous [2].
  See other NON-METAL HALIDES, OXIDANTS, XENON COMPOUNDS

4378. Iodine heptafluoride
[16921-96-3]

F_7I

$$IF_7$$

Carbon, or Combustible gases
  Booth, H. S. *et al.*, *Chem. Rev.*, 1947, **41**, 428

Activated carbon ignites immediately in the gas, mixtures with methane ignite, and those with carbon monoxide ignite on warming, while those with hydrogen explode on heating or sparking.

Metals
Booth, H. S. *et al., Chem. Rev.*, 1947, **41**, 427
Interaction with barium, potassium and sodium is immediate, accompanied by evolution of light and heat. Aluminium, magnesium, and tin are passivated on contact, but on heating react similarly to the former metals.

Organic solvents
Booth, H. S. *et al., Chem. Rev.*, 1947, **41**, 428
Benzene, light petroleum, ethanol and ether ignite in contact with the gas, while the exotherm with acetic acid, acetone or ethyl acetate causes rapid boiling. General organic materials (cellulose, grease, oils) ignite if excess heptafluoride is present.
See other INTERHALOGENS

4379. Potassium heptafluorotantalate(V)
[16924-00-8] F_7K_2Ta
$$K_2[TaF_7]$$

Water
Ephraim, 1939, 353
Hydrolysis is less violent than that of tantalum pentafluoride.
See related METAL HALIDES

4380. Difluoroammonium hexafluoroantimonate
[56533-31-4] F_8H_2NSb
$$F_2N^+H_2SbF_6^-$$

See Difluoroammonium hexafluoroarsenate
See related N-HALOGEN COMPOUNDS

4381. Bis(S,S-difluoro-N-sulfimido)sulfur tetrafluoride
[52795-23-0] $F_8N_2S_3$
$$F_2S:NSF_4N:SF_2$$

Water
Hofer, R. *et al., Angew, Chem. (Intern. Ed.)*, 1973, **12**, 1000
Either the liquid tetrafluoride or its viscous polymer decomposes explosively in contact with water.
See other NON-METAL HALIDES, N–S COMPOUNDS

4382. Xenon(II) pentafluoroorthoselenate
[38344-58-0] $F_{10}O_2Se_2Xe$

$$Xe(OSeF_5)_2$$

Oxidisable materials
    Seppelt, K., *Angew. Chem. (Intern. Ed.)*, 1972, **11**, 724
    Interaction is explosive.
    See other OXIDANTS, XENON COMPOUNDS

4383. Xenon(II) pentafluoroorthotellurate
[25005-56-5] $F_{10}O_2Te_2Xe$

$$Xe(OTeF_5)_2$$

Organic solvents
    Sladky, F., *Angew. Chem. (Intern. Ed.)*, 1969, **8**, 523; *Inorg. Synth.*, 1986, **24**, 36
    Explosive or very vigorous reactions occur on contact with acetone, benzene or
    ethanol.
    See other OXIDANTS, XENON COMPOUNDS

4384. Tetrafluoroammonium hexafluoromanganate
[74449-37-9] $F_{14}MnN_2$

$$[F_4N]_2[MnF_6]$$

Water
    Christe, K. O. *et al., Inorg. Chem.*, 1980, **19**, 3254
    Reaction of the powerful oxidant with water is extremely violent and needs proper
    precautions. In absence of fuels, it is not shock-sensitive.
    See other N-HALOGEN COMPOUNDS, OXIDANTS

4385. Tetrafluoroammonium hexafluoronickelate
[63105-40-8] $F_{14}N_2Ni$

$$[F_4N]_2[NiF_6]$$

    Christe, K. O., US Pat. 4 108 965, 1978
    This high energy oxidant is useful in propellants.

Water
    Christe, K. O., *Inorg. Chem.*, 1977, **16**, 2238
    The hydrolysis of this powerful oxidant is very violent, and may be explosive if
    attempted appreciably above $-180°C$.
    See other N-HALOGEN COMPOUNDS, OXIDANTS

4386. Tetrafluoroammonium octafluoroxenate
[82963-15-3] $F_{16}N_2Xe$

$$[F_4N]_2[XeF_8]$$

Christe, K. O. *et al.*, US Pat. 4 447 407, 1984

It produces the highest theoretical detonation pressures in explosive formulations (maximal yields of fluorine and nitrogen trifluoride).

See other N-HALOGEN COMPOUNDS, OXIDANTS, XENON COMPOUNDS

4387. Octakis(trifluorophosphine)dirhodium
[14876-96-1] $F_{24}P_8Rh_2$

$$[(F_3P)_4RhRh(PF_3)_4]$$

Acetylenic esters

Bennett, M. A. *et al.*, *Inorg. Chem.*, 1976, **15**, 107–108

Formation of complexes with excess methyl propiolate or dimethyl acetylenedicarboxylate must not be allowed to proceed at above +20°C, or violently explosive polymerisation of the acetylene esters will occur.

See other POLYMERISATION INCIDENTS

4388. Iron
[7439-89-6]   Fe

Fe

1. Bailar, 1973, Vol. 3, 985
2. Gusein, M. A. *et al.*, *Chem. Abs.*, 1974, **81**, 124774
3. Jakusch, H. *et al.*, *Chem. Abs.*, 1983, **98**, 130471
4. Brinza, V. N. *et al.*, *Chem. Abs.*, 1983, **99**, 7872
5. Bandyopadhyay, A. *et al.*, *Steel India*, 1987, **10**, 58–63
6. Ganguly, A. *et al.*, *Tool Alloy Steels*, 1988, **22**(1), 19–26

The known pyrophoric [1] and explosive properties of ultrafine iron powder were examined in detail [2]. Pyrophoric iron particles may be stabilised by heat treatment with oxygen-containing gases in 2 stages at 25–45°, then 50–70°C [3]. The ignition temperature of sponge iron (183–203°C) is independent of heating rate and coincides with the dissociation temperature of iron hydroxides formed on the surface [4]. An incident involving spontaneous combustion of hydrogen-reduced sponge iron (Direct Reduced Iron, DRI) is analysed. The surface area and porosity of coal-reduced DRI prepared at over 1000°C is lower, and the product, a more dense, sintered DRI has greater oxidation resistance. Fire control and salvage techniques are discussed [5]. Other aspects of storing, handling and shipping of sponge iron are reviewed [6].

See entry HIGH SURFACE-AREA SOLIDS *See other* PYROPHORIC METALS

Air, Oil

Glaser, A., *Arbeitschutz*, 1941, 134–135

Oxidative heating of oily iron dust in a collecting vessel caused vaporisation of oil and subsequent ignition, causing an explosion.
See other SELF-HEATING AND IGNITION INCIDENTS

Air, Water
1. Brimelow, H. C., private comm., 1972
2. Unpublished observations, 1949
2-Nitrophenylpyruvic acid was reduced to oxindole using iron pin-dust−ferrous sulfate in water. The iron oxide−iron residues, after filtering and washing with chloroform, rapidly heated in contact with air and shattered the Buchner funnel [1]. Previously, rapid heating effects had been observed on sucking air through the iron oxide residues from hot filtration of aqueous liquor from reduction of a nitro compound with reduced iron powder [2].
See Water, below

Acetaldehyde
See Acetaldehyde: Metals

Chloric acid
See Chloric acid: Metals, etc.

Chloroformamidinium nitrate
See Chloroformamidinium nitrate: Alone, etc.

Halogens or Interhalogens	MRH values show % of halogen
See Bromine pentafluoride: Acids, etc.	MRH 4.23/54
Chlorine: Metals	MRH 2.68/56
Chlorine trifluoride: Metals	
Fluorine: Metals	MRH 7.49/40

Other reactants
Yoshida, 1980, 233
MRH values calculated for 19 combinations with oxidants are given.

Oxidants	MRH values show % of oxidant
See Halogens or Interhalogens above	
Ammonium nitrate: Metals	MRH 3.35/68
Ammonium peroxodisulfate: Iron	MRH 0.92/78
Dinitrogen tetraoxide: Metals	MRH 4.43/38
Hydrogen peroxide: Metals	MRH 4.60/48
Nitryl fluoride: Metals	
Peroxyformic acid	MRH 5.69/99+
Potassium dichromate: Iron	MRH 0.42/66
Potassium perchlorate: Metal powders	
Sodium peroxide: Metals	MRH 1.55/68

Polystyrene
Unpublished observation, 1971

1548

Iron flake powder and polystyrene beads had been blended in a high-speed mixer. The mixture ignited and burned rapidly when discharged into a polythene bag. Rapid oxidation of the finely divided metal and/or static discharge may have initiated the fire. No ignition occurred when the iron powder was surface coated with stearic acid.
See other STATIC INITIATION INCIDENTS

Sodium acetylide
See Sodium acetylide: Metals

Water
1. Asakura, S. *et al., Chem. Abs.*, 1983, **99**, 110033
2. Anon, *Safety Digest Univ. Safety Assoc.*, 1991, **41**, 10
3. Clancey, V. J., *Proc. 41st Ironmaking Conf.*, 1982, 244–250
In a study of the spontaneous exothermic reaction of iron powder or turnings with water in absence of chlorides, rust was found to catalyse the reaction [1]. On drilling holes into steel tubes filled with possibly wet ferrous scrap and sealed some years, they were found to be pressurised. The third tube produced an explosion and jet of flame causing some injury to the driller. This was probably attributable to hydrogen generation [2]. Hazards associated with shipping of direct-reduced iron and arising from reactivity with condensation etc. in ship's holds, with liberation and ignition of hydrogen, are discussed [3].
See Air, Water, above
See other CORROSION INCIDENTS, GAS EVOLUTION INCIDENTS *See other* METALS

4389. Ferromanganese (Iron–manganese alloy)
[73202-12-7]

Fe—Mn

Fe—Mn

See entry FERROALLOY POWDERS *See other* ALLOYS

4390. Ferrosilicon (Iron–silicon alloy)
[50645-52-8]

Fe–Si

Fe—Si

1. Delyan, V. I. *et al., Chem. Abs.*, 1983, **99**, 25024
2. Babaitsev, I. V. *et al., Chem. Abs.*, 1987, **107**, 241804
The flammability and explosive hazard of ferrosilicon powder is increased substantially during grinding in a vibratory mill [1]. Explosion hazards from air–hydrogen, — acetylene, or — propane mixtures formed during preparation of ferrosilicon containing alkaline earth additives are attributed to contact of barium or magnesium carbide or silicide additive with atmospheric moisture [2].
See FERROALLOY POWDERS

Sodium hydroxide, Water
Kirk-Othmer (3rd edn.), 1982, Vol. 19, 494

Intimate mixtures of ferrosilicon with solid sodium hydroxide incandesce when moistened.

Water
1. Anon., *Chem. Trade J.*, 1956, **139**, 1180
2. Horn, Q.C. *et al.*, *Chem. Abs.* 1998, **128**, 37646c; 37653c
Ferrosilicon containing 30–75% of silicon is hazardous, particularly when finely divided, and must be kept in a moisture-tight drum. In contact with water, the impurities present (arsenide, carbide, phosphide) evolve extremely poisonous arsine, combustible acetylene and spontaneously flammable phosphine [1 & 2].
See related METAL NON-METALLIDES

4391. Ferrotitanium (Iron–titanium alloy)
[87490-22-0] $Fe-Ti$
$$Fe-Ti$$

See entry FERROALLOY POWDERS *See other* ALLOYS

4392. Iron(II) hydroxide
[18624-44-7] FeH_2O_2
$$Fe(OH)_2$$

Gibson, 1969, 121
Prepared under nitrogen, it is pyrophoric in air, producing sparks.
See other PYROPHORIC MATERIALS, REDUCANTS

4393. Ammonium iron(III) sulfate
[7783-83-7] $FeH_4NO_8S_2.12H_2O$
$$NH_4Fe(SO_4)_2.12H_2O$$

Sulfuric acid
See Sulfuric acid: Ammonium iron(III) sulfate
See related METAL OXONON-METALLATES

4394. Iron(III) phosphinate
[7783-84-8] $FeH_6O_6P_3$
$$Fe[OP(H)OH]_3$$

Mellor, 1981, Vol. 8, Suppl. 3, 626
It is used in impact-sensitive priming conmpositions.
See other METAL PHOSPHINATES, REDOX COMPOUNDS

4395. Iron(II) iodide
[7783-86-0] FeI_2

$$FeI_2$$

Alkali metals
    See Potassium: Metal halides
        Sodium: Metal halides
    See other METAL HALIDES

4396. Potassium peroxoferrate(2−)
[ ] FeK_2O_5

$$K_2[(O_2)FeO_3]$$

Alone, or Non-metals, or Sulfuric acid
    Goralevich, D. K., *J. Russ. Phys. Chem. Soc.*, 1926, **58**, 1115
    It explodes on heating or impact, or in contact with charcoal, phosphorus, sulfur
    or sulfuric acid.
    See other PEROXOACID SALTS

4397. Iron(III) nitrate
[10421-42-4] FeN_3O_9

$$Fe(ONO_2)_3$$

Dimethyl sulfoxide
    See Dimethyl sulfoxide: Metal oxosalts
    See other METAL NITRATES

4398. Iron(II) oxide
[1345-25-1] FeO

$$FeO$$

    1. Mellor, 1941, Vol. 13, 715
    2. Bailar, 1973, Vol. 3, 1008–1009
    The oxide (prepared at 300°C) burns in air above 200°C, while the finely divided
    oxide prepared by reduction may be pyrophoric at ambient temperature [1]. That
    prepared by thermal decomposition under vacuum of iron(II) oxalate is also
    pyrophoric [2].

Oxidants
    See Nitric acid: Iron(II) oxide
        Hydrogen peroxide: Metals, etc.

Sulfur dioxide
    Mellor, 1941, Vol. 13, 715

The oxide incandesces when heated in sulfur dioxide.
See other METAL OXIDES, REDUCANTS

4399. Iron(II) sulfate
[7720-78-7]   FeO$_4$S

$$FeSO_4$$

HCS 1980, 502

Arsenic trioxide, Sodium nitrate
See Sodium nitrate: Arsenic trioxide, Iron(II) sulfate

Methyl isocyanoacetate
See Methyl isocyanoacetate: Heavy metals
See other METAL OXONON-METALLATES, REDUCANTS

4400. Iron(II) sulfide
[1317-37-9]   FeS

$$FeS$$

1. Mellor, 1942, Vol. 14, 157
2. Anon., *Chem. Age.*, 1939, **40**, 267
3. Davie, F. M. *et al., J. Loss Prev.*, 1993, **6**(3), 139 & 145
4. Walker, R. *et al., Ind. Eng. Chem. Res.*, 1996, **35**(5), 1736

The moist sulfide readily oxidises in air exothermally, and may reach incandescence. Grinding in a mortar hastens this [1]. The impure sulfide formed when steel processing equipment is used with materials containing hydrogen sulfide or volatile sulfur compounds is pyrophoric, and has caused many fires and explosions when such equipment is opened without effective purging. Various methods of purging are discussed [2]. Formation of pyrophoric FeS in bitumen tanks is considered as a cause of spontaneous ignition and explosion in the head space [3]. A detailed study of formation of possibly pyrophoric sulphides from rust in crude oil tankers has been made [4].

Lithium
See Lithium: Metal oxides, etc.
See other IGNITION SOURCES, METAL SULFIDES

4401. Iron disulfide (Iron pyrites)
[1309-36-0]   FeS$_2$

$$FeS_2$$

1. Anon., *Angew. Chem. (Nachr.)*, 1954, **2**, 219
2. Bowes, P. C., *Ind. Chemist*, 1954, **30**, 12–14

3. Ruiss, I. G. *et al.*, *J. Chem. Ind.* (Moscow), 1935, **12**, 692–696
4. Guedes de Carvalho, R. A. *et al.*, *Chem. Abs.*, 1980, **93**, 223677

Finely powdered pyrites, especially in presence of moisture, will rapidly heat spontaneously and ignite, particularly in contact with combustible materials [1]. Inert gas blanketing will prevent this [2]. Precautions to reduce the self-ignition hazards of powdered pyrites, and the explosion hazards of pyrites air mixtures in the furnaces of sulfuric acid plants have been detailed and discussed [3]. Further studies on minimum moisture content of Portuguese pyrites for safe transportation and storage are reported [4].

See other SELF-HEATING AND IGNITION INCIDENTS

Carbon

Ruiss, I. G. *et al.*, *J. Chem. Ind.* (Moscow), 1935, **12**, 696

The presence of carbon in pyrites lowers the ignition temperature to 228–42°C and increases the explosivity of dust suspensions in air.

See other METAL SULFIDES

4402. Di-μ-iodotetranitrosyldiiron
[15002-08-1] $Fe_2I_2N_4O_4$

$$I(ON)_2FeFe(NO)_2I$$

Kegzdins, P. *et al.* US Pat. 5,631,284, 1997; *Chem. Abs.*, 1997, **127**, 50793x

This compound may decompose explosively during sublimation, if overheated.

4403. Iron(III) oxide
[1309-37-1] Fe_2O_3

$$Fe_2O_3$$

Aluminium

1. Mellor, 1946, Vol. 5, 217
2. Bodur, G. M., *J. Chem. Educ.*, 1985, **62**, 1107

An intimately powdered mixture, usually ignited by magnesium ribbon as a high-temperature fuse, reacts with an intense exotherm to produce molten iron and was used formerly (before the advent of gas or arc welding) in the commercial 'thermite' welding process. Incendive particles have been produced by this reaction on impact between aluminium and rusty iron. (The term 'thermite reaction' has now been extended to include many combinations of reducing metals and metal oxides) [1]. Some accidents in demonstrating the thermite reaction are described [2].

See Calcium disilicide, below
See Aluminium: Metal oxides, etc. *also* Magnesium: Metal oxides
See also LIGHT ALLOYS
See other THERMITE REACTIONS

Aluminium, Propene

Batty, G. F., private comm., 1972

Use of a rusty iron tool on an aluminium compressor piston caused incendive sparks which ignited residual propene–air mixture in the cylinder.
See Aluminium, above

Aluminium–magnesium alloy, Water
See Aluminium–magnesium alloy: Iron(III) oxide, etc.

Aluminium–magnesium–zinc alloys
See Aluminium–magnesium–zinc alloy: Rusted steel

Calcium disilicide
Berger, E., Compt. rend., 1920, 170, 29
The mixture ('silicon thermite') attains a very high temperature when heated, producing molten iron like the usual thermite mixture.
See Aluminium, above

Carbon monoxide
Othen, C. W., *School, Sci. Rev.*, 1964, **45**(156), 459
The reason for a previously reported explosion during reduction of iron oxide with carbon monoxide is given as the formation of pentacarbonyliron at temperatures between 0 and 150°C. Suitable heating arrangements and precautions will eliminate this hazard.
See Pentacarbonyliron

Ethylene oxide
See Ethylene oxide: Contaminants

Guanidinium perchlorate
See Guanidinium perchlorate: Iron(III) oxide

Hydrogen peroxide
See Hydrogen peroxide: Metals, etc.

Magnesium
See Magnesium: Metal oxides

Metal acetylides
See Calcium acetylide: Iron(III) chloride, etc.
   Caesium acetylide: Iron(III) oxide
   Rubidium acetylide: Metal oxides
See other METAL OXIDES

4404. Iron(III) sulfide
   [12063-27-3] Fe_2S_3
                           Fe_2S_3

Moore, F. M., *Proc. Gas Cond. Conf.*, 1976, **26**, 1

Hydrogen sulfide is removed from natural gas by passage over iron sponge, when flammable iron sulfide is produced. Handling precautions during regeneration of the reactor beds are detailed.

See other METAL SULFIDES, PYROPHORIC IRON–SULFUR COMPOUNDS

4405. Iron(II,III) oxide ('Magnetite')
[1317-61-9]   **Fe₃O₄**

$$FeO.Fe_2O_3$$

Aluminium, Calcium silicide, Sodium nitrate.
    Schierwater, F-W., *Sichere Chemiearb.*, 1976, **28**, 30–31
    During the preparation of a foundry mixture of the finely divided oxide with aluminium powder and small amounts of sodium nitrate and calcium silicide with calcium fluoride as flux, a violent explosion occurred in the conical mixer. It had been established previously that the mixture could not be ignited by impact or friction and that if ignited by a very high-energy source (magnesium ribbon), it burned rather slowly to a glowing liquid. The possibility of ignition of the solid mixture by silane produced from water acting on the silicide content was discounted, but an aluminium dust explosion may have been involved with this ignition source.

Aluminium, Sulfur
    Crozier, T. H., *HM Insp. Expl. Spec. Rept. 237*, London, HMSO, 1929
    A 20 t quantity of an incendiary bomb mixture of the finely powdered oxide, aluminium and sulfur became accidentally ignited and burned with almost explosive violence. It is similar to thermite mixture.
    See other THERMITE REACTIONS

Hydrogen trisulfide
    See Hydrogen trisulfide: Metal oxides
    See other METAL OXIDES

4406. Gallium
[7440-55-3]   **Ga**

$$Ga$$

Aluminium alloys
    Marshall, C., private comm., 1981
    Serious problems may arise if gallium or its liquid alloys contact aluminium alloy structural components in aircraft, when rapid 'amalgamation' and weakening occurs.
    See other CORROSION INCIDENTS

Halogens
    1. Walker, H. L., *School Sci. Rev.*, 1956, **37**(132), 196
    2. Bailar, 1973, Vol. 1, 1084

The metal reacts with cold chlorine strongly exothermically, and the compact metal with bromine even at $-33°C$, reaction being violent at ambient temperature [1]. Interaction of gallium with liquid bromine at $0°C$ proceeds with a flash, resembling the action of alkali metals with water [2].

Hydrogen peroxide
See Hydrogen peroxide: Gallium, etc.
See other METALS

4407. Lithium tetrahydrogallate
[17836-90-7]   GaH$_4$Li

Li[H$_4$Ga]

Gaylord, 1956, 26
Though of lower stability than the analogous aluminate, its reactivity is generally similar to that of the latter.
See other COMPLEX HYDRIDES

4408. Sodium tetrahydrogallate
[32106-51-7]   GaH$_4$Na

Na[H$_4$Ga]

Water
McKay, 1966, 169
It is explosively hydrolysed by water.
See other COMPLEX HYDRIDES

4409. Gallium azide (Triazidogallane)
[ ]   GaN$_9$

Ga(N$_3$)$_3$

Fischer, R. A. et al., Chem. Eur. J., 1996, **2**(11), 1353
The solid detonates violently when heated rapidly to above $280°C$.
See other METAL AZIDES

4410. Digallane
[12140-58-8]   Ga$_2$H$_6$

H$_3$GaGaH$_3$

Leleu, Cahiers, 1976, (85), 585
According to some authors it ignites in air.
See other METAL HYDRIDES

1556

4411. Gallium(I) oxide
[12024-20-3]   Ga$_2$O

$$Ga_2O$$

Bromine
   Bailar, 1973, Vol. 1, 1091
   It is a strong reducant, reacting violently with bromine.
   See other METAL OXIDES, REDUCANTS

4412. Germanium
[7440-56-4]   Ge

$$Ge$$

Halogens
   Mellor, 1941, Vol. 7, 260
   The powdered metal ignites in chlorine, and lumps will ignite on heating in chlorine
   or bromine.

Oxidants
   Mellor, 1941, Vol. 7, 260–261
   The powdered metal reacts violently with nitric acid, and mixtures with potassium
   chlorate or nitrate explode on heating. Heated germanium burns with incandescence
   in oxygen.
   See Potassium hydroxide: Germanium
        Sodium peroxide: Metals (reference 2)
   See other METALS

4413. Poly(germanium monohydride)
[13572-99-1]   (GeH)$_n$

$$(GeH)_n$$

Jolly, W. L. *et al.*, *Inorg. Synth.*, 1963, **7**, 39
The solid polymeric hydride sometimes decomposes explosively into its elements
on exposure to air.
See other METAL HYDRIDES

4414. Germanium imide
[26257-00-1]   GeHN

$$Ge:NH$$

Air, or Oxygen
   Johnson, O. H., *Chem. Rev.*, 1952, **51**, 449
   On exposure to air it reacts violently, and in oxygen incandescence occurs.
   See other N-METAL DERIVATIVES

4415. Poly(germanium dihydride)
[32028-94-7] $(GeH_2)_n$

$(-GeH_2-)_n$

Bailar, 1973, Vol. 2, 13
Impact may cause explosive decomposition to the elements, with ignition of the liberated hydrogen.
See other METAL HYDRIDES

4416. Azidogermane
[21138-22-7] GeH_3N_3

N_3GeH_3

Fluorosilane
Anon., *Angew. Chem. (Nachr.)*, 1970, **18**, 272
An attempt to prepare azidosilane by interaction of azidogermane and fluorosilane exploded.
See related METAL AZIDES, METAL HYDRIDES

†4417. Germane
[7782-65-2] GeH_4

GeH_4

ΔH°_f 90 kJ/mol

1. Brauer, 1963, Vol. 1, 715
2. *Chem. Abs.*, 1994, **120**, 80876
Germane and its higher homologues decompose in air, often igniting [1]. The Japanese have studied the explosive decomposition of germane, the minimum pressure for this was a little more than 0.1 bar [2].

Bromine
See Bromine: Germane
See other METAL HYDRIDES

4418. Sodium germanide
[12265-93-9] $GeNa$

$NaGe$

Air, or Water
Johnson, O. H., *Chem. Rev.*, 1952, **51**, 452
The binary alloy is pyrophoric and may ignite in contact with water, as do other alkali metal germanides.
See other ALLOYS, PYROPHORIC MATERIALS

4419. Germanium(II) sulfide
    [12025-32-0]
$$GeS$$

GeS

Potassium nitrate
    See Potassium nitrate: Metal sulfides
    See other METAL SULFIDES

†4420. Digermane
    [13819-89-8]
$$H_3GeGeH_3$$

Ge_2H_6

1. Mellor, 1941, Vol. 7, 264
2. Brauer, 1963, Vol. 1, 715
3. McKay, K. M., *Inorg. Synth.*, 1974, **15**, 170
It may ignite in air [1], particularly if air is admitted suddenly into the gas at reduced pressure [2]. Although digermane and its homologues do not usually ignite on exposure to air, their autoignition temperatures appear to be about 50°C and combustion is rapid or explosive [3].
See other METAL HYDRIDES

4421. Trigermane
    [14691-44-2]
$$H_3GeGeH_2GeH_3$$

Ge_3H_8

Brauer, 1963, Vol. 1, 715
Air-sensitive, it may ignite.
See other METAL HYDRIDES

4422. Poly(dimercuryimmonium hydroxide) ('Millon's base anhydride')
    [12529-66-7]
$$(Hg:N^+:Hg\ OH^-)_n$$

$(HHg_2NO)_n$

Sidgwick, 1950, 318
The anhydride of Millon's base explodes if touched or heated to 130°C.
See Mercury: Ammonia
See other FULMINATING METALS, MERCURY COMPOUNDS, POLY(DIMERCURYIM-MONIUM) COMPOUNDS, *N*-METAL DERIVATIVES

4423. Hydriodic acid
    [10034-85-2]
$$HI$$

HI

Hydrogen iodide is mildly endothermic (ΔH_f° (g) +25.9 kJ/mol, 0.20 kJ/g).

Muir, G. D., private comm., 1968

During preparation of hydriodic acid by distillation of phosphorus and wet iodine, the condenser became blocked with by-product phosphonium iodide, and an explosion, possibly also involving phosphine, occurred. There is also a purification hazard.

See Phosphorus: Hydriodic acid

Metals
See Magnesium: Hydrogen halides
Potassium: Hydrogen halides

Oxidants
Leleu, *Cahiers*, 1974, (75), 271

Hydrogen iodide ignites in contact with fluorine, dinitrogen trioxide, dinitrogen tetraoxide and fuming nitric acid.

See Ethyl hydroperoxide: Hydriodic acid
Perchloric acid: Iodides
Potassium chlorate: Hydrogen iodide

Phosphorus
See Phosphorus: Hydriodic acid
See other INORGANIC ACIDS, NON-METAL HALIDES, NON-METAL HYDRIDES

4424. Iodic acid
[7782-68-5] HIO_3

$$HOIO_2$$

Non-metals
1. Mellor, 1946, Vol. 5, 15
2. Partington, 1967, 813

Interaction with boron below 40°C is vigorous, attaining incandescence [1]. Charcoal, phosphorus and sulfur deflagrate on heating [2].

See other INORGANIC ACIDS, OXIDANTS, OXOHALOGEN ACIDS, IODINE COMPOUNDS

4425. Periodic acid
[13444-71-8] HIO_4

$$HOIO_3$$

Dimethyl sulfoxide
Rowe, J. J. M. *et al.*, *J. Amer. Chem. Soc.*, 1968, **90**, 1924

Although 1.5 M solutions of periodic acid in dimethyl sulfoxide explode after a few min, 0.15 M solutions appear stable.

See Dimethyl sulfoxide: Metal oxosalts, or: Perchloric acid

Tetraethylammonium hydroxide
1. Cookson, R. C. *et al., Chem. Brit.*, 1979, **15**, 329
2. Kirby, G. W. *et al,., J. Chem. Soc., Perkin Trans.*, 1981, 3252
During the preparation of the quaternary oxidant tetraethylammonium iodate from the aqueous reagents, the residue after vacuum evaporation of most of the water exploded, breaking the flask. This was attributed to possible presence of excess periodic acid in the reaction mixture [1]. Further details and precautions to avoid heating the salt, normally stable in storage, are given [2].
See QUATERNARY OXIDANTS
See other INORGANIC ACIDS, OXIDANTS, OXOHALOGEN ACIDS, IODINE COMPOUNDS

4426. Diiodamine (Iodimide)
[15587-44-7]   HI$_2$N

$$I_2NH$$

Mellor, 1940, Vol. 8, 607
Explosive, formed on prolonged contact of nitrogen triiodide with water.
See Nitrogen triiodide–ammonia *See other* N-HALOGEN COMPOUNDS

4427. Potassium hydride
[7693-26-7]   HK

$$KH$$

1. Sorbe, 1968, 67
2. Brown, C. A., *J. Org. Chem.*, 1974, **39**, 3913–3918
It ignites on exposure to air [1], and the hydride dispersed in oil is much more highly reactive than sodium hydride dispersions, and rather more careful handling is necessary for safe working. Such precautions are detailed. Contact with water of even traces of the dispersion in flammble solvents will lead to ignition [2].
See Potassium hexahydroaluminate

Bromopentaborane(9), Dimethyl ether
Anon., *Lab. Accid. Higher Educ.*, Item 11, HSE, Barking, 1987
During the reaction of bromopentaborane and potassium hydride in dimethyl ether at −78°C, the reaction became uncontrollable, shattering the glassware and igniting. Cause may have been contamination or effect of scale-up.

O-2,4-Dinitrophenylhydroxylamine
See O-2,4-Dinitrophenylhydroxylamine: Potassium hydride

Fluoroalkene (unspecified)
MCA Case History No. 2134
After a few minutes' reflux at 12°C, a mixture of the hydride (0.01 mol) and a fluoroalkene (0.02 mol) exploded violently. This was attributed to possible presence of metallic potassium in the hydride, causing polymerisation or formation of a fluoroacetylene.

Oxidants
See Fluorine: Metal hydrides
Oxygen (Gas): Metal hydrides
See other METAL HYDRIDES

4428. Potassium hydroxide
[1310-58-3]   HKO
KOH

HCS 1980, 768

Acids
MCA Case History No. 920
Incautious addition of acetic acid to a vessel contaminated with potassium
hydroxide caused eruption of the acid.
See other NEUTRALISATION INCIDENTS

Ammonium hexachloroplatinate
See Ammonium hexachloroplatinate: Potassium hydroxide

1,4-Bis(1,2-dibromoethyl)benzene
See 1,4-Bis(1,2-dibromoethyl)benzene: Potassium hydroxide

Bromoform, Cyclic polyethylene oxides
See Bromoform: Cyclic polyethylene oxides, etc.

2-Bromo-2,5,5-trimethylcyclopentanone, Triethylamine
See 2-Bromo-2,5,5-trimethylcyclopentanone: Potassium hydroxide, etc.

Chlorine dioxide   MRH 1.88/55
See Chlorine dioxide: Potassium hydroxide

Cyclopentadiene
See Cyclopentadiene: Potassium hydroxide

Diamond
See Carbon: Potassium hydroxide

Germanium
Partington, 1967, 181
Germanium is oxidised by the fused hydroxide with incandescence.

Glass
Lloyd, A., *Chem. Brit.*, 1987, **23**, 208
During the distillation of hexane from potassium hydroxide pellets, it was found
that the 2 l flask had become perforated by alkaline attack, and that hexane was
leaking through two pinholes, generating much flammable vapour. The high rate

of corrosion of glass is probably associated with the fact that fresh pellets of potassium hydroxide already contain some 14% of water, and absorption of further water from solvents during drying over potash pellets leads to formation of drops of very concentrated aqueous potassium hydroxide solution.
See other CORROSION INCIDENTS, GLASS INCIDENTS

Glycols
There are no reports of mishap, but:
See Sodium hydroxide: Glycols

Hyponitrous acid
See Hyponitrous acid: Alone, or Potassium hydroxide

Maleic anhydride   MRH 1.67/99+
See Maleic anhydride: Bases, etc.

Nitroalkanes   MRH Nitromethane 5.69/99+, nitroethane 4.52/99+
See Nitromethane: Acids, etc.
See NITROALKANES: inorganic bases

Nitroaryl compounds
Solid mixtures may deflagrate readily.
See 4-Chloronitrobenzene: Potassium hydroxide
    Nitrobenzene: Alkali
    4-Nitrobenzoic acid: Potassium hydroxide
    2-Nitrophenol: Potassium hydroxide
    4-Nitrophenol: Potassium hydroxide
See DEFLAGRATION INCIDENTS

Nitrogen trichloride
See Nitrogen trichloride: Initiators

Other reactants
Yoshida, 1980, 212
MRH values calculated for 10 combinations, largely with catalytically decomposed materials, are given.

Potassium peroxodisulfate
See Potassium peroxodisulfate: Potassium hydroxide

Sugars
See SUGARS

2,2,3,3-Tetrafluoropropanol
See 2,2,3,3-Tetrafluoropropanol: Potassium hydroxide

Tetrahydrofuran
See Tetrahydrofuran: Caustic alkalies

Thorium dicarbide
See Thorium dicarbide: Non-metals, etc.

2,4,6-Trinitrotoluene
See 2,4,6-Trinitrotoluene: Added impurities

Water
Anon., *CISHC Chem. Safety Summ.*, 1976, **46**, 8–9
A mixture of flake potassium hydroxide and sodium hydroxide was added to a reaction mixture without the agitator running. When this was started the batch erupted, owing to the sudden solution exotherm. Although this is a physical hazard rather than a chemical hazard, similar incidents have occurred frequently.
See Sodium hydroxide: Water.
See other AGITATION INCIDENTS
See other INORGANIC BASES

4429. Potassium hydrogen xenate
[[73378-54-8], sesquihydrate] HKO_4Xe
$KOXe(O)_2OH.1.5H_2O$

See other XENON COMPOUNDS

4430. Potassium hydrogen peroxomonosulfate
[10058-23-8] HKO_5S
$KOSO_2OOH$

Alone, or Organic matter
Castrantas, 1965, 5
It melts with decomposition at 100°C, and forms explosive mixtures with as little as 1% of organic matter.
See other PEROXOACID SALTS

4431. Dipotassium phosphinate
[13492-26-7] HK_2O_2P
$KHP(O)OK$

Water
Mellor, 1971, Vol. 8, Suppl. 3, 623
The salt ignites in contact with a little water.
See other METAL PHOSPHINATES, REDUCANTS

4432. Lithium hydride
[60380-67-8]   **HLi**

LiH

1. Bailar, 1973, Vol. 1, 344
2. Brinza, V. A. *et al., Chem. Abs.*, 1979, **91**, 96308
The powdered material burns readily on exposure to air [1], and the rate of combustion of air suspensions has been studied with respect to concentrations of hydride and moisture in the air [2].

Dinitrogen oxide
See Dinitrogen oxide: Lithium hydride

Oxygen
See Oxygen (Gas): Metal hydrides
Oxygen (Liquid): Lithium hydride
See other METAL HYDRIDES

4433. Manganese diazide hydroxide
[ ]   **HMnN$_6$O**

(N$_3$)$_2$MnOH

Mellor, 1940, Vol. 8, 354
It explodes at 203°C.
See related METAL AZIDES

4434. Permanganic acid
[13465-41-3]   **HMnO$_4$**

HOMnO$_3$

Organic materials
1. Frigerio, N. A., *J. Amer. Chem. Soc.*, 1969, **91**, 6201
2. von Schwartz, 1918, 327
The crystalline acid and its dihydrate are very unstable, often exploding at about 3° and 18°C, respectively, but they may be stored virtually unchanged at −75°C. The anhydrous acid ignited explosively every organic compound with which it came into contact except mono-, di- or tri-chloromethanes [1]. The solution of permanganic acid (or its explosive anhydride, dimanganese heptoxide) produced by interaction of permanganates and sulfuric acid, will explode on contact with benzene, carbon disulfide, diethyl ether, ethanol, flammable gases, petroleum or other organic substances [2].
See Dioxonium hexamanganato(VII)manganate
See other INORGANIC ACIDS, OXIDANTS

4435. Nitrous acid
    [7782-77-6]   **HNO$_2$**

                              HON:O

2-Amino-5-nitrophenol
  See 2-Amino-5-nitrophenol: Nitrous acid

Ammonium decahydrodecaborate(2−)
  See Ammonium decahydrodecaborate(2−): Nitrous acid

Anilines
  Diazonium compounds or solutions produced by action of nitrous acid on the
  anilines below have been involved in various incidents.
  See Aniline, 4-Bromoaniline, 2-Chloroaniline, 3-Chloroaniline, 2-Nitroaniline, 3-
  Nitroaniline, 4-Nitroaniline.

A semicarbazone, Silver nitrate
  Mitchell, J. J., *Chem. Eng. News*, 1956, **34**, 4704
  Use of nitrous acid to liberate a free keto-acid from its semicarbazone caused
  formation of hydrogen azide which was co-extracted into ether with the product.
  Addition of silver nitrate to precipitate the silver salt of the acid also precipitated
  silver azide, which later exploded on scraping from a sintered disc. The possibility
  of formation of free hydrogen azide from interaction of nitrous acid and hydrazine
  or hydroxylamine derivatives is stressed.

Phosphine
  See Phosphine (reference 3)

Phosphorus trichloride
  Mellor, 1940, Vol. 8, 1004
  The trichloride explodes with nitrous (or nitric) acid.
  See other INORGANIC ACIDS, OXIDANTS

4436. Nitric acid
    [7697-37-2]   **HNO$_3$**

                              HONO$_2$

  (*MCA SD-5*, 1961); *FPA H23*, 1974; *HCS 1980*, 678 (<70%), 679 (>70%); *RSC
  Lab. Hazards Data Sheet No. 32*, 1985

  1. *MCA SD-5*, 1961
  2. Bretherick, L., *CISHC Chem. Safety Summ.*, 1978, **49**(197), 3
  3. Merck, 11th edn., 6497
  4. Kirk Othmer, 3rd edn., Vol. 6, 866
  5. Editor's comments, 1999
  The oxidising power and hazard potential of nitric acid increases progressively with
  increase in strength from the conc. acid (70 wt%) through fuming acids (above

85 wt%) to the anhydrous 100% acid. Nitric acid is the common chemical most frequently involved in reactive incidents (note the extent of this compound entry), and this is a reflection of its exceptional ability to function as an effective oxidant even when fairly dilute (unlike sulfuric acid) or at ambient temperature (unlike perchloric acid). Its other notable ability to oxidise most organic compounds to gaseous carbon dioxide, coupled with its own reduction to gaseous 'nitrous fumes' has been involved in many of the incidents in which closed, or nearly closed, reaction vessels or storage containers have failed from internal gas pressure [2]. Screw-capped winchesters of fuming acid may therefore develop internal pressure if the inert liner to the plastic cap fails or is missing.

There is a potentially dangerous confusion over the meaning of fuming nitric acid. Once, and sometimes still [3], this meant the red fuming acid which consists of nitrogen dioxide dissolved in nitric acid. This is a potent oxidising agent [1] and best avoided for nitrations, where oxidation is an unwanted and dangerous side-reaction. Increasingly, however, the white fuming acid of American Chemical Society analytical specification [4] is meant. This is a good nitrating agent but a poor oxidant, showing induction periods of hours in length followed by auto-catalytic runaway. When following previous literature it is wise to consider which was meant. Nitric acid generally is becoming whiter and less oxidant, which can be partially rectified by adding sodium nitrite.

Anhydrous nitric acid is miscible even with hydrocarbons, and the increased availability of high quality white grades increases the danger of inadvertently forming such mixtures. If anywhere near stoichiometric composition, an homogeneous mixture of nitric acid and virtually any organic is a sensitive high explosive. So much was obvious to learned men of the last century, and many such mixtures were patented as explosives, by Sprengel and others, from 1871 onwards. Recent correspondence makes it apparent that this is neither known nor obvious to many Professors of Chemistry in the late 1990s: though learning has departed, danger has not [5]!

See Polyalkenes, below
See other GAS EVOLUTION INCIDENTS

Preparative hazard
Crawley, F. K., *Loss Prev. Bull.*, 1992, (104), 4
Corrosion permitted leak of nitric acid into the ammonia feed passing through the cooling tubes of a nitric acid plant. An ammonium nitrate deposit built up and exploded when the tubes were cut out with a welding torch for replacement.

Acetic acid
1. Raikova, V. M., *Chem. Abs.*, 1983, **98**, 5958
2. Vidal. P. *et al., C. R. Acad. Sci. Ser. II*, 1991, **313**(12), 1383; *J. Energ. Mater.*, 1993, **11**(21), 135.
Critical detonation diameters for mixtures have been determined.

Acetic acid, Sodium hexahydroxyplatinate(IV)
1. Davidson, J. M. *et al., Chem. & Ind.*, 1966, 306
2. Malerbi, B. W., *Chem. & Ind.*, 1970, 796

During preparation of diacetatoplatinum(II) by alternative procedures, the hexahy-droxyplatinate in mixed nitric–acetic acids was evaporated to a syrup and several explosions were experienced [1], possibly owing to formation of acetyl nitrate. On one occasion a brown solid was isolated and dried, but subsequently exploded with great violence when touched with a glass rod. The material was thought to be a mixture of platinum (IV) acetate–nitrate species [2].

Acetic anhydride   MRH 4.94/34

1. Brown, T. A. *et al., Chem. Brit.*, 1967, **3**, 504
2. Dubar, J. *et al., Compt. rend. C*, 1968, **266**, 1114
3. Dingle, L. E. *et al., Chem. Brit.*, 1968, **4**, 136
4. Ullmann, A10, 157
5. *MCA Case History No. 103*
6. Venters, K. *et al., Chem. Abs.*, 1982, **97**, 45372
7. Raikova, V. M., *Chem. Abs.*, 1983, **98**, 5958

Mixtures containing 50–85 wt% of fuming nitric acid are detonable and very sensitive to initiation by friction or shock (possibly owing to formation of acetyl nitrate and/or tetranitromethane). For preparation of mixtures outside these limits, the order of mixing is important (below 50%, acid into anhydride; above 85%, vice versa; and all below 10°C) [1]. Similar information is also presented diagrammat-ically [2]. Mixtures containing less than 50% of nitric acid are also dangerous in that addition of small amounts of water (or water-containing mineral acids) readily initiates an uncontrollable exothermic fume-off, which will evaporate most of the liquid present. Equimolar mixtures of 38% nitric acid with acetic anhydride can be detonated at room temperature after aging for a few h [3]. Others put the lower limit of detonability at 30% [4]. Accidental contact of the 2 materials caused a violent explosion [5]. Thermodynamic calculations show maximum values for heats of decomposition for acetyl nitrate or its equivalent mixtures of the components [6]. Critical detonation diameters for mixtures have been determined [7].
See 1,3,5-Triacetylhexahydrotriazine, etc., below

Acetic anhydride, Hexamethylenetetramine acetate
Leach, J. T., *J. Haz. Mat.*, 1981, **4**, 271–281
The military explosives RDX and HMX are manufactured from the 3 components using the Bachman process. Some of the possible mixtures may lead to fires in open vessels and explosions under confinement, and the exothermic and other effects (some calculated by the CHETAH program) for a wide range of mixtures are presented as ternary diagrams. It was also found that acetic anhydride layered onto solutions of ammonium nitrate in nitric acid exploded, owing to formation of acetyl nitrate.

Acetone   MRH 5.86/23

1. Kennedy, R., private comm., 1975
2. Anon., *Safety Digest Univ. Safety Assoc.*, 1989 **34**, 14

A winchester of fuming nitric acid with a plastics cap burst, probably owing to internal pressure build-up and uneven wall thickness. The explosion fractured

an adjacent bottle of acetone which ignited on contact with the oxidant [1]. Segregation of oxidants and fuels in storage is essential to prevent such incidents, which have occurred previously elsewhere. Another explosion resulted, when attempting to clear a jammed glass stopper by successive application of acetone and nitric acid [2].
See Polyalkenes, below
See also SAFE STORAGE OF CHEMICALS

Acetone, Acetic acid
1. Frant, M. S., *Chem. Eng. News*, 1960, **38**(43), 56
2. Secunda, W. J., *Chem. Eng. News*, 1960, **38**(46), 5
A mixture of equal parts of nitric acid, acetone and 75% acetic acid, used to etch nickel, will explode 1.5–6 h after mixing if kept in a closed bottle. The presence of the diluted acetic acid would probably slow the known violent oxidation of acetone by nitric acid [1] (but would not prevent slow oxidation with gas evolution). Alternatively, the formation of tetranitromethane and subsequent oxidation of acetone was postulated [2], but this is perhaps unlikely, because presence of acetic anhydride is normally necessary for formation of tetranitromethane.

Acetone, Sulfuric acid
Fawcett, H. H., *Ind. Eng. Chem.*, 1959, **51**, 89A
Acetone is oxidised violently by mixed nitric–sulfuric acids, and if the mixture is confined in a narrow-mouthed vessel, it may be ejected or explode.

Acetonitrile   MRH 6.15/23
Andrussow, L., *Chim. Ind.* (Paris), 1961, **86**, 542
Mixtures of fuming nitric acid and acetonitrile are high explosives.

4-Acetoxy-3-methoxybenzaldehyde
Legrand, M. *et al., J. Chem. Educ.*, 1978, **55**, 769
The dangerously exothermic nitration of the aldehyde with conc. nitric acid is adequately controlled on a several hundred g scale by an automated assembly of laboratory apparatus.

Acetylene, Mercury(II) salts, Sulfuric acid
Orton, K. J. P. *et al., J. Chem. Soc.*, 1920, **117**, 29
Contact of acetylene with the conc. acid in presence of mercury salts forms trinitromethane, explosive above its m.p., 15°C. Subsequent addition of sulfuric acid produces tetranitromethane, a powerful oxidant of limited stability, in high yield.
See Trinitromethane, *also* Tetranitromethane

Acrylonitrile
See Acrylonitrile: Acids

Acrylonitrile–methacrylate copolymer
Caprio, V. *et al., J. Haz. Mat.*, 1983, **7**, 383–391

The Asahi process for wet-spinning the copolymer involves water dilution at below 0°C of a solution of the copolymer in aqueous 68 wt% nitric acid (the azeotropic composition). The potential for slow self heating and decomposition have been investigated experimentally with variations in several parameters in a Sikarex safety calorimeter. At 20% polymer content, the slow self heating starts even at ambient temperature, and later involves evolution of 30 mol of uncondensable gas per kg of copolymer with comcommittant boiling of the nitric acid. A 2 step mechanism has been proposed.

See other GAS EVOLUTION INCIDENTS

Alcohols   MRH Ethanol 5.56/24, furfuryl alcohol 5.86/28
1. Fawcett, H. H., *Chem. Eng. News*, 1949, **27**, 1396
2. Fromm, F. *et al.*, *Chem. Eng. News*, 1949, **27**, 1956
3. Unpublished observations, 1956
4. *MCA Case History No. 1152*
5. Potter, C. R., *Chem. & Ind.*, 1971, 501
6. Spengler, G. *et al.*, *Brennst. Chem.*, 1965, **46**, 117
7. Long, L. A., private comm., 1972
8. Kurbangalina, R. K., *Zh. Prikl. Khim.*, 1959, **32**, 1467–1470
9. Albright, Hanson, 1976, 341–343
10. Anon., *CIHSC Chem. Safety Summ.*, 1980, **51**, 4
11. Anon., *Loss Prev. Bull.*, 1980, (035), 7–9
12. Anon., *Univ. Safety Assoc. Safety News*, 1982, (16), 4
13. Mabbott, D. J., 1996, personal communication

A 15% solution of nitric acid in ethanol was used to etch a bismuth crystal. After removing the metal, the mixture decomposed vigorously [1]. Mixtures of nitric acid and alcohols ('Nital') are quite unstable when the concentration of acid is above 10%, and mixtures containing over 5% should not be stored [2]. The use of a little alcohol and excess nitric acid to clean sintered glassware (by 'nitric acid fizzing') is not recommended. At best it is a completely unpredictable approximation to a nitric acid–alcohol rocket propulsion system. At worst, if heavy metals are present, fulminates capable of detonating the mixture may be formed [3]. Chromic acid mixture is less hazardous for such cleaning operations. The Case History describes a violent explosion caused by addition of conc. acid to a tank car contaminated with a little ethanol [4]. During oxidation of cyclohexanol to the 1,2-dione by an established process, a violent explosion occurred. Two intermediates are possible suspects [5]. Furfuryl alcohol is hypergolic with high-strength nitric acid [6] and methanol has been used as a propellant fuel. It also readily forms the explosive ester, methyl nitrate [7]. Mixtures with ethylene glycol are easily detonated by heat, impact or friction [8]. The injector process for safe continuous nitration of glycerol and other alcohols is reviewed [9]. Use of the wrong proportions of denatured ethanol and nitric acid (5:95 instead of 95:5) led to an explosion [10]. During attempts to clean a reactor, unclear instructions led to 120 l of conc. nitric acid, rather than the aqueous 2.6% solution intended, being charged via a pump which contained some 5 l of 2-propanol from previous use. Rapid oxidation of the alcohol, (probably accompanied by formation of explosive isopropyl nitrate) and gas evolution burst the reactor [11]. A stock solution of conc. nitric acid in methanol for cleaning glassware was stored in a screw-capped bottle which later burst from internal gas pressure. Such mixtures are dangerously unstable and should not be used or stored [12]. A 33% by volume solution of nitric acid in ethanol was mixed for metallurgical purposes in a large open bottle, after two hours the still warm mixture was capped. Perhaps 7 hours after mixing, there was an exceedingly violent explosion. It was decided not to use this mixture in future but to substitute one containing perchloric acid, (which may be no less hazardous) [13].

See Methanol, Sulfuric acid, below

See other GAS EVOLUTION INCIDENTS, NITRATION INCIDENTS, PLANT CLEANING INCIDENTS

Alcohols, Disulfuric acid
Rastogi, R. P. *et al., Amer. Inst. Aero. Astronaut. J.*, 1966, **4**, 1083–1085
Ignition delays were determined for contact of various aliphatic alcohols with mixtures of disulfuric acid (*d*, 1.9) and red fuming nitric acid (*d*, 1.5). With ethanol, a minimum of 30 wt% of disulfuric acid was required for ignition.

Alcohols, Potassium permanganate
See Potassium permanganate: Alcohols, Nitric acid

Aliphatic amines
1. Schalla, R. L. *et al., Amer. Rocket Soc. J.*, 1959, **29**, 33–39
2. Schneebeli, P. *et al., Rech. Aeron.*, 1962, **87**, 33–46
In contact with white fuming nitric acid, ignition delays for tributylamine, tripropylamine and triethylamine were less than those of the corresponding dialkylamines, and under good mixing conditions, butylamine did not ignite [1]. Triethylamine ignites on contact with the conc. acid [2].
See Aromatic amines (reference 2); Cyclohexylamine; and Dinitrogen tetraoxide: Nitrogenous fuels, all below

Alkanethiols
1. Filby, W. G., *Lab. Practice*, 1974, **23**, 355
2. McCullough, F. *et al., Proc. 5th Comb. Symp.*, 181, New York, Reinhold, 1955
3. Illingworth, B., private comm., 1980
4. Quam, G. N., *Safety Practice for Chemical Laboratories*, 52–53, Villanova (Pa.), Villanova Press, 1963
5. Reychler, A., *Bull. Soc. Chim. Belg.*, 1913, **110**, 217, 300
Oxidation of alkanethiols to alkanesulfonic acids with excess conc. acid as usually described is potentially hazardous, the exotherm often causing ignition of the thiol. A modified method involving oxidation under nitrogen and at temperatures 1–2°C above the m.p. of the thiol is safer and gives purer products [1]. Technical butanethiol (containing 28% of propane- and 7% of pentane-thiols) is hypergolic with 96% acid [2]. Oxidation of several thiols to the sulfonic acids by addition to stirred conc. acid had been effected normally, but when 2 new

1571

batches of pentanethiol were used, flame was observed in the vapour phase a few s after addition. No unusual impurities were detected. The oxidation can be effected safely with nitric acid–water (2:3) with rapid stirring at below 35°C [3]. Oxidation of dodecanethiol [4] and hexadecanethiol [5] with fuming nitric acid proceeded explosively.

2-Alkoxy-1,3-dithia-2-phospholane ('O-Alkyl ethylene dithiophosphate')
  Arbuzov, A. E. et al., Chem. Abs., 1953, 47, 4833e,f
  The methoxy and ethoxy derivatives ignite with the conc. acid.
  See Phosphorus compounds, below

Alkylamines
  1. Vouillamox, L., Chimia, 1995, 49(11), 439
  During propionitrile manufacture by oxidative amination of propanol, nitric acid was used for neutralisation. This eventually led to concentration of nitrates, which exploded, at the bottom of the final distillation column [1]. Many amines form highly crystalline nitrates, which are usually explosive once isolated.
  See AMINIUM NITRATES

2-Aminothiazole
  Biasutti, G. S. et al., Loss. Prev., 1974, 8, 123–125
  In an attempt to produce the nitrate salt, 2-aminothiazole was added to the required amount of nitric acid and the mixture stirred and heated: the mixture exploded without warning.
  See other NITRATION INCIDENTS

2-Aminothiazole, Sulfuric acid
  Silver, L., Chem. Eng. Progr., 1967, 63(8), 43
  Nitration of 2-aminothiazole with nitric–sulfuric acids was normally effected by mixing the reactants at low temperature, heating to 90°C during 30 min and then applying cooling. When the cooling was omitted, a violent explosion occurred. Experiment showed that this was due to a slow exothermic reaction accelerating out of control under the adiabatic conditions. N-Nitroamines were not involved.
  See other NITRATION INCIDENTS, SELF-ACCELERATING REACTIONS

Ammonia
  Mellor, 1940, Vol. 8, 219
  A jet of ammonia will ignite in nitric acid vapour.

Ammonium nitrate
  Biasutti, 1981, 135
  Decomposition of a 70% nitric acid–ammonium nitrate slurry explosive led to overflow, contact with wood and a fire. This spread to detonators, which initiated detonation of the slurry.

Anilinium nitrate
  Andrussow, L., Chim. Ind . (Paris), 1961, 86, 542

1572

Although aniline may be hypergolic with nitric acid (below), anilinium nitrate dissolves unchanged in 98% acid and can be stored for long periods, though the solution has explosive properties.
See Cyclohexylamine, below

Anilinium nitrate, Inorganic additives
Munjal, M. N. *et al., Indian J. Chem.*, 1967, **5**, 320–322
The effect of inorganic additives upon ignition delay in anilinium nitrate–red fuming nitric acid systems was examined. The insoluble compounds copper(I) chloride, potassium permanganate, sodium pentacyanonitrosylferrate and vanadium(V) oxide were moderately effective promoters, while the soluble ammonium or sodium metavanadates were very effective, producing vigorous ignition.

Aromatic amines   MRH Aniline 5.98/19
1. Kit and Evered, 1960, 239, 242
2. *Aniline*, 85, Allied Chem. Corp., New York, 1964
3. Miller, R. O., *Tech. Note No. 3884*, 1–32, Washington, Nat. Advisory Comm. Aeronaut., 1956
4. Spengler, G. *et al., Brennst. Chem.*, 1965, **46**, 1172
5. Bernard, M. L. *et al., Compt. rend. C*, 1966, **263**, 24–26
6. Schalla, R. L. *et al., Amer. Rocket Soc. J.*, 1959, **29**, 38
Many aromatic amines (aniline, *N*-ethylaniline, *o*4-toluidine, xylidine, etc., and their mixtures) are hypergolic with red fuming nitric acid [1]. When the amines are dissolved in triethylamine, ignition occurs at −60°C and below [2]. Addition of a mixture of aniline, dimethylaniline, xylidine and pentacarbonyliron renders hydrocarbons hypergolic with conc. nitric acid [3]. Although aniline is not hypergolic with 96% acid, presence of sulfuric acid (5% or above) renders it so. Presence of dinitrogen tetraoxide further reduces ignition delay [4]. The mechanism of ignition of *p*-phenylenediamine (or furaldehyde azine) on contact with red fuming nitric acid is quantified [5]. Traces of sulfuric acid may be essential for ignition to occur [6].
See Benzidine, below

Aromatic amines, Metal compounds
Rastogi, R. P. *et al., Indian J. Chem.*, 1964, **2**, 301–307
The effects of various metal oxides and salts which promote ignition of amine–red fuming nitric acid systems were examined. Among soluble catalysts, copper(I) oxide, ammonium metavanadate, sodium metavanadate, iron(III) chloride (and potassium hexacyanoferrate(II) with *o*-toluidine) are most effective. Of the insoluble materials, copper(II) oxide, iron(III) oxide, vanadium(V) oxide, potassium chromate, potassium dichromate, potassium hexacyanoferrate(III) and sodium pentacyanonitrosylferrate(II) were effective.
See Anilinium nitrate, etc., *also* Aromatic amines (reference 3), both above

Aromatic compound
Anon., *Safety Digest Univ. Safety Assoc.*, 1989, **34**, 13

An explosion resulted on nitrating an unrevealed aromatic, possibly a phenethyl alcohol derivative, with fuming nitric at −40°C, adding the acid to substrate. Probably too cold and too basic for initial reaction, leading to runaway (oxidation?) when almost all the nitric acid had been charged. Nitration only proceeds when the acidity is sufficient to protonate much of the nitric acid present, in this context any oxygen in the substrate is a base. The editor's experience is that, for monoalkylbenzenes, nitration is very slow, even with substantial sulfuric acid also present as catalyst and dehydrating agent, at −20°C and below. Red fuming nitric acid, which contains dissolved oxides, encourages oxidation (much more exothermic, autocatalytic and gas evolving) in place of nitration.
See Benzeneethanol, below
See other NITRATION INCIDENTS

Arsine–boron tribromide
See Arsine–boron tribromide: Oxidants

Benzeneethanol, Sulfuric acid
Rashevskaya, S. K. *et al., Zh. Obshch. Khim., 1963,* **33**(12), 3998; *Chem. Abs.,* 1964, 9178f
Indifferent yields of mononitro-products resulted from nitration in 1:1 nitric and sulfuric acid at 0°C. At 18°C there was an explosion. (The editor has dinitrated this compound with no trouble, but he used far more sulfuric acid and added the nitric acid to the other components).
See 2-Methoxyethylbenzene, Sulfuric acid; below
See other NITRATION INCIDENTS

Benzidine
Krishna, P. M. M. *et al., Chem. Abs.,* 1981, **94**, 124107
Benzidine is hypergolic with red fuming nitric acid.
See Aromatic amines, above

Benzonitrile, Sulfuric acid
Anon., *Loss. Prev. Bull.,* 1978, (024), 168
Nitration of benzonitrile by addition to mixed nitrating acid had been effected uneventfully in 250 batches. Plant 'improvements' led to the same dosing vessel being used alternately for the nitrile and mixed acids. Some nitration occurred in the uncooled and unstirred dosing vessel containing a little residual nitrile and an explosion occurred during the second batch in the modified plant.

Benzo[*b*]thiophene derivatives
Brown, I. *et al., Chem. & Ind.,* 1962, 982
During nitration of several amino derivatives, diazotisation and oxidation occurred to produce internal diazonium phenoxide compounds. 5-Acetylamino-3-bromobenzo[*b*]thiophene unexpectedly underwent hydrolysis, diazotisation and oxidation to the explosive compound below.
See 3-Bromo-2,7-dinitro-5-benzo[*b*]thiophenediazonium-4-oxide

See 4-Chloro-2-nitroaniline, below
See other NITRATION INCIDENTS, PLANT CLEANING INCIDENTS

N-Benzyl-*N*-ethylaniline, Sulfuric acid
  Anon., *Loss. Prev. Bull.*, 1978, (024), 169; 1980, (038), 27
  Addition of 2–3 l of the amine to a dosing vessel not completely free of traces of
  mixed acids led to a vigorous decomposition which burst a vent line.
  See other NITRATION INCIDENTS, PLANT CLEANING INCIDENTS

1,4-Bis(methoxymethyl)2,3,5,6-tetramethylbenzene
  Anon., *Jahresber.*, 1972, 84
  Oxidation of the durene derivative to benzenehexacarboxylic acid (mellitic acid)
  in an autoclave is normally effected in stages, initially by heating at 80–104°C
  with the vent open to allow escape of the evolved gases. Subsequent heating to a
  higher temperature with the vent closed completes the reaction. On one occasion
  omission of the first vented heating phase led to explosive rupture of the autoclave
  at 80°C.
  See other GAS EVOLUTION INCIDENTS
  See Hydrocarbons, below

1,3-Bis(trifluoromethyl)benzene, Sulfuric acid
  Grewer, T., *Proc. 1st Int. Symp. Prev. Occ. Risks in Chem. Ind.*, 153–160, Heidelberg, ISSA, 1970
  The fluoro compound is resistant to nitration and an operating temperature of 90°C
  is necessary to ensure formation of the 5-nitro derivative. Under these conditions,
  the atmosphere (containing the fluoro compound, its nitro derivative and nitric
  acid vapours) in the nitration vessel is explosive and above the flash point. An
  unknown ignition source led to an explosion and rupture of the 3 cu. m vessel,
  and a maximum explosion pressure of 50 bar was confirmed experimentally. Such
  explosive atmospheres are not found in low temperature nitration reactions.
  See Chlorobenzene, below
  See other NITRATION INCIDENTS

Bromine pentafluoride
  See Bromine pentafluoride: Acids, etc.

tert-Butyl-*m*-xylene, Sulfuric acid
  Kotoyori, T., private comm., 1983
  Normally, the hydrocarbon (360 kg) was tri-nitrated by slow addition during 15 h
  to premixed 97% nitric–98% sulfuric acids (wt ratio 1.04:1, 1470 kg) agitated at
  35–40°C, maintaining the temperature (checked at 30 min intervals) by controlling
  the jacket cooling and rate of addition. Because of a confusing and inadequate
  control panel layout, stirring was accidentally stopped after 3.3 h and addition
  was continued for a further 2 h before the absence of agitation was noticed, when
  the temperature was 34°C, and addition was stopped. At this stage, a layer of
  the liquid hydrocarbon (45 kg, $d = 0.8$) would be on top of the solid trinitro
  compound (150 kg, m.p. 112°C), itself floating on the mixed acid (d about 1.75).

As the operator realised it would cause an immediate violent exotherm if full agitation were restored, he attempted to start mild agitation by flicking the agitator on for 1 s only. After some 20 s, yellow-brown fumes began to escape from the vent pipe, indicating the onset of exothermic decomposition, so he went to open the dump valve but was prevented by the thick fumes. During the next couple of minutes the velocity of the fumes (now white) and the associated noise increased to a very high level and the vessel finally exploded, the lid being blown off and the contents becoming ignited.
See other AGITATION INCIDENTS, NITRATION INCIDENTS

Cadmium phosphide
Juza, R. *et al., Z. Anorg. Chem.*, 1956, **283**, 230
Reaction with conc. acid is explosive.

Cellulose
MCA SD-5, 1961
Cellulose may be converted to the highly flammable nitrate ester on contact with the vapour of nitric acid, as well as by the liquid acid.

Chlorobenzene
Anon., *Jahresber.*, 1974, 86
In a plant for the continuous nitration of chlorobenzene, maloperation during start-up caused the addition of substantial amounts of reactants into the reactor before effective agitation and mixing had been established. The normal reaction temperature of 60°C was rapidly exceeded by at least 60° and an explosion occurred. Subsequent investigation showed that at 80°C an explosive atmosphere was formed above the reaction mixture, and that the adiabatic vapour-phase nitration would attain a temperature of 700°C and ignite the explosive atmosphere in the reactor.
See 1,3-Bis(trifluoromethyl)benzene, above
See other AGITATION INCIDENTS, NITRATION INCIDENTS, SELF-HEATING AND IGNITION INCIDENTS

4-Chloro-2-nitroaniline
1. Elderfield, R. C. *et al., J. Org. Chem.*, 1946, **11**, 820
2. *MCA Case History No. 1489*
The literature procedure for preparation of 4-chloro-2,6-dinitroaniline [1], involving direct nitration in 65% nitric acid, was modified by increasing the reaction temperature to 60°C 1 h after holding at 30−35°C as originally specified. This procedure was satisfactory on the bench scale, and was then scaled up into a 900 l reactor. After the temperature had reached 30°C, heating was discontinued, but the temperature continued to rise to 100−110°C. Decomposition set in with copious evolution of nitrous fumes and production of a very shock-sensitive explosive solid. This was identified as 4-chloro-2,5-dinitrobenzenediazonium-6-oxide, produced by hydrolysis of a nitro group in the expected product by the diluted nitric acid at high temperature, diazotisation of the free amino group by the nitrous acid produced in the hydrolysis (or by the nitrous fumes), and introduction of a further nitro

group under the prevailing reaction conditions. It is recommended that primary aromatic amines should be protected by acetylation before nitration, to prevent the possibility of accidental diazotisation.
See Benzo[*b*]thiophene derivatives, above
See also ARENEDIAZONIUM OXIDES
See other NITRATION INCIDENTS

Coal
Green, J. B. *et al., Anal. Chem.*, 1979, **51**, 1126
The mixtures of bituminous coal and conc. nitric acid used to prepare nitrohumic acids are potentially explosive, and appropriate care is necessary.
See Organic materials, below

Contaminants
NSC Safety Newsletter, Chem. Sect., July, 1977
A 50% nitric acid solution was used to clean glass containers and then stored in a screw-capped bottle. Some time later the bottle burst, owing to gas generation by the dissolved contaminant(s).
See other GAS EVOLUTION INCIDENTS

Copper
1. Anon., *Loss Prev. Bull.*, 1987, (074), 23–27
2. Anon., *Loss Prev. Bull.*, 1988, (080), 31–34
A 'canned' electric pump was being used to pump 89% nitric acid during a plant malfunction when a violent explosion occurred. This was attributed to penetration of nitric acid into the sealed copper windings of the motor and subsequent generation of nitric oxide which pressurised the stator enclosure to some 60–280 bar, causing it to burst. Preventive precautions are detailed [1]. The pump manufacturers, however, subsequently challenged this account and explanation and revealed that the pump, which had been returned for service and repair on previous occasions, had shown signs of reverse rotation and presence of nitric acid in the windings. Such pressure build-up was most unlikely, since the stator enclosure had both an external vent and a separate pressure relief valve. The remains of the exploded pump also showed signs of reverse rotation and the impellor appeared to have come loose on the shaft prior to the explosion. The latter occurred when the motor was energised via replacement fuses 50% above recommended capacity. The precise cause of the explosion is not clear, but may well have been spark-initiated [2].
See IGNITION SOURCES
See other GAS EVOLUTION INCIDENTS

Copper(I) nitride
Mellor, 1940, Vol. 8, 100
Interaction with conc. acid is very violent.

Cotton, Rubber, Sulfuric acid, Water
Biasutti, 1981, 76

A rubber glove and a box of cotton waste fell into a tank of spent nitrating acid from the preparation of TNT. The total contents of the tank ignited, leading to a series of explosions which caused severe damage and some 900 casualties, including 82 killed.

See other NITRATION INCIDENTS

Crotonaldehyde

Andrussow, L., *Chim. Ind* . (Paris), 1961, **86**, 542
Crotonaldehyde is hypergolic with conc. nitric acid, ignition delay being 1 ms.
See ROCKET PROPELLANTS

Cycloalkanones

1. Dye, W. T., *Chem. Eng. News*, 1959, **37**, 48
2. Field, K. W. *et al., J. Chem. Educ.*, 1985, **62**, 637

Oxidation of 4-methylcyclohexanone by addition of nitric acid at about 75°C caused a detonation to occur. These conditions had been used previously to oxidise the corresponding alcohol, but although the ketone is apparently an intermediate in oxidation of the alcohol, the former requires a much higher temperature to start and maintain the reaction. An OTS report, PB73591, mentions a similar violent reaction with cyclohexanone [1]. Presence of nitrous acid is essential for the smooth oxidation of cycloalkanones with nitric acid to α, ω-hexanedioic acids. Because high-purity nitric acid (free of nitrous acid) is now commonly available, addition of a little sodium or potassium nitrite to the acid is necessary before its use to oxidise cycloalkanones [2].
See Acetone, etc., above

Cyclohexanol, Cyclohexanone

1. Wilson, B. J. et al., *Loss. Prev.*, 1979, **12**, 27–29
2. Vidal, P. *et al., Thermochim. Acta*, 1993, **225**(2), 223

The process for manufacturing adipic acid by nitric acid oxidation of KA (the ketone–alcohol mixture produced by air oxidation of cyclohexane in the liquid phase) at 68°C is potentially hazardous, in that reactor or cooling failure could lead to a large and vigorous evolution of hot toxic nitrous fumes. The overall process has been studied and modified to reduce these potential hazards [1]. Solutions of organic diacids in nitric acid, such as result in adipic acid manufacture, were studied for exotherm, gas evolution, deflagration and detonation. Critical diameters were plotted against composition. Recommendations for plant design result [2].

Cyclohexylamine   MRH 6.07/18

Andrussow, L., *Chim. Ind*. (Paris), 1961, **86**, 542
Although cyclohexylamine has been used as a fuel with nitric acid in rocket motors, cyclohexylammonium nitrate dissolves unchanged in fuming nitric acid to give a solution stable for long periods.
See Anilinium nitrate, above
See also ROCKET PROPELLANTS

1,3-Diacetoxybenzene
See 4,6-Dinitro-1,3-benzenediol
See other INDUCTION PERIOD INCIDENTS, NITRATION INCIDENTS

1,2-Diaminoethanebis(trimethylgold)
See 1,2-Diaminoethanebis(trimethylgold)

Di-2-butoxyethyl ether
   Hanson, C. *et al., Proc. Int. Solvent Extr. Conf.* (Liege), 1980, 2, paper 80–70
   The solvent ('Butex'), an extractant in nuclear reprocessing may decompose
   violently in contact with nitric acid. Hydrolysis to butanol, followed by violent
   oxidation, catalysed by nitrous acid, is involved.
   See Alcohols, above

1,2-Dichloroethane
   Kurbangalina, R. K., *Zh. Prikl. Khim.*, 1959, **32**, 1467
   Mixtures are easily detonated by heat, impact or friction.

Dichloroethylene
   Raikova, V. M., *Chem. Abs.*, 1983, **98**, 5958
   Critical detonation diameters for mixtures have been determined.

Dichloromethane
   Andrussow, L., *Chim. Ind* . (Paris), 1961, **86**, 542
   Dichloromethane dissolves endothermically in conc. nitric acid to give a detonable
   solution.

Dichromates, Organic fuels
   Munjal, N. L. *et al., Amer. Inst. Aero. Astronaut. J.*, 1972, **10**, 1722–1723
   The effects of dichromates in promoting ignition of non-hypergolic mixtures of
   red fuming nitric acid with cyclohexanol, 2-cresol or 3-cresol, and furfural were
   studied. Ammonium dichromate was most effective in all cases, and the only
   effective catalyst for cyclohexanol. Potassium chromate and potassium dichromate
   were also examined.

Diethylaminoethanol
   See 2-Diethylammonioethyl nitrate nitrate

Diethyl ether   MRH 5.94/21
   1. Foote, C. S., private comm., 1965
   2. Holt, C., *School. Sci. Rev.*, 1976, **58**(202), 159–160
   3. Fischer, E., *Ber.*, 1902, **35**, 3794
   4. Anon., *Sichere Chemiearb.*, 1981, **33**, 94–95
   5. Anon., *Univ. Saf. Assoc., Saf. Digest*, 1988, (25), 6–8
   Addition of ether to a nitration mixture (2-bromotoluene and conc. nitric acid) diluted
   with an equal volume of water in a separating funnel led to a low order explosion.
   This was attributed to oxidation of the ether (possibly containing alcohol) by the

acid. Addition of more water before adding ether was recommended [1]. Attention is drawn to the delayed vigorous decomposition reaction which occurs some time after the initially homogeneous mixture of conc. nitric acid and ether has separated into 2 phases (possibly involving formation of ethyl nitrate?) [2]. In the preparation of 3-chlorolactic acid by oxidation of epichlorhydrin with nitric acid [3], the product was extracted into ether. Vacuum evaporation of solvent at 40°C led to a violent explosion, wrongly attributed to formation of peroxides. Experiment showed that extraction of 200 ml of a 30% aqueous solution of nitric acid with 8×200 ml portions of ether extracted 88% of the acid into the ether [4]. Upon concentration, this would produce very concentrated warm acid in contact with organic material, which would be oxidised violently or explosively (and could form ethyl nitrate). It is dangerous to extract nitric acid solutions with ether or other organic solvents without first removing the excess acid, either by neutralisation or ion exchange.

Nitric acid was being used to clean laboratory glassware to a very high state of surface cleanliness by soaking in a bucket of conc. acid for a week. During this time, one of the empty winchesters was inadvertently used as a solvent residues bottle and some 250 ml of diethyl ether was put in. The acid was to be recovered for re-use and it was put back into the original bottles. When the bottle containing ether was refilled and screw capped, after a few seconds it exploded violently [5].
See Glassware, below

Diethyl ether, Sulfuric acid
  1. Kirk-Othmer, 1965, Vol. 8, 479
  2. Van Alphen, J., *Rec. Trav. Chim.*, 1930, 492–500
Interaction of ether with anhydrous nitric acid to produce ethyl nitrate may proceed explosively [1], and in presence of conc. sulfuric acid, ether and conc. nitric acid explode violently [2].
See Dimethyl ether, below

3,6-Dihydro-1,2,2*H*-oxazine
  Wichterle, O. *et al., Coll. Czech. Chem. Comm.*, 1950, **15**, 309–321
The product of interaction (possibly an *N*-nitro and/or ring-opened nitrate ester derivative) is explosive.

Dimethyl 4-acetamidophthalate
  See Dimethyl 4-acetamidophthalate: Nitric acid

Dimethylaminomethylferrocene, Water
  Koch-Light Ltd., private comm., 1976
In an assay procedure the amine was heated with diluted nitric acid, and near-explosive decomposition occurred.

Dimethyl ether
  Andrussow, L., *Chim. Ind.* (Paris), 1961, **86**, 542–545
The latent hazards in storing and handling the explosive mixtures with the conc. acid are discussed (methyl nitrate may be formed).
  See Diethyl ether, above

1580

1,1-Dimethylhydrazine
See Hydrazine and derivatives, below

1,1-Dimethylhydrazine, Organic compounds
Spengler, G. *et al., Brennst. Chem.*, 1965, *46*, 117
Contact of nitric acid (or dinitrogen tetraoxide) with dimethylhydrazine is hypergolic and well described in rocket technology. While hydrocarbons and several other classes of organic compounds are not hypergolic with these oxidisers, addition of a proportion of dimethylhydrazine to a wide range of hydrocarbons, alcohols, amines, esters and heterocyclic compounds renders them hypergolic in contact with nitric acid or dinitrogen tetraoxide.

Dimethyl sulfide, 1,4-Dioxane
Rudakov, E. S. *et al., Chem. Abs.*, 1972, **76**, 13515
The mechanism which leads to delayed explosion in the system with nitric acid, even when cooled in liquid nitrogen, was investigated.

Dimethyl sulfoxide, Water
Fiola, J. *et al.*, Czech. Pat. 170 169, 1977
During the oxidation of the sulfoxide to the sulfone, the concentration of water in the reaction mixture must be maintained above 14% to prevent the risk of detonation.

1,3-Dinitrobenzene   MRH 5.02/99+
Biasutti, 1981, 92
After a detonation during manufacture of an explosive solution of the nitro compound in conc. nitric acid, manufacture was discontinued as being too dangerous.

Dinitrogen tetraoxide, Nitrogenous fuels
1. Durgapal, U. C. *et al., Amer. Inst. Aero. Astronaut. J.*, 1974, **12**, 1611
2. Munjal, N. L. *et al., J. Spacecr. Rockets*, 1974, **11**, 4286–430
3. Durgapal, U. C. *et al., J. Spacecr. Rockets*, 1974, **11**, 447–448
The effect on decrease in hypergolic ignition delay of increasing concentrations of dinitrogen tetraoxide in red fuming nitric acid was studied with triethylamine, dimethylhydrazine or *mixo*-xylidine as fuels [1]. The effect of various catalysts on ignition delay after contact of red fuming nitric acid with various arylamine–formaldehyde condensation products was also studied [2,3].

Dioxane, Perchloric acid
See Perchloric acid: Dioxane, Nitric acid

Diphenyldistibene
See Diphenyldistibene: Air, etc.

Diphenylmercury
Whitmore, 1921, 43, 168
Interaction in carbon disulfide, even at −15°C, is violent.

Diphenyltin
Krause, E. *et al., Ber.*, 1920, **53**, 177
Ignition occurs on contact with fuming nitric acid.

Disodium phenyl orthophosphate
Muir, G. D., private comm., 1968
Concentration by evaporation of the nitration product of the phosphate ester caused a violent explosion, possibly involving picric acid derivatives from over-nitration.

Divinyl ether   MRH 6.44/22
Andrussow, L., *Chim. Ind* . (Paris), 1961, **86**, 542
Divinyl ether is hypergolic with conc. nitric acid, ignition delay being 1 ms.
See ROCKET PROPELLANTS

Ethanesulfonamide
Suter, C. M., *Organic Chemistry of Sulfur*, 111, London, Wiley, 1944
Reaction with nitric acid at a moderate temperature is explosive, dinitrogen oxide being evolved.

5-Ethyl-2-methylpyridine
1. Frank, R. L., *Chem. Eng. News*, 1952, **30**, 3348
2. Rubinstein, H. *et al., J. Chem. Eng. Data*, 1967, **30**, 3348
3. Elam. E. V. *et al., Chem. Eng. News*, 1973, **51**(34), 42
Following a patented procedure, the two reactants were being heated together at 145°C/14.5 bar to produce 2,5-pyridinedicarboxylic acid. The temperature rose to 160°/43.5 bar, and the autoclave was vented and cooled, but 90 s later a violent explosion occurred, although both rupture disks (rated 105 and 411 bar) had relieved. General precautions are discussed [1]. Up to 20% of 5-(1,1-dinitroethyl)-2-methylpyridine, probably an explosive compound, is produced in this reaction. However, in presence of added water, no instability was seen in a series of reactions at temperatures up to 160°C and pressures up to 102 bar [2]. A further incident in an electrically heated titanium-lined autoclave involved heating the pyridine with excess 70% acid to 140°C during 1 h. When heating was discontinued, the temperature continued to rise in spite of air cooling, and an explosion occurred at 172°C which bulged the autoclave [3]. It is remotely possible that reaction of the titanium liner with the pressurised hot acid may have contributed to the incident.
See Metals (reference 1), below *also* Titanium: Nitric acid

Fat, Sulfuric acid
1. Tyler, L. J., *Chem. Eng. News*, 1973, **51**(31), 32
2. Engan, W. L., *Chem. Eng. News*, 1973, **51**(51), 37
A wet-ashing procedure for analysis of fatty animal tissue was modified by using Teflon-lined bombs rated for use at 340 bar instead of open crucibles. Bombs cooled to well below 0°C were charged with fuming nitric and fuming sulfuric acids (1 ml of each) and adipose tissue (0.5 g), removed from the cooling bath and sealed. After 10 min delay, the bombs exploded, probably owing to development of

high internal temperature and pressure (calculated as 4000 K and 1000 bar max.) from complete oxidation of all the organic material [1]. Formation of glyceryl nitrate from the lipid content may have contributed to the violence of the explosion. The presence of 2–5% of water in the mixed acids is recommended to reduce the nitrating potential when in contact with organic fuels [2].

See Organic materials
See Organic materials
See Sulfuric acid

Formaldehyde

1. Pirie, C. J. *et al., Chem. Brit.*, 1979, **15**, 11–12
2. Bretherick, L., *CISHC Chem. Safety Summ.*, 1979, **50**(197), 3
3. Vergnano, L. P., *CISHC Chem. Safety Summ.*, 1979, **50**(197), 2
4. Thurger, E. T., *CISHC Chem. Safety Summ.*, 1979, **50**(197), 3
5. Walkcr, J. F., *Formaldehyde, ACS Monograph 159*, 254–255, New York, Reinhold, 3rd ed. 1964
6. Winsor, L., *Chem. Brit.*, 1979, **15**, 281–282
7. Anon., *Loss. Prev. Bull.*, 1978, (020), 42
8. Haas, P. A., *Chem. Eng. News*, 1987, **65**(16), 2

A closed bottle of decalcifying fluid (a 10% nitric acid–formalin mixture) exploded a few h after the contents had been used and returned to the bottle for storage [1]. The incident was attributed to the interaction of an oxidant with a reducant with evolution of gas, promoted by dissolved impurities arising from its use, leading to pressurisation of the bottle to its bursting point [2,3,4]. This reaction is discussed in detail [5]. 5–10% Solutions of nitric acid or formic acid are preferred decalcifying fluids [6]. After failure of a pump which continuously fed nitric acid into circulating formaldehyde solution (used as a plant cleaning mixture), 30 l of 60% nitric acid was batch-charged into a filter vessel in the fluid circuit. When circulation was restarted to mix in the acid and continue the cleaning operation, a violent redox reaction blew off the lid of the filter vessel [7]. Process development wastes containing urea and hexamethylenetetramine were acidified with nitric acid to dissolve hydrated metal oxides. Under these conditions, hexamine hydrolyses to formaldehyde and ammonium nitrate, and it was calculated that the composition of the acidified waste would be 2M nitric acid, 2M formaldehyde, 1M ammonium nitrate and 0.5M metal nitrate. Such mixtures stored for 2 months in a loosely capped polythene bottle decomposed with vigorous gas evolution which ruptured the container [8].

See other GAS EVOLUTION INCIDENTS, PLANT CLEANING INCIDENTS, REDOX REACTIONS

Formic acid

1. Halze, K. *et al., Proc. 3rd. Int. Symp. Loss. Prev. Safety Prom. Proc. Ind.*, 1413–1424, Basle, SSCI, 1980

The use of excess formic acid to destroy excess nitric acid (5M) in nuclear fuel reprocessing waste solutions at 100°C is potentially hazardous because of an induction period, high exothermicity and the evolution of large amounts of gas, mainly carbon dioxide, dinitrogen oxide and nitrogen oxide, with some nitrogen and dinitrogen tetraoxide. The system has been studied thermokinetically, and the effects of various salts (which decrease the reaction rate) and sulfuric acid (which increases the rate) were determined [1].

See other GAS EVOLUTION INCIDENTS, INDUCTION PERIOD INCIDENTS, REDOX REACTIONS

Formic acid, Oxalic acid

Cardillo, P. *et al., Chim. Ind.* (Milan) 1989 **71**(4) 61

A study of, and details of how to control, the stop–start nature of the oxidation of formic acid. Probably relevant to other nitric acid oxidations, especially preparation of glyoxylic acid from glyoxal.

Formic acid, Urea

Mejdell, G. T., *Proc. 3rd. Int. Symp. Loss. Prev. Safety Prom. Proc. Ind.*, 1325–1338, Basle, SSCI, 1980

In the production of formic acid, a slurry of calcium formate in 50% aqueous formic acid containing urea is acidified with strong nitric acid to convert the calcium salt to free acid, and interaction of formic acid (reducant) with nitric acid (oxidant) is inhibited by the urea. When only 10% of the required amount of urea had been added (unwittingly, because of a blocked hopper), addition of the nitric acid caused a thermal runaway (redox) reaction to occur which burst the (vented) vessel. A small-scale repeat indicated that a pressure of 150–200 bar may have been attained. A mathematical model was developed which closely matched experimental data.

See other REDOX REACTIONS, RUNAWAY REACTIONS

2-Formylamino-1-phenyl-1,3-propanediol

1. Biasutti, G. S. *et al., Loss Prev.*, 1974, **8**, 123
2. Biasutti, 1981, 131

Large scale nitration to the *p*-nitro derivative in a stirred and pump-circulated reactor had been effected uneventfully for 15 years when a violent explosion occurred. Subsequent investigation showed that with sufficient priming energy, the reaction mixture could be caused to detonate with an energy release of 3.77 kJ/g [1]. It was supposed that the priming energy originated from a bolt from the agitation gear being sucked into the reactor circulating pump [2].

See other AGITATION INCIDENTS, NITRATION INCIDENTS

Fluorine   MRH 0.50/47

See Fluorine: Nitric acid

Furfural

Giller, S. A. *et al.*, Brit. Pat., 1 466 043, 1977

A safe method of nitrating furfuraldehyde or its *O,O*-diacetate to 5-nitrofurfural diacetate is described which minimises the formation of acetyl nitrate.

Furfurylidene ketones

Panda, S. P. *et al., J. Armam. Stud.*, 1974, **10**(2), 44–45; 1976, **12**, 138–139

In 10 out of 12 cases, furfurylidene derivatives of ketones were hypergolic with red fuming nitric acid. Difurfurylidene hydrazine behaves similarly, and a mechanism involving polymerisation of the compounds is proposed.

Glassware
Anon., *Univ. Saf. Assoc., Saf. Digest*, 1988, (25), 6–8
Nitric acid has long been used to ensure absolute surface cleanliness on laboratory glassware where this is essential for technical reasons. Chromic acid cannot be used for this purpose because a surface film of adsorbed chromium ions remains after cleaning. Following an accident in which some diethyl ether was inadvertently put into the bottle used to store the conc. nitric acid after use and which led to an explosion, a Code of Practice covering the use of acid cleaning procedures for glassware was developed. This proposed that formulated detergents should be used for routine cleaning, and that use of acid processes would only be permitted if it could be practically demonstrated that they were essential for the proposed work. The Code covers the preparation of written procedures dealing with all aspects of the use, storage and disposal of acids, and personal protection requirements.
See Diethyl ether, above;
See CLEANING BATHS FOR GLASSWARE
See other GLASS INCIDENTS

Glycerol
Sanders, R. E., *Chem. Eng. Progress*, 1992, 88(5), 78
Glycerol was inadvertently used as the fluid in a gas bubbler on a nitric acid plant. It nitrated and detonated two days later.

Glycerol, Hydrofluoric acid
1. Buck, R. H., *J. Electrochem. Soc.*, 1966, **113**, 1352–1353
2. Wright, J. L., private comm., 1984
3. Anon., *Safety Digest Univ. Safety Assoc.*, 1992, **44**, 25
A chemical polishing solution consisting of nitric acid and hydrofluoric acid (1 Vol. each) and glycerol (2 vols.) generated enough pressure during storage for 4 h to rupture the closed plastics container. This was caused by gas evolution from oxidation of glycerol by the strongly oxidising mixture [1]. A mixture of nitric acid (80 ml), hydrofluoric acid (80 ml) and glycerol (240 ml) was used immediately for etching metal, again the next day, and then stored in a stoppered flask. After some 2–3 days, the stopper was ejected and 300 ml was sprayed around the fume cupboard containing the flask [2]. The metals dissolved during use further destabilise the mixture, which should not be stored under any circumstances. A freshly made etching mixture with hydrochloric acid, which one would expect to catalyse the oxidant power of nitric acid, burst its bottle in 30 minutes [3].
See Alcohols, above; Hydrofluoric acid, Lactic acid, below
See other GAS EVOLUTION INCIDENTS

Glycerol, Sulfuric acid
Dunn, B. W., *Bur. Expl. Accid. Bull.*, 35, Pennsylvania, Amer. Rail Assoc., 1916

Charging mixed nitrating acid into an insufficiently cleaned glycerol drum led to a violent explosion. Formation and detonation of glyceryl nitrate may have added to the oxidation energy release.

Glyoxal
Urben, P. G., personal experience
The oxidation of aqueous ethanedial to glyoxylic acid is inclined to induction periods and then runaway, cessation and renewed runaway, as the nitric acid is progessively added. Probably this is the same problem as with formic acid, oxalic acid (above).

Hexalithium disilicide
See Hexalithium disilicide: Acids

2,2,4,4,6,6-Hexamethyltrithiane
Baumann, E. *et al., Ber.*, 1889, **22**, 2596
Oxidation of 'tri(thioacetone)' by conc. nitric acid is explosively violent.
See Thioaldehydes, etc., below

2-Hexenal
Lobanov, V. I., *Chem. Abs.*, 1966, **64**, 14013c
Hexenal is determined by photocolorimetry after oxidation with nitric acid. The yellow-orange oxidation product explodes on heating.

Hydrazine and derivatives
1. Andrussow, L., *Chim. Ind* . (Paris), 1961, **86**, 542
2. Jain, S. R., *Combust. Flame*, 1977, **28**, 542
3. Jain, S. R. *et al., J. Spacecr. Rockets*, 1979, **16**, 69–73
Hydrazine is hypergolic with conc. nitric acid [1]. Of a series of hydrazones and azines derived from aldehydes and ketones, only those which decomposed when heated alone were hypergolic with the acid when heated at 12.5°C/min [2]. Solid hydrazones formed from various aldehydes with dimethylhydrazine or phenylhydrazine are hypergolic with the acid [3].
See ROCKET PROPELLANTS

Hydrocarbons
1. Andrussow, L., *Chim. Ind.* (Paris), 1961, **86**, 542
2. Wilson, P. J. *et al., Chem. Rev.*, 1944, **34**, 8
3. Trent, C. H. *et al., J. Amer. Rocket Soc.*, 1951, **21**, 129–131
4. Ingolgaonkar, M. B. *et al., J. Armam. Stud.*, 1979, **15**(2), 33–42
5. Sykes, W. G. *et al., Chem. Eng. Progr.*, 1963, **59**(1), 70–71
6. Mason, C. M. *et al., J. Chem. Eng. Data*, 1965, **10**, 173
7. Urbanski, 1961, Vol. 1, 140
8. Derbyshire, D. H., *Proc. 1st Symp. Chem. Process Haz. Ref. Plant Des.*, 37–41, Symp. Ser. No. 7, London, IChE, 1960
9. Powell, G. *et al., Org. Synth.*, 1943, Coll. Vol. 2, 450
10. Wilms, H. *et al., Angew. Chem.*, 1962, **74**, 465

11. Chaigneau, M. *Compt. rend.*, 1951, **223**, 657–659
12. Anon., *Chemisches Ind.*, 1914, **37**, 337–342
13. Biasutti, 1981, 48
14. Bryant, T. J. *et al., Rept. AD-B-000447*, Richmond (Va.), USNTIS, 1974
15. Panda, S. P. *et al., Chem. Abs.*, 1981, **94**, 194475
16. Dutta, P. K. *et al., J. Armam. Stud.*, 1976, **12**, 108–117
17. Mellor, 1964, Vol. 8, Suppl. 2, 341

MRH Benzene 8.15/18, cyclopentadiene 6.44/18, divinylbenzene 6.23/19, hexane 6.07/18, toluene 6.07/18, *o*-xylene 6.07/17

Dienes and acetylene derivatives are hypergolic in contact with conc. nitric acid, ignition delay being 1 ms [1]. Cyclopentadiene reacts explosively with fuming nitric acid [2], igniting under nitrogen [3], and dicyclopentadiene is highly hypergolic with red or white fuming nitric acid, also in presence of sulfuric acid [4]. Burning fuel oil and other petroleum products detonate immediately on contact with conc. nitric acid [5]. Very high sensitivity is shown by mixtures with benzene close to the stoicheiometric proportions of around 84% acid [6]. Lack of proper control in nitration of toluene with mixed acids may lead to runaway or explosive reaction. A contributory factor is the oxidative formation, and subsequent nitration and decomposition, of nitrocresols [7]. Oxidation of *p*-xylene with nitric acid under pressure in manufacture of terephthalic acid carries explosion hazards in the autoclaves and condensing systems [8]. During nitration of mesitylene in acetic acid–acetic anhydride solution, fuming nitric acid must be added slowly to the cooled mixture to prevent the temperature exceeding 20°C, when an explosive reaction may occur [9]. During oxidation of mesitylene with nitric acid in an autoclave at 115°C to give 3,5-dimethylbenzoic acid, a violent explosion occurred. Ths was attributed to local overheating, formation of 1,3,5-tris(nitromethyl)benzene and violent decomposition of the latter. Smaller scale preparations with better temperature control were uneventful [10].

The explosive hazards involved in the preparation of mellitic acid by fuming nitric acid oxidation of hexamethylbenzene make the procedure only suitable for small scale operation [11]. Large-scale addition of too-cold nitrating acid to benzene without agitation later caused an uncontrollably violent reaction to occur when stirring was started. The vapour–air mixture produced was ignited by interaction of benzene and nitric acid at 100–170°C and caused an extremely violent explosion [12]. Failure to start the agitator or to apply water cooling in nitration of benzene led to a pressure burst of the nitrating vessel [13]. In the impact-initiated combustion of hydrocarbons in acid at high pressure, the effects of impact, hydrocarbon structure and stoicheiometry upon ignition were studied [14]. The pure terpene 3-carene is more hypergolic with red fuming nitric acid than the mixture with other terpenes in turpentine [15]. The use of turpentine and cashew nutshell oil as bases for hypergolic rocket fuels with red fuming nitric acid has been evaluated [16]. The nitric acid–gasoline propellant system is non-hypergolic, but can be made so by addition of 15–25% of unsaturated hydrocarbons to the gasoline [17].

See other AGITATION INCIDENTS, NITRATION INCIDENTS

Hydrocarbons, 1,1-Dimethylhydrazine
Ingolgaonkar, M. B. *et al., J. Armam. Stud.*, 1979, **15**(2), 33–42
Presence of the hydrazine confers hypergolicity upon mixtures of several non-hypergolic hydrocarbons with red fuming nitric acid.

Hydrofluoric acid, Lactic acid
Bubar, S. F. *et al., J. Chem. Educ.*, 1966, **43**, A956
Mixtures of the 3 acids, used as metal polishing solutions, are unstable and should not be stored. Lactic acid and nitric acid react autocatalytically after a quiescent period, attaining a temperature of about 90°C with vigorous gas evolution after about 12 h. Prepare freshly, discard after use and handle carefully.
See Glycerol, etc., above
See other INDUCTION PERIOD INCIDENTS

Hydrofluoric acid, Propylene glycol, Silver nitrate
Leleu, *Cahiers*, 1979, (94), 125
A chemical polishing mixture was put into a closed glass bottle which burst 30 min later, and formation of silver fulminate was suggested. However, in absence of the silver salt such mixtures evolve gas and should not be stored in any event, especially after use for metal polishing, when the dissolved metal(s) tend to further destabilise the mixture.
See Glycerol, etc., above

Hydrogen iodide, or Hydrogen selenide, or Hydrogen sulfide
See Non-metal hydrides, below

Hydrogen peroxide, Ketones
See Hydrogen peroxide, Ketones, etc.

Hydrogen peroxide, Mercury(II) oxide
See Hydrogen peroxide: Mercury(II) oxide, etc.

Hydrogen peroxide, Soils
Krishnamurty, K. V. *et al., At. Abs. Newslett.*, 1976, **15**, 68–70
When preparing soil and sediment samples for atomic absorption spectral analysis for trace metals, pre-oxidation with nitric acid before addition of hydrogen peroxide eliminates the danger of explosion.

Ion exchange resins
1. Barghusen, J. *et al., Reactor Fuel Process.*, 1964, **7**, 297
2. McBride, J. A. *et al., 3rd Geneva Conf. Peaceful Uses At. En. A/CONF. 28 28/P/278*, 1965
3. Kalkworf, D. R., *Rept. BNWL-2931*, Richland (Wa.), Battelle, 1977
4. Colman, C., *Chem. Eng.* (New York), 1980, **87**, 271–274
Several cases of interaction between anion exchange resins and nitric acid causing rapid release of energy or explosion have occurred [1]. The cause has been attributed to oxidative degradation of the organic resin matrix and/or nitration of

the latter. Suggested precautions include control of temperature, acid concentration and contact time. Presence of heavy ions (Pu) or oxidising agents (dichromates) tends to accelerate the decomposition [2]. The present state of knowledge was reviewed and relevant process parameters identified [3]. Several case histories are given, and safety aspects of regeneration of large-volume beds of strong-base anion exchange resins or cationic exchangers with nitric acid are discussed, but weak-base resins should not be treated with nitric acid. Storage of nitric acid-containing resins may lead to ignition [4].

See other SELF-HEATING AND IGNITION INCIDENTS

Iron(II) oxide
Mellor, 1941, Vol. 13, 716
The finely divided (pyrophoric) oxide incandesces with nitric acid.

Lead-containing rubber
Johnson, T. C. *et al., Rep. RFP-1354*, 1–8, Washington (DC), USAEC, 1969
Radiation resistant lead-containing dry-box gloves may ignite in nitric acid environments.

Magnesium, 2-Nitroaniline
Gune, S. G. *et al., J. Armam. Stud.*, 1983, **18–19**, 39–41
Though neither of the components is individually hypergolic with red fuming nitric acid, mixtures with magnesium containing 20% and 30% of 2-nitroaniline ignite with 67 and 23 ms delay, respectively.

Magnesium silicide
Mayes, R. B. *et al., School Sci. Rev.*, 1975, **56**(197), 819–820
When the residue from combustion of magnesium in air was removed from the porcelain crucible, a grey stain remained. Addition of cold conc. nitric acid to remove the stain led to a violent reaction. This was found to be caused by the presence of magnesium silicide in the stain.

Magnesium phosphide
Mellor, 1940, Vol. 8, 842
Oxidation of the phosphide proceeds with incandescence.

Manufacture hazard
Lawrence, G. M. , *Plant/Oper. Progr.*, 1989, **8**(1), 33
An explosion in the vent of an ammonia combustion plant was attributed to deposition of ammonium nitrite/nitrate crystals. It is considered that the very unstable nitrite acts as a sensitiser to the nitrate, and that explosion is triggered by contact with acid.
See Ammonium nitrate, Ammonium nitrite

Metal acetylides
Mellor, 1946, Vol. 5, 8482

Caesium acetylide and rubidium acetylide explode in contact with nitric acid, and the sodium and potassium analogues probably react violently.

Metal cyanides
491M, 1975, 268
Explosive reactions.

Metal hexacyanoferrates
Sidgwick, 1950, 1344
The action of 30% nitric acid on hexacyanoferrates(II) or (III) to produce penta-cyanonitrosylferrate(II) ('nitroprusside') is violent.

Metals
1. Mellor, 1940, Vol. 2, 470; 1940, Vol. 4, 270, 483; 1941, Vol. 7, 260; 1940, Vol. 9, 627; 1942, Vol. 12, 32, 188
2. Pascal, 1956, Vol. 10, 504
3. Condike, G. F., *Chemistry*, 1974, **47**(4), 27–28
4. McGarman, A. R., *Chemistry*, 1974, **47**(11), 27–28
MRH Calcium 9.33/61, lithium 15.27/40, magnesium 11.96/48, manganese 4.88/59, nickel 2.64/70, sodium 5.48/64, zinc 3.60/72

Bismuth powder glows red-hot in contact with fuming nitric acid, while the molten metal (271°C) explodes in contact with conc. acid. Powdered germanium reacts violently with the latter, and lithium ignites. Manganese powder incandesces and explodes feebly with nitric acid, and sodium ignites with nitric acid of density above 1.056. Titanium alloys form an explosive deposit with fuming nitric acid. Although uranium powder reacts vigorously with red fuming nitric acid, under some conditions explosive deposits may be formed. Addition of conc. acid to molten zinc (419°C) causes it to incandesce, and magnesium burns brilliantly in nitric acid vapour [1]. Antimony may be attacked violently by fuming nitric acid [2]. The experimental quantities of tin and nitric acid specified for a laboratory demonstration [3] are likely to lead to formation of the explosive nitrate oxide [4].
See Tetrahydroxotritin(2+) nitrate, Silver nitrate: Titanium
See Copper, above

Metal salicylates
Belcher, R., private comm., 1968
Metal salicylates are occasionally incorporated into mixtures of 'unknowns' for qualitative inorganic analysis. During the conventional group separation, organic radicals are removed by evaporation with nitric acid. When salicylates are present, this can lead to formation of trinitrophenol through nitration and decarboxylation. This may react with any heavy metal ions present to form unstable or explosive picrates, if the evaporation is taken to dryness. The MAQA alternative scheme of analysis obviates this danger.
See Lead(II) picrate

Metal thiocyanate   MRH Potassium thiocyanate 4.31/42
MCA Case History No. 853

When the (unspecified) thiocyanate solution was pumped through an 80 mm pipeline containing nitric acid, a violent explosion occurred. This was confirmed experimentaly and attributed to the redox reaction between the reducing solution and the oxidant acid.

See other REDOX REACTIONS

Methanol, Sulfuric acid
Anon., *Chem. & Ind.*, 1986, 725
A nitration reactor was charged with methanol which was wrongly labeled as xylene. Addition of nitrating acid mixture led to a runaway exothermic reaction which eventually ruptured the reactor as well as the bursting disk, allowing 2700 l of mixed acid to spray out.

See Alcohols, above
See related NITRATION INCIDENTS

2-Methoxyethylbenzene, Sulfuric acid
Urben, P. G., personal experience
This substrate, dissolved in 3 molar equivalents of sulfuric acid, was batch nitrated by slow addition of 1.05 moles of concentrated nitric acid at 5–20°C followed by a stir-out and quenching into water. No problems were seen in the laboratory but on pilot plant ($\frac{1}{2}$ kilomole) there was twice a sharp exotherm $2\frac{1}{2}$ hours after completion of addition, blowing the bursting disk with gas evolution, leaving 4-nitrobenzoic acid as the (minor) isolable product. With good thermal insulation and poor agitation it was possible to reproduce this in the lab. No evidence of dinitration could be found and mononitration appeared almost instantaneous. Reduction of the nitric acid to 0.99 equivalents prevented recurrence, even on a much larger scale with day long delays, while not changing the near quantitative yield.

The end point corresponded to generation of sulfuric acid monohydrate; the experimenter subsequently adopted the rules of thumb that oxidative side reactions do not occur while the availability of water is less than this, and that dinitration side-products are a consequence of charging substrate, traditionally but irrationally, to nitration mixture.

See related NITRATION INCIDENTS

2-Methylbenzimidazole, Sulfuric acid
Zmojdzin, A. *et al.*, Ger. Offen. 2 310 414, 1974
A safe method of preparing 5-nitro-2-methylbenzimidazole involves preliminary addition of the heterocycle to nitric acid (*d* 1.40) with subsequent addition of sulfuric acid, keeping the temperature below 110°C.

See related NITRATION INCIDENTS

4-Methylcyclohexanone
Dye, W. T., *Chem. Eng. News*, 1959, **37**, 48
Oxidation of 4-methylcyclohexanone by addition of nitric acid at about 75°C caused a detonation to occur. These conditions had been used previously to oxidise the corresponding alcohol, but although the ketone is apparently an intermediate in oxidation of the alcohol, the former requires a much higher temperature to start

and maintain the reaction. An OTS report, PB73591, mentions a similar violent reaction with cyclohexanone.

See Acetone, etc., above

Methylthiophene
Rüst, 1948, 318

During the nitration of methylthiophene, direct liquid contact caused ignition to occur, so air saturated with the organic vapour was passed into the cooled conc. acid. After a while, the whole apparatus exploded violently, probably owing to ignition of the air–fuel mixture being passed in.

See other NITRATION INCIDENTS *SEE* THIOPHENE, BELOW

Molybdenum nitride
See Molybdenum nitride: Nitric acid

Naphthalene-1-sulfonic acid, Sulfuric acid
Anon. *Loss Prev. Bull.*, 1994, (116), 21

Production of dye intermediates by nitration of the sulfonic acid was conducted under automatic control, nitric acid being added automatically so long as the temperature was below 30°C. There was no other control on nitric acid addition. The agitation and probably also the thermocouple reading being inadequate, excess nitric acid appears to have charged, layered, then reacted, blowing apart the vessel.

See other AGITATION INCIDENTS

Nitroaromatics
1. Urbanski, 1967, Vol. 3, 290
2. Raikova, V. M., *Chem. Abs.*, 1983, **98**, 5958

MRH Nitrobenzene 6.32/29, nitroxylene 6.23/25, dinitrobenzene 5.02/99+, 2,4-dinitrotoluene 4.56/99+

A series of mixtures of nitric acid with one or more of mono- and di-nitrobenzenes, di- and tri-nitrotoluenes have been shown to possess high-explosive properties [1]. Critical detonation diameters for mixtures of nitric acid with dinitrotoluene and TNT have been determined [2].

Nitrobenzene, Sulfuric acid
Fritz, E. J., *Loss Prev.*, 1969, **3**, 41–44

Failure of the agitator during addition of mixed nitrating acid allowed unreacted reagents to build up in the reaction system which, owing to absence of the usual reaction exotherm, became under-cooled. Application of heat and agitation caused a runaway reaction, terminating in explosion, to occur. Laboratory-scale repetition of this sequence showed an exotherm to 200°C in 0.1 s, a rate of almost 2000°/s. Operating improvements are detailed.

See other AGITATION INCIDENTS, NITRATION INCIDENTS, RUNAWAY REACTIONS

Nitrobenzene, Water
1. Anon., *J. R. Inst. Chem.*, 1960, **84**, 451
2. Van Dolah, R. W., *Loss. Prev.*, 1969, **3**, 32

A plant explosion involved a mixture of nitrobenzene, nitric acid and a substantial quantity of water. Detonation occurred with a speed and power comparable to TNT. This was unexpected in view of the presence of water in the mixture [1]. The later reference deals with a detailed practical and theoretical study of this system and determination of the detonability limits and shock-sensitivity. The limits of detonability coincided with the limits of miscibility over a wide portion of the ternary composition diagram. In absence of water, very high sensitivity (similar to that of glyceryl nitrate) occurred between 50 and 80% nitric acid, the stoicheiometric proportion being 73% [2].
See Nitroaromatics, above

Nitromethane   MRH 6.19/62
Olah, G. A. *et al., Org. Synth.*, 1957, **47**, 60
Mixtures are extremely explosive.

1-Nitronaphthalene, Sulfuric acid
Biasutti, 1981, 83
Towards the end of nitration of mononitronaphthalene to trinitronaphthalene, a drain valve became blocked with tarry solid. Raising the temperature to 60°C to melt the obstruction led to separation of more solid, failure of the agitator, then a runaway reaction and detonation.
See other AGITATION INCIDENTS, NITRATION INCIDENTS, RUNAWAY REACTIONS

Non-metal hydrides
1. Mellor, 1946, Vol. 5, 36; 1940, Vol. 6, 814; 1939, Vol. 9, 56, 397
2. Hofmann, A. W., *Ber.*, 1870, **3**, 658–660
3. Pascal, 1956, Vol. 10, 505
Arsine, phosphine and tetraborane(10) are all oxidised explosively by fuming nitric acid, while stibine behaves similarly with the conc. acid [1]. Hydrogen iodide, hydrogen selenide, hydrogen sulfide and phosphine all ignite when the fuming acid is dripped into the gas [2]. Hydrogen telluride ignites with cold conc. acid, sometimes exploding [3].
See Ammonia, above; and Phosphine derivatives, below

Non-metals
1. Mellor, 1946, Vol. 5, 16; 1940, Vol. 8, 787, 845
2. Bailar, 1973, Vol. 1, 1337
3. Pascal, 1956, Vol. 10, 504
4. *MCA Case History No. 1969*
Boron (finely divided forms) reacts violently with conc. acid and may attain incandescence. The vapour of phosphorus, heated in nitric acid in presence of air, may ignite. Boron phosphide ignites with the conc. acid [1]. Silicon crystallised from its eutectic with aluminium reacts violently with conc. acid [2], arsenic may react violently with the fuming acid, and finely divided carbon similarly with conc. acid [3]. Use of conc. acid to clean a stainless steel hose contaminated with phosphorus led to an explosion [4].

Oleoyl chloride

Nolan, 1983, Case history 62

3-Nitro-4-cresol was to be manufactured by adding 4-toluidine to nitric acid. One carboy of nitric acid was charged into the reactor, but the next one charged actually contained oleoyl chloride in error. The violent reaction ensuing ruptured the reactor and there was a fire.

See related NITRATION INCIDENTS

Organic diacids

See Cyclohexanol, Cyclohexanone, above

Organic materials

1. Bowen, H. J. M., private comm., 1968
2. Turnbull, A. J., *Chem. Brit.*, 1979, **15**, 377
3. Williams, E. V., *J. Food Technol.*, 1978, **13**, 167
4. Everson, R. J., *Chem. Eng. News*, 1980, **58**(11), 5
5. Coleman, C. B., *Chem. Eng. News*, 1980, **58**(11), 5
6. Katz, S. A., *Chem. Eng. News*, 1980, **58**(17), 51
7. Borgaard, O. K. *et al.*, *J. Assoc. Offic. Anal. Chem.*, 1982, **65**, 762–763
8. Anon., *Inst. Univ. Safety Officers Bull.*, 1987, (8), 9
9. Ward, N. I., *Internat. Analyst*, 1987, **1**(2), 7

When 16 M (70%) nitric acid was poured down a sink without diluting water, interaction with (unspecified) organic material in the trap caused a delayed explosion [1]. Digestion of biological materials with nitric acid in Teflon-lined bombs has led to several violent explosions. A suggested improvement was to leave the sample in contact with nitric acid overnight before sealing and heating the bomb [2]. An alternative procedure of digesting samples with nitric acid at 20°C for 24 h in polystyrene tubes with hand-tightened polythene screw caps [3] also led to explosions when generous samples of dry-powdered food products were digested [2]. Digestion of 0.8 g of dry dog-food with 5 cc of nitric acid at 130°C in Teflon lined stainless bombs led to explosions [4], or of 0.2 g of sewage sludge with 2.5 cc of acid [5]. The amount of organic material is normally restricted to 0.1 g in the bomb manufacturer's instructions [6]. Note that 0.1 g of carbon on oxidation with nitric acid gives a minimum volume of 540 cc of gas measured at 1 bar/120°C, so a static pressure of many bar would be attained in a 10–20 cc bomb. The hazards of digesting organic samples with nitric acid in closed systems for elemental analysis have been stressed. Digestion in an open flask may be a safer alternative in many cases [7]. In preparation for the analysis of powdered milk for trace metals, 1 g portions were heated at 80°C with 5 ml of conc. nitric acid in 25 ml Teflon lined stainless digestion vessels made to a design maximum of 500 bar, with maximum working pressure limited to 125 bar. With all 3 samples explosions occurred, rupturing the digestion vessels and causing local damage. In one account of the incident, the explosions were attributed to formation of glyceryl trinitrate [8], but this is extremely unlikely, as its formation normally requires presence of conc. sulfuric acid as dehydrating agent. In a second account the correct explanation of a runaway exothermic oxidation reaction of an excessive sample

by undiluted nitric acid accompanied by gas formation was deduced. The strong recommendation to limit sample size to 0.1 g maximum [9] coincides with that made 7 years previously [6].
See other GAS EVOLUTION INCIDENTS

Organic materials, Perchloric acid
See Perchloric acid: Nitric acid, etc.

Organic materials, Potassium chlorate
See Potassium chlorate: Nitric acid, etc.

Organic materials, Sulfuric acid
Anon., *ABCM Quart. Safety Summ.*, 1934, **5**, 17
Use of the mixed conc. acids to dissolve an organic residue caused a violent explosion. Nitric acid is a very powerful and rapid oxidant and may form unstable fulminic acid or polynitro compounds under these conditions.

Other reactants
Yoshida, 1980, 188–189
MRH values calculated for 28 combinations, largely with oxidisable materials, are given.

Phenylacetylene, 1,1-Dimethylhydrazine
Spengler, G. *et al., Sci. Tech. Aerospace Rept.*, **2**(17), 2392, Washington, NASA, 1964
Phenylacetylene does not itself ignite on contact with nitric acid, but addition of 1,1-dimethylhydrazine renders it hypergolic.
See Hydrocarbons, above

Phosphine derivatives   MRH Ethylphosphine 6.15/23
1. von Schwartz, 1918, 325
2. Mellor, 1947, Vol. 8, 827, 1041
3. Graham, T., *Trans. Roy. Soc. Edinburgh*, 1835, **13**, 88
Phosphine ignites in conc. nitric acid and addition of warm fuming nitric acid to phosphine causes explosion [1]. Phosphonium iodide ignites with nitric acid, and ethylphosphine explodes with fuming acid [2]. Tris(iodomercuri)phosphine is violently decomposed by nitric acid or aqua regia [3].

Phosphorus compounds
Mellor, 1947, Vol. 8, 1061; 1971, Vol. 8, Suppl. 3, 348, 335, 373
Tetraphosphorus tetraoxide trisulfide or neodymium phosphide are violently oxidised, nickel tetraphosphide ignites with the fuming acid, and tetraphosphorus diiodo triselenide reacts explosively with nitric acid.
See 2-Alkoxy-1,3-dithia-2-phospholane, above

Phosphorus halides
Mellor, 1947, Vol. 8, 827, 1004, 1038

Tetraphosphorus iodide ignites in contact with cold conc. nitric acid, and phosphorus trichloride explodes with nitric (or nitrous) acid.

Phthalic anhydride, Sulfuric acid
1. Tyman, J. H. P. *et al., Chem. & Ind.*, 1972, 664
2. Bentley, R. K., *Chem. & Ind.*, 1972, 767
3. Bretherick, L., *Chem. & Ind.*, 1972, 790
Attempts to follow a published method for nitrating phthalic anhydride in sulfuric acid at 80–100°C with fuming nitric acid caused an eruptive decomposition to occur after 2 h delay [1]. The hazard can be eliminated by use of a smaller excess of nitrating acid at 55–65°C [2]. Acyl nitrates were suggested as a possible cause of the delayed eruption [3].
See Sulfuric acid, Terephthalic acid, below

Polyalkenes
1. Marsh, J. R., *Chem. & Ind.*, 1968, 1718
2. Dabeka, R. W. *et al., Anal. Chem.*, 1976, **48**, 2048
3. Kaufman, J. A., private comm., 1987
4. Weker, R. A. *et al., Chem. Eng. News*, 1995, **73**(25), 4
Fuming nitric acid had seeped past the protective polytetrafluoroethylene liner inside the polyethylene or polypropylene screw cap and attacked the latter, causing pressure build-up in the glass bottle [1]. Polypropylene bottles are unsuitable for long term storage of nitric acid, because slow embrittlement and cracking occur [2]. A plastic bottle (polythene?) used to store conc. nitric acid collapsed while being carried [3]. Digestion of airborne particulate metals caught on a PTFE filter by nitric acid in a microwave oven led to explosion. The filter membrane was mounted in a poly(methylpentene) support, which was the cause of the violent reaction. Use of either an all PTFE filter, or cutting away the support ring, is advisable [4].

Poly(dibromosilylene)
See Poly(dibromosilylene): Oxidants

Poly(ethylene oxide) derivatives
1. Corby, M. P., *Chem. Brit.*, 1975, **11**, 334; private comm., 1975
2. Corby, M. P., *Chem. Brit.*, 1975, **11**, 456
A mixture of nitric and phosphoric acids (50, 17%, respectively) with a primary alcohol ethoxylate surfactant (0.1%) and water exploded after storage for 7 months in a glass bottle [1]. Progressive hydrolysis under these conditions would be expected to lead to production of the readily oxidised ethylene glycol, and gaseous decomposition products leading to pressure build-up. A general warning against mixing surfactants and oxidising acids is given [2]. The formation of ethylene glycol dinitrate is perhaps unlikely in view of the 33% water content in the mixture.

Polymer (unspecified)
Anon., *Safety Digest Univ. Safety Assoc.*, 1989, **34**, 14

1596

Throat cut, almost fatally, by shards of a test tube from which a polymeric residue was being cleared. Explosion apparently almost immediate; were there also miscible solvent residues?

Poly(silylene)
See Poly(silylene): Oxidants

Polyurethane foam
Anon., *Loss Prev. Bull.*, 1980, (035), 10–11
Mechanical cleaning of a multitubular stainless steel condenser (blocked by a rigid polyurethane foam) by rodding the tubes was laborious, so chemical cleaning with conc. nitric acid was attempted. When the initial vigorous reaction ('fireworks') subsided, owing to crust formation, the rod was again inserted, but a sudden explosion occurred which ruptured the condenser.
See other PLANT CLEANING INCIDENTS

Propiophenone, Sulfuric acid
1. Baker, J. W., *J. Chem. Soc.*, 1932, 1155
2. Dowell, R. I., *Chem. Brit.*, 1981, **17**, 56–57
3. Keneford, J. R., *et al., J. Chem. Soc.*, 1948, 356
During the nitration of propiophenone following a published procedure [1] by adding the ketone to the mixed acids at −5 to 0°C, an uncontrollable exotherm developed and finally accelerated to explosion [2]. The nitration was subsequently effected by an alternative procedure operated at −10 to −5°C [3].
See other NITRATION INCIDENTS

Pyrocatechol
Andrussow, L., *Chim. Ind* . (Paris), 1961, **86**, 542
The phenol is hypergolic with conc. nitric acid, with a 1 ms ignition delay.

Pumps
Anon. *Loss Prev. Bull.*, 1994, 116, 21
Two incidents of pump explosions in chemical plants are reported, both occasioned by nitric acid reaching parts of a pump it should not and there corroding metal or contacting electrics. Proper pump selection, installation and maintenance would have avoided these incidents.

Reducants
A variety of reducants ignite or explode with nitric acid.
See Hydrazine and derivatives, above
       Hydrogen sulfide: Oxidants
       Non-metal hydrides, above
       Potassium phosphinate: Air, or Nitric acid
       Sulfur dioxide, below

Resorcinol
Biasutti, 1981, 104

During the preparation of dinitroresorcinol, too low a concentration of nitric acid (82%) led to formation of tarry material, some of which remained in the nitrator discharge line. In the following batch, the tar decomposed in contact with higher strength acid, leading to an explosion.
See other NITRATION INCIDENTS, PLANT CLEANING INCIDENTS

Rubber
1. Read, T. R., *School Sci. Rev.*, 1976, **57**(200), 592
2. Mishra, B. *et al., J. Hazard. Mater.*, 1997, 56, 107
In a demonstration of the corrosive effect of fuming nitric acid on rubber tubing, soon after the initial vigorous reaction had subsided, a small explosion followed by ignition occurred. (The sulfur compounds used to compound the rubber were probably involved) [1]. Leaded neoprene rubber gloves, (the lead present as red lead, Pb_3O_4) used in the nuclear industry, may form explosive compounds in contact with nitric acid. It is hypothesised this could be lead fulminate formation. Since mixtures of lead nitrate with organics, or with sulfur from vulcanisers, would be explosive, this seems unduly complicated [2]
See other CORROSION INCIDENTS

Salicylic acid
See Lead(II) picrate

Silicone oil
Phillips, C. V. *et al., Lab. Pract.*, 1981, **30**, 598
A mineral was being leached by heating with 7 M nitric acid in a PTFE-lined bomb heated by immersion in a silicone oil bath, and at 195°C a violent explosion occurred. This was attributed to prior leakage of oil into the pressure vessel, which had been immersed in the oil at 120°C, then allowed to cool, before being heated to the higher temperature. Appropriate precautions are recommended.

Silver buten-3-ynide
See Silver buten-3-ynide: Alone, or Ammonia, etc.

Steel gas cylinder
Anon., *CISHC Chem. Safety Summ.*, 1974–5, **45–46**, 3–4
Conc. nitric acid leaking from a faulty road tanker became partially diluted with water and was prevented from running away along the roadside gully by a full oxygen cylinder in the horizontal position. After several h, the cylinder was sufficiently weakened by corrosion to split open under the internal gas pressure.
See other CORROSION INCIDENTS

Sucrose
Poncini, L., *School Sci. Rev.*, 1981, **62**(200), 515–516
Use of aqueous solutions of sucrose rather than the solid in class demonstrations of preparation of oxalic acid is described as safer.

Sulfur dioxide
Coleman, G. H. *et al., Inorg. Synth.*, 1939, **1**, 55

Presence of dinitrogen tetraoxide appears to be essential to catalyse smooth formation of nitrosylsulfuric acid from sulfur dioxide and nitric acid. In its absence, reaction may be delayed and then proceed explosively.
See other CATALYTIC IMPURITY INCIDENTS

Sulfur halides
Mellor, 1947, Vol. 10, 646, 651–652
Interaction with sulfur dichloride, sulfur dibromide or disulfur dibromide is violent, the hydrogen halide being liberated.
See other GAS EVOLUTION INCIDENTS

Sulfuric acid
1. Anon., *BCISC Quart. Safety Summ.*, 1964, **35**, 3
2. *FPA H50*, 1976; *HCS 1980*, 666
The gland of a centrifugal pump being used to pump nitrating acid (nitric:sulfuric, 1:3) exploded after 10 min use. This was attributed to nitration of the gland packing, followed by frictional detonation. Inert shaft sealing material is advocated [1]. General handling precautions for nitrating acid are detailed [2].
See 2,2-Oxybis(ethyl nitrate)

Sulfuric acid, Terephthalic acid
Withers, C. V., *Chem. & Ind.*, 1972, 821; private comm., 1972
During nitration of the acid with fuming nitric acid in oleum, a delayed exotherm increased the temperature after 2 h from 100° to 160°C, causing eruption of the contents. At 120° the delay was 1 h and at 130°C, 30 min.
See Phthalic anhydride, etc., above
See other INDUCTION PERIOD INCIDENTS

Thioaldehydes, or Thioketones
Campaigne, E., *Chem. Rev.*, 1946, **39**, 57
Nitric acid generally reacts too violently with thials or thiones for the reactions to be of preparative interest.

Thiols
See Alkanethiols, above

Thiophene
1. Meyer, V., *Ber.*, 1883, **16**, 1472
2. Babasinian, V. S., *Org. Synth.*, 1944, Coll. Vol. 2, 467
Interaction of thiophene with fuming nitric acid is very violent if uncontrolled, extensive oxidation occurring [1]. Use of a diluent and close control of temperature is necessary for preparation of nitrothiophene [2].
See Methylthiophene, above

1,3,5-Triacetylhexahydro-1,3,5-triazine, Trifluoroacetic anhydride
1. Albright and Hanson, 1976, 327
2. Bedford, C. D., *Chem. Eng. News*, 1980, **58**(35), 33, 43

3. Gilbert, E. E., *Chem. Eng. News*, 1980, **58**(40), 5
4. Bretherick, L., *Chem. Eng. News*, 1981, **59**(14), 59
During the attempted nitrolysis of the triacetyl compound to the 1-acetyl-3,5-dinitro compound with 99% nitric acid in trifluoroacetic anhydride at 30°C, following a general procedure [1], a violent explosion occurred on the 1 g scale [2]. This was ascribed to the formation of acetyl nitrate, expected to be formed under the reaction conditions [3,4]. Caution with nitrolysis of any acetyl compound is urged [3].
See Acetic anhydride, above; Triazine, etc., below

Triazine, Trifluoroacetic anhydride
491M, 1975, 272
N-Nitration of triazine with 99% nitric acid in trifluoroacetic anhydride at 36°C proceeded explosively.

Tributyl phosphate
See NUCLEAR WASTES

Triethylgallium etherate
See Triethylgallium

2,4,6-Trimethyltrioxane
Muir, G. D., private comm., 1968
Oxidation of the trioxane ('paraldehyde') to glyoxal by action of nitric acid is subject to an induction period, and the reaction may become violent if addition of the trioxane is too fast. Presence of nitrous acid eliminates the induction period.
See other INDUCTION PERIOD INCIDENTS

Turpentine
Hermoni, A., *Appl. Chem.*, 1958, **8**, 670–672
Turpentine and fuming nitric acid do not ignite on contact in absence of added catalysts (fuming sulfuric acid, iron(III) chloride, ammonium metavanadate or copper(II) chloride).
See Hydrocarbons, above

Uranium disulfide
Sidgwick, 1950, 1081
Interaction is violent.

Wood
1. Personal experience, 1974
2. Anon., *Chem. Eng. News*, 1975, **53**(11), 14
3. *MCA Case History No. 1797*
4. Anon., *Chem. & Ind.*, 1994, (6), 212
A cracked winchester of conc. acid leaked into sawdust packing and caused a fire [1]. A similar incident was involved in a freight-plane crash [2]. Fuming acid, leaking from a cracked bottle, ignited a wooden truck [3]. A retrospective survey on nitric acid packed in sawdust reveals an incident before 1894 [4].
See other PACKAGING INCIDENTS

Zinc ethoxide
See Zinc ethoxide: Nitric acid
See other INORGANIC ACIDS, OXIDANTS

4437. Peroxonitric acid
[26604-66-0]   HNO$_4$

$$O_2NOOH$$

1. Schwarz, R., *Z. Anorg. Chem.*, 1948, **256**, 3
2. Kenley, R. A. *et al.*, *J. Amer. Chem. Soc.*, 1981, **103**, 2205

The pure peroxoacid, prepared at −80°C, decomposes explosively at −30°C. Solutions in acetic acid or water of below the limiting concentration (corresponding to a stoicheiometric mixture of 70% aqueous nitric acid and 100% hydrogen peroxide) are stable, while those above the limit decompose autocatalytically, eventually exploding [1]. Explosion of the vapour when passed into a mass spectrometer inlet at 427°C (but not at 327°) was noted [2].
See other PEROXOACIDS

4438. Nitrosylsulfuric acid
[7782-78-7]   HNO$_5$S

$$O:NOSO_2OH$$

Preparative hazard
See Nitric acid: Sulfur dioxide

6-Chloro-2,4-dinitroaniline
See 2-Chloro-4,6-dinitroaniline: Nitrosylsulfuric acid

Dinitroaniline
Anon., *BCISC Quart. Safety Summ.*, 1970, **41**, 28
During plant-scale diazotisation of a dinitroaniline hydrochloride, local increase in temperature, owing to high concentration of the reaction mixture, caused a violent explosion.
See Nitric acid: 4-Chloro-2-nitroaniline
See other INORGANIC ACIDS, NITROSO COMPOUNDS, OXIDANTS

4439. Lead(II) imide
[12397-26-1]   HNPb

$$Pb:NH$$

Alone, or Acids, or Water
Mellor, 1940, Vol. 8, 265
It explodes on heating, or in contact with water or dilute acids.
See other N-METAL DERIVATIVES

4440. Phospham
[33849-97-1]

(N:PNH)$_n$

HN$_2$P

Hydrogen sulfide
Mellor, 1940, Vol. 8, 270
The solid produced by interaction of phospham and hydrogen sulfide at red heat
is probably a trimeric triphosphatriazine such as phospham. The solid ignites in
slightly warm air or in dinitrogen tetraoxide, and is violently oxidised by nitric
acid.

Oxidants
Mellor, 1940, Vol. 8, 269–270
Interaction with copper(II) oxide or mercury(II) oxide proceeds incandescently.
Mixtures with a chlorate or nitrate explode on heating. Phospham ignites in dini-
trogen tetraoxide.
See related NON-METAL HYDRIDES

4441. Hydrogen azide (Hydrazoic acid)
[7782-79-8]

HN$_3$

HN$_3$

1. Smith, 1966, Vol. 2, 214
2. Bowden, F. P. *et al., Endeavour*, 1962, **21**, 121
3. Audreith, L. F. *et al., Inorg. Synth.*, 1939, **1**, 77
4. Kemp, M. D., *J. Chem. Educ.*, 1960, **37**, 142
5. Birkofer, L. *et al., Org. Synth.*, 1970, **50**, 109
6. Shapiro, E. L., *Chem. Eng. News*, 1974, **52**(2), 5
7. Chandross, E. A. and Gunderson, H., *Chem. Eng. News*, 1974, **52**(24), 5
8. Sood, R. K. *et al., J. Therm. Anal.*, 1981, **20**, 491–493

Hydrogen azide is quite safe in dilute solution, but is violently explosive and
of variable sensitivity in the concentrated (17–50%) or pure states. Wherever
possible, a low-boiling solvent (ether, pentane) should be added to its solutions to
prevent inadvertent concentration by evaporation and recondensation. If this is not
possible, no unwetted part of apparatus containing its solutions should be kept at a
temperature appreciably below the boiling point (35°C) of the pure acid. The pure
acid has often been isolated by distillation, but appears to undergo rapid sensiti-
sation on standing, so that after an hour, faint vibrations or speech are enough to
initiate detonation [1]. The solid acid (−80°C) is also very unstable [2]. Preparative procedures have been detailed [3]. It is readily formed on contact of hydrazine
to its salts with nitrous acid or its salts. A safe procedure for the preparation of
virtually anhydrous hydrogen azide has been described [4]. Trimethylsilyl azide
serves as a safe and stable substitute for hydrogen azide in many cases [5]. A
laboratory explosion [6] seems likely to have been caused by the use of a huge
(90-fold) excess of azide in too-concentrated solution and at too low an ambient
temperature, leading to condensation of highly concentrated hydrogen azide. This

penetrated into a ground-glass joint, and explosion was initiated on removing the stopper [7]. It is claimed that it may be distilled without fear of explosion if the trapping of the liquid azide in ground joints is prevented by thorough pregreasing of all joints [8].

As the lowest MW azide, hydrogen azide is extremely endothermic (ΔH°_{f} (g) +294.1 kJ/mol, 6.83 kJ/g).

See Methyl azide, *also* 4-Chlorophenylisocyanate, *also* Nitrous acid: A semicarbazone, etc.

See other GLASS INCIDENTS

Heavy metals
1. Napier, D. H., private comm., 1972
2. Cowley, B. R. *et al., Chem. & Ind.*, 1973, 444
3. Mellor, 1967, Vol. 8, Suppl. 2, 4

Great care is necessary to prevent formation of explosive heavy metal azides from unsuspected contact of hydrogen azide with heavy metals. Interaction of hydrazine and nitrite salts in a copper drainage system caused formation and explosion of copper azide deposits [1]. Use of a brass water-pump and vacuum gauge during removal of excess hydrogen azide under vacuum formed deposits which exploded when the pump and gauge were handled later [2]. Raney nickel catalyses the vigorous decomposition of solutions of hydrogen azide [3].

See Cadmium azide (reference 2), Sodium azide: Heavy metals
See other ENDOTHERMIC COMPOUNDS, INORGANIC ACIDS, NON-METAL AZIDES

4442. Dinitramine (Dinitramide)
[115045-20-4] $HN_3 O_4$
$$HN(NO_2)_2$$

1. Borman, S., *Chem. Eng. News*, 1994, **72**(3), 18
2. Dawe, J. R. *et al., Proc. Int. Pyrotech. Semin.*, 1998, **24th**, 789

Salts of the acidic dinitramine find use as oxidants in propellant formulations. The dinitramine itself is explosively unstable, though more stable than nitramide [1]. Although described as relatively stable the sodium salt has an autoignition temperature of 123°C, the potassium 140°C [2].

See Nitric amide
See other N-NITRO COMPOUNDS

4443. Pentazole
[289-19-0] HN_5
Sorbe, 1968, 140
Pentazole (98.7% nitrogen) and its compounds are explosive.
See other HIGH-NITROGEN COMPOUNDS
See related TETRAZOLES

4444. Sodium hydride
[7646-69-7]   **HNa**
                                        NaH

1. Plesek, J. *et al., Sodium hydride* (Engl. transl., Jones, G.) 5, 8, London, Iliffe, 1968
2. Bigou, F., personal communication, 1993
3. Gibb, T. R. P., US Pat. 2 702 281, 1955
4. Urben, P. G., personal experience

Commercial sodium hydride may contain traces of sodium which render it spontaneously flammable in moist air, or air enriched with carbon dioxide. The very finely divided dry powder ignites in dry air. Dispersions of the hydride in mineral oil are safe to handle. All normal extinguishers are unsuitable for solid sodium hydride fires: powdered sand, ashes or sodium chloride are suitable [1]. Commercial oil-coated sodium hydride was washed with dichloromethane to remove the oil. During later charging to a reactor containing more dichloromethane and an involatile substrate the hydride ignited. Ignition was probably by traces of moisture, but combustion seems to have continued in an atmosphere of hydrogen and dichloromethane vapour. It seems inadvisable to mix sodium hydride with most chlorinated solvents. Although a protective coating of non-hydridic material may prevent immediate reaction, it is thermodynamically capable of reacting energetically with halocarbons [2].

Many acylation reactions of esters using sodium hydride as base appear autocatalytic, with considerable potential for runaway, since the active base in solution is an alkoxide and the alcohol is a product of reaction [4]. A safe form of sodium hydride (as a solid solution in a halide) for large-scale industrial use has been claimed [3].

See Sodium: Halocarbons

Acetylene
Mellor, 1940, Vol. 2, 483
Dry acetylene does not react with sodium hydride below 42°C, but in presence of moisture, reaction is vigorous even at −60°C.

Diethyl succinate, Ethyl trifluoroacetate
Harris, A. R. *et al., Chem. Brit.*, 1983, **19**, 645
Acylation of diethyl succinate by ethyl trifluoroacetate in presence of sodium hydride and in absence of a solvent is hazardous, fire or explosion occurring on 2 occasions some 10−20 min after adding a little of the succinate to the hydride−trifluoroacetate premixture at 60°C. Presence of a solvent appears to eliminate the hazard.
See Ethyl trifluoropropionate, below

Dimethylformamide
1. Buckley, J. *et al., Chem. Eng. News*, 1982, **60**(28), 5
2. De Wall, G. *Chem. Eng. News*, 1982, **60**(37), 5, 43

A mixture of the hydride and solvent was heated to, and held at 50°C. An exothermic reaction, which increased the temperature to 75°C, could not be controlled by external cooling, and the pilot-scale reactor contents erupted. It was later found that an exotherm began to develop at 26° (or at 40–50° with carefully dried solvent), and the subsequent reaction accelerated rapidly. Avoidance of holding mixtures hot is recommended, particularly when scaling up reactions. Similar behaviour was seen with dimethylacetamide [1]. A further similar plant-scale incident was reported, with onset of the exotherm at 40°C, followed by self-heating to 100°C in 10 min, even with maximum cooling applied [2].
See other SELF-ACCELERATING REACTIONS

Dimethyl sulfoxide
See Dimethyl sulfoxide: Sodium hydride

Ethyl 2,2,3-trifluoropropionate
Bagnall, R. D., private comm., 1972
The ester decomposes violently in presence of sodium hydride, probably owing to hydride-induced elimination of hydrogen fluoride and subsequent exothermic polymerisation.
See Methyl trichloroacetate: Trimethylamine

Glycerol
Unpublished observations, 1956
Exothermic interaction of granular hydride with undiluted (viscous) glycerol with inadequate stirring caused charring to occur. Dilution with THF to reduce viscosity and improve mixing prevented local overheating during formation of monosodium glyceroxide.

Halogens   MRH Chlorine 3.10/30
Mellor, 1940, Vol. 2, 483
Interaction with chlorine or fluorine is incandescent at ambient temperature, and with iodine at 100°C.

Other reactants
Yoshida, 1980, 215
MRH values calculated for 8 combinations with various materials are given.

Oxygen   MRH 2.59/31
See Oxygen (Gas): Metal hydrides

Sulfur   MRH 1.25/16
See Sulfur: Sodium hydride

Sulfur dioxide   MRH 1.21/43
Moissan, H., *Compt. rend.*, 1902, **135**, 647
Sulfur dioxide reacts explosively in contact with sodium hydride unless diluted with hydrogen.

Water
1. Braidech, M. M., *J. Chem. Educ.*, 1967, **44**, 321
2. *MCA Case History No. 1587*
Addition of sodium hydride to a damp reactor which had not been purged with
inert gas caused evolution of hydrogen and a violent explosion. Solid dispersions
of the hydride in mineral oil are more easily and safely handled [1]. When an
unprotected polythene bag containing the hydride was moved, some of the powder
leaked from a hole, contacted moisture and immediately ignited. Such materials
should be kept in tightly closed containers in an isolated, dry location [2].
See other GAS EVOLUTION INCIDENTS
See other METAL HYDRIDES, REDUCANTS

4445. Sodium hydroxide
    [1310-73-2]   **HNaO**
                            NaOH

(*MCA SD-9*, 1968); *NSC 373*, 1982; *HCS 1980*, 845 (solid), 846 (33% solution);
RSC Lab. Hazards Safety Data Sheet No. 36, 1985

Aluminium, Arsenical materials
    See Aluminium: Arsenic trioxide, etc.

Ammonia, Silver nitrate
    See Silver nitrate: Ammonia, Sodium hydroxide

1,4-Benzenediol
    See 1,4-Benzenediol: Sodium hydroxide

N,N'-Bis(2,2,2-trinitroethyl)urea
    See N,N'-Bis(2,2,2-trinitroethyl)urea: Sodium hydroxide

Bromine
    See Bromine: Sodium hydroxide

4-Chlorobutyronitrile
    See 4-Chlorobutyronitrile: Sodium hydroxide

Chloroform, Methanol   MRH Chloroform 2.05/43
    See Chloroform: Sodium hydroxide, etc.

4-Chloro-2-methylphenol
    See 4-Chloro-2-methylphenol: Sodium hydroxide

Cinnamaldehyde
    See Cinnamaldehyde: Sodium hydroxide

1606

Cyanogen azide
See Cyanogen azide: Sodium hydroxide

Diborane
See Diborane: Octanal oxime, etc.

2,2-Dichloro-3,3-dimethylbutane
See 2,2-Dichloro-3,3-dimethylbutane: Sodium hydroxide

Glycols
Cardillo, P. *et al., Chim. e Ind.* (Milan), 1982, **64**, 781–784
Various mixtures of the base with ethylene glycol or diethylene glycol when heated in DSC capsules show exothermic decomposition around 230°C with rapid evolution of hydrogen. The exotherms increase with the base:glycol ratio, and the principal hazard arises from the rapid increase of pressure in closed systems caused by hydrogen evolution.
See Sodium 2-hydroxyethoxide
See entries GAS EVOLUTION INCIDENTS, PRESSURE INCREASE IN EXOTHERMIC DECOMPOSITION

Maleic anhydride   MRH 1.67/99+
See Maleic anhydride: Bases, etc.

4-Methyl-2-nitrophenol
See 4-Methyl-2-nitrophenol: Sodium hydroxide

3-Methyl-2-penten-4-yn-1-ol
See 3-Methyl-2-penten-4-yn-1-ol: Sodium hydroxide

Nitrobenzene
See Nitrobenzene: Alkali

Other reactants
Yoshida, 1980, 213
MRH values calculated for 12 combinations with various materials are given.

Sodium tetrahydroborate
See Sodium tetrahydroborate: Alkali

Succinic anhydride
See Succinic anhydride: Sodium hydroxide

Sugars
See SUGARS

1,2,4,5-Tetrachlorobenzene
See 1,2,4,5-tetrachlorobenzene: Sodium hydroxide, Solvent

12,2,2-Trichloroethanol
See 2,2,2-Trichloroethanol: Sodium hydroxide

Trichloroethylene   MRH 2.26/45
See Trichloroethylene: Alkali

Trichloronitromethane   MRH 1.80/46
See Trichloronitromethane: Sodium hydroxide

Water
1. *MCA SD-9*, 1968
2. *Haz. Chem. Data*, 1975, 267
3. *MCA Case History No. 2166*
4. Anon., *Loss Prev. Bull.*, 1987, (078), 26

The heat of solution is very high, and with limited amounts of water, violent
boiling or even ignition of adjacent combustibles may occur [1,2]. The case history
describes the disposal of open ended drums of waste caustic soda by putting into
an empty hopper and sluicing out the solid alkali with a hot water jet. When the
outlet became blocked, the hopper partly filled with hot water, and addition of a
further drum caused a violent eruption to occur [3]. A tank holding 73% caustic
soda solution at 121°C was found to be leaking and some of the contents were
pumped out. A residue of 50 m^3 was to be diluted with water before removal and
100 m^3 of hot water was introduced via a hose into the top of the tank, where
it formed a top layer. A compressed air lance was then introduced to mix the
contents. After 30 min. a violent eruption occurred as the layers suddenly mixed
to give a very large exotherm, and the steam so produced buckled the top of the
tank [4].
See Potassium hydroxide: Water

Zinc
See Zinc: Sodium hydroxide

Zirconium
See Zirconium: Oxygen-containing compounds
See other INORGANIC BASES

4446. Sodium hydrogen sulfate
[7681-38-1]   HNaO$_4$S
NaOSO$_2$OH

Acetic anhydride, Ethanol
See Acetic anydride: Ethanol etc.

Calcium hypochlorite
See Calcium hypochlorite: Sodium hydrogen sulfate
See other METAL OXONON-METALLATES *See related* INORGANIC ACIDS

4447. Sodium hydrogen xenate
[73378-53-7], sesquihydrate   HNaO$_4$Xe
$$NaOXe(O)_2OH.1.5H_2O$$

See other XENON COMPOUNDS

4448. Rubidium hydrogen xenate
[73378-55-9], sesquihydrate   HO$_4$RbXe
$$RbOXe(O)_2OH.1.5H_2O$$

See other XENON COMPOUNDS

4449. 'Solid Phosphorus hydride'
[ ]   HP$_2$
(Indeterminate structure)

1. Mellor, 1940, Vol. 8, 851
2. Sidgwick, 1950, 730
This material (possibly phosphine adsorbed on phosphorus, and produced by decomposition of diphosphane in light) ignites in air, on impact, or on sudden heating to 100°C [1,2].
See other NON-METAL HYDRIDES

4450. Rubidium hydride
[13446-75-8]   HRb
RbH

Acetylene
   Mellor, 1940, Vol. 2, 483
   In presence of moisture, interaction of the hydride and acetylene is vigorous at −60°C, while in dry acetylene, reaction only occurs above 42°C.

Oxygen
   See Oxygen (Gas): Metal hydrides

Water
   Mellor, 1963, Vol. 2, Suppl. 2.2, 2187
   Interaction with water is too violent to permit of safe use of the hydride as a drying agent. When dispersed as a solid solution in a metal halide, it can be used as a drying or reducing agent.
   See other METAL HYDRIDES

4451. Silicon monohydride (Silylidyne)
[13774-94-2] $(HSi)_n$

$$(SiH)_n$$

Alkali

Stock, G. *et al., Angew. Chem.*, 1956, **68**, 213
The polymeric hydride is relatively stable to water, but reacts violently with alkali, evolving hydrogen.
See other NON-METAL HYDRIDES

4452. 'Unsaturated' Silicon hydride
[ ] $(H_{1.5}Si)_n$

$$(SiH_{1.5})_n$$

Bailar, 1973, Vol. 1, 1350
The polymeric material (composition varies from $SiH_{1.42}$ to $SiH_{1.56}$) burns with a shower of sparks if heated in air.
See other NON-METAL HYDRIDES

4453. Hydrogen (Gas)
[1333-74-0] H_2

$$H_2$$

FPA H20, 1974; *HCS 1980*, 544 (cylinder)

1. Pignot, A., *Chaleur Ind.*, 1939, **20**, 251–259
2. Anon., *Loss Prev. Bull.*, 1977, (015), 2
3. Fenning, R. W. *et al., Engineering*, 1930, **130**, 252
4. Messelt, W., *Engineering*, 1922, **113**, 502
5. Neer, M. E., *Diss. Abstr. Int. B*, 1972, **33**, 686
6. Carhart, H. W., *et al., AD Rep. No. 781862/1GA*, 1–39, Richmond (Va.), USNTIS, 1974
7. Neer, M. E., *Amer. Inst. Aero. Astron. Paper 74-1159*, 1974, 1–7
8. *NASA Tech. Memo. TMX-71565*, Richmond (Va.), USNTIS, 1971
9. Andrew, H. G., *School Sci. Rev.*, 1979, **61**(215), 292–294
10. Boeck, H., *Chem. Sich.*, 1982, **29**(4), 158–160
11. McCue, J. C., *Altern. Energ. Sources, 1980*, 1983, **3**, 419–427
12. Hydrogen Safety Manual, Balthasar, W. *et al.*, Luxembourg, Comm. Euro. Commun., 1983
13. Angus, H. C., *Chem. & Ind.*, 1984, 68–72
14. Lee, J. H. S. *et al., Adv. Heat Transfer*, 1997, **25**, 59

Several of the combustion-related properties of hydrogen in air, such as its wide flammability limits (4–75 vol%), wide detonation range (20–65 vol%), very low spark ignition energy (0.02 mJ), high heat of combustion (121 kJ/g) and high flame temperature (2050°C) combine to emphasise the high fire-related hazards of the lightest element. Sudden release of hydrogen into the atmosphere from storage at above 79 bar may cause spontaneous ignition, owing to the inverse Joule-Thompson effect [1]. This, however, does not always occur in practice, largely because of the low value of the effect (1.8°C/110 bar drop at 20°C; 3.9°/100 bar at 100°), so that ignition is unlikely unless the gas temperature is close to the auto-ignition temperature of 530–590°C [2] (or unless catalytic impurities are involved). Release of hydrogen at 47.5 bar into a vented 17.5 l chromium-plated sphere had caused explosive ignition [3]. Earlier, it had been found that a release of hydrogen for filling balloons would ignite under certain circumstances when the aperture was rusted, a brush discharge then being visible [4]. Spontaneous ignition of flowing hydrogen–air mixtures has been studied [5]. All aspects of hazards involved in the production, storage, handling and use of hydrogen as a Navy fuel are discussed [6]. In a study of the causes of auto-ignition of fast flowing hydrogen–air mixtures, at velocities above 750 m/s ignition occurred at temperatures 400°C below those established for static mixtures [7]. All recorded incidents and accidents involving hydrogen ignition have been reviewed and analysed [8]. Safety aspects of the generation and use of hydrogen in school demonstrations are detailed [9,10]. A compact portable ultrasafe storage unit for hydrogen for laboratory use has been developed, based on a lanthanum–nickel alloy hydride storage capsule, to elim-inate hazards associated with use of high pressure storage of hydrogen [11]. A safety manual devoted to hydrogen has been published [12], and safety matters are included in a review of current technological aspects of storage, distribution and compression of hydrogen [13]. A detailed review of combustion and explo-sion properties of hydrogen, and of associated parameters, with a particular eye to nuclear reactor safety, has been published more recently [14].

Acetylene, Ethylene
Anon., *BCISC. Quart. Safety Summ.*, 1974, **45**, 2–3
In a plant producing 200 kt/a of ethylene from cracked naphtha, acetylene in the product was hydrogenated to ethylene in a catalytic unit operated under condi-tions mild enough not to hydrogenate ethylene. During a temporary shut-down and probably owing to operating error, the internal temperature in the catalytic unit rose to about 400°C, though the external wall temperature was recorded as excessive at 120°C. Attempts to reduce the temperature by passing in additional ethylene were unsuccessful, as the conditions were now severe enough to hydro-genate ethylene. This exothermic reaction increased the temperature, finally to 950°C, and the extensive cracking to methane, carbon and hydrogen now occur-ring was accompanied by further pressure increase. Plant failure was followed by an explosion and fire which took 4 days to extinguish, and damage totalled 6 M sterling.
See other HYDROGENATION INCIDENTS

Air
1. Berman, M., *Nucl. Sci. Eng.*, 1986, **93**(4), 321–347
2. Edeskuty, F. J. *et al., Adv. Hydrogen Energy*, 1986, **5** (Hydrogen Energy Prog. 6, Vol. 1), 147–158

3. Cummings, J. C. *et al., Sandia Nat. Lab. Report, SAND-86-0173*, 1987, DE87011907 (USNTIS. Richmomd, Va)
4. Mottin, M. C., *Proc. 11th Int. Sympos. Prev. Occup. Risks Chem. Ind.*, 383–385, Heidelberg, ISSA, 1987

Recent large-scale experimental work indicating that transition of hydrogen–air mixtures from deflagration to detonation is considerably affected by geometrical factors and scale, and that transition to detonation may be rather more likely in large-scale incidents than previously supposed [1]. Safety aspects of large-scale combustion of hydrogen–air mixtures, especially in confined volumes, have been investigated [2]. Factors involved in the flame acceleration observed in combustion of hydrogen–air mixtures were investigated experimentally in closed and vented vessels, and the results cast doubt on the validity of existing explosion-venting guidelines [3]. Based on several accidents and on large-scale experimental investigations of unconfined explosions of hydrogen–air mixtures, it was concluded that 0.1% of the energy potentially available was effectively released [4].

Air, Catalysts
Mellor, 1942, Vol. 1, 325, 1937, Vol. 16, 146
Catalytically active platinum and similar metals containing adsorbed oxygen or hydrogen will heat and cause ignition in contact with hydrogen or air, respectively. Nitrogen purging before exposure to atmosphere will eliminate the possibility.

Air, Various vapours
Schumacher, H. J., *Angew. Chem.*, 1951, **63**, 560–561
The effects of the presence of 44 gaseous or volatile materials upon the upper explosion limits of hydrogen–air mixtures have been tabulated.

Benzene, Raney nickel catalyst
See Benzene: Hydrogen, Raney nickel

Calcium carbonate, Magnesium
See Magnesium: Calcium carbonate, Hydrogen

Carbon monoxide, Oxygen
See Oxygen: Carbon monoxide, Hydrogen

Catalyst
Bretherick, L., *CISHC Chem. Safety Summ.*, 1980, **51**, 111
Ignition of hydrogen leaking from a hydrogenation autoclave stirrer gland was attributed to traces of hydrogenation catalyst dust outside the reactor.

Catalyst, 3,4-Dichloronitrobenzene
See 3,4-Dichloronitrobenzene: Catalyst, Hydrogen

Catalysts, Vegetable oils
Smirnov, V. M., *Chem. Abs.*, 1938, **32**, 4368$_6$

Flash fires and explosions which frequently occurred on discharge of the hot products of catalytic hydrogenation of vegetable oils were attributed to formation of phosphine from the phosphatides present to a considerable extent in, e.g., rape-seed and linseed oils.
See other HYDROGENATION INCIDENTS

Ethylene, Nickel catalyst
Anon., *Loss Prev. Bull.*, 1977, (015), 4–5
Hydrogen from a naphtha cracker normally contains 10% of methane, practically no ethylene and a little carbon monoxide, and the latter is removed by passage over a heated nickel catalyst when it is hydrogenated to methane. Failure of the ethylene refrigeration compressor led to the presence of 8.5% of ethylene in the hydrogen. Presence of catalyst dust in the inlet tube caused rapid hydrogenation of ethylene to occur in this confined volume and the temperature rapidly increased. A temperature monitor set at 500°C also failed, and when the inlet tube reached 800°C it burst under the prevailing pressure and the hydrogen released immediately ignited. Preventive measures are detailed.
See other HYDROGENATION INCIDENTS

Halogens, or Interhalogens   MRH values show % of halogen
See Bromine: Hydrogen 0.45/99
Bromine fluoride: Hydrogen
Bromine trifluoride: Halogens, etc.
Chlorine: Hydrogen 2.55/97
Chlorine trifluoride: Hydrogen-containing materials 6.82/97
Fluorine: Hydrogen 13.39/95
Iodine heptafluoride: Carbon, etc.

Hydrogen peroxide, Palladium catalyst
See Hydrogen peroxide: Hydrogen, Palladium catalyst

Liquid nitrogen
1. Mel'nik, B. D. *et al.*, *Chem Abs.*, 1963, **59**, 7309g
2. Bohlken, S. F., *Chem. Eng. Progr.*, 1961, **57**(4), 49–52
During the purification of washed hydrogen from cracking of natural gas, cooling with liquid nitrogen can lead to trapping of explosive products from interaction of alkenes in the gas with oxides of nitrogen arising from biologically derived ammonium nitrate or nitrite in the scrubbing water. Various measures to prevent this are discussed. Similar effects may be observed when alkenes are oxidised in presence of nitrogen [1]. Analysis of a similar incident involving hydrogen derived from fuel oil implicated resins derived from interaction of nitrogen oxide (and possibly dinitrogen oxide) in a low-temperature heat exchanger section operating at −130 to −145°C [2].
See Nitrogen oxide: Dienes, Oxygen

Metals
Mellor, 1942, Vol. 1, 327

Lithium, calcium, barium and strontium react readily, sometimes igniting, in hydrogen above 300°C, while sodium and potassium react more slowly to form the hydrides.

Nickel, Oxygen
See Nickel: Hydrogen, Oxygen

2-Nitroanisole
See 2-Nitroanisole: Hydrogen

Oxidants
See Chlorine dioxide: Hydrogen
    Copper(II) oxide: Hydrogen
    Dichlorine oxide: Oxidisable materials
    Difluorodiazene: Hydrogen
    Dinitrogen oxide: Hydrogen, etc.
    Dinitrogen tetraoxide: Hydrogen, etc.
    Fluorine: Hydrogen, or: Hydrogen, Oxygen
    Fluorine perchlorate: Hydrogen
    Iodine heptafluoride: Carbon, etc.
    Nitrogen oxide: Hydrogen, etc.
    Nitryl fluoride: Non-metals
    Oxygen (Gas): Hydrogen
    Palladium(II) oxide: Hydrogen
    Xenon hexafluoride: Hydrogen

Oxygen, Sulfuric acid
Tabata, Y. et al., J. Haz. Mat., 1987, 17, 47–59
Three PVC chlorine drying towers in a mercury amalgam cell chlorine plant suddenly exploded violently some time after the AC auxiliary power supply to the plant had failed, while the DC cell and brine feed supply had remained on. The failure of the pumped mercury circulation (and its flow alarm) through the electrolysis cells caused hydrogen (instead of sodium) to be liberated at the cathodes and oxygen at the anodes, so the chlorine gas output fed into the PVC drying towers trickle-fed with conc. sulfuric acid contained hydrogen and oxygen also. Ignition of the explosive mixture in the towers was caused by static sparks generated by the falling sulfuric acid drops: this was confirmed by a measurement of a static negative potential of 5kV inside and near the base of the towers. Other possible hazardous malfunctions in amalgam chlorine cells and in non-conductive drying towers are described, and precautions to avoid such hazards discussed.
See other STATIC INITIATION INCIDENTS

Palladium, 2-Propanol
491M, 1975, 205
A stream of hydrogen containing entrained 2-propanol vapour and catalyst particles ignited in contact with air.

See other HYDROGENATION INCIDENTS
See HYDROGENATION CATALYSTS

Palladium trifluoride
See Palladium trifluoride: Hydrogen

Platinum catalyst
Dimitrov, I. *et al., Chem. Abs.*, 1981, **94**, 17876
Some characteristics of ignition of hydrogen on platinum catalyst (title only translated).

Poly(carbon monofluoride)
See Poly(carbon monofluoride): Hydrogen

1,1,1-Tris(azidomethyl)ethane, Palladium catalyst
See 1,1,1-Tris(azidomethyl)ethane: Hydrogen, etc.

Tris(hydroxymethyl)nitromethane
See other NON-METALS, REDUCANTS
See Tris(hydroxymethyl)nitromethane: Hydrogen etc.

4454. Hydrogen (Liquid)

1. Kit and Evered, 1960, 123
2. *Rept. UCRL-3072*, Univ. of California, Berkeley, 1955
3. Weintraub, A. A. *et al., Health Phys.*, 1962, **8**, 11
4. Edutsky, F. J., *Progr. Refrig. Sci. Technol., Proc. 12th Int. Congr. Refrig.*, 1969, **1**, 283–300
5. Lodini, M. A. K. *et al., Int. J. Hydrogen Energy*, 1989, **14**, 35–43
6. *NFPA 50B*, Quincy (Ma), Natl. Fire Prot. Assoc., 1989

The main precaution necessary for use of liquid hydrogen is to prevent air leaking into the system, where it will be condensed and solidified. Fracture of a crystal of solid air or oxygen could produce a spark to initiate explosion [1]. Procedures for the safe handling of liquid hydrogen in the laboratory [2] and in liquid hydrogen bubble chambers [3] have been detailed. Safety problems in large-scale handling and transport of liquid hydrogen have been discussed and safety codes described [4]. Sudden rupture of a 34 kl vacuum jacketed tank of liquid hydrogen appears to have been caused by accidental admission of air to the jacket space, causing excessive boil-off of gas beyond the capacity of the pressure relief [5]. A new US National Fire Code covers all aspects of construction, siting, piping, components and safety devices in consumer systems for liquid hydrogen [6].

Oxygen
1. Cai, T.-J., *Chem. Abs.*, 1981, **94**, 158861
2. Gamezo, V. N., *et al., Fizika Goreniya i Vzryva*, 1992, **28**(1), 106

Procedures to remove solid oxygen from liquefied hydrogen before storage are described [1]. Solid oxygen–liquid hydrogen mixtures are very detonable, with low critical diameters and anomalously high detonation velocities [2].

Ozone
See Ozone: Hydrogen (Liquid)
See other CRYOGENIC LIQUIDS

4455. Poly(dimercuryimmonium iodide hydrate)
[ ] $(H_2Hg_2INO)_n$

$$(Hg:N^+:HgI^-H_2O)_n$$

Sorbe, 1968, 97
It explodes on heating.
See other IODINE COMPOUNDS, POLY(DIMERCURYIMMONIUM) COMPOUNDS

4456. Potassium amide
[17242-52-3] H_2KN

$$KNH_2$$

1. Brandsma, 1971, 20–21
2. Sorbe, 1968, 68
3. Sanders, D. R., *Chem. Eng. News*, 1986, **64**(21), 2

It has similar properties to the much more widely used and investigated sodium amide, but may be expected on general grounds to be more violently reactive than the former. The frequent fires or explosions observed during work-up of reaction mixtures involving the amide were attributed to presence of unreacted (oxide-coated) particles of potassium in the amide solution in liquid ammonia. A safe filtration technique for removal of the particles is described [1]. It also ignites on heating or friction in air [2]. After preparation from liquid ammonia and potassium, the dry product after evaporation exploded while being chiselled out of the evaporator flask [3].

Ammonia, Copper(II) nitrate
See Copper(II) nitrate: Ammonia, Potassium amide

Potassium nitrite
See Potassium nitrite: Potassium amide

Tetraphenyllead
Houben-Weyl, 1975, Vol. 13.3, 241
One of the by-products of interaction is a highly explosive lead(IV) compound.

Water
Mellor, 1940, Vol. 8, 255
Interaction is violent and ignition may occur, even in contact with humid air. Old samples may explode with considerable delay after contact with water.
See Sodium amide: Water
See other N-METAL DERIVATIVES

4457. Potassium amidosulfate
[13823-50-2]   H₂KNO₃S

$$KOSO_2NH_2$$

Metal nitrates or nitrites
See entry METAL AMIDOSULFATES

4458. Potassium hydroxylamine-*O*-sulfonate
[49559-20-8]   H₂KNO₄S

$$KOSO_2ONH_2$$

1. Anon., private comm., 1985
2. Urben, P. G., private comm., 1989
The salt was isolated (possibly as a hydrate) by unheated vacuum rotary evaporation. The stopper was later ejected from the flask by an exothermic decomposition. It was thought that the salt had been hydrolysed by the water of hydration to give potassium hydrogen sulfate and anhydrous hydroxylamine which is unstable at ambient temperature [1]. It seems more likely that the warm salt would undergo a bimolecular disproportionation to give potassium hydrogen sulfate and the *N*-amino derivative of the title compound, which would decompose to give more potassium hydrogen sulfate and the very reactive and unstable diazene. Hydroxylamine is now thought to be stable when pure [2].
See Diazene
See other N–O COMPOUNDS

4459. Potassium phosphinate ('Potassium hypophosphite')
[7782-87-8]   H₂KO₂P

$$KOP(O)H_2$$

Air, or Nitric acid
Mellor, 1940, Vol. 8, 882
The salt burns (owing to evolution of phosphine) when heated in air, and explodes when evaporated with nitric acid.
See other REDOX REACTIONS
See other METAL PHOSPHINATES, REDUCANTS

4460. Potassium dihydrogenphosphide
[13659-67-1]   H₂KP

$$KPH_2$$

Mellor, 1971, Vol. 8, Suppl. 3, 283
The solid ignites in air.
See related PHOSPHINES

4461. Lanthanum dihydride
[13823-36-4] H_2La

Kirk-Othmer, 1966, Vol. 11, 207
The very reactive hydride ignites in air.
See other METAL HYDRIDES

4462. Lithium amide
[7782-89-0] H_2LiN

$LiNH_2$

Water
Bergstrom, F. W. *et al., Chem. Rev.*, 1933, **12**, 61
It reacts readily with water with a potentially dangerous exotherm.
See other N-METAL DERIVATIVES

4463. Magnesium hydride
[7693-27-8] H_2Mg

MgH_2

Air, or Water
1. Bailar, 1973, Vol. 1, 34–35
2. Ashby, E. C., *Inorg. Synth.*, 1977, **17**, 4
3. Anon., *Chem. Brit.*, 1985, **21**, 711
4. Vigeholm, B., *Chem. Abs.*, 1987, **106**, 165264

The finely divided hydride produced by pyrolysis is pyrophoric in air, while synthesis from the elements produces a substantially air-stable product [1]. That prepared by reduction of butylmagnesium bromide with lithium tetrahydroalumi-nate is pyrophoric and reacts violently with water and other protic compounds [2]. The hydride produced from magnesium anthracene has a very large specific surface area and is pyrophoric [3]. In the context of use of the hydride for energy storage purposes, ignition and combustion behaviour of 100–400 g portions were studied, as well as the reaction with water [4].
See other PYROPHORIC MATERIALS

Oxygen (Gas)
See Oxygen (Gas): Metal hydrides
See other METAL HYDRIDES

4464. Magnesium–nickel hydride
[67016-28-8] $H_2Mg.H_2Ni$

$MgH_2.NiH_2$

Hariguchi, S. *et al., Chem. Abs.*, 1981, **94**, 159191
Of several mixed hydrides, the magnesium–nickel hydrides were the most hazardous in terms of dust explosions.
See other DUST EXPLOSION INCIDENTS, METAL HYDRIDES

4465. Sodium amide
[7782-92-5]

H_2NNa

NaNH₂

FPA H102, 1981; *HCS 1980*, 826

1. Bergstrom, F. W. *et al., Chem. Rev.*, 1933, **12**, 61
2. Brauer, 1963, Vol. 1, 467
3. Krüger, G. R. *et al., Inorg. Synth.*, 1966, **8**, 15
4. Shreve, R. N. *et al., Ind. Eng. Chem.*, 1940, **32**, 173
5. Sandor, S., *Munkavédelem*, 1960, **6**(4–6), 20
6. Rüst, 1948, 283
7. Anon., *Univ. Safety Assoc. Safety News*, 1977, (8), 16
8. Ullmann, 1993, **A24**, 272
It frequently ignites or explodes on heating or grinding in air, particularly if previously exposed to air or moisture to produce degradation products (possibly peroxidic) [1,2,3]. Only one explosion not involving exposure to air has been recorded, during pulverisation [4]. The following oxidation products, all explosively unstable, have been identified: sodium hyponitrite, sodium trioxodinitrate, sodium tetraoxodinitrate, sodium pentaoxodinitrate and sodium hexaoxodinitrate [5]. Several cases of explosive incidents during packing or use of air-exposed material are described [6]. When a half-used 500 g bottle of sodamide was opened, sparks, fumes and intense heat were produced, leading to collapse of the bottle with subsequent explosion [7]. A test for safety of use consists of burning a gram on a spoon, if there are crepitations it is unsafe to use and should be cautiously destroyed [8].

Aryl halide, 1-Alkoxy-1-(trimethylsilyloxy)alkenes
Ferguson, J. R. *et al., Chem. Brit.*, 1997, **33**(6), 21
A mixture of sodamide, bromoanisole, and 1-methoxy-2-methyl-1-(trimethylsilyloxy)-1-butene were reacted at room temperature at a 50 millimolar scale. After some two hours, with slight emission of ammonia, the reaction suddenly became exothermic, with violent gas emission and on one occasion a fire. This was a modest scale-up of a literature procedure for synthesis of 2-alkylbenzoic acids, via a benzyne intermediate. It is advised that this reaction be employed only on smaller scale, with safety precautions. The reaction must pass through a benzocyclobutane intermediate, this, or another, high energy species might accumulate and then decompose.

Halocarbons   MRH Carbon tetrachloride 3.31/50
Sorbe, 1968, 85
Interaction is explosively violent.

Other reactants
Yoshida, 1980, 264
MRH values calculated for 14 combinations with various materials are given.

Oxidants

MRH values show % of oxidant

See Chromium trioxide: Sodium amide

Dinitrogen tetraoxide: Sodium amide
MRH 4.31/48

Potassium chlorate: Sodium amide
MRH 2.64/76

Sodium nitrite: Sodium amide

Water
1. Mellor, 1940, Vol. 2, 255
2. Personal experience
Fresh material behaves like sodium with water, hissing, forming a diminishing floating globule, and often finally exploding [1]. Old, degraded (yellow) samples may be immersed in water for appreciable periods with little action, and then explode very violently. Disposal by controlled burning is safer [2].
See other N-METAL DERIVATIVES

4466. Sodium O-hydroxylamide
[22755-22-2]

H_2NNaO

NaONH$_2$

See Hydroxylamine: Metals

4467. Sodium amidosulfate
[13845-18-6]

H_2NNaO_3S

NaOSO$_2$NH$_2$

Metal nitrates or nitrites
See entry METAL AMIDOSULFATES

4468. Lead(II) nitrate phosphinate
[ ]

H_2NO_5PPb

O$_3$NPbOP(O)H$_2$

Bailar, 1973, Vol. 2, 131
It is powerfully explosive.
See other HEAVY METAL DERIVATIVES, REDOX COMPOUNDS

4469. Diazene
[3618-05-1]

H_2N_2

HN=NH

Its extreme instability is matched by its extremely high endothermicity (ΔH_f° (g) +212.1 kJ/mol, 7.07 kJ/g)
See Potassium hydroxylamine-O-sulfonate (reference 2)
See other ENDOTHERMIC COMPOUNDS

1620

4470. Hyponitrous acid
[14448-38-5] $H_2N_2O_2$

HON:NOH

Alone, or Potassium hydroxide
1. Sidgwick, 1950, 693
2. Mellor, 1940, Vol. 8, 407

An extraordinarily explosive solid, of which the sodium salt also explodes on heating to 260°C [1]. An attempt to prepare the acid by treating its silver salt with hydrogen sulfide caused explosive decomposition. Contact with solid potassium hydroxide caused ignition [2].
See Lead hyponitrite
See other N–O COMPOUNDS, REDUCANTS

4471. Nitric amide (Nitramide)
[7782-94-7] $H_2N_2O_2$

O_2NNH_2

Canis, C., *Rev. Chim. Minerale*, 1964, **1**, 521
Nitramide is quite unstable and various reactions in which it is formed are violent. Attempts to prepare it by interaction of various nitrates and sulfamates showed that the reactions became explosive at specific temperatures.

Alkalies
Thiele, J. *et al., Ber.*, 1894, **27**, 1909
A drop of conc. alkali solution added to solid nitramide causes a flame and explosive decomposition.

Sulfuric acid
Urbanski, 1967, Vol. 3, 16
Nitramide decomposes explosively on contact with conc. sulfuric acid.
See other N-NITRO COMPOUNDS

4472. Sulfamoyl azide (Amidosulfuryl azide)
[13449-16-6] $H_2N_4O_2S$

$H_2NSO_2N_3$

Shozda, R. J. *et al., J. Org. Chem.*, 1967, **32**, 2876
It is a low-melting solid, as shock-sensitive as glyceryl nitrate.
See other ACYL AZIDES

4473. Sodium phosphinate ('Sodium hypophosphite')
[7681-53-0] H_2NaO_2P

$$NaOP(O)H_2$$

Mellor, 1940, Vol. 8, 881
Evaporation of aqueous solutions by heating may cause an explosion, phosphine being evolved.

Other reactants
Yoshida, 1980, 345
MRH values calculated for 6 combinations, largely with oxidants, are given.

Oxidants MRH Sodium chlorate 2.88/49, sodium nitrate 1.00/44
1. Costa, R. L., *Chem. Eng. News*, 1947, **25**, 3177
2. Mellor, 1940, Vol. 8, 881
3. Mellor, 1971, Vol. 8, Suppl. 3, 624
Evaporation of a moist mixture of sodium phosphinate with a trace of sodium chlorate by slow heating caused a violent explosion. It was concluded that, once started, the decomposition of the phosphinate proceeds spontaneously [1]. Similar reactions have been reported with nitrates instead of chlorates. Such mixtures had previously been proposed as explosives [2]. Interaction of iodine with the anhydrous salts is violently exothermic, causing ignition [3].
See Perchloric acid: Sodium phosphinate
See other REDOX REACTIONS
See other METAL PHOSPHINATES, REDUCANTS

4474. Sodium dihydrogen phosphide
[24167-76-8] H_2NaP

$$NaPH_2$$

Albers, H. *et al., Ber.*, 1943, **76**, 23
It ignites in air.
See related PHOSPHINES

4475. Diprotium monoxide (Water)
[7732-18-5] H_2O

$$HOH$$

1. Carson, P. J. *et al., Loss Prev. Bull.*, 1992, **102**, 7
2. Anon. *Safety Digest Univ. Safety Assoc.*, 1994, (50), 22 & 25
3. Editor's comments, 1999
An extremely reactive liquid, solid or vapour with a dangerously high thermal capacity in both liquid and vapour state. When heated, the commonest cause of mid-19th century industrial explosions; it also bursts containers on cooling to low ambient temperatures. Still a frequent cause of vapour explosions today.

Reacts with many metals to give hydrogen, sometimes violently. With non-metals pyrophoric hydrides may result. Frequently initiates explosive reactions between other substances. Violent reactions with many non-metal and some metal halides and oxyhalides, also with many organometallic compounds. Many metal nonmetallides produce toxic, flammable or pyrophoric gases on contact with diprotium monoxide.

The oxide interacts exothermically with strong acids and bases. When heated to decomposition, diprotium monoxide evolves the dangerous materials hydrogen, oxygen and hydrogen peroxide (all have individual entries in this Handbook). For a few incidents involving diprotium monoxide reactions see [1]. An opposed pair of explosions from careless handling of diprotium monoxide are reported. A student cooled some glass bottles of impure material in liquid nitrogen, they exploded surprisingly violently, causing lacerations. A technician removed a heated carboy of the pure material from an autoclave, a few seconds later it exploded (having presumably been superheated) causing severe burns (these were lessened because she was not wearing safety type footwear, which might have filled with the hot liquid, but open shoes able to drain) [2]. The editor has known injury result from safety footwear when diprotium monoxide at only 70°C ran into it. He has himself emerged almost unscathed in sandals when it was poured on his ankle at 100°C; a similar accident to a colleague wearing shoes immobilised the victim for a week. Reading those Internet sites which cover chemical safety, anecdotally, it is apparent that diprotium monoxide is almost level with nitric acid as a cause of memorable mishap, usually by contact with unsuspected metal or metal hydride traces, sometimes as a catalyst to solid mixtures of an oxidant and a reductant, and occasionally through explosive boiling when heated [3]. In terms of human deaths caused, the most deadly material in this Handbook, also the initiator of the single chemical accident with the highest death toll. Too many potential cross-references to list

See Deuterium oxide
See Methyl isocyanate

Aluminium, Sodium dithionite
Kirschner, E., *Chem. Eng. News*, 1995, **73**(18), 6
During attempts to clear a blocked benzaldehyde pipe, water was allowed to flow into a vessel containing 400 kg of aluminium powder and 2,500 kg of 'sodium hydrosulfite'. The water reacted slowly with the hydrosulfite, evolving heat and hydrogen sulfide. Little attempt was made to deal with the reaction for 10 h. About 2 h later there was an explosion and fire, killing four operators who were attempting to nitrogen blanket the reactor. This was attributed to the aluminium having attained a temperature to react (in which case nitrogen would be useless, since air is unlikely to have been involved).
See Aluminium, Sodium dithionite

Incinerator dust
Takatsuki, H., *Chem. Abs.*, 1995, **122**, 168826k
Evolution of hydrogen, attributed to presence of aluminium particles, formed an explosive atmosphere over wetted incinerator residues.

Metals
U.S. Dept of Energy, 1994, HDBK-1081-94, (*Spontaneous Heating and Pyrophoricity*), 51

Several incidents in which moisture has contributed to fires or explosions, in some of which water was definitely the sole oxidant, in zirconium, magnesium, uranium and thorium scraps or powders are retailed, largely sourced from an earlier paper. However, a plutonium fire was extinguished with water.

See also WATER-REACTIVE COMPOUNDS

4476. Oxosilane
[22755-01-7] H_2OSi

$$O:SiH_2$$

Kautsky, K., *Z. Anorg. Chem.*, 1921, **117**, 209
It ignites in air.
See related SILANES

4477. Hydrogen peroxide
[7722-84-1] H_2O_2

HOOH

(*MCA SD-53*, 1969); *FPA H3*, 1972; *HCS 1980*, 549 (50%), 550 (30%), 551 (10%); *RSC Lab. Hazard Safety Data Sheet No. 57*, 1987

1. Shanley, E. S. *et al.*, *Ind. Eng. Chem.*, 1947, **39**, 1536
2. Naistat, S. S. *et al.*, *Chem. Eng. Progr.*, 1961, **57**(8), 76
3. Kirk-Othmer, 1966, Vol. 11, 407; *MCA SD-53*, 1969
4. Anon., *Fire Prot. Ass. J.*, 1954, 215
5. Smith, I. C. P., private comm., 1973
6. *MCA Case History No. 1121*
7. McClure, J. D., *J. Org. Chem.*, 1962, **27**, 627
8. Campbell, G. A. *et al.*, *Proc. 4th Symp. Chem Process Haz. Ref. Plant Des.*, 37–43, London, IChE, 1972
9. Clark, M. C. *et al.*, *Chem. & Ind.*, 1974, 113
10. Cookson, P. G. *et al.*, *J. Organomet. Chem.*, 1975, **99**, C31–32
11. *Manufacture of Organic Chemicals using Hydrogen Peroxide and Related Compounds*, AO 1.2, Widnes, Interox Chemicals Ltd., 1979
12. Berthold, W. *et al.*, *Proc. 3rd Int. Symp. Loss Prev. Safety Prom. Chem. Ind.*, 1431–1434, Basle, SSCI, 1980
13. Clemens, D., *Chem. Eng. News*, 1986, **64**(50), 2
14. Manning, D., *Chem. Eng. News*, 1987, **65**(9), 4
15. Merrifield, 1988, 23 pp.
16. '*Hydrogen Peroxide in Organic Chemistry*', Schirmann, J. P., Delavarenne, S. T., Paris, Informations Chimie, 1987 (English edn.)
17. Ullmann, 1989, **A13**, 462
18. Mackenzie, J., *Chem. Eng.*, 1990, **97**(6), 84

19. Tojo, G. INTERNET, 1997
20. Davies, A. G. *et al., J. Chem. Soc., Perkin Trans,* 2, 1981, 1512
The hazards attendant upon use of conc. hydrogen peroxide solutions have been reviewed [1,2,3]. Salient points include:

Release of enough energy during catalytic decomposition of 65% peroxide to evaporate all water present and formed, and subsequent liability of ignition of combustible materials.

Most cellulosic materials contain enough trace-metal catalysts to cause spontaneous ignition with 90% peroxide.

Contamination of conc. peroxide causes possibility of explosion. Readily oxidisable materials, or alkaline substances containing heavy metals, may react violently.

Soluble fuels (acetone, ethanol, glycerol) will detonate on admixture with peroxide of over 30% concentration, the violence increasing with concentration.

Handling systems must exclude fittings of iron, brass, copper, Monel, and screwed joints caulked with red lead preparations.

Concentrated peroxide may decompose violently in contact with iron, copper chromium and most other metals and their salts, and dust (which frequently contains rust). Absolute cleanliness, suitable equipment (PVC, butyl or Neoprene rubber, Teflon) and personal protection are essential for safe handling [4].

During concentration under vacuum of aqueous [5] or of aqueous–alcoholic [6] solutions of hydrogen peroxide, violent explosions occurred when the concentration was sufficiently high (probably over 90%) [3]. It is possible to accidentally concentrate 30% hydrogen peroxide to explosion on a laboratory rotary evaporator [19]. A suspension of 90% hydrogen peroxide in dichloromethane is shock-sensitive [7]. Detonation of hydrogen peroxide vapour has been studied experimentally [8]. Explosion of a screw capped winchester of 35% peroxide solution after 2 years owing to internal pressure of liberated oxygen emphasises the need to date-label materials of limited stability, and to vent the container automatically by fitting a Bunsen valve or similar device [9]. It has been suggested that hazards in use and handling of concentrated hydrogen peroxide can be avoided by using the solid 2:1 complex of hydrogen peroxide with 1,4-diazabicyclo[2.2.2]octane. This is hygroscopic but supposedly stable for at least several months in storage, although it has been reported to decompose above 60°C [10]. The 1:1 complex, which is of considerably lower energy, has proved explosive, however [20].

A survey, with many references, of 14 classes of preparative reactions involving hydrogen peroxide or its derivatives emphasises safety aspects of the various procedures [11]. Following the decomposition of 100 1 of 50% aqueous hydrogen peroxide which damaged the 630 1 stainless vessel rated at 6 bar, the effect of added contaminants and variations in temperature and pH on the adiabatic decomposition was studied in a 1 1 pressure vessel, where a final temperature of 310°C and a pressure around 200 bar were attained. Rust had little effect, but addition of a little ammonia (pH increased from 1.8 to 6.0) caused the induction period to fall dramatically, effectively from infinity to a few h at 40°C and a few min at 80°C. Addition of sodium hydroxide to pH 7.5 reduced the induction period at 24°C from infinity to about 4 min [12].

Several 0.5–1 l bottles of an unspecified plastic containing 30% hydrogen peroxide which had been stored for over a year became brittle and developed cracks during normal handling operations. One bottle stored for more than 4 years broke into small fragments when squeezed by hand [13]. However, these bottles were undoubtedly of the wrong material, because storage of up to 54% hydrogen peroxide solutions in suitable plastic bottles with no signs of container deterioration is a well-established commercial practice [14]. A recent report and review update and extend general information on storage and handling of the powerful oxidant. The second pays considerable attention to headspace explosions caused by decomposition in the presence of volatile fuels. It is calculated that during adiabatic decomposition of 90% peroxide solution, a temperature of 740°C could be attained, accompanied by the release of 5233 volumes of oxygen and steam [15]. Detonation limit diagrams for hydrogen peroxide vapours against liquid phase composition, temperature and pressure are given [17]. Each chapter of a recent monograph is devoted to a specific functional group, and the final chapter is concerned with safety aspects of the use of hydrogen peroxide as oxidant in mixtures with organic compounds [16].
See other INDUCTION PERIOD INCIDENTS

Acetal, Acetic acid
Ashley, J. N. *et al., Chem. & Ind.*, 1957, 702
An organic sulfur compound containing an acetal function had been oxidised to the sulfone with 30% hydrogen peroxide in acetic acid. After the liquor had been concentrated by vacuum distillation at 50–60°C, the residue exploded during handling. This was attributed to formation of the peroxide of the acetal (formally a *gem*-diether) or of the aldehyde formed by hydrolysis, but formation and explosion of peracetic acid seems a more likely explanation.

Acetaldehyde, Desiccants
Karnojitzky, V. J., *Chim. Ind* . (Paris), 1962, **88**, 235
Interaction gives the extremely explosive poly(ethylidene) peroxide, also formed on warming peroxidised diethyl ether.

Acetic acid
1. Grundmann, C. *et al., Ber.*, 1939, **69**, 1755
2. *491M*, 1975, 207
During preparation of peracetic acid, the temperature should not be too low to prevent reaction as the reagents are mixed, because reaction may begin later with explosive violence [1]. Heating well-diluted peroxide and acid will exothermically produce the acid which may explode at 110°C [2].

Acetic acid, *N*-Heterocycles
1. Wommack, J. B., *Chem. Eng. News*, 1977, **55**(50), 5
2. Dholakia, S. *et al., Chem. & Ind.*, 1977, 963
3. Daisley, R. W. *et al., Org. Prep. Proc. Int.*, 1983, **15**, 281
During isolation of the di-*N*-oxide of 2,5-dimethylpyrazine [1] and of the mono-*N*-oxide of 2,2-bipyridyl [1], prepared by action of hydrogen peroxide in acetic

acid on the heterocycles, violent explosions occurred on evaporation of the excess peroxyacetic acid. In the preparation of 3-bromopyridine *N*-oxide, the necessity to take adequate precautions against the possibility of explosion while vacuum distilling off excess reagents from a water bath maintained at 50°C is stressed [3]. For a method not involving this considerable explosion risk.
See N-OXIDES
See 3-Methyl-4-nitropyridine *N*-oxide

Acetic acid, Jute
De Forest, P. *et al., Chem. Eng. News*, 1987, **65**(31), 2; *J. Chem. Educ.*, 1988, **65**(8), 728–729
Jute fibre (0.1 g) was being prepared for a standard forensic examination by treatment in glacial acetic acid (20 ml) with hydrogen peroxide solution (20 ml, but in error 30% peroxide was used in place of the specified 6% '20-vol') strength. After heating on a water bath for 4 h, there was a violent explosion which shattered the ceramic hotplate below the bath with formation of 'ceramic shrapnel'. The available hydrogen peroxide would lead to formation of 14.7 g of peracetic acid as a 34 wt% solution in aqueous 33% acetic acid. This would be expected to become more concentrated by evaporation during prolonged heating, and concentrated solutions are known to be thermally unstable.
See Perchloric acid: Clay

Acetic acid, 3-Thietanol
1. Dittmer, D. C. *et al., J. Org. Chem.*, 1961, **26**, 1324
2. Dittmer, D. C. *et al., J. Org. Chem.*, 1971, **36**, 1324
3. Chang, P. *et al., J. Org. Chem.*, 1969, **34**, 2791
4. Lamm, B. *et al., Acta Chem. Scand. Ser. B*, 1974, **28**, 701–703
When following the original route [1] to 3-hydroxythietane-1,1-dioxide, it is essential to dilute and evaporate the hydrogen peroxide–acetic acid reaction mixture slowly from a large dish, to prevent explosions arising from concentration of peroxyacetic acid [2]. Alternative routes to avoid this hazard are available [3,4].
See Peroxyacetic acid

Acetic anhydride
1. Prett, K., *Textilveredlung*, 1966, **1**, 288–290
2. Swern, D., *Chem. Rev.*, 1945, **45**, 5
During preparation of peracetic acid solutions for textile bleaching operations, the reaction mixture must be kept acid. Under alkaline conditions, highly explosive diacetyl peroxide separates from solution [1]. An excess of the anhydride has the same effect [2].

Acetone, Other reagents   MRH Acetone 6.36/19
1. Anon., *Angew. Chem. (Nachr.)*, 1970, **18**, 3
2. *MCA Case History No. 233*
3. *MCA Case History No. 223*
4. Stirling, C. J. M., *Chem. Brit.*, 1969, **5**, 36

5. Seidl, H. *Angew. Chem. (Intern. Ed.)*, 1964, **3**, 640
6. Treibs, W., *Angew. Chem. (Intern. Ed.)*, 1963, **3**, 802
7. Brewer, A. D., *Chem. Brit.*, 1975, **11**, 335
8. Schurz, J. *et al.*, *Angew. Chem.*, 1956, **68**, 182
9. Ghilardi, C. A. *et al.*, *J. Chem. Soc., Dalton Trans.*, 1986, 12075–2081
10. Anon., *Sichere Chemiearb.*, 1987, **39**, 70
11. Ullmann, 1989, **A13**, 462

Acetone and hydrogen peroxide readily form explosive dimeric and trimeric cyclic peroxides, particularly during evaporation of the mixture. Many explosions have occurred during work-up of peroxide reactions run in acetone as solvent, including partial hydrolysis of a nitrile [1] and oxidation of 2,2-thiodiethanol [2] and of an unspecified material [3]. The reaction mixture from oxidation of a sulfide with hydrogen peroxide in acetone exploded violently during vacuum evaporation at 90°C. On another occasion, oxidation of a sulfide in acetone in presence of molybdate catalyst proceeded with explosive violence. A general warning against using acetone as a solvent for peroxide oxidations is given [4]. During the isolation of 1-tetralone, produced by oxidation of tetralin with hydrogen peroxide in acetone, a violent explosion occurred which was attributed to acetone peroxide [5]. The originator of the method later gave detailed instructions for a safe procedure, which feature the exclusion of mineral acids, even in traces [6].

During oxidation of an unspecified sulfur heterocycle in acetone with excess 35% hydrogen peroxide, a white solid separated during 3 days standing in a cool place. The solid (20 g) appeared to have been acetone peroxide, because it exploded with great violence during drying in a vacuum oven. The previous warning [4] on incompatibility of acetone and hydrogen peroxide was repeated [7]. A cyclic diketone was oxidised with alkaline peroxide in acetone solution and the product isolated by a solvent extraction procedure. During vacuum evaporation of the ethyl acetate extract, gassing was noted and the concentrate was treated twice with platinum catalyst to decompose excess hydrogen peroxide. After filtration and further vacuum evaporation, further gassing was noted and heating was discontinued, but the residue exploded [8]. This seems likely to have happened because the cyclic acetone peroxides, as dialkyl peroxides (and unlike hydroperoxides) would largely survive the catalytic decomposition treatment.

During the oxidation of the ligand tris(diphenylphosphinylethyl)amine with hydrogen peroxide in acetone to give tris(diphenylphosphonylethyl)amine, overheating must be avoided to prevent damaging explosions (of acetone peroxide). A Teflon beaker is recommended for the reaction [9]. A thiol was being oxidised in acetone to the disulfide with hydrogen peroxide. No product separated on standing, so the solution was concentrated to one fifth of its bulk in a rotary vacuum evaporator, and a white solid separated. During subsequent operations, the solid (trimeric acetone peroxide) exploded with great violence [10].

Detonation limit diagrams for hydrogen peroxide, water, acetone systms are given [11].

See 3,3,6,6,9,9-Hexamethyl-1,2,4,5,7,8-hexoxonane
See also KETONE PEROXIDES

Aconitic acid

Anon., private communication, 1993

An attempt was made to convert aconitic acid (2 g) to isocitric acid using 25 ml. of 30% aqueous hydrogen peroxide, time and temperature unspecified. A crystalline product precipitated which exploded on grinding in a mortar.

Alcohols   MRH Ethanol 6.19/19

1. *MCA SD-53*, 1969
2. Spengler, G. *et al., Brennst. Chem.*, 1965, **46**, 117
3. Heemskerk, A. H. *et al., Loss Prevention and Safety Promotion in the Process Industries*, Vol I, (Mewis, J. J., Pasman, H. J. & De Rademaker, E. E. Eds.), 411, Amsterdam, Elsevier, 1995

Homogeneous mixtures of concentrated peroxide with alcohols or other peroxide-miscible organic liquids are capable of detonation by shock or heat [1]. Furfuryl alcohol ignites in contact with 85% peroxide within 1 s [2]. Detonability limits of mixtures with 2-propanol have been measured. Approximately stoichiometric combinations of 50% hydrogen peroxide and the alcohol could be made to detonate, as could a wider range of mixtures with higher test peroxide [3].

See Oxygenated compounds, etc., below

Alcohols, Sulfuric acid.

1. Hedaya, E. *et al., Chem. Eng. News*, 1967, **45**(43), 73
2. Merrifield, 1988, 5–6, 16

During conversion of alcohols to hydroperoxides, the order of mixing of reagents is important. Addition of conc. acid to mixtures of an alcohol and conc. peroxide almost inevitably leads to explosion, particularly if the mixture is inhomogeneous and the alcohol is a solid [1]. In presence of concentrated acid, hydrogen peroxide forms peroxymonosulfuric acid (Caro's acid), especially if the rate of conversion of the alcohol is low. The peroxyacid is a very powerful and unstable oxidant, and its reactions are often violent. Thus in the hydrogen peroxide–sulfuric acid–alcohol ternary system, an autodetonation region may exist. This is shown in a triangular diagram for 2-propanol and *tert*-butanol [2].

See tert-Butanol, etc.; 2-Phenyl-1,1-dimethylethanol; 3,5-Dimethyl-3-hexanol, all below

Aluminium isopropoxide, Heavy metal salts

Ward, D. S., unpublished information, 1974

During preparation of an alumina catalyst, the isopropoxide was stirred with 6% hydrogen peroxide to generate a slurry of pseudo-boehmite alumina, to which was added a solution of heavy metal salts. The heating and foaming of the mixture was excessive and foam overflowed into a safety tray, then ignited. The incident was attributed to either spontaneous or static-induced ignition of the foam consisting of oxygen-rich bubbles in the isopropanol-containing liquid medium. The heavy metal salts would catalyse decomposition of the hydrogen peroxide and may also have been involved in the ignition process.

See Metals, etc., below

2-Amino-4-methyloxazole
Rapi, G. *et al., J. Chem. Soc., Chem. Comm.*, 1982, 1339–1340
Oxidative rearrangement of the oxazole to 4-hydroperoxy-5-hydroxy-4-methylimidazolidin-2-one in presence of iron(II) catalysts at ambient temperature may become explosive if not controlled effectively.

Ammonia
Matlow, S. L., *Chem. Eng. News*, 1990, **68**(30), 2
A sealed container of mixed 30% aqueous solution and aqueous ammonia exploded on storage, driving shards into a wall 18 m away.
See Oxygen(Gas): Ammonia and Oxygen(Liquid): Ammonia

Aromatic hydrocarbons, Trifluoroacetic acid   MRH Toluene 6.62/15
Demo, N. C. *et al., Tetrahedron Lett.*, 1977, 1703–1704
A solution of 30% aqueous hydrogen peroxide in trifluoroacetic acid is useful for destructive oxidation of the aromatic ring in preference to the side chains as is usual with most oxidants. During work-up operations, the excess peroxide must be catalytically decomposed with manganese dioxide before removal of solvent to prevent explosions.

Azelaic acid, Sulfuric acid
Berkowitz, S., US Pat. 4 147 720, 1979
Conversion of the acid to diperoxyazelaic acid in hydrogen peroxide/sulfuric acid medium at 45–50°C was uncontrollably exothermic and led to explosion. Use of a peroxide/phosphoric acid medium gives a safe and effective conversion of C_6–C_{16} diacids to the diperoxy acids, though up to 5% of sulfuric acid is needed as catalyst for C_{12}–C_{16} acids.
See Carboxylic acids, below

Benzene
Merrifield, 1988, 13
A triangular diagram shows the range of explosive mixtures in the hydrogen peroxide–benzene–water ternary system, and also includes data for various oxygenated water-soluble organic compounds.
See Oxygenated compounds, below

Benzenesulfonic anhydride
Fichter, F., *Helv. Chim. Acta*, 1924, **7**, 1072
Attempts to prepare benzeneperoxysulfonic acid by interaction of the anhydride and 90–95% peroxide led to explosively violent decomposition within a few s. When diluted with acetic acid, the reaction mixture soon began to decompose, leading to violent boiling. It was concluded that the peroxyacid is too unstable for more than transitory existence.

tert-Butanol, Sulfuric acid
Schenach, T. A., *Chem. Eng. News*, 1973, **51**(6), 39

Preparation of di-*tert*-butyl peroxide by addition of *tert*-butanol to 50% hydrogen peroxide–78% sulfuric acid (1:2 by wt) is a dangerously deceptive procedure. On the small scale, and with adequate cooling capacity it may be possible to prevent the initial stage (exothermic formation of *tert*-butyl hydroperoxide) getting out of control and initiating violent or explosive decomposition of the peroxide–peroxomonosulfuric acid mixture. This hazard diminishes as the reaction proceeds with consumption of hydrogen peroxide and dilution by the water of reaction. On the plant scale several severe explosions occurred, preceded only by a gradual temperature increase, during attempted process development work.
See Alcohols, above

Carbon   MRH 6.19/15
  1. Mellor, 1939, Vol. 1, 936–938
  2. Schumb, 1955, 402, 478
The violent decomposition observed on adding charcoal to conc. hydrogen peroxide is mainly owing to catalysis by metallic impurities present and the active surface of the charcoal, rather than to direct oxidation of the carbon [1]. Charcoal mixed with a trace of manganese dioxide ignites immediately on contact with conc. peroxide [2].

Carboxylic acids   MRH Formic acid 3.30/58
  Kuchta, J. M. *et al., Rep. Invest. No. 5877*, Washington, US. Bur. Mines, 1961
Admixture produces peroxyacids, some of which are unstable and explosive. Aqueous peroxide solutions containing formic acid, acetic acid or tartaric acid above certain concentrations can be caused to detonate by a severe explosive shock.
See Azelaic acid, etc., above; Oxygenated compounds, etc., below
See PEROXYACIDS

Catalyst (unspecified)
  1. Anon., *Sichere Chemiearb.*, 1979, **31**, 63
  2. Merrifield, 1988, 16
Aqueous cyanide residues were being detoxified by treatment with a catalyst and then 50% peroxide solution to oxidise cyanide to cyanate at pH 10 ± 0.5 at above 40°C. Apparently all the alkaline catalyst solution had not been completely drained from the 100 l dosing flask before the peroxide solution (28 kg) was charged. The latter solution began to decompose and this accelerated to explosion, destroying the flask [1]. In a continuous bleaching plant, a 200 l stainless dosing vessel, unrelieved except for a small overflow line, was connected directly to a supply of 35 wt% peroxide solution and to the bleaching process vessel. Back-contamination of the peroxide solution in the dosing vessel led to exothermic decomposition, and the evolved gases caused the vessel to pressure-burst [2].
See other GAS EVOLUTION INCIDENTS

Coal
  Wood, C. R., *Fuel*, 1974, **53**, 220

Coal (72 mesh) slurried with hydrogen peroxide solution (30 wt%) reacts steadily on warming to 85°C, but if much pyrites is present, the oxidation may accelerate to a dangerous extent.

See other CATALYTIC IMPURITY INCIDENTS

Copper(II) chloride

Merrifield, 1988, 7

In a printed circuit board etching line using copper(II) chloride solution, 45 wt% hydrogen peroxide solution was used to recover the copper salts. The peroxide header tank became contaminated with trace amounts of the etching solution, and catalytic decomposition of the peroxide led to a pressure burst of the tank.

See other CATALYTIC IMPURITY INCIDENTS, GAS EVOLUTION INCIDENTS

Cotton waste

Merrifield, 1988, 6

Spillage of a weak (5%) aqueous peroxide solution onto cotton waste led, after some time, to ignition. This probably involved concentration of the peroxide by evaporation/absorption of the water, the heat insulating effect of the fibrous mass preventing heat loss as oxidation proceeded, and possibly the presence of trace metals in the waste promoting catalytic decomposition.

See Wood, below

Diethyl ether

Bruhl, J. W., Ber., 1895, 28, 2856–2857

Evaporation of an ethereal solution of hydrogen peroxide gave a residue of which a drop on a platinum spatula exploded weakly on exposure to flame. When the sample (1–2 g) was stirred with a glass rod (not fire polished), an extremely violent detonation occurred.

See other GLASS INCIDENTS
See ETHERS

3,5-Dimethyl-3-hexanol, Sulfuric acid

MCA Guide, 1972, 315

The alcohol was treated with 90% hydrogen peroxide and a trace of sulfuric acid at 0°C. While warming to ambient temperature overnight it exploded violently. Some 3,5-dihydroperoxide may have formed from autoxidation at the 5-position.

See Alcohols, etc.; also tert-Butanol, etc.; both above, and 2-Methyl-1-phenyl-2-propanol, etc., below

Dimethylphenylarsine

Anon., Lab. Accid. Higher Educ., item 8, Barking, HSE, 1987

During the preparation of the As-oxide in glassware by oxidation with hydrogen peroxide (conditions unknown), an explosion occurred, possibly due to presence of impurities.

See Dimethylphenylphosphine, next below

Dimethylphenylphosphine

Dennister, M. L. et al., Inorg. Synth., 1977, 17, 185

Oxidation of the phosphine to the oxide by adding its solution in ether to stirred 15% aqueous peroxide may become very violent if the rate of addition is too fast.
See Dimethylphenylarsine, next above

Diphenyl diselenide
Bloodworth, A. J., *et al., J. Chem. Soc., Perkin Trans. 1*, 1983, 471–473
Attempted thermal dehydration of benzeneseleninic acid, formed by oxidation of diphenyl diselenide with hydrogen peroxide, gave a solid which exploded at 53–55°C. The solid may have been the complex of the acid with hydrogen peroxide.

2-Ethoxyethanol, Polyacrylamide gel, Toluene
1. Darnall, D. W., *Chem. Eng. News*, 1978, **56**(47), 47
2. Shanley, E. S., *Chem. Eng. News*, 1979, **57**(21), 43
Polyacrylamide gels were dissolved in 30% peroxide solution and added to a scintillation mixture in 1:1 2-ethoxyethanol–toluene. After counting, the mixtures were bulked and evaporated intermittently with heat during a 4 week period, and the accumulated peroxidised residues eventually exploded violently [1]. A subsequent comment indicated that peroxidised materials may not necessarily have been formed, because solutions of organic materials in aqueous peroxide are themselves potentially explosive [2].
See Oxygenated compounds, below

Ethyl acetate
Schierwater, F.-W., *Jahresber.*, 1981, 77
Organic material was extracted from a sample with methanol, and after evaporation the small residue was refluxed with aqueous peroxide. Extraction with ethyl acetate, separation and evaporation gave a residue which was scarcely visible but which exploded violently when the flask was moved. Several peroxidic species, including methyl hydroperoxide, peroxyacetic acid or its ethyl ester, may have been involved.
See Ethyl acetate

Fatty acid
Merrifield, 1988, 7
In a continuous operation to bleach a fat-derived acid, the latter was stirred and treated at 80°C with a slow stream of 50 wt% hydrogen peroxide. Un-noticed failure of the agitator led to peroxide build-up, layer formation and eventual formation of an explosive mixture. When this passed into a centrifugal pump, it detonated.
See other AGITATION INCIDENTS

Formic acid, Metaboric acid
See Peroxyformic acid

Gadolinium hydroxide
Bogdanov, G. A. *et al., Chem. Abs.*, 1976, **85**, 52253

Interaction gives a hydrated basic peroxide which decomposes explosively at 80–90°C.

Gallium, Hydrochloric acid
Titze, H., *Mikrochem. Acta*, 1977, 480
During dissolution for analysis of 1–9 g samples of gallium in conc. hydrochloric acid containing small portions of 30% hydrogen peroxide solution, cooling may be necessary to prevent development of explosively violent reactions at higher temperatures.

Hexamethylenetetramine
See 1,6-Diaza-3,4,8,9,12,13-hexaoxabicyclo[4.4.4]tetradecane

Hydroboration product of 2-methyl-2-propenyl tetrahydropyranyl ether
See TETRAHYDROPYRANYL ETHER DERIVATIVES

Hydrogen, Palladium catalyst
Anon., *Jahresber.*, 1982, 65–66
In the cyclic anthraquinone process for preparing hydrogen peroxide, a hydrocarbon solution of a 2-alkylanthraquinone is catalytically reduced with hydrogen to the 9,10-diol, which is then air-oxidised to the quinone with formation of hydrogen peroxide. The latter is removed from the process solution by water washing and phase separation, and the process stream, now virtually free of peroxide is recycled to the reactor. Failure of the compressed air supply prevented the separation of the aqueous hydrogen peroxide from the hydrocarbon phase, and both layers were recycled into the process. Contact of the peroxide solution with the catalyst released oxygen, which mixed with hydrogen and was ignited by the catalyst. The initial explosion released some of the flammable process solution which ignited, and eventually some 70 m^3 of the latter was involved in a major fire, with a further explosion in the hydrogenation plant.
See other HYDROGENATION INCIDENTS

Hydrogen selenide
Mellor, 1939, Vol. 1, 941
Interaction is very fast.

Immiscible organic liquids
Merrifield, 1988, 5
Emulsions formed from immiscible organic liquids in aqueous peroxide mixtures may behave in the same way as miscible organic liquids, but if the emulsion breaks and separation of the organic phase occurs, passage into an explosive region of the peroxide–water–organic liquid ternary system may occur, and this is potentially very dangerous.

Iron(III) chloride, Hydrocarbons
Merrifield, 1988, 7–8

A 6000 m^3 aqueous refinery effluent containing sulfides and traces of hydrocarbons was to be treated to remove sulfides before discharge. Aqueous ferric chloride was added, followed by 1800 l of 50 wt% peroxide added over 40 min, and after a further 30 min an explosion occurred which blew off the lid of the treatment vessel. This was attributed to ignition of the explosive mixture of hydrocarbon vapours and oxygen (from iron-catalysed decomposition of peroxide) formed above the liquid surface.
See other GAS EVOLUTION INCIDENTS

Iron(II) sulfate, 2-Methylpyridine, Sulfuric acid
Mond. Div., ICI, private. comm., 1969
Addition of 30% peroxide and sulfuric acid to 2-methylpyridine and iron(II) sulfate caused a sudden exotherm, followed by a vapour phase explosion and ignition. Lack of stirring is thought to have caused local overheating, vaporisation of the base and its ignition in the possibly oxygen-enriched atmosphere.
See other AGITATION INCIDENTS

Iron(II) sulfate, Nitric acid, Sodium carboxymethylcellulose
1. Sansoni, B. *et al., Z. Anal. Chem.*, 1968, **243**, 209
2. Cela Torrijos, R. *et al., Chem. & Eng. News*, 1978, **56**(49), 51
A published analytical procedure [1] for decomposing the sodium salt with ferrous-sulfate/peroxide in nitric acid at pH 2 led to deflagration or explosion during the evaporation stage when applied to 10 g samples, but not with 5 g samples. Presence of sulfuric acid avoids the problem [2].

Ketene
Swern, D., Chem. Rev., 1949, 45, 7
Interaction with excess ketene rapidly forms explosive diacetyl peroxide.

Ketones, Nitric acid
Bjorklund, G. H. *et al., Trans. R. Soc. Can., (Sect. III)*, 1950, **44**, 25
Unless the temperature and concentrations of reagents were carefully controlled, mixtures of hydrogen peroxide, nitric acid and acetone overheated and exploded violently. Under controlled conditions, the explosive dimeric or trimeric acetone peroxides were produced. 2-Butanone and 3-pentanone gave shock- and heat-sensitive oily peroxides. Cyclopentanone reacts vigorously, giving a solid which soon produces a series of explosions if left in contact with the undiluted reaction liquor. The isolated trimeric peroxide is very sensitive to shock, slight friction or rapid heating, and explodes very violently. Cyclohexanone and 3-methylcylohexanone gave oily, rather explosive peroxides.
See CYCLIC PEROXIDES, KETONE PEROXIDES

Lead, Trioxane
Bamberger, M. *et al., Z. Ges. Schiess-u. Sprengstoffw.*, 1927, **22**, 125–128
Mixtures of trioxane with 60% hydrogen peroxide solution are detonable by heat or shock, or spontaneously after contact with metallic lead. The latter may be owing to the heat of oxidation of lead.
See Metals *also* Oxygenated compounds, etc., both below

1635

Lithium tetrahydroaluminate
   Osmon, R. V., *Chem. Eng. Progr. Symp. Ser.*, 1966, **62**(61), 92–102
   This oxidiser–fuel combination showed promise in rocketry.
   See other REDOX REACTIONS

Mercury(II) oxide, Nitric acid   MRH Mercury(II) oxide 1.59/tr.
   Mellor, 1940, Vol. 4, 781
   Although red mercuric oxide usually vigorously decomposes hydrogen peroxide,
   the presence of traces of nitric acid inhibits decomposition and promotes formation
   of red mercury(II) peroxide. This explodes on impact or friction, even when wet,
   if the mercury oxide was finely divided.

Metals, or Metal oxides, or Metal salts
   1. Mellor, 1939, Vol. 1, 936–944
   2. Schumb, 1955, 480
   3. Kit and Evered, 1960, 136
   4. *Hydrogen Peroxide Data Manual*, Laporte Chemicals Ltd., Luton, 1960
   5. Hardin, M. C. *et al., Chem. Abs.*, 1963, **59**, 2583h
   6. Winnacker, 1970, Vol. 2, 552
   MRH Aluminium 12.76/35, iron 4.60/52, lithium 13.64/29, magnesium 12.68/41,
   silver, 1.59/tr., sodium 5.81/57; manganese dioxide, mercury(I) oxide, mercury(II)
   oxide, calcium permanganate, all 1.59/tr.
   The noble metals are all very active catalysts, particularly when finely divided
   or dispersed on high-area surfaces, for the decomposition of hydrogen peroxide,
   silver being used for this purpose in peroxide-powered rocket motors. Gold and
   the platinum group metals behave similarly [2,3,4]. Addition of platinum black to
   conc. peroxide solution may cause an explosion, and powdered magnesium and
   iron, promoted by traces of manganese dioxide, ignite on contact [1]. Potassium
   and sodium are oxidised violently by conc. peroxide solutions [1], and interaction
   of lithium and hydrogen peroxide is hypergolic and controllable under rocket
   motor conditions [5]. The effectiveness of metals for catalytic decomposition of
   peroxide solutions decreases in the order: osmium, palladium, platinum, iridium,
   gold, silver, manganese, cobalt, copper, lead [6]. Oxides of cobalt, iron (especially
   rust), lead (also the hydroxide), manganese, mercury and nickel are also very active
   and the parent metals and their alloys must be rigorously excluded from peroxide
   handling systems [1,4]. Soluble derivatives of many other metals, particularly under
   alkaline conditions, will also catalyse the exothermic decomposition, even at low
   concentrations [4]. Calcium permanganate has been used, either as a solid or in
   conc. solution, to ignite peroxide rocket motors [3]. Comprehensive data on all
   aspects of handling and use of conc. peroxide are available [4].
   See Carbon, above

Methanol, *tert*-Amine, Platinum catalyst
   Laughlin, R. G., *Chem. Eng. News*, 1978, **56**(19), 38
   To solubilise a dialkylmethylamine during oxidation to the *N*-oxide with hydrogen
   peroxide, methanol was used as diluent in place of water. After oxide formation

was complete, platinum black was added to decompose excess peroxide, and an explosion occurred. This was attributed to ignition of the methanol vapour–oxygen mixture by glowing catalyst particles.
See Oxygenated compounds, below

Methanol, Copper sulfate, Cyanide residues
Merrifield, 1988, 8
Aqueous cyanide effluent containing a little methanol in a 2 m^3 open tank was being treated to destroy cyanide by oxidation to cyanate with hydrogen peroxide in the presence of copper sulfate as catalyst. The tank was located in a booth with doors. Addition of copper sulfate (1 g/l) was followed by the peroxide solution (27 l of 35 wt%), and after the addition was complete an explosion blew off the doors of the booth. This was attributed to formation of a methanol vapour–oxygen mixture above the liquid surface, followed by spontaneous ignition. It seems remotely possible that unstable methyl hydroperoxide may have been involved in the ignition process.
See Waste treatment, below.
See other GAS EVOLUTION INCIDENTS

Methanol, Phosphoric acid
1. Anon., *J. Electrochem. Soc.*, 1965, **112**, 251C
2. King, C. V., *J. Electrochem. Soc.*, 1965, **112**, 1057; 1966, **113**, 519
In mixtures of the 3 components intended for use as chemical polishing or etching solutions [1], conc. peroxide must not be used [2].
See Oxygenated compounds, below

4-Methyl-2,4,6-triazatricyclo[5.2.2.0^{2,6}]undeca-8-ene-3,5-dione, Potassium hydroxide
1. Olsen, H. *et al., J. Amer. Chem. Soc.*, 1977, **99**, 1533
2. Engel, P. S. *et al., J. Amer. Chem. Soc.*, 1983, **105**, 7107
Treatment of the dione dissolved in ethylene glycol–30% hydrogen peroxide with potassium hydroxide leads to a violently exothermic hydrolysis — N-oxidation reaction [1]. Simultaneous addition of peroxide and alkali solutions from separate funnels gives a controllable and higher yielding reaction [2].

Nitric acid, Soils
See Nitric acid: Hydrogen peroxide, Soils

Nitric acid, Thiourea
Bjorklund, G. H. *et al., Trans. R. Soc. Can., (Sect. III)*, 1950, **44**, 28
The solid peroxide produced by action of hydrogen peroxide and nitric acid on thiourea (and possibly a hydrogen peroxidate of thiourea dioxide) decomposed violently on drying in air, with evolution of sufur dioxide and free sulfur.
See CRYSTALLINE HYDROGEN PEROXIDATES

Nitriles
Wakakura, M. *et al., Netsu Sokutsei*, 1994, **21**(2), 68; *Chem. Abs.*, 1994, **121**, 57411f

A calorimetric study of the alkaline hydrogen peroxide conversion of organic nitriles to amides was undertaken. The reaction showed potential for runaway if carelessly scaled up.

Nitrogenous bases   MRH values below references
1. Stone, F. S. *et al., J. Chem. Phys.*, 1952, **20**, 1339
2. Urbanski, 1967, Vol. 3, 306
3. *Haz. Chem. Data*, 1975, 296
MRH Ammonia, 5.86/25, aniline 6.44/17, dimethylhydrazine 6.69/19
Ammonia dissolved in 99.6% peroxide gave an unstable solution which exploded violently [1]. In the absence of catalysts, conc. peroxide does not react immediately with hydrazine hydrate. This induction period has caused a number of explosions and accidents owing to sudden reaction of accumulated materials [2]. 1,1-Dimethylhydrazine is hypergolic with high-test peroxide [3].
See other INDUCTION PERIOD INCIDENTS

Organic compounds   MRH Acetic acid 5.10/31, ethanol 6.19/19
1. Hutton, E., *Chem. Brit.*, 1969, **5**, 287
2. *The Manufacture of Organic Chemicals using Hydrogen Peroxide*, AO 1.2, 36–37, Widnes, Interox Chemicals Ltd., 1979
3. Nolan, 1983, Case history 156
4. Merrifield, 1988, 8
5. Mackenzie, J., *Plant/Oper. Progr.*, 1991, **10**(3), 143
6. Klais, O., *Thermochim. Acta*, 1993, **225**(2), 213
Although under certain circumstances mixtures of hydrogen peroxide and organic compounds are capable of developing more explosive power than an equivalent wt of TNT, in many cases interaction can be effected safely and well under control by applying well-established procedures, to which several references are given [1]. A triangular diagram specifically for the system hydrogen peroxide–glycerol–water, but also generally applicable to other organic compounds, such as acetic acid, ethanol, aniline or quinoline, shows the range of detonable compositions, and gives guidance and examples on avoiding the hazardous regions in reaction systems [2]. A polyester was made from propylene glycol, adipic acid and aluminium stearate catalyst by distilling out the water of reaction at 230°C under nitrogen. A portion of hydrogen peroxide, added to bleach the product, failed to do this, so a second portion was added. The explosion which occurred later was attributed to accumulation of highly concentrated excess peroxide and traces of organic material (either volatile or entrained) in the reflux dividing head [3]. Only a few ppm of impurities are necessary to cause autodetonation of very concentrated hydrogen peroxide [4]. A survey of explosion hazards likely to be encountered from reagent use of hydrogen peroxide in organic chemistry is given [5]. Charging too concentrated peroxide to organic materials in surfactant manufacture caused a pressure increase which burst a pipe, starting a catalogue of mishap which destroyed the plant. Laboratory investigations showed that 10% peroxide by weight with 'organic' material could cause pressure surges in excess of the plant rating. The editor suspects this is base catalysed peroxide decomposition, the organic material being tertiary amines for amine oxide surfactants [6].

Organic materials, Sulfuric acid

Analytical Methods Committee, *Analyst*, 1976, **101**, 62–66

Advantages and potential hazards in the use of mixtures of 50% hydrogen peroxide solution and conc. sulfuric acid to destroy various types of organic materials prior to analysis are discussed in detail. The method is appreciably safer than those using perchloric and/or nitric acids, but the use of an adequate proportion of sulfuric acid with a minimum of peroxide is necessary to avoid the risk of explosive decomposition. The method is not suitable for use in pressure-digestion vessels (PTFE lined steel bombs), in which an explosion occurred at 80°C.

Other reactants

Yoshida, 1980, 85–86

MRH values calculated for a wide range of 25 oxidisable or catalytically active materials are given.

Oxygenated compounds, Water

1. Monger, J. M. *et al., J. Chem. Eng. Data*, 1961, **6**(1), 23
2. Shanley, E. S., *Chem. Eng. News*, 1979, **57**(21), 43

MRH Acetaldehyde 6.27/21, acetic acid 5.27/31, acetone 6.26/19, ethanol 6.19/19, formaldehyde 6.44/31, formic acid 3.30/58, methanol 5.98/34, 2-propanol 6.32/16, propionaldehyde 6.44/19

The explosion limits have been determined for liquid systems containing hydrogen peroxide, water and acetaldehyde, acetic acid, acetone, ethanol, formaldehyde, formic acid, methanol, 2-propanol or propionaldehyde, under various types of initiation [1]. In general, explosive behaviour is noted where the ratio of hydrogen peroxide to water is >1, and if the overall fuel–peroxide composition is stoicheiometric, the explosive power and sensitivity may be equivalent to those of glyceryl nitrate [2].

See Alcohols, above

2-Phenyl-1,1-dimethylethanol, Sulfuric acid

1. Winstein, S. *et al., J. Amer. Chem. Soc.*, 1967, **89**, 1661
2. Hedaya, E. *et al., Chem. Eng. News*, 1967, **45**(43), 73
3. Hiatt, R. R. *et al., J. Org. Chem.*, 1963, **28**, 1893

Directions given [1] for the preparation of 2-phenyl-1,1-dimethylethyl hydroperoxide by adding sulfuric acid to a mixture of the alcohol and 90% hydrogen peroxide are wrong [2] and will lead to explosion [3]. The acidified peroxide (30–50% solution is strong enough) is preferably added to the alcohol, with suitable cooling and precautions [2].

See Alcohols, above, *also* Oxygenated compounds, etc., below

See 2-(4-Chlorophenyl)-1,1-dimethylethyl hydroperoxide

α-Phenylselenoketones

Reich, H. J. *et al., J. Amer. Chem. Soc.*, 1975, **97**, 5442

During conversion of the selenoketone to the nor-phenylselenoenones with hydrogen peroxide, care is necessary to control the strongly exothermic reaction.

Under no circumstances should oxidation of amounts above 5 mmol be effected by adding the full amount of peroxide before oxidation has started.

Phosphorus   MRH (Yellow) 7.78/27
Anon., *J. R. Inst. Chem.*, 1957, **81**, 473
If yellow or red phosphorus is incompletely immersed while undergoing oxidation in hydrogen peroxide solutions, heating at the air–solution interface can ignite the phosphorus and lead to a violent reaction. Such behaviour has been observed with peroxide solutions above 30% (110 vol) concentration.

Phosphorus(V) oxide   MRH 1.59/tr.
Toennies, G., *J. Amer. Chem. Soc.*, 1937, **59**, 555
The extremely violent interaction of phosphorus(V) oxide and conc. hydrogen peroxide to give peroxomonophosphoric acid may be moderated by using acetonitrile as a diluent.

Poly(acetoxyacrylic acid lactone), Poly(2-hydroxyacrylic acid)
Anon., *Jahresber.*, 1978, 75
The complex peroxide formed by stirring the polymeric compounds with 30% peroxide solution and sodium hydroxide at 5°C for an hour, according to a patented procedure, was precipitated with methanol. After intensive vacuum drying at 1.3 mbar without heating, the product suddenly exploded.
See other POLYPEROXIDES

Polymer residues
Merrifield, 1988, 7
A jacketed polymerisation vessel had become coated internally by a build up of polymer residues, and the vessel was being cleaned by treatment with aqueous hydrogen peroxide. To 5000 l of water in the vessel was added 150 kg of 27 wt% peroxide solution, and the vessel was heated by application of 10 bar steam (180°C) to the jacket. After a few minutes an explosion occurred, attributed to spontaneous ignition of a mixture of oxygen from the decomposing peroxide and monomer vapours produced by depolymerisation of the residue on the heated walls of the vessel.
See other GAS EVOLUTION INCIDENTS

2-Propanol
Merrifield, 1988, 7
A large volume (11.25 m^3) of mixed fatty acids was to be bleached by treatment with successive portions of 50 wt% hydrogen peroxide. 2-Propanol (450 l) was added to the acids (to improve the mutual solubility of the reactants). The first 20 l portion of peroxide (at 51°C) was added, followed after 1 min by a second portion. Shortly afterwards an explosion occurred, which was attributed to spontaneous ignition of a 2-propanol vapour–oxygen mixture formed above the surface of the liquid. Oxygen is almost invariably evolved from hydrogen peroxide reactions, and volatile flammable solvents are therefore incompatible components in peroxide systems.

See Alcohols, above
See other GAS EVOLUTION INCIDENTS

Sulfides
Editor's personal experience
Addition of 30% hydrogen peroxide to moderately strong sodium sulfide solutions produced superheating below the surface, followed by cavitational explosions sufficient to break laboratory glass beakers.

Sulfuric acid
1. Analytical Methods Committee, *Analyst*, 1967, **92**, 404
2. Dobbs, D. A. *et al., Chem. Eng. News*, 1990, **68**(17), 2
3. Erickson, T. V., *Chem. Eng. News*, 1990, **68**(33), 2
4. Anon., *Environment, Safety & Health Bull.*, 1993, **93**(2), 1
5. Stuart, R., Internet, 1991
Evaporation of mixtures of excess 50% hydrogen peroxide solutions with sulfuric acid (10:1) leads to loud but non-shattering explosions of the peroxomonosulfuric acid formed [1]. A freshly-made mixture of equal volumes of sulfuric acid and 30% hydrogen peroxide, "Piranha Solution" used for cleaning glass frits, exploded violently on storage. Explosions have also been experienced when cleaning frits, when contamination with organic solvent was suspected. The mixture will, in any case, slowly evolve oxygen to pressurise sealed containers [2,3]. Piranha solution also burst its container after addition of nitric acid [4]. A similar mixture, also containing small quantities of lithium sulfate and selenious acid and intended as a digesting agent for soil samples, burst its container after twelve days undisturbed in a refrigerator. Such solutions are best not stored and, if stored, should not be sealed [5].
See Organic materials, Sulfuric acid, above

Tetrahydrothiophene
Koppel, H. C., *Chem. Eng. News*, 1974, **52**(39), 3
Preparative reactions involving oxidation of tetrahydrothiophene to the sulfoxide by slow addition of 37% peroxide solutions exploded violently on 3 occasions. No explanation is apparent, and similar reactions had been run uneventfully over a period of 10 years.

Tin(II) chloride   MRH 1.21/85
Vickery, R. C. *et al., Chem. & Ind.*, 1949, 657
Interaction is strongly exothermic, even in solution. Addition of peroxide solutions of above 3 wt% strength causes a violent reaction.

Unsaturated compounds
Swern, 1971, Vol. 2, 432
In commercial-scale epoxidation of unsaturated organic compounds, 50–70% hydrogen peroxide is added during 2–8 h to the stirred mixture of unsaturated compound, aliphatic acid and acid catalyst at 50–80°C, the rate of addition being

dependent on heat-transfer capacity of the reaction system. When 70% peroxide is being used, care must be taken to avoid the dangerous detonable region by ensuring that an exotherm has occurred before more than 25% of the hydrogen peroxide is added.

See Carboxylic acids, *also* Organic compounds, *also* Oxygenated compounds, all above

Vinyl acetate

Anon., *ABCM Quart. Safety Summ.*, 1948, **19**, 18

Vinyl acetate had been hydroxylated by treatment with excess hydrogen peroxide in presence of osmium tetraoxide catalyst. An explosion occurred while excess vinyl acetate and solvent were being removed by vacuum distillation. This was attributed to the presence of peracetic acid, formed by interaction of excess hydrogen peroxide with acetic acid produced by hydrolysis of the vinyl acetate.

Waste treatment

1. Anon., 1989, private communication.
2. Anon., *Eur. Chem. News*, 1992, **58**(1537), 34
3. Wehrum, W. L., *Process Safety Progr.*, 1993, **12**(4), 199
4. Pratt, T. H., *Process Safety Progr.*, 1993, **12**(4), 203
5. Anon., *Loss Prev. Bull.*, 1991, (102), 17

In the interests of greater safety and environmental consciousness, a chemical factory which had previously used sodium hypochlorite to treat alkaline aqueous wastes containing methanethiol, other volatile sulfur compounds, methanol and possibly dimethyl ether, changed to 30% hydrogen peroxide. Shortly after, an explosion in the headspace shattered the vessel, killing a worker and breaking a pipe containing flammable solvent, leading to a fire which destroyed the factory. It appears that decomposition of peroxide in alkali had produced an oxygen enriched headspace, allowing the organic vapours therein to enter explosive limits despite a nitrogen purge. The actual source of ignition is unknown [1]. Another fatal explosion when treating effluent, NO containing off gas, with peroxide is reported, presumably by creation and ignition of an oxygen rich atmosphere containing organic vapours [2]. A third waste treatment incident, possibly fuelled by acetone is reported [3]. An ignition mechanism, by means of electrostatic effects from bursting oxygen bubbles, is postulated [4]. A related accident which killed 17 men was due to an explosion in a waste tank from a propylene oxide/styrene plant. This contained hydrocarbons over aqueous waste and evolved oxygen from decomposition of 'light peroxide' by-product decomposition. The headspace was nitrogen flushed. During maintenance the nitrogen flow was cut off for 34 hours, but oxygen readings remained low. Minutes after nitrogen purging was restarted the tank exploded. It is thought the oxygen meter may have been ill positioned [5].

See Methanol, Copper sulfate, Cyanide residues, above

Wood

1. *MCA Case History No. 1626*
2. *MCA Case History No. 1648*

Leakage from drums of 35% hydrogen peroxide onto a wooden pallet caused ignition of the latter when it was moved. Combustion, though limited in area, was fierce and took some time to extinguish [1]. Leakage of 50% peroxide onto supporting pallets under polythene sheeting led to spontaneous ignition and a fierce fire. Contact of 50% peroxide with wood does not usually lead to spontaneous ignition, but hot weather, dry wood (possibly catalytically contaminated) and the thermal insulation of the cover may have contributed to ignition [2].
See other PACKAGING INCIDENTS

Yeast
Bodenbaugh, J. H. *et al., J. Chem. Educ.*, 1988, **66**(5), 461–463
Hydrogen peroxide solution may be catalytically decomposed to give oxygen gas safely using dry yeast powder. (The active component is enzymatic.)
See other GAS EVOLUTION INCIDENTS
See other OXIDANTS, PEROXIDES

4478. Zinc hydroxide
[20427-58-1] H_2O_2Zn
$$Zn(OH)_2$$

Chlorinated rubber
See CHLORINATED RUBBER: metal oxides or hydroxides

4479. Sulfuric acid
[7664-93-9] H_2O_4S
$$HOSO_2OH$$

(*MCA SD-20*, 1963); *NSC 325*, 1982; *NSC 210*, 1967 (oleum); *FPA H31*, 1974; *HCS 1980*, 878, 879 (solutions), 702–704 (oleum); *RSC Lab. Hazards Safety Data Sheet No. 33*, 1985

The MCA Data Sheet covers safe handling procedures for all grades of sulfuric acid from fuming (oleum) through conc. to diluted acid. Coverage of the other series is as shown.
For an unusual incident involving static charge generation by sulfuric acid drops
See Hydrogen: Oxygen, Sulfuric acid

Acetaldehyde   MRH 1.34/57
See Acetaldehyde: Sulfuric acid

Acetone, Nitric acid
See Nitric acid: Acetone, etc.

Acetonitrile, Sulfur trioxide
See Acetonitrile: Sulfuric acid

Acrylonitrile   MRH 2.26/52
  See Acrylonitrile: Acids

Alkyl nitrates   MRH Methyl nitrate 6.53/99+
  See ALKYL NITRATES: LEWIS ACIDS

2-Aminoethanol
  Nolan, 1983, Case history 101
  The two components are immiscible and must be stirred to effect conversion at
  110°C to 2-aminoethyl hydrogen sulfate. A shift change led to the base being
  added to the acid without agitation. When the agitator was started later, reaction
  of the two cold and viscous layers of acid and base proceeded explosively.
  See other AGITATION INCIDENTS

Ammonium iron(III) sulfate dodecahydrate
  Clark, R. E. D., private comm., 1973
  A few dense crystals heated with sulfuric acid exploded, owing to the exotherm
  in contact with water liberated as the crystals disintegrated.

Aniline, Glycerol, Nitrobenzene
  See Quinoline

Benzyl alcohol   MRH 1.13/52
  Grignard, 1960, Vol. 5, 1005
  A mixture of the alcohol with 58% sulfuric acid decomposes explosively at about
  180°C.

Bromine pentafluoride
  See Bromine pentafluoride: Acids, etc.

tert-Butyl-*m*-xylene, Nitric acid
  See Nitric acid: *tert*-Butyl-*m*-xylene, etc.

1-Chloro-2,3-epoxypropane
  See 1-Chloro-2,3-epoxypropane: Sulfuric acid

4-Chloronitrobenzene, Sulfur trioxide
  1. Grewer, T., *Chem. Ing. Tech.*, 1975, **47**, 233
  2. Vervalin, C. H., *Hydrocarbon Process.*, 1976, **55**(9), 323
  3. *See entry* SELF-ACCELERATING REACTIONS
  4. Nolan, 1983, Case history 100
  DTA shows that the reaction mixture from sulfonation of the nitro compound in
  20% oleum, containing 35 wt% of 2-chloro-5-nitrobenzenesulfonic acid, exhibits
  2 exothermic stages at 100 and 220°C, respectively, the latter being violently
  rapid. The adiabatic reaction mixture, initially at 89° attained 285°C with boiling
  after 17 h. At 180° the induction period was about 20 min [1]. Sulfonation of
  4-chloronitrobenzene with 65% oleum at 46°C led to a runaway decomposition

reaction in a 2000 l vessel. The original process using 20% oleum was less sensitive to heating rate and temperature. Knowledge that the reaction could be dangerous above 50°C had not been applied [2]. More thermoanalytical work has established safer operating conditions with a greater margin for remedial action [3]. In a further incident, the sulfonation (an established process) had been effected with 65% oleum at below 80°C, and the reaction mixture had been heated to, and held at 115°C, and then cooled to 80°C. Half of the batch was quenched in water with cooling to crystallise the product, but the remaining half in the reactor overheated and exploded. This was attributed to backflow of water from the quench tank into the reactor, and the exotherm from reaction of water with oleum led to exothermic decomposition of the sulfonic acid at around 160°C [4].

See THERMAL STABILITY OF REACTION MIXTURES AND SYSTEMS
See other INDUCTION PERIOD INCIDENTS, SULFONATION INCIDENTS

Copper
Campbell, D. A., *School Sci. Rev.*, 1939, **20**(80), 631
The generation of sulfur dioxide by reduction of sulfuric acid with copper is considered too dangerous for a school experiment.

2-Cyano-4-nitrobenzenediazonium hydrogen sulfate
See 2-Cyano-4-nitrobenzenediazonium hydrogen sulfate: Sulfuric acid

2-Cyano-2-propanol
1. *Occupancy Fire Record, FR 57-5*, 5, Boston, NFPA, 1957
2. Kirk-Othmer, 1967, Vol. 13, 333
Addition of sulfuric acid to the cyano-alcohol caused a vigorous reaction which pressure-ruptured the vessel [1]. This seems likely to have been caused by insufficient cooling to prevent dehydration of the alcohol to methacrylonitrile and lack of inhibitors to prevent exothermic polymerisation of the nitrile [2].

Cyclopentadiene   MRH 1.75/40
Wilson, P. J. *et al.*, *Chem. Rev.*, 1944, **34**, 8
It reacts violently with charring, or explodes in contact with conc. sulfuric acid.
See Nitric acid: Hydrocarbons

Cyclopentanone oxime
See Cyclopentanone oxime: Sulfuric acid

1,3-Diazidobenzene
See 1,3-Diazidobenzene: Acids

Dichloromethane, Ethanol, Nitrate or nitrite
Dickie, R. J., private comm., 1981
The residues from a cortisol assay procedure (5 cc dichloromethane, 2.5 cc of a fluorescent reagent in 15:85 ethanol–sulfuric acid) were added to a 500 cc bottle and screw capped. After a 90 s delay, the bottle burst violently and brown fumes were seen. It was surmised that a nitrate or nitrite contaminant in the bottle had

liberated oxides of nitrogen or nitric acid which had oxidised the organics exother-mically.
See Nitric acid: Alcohols, *also* Dinitrogen tetraoxide: Halocarbons

Diethylamine
Anon., *CISHC Chem. Safety Summ.*, 1984, **55**(218), 34–35
Diethylamine fumes from a reactor were usually absorbed in a glass scrubber through which sulfuric acid was circulated, but an unresolved fault in the level sensor caused the acid circulation pump to operate intermittently. While the pump was not running, amine fumes condensed in the dip pipe, forming a solid crust (of the sulfate) which allowed a quantity of condensed amine to accumulate out of contact with the acid. When the pump was restarted, the neutralisation exotherm was sufficient to shatter the scrubber and distort the mesh guard around it.
See other NEUTRALISATION INCIDENTS

Dimethoxydinitroanthraquinone
See mixo-Dimethoxydinitroanthraquinone: Sulfuric acid

4-Dimethylaminobenzaldehyde
MCA Case History No. 2101
During the preparation of a solution of the aldehyde in dilute sulfuric acid, the latter should be prepared before addition of the aldehyde. An attempt to prepare the solution by adding a slurry of the basic aldehyde in a little water to conc. sulfuric acid caused the stoppered flask to burst from the large exotherm generated by this procedure.

2,5-Dinitro-3-methylbenzoic acid, Sodium azide
See 2,5-Dinitro-3-methylbenzoic acid: Oleum, Sodium azide

1,5-Dinitronaphthalene, Sulfur
See 1,5-Dinitronaphthalene: Sulfur, etc.

Ethoxylated nonylphenol
Anon., *CISHC Chem. Safety Summ.*, 1978, **49**, 73
Breakage of glassware allowed 92% sulfuric acid into a heating bath containing hot Synperonic NX (nearly anhydrous ethoxylated nonylphenol, a liquid surfactant). Hydrolysis of the latter lowered the flashpoint and the bath ignited (possibly from formation of dioxane, flash point 12°C).

Hexalithium disilicide
See Hexalithium disilicide: Acids

Hydrofluoric acid
Bentzinger, von R. *et al., Praxis Naturwiss. Chem.*, 1987, 36, 37
A student was attempting to prepare anhydrous hydrogen fluoride by dehydrating aqueous 60% hydrogen fluoride solution with conc. sulfuric acid. Addition of 200 ml of sulfuric acid to 500 ml of hydrofluoric acid in a 1 l copper flask led to

a rumbling noise, then a fountain from the flask neck of hot mixed acids which severely corroded the window glass and the floor tiles.
See Water, below
See other CORROSION INCIDENTS, GLASS INCIDENTS

Hydrogen peroxide   MRH 1.59/99+
See Hydrogen peroxide: Sulfuric acid

Mercury nitride
See Mercury nitride: Alone, or Sulfuric acid

Metal acetylides or carbides   MRH Calcium carbide 4.35/56
   1. Mellor, 1946, Vol. 5, 849
   2. *MCA SD-50*, 1963
Monocaesium and monorubidium acetylides ignite with conc. sulfuric acid [1]. Other carbides are hazardous in contact [2].

Metal chlorates   MRH Sodium chlorate 1.30/99+
   See METAL CHLORATES: ACIDS, *ALSO* METAL HALOGENATES: METALS, ETC.

Metal perchlorates
See METAL PERCHLORATES: SULFURIC ACID

4-Methylpyridine
   Anon., *Loss. Prev. Bull.*, 1979, (029), 130–131
The base was added to 60 l of acid to convert it to the salt, but malfunction of the swivelling blade glass stirrer caused no agitation and the liquid base remained layered above the conc. acid. Switching off the stirrer caused it to become effective and the whole of the contents erupted from the 100 l flask as an acid mist.
See other AGITATION INCIDENTS, NEUTRALISATION INCIDENTS

Nitramide
See Nitric amide: Sulfuric acid

Nitric acid, Organic materials
See Nitric acid: Organic materials, Sulfuric acid

Nitric acid, Toluene
See Nitric acid: Hydrocarbons (reference 7)

Nitroaryl bases and derivatives   MRH Nitroaniline 3.30/99+
   1. Hodgson, J. F., *Chem. & Ind.*, 1968, 1399; private comm., 1973
   2. Poshkus, A. C. *et al., J. Appl. Polym. Sci.*, 1970, **14**, 2049–2052
A series of 2- and 4-nitroaniline derivatives and analogues when heated with conc. sulfuric acid to above 200°C undergo, after an induction period, a vigorous reaction. This is accompanied by gas evolution which produces up to a 150-fold increase in volume of a solid foam, and is rapid enough to be potentially hazardous if confined.

2-Nitroaniline reacts almost explosively [1], and 4-nitroaniline, 4-nitroacetanilide, aminonitrobiphenyls, aminonitronaphthalenes and their various derivatives [2], as well as some nitro-N-heterocycles [1,2], also react vigorously. 4-Nitroanilinium sulfate and 4-nitroaniline-2-sulfonic acid and its salts also generate foams when heated without sulfuric acid. The mechanism is not clear, but involves generation of a polymeric matrix foamed by sulfur dioxide and water eliminated during the reaction [1].

See Diethyl sulfate: 2,7-Dinitro-9-phenylphenanthridine
See other INDUCTION PERIOD INCIDENTS, GAS EVOLUTION INCIDENTS

Nitrobenzene
See Nitrobenzene: Sulfuric acid

3-Nitrobenzenesulfonic acid
See 3-Nitrobenzenesulfonic acid: Sulfuric acid

Nitro-explosives
Gamezo, V. N. *et al., J. Phys. IV*, 1995, **5** (C4), 395
The effect of sulphuric acid and oleum on the explosive properties of nitromethane, dinitrotoluene, trinitrobenzene and trinitrotoluene was studied. They are considerably sensitised.
See OTHER NITRO-DERIVATIVES, ABOVE AND BELOW

Nitromethane
See Nitromethane: Acids

N-Nitromethylamine
See N-Nitromethylamine: Sulfuric acid

4-Nitrotoluene
See 4-Nitrotoluene: Sulfuric acid

Other reactants
Yoshida, 1980, 393
MRH values calculated for 15 combinations with various materials are given.

Permanganates   MRH Potassium permanganate 0.50/76
Interaction produces the powerful oxidant, permanganic acid.
See Permanganic acid: Organic materials
Potassium permanganate: Sulfuric acid

Phosphorus   MRH (Red) 2.64/28
1. Mellor, 1940, Vol. 8, 786
2. Zschalich, A. *et al., Chem. Abs.*, 1993, 11532b
3. Urben, P. G., personal experience
White phosphorus ignites in contact with boiling sulfuric acid or its vapour [1]. An explosion hazard exists at industrial scale under milder conditions [2]. Even

red phosphorus reacts vigorously with sulfuric acid in the presence of iodine or iodides, to produce hydrogen sulfide, from about 80°C [3].

Phosphorus(III) oxide
See Tetraphosphorus hexaoxide: Sulfuric acid

Poly(silylene)
See Poly(silylene): Oxidants

Potassium
See Potassium: Sulfuric acid

Potassium *tert*-butoxide
See Potassium *tert*-butoxide: Acids

3-Propynol
See 3-Propynol: Sulfuric acid

Silver permanganate
See Silver permanganate: Sulfuric acid

Silver peroxochromate

Sodium   MRH 3.89/31
See Sodium: Acids

Sodium carbonate
See Sodium carbonate: Sulfuric acid

Sodium tetrahydroborate
See Sodium tetrahydroborate: Acids

Sodium thiocyanate
Meyer, 1977, 39
Violent exotherm on contact, carbonyl sulfide being evolved.

Steel
MARS Database, 1998, short report 021
Steel is normally considered a safe material for contact with (reasonably pure) sulphuric acid. However this report is of a fatal explosion when working on a sulphuric acid tank with cutting torches, attributed to hydrogen in the headspace arising from corrosion of the metal ascribed to poor maintenance.

Sucrose
Meeks, E. G., *School Sci. Rev.*, 1979, **61**(215), 281–283
In the demonstration of the foaming carbonisation of sucrose by sulfuric acid, it has been found that about half of the gas produced is carbon monoxide and not

carbon dioxide as previously supposed. (Other sugars and carbohydrates may be expected to behave similarly.)

Tetramethylbenzenes
1. Birch, S. F. *et al., J. Amer. Chem. Soc.*, 1949, **71**, 1364
2. Smith, L. I., *Org. Reactions*, 1942, Vol. 1, 382
Sulfonation of the mixed isomers of 1,2,3,5- and 1,2,4,5-tetramethylbenzenes was too violent [1], for shaking in a closed glass vessel as originally described [2].
See other SULFONATION INCIDENTS

1,2,4,5-Tetrazine
See 1,2,4,5-Tetrazine: Sulfuric acid

Thallium(I) azidodithiocarbonate
See Thallium(I) azidodithiocarbonate: Sulfuric acid

1,3,5-Trinitrosohexahydro-1,3,5-triazine
See 1,3,5-Trinitrosohexahydro-1,3,5-triazine: Sulfuric acid

Water
Mellor, 1947, Vol. 120, 405–408
Dilution of conc. sulfuric acid by water is vigorously exothermic and must be effected by adding acid to water to avoid local boiling. Mixtures of sulfuric acid and excess snow form powerful freezing mixtures. Fuming sulfuric acid (oleum, containing sulfur trioxide) reacts violently with water.
See 3-Nitrobenzenesulfonic acid: Sulfuric acid

Zinc iodide
Pascal, 1962, Vol. 5, 168
Interaction with the conc. acid is violent.
See other INORGANIC ACIDS, OXIDANTS

4480. Poly(dihydroxydioxodisilane)

[ ] $(H_2O_4Si_2)_n$

$$(Si(O)(OH)Si(O)(OH))_n$$

1. Sorbe, 1968, 127
2. Mellor, 1940, Vol. 6, 230
The dry compound explodes on impact or heating [1]. This polymeric product of hydrolysis of hexahalo- or hexaethoxy-disilane decomposes with more or less violence if heated in air (when it ignites) or in a test tube (when it explodes) [2].
See related NON-METAL OXIDES

4481. Peroxomonosulfuric acid
[7722-86-3]

$H_2O_5S_2$

HOOSO$_2$OH

1. Edwards, J. O., *Chem. Eng. News*, 1955, **33**, 3336
2. Brauer, 1963, Vol. 1, 389

A small sample had been prepared from chlorosulfuric acid and 90% hydrogen peroxide; the required acid phase was separated and stored at 0°C overnight. After warming slightly, it exploded [1]. The handling of large amounts is dangerous owing to the possibility of local overheating (e.g., from contact with moisture) and explosive decomposition [2].

Acetone
1. *MCA Case History No. 662*
2. Bayer, A., *Ber.*, 1900, **33**, 858

Accidental addition of a little acetone to the residue from wet-ashing a polymer with mixed nitric–sulfuric acids and hydrogen peroxide caused a violent explosion [1]. The peroxoacid would be produced under these conditions, and is known to react with acetone to produce the explosive acetone peroxide [2], but direct oxidation of acetone may have been responsible.

Alcohols
1. Toennies, G., *J. Amer. Chem. Soc.*, 1937, **59**, 552
2. Merrifield, 1988, 6, 16

Contact of the acid with *sec-* or *tert*-alcohols, even with cooling, may lead to violent explosions [1]. The autodetonation region which exists for the hydrogen peroxide–sulfuric acid −2-propanol or–*tert*-butanol system is shown on a triangular diagram [2].

Aromatic compounds
Sidgwick, 1950, 939

Mixtures with aniline, benzene, phenol etc., explode.

Catalysts
Mellor, 1946, Vol. 5, 483–484

The 92% acid is decomposed explosively on contact with massive or finely divided platinum, manganese dioxide or silver. Neutralised solutions of the acid also froth violently on treatment with silver nitrate, lead dioxide or manganese dioxide.

Fibres
Ahrle, H., *Z. Angew. Chem.*, 1909, **23**, 1713

Wool and cellulose are rapidly carbonised by the 92% acid, while cotton ignites after a short delay.

See other INORGANIC ACIDS, OXIDANTS, PEROXOACIDS

4482. Peroxodisulfuric acid
  [13445-49-3] $H_2O_8S_2$
                              $HOSO_2OOSO_2OH$

Aromatic amines, Sulfuric acid
  Nielsen, A. T. *et al., J. Org. Chem.*, 1980, **45**, 2341–2347
  The acid, prepared from 90–98% hydrogen peroxide and oleum or 100% sulfuric acid, is one of the most powerful known oxidants and its use for oxidising aromatic amines to nitro compounds has been studied. Some mono- di- and tri-amines are destroyed exothermically with violent fume-off. Precautions for use are detailed.

Organic liquids
  1. D'Ans, J. *et al., Ber.*, 1910, **43**, 1880; *Z. Anorg. Chem.*, 1911, **73**, 1911
  2. Sidgwick, 1950, 938
  A very powerful oxidant; uncontrolled contact with aniline, benzene, ethanol, ether, nitrobenzene or phenol may cause explosion [1]. Alkanes are slowly carbonised [2].
  See other OXIDANTS, PEROXOACIDS

†4483. Hydrogen sulfide
  [7783-06-4] H_2S
                                  H_2S

(*MCA SD-36*, 1968); *NSC 284*, 1977; *FPA H78*, 1979; *HCS 1980*, 553 (cylinder); *RSC Lab. Hazards Safety Data Sheet No. 35*, 1985

1,2-Bis(2-azidoethoxy)ethane, Ethanol
  See 1,2-Bis(2-azidoethoxy)ethane: Ethanol, etc.

4-Bromobenzenediazonium chloride
  See DIAZONIUM SULFIDES AND DERIVATIVES

Copper, Oxygen
  Merz, V. *et al., Ber.*, 1880, **13**, 722
  Addition of powdered copper to a 1:2 mixture of hydrogen sulfide and oxygen causes the metal to become incandescent and ignite the explosive mixture.
  See Metals, below

Metal oxides   MRH Copper oxide 1.05/70, chromium trioxide 2.05/88, lead dioxide 0.54/78, manganese dioxide 0.84/94, sodium peroxide 3.01/70
  Mellor, 1947, Vol. 10. 129, 141
  Hydrogen sulfide is rapidly oxidised, and may ignite in contact with a range of metal oxides, including barium peroxide, chromium trioxide, copper oxide, lead dioxide, manganese dioxide, nickel oxide, silver(I) oxide, silver(II) oxide, sodium peroxide, and thallium(III) oxide. In the presence of air, contact with mixtures of calcium oxide or barium oxide with mercury oxide or nickel oxide may cause vivid incandescence or explosion.
  See Rust, below

Metals
>Mellor, 1947, Vol. 10, 140; 1943, Vol. 11, 731
A mixture with air passed over copper powder may attain red heat. Finely divided tungsten glows red hot in a stream of hydrogen sulfide.
See Copper, Oxygen, above *See also* Sodium: Sulfides

Nitrogen trichloride
See Nitrogen trichloride: Initiators

Other reactants
Yoshida, 1980, 390
MRH values calculated for 11 combinations with oxidants are given.

Oxidants   MRH values show % of oxidant
See Metal oxides, above; Oxygen, and Rust, both below

Bromine pentafluoride: Hydrogen-containing materials	MRH 3.01/70
Chlorine trifluoride: Hydrogen-containing materials	MRH 2.05/88
Chromium trioxide: Hydrogen sulfide	
Copper chromate oxide: Hydrogen sulfide	
Dibismuth dichromium nonaoxide: Hydrogen sulfide	
Dichlorine oxide: Oxidisable materials	MRH 3.22/91
Fluorine: Hydrogen sulfide	MRH 9.29/72
Heptasilver nitrate octaoxide: Alone, or Sulfides	
Lead(II) hypochlorite: Hydrogen sulfide	
Lead(IV) oxide: Hydrogen sulfide	MRH 0.54/78
Mercury(I) bromate: Hydrogen sulfide	
Nitric acid: Non-metal hydrides	MRH 4.43/74
Oxygen difluoride: Combustible gases	
Perchloryl fluoride: Hydrocarbons, etc.	
Silver bromate: Sulfur compounds	
Sodium peroxide: Hydrogen sulfide	MRH 3.01/70

Oxygen
Gray, P. *et al., J. Chem. Soc., Faraday Trans. 1*, 1974, **70** 2338–2350
During an investigation of the spontaneously explosive oxidation of near-stoicheiometric gas mixtures in the range 280–360°C, extensive self-heating was observed before ignition occurred. The second and third ignition limits were also investigated.
See Oxidants, above

Rust
Mee, A. J., *School Sci. Rev.*, 1940, **22**(85), 95
Hydrogen sulfide may ignite if passed through rusty iron pipes.
See Metal oxides, above

Silver fulminate
See Silver fulminate: Hydrogen sulfide

Soda-lime
1. Mellor, 1947, Vol. 10, 140
2. Bretherick, L., *Chem. & Ind.*, 1971, 1042
Interaction is exothermic, and if air is present, incandescence may occur with freshly prepared granular material. Admixture with oxygen causes a violent explosion [1]. Soda-lime, used to absorb hydrogen sulfide, will subsequently react with atmospheric oxygen and especially carbon dioxide (from the solid coolant) with a sufficient exotherm in contact with moist paper wipes (in a laboratory waste bin) to cause ignition [2]. Spent material should be saturated with water before separate disposal. Mixture analogous to soda-lime, such as barium hydroxide with potassium or sodium hydroxides, also behave similarly [1].
See other SELF-HEATING AND IGNITION INCIDENTS
See other NON-METAL HYDRIDES, NON-METAL SULFIDES

†4484. Hydrogen disulfide
[13465-07-1]   H₂S₂

H_2S_2

HSSH

Alkalies
Bloch, I. *et al., Ber.*, 1908, **41**, 1977
Rather more flammable than hydrogen trisulfide, it is decomposed violently by alkalies.
See other NON-METAL HYDRIDES, NON-METAL SULFIDES

4485. Hydrogen trisulfide
[13845-23-3] H_2S_3

HSSSH

Benzenediazonium chloride
King, W. B. *et al., J. Amer. Chem. Soc.*, 1932, **54**, 3073
Addition of the dry diazonium salt to the crude liquid sulfide causes explosively violent interaction. Slow addition of the sulfide to the cooled salt, or dilution with a solvent, moderates the reaction.
See other DIAZONIUM SULFIDES AND DERIVATIVES

Metal oxides
Mellor, 1947, Vol. 10, 159
Contact with copper oxide, lead(II) oxide, lead(IV) oxide, mercury(II) oxide, tin(IV) oxide or iron(II,III) oxide causes violent decomposition and ignition. Dry powdered silver oxide causes an explosion.

Nitrogen trichloride
See Nitrogen trichloride: Initiators

Pentanol
Mellor, 1947, Vol. 10, 158
Interaction is explosively violent.

Potassium permanganate
See Potassium permanganate: Hydrogen trisulfide
See other NON-METAL HYDRIDES, NON-METAL SULFIDES

†4486. Hydrogen selenide
[7783-07-5]

H_2Se

H_2Se

It is moderately exothermic (ΔH°_f (g) +85.8 kJ/mol, 1.06 kJ/g).

Oxidants
See Hydrogen peroxide: Hydrogen selenide
  Nitric acid: Non-metal hydrides
See other ENDOTHERMIC COMPOUNDS, NON-METAL HYDRIDES
See related NON-METAL SULFIDES

4487. Poly(silylene)
[32078-95-8]

$(H_2Si)_n$

$(-SiH_2-)_n$

Brauer, 1963, Vol. 1, 682
The dry solid ignites in air.

Oxidants
Bailar, 1973, Vol. 1, 1352
It ignites in contact with conc. nitric acid and explodes with sulfuric acid.
See related SILANES

†4488. Hydrogen telluride
[7783-09-7]

H_2Te

H_2Te

It is moderately endothermic (ΔH°_f (g) +154.4 kJ/mol, 1.19 kJ/g).

Nitric acid
See Nitric acid: Non-metal hydrides
See other ENDOTHERMIC COMPOUNDS, METAL HYDRIDES *See related* NON-METAL
HYDRIDES

4489. Thorium dihydride
[16689-88-6]

H_2Th

ThH_2

1. Hartmann, I. *et al., Rept. Invest. No. 4835*, Washington, US Bur. Mines, 1951
2. Grossel, S. S., *J. Loss Prev. Proc. Ind.*, 1988, **1**, 62–74

Extremely pyrophoric when powdered [1], and is also a significant dust exploion hazard [2].
See other METAL HYDRIDES, PYROPHORIC MATERIALS

4490. Titanium dihydride
[7704-98-5]   H₂Ti

$$TiH_2$$

1. Alekseev, A. G. *et al., Chem. Abs.*, 1974, **81**, 123688
2. Grossel, S. S., *J. Loss Prev. Proc. Ind.*, 1988, **1**, 62–74
Pyrophoricity and detonation behaviour of titanium hydride powders of various particle sizes were studied in comparison with those of titanium metal powders [1]. Maximum dust explosion pressures of 8.2 bar, with a maximum rate of rise of 816 bar/s have been recorded [2].
See Zirconium (reference 6)
See other METAL HYDRIDES, PYROPHORIC MATERIALS

4491. Titanium–zirconium hydride
[57425-12-4]   H₂Ti.H₂Zr

$$TiH_2.ZrH_2$$

See Zirconium (reference 6)
See other METAL HYDRIDES

4492. Zinc hydride
[14018-82-7]   H₂Zn

$$ZnH_2$$

Barbaras, G. D. *et al., J. Amer. Chem. Soc.*, 1951, **73**, 4587
Fresh samples reacted slowly with air, but aged and partly decomposed samples (containing finely divided zinc) may ignite in air.

Acids, Water
Ashby, E. C. *et al., Inorg. Synth.*, 1977, **17**, 9
Hydrolysis with water occurs slowly, but is violent with aqueous acids.
See other METAL HYDRIDES

4493. 'Zirconium hydride'
[7704-99-6]   H₂Zr

$$ZrH_2$$

1. Mellor, 1941, Vol. 7, 114
2. Sidgwick, 1950, 633

3. Baumgart, H., *Chem. Abs.*, 1985, **10** 3, 73412
4. Kuroda, E. *et al., Chem. Abs.*, 1997, **127**, 150727x

The product of sorbing hydrogen on to hot zirconium powder burns with incandescence when heated in air [1,2]. Ignition and combustion properties of pyrophoric zirconium and zirconium hydride powder (<10 μ), and safety measures are discussed [3]. Ignition energies for zirconium hydrides, ZrH_x; $x = 0 - 2$, in air have been studied in Japan [4].
See Zirconium (reference 6)
See other METAL HYDRIDES

4494. Dipotassium triimidotellurite

[ ] $H_3K_2N_3Te$

$$2K^+ \ (HN{=})_3Te^{2-}$$

1. Schmitz-Du Mont, O. *et al., Angewand. Chem. (Int.)*, 1967, **6**, 1071
2. Chivers, T. *et al., Angewand. Chem. (Int.)*, 1995, **34**(22), 2549

Colourless crystals, exploding on contact with water unless benzene damp. Ammonolysis converts it to the still more explosive tritellurium tetranitride [1]. Highly explosive [2].
See also N–S COMPOUNDS

4495. Lanthanum hydride
[13864-01-2] H_3La

$$LaH_3$$

1. Mackay, 1966, 66

It ignites in air.
See other METAL HYDRIDES

4496. Lithium hydrazide
[37067-42-8] H_3LiN_2

$$Li^+HN^- {-}NH_2$$

Hydrazine and its Derivatives, Schmidt, F. W., New York, Wiley, 1984, 75
Merely smouldering on contact with air, it is claimed to be a safe metal hydrazide.
See other METAL HYDRAZIDES

†4497. Ammonia
[7664-41-7] H_3N

$$NH_3$$

(*MCA SD-8*, 1960); *NSC 251*, 1979; *FPA H13*, 1973; *HCS 1980*, 139 (cylinder), 140 (25% solution); *RSC Lab. Hazards Safety Data Sheet No. 37*, 1985

1. *MCA SD-8*, 1960
2. Anon., *ABCM Quart. Safety Summ.*, 1950, **45**, 14
3. Anon., *ABCM Quart. Safety Summ.*, 1974, **45**, 14
4. Unpublished observations, 1968–1974
5. Henderson, D., *Ammonia Plant Safety*, 1975, **17**, 132–134
6. Krivulin, V. N. *et al., Chem. Abs.*, 1976, **84**, 124132
7. Harris, G. F. P. *et al., Proc. 6th Symp. Chem. Proc. Haz. Ref. Plant Des.*, 29–38, Symp. Ser. 49, Rugby, IChE, 1977
8. McRae, M. H., *Plant/Oper. Progr.*, 1987, **6**(1), 17–19
9. Klem, T. J., *Ammonia Plant Saf.*, 1987, (26), 145–149
10. Andersson, B. O., *Hazards XI*, 15, Symp. Ser. **124**, Rugby(UK), IChE, 1991.
11. Andersson, B. O., *Ammonia Plant Safety*, 1991, **31**, 5
12. Carlsson, T. O. *et al., Sci. Tech Froid*, 1996, (3), 277

Although there is a high lower explosive limit in air and ignition is not easy, there is a long history of violent gas–air explosions in refrigeration practice, in which ammonia previously was used widely [1]. Mixtures of ammonia and air lying within the explosive limits can occur above aqueous solutions of certain strengths. Welding operations on a vessel containing aqueous ammonia caused a violent explosion [2]. Several incidents involving sudden 'boiling' of conc. ammonia solution (d, 0.880, 35 wt%) have occured when screw-capped winchesters are opened [3,4]. These are attributable to supersaturation of the solution with gas, caused by increase in temperature subsequent to preparation and bottling. The effect is particularly marked with winchesters filled in winter and opened in summer. Usually boiling-off begins immediately the cap is loosened, and if this is done carefully using gloves and eye-protection in a fume cupboard or under local extract ventilation, no great problems arise [4]. However a bottle which had been gently and successfully opened in this way without boiling-off subsequently erupted violently when the surface of the liquid was disturbed by a pipette [3]. This hazard may be avoided by using the slightly less concentrated solution (d, 0.990, 25 wt%) where this is technically acceptable.

There is an account of an incident involving combustion/explosion in the free space of a weak ammonia liquor tank where detailed examination revealed no evidence for an ignition source [5]. The existence of ammonia–air mixtures able to burn only in sufficiently large enclosed volumes was established [6]. Previous data on flammability has been summarised and extended by determination of flammability characteristics of ammonia admixed with dry air and oxygen-deficient air at temperatures from ambient up to 400°C. Similar work with added water vapour at 80°C shows that aqueous solutions of below 5% ammonia content do not produce flammable vapours at any temperature, and that above 49°C no flammable vapours are produced by ammonia solution of any concentration. Some data on ignition energy and explosibility are also given [7]. Two recent US incidents serve as a reminder of the considerable explosive potential of ammonia–air mixtures in confined volumes. An ice-cream factory with ammonia refrigeration plant which had been installed over 50 years previously had not been upgraded to meet modern engineering practice and electrical fire codes. There were no check valves or oil separators in the compressor discharge lines, and the plant was lit by unprotected tungsten filament lamps. During the night an ammmonia leak developed in the basement, and a few minutes later there was a violent explosion which largely destroyed the building. The major devastation prevented identification of the cause of the leak or of the ignition source, and it was postulated that oil mist may have been involved in initiating the explosion [8]. During an attempt to isolate a leak in the anhydrous ammonia refrigeration system in a cold store warehouse, an electric fork lift truck was to be used to replace a faulty valve. The truck crashed into an internal wall and the large explosion which followed no doubt involved spark ignition of the flammable atmosphere [9]. Both incidents revealed a widespread belief that ammonia is non-flammable, possibly arising from the US classification of anhydrous ammonia in cylinders as 'a non-flammable gas'. In Europe, it is realistically classified as flammable, as also are highly concentrated aqueous solutions. With returning interest in it as a domestic refrigerant, studies of ignition energies and flame propagation in mixtures with air have been performed [12]

A report is given of explosion of a tank containing 7000 t of liquid ammonia, in Lithuania, consequent upon thermal disequilibrium and 'bumping' of the liquid [10,11]. The liberated ammonia caught fire in the open air, the first time this is thought to have been observed.
See Nitric acid: Manufacture hazard

Air, Hydrocarbons
Kalkert, N. *et al.*, *Chem. Ing. Tech.*, 1979, **51**, 895
Explosion limits have been estimated for mixtures containing $C_1 - C_3$ hydrocarbons.

Air, Rust
Verduijn, W. D., *Process Safety Progress*, 1996, **15**(2), 89; Verduijn, W. D. *Ammonia Plant Safety*, 1996
An account is given of a sequence of mishaps starting with a burst pipe in a nitric acid plant, apparently due to pre-ignition by rust catalysis of an ammonia/air mixture at 220°C and 11 bar.

Boron halides
Sidgwick, 1950, 380
The boron halides react violently with ammonia.

Calcium
See Calcium: Ammonia

1-Chloro-2,4-dinitrobenzene
See 1-Chloro-2,4-dinitrobenzene: Ammonia

Chlorine azide
See Chlorine azide: Alone, or Ammonia

Chloroformamidinium nitrate
See Chloroformamidinium nitrate: Alone, or Amines, etc.

2-, or 4-Chloronitrobenzene
See 2-Chloronirobenzene: Ammonia or 4-Chloronitrobenzene: Ammonia

1,2-Dichloroethane
491M, 1975, 36
Liquid ammonia and the solvent may explode when mixed. (It is possible this was a liquefied gas (physical) explosion, rather than an exothermic chemical reaction.)
See LIQUEFIED GASES

Ethanol, Silver nitrate
See Silver nitrate: Ammonia, Ethanol

Ethylene oxide
See Ethylene oxide: Ammonia

Germanium derivatives
Paschenko, I. S. *et al., Chem. Abs.*, 1976, **84**, 8568
'Causes of the Explosiveness of Gas–Liquid Mixtures during Neutralisation of Germanium-containing Solutions by Ammonia' (title only translated).
See Germanium imide

Gold(III) chloride
See Gold(III) chloride: Ammonia, etc.

Halogens, or Interhalogens
Ammonia either reacts violently, or produces explosive products, with all 4 halogens and some of the interhalogens.
See Bromine: Ammonia
    Bromine pentafluoride: Hydrogen-containing materials
    Chlorine: Nitrogen compounds
    Chlorine trifluoride: Hydrogen-containing materials
    Fluorine: Hydrides
    Iodine: Ammonia

Heavy metals
Ammonia is capable of reacting with some heavy metal compounds (silver, gold, mercury) to produce materials, some of uncertain constitution, which may explode violently when dry.
See Gold(III) chloride: Ammonia
    Mercury: Ammonia
    Potassium triamidothallate ammoniate
    Silver azide: Ammonia
    Silver chloride: Ammonia
    Silver nitrate: Ammonia
    Silver(I) oxide: Ammonia
See N-METAL DERIVATIVES

Iodine, Potassium
Staley, S. W. *et al., Tetrahedron*, 1975, **31**, 1133
During the reductive cleavage of cyclopolyenes with potassium in liquid ammonia, the intermediate anionic species are quenched with iodine–pentane mixtures. The possibility of formation of explosive nitrogen triiodide and the need for precautions are stressed.

Magnesium perchlorate
See Magnesium perchlorate: Ammonia

Mercury
1. Marsch, H. D., *Ammonia Plant Safety*, 1991, **31**, 8
2. Wilhelm, S. M., *Ammonia Plant Safety*, 1991, **31**, 20
An explosion rupturing an ammonia separator (still) in an ammonia production unit, probably because mercury vapour from geological sources entered with hydrogen syngas originating from natural gas and reacted to give explosive nitride deposits. The separator remains crackled when scraped [1]. For a more academic study of the effects of mercury on ammonia plants, including embrittlement and corrosion, as well as explosive deposits [2].
See other CORROSION INCIDENTS
See Mercury: Ammonia

Nitrogen trichloride
See Nitrogen trichloride: Initiators

Other reactants
Yoshida, 1980, 36
MRH values calculated for 16 combinations, largely with oxidants, are given.

Oxidants   MRH values show % of oxidant
Maslonka, F., *Chem. Abs.*, 1978, **88**, 193935
The flammability and explosion susceptibility of ammonia is reviewed, with thermochemical data, explosion limits for various oxidants, and explosion parameters, both for the gas and its conc. aqueous solutions.
See Ammonium peroxodisulfate: Ammonia, etc.
　　Chromium trioxide: Ammonia
　　Chromyl chloride: Ammonia
　　Dichlorine oxide: Oxidisable materials
　　Dinitrogen tetraoxide: Ammonia
　　Hydrogen peroxide: Nitrogenous bases　　　　　　MRH 5.86/75
　　Nitric acid: Ammonia　　　　　　　　　　　　MRH 5.10/69
　　Nitrogen oxide: Ammonia
　　Nitryl chloride: Inorganic materials
　　Oxygen (Liquid): Ammonia, etc.
　　Potassium chlorate: Ammonia　　　　　　　　　MRH 4.31/78
　　'Trioxygen difluoride': Various materials

Oxygen, Platinum
    See Oxygen (Gas): Ammonia, Platinum

Ozone
    See Ozone: Ammoniacal vapours

Pentaborane(9)
    See Pentaborane(9): Ammonia

Silver compounds
    See Silver azide: Ammonia
        Silver chloride: Ammonia
        Silver nitrate: Ammonia
        Silver(I) oxide: Ammonia, etc.

Stibine
    See Stibine: Ammonia

Tellurium halides
    See Tellurium tetrabromide: Ammonia
        Tellurium tetrachloride: Ammonia

Tetramethylammonium amide
    See Tetramethylammonium amide: Ammonia

Thiocarbonyl azide thiocyanate
    See Thiocarbonyl azide thiocyanate: Ammonia, etc.

Sulfinyl chloride
    See Sulfinyl chloride: Ammonia

Thiotrithiazyl chloride
    See Thiotrithiazyl chloride: Ammonia
    See other INORGANIC BASES, NON-METAL HYDRIDES

4498. Hydroxylamine
[7803-49-8]   **H₃NO**

$$HONH_2$$

HCS 1980, 557

1. Ashford, J. S., private comm., 1967
2. Brauer, 1963, Vol. 1, 502
3. Vervalin, C. H., *Hydrocarbon Proc. Petr. Ref.*, 1963, **42**(2), 174
4. Rüst, 1948, 302
5. Bailar, 1973, Vol. 2, 272

The base was being prepared by distilling a mixture of hydroxylamine hydrochloride and sodium hydroxide in methanol under reduced pressure, and a violent explosion occurred towards the end of distillation [1], probably owing to an increase in pressure above 53 mbar. It explodes when heated under atmospheric pressure [2]. Traces of hydroxylamine remaining after reaction with acetonitrile to form acetamide oxime caused an explosion during evaporation of solvent. Traces can be removed by treatment with diacetyl monoxime and ammoniacal nickel sulfate, forming nickel dimethylglyoxime [3]. An account of an extremely violent explosion towards the end of vacuum distillation had been published previously [4]. Anhydrous hydroxylamine is usually stored at 10°C to prevent internal oxidation–reduction reactions which occur at ambient temperature [5].
See other REDOX REACTIONS

Aldehyde (unspecified), Pyridine
MCA Guide, 1972, 300
During the preparation of an unspecified aldoxime, the ethanolic reaction mixture was extracted with ether and the extract was concentrated by vacuum distillation from a bath at 70–80°C. Towards the end of distillation violent decomposition occurred (probably of traces of unreacted hydroxylamine in the extract).
See above

Barium oxide   MRH 3.86/tr.
See Oxidants, below

Carbonyl compound (unspecified), Pyridine
Anon., *Chem. Processing* (Chicago), 1963, **26**(24), 30
During preparation of an unspecified oxime, the carbonyl compound, pyridine, hydroxylamine hydrochloride and sodium acetate were heated in a stainless steel autoclave. At 90°C a sudden reaction caused a pressure surge to 340 bar, when the bursting disk failed. The reaction had been run previously and uneventfully on one-tenth scale in a glass lined autoclave.
See Metals, below

Copper(II) sulfate
Mellor, 1940, Vol. 8, 290
Hydroxylamine ignites with the anhydrous salt (possibly owing to heat of coordination).

Hydrogenation catalyst, Acid, Fuel
Riesthuis, P. *et al., J. Loss Prev. Process Ind.*, 1997, **10**(10), 67
In the presence of precious metal hydrogenation catalyst, hydroxylamine salts may disproportionate and form dinitrogen monoxide. Such a mixture is present in a process whereby the hydroxyamine is formed by hydrogenation of nitrate. An explosion in the degassing line, after a period of abnormal operation, was attributed to nitrous oxide build-up. Fuel, in the form of hydrogen and methane diluent, was already present.
See Dinitrogen oxide

Metals   MRH Calcium, sodium, zinc all 3.85/tr.
1. Mellor, 1940, Vol. 8, 292
2. Sorbe, 1968, 158
Sodium ignites in contact with hydroxylamine alone, but reacts smoothly in ether solution to give *N*-sodium hydroxylamide which may be pyrophoric in air. Calcium reacts to give the bis(hydroxylamide) which explodes at 180°C. Finely divided zinc either ignites or explodes when warmed in contact with hydroxylamine [1]. Zinc and hydroxylamine form the bis(hydroxylamide) solvated with hydroxylamine (3 mol), an explosive compound [2].
See also OXIMES

Other reactants
Yoshida, 1980, 286
MRH values calculated for 16 combinations with various materials are given.

Oxidants   MRH values below references show % of oxidant
1. Brauer, 1963, Vol. 1, 502
2. Mellor, 1930, Vol. 3, 670; 1940, Vol. 8, 287–294
MRH Barium peroxide 3.85/tr., calcium hypochlorite 3.77/52, lead dioxide 1.92/64, potassium dichromate 3.85/tr., potassium permanganate 2.80/49, sodium hypochlorite 3.97/53
Hydroxylamine is a powerful reducant, particularly when anhydrous, and if exposed to air on a fibrous extended surface (filter paper) it rapidly heats by aerobic oxidation. It explodes in contact with air above 70°C [1]. Barium peroxide will ignite aqueous hydroxylamine, while the solid ignites in dry contact with barium oxide, barium peroxide, lead dioxide and potassium permanganate, but with chlorates, bromates and perchlorates only when moistened with sulfuric acid. Contact of the anhydrous base with potassium dichromate or sodium dichromate is violently explosive, but less so with ammonium dichromate or chromium trioxide. Ignition occurs in gaseous chlorine, and vigorous oxidation occurs with hypochlorites.
See other REDOX REACTIONS

Phosphorus chlorides   MRH Phosphorus(III) & (V) chlorides both 3.85/tr.
Mellor, 1940, Vol. 8, 290
The base ignites in contact with phosphorus trichloride and phosphorus pentachloride.
See also HYDROXYLAMINIUM SALTS
See other N–O COMPOUNDS, REDUCANTS

4499. Amidosulfuric acid (Sulfamic acid)
[5329-14-6] H_3NO_3S

$$H_2NSO_2OH$$

HCS 1980, 869

Chlorine
See Chlorine: Nitrogen compounds (reference 7)

Metal nitrates or nitrites
See METAL AMIDOSULFATES
See other INORGANIC ACIDS, N–S COMPOUNDS

4500. Sodium hydrazide
[13598-47-5] H_3N_2Na

$$NaNHNH_2$$

Air, or Ethanol, or Water
1. Mellor, 1940, Vol. 8, 317
2. Kauffmann, T., *Angew. Chem. (Intern. Ed.)*, 1964, **3**, 342
Contact with traces of air, ethanol or water causes a violent explosion [1], as does heating to 100°C [2].
See other METAL HYDRAZIDES

4501. Azidosilane
[13847-60-4] H_3N_3Si

$$N_3SiH_3$$

Preparative hazard
See Azidogermane: Fluorosilane
See other NON-METAL AZIDES *See related* SILANES

4502. Hydrogen hexaazidothallate(III)
[63338-74-9] $H_3N_{18}Tl$

$$H_3[Tl(N_3)_6]$$

Krommes, P. *et al., J. Organomet. Chem.*, 1977, **131**, 418
Not more than 200–500 mg portions of this extremely heat- and impact-sensitive explosive should be prepared.
See related METAL AZIDES *See other* N-METAL DERIVATIVES

4503. Phosphinic acid ('Hypophosphorous acid')
[6303-21-5] H_3O_2P

$$HOP(O)H_2$$

1. Mellor, 1940, Vol. 8, Suppl. 3, 614
2. Albright and Wilson (Manufacturer's safety sheet)
3. *MARS Database*, 1998, short report 181
The explosion hazard associated with the usual laboratory preparation from white phosphorus and alkali may be avoided by an alternative method involving oxidation of phosphine with an aqueous iodine solution [1]. The commercial 50% solution reacts violently with oxidants. On heating, it decomposes rapidly above 100°C evolving phosphine, which is liable to explode with air. It is recommended it

be kept below 50°C [2]. During vacuum concentration of a dilute solution on small industrial scale, a crust formed, overheated and decomposed, pressurising the system with phosphine which escaped and produced two explosions on mixing with air, causing modest damage to the factory installations [3].

Mercury(II) nitrate
See Mercury(II) nitrate: Phosphinic acid

Mercury(II) oxide   MRH 0.71/86
  Mellor, 1940, Vol. 4, 778
  The redox reaction is explosive.
  See other REDOX REACTIONS

Other reactants
  Yoshida, 1980, 343
  MRH values calculated for 6 combinations with various other materials are given.
  See other INORGANIC ACIDS, REDUCANTS

4504. Phosphonic acid (Phosphorous acid)
[13598-36-2]   H₃O₃P

$$P(OH)_3 \text{ or } HP(O)(OH)_2$$

1. Partington, 1946, 614
2. Albright and Wilson, (Manufacturer's safety sheet)
It decomposes at around 200°C with evolution of phosphine which may ignite [1]. Yellow deposits may form in aqueous solution which are spontaneously flammable on drying [2].
See Phosphorus trichloride: Carboxylic acids
See other INORGANIC ACIDS, REDUCANTS

4505. Orthophosphoric acid
[7664-38-2]   H₃O₄P

$$P(O)(OH)_3$$

(*MCA SD-7*, 1958); *NSC 674*, 1978; *HCS 1980*, 742, 743, *MCA SD-70*, 1958

Although it is not an oxidant, it is an active acid which, because of its low equivalent wt. and availability in high concentration and density (90% soln. is 50 N), can develop a large exotherm on neutralisation. Handling precautions are detailed.

Chlorides, Stainless steel
  Piekarz, J., *Chem. in Can.*, 1961, **13**(4), 40–41
  Presence of traces of chloride ion in technical 75% phosphoric acid in a closed storage tank caused corrosion and liberation of hydrogen which later exploded at a sparking contact.
  See other CORROSION INCIDENTS, GAS EVOLUTION INCIDENTS

Nitromethane
See Nitromethane: Acids, etc.

Sodium tetrahydroborate
See Sodium tetrahydroborate: Acids
See other INORGANIC ACIDS

4506. Peroxomonophosphoric acid
[13598-52-5] H_3O_5P

HOOP(O)(OH)$_2$

Preparative hazard
See Hydrogen peroxide: Phosphorus(V) oxide

Coal, Potassium permanganate
See Potassium permanganate: Coal, etc.

Organic material
Castrantas, 1965, 5
The 80% solution causes ignition when dropped onto organic material.
See other INORGANIC ACIDS, PEROXOACIDS

4507. Hydroxyaqua(oxo)diperoxorhenium VII
[ ] H_3O_7Re

Herrmann, W. A. et al., Chem. Eur. J., 1996, **2**(2), 168
A highly explosive red/orange solid, obtained from rhenium trioxide and hydrogen
peroxide reacting in hexamethylphosphoramide as solvent, was tentatively assigned
this structure.
See other INORGANIC PEROXIDES

†4508. Phosphine
[7803-51-2] H_3P

PH$_3$

HCS 1980, 745 (cylinder); RSC Lab. Hazards Safety Data Sheet No. 49 [0], 1986

1. Rüst, 1948, 301
2. Sidgwick, 1950, 729
3. Mellor, 1940, Vol. 8, 811, 814
4. McKay, H. A. C., Chem. & Ind., 1964, 1978
5. Anon., Lab. Haz. Bull., 1982, (3), item 120

6. Green, A. R. *et al., Dev. Agric. Eng.*, 1984, **5** (Controlled Atmos. Fumig. Grain Storage), 433–449
7. Ethenakis, V. M. *et al., Sol. Cells*, 1987, **22**, 305–317

Liquefied phosphine (a mildly endothermic compound, $\Delta H°_f$ (g) +22.8 kJ/mol, 0.67 kJ/g) can be detonated by powerful initiation [1]. Pure phosphine does not spontaneously ignite in air below 150°C unless it is thoroughly dried, when it ignites in cold air. The presence of traces (0.2%) of diphosphane in phosphine as normally prepared causes it to ignite spontaneously in air, even at below −15°C [2]. Traces of oxidants promote pyrophoricity [3]. Phosphine, generated by action of water on calcium phosphide, was dried by passage through towers packed with the latter. Soon after refilling the generator (but not the towers) and starting purging with argon, a violent explosion occurred. This was attributed to the air, displaced from the generator by argon, reacting explosively with dry phosphine present in the drying towers, possibly catalysed by the orange-yellow polyphosphine formed on the surface of calcium phosphide. Fresh calcium phosphide in both generator and drying towers, with separate purging, is recommended [4]. During vacuum transfer of the supposedly pure gas, air leaked into the cold-trap system and caused an explosion [5]. Lower flammability limit of pure phosphine in moist air (0.39 vol% water vapour) at 1037 mbar was 2.1% at 10° and 1.85% at 50°C. Presence of ammonia may tend to reduce the limit value [6]. Potential hazards involved in use of phosphine in manufacture of photovoltaic cells are reviewed [7].
See Oxygen, below, *also* Phosphonium iodide: Potassium hydroxide

Boron trichloride
See Boron trichloride: Phosphine

Dichlorine oxide   MRH 5.36/84
See Dichlorine oxide: Oxidisable materials

Halogens   MRH Chlorine 2.13/89
Mellor, 1940, Vol. 8, 812
Ignition occurs on contact with chlorine or bromine or their aqueous solutions.

Metal nitrates   MRH Silver nitrate 3.77/87
Mellor, 1941, Vol. 3, 471; 1940, Vol. 4, 993
Passage of phosphine into silver nitrate solution causes ignition or explosion, depending on the gas rate. Mercury(II) nitrate solution gives a complex phosphide, explosive when dry.

Nitric acid   MRH 7.78/74
See Nitric acid: Non-metal hydrides

Nitrogen trichloride   MRH 3.64/91

Other reactants
Yoshida, 1980, 342
MRH values calculated for 10 combinations with oxidants are given.

Oxidants   MRH values below reference
    Mellor, 1940, Vol. 8, 811, 814
    MRH Dinitrogen monoxide 6.99/84, dinitrogen trioxide 9.25/75, nitrogen oxide
    9.71/78, nitrous acid 7.11/79
        Pure phosphine is rendered pyrophoric by traces of dinitrogen trioxide, nitrous
    acid, or similar oxidant.
    See Dichlorine oxide; Halogens; Metal nitrates; Nitric acid, all above, Oxygen,
    below

Oxygen
    1. Fischer, E. O. *et al., Angew. Chem. (Intern. Ed.)*, 1968, **7**, 136
    2. Mellor, 1971, Vol. 8, Suppl. 3, 281–282
    Even small amounts of oxygen present in phosphine give an explosive mixture,
    in which autoignition occurs at low pressures [1]. The effects of other materials
    upon the explosive interaction have been studied [2].
    See other ENDOTHERMIC COMPOUNDS, NON-METAL HYDRIDES

4509. Plutonium(III) hydride
[15457-77-9] H_3Pu

$$PuH_3$$

Stout, E. L., *Chem. Eng. News*, 1958, **36**(8), 64
The hydride (and the metal) when finely divided are spontaneously flammable,
and burning causes a specially dangerous contamination problem, in view of the
radioctive and toxic hazards.

Water
    Bailar, 1973, Vol. 5, 150
    Rapid addition of water often causes ignition of the hydride.
    See other METAL HYDRIDES

†4510. Stibine
[7803-52-3] H_3Sb

$$SbH_3$$

Mellor, 1939, Vol. 9, 394
During evaporation of liquid stibine at $-17°C$, a relatively weak and isothermal
explosive decomposition may occur. Gaseous stibine at ambient temperature may
propagate an explosion from a hot spot on the retaining vessel wall, and it autocatalytically decomposes, sometimes explosively, at 200°C.
    It is moderately endothermic ($\Delta H^°_f$ (g) $+145.2$ kJ/mol, 1.16 kJ/g).

Ammonia
    Mellor, 1939, Vol. 9, 397
    A heated mixture explodes

Oxidants
See Chlorine: Non-metal hydrides
Nitric acid: Non-metal hydrides
Ozone: Stibine
See related NON-METAL HYDRIDES See other ENDOTHERMIC COMPOUNDS, METAL HYDRIDES

4511. Uranium(III) hydride
[13598-56-6] H_3U

$$UH_3$$

Stout, E. L., *Chem. Eng. News*, 1958, **36**(8), 64
The dry powdered hydride readily ignites in air.

Halocarbons
Bailar, 1973, Vol. 5, 150
Contact of the hydride with halogen-containing solvents may lead to explosive interaction.

Water
Bailar, 1973, Vol. 5, 150
Rapid addition of water often causes ignition of the hydride.
See other METAL HYDRIDES

4512. Ammonium iodide
[12027-06-4] H_4IN

$$NH_4I$$

Bromine trifluoride
See Bromine trifluoride: Ammonium halides
See related METAL HALIDES

4513. Ammonium iodate
[13446-09-8] H_4INO_3

$$NH_4IO_3$$

Anon., *ABCM Quart. Safety Summ.*, 1955, **26**, 24
Violent decomposition occurred on touching with a scoop (which presumably was contaminated). A similar batch of material contained less than 100 ppm of periodate.

Other reactants
Yoshida, 1980, 375
MRH values calculated for 15 combinations with oxidisable materials are given.
See related METAL OXOHALOGENATES See other IODINE COMPOUNDS

4514. Ammonium periodate
[13446-09-8] H_4INO_4

$$NH_4IO_4$$

1. Anon., *Chem. Eng. News*, 1951, **29**, 1770
2. Mellor, 1940, Vol. 2, 408

It exploded while being transferred by scooping [1]. The sensitivity towards heat, but not abrasive impact, was known previously [2].

See related METAL OXOHALOGENATES
See other IODINE COMPOUNDS

4515. Phosphonium iodide
[12125-09-6] H_4IP

$$PH_4I$$

Oxidants
   Mellor, 1947, Vol. 8, 827
   Bromates, chlorates or iodates ignite in contact with phosphonium iodide at ambient temperature if dry, or in presence of acid to generate bromic acid, etc. Ignition also occurs with nitric acid, and reaction with dry silver nitrate is very exothermic. Interaction with antimony pentachloride at ambient temperature proceeds explosively.
   See other REDOX REACTIONS

Potassium hydroxide
   Anon., *Univ. Safety Assoc., Safety Newsletter*, 1983–1984
   Phosphine, generated by action of the alkali on phosphonium iodide, was shown to be pure by mass spectrometry. During a second preparation, an air leak developed during cold-trap transfer and an explosion occurred. Presence of diphosphane was suspected.
   See Phosphine (reference 2)

Water
   See Phosphorus: Hydriodic acid
   See other REDUCANTS

4516. Potassium tetrahydrozincate
[34324-17-9] H_4K_2Zn

$$K_2[H_4Zn]$$

Ashby, E. C. *et al., Inorg. Chem.*, 1971, **10**, 2486
Extremely reactive, pyrophoric in air.
See other COMPLEX HYDRIDES

4517. Magnesium phosphinate
[7783-17-7] $H_4MgO_4P_2$

$$Mg[OP(O)H_2]_2$$

Mellor, 1940, Vol. 8, 885
On heating it evolves phosphine which ignites in air.
See other METAL PHOSPHINATES, REDUCANTS

4518. Ammonium permanganate
[13446-10-1] H_4MnNO_4

$$NH_4MnO_4$$

1. Mellor, 1942, Vol. 12, 302
2. Pavlychenko, M. M. *et al., Chem. Abs.*, 1962, **57**, 2897e
The dry material is friction-sensitive and explodes at 60°C in air [1]. Under vacuum, decomposition becomes explosive above 100°C, and sparks will also initiate decomposition.
See other OXOSALTS OF NITROGENOUS BASES (reference 1), OXIDANTS

4519. Manganese(II) phosphinate
[10043-84-2] $H_4MnO_4P_2$

$$Mn[OP(O)H_2]_2$$

Mellor, 1940, Vol. 8, 889
It explodes above 200°C.
See other HEAVY METAL DERIVATIVES, METAL PHOSPHINATES, REDUCANTS

†4520. Hydrazine
[302-01-2] H_4N_2

$$H_2NNH_2$$

HCS 1980, 536 (anhydrous), 537 (hydrate); *RSC Lab. Hazard Data Sheet No. 14*, 1983

1. Troyan, J. E., *Ind. Eng. Chem.*, 1953, **45**, 2608
2. Day, A. C. *et al., Org. Synth.*, 1970, **50**, 42
3. *491M*, 1975, 199
4. Mitchell, J. W., *Anal. Chem.*, 1980, **52**, 774–776
5. Bellerby, J. M. *et al., Chem. Abs.*, 1982, **96**, 222535
6. Bunker, R. L. *et al., Prog. Astronaut. Aeronaut.*, 1991, **133**, 324
7. *Hydrazine and its Derivatives*, Schmidt, F. W., New York, Wiley, 1984

It is rather endothermic (ΔH_f° (l) +50.4 kJ/mol, 1.57 kJ/g) and readily decomposed.

Hydrazine is an endothermic compound and the vapour is exceptionally hazardous in that once it is ignited, it will continue to burn by exothermic decomposition in complete absence of air or other oxidant [1]. Distillation of anhydrous hydrazine (prepared by dehydrating hydrazine hydrate with solid sodium

hydroxide) must be carried out under nitrogen to avoid the possibility of an explosion if air is present [2]. If spread with an extended surface (as a thin film, or absorbed on porous or fibrous material), anhydrous hydrazine may autoxidise with sufficient heat generation and retention to cause ignition. Presence of traces of metal catalysts (as in some cellulosic materials) will greatly accelerate this effect [3]. Procedures for safe handling and purification of aqueous hydrazine solutions are detailed [4], and autoxidation processes of hydrazine and its methylated derivatives have been reviewed [5]. The mechanism by which hydrazine can be detonated during pumping operations has been studied [6]. A comprehensive monograph includes sections on decomposition and other hazard-related aspects of the preparation and applications of hydrazine [7].
See Barium oxide; Cotton waste, Heavy metals; Metal catalysts, all below

Barium oxide, or Calcium oxide
1. Gmelin, 1935, Syst. 4, 318
2. *ABCM Quart. Safety Summ.*, 1950, **21**, 18
The residue from dehydrating hydrazine with barium oxide slowly decomposes exothermically in daylight and finally explodes [1]. Dehydration of 95% hydrazine by boiling with calcium oxide under nitrogen caused violent explosions on 2 occasions. It was concluded that use of calcium or barium oxides for this purpose is potentially dangerous, and that boiling under reduced pressure may be advisable to lower the liquid temperature. General precautions in handling hydrazine are discussed [2].
See Sodium hydroxide, below

Benzeneseleninic acid or anhydride
See Benzeneseleninic acid: Hydrazine derivatives

Calcium
Hydrazine and its Derivatives, Schmidt, F. W., New York, Wiley, 1984, 75
Use of excess calcium metal, or its solution in ammonia (presumable calcamide) to dry hydrazine leads to formation of calcium hydrazide and explosion. Calcium hydride and carbide are, apparently, safe drying agents
See METAL HYDRAZIDES

Carbon dioxide, Stainless steel
Bellerby, J. M., *J. Haz. Mat.*, 1983, **7**, 187–197
The rate of decomposition of hydrazine in stainless steel vessels (which is accompanied by corrosion) is directly proportional to carbon dioxide concentration over 20 ppm and below 250 ppm. The species responsible for the catalytic decomposition is not one of the expected corrosion products, iron(II) carbazate or its nickel or chromium(II) analogues.
See other CATALYTIC IMPURITY INCIDENTS, CORROSION INCIDENTS

1-Chloro-2,4-dinitrobenzene
See 1-Chloro-2,4-dinitrobenzene: Hydrazine hydrate

2-Chloro-5-methylnitrobenzene
See 2-Chloro-5-methylphenylhydroxylamine

Cotton waste, Heavy metals
Orme, R. J., private comm., 1980
In a factory manufacturing organo-iron and organo-manganese catalysts, use of cotton waste to mop up a spill of dilute hydrazine solution led to a spontaneous fire later. The fire was attributed to onset of rapid metal-catalysed decomposition of the hydrazine after sufficient water had evaporated from the waste, the traces of heavy metals originating from dust contamination, etc.
See Metal catalysts, below

Dicyanofurazan
See Dicyanofurazan: Nitrogenous bases

(Difluoroamino)difluoroacetonitrile
See (Difluoroamino)difluoroacetonitrile: Hydrazine

Dinitrogen oxide
See Dinitrogen oxide: Hydrazine

Iodine pentoxide
Speeds, S. J. *et al.*, US Pat. Appl. 844 082, 1977
The oxide is used to effect rapid decomposition of hydrazine in a high-temperature gas generator which produces exit pressures (of nitrogen, hydrogen and ammonia) of 550 bar.

Metal catalysts
1. Mellor, 1940, Vol. 8, 317: 1967, Vol. 8, Suppl. 2.2, 83
2. Audrieth, L. F. *et al., Ind. Eng. Chem.*, 1951, **43**, 1774
3. Troyan, J. E., *Ind. Eng. Chem.*, 1953, **45**, 2608
4. Blumenthal, J. L. *et al.*, US Pat. 3 846 339, 1974
5. Rao, K. R. *et al., Trans. Powder Metall. Assoc. India*, 1975, **2**, 64–66
6. Armstrong, W. E. *et al.*, US Pat. 4 124 538, 1978
7. Bellerly, J. M., *J. Haz. Mat.*, 1986, **13**(1), 57–60
In contact with metallic catalysts (platinum black, Raney nickel) hydrazine is catalytically decomposed, yielding ammonia, hydrogen and nitrogen. With conc. hydrazine the reaction may be violent, and the ammonia and hydrogen evolved could be ignited by particles of dry catalyst [1]. Measures to prevent autoxidation of hydrazine catalysed by traces of copper are discussed, including displacement of air from transportation containers with nitrogen [2]. Catalytic decomposition is effected by iron oxide, molybdenum-stabilised stainless steel, molybdenum and its oxides, and finely divided solids. Handling procedures are discussed [3]. A rhenium–alumina catalyst causes immediate ignition of undiluted hydrazine [4]. The effect upon ignition delay of thermal treatment and particle size of powdered transition metal catalysts has been studied [5]. Iridium or iridium–ruthenium on alumina also effect catalytic decomposition of hydrazine [6]. Soluble chromium and/or manganese species do not increase the rate of decomposition of anhydrous hydrazine, but presence of traces of acid as well leads to significantly increased rates [7].

Metal salts

1. Mellor, 1941, Vol. 7, 430; 1940, Vol. 8, 318; 1967, Vol. 8, Suppl. 2.2, 882
2. Traynham, J. G. *et al., J. Org. Chem.*, 1967, **32**, 3285
3. McCoy, P. O., *Chem. Eng. News*, 1970, **48**(48), 9

Several of its complexes with metallic salts are unstable, including those of basic cadmium perchlorate (highly explosive), cadmium nitrite (explosive), copper chlorate (explodes on drying without heat), manganese nitrate (ignites at 150°C), mercury(I) chloride, mercury(II) chloride, mercury(I) nitrate, mercury(II) nitrate (all are explosive) and tin(II) chloride (explodes on heating): other examples are given [1]. There is a published method for reducing *cis-trans*-cyclododecadiene to the *cis*-monoene with diazene ('diimide') generated in situ from hydrazine, copper(II) salts and air [2]. During a modified reaction sequence using crude diene, much sludge containing polymer, copper(II) acetate and hydrazine was produced. When filtered off in air, the sludge heated and eventually glowed. Use of purified diene and filtration under nitrogen avoids the problem [3].

See Trihydrazinealuminium perchlorate

See also AMMINEMETAL OXOSALTS

Methanol, Nitromethane
See Nitromethane: Hydrazine, Methanol

Other reactants
Yoshida, 1980, 283
MRH values calculated for 16 combinations with oxidants are given.

Oxidants   MRH values below references show % of oxidant

1. *Haz. Chem Data*, 1975, 166
2. Mellor, 1941, Vol. 3, 137; 1940, Vol. 8, 313–319; 1943, Vol. 11, 234
3. *ASESB Op. Incid. Rept. No. 105*
4. Troyan, J. E., *Ind. Eng. Chem.*, 1953, **45**, 4608
5. Wannagat, U. *et al., Monatsh. Chem.*, 1969, **97**, 1157–1162
6. Ephraim, 1939, 651

MRH Ammonium nitrate 3.97/84, chromium trioxide 3.01/81, hydrogen peroxide 6.36/68, nitric acid 5.84/61, potassium dichromate 1.59/tr., sodium chlorate 6.07/69, sodium nitrate 3.18/68

Hydrazine is a powerful, endothermic reducing agent, and its interaction with oxidants may be expected to be violent if unmoderated, as in rocket propulsion systems (often hypergolic). Mixtures of hydrazine vapour with air are explosive over a very wide range (4.7 — 100%) and ignition temperature can be very low (24°C on a rusty iron surface). When mixed with, or spilt on, highly porous materials (asbestos, cloth, earth, wood), autoxidation in air may proceed fast enough to cause ignition. Contact with hydrogen peroxide or nitric acid may cause ignition with conc. reagents and with dinitrogen tetraoxide is hypergolic [1]. Contact with chromate salts or chromium trioxide causes explosive decomposition. Copper oxide or lead oxide cause a vigorous decomposition, while dropping hydrazine hydrate on to mercury oxide can cause an explosion. This may be owing to known oxidation of hydrazine to hydrogen azide by two electron transfer, and

formation of explosive mercury azide [2]. Contact of hydrazine with the explosive
N, 2,4,6-tetranitroaniline caused ignition [3]. Contact of iron oxide, chlorates or
peroxides may lead to violent reactions [4]. The ignition delay with fuming nitric
acid was 8 ms, explosion also occurring [5]. Contact with potassium permanganate
leads to explosion [6].
See Lithium perchlorate: Hydrazine
    Potassium peroxodisulfate: Hydrazine, etc.
    Silver(I) oxide: Ammonia, etc.
    Sodium perchlorate: Hydrazine   MRH 5.86/65
    'Trioxygen difluoride': Various materials
See N-HALOIMIDES
See ROCKET PROPELLANTS
See also SILYLHYDRAZINES: OXIDANTS

Potassium
    Gmelin, 1935, Syst. 4, 318
    Explosive interaction.
    See Sodium hydrazide

Rust
    MCA Case History No. 1893
    Use of rusty tweezers, rather than the glass rod specified to handle specimens
    being immersed in hot 64% hydrazine, caused the hydrazine vapour to ignite.

Ruthenium(III) chloride
    Allen, A. D. et al., Inorg. Synth., 1970, 12, 3
    During the preparation of pentaamminedinitrogenruthenium(II) solutions, the initial
    gas producing reaction is so vigorous that increase in scale above that described
    (9 mmol) is not advised.
    See Metal catalysts, above
    See Pentaamminechlororuthenium chloride: Sodium azide

Silver compounds   MRH Silver nitrate 3.10/78
    Anon., CISHC Chem. Safety Summ., 1978, 49(194), 29
    For analysis, a silver-containing solution was made alkaline with 25% sodium
    hydroxide solution and filtered, then the precipitate was washed with ammonium
    hydroxide to redissolve the silver. Hydrazine (as sulfate) was added to precip-
    itate the silver, and when the mixture was heated, an explosion occurred. This
    could have been caused by precipitation of explosive silver nitride, rapid catalytic
    decomposition by silver compounds of the hydrazine salt, and/or ignition of the
    hydrogen evolved.

Sodium
    See Sodium: Hydrazine

Sodium hydroxide
    1. Anon., Univ. Safety Assoc., Safety News, 1978, (10), 12
    2. Day, A. C. et al., Org. Synth., 1970, 50, 42

While drying hydrazine over sodium hydroxide pellets, explosive decomposition occurred, spraying molten sodium hydroxide around. This was attributed to possible leakage of air into the evacuated flask.

Thiocarbonyl azide thiocyanate
See Thiocarbonyl azide thiocyanate: Ammonia, etc.

Titanium compounds
Bains, M. S., *Can. J. Chem.*, 1966, **44**, 534–538
Interaction of anhydrous hydrazine and titanium isopropoxide is explosive at 130°C in absence of solvent. Evaporation of solvent ether from the reaction product of tetrakis(dimethylamino)titanium and anhydrous hydrazine caused an explosion, attributed to formation and ignition of dimethylamine. *N*-Metal derivatives may also have been formed.
See other NON-METAL HYDRIDES, REDUCANTS

4521. Ammonium nitrite
[13446-48-5]   H₄N₂O₂
NH₄NO₂

See Nitric Acid: Manufacture Hazard
See entry NITRITE SALTS OF NITROGENOUS BASES

4522. Ammonium nitrate
[6484-52-2]   H₄N₂O₃
NH₄NO₃

FPA H33, 1975; *HCS 1980*, 147

1. Federoff, 1960, 35
2. Urbanski, 1965, Vol. 2, 460
3. Popper, H., *Chem. Eng.*, 1962, **70**, 91
4. Sykes, W. G. *et al.*, *Chem. Eng. Progr.*, 1963, **59**(1), 66
5. Yoshida, T., private comm., 1983
6. Urben, P. G., *Loss Prev. Bull.*, 1992, **108**, 31
7. Mellor, 1964, Vol. 8, Suppl. 1, 543
8. *MCA Guide*, 1972, 308
9. Croysdale, L. G. *et al.*, *Chem. Eng. Progr.*, 1965, **61**(1), 76
10. *MCA Case History No. 873*
11. Bailar, 1973, Vol. 2, 317
12. Vakhrushev, Yu. A. *et al.*, *Chem. Abs.*, 1975, **82**, 113745
13. Pascal, 1956, Vol. 10. 216
14. Rozman, B. Yu., *Chem. Abs.*, 1960, **54**, 12587c
15. Lee, P. R., *J. Appl. Chem.*, 1969, **19**, 345–351
16. Freeman, R., *Chem. Eng. Progr.*, 1975, **71**(11), 71–74
17. David, A. *et al.*, *Hydrocarbon Process.*, 1978, **57**(11), 149–175
18. Shah, K. D., *Fert. Sci. Tech. Ser.*, 1985, **4**, 261–273

19. Kirschner, E. *Chem. Eng. News*, 1994, **72**(51), 4; *ibid.*, **73**(37), 7
20. Anon., *Chem. Eng. News*, 1996, **74**(5), 11
21. Wiswesser, W. J., private comm., 1988
22. Anon., *Fire Prevention*, 1990, (231), 40
23. Li, R. *et al., Arch. Combust.*, 1988, **8**(1), 65–70
24. Buczkowski, D. *et al., Chem Abs.*, 1996, **125**, 176043e

The decomposition, fire and explosion hazards of this salt of positive oxygen balance have been adequately reviewed [1]. In the absence of impurities it is difficult, but not impossible, to cause ammonium nitrate to detonate. Use of explosives to break up the caked double salt, ammonium nitrate–ammonium sulfate (2:1), caused a 4.5 Mkg dump to detonate [4], even though some 45% of ammonium sulfate was present, effectively as a diluent, in the salt. Although the explosion tests available in 1921 did not give positive results for the nitrate–sulfate salt, it has recently been shown, using the BAM 50/60 steel tube test, that it will explode [5]. The ammonium sulphate is unlikely to be totally inert, having a negative enthalpy of decomposition to steam, nitrogen and sulfur, which last is capable of oxidation to sulfur dioxide, bringing the double salt into oxygen balance [6]. The effect of various impurities and additives on the thermal stability of ammonium nitrate has been widely studied [1,2,7]. Impure ammonium nitrate now constitutes more than 95% of all civil explosives used in the USA. A few incidents involving explosive decomposition of aqueous solutions of the salt during evaporation [8], or transfer operations (some in presence of oil) have been recorded [9]. Flame cutting a mild steel pipe blocked by the solid (impure) salt caused it to explode [10]. During preparation of dinitrogen oxide by the exothermic pyrolysis of the nitrate at 170°C, the temperature must not be allowed to exceed 250°C, as explosion may occur [11].

The thermal stability of ammonium nitrate solutions is decreased by presence of sodium 1,1′-methylenebis(6-naphthalenesulfonate) at 0.15%, but zinc oxide or zinc sulfate–apatite mixtures act as stabilisers [12]. Presence of ammonium chloride (0.1%) causes thermal decomposition to become violent around 175°C, and some chlorine is evolved [13]. Magnesium nitrate is claimed to significantly desensitise ammonium nitrate [24]. The parameters involved in the self-ignition of the nitrate during storage or transport are described, and a mechanism for instantaneous decomposition is proposed [14]. The factors affecting potentially explosive decomposition in bulk storage of the salt are discussed, and steady state thermal explosion theory is applied to the prediction of critical masses in relation to temperature in bulk storage. Application of the results to current storage practice is also discussed [15]. A detailed account of the investigation of a fire and explosion (basically unexplained) involving some 5 t of nitrate of a 14 kt warehouse stock has been published [16]. A 2-stage production process is designed to minimise danger of decomposition of the salt [17]. Accidents involving ammonium nitrate and the factors causing them are reviewed, and procedures for manufacturing, handling, storage and transportation are discussed [18]. A large explosion, probably of some tons TNT equivalent with several fatalities, during concentration of ammonium nitrate solution for fertiliser manufacture is discussed. There had been 750 tons of the nitrate present, the initiation mechanism is disputed though acidity, impurities, heat and lack of supervision were involved [19,20]. Road trailers laden with 14 t and 7 t of nitrate were involved in a fire and 3 subsequent detonations [21]. Another account makes it clear that it was, for the most part, ammonium nitrate/fuel oil (ANFO) explosive rather than the pure nitrate which exploded [22]. During pneumatic pipeline conveyance of the powdered salt, electrostatic generation is unlikely to be a problem under smooth flow conditions. However, the possibility of fast charge accumulation if flow discontinuities arise must not be overlooked [23].

Acetic acid   MRH 3.30/16
von Schwartz, 1918, 322
Conc. mixtures ignite on warming.

Acetic anhydride, Hexamethylenetetramine acetate, Nitric acid
See Nitric acid: Acetic anhydride, Hexamethylenetetramine acetate

Aluminium, Calcium nitrate, Formamide
See Calcium nitrate: Aluminium, etc.

Ammonia
Kohczkowski, A. *et al., J. Chem. Technol. Biotechnol.*, 1981, **31**, 327–332
Depending on the conditions, presence of free ammonia in ammonium nitrate may either stabilise, or tend to destabilise, the salt.

Ammonium chloride, (Barium nitrate), Water, Zinc
1. Bailey, P. S. *et al., J. Chem. Educ.*, 1975, **52**, 525
2. Jackson, H., *Spectrum*, 1969, **7**(2), 82
3. Bodner, G. M., *J. Chem. Educ.*, 1985, **62**(12), 1106
Addition of water to an intimate mixture of zinc powder and the salts causes spontaneous ignition [1], and a mixture of ammonium nitrate and ammonium chloride (9:1) sprinkled with zinc dust ignites vigorously when moistened [2]. Premature spontaneous ignition of the mixture was attributed to absorption of moisture by the hygroscopic salt mixture [3].
See Metals, below
See other REDOX REACTIONS

Ammonium phosphate, Potassium sulfate
Gryzlov, L. D. *et al., Chem. Abs.*, 1986, **104**, 185477
During manufacture of 26/16/10 N/P/K fertiliser by ammoniation of nitric–phosphoric acid mixtures, followed by concentration, addition of potassium sulfate, then granulation at up to 250°C, the possibility of a thermal explosion exists. Kinetic studies showed that 240°C appears to be a safe granulation temperature, but that changes in the composition and pH of the mixture may decrease this critical temperature.

Ammonium sulfate, Potassium
Staudinger, H., *Z. Elektrochem.*, 1925, **31**, 549
Ammonium nitrate containing the sulfate readily explodes on contact with potassium or its alloy with sodium.

Alkali metals   MRH Sodium 4.15/36
  Mellor, 1964, Vol. 8, Suppl. 1, 546
  Sodium progressively reduces the nitrate, eventually forming a yellow explosive
  solid, probably sodium hyponitrite.

Calcium superphosphate, Lignin, Phosphoric acid, Urea
  Dobreva, Ts. *et al., Chem. Abs.*, 1988, **109**, 37213
  The products of thermal decomposition of organic–mineral fertiliser mixtures were
  studied by IR spectroscopy, and implications for fire and explosion hazards consid-
  ered.
  See Urea, below

Charcoal, Metal oxides   MRH Carbon 3.64/7
  Herbst, H., *Chem. Ztg.*, 1935, **59**, 744–745
  The pelleted explosive ('ammonpulver', containing 10% charcoal) normally ignites
  at 160–165°C, but presence of rust, or copper oxide or zinc oxide lowers the
  temperature to 80–120°C.
  See Non-metals, below

Chloride salts
  1. Pascal, 1956, Vol. 10, 216
  2. Pany, V. *et al., Chem. Abs.*, 1976, **85**, 56018
  The nitrate containing 0.1% of ammonium chloride decomposes vigorously below
  175°C [1]. Presence of 0.1% of calcium chloride or iron(III) chloride in the nitrate
  lowers its initiation temperature sufficiently to give violent or explosive decomposition. Thermal analysis plots for aluminium chloride, calcium chloride and
  iron(III) chloride are given [2].

Copper iron(II) sulfide
  Kuznetsov, G. V. *et al., Chem. Abs.*, 1975, **82**, 75133
  During preparations for blasting the sulfide mineral copper pyrites, ammonium
  nitrate-based blasting cartridges exploded prematurely in the blast holes. This was
  attributed to exothermic interaction of acid ground-water with the sulfide–oxidant
  combination.
  See Metal sulfides, below

Cyanoguanidine
  See Cyanoguanidine: Oxidants

Fertiliser materials
  Davies, R. O. E. *et al., Ind. Eng. Chem.*, 1945, **37**, 59–63
  Mixtures of ammonium nitrate, superphosphate and organic materials stored in
  bulk may ignite if the internal temperature exceeds 90°C. This is owing to the free
  acid in the superphosphate, and may be prevented by neutralisation with ammonia.

Hydrocarbon oils   MRH Hexane 3.80/5, Toluene 3.80/7
  Goffart, P., *Chem. Abs.*, 1975, **82**, 61420
  Detonability of several ammonium nitrate-based fertilisers, with or without the
  addition of fuel oil, was studied.

Metals

1. Mellor, 1964, Vol. 8, Suppl. 1, 543–546
2. Soda, N. *et al., Chem. Abs.*, 1969, **70**, 49151
3. Peterson, W. S. *Light Met* . (Warrendale, Pa) 1995, 899

MRH Aluminium 8.49/25, lead 1.38/56, magnesium 8.79/31, sodium 4.15/36, zinc 3.22/tr.

Many of the following powdered metals reacted violently or explosively with fused ammonium nitrate below 200°C: aluminium, antimony, bismuth, cadmium, chromium, cobalt, copper, iron, lead, magnesium, manganese, nickel, tin, zinc; also brass and stainless steel. Mixtures with aluminium powder are used as the commercial explosive Ammonal. Sodium reacts to form the yellow explosive compound sodium hyponitrite, and presence of potassium sensitises the nitrate to shock [1]. Shock-sensitivity of mixtures of ammonium nitrate and powdered metals decreases in the order titanium, tin, aluminium, magnesium, zinc, lead, iron, antimony, copper [2]. Contact between molten aluminium and the salt is violently explosive, apparently there is a considerable risk of this happening in scrap remelting [3].

See Ammonium sulfate, etc., above
Aluminium: Ammonium nitrate
Potassium: Nitrogen-containing explosives

Metal salts

1. Glazkova, A. P., *Chem. Abs.*, 1970, **73**, 132646
2. Rosser, W. A. *et al., Trans. Faraday Soc.*, 1964, **60**, 1618–1625

Catalytic effects on the thermal decomposition and burning under nitrogen of the nitrate were determined for ammonium dichromate, potassium dichromate, potassium chromate, barium chloride, sodium chloride and potassium nitrate. Chromium(VI) salts are most effective in decomposition, and the halides salts during burning of the nitrate [1]. The effect of chromium compounds soluble in the molten nitrate, all of which promote decomposition of the latter, was studied (especially using ammonium dichromate) in kinetic experiments [2].

Metal sulfides

Proc. 13th Symp. Explos. Pyrotech., 1986, **VII**, 1–7; (*Chem. Abs.*, 1989 **110**, 98227s)

In presence of sulfide ores, specifically pyrites, explosives containing ammonium nitrate may undergo runaway reaction, leading to detonation at temperatures below 40°C if pH is less than 2. The reaction is acid catalysed.

Nitric acid

Rubtsov, Yu. I. *et al., Chem. Abs.*, 1987, **106**, 158775

In the context of safety of the process of neutralisation of nitric acid with ammonia, the effects of temperature (160–230°C), pressure (2.3–9.8 bar), and concentrations of ammonium nitrate (86–94%) and of nitric acid (0–4%) upon decomposition rate were studied.

See Nitric acid: Manufacture Hazard

Non-metals   MRH Carbon 3.64/7, phosphorus (y) 4.60/14
Mellor, 1940, Vol. 2, 841–842; 1946, Vol. 5, 830
The powdered salt in admixture with charcoal explodes at 170°C, or the solid salt
on contact with glowing charcoal. Phosphorus ignites on the fused salt.
See Sulfur, below, *also* Phosphorus: Nitrates

Organic fuels   MRH values below references
1. Sykes, W. G. *et al., Chem. Eng. Progr.*, 1963, **59**(1), 66
2. Urbanski, 1965, Vol. 2, 461
3. Bernoff, R. A. *et al.*, US Pat. 3 232 940, 1969
MRH Acetone 3.77/8, aniline 3.77/8, carbon disulfide 3.26/11, ethanol 3.68/9,
ethylene glycol 3.56/14
In a review of ammonium nitrate explosion hazards, several incidents involving
mixtures of this positive oxygen balance material with various organic materials
were analysed [1]. In general, fire incidents involving ammonium nitrate admixed
with 1% of wax, oil, or stearates (as anti-caking additives) tended more towards
explosion than the pure salt, although the degree of confinement is also important
[1,2]. The ease of detonation of mixtures is much greater when 2–4% of oil is
present, and such mixtures are used as commercial explosives [2]. The preparation
of ammonium nitrate–fuel systems at a molecular level is covered by the patent,
which describes several alkylammonium nitrates as high-energy materials [3].

Other reactants
Yoshida, 1980, 190
MRH values calculated for 19 combinations with oxidisable materials are given.

Potassium nitrite
Mellor, 1940, Vol. 2, 842
Contact of the solid nitrate with fused nitrite causes incandescence.
See NITRITE SALTS OF NITROGENOUS BASES

Potassium permanganate
See Potassium permanganate: Ammonium nitrate

Sawdust
Biasutti, 1981, 151
A violent explosion in an ammonium nitrate store with sawdust-covered floors
(to absorb spillage and prevent sparks) was attributed to local decomposition of
the moist nitrate-containing sawdust, leading to temperature rise and spontaneous
ignition. The observation of red-brown fumes just before the explosion supports
this hypothesis.

Sugar
Gonzalez, M. M., *Chem. Abs.*, 1981, **65**, 83230
Mixtures of the salt with sugar syrup are effective explosives.

Sulfide ores
Liao, M. *et al., Proc. 13th Symp. Explos. Pyrotech.*, 1988, VII.1–VII.7

The main factors leading to spontaneous explosions of blasting explosives in sulfide ore mines are identified as presence of ammonium nitrate, presence of pyrites and their weathering products, as well as the pH and temperature. At pH below 2, the critical temperature for spontaneous explosion may be lower than 40°·

Sulfur
1. Mason, C. M. *et al.*, *J. Agric. Food Chem.*, 1967, **15**, 954
2. Prugh, R. W., *Chem. Eng. Progr.*, 1967, **63**(11), 53–55

The fire risks of nitrate–sulfur mixtures have been discussed [1], and explosion risks assessed by DTA [2].

Trinitroanisole
Biasutti, 1981, 54

A mixture of ammonium nitrate and 2,4,6-trinitroanisole, prepared as an explosive by mixing the hot components, ignited spontaneously and later exploded violently.

Urea
1. Croysdale, L. C. *et al.*, *Chem. Eng. Progr.*, 1965, **61**(1), 72
2. Anderson, J. F., *Safety Air Ammon. Plants*, 1967, **9**, 70–71
3. Leach, J. W., *Rept. AD-A075338*, Richmond (Va.), USNTIS, 1979
4. Perbal, G., *Proc. 4th Int. Symp. Loss Prev. Safety Prom. Proc. Ind.*, Vol. 2, M1–M19, Rugby, IChE, 1983
5. Rubstov, Yu. I. *et al.*, *Chem. Abs.*, 1988, **108**, 137111

Conc. solutions of ammonium nitrate and urea exploded during large-scale mixing operations. Although the cause was not established, the hazards of these operations were discussed in relation to the circumstances [1]. A further explosion under similar circumstances was reported [2]. Some of the liquid streams involved in preparation of nitroguanidine, which are somewhat similar to those in the present process, are detonable if less than 25% of water is present in the solutions. Thermal characteristics and detonation data were developed for the nitroguanidine process and that to produce guanidinium nitrate [3]. The circumstances of the 3 explosions [1,2] have been reviewed and detailed experimental work has confirmed that the solutions (54–46% ammonium nitrate, 35–39% urea and 15–20% water) cannot be detonated. However, dehydration of the solution gives a solid residue capable of deflagration at 240°C, and the residue from prolonged heating at 158°C is capable of detonation. It is concluded that the explosions were most likely caused by leakage and dehydration in the lagging of the steam-traced pipelines at 158°C, slow development of the unstable product and of a hot spot, with eventual initiation of detonation. Any traces of urea nitrate would not have contributed significantly to the events [4]. There is no danger of violent reaction as long as the components remain in solution. Processing temperatures should not exceed 120°C [5].

Water
Kolaczkowski, A. *et al.*, *Chem. Abs.*, 1982, **97**, 78074

Hot aqueous solutions of the nitrate of above 50% concentration may decompose explosively under adiabatic conditions and under conditions of confinement (small, or no vents). A 9 m^3 tank of 85% solution also containing 0.9% of nitric acid and 0.2% of chloride ion (which catalyses decomposition) at 110°C exploded powerfully.

See related METAL NITRATES
See other OXIDANTS, OXOSALTS OF NITROGENOUS BASES

4523. Ammonium nitridoosmate
[22493-01-2] $H_4N_2O_3Os$

$$NH_4[NOsO_3]$$

Mellor, 1942, Vol. 15, 727
It decomposes explosively at 150°C, like the barium salt.
See related NITRIDES

4524. Hydroxylaminium nitrate
[13465-08-2] $H_4N_2O_4$

$$HON^+H_3\ NO_3^-$$

1. Mellor, 1940, Vol. 8, 303
2. Kirk-Othmer, 1981, **16**, 179
3. Asaki, A., *Chem. Abs.*, 1983, **98**, 48238
4. Van Disk, C. A., *Chem. Abs.*, 1984, **100**, 70566

The redox salt which decomposes at 100°C, evolving nitrogen and oxygen [1] is overall an oxidant but has been used as a reducant in actinide processing [2]. It has been used recently as a missile monopropellant [3], and also as an oxidant, with its isopropyl derivative as the fuel [4].
See other HYDROXYLAMINIUM SALTS

Amine nitrates
  Klein, N., *Progr. Astronaut. Aeronaut.*, 1988, **109** (Gun Propell. Technol.), 473–497
  The use of aqueous mixtures of hydroxylammonium nitrate with various butylammonium nitrates as gun propellants is reviewed.
See other OXOSALTS OF NITROGENOUS BASES, REDOX COMPOUNDS

4525. Tetrahydroxotritin(2+) nitrate ('Basic stannous nitrate')
[73593-33-6] (ion) $H_4N_2O_{10}Sn_3$

$$[Sn_3(OH)_4][NO_3]_2$$

1. Anon., *J. Soc. Chem. Ind.*, 1922, **41**, 423–434R
2. Donaldson, J. D. *et al.*, *J. Chem. Soc.*, 1961, 1996

An explosion in flour-bleaching operations was attributed to violent decomposition of the basic nitrate [1], which is an impact-, friction- and heat-sensitive explosive [2]. The instability is associated with the presence of reducant and oxidant functions in the same molecule. The previous formulation as tin(II) nitrate oxide is revised to that above.
See other REDOX COMPOUNDS
See related METAL NITRATES

1684

4526. Ammonium azide
[12164-94-2] H_4N_4

$$NH_4N_3$$

1. Mellor, 1940, Vol. 8, 344
2. Sorbe, 1968, 129
3. Obenland, C. O. *et al., Inorg. Synth.*, 1966, **8**, 53

It explodes on rapid heating [1] and contains ~93% of nitrogen. It is also friction-
and impact-sensitive [2], and preparative precautions have been detailed [3].
See other HIGH-NITROGEN COMPOUNDS, NON-METAL AZIDES

4527. Tetraimide (Tetraazetidine)
[58674-00-3] H_4N_4

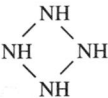

Rice, F. O. *et al., J. Amer. Chem. Soc.*, 1957, **79**, 1880–1881

This blue solid (a cyclic dimer of diazene, 93.3% of nitrogen), obtained by freezing
out at −195°C the pyrolysis products of hydrogen azide, is extremely explosive
above this temperature. An explosion at −125°C destroyed the apparatus.
See other HIGH-NITROGEN COMPOUNDS

4528. Ammonium dinitramide
[140456-78-6] $H_4N_4O_4$

$$H_4N^+ \ {}^-N(NO_2)_2$$

See Dinitramine
See other N-NITRO COMPOUNDS

4529. 'Sodium perpyrophosphate'
[18178-05-7] $H_4Na_4O_{11}P_2$

$$(NaO)_2P(O)OP(O)(ONa)_2.2H_2O_2$$

See Sodium pyrophosphate hydrogen peroxidate

4530. Oxodisilane
[22755-00-6] H_4OSi_2

$$H_3SiSi(O)H$$

Kautsky, K., *Z. Anorg. Chem.*, 1921, **117**, 209
It ignites in air.
See related SILANES

4531. Lead(II) phosphinate
[10294-58-3] $H_4O_4P_2Pb$

$$Pb(OP(O)H_2)_2$$

Mellor, 1971, Vol. 8, Suppl. 3, 623
Impact-sensitive, used as priming explosive.
See other HEAVY METAL DERIVATIVES, METAL PHOSPHINATES, REDUCANTS

4532. Ruthenium(IV) hydroxide
[12181-34-9] H_4O_4Ru

$$Ru(OH)_4$$

Mellor, 1942, Vol. 15, 516
It deflagrates incandescently above 300°C.

4533. Xenon(IV) hydroxide
[ ] H_4O_4Xe

$$Xe(OH)_4$$

Sorbe, 1968, 158
Very unstable, it explodes violently on heating.
See other NON-METAL OXIDES, XENON COMPOUNDS

4534. Zirconium hydroxide
[14475-63-9] H_4O_4Zr

$$Zr(OH)_4$$

Zirconium
See Zirconium: Oxygen-containing compounds

4535. Lead(II) phosphite
[15521-60-5] $H_4O_6P_2Pb$

$$Pb(OPH(O)OH)_2$$

491M, 1975, 228

A fibre drum of the salt ignited spontaneously. This was attributed to slow decomposition with formation of phosphine which accumulated and eventually ignited inside the drum.
See other HEAVY METAL DERIVATIVES, REDUCANTS

4536. Tetrahydroxydioxotrisilane
[ ] $H_4O_6Si_3$

$$HOSi(O)Si(OH)_2Si(O)OH$$

1. Mellor, 1940, Vol. 6, 230
2. Müller, R. *et al., J. Prakt. Chem.*, 1966, **31**, 1–6

The hydrolysis product of octachlorotrisilane (and probably polymeric), it decomposes violently or explosively when heated in air, or sometimes spontaneously [1], and ignites when subjected to friction [2].

See Poly(dihydroxydioxodisilane)
See related NON-METAL OXIDES

4537. Oxobis[aqua(oxo)diperoxorheniumVII]
[174753-06-1] $H_4O_{13}Re_2$

Herrmann, W. A. *et al.*, *Chem. Eur. J.*, 1996, **2**(2), 168

An explosive red/orange solid, obtained from dirhenium heptoxide and hydrogen peroxide reacting in ether as solvent.

See other INORGANIC PEROXIDES

†4538. Diphosphane
[13445-50-6] H_4P_2

$$H_2PPH_2$$

Mellor, 1940, Vol. 8, 829

It ignites in air, and will cause other flammable gases to ignite when present at 0.2 vol% concentration.

See other PHOSPHINES

†4539. Silane
[7803-62-5] H_4Si

$$SiH_4$$

1. Mellor, 1940, Vol. 6, 220–221
2. Braker, 1980, 632
3. Gale, H. J., *Lab. Haz. Bull.*, 1984, (3), item 140
4. Britton, L. G., *Plant/Oper. Progr.*, 1990, **9**(1), 16
5. Katou, Y., *Chem. Abs* . 1990, **113**, 84061h
6. Kondo, S. *et al.*, *Combust. Flame*, 1994, **97**(3–4), 296.
7. Horiguchi, S., *Chem. Abs* . 1997, **127**, 143180k

It is mildly endothermic (ΔH°_f (g) +32.5 kJ/mol, 1.01 kJ/g).

Pure material is said not to ignite in air unless the temperature be increased or the pressure reduced. Presence of other hydrides as impurities causes ignition always to occur on contact [1]. However, 99.95% pure material, even at concentrations down to 1% in hydrogen and/or nitrogen, ignites in contact with air unless emerging at

very high gas velocity, when mixtures with up to 10% silane content may not ignite [2]. Silane from a cylinder at 24 bar contained inside a ventilated cabinet inadvertently flowed back to the low pressure side of a nitrogen cylinder regulator outside the cabinet. When the regulator failed, the leaking silane ignited and burned outside the cabinet, preventing access to the main silane valve, and the leak reignited after extinguishing the fire. The system has been redesigned to eliminate the problems [3]. An exhaustive review of combustion and associated hazards is found in [4]. This also reveals potentially dangerous deflagration with the chlorofluoromethanes used in some fire extinguishers. A detailed study of silane flammability and blast effects in mixtures with nitrogen discharged to air [5]. The lower flammable limit of silane in air declines with decreasing oxygen content [6]. A Japanese review of explosion hazards with silane has been published . Spontaneous combustion in (somewhat oxygen depleted) air has been observed down to −162 C. In addition to the materials listed separately below, trifluoramine, interhalogens and a range of fluorocarbons and halofluorocarbons form explosive mixtures, as do sulfur tetra- and hexa-fluorides. Most of these are described as pyrophoric with silane, which may be a mistranslation [7].
See other IGNITION SOURCES

Air, *cis*-2-Butene
Urtiew, P. A. *et al., Rept. UCRL-52007*, Richmond (Va.), USNTIS, 1976
Presence of the alkene delays or prevents spontaneous ignition of silane–air mixtures.

Halogens, or Covalent chlorides   MRH values below references
1. Mellor, 1940, Vol. 6, 220–221
2. Braker, 1980, 632
MRH Bromine 1.21/91, carbon tetrachloride 2.89/83, chlorine 3.85/82
Silane burns in contact with bromine, chlorine or covalent chlorides (carbonyl chloride, antimony pentachloride, tin(IV) chloride, etc.) [1]. Extreme caution is necessary when handling silane in systems with halogenated compounds, as a trace of free halogen may cause violent explosions [2].

Nitrogen oxides
Horiguchi, S. *et al., Chem. Abs.*, 1990, **112** 121711
Mixtures of silane and nitrogen oxides have been studied, that with nitrous oxide detonates very easily.

Other reactants
Yoshida, 1980, 371
MRH values calculated for 4 combinations with the materials indicated are given.

Oxygen   MRH 14.31/66
1. Bailar, 1973, Vol. 1, 1366
2. Ring, M. A. *et al., AIP Conf. Proc.*, 1988, **166** (Photovolt. Safety), 175–182
Very pure silane does not immediately explode with oxygen, but the decomposition products may ignite after a delay [1]. Mixtures of silane with 30% oxygen are metastable and potentially explosive under all pressure conditions studied, and

become explosive at above 80°C. Addition of a few % of disilane renders the mixtures much more explosive, and mechanisms are proposed [2].
See other SILANES

4540. Thorium hydride
[15457-87-1] H_4Th

ThH_4

1. Mellor, 1941, Vol. 7, 207
2. Stout, E. L., *Chem. Eng. News*, 1958, **36**(8), 64
3. Hartmann, I. *et al., Rept. Invest. No. 4835*, Washington, US Bur. Mines, 1951
Thorium hydride explodes on heating in air [1], and the powdered hydride readily ignites on handling in air [2]. Layers of thorium or uranium hydrides ignited spontaneously after exposure to ambient air for a few min [3].
See other METAL HYDRIDES

4541. Uranium(IV) hydride
[51680-55-8] H_4U

UH_4

1. Hartmann, I. *et al., Rept. Invest. No. 4835*, Washington, US Bur. Mines, 1951
2. Starks, D. F. *et al., J. Amer. Chem. Soc.*, 1973, **95**, 3423
Layers of uranium or thorium hydrides ignited spontaneously after exposure to ambient air for a few min [1]. Thermal decomposition yields pyrophoric uranium [2].

Oxygen
See Oxygen (Gas): Metal hydrides
See other METAL HYDRIDES

4542. Orthoperiodic acid
[10450-60-9] H_5IO_6

$O{:}I(OH)_5$

Azo-pigment, Perchloric acid
See Perchloric acid: Azo-pigment, etc.
See other INORGANIC ACIDS, OXOHALOGEN ACIDS

4543. Iododisilane
[14380-76-8] H_5ISi_2

$ISiH_2SiH_3$

Bailar, 1973, Vol. 1, 1371
It ignites in air.
See other HALOSILANES, IODINE COMPOUNDS

4544. Ammonium hydroxide
[1336-21-6]   H₅NO

$$NH_4OH$$

(*MCA SD-13*, 1947); *NSC 701*, 1983; *HCS 1980*, 140

Nitromethane
See Nitromethane: Acids, etc.

Other reactants
Yoshida, 1980, 211
MRH values calculated for 11 combinations with oxidants are given.
See other INORGANIC BASES *See related* NON-METAL HYDRIDES

4545. Ammonium hydrogen sulfite
[10192-30-0]   H₅NO₃S

$$NH_4OS(O)OH$$

Self accelerating decomposition ($Q = 0.43$ kJ/g) occurs at 130°C in closed systems, such as are used for the Bucherer reaction.
See entry SELF-ACCELERATING REACTIONS

4546. Hydrazinium salts
[ ]   H₅N₂⁺ Z⁻

$$H_2NN^+H_3Z^-$$

Metal nitrites
See Hydrogen azide
See also NITRITE SALTS OF NITROGENOUS BASES

4547. Diamidophosphorous acid
[22750-67-0]   H₅N₂OP

$$(H_2N)_2P(O)H$$

Water
Mellor, 1940, Vol. 8, 704
Interaction is violent with incandescence.
See other INORGANIC ACIDS

4548. Hydrazinium nitrite
[ ]   H₅N₃O₂

$$H_2NN^+H_3NO_2^-$$

See entries HYDRAZINIUM SALTS, NITRITE SALTS OF NITROGENOUS BASES

1690

4549. Hydrazinium nitrate
[13464-97-6] $H_5N_3O_3$

$$H_2NN^+H_3NO_3^-$$

Alone, or Metals, or Metal compounds
   Mellor, 1940, Vol. 8, 327; 1967, Vol. 8, Suppl. 2.2, 84, 96
   It is an explosive of positive oxygen balance, less stable than ammonium nitrate,
   and has been studied in detail. Stable on slow heating to 300°C, it decomposes
   explosively on rapid heating or under confinement. Presence of zinc, copper, most
   other metals and their acetylides, nitrides, oxides or sulfides cause flaming decomposition above the m.p. (70°C). Commercial cobalt (cubes) causes an explosion
   also.
   See entry HYDRAZINIUM SALTS

2-Hydroxyethylamine
   Fujihara, S. *et al., Chem. Abs.*, 1976, **84**, 7212
   Mixtures of the salt (80%) and amine (15%) with water are useful as an impact-
   insensitive but powerful liquid explosive.

Potassium dichromate
   Shidlovskii, A. A. *et al., Chem. Abs.*, 1960, **54**, 22132g
   Thermal decomposition becomes explosive above 270°C, or above 100°C in pres-
   ence of 5% of potassium dichromate.
   See other OXOSALTS OF NITROGENOUS BASES, REDOX COMPOUNDS

4550. Hydrazinium azide
[38551-29-0] H_5N_5

$$H_2NN^+H_3N_3^-$$

It contains some 94% of nitrogen.
See other HIGH-NITROGEN COMPOUNDS, HYDRAZINIUM SALTS, NON-METAL AZIDES

4551. Potassium triamidothallate ammoniate
[ ] $H_6K_2N_3Tl.xNH_3$

$$K_2[(H_2N)_3Tl].xNH_3$$

Alone, or Acids, or Water
   Mellor, 1940, Vol. 8, 262
   Like thallium(I) amide from which it is derived by treatment with potassium amide
   in liquid ammonia, the ammoniated salt ($x = 2$ or less) explodes violently on
   heating, friction, or contact with dilute acids or water.
   See other N-METAL DERIVATIVES

4552. Lanthanum pentanickel hexahydride
[54847-21-1] H_6LaNi_5

$$LaNi_5H_6$$

Imamoto, T. *et al., J. Org. Chem.*, 1987, **52**, 5695–5699
The alloy hydride has been investigated as a useful hydrogenation catalyst for a wide variety of substrates under mild conditions. It is however pyrophoric in air, and an experimental procedure has been developed to avoid this hazard. A related hydride, $LaNi_{4.5}Al_{0.5}H_5$ has similar properties.
See Lanthanum–nickel alloy
See LANTHANIDE–TRANSITION METAL ALLOY HYDRIDES

4553. Dioxonium hexamanganato(VII)manganate
[51184-04-4] $H_6Mn_{14}O_{52}\cdot11H_2O$
[54065-28-0] (new unsolv. structure $H_2Mn_7O_{24}$)
$$[H_3O^+]_2[Mn(MnO_4)_6]_2^-\cdot11H_2O$$

Krebs, B. *et al., Angew. Chem. (Intern. Ed.)*, 1974, **13**, 603
It has been shown that 'solid permanganic acid dihydrate', the powerful and unstable oxidant, is an undecahydrate as formulated above.
See Permanganic acid
See other INORGANIC ACIDS, OXIDANTS

4554. Ammonium phosphinate
[7803-65-8] H_6NO_2P

$$NH_4OP(O)H_2$$

Mellor, 1940, Vol. 8, 880
The salt evolves spontaneously flammable phosphine around 240°C.
See other REDUCANTS *See related* METAL PHOSPHINATES

4555. Hydroxylaminium phosphinate
[ ] H_6NO_3P

$$HON^+H_3\ O^-P(O)H_2$$

Mellor, 1940, Vol. 8, 880
The salt detonates above its m.p., 92°C.
See other HYDROXYLAMINIUM SALTS, REDUCANTS *See related* METAL PHOSPHI-NATES

4556. Ammonium amidosulfate (Ammonium sulfamate)
[7773-06-0] $H_6N_2O_3S$

$$NH_4OSO_2NH_2$$

1. Hunt, J. K., *Chem. Eng. News*, 1952, **30**, 707
2. Rogers, M. G., private comm., 1973

A 60% solution of ammonium sulfamate (pH above 4.5) will not undergo rapid hydrolysis below 200°C. Addition of acid (to pH 2) causes a runaway exothermic hydrolysis to set in at 130°C. Superheating and vigorous boiling can occur under appropriate physical conditions [1]. The use of urea–formaldehyde resins as temporary binders in the firing of refractories and ceramics at high temperatures can lead to the formation of substantial deposits of ammonium sulfamate in the cooler parts of kilns, should a fuel oil containing appreciable amounts of sulfur be used for firing. Ammonia from decomposition of the resin combines with sulfur trioxide to form ammonium sulfamate which accumulates as either a solidified deposit in flues or as a white deposit on walls, causing corrosion and handling problems [2].
See other CORROSION INCIDENTS, N–S COMPOUNDS, OXOSALTS OF NITROGENOUS BASES

4557. Ammonium amidoselenate
[13767-10-0] $H_6N_2O_3Se$

$$NH_4OSeO_2NH_2$$

Dostal, K. *et al., Z. Anorg. Chem.*, 1958, **296**, 29–35
It explodes if rapidly heated to 120°C.
See related N–S COMPOUNDS *See other* OXOSALTS OF NITROGENOUS BASES

4558. Hydrazinium hydrogen selenate
[ ] $H_6N_2O_4Se$

$$H_2NN^+H_3 \ O^-SeO_2OH$$

Meyer, J. *et al., Ber.*, 1928, **61**, 1839
The salt explodes in contact with a hot glass rod.
See other HYDRAZINIUM SALTS, OXOSALTS OF NITROGENOUS BASES

4559. Diamminepalladium(II) nitrite
[28068-05-5] $H_6N_4O_4Pd$

$$[(H_3N)_2Pd] \ [NO_2]_2$$

Fountain, N. O., private comm., 1983
The dry salt is not readily ignitable, but is marginally more sensitive to impact than dinitrobenzene and can be ignited by impacted friction. The deflagration rate, once confined ignition had begun, was just greater than that of typical deflagrating explosives.
See other AMMINEMETAL OXOSALTS *See related* PLATINUM COMPOUNDS

4560. *cis*-**Diammineplatinum(II) nitrite**
[14286-02-3] $H_6N_4O_4Pt$

$$[(H_3N)_2Pt][NO_2]_2$$

1. Brauer, 1965, Vol. 2, 1560
2. Holifield, P. J., private comm., 1974
3. Fountain, N. O., private comm., 1983

It decomposes explosively at 200°C [1]. Dry material stored in clear bottles in sunlight for several weeks became sensitive and exploded violently on slight mechanical shock. The material is now supplied commercially moistened with water [2]. The dry salt may be readily ignited, and deflagration under confinement is similar to that of gunpowder. Presence of 10% of water reduced ignitability but did not slow the deflagration rate thereafter. Sensitivity to impact or friction was greater than that of nitrobenzene [3].

See other AMMINEMETAL OXOSALTS, PLATINUM COMPOUNDS

4561. Hydrazinium dinitrate
[13464-98-7] $H_6N_4O_6$

$$H_3N^+N^+H_3 \ (NO_3^-)_2$$

Sorbe, 1968, 130
Like the mono-salt, it explodes on heating or impact.
See other HYDRAZINIUM SALTS, OXOSALTS OF NITROGENOUS BASES

4562. Diamminepalladium(II) nitrate
[34090-17-0] $H_6N_4O_6Pd$

$$[(H_3N)_2Pd][NO_3]_2$$

White, J. H., private comm., 1965
There is a danger of explosion if the nitrate is dried.
See other AMMINEMETAL OXOSALTS *See related* PLATINUM COMPOUNDS

4563. Zinc dihydrazide
[25546-98-9] H_6N_4Zn

$$Zn(NHNH_2)_2$$

Mellor, 1940, Vol. 8, 315
It explodes at 70°C.
See other METAL HYDRAZIDES

4564. Sodium hexahydroxyplatinate(IV)
  [12325-31-4] $H_6Na_2O_6Pt$

$$Na_2[Pt(OH)_6]$$

Acetic acid, Nitric acid
  See Nitric acid: Acetic acid, Sodium hexahydroyxyplatinate(IV)
  See other PLATINUM COMPOUNDS

4565. Disilyl oxide
  [13597-73-4] H_6OSi_2

$$H_3SiOSiH_3$$

Chlorine
  See Chlorine: Disilyl oxide
  See related SILANES

4566. 2,4,6-Trisilatrioxane (Cyclotrisiloxane)
  [291-50-9] $H_6O_3Si_3$

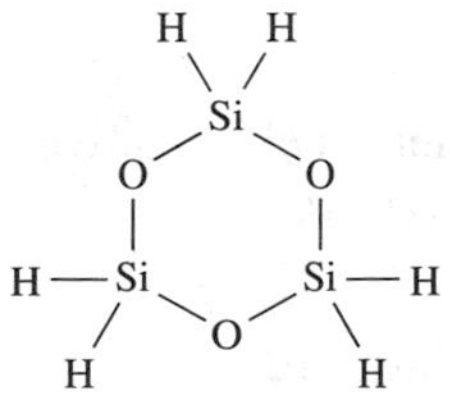

Air, or Chlorine
  Mellor, 1940, Vol. 6, 234
  The solid polymer (approximating to the trimer) ignites in air or chlorine.
  See related SILANES

4567. Tetraphosphoric acid
  [13813-62-2] $H_6O_{13}P_4$

$$HO[HOP(O)O]_3P(O)(OH)_2$$

Water
  Mellor, 1971, Vol. 8, Suppl. 3, 736
  Dilution of polyphosphoric acids with water in absence of cooling may lead to
  a large exotherm. Thus, tetraphosphoric acid diluted from 84 to 54% phosphorus
  pentoxide content rapidly attains a temperature of 120–140°C.
  See other INORGANIC ACIDS

4568. Disilyl sulfide
  [16544-95-9] H_6SSi_2

$$H_3SiSSiH_3$$

Sorbe, 1968, 127
  It ignites in moist air.
  See related NON-METAL SULFIDES, SILANES

†4569. Disilane
[1590-87-0] H_6Si_2

$$H_3SiSiH_3$$

Bromine
See Bromine: Non-metal hydrides

Non-metal halides, or Oxygen
Mellor, 1940, Vol. 6, 220–224
It explodes on contact with carbon tetrachloride or sulfur hexafluoride, and contact with chloroform causes incandescence. Disilane ignites spontaneously in air, even when pure, and ingress of air or oxygen into a volume of disilane causes explosion.
See other SILANES

4570. Dihydrazinemanganese(II) nitrate
[39957-12-5] $H_8MnN_6O_6$

$$[(H_4N_2)_2Mn][NO_3]_2$$

Bailar, 1973, Vol. 3, 827
It ignites at 150°C, but is not shock-sensitive.
See other AMMINEMETAL NITRATES

4571. Amminepentahydroxyplatinum
[ ] H_8NO_5Pt

$$[H_3NPt(OH)_5]$$

Jacobsen, J., *Compt. rend.*, 1909, **149**, 575
It explodes fairly violently above 250°C, as does the pyridine analogue.
See other N-METAL DERIVATIVES, PLATINUM COMPOUNDS

4572. Diamminedihydroxyosmium
[ ] $H_8N_2O_2Os$

$$[(H_3N)_2Os(OH)_2]$$

Ephraim, 1939, 463
It decomposes vigorously on heating.
See other N-METAL DERIVATIVES *See related* PLATINUM COMPOUNDS

4573. Ammonium thiosulfate
[7783-18-8] $H_8N_2O_3S_2$

$$H_4NSSO_2ONH_4$$

Sodium chlorate
See Sodium chlorate: Ammonium salts, etc.
See related N–S COMPOUNDS

4574. Ammonium sulfate
[7783-20-2]   H₈N₂O₄S

$$(NH_4)_2SO_4$$

Other reactants
Yoshida, 1980, 394
MRH values calculated for 14 combinations with various materials are given.

Ammonium nitrate
See Ammonium Nitrate

Potassium chlorate
See Potassium chlorate: Ammonia, etc.

Sodium hypochlorite
See Sodium hypochlorite: Ammonium salts
See related METAL OXONON-METALLATES

4575. Hydroxylaminium sulfate
[10039-54-0]   H₈N₂O₆S

$$(HONH_3)_2SO_4$$

FPA H66, 1978; HCS 1980, 559

1. Grewer, T. *et al., Hazards from Pressure*, IChE Symp. Ser. No. 102, 1–9, Oxford, Pergamon, 1987
2. Grewer, T. *et al., Exothermic Decomposition*, Technical Report 01VD 159/0329 for Federal German Ministry for Res. Technol., Bonn. 1986

The solid deflagrates at 5.6 cm/min, and will attain a maximum of 250 bar pressure in a closed system [1]. Energy of decomposition (in range 180–280°C) measured as 1.6 kJ/g [2].
See other DEFLAGRATION INCIDENTS

Aniline, Chloral hydrate
See 2,2,2-Trichloro-1,1-ethanediol
See entry HYDROXYLAMINIUM SALTS

4576. Ammonium peroxodisulfate (Ammonium persulphate)
[7727-54-0]   H₈N₂O₈S₂

$$H_4NOSO_2OOSO_2ONH_4$$

HCS 1980, 149

Energy of decomposition (in range 140–210°C) measured as 0.323 kJ/g.
See entry THERMOCHEMISTRY AND EXOTHERMIC DECOMPOSITION (reference 2)

Aluminium, Water   MRH Aluminium 5.90/24
    Pieters, 1957, 30
    A mixture including the powdered metal may explode.

Ammonia, Silver salts
    Mellor, 1947, Vol. 10, 466
    In conc. solutions the silver catalysed oxidation of ammonia to nitrogen may be
    very violent.

Azoisobutyronitrile
    1. Anon., *Chemical Engineer*, 1993, **536**, 4
    2. *The Fire at Allied Colloids Ltd.*, HSE Books, Sudbury UK, 1994
    The azonitrile (AZDN) and ammonium persulphate were stored together as poly-
    merisation initiators. The AZDN was warmed by a supposedly blanked-off steam
    pipe and partially decomposed, rupturing its kegs and falling onto the persulphate
    in sacks below. Some degree of mixing was effected by a falling keg lid piercing
    a sack and carrying the AZDN with it. Subsequent ignition of the explosive mix
    (exactly how is uncertain) caused explosion and a major fire. Though they serve
    the same purpose, oxidants should not be stored with reducants.
    See Azoisobutyronitrile
    See RADICAL INITIATORS

Iron   MRH 0.92/22
    Mellor, 1947, Vol. 10, 470
    Iron exposed to the action of a slightly acid conc. solution of ammonium perox-
    odisulfate dissolves violently.

Other reactants
    Yoshida, 1980, 328
    MRH values calculated for 11 combinations, largely with oxidisable materials, are
    given.

Sodium peroxide   MRH 2.72/40
    See Sodium peroxide: Ammonium peroxodisulfate

Sodium sulphide
    Zhang, G-S. *et al., Loss Prevention and Safety Promotion in the Process Industries*,
    Vol I, (Mewis, J. J., Pasman, H. J. & De Rademaker, E. E. Eds.), 277, Amsterdam,
    Elsevier, 1995
    An account of a serious warehouse explosion (15 dead, 141 injured). The two prin-
    cipal detonations were mostly due to ammonium nitrate, of which some hundred
    tonnes had been present, but the initiating fire was first observed in ammonium
    persulfate. This had been promiscuously stored alongside potassium permanganate,
    matches, potassium nitrate and sodium sulphide (or possibly sulphite), inter alia.
    None of these would improve the safety of ammonium persulfate. It was shown
    that the persulphate gives an immediate exothermic reaction with the sulphide.
    This was ascribed as the ultimate initiation. It was concluded that oxidants and

reducants should not be mixed in storage [many other oxidants, such as permanganates, should not be trusted near a peroxy compound—Ed.] and that storage areas should be better separated one from another.

See also POTASSIUM PERMANGANATE: AMMONIUM NITRATE

Sulfuric acid
See CLEANING BATHS FOR GLASSWARE

Zinc, Ammonia
Rept. HM Insp. Explos., 33, London, HMSO, 1950
The salt exploded during drying, but no cause was determined. However, the reputed explosive character of tetraamminezinc peroxodisulfate, possibly formed from interaction of the ammonium salt, galvanised iron and ammonia, was mentioned as a possible cause.
See Tetraamminezinc peroxodisulfate
See other OXIDANTS, PEROXOACID SALTS

4577. Ammonium sulfide
[12135-76-1] : polysulfide is [12259-52-6] H_8N_2S
$(NH_4)_2S$

Zinc
Anon., *Sichere Chemiearb.*, 1960, **12**(4), 29
A closed zinc container filled with a conc. solution of ammonium sulfide exploded, owing to liberation of hydrogen sulfide and hydrogen, accompanying the formation of zinc ammonium sulfide.
See related METAL SULFIDES

4578. Ammonium tetranitroplatinate(II)
[22289-82-3] (ion) $H_8N_6O_8Pt$
$[NH_4]_2[Pt(NO_2)_4]$

Nilson, L. F., *J. Prakt. Chem. [2]*, 1877, **16**, 249
It decomposes explosively on heating.
See other OXOSALTS OF NITROGENOUS BASES, PLATINUM COMPOUNDS

†4579. Trisilane
[7783-26-8] H_8Si_3
$H_3SiSiH_2SiH_3$

Air, or Carbon tetrachloride, or Oxygen
Mellor, 1940, Vol. 6, 224
It ignites and explodes in air or oxygen, and reacts vigorously with carbon tetrachloride.
See other SILANES

4580. Trisilylamine
[13862-16-3]   H₉NSi₃

$$(H_3Si)_3N$$

Air, or Ammonia, or Hydrogen
  Mellor, 1940, Vol. 8, 262
  The liquid ignites in air, and reacts vigorously with ammonia, hydrogen or moisture.
See related SILANES

4581. Sodium triammine
[84937-00-8]   H₉N₃Na

$$Na(NH_3)_3$$

Winter, R. *et al., Ber. Bunsenges. Phys. Chem.*, 1982, **86**, 1093–1097
The new crystalline solid, metastable in liquid nitrogen, dissociates at ambient temperature and caused several sample tubes to explode from internal pressure of ammonia.
See other GAS EVOLUTION INCIDENTS
See related N-METAL DERIVATIVES

4582. Triamminenitratoplatinum(II) nitrate
[17524-18-4]   H₉N₅O₆Pt

$$[(H_3N)_3PtNO_3]NO_3$$

Mellor, 1942, Vol. 16, 409
It decomposes violently on heating.
See other AMMINEMETAL OXOSALTS, PLATINUM COMPOUNDS

4583. Trisilylphosphine
[15110-33-5]   H₉PSi₃

$$(H_3Si)_3P$$

Amberger, E. *et al., Angew. Chem.*, 1962, **74**, 32–33
The liquid ignites in air.
See related PHOSPHINES, SILANES

4584. Tetrasilane
[7783-29-1]   H₁₀Si₄

$$H(SiH_2)_4H$$

Air, or Carbon tetrachloride, or Oxygen
  Mellor, 1940, Vol. 6, 224

It ignites and explodes in air or oxygen, and reacts vigorously with carbon tetrachloride.
See other SILANES

4585. Tetrasilylhydrazine
[25573-59-5] $H_{12}N_2Si_4$

$$(H_3Si)_2NN(SiH_3)_2$$

Bailar, 1973, Vol. 1, 1377
It explodes in air.
See SILYLHYDRAZINES *See related* SILANES

4586. Tetraamminezinc peroxodisulfate
[39733-13-6] $H_{12}N_4O_8S_2Zn$

$$[(H_3N)_4Zn]S_2O_8$$

Barbieri, G. A. *et al., Z. Anorg. Chem.*, 1911, **71**, 347
It explodes on heating or impact.
See other AMMINEMETAL OXOSALTS

4587. Tetraamminenickel(II) nitrate
[15651-35-1] $H_{12}N_6NiO_6$

$$[(H_3N)_4Ni][NO_3]_2$$

Explosive.
See other AMMINEMETAL NITRATES

4588. Tetraamminepalladium(II) nitrate
[13601-08-6] $H_{12}N_6O_6Pd$

$$[(H_3N)_4Pd][NO_3]_2$$

White, J. H., private comm., 1965
Evaporation of a solution of the salt used for plating gave a moist residue which ignited and burned violently. Reclamation of palladium from such solutions by direct reduction is recommended. It was subsequently shown to be explosive
See other AMMINEMETAL NITRATES *See related* PLATINUM COMPOUNDS

4589. Tetraammineplatinum(II) nitrate (Tetraamminedinitratoplatinum)
[20634-12-2] tetrahedral $H_{12}N_6O_6Pt$

$$[(H_3N)_4Pt][NO_3]_2$$

Similar to the palladium complex (above)
See other AMMINEMETAL NITRATES, PLATINUM COMPOUNDS

4590. Tetraammineplatinum(II) nitrate (Tetraamminedinitratoplatinum)
[133163-34-5] octahedral $H_{12}N_6O_6Pt$

$$O_2NOPt(NH_3)_4ONO_2$$

Similar to the palladium complex (above)
See other AMMINEMETAL NITRATES, PLATINUM COMPOUNDS

4591. Ammonium 2,4,6-tris(dioxoselena)hexahydrotriazine-1,3,5-triide
('Ammonium triselenimidate')
[ ] $H_{12}N_6O_6Se_3$

Explosive.
See Selenium difluoride dioxide: Ammonia
See related N–S COMPOUNDS

4592. Trihydrazinenickel(II) nitrate
[69101-54-8] $H_{12}N_8NiO_6$

$$[(H_4N_2)_3Ni][NO_3]_2$$

Ellern, H. *et al., J. Chem. Educ.*, 1955, **32**, 24
A dry sample exploded violently and a moist sample spontaneouly deflagrated.
See other AMMINEMETAL NITRATES

4593. Amminedecahydroxydiplatinum
[ ] $H_{13}NO_{10}Pt$

$$[H_3NPt_2(OH)_{10}]$$

Jacobsen, J., *Compt. rend.*, 1909. **149**, 574–577
The compound, of unknown structure, explodes violently on heating.
See other PLATINUM COMPOUNDS

4594. Tetraamminehydroxynitratoplatinum(IV) nitrate
[ ] $H_{13}N_7O_{10}Pt$

$$[(H_3N)_4Pt(OH)NO_3][NO_3]_2$$

Mellor, 1942, Vol. 16, 411
It explodes violently on heating.
See other AMMINEMETAL OXOSALTS, PLATINUM COMPOUNDS

4595. Tetraamminelithium dihydrogenphosphide
[44023-01-6] $H_{14}LiN_4P$

$$[(H_3N)_4Li]PH_2$$

Water
Legoux, C., *Compt. rend.*, 1938, **207**, 634
It reacts vigorously with water, evolving phosphine and ammonia which may ignite.
See related METAL NON-METALLIDES

4596. Pentaamminedinitrogenruthenium(II) salts
[19504-40-6] (ion) $H_{15}N7Ru^{2+}$

$$[(H_3N)_5RuN_2]^{2+}$$

Preparative hazard
See Pentaamminechlororuthenium chloride: Sodium azide
Hydrazine: Ruthenium(III) chloride
See other N-METAL DERIVATIVES

4597. Octaammine-μ-hydroxy[μ-superoxido-*O*,*O'*)]dirhodium(4+) nitrate
[113753-12-1] $H_{25}N_{12}O_{15}Rh_2$

$$[(H_3N)_4Rh:OH,O_2:Rh(NH_3)_4] \ [NO_3]_4$$

Springborg, J. *et al., Acta Chem Scand. Sec. A*, 1987, **A41**, 485
The complex may detonate when subjected to mechanical shock.
See other AMMINEMETAL OXOSALTS

4598. Undecaamminetetraruthenium dodecaoxide
[ ] $H_{33}N_{11}O_{12}Ru_4$

$$[H_3N)_{11}Ru_4O_{12}]$$

Preparative hazard
See Ruthenium(VIII) oxide: Ammonia
See related METAL OXIDES

4599. Hafnium
[7440-58-6]   Hf

$$Hf$$

Alone, or Non-metals, or Oxidants
(*MCA SD-52*, 1966)

Although the massive metal is relatively inert, when powdered it becomes very reactive. The dry powder may react explosively at elevated temperatures with nitrogen, phosphorus, oxygen, sulfur and other non-metals. The halogens react similarly, and in contact with hot conc. nitric acid and other oxidants it may explode (often after a delay with nitric acid). The powder is pyrophoric and readily ignitable by friction, heat or static sparks, and if dry burns fiercely. Presence of

water (5–10%) slightly reduces the ease of ignition, but combustion of the damp powder proceeds explosively (the oxygen content of water, 89%, being much higher than that of air). A minimum of 25% water is necessary to reduce handling hazards to a minimum. Full handling precautions are detailed.
See other METALS, PYROPHORIC METALS

4600. Mercury
[7439-97-6]

Hg

Hg

NSC 203, 1976; *HCS 1980*, 620; *RSC Lab. Hazards Data Sheet No. 47*, 1986

Acetylenic compounds
See 3-Bromopropyne: Metals 2.18/99+
2-Butyne-1,4-diol: Alkalies, etc.
Sodium acetylide: Metals
See ACETYLENIC COMPOUNDS: metals

Alkynes, Silver perchlorate
See Silver perchlorate: Alkynes, etc.

Ammonia
1. Sampey, J. J., *Chem. Eng. News.*, 1947, **25**, 2138
2. Braidech, M. M., *J. Chem. Educ.*, 1967, **44**, A324
3. Thodos, G., *Amer. Inst. Chem. Engrs. J.*, 1964, **10**, 274
4. Henderson, L. M., *Ind. Eng. Chem. (News Ed.)*, 1932, **10**, 73
5. Brunt, C. Van, *Science*, N. Y., 1927, **65**, 63–64
A mercury manometer used with ammonia became blocked by deposition of a grey-brown solid, which exploded during attempts to remove it mechanically or on heating. The solid appeared to be a dehydration product of Millon's base and was freely soluble in sodium thiosulfate solution. This method of cleaning is probably safer than others, but the use of mercury manometers with ammonia should be avoided as intrinsically unsafe [1,2]. Although pure dry ammonia and mercury do not react even under pressure at 340 kbar and 200°C, the presence of traces of water leads to the formation of an explosive compound, which may explode during depressurisation of the system [3]. Explosions in mercury–ammonia systems had been reported previously [4,5].
See Poly(dimercuryimmonium hydroxide)
See also FULMINATING METALS
See other CORROSION INCIDENTS

Boron diiodophosphide
See Boron diiodophosphide: Metals

Ethylene oxide   MRH 3.72/99+
See Ethylene oxide(reference 4)

Metals

Bretherick, L., *Lab. Pract.*, 1973, **22**, 533

The high mobility and tendency to dispersion exhibited by mercury, and the ease with which it forms alloys (amalgams) with many laboratory and electrical contact metals, can cause severe corrosion problems in laboratories. A filter–cyclone trap is described to contain completely mercury ejected accidentally by overpressuring of mercury manometers and similar items.

See Aluminium: Mercury
     Calcium: Mercury
     Lithium: Mercury
     Potassium: Mercury
     Rubidium: Mercury
     Sodium: Mercury
See other CORROSION INCIDENTS

Methyl azide
See Methyl azide: Mercury

Methysilane, Oxygen
See Methylsilane: Mercury, etc.

Other reactants

Yoshida, 1980, 210

MRH values calculated for 6 combinations, largely with materials showing catalytic decomposition, are given.

Oxidants   MRH 5.69/99+

See Bromine: Metals
     Chlorine: Metals (reference 1)
     Chlorine dioxide: Mercury
     Nitric acid: Alcohols (reference 2)
     Peroxyformic acid: Metals, etc.

Tetracarbonylnickel, Oxygen   MRH Tetracarbonylnickel 0.88/99+
See Tetracarbonylnickel: Mercury, etc.
See other METALS

4601. Zinc amalgam
    [72780-86-0]   **Hg−Zn**

                              Hg−Zn

Brimelow, H. C., private comm., 1972

Amalgamated zinc residues isolated from Clemmensen reduction of an alkyl aryl ketone in glacial acetic acid were pyrophoric, and had to be immediately dumped into water after filtration to prevent ignition.

See other PYROPHORIC ALLOYS

4602. Mercury(II) iodide
[7774-29-0]

HgI$_2$

HgI$_2$

Chlorine trifluoride
See Chlorine trifluoride: Metals, etc.
See other MERCURY COMPOUNDS, METAL HALIDES

4603. Mercury(II) nitrate
[10045-94-0]

HgN$_2$O$_6$

Hg(NO$_3$)$_2$

HCS 1980, 616

Acetylene
Mellor, 1940, Vol. 4, 993
Contact of acetylene with the nitrate solution gives mercury acetylide, an explosive sensitive to heat, friction or contact with sulfuric acid.

Ethanol
Mellor, 1940, Vol. 4, 993
Addition of mercury(II) nitrate solution to ethanol gives mercury fulminate.

Ferrocene
Sallott, G. P. et al., Proc. Int. Pyrotech. Semin., 1984, 589–602
Compositions prepared from mercury(II) nitrate and ferrocene or its derivatives show promise as explosive priming mixtures, but such mixtures are fairly sensitive to electrostatic initiation and should be handled in the wet state.

Isobutene
See 2-Methyl-1-nitratodimercurio-2-nitratomercuriopropane

Petroleum hydrocarbons
1. Mixer, R. Y., Chem. Eng. News, 1948, 26, 2434
2. Ball, J., Chem. Eng. News, 1948, 26, 3300
Gas oil was stirred vigorously with the finely divided solid nitrate to complete the removal of sulfur compounds. After the mixture had congealed, preventing further stirring, a violent reaction set in which reached incandescence, accompanied by vigorous evolution of nitrous fumes. This was attributed to the self-catalysed nitrating action of mercury nitrate in a semi-solid environment unable to lose heat effectively. A similar occurrence was observed when crushed nitrate was just covered with cracked naphtha [1]. The second publication reveals that this type of hazard was known in cracked distillates containing a high proportion of unsaturates and aromatics, when allowed to stand in prolonged contact with mercury(II) nitrate. The hazard may be avoided by using several small portions of the salt sequentially, and working with 100 g portions of hydrocarbons [2].

Phosphine
    Mellor, 1940, Vol. 9, 993
    Phosphine reacts with the aqueous salt solution to give a complex nitrate–phosphide, which when dry explodes on heating or impact.

Phosphinic acid
    Mellor, 1940, Vol. 4, 993
    Phosphinic ('hypophosphorous') acid violently reduces the salt to the metal.

Potassium cyanide
    Rüst, 1948, 337
    Mixtures exploded when heated, but only if contained in narrow ignition tubes. Formation of nitrite, a more powerful oxidant than nitrate, may have been involved.
    See other MERCURY COMPOUNDS, METAL NITRATES, OXIDANTS

4604. Mercury(II) azide
[14215-33-9]   HgN$_6$

$$Hg(N_3)_2$$

1. Mellor, 1940, Vol. 8, 351; 1967, Vol. 8. Suppl. 2, 43, 50
2. Barton, A. F. M. *et al., Chem. Rev.*, 1973, **73**, 138
3. Heathcock, C. H., *Angew. Chem. (Intern. Ed.)*, 1969, **8** 134
The very high friction-sensitivity, particularly of large crystals, and brisance on explosion are to be expected from the thermodynamic properties of the salt. Its great sensitivity, even under water, renders it unsuitable as a practical detonator [1]. Spontaneous explosions during intercrystalline transformations have been observed, or on crystallisation from hot water [2]. A safe method of preparing solutions in aqueous THF for synthetic purposes is available [3].
    It is strongly endothermic (ΔH°_f (s) +556.5 kJ/mol, 1.96 kJ/g).
See other MERCURY COMPOUNDS, METAL AZIDES

4605. Mercury(II) oxide
[21908-53-2]   HgO

$$HgO$$

HCS 1980, 617

Under appropriate conditions, it can function as a powerful oxidant and/or catalyst, owing to the tendency to dissociate to metal and oxygen.

Acetyl nitrate
    See Acetyl nitrate: Mercury(II) oxide

Butadiene, Ethanol, Iodine
    See 2-Ethoxy-1-iodo-3-butene

Chlorine, Hydrocarbons
See Methane: Halogens, etc.
Ethylene: Chlorine

Diboron tetrafluoride
See Diboron terafluoride: Metal oxides

Disulfur dichloride
See Disulfur dichloride: Mercury oxide

Hydrogen peroxide
See Hydrogen peroxide: Mercury(II) oxide, or: Metals, etc.

Hydrogen trisulfide
See Hydrogen trisulfide: Metal oxides

Chlorine
1. Tabata, Y. et al., J. Haz. Mat., 1987, **17**, 55
2. Shilov, E. A., Compt. Rend. Acad. Sci. USSR, 1945, **46**, 64–65
Mercury(II) oxide reacts with chlorine to form dichlorine oxide which can explode at temperatures close to ambient [1]. This confirms a previous report [2].

Metals
1. Staudinger, H., Z. Elektrochem., 1925, **312**, 549
2. Mellor, 1940, Vol. 4, 272
Mixtures of the red or yellow oxides with sodium–potassium alloy explode violently on impact, the yellow (more finely particulate) oxide giving the more sensitive mixture [1]. Mixtures with magnesium or potassium may explode on heating [2].

Methanethiol
Klason, P., Ber., 1887, **20**, 3410
Interaction is rather violent in absence of diluent.

Non-metals
Mellor, 1940, Vol. 4, 777–778
Mixtures with phosphorus explode on impact or on boiling with water. A mixture with sulfur explodes on heating.

Phospham
See Phospham: Oxidants

Reducants
Mellor, 1940, Vol. 4, 778; Vol. 8, 318
Hydrazine hydrate and phosphinic acid both explosively reduce the oxide when dropped on to it.
See other MERCURY COMPOUNDS, METAL OXIDES, OXIDANTS

4606. Mercury peroxide
[12298-67-8] HgO_2

$$HgO_2$$

Preparative hazard
See Hydrogen peroxide: Mercury(II) oxide, etc.
See other MERCURY COMPOUNDS, METAL PEROXIDES

4607. Mercury(II) sulfide
[1344-48-5] HgS

$$HgS$$

Oxidants
Mellor, 1941, Vol. 2, 242; 1941, Vol. 3, 377; 1940, Vol. 4, 952
The sulfide causes dichlorine oxide to explode, and it incandesces in chlorine.
Grinding with silver oxide ignites the mixture.
See other MERCURY COMPOUNDS, METAL SULFIDES

4608. Poly(dimercuryimmonium permanganate)
[ ] $(Hg_2MnNO_4)_n$

$$(Hg{:}N^+{:}Hg\ MnO_4^-)_n$$

Sorbe, 1968, 97
It is highly explosive.
See other POLY(DIMERCURYIMMONIUM) COMPOUNDS

4609. Mercury(I) nitrate
[10415-75-5] $Hg_2N_2O_6$

$$Hg_2(NO_3)_2$$

Carbon
Mellor, 1940, Vol. 4, 987
Contact with red-hot carbon causes a mild explosion.

Phosphorus
See Phosphorus: Nitrates
See other MERCURY COMPOUNDS, METAL NITRATES

4610. Mercury(I) thionitrosylate
[ ] $(Hg_2N_2S_2)_n$

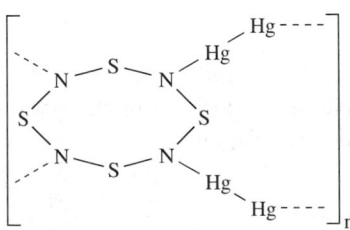

Goehring, M. *et al.*, *Z. Anorg. Chem.*, 1956, **285**, 702

It explodes on heating in a flame. The structure may be the indicated linear polymer of tetrasulfur tetraimidate rings linked by pairs of dimercury bonds.

See other MERCURY COMPOUNDS, N–S COMPOUNDS

4611. Poly(dimercuryimmonium azide)

[ ] $(Hg_2N_4)_n$

$$(Hg{:}N^+{:}Hg\ N_3^-)_n$$

Sorbe, 1968, 97

It is highly explosive.

See other POLY(DIMERCURYIMMONIUM) COMPOUNDS *See related* METAL AZIDES

4612. Mercury(I) azide

[38232-63-2] Hg_2N_6

$$Hg_2(N_3)_2$$

Mellor, 1940, Vol. 8, 351; 1967, Vol. 8, Suppl. 2, 25, 50

It is less sensitive and a less powerful explosive than silver azide or lead azide. It explodes on heating in air to above 270°C, or after an induction period at 140°C in the dark under vacuum.

See other INDUCTION PERIOD INCIDENTS, MERCURY COMPOUNDS, METAL AZIDES

4613. 'Mercury(I) oxide'

[15829-53-5] Hg_2O

$$Hg_2O$$

Sidgwick, 1950, 292

The material is known to be an intimate mixture of mercury(II) oxide and finely dispersed metallic mercury.

Alkali metals
Mellor, 1940, Vol. 4, 771
Interaction with molten potassium or sodium is accompanied by a brilliant light and a light explosion.

Chlorine, Ethylene
See Ethylene: Chlorine

Hydrogen peroxide   MRH 1.59/99+
Antropov, V., *J. Prakt. Chem.*, 1908, **77**, 316
Contact cause explosive decomposition of the peroxide.
See Hydrogen peroxide: Metals, etc.

Non-metals
Mellor, 1940, Vol. 4, 771

Mixtures with phosphorus explode on impact, and that with sulfur ignites on frictional initiation.

See other MERCURY COMPOUNDS, METAL OXIDES

4614. Tris(iodomercurio)phosphine

[ ] Hg_3I_3P

$$(IHg)_3P$$

Nitric acid
   See Nitric acid: Phosphine derivatives
   See related MERCURY COMPOUNDS, PHOSPHINES

4615. Mercury nitride

[12136-15-1] Hg_3N_2

$$Hg_3N_2$$

Alone, or Sulfuric acid.
   1. Mellor, 1940, Vol. 8, 108
   2. Fischer, F. *et al., Ber.*, 1910, **43**, 1469
   3. Wilhelm, S. M., *Plant/Oper. Progr.*, 1991, **10**(4), 192
   It explodes on friction, impact, heating or in contact with sulfuric acid [1]. A sample at below −40°C exploded when disturbed [2]. It may form from the metal and ammonia in some circumstances [3]
   See other FULMINATING METALS, MERCURY COMPOUNDS, *N*-METAL DERIVATIVES

4616. Trimercury tetraphosphide

[12397-29-4] Hg_3P_4

$$Hg_3P_4$$

Oxidants
   Mellor, 1940, Vol. 8, 844
   It ignites when warmed in air, or cold on contact with chlorine. A mixture with potassium chlorate explodes on impact.
   See Potassium chlorate: Metal phosphides
   See other MERCURY COMPOUNDS, METAL NON-METALLIDES

4617. Mercury(I) hypophosphate

[ ] $Hg_4O_6P_2$

Mellor, 1971, Vol. 8, Suppl. 3, 651
Explosive decomposition occurs on heating.
See other MERCURY COMPOUNDS

4618. Potassium iodide
[7861-11-0]   IK

KI

HCS 1980, 770

Charcoal, Ozone
See Ozone: Charcoal, Potassium iodide

Diazonium salts
See DIAZONIUM TRIIODIDES

Diisopropyl peroxydicarbonate
See Diisopropyl peroxydicarbonate (reference 2)

Oxidants
See Bromine pentafluoride: Acids, etc.
    Chlorine trifluoride: Metals, etc.
    Fluorine perchlorate: Alone, etc.
    Trifluoroacetyl hypofluorite
See other METAL HALIDES

4619. Potassium iodate
[7758-05-6] IKO_3

$KOIO_2$

HCS 1980, 769

Charcoal, Ozone
See Ozone:Charcoal, Potassium iodide

Metals and oxidisable derivatives
See METAL HALOGENATES: METALS AND OXIDISABLE DERIVATIVES
Phosphorus: Metal halogenates

Other reactants
Yoshida, 1980, 376
MRH values calculated for 27 combinations with oxidisable materials are given.

Sodium disulfite, Water
Association for Science Education, Internet, 1997
A student, who should have mixed solutions of these two salts, mixed the solids
and added water, which provoked a violent reaction. It is suggested that pupils
should not be allowed access to solid, undiluted, oxidants and reductants at the
same time.
See other IODINE COMPOUNDS, METAL HALOGENATES, OXIDANTS

4620. Potassium periodate
[7790-21-8]   IKO$_4$

KOIO$_3$

HCS 1980, 773

Ammonium perchlorate
See Ammonium perchlorate: Impurities
See other IODINE COMPOUNDS, METAL OXOHALOGENATES, OXIDANTS

4621. Iodine azide
[14696-82-3]   IN$_3$

IN$_3$

1. Hantzsch, A., *Ber.*, 1900, **33**, 525
2. Dehnicke, K., *Angew. Chem. (Intern. Ed.)*, 1976, **15**, 553
3. Dehnicke, K., *Angew. Chem. (Intern. Ed.)*, 1979, **18**, 507–514
4. Klapötke, T. M. *et al., Angew. Chem. (Int.)*, 1993, **32**(2), 275
The isolated solid is a very shock- and friction-sensitive explosive [1], but the preparation and safe handling of dilute solutions in solvents other than ether have been described [2]. The need to use appropriate techniques and precautions when using iodine azide as a reagent is stressed [3]. The purer the more explosive; explosive properties are characterised (lead-block test, etc.) in a footnote to [4].

Sulfur-containing alkenes
Hassner, A. *et al., J. Org. Chem.*, 1968, **33**, 2690
Interaction is accompanied by violent decomposition of the azide.
See other HALOGEN AZIDES

4622. Iodinated poly(sulfur nitride)
[ ]   (IN$_{4.2}$S$_6$)$_n$

(Indeterminate structure)

Banister, A. J. *et al., J. Chem. Soc., Dalton Trans.*, 1980, 937
The polymeric product is potentially explosive.
See other IODINE COMPOUNDS, N–S COMPOUNDS

4623. Sodium iodide
[7681-82-5]   INa

NaI

HCS 1980, 849

Oxidants
See Perchloryl fluoride: Calcium acetylide, etc.
     Perchloric acid: Iodides
See other METAL HALIDES

4624. Sodium iodate
[7681-55-2]   INaO$_3$

NaOIO$_2$

Metals   MRH Magnesium 5.98/27
Webster, H. A. *et al., Rept. AD-782510/2GA*, Springfield (Va.), USNTIS, 1974
The use of mixtures with magnesium in pyrotechnic flares is discussed.
See Potassium: Oxidants   MRH 1.42/31
     Sodium: Iodates   MRH 3.10/42

Other reactants
Yoshida, 1980, 381–382
MRH values calculated for 27 combinations with oxidisable materials are given.
See other IODINE COMPOUNDS, METAL HALOGENATES, OXIDANTS

4625. Iodine
[7553-56-2]   I$_2$

I$_2$

(*MCA SD-43*, 1952); *NSC 457*, 1968; *HCS 1980*, 560

Acetaldehyde
MCA SD-43, 1952
Interaction may be violent.

Acetylene   MRH 8.70/99+
See Acetylene: Halogens

Aluminium, Diethyl ether
Garcia, F. J. A., *Inorg. Chem.*, 1980, **20**, 83
In the preparation of aluminium iodide etherate from the elements in ether, the
aluminium must be as turnings, rather than fine powder, to keep the reaction under
control.
See Aluminium: Halogens

Ammonia
Mellor, 1940, Vol. 8, 605; 1967, Vol. 8, Suppl. 2.2, 416
Ammonia solutions react with iodine (or potassium iodide) to produce highly
explosive addition compounds of nitrogen triiodide and ammonia.
See Nitrogen triiodide–ammonia

Ammonia, Lithium 1-heptynide
  Houben-Weyl, 1977, Vol. 5.2a, 605
  Reaction of iodine with the lithium heptynide in liquid ammonia to give (explosive)
  1-iodoheptyne may lead to formation of nitrogen triiodide as a black precipitate.
  Low temperatures minimise the formation, and it may be destroyed by adding
  sodium ethoxide solution.

Ammonia, Potassium
  See Ammonia: Iodine, Potassium

Butadiene, Ethanol, Mercuric oxide
  See 2-Ethoxy-1-iodo-3-butene

Caesium oxide
  See Caesium oxide: Halogens

Dipropylmercury
  See Dipropylmercury: Iodine

Ethanol, Phosphorus
  1. Read, C. W. W., *School Sci. Rev.*, 1940, **21**(83), 967
  2. Llowarch, D., *School Sci. Rev.*, 1956, **37**(133), 434
  Interaction of ethanol, phosphorus and iodine to form iodoethane was considered
  too dangerous for a school experiment [1]. A safer modification is now available [2].
  See Phosphorus: Halogens

Formamide, Pyridine, Sulfur trioxide
  See Formamide: Iodine, etc.

Halogens, or Interhalogens
  See Bromine trifluoride: Halogens
       Bromine pentafluoride: Acids, etc.
       Chlorine trifluoride: Iodine
       Fluorine: Halogens, etc.

Metal acetylides or carbides   MRH values show % of those
  Several react very exothermally with iodine.
  See Barium acetylide: Halogens
       Calcium acetylide: Halogens   MRH 1.46/20
       Caesium acetylide: Halogens
       Dicopper(I) acetylide: Halogens
       Lithium acetylide: Halogens
       Rubidium acetylide: Halogens
       Strontium acetylide: Halogens   MRH 1.34/31
       Zirconium dicarbide: Halogens

Metals   MRH Lithium 2.01/5, potassium 1.97/24
1. Mellor, 1941, Vol. 2, 469; 1963, Vol. 2, Suppl. 2.2, 1563; 1939, Vol. 9, 379
2. Tarnqvist, E. G. M. *et al., Inorg. Chem.*, 1979, **18**, 1793
Iodine and antimony powder react so violently as to cause ignition or explosion
of the bulk of the mixture. A mixture of potassium and iodine explodes weakly
on impact, while potassium ignites in contact with molten iodine [1]. Interaction of molten iodine with titanium above 113°C under vacuum to form titanium
tetraiodide is highly exothermic and sparks are produced. The preparative technique described permits the progressive reaction of 0.5 g portions of the titanium
powder charged (7.2 g) to minimise hazard [2].
See Aluminium: Oxidants
    Aluminium–titanium alloys: Oxidants
    Hafnium: Alone, etc.
    Sodium: Halogens (reference 5)

Metals, Water
Jackson, H., *Spectrum*, 1969, **7**, 82
Flash-ignition occurs when mixtures of iodine with powdered aluminium, magne-
sium or zinc are moistened with a drop of water.

Non-metals
See Boron: Halogens
    Phosphorus: Halogens

Other reactants
Yoshida, 1980, 372
MRH values calculated for 8 combinations, largely with oxidisable materials, are
given.

Oxygen difluoride
See Oxygen difluoride: Halogens, etc.

Poly(acetylene)
Pekker, S. *et al., Chem. Abs.*, 1983, **99**, 213009
In TGA studies on the decomposition of iodine-doped polyacetylene, at high
heating rates (30°C/min), decomposition becomes explosive at the m.p. of iodine,
113°C. This was attributed to exothermic reaction of liquid iodine with polyacetylene.

Silver azide   MRH 2.05/99+
See Silver azide: Halogens

Sodium phosphinate
See Sodium phosphinate: Oxidants

Tetraamminecopper(II) sulfate, Ethanol
Schwarzenbach, V., *Ber.*, 1875, **8**, 1233
1716

Addition of ethanol to a mixture of iodine and the salt soon led to explosions of variable intensity, involving formation of *N*-iodo derivatives.
See Mercury(II) amide chloride: Halogens

Trioxygen difluoride
See 'Trioxygen difluoride': Various materials
See other HALOGENS, OXIDANTS

4626. *N*,*N*'-Diiodosulfurdiimide
[ ]   I₂N₂S
$$IN=S=NI$$

Seppelt, K. *et al., Angew. Chem. (Int.)*, 1969, **8**, 771
This explodes both on melting and impact. The bromo-analogue is more sensitive, the chloro-compound is not known.
See other N-HALOGEN COMPOUNDS, N–S COMPOUNDS, IODINE COMPOUNDS

4627. Iodine(V) oxide
[12029-98-0]   I₂O₅
$$I_2O_5$$

Aluminium
Ivanov, V. G. *et al., Chem. Abs.*, 1981, **95**, 9423
Combustion of compressed mixtures of metal powder and the oxide is very vigorous, a layer of molten aluminium being produced initially, followed by a jet of flame.

Bromine pentafluoride
See Bromine pentafluoride: Acids, etc.

Hydrazine
See Hydrazine: Iodine pentoxide

Non-metals
Mellor, 1941, Vol. 2, 295
Iodine pentaoxide reacts explosively with warm carbon, sulfur, rosin, sugar or powdered, easily combustible elements
See other HALOGEN OXIDES, IODINE COMPOUNDS, OXIDANTS

4628. Iodine(VII) oxide
[12055-24-2]   I₂O₇
$$I_2O_7$$

Diethyl ether
Mishra, H. C. *et al. J. Chem. Soc.*, 1962, 1195–1196

Washing of the (incompletely characterised) solid with ether occasionally led to explosive decomposition.

See other HALOGEN OXIDES, IODINE COMPOUNDS, OXIDANTS

4629. Phosphorus diiodide triselenide
[30911-12-7] $I_2P_4Se_3$

$$P_4I_2Se_3$$

Nitric acid
  See Nitric acid: Phosphorus compounds
  See related IODINE COMPOUNDS, NON-METAL HALIDES, NON-METAL SULFIDES

4630. Titanium diiodide
[13783-07-8] I_2Ti

$$TiI_2$$

Preparative hazard
  Gibson, 1969, 63
  It may ignite in moist air.
  See Titanium: Halogens (reference 3)
  See other METAL HALIDES, PYROPHORIC MATERIALS

4631. Tungsten diiodide
[13470-17-2] I_2W

$$IWI$$

Preparative hazard
  Mussell, R. D. *et al., Inorg. Chem.*, 1990, **29**(19), 3711
  Explosions sometimes resulted on opening sealed tubes in which complex mixed halide salts of this compound had been prepared. There is no obvious source of pressure in the reaction mixture of dipotassium tetradecachlorotungstate, lithium iodide and potassium iodide.
  See also Tungsten dichloride *See other* METAL HALIDES

4632. Zinc iodide
[10139-47-6] I_2Zn

$$ZnI_2$$

Sulfuric acid
  See Sulfuric acid: Zinc iodide
  See other METAL HALIDES

4633. Nitrogen triodide (Nitrogen iodide)
[13444-85-4]   I₃N

$$NI_3$$

Klapötke, T. M. *et al., Angew. Chem. (Int.)*, 1990, **29**(6), 677
The uncomplexed material has been prepared and is very explosive, decomposing spontaneously at 0°C or below. Like its bromine and chlorine analogues, it is highly endothermic.
See next two below *See other* N-HALOGEN COMPOUNDS, ENDOTHERMIC COMPOUNDS, IODINE COMPOUNDS

4634. Nitrogen triiodide–silver amide
[ ]   I₃N.AgH₂N

$$NI_3.AgNH_2$$

Mellor, 1967, Vol. 8. Suppl. 2, 418
The dry complex may explode.
See other N-HALOGEN COMPOUNDS, IODINE COMPOUNDS, N-METAL DERIVATIVES, SILVER COMPOUNDS

4635. Nitrogen triiodide–ammonia
[34641-74-2], [14014-86-9] (1:1)   I₃N.H₃N

$$NI_3.NH_3$$

Alone, or Halogens, or Oxidants, or Concentrated acids
  1. Mellor, 1940, Vol. 8, 607; 1967, Vol. 8, Suppl. 2, 418
  2. Garner, W. E. *et al., Nature*, 1935, **135**, 832
  3. Taylor, G. N., *Chem. Brit.*, 1981, **17**, 107
  4. Houben-Weyl, 1977, Vol. 5.2a, 605
  5. Bodner, G. M., *J. Chem. Educ.*, 1985, **62**(12), 1107
  6. Bentzinger, von R. *et al., Praxis Naturwiss. Chem.*, 1987, **36**, 37
Readily formed in systems containing ammonia and iodine or some of its derivatives, the addition compound of nitrogen triiodide and ammonia when dry is an extremely sensitive, unstable detonator capable of initiation by minimal amounts of any form of energy (light, heat, sound, nuclear radiation, mechanical vibration), even at sub-zero temperatures, and occasionally even with moisture present. It may be handled cautiously when wet, but heavy friction will still initiate it. It explodes in boiling water and is decomposed by cold water to explosive diiodamine. It explodes, possibly owing to heat-initiation on contact with virtually any conc. acid, with chlorine or bromine, ozone or hydrogen peroxide solution [1]. Crystals desiccated in vacuum spontaneously explode when dry [2]. It has been formed in relatively large amounts during preparation of iodoacetylene derivatives in liquid ammonia [3,4], and may be destroyed by addition of sodium ethoxide [4]. Accidents involving demonstration of the explosive properties have been summarised [5]. A student prepared some to spread about on the floor as a joke, but it exploded

prematurely while still in contact with the solution, causing eye injuries to the joker [6].

See other N-HALOGEN COMPOUNDS, IODINE COMPOUNDS

4636. Phosphorus triiodide
[13455-01-1] I_3P

$$PI_3$$

Hydroxylic compounds, or Oxygen
Leleu, *Cahiers*, 1974, (75), 273
Interaction with water or glycerol is violent, and the iodide ignites in oxygen.

Sulfur acids
Dillon, K. B. *et al., J. Chem. Soc., Dalton Trans.*, 1979, 887–888
Contact with 25 or 50% oleum led to violently exothermic reactions, and vigorous reaction with chlorosulfuric acid.
See Phosphorus tribromide: Sulfur acids
   Phosphorus trichloride: Sulfur acids
See other IODINE COMPOUNDS, HALOPHOSPHINES, NON-METAL HALIDES

4637. Tetraiododiphosphane
[13455-00-0] I_4P_2

$$I_2PPI_2$$

Preparative hazard
See Phosphorus: Halogens (references 3,4)
See other HALOPHOSPHINES, IODINE COMPOUNDS, NON-METAL HALIDES

4638. Titanium tetraiodide
[7720-83-4] I_4Ti

$$TiI_4$$

Preparative hazard
Denes, G., *J. Solid State Chem.*, 1988, **77**(1), 54–59
Use of a 'bent copper tube' reactor to prepare the iodide is hazardous.
See Titanium: Halogens
See other METAL HALIDES

4639. Zirconium tetraiodide
[13986-26-0] I_4Zr

$$ZrI_4$$

Ethanol
Pascal, 1963, Vol. 9, 565

Interaction is very violent.
See other METAL HALIDES

4640. Indium
[7440-74-6]   In

In

Acetonitrile, Dinitrogen tetraoxide
See Dinitrogen tetraoxide: Acetonitrile, etc.

Mercury(II) bromide
See Mercury(II) bromide: Indium

Sulfur
  See Sulfur: Metals
  See other METALS

4641. Indium(II) oxide
[12136-26-4]   InO

InO

Ellern, 1968, 33
The lower oxide, prepared by hydrogenation, incandesces on exposure to air.
See other METAL OXIDES, PYROPHORIC MATERIALS

4642. Indium phosphide
[22398-80-7]   InP

InP

Preparative hazard
  See Phosphorus: Metals
  See METAL PNICTIDES

4643. Iridium
[7439-88-5]   Ir

Ir

The finely divided catalytic metal may be pyrophoric.
See Zinc: Catalytic metals *See entry* HYDROGENATION CATALYSTS

Interhalogens
  See Bromine pentafluoride: Acids, etc.
    Chlorine trifluoride: Metals
  See other METALS, PYROPHORIC METALS

4644. Iridium(IV) oxide
[12030-49-8] **IrO_2**

IrO_2

Peroxyformic acid
See Peroxyformic acid: Metals, etc.
See other METAL OXIDES

4645. Potassium
[7440-09-7] **K**

K

FPA H38, 1975; *HCS 1980*, 759

1. Gilbert, H. N., *Chem. Eng. News*, 1948, **26**, 2604
2. Johnson, W. S. *et al., Org. Synth.*, 1963, Coll. Vol. 4, 134–135
3. *MCA Case History No. 1891*
4. Taylor, D. A. H., *Chem. Brit.*, 1974, **10**, 101
5. Diaper, D. G. M., *Chem. Brit.*, 1974, **10**, 312
5a. Anon., *Fire Precaut.*, 1988, (213), 46
6. Houben-Weyl, 1970, Vol. 13.1, 264
7. Burfield, D. R. *et al., Chem. & Ind.*, 1979, 89
8. Davis, A. C., *Chem. Brit.*, 1979, **15**, 179
9. Levy, J. *et al., Angew. Chem. (Intern. Ed.)*, 1981, **20**, 1033
10. Mellor, 1941, Vol. 2, 468; 1963, Vol. 2, Suppl. 2.2, 1559
11. Mellor, 1941, Vol. 2, 493
12. Short, J. F., *Chem. & Ind.*, 1964, 2132
13. Brazier, A. D., *Chem. & Ind.*, 1965, 220; Balfour, A. E., ibid., 353; Bil, M. S., ibid., 812; Cole, R. J., ibid., 944
14. March, R. G., *Chem. Brit.*, 1979, **15**, 65
15. Brock, T. H. *et al., Nachr. Chem. Tech. Lab.*, 1998, **46**(1), 16
16. Brandsma, 1971, 10, 21

In a review of the comparative properties of sodium and potassium, the latter is rated as invariably the more hazardous [1]. Laboratory procedures for safe handling of potassium have been detailed [2]. A safe method for disposal of potassium residues in bulk storage and processing vessels has been developed. Basically the method involves reaction of the metallic residues with dry (condensate-free) steam under closely controlled conditions. It may be used where small-scale techniques (dissolution in higher alcohols) are inapplicable [3]. Hazards associated with storage of potassium in aluminium containers are discussed. The severe corrosion leading to perforation of such commercially supplied containers [4] was attributed to the deliquescent and subsequently corrosive nature of potassium carbonate formed from atmospheric carbon dioxide inside the container. Sodium may be safely stored in aluminium cans because the derived sodium carbonate is not deliquescent and causes no corrosion [5]. A serious laboratory fire originating in a store cupboard containing potassium in an aluminium container may have arisen through corrosive failure of the latter [5]. There is a considerable risk of fire if powdered potassium ('potassium sand') dispersed in benzene is exposed to air [6]. A practical survey of methods of safe handling and disposal of potassium covers the separation of metal from adhering oxide by melting under xylene containing 1% of isopropanol, disposal of small metal residues by treatment with 1:1 ethyl acetate–hydrocarbon diluent mixtures, and disposal of 30 g blocks by dropping into a 2 m deep hole in the ground containing 0.5 m of water [7]. Several fires (but not explosions) occurred when potassium was 'blotted' free of solvent (used to remove traces of oil) with previously used filter paper. The fires were attributed to rapid formation of potassium hydroxide solution from traces of potassium on the filter paper under humid conditions, and ignition of the dry metal on contact with the wet alkali [8]. Potassium dispersed on silica is relatively air stable [9], unlike the dispersions on sodium carbonate or potassium carbonate which are pyrophoric.

It is convenient to consider interaction of potassium and air under two headings.

Rapid oxidation

Reaction with air or oxygen in complete absence of moisture does not occur, even on heating to the boiling point. However, in contact with normal (moist) air, oxidation may become so fast that melting or ignition occurs, particularly if pressure is applied locally to cause melting and exposure of a fresh surface, as when potassium is pressed through a die to form potassium wire [10].

Slow oxidation

Metallic potassium on prolonged exposure to air forms a coating of yellowish potassium superoxide (KO_2) under which is a layer of potassium oxide in contact with the metal [11]. The previous statements that normal contact of potassium with the superoxide causes ignition to occur [11], and that if the layer of superoxide is impacted into the underlying metal by dry cutting operations or a hammer blow an explosion occurs [1,12] are now known to be incorrect. The explosions are caused by interaction of residual traces of mineral oil or other organic contaminant, rather than potassium metal, with the surface layer of superoxide, initiated by the blade pressure or impact. Cases of flashes of light and fires [13] or explosion [14] when potassium (some were old samples) was cut under oil have been reported. A review with references on explosion hazards of potassium, gives particular attention to oxidised crusts [15].

Fresh potassium should be stored under dry xylene in airtight containers to prevent oxidation [11]. Old stocks, where the coating is orange or yellow, should not be cut, but destroyed by burning on an open coke fire, or by addition of *tert*-butanol to small portions under xylene in a hood. It is dangerous to use methanol or ethanol (either wet or dry) as a replacement for *tert*-butanol. Recommended procedures include cutting and handling the metal with forceps under dry xylene, and disposal of scraps in xylene by addition of *tert*-butanol [1,12,13], even the latter being capable of violent reaction [16]. Alternatively, ethyl acetate–hydrocarbon mixtures may be used for small disposals [7].

See Potassium dioxide (reference 3)
See Potassium–sodium alloy

Acetylene
See Acetylene: Potassium
See other CORROSION INCIDENTS

Alcohols
Pratt, E. F. *et al., J. Amer. Chem. Soc.*, 1954, **75**, 54
Interaction with a range of alcohols (n-propanol to n-octanol, benzyl alcohol, cyclo-hexanol) to form the alkoxides usually led to explosions unless air in the containing vessel was displaced by nitrogen before addition of potassium in small portions with stirring.
See Slow oxidation, above

Aluminium, Air
Anon., *Fire Prevention*, 1988, (213), 46
After a fire which started in a disused fume cupboard employed as a storage space for various inorganic chemicals, including potassium, it is suggested that the potassium oxides which inevitably coat aged potassium may have corroded the aluminium can, releasing the oil in which the metal is stored and exposing it to accelerated oxidation. This argument is thermodynamically sound, but kinetically questionable as it involves a solid/solid interaction of potassium and aluminium oxides.

Carbon
1. Mellor, 1963, Vol. 2, Suppl. 2.2, 1566
2. Mellor, D. P., *Chem. & Ind.*, 1965, 723
3. *Alkali metals*, 1957, 169
Reaction of various forms of carbon (soot, graphite or activated charcoal) is exothermic and vigorous at elevated temperatures, and if the carbon is finely divided and air is present, ignition may occur leading to explosions, possibly owing to the potassium superoxide which would be formed [1]. Explosions caused by attempts to extinguish potassium fires with graphite powder have been so attributed [2]. Potassium cannot be produced by electrolysis of potassium chloride because of interaction of the metal with graphite electrodes and formation of explosive 'carbonylpotassium' (potassium benzenehexoxide) [3].

Carbon disulfide
See Carbon disulfide: Alkali metals

Dimethyl sulfoxide
See Dimethyl sulfoxide: Potassium

Ethylene oxide   MRH 3.72/99+
See Ethylene oxide: Contaminants

Halocarbons   MRH values below references
1. Staudinger, H., *Z. Angew. Chem.*, 1922, **35**, 657
2. Staudinger, H., *Z. Elektrochem.*, 1925, **31**, 549

1724

3. Lenze, F. *et al. Chem. Ztg.*, 1932, **56**, 921
4. Rampino, L. D., *Chem. Eng. News*, 1958, **36**(32), 62
MRH Bromoform 3.14/68, carbon tetrachloride 5.19/50, chloroform 4.93/50, pentachloroethane 4.98/51, tetrachloroethane 4.77/52, tetrachloroethylene 5.23/51
Although apparently stable on contact, mixtures of potassium (or its alloys) with a wide range of halocarbons are shock-sensitive and may explode with great violence on light impact. Chloroethane, dichloroethane, trichloroethane, pentachloroethane, bromoform, dibromomethane and diiodomethane are among those investigated. Sensitivity increases generally with the degree of substitution, and potassium–sodium alloy gives extremely sensitive mixtures. The mixture with carbon tetrachloride is 150–200 times as shock-sensitive as mercury fulminate, and a mixture of potassium with bromoform was exploded by a door slamming nearby [1,2]. Mixtures with tetrachloroethane and pentachloroethane will often explode after a short delay during which a voluminous solid separates out [3]. When heated together, potassium and tetrachloroethylene exploded at 97–99°C, except when the metal had been very recently freed of its usual oxide film. Sodium did not explode under the same conditions [4].
See Oxalyl dihalides, below

Halogens or Interhalogens   MRH Bromine 3.22/68, chlorine 5.77/48
1. Mellor, 1941, Vol. 2, 114, 469; 1963, Vol. 2, Suppl. 2.2, 1563
2. Pascal, 1960, Vol. 16, 578
Potassium ignites in fluorine and in dry chlorine (unlike sodium). In bromine vapour it incandesces, and explodes violently in liquid bromine. Mixtures with iodine incandesce on heating, and explode weakly on impact. Potassium reacts explosively with molten iodine bromide and iodine, and a mixture with the former is shock-sensitive and explodes strongly. Molten potassium reacts explosively with iodine pentafluoride [1]. Contact with iodine trichloride causes ignition [2].

Hydrazine
See Hydrazine: Potassium

Hydrogen halides   MRH Hydrogen chloride 3.72/48
1. Cueilleron, J., *Bull. Soc. Chim. Fr.*, 1945, **12**, 88
2. Pascal, 1963, Vol. 2.2, 31
Impact causes a mixture of potassium and anhydrous hydrogen chloride to explode very violently [1]. Molten potassium ignites in contact with hydrogen chloride, hydrogen bromide or hydrogen iodide [2].

Magnesium halides, Potassium iodide
Rieke, R. D. *et al., Org. Synth.*, 1979, **59**, 85–94
Interaction of potassium with magnesium bromide, magnesium chloride or magnesium iodide in refluxing THF produces very finely divided and highly reactive magnesium which will ignite if exposed long to air.
See PYROPHORIC METALS

Maleic anhydride
See Maleic anhydride: Bases, etc.

Mercury
Mellor, 1941, Vol. 2, 469
Interaction to form amalgams is vigorously exothermic and may become violent if too much potassium is added at once.

Metal halides
1. Staudinger, H., *Z. Elektrochem.*, 1925, **31**, 549
2. Cueilleron, J., *Bull. Soc. Chim. Fr.*, 1945, **12**, 88
3. Mellor, 1963, Vol. 2, Suppl. 2.2, 1571
4. Rieke, R. D., *Accts. Chem. Res.*, 1977, **10**, 301–305
Cobalt(II) chloride 2.59/62, iron(II) bromide 1.80/74, iron(III) bromide 1.88/72, iron(III) chloride 3.22/58, iron(II) iodide 1.38/80, aluminium chloride 2.43/53, copper(I) chloride 2.18/72, copper(II) chloride 3.10/63, silver fluoride 2.18/76
Mixtures of potassium with metal halides are sensitive to mechanical shock, and the ensuing explosions have been graded. Very violent explosions occurred with calcium bromide, iron(III) bromide, iron(III) chloride, iron(II) bromide, iron(II) iodide or cobalt(II) chloride. Strong explosions occurred with silver fluoride, mercury(II) bromide, mercury(II) chloride, mercury(II) fluoride, mercury(II) iodide, copper(I) bromide, copper(I) chloride, copper(I) iodide, copper(II) bromide, copper(II) chloride, and ammonium tetrachlorocuprate; zinc and cadmium chlorides, bromides and iodides; aluminium fluoride, chloride, or bromide, thallium(I) bromide; tin(II) or (IV) chloride, tin(IV) iodide (with sulfur), arsenic trichloride and triiodide, antimony and bismuth tribromides, trichlorides and triiodides; vanadium(V) chloride, chromium(IV) chloride, manganese(II) and iron(II) chlorides, and nickel bromide, chloride or iodide. Weak explosions occurred with a wide variety of other halides [1,2,3]. These reactions may be moderated by the use of ether or hydrocarbon solvents. The reactions are still highly exothermic but controllable, and give finely divided and highly reactive metals, some pyrophoric in air, such as magnesium and aluminium [4].
See Magnesium halides, etc., above

Metal oxides
1. Mellor, 1963, Vol. 2, Suppl. 2.2, 1571
2. Mellor, 1941, Vol. 3, 138
3. Mellor, 1940, Vol. 4, 770, 779
4. Mellor, 1941, Vol. 7, 401
5. Mellor, 1943, Vol. 11, 237, 542
6. Mellor, 1941, Vol. 9, 649
Lead peroxide reacts explosively [1] and copper(II) oxide incandescently [2] with warm potassium. Mercury(II) and (I) oxides react with molten potassium with incandescence and explosion, respectively [3]. Tin(IV) oxide is reduced incandescently on warming [4] and molybdenum(III) oxide on heating [5]. Warm bismuth trioxide is reduced with incandescence [6].
See 'Mercury(I) oxide': Alkali metals

Nitric acid
Pascal, 1963, Vol. 2.2, 31
Interaction with conc. nitric acid is explosive.

Nitrogen-containing explosives
Staudinger, H., *Z. Elektrochem.*, 1925, **31**, 549
Nitro or nitrate explosives, normally shock-insensitive, are rendered extremely sensitive by addition of traces of potassium or potassium–sodium alloy. Ammonium nitrate, and nitrate–sulfate mixtures, picric acid, and even nitrobenzene respond in this way.

Non-metal halides
1. Mellor, 1963, Vol. 2, Suppl. 2.2, 1564–1568
2. Mellor, 1940, Vol. 8, 1006
3. Mellor, 1947, Vol. 10, 642–646, 908, 912
Diselenium dichloride and seleninyl chloride both explode on addition of potassium [1,3], while the metal ignites in contact with phosphorus trichloride vapour or liquid [2]. Mixtures of potassium with sulfur dichloride or sulfur dibromide, phosphorus tribromide or phosphorus trichloride, and with phosgene are shock-sensitive, usually exploding violently on impact. Potassium also explodes violently on heating with disulfur dichloride, and with sulfur dichloride or seleninyl bromide without heating [3].

Non-metal oxides
1. Gilbert, H. N., *Chem. Eng. News*, 1948, **26**, 2604
2. Mellor, 1941, Vol. 2, 241
3. Mellor, 1940, Vol. 8, 436, 544, 554, 945
4. Pascal, 1963, Vol. 2.2, 31
MRH Dinitrogen oxide 3.72/54, nitrogen oxide 4.60/61
Mixtures of potassium and solid carbon dioxide are shock-sensitive and explode violently on impact, and carbon monoxide readily reacts to form explosive 'carbonylpotassium' (potassium benzenehexoxide) [1]. Dichlorine oxide explodes on contact with potassium [2]. Potassium ignites in dinitrogen tetraoxide or dinitrogen pentaoxide at ambient temperature and incandesces when warmed with nitrogen oxide or phosphorus(V) oxide [3]. At −50°C, potassium and carbon monoxide react to give dicarbonylpotassium, which explodes in contact with air or water, or at 100°C. At 150°C, the product is a trimer of this, potassium benzenehexoxide. The just-molten metal ignites in sulfur dioxide [4].

Organic samples for qualitative analysis
See LASSAIGNE TEST (reference 2)

Other reactants
Yoshida, 1980, 96–97
MRH values calculated for 49 combinations with a wide variety of other materials are given.

Oxalyl dihalides
Staudinger, H., *Z. Angew. Chem.*, 1922, **35**, 657; *Ber.*, 1913, **46**, 1426
In absence of mechanical disturbance, potassium or potassium–sodium alloy appears to be stable in contact with oxalyl dibromide or oxalyl dichloride, but the mixtures are very shock-sensitive and explode very violently.
See Halocarbons, above

Oxidants
Mellor, 1963, Vol. 2, Suppl. 2.2, 1571
The potential for violence of interaction between the powerful reducing agent potassium and oxidant classes has been well described. Other miscellaneous oxygen-containing substances which react violently or explosively include sodium iodate, silver iodate, lead sulfate and boric acid.
See Metal oxides, *also* Non-metal oxides, both above
Selenium, etc., *also* Sulfur, *and* Sulfuric acid, all below
Ammonium nitrate: Ammonium sulfate, etc.
Chlorine trifluoride: Metals
Chromium trioxide: Alkali metals
Dichlorine oxide: Oxidisable materials
Nitryl fluoride: Metals

Selenium, or Tellurium
Mellor, 1947, Vol. 10, 767; 1943, Vol. 11, 40
Interaction of selenium and potassium is exothermic and ignition occurs, or, with excess potassium, a slight explosion. Tellurium and potassium become incandescent when warmed in a hydrogen atmosphere to prevent aerobic oxidation.

Sulfur
Pascal, 1963, Vol. 2.2, 30
Interaction is violent on warming.

Sulfuric acid   MRH 3.57/56
Kirk-Othmer, 1961, Vol. 16, 362
Interaction is explosive.

Water
1. Mellor, 1941, Vol. 2, 469; 1963, Vol. 2, Suppl. 2.2, 1560
2. Angus, L. H., *School Sci. Rev.*, 1950, **31**(115), 402
3. Markowitz, M. M., *J. Chem. Educ.*, 1963, **40**, 633–636
4. Ashman, A., *School Sci. Rev.*, 1983, **65**(231), 387
Interaction is violently exothermic, and the heat evolved with water at 20°C is enough to ignite evolved hydrogen. Larger pieces of potassium invariably explode in water and scatter burning potassium particles over a wide area. Aqueous alcohols should not be used for waste metal disposal [1]. Small pieces of potassium will also explode with a restricted amount of water [2]. The reactivity of potassium and other alkali metals with water has been discussed in detail. The vigour of

reaction is scarcely reduced by contact with ice–water slurries [3]. A convenient demonstration of ignition of potassium by placing a small piece on an ice cube is due to Faraday [4].

See ALKALI METALS *See other* METALS, REDUCANTS

4646. Potassium–sodium alloy
[12532-69-3]   K–Na

K–Na

1. *Health and Safety Information*, 251, Washington, USAEC, 1967
2. Ellis, J. E. *et al., Inorg. Synth.*, 1976, **16**, 70
3. Mellor, 1961, Vol. 2, Suppl. 2.1, 562
4. Anon., *Fire Prot. Assoc. J.*, 1965, **68**, 1402

Appropriate precautions in handling the alloy, used as a thermally stable and radiation-resistant liquid coolant for reactor cores, are described. It is generally more hazardous than either of the component metals, because alloys in the range 50–80 wt% of potassium are liquid at ambient temperature and can therefore come into more intimate contact with reagents than the solid metals. Most of the entries under Potassium above, and some of those under Sodium may be expected to apply to their alloys, with allowance for the composition [1]. For the destruction of residual small amounts of the alloy, treatment with a 1:1 mixture of isopropanol and heptane over a safety tray is recommended [2]. The alloy usually ignites on exposure to air, with which it reacts in any case to form potassium dioxide (or superoxide), a very powerful oxidant [3]. Fires may be extinguished with dried sodium carbonate, calcium carbonate, sand or resin-coated sodium chloride. Graphite is not suitable as violent reaction with the superoxide is possible. Carbon dioxide or halocarbon extinguishers must not be used, to avoid explosions [4].

See Carbon dioxide, *also* Halocarbons, both below
See Potassium dioxide: Carbon

tert-Butanol
Renfrew, M. M., *J. Chem. Educ.*, 1983, **60**, A229
Contact of the alloy with *tert*-butanol caused ignition.
See reference 2, main entry above

Carbon dioxide
Staudinger, H., *Z. Elektrochem.*, 1925, **31**, 549
Mixtures of the alloy and solid carbon dioxide are powerful explosives, some 40 times more sensitive to shock than mercury fulminate.

Carbon disulfide
See Carbon disulfide: Alkali metals

Fluoropolymers
1. *491M*, 1975, 394
2. Kimura, T. *et al., J. Chem. Educ.*, 1973, **50**, A85

Poly-tetrafluoroethylene or -hexafluoropropylene sealing tapes burned vigorously in contact with the alloy in a helium atmosphere [1]. A teflon-coated magnetic stirrer bar used to stir the alloy under propane atmosphere ignited when the speed was increased and generated enough heat to melt the glass. Triboelectric initiation was postulated [2].
See other GLASS INCIDENTS

Halocarbons
1. Staudinger, H., *Z. Elektrochem.*, 1925, **31**, 550
2. *Inform. Exch. Bull.*, Lawrence Radiation Lab., 1961, **1**, 22
3. Schmidt, E. *et al., Chem. Abs.*, 1975, **83**, 163034
4. *491M*, 1975, 394
The liquid alloy gives mixtures with halocarbons even more shock-sensitive than those with potassium. Highly chlorinated methane derivatives are more reactive than those of ethane, often exploding spontaneously after a delay [1]. Contact of 1,1,1-trichloroethane with a trapped alloy residue in a valve caused an explosion [2]. It is to be expected that chlorofluorocarbons will also form hazardous mixtures in view of their reactivity with barium. Precautionary measures for demonstrating the explosion of the alloy with chloroform are detailed [3]. Addition of 2 drops of 1,1,2-trichlorotrifluoroethane to the alloy caused a violent explosion [4].
See Fluoropolymers, above
See other METAL–HALOCARBON INCIDENTS

Metal halides
1. Rieke, R. D., *Accts. Chem. Res.*, 1977, **10**, 302
2. Ellern, 1968, 43
Use of the alloy to reduce metal halides in solvents to the finely divided and highly reactive metals is not recommended for cases where the halide is highly soluble in the solvent (e.g. zinc chloride or iron(III) chloride in THF). Explosive reaction may ensue [1]. The alloys explode violently in contact with silver halides.
See Potassium: Metal halides

Metal oxides
Staudinger, H., *Z. Elektrochem.*, 1925, **31**, 551
Mixtures of the alloy with silver oxide or mercury oxide are shock-sensitive powerful explosives. The red form of mercury(I) oxide gives mixtures 40 times, and the yellow form 140 times as sensitive as mercury fulminate.

Nitrogen-containing explosives
See Potassium: Nitrogen-containing explosives

Water
Anon., *Loss Prev. Bull.*, 1977, (018), 18
The liquid alloy was used in an unlabelled Dreschel bottle to dry inert gas. A violent explosion occurred when the bottle was being cleaned, owing to contact of the alloy with water. Other desiccants are much safer (and would avoid the possibility of mistaking the alloy for mercury).
See other ALLOYS, PYROPHORIC MATERIALS

4647. Potassium permanganate
  [7722-64-7]   KMnO$_4$
                              KMnO$_4$

FPA H28, 1974; *HCS 1980*, 774

Acetic acid   MRH 2.34/84
or Acetic anhhydride   MRH 2.55/86
  von Schwartz, 1918, 34
  Cooling is necessary to prevent possible explosion from contact of potassium
  permanganate (or the sodium or calcium salts) with acetic acid or acetic anhydride.
  See Oxygenated organic compounds, below

Acetone, *tert*-Butylamine   MRH Acetone 2.76/8
  1. Kornblum, N. *et al., Org. Synth.*, 1973, Coll. Vol. 5, 847
  2. Kornblum, N. *et al., Org. Synth.*, 1963, **43**, 89
  3. Haynes, R. K. *et al., Lab. Equip. Digest*, 1974, **12**(6), 98; private comm., 1974
  In the reprinted description [1] of a general method of oxidising *tert*-alkylamines
  to the corresponding nitroalkanes, a superscript reference indicating that *tert*-
  butylamine should be oxidised in water alone, rather than in acetone containing
  20% of water, is omitted, although it was present in the original description [2].
  This appears to be important, because running the reaction in 20% aqueous acetone
  led to a violent reaction with eruption of the flask contents. This was attributed to
  caking of the solid permanganate owing to inadequate agitation, and onset of an
  exothermic reaction between oxidant and solvent [3].

Alcohols, Nitric acid
  Numjal, N. L. *et al., Amer. Inst. Aero. Astronaut. J.*, 1972, **10**, 1345
  Methanol, ethanol, isopropanol, pentanol or isopentanol do not ignite immediately
  upon mixing with red fuming nitric acid, but addition of potassium permanganate
  (20%) to the acid before mixing causes immediate ignition.
  See Nitric acid: Alcohols

Aluminium carbide
  Mellor, 1946, Vol. 5, 872
  Incandescence on warming.

Ammonia, Sulfuric acid
  Mellor, 1941, Vol. 1, 907
  Ammonia is oxidised with incandescence in contact with the permanganic acid
  formed in the mixture.

Ammonium nitrate
  Urbanski, 1965, Vol. 2, 491
  A mixture of 0.5% of potassium permanganate with an ammonium nitrate explosive caused an explosion 7 h later. This was owing to formation and exothermic
  decomposition of ammonium permanganate, leading to ignition.

Ammonium perchlorate
See Ammonium perchlorate: Impurities

Antimony, or Arsenic   MRH 0.25/86, 0.25/91, resp.
Mellor, 1942, Vol. 12, 322
Antimony ignites on grinding in a mortar with the solid oxidant, while arsenic explodes.

Coal, Peroxomonosulfuric acid
Rawat, N. S., *Chem. & Ind.*, 1976, 743
In a new method for determination of sulfur in coal, the samples are oxidised with an aqueous mixture of permanganate and the peroxoacid. During the digestion, a reflux condenser is essential to prevent loss of water, which could lead to explosively violent oxidation.

Dichloromethylsilane
See Dichloromethylsilane: Oxidants

Dimethylformamide
Finlay, J. B., *Chem. Eng. News*, 1980, **58**(38), 65
Addition of potassium permanganate to dimethylformamide to give a 20% (approx. saturated) solution led to an explosion after 5 min. Subsequent tests on 1 g of oxidant with 5 g of solvent showed a rapid exotherm after 3–4 min, accompanied by popping noises from undissolved oxidant.
See Oxygenated organic compounds, below; *also* Dimethylformamide: Oxidants

Dimethyl sulfoxide   MRH 3.01/86
Kirk-Othmer (3rd ed.), 1982, Vol. 19, 494
The solvent ignites in contact with the solid oxidant.
See Oxygenated organic compounds, below

Ethanol, Sulfuric acid
Koch, K. R., *J. Chem. Educ.*, 1982, **58**, 973–974
In a demonstration of the powerful oxidant effect of manganese heptoxide on ethanol layered on top of sulfuric acid, it is essential to observe all the precautions given to prevent violent exothermic reactions.

Formaldehyde
1. Piefel, W. *et al., Chem. Abs.*, 1977, **86**, 60508
2. Robinson, P. J., *Chem. & Ind.*, 1978, 723
3. Wainmann, H. E., *Chem. & Ind.*, 1978, 744
4. Craven, J. R., *Chem. Brit.*, 1979, **15**, 66
5. Young, J. A., *CHAS Notes*, 1988, **6**(3), 6
Formaldehyde gas for disinfection purposes may be released from the aqueous solution ('formalin') by treatment with potassium permanganate (the heat for evaporation arising from the redox reaction), but the quantities must be limited to avoid

the risk of fire or explosion [1]. There is an account of an incident involving an extremely exothermic reaction overnight after addition of 0.1 l of formalin to 50 g of oxidant in a plastic beaker, which was melted [2]. Electrically heated evaporation of formalin is a safer fumigation technique [3]. The wisdom of using plastic containers for any unsupervised exothermic reaction is questioned [4]. An attempt to disinfect a large building and its hundreds of incubators all at once by running formalin from a 180 l drum onto several kg of potassium permanganate in a small waste container led, predictably, to a large fire which destroyed the building and its contents [5].

Glycerol

1. 'Leaflet No. 5', London, Inst. of Chem., April, 1940
2. Anon., *BCISC Quart. Safety Summ.*, 1974, **45**, 11–122
3. Bozzelli, J. W. *et al., J. Chem. Educ.*, 1979, **56**, 675–676
4. Kauffman, G. B. *et al., J. Chem. Educ.*, 1981, **58**, 802

Contact of glycerol with solid potassium permanganate caused a vigorous fire [1]. During the preparation of a solution for the decontamination of tetramethyllead spills, addition of the solid oxidant to a bucket contaminated with glycerol caused ignition to occur after a few s [2]. The trifunctionality and high viscosity of glycerol give high heat release and poor dissipation of reaction heat, respectively. The combination is a reliable method of igniting thermite mixture demonstrations [3], which should be conducted out of doors [4].
See Oxygenated organic compounds, below

Hydrochloric acid   MRH 0.75/63

1. Curry, J. C., *School Sci. Rev.*, 1965, **46**(160), 770
2. Adair, A., *Chem. & Ind.*, 1965, 1723
3. Jenkins, E. W., private comm., 1976
4. Ephraim, 1939, 162

During preparation of chlorine by addition of the conc. acid to solid permanganate, a sharp explosion occurred on one occasion [1]. Sulfuric acid was not used in error, nor was tube blockage involved [2]. A similar incident was reported later [3]. It appears remotely possible that permanganate may be able to oxidise chlorine to chlorine oxide [4], which as a dilute mixture with chlorine would be mildly explosive in the gas phase.

Hydrofluoric acid

Black, A. M. *et al., J. Chem. Soc., Dalton Trans.*, 1974, 977

During preparation of potassium hexafluoromanganate(IV), addition of the solid oxidant to exceptionally conc. hydrofluoric acid (60–90%, rather than 40% previously used) caused a violent exotherm with light emission.

Hydrogen peroxide

Anon., *J. Pharm. Chim.*, 1927, **6**, 410

Contact of hydrogen peroxide from a broken bottle with pervious packages of permanganate caused a violent reaction and fire.
See Hydrogen peroxide: Metals, etc. (reference 3)

Hydrogen trisulfide
Mellor, 1947, Vol. 10, 159
Contact with solid permanganate ignites the liquid sulfide.

Hydroxylamine   MRH 3.85/99+
See other REDOX REACTIONS

Non-metals   MRH values below reference
Mellor, 1942, Vol. 12, 319–323
MRH Carbon 2.59/7, phosphorus (yellow) 6.88/62, (red) 2.34/16, sulfur 2.80/12
    A mixture of carbon and potassium permanganate is not friction-sensitive, but
burns vigorously on heating. Mixtures with phosphorus or sulfur react explosively
on grinding or heating, respectively.

Organic nitro compounds
Blinov, I., J. Chem. Ind. USSR, 1937, 14, 1151–1153
Mixtures ignite easily on heating, shock, or contact with sulfuric acid.

Other reactants
Yoshida, 1980, 90
MRH values calculated for 26 combinations with oxidisable materials are given.

Oxygenated organic compounds
    1. Rathsbury, H. et al., Chem. Ztg., 1941, 65, 426–427
    2. Gallais, 1957, 697
    3. Partington, 1967, 830–831
Contact of several liquid organic compounds, or of aqueous solutions of solids,
with the powdered oxidant leads to ignition. Presence of hydroxy and /or
keto groups causes an extraordinary increase in sensitivity. Such compounds
include ethylene glycol, propane-1,2-diol, erythritol, mannitol, triethanolamine, 3-
chloropropane-1,2-diol; acetaldehyde, isobutyraldehyde, benzaldehyde; acetylace-
tone; esters of ethylene glycol, lactic acid, acetic acid, oxalic acid. The necessary
presence of hydroxy or keto groups may be connected with solubility factors [1].
Mixtures of the solid oxidant with solid reducing sugars may react violently or
explosively [2]. Powdered oxalic acid and the oxidant ignite soon after mixing [3].
See Ethylene glycol: Oxidants

Polypropylene
MCA Case History No. 1842
While using a screw conveyor to move the solid oxidant, ignition of a polypropylene tube in the feed system occurred. This could not be reproduced, even when
likely contaminants were present.

Potassium chloride, Sulfuric acid
491M, 1975, 347
An attempt to prepare permanganyl chloride by cautiously adding conc. sulfuric
acid to an intimate mixture of the salts at 0°C in clean glass apparatus caused a
violent explosion.

Reducants
Interaction may be violent or explosive.
See Formaldehyde, above
 Hydrazine: Oxidants
 Hydroxylamine: Oxidants

Slag wool
Hallam, B. J. *et al., School Sci. Rev.*, 1975, **56**(197), 820
Generation of oxygen by heating a layer of potassium permanganate retained in a test tube by a plug of 'Rocksil' slag wool led to minor explosions. The wool liberated an organic distillate on heating, and the explosions were attributed to combustion of the distillate in the liberated oxygen. Roasting the wool before using it with oxidants is recommended.

Sulfuric acid, Water
1. Archer, J. R., *Chem. Eng. News*, 1948, **26**, 205
2. Anon., *ABCM Quart. Safety Summ.*, 1946, **17**, 2
3. Stevens, G. C., private comm., 1971
4. Haight, G. P. *et al., Chem. Eng. News*, 1980, **58**(13), 3
5. Mann, L. T., *Chem. Eng. News*, 1980, **58**(18), 2
6. Gulick, W. M., *Chem. Eng. News*, 1980, **58**(28), 5
7. Olley, R. H., *J. Polym. Sci. Polym. Phys. Edn.*, 1979, **176**, 627
8. Olley, R. H., *Chem. Brit.*, 1988, **24**, 31
9. Anon., *Polymer*, 1982, **23**, 1707
Addition of conc. sulfuric acid to the slightly damp permanganate caused an explosion. This was attributed to the formation of permanganic acid, dehydration to dimanganese heptoxide and explosion of the latter, caused by heat liberated from interaction of sulfuric acid and moisture [1]. A similar incident was reported previously when a solution of potassium permanganate in sulfuric acid, prepared as a cleaning agent, exploded violently [2]. There is, however a reputedly safe procedure for the final ultra-cleaning of glassware in cases where the adsorbed chromium film left by chromic acid cleaning would be intolerable. This involves treatment of the glassware with conc. sulfuric acid (10 ml) to which one or two small crystals (not more) of permanganate have been added. If the solution changes to a brown colour, it must immediately be discarded into excess water [3]. A lecture demonstration of the 'permanganate volcano' involving preparation of manganese heptoxide exploded with great violence. It was concluded that the demonstration was too dangerous to use [4]. Similar comments followed [5,6]. When preparing a 7% solution of potassium permanganate in conc. sulfuric acid as an etchant for polyalkenes, it is essential to use only small quantities and to stir the acid rapidly while adding the oxidant and until it is completely dissolved [7]. An attempt to do this with inadequate instructions (3 l of acid; no agitation) led to a violent explosion which demolished a fume cupboard [8]. A safer, more dilute etchant solution has been described [9].

Titanium
Mellor, 1941, Vol. 7, 20
A mixture of powdered metal and oxidant explodes on heating.

Trifluoroacetic anhydride
1. Perrin, D. D. *et al.*, *Purification of Laboratory Chemicals*, 448, Sydney, Pergamon, 2nd ed., 1980
2. Mook, R. A., *Chem. Eng. News*, 1987, **65**(13), 2
3. Author's comments
A bottle of the anhydride had been purified uneventfully by distillation from potassium permanganate according to a published procedure [1]. When a second batch was attempted, a brown colour was observed in the distillation flask and the distillate was green, so distillation was stopped. Several hours later the distillation flask exploded spontaneously and with great violence. This was attributed to the formation of dimanganese heptaoxide (which caused the brown colouration). Trifluoroacetic acid may also explode if the same purification procedure is followed [2]. It seems likely that the first bottle of anhydride had a low content of trifluoroacetic acid, so that little of the dimanganese heptaoxide would be produced by liberation of permanganic acid and its subsequent dehydration by the anhydride. If the second sample of anhydride had an appreciably higher acid content, a substantial amount of the explosive heptaoxide could have been produced during the several hours standing by the progressive reaction sequence below. [3]

$$2F_3CCO.OH + 2KMnO_4 \rightarrow 2HMnO_4 + 2F_3CO.OK$$
$$2HMnO_4 + (F_3CCO.)_2O \rightarrow Mn_2O_7 + 2F_3CCO.OH$$

3,4,4'-Trimethyldiphenyl sulfone
Adduci, J. M. *et al.*, *Macromol. Synth.*, 1982, **8**, 53–56
During oxidation of the sulfone to the tricarboxylic acid, addition of oxidant must be in small portions to avoid an exothermic vigorous reaction.

Wood   MRH Cellulose 2.68/15
1. Anon., *Chem. Ztg.*, 1927, **51**, 221
2. Dunn, B. W., *Amer. Rail Assoc., Bur Explos., Rept. No. 10*, 1917
Contact between the solid oxidant and wood, in presence of either moisture [1] or mechanical friction [2] may cause a fire.
See other METAL OXOMETALLATES, OXIDANTS

4648. Potassium thiazate
[73400-02-9]   **KNOS**
                                    KN:S:O

Alone, or Water
Armitage, D. A. *et al.*, *J. Chem. Soc., Chem. Comm.*, 1979, 1078–1079
It explodes upon ignition dry, and ignites in contact with water.
See other N-METAL DERIVATIVES, N–S COMPOUNDS

4649. Potassium nitrite
[7758-09-0]   KNO$_2$

KON:O

HCS 1980, 772

Ammonium salts
  Mellor, 1941, Vol. 2, 702
  Addition of ammonium sulfate to the fused nitrite causes effervescence and ignition.
  See NITRITE SALTS OF NITROGENOUS BASES

Boron
  Mellor, 1946, Vol. 5, 16
  Addition of boron to the fused nitrite causes violent decomposition.

Potassium amide
  Bergstrom, F. W., *Chem. Rev.*, 1933, **12**, 64
  Heating a mixture of the solids under vacuum causes a vigorous explosion.

Potassium cyanide
  See Sodium nitrite: Metal cyanides
  See other METAL NITRITES, METAL OXONON-METALLATES, OXIDANTS

4650. Potassium nitrate
[7757-79-1]   KNO$_3$

KONO$_2$

FPA H74, 1978; *HCS 1980*, 771

Aluminium, Barium nitrate, Potassium perchlorate, Water
  See Aluminium: Metal nitrates, etc.

1,3-Bis(trichloromethyl)benzene
  See 1,3-Bis(trichloromethyl)benzene: Oxidants

Boron, 'Laminac', Trichloroethylene
  MCA Case History No. 745
  A 14 kg batch of the mixture, mainly of boron and potassium nitrate, with a
  minority of the synthetic resin adhesive and solvent, exploded 5 min after blending
  had started. Several possibilities of a frictional initiation source were considered.

Calcium silicide
  1. Berger, F., *Compt. rend.*, 1920, **170**, 1492
  2. Smolin, A. O., *Chem. Abs.*, 1977, **86**, 173758
  A mixture of potassium nitrate (or sodium nitrate) and calcium silicide (60:40) is
  a readily ignited primer which burns at a very high temperature. It is capable of

initiating many high-temperature reactions [1]. The topic has been discussed more recently but no details were translated [2].

Calcium hydroxide, Polychlorinated phenols
See Calcium hydroxide: Polychlorinated phenols, etc.

Cellulose
Ishida, H. *et al., Chem. Abs.*, 1988, **109**, 56797
Thermal reaction hazards of potassium nitrate–cellulose mixtures were evaluated by ARC. Stoicheiometric mixtures (zero oxygen balance) showed the lowest ignition temperatures.

Chromium nitride
Partington, 1967, 744
The nitride deflagrates with the molten nitrate.

Honey, Sulfur
Winfield, J., *The Gunpowder Mills of Fernilee*, 20 (Private publ. by Author: 21 New Rd, Whaley Bridge, SK23 7JG, UK), 1996
An early Chinese manuscript warns that seething this mixture endangers the experimenter's beard. It is a probable precursor of gunpowder.

Lactose
See Lactose: Oxidants

Metals
Mellor, 1941, Vol. 7, 20, 116, 261; 1939, Vol. 9, 382
MRH Aluminium 7.15/35, iron 1.55/53, magnesium 7.57/42, sodium 3.10/58
   Mixtures of potassium nitrate and powdered titanium, antimony or germanium explode on heating, and with zirconium at the fusion temperature of the mixture.

Metal sulfides
1. Mellor, 1939, Vol. 9, 270, 524
2. Mellor, 1941, Vol. 3, 745; Vol. 7, 91, 274
3. Mellor, 1943, Vol. 11, 647
4. Pascal, 1963, Vol. 8.3, 404
MRH Antimony trisulfide 2.30/37, titanium disulfide 3.42/26
   Mixtures of potassium nitrate with antimony trisulfide [1], barium sulfide, calcium sulfide, germanium monosulfide or titanium disulfide all explode on heating [2]. The mixture with arsenic disulfide is detonable, and addition of sulfur gives a pyrotechnic composition [2]. Mixtures with molybdenum disulfide are also detonable [3]. Interaction with sulfides in molten mixtures is violent [4].

Non-metals
1. *MCA Case History No. 1334*
2. Mellor, 1946, Vol. 5, 16
3. Brede, U., *Chem. Abs.*, 1981, **94**, 86603

4. Leleu, *Cahiers*, 1980, (99), 278
5. Mellor, 1941, Vol. 2, 820, 825; 1963, Vol. 2, Suppl. 2.2, 1939
6. Mellor, 1940, Vol. 8, 788
7. Mellor, 1939, Vol. 9, 35
MRH Carbon 3.26/13, Phosphorus (yellow) 3.14/27, sulfur 2.97/21

A pyrotechnic blend of a finely divided mixture with boron ignited and exploded when the aluminium container was dropped [1]; (the aluminium container also may have been involved). Boron is not attacked at below 400°C, but is at fusion temperature or at lower temperatures if decomposition products (nitrites) are present [2]. The mixture has also been evaluated as a propellant [3]. Contact of powdered carbon with the nitrate at 290°C causes vigorous combustion [4] and a mixture explodes on heating. Gunpowder is the oldest known explosive and contains potassium nitrate, charcoal and sulfur, the latter to reduce ignition temperature and to increase the speed of combustion [5]. Mixtures of white phosphorus and potassium nitrate explode on percussion, and a mixture with red phosphorus reacts vigorously on heating [6]. Mixtures of potassium nitrate with arsenic explode vigorously on ignition [7].

Organic materials

1. Smith, A. J., *Quart. Natl. Fire Prot. Assoc.*, 1930, **24**, 39–44
2. Unpublished information, 1979
MRH Aniline 3.51/13, acetone 3.47/15, ethanol 3.31/16, toluene 3.56/12

Potassium nitrate in cloth sacks stowed next to baled peat moss became involved in a ship fire and caused rapid flame spread and explosions [1]. Heat transfer salt from a new supplier was added to a pilot plant reactor salt bath. Some 12 h after start of heating to melt the bath contents a muffled explosion occurred, attributed to presence of organic impurities in the new salt [2].
See Cellulose, above.
See Sodium nitrate: Jute, etc.

Other reactants

Yoshida, 1980, 192
MRH values calculated for 19 combinations with oxidisable materials are given.

Phosphides

Mellor, 1940, Vol. 8, 839, 845
Boron phosphide ignites in molten nitrates; mixtures of the nitrate with copper(II) phosphide explode on heating, and that with copper monophosphide explodes on impact.

Reducants

Mellor, 1941, Vol. 2, 820
Mixtures of potassium nitrate with sodium phosphinate and sodium thiosulfate are explosive, the former being rather powerful.
See other REDOX REACTIONS

Sodium acetate   MRH 2.55/34
  Pieters, 1957, 30
  Mixtures may be explosive.

Thorium dicarbide
  See Thorium dicarbide: Non-metals, etc.
  See other METAL NITRATES, OXIDANTS

4651. Potassium nitridoosmate
  [21774-03-8]   KNO₃Os

$$K[NOsO_3]$$

  Clifford, A. F. *et al., Inorg. Synth.*, 1960, **6**, 205
  Heating above 180°C at ambient pressure causes it to explode.
  See related NITRIDES

4652. Potassium azide
  [20762-60-1]   KN₃

$$KN_3$$

  Mellor, 1940, Vol. 8, 347
  It is insensitive to shock; on heating progressively it melts, then decomposes
  evolving nitrogen, and the residue ignites with a feeble explosion.

Carbon disulfide
  See Carbon disulfide: Metal azides

Manganese dioxide
  See Manganese(IV)oxide: Potassium azide

Sulfur dioxide
  Mellor, 1940, Vol. 8, 347
  Potassium azide explodes at 120°C when heated in liquid sulfur dioxide.
  See other METAL AZIDES

4653. Potassium azidosulfate
  [ ]   KN₃O₃S

$$KOSO_2N_3$$

  Mellor, 1940, Vol. 8, 314
  It explodes on heating.
  See other ACYL AZIDES

4654. Potassium dinitramide
[140456-79-7]   KN₃O₄

$$K^+ \ ^-N(NO_2)_2$$

See Dinitramine
See other N-NITRO COMPOUNDS

4655. Potassium azidodisulfate
[67880-14-2]   KN₃O₆S

$$KOSO_2OSO_2N_3$$

Alone, or Water
Mellor, 1967, Vol. 8, Suppl. 2.2, 36
On keeping or slow heating, the salt produces explosive disulfuryl azide, and the salt reacts explosively with water.
See Sulfuryl diazide *See other* ACYL AZIDES

4656. Potassium dioxide (Potassium superoxide)
[12030-88-5]   KO₂

$$KO_2$$

1. *Alkali Metals*, 1957, 174
2. Gilbert, H. N., *Chem. Eng. News*, 1948, **26**, 2604
3. Madaus, private comm., 1976
4. Kirk-Othmer, 1968, Vol. 16, 365–366
5. Commander, J. C., *ERDA Rep. ANCR 1217; TID-4500*, Springfield (Va.), USNTIS, 1975
6. Sloan, S. A., *Proc. Int. Conf. Liquid Met. Technol.*, Champion (Pa.), Int. At. En. Auth., 1976

The earlier references, which state that this powerful oxidant is stable when pure, but explosive when formed as a layer on metallic potassium [1,2], are not wholly correct [3], because the superoxide is manufactured uneventfully by spraying the molten metal into air to effect oxidation [4]. Previous incidents appear to have involved explosive oxidation of unsuspected traces of mineral oil or solvents [3]. However, mixtures of the superoxide with liquid or solid potassium–sodium alloys will ignite spontaneously after an induction period of 18 min, but combustion while violent is not explosive [3]. The additional presence of water (which reduces the induction period) or hydrocarbon contaminant did produce explosion hazards under various circumstances [5]. Contact of liquid potassium with the superoxide gives no obvious reaction below 117°C and a controlled reaction between 117 and 177°C, but an explosive reaction occurs above 177°C. Heating at 100°C/min from 77° caused explosion at 208°C [6].
See other INDUCTION PERIOD INCIDENTS
See Potassium:- *Slow oxidation*
See Dipotassium μ cyclooctatetraene

2-Aminophenol, Tetrahydrofuran

    1. Crank, G. *et al., Aust. J. Chem.*, 1984, **37**, 845–855

    2. Crank, G. *et al., Chem. Eng. News*, 1988, **68**(35), 2

    2-Aminophenol was being oxidised in THF solution at 65°C using a larger than normal proportion of potassium dioxide. When stirring was stopped after 6 h, a violent explosion occurred. This was attributed to formation of tetrahydrofuranyl hydroperoxide by the excess dioxide. THF was described as an unsafe solvent for superoxide reactions [1]. A later attempt at the same reaction in toluene also led to explosion, now blamed on the substrate [2].

    See Hydrocarbons, below.

Carbon

    Mellor, D. P., *Chem. & Ind.*, 1965, 723

    Residues from extinguishing small sodium–potassium alloy fires with graphite were accumulated in an airtight drum. Later, burning alloy fell into the drum and caused a violent explosion. This was attributed to formation of potassium superoxide during storage, and explosive reaction of the latter with graphite, initiated by the burning alloy. Graphite is not a suitable extinguishant for fires involving potassium or its alloys.

Diselenium dichloride

    Mellor, 1947, Vol. 10, 897

    Interaction is very violent.

Ethanol

    Health & Safety Inf., 251, Washington, USAEC, 1967

    Disposal of a piece of potassium–sodium alloy under argon in a glove box by addition of ethanol caused violent gas evolution which burnt a glove and produced a flame. A piece of highly oxidised potassium exploded when dropped into ethanol. Both incidents were attributed to violent interaction of potassium superoxide and ethanol.

Hydrocarbons

    Health & Safety Inf., 251, Washington, USAEC, 1967

    Residues of potassium–sodium alloy in metal containers were covered with oil prior to later disposal. When a lid was removed later, a violent explosion occurred. This was attributed to frictional initiation of the mixture of potassium superoxide (formed on long standing of the alloy) and oil.

Metals

    Mellor, 1941, Vol. 2, 493

    Oxidation of arsenic, antimony, copper, potassium, tin or zinc proceeds with incandescence.

    See reference 6 above

Organic compounds

    Frimer, A. A. *et al., Photochem. Photobiol.*, 1978, **28**(4–5. Singlet oxygen), 718

Potassium superoxide must not be added to neat oxidation substrates, or ignition may occur, and weighing it out on filter paper is also hazardous.

Potassium–sodium alloy
Sloan, S. A. *et al., Microscope*, 1977, **25**, 237–243
A runaway self heating reaction occurs 3.8 s after mixing at 201°C or above, and at that temperature an exotherm to 304°C at a rate above 2000°/s was observed, but no pressure effects were seen.
See related METAL PEROXIDES *See other* METAL OXIDES, OXIDANTS

4657. Potassium trioxide (Potassium ozonide)
[12030-89-6] KO_3

$$KO_3$$

See Caesium trioxide, *also* Fluorine: Potassium hydroxide
See related METAL PEROXIDES *See other* METAL OXIDES

4658. Potassium silicide
[16789-24-5] KSi

$$KSi$$

Bailar, 1973, Vol. 1, 1357
It ignites in air.
See other METAL NON-METALLIDES

4659. Potassium diperoxomolybdate
[58412-05-8] K_2MoO_6

$$K_2[(O_2)_2MoO_2]$$

Mellor, 1943, Vol. 11, 607
It explodes on grinding.
See other PEROXOACID SALTS

4660. Potassium tetraperoxomolybdate
[56094-15-6] K_2MoO_8

$$K_2[Mo(O_2)_4]$$

Jahr, K. F., *FIAT Rev. Ger. Sci: Inorg. Chem., Pt. III*, 1948. 170
It is explosive.
See other PEROXOACID SALTS

4661. Potassium nitrosodisulfate
[14293-70-0] $K_2NO_7S_2$
$$(KOSO_2)_2NO\cdot$$

1. Zimmer, H. *et al., Chem. Rev.*, 1971, **71**, 230, 243
2. Wehrli, P. A. *et al., Org. Synth.*, 1972, **52**, 86
The observed instability of the solid radical on storage, ranging from slow decomposition to violent explosion (probably depending on the degree of confinement), depends largely upon the degree of contamination by nitrite ion. Storage conditions to enhance stability are detailed [1]. Synthetic use of solutions of the salt has been preferred, and the storage of any isolated salt as an alkaline slurry is recommended [2].
See related NITROSO COMPOUNDS

4662. Potassium *N*-nitrosohydroxylamine-*N*-sulfonate ('Potassium dinitrososulfite')
[26241-10-1] $K_2N_2O_5S$
$$KOSO_2N(NO)OK$$

Weitz, E. *et al., Ber.*, 1933, **66**, 1718
The salt (formerly named a dinitrososulfite) explodes on heating.
See other N–O COMPOUNDS, N–S COMPOUNDS

4663. Potassium azodisulfonate
[ ] $K_2N_2O_6S_2$
$$KOSO_2N:NSO_2OK$$

Alone, or Water
Konrad, E. *et al., Ber.*, 1926, **59**, 135
There is a possibility of explosion during vacuum desiccation of the salt, which evolves nitrogen violently in contact with water.
See related AZO COMPOUNDS *See other* N–S COMPOUNDS

4664. Potassium sulfurdiimidate
[[79796-14-8] (ion)] K_2N_2S
$$KN:S:NK$$

Alone, or Solvents, or Water
1. Goehring, 1957, 39, 6
2. Hornemann, K. *et al., Angew. Chem. (Intern. Ed.)*, 1982, **21**, 633
It ignites in air, and reacts violently and may explode with traces of water [1]. It is stable up to 180°C, but reacts explosively and ignites with water, methanol or chloromethane, methylene chloride or carbon tetrachloride [2].
See other *N*-METAL DERIVATIVES, N–S COMPOUNDS

4665. Potassium hexaazidoplatinate(IV)
[41909-62-0] (ion) $K_2N_{18}Pt_2$
$$K_2[Pt(N_3)_6]$$

Mellor, 1940, Vol. 8, 355
Evaporation of a solution of hexachloroplatinic acid with a deficiency of potassium azide, or with an equivalence of ammonium azide gives explosive residues. Evaporation of a solution of the acid with an equivalence (8 mol) of potassium azide leads to explosion of the conc. solution of the title compound.
See other PLATINUM COMPOUNDS *See related* METAL AZIDES

4666. Potassium peroxide
[17014-71-0] K_2O_2
$$KOOK$$

Other reactants
Yoshida, 1980, 83
MRH values calculated for 15 combinations with oxidisable materials are given.

Water
Pascal, 1963, Vol. 2.2, 59
Interaction is violent or explosive.
See other METAL PEROXIDES, OXIDANTS

4667. Potassium diperoxoorthovanadate
[ ] K_2O_6V
$$K_2[(O_2)_2VO_2]$$

1. Mellor, 1939, Vol. 9, 795
2. Eméleus, 1960, 425
It explodes on heating [1,2].
See other PEROXOACID SALTS

4668. Potassium peroxodisulfate
[7727-21-1] $K_2O_8S_2$
$$KOSO_2OOSO_2OK$$

FPA H106, 1981; *HCS 1980*, 775

Energy of decomposition (in range 190–840°C) measured as 0.29 kJ/g.
See entry THERMOCHEMISTRY AND EXOTHERMIC DECOMPOSITION (reference 2)

Alkali, Hydrazine salts
Mellor, 1947, Vol. 10, 466

Addition of alkali to the mixed salts liberates hydrazine which is vigorously oxidised to nitrogen gas.

Potassium hydroxide
Anon., *BCISC Quart. Safety Summ.*, 1965, **36**, 41
Surface contamination of 2 kg of the dry salt with as little as 2 flakes of moist potassium hydroxide caused a vigorous self-sustaining fire, which was extinguished with water, but not by carbon dioxide or dry powder extinguishers.

Water
Castrantas, 1965
The salt rapidly liberates oxygen above 100°C when dry, but at only 50°C when wet.
See other PEROXOACID SALTS, OXIDANTS

4669. Potassium tetraperoxotungstate
[37346-96-6] K_2O_8W

$$K_2[W(O_2)_4]$$

Mellor, 1943, Vol. 11, 836
It explodes on friction or rapid heating to 80°C.
See other PEROXOACID SALTS

4670. Potassium sulfide
[1312-73-8] K_2S
                                   KSK

HCS 1980, 387

Merck Index, 1983, p1104
The anhydrous sulfide is unstable, and may explode on percussion or rapid heating.

Chloroform: Ethanol (or Methanol)
See Potassium dithioformate

Nitrogen oxide
See Nitrogen oxide: Potassium sulfide

Other reactants
Yoshida, 1980, 83
MRH values calculated for 13 combinations with various reagents are given.
See other METAL SULFIDES

4671. Potassium nitride
[29285-24-3]

K_3N

K_3N

Mellor, 1940, Vol. 8, 99
It usually ignites in air.

Phosphorus
See Phosphorus: Potassium nitride

Sulfur
See Sulfur: Potassium nitride
See other NITRIDES

4672. Potassium 2,4,6-tris(dioxoselena)hexahydrotriazine-1,3,5-triide ('Potassium triselenimidate')
[51593-99-8]

$K_3N_3O_6Se_3$

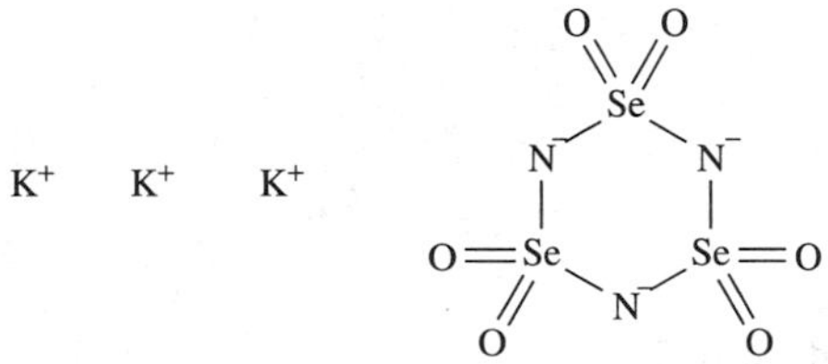

Explosive.
See Selenium difluoride dioxide: Ammonia
See other N-METAL DERIVATIVES

4673. Potassium antimonide
[16823-94-2]

K_3Sb

K_3Sb

1. Mellor, 1939, Vol. 9, 403
2. Rüst, 1948, 342

It usually ignites when broken in air [1] and explodes on exposure to moisture (in breath) [2].
See other ALLOYS, PYROPHORIC MATERIALS

4674. Potassium hexaoxoxenonate–xenon trioxide
[12273-50-6]

$K_4O_6Xe.2O_3Xe$

$K_4[O_6Xe].2XeO_3$

1. Grosse, A. V., *Chem. Eng. News*, 1964, **42**(29), 42
2. Spittler, T. M. *et al.*, *J. Amer. Chem. Soc.*, 1966, **88**, 2942
3. Appelman, E. H. *et al.*, *J. Amer. Chem. Soc.*, 1964, **86**, 2141

The complex salt is very sensitive to mechanical shock and explodes violently [1,2], even when wet [3].
See other XENON COMPOUNDS
See related NON-METAL OXIDES

4675. Lithium tripotassium tetrasilicide
[102210-64-0] K_3LiSi_4

$$LiK_3Si_4$$

von Schering, H. G. *et al., Angew. Chem. (Intern. Ed.)*, 1986, **25**, 566
It ignites in air.
See other METAL NON-METALLIDES

4676. Lithium heptapotassium di(tetrasilicide)
[102210-65-1] K_7LiSi_8

$$LiK_7(Si_4)_2$$

von Schering, H. G. *et al., Angew. Chem. (Intern. Ed.)*, 1986, **25**, 566
It ignites in air.
See other METAL NON-METALLIDES

4677. Lanthanum
[7439-91-0]   La

$$La$$

Nitric acid
   Mellor, 1946, Vol. 5, 603
   Oxidation is violent.

Phosphorus
See Phosphorus: Metals
See other METALS

4678. Lanthanum–nickel alloy
[12196-72-4] $LaNi_5$

$$LaNi_5$$

Hydrogen
See Lanthanum pentanickel hexahydride

Hydrogen, Poly(tetrafluoroethylene)
See LANTHANIDE–TRANSITION METAL ALLOY HYDRIDES
See Poly(tetrafluoroethylene): Metal hydrides
See other ALLOYS, HYDROGENATION CATALYSTS

4679. Lanthanum oxide
[1312-81-8] La_2O_3

$$La_2O_3$$

Chlorine trifluoride
See Chlorine trifluoride: Metals, etc.

Water

Sidgwick, 1950, 441

Interaction is vigorously exothermic, accompanied by hissing, as for calcium oxide.

See Calcium oxide: Water

See other METAL OXIDES

4680. Lithium
[7439-93-2]   Li

Li

NSC 566, 1978; *HCS 1980*, 592

1. Mellor, 1941, Vol. 2, 468: 1961, Vol. 2, Suppl. 2.1, 71
2. *Haz. Chem. Data*, 1975, 187
3. Bullock, A., *Chem. Brit.*, 1975, **11**, 115
4. Bullock, A., *School Sci. Rev.*, 1975, **57**(119), 311–314; private comm., 1975
5. Lloyd, W. H., and Emley, E. F., *Chem. Brit.*, 1975, **11**, 334
6. Ireland, R. E. *et al.*, *Org. Synth.*, 1977, **56**, 47
7. Fritsch, *Proc. 1st Int. Symp. Prev. Occup. Risks Chem. Ind.*, 208–211, Heidelberg, ISSA, 1970

MRH Oxygen 19.95/54

Finely divided metal may ignite in air at ambient temperature and massive metal above the m.p., 180°C, especially if oxide or nitride is present. Since lithium will burn in oxygen, nitrogen or carbon dioxide, and when alight it will remove the combined oxygen in sand, sodium carbonate, etc., it is difficult to extinguish once alight [1]. (*Note the very high MRH value above.*) Molten lithium is extremely reactive and will attack concrete and refractory materials. Use of normal fire extinguishers (containing water, foam, carbon dioxide, halocarbons, dry powders) will either accelerate combustion or cause explosion. Powdered graphite, lithium chloride, potassium chloride or zirconium silicate are suitable extinguishants [2]. A well-tried and usually uneventful demonstration of atmospheric oxidation of molten lithium led to an explosion [3]. A high degree of correlation of incidence of explosions with high atmospheric humidity was demonstrated, with the intensity of explosion apparently directly related to the purity of the sample of metal [4]. Other possible factors were also identified [5]. While cleaning lithium wire by washing with hexane, the wire must be dried carefully with a paper towel. Too-vigorous rubbing will cause a fire [6].

Lithium blocks containing traces of nitride but which had been supplied under argon in sealed tins were cut into 1 cm strips and stored under air in closed tins overnight until used. Lithium containing some nitride reacts slowly with nitrogen at ambient temperature to form more nitride, which autocatalyses the reaction which progressively accelerates and becomes exothermic. The strips of lithium reacted with the nitrogen of the air in the closed tins, causing a partial vacuum and an oxygen-enriched atmosphere. When the tins collapsed, the impact and/or compression of the oxygen-enriched atmosphere caused ignition and fierce burning of the lithium, which was very difficult to extinguish inside the crushed tins [7].

See Metal chlorides, *also* Water, both below

Acetonitrile, Sulfur dioxide

Ebner, W. B. *et al., Proc. 8th Power Sources Symp.*, 119–124, 1982

An ARC study of the thermal and pressure behaviour of actual electric batteries under various atypical conditions showed the major contributions to the exothermic behaviour as the reactions between lithium and acetonitrile, lithium and sulfur and the decomposition of lithium dithionite. The first reaction can generate enough heat to trigger other exothermic rections. The hazards associated with the various parameters are quantified.

See Sulfur dioxide, below

Bromine pentafluoride

See Bromine pentafluoride: Acids, etc.

Bromobenzene

Koch-Light Labs., Ltd., private comm., 1976

In a modified preparation of phenyllithium, bromobenzene was added to finely powdered lithium (rather than coarse particles) in ether. The reaction appeared to be proceeding normally, but after about 30 min it became very vigorous and accelerated to explosion. It was thought that the powdered metal may have been partially coated with oxide or nitride which abraded during stirring, exposing a lot of fresh metal surface on the powdered metal.

See Sodium: Halocarbons (reference 8) *See also* Halocarbons, below

Carbon, Lithium tetrachloroaluminate, Sulfinyl chloride

Kilroy, W. P. *et al., J. Electrochem. Soc.*, 1981, **128**, 934–935

In electric battery systems, lithium is inert to the electrolyte components in absence of carbon, but in presence of over 10% of carbon (pre-mixed by grinding with the metal), contact with the electrolyte mixture leads to ignition or explosion.

See Sulfinyl chloride, below

Carbon, Sulfinyl chloride

James, S. D. *et al., J. Electrochem. Soc.*, 1983, **130**, 2037–2040

Pregrinding lithium with carbon leads to ignition on contact with sulfinyl chloride in electric battery systems. The effect of moisture and purity of the carbon on reactivity was studied.

Carbon disulfide

See Carbon disulfide: Alkali metals

Chlorine tri- or penta-fluorides

See Chlorine trifluoride: Metals, etc.

1-Chloro-3-dimethylaminopropane

1. Eisenbach, A. *et al., Euro. Polym. J.*, 1975, **11**, 699
2. Service, D., *Chem. Brit.*, 1987, **23**, 27
3. Bell, J. A. *et al., J. Org. Chem.*, 1959, **24**, 2036
4. Kamienskis, C. W., *Chem. Brit.*, 1987, **23**, 531–532

The preparation of 1-lithio-3-dimethylaminopropane according to a published procedure [1] involved reaction of the chloro compound with a mineral oil dispersion of lithium (30%) in hexane at 0°C. It had previously been found that lithium with a sodium content of some 0.3% reacted slowly, even at temperatures above 15°C. When a new batch of lithium (later found to contain 1.9% of sodium) was used, a vigorous reaction which set in at between 0° and −35°C led to ignition of the reaction mixture [2]. Although it was known that the presence of sodium increased the reactivity of lithium towards organohalides up to a sodium content of 2% [3], it had not been appreciated that such wide variations in reactivity were likely, or that a hazardous situation could develop [1,2]. However, a more likely cause proposed for the runaway exothermic reaction was the fact that all the reagents were mixed at 0°C, rather than the more usual course of adding the halide slowly to the lithium dispersion in a hydrocarbon solvent lower-boiling than hexane, so as to maintain gentle reflux at 35–50°C. Some general precautions for this type of reaction are given [4].
See Halocarbons, below

1,2-Diaminoethane, Tetralin
Reggel, A. *et al., J. Org. Chem.*, 1957, **22**, 892
Reduction of tetralin to octalin with lithium and ethylenediamine proceeds slowly, but if heated to 85°C it becomes violent, with rapid evolution of hydrogen.
See other GAS EVOLUTION INCIDENTS

Diazomethane
See Diazomethane: Alkali metals

Diborane   MRH 1.13/99+
See Diborane: Metals

Ethylene
Pascal, 1966, Vol. 2.1, 38
Passage of the gas over heated lithium causes the latter to incandesce, producing a mixture of lithium hydride and lithium acetylide.
See Metal chlorides, etc., below

Halocarbons   MRH Bromoform 3.77/92, carbon tetrabromide 3.89/92,
                               carbon tetrachloride 8.24/85, carbon tetraiodide 1.42/95,
                               chloroform 7.78/85, fluorotrichloromethane 9.29/83,
                               tetrachloroethylene 8.16/86
1. Staudinger, H., *Z. Elektrochem.*, 1925, **31**, 549
2. *Pot. Incid. Rep.*, 1968, **39**, ASESB, Washington
3. Pittwell, L. R., *J. R. Inst. Chem.*, 1959, **80**, 552
4. BDH Catalogue Safety Note, 969DD/14.0/0773, 1973
5. *491M*, 1975, 230
Mixtures of lithium shavings and several halocarbon derivatives are impact-sensitive and will explode, sometimes violently [1,2]. Such materials include:

bromoform, carbon tetrabromide, carbon tetrachloride, carbon tetraiodide, chloroform, dichloromethane, diiodomethane, fluorotrichloromethane, tetrachloroethylene, trichloroethylene and 1,1,2-trichlorotrifluoroethane. In an operational incident, shearing samples off a lithium billet immersed in carbon tetrachloride caused an explosion and continuing combustion of the immersed metal [3]. Lithium which had been washed in carbon tetrachloride to remove traces of oil exploded when cut with a knife. Hexane is recommended as a suitable washing solvent [4]. A few drops of carbon tetrachloride on burning lithium was without effect, but a 25 cc portion caused a violent explosion [5].

See 1-Chloro-3-dimethylaminopropane, above
Poly(1,1-difluoroethylene–hexafluoropropylene), below
See entry HALOCARBONS: metals

Halogens   MRH Chlorine 9.62/83
1. Staudinger, H., *Z. Elektrochem.*, 1925, **31**, 549
2. Mellor, 1961, Vol. 2, Suppl. 2.1, 82
Mixtures of lithium and bromine are unreactive unless subject to heavy impact, when explosion occurs [1]. Lithium and iodine react above 200°C with a large exotherm [2].

Hydrogen
See Hydrogen: Metals

Mercury
1. Smith, G. McP. *et al., J. Amer. Chem. Soc.*, 1909, **31**, 799
2. Alexander, J. *et al., J. Chem. Educ.*, 1970, **47**, 277
Interaction to form lithium amalgam is violently exothermic and may be explosive if large pieces of lithium are used [1]. An improved technique, using *p*-cymene as inerting diluent, is described in the later reference [2].

Metal chlorides, Nitrogen
Anon., *BCISC Quart. Safety Summ.*, 1969, **40**, 16
Accidental contamination of lithium strip with anhydrous chromium trichloride or zirconium tetrachloride caused it to ignite and burn vigorously in the nitrogen atmosphere of a glove box.

Metal oxides and chalcogenides   MRH Chromium trioxide 7.20/83,
                                                                    sodium peroxide 13.64/71
1. Mellor, 1967, Vol. 2, Suppl. 2.1, 81–82
2. *Alkali Metals*, 1957, 11
Lithium is used to reduce metallic oxides in metallurgical operations, and the reactions, after initiation at moderate temperatures, are violently exothermic and rapid. Chromium(III) oxide reacts at 185°C, reaching 965°; similarly molybdenum trioxide (180 to 1400°), niobium pentoxide (320 to 490°), titanium dioxide (200–400 to 1400°), tungsten trioxide (200 to 1030°), vanadium pentoxide (394 to 768°); also iron(II) sulfide (260 to 945°), and manganese telluride (230 to 600°C)

[1]. Residual mixtures from lithium production cells containing lithium and rust sometimes ignite when left as thin layers exposed to air [2].

Metals
1. Bailar, 1973, Vol. 1, 343–344
2. Popov, E. I. *et al., Chem. Abs.*, 1989, **110**, 100000
Formation of various intermetallic comounds of lithium by melting with aluminium, bismuth, calcium, lead, mercury, silicon, strontium, thallium or tin may be very vigorous and dangerous to effect [1]. Ignition and combustion hazards of alloys of lithium with aluminium or magnesium have been studied, the latter being more reactive than the former. Use of nitrogen as a protective medium for alloy powders is ineffective, mixed nitrides being formed [2].
See Mercury, above; Platinum, below

Nitric acid   MRH 15.27/60
Accident & Fire Prev. Info., March, 1974, Washington, USAEC
Lithium ignited on contact with nitric acid and the reaction became violent, ejecting burning lithium.

Nitryl fluoride
See Nitryl fluoride: Metals

Non-metal oxides
Mellor, 1961, Vol. 2, Suppl. 2.1, 74, 84, 88
Although carbon dioxide reacts slowly with lithium at ambient temperature, the molten metal will burn vigorously in the gas, which cannot be used as an extinguisher on lithium fires. Carbon monoxide reacts in liquid ammonia to give the carbonyl which reacts explosively with water or air. Lithium rapidly attacks silica or glass at 250°C.
See Sulfur dioxide, below
See other GLASS INCIDENTS

Other reactants
Yoshida, 1980, 385–386
MRH values calculated for 24 combinations with various reducible materials are given.

Platinum
See Platinum: Lithium

Poly(1,1-difluoroethylene–hexafluoropropylene) (Viton)
Markowitz, M. M. *et al., Chem. Eng. News*, 1961, **39**(32), 4
Mixtures of finely divided metal and shredded polymer ignited in air on contact with water, or on heating to 369°C, or at 354° under argon.
See Halocarbons, above

Sodium carbonate   MRH Sodium carbonate 4.27/73
or Sodium chloride
    Mellor, 1961, Vol. 2, Suppl. 2.1, 25
    Sodium carbonate and sodium chloride are unsuitable to use as extinguishers for
    lithium fires, since burning lithium will liberate the more reactive sodium in contact
    with them.

Sulfur
    Mellor, 1961, Vol. 2, Suppl. 2.1, 75
    Interaction when either is molten is very violent and, even in presence of inert
    diluent, the reaction begins explosively. Reaction of sulfur with lithium dissolved
    in liquid ammonia at −33°C is also very vigorous.

Sulfinyl chloride
    1. Abraham, K. M. *et al., Proc. 28th Power Sources Symp.*, 1978, 255–257
    2. Subbarao, S. *et al., Proc. Electrochem. Soc.*, 1988, **88-6**, 187–200
    3. Ducatman, A. M. *et al., J. Occup. Med.*, 1988, **30**, 309–311
    4. Butakov, A. A. *et al., J. Power Sources*, 1992, **39**(3), 375
    An experimental investigation of explosion hazards in lithium–sulfinyl chloride
    cells on forced discharge showed cathode limited cells are safe, but anode limited
    cells may explode without warning signs [1]. Extended reversal at −40°C caused
    explosion on warming to ambient temperature, owing to thermal runaway caused
    by accelerated corrosion of lithium [2]. The violent explosion of a large pris-
    matic cell of a battery is described [3]. Another study of explosion mechanisms
    in lithium/thionyl chloride batteries is reported [4]
    See Carbon, etc., above

Sulfur dioxide
    1. Dey, A. N. *et al., Chem. Abs.*, 1980, **93**, 10716, 10717
    2. Crofts, C. C., *Chem. Abs.*, 1983, **99**, 56425
    3. Subbarao, S. *et al., J. Power Sources*, 1987, **21**(3–4), 221–237
    4. Donaldson, G. J., *et al., Power Sources*, 1991, **13**, 363
    Two reports cover safety studies on lithium–sulfur dioxide batteries [1]. The cause
    of violent venting of discharged lithium–sulfur dioxide cells was ascribed to
    corrosion in a glass to metal seal and formation of lithium–aluminium alloys
    and other cathode reaction product(s) which are both shock-sensitive [2] The
    pyrophoric charged anodes of lithium–sulfur dioxide batteries are covered with
    smooth crystalline platelets, but partially discharged anodes are covered with
    a rough, non-adherent layer of lithium dithionite. The explosions which may
    occur during charging are attributed to thermal runaway reactions of lithium and
    sulfur dioxide to form lithium dithionite, $LiOSO.SO.OLi$ [3]. A study of battery
    explosions consequent upon anode breakage in lithium–sulfur dioxide cells is
    reported [4].
    See Acetonitrile, etc., *also* Non-metal oxides, both above

Trifluoromethyl hypofluorite
    See Trifluoromethyl hypofluorite: Lithium

Water
1. Mellor, 1961, Vol. 2, Suppl. 2.1, 72
2. Bailar, 1973, Vol. 1, 337
3. Markowitz, M. M., *J. Chem. Educ.*, 1963, **40**, 633–636
Reaction with cold water is of moderate vigour, but violent with hot water, and the liberated hydrogen may ignite [1]. The powdered metal reacts explosively with water [2]. The reactivity of lithium and other alkali metals with various forms of water has been discussed in detail. Prolonged contact with steam forms a thermally insulating layer which promotes overheating of the metal and may lead to a subsequent explosion as the insulating layer breaks up [3].
See ALKALI METALS
See other METALS

4681. Lithium–magnesium alloy
[12384-02-0]   Li—Mg

$$Li-Mg$$

Danyshova, T. A. *et al., Chem. Abs.*, 1982, **97**, 167510
Fire hazard (title only translated).
See other ALLOYS

4682. Lithium–tin alloys
[12359-06-7]   Li—Sn

$$Li-Sn$$

Alkali metals, 1957, 13

Lithium–tin alloys are pyrophoric.
See other ALLOYS, PYROPHORIC MATERIALS

4683. Lithium sodium nitroxylate
[ ] $LiNNaO_2$

$$Li[O:NONa]$$

Mellor, 1961, Vol. 2, Suppl. 2.1, 78
It decomposes violently at 130°C.
See other N–O COMPOUNDS

4684. Lithium nitrate
[7790-69-4] $LiNO_3$

$$LiONO_2$$

Ethanol
See VAPOUR EXPLOSIONS (reference 3)

Other reactants
    Yoshida, 1980, 208
    MRH values calculated for 15 combinations with oxidisable materials are given.

Propene, Sulfur dioxide
    See Propene: Lithium nitrate, etc.
    See other METAL NITRATES

4685. Lithium azide
[19597-69-4]   LiN$_3$

LiN$_3$

Mellor, 1940, Vol. 8, 345
Insensitive to shock, the moist or dry salt decomposes explosively at 115–298°C, depending on the rate of heating.

Alkyl nitrates, Dimethylformamide
    Polansky, O. E., *Angew. Chem. (Intern. Ed.)*, 1971, **10**, 412
    Although these reaction mixtures are stable at 25°C during preparation of *tert*-alkyl azides, above 200°C the mixtures are shock-sensitive and highly explosive.
    See other METAL AZIDES

4686. Lithium oxide
[12057-24-8]   Li$_2$O

LiOLi

Metal halide hydrates
    Rowley, A. T. *et al., Inorg. Chem. Acta*, 1993, **211**(1), 77
    Preparation of metal oxides by fusing metal halides with lithium oxide in a sealed tube leads to explosions if halide hydrates are employed, particularly lanthanide trihalide hydrates. The preparation succeeds with anhydrous halides. This will be purely a question of vapour pressure above an exothermic reaction; the question is whether the vapour is water, or metal halide, and the reaction oxide formation, or hydration of lithium oxide. Like other alkali metal oxides, hydration is extremely energetic.
    See Diprotium monoxide

4687. Lithium dithionite
[59744-77-3]   Li$_2$O$_4$S$_2$

LiOSO.SO.OLi

See Lithium: Sulfur dioxide (reference 3)
See other METAL OXONON-METALLATES, REDUCANTS

4688. Lithium nitride
[26134-62-3]

Li₃N

Li_3N

Air, or Water
1. 'Data Sheet TD 121', Exton, Foote Mineral Co., 1965
2. Schönher, E. *et al.*, *Inorg. Synth.*, 1983, **22**, 48–55
The finely divided powder may ignite if sprayed into moist air [1]. Procedures for preparing polycrystalline and single crystal materials are detailed, with precautions to prevent ignition of material deposited on the walls of the reaction chamber when it is opened to air, or cleaned with water [2].

Copper(I) chloride
Mellor, 1940, Vol. 8, 100
Heated interaction to produce metallic copper is violent.

Silicon tetrafluoride
Porritt, L. J., *Chem. Brit.*, 1979, **15**, 282–283
Interaction is violently exothermic, causing sealed glass tubes at 350°C to explode, and a flow reaction system partially to melt.
See other GLASS INCIDENTS

Transition metal halides
Hector, A. *et al.*, *J. Chem. Soc., Chem. Comm.*, 1993, (13), 1095; *J. Chem. Soc., Dalton Trans.*, 1993, (16), 2435
Titanium and vanadium nitrides may be prepared by a metathesis reaction of their tetrachlorides with the nitride, initiated by heat or friction. The reaction is potentially explosive. Other transition metal halides may cause ampules to explode after thermal initiation when anhydrous and were invariably found to do so when the hydrates were used.
See METAL PNICTIDES, METATHESIS
See other NITRIDES

4689. Hexalithium disilicide
[12136-61-7]

Li_6Si_2

Li₃SiSiLi₃

Acids
Mellor, 1940, Vol. 6, 170
The silicide incandesces in conc. hydrochloric acid, and with dilute acid evolves silanes which ignite. It explodes with nitric acid and incandesces when floated on sulfuric acid.

Halogens
Mellor, 1940, Vol. 6, 169
When warmed, it ignites in fluorine, but higher temperatures are necessary with chlorine, bromine and iodine.

Non-metals
Mellor, 1940, Vol. 6, 169
Interaction with phosphorus, selenium or tellurium causes incandescence. Sulfur also reacts vigorously.

Water
Mellor, 1940, Vol. 6, 170
Reaction with water is very violent, and the silanes evolved ignite.
See other METAL NON-METALLIDES

4690. Magnesium
[7439-95-4]   Mg

Mg

NSC 426, 1981; *FPA H21*, 1974 (not powder); *HCS 1980*, 598

Preparative hazard
  1. *Dust Explosion Prevention: Magnesium Powder, NFPA Standard Codes 480, 651*, both 1987
  2. Popov, E. I., *Chem. Abs.*, 1975, **82**, 61415
  3. Grossel, S. S., *J. Loss Prev. Proc. Ind.*, 1988, **1**, 62–74
All aspects of prevention of magnesium (and aluminium) dust explosions in storage, handling or processing operations are covered in two recent US National Fire Codes [1]. Effects of various parameters on ignition of magnesium powders were studied [2]. Maximum explosion pressures of 7.9 bar, with maximum rate of rise of 884 bar/s have been recorded [3].
See Potassium: Magnesium chloride, etc., *also* Metal halides, below

Acetylenic compounds
See ACETYLENIC COMPOUNDS: metals

Aluminium, Rusted steel, Zinc
See Aluminium–magnesium–zinc alloy: Rusted steel

Ammonium salts, Chlorate salts   MRH Ammonium chlorate 9.07/74
Kirk-Othmer (3rd ed.), 1982, Vol. 19, 494
The mixture ignites when wet owing to formation of unstable ammonium chlorate, and combustion of the metal is very intense.

Atmospheric gases, Water
Darras, R. *et al., Chem. Abs.*, 1961, **55**, 13235
The temperature at which massive magnesium and its alloys will ignite in air depends on the heating programme and the presence or absence of moisture.

Barium carbonate, Water
MCA Case History No. 1849

1758

Fusion of the metal and salt formed barium acetylide, to which water was added without effective cooling. The vigorous evolution of acetylene blew off the reactor lid and the hot acetylene ignited in air.
See other GAS EVOLUTION INCIDENTS

Beryllium fluoride
Walker, H. L., *School Sci. Rev.*, 1954, **35**(127), 348
Interaction is violently exothermic.

Boron diiodophosphide
See Boron diiodophosphide: Metals

Calcium carbonate, Hydrogen
Mellor, 1940, Vol. 4, 271
Heating an intimate mixture of the powdered metal and carbonate in a stream of hydrogen leads to a violent explosion.

Carbon dioxide   MRH 8.66/49
  1. Partington, 1967, 475
  2. Driscoll, J. A., *J. Chem. Educ.*, 1978, **55**, 450–451
  3. Darras, R. *et al., Chem. Abs.*, 1961, **55**, 13235
A mixture of solid carbon dioxide and magnesium burns very rapidly and brilliantly when ignited [1]. Details for a safe demonstration of the violent combustion have been published [2]. The temperature at which massive magnesium and its alloys will ignite carbon dioxide at 1 bar or 15 bar pressure depends on the heating programme and the presence or absence of moisture. Steady progressive heating caused ignition in carbon dioxide at 900°C (780° if moist), or at 800°C after 4 h (650°/40 min if moist), combustion being violent [3].

Carbon tetrachloride   MRH 5.65/76
See Methanol: Carbon tetrachloride, Metals

Chloroformamidinium nitrate
See Chloroformamidinium nitrate: Metals

Copper compounds
Gorbunov, V. V. *et al., Combustion, Expls, & Shockwaves, 1986*, **22**, 726
Not merely cupric oxide, but the fluoride, chloride and sulfide all form mixtures with heats of combustion above 2 kJ/g. The equivalent lead compounds are less active.

Ethylene oxide   MRH 6.82/79
See Ethylene oxide(reference 3)

Fluorocarbon polymers
Kirk-Othmer (3rd ed.), 1982, Vol. 19, 495
Compressed mixtures are used as special igniters.
See Halocarbons, below

Glass powders
See Metal oxides, below (reference 3)

Halocarbons   MRH Carbon tetrachloride 5.65/76, trichloroethylene 5.48/78
1. Clogston, C. C., *Bull. Res. Underwriters Lab.*, 1945, **34**, 5
2. *Pot. Incid. Rept.*, 39, ASESB, Washington, 1968
3. Mond. Div. ICI, private comm., 1968
4. Hartmann, I., *Ind. Eng. Chem.*, 1948, **40**, 756
5. *Haz. Chem. Data.*, 1975, 115
The powdered metal reacts vigorously and may explode on contact with chloromethane, chloroform, or carbon tetrachloride, or mixtures of these [1]. Mixtures of powdered metal with carbon tetrachloride or trichloroethylene will flash on heavy impact [2]. Violent decomposition with evolution of hydrogen chloride can occur when 1,1,1-trichloroethane comes into contact with magnesium or its alloys with aluminium [3]. Magnesium dust ignited at 400°C in dichlorodifluoromethane (used as an extinguishant for hydrocarbon fires) and suspensions of the dust exploded violently on sparking in the vapour [4]. Interaction with 1,2-dibromoethane may become violent, and gives air-sensitive Grignard compounds [5].
See Fluorocarbon polymers, above; Polytetrafluoroethylene, below
See Bromomethane: Metals
See other CORROSION INCIDENTS

Halogens, or Interhalogens   MRH Chlorine 6.78/75, fluorine 17.78/61
Mellor, 1940, Vol. 4, 267
It ignites if moist and burns violently in fluorine, and ignites in moist or warm chlorine. It burns not very readily in bromine vapour and may ignite if finely divided on heating in iodine vapour.
See Chlorine trifluoride: Metals
     Iodine heptafluoride: Metals
     Iodine: Metals, Water

Hydrogen iodide
Mellor, 1941, Vol. 2, 206
Contact causes momentary ignition.

Lead compounds
See Copper compounds, above

Magnesium sulfate
See Magnesium sulfate: Magnesium

Metal cyanides
Mellor, 1940, Vol. 4, 271
Magnesium reacts with incandescence on heating with cadmium cyanide, cobalt cyanide, copper cyanide, lead cyanide, nickel cyanide and zinc cyanide. With gold

cyanide or mercury cyanide, the cyanogen released by thermal decomposition of these salts reacts explosively with magnesium.

Metal oxides   MRH Chromium trioxide 6.44/73, lead dioxide 3.22/83
1. Mellor, 1941, Vol. 3, 138, 378; 1940, Vol. 4, 272; 1941, Vol. 7, 401
2. Stout, E. L., *Chem. Eng. News*, 1958, **36**(8), 64
3. Lawrence, K. D. *et al., Chem. Abs.*, 1979, **91**, 41576
Magnesium will reduce violently metal oxides on heating, similarly to aluminium powder in the thermite reaction. Beryllium oxide, cadmium oxide, copper oxide, mercury oxide, molybdenum oxide, tin oxide and zinc oxide are all reduced explosively on heating. Silver oxide reacts with explosive violence when heated with magnesium powder in a sealed tube [1]. Interaction of molten magnesium and iron oxide scale is violent [2]. Use of glass powders as suppressants for burning magnesium showed that some lithium oxide–magnesium oxide frits reacted violently [3].
See other GLASS INCIDENTS, THERMITE REACTIONS

Metal oxosalts   MRH Ammonium nitrate 8.79/69, potassium nitrate 7.57/58, silver
                    nitrate 6.95/70, barium sulfate 4.27/71, copper(II) sulfate 7.03/58,
                    sodium sulfate 5.86/60, potassium carbonate, 3.10/66.
1. Mellor, 1940, Vol. 4, 272; Vol. 8, Suppl. 2.1, 545
2. Pieters, 1957, 30
3. Shimizu, T., *Chem. Abs.*, 1991, **114**, 9175
Interaction with fused ammonium nitrate or with metal nitrates, phosphates or sulfates may be explosively violent [1]. Lithium and sodium carbonates may also react vigorously [2]. The mixture with magnesium sulfate has been described as a noisy but low power bursting charge for pyrotechny [3].
See Silver nitrate: Magnesium
      Potassium perchlorate: Metal powders
See other THERMITE REACTIONS

Methanol   MRH 5.02/73
1. Vogel, 1957, 169
2. Personal experience, 1952
3. Shidlovskii, A. A., *Chem Abs.*, 1947, **41**, 1105d
The reaction of magnesium and methanol to form magnesium methoxide and used to prepare dry methanol [1] is very vigorous, but often subject to a lengthy induction period. Sufficient methanol must be present to absorb the violent exotherm which sometimes occurs [2]. Mixtures of powdered magnesium (or aluminium) and methanol are capable of detonation and are more powerful than military explosives [3].
See other INDUCTION PERIOD INCIDENTS

Methanol, 3-Methyl-2-butenoylanilide
Brettle, R. *et al., Tetrahedron Lett.*, 1980, **21**, 2915
Reduction of the double bond in the anilide by magnesium in methanol is vigorously exothermic after an induction period, and efficient cooling is then necessary.

See Methanol, above.
See other INDUCTION PERIOD INCIDENTS

Molybdenum disulfide
See Molybdenum disulfide, Magnesium

Nitric acid, 2-Nitroaniline
See Nitric acid: Magnesium, 2-Nitroaniline

Nitrogen
See Nitrogen (Liquid): Magnesium

Other reactants
Yoshida, 1980, 347–348
MRH values calculated for 31 combinations, largely with oxidants, are given.

Oxidants   MRH values show % of oxidant
See Carbon dioxide; Halogens etc.; Metal oxides; Metal oxosalts, all above;
    and Sulfur, etc., below

Ammonium nitrate: Metals	MRH 8.79/69
Barium peroxide: Metals	
Dinitrogen tetraoxide: Metals	MRH 12.97/50
Hydrogen peroxide: Metals	MRH 12.68/59
Lead(IV)oxide: Metals	MRH 3.22/83
Nitric acid: Metals	MRH 11.97/52
Oxygen (Liquid): Metals	
Potassium chlorate: Metals	MRH 9.50/63
Potassium perchlorate: Powdered metals	
Sodium iodate: Metals	
Sodium nitrate: Magnesium	MRH 8.58/58

Poly(tetrafluoroethylene)
1. Anon., *Indust. Res.*, 1969(9), 15
2. Diercks, B. V., *J. Haz. Mat.*, 1986, **12**(1), 3
3. Potter, L. J., *Proc. 13th Int. Pyrotech. Semin.*, 1988, 639–659
Finely divided magnesium and Teflon are described as a hazardous combination
of materials [1]. The compressed pellets of powdered metal and polymer used as
infra-red decoy flares have been found sensitive to ignition by electrostatic sparks
during manufacture [2]. The ignition sensitivity of igniter mixtures containing
magnesium, Teflon and Viton was studied, following two fires [3].
See Fluoropolymers, *also* Halocarbons, both above

Potassium carbonate   MRH 3.10/66
Winckler, C., *Ber.*, 1890, **23**, 442
The mixture of magnesium and potassium carbonate recommended by Castellana
as a safe substitute for molten sodium in the Lassaigne test can itself be hazardous,

as an equimolar mixture gives an explosive substance (possibly 'carbonylpotassium', potassium benzenehexoxide) on heating.

Rusty steel

Billinge, K., *Fire Prev. Sci. Technol.*, 1981, (24), 13–19

Impact of a magnesium anode on rusty steel led to ignition and explosion in a tanker.

See other THERMITE REACTIONS

Silicon dioxide

1. Barker, W. D., *School Sci. Rev.*, 1938, **20**(77), 150
2. Marle, E. R., *Nature*, 1909, **82**, 428

Heating a mixture of powdered magnesium and silica (later found not to be absolutely dry) caused a violent explosion rather than the vigorous reaction anticipated [1]. A warning had been published previously [2].

See also FLASH POWDER

Sulfur, or Tellurium   MRH Sulfur 6.15/57

1. Brauer, 1963, Vol. 1. 913
2. Hutton, K., *School Sci. Rev.*, 1950, **31**(114), 265
3. Bodner, G. M., *J. Chem. Educ.*, 1985, **62**(12), 1107

Interaction of magnesium with sulfur or tellurium at elevated temperatures may be violent [1] or even explosive [2], even with tiny amounts [3].

See Sulfur: Metals

Water

1. Shidlovskii, A. A., *Chem. Abs.*, 1947, **41**, 1105d
2. Taylor, F. R. *et al.*, *Rept. ARAED-T-87022*, 1–26, 1988; O/No. AD A187158, NTIS, Richmond (Va.)

Mixtures of magnesium (or aluminium) powder with water can be caused to explode powerfully by initiation with a boosted detonator [1]. The use of organic coatings on magnesium or aluminium powder in pyrotechnic compositions prevents reaction with atmospheric moisture and problems resulting from hydrogen evolution [2].

See Methanol, above; *also* Aluminium–magnesium alloy: Water

Water, Uranium

1. Anon., *Loss Prev. Bull.*, 1992, **103**, 19
2. Pitt, M. J., personal communication, 1993

A drum of graphite with magnesium and uranium, stored underwater with a 6mm diameter relief hole, burst and scattered its contents. This was attrributed to insufficient venting of hydrogen evolved by reaction of magnesium and water [1]. If the uranium was present as metal, it seems likely to have been an even more potent source of hydrogen [2].

See other CORROSION INCIDENTS
See other METALS, REDUCANTS

4691. Magnesium permanganate
[10377-62-5]   MgMn$_2$O$_8$

$$Mg(MnO_4)_2$$

Organic compounds
See Zinc permanganate: Organic compounds
See other METAL OXOMETALLATES, OXIDANTS

4692. Magnesium nitrite
[15070-34-5]   MgN$_2$O$_4$

$$Mg(NO_2)_2$$

Ephraim, 1939, 686–687
Less stable, more hygroscopic and readily hydrolysed than alkali metal nitrites, it
may function as a more powerful oxidant.
See other METAL NITRITES, OXIDANTS

4693. Magnesium nitrate
[10377-60-3]   MgN$_2$O$_6$

$$Mg(NO_3)_2$$

HCS 1980, 601

Dimethylformamide   MRH 5.06/22
'DMF Brochure', Billingham, ICI Ltd., 1965
Magnesium nitrate has been reported to undergo spontaneous decomposition in
DMF, (possibly as a result of hydrolysis of the hexahydrate above its m.p., 90°C
to liberate nitric acid). Although this effect has not been observed with other
nitrates, reaction mixtures with hydrolysable nitrates should be treated with care.
See Sodium nitrate: Jute, Magnesium chloride

Other reactants
Yoshida, 1980, 205
MRH values calculated for 16 combinations with oxidisable materials are given.

Tin(II) fluoride
Denes, G., *Chem. Eng. News*, 1988, **66**(14), 2
Addition of an aqueous solution of the fluoride to one of the nitrate unexpectedly
gave no precipitate of magnesium fluoride, and the solution was allowed to evaporate, first at ambient temperature, and then in a warm place, giving a wet white
solid. While being manipulated with a spatula, this exploded with some violence,
brown fumes were evolved and the solid was transformed into a dry tan-coloured
powder. It was thought that oxidation of tin(II) to (IV) had occurred, the nitrate
being reduced to nitrogen oxide.
See Tetrahydroxotritin(2+) nitrate

See other REDOX REACTIONS
See other METAL NITRATES, OXIDANTS

4694. Magnesium azide
[39108-12-8]   MgN$_6$

$$Mg(N_3)_2$$

Mellor, 1940, Vol. 8, 350
It explodes on heating.
See other METAL AZIDES

4695. Magnesium oxide
[1309-48-4]   MgO

$$MgO$$

HCS 1980, 602

Interhalogens
See Bromine pentafluoride: Acids, etc., Chlorine trifluoride: Metals, etc.

Phosphorus pentachloride
See Phosphorus pentachloride: Magnesium oxide
See other INORGANIC BASES, METAL OXIDES

4696. Magnesium sulfate
[18939-43-0]   MgO$_4$S

$$MgSO_4$$

HCS 1980, 603

Ethoxyethynyl alcohols
See ETHOXYETHYNYL ALCOHOLS See other METAL OXONON-METALLATES

Magnesium
Shimizu, T., Chem. Abs., 1991, **114**, 9175
The mixture with magnesium powder has been described as a pyrotechnic explosive.
See Magnesium: Metal oxosalts

4697. Magnesium silicide
[22831-39-6]   Mg$_2$Si

$$Mg:Si:Mg$$

Acids, Water
Sorbe, 1968, 84

Contact with moisture under acidic conditions generates silanes which ignite in air.
See other METAL NON-METALLIDES

4698. Magnesium nitride
[12057-71-5] Mg_3N_2

$$Mg_3N_2$$

See entry REFRACTORY POWDERS
See other NITRIDES

4699. Magnesium phosphide
[12057-74-8] Mg_3P_2

$$Mg_3P_2$$

Oxidants
   Mellor, 1940, Vol. 8, 842
   It ignites on heating in chlorine, or in bromine or iodine vapours at higher temper-
   atures. Reaction with nitric acid causes incandescence.

Water
   Mellor, 1940, Vol. 8, 842
   Phosphine is evolved and may ignite.
   See other METAL NON-METALLIDES

4700. Manganese
[7429-96-5] Mn

$$Mn$$

NSC 306, 1982

Aluminium
   Occup. Hazards, 1966, **28**(11), 44
   The finely divided metal is pyrophoric, and a mixture of manganese and aluminium
   dusts accidentally released from a filter bag exploded violently.

Carbon dioxide
   See Carbon dioxide: Metals

Other reactants
   Yoshida, 1980, 352
   MRH values calculated for 16 combinations with oxidants are given.

Oxidants   MRH Chlorine 3.85/56, hydrogen peroxide 5.36/48, nitric acid 4.85/41
   Mellor, 1942, Vol. 12. 186–188
   Powdered manganese ignites and becomes incandescent in fluorine or on warming
   in chlorine. Contact with conc. hydrogen peroxide causes violent decomposition

and/or ignition, and with nitric acid incandescence and a feeble explosion were observed.

See Ammonium nitrate: Metals   MRH 3.80/69
    Bromine pentafluoride: Acids, etc.   MRH 4.52/65
    Nitryl fluoride: Metals

Phosphorus
    See Phosphorus: Metals

Sulfur dioxide
    Mellor, 1942, Vol. 12, 187
    Pyrophoric manganese burns brilliantly on warming in sulfur dioxide.
    See other METALS, PYROPHORIC METALS

4701. Manganese(II) nitrate
[10377-66-9]   MnN₂O₆
                              Mn(NO₃)₂

Other reactants
    Yoshida, 1980, 206
    MRH values calculated for 16 combinations with oxidisable materials are given.

Urea
    Novikov, A. V. et al., Chem. Abs., 1975, **82**, 48151
    The anhydrous complex with urea finally decomposes at 240°C with a light explosion.
    See also AMMINEMETAL OXOSALTS
    See other METAL NITRATES

4702. Manganese(III) azide
[[ ] (not isolated)]   MnN₉
                              Mn(N₃)₃

Alkenes
    1. Fristad, W. F. et al., J. Org. Chem., 1985, **50**, 3647–3649
    2. Chem. Abs., **109**, 82979g
    The preparation and immediate use of manganese(III) azide species generated slowly in situ by refluxing manganese(III) acetate and sodium azide in acetic acid in presence of alkene reaction substrates to prepare 1,2-diazidoalkanes avoids the need to isolate manganese(III) azide which has a high probability of explosive instability [1]. The (II) azide is known [20260-90-6], isolable, detonable and has been patented as a power source for lasers [2].

Styrene
    See Sodium azide: Manganese(III) salts, Styrene
    See related METAL AZIDES

4703. Sodium permanganate
[10101-50-5]   MnNaO$_4$

$$NaMnO_4$$

Acetic acid, or Acetic anhydride
  See Potassium permanganate: Acetic acid

Methanol
  US Governement Source, Internet (www.dne.bnl.gov/etd/csc), 1996
  An exothermic reaction from these two is reported, causing boiling and spraying
  of the mixture. It is not clear if anything else (acid?) was involved.

Other reactants
  Yoshida, 1980, 93
  MRH values calculated for 20 combinations with oxidisable materials are given.
  See other METAL OXOMETALLATES, OXIDANTS

4704. Manganese(II) oxide
[1344-43-0]   MnO

$$MnO$$

Hydrogen peroxide
  See Hydrogen peroxide: Metals, etc.
  See other METAL OXIDES

4705. Manganese(IV) oxide (Manganese dioxide)
[1313-13-9]   MnO$_2$

$$MnO_2$$

  HCS 1980, 609

Aluminium
  Sidgwick, 1950, 1265
  Interaction on heating is very violent (a thermite reaction).
  See Aluminium: Metal oxides
  See other THERMITE REACTIONS

Anilinium perchlorate
  See Anilinium perchlorate: Metal oxides

Calcium hydride
  Pascal, 1958, Vol. 4, 304
  Interaction becomes incandescent on warming.
  See other REDOX REACTIONS

Carbon black, Chlorinated paraffin, Lead(IV) oxide, Manganese(IV) oxide
  See Lead(IV) oxide: Carbon black, etc.

Diboron tetrafluoride
See Diboron tetrafluoride: Metal oxides

Rubidium acetylide
See Rubidium acetylide: Metal oxides

Hydrogen sulfide
See Hydrogen sulfide: Metal oxides

Hydroxylaminium chloride
See Hydroxylaminium chloride: Manganese dioxide

Lithium
Hamportzumian, K. *et al., J. Power Sources*, 1982, **8**, 35–40
Manganese dioxide–lithium batteries arc generally safer than sulfur dioxide–lithium cells.
See Lithium: Sulfur dioxide

Oxidants   MRH Chlorine trifluoride 1.80/52, hydrogen peroxide 1.59/99+
1. Mellor, 1942, Vol. 12, 254
2. Mellor, 1941, Vol. 1, 936
3. Ahrle, H., *Z. Angew. Chem.*, 1909, **22**, 713
4. von Schwartz, 1918, 323
5. Blanco Prieto, J., *Chem. Abs.*, 1976, **84**, 46791
Action of chlorine trifluoride causes incandescence [1]. Manganese dioxide catalytically decomposes powerful oxidising agents, often violently. Dropped into conc. hydrogen peroxide, the powdered oxide may cause explosion [2]. Either the massive or the powdered oxide explosively decomposes 92% peroxomonosulfuric acid [3], and mixtures with chlorates ('oxygen mixture', heated to generate the gas) may react with explosive violence [4]. Cuban pyrolusite can be used in place of potassium dichromate to promote thermal decomposition of potassium chlorate in match-head formulations [5].
See Peroxyformic acid: Metals, etc.

Potassium azide
Mellor, 1940, Vol. 8, 347
On gentle warming, interaction is violent.
See other METAL OXIDES, OXIDANTS

4706. Manganese(II) sulfide
[18820-29-6]   MnS

                                        MnS

Mellor, 1942, Vol. 12, 394
The vacuum-dried red sulfide becomes red-hot on exposure to air.
See other METAL SULFIDES, PYROPHORIC MATERIALS

4707. Manganese(IV) sulfide
[12125-23-4]   MnS$_2$

$$MnS_2$$

Sorbe, 1968, 84
It explodes at 580°C.
See other METAL SULFIDES

4708. Manganese(II) telluride
[12032-88-1]   MnTe

$$MnTe$$

Lithium
   See Lithium: Metal oxides, etc.
   See other METAL NON-METALLIDES

4709. Manganese(VII) oxide
[12057-92-0]   Mn$_2$O$_7$

$$O_3MnOMnO_3$$

1. Mellor, 1942, Vol. 12, 293
2. Bailar, 1973, Vol. 3, 805
An unstable powerful oxidant, it explodes between 40 and 70°C, or on friction
or impact, sensitivity being as great as that of mercury fulminate [1]. Detonation
occurs at 95°C, and under vacuum explosive decomposition occurs above 10°C [2].
See Potassium permanganate: Sulfuric acid

Organic material
   1. Patterson, A. M., *Chem. Eng. News*, 1948, **26**, 711
   2. Mellor, 1940, Vol. 12, 293
   A sample of the heptoxide exploded in contact with the grease on a stopcock
   [1], and explosion or ignition has been noted with various solvents, oils, fats and
   fibres [2].
   See other METAL OXIDES, OXIDANTS

4710. Zinc permanganate
[23414-72-4]   Mn$_2$O$_8$Zn

$$Zn(MnO_4)_2$$

Cellulose
   Sörgel, U. C., *Pharm. Prax., Beil. Pharmaz.*, 1960, (2), 30
   An explosion involved contact of the permanganate with cellulose.

Organic compounds
   Wolfe, S. *et al., J. Amer. Chem. Soc.*, 1983, **105**, 7755–7757

Solid zinc permanganate and magnesium permanganate react violently with many classes of organic compounds. The reactions may be controlled successfully by supporting the oxidants on silica. The zinc salt is more effective than the magnesium salt, but both are much more reactive than potassium permanganate.

See other METAL OXOMETALLATES, OXIDANTS

4711. Manganese phosphide
[12263-33-1]

$$Mn_3P_2$$

$$Mn_3P_2$$

Chlorine
   See Chlorine: Phosphorus compounds
   See other METAL NON-METALLIDES

4712. Molybdenum
[7439-98-7]

$$Mo$$

$$Mo$$

Oxidants   MRH value shows % of oxidant
   See Bromine pentafluoride: Acids, etc.
      Bromine trifluoride: Halogens, etc.
      Chlorine trifluoride: Metals
      Fluorine: Metals   MRH 7.57/54
      Iodine pentafluoride: Metals
      Lead(IV) oxide: Metals
      Nitryl fluoride: Metals
      Potassium perchlorate: Metal powders
      Sodium peroxide: Metals
   See other METALS

4713. Sodium molybdate
[10102-40-6]

$$MoNa_2O_4$$

$$Na_2MoO_4$$

Aniline, Nitrobenzene
   Nolan, 1983, Case history 144
   In the preparation of a dyestuff from aniline, nitrobenzene (as oxidant), hydrochloric acid and sodium hydroxide, ferric chloride is often used as catalyst, but sodium molybdate was substituted as a more effective catalyst. The materials were charged into a 4.5 m^3 reactor and heating was started after addition of nitrobenzene, but the temperature controller was mis-set, and overheating at a high rate ensued. The exotherm was much higher than normal because of the more effective catalyst, and partial failure of the cooling water led to an uncontrollable exotherm.
   See other METAL OXOMETALLATES, RUNAWAY REACTIONS

4714. Molybdenum nitride
[37245-81-1] $\qquad\qquad$ Mo_nN
[12033-19-1] MoN; [12033-31-7] Mo_2N; [97953-19-0] Mo_4N
Complex structures

Nitric acid
Chorley, R. W., *Chem. Brit.*, 1993, **29**(10), 867
High surface area samples may be pyrophoric, but can be passivated by exposure to oxygen. A passivated sample (50 mg) caused a small explosion when concentrated nitric acid was added to it.
See other NITRIDES
See also MIXING

4715. Sodium tetraperoxomolybdate
[42489-15-6] (ion) $\qquad\qquad$ $MoNa_2O_8$
$$Na_2[Mo(O_2)_4]$$

Castrantas, 1965, 5
It decomposes explosively under vacuum.
See other PEROXOACID SALTS

4716. Molybdenum(IV) oxide
[18868-43-4] $\qquad\qquad$ MoO_2
$$MoO_2$$

Gibson, 1969, 102
It becomes incandescent in contact with air.
See other METAL OXIDES, PYROPHORIC MATERIALS

4717. Molybdenum(VI) oxide
[1313-27-5] $\qquad\qquad$ MoO_3
$$MoO_3$$

Graphite
1. Young, J. A., private comm., 1989; *CHAS Notes*, 1992, **X**(6), 2
2. Partington, 1967, 753
During preparation of an alloy steel, molybdenum trioxide (372 kg), and then graphite (340 kg, 1.56 equiv., in 4 portions), were charged into an electric furnace containing a charge of melted scrap steel at 1500–1600°C, using a charging ram and rotatable box for the additions through the furnace door. Immediately after the last graphite addition, as the ram and box were being withdrawn, a large fireball emerged through the furnace door and fatally burned the ram operator. This was caused by the very rapid exothermic redox reaction between the oxide and graphite to give molybdenum and a huge volume (some 800 m^3) of very

hot carbon monoxide) which emerged at high velocity from the 27 m³ furnace and ignited, giving a flame-jet some 15 m long and 2–3 m in diameter [1]. It is possible that some of the graphite particles were ejected in the gas jet and augmented the combustion intensity. Reduction of the trioxide to molybdenum by strongly heating with carbon has long been a commercial metallurgical process [2].
See other REDOX REACTIONS

Interhalogens
See Bromine pentafluoride: Acids, etc.
    Chlorine trifluoride: Metals. etc.

Metals
Mellor, 1943, Vol. 11, 542
Reduction of the oxide by heated sodium or potassium proceeds with incandescence, and explosion occurs in contact with molten magnesium.
See Lithium: Metal oxides
See other METAL OXIDES

4718. Oxodiperoxomolybdenum–hexamethylphosphoramide
[53474-95-6 + 680-31-9] $MoO_5.C_6H_{18}N_3OP$
$MoO(O_2)_2.O:P(NMe_2)_3$

Fieser, 1974, Vol. 4, 203
The monohydrated complex exploded after storage at ambient temperature for a month. Refrigerated storage is recommended.
See related METAL PEROXIDES, AMMINECHROMIUM PEROXOCOMPLEXES

4719. Molybdenum(IV) sulfide
[1317-33-5] MoS_2
MoS_2

Magnesium
Danielyan, N. G. *et al., Mod. Phys. Lett. B.*, 1991, **5**(19), 1301
Molybdenum disulphide is capable of forming a thermite type mixture with magnesium, which may explode on grinding.
See also THERMITE REACTIONS

Potassium nitrate
See Potassium nitrate: Metal sulfides
See other METAL SULFIDES

4720. Sodium nitrite
[7632-00-0] $NNaO_2$
NaON:O

FPA H133, 1984; *HCS 1980*, 852

Aminoguanidine salts
  Urbanski, 1967, Vol. 3, 207
  Interaction, without addition of acid, produces tetrazolylguanidine ('tetrazene'), a primary explosive of equal sensitivity to mercury(II) azide, but more readily initiated.

Ammonium salts   MRH Ammonium chloride 1.80/44, nitrate 1.46/99+
  1. von Schwartz, 1918, 299
  2. Mellor, 1967, Vol. 8, Suppl. 2.1, 388
  3. *RoSPA Occ. Safety and Health Suppl.*, 1972, **2**(10), 32
  Heating a mixture of an ammonium salt with a nitrite salt causes a violent explosion on melting [1], owing to formation and decomposition of ammonium nitrite. Salts of other nitrogenous bases behave similarly. Mixtures of ammonium chloride and sodium nitrite are used as commercial explosives [2]. Accidental contact of traces of ammonium nitrate with sodium nitrite residues caused wooden decking on a truck to ignite [3].
  See wood, below; *also* NITRITE SALTS OF NITROGENOUS BASES

1,3-Butadiene
  1. Beer, R. N., private comm., 1972
  2. Keister, R. G. *et al., Loss. Prev.*, 1971, **5**, 69
  Sodium nitrite solution is used to inhibit 'popcorn' polymerisation of butadiene in processing plants. If conc. nitrite solutions (5%) are used, a black sludge is produced, which when dry, will ignite and burn when heated to 150°C, even in absence of air [1]. The sludge produced from use of nitrite solutions to scavenge oxygen in butadiene distillation systems contained 80% of organic polymer and a nitrate:nitrite ratio of 2:1. Use of dilute nitrite solution (0.5%) or pH above 8 prevents sludge formation [2].
  See Buten-3-yne: 1,3-Butadiene, *and* Nitrogen oxide: Dienes

Hydroxylamine hydrochloride
  Scott, E. S., *J. Chem. Ed.*, 1992, **69**(12), 1028
  If hydroxylamine hydrochloride and sodium nitrite, of normal laboratory particle size, are swiftly shaken together and left to stand, a vigorous gas evolving reaction will start after a short induction period. This is a rare example of spontaneous reaction of solids
  See Ammonium salts, above

Maize husks, Sodium nitrate
  See Sodium nitrate: Maize husks, etc.

Metal amidosulfates
  Heubel, J. *et al., Compt. rend.*, 1963, **257**, 684–686
  Interaction of nitrites when heated with metal amidosulfates (sulfamates) may become explosively violent owing to liberation of nitrogen and steam. Mixtures with ammonium sulfamate form ammonium nitrite which decomposes violently around 80°C.
  See Sodium nitrate: Metal amidosulfates

1774

Metal cyanides
1. von Schwartz, 1918, 299
2. Mellor, 1940, Vol. 8, 478
3. Greenwood, P. H. S., *J. Proc. R. Inst. Chem.*, 1947, 137
4. Elson, C. H. R., *J. Proc. R. Inst. Chem.*, 1947, 19
5. Eiter, K. *et al.*, Austrian Pat. 176 784, 1953

Mixtures of sodium nitrite and various cyanides [1] explode on heating, including potassium cyanide [2], potassium hexacyanoferrate(III), sodium pentacyanonitrosylferrate(II) [3], potassium hexacyanoferrate(II) [4], or mercury(II) cyanide [5]. Such mixtures have been proposed as explosives, initiable by heat or a detonator [5].

Other reactants
Yoshida, 1980, 13
MRH values calculated for 19 combinations with oxidisable materials are given.

Paper, Sulfur
Anon., *CIHSC Chem. Safety Summ.*, 1984, **55**(218), 36
Empty paper sacks, some of which had contained sulfur or inorganic salts but mostly sodium nitrite, were dumped (with general rubbish) into a skip located inside by windows in direct sunlight. The fire which resulted was probably caused by exothermic interaction of the sodium nitrite with sulfur and damp paper, with further self heating aided by the generally insulating nature of the mound of paper sacks in the draught-free skip warmed by sunlight.
See Wood, below
See other SELF-HEATING AND GNITION INCIDENTS

Phenol
Hatton, J. P., private comm., 1976
A mixture exploded violently on heating in a test tube.

Phthalic acid, or Phthalic anhydride
Hawes, B. V. W., *J. R. Inst. Chem.*, 1955, **79**, 668
Mixtures of sodium nitrite and phthalic acid or phthalic anhydride explode violently on heating. A nitrite ester may have been produced.
See ACYL NITRITES

Potassium thiocyanate   MRH (sodium salt) 3.57/28
1. Mellor, 1940, Vol. 8, 478
2. Baum, R., *Chem. Eng. News*, 1982, **60**(41), 29
3. Fawcett, H. H., *Chem. Eng. News*, 1982, **60**(46), 4
4. Shriver, D. F., also Coleman, L. F., *Chem. Eng. News*, 1982, **60**(47), 2
5. Pinsky, B., *Chem. Eng. News*, 1982, **60**(47), 2
6. Viola, A., *Chem. Eng. News*, 1983, **61**(3), 76

A mixture of the nitrite with sodium thiocyanate explodes on heating [1]. Preparation of a molten salt bath from 0.45 kg of potassium thiocyanate (reducant) and 1.35 kg of sodium nitrite (oxidant) led to a violent explosion on melting, which caused severe structural damage to the laboratory [2]. It was claimed that this could not have been foreseen from available information [2], but this was vigorously refuted [3,4]. Use of sand baths, rather than salt baths for laboratory heating purposes is to be preferred [5]. The relative lack of descriptive chemistry in modern curricula is cited as the major contributory factor to the general ignorance which led to the explosion [6].

See METAL THIOCYANATES: oxidants, MOLTEN SALT BATHS
See other REDOX REACTIONS

Reducants
    See Potassium thiocyanate, above
        Sodium disulfite; Sodium thiosulfate, both below
    See Sodium nitroxylate

Sodium amide
    Bergstrom, F. W. *et al., Chem. Rev.*, 1933, **12**, 64
    Addition of solid sodium nitrite to the molten amide caused immediate gas evolution, followed by a violent explosion.

Sodium disulfite   MRH 1.84/68
    MCA Case History No. 183
    Large-scale addition of solid sodium disulfite to an unstirred and too-concentrated solution of sodium nitrite caused a vigorous exothermic reaction.
    See other REDOX REACTIONS

Sodium nitrate, Sodium sulfide
    Janz, 1976, Table 3, 22
    Accidental mixing of the 3 molten salts caused a violent explosion.
    See other REDOX REACTIONS

Sodium thiocyanate
    See Potassium thiocyanate, above
    See other REDOX REACTIONS

Sodium thiosulfate
    Mellor, 1940, Vol. 8, 478; 1947, Vol. 10, 501
    There is no interaction between solutions, but evaporation of the mixture gave a residue which explodes on heating. The mixed solids behave similarly.
    See other REDOX REACTIONS

Urea   MRH 2.80/30
    Bucci, F., *Ann. Chim.* (Rome), 1951, **41**, 587
    Fusion of urea (2 mol) with sodium nitrite (or potassium nitrite, 1 mol of either) to give high yields of the cyanate must be carried out exactly as described to avoid the risk of explosion.

Wood
    Anon., *ABCM Quart. Safety Summ.*, 1944, **15**, 30

Wooden staging, which had become impregnated over a number of years with sodium nitrite, became accidentally ignited and burned as fiercely as if impregnated with potassium chlorate. Although the effect of impregnating cellulosic material with sodium nitrate is well known, that due to sodium nitrite was unexpected.

See Ammonium salts (reference 3), *also* Paper, Sulfur, both above
See other METAL NITRITES, METAL OXONON-METALLATES, OXIDANTS

4721. Sodium nitrate
[7631-99-4]   NNaO$_3$

NaNO$_3$

FPA H26, 1983; *HCS 1980*, 851

Acetic anhydride
See Acetic anhydride: Metal nitrates

Aluminium   MRH Aluminium 4.43/27
1. Anon., *Fire*, 1935, **28**, 30
2. Farnell, P. L. *et al., Chem. Abs.*, 1972, **76**, 129564
3. Gerling, H. W., *Metall* (Berlin), 1982, **36**, 698–699
Mixtures of the nitrate with powdered aluminium or its oxide (the latter seems unlikely) were reported to be explosive [1], and the performance characteristics of flares containing compressed mixtures of the metal and nitrate have been evaluated [2]. A violent explosion in a copper smelting works was caused mainly by reaction of aluminium with sodium nitrate [3].

Aluminium, Water
Jackson, B. *et al., Chem. Abs.*, 1976, **84**, 46798
During investigation of pyrotechnic flare formulations, it was found that mixtures of the metal powder and oxidant underwent a low-temperature exothermic reaction at 70–135°C in presence of moisture.

Antimony
Mellor, 1939, Vol. 9, 382
Powdered antimony explodes when heated with an alkali metal nitrate.

Arsenic trioxide, Iron(II) sulfate
Dunn., B. W., *Bur. Expl. Rept. 13*, Pennsylvania, Amer. Rail Assoc., 1920
A veterinary preparation containing the oxidant and reducant materials (possibly with some additional combustibles) ignited spontaneously.
See other REDOX REACTIONS

Barium thiocyanate
Pieters, 1957, 30
Mixtures may explode
See Sodium thiocyanate, below

Bitumen

Xu, Y.-C. *et al.*, *Chem. Abs.*, 1980, **93**, 209761

The induction periods for the reaction of sodium nitrate–bitumen mixtures (43:57 wt) heated at 195, 234 or 260°C are 44, 2 and 0.5 h, respectively. Further study of sodium nitrate/bitumenised waste systems held at these temperatures showed an initial weak exotherm around 260°C and a larger exotherm (0.96–1.21 kJ/g) accompanied by 50% wt loss around 430°C.

Boron phosphide

Mellor, 1940, Vol. 8, 845

Deflagration occurs in contact with molten alkali metal nitrates.

Calcium–silicon alloy

Smolin, A. O., *Chem. Abs.*, 1974, **81**, 138219

Combustion of the alloy in admixture with sodium nitrate is mentioned in an explosives context, but no details are translated.

Fibrous material   MRH Cellulose 3.85/30

1. Mellor, 1961, Vol. 2, Suppl. 2.1, 1244
2. Anon., *ABCM Quart. Safety Summ.*, 1944, **15**, 30

Fibrous organic material (jute storage bags) is oxidised in contact with sodium nitrate above 160°C and will ignite below 220°C [1]. Wood and similar cellulosic materials are rendered highly combustible by nitrate impregnation [2].

Jute, Magnesium chloride

van Hoogstraten, C. W., *Chem. Abs.*, 1947, **41**, 3692b

Solid crude sodium nitrate packed in jute bags sometimes ignited the latter in storage. Normally ignition did not occur below 240°C, but in cases where magnesium chloride (up to 16%) was present, ignition occurred at 130°C. This was attributed to formation of magnesium nitrate hexahydrate, which hydrolyses above its m.p. (90°C) liberating nitric acid. The latter was thought to have caused ignition of the jute bags under unusual conditions of temperature and friction.

See Magnesium nitrate: Dimethylformamide

Magnesium   MRH 8.58/42

Bond, B. D. *et al.*, *Combust. Flame*, 1966, **10**, 349–354

A study of the kinetics in attack of magnesium by molten sodium nitrate indicates that decomposition of the nitrate releases oxygen atoms which oxidise the metal so exothermally that ignition ensues.

See Magnesium: Metal oxosalts

Maize husks, Sodium nitrite

Anon., *CISHC Chem. Safety Summ.*, 1985, **52**(222), 66

A rotary drum had been used previously to dry metal components which had been heat-treated in nitrate–nitrite molten salt baths, washed, then tumble dried with ground maize husks to absorb adhering water. When the drum was taken out of

service, it was not cleaned out. After some 10 months it was recommissioned, but while being heated up to operating temperature, an explosion occurred which ejected flame jets for several m. This was attributed to presence of considerable contamination of the maize husks by metal nitrate–nitrite residues, and ignition on heating of such material was confirmed experimentally.
See MOLTEN SALT BATHS

Metal amidosulfates
Heubel, J. *et al., Compt. rend.*, 1962, **255**, 708–709
Interaction of nitrates when heated with amidosulfates (sulfamates) may become explosively violent owing to liberation of dinitrogen oxide and steam.
See Sodium nitrite: Metal amidosulfates

Metal cyanides   MRH Sodium cyanide 4.31/37
See MOLTEN SALT BATHS

Non-metals   MRH Phosphorus (r) 4.43/27, sulfur 3.39/24
1. Mellor, 1941, Vol. 2, 820
2. Leleu, *Cahiers*, 1980, (99), 279
Contact of powdered charcoal with the molten nitrate, or of the solid nitrate with glowing charcoal, causes vigorous combustion of the carbon. Mixtures with charcoal and sulfur have been used as black powder [1]. Charcoal powder–nitrate mixtures burn briskly at 200°C [2].

Other reactants
Yoshida, 1980, 199
MRH values calculated for 19 combinations with oxidisable materials are given.

Peroxyformic acid
See Peroxyformic acid: Sodium nitrate

Phenol, Trifluoroacetic acid
Spitzer, U. A. *et al., J. Org. Chem.*, 1974, **39**, 3936
When the salt–acid nitration mixture was applied to phenol, a potentially hazardous rapid exothermic reaction occurred producing tar.
See other NITRATION INCIDENTS

Sodium   MRH 3.85/57
Mellor, 1961, Vol. 2, Suppl. 2.1, 518
Interaction of sodium nitrate and sodium alone, or dissolved in liquid ammonia, eventually gives a yellow exlosive compound.
See Sodium nitroxylate

Sodium nitrite, Sodium sulfide
See Sodium nitrite: Sodium nitrate, Sodium sulfide

Sodium phosphinate   MRH 1.00/56
Mellor, 1941, Vol. 2, 820

A mixture exploded violently on warming.
See other REDOX REACTIONS

Sodium thiocyanate
Anon., *Sichere Chemiearbeit*, 1994, **46**, 80
Sodium nitrate and sodium thiocyanate were being blended in a ploughshare mixer to produce a plating formulation. The blend deflagrated, burning four workers and blowing apart the mixer. Subsequent laboratory investigation showed such mixtures to be friction sensitive. The firm has ceased to prepare such mixtures. Caution is advised when mixing oxidants and reducants.

Sodium thiosulfate
Mellor, 1941, Vol. 2, 820
A mixture is explosive when heated.
See other REDOX REACTIONS

Tris(cyclopentadienyl)cerium
1. Jakob, K., *Z. Anorg. Allg. Chem.*, 1988, **556**, 170
2. Grodett, P. S., *Chem. Eng. News*, 1988, **66**(31), 2
Reaction of ammonium hexanitrocerate and cyclopentadienylsodium under inert conditions gives tris(cyclopentadienyl)cerium and sodium nitrate, removed by filtration before evaporation of solvent [1]. When the filtration step was omitted, and the evaporated solid mixture was heated to 75°C, a violent explosion occurred. This may have involved complexes of the type $Ce(NO_3)Cp_2 \cdot NaNO_3$[2], but a direct redox reaction between the reactive $CeCp_3$ and the oxidant is also possible.
See other REDOX REACTIONS

Wood
See Fibrous material, above
See other METAL NITRATES, OXIDANTS

4722. Sodium nitroxylate
[13968-14-4] NNa_2O_2
                                        Na[O:NONa]

Air, or Carbon dioxide, or Water
Mellor, 1963, Vol. 2, Suppl. 2.2, 1566
The solid is very reactive towards air, moisture or carbon dioxide, and tends to explode readily, also decomposing violently on heating. It is produced from sodium nitrate or sodium nitrite by electrolytic reduction, or action of sodium.
See other N–O COMPOUNDS

4723. Sodium nitride
[12136-83-3]

NNa_3

Na_3N

It decomposes explosively on gentle warming.
See other NITRIDES

4724. Nitrogen oxide ('Nitric oxide')
[10102-43-9]

NO

NO

1. *UK Sci. Mission Rept. 68/79*, Washington, UKSM, 1968
2. Miller, R. O., *Ind. Eng. Chem., Proc. Res. Dev.*, 1968, **7** 590–593
3. Ribovich, J. *et al., J. Haz. Mat.*, 1977, **1**, 275–287
4. Ramsay, J. B. *et al., Proc. 6th Int. Symp. Detonation*, 1976, 723–728

Liquid nitrogen oxide and other cryogenic oxidisers (ozone, fluorine in presence of water) are very sensitive to detonation in absence of fuel, and can be initiated as readily as glyceryl nitrate [1,2]. Detonation of the endothermic liquid oxide close to its b.p. (−152°C) generated a 100 kbar pulse and fragmented the test equipment. It is the simplest molecule that is capable of detonation in all 3 phases [3]. The liquid oxide is sensitive and may explode during distillation [4].

It is highly endothermic (ΔH°_f (g) +90.4 kJ/mol, 3.01 kJ/g) and an active oxidant.

Acetylene, Perchloric acid
See Perchloric acid: Acetylene, Nitric oxide

Alkenes, Oxygen
See Nitrogen dioxide: Alkenes

Ammonia
Checkel, M. D. *et al., J. Loss Prevention*, 1995, **8**(4), 215
Flammability limits of mixtures of nitric oxide and ammonia were studied. The limits are 15–70% ammonia; tenfold pressure rise and explosion are possible. Introduction of oxygen into flammable compositions will produce autoignition, via nitrogen dioxide formation, contamination with air might.

Carbon, Potassium hydrogen tartrate
Pascal, P. *et al., Mém. Poudres*, 1953, **35**, 335–347
The ignition temperature of 400°C for carbon black in nitrogen oxide was reduced by the presence of the tartrate.

Carbon disulfide
Winderlich, R., *J. Chem. Educ.*, 1950, **27**, 669
A demonstration of combustion of carbon disulfide in nitrogen oxide (both endothermic compounds) exploded violently.

Dichlorine oxide
See Dichlorine oxide: Nitrogen oxide

Dienes, Oxygen
1. Haseba, S. *et al., Chem. Eng. Progr.*, 1966, **62**(4), 92
2. Schuftan, P. M., *Chem. Eng. Progr.*, 1966, **62**(7), 8
Violent explosions which occurred at −100 to −180°C in ammonia synthesis gas units were traced to the formation of explosive addition products of dienes and oxides of nitrogen, produced from interaction of nitrogen oxide and oxygen. Laboratory experiments showed that the addition products from 1,3-butadiene or cyclopentadiene formed rapidly at about −150°C, and ignited or exploded on warming to −35 to −15°C. The unconjugated propadiene, and alkenes or acetylene reacted slowly and the products did not ignite until +30 to +50°C [1]. This type of derivative ('pseudo-nitrosite') was formerly used (Wallach) to characterise terpene hydrocarbons. Further comments were made later [2].
See Sodium nitrite: 1,3-Butadiene

Fluorine
See Fluorine: Non-metal oxides

Hydrogen, Nitrous oxide
Glowinski, J. *et al., Przem. Chem.*, 1996, **75**(2), 63
Explosive limits for this combination have been studied from 1 to 20 bar at 50°C.

Hydrogen, Oxygen
Chanmugam, J. *et al., Nature*, 1952, **170**, 1067
Pre-addition of nitrogen oxide (or nitrosyl chloride as its precursor) to stoicheiometric hydrogen−oxygen mixtures at 240 mbar/360°C will cause immediate ignition under a variety of circumstances.

Metal acetylides or carbides
Mellor, 1946, Vol. 5, 848, 891
Rubidium acetylide ignites on heating, and uranium dicarbide incandesces in the gas at 370°C.

Metals
Mellor, 1940, Vol. 8, 436; 1943, Vol. 11, 162; 1942, Vol. 12, 32
Pyrophoric chromium attains incandescence in the oxide, while calcium, potassium and uranium need heating before ignition occurs, when combustion is brilliant in the 53% oxygen content.

Methanol
Partel, G., *Riv. Ing.*, 1965, **15**, 969–976
This oxidant−fuel system was evaluated as a rocket propellant combination.

Nitrogen trichloride
See Nitrogen trichloride: Initiators

Non-metals
    Mellor, 1940, Vol. 8, 109, 433, 435
    Amorphous (not crystalline) boron reacts with brilliant flashes at ambient temperature, and charcoal or phosphorus continue to burn more brilliantly than in air (which has a much lower oxygen content).

Ozone
    See Ozone: Nitrogen oxide

Pentacarbonyliron
    Manchot, W. *et al., Ann.*, 1929, **470**, 275
    Rapid heating to above 50°C in an autoclave caused an explosive reaction.

Perchloryl fluoride
    See Perchloryl fluoride: Hydrocarbons, etc.

Phosphine, Oxygen
    Mellor, 1940, Vol. 8, 435
    Addition of oxygen to a mixture of phosphine and nitrogen oxide causes explosion.
    See Phosphine: Oxygen

Potassium sulfide
    Mellor, 1940, Vol. 8, 434
    The pyrophoric sulfide ignites in the oxide.

Sodium diphenylketyl
    1. Mendenhall, D., *J. Amer. Chem. Soc.*, 1974, **96**, 5000
    2. Mendenhall, D., *Chem. Eng. News*, 1982, **60**(20), 2
    A procedure for preparation of sodium hyponitrite involving reduction of nitrogen oxide by the ketyl, followed by extraction into water [1], had been operated routinely on the small scale. A 41-fold scaled up run exploded and ignited after 200 ml of water had been added as part of the work-up. This was attributed to the presence of an unusually large proportion of coagulated unreacted sodium in the ketyl. Small-scale operation with precautions is urged [2].

Vinyl chloride
    See Vinyl chloride: Oxides of nitrogen
    See other ENDOTHERMIC COMPOUNDS, NON-METAL OXIDES, OXIDANTS

4725. Nitrogen dioxide (Dinitrogen tetraoxide)
    [10102-44-0] NO_2
                                    NO_2

    At ambient temperatures the title gases exist as an equilibrium mixture; for details
    See Dinitrogen tetraoxide

Alkenes
  Rozlovskii, A. I., *Chem. Abs.*, 1975, **83**, 118122
  The mechanisms of explosions in solidified gas mixtures at low temperatures
  containing unsaturated hydrocarbons and oxides of nitrogen is discussed. Fast
  radical addition of nitrogen dioxide to double bonds is involved, and with dienes
  it is a fast reaction of very low energy of activation. Possibilities of preventing
  explosions are discussed.
  See Nitrogen oxide: Dienes, Oxygen
  See other NON-METAL OXIDES, OXIDANTS

4726. Plutonium nitride
  [12033-54-4]   NPu

PuN

Oxygen, Water
  Bailar, 1973, Vol. 5, 343
  The ignition temperature of the nitride in oxygen is lowered from 300° to below
  100°C by the presence of moisture.
  See other NITRIDES

4727. Rubidium nitride
  [12136-85-5]   NRb₃

Rb₃N

Mellor, 1940, Vol. 8, 99
  The alkali nitrides burn in air.
  See other N-METAL DERIVATIVES

4728. Poly(sulfur nitride)
  [56422-03-8]   (NS)ₙ
  (**Complex structure**)

Kennet, F. A. *et al., J. Chem. Soc., Dalton Trans.*, 1982, 851−857
  Hazards involved in various routes to the polymer are detailed.
  See Caesium azide: Sulfur dioxide, *also* Poly(selenium nitride)
  See other NITRIDES, N–S COMPOUNDS

4729. Antimony(III) nitride
  [12333-57-2]   NSb

SbN

Alone, or Water
  1. Fischer, F. *et al., Ber.*, 1910, **43**, 1471
  2. Franklin, E. C., *J. Amer. Chem. Soc.*, 1905, **27**, 850

Explosive decomposition occurs on warming under vacuum [1] and impure material explodes mildly on heating in air, or on contact with water or dilute acids.
See other NITRIDES

4730. Poly(selenium nitride)
[12033-59-9] $(NSe)_n$

(Complex structure)

Preparative hazard
See Trimethylsilyl azide: Selenium halides
See also Poly(sulfur nitride) *See other* NITRIDES
See related N–S COMPOUNDS

4731. Thallium(I) nitride
[12033-67-9] NTl

Tl N

Alone, or Water
Mellor, 1940, Vol. 8, 262
The nitride explodes violently on exposure to shock, heat, water or dilute acids.
See other NITRIDES

4732. Uranium(III) nitride
[25658-43-9] NU

UN

Bailar, 1973, Vol. 5, 340
Finely powdered material is pyrophoric on exposure to air at ambient temperature.
See other NITRIDES, PYROPHORIC MATERIALS

4733. Zirconium nitride
[25658-42-8] NZr

ZrN

See entry REFRACTORY POWDERS *See other* NITRIDES

4734. Nitrogen (Gas)
[7727-37-9] N_2

N_2

Kletz, T. A., *Proc. 3rd Int. Symp. Loss Prev. Safety Prom. Proc. Ind.*, 1519–1527, Basle, SSCI, 1980
The benefits of large scale use of nitrogen as an inerting gas to prevent fire or explosion have been at the expense of many fatal asphyxiations arising from its

accidental use in place of air or from its unexpected presence in confined spaces. At an oxygen concentration below 10%, collapse is almost instantaneous. Various accidents are described with appropriate precautions, and it was concluded that nitrogen may have killed more people than any other single substance in industry.

Lithium
See Lithium(reference 1), *also* Lithium: Metal chlorides

Lithium tetrahydroaluminate
See Lithium tetrahydroaluminate: Nitrogen

Ozone
See Ozone: Nitrogen

Titanium
See Titanium: Nitrogen

4735. Nitrogen (Liquid)

1. Anon., *BCISC Quart. Safety Summ.*, 1964, **35**, 25
2. Anon., *BCISC Quart.Safety Summ.*,1974, **45**, 11
3. Lindsay, W. N., *Chem. Eng. News*, 1987, **65**(24), 2

Although liquid nitrogen is inherently safer than liquid oxygen or liquid air as a coolant, its ability to condense liquid oxygen out of the atmosphere can create hazards. A distillation tube containing a little solvent was cooled in liquid nitrogen while being sealed off in a blowpipe flame. A few minutes late the tube exploded, probably owing to high internal pressure caused by evaporation of liquid oxygen which had condensed into the tube during sealing. It is possible the atmosphere close to the blowpipe was oxygen-enriched. Open vessels which contain organic materials, or which are to be hermetically sealed should not be cooled in liquid nitrogen but in a coolant at a higher temperature. Liquid nitrogen should normally be used only for cooling evacuated or closed vessels where extreme cooling is necessary, and it should be removed from around the vacuum trap or other vessel before opening them to atmosphere [1]. The later reference describes 2 incidents of forceful explosion of 75 l cryogenic containers of liquid nitrogen. The cause of the first appeared to be blockage of the 19 mm neck by an ice-plug formed in a very humid environment. The cause of the second was probably 'roll-over' of a surface layer of superheated liquid, and fracture of the vessel by the ensuing violent boil-off of gas. Superheating of $10-15°C$ is known to occur in high-purity liquid nitrogen, but much higher levels of superheating may be possible. Various remedies (stirring, generation of a warm spot or vertical submerged conductive gauzes) are discussed [2]. With the current heightened interest in new materials which become superconducting at liquid nitrogen temperature ($-195.8°C$), the attention of workers (not necessarily chemists) in this area is drawn to the fact that containers of liquid nitrogen exposed to atmosphere inevitably will condense some oxygen (b.p. $-183°C$) into the nitrogen coolant. Such mixtures may behave as an oxidant and react violently with traces of oil, grease or other organic material. Vacuum flasks of liquid nitrogen should not be left about unused or uncovered [3].

Argon
See Argon: Liquid nitrogen

Fatty materials
Anon., *CISHC Chem Safety Summ.*, 1979, **50**, 91
Use of liquid nitrogen in cryogenic grinding of fatty materials led to an explosion. Condensation of liquid oxygen onto the fatty material, with initiation by the grinding friction seems a likely causative sequence.

Hydrogen
See Hydrogen (Gas): Liquid nitrogen

Magnesium
Ephraim, 1939, 625
A mixture of magnesium powder and liquid nitrogen reacts very violently when lit with a fuse, forming magnesium nitride.

Oxygen, Radiation
1. Chen, C. W. *et al., Cryogenics*, 1969, **9**, 131–132
2. Takehisa, M. *et al., Chem. Abs.*, 1977, **87**, 30593
3. Perdue, P. T., *Health Phys.*, 1972, **23**, 116–117
4. Watanabe, H. *et al., Chem. Abs.*, 1980, **92**, 30294
Liquid nitrogen subject to nuclear radiation (high neutron and gamma fluxes) must be kept free of oxygen to prevent explosions occurring in reactor cryostats. The explosive species generated is not nitrogen oxide or nitrogen dioxide, because solutions of these will not cause explosion on contact with a drop of acrylonitrile, whereas irradiated oxygen-containing nitrogen does so. Ozone was thought to be responsible [1], and this was confirmed by a detailed experimental investigation [2], though trace organic impurities may also have been involved [1]. The subject has been reviewed, and oxygen must also be eliminated from liquid helium, hydrogen or noble gases before irradiation [3]. Similar effects and conclusions were associated with an explosion which occurred when liquid nitrogen was used to cool a target during electron-beam irradiation [4].
See other CRYOGENIC LIQUIDS, IRRADIATION DECOMPOSITION INCIDENTS

4736. Sodium hyponitrite
[13517-64-7] $N_2Na_2O_2$
                          NaON:NONa

Preparative hazard
See Nitrogen oxide: Sodium diphenylketyl
See Hyponitrous acid, *also* Sodium amide (reference 4)

4737. Sodium trioxodinitrate
[13826-64-7] $N_2Na_2O_3$

NaON(O):NONa

See Sodium amide(reference 4)
See other METAL NITRATES

4738. Sodium tetraoxodinitrate
[ ] $N_2Na_2O_4$

NaON(O):N(O)ONa

See Sodium amide (reference 4)
See other METAL NITRATES

4739. Sodium pentaoxodinitrate
[59795-18-5] $N_2Na_2O_5$

NaON(O$_2$):N(O)ONa

See Sodium amide(reference 4)
See other METAL NITRATES

4740. Sodium hexaoxodinitrate
[ ] $N_2Na_2O_6$

NaON(O$_2$):N(O$_2$)ONa

See Sodium amide(reference 4)
See other METAL NITRATES

4741. Dinitrosylnickel
[ ] N_2NiO_2

ONNiNO

Mond, R. L. *et al., J. Chem. Soc.*, 1922, **121**, 38
A blue, impure and probably polymeric solid, produced from reaction of nitric oxide and nickel carbonyl, decomposed with incandescence at 90°C. The structure is very doubtful but a dinitrosyl was tentatively postulated. A trinitrosyl, [115380-62-6], has been listed recently.
See other HEAVY METAL DERIVATIVES, NITROSO COMPOUNDS

4742. Nitritonitrosylnickel
[ ] N_2NiO_3

O:NNiON:O

Water
Bailar, 1973, Vol. 3, 1119

It is probably polymeric and ignites on contact with water.
See other NITROSO COMPOUNDS
See related METAL NITRITES

4743. Nickel(II) nitrate
[13138-45-9] N_2NiO_6

$$Ni^{2+} \ (NO_3^-)_2$$

Oxidant. The tetrammine and tetrahydrazine complexes are explosive.
See AMMINEMETAL NITRATES, METAL NITRATES: organic matter

1-Methylbenzotriazole
Diamantopoulou, E. *et al., Polyhedron*, 1994, **13**(10), 1593
A study of the 1-methylbenzotriazole complexes, of various stoicheiometry and
hydration, showed several to be explosive.
See other METAL NITRATES, OXIDANTS

4744. Dinitrogen oxide ('Nitrous oxide')
[10204-97-2] N_2O

$$N_2O$$

FPA H111, 1982 (cylinder); *HCS 1980*, 692 (cylinder)

1. Rüst, 1948, 278
2. Conrad, D. *et al., Chem. Abs.*, 1983, **99**, 55875
3. Editorial feature, *Reading Times* (PA), 14th May, 1987
4. Bretherick, L., *Chem. Eng. News*, 1987, **65**(31), 3
5. Hamsen, B. *et al., Chem. Eng. News*, 1991, **69**(29), 2
During transfer of the liquefied gas from a stock steel cylinder into smaller cylinders, the effective expansion caused cooling of the stock cylinder and a fall in
pressure to occur. Application of a flame to the stock cylinder of the endothermic
oxide led to decomposition and explosive rupture of the cylinder [1]. The explosive decomposition of the gas at 0.05−8 bar at ambient temperature after initiation
by electric discharge can be prevented by addition of 30 vol% of air, nitrogen or
oxygen [2]. Oxyacetylene welding repair work on or near to a 6 t tank of liquefied
oxide led to a violent explosion [3], effectively a large scale rerun 85 years after
the original incident described above [4]. When using nitrous oxide, containing
organic solutes, as a supercritical fluid for hplc purposes, a 1 ml sample exploded,
causing considerable shrapnel damage. It is recommended that it be not used as a
supercritical fluid [5].
    It is highly endothermic ($\Delta \ H_f^\circ$ (g) +90.4 kJ/mol, 2.01 kJ/g) and an active
oxidant.
See Fuels, below
See other ENDOTHERMIC COMPOUNDS
See also INDIGESTION

Preparative hazard
Bailar, 1973, Vol. 2, 317

Acetylene
See ATOMIC ABSORPTION SPECTROSCOPY

Boron
Mellor, 1940, Vol. 8, 109
Amorphous boron (not crystalline) ignites on heating in the dry oxide.

Carbon monoxide
Wilton, C. *et al., Chem. Abs.*, 1975, **82**, 113724
Possible blast hazards associated with use of the propellant combination liquid dinitrogen oxide–liquid carbon monoxide have been evaluated.

Combustible gases
Sorbe, 1968, 131
The oxide (with an oxygen content 1.7 times that of air) forms explosive mixtures with ammonia, carbon monoxide, hydrogen, hydrogen sulfide and phosphine.

Fuels
1. L. Bretherick's comments
2. Young, J. A., *CHASNotes*, 1993, **XI**(6), 2
Although the oxygen in dinitrogen oxide is chemically (but endothermically) bound to nitrogen, it is present in a much higher concentration (36.4%) than the 21% present in air in admixture with nitrogen, so combustion or oxidation in the oxide will be much faster than in air. Advantage of this has been taken by feeding the oxide together with air into carburettor intakes to boost internal combustion engine performance, but strict control is necessary to prevent premature ignition. Nitrogen oxide, though of even higher oxygen content (53.3%), is too reactive and corrosive for this application [1].
   A man habitually enjoyed the euphoric effects of inhaling whiffs of nitrous oxide in seclusion, and kept a cylinder of the gas in his sedan for that purpose. He decided to spray the faded car seats with an aerosol can of vinyl dressing (propane/butane propellant) with the windows closed. Then he had a whiff of gas from the briefly opened cylinder, and settled back to enjoy the euphoria and a cigarette. He was lucky to survive the resulting explosion of the fuel/oxidant mixture in a closed vessel [2].
See OXYGEN ENRICHMENT

Hydrazine
Mellor, 1967, Vol. 8, Suppl. 2.2, 214
Ignition occurs on contact.
See other REDOX REACTIONS

Hydrogen, Nitric oxide
See Nitrogen oxide: Hydrogen, etc.

Hydrogen, Oxygen
Bailar, 1973, Vol. 2, 321–322
The ignition temperature of mixtures of hydrogen with dinitrogen oxide is lower than that of hydrogen admixed with air or oxygen. The oxide also sensitises mixtures of hydrogen and oxygen, so that addition of oxygen to a hydrogen–oxide mixture will cause instancous ignition or explosion. Explosive limits are extremely wide.

Lithium hydride
Mellor, 1967, Vol. 8, Suppl. 2.2, 214
The hydride ignites in the gas.

Phosphine
Thénard, J., *Compt. rend.*, 1844, **18**, 652
A mixture with excess oxide can be exploded by sparking.

Plastic tubes
See Oxygen (Gas): Plastic tubes

Silane
See Silane: Nitrogen oxides

Tin(II) oxide
See Tin(II) oxide: Non-metal oxides

Tungsten carbides
See Tungsten carbide: Nitrogen oxides
Ditungsten carbide: Oxidants
See other NON-METAL OXIDES, OXIDANTS

4745. Lead hyponitrite
[19423-89-3] N_2O_2Pb

$$(-PbON:NO-)_n$$

Partington, J. R. *et al., J. Chem. Soc.*, 1932, 135, 2589
The lead salt decomposes with explosive violence at 150–160°C and should not be dried by heating. The salt prepared from lead acetate explodes more violently than that from the nitrate.
See other HEAVY METAL DERIVATIVES, N–O COMPOUNDS

4746. Dinitrogen trioxide
[10544-73-7] N_2O_3

$$O:NON:O$$

Acetic acid, 6-Hexanelactam
See 6-Hexanelactam: Acetic acid, etc.

Chloroprene
See 2-Chloro-1-nitro-4-nitroso-2-butene

Phosphine
See Phosphine (reference 3)

Phosphorus
See Phosphorus: Non-metal oxides
See other NON-METAL OXIDES, OXIDANTS

4747. Dinitrogen tetraoxide (Nitrogen dioxide)
[10544-72-6] N_2O_4
$$N_2O_4$$

HCS 1980, 675 (cylinder)

The equilibrium mixture of nitrogen dioxide and dinitrogen tetraoxide is completely associated at −9°C to the latter form which is marginally endothermic (ΔH°_f (g) +9.7 kJ/mol, 0.10 kJ/g). Above 140°C it is completely dissociated to nitrogen dioxide, which is moderately endothermic (ΔH°_f (g) +33.8 kJ/mol, 0.74 kJ/g).

Acetonitrile, Indium   MRH Acetonitrile 7.87/25
Addison, C. C. *et al., Chem. & Ind.*, 1958, 1004
Shaking a slow-reacting mixture caused detonation, attributed to indium-catalysed oxidation of acetonitrile.

Alcohols
Daniels, F., *Chem. Eng. News*, 1955, **33**, 2372
A violent explosion occurred during the ready interaction to produce alkyl nitrates.

Ammonia   MRH 6.61/33
Mellor, 1940, Vol. 8, 541
Liquid ammonia reacts explosively with the solid tetraoxide at −80°C, while aqueous ammonia reacts vigorously with the gas at ambient temperature.

Barium oxide
Mellor, 1940, Vol. 8, 545
In contact with the gas at 200°C the oxide suddenly reacts, reaches red heat and melts.

Boron trichloride
Mellor, 1946, Vol. 5, 132
Interaction is energetic.

Carbon disulfide
1. Mellor, 1940, Vol. 8, 543
2. Sorbe, 1968, 132

Liquid mixtures proposed for use as explosives are stable up to 200°C [1], but can be detonated by mercury fulminate, and the vapours by sparking [2].

Carbonylmetals
Cloyd, 1965, 74
Combination is hypergolic.

Cellulose, Magnesium perchlorate
See Magnesium perchlorate: Cellulose, etc.

Cycloalkenes, Oxygen
Lachowicz, D. R. *et al.*, US Pat. 3 621 050, 1971
Contact of cycloalkenes with a mixture of dinitrogen tetraoxide and excess oxygen at temperatures of 0°C or below produces nitroperoxonitrates of the general formula–$CHNO_2$–$CH(OONO_2)$ — which appear to be unstable at temperatures above 0°C, owing to the presence of the peroxonitrate group.
See Hydrocarbons, below

Difluorotrifluoromethylphosphine
Mahler, W., *Inorg. Chem.*, 1979, **18**, 352
A reaction, to produce the phosphine oxide on 12 mmol scale, ignited.

Dimethyl sulfoxide   MRH 6.99/36
See Dimethyl sulfoxide: Dinitrogen tetraoxide

Formaldehyde
1. Pollard, F. H. *et al.*, *Trans. Faraday Soc.*, 1949, **45**, 767–770
2. Rastogi, R. P. *et al.*, *Chem. Abs.*, 1975, **83**, 12936
The slow (redox) reaction becomes explosive around 180°C [1], or even lower [2].
See other REDOX REACTIONS

Halocarbons   MRH Chloroform 2.38/67, 1,2-dichloroethane 5.06/42,
                                      1,1-dichloroethylene 5.06/46, trichloroethylene 3.97/56
1. Turley, R. E., *Chem. Eng. News*, 1964, **42**(47), 53
2. Benson, S. W., *Chem. Eng. News*, 1964, **42**(51), 4
3. Shanley, E. S., *Chem. Eng. News*, 1964, **42**(52), 5
4. Kuchta, J. M. *et al.*, *J. Chem. Eng. Data*, 1968, **13**, 421–428
Mixtures of the tetraoxide with dichloromethane, chloroform, carbon tetrachloride, 1,2-dichloroethane, trichloroethylene and tetrachloroethylene are explosive when subjected to shock of 25 g TNT equivalent or less [1]. Mixtures with trichloroethylene react violently on heating to 150°C [2]. Partially fluorinated chloroalkanes were more stable to shock. Theoretical aspects are discussed in the later reference [2,3]. The effect of pressure on flammability limits has been studied [4].
See Vinyl chloride: Oxides of nitrogen
See Uranium: Nitric acid

Heterocyclic bases   MRH Pyridine 7.82/22, quinoline 7.87/22
Mellor, 1940, Vol. 8, 543
Pyridine and quinoline are attacked violently by the liquid oxide.

Hydrazine derivatives
1. Cloyd, 1965, 74
2. Miyajima, H. *et al., Combust. Sci. Technol.*, 1973, **8**, 199–200
Combinations with hydrazine, methylhydrazine, 1,1-dimethylhydrazine or mixtures
thereof are hypergolic and used in rocketry [1]. The hypergolic gas-phase ignition
of hydrazine at 70–160°C/53–120 mbar has been studied [2].
See ROCKET PROPELLANTS

Hydrocarbons   MRH values below references
1. Mellor, 1967, Vol. 8, Suppl. 2.2, 264
2. Fierz, H. E., *J. Soc. Chem. Ind.*, 1922, **41**, 114R
3. Raschig, F., *Z. Angew. Chem.*, 1922, **35**, 117–119
4. Berl, E. *Z. Angew. Chem.*, 1923, **36**, 87–91
5. Schaarschmidt, A., *Z. Angew. Chem.*, 1923, **36**, 533–536
6. Berl, E., *Z. Angew. Chem.*, 1924, **37**, 164–165
7. Schaarschmidt, A., *Z. Angew. Chem.*, 1925, **38**, 537–541
8. *MCA Case History No. 128*
9. Folecki, J. *et al., Chem. & Ind.*, 1967, 1424
10. Cloyd, 1965, 74
11. Urbanski, 1967, Vol. 3, 289
12. Biasutti, 1981, 50
13. Biasutti, 1981, 53–54
MRH Benzene 7.99/19, hexane 7.91/17, isoprene 8.28/18, methylcyclohexane
7.87/17
   A mixture of the tetraoxide and toluene exploded, possibly initiated by
unsaturated impurities [1]. During attempted separation by low temperature
distillation of an accidental mixture of light petroleum and the oxide, a
large bulk of material awaiting distillation became heated by unusual climatic
conditions to 50°C and exploded violently [2]. Subsequently, discussion of
possible alternative causes involving either unsaturated or aromatic compounds
was published [3] [4] [5] [6] [7]. Erroneous addition of liquid in place of gaseous
nitrogen tetraoxide to hot cyclohexane caused an explosion [8]. During kinetic
studies, one sample of a 1:1 molar solution of tetraoxide in hexane exploded during
(normally slow) decomposition at 28°C [9]. Cyclopentadiene is hypergolic with the
oxide [10]. These incidents are understandable because of their similarity to rocket
propellant systems and liquid mixtures previously used as bomb fillings [11]. The
liquid oxide leaking from a ruptured 6 t storage tank ran into a gutter containing
toluene and a violent explosion ensued [12]. An alternative account describes the
hydrocarbon as benzene [13].
See Cycloalkenes, above; Unsaturated hydrocarbons, below

Hydrogen, Oxygen
Lewis, B., *Chem. Rev.*, 1932, **10**, 60

The presence of small amounts of the oxide in non-explosive mixtures of hydrogen and oxygen renders them explosive.

Isopropyl nitrite, Propyl nitrite
Safety in the Chemical Laboratory, Vol. 1, 121, Steere, N. V. (Ed.), Easton (Pa.) *J. Ch. Ed.*, 1967
A pressurised mixture of the cold components exploded very violently during a combustion test run. The mixture was known to be autoexplosive at ambient temperature, and both of the organic components are capable of violent decomposition in absence of added oxidant.

Laboratory grease
Arapava, L. D. *et al., Chem. Abs.*, 1985, **102**, 169310
Contact of the lubricating grease Litol-24 with the oxidant at below 80°C led to explosion on subsequent impact. This involved nitration products of the antioxidant present, 4-hydroxydiphenylamine. Above 80°C decomposition superceded nitration, and no explosion occurred.
See other NITRATION INCIDENTS

Metal acetylides or carbides   MRH values show % of oxidant
Mellor, 1946, Vol. 5, 849
Caesium acetylide ignites at 100°C in the gas.
See Tungsten carbide: Nitrogen oxides   MRH 4.02/63
Ditungsten carbide: Oxidants   MRH 3.85/67

Metals   MRH Magnesium 12.97/50, potassium 3.72/46
1. Mellor, 1940, Vol. 8, 544–545; 1942, Vol. 13, 342
2. Pascal, 1956, Vol. 10, 382; 1958, Vol. 4, 291
Reduced iron, potassium and pyrophoric manganese all ignite in the gas at ambient temperature. Magnesium filings burn vigorously when heated in the gas [1]. Slightly warm sodium ignites in contact with the gas, and interaction with calcium is explosive [2].
See Aluminium: Oxidants

Nitroaniline
Anon., *CISHC Chem. Safety Summ.*, 1978, **49**, 3–4
Process errors led to discharge of copious amounts of nitrous fumes into the glass reinforced plastic ventilation duct above a diazotisation vessel. On two occasions fires were caused in the duct by vigorous reaction of the dinitrogen tetraoxide with nitroaniline dusts in the duct. Laboratory tests confirmed this to be the cause of the fires, and precautions are detailed.

Nitroaromatics
1. Urbanski, 1967, Vol. 3, 288
2. Kristoff, F. T. *et al., J. Haz. Mat.*, 1983, **7**, 199–210
Mixtures with nitrobenzene were formerly used as liquid high explosives, with addition of carbon disulfide to lower the freezing point, but high sensitivity to mechanical stimulus was disadvantageous [1]. During the recovery of acids from nitration of toluene, mixtures of the oxide with nitrotoluene or dinitrotoluene may

be isolated under certain process conditions. While such mixtures are not unduly sensitive to impact, friction or thermal initiation, when oxygen-balanced they are extremely sensitive to induced shock and are capable of explosive propagation at film thicknesses below 0.5 mm. It is suspected that many explosions in TNT acid recovery operations, previously attributed to tetranitromethane, may have been caused by such mixtures [2].

Nitrogen trichloride
See Nitrogen trichloride: Initiators

Organic compounds
Riebsomer, J. L., *Chem. Rev.*, 1945, **36**, 158
In a review of the interaction of the oxidant with organic compounds, attention is drawn to the possibility of formation of unstable or explosive products.

Other reactants
Yoshida, 1980, 269
MRH values calculated for 18 combinations with oxidisable materials are given.

Ozone
See Ozone: Nitrogen oxide

Phospham
See Phospham: Oxidants

Phosphorus   MRH 9.12/35
See Phosphorus: Non-metal oxides

Sodium amide
Beck, G., *Z. Anorg. Chem.*, 1937, **233**, 158
Interaction with the oxide in carbon tetrachloride is vigorous, producing sparks.

Tetracarbonylnickel
Bailar, 1973, Vol. 3, 1130
Interaction of the liquids is rather violent.
See Carbonylmetals, above

Tetramethyltin
Bailar, 1973, Vol. 2, 355
Interaction is explosively violent even at $-80°C$, and dilution with with inert solvents is required for moderation.

2-Toluidinium nitrate
Rastogi, R. P. *et al., Indian J. Chem., Sect. A*, 1980, **19** A, 317–321
Reaction in this hybrid rocket propellant system is enhanced by presence of ammonium vanadate.

Triethylamine
Davenport, D. A. *et al., J. Amer. Chem. Soc.*, 1953, **75**, 4175
The complex, containing excess oxide over amine, exploded at below $0°C$ when free of solvent.

Triethylammonium nitrate

Addison, C. C. *et al.*, *Chem. & Ind.*, 1953, 1315

The two component form an addition complex with diethyl ether, which exploded violently after partial desiccation: an ether-free complex is also unstable.

See Triethylamine, above

Unsaturated hydrocarbons   MRH Isoprene 8.28/18

1. Sergeev, G. P. *et al.*, *Chem. Abs.*, 1966, **65**, 3659g
2. Biasutti, 1981, 123

Dinitrogen tetraoxide reacts explosively between $-32°$ and $-90°C$ with propene, 1-butene, isobutene, 1,3-butadiene, cyclopentadiene and 1-hexene, but 6 other unsaturates failed to react [1]. Reaction of propene with the oxide at 2 bar/$30°C$ to give lactic acid nitrate was proceeding in a pump-fed tubular reactor pilot plant. A violent explosion after several hours of steady operation was later ascribed to an overheated pump gland which recently had been tightened. A similar pump with a tight gland created a hot-spot at $200°C$ [2].

See Nitrogen dioxide: Alkenes

Vinyl chloride

See Vinyl chloride: Oxides of nitrogen

Xenon tetrafluoride oxide

Christe, K. O., *Inorg. Chem.*, 1988, **27**, 3764

In the reaction of the pentaoxide with xenon tetrafluoride oxide to give xenon difluoride dioxide and nitryl fluoride, the xenon tetrafluoride oxide must be used in excess to avoid formation of xenon trioxide, which forms a sensitive explosive mixture with xenon difluoride dioxide.

See Xenon tetrafluoride oxide: Caesium nitrate

See other ENDOTHERMIC COMPOUNDS, NON-METAL OXIDES, OXIDANTS

4748. Dinitrogen pentaoxide (Nitryl nitrate)
[10102-03-1] N_2O_5

$$NO_{2+}NO_{3-}$$

Partington, 1967, 569

If suddenly heated, the solid (m.p. $30°C$) explodes. The structure is now considered to be nitryl nitrate.

Acetaldehyde

Kacmarek, A. J. *et al.*, *J. Org. Chem.*, 1975, **40**, 1853

Direct combination to produce ethylidene dinitrate at $-196°C$ is violently explosive, but uneventful when the acetaldehyde is diluted with nitrogen.

Aromatic compounds

Editor's comments, 1999

Increased availability of this compound means it is finding use as a clean and mild nitrating agent. A stoichiometric byproduct in many procedures is anhydrous nitric acid, which is often blithely ignored. Anhydrous nitric acid is a very hazardous

material in its own right, being surprisingly unreactive — except as an oxidant during detonations! In particular, work-up by distillation is virtually certain to form detonable compositions in the condenser at some stage. The editor is not yet aware of any accidents during such distillation, but counsels forethought.
See Nitric acid

tert-Butylhydroperoxide, Toluene
See tert-Butylhydroperoxide: Toluene, Dinitrogen pentaoxide

Dichloromethane
See NITRATING AGENTS

Ethylene
Nitro Compounds, Feuer H. & Nielsen A. T. (Eds), New York, VCH, 1990, 315
The crude product is too explosive to be worked up. Analogy with other olefins indicates that the final product will be ethylene glycol dinitrate, a known explosive though safe enough to have largely replaced nitroglycerine. That will be preceded by 1,2-dinitroethane and nitratonitroethane, more sensitive if less powerful.

Metals
Mellor, 1940, Vol. 8, 554
Potassium and sodium burn brilliantly in the gas, while mercury and arsenic are vigorously oxidised.

Naphthalene
Mellor, 1940, Vol. 8, 554
Naphthalene explodes and other organic materials react vigorously with the powerful oxidant.

Sodium acetylide
See Sodium acetylide: Oxidants

Strained ring heterocycles
Golding, P. *et al., Tetrahedron Lett.*, 1988, **29**, 2731–2734; 2735–2736
Strained ring oxygen heterocycles such as epoxides and oxetanes react with the oxide in dichloromethane at sub-ambient temperatures to form high yields of the powerfully explosive glycol dinitrate esters. Epoxides give 1,2-dinitrates (such as those from ethylene, propene, butene and 4 other oxides), while oxetanes (6 examples) give 1,3-dinitrates. Strained ring nitrogen heterocycles give explosive *N*-nitroaminoalkyl nitrates, and several derived from aziridines and 3 derived from azetidines are exemplified.
See other STRAINED-RING COMPOUNDS
See NITRATING AGENTS

Sulfur dichloride, or Sulfuryl chloride
Schmeisser, M., *Angew. Chem.*, 1955, **67**, 495, 499
Interaction is explosively violent.
See other NON-METAL OXIDES, OXIDANTS
See related ACID ANHYDRIDES

4749. Lead(II) nitrate
[10099-74-8]　　　　　　　　　　　　　　　　　　　　　　　　　　　　　　　N_2O_6Pb

$$Pb(NO_3)_2$$

HCS 1980, 589

Carbon　　　　　　　　　　　　　　　　　　　　　　　　　　　　　　　　　MRH 1.76/7
　　Mellor, 1941, Vol. 7, 863
　　Contact with red hot carbon causes an explosion with showers of sparks.

Calcium–silicon alloy
　　Smolin, A. O., *Chem. Abs.*, 1975, **85**, 7591
　　'Combustion of Calcium–Silicon in a Mixture with Lead Nitrate' (title only translated)

Cyclopentadienylsodium
　　Houben-Weyl, 1975, Vol. 13.3, 200
　　If lead nitrate is used rather than the chloride or acetate as the source of divalent lead, the crude dicyclopentadienyllead may explode violently during purification by high-vacuum sublimation at 100–130°C.

Other reactants
　　Yoshida, 1980, 200
　　MRH values for 17 combinations with oxidisable materials are given.

Potassium acetate
　　Mee, A. J., *School Sci. Rev.*, 1940, **28**(86), 95
　　A heated mixture of the oxidant and organic salt exploded violently.
　　See other METAL NITRATES, OXIDANTS

4750. Tin(II) nitrate
[22755-27-7]　　　　　　　　　　　　　　　　　　　　　　　　　　　　　　N_2O_6Sn

$$Sn(NO_3)_2$$

Bailar, 1973, Vol. 2, 74
During attempted isolation of the nitrate by evaporation of its aqueous solutions, a number of explosions have occurred.
See other METAL NITRATES, REDOX COMPOUNDS

4751. Zinc nitrate
[7779-88-6]　　　　　　　　　　　　　　　　　　　　　　　　　　　　　　　N_2O_6Zn

$$Zn(NO_3)_2$$

HCS 1980, 967

Carbon
　　Mellor, 1940, Vol. 4, 655
　　When the nitrate is sprinkled on to hot carbon, an explosion occurs.
　　See other METAL NITRATES, OXIDANTS

4752. Uranyl nitrate
[10102-06-4] N_2O_8U

$$O_2U(NO_3)_2$$

Cellulose
Clinton, T. G., *J. R. Inst. Chem.*, 1958, **82**, 633
The analytical use of cellulose fibre to absorb uranyl nitrate solution prior to ignition has led to explosions during ignition, owing to formation of cellulose nitrate. An alternative method is described.

Diethyl ether
1. Ivanov, W. N., *Chem. Ztg.*, 1912, **36**, 297
2. Andrews, L. W., *J. Amer. Chem. Soc.*, 1912, **34**, 1686–1687
3. Müller, A., *Chem. Ztg.*, 1916, **40**, 38; 1917, **41**, 439
The mild detonations reported when the crystalline salt was disturbed [1] were thought to have been caused by presence of solvent ether in the crystals (6 mols of water may be replaced by 2 of ether) [2]. This was later confirmed [3], and the formation of ethyl nitrate or diethyloxonium nitrate may have been involved, as the anhydrous salt functions as a powerful nitrator. Solutions of the nitrate in ether should not be exposed to sunlight to avoid the possibility of explosions.
See Organic solvents, below
See other IRRADIATION DECOMPOSITION INCIDENTS

Ethylene glycol
See Ethylene glycol: Oxidants

Organic solvents
Burriss, W. H., *Nucl. Sci. Abs.*, 1976, **33**, 19790
During routine operations to reduce the hexahydrate to uranium trioxide, an excessive amount of organic solution entered the denitrator pots and ignited.
See Diethyl ether, above
See also NITRATING AGENTS
See other METAL NITRATES

4753. Lead nitride
[58572-21-7] N_2Pb_3

$$Pb_3N_2$$

Fischer, F. *et al., Ber.*, 1910, **43**, 1470
Very unstable, it decomposes explosively during vacuum degassing.
See other NITRIDES

4754. Disulfur dinitride
[25474-92-4] N_2S_2

1. Brauer, 1963, Vol. 12, 410
2. Williams, J. D., *Polyhedron*, 1987, **6**(5), 939–941
Explosive, initiated by shock, friction, pressure or temperatures over 30°C [1]. It
may also detonate spontaneously at ambient temperature [2].

Metal chlorides
Roesky, H. W. *et al., Inorg. Chem.*, 1984, **23**, 75
The 1:1 complexes with beryllium chloride or titanium tetrachloride may explode
violently.
See other NITRIDES, N–S COMPOUNDS

4755. Tetrasulfur dinitride
[32607-15-1]   N₂S₄

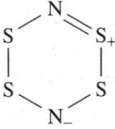

1. Brauer, 1963, Vol. 1, 408
2. Heal, H. G. *et al., J. Inorg. Nucl. Chem.*, 1975, **37**, 286
It decomposes explosively to its elements at 100°C. Previously prepared from the
explosive tetranitride [1], it may now safely be prepared from the stable intermediate, heptasulfur imide [2].
See other NITRIDES, N–S COMPOUNDS

4756. Tetraselenium dinitride
[132724-39-1]   N₂Se₄

Dehnicke, K. *et al., Angew. Chem. (Int.)*, 1991, **30**(5), 577
Made from diselenium dichloride and trimethylsilyl azide, the crystalline nitride
explodes under a hammer blow, or on heating to 80°C. Can apparently be handled
safely at room temperature.
See related N–S COMPOUNDS *See other* NITRIDES

4757. Poly(disilicon nitride)
[12438-96-9]   (N₂Si₂)ₙ
(Complex Structure with Si-Si bonds)

Alone, or Water
Gmelin, 1959, Vol. 15.B, 610
It ignites in air, explodes on heating and contact with drops of water generates
sparks.
See other NITRIDES

4758. Sodium azide
[26628-22-8]

NaN$_3$

N$_3$Na

HCS 1980, 827

1. Mellor, 1940, Vol. 8, 345
2. Lambert, B., *School Sci. Rev.*, 1927, **8**(31), 218
3. Anon., *Jahresber.*, 1981, 77–79
4. Anon., *Safety*, 1982, (3), 2
5. Grewer, T. *et al., Exothermic Decomposition*, Technical Report 01VD 159/0329 for Federal German Ministry for Res. Technol., Bonn. 1986

Insensitive to impact, it decomposes, sometimes explosively, above its m.p. [1], particularly if heated rapidly [2]. Although used in aqueous solutions as a preservative in pharmaceutical preparations, application of freeze-drying techniques to such solutions has led to problems arising from volatilisation of traces of hydrazoic acid from non-neutral solutions, condensation in metal lines, traps or filters, and formation of heavy metal azides in contact with lead, copper or zinc components in the drying plant [3,4].

Energy of exothermic decomposition in range 230–260°C was measured as 0.76 kJ/g by DSC, and T_{ait24} was determined as 253°C by adiabatic Dewar tests, with an apparent energy of activation of 145 kJ/mol [5].

See Heavy metals, below

Acids
See Hydrogen azide

Ammonium chloride, Trichloroacetonitrile
See 5-Trichloromethyltetrazole

Barium carbonate
Henneburg, G. O. *et al., Can. J. Res.*, 1950, **28B**, 345
Interaction to form cyanide ion requires careful control of temperature at 630°C to prevent explosions.

Bromine
See Bromine: Metal azides

Carbon disulfide
Stohlmeier, M. *et al., Chem. Ind. Dig.*, 1993, **6**(2), 124
A review on all the more or less explosive products which may be prepared from the above, with advice on safe handling and preparation.
See also Carbon disulfide: Metal azides

Carbonyl dichloride (Phosgene)
See tert-Butyl azidoformate (reference 4)

Chloroform
See Dichloromethane: Quaternary ammonium azides, etc.

Cyanuric chloride
See 2,4,6-Triazido-1,3,5-triazine (reference 2)

Chromyl chloride
See Chromyl chloride: Sodium azide

Dichloromethane, Dimethyl sulfoxide
1. Peet, N. P. *et al., Chem. Eng. News*, 1993, **71**(16), 4
2. Hruby, V. J. *et al., Chem. Eng. News*, 1993, **71**(41), 2
An explosion was experienced during work up of an epoxide opening reaction involving acidified sodium azide in a dichloromethane/dimethyl sulfoxide solvent. The author ascribes this to diazidomethane formation from dichloromethane [1]. A second report of an analoguous accident, also attributed to diazidomethane, almost certainly involved hydrogen azide for the cold traps of a vacuum pump on a rotary evaporator were involved; this implies an explosive more volatile than dichloromethane. It is recommended that halogenated solvents be not used for azide reactions [2].
See Dichloromethane: Quaternary ammonium azides, etc.

2,5-Dinitro-3-methylbenzoic acid, Oleum
See 2,5-Dinitro-3-methylbenzoic acid: Oleum, Sodium azide

Heavy metals
1. Wear, J. O., *J. Chem. Educ.*, 1975, **52**, A23–25
2. Becher, H. H., *Naturwiss.*, 1970, **57**, 671
3. Anon., *Chem. Eng. News*, 1976, **54**(36), 6
4. Pobiner, H., *Chem. Eng. News*, 1982, **60**(125), 4
5. Anon., *Safety Digest Univ. Safety Assoc.*, 1992, **44**, 22
6. Kurtz, L., Internet, 1995
7. Mellor, 1967, Vol. 8, Suppl. 2, 4
The effluent from automatic blood analysers in which 0.01–0.1% sodium azide solutions are used, may lead over several months to formation of explosive heavy metal azides in brass, copper or lead plumbing lines, especially if acids are also present. Several incidents during drain-maintenance are described, and preventative flushing and decontamination procedures are discused [1]. Brass plates, exposed to sodium azide solution during several months in soil-percolation tests and then dried, caused explosions owing to formation of copper and/or zinc azides [2]. During repairs to a metal thermostat bath in which sodium azide had been used as a preservative, a violent explosion occurred [3]. Use of sodium azide in automatic sulfur titrators led to explosions arising from formation of copper azide in a copper/brass valve assembly [4]. Mixing sodium azide solutions with heavy metal salts will produce the metal azides with greater facility; an explosion with mercuric chloride is reported [5]. A similar incident is reported later, apparently

the mixture has been recommended as a biocide for aqueous solutions. In this case an explosion occurred while stirring water containing perhaps 5% of each [6]. Raney nickel catalyses the vigorous decomposition of solutions of hydrogen azide or azide salts [7].

Manganese(III) salts, Styrene
Fristad, W. E. *et al., J. Org. Chem.*, 1985, **50**, 3468
After applying an apparently safe procedure for diazidation of alkenes to styrene, an explosion occurred as the clear solution was cooling.

Nitryl fluoride
Klapötke, T. M. *et al., Chem. Ber.*, 1994, **127**(11), 2181
Reaction with sodium azide even at −78°C leads to explosions attributed to fluorine azide formation.
See Fluorine azide

Sulfuric acid
Ross, F. F., *Water & Waste Treatment*, 1964, **9**, 528; private comm., 1966
One of the reagents required for the determination of dissolved oxygen in polluted water is a solution of sodium azide in 50% sulfuric acid. It is important that the diluted acid should be quite cold before adding the azide, since hydrogen azide boils at 36°C and is explosive in the condensed liquid state.

Trifluoroacryloyl fluoride
See Trifluoroacryloyl fluoride: Sodium azide

Water
Anon., *Angew. Chem.*, 1952, **64**, 169
Addition of water to sodium azide which had been heated strongly caused a violent reaction. This was attributed to formation of metallic sodium or sodium nitride.
See other METAL AZIDES

4759. Sodium azidosulfate
[67880-15-3] N_3NaO_3S

$$NaOSO_2N_3$$

Shozda, R. J. *et al., J. Org. Chem.*, 1967, **32**, 2876
It is a weak explosive with variable sensitivity to mechanical shock and heating.
See other ACYL AZIDES

4760. Sodium trisulfurtrinitridate
[65107-36-0] (ion) N_3NaS_3

Bojes, J. *et al.*, *Inorg. Chem.*, 1978, **17**, 319
Like tetrasulfur tetranitride, salts of trisulfur trinitride and of pentasulfur tetran-
itride (particularly alkali-metal salts) are heat- and friction-sensitive explosives.
Preparation on a scale limited to 1 g, and care in use of spatulae or in preparation
of samples for IR examination is recommended.
See Sodium tetrasulfur pentanitridate
See related NITRIDES
See other N–S COMPOUNDS

4761. Thallium 2,4,6-tris(dioxoselena)hexahydrotriazine-1,3-5-triide ('Thallium triselenimidate')

[ ] $N_3O_6Se_3Tl_3$

Explosive.
See Selenium difluoride dioxide: Ammonia
See other N-METAL DERIVATIVES

4762. Thallium(III) nitrate

[13746-98-0] N_3O_9Tl

$$Tl(NO_3)_3$$

Formic acid, 4-Hydroxy-3-methoxybenzaldehyde
491M, 1975, 416
Addition of a little of the aldehyde (vanillin) to a strong solution of the trihydrated
nitrate in 90% formic acid led to a violent (redox) reaction.
See other METAL NITRATES, REDOX REACTIONS

4763. Vanadium trinitrate oxide

[16017-37-1] $N_3O_{10}V$

$$VO(NO_3)_3$$

Organic materials
1. Schmeisser, M., *Angew. Chem.*, 1955, **67**, 495
2. Harris, A. D. *et al.*, *Inorg. Synth.*, 1967, **9**, 87
Many hydrocarbons and organic solvents ignite on contact with this powerful
oxidant and nitrating agent [1], which reacts like fuming nitric acid with paper,
rubber or wood [2].
See related METAL NITRATES
See other OXIDANTS

4764. Tris(thionitrosyl)thallium

[ ] N_3S_3Tl

$(S{:}N)_3Tl$

Bailar, 1973, Vol. 1, 1162
The compound and its adduct with ammonia explode very easily with heat or shock.

See other N-METAL DERIVATIVES, N–S COMPOUNDS

4765. Thallium(I) azide

[13847-66-0] N_3Tl

TlN_3

Mellor, 1940, Vol. 8, 352
A relatively stable azide, it can be exploded on fairly heavy impact, or by heating at 350–400°C.

See other METAL AZIDES

4766. Nitrosyl azide

[62316-46-5] N_4O

$O{:}NN_3$

1. Lucien, H. W., *J. Amer. Chem. Soc.*, 1958, **80**, 4458
2. Klapötke, T. M., *et al., Angew. Chem. (Int.)*, 1993, **32**(11), 1610
3. Tornieporth-Oetting, I. C. *et al., Angewand. Chem. (Int.)*, 1995, **34**(5), 511

Explosions were experienced on several occasions during preparation of nitrosyl azide by various methods [1]. A yellow solid at −85°C, it decomposes even at −50°C [2]. It has an enthalpy of detonation of 6.3 kJ/g, the same as nitroglycerine, and a cavity volume greater than 600ml in the lead block test (TNT = 300; Nitroglycerine = 520 ml) [3].

See other HIGH-NITROGEN COMPOUNDS, NON-METAL AZIDES

4767. Thiotrithiazyl nitrate

[79796-40-0] (ion) $N_4O_3S_4$

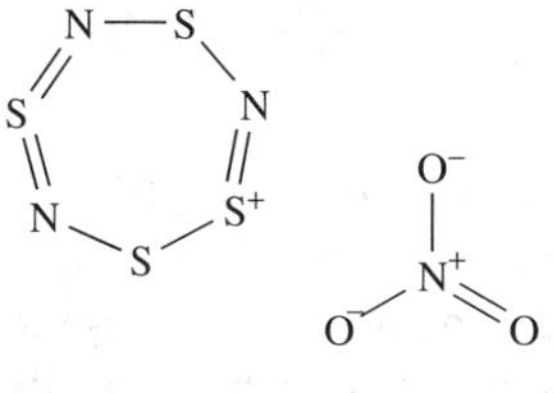

Goehring, 1957, 74
It explodes on friction or impact.

See other N–S COMPOUNDS

4768. Plutonium(IV) nitrate
[13823-27-2]   N₄O₁₂Pu

$$Pu(NO_3)_4$$

MCA Case History No. 1498

Polythene bottles are not suitable for long-term storage of plutonium nitrate solutions as radiation-induced stress cracks appeared in the bases of several 10 l bottles during 6 months' storage. Short-term storage and improved venting are recommended.
See Americium trichloride
See other IRRADIATION DECOMPOSITION INCIDENTS, METAL NITRATES

4769. Diseleniumdisulfur tetranitride
[   ]   N₄S₂Se₂

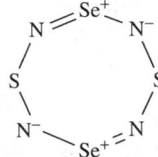

Maaninen, A. *et al., Phosphorus, Sulfur, Silicon Relat. Elem.*, 1997, (124–125), 457
Like the homlogues below, this compound is explosive on heating or mechanical stress.
See other SULFUR, SELENIUM AND TELLURIUM TETRANITRIDES IN IMMEDIATELY SUBSEQUENT ENTRIES
See other N–S COMPOUNDS, NITRIDES

4770. Tetrasulfur tetranitride
[28950-34-7]   N₄S₄

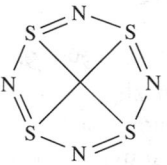

1. Mellor, 1940, Vol. 8, 625
2. Brauer, 1963, Vol. 1, 406
3. Villema-Blanco, M. *et al., Inorg. Synth.*, 1967, **9**, 101
4. Banister, A. J., *Inorg. Synth.*, 1977, **17**, 197–198
5. Witt, M. W. R. *et al., J. Amer. Chem. Soc.*, 1983, **105**, 1668
6. Mark, H. *et al., Chem. Eng. News*, 1984, **62**(5), 4
7. Buntain, G. A., *Chem. Eng. News*, 1988, **66**(18), 2

The endothermic nitride is susceptible to explosive decomposition on friction, shock or heating above 100°C [1]. Explosion is violent if initiated by a detonator [2]. Sensitivity toward heat and shock increases with purity. Preparative precautions have been detailed [3], and further improvements in safety procedures and handling described [4]. An improved plasma pyrolysis procedure to produce poly(sulfur nitride) films has been described [5]. Light crushing of a small sample of impure material (m.p. below 160°C, supposedly of relatively low sensitivity) prior to purification by sublimation led to a violent explosion [6] and a restatement of the need [4] for adequate precautions. Explosive sensitivity tests have shown it to be more sensitive to impact and friction than is lead azide, used in detonators. Spark-sensitivity is, however, relatively low [7].

Metal chlorides
Roesky, H. W. *et al., Inorg. Chem.*, 1984, **23**, 75–76
The 1:1 complex with titanium chloride or the 1:2 complex with beryllium chloride may explode violently. Careful pyrolysis of these gives mixtures of disulfur dinitride and its 1:1 complexes with the metal chlorides, which may also explode, so screening is essential.

Oxidants
Pascal, 1956, Vol. 10, 645–646
Contact with fluorine leads to ignition, and interaction with metal chlorates or oxides is violent.
See other ENDOTHERMIC COMPOUNDS, NITRIDES, N–S COMPOUNDS

4771. Tetraselenium tetranitride
  [12033-88-6] N_4Se_4

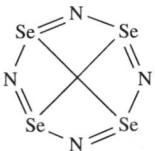

Alone, or Halogens and derivatives
1. Mellor, 1947, Vol. 10, 789
2. Kelly, P. F. *et al., Polyhedron*, 1990, **9**(13), 1567
The dry material explodes on slight compression, or on heating at 130–230°C. Contact with bromine, chlorine or a little fuming hydrochloric acid also causes explosion [1]. Explosions when manipulated with a metal spatula are reported [2].
See related N–S COMPOUNDS
See other NITRIDES

4772. Tritellurium tetranitride
  [   ] N_4Te_3
                        Structure unknown

Schmitz-Du Mont, O. *et al., Angew. Chem. (Int.)*, 1967, **6**, 1071

This yellow compound, prepared by ammonolysis of dipotassium triimidotellurite, was too explosive to be able to weigh it out for analysis.
See Tetratellurium tetranitride
See other NITRIDES

4773. Tetratellurium tetranitride
[12164-01-1]   N$_4$Te$_4$

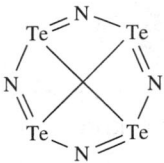

1. Fischer, F. *et al., Ber.*, 1910, **43**, 1472
2. Garcia-Fernandez, H., *Bull. Soc. Chim. Fr. [4]*, 1973, 1210–1215

The compound, previously formulated as tritellurium tetranitride is now shown to have the title structure (though probably polymeric, with inter-ring bonding). Two forms were originally described, one black which explodes on impact, and one yellow which explodes at 200°C [1]. It is explosive when dry, but may be stored safely under carbon tetrachloride [2].
See related N–S COMPOUNDS
See other NITRIDES

4774. Trithorium tetranitride
[12033-90-8]   N$_4$Th$_3$

Th$_3$N$_4$

Air, or Oxygen
Mellor, 1940, Vol. 8, 122
It burns incandescently on heating in air, and very vividly in oxygen.
See other NITRIDES

4775. Sodium tetrasulfur pentanitridate
[60241-33-0]   N$_5$NaS$_4$

Na$^+$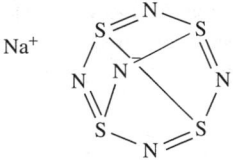

Scherer, O. J. *et al., Angew. Chem. (Intern. Ed.)*, 1975, **14**, 485
It explodes at about 180°C, or under pressure, or by friction of a spatula on a sintered glass filter.
See other NITRIDES, *N*-METAL DERIVATIVES, N–S COMPOUNDS

4776. Triphosphorus pentanitride
[12316-91-3] N_5P_3

$$P_3N_5$$

Metals
Mellor, 1971, Vol. 8, Suppl. 3, 370–371
Interaction with molten magnesium at 651°C is explosive, and more so with calcium at 200°C.

Oxidants
Partington, 1946, 618
It ignites on heating in chlorine or oxygen.
See also 'Tetraphosphorus hexanitride' *See other* NITRIDES

4777. Nickel azide
[59865-91-7] N_6Ni

$$Ni(N_3)_2$$

Mellor, 1940, Vol. 8, 355
It explodes at 200°C.
See other METAL AZIDES

4778. Sulfinyl azide
[ ] N_6OS

$$O{:}S(N_3)_2$$

Pascal, 1961, Vol. 10, 634
An explosive liquid.
See other ACYL AZIDES, NON-METAL AZIDES

4779. Sulfuryl diazide
[72250-07-8] N_6O_2S

$$N_3SO_2N_3$$

1. Curtius, T. *et al., Ber.*, 1922, **55**, 1571
2. Nojima, M., *J. Chem. Soc., Perkin Trans. 1*, 1979, 1813
It explodes violently when heated and often spontaneously at ambient temperature [1]. A safe method of preparation in solution has been described [2].
See other ACYL AZIDES, NON-METAL AZIDES

4780. Disulfuryl diazide
[73506-23-7] $N_6O_5S_2$

$$N_3SO_2OSO_2N_3$$

Alone, or Alkali
Lehmann, H. A. *et al., Z. Anorg. Chem.*, 1957, **293**, 314

The azide decomposes explosively below 80°C and should only be stored in 1 g quantities. In contact with dilute alkali at 0°C an explosive deposit is formed (? sodium azidosulfate?).

See other ACYL AZIDES, NON-METAL AZIDES

4781. 'Tetraphosphorus hexanitride'
[ ]   'N₆P₄'

Unknown structure

Mellor, 1971, 8, Suppl. 3, 371
Supposedly impure triphosphorus pentanitride, it ignites in air.
See related NITRIDES

4782. Lead(II) azide
[13424-46-9]   N₆Pb

Pb(N₃)₂

1. Mellor, 1940, Vol. 8, 353;, 1967, Vol. 8, Suppl. 2, 21, 50
2. Taylor, G. W. C. *et al., J. Crystal Growth*, 1968, **3**, 391
3. Barton, A. F. M. *et al., Chem. Rev.*, 1973, **73**, 138
4. *MCA Case History No. 2053*

As a heavy metal azide, it is rather endothermic (ΔH°_{f} (s) +436.4 kJ/mol, 1.50 kJ/g).

As a widely used detonator, its properties have been studied in great detail. Although quantitatively inferior to mercury fulminate as a detonator, it has proved to be more reliable in service. The pure compound occurs in 2 crystalline forms, one of which appears to be much more sensitive to initiation [1]. Factors which suppressed spontaneous explosions of lead azide during crystallisation were vigorous agitation and use of hydrophilic colloids [2]. These aspects have been reviewed [3]. A vacuum-desiccated sample of 5 g of the azide exploded violently when touched with a metal spatula [4] (possibly owing to static charge, generated by rapid evaporation of the solvent from the glass-insulated system, being earthed by the spatula).

See Sodium azide: Heavy metals (reference 1)

Calcium stearate

MCA Case History No. 949
An explosion occurred during blending and screening operations on a mixture of lead azide and 0.5% of calcium stearate. If free stearic acid were present as impurity in the calcium salt, free hydrogen azide may have been involved.

Copper, or Zinc

Federoff, 1960, A532, 551
Lead azide, on prolonged contact with copper, zinc or their alloys, forms traces of the extremely sensitive copper or zinc azides which may initiate detonation of the whole mass of azide.

See other ENDOTHERMIC COMPOUNDS, METAL AZIDES

4783. Palladium(II) azide
[13718-25-7]   N$_6$Pd

$$Pd(N_3)_2$$

Ukhin, L. Yu. *et al., J. Organomet. Chem.*, 1981, **210**, 270
The explosive precipitate needs careful handling.
See other METAL AZIDES

4784. Pentasulfur hexanitride
[68993-02-2]   N$_6$S$_5$

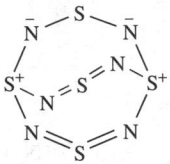

Sheldrick, W. S. *et al., Inorg. Chem.*, 1980, **19**, 539
Powerfully explosive, like its dimethyl homologue. A m.p. apparatus was wrecked,
and a large glass funnel containing 50 mg was pulverised.
See other NITRIDES, N–S COMPOUNDS

4785. Strontium azide
[19465-89-5]   N$_6$Sr

$$Sr(N_3)_2$$

It is somewhat endothermic (ΔH°_f (s) +204.6 kJ/mol, 1.19 kJ/g).
See Barium azide, *also* Calcium azide
See other ENDOTHERMIC COMPOUNDS, METAL AZIDES

4786. Zinc azide
[14215-28-2]   N$_6$Zn

$$Zn(N_3)_2$$

Bailar, 1973, Vol. 3, 217
Zinc azide is said to explode easily.
See Sodium azide: Heavy metals *See other* METAL AZIDES

4787. Poly(diazidophosphazene)
[ ]   (N$_7$P)$_n$

$$[-N=P(N_3)_2-]_n$$

Dillon, K. B. *et al., J. Chem. Soc., Dalton Trans.*, 1980, 1037
This polymer of phosphonitrilic diazide is a viscous explosive oil.
See 1,1,3,3,5,5-Hexaazido-2,4,6-triaza-1,3,5-triphosphorine
See other NON-METAL AZIDES

4788. Phosphorus triazide oxide
[4635-46-5] N_9OP

$$O:P(N_3)_3$$

Buder, W. *et al., Z. Anorg. Chem.*, 1975, **415**, 263
The liquid may explode violently on warming from 0°C to ambient temperature, but can be used safely in solution.
See related ENDOTHERMIC COMPOUNDS, NON-METAL AZIDES

4789. Phosphorus triazide
[56280-76-3] N_9P

$$P(N_3)_3$$

1. Lines, E. L. *et al., Inorg. Chem.*, 1972, **11**, 2270
2. Buder, W. *et al., Z. Anorg. Chem.*, 1975, **415**, 263
It is a highly explosive liquid [1], which may explode on warming from 0°C to ambient temperature, but can be used safely in solution [2].
See other NON-METAL AZIDES

4790. Lead(IV) azide
[73513-16-3] $N_{12}Pb$

$$Pb(N_3)_4$$

1. Mellor, 1967, Vol. 8, Suppl. 2, 22
2. Möller, H., *Z. Anorg. Chem.*, 1949, **26** 0, 10
3. Zbirol, E., *Synthesis*, 1972, (6), 285–302
The crystalline product appears less stable than the diazide, spontaneously decomposing, sometimes explosively [1]. It was rated as too unstable for use as a practical detonator or explosive [2]. Lead(IV) acetate azide (probably the triacetate azide) is also rather unstable, evolving nitrogen above 0°C with precipitation of lead(II) azide [3]. Lead(IV) azide will be considerably more endothermic than the lead(II) salt.
See other ENDOTHERMIC COMPOUNDS, METAL AZIDES

4791. Silicon tetraazide
[27890-58-0] $N_{12}Si$

$$Si(N_3)_4$$

1. Mellor, 1967, Vol. 8, Suppl. 2, 20
2. Wiberg, E. *et al., Z. Naturforsch. B*, 1954, **9B**, 500
3. Anon., *Angew. Chem. (Nachr.)*, 1970, **18**, 27
Spontaneous explosions have been observed [1] with this dangerously explosive material, especially when pure. A sample at 0°C exploded during removal of traces of benzene under high vacuum [2]. A residue containing the tetraazide, silicon chloride triazide and probably silicon dichloride diazide, exploded on standing for 2 or 3 days, possibly owing to hydrazoic acid produced by hydrolysis.
See other ENDOTHERMIC COMPOUNDS, NON-METAL AZIDES

4792. Thallium(I) tetraazidothallate
[ ] $N_{12}Tl_2$

$$Tl[(N_3)_4Tl]$$

Bailar, 1973, Vol. 1, 1152
A heat-, friction- and shock-sensitive explosive.
See related ENDOTHERMIC COMPOUNDS, HEAVY METAL DERIVATIVES, METAL AZIDES

4793. Pentaazidophosphorane
[56295-87-5] $N_{15}P$

$$P(N_3)_5$$

See Sodium hexaazidophosphate (next below)

4794. Sodium hexaazidophosphate
[60955-41-1] $N_{18}NaP$

$$Na[P(N_3)_6]$$

1. Volgnandt, P. *et al., Z. Anorg. Chem.*, 1976, **425**, 189–192
2. Buder, W. *et al., Z. Anorg. Chem.*, 1975, **415**, 263
The product of interaction of sodium azide and phosphorus trichloride occasionally exploded on warming from 0°C to ambient temperature, but was examined safely in solution. The structure of the explosive product is determined as the title compound [1], rather than pentaazidophosphorane as originally reported [2]. It contains some 82% of nitrogen.
See related NON-METAL AZIDES
See other ENDOTHERMIC COMPOUNDS, HIGH-NITROGEN COMPOUNDS

4795. 1,1,3,3,5,5-Hexaazido-2,4,6-triaza-1,3,5-triphosphorine
[22295-99-4] $N_{21}P_3$

1. Grundmann, C., *Z. Naturforsch. B*, 1955, **10B**, 116
2. Tornieporth-Oetting, I. C. *et al., Angewand. Chem. (Int.)*, 1995, **34**(5), 511

This viscous liquid (containing 73.5% N) is violently explosive when subjected to shock, friction [1] or electric discharge [2].

See Poly(diazidophosphazene)

See other ENDOTHERMIC COMPOUNDS, HIGH-NITROGEN COMPOUNDS, NON-METAL AZIDES

4796. Sodium

[7440-23-5]

Na

(*MCA SD-47*, 1952); *NSC 231*, 1978; *FPA H4*, 1972; *HCS 1980*, 823

1. Hawkes, A. S. *et al., J. Chem. Educ.*, 1953, **30**, 467
2. *Alkali Metals*, 1957
3. Houben-Weyl, 1979, Vol. 13.1, 287–289
4. Muir, G. D., private comm., 1968
5. Reuillon, M. *et al., Proc. 2nd Int. Symp. Loss Prev. Safety Prom. Proc. Ind.*, 481–486, Frankfurt, Dechema, 1978
6. Anon., *Fire Prevention*, 1995, (May), 39

In a discussion of safe methods for laboratory use of sodium, disposal of small quantities (up to 5–10 g) by immersion in isopropanol, which may contain up to 2% of water to increase the rate of reaction, is recommended. Quantities up to 50 g should be burned in a heavy metal dish, using a gas flame [1], (with suitable arrangements for dispersion or absorption of the alkaline particulate smoke so produced). Handling techniques and safety precautions for large-scale operations have also been detailed for this reactive metal [2]. Techniques for the preparation and precautions in use of finely dispersed sodium ('sodium sand') are given. It is to some extent pyrophoric [3]. Dispersions of sodium in volatile solvents may become pyrophoric if the solvent evaporates round the neck of a container. Serum cap closures are safer than screw caps [4]. Improved methods for extinction of liquid sodium (and other metal) fires are described which involve application of low-melting inorganic powder mixtures which melt and form a protective crust. Mixtures of sodium carbonate or graphite, with lithium carbonate, sodium chloride or sodium fluoride are very effective for sodium [5]. A fire involving a leak of bulk liquid sodium, which ignited, is reported. Autoignition in air can be as low as 100°C [6].

Acids   MRH values below reference
Mellor, 1941, Vol. 2, 469–470
MRH Hydrogen chloride 5.36/61, hydrogen fluoride 6.95/47, nitric acid 5.48/64, sulfuric acid 3.89/69
Anhydrous hydrogen chloride, hydrogen fluoride or sulfuric acid react slowly with sodium, while the aqueous solutions react explosively. Nitric acid of density above 1.056 causes ignition of sodium.

Ammonia, Aromatic hydrocarbons
Stevenson, J. R. *et al., J. Org. Chem.*, 1984, **49**, 3444
When naphthalene is reduced in liquid ammonia by metallic potassium, evaporation of excess ammonia gives the solvated solid potassium naphthalide ($K_2 \cdot$

$NH_3 \cdot Naphth_2$). If sodium is used in place of potassium, the product detonates as crystallisation starts. This is attributed to energetic expulsion of ammonia held endothermically in the growing crystal lattice. The same also occurs with anthracene and sodium, and nitrobenzene and barium. Caution in preparing and using these compounds is urged.

Ammonia, 1,2-Dimethoxyethane, Pyridine
Nolan, 1983, Case history 215
Pyridine is converted to a slurry of a dihydrobipyridyl by treatment with sodium in liquid ammonia at $-33°C$, then warming to $10°C$ to evaporate excess ammonia. The slurry was then transferred to a decomposition vessel (for subsequent aerobic oxidation to liberate bipyridyl), the transfer pipe being flushed through with cold dimethoxyethane. When warm solvent was used for this purpose on one occasion, oxidation proceeded so fast that ignition occurred.

Aqua Regia
See AQUA REGIA: SODIUM

Calcium, Mixed oxides
See Calcium: Sodium, etc.

Chlorobenzene, Phosphorus trichloride
Venczel, A., Rom. Pat. RO 72 608, 1980
In the preparation of triphenylphosphine from chlorobenzene, phosphorus trichloride and sodium dispersed in toluene or xylene, the possibility of explosion is avoided by adding about 1 mol% of a lower alcohol, based on sodium usage.

1-Chlorobutane, Light petroleum
Shackelford, S. A., J. Org. Chem., 1979, **44**, 3492, footnote 30
In formation of butylsodium, the temperature of interaction of the chloro compound with the dispersion of sodium in light petroleum must not be too low. At $-23°C$, reaction is smooth, but at $-78°$ excess of chlorobutane accumulated, and when reaction did start, it accelerated to violent explosion.
See Halocarbons (reference 7), below

Chloroform, Methanol
See Chloroform: Sodium, Methanol

Diazomethane
See Diazomethane: Alkali metals

Diethyl ether
Hey, P., private comm., 1965
While sodium wire was being pressed into ether, the die-hole blocked. Increasing the pressure on the ram to free it caused ignition of the ejected sodium and explosion of the flask of ether. Pressing the sodium into less flammable xylene or toluene and subsequent replacement of solvent with ether was recommended.

1816

Dimethylformamide

'DMF Brochure', Billingham, ICI, 1965

A vigorous reaction occurs on heating DMF with sodium metal.

Ethanol

Brändstrom, A., *Acta Chem. Scand.*, 1951, **4**, 1608–1610

Air must be excluded during exothermic interaction of ethanol with sodium finely dispersed in hydrocarbons to avoid the possibility of hydrogen–air mixture explosions.

See Potassium: Alcohols

Fluorinated alkanols

1. Anon., personal communication, 1996
2. Editor's comments

By mistake, sodium was added to excess liquid perfluoroalkylmethanol in the absence of the diluent which had previously moderated the reaction. Vigorous initial reaction culminated in an explosion which destroyed the flask and attached glassware and cracked the fume cupboard window, this was attributed to ignition of evolved hydrogen [1]. Since air would rapidly be swept from the flask by the gas flow, it seems probable that a hydrogen explosion sufficient to damage the hood would be external to the reaction flask. The desired fluoroalkoxide is undoubtedly thermodynamically unstable with respect to sodium fluoride formation and a number of other metal derivatives of fluorinated organics have proved explosively metastable [2].

See METAL DERIVATIVES OF ORGANOFLUORINE COMPOUNDS

Fluorinated compounds

Herring, D. E., private comm., 1964

Fusion of fluorinated compounds with sodium for qualitative analysis requires a high temperature for reaction because of their unreactivity. When reaction does occur, there is often an explosion and precautions must be taken.

Halocarbons   MRH values below references

1. Staudinger, H., *Z. Elektrochem.*, 1925, **31**, 549
2. Lenze, F. *et al.*, *Chem. Ztg.*, 1932, **56**, 921
3. Ward, E. R., *Proc. Chem. Soc.*, 1963, 15
4. *MCA SD-34*, 1949
5. Rayner, P. N. G., private comm., 1968
6. Anon., *Angew. Chem. (Nachr.)*, 1970, **18**, 397
7. Read, R. R. *et al.*, *Org. Synth.*, 1955, Coll. Vol. 3, 158
8. Houben-Weyl, 1970, Vol. 13.1, 394–395
9. Firat, Y., *Makromol. Chem.*, 1975, Suppl. 1, 207
10. Grattan, D. W. *et al.*, *J. Chem. Soc., Dalton Trans.*, 1977, 1734–1744

MRH Carbon tetrachloride 6.10/63, chloroform 5.82/63, 1,2-dichloroethylene 5.52/68, trichloroethylene 5.94/66

Although apparently stable standing in contact, mixtures of sodium with a range of halogenated alkane solvents are metastable and capable of initiation to explosion by shock or impact. Carbon tetrachloride [1,2,3], chloroform, dichloromethane

and chloromethane [1,2], tetrachloroethane [1,2,4] have been investigated, among others. Generally the sensitivity to initiation and the force of the explosion increase with the degree of halogen substitution; the two former less than in the corresponding systems with potassium or potassium–sodium alloy. Any aliphatic halocarbon (except fully fluorinated alkanes) may be expected to behave in this way. Small portions of sodium and hexachlorocyclopentadiene were mixed in preparation for a sodium fusion test. On shaking a few min later, the tube exploded [5]. On the small scale, no reaction occurred on boiling perfluorohexyl iodide in contact with metallic sodium. With 140 g of iodide and 7 g of sodium an explosion occurred after 30 min [6]. The temperature range for smooth interaction of bromobenzene, 1-bromobutane and sodium in ether to give butylbenzene is critical. Below 15°C reaction is delayed but later becomes vigorous, and above 30°C the reaction becomes violent [7]. Sodium wire and chlorobenzene react exothermally in benzene under nitrogen to give phenylsodium, and the reaction must be controlled by cooling. Use of finely divided sodium will lead to an uncontrollable, explosive reaction [8]. The hazards of contacting dichloromethane with a sodium film during an ultra-drying procedure are stressed. Thorough pre-drying and operations under vacuum are essential precautions [9]. Purification of chlormethane, chloroethane, or dichloromethane by condensation onto a fresh sodium mirror is extremely dangerous unless the halocarbons are very dry and oxygen-free [10].
See Chlorobenzene, etc.
   1-Chlorobutane
   Chloroform, etc.
   Chloroform: Metals
   Iodomethane: Sodium
   Lithium: Bromobenzene
See entry METAL–HALOCARBON INCIDENTS

Halogen azides
   See HALOGEN AZIDES: METALS

Halogens, or Interhalogens   MRH Chlorine 6.90/61, Fluorine 13.43/45
   1. Mellor, 1961, Vol. 2, Suppl. 2.1, 450–452
   2. Mellor, 1941, Vol. 2, 114, 469
   3. Mellor, 1941, Vol. 2, 92, 469
   4. Staudinger, H., *Z. Elektrochem.*, 1925, **31**, 549
   5. Booth, H. S. *et al., Chem. Rev.*, 1947, **41**, 427
   6. Pascal, 1966, Vol. 2.1, 221
Sodium ignites in fluorine gas but is inert in the liquefied gas [1]. Cold sodium ignites in moist chlorine [2] but may be distilled unchanged in the dry gas [1]. Sodium and liquid bromine appear to be unreactive on prolonged contact [3], but mixtures may be detonated violently by mechanical shock [4]. Finely divided sodium luminesces in bromine vapour [1]. Iodine bromide or iodine chloride react slowly with sodium, but mixtures will explode under a hammer-blow [1]. Interaction of iodine pentafluoride with solid sodium is initially vigorous, but soon slows with film-formation, while that with molten sodium is explosively violent

[2]. Sodium reacts immediately and incandescently with iodine heptafluoride [5]. Mixtures of solid sodium and iodine explode lightly when initiated by shock [6].

Hydrazine
Mellor, 1940, Vol. 8, 316
Anhydrous hydrazine and sodium react in ether to form sodium hydrazide, which explodes in contact with air. Hydrazine hydrate and sodium react very exother mally, generating hydrogen and ammonia.

Hydroxylamine   MRH 3.85/99+
See Hydroxylamine: Metals

Iodates
Cueilleron, J., *Bull. Soc. Chim. Fr.*, 1945, **12**, 88–89
Mixtures of sodium with silver iodate or sodium iodate explode when initiated by shock.
See Oxidants, below

Mercury
Brauer, 1965, Vol. 2, 1802
Interaction of sodium and mercury to form sodium amalgam is violently exothermic, and moderation of the reaction with an inert liquid, or by adding mercury slowly to the sodium is necessary. Even so, temperatures of 400°C may be attained.

Metal halides
1. *Alkali Metals*, 1957, 129
2. Mellor, 1961, Vol. 2, Suppl. 2.1, 494–496
3. Staudinger, H., *Z. Elektrochem.*, 1925, **31**, 549
4. Cueilleron, J., *Bull. Soc. Chim. Fr.*, 1945, **12**, 88
MRH Ammonium chloride 2.64/66, iron(III) chloride, 3.60/70, manganese(II) chloride 1.97/73, tin(II) chloride 1.38/84, vanadium(III) chloride 2.38/83, vanadyl chloride 2.30/82
Sodium dispersions will reduce many metal halides exothermically. Iron(III) chloride is reduced fairly smoothly at ambient temperature or below in presence of 1,2-dimethoxyethane. Nickel chloride, cobalt chloride, lead chloride or calcium chloride require higher temperatures to initiate the reaction, the exotherm with cobalt chloride increasing the temperature from 325 to 375°C and causing evaporation of most of the dispersing oil [1]. The finely powdered metals produced by reduction of halides of cadmium, chromium, cobalt, copper, iron, manganese, molybdenum, nickel, silver, tin or zinc with sodium dispersed in ether or toluene, are all pyrophoric [2]. Interaction of sodium with vanadyl chloride at 180°C is violent [2]. Mixtures of sodium with metal halides are sensitive to mechanical shock [3]; the ensuing explosions have been graded for intensity. Very violent explosions occurred with iron(III) chloride, iron(III) bromide, iron(II) bromide, iron(II) iodide, cobalt(II) chloride or cobalt(II) bromide. Strong explosions occurred with the various halides of aluminium, antimony, arsenic, bismuth, copper(II), mercury, silver, and tin (including a mixture of tin(IV) iodide and

sulfur), also vanadium(V) chloride and ammonium tetrachlorocuprate. Ammonium, copper(I), cadmium and nickel halides generally gave weak explosions, and most of the alkali- and alkaline earth-metals were insensitive [4].
See Molybdenum pentachloride: Sodium, or Sodium sulfide
See Tungsten hexachloride: Sodium sulfide

Metal oxides
1. Mellor, 1939, Vol. 9, 649
2. Mellor, 1943, Vol. 11, 237, 542
3. Mellor, 1941, Vol. 3, 138
4. Mellor, 1941, Vol. 7, 658, 401
5. Mellor, 1940, Vol. 4, 770
6. Brewer, L. *et al., Rept. UCRL 1864*, 7, Washington, USAEC, 1952
7. Mellor, 1961, Vol. 2, Suppl. 2.1, 474
MRH Chromium trioxide 3.39/79, copper(II) oxide 2.05/63, sodium peroxide 2.55/63, vanadium(V) oxide 1.30/84

Bismuth(III) oxide[1], chromium trioxide [2], and copper(II) oxide [3], are reduced with incandescence on heating with sodium. Finely divided sodium ignites on admixture with fine lead oxide without heating, while the coarse material reacts vigorously with molten sodium [4]. The latter reduces mercury(I) oxide [5] or molybdenum trioxide [2] with incandescence, the former producing a light explosion also. Sodium reduces sodium peroxide vigorously at 500°C [6] and tin(IV) oxide incandescently on gentle heating [4]. Finely dispersed sodium reduces metal oxides on heating at temperatures between 100 and 300°C, producing, e.g., pyrophoric iron, nickel and zinc [4].

Nitrogen-containing explosives
Leleu, *Cahiers*, 1975, (79), 266
Mixtures of sodium (or its alloy with potassium) and nitromethane, trichloronitromethane, nitrobenzene, dinitrobenzene, dinitronaphthalene, ethyl nitrite, ethyl nitrate or glyceryl nitrate are shock-sensitive, the sensitivity increasing with the number of nitro groups.

Non-metal halides   MRH values below references
1. Mellor, 1940, Vol. 8, 1033
2. Mellor, 1941, Vol. 2, 470
3. Mellor, 1940, Vol. 8, 1016
4. Mellor, 1961, Vol. 2, Suppl. 2.1, 455, 460
5. Mellor, 1940, Vol. 8, 1073
6. Cueilleron, J., *Bull. Soc. Chem. Fr.*, 1945, **12**, 88
7. Mellor, 1947, Vol. 10, 912
8. Kirk-Othmer, 1969, Vol. 18, 438
9. Leleu, *Cahiers*, 1975, (79), 270
MRH Phosphorus trichloride 3.10/67, phosphorus pentachloride 4.89/65, phosphoryl chloride 4.27/69, sulfinyl chloride 5.27/67

Sodium floats virtually unchanged on phosphorus tribromide, but added drops of water caused a violent explosion [1]. Molten sodium explodes with phosphorus trichloride [2] and may ignite or explode with phosphorus pentachloride

[3]. Diselenium dichloride reacts vigorously with sodium on heating, emitting light and heat [4]. Sodium ignites in sulfinyl chloride vapour at 300°C [4], or in a stream of thiophosphoryl fluoride [5]. The shock-sensitive mixtures of sodium with phosphorus pentachloride, phosphorus tribromide or sulfur dichloride gave violent explosions on impact, while those with boron tribromide or sulfur dibromide gave strong explosions [6]. Sodium and phosphoryl chloride interact explosively on heating [8], and mixtures of sodium with sulfinyl fluoride or silicon tetrachloride or silicon tetrafluoride are shock-sensitive explosives [9].

Non-metal oxides   MRH values below references
1. Mellor, 1940, Vol. 6, 70; 1961, Vol. 2, Suppl. 2.1, 468
2. Gilbert, H. N., *Chem. Eng. News*, 1948, **26**, 2604
3. Mellor, 1940, Vol. 8, 554, 945; 1961, Vol. 2, Suppl. 2.1, 458
4. Leleu, *Cahiers*, 1975, (79), 269
5. Weiss, E., *Angew. Chem. (Int.)*, 1993, **32**(11), 1518
MRH Carbon monoxide 6.15/65, chlorine dioxide 5.69/75, sulfur dioxide 4.56/74, sulfur trioxide 5.06/70

Sodium and carbon dioxide are normally unreactive till red-heat has been attained [1], but mixtures of the 2 solids are impact-sensitive and explode violently [2]. Carbon dioxide is unsuitable as an extinguishant for the burning metal alone, as the intensity of combustion is increased by replacing air with carbon dioxide (72.7% oxygen content). However it has been used successfully to extinguish solvent fires where sodium was also present, since it often fails to ignite because of the blanketing effect of solvent vapour. Conversely addition of kerosene to burning sodium enables the whole to be extinguished with carbon dioxide [1]. Finely divided silica (sand) will often react with burning sodium, so is not entirely suitable as an extinguishant [1,2]. (Anhydrous sodium carbonate and sodium chloride are suitably unreactive for this purpose.) Solid sodium is inert to dry liquid or gaseous sulfur dioxide, but molten sodium reacts violently [2]. The moist gas reacts almost as vigorously as water [1]. Phosphorus pentaoxide becomes incandescent on warming with sodium, which also ignites in dinitrogen pentaoxide [3]. The product of interaction of carbon monoxide with sodium (not sodium carbonyl as previously thought [2], but the hexamer), sodium benzenehexoxide, is shock sensitive, explodes at 90°C and contact with water causes explosion or ignition to occur [4]. Later work [5] makes it likely that this compound is, in fact, the dimer disodium ethynediolate.

Non-metals   MRH Sulfur 4.81/41
1. Mellor, 1961, Vol. 2, Suppl. 2.1, 466–467, 454–455
2. Mellor, 1941, Vol. 2, 469
3. Mellor, 1947, Vol. 10, 766
Explosions have occasionally occurred when carbon powder is in contact with evaporating sodium and air [1]. The violent interaction of ground or heated mixtures of sodium and sulfur may be moderated by the presence of sodium chloride or boiling toluene [1,2]. Selenium reacts incandescently with sodium when heated [3], and molten tellurium reacts vigorously when poured on to solid sodium [1].

Organic samples for qualitative analysis
See LASSAIGNE TEST

Other reactants
Yoshida, 1980, 262–263
MRH values calculated for 19 combinations with various materials are given.

Oxidants   MRH values show % of oxidant
See Halogens or Interhalogens; Metal halides;
    Metal oxides; Non-metal oxides, all above
    Ammonium nitrate: Metals   MRH 4.15/64
    Nitric acid: Metals   MRH 5.84/32
    Nitrosyl fluoride: Sodium
    Nitryl fluoride: Metals
    Sodium nitrate: Sodium   MRH 3.85/43
    Sodium nitrite: Reducants   MRH 3.42/50
    Sodium peroxide: Metals   MRH 2.55/63

Oxygenated compounds
Leleu, *Cahiers*, 1975, (79), 267
Mixtures of inorganic oxygenated compounds (halide oxides or oxide sulfides) or
oxygen-rich organic compounds (alkyl oxalates) with sodium (or its alloy with
potassium) are shock-sensitive explosives.
See Non-metal oxides, above

Sulfides
Leleu, *Cahiers*, 1975, (79), 268, 270
Passage of moist hydrogen sulfide over unheated sodium causes melting and then
usually ignition of the metal. Mixtures of sodium (or its alloy with potassium) and
carbon disulfide are shock-sensitive explosives.

Sulfur
Hames, D. *et al., Proc. Inst. Electr. Eng.*, 1979, **126**, 1157–1161
Safety aspects of large arrays of sodium–sulfur batteries for rail traction are consid-
ered among technical aspects.

2,2,3,3-Tetrafluoropropanol
See 2,2,3,3-Tetrafluoropropanol: Potassium hydroxide, etc.

Tetrahydrofuran, Water
Anon., *Lab. Accid. Higher Educ.*, item 5, Barking, HSE, 1987
THF was being dried by refluxing over metallic sodium, and after the reflux period,
heat was turned off and a stopper was put in the top of the condenser. The presence of the stopper was forgotten, and when heating was restarted, the condenser
was ejected from the flask and broke. Escaping water came into contact with the
sodium, evolving hydrogen which ignited and then set fire to the escaping THF
vapour in a fume cupboard, and an explosion ensued.
See Water, below

Toluene
See Xylene, below

Water

1. Mellor, 1941, Vol. 2, 469
2. Mellor, 1961, Vol. 2, Suppl. 2.1, 362
3. *MCA Case History No. 456*
4. Markowitz, M. M., *J. Chem. Educ.*, 1963, **40**, 633–636
5. *MCA Case History No. 1653*
6. *MCA Case History No. 2082*
7. Peacocke, T. A. H., *Chem. Brit.*, 1979, **15**, 233

Small pieces of sodium in contact with water react vigorously but do not usually ignite the evolved hydrogen unless the water is above 40°C, or if rapid dissipation of heat is prevented by immobilising the sodium by use of a viscous aqueous solution (starch paste, etc.) or wet filter paper [1]. In contact with ice, sodium explodes violently [2]. Small, hot particles of sodium (remaining from dissolution of larger pieces) may finally explode as do large lumps of the metal in water [1]. Sodium residues from a Wurtz reaction were treated with alcohol to destroy sodium, but later accidental contact with water caused a fire, showing that the alcohol treatment was incomplete. This may have been owing to a crust of alcohol-insoluble halide over the residual sodium [3]. Reactivity of sodium and other alkali-metals with water or steam has been discussed in detail [4]. Rolling a drum containing a 12 cm layer of sodium sludge mixed with soda ash led to a mild pressure explosion. Moisture present in the drum contacted the sodium residue when rolled [5]. Preparing to dispose of a few l of old sodium dispersion by adding a solvent and then burning it, an operator opened the container under cover but in an atmosphere rendered humid by very recent rain. The dispersion immediately ignited and exploded [6]. An explosion of a small piece of sodium retained under water (out of contact with air) was attributed to formation of organic peroxides from arylsodium derivatives produced in storage of sodium under naphtha [7].

Xylene

Anon., *Sichere Chemiarb.*, 1989, **41**(5), 58

After storage for 16 years in a tin, a sealed bottle originally holding sodium dispersed in xylene was found to contain a yellow/white solid layer in place of the expected supernatant xylene. Scraping the solid out caused a violent explosion. The force of the explosion leads to a suspicion of peroxide formation, but there is no obvious explanation. Reactive materials like alkali-metal dispersions in volatile solvents should not be stored indefinitely, but clearly labeled after receipt or preparation to show the disposal date. Disposal of such dispersions by burning is recommended. Sodium dispersed in toluene might behave in the same way.

See ALKALI METALS
See other METALS

4797. Sodium–antimony alloy
[[12300-03-7] (1:1 alloy)]   Na—Sb

Na—Sb

See Nitrosyl tribromide: Sodium–antimony alloy
See other ALLOYS

4798. Sodium–zinc alloy
 [[39422-73-6] (1:6 alloy)]   Na–Zn
                                     Na–Zn

Houben-Weyl, 1973, Vol. 13.2a, 571
Preparation by addition of sodium to molten zinc (1:4) in absence of air is rather
violent.
See other ALLOYS

4799. Sodium dioxide (Sodium superoxide)
 [12034-12-7]   NaO$_2$
                                     NaO$_2$

Sodium peroxide
  Gausing, W. *et al., Angew. Chem. (Intern. Ed.)*, 1978, **17**, 372
  A by-product mixture of sodium dioxide and sodium peroxide separated by filtra-
  tion was sensitive to warmth and shock when dry.
  See related METAL PEROXIDES *See other* METAL OXIDES, OXIDANTS

4800. Sodium trioxide
 [12058-54-7]   NaO$_3$
                                     NaO$_3$

Petrocelli, A. W. *et al., J. Chem. Educ.*, 1962, **39**, 557
This is the least stable of the alkali-metal ozonides.
See related METAL PEROXIDES *See other* METAL OXIDES, OXIDANTS

4801. Sodium silicide
 [12164-12-4]   NaSi
                                     NaSi

1. Mellor, 1961, Vol. 2, Suppl. 2.1, 564
2. Bailar, 1973, Vol. 1, 446
Sodium silicide ignites in air [1], and like its potassium, rubidium and caesium
analogues, ignites explosively on contact with water or dilute acids [2].
See other METAL NON-METALLIDES

4802. Sodium oxide
 [1313-59-3]   Na$_2$O
                                     Na$_2$O

2,4-Dinitrotoluene
  See 2,4-Dinitrotoluene: Sodium oxide

Phosphorus(V) oxide
  See Tetraphosphorus decaoxide: Inorganic bases

Water
  Interaction is likely to be violently exothermic.
  See other INORGANIC BASES, METAL OXIDES

1824

4803. Sodium peroxide
[1313-60-6]   Na$_2$O$_2$

NaOONa

FPA H104, 1981; *HCS 1980*, 854

1. Allen, C. F. H. *et al., J. Chem. Educ.*, 1942, **19**, 72
2. Rüst, 1948, 297

Hazards attendant on the use of this powerful oxidant may in many cases be eliminated by substitution with 'sodium perborate' (actually sodium borate hydrogen peroxidate) [1]. One of several wooden boxes of the peroxide (not clearly labelled as such) exploded with great violence during handling operations. It seems likely that contamination with a combustible material, or possibly with moisture, had occurred [2].

See Fibrous materials, *also* Wood, both below

Acetic acid   MRH 2.80/16
von Schwartz, 1918, 321
Admixture causes explosion, owing either to direct oxidation of acetic acid by the highly concentrated hydrogen peroxide produced, or perhaps to formation of concentrated peroxyacetic acid.

Aluminium   MRH 4.56/19

Aluminium chloride   MRH 2.30/53
Anon., *Chem. Eng. News*, 1954, **32**, 258
A mixture of the three slowly reacted, creating a pressure of 122 bar (? of chlorine and possibly hydrogen chloride) in 41 days, and the residue reacted spontaneously on exposure to air.
See Metals, etc., below

Ammonium peroxodisulfate   MRH 2.72/60
Mellor, 1947, Vol. 10, 464
A mixture exploded on being subjected to friction in a mortar, heating above 75°C, or exposure to carbon dioxide or drops of water.

Boron nitride
Mellor, 1940, Vol. 8, 111
Addition of powdered nitride to the molten peroxide caused incandescence.

Calcium acetylide   MRH 3.01/14
Mellor, 1941, Vol. 2, 490
A mixture is explosive.

Carbon dioxide   MRH 1.84/36

Cotton wool   MRH Cellulose 1.97/99+
1. Bulavin, Yu. I., *Chem. Abs.*, 1978, **89**, 75003
2. Anon., *Chemist-Analyst*, 1978, **67**(3), 7

Cotton wool sprinkled with the peroxide (used in guard tubes) will ignite a few s after exposure to moist carbon dioxide. Exothermic interaction of sodium peroxide and carbon dioxide heats the cotton which then ignites in the liberated oxygen [1]. Cotton contaminated with peroxide should not be stored under any circumstances as it may ignite with atmospheric carbon dioxide [2].

Fibrous materials, Water   MRH Cellulose 1.97/99+
1. von Schwartz, 1918, 328
2. Betteridge, D., private comm., 1973
Contact of the solid peroxide with moist cloth, paper or wood often causes ignition [1], and addition of water to intermixed cotton wool and peroxide causes violent ignition [2].

Glucose, Potassium nitrate   MRH 2.93/15
491M, 1975, 391
A micro-bomb calorimeter exploded when the wrong proportions of sample and oxidants were used. Instead of 4 g of peroxide and 0.2 g of nitrate for 0.2 g of the sugar sample, 0.35 g of peroxide and 2.6 g of dextrose were used. The deficiency of peroxide to absorb the decomposition gases and excess of organic matter led to a rapid rise in temperature and pressure, which burst the bomb calorimeter.

Hexamethylenetetramine   MRH 2.47/10
Ellern, 1968, 46
A mixture ignites when moistened with water.

Hydrogen sulfide   MRH 3.01/30
1. Barrs, C. E., *J. R. Inst. Chem.*, 1955, **79**, 43
2. Mellor, 1947, Vol. 10, 129
Solid sodium peroxide causes immediate ignition in contact with gaseous hydrogen sulfide [1]. Barium peroxides and other peroxides behave similarly [2].

Hydroxy compounds   MRH values below reference
Castrantas, 1965, 4
MRH Ethanol 2.43/9, ethylene glycol 2.68/14, glycerol 2.93/17, allose 2.93/17, acetic acid 2.80/16
   The exothermic oxidation of ethanol, ethylene glycol, glycerol, sugar or acetic acid may lead to fire or explosion.
See Water, below, *also* Ethylene glycol: Oxidants

Magnesium
Bentzinger, von R. *et al., Praxis Naturwiss. Chem.*, 1987, **36**, 38
A priming mixture of magnesium powder and barium peroxide to ignite thermite mixture was to be prepared, but as no barium peroxide was available, sodium peroxide was used instead. Some time after preparation, the mixture ignited spontaneously, because sodium peroxide, unlike barium peroxide, is very hygroscopic and forms hot conc. hydrogen peroxide from contact with atmospheric moisture.
See Carbon dioxide, etc., above

Metal halides
Wiley, J. B. *et al., Mater. Res. Bull.*, 1993, **28**(9), 893
Metathesis reaction to prepare crystalline transition and B metal oxides from halides and sodium peroxide is initiated by a hot wire and can be highly exothermic and explosive in some instances.
See METATHESIS

Metals   MRH Aluminium 4.56/19, iron 1.55/32, magnesium 4.945/24, sodium 2.55/37, tin 1.38/44
Bunzel, E. G. *et al., Z. Anorg. Chem.*, 1947, **254**, 20
At 240°C mixtures of finely divided metals (aluminium, iron, tungsten) with the peroxide ignite under high friction, and molybdenum powder reacts explosively.

Metals, Carbon dioxide, Water
Mellor, 1941, Vol. 2, 490; 1961, Vol. 2, Suppl. 2.1, 633; 1946, Vol. 5, 217
Intimate mixtures of sodium peroxide with aluminium, magnesium or tin powders ignite on exposure to moist air and become incandescent on heating in air or on moistening with water. Exposure of such mixtures to carbon dioxide causes an explosion. (Interaction of the peroxide and carbon dioxide is highly exothermic.) Sodium is oxidised vigorously at 500°C.

Non-metal halides
1. Mellor, 1947, Vol. 10, 897; 1961, Vol. 2, Suppl. 2.1, 634
2. Comanducci, E., *Chem. Ztg.*, 1911, **15**, 706
Violent interactions occur with diselenium dichloride or disulfur dichloride, the latter emitting light and heat [1]. The very exothermic reaction with phosphorus trichloride may accelerate to explosion [2].

Non-metals   MRH Carbon 3.05/7, phosphorus 2.30/14, sulfur 3.35/29
1. von Schwartz, 1918, 328
2. Mellor, 1941, Vol. 2, 490
3. Anon., *ABCM Quart. Safety Summ.*, 1948, **19**, 13
Intimate mixtures with carbon or phosphorus may ignite or explode [1]. Other readily oxidisable materials (probably antimony, arsenic, boron, sulfur, selenium) also form explosive mixtures [2]. Use of finely powdered carbon, rather than the granular carbon specified for a reagent, mixed with sodium peroxide caused an explosion [3].

Organic liquids, Water   MRH Aniline 2.76/7, benzene 2.76/7, diethyl ether 2.55/8, glycerol 2.72/15
von Schwartz, 1918, 328
Simultaneous contact of sodium peroxide with water and aniline, or benzene, diethyl ether or glycerol, causes ignition: (equivalent to contact with conc. hydrogen peroxide).

Organic materials
Rüst, 1948, 337

A 'medicinal' mixture of the peroxide (30%) with liquid paraffin (18%), dried soap (42%) and almond oil (10%) ignited explosively during preparation.

Organic material, Water
Bentzinger, von R. *et al., Praxis Naturwiss. Chem.*, 1987, **36**, 40
In a combustion demonstration, 50 g of sodium peroxide and half its volume of sago flour are cautiously mixed with a feather. Addition of a few drops of hot water immediately ignites the mixture, which burns vigorously. (The scale of the experiment, using 0.64 mol of sodium peroxide, seems excessive.)

Other reactants
Yoshida, 1980, 87–88
MRH values calculated for 30 combinations with oxidisable materials are given.

Peroxyformic acid
See Peroxyformic acid: Metals, etc.

Silver chloride   MRH 0.21/78

Charcoal   MRH 3.05/7
Mellor, 1941, Vol. 3, 401
An intimate mixture ignites after a short delay (but the same is probably true if silver chloride is omitted).
See Non-metals, above

Sodium dioxide
See Sodium dioxide: Sodium peroxide

Water
1. *Haz. Chem. Data*, 1975, 269
2. Friend, J. N. *et al., Nature*, 1934, **134**, 778
3. Cheeseman, G. H. *et al., Nature*, 1934, **134**, 971
Reaction with water is vigorous, and with large amounts of peroxide it may be explosive. Contact of the peroxide with combustibles and traces of water may cause ignition [1]. Violent explosions on two occasions during attempted preparation of oxygen were attributed to traces of sodium in the peroxide. The former would liberate hydrogen and ignite the detonable mixture [2,3].

Wood   MRH Cellulose 1.97/99+
Dupré, A., *J. Soc. Chem. Ind.*, 1897, **16**, 492
Friction of the peroxide between wooden surfaces caused ignition of the latter.
See other METAL PEROXIDES, OXIDANTS

4804. Sodium thiosulfate
[7772-98-7] $Na_2O_3S_2$

NaSSO$_2$ONa

HCS 1980, 863

Widmann, G. *et al., Thermochim. Acta*, 1988, **134**, 451–455
Energy of decomposition (320–520°C) was determined by DSC as 2.42 kJ/g,
peaking at 427°C.

Metal nitrates   MRH values show % of nitrate
  See Potassium nitrate: Reducants   MRH 2.22/50
     Sodium nitrate: Sodium thiosulfate   MRH 2.38/46

Other reactants
  Yoshida, 1980, 230
  MRH values calculated for 9 combinations with various materials are given.

Sodium nitrite   MRH 2.26/47
  Stevens, H. P., *J. Proc. R. Inst. Chem.*, 1946, 285
  A mixture of these oxidising and reducing salts will explode violently after most
  of the water of crystallisation has been driven off by heating.
  See other REDOX REACTIONS
  See other METAL OXONON-METALLATES, REDUCANTS

4805. Sodium metasilicate
[6834-92-0] $(Na_2O_3Si)_n$

(-OSi(ONa)$_2$-)$_n$

Fluorine
  See Fluorine: Metal salts
  See other METAL OXONON-METALLATES

4806. Sodium sulfate
[7681-38-1] Na_2O_4S

NaOSO$_2$ONa

Aluminium
  See Aluminium: Sodium sulfate
  See other METAL OXONON-METALLATES

4807. Sodium dithionite ('Sodium hydrosulfite')
[7775-14-6]

NaOS(O)S(O)ONa

$Na_2O_4S_2$

1. *MCA Case History No. 882*
2. Tyler, B. J. *et al., Hazards from Pressure*, IChE Symp. Ser. No. 102, 45–59, Oxford, Pergamon, 1987
3. Henderson, D. K. *et al., J. Haz. Mat.*, 1988, **19**, 155–159
A batch decomposed violently during drying in a graining bowl. No explanation was offered but contamination with water and/or an oxidant seems likely. Thermal decomposition of pure material occurs violently at 190°C [1]. Spontaneous ignition of thin layers of sodium dithionite dust was studied experimentally. Ignition temperatures were measured for 5–40 mm layers of mixtures with less than 75% of inert dust. The overall importance of the effect of layer depth on heat transfer parameters at the cold surface was established, and the results were compared with theoretical predictions [2]. Using increasing temperature regimes to determine ignition temperature of dithionite dust layers confirmed the previous value of 190°C, but when decreasing regimes were used the much higher figure of 400°C was found. This was explained on the basis of a 2 stage decomposition characteristic [3].

Aluminium, Water

Kirschner, E., *Chem. Eng. News*, 1995, **73**(18), 6
As part of the process for manufacture of a gold-precipitant, 0.5 tonnes of aluminium powder and 3.5 tonnes of sodium dithionite were in a blender to which benzaldehyde was to be added (other sources also claim presence of potassium carbonate). Either from a leaking seal, or in trying to clear a benzaldehyde feed line, some water was admitted, starting a slow reaction emitting sulphurous fumes. After about eleven hours an attempt was made to blanket with nitrogen (if air had initially been involved, it seems probable the blender had long since purged itself with sulphur dioxide). There was an explosion, followed by fire, killing five workers. In view of the known properties of 'hydros', admission of water might warm localised regions to a temperature where self-sustaining decomposition would ensue in the absence of air. Although a reducing agent in wide use, where aluminium is concerned sodium dithionite and its decomposition products are most definitely oxidants (as is potassium carbonate). Aluminium oxidation was probably responsible for the main blast.
See Aluminium: Metal oxides, or Oxosalts, or Sulfides
See also Water, below

Sodium chlorite
See Sodium chlorite: Sodium dithionite

Water
1. Anon., *Chem. Trade J.*, 1939, **104**, 355
2. *MCA Case History No. 2292*
3. Ullmann, 1994, Vol A25, 483

Addition of 10% of water to the solid anhydrous material caused a vigorous exotherm and spontaneous ignition. Bulk material may decompose at 135°C [1]. A loose drum lid allowed ingress of water and smouldering started. As the drum was tipped over for disposal, ignition occurred [2]. Air is usually involved in the exotherm leading to fire after water contamination. The dihydrate may be pyrophoric when finely divided. In the absence of air, exothermic decomposition can occur from 90°C after prolonged heating [3].

See other METAL OXONON-METALLATES, REDUCANTS

4808. Sodium disulfite ('Sodium metabisulfite')
[7681-57-4] $Na_2O_5S_2$

$$NaOS(O)OS(O)ONa$$

Potassium iodate
See Potassium iodate: Sodium disulfite, Water

Sodium nitrite
See Sodium nitrite: Sodium disulfite
See other METAL OXONON-METALLATES, REDUCANTS

4809. Sodium peroxodisulfate
[7775-27-1] $Na_2O_8S_2$

$$NaOSO_2OOSO_2ONa$$

T_{ait24} was determined as 148°C by adiabatic Dewar tests, with an apparent energy of activation of 75 kJ/mol.
See entry THERMOCHEMISTRY AND EXOTHERMIC DECOMPOSITION (reference 2)

2,2'-Azobis(2-amidiniopropane) chloride
See 2,2'-Azobis(amidiniopropane) peroxodisulfate
See other PEROXOACID SALTS, OXIDANTS

4810. Sodium tetraperoxotungstate
[12501-01-8] Na_2O_8W

$$Na_2[W(O_2)_4]$$

Mellor, 1943, Vol. 11, 835
It explodes feebly on warming.
See other PEROXOACID SALTS

4811. Sodium sulfide
[1313-82-2, 27610-45-2 (anhydr., 9H$_2$O, resp.)]   Na$_2$S
Na$_2$S

FPA H115, 1982; *HCS 1980*, 860

1. Anon., *ABCM. Quart. Safety Summ.*, 1942, **13**, 5
2. Dunn, B. W., *Bur. Explos. Accid. Bull., Amer. Rail Assoc.*, 1917, **25**, 36
Fused sodium sulfide in small lumps is liable to spontaneous heating from aerobic oxidation, temperatures of up to 120°C being observed after exposure to moisture and air. Packing in hermetically closed containers is essential [1]. Previously, similar material packed in wooden barrels had ignited in transit [2].
See also SULFUR BLACK

Ammonium persulfate
See Ammonium peroxodisulfate: Sodium sulfide

Carbon   MRH 0.041/tr.
Creevey, J., *Chem. Age*, 1941, **44**, 257
Mixtures of sodium sulfide and finely divided carbon exhibit an exotherm on exposure to air. As indicated by the low MRH value, this is probably not a direct interaction, but arises from co-promotion of aerobic oxidation of the individual components.

Diazonium salts
See DIAZONIUM SULFIDES

N,N-Dichloromethylamine
See N,N-Dichloromethylamine: Calcium hypochlorite, etc.

Glass
L.B., Personal experience, 1964
An old soda glass bottle of the hygroscopic nonahydrate had become severely etched and corroded internally by the contents, which had leaked through the perforated wall of the bottle. It is probable that the old material had become oxidised by air to a strongly alkaline mixture (below), corrosive to the soft glass.
$$2Na_2S + O_2 + H_2O \rightarrow Na_2S_2O_3 + 2NaOH$$
See other CORROSION INCIDENTS, GLASS INCIDENTS

Hydrogen peroxide
See Hydrogen peroxide: Sulfides

Metal halides
See Tungsten hexachloride: Sodium sulfide

Other reactants
Yoshida, 1980, 392
MRH values calculated for 18 combinations, largely with oxidants, are given.

Sodium carbonate, Water
See SMELT: WATER
See other METAL SULFIDES

4812. Sodium disulfide
[22868-13-9] Na_2S_2

NaSSNa

Diazonium salts
See DIAZONIUM SULFIDES *See other* METAL SULFIDES

4813. Sodium polysulfide
[1344-08-7] Na_2S_x

NaS$_x$Na

Diazonium salts
See DIAZONIUM SULFIDES AND DERIVATIVES *See other* METAL SULFIDES

4814. Sodium orthophosphate
[7601-54-9] Na_3O_4P

O:P(ONa)$_3$

Sugars
See SUGARS

4815. Sodium phosphide
[12058-85-4] Na_3P

Na$_3$P

Water
Mellor, 1940, Vol. 8, 834
Sodium phosphide (or potassium phosphide) is decomposed by moist air or water,
evolving phosphine, which often ignites.
See other METAL NON-METALLIDES

4816. Sodium pyrophosphate hydrogen peroxidate
[15039-07-3] $Na_4O_7P_2.2H_2O_2$

(NaO)$_2$P(O)OP(O)(ONa)$_2$.2H$_2$O$_2$

See entry CRYSTALLINE HYDROGEN PEROXIDATES

4817. Niobium
[7440-03-1]   Nb

                                Nb

Halogens, or Interhalogens
Mellor, 1939, Vol. 9, 849; 1956, Vol. 2, Suppl. 1, 165
Niobium ignites in cold fluorine, and in chlorine at 205°C, and incandesces in
contact with bromine trifluoride.
See other METALS

4818. Niobium(V) oxide
[1313-96-8]   Nb$_2$O$_5$

                          O$_2$NbONbO$_2$

Lithium
    See Lithium: Metal oxides
    See other METAL OXIDES

4819. Neodymium
[7440-00-8]   Nd

                                Nd

Phosphorus
    See Phosphorus: Metals *See other* METALS

4820. Nickel
[7440-02-0]   Ni

                                Ni

RSC Lab. Hazards Data Sheet No. 60, 1987 (Ni and compounds)

1. Sasse, W. H. F., *Org. Synth.*, 1966, **46**, 5
2. Whaley, T. P., *Inorg. Synth.*, 1957, **5**, 197
3. Brown, C. A. *et al., J. Org. Chem.*, 1973, **38**, 2226
4. Savel'eva, O. N., *Chem. Abs.*, 1982, **96**, 165547
5. Anon., *Jahresber.*, 1987, 65

Raney nickel catalysts must not be degassed by heating under vacuum, as large
amounts of heat and hydrogen may be evolved suddenly and dangerous explosions
may be caused [1]. Nickel powder prepared by several alternative methods may
be pyrophoric if particles are fine enough [2]. A non-pyrophoric highly active
colloidal hydrogenation catalyst (P-2 nickel) is produced by reduction of nickel
acetate in ethanol by sodium tetrahydroborate [3]. Catalytic Raney nickel electrodes
are rendered non-pyrophoric by aqueous metal nitrate treatment [4]. After storage
for some time, a drum containing Raney nickel catalyst under water was opened for
inspection. As the lid was being removed, there was an explosion which blew it up

into the air. This is attributed to desorption of hydrogen into the air-space to form an explosive mixture, and ignition of the latter by traces of dry (and pyrophoric) catalyst in the top of the drum which sparked in contact with the ingress of air as the lid was opened. It was suggested that such stored drums should be shaken to wet any catalyst at the top of the drum before opening. Displacing any air with carbon dioxide before sealing would seem also to be advisable [5].

Aluminium
See Aluminium–nickel alloys

Aluminium chloride, Ethylene
See Ethylene: Aluminium chloride

1,4-Dioxane   MRH 1.46/99+
See 1,4-Dioxane: Nickel

Hydrogen
  1. Adkins, H. et al., J. Amer. Chem. Soc., 1948, **70**, 695
  2. Adkins, H. et al., Org. Synth., 1955, Coll. Vol. 3, 176
  3. Chadwell, A. J. et al., J. Phys. Chem., 1956, **60**, 1340
During hydrogenation of an unspecified substrate (possibly 4-nitrotoluene) at high pressure with the highly active W6 type of Raney nickel catalyst at 150°C, a sudden exotherm caused the initial pressure to double rapidly to 680 bar [1]. This does not happen at 100°C or below. Care is necessary with selection of reaction conditions for highly active catalysts [2]. Hydrogen-laden Raney nickel catalyst when heated strongly under vacuum undergoes explosive release of the sorbed hydrogen [3].

Hydrogen, Oxygen
Lee, W. B., Ind. Eng. Chem., Prod. Res. Dev., 1967, **6**(1), 59–64
Raney nickel powder entrained in cryogenic hydrogen gas causes ignition on contact with liquid or cryogenic gaseous oxygen.
See other CATALYTIC IMPURITY INCIDENTS

Magnesium silicate
Blake, E. J., private comm., 1974
The pyrophoricity of nickel-on-sepiolite catalysts after use in petroleum processing operations may be caused by the presence of finely divided nickel and/or carbon.
See other PYROPHORIC CATALYSTS

Methanol
  1. MCA Case History No. 1225
  2. Anon., Jahresber., 1983, 77–78
  3. Anon., Sichere Chemiearbeit, 1990, **42**, 45
Ignition occurred when methanol was poured through the open manhole of a 230 l reactor containing Raney nickel catalyst. Although the reactor had been purged thoroughly with nitrogen before opening, it was later shown that air was entrained

during pouring operations, and this would cause nickel particles round the manhole opening to glow. A closed charging system was recommended [1]. A slurry of Raney nickel catalyst in methanol was to be added to a nitrogen-filled reactor containing a methanol solution. A valve connecting the reactor to a closed circuit acid-scrubber was opened to equalise the pressure, but the scrubber was at a rather lower pressure and reduced the pressure in the total system below ambient. When the charging port was opened, air was drawn in carrying residual catalyst particles into the vessel and ignition occurred. Remedial measures are outlined [2]. An identical ignition injuring two workers is reported subsequently [3].

Non-metals
Mellor, 1942, Vol. 15, 148, 151
On heating, mixtures of powdered nickel with sulfur or selenium react incandescently.
See Sulfur compounds, below

Organic solvents
Hotta, K. *et al.*, *Chem. Abs.*, 1969, **70**, 81308
Raney nickel catalyst evaporated with small amounts of methanol, ethanol, isopropanol, pentanol, acetone, benzene, cyclohexane or *p*-dioxane and then heated towards 200°C eventually explodes
See 1,4-Dioxane: Nickel

Other reactants
Yoshida, 1980, 270
MRH values calculated for 14 combinations, largely with oxidants, are given.

Oxidants
values show % of oxidant
See Ammonium nitrate: Metals   MRH 2.59/58
      Bromine pentafluoride: acids, etc.   MRH 3.64/55
      Chlorine: Metals   MRH 2.34/55
      Nitryl fluoride: Metals
      Peroxyformic acid: Metals   MRH 5.69/99+
      Potassium perchlorate: Metal powders   MRH 2.64/37

Sulfur compounds
Kornfeld, E. C., *J. Org. Chem.*, 1951, **16**, 137
Raney nickel catalyst, containing appreciable amounts of the sulfide (after use to desulfurise thioamides) is rather pyrophoric.
See other HYDROGENATION CATALYSTS, METALS, PYROPHORIC METALS

4821. Nickel(II) oxide
[1313-99-1]

NiO

NiO

Anilinium perchlorate
See Anilinium perchlorate: Metal oxides

Fluorine
See Fluorine: Metal oxides

Hydrogen peroxide
See Hydrogen peroxide: Metals, etc.

Hydrogen sulfide
See Hydrogen sulfide: Metal oxides
See other METAL OXIDES

4822. Nickel(IV) oxide
[12035-36-8]

NiO₂

NiO_2

Fluorine
See Fluorine: Metal oxides
See other METAL OXIDES

4823. Nickel(III) oxide
[1314-06-3]

Ni_2O_3

Ni_2O_3

Hydrogen peroxide
See Hydrogen peroxide: Metals, etc.

Nitroalkanes
See NITROALKANES: metal oxides
See other METAL OXIDES

4824. Lead(II) oxide
[1317-36-8]

OPb

PbO

Aluminium carbide
Mellor, 1946, Vol. 5, 872
The carbide is oxidised with incandescence on warming with lead oxide.
See Metal acetylides, below

Chlorinated rubber
See CHLORINATED RUBBER: metal oxides

Chlorine, Ethylene
See Chlorine: Hydrocarbons (references 1,4)

Dichloromethylsilane
See Dichloromethylsilane: Oxidants

Fluorine, Glycerol
See Fluorine: Miscellaneous materials (reference 1)

Fluoroelastomers
Simpson, M. B. H., *Kautsch. Gummi, Kunstst.*, 1980, **33**, 83–85
Dispersions of lead oxide in fluoroelastomers underwent severe exothermic decomposition when heated above the normal mixing temperature of below 200°C, forming elemental lead.

Glycerol, Perchloric acid
See Perchloric acid: Glycerol, etc.

Hydrogen trisulfide
See Hydrogen trisulfide: Metal oxides

Linseed oil
Stolyarov, A. A., *Chem. Abs.*, 1935, **29**, 3860$_3$
Spontaneous ignition occurring during the early stages of mixing and grinding the components was traced to individual grades of lead oxide and the presence of undispersed lumps in the mixture.

Metal acetylides
Mellor, 1945, Vol. 5, 849
Interaction at 200°C with rubidium acetylide is explosive, and with lithium acetylide, incandescent.

Metals
Mellor, 1946, Vol. 5, 217; 1941, Vol. 7, 116, 656–658
Mixtures of the oxide with aluminium powder give a violent or explosive 'thermite' reaction on heating. Finely divided sodium ignites on admixture with the oxide, and a mixture of the latter with zirconium explodes on heating. Titanium is also oxidised violently on warming.

Non-metals
1. Mellor, 1941, Vol. 7, 657
2. Goodfield, J. A. C. *et al.*, *Fuel*, 1982, **61**, 843–847
A mixture with boron incandesces on heating, and with silicon the reaction is vigorous. If aluminium is present the mixture explodes on heating (but the same

is true if silicon is absent) [1]. The pyrotechnic reaction of boron and lead oxide mixtures has been studied by DSC [2].

Peroxyformic acid
  See Peroxyformic acid: Metals, etc.

Seleninyl chloride
  See Seleninyl chloride: Metal oxides
  See other METAL OXIDES, OXIDANTS

4825. Palladium(II) oxide
[1314-08-5]   OPd
                                     PdO

Hydrogen
  Sidgwick, 1950, 1558
  It is a strong oxidant and glows in contact with hydrogen at ambient temperature.
  See other METAL OXIDES, OXIDANTS

4826. Thorium oxide sulfide
[12218-77-8]   OSTh
                                   O:Th:S

Mellor, 1941, Vol. 7, 240
It ignites in contact with air.
See related METAL OXIDES, METAL SULFIDES *See other* PYROPHORIC MATERIALS

4827. Zirconium oxide sulfide
[12164-95-3]   OSZr
                                   O:Zr:S

Sorbe, 1968, 160
It ignites in air.
See related METAL OXIDES, METAL SULFIDES *See other* PYROPHORIC MATERIALS

4828. Silicon oxide
[10097-28-6]   OSi
                                     SiO

Zintl, E. et al., *Z. Anorg. Chem.*, 1940, **245**, 1
The freshly prepared material ignites in air.
See other NON-METAL OXIDES, PYROPHORIC MATERIALS

4829. Tin(II) oxide
  [21651-19-4]   OSn

SnO

Bailar, 1973, Vol. 2, 64
On heating at 300°C in air, oxidation proceeds incandescently.

Non-metal oxides
  Mellor, 1941, Vol. 7, 388
  The oxide ignites in nitrous oxide at 400°C, and incandesces when heated in sulfur
  dioxide.
  See other METAL OXIDES

4830. Zinc oxide
  [1314-13-2]   OZn

ZnO

NSC 267, 1978; *HCS 1980*, 968

Aluminium, Hexachloroethane
  See Hexachloroethane: Metals

Chlorinated rubber
  See CHLORINATED RUBBER: metal oxides

Linseed oil
  Anon., *Chem. Trade J.*, 1933, **92**, 278
  Slow addition of zinc white (a voluminous oxide containing much air) to cover the
  surface of linseed oil varnish caused generation of heat and ignition. Lithopone, a
  denser (air-free) grade of oxide did not cause heating.

Magnesium
  See Magnesium: Metal oxides
  See other METAL OXIDES

4831. Oxygen (Gas)
  [7782-44-7]   O₂

O₂

NSC 472, 1977; *FPA H12*, 1973; *HCS 1980*, 708

1. Anon., *Sichere Chemiearbeit*, 1994, **46**(8), 94
  Even a slight increase in the oxygen content of air above its usual 21 vol.% will
  greatly increase the rate of oxidation or combustion of many substances, including
  the human anatomy. An autoclave for hot oxygen treatment (900°C, 406 bar) was
  located in a bunker. To purge it with argon for cooling it was necessary to enter

the bunker and open a valve. This discharged hot oxygen into a pipe and ignited a plastic seal, thence the metal of the pipe, then combustible contents of the bunker burnt explosively in the escaped oxygen. The operator was killed immediately [1]. Low concentrations of oxygen in other gases may be concentrated as liquid when working at low temperatures.
See Oxygen (Liquid)
See OXYGEN ENRICHMENT

Acetaldehyde
See Acetaldehyde (references 2,4)

Acetone
Anon., private comm., 1988
To 'ensure proper ignition' during oxygen-flask analysis, a technician put a single drop of acetone onto the filter paper enclosing the sample, before igniting the tail and insertion into the oxygen-filled flask. An explosion followed which shattered the flask, fortunately enclosed in a mesh screen which retained the fragments. It was calculated that 0.07 ml of acetone would have formed an explosive vapour–air mixture in the flask above the lower explosive limit, so considerably less acetone would have given an explosive mixture in the oxygen atmosphere.
See Barium acetate: Copper(II) oxide, etc.

Acetone, Acetylene
Anon., *CISHC Chem. Safety Summ.*, 1976, **47**, 33–34
A combination of faulty equipment and careless working led to an extremely violent explosion during oxy-acetylene cutting work. The oxygen cylinder was nearly empty and the regulator had a cracked diaphragm. The acetylene cylinder was lying on its side and was feeding a mixture of liquid acetone and acetylene gas to the burner head. When the oxygen ran out, the excess pressure from the acetylene line forced the acetone–acetylene mixture back up the oxygen line and into the cylinder via the cracked diaphragm. The explosion destroyed the whole plant.
See Acetylene: Oxygen

Acetylene
Skobelar, V. F., *Chem. Abs.*, 1981, **94**, 120243
A safe method for demonstrating explosive combustion of acetylene–oxygen mixtures in bubbles is described.

sec-Alcohols
1. Davies, 1961, 80
2. Redemann, E. G., *J. Amer. Chem. Soc.*, 1942, **64**, 3049
Secondary alcohols are readily autoxidised in contact with oxygen or air, forming ketones and hydrogen peroxide [1]. A partly full bottle of 2-propanol exposed to sunlight for a long period became 0.36 M in peroxide and potentially explosive [2].
See 2-Propanol: 2-Butanone

See 2-Butanol
See Hydrogen peroxide: Acetone, or: Alcohols

Alkali metals
1. Mellor, 1941, Vol. 2, 469
2. Sidgwick, 1950, 65
Reactivity towards air or oxygen increases from lithium to caesium, and the intensity depends on state of subdivision and on presence or absence of moisture. Lithium normally ignites in air above its m.p., while potassium may ignite after exposure to atmosphere, unless it is unusually dry. Rubidium and caesium ignite immediately on exposure [1]. It is reported that sodium and potassium may be distilled unchanged under perfectly dried oxygen [2].

Alkaline earth metals
Mellor, 1941, Vol. 3, 637–638
Finely divided calcium may ignite in air, and the massive metal ignites on heating in air, and burns vigorously at 300°C in oxygen. Strontium and barium behave similarly.

Aluminium–titanium alloys
See Aluminium–titanium alloys: Oxidants

Ammonia
Rüst, 1948, 332–333
Accidental connection of an oxygen cylinder to top-up an ammonia-containing refrigeration system led to explosive destruction of the compressor.

Ammonia, Platinum
1. Nealy, D., and Williams, E. J. *et al., School Sci. Rev.*, 1935, **16**(63), 410
2. Long, G. C., *Spectrum* (Pretoria), 1980, **18**, 30
In school demonstrations of the oxidation of ammonia to nitric acid over platinum catalysts, substitution of oxygen for air causes fairly vigorous explosions to occur [1]. Practical details are given [2].

1,4-Benzenediol, 1-Propanol
1. Burgiel, J. C. *et al., J. Chem. Phys.*, 1965, **43**, 4291
2. Anon., *Univ. Safety Assoc. Safety News*, 1978, (9), 11–12; Peacock, C. J., private comm., 1978
The solid inclusion (clathrate) complex of oxygen with hydroquinone had been prepared twice previously by a published method which involved saturating a solution of hydroquinone in propanol in an autoclave at 70°C with oxygen at 20–150 bar, followed by slow cooling under oxygen pressure (and most probably without stirring to allow large crystals of the inclusion complex to form) [1]. In a third attempt, (to produce smaller crystals free of the 'quinhydrone complex' impurity) a solution was prepared and saturated with oxygen at 80 bar, then heated to 90°C (when the partial pressure of oxygen would increase to around 100 bar),

then allowed to cool with agitation. After 20 min the bursting disk failed at 450 bar and an explosion damaged the autoclave and premises [2]. Without agitation, the rate of oxidation of the solvent and hydroquinone would be controlled by diffusion of oxygen through the (small) gas liquid interface. With agitation, the rate of oxidation would be expected to increase greatly, and lead to runaway oxidation with very fast increase in temperature and pressure and explosive combustion when the autoignition temperature (371°C for propanol in air at 1 bar, less in oxygen) were attained.

See Hydrocarbons (reference 3), below

Benzoic acid
1. Corbett, R. A., private comm., 1976
2. Wilson, L. Y. et al., *J. Chem. Educ.*, 1985, **62**(10), 902
Benzoic acid is burned in oxygen as a primary thermochemical standard to calibrate oxygen bomb calorimeters used in the IP12/ASTM D240 standard tests for determination of calorific value of liquid hydrocarbon fuels. If the benzoic acid is powdered (rather than pelleted as IP12 recommends), very rapid combustion occurs and the flame front may burn through the non-metallic (Teflon) seals on valve seats and the bomb may be destroyed [1]. Ignition of a pelleted sample in a bomb with a faulty closure valve led to sudden venting of combustion gases which blew off the insulating cover and thermometer. Safety precautions are listed for bomb calorimetry work [2].

Biological material, Ether
Napier, D. H., private comm., 1972
Biological material in a polythene bag filled with oxygen and being prepared for analytical combustion exploded. Diethyl ether used to anaesthetise the experimental animal from which the sample was derived may have still been present, and ignition from static charge on the plastics bag may have been involved.

See related STATIC INITIATION INCIDENTS

Boron tribromide
See Arsine–boron tribromide

Bromine, Chlorotrifluoroethylene
See Bromine: Chlorotrifluoroethylene, Oxygen

Calcium phosphide
See Calcium phosphide: Oxygen

Carbon disulfide
Dudkin, V. A., *Chem. Abs.*, 1982, **97**, 203762
The lower limit for spontaneous ignition of mixtures with oxygen has been studied.

Carbon disulfide, Mercury, Anthracene
Anon., *ABCM Quart. Safety Summ.*, 1953, **25**, 2

Shortly after mercury was accidentally introduced into a system containing a solution of anthracene in carbon disulfide under an atmosphere of oxygen, an explosion occurred. Presence of mercury may have catalysed rapid oxidation of carbon disulfide.
See other CATALYTIC IMPURITY INCIDENTS

Carbon monoxide, Hydrogen
    Schierwater, F.-W., Sichere Chemiearb., 1985, 37(9), 101–102
    Synthesis gas ($CO + H_2$) at 40 bar containing a low level of hydrogen sulfide was to be freed of the latter impurity by adding the theoretical quantity of oxygen and passing the mixture over a catalyst. Introduction of oxygen (from a supply at 60 bar) via a simple T-piece (instead of through the recommended small bore coaxial injection nozzle to ensure thorough mixing with the gas stream) caused development of an intense inverse flame in the locally very high oxygen concentration which burned through the reactor side wall opposite the oxygen inlet and ejected a metre-long flame-jet.

Copper, Hydrogen sulfide
    See Hydrogen sulfide: Copper, Oxygen

Cyclohexane-1,2-dione bis(phenylhydrazone)
    1. Schulze, M. et al., Z. Chem., 1979, 19, 210–211
    2. Adam, W. A. et al., Tetrahedron, 1985, 41, 2045–2056
    Autoxidation of the hydrazone (O_2/C_6H_6/UV) gives the explosive isomeric 1,2-dihydroperoxy-1,2-bis(benzeneazo)cyclohexane [1], and the same is true for COT derivatives [2].
    See 1,2-Dihydroperoxy-1,2-bis(benzeneazo)cyclohexane
    See α-PHENYLAZO HYDROPEROXIDES

Cyclooctatetraene
    Adam, W. A. et al., Tetrahedron Lett., 1982, 23, 3156
    In a procedure to oxygenate photochemically COT to the 1,4-endoperoxide, highly explosive undefined polymeric peroxides are formed as by-products.
    See other POLYPEROXIDES

Diborane
    See Diborane: Oxygen

Diboron tetrafluoride
    See Diboron tetrafluoride: Oxygen

Dimethoxymethane
    See Dimethoxymethane: Oxygen

Dimethylketene
    See Poly(peroxyisobutyrolactone)

Dimethyl sulfide

Harkness, A. C. *et al., Atmos. Environ.*, 1967, **1**, 491–497

Interaction is explosive at 210°C or above, and the mechanism has been studied.

Ethers

Davies, 1961, 79

Many ethers, either of open chain (diethyl or diisopropyl ether) or cyclic type (tetrahydrofuran, dioxane), are readily autoxidised on exposure to air or oxygen in presence of light. The hydroperoxides formed are less volatile than the parent ether and may be concentrated to a dangerous extent if distillation of peroxidised material is attempted.

See AUTOXIDATION, PEROXIDES IN SOLVENTS

Fibrous materials

1. Anon., *Loss Prev. Bull.*, 1978, (020), 49
2. Anon., *Sichere Chemiearbeit*, 1991, **43**(10), 112
3. Anon., *Loss Prev. Bull.*, 1978, (023), 141–143

Most fibrous fabrics will sorb oxygen when exposed to concentrations greater than the normal 21% in air, and this is retained for a long time after excess oxygen is no longer present, greatly increasing the possibility of ignition. Hose leakage led to oxygen enrichment inside a vessel in which welding was taking place. A welder lit a cigarette but did not appreciate the significance of its rapid combustion and the unusually long lighter flame. Sparks from welding later ignited clothing, leading to fatal burns [1]. Accidental connection of a pneumatic grinder to an oxygen line instead of a compressed air line led to ignition of the apprentices' clothing when the grinder was started. Another grinding incident occurred in the open air near the vent of a liquid oxygen tank, not only clothing but the worker's hair and beard ignited [2]. Accidental connection of a breathing set to an oxygen cylinder rather than to a medical air cylinder caused ignition of the facepiece [3].

See OXYGEN ENRICHMENT

Fluorine, Hydrogen

See Fluorine: Hydrogen, Oxygen

Fluorocarbons (including perfluorocarboms)

See FLUOROCARBONS: OXYGEN

Fuel gases

NFPA 51, Quincy (Ma), Natl. Fire Prot. Assocn., 1984

The US fire code covers installation and use of gaseous oxygen–fuel gas systems for welding and cutting, for the thermodynamically unstable fuels acetylene, MAPP (methylacetylene–allene–propene–propane mixtures), and the stable hydrocarbons propane or butane.

Fuels

1. Ivanov, B. A. *et al., Chem. Abs.*, 1975, **82**, 75130

2. Rasbash, D. J., *Fire Saf. J.*, 1986, **11**, 113–125
3. Plewinsky, B. *et al., Prax. Sicherheitstech.*, 1997, **4**, 95
Possible reasons for explosions of 40 1 steel cylinders containing oxygen at or below 150 bar are discussed. Presence of a substantial (0.1–0.2 mm) layer of mineral oil, or more likely, ingress of a combustible gas from a higher-pressure source, are possibilities [1]. The major factors affecting the occurrence and progress of explosions in mixtures of air with flammable gases, vapours, mists and dusts are discussed. These include flammability limits, burning velocity in laminar and turbulent conditions, possible onset of detonation, pressure piling, etc. [2]. A review, with references, of detonation in systems with gaseous oxidants and condensed fuels, with particular relevance to the design of oxidation reactors, has been published [3].
See Hydrocarbons, *and* Oil films, both below

Glycerol
   Smith, B. S., *Chem. Abs.*, 1978, **89**, 47649
Glycerol is oxidised at an appreciable rate by high-pressure oxygen in the presence of copper alloys to give glyceric acid, oxalic acid and formic acid, which lead to severe corrosion of the alloy. Malfunction of aircraft oxygen gauges operating at 125 bar was attributed to blockage of the brass Bourdon tubes by copper salts of the acids produced by long-term oxidation of residual traces of glycerol–water solutions used for gauge testing.
See other CORROSION INCIDENTS

Halocarbons
   1. Fawcett, H. H., *Chem. Eng. News*, 1957, **35**(43), 60
   2. Weber, U., *Chim. Ind* . (Paris), 1963, **90**, 178
   3. Bashford, L. A. *et al., J. Chem. Soc.*, 1938, 1064
   4. Haszeldine, R. N. *et al., J. Chem. Soc.*, 1959, 1085–1086
1,1,1-Trichloroethane exploded after heating under oxygen at 54 bar and 100°C for 3 h [1]. Trichloroethylene, remaining in a pipe after cleaning operations, exploded under 27 bar pressure of oxygen at ambient temperature. It was later found possible to explode stoicheiometric mixtures [2]. Chlorotrifluoroethylene and bromotrifluoroethylene each react explosively with oxygen at ambient temperature [3], but controlled oxidation of the former produces an explosive cyclic peroxide, 4.5-dichloro-3,3,4,5,6,6-hexafluoro-1,2-dioxane [4].
See Tetrafluoroethylene
See Oxygen (Liquid): Halocarbons
Iodomethane: Oxygen

Hydrocarbons
   1. Davies, 1961, 11
   2. Staudinger, H., *Z. Elektrochem.*, 1925, **31**, 549
   3. Unpublished information, 1968
   4. Burgess, G. K., *Ind. Eng. Chem.*, 1930, **22**, 473
   5. Francis, C. K., *Ind. Eng. Chem.*, 1930, **22**, 896

6. Hunn, E. B., *Ind. Eng. Chem.*, 1930, **22**, 804
7. Dunn., B. W., *Bur. Expl. Accid. Bull., Amer. Rail Assoc.*, 1924, **32**(1), 1–2
8. Anon., *CISHC Chem. Safety Summ.*, 1977, **48**, 8
9. Anon., *Loss Prev. Bull.*, 1978, (020), 49
10. Davies, R., *Loss Prev.*, 1979, **12**, 103–110
11. Kletz, T. A., *Loss Prev.*, 1979, **12**, 96–102
12. Herrmann, G. *et al.*, Ger. Offen. DE 3 514 924, 1986

Interaction of hydrocarbons with gaseous oxygen may be slow or fast, depending on conditions and substrate, but peroxidic products are always involved, and the rate of formation is highest at a C–H link adjacent to an aromatic ring or a double bond [1]. Predictable conditions for explosion are those relevant to the internal combustion engine (Otto or Diesel types). Unpredicted conditions include the following cases. During studies on autoxidation of 1,1-diphenylethylene with oxygen at high pressure and low temperature an explosion occurred [2]. Partial oxidation of a gasoline fraction in an autoclave under oxygen, initially at 22 bar and 100°C, ran wild and exploded; several smaller reactions had proceeded uneventfully [3]. Previously oxidation reactions of gasoline at 20 bar/105°C had exploded [4], and a severe explosion occurred while admitting oxygen, virtually at 1 bar pressure, to an autoclave containing a glass bottle of gasoline [5]. Precautions were detailed later [6].

An attempt to decarbonise a gasoline motor cycle engine by inserting a lighted match through the spark plug hole into the cylinder, and then blowing in oxygen, caused a violent explosion to occur [7]. Comment seems superfluous. Lack of a duplicate oxygen monitor on a plant in which oxygen and hydrocarbon were mixed led to a violent explosion during the hour-long daily monitor-checking procedure [8]. Use of oxygen instead of compressed air to start a diesel engine, or to clear a blocked petrol pipe, led to violent explosions [9]. In an overview of hazards of industrial air-oxidation processes, they are all identified as inherently hazardous, and the majority extremely so. Particular hazards are detailed, and major incidents involving liquid phase oxidation of cyclohexane or cumene, and vapour phase oxidation of ethylene to the oxide are listed, together with the causes [10]. Further examples of the causes of fires in the last 3 processes, and in the liquid phase oxidation of *p*-xylene to terephthalic acid are given [11]. A technique for liquid-phase oxidation of hydrocarbons in a vertical column containing a bottom water layer and no gas-phase above the hydrocarbon layer is claimed to be free of the danger of detonation [12].

See Fuel gases, Fuels, *also* Hydrogen, all above
and Oil films, below
Buten-3-yne: Oxygen
See Tetracarbonylnickel: Oxygen
See also BITUMEN

Hydrocarbons, Promoters
Barisov, A. A. *et al., Chem. Abs.*, 1984, **101**, 93812

Among other compounds, methyl nitrate, nitromethane, ethyl nitrate and tetrafluorohydrazine function as promoters of spontaneous ignition of mixtures of methane or propane with oxygen and argon.

Hydrogen
1. Bailey, D. J. C., *Chem. & Ind.*, 1954, 492
2. Cook, M. A. *et al.*, *Chem. Eng. News*, 1956, **34**, 4436
3. Weissveiler, A., *Z. Elektrochem.*, 1936, **42**, 499
4. Carty, P., *School Sci. Rev.*, 1978, **59**(208), 517–518
5. Loubeyre, P. *et al.*, *Nature*, 1995, 378, 44
6. Editor's comments

When oxygen was passed through a drying tower containing activated alumina previously used to dry hydrogen, explosions occurred. Nitrogen purging between changing gases would prevent this [1]. During preparation of stoicheiometric mixtures in a steel mixing tank (at 17–82 bar, with or without 15% argon), several spontaneous explosions occurred during valve manipulation at 30 min after mixing, but not 10 min after mixing. A possible catalytic effect of the surface of the steel tank was eliminated by coating it thinly with silcone grease [2]. There is a narrow range of concentrations in which the mixture is supersensitive to initiation [3]. A simple and safe method of demonstrating the explosive combustion of stoicheiometric hydrogen–oxygen mixtures with spark ignition in a polythene bottle is described [4]. It is claimed that, at ultra-high pressure, hydrogen and oxygen mixtures become non-explosive [5]. Since an extremely small test apparatus is involved and no consideration given to critical diameter or other surface effects, this is open to doubt [6].

Hydrogen sulfide
See Hydrogen sulfide: Oxygen

Lithiated dialkylnitrosamines
Mochizuki, M., *Tetrahedron Lett.*, 1980, **21**, 1766
Oxygenation of lithiated dialkylnitrosamines in THF at $-78°C$ is fast and gives good yields of the N-alkyl-N-(1-hydroperoxyalkyl)nitrosamines. If oxygenation is too prolonged, poor yields and explosive by-products result.

Mercury, Tetracarbonylnickel
See Tetracarbonylnickel: Mercury, etc.

Metal hydrides
1. Mellor, 1941, Vol. 2, 483
2. Mackay, 1966, 30, 67
Sodium hydride ignites in oxygen at 230°C, and finely divided uranium hydride ignites on contact. Lithium hydride, sodium hydride and potassium hydride react slowly in dry air, while rubidium and caesium hydrides ignite. Reaction is accelerated in moist air, and even finely divided lithium hydride ignites then [1]. Finely divided magnesium hydride, prepared by pyrolysis, ignites immediately in air [2].
See also COMPLEX HYDRIDES

Metals
1. Heinicke, G., *Technik*, 1969, **24**, 313–319
2. Hirano, T. *et al., J. Loss Prev.*, 1993, **6**(3), 151
General data on fire hazards present in industrial oxygen-producing plants are given with details of experimental study of factors involved in combustion of various metals and alloys used in such plant. Safety in selection of materials is discussed [1]. A study of combustion of structural metals (Fe, Al) in oxygen is given [2].
See Oxygen (Liquid): Metals

Methoxy-1,3,5,7-cyclooctatetraene
See Methoxy-1,3,5,7-cyclooctatetraene: Oxygen

4-Methoxytoluene
Imamura, J. *et al.*, Brit. Pat. 1 546 397, 1979
Liquid phase oxidation of 4-methoxytoluene to anisaldehyde at 60 bar/115°C in acetic acid containing heavy metal salts is violent, and must be controlled by the rate of addition of oxygen gas.

Non-metal hydrides
1. Mellor, 1946, Vol. 5, 36
2. Sidgwick, 1950, 344, 553
3. Mellor, 1940, Vol. 6, 220–225
The reported ignition of diborane and tetraborane(10) in contact with air or oxygen is due to the presence of traces of silicon hydrides [1]. The lower members of the latter class ignite or explode in air or oxygen, especially at reduced pressure [2]. Phosphine is also sensitive [3].
See Aluminium tetrahydroborate: Alkenes, Oxygen
　　Decaborane(14)
　　Pentaborane(11)
　　Phosphine: Oxygen

Oil films
Werley, B. M., *ASTM Spec. Tech. Publ.*, 1983, (812), 108–125
The hazards arising from presence of oil films in oxygen-handling systems are reviewed, with consideration of cleanliness specifications, film flammability, film migration and hazard mechanisms.
See Hydrocarbons, above

Organic analytical samples
1. Pack, D. E., *Chem. Eng. News*, 1966, **44**(2), 4
2. Devereaux, H. D., *Chem. Eng. News*, 1966, **44**(6), 4
3. Heysel, R. M. *et al., Chem. Eng. News*, 1966, **44**(10), 4
4. MacDonald, A. M. G., *Chem. Eng. News*, 1966, **44**(19), 7
5. Jenkin, J. B., *Chem. & Ind.*, 1974, 585
6. Guthrie, R. D. *et al., Chem. Brit.*, 1979, **15**, 614; *J. Chem. Educ.*, 1980, **57**(3), 226

Hazards involved in the use of the oxygen flask combustion technique are discussed and illustrated with several examples [1]. Some explosions occurred after completion of combustion, when the oxygen concentration in the flask is still some 75% [2]. Shielding of the flask [3], minimally with wire gauze [4] seems advisable. Explosion during a blank combustion (of a folded filter paper) was attributed possibly to thermal strains. Similar occurrences had been noted previously [5]. Fitting a simple pressure relief (Bunsen) valve to the flask improves safety aspects, and the oxygen flask method is also to be preferred (for qualitative work) to the more hazardous sodium fusion method [6].
See Sodium: Halocarbons, or: Organic analytical samples

Organic materials
Burnett, R. J. *et al., J. Chem. Educ.*, 1985, **62**(5), A157, A159
Facilities and procedures necessary to encourage safety in laboratory work involving oxidation of organic materials with pressurised oxygen are reviewed.

Ozone, Rubber
See Ozone: Oxygen, Rubber

Pentaborane(9)
See Pentaborane(9): Oxygen

Perfluorocarbons
See FLUOROCARBONS: OXYGEN

Phosphine
See Phosphine: Oxygen

Phosphorus(III) oxide
See Tetraphosphorus hexaoxide: Oxygen

Phosphorus tribromide
Bailar, 1973, Vol. 2, 426
Oxidation of the bromide with gaseous oxygen is not easily controlled and becomes explosive.

Plastic cylinder caps
Anon., *Loss Prev. Bull.*, 1978, (020), 49
The plastic caps (or tape) which cover the outlets on refilled oxygen cylinders will burn readily in oxygen and must be completely removed and no bits trapped in the fittings attached to the cylinder outlet.
See Organic materials, above

Plastic tubes
1. Wolf, G. L. *et al., Anaesthesiology*, 1987, **67**, 236–239
2. Davies, P., Personal communication, 1993

Following fires in which endotracheal tubes became ignited by surgical lasers or electrocautery in atmospheres enriched by oxygen and/or nitrous oxide, the flammability of PVC, silicone rubber and red rubber tubes in enriched atmospheres was studied [1]. Ozonised oxygen was reacted with hydrogen at low pressure to generate hydroxyl radicals. Pressure in the apparatus was maintained by a vacuum pump protected from ozone by a tube of heated silver foil. On two occasions there was an explosion in the plastic vent pipe from the vacuum pump. The vent gas should have been outside explosive limits and the exact cause is not clear; the editor suspects peroxide formation.

Polymers
1. *MCA Case History No. 395*
2. *MCA Case History No. 1111*
3. *491M*, 1975, 294
A foam rubber sample, being tested for oxidation resistance under oxygen at 34 bar at 90°C exploded with extreme violence after 4 days [1]. Use of a Neoprene-lined hose in a high-pressure oxygen manifold caused failure and ignition of the burst hose. Possible ignition sources include adiabatic compressive heating, and friction from vibration of metal reinforcing fibres in the high velocity stream of escaping oxygen [2]. Polytetrafluoroethylene (Teflon) ignited at 705°C in oxygen at 0.34 bar pressure when used as wire insulation, while polyolefine insulation ignited at 593°C under the same conditions [3].

Polytetrafluoroethylene, Stainless steel
Anon., *Petroleum Rev.*, 1979, 48
In two test methods (IP40, IP38) for the oxidative stability of hydrocarbons, samples are heated in a valve-sealed stainless bomb under an oxygen atmosphere. On several occasions minor explosions occurred when valves with Teflon sealing discs (rather than the older metal needle closures) were operated during oxygen purging or venting operations. (This was associated with presence of metal particles and traces of grease being embedded in the Teflon discs). Only stainless needle valves, and a controlled rate of pressure release, are recommended for these tests.
See PERFLUOROCARBONS

Propylene oxide
See Propylene oxide: Oxygen

Rhenium
Mellor, 1942, Vol. 12, 471
The metal ignites in oxygen at 300°C.

Rubberised fabric
Anon., *Loss Prev. Bull.*, 1978, (020), 52
Accidental connection of a cylinder of oxygen instead of medical air to an air breathing set led to ignition of the face-piece.

Tetrafluoroethylene

1. Pajaczkowski, A., *Chem. & Ind.*, 1964, 659
2. Ger. Pat. 773 900, 1971

Accidental admixture of oxygen gas with unstabilised liquid tetrafluoroethylene produced a polymeric peroxide which was powerfully explosive, and sensitive to heat, impact or friction [1]. Removal of oxygen by treatment with pyrophoric copper to prevent explosion of tetrafluoroethylene has been claimed [2].

See other POLYPEROXIDES

Tetramethyldisiloxane

1. Anon., *Jahresber.*, 1983, 71–72
2. Plewinsky, B. *et al.*, *Thermochim. Acta*, 1985, **94**, 33–42
3. Plewinsky, B., *Chem. Abs.*, 1990, **113**, 234231

A 4.9 g sample of the liquid siloxane in a glass dish was put into a bomb calorimeter (on an open bench) containing 5 ml of sodium hydroxide solution to absorb combustion gases. The electric igniter system consisted of a metal wire in contact with a cotton-wool wick which dipped into the siloxane sample. The bomb was sealed, pressured up to 39–44 bar with oxygen, and the igniter was fired. A violent explosion blew the lid off the bomb (rated at 190 bar working, 250 bar test), and examination of the deformed bomb indicated that a maximum detonation pressure of around 900 bar had been attained.

Detailed examination of the reaction showed a 2 stage mechanism. Hydrolysis of the volatile partially methylated siloxane by the alkaline solution liberated hydrogen which formed an explosive mixture with oxygen. When the igniter was fired, the hydrogen–oxygen explosion atomised the remaining liquid siloxane leading to a very violent secondary explosion. The wick also played a decisive separate role, as the siloxane-soaked cotton also underwent extremely rapid detonative decomposition on ignition. The rate of pressure rise was measured as 10 kbar/s, with a maximum pressure calculated for a 4.9 g sample of 800 bar. Hexamethyldisiloxane (which can't produce hydrogen by hydrolysis) burned smoothly under the same conditions. It is recommended that no more than 0.5 g samples should be used, with oxygen pressure limited to 20 bar and protective enclosure of the bomb equipment [1]. A more detailed investigation of the combustion of tetramethyldisiloxane in oxygen showed that, even in the absence of hydrogen, combustion was still extremely violent, the rate of pressure increase being inversely proportional to the pressure of oxygen, as expected for a homogeneous gas-phase combustion reaction. The unusually high rate of pressure rise (265 kbar/s, to max 113 bar for 10 bar oxygen pressure) is indicative of detonation. Under heterogeneous conditions, liquid tetramethyldisiloxane gave a maximum rate of pressure rise of 90 kbar/s to max 160 kbar with 30 bar oxygen pressure, with an abrupt change to the higher values between 15 and 20 bar oxygen pressure [2]. Further study showed that detonation-like transmission of combustion across the surface of either a soaked wick, or bulk liquid, could take place with a gas-phase composition below the lower flammability limit [3]. Pentamethyldisiloxane showed generally similar behaviour (maximum rate 38 kbar/s to max 205 bar for 30 bar oxygen pressure), with an abrupt increase between 25 and 30 bar pressure of oxygen [2].

See SILANES; OXYGEN BOMB CALORIMETRY
See entry PRESSURE INCREASE IN EXOTHERMIC DECOMPOSITION

Titanium
Anon., *ABCM Quart. Safety Summ.*, 1962, **33**, 24
Passage of oxygen through a titanium feed pipe into a titanium autoclave caused a titanium–oxygen fire and explosion at 44 bar. When the surface oxide film is damaged, titanium can ignite at 24 bar under static conditions and at 3.4 bar under dynamic conditions, with oxygen at ambient temperature.
See Aluminium–titanium alloys: Oxidants

Titanium trichloride, Vapours of low ignition temperatures
Huneker, K. H., *Chem. Abs.*, 1977, **87**, 15458
The Dräger oxygen-measuring tube contains titanium trichloride as the indicating substance, and a possible hazard in using these tubes to measure oxygen contents above 25 vol% in gas mixtures has been noted. The temperature in the tube can reach 120°C during oxidation of titanium trichloride to titanium dichloride oxide, so the test should not be used if compounds with ignition points below 135°C (e.g. carbon disulfide) are present.

Trirhenium nonachloride
1. Edwards, P. G. *et al., J. Chem. Soc., Dalton Trans.*, 1980, 2467, 2472
2. Edwards, P. G., *Lab. Haz. Bull.*, 1984(4), 3, item 212
During preparation of perrhenyl chloride by combustion of rhenium nonachloride in an oxygen stream [1], a violent explosion, possibly involving chlorine oxides, occurred on heating the chloride to 250°C. A safe procedure involving heating the chloride to 250°C under nitrogen, and then introducing oxygen, is proposed [2].
See other NON-METALS, OXIDANTS

4832. Oxygen (Liquid)
[7782-44-7]   **O$_2$**

$$O_2$$

HCS 1980, 709

NFPA 50, Quincy (Ma), Natl. Fire Prot. Assoc., 1985
The recent US National Fire Code covers all aspects of equipment, installation and safe operational practices necessary for bulk oxygen storage at consumer sites.

Preparative hazard
Anon., *Sichere Chemiearbeit*, 1994, **46**(8), 94
During regeneration of an adsorber in the cold box of an air separation plant, regeneration gas exited via the adsorber drain pipe. This still contained some liquid oxygen. Very low regeneration-gas flows permitted desorption of acetylene and other hydrocarbons at above normal (ppm) concentrations, these condensed or crystallised into the liquid oxygen and the mixture finally detonated, causing

limited damage. It is recommended that regeneration gas should never be vented via the drain pipe.
See Hydrocarbons, below

Author's comments
Although the entries below refer to the specific hazards observed with a few organic or oxidisable materials, it is probable that most organic materials, inorganic reducing agents, and many unoxidised inorganic materials may be highly hazardous with liquid oxygen under approprate conditions of contact and initiation. Liquid oxygen also implies a region of oxygen enriched atmosphere near the vent of a storage vessel, or during transfer; this has its own hazards.
See CRYOGENIC LIQUIDS, OXYGEN ENRICHMENT

Acetone
Blau, K., private comm., 1965
Accidental addition of liquid oxygen to vacuum jars containing acetone residues from trap-cooling use caused a violent explosion. Liquid nitrogen is less hazardous as a trap coolant, but only under controlled conditions.
See Nitrogen (Liquid)

Acetylene, Oil
Basyrov, Z. B. *et al., Chem. Abs.*, 1960, **54**, 12587a
The detonation capacity of mixtures of acetylene and liquid oxygen is increased by the presence of organic material (oils) in the oxygen. Hazards of accumulation of oil in air-liquefaction and -fractionation plants are emphasised.
See Hydrocarbons, below

Ammonia, Ammonium nitrate, Diphenyl carbonate
Buzard, J. *et al., J. Amer. Chem. Soc.*, 1952, **74**, 2925–2926
During the synthesis of ^{15}N-labelled urea by interaction of labelled ammonium nitrate, liquid ammonia and diphenyl carbonate in presence of copper powder, a series of explosions of the refrigerated sealed tubes was encountered. This was almost certainly caused by condensation of traces of oxygen in the tubes cooled to −196°C during condensation of ammonia before sealing the tubes. Cooling to −80°C would have been adequate and have avoided the hazard.
See Nitrogen (Liquid)

Asphalt
1. Weber, U., *Chim. et Ind.* (Paris), 1963, **90**, 178
2. English, W. D. *et al., Chem. Eng. News*, 1992, **70**(10), 18
3. Gayle, B., *Rev. Tech. Feu*, 1979, **20**(189), 62
Mechanical impact on a road surface on to which liquid oxygen had leaked caused a violent explosion [1]. Anecdotal evidence that there have been several like incidents is given [2]. Mixtures of asphalt and liquid oxygen were shown to be impact-sensitive on the small scale, but on the larger scale a detonator was necessary to initiate mild explosion of liquid oxygen on a layer of asphalt [3]. Oil,

rubber or other impurities may have been present on the road surface in the first incident.

Carbon

1. Kirshenbaum, 1965, 4
2. Stettbacher, A., *Spreng-u. Schiess-Stoffe*, 71, Zurich, Racher, 1948
3. Angot, A., *Ann. Mines [13]*, 1934, **5**, 46–51
4. Hempseed, J. W. *et al.*, *Plant/Oper. Progr.*, 1991, **10**(3), 184

Mixtures of carbon and liquid oxygen have been used as blasting explosives for some time [1,2], Mixtures with carbon black appear unusually sensitive to impact, and a blasting cartridge exploded when dropped [3]. Purification of helium containing only 1.4% oxygen and 0.7% nitrogen by passing through a carbon absorber at −195°C led to an explosion of 20 kg TNT equivalent. It was found that a liquid phase of 85% oxygen content was capable of condensing onto the carbon at the entry.

Carbon, Iron(II) oxide

Leleu, *Cahiers*, 1975, (78), 121

Carbon containing 3.5% of the oxide explodes on contact with liquid oxygen.
See other CATALYTIC IMPURITY INCIDENTS

Halocarbons

Anon., *Chem. Eng. News*, 1965, **43**(24), 41

Mixtures of liquid oxygen with dichloromethane, 1,1,1-trichloroethane, trichloroethylene and 'chlorinated dye penetrants 1 and 2' exploded violently when initiated with a blasting cap. Carbon tetrachloride exploded only mildly, and a partly fluorinated chloroalkane not at all. Trichloroethylene has been used for degreasing metallic parts before use with liquid oxygen, but is not safe.
See Oxygen (Gas): Halocarbons

Hexafluoropropene, Oxygen difluoride
See Oxygen difluoride: Hexafluoropropene, etc.

Hydrocarbons

1. Mellor, 1941, 1, 379
2. Kirshenbaum, 1956, 17, 22
3. *MCA Case History No. 865*
4. Wright, G. T., *Chem. Eng. Progr.*, 1961, **57**(4), 45–48
5. Rotzler, R. W., *et al.*, *Chem. Eng. Progr.*, 1960, **56**(6), 68–73
6. Matthews, L. G., *Chem. Eng. Progr.*, 1961, **57**(4), 48–49
7. Karwat., E., *Chem. Eng. Progr.*, 1958, **54**(10), 96–101
8. Streng, A. G. *et al.*, *J. Chem. Eng. Data*, 1959, **4**, 127–131
9. Pelman, M., *AD Rep. 768744/5GA*, Springfield (Va.), USNTIS, 1973
10. Gustov, V. F. *et al.*, *Chem. Abs.*, 1984, **100**, 12006

Mixtures of hydrocarbons with liquid oxygen are highly dangerous explosives, not always requiring external initiation. The earliest example must be that of Claudé, who accidentally knocked a lighted candle into a bucket of liquid oxygen in 1903 and suffered from the violent explosion ensuing [1]. Mixtures with liquid methane or benzene are specifically described as explosive [2]. Traces of oil in a liquid oxygen transfer pump caused the explosion described in the Case History [3]. Mixtures with petroleum and absorbent charcoal have been used experimentally as blasting explosives [1]. Addition of a little aluminium powder to liquid methane−oxygen mixtures increases the explosive power [2]. An explosion in a liquid oxygen evaporator was attributed to the presence of acetylene, arising from unusual plant conditions and higher than usual hydrocarbon concentrations in the atmospheric air taken in for liquefaction [4]. Two similar explosions in main condensers were attributed to presence of acetylene [5] or unspecified hydrocarbons [6]. Suitable precautions are detailed in all cases. Experimental work appeared to implicate ozone as a major contributory factor [7].

The explosive properties of liquid methane−oxygen mixtures were determined [8]. During investigation of an explosion in a portable air liquefaction−separation plant, hydrocarbon oil was found in a silica filtration bed [9]. The mechanism of slow heterogeneous accumulation of hydrocarbons dissolved in trace amounts in liquid oxygen on the liquid evaporator surfaces is discussed. It was concluded that months of continuous evaporation would be required to attain explosion-hazardous levels in real evaporators [10].

Hydrogen (Liquid)
See Hydrogen (Liquid): Oxygen

Liquefied gases
1. Kirshenbaum, 1956, 17, 30, 34
2. Lester, D. H., *Rept. NASA-CR-148711*, Richmond (Va), USNTIS, 1976
Mixtures with liquid carbon monoxide, cyanogen (solidified), and methane are highly explosive [1]. Autoignition in liquid oxygen−hydrogen propellant systems has been reviewed [2].

Lithium hydride
Kirshenbaum, 1956, 44
Mixtures of lithium hydride powder and liquid oxygen are detonable explosives of greater power than TNT.

Metals
1. Anon., *Chem. Eng. News*, 1957, **35**(24), 90
2. Anon., *Chem. Eng. News*, 1957, **35**(31), 10
3. Austin, C. M. *et al.*, *J. Chem. Educ.*, 1959, **36**, 54, 308
4. *MCA Case History No. 824*
5. Kirshenbaum, 1956, 4, 16
6. *MCA Case History No. 988*
7. Strauss, W. A., *Amer. Inst. Aeronaut. Astronaut. J.*, 1968, **6**, 1753
8. Rozovskii, A. S. *et al.*, *Chem. Abs.*, 1979, **91**, 23447
9. Lewis, D. J., *Haz. Cargo. Bull.*, 1983, **4**(11), 28−29
10. Bodner, G. M., *J. Chem. Educ.*, 1985, **62**(12), 1106

A demonstration of combustion of aluminium powder in oxygen exploded violently, probably owing to presence of unevaporated liquid at ignition [1]. Stoicheiometric mixtures of the two are explosives much more powerful than TNT [2]. Further details and comments were given later [3]. An aluminium filter in a high-capacity liquid oxygen transfer line exploded violently, possibly owing to friction or impact ignition of an aluminium component in contact with an abrasive particle, which would penetrate the protective oxide layer [4]. Powdered magnesium, titanium or zirconium mixed with liquid oxygen are detonable [5]. A nitrogen-pressurised liquid oxygen dispenser made of titanium alloy failed during nitrogen pressurising. The vessel failed because of reaction of the titanium alloy with liquid oxygen. Titanium is more reactive towards oxygen than either stainless steel or aluminium, and should therefore not be used for oxygen service, either gas or liquid [6]. Mixtures of liquid oxygen with 48–64 wt% of fine aluminium powder are detonable, and the parameters were investigated [7]. Detonation of mixtures with powdered aluminium, iron, titanium and chromium–nickel alloy powders was studied [8]. A new liquid oxygen road tanker constructed of aluminium ruptured violently during manoeuvring operations after a delivery, when the contents had been pressured up with gaseous oxygen (evaporated from the liquid) to expel the liquid. The rupture was traced to erosion and thinning of the tank wall adjacent to an internal baffle joint, where aluminium particles, oil and a halocarbon degreasing solvent were trapped in a cavity in the weld. These had reacted with the pressurising oxygen atmosphere, raising the wall temperature in the vapour space to that where direct reaction of the aluminium container with gaseous oxygen was possible. Some 70 kg of metal was missing in all [9]. Accounts of many further accidents in lecture demonstrations of rapid combustion of aluminium powder–liquid oxygen mixtures have been published [10].
See Aluminum: Oxidants, *also* Magnesium: Oxidants

Organic materials
1. Hauser, R. L. *et al., Adv. Cryog. Eng.*, 1962, **8**, 242–250
2. Clippinger, D. E. *et al., Aircraft. Eng.*, 1959, **32**(365), 204–205
Many common polymers, polymeric additives and lubricants oxidise so rapidly after impact in liquid oxygen that they are hazardous. Of those tested, only acrylonitrile–butadiene, poly(cyanoethylsiloxane), poly(dimethylsiloxane) and polystyrene exploded after impact of 6.8–95 J intensity (5–70 ft.lbf). All plasticisers (except dibutyl sebacate) and antioxidants examined were very reactive. A theoretical treatment of rates of energy absorption and transfer is included [1]. Previously, many resins and lubricants had been examined similarly, and 35 were found acceptable in liquid oxygen systems [2].

1,3,5-Trioxane
Kirshenbaum, 1956, 31
Mixtures with liquid oxygen are highly explosive.

Wood (+ charcoal)
1. Weber, U., *Chim. et Ind.* (Paris), 1963, **90**, 178
2. Lang, A., *Chem. Eng. Progr.*, 1962, **58**(2), 71

Several major accidents during handling of tonnage amounts of liquid oxygen are described [1]. One involved a violent explosion caused by leakage of liquid oxygen on to metal-encased timber (which previously had been charred) and spark ignition from welding in the oxygen-enriched atmosphere of an air-separation plant [2].
See other CRYOGENIC LIQUIDS, NON-METALS, OXIDANTS

4833. Osmium(IV) oxide
[12036-02-1] O_2Os

OsO_2

Preparative hazard
Sidgwick, 1950, 1493
The amorphous form, prepared by dehydration at low temperature, is a pyrophoric powder.
See Sodium chlorate: Osmium
See other METAL OXIDES, PYROPHORIC MATERIALS

4834. Lead(IV) oxide (Lead dioxide)
[1209-60-0] O_2Pb

PbO_2

HCS 1980, 590

Carbon black, Chlorinated paraffin, Manganese(IV) oxide
1. Rasmussen, B., *Unwanted chemical reactions; Appendix B*, Risøl-279, RisøNational Laboratory, Roskilde (DK), 1987; *J. Haz. Mat.*, 1988, **19**, 279–288
2. Author's comments to Dr Rasmussen, 1988
During the established manufacture of an accelerator mixture for a joint-filling composition, carbon black, lead(IV) oxide and manganese(IV) oxide were dispersed in a mixture of viscous chlorinated paraffins in presence of a silicone surfactant by agitation in a 1200 l reactor. The mixture was subsequently milled in a continuously fed shear disk bead mill to reduce the particle size and increase the degree of dispersion to a predetermined value. The bulk temperature of the material was increased to 55°C by the milling operation, and it was returned to the reactor for storage before re-milling, as the particle size was too coarse. After 2 days, exothermic decomposition set in with evolution of hydrogen chloride and probably chlorine also, and the mixture in the reactor eventually ignited. The reactor was eventually quenched with water and was found to contain a solid resiidue resembling clinker.

Some variations from the normal routine had occurred in the course of preparing this batch, including:- addition of a major portion of the chlorinated paraffin to the reactor which still contained 100–150 kg of a previous batch of accelerator, followed by standing for 16 days over a winter holiday break; use of 20% more manganese dioxide than usual; the mixture had not dispersed properly during the normal single milling pass, so had to be stored.

Subsequent investigation of the incident was undertaken both by DSC and by both 5 kg and 1 kg simulations of the incident in electrically heated glass or metal reaction vessels. DSC showed that the stabilised chlorinated paraffin began to decompose exothermally only above 300°C, with $\Delta H = 1.25$ kJ/g, while the reaction mixture showed an initial exotherm at 91°C, with the main exotherms above 200°C, and $\Delta H = 3.64$ kJ/g. The 5 kg process simulation experiment showed that non-acidic gas began to be evolved at about 150°C, and no hydrogen chloride was evolved until after the reaction mixture had been heated for several hours at temperatures up to about 200°C (during which period the inhibitor had probably been consumed). Further heating to 220°C initiated a thermal runaway reaction, with temperatures exceeding 300°C and violent evolution of white fumes, which continued (without external heating) for over an hour. The residue was a grey clinker, samples of which showed a relatively much lower lead content and higher manganese content than the original mixture. The addition of ferric chloride to the chlorinated paraffin cused considerable reduction of the thermal stability of the latter.

It was concluded that the reaction mixture was thermally unstable if subjected to thermal initiation, and that temperatures of over 300°C were capable of attainment, when direct reaction between the metal oxides amd the chlorinated paraffins or their degradation products were possible. The origin of the heat energy necessary to increase the temperature of the milled batch from 55 to 91°C (the DSC-identified exotherm teperature) could not be identified [1].

However, a reasonable and coherent explanation seems possible if the following assumptions are made.

1. During the 16-day contact of the first 250 kg portion of chlorinated paraffin with the 150 kg residue of previously processed accelerator, the 1% inhibitor content of the former became depleted or exhausted by adsorption and/or reaction, reducing the thermal stability of the liquid phase.

2. During the 2 day storage (apparently without agitation) of the milled reaction mixture, initially at 55°C, the oxidants (mainly lead dioxide) began to react with the carbon black (effectively a very finely divided organic fuel), the 12.5 kg of which could release up to 373 MJ of heat if oxidised fully to carbon dioxide. The heat release may well have not been uniform throughout the batch, probably being concentrated in the bottom of the unstirred batch, where the not fully dispersed lead dioxide ($d = 9.4$) would settle out. This heat release may have been sufficient to increase the temperature, at least locally, to 91°C, when the total exotherm would then have been increased.

3. Once the thermal runaway had become established, the high temperature could allow a 'smelting reaction' (reduction of lead oxides by carbon to metallic lead) to occur, and the low m.p. and high density (327°C, 10.6, respectively) would cause the molten lead to concentrate at the base of the reactor, decreasing the lead content and raising the manganese content of the bulk of the clinker residue, as was observed [2].

See other AGITATION INCIDENTS, INDUCTION PERIOD INCIDENTS, REDOX REACTIONS

Chlorine trifluoride
See Chlorine trifluoride: Metals, etc.

Hydrogen sulfides   MRH Hydrogen sulfide 0.56/82
Mellor, 1941, Vol. 7, 689; 1947, Vol. 10, 129, 159
Contact of hydrogen sulfide with dry or moist lead dioxide causes attainment of
red heat and ignition. Contact of hydrogen trisulfide with the dioxide causes violent
decomposition and ignition.

Metal acetylides or carbides
Mellor, 1946, Vol. 5, 850, 872
Interaction with aluminium carbide is incandescent, and with caesium acetylide
at 350°C is explosive. Other related compounds may be expected to be oxidised
violently if unmoderated.

Metals   MRH Magnesium 3.22/17
1. Mellor, 1963, Vol. 2, Suppl. 2.2, 1571
2. Mellor, 1940, Vol. 4, 272
3. Mellor, 1946, Vol. 5, 217
4. Mellor, 1941, Vol. 7, 658, 691
5. Collvin, P., *Chem. Abs.*, 1982, **96**, 87999
Warm potassium reacts explosively with lead dioxide, and sodium probably
behaves similarly [1]. Magnesium reacts violently [2], and powdered aluminium
probably does also, as it reacts violently with the monoxide [3]. Mixtures of
powdered molybdenum or tungsten with the dioxide incandesce on heating [4].
The combination with zirconium is rated highest on a scale for sensitivity to
deflagration, friction, ignition and static electrical initiation [5].
See other STATIC INITIATION INCIDENTS

Metal sulfides
Mellor, 1941, Vol. 3, 745
Calcium sulfide, strontium sulfide or barium sulfide react vigorously with lead
dioxide on heating.

Nitroalkanes
See NITROALKANES: metal oxides

Nitrogen compounds   MRH Hydroxylamine 1.92/36
1. Mellor, 1941, Vol. 7, 637
2. Mellor, 1940, Vol. 8, 291
Hydroxylamine ignites in contact with lead dioxide [1], while phenylhydrazine
immediately reacts vigorously [2].

1. Mellor, 1941, Vol. 7, 690
2. Mellor, 1947, Vol. 10, 676, 909

Warm phosphorus trichloride reacts with incandescence [1] and seleninyl chloride vigorously. Sulfonyl dichloride may react explosively [2].

Non-metals
1. Mellor, 1946, Vol. 5, 17
2. Mellor, 1941, Vol. 7, 689–690
3. Gibson, N., *Runaway Reactions*, 1981, Paper 3/R, 6

Boron or yellow phosphorus explode violently on grinding with lead dioxide, while red phosphorus ignites [1]. Mixtures with sulfur ignite on grinding or addition of sulfuric acid [2]. An initiating mixture of silicon and lead dioxide (2:1) attains a temperature around 1100°C after ignition by a small flame [3].

Peroxyformic acid
See Peroxyformic acid: Metals, etc.

Potassium
See Potassium: Metal oxides

Sulfur dioxide
Mellor, 1941, Vol. 7, 689
Interaction is incandescent.
See other METAL OXIDES, OXIDANTS

4835. Palladium(IV) oxide
[12036-04-3] O_2Pd

PdO_2

Bailar, 1973, Vol. 3, 1279
The hydrated oxide is a strong oxidant, and slowly evolves oxygen at ambient temperature.
See other METAL OXIDES, OXIDANTS *See related* PLATINUM COMPOUNDS

4836. Platinum(IV) oxide
[1314-15-4] O_2Pt

PtO_2

Acetic acid, Hydrogen
Adams, R. *et al., Org. Synth.*, 1941, Coll. Vol. 1, 66
Addition of fresh platinum oxide catalyst to a hydrogenation reaction in acetic acid caused immediate explosion. Several similar incidents, usually involving acetic acid as solvent, are known to the author.
See other METAL OXIDES, OXIDANTS, PLATINUM COMPOUNDS

4837. Sulfur dioxide
[7446-09-5]   O₂S
                              SO₂

(*MCA SD-52*, 1953); *HCS 1980*, 876 (cylinder)

Preparative hazard
  See Sulfuric acid: Copper

Barium peroxide
  See Barium peroxide: Non-metal oxides

Boranes
  See BORANES: aluminium chloride, sulfur dioxide

Caesium azide
  See Caesium azide: Sulfur dioxide

Diethylzinc
  See Diethylzinc: Sulfur dioxide

Halogens, or Interhalogens   MRH values show % of interhalogen
  See Bromine pentafluoride: Acids, etc.   MRH 1.72/53
       Chlorine trifluoride: Metals, etc.   MRH 2.43/74
       Fluorine: Non-metal oxides   MRH 5.10/64

Lithium acetylide–ammonia
  See Monolithium acetylide–ammonia: Gases, etc.

Lithium nitrate,   MRH 1.00/30

Propene   MRH 2.18/30
  See Propene: Lithium nitrate, etc.

Metal acetylides
  Mellor, 1946, Vol. 5, 848–849
  Monocaesium acetylide or monopotassium acetylide, and the ammoniate of monolithium acetylide all ignite and incandesce in unheated sulfur dioxide. The dimetal
  derivatives including sodium acetylide appear to be less reactive, needing heat
  before ignition occurs.

Metal oxides   MRH Iron(II) oxide 1.21/40
  1. Mellor, 1941, Vol. 2, 487
  2. Mellor, 1942, Vol. 13, 715
  3. Mellor, 1941, Vol. 7, 388, 689
  Caesium oxide [1], iron(II) oxide [2], tin oxide and lead(IV) oxide [3] all ignite
  and incandesce on heating in the gas.

Metals   MRH Chromium 3.39/52, manganese 2.89/46, sodium 4.56/26
  1. Mellor, 1943, Vol. 11, 161
  2. Mellor, 1942, Vol. 12, 187
  3. Gilbert, H. N., *Chem. Eng. News*, 1948, **26**, 2604
  4. Mellor, 1961, Vol. 2, Suppl. 2.1, 468
Finely divided (pyrophoric) chromium incandesces in sulfur dioxide [1], while pyrophoric manganese burns brilliantly on heating in the gas [2]. Molten sodium reacts violently with the dry gas or liquid, while the moist gas reacts as vigorously as water with cold sodium [3].

Other reactants
  Yoshida, 1980, 267
  MRH values calculated for 12 combinations with various materials are given.

Peat
  Byrne, P. J., *Chem. Abs.*, 1976, **104**, 21731
  Treatment of peat stockpiles with sulfur dioxide surprisingly accelerated the self-heating process.

Polymeric tubing
  MCA Case History No. 1044
  Plastics tubing normally capable of withstanding an internal pressure of 7 bar failed below 2 bar when used to convey gaseous sulfur dioxide.

Potassium chlorate   MRH 1.80/44
  See Potassium chlorate: Sulfur dioxide

Silver azide
  See Silver azide: Sulfur dioxide

Sodium hydride   MRH 3.72/23
  See Sodium hydride: Sulfur dioxide
  See other NON-METAL OXIDES, REDUCANTS

4838. Selenium dioxide
  [7446-08-4] O_2Se

$$SeO_2$$

  HCS 1980, 814

Preparative hazard
  See Selenium: Oxygen, Organic matter

1,3-Bis(trichloromethyl)benzene
  See 1,3-Bis(trichloromethyl)benzene: Oxidants

Phosphorus trichloride
  Mellor, 1940, Vol. 8, 1005

A mixture of the cold components attains red-heat.
See related METAL OXIDES *See other* OXIDANTS

4839. Silicon dioxide (Silica)
[7631-86-9] O_2Si

$$SiO_2$$

HCS 1980, 819

Hydrochloric acid
MCA Case History No. 1857
A glass drying trap on a hydrochloric acid storage tank was filled with silica gel instead of the calcium sulfate specified. The glass trap fractured, probably owing to thermal shock from the much higher heat of adsorption of water and hydrogen chloride on silica gel.
See other GLASS INCIDENTS

Metals
See Magnesium: Silicon dioxide
Sodium: Non-metal oxides

Oxygen difluoride
See Oxygen difluoride: Adsorbents

Ozone
See Ozone: Silica gel

Silicone liquid, Unstated salt
See SILICONE LIQUID: SILICA, ETC.

Steel
These two anciently served as a means of ignition. They still can. Although the sparking of flint and steel is strictly a mechanical phenomenon, any subsequent fires will not be. Other forms of silica still serve as igniters to this day, via the piezoelectric effect.
See GRAVEL

Vinyl acetate
See Vinyl acetate: Desiccants

Xenon hexafluoride
See Xenon hexafluoride: Silicon dioxide
See other NON-METAL OXIDES

4840. Tin(IV) oxide
[18282-10-5] O_2Sn

$$SnO_2$$

Chlorine trifluoride
See Chlorine trifluoride: Metals, etc.

Hydrogen trisulfide
See Hydrogen trisulfide: Metal oxides

Metals
See Aluminium: Metal oxides
Magnesium: Metal oxides
Potassium: Metal oxides
Sodium: Metal oxides
See other METAL OXIDES

4841. Strontium peroxide
[1314-18-7] O_2Sr

$$SrO_2$$

Organic materials
Haz. Chem. Data, 1975, 272
Mixtures of the peroxide with combustible organic materials very readily ignite on friction or contact with moisture.
See other METAL PEROXIDES

4842. Titanium(IV) oxide (Titanium dioxide)
[13463-67-7] O_2Ti

$$TiO_2$$

HCS 1980, 903

Metals
Mellor, 1941, Vol. 7, 10, 44; 1961, Vol. 2, Suppl. 2.1, 81
Reduction of the oxide by aluminium, calcium, magnesium, potassium, sodium or zinc is accompanied by more or less incandescence (lithium, magnesium and zinc especially).
See other METAL OXIDES

4843. Uranium(IV) oxide
[1344-57-6] O_2U

$$UO_2$$

Mellor, 1942, Vol. 12, 42

The finely divided oxide prepared at low temperature is pyrophoric when heated in air, and burns brilliantly.

See other METAL OXIDES, PYROPHORIC MATERIALS

4844. Tungsten(IV) oxide
[12036-22-5] O_2W

$$WO_2$$

Chlorine
See Chlorine: Tungsten dioxide
See other METAL OXIDES

4845. Zinc peroxide
[1314-22-3] O_2Zn

$$ZnO_2$$

Alone, or Metals
1. Mellor, 1940, Vol. 4, 530
2. Sidgwick, 1950, 270
The hydrated peroxide (of indefinite composition) explodes at 212°C [1], and mixtures with aluminium or zinc powders burn brilliantly [2].
See other METAL PEROXIDES

4846. Ozone (Trioxygen)
[10028-15-6] O_3

$$O_3$$

HCS 1980, 710 (cyl., dissolved gas); *RSC Lab. Hazards Data Sheet No. 66*, 1987

1. Sidgwick, 1950, 860
2. Adley, F. E., *Nucl. Sci. Abs.*, 1962, **17**, 18
3. Streng, A. G., *Explosivstoffe*, 1960, **8**, 225
4. Clough, P. N. *et al., Chem. & Ind.*, 1966, 1971
5. Gatwood, G. T. *et al., J. Chem. Educ.*, 1969, **46**, A103
6. Loomis, J. S. *et al., Nucl. Instr. Methods*, 1962, **15**, 243
7. Brereton, S. J., *Fusion Technol.*, 1988, **15**(2 pt.23), 833
8. *Ozone Reactions with Organic Compounds*, ACS 112, Washington, ACS, 1972
9. Walker, K. G., *Chem. Brit.*, 1987, **23**, 839
Ozone is strongly endothermic (ΔH°_f (g) +142.2 kJ/mol, 2.96 kJ/g) and the pure solid or liquid materials are highly explosive. Evaporation of a solution of ozone in liquid oxygen causes ozone enrichment and ultimately explosion [1]. Organic liquids and oxidisable materials dropped into liquid ozone will also cause explosion of the ozone [2]. Ozone technology and hazards have been reviewed [3], a safe process to concentrate ozone by selective adsorption on silica gel at low

temperatures has been described [4], and safe techniques for laboratory generation and handling of ozone have been detailed [5]. Ozone has become available as a dissolved gas in cylinders. Explosive hazards involved in use of liquid nitrogen as coolant where ozone may be incidentally produced during operation of van der Graaff generators [6] and radiation sources [7], etc. have been discussed. The chemistry of interaction with organic compounds has been extensively reviewed [8]. Methods and equipment for generation of ozone (up to 13%) in oxygen and storage at high pressure and low temperature for extended periods are described [9].
See Silica gel, below

Acetylene
Grignard, 1935, Vol. 3, 166
Passage of ozone (endothermic oxidant) into acetylene (endothermic reducant) leads to a violent explosion when 50 mg/l of ozone is present.
See other REDOX REACTIONS

Alkenes
1. Sidgwick, 1950, 862
2. Davies, 1961, 97
3. Stull, 1977, 21
Interaction of alkenes with ozonised oxygen tends to give several types of products or their polymers, some of which show more pronounced explosive tendencies than others [1]. The cyclic *gem*-diperoxides are more explosive than the true ozonides [2]. It has been calculated that ozonisation of the endothermic *trans*-stilbene (ΔH°_f +135.4 kJ/mol, 0.78 kJ/g) would give, in the event of decomposition of the unstable ozonide, an exothermic release of 1.41 kJ/g which would attain an adiabatic decomposition temperature approaching 750°C with a 27-fold pressure increase in a closed vessel [3].
See CYCLIC PEROXIDES

Alkylmetals
Lee, H. U. *et al.*, *Combust. Flame*, 1975, **24**, 27–34
The spectra of flames from spontaneous interaction of ozone with dimethylzinc and diethylzinc in a flow system have been studied.

Ammoniacal vapours
1. Bond, 1991, 31
2. Anon., *Fire Prevention*, 1986, (191), 45
Vent air from an animal facility was deodourised by ozone treatment; an extensive fire was started by cutting into the polypropylene ducting carrying the treated air to atmosphere. This was attributed to formation of ammonium nitrate by ammonia oxidation, which mixed with organic dusts also present in the vent-line [1]. A more detailed account of the incident reveals that initial ignition was probably by sawing into metal. Ozone was also found to have degraded the polypropylene pipework, increasing flammability (peroxide formation?) [2].
See Ammonium nitrate

Aromatic compounds
    Mellor, 1941, Vol. 1, 911
    Benzene, aniline and other aromatic compounds give explosive gelatinous
    ozonides, among other products, on contact with ozonised oxygen.

Benzene, Oxygen, Rubber
    Morrell, S. H., private comm., 1968
    During ozonisation of rubber dissolved in benzene, an explosion occurred. This
    seems unlikely to have been owing to formation of benzene triozonide (which
    separates as a gelatinous precipitate after prolonged ozonisation), since the
    solution remained clear. A rubber ozonide may have been involved, but the
    benzene–oxygen system itself has high potential for hazard.

Bromine
    Lewis, B. et al., J. Amer. Chem. Soc., 1931, 53, 2710
    Interaction becomes explosive above 20°C and a minimum critical pressure.
    See Hydrogen bromide, below, also Bromine trioxide

Charcoal, Potassium iodide
    Anon., Lab. Haz. Bull., 1983(9), 3, item 660
    Carbon impregnated with potassium iodide was used as an ozone-scrubbing filter
    in a chemiluminescence NO_x analyser. When the level of iodide was inadvertently
    increased to the high level of 40%, the filter exploded violently during replace-
    ment after use. This was attributed to oxidation of iodide to iodate by ozone, and
    frictional initiation of the iodate–carbon mixture when the filter was dismantled.

Citronellic acid
    Yost, Y., Chem. Eng. News, 1977, 55(21), 4
    During work-up of the products of ozonolysis of R- and S-citronellic acids, a
    substantial quantity of the highly explosive trimeric acetone peroxide (3,3,6,6,9,9-
    hexamethyl-1,2,4,5,7,8-hexoxonane) was unwittingly isolated by distillation at
    105–135°C to give the solid m.p. 95°C. The peroxide appears to have been
    produced by ozonolysis of the isopropylidene group in citronellic acid, and
    presumably the same could occur when any isopropylidene group is ozonised.
    Appropriate care is advised.

Combustible gases
    1. Streng, A. G., Chem. Abs., 1964, 55, 8862c
    2. Nomura, Y. et al., Chem. Abs., 1952, 46, 4234a
    Carbon monoxide and ethylene admitted into contact with ozone via an aluminium
    tip ignited and burned smoothly, while normal contact of carbon monoxide,
    nitrogen oxide, ammonia or phosphine causes immediate explosion at 0° or −78°C
    [1]. At pressures below 10 mbar, contact with ethylene is explosive at −150°C [2].
    See Nitrogen oxide, below

trans-2,3-Dichloro-2-butene
    1. Griesbaum, K. et al., J. Amer. Chem. Soc., 1976, 98, 2880
    2. Gäb, S. et al., J. Org. Chem., 1984, 49, 2711–2714

The crude products of ozonolysis at $-30°C$ of the chloroalkene tended to decompose explosively on warming to ambient temperature, particularly in absence of solvents. The products included the individually explosive compounds acetyl 1,1,-dichloroethyl peroxide, 3,6-dichloro-3,6-dimethyl-2,3,5,6-tetraoxane and diacetyl peroxide [1]. Ozonolysis in ethyl formate saturated with hydrogen chloride gives a high yield of 1,1-dichloroethyl hydroperoxide as a further unstable intermediate product [2].
See OZONIDES

Dicyanogen
  Anon., *Chem. Eng. News*, 1957, **35**(21), 22
  This hypergolic combination of two extremely endothermic compounds appears to have rocket-propellant capabilities.
  See ROCKET PROPELLANTS

Dienes, Oxygen
  Griesbaum, K. *et al.*, *Chem. Eng. News*, 1982, **60**(8), 63
  Treatment of isoprene, 2,3-dimethylbutadiene or cyclopentadiene in pentane at $-78°C$ with a high ratio ozone/oxygen stream led to immediate ignition and flames in both liquid and gas phases.

Diethyl ether
  Mellor, 1941, Vol. 1, 911
  Contact of ether with ozonised oxygen produces some of the explosive diethyl peroxide.

1,1-Difluoroethylene
  1. Gillies, C. W. *et al.*, *J. Phys Chem.*, 1979, **83**, 1545
  2. Agopovich, J. W. *et al.*, *J. Amer. Chem. Soc.*, 1983, **105**, 5049
  The product of ozonolysis of 1,1-difluoroethylene is a shock-sensitive peroxidic species [1], which reacts vigorously with aqueous potassium iodide [2].

Ethylene
  Mellor, 1941, Vol. 1, 911
  Interaction may be explosive, owing to the very low stability of ethylene ozonide.
  See Combustible gases, above, *also* Ethylene ozonide

Ethylene, Formyl fluoride
  Mazur, U. *et al.*, *J. Org. Chem.*, 1979, **44**, 3188
  Two attempts to ozonise mixtures of ethylene and formyl fluoride in absence of solvent led to explosions.

Fluoroethylene
  See Fluoroethylene ozonide

Hydrogen (liquid)
  Schwab, G. M., *Umschau*, 1922, 538–539
  Liquid hydrogen and solid ozone form very powerfully explosive mixtures.

Hydrogen, Oxygen difluoride
Boehm, R., *Chem. Abs.*, 1975, **83**, 82254
The use of liquid oxygen difluoride (40 or 90%) to stabilise liquid ozone as oxidant for gaseous hydrogen in a rocket motor was not entirely successful, explosions occurring at both concentrations.
See Oxygen fluorides, below

Hydrogen bromide
Lewis, B. *et al., J. Amer. Chem. Soc.*, 1931, **53**, 3565
Interaction is rapid and becomes explosive above a total pressure of ∼40 mbar, even at −104°C.
See Bromine, above

4-Hydroxy-4-methyl-1,6-hepatadiene
See 4-Hydroxy-4-methyl-1,6-heptadiene: Ozone

3-Hydroxy-2,2,4-trimethyl-3-pentenoic acid lactone
Perrin, C. L. *et al., J. Org. Chem.*, 1980, **45**, 1706
During work-up after ozonolysis of the lactone to give dimethylmaleic anhydride, removal of excess ozone by a stream of oxygen leads to the precipitation of peroxidic material at the concentration used in the preparation.

Isopropylidene compounds
See Citronellic acid, above

2-Methyl-1,3-butadiene
See Dienes, above, *also* 2-Methyl-1,3-butadiene: Ozone

Nitrogen
Strakhov, B. V. *et al., Chem. Abs.*, 1963, **58**, 5069a
The explosive oxidation of nitrogen in admixture with ozone in metallic vessels has been studied.

Nitrogen oxide
Nomura, Y. *et al., Sci. Reps. Tohoku Univ.*, 1950[A], **2**, 229–232
Violent explosions occurred in mixtures at −189°C.
See Combustible gases, above

Nitrogen trichloride
See Nitrogen trichloride: Initiators

Oxygen, Rubber powder
Anon., *Jahresber.*, 1983, 72
In an attempt to introduce carboxyl groups to improve the adhesive properties of rubber, 6 l of powdered rubber (0.1–0.5 mm particle size) in an 8 l flask was treated with ozonised oxygen. The gas stream, containing 5% of ozone was led to the bottom of the flask via a dip tube at the rate of 4 l/min for 2 min, when

treatment was discontinued and the flask closed. After 5 min a violent explosion occurred. This was attributed to formation of ozonides, but exothermic interaction of the high surface area rubber with almost pure oxygen under virtually adiabatic conditions may also have been involved.
See Oxygen: Polymers

Oxygen fluorides
Streng, A. G., *Chem. Rev.*, 1963, **63**, 613, 619
Mixtures with dioxygen difluoride explode at $-148°C$, and with 'dioxygen trifluoride' (probably mixed dioxygen di- and tetra-fluorides) at $-183°C$.

Silica gel
Cohen, Z. *et al., J. Org. Chem.*, 1975, **40**, 2142, footnote 6
Silica gel at $-78°C$ adsorbs 4.5 wt% of ozone, and below this temperature the concentration increases rapidly. At below $-112°C$ ozone liquefies and there is a potential explosion hazard at temperatures below $-100°C$ if organic material is present.
See reference 3, main entry

Stibine
Stock, A. *et al., Ber.*, 1905, **38**, 3837
Passage of oxygen containing 2% of ozone through stibine at $-90°C$ caused an explosion. On standing, a suspension of solid stibine in a liquid oxygen solution of ozone eventually exploded, as oxygen evaporated increasing the concentration of ozone and the temperature.

Tetrafluorohydrazine
See Tetrafluorohydrazine: Ozone

Tetramethylammonium hydroxide
Solomon, I. J. *et al., J. Amer. Chem. Soc.*, 1960, **82**, 5640
During interaction to form tetramethylammonium ozonide, constant agitation and slow ozonisation were necessary to prevent ignition.

Trifluoroethylene
Agopovich, J. W. *et al., J. Amer. Chem. Soc.*, 1983, **105**, 5048
After ozonisation of trifluoroethylene in chlorotrifluoromethane solution at $-95°C$, the reactor must only be allowed to warm slightly (to $-88°C$) during subsequent cryogenic distillation. When allowed to warm towards ambient temperature the reactor exploded violently. The residue from distillation is peroxidic and probably sensitive to sudden temperature increases.

Unsaturated acetals
Schuster, L., Ger. Offen. 2 514 001, 1976
Ozonolysis of unsaturated acetals at -70 to $0°C$ to give glyoxal monoacetals is uneventful in organic solvents, but leads to explosions in aqueous solutions.
See other NON-METALS, OXIDANTS

4847. Phosphorus(III) oxide
[1314-24-5] O_3P_2

$$P_2O_3$$

See Tetraphosphorus hexaoxide

4848. Palladium(III) oxide
[80680-07-5] O_3Pd_2

$$Pd_2O_3$$

Sidgwick, 1950, 1573
If the hydrated oxide is heated to remove water, it incandesces or explodes, giving
the monoxide.
See other METAL OXIDES *See related* PLATINUM COMPOUNDS

4849. Sulfur trioxide
[7446-11-9] O_3S

$$SO_3$$

HCS 1980, 880

The metastable liquid form boils at 44°C, the more stable solid alpha form melts
at 62°C and has 25–30 J/g heat of fusion. This permits pressure generation when
proceeding in either direction through the phase change.

Acetonitrile, Sulfuric acid
See Acetonitrile: Sulfuric acid, etc.

Chlorinated solvents
Gilbert, E. E. *Chem. Eng. News* 1989 **67**(41), 2
Details of exothermic and gas-evolving reactions are given, storage of such
mixtures is not recommended.

Cyclohexanone oxime, Sulfuric acid
See Cyclohexanone oxime: Oleum

Dimethylformamide
See Sulfur trioxide–dimethylformamide

Dimethyl sulfoxide
Seel, F., *Inorg. Synth.*, 1947, **2**, 174
Dissolution of sulfur trioxide in the sulfoxide is very exothermic and must be done
slowly with cooling to avoid decomposition.

Dioxane
Sisler, H. H. *et al., Inorg. Synth.*, 1947, **2**, 174

Since the 1:1 addition complex sometimes decomposes violently on storing at ambient temperature, it should only be prepared immediately before use (if refrigerated storage is not possible).

Dioxygen difluoride
See Dioxygen difluoride: Sulfur trioxide

Diphenylmercury
See Diphenylmercury: Chlorine monoxide, etc.

Ethylene oxide
See Ethylene oxide: Pyridine, Sulfur dioxide

Formamide, Iodine, Pyridine
See Formamide: Iodine, etc.

Metal oxides
1. Partington, 1967, 706
2. Mellor, 1941, Vol. 7, 654
The violent interaction with sulfur trioxide causes incandescence with barium oxide [1] and lead oxide [2].

Nitryl chloride
See Nitryl chloride: Inorganic materials

Organic materials, Water
Unremembered suppliers brochure, 1979
Organic materials, particularly if fibrous with adsorbed or absorbed moisture present, may char or ignite in contact with the stabilised liquid form because of the very high heat of hydration (2.1 kJ/g) and formation of hot oleum which then functions as an oxidant.
See ACID ANHYDRIDES

Other reactants
Yoshida, 1980, 128
MRH values calculated for 18 combinations with various materials are given.

Phosphorus
See Phosphorus: Non-metal oxides

Sulfuric acid
HCS 1980, 702 (65%), 703 (30%), 704 (20%)
Solutions of sulfur trioxide in conc. sulfuric acid (oleum) are available with various proportions of dissolved sulfur trioxide, and the 3 entries in *Handling Chemicals Safely 1980* deal with the properties and hazards of solutions with 65, 30 and 20% of sulfur trioxide, respectively. As expected, the reactivity and oxidising power increase with the content of sulfur trioxide.

Tetrafluoroethylene

1. Anderson, A. W., *Chem. Eng. News*, 1971, **49**(22), 3
2. Xu, B. *et al., Chem. Abs.*, 1982, **97**, 182248
3. Aihara, R. *et al., Chem. Abs.*, 1986, **101** 05, 99522

Reaction to give tetrafluorooxathietane 2,2-dioxide (tetrafluoroethane sultone) had been used industrially and uneventfully, but reaction with excess sulfur trioxide may cause explosive decomposition to carbonyl fluoride and sulfur dioxide [1]. An incident involving the same explosion hazard was reported 11 years later [2]. Use of inert gas to prevent explosion has been patented [3].

Water

1. Mellor, 1947, Vol. 10, 344
2. Willis, P. B., *Chem. Eng. News*, 1986, **64**(38), 2
3. Eccles, E. J., *Chem. Eng. News*, 1987, **65**(12), 2
4. Rosin, J., *Chem. Eng. News*, 1987, **65**(24), 2

Interaction is vigorously exothermic, sometimes explosive, with evolution of light and heat. A 4:1 mixture of oxide and water completely vaporises, with simultaneous emission of light. Unused portions of ampoules of sulfur trioxide were prepared for disposal by dripping a little water each day during several weeks into the ampoules. When inverted to empty the contents, one such ampoule exploded, scattering conc. sulfuric acid around [2]. This technique probably produces a layer of water above one of conc. sulfuric acid, and when the layers mixed the exotherm burst the narrow-necked ampoule. Moderate quantities of liquid sulfur trioxide (100–500 g) may be disposed of into a strong water stream discharging into an acid-proof drain, using personal protection [3]. It is, however, stressed that all disposal methods involving addition to water are inherently unsafe, because sulfur trioxide is insoluble in water and also highly hydrophobic, so the potential for sudden and violent reaction always exists. The safest method is to dissolve sulfur trioxide in a large volume of conc. sulfuric acid, and then dispose of the weak oleum by slow addition to a large and well stirred volume of water [4].

See other ACID ANHYDRIDES, NON-METAL OXIDES, OXIDANTS

4850. Sulfur trioxide–dimethylformamide

[29584-42-7] $O_3S.C_3H_7NO$

$SO_3.Me_2NCO.H$

Koch-Light Laboratories Ltd., private comm., 1976

A bottle of the complex exploded in storage. No cause was established, but diffusive ingress of moisture to form sulfuric acid, and subsequent hydrolysis of the solvent with formation of carbon monoxide appear to be possible contributory factors.

See related ACID ANHYDRIDES, NON-METAL OXIDES

4851. Antimony(III) oxide
[1309-64-4] O_3Sb_2

O:SbOSb:O

HCS 1980, 164

Mellor, 1939, Vol. 9, 425
The powdered oxide ignites on heating in air.

Bromine trifluoride
See Bromine trifluoride: Antimony trichloride oxide

Chlorinated rubber
See CHLORINATED RUBBER: metal oxides
See other METAL OXIDES

4852. Selenium trioxide
[13768-86-0] O_3Se

SeO₃

Organic materials
Sorbe, 1968, 124
A powerful oxidant which forms explosive mixtures with organic materials.
See other NON-METAL OXIDES

4853. Tellurium trioxide
[13451-18-8] O_3Te

TeO₃

Metals, or Non-metals
Bailar, 1973, Vol. 2, 971
The yellow α-form is a powerful oxidant, reacting violently when heated with a
variety of metallic and non-metallic elements.
See other NON-METAL OXIDES

4854. Thallium(III) oxide
[1314-32-5] O_3Tl_2

O:TlOTl:O

Sulfur, or Sulfur compounds
Mellor, 1946, Vol. 5, 434
Mixtures of the oxide with sulfur or antimony trisulfide explode on grinding in a
mortar. Dry hydrogen sulfide ignites, and sometimes feebly explodes, over thallium
oxide.
See other METAL OXIDES

4855. Vanadium(III) oxide
[1314-34-7] O_3V_2

O:VOV:O

Sidgwick, 1950, 825
It ignites on heating in air.
See other METAL OXIDES

4856. Tungsten(VI) oxide
[1314-35-8] O_3W

WO_3

Boron tribromide
See Boron tribromide: Tungsten trioxide

Bromine, Tungsten
See Bromine: Tungsten, etc.

Interhalogens
See Bromine pentafluoride: Acids, etc.
Chlorine trifluoride: Metals, etc.

Lithium
See Lithium: Metal oxides
See other METAL OXIDES

4857. Xenon trioxide
[13776-58-4] O_3Xe

XeO_3

1. Malm, J. G. *et al., Chem. Rev.*, 1965, **65**, 199
2. Chernick, C. L. *et al., J. Chem. Educ.*, 1966, **43**, 619; *Inorg. Synth.*, 1968, **11**, 206, 209
3. Anon., *Chem. Eng. News*, 1963, **41**(13), 45
4. Holloway, J. H., *Talanta*, 1967, **14**, 871–873
5. Shackelford, S. A. *et al., Inorg. Nucl. Chem. Lett.*, 1973, **9**, 609, footnote 9
6. Shackelford, S. A. *et al., J. Org. Chem.*, 1975, **40**, 1869–1875
7. Foropoulos, J. *et al., Inorg. Chem.*, 1982, **21**, 2503–2504

It is a powerful explosive [1] produced when xenon tetrafluoride or xenon hexafluoride are exposed to moist air and hydrolysed. Some tetrafluoride is usually present in xenon difluoride, so the latter is potentially dangerous. Although safe to handle in small amounts in aqueous solution, great care must be taken to avoid solutions drying out, e.g. around ground stoppers [2]. Full safety precautions have been discussed [2,3,4]. Precautions necessary for use of aqueous solutions of the trioxide as an epoxidation reagent are detailed [5,6]. A safe method of preparation, so far free from explosions, from xenon hexafluoride and difluorophosphoric acid at 0°C has been described [7].

See other NON-METAL OXIDES, OXIDANTS, XENON COMPOUNDS

4858. Osmium(VIII) oxide
[20816-12-0] O_4Os

$$OsO_4$$

Hydrogen peroxide
See Hydrogen peroxide: Metals, etc.

1-Methylimidazole
Nielsen, A. J., *J. Chem. Soc., Dalton Trans.*, 1979, 1084–1088
An explosion occurred when the oxidant was added to the imidazole.
See other METAL OXIDES, OXIDANTS

4859. Tetraphosphorus tetraoxide trisulfide
[ ] $O_4P_4S_3$

Water
Mellor, 1971, Vol. 8, Suppl. 3, 435
It ignites if moistened with water.
See related NON-METAL OXIDES, NON-METAL SULFIDES

4860. Lead sulfate
[7446-14-2] O_4PbS

$$PbSO_4$$

Potassium
See Potassium: Oxidants
See other METAL OXONON-METALLATES

4861. Dilead(II)lead(IV) oxide
[1314-41-6] O_4Pb_3

$$2PbO.PbO_2$$

HCS 1980, 808

Dichloromethylsilane
See Dichloromethylsilane: Oxidants

Eosin lead salts
Anon., *Chem. & Ind.*, 1995, (22), 912

The mixture of the lead salts of eosin and red lead formerly used as a pigment, proved to be a very easily ignited pyrotechnic mix.

Peroxyformic acid
See Peroxyformic acid: Metals, etc.

Seleninyl chloride
See Seleninyl chloride: Metal oxides

2,4,6-Trinitrotoluene
See 2,4,6-Trinitrotoluene: Added impurities
See other METAL OXIDES, OXIDANTS

4862. Ruthenium(VIII) oxide
[20427-56-9] O_4Ru
$$RuO_4$$

1. Mellor, 1942, Vol. 15, 520
2. Brauer, 1965, Vol. 2, 1600
3. Bailar, 1973, Vol. 3, 1194
4. Ullmann, 1992, Vol. A21, 125/6
The liquid (which shows no true b.p.) is said to decompose explosively either above 106° or above 180°C [1,2,3]. The oxide should not be stored in bulk, either as liquid or solid [4].

Ammonia
Watt, J., *J. Inorg. Nucl. Chem.*, 1965, **27**, 262
Rapid interaction at −70°C/1 mbar causes ignition. Slow interaction produced a solid which exploded at 206°C, probably owing to formation of ruthenium amide oxide or ruthenium nitride.

Hydriodic acid
Mellor, 1942, Vol. 15, 520
Reaction with the acid to give ruthenium triiodide is explosively violent.

Organic materials
1. Mellor, 1942, Vol. 15, 517, 520
2. Brauer, 1965, Vol. 2, 1600
The solid oxide, or its concentrated solutions or vapour, tends to oxidise ethanol, cellulose fibres, etc., explosively [1,2]. Ethanol also gives an explosive by-product, probably an organoruthenium derivative [1].

Phosphorus tribromide
Mellor, 1971, Vol. 8, Suppl. 3, 521
Interaction is vigorously exothermic.

Sulfur
    Mellor, 1942, Vol. 15, 520
    The vapour explodes in contact with sulfur.
    See other METAL OXIDES, OXIDANTS

4863. Xenon tetraoxide
    [12340-14-6]   O₄Xe

$$XeO_4$$

    Selig, H. *et al., Science* (New York), 1964, **143**, 1322
    Decomposition may be explosive, even at −40°C.
    See other NON-METAL OXIDES, OXIDANTS, XENON COMPOUNDS

4864. Phosphorus(V) oxide
    [1314-56-3]   O₅P₂

$$P_2O_5$$

    See Tetraphosphorus decaoxide

4865. Tantalum(V) oxide
    [1314-61-0]   O₅Ta₂

$$O_2TaOTaO_2$$

Chlorine trifluoride
    See Chlorine trifluoride: Metals, etc.
    See other METAL OXIDES

4866. Vanadium(V) oxide
    [1314-62-1]   O₅V₂

$$O_2VOVO_2$$

    HCS 1980, 950

Calcium, Sulfur
    491M, 1975, 81
    A mixture of the metal and oxidant, contaminated with sulfur and some water, led
    to a serious fire.

Chlorine trifluoride
    See Chlorine trifluoride: Metals, etc.

Lithium   MRH 2.43/7
    See Lithium: Metal oxides

Other reactants
    Yoshida, 1980, 118
    MRH values calculated for 9 combinations, largely with oxidisable materials, are
    given.

Peroxyformic acid   MRH 5.69/99+
    See Peroxyformic acid: Metals, etc.
    See other METAL OXIDES, OXIDANTS

4867. Tetraphosphorus hexaoxide (Phosphorus(III) oxide)
   [10248-58-5] O_6P_4

Ammonia
    Thorpe, T. E. *et al., J. Chem. Soc.*, 1891, **59**, 1019
    Interaction of the molten oxide and ammonia under nitrogen is rather violent and
    the mixture ignites. Use of a solvent and cooling controls the reaction to produce
    diamidophosphinic acid.

Disulfur dichloride
    Mellor, 1940, Vol. 8, 898
    Interaction is very violent.

Halogens
    1. Thorpe, T. E. *et al., J. Chem. Soc.*, 1891, **59**, 1019
    2. Mellor, 1940, Vol. 8, 897
    The oxide ignites in contact with excess chlorine gas [1], and reacts violently,
    usually igniting, with liquid bromine [2].

Organic liquids
    1. Thorpe, T. E. *et al., J. Chem. Soc.*, 1890, **57**, 569–573
    2. Mellor, 1971, Vol. 8, Suppl. 3, 382
    The oxide ignites immediately with ethanol at ambient temperature [1]. The
    liquid oxide (above 24°C) reacts very violently with methanol, dimethylformamide,
    dimethyl sulfite or dimethyl sulfoxide (also with arsenic trifluoride) and charring
    may occur [2].

Oxygen
    Mellor, 1940, Vol. 8, 897–898; 1971, Vol. 8, Suppl. 3, 380–381
    Interaction with air or oxygen is rapid and, at slightly elevated temperatures in air
    or at high concentration of oxygen, ignition is very probable, particularly if the
    oxide is molten (above 22°C) or distributed as a thin layer. The solid in contact
    with oxygen at 50–60°C ignites and burns very brilliantly. During distillation of
    phosphorus(III) oxide, air must be rigorously excluded to avoid the possibility of

explosion. The later reference states that the carefully purified oxide does not ignite in oxygen, and that the earlier observations were on material containing traces of white phosphorus.

Phosphorus pentachloride
See Phosphorus pentachloride: Phosphorus(III) oxide

Sulfur
1. Thorpe, T. E. *et al., J. Chem. Soc.*, 1891, **59**, 1019
2. Pernert, J. C. *et al., Chem. Eng. News*, 1949, **27**, 2143
Interaction of a mixture of sulfur and the oxide under inert atmosphere above 160°C to form tetraphosphorus hexaoxide tetrasulfide is violent and dangerous on scales of working other than small [1]. A safer procedure involving distillation of phosphorus(V) oxide and phosphorus(V) sulfide is described [2].

Sulfuric acid
Mellor, 1940, Vol. 8, 898
Addition of sulfuric acid to the oxide causes violent oxidation, and ignition if more than 1–2 g is used.

Water
Mellor, 1940, Vol. 8, 897
Reaction with cold water is slow, but with hot water, violent, the evolved phosphine igniting. With more than 2 g of oxide, violent explosions occur.
See other NON-METAL OXIDES, REDUCANTS

4868. Tetraphosphorus hexaoxide–bis(borane)
[14940-94-4] $O_6P_4.B_2H_6$

$$P_4O_6.2BH_3$$

Water
Mellor, 1971, Vol. 8, Suppl. 3, 382
The compound ignites in contact with a little water.
See related BORANES, NON-METAL OXIDES

4869. Tetraphosphorus hexaoxide tetrasulfide
[15780-31-1] $O_6P_4S_4$

Preparative hazard
See Tetraphosphorus hexaoxide: Sulfur
See related NON-METAL OXIDES, NON-METAL SULFIDES

4870. Disulfur heptaoxide
    [12065-85-9] O_7S_2

$$(OSO_2OOSO_2)_n$$

Meyer, F. *et al., Ber.*, 1922, **55**, 2923
The crystalline material ('peroxydisulfuric anhydride', of unknown structure) soon
explodes on exposure to moist air.
See other ACID ANHYDRIDES, NON-METAL OXIDES

4871. Triuranium octaoxide
    [1344-59-8] O_8U_3

$$U_3O_8$$

Barium oxide
    Mellor, 1942, Vol. 12, 49
    Interaction below 300°C is vigorous and very exothermic.
    See other METAL OXIDES

4872. Tetraphosphorus decaoxide (Phosphorus(V) oxide)
    [16752-60-6] $O_{10}P_4$

(*MCA SD-28*, 1948); *HCS 1980*, 747

Barium sulfide
    See Barium sulfide: Phosphorus(V) oxide

Formic acid   MRH 0.29/99+
    See Formic acid: Phosphorus pentaoxide

Hydrogen fluoride
    Gore, G., *J. Chem Soc.*, 1869, **22**, 368
    Interaction is vigorous below 20°C.

Inorganic bases
    1. Mellor, 1940, Vol. 8, 945
    2. Van Wazer, 1958, Vol. 1, 279
    Dry mixtures with calcium or sodium oxides do not react in the cold, but interact
    violently if warmed or moistened, evolving phosphorus(V) oxide vapour [1]. A

mixture of the oxide and sodium carbonate may be initiated by strong local heating, when the whole mass will suddenly become hot [2].

Iodides
Pascal, 1960, Vol. 16.1, 538
Interaction is violent

Metals   MRH Aluminium 4.89/39, magnesium 5.48/45
Mellor, 1940, Vol. 8, 945
Interaction with warm sodium or potassium is incandescent and explosive with heated calcium.

Methyl hydroperoxide
See Methyl hydroperoxide: Alone, etc.

Other reactants
Yoshida, 1980, 119
MRH values calculated for 5 combinations with other materials are given.

Oxidants
See Bromine pentafluoride: Acids, etc.
Chlorine trifluoride: Metals, etc.
Hydrogen peroxide: Phosphorus(V) oxide
Oxygen difluoride: Phosphorus(V) oxide
Perchloric acid: Dehydrating agents

3-Propynol
See 3-Propynol: Phosphorus pentaoxide

Water
1. Mellor, 1940, Vol. 8, 944
2. *MCA SD-28*, 1948
Phosphorus(V) oxide is a very powerful desiccant and its interaction with liquid water is very energetic and highly exothermic [1]. The increase in temperature may be enough to ignite combustible materials if present and in contact [2].
See other ACID ANHYDRIDES, NON-METAL OXIDES

4873. Osmium
[7440-04-2]   **Os**
                                    Os

Chlorine trifluoride
See Chlorine trifluoride: Metals

Fluorine
See Fluorine: Metals

Phosphorus
See Phosphorus: Metals
See other METALS

†4874. Phosphorus
[7723-14-0]   P

(*MCA SD-16*, 1947); *FPA H19*, 1973; *HCS 1980*, 748 (red), 750 (white); *RSC Lab. Hazard Data Sheet No. 79*, 1989 (white P)

1. Albright and Wilson, (Manufacturer's safety sheet)
2. Anon., *Chem. Trade J.*, 1936, **98**, 522
3. Anon., *CISHC Chem. Safety Summ.*, 1974–1975, **45–46**, 3
4. Anon., *Jahresber.*, 1979, 73
5. Anon., *Fire Prev.*, 1987, (197), 42–43
6. Anon., *Fire Prevention*, 1992, (249), 45

White phosphorus has an autoignition temperature only slightly above ambient, dispersed it will soon heat itself to that by the slow oxidation responsible for its glow. Red is not spontaneously combustible, however if it does catch fire white will be produced, so that the fire, once extinguished, may spontaneously re-ignite. Both can produce phosphine, among other products, by slow reaction with water. Sealed containers of damp phosphorus (white is often stored under water) may pressurise with highly toxic, pyrophoric, gas mixtures [1].

Red phosphorus powder being charged into a reaction vessel exploded when the lid of the tin fell in and was struck by the agitator [2]. A further case of ignition of red phosphorus by impact from its metal container is reported [3]. Impact of a rubber hammer on a nitrogen-purged stirred metal feed container of red phosphorus powder led to ignition and a jet of flame [4]. A rail tanker containing 44 kl of molten white phosphorus at about 50°C was damaged in a derailment, and the contents on exposure to ambient air immediately ignited, and burned for 5 days before being brought under control [5]. A very similar rail tankwagon fire occurred a few years later [6].

See Oxygen, below
See other FRICTIONAL INITIATION INCIDENTS

Alkalies
Mellor, 1940, Vol. 8, 802
Contact of phosphorus with boiling caustic alkalies or with hot calcium hydroxide evolves phosphine, which usually ignites in air.

Chlorosulfuric acid   MRH 2.51/75
Heumann, K. *et al.*, *Ber.*, 1882, **15**, 417

1884

White phosphorus begins to reduce the acid at 25–30°C and the vigorous reaction accelerates to explosion. Red phosphorus is similar at a higher initial temperature.

Cyanogen iodide
Mellor, 1940, Vol. 8, 791
Molten phosphorus reacts incandescently.

Halogen azides
See HALOGEN AZIDES: metals, or phosphorus

Halogen oxides   MRH values show % of oxidant
See Chlorine dioxide: Non-metals   MRH 7.87/73
      Dichlorine oxide: Oxidisable materials   MRH 0.92/92
      Oxygen difluoride: Non-metals
      Trioxygen difluoride: Various materials

Halogens, or Interhalogens   MRH Chlorine 1.92/85
1. Mellor, 1940, Vol. 8, 785; 1956, Vol. 2, Suppl. 1, 379
2. Christomanos, A. C., *Z. Anorg. Chem.*, 1904, **41**, 279
3. Kuhn, R. *et al., Helv. Chim. Acta*, 1928, **11**, 107
4. Newkome, G. R. *et al., J. Chem. Soc., Chem. Comm.*, 1975, 885–886
Both yellow and red phosphorus ignite on contact with fluorine and chlorine; red ignites in liquid bromine or in a heptane solution of chlorine at 0°C. Yellow phosphorus explodes in liquid bromine or chlorine, and ignites in contact with bromine vapour or solid iodine [1]. Interaction of bromine and white phosphorus in carbon disulfide gives a slimy by-product which explodes violently on heating [2]. Interaction of phosphorus and iodine in carbon disulfide is rather rapid [3]. A less hazardous preparation of diphosphorus tetraiodide from phosphorus trichloride and potassium iodide in ether is recommended [4].
See Bromine pentafluoride: Acids, etc.
      Bromine trifluoride: Halogens, etc.
      Chlorine trifluoride: Metals, etc.
      Iodine pentafluoride: Metals, etc.
      Iodine trichloride: Phosphorus

Hexalithium disilicide
Mellor, 1940, Vol. 6, 169
Interaction is incandescent.

Hydriodic acid
1. Blau, K., private comm., 1965
2. Agranat, I. *et al., J. Chem. Soc., Perkin Trans. 1*, 1974, 1159
3. Villain, F. J. *et al., J. Med. Chem.*, 1964, **7**, 457, footnote 9
During removal of free iodine from hydriodic acid by distillation from red phosphorus, phosphine was produced. When air was admixed by changing the receiver flask, an explosion occurred. Omission of distillation by boiling the reactants in inert atmosphere, and separating the phosphorus by fitration through a sintered

glass funnel containing solid carbon dioxide gave a colourless product [1]. Potential hazards incidental to the use of red phosphorus and hydriodic acid as a reducing agent for organic carbonyl compounds have been noted. Accumulation in a reflux condenser of phosphonium iodide which will react violently with water [2], and 2 unexplained explosions [3], possibly owing to the same cause, have been reported. *See* Phosphonium iodide

Hydrogen peroxide   MRH 7.78/73
   See Hydrogen peroxide: Phosphorus

Magnesium perchlorate
   1965 Summary of Serious Accidents, Washington, USAEC, 1966
   Phosphorus and magnesium perchlorate exploded violently while being mixed. It seems likely that this was initiated by interaction of phosphorus and traces of perchloric acid to form magnesium phosphate. The acid may have been present as impurity in the magnesium perchlorate, or might have been formed from traces of phosphoric acid in the phosphorus. Once formed, the perchloric acid would be rendered anhydrous by the powerful dehydrating action of magnesium perchlorate.

Metal acetylides
   Mellor, 1946, Vol. 5, 848–849
   Monorubium acetylide and monocaesium acetylide incandesce with warm phosphorus. Lithium acetylide and sodium acetylide burn vigorously in phosphorus vapour, and the potassium, rubidium and caesium analogues should react with increasing violence.

Metal halogenates   MRH values below references
   Mellor, 1941, Vol. 2, 310; 1940, Vol. 8, 785–786
   MRH Barium chlorate 5.06/83, calcium chlorate 5.61/77, potassium chlorate 6.07/76, sodium bromate 4.98/80, sodium chlorate 7.32/75, zinc chlorate 6.11/76
   Dry finely divided mixtures of red (or white) phosphorus with chlorates, bromates or iodates of barium, calcium, magnesium, potassium, sodium or zinc will readily explode on initiation by friction, impact or heat. Fires have been caused by accidental contact in the pocket between the red phosphorus in the friction strip on safety-match boxes and potassium chlorate tablets. Addition of a little water to a mixture of white or red phosphorus and potassium iodate causes a violent or explosive reaction. Addition of a little of a solution of phosphorus in carbon disulfide to potassium chlorate causes an explosion when the solvent evaporates. The extreme danger of mixtures of red phosphorus (or sulfur) with chlorates was recognised in the UK some 50 years ago when unlicenced preparation of such mixtures was prohibited by Orders in Council.

Metal halides
   Mellor, 1939, Vol. 9, 467; 1943, Vol. 11, 234
   Phosphorus ignites in contact with antimony pentachloride and explodes with chromyl chloride in presence of moisture at ambient temperature.

Metal oxides or peroxides   MRH values below references
  Mellor, 1941, Vol. 2, 490; Vol. 7, 690; 1940, Vol. 8, 792; 1943, Vol. 11, 234
  MRH Chromium trioxide 3.68/84, lead peroxide 1.30/90, potassium peroxide
  5.90/81, sodium peroxide 2.30/86
    Red phosphorus reacts vigorously on heating with copper oxide or manganese
  dioxide; and on grinding with lead oxide, mercury oxide or silver oxide, ignition
  may occur. Red phosphorus ignites in contact with lead peroxide, potassium
  peroxide or sodium peroxide, while white phosphorus explodes, and also in contact
  with molten chromium trioxide at 200°C.

Metals
  1. Mellor, 1941, Vol. 7, 115; 1940, Vol. 8, 842, 847, 853; 1971, Vol. 8, Suppl. 3,
     228
  2. Van Wazer, 1958, Vol. 1, 159
  3. Mellor, 1942, Vol. 15, 696; 1937, Vol. 16, 160
  4. Iseler, G. W. *et al., Int. Conf. Indium Phosphide Relat. Mater.*, 1992, 266
  Reaction of beryllium, copper, manganese, thorium or zirconium is incandescent
  when heated with phosphorus [1] and that of cerium, lanthanum, neodymium and
  praseodymium is violent above 400°C [2]. Osmium incandesces in phosphorus
  vapour, and platinum burns vividly below red-heat [3]. Red phosphorus shows very
  variable vapour pressure between batches (not surprising, it is an indeterminate
  material). This leads to explosions when preparing indium phosphide by reactions
  involving fusion with phosphorus in a sealed tube [4].
  See Aluminium: Non-metals

Metal sulfates
  Berger, E., *Compt. rend.*, 1920, **170**, 1492
  Excess red phosphorus will burn admixed with barium or calcium sulfates if primed
  at a high temperature with potassium nitrate–calcium silicide mixture.
  See Potassium nitrate: Calcium silicide

Nitrates   MRH Ammonium nitrate 4.60/86, potassium nitrate 3.14/73,
                                        silver nitrate 3.89/82, sodium nitrate 4.43/73
  1. Mellor, 1941, Vol. 3, 470; 1940, Vol. 4, 987; Vol. 8, 788
  2. Anon., *Jahresber.*, 1975, 83–84
  Yellow phosphorus ignites in molten ammonium nitrate, and mixtures of phosphorus with ammonium nitrate, mercury(I) nitrate or silver nitrate explode on
  impact. Red phosphorus is oxidised vigorously when heated with potassium nitrate
  [1]. During development of new refining agents for aluminium manufacture, a
  mixture containing red phosphorus (16%) and sodium nitrate (35%) was being
  pressed into 400 g tablets. When the die pressure was increased to 70 bar, a violent
  explosion occurred [2].

Nitric acid   MRH 7.66/67
  See Nitric acid: Non-metals

Nitrogen halides
  See Nitrogen tribromide hexaammoniate: Initiators
        Nitrogen trichloride: Alone, or Non-metals   MRH 3.35/87

Nitrosyl fluoride
See Nitrosyl fluoride: Metals, etc.

Nitryl fluoride
See Nitryl fluoride: Non-metals

Non-metal halides
1. Pascal, 1960, Vol. 13.2, 1162
2. Mellor, 1946, Vol. 5, 136; 1940, Vol. 8, 787; 1947, Vol. 10, 906
Either the white or red form incandesces with boron triiodide [1]. Red phosphorus incandesces in seleninyl choride while white explodes. Red phosphorus reacts vigorously on warming with sulfuryl chloride or disulfuryl chloride, and violently with disulfur dibromide [2].

Non-metal oxides   MRH Dinitrogen tetraoxide 9.12/65
Mellor, 1940, Vol. 8, 786–787
Warm or molten white phosphorus burns vigorously in nitrogen oxide, dinitrogen tetraoxide or dinitrogen pentaoxide. White phosphorus ignites after some delay in contact with the vapour of sulfur trioxide, but immediately in contact with the liquid if a large portion is used.

Other reactants
Yoshida, 1980, 73–75
MRH values calculated for 36 combinations, largely with oxidants, are given.

Oxidants
Inoue, Y. *et al., Chem. Abs.*, 1988, **108**, 134435
Impact sensitivities of mixtures of red phosphorus with various oxidants were determined in a direct drop-ball method, which indicated higher sensitivities than those determined with an indirect striker mechanism. Mixtures with silver chlorate were most sensitive, those with bromates, chlorates and chlorites were extremely sensitive, and mixtures with sodium peroxide and potassium superoxide were more sensitive than those with barium, calcium, magnesium, strontium or zinc peroxides. Mixtures with perchlorates or iodates had sensitivities comparable to those of unmixed explosives, such as lead azide, 3,5-dinitrobenzenediazonium-2-oxide etc.

Oxygen
1. Mellor, 1940, Vol. 8, 771
2. Van Wazer, 1958, 1, 97
3. Mellor, 1971, Vol. 8, Suppl. 3, 237–245
The reactivity of phosphorus with air or oxygen depends on the allotrope of phosphorus involved and the conditions of contact, white (yellow) phosphorus being by far the more reactive. White phosphorus readily ignites in air if warmed, finely divided (e.g. from an evaporating solution) or under conditions where the slow oxidative exotherm cannot be dissipated. Contact with finely divided charcoal or lampblack promotes ignition, probably by adsorbed oxygen [1]. Contact with amalgamated aluminium also promotes ignition [2]. Oxidation of phosphorus under both explosive and non-explosive conditions has been reviewed in detail [3].

Peroxyformic acid   MRH 6.65/72
    See Peroxyformic acid: Non-metals

Potassium nitride
    Mellor, 1940, Vol. 8, 99
    Potassium and other alkali-metal nitrides react on heating with phosphorus to give
    a highly flammable mixture which evolves ammonia and phosphine with water.

Potassium permanganate   MRH 2.34/84
    Mellor, 1942, Vol. 12, 322
    Grinding mixtures of phosphorus and potassium permanganate causes explosion,
    more violent if also heated.

Selenium
    See Selenium: Phosphorus

Sodium chlorite   MRH 6.23/79
    See Sodium chlorite: Phosphorus

Sulfur
    See Sulfur: Non-metals

Sulfuric acid   MRH 2.64/72
    See Sulfuric acid: Phosphorus
    See other NON-METALS, PYROPHORIC MATERIALS

4875. Phosphorus(V) sulfide
[1314-80-3] P_2S_5
$$P_2S_5$$

    See Tetraphosphorus decasulfide

4876. Zinc phosphide
[1314-85-7] P_2Zn_3
$$Zn_3P_2$$

Perchloric acid
    See Perchloric acid: Zinc phosphide

Water
    Fehse, W., *Feuerschutz*, 1938, **18**, 17
    Contact of one drop of water with a zinc phosphide rodenticide preparation caused
    ignition.
    See other METAL NON-METALLIDES

4877. Tetraphosphorus trisulfide
[1314-85-8]   P₄S₃

Gibson, 1969, 118–119
It ignites in air above 100°C.

Other reactants
Yoshida, 1980, 131
MRH values calculated for 16 combinations, largely with oxidants, are given.
See other NON-METAL SULFIDES

4878. Tetraphosphorus decasulfide (Phosphorus(V) sulfide)
[15857-57-5]   P₄S₁₀

1. (*MCA SD-71*, 1958)
2. Grossel, S. S., *J. Loss Prev. Proc. Ind.*, 1988, **1**, 62–74
3. Albright and Wilson, (Manufacturer's safety sheet)
The general reactivity of the sulfide depends markedly on the physical form, and differences of a factor of 10 may be involved. It is ignitable by friction, sparks or flames, and ignites in dry air if heated close to the m.p., 275–280°C. The dust (200 mesh) forms explosive mixtures in air above a concentration of 0.5% w/v [1], and maximum explosion pressures of 4.35 bar, with maximum rate of rise exceeding 680 bar/s have been determined [2]. The dust can acquire sufficient static electricity from movement for ignition to occur [3].

Alcohols
1. *MCA Case History No. 227*
2. Ashford, J. S., private comm., 1964
A mixture of the sulfide, ethylene glycol and hexane in a mantle-heated flask spontaneously overheated and exploded at an internal temperature of around 180°C. It had been intended to maintain the reaction temperature at 60°C, but since alcoholysis of the sulfide is exothermic, presence of the heating mantle prevented the dissipation of heat, and the reaction accelerated continuously until explosive decomposition occurred [1]. An incident in similar circumstances involving interaction of the sulfide with 4-methyl-2-pentanol also led to violent eruption of the flask contents.

1890

Diethyl maleate, Methanol

Nolan, 1983, Case history 152

In manufacture of Malathion, the phosphorus sulfide dissolved in toluene is treated first with methanol, then diethyl maleate. Use of insufficient toluene to dissolve the sulfide led to a runaway exothermic reaction and explosion.

See Alcohols, above

Water

Haz. Chem. Data, 1975, 239

It heats and may ignite in contact with limited amounts of water, hydrogen sulfide also being evolved.

See other NON-METAL SULFIDES

4879. Tetraphosphorus decasulfide . pyridine complex
[71257-31-3] $P_4S_{10}.4C_5H_5N$

$$P_4S_{10}.4Py$$

Riedel-de Haën catalogue, 1986

Although convenient to handle, the complex is classed as highly flammable.

See related NON-METAL SULFIDES

4880. Tetraphosphorus triselenide
[1314-86-9] P_4Se_3

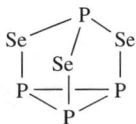

1. Mellor, 1947, Vol. 10, 790
2. Stoppioni, P. *et al., Gazz. Chim. Ital.*, 1988, **118**, 581–582

It ignites when heated in air [1]. A new and safer synthesis from red phosphorus and grey selenium has been described [2].

See related NON-METAL SULFIDES

4881. Lead pentaphosphide
[ ] P_5Pb

$$PbP_5$$

Bailar, 1973, Vol. 2, 117

It ignites in air

See other METAL NON-METALLIDES, PYROPHORIC MATERIALS

4882. Lead
[7439-92-1]   Pb

Pb

HCS 1980, 534 (powder)

Mézáros, L., *Tetrahedron Lett.*, 1967, 4951
The finely divided lead produced by reduction of the oxide with furfural vapour
at 290°C is pyrophoric and chemically reactive.

Hydrogen peroxide, Trioxane
    See Hydrogen peroxide: Lead, Trioxane

Nitric acid, Rubber
    See Nitric acid: Lead-containing rubber

Oxidants
    See Ammonium nitrate: Metals
        Chlorine trifluoride: Metals
        Hydrogen peroxide: Metals

Sodium acetylide
    See Sodium acetylide: Metals
    See other METALS, PYROPHORIC METALS

4883. Lead–tin alloys
[12333-97-0] (1:1 alloy)   Pb—Sn

Pb—Sn

Palm oil, Steel sheets
    Nedelcu, C. *et al.*, Rom. Pat. 68 602, 1979
    The explosion hazards associated with use of palm oil when coating steel sheets
    with molten alloys (8–12% tin) is eliminated by use of molten sodium acetate in
    place of the oil.
    See other ALLOYS

4884. Lead–zirconium alloys
[12412-94-1] (3:5 alloy)   Pb—Zr

Pb—Zr

Alexander, P. P., US Pat. 2 611 316. 1952
Alloys containing 10–70% of zirconium will pulverise and ignite on impact.
See other ALLOYS, PYROPHORIC ALLOYS

1892

4885. Palladium
[7440-05-3]   Pd

Pd

Finely divided palladium (palladium-black) used as a hydrogenation catalyst is usually pyrophoric and needs appropriate handling precautions.

Aluminium
See Aluminium: Catalytic metals

Arsenic
See Arsenic: Metals

Carbon
1. Mozingo, R., *Org. Synth.*, 1955, Coll. Vol. 3, 687
2. Rachlin, A. I. *et al., Org. Synth.*, 1971, **51**, 687
3. Fraser, R., private comm., 1973

Palladium-on-carbon catalysts prepared by formaldehyde reduction are less pyrophoric than those reduced with hydrogen [1]. Such catalysts become extremely pyrophoric on thorough vacuum drying [2]. Those prepared on high surface-area supports (up to 2000 m^2/g) are highly active and readily will cause catalytic ignition of hydrogen–air or solvent–air mixtures. Methanol is noted for easy ignition because of its high volatility [3].

Formic acid
See Formic acid: Palladium–carbon catalyst

Hydrogen, Hydrogen peroxide
See Hydrogen peroxide: Hydrogen, Palladium

Hydrogen, Poly(tetrafluoroethylene)
See Poly(tetrafluoroethylene): Metal hydrides

Ozonides
See OZONIDES (reference 4)

Sodium tetrahydroborate
Augustine, 1968, 75

In preparation for the reduction of a nitro compound, the tetrahydroborate solution is added to an aqueous supension of palladium-on-charcoal catalyst. The reversed addition of dry catalyst to the tetrahydroborate solution may cause ignition of liberated hydrogen.

Sulfur
See Sulfur: Metals
See other HYDROGENATION CATALYSTS, METALS, PYROPHORIC METALS

4886. Praseodymium
  [7440-10-0]   **Pr**

                                          Pr

Bromine
  See Bromine: Metals,
  See also LANTHANIDE METALS *See other* METALS

4887. Platinum
  [7440-06-4]   **Pt**
                                          Pt

  1. van Campen, M. G., *Chem. Eng. News*, 1954, **32**, 4698
  2. Tilford, C. H., private comm., 1965
  Finely divided (catalytic) forms of platinum are hazardous to handle if allowed to
  dry. Used Adams' hydrogenation catalyst exploded while being sieved in air. It
  had been well washed, finally with water, and then air dried [1]. Manipulation of
  used catalyst on filter paper caused a violent explosion [2]. The explosive catalysed
  oxidation of adsorbed hydrogen is suspected here.
  See Zinc: Catalytic metals

Aluminium
  See Aluminium: Catalytic metals

Acetone, Nitrosyl chloride
  See Nitrosyl chloride: Acetone, etc.

Arsenic
  Wöhler, L., *Z. Anorg. Chem.*, 1930, **186**, 324
  During interaction to give platinum diarsenide in a sealed tube at 270°C, the highly
  exothermic reaction may burst the tube.

Carbon, Methanol
  Unpublished information, 1982
  Ignition occurred when methanol was charged into an unpurged autoclave
  containing platinum-on-carbon catalyst.

Dioxygen difluoride
  See Dioxygen difluoride: Various materials

Ethanol
  Elkins, J. S., private comm., 1968
  Addition of platinum-black catalyst to ethanol caused ignition. Pre-reduction with
  hydrogen and/or nitrogen purging of air prevented this.

Hydrazine
  See Hydrazine: Metal catalysts

Hydrogen, Air
  Coleman, J. W., private comm., 1965

A platinised alumina catalyst had been treated with flowing hydrogen at ambient temperature. Purging with air caused an explosion. Inert gas purging is essential after treating catalysts or adsorbent with hydrogen before admitting air.

Hydrogen peroxide
See Hydrogen peroxide: Metals

Lithium
Nash, C. P., *Chem. Eng. News*, 1961, **39**(5), 42
Interaction to form an intermetallic compound sets in violently at 540°C±20°C.

Methyl hydroperoxide
See Methyl hydroperoxide: Alone, etc.

Ozonides
See OZONIDES (reference 4)

Peroxomonosulfuric acid
See Peroxomonosulfuric acid: Catalysts

Phosphorus
See Phosphorus: Metals

Selenium, or Tellurium
Mellor, 1937, Vol. 16, 158
Finely divided or spongy platinum reacts incandescently when heated with selenium or tellurium.

Vanadium dichloride, Water
See Vanadium dichloride: Platinum, Water
See other HYDROGENATION CATALYSTS, METALS, PYROPHORIC METALS

4888. Plutonium
[7440-07-5]   Pu

Pu

1. Bailar, 1973, Vol. 5, 44–45
2. Williamson, G. K., *Chem. & Ind.*, 1960, 1384
The reactive metal is pyrophoric in air at elevated temperatures; lathe turnings have ignited at 265°C and 140-mesh powder at 135°C [1]. This most extremely toxic metal may be worked under nitrogen, argon, or helium (but not carbon dioxide) at temperatures above 600°C [2].

Carbon tetrachloride
Serious Accid. Series, No. 246, Washington, USAEC, 1965

While draining after degreasing in the solvent, plutonium chips ignited, and when accidentally dropped into the solvent, caused an explosion.

See other METAL–HALOCARBON INCIDENTS

Water

MCA Case History No. 1212

Plutonium components, normally kept under argon, were accidentally exposed to air and moisture, probably forming plutonium hydride and hydrated oxides. When the plastics containing bag was disturbed, ignition occurred, causing widespread radiation contamination.

See Plutonium(III) hydride

See other METALS, PYROPHORIC METALS

4889. Rubidium
[7440-17-7]   **Rb**

Rb

Mellor, 1941, Vol. 2, 468; 1963, Vol. 2, Suppl. 2.2, 2172

Rubidium is a typical but very reactive member of the series of alkali metals.It is appreciably more reactive than potassium, but less so than caesium, and so would be expected to react more violently with those materials that are hazardous with potassium or sodium. Rubidium ignites on exposure to air or dry oxygen, largely forming the oxide.

Halogens

Mellor, 1941, Vol. 2, 13; 1963, Vol. 2, Suppl. 2.2, 2174

Ignition occurs in fluorine, chlorine and bromine or iodine vapours. Contact with liquid bromine would be expected to cause a violent explosion.

Mercury

Mellor, 1941, Vol. 2, 469

Interaction is exothermic and may be violent.

Non-metals

Mellor, 1963, Vol. 2, Suppl. 2.2, 2174–2175

The molten metal ignites in sulfur vapour, and reacts with various forms of carbon exothermically, one product, the interstitial rubidium octacarbide, being pyrophoric.

Vanadium trichloride oxide

Mellor, 1963, Vol. 2, Suppl. 2.2, 2176

Interaction is violent at 60°C.

Water

1. Mellor, 1941, Vol. 2, 469
2. Markowitz, M. M., *J. Chem. Educ.*, 1963, **40**, 633–636

Contact with cold water is exothermic enough to ignite the hydrogen evolved [1]. Reactivity of rubidium and other alkali metals with water has been discussed in detail [2].

See ALKALI METALS *See other* METALS, PYROPHORIC METALS

4890. Rhenium
[7440-15-5]

Re

Re

Oxygen (Gas)
See Oxygen (Gas): Rhenium
See other METALS

4891. Rhenium(VII) sulfide
[12038-67-4]

Re_2S_7

Re_2S_7

Sidgwick, 1950, 1299
It is readily oxidised in air, sometimes igniting.
See other METAL SULFIDES

4892. Rhodium
[7440-16-6]

Rh

Rh

Sidgwick, 1950, 1513
Metallic rhodium prepared by heating its compounds in hydrogen must be allowed to cool in an inert atmosphere to prevent catalytic ignition of the sorbed hydrogen on exposure to air.
See Zinc: Catalytic metals

Interhalogens
See Bromine pentafluoride: Acids, etc.
    Chlorine trifluoride: Metals
See other HYDROGENATION CATALYSTS, METALS, PYROPHORIC METALS

4893. Radon
[10043-92-2]

Rn

Rn

Water
John, G., *Health Phys.*, 1985, **49**, 977–979
A 22 year old glass ampoule containing an aqueous solution of radon was under considerable pressure when opened and ejected the top vigorously. This was attributed to formation of hydrogen by radiolysis of the solvent water.

See Americium trichloride
See other IRRADIATION DECOMPOSITION INCIDENTS

4894. Ruthenium
[7440-18-8]   **Ru**

Ru

1. Mellor, 1942, Vol. 15, 502
2. Ullmann, 1992, Vol. A21, 125/6
Dissolution of a zinc–ruthenium alloy in hydrochloric acid leaves an explosive residue of finely divided ruthenium [1]. More probably this is the hydride, which may decompose on slight stimulus, the evolved hydrogen probably igniting because of the catalytic activity of the metal. Ruthenium prepared from its compounds by borohydride reduction is especially dangerous in this respect [2].

Aqua regia, Potassium chlorate
See Potassium chlorate: Aqua regia, etc.
See other HYDROGENATION CATALYSTS, METALS

4895. Ruthenium(IV) sulfide
[12166-20-0]   **RuS₂**

RuS₂

Air, or Carbon dioxide
Mellor, 1942, Vol. 15, 541
The sulfide decomposes explosively if heated in air or air–nitrogen mixtures, and incandesces, then explodes if heated under carbon dioxide.
See other METAL SULFIDES

4896. Ruthenium salts
[22541-59-9] (2+)   **RuZ₂**
[22541-88-4] (3+)   **RuZ₃**

RuZ₂, etc.

Sodium tetrahydroborate
See Sodium tetrahydroborate: Ruthenium salts

†4897. Sulfur
[7704-34-9]   **S**

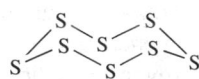

(*MCA SD-74*, 1959); *NSC 612*, 1979 (solid); *NSC 592*, 1978 (liquid); *FPA H14*, 1973; *HCS 1980*, 871 (liquid), 872 (solid)
NFPA 655, Quincy (Ma), Natl. Fire Prot. Assoc., 1988

1. *MCA SD-74*, 1959
2. Anon., *Loss Prev. Bull.*, 1987, (078), 23–25
3. Ayers, P. G., *Loss. Prev. Bull.*, 1988, (094), 34
4. Poole, G. A., *Loss. Prev. Bull.*, 1988, (084), 35

The new US National Fire Code covers the precautions necessary to prevent fires or explosions during crushing of the solid, or during handling of the molten element.

It may become ignited through frictional heat or incendive sparks, particularly when suspended as dust in air, for which the approximate explosive limits are 35–1400 g/m^3. Handling precautions for the solid and molten element are detailed. In the hazardous reactions listed below, the dual capacity of sulfur to oxidise or be oxidised is apparent [1]. Molten sulfur at '150–200°C' was being transported in a truck with two tank trailers each holding 12 t (57% of capacity). A traffic accident on a bridge caused both tanks to vault over the central barrier (thought to be caused by surging of the tank contents), and the impact caused tank rupture and ignition of the spilled hot sulfur. The fire led to dense clouds of sulfur dioxide and smoke which greatly hampered emergency services. The flash point of pure sulfur is 188°C, but impure sulfur may ignite at 168°C [2]. Later correspondents corrected the temperature limits for transportation of molten sulfur to 130–140° or 132–149°C [3,4]. Several values have been quoted in the literature for both flash point and autoignition temperature of molten sulfur, e.g. fl.p. 168, 188; a.i.t. 190, 232°C.
See entry HIGH SURFACE-AREA SOLIDS

Aluminium, Copper
See Aluminium: Copper, Sulfur

Aluminium, Starch
Biasutti, 1981, 138
Ignition of a dry mixture of the 3 powdered materials being mixed in a screw extruder led to a minor explosion, then a major dust explosion.

Calcium, Vanadium pentaoxide
See Vanadium(V) oxide: Calcium, etc.

Diethyl ether
Taylor, H. F., *J. R. Inst. Chem.*, 1955, **79**, 43
Evaporation of an ethereal extract of sulfur caused an explosion of great violence. Experiment showed that evaporation of wet, peroxidised ether gave a mildly explosive residue, which became violently explosive on addition of sulfur.

Fibreglass, Iron
Anon., *Info. Circ. No. 8272(10)*, Washington, US Bur. Mines, 1964

A mixture of sulfur with iron filings and fibreglass reacts exothermally at 125–145°C.
See Metals, below (reference 12)

Fluorine
See Fluorine: Non-metals

Halogen oxides　　　　　　MRH Chlorine dioxide 4.64/76, chlorine trioxide 7.57/60,
　　　　　　　　　　　　　　dichlorine oxide 3.01/89, trioxygen difluoride 6.82/67
Mellor, 1941, Vol. 2, 289, 241, 295; 1956, Vol. 2, Suppl. 1, 540, 542
Sulfur ignites in chlorine dioxide gas, and may cause an explosion, which usually happens on contact of sulfur with the very unstable dichlorine monoxide. Iodine(V) oxide reacts explosively on warming with sulfur. Chlorine trioxide would be expected to oxidise sulfur violently but dichlorine heptaoxide does not react.
See 'Trioxygen difluoride': Various materials

Interhalogens
See Bromine pentafluoride: Acids, etc.
　　　Bromine trifluoride: Halogens, etc.
　　　Chlorine trifluoride: Metals, etc.　　　　　　　　　　MRH 4.10/85
　　　Iodine pentafluoride: Metals, etc.

Limonene
Nolan, 1983, Case history 216
Addition of molten sulfur to limonene in a 9 kl reactor led to a violent runaway exothermic reaction. Small scale pilot runs had not shown the possibility of this. Heating terpenes strongly with sulfur usually leads to formation of benzene derivatives with evolution of hydrogen sulfide.
See other GAS EVOLUTION INCIDENTS

Metal acetylides or carbides
Mellor, 1946, Vol. 5, 849, 862, 886, 891
Monorubidium acetylide ignites in molten sulfur; barium carbide ignites in sulfur vapour at 150°C and incandesces; while calcium, strontium, barium and uranium carbides need a temperature around 500°C to ignite.

Metal halogenates
See METAL HALOGENATES: non-metals
See also METAL CHLORATES: phosphorus, etc.

Metal oxides　　　　　　　　　　　　　　MRH values show % of oxide
See Chromium trioxide: Sulfur　　　　　　　　　　MRH 1.76/86
　　　Lead(IV) oxide: Non-metals
　　　'Mercury(I) oxide': Non-metals
　　　Mercury(II) oxide: Non-metals
　　　Silver(I) oxide: Non-metals

1900

Sodium peroxide: Non-metals   MRH 3.35/71
Thallium(III) oxide: Sulfur

Metals   MRH Aluminium 4.81/36, calcium 6.69/56, magnesium 6.15/43,
                                    rubidium 1.76/84, sodium 4.81/59, zinc 2.09/67

1. Mellor, 1940, Vol. 4, 268, 480
2. Mellor, 1941, Vol. 3, 639
3. Mellor, 1946, Vol. 5, 210, 393
4. Mellor, 1942, Vol. 15, 149, 527, 627, 696
5. Mellor, 1941, Vol. 7, 208, 328
6. Mellor, 1942, Vol. 12, 31
7. Rothman, R., *Philad. Astronaut. Soc. Bull.*, 1953, **1**, 7
8. Sleight, A. W. *et al.*, *Inorg. Synth.*, 1973, **14**, 152–153
9. Revelli, J. F., *Inorg. Synth.*, 1979, **19**, 35
10. Kotoyori, T. *et al.*, *Chem. Abs.*, 1984, **100**, 197111
11. Partington, 1946, 552–853
12. Ephraim, 1939, 519
13. Chakurov, Kh. *et al.*, *J. Solid State Chem.*, 1987, **71**, 522–529

A mixture of sulfur with powdered zinc explodes on warming, and that with cadmium reacts rather less vigorously [1]. Mixtures of calcium and sulfur explode on ignition and calcium burns in sulfur vapour at 400°C [2]. Indium [3], palladium and rhodium [4], thorium and tin [5] all ignite and incandesce on heating admixed with sulfur. Uranium [6], osmium and nickel [4] (powders) ignite and incandesce in boiling sulfur or its vapours at 600°C, although more finely divided (catalytic) forms of nickel should react more readily. Magnesium will react when red-hot or molten, with molten sulfur or its vapour [1] and a mixture of aluminium powder and sulfur will burn violently or explode if ignited with a high temperature (magnesium) fuse [3]. Intimate mixtures of micro-grained zinc and sulfur were used as a rocket propellant [7]. In the reaction of gadolinium and sulfur in a heated quartz ampoule to prepare gadolinium sulfide, good temperature control is necessary to prevent violent interaction, with attack and possible rupture of the ampoule [8]. The same is true of the reaction of tantalum with sulfur [9]. Spontaneous combustion occurred in a copper powder–sulfur mixture used in manufacture of brake pads. Thermal analysis showed that the mixture ignited at 115°C. Preventive measures are discussed [10]. Iron filings burn when sprinkled into sulfur vapour, and a mixture with powdered sulfur attains incandescence when heated. Addition of water to the dry mixture will initiate an exothermic reaction [11]. Mercury, iron, magnesium or copper will react exothermally with sulfur at ambient temperature if the state of subdivision is fine enough [12]. In the preparation of the sulfides of cadmium, tin and zinc by 'mechanochemical reaction' (grinding?), the violent, often explosive reactions can be rendered non-explosive by the presence of a critical amount of inert material, and a mechanism is proposed [13].

See Aluminium, Copper; also Aluminium, Starch; and Fibreglass, Iron, all above
    Lithium: Sulfur
    Potassium: Sulfur

Rubidium: Non-metals
Sodium: Non-metals
See Selenium: Metals

Non-metals
1. Mellor, 1946, Vol. 5, 15
2. von Schwartz, 1918, 328
3. Mellor, 1940, Vol. 8. 786
4. Phillips, R., *Org. Synth.*, 1943, Coll. Vol. 2, 579

Boron reacts with sulfur at 600°C becoming incandescent [1]. Mixtures of sulfur with lamp black or freshly calcined charcoal ignite spontaneously, probably owing to adsorbed oxygen on the catalytic surface [2]. Mixtures of yellow phosphorus and sulfur ignite and/or explode on heating [3]. Ignition of an intimate mixture of red phosphorus and sulfur causes a violent exothermic reaction [4].

Other reactants
Yoshida, 1980, 37–38
MRH values calculated for 32 combinations, mainly with oxidants, are given.

Oxidants   MRH values show % of oxidant
1. Ivanov, V. G. *et al., Chem. Abs.*, 1980, **93**, 116708
2. Ivanov, V. G. *et al., Chem. Abs.*, 1981, **94**, 86592

The self-ignition of sulfur with potassium chlorate or iodine(V) oxide at 145–160°C, and with potassium perchlorate at 385°C was studied using DTA [1], and combustion characterisitics of the mixtures were determined [2].

See Halogen oxides, Interhalogens, Metal oxides, all above

Ammonium nitrate: Sulfur	MRH 3.01/88
Barium bromate: Sulfur	MRH 2.26/88
Chromyl chloride: Non-metals	
Fluorine: Non-metals	MRH 8.28/78
Heptasilver nitrate octaoxide: Alone, or Sulfides, etc.	
Lead(II) chlorite: Non-metals	
Lead chromate: Sulfur	
Phosphorus(III) oxide	
Potassium bromate: Non-metals	MRH 2.55/86
Potassium chlorate: Charcoal, etc., or: Non-metals	MRH 3.22/82
Potassium chlorite: Sulfur	
Potassium perchlorate: Sulfur	MRH 3.39/79
Potassium permanganate: Non-metals	MRH 1.63/90
Silver bromate: Sulfur compounds	MRH 2.00/88
Silver chlorite: Hydrochloric acid, etc.	
Silver nitrate: Non-metals	MRH 1.67/84
Sodium chlorate: Ammonium salts, etc.	MRH 4.27/80
Tetraphosphorus hexaoxide: Sulfur	

Potassium nitride
Mellor, 1940, Vol. 8, 99

Potassium nitride and other alkali metal nitrides react with sulfur to form a highly flammable mixture, which evolves ammonia and hydrogen sulfide in contact with water.
See Caesium nitride

Sodium hydride
Mellor, 1941, Vol. 2, 483
Interaction with sulfur vapour is vigorous.

Static discharges
Enstad, G., *A Reconsideration of the Concept of Minimum Ignition Energy*, paper at Euro. Fed. Chem. Engrs. Wkg. Party meeting, 16th March, 1975.
Sulfur dust has a very low minimum ignition energy (3 mJ), so particular care is necessary in formulating protective measures.

Tetraphenyllead
See Tetraphenyllead: Sulfur
See other NON-METALS, OXIDANTS

4898. Silicon monosulfide
[12504-41-5]   SSi
                      SiS

Water
Bailar, 1973, Vol. 1, 1355
It reacts rapidly with water and may ignite.
See other NON-METAL SULFIDES

4899. Samarium sulfide
[29678-92-0]   SSm
                      SmS

Gadolinium sulfide
Jayaraman, A. *et al., Phys. Rev. Lett.*, 1973, **31**, 700
Cooling of samples of crystalline samarium sulfide containing 22 atom% of gadolinium in the lattice to below −153°C causes a near-explosive rearrangement to a powder form. This was attributed to changes in the lattice constants corresponding to a volume increase of 7.5%.
See other METAL SULFIDES

4900. Tin(II) sulfide
[1314-95-0]   SSn
                      SnS

See entry REFRACTORY POWDERS

Oxidants
    See Chloric acid: Metal sulfides
        Dichlorine oxide: Oxidisable materials
    See also METAL HALOGENATES: metals, etc.
    See other METAL SULFIDES

4901. Strontium sulfide
[1314-96-1]
SSr

$$SrS$$

Lead dioxide
    See Lead(IV) oxide: Metal sulfides
    See other METAL SULFIDES

4902. Tin(IV) sulfide
[1315-01-1]
S_2Sn

$$SnS_2$$

Oxidants
    See Chloric acid: Metal sulfides
        Dichlorine oxide: Oxidisable materials
    See METAL HALOGENATES: metals, etc.
    See other METAL SULFIDES

4903. Tantalum(IV) sulfide
[12143-72-5]
S_2Ta

$$TaS_2$$

Preparative hazard
    See Sulfur: Metals (reference 9)
    See other METAL SULFIDES

4904. Titanium(IV) sulfide
[12039-13-3]
S_2Ti

$$TiS_2$$

Potassium nitrate
    See Potassium nitrate: Metal sulfides
    See other METAL SULFIDES

4905. Uranium(IV) sulfide
[12039-14-4]

US_2

S_2U

Nitric acid
See Nitric acid: Uranium disulfide
See other METAL SULFIDES

4906. Antimony trisulfide
[1345-04-6]

Sb_2S_3

S_3Sb_2

Other reactants
Yoshida, 1980, 130
MRH values calculated for 11 combinations with various materials are given.

Oxidants   MRH values show % of oxidant.
See Chloric acid: Metal sulfides
    Dichlorine oxide: Oxidisable materials   MRH 3.72/49
    Fluorine: Sulfides   MRH 6.40/56
    Heptasilver nitrate octaoxide: Alone, or Sulfides
    Lead(II) chlorite: Antimony sulfide
    Potassium chlorate: Metal sulfides   MRH 2.84/61
    Potassium nitrate: Metal sulfides   MRH 2.30/63
    Silver(I) oxide: Metal sulfides
    Thallium(III) oxide: Sulfur, etc.
See other METAL SULFIDES

4907. Antimony
[7440-36-0]

Sb

Sb

NSC 408, 1978 (metal and compounds)

Krebs, H. *et al., Z. Anorg. Chem.*, 1955, **282**, 177
Electrolysis of acidified, stirred antimony halide solutions at low current density
produces explosive antimony which contains substantial amounts of halogen.

Alkali nitrates
Mellor, 1947, Vol. 9, 382
Mixtures detonate on heating, forming antimonates.

Aluminium
See Aluminium: Antimony, etc.

Disulfur dibromide
See Disulfur dibromide: Metals

Halogens   MRH Chlorine 1.67/47, fluorine 5.10/32
    Mellor, 1947, Vol. 9, 379
    Antimony ignites in fluorine, chlorine and bromine. With iodine, ignition or explosion may occur if quantities are large enough.

Other reactants
    Yoshida, 1980, 34
    MRH values calculated for 9 combinations with oxidants are given.

Oxidants   MRH values show % of oxidant
    See Halogens, above
        Ammonium nitrate: Metals   MRH 2.30/62
        Bromine pentafluoride: Acids, etc.   MRH 2.76/46
        Bromine trifluoride: Halogens, etc.   MRH 2.34/53
        Chloric acid: Metals, etc.
        Chlorine trifluoride: Metals, etc.   MRH 3.47/36
        Dichlorine oxide: Oxidisable materials   MRH 2.64/41
        Iodine pentafluoride: Metals
        Iodine: Metals
        Nitrosyl fluoride: Metals
        Perchloric acid: Antimony(III) compounds
        Potassium dioxide: Metals
        Potassium permanganate: Antimony, etc.   MRH 0.25/86
        Seleninyl chloride: Antimony
        Sodium nitrate: Antimony
    See other METALS

4908. Selenium
    [7782-49-2]   **Se**

                                    Se

    NSC 578, 1982 (element and compounds)

Hexalithium disilicide
    See Hexalithium disilicide: Non-metals

Metal acetylides or carbides
    Mellor, 1946, Vol. 5, 862, 886
    Metal acetylides incandesce on heating in selenium vapour; barium acetylide at 150°C, calcium acetylide and strontium acetylide at 500°C; and thorium carbide at an unstated temperature.
    See Rubidium acetylide: Non-metals

Metal amides
    Pascal, 1960, Vol. 16.2, 1724
    Interaction of selenium with alkali metal amides and alkaline earth metal amides gives explosive products.

Metal chlorates

Mellor, 1956, Vol. 2, Suppl. 1, 583

A slightly moist mixture of selenium with any chlorate (except of alkali metals) becomes incandescent.

See METAL HALOGENATES: non-metals

Metals

1. Mellor, 1942, Vol. 15, 151
2. Mellor, 1947, Vol. 10, 766–777
3. Mellor, 1942, Vol. 12, 31
4. Mellor, 1940, Vol. 4, 480
5. Mellor, 1942, Vol. 16, 158
6. Reisman, A. *et al., J. Phys. Chem.*, 1963, **67**, 22
7. Bailar, 1973, Vol. 2, 79

Nickel and selenium interact with incandescence on gentle heating [1], as do also sodium and potassium, the latter mildly explosively [2]. Uranium [3] and zinc [4] also incandesce when their mixtures with selenium are heated, and platinum sponge incandesces vividly [5]. The particle size of cadmium and selenium must be below a critical size to prevent explosions during synthesis of cadmium selenide by heating the elements together. Similar considerations also apply to interaction of cadmium or zinc with sulfur, selenium or tellurium [6]. Interaction of powdered tin and selenium at 350°C is extremely exothermic [7].

Nitrogen trichloride

See Nitrogen trichloride: Initiators

Oxidants

See Metal chlorates, above; Oxygen, etc., below
    Barium peroxide: Selenium
    Bromine pentafluoride: Acids, etc.
    Chlorine trifluoride: Metals, etc.
    Chromium trioxide: Selenium
    Fluorine: Non-metals
    Potassium bromate: Non-metals
    Silver(I) oxide: Non-metals
    Sodium peroxide: Non-metals

Oxygen, Organic matter

1. Astin, S. *et al., J. Chem. Soc.*, 1933, 391
2. Watkins, C. R. *et al., Chem. Rev.*, 1945, **36**, 235

During attempted conversion of recovered selenium metal to the dioxide by heating in oxygen, a vigorous explosion occurred. This was attributed to selenium-catalysed oxidation of traces of organic impurities in the selenium. Oxidation of recovered selenium with nitric acid is rendered vigorous by presence of organic impurities, but is a safe procedure [1]. The later reference incorrectly attributed the explosions to the nitric acid procedure [2].

Phosphorus
Pascal, 1956, Vol. 10, 724
Interaction warm attains incandescence.
See related METALS, NON-METALS

4909. Silicon
[7440-21-3]
Si

Si

1. Eckhoff, R. K. *et al.*, *J. Electrochem. Soc.*, 1986, **133**, 2631–2637
2. Grossel, S. S., *J. Loss Prev. Proc. Ind.*, 1988, **1**, 62–74
3. See entry Silane (reference 4)

Finely powdered silicon can give significant dust explosion hazards. Relationships of sensitivity to spark ignition and of explosibility to particle size are studied [1]. Maximum explosion pressures of 6.4 bar, with maximum rate of rise of 884 bar/s have been determined [2]. Silicon dust is likely to result from processes using silanes in the gas phase [3].

Calcium
See Calcium: Silicon

Metal acetylides
See Caesium acetylide: Non-metals
Rubidium acetylide: Non-metals

Metal carbonates
Mellor, 1940, Vol. 6, 164
Amorphous or crystalline silicon both react exothermally when heated with alkali-metal carbonates, attaining incandescence and evolving carbon monoxide.

Metal hexafluorides
1. Paine, R. T. *et al.*, *Inorg. Chem.*, 1975, **14**, 1111
2. Paine, R. T. *et al.*, *Inorg. Synth.*, 1979, **19**, 138

During the reduction of iridium hexafluoride, osmium hexafluoride or rhenium hexafluoride with silicon to the pentafluorides, the hexafluorides must not be condensed directly onto undiluted silicon powder, or explosions may result [1]. The same is true for molybdenum hexafluoride and uranium hexafluoride [2].

Oxidants
See Metal hexafluorides, above
Bromine trifluoride: Halogens, etc.
Chlorine trifluoride: Metals, etc.
Chlorine: Non-metals
Cobalt trifluoride: Silicon
Fluorine: Non-metals
Iodine pentafluoride: Metals, etc.

Lead(II) oxide: Non-metals
Nitrosyl fluoride: Metals, etc.
Oxygen difluoride: Non-metals
Peroxyformic acid: Non-metals
Silver fluoride: Non-metals

Water
See Boron: Water
See related METALS, NON-METALS

4910. Silicon–zirconium alloys
    [12138-26-0] (1:1)   Si–Zr
    [12211-03-9] (1:2)

                                    Si–Zr

See Zirconium (reference 6)
See other ALLOYS

4911. Samarium
    [7440-31-5]   Sm

                                    Sm

1,1,2-Trichlorotrifluoroethane
    Anon., *NSC Newsletter, R & D Sect.*, 1970, (8)
    An attempt to mill a slurry of metal in the solvent caused a violent explosion.
    Experiment showed violent interaction of the freshly ground metal surfaces with
    the halocarbon solvent.
    See other METAL–HALOCARBON INCIDENTS *See other* METALS

4912. Tin
    [7440-31-5]   Sn

                                    Sn

HCS 1980, 901 (powder)

Carbon tetrachloride, Water
    Pascal, 1963, Vol. 8.3, 309
    In the presence of water vapour, interaction is violent
    See other METAL–HALOCARBON INCIDENTS

Disulfur dichloride
    See Disulfur dichloride: Tin

Other reactants
    Yoshida, 1980, 217
    MRH Values calculated for 9 combinations with oxidants are given.

Oxidants
MRH values show % of oxidant

 See Ammonium nitrate: Metals   MRH 2.92/58
 Chlorine: Metals   MRH 2.09/55
 Chlorine trifluoride: Metals
 Copper(II) nitrate: Tin   MRH 2.68/38
 Fluorine: Metals
 Iodine bromide: Metals
 Iodine heptafluoride: Metals
 Potassium dioxide: Metals
 Sodium peroxide: Metals   MRH 1.38/56
 Sulfur: Metals
 Tellurium: Tin

Water
1. Board, S. J. *et al., Int. J. Heat Mass Transf.*, 1974, **17**, 331–339
2. Board, S. J. *et al., Chem. Abs.*, 1975, **83**, 12950
3. Frost, D. L. *et al., Progr. Astronaut. Aeronaut.*, 1988, **114** (Dynam. Explos.), 451–473

Experiments involving explosions of molten tin and water are described [1], and the mechanism of propagation of the thermal explosions was studied [2]. The dynamics of explosive interaction between multiple drops of molten tin and water has been examined experimentally [3].
See MOLTEN METAL EXPLOSIONS *See other* METALS

4913. Strontium
[7440-25-7]   Sr

$$Sr$$

Sorbe, 1968, 154
The finely divided metal may ignite in air.

Halogens
Pascal, 1958, Vol. 4, 579
At 300°C strontium incandesces in chlorine, and at 400°C it ignites in bromine vapour.

Water
Pascal, 1958, Vol. 4, 579
Interaction is more vigorous than with calcium.
See other METALS, PYROPHORIC METALS

4914. Tantalum
[7440-25-7]   Ta

$$Ta$$

Although very unreactive in massive forms, the finely divided metal may be pyrophoric.
See entry PYROPHORIC METALS (references 3,9)

Bromine trifluoride
 See Bromine trifluoride: Halogens, etc.

Fluorine
 See Fluorine: Metals

Lead chromate
 See Lead chromate: Tantalum
 See other METALS

4915. Technetium
[7440-26-8]

Tc

Tc

Bailar, 1973, Vol. 3, 881
In finely divided forms (sponge or powder) it readily burns in air to the heptaoxide.
See other METALS, PYROPHORIC METALS

4916. Tellurium
[13494-80-9]

Te

Te

Halogens, or Interhalogens
 See Bromine pentafluoride: Acids, etc.
  Chlorine: Non-metals
  Chlorine fluoride: Tellurium
  Chlorine trifluoride: Metals
  Fluorine: Non-metals

Hexalithium disilcide
 See Hexalithium disilicide: Non-metals

Metals
 See Cadmium: Selenium, etc.
  Platinum: Selenium, etc.
  Potassium: Selenium, etc.
  Sodium: Non-metals

Silver iodate
 See Silver iodate: Tellurium

Tin
 Bailar, 1973, Vol. 2, 80
 Interaction hot to give tin telluride attains incandescence.
 See Selenium: Metals
 See related METALS, NON-METALS

4917. Thorium
    [7440-29-1]   **Th**
                                     Th

1. Mellor, 1941, Vol. 7, 207
2. Stout, E. L., *Chem. Eng. News*, 1958, **36**(8), 64
3. *Fire Record Bull. FR58-3*, Boston, NFPA, 1958

The massive metal ignites below red-heat when heated in air, and the powdered
metal ignites when rubbed or crushed [1], or on pouring as a stream in air, owing to
intergranular friction. Storage and handling precautions are discussed [2]. During
destruction by burning of radioactive thorium scrap, addition of a 4 cm lump of
the metal waste to burning thorium wafers led to a series of violent explosions.
This was attributed to presence of water or other impurities in the lump [3].
See THORIUM FURNACE RESIDUES

Carbon dioxide, Nitrogen
   See Carbon dioxide: Metals, Nitrogen

Halogens
   Mellor, 1941, Vol. 7, 207
   Thorium incandesces in chlorine, bromine or iodine, and would be expected to
   ignite readily in fluorine.

Nitryl fluoride
   See Nitryl fluoride: Metals

Oxygen
   Mellor, 1941, Vol. 7, 207
   The massive metal ignites on heating in oxygen.

Peroxyformic acid
   See Peroxyformic acid: Metals

Phosphorus
   See Phosphorus: Metals

Silver
   See Silver–thorium alloy

Sulfur
   See Sulfur: Metals
   See other METALS, PYROPHORIC METALS

4918. Thorium salts
[16065-92-2]   Th^{4+}

$$Th^{4+}$$

'Cupferron'
See Ammonium *N*-Nitrosophenylaminooxide: Thorium salts

4919. Titanium
[7440-32-6]   Ti

$$Ti$$

NSC 485, 1968; *FPA H87*, 1980

1. Mellor, 1941, Vol. 7, 19
2. Stout, E. L., *Chem. Eng. News*, 1958, **36**(8), 64
3. *NSC Data Sheet 485*, 1968
4. Lowry, R. N. *et al.*, *Inorg. Synth.*, 1967, **10**, 3
5. *Standard for Production, Processing, Handling and Storage of Titanium*, NFPA 481, Quincy (Ma), 1987
6. Alekseev, A. G. *et al.*, *Chem. Abs.*, 1975, **83**, 83527
7. Popov, E. I. *et al.*, *Chem. Abs.*, 1975, **83**, 150922
8. Elrod, C. W., *Chem. Abs.*, 1982, **97**, 184846
9. Poyarkov, V. G. *et al.*, *Chem. Abs.*, 1983, **98**, 37077
10. Grossel, S. S., *J. Loss Prev. Proc. Ind.*, 1988, **1**, 62–74
11. Prine, B. A., *Plant/Oper. Progr.*, 1992, **11**(2), 113
12. Mahnken, G. E. *et al.*, *Process Safety Progress*, 1997, **16**(1), 54

While massive titanium ignites in air only at 1200°C [1], the finely divided metal is very pyrophoric and should be stored damp in metal containers [2]. Titanium fires are difficult to extinguish as it burns brilliantly in nitrogen above 800°C, in carbon dioxide above 550°C [3], and may react explosively with metal carbonates [1]. Residual titanium sponge after vapour phase reaction with iodine at 400–425°C is often pyrophoric [4]. All aspects of preventing hazards in use of titanium are covered in a US Standard [5]. Powdered metal made by calcium hydride reduction shows a greater explosion hazard than electrolytically reduced metal. The effect of inert gases was also examined [6]. The relationship between spontaneous ignition temperatures (250–600°C) and lower explosive limits (40–300 mg/l) of titanium powders was studied [7]. Incidents involving combustion of titanium components in aviation gas turbines are reviewed [8]. Ignition parameters for titanium powder of <100 μ particle size were determined. Ignition temperature decreases from 640 to 508°C with increase in specific surface area from 960 to 1520 cm^2/g [9]. Maximum explosion pressures of 5.8 bar, with maximum rates of rise of 816 bar/s have been determined [10]. Titanium combustion as a result of welding operations appears to be potentiated by carbon steel, there are several recorded cases, mechanism not established [11]. Another fire when cutting into a steel-clad Ti heat-exchanger, with explosions when sprayed with water is reported [12].

Aluminium
See Aluminium–titanium alloys

Bromine trifluoride
See Bromine trifluoride: Halogens, etc.

Carbon black
Afanas'eva, L. F. *et al., Chem. Abs.*, 1981, **95**, 64482
Fire hazards and pyrophoricity of mixed charges for preparing titanium carbide are discussed.

Carbon dioxide, or Metal carbonates
1. Stout, E. L., *Chem. Eng. News*, 1958, **36**(8), 64
2. Mellor, 1941, Vol. 7, 20
Titanium powder has ignited as a thin layer under carbon dioxide, and burns in the gas above 550°C [1]. Contact of titanium with fused alkali-metal carbonates causes incandescence [2].

Carbon dioxide, Nitrogen
See Carbon dioxide: Metals, or: Metals, Nitrogen

Halocarbons
Pot. Incid. Rep. 39, ASESB, Washington, 1968
Mixtures of powdered titanium and trichloroethylene or 1,1,2-trichlorotrifluoro-ethane flash or spark under heavy impact.
See other METAL–HALOCARBON INCIDENTS

Halogens
1. Mellor, 1941, Vol. 7, 192
2. Kirk-Othmer, 1969, Vol. 20, 372
3. Denes, G., *Chem. Eng. News*, 1987, **65**(7), 2
4. Fast, J. D., *Rec. Trav. Chim.*, 1939, **58**, 174
5. Cobb, C. M., *Chem. Eng. News*, 1987, **65**(14), 2
6. Sisler, H. H., *Chem. Abs.*, 1988, **108**, 15143
Reaction of heated titanium with the halogens usually causes ignition and incandescence (chlorine at 350°C, bromine at 360°C, iodine higher, but weak combustion) [1], especially in absence of moisture [2]. Attempts to prepare titanium tetraiodide (5 g) by heating titanium powder with iodine in sealed copper tubes heated in a water bath led to explosions when the bath temperature reached 90°C. Rupture of the tubes also occurred with 0.5 g preparations [3]. Attempts to prepare titanium diiodide according to an earlier high-vacuum procedure [4] also led to violent explosions after two days [5]. A procedure to prepare the tetraiodide by heating a mixture of titanium and excess iodine in a Pyrex glass apparatus is reported [6].
See other HALOGENATION INCIDENTS
See Chlorine: Metals (references 2,3), *also* Fluorine: Metals

Metal oxides   MRH Chromium trioxide 4.85/73
    Mellor, 1961, Vol. 7, 20
    Interaction with copper(II) oxide or lead(II,IV) oxide is violent.

Metal oxosalts
    Mellor, 1941, Vol. 7, 20
    Contact of powdered titanium with molten potassium chlorate, alkali-metal carbonates or mixed potassium carbonate/nitrate causes vivid incandescence.

Nitric acid   MRH 8.79/51
    1. Anon., *Occ. Health and Safety*, 1957, **7**, 214
    2. Anon., *Chem. Eng. News*, 1953, **31**, 3320
    Investigation showed that commercial titanium alloys in contact with acid containing less than 1.34% water and more than 6% of dinitrogen tetraoxide may become sensitive to impact, and react explosively with the acid. Possible causes are discussed [1]. The spongy residue formed by prolonged corrosion of titanium–manganese alloys by red fuming nitric acid will explode on exposure to friction or heat [2].
    See other CORROSION INCIDENTS

Nitrogen
    Bailar, 1973, Vol. 3, 359
    Titanium is claimed to be the only element that will burn in nitrogen.
    See Lithium, Nitrogen: Magnesium

Nitryl fluoride
    See Nitryl fluoride: Metals

Other reactants
    Yoshida, 1980, 231
    MRH values for 13 combinations with oxidants are given.

Oxidants
    Mellor, 1941, Vol. 7, 20
    MRH Ammonium nitrate 5.48/77, hydrogen peroxide 8.74/59, potassium permanganate 3.97/63, sodium bromate 6.19/68, sodium chlorate 8.45/60, sodium chlorite 7.36/65, sodium nitrate 6.07/58, sodium perchlorate 8.37/57, sodium peroxide 3.51/76
    Mixtures of the powdered metal with potassium chlorate, potassium nitrate or potassium permanganate react explosively on heating. Addition of the powdered metal to fused potassium chlorate at 370°C leads to incandescence.
    See Metal oxides; Nitric acid, both above
    Iodine: Metals
    Potassium perchlorate: Metal powders (references 1,3)
    Sodium chlorate, below

Oxygen
    1. Kirk-Othmer, 1969, Vol. 20, 372
    2. Komamiya, K., *Chem. Abs.*, 1982, **96**, 90906

3. Shenoy, V. R., *Chem. Eng. News*, 1992, **70**(40), 4
4. Anon., *Sichere Chemiearbeit*, 1994, **46**(6), 67
Ignition may occur in liquid or gaseous oxygen and in gas mixtures containing over 46% oxygen, under impact or abrasion. Mixtures of powdered titanium and liquid oxygen are detonable [1]. In the absence of a surface film of grease, etc. on a freshly abraded surface, titanium will not burn under high pressure oxygen at high temperature [2]. Two explosions were experienced when studying treatment of ppm levels of organics in waste under 3-4 bar of oxygen at around 200°C. In each case the problem seems to have started at joints where abrasion, flakes of metal, grease and/or polymeric seals might have been present [3]. An inaccessible valve of partly titanium construction in a line filled with oxygen at 140°C (pressure unspecified) ignited for no apparent reason and burnt out. Titanium is not a suitable material of construction for oxygen lines, being even less protected by its oxide coating than is aluminium [4].
See Oxygen (Gas): Titanium, *and* Oxygen (Liquid): Metals, *also* Oxygen (Gas): Polytetrafluoroethylene, Stainless steel

Platinum group metals
Ullmann, 1992, Vol. A21, 125/6
Alloy formation between the powdered metals may be violent

Silver fluoride
See Silver fluoride: Titanium

Silver nitrate
See Silver nitrate: titanium

Sodium chlorate, Water   MRH 8.45/60
Scribner, H. C., *Tappi J.*, 1986, **69**(12), 16
A steam-heated titanum heat exchanger was used to heat sodium chlorate solution passing through its tubes, some of which were blocked by polypropylene film and wood chip debris. This led to local boiling and separation of solid sodium chlorate, which ignited the organic material and heated the titanium tube to its ignition point (427°C). It then started to burn, reacting with steam to liberate hydrogen which led to a violent explosion.
See other GAS EVOLUTION INCIDENTS

Water
1. Mellor, 1941, Vol. 7, 19
2. Anon., *Loss Prev. Bull.*, 1987, (075), 22
Finely divided titanium reacts with steam at 700°C, and in presence of air the evolved hydrogen may ignite or explode [1]. A heat exchanger with titanium tubes was being dismantled with flame cutting equipment, which caused ignition of the hot titanium in air. Application of water to extinguish the violent fire led to evolution of hydrogen and then an explosion [2].
See other METALS, PYROPHORIC METALS

4920. Titanium salts
[16043-45-1] (4+)
$$TiZ_4$$

TiZ₄ (right margin) — TiZ_4

'Cupferron'
See Ammonium *N*-nitrosophenylaminooxide: Thorium salts

4921. Titanium–zirconium alloys
[50646-37-2] (1:1)
$$Ti-Zr$$

Ti–Zr (right margin)

See Zirconium (reference 6)
See other ALLOYS

4922. Thallium
[7440-28-0]
$$Tl$$

Tl (right margin)

Fluorine
  See Fluorine: Metals
  See other METALS

4923. Uranium
[7440-61-1]
$$U$$

U (right margin)

1. *MCA Case History No. 1296*
2. Bailar, 1973, Vol. 5, 40
3. Rieke, R. D. *et al., J. Org. Chem.*, 1979, **44**, 3446
4. Starks, D. F. *et al., J. Amer. Chem. Soc.*, 1973, **95**, 3423

Storage of uranium foil in closed containers in presence of air and water may produce a pyrophoric surface [1]. Uranium must be machined in a fume hood because, apart from the radioactivity hazard, the swarf is easily ignited. The massive metal ignites at 600–700°C in air [2]. The finely divided reactive form of uranium produced by pyrolysis of the hydride is pyrophoric [3], while that produced as a slurry by reduction of uranium tetrachloride in dimethoxyethane by potassium–sodium alloy is not [4].

Ammonia
  Mellor, 1942, Vol. 12, 31
  At dull red heat the metal incandesces in ammonia.

Bromine trifluoride
  See Bromine trifluoride: Uranium, etc.

Carbon dioxide
 Bailar, 1973, Vol. 5, 42
 At 750°C interaction is so rapid that ignition will occur with the finely divided metal, and at 800°C the massive metal will ignite.

Carbon dioxide, Nitrogen
 See Carbon dioxide: Metals, Nitrogen

Carbon tetrachloride
 491M, 1975, 435
 Use of a carbon tetrachloride fire extinguisher on a small uranium fire led to an explosion.
 See other METAL–HALOCARBON INCIDENTS

Halogens
 Mellor, 1942, Vol. 12, 31
 Uranium powder ignites in fluorine at ambient temperature, in chlorine at 150–180°, in bromine vapour at 210–240° and in iodine vapour at 260°C.

Nitric acid, Trichloroethylene
 MCA Case History No. 1104
 Uranium scrap containing rather fine turnings was being pickled, after degreasing with trichloroethylene, by treatment with 4–6 M nitric acid solution. During subsequent rinsing with cold water, an explosion ejected the metal and acid liquor from the beaker. It has been reported in the literature that uranium reacts with nitric acid or dinitrogen tetraoxide either explosively, or with the formation of an explosive surface coating or residue. This corrosive effect is inhibited by presence of fluoride ion during pickling. Preventive measures proposed include separate treatment of finely divided uranium scrap, thorough pre-mixing of the acid solution and addition of ammonium hydrogen fluoride. An alternative, or contributory, cause could have been the violent reaction between dinitrogen tetraoxide and residual trichloroethylene.
 See Dinitrogen tetraoxide: Halocarbons
 See other CORROSION INCIDENTS

Nitrogen oxide
 See Nitrogen oxide: Metals

Nitryl fluoride
 See Nitryl fluoride: Metals

Selenium, or Sulfur
 Mellor, 1942, Vol. 12, 31
 Uranium incandesces in sulfur vapour, and with selenium also.

Water
 See Magnesium: Water, Uranium
 See other METALS, PYROPHORIC METALS

1918

4924. Vanadium
[7440-68-2]   V

V

Pascal, 1958, Vol. 12, 45
Some samples of vanadium prepared by magnesium reduction of vanadium dichloride or vanadium trichloride are pyrophoric.

Oxidants
   See Bromine trifluoride: Halogens, etc.
      Chlorine: Metals
      Nitryl fluoride: Metals
   See other METALS, PYROPHORIC METALS

4925. Tungsten
[7440-33-7]   W

W

Mellor, 1943, Vol. 11, 729
The finely divided metal may ignite on heating in air.

Oxidants
   1. Mellor, 1943, Vol. 11, 729
   2. Rose, J. E., *Chem. Abs.*, 1980, **92**, 8483
The finely divided metal may ignite in contact with a range of oxidants, usually on heating [1]. Mixtures of the metal powder with barium chromate and potassium chlorate are used as a pyrotechnic fuse [2].
   See Bromine: Tungsten, etc.
      Bromine pentafluoride: Acids, etc.
      Chlorine trifluoride: Metals
      Fluorine: Metals
      Hydrogen sulfide: Metals
      Iodine pentafluoride: Metals
      Lead(IV) oxide: Metals
      Nitryl fluoride: Metals
      Oxygen difluoride: Metals
      Potassium dichromate: Tungsten
      Potassium perchlorate: Barium chromate, etc.
      Sodium peroxide: Metals
   See other METALS, PYROPHORIC METALS

4926. Xenon
[7440-63-3]   Xe

Xe

Sponsles, M. *et al., J. Amer. Chem. Soc.*, 1989, **111**(17), 6841

1919

Liquid xenon, used as a solvent, will burst sealed apparatus if allowed to warm up. Pressure relief valves are recommended.

Fluorine
See Fluorine: Xenon
See other NON-METALS

4927. Zinc
  [7440-66-6]   **Zn**

                                   Zn

NSC 267, 1978; *FPA H96*, 1980; *HCS 1980*, 966 (dust); *RSC Lab. Hazard Data Sheet No. 76*, 1988 (zinc and compounds)

1. Anon., *Chem. Ztg.*, 1940, 289
2. Anon., *Ind. Eng. Chem.*, 1944, **20**, 330
3. Robson, S., *School Sci. Rev.*, 1934, **16**(62), 165
4. Anon., *Jahresber.*, 1978, 72
A cloud of zinc dust generated by sieving the hot dried material exploded violently, apparently after initiation by a spark from the percussive sieve-shaking mechanism [1]. Precautions recommended include use of cold zinc and total enclosure of such processes [2]. The possibility of explosions of zinc dust suspended in air is presented as a serious hazard [3]. A serious dust explosion in the air filter unit of a zinc grinding mill was initiated by a spark from an explosion relief panel [4].
See Zinc chloride, below

Acetic acid
Wehrli, S., *Chem. Fabrik*, 1940, 362–363
The residues from zinc dust–acetic acid reduction operations may ignite after a long delay if discarded into waste-bins with paper. Small amounts appear to ignite more rapidly than larger portions.
See other PYROPHORIC MATERIALS

Aluminium
See Aluminium: Zinc

Aluminium, Magnesium, Rusted steel
See Aluminium–magnesium–zinc alloy: Rusted steel

Ammonium nitrate,   MRH 3.22/5

Ammonium chloride, Water
1. Sorbe, 1968, 158
2. Hanson, R. M., *J. Chem. Educ.*, 1976, **53**, 578
A mixture of zinc dust with ammonium nitrate [1], or mixed nitrate and chloride [2] ignites when moistened.
See Ammonium nitrate: Metals

Ammonium sulfide
  See Ammonium sulfide: Zinc

Arsenic alloy, Water
  Carson, P. A. *et al., Loss Prev. Bull.* 1991, **102**, 7
  An arsenic alloy in a galvanised container with water generated the extremely toxic gas arsine.
  See Aluminium: Arsenic trioxide, Sodium arsenate, Sodium hydroxide
  See Water, below

Arsenic trioxide
  Mellor, 1940, Vol. 4, 486
  A mixture with excess zinc filings explodes on heating

Calcium chloride
  See Calcium chloride: Zinc

Carbon disulfide   MRH 2.43/37
  Mellor, 1940, Vol. 4, 4
  Carbon disulfide and zinc powder interact with incandescence.

Carbon tetrachloride,   MRH 2.43/54

Methanol
  See Methanol: Carbon tetrachloride, Metals

Chlorinated rubber
  See Halocarbons (reference 2), below

Catalytic metals, Acids
  Sidgwick, 1950, 1530, 2578
  Alloys of zinc with iridium, platinum or rhodium, after extraction with acid, leave residues which explode on warming in air, owing to the presence of occluded hydrogen (or oxygen) in the catalytic metal powders so produced.

Electrolytes, Silver
  See Silver: Electrolytes, etc.

Ethyl acetoacetate, Tris(bromomethyl)ethanol
  See Ethyl acetoacetate: 2,2,2-Tris(bromomethyl)ethanol, Zinc

Halocarbons   MRH Carbon tetrachloride 2.43/54
  1. Berger, E., *Compt. rend.*, 1920, **170**, 29
  2. Anon., *ABCM Quart. Safety Summ.*, 1963, **34**, 12
  3. Lamouroux, A. *et al., Mém. Poudres*, 1957, **39**, 435–445
  A paste of zinc powder and carbon tetrachloride (with kieselguhr as thickener) will readily burn after ignition by a high-temperature primer [1]. Intimate mixtures of

chlorinated rubber with powdered zinc (or its oxide) in presence or absence of hydrocarbon or halocarbon solvents react violently or explosively at about 216°C [2]. Powdered zinc initially reacts more violently with hexachloroethane in ethanol than does aluminium [3].

See Bromomethane: Metals *also* Chloromethane: Metals
See also CHLORINATED RUBBER: metal oxides
See other METAL–HALOCARBON INCIDENTS

Halogens, or Interhalogens   MRH Chlorine 3.05/52, fluorine 7.45/47
Mellor, 1940, Vol. 4, 476
Warm zinc powder incandesces in fluorine, and zinc foil ignites in cold chlorine if traces of moisture are present.
See Bromine pentafluoride: Acids, etc.
    Chlorine trifluoride: Metals
    Iodine: Metals, Water

Lead azide   MRH 1.63/99+
See Lead(II) azide: Copper, etc.

Manganese dichloride
Terreill, A., *Bull. Soc. Chim. Fr. [2]*, 1874, **21**, 289
Zinc foil reacts explosively when heated with anhydrous manganese dichloride.

Metal oxides   MRH Chromium trioxide 2.51/50, potassium peroxide 1.72/45
See Potassium dioxide: Metals
    Titanium dioxide: Metals
    Zinc peroxide: Alone, etc.

2-Nitroanisole, Sodium hydroxide
See 2-Nitroanisole: Sodium hydroxide, etc.

Nitrobenzene   MRH 4.52/99+
Muir, G. D., private comm., 1968
Zinc residues from reduction of nitrobenzene to *N*-phenylhydroxylamine are often pyrophoric and must be kept wet during disposal.

Nitrogen compounds   MRH values show % of nitrogen compound
    See Hydrazinium nitrate: Alone, etc.   MRH 1.67/50
    Hydroxylamine: Metals   MRH 3.85/99+

Non-metals   MRH Sulfur 2.09/33
1. Mellor, 1940, Vol. 4, 480, 485
2. Read, C. W. W., *School Sci. Rev.*, 1940, **21**(83), 976
Arsenic, selenium and tellurium all react with incandescence on heating [1]. Interaction on heating powdered zinc and sulfur is considered to be too violent for use as a school experiment [2].

Other reactants
   Yoshida, 1980, 1–2
   MRH values calculated for 31 combinations, mainly with oxidants, are given.

Oxidants   MRH values show % of oxidant
   See Halogens; Interhalogens; Metal oxides, all above
       Ammonium nitrate: Metals   MRH 3.22/55
       Nitric acid: Metals   MRH 3.59/28
       Nitryl fluoride: Metals
       Peroxyformic acid: Metals   MRH 5.69/99+
       Potassium chlorate: Metals   MRH 3.63/38

Paint primer base
   Muller, K. A., *Spontaneous Combustion of Zinc-rich Paint Mixtures*, Washington, API, 1969
   Zinc residues from poorly mixed priming paint spontaneously inflamed after prolonged exposure to atmosphere. Aluminium residues may behave similarly.

Pentacarbonyliron, Transition metal halides
   See Pentacarbonyliron: Transition metal halides, etc.

Seleninyl bromide
   See Seleninyl bromide: Metals

Sodium hydroxide
   1. Anon., *ABCM Quart. Safety Summ.*, 1963, **34**, 14
   2. Jackson, H., *Spectrum*, 1969, **7**(2), 82
   3. *Summerlin, L. unpublished letter to Chem. Eng. News*, 1988
   Accidental contamination of a metal scoop with flake sodium hydroxide, prior to its use with zinc dust, caused ignition of the latter [1]. A stiff paste prepared from zinc dust and 10% sodium hydroxide solution attains a temperature above 100°C after exposure to air for 15 min [2]. The residue of zinc dust and sodium hydroxide solution from a lecture demonstration involving zinc plating a copper coin presents a high fire risk if discarded onto paper in a waste bin, ignition of the paper having occurred on many occasions. Dissolve the residue in dilute acid before flushing away with water [3].
   See PAPER TOWELS
   See Water, next below

Water
   1. Anon., *Metallwirtschaft*, 1941, **20**, 475
   2. Anon., *Chem. Fabrik*, 1940, 362
   3. Mellor, 1940, Vol. 4, 474
   4. Carson, P. J. *et al., Loss Prev. Bull.*, 1992, **102**, 15
   In contact with atmospheric oxygen and limited amounts of water, zinc dust will generate heat, and may become incandescent [1]. Presence of acetic acid and copper shortens the induction period. Store the metal dry, and keep residues

thoroughly wet until disposal [2]. Hydrogen is evolved, especially under acid or alkaline conditions [3]. An incident of a burst galvanised drum from hydrogen evolution is recorded [4].

See other CORROSION INCIDENTS, GAS EVOLUTION INCIDENTS, INDUCTION PERIOD INCIDENTS

Zinc chloride

Bylo, Z. *et al., Chem. Abs.*, 1983, **98**, 146031

Flammability of some commercial zinc dusts or powders is attributed to presence of zinc chloride.

See other METALS, PYROPHORIC METALS, REDUCANTS

4928. Zirconium
[7440-67-7]   Zr

Zr

(*MCA SD-92*, 1966); *NSC 382*, 1968 (powder)

1. *Standard for Production, Processing, Handling and Storage of Zirconium*, NFPA 482M, 1974
2. *Zirconium Fire and Explosion Incidents*, TID-5365, Washington, USAEC, 1956
3. Gleason, R. P. *et al., J. Amer. Soc. Safety Eng.*, 1964, **9**(3), 15
4. Bulmer, G. H. *et al., Rept. AHSB(S) R59*, London, UKAEA, 1964
5. *MCA Case History No. 1234*
6. Ioffe, V. G., *Chem. Abs.*, 1967, **66**, 12614d
7. Karlowicz, P. *et al., J. Electrochem. Soc.*, 1961, **108**, 659–663
8. Leuschke, G. *et al., Chem. Abs.*, 1983, **98**, 221174
9. Baumgart, H., *Chem. Abs.*, 1985, **103**, 73412
10. Cooper, T D., *Chem. Abs.*, 1986, **104**, 65449
11. *See entry* DUST EXPLOSION INCIDENTS (reference 22)
12. Bond, 1991, 60

All aspects of hazards in the production, use and storage of zirconium are covered in a US Standard [1], and there is a compilation of 43 abstracts of unusual zirconium fire and explosion incidents, most of which involved the pyrophoric finely divided metal, moisture and friction, occasionally accompanied by the possibility of static sparks from polythene bags [2]. The increasing use of zirconium necessitated a review of methods of controlling the pronounced pyrophoric properties of zirconium powder. Five considerations for the safe use of zirconium include: exclusion of air or oxygen by use of inert gases; exclusion of water, its vapour and other contaminants or oxidants; control of particle size; limitation of amount of powder handled or exposed; and limitation of exposure of personnel [3]. Storage and disposal of irradiated metal wastes in various forms has also been discussed [4]. Ignition may not be immediate, however, as fine zirconium dust left uncleared from previous grinding operations was ignited violently by a spark from subsequent grinding operations [5]. The lower ignition limits of air suspensions of zirconium powder, of its alloys with silicon and titanium, and of zirconium–titanium hydride have been studied in relation to the history of exposure to air [6]. Treatment of zirconium powder with 1% hydrogen fluoride solution considerably desensitises it to electrostatic ignition [7]. Particles of Zircaloy (98–99% Zr) blown from tubes by compressed air ignited by friction or impact, even at 2 bar blow-off pressure, and could initiate dust explosions [8]. Ignition and combustion properties of pyrotechnic zirconium and zirconium hydride powders (<10 μ) and safety measures are discussed [9]. In a review of pyrophoricity of zirconium and its alloys, it emerged that the maximum particle size for pyrophoric behaviour is 54 μ, and that smaller particles may also explode as well as ignite in air. Factors influencing ignition temperature were explored, and recommendations made for safe handling, transportation and storage [10]. In dust explosions, maximum pressures of 5.1 bar, with maximum rates of rise of 748 bar/s have been determined [11]. Flakes of zirconium came away when a manhole cover was opened on a zirconium-distillation column containing methanol. The zirconium particles ignited, setting fire to the methanol vapour [12].

See Lead–zirconium alloys
See METAL DUSTS, PYROPHORIC METALS
See Hafnium

Carbon dioxide, Nitrogen
See Carbon dioxide: Metals, or: Metals, Nitrogen

Carbon tetrachloride
1. *Zirconium Fire and Explosion Incidents*, TID-5365, Washington, USAEC, 1956
2. *491M*, 1975, 446
A mixture of powdered zirconium and carbon tetrachloride exploded violently while being heated [1], and immersion of zirconium sponge in the unheated solvent also caused explosion [2].
See other METAL–HALOCARBON INCIDENTS

Nitryl fluoride
See Nitryl fluoride: Metals

Oxygen-containing compounds
1. Mellor, 1941, Vol. 7, 116
2. Ellern, 1968, 249
The affinity of zirconium for oxygen is great, particularly when the metal is finely divided, and extends to many oxygen-containing compounds. The vigorous reactions with potassium chlorate, potassium nitrate, copper(II) oxide or lead oxide might be anticipated, but there are also explosive reactions with the combined oxygen when zirconium is heated with alkali-metal hydroxides or carbonates, or even zirconium hydroxide. Hydrated sodium tetraborate behaves similarly [1]. Several alkali-metal oxosalts (chromates, dichromates, molybdates, sulfates, tungstates) react violently or explosively when reduced to the parent metals by heating with zirconium [2].
See Copper(II) oxide: Metals
    Lead(II) oxide: Metals
    Lithium chromate: Zirconium

Potassium chlorate: Metals
Potassium nitrate : Metals

Phosphorus
See Phosphorus: Metals

Sulfuric acid, Potassium bisulfate
Ullmann, 1996, A28, 554,
Zirconium powder mixed with an approximately equal quantity of 17% potassium bisulfate in concentrated sulfuric acid exploded violently.
See Oxygen-containing compounds, above

Water
1. *MCA SD-92*, 1966
2. Muir, G. D., private comm., 1968
3. *Haz. Chem Data*, 1975, 305
4. Hall, A. N. *et al., Loss Prev. Bull.*, 1987, (076), 22–29
Powder damp with 5–10% of water may ignite, and although 25% of water is regarded as a safe concentration [1], ignition of a 50% paste on breakage of a glass container has been observed [2]. Although water is used to prevent ignition, the powder, once ignited, will burn under water (88.8% oxygen) more violently than in air (21% oxygen) [3]. The relatively minor part played by steam and zircalloy fuel containment cans (ruptured by the steam explosion) to form hydrogen towards the end of the catastrophic sequence involved in the Chernobyl nuclear reactor accident is noted in a comprehensive account of the causes of the disaster [4].
See other METALS, PYROPHORIC METALS

4929. Zirconium salts
[15543-40-5] Zr^{4+}

$$Zr^{4+}$$

'Cupferron'
See Ammonium *N*-nitrosophenylaminooxide: Thorium salts

Appendix 1

Source Title Abbreviations used in Handbook References

The abbreviations used in the references for titles of journals and periodicals are those used in BP publications practice and conform closely to the recommendations of the *Chemical Abstracts* system. Abbreviations which have been used to indicate textbook and reference book sources of information are set out below with the full titles and publication details.

ABCM Quart. Safety Summ.,	*Quarterly Safety Summaries*, London, Association of British Chemical Manufacturers, 1930–1964
ACS 54, 1966	*Advanced Propellant Chemistry*, ACS 54, Washington, American Chemical Society, 1966
ACS 88, 1969	*Propellants Manufacture, Hazards and Testing*, ACS 88, Washington, American Chemical Society, 1969
Albright, Hanson, 1976	*Industrial and Laboratory Nitrations*, ACS Symposium Series 22, Albright, L. F., Hanson, C., (Eds.), Washington, American Chemical Society, 1976
Alkali Metals, 1957	*Alkali Metals*, ACS 19, Washington, American Chemical Society, 1957
ASHRAE. J.	*American Society of Heating, Refrigeration and Air-conditioning Engineers Journal*, New York.
Augustine, 1968	*Reduction, Techniques and Applications in Organic Chemistry*, Augustine, R. L., London, Edward Arnold, 1968

Augustine, 1969, 1971

Oxidation, Techniques and Applications in Organic Synthesis, Augustine, R. L., (Ed.), New York, Marcel Dekker, Vol. 1, 1968, Vol. 2 (with Trecker, D. J.), 1971

Bahme, 1972

Fire Officer's Guide to Dangerous Chemicals, FSP-36, Bahme, C. W., Boston, National Fire Protection Association, 1972

Bailar, 1973

Comprehensive Inorganic Chemistry, Bailar, J. C., Eméleus, H. J., Nyholm, R. S., Trotman-Dickenson, A. F., (Eds.), Oxford, Pergamon, 5 vols., 1973

BCISC Quart. Safety Summ.,

Quarterly Safety Summaries, London, British Chemical Industries Safety Council, 1965–1973

Barton & Ollis, 1979

Comprehensive Organic Chemistry, Barton, D. H. R.,Ollis, W. D., (Eds), Oxford, Pergamon, 1979.

Biasutti, 1981

History of Accidents in the Explosives Industry, Biasutti, G. S. (Author and publisher), Vevey, 1981

Bond, 1991

Sources of Ignition, Bond, J., Oxford, Butterworth-Heinemann, 1991

Braker, 1981

Matheson Gas Data Book, Braker, W., Mossman, A. L., Lyndhurst (NJ), Matheson Div. Searle, 6th edn. 1981

Brandsma, 1971

Preparative Acetylenic Chemistry, Brandsma, L., Barking, Elsevier, 1971

Brauer, 1961, 1965

Handbook of Preparative Inorganic Chemistry, Brauer, G., (Translation Ed. Riley, R. F.), London, Academic Press, 2nd edn. Vol. 1, 1963; Vol. 2, 1965

Bretherick, 1986

Hazards in the Chemical Laboratory, Bretherick, L. (Ed.), London, Royal Society of Chemistry, 4th edn., 1986

Castrantas, 1965

Fire and Explosion Hazards of Peroxy Compounds, Special Publication No. 394, Castrantas, H. M., Banerjee, D. K., Noller, D. C., Philadelphia, ASTM, 1965

Castrantas, 1970

Laboratory Handling and Storage of Peroxy Compounds, Special Publication No. 491, Castrantas, H. M., Banerjee, D. K., Philadelphia, ASTM, 1970

CHAS Notes	*Chemical Health and Safety Notes*, ACS Division of Chemical Health and Safety, Washington, 1982 to date (4 issues/yr)
Chemiarbeit	*Chemiearbeit, Schutzen und Helfen*, Dusseldorf, Berufsgenossenschaft der Chemische Industrie, 1949–1962 (supplement to Chemische Industrie)
Chem. Hazards Ind.	*Chemical Hazards in Industry*, Royal Society of Chemistry Information Group, Cambridge, 1984 to date (monthly abstracts)
CISHC Chem. Safety Summ.,	*Chemical Safety Summaries*, London, Chemical Industry Safety and Health Council of the Chemical Industries Association, 1974 to 1986 (terminated)
CHETAH 1990	*The ASTM Chemical Thermodynamic and Energy Release Evaluation Programme*, Version 4.4, 2nd Edn (ASTM Data Series Publication 51A), Davies C. A. *et al.*, Philadelphia, ASTM, Appendix V. Upgrades continue to be issued (1998)
Cloyd, 1965	*Handling Hazardous Materials*, Technical Survey No. SP-5032, Cloyd, D. R., Murphy, W. J., Washington, NASA, 1965
Coates, 1960	*Organometallic Compounds*, Coates, G. E., London, Methuen, 1960
Coates, 1967, 1968	*Organometallic Compounds*, Coates, G. E., Green, M. L. H., Wade, K., London, Methuen, Vol. 1, 1967, Vol. 2, 1968
Compendium 1985, 1987	*Compendium of Safety Data Sheets, for Research and Industrial Chemicals*, Keith, L. H., Walters, D. B., Deerfield Beach/Weinheim, VCH, vols. 1–3 of 867 sheets 1985, vols 4–6 of 723 sheets, 1987
Dangerous Loads, 1972	*Dangerous Loads*, Leicester, Institution of Fire Engineers, 1972
Dangerous Substances, 1972	*Dangerous Substances: Guidance on Fires and Spillages* (Section 1, 'Inflammable liquids'), London, HMSO, 1972
Davies, 1961	*Organic Peroxides*, Davies, A. G., London, Butterworths, 1961
Davis, 1943	*Chemistry of Powder and Explosives*, Davis, T. L., New York, Wiley, 1943

DOC 5, 1982 — *Dictionary of Organic Compounds*, Buckingham, J., (Ed.), London, Chapman & Hall, 5th edn. in 5 vols. + indices, 1982, annual supplementary vols. to date

Dunlop, 1953 — *The Furans*, ACS 119, Dunlop, A. P., Peters, F. N., New York, Reinhold, 1953

Ellern, 1968 — *Military and Civilian Pyrotechnics*, Ellern, H., New York, Chemical Publishing Co., 1968

Emeléus, 1960 — *Modern Aspects of Inorganic Chemistry*, Emeléus, H. J., Anderson, J. S., London, Rutledge and Kegan Paul, 1960

Ephraim, 1939 — *Inorganic Chemistry*, Ephraim, F., (3rd Engl. edn., Thorne, P. C. L., Ward, A. M.), London, Gurney & Jackson, 1939

Federoff, 1960 — *Encyclopaedia of Explosives and Related Compounds*, Federoff, B. T. (Ed.), Dover (NJ), Piccatinny Arsenal, Vol. 1, 1960; Vol. 10, 1984 (from USNTIS)

Fieser, 1967–1986 — *Reagents for Organic Synthesis*, Fieser, L. F. and Fieser, M., New York, Wiley, Vol. 1, 1967, The series continues after the death of the originator.

Freifelder, 1971 — *Practical Catalytic Hydrogenation*, Freifelder, M., New York, Wiley-Interscience, 1971

Gallais, 1957 — *Chimie Minérale Théoretique et Expérimental (Chimie Électronique)*, Gallais, F., Paris, Masson, 1957

Gaylord, 1956 — *Reductions with Complex Metal Hydrides*, Gaylord, N. G., New York, Interscience, 1956

Gibson, 1969 — *Handbook of Selected Properties of Air-and Water-Reactive Materials*, Gibson, H., Crane (In.), US Naval Ammunition Depot, 1969

Goehring, 1957 — *Ergebnisse und Probleme der Chemie der Schwefelstickstoff-verbindung, Scientia Chemica*, Vol. 9, Goehring, M., Berlin, Akademie Verlag, 1957

Grignard, 1935–54 — *Traité de Chimie Organique*, Grignard, V. and various volume Eds., Paris, Masson, Vol. 1, 1935, Vol. 23, 1954

Guide for Safety, 1972	*Guide for Safety in the Chemical Laboratory*, MCA, New York, Van Nostrand Reinhold, 2nd edn., 1972
Harmon, 1974	*A Review of Violent Monomer Polymerisation*, Harmon, M., King, J., Rept. AD-017443, Springfield (Va.), NTIS, 1974
Haz. Chem. Data, 1975	*Hazardous Chemical Data*, NFPA 49, Boston, National Fire Protection Association, 1975
HCS 1980	*Handling Chemicals Safely*, Dutch Assoc. of Safety Experts, Dutch Chemical Industry Assoc., Dutch Safety Inst., 1980 (Engl. transl. of 2nd Dutch edn.)
Houben-Weyl, 1953 to date	*Methoden der Organischen Chemie*, Müller, E. (Ed.), Stuttgart, Georg Thieme, 4th edn., published unsequentially from 1953.
Inorg. Synth., 1939–86	*Inorganic Syntheses*, various Eds., Maidenhead, McGraw-Hill, Vol. 1, 1939, Vol. 24, 1986
Jahresber. 1960 to date	*Jahresberichte*, Berufsgenossenschaft der Chemischen Industrie, Heidelberg, annually since 1960
JANAF, 1971	*JANAF Thermochemical Tables*, Stull, D. R., Prophet, H., National Bureau of Standards, 2nd edn., 1971
Karrer, 1950	*Organic Chemistry*, Karrer, P., London, Elsevier, 1950
Janz, 1976	*Eutectic Data, Table 2 (Safety Hazards, etc.)* Janz, G. J., Allen, C. B., Molten Salts Data Center, ERDA Rept. TID 27163-21, 1976; subsequently publ. in J. Haz. Mat.,1980–81, **4** (1), 145–176
Kharasch and Reinmuth, 1954	*Grignard Reactions of Non-Metallic Substances*, Kharasch, M. S. and Reinmuth, O., London, Constable, 1954
King, 1990	*Safety in the Process Industries*, King, R., London, Butterworth-Heinemann, 1990
Kirk-Othmer, 1963–71	*Encyclopaedia of Chemical Technology*, (3rd edn., 1978–83) Kirk, R. E., Othmer, D. F., London, Wiley-Interscience, Vol. 1, 1963, Vol. 22, 1970, Suppl. Vol. 1971. (4th edn., from 1991) 1993;

1931

Kirshenbaum, 1956	*Final Report on Fundamental Studies of New Explosive Reactions*, Kirshenbaum, A. D., Philadelphia, Research Institute of Temple University, 1956
Kit and Evered, 1960	*Rocket Propellant Handbook*, Kit, B., Evered, D. S., London, MacMillan, 1960
Lab. Hazards Bull.	*Laboratory Hazards Bulletin*, Royal Society of Chemistry Information Group, Cambridge, 1981 to date (monthly abstracts)
Lawless, 1968	*High Energy Oxidisers*, Lawless, E. W. and Smith, I. C., New York, Marcel Dekker, 1968
Leleu, Cahiers, 1972–80	*Les Réactions Chimique Dangereuse*, Leleu, J., published serially in *Cahiers de Notes Documentaires*, from 1972 **(68)**, 319 to 1980, **(100)**, 426
Loss Prev. Bull., 1974 to date	*Loss Prevention Bulletin*, Rugby, Institution of Chemical Engineers, 6 issues annually since 1974
Mackay, 1966	*Hydrogen Compounds of the Metallic Elements*, Mackay, K. M., London, Spon, 1966
Major Loss Prevention, 1971	*Major Loss Prevention in the Process Industries, Symposium Series* No. 34, London, Institution of Chemical Engineers, 1971
Martin, 1971,	*Dimethylsulfoxid*, Martin, D., Hauthal, H. G., Berlin, Akademie Verlag, 1971
Matheson Guide, 1982	*Guide to Safe Handling of Compressed Gases*, Matheson Div. Searle Medical Products, Lyndhurst (NJ), 1982
MCA Case Histories, 1950–78	*Case Histories of Accidents in the Chemical Industry*, Washington, MCA, (published monthly, republished as 4 ndexed collected volumes 1953–75, discontinued 1978)
(MCA SD-Series) 1947–76	*Safety Data Sheets*, MCA, Washington. The whole series of 101 sheets was withdrawn in 1980, but remains a useful source of collected information on basic industrial chemicals.
Meidl, 1972	*Hazardous Materials Handbook*, Meidl, J. H., Beverley Hills, Glencoe Press, 1972

Mellor, 1947–71   *Comprehensive Treatise on Inorganic and*
                                 Theoretical Chemistry, Mellor, J. W.,
                                 London, Longmans Green, Vol. 1, repr. 1941,
                                 Vol. 16, publ. 1937; and isolated
                                 Supplementary vols. up to Vol. 8 Suppl. 3,
                                 1971

Mellor MIC, 1961   *Mellor's Modern Inorganic Chemistry*,
                                 Parkes, G. D. (Revision Ed.), London,
                                 Longmans Green, 1961

Merck, 1976, 1983, 1989, 1996   *The Merck Index*, Chapman and Hall,
                                 London., 9th edn. 1976, 10th edn. 1983, 11th
                                 edn. 1989, 12th edn. 1996

Merrifield, 1988   *Fire and Explosion Hazards associated with*
                                 Storage and Handling of Hydrogen Peroxide,
                                 Merrifield, R., Specialist Inspectors Report
                                 No. 19, Bootle, HSE. 1988

Meyer, 1977   *Chemistry of Hazardous Materials*,
                                 Meyer, E., Englewood Cliffs (NJ),
                                 Prentice-Hall, 1977

Miller, 1965–66   *Acetylene, its Properties, Manufacture and*
                                 Uses, Miller, S. A., London, Academic Press,
                                 Vol. 1, 1965, Vol. 2, 1966

Moissan, 1900   *Le Fluor et ses Composés*, Moissan, H.,
                                 Paris, Steinheil, 1900

NFPA 491M, 1975   *Manual of Hazardous Chemical Reactions*,
                                 NFPA 491M, Quincy (MA), National Fire
                                 Potection Association, 5th edn., 1975 (larger
                                 format of same text, 1985)

Nolan, 1983   *Case Histories of Runaway Reactions*,
                                 Nolan, P. F., London, Polytechnic of the
                                 South Bank, 1983. A summary of the
                                 conclusions on the causes was published in
                                 J. Haz. Materials, 1987, **14**, 233–239

Org. Synth., 1941 to date   *Organic Syntheses*, various Eds., New York,
                                 Wiley, Collective Vol. 1, 1944; Coll. Vol. 2,
                                 1944; Coll. Vol. 3, 1955; Coll. Vol. 4, 1962;
                                 Coll. Vol. 5, 1973; Coll. Vol. 6, 1987; Coll.
                                 Vol. 7, 1990; Coll. Vol. 8, 1993; annual
                                 volumes (from 70) thereafter.

Partington, 1967   *General and Inorganic Chemistry*,
                                 Partington, J. R., London, MacMillan, 4th
                                 edn., 1967

Pascal, 1956–70	*Nouveau Traité de Chimie Minérale*, Pascal, P. (Ed.), Paris, Masson, Vol. 1, 1956; Vol. 20.3, 1964; Vol. 15.5, 1970
Pieters, 1957	*Safety in the Chemical Laboratory*, Pieters, H. A. J., Creyghton, J. W., London, Academic Press, 2nd edn., 1957
Plant/Oper. Progress	*Plant Operations Progress* continues as *Process Safety Progress* 1993-
Pot. Incid. Rept.	*ASESB Potential Incident Report*, Washington, Armed Services Explosives Safety Board
Proc. nth Combust. Symp.	*Proceedings of Combustion Symposia*, various Eds. and Publishers, dated year following the event. 1st Symposium 1928; 19th Symposium, 1982
Renfrew, 1981	*Safety in the Chemical Laboratory*, Collected reprinted articles from *J. Chem. Educ.*, Renfrew, M. M. (Ed.), Vol. IV, 1981
Rodd, 1951–62	*The Chemistry of Carbon Compounds*, Rodd, E. H. (Ed.), London, Elsevier, Vol. 1A, 1951, Vol. 5, 1962; 2nd edn. began in 1979
RSC Data Sheets, 1982 to date	*Laboratory Hazard Data Sheet Series*, Cambridge, Royal Society of Chemistry, No. 1, 1982; No. 89, 1989, issued monthly, ceased 1993
RSC Safety Data Sheets, 1989	*Chemical Safety Data Sheets*, Cambridge, Royal Society of Chemistry, 1989 Generally similar to the monthly Laboratory Hazards Safety Data Sheet series, but the information is presented under headings reflecting legislative requirements. Volume 1 deals with some 100 largely organic liquid solvents and reagents, while Volume 2 covers main group metals and their compounds, 122 items in all. Vol. 3; Corrosives and Irritants, 1990. Vol 4; Toxic Chemicals, Pt. I 1990, Pt. II 1991. Vol 5, Flammables, 1989. There is considerable overlap among the volumes
Runaway Reactions, 1981	*Runaway Reactions, Unstable Products and Combustible Powders*, Symposium Series No. 68 (Chester 1981), Rugby, Institution of Chemical Engineers, 1981

Rüst, 1948

Unfälle beim Chemischen Arbeiten, Rüst, E., Ebert, A., Zurich, Rascher Verlag, 2nd edn., 1948

Rutledge, 1968

Acetylenic Compounds, Rutledge, T. F., New York, Reinhold, 1968

Schmidt, 1967

Handling and Use of Fluorine and Fluorine-Oxygen Mixtures in Rocket Systems, SP-3037, Schmidt, H. W., Harper, J. T., Washington, NASA, 1967

Schumacher, 1960

Perchlorates, their Properties, Manufacture and uses, ACS 146, Schumacher, J. C., New York, Reinhold,1960

Schumb, 1955

Hydrogen Peroxide, ACS 128, Schumb, W. C., Satterfield, C. N., Wentworth, R. L., New York, Reinhold, 1955

Shriver, 1986

Manipulation of Air-Sensitive Compounds, Shriver, D. F. and Drozdzon, M. A. (Eds.), New York, Wiley, 2nd edn.,1986

Sichere Chemiearb.

Sichere Chemiarbeit, Heidelberg,Berufsgenossenschaft der Chemischen Industrie, 1963 to date

Sidgwick, 1950

The Chemical Elements and their Compounds, Sidgwick, N. V., Oxford, Oxford University Press, 1950

Sittig, 1981

Handbook of Toxic and Hazardous Chemicals, Sittig, M., Park Ridge (NJ), Noyes Data Corp., 2nd ed., 1985

Smith, 1965–6

Open-Chain Nitrogen Compounds, Smith, P. A. S., New York, Benjamin, Vol. 1, 1965; Vol. 2, 1966

Sorbe, 1968

Giftige und Explosive Substanzen, Sorbe, G., Frankfurt, Umschau Verlag, 1968

Steere, 1971

Handbook of Laboratory Safety, Steere, N. V. (Ed.), Cleveland, Chemical Rubber Publishing Co., 2nd edn., 1971. 3rd edn., Furr, A. K. (Ed.), 1989

Steere, 1967, '71, '74

Safety in the Chemical Laboratory, N. V. Steere (Ed.), Collected reprinted articles from *J. Chem. Educ.*, Vol. I, 1967; Vol. II, 1971; Vol. III, 1974

Stull, 1977
Fundamentals of Fire and Explosion, AIChE Monograph Series No. 10, Stull, D. R., New York, American Institute of Chemical Engineers, 1977

Swern, 1970-52
Organic Peroxides, Swern, D. (Ed.), London, Wiley-Interscience, Vol. 1, 1970; Vol. 2, 1971; Vol. 3, 1972

Ullmann, 1998
Ullmann's Encyclopedia of Industrial Chemistry, Weinheim, 6th edn. 1998 (CD update 1999)

Urbanski, 1964–84
Chemistry and Technology of Explosives, Urbanski, T., London, MacMillan, Vol. 1, 1964; Vol. 2, 1965; Vol. 3, 1967; Oxford, Pergamon, Vol. 4, 1984

Van Wazer, 1958–61
Phosphorus and its Compounds, Van Wazer, J. R., New York, Interscience, Vol. 1, 1958; Vol. 2, 1961

Vervalin, 1973
Fire Protection Manual for Hydrocarbon Processing Plants, Vervalin, C. H. (Ed.), Houston, Gulf Publishing Co., 2nd edn., 1973

Vogel, 1957
A Textbook of Practical Organic Chemistry, Vogel, A. I., London, Longmans Green, 3rd edn., 1957

von Schwartz, 1918
Fire and Explosion Risks, von Schwartz, E., London, Griffin, 1918

Weast, 1988
Handbook of Chemistry and Physics, Weast, R. C. (Ed.), Boca Raton (Fa.), CRC Press, 1988; continued under the editorship of Lide, D. R.

Whitmore, 1921
Organic Compounds of Mercury, Whitmore, F. C., New York, Chemical Catalog Co. Inc., 1921

Yoshida, 1980
Handbook of Hazardous Reactions with Chemicals, Yoshida, T., Tokyo, Tokyo Fire Department, 1980 (copy in British Library, Boston Spa)

Zabetakis, 1965
Flammability Characteristics of Combustible Gases and Vapours, Zabetakis, M. G., Washington, US Bureau of Mines, 1965

Tabulated Fire-related Data

Compounds which are considered to be unusually hazardous in a fire context because of their low flash points (below 25°C) or auto-ignition temperatures (below 225°C) are included in the table. The names used are those titles in the text of Section 1 which are prefixed with a dagger. Synonyms may be found either in Section 1 or in the alphabetical index of chemical names and synonyms in Appendix 4. Boiling points are given for those compounds boiling below 50°C.

The figures for flash points are closed-cup values except where a suffix (o) indicates the (usually higher) open-cup value. The figures for explosive limits (or flammability limits) are % by volume in air at ambient temperature except where indicated otherwise. Where no figure has been found for the upper limit, a query has been inserted. Figures for auto-ignition temperatures are usually those determined in glass (without catalytic effects) except where stated.

Most of the values are those quoted in the references given in the topic entries:

AUTOIGNITION TEMPERATURES
FIRE
FLAMMABILITY
FLASH POINTS

Name	Formula	B.P./°C	Fl.P./°C	E.L./%	A.I.T./°C
Arsine	AsH_3	−62	—	flammable	—
Diborane	B_2H_6	−93	−90	0.9–98	38–52
Pentaborane(9)	B_5H_9	—	30	0.4–?	35
Bromosilane	BrH_3Si	2	< 0	—	ambient
Tribromosilane	Br_3HSi	112	—	flammable	ambient
Hydrogen cyanide	CHN	25.7	−18	6.0–41	—
Dichloromethane	CH_2Cl_2	39.7	calc −9	12.0–19.0	605
Formaldehyde	CH_2O	−19	−19	7.0–73	—
Bromomethane	CH_3Br	3.5	none	see text	—
Chloromethane	CH_3Cl	−24	< 0	8.1–17	—
Methyltrichlorosilane	CH_3ClSi	—	8	7.6–?	—
Fluoromethane	CH_3F	−78	—	flammable	—
Methyl nitrite	CH_3NO_2	−12	—	flammable	explodes
Methane	CH_4	−161	−187	5.3–15	—
Dichloromethylsilane	CH_4Cl_2Si	—	−32	—	ambient
Methanol	CH_4O	—	10	6–36.5/60°	—
Methanethiol	CH_4S	6	−18	3.9–21.8	—

Name	Formula	B.P./°C	Fl.P./°C	E.L./%	A.I.T./°C
Methylamine	CH_5N	−6	−18	4.5−21	—
35%w/v in water		—	7.5	—	—
Methylhydrazine	CH_6N_2	—	18	2.5−97	194
Carbon monoxide	CO	−192	—	12.5−74	—
Carbonyl sulfide	COS	−50	—	6.5−28.2	—
Carbon disulfide	CS_2	46	−30	0.6−50	125 (below 100 if rust present) ambient
Bromotrifluoroethylene	C_2BrF_3	−3	—	—	
Chlorotrifluoroethylene	C_2ClF_3	−28	—	8.4−38.7	—
Tetrafluoroethylene	C_2F_4	−76	—	11−60	180
Trichloroethylene	C_2HCl_3	—	32 (above 30°C)	12.5−90	410
Trifluoroethylene	C_2HF_3	—	—	15.3−27	—
Acetylene	C_2H_2	−83	−18	2.5−82	—
1,1-Dichloroethylene	$C_2H_2Cl_2$	32	−15(o)	7.3−16	—
cis-1,2-Dichloroethylene	$C_2H_2Cl_2$	—	6	3.3−15	—
trans-1,2-Dichloroethylene	C_2HCl_2	4	2	9.7−12.8	—
1,1-Difluoroethylene	$C_2H_2F_2$	−86	—	5.5−21.3	—
Ethanedial	$C_2H_2O_2$	50	—	—	—
Bromoethylene	C_2H_3Br	16	< −8	6−16	—
Chloroethylene	C_2H_3Cl	−14	−8	4−22 (or 33)	—
1-Chloro-1,1-difluoroethane	$C_2H_3ClF_2$	−9	—	9−14.8	—
Acetyl chloride	C_2H_3ClO	—	4	5.0−?	—
Methyl chloroformate	$C_2H_3ClO_2$	—	12	—	—
1,1,1-Trichloroethane	$C_2H_3Cl_3$	—	none	8−10.5	500
Trichlorovinylsilane	$C_2H_3Cl_3Si$	—	−9	—	—
Fluoroethylene	C_2H_3F	−73	—	2.6−22	—
1,1,1-Trifluoroethane	$C_2H_3F_3$	−47	—	9.2−18.4	—
Acetonitrile	C_2H_3N	—	6(o)	4−16	—
Methyl isocyanate	C_2H_3NO	39	−7	—	—
Ethylene	C_2H_4	−104	−136	3.0−34	—
1,1-Dichloroethane	$C_2H_4Cl_2$	—	−12	5.6−11.4	—
1,2-Dichloroethane	$C_2H_4Cl_2$	—	13	6.2−15.9	—
Bis(chloromethyl) ether	$C_2H_4Cl_2O$	—	< 19	—	—
1,1-Difluoroethane	$C_2H_4F_2$	−25	—	3.7−18	—
Acetaldehyde	C_2H_4O	20	−38	4.0−57	204 (140 by DIN 51794)
Ethylene oxide	C_2H_4O	11	−20	3.0−100	—
Thiolacetic acid	C_2H_4OS	—	11	—	—
Methyl formate	$C_2H_4O_2$	32	−32	5.9−20	—
Thiiran	C_2H_4S	—	10	—	—
Bromoethane	C_2H_5Br	38	−20	6.7−11.3	—
Chloroethane	C_2H_5Cl	12	−50	3.8−15.4	—
Chloromethyl methyl ether	C_2H_5ClO	—	−8	—	—
Trichloroethylsilane	$C_2H_5Cl_3Si$	—	14	—	—
Fluoroethane	C_2H_5F	−38	—	—	—
Aziridine	C_2H_5N	—	−11	3.3−54.8	—
Acetaldehyde oxime	C_2H_5NO	—	< 22	—	—
Ethyl nitrite	$C_2H_5NO_2$	17	35	3.1−> 50	expl. > 90
Ethyl nitrate	$C_2H_5NO_3$	—	10	3.8−?	expl. 85
Ethane	C_2H_6	−89	−93	3−12.5	—
Dichlorodimethylsilane	$C_2H_6Cl_2Si$	−9	—	3.4−9.5	—
Dichloroethylsilane	$C_2H_6Cl_2Si$	—	−1	—	—
Dimethyl ether	C_2H_6O	−24	−41	3.4−18 (27)	—
Ethanol	C_2H_6O	—	12	3.3−19/60°C	—
Dimethyl sulfoxide	C_2H_6OS	—	—	—	215
Dimethyl sulfate	$C_2H_6O_4S$	—	—	—	188
Dimethyl sulfide	C_2H_6S	37	−37	2.2−19.7	206
Ethanethiol	C_2H_6S	36	−17	2.8−18	—
Dimethyl disulfide	$C_2H_6S_2$	—	7	1.1−?	—
Dimethylamine	C_2H_7N	7	−50	2.8−14.4	—
40%w/v in water		—	−16	—	—
25%w/v in water		—	6	—	—

Name	Formula	B.P./°C	Fl.P./°C	E.L./%	A.I.T./°C
Ethylamine	C_2H_7N	17	−16	3.5−14	—
70%w/v in water	—	—	< 22	—	—
1,1-Dimethylhydrazine	$C_2H_8N_2$	—	1	2.1−95	—
1,2-Dimethylhydrazine	$C_2H_8N_2$	—	−17	—	—
1,1-Dimethyldiborane	$C_2H_{10}B_2$	−3	< −10	—	—
1,2-Dimethyldiborane	$C_2H_{10}B_2$	−49	< −55	—	—
Dicyanogen	C_2N_2	−21	—	6−32	—
Hexafluoropropene	C_3F_6	—	—	15−20	—
2-Chloroacrylonitrile	C_3H_2ClN	—	8	—	—
3-Bromopropyne	C_3H_3Br	—	10	3.0−?	—
3-Chloropropyne	C_3H_3Cl	—	−16	—	—
Acryloyl chloride	C_3H_3ClO	—	18	—	—
3,3,3-Trifluoropropene	$C_3H_3F_3$	−18	< −30	4.7−23.5	—
1,1,1-Trifluoroacetone	$C_3H_3F_3O$	22	−30	—	—
Acrylonitrile	C_3H_3Ne	—	−1	3.0−17	—
5%w/v in water	—	—	< 9	—	—
Isoxazole	C_3H_3NO	—	8	—	—
Oxazole	C_3H_3NO	—	18	—	—
Propadiene	C_3H_4	−34	—	1.7−12	—
Propyne	C_3H_4	−23	—	2.15−12.5	—
1-Chloro-3,3,3-trifluoropropane	$C_3H_4ClF_3$	—	—	6−14	—
2,3-Dichloropropene	$C_3H_4Cl_2$	—	10	—	—
Acrylaldehyde	C_3H_4O	—	−26	2.8−31	—
Methoxyacetylene	C_3H_4O	23	< −20	—	—
Vinyl formate	$C_3H_4O_2$	46	< 0	—	—
3-Bromo-1-propene	C_3H_5Br	—	−1	4.4−7.3	—
1-Chloro-1-propene	C_3H_5Cl	35	< −6	4.5−16	—
2-Chloropropene	C_3H_5Cl	22	−37	4.5−16	—
3-Chloropropene	C_3H_5Cl	45	−32	3.3−11.2	—
1-Chloro-2,3-epoxypropane	C_3H_5ClO	—	21	—	—
Propionyl chloride	C_3H_5ClO	—	12	—	—
Ethyl chloroformate	$C_3H_5ClO_2$	—	16	—	—
1-Fluoro-2,3-epoxypropane	C_3H_5FO	—	4	—	—
3-Iodopropene	C_3H_5I	—	5	—	—
Propiononitrile	C_3H_5N	—	6	3.1−?	—
Cyclopropane	C_3H_6	−33	—	2.4−10.4	—
Propene	C_3H_6	−48	−108	2.4−11.1 (see text)	—
1,1-Dichloropropane	$C_3H_6Cl_2$	—	21	3.1−?	—
1,2-Dichloropropane	$C_3H_6Cl_2$	—	15	3.4−14.5	—
Dichloromethylvinylsilane	$C_3H_6Cl_2Si$	—	−1	—	—
Acetone	C_3H_6O	—	−17	2.6−12.8	—
Methyl vinyl ether	C_3H_6O	5.5	−56(o)	2.6−39.0	—
Oxetane	C_3H_6O	50	−28	—	—
2-Propen-1-ol	C_3H_6O	—	21	3.0−18.0	—
Propionaldehyde	C_3H_6O	49	−9	2.9−17	207
Propylene oxide	C_3H_6O	34	−37	3.1−27.5	—
1,3-Dioxolane	$C_3H_6O_2$	—	2	—	—
Ethyl formate	$C_3H_6O_2$	—	−34	2.7−13.5	—
Methyl acetate	$C_3H_6O_2$	—	−9	3.1−16	—
Dimethyl carbonate	$C_3H_6O_3$	—	18	—	—
2-Propene-1-thiol	C_3H_6S	—	−10	—	—
1-Bromopropane	C_3H_7Br	—	−1	4.6−?	—
2-Bromopropane	C_3H_7Br	—	1	—	—
1-Chloropropane	C_3H_7Cl	47	< −18	2.6−11.1	—
2-Chloropropane	C_3H_7Cl	36	−32	2.8−10.7	—
Chloromethyl ethyl ether	C_3H_7ClO	—	15	—	—
2-Iodopropane	C_3H_7I	—	20	—	—
3-Aminopropene	C_3H_7N	—	−29	2.2−22	—
Azetidine	C_3H_7N	61	−20	—	—
Cyclopropylamine	C_3H_7N	—	1	—	—
2-Methylaziridine	C_3H_7N	—	−10	—	—
Isopropyl nitrite	$C_3H_7NO_2$	40	< 10	—	—
Propyl nitrite	$C_3H_7NO_2$	47	< 20	—	—
Isopropyl nitrate	$C_3H_7NO_3$	—	11	?−100	175

Name	Formula	B.P./°C	Fl.P./°C	E.L./%	A.I.T./°C
Propyl nitrate	$C_3H_7NO_3$	—	20	2–100	—
Propane	C_3H_8	7–45	−104	2.2–9.5	—
Ethyl methyl ether	C_3H_8O	11	−37	2–10.1	190
Propanol	C_3H_8O	—	15	2.5–13.5	—
2-Propanol	C_3H_8O	—	12	2.3–12.7	—
Dimethoxymethane	$C_3H_8O_2$	44	−18(o)	—	—
Ethyl methyl sulfide	C_3H_8S	—	−15	—	—
Propanethiol	C_3H_8S	—	−20	—	—
2-Propanethiol	C_3H_8S	—	−35	—	—
Trimethyl borate	$C_3H_9BO_3$	—	−8	—	—
Chlorotrimethylsilane	C_3H_9ClSi	—	−20	—	—
Iodotrimethylsilane	C_3H_9ISi	106	−31	—	—
Isopropylamine	C_3H_9N	32	−50	2.3–10.4	—
Propylamine	C_3H_9N	49	−37	2.0–10.4	—
Trimethylamine	C_3H_9N	3	−5	2.0–11.6	190
25%w/v in water		—	5	—	—
2-Methoxyetylamine	C_3HnNO	—	9	—	—
Trimethylsilyl azide	$C_3H_9N_3Si$	—	23	—	—
1,2-Diaminopropane	$C_3H_{10}N_2$	—	24	—	—
1,3-Diaminopropane	$C_3H_{10}N_2$	—	24(o)	—	—
Propadiene-1,3-dione	C_3O_2	7	—	6–30	—
1,3-Butadiyne	C_4H_2	10	—	1.4–100	—
Buten-3-yne	C_4H_4	5	< −5	1.8–40(-93% in oxygen)	—
Furan	C_4H_4O	32	−36	2.3–14.3	—
Diketene	$C_4H_4O_2$	—	33	2–11.7	200
Methyl propiolate	$C_4H_4O_2$	—	10	—	—
Thiophene	C_4H_4S	—	−6	—	—
2-Chloro-1,3-butadiene	C_4H_5Cl	—	−20	2.5–20	—
Ethyl trifluoroacetate	$C_4H_5F_3O_2$	—	−17	—	—
1-Cyanopropene	C_4H_5N	—	16	—	—
3-Cyanopropene	C_4H_5N	—	19	—	—
1,2-Butadiene	C_4H_6	19	< 0	—	—
1,3-Butadiene	C_4H_6	−4	< −17	2–11.5	—
1-Butyne	C_4H_6	8	< −7	—	—
2-Butyne	C_4H_6	28	−31	1.4–?	—
Cyclobutene	C_4H_6	2	< −10	—	—
1-Buten-3-one	C_4H_6O	—	−7	2.1–15.6	—
Crotonaldehyde	C_4H_6O	—	13	2.1–15.5/60°	207
2,3-Dihydrofuran	C_4H_6O	54	−24	—	—
Divinyl ether	C_4H_6O	29	< −30	1.7–27	—
3,4-Epoxybutene	C_4H_6O	—	< −50	—	—
Ethoxyacetylene	C_4H_6O	50	< −7	—	—
Methacrylaldehyde	C_4H_6O	—	−15	—	—
Allyl formate	$C_4H_6O_2$	—	< 22	—	—
Butane-2,3-dione	$C_4H_6O_2$	—	7	—	—
1,2:3,4-Diepoxybutane	$C_4H_6O_2$	—	−10	—	—
Methyl acrylate	$C_4H_6O_2$	—	−3(o)	2.8–25	—
Vinyl acetate	$C_4H_6O_2$	—	−8	2.6–13.4	—
1-Bromo-2-butene	C_4H_7Br	—	13	—	—
4-Bromo-1-butene	C_4H_7Br	—	1	—	—
2-Chloro-2-butene	C_4H_7Cl	—	−25	2.3–9.3	—
3-Chloro-1-butene	C_4H_7Cl	—	−27	2.2–?	—
3-Chloro-2-methyl-1-propene	C_4H_7Cl	—	−19	2.3–9.3	—
Butyryl chloride	C_4H_7ClO	—	18	—	—
2-Chloroethyl vinyl ether	C_4H_7ClO	—	25	—	—
Isobutyryl chloride	C_4H_7ClO	—	8	—	—
Isopropyl chloroformate	$C_4H_7ClO_2$	—	11	—	—
Butyronitrile	C_4H_7N	—	21	—	—
Isobutyronitrile	C_4H_7N	—	8	—	—
2-Methyl-2-oxazoline	C_4H_7NO	—	20	—	—
1-Butene	C_4H_8	−6	−80	1.6–9.3	—
cis-2-Butene	C_4H_8	4	−12	1.7–9.7	—
trans-2-Butene	C_4H_8	3	< −6	1.8–9.7	—
Cyclobutane	C_4H_8	13	< 10	1.8–?	—

1940

Name	Formula	B.P./°C	Fl.P./°C	E.L./%	A.I.T./°C
Methylcyclopropane	C_4H_8	4	< 0	—	—
2-Methylpropene	C_4H_8	−7	< −10	1.8–8.8	—
mixo-Dichlorobutane	$C_4H_8Cl_2$	—	21	—	—
2-Butanone	C_4H_8O	—	−1	1.8–11.5	—
Butyraldehyde	C_4H_8O	—	−6	2.5–?	230
Cyclopropyl methyl ether	C_4H_8O	—	< 10	—	—
1,2-Epoxybutane	C_4H_8O	44.7	−15	1.5–18.3	—
Ethyl vinyl ether	C_4H_8O	36	−46	1.7–29	202
Isobutyraldehyde	C_4H_8O	—	−25	1.6–10.6	223
Tetrahydrofuran	C_4H_8O	—	−17	1.8–11.8	224
1,3-Dioxane	$C_4H_8O_2$	—	2(o)	2–22	—
1,4-Dioxane	$C_4H_8O_2$	—	12	2–22	180
Ethyl acetate	$C_4H_8O_2$	—	−4	2.2–11.5	—
Isopropyl formate	$C_4H_8O_2$	—	−6	—	—
Methyl propionate	$C_4H_8O_2$	—	−2	2.5–13	—
Propyl formate	$C_4H_8O_2$	—	−3	2.3–?	—
Tetrahydrothiophene	C_4H_8S	—	13	—	—
1-Bromobutane	C_4H_9Br	—	18	2.8–6.6/100	—
2-Bromobutane	C_4H_9Br	—	21	—	—
1-Bromo-2-methylpropane	C_4H_9Br	—	22	—	—
2-Bromo-2-methylpropane	C_4H_9Br	—	−18	—	—
2-Bromoethyl ethyl ether	C_4H_9BrO	—	21	—	—
1-Chlorobutane	C_4H_9Cl	—	−12	1.9–10.1	—
2-Chlorobutane	C_4H_9Cl	—	−10	—	—
1-Chloro-2-methylpropane	C_4H_9Cl	—	−6	2.0–8.7	—
2-Chloro-2-methylpropane	C_4H_9Cl	—	0	—	—
2-Iodobutane	C_4H_9I	—	−10	—	—
1-Iodo-2-methylpropane	C_4H_9I	—	0	—	—
2-Iodo-2-methylpropane	C_4H_9I	—	−10	—	—
Pyrrolidine	C_4H_9N	—	3	—	—
Butyl nitrite	$C_4H_9NO_2$	—	10	—	—
tert-Butyl nitrite	$C_4H_9NO_2$	—	−13	—	—
Cyanotrimethylsilane	C_4H_9NSi	114	1	—	—
Butane	C_4H_{10}	1	−60	1.9–8.5	—
Isobutane	C_4H_{10}	−12	−81	1.9–8.5	—
Dichlorodiethylsilane	$C_4H_{10}Cl_2Si$	—	24	—	—
2-Butanol	$C_4H_{10}O$	—	14	(1.7–9.8/100°)	—
tert-Butanol	$C_4H_{10}O$	—	10	2.4–8	—
Diethyl ether	$C_4H_{10}O$	36	−45	1.8–36.5	180
Methyl propyl ether	$C_4H_{10}O$	39	−20	2.0–14.8	—
1,1-Dimethoxyethane	$C_4H_{10}O_2$	—	1(o)	—	—
1,2-Dimethoxyethane	$C_4H_{10}O_2$	—	1	—	—
Trimethyl orthoformate	$C_4H_{10}O_3$	—	15	—	—
1-Butanethiol	$C_4H_{10}S$	—	2	—	—
2-Butanethiol	$C_4H_{10}S$	—	−23	—	—
Diethyl sulfide	$C_4H_{10}S$	—	−10	—	—
2-Methylpropanethiol	$C_4H_{10}S$	—	−10	—	—
2-Methyl-2-propanethiol	$C_4H_{10}S$	< −29	—	—	—
Butylamine	$C_4H_{11}N$	—	−12	1.7–9.8	—
2-Butylamine	$C_4H_{11}N$	—	−9	—	—
tert-Butylamine	$C_4H_{11}N$	45	−7	(1.7–9.8/100°)	—
Diethylamine	$C_4H_{11}N$	—	−39	1.8–10.1	—
Ethyldimethylamine	$C_4H_{11}N$	—	−36	—	—
Isobutylamine	$C_4H_{11}N$	—	−9	—	—
2-Dimethylaminoethylamine	$C_4H_{12}N_2$	—	11	—	—
Tetramethyl orthosilicate	$C_4H_{12}O_4Si$	—	20	—	—
Tetramethylsilane	$C_4H_{12}Si$	26	< 0	—	—
Tetramethyltin	$C_4H_{12}Sn$	—	−12	1.9–?	—
Dicyanoacetylene	C_4N_2	76	—	—	130
Tetracarbonylnickel	C_4NiO_4	43	< −20	2.0–?	—
Pentacarbonyliron	C_5FeO_5	—	−15	—	ambient
Pyridine	C_5H_5N	—	20	1.8–12.4	—
Cyclopentadiene	C_5H_6	42	−3	—	—
2-Methyl-1-buten-3-yne	C_5H_6	—	< −7(o)	—	—
2-Methylfuran	C_5H_6O	—	−30	—	—

Name	Formula	B.P./°C	Fl.P./°C	E.L./%	A.I.T./°C
2-Methylthiophene	C_5H_6S	—	8	—	—
1-Methylpyrrole	C_5H_7N	—	16	—	—
Cyclopentene	C_5H_8	44	−29	—	—
3-Methyl-1,2-butadiene	C_5H_8	40	−12	—	—
2-Methyl-1,3-butadiene	C_5H_8	34	−53	2.0−8.9	220
3-Methyl-1-butyne	C_5H_8	29.3	—	—	—
1,2-Pentadiene	C_5H_8	44.9	—	—	—
1,3-Pentadiene	C_5H_8	42	−43	2−8.3	—
1,4-Pentadiene	C_5H_8	26	4	—	—
1-Pentyne	C_5H_8	—	−20	—	—
2-Pentyne	C_5H_8	56	—	—	—
Allyl vinyl ether	C_5H_8O	—	< 21(o)	—	—
Cyclopentene oxide	C_5H_8O	—	10	—	—
Cyclopropyl methyl ketone	C_5H_8O	—	13	—	—
2,3-Dihydropyran	C_5H_8O	—	−16	—	—
2-Methyl-3-butyn-2-ol	C_5H_8O	—	20	—	—
Methyl isopropenyl ketone	C_5H_8O	—	21	(1.8−9.0/50°)	—
Allyl acetate	$C_5H_8O_2$	—	22	—	—
Ethyl acrylate	$C_5H_8O_2$	—	9	1.8−?	—
Methyl crotonate	$C_5H_8O_2$	—	−1	—	—
Methyl cyclopropanecarboxylate	$C_5H_8O_2$	—	17	—	—
Methyl methacrylate	$C_5H_8O_2$	—	8	2.1−12.5	—
Isopropenyl acetate	$C_5H_8O_2$	—	16	1.9−?	—
Vinyl propionate	$C_5H_8O_2$	—	1	—	—
Chlorocyclopentane	C_5H_9Cl	—	16	—	—
Pivaloyl chloride	C_5H_9ClO	—	8	—	—
Pivalonitrile	C_5H_9N	—	21	—	—
1,2,3,6-Tetrahydropyridine	C_5H_9N	—	16	—	—
Butyl isocyanate	C_5H_9NO	—	11	—	—
Cyclopentane	C_5H_{10}	49	−25	—	—
1,1-Dimethylcyclopropane	C_5H_{10}	20	—	—	—
Ethylcyclopropane	C_5H_{10}	36	< 10	—	—
2-Methyl-1-butene	C_5H_{10}	39	−20	—	—
2-Methyl-2-butene	C_5H_{10}	38	−20	—	—
3-Methyl-1-butene	C_5H_{10}	20	−7	1.5−9.1	—
Methylcyclobutane	C_5H_{10}	41	< 10	—	—
1-Pentene	C_5H_{10}	30	−18	1.5−8.7	—
2-Pentene	C_5H_{10}	30	−18	—	—
Allyl ethyl ether	$C_5H_{10}O$	—	−20	—	—
Ethyl propenyl ether	$C_5H_{10}O$	—	< −5	—	—
Isopropyl vinyl ether	$C_5H_{10}O$	—	−32	—	—
Isovaleraldehyde	$C_5H_{10}O$	—	−5	—	—
3-Methyl-2-butanone	$C_5H_{10}O$	—	−3	—	—
2-Methyl-3-butene-2-ol	$C_5H_{10}O$	—	18	—	—
2-Methyltetrahydrofuran	$C_5H_{10}O$	—	−12	—	—
2-Pentanone	$C_5H_{10}O$	—	7	1.6−7.2	—
3-Pentanone	$C_5H_{10}O$	—	13	1.6−?	—
4-Penten-2-ol	$C_5H_{10}O$	—	25	—	—
Tetrahdropyran	$C_5H_{10}O$	—	−20	—	—
Valeraldehyde	$C_5H_{10}O$	—	12	—	—
Butyl formate	$C_5H_{10}O_2$	—	18	1.7−8	—
3,3-Dimethoxypropene	$C_5H_{10}O_2$	—	19	—	—
2,2-Dimethyl-1,3-dioxolane	$C_5H_{10}O_2$	—	−1	—	—
Ethyl propionate	$C_5H_{10}O_2$	—	12	1.9−11	—
Isobutyl formate	$C_5H_{10}O_2$	—	5	2.0−8.9	—
Isopropyl acetate	$C_5H_{10}O_2$	—	4	1.7−7.8	—
2-Methoxyethyl vinyl ether	$C_5H_{10}O_2$	—	18(o)	—	—
Methyl butyrate	$C_5H_{10}O_2$	—	14	—	—
4-Methyl-1,3-dioxane	$C_5H_{10}O_2$	—	16	—	—
Methyl isobutyrate	$C_5H_{10}O_2$	—	13(o)	—	—
Propyl acetate	$C_5H_{10}O_2$	—	14	1.7−8	—
Diethyl carbonate	$C_5H_{10}O_3$	—	25	—	—
1-Bromo-3-methylbutane	$C_5H_{11}Br$	—	21	—	—
2-Bromopentane	$C_5H_{11}Br$	—	20	—	—
1-Chloro-3-methylbutane	$C_5H_{11}Cl$	—	< 21	1.5−7.4	—

Name	Formula	B.P./°C	Fl.P./°C	E.L./%	A.I.T./°C
2-Chloro-2-methylbutane	C$_5$H$_{11}$Cl	—	3	1.5–7.4	—
1-Chloropentane	C$_5$H$_{11}$Cl	—	1	1.6–8.6	—
2-Iodopentane	C$_5$H$_{11}$I	—	< 23	—	—
Cyclopentylamine	C$_5$H$_{11}$N	—	13	—	—
1-Methylpyrrolidine	C$_5$H$_{11}$N	—	3	—	—
Piperidine	C$_5$H$_{11}$N	—	3	—	—
4-Methylmorpholine	C$_5$H$_{11}$NO	—	24	—	—
Isopentyl nitrite	C$_5$H$_{11}$NO$_2$	—	−18	1–?(calc.)	209
Pentyl nitrite	C$_5$H$_{11}$NO$_2$	—	< 23	—	—
2,2-Dimethylpropane	C$_5$H$_{12}$	9	< −7	1.4–7.5	—
2-Methylbutane	C$_5$H$_{12}$	28	−51	1.4–7.6	—
Pentane	C$_5$H$_{12}$	36	−40	1.4–8.0	—
Butyl methyl ether	C$_5$H$_{12}$O	—	−10	—	—
tert-Butyl methyl ether	C$_5$H$_{12}$O	55	−28	2.5–15.1	—
Ethyl isopropyl ether	C$_5$H$_{12}$O	—	< −15	—	—
Ethyl propyl ether	C$_5$H$_{12}$O	—	< −20	1.7–9	—
tert-Pentanol	C$_5$H$_{12}$O	—	19	1.2–9	—
Diethoxymethane	C$_5$H$_{12}$O$_2$	—	(55	—	—
1,1-Dimethoxypropane	C$_5$H$_{12}$O$_2$	—	< 10	—	—
2,2-Dimethoxypropane	C$_5$H$_{12}$O$_2$	—	−7	—	—
Tetramethyl orthocarbonate	C$_5$H$_{12}$O$_4$	—	6	—	—
2-Methyl-2-butanethiol	C$_5$H$_{12}$S	—	−1	—	—
3-Methylbutanethiol	C$_5$H$_{12}$S	—	18	—	—
Pentanethiol	C$_5$H$_{12}$S	—	18	—	—
1,1-Dimethylpropylamine	C$_5$H$_{13}$N	—	−1	—	—
1,2-Dimethylpropylamine	C$_5$H$_{13}$N	—	−20	—	—
2,2-Dimethylpropylamine	C$_5$H$_{13}$N	—	25	—	—
N,N-Dimethylpropylamine	C$_5$H$_{13}$N	—	−11	—	—
Isopentylamine	C$_5$H$_{13}$N	—	−1	2.3–22	—
N-Methylbutylamine	C$_5$H$_{13}$N	—	2	—	—
Pentylamine	C$_5$H$_{13}$N	—	−1	—	—
2-Dimethylamino-N-methyl-ethylamine	C$_5$H$_{14}$N$_2$	—	14	—	—
Dimethylaminotrimethylsilane	C$_5$H$_{15}$NSi	—	−19	—	—
Hexafluorobenzene	C$_6$F$_6$	—	10	—	—
1,2,4,5-Tetrafluorobenzene	C$_6$H$_2$F$_4$	—	4(o)	—	—
1,2,4-Trifluorobenzene	C$_6$H$_3$F$_3$	—	−5(o)	—	—
2-, 3-, or 4-Chlorofluorobenzene	C$_6$H$_4$ClF	—	all 18	—	—
1,3-Difluorobenzene	C$_6$H$_4$F$_2$	—	< 0	—	—
1,4-Difluorobenzene	C$_6$H$_4$F$_2$	—	−5(o)	—	—
Fluorobenzene	C$_6$H$_5$F	—	−15	—	—
Benzene	C$_6$H$_6$	—	−11	1.4–8.0	—
1,5-Hexadien-3-yne	C$_6$H$_6$	—	< −20	1.5–?	—
1,3-Cyclohexadiene	C$_6$H$_8$	—	−6	—	—
1,4-Cyclohexadiene	C$_6$H$_8$	—	−11	—	—
2,5-Dimethylfuran	C$_6$H$_8$O	—	16	—	—
Cyclohexene	C$_6$H$_{10}$	—	−12	1.2–?/100°	310
1,3-Hexadiene	C$_6$H$_{10}$	—	−3	—	—
1,4-Hexadiene	C$_6$H$_{10}$	—	−21	2.0–6.1	—
1,5-Hexadiene	C$_6$H$_{10}$	—	−46	—	—
c- or t-2-trans-4-Hexadiene	C$_6$H$_{10}$	—	−7	—	—
1-, 2- or 3-Hexyne	C$_6$H$_{10}$	—	−10	—	—
2-Methyl-1,3-pentadiene	C$_6$H$_{10}$	—	−12	—	—
4-Methyl-1,3-pentadiene	C$_6$H$_{10}$	—	−34	—	—
Diallyl ether	C$_6$H$_{10}$O	—	−6	—	—
3-Methyl-2-methylenebutanal	C$_6$H$_{10}$O	—	11	—	—
Ethyl crotonate	C$_6$H$_{10}$O$_2$	—	2	—	—
Ethyl cyclopropanecarboxylate	C$_6$H$_{10}$O$_2$	—	18	—	—
Ethyl methacrylate	C$_6$H$_{10}$O$_2$	—	21	1.8–?	—
Vinyl butyrate	C$_6$H$_{10}$O$_2$	—	21(o)	1.4–8.8	—
Diallylamine	C$_6$H$_{11}$N	—	15	—	—
Cyclohexane	C$_6$H$_{12}$	—	−20	1.3–8.4	—
Ethylcyclobutane	C$_6$H$_{12}$	—	−15	1.2–7.7	210
1-Hexene	C$_6$H$_{12}$	—	−26	1.2–6.9	—
2-Hexene	C$_6$H$_{12}$	—	−21	—	—

Name	Formula	B.P./°C	Fl.P./°C	E.L./%	A.I.T./°C
Methylcyclopentane	C_6H_{12}	—	−29	—	—
2-Methyl-1-pentene	C_6H_{12}	—	−28	—	—
4-Methyl-1-pentene	C_6H_{12}	—	−7	—	—
cis 8–4-Methyl-2-pentene	C_6H_{12}	—	−32	—	—
trans 8-4-Methyl-2-pentene	C_6H_{12}	—	−29	—	—
Butyl vinyl ether	$C_6H_{12}O$	—	−1	—	—
3,3-Dimethyl-2-butanone	$C_6H_{12}O$	—	12	—	—
2-Ethylbutyraldehyde	$C_6H_{12}O$	—	21	1.2–7.7	—
2-Hexanone	$C_6H_{12}O$	—	23	1.2–8.0	—
3-Hexanone	$C_6H_{12}O$	—	14	—	—
Isobutyl vinyl ether	$C_6H_{12}O$	—	−9	—	—
2-Methylpentanal	$C_6H_{12}O$	—	20	—	195
3-Methylpentanal	$C_6H_{12}O$	—	< 23	—	—
2-Methyl-3-pentanone	$C_6H_{12}O$	—	1	—	—
3-Methyl-2-pentanone	$C_6H_{12}O$	—	15	—	—
4-Methyl-2-pentanone	$C_6H_{12}O$	—	17	1.2–8	—
Butyl acetate	$C_6H_{12}O_2$	—	23	1.4–7.5	—
2-Butyl acetate	$C_6H_{12}O_2$	—	18	1.3–7.5	—
2,6-Dimethyl-1,4-dioxane	$C_6H_{12}O_2$	—	24	—	—
Ethyl isobutyrate	$C_6H_{12}O_2$	—	13	—	—
2-Ethyl-2-methyl-1,3-dioxolane	$C_6H_{12}O_2$	—	23(o)	—	—
4-Hydroxy-4-methyl-2-pentanone	$C_6H_{12}O_2$	—	9	—	—
Isobutyl acetate	$C_6H_{12}O_2$	—	17	2.4–10.5	—
Isopentyl formate	$C_6H_{12}O_2$	—	23	1.2–8.0	—
Isopropyl propionate	$C_6H_{12}O_2$	—	21	—	—
Methyl isovalerate	$C_6H_{12}O_2$	—	20	—	—
Methyl pivalate	$C_6H_{12}O_2$	—	7	—	—
Methyl valerate	$C_6H_{12}O_2$	—	< 25	—	—
Isobutyl peroxyacetate	$C_6H_{12}O_3$	—	< 25	—	—
2,4,6-Trimethyltrioxane	$C_6H_{12}O_3$	—	17	1.3–?	—
Cyclohexylamine	$C_6H_{13}N$	—	21	—	—
1-Methylpiperidine	$C_6H_{13}N$	—	3	—	—
2-Methylpiperidine	$C_6H_{13}N$	—	10	—	—
3-Methylpiperidine	$C_6H_{13}N$	—	8	—	—
4-Methylpiperidine	$C_6H_{13}N$	—	9	—	—
Perhydroazepine	$C_6H_{13}N$	—	22	—	—
2,2-Dimethylbutane	C_6H_{14}	50	−48	(1.2–7.0/100°)	—
2,3-Dimethylbutane	C_6H_{14}	—	−29	1.2–7.0	—
Hexane	C_6H_{14}	—	−23	1.1–7.5	225
Isohexane	C_6H_{14}	—	−7	1.2–7.0	—
3-Methylpentane	C_6H_{14}	—	−7	1.2–7.0	—
Butyl ethyl ether	$C_6H_{14}O$	—	−1	—	—
tert-Butyl ethyl ether	$C_6H_{14}O$	—	−19	—	—
Diisopropyl ether	$C_6H_{14}O$	—	−36	1.4–21	—
Dipropyl ether	$C_6H_{14}O$	—	−21	—	215
2-Methyl-2-pentanol	$C_6H_{14}O$	—	21	—	—
1,1-Diethoxyethane	$C_6H_{14}O_2$	—	−21	1.7–10.4	—
1,2-Diethoxyethane	$C_6H_{14}O_2$	—	—	—	205
Triethylaluminium	$C_6H_{15}Al$	—	< −53	—	< −53
Triethyl borate	$C_6H_{15}BO_3$	—	11	—	—
Butylethylamine	$C_6H_{15}N$	—	18(o)	—	—
Diisopropylamine	$C_6H_{15}N$	—	−7	—	—
1,3-Dimethylbutylamine	$C_6H_{15}N$	—	13	—	—
Dipropylamine	$C_6H_{15}N$	—	17	—	—
Triethylamine	$C_6H_{15}N$	—	−7	1.2–8.0	—
1,2-Bis(dimethylamino)ethane	$C_6H_{16}N_2$	—	18	—	—
Diethoxydimethylsilane	$C_6H_{16}O_2Si$	—	11	—	—
Bis(trimethylsilyl) oxide	$C_6H_{18}OSi_2$	—	−1	—	—
Hexamethyldisilazane	$C_6H_{19}NSi_2$	—	14	—	—
Benzaldehyde	C_7H_6O	—	—	—	190
2-Fluorotoluene	C_7H_7F	—	8	—	—
3-Fluorotoluene	C_7H_7F	—	12	—	—
4-Fluorotoluene	C_7H_7F	—	10	—	—
Bicyclo[2.2.1]-2,5-heptadiene	C_7H_8	—	−21	—	—

Name	Formula	B.P./°C	Fl.P./°C	E.L./%	A.I.T./°C
1,3,5-Cycloheptatriene	C_7H_8	—	4	—	—
Toluene	C_7H_8	—	4	1.3–7.0	—
Cycloheptene	C_7H_{12}	—	−6	—	—
1-Heptyne	C_7H_{12}	—	−2	—	—
4-Methylcyclohexene	C_7H_{12}	—	−1(o)	—	—
Cycloheptane	C_7H_{14}	—	6	—	—
Ethylcyclopentane	C_7H_{14}	—	15	1.1–6.7	—
1-Heptene	C_7H_{14}	—	−1	—	—
2-Heptene	C_7H_{14}	—	< −1	—	—
3-Heptene	C_7H_{14}	—	< −7	—	—
Methylcyclohexane	C_7H_{14}	—	−4	1.2–6.7	—
2,3,3-Trimethylbutene	C_7H_{14}	—	< 0	—	—
2,4-Dimethyl-3-pentanone	$C_7H_{14}O$	—	15	—	—
3,3-Diethoxypropene	$C_7H_{14}O_2$	—	4	—	—
Ethyl isovalerate	$C_7H_{14}O_2$	—	25	—	—
Isobutyl propionate	$C_7H_{14}O_2$	—	18	—	—
Isopentyl acetate	$C_7H_{14}O_2$	—	25	(1.1–7.0/100°)	—
Isopropyl butyrate	$C_7H_{14}O_2$	—	24	—	—
Isopropyl isobutyrate	$C_7H_{14}O_2$	—	20	—	—
Pentyl acetate	$C_7H_{14}O_2$	—	25	1.1–7.5	—
2-Pentyl acetate	$C_7H_{14}O_2$	—	23	1.1–7.5	—
2,6-Dimethylpiperidine	$C_7H_{15}N$	—	16	—	—
1-Ethylpiperidine	$C_7H_{15}N$	—	19	—	—
2-Ethylpiperidine	$C_7H_{15}N$	—	−12	—	—
2,3-Dimethylpentane	C_7H_{16}	—	−56	1.1–6.7	—
2,4-Dimethylpentane	C_7H_{16}	—	−12	—	—
Heptane	C_7H_{16}	—	−4	1.2–6.7	223
2-Methylhexane	C_7H_{16}	—	−1	1.0–6.0	—
3-Methylhexane	C_7H_{16}	—	−1	1.0–6.0	—
2,2,3-Trimethylbutane	C_7H_{16}	—	−20	1.0–?	—
Phenylglyoxal	$C_8H_6O_2$	—	< 23	—	—
1,3,5,7-Cyclooctatetraene	C_8H_8	—	21	—	—
Ethylbenzene	C_8H_{10}	—	15	1.2–6.8	—
o-Xylene	C_8H_{10}	—	17	1.0–6.0	—
m- or p-Xylene	C_8H_{10}	—	25	1.1–7.0	—
4-Vinylcyclohexene	C_8H_{12}	—	16	—	—
1,7-Octadiene	C_8H_{14}	—	25	—	—
1-Octyne	C_8H_{14}	—	16	—	—
2-, 3-, or 4-Octyne	C_8H_{14}	—	all < 23	—	—
Vinylcyclohexane	C_8H_{14}	—	16	—	—
cis-1,2-Dimethylcyclohexane	C_8H_{16}	—	11	—	—
trans-1,2-Dimethylcyclohexane	C_8H_{16}	—	7	—	—
1,3-Dimethylcyclohexane	C_8H_{16}	—	6	—	—
cis-1,4-Dimethylcyclohexane	C_8H_{16}	—	10	—	—
trans-1,4-Dimethylcyclohexane	C_8H_{16}	—	16	—	—
1-Octene	C_8H_{16}	—	8	—	—
2-Octene	C_8H_{16}	—	14	—	—
2,3,4-Trimethyl-1-pentene	C_8H_{16}	—	5	—	—
2,4,4-Trimethyl-1-pentene	C_8H_{16}	—	−29	—	—
2,3,4-Trimethyl-2-pentene	C_8H_{16}	—	1	—	—
2,4,4-Trimethyl-2-pentene	C_8H_{16}	—	2	—	—
3,4,4-Trimethyl-2-pentene	C_8H_{16}	—	< 21	—	—
2-Ethylhexanal	$C_8H_{16}O$	—	—	—	197
2,3-Dimethylhexane	C_8H_{18}	—	7(o)	—	—
2,4-Dimethylhexane	C_8H_{18}	—	10(o)	—	—
3-Ethyl-2-methylpentane	C_8H_{18}	—	< 21	—	—
2-Methylheptane	C_8H_{18}	—	4	—	—
3-Methylheptane	C_8H_{18}	—	−23	—	—
Octane	C_8H_{18}	—	13	1.0–4.7	220
2,2,3-Trimethylpentane	C_8H_{18}	—	< 21	—	—
2,2,4-Trimethylpentane	C_8H_{18}	—	−12	1.1–6.0	—
2,3,3-Trimethylpentane	C_8H_{18}	—	<21	—	—
2,3,4-Trimethylpentane	C_8H_{18}	—	4	—	—
Dibutyl ether	$C_8H_{18}O$	—	25	1.5–7.6	194
Di-tert-butyl peroxide	$C_8H_{18}O_2$	—	12	—	—

Name	Formula	B.P./°C	Fl.P./°C	E.L./%	A.I.T./°C
Bis(2-ethoxyethyl) ether	$C_8H_{18}O_3$	—	—	—	205
2-(2-Butoxyethoxy)ethanol	$C_8H_{18}O_3$	—	—	—	225
Di-2-butylamine	$C_8H_{19}N$	—	24	—	—
Diisobutylamine	$C_8H_{19}N$	—	21	—	—
2,6-Dimethyl-3-heptene	C_9H_{18}	—	21(o)	—	—
1,3,5-Trimethylcyclohexane	C_9H_{18}	—	19	—	—
3,3-Diethylpentane	C_9H_{20}	—	< 21	0.7–7.7	—
2,5-Dimethylheptane	C_9H_{20}	—	23	—	—
3,5-Dimethylheptane	C_9H_{20}	—	23	—	—
4,4-Dimethylheptane	C_9H_{20}	—	21	—	—
3-Ethyl-2,3-dimethylpentane	C_9H_{20}	—	8	—	—
3-Ethyl-4-methylhexane	C_9H_{20}	—	24	—	—
4-Ethyl-2-methylhexane	C_9H_{20}	—	< 21	—	—
2- or 3-Methyloctane	C_9H_{20}	—	—	—	220
4-Methyloctane	C_9H_{20}	—	—	—	225
Nonane	C_9H_{20}	—	—	—	205
2,2,5-Trimethylhexane	C_9H_{20}	—	13	—	—
2,2,3,3-Tetramethylpentane	C_9H_{20}	—	< 21	0.8–4.9	—
2,2,3,4-Tetramethylpentane	C_9H_{20}	—	< 21	—	—
Dicyclopentadiene	$C_{10}H_{12}$	—	−7(o)	—	—
2-Ethylhexyl vinyl ether	$C_{10}H_{20}O$	—	—	—	202
Isopentyl isovalerate	$C_{10}H_{20}O_2$	—	24	—	—
Decane	$C_{10}H_{22}$	—	—	—	210
2-Methylnonane	$C_{10}H_{22}$	—	—	—	210
Dipentyl ether	$C_{10}H_{22}O$	—	—	—	170
tert-Butyl peroxybenzoate	$C_{11}H_{14}O_3$	—	19	—	—
Dihexyl ether	$C_{12}H_{26}O$	—	—	—	185
Triisobutylaluminium	$C_{12}H_{27}Al$	—	< 0	—	< 4
Tributylphosphine	$C_{12}H_{27}P$	—	—	—	200
Tetradecane	$C_{14}H_{20}$	—	—	—	200
Hexadecane	$C_{16}H_{34}$	—	—	—	205
Dichlorosilane	Cl_2H_2Si	8.3	—	—	58 ± 5°
Trichlorosilane	Cl_3HSi	32	−50	—	104
Deuterium	D_2	−249	—	5–75	—
Trifluorosilane	F_3HSi	−95	—	flammable	—
Tetrafluorohydrazine	F_4N_2	−73	—	—	explosive
Germane	GeH_4	−90	—	—	ambient
Digermane	Ge_2H_6	29	—	—	—
Hydrogen	H_2	−253	—	4.1–74.2	—
Hydrogen sulfide	H_2S	−62	—	4–44	—
Hydrogen disulfide	H_2S_2	—	< 22	—	—
Hydrogen selenide	H_2Se	−42	—	—	—
Hydrogen telluride	H_2Te	−2	—	—	50°
Ammonia	H_3N	−33	—	15.8–25	—
Phosphine	H_3P	−88	—	1.79–?	see text
Stibine	H_3Sb	−17	—	—	—
Hydrazine	H_4N_2	—	38	4.7–100	23 on rust 132 on iron 156 on st. steel
Diphosphane	H_4P_2	—	—	—	ambient
Silane	H_4Si	−112	—	1.4–?	−162
Disilane	H_6Si_2	−14.5	—	—	ambient
Trisilane	H_8Si_3	< 0	< 0	—	ambient
Phosphorus (white) P		—	—	—	38.5
Sulfur (hot solid) S		—	see text	see text	see text
Sulfur (hot liquid) S		—	see text	—	see text

Glossary of Abbreviations and Technical Terms

Aerobic	In presence of air
AIT	Autoignition temperature
Alloy	Mixture of 2 or more metals
Amalgam	Alloy of a metal with mercury
Ambient	Usual or surrounding
Anaerobic	In absence of air
Analogue	Compound of the same structural type
AO	Active oxygen content of peroxides
Aprotic	Without labile hydrogen atom(s)
Aqua regia	Mixture of nitric and hydrochloric acids
ARC	Accelerating rate calorimetry
ASTM	American Society for Testing and Materials
Autoxidation	Slow reaction with air
BAM	Bundes Anstalt fur Materialsprufung (similar to ASTM)
Basic	Fundamental, or, alkaline (acting as a base)
BLEVE	Boiling liquid expanding vapour explosion
Blowing agent	Material producing much gas on decomposition
b.p.	Boiling point
BSC	Bench scale calorimeter
Carbonaceous	Containing elemental carbon (as opposed to organic, containing combined carbon)
CHETAH	A computer program to predict energy release hazards
CIT	Critical ignition temperature
Class	Collection of related chemical groups or topics
COI	Critical oxygen index
conc.	concentrated
Congener	Compound with related but not identical structure
COT	Cyclooctatetraene
Critical diameter	Minimum diameter of an explosive charge capable of maintaining detonation
Cryogenic	At a very low (freezing) temperature
CVD	Chemical vapour deposition

Deflagration	Self sustaining internal combustion propagating by means of molecular heat transfer slower than the speed of sound (the explosion mechanism gunpowder and other 'low' explosives)
DH°f	Standard heat of formation
Desiccate	Dry intensively
Detonable	Capable of detonation
Detonation	A self sustaining decomposition reaction propagating faster than the speed of sound by means of a shock wave (the characteristic property of 'high' as opposed to 'low' explosives).
Diglyme	Diethyleneglycol dimethyl ether
Digol	Diethyleneglycol
DMF	Dimethylformamide
DMSO	Dimethyl sulfoxide
DSC	Differential scanning calorimetry
DTA	Differential thermal analysis
EL	Explosive limits (or Flammable limits), vol% in air
Endotherm	Absorption of heat
ESCA	Electron scanning chemical analysis
Exotherm.	Liberation of (reaction) heat
Freeze drying	Drying without heat by vacuum evaporation of frozen solvent
GLC	Gas-liquid chromatography
Glyme	Diethyleneglycol monomethyl ether
Halocarbon	Partially or fully halogenated hydrocarbon
HMPA	Hexamethylphosphoramide
HMSO	Her Majesty's Stationery Office
Homologue	Compound of the same (organic) series
Hypergolic	Ignites on contact
Ignition source	A source of energy which ignites flammables
IMS	Industrial methylated spirit (ethanol)
Induction period	Delay in reaction starting, caused by inhibitors
Inorganic	Not containing combined carbon, of mineral origin
Intermolecular	Between different molecules
Intramolecular	Within the same molecule
IR	Infrared spectroscopy
Initiation	Triggering of explosion or decomposition
LAH	Lithium tetrahydroaluminate
Lanthanide	Of the group of rare-earth metals
LRS	Laser Raman spectroscopy
LPG	Liquefied petroleum gas
MAPP	Methylacetylene/propadiene/propene mixture
Molecular sieve	A zeolite lattice with micropores of specific sizes, useful for molecular separations

m.p.	Melting point
MRH	Maximum reaction heat
NMR	Nuclear magnetic resonance spectrosopy
Nomenclature	System of naming chemicals
Off-spec.	Off-specification (low quality)
Oleum	Sulfar trioxide dissolved in sulfuric acid
Organic	A compound containing combined carbon
Organometallic	Containing carbon to metal bonding
Oxidant	Oxidising agent (electron sink)
PCB	Polychlorinated biphenyl
PTFE	Polytetrafluoroethylene
Propagation	Spread or transmission of decomposition, flame or explosion
Propellant	Energetic composition used in ballistics
PVC	Polyvinyl chloride
Pyrophoric	Igniting on contact with air (or frictional sparking)
Q	Heat of (exothermic) reaction or polymerisation
Quaternary salt	Tetra-substituted ammonium salt etc.
RCHD	Reactive Chemical Hazards Database, which you are using
Redox compound	Compound with reducing and oxidising features
Reducant	Reducing agent (electron source)
Refractory	Heat resisting
REITP2	A computer program to calculate MRH
RSC	Reaction Safety Calorimeter, or Royal Society of Chemistry
Runaway	Reaction out of control
Self-accelerating	Reaction catalysed by its own products
Self-heating	When substance generates heat faster than it is dissipated
Silicones	Organic derivatives of a polysiloxane chain
Slurry	Pourable mixture of solid and liquid
Smelt	Molten sodium sulfide and carbonate from evaporated sulfite liquor
Substance	Not a single chemical species, often of natural origin
Superheated	A liquid at a temperature above its boiling point
Tait24	Temperature of adiabatic storage which gives an induction time to exothermic decomposition of 24 hours
Thermite reaction	Reaction of aluminium powder and iron oxide producing molten iron (and analoguous reactions)
THF	Tetrahydroffiran
TGA	Thermogravimetric analysis
Thermochemistry	Study of heat effects of chemical reactions
TLC	Thin layer chromatography
TNT	Trinitrotoluene
U	Heat of (exothermic) decomposition

Unit operation	A single operational stage of a chemical process sequence
Unit process	A single chemical reaction stage in a process sequence
USNTIS	US National Technical Information Service
UV	Ultraviolet spectroscopy

Appendix 4

Index of Chemical Names and Synonyms used in Section 1

The majority of the names for chemicals in this alphabetically arranged index conform to one of the systematic series permitted under various sections of the IUPAC Definitive Rules for Nomenclature. Where there is a marked difference between these names and the alternative names recommended in the IUPAC-based BS2472:1983 or ASE 1985 nomenclature lists, or long established traditional names, these are given as synonyms in parentheses after the main title. These synonyms also have their own index entry, cross-referenced back to the IUPAC-based names used as bold titles in the text of Volume 1.

Those titles prefixed in the text with a dagger to show high flammability risks are prefixed similarly in this index. The four character numbers following the names or synonyms are the serial numbers of the entries, not page numbers.

It should be noted that italicised hyphenated prefixes which indicate structure, such as *cis-*, *o-*, *m-*, *tert-*, *mixo-*, *N-*, *O-*, etc., have been ignored during the alpha-sorting routine used on this index and the group-lists, while the roman character structural prefixes iso and neo, and roman multiplying prefixes such as di, tris, tetra and hexakis, have been included in the indexing procedure.

† Acetaldehyde, 0828
† Acetaldehyde oxime, 0865
Acetic acid, 0833
Acetic anhydride, 1534
Acetoacet-4-phenetidide, 3529
Acetohydrazide, 0912
† Acetone, 1220
Acetone oxime, 1258
† Acetonitrile, 0758
Acetonitrileimidazole-5,7,7,12,14,14-hexamethyl-1,4,8,11-tetraaza-
    4,11-cyclotetradecadieneiron(II) perchlorate, 3828
Acetonitrilethiol, *see* Mercaptoacetonitrile, 0767
Acetoxydimercurio(perchloratodimercurio)ethenone, 1413
1-Acetoxy-1-hydroperoxy-6-cyclodecanone, 3669
3-Acetoxy-4-iodo-3,7,7-trimethylbicyclo[4.1.0]heptane, 3544
Acetoxymercurio(perchloratomercurio)ethenone, 1412
4-Acetoxy-3-methoxybenzaldehyde, 3290
4-Acetoxy-3-methoxy-2-nitrobenzaldehyde, 3266
1-Acetoxy-6-oxo-cyclodecyl hydroperoxide, 3547
3-Acetoxy-1,2,4-trioxolane, *see* Vinyl acetate ozonide, 1543
2-Acetylamino-3,5-dinitrothiophene, 2280
Acetyl azide, 0771
Acetyl bromide, 0728
† Acetyl chloride, 0735
Acetyl cyclohexanesulfonyl peroxide, 3033
Acetyldimethylarsine, 1628
† Acetylene, 0686
Acetylenebis(triethyllead), 3672
Acetylenebis(triethyltin), 3674
Acetylenedicarboxaldehyde, 1403
Acetylenedicarboxylic acid, 1405
Acetyl hypobromite, 0729
Acetyl hypofluorite, 0751
2-Acetyl-3-methyl-4,5-dihydrothiophen-4-one, 2807
Acetyl nitrate, 0765
Acetyl nitrite, 0762
Acetyl nitro peroxide, *see* Peroxyacetyl nitrate, 0766
O-Acetylsalicylic acid, 3137
1-Acetyl-4-(4′-sulfophenyl)-3-tetrazene, 2982
4-(4-Acetyl-1-tetrazenyl)benzenesulfonic acid, *see* 1-Acetyl-4-(4′-
    sulfophenyl)-3-tetrazene, 2982
Aconitic acid, *see* *E*-Propene-1,2,3-tricarboxylic acid, 2342
† Acrylaldehyde, 1145
† Acrylamide, 1180
† Acrylic acid, 1148
† Acrylonitrile, 1107

† Acrylonitrile, 1107
† Acryloyl chloride, 1093
1,2-Cyclohexanedione, 2385a
Adipic acid, 2441
Alane–*N*,*N*-dimethylethylamine, 0071
† Allene, *see* Propadiene, 1124
all isomers except 3,5-, *see* Dinitroanilines, 2276
† 'Alloxan', *see* Pyrimidine-2,4,5,6-(1*H*,3*H*)-tetrone 1398
† Allyl acetate, 1912
† Allyl alcohol, *see* 2-Propen-1-ol, 1223
† Allylamine, *see* 3-Aminopropene, 1254
† Allyl azide, *see* 3-Azidopropene, 1188
Allyl benzenesulfonate, 3155
† Allyl bromide, *see* 3-Bromo-1-propene, 1153
† Allyl chloride, *see* 3-Chloropropene, 1158
Allyldimethylarsine, 1983
† Allyl ethyl ether, 1955
† Allyl formate, 1524
Allyl glycidyl ether, 2-Propenyloxymethyloxirane, *see* 1-Allyloxy-2,3-
    epoxypropane, 2434
Allyl hydroperoxide, 1226
† Allyl iodide, *see* 3-Iodopropene, 1174
Allyl isothiocyanate, 1471
Allyllithium, 1177
† Allyl mercaptan, *see* 2-Propene-1-thiol, 1239
Allylmercury(II) iodide, 1173
1-Allyloxy-2,3-epoxypropane, 2434
Allyl phosphorodichloridite, 1169
N-Allylthiourea, 1600
Allyl 4-toluenesulfonate, 3315
† Allyl vinyl ether, 1904
Allylzinc bromide, 1155
'Alumina', *see* Aluminium oxide 0087
Aluminium, 0048
Aluminium abietate, 3917
Aluminium amalgam, 0051
Aluminium azide, 0082
Aluminium bromide, 0060
Aluminium carbide, 1031
Aluminium chlorate, 0065
Aluminium chloride, 0062
Aluminium chloride–nitromethane, 0063
Aluminium chloride–trimethylaluminium complex, *see*
    Trimethyldialuminium trichloride, 1292
Aluminium–cobalt alloy, 0049
Aluminium copper(I) sulfide, 0084

Aluminium–copper–zinc alloy, 0050
Aluminium dichloride hydride diethyl etherate, 0061
Aluminium formate, 1089
Aluminium hydride, 0070
Aluminium hydride–diethyl ether, 0073
Aluminium hydride–trimethylamine, 0072
Aluminium hydroxide, 0074
Aluminium iodide, 0079
Aluminium isopropoxide, 3216
Aluminium–lanthanum–nickel alloy, 0080
Aluminium–lithium alloy, 0052
Aluminium–magnesium alloy, 0053
Aluminium–magnesium–zinc alloy, 0054
Aluminium–mercury alloy, *see* Aluminium amalgam, 0051
Aluminium–nickel alloys, 0055
Aluminium oxide, 0087
Aluminium perchlorate, 0066
Aluminium phosphide, 0083
Aluminium phosphinate, 0078
Aluminium stearate, 3914
Aluminium tetraazidoborate, 0059
Aluminium tetrahydroborate, 0058
Aluminium–titanium alloys, 0056
Americium trichloride, 0090
Amidosulfuric acid, 4499
Amidosulfuryl azide, *see* Sulfamoyl azide, 4472
4-Aminobenzenediazonium perchlorate, 2304
† 1-Aminobutane, *see* Butylamine, 1723
† 2-Aminobutane, *see* 2-Butylamine, 1724
2-Amino-4,6-dihydroxy-5-nitropyrimidine, 1436
2-Amino-4,6-dinitrophenol, 2278
2-Amino-3,5-dinitrothiophene, 1421
3-Amino-2,5-dinitrotoluene, 2783
2-(2-Aminoethyl)aminoethanol, *see* N-2-Hydroxyethyl-1,2-diaminoethane, 1756
2-(2-Aminoethylamino)-5-methoxynitrobenzene, 3172
2-Aminoethylammonium perchlorate, 0958
5-Amino-2-ethyl-2H-tetrazole, 1268
Aminoguanidine, 0507
Aminoguanidinium nitrate, 0512
3-Aminoisoxazole, 1136
5-Aminoisoxazole-3-carbonamide, 1476
4-Amino-3-isoxazolidinone, 1137
3-Amino-5-methylisoxazole, 1501
5-Amino-3-methylisoxazole, 1502
2-Amino-4-methyloxazole, 1500

Ammonium iodide, 4512
Ammonium iron(III) sulfate, 4393
Ammonium 3-methyl-2,4,6-trinitrophenoxide, 2803
Ammonium nitrate, 4522
Ammonium nitridoosmate, 4523
Ammonium nitrite, 4521
Ammonium *aci*-nitromethanide, 0504
Ammonium *N*-nitrosophenylaminooxide, 2399
Ammonium pentaperoxodichromate(2−), 4247
Ammonium perchlorate, 4004
Ammonium perchlorylamide, 4009
Ammonium periodate, 4514
Ammonium permanganate, 4518
Ammonium peroxoborate, 0146
Ammonium peroxodisulfate, 4576
Ammonium persulphate, *see* Ammonium peroxodisulfate, 4576
Ammonium phosphinate, 4554
Ammonium picrate, 2322
Ammonium sulfamate, *see* Ammonium amidosulfate, 4556
Ammonium sulfate, 4574
Ammonium sulfide, 4577
Ammonium tetrachromate(2−), 4253
Ammonium tetranitroplatinate(II), 4578
Ammonium tetraperoxochromate(3−), 4232
Ammonium thiocyanate, 0479
Ammonium thiosulfate, 4573
Ammonium trichromate(2−), 4252
Ammonium 2,4,5-trinitroimidazolide, 1144
Ammonium 2,4,6-tris(dioxoselena)hexahydrotriazine-1,3,5-triide, 4591
'Ammonium triselenimidate', *see* Ammonium 2,4,6-tris(dioxoselena)
    hexahydrotriazine-1,3,5-triide, 4591
Aniline, 2354
Anilinium chloride, 2365
Anilinium nitrate, 2379
Anilinium perchlorate, 2367
Anisaldehyde, *see* 4-Methoxybenzaldehyde, 2956
m-Anisidine, *see* 3-Methoxyaniline, 2817
o-Anisidine, *see* 2-Methoxyaniline, 2816
p-Anisidine, *see* 4-Methoxyaniline, 2818
Anisoyl chloride, *see* 4-Methoxybenzoyl chloride, 2930
Anisyl chloride, *see* 4-Methoxybenzyl chloride, 2962
Anthracene, 3632
Antimony, 4907
Antimony(III) chloride oxide, 4041
Antimony(III) nitride, 4729
Antimony(III) oxide, 4851

Antimony(III) oxide perchlorate, 4046
Antimonyl chloride, *see* Antimony trichloride oxide, 4150
Antimony pentachloride, 4184
Antimony trichloride, 4157
Antimony trichloride oxide, 4150
Antimony trisulfide, 4906
Aqua-1,2-diaminoethanediperoxochromium(IV), 0965
Aqua-1,2-diaminopropanediperoxochromium(IV) dihydrate, 1336
trans-Aquadioxo(terpyridine)ruthenium(2+) diperchlorate, 3687
Aquafluorobis(1,10-phenanthroline)chromium(III) perchlorate, 3846
Aquaglycinatonitratocopper, *see* Copper(II) glycinate nitrate, 0905
Argon, 0091
Arsenic, 0092
Arsenic pentafluoride, 0095
Arsenic pentaoxide, 0106
Arsenic trichloride, 0094
Arsenic trioxide, 0105
Arsenic trisulfide, 0108
Arsenous triazide, *see* Triazidoarsine 0103
† Arsine, 0100
Arsine–boron tribromide, 0101
Ascaridole, *see* 1,4-Epidioxy-2-*p*-menthene, 3347
2-Aza-1,3-dioxolanium perchlorate, 0898
2-Azatricyclo[2.2.1.0^{2,6}]hept-7-yl perchlorate, 2368
AZDN, *see* Azoisobutyronitrile, 3011
Azelaic acid, 3187
† Azetidine, 1255
Azidoacetaldehyde, 0772
Azidoacetic acid, 0774
† Azidoacetone, 1190
Azidoacetone oxime, 1215
Azidoacetonitrile, 0714
Azidoacetyl chloride, 0693
4-Azidobenzaldehyde, 2697
Azidobenzene, *see* Phenyl azide, 2271
7-Azidobicyclo[2.2.1]hepta-2,5-diene, 2777
2-Azido-1,3-butadiene, 1472
1-Azidobutane, 1666
Azido-2-butyne, 1473
N-Azidocarbonylazepine, 2728
Azidocarbonyl fluoride, 0339
Azidocarbonylguanidine, 0820
4-Azidocarbonyl-1,2,3-thiadiazole, 1069
2-*trans*-1-Azido-1,2-dihydroacenaphthyl nitrate, 3466
N-Azidodimethylamine, 0915
4-Azido-*N*,*N*-dimethylaniline, 3331

4,4'-Biphenylenebis(diazonium) perchlorate, 3457
4,4'-Biphenylene-bis-sulfonylazide, 3468
2,2'-Bipyridine N,N'-dioxide-dicarbonylrhodium(I) perchlorate, 3454
4,4'-Bipyridyl-bis(pentaammineruthenium(III) perchlorate), 3385
2,2'-Bipyridyldichloropalladium(IV) perchlorate, 3255
2,2'-Bipyridyl 1-oxide, 3258
1,2-Bis(acetoxyethoxy)ethane, 3355
Bis(acrylonitrile)nickel(0), 2312
Bis(2-aminoethyl)amine, 1777
Bis(2-aminoethyl)aminecobalt(III) azide, 1775
Bis(2-aminoethyl)aminediperoxochromium(IV), 1776
Bis(2-aminoethyl)aminesilver nitrate, 1773
N,N'-Bis(2-aminoethyl)1,2-diaminoethane, 2599
N,N-Bis(3-aminopropyl)-1,4-diazacycloheptanenickel(II) perchlorate, 3415
1,3-Bis(5-amino-1,3,4-triazol-2-yl)triazene, 1576
Bis(2-azidobenzoyl) peroxide, 3628
cis-1,2-Bis(azidocarbonyl)cyclobutane, 2323
1,2-Bis(azidocarbonyl)cyclopropane, 1835
1,2-Bis(2-azidoethoxy)ethane, 2479
Bis(2-azidoethoxymethyl)nitramine, 2481
N,N-Bis(azidomethyl)nitric amide, see 1,3-Diazido-2-nitroazapropane, 0824
3,3-Bis(azidomethyl)oxetane, 1902
Bis(azidothiocarbonyl) disulfide, 1016
Bis(benzeneazo) oxide, 3490
Bis(η-benzene)chromium(0), 3511
Bis(benzene)chromium dichromate, 3851
Bis(benzenediazo) sulfide, 3491
Bis(η-benzene)iron(0), 3512
Bis(benzeneiron)–fulvalenediyl complex, 3834
Bis(η-benzene)molybdenum(0), 3514
Bis(1-benzo[d]triazolyl) carbonate, 3598
Bis(1-benzo[d]triazolyl) oxalate, 3629
1,1-Bis(benzoylperoxy)cyclohexane, 3804
Bis(benzyl 1-methylhydrazinocarbodithioate N^2,S')(perchlorato-O,O')
    copper(1+) perchlorate, 3766
Bis[1,5-bis(4-methylphenyl)-1,3-pentaazadienato-N3,N5]-(T-4) cobalt, 3868
Bis(borane)–hydrazine, 0143
Bis(bromobenzenesulfonyl) peroxide, 3453
N,N-Bis(bromomercurio)hydrazine, 0268
1,2-Bis($tert$-butylphosphino)ethane, 3378
Bis(3-carboxypropionyl) peroxide, 2990
1,4-Bis(chlorimido)-2,5-cyclohexadiene, see Benzoquinone
    1,4-bis(chloroimine), 2159
Bis(4-chlorobenzenediazo) oxide, 3456
Bis(4-chlorobenzenesulfonyl) peroxide, 3458
I,I-Bis(3-chlorobenzoylperoxy)-4-chlorophenyliodine, 3793

N-Bromotetramethylguanidine, 2002
α-Bromo-γ-thiobutyrolactone, 1450
Bromotrichloromethane, 0310
3-Bromo-1,1,1-trichloropropane, 1126
† Bromotrifluoroethylene, 0579
Bromotrifluoromethane, 0311
N-Bromotrimethylammonium bromide(?), 1300
2-Bromo-2,5,5-trimethylcyclopentanone, 3018
Bromotrimethyl-*N,N,N',N'*-tetramethylethanediamine-*N,N'*-palladium, *see*
    N,N,N',N'-Tetramethylethane-1,2-diamine, trimethylpalladium(IV)
    bromide, 3225
Bromyl fluoride, 0239
Buckminsterfullerene, 3916
† 1,2-Butadiene, 1479
† 1,3-Butadiene, 1480
† 1,3-Butadiyne, 1385
    4,4'-(Butadiyne-1,4-diyl)bis(2,2,6,6-tetramethyl-4-hydroxypiperidine-1-oxyl),
    3838
† Butanal, *see* Butyraldehyde, 1607
† Butane, 1668
Butane-1,4-diisocyanate, 2376
1,4-Butanedinitrile, *see* Succinodinitrile, 1433
† Butane-2,3-dione, 1525
Butane-2,3-dione dioxime, 1595
2,3-Butanedione monoxime, 1570
† 2-Butanethiol, 1713
† Butanethiol, 1712
Butanoic acid, *see* Butyric acid, 1614
1-Butanol, 1694
† 2-Butanol, 1695
† *tert*-Butanol, 1696
† 2-Butanone, 1606
2-Butanone oxime, 1654
2-Butanone oxime hydrochloride, 1678
† Butanonitrile, *see* Butyronitrile, 1563
† Butanoyl chloride, *see* Butyryl chloride, 1555
† 2-Butenal, *see* Crotonaldehyde, 1516
† 1-Butene, 1577
† *cis*-2-Butene, 1578
† *trans*-2-Butene, 1579
trans-2-Butene-1,4-dioic acid, *see* Fumaric acid, 1446
trans-2-Butene ozonide, 1623
† 1-Buten-3-one, 1515
† 2-Butenonitrile, *see* 1-Cyanopropene, 1464
† 3-Butenonitrile, *see* 3-Cyanopropene, 1465
2-Buten-1-yl benzenesulfonate, 3316

Carbon tetrachloride, 0332
Carbon tetrafluoride, 0349
Carbon tetraiodide, 0525
Carbonyl azide, *see* Carbonic diazide, 0550
Carbonyl-bis(triphenylphosphine)iridium–silver diperchlorate, 3898
Carbonyl dichloride, 0329
Carbonyl dicyanide, *see* Oxopropanedinitrile, 1341
Carbonyl difluoride, 0343
Carbonyl diisothiocyanate, 1342
Carbonyl(pentasulfur pentanitrido)molybdenum, 0535
† Carbonyl sulfide, 0556
Carboxybenzenesulfonyl azide, 2699
2-Carboxy-3,6-dimethylbenzenediazonium chloride, 3138
Carboxymethyl carbamimoniothioate chloride, *see* S-Carboxymethyl-
    isothiouronium chloride, 1245
S-Carboxymethylisothiouronium chloride, 1245
'Carbyl sulfate', *see* 2,4-Dithia-1,3-dioxane-2,2,4,4-tetraoxide, 0839
Δ3-Carene, 3336
Ceric ammonium nitrate, Ammonium cerium(IV) nitrate, *see* Ammonium
    hexanitrocerate, 3964
Cerium, 3961
Cerium azide, 3966
Cerium carbide, 0590
Cerium dihydride, 3962
Cerium(III) tetrahydroaluminate, 0089
Cerium nitride, 3965
Cerium trihydride, 3963
Cerium trisulfide, 3967
Cetyltrimethylammonium permanganate, *see* Hexadecyltrimethylammonium
    permanganate, 3785
Chloral hydrate, *see* 2,2,2-Trichloro-1,1-ethanediol, 0745
Chloramide, *see* Chloramine, 4000
Chloramine, 4000
'Chloramine B', *see* Sodium N-chlorobenzenesulfonamide, 2229
Chloratomercurio(formyl)methylenemercury(II), 0654
Chloric acid, 3996
Chlorimide, *see* Dichloramine, 4063
3-Chlorimino-1-phenylpropene, *see* N-Chlorocinnamaldimin, 3126
Chloriminovanadium trichloride, 4165
Chlorine, 4047
Chlorine azide, 4030
Chlorine dioxide, 4042
Chlorine dioxygen trifluoride, 3983
Chlorine fluoride, 3971
Chlorine fluorosulfate, 3975
Chlorine(I) trifluoromethanesulfonate, 0320

'Chlorine nitrate', *see* Nitryl hypochlorite, 4026
Chlorine pentafluoride, 3989
Chlorine perchlorate, 4101
Chlorine trifluoride, 3981
Chlorine trifluoride oxide, 3982
Chlorine trioxide, 4044
Chloroacetaldehyde oxime, 0787
N-Chloroacetamide, 0789
Chloroacetamide, 0788
Chloroacetamide oxime, *see* 2-Chloro-*N*-hydroxyacetamidine, 0849
† Chloroacetone, 1161
4-Chloroacetophenone, 2929
Chloroacetylene, 0652
† 2-Chloroacrylonitrile, 1074
† *N*-Chloroallylamine, 1202
4-Chloro-2-aminophenol, 2303
† *N*-Chloro-3-aminopropene, *see* *N*-Chloroallylamine, 1202
N-Chloro-3-aminopropyne, 1129
2-Chloroaniline, 2300
3-Chloroaniline, 2301
4-Chloroaniline, 2302
1-Chloroaziridine, 0786
Chlorobenzene, 2228
2-Chloro-1,4-benzenediamine, 2349
4-Chloro-1,2-benzenediamine, 2347
4-Chloro-1,3-benzenediamine, 2348
2-Chlorobenzenediazonium salts, 2148
3-Chlorobenzenediazonium salts, 2149
4-Chlorobenzenediazonium triiodide, 2137
4-Chlorobenzenesulfonyl azide, 2152
1-Chlorobenzotriazole, 2150
4-Chlorobenzoyl azide, 2654
2-Chlorobenzylidenemalononitrile, 3333
N-Chloro-bis(2-chloroethyl)amine, 1590
2-Chloro-1,1-bis(fluorooxy)trifluoroethane, 0595
N-Chlorobis(trifluoromethanesulfonyl)imide, 0596
† 2-Chloro-1,3-butadiene, 1451
† 1-Chlorobutane, 1637
† 2-Chlorobutane, 1638
1-Chloro-2-butanone, 1556
4-Chlorobutanonitrile, *see* 4-Chlorobutyronitrile, 1487
† 2-Chloro-2-butene, 1551
† 3-Chloro-1-butene, 1552
1-Chloro-1-buten-3-one, 1454
4-Chloro-2-butynol, 1455
4-Chlorobutyronitrile, 1487

Chromium(II) sulfide, 4245
Chromium nitride, 4237
Chromium pentafluoride, 4226
Chromium trioxide, 4242
Chromocene, *see* Dicyclopentadienylchromium, 3273
Chromyl acetate, 1494
Chromyl azide, 4239
Chromyl azide chloride, 3968
Chromyl chloride, 4054
'[(Chromyldioxy)iodo]benzene', *see* Phenyliodine(III) chromate, 2247
Chromyl fluorosulfate, 4225
Chromyl isocyanate, 0613
Chromyl isothiocyanate, 0612
Chromyl nitrate, 4238
Chromyl perchlorate, 4055
Cinnamaldehyde, 3134
Citral, *see* 3,7-Dimethyl-2,6-octadienal, 3346
Citric acid, 2389
Citronellic acid, 3352
Cobalt, 4199
Cobalt(II) acetate tetrahydrate, *see* Diacetatotetraaquocobalt, 1780
Cobalt(II) azide, 4216
Cobalt(II) bromide, 0263
Cobalt(II) chelate of bi(1-hydroxy-3,5-diphenylpyrazol-4-yl *N*-oxide), 3872
Cobalt(II) chelate of 1,3-bis(*N*-nitrosohydroxylamino)benzene, 2164
Cobalt(II) chloride, 4048
Cobalt(III) amide, 4201
Cobalt(III) chloride, 4122
Cobalt(III) nitride, 4214
Cobalt(III) oxide, 4221
Cobalt(II) nitrate, 4215
Cobalt(II) oxide, 4217
Cobalt(II) perchlorate hydrates, 4051
Cobalt(II) picramate, 3460
Cobalt(II) sulfide, 4218
Cobalt trifluoride, 4200
Cobalt tris(dihydrogenphosphide), 4203
Columbium, 3948
Copper, 4267
Copper bis(1-benzeneazothiocarbonyl-2-phenyl-2-hydrazide), 3861
Copper chromate oxide, 4223
Copper diphosphide, 4284
Copper(I) azide, 4287
Copper(I) benzene-1,4-bis(ethynide), 3236
Copper(I) bromide, 0265
Copper(I) chloride, 4056

Cyanodimethylarsine, 1199
† 2-Cyanoethanol, 1181
N-(2-Cyanoethyl)ethanolamine, *see* 3-(2-Hydroxyethylamino)propionitrile, 1952
N-(2-Cyanoethyl)ethylhydroxylamine, 1951
3-Cyanoethylhydrazine, *see* 3-Hydrazinopropanenitrile, 1275
Cyanoform, 1383
Cyanoformyl chloride, 0600
Cyanogen azide, 0544
Cyanogen bromide, 0313
Cyanogen chloride, 0323
Cyanoguanidine, 0813
Cyanohydrazonoacetyl azide, 1083
5-Cyano-2-methyltetrazole, 1120
Cyanonitrene, 0540
2-Cyano-4-nitrobenzenediazonium hydrogen sulfate, 2668
4-Cyano-3-nitrotoluene, 2917
3(3-Cyano-1,2,4-oxadiazol-5-yl)-4-cyanofurazan 2-(or 5-) oxide, 2630
2-Cyano-2-propanol, 1566
† 1-Cyanopropene, 1464
† 3-Cyanopropene, 1465
1-Cyano-2-propen-1-ol, 1467
2-Cyano-2-propyl nitrate, 1506
3-Cyanopropyne, 1416
2(5-Cyanotetrazole)pentaamminecobalt(III) perchlorate, 0974
† Cyanotrimethylsilane, 1665
2-Cyano-1,2,3-tris(difluoroamino)propane, 1432
† Cyanuric acid, *see* 2,4,6-Trihydroxy-1,3,5-triazine, 1118
Cyanuric chloride, *see* 2,4,6-Trichloro-1,3,5-triazine, 1038
Cyclic poly(ethylene oxides), 0830
Cyclo-*oligo*-butadiynyl-2,3-dioxocycloutenylidene, *see* Oligo(octacarbon-dioxide), 3108
† Cyclobutane, 1580
1,2-Cyclobutanedione, 1440
† Cyclobutene, 1483
Cyclobutyl 4-methylbenzenesulfonate, 3404
Cyclobutylmethyl methanesulfonate, 2510
cis-Cyclododecene, 3351
† Cycloheptane, 2849
† 1,3,5-Cycloheptatriene, 2790
† Cycloheptene, 2837
† 1,3-Cyclohexadiene, 2361
† 1,4-Cyclohexadiene, 2362
Cyclohexadiene-1,4-dione, *see* 1,4-Benzoquinone, 2214
2,5-Cyclohexadienone-4-imine, *see* 1,4-Benzoquinone monoimine, 2260
† Cyclohexane, 2457

Diazido(trifluoromethyl)arsine, 0307
Diazirine, 0405
Diazirine-3,3-dicarboxylic acid, 1080
Diazoacetaldehyde, 0710
Diazoacetonitrile, 0675
Diazoacetyl azide, 0679
'Diazoaminomethane', *see* 1,3-Dimethyltriazene, 0945
2-Diazocyclohexanone, 2375
Diazocyclopentadiene, 1832
Diazodicyanoimidazole, *see* 2-Diazonio-4,5-dicyanoimidazolide, 2050
Diazodicyanomethane, *see* Diazomalononitrile, 1344
2-Diazo-2,6-dihydro-6-oxo-1.3-azulenedicarbonitrile, *see* 1,3-Dicyano-
     2-diazo-2,6-azulene quinone, 3434
2-Diazo-2*H*-imidazole, 1082
1-Diazoindene, 3116
Diazomalonic acid, 1081
Diazomalononitrile, 1344
Diazomethane, 0406
5-(Diazomethylazo)tetrazole, 0719
Diazomethyldimethylarsine, 1240
5-(4-Diazoniobenzenesulfonamido)thiazole tetrafluoroborate, 3119
2-Diazonio-4,5-dicyanoimidazolide, 2050
1-(4-Diazoniophenyl)-1,2-dihydropyridine-2-iminosulfinate, 3392
1(2'-Diazoniophenyl)2-methyl-4,6-diphenylpyridinium diperchlorate, 3847
3-Diazoniopyrazolide-4-carboxamide, 1399
3-Diazoniopyrazolide-4-carboxamide, 1422
5-Diazoniotetrazolide, 0548
3-Diazo-5-phenyl-3*H*-pyrazole, 3118
3-Diazo-5-phenyl-3*H*-1,2,4-triazole, 2909
4-Diazo-5-phenyl-1,2,3-triazole, 2910
† 3-Diazopropene, 1135
Diazopropyne, 1070
α-Diazo-γ-thiobutyrolactone, 1434
3-Diazo-3*H*-1,2,4-triazole, 0677
4-Diazo-1,2,3-triazole, 0678
Dibenzenesufonyl peroxide, 3499
2,3:5,6-Dibenzobicyclo[3.3.0]hexane, 3633
Dibenzocyclododeca-1,7-dienetetrayne, 3786
Dibenzo-d,f-1,2-dioxocin-3,8-dione, *see* 2,2-Biphenyldicarbonyl peroxide,
     3631
1,5-Dibenzoylnaphthalene, 3845
Dibenzoyl peroxide, 3639
'Dibenzyl chlorophosphate', *see* Dibenzyl phosphorochloridate, 3650
Dibenzyl ether, 3655
Dibenzyl phosphite, 3658
Dibenzyl phosphorochloridate, 3650

Dibutyl hyponitrite, 3069
trans-Di-*tert*-butyl hyponitrite, 3070
Dibutylmagnesium, 3068
Dibutyl-3-methyl-3-buten-1-ynlborane, 3618
2,6-Di-*tert*-butyl-4-nitrophenol, 3666
† Di-*tert*-butyl peroxide, 3074
2,2-Di(*tert*-butylperoxy)butane, 3565
Dibutyl phthalate, 3714
2,4-Di-*tert*-butyl-2,2,4,4-tetrafluoro-1,3-dimethyl-1,3,2,4-diazadiphos-
    phetidine, 3377
Dibutylthallium isocyanate, 3193
Dibutylzinc, 3080
Di-3-camphoroyl peroxide, 3807
Dicarbonyl-π-cycloheptatrienyltungsten azide, 3123
cis-Dicarbonyl(cyclopentadienyl)cyclooctenemanganese, 3689
Dicarbonylmolybdenum diazide, 0995
Dicarbonyl(phenanthroline *N*-oxide)rhodium(I) perchlorate, 3626
Dicarbonylpyrazinerhodium(I) perchlorate, 2147
Dicarbonyl-η-trichloropropenyldinickel chloride dimer, 3233
Dicarbonyltungsten diazide, 1013
Dichloramine, 4063
Dichlordifluoromethane, 0326
Dichlorine heptaoxide, *see* Perchloryl perchlorate, 4107
Dichlorine oxide, 4095
Dichlorine trioxide, 4100
Dichloroacetylene, 0602
† *N*,*N*-Dichloro-β-alanine, 1168
† 2-Dichloroaminopropanoic acid, *see* *N*,*N*-Dichloro-β-alanine, 1168
1-Dichloroaminotetrazole, 0371
N,*N*-Dichloroaniline, 2242
2,3-Dichloroaniline, 2237
2,4-Dichloroaniline, 2238
2,5-Dichloroaniline, 2239
2,6-Dichloroaniline, 2240
3,4-Dichloroaniline, 2241
1,2-Dichlorobenzene, 2156
N,*N*-Dichlorobenzenesulfonamide, 2244
2,6-Dichlorobenzoquinone-4-chloroimide, 2078
cis-Dichlorobis(2,2′-bipyridyl)cobalt(III) chloride, 3799
trans-Dichlorobis[1,2-phenylenebis(dimethylarsine)]palladium(IV)
    diperchlorate, 3808
N,*N*′-dichlorobis(2,4,6-trichlorophenyl)urea, 3596
Dichloroborane, 0124
1,4-Dichloro-1,3-butadiyne, 1353
† *mixo*-Dichlorobutane, 1587
1,4-Dichloro-2-butyne, 1428

Dimethyl 4-nitrophenyl thionophosphate, 2979
N,N-Dimethyl-4-nitrosoaniline, 2980
1,2-Dimethylnitrosohydrazine, 0946
3,7-Dimethyl-2,6-octadienal, 3346
O,O-Dimethyl-S-[(4-oxo-1,2,3-benxotriazin-3[4H]-yl)methyl]
    phosphorodithioate, 3307
† 2,3-Dimethylpentane, 2871
† 2,4-Dimethylpentane, 2872
† 2,4-Dimethyl-3-pentanone, 2857
S,S-Dimethylpentasulfur hexanitride, 0917
Dimethyl peroxide, 0923
Dimethyl peroxydicarbonate, 1544
Dimethylphenylarsine, 2991
Dimethyl(phenylethynyl)thallium, 3298
Dimethylphenylphosphine, 3003
Dimethylphenylphosphine oxide, 3002
3,3-Dimethyl-1-phenyltriazene, 2996
Dimethylphosphine, 0948
Dimethyl phosphoramidate, 0952
Dimethylphosphorohydrazine, see Dimethyl hydrazidophosphate, 0959
† 2,6-Dimethylpiperidine, 2868
† 2,2-Dimethylpropane, 1999
† 1,1-Dimethylpropanol, see tert-Pentanol, 2015
† N,N-Dimethylpropylamine, 2031
† 1,1-Dimethylpropylamine, 2028
† 1,2-Dimethylpropylamine, 2029
† 2,2-Dimethylpropylamine, 2030
Dimethylpropyleneurea, see 1,3-Dimethylhexahydropyrimidone, 2475
Dimethyl-1-propynylthallium, 1938
2,5-Dimethylpyrazine 1,4-dioxide, 2378
2,6-Dimethylpyridine N-oxide, 2819
3,3-Dimethyl-1-(3-quinolyl)triazene, 3398
Dimethyl selenate, 0930
(Dimethylsilylmethyl)trimethyllead, 2605
† Dimethyl sulfate, 0929
† Dimethyl sulfide, 0932
Dimethyl sulfite, 0927
† Dimethyl sulfoxide, 0921
Dimethyl terephthalate, 3292
3,6-Dimethyl-1,2,4,5-tetraoxane, 1625
Dimethylthallium fulminate, 1211
Dimethylthallium N-methylacetohydroxamate, 2008
Dimethyltin dinitrate, 0914
2,2-Dimethyltriazanium perchlorate, 0963
1,3-Dimethyltriazene, 0945
1,2-Dimethyl-2-trimethylsilylhydrazine, 2047

Dinitrogen oxide, 4744
Dinitrogen pentaoxide, 4748
Dinitrogen tetraoxide, 4747
Dinitrogen tetraoxide, *see* Nitrogen dioxide, 4725
Dinitrogen trioxide, 4746
3,5-Dinitro-4-hydroxybenzenediazonium 2-oxide, 2090
1,3-Dinitro-2-imidazolidinone, 1142
Dinitromethane, 0410
N,N'-Dinitromethanediamine, lead(II) salt, *see* Lead methylenebis(nitramide), 0412
3,5-Dinitro-2-methylbenzenediazonium-4-oxide, 2666
3,5-Dinitro-6-methylbenzenediazonium-2-oxide, 2667
2,5-Dinitro-3-methylbenzoic acid, 2919
N,N'-Dinitro-N-methyl-1,2-diaminoethane, 1276
1,5-Dinitronaphthalene, 3242
1,4-Dinitropentacyclo[4.2.0.0^{2,5}.0^{3,8}.0^{4,7}]octane, 2918
2,4-Dinitropentane isomers, 1953
2,6-Dinitro-4-perchlorylphenol, 2101
2,4-Dinitrophenol, 2197
2,4-Dinitrophenylacetyl chloride, 2902
2,4-Dinitrophenylhydrazine, 2321
2,4-Dinitrophenylhydrazinium perchlorate, 2351
O-(2,4-Dinitrophenyl)hydroxylamine, 2279
3,8-Dinitro-6-phenylphenanthridine, 3780
3,5-Dinitro-2-(picrylazo)pyridine, 3387
4,6-Dinitro-1-picrylbenzotriazole, 3438
5,7-Dinitro-1-picrylbenzotriazole, 3439
2,2-Dinitropropylhydrazine, 1277
1-(3,5-Dinitro-2-pyridyl)-2-picrylhydrazine, 3389
4,6-Dinitroresorcinol, *see* 4,6-Dinitro-1,3-benzenediol, 2198
3,5-Dinitrosalicylic acid, 2665
3,7-Dinitroso-1,3,5,7-tetraazabicyclo[3.3.1]nonane, 1954
Dinitrosylnickel, 4741
3,5-Dinitro-2-toluamide, 2941
2,4-Dinitrotoluene, 2726
4,6-Dinitro-2-(2,4,6-trinitrophenyl)benzotriazole N-oxide, 3441
7,8-Dioxabicyclo[4.2.2]-2,4,7-decatriene, 2955
† 1,3-Dioxane, 1616
† 1,4-Dioxane, 1617
cis-1,4-Dioxenedioxetane, 1538
1,4-Dioxobut-2-yne, *see* Acetylenedicarboxaldehyde, 1403
3,6-Dioxo-1,2-dioxane, 1445
† 1,3-Dioxolane, 1228
† 1,3-Dioxol-4-en-2-one, 1087
Dioxonium hexamanganato(VII)manganate, 4553

5,12-Dioxo-4,4,11,11-tetrahydroxy-14*H*-[1,2,5]oxadiazolo[3,4-*e*]
  [1,2,5]oxadiazolo[3′,4′:4,5]benzotriazolo[2,1-*a*]benzotriazolo-*b*-ium
  1,8-dioxide, 3440
1,1′-Dioxybiscyclohexanol, *see* Bis(1-hydroxycyclohexyl) peroxide, 3553
Dioxydicyclohexylidenebishydroperoxide, *see* Bis(1-hydroperoxycyclohexyl)
  peroxide, 3556
Dioxygen difluoride, 4320
Dioxygenyl tetrafluoroborate, 0132
† Dipentyl ether, 3369
2,6-Diperchloryl-4,4-diphenoquinone, 3444
Diperoxyazelaic acid, 3189
Diperoxyterephthalic acid, 2925
Diphenylacetylene, *see* Diphenylethyne, 3634
Diphenylamine, 3504
Diphenyl azidophosphate, 3489
2,5-Diphenyl-3,4-benzofuran-2,5-endoperoxide (1,4-Epoxy-1,4-dihydro-
  1,4-diphenyl-2,3-benzodioxin), 3798
4,4′-Diphenyl-2,2′-bi(1,3-dithiol)-2′-yl-2-ylium perchlorate, 3746
1,6-Diphenyl-2,5-bis(diazo)-1,3,4,6-tetraoxohexane, 3744
Diphenylcyclopropenylium perchlorate, 3680
Diphenyldiazene, *see* Azobenzene, 3483
Diphenyl diselenide, 3501
Diphenyldistibene, 3500
Diphenyl ether, 3492
1,1-Diphenylethylene, 3642
trans-1,2-Diphenylethylene, 3643
Diphenylethyne, 3634
1,2-Diphenylhydrazine, 3517
1,1-Diphenylhydrazinium chloride, 3519
Diphenylmagnesium, 3482
Diphenylmercury, 3480
1,8-Diphenyloctatetrayne, 3788
1,5-Diphenylpentaazadiene, 3507
N,*N*-Diphenyl-3-phenylpropenylidenimmonium perchlorate
Diphenylphosphine, 3508
Diphenylphosphorus(III) azide, 3488
Diphenylselenone, 3494
Diphenyl sulfoxide, 3493
2,2-Diphenyl-1,3,4-thiadiazoline, 3647
Diphenyltin, 3502
1,3-Diphenyltriazene, 3506
5,6-Diphenyl-1,2,4-triazine-3-diazonium tetrafluoroborate 2-oxide, 3677
† Diphosphane, 4538
1,2-Diphosphinoethane, 0957
Diphosphoryl chloride, 4168
Dipotassium μ-cyclooctatetraene, 2947

2,3-Epoxypropionaldehyde 2,4-dinitrophenylhydrazone, 3130
2,3-Epoxypropionaldehyde oxime, 1182
4(2,3-Epoxypropoxy)butanol, 2866
† 2,3-Epoxypropyl nitrate, 1186
3(2,3-Epoxypropyloxy)2,2-dinitropropyl azide, 2403
Erbium perchlorate, 4132
† Ethanal, *see* Acetaldehyde, 0828
† Ethanamine, *see* Ethylamine, 0942
† Ethane, 0881
† Ethanedial, 0723
† Ethanedinitrile, *see* Dicyanogen, 0996
Ethanedioic acid, *see* Oxalic acid, 0725
1,2-Ethanediol, *see* Ethylene glycol, 0924
'Ethane hexamercarbide', *see* 1,2-Bis(hydroxomercurio)-1,1,2,2-
    bis(oxydimercurio)ethane, 0708
† Ethanenitrile, *see* Acetonitrile, 0758
Ethaneperoxoic acid, *see* Peroxyacetic acid, 0837
† Ethanethiol, 0933
Ethanoic acid, *see* Acetic acid, 0833
† Ethanol, 0920
Ethanolamine, *see* 2-Hydroxyethylamine, 0943
Ethanolamine perchlorate, *see* 2-Hydroxyethylaminium perchlorate, 0951
† Ethanoyl chloride, *see* Acetyl chloride, 0735
Etheneamine, *see* Vinylamine, 0864
† Ethene, *see* Ethylene, 0781
Ethenone, *see* Ketene, 0722
† 3-Ethenoxypropene, *see* Allyl vinyl ether, 1904
Ethenylbenzene, *see* Styrene, 2945
† Ethenyl ethanoate, *see* Vinyl acetate, 1532
Ethenylmagnesium chloride, 0732
† Ethenyl methanoate, *see* Vinyl formate, 1149
† Ethoxyacetylene, 1521
4-Ethoxybutyldiethylaluminium, 3372
3-Ethoxycarbonyl-4,4,5,5-tetracyano-3-trimethylplumbyl-4,5-dihydro-
    3*H*-pyrazole, 3611
Ethoxydiisobutylaluminium, 3373
† Ethoxyethane, *see* Diethyl ether, 1697
2-Ethoxyethanol, 1702
† Ethoxyethene, *see* Ethyl vinyl ether, 1610
2-Ethoxyethyl acetate, 2511
† Ethoxyethyne, *see* Ethoxyacetylene, 1521
2-Ethoxy-1-iodo-3-butene, 2449
4-Ethoxy-2-methyl-3-butyn-2-ol, 2844
3-Ethoxymethylene-2,4-pentanedione, 3014
† 3-Ethoxypropene, *see* Allyl ethyl ether, 1955
† 1-Ethoxypropene, *see* Ethyl propenyl ether, 1956

2-Furoyl azide, 1821
Furoyl chloride, 1818
2-Furylchlorodiazirine, 1817

Gallium, 4406
Gallium azide, 4409
Gallium(I) oxide, 4411
Gallium perchlorate, 4135
† Germane, 4417
Germanium, 4412
Germanium(II) sulfide, 4419
Germanium imide, 4414
Germanium isocyanate, 1377
Germanium tetrachloride, 4162
Glucose, 2518
Glutarodinitrile, 1870
Glutaryl diazide, 1874
Glycerol, 1286
† Glyceryl trinitrate, 1196
Glycidyl azide, 1191
Glycidyl phenyl ether, *see* 3-Phenoxy-1,2-epoxypropane, 3150
Glycolonitrile, 0760
† Glyoxal, *see* Ethanedial, 0723
Glyoxylic acid, 0724
Gold, 0110
Gold(I) acetylide, 0573
Gold(I) cyanide, 0308
Gold(III) chloride, 0111
Gold(III) hydroxide–ammonia, 0112
Gold(III) nitride trihydrate, 0118
Gold(III) oxide, 0115
Gold(III) sulfide, 0116
Gold(I) nitride–ammonia, 0117
Graphite hexafluorogermanate, 3422
Guanidinium dichromate, 0971
Guanidinium nitrate, 0509
Guanidinium perchlorate, 0502

Hafnium, 4599
Hafnium carbide, 0521
Hafnium(IV) tetrahydroborate, 0182
Hafnium tetrachloride, 4163
Heptafluorobutyramide, 1392
Heptafluorobutyryl hypochlorite, 1352
Heptafluorobutyryl hypofluorite, 1369

Heptafluorobutyryl nitrite, 1368
Heptafluoroisopropyl hypochlorite, 1035
2-Heptafluoropropyl-1,3,4-dioxazolone, 1812
Heptafluoropropyl hypofluorite, 1060
Heptakis(dimethylamino)trialuminium triboron pentahydride, 3676
† Heptane, 2873
1,1,1,3,5,5,5-Heptanitropentane, 1854
Heptasilver nitrate octaoxide, 0047
Heptaspiro[2.4.2.4.2.4.2.4.2.4.2.4.2.4]nonatetraconta-
    4,6,11,13,18,20,25,27,32,34,39,41,46,48-tetradecayne, 3912
Hepta-1,3,5-triyne, 2641
† 1-Heptene, 2851
† 2-Heptene, 2852
† 3-Heptene, 2853
1-Heptene-4,6-diyne, 2707
† 1-Heptyne, 2838
2-Heptyn-1-ol, 2841
Hexaamminecalcium, 3933
Hexaamminechromium(III) nitrate, 4234
Hexaamminechromium(III) perchlorate, 4129
Hexaamminecobalt(III) chlorate, 4125
Hexaamminecobalt(III) hexanitrocobaltate(3−), 4219
Hexaamminecobalt(III) iodate, 4208
Hexaamminecobalt(III) nitrate, 4210
Hexaamminecobalt(III) perchlorate, 4126
Hexaamminecobalt(III) permanganate, 4209
Hexaamminenickel chlorate, 4074
Hexaamminenickel perchlorate, 4075
1,1,3,3,5,5-Hexaazido-2,4,6-triaza-1,3,5-triphosphorine, 4795
Hexaborane(10), 0191
Hexaborane(12), 0192
Hexabromoethane, 0584
Hexacarbonylchromium, 2056
Hexacarbonylmolybdenum, 2628
Hexacarbonyltungsten, 2636
Hexacarbonylvanadium, 2635
Hexachlorobenzene, 2055
1,2,3,4,5,6-Hexachlorocyclohexane, 2306
Hexachlorocyclopentadiene, 1808
Hexachlorocyclotriphosphazine, 4189
Hexachlorodisilane, 4191
1,2,3,4,10,10-Hexachloro-6,7-epoxy-1,4,4a,5,6,7,8,8a-octahydro-
    endo,endo-1,4:5,8-dimethanonaphthalene), *see* Endrin, 3459
Hexachloroethane, 0611
Hexachloromelamine, *see* 2,4,6-Tris(dichloroamino)-1,3,5-triazine, 1043

Hexacyclohexyldilead, 3894
† Hexadecane, 3726
Hexadecanethiol, 3727
N-Hexadecylpyridinium permanganate, 3827
Hexadecyltrimethylammonium permanganate, 3785
2,4-Hexadienal, 2383
† 1,3-Hexadiene, 2409
† 1,4-Hexadiene, 2410
† 1,5-Hexadiene, 2411
† cis-2-trans-4-Hexadiene, 2412
† trans-2-trans-4-Hexadiene, 2413
2,4-Hexadienoic acid, 2385b
1,3-Hexadien-5-yne, 2290
† 1,5-Hexadien-3-yne, 2291
4,5-Hexadien-2-yn-1-ol, 2328
1,5-Hexadiyne, 2292
2,4-Hexadiyne, 2293
2,4-Hexadiyne-1,6-dioic acid, 2094
2,5-Hexadiyn-1-ol, 2329
1,5-Hexadiyn-3-one, 2213
2,4-Hexadiynylene chloroformate, 2890
2,4-Hexadiynylene chlorosulfite, 2162
Hexaethyltrialuminium trithiocyanate, 3695
† Hexafluorobenzene, 2060
† 2,2,2,4,4,4-Hexafluoro-1,3-dimethyl-1,3,2,4-diazadiphosphetidine, 0906
Hexafluoroglutaryl dihypochlorite, 1806
† Hexafluoroisopropylideneamine, 1067
Hexafluoroisopropylideneaminolithium, 1052
Hexafluorophosphoric acid, see Hydrogen hexafluorophophosphate, 4360
† Hexafluoropropene, 1051
Hexahydro[18]annulene, see 1,7,13-Cyclooctadecatriene-3,5,9,11,15,17-
    hexayne, 3740
Hexahydropyrazine, see Piperazine, 1689
Hexahydroxylaminecobalt(III) nitrate, 4211
Hexakis(pyridine)iron(II) tridecacarbonyltetraferrate(2−), 3906
Hexalithium disilicide, 4689
Hexamethyldiplatinum, 2606
Hexamethyldisilane, 2608
† Hexamethyldisilazane, 2610
5,6,14,15,20,21-Hexamethyl-1,3,4,7,8,10,12,13,16,17,19,22-
    dodecaazatetracyclo[8.8.4.1^{3,17}.1^{8,12}]tetracosa-4,6,13,15,19,21-
    hexaene-N^4,N^7,N^{13},N^{16},N^{19},N^{22}cobalt(II) perchlorate, 3768
5,6,14,15,20,21-Hexamethyl-1,3,4,7,8,10,12,13,16,17,19,22-
    dodecaazatetracyclo[8.8.4.1^{3,17} 1^{8,12}]tetracosa-4,6,13,15,19,21-
    hexaene-N^4,N^7,N^{13},N^{16},N^{19},N^{22}iron(II) perchlorate, 3769

5,6,14,15,20,21-Hexamethyl-1,3,4,7,8,10,12,13,16,17,19,22-
    dodecaazatetracyclo[8.8.4.1^{3,17}.1^{8,12}]tetracosa-4,6,13,15,19,21-
    hexaene-N^4,N^7,N^{13},N^{16},N^{19},N^{22}nickel(II) perchlorate, 3770
† Hexamethyleneimine, *see* Perhydroazepine, 2526
Hexamethylenetetramine, 2477
Hexamethylenetetramine tetraiodide, 2471
Hexamethylenetetrammonium tetraperoxochromate(V)? 3779
'Hexamethylenetriperoxydiamine', *see* 1,6-Diaza-3,4,8,9,12,13-
    hexaoxabicyclo[4.4.4]tetradecane, 2476
Hexamethylerbium–hexamethylethylenediaminelithium complex, 2592
2,4,6,8,9,10-Hexamethylhexaaza-1,3,5,7-tetraphosphaadamantane, 2600
3,3,6,6,9,9-Hexamethyl-1,2,4,5,7,8-hexoxonane, 3195
2,7,12,18,23,28-Hexamethyl-3,7,11,19,23,27,33,34-octaazatricyclo
    [27.3.1.1(13,17)]tetratriaconta-1(33),2,11,13,15,17(34),
    18,27,29,31-decaenediaquadichlorodiiron(II) diperchlorate, 3880
Hexamethylrhenium, 2607
5,7,7,12,14,14-Hexamethyl-1,4,8,11-tetraaza-4,11,-cyclotetradecadiene-
    1,10-phenanthrolineiron(II) perchlorate, 3869
1,1,1,3,3,3-Hexamethyl-2-(trimethylsilyl)trisilane, *see*
    Tris(trimethylsilyl)silane, 3231
Hexamethyltrisiloxane, 2617
2,2,4,4,6,6-Hexamethyltrithiane, 3197
Hexamethyltungsten, 2609
† Hexane, 2531
Hexane-1,6-dioic acid, *see* Adipic acid, 2441
1,6-Hexanediyl perchlorate, 2470
6-Hexanelactam, *see* Caprolactam, 2451
Hexanitrobenzene, 2632
Hexanitroethane, 1014
Hexanitrohexaazaisowurtzitane, 2325
Hexanitrosobenzene, *see* Benzotri(furazan *N*-oxide), 2631
† 2-Hexanone, 2488
† 3-Hexanone, 2489
1,4,7,10,13,16-Hexaoxacyclooctadecane, 3561
Hexaoxygen difluoride, 4327
Hexaphenylhexaarsane, 3891
Hexaspiro[2.4.2.4.2.4.2.4.2.4.2.4]dotetraconta-4,6,11,13,18,20,25,
    27,32,34,39,41-dodecayne, 3905
1,3,5-Hexatriyne, 2074
Hexaureachromium(III) nitrate, 2623
Hexaureagallium(III) perchlorate, 2620
2-Hexenal, 2432
† 1-Hexene, 2459
† 2-Hexene, 2460
trans-2-Hexene ozonide, 2513
Hexyl perchlorate, 2520

Hydroperoxymethyl formate, 0838
2-Hydroperoxy-2-methylpropane, *see* *tert*-Butyl hydroperoxide, 1698
2-Hydroperoxypropane, *see* Isopropyl hydroperoxide, 1283
Hydroquinone, *see* 1,4-Benzenediol, 2333
Hydroquinone–oxygen clathrate, *see* 1,4-Benzenediol–oxygen complex, 2334
† Hydroximinoethane, *see* Acetaldehyde oxime, 0865
Hydroxobis[5,7,7,12,14,14-hexamethyl-1,4,8,11-tetraaza-
    4,11-cyclotetradecadieneiron(II)] triperchlorate, 3881
Hydroxyacetone, 1231
Hydroxyacetonitrile, *see* Glycolonitrile, 0760
Hydroxyaqua(oxo)diperoxorheniumVII, 4507
N-Hydroxybenzeneamine, *see* *N*-Phenylhydroxylamine, 2356
4-Hydroxybenzenediazonium-3-carboxylate, 2663
Hydroxybenzene, *see* Phenol, 2330
1-Hydroxy-1,2-benziodoxol-3-one-1-oxide, *see* 2-Iodylbenzoic acid, 2681
1-Hydroxybenzotriazole, 2272
2-Hydroxy-4,6-bis(nitroamino)-1,3,5-triazine, 1122
2-Hydroxy-3-butenonitrile, *see* 1-Cyano-2-propen-1-ol, 1467
1-Hydroxy-3-butyl hydroperoxide, 1706
4-Hydroxy-*trans*-cinnamic acid, 3136
Hydroxycopper(II) glyoximate, 0799
4-Hydroxy-3,5-dimethyl-1,2,4-triazole, 1575
4-Hydroxy-3,5-dinitrobenzenearsonic acid, 2220
2-Hydroxy-3,5-dinitrobenzoic acid, *see* 3,5-Dinitrosalicylic acid, 2665
2-Hydroxy-3,5-dinitropyridine, 1822
N-Hydroxydithiocarbamic acid, 0454
2(2-Hydroxyethoxy)ethyl perchlorate, 1643
2-Hydroxyethylamine, 0943
2-Hydroxyethylaminium perchlorate, 0951
3-(2-Hydroxyethylamino)propionitrile, 1952
N-2-Hydroxyethyl-1,2-diaminoethane, 1756
N-2-Hydroxyethyldimethylamine, 1729
O-(2-Hydroxyethyl)hydroxylamine, 0944
1-Hydroxyethylidene-1,1-diphosphonic acid, 0956
2-Hydroxyethylmercury(II) nitrate, 0857
1-Hydroxyethyl peroxyacetate, 1626
4(1-Hydroxyethyl)pyridine *N*-oxide
'Hydroxyethyltriethylenetetramine', *see* 3,6,9-Triaza-11-aminoundecanol,
    3100
2-Hydroxyethyltrimethylstibonium iodide, 2037
1-Hydroxy-2-hydroxylamino-1,3,4-triazole, 0817
1-Hydroxyimidazole-2-carboxaldoxime 3-oxide, 1477
1-Hydroxyimidazole *N*-oxide, 1134
1-Hydroxyiminobutane, *see* Butyraldehyde oxime, 1655
2-Hydroxyiminopropane, *see* Acetone oxime, 1258
Hydroxylamine, 4498

Hydroxylaminium chloride, 4002
Hydroxylaminium nitrate, 4524
Hydroxylaminium perchlorate, 4005
Hydroxylaminium phosphinate, 4555
Hydroxylaminium sulfate, 4575
2-Hydroxyliminiobutane chloride, *see* 2-Butanone oxime hydrochloride, 1678
4-Hydroxy-3-methoxybenzaldehyde, 2958
N-Hydroxymethylacrylamide, 1571
2-Hydroxy-2-methylglutaric acid, 2445
4-Hydroxy-4-methyl-1,6-heptadiene, 3029
4-Hydroxy-4-methyl-1,6-heptadiene diozonide, 3036
Hydroxymethyl hydroperoxide, 0487
Hydroxymethyl methyl peroxide, 0926
2-Hydroxymethyl-2-nitropropane-1,3-diol, *see* Tris(hydroxymethyl)
    nitromethane, 1664
† 4-Hydroxy-4-methyl-2-pentanone, 2501
N-Hydroxymethyl-2-propenamide, *see* *N*-Hydroxymethylacrylamide, 1571
3-Hydroxy-4-nitrobenzaldehyde, 2690
4-Hydroxy-3-nitrobenzaldehyde, 2691
5-Hydroxy-2-nitrobenzaldehyde, 2692
4-Hydroxy-3-nitrobenzenesulfonyl chloride, 2146
2-Hydroxy-6-nitro-1-naphthalenediazonium-4-sulfonate, 3240
(Hydroxy)(oxo)(phenyl)-λ^3-iodanium perchlorate, 2299
† 3-Hydroxypropanenitrile, *see* 2-Cyanoethanol, 1181
† 1-Hydroxypropane, *see* Propanol, 1279
2-Hydroxypropane-1,2,3-tricarboxylic acid, *see* Citric acid, 2389
2-Hydroxypropanoic acid, *see* Lactic acid, 1234
N-Hydroxysuccinimide, 1469
Hydroxyurea, 0477
Hypochlorous acid, 3995
Hyponitrous acid, 4470
'Hypophosphorous acid', *see* Phosphinic acid, 4503

† Imidazoline-2,4-dithione, 1141
Imidodicarbonic diamide, 0874
Iminobisacetonitrile, 1474
3,3′-Iminobispropylamine, 2587
† 2-Iminohexafluoropropane, *see* Hexafluoroisopropylideneamine, 1067
2-Indanecarboxaldehyde, 3288
Indium, 4640
Indium bromide, 0289
Indium(II) oxide, 4641
Indium(I) perchlorate, 4014
Indium phosphide, 4642
Iodic acid, 4424
Iodimide, *see* Diiodamine, 4426

† Isopropyl formate, 1619
Isopropyl hydroperoxide, 1283
Isopropyl hypochlorite, 1247
6-Isopropylidenethiacyclodeca-3,8-diyne, 3527
† Isopropyl iodide, *see* 2-Iodopropane, 1252
† Isopropyl isobutyrate, 2863
Isopropylisocyanide dichloride, 1561
5-Isopropyl-2-methyl-1,3-cyclohexadiene, *see* 1,5-*p*-Menthadiene, 3338
† Isopropyl nitrate, 1266
† Isopropyl nitrite, 1262
† Isopropyl propionate, 2504
† Isopropyl vinyl ether, 1957
Isosorbide dinitrate, 2380
Isosorbide mononitrate, *see* 1,4,3,6-Dianhydroglucitol 2-nitrate, 2396
3-Isothiocyanatopropene, *see* Allyl isothiocyanate, 1471
† Isovaleraldehyde, 1958
† Isoxazole, 1110

Ketene, 0722
Krypton difluoride, 4313

Lactic acid, 1234
Lactose, 3557
Lanthanum, 4677
Lanthanum carbide, 0991
Lanthanum dihydride, 4461
Lanthanum hexaboride, 0193
Lanthanum hydride, 4495
Lanthanum–nickel alloy, 4678
Lanthanum 2-nitrobenzoate, 3815
Lanthanum oxide, 4679
Lanthanum pentanickel hexahydride, 4552
Lanthanum picrate, 3741
Lead, 4882
Lead abietate, 3902
Lead acetate–lead bromate, 1540
Lead 2-amino-4,6-dinitrophenoxide, 3469
Lead 5,5'-azotetrazolide, 1020
Lead bis(1-benzeneazothiocarbonyl-2-phenyl-2-hydrazide), 3862
Lead bromate, 0278
Lead carbonate, 0558
Lead carbonate–lead hydroxide, 0726
Lead chloride, 4112
Lead chromate, 4243
Lead dioxide, *see* Lead(IV) oxide, 4834

Lead hyponitrite, 4745
Lead(II) azide, 4782
Lead(II) chlorate, 4105
Lead(II) chlorite, 4102
Lead(II) cyanide, 0999
Lead(II) fluoride, 4329
Lead(II) hypochlorite, 4098
Lead(II) imide, 4439
Lead(II) nitrate, 4749
Lead(II) nitrate phosphinate, 4468
Lead(II) oxide, 4824
Lead(II) phosphinate, 4531
Lead(II) phosphite, 4535
Lead(II) picrate, 3436
Lead(II) thiocyanate, 1000
Lead(II) trinitrosobenzene-1,3,5-trioxide, 3594
Lead(IV) acetate azide, 2402
Lead(IV) azide, 4790
Lead(IV) oxide, 4834
Lead methylenebis(nitramide), 0412
Lead nitride, 4753
Lead oleate, 3895
Lead pentaphosphide, 4881
Lead perchlorate, 4108
'Lead styphnate', *see* Lead 2,4,6-trinitroresorcinoxide, 2071
Lead sulfate, 4860
Lead tetrachloride, 4172
Lead−tin alloys, 4883
Lead 2,4,6-trinitroresorcinoxide, 2071
Lead−zirconium alloys, 4884
Limonene, 3337
Linolenic acid, *see* 9,12,15-Octadecatrienoic acid, 3771
Lithium, 4680
Lithium acetylide, 0992
Lithium aluminium hydride, *see* Lithium tetrahydroaluminate, 0075
Lithium amide, 4462
Lithium azide, 4685
Lithium benzenehexoxide, 2627
Lithium bis(trimethylsilyl)amide, 2594
Lithium borohydride, *see* Lithium tetrahydroborate, 0145
Lithium bromoacetylide, 0580
Lithium carbonate, 0533
Lithium chlorite, 4020
Lithium chloroacetylide, 0599
Lithium chloroethynide, *see* Lithium chloroacetylide, 0599
Lithium chromate, 4236

Lithium diazomethanide, 0379
Lithium diethylamide, 1686
Lithium dihydrocuprate, 4271
Lithium 2,2-dimethyltrimethylsilylhydrazide, 2041
Lithium diphenylhydridotungstate(2−), 3503
Lithium dithionite, 4687
Lithium ethynediolate, 0993
Lithium heptapotassium di(tetrasilicide), 4676
Lithium 1-heptynide, 2834
Lithium hexaazidocuprate(4−), 4278
Lithium hexamethylchromate(3−), 2590
Lithium hexaphenyltungstate(2−), 3892
Lithium hydrazide, 4496
Lithium hydride, 4432
Lithium−magnesium alloy, 4681
Lithium nitrate, 4684
Lithium nitride, 4688
Lithium 4-nitrothiophenoxide, 2177
Lithium octacarbonyltrinickelate, 3109
Lithium oxide, 4686
Lithium pentahydrocuprate(4−), 4273
Lithium pentamethyltitanate−bis(2,2′-bipyridine), 2042
Lithium perchlorate, 4021
Lithium sodium nitroxylate, 4683
Lithium tetraazidoaluminate, 0081
Lithium tetraazidoborate, 0151
Lithium tetradeuteroaluminate, 0069
Lithium 3-(1,1,2,2-tetrafluoroethoxy)propynide, 1820
Lithium tetrahydroaluminate, 0075
Lithium tetrahydroborate, 0145
Lithium tetrahydrogallate, 4407
Lithium tetramethylborate, 1740
Lithium tetramethylchromate(II), 1744
Lithium 1,1,2,2-tetranitroethanediide, 0994
Lithium thiocyanate, 0532
Lithium−tin alloys, 4682
Lithium triethylsilylamide, 2578
Lithium trifluoropropynide, 1046
Lithium tripotassium tetrasilicide, 4675

Magnesium, 4690
Magnesium azide, 4694
Magnesium boride, 0168
Magnesium carbonate hydroxide, 0534
Magnesium chlorate, 4083
Magnesium chloride, 4081

Magnesium hydride, 4463
Magnesium hypochlorite, 4082
Magnesium−nickel hydride, 4464
Magnesium nitrate, 4693
Magnesium nitride, 4698
Magnesium nitrite, 4692
Magnesium oxide, 4695
Magnesium perchlorate, 4084
Magnesium permanganate, 4691
Magnesium phosphide, 4699
Magnesium phosphinate, 4517
Magnesium silicide, 4697
Magnesium sulfate, 4696
Magnesium tetrahydroaluminate, 0085
'Magnetite', *see* Iron(II,III) oxide, 4405
Maleic anhydride, 1404
Maleic anhydride ozonide, 1406
Maleic hydrazide, 1435
Maleimide, 1418
Malonic acid, 1151
† Malononitrile, 1078
Manganese, 4700
Manganese abietate, 3901
Manganese chloride trioxide, 4022
Manganese diazide hydroxide, 4433
Manganese dichloride dioxide, 4086
Manganese dioxide, *see* Manganese(IV) oxide, 4705
Manganese fluoride trioxide, 4301
Manganese(II) bis(acetylide), 1396
Manganese(II) chlorate, 4087
Manganese(II) chloride, 4085
Manganese(II) N,N-diethyldithiocarbamate, 3358
Manganese(III) azide, 4702
Manganese(II) nitrate, 4701
Manganese(II) oxide, 4704
Manganese(II) perchlorate, 4088
Manganese(II) phosphinate, 4519
Manganese(II) sulfide, 4706
Manganese(II) telluride, 4708
Manganese(II) tetrahydroaluminate, 0086
Manganese(IV) oxide, 4705
Manganese(IV) sulfide, 4707
Manganese phosphide, 4711
Manganese picrate hydroxide, 3742
Manganese tetrafluoride, 4343
Manganese trichloride oxide, 4141

Manganese trifluoride, 4335
Manganese(VII) oxide, 4709
1,5-*p*-Menthadiene, 3338
Mercaptoacetonitrile, 0767
Mercury, 4600
Mercury 5,5'-azotetrazolide, 0980
Mercury bis(chloroacetylide), 1356
'Mercury diazocarbide', *see* Poly(diazomethylenemercury), 0522
Mercury(I) azide, 4612
Mercury(I) bromate, 0271
Mercury(I) chlorite, 4080
Mercury(I) cyanamide, 0523
Mercury(I) fluoride, 4312
Mercury(I) hypophosphate, 4617
Mercury(II) acetylide, 0975
Mercury(II) amide chloride, 3999
Mercury(II) azide, 4604
Mercury(II) bromate, 0270
Mercury(II) bromide, 0269
Mercury(II) chloride, 4076
Mercury(II) chlorite, 4077
Mercury(II) cyanate, 0977
Mercury(II) cyanide, 0976
Mercury(II) *aci*-dinitromethanide, 0707
Mercury(II) formohydroxamate, 0804
Mercury(II) fulminate, 0978
Mercury(II) iodide, 4602
Mercury(II) nitrate, 4603
Mercury(II) 5-nitrotetrazolide, 0981
Mercury(II) oxalate, 0982
Mercury(II) oxide, 4605
Mercury(II) oxycyanide, *see* Dimercury dicyanide oxide, 0983
Mercury(II) perchlorate, 4078
Mercury(II) perchlorate . 6 (or 4)dimethyl sulfoxide, 4079
Mercury(II) *N*-perchlorylbenzylamide, 3651
Mercury(II) peroxybenzoate, 3637
Mercury(II) picrate, 3433
Mercury(II) sulfide, 4607
Mercury(II) thiocyanate, 0979
Mercury(I) nitrate, 4609
'Mercury(I) oxide' 4613
Mercury(I) thionitrosylate, 4610
Mercury nitride, 4615
Mercury peroxide, 4606
Mesitylene, 3162
O-Mesitylenesulfonylhydroxylamine, 3170

2028

N-Methylmorpholine oxide, 1997
4-Methylmorpholine-4-oxide, *see* N-Methylmorpholine oxide, 1997
Methyl nitrate, 0457
2-Methyl-1-nitratodimercurio-2-nitratomercuriopropane, 1592
† Methyl nitrite, 0455
N-Methyl-4-nitroaniline, 2800
2-Methyl-5-nitroaniline, 2801
N-Methyl-*p*-nitroanilinium 2(N-methyl-N-*p*-nitrophenylaminosulfonyl)
  ethylsulfate, 3713
3′-Methyl-2-nitrobenzanilide, 3645
Methyl 2-nitrobenzenediazoate, 2782
2-Methyl-5-nitrobenzenesulfonic acid, 2772
4-Methyl-3-nitrobenzenesulfonic acid, 2773
2-Methyl-5-nitrobenzimidazole, 2940
3-Methyl-2-nitrobenzoyl chloride, 2913
3-Methyl-4-nitro-1-buten-3-yl acetate, 2835
3-Methyl-4-nitro-2-buten-1-yl acetate, 2836
1-Methyl-3-nitroguanidinium nitrate, 0947
1-Methyl-3-nitroguanidinium perchlorate, 0939
2-Methyl-4-nitroimidazole, 1475
'Methylnitrolic acid', *see* Nitrooximinomethane, 0409
1-Methyl-3-nitro-1-nitrosoguanidine, 0876
3-Methyl-4-nitrophenol, 2766
4-Methyl-2-nitrophenol, 2767
2-Methyl-2-nitropropane, *see* *tert*-Nitrobutane, 1660
3-Methyl-4-nitropyridine N-oxide, 2317
N-Methyl-N-nitrosourea, 0875
Methylnitrothiophene, 1850
† 2-Methylnonane, 3366
S-7-Methylnonylthiouronium picrate, 3737
† 2-Methyloctane, 3206
† 3-Methyloctane, 3207
† 4-Methyloctane, 3208
2-Methyl-3,5,7-octatriyn-2-ol, 3135
† 2-Methyl-2-oxazoline, 1569
† Methyloxirane, *see* Propylene oxide, 1225
Methyloxobismuthine, *see* Methylbismuth oxide, 0428
† 2-Methyl-1,3-pentadiene, 2416
† 4-Methyl-1,3-pentadiene, 2417
† 2-Methylpentanal, 2491
† 3-Methylpentanal, 2492
† 3-Methylpentane, 2533
† 2-Methylpentane, *see* Isohexane, 2532
† 2-Methyl-2-pentanol, 2544
† 2-Methyl-3-pentanone, 2493
† 3-Methyl-2-pentanone, 2494

†   2-Methyltetrahydrofuran, 1961
2-Methyltetrazole, 0814
S-Methylthioacetohydroximate, 1260
†   2-Methylthiophene, 1882
4-Methyl-2,4,6-triazatricyclo[5.2.2.0^{2,6}]undeca-8-ene-3,5-dione, 3160
1-Methyl-1,2,3-triazole, 1189
Methyl trichloroacetate, 1096
†   Methyltrichlorosilane, 0439
Methyltrifluoromethyldioxirane, 1102
Methyltrifluoromethyltrichlorophosphorane, 0742
Methyl trifluorovinyl ether, 1101
3-Methyl-2,4,6-trinitrophenol, 2703
3-Methyl-1,2,4-trioxolane, 1235
3-(3-Methyl-1,2,4-trioxolan-3-yl)-1,2,4-trioxolane, *see* Isoprene diozonide, 1920
Methyltriphenylphosphonium permanganate, 3784
†   Methyl valerate, 2507
†   Methyl vinyl ether, 1221
†   Methyl vinyl ketone, *see* 1-Buten-3-one, 1515
2-Methyl-5-vinyltetrazole, 1511
Methylzinc iodide, 0447
†   2-Methypropylamine, *see* Isobutylamine, 1728
'Millon's base anhydride', *see* Poly(dimercuryimmonium hydroxide), 4422
Molybdenum, 4712
Molybdenum azide tribromide, 0288
Molybdenum diazide tetrachloride, 4164
Molybdenum hexafluoride, 4365
Molybdenum hexamethoxide, 2596
Molybdenum(IV) oxide, 4716
Molybdenum(IV) sulfide, 4719
Molybdenum nitride, 4714
Molybdenum pentachloride, 4180
Molybdenum(VI) oxide, 4717
Monocaesium acetylide, 0661
Monofluoroxonium hexafluoroarsenate, 0097
Monolithium acetylide, 0673
Monolithium acetylide–ammonia, 0674
Monoperoxysuccinic acid, 1542
Monopotassium acetylide, 0671
Monopotassium perchlorylamide, 3994
Monorubidium acetylide, 0685
Monosilver acetylide, 0650
Monosodium acetylide, 0684
Monosodium cyanamide, 0382
Morpholine, 1657

4-Morpholinesulfenyl chloride, 1586
Morpholinium perchlorate, 1680

Naphthalene, 3254
1-Naphthalenediazonium perchlorate, 3246
2-Naphthalenediazonium perchlorate, 3247
1-Naphthalenediazonium salts, 3251
2-Naphthalenediazonium salts, 3252
2-Naphthalenediazonium trichloromercurate, 3248
1-Naphthylamine, 3264
2-Naphthylamine, 3265
1-Naphthyl isocyanate, 3390
Naphthylsodium, 3253
1-(2-Naphthyl)-3-(5-tetrazolyl)triazene, 3394
(1-Napththyl)-1,2,3,4-thiatriazole, 3391
Neodymium, 4819
Neodymium perchlorate . 2acetonitrile, 4148
† Neopentane, *see* 2,2-Dimethylpropane, 1999
† Neopentylamine, *see* 2,2-Dimethylpropylamine, 2030
Neptunium hexafluoride, 4366
Nickel, 4820
Nickel azide, 4777
Nickel chlorite, 4093
Nickel 2,4-dinitrophenoxide hydroxide, 3745
Nickel(II) cyanide, 0997
Nickel(III) oxide, 4823
Nickel(II) nitrate, 4743
Nickel(II) oxide, 4821
Nickel(IV) oxide, 4822
Nickel 2-nitrophenoxide, 3464
Nickelocene, 3287
Nickel perchlorate, 4094
Nickel picrate, 3435
† Nickel tetracarbonyl, *see* Tetracarbonylnickel, 1805
Niobium, 4817
Niobium(V) oxide, 4818
Nitric acid, 4436
Nitric amide (Nitramide), 4471
'Nitric oxide', *see* Nitrogen oxide, 4724
Nitrilotris(oxiranemethane), 3181
Nitritonitrosylnickel, 4742
Nitroacetaldehyde, 0763
2-Nitroacetaldehyde oxime, 0809
4-Nitroacetanilide, 2950
† Nitroacetone, 1184
Nitroacetonitrile, *see* 2-Nitroethanonitrile, 0711

Oleic acid, *see cis*-9-Octadecenoic acid, 3774
Oleoyl chloride, 3772
Oligo(octacarbondioxide), 3108
1,5-(or 1,8-)Bis(dinitrophenoxy)-4,8-(or 4,5-)dinitroanthraquinone, 3860
1(2′-, 3′-, or 4′-Diazoniophenyl)-2,4,6-triphenylpyridinium diperchlorate, 3870
3- or 4-Methoxy-5,6-benzo-6*H*-1,2-dioxin, 3154
Orthoperiodic acid, 4542
Orthophosphoric acid, 4505
2-, 3- or 4-Trifluoromethylphenylmagnesium bromide, 2643
Osmium, 4873
Osmium hexafluoride, 4370
Osmium(IV) oxide, 4833
Osmium(VIII) oxide, 4858
† 6-Oxabicyclo[3.1.0]hexane, *see* Cyclopentene oxide, 1906
Oxalic acid, 0725
Oxalyl chloride, *see* Oxalyl dichloride, 0605
Oxalyl dibromide, 0583
Oxalyl dichloride, 0605
1,4-Oxathiane, 1613
† Oxazole, 1111
Oxepin-3,6-endoperoxide, *see* endo-2,3-Epoxy-7,8-dioxabicyclo
    [2.2.2]oct-5-ene, 2336
† Oxetane, 1222
2-Oximinobutane, *see* 2-Butanone oxime, 1654
4-Oximino-4,5,6,7-tetrahydrobenzofurazan *N*-oxide, 2357
Oxiranecarboxaldehyde oxime, *see* 2,3-Epoxypropionaldehyde oxime, 1182
† Oxirane, *see* Ethylene oxide, 0829
Oxiranemethanol, *see* 2,3-Epoxypropanol, 1229
† Oxiranemethanol nitrate, *see* 2,3-Epoxypropyl nitrate, 1186
Oxoacetic acid, *see* Glyoxylic acid, 0724
Oxobis[aqua(oxo)diperoxorheniumVII], 4537
μ-Oxo-*I,I*-bis(trifluoroacetato-*O*)-*I,I*-diphenyldiodine(III), 3698
Oxodiperoxodipiperidinechromium(VI), 3367
Oxodiperoxodipyridinechromium(VI), 3274
Oxodiperoxodi(pyridine *N*-oxide)molybdenum, 3279
Oxodiperoxodi(pyridine *N*-oxide)tungsten, 3283
Oxodiperoxodiquinolinechromium(VI), 3748
Oxodiperoxomolybdenum–hexamethylphosphoramide, 4718
Oxodiperoxopyridinechromium *N*-oxide, 1844
(Oxodiperoxy(1,3-dimethyl-2,4,5,6-tetrahydro-2-1*H*)-
    pyrimidinone)molybdenum, 2472
Oxodiperoxy(pyridine)(1,3-dimethyl-2,4,5,6-tetrahydro-2-1*H*)-
    (pyrimidinone)molybdenum, 3410
Oxodisilane, 4530
Oxopropanedinitrile, 1341
2-Oxopropanoic acid, *see* Pyruvic acid, 1150

Oxosilane, 4476
1-Oxo-1,2,3,4-tetrahydronaphthalene, 3289
Oxybis[aqua-5,7,7,12,14,14-hexamethyl-1,4,8,11-tetraaza-
    4,11-cyclotetradecadieneiron(II)] tetraperchlorate, 3882
1,1'-Oxybis-2-azidoethane, 1604
4,4'-Oxybis(benzenesulfonylhydrazide), 3525
Oxybis[bis(cyclopentadienyl)titanium] 3803
Oxybis(N,N-dimethylacetamidetriphenylstibonium) perchlorate, 3911
[1,1'-Oxybis[ethane]]tris(pentafluorophenyl)aluminium, 3829
[1,1'-Oxybis[ethane]]tris(trimethyl silyl)aluminium, *see*
    Tris(trimethylsilyl)aluminium etherate, 3620
† 1,1'-Oxybisethene, *see* Divinyl ether, 1519
2,2'-Oxybis(ethyl nitrate), 1599
2,2'-Oxybis(iminomethylfuran) mono-*N*-oxide, 3260
† Oxybismethane, *see* Dimethyl ether, 0919
1,1-Oxybis(4-methylphenyldiazene), *see* Bis(toluenediazo) oxide, 3654
Oxybisphenyliodonium bistetrafluoroborate, 3475
Oxygen difluoride, 4317
Oxygen (Gas), 4831
Oxygen (Liquid), 4832
Oxynitrotriazole, *see* 3-Nitro-1,2,4-triazolone, 0716
Ozobenzene, *see* Benzene triozonide, 2343
Ozone, 4846

Palladium, 4885
Palladium(II) acetate, 1541
Palladium(II) azide, 4783
Palladium(II) azidodithioformate, 1015
Palladium(III) oxide, 4848
Palladium(II) oxide, 4825
Palladium(IV) oxide, 4835
Palladium tetrafluoride, 4347
Palladium trifluoride, 4341
Paraformaldehyde, 0417
Pentaammineaquacobalt(III) chlorate, 4124
Pentaammineazidoruthenium(III) chloride, 4073
Pentaamminechlorocobalt(III) perchlorate, 4123
Pentaamminechlororuthenium chloride, 4137
Pentaamminedinitrogenosmium(II) perchlorate, 4072
Pentaamminedinitrogenruthenium(II) salts, 4596
Pentaamminenitratocobalt(III) nitrate, 4207
Pentaamminenitrochromium(III) nitrate, 4233
Pentaamminephosphinatochromium(III) perchlorate, 4053
Pentaamminephosphinatocobalt(III) perchlorate, 4050
Pentaamminepyrazineruthenium(II) perchlorate, 1792
Pentaamminepyridineruthenium(II) perchlorate, 2048

Permanganic acid, 4434

μ-Peroxobis[ammine(2,2′,2″-triaminotriethylamine)cobalt(III)](4+) perchlorate, 3591

Peroxodisulfuric acid, 4482

Peroxodisulfuryl difluoride, 4328

Peroxomonophosphoric acid, 4506

Peroxomonosulfuric acid, 4481

Peroxonitric acid, 4437

Peroxyacetic acid, 0837

Peroxyacetyl nitrate, 0766

Peroxyacetyl perchlorate, 0737

Peroxybenzoic acid, 2733

Peroxy-2-butenoic acid, *see* Peroxycrotonic acid, 1535

3-Peroxycamphoric acid, 3348

Peroxycrotonic acid, 1535

Peroxyformic acid, 0420

Peroxyfuroic acid, 1837

Peroxyhexanoic acid, 2514

Peroxypropionic acid, 1236

Peroxypropionyl nitrate, 1187

Peroxypropionyl perchlorate, 1167

Peroxytrifluoroacetic acid, 0666

Perrhenyl chloride, *see* Rhenium chloride trioxide, 4045

Perylenium perchlorate, 3790

1,10-Phenanthroline-5,6-dione, 3578

Phenethyl alcohol, *see* Benzeneethanol, 2983

Phenol, 2330

Phenoxyacetylene, 2920

3-Phenoxy-1,2-epoxypropane, 3150

Phenylacetonitrile, 2935

Phenylacetylene, 2912

Phenylaminothia-2,3,4-triazole, *see* N-Phenyl-1,2,3,4-thiatriazolamine, 2729

Phenyl azide, 2271

α-Phenylazobenzyl hydroperoxide, 3609

α-Phenylazo-4-bromobenzyl hydroperoxide, 3607

1-Phenylazocyclohexyl hydroperoxide, 3539

α-Phenylazo-4-fluorobenzyl hydroperoxide, 3608

N-Phenylazopiperidine, 3407

1-Phenylbiguanidinium hydrogen dichromate, 3019

1-Phenylboralane, 3317

1-Phenyl-3-*tert*-butyltriazene, 3335

Phenylchlorodiazirine, 2673

Phenyldiazidomethane, 2730

Phenyldiazomethane, 2725

2-Phenyl-1,1-dimethylethyl hydroperoxide, 3332

1-(4-Phenyl-1,3-diselenolylidene)piperidinium perchlorate, 3659

Poly[bis(2,2,2-trifluoroethoxy)phosphazene] 1431
Poly([7,8-bis(trifluoromethyl)tetracyclo [4.2.0.0^{2,8}.0^{5,7}]octane-3,4-diyl]-
  1,2-ethenediyl), 3463
Poly[borane(1)], 0134
cis-Poly(butadiene), 1484
Poly(1,3-butadiene peroxide), 1533
Poly(butadiyne), 1386
Poly(carbon monofluoride), 0337
Poly(chlorotrifluoroethylene), 0592
Poly(1,3-cyclohexadiene peroxide), 2386
Poly(cyclopentadienyltitanium dichloride), 1843
Poly(diazidophosphazene), 4787
Poly(diazomethylenemercury), 0522
Poly(dibromosilylene), 0283
Poly(difluorosilylene), 4330
Poly(dihydroxydioxodisilane), 4480
Poly(dimercuryimmonium acetylide), 0669
Poly(dimercuryimmonium azide), 4611
Poly(dimercuryimmonium bromate), 0253
Poly(dimercuryimmonium hydroxide), 4422
Poly(dimercuryimmonium iodide hydrate), 4455
Poly(dimercuryimmonium perchlorate), 4012
Poly(dimercuryimmonium permanganate), 4608
Poly(dimethylketene peroxide), *see* Poly(peroxyisobutyrolactone), 1536
Poly(dimethylsiloxane), 0922
Poly(disilicon nitride), 4757
Poly(ethenyl nitrate), *see* Poly(vinyl nitrate), 0764
Poly(ethylene), 0782
Poly(ethylene terephthalate), 3262
Poly(ethylidene peroxide), 0835
Poly(furan-2,5-diyl), 1402
Poly(germanium dihydride), 4415
Poly(germanium monohydride), 4413
Poly(2,4-hexadiyne-1,6-ylene carbonate), 2669
Poly(isobutene), 1583
Poly(methyl methacrylate peroxide), 1919
Poly(oxycarbonyl-2,4-hexadiyne-1,6-diyl), *see* Poly(2,4-hexadiyne-1,6-ylene
  carbonate), 2669
Poly[oxy(methyl)silylene], 0485
Poly(1-pentafluorothio-1,2-butadiyne), 1379
Poly(peroxyisobutyrolactone), 1536
Poly(pyrrole), 1417
Poly(selenium nitride), 4730
Poly(silylene), 4487
Poly(styrene peroxide), 2957
Poly(sulfur nitride), 4728

Poly(tetrafluoroethylene), 0629

Poly(thiophene), 1407

Poly(vinyl acetate peroxide), 1539

Poly(vinyl alcohol), 0831

Poly(vinyl butyral), 3030

Poly(vinyl nitrate), 0764

Potassium, 4645

Potassium acetylene-1,2-dioxide, 0990

Potassium acetylide, 0987

Potassium amide, 4456

Potassium amidosulfate, 4457

Potassium antimonide, 4673

Potassium azide, 4652

Potassium azidodisulfate, 4655

Potassium azidopentacyanocobaltate(3−), 1809

Potassium azidosulfate, 4653

Potassium azodisulfonate, 4663

Potassium benzenehexoxide, 2626

Potassium benzenesulfonylperoxosulfate, 2257

Potassium $O-O$-benzoylmonoperoxosulfate, 2684

Potassium bis(8-(4-nitrophenylthio)undecahydrodicarbaundecaborato)
   cobaltate(1−), 3719

Potassium bis(8-(4-nitrophenylthio)undecahydrodicarbaundecaborato)
   ferrate(1−), 3720

Potassium bis(phenylethynyl)palladate(2−), 3699

Potassium bis(phenylethynyl)platinate(2−), 3700

Potassium bis(propynyl)palladate, 2308

Potassium bis(propynyl)platinate, 2309

Potassium bromate, 0255

Potassium *tert*-butoxide, 1650

Potassium carbonate, 0531

Potassium chlorate, 4017

Potassium chloride, 4015

Potassium chlorite, 4016

Potassium citrate tri(hydrogen peroxidate), 2258

Potassium cyanate, 0528

Potassium cyanide, 0526

Potassium cyanide–potassium nitrite, 0527

Potassium cyclohexanehexone 1,3,5-trioximate, 2625

Potassium cyclopentadienide, 1846

Potassium [(7,8,9,10,11-η)-1,2,3,4,5,6,7,8,9,11-decahydro-10-(4-
   nitrobenzenethiolato-S)-7,8-dicarbaundecaborato(2−)][(7,8,9,10,11-
   η)undecahydro-7,8-dicarbadecaborato(2−)cobaltate(1−)], *see* Potassium
   [8-(4-nitrophenylthio)undecahydrodicarbaundecaborato]undecahydro-
   dicarbaundecaboratocobaltate(1−) 3380

Potassium 1,3-dibromo-1,3,5-triazine-2,4-dione-6-oxide, 1032

Potassium dichromate, 4248
Potassium diethylamide, 1685
Potassium diethynylpalladate(2−), 1394
Potassium diethynylplatinate(2−), 1395
Potassium dihydrogenphosphide, 4460
Potassium dinitramide, 4654
Potassium dinitroacctamide, 0709
Potassium 2,5-dinitrocyclopentanonide, 1847
Potassium dinitrogentris(trimethylphosphine)cobaltate(1−), 3228
Potassium dinitromethanide, 0377
Potassium dinitrooxalatoplatinate(2−), 0988
Potassium 1,1-dinitropropanide, 1175
'Potassium dinitrososulfite', *see* Potassium *N*-nitrosohydroxylamine-
    N-sulfonate, 4662
Potassium 3,5-dinitro-2(1-tetrazenyl)phenoxide, 2256
Potassium 2,6-dinitrotoluene, 2721
Potassium dioxide, 4656
Potassium diperoxomolybdate, 4659
Potassium diperoxoorthovanadate, 4667
Potassium *O,O*-diphenyl dithiophosphate, 3481
Potassium dithioformate, 0378
Potassium ethoxide, 0861
† Potassium *O*-ethyl dithiocarbonate, 1176
'Potassium ethyl xanthate', *see* Potassium *O*-ethyl dithiocarbonate
Potassium ethynediolate, *see* Potassium acetylene-1,2-dioxide, 0990
'Potassium ferricyanide', *see* Potassium hexacyanoferrate(III), 2063
'Potassium ferrocyanide', *see* Potassium hexacyanoferrate(II), 2064
Potassium fluoride hydrogen peroxidate, 4300
Potassium graphite, 3107
Potassium heptafluorotantalate(V), 4379
Potassium hexaazidoplatinate(IV), 4665
Potassium hexachloroplatinate, 4187
Potassium hexacyanoferrate(II), 2064
Potassium hexacyanoferrate(III), 2063
Potassium hexaethynylcobaltate(4−), 3446
Potassium hexaethynylmanganate(3−), 3448
Potassium hexafluorobromate, 0244
Potassium hexafluoromanganate(IV), 4363
Potassium hexafluorosilicate(2−), 4364
Potassium hexahydroaluminate(3−), 0077
Potassium hexanitrocobaltate(3−), 4213
Potassium hexaoxoxenonate−xenon trioxide, 4674
Potassium hydride, 4427
Potassium hydrogen acetylenedicarboxylate, 1382
Potassium hydrogen diazirine-3,3-dicarboxylate, 1068
Potassium hydrogen peroxomonosulfate, 4430

Radon, 4893
Raney cobalt alloy, *see* Aluminium−cobalt alloy, 0049
Raney nickel alloys, *see* Aluminium−nickel alloys, 0055
Resorcinol, *see* 1,3-Benzenediol, 2332
Resorcinol diacetate, *see* 1,3-Diacetoxybenzene, 3291
Rhenium, 4890
Rhenium chloride trioxide, 4045
Rhenium hexafluoride, 4373
Rhenium hexamethoxide, 2603
Rhenium nitride tetrafluoride, 4344
Rhenium tetrachloride oxide, 4167
Rhenium(VII) sulfide, 4891
'Rhodanine', *see* 2-Thioxo-4-thiazolidinone, 1112
Rhodium, 4892
Rhodium(III) chloride, 4155
Rhodium tetrafluoride, 4349
Rubidium, 4889
Rubidium acetylide, 1025
Rubidium fluoroxysulfate, 4309
Rubidium graphite, 3110
Rubidium hexafluorobromate, 0245
Rubidium hydride, 4450
Rubidium hydrogen xenate, 4448
Rubidium nitride, 4727
Rubidium tetrafluorochlorate(1−), 3988
Ruthenium, 4894
Ruthenium(III) chloride, 4156
Ruthenium(IV) hydroxide, 4532
Ruthenium(IV) sulfide, 4895
Ruthenium salts, 4896
Ruthenium(VIII) oxide, 4862

Samarium, 4911
Samarium sulfide, 4899
Scandium 3-nitrobenzoate, 3816
Sebacoyl chloride, 3341
Seleninyl bis(dimethylamide), 1758
Seleninyl bromide, 0275
Seleninyl chloride, 4097
Selenium, 4908
Selenium difluoride dioxide, 4321
Selenium dioxide, 4838
Selenium tetrabromide, 0295
Selenium tetrafluoride, 4351
Selenium trioxide, 4852
† Silane, 4539

Silica, *see* Silicon dioxide, 4839
Silicon, 4909
Silicon dibromide sulfide, 0281
Silicon dioxide, 4839
Silicon monohydride (Silylidyne), 4451
Silicon monosulfide, 4898
Silicon oxide, 4828
Silicon tetraazide, 4791
Silicon tetrachloride, *see* Tetrachlorosilane, 4173
Silicon tetrafluoride, 4352
Silicon–zirconium alloys, 4910
Silver, 0001
Silver acetylide, 0568
Silver acetylide–silver nitrate, 0569
Silver–aluminium alloy, 0002
Silver amide, 0015
Silver 5-aminotetrazolide, 0392
Silver azide, 0023
Silver azide chloride, 0009
Silver 2-azido-4,6-dinitrophenoxide, 2075
Silver azidodithioformate, 0303
Silver 1-benzeneazothiocarbonyl-2-phenylhydrazide, 3605
Silver benzeneselenonate, 2219
Silver benzo-1,2,3-triazole-1-oxide, 2127
Silver bromate, 0007
Silver buten-3-ynide, 1408
Silver chlorate, 0011
Silver chloride, 0008
Silver chlorite, 0010
Silver chloroacetylide, 0566
Silver cyanate, 0300
Silver cyanide, 0299
Silver cyanodinitromethanide, 0567
Silver 3-cyano-1-phenyltriazen-3-ide, 2670
Silver cyclopropylacetylide, 1838
Silver difluoride, 0014
Silver dinitroacetamide, 0689
Silver 3,5-dinitroanthranilate, 2642
Silver 1,3-di(5-tetrazolyl)triazenide, 0690
Silvered copper, 0003
Silver fluoride, 0013
Silver fulminate, 0301
Silver hexahydrohexaborate(2−), 0027
Silver hexanitrodiphenylamide, 3429
Silver 1,3,5-hexatriynide, 2052
Silver 3-hydroxypropynide, 1088

Sodium amide, 4465
Sodium amidosulfate, 4467
Sodium–antimony alloy, 4797
Sodium arachidonate, *see* Sodium 5,8,11,14-eicosatetraenoate, 3809
Sodium azide, 4758
Sodium azidosulfate, 4759
Sodium 5-azidotetrazolide, 0551
Sodium benzenehexoxide, 2634
Sodium 2-benzothiazolylthiolate, 2658
Sodium 1,4-bis(*aci*-nitro)-2,5-cyclohexadienide, 2188
Sodium 3,5-bis(*aci*-nitro)cyclohexene-4,6-diiminide, 2203
Sodium borate hydrogen peroxidate, 0154
Sodium borohydride, *see* Sodium tetrahydroborate, 0147
Sodium bromate, 0257
Sodium bromoacetylide, 0581
Sodium carbonate, 0552
Sodium carbonate hydrogen peroxidate, 0553
Sodium chlorate, 4039
Sodium chloride, 4036
Sodium chlorite, 4038
Sodium chloroacetate, 0694
Sodium 4-chloroacetophenone oximate, 2927
Sodium chloroacetylide, 0601
Sodium *N*-chlorobenzenesulfonamide, 2229
Sodium chloroethynide, *see* Sodium chloroacetylide, 0601
Sodium 4-chloro-2-methylphenoxide, 2720
Sodium *N*-chloro-4-toluenesulfonamide, 2739
Sodium 4-cresolate, *see* Sodium 4-methylphenoxide, 2788
Sodium cyanide, 0536
Sodium diazomethanide, 0383
Sodium 1,3-dichloro-1,3,5-triazine-2,4-dione-6-oxide, 1037
Sodium dichromate, 4250
Sodium diethylamide, 1688
Sodium diformylnitromethanide hydrate, 1076
Sodium dihydrobis(2-methoxyethoxy)aluminate, 2575
Sodium dihydrogen phosphide, 4474
Sodium 1,3-dihydroxy-1,3-bis(*aci*-nitromethyl)-2,2,4,4-tetramethylcyclo-
  butandiide, 3344
Sodium 4,4-dimethoxy-1-*aci*-nitro-3,5-dinitro-2,5-cyclohexadienide, 2951
Sodium dimethylsulfinate, 0879
Sodium dinitroacetamide, 0712
Sodium dinitromethanide, 0384
Sodium 5-(dinitromethyl)tetrazolide, 0681
Sodium 2,4-dinitrophenoxide, 2115
Sodium dioxide, 4799
Sodium diphenylketyl, 3604

Sodium disulfide, 4812
Sodium disulfite, 4808
Sodium dithionite, 4807
Sodium 5,8,11,14-eicosatetraenoate, 3809
Sodium ethaneperoxoate, *see* Sodium peroxyacetate, 0780
Sodium ethoxide, 0878
Sodium ethoxyacetylide, 1478
Sodium ethylenediaminetetraacetate, 3305
Sodium ethynediolate, 1023
Sodium ethynide, *see* Sodium acetylide, 1022
Sodium fulminate, 0537
Sodium germanide, 4418
Sodium hexaazidophosphate, 4794
Sodium hexahydroxyplatinate(IV), 4564
Sodium hexakis(propynyl)ferrate(4−), 3759
Sodium hexaoxodinitrate, 4740
Sodium hydrazide, 4500
Sodium hydride, 4444
Sodium hydrogen carbonate, 0390
Sodium hydrogen sulfate, 4446
Sodium hydrogen xenate, 4447
'Sodium hydrosulfite', *see* Sodium dithionite, 4807
Sodium hydroxide, 4445
Sodium 2-hydroxyethoxide, 0880
Sodium *O*-hydroxylamide, 4466
Sodium 3-hydroxymercurio-2,6-dinitro-4-*aci*-nitro-2,5-cyclohexa-
    dienonide, 2083
Sodium 2-hydroxymercurio-4-*aci*-nitro-2,5-cyclohexadienonide, 2171
Sodium 2-hydroxymercurio-6-nitro-4-*aci*-nitro-2,5-cyclohexadienonide, 2111
Sodium 5(5′-hydroxytetrazol-3′-ylazo)tetrazolide, 0682
Sodium hypoborate, 0164
Sodium hypochlorite, 4037
Sodium hyponitrite, 4736
'Sodium hypophosphite', *see* Sodium phosphinate, 4473
Sodium iodate, 4624
Sodium iodide, 4623
Sodium isopropoxide, 1270
'Sodium metabisulfite', *see* Sodium disulfite, 4808
Sodium metasilicate, 4805
Sodium methoxide, 0464
Sodium methoxyacetylide, 1123
Sodium 3-methylisoxazolin-4,5-dione-4-oximate, 1419
Sodium 4-methylphenoxide, 2788
Sodium molybdate, 4713
Sodium monoperoxycarbonate, 0554
Sodium nitrate, 4721

Sodium tetraperoxomolybdate, 4715
Sodium tetraperoxotungstate, 4810
Sodium tetrasulfur pentanitridate, 4775
Sodium 1-tetrazolacetate, 1119
Sodium thiocyanate, 0538
Sodium thiosulfate, 4804
Sodium triammine, 4581
Sodium triazidoaurate(?), 0113
Sodium tricarbonylnitrosylferrate, 1064
Sodium trichloroacetate, 0608
Sodium 2,2,2-trifluoroethoxide, 0705
Sodium 2,2,2-trinitroethanide, 0713
Sodium 2,4,6-trinitrophenoxide, *see* Sodium picrate, 2086
Sodium trioxide, 4800
Sodium trioxodinitrate, 4737
Sodium tris(*O,O*'-1-oximatonaphthalene-1,2-dione)ferrate (Pigment
    green B), 3871
Sodium trisulfurtrinitridate, 4760
Sodium–zinc alloy, 4798
'Solid Phosphorus hydride' 4449
Sorbic acid, *see* 2,4-Hexadienoic acid, 2385b
Spiro[3*H*-diazirine-3,9'pentacyclo[4,3,0,0,^{2,5}0,^{3,8}0,^{4,7}]nonane], *see* Spiro-
    (homocubane-9,9'-diazirine), 3127
Spiro(homocubane-9,9'-diazirine), 3127
† Stibine, 4510
Stibobenzene, *see* Diphenyldistibene, 3500
Stilbene, *see* trans-1,2-Diphenylethylene, 3643
Strontium, 4913
Strontium acetylide, 1026
Strontium azide, 4785
Strontium peroxide, 4841
Strontium sulfide, 4901
Styrene, 2945
Succinic anhydride, 1443
Succinodinitrile, 1433
Succinoyl diazide, 1438
Succinyl peroxide, *see* 3,6-Dioxo-1,2-dioxane, 1445
Sucrose, 3558
Sulfamic acid, *see* Amidosulfuric acid, 4499
Sulfamoyl azide, 4472
Sulfinyl azide, 4778
† Sulfinylbismethane, *see* Dimethyl sulfoxide, 0921
Sulfinyl bromide, 0274
Sulfinyl chloride, 4096
Sulfinylcyanamide, 0542
Sulfinyl fluoride, 4318

Sulfinylsulfamoyl fluoride, *see* Sulfur oxide-(*N*-fluorosulfonyl)imide, 4305
Sulfolane, *see* Tetrahydrothiophene-1,1-dioxide, 1622
Sulfonyl chloride, 4099
† Sulfur, 4897
Sulfur dibromide, 0280
Sulfur dichloride, 4113
Sulfur dioxide, 4837
Sulfur hexafluoride, 4374
Sulfuric acid, 4479
Sulfur oxide-(*N*-fluorosulfonyl)imide, 4305
Sulfur tetrafluoride, 4350
Sulfur thiocyanate, 1002
Sulfur trioxide, 4849
Sulfur trioxide–dimethylformamide, 4850
Sulfuryl azide chloride, 4031
Sulfuryl chloride, *see* Sulfonyl chloride, 4099
Sulfuryl diazide, 4779

Tantalum, 4914
Tantalum(IV) sulfide, 4903
Tantalum pentachloride, 4185
Tantalum(V) oxide, 4865
Tartaric acid, 1545
Technetium, 4915
Tellurane-1,1-dioxide, 1980
Tellurium, 4916
Tellurium tetrabromide, 0296
Tellurium tetrachloride, 4175
Tellurium trioxide, 4853
Terephthalic acid, 2924
Terephthaloyl chloride, 2889
Tetraacrylonitrilecopper(I) perchlorate, 3510
Tetraallyl-2-tetrazene, 3545
Tetraallyluranium, 3548
Tetra(3-aminopropanethiolato)trimercury perchlorate, 3582
Tetraamminebis(dinitrogen)osmium(II) perchlorate, 4071
Tetraamminebis(5-nitro-2*H*-tetrazolato)cobalt(1$^+$) perchlorate, 0970
Tetraammine-2,3-butanediimineruthenium(III) perchlorate, 1794
Tetraamminecadmium permanganate, 3956
Tetraamminecopper(II) azide, 4277
Tetraamminecopper(II) bromate, 0264
Tetraamminecopper(II) nitrate, 4276
Tetraamminecopper(II) nitrite, 4275
Tetraamminecopper(II) sulfate, 4274
trans Tetraamminediazidocobalt(III) *trans*-diamminetetraazido-
    cobaltate(1−), 4220

Tetrafluorourea, 0352

1.3a,4,6a-Tetrahydro-*N*,*N*′-dinitroimidazo[4,5-*d*]imidazole-2,5-diamine, *see*
Octahydro-2,5-bis(nitroimino)imidazo[4,5-*d*]imidazole, 1514

1,4,5,6-Tetrahydro-*N*,5-dinitrotriazin-2-amine, *see* 5-Nitro-2-
(nitroimino)hexahydro-1,3,5-triazine, 1218

† Tetrahydrofuran, 1612

2-Tetrahydrofuranylidene(dimethylphenylphosphine-
trimethylphosphine)-2,4,6-trimethylphenylnickel perchlorate, 3854

Tetrahydrofurfuryl alcohol, 1978

2-Tetrahydrofuryl hydroperoxide, 1624

4a,5,7a,8-Tetrahydro-4,8-methano-4*H*-indeno[5,6-c]-1,2,5-oxadiazole 1-oxide,
3281

4a,5,7a,8-Tetrahydro-4,8-methano-4*H*-indeno[5,6-c]-1,2,5-oxadiazole 3-oxide,
3282

Tetrahydronaphthalene, 3300

1,2,3,4-Tetrahydro-1-naphthyl hydroperoxide, 3312

Tetrahydro-1,4-oxazine, *see* Morpholine, 1657

† Tetrahydropyran, 1965

2-Tetrahydropyranylidene-bis(dimethylphenylphosphine)-
3,4,6-trimethylphenylnickel perchlorate, 3877

† 1,2,3,6-Tetrahydropyridine, 1930

1,4,5,8-Tetrahydro-1,4,5,8-tetrathiafulvalene, *see* 2,2′-Bi-1,3-dithiole, 2215

† Tetrahydrothiophene, 1627

Tetrahydrothiophene-1,1-dioxide, 1622

Tetrahydroxotritin(2+) nitrate, 4525

Tetrahydroxydioxotrisilane, 4536

Tetraimide (Tetraazetidine), 4527

Tetraiodoarsonium tetrachloroaluminate, 0057

Tetraiododiphosphane, 4637

Tetraiodoethylene, 0986

Tetraiodomethane, *see* Carbon tetraiodide, 0525

Tetraisocyanatogermane, *see* Germanium isocyanate, 1377

Tetraisopropylchromium, 3577

1,1,6,6-Tetrakis(acetylperoxy)cyclododecane, 3810

Tetrakis(μ_3-2-amino-2-methylpropanolato)tetrakis(μ_2-2.amino-
2-methylpropanolato)hexacopper(II) perchlorate, 3883

3,3,6,6-Tetrakis(bromomethyl)-9,9-dimethyl-1,2,4,5,7,8-hexoxonane, 3173

Tetrakis(butylthio)uranium, 3731

Tetrakis(chloroethynyl)silane, 2879

Tetrakis(*N*,*N*-dichloroaminomethyl)methane, 1900

Tetrakis(diethylphosphino)silane, 3733

Tetrakis(dimethylamino)titanium, 3104

Tetrakis(ethylthio)uranium, 3096

1,1,4,4-Tetrakis(fluoroxy)hexafluorobutane, 1374

Tetrakis(hydroxymercurio)methane, 0471

Tetrakis(hydroxymethyl)phosphonium nitrate, 1754

Tetrakis(3-methylpyrazole)cadmium sulfate, 3716

Tetrakis(3-methylpyrazole)manganese(II) sulfate, 3717

Tetrakis(4-*N*-methylpyridinio)porphinecobalt(III)(5+) perchlorate, 3908

Tetrakis(4-*N*-methylpyridinio)porphineiron(III)(5+) perchlorate, 3909

1,3,4,6-Tetrakis(2-methyltetrazol-5-yl)-hexaaza-1,5-diene, 3013

5,10,15,20-Tetrakis(2-nitrophenyl)porphine, 3907

Tetrakis(pentafluorophenyl)titanium, 3842

Tetrakis(pyrazole)manganese(II) sulfate, 3538

Tetrakis(pyridine)bis(tetracarbonylcobalt)magnesium, 3867

Tetrakis(thiourea)manganese(II) perchlorate, 1785

Tetrakis(trimethylsilyl)diaminodiphosphene, 3588

Tetrakis(2,2,2-trinitroethyl)orthocarbonate, 3132

Tetramethoxyethylene, 2516

† Tetramethoxymethane, *see* Tetramethyl orthocarbonate, 2022

† Tetramethoxysilane, 1764

Tetramethylammonium amide, 1782

Tetramethylammonium azidocyanatoiodate(I), 2006

Tetramethylammonium azidocyanoiodate(I), 2005

Tetramethylammonium azidoselenocyanatoiodate(I), 2007

Tetramethylammonium chlorite, 1742

Tetramethylammonium diazidoiodate(I), 1750

Tetramethylammonium monoperchromate, 1745

Tetramethylammonium ozonate, 1753

Tetramethylammonium pentafluoroxenoxide, 1747

Tetramethylammonium pentaperoxodichromate, 3103

Tetramethylammonium periodate, 1749

Tetramethylammonium superoxide, 1751

Tetramethylbis(trimethysilanoxy)digold, 3382

2,4,6,8-Tetramethylcyclotetrasiloxane, 1790

Tetramethyldialuminium dihydride, 1778

Tetramethyldiarsane, 1736

Tetramethyldiarsine, *see* Tetramethyldiarsane, 1736

Tetramethyldiborane, 1779

Tetramethyldigallane, 1748

Tetramethyldigold diazide, 1739

Tetramethyl-1,2-dioxetane, 2508

Tetramethyldiphosphane, 1765

Tetramethyldiphosphane disulfide, 1766

Tetramethyldiphosphine, *see* Tetramethyldiphosphane, 1765

Tetramethyldisiloxane, 1783

Tetramethyldistibane, 1769

Tetramethyldistibine, *see* Tetramethyldistibane, 1769

N,N,N′,N′-Tetramethylethane-1,2-diamine, trimethylpalladium(IV) bromide, 3225

N,N,N,N-Tetramethylethane-1,2-diamine, trimethylpalladium(IV) iodide, 3226

N,N,N′,N′-Tetramethylformamidinium perchlorate, 2026

2063

1,2,4-Trioxolane, *see* Ethylene ozonide, 0836
'Trioxygen difluoride' 4323
Trioxygen, *see* Ozone, 4846
Triphenylaluminium, 3749
Triphenylchromium tetrahydrofuranate, 3751
3,3,5-Triphenyl-4,4-dimethyl-5-hydroperoxy-4,5-dihydro(3*H*)pyrazole, 3841
Triphenyllead nitrate, 3753
Triphenylmethyl azide, 3783
Triphenylmethyl nitrate, 3782
Triphenylmethylpotassium, 3781
1,3,5-Triphenyl-1,4-pentaazadiene, 3754
Triphenylphosphine, 3756
Triphenylphosphine oxide hydrogen peroxidate, 3755
Triphenylphosphine oxide-oxodiperoxochromium(VI), 3752
2,4,6-Triphenylpyrilium perchlorate, 3839
Triphenylsilyl perchlorate, 3750
Triphenyltin hydroperoxide, 3758
† Triphosgene, *see* Trichloromethyl carbonate, 1040
Triphosphorus pentanitride, 4776
Tripotassium hexafluoroferrate(3−), 4359
Tri-2-propenylborane, 3177
Tripropylaluminium, 3215
Tripropylantimony, 3221
Tripropylborane, 3217
Tripropylindium, 3219
Tripropyllead fulminate, 3363
Tripropylsilyl perchlorate, 3218
Trirhenium nonachloride, 4196
1,1,1-Tris(aminomethyl)ethane, 2044
Tris(2-azidoethyl)amine, 2482
1,1,1-Tris(azidomethyl)ethane, 1937
1,3,5-Tris(4-azidosulfonylphenyl)-1,3,5-triazinetrione, 3818
Tris(2,2′-bipyridine)chromium(0), 3875
Tris(2,2′-bipyridine)chromium(II) perchlorate, 3874
Tris(2,2′-bipyridine)silver(II) perchlorate, 3873
Tris(bis-2-methoxyethyl ether)potassium hexacarbonylniobate(1−), 3856
2,4,6-Tris(bromoamino)-1,3,5-triazine (Tribromomelamine), 1091
2,2,2-Tris(bromomethyl)ethanol, 1922
1,1,1-Tris(bromomethyl)methane, 1550
† 2,4,6-Tris(chloroamino)-1,3,5-triazine, 1095
Tris(cyclopentadienyl)cerium, 3683
Tris(cyclopentadienyl)plutonium, 3684
Tris(cyclopentadienyl)uranium, 3685
Tris(2,3-diaminobutane)nickel(II) nitrate, 3589
Tris(1,2-diaminoethane)chromium(III) perchlorate, 2619
Tris(1,2-diaminoethane)cobalt(III) nitrate, 2622

Trithorium tetranitride, 4774
Tri(4-tolyl)ammonium perchlorate, 3824
Triuranium octaoxide, 4871
Trivinylantimony, 2405
Trivinylbismuth, 2391
Tropylium perchlorate, 2742
Tungsten, 4925
Tungsten azide pentabromide, 0297
Tungsten azide pentachloride, 4182
Tungsten carbide, 0563
Tungsten dichloride, 4119
Tungsten diiodide, 4631
Tungsten hexachloride, 4193
Tungsten hexafluoride, 4376
Tungsten hexamethoxide, 2604
Tungsten(IV) oxide, 4844
Tungsten tetrabromide oxide, 0294
Tungsten(VI) oxide, 4856

Undecaamminetetraruthenium dodecaoxide, 4598
'Unsaturated' Silicon hydride, 4452
Uranium, 4923
Uranium azide pentachloride, 4181
Uranium carbide, 0562
Uranium dicarbide, 1028
Uranium hexachloride, 4192
Uranium hexafluoride, 4375
Uranium(III) hydride, 4511
Uranium(III) nitride, 4732
Uranium(III) tetrahydroborate, 0178
Uranium(IV) hydride, 4541
Uranium(IV) oxide, 4843
Uranium(IV) sulfide, 4905
Uranium(IV) tetrahydroborate etherates, 0183
Uranyl nitrate, 4752
Uranyl perchlorate, 4111
Urea, 0475
Urea hydrogen peroxidate, 0476
Urea nitrate, *see* Uronium nitrate, 0494
Urea perchlorate, *see* Uronium perchlorate, 0491
Uronium nitrate, 0494
Uronium perchlorate, 0491

† Valeraldehyde, 1966
Vanadium, 4924

Xenon(II) fluoride perchlorate, 3977
Xenon(II) fluoride trifluoroacetate, 0634
Xenon(II) fluoride trifluoromethanesulfonate, 0356
Xenon(II) pentafluoroorthoselenate, 4382
Xenon(II) pentafluoroorthotellurate, 4383
Xenon(II) perchlorate, 4110
Xenon(IV) hydroxide, 4533
Xenon tetrafluoride, 4353
Xenon tetrafluoride oxide, 4346
Xenon tetraoxide, 4863
Xenon trioxide, 4857
† *m*-Xylene, 2971
† *o*-Xylene, 2970
† *p*-Xylene, 2972
† *mixo*-Xylene, 2969

Yttrium 4-nitrobenzoate trihydrate, 3817

Zinc, 4927
Zinc abietate, 3903
Zinc amalgam, 4601
Zinc azide, 4786
Zinc bis(1-benzeneazothiocarbonyl-2-phenyl-2-hydrazide), 3863
Zinc bromate, 0279
Zinc chlorate, 4106
Zinc chloride, 4120
Zinc cyanide, 1004
Zinc dihydrazide, 4563
Zinc ethoxide, 1704
Zinc ethylsulfinate, 1711
Zinc hydride, 4492
Zinc hydroxide, 4478
Zinc iodide, 4632
Zinc nitrate, 4751
Zinc oxide, 4830
Zinc permanganate, 4710
Zinc peroxide, 4845
Zinc phosphide, 4876
Zinc picrate, 3437
Zinc stearate, 3897
Zirconium, 4928
Zirconium carbide, 0565
Zirconium dibromide, 0285
Zirconium dicarbide, 1029
'Zirconium hydride', 4493

Appendix 5

Index of CAS Registry Numbers *vs* Serial Numbers in Section 1

The CAS Registry numbers are the key to the full structural and stereochemical information held within the CAS chemical identification system, available both in printed form and on-line via several hosts worldwide.

The Index is arranged in numerically increasing order of the registry numbers. The serial numbers of those compounds for which no registry number has been found are listed separately at the end of the Index.

[50-00-0], 0416	[62-23-7], 2695	[74-82-8], 0466	[75-09-2], 0397	[75-47-8], 0376
[50-71-5], 1398	[62-53-3], 2354	[74-83-9], 0429	[75-11-6], 0400	[75-50-3], 1310
[50-78-2], 3137	[62-56-6], 0480	[74-84-0], 0881	[75-12-7], 0453	[75-52-5], 0456
[50-99-7], 2518	[63-42-3], 3557	[74-85-1], 0781	[75-15-0], 0560	[75-54-7], 0470
[51-17-2], 2724	[64-17-5], 0920	[74-86-2], 0686	[75-18-3], 0932	[75-55-8], 1257
[51-28-5], 2197	[64-18-6], 0418	[74-87-3], 0432	[75-19-4], 1197	[75-56-9], 1225
[51-79-6], 1261	[64-19-7], 0833	[74-88-4], 0445	[75-20-7], 0585	[75-62-7], 0310
[54-87-5], 2908	[64-67-5], 1710	[74-89-5], 0493	[75-21-8], 0829	[75-63-8], 0311
[55-63-0], 1196	[65-85-0], 2732	[74-90-8], 0380	[75-24-1], 1291	[75-64-9], 1725
[56-18-8], 2587	[67-56-1], 0484	[74-93-1], 0489	[75-25-2], 0368	[75-65-0], 1696
[56-23-5], 0332	[67-63-0], 1280	[74-94-5], 0746	[75-26-3], 1242	[75-66-1], 1716
[56-81-5], 1286	[67-64-1], 1220	[74-95-3], 0395	[75-28-5], 1669	[75-68-3], 0731
[57-06-7], 1471	[67-66-3], 0372	[74-96-4], 0846	[75-29-6], 1244	[75-69-4], 0330
[57-13-6], 0475	[67-72-1], 0611	[74-98-6], 1271	[75-30-9], 1252	[75-71-8], 0326
[57-14-7], 0954	[68-12-2], 1259	[74-99-7], 1125	[75-31-0], 1308	[75-73-0], 0349
[57-50-1], 3558	[68-39-3], 1137	[75-00-3], 0848	[75-33-2], 1290	[75-74-1], 1767
[57-55-6], 1285	[69-68-5], 0921	[75-01-4], 0730	[75-34-3], 0790	[75-75-2], 0488
[57-71-6], 1570	[70-25-7], 0876	[75-02-5], 0747	[75-35-4], 0695	[75-76-3], 1770
[60-09-3], 3487	[70-34-8], 2108	[75-03-6], 0858	[75-36-5], 0735	[75-77-4], 1304
[60-12-8], 2983	[71-23-8], 1279	[75-04-7], 0942	[75-37-6], 0801	[75-78-5], 0902
[60-29-7], 1697	[71-43-2], 2288	[75-05-8], 0758	[75-38-7], 0700	[75-79-6], 0439
[60-34-4], 0503	[71-55-6], 0740	[75-07-0], 0828	[75-44-5], 0329	[75-83-2], 2529
[60-51-5], 2009	[72-20-8], 3459	[75-08-1], 0933	[75-45-6], 0369	[75-85-4], 2015

[106-95-6], 1153

[106-96-7], 1090

[106-97-8], 1668

[106-98-9], 1577

[106-99-0], 1480

[107-00-6], 1481

[107-02-8], 1145

[107-03-9], 1289

[107-05-1], 1158

[107-06-2], 0791

[107-10-8], 1309

[107-11-9], 1254

[107-12-0], 1179

[107-13-1], 1107

[107-15-3], 0953

[107-16-4], 0760

[107-18-6], 1223

[107-19-7], 1147

[107-21-1], 0924

[107-22-2], 0723

[107-25-5], 1221

[107-29-9], 0865

[107-30-2], 0850

[107-31-3], 0834

[107-32-4], 0420

[107-39-1], 3043

[107-40-4], 3045

[107-46-0], 2601

[107-71-1], 2509

[107-81-3], 1985

[107-82-4], 1984

[107-83-5], 2532

[107-84-6], 1986

[107-85-7], 2032

[107-87-9], 1962

[107-92-6], 1614

[108-00-9], 1755

[108-01-0], 1729

[108-03-2], 1263

[108-05-4], 1532

[108-08-7], 2872

[108-09-8], 2567

[108-10-1], 2495

[108-18-9], 2566

[108-19-0], 0874

[108-20-3], 2542

[108-21-4], 1972

[108-23-6], 1560

[108-24-7], 1534

[108-30-5], 1443

[108-31-6], 1404

[108-38-3], 2971

[108-40-7], 2809

[108-42-9], 2301

[108-44-1], 2814

[108-46-3], 2332

[108-58-7], 3291

[108-64-5], 2859

[108-67-3], 3162

[108-72-5], 1916

[108-77-0], 1038

[108-80-5], 1118

[108-87-2], 2854

[108-88-3], 2791

[108-90-7], 2228

[108-91-8], 2521

[108-93-0], 2485

[108-94-1], 2429

[108-95-2], 2330

[108-98-5], 2344

[109-02-4], 1995

[109-05-7], 2523

[109-06-8], 2355

[109-53-5], 2490

[109-54-6], 2003

[109-55-7], 2039

[109-60-4], 1977

[109-63-7], 1674

[109-65-9], 1631

[109-66-0], 2001

[109-69-3], 1637

[109-71-7], 1073

[109-72-8], 1651

[109-73-9], 1723

[109-74-0], 1563

[109-75-1], 1465

[109-76-1], 1946

[109-76-2], 1326

[109-77-3], 1078

[109-78-4], 1181

[109-79-5], 1712

[109-85-3], 1312

[109-86-4], 1284

[109-87-5], 1281

[109-89-7], 1726

[109-92-2], 1610

[109-93-3], 1519

[109-94-4], 1230

[109-95-5], 0867

[109-97-7], 1466

[109-99-9], 1612

[110-00-9], 1439

[110-01-0], 1627

[110-02-1], 1448

[110-05-4], 3074

[110-17-8], 1446

[110-18-9], 2579

[110-19-0], 2502

[110-22-2], 1537

[110-44-1], 2385b

[110-45-2], 2503

[110-46-3], 1996

[110-51-0], 0139

[110-54-3], 2531

[110-58-7], 2034

[110-61-2], 1433

[110-62-3], 1966

[110-65-6], 1526

[110-66-7], 2025

[110-68-9], 2033

[110-71-4], 1701

[110-74-7], 1621

[110-75-8], 1557

[110-80-5], 1702

[110-82-7], 2457

[110-83-8], 2406

[110-85-0], 1689

[110-86-1], 1848

[110-87-2], 1908

[110-88-3], 1237

[110-89-4], 1993

[110-91-8], 1657

[110-96-3], 3085

[111-15-9], 2511

[111-19-3], 3341

[111-21-7], 3355

[111-34-2], 2484

[111-36-4], 1931

[111-40-0], 1777

[111-41-1], 1756

[111-43-3], 2543

[111-46-6], 1705

[111-49-9], 2526

[111-65-9], 3059

[111-66-0], 3040

[111-67-1], 3041

[111-77-3], 2019

[111-84-2], 3209

[111-92-2], 3083

[111-94-4], 2397

[111-96-6], 2549

[112-24-3], 2599

[112-34-5], 3077

[112-36-7], 3076

[112-57-2], 3101

[112-58-3], 3563

[112-73-2], 3564

[112-77-6], 3772

[112-80-1], 3774

[115-00-6], 3465

[115-07-1], 1198

[115-10-6], 0919

[115-11-7], 1582

[115-18-4], 1960

[115-19-5], 1910

[115-20-8], 0744

[115-21-9], 0854

[115-77-5], 2021

[116-09-6], 1231

[116-14-3], 0628

[116-15-4], 1051

[118-52-5], 1865

[118-74-1], 2055

[118-96-7], 2701

[119-26-6], 2321

[119-33-5], 2767

[119-64-2], 3300

[119-90-4], 3660

[120-12-7], 3632

[120-37-0], 3691

[120-61-6], 3292

[120-80-9], 2331

[120-92-3], 1905

[120-94-5], 1992

[121-03-9], 2772

[121-14-2], 2726

[121-33-5], 2958

[121-43-7], 1296

[121-44-8], 2569

[121-45-9], 1315

[121-46-0], 2789

[121-66-4], 1117

[121-69-7], 2993

[121-82-4], 1219

[121-86-8], 2711

[121-88-0], 2316

[121-92-6], 2694

[122-39-4], 3504

[122-40-7], 3663

[122-56-5], 3569

[122-60-1], 3150

[122-82-7], 3529

[122-97-4], 3165

[123-05-7], 3050

[123-11-5], 2956

[123-15-9], 2491

[123-20-6], 2439

[123-23-9], 2990

[123-25-1], 3031

[123-31-9], 2333

[123-33-1], 1435

[123-38-6], 1224

[123-39-7], 0866

[123-42-2], 2501

[123-51-3], 2014

[123-63-7], 2515

[123-72-8], 1607

[123-75-1], 1653

[123-77-3], 0816

[123-86-4], 2496

[123-91-1], 1617

[123-92-2], 2861

[123-99-9], 3187

[124-02-7], 2450

[124-04-9], 2441

[124-18-5], 3365

[124-38-9], 0557

[124-40-3], 0941

[124-41-4], 0464

[124-43-6], 0476

2085

[1605-58-9], 2036	[1873-77-4], 3231	[2386-62-1], 0842	[2809-21-4], 0956	[3221-61-2], 3206
[1607-30-3], 3354	[1884-64-6], 0540	[2397-67-3], 3214	[2809-67-8], 3023	[3236-56-4], 3194
[1609-47-8], 2444	[1886-68-6], 3656	[2400-59-1], 3554	[2809-69-0], 2293	[3242-56-6], 2375
[1618-08-2], 1344	[1886-69-7], 3458	[2402-95-1], 1829	[2829-31-4], 3609	[3248-28-0], 2442
[1623-99-0], 2287	[1890-04-6], 1860	[2402-97-3], 1828	[2845-00-3], 1173	[3251-23-8], 4279
[1624-19-4], 3184	[1931-62-0], 3015	[2404-52-6], 4340	[2851-55-0], 0990	[3264-86-6], 2626
[1630-94-0], 1940	[1941-24-8], 0506	[2406-51-1], 0807	[2868-37-3], 1917	[3272-96-6], 0849
[1631-82-9], 2810	[1941-79-3], 3189	[2407-94-5], 3553	[2875-36-7], 3053	[3277-26-7], 1783
[1634-04-4], 2011	[1942-45-6], 3025	[2417-93-8], 1253	[2881-62-1], 1022	[3279-95-6], 0944
[1637-31-6], 1389	[1944-83-8], 3332	[2424-09-1], 1758	[2893-78-9], 1037	[3282-30-2], 1925
[1640-89-7], 2850	[1956-39-4], 2927	[2425-74-3], 1994	[2899-02-7], 3596	[3296-62-3], 0694
[1645-75-6], 1067	[1972-28-7], 2425	[2425-79-8], 3353	[2916-31-6], 1969	[3312-92-6], 1024
[1653-57-2], 3364	[1984-04-9], 3390	[2428-04-8], 1043	[2917-26-2], 3727	[3315-16-0], 0300
[1656-16-2], 2073	[2003-31-8], 1036	[2446-84-6], 1528	[2918-51-8], 2682	[3315-89-7], 0972
[1656-44-6], 2100	[2028-76-4], 2151	[2506-80-1], 2026	[2923-28-6], 0444	[3319-45-7], 3755
[1663-35-0], 1973	[2032-04-4], 1135	[2524-02-9], 3980	[2925-21-5], 0657	[3327-66-1], 0586
[1666-13-3], 3501	[2036-15-9], 2552	[2536-18-7], 1142	[2938-46-7], 0423	[3329-56-4], 2428
[1675-54-3], 3825	[2041-76-1], 2844	[2551-62-4], 4374	[2938-73-0], 0844	[3333-52-6], 3007
[1679-09-0], 2023	[2045-70-7], 1421	[2554-06-5], 3560	[2944-05-0], 0559	[3356-63-6], 1170
[1686-88-3], 1472	[2083-91-2], 2043	[2567-83-1], 3088	[2944-66-3], 2065	[3385-78-2], 1307
[1711-42-8], 2925	[2092-17-3], 0576	[2581-34-2], 2766	[2958-89-6], 1586	[3400-09-7], 4063
[1712-64-7], 1266	[2152-13-7], 3267	[2589-01-7], 3533	[2980-33-8], 1822	[3437-84-1], 3032
[1717-00-6], 0738	[2154-71-4], 0638	[2592-85-0], 2178	[2997-92-4], 3089	[3444-13-1], 0982
[1718-18-9], 0642	[2155-71-7], 3715	[2592-95-2], 2272	[2999-74-8], 0908	[3445-52-1], 1476
[1719-53-5], 1683	[2156-66-3], 0915	[2611-42-9], 1023	[3003-15-4], 1321	[3454-11-3], 0756
[1730-69-4], 3065	[2156-71-0], 1948	[2612-33-1], 1160	[3009-34-5], 2260	[3463-17-0], 3604
[1739-53-3], 0845	[2167-23-9], 3565	[2614-76-8], 1287	[3011-34-5], 2691	[3468-63-1], 3702
[1746-09-4], 2591	[2169-38-2], 1740	[2629-78-9], 1361	[3013-38-5], 2672	[3469-17-8], 3638
[1750-42-1], 1136	[2175-91-9], 2967	[2643-84-5], 2960	[3015-98-3], 3219	[3470-17-5], 2631
[1762-95-4], 0479	[2198-20-1], 3351	[2644-70-4], 4007	[3017-85-4], 0882	[3495-54-3], 0063
[1779-25-5], 3064	[2200-13-7], 3876	[2651-85-6], 2572	[3021-39-4], 0977	[3504-13-0], 1542
[1781-66-4], 1884	[2203-57-8], 1060	[2675-89-0], 1585	[3031-73-0], 0486	[3514-82-7], 3481
[1789-58-8], 0903	[2216-12-8], 3472	[2691-41-0], 1605	[3031-74-1], 0925	[3522-94-9], 3210
[1792-40-1], 2940	[2216-30-0], 3200	[2696-92-6], 4023	[3031-75-2], 1283	[3525-44-8], 1667
[1794-84-9], 0396	[2216-33-3], 3207	[2697-42-9], 0952	[3034-94-4], 2905	[3531-19-9], 2153
[1795-27-3], 3191	[2216-34-4], 3208	[2698-41-1], 3333	[3052-45-7], 1177	[3536-96-7], 0732
[1809-19-4], 3086	[2217-55-2], 3449	[2698-69-3], 3266	[3054-95-3], 2858	[3558-60-9], 3164
[1809-53-6], 3020	[2235-08-7], 1416	[2699-12-9], 3556	[3058-38-0], 2324	[3570-93-2], 2400
[1817-47-6], 3157	[2245-68-3], 1685	[2712-78-9], 3239	[3071-32-7], 2985	[3585-32-8], 0945
[1817-73-8], 2132	[2252-95-1], 1363	[2713-09-9], 0662	[3074-75-7], 3205	[3607-48-5], 0524
[1838-59-1], 1524	[2278-22-0], 0766	[2735-04-8], 2998	[3074-77-9], 3204	[3618-05-1], 4469
[1840-42-2], 0340	[2294-47-5], 2208	[2738-18-3], 3190	[3086-29-1], 1303	[3676-91-3], 1765
[1850-14-2], 2022	[2306-67-0], 0888	[2749-79-3], 2328	[3159-98-6], 3542	[3676-97-9], 1766
[1854-19-9], 3081	[2370-88-9], 1790	[2783-96-2], 0681	[3161-99-7], 2074	[3686-43-9], 1567
[1871-52-9], 2966	[2372-21-6], 3051	[2786-01-8], 2657	[3179-56-4], 3033	[3692-81-7], 3571
[1872-00-0], 1722	[2378-02-1], 1380	[2802-70-2], 1059	[3188-13-4], 1246	[3698-54-2], 2122

[3698-83-7], 2077
[3706-77-2], 1094
[3710-30-3], 3021
[3751-44-8], 1576
[3768-60-3], 1757
[3802-95-7], 0347
[3811-04-9], 4017
[3823 94-7], 1101
[3851-22-7], 1704
[3855-45-6], 0348
[3880-04-4], 0906
[3884-87-0], 0305
[3917-15-5], 1904
[3946-86-9], 1751
[3958-60-9], 2710
[3982-91-0], 4154
[3999-10-8], 2727
[4000-16-2], 0517
[4023-65-8], 2342
[4039-32-1], 2594
[4100-38-3], 3309
[4104-85-2], 0515
[4109-96-0], 4066
[4112-22-5], 1494
[4113-57-9], 0655
[4150-34-9], 3758
[4154-69-2], 1678
[4170-30-3], 1516
[4185-47-1], 1602
[4209-91-0], 3073
[4212-43-5], 1236
[4222-21-3], 0733
[4222-26-8], 2928
[4222-27-9], 0734
[4230-63-1], 3488
[4254-22-2], 0729
[4264-83-9], 2185
[4279-76-9], 2920
[4312-28-1], 3848
[4328-17-0], 2120
[4331-98-0], 0998
[4343-68-4], 0541
[4377-73-5], 2310
[4388-03-8], 1593
[4394-93-8], 0362
[4403-63-8], 3581

[4415-23-0], 1007
[4417-80-5], 2433
[4418-61-5], 0461
[4419-11-8], 3668
[4437-18-7], 1840
[4452-58-8], 0554
[4457-90-3], 2053
[4457-91-4], 3418
[4460-46-2], 2673
[4461-41-0], 1551
[4461-48-7], 2465
[4472-06-4], 0386
[4474-17-3], 0600
[4504-27-2], 1190
[4531-35-5], 2895
[4538-37-8], 2376
[4543-33-3], 3442
[4547-68-6], 2152
[4547-69-7], 2210
[4549-46-6], 1690
[4572-12-7], 3788
[4591-46-2], 2123
[4606-07-9], 2437
[4635-46-5], 4788
[4648-54-8], 1314
[4656-04-6], 2593
[4676-82-8], 1624
[4682-03-5], 2087
[4702-38-9], 0407
[4711-74-4], 2606
[4711-95-9], 2866
[4712-19-0], 3468
[4733-52-2], 2900
[4744-10-9], 2017
[4747-90-4], 1079
[4749-28-4], 1183
[4755-77-5], 1456
[4756-66-5], 1377
[4774-73-6], 0918
[4784-77-4], 1548
[4806-61-5], 2458
[4845-05-0], 2435
[4873-09-0], 3315
[4884-18-8], 3693
[4887-24-5], 2390
[4888-97-5], 3010

[4897-21-6], 1453
[4904-55-6], 3667
[4911-55-1], 1825
[4922-98-9], 2952
[4972-29-6], 2234
[4981-48-0], 1935
[4984-82-1], 1855
[5071-96-5], 2997
[5106-46-7], 2514
[5120-07-0], 3672
[5128-28-9], 2089
[5131-60-2], 2348
[5131-88-4], 0630
[5158-50-9], 0887
[5162-44-7], 1549
[5164-35-2], 1856
[5194-50-3], 2412
[5194-51-4], 2413
[5256-74-6], 2828
[5274-48-6], 2217
[5314-83-0], 1673
[5314-85-2], 1335
[5329-14-6], 4499
[5332-73-0], 1730
[5332-96-7], 3142
[5345-54-0], 2795
[5366-84-7], 3240
[5388-62-5], 2154
[5390-07-8], 2236
[5392-40-5], 3346
[5425-78-5], 1245
[5428-37-5], 3259
[5434-72-0], 3902
[5470-11-1], 4002
[5492-69-3], 3090
[5500-21-0], 1463
[5518-62-7], 0957
[5573-38-3], 3229
[5588-74-9], 2745
[5593-70-4], 3730
[5610-59-3], 0301
[5613-68-3], 2405
[5613-69-4], 3221
[5637-83-2], 1348
[5653-21-4], 0809
[5702-61-4], 3500

[5789-17-3], 1141
[5796-89-4], 1187
[5796-89-4], 1574
[5797-06-8], 1837
[5809-59-6], 1467
[5813-49-0], 0762
[5813-64-9], 2030
[5813-77-4], 1535
[5888-61-9], 3822
[5960-88-3], 0853
[5989-27-5], 3337
[6044-68-4], 1968
[6066-82-6], 1469
[6069-42-7], 1226
[6088-91-1], 1381
[6109-04-2], 3631
[6130-87-6], 1759
[6147-53-1], 1780
[6180-21-8], 0599
[6232-88-8], 2926
[6269-50-7], 2220
[6303-21-5], 4503
[6361-21-3], 2649
[6361-22-4], 2650
[6381-06-2], 3502
[6407-56-3], 3701
[6410-13-5], 3697
[6411-21-8], 3379
[6443-91-0], 1146
[6470-09-3], 1342
[6476-36-4], 3220
[6477-64-1], 3436
[6484-52-2], 4522
[6488-00-2], 2976
[6522-40-3], 2369
[6540-76-7], 2394
[6569-51-3], 0176
[6592-86-1], 2129
[6610-25-9], 3809
[6628-86-0], 2652
[6630-99-5], 0413
[6659-62-7], 1186
[6713-82-2], 3894
[6728-26-3], 2432
[6786-32-9], 2689
[6798-76-1], 3903

[6824-18-6], 1444
[6832-13-9], 0710
[6832-16-2], 1138
[6834-92-0], 4805
[6865-68-5], 2622
[6866-10-0], 3147
[6890-38-6], 2378
[6918-51-0], 0930
[6921-27-3], 2327
[6925-01-5], 2189
[6926-39-2], 2004
[6926-40-5], 2002
[6928-74-1], 1498
[6996-92-5], 2335
[7119-27-9], 1454
[7154-79-2], 3211
[7176-89-8], 3149
[7182-87-8], 0374
[7203-21-6], 3786
[7226-23-5], 2475
[7227-91-0], 2996
[7245-18-3], 1317
[7279-42-0], 2179
[7289-92-1], 2598
[7318-34-5], 3270
[7332-00-5], 1666
[7360-53-4], 1089
[7388-28-5], 0880
[7400-08-0], 3136
[7416-48-0], 1626
[7429-90-5], 0048
[7429-96-5], 4700
[7436-07-9], 3140
[7439-88-5], 4643
[7439-89-6], 4388
[7439-91-0], 4677
[7439-92-1], 4882
[7439-93-2], 4680
[7439-95-4], 4690
[7439-97-6], 4600
[7439-98-7], 4712
[7440-00-8], 4819
[7440-02-0], 4820
[7440-03-1], 3948
[7440-03-1], 4817
[7440-04-2], 4873

[7440-05-3], 4885	[7446-08-4], 4838	[7681-57-4], 4808	[7773-06-0], 4556	[7783-75-7], 4362
[7440-06-4], 4887	[7446-09-5], 4837	[7681-82-5], 4623	[7774-29-0], 4602	[7783-77-9], 4365
[7440-07-5], 4888	[7446-11-9], 4849	[7693-26-7], 4427	[7775-14-6], 4807	[7783-81-5], 4375
[7440-09-7], 4645	[7446-14-2], 4860	[7693-27-8], 4463	[7775-27-1], 4809	[7783-82-6], 4376
[7440-10-0], 4886	[7446-70-0], 0062	[7697-37-2], 4436	[7775-41-9], 0013	[7783-83-7], 4393
[7440-15-5], 4890	[7447-40-7], 4015	[7704-34-9], 4897	[7778-18-9], 3939	[7783-84-8], 4394
[7440-16-6], 4892	[7529-22-8], 1997	[7704-98-5], 4490	[7778-50-9], 4248	[7783-86-0], 4395
[7440-17-7], 4889	[7544-40-3], 3027	[7704-99-6], 4493	[7778-54-3], 3924	[7783-89-3], 0007
[7440-18-8], 4894	[7550-45-0], 4176	[7705-07-9], 4158	[7778-74-7], 4018	[7783-90-6], 0008
[7440-21-3], 4909	[7553-56-2], 4625	[7705-08-0], 4133	[7779-88-6], 4751	[7783-91-7], 0010
[7440-22-4], 0001	[7568-37-8], 0431	[7705-12-6], 1393	[7782-40-3], 0298	[7783-92-8], 0011
[7440-23-5], 4796	[7570-25-4], 0770	[7713-79-3], 3122	[7782-41-4], 4310	[7783-93-9], 0012
[7440-25-7], 4913	[7572-29-4], 0602	[7718-98-1], 4159	[7782-42-5], 0298	[7783-95-1], 0014
[7440-25-7], 4914	[7575-57-7], 3155	[7719-09-7], 4096	[7782-44-7], 4831	[7783-97-3], 0020
[7440-26-8], 4915	[7601-54-9], 4814	[7719-12-2], 4153	[7782-44-7], 4832	[7783-98-4], 0021
[7440-28-0], 4922	[7601-87-8], 1512	[7720-78-7], 4399	[7782-49-2], 4908	[7784-04-5], 0031
[7440-29-1], 4917	[7601-89-0], 4040	[7720-83-4], 4638	[7782-50-3], 4047	[7784-21-6], 0070
[7440-31-5], 4911	[7601-90-3], 3998	[7721-01-9], 4185	[7782-63-9], 3175	[7784-23-8], 0079
[7440-31-5], 4912	[7616-83-3], 4078	[7722-64-7], 4647	[7782-65-2], 4417	[7784-34-1], 0094
[7440-32-6], 4919	[7616-94-6], 3974	[7722-84-1], 4477	[7782-68-5], 4424	[7784-36-3], 0095
[7440-33-7], 4925	[7631-86-9], 4839	[7722-86-3], 4481	[7782-77-6], 4435	[7784-42-1], 0100
[7440-36-0], 4907	[7631-99-4], 4721	[7723-14-0], 4874	[7782-78-7], 4438	[7786-30-3], 4081
[7440-37-1], 0091	[7632-00-0], 4720	[7726-95-6], 0261	[7782-79-8], 4441	[7787-47-5], 0221
[7440-38-2], 0092	[7632-04-4], 0155	[7727-15-3], 0060	[7782-87-8], 4459	[7787-49-7], 0223
[7440-39-3], 0200	[7632-51-1], 4177	[7727-18-6], 4151	[7782-89-0], 4462	[7787-52-2], 0224
[7440-41-7], 0220	[7637-07-2], 0129	[7727-21-1], 4668	[7782-92-5], 4465	[7787-62-4], 0227
[7440-42-8], 0119	[7646-69-7], 4444	[7727-37-9], 4734	[7782-94-7], 4471	[7787-70-4], 0265
[7440-43-9], 3949	[7646-78-8], 4174	[7727-43-7], 0217	[7783-06-4], 4483	[7787-71-5], 0241
[7440-44-0], 0298	[7646-79-9], 4048	[7727-54-0], 4576	[7783-07-5], 4486	[7789-09-5], 4246
[7440-45-1], 3961	[7646-85-7], 4120	[7731-34-2], 3406	[7783-09-7], 4488	[7789-18-6], 4261
[7440-46-2], 4254	[7647-01-0], 3993	[7732-18-5], 4475	[7783-17-7], 4517	[7789-20-0], 4291
[7440-47-3], 4222	[7647-14-5], 4036	[7738-94-5], 4229	[7783-18-8], 4573	[7789-25-5], 4302
[7440-48-4], 4199	[7647-18-9], 4184	[7740-53-1], 4292	[7783-20-2], 4574	[7789-26-6], 4304
[7440-50-8], 4267	[7651-91-4], 0437	[7757-79-1], 4650	[7783-26-8], 4579	[7789-30-2], 0243
[7440-55-3], 4406	[7659-31-6], 0568	[7758-01-2], 0255	[7783-29-1], 4584	[7789-31-3], 0248
[7440-56-4], 4412	[7664-38-2], 4505	[7758-05-6], 4619	[7783-41-7], 4317	[7789-33-5], 0254
[7440-57-5], 0110	[7664-39-3], 4294	[7758-09-0], 4649	[7783-42-8], 4318	[7789-38-0], 0257
[7440-58-6], 4599	[7664-41-7], 4497	[7758-19-2], 4038	[7783-44-0], 4320	[7789-41-5], 0262
[7440-61-1], 4923	[7664-93-9], 4479	[7758-89-6], 4056	[7783-46-2], 4329	[7789-43-7], 0263
[7440-63-3], 4926	[7673-09-8], 1095	[7758-94-3], 4061	[7783-47-3], 4331	[7789-46-0], 0266
[7440-66-6], 4927	[7677-24-9], 1665	[7758-95-4], 4112	[7783-53-1], 4335	[7789-47-1], 0269
[7440-67-7], 4928	[7681-38-1], 4446	[7758-97-6], 4243	[7783-54-2], 4336	[7789-51-7], 0275
[7440-68-2], 4924	[7681-38-1], 4806	[7761-88-8], 0022	[7783-55-3], 4339	[7789-57-3], 0287
[7440-69-9], 0226	[7681-52-9], 4037	[7772-98-7], 4804	[7783-60-0], 4350	[7789-60-8], 0293
[7440-70-2], 3922	[7681-53-0], 4473	[7772-99-8], 4116	[7783-61-1], 4352	[7789-65-3], 0295
[7440-74-6], 4640	[7681-55-2], 4624	[7773-01-5], 4085	[7783-66-6], 4355	[7789-78-8], 3927

[7789-79-9], 3931
[7790-21-8], 4620
[7790-69-4], 4684
[7790-89-8], 3971
[7790-91-2], 3981
[7790-92-3], 3995
[7790-93-4], 3996
[7790-94-5], 3997
[7790-98-9], 4004
[7790-99-0], 4013
[7791-03-9], 4021
[7791-08-4], 4041
[7791-21-1], 4095
[7791-23-3], 4097
[7791-25-5], 4099
[7791-27-7], 4103
[7796-16-9], 0659
[7803-49-8], 4498
[7803-51-2], 4508
[7803-52-3], 4510
[7803-62-5], 4539
[7803-65-8], 4554
[7847-94-7], 4076
[7861-11-0], 4618
[7958-20-3], 2550
[8049-11-4], 0050
[9002-81-7], 0417
[9002-83-9], 0592
[9002-84-0], 0629
[9002-88-4], 0782
[9002-89-5], 0831
[9003-17-2], 1484
[9003-27-4], 1583
[9004-73-3], 0485
[9016-00-6], 0922
[9018-45-5], 3836
[10022-31-8], 0212
[10022-50-1], 4303
[10025-67-9], 4114
[10025-68-0], 4115
[10025-73-7], 4127
[10025-78-2], 4136
[10025-85-1], 4143
[10025-87-3], 4149
[10025-91-9], 4157
[10026-04-7], 4173

[10026-07-0], 4175
[10026-11-6], 4178
[10026-13-8], 4183
[10026-18-3], 4200
[10026-22-9], 4215
[10027-74-4], 1707
[10028-15-6], 4846
[10031-26-2], 0286
[10031-27-3], 0296
[10031-43-3], 4279
[10034-81-8], 4084
[10034-85-2], 4423
[10035-10-6], 0247
[10036-47-2], 4345
[10038-98-9], 4162
[10039-54-0], 4575
[10043-11-5], 0152
[10043-35-3], 0144
[10043-52-4], 3923
[10043-84-2], 4519
[10043-92-2], 4893
[10045-94-0], 4603
[10049-03-3], 3976
[10049-04-4], 4042
[10049-05-5], 4052
[10049-06-6], 4117
[10049-07-7], 4155
[10049-08-8], 4156
[10049-17-9], 4373
[10051-06-6], 0359
[10058-23-8], 4430
[10097-28-6], 4828
[10099-74-8], 4749
[10101-50-5], 4703
[10102-03-1], 4748
[10102-06-4], 4752
[10102-40-6], 4713
[10102-43-9], 4724
[10102-44-0], 4725
[10112-08-0], 0492
[10118-76-0], 3934
[10124-37-5], 3935
[10124-48-8], 3999
[10137-74-3], 3952
[10138-17-7], 2498
[10139-47-6], 4632

[10158-43-7], 3298
[10192-29-7], 4003
[10192-30-0], 4545
[10204-97-2], 4744
[10210-68-1], 2881
[10218-83-4], 0430
[10229-09-1], 1508
[10230-68-9], 1184
[10233-03-1], 4082
[10241-03-9], 4160
[10241-04-4], 4122
[10241-05-1], 4180
[10248-58-5], 4867
[10256-92-5], 0352
[10272-07-8], 2999
[10294-33-4], 0122
[10294-34-5], 0127
[10294-46-9], 4057
[10294-47-0], 4105
[10294-48-1], 4107
[10294-58-3], 4531
[10308-82-4], 0512
[10308-83-5], 0514
[10308-84-6], 0502
[10308-90-4], 1276
[10308-93-7], 3048
[10308-95-9], 3092
[10311-08-7], 3002
[10325-39-0], 0124
[10326-21-3], 4083
[10361-95-2], 4106
[10369-17-2], 2600
[10377-60-3], 4693
[10377-62-5], 4691
[10377-66-9], 4701
[10397-28-1], 2916
[10405-27-3], 4311
[10415-75-5], 4609
[10420-90-3], 2290
[10421-42-4], 4397
[10437-85-1], 3404
[10450-60-9], 4542
[10467-10-4], 0859
[10504-99-1], 3494
[10534-86-8], 4210
[10544-72-6], 4747

[10544-73-7], 4746
[10545-99-0], 4113
[10557-85-4], 1868
[10567-69-8], 0211
[10578-16-2], 4314
[10580-52-6], 4118
[10588-01-9], 4250
[10599-90-3], 4000
[11079-26-8], 3708
[11144-29-9], 0002
[11713-32-0], 0631
[11802-21-2], 3469
[12001-65-9], 2471
[12001-89-7], 3767
[12003-23-5], 0084
[12003-69-9], 0051
[12003-96-2], 0056
[12007-25-9], 0168
[12007-33-9], 0172
[12007-46-4], 0177
[12008-21-8], 0193
[12008-61-6], 0199
[12010-53-6], 0231
[12011-67-5], 0365
[12012-32-7], 0590
[12013-55-7], 3943
[12013-56-8], 3944
[12013-82-0], 3946
[12014-93-6], 3967
[12015-30-4], 4033
[12015-76-8], 4269
[12018-00-7], 4241
[12018-06-3], 4245
[12019-11-3], 4284
[12019-27-1], 4268
[12020-65-4], 4293
[12024-20-3], 4411
[12025-32-0], 4419
[12027-06-4], 4512
[12029-98-0], 4627
[12030-49-8], 4644
[12030-88-5], 4656
[12030-89-6], 4657
[12032-88-1], 4708
[12033-19-1], 4714
[12033-31-7], 4714

[12033-54-4], 4726
[12033-59-9], 4730
[12033-67-9], 4731
[12033-88-6], 4771
[12033-90-8], 4774
[12034-12-7], 4799
[12035-36-8], 4822
[12036-02-1], 4833
[12036-04-3], 4835
[12036-22-5], 4844
[12038-67-4], 4891
[12039-13-3], 4904
[12039-14-4], 4905
[12042-38-5], 0053
[12043-56-0], 0049
[12044-52-9], 0107
[12045-01-1], 0128
[12047-79-9], 0219
[12051-08-0], 4028
[12053-67-7], 4263
[12055-24-2], 4628
[12057-24-8], 4686
[12057-71-5], 4698
[12057-74-8], 4699
[12057-92-0], 4709
[12058-54-7], 4800
[12058-85-4], 4815
[12063-27-3], 4404
[12065-85-9], 4870
[12069-32-8], 0309
[12069-85-1], 0521
[12070-08-5], 0561
[12070-09-6], 0562
[12070-12-1], 0563
[12070-13-2], 0564
[12070-14-3], 0565
[12070-14-3], 1029
[12070-27-8], 0575
[12071-15-7], 0991
[12071-29-3], 1026
[12071-33-9], 1028
[12075-68-2], 2556
[12079-66-2], 2882
[12081-54-8], 3098
[12081-88-8], 3107
[12082-46-1], 3180

[12097-97-1], 3830 [12265-93-9], 4418 [13093-00-0], 4111 [13453-07-1], 0111 [13537-24-1], 4134
[12116-83-5], 3296 [12266-58-9], 2312 [13101-58-1], 0548 [13453-57-1], 4102 [13537-32-1], 4297
[12124-97-9], 0251 [12273-50-6], 4674 [13138-21-1], 0675 [13455-00-0], 4637 [13537-33-2], 4298
[12125-02-9], 4001 [12275-58-0], 2951 [13138-45-9], 4743 [13455-01-1], 4636 [13551-73-0], 1864
[12125-09-6], 4515 [12279-90-2], 0109 [13147-25-6], 1691 [13455-15-7], 4348 [13556-31-5], 3172
[12125-23-4], 4707 [12281-65-1], 0567 [13154-66-0], 1331 [13463-30-4], 4172 [13556-50-8], 2207
[12129-68-9], 3514 [12298-67-8], 4606 [13168-78-0], 2263 [13463-39-3], 1805 [13569-43-2], 0246
[12133-31-2], 3933 [12300-03-7], 4797 [13172-31-1], 0282 [13463-40-6], 1814 [13569-50-1], 3962
[12133-60-7], 3983 [12312-33-3], 3269 [13196-64-7], 3188 [13463-67-7], 4842 [13572-99-1], 4413
[12134-29-1], 4266 [12316-91-3], 4776 [13213-29-1], 3804 [13463-98-4], 3900 [13596-23-1], 4194
[12134-35-9], 4290 [12325-31-4], 4564 [13214-58-9], 0748 [13464-46-5], 0090 [13596-24-2], 4198
[12135-76-1], 4577 [12331-76-9], 4235 [13223-78-4], 1010 [13464-97-6], 4549 [13597-73-4], 4565
[12136-15-1], 4615 [12333-57-2], 4729 [13232-74-1], 2632 [13464-98-7], 4561 [13597-95-0], 0222
[12136-26-4], 4641 [12333-97-0], 4883 [13235-16-0], 3940 [13465-07-1], 4484 [13598-36-2], 4504
[12136-61-7], 4689 [12340-14-6], 4863 [13270-95-6], 2578 [13465-08-2], 4524 [13598-47-5], 4500
[12136-83-3], 4723 [12354-85-7], 3806 [13271-93-7], 2047 [13465-33-3], 0271 [13598-52-5], 4506
[12136-85-5], 4727 [12359-06-7], 4682 [13271-97-1], 3583 [13465-41-3], 4434 [13598-56-6], 4511
[12138-26-0], 4910 [12380-95-9], 3960 [13271-98-2], 3777 [13465-66-2], 4351 [13600-88-9], 4204
[12139-70-7], 4214 [12384-02-0], 4681 [13272-02-1], 3230 [13465-71-9], 4333 [13601-08-6], 4588
[12140-58-8], 4410 [12397-26-1], 4439 [13272-03-2], 3586 [13465-73-1], 0250 [13620-57-0], 3653
[12143-72-5], 4903 [12397-29-4], 4616 [13272-04-3], 3696 [13465-77-5], 4191 [13637-63-3], 3989
[12154-56-2], 3605 [12412-94-1], 4884 [13272-05-4], 3852 [13465-95-7], 0206 [13637-65-5], 3992
[12164-01-1], 4773 [12430-27-2], 0194 [13280-07-4], 1455 [13466-78-9], 3336 [13637-71-3], 4094
[12164-12-4], 4801 [12438-96-9], 4757 [13283-01-7], 4193 [13469-09-3], 0289 [13637-76-8], 4108
[12164-94-2], 4526 [12442-63-6], 4104 [13283-31-3], 0135 [13470-12-7], 4119 [13637-87-1], 3978
[12164-95-3], 4827 [12501-01-8], 4810 [13292-87-0], 0136 [13470-17-2], 4631 [13637-88-2], 3987
[12166-20-0], 4895 [12504-41-5], 4898 [13298-76-5], 3009 [13477-00-4], 0205 [13659-67-1], 4460
[12171-97-0], 3840 [12505-77-0], 0169 [13360-63-9], 2565 [13477-09-3], 0207 [13693-05-5], 4371
[12181-34-9], 4532 [12517-41-8], 4283 [13400-13-0], 4255 [13478-33-6], 4051 [13693-06-6], 4372
[12191-80-9], 4327 [12529-66-7], 4422 [13422-81-6], 1782 [13478-38-1], 4279 [13693-09-9], 4377
[12193-29-2], 3110 [12532-69-3], 4646 [13424-46-9], 4782 [13483-59-5], 0252 [13701-67-2], 0161
[12196-72-4], 4678 [12540-13-5], 0615 [13444-71-8], 4425 [13492-26-7], 4431 [13706-14-4], 3986
[12206-14-3], 4240 [12542-85-7], 1292 [13444-85-4], 4633 [13494-80-9], 4916 [13709-32-5], 4328
[12211-03-9], 4910 [12576-63-5], 3790 [13444-89-8], 0291 [13497-91-1], 4171 [13709-36-9], 4332
[12212-56-5], 3705 [12635-29-9], 0055 [13444-90-1], 4025 [13497-97-7], 0061 [13709-55-2], 4347
[12213-13-7], 3861 [12674-40-7], 1027 [13445-49-3], 4482 [13498-06-1], 4148 [13709-61-0], 4353
[12216-68-9], 3684 [12774-81-1], 1362 [13445-50-6], 4538 [13498-14-1], 4168 [13718-25-7], 4783
[12218-77-8], 4826 [12785-36-3], 0004 [13446-09-8], 4513 [13499-05-3], 4163 [13730-91-1], 0143
[12228-13-6], 0132 [13007-92-6], 2056 [13446-09-8], 4514 [13517-00-5], 4286 [13746-66-2], 2063
[12228-28-3], 0149 [13036-75-4], 4326 [13446-10-1], 4518 [13517-10-7], 0150 [13746-98-0], 4762
[12232-97-2], 0230 [13078-30-3], 2729 [13446-48-5], 4521 [13517-64-7], 4736 [13749-37-6], 1126
[12244-59-6], 2113 [13084-45-2], 1496 [13446-75-8], 4450 [13520-74-6], 0281 [13749-94-5], 1260
[12245-12-4], 3451 [13084-46-3], 1497 [13449-15-5], 4031 [13520-77-9], 0294 [13762-26-0], 4121
[12258-22-9], 0047 [13084-47-4], 0802 [13449-16-6], 4472 [13520-90-6], 0292 [13762-80-6], 4011
[12259-52-6], 4577 [13086-63-0], 0366 [13449-22-4], 1693 [13529-75-4], 2041 [13763-19-4], 4197
[12263-33-1], 4711 [13092-75-6], 0650 [13451-18-8], 4853 [13536-85-1], 4324 [13763-23-0], 4192

[13766-26-2], 0134
[13767-10-0], 4557
[13768-38-2], 4370
[13768-86-0], 4852
[13769-36-3], 0267
[13770-16-6], 4088
[13770-86-0], 0209
[13770-96-2], 0076
[13772-47-9], 4258
[13773-81-4], 4313
[13774-85-1], 4346
[13774-94-2], 4451
[13776-58-4], 4857
[13780-64-8], 4319
[13782-01-9], 4213
[13783-04-5], 0284
[13783-07-8], 4630
[13812-39-0], 4068
[13813-62-2], 4567
[13814-76-1], 4167
[13815-21-9], 3985
[13815-30-0], 4359
[13816-48-3], 3926
[13819-89-8], 4420
[13820-81-0], 4124
[13820-83-2], 4126
[13823-27-2], 4768
[13823-36-4], 4461
[13823-50-2], 4457
[13825-86-0], 4244
[13826-64-7], 4737
[13826-86-3], 0131
[13842-82-5], 4341
[13845-18-6], 4467
[13845-23-3], 4485
[13847-60-4], 4501
[13847-65-9], 4337
[13847-66-0], 4765
[13862-16-3], 4580
[13863-59-7], 0238
[13863-88-2], 0023
[13864-01-2], 4495
[13864-02-3], 3963
[13875-06-4], 4322
[13932-10-0], 4044
[13933-23-8], 4062

[13939-06-5], 2628
[13943-58-3], 2064
[13952-84-6], 1724
[13957-21-6], 2449
[13965-73-6], 0162
[13967-25-4], 4312
[13967-90-3], 0202
[13968-14-4], 4722
[13973-87-0], 0256
[13973-88-1], 4030
[13980-04-6], 1217
[13985-87-0], 3801
[13986-26-0], 4639
[13997-90-5], 3975
[14014-86-9], 4635
[14017-53-9], 4131
[14017-55-1], 4132
[14018-82-7], 4492
[14024-00-1], 2635
[14024-50-1], 3326
[14038-43-8], 3739
[14040-11-0], 2636
[14044-65-6], 0138
[14090-22-3], 1629
[14104-20-2], 0005
[14128-54-2], 0069
[14168-44-6], 0483
[14215-28-2], 4786
[14215-29-3], 3957
[14215-30-6], 4280
[14215-31-7], 4216
[14215-33-9], 4604
[14222-60-7], 3163
[14233-86-4], 1645
[14243-51-7], 2558
[14259-65-5], 4188
[14259-66-6], 4164
[14259-67-7], 3968
[14268-23-6], 0530
[14283-05-7], 4274
[14284-90-3], 3690
[14285-68-8], 3386
[14286-02-3], 4560
[14293-70-0], 4661
[14306-88-8], 3838
[14307-35-8], 4236

[14309-25-2], 3783
[14312-20-0], 0280
[14314-27-3], 4016
[14322-50-0], 4075
[14335-40-1], 3979
[14336-80-2], 4287
[14351-66-7], 3805
[14353-90-3], 2058
[14355-21-6], 0120
[14362-68-6], 0364
[14362-70-0], 0350
[14368-49-1], 2202
[14380-76-8], 4543
[14391-39-6], 0140
[14391-40-9], 0142
[14399-53-2], 0805
[14402-89-2], 1813
[14404-36-5], 4207
[14435-92-8], 0550
[14448-38-5], 4470
[14451-00-4], 3447
[14452-39-2], 0066
[14459-54-2], 4150
[14475-63-9], 4534
[14477-33-9], 1329
[14481-29-9], 2577
[14488-49-4], 0950
[14516-66-6], 3890
[14519-07-4], 0279
[14519-10-9], 0249
[14521-05-2], 4366
[14522-39-5], 3799
[14524-44-8], 3829
[14545-08-5], 0670
[14545-72-3], 4026
[14548-58-4], 3132
[14548-59-5], 2786
[14550-84-6], 0260
[14589-65-2], 4208
[14597-45-6], 2776
[14604-29-6], 2774
[14621-84-2], 0336
[14642-66-1], 2877
[14649-03-7], 3139
[14652-46-1], 4206
[14666-77-4], 3356

[14666-78-5], 2446
[14668-82-7], 1109
[14674-72-7], 3925
[14674-74-9], 0204
[14678-02-5], 1502
[14691-44-2], 4421
[14696-82-3], 4621
[14705-99-8], 1809
[14746-03-3], 2377
[14751-89-4], 3875
[14760-99-7], 1513
[14774-78-8], 2139
[14781-32-9], 1784
[14781-33-0], 1784
[14796-11-3], 2367
[14848-01-2], 2654
[14856-86-1], 1046
[14871-79-5], 0210
[14871-92-2], 3255
[14876-96-1], 4387
[14877-32-9], 0283
[14878-31-0], 1376
[14884-42-5], 4226
[14913-74-7], 0496
[14917-59-0], 2644
[14931-97-6], 2590
[14932-06-0], 1787
[14940-94-4], 4868
[14973-59-2], 4196
[14976-54-6], 3070
[14977-61-8], 4054
[14984-81-7], 4321
[14986-53-9], 4295
[14986-60-8], 4307
[14989-38-9], 4165
[14989-89-0], 1114
[15002-08-1], 4402
[15039-07-3], 4816
[15061-57-1], 4058
[15070-34-5], 4692
[15100-34-6], 0102
[15110-33-5], 4583
[15114-46-2], 1032
[15114-92-8], 2595
[15156-18-0], 4123
[15179-32-5], 4367

[15192-14-0], 4325
[15192-76-4], 0618
[15195-58-1], 4343
[15203-80-2], 4129
[15213-49-7], 0988
[15219-34-8], 0583
[15220-72-1], 3376
[15232-76-5], 3024
[15245-44-0], 2071
[15246-55-6], 2619
[15263-28-7], 4234
[15284-39-6], 1511
[15285-42-4], 2765
[15320-25-9], 4019
[15321-10-5], 3988
[15321-67-7], 3969
[15336-58-0], 0569
[15383-68-3], 1408
[15388-46-2], 3874
[15411-45-7], 1544
[15457-77-9], 4509
[15457-87-1], 4540
[15477-33-5], 0065
[15511-25-8], 3251
[15521-60-5], 4535
[15529-45-0], 3382
[15538-67-7], 0175
[15543-40-5], 4929
[15576-35-9], 1322
[15584-65-3], 3682
[15587-44-7], 4426
[15588-62-2], 4005
[15598-34-2], 1863
[15605-27-3], 4022
[15605-28-4], 4027
[15630-89-4], 0553
[15640-93-4], 0133
[15651-35-1], 4587
[15663-42-0], 0519
[15676-66-1], 3573
[15677-44-8], 3572
[15685-17-3], 3358
[15712-13-7], 1358
[15713-42-1], 1732
[15718-71-5], 0964
[15721-33-2], 2535

[15727-43-2], 3476
[15736-98-8], 0537
[15742-33-3], 4219
[15755-09-6], 0897
[15760-45-9], 3019
[15760-46-0], 3182
[15769-72-9], 3373
[15780-31-1], 4869
[15790-54-2], 1269
[15829-53-5], 4613
[15842-89-4], 3005
[15857-57-5], 4878
[15861-05-9], 4296
[15865-57-3], 2915
[15873-50-4], 2350
[15875-44-2], 0500
[15876-39-8], 3787
[15877-57-3], 2492
[15930-75-3], 4338
[15932-89-5], 0487
[15980-15-1], 1613
[15995-42-3], 2044
[16007-16-2], 3173
[16016-20-0], 2396
[16017-37-1], 4763
[16017-38-2], 4238
[16029-98-4], 1306
[16043-45-1], 4920
[16065-92-2], 4918
[16066-38-9], 3035
[16099-75-5], 4299
[16102-24-2], 2824
[16130-58-8], 1075
[16143-80-9], 3871
[16166-61-3], 3400
[16176-02-6], 1432
[16187-03-4], 1982
[16187-15-8], 1623
[16260-59-6], 2157
[16282-67-0], 0355
[16329-92-3], 0641
[16329-93-4], 1062
[16408-92-7], 0649
[16408-93-8], 0647
[16408-94-9], 0637
[16413-88-0], 2901

[16438-87-3], 0542
[16462-53-6], 3751
[16544-95-9], 4568
[16588-34-4], 2651
[16607-77-5], 2944
[16664-33-8], 3550
[16668-69-2], 3740
[16681-65-5], 1189
[16681-78-0], 0814
[16689-88-6], 4489
[16694-46-5], 1273
[16714-23-1], 2939
[16747-33-4], 3203
[16752-60-6], 4872
[16774-21-3], 3964
[16789-24-5], 4658
[16823-94-2], 4673
[16824-78-5], 3430
[16824-81-0], 3437
[16825-72-2], 2754
[16825-74-4], 2755
[16829-28-0], 4323
[16829-30-4], 4305
[16853-85-3], 0075
[16871-90-2], 4364
[16872-11-0], 0130
[16875-44-2], 0501
[16903-32-5], 0164
[16921-96-3], 4378
[16924-00-8], 4379
[16924-32-6], 0960
[16940-66-2], 0147
[16940-81-1], 4360
[16949-15-8], 0145
[16962-07-5], 0058
[16962-31-5], 4363
[16968-06-2], 1345
[16968-19-7], 3646
[16971-92-9], 3994
[16978-76-0], 3407
[16986-83-7], 2418
[17003-75-7], 0750
[17003-82-6], 0341
[17013-07-9], 0072
[17014-71-0], 4666
[17043-56-0], 0436

[17068-96-0], 0158
[17082-12-1], 3484
[17083-63-5], 0077
[17088-37-8], 3195
[17088-73-2], 0928
[17097-12-0], 0146
[17117-12-3], 2297
[17122-97-3], 2730
[17156-85-3], 0179
[17156-88-6], 0961
[17168-82-0], 0965
[17168-82-0], 0968
[17168-83-1], 1781
[17168-85-3], 4231
[17176-77-1], 3658
[17185-68-1], 1336
[17194-00-2], 0208
[17206-00-7], 2527
[17224-08-7], 0358
[17224-09-8], 0351
[17242-52-3], 4456
[17292-62-5], 0382
[17300-62-8], 0085
[17333-74-3], 2775
[17333-82-2], 2131
[17333-83-4], 2148
[17333-84-5], 2149
[17333-86-7], 2659
[17333-87-8], 2660
[17333-88-9], 2661
[17408-16-1], 3141
[17417-38-8], 4059
[17421-82-8], 3603
[17440-85-6], 0157
[17455-13-9], 0830
[17455-13-9], 3561
[17495-81-7], 4029
[17496-59-2], 4100
[17498-10-1], 1692
[17508-17-7], 2279
[17510-52-0], 1159
[17524-18-4], 4582
[17530-57-3], 3599
[17548-36-6], 3695
[17596-45-1], 0141
[17607-20-4], 1902

[17658-52-8], 3428
[17683-09-9], 3657
[17697-12-0], 3495
[17702-41-9], 0198
[17761-23-8], 1070
[17814-73-2], 0448
[17814-74-3], 3443
[17835-81-3], 4135
[17836-90-7], 4407
[17891-55-3], 3222
[17927-57-0], 0160
[17933-16-2], 0426
[18139-02-1], 0800
[18139-03-2], 1459
[18178-05-7], 4529
[18204-79-0], 1302
[18230-75-6], 1330
[18244-91-2], 2562
[18264-75-0], 0495
[18273-30-8], 0803
[18282-10-5], 4840
[18283-93-7], 0181
[18356-02-0], 0463
[18365-12-3], 1251
[18380-68-2], 0900
[18428-81-4], 4145
[18433-84-6], 0189
[18501-44-5], 4211
[18523-48-3], 0774
[18532-87-1], 4137
[18558-16-4], 0415
[18608-30-5], 1041
[18624-44-7], 4392
[18649-64-4], 2212
[18727-07-6], 0491
[18810-58-7], 0214
[18815-16-2], 4179
[18815-73-1], 0447
[18820-29-6], 4706
[18820-63-8], 4306
[18864-05-6], 3549
[18868-43-4], 4716
[18880-14-3], 1601
[18902-42-6], 4144
[18918-17-7], 3637
[18922-71-9], 2893

[18925-10-5], 1155
[18925-69-4], 3321
[18939-43-0], 4696
[18939-61-2], 4282
[19155-52-3], 0773
[19162-21-1], 3223
[19162-94-8], 0360
[19195-69-8], 3984
[19212-11-4], 3420
[19232-39-4], 2404
[19267-68-6], 1791
[19273-47-3], 1562
[19273-49-5], 1458
[19287-45-7], 0166
[19372-47-2], 3325
[19416-93-4], 0449
[19419-98-8], 0672
[19423-89-3], 4745
[19465-88-8], 3936
[19465-89-5], 4785
[19482-31-6], 2048
[19504-40-6], 4596
[19521-84-7], 2227
[19528-48-4], 3433
[19597-69-4], 4685
[19624-22-7], 0188
[19708-47-5], 0335
[19733-68-7], 1584
[19792-18-8], 1360
[19808-30-1], 1410
[19816-89-8], 0783
[19899-80-0], 1122
[19932-64-0], 0679
[19987-14-5], 2845
[20039-37-6], 3304
[20205-91-8], 0156
[20281-00-9], 4264
[20394-65-4], 4009
[20421-00-5], 3679
[20427-56-9], 4862
[20427-58-1], 4478
[20436-27-5], 0165
[20442-99-3], 3628
[20443-67-8], 3316
[20446-69-9], 0963
[20471-57-2], 2334

2094

[20519-92-0], 1332
[20539-85-9], 2978
[20548-54-3], 3941
[20611-53-4], 4072
[20634-12-2], 4589
[20667-12-3], 0032
[20720-71-2], 0685
[20731-44-6], 2143
[20737-02-4], 0038
[20738-31-2], 3891
[20762-60-1], 4652
[20762-98-5], 1821
[20816-12-0], 4858
[20829-66-7], 0966
[20854-03-9], 2743
[20854-05-1], 2744
[20859-73-8], 0083
[20914-03-0], 3712
[20919-99-7], 1854
[20944-65-4], 3528
[20972-64-9], 1450
[20982-74-5], 3111
[20991-79-1], 0539
[21041-85-0], 0717
[21050-95-3], 2150
[21107-27-7], 1334
[21109-95-5], 0218
[21138-22-7], 4416
[21230-20-6], 2273
[21245-43-2], 3692
[21249-40-0], 3811
[21251-20-7], 1403
[21254-73-9], 1839
[21255-83-4], 0258
[21261-07-4], 3371
[21299-20-7], 3455
[21372-60-1], 1172
[21380-82-5], 1628
[21473-42-1], 0401
[21482-59-7], 1779
[21483-09-8], 3753
[21519-20-0], 3862
[21548-73-2], 0026
[21548-73-2], 0035
[21572-61-2], 1807
[21645-51-2], 0074

[21651-19-4], 4829
[21774-03-8], 4651
[21790-54-5], 3863
[21844-15-1], 0153
[21892-31-9], 2069
[21908-53-2], 4605
[21916-66-5], 1883
[21958-87-5], 2093
[21961-22-8], 1365
[21981-95-3], 3552
[22086-53-9], 2534
[22113-51-5], 2788
[22143-20-0], 4301
[22205-57-8], 4260
[22281-49-8], 3616
[22289-82-3], 4578
[22295-11-0], 1768
[22295-99-4], 4795
[22388-72-3], 4209
[22392-07-0], 1682
[22398-80-7], 4642
[22410-18-0], 1374
[22432-68-4], 1042
[22471-42-7], 2623
[22480-64-4], 2130
[22493-01-2], 4523
[22524-35-2], 1238
[22524-45-4], 2510
[22541-59-9], 4896
[22541-88-4], 4896
[22575-95-7], 3037
[22585-64-4], 0239
[22620-90-2], 0617
[22650-46-0], 4060
[22653-19-6], 1739
[22675-68-9], 1035
[22688-10-6], 1979
[22691-91-4], 2395
[22692-30-4], 2934
[22710-73-2], 2954
[22722-03-8], 1327
[22722-98-1], 2575
[22750-47-6], 0228
[22750-53-4], 3954
[22750-56-7], 0614
[22750-57-8], 4262

[22750-67-0], 4547
[22750-69-2], 1143
[22750-93-2], 0852
[22751-18-4], 2705
[22751-23-1], 2914
[22751-24-2], 2155
[22754-96-7], 0987
[22754-97-8], 1025
[22755-00-6], 4530
[22755-01-7], 4476
[22755-07-3], 3491
[22755-09-5], 2366
[22755-14-2], 1249
[22755-22-2], 4466
[22755-25-5], 2182
[22755-27-7], 4750
[22755-34-6], 1091
[22755-36-8], 1323
[22784-01-6], 3580
[22826-61-5], 2633
[22831-39-6], 4697
[22834-73-8], 3907
[22868-13-9], 4812
[22887-15-6], 4270
[22978-89-8], 3640
[22996-18-5], 2716
[23097-77-0], 4141
[23115-33-5], 2602
[23129-50-2], 3906
[23143-88-6], 0339
[23273-02-1], 0967
[23292-52-6], 0627
[23303-78-8], 4342
[23326-27-4], 1881
[23362-09-6], 0499
[23380-38-3], 3799
[23411-89-4], 3587
[23414-72-4], 4710
[23418-85-1], 3399
[23679-20-1], 2456
[23705-25-1], 2880
[23753-67-5], 0186
[23777-55-1], 0513
[23777-63-1], 0167
[23777-80-2], 0191
[23784-96-5], 1830

[23834-96-0], 0159
[23840-95-1], 0184
[23936-60-9], 2585a
[24094-93-7], 4237
[24115-83-1], 3124
[24156-53-4], 0771
[24167-76-8], 4474
[24173-36-2], 2697
[24264-08-2], 2430
[24308-92-7], 1711
[24345-74-2], 1604
[24356-04-5], 3865
[24423-68-5], 3368
[24516-71-0], 3462
[24573-95-3], 3331
[24621-17-8], 0285
[24704-64-1], 0078
[24748-23-0], 3562
[24748-25-2], 1754
[24869-88-3], 1564
[24903-95-5], 3175
[25005-56-5], 4383
[25014-41-9], 1108
[25021-40-5], 3029
[25038-59-9], 3262
[25067-58-7], 0687
[25077-87-6], 2138
[25081-01-0], 4024
[25136-85-0], 0337
[25147-05-1], 0743
[25167-31-1], 0786
[25167-81-1], 2161
[25196-24-9], 3378
[25230-72-2], 2742
[25233-34-5], 1407
[25238-02-2], 1128
[25240-93-1], 0545
[25251-03-0], 0240
[25253-54-7], 4109
[25284-83-7], 0490
[25382-52-9], 0402
[25393-98-0], 3268
[25395-83-9], 0213
[25398-08-7], 3253
[25402-50-0], 4334
[25455-73-6], 0033

[25474-92-4], 4754
[25480-76-6], 1488
[25483-10-7], 0242
[25496-08-6], 2746
[25523-79-9], 4110
[25546-98-9], 4563
[25573-59-5], 4585
[25573-61-9], 0125
[25582-86-9], 3977
[25596-24-1], 1299
[25604-70-0], 0578
[25604-71-1], 0598
[25639-45-6], 3245
[25658-42-8], 4733
[25658-43-9], 4732
[25682-07-9], 0958
[25710-89-8], 0634
[25737-28-4], 3328
[25764-08-3], 3965
[25854-38-0], 0403
[25854-41-5], 0384
[25869-93-6], 0182
[25870-02-4], 0028
[25875-18-7], 1064
[25910-37-6], 2201
[25987-94-4], 0302
[26028-46-6], 3183
[26134-62-3], 4688
[26156-56-9], 4125
[26157-96-0], 2054
[26163-27-9], 0689
[26163-27-9], 0712
[26222-92-4], 0236
[26241-10-1], 4662
[26249-01-4], 2658
[26257-00-1], 4414
[26283-13-6], 3970
[26299-14-9], 1862
[26351-01-9], 0073
[26386-83-9], 3489
[26388-78-3], 1786
[26445-82-9], 0174
[26464-99-3], 2045
[26493-63-0], 1775
[26519-91-5], 2363
[26522-91-8], 0270

[37067-42-8], 4496 · [39422-73-6], 4798 · [43135-83-7], 3363 · [51593-99-8], 4672 · [53474-95-6], 4718
[37150-27-9], 2191 · [39422-94-1], 3461 · [44023-01-6], 4595 · [51655-36-8], 1636 · [53477-43-3], 3000
[37205-27-9], 3706 · [39437-99-5], 1743 · [44397-85-5], 0763 · [51680-55-8], 4541 · [53490-67-8], 2035
[37218-25-0], 0003 · [39597-90-5], 2336 · [44750-17-6], 2173 · [51688-25-6], 3397 · [53496-15-4], 2865
[37235-82-8], 0232 · [39687-95-1], 1470 · [45101-14-2], 1760 · [51819-69-3], 4086 · [53504-62-4], 3461
[37245-81-1], 4714 · [39729-81-2], 4275 · [45722-89-2], 2388 · [51821-32-0], 0857 · [53504-66-8], 3461
[37271-59-3], 0049 · [39733-13-6], 4586 · [48114-68-7], 2822 · [51832-56-5], 3520 · [53504-80-6], 3285
[37279 26-8], 3424 · [39753-42-9], 2840 · [49558-46-5], 2700 · [51877-12-4], 1387 · [53534-20-6], 0938
[37297-87-3], 0975 · [39957-12-5], 4570 · [49559-20-8], 4458 · [52003-45-9], 0321 · [53535-47-0], 3434
[37298-41-2], 3324 · [40089-12-1], 3133 · [49571-31-5], 2655 · [52096-16-9], 0777 · [53578-07-7], 1247
[37346-96-6], 4669 · [40237-34-1], 0914 · [49716-47-4], 3689 · [52096-18-1], 2283 · [53654-97-0], 0318
[37388-50-4], 4315 · [40330-52-7], 4227 · [50274-95-8], 2678 · [52096-20-5], 2282 · [53808-72-3], 3174
[37414-44-1], 3288 · [40330-52-7], 4228 · [50277-65-1], 1065 · [52096-22-7], 2704 · [53808-74-5], 3340
[37436-17-2], 2374 · [40330-52-7], 4230 · [50311-48-3], 0346 · [52106-86-2], 2720 · [53808-76-7], 3659
[37480-75-4], 3591 · [40358-04-1], 1388 · [50327-56-5], 3813 · [52118-12-4], 3844 · [53847-48-6], 2723
[37609-69-1], 2948 · [40428-75-9], 1438 · [50424-93-6], 2913 · [52157-57-0], 2352 · [53852-44-1], 2204
[37647-36-2], 3680 · [40560-76-7], 2780 · [50577-64-5], 0392 · [52324-65-9], 4212 · [53939-55-2], 3701
[37759-21-0], 3842 · [40561-27-1], 1506 · [50623-64-8], 3645 · [52412-56-3], 3515 · [53966-09-9], 3478
[37895-51-5], 1415 · [40575-26-6], 3235 · [50645-52-8], 4390 · [52470-25-4], 0509 · [54007-00-0], 3685
[38005-27-5], 4308 · [40575-26-6], 3236 · [50646-37-2], 4921 · [52716-12-8], 1924 · [54065-28-0], 4553
[38005-31-1], 4140 · [40661-97-0], 0433 · [50650-92-5], 0709 · [52732-14-6], 3590 · [54086-40-7], 0306
[38078-09-0], 1684 · [40898-91-7], 4043 · [50662-24-3], 2042 · [52762-90-0], 3611 · [54086-41-8], 0574
[38092-76-1], 0951 · [40953-35-3], 2050 · [50707-40-9], 3171 · [52781-75-6], 3452 · [54153-83-2], 4253
[38115-19-4], 4316 · [40985-41-9], 0469 · [50744-08-6], 3179 · [52795-23-0], 4381 · [54196-34-8], 0451
[38139-15-0], 0520 · [41029-45-2], 1461 · [50764-78-8], 1214 · [52851-26-0], 2398 · [54217-98-0], 3464
[38232-63-2], 4612 · [41084-90-6], 0621 · [50831-29-3], 1785 · [52913-14-1], 2821 · [54217-99-1], 3464
[38293-27-5], 1844 · [41334-40-1], 3119 · [50846-98-5], 1082 · [52936-24-0], 2520 · [54222-33-2], 3359
[38344-58-0], 4382 · [41337-62-6], 3798 · [50859-36-2], 0096 · [52936-25-1], 0795 · [54222-34-3], 3707
[38361-85-2], 3100 · [41409-50-1], 1371 · [50892-86-9], 3438 · [52936-53-1], 1250 · [54448-39-4], 0472
[38386-55-9], 3038 · [41422-43-9], 1769 · [50892-90-5], 3439 · [52999-19-6], 3545 · [54514-12-4], 2736
[38551-29-0], 4550 · [41463-64-3], 1018 · [50892-99-4], 1872 · [53004-03-8], 3161 · [54514-15-7], 2751
[38682-34-7], 0123 · [41481-90-7], 1792 · [50906-29-1], 2898 · [53055-05-3], 2937 · [54514-15-7], 2752
[38711-69-2], 3577 · [41570-79-0], 3742 · [51006-26-9], 3867 · [53078-10-7], 4161 · [54524-31-1], 0767
[38787-96-1], 1235 · [41909-62-0], 4665 · [51027-90-8], 2984 · [53088-06-5], 3600 · [54567-24-7], 0820
[38870-89-2], 1165 · [41924-27-0], 3584 · [51104-87-1], 1353 · [53144-64-2], 0825 · [54573-23-8], 1871
[38892-09-0], 2115 · [42017-07-2], 2834 · [51138-16-0], 3001 · [53149-77-2], 3908 · [54675-76-2], 3901
[38910-25-8], 2474 · [42081-47-0], 4257 · [51184-04-4], 4553 · [53151-88-5], 3566 · [54689-17-7], 4220
[38989-47-8], 4354 · [42087-90-1], 3477 · [51249-08-2], 2280 · [53182-16-4], 3178 · [54747-08-9], 3612
[39003-82-2], 0093 · [42238-29-9], 2205 · [51269-39-7], 3803 · [53201-99-3], 4271 · [54847-21-1], 4552
[39047-21-7], 1795 · [42246-25-3], 4045 · [51286-83-0], 1106 · [53213-77-7], 3746 · [55011-46-6], 0387
[39108-12-8], 4694 · [42294-95-6], 0589 · [51286-84-1], 1119 · [53213-78-8], 2589 · [55042-15-4], 4181
[39108-14-0], 0082 · [42404-59-1], 3543 · [51317-78-3], 3322 · [53214-97-4], 1799 · [55044-04-7], 0811
[39254-48-3], 2338 · [42454-06-8], 2692 · [51325-04-3], 1402 · [53346-21-7], 3303 · [55058-76-9], 3664
[39254-49-4], 2219 · [42489-15-6], 4715 · [51329-17-0], 3099 · [53378-72-6], 2046 · [55060-79-2], 3717
[39274-39-0], 0356 · [42494-76-1], 1200 · [51451-05-9], 0787 · [53421-36-6], 0622 · [55060-80-5], 3540
[39409-82-0], 0534 · [43008-25-9], 2160 · [51470-74-7], 2480 · [53436-80-9], 0067 · [55060-82-7], 3716

[55106-91-7], 3238   [57508-69-7], 3115   [59598-02-6], 2613   [63104-54-1], 3232   [65343-66-0], 3227

[55106-92-8], 3389   [57530-25-3], 1194   [59643-77-5], 3625   [63105-40-8], 4385   [65381-02-4], 3473

[55106-93-9], 3387   [57664-97-8], 2887   [59675-70-6], 4130   [63128-08-5], 3770   [65521-60-0], 1044

[55106-95-1], 3735   [57670-85-6], 1002   [59744-77-3], 4687   [63148-65-2], 3030   [65579-06-8], 0089

[55106-96-2], 3734   [57678-14-5], 3143   [59795-18-5], 4739   [63160-33-8], 2830   [65579-07-9], 0183

[55124-14-6], 0394   [57707-64-9], 0714   [59865-91-7], 4777   [63181-09-9], 3228   [65597-24-2], 0320

[55153-36-1], 2512   [57846-03-4], 2346   [59871-26-0], 1240   [63246-75-3], 3335   [65597-25-3], 0594

[55341-15-6], 1554   [57891-85-7], 2114   [60241-33-0], 4775   [63338-74-9], 4502   [65651-91-4], 2586

[55358-23-1], 3870   [57965-32-9], 2250   [60345-95-1], 0981   [63354-77-8], 2669   [65664-23-5], 2896

[55358-25-3], 3847   [58249-49-3], 3990   [60380-67-8], 4432   [63382-64-9], 2467   [65776-39-8], 0086

[55358-29-7], 3870   [58281-24-6], 3470   [60489-76-1], 3808   [63609-41-6], 3148   [65901-39-5], 4053

[55358-33-3], 3870   [58302-41-3], 3627   [60553-18-6], 0752   [63618-84-8], 3636   [66080-22-6], 3423

[55374-98-6], 4070   [58302-42-4], 2642   [60580-20-3], 4170   [63643-78-7], 0799   [66217-76-3], 2799

[55644-07-0], 1803   [58412-05-8], 4659   [60617-65-4], 4349   [63672-69-3], 4152   [66326-45-2], 4008

[55644-07-0], 1804   [58426-27-0], 1206   [60633-76-3], 0393   [63706-85-4], 0163   [66326-46-3], 4010

[55761-75-8], 3602   [58430-57-2], 0879   [60672-60-8], 0755   [63839-60-1], 1144   [66427-01-8], 0454

[55800-08-3], 3835   [58501-09-0], 4224   [60784-40-9], 1131   [63915-60-6], 1590   [66459-02-7], 0080

[55864-39-6], 1192   [58572-21-7], 4753   [60955-41-1], 4794   [64010-65-7], 4273   [66486-68-8], 2641

[55870-36-5], 1793   [58609-72-6], 3614   [61014-23-1], 3181   [64057-57-4], 0780   [66599-67-5], 3589

[55912-20-4], 2717   [58656-11-4], 3377   [61033-15-6], 2359   [64297-36-5], 3233   [66649-16-9], 3275

[55999-69-4], 1749   [58659-15-7], 3028   [61204-16-8], 2095   [64297-64-9], 2681   [66737-76-7], 2213

[56090-02-9], 2607   [58674-00-3], 4527   [61224-59-7], 3392   [64297-75-2], 3319   [66793-67-7], 1373

[56092-91-2], 2630   [58675-02-8], 1452   [61447-07-2], 2835   [64365-00-0], 3909   [66862-11-1], 2592

[56094-15-6], 4660   [58696-86-9], 2803   [61514-68-9], 2218   [64445-49-4], 2206   [66946-48-3], 3263

[56139-33-4], 1926   [58776-08-2], 2251   [61565-16-0], 1386   [64624-44-8], 1874   [66955-43-9], 0737

[56280-76-3], 4789   [58816-51-8], 0959   [61695-12-3], 3576   [64771-59-1], 1356   [66955-44-0], 1167

[56295-87-5], 4793   [58911-30-1], 0593   [61832-41-5], 1596   [64781-77-7], 2910   [66986-75-2], 3699

[56370-81-1], 1794   [58941-14-3], 0739   [62072-11-1], 2831   [64781-78-8], 0677   [67016-28-8], 4464

[56386-22-2], 3859   [58941-15-4], 1168   [62072-18-8], 3118   [64825-48-5], 2614   [67165-30-4], 4232

[56391-49-2], 3789   [59034-32-1], 1414   [62116-25-0], 1201   [64867-41-0], 0029   [67187-50-2], 2621

[56411-66-6], 2298   [59034-33-2], 1866   [62126-20-9], 0068   [65036-47-7], 2097   [67312-43-0], 0547

[56411-67-7], 2339   [59034-34-3], 1411   [62127-48-4], 1550   [65037-43-6], 2674   [67362-62-3], 0824

[56422-03-8], 4728   [59104-93-7], 0680   [62133-36-2], 2588   [65039-20-5], 2796   [67400-85-2], 0408

[56433-01-3], 2671   [59183-17-4], 1492   [62316-46-5], 4766   [65082-64-6], 3412   [67401-50-7], 3305

[56522-42-0], 0683   [59183-18-5], 1491   [62375-91-1], 2782   [65107-36-0], 4760   [67405-47-4], 2536

[56533-30-3], 0098   [59261-17-5], 1538   [62497-91-0], 3224   [65232-69-1], 2555   [67517-99-7], 3665

[56533-31-4], 4380   [59320-16-0], 2977   [62554-90-9], 2133   [65235-78-1], 3372   [67536-44-1], 0708

[56581-58-9], 3885   [59327-98-9], 2708   [62597-99-3], 4055   [65235-79-2], 0015   [67711-62-0], 2387

[56743-33-0], 2699   [59348-62-8], 1081   [62763-56-8], 4014   [65313-31-7], 1272   [67722-96-7], 1182

[56892-92-5], 0154   [59391-85-4], 2947   [62785-41-5], 0425   [65313-32-8], 0101   [67772-28-5], 1536

[56929-36-3], 0778   [59419-71-5], 1776   [62861-56-7], 1426   [65313-33-9], 0892   [67849-01-8], 0059

[56943-26-1], 0173   [59420-01-8], 3421   [62861-57-8], 2353   [65313-34-0], 1676   [67849-02-9], 0081

[57284-58-7], 2121   [59483-61-3], 2231   [62865-97-6], 2337   [65313-35-1], 2391   [67870-97-7], 2576

[57304-94-6], 3856   [59544-89-7], 0099   [62905-84-4], 2847   [65313-36-2], 0798   [67870-99-9], 3766

[57421-56-4], 1071   [59559-04-7], 2479   [62909-81-3], 4079   [65313-37-3], 1748   [67880-11-9], 0772

[57425-12-4], 4491   [59589-13-8], 3123   [63085-06-3], 3435   [65321-68-8], 2783   [67880-13-1], 0009

[67880-14-2], 4655	[68979-48-6], 1835	[71178-86-4], 3534	[73452-31-0], 2612	[76695-67-5], 3796
[67880-15-3], 4759	[68993-02-2], 4784	[71231-59-9], 3102	[73452-32-1], 2878	[76695-68-6], 3743
[67880-17-5], 0720	[69042-75-7], 3385	[71250-00-5], 0518	[73469-45-1], 2624	[76710-75-3], 3342
[67880-19-7], 4092	[69101-54-8], 4592	[71257-31-3], 4879	[73506-23-7], 4780	[76710-76-4], 3899
[67880-20-0], 0818	[69167-99-3], 3018	[71356-36-0], 3018	[73506-39-5], 1478	[76710-77-5], 3827
[67880-21-1], 1347	[69180-59-2], 2753	[71359-61-0], 1034	[73513-15-2], 3120	[76828-34-7], 2403
[67880-22-2], 1017	[69338-55-2], 3521	[71359-62-1], 1352	[73513-16-3], 4790	[77486-94-3], 3395
[67880-25-5], 4035	[69517-30-2], 0742	[71359-63-2], 0653	[73513-17-4], 4077	[77625-37-7], 3522
[67880-26-6], 2737	[69587-03-5], 3247	[71359-64-3], 1806	[73526-98-4], 2421	[77628-04-7], 3192
[67880-27-7], 0379	[69597-04-6], 2740	[71369-69-2], 2343	[73566-71-5], 3414	[77664-36-9], 3820
[67880-28-8], 0383	[69681-97-6], 3154	[71404-13-2], 3762	[73593-33-6], 4525	[77695-66-4], 3797
[67922-18-3], 1742	[69681-98-7], 3154	[71799-32-1], 2145	[73680-23-6], 3279	[77812-44-3], 3846
[68206-59-5], 3466	[69681-99-8], 3313	[71901-54-7], 0917	[73730-53-7], 0049	[78114-26-8], 3417
[68348-85-6], 0995	[69682-00-4], 3313	[71939-10-1], 1210	[73771-13-8], 2086	[78141-00-1], 3243
[68379-32-8], 1013	[69682-01-5], 3314	[72040-09-6], 1490	[73954-59-3], 4051	[78343-32-5], 1058
[68436-99-7], 0180	[69720-86-1], 3864	[72044-13-4], 0523	[74087-85-7], 1227	[78350-94-4], 2177
[68448-47-5], 3248	[69833-15-4], 3585	[72151-31-6], 4356	[74093-43-9], 0303	[78398-48-8], 3834
[68499-65-0], 3394	[69932-17-8], 0703	[72172-64-6], 3953	[74167-05-8], 3425	[78657-29-1], 1753
[68533-38-0], 0148	[70142-16-4], 0312	[72248-72-7], 4093	[74190-87-7], 3559	[78790-57-5], 3160
[68560-64-3], 2209	[70187-32-5], 1997	[72250-07-8], 4779	[74273-75-9], 1400	[78925-61-8], 3535
[68574-13-0], 1750	[70212-12-3], 3415	[72385-44-5], 1927	[74311-45-7], 4128	[78937-12-9], 1811
[68574-15-2], 2006	[70247-32-4], 0974	[72399-85-0], 1869	[74386-96-2], 0391	[78937-14-1], 2049
[68574-17-4], 2005	[70247-49-3], 3114	[72437-42-4], 3606	[74415-60-4], 3234	[78948-09-1], 0751
[68594-17-2], 0371	[70247-50-6], 1419	[72443-10-8], 2216	[74449-37-9], 4384	[79151-63-6], 3853
[68594-19-4], 0827	[70247-51-7], 1409	[72447-41-7], 3156	[74542-20-4], 4357	[79251-56-2], 2548
[68594-24-1], 1012	[70277-99-5], 2974	[72644-04-3], 3768	[74637-34-6], 3193	[79251-57-3], 2548
[68596-88-3], 3128	[70278-00-1], 2242	[72780-86-0], 4601	[74936-21-3], 0815	[79263-47-1], 3146
[68596-89-4], 2663	[70299-48-8], 1282	[72848-18-1], 0721	[74999-19-2], 0826	[79403-75-1], 3633
[68596-91-8], 2758	[70324-20-8], 2670	[72957-64-3], 2364	[75027-71-3], 3872	[79472-87-0], 1434
[68596-92-9], 2757	[70324-23-1], 3398	[72985-54-7], 1061	[75038-71-0], 1800	[79723-21-0], 0427
[68596-93-0], 2137	[70324-26-4], 3757	[72985-56-9], 0595	[75109-11-4], 3271	[79796-14-8], 4664
[68596-94-1], 2756	[70324-27-5], 2982	[73017-74-2], 2867	[75142-81-3], 3662	[79796-40-0], 4767
[68596-95-2], 2254	[70324-35-5], 2256	[73033-58-6], 2718	[75441-10-0], 3620	[79999-60-3], 3647
[68596-99-6], 2084	[70343-15-6], 2919	[73132-51-5], 1921	[75476-36-7], 3310	[80005-73-8], 3761
[68597-02-4], 3246	[70343-16-7], 2666	[73202-12-7], 4389	[75516-37-9], 3769	[80044-09-3], 1191
[68597-05-7], 2102	[70343-16-7], 2667	[73257-07-5], 3785	[75524-40-2], 0467	[80044-10-6], 1191
[68597-06-8], 2106	[70348-66-2], 1594	[73335-41-8], 3784	[75900-58-2], 4358	[80410-45-3], 3855
[68597-08-0], 2784	[70473-76-6], 2307	[73347-64-5], 4309	[75952-58-8], 1297	[80466-56-4], 1436
[68597-10-4], 2668	[70664-49-2], 1384	[73378-53-7], 4447	[75993-65-6], 0421	[80605-27-2], 3509
[68602-57-3], 1057	[70682-07-4], 1816	[73378-54-8], 4429	[76002-04-5], 0817	[80605-36-3], 3509
[68674-44-2], 0603	[70806-67-6], 4256	[73378-55-9], 4448	[76321-00-0], 3422	[80670-36-6], 2909
[68795-10-8], 0749	[70843-45-7], 3345	[73378-56-0], 4259	[76371-73-8], 3408	[80680-07-5], 4848
[68825-98-9], 0288	[70865-40-6], 0196	[73399-68-5], 1412	[76429-97-5], 1340	[80841-08-3], 3505
[68844-25-7], 3617	[70950-00-4], 0573	[73399-71-0], 1413	[76429-98-6], 1080	[80866-80-4], 2715
[68868-87-1], 3512	[70992-03-9], 4277	[73399-75-4], 3721	[76556-13-3], 0823	[80913-65-1], 1997
[68897-47-2], 0505	[71155-12-9], 3405	[73400-02-9], 4648	[76695-64-2], 3744	[80913-66-2], 1997

2100

[126623-64-1], 2792   [133610-01-2], 3841   [153757-93-8], 0822   [159148-53-5], 3832   [168848-75-7], 3920
[126623-77-6], 3403   [134927-04-1], 2381   [153757-95-0], 0718   [159793-62-1], 1328   [168848-76-8], 3921
[127510-65-0], 1045   [135285-90-4], 2325   [155147-41-4], 3720   [159793-64-3], 1324   [171565-26-7], 0104
[127526-90-3], 2892   [135840-44-7], 2596   [155147-43-6], 3380   [159793-67-6], 1741   [174753-06-1], 4537
[128568-96-7], 2472   [139394-51-7], 1514   [155147-44-7], 3719   [160693-88-2], 2381   [174872-99-2], 3256
[128575-71-3], 3410   [140456-78-6], 4528   [155793-91-2], 2483   [161535-67-6], 1747   [180570-35-8], 0624
[129103-51-1], 3261   [140456-79-7], 4654   [155797-52-7], 3905   [161616-67-7], 3308   [180570-38-1], 0701
[129215-84-5], 3127   [144077-02-2], 3802   [155797-58-3], 3918   [164077-35-4], 3887   [187585-04-6], 0389
[129660-17-9], 1195   [144673-73-4], 3012   [156545-33-4], 2761   [164077-36-5], 3915
[130400-13-4], 1218   [148739-67-7], 2040   [157066-32-5], 2680   [165610-13-9], 3381
[132724-39-1], 4756   [153498-61-4], 1008   [157951-76-3], 0307   [167771-41-7], 0103
[133163-34-5], 4590   [153757-93-8], 0821   [158668-93-0], 2581   [168848-73-5], 3912

Serial numbers with no Registry number

0016	0334	0904	1591	2094	2539	3129	3507	3807
0017	0338	0911	1592	2101	2570	3130	3510	3810
0018	0367	0926	1643	2103	2571	3135	3513	3814
0024	0422	0939	1644	2111	2583	3144	3523	3819
0027	0428	0947	1663	2112	2603	3145	3524	3823
0030	0443	0962	1706	2117	2605	3153	3527	3849
0034	0471	0984	1744	2128	2625	3176	3538	3851
0036	0497	0994	1745	2146	2639	3185	3544	3854
0037	0504	1003	1746	2147	2683	3196	3547	3860
0039	0516	1006	1752	2162	2684	3218	3592	3866
0040	0527	1011	1761	2171	2707	3257	3593	3873
0041	0535	1016	1762	2175	2709	3260	3594	3886
0042	0543	1019	1763	2180	2721	3274	3595	3888
0043	0549	1021	1773	2188	2728	3281	3597	3892
0044	0566	1030	1774	2200	2804	3282	3601	3898
0045	0571	1048	1788	2203	2805	3283	3618	3904
0046	0580	1056	1796	2211	2807	3330	3619	3919
0054	0581	1068	1798	2221	2829	3334	3621	3929
0088	0601	1069	1802	2233	2833	3343	3622	3930
0097	0612	1077	1810	2247	2846	3344	3641	3932
0112	0613	1084	1815	2255	2856	3362	3669	3945
0113	0654	1088	1817	2257	2879	3367	3670	3950
0114	0668	1123	1823	2270	2883	3370	3681	3951
0117	0669	1211	1834	2281	2884	3374	3687	3966
0118	0674	1277	1838	2284	2890	3384	3700	3972
0126	0682	1300	1845	2285	2897	3388	3703	4012
0137	0688	1333	1850	2286	2904	3396	3704	4032
0151	0690	1337	1859	2304	2933	3413	3711	4046
0171	0691	1338	1900	2308	2963	3419	3713	4049
0178	0692	1354	1919	2309	2964	3426	3718	4050
0187	0707	1355	1938	2320	2965	3427	3731	4064
0197	0711	1369	1951	2323	2988	3431	3733	4065
0229	0713	1375	1980	2326	3017	3440	3737	4069
0237	0719	1390	1989	2341	3036	3444	3741	4071
0253	0764	1394	2007	2357	3075	3445	3745	4074
0264	0796	1395	2008	2358	3079	3446	3748	4089
0268	0804	1396	2027	2370	3091	3448	3759	4090
0272	0806	1420	2052	2386	3094	3450	3760	4091
0276	0835	1424	2066	2393	3096	3456	3763	4098
0277	0838	1445	2067	2402	3103	3457	3765	4142
0290	0889	1495	2068	2453	3104	3467	3775	4146
0297	0891	1510	2072	2454	3105	3471	3779	4147
0304	0893	1539	2083	2513	3112	3490	3793	4166
0325	0898	1540	2088	2519	3126	3503	3800	4169

4190	4361	4468	4551	4598	4634	4740	4781
4195	4396	4480	4555	4608	4653	4741	4787
4201	4409	4494	4558	4610	4663	4742	4792
4203	4433	4507	4571	4611	4667	4761	4859
4239	4449	4533	4572	4614	4683	4764	4881
4247	4452	4536	4591	4617	4702	4769	
4249	4454	4546	4593	4622	4735	4772	
4278	4455	4548	4594	4626	4738	4778